WORLD ECONOMIC AND BUSINESS REVIEW

1990

WORLD ECONOMIC AND BUSINESS REVIEW

1990

Compiled by

WORLD OF INFORMATION

Blackwell Reference

First published 1990

Basil Blackwell Ltd
108 Cowley Road, Oxford, OX4 1JF, UK

Basil Blackwell, Inc
3 Cambridge Center
Cambridge
Massachusetts 02142, USA

British Library Cataloguing in Publication Data
World economic and business review.
1990-.
1. Economic conditions.
2. World of information
3309005

ISBN 0-631-17345-5
ISSN 0957-8099

Printed in Great Britain by The Bath Press

Contents

Contributors

Barbara Annis is a Washington-based journalist who writes about Latin America. She is editor of the bi-weekly *Latin American Index* and regularly corresponds for Latin American Newsletters and *America Economia*. She frequently travels to Latin America.

Hilary Barnes is Copenhagen correspondent of the London *Financial Times* and Nordic correspondent of *The Economist*, and publisher and editor of a monthly review of the Nordic countries, *Scandinavian Economics*. He has lived and worked in Scandinavia for more than 20 years.

Jacques Bekaert is a Belgian journalist based in Bangkok and is a columnist on Indochinese affairs for the *Bangkok Post*. He also writes for French daily *Le Monde* and for the BBC Far Eastern Service.

Adotey Bing is an economist and writer on African affairs. He is senior research editor of Africa Books.

Sue Branford is a talks writer in the Latin American Service of the BBC External Services and is currently based in London.

George Blazyca is an economist and Deputy Head of the School of Social Sciences at Thames Polytechnic in London and has published widely on Polish and East European economic affairs. He is author of the quarterly *Polish Country Report* published by the Economist Intelligence Unit Ltd.

Jim Bodgener is the *Financial Times* correspondent in Ankara. Until 1988 he was a staff writer with *Middle East Economic Digest*, where he had covered Turkey since 1984. Before joining *MEED* he reported on overseas business for the UK weekly *Construction News*.

Gerald Bourke is a freelance journalist based in Abidjan, Côte d'Ivoire, covering West Africa. He reports for the BBC, *The Independent, Africa Economic Digest* and *South Magazine*.

David Bradshaw has specialised in Middle East affairs since the mid-1970s. After research in Iran on tribal nomadism, he joined the BBCs Arabic Service. In 1984-86 he was based in Algiers as the BBC's North Africa correspondent. In 1989 he joined *The Economist* foreign department.

Jimmy Burns is on the staff of the *Financial Times* with a special responsibility for Latin American affairs. His book on Argentina, 'The Land that lost its Heroes: The Falklands, The Post-War and Alfonsin' (Bloomsbury), won the 1988 Somerset Maugham Award. A further book, 'Beyond the Silver River: South American Encounters' was published in 1989. He is researching a book on Christopher Columbus.

Tom Burns is Spain correspondent for *The Washington Post* and *Newsweek*. He is a director of the English language business newsletter *Spanish Trends* and is a regular contributor to the *Financial Times* and *The Observer*.

John Burstein is an associate of the Washington Office on Latin America (WOLA) specialising in Central America, and writes frequently on political and economic issues and US foreign policy.

Jane Carroll is a freelance writer and journalist and a commissioning editor for the *Economist Guides*. She was formerly Deputy Senior Editor of Europa Publications Ltd and East Africa editor for *Africa Economic Digest*.

Greg Chamberlain is the Caribbean specialist of the London *Guardian* and has written about the region for newspapers and magazines in Britain, France, the USA and the Caribbean for the past 20 years. He also follows Albanian affairs.

Nicholas Childs is the senior talks writer for the BBC Arabic Service specialising in the affairs of the Arabian Gulf.

Julia Crawford is a Paris-based freelance journalist specialising in French and European affairs. She is a correspondent for *EuroBusiness, Business International* (the Economist Group) and Lafferty Group newsletters. Before moving to Paris, she worked as a business journalist in London and as an editor for the Dutch national news agency in The Hague.

Gillian Cronje is a writer on African political and economic affairs and co-author of 'Lonrho: Portrait of a Multinational' and 'The Workers of Namibia' with Suzanne Cronje.

Suzanne Cronje (who also writes under the name of Susi Pollak) is a freelance writer on African affairs and author of several books including two on Nigeria. With Gillian Cronje she wrote 'The Workers of Namibia' (IDAP, 1978) and, with Gillian Cronje and Margaret Lind, 'Lonrho: Portrait of a Multinational' (Penguin Books, 1976).

Nicholas Cumming-Bruce is a Bangkok-based correspondent for London's *The Guardian* in South-East Asia.

Christopher Cviic is a Croat from Yugoslavia. After a number of years in the BBC External Services where his last post was assistant head of talks, European English Service, in 1969 he joined *The Economist* as a correspondent and leader-writer specialising in Eastern Europe. Since 1983 he has also been editing *The World Today*, the monthly journal of the Royal Institute of International Affairs in London.

John Davies is on the foreign desk of the *Financial Times* in London and was formerly an *FT* correspondent in Frankfurt. A graduate of the University of Sydney, he has worked as a journalist in Australia, the UK and France, as well as West Germany.

Judy Dempsey is the Vienna correspondent for the *Financial Times*.

Pam Dougherty is a freelance journalist based in Amman.

Vaudine England is a talks writer for the Far Eastern Service of the BBC, arrived in Britain from Indonesia, where she wrote for the *Far Eastern Economic Review*. Her work as a journalist began in Hong Kong but has since taken her on extensive working trips throughout South East Asia for the last eight years.

Teresa English is a staff writer for *MEED* and currently writes on Qatar, UAE and Bahrain.

Roberto Espíndola is Senior Lecturer in Politics and Co-Director of the Research Unit on European-Latin American Relations at the University of Bradford. He is also a consultant on Central American and Caribbean affairs for *The Independent* and writes for several specialised publications.

George Fodor is a freelance journalist who writes on domestic and foreign developments in all East European countries as well as on East-West relations. He is a regular contributor to the South East-European Department of the BBC World Service.

Alfredo W. Forti, a native of Argentina, has worked extensively in Latin America, and is presently the Program Director at the Commission on US Latin American Relations, a private Washington DC-based organisation.

Aidan Foster-Carter is Senior Lecturer in Sociology at the University of Leeds, UK, where he is also Director of the Leeds University Korea Project. He is the author of many scholarly and journalistic articles on Korean affairs, especially North Korea; and has visited both halves of Korea.

Colm Foy is the author of 'Cape Verde: Politics, Economics and Society', published by Frances Pinter, London. He was formerly assistant editor of *AfricAsia* magazine and currently works as an independent journalist in Paris. He is an established expert on the Portuguese-speaking nations of Africa.

Michael Gawenda has been a journalist since 1968. He was London correspondent of the Australian *Herald* and *Weekly Times* between 1978 and 1981. He read economics at Monash University and is currently assistant editor for *Time Australia* magazine.

Fay Gjester was with Reuters in London from 1953 to 1955, and then Oslo correspondent of the *Financial Times* (1955-86). Since returning to England she has worked as a freelance journalist specialising in Norwegian affairs, particularly energy topics. She writes regularly for *FT* energy newsletters, *European Energy Report* and *Power in Europe*, and is Scandinavian editor of *Offshore Engineer*.

Michael Griffin is a freelance journalist and photographer specialising in East Africa and Indian Ocean affairs but is currently based in London.

Jón Hákon Magnússon is an Icelandic journalist, who was the *Financial Times* correspondent in Reykjavík before establishing a public relations consultancy, Kynning Og Markadur HF.

Anthony Hawkins is Professor of Business Studies at the University of Zimbabwe.

Stephen Hays was formerly editor of the Latin American Monitor's Southern Cone Report. He has travelled as a freelance correspondent in South America writing for a number of publications and was also on the staff of the *Buenos Aires Herald*. He now works for Reuters.

Adrian Hewitt is deputy director of the UK's Overseas Development Institute, and research adviser to the British All-Party Parliamentary Group on Overseas Development. He has specialised at ODI on trade and aid policy and relations between developing countries and the European Community.

Roland Hill has been London correspondent of leading West German and Austrian newspapers for 30 years. Brought up in Germany and Austria, he studied history at Milan and London Universities. He is the author of several books and has been president of the Foreign Press Association in London. He is now working on a biography of the historian Lord Acton.

Lindsey Hilsum is a freelance journalist. She was based in Nairobi for six years, and covered East Africa for the BBC and *The Guardian* from 1986 to 1989. She specialises in news and current affairs, but has also written about development issues, health and travel.

Kerin Hope is the *Financial Times* correspondent in Greece.

Vernon Hough has worked as a physicist in the development of nuclear power, edited the journals *Nuclear Engineering* and *Petroleum Times*, and spent 12 years with a major oil company. He also writes for *Petroleum Economist* and The Economist Intelligence Unit.

Diana Hubbard is a journalist and economist with a special interest in francophone Africa. She regularly writes for the Economist Intelligence Unit and *Africa Economic Digest* and also does consultancy work for UNIDO.

Gregg Huff is a lecturer in economics at the University of Durham where he teaches macroeconomics.

T. R. Hughes is a journalist specialising, for more than 30 years, in foreign and Pacific affairs. He was on the staff of Reuters, London's *Daily Mail* and London's *Daily Telegraph*. He was previously Senior Information Officer in the Solomon Islands and author of the country's first official *Handbook*. He currently edits the United Kingdom Pacific Islands Society's newsletter, *The Outrigger*.

Anthony Hyman is a freelance writer for the press and radio on Middle East and South Asian affairs. A new edition of his book 'Afghanistan under Soviet Domination 1964-89' will be published by Macmillan in 1990.

Canute James is a journalist based in Jamaica. He is the Caribbean correspondent of the London *Financial Times*.

Michael Jansen is a writer on Middle East affairs who contributes to *Middle East International* (London) and the *Irish Times*. She is author of several books on Palestine, the latest being 'Dissonance in Zion' published by Zed in 1987, and a novel set in Cyprus. She lives in Nicosia.

Luis Kaffman was for many years the Economic and Financial Editor at *Latin American Newsletters*. He now works at *The Economist Intelligence Unit* as the Latin American Editor of the Forecasting Unit.

Bob King was the *Financial Times* Taipei correspondent from 1980 to 1989, and now freelances extensively. He is a regular contributor to the BBC.

Patrick Knight has returned to live in Brazil after three years based in London. Previously there from 1974 to 1985 as a correspondent of *The Times* he is now working for a wide range of publications specialising on the economy, including *Lloyds List, Mining Journal, Commodity Week, Pulp and Paper International.*

Nicolas Kulibaba is a Washington-based political analyst and consultant. He advises international financial institutions, development agencies, and multinational businesses operating in Africa, Latin America and the Caribbean.

David Lane is a freelance business journalist living in Rome.

Xueling Lin is a Copenhagen-based journalist, specialising in the Scandinavian economies.

Margie Lindsay is editor of the *Financial Times* fortnightly newsletter, *East European Markets*. Her book, 'International Business in Gorbachev's Soviet Union', was published in 1989 by Pinter Publishers. She is now writing a book on capital market formation in Eastern Europe.

Simon Long is a talks writer with the Far Eastern Service of the BBC. He writes and broadcasts about Asia and the Pacific. He has travelled widely in the region and in 1988-89 he moved to Peking as a correspondent for the BBC World Service and the *Sunday Correspondent*.

W. L. Luetkens was with Reuters in London from 1943 to 1953, and in Frankfurt and Bonn from 1953 to 1958. He was then Bonn correspondent of the *Financial Times* (1959-69), before returning to London as a foreign leader writer.

Andrew Lycett is a freelance journalist and consultant specialising in international politics and economics. He writes regularly for *The Times* and other British newspapers. He is co-author of 'Qaddafi and the Libyan Revolution' published in 1987. His new book 'Destination Jeddah' will be published this year.

John Madeley publishes and edits *International Agricultural Development* magazine. He has travelled widely in Africa and writes and broadcasts extensively on African issues.

Jon Marks is North Africa writer for *MEED* and writes, broadcasts and lectures on Maghreb politics, economics and history. He also writes for *Business International*, the Economist Intelligence Unit and is associate editor of the *Financial Times* fortnightly *International Trade Finance*.

Jan Marsh is a freelance author, formerly research worker with Mozambique, Angola and Guinea Information Centre and International Defence and Aid Fund for Southern Africa.

Patrick Middleton is a lecturer in the Department of Marketing at the Institute of Technology in the University of Nice. He is Deputy Editor of *Riviera Reporter* magazine and presents a weekly three-hour magazine and talk back show on Riviera Radio, at Capital Radio, Monaco's subsidiary. He has lived in the south of France since 1974.

Michelle Misquitta/Rafferty is an Indian journalist who has travelled widely in Asia and the Pacific and has lived in India, Hong Kong and Japan.

Jane Monahan is a freelance journalist specialising in Spain and Latin America.

John Murray-Brown is a correspondent in Jakarta for a number of publications including the *Financial Times*.

Peter Nares lives in Colombia, writes for a number of international publications and is a correspondent for the BBC.

Edmund Neville-Rolfe was director of the Bureau Européen de Recherches, Brussels, from 1973 to 1988.

Andrew Nickson teaches development economics at the School of Public Policy, University of Birmingham.

Nicholas Nugent is a senior commentator on South Asia with the BBC World Service. He regularly visits and reports from the region.

Elizabeth Obadina is a Lagos-based freelance journalist. She corresponds regularly for the BBC, and *The Times* (London). She has acted as a consultant to UNICEF and contributes to the IPPF magazine *People*.

Tim Owen is a freelance writer and lecturer on Middle East affairs who has contributed to the British press and Middle East publications. In government service for many years, he has travelled widely in the Middle East and elsewhere. He has worked in Lebanon, Aden, the Gulf and Oman. Until recently he was a consultant to the Middle East Association, and continues to visit the Middle East.

Thomas Pennant taught at the University of Malawi from 1977 to 1983. He has written various articles on Malawi, including a contribution to Amis, P & Lloyd, P (eds), 'Housing Africa's Urban Poor'. (Manchester University Press for the International African Institute, forthcoming).

Trevor Petch is a freelance writer and broadcaster specialising in Caribbean affairs. He is Deputy Editor of *Caribbean Insight*, and edits and publishes the newsletter *The Belize Briefing*.

George Philip is reader in Latin American politics, jointly at the London School of Economics and the Latin American Institute, with a special interest in Mexico.

Kevin Rafferty is managing editor of *International Media Partners*, New York. He is also author of the recently published 'City on the Rocks: Hong Kong's Uncertain Future', Viking. He was previously in charge of the Asia coverage of the *Financial Times*, founder editor of *Business Times Malaysia* and of *Asia and Pacific Review*.

Stephen Riley is a senior lecturer in politics at Staffordshire Polytechnic in Stoke on Trent. He has also taught at the University of Sierra Leone, and has written many articles on Sierra Leonean politics and development issues. He is the co-author or editor of books on Sierra Leone, African debt and Third World politics, and is currently compiling a general bibliography of Sierra Leone.

David Robie is a New Zealand journalist and author specialising in South Pacific politics and development issues. He is author of 'Eyes of Fire' and 'Blood on their Banner'; 'Nationalist Struggles in the South Pacific', and a specialist contributor to *Pacific Islands Monthly*, the *Dominion*, *Gemini News Service* and other publications.

Anthony Robinson is the Johannesburg correspondent of the London *Financial Times*. Before moving to South Africa in September 1984, he covered the communist world for six years both as East Europe correspondent and Moscow correspondent for the *Financial Times* after spending eight years with Reuters and the *Financial Times* in Italy.

Mike Rose has an MA in Latin American politics and is currently researching in the field. He has broadcast on Latin America for national and local radio in the UK and he has written extensively about the region. He is a Latin America correspondent for *Gemini News Service* and has worked in Mexico on the staff of the *Mexico City News* and now works for the BBC World Service in Reading.

Alan J. K. Sanders is a writer and broadcaster on Soviet, Central Asian and Mongolian affairs. A Fellow of the Institute of Linguists, London, he is author of 'The People's Republic of Mongolia' published by Oxford University Press and 'Mongolia: Politics, Economics and Society', published by Frances Pinter, London, and numerous articles in specialist journals.

Howard Schissel is a Paris-based writer and consultant on African affairs, with a special interest in francophone Africa and African natural resources.

Robert Shaw is a Nairobi-based freelance journalist and a businessman. He writes mainly about Kenyan economic and development issues and contributes to several domestic and international publications.

Michael Simmons is the East Europe correspondent of the *Guardian*. His book on East Germany called 'The Unloved Country' was published by Abacus in 1989. He has also written a full length political history of the city of Berlin.

Bernard Simon is a former correspondent in Canada for the *Financial Times* of London and is presently managing editor of the *Financial Post* of Canada in Toronto.

Geoffrey Stork, who sadly died on December 2nd 1989, was a well known business consultant specialising on the Southern Cone countries of Argentina, Chile, Paraguay and Uruguay. In the 1960s he ran the Economist Intelligence Unit's Buenos Aires office before going into business consultancy himself in 1969.

Janet Sutherland was English editor of the magazine *UR* which was produced by the Iraqi Cultural Centre in London from 1979 to 1981, when she joined the staff of the *Baghdad Observer*, contributing features as well as working as sub-editor for Naji el-Hadithi, editor-in-chief of the English language daily.

Richard Synge is editor of the London-based weekly *Africa Economic Digest* and a specialist on the Nigerian economy. He is also active in contributing to publications covering the African continent, and was founding editor of *Africa Review*.

Paul Tansey is an economic consultant and commentator on economic affairs. A former assistant editor of the *Irish Times*, he now writes a weekly economic commentary in *The Sunday Tribune*. He is a partner in Tansey, Webster & Associates, economic consultants, Dublin.

Ian Templeton is a political correspondent of the *Auckland Star* and New Zealand correspondent of the London *Guardian*. He edits *Trans-Tasman*, a weekly newsletter which informs business leaders of political and economic developments in the making, with special emphasis on relations between Australia and New Zealand.

Enrique Tessieri is a freelance journalist who writes regularly for the national as well as international press. His byline appears frequently in the *Financial Times* and in other specialised economic publications that deal with energy, trade and finance topics.

Tony Thorndike is professor of international relations at the Staffordshire Polytechnic, England. A teacher of Caribbean politics and international affairs, he has travelled widely in the Caribbean and Central America and written extensively on issues affecting the region since the early 1970s.

Susan Traill is a New York-based freelance journalist specialising in US, European and Third World affairs.

Ron van de Krol is a freelance journalist in Amsterdam. He formerly worked for Reuters news agency in London, Amsterdam and Bonn.

Richard Vokes is a lecturer in Economics and South-East Asian Studies at the University of Kent at Canterbury with teaching and research interests in socialist economic transformation in Indo-China. He has lived and worked in South-East Asia and is a regular visitor to the region.

Anthony Walker is a correspondent in the Middle East for *The Age* of Melbourne and the *Financial Times*. Since 1985 he has been based in Cairo. From 1979 to 1983 he was in Beijing.

Catharine Watson is a journalist based in Uganda. She is correspondent for *The Independent, The Observer, Africa Report* and other publications. She also strings for the BBC.

David Watts is assistant news editor of *The Times*. He was previously Tokyo bureau chief, Middle East correspondent and South East Asia correspondent for *The Times*.

Sara Webb is Stockholm correspondent for the *Financial Times*. She has worked for Swedish newspapers as well as contributing to various British publications.

Peter West is a British economist who lived in Caracas for ten years, where he worked as Economic Advisor to a major Venezuelan commercial bank. Since 1986 he has been working as a senior economist specialising in Latin America in an international organisation in Washington.

Kaye Whiteman is the editor-in-chief of *West Africa* magazine. He has written and broadcast extensively on African affairs for the last 25 years, and has a special interest in francophone Africa.

Andrew Whitley recently concluded a three and a half year assignment in Israel as the staff correspondent for the *Financial Times*. He is currently a Visiting Fellow of the Kevorkian Centre for Near Eastern Studies at New York University, where he is engaged in writing a biography of the late Shah of Iran.

John Wicks is Zurich correspondent of the *Financial Times* and contributes on economic and financial matters to a number of other British, Swiss and American publications. He is a frequent visitor in Liechtenstein and has written numerous articles on the Principality in the *Financial Times* and other newspapers and journals.

Nalin Wijesekera was successively a journalist on the staff of the *Sri Lanka Observer, Times* and *Daily Mirror*, and for more than ten years now has been a freelance broadcaster and contributer to over 15 international publications, including *Asian Business*. He is editor of the *Clothing Magazine* of the Clothing Industry Training Institute.

Michael Williams is the BBC's senior commentator on the Far East and South-East Asia. He is the author of several studies and articles on the region's history and politics, and a frequent visitor to the region.

Peter Wise is a freelance journalist based in Lisbon. He contributes to several publications including *The Washington Post, EuroBusiness, Business International, The Far Eastern Economic Review* and the *Financial Times*.

Michael B. Yahuda is a reader in international relations at the London School of Economics. He is well known as a China specialist, and is the author of 'China's Role in World Affairs, Towards the End of Isolationism: China's Foreign Policy after Mao' and editor of 'New Directions in the Social Sciences and Humanities in China.'

Sam Younger is head of the BBC Arabic Service and a former assistant editor of *Middle East International Magazine*.

John M. Zindar writes and lectures on the politics and economics of Latin America and Eastern Europe. He currently works as Project Director at the International Center for Development Policy in Washington DC.

World of Information

Anthony Axon	Managing Director
Julie Clay	Trade Sales
Daughn Eva	Production Coordinator
Richard Green	Editorial Director
Belinda Isaac	Editorial Coordinator
Joan Jones	Administration
Marianne Keating	Database Coordinator
Sue Lambert	Typesetting and Layout
Andrew Page	Financial Controller
Ailsa Shelley	Administration
Sue Staniforth	Subscriptions

Principal statistical sources

Organisation for International Co-operation and Development
International Monetary Fund (IMF): International Financial Statistics
IMF: Direction of Trade Statistics
World Bank Atlas
World Bank: World Development Report
United Nations Industrial Development Organisation
BP Statistical Review of World Energy
ABECOR Country Reports
National Sources

World currencies

	Currency Unit	Units per Dollar January 1987	January 1988	January 1989	November 1989	Units per £ Sterling November 1989
Afghanistan	Afghani	50.6	50.6	56.3	62.8	99.3
Albania	Lek	6.84	5.53	5.60	6.40	10.10
Algeria	Dinar	4.63	4.96	6.40	8.10	12.80
American Samoa	US Dollar	1.00	1.00	1.00	1.00	1.58
Angola	Kwanza	29.9	29.9	31.1	29.9	47.0
Anguilla	EC Dollar	2.70	2.70	2.70	2.68	4.24
Antigua	EC Dollar	2.70	2.70	2.70	2.68	4.24
Argentina	Peso/Austral	1.25	3.78	16.70	652.50	1028.28
Australia	Dollar	1.50	1.41	1.16	1.28	2.01
Austria	Schilling	13.6	11.5	12.8	13.0	20.6
Bahamas	Dollar	1.00	1.00	1.00	1.00	1.58
Bahrain	Dinar	0.38	0.38	0.38	0.37	0.59
Bangladesh	Taka	31.0	31.1	31.3	31.0	49.0
Barbados	Dollar	2.01	2.01	2.01	2.00	3.16
Belgium	Belgian Franc	40.4	34.1	38.3	38.8	61.4
Belize	Dollar	2.00	2.00	2.00	2.00	3.15
Benin	CFA Franc	321	276	312	314	496
Bermuda	Dollar	1.00	1.00	1.00	1.00	1.58
Bhutan	Ngultrum	13.1	13.0	15.0	16.7	26.4
Bolivia	Peso/Boliviano	1.91	2.18	2.57	2.88	4.55
Botswana	Pula	1.85	1.59	1.95	1.95	3.07
Brazil	Cruzeiro/Cruzado	14.9	51.7	810.7	5.2	8.2
British Virgin Islands	US Dollar	1.00	1.00	1.00	1.00	1.58
Brunei	Dollar	2.18	2.10	1.95	1.96	3.09
Bulgaria	Lev	0.90	0.83	0.85	0.84	1.33
Burkina Faso	CFA Franc	321	276	312	314	496
Burundi	Franc	100	116	154	159	251
Cambodia	Riel	n.a.	100	100	150	236
Cameroon	CFA Franc	321	276	312	314	496
Cape Verde Islands	Escudo	89.3	73.4	73.7	77.3	122.2
Cayman Islands	Dollar	0.84	0.84	0.83	0.83	1.30
Central African Republic	CFA Franc	321	276	312	314	496
Chad	CFA Franc	321	276	312	314	496
Chile	Peso	201	244	251	272	429
China	Renminbi Yuan	3.72	3.72	3.79	3.71	5.86
Colombia	Peso	221	264	339	141	223
Comoros	CFA Franc	321	276	312	314	496
Congo	CFA Franc	321	276	312	314	496
Cook Islands	NZ Dollar	1.91	1.51	1.58	1.70	2.70
Costa Rica	Colon	58.9	69.3	79.6	83.0	131.1
Côte d'Ivoire	CFA Franc	321	276	312	314	496
Cuba	Peso	0.80	0.73	0.76	0.79	1.25
Cyprus	Cyprus Pound	0.51	0.44	0.47	0.49	0.78
Czechoslovakia	Koruna	5.75	5.15	5.38	15.00	23.70
Denmark	Danish Krone	7.34	6.27	7.05	7.18	11.34
Djibouti	Franc	177	177	177	176	278
Dominica	EC Dollar	2.70	2.70	2.70	2.70	4.25
The Dominican Republic	Peso	3.03	4.65	6.35	6.70	10.60
Easter Islands	Chilean Peso	201	226	251	272	429
Ecuador	Sucre	147	253	415	597	943

World currencies

	Currency Unit	Units per Dollar January 1987	January 1988	January 1989	November 1989	Units per £ Sterling November 1989
Egypt	Pound	1.36	2.20	2.36	2.58	4.10
El Salvador	Colon	5.00	5.00	5.00	4.99	7.90
Equatorial Guinea	CFA Franc	321	276	312	314	496
Ethiopia	Birr	2.05	2.07	2.05	2.04	3.23
Fiji	Dollar	1.14	1.45	1.40	1.51	2.38
Finland	Markka	4.79	4.00	4.25	4.27	6.74
France	Franc	6.42	5.51	6.23	6.28	9.90
French Guiana	Franc	6.42	5.51	6.23	6.28	9.90
French Polynesia	CFP Franc	117	110	112	113	178
Gabon	CFA Franc	321	276	312	314	496
Gambia	Dalasi	7.58	2.27	6.73	7.63	12.04
German Democratic Republic	DDR Mark	1.94	1.63	1.83	1.85	2.93
German Federal Republic	Deutsche Mark	1.94	1.63	1.83	1.85	2.93
Ghana	Cedi	90	176	231	288	455
Gibraltar	Gibraltar Pound	0.68	0.61	0.57	0.63	1.00
Greece	Drachma	140	140	152	165	261
Greenland	Danish Krone	7.34	7.01	7.06	7.17	11.33
Grenada	EC Dollar	2.70	2.70	2.70	2.70	4.25
Guadeloupe and Martinique	Franc	6.42	5.51	6.23	6.28	10.00
Guam	US Dollar	1.00	1.00	1.00	1.00	1.58
Guatemala	Quetzal	1.00	1.00	1.00	1.00	1.58
Guinea Republic	Franc	340	440	300	299	472
Guinea Bissau	Peso	170	650	651	650	1023
Guyana	Dollar	4.41	10.00	9.00	29.90	47.30
Haiti	Goude	5.00	5.00	5.00	5.00	7.87
Hawaii	US Dollar	1.00	1.00	1.00	1.00	1.58
Honduras	Lempira	2.10	2.00	2.00	2.00	3.15
Hong Kong	Dollar	7.79	7.79	7.83	7.80	12.31
Hungary	Forint	45.9	47.9	54.1	58.9	93.0
Iceland	Icelandic Krona	40.2	39.0	49.5	62.2	98.3
India	Rupee	13.1	13.0	15.0	16.7	26.4
Indonesia	Rupiah	1641	1656	1757	1786	2820
Iran	Rial	75.4	66.6	70.4	71.7	113.8
Iraq	Dinar	0.31	0.31	0.31	0.30	0.49
Ireland	Punt	0.71	0.68	0.68	0.70	1.10
Israel	Shekel/New Shekel	1.49	1.61	1.83	1.98	3.12
Italy	Lira	1351	1317	1340	1357	2144
Jamaica	Dollar	5.50	5.50	5.40	6.00	9.40
Japan	Yen	160	127	127	144	227
Jordan	Dinar	0.34	0.34	0.49	0.66	1.04
Kenya	Shilling	15.7	15.6	18.5	21.5	34.0
Kiribati	Aust. Dollar	1.50	1.39	1.16	1.28	2.00
Korea DPR	Won	0.94	0.94	0.97	0.97	1.52
Republic of Korea	Won	861	792	691	668	1055
Kuwait	Dinar	0.29	0.27	0.28	0.30	0.47
Laos	Kip/New Kip	35.0	350.0	430.4	580.8	917.4
Lebanon	Pound	91	492	542	429	678
Lesotho	Maloti	2.18	1.96	2.40	2.60	4.16
Liberia	Dollar	1.00	1.00	1.00	1.00	1.58
Libya	Dinar	0.31	0.28	0.29	0.30	0.47

World currencies

	Currency Unit	Units per Dollar January 1987	January 1988	January 1989	November 1989	Units per £ Sterling November 1989
Liechtenstein	Swiss Franc	1.62	1.52	1.56	1.62	2.57
Luxembourg	Luxembourg Franc	40.4	37.9	38.3	38.8	61.4
Macao	Pataca	8.10	8.02	8.05	8.04	12.70
Madagascar	Franc	724	1234	1423	1395	2203
Malawi	Kwacha	1.96	2.10	2.61	2.73	4.31
Malaysia	Ringgit	2.60	2.53	2.74	2.70	4.26
Maldives	Rufiya	7.00	10.10	8.71	8.98	14.17
Mali	CFA Franc	321	276	312	314	496
Malta	Lira	0.37	0.35	0.34	0.35	0.55
Marshall Islands	US Dollar	1.00	1.00	1.00	1.00	1.58
Mauritania	Ouguiya	74.8	72.4	75.1	85.2	134.6
Mauritius	Rupee	13.1	12.5	14.0	15.4	24.4
Mexico	Peso	916	2231	2294	2630	4146
Monaco	French Franc	6.42	6.07	6.23	6.28	10.00
Mongolia	Tugrik	3.36	3.36	3.40	3.40	5.28
Montserrat	EC Dollar	2.70	2.70	2.70	2.70	4.25
Morocco	Dirham	8.68	7.79	8.25	8.33	13.16
Mozambique	Metical	39	455	652	805	1271
Myanma	Kyat	6.96	6.14	6.42	6.82	10.78
Namibia	South African Rand	2.18	1.96	2.40	2.63	4.16
Nauru	Australian Dollar	1.50	1.39	1.16	1.28	2.00
Nepal	Rupee	21.4	21.0	24.0	25.0	39.0
The Netherlands	Guilder	2.19	2.05	2.06	2.10	3.30
Netherlands Dependencies	Guilder	1.80	1.79	2.06	2.10	3.30
New Caledonia	CFP Franc	117	110	112	113	178
New Zealand	Dollar	1.91	1.54	1.58	1.70	2.70
Nicaragua	Cordoba1	500	2190	923	21256	33573
Niger Republic	CFA Franc	321	276	312	314	496
Nigeria	Naira	3.28	4.14	5.30	7.38	11.64
Niue	NZ Dollar	1.91	1.51	1.58	1.70	2.70
Norfolk Island	Australian Dollar	1.50	1.39	1.16	1.28	2.00
N Marianas	US Dollar	1.00	1.00	1.00	1.00	1.58
Norway	Norwegian Krone	7.39	6.67	6.67	6.92	10.90
Oman	Rial	0.38	0.39	0.39	0.38	0.60
Pakistan	Rupee	17.3	17.5	18.6	21.0	33.0
Palau	US Dollar	1.00	1.00	1.00	1.00	1.58
Panama	Balboa	1.00	1.00	1.00	1.00	1.58
Papua New Guinea	Kina	0.96	0.90	0.83	0.86	1.40
Paraguay	Guarani	240	320	321	1180	1863
Peru	Sol/Inti	14.0	33.0	500.4	5603.3	8850.4
Philippines	Peso	20.6	20.7	20.6	21.1	33.4
Poland	Zloty	199	297	519	2814	4444
Portugal	Escudo	146	144	149	158	250
Puerto Rico	US Dollar	1.00	1.00	1.00	1.00	1.58
Qatar	Riyal	3.64	3.64	3.67	3.61	5.71
Romania	Leu	4.30	10.10	8.38	9.10	14.38
Rwanda	Franc	83.9	74.1	76.6	82.6	130.4
St Kitts Nevis	EC Dollar	2.70	2.70	2.70	2.70	4.80
St Lucia	EC Dollar	2.70	2.70	2.70	2.70	4.80
St Vincent	EC Dollar	2.70	2.70	2.70	2.70	4.80

World currencies

	Currency Unit	Units per Dollar January 1987	January 1988	January 1989	November 1989	Units per £ Sterling November 1989
San Marino	San Marino Lira	1351	1317	1340	1357	2144
São Tomé & Príncipe	Dobra	37.0	32.6	101.8	106.9	168.9
Saudi Arabia	Riyal	3.75	3.75	3.76	3.73	5.90
Senegal	CFA Franc	321	276	312	314	496
Seychelles	Rupee	5.91	5.22	5.36	5.53	8.74
Sierra Leone	Leone	31.0	22.0	31.3	62.9	10.0
Singapore	Dollar	2.18	2.02	1.95	1.96	3.10
Solomon Islands	Dollar	1.98	2.00	2.11	2.40	3.77
Somalia	Shilling	113	100	244	408	645
South Africa	Rand	2.18	1.96	2.40	2.63	4.15 (c)
		4.67	3.15	4.01	4.01	6.34 (r)
Spain	Peseta	132	121	115	117	184
Sri Lanka	Rupee	28.5	30.8	32.9	40.0	62.4
Sudan	Pound	2.45	4.50	4.50	4.50	7.10
Suriname	Guilder	1.79	1.79	1.79	1.78	2.80
Swaziland	Lilangeni	2.18	1.96	2.40	2.63	4.16
Sweden	Swedish Krone	6.78	6.41	6.25	6.42	10.20
Switzerland	Franc	1.62	1.52	1.56	1.62	2.57
Syria	Pound	3.92	3.92	21.00	21.00	33.04
Taiwan	Dollar	35.5	28.5	28.3	25.8	40.7
Tanzania	Shilling	51.4	87.2	127.1	147.3	232.7
Thailand	Baht	26.1	25.2	25.0	25.5	40.2
Togo	CFA Franc	321	276	312	314	496
Tokelau	NZ Dollar	1.91	1.51	1.58	1.70	2.70
Tonga	Pa'anga	1.50	1.39	1.16	1.28	2.00
Trinidad and Tobago	Dollar	2.41	3.60	4.25	4.23	6.70
Tunisia	Dinar	0.84	0.78	0.92	0.95	1.47
Turkey	Lira	758	1033	1868	2290	3616
Turks and Caicos Islands	US Dollar	1.00	1.00	1.00	1.00	1.58
Tuvalu	Australian Dollar	1.50	1.39	1.16	1.28	2.00
Uganda	Shilling/New Shilling	1384	62	166	339	535
United Arab Emirates	Dirham	3.67	3.67	3.68	3.67	5.76
United Kingdom	Pound	0.68	0.61	0.57	0.63	1.00
Uruguay	Peso	179	282	456	725	1145
USSR	Rouble	0.68	0.64	0.62	0.62	0.98
Vanuatu	Vatu	117	111	104	121	191
Vatican City	Lira	1351	1317	1340	1357	2144
Venezuela	Bolivar	14.5	14.5	14.5	42.4	67.0
Vietnam	Dong	80.0	368.0	2802.4	4482.9	7080.8
Wallis & Futuna	CFP Franc	117	110	112	113	178
Western Samoa	Tala	2.24	2.05	2.15	2.30	3.64
Yemen Arab Republic	Rial	10.50	9.88	10.32	9.84	15.34
Yemen PDR	Dinar	0.34	0.34	0.34	0.34	0.54
Yugoslavia	Dinar	457	880	5253	50811	80256
Zaïre	Zaïre	68.6	130.0	245.3	438.4	692.5
Zambia	Kwacha	12.71	8.32	9.70	9.52	17.63
Zimbabwe	Dollar	1.68	1.71	1.97	2.23	3.52

(c) commercial rate (r) financial rate

World indicators 1988/89

	Population (m)		Area ('000 sq km)	GDP per capita ($)		Inflation (%)		GDP real growth (%)		Balance of trade ($m)	
Afghanistan	15.50		647.5	170	(1,e)	–		3.5	(3,e)	196	
Albania	3.20		27.4	850	(2,e)	–		–		42	(2,e)
Algeria	23.84		2381.7	2271		6.0		1.8		640	
Andorra	0.05		0.5	9834	(1)	–		–		–	
Angola	9.20	(1)	1246.7	–		–		–		118	(1)
Anguilla	0.01		0.1	3562	(1)	5.0		8.2		–	
Antigua	0.08		0.4	3399	(1)	1.2	(1)	6.6		230	(1)
Argentina	32.00	(1)	2766.9	2862	(1)	4427.0		3.1		4234	(1)
Aruba	0.07	(1)	0.2	–		–		–		–	
Australia	16.50		7682.3	15097		7.6		3.7		1136	
Austria	7.60		83.9	13900	(1)	1.9		1.0	(1)	4700	
Bahamas	0.23		13.9	11317		4.2	(1)	2.0		774	(1)
Bahrain	0.46		0.7	7040	(1,e)	0.3		2.8	(1)	13	
Bangladesh	104.50		144.0	170	(n)	9.4		2.6		1444	
Barbados	0.25		0.4	4233		4.8		3.5		181	(1)
Belgium	9.92		30.5	15345		1.2		4.2		1105	
Belize	0.17		23.0	1138	(1)	0.3		7.6		61	
Benin	4.30	(1)	112.6	351	(2)	5.0	(2)	8.0	(1,e)	385	(1)
Bermuda	0.06		0.1	18000	(1,e)	5.3		4.0	(1)	530	(2)
Bhutan	1.35	(1)	47.0	150	(1)	–		6.7	(4)	63	(1,e)
Bolivia	6.99		1098.6	724		16.0		2.8		57	
Botswana	1.21		582.0	973		8.4		1.3		386	
Brazil	144.00		8512.0	2449		683.0		0.0		19182	
British Virgin Islands	0.02		0.2	9491	(1)	5.2		15.1	(1)	35	(1)
Brunei	0.30		5.8	15390	(1)	2.0		2.2		800	
Bulgaria	9.00		110.9	2750	(1,m)	–		5.1	(1,s)	161	(1)
Burkina Faso	8.30	(1)	274.2	216	(1)	4.2	(1)	3.2	(1)	418	(2)
Burundi	5.15		27.8	232	(1)	4.5		2.0	(1)	78	(1)
Cambodia	6.70		181.0	–		–		–		200	(2,e)
Cameroon	10.70		475.4	1185		8.6		2.0	(1)	398	(e)
Canada	26.10		9976.0	–		4.0		4.5	(e)	8881	
Cape Verde Islands	0.36		4.0	500	(1,n)	–		6.0	(2,n,e)	124	(e)
Cayman Islands	0.03		0.3	18239	(1)	5.2		10.0		10	(1)
Central African Republic	3.00		623.0	330	(1,n)	4.0		0.4	(1,n,e)	71	(1)
Chad	5.40		1284.0	130	(1)	6.5	(1,e)	0.5	(1,n)	150	(1)
Chile	12.80		756.6	2518		14.7		7.4		2219	
China	1100.00		9597.0	227	(n,e)	18.5		11.2		5500	(e)
Colombia	30.20		1138.9	1739		28.1		3.7		648	
Comoros	0.42	(1)	2.2	370	(1)	–		1.7	(1,n,e)	34	(3)
Congo	2.00		342.0	842	(n,e)	16.5		0.6		373	
Costa Rica	2.90		51.1	2235		20.8		3.0		86	
Côte d'Ivoire	11.60		322.5	857		7.0		3.9	(1,n)	815	
Cuba	10.00	(2)	114.5	2238	(3,n)	–		3.7	(1,n)	–	
Cyprus	0.69		9.3	5985		3.5		6.9		1030	
Czechoslovakia	15.61		127.9	6045	(2,m)	2.3	(2)	2.0	(1,m)	–	
Denmark	5.13		43.1	21060		4.5		0.4		1300	

World indicators 1988/89

	Population (m)	Area ('000 sq km)	GDP per capita ($)	Inflation (%)	GDP real growth (%)	Balance of trade ($m)
Djibouti	0.50	23.2	430 (1,e)	–	–	182 (1)
Dominica	0.08	0.8	1550 (1)	1.7	5.6	24
The Dominican Republic	6.90	48.4	354	46.0	0.8	715
Ecuador	10.20	270.7	1477	58.3	12.8	589
Egypt	54.00	1101.5	731 (1,e)	17.6	1.5 (1)	6800
El Salvador	5.10	21.4	955	19.7	1.6	305 (1,e)
Equatorial Guinea	0.41	28.1	399 (1)	–	–	–
Ethiopia	47.90	1251.0	113	7.0	8.0 (1)	1125 (1,e)
Fiji	0.74	18.3	908	11.7 (e)	2.5	54
Finland	4.95	338.1	18163 (1)	5.1	5.2	1390
France	55.90	544.0	16859	3.1	3.7	8100
French Guiana	0.09	91.0	–	4.0 (1)	–	337 (1,e)
Gabon	1.09	267.7	2988	6.0	12.3 (e)	404
Gambia	0.81	11.3	290 (1)	11.7	–	31
G D R	16.70	108.3	8659 (s)	–	8.2 (2,s)	1613
German Federal Republic	61.20	248.7	19261 (n)	1.1	3.6 (n)	68760
Ghana	14.10	239.5	624 (1)	31.3	6.0 (e)	125 (1)
Gibraltar	0.03	0.0	7346 (1,n)	4.4	–	167
Greece	10.00	132.0	5242	13.5	1.0	6702
Greenland	0.05	2175.6	–	–	–	157 (1)
Grenada	0.10 (1)	0.4	1346 (1)	5.8	5.8	60
Guadeloupe	0.34	1.7	3600 (3)	3.5 (1)	–	938 (1)
Guatemala	8.70	108.9	1502	10.9	3.5	396 (1)
Guinea Republic	6.50 (1)	246.0	320 (1)	–	5.9 (1,e)	116 (1)
Guinea Bissau	0.92 (1)	36.1	170 (1,n)	30.0 (2)	–	30 (1)
Guyana	0.72	215.0	995	40.0	3.7	21
Haiti	6.30	27.8	319 (e)	4.0	0.2 (e)	128
Honduras	4.80	112.1	851 (1)	4.6	3.0	24
Hong Kong	5.68	1.1	1058 (1)	7.4	6.0 (n)	5729
Hungary	10.60	93.0	2471	15.7	0.4	–
Iceland	0.25	102.9	22976	23.5	1.9 (n)	153
India	796.60	3287.6	330 (n)	9.4 (e)	1.7 (n,e)	7900
Indonesia	175.20	1919.4	450 (1,e)	8.0 (e)	2.1 (1,e)	5726
Iran	52.70	1648.0	3408 (n)	15.0 (1,e)	0.2 (1,e)	144
Iraq	17.20	434.9	2943 (1,e)	25.0 (1,e)	1.7 (1,e)	879
Ireland	3.54	70.3	8323	2.1	3.0	2983
Israel	4.50	20.8	7677	16.3	1.8	3144
Italy	57.40	301.3	14430	5.0	3.9	9109
Jamaica	2.40	11.0	1388	8.8	1.9	395
Japan	121.80 (1)	377.7	16136 (1,n)	0.7	3.8 (n)	94990
Jordan	2.90	95.4	1552	14.0	3.5	1705
Kenya	23.90	582.7	360	8.2	6.1	785
Korea DPR	21.90	123.0	–	–	2.5 (e)	530 (2)
Republic of Korea	42.00	99.1	4082	7.1	12.0	11450
Kuwait	1.96	17.8	13160 (n)	1.5	–	1907
Laos	3.90	236.8	141 (n)	–	8.0 (4,e)	130 (e)

World indicators 1988/89

	Population (m)	Area ('000 sq km)	GDP per capita ($)	Inflation (%)	GDP real growth (%)	Balance of trade ($m)
Lebanon	2.80	10.5	–	700.0 (2)	–	1540 (2,e)
Lesotho	1.70	30.4	234 (1)	12.2 (1)	5.8 (2)	409 (1)
Liberia	2.50	111.4	450 (1,n)	5.0 (1)	1.0 (1,n,e)	63 (1)
Libya	4.30	1761.1	5500 (1)	–	–	433 (2)
Liechtenstein	0.03	0.2	35600 (1)	1.9	0.0	472
Luxembourg	0.38	2.6	18486 (1)	1.6	5.2	–
Macao	0.44	0.0	5242	7.8	7.5 (e)	–
Madagascar	10.90	587.0	190 (2)	15.0 (1,e)	1.0 (1,n,e)	17 (2)
Malawi	8.40	118.5	182 (1)	33.9	3.6	139 (1)
Malaysia	16.92	330.4	2046 (e)	2.5	7.4 (e)	5559
Maldives	0.20	0.3	405 (1,e)	4.0 (2)	8.7 (e)	65
Mali	8.70 (1)	1240.2	222 (1,n)	–	2.0 (2,n)	80 (1)
Malta	0.35	0.3	5168	0.9	6.2 (1)	644
Martinique	0.34	1.1	4761 (3)	3.6 (1)	–	905 (1)
Mauritania	2.00	1030.7	503 (1,n)	8.2 (1)	0.5 (2,n)	146 (2)
Mauritius	1.08	1.9	1714 (1)	9.1	7.4 (e)	147
Mexico	82.70	1958.2	2588	114.0	1.1 (e)	1600
Monaco	0.03	0.0	–	–	–	–
Mongolia	2.04	1565.0	2140 (2,n)	–	–	–
Montserrat	0.01 (1)	0.1	3997 (1)	1.6	12.4	17
Morocco	23.30	711.0	720 (1)	2.3	8.0	752
Mozambique	14.70	799.4	95 (n)	–	3.1 (1)	549 (1,e)
Myanma	38.30	676.6	254 (1)	16.1	3.7 (1)	487
Namibia	1.70 (1)	823.6	1014 (2)	16.0 (2)	1.0 (2,e)	65 (2,e)
Nepal	18.30	147.2	159	9.0	4.5	96
The Netherlands	14.80	41.5	14558	0.7	3.0	8171
Netherlands Dependencies	0.20 (2)	0.8	7500 (1,n)	2.6	2.5 (1,n)	195 (1)
New Zealand	3.29	268.7	9389	6.4	0.6 (n,e)	2015 (1)
Nicaragua	3.62	147.9	819	33000.0	8.0	612
Niger Republic	6.80 (1)	1267.0	318 (1,n)	6.7 (1)	6.6 (2,n)	159 (1)
Nigeria	105.00	923.8	287 (1)	3.2	4.2	2419 (1)
Norway	4.20	324.0	22457	6.7	1.1	1079
Oman	1.38	320.0	5780 (1,n)	0.0 (1)	1.6 (2,n)	439
Pakistan	105.41	804.0	362	8.8	5.8	2607
Panama	2.32	77.1	2229	0.3	25.0	177
Papua New Guinea	3.60	461.7	721 (1)	5.4	3.9 (e)	1198
Paraguay	4.04	406.7	1503	23.4	6.2	69
Peru	21.30	1285.2	1503	1191.0	7.1	122
Philippines	60.00	300.4	668 (e)	9.0	6.1 (e)	1085 (e)
Poland	37.80	312.7	1363 (m,e)	60.0	4.7 (m)	1089
Portugal	10.40	92.1	3465 (1)	9.6	4.7 (1)	5100
Puerto Rico	3.30	8.9	7762	2.8	5.2	2000
Qatar	0.35	11.4	15645	1.6 (1)	0.4 (1)	619
Romania	23.10	237.5	3456	–	3.6 (1)	1900 (2)
Rwanda	6.80	26.3	319 (1)	2.9	4.5 (2)	161
St Kitts Nevis	0.05	0.3	2119 (1)	1.0	4.0	24 (1)

World indicators 1988/89

	Population (m)	Area ('000 sq km)	GDP per capita ($)	Inflation (%)	GDP real growth (%)	Balance of trade ($m)
St Lucia	0.14	0.6	1400 (1)	1.0	5.0	82
St Vincent	0.11	0.4	1210 (1)	2.0	3.0	99
San Marino	0.02	0.1	7001 (3)	–	–	–
São Tomé & Príncipe	0.11	1.0	280 (1)	–	5.1	10
Saudi Arabia	14.00	2149.7	5535	1.0	0.0 (1)	1952
Senegal	7.00	196.2	510 (1,n)	6.4 (2)	1.3 (2,n)	447 (1)
Seychelles	0.07	0.5	3723 (1)	1.7	2.5 (n)	113
Sierra Leone	4.00	72.3	300 (1,n)	179.0 (1)	0.4 (1,n)	25 (1)
Singapore	2.65	0.6	8932	1.5	9.0	2345
Somalia	5.70	738.0	290 (1,n)	125.0	12.4 (1,n)	382 (1)
South Africa	33.80	1221.0	2574	12.8	3.2	5200
Spain	39.10	504.8	4236 (2)	4.8	5.0	18000
Sri Lanka	16.60	65.6	423 (n,e)	14.0	3.5 (n,e)	544
Sudan	23.10 (1)	2505.8	482 (1,n)	24.4 (2)	3.2 (1,n)	430 (1)
Suriname	0.40	163.3	3420	30.0	2.2 (1)	119 (2)
Swaziland	0.70 (1)	17.4	773 (1,n)	12.4 (1)	0.9 (2,n)	27 (1)
Sweden	8.40	449.0	21289	5.8	2.0	3903
Switzerland	6.51	41.3	37361	1.9	3.2	10900
Syria	11.34	185.2	1820 (1,n)	59.0 (1)	9.3 (1)	869 (1)
Taiwan	19.80	36.0	5696	5.0	7.0	10760
Tanzania	23.20	945.1	153 (1)	35.0	3.9 (1)	860
Thailand	54.50	514.0	841 (1)	3.8	8.4 (1)	2074
Togo	3.30	56.6	300 (1,n)	0.1	1.4 (1)	28
Trinidad and Tobago	1.20	5.1	3636	7.7	4.0	25
Tunisia	7.80	164.2	1286	6.3	1.5	1090
Turkey	54.10	779.5	1342 (n)	75.4	3.4 (n)	1800
Turks and Caicos Islands	0.01	0.4	3500 (n,e)	–	–	24 (3)
Uganda	17.20	236.0	260 (1,n)	60.0	5.0	316
United Arab Emirates	1.50	90.6	15527	6.0 (1,e)	2.1	–
United Kingdom	57.10	244.1	7937	4.8	2.6	24937
United States of America	246.30	9300.0	–	4.0	3.9 (n)	126780
Uruguay	3.10 (1)	176.2	2989	62.6	0.5	292
USSR	287.50	22402.2	–	–	4.4 (m)	2725
US Virgin Islands	0.11	0.4	9724 (2,n,e)	–	6.2 (2,n)	521 (2)
Venezuela	18.80	916.5	4544	35.5	5.7	1347
Vietnam	65.20	329.6	150 (2,n,e)	700.0 (e)	–	1200 (2,e)
Yemen Arab Republic	9.30	195.0	576 (1,n)	–	2.4 (2,n)	1141 (1)
Yemen PDR	2.48	333.0	420 (1,n)	–	1.2 (2,n)	–
Yugoslavia	23.60	255.8	2355 (2,m)	194.0	3.6 (2,m)	779
Zaïre	34.00	2345.4	90 (1)	100.0	2.6 (1)	563
Zambia	7.50	752.6	250 (1,n)	55.7	2.7	497
Zimbabwe	8.90	391.1	580 (1,n)	12.5 (1)	2.6 (2)	381 (1)

(1) 1987 (2) 1986 (3) 1985 (4) 1984 (e) estimated figure (n) gross national product (GNP) (m)net material product (NMP) (s)national income (NI)

The year of liberal revolution
A survey of 1989

John Toye

The year 1989 began with nothing more than a keen anticipation of the historical anniversaries which would be marked during its course. It would bring the bicentennial of the outbreak of the French Revolution, followed quickly by the 75th anniversary of the outbreak of the First World War and the 50th anniversary of the start of the Second World War. All these were indeed celebrated; the first by a competition in public extravagance between President Mitterrand and Mr Chirac, the Mayor of Paris, the second and third by the sombre reconstruction in newspapers, television documentaries and old newsreels of the tragic steps that, twice this century, took Europe and then the world to the depths of disaster and destruction. In a quiet year these pageants would themselves have been the outstanding events. But in 1989 the reality of liberal revolution transcended the memory of it, and the outbreak of peace in Europe and the world triumphed over the ritual re-enactment of the outbreak of war.

Poland: government by Solidarity

The old news footage served as a sharp reminder of the continuing centrality of Poland in European and world affairs. The rise of the Solidarity Union in the early 1980s began the process of challenging Soviet-backed state socialism in Eastern Europe even before the accession to power in the USSR of Mikhail Gorbachev in 1985. Various strategies of repression and attempted outmanoeuvring by the martial law government of General Jarazelski had failed to prevent the political re-emergence of Solidarity by late 1988 . The union was invited to a round-table negotiation with the government and other sections of Polish society, in the wake of mine and shipyard strikes in August 1988. The result of this conference, which took place from 6 February to 5 April 1989, was a bargain whereby Solidarity was to become legal again in exchange for its promise to participate in an election in which, by prior agreement, 65 per cent of seats in the lower house of parliament would be reserved for the Communist Party and its allies.

To the amazement of both the government and Solidarity, the latter won every seat but one that it contested in this curious election. A period of political shadow-boxing followed, as Lech Walesa played a very cautious hand – effectively deceiving the government about how much of its political strength remained. But by August he had succeeded in detaching the government's non-communist allies, the Democrats and the Peasants' Party, and involving them in the 'Little Coalition' with Solidarity. With this support, Walesa was able to engineer the appointment of Tadeusz Mazowiecki, editor of the union journal, as the first non-communist prime minister of an Eastern European country for 40 years. Once this had happened on 24 August, the long overture to the 1989 political revolution in Eastern Europe was completed, and the drama itself began.

Hungary: pre-empting the liberal revolution

The main protagonist was the Hungarian Government. On 10 September the border restrictions in force between Hungary and Austria were changed. This permitted tens of thousands of East Germans, who had come to Hungary 'on holiday' in anticipation of this change, to cross into West Germany via Austria. The communist government in Hungary had, ever since the 1956 rising, followed a more liberal path than other Eastern European communist regimes, and the long rule of Janos Kadar had already given way to a reformist regime in late 1988. Now the reformist government in Hungary made the highly significant decision to put its obligations under the 1972 Helsinki Agreement above its agreements with its CMEA partners on travel restrictions. East Germans, particularly young ones who said they saw no future for themselves under the communist regime, began to vote with their feet, using the first opportunity to do so since the sealing of the Berlin Wall in 1961. This haemorrhage of young workers put a pressure, which soon became intolerable, on the regime in East Germany.

Meanwhile, in Hungary itself, the reformist Communist government decided not to await a popular rebuff, but to transform itself into a Western European-style Social Democratic party. The repudiation of the monopoly of power by the Communist Party on 8 October was followed on the 19th by the establishment of a multi-party political system in which free elections were scheduled for the late spring of 1990. This adroit move by the new Hungarian Socialist Party left it in better shape for an electoral contest than any of its sister parties. Internationally it gained valuable recognition when it was admitted to the Socialist International. Internally it was able to focus on the strong current of Hungarian nationalism, as it completed the transformation in time for the 33rd anniversary of the 1956 Hungarian uprising on 24 October. By the end of the year, Hungary had approved a tough austerity budget, paving the way for a $350m IMF stand-by credit.

East Germany: last throes of a political dinosaur

The regime of Erich Honecker in the GDR had been in power for 15 years when Mikhail Gorbachev emerged in 1985 as the reforming leader of Soviet communism. Honecker made no concessions to the new climate of glasnost and perestroika, and succeeded in maintaining his old-style Stalinist system until Hungary (and then Czechoslovakia on 1 October) provided exit routes for hundreds of thousands of young emigrants. The spectacle of mass emigration was sufficient to dislodge Honecker in an internal coup, and to begin a brief attempt by his successor, Egon Krenz, to salvage the position of the Communist Party. This involved the dramatic opening of the Berlin Wall (9 November), the announcement of free elections and a move to a coalition with non-communist groups. But even the abolition of the party's leading role (1 December) failed to quieten the large street demonstrations in Leipzig, East Berlin and other major cities. The issue that would not go away was the alleged corruption of the Honecker regime (of which Krenz himself, as security chief, had been part). Krenz resigned on 3 December, having lost the confidence of everyone, including his own Communist Party, which thereupon entered a severe internal crisis. Government continued under the moderate reformer, Hans Modrow, who welcomed Chancellor Helmut Kohl on his first visit to East Germany and began a series of moves aimed at rapid development of relations between the GDR and the FRG.

Czechoslovakia: power to the people

Czechoslovakia, by providing transit for young East German emigrants during October, increased the pressure on the Honecker regime; but its collapse, in turn, suddenly switched attention and political pressure on to Czechoslovakia itself. After the opening of the Berlin Wall, demonstrations for change suddenly grew in Prague and Bratislava against Milos Jakes, the party chief of the only remaining Stalinist regime (except Romania and Albania). The initial decision was to repress this upsurge of opposition. On 17 November riot police in Prague beat up students among a crowd of 50,000 protesters in Wenceslas Square. This action only stimulated larger demonstrations. The Jakes leadership drew back from larger-scale repression, instead resigning with the entire Communist Party praesidium and secretariat on 24 November. As in East Germany, a brief interregnum followed, under Ladislav Adamec, but his move towards a coalition government was too half-hearted to satisfy the opposition led by the Civic Forum of Vaclav Havel, the playwright. Adamec resigned in the face of continuing strikes and demonstrations. So did President Gustav Husak who had been instrumental in ruling the country for the 21 years since the crushing by Soviet tanks of the liberal episode of the Prague Spring in 1968. Czechoslovakia now also looks forward to free elections in 1990 within a multi-party system. The Czech Communist Party held an extraordinary party congress which began its transformation to a radical socialist party, and Havel himself was elected to replace Husak as the Czech president.

Bulgaria: change from above

Almost unnoticed in the jubilation over the Berlin Wall opening, Todor Zhivkov, who had presided for 35 years over the communist regime in Bulgaria, was ousted by Petar Mladenov, a liberal, who was able to capitalise on anti-Zhivkov dissent within the Politburo. Some of this dissent was due to economic mismanagement, but some also arose from Zhivkov's policy of attempting to assimilate Bulgaria's ethnically Turkish Islamic minority and the international condemnation which this policy aroused after more than 300,000 (some quarter of the total) fled to Turkey during the summer. The new regime will try to edge away from active assimilation while not antagonising the anti-Turkish sentiments which Zhivkov stirred up. It will also make moves towards a mixed economy and greater political and human rights.

Romania: the repressive option fails

Until December it seemed as if Nicolae Ceausescu's regime in Romania would, because of its isolation and much-feared army and secret police, be able to ride out the upheavals in the rest of Eastern Europe. At the end of November Ceausescu was re-elected leader at a Communist Party congress which turned its face against democratic reforms. But shortly afterwards he came forward with some uncharacteristic criticisms of the regime which he had led for 24 years. These admitted economic failures as well as political mistakes in policy-making and implementation, but they did not countenance any dilution of Communist Party control. But as in Bulgaria, ethnic tensions were at work beneath the surface. In Romania's case, these centred on ethnic Hungarians in western Romania. On 17 December attempts to defend a pro-Hungarian pastor developed into more general demonstrations against the Ceausescu regime in Timisoara. According to eye-witnesses, military force was deployed against the demonstrators, at least 400 of whom were seriously injured or killed. Ceausescu himself was visiting Iran at the time. However, it subsequently became clear that repression could not overcome the wave of popular revolutionary anger and the Ceausescu regime collapsed, after only a week's fighting, on 25 December. The new National Salvation Front executed the Ceausescus and abolished the Communist Party. Elections have been called for spring 1990.

The Soviet Union: the Gorbachev revolution

The sudden domino-like collapse of most old-style communist regimes in Eastern Europe during the autumn and winter of 1989 came as a great surprise, both in terms of its scope and its speed. Taking the longer view, however, it is the mirror image of the sudden proliferation of Stalinist communist regimes 40 years ago. Both the rise and the fall of these regimes reflect the very powerful Soviet influence over Eastern European politics, although at the end of the 1980s that influence is exercised benignly by Mikhail Gorbachev. Evidence of consultation with Moscow during the regime changes has been plentiful. Such changes as have come about are in line with Mr Gorbachev's vision of a 'common European home', encompassing regimes which are more democratic, better respecters of human rights and economically restructured.

The moves towards glasnost and perestroika in the Soviet Union itself since 1985 have created quite formidable internal problems for the Soviet leadership under its new executive President Gorbachev. A common perception is that glasnost has made more progress than perestroika. Access to information has dramatically increased, not least with the televising of the proceedings of the newly elected Congress of People's Deputies since April. Transparency is an important source of strength for the Soviet reformers. When, for example, nationalist demonstrators in Tbilisi, Georgia, were attacked by Soviet troops and about 20 of them killed (9 April), secrecy was impossible. Defenders of the army's action had to identify themselves publicly, and the subsequent adverse publicity about this episode has tended to discredit them. If it was a planned provocation by conservatives in the Politburo, glasnost has helped it to back-fire.. The reshuffle of the Politburo which followed saw the position of the conservatives weakened, on balance, and their leader, Yegar Ligachev, switched to the difficult agriculture portfolio.

Economic restructuring, or perestroika, by contrast, has proved a much more difficult task. The 1987 Law of State Enterprises was due to be fully implemented by the end of 1989, but the degree of decentralisation for which it provides is not sufficiently radical to prevent the Soviet bureaucracy commandeering the bulk of production, much as usual, for distribution through administrative channels. Leonid Abalkin, the deputy prime minister in charge of economic reform, warned against the dangers of creeping back to the old system of command economy. At the same time he pointed out that Soviet public opinion was largely sceptical about the likely success of reform. The government's plan for decentralisation and price reform relies heavily on the idea of a prices and incomes policy, but there is little clear understanding of how such a policy should be operated.

Apart from the poor progress with economic reform, the Gorbachev revolution faces further dangers from the sudden liberalisation of most of Eastern Europe. Gorbachev has already marked out the limits of the changes which he is prepared to countenance. In brief, his position involves complete opposition to any changes in political boundaries; the maintenance of Communist Party hegemony in the Soviet Union (though democratised); and the maintenance of a socialist economy (though one effectively decentralised and re-invigorated). The Eastern European changes make it more difficult for him to hold to all of these policies.

Events in East Germany have suddenly revived the old question of German reunification. Although most parties recognise that even economic union is not now, and cannot be for quite a few years, a practical option, 1990 is bound to see a succession of plans for closer association or even some form of confederation between the FRG and the GDR. Other pressures for re-drawing the map will arise from the strong nationalism which recent changes have provoked in Eastern Europe and the existence of many ethnic minorities in the newly liberalised countries. This is bound to feed back into the USSR itself, where it is not clear that the stasis of nationalities implied by the 'no change' decision on Nagorno-Karabakh will hold. For example, with the fall of the Ceausescu regime, unrest in Moldavia (which is ethnically Romanian) is bound to increase.

The hegemony of the Communist Party within the USSR has already been challenged by the parliaments of Lithuania and Estonia, which have voted, against directives of the Soviet Government, to permit non-communist political parties and multi-party elections. If this becomes the norm in Eastern Europe in 1990, it will become harder to hold the line within all the republics of the USSR. (The Lithuanian Communist Party has also broken with the CPSU, in the first split since 1917.) Finally, moves by Poland and Hungary towards taking a substantial IMF loan on the condition of undertaking a package of orthodox economic stabilisation measures may – depending on its success – make it more difficult for the Soviet Union to persevere with its version of a socialist economy. If the magic of the market is seen to work in Poland or Hungary, a more radical approach to the Soviet economy may become irresistible.

China: confusion, brutality and stagnation

The Soviet Union's long-matured attempt to normalise its relations with the People's Republic of China in 1989 succeeded in formal terms, but also went dramatically wrong. President Gorbachev's visit to Beijing in May was designed to heal – after many failed previous attempts – the 30-year-long rupture between the two communist superpowers. It did so, and full diplomatic relations were restored. But it also sparked off a great wave of youthful adulation by better-off Chinese students for Gorbachev and the 'Gorbachev phenomenon' of openness, democratisation and intra-party renewal. This was symbolised by the gathering in Tiananmen Square in Beijing in May and early June of students demonstrating for democracy and liberty. The Tiananmen Square students not only disrupted the pomp and circumstance of the Gorbachev visit, they inflicted severe loss of face on the Chinese Communist Party leadership, and, through the support they enjoyed in many provinces, succeeded in splitting the leadership into pro-reform and anti-reform factions.

After a period of hesitation, Deng Xiaoping, as chairman of the military commission, put together a hardline coalition, which proceeded to massacre about 2,000 students and their sympathisers in Tiananmen Square on 4 June (oddly coinciding with Solidarity's triumph in the Polish elections). Since the massacre was televised by Western media, the international outrage was tremendous, although as early as July, President Bush sent his national security adviser to Beijing in an attempt to salvage US-PRC good relations. In the aftermath of the crackdown, the Chinese Communist Party leader, Zhao Ziyang, was relieved of his post and anti-reform Premier Li Peng emerged victorious from the power struggle. On 1 October, the 40th anniversary of the Communist Revolution was celebrated in Tiananmen Square before a very carefully invited audience.

It is probably true that the repression which has followed the massacre is less severe than previous such episodes in China's post-war history. This is in part because China is suffering an increasing power vacuum. Deng Xiaoping, at 85, has been succeeded by Jiang Zenin as Chairman of the Military Commission, but he is essentially a bureaucrat in a system which is manipulated through powerful family and kin loyalties. It is not at all clear how much power Deng retains, or how far the rise of the Yang family dynasty will be able to survive the current period of political and economic stagnation. Politically, China at the end of 1989 was paralysed by a whole gamut of cross-cutting conflicts. Economically the process of structural change has been halted, while the government struggles to get inflation and the foreign balance under control.

China has once again chosen to differ from the Soviet line, denying itself thereby any interesting options for progress. Meanwhile, the carefully negotiated agreement with the UK for the return of Hong Kong has been placed in jeopardy. Confidence in the future of the colony after 1997 has ebbed dramatically, and the UK is placed in the position of having to guess how few British passports it can grant to Hong Kong residents without further precipitating the collapse of the territory's now fragile economy. The US administration has decided to gamble on radical changes occurring sooner rather than later, but at the cost of significant domestic approval for President Bush. China's xenophobia prevents it from seeing how far the Western allies are prepared to go, in search of longer-term economic advantages, to forgive and forget the massacre in Tiananmen Square.

The United States: deficits, drugs and debt

As the socialist countries decided whether to embrace or revile democracy, in other countries the democratic process went on in its accustomed way. In the United States George Bush was elected as president in November 1988, after a campaign which he won by pinning the obnoxious label of 'liberal' on his opponent, Michael Dukakis. Once elected, however, he declared his aim to be to bring about a kinder, gentler America. In the USA not only are democracy and liberalism awkward bedfellows, but the voters do not necessarily get what they vote for – two cautionary tales which should be noted in Berlin, Warsaw, Budapest and Prague.

President Bush immediately ran into trouble with Congress when it overturned his nomination for defence secretary (John Tower) on grounds of ethics violations. This defeat reflected increased Democrat strength in Congress, and its low estimation of Bush's political skills. Although Bush triumphed on his first big international occasion, the NATO summit in late May, by recognising the strength of West German opposition to the modernisation of short-range nuclear weapons, he cut a poor figure on his visit to Poland in July because domestic budget pressures left him with little financial aid to offer the economically beleaguered Poles. A mere $60m was seen as miserly, but the size of the proposed US financial support was greatly increased after Lech Walesa's visit to the USA at the end of the year.

This embarrassment arose directly from Bush's 'watch my lips' pledge during his election campaign of no new taxes, combined with a promise to cut capital gains tax. Congressional opposition to the latter prevented more than a temporary cut, but absorbed legislative time that should have gone to serious attempts to reduce the government budget deficit. With Congress opposed to spending cuts and the president firm against new taxes, it was almost inevitable that the Gramm–Rudman–Hollings Act would come into force, as it did in October, with mandatory across-the-board spending cuts. The act has forced renewed negotiations between president and Congress to come up with a budgetary package. This has been achieved, but only after inflicting much arbitrary disruption on programmes. Moreover, some measures in the new package are window-dressing rather than real reductions in the resources absorbed by public activities. Real cuts amount to about $14bn, while some $5.5bn will be raised as additional revenue. The overall deficit will be reduced by about $17bn in the 1990 fiscal year. But a number of awkward fiscal problems have merely been postponed. President Bush should be looking keenly for expenditure savings that can be made from negotiated defence cuts.

The prospects for continuation of the arms-cutting process after the INF Treaty of 1988 were strengthened by the December summit meeting between Bush and Gorbachev in Malta. Bush was evidently pleased to be able to fix the meeting at all. He insisted that it should be a purely exploratory discussion, but then agreed to Gorbachev's proposal of an international conference on the future of Europe in 1990. This might provide a framework for further mutual reductions of personnel and equipment by NATO and the Warsaw Pact.

The war against drugs was another important plank in Bush's election platform, but action here has also been constrained by the budgetary position. Bush's approach was to be largely reactive rather than pro-active on the drugs issue. He waited until the Colombian drug traffickers murdered Luis Carlos Galan, an anti-drugs presidential candidate and the police chief of Medellin, the city which was their base, before coming forward with an offer of support for the Colombian Government in the shape of advisers and equipment. The drug barons have put up a very vigorous fight to avoid their extradition to the USA on drugs charges. General Manuel Noriega, former head of the government of Panama, was also (since February 1988) wanted for extradition to the USA on drug-smuggling and money-laundering charges. Having lost an election in May to Guillermo Enrada, Noriega declared the result invalid. On 3 October the CIA was involved in a coup against Noriega that failed. Eventually, on 20 December, a US force of 9,500 (plus the 13,000 already stationed in the Canal Zone) invaded Panama City, to widespread international protests, but failed to capture Noriega, who later surrendered, having agreed to unknown terms for doing so. All this amounted to much less than a triumph in what is admittedly an extraordinarily difficult task. One bright spot was that the drugs issue encouraged some improved co-operation between the USA and Cuba on narcotics enforcement, especially after Fidel Castro discovered and harshly punished drug-related corruption among his own colleagues.

One other aspect of US relations with Latin America was hindered by the US Government's financial constraints. In recognition that the Baker Plan to end the Third World (and particularly the Latin American) debt crisis had achieved very little, US Treasury Secretary Nicholas Brady launched a new Brady Plan very soon after the president's inauguration. The new debt plan faced many inherent difficulties, not least that commercial banks were still expected to participate voluntarily in the reduction of old debt or the provision of new money, and that banking regulations sometimes make that impossible. The main result so far has been a new debt-relief agreement for Mexico. Cynics say that this was all the Brady Plan was ever intended to do, and other deeply indebted Latin American countries cannot expect much from it. However, the USA may have to be more forthcoming on debt if it wishes to sustain the new civilian democratic regimes which emerged in 1989 in Chile, Brazil and Argentina.

The US balance-of-payments deficit has been financed in part by Japan's use of her surpluses to invest in US assets. This has caused some noticeable ill-feeling of a sort that, for example, European foreign investment does not normally provoke. The US Government has at the same time decided to take a much stronger line against Japanese trade and commercial practices which are alleged to be damaging US exports, particularly in the high-technology field.

Western Europe: watching from the wings

Like the United States, and possibly even more so, Western Europe has been largely an observer in 1989 of the liberal political revolution. With few elections, Western Europe has enjoyed political stability: even where elections have taken place, as in Portugal, they did not generally bring major change. Greece is an exception, where the PASOK Party was finally rejected by the electorate. Much sympathy has been expressed with the liberal advances in the East. Some practical help has also been given – particularly by the FRG, which has absorbed hundreds of thousands of refugees from the East. More has been promised, in the form of financial aid or investment once this is permitted by the new Eastern European regimes. Less directly, the common view has been taken that the further strengthening of the European Community would be helpful to the new regimes in the East by eventually providing forms of association that would strengthen their liberal tendencies.

The leading role of West Germany has been emphasised in 1989, and especially Helmut Kohl's opposition to the modernisation of short-range nuclear weapons at the spring NATO summit and his skilful handling of the arrival of East German refugees. Since the opening of the Berlin Wall, he has been faced with strong emotional demands from both sides for the unification of Germany. His response has been affirmative with the important riders of 'not yet' and 'only within a European framework'. This is such a crucial and sensitive issue for Europe that it is difficult to exaggerate its importance. German nationalism is bound to flourish, but even after 50 years the old fears in France, Poland and the Soviet Union lie only just below the surface. The aim of unification will in practice prove very difficult, if only because for the last 40 years the two Germanies have been deliberately developed as models of very different economic and social systems, and their populations have been segregated from each other to an almost unheard of extent among neighbouring states. Whether Helmut Kohl can manage to handle the inevitable frustration which German nationalism must suffer in the early 1990s will be one of the key issues of the decade.

Apart from celebrating their own 1789 Revolution, the French have contributed mainly through their leadership in the European Community's plans for further integration. Progress towards the single market in 1992 requires the solution of several outstanding problems – for example, the harmonisation of public procurement practices and policies over the control of industrial mergers. The French hoped to have these problems resolved before the end of their EC presidency on 31 December. President François Mitterrand proposed a new treaty to amend the Treaty of Rome to permit the completion of European economic and monetary union. This proposal, along with a rather weak Social Charter for Europe, was welcomed by all EC member states at Strasbourg in December, with the exception of the UK.

The UK fought a rearguard action against the programme of further European integration, preferring instead the concept of co-operation between member nations. The pound sterling has not joined the European Exchange Rate Mechanism, and the internal debate in the ruling Conservative Party about when and on what conditions it should do so led to the resignation of the Chancellor of the Exchequer, Nigel Lawson. This was a major factor in the downturn of Margaret Thatcher's political fortunes after celebrating 10 years as prime minister on 4 May. Even the attempt to switch to 'green' or environmentally sound policies has not stemmed her loss of support: the unfavourable economic situation influenced voters more, and looks set to do so throughout 1990.

Japan: Liberal Democrats in the doldrums

Japan has a liberal and democratic system, and indeed has been ruled since the end of the Second World War by the Liberal Democratic Party. The LDP's 34 years of unbroken rule have been accompanied by a growing number of corruption scandals. In an even more widespread and murkier deal than the Lockheed bribery scandal which drove Prime Minister Kakuei Tanaka out of office, Prime Minister Noboru Takeshita was finally forced to resign in April as the Recruit shares-for-favours affair implicated more and more of the leading LDP figures. His hapless successor, Sosuke Uno, who almost entirely lacked political support, did not last long, being driven from office by a sex scandal. In the July election, the Socialist Party, led since 1986 by Takako Doi, focused public anger about the new consumption tax and scored a remarkable triumph in more than doubling its number of seats in the upper House of Representatives. The lower house remains overwhelmingly controlled by the LDP (with a 295-seat majority), but new elections are due soon. A socialist win in the lower house looks rather unlikely, not least because the Socialist Party itself has reservations about Ms Doi's strategy of going all out to mobilise women voters against the corruption and sexism of the LDP, but the LDP cannot now take anything for granted.

Despite the political setbacks of the government party, Japan continues to flourish economically, reinvesting its huge balance of payments surpluses abroad, and particularly in the USA. The policy of trying to reduce the trade surplus has not had much success. Against this background, the USA has taken much stronger action against Japan in relation to unfair practices against US exports. Questions about how Japan and the USA compare in economic strength and which is the greatest economic power in the world have soured the atmosphere, and Japanese complaints of US 'racism' are not entirely without foundation, even if, coming from a xenophobic culture, they are hard to accept. The ability of the new Prime Minister, Toshiki Kaifu, to project a strong international image for Japan is weakened by the LDP's domestic electoral problems. This is the key to Japan's difficulties in aiding financially the new governments in Eastern Europe.

India: elections bring a new government

India, like Japan, is a democracy which in 1989 saw its conventional assumptions about a dominant political party rocked by election results. The Congress Party has been in power in the central government for all but two (1977-79) years since independence in 1947. It was always likely that its leader, Rajiv Gandhi, would poll worse than he did in 1984, when voters gave him a landslide victory in sympathy for the Sikh assassination of his mother. Much of the Indian politics of 1989 was a lead-in to what many saw as a possibly epoch-making contest. Mr Gandhi announced radical populist programmes of rural decentralisation. His opponents strained to tar him with the alleged Bofors corruption scandal, and to compose their own differences. Both sides flirted with the politics of Hindu – Muslim tensions, and Muslims were alienated from their traditional support of Congress.

In the event, Congress lost massively and could not form a majority government unaided. The opposition combined to do so and Mr V.P. Singh emerged as its leader and the new prime minister. Although Congress remained the largest single party, with just over 200 seats, both the Hindu BJP party and the various marxist parties refused to back a Congress government, and instead opted to support a minority government of the National Front. The new government's first actions have been to appoint an ombudsman with powers to investigate cases of high-level corruption, and to facilitate the writing-off of debts on rural loans. Much now depends on whether the Congress rump will split, with discontented groups shifting allegiance to the National Front. Mr Gandhi's decision to remain as Congress leader is said to increase the likelihood of that happening. On economic policy, Mr V.P. Singh's record as finance minister in 1985-87 shows him as favouring the liberalising of trade and industrial controls. He has yet to show whether he will push that strategy vigorously as prime minister.

The Middle East: pragmatism and peace hopes

After 10 years of theocratic power, the Ayatollah Khomeini departed from Iran's political scene, with the same large and frenzied crowds mourning his death as welcomed him back from exile after the Shah's overthrow in 1979. His spiritual successor, Ayatollah Ali Khameni, is a much younger man who cannot expect to exercise the same kind of spiritual authority as the leader of the clerical revolution. A reappraisal of policies was therefore easier for the incoming president to make, and the presidential position was itself strengthened so that its holder exercised full executive power. The overwhelming victor in the August presidential election was the former speaker of the parliament, Ali Akbar Hashemi Rafsanjani. He presents a markedly more technocratic and pragmatic approach to politics than has been usual in the last decade.

On the economic front, a start in the direction of economic liberalisation ran quickly into the opposition of business interests which benefit from the current set of controls. Suggestions that Iran may need to seek foreign loans or foreign investment to help it through an adjustment process were not well regarded by the clergy, who still maintain the ideology of 'foreign devils'. While economic reforms are stalled, the population continues to grow very rapidly. Corruption, theft and drug abuse remain widespread and there is a strong undercurrent of popular frustration that things are getting no better. Iran still relies very heavily on oil exports for foreign exchange, and is thus tied into the soft oil market with only limited ability to increase its own rate of production. Despite the Salman Rushdie death threat affair, and the occasional terrorist atrocities which are claimed to be traceable back to Iran, it would be hard to interpret 1989 as a year which saw significant advances for Islamic fundamentalism.

As far as the Arab–Israeli conflict over the West Bank and Gaza Strip is concerned, 1989 brought some slightly hopeful signs. The Palestinian uprising (*intifada*) was contained by the Israeli military, but only with difficulty: there is a growing recognition in Israel that military containment cannot be a permanent solution.

The PLO now recognises the existence of Israel, but expects more states to recognise the Palestine National Council. The Soviet Union seems to have been working to influence President Assad of Syria to fall in line behind Egypt as leader of the Arab world, and to end the 12 years of Syrian–Egyptian hostility which followed Egypt's acceptance of the Camp David accords of 1977. If the Soviet Union can bring about a much greater Arab unity behind an Arab–Israeli peace process, it will only remain for the USA to exert real pressure on its ally, Israel, for greater progress towards peace to be made in the region. That even this most recalcitrant of all regional conflicts shows some small signs of yielding is another important measure of how far, in 1989, peace has broken out.

John Toye
Institute of Development Studies
Brighton, UK
December 1989

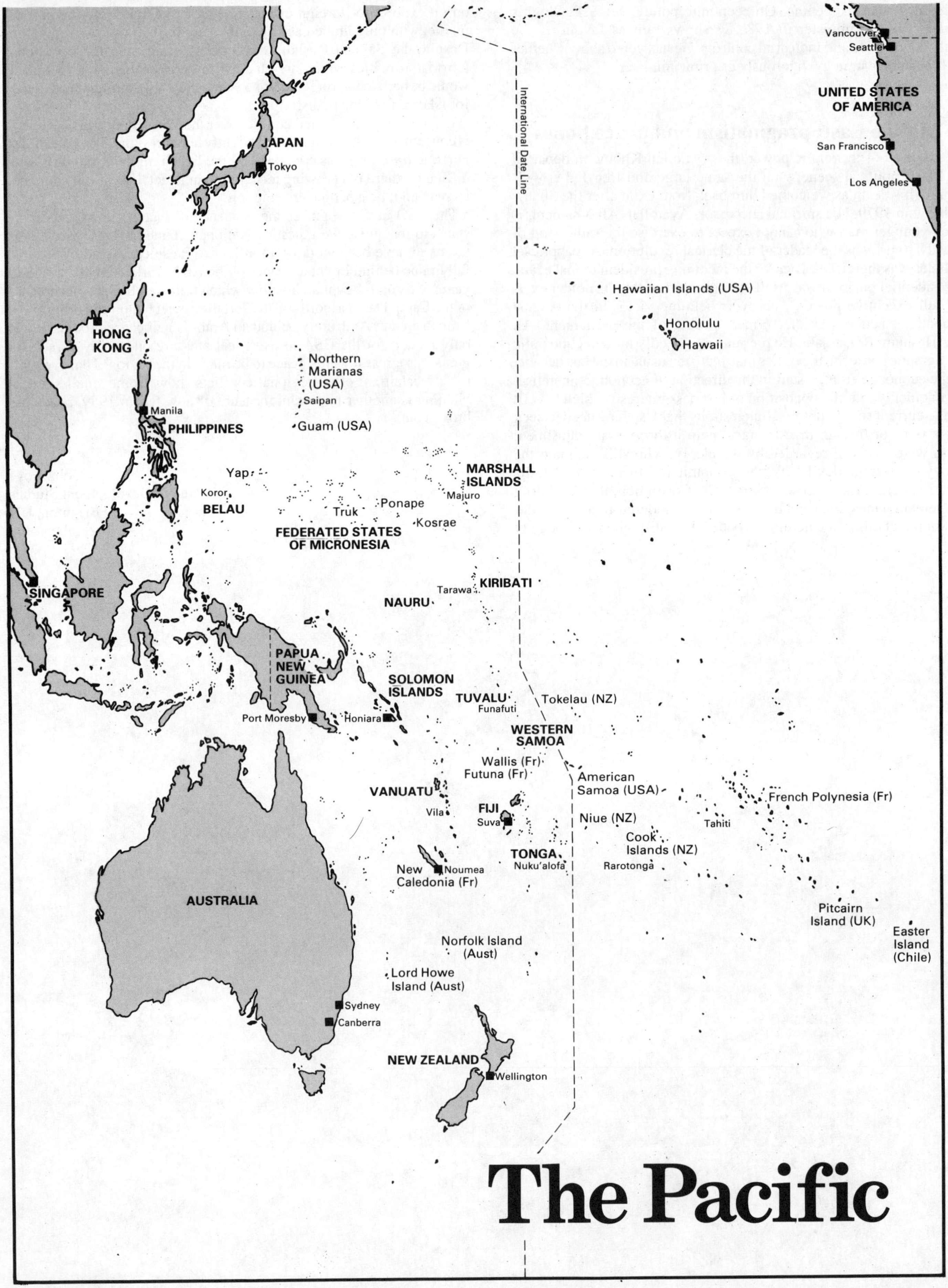
Vancouver
Seattle
UNITED STATES OF AMERICA
San Francisco
Los Angeles
International Date Line
JAPAN
Tokyo
Hawaiian Islands (USA)
Honolulu
Hawaii
HONG KONG
Northern Marianas (USA)
Saipan
Manila
PHILIPPINES
Guam (USA)
Yap
MARSHALL ISLANDS
Majuro
Koror
BELAU
Truk
Ponape
Kosrae
FEDERATED STATES OF MICRONESIA
SINGAPORE
Tarawa
KIRIBATI
NAURU
PAPUA NEW GUINEA
SOLOMON ISLANDS
TUVALU
Funafuti
Tokelau (NZ)
Port Moresby
Honiara
WESTERN SAMOA
Wallis (Fr)
Futuna (Fr)
American Samoa (USA)
VANUATU
French Polynesia (Fr)
FIJI
Vila
Suva
Niue (NZ)
Tahiti
Cook Islands (NZ)
TONGA
Nuku'alofa
Rarotonga
New Caledonia (Fr)
Noumea
AUSTRALIA
Pitcairn Island (UK)
Easter Island (Chile)
Norfolk Island (Aust)
Lord Howe Island (Aust)
Sydney
Canberra
NEW ZEALAND
Wellington

The Pacific

Afghanistan

Anthony Hyman

The Soviet troops' withdrawal from Afghanistan was completed by February 1989, but it did not bring peace. Instead, the Afghan war looked set to go on indefinitely, with little hope of a compromise political settlement or more than lip-service being paid by the superpowers to the main points in the UN-sponsored Geneva accords. Neither the Kabul government nor the guerrilla resistance forces seemed in a position to win a decisive victory in this bitter struggle, which developed massively in scale from 1980, after the Soviet military invasion.

President Najibullah moved to declare a state of emergency only four days after the last Soviet troops left, suspending the constitution and setting up a supreme council with a large military element. This replaced most of the non-PDPA (People's Democratic Party) members of the cabinet, who had been inducted supposedly to broaden the government. At the same time, some well-known communist figures who had earlier been retired returned to senior posts, notably Soltan Ali Keshtmand, (prime minister 1981-88), and Mahmud Barialay, a half-brother of deposed President Babrak Karmal.

Predictions that the PDPA regime would rapidly lose control of the towns and then conveniently disintegrate in the wake of Soviet military withdrawal proved wishful thinking. Though tensions continued between Khalq and Parcham factions of the party, with many arrests made in Kabul from February 1989, it held together. The army, militias and air force in fact fought creditably, proving they were no longer so dependent on Soviet forces or advisers. Thousands of rocket attacks on Kabul killed mainly civilians, though some military targets were also hit in 1989.

A massed guerrilla assault on the town of Jalalabad from February 1989 proved to be badly led and uncoordinated. It caused great damage to the town, but completely failed in its objective. The garrison town of Jalalabad, lying strategically at the halfway point on the highway between the Afghan capital and the Pakistan border, would have made an ideal base to install the opposition HQ, and from which to go on to challenge the regime's control of Kabul. But the failure to take it, and heavy losses among mujaheddin (fighters for the faith) exposed to devastating air attacks, badly demoralised the opposition.

Political manoeuvres

The PDPA government made much of Pakistani involvement in the fighting around Jalalabad, claiming the entire operation was directed and ordered by Pakistan. The regime tried hard to drive a wedge between field commanders of the guerrillas and leaders of exile parties. Agents of KHAD, the KGB-trained Afghan secret police, were actively trying to make deals with local commanders, and to exploit traditional ethnic and tribal rivalries. The main strength of the PDPA regime lay in its organisational capacity, especially when contrasted with the largely chaotic resistance.

Opposition leaders of the seven main Afghan parties based in exile proved quite inadequate to exploit the delicate political situation in February 1989. Attempts to produce opposition unity by declaring an alternative government from exile proved a fiasco. The so-called *Shura* (consultative council) in Rawalpindi was dominated by nominated representatives of the seven Peshawar-based parties. They failed to reach agreement between themselves, let alone with the Shi'a Muslim parties on their representation in the government, as had been hoped.

Blatant interference in the Shura by Pakistan's ISI, (Inter-Services Intelligence, a powerful military agency), provoked a strong counter-reaction in Afghan nationalist circles. As a result, few guerrilla commanders were willing to attend the Shura. However, an interim government was formed, and it was recognised by Saudi Arabia, Bahrain, Sudan and Malaysia. Its representatives took up the vacant seat of Afghanistan at the Islamic Conference Organisation in March 1989, though wider international recognition looked unlikely unless mujaheddin forces succeeded in taking the capital.

Both superpowers appeared committed to continue supplying plentiful supplies of weapons – at least until the beginning of 1990 – to their respective sides in this proxy war. The key regional states, Pakistan and Iran, continued to press the cause of the resistance, though with emphasis on different parties. Though Benazir Bhutto's new government in Pakistan was under considerable domestic pressure to reach a settlement with Kabul and to get a return of Afghan refugees, it was far from certain if Islamabad's policy had altered and become more independent of Washington. The prospect for 1990 was more war.

Reports from independent observers strongly suggested, though, that disillusionment was spreading among the Afghan population at the in-fighting between guerrillas and the sheer ineptitude of the opposition leaders. Political rivalries within the resistance caused a well-publicised ambush and killing in the northern province of Takhar in July of 30 commanders and fighters of Jamiat-i-Islami by guerrillas of Hizb-i-Islami of Gulbuddin Hekmatyar, the party

KEY FACTS

Official title: Democratic Republic of Afghanistan

Head of state: President of Revolutionary Council (acting): Dr Najibullah

Head of government: Chairman of Council of Ministers, Soltan Ali Keshtmand

Ruling party: People's Democratic Party; PDP leader: Dr Najibullah

Capital: Kabul

Official Languages: Pushtu, Dari

Currency: Afghani (Af = 100 puls)

Exchange rate: Af62.8 per $ (Nov 1989)

Area: 647,497 sq km

Population: 15.5m* (1988)

GDP per capita: $170* (1987)

GDP real growth: 3.5%* (1985)

Trade balance: –$196.0m (1988)

Foreign debt: $1,499m (1987)

* estimated figure

consistently favoured by Pakistan's ISI. Jamiat leader Professor Burhanuddin Rabbani demanded the suspension of his colleague in the interim government, also demanding that the Hizb local commander responsible for the massacre be handed over to an Islamic court for trial. The factor of war-weariness was bringing pressure to wind down the fighting. Many Afghans living inside the country, as well as 5m refugees in exile, were anxious for a compromise peace. Some clearly believed this war was no longer a true *jihad* (or holy war), since the departure of Soviet troops. There were reported difficulties in finding volunteers from the refugee camps in Pakistan for the Jalalabad front, where hundreds of children were seen in mujaheddin ranks.

Rebuilding Afghanistan

Under UN auspices, a programme was being prepared in 1989 to help pay for rebuilding Afghanistan's shattered economy. The chief UN co-ordinator for the return of Afghan refugees, Prince Sadruddin Aga Khan, launched the international appeal for $2,000m in October 1988. The response by August 1989 was poor, with aid commitments almost entirely in kind or in roubles, as in the case of the USSR. Many potential donor states preferred to stick to their bilateral projects, or support independent humanitarian organisations already working for Afghan relief.

Total damage is claimed to be many times this figure. (Soviet officials in Kabul in August 1988 calculated the cost of rebuilding the shattered economy at $10bn.) Reconstruction will inevitably be a painfully slow process. It can only really begin with peace, and after millions of Afghan refugees presently in Pakistan, Iran and elsewhere start to return home. In 1989, small groups of refugees returned across the Afghan borders from Iran and Pakistan, but at the same time scores of thousands of fresh refugees fled to Pakistan, after heavy fighting in border areas.

Unless a genuine coalition government is formed in Kabul, bringing with it peace together with some form of stability over the country as a whole, the vital but daunting task of reconstruction will continue to be hindered by international politics. While the USSR showed its determination to back up its client regime in Kabul, the common policy of the Western-bloc states appeared to be to block any financial help which could be of assistance to its survival. Thus the genuine humanitarian needs of the Afghan people risk being ignored.

The 11-year war ravaged the Afghan countryside. Thousands of villages and much of the country's livestock were destroyed, along with much of the infrastructure built over previous decades of steady development. Afghan agriculture is in sharp decline, with crops down by an estimated two-thirds in worst affected areas. Agronomists vary in their estimates of the precise impact of war conditions on crop production in the regions, but they are in agreement on the urgency of reconstruction.

Agriculture was traditionally by far the most important part of the Afghan economy, providing around 55 per cent of GDP. Serious damage has occurred to the country's vital irrigation system of *qanats* (underground water channels dug and maintained by communal effort). Since 1979, at least 1m Afghans have died from the war, directly and indirectly, according to conservative estimates.

The exodus since 1979 of some 4.5m refugees across the borders to Pakistan and Iran – together with over 1m more displaced persons fleeing to safety in Afghan cities or remote rural areas – has resulted in thousands of villages becoming deserted or acutely short of labour and livestock.

This damage was aggravated by war operations, scorched earth tactics and reprisals by Soviet-Afghan forces. Destruction of crops and livestock by bombardments and air raids was particularly severe in

KEY INDICATORS

	Unit	1984	1985	1986	1987	1988
Population	'000	16,000*	15,000*	15,000*	15,220	15,510
Gross domestic product (GDP)	Af bn	150.3	155.6	160.0*	–	–
GDP per capita	Af	8,383	10,120*	10,666*	–	–
Consumer prices	1980=100	116.0	126.6	–	–	–
Manufacturing production						
Cement	m tonnes	0.12	–	–	–	–
Wheat flour	m tonnes	0.15	–	–	–	–
Casing	m coils	1.55	–	–	–	–
Agricultural production						
Wheat	m tonnes	2.86	2.85	2.75	2.80	2.90
Maize	m tonnes	0.80	0.80	0.75	0.03	0.03
Barley	m tonnes	0.33	0.33	0.33	0.33	0.35
Rice paddy	m tonnes	0.48	0.48	0.48	0.48	0.49
Potatoes	m tonnes	0.32	0.31	0.34	0.35	0.36
Sugar beet	m tonnes	0.04	0.005	0.02	0.02	0.02
Sugar-cane	m tonnes	0.07	0.07	0.07	0.08	0.08
Energy production						
Coal	m tonnes	–	–	–	–	–
Natural gas	m cu metres	2,900	2,900	2,900	2,790	3,000
Electricity	bn kWh	–	1375	–	–	–
Exports	$m	632.9	556.8	551.9	511.9	403.0
Fruits and nuts	$m	181.9	120.0	134.9	128.6	–
Karakul skins	$m	9.2	10.5	10.5	8.3	–
Natural gas	$m	314.3	309.4	259.6	204.5	–
Carpets	$m	42.2	58.4	39.5	48.9	–
Cotton	$m	22.4	19.2	9.8	8.9	–
Imports	$m	1,389.5	1,194.2	1,403.5	995.9	599.0
Balance of trade	$m	–575.1	–637.2	–851.6	–484.0	–196.0
External debt	$m	872.0	1,122.0	1,289.0	1,499.0	–
Debt service ratio	%	4.9	4.2	3.6	3.4	–
Exchange rate	Af per $	50.60	50.60	50.60	50.6	50.6

* estimated figure

Afghan provinces lying close to the Pakistan border, around the capital Kabul, in Herat and other fertile northern regions.

At the same time there was through the 1980s a drying-up of access to improved seed and fertiliser from state agencies for the great majority of Afghan farmers, living in areas outside the control of the Kabul regime. Production of fertiliser is, however, continuing. The modern fertiliser and thermal power plant at Mazar-e-Sharif, capital of Balkh province in northern Afghanistan, reportedly increased its annual fertiliser production to 120,000 tonnes in 1987.

This is the only petrochemical plant in the country, though another larger one was planned in 1983. It employs 2,216 workers. The Mazar-e-Sharif plant is currently being expanded with Soviet aid. It produced 61,500 tonnes of fertiliser and 120,370,000 kwh in the first half of the Afghan year from 21 March 1988.

A Five-Year-Plan announced for 1986-91 is effectively a dead letter. In reality, the 'Plan' was more a list of ideal targets for peacetime, for the weakly developed heavy industry and energy sectors. Development has continued in the 1980s, against all the odds, with aid from the USSR and other CMEA-bloc states, but actual performance of most industrial and manufacturing concerns is at a very low ebb.

Two parallel economies have come into being. That outside the control of the Soviet-backed government is largely self-sufficient but fragile, with past trading links with Kabul and the other cities often disrupted by war. Wheat is commonly cultivated in the plains and river valleys so as to feed the local population with the staple food of bread. In some northern areas, which used to grow cash crops of cotton, commanders of the mujaheddin ordered farmers to grow wheat instead.

Kabul and other towns under the Kabul government's control are now acutely dependent upon imported foodstuffs and other supplies transported by trucks across the Hindu Kush mountains along the Salang highway route from the USSR border. Soviet wheat forms an important part of the diet of Afghans living in the capital. Estimates of the annual imports of wheat from the USSR needed to feed the swollen population of the capital are as much as 500,000 tonnes.

The population of the Afghan capital, according to a Kabul Radio report of February 1989, was 1,420,000, living in 215,000 households. A census carried out in Kabul at the end of 1986 had placed the population at 1.287m – close on double the pre-war estimated figure, but still considerably less than many independent estimates, some of which place it at over 2m.

Trade patterns

Afghanistan's trade balance, still in modest surplus in the early years of the war, slipped into regular deficit. The private sector was responsible for a large part of foreign trade, earning the bulk of the state's foreign currency, in addition to 80 per cent of domestic trade. According to Syed Amanuddin Amin, a deputy prime minister, the private sector accounted in 1987-88 for 53.5 per cent of exports and 49.9 per cent of imports into Afghanistan. Mr Amin also admitted in 1988 that the value of exports had fallen during the previous four years.

This was mainly due to the PDPA regime's increased dependence upon the USSR. The economies of the two countries have become inextricably linked. In return for 97 per cent of Afghanistan's natural gas production, the USSR provides most of Afghanistan's energy needs and much of its machinery, consumer goods, wheat and other food items. Soviet officials stated in Kabul in August 1988 that Afghanistan owed $12bn to the USSR.

Afghanistan was almost totally dependent for its industrial progress upon the USSR, Czechoslovakia and other CMEA states of eastern Europe. All Afghanistan's main Western aid-givers – West Germany, France, the USA – plus China, Japan and all non-CMEA states (with the exception of India) cut off their aid programmes in 1980, after the Soviet military invasion. Western funds were diverted to UNHCR and other agencies providing humanitarian relief for Afghan refugees in Pakistan.

Soviet loans and aid in the five years to 1987, at a reported total of $573m, represented 74 per cent of all foreign aid. This does not take account of Soviet military loans, to a value estimated at $6-8bn.

The economies and infrastructure of Afghanistan were becoming steadily merged with those of Central Asia. Most of the mineral resources so far discovered lie north of the Hindu Kush mountain barrier, and most industrial centres also lie in the north. Over 320 large and small industrial projects with Soviet economic and technical assistance had been built or signed for by 1988. Of these, almost half were already in operation. Plants built with Soviet aid accounted for 60 per cent of manufacturing output and 75 per cent of the state sector's output.

Permanent links with the USSR by road and rail are of obvious geo-strategic and military as well as of economic significance. New airports have been built close to the Afghan borders, or greatly expanded, especially for the capital Kabul. The first bridge across the Amu Darya (Oxus) river border was opened in 1982, and a railhead established on the Afghan side of the border. However, progress with construction work for the projected railway, as well as pipelines and some industrial projects were hindered by guerrilla attacks. A big planned expansion of power generation will be by means of a link-up to the USSR's electric system grid.

Exploitation and exploration of Afghanistan's considerable mineral riches in natural gas, oil, coal and copper is going on steadily, in spite of wartime dangers. In the past decade the gas industry became by far the most important item for state revenues, accounting for some 90 per cent of the total in recent years. A full 97 per cent of the country's annual gas production is exported by pipeline to Soviet Central Asia.

Domestic gas production was expanded by the opening in 1986 of the Jarquduq-e-Shebergan gas condensate refinery, built with Soviet aid. Its annual production target was 4,000 tonnes condensate, 3,000 tonnes diesel and 1,000 tonnes fuel. A three-year programme worth R33m for geological oil and gas surveys in the northern provinces with the USSR was announced for 1987-89. Natural gas production has remained little changed through the 1980s, at between 2.3 and 2.6bn cubic metres.

Oil exploration is firmly under Soviet direction, and geologists believe oil reserves are relatively large. All petroleum products, though, are imported from the USSR. Afghanistan agreed to purchase 547,000 tonnes of petroleum, kerosene and diesel fuel during 1987.

Problems in coal production, centred at Pul-e-Khumri, were acknowledged indirectly in official statistics, which gave production for the last nine months of 1988 as just 30,000 tonnes. Coal mining had reportedly reached 150,000 tonnes in the Afghan year 1986-87. Coal-mining equipment has been supplied mainly by Czechoslovakia, the most important and by far the best-established of the other East European states trading and investing in Afghan industries. Bulgaria, East Germany and Poland were increasingly active in industrial as well as agricultural sectors of the Afghan economy.

Dried fruit and nut exports are the main hard-currency earners, and the second largest export item, after gas. Sales in western markets fell sharply after 1983-84 (when exports were worth $192m), but had recovered ground by 1987. 40,000 tonnes of dried fruit were exported to a value of $59m in the half year from 21 March 1986. The market trend was for sales of dried fruit (mainly raisins) to increase in the USSR, eastern Europe and Pakistan rather than in western Europe. Fresh fruits went mainly to the USSR and India.

Cotton, wool and karakul (Persian lambskins) are all traditional exports which languished in the 1980s. The export potential of carpets and the handicraft industry was recognised in a number of state initiatives. The Industrial Development Bank of Afghanistan was set to provide a total of Afs2,660m ($53m) to promote the handicraft industry, over the period 1986-91. Successive annual falls in export levels of carpets and rugs were halted by 1987. Over 800 private traders are grouped in the Afghan Carpet Exporters' Guild. Carpet weaving is an important source of rural employment, mainly of women and girls, providing extra cash for many families in northern Afghanistan.

COUNTRY PROFILE

Political system: Politics in Afghanistan is complicated by the bloody ongoing civil war between a Soviet-backed government and fundamentalist Muslim guerillas. More than five million of the 17 million people fled across the borders to Pakistan and Iran. The position became more complicated in 1989 after the 115,000 Russian troops completed their withdrawal and United Nations negotiators failed to form a broadbased national government that would be neutral politically. The Kabul government under Sayid Mohammad Najibullah had a Soviet-style politburo elected by the 36-member Central Committee of the People's Democratic Party. In November 1987 the Soviet-backed government held a loya jirga (grand tribal council) which set up a new constitution with a president (Najibullah) elected for seven years. Elections were held in 1988 for the assembly dominated by Najibullah's party. A Shura (Assembly) was held in the Pakistani city of Rawalpindi in February 1989, attended by 60 representatives of each of the seven Afghan opposition parties based in Pakistan, together with representatives of Shi'a Muslim parties and a few independent non-party Afghans. It failed to unite the rival parties, but the interim Afghan government declared at the end of the Shura was recognised by four Muslim states and it took the seat held vacant for Afghanistan since 1980 in the ICO (Islamic Conference Organisation). UN proposals were for a ceasefire and loya jirga in March 1989 which would elect a broadbased government. But Muslim guerrilla parties were opposed to co-operating with the Najibullah regime.

Main city: Kabul (population 2m* in 1989).

Media
Radio: There are domestic and external radio broadcasts in a number of languages.
Television: There is a national TV network.

Domestic economy: The economy particularly has been badly disrupted by warfare and the flight of refugees. Even before the disruption, Afghanistan was a traditionally pastoral country with a large nomadic population (estimated about 15 per cent). Being landlocked, with much of the territory consisting of high plateau desert and mountains, communications have always been an impediment to economic progress. But there were hopes for development of agriculture in the fertile valleys and investment in mineral and hydro-electric resources. With the Soviet invasion in 1979, Afghanistan began to rely almost exclusively on aid from the Soviet Union and CMEA. According to figures supplied to the ADB, growth of gross domestic product was an estimated 0.9 per cent a year between 1983 and 1987. Agriculture accounts for almost 70 per cent of GDP and industry for 17 per cent. The government claimed that GDP in 1987 was 11 per cent higher than in 1978.

Employment: Lack of skilled labour has been an impediment to development.

External trade: Trade is controlled by permits and much trade is undertaken on a countertrade basis, which is difficult to quantify. Changes have occurred in the direction of trade and the importance of different sectors in recent years. In 1988 the government announced freeing of trade allowed through private concerns.
Exports: Total exports were valued at $483m in 1987. Main exports are natural gas (typically, 36% of the total), fruits and nuts (22%), carpets (14%), Karakul skins (5%) and cotton. Main destinations are USSR, CMEA countries, UK, German FR, USA and the Netherlands.
Imports: Total imports were valued at $689m in 1987. Main imports are food (wheat, sugar, beverages), petroleum, fertilisers, basic manufactured goods and vehicles. Main sources are USSR and CMEA countries.

Agriculture: Agriculture contributes about 60 per cent of GNP and 77 per cent of export income, although much food is imported. Over 65 per cent of the labour force is employed on the land, mainly cultivating smallholdings or herding sheep, cattle and goats. Animal husbandry was up 9 per cent in 1987. 12 per cent of total area is cultivated, another 10 per cent is pasture, further 5-6 per cent considered by some sources to have agricultural potential. Most cultivated land situated in river valleys or plains which are often fertile; irrigation is carried out on an estimated two-thirds of cultivated land and yield per ha has been good. But food output was 15 per cent below what was required to feed the existing population. In total more than 12,000 farming villages had been abandoned or destroyed out of 22,000. Main crops are wheat, fruit and vegetables, maize, rice, barley, cotton, sugar-beet, sugar-cane, oil seeds. Livestock raised include sheep, cattle, goats and poultry, with donkeys, horses, camels, mules and buffaloes kept as draught animals. Sheep provide a major source of protein and animal fat. Some 70 per cent of wool production, along with hides from Karakul sheep, is exported. Most exploitation of forest resources is for fuel wood.

Industry: Manufacturing relies on the agricultural sector for much of its raw materials. Industrial production was up 15 per cent in 1987. Major products include cotton, wool and synthetic textiles, carpets, footwear, leather goods, dried fruit, cottonseed oil, furniture, prefabricated buildings, glass, bicycles, mechanical spares, fertilisers and other industrial chemicals and plastics. Czechoslovakian aid is committed to developing a cement factory. A new oil refinery is planned in Jowzjan.

COUNTRY PROFILE

Mining: Afghanistan's potential mineral wealth is probably great, and there has been considerable prospecting in recent years. Resources currently exploited include coal, natural gas, salt, lapis lazuli, talc and barium sulphate. Czechoslovakian aid is committed to developing various mining projects. Long-term mineral development projects include copper mining and smelting at Ainak and high-grade iron ore mining at Hajigak.

Hydrocarbons: Natural gas production has increased to 2.4bn cubic metres with about 97 per cent exported to the USSR via a 180 km pipeline. There are reports of additional finds of natural gas and oil.

Energy: Hydroelectric power is generated in small but increasing quantities. There is a distribution point at Hairaton for electric power from the USSR. A tiny percentage of natural gas production is consumed domestically.

Membership of international organisations: Asian Development Bank, CCC, Colombo Plan, ESCAP, FAO, G-77, IAEA, IBRD, ICAO, IDA, IFAD, IFC, ILO, IMF, INTELSAT, Islamic Development Bank, ITU, NAM, UN, UNESCO, UPU, WFTU, WHO, WMO, WTO, WSG.

BUSINESS GUIDE

Time: GMT + 4.5 hrs.

Climate: The climate is generally dry and varies according to altitude. In Kabul (altitude 1800 metres) winter temperatures can fall below -20°C, with average temperatures of 32°C in Jul-Aug. Lower regions are warmer; summer temperatures can reach 40°C plus. Most rainfall from Mar to May (Kabul average 335 mm).

Entry requirements
Passport: Required by all.
Visa: Visas are required by all with few exceptions. It is not possible to obtain a tourist visa. Exit permits required.
Prohibited entry: Nationals of Israel and South Africa are forbidden to enter.
Currency: The currency has not been convertible overseas. Foreign currency may be taken into the country, but controls are imposed on entry and exit. Up to Afghani 500 may be exported. All foreign exchange transactions must be through the Central Bank of Afghanistan.

Health precautions
Mandatory: Vaccination certificates for cholera if travelling from an infected zone.
Advisable: Typhoid inoculations and anti-malarial precautions. Water should be boiled before drinking.

Representation overseas: Ankara, Baghdad, Beijing, Belgrade, Berlin, Bonn, Dacca, Havana, Jakarta, Jeddah, Kuwait City, London, Moscow, New Delhi, Paris, Prague, Rome, Sofia, Tokyo, Warsaw, Washington.

Air access
National airline: Ariana Afghan.
Other airlines: Aeroflot.
Tax: An airport tax of Af 200 is levied on departure.
International airport: Kwaja Rawash (KBL), 16 km from Kabul, bank, buffet, post office, restaurant.

Surface access
Main ports: There are no sea ports.
Road: There are links to Iran and Pakistan via the Asia Highway, with a bus service, and to the USSR via road and rail. Hostilities have periodically closed the Pakistan route; check before travelling.
Rail: Links between Kabul and the Soviet Union.

Hotels: All hotel booking is through the Afghan Tourist Organisation. Tipping is not usual. A five per cent government tax is levied.

Car hire: International driving licences are required if cars are available, possibly through Afghan Tourist Organisation.

City transport
Taxis: Tipping is not usual; a fixed rate system has been operated.

National transport
Air: The domestic airline, Bakhtar Afghan Airlines, serves main cities (Kabul, Herat, Kandahar, Mazar-i-sharif).
Road: Main centres are linked by paved roads but secondary roads vary in condition and by season. There were 22,000 km of roads in 1986.
Buses: Bus service is available, but unreliable and dangerous for internal travel.
Water: There are 1,200 km of navigable inland waterways, including the Amu Darya River.

Public holidays
Fixed dates: 27 Apr (Revolution Day), 27 May (Independence Day), 19 Aug.
Variable dates: Lunar New Year, Ramadan, Id al-Fitr, Id al-Adha, Ashoura.

Working hours
Government and business: (Mon-Wed, Sat) 0830 – 1200, 1300 – 1630; (Thu) 0830 – 1330.
Banking: (Sat-Tue, Sat) 0800 – 1330/1400; (Thu) 0800 – 1100/1200. Wednesday opening is erratic.

Telecommunications
Telephone: Dialling for Afghanistan is via the International Operator. There were approximately 31,000 telephones in 1979, with less than 60 per cent automatic service and approximately 14,000 business lines.
Telex: Available at the central post office in Kabul.

Banking: *Central bank:* Da Afghanistan Bank.
Head offices: Afghanistan Bank, Agricultural Development Bank of Afghanistan, Bank Millie Afghan, Export Promotion Bank of Afghanistan, Industrial Development Bank of Afghanistan, Mortgage & Construction Bank, Pashtany Tejarty Bank (The Afghan Commercial Bank).

Weights and measures: Metric system (local units also in use).

Representation in capital: Austria, Bangladesh, Bulgaria, China PR, Cuba, Czechoslovakia, France, German DR, German FR, Hungary, India, Indonesia, Iran, Iraq, Italy, Japan, Korea PDR, Libya, Mongolia, Pakistan, Poland, Turkey, USSR, UK, USA, Vietnam, Yugoslavia.

BUSINESS DIRECTORY

Hotels: Kabul
Inter-Continental Kabul, Bagh-e-Bala Road, P.O. Box: 562 (tel: 31851/5).

Kabul, Avicina Street (tel: 24750).

Park, Mohd Jan Khan Watt, P.O. Box: 3201 (tel: 25202).

Spinzar, Mohd Jan Khan Watt, P.O. Box: 320 (tel: 22961/3).

Yama, Timur Shah Park (tel: 26509).

Business advice and information

Afghan Chambers of Commerce and Industry, Mohd Jan Khan Watt (tel: 26796, 25201; tx: 45; cable AFCHAMB).

Afghan Tourist Organisation (ATO), Ansari Watt, Shar-i-Nau, Kabul (tel: (93)30323).

Ariana Afghan Airlines (tel: 63601).

Central Statistics Authority, Block 4, Macroraion, Kabul (tel: 24883, 24188).

Federation of Afghan Chambers of Commerce and Industry, Daraulaman Watt, Kabul.

State Planning Committee, Chara-e-Malik Asghar, Kabul.

Banking
Da Afghanistan Bank (central bank), Ibni Sina Watt, Kabul (tel: 24075; tx: 223).

Agricultural Development Bank of Afghanistan, Jadah Maiwand, Pamir Bldg, PO Box 414, Kabul (tel: 22284; tx: 274).

Bank Millie Afghan, Jada Ibn Suna, Kabul (tel: 25451; tx: 31).

Pashtany Tajarahy Bank, Mohd Jan Khan Watt, Kabul (tel: 26551/54; tx: 243).

Albania

Greg Chamberlain

As the rest of Eastern Europe washed its hands of half a century of communism, Albania's neo-Stalinist rulers denounced the upheavals as 'lethal to socialism' and accused the defeated communist parties of 'laying the red carpet to the counter-revolution.' But at the same time, in other language, they urged a pragmatism and efficiency on their subjects in response to the same urgent economic pressures that brought down those regimes.

By outside standards, little stirred in Europe's most secretive and isolated country over the past year. Cut off by high mountains and a fortress mentality honed by 2,000 years of foreign occupation, Albania's three million people are still virtually the poorest in Europe. They are still not allowed to own cars (officially 'bourgeois egotism'), practise religion or have any but the merest of contacts with foreigners, only a carefully-scrutinised few thousand of whom are allowed to visit Albania annually and then only in supervised groups.

Yet returning travellers reported a more relaxed mood. While firmly rejecting the changes beyond the country's borders, party leader and head of state Ramiz Alia conceded in November 1989 that Albania was 'not an extra-terrestrial paradise where everything is solved for the best and where life is full of blessings.'

But, Alia warned, the changes could lead only to 're-establishment of class oppression and exploitation.' Albania was the 'only country building socialism,' he said, in the face of a 'permanent imperialist-revisionist encirclement' by 'enemies fighting to eradicate socialism from the face of the earth' and seeking to blame it for 'every evil in world history.' Change in Albania would be 'without haste and without force.' Added foreign minister Reis Malile: 'What is happening in Eastern Europe has nothing to do with us.'

The continued adoration of Enver Hoxha (the founder of communist Albania and its ruler for 41 years) which Alia has promoted as a means to revolutionary legitimacy is one measure of this conservatism. So is the sixth of the adult population estimated to work for the secret police. The military share of the national budget rose 16 per cent in 1989 to nearly a fifth of the total. The number of political prisoners, impossible to estimate according to Amnesty International, was put at 4-5,000 by the Greek press in April 1989.

Yet even as the UN Human Rights Commission condemned Albania for religious intolerance, soon after the reported death of 87-year-old Father Pjeter Meshkalla after a total of 31 years in a labour camp, the regime seemed to be softening its sometimes-ferocious pursuit of the religious. Mother Teresa, the Yugoslav-born Albanian missionary and Nobel Peace laureate, was even allowed to visit in August 1989 and was received by Malile and Hoxha's influential widow Nexhmije, though the price of the visit was her silence about repression of religion. A few weeks later, however, Alia stressed that there would be 'no concession to religious ideology in any of its various forms.'

Hints of official worry about revived religious feeling have been expressed along with concern at the restiveness of one of the youngest populations in Europe. A remarkably outspoken Union of Albanian Youth conference in August 1989 criticised lack of entertainment and fashionable clothes and even praised normally taboo rock music. Cultural shortcomings were fertile ground for unhealthy foreign influences, delegates warned.

Though few Albanians were inclined to take him at his word, Foto Cami, who is the regime's number two figure, called for experiment and open discussion to replace what he said was the present cultural mediocrity. His words were echoed at an April 1989 congress of the official Writers' and Artists' Union, where tolerance, 'emancipation' and a 'fight for civilised debate' was urged by senior figures, backed by the country's influential leading novelist, Ismail Kadare.

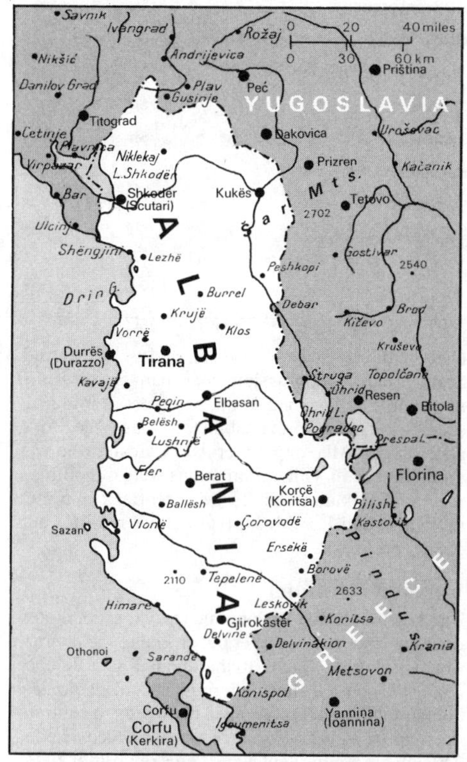

Economy

Both Cami and Alia extended this kind of plea for flexibility to the economy, which is plagued by outdated equipment for lack of foreign contacts and by inefficiency and lack of worker motivation springing from over-centralisation and excessive planning. Cami, at a March 1989 seminar entitled 'Perestroika – anti-socialist theory and practice,' roundly denounced Gorbachev's reforms, but at the same time called for new ideas and 'arguments, not slogans' to solve problems. He attacked rigidity, dogma and 'politicising everything,' and said the teachings of Hoxha should serve only as a 'compass' and not be 'mechanically adopted.' Six months later Alia seemed to take a step backwards by stressing that there would be no 'concessions to bourgeois ideology in any field –politics, art and culture or economic relations'. Yet in the same speech, he conceded that thorough investigation should be made into why socialism had been breaking down and astonished many by saying party membership should not be a requirement for appointment in a post. He firmly ruled out a multiparty system, however.

Earlier, Alia had publicly admitted that 1988 had been a 'very bad year' for agriculture, the economy's mainstay. The 1988-89 winter had been the driest for 40 years and hit livestock and nearly all crops except wheat. In a country which boasts about its abolition of private enterprise, that prices have not risen for nearly half a century and where wages have a maximum 2-1 differential, he called for bold economic experiments, less central control and 'economic solutions for economic problems.' He also urged farm co-operatives to sell their surpluses and to promote initiative and pay differentials as a way to raise production. Alia's increasingly frequent talk of giving priority to consumer goods and, in November 1988, the need to import 'a few million dollars' worth' of food to meet shortages give the lie to Albania's boast of being self-sufficient in food and almost so in industrial production.

Alia brought matters to a head at a special party plenum in February 1989 at which he warned of 'serious political repercussions' if the farm crisis continued. He announced increased investment in agriculture, a campaign for land reclamation and substantial imports of foreign agricultural machinery. The plenum also formalised the earlier calls for decentralisation and wage incentives.

The changes were accompanied by the first major government shake-up of Alia's four-year rule. The party chief stressed that the changes were not political, but simply in the interests of economic efficiency, perhaps a sign of the strength of the conservatives opposed to the relatively liberal line of Alia and Cami. A major victim was hardliner Simon Stefani, who lost his key job as central committee secretary in charge of military and security matters to become interior minister and a deputy premier. Another central committee secretary, Hajredin Celiku, economist Besnik Bekteshi and Vangjel Cerrava (both deputy premiers) and ministers Themie Thomai (agriculture), Llambi Gegprifti (industry and mining) and the transport and

KEY FACTS

Official title: The People's Socialist Republic of Albania

Head of state: President Ramiz Alia

Head of government: Prime Minister Adil Carcani

Ruling party: Albanian Party of Labour; Party leader: Ramiz Alia

Capital: Tirana

Official Languages: Albanian (dialects: Gheg and Tosk)

Currency: Lek = 100 Quindarka

Exchange rate: Lek 6.35 per $ (Dec 1989)

Area: 27,398 sq km

Population: 3.2m (1989)

Trade balance: –$42m* (1986)

* estimated figure

KEY INDICATORS

	Unit	1984	1985	1986	1987	1988
Population	m	2.9	3.0	3.0	3.1	3.2
Gross domestic product (GDP)	$m	2,360	2,400*	2,450*	–	–
GDP real growth	%	4.4	4.0*	3.1*	–	–
GDP per capita	$	833	800	766	–	–
Agricultural production	1975=100	132	–	–	–	–
Imports	$m	280*	208*	–	–	–
Exports	$m	200*	158*	–	–	–
Balance of Trade	$m	–81*	–50*	–42*	–	–
Exchange rate	lek per $	7.2	7.7	6.7	–	6.0

* estimated figure

energy ministers were all dismissed or transferred. State control commission chairman Enver Halili meanwhile was named to the new post of minister-secretary-general to the cabinet, to 'ease' the work, according to Alia, of premier Adil Carcani, who was thought to be ill. Other promotions were of politburo members Hekuran Isai (to be a central committee secretary) and Pali Miska (agriculture minister and deputy premier).

Foreign relations

The key to Albania's economic revival is not a government reshuffle however but an end to the international isolation which Hoxha created as a defence against supposedly hostile foreigners and ideological contamination. Alia continued the steady, if cautious opening-up he has led since he succeeded Hoxha. Though the regime kept up its fierce opposition to the USA and the 'social imperialist' Soviet Union as war-mongering great powers heading ideological blocs, Albania broke its long boycott of most international gatherings by attending the January 1989 chemical weapons ban conference in Paris. Yet still no foreign head of government has visited Tirana since Nikita Khrushchev, just before Albania's 1961 break with the Soviet Union. And foreign minister Malile is the most senior Albanian official to have stepped into the outside world since then. Malile, who has visited only a few countries, went to France, Spain and Turkey for the first time in 1989.

The Albanians are cultivating capitalist countries rather than the Eastern bloc in their quest for help with modernisation and top of the list is West Germany, which gave Albania an $11m grant in March 1989, half as financial support and half as technical aid for the mining, steel and oil sectors. Supply of West German textile and metallurgical equipment was being negotiated and Albania bought more West German goods. But trade with all countries was still hampered by Albania's constitutional ban on foreign credits, limiting all exchanges to barter.

Air France followed Lufthansa in starting services to Tirana in 1989, soon after Malile's visit to France, with which Albania is negotiating a deal involving aid for food production and for the country's hydropower industry. Ties with Greece, with whom Tirana has a quarrel over the Greek minority in southern Albania, advanced with more ministerial visits, agreements on border crossings and the start of a day-trippers' ferry from Corfu to the nearby Albanian port of Saranda.

Links with another important neighbour, Italy, were still snarled up by the presence in the Italian embassy in Tirana for the fourth consecutive year of an Albanian refugee family which the regime will not allow to leave. The Italian parliament has vetoed their handing over to the authorities. But Tirana has hinted it might break the deadlock soon.

Albania's long-frosty ties with Eastern Europe, a legacy of its break with Moscow, continued to thaw. In June 1989, East German Foreign Minister Oskar Fischer became the most senior Eastern European official since 1961 to go to Tirana. A wide-ranging economic, scientific and technical agreement was signed but the Albanians declined an offer of help to develop tourism.

Relations also improved with Czechoslovakia (ambassadorial ties were restored) and Bulgaria, where state planning commission chairman Niko Gjyzari became the most senior Albanian official to go to Eastern Europe for 28 years when he visited Sofia for talks with Todor Zhivkov. Alia however later denounced Bulgaria's repression of its Turkish minority. An ambassador was also appointed to Hungary only for the regime to denounce the rehabilitation of 1956 uprising hero Imre Nagy in June 1989. China too – Albania's political and economic prop from 1961-78 – got a tongue-lashing for its June 1989 repression of students three months after deputy foreign minister Muhamet Kapllani became the most senior Albanian to visit Beijing since the dramatic 1978 break. Only Poland remains in Eastern Europe to restore ambassadorial ties with Tirana.

Albania's perestroika efforts in foreign relations were crowned by its promotion of co-operation among the ideologically various Balkan community it is keen to rejoin. Albania hosted a meeting of six deputy foreign ministers in January 1989 and was due to chair a meeting of full ministers in Tirana in October 1990. It was the first time communist Albania had hosted any such international gathering. The participants avoided the many prickly ethnic and ideological issues which divide them and focused on such things as drug problems, tourism, arms trafficking and technical exchanges. But in September 1989, a disillusioned Alia said regional nationality conflicts had 'badly compromised Ballcan co-operation.

Indeed, only weeks after the January meeting, Albania's relations with neighbouring Yugoslavia had taken another nosedive over the perennial problem of the Albanian population of the poverty-stricken border province of Kosovo. As a general strike by the Albanians against repressive rule by Serbia, of which Kosovo is a part, led to a state of emergency and then in March clashes which killed 22 Albanians, the Yugoslavs accused Tirana of helping to foment the trouble. Albania denied interfering but warned that the rights of the 1.7m ethnic Albanians there were threatened by rising Serbian nationalism, as expressed by new direct rule over Kosovo and constitutional changes to reduce Kosovo's autonomy. Alia also spoke of the danger of foreign invasion by the major powers if things got out of hand.

Behind the rhetoric however, calmer counsels prevailed. Foreign Minister Malile spoke of overcoming the 'sad legacy of the past' and said Albania's desire, possible only if Kosovo's rights were respected, was a stable Yugoslavia. The two countries' trade fell by 40 per cent in 1988, to some $50m, with Yugoslavia buying less Albanian chrome, electricity and iron. But a new joint economic co-operation committee resolved to almost double the figure in 1989. The Yugoslavs agreed to subsidise for a little longer the new freight-only railway which is Albania's sole rail link with the outside world. Belgrade complains that Tirana has broken its promise to use the line, which is operating at only 15 per cent capacity.

COUNTRY PROFILE

Political system: Communist state with legislative power vested in 250-member unicameral People's Assembly, elected every four years. Executive authority held by Party of Labour politbureau and council of ministers, which is elected by Assembly. Chairman of council is head of government.

Main cities/towns: Tirana (population 220,000), Durres, Shkodër, Elbasan, Vlorë, Korce.

Media
Press: State news agency, ATA (Albanian Telegraphic Agency). Newspapers all officially controlled. *Dailies:* Most influential – *Zeri i Popullit* (organ of the Party of Labour) with circulation of 105,000, and *Bashkimi* (organ of the Democratic Front), circulation 30,000. *Weeklies:* Mostly controlled by local party/committees.
Radio: Approximately 200,000 receivers in use, served by one national and seven regional stations. Wire relay service covers whole country. Radio Tirana broadcasts in several European languages.
Television: Approximately 10,000 receivers in use served by stations at Tirana, Kukes, Berat, Pogradec. Greek and Austrian programmes are to be broadcast. Italian TV is relayed in censored form.

Domestic economy: Centrally planned economy. Emphasis on industrialisation and self-sufficiency in agriculture. Despite claimed economic growth of six per cent per annum 1981-85, Albania ranks as one of the poorest countries in Europe. Low levels of productivity and capital investment combined with shortages of skilled labour are major constraints on growth.

External trade: Totally state-controlled. Major exports include chrome, copper wire, ferro-nickel ore, bitumen, oil, hydro-electricity. Agriculture (tobacco, wood products, fruit and vegetables) accounted for 40 per cent of exports in the late 1980s. Imports include chemicals, machinery, textiles, vehicles, lubricants and various consumer goods. Principal trading partners are Yugoslavia (trade turnover around $93m) and Greece (trade turnover around $70m). Other trading partners include GDR, Italy, Romania, FRG, Hungary, Poland, China, France, Czechoslovakia and Bulgaria.

Agriculture: Mainstay of the economy, employing over 55 per cent of the workforce. Some 460 co-operatives produce around 75 per cent of the output. Emphasis on production of cereals. Despite rapid mechanisation and extensive irrigation programme, sector remains under-developed. Major crops include sugar-beet, cotton, grains, beans and sunflower seeds. Livestock sector.

COUNTRY PROFILE

Industry: Predominantly small-scale with bias towards engineering, chemicals, metals, construction materials, food processing and other agro-allied industries. Employs 25 per cent of workforce and accounts for 55 per cent of GDP.

Mining: World's third largest producer of chromium ore although extraction fell well short of target in 1987. Extensive reserves of copper, lignite and ferro-nickel. Also some coal.

Energy: Crude oil is major source of primary energy. Oil production largely from two small fields at Patos and Morinza. Refining capacity 1.5 million tonnes per annum. Major gas fields at Diviak and Bubuline. Hydro-electricity developed, with surplus exported.

Legal system: Justice administered under the constitution by the Supreme Court, and by district, village and county courts.

Membership of international organisations: CCC, CEMA, FAO, IAEA, IPU, ITU, UN, UNESCO, UPU, WFTU, WHO, WMO.

BUSINESS GUIDE

Time: GMT + 1 (GMT + 2 summer).

Climate: Mediterranean; average annual temperature 16°C. Warmest in south-west, coldest in north-east.

Entry requirements
Visa: Must be obtained in advance. Available from Albanian embassies in Paris, Rome, Belgrade and Stockholm. US and Soviet nationals refused entry, others may encounter difficulties.

Air access: From Athens, Bucharest, Budapest, East Berlin, Frankfurt, Ioannina, Paris, Rome, Vienna, Zurich.
National airline: Albtransport.
International airport: Rinas (TIA), 29 km from Tirana. Airport tax: 30 lek.

Surface access: Border crossing with Greece at Kakavia and Kristalopigi, and with Yugoslavia at Han-i-Hotit, Vrbnica and Cafasan. Rail link with Yugoslavia completed in 1985, carrying freight only, but closed in 1989.
Main ports: Durres, Sarande, Vlore.

Hotels: All bookings handled by Albturist, except for business visitors. Modern hotels in main towns.

Car hire: Private cars banned. Cars only available for party and government leaders and foreign diplomats.

City transport
Taxis: Taxi service available only in Tirana.

National transport
Air: No regular internal air service.
Road: Network of about 21,000 km (3,000 km paved); mountain roads often impassable.
Buses: Flat-fare bus service in Tirana.
Rail: Rail network of about 350 km.
Water: Ferry service connects Durres with Trieste (Italy).

Public holidays
Fixed dates: 1 Jan, 11 Jan, 1 May, 8 Nov, 28 Nov, 29 Nov.

Telecommunications
Telephone: Approximately 32,000 telephones in use. Country code for telex: 604 AB.

Representation in capital: Bulgaria, China, Cuba, Czechoslovakia, Egypt, France, Federal Republic of Germany, German Democratic Republic, Greece, Guyana, Hungary, Italy, Democratic People's Republic of Korea, Poland, Romania, Turkey, Vietnam, Yugoslavia.

BUSINESS DIRECTORY

Academic/research institutes
University of Tiranë.

Business advice and information
Chamber of Commerce, Konferenca e Pezës Street 6, Tirana (tel: 42-46, 29-34).

Banking
Drejbanka – Banque de L'Etat Albanais, (State Bank of Albania), Sheshi Skënderbej 1, Tirana (tel: 24-35; tx: 2153 BANKA AB).

Principal newspapers
Bashkimi, Bulevardi Stalin, Tirana (tel: 81-10).

Zeri i Popullit, Bulevardi Stalin, Tirana (tel: 78-13).

Other useful addresses

Agroeksport – State Trade Organisation for the Export of Agricultural and Food Products, 4 Shkurti Street 6, Tirana (tel: 52-57, 52-59, 51-29, 31-28, 25-33; tx: 2137 AGREKS AB).

Albanian Telegraphic Agency (ATA), Boulevard Marcel Cachin 23, Tirana (tel: 44-12; tx: 2142).

Albimpeks – State Trading Organisation for the Import of Materials, Rue 4, Shurti Street 6, Tirana (tel: 27-11, 61-23, 45-40, 40-51, 54-90, 73-53, 73-58; tx: 2112 EXIMP AB).

Albkontrol – Organisation for Inspection of Exported and Imported Goods, Enver Hoxha Street, Durrës (tel: 23-54).

Albtransport – State Organisation for Air Transport, Kongresii Permetit Street 202, Tirana (tel: 30-26; tx: 2124 AERIN AB). Also at: Rinas Airport, Tirana (tel: 31-07; tx: 2445 AJTRAN AB).

Albturist – State Tourist Organisation, Deshmore e Kombit, Boulevard 6, Tirana (tel: 38-60, 48-53; tx: 2148 HODAJT AB).

Artimpex – State Organisation for Export, 4 Shkurti Street 6, Tirana.

Bureau for the Registration of Patents & Trade Marks, Konferenca e Pezes Street 6, Tirana.

Centrocoop – Central Union of Consumers Co-operative for Export and Import, Rue Mujo Ulquinaku 2, Tirana (tel: 32-28).

Drejtkursig – Savings and Insurance Offices of Albania (Head Office), Tirana (tel: 32-06).

Industrialeksport – Rue 4 Shkurti Street 6, Tirana (tel: 45-50; tx: 2127 MAKIMP AB).

Kinostudio – 'Shquiperia e Rue', Cinematographic Studios 'New Albania' Rue Aleksander Moisiu 76, Tirana (tel: 33-38, 27-47).

Makinaimport – State Trade Organisation for the Import of Machinery, Rue 4 Shkurti Street 6, Tirana (tel: 32-67, 52-20; tx: 2127 MAKIMP AB).

Mineralimpex – State Organisation for Export of Minerals, Rue 4 Shkurti Street 6, Tirana (tel: 33-70; tx: 2116 MIMPEX AB).

Ministry of Foreign Trade, Tirana (tx: 2152).

Ministry of Light Industry, Tirana.

Ministry of Industry & Mining, Tirana.

Ministry of the Foodstuff Industry, Tirana.

Ndermarria E Librit – State Organisation for the Book Trade, Konferenca e Pezes Street, Tirana (tel: 33-23).

Transshqip – State Organisation for the Transport of Goods in Foreign Trade, Rue 4 Shkurti Street 6, Tirana (tel: 30-76, 46-59, 74-29; tx: 2131 TRANSH AB).

Algeria

Jon Marks

An unprecedented crisis of public confidence in the regime which followed riots throughout northern Algeria in October 1988 forced President Chadli Bendjedid to speed the pace of political change to accompany economic reforms already under way. A new government was formed to accelerate economic liberalisation while also coping with the debt burden during a period when annual debt service payments were in excess of $6bn. Underlining the increasing pragmatism of the Algerian leadership, in the first half of 1989 it took a previously unthinkable decision to borrow from the IMF. Parallel to economic reforms, two referenda on constitutional changes were held, opening the way for a radical overhaul of the political system erected in the period since independence. This activity opened the way for the creation of a less rigid polity, responsive to popular demands. However, the political elite remained divided over the pace and extent of reforms and despite the positive developments of the year after the riots, the political environment remained in a state of volatility and tension which was not apparent before the explosion of October 1988.

October 1988 riots

The riots which erupted in early October 1988 were suppressed with unprecedented ferocity by the military. The riots were accompanied by appeals for more social justice – reflecting public concern that economic liberalisation had been accompanied by unprecedented divisions of wealth – and demands for a more open, responsive political system. The shock of the riots, the regime's response (with the shooting and torture of civilians which the majority of Algerians found totally unacceptable) and the ensuing political crisis undermined the ruling legitimacy of two main pillars of the Algerian establishment, military and the Front de Liberation Nationale (FLN).

Even before the October riots broke the surface calm of Algerian politics, tensions had been rising with several weeks of social tensions and labour disputes focusing on heated negotiations over national salary scales, and widespread shortages of water and basic goods. Clashes between citizens and the authorities were reported in the late summer of 1988 – and were to continue through the following year – and a serious confrontation between organised labour and riot police was threatened in late September following a dispute over new pay scales and the streamlining of terms of employment (ending traditional benefits including an AD100 per child payment to workers with families) at the Rouiba heavy vehicles plant.

When they came the riots were more serious than any recent outbreak, affecting Algiers, Oran, Annaba, Blida and a number of smaller towns in the north and high plateaux. News agencies and other foreign media published estimates of the total number of dead ranging from 200 to 600. Official figures for losses were presented by Interior Minister El-Hadi Khediri on 20 October: 159 dead (5 of whom were members of the security services) and 154 wounded. There were also reports that youths were tortured by elements of the security forces – a charge which has a special resonance in Algeria because of its bitter anti-colonial struggle.

In many respects the riots were a spontaneous explosion of violence involving mainly younger Algerians, frustrated at a lack of opportunities in a rigid system suffering the effects of economic austerity. After the initial troubles, which were started on 4–5 October by lyceens and other youths, radical Islamist groups entered the fray, playing a significant role as the most visible and vocal force acting against the military intervention. Strongholds of radical Islam like the Algiers neighbourhoods of Kouba, El Biar and Bab el-Oued suffered some of the heaviest casualties. Islamist groups organised the largest demonstrations and opposition imams emerged to articulate popular concern at social and political inequalities.

This underlined the fact that after a period when the Islamic opposition had been fragmented and suppressed it was now re-establishing itself as a force in local politics. On 10 October Chadli held unprecedented talks with the popular imam Ali Belhadj, and the more established Shaikh Sahnoun of Kouba and Shaikh Mahfoud of Blida. They demanded reforms that would appeal to Islamic sentiment and were then instrumental in calming tempers.

But despite the significance of the Islamists' role, the events of October did not constitute a fundamentalist uprising. Rather, the Islamists tried to take advantage of the troubles which they did not start and, perhaps more importantly for the future, showed they had an authority which could influence events which had gone out of the control of the leadership.

There is some evidence to show that the riots were more than a spontaneous protest, with some degree of trouble-making coming from disaffected elements

KEY FACTS

Official title: Democratic and Popular Republic of Algeria

Head of state: President Chadli Bendjedid

Ruling party: Front de Libération Nationale (FLN)

Capital: Algiers

Official Languages: Arabic (Berber, French widely spoken)

Currency: Algerian Dinar (AD) = 100 centimes

Exchange rate: AD8.06 per $ (Nov 1989)

Area: 2,381,741 sq km

Population: 23.84m* (1988)

GDP per capita: $2,271 (1988)

GDP real growth: –1.8% (1988)

Inflation: 6.0% (1988)

Trade balance: $640m (1988)

Foreign debt: $23bn (1988)

* estimated figure

KEY INDICATORS

	Unit	1984	1985	1986	1987	1988
Population	m	21.04	21.85	22.52	23.10	23.84
Gross domestic product (GDP)	AD bn	259.9	287.4	286.5	307.9	320.0
GDP per capita	AD '000	12,353	13,153	12,722	13,329	13,422
GDP real growth	%	4.0	4.0	–1.5	–1.1	–1.8
Inflation	%	6.3	10.5	12.4	7.4	6.0
Agricultural production	1979-81=100	107.5	125.0	120.1	122.4	121.7
Food production	1979-81=100	106.8	124.2	119.3	121.6	120.7
Oil production	m barrels/d	695.4	672.4	673.9	648.2	–
Exports	AD m	63,764	64,564	34,935	41,736	45,390
Petroleum	AD m	45,793	33,678	28,613	40,694	42,965
Imports	AD m	51,257	49,491	43,393	34,153	43,959
Foodstuffs	AD m	9,259	11,393	6,926	7,502	10,605
Raw materials	AD m	18,311	15,352	13,451	11,233	13,599
Consumer goods	AD m	7,669	6,604	4,991	5,000	5,165
Balance of trade	$m	1,258	2,032	–1,524	1,202	640
Current account	$m	74	1,105	–2,230	141	–760
Total external debt	$m	13,865	15,330	19,300	22,881	–
Debt service ratio	%	32.8	32.5	51.0	46.8	–
Exchange rate	AD per $	4.98	5.03	4.70	4.85	5.91

* estimated figure

within the FLN establishment itself. This followed at least two years in which President Chadli Bendjedid and his supporters had promoted reforms in the face of opposition from conservative 'old guard' and radical socialist opinion within the FLN. In the run up to the sixth party congress, which was scheduled for December 1988, the president and his reformist allies had failed to achieve a consensus with conservatives within the party over how far reforms (and the attack on vested interests within the FLN apparatus) should go. Representative of these hardening attitudes was the Rouiba dispute, in which the pay and company reform policies of an FLN government were opposed by senior members of the Union Generale des Travailleurs Algeriens (UGTA), the mass trade union organisation affiliated to the FLN, as well as workers fearful for their jobs and living standards.

During a difficult summer rumours of policy struggles were rife and the Algiers rumour mill worked overtime with reports of high-level corruption, some of which may have been initiated by old guard elements. It was also claimed that a spate of shortages and the import of poor quality goods into state shops was the result of maladministration rather than a lack of foreign exchange to cover the cost of essential imports. These factors helped fuel the popular explosion when it came, giving the timing and nature of the October events a pattern. The rioters' targets were highly selective and well-timed. Mainly destroyed were government and party offices, Souk el Fellah and Galeries Algeriennes supermarkets and shops, and other official property. Marches and protests started at roughly the same time in several locations. Areas which might have been expected to rise in such a widespread movement, most notably the Berber heartland of Kabylie, did not follow.

If old guard elements within the FLN did whip up discontent, the troubles had the opposite effect from that intended. After initial confusion – in which the government effectively lost control of the country, forcing Chadli to call in the army – the troubles were used by Chadli and the party's reformist wing to gain the upper hand, to outflank old guard rivals and force through reforms (including the retirement of older generation and left-wing officials within the FLN hierarchy and military establishment). A clear commitment to reform was welcomed in the country, effectively keeping Chadli in power, and also abroad.

International opinion welcomed Chadli's promise to reform the political as well as the economic system. However, the violence of October 1988 had done much to undermine perceptions of Algeria's political stability, with potentially damaging consequences for the country's credit rating and for the introduction of reforms. Even though many banks and export credit agencies reassessed their attitudes to Algerian risk, bilateral supporters, led by France, Spain and Italy, and multilateral agencies mobilised new support, allowing the government to put the economy on a more even keel and ride out the political storm.

Constitutional reforms

The riots provoked loud calls for reform to which Chadli quickly responded with the offer of a new constitution. This was to be put to Algerians for approval in two referenda, one in November 1988, the other in February 1989. These changes were, by mid-1989, to lead to the introduction of a form of multi-party system, ending the FLN's formal monopoly of political power, held since independence.

Not long before Algerians were to vote in the first referendum, on 3 November 1988, Chadli clearly signalled his commitment to change by removing the figure most clearly associated with the FLN's old guard, Mohamed Cherif Messaadia, and replacing him with Abdelhamid Mehri as head of the party's permanent secretariat. Mehri's brief was to reform the party. It was one of several initiatives which allowed Chadli to win a substantial 'yes' vote in the first referendum. Few Algerians were going to vote against change, even if they were revolted by the October events; in the event Chadli came out well from a referendum which was widely seen as a vote of confidence in his leadership. There was a 92.27 per cent yes vote, approving constitutional amendments which would transfer power away from the presidency to a government responsible to parliament. The national turnout was 83.1 per cent, with voting highest in provincial towns and rural areas and support considerably lower in urban centres.

A new head of government was announced two days later, Kasdi Merbah, who was responsible for naming a government which under the new constitutional terms would be relatively independent of the executive but answerable to parliament. Merbah was a prominent member of several former governments, a former security chief and confidant of the late president Boumediene who had established a new reputation as a pragmatic and successful minister of heavy industry, agriculture and health. His government comprised economic liberals and 'technocrats': it included as finance minister Sid Ahmed Ghozali, who for several years had been out of favour for his liberal ideas of economic management, which included selling hydrocarbons at market sensitive rather than higher 'political' prices.

On 22 December 1988 Chadli consolidated his position in a presidential election where he was the only candidate but was competing against public apathy or even opposition. Turnout was high compared to the November and February referenda with 89.07 per cent of the electorate voting. Chadli gained 81.17 per cent of the votes cast.

While the government got on with the business of implementing reforms there were changes elsewhere in the political system, most notably when on 3 March 1989 it was announced that the military would withdraw from the FLN central committee – and in effect from everyday political life. This was announced by a younger generation officer, Colonel Yahia Rahal, and supported by modernisers from within the military establishment including the Chief-of-Staff General Khaled Nezzer and the First Military Region Commander General El-Hashemi Hadjeres. There were reports of resistance to this de-militarisation of politics from within the military establishment, despite an acceptance in recent years that the armed forces would be better occupied modernising, rather than dabbling in politics. However, this did not provoke open provocation: the evidence of subsequent months suggested that although military men were less active in politics, their influence remained strong and the army was, as ever, a factor which could be ignored in Algerian politics.

In the first months of the new constitutional arrangement the president took on the role of arbiter between conflicting groups within a divided society. In a speech to the FLN central committee on 28 March 1989 Chadli told the party that it had to accept the opposition which 'existed in the sixties, the seventies and the eighties; why deny the reality?' The difference, he said, was that before 'we employed other means against them which you all know about – prison and the various methods'. At this stage the introduction of a multi-party system was apparently ruled out: 'I do not believe in current circumstances in the multiplicity of parties', Chadli told the central committee, adding the rider that 'if the Algerian people want this I cannot object to the wish of the nation'. Opposition groups were to be given a public voice, being allowed to organise as 'associations of a political nature', while the FLN would revert to its pre-independence status as a broad-based front representing a variety of national interests.

Despite these widely welcomed developments, opposition to reforms came from a variety of directions including the UGTA, which attacked constitutional changes which would disaffiliate it from the FLN and allow independent unions to form. Trade unionists were deeply concerned with constitutional clauses which apparently ended Algeria's commitment to socialism (a charge denied by the leadership) and the prospect of much-increased foreign influence on the economy. These concerns were reflected in a wave of strikes which affected manufacturing and service industries. Also protesting were previously passive groups such as consumers, with consumer boycotts over price rises during Ramadhan in Algiers, Oran, Tizi Ouzou and other towns. In some cases violence was reported, as was also the case in a spate of protests against allocation of housing, a major social issue and problem for the government.

Articulating opposition views were more than 20 political 'associations' – in effect parties – including groups representing Islamist, Kabyle and left-wing sentiment. Among the most notable were the Front

Islamique du Salut (FIS) and the Ligue de la Da'wa Islamique who might eventually form the core of a moderate, legal Islamic opposition. These and other groups inspired by Islam have a large, but still minority, following. The existence of several Islamist groupings reflects divisions over strategy.

Other opposition groups also have support, not least the communist Parti de l'Avant Garde Socialiste (PAGS) and the strongly Kabyle Rasssemblement pour la Culture et la Democratie. But in the case of leftists and Berber activists, as with other tendencies, groups divided by internal differences would have problems coming together to form a united front against the still dominant FLN.

A commitment to genuine change – and the fear of the majority of Algerians of what further political turbulence might bring – bought time for Chadli whose position was vulnerable after the riots. To maintain the pace of change a second referendum on constitutional reforms was held on 23 February 1989. The result highlighted a degree of alienation within Algerian society but also provided a further mandate for change: 7.3m voting for the new constitution and an unprecedented 2.6m voting against. Turnout was 79 per cent – just enough to give credibility to the 'Second Republic'.

The new constitution promised a reduction in the FLN's power and the emergence of political 'platforms', the precursors of a genuine multi-party system. When legislation allowing opposition 'associations of a political nature' and a revised electoral code reached parliament in the summer of 1989 it was clear that the move to a multi-party system had been more radical than originally planned – with popular protests at local grievances a constant reminder of the need to prove the commitment to change.

In July 1989 legislation approving opposition politics and a competitive electoral system was passed by a parliament that was still 100 per cent FLN. However, while the new laws represented a major break with the past it was widely noted that the opposition would remain widely constrained and that the FLN was expected to do best in elections, remaining the country's dominant party – helped by an electoral law weighted towards the largest, best-organised party in

COUNTRY PROFILE

Political system: The president is elected by universal suffrage for five years. He appoints a prime minister, who in turn appoints a government. The unicameral, popularly elected, 281-member National Popular Assembly holds legislative power. The president can dissolve it, and call elections. Independence gained from France in 1962. A revised version of the constitution was approved by national referenda in November 1988 and February 1989.

Population: 99 per cent of the population are Berber Arabs, with the remainder mainly European. Islam is the dominant faith.

Main cities/towns: Algiers (population 2.4m), Oran (pop 1.2m), Constantine (pop 1m), Annaba (pop 800,000).

Language: Arabic is the official language; French and Berber languages also spoken.

Media
Press: Leading national dailies: *Al Chaab* (Arabic), *El Moudjahid* (French), *El Massaa* (Arabic), *Horizons* (English section). Main provincial dailies: *An Nasr* (Arabic), *Al Joumbouria* (Arabic). *Broadcasting:* Radio services (four stations) are in Arabic, French, Berber, English and Spanish. Television services are in Arabic and French. Network operated by Entreprise Nationale de Télévision (ENTV).

External trade: Strict control of imports combined with broadening of hydrocarbons exports helped restore balance of trade surplus: $1.5bn in 1987. Exports are heavily dependant on oil and gas. Increasing sales of Liquefied Natural Gas (LNG) and condensates have offset falling oil revenues due to the oil glut. The sharp fall in oil prices in 1986 sharply reduced both gas and oil earnings, but the situation stabilised in 1987 with the improvement in prices. The government has introduced a series of measures designed to boost non-hydrocarbon exports, while cutting on imports of consumer goods. Invisible account has remained heavily in deficit owing to high loan repayments and slump in invisible earnings (tourism, workers' remittances). In the late 1980s this deficit together with a reduced trade surplus has left the current account in deficit. Countertrade and compensation trade agreements are now common.
Exports: Hydrocarbons (total around 96-98% – natural gas 25%, condensates 25%, crude petroleum 24%, petroleum products 20%), wine, tobacco, foodstuffs. Main destinations: France (22% of total value in 1986), Italy (20%), USA (17.4%), Netherlands (15%), Benelux (6%), UK, Japan.
Imports: Capital goods (typically around 24-26% of total value), semi-finished goods (24-26%), food and tobacco (20% in 1985), consumer goods (15-17%), transport equipment (14-16%), raw materials (5-8%). Main sources: France (24% of total value in 1986), German FR (11%), Italy (10%), USA (8%), Japan, Spain (5%), Belgium, Canada, UK.

Agriculture: Although still the major employer (22 per cent of the labour force), the sector has declined steadily during the 1980s and currently contributes less than six per cent to GDP. Good rainfall produced a record harvest in 1985, and reforms helped increase productivity. Around 30 per cent of total area is potentially cultivable with irrigation, with about five per cent currently under cultivation, mostly in the northern coastal strip. Despite extensive irrigation programme and the dividing of state holdings into smaller units, agricultural output has failed to keep pace with the rate of population growth. Imports, at AD8.5bn in 1987, now provide around 60 per cent of domestic food requirements. Underlying constraints to growth include soil erosion, desert encroachment, inefficient management in the state sector, poor marketing and recurrent droughts. The government plans to increase output and reduce dependence on imports by a series of measures including investing in new technology, financial incentives for state and private sector farms to buy equipment, less interference in the private sector and tree planting to arrest desertification. Reforms in October 1987 included the renting or leasing of 3,200 state farms to the private sector. Main cash crops are grapes, oranges, olives, dates, tobacco, sugar-beet and tomatoes. Hard and soft wheat and barley are grown for the home market, as are vegetables, pulses and citrus fruit. Livestock farming, mainly in the north-east, and fishing (sardines, anchovies, sprats, tuna, shellfish) also important.

Industry: Manufacturing contributes around 13 per cent to GDP. Production dominated by state-owned heavy industries such as steel, petrochemicals, fertilisers and cement. Traditional agro-allied industries also important, particularly textiles, food processing and tobacco and cigarette processing, which are being expanded, and public companies are now allowed to set prices and invest profits. Also opening up of industry to more foreign involvement, particularly in large-scale projects such as motor vehicle assembly (one 30,000-vehicles per annum planned). Industrial output is now rising after stagnating 1980-1983. Main constraints to development are shortages of vital inputs and skilled labour, high production and transport costs and maintenance problems. Industrial development is centralised in the northern coastal strip, but plans exist to extend industry to the high plateaux in the south.

Mining: Rich in minerals, including iron-ore, uranium, zinc, phosphates, gold, antimony, bituminous coal, tungsten, manganese, lead, mercury and salt. The mining of iron-ore and phosphate for feedstocks (for local steel and fertilizer production respectively) and for export are the most important. Developments in the Hoggar region in the south of the country, where reserves of uranium, gold and tungsten have been pinpointed, were suspended because of cost. Also located near the Moroccan border are iron ore reserves estimated at 2bn tonnes. The remote location and Western Sahara/Morocco conflict have so far prevented exploitation.

Hydrocarbons: Sector is of fundamental importance and currently accounts for 95 per cent of export earnings (1988), 23 per cent of government revenues and around 25 per cent of GDP. Unlike most members of OPEC, Algeria's hydrocarbons sector is relatively diversified and thus partially protected from world price shifts. Current emphasis on cutting production of crude oil (approx 640,000 b/d in 1988), to conserve reserves, and expanding output of natural gas, condensates and LPG. The aim is to maximise domestic natural gas consumption and to replace crude oil as the main foreign currency earner by LNG, condensates and LPG (which are expected to make up 90 per cent of hydrocarbon exports by the end of the century). Production and exploration of hydrocarbons are controlled by state-owned SONATRACH.

Energy: Oil: High quality, low sulphur which sells at a premium. Reserves estimated at around 6.6bn barrels, sufficient at current rates of extraction to last the next 25 years. Owing to a policy of crude oil conservation and OPEC quotas, production has fallen consistently since late 1970s to current level of 600,000 b/d – about half of normal capacity. Main fields situated in eastern regions near Edjele and Hassi Messaoud, and In Amenas. In 1985 small but promising new finds of oil were made in the Hassi-Messaoud region and near Adrar in the west. Legislation was approved in August 1986 to encourage foreign companies into joint-venture oil exploration in the hope that commercially exploitable reserves will be found. Natural gas: Proven reserves amount to over 3,000bn cu m, fourth largest in the world. Expansion of LNG export capacity boosted to over 40bn cu m, although 1985 exports amounted to about half that. LNG exports (mainly to Europe) account for 25 per cent of total exports. Sales of gas to Italy through the TransMed pipeline are running at 12bn cu metres a year. Greece, Turkey, UK and Yugoslavia agreed to start taking natural gas even though the proposed increase in LNG exports was hampered by a weak world market and disputes between SONATRACH and customers over volumes and prices. LPG production is being increased from current 1m tonnes a year, for domestic use and export.

Membership of international organisations: African Development Bank, Arab Fund for Economic and Social Development, Arab League, GATT, ILO, IMF, Islamic Conference, Islamic Development Bank, OAPEC, OAU, OPEC, UN, WHO, World Bank.

the country (the FLN) and forcing moderation on opposition groups; by the FLN's hold on government and the administration; and also by divisions among the opposition. Thus, even with a previously unimaginable multi-party system in place, the FLN remained in control and the opposition would find it very difficult to get a foothold in positions of power.

In September 1989, with the October anniversary looming and further in-fighting within the FLN manifesting itself, Chadli and his senior officials acted again to push through political reforms. There was concern that laws on elections and information had been amended by FLN traditionalists in parliament, partly due to a lack of concerted support for these reforms from the Merbah government. Thus less than a year after its appointment, the government – which had done much to push through part of the economic reform programme and stabilise the external payments situation – was dismissed. This prompted an unprecedented public outburst by the premier, who complained that his dismissal was unconstitutional.

A close presidential aide, Mouloud Hamrouche, was appointed prime minister with a brief to accelerate reforms. Hamrouche – a leading reformist and secretary-general to the presidency – made it clear that legislation favoured by the presidency would be pushed through, or parliament would be dissolved. The September 1989 government reshuffle was to be followed by another party congress in late October and local government elections were scheduled for 1990. These events would show the relative strengths of reformers over conservatives in the FLN establishment, and just how far the presidency had asserted its vision of change over the country. Elections would prove a first-ever competitive test for the FLN, with the prospect of introducing a new generation of FLN and opposition politicians to politics at grass roots level.

Their success would be closely watched as an indicator of how far the reform process would go. Also closely watched would be personnel changes within the elite: despite a year spent institutionalising democracy in Algeria, older traditions of faction and internal conflict were not expected to disappear overnight.

Maghreb unity

While internal politics dominated Algeria during the year, efforts to build a united Maghreb once again played a leading role in foreign policy. Rapprochement with Morocco continued, with Chadli paying the first visit by an Algerian president to Morocco in more than a decade. Wider regional unity initiatives also bore fruit. In February 1989 a second summit of Maghreb leaders was held in Marrakesh and a regional grouping, the Arab Maghreb Union (known by its emotive French acronym UMA) was established.

Crucial to building regional unity was the promotion of relations with Morocco, which involved mending fences after years of conflict over the Western Sahara. Helping the process along, on 4 January 1989 King Hasan met Polisario leaders in Marrakesh for talks which were applauded by Algiers. Despite its rapprochement with Rabat following the restoration of diplomatic ties in May 1988, Algeria remained publicly committed to Polisario and the cause of an independent Sahara. In private, however, the majority of Algerian leaders were now more committed to the wider goal of Maghreb unity, even if this meant giving ground to Rabat on the issue of Sahrawi independence.

Chadli visited Morocco on 6–8 February 1989, while UN and African officials continued efforts to bring peace in the Western Sahara. Hasan was expected to hold further talks with Polisario leaders soon after, but did not. However, with the Saharan peace initiative under way, the way was clear for Algeria and Morocco to concentrate on Maghreb unity and bilateral economic ties.

On the agenda for talks was a major scheme to build a gas pipeline linking Algeria and Spain. This would pass via Morocco, which would take substantial quantities of gas. It was announced that a 50:50 Algerian–Moroccan venture would carry out marketing surveys, feasibility studies and examine financing for a pipeline running from Hassi R'Mel to Oujda, from where it would take a circuitous route across Morocco before crossing the Strait of Gibraltar to Spain. A gas deal would hugely increase the level of bilateral trade with Morocco and Spain, but despite an upsurge in interest in the scheme, cost remained a major problem. According to Spanish estimates, the second Transmediterranean pipeline would cost over $4bn.

Parallel to the Algerian–Moroccan rapprochement was a wider programme of shuttle diplomacy, aimed at bringing together the five established north-west African states (Algeria, Tunisia, Mauritania, Morocco and Libya). Their leaders met for a second time (following the June 1988 Algiers summit) in Marrakesh on 17–18 February 1989 to establish the UMA, a regional organisation with the administrative structures necessary to bring North African economies closer together and rationalise legal and political systems to speed regional integration. The UMA was also intended to allow Maghreb leaders to meet on a regular basis, which held the promise of reducing tensions which for so long had led to bilateral conflicts in the region. The supreme body was a council of heads of state with an informal secretariat which was to move with the revolving presidency.

Regional unity was seen as essential so that North African states could confront economic challenges, not least 1992 and the EC single market which some UMA member feared would cause further disequilibrium of trade in the Mediterranean basin. It was also hoped that Maghreb unity would act as a stimulus to export-led growth within the UMA, forming a basis for integrated economic development which the region's hard-pressed economies badly need.

A vulnerable economy

The social and economic tensions which found expression in the October 1988 riots stemmed from structural weaknesses in an economy still dependent on hydrocarbons exports for its hard currency earnings. Algerian customs figures for 1988 show that hydrocarbons accounted for around 95 per cent of all exports (AD45.6bn out of AD48.1bn), while a 2m tonnes cereals harvest in the summer of 1989 showed that the economy remains vulnerable to drought.

The balance of payments situation was controlled by cutting imports to the bone and bringing the economy slowly more into line with world market values – which did not ease the situation for companies and consumers dependent on imported goods. These problems were reflected by inflation in the formal economy and in the black market, which had become the only source for some goods. The black market dinar now traded at three to five times its official rate.

The riots served to highlight the limits of a balance of payments strategy centred on compressing imports: it was now clear that to build a stronger economy the problems of inefficient management and corrupt practices had to be challenged even more vigorously. This was of central importance to the Merbah government and was emphasised as of critical importance by Hamrouche on his takeover.

Just how successful reforms were is questionable. Data on performance in agriculture, where changes were radical to encourage private enterprise, are clouded by drought conditions in parts of the country which would have reduced production, reforms or no. Other data is difficult to interpret given traditional problems of conflicting figures and politically-motivated statements on economic performance.

The economy grew by 2.7 per cent in 1988 according to figures released with the budget and annual plan. However, according to authoritative independent estimates – later corroborated by official statements – there was negative GDP growth (of around 1.8 per cent) that year. A higher level of growth was expected for 1989, with relatively high oil prices and the benefits of reforms then under way coming into effect. Not long before his dismissal, Merbah was forecasting 4–5 per cent growth in 1989: on his takeover Hamrouche described this forecast as 'optimistic' and a figure of nearer 3 per cent is expected for the year, after two years of negative growth.

Growth was essential to help tackle some of the country's most pressing problems, led by shortages of housing and unemployment. In an increasingly open political atmosphere such problems were being discussed more openly – a fact recognised by the publication of official figures which showed the total of unemployed to be 1.2m, about 22 per cent of the total workforce.

The Merbah government was committed to tackling these problems but followed the budgetary policies of its predecessor. Reconciling a conservative financial policy with the need to ease social pressures was to prove an unenviable task. The 1989 budget was based on tight controls on public spending to avoid precipitating a financial crisis which might end in rescheduling. Budget estimates looked to a reduction in the budget deficit rather than substantially increased investment spending: the government was also committed to increasing production in sectors that could boost exports and substitute imports, but this was not to be at the expense of over-spending. Current spending was set to rise by 8.9 per cent (an AD6.4bn increase), but a public sector pay rise awarded in January 1989 meant that wages were to rise by an estimated AD7.2bn in 1989. Government investment spending was held to 4.2 per cent growth.

Economic policy remained relatively constant, even though it was not untouched by the wider political mood of the country. The 1989 budget and annual plan were passed in late December only after unprecedented public parliamentary debate. The bulk of the draft budget was approved as planned but the government was forced to remove some of the most unpopular details.

Later in the year parliament was to be more vocal in opposing policies – and the government's failure to deal with this challenge was to be given as a reason for Merbah's dismissal. In the budget debates more than half the deputies in the national assembly spoke – an unprecedented response: in subsequent voting, the majority voted against planned petrol price rises. Rumours of higher prices had already caused panic buying and shortages.

Political crises and rising concern over trade-related debt caused problems as Western export credit agencies reassessed their attitude to Algerian risk. After the riots several agencies, including the Netherlands' NCM, decided to downgrade cover for exporters. NCM, for example, reduced its ceiling for cover on medium-term transactions from 100m guilders to 50m guilders. The UK export credit agency, the Export Credits Guarantee Department (ECGD), downgraded Algeria to its D category, in part to counteract recent heavy insurance payouts. The Export-Import Bank of the USA (Eximbank) took similar measures. Other agencies – most notably Coface of France and Italy's SACE – maintained their support (see below), but given the importance of trade credits to Algeria's economic strategy, increased reluctance to provide funds at favourable rates was a major problem.

Also problematic was a growing feeling among commercial banks that (short-term lending apart) they were reaching the limits for their Algerian lending. In 1988 syndicated medium to long-term commercial bank loans provided over $1bn, with Japanese banks taking a lead. In 1989 it was clear that borrowing from this source would be much-reduced, with multilateral agencies and bilateral borrowers providing a greater proportion of funds needed to make up the financing gap. This was, however, to be at the cost of a cornerstone of traditional Algerian ideology.

IMF agreement

Under pressure from its creditors, Algeria in early 1989 decided to break with its past aloofness (and open criticism) towards the IMF – with whom relations had been improving for some time – and reached an agreement putting in place a standby credit. The standby agreement represented a compromise, opening the way to securing sufficient new finance to avoid the need to reschedule even if the economic outlook deteriorated sharply over the next year.

The decision nonetheless represented a bold break with the past: the FLN had always rejected IMF intervention on the grounds that it would take policy-making away from Algiers, thereby infringing national sovereignty. The avoidance of a rescheduling – which would close Algiers' access to international money markets and further constrain its autonomy of action – remained a centrepiece of Algerian policy.

The burden of external debt remained heavy in 1989, with debt service estimated at around $6bn and (before the standby agreement) multilateral as well as commercial bank finance becoming more difficult to obtain. The external situation was eased by strengthening oil prices and consolidated when in late May the IMF board approved a loan package worth around $620m. This included the 12-month standby loan, worth SDR155.7m, and a SDR315.2m compensatory and contingency financing facility (CCFF). The CCFF, a three-year loan, was intended to help finance shortfalls in merchandise exports and the rising cost of cereals imports.

The IMF accord unlocked funds from other donors, including the World Bank, which approved loans worth $341m in its fiscal year to 30 June 1989 and was expected to disburse about $400m in the next year. The Bank not only accelerated its programme of project lending but, in September 1989, also approved a first structural adjustment loan – called an economic reform support loan – worth $300m. There were suggestions that new World Bank funding may have been in jeopardy if Algiers had not worked out a structured policy for managing its external debt.

Given the political sensitivities surrounding the standby's approval, it was not accompanied by a classic IMF programme – and all of those involved minimised the aspects of conditionality that might be involved. Rather, the standby was said to indicate that the IMF had given its blessing to the next round of the government's reform strategy. This was to be based around an unpublished three-year programme of reforms, affecting such sensitive areas as subsidies and exchange rate policy (with the government privately planning to bring the official exchange rate and the black market rates of the dinar roughly into line by 1993). Further devaluation of the dinar was expected to follow.

Sonatrach settles gas disputes

A key element in the government's strategy to put the economy on an even keel was to maximise hydrocarbons earnings, through offering more competitive prices than in the past and, where necessary, investing in facilities. The new government – which included a young former head of state hydrocarbons concern Sonatrach, Sadek Boussena, as energy minister – argued that while the promotion of non-hydrocarbons industries was essential for the longer term, in the short term it was essential for Algeria to maximise its exports where it could, namely through sales of oil and especially gas. This meant offering gas at competitive prices to potential new customers in Europe, Asia and the USA, and also bringing to an end protracted disputes between Sonatrach and three of its four key European customers for natural gas. The process of reconciliation with Gaz de France (GdF), Spain's Enagas and Belgium's Distrigaz was helped by the concern of the governments of France, Spain and Belgium to support Chadli during a difficult period.

The damaging dispute over long-term natural gas supply contracts with GdF – which had led to a wider trade dispute damaging French interests in what remained a major market for France – was settled in January 1989 with a compromise agreement, involving a new pricing arrangement, the possibility of wide-ranging co-operation between Sonatrach and GdF (including French participation in the planned gas pipeline passing through Morocco) and a compensation payment for arrears claimed by Algeria after GdF began a unilateral pricing policy in January 1987.

The four-year arrangement kept three existing long-term contracts (signed in 1965, 1972 and 1982) in place, calling for GdF to lift 9.15bn cu metres of LNG a year, but meant that GdF must pay more than it previously wanted. It had previously argued for a price around $2.14 per million British thermal units (Btu) but finally agreed to a price around $2.28 per million Btu (and also said it would contribute towards liquifaction costs) – in effect providing a price above the rival Soviet and Dutch prices but below what Algiers was demanding. Sonatrach was invoicing GdF at around $2.79 per million Btu, while GdF was paying about $1.94 per million Btu. To soften the compromise GdF agreed to pay arrears and an initial payment of Fr850m was made.

In some respects the new agreement represented a victory for Sonatrach, overcoming GdF's demands to break with rigid contract terms. In return Sonatrach showed flexibility on the question of pricing structures (as it has already done in its negotiations with new customers) by no longer demanding that prices should be fixed according to fluctuations within a basket of OPEC crudes. Instead there was to be a realistic index based on the market prices of 10 Middle East and African crudes. Both sides were anxious to stress that the gas agreement was a matter for the two companies, rather than their governments. However, the GdF accord was accompanied by intense government activity and ran parallel to a major agreement providing French credits.

Algeria had publicly rejected offers of support immediately after the October 1988 riots. However, the appointment of a more pragmatic government in Algiers and a view in Paris that Algeria was an important country which should be given full support – reflecting a growing belief in Europe that turbulence in the southern Mediterranean could affect the internal politics of EEC member states such as France, where more than 1m Algerians live and work – combined to produce the offer of unprecedented French credits. In early January 1989 it was announced that agreement in principle had been reached on a Fr7bn ($1.2bn) mixed credit package. This was followed three days later by the announcement of the GdF settlement.

This package, it was hoped, would reverse a slump in French trade with Algeria: French exports had fallen to Fr4.63bn in January–June 1988 from Fr5.59bn in the first six months of 1987. The Fr7bn mixed credit package had much in common with the annual financial protocols agreed by France with many states but previously shunned by Algiers. Fr4bn was offered in the form of financial aid, for balance of payments support and projects. Half would come from government loans (repayable over 30 years including 10 years' grace at 2 per cent interest) and the rest from commercial credits. Fr3bn in revolving credits would be guaranteed by France's export credit agency Coface, and used to finance imports of spare parts, tyres and small-scale equipment purchases, rather than capital goods or food. It was intended as quick-disbursing support during a critical period of peak payments and economic restructuring.

This was not to be the limit of French support during the year: in August 1989 the French Economy, Budget and Finance Minister Pierre Beregovoy visited Algiers and offered another Fr1bn in credits to support the Fr3bn line. He also promised at least Fr1bn in financial credits, which would help to roll over payments due on trade debt contracted in 1985–88. Algiers asked for Fr2bn in financial credits, which the French Government agreed to, but which was initially opposed by the banks who would have to raise the money. Whereas trade credits are guaranteed by Coface, financial credits would have to be provided by the banks as sovereign lending, without official support. Banks were unenthusiastic about taking on more of this form of Algerian risk, but following intense political pressure were expected to do so.

The settlement of the GdF dispute set the pace for talks with Enagas and Distrigaz. Enagas' contract was subsequently renegotiated with pricing similar to the GdF deal at around $2.28 per million Btu. Under the terms of the accord Enagas is to take 3bn cubic metres of LNG a year until 2002, after which it will take 3.5bn cubic metres and by 2004 3.8bn cubic metres. An agreement with Distrigaz was formally signed on 22 June 1989, following a dispute clouded by the Belgian firm wanting to reduce its contracted volume of liftings as well as its prices. A compromise was reached with Distrigaz committing itself to import 4.5bn cubic metres of LNG a year, up from the 3–4bn cubic metres then being imported but lower than the 5bn cubic metres which Distrigaz was contracted to take under its previous agreement. Pricing was set at around the $2.28 per million Btu level. Distrigaz also agreed to pay arrears, put at BFr20bn ($490m) according to reports from Brussels, of which BFr11bn would come from provisions already made by Distrigaz.

Sonatrach also devoted new energy to attracting new customers. Among these, Japan's Osaka Gas Company took its first delivery of LNG in 1989, with the gas reportedly priced at rates below the rival Indonesian or Brunei gas. Although this seemed to be little more than a marketing ploy – and the long distance Algerian gas has to travel to reach Japan would seem to preclude much larger contracts – it was reported that Osaka Gas was negotiating a long-term supply contract. More important still, the trans-Atlantic gas trade was moving again, through revised agreements with Panhandle and Distrigas and the prospect of supply contracts with at least two other East Coast utilities. This would do much to secure much-needed export revenues and was well-received by Algeria's creditors.

In order to maximise this export potential, a major expansion of LNG facilities was planned for the 1990s. Four international companies were prequalified to study the revamping of LNG facilities at the Skikda and Arzew complexes – with the prospect of a revamping contract and the extension of all existing LNG plants, which would be essential to bring capacity to a level where it could cope with the 40bn cubic metres a year sales Boussena has forecast by 1994.

Other hydrocarbons facilities are also to be upgraded, and after years of troubled relations with international oil companies, the government is trying to encourage more firms to involve themselves in exploration. Currently only France's Total, Spain's Cepsa, Italy's Agip, the US Anadarko Petroleum Corporation and Australia's BHP Petroleum have commitments in the country.

European credits

It was not only France that provided new financial support in early 1989. While many OECD creditors took stock of their position, several Mediterranean partners promised much-increased support. These included Spain, who after three years of talks offered a credit package worth up to Pts140bn ($1.2bn) for the 1989–91 period. The package, providing up to $380m a year, consists of 50 per cent soft credits from Spain's concessionary lending agency Fondo de Ayuda al Desarrollo and the rest from commercial credits, repayable at the OECD consensus rate. Notably Madrid's credit offer for Algeria was even larger than its record Pts125bn credits offered to Morocco in 1988. Italy weighed in with a IL320bn credit package for 1988–90, agreed in December 1988 during the visit of the Italian Minister of Foreign Affairs, Giulio Andreotti. The package mainly consisted of soft government loans and covered priority projects in sectors such as agriculture and industry, including support for the planned Fiat car plant and other motor industry projects.

While Japanese banks proved less keen to lend in 1989 than in 1988, Tokyo maintained its strong support, with Exim Japan providing three co-financing loans with the World Bank worth a total of $243.5m and also more credits for Sonatrach, which backed work by Japanese firms on the upgrading of hydrocarbons facilities.

Reforms go ahead

Fresh credits are essential to allow Algeria to pull through a very difficult period. Just as essential is the need for further reforms in all sectors of the economy. Changes made since 1983 have had beneficial effects but it is widely felt that much more still is needed.

Reflecting the need for change, the planning system is being overhauled with central and local government, public sector organisations and autonomous Entreprises Publiques Economiques (EPEs) working together to prepare a third Five-Year Plan for the period 1990–94. The new Plan will be a major departure from its predecessors with some 40 committees, each responsible for a sector, formulating policy in the form of a medium-term plan for its sector, to be incorporated into the Five-Year Plan.

The 1990–94 Plan is intended not to give detailed policy directives but to define broad national objectives. If a company thinks the direction the Plan is leading it is wrong it can object and go its own way – in theory. The aim is to give the economy direction, but to avoid the political interference that in the past has taken initiative away from managers and inflated the bureaucracy to unacceptable levels.

If Algeria is to succeed in facing the challenges of the 1990s it is by overcoming unnecessary state control and freeing individuals to stimulate growth that the regime will depend for its success. While it is likely that Algeria will remain a heavily centralised state, a liberal trend is establishing itself that offers the promise of a substantial break with the past traditions of a corporate state. As the October 1988 riots clearly showed, this is more than an argument over ideological niceties.

BUSINESS GUIDE

Time: GMT +1.

Climate: (Coastal region) Mediterranean with warm and humid summers from May to Sep, temperatures reaching 30-38°C. From Oct to May rainfall can be heavy, especially from Nov to Jan, and temperatures vary from 22-30°C – coolest months Dec and Jan. (Inland) Hot and dry all the year round with daytime summer temperature up to 43°C, cool/cold nights.

Entry requirements
Passport: Required by all foreign nationals.
Visa: Required by all except nationals of Denmark, Italy, Spain, UK, Finland, Norway, Sweden, Switzerland. Visas cannot be issued at the border. Extensions can be obtained from the Service des Etrangers. Entry visas usually valid for 90 days.
Prohibited entry: Nationals of Israel, Rep of Korea, Malawi, South Africa, Taiwan and any visitor whose passport shows evidence of a visit to Israel or South Africa. Visitors without a confirmed return date of travel indicated in their ticket may be refused entry.
Currency: Visitors entering the country must change a fixed sum (AD1000) of foreign exchange into Algerian dinars at point of entry. Unlimited amounts of foreign currency may be imported but must be declared. To enable re-export of any unchanged foreign currency, declaration forms issued on arrival should be kept, used at each successive currency change and surrendered on departure. Failure to comply with these regulations may mean visitors are liable to forfeit the currency. Import/export of local currency limited to AD50. Note: it can sometimes be difficult to reconvert dinars to foreign currency.

Health precautions
Mandatory: Yellow fever and/or cholera certificate if arriving from infected or endemic areas.
Advisable: Cholera, typhoid, polio vaccinations. Malaria risk in some areas, therefore prophylaxis recommended. Water precautions should be taken throughout the country.

Representation overseas: Abidjan, Abu Dhabi, Accra, Addis Ababa, Amman, Antananarivo, Athens, Baghdad, Bamako, Beirut, Berlin, Beijing, Belgrade, Berne, Bissau, Bonn, Brasilia, Brazzaville, Brussels, Buenos Aires, Caracas, Conakry, Cotonou, Dakar, Damascus, Dar es Salaam, The Hague, Hanoi, Harare, Havana, Islamabad, Jakarta, Jeddah, Khartoum, Kinshasa, Kuwait City, Lagos, Lima, Lisbon, London, Luanda, Madrid, Maputo, Mexico City, Mogadishu, Monrovia, Moscow, Nairobi, New Delhi, Naimey, Nouakchott, Ottawa, Paris, Prague, Rome, Sana'a, Sofia, Stockholm, Tehran, Tokyo, Tripoli, Tunis, Vienna, Warsaw, Washington, Yaoundé.

Air access
National airline: Air Algérie.
Other airlines: Aeroflot, Air France, Alitalia, Balkan, CSA, Egyptair, Iberia, Interflug, JAT, Libyan Arab LOT, Lufthansa, Malev, Royal Air Maroc, Sabena, Saudi, Swissair, Syrian Arab Airlines, Tarom, THY Turkish, Tunis Air, Tunisavia.
International airports: Algiers-Houari Boumédienne (Code: ALG), 19 km from city, duty-free shop, restaurant, buffet, bank, post office, shops, car hire; Annaba-Les Salines (Code: AAE), 12 km from city, restaurant, buffet, bank, post office, shops, car hire; Constantine-Ain el Bey (Code: CZL), 9 km from city, restaurant; Oran-Es Senia (Code: ORN), 10 km from city, bank, car hire, buffet.

Surface access
Main ports: Algiers, Annaba, Arzew, Bejaia, Oran. Regular ferry services connect Algiers with Marseilles and other Mediterranean ports.
Overland access: The National Travel and Transport Company (SNTV) operates bus services linking Libya, Niger, Morocco and Tunisia with Algeria. Road access also possible from Mali by one of the trans-Saharan highways. A daily train service (the Transmaghreb) links Tunis with Algiers and Oran.

Hotels: Classified from one to five stars. Major hotels in the business centres tend to be expensive. The government travel agency, Altour, operates many of the hotels used by business visitors – these are generally of a good standard. Advisable to book well in advance as accommodation in Algiers is difficult to obtain. Service charge usually 15 per cent.

Credit cards: Use of credit cards is restricted.

Car hire: Available in most main towns and at airports. International driving licence and third party insurance required. Maximum speed limits 50 kph in town, 100 kph on main roads.

City transport
Taxis: Widely available in main centres. Identified by local colour code. Radio taxis in Algiers. Taxis are metered and a minimum fare system operates. Surcharges after dark. Usual tip 10 per cent.
Buses: State-owned service operates in Algiers. Can be overcrowded during rush hours. Daily and longer duration tickets are available.

National transport
Air: Frequent services from Algiers to Annaba, Constantine and Oran are operated by Inter Air Services (internal branch of Air Algérie). Regular flights also link these towns with other principal centres. Fares are generally low and overbooking can occur, especially in summer.
Road: Total length 82,000 km, of which 24,000 main, 19,000 secondary roads. Main roads are in good condition generally.
Buses: Long-distance coach services operated by SNTV and Altour. Bookings for long trips should be made well in advance. Routes include: Algiers-Beskra, Algiers-Oran, Algiers-Djelfa, Algiers-Metlili, Algiers-Tamanghest, Adrar-Béchar.
Rail: Service operated by SNTF. Two classes; some services are air-conditioned and/or have couchettes. Routes include: Algiers-Oran, Béchar-Mohammadia, Mohammadia-Mostaganem, Algiers-Annaba, Constantine-Touggourt, Constantine-Skikda, Algiers-Tizi Ouzou, Annaba-Tébessa.

Public holidays
Fixed dates: 1 Jan, 1 May, 19 Jun (Anniversary Day), 5 Jul (Independence), 1 Nov (Anniversary of the Revolution). Also Muslim holidays.
Variable dates: Id el-Fitr, Id el-Adha, Hegira, Ashura, Mawlid-al-Nabi.

Working hours: The Muslim weekend is observed – Thursday and Friday.
Government and business: (Sat-Tue) 0800 – 1200 and 1300 – 1700, (Wed) 0800 – 1200 and 1300 – 1600.
Banking: (Sun-Thur) 0845/0900 – 1530/1600.
Shops: 0800 – 1230 and 1430 – 1800.

Telecommunications
Telephone: Public call boxes in all post offices, some streets and main hotels. Automatic dialling available. For enquiries dial 19, for telegrams 13, for police 17. Dialling code for Algeria: IDD access code + 213 followed by area code (2 for Algiers) followed by subscriber's number.
Telex: International telex facilities at main post office in Algiers and at El Aurassi, Algiers and El Safir hotels. Service is generally poor and restricted to certain hours. International telex code: DZ. The main post office in Algiers maintains a 24-hr telegraph service.

Postal service: All business correspondence should be sent air mail and marked 'Documentation Technique – Sans Valeur Commerciale'. Advisable to rent a post box.

Banking
Central bank: Banque Centrale d'Algérie. *Head offices:* Banque Algérienne de Développement, Banque Algérienne de l'Agriculture et du Développement Rural, Banque Extérieure d'Algérie, Banque Nationale d'Algérie, Caisse Centrale de Coopération Économique, Caisse Nationale d'Epargne et de Prévoyance, Crédit Populaire d'Algérie. *Commercial banks permitted to transact foreign business:* Banque Extérieure, Banque Nationale, Crédit Populaire, Banque

BUSINESS GUIDE

Algérienne de l'Agriculture et du Développement Rural.

Trade fairs: International Trade Fair held every May-June in Algiers. Several more specialised exhibitions take place every year.

Electricity supply: Electricity supply varies from 127-220 V. All types of plug fittings in use.

Representation in capital: Albania, Angola, Argentina, Australia, Austria, Bangladesh, Belgium, Benin, Brazil, Bulgaria, Burundi, Cameroon, Canada, Cape Verde, Chad, China PR, Congo, Cuba, Czechoslovakia, Denmark, Finland, France, Gabon, German DR, German FR, Ghana, Greece, Guinea, Guinea-Bissau, Hungary, India, Indonesia, Iran, Iraq, Ireland, Italy, Côte d'Ivoire, Japan, Jordan, Korea PDR, Kuwait, Lebanon, Libya, Madagascar, Mali, Mauritania, Mexico, Mongolia, Netherlands, Niger, Nigeria, Oman, Pakistan, Philippines, Poland, Portugal, Qatar, Romania, Saudi Arabia, Senegal, Somalia, Spain, Sweden, Switzerland, Syria, Tunisia, Turkey, USSR, UAE, UK, USA, Vatican, Venezuela, Vietnam, Yemen AR, Yemen PDR, Yugoslavia, Zaïre, Zimbabwe.

Hotels: Algiers
Albert, 5 Avenue Pasteur (tel: 63-00-20, 63-00-22; tx: 53556).

Aurassi, Avenue Frantz Fanon (tel: 64-82-52; tx: 52475).

El Djazair, 24 Avenue Souidani Boudjemaa (tel: 60-10-00; tx: 661115).

Safin, 1 Rue Asselah Hocine (tel: 63-50-40; tx: 52142).

Suisse, 6 Re Drouillet (tel: 63-49-52, 63-49-11).

Car hire: Algiers Algérie Auto-Tourisme (AAT): 5 Rue du Professeur Curtillet (tel: 59-19-99, 59-19-95; tx: 53423 aat).

Alger Location de Voitures: 14 Rue Docteur Saadawe (tel: 63-69-06).

Office National Algérien du Tourisme: 5 Boulevard Ben Boulaid (tel: 63-13-87, 63-63-88; tx: 53960).

Other useful addresses
Algiers: Agence Nationale d'Edition et de Publicité (ANEP), 1 Boulevard Pasteur.

Air Algérie, 1 Place Maurice Audin, PO Box 858 (tel: 63-12-82/89; tx: 52436).

Algérie Presse Service (APS), 7 Boulevard Ché Guévara.

Banque Centrale d'Algérie, 8 Boulevard Zirout Yousef (tel: 64-75-00; tx: 52046).

Central Post Office, 5 Boulevard Mohamed Khémisti.

COMEX (Chamber of Commerce), 6 Boulevard Amilcar Cabral (tel: 62-70-44).

Directorate of Statistics, Ibn Badis Al-Mouiz Chemin, El-Biar (tel: 78-03-23; tx: 52560).

Entreprise Nationale d'Engineering Petrolier, 2 Boulevard Mohamed V (tel: 63-08-92).

Entreprise Nationale de Raffinage et de Distribution des Produits Petroliers, Avenue Ahmed Ghermoul.

Institut National Algérien du Commerce Extérieur, Boulevard Amilcar Cabral.

Office Algérien Interprofessionel des Céréales, Rue Ferhat Bousaad.

Office National Algérien du Tourisme, 27 Rue Khelifa Boukhalfa (tel: 61-26-55, 61-29-86; tx: 66383, 66339).

Office National des Foires et Expositions, PO Box 656.

Radiodiffusion Télévision Algérienne (RTA), 21 Boulevard des Martyrs (tel: 60-23-00; tx: 52042).

Service des Étrangers, Boulevard Zeroud Yousef 19A.

SNTF, 21 Boulevard Mohamed V (tel: 61-15-10).

Société Centrale pour l'Equipement du Territoire, Rue Sergent Addoun.

Société Nationale d'Edition et de Diffusion, Boulevard Zeroud Yousef (deals with books, stationery, advertising).

Société Nationale de Constructions Mécaniques, Route Nationale, Bukhadem (deals with vehicles, agricultural equipment etc).

Société Nationale de Gestion et de Développement des Industries Alimentaires, Avenue Mustapha El Ouali Sayed.

Société Nationale des Matériaux de Construction, Rue Didouche Mourad.

Société Nationale d'Etudes de Gestion, de Réalisations et d'Exploitations Industrielles, Rue Khélifa Boukhalfa.

American Samoa

T. R. Hughes

(map on page xxiv)

KEY FACTS

Official title: Territory of American Samoa

Head of state: President of the USA

Head of government: Governor Peter Tali Coleman

Capital: Fagatogo (on Tutuila)

Official Languages: English, Samoan

Currency: US Dollar ($ = 100 cents)

Area: 196 sq km (5 islands) – Tutuila: 135 sq km

Population: 36,900 (1988)

Foreign debt: $52m (1987) (US Pacific Islands)

The resignation of Fofo If Sunia (D), American Samoa's non-voting delegate to the US House of Representatives, for misusing congressional funds and the subsequent close fight in November 1988 to fill his Washington seat overshadowed everything else on the political scene. Sunia's successor Faleomaveaga Eni F. Hunkin Jr kept the Democratic flag flying but only just. He failed at first to get a majority in the four-cornered contest but secured a narrow win in the runoff. Like Americans elsewhere, American Samoans split their ballots. The election also saw the comeback of the veteran politician Peter Tali Coleman, American Samoa's first governor and local Republican Party chairman, to the gubernatorial post, signalling that he is still a power in the island although he is nearly 70.

American Samoa, an unincorporated territory, is the United States' only possession in the southern hemisphere. Its people are US nationals and can move freely within US territory. The absentee vote (almost 20 per cent) served to emphasise how much the American Samoans value this arrangement: more than double the resident population of about 36,900 are estimated to be living in the US mainland or Hawaii.

Sunia resigned from the House in September 1988 before it could take action on his guilty plea in a payroll-padding case. He admitted a single count of conspiracy after being charged with defrauding the US Government of more than $130,000 in payments to Samoans falsely listed as working on his staff and was sentenced to 5 and 15 months in prison. He also paid a steep fine: $65,000. Sunia's long-time administrative aide, Matthew Iuli, was jailed for three months.

Quite apart from the public indignity, the case was a blow to American Samoa's cause as Sunia had set a precedent by becoming chairman of a House subcommittee and was well versed in using his Capitol Hill powers to find new and legitimate ways of funding projects in his homeland. A high chief and matai, Sunia was the first Samoan to hold the delegate's seat in the House when elected under a revised constitution in 1980. By resigning so promptly, he hoped that Governor A. P. Lutali, a Democratic ally, could fill his seat quickly in a snap election so that a successor would have seniority over other newcomers to be seated in the House in January 1989. But constitutional lawyers ruled out this tactic.

Sunia's successor cut a popular profile in Pago Pago despite his uphill struggle at the polls. At 45 and with excellent credentials, Faleomaveaga (a chiefly title but one that Eni Hunkin has been using as a surname) is no stranger to Washington either. As the son of a US Navy petty officer, he spent many years of his life on the US mainland before being elected American Samoa's lieutenant-governor in 1984. He is a Vietnam veteran, lawyer and sportsman.

In the fourth gubernatorial combat since the territory was granted a measure of autonomy in 1977, Peter Coleman's win over the incumbent A. P. Lutali was achieved in an atmosphere akin to a return world heavyweight prize fight. Coleman had served for two terms, each time defeating Lutali, but was barred by law from offering himself for a third consecutive spell. This time the roles of the two political giants were reversed with challenger Coleman overcoming fears that he might not be able to win over the young new voters registering for the first time.

Faleomaveaga's priorities include pressing for an improvement in the island's 89-year-old relationship with the United States to get rid of the 'colonial' label together with a plea for an economic deal to upgrade its electricity supply, water and roads and making the island less dependent on Washington. In fact, the island has one of the best infrastructures in the Pacific. Its international airport can handle aircraft of any size. The harbour is one of the deepest in the region and can handle most vessels. Telecommunications are adequate.

Economic link

Economically as well as politically, the island is closely enmeshed with the US mainland despite cultural differences and the vast distances between them. American Samoa has relatively well-developed Western consumer tastes even though the per capita income is about one-fifth of that of Hawaii. About 80 per cent of its imports originate in the United States with an increasing flow of goods from Japan, Australia and New Zealand; but there is very little inter-island Pacific trade.

The government is the biggest single employer giving jobs to some 4,000 out of a labour force of 11,000. Total US aid is about $45m a year which is the equivalent of half the local budget but the US contribution has remained static for a decade – an effective drop in real terms of 46 per cent. The gap has been made up of local taxes and revenue from the expanding private sector. The number of private businesses registered in the territory in 1986 was put at 1,388, up 13.6 per cent over 1985. Due largely to the election, however, the government expects a budget shortfall of around $5m in the 1989 financial year.

Imports far exceed exports which are almost exclusively generated by two tuna canneries in Pago Pago employing altogether 3,600 workers and contributing upwards of $20m a year to the local economy. The two canneries in American Samoa and six others make up the entire US tuna canning industry: the territory's share of overall US production is about a quarter.

While the local government has provided long-term leases and tax incentives to persuade the canneries to stay, it is equally aware of the need for diversification. Tourism, stagnant for some years, is thought to be a good option provided it does not impinge on the local way of life.

Cultural ties

American Samoans have strong cultural ties which have kept them nationalistic and cautious about changes. Local government has its roots in the traditional Samoan political structure, the matai or village chief system which is a complex hierarchy beginning at village level and leading up to a paramount chief who would be the equivalent of a Western head of state. Under United States influence, the system has been modified to fit a Western-style democracy. But, in essence, it works as it has for centuries.

Communal ownership of land is a disincentive to foreign investors and efforts are under way to modify the land tenure system. The exodus of young, skilled workers is another drag on progress. But the Bank of Hawaii, in an encouraging review, says that with careful accommodation of Samoan customs and meeting Western requirements for investment, the potential for economic development appears good, especially as the island has unfettered access to the US market.

COUNTRY PROFILE

Political system: American Samoa is a flag territory of the US, administered by the US Department of the Interior and is represented by a non-voting delegate to the US House of Representatives. Local executive power rests with an elected governor. The legislature has 18 senators elected according to Samoan custom from local chiefs or matai and 20 representatives chosen by popular vote, plus one non-voting member in the lower house.

Main cities/towns: Pago Pago (population 5,000* in 1987; on Tutuila which has a total population of about 35,000).

Language: English is the official language and is used for business and commerce, but Samoan (a Polynesian language) is more common among the local population.

Media
Press: Newspapers: *Samoan News* three times a week and *Samoa Journal Advertiser* weekly.
Broadcasting: Radio Samoa broadcasts in English and Samoan transmitted 24 hrs/day. KSBS FM broadcasts in English and Samoan 0600 – 2400. Three TV channels broadcast at least 7.5 hrs/day in colour.

Domestic economy: American Samoa's subsistence agricultural economy has been transformed during its association with US; Western-style consumer values supported by US aid, about $2,600 per capita p.a. There is a programme to accelerate privatisation of services, transferred garbage collection, ambulance services, a motor pool and hospital maintenance to private management. Certain state-owned hotels are expected to follow.

Employment: US administration, civil administration and the business sector support a wage economy on main islands; many Samoans work temporarily overseas in USA and elsewhere.

Agriculture: Volcanic soil, about one-third of land area cultivable. Smallholdings produce coconuts, bananas, breadfruit, cassava, yams, pineapples, vegetables, fruit and other produce, mainly for local consumption. Tuna and deep-sea fishing is important to the economy.

Industry: Two US-owned fish canneries employ at least 35 per cent of labour force. Other industries include meat canning, dairy produce, jewellery, handicrafts and tourism. The US Government is trying to encourage joint ventures and other foreign investment for any product with a 30 per cent local content. Three new factories, processing soap, liquor and perfume, were opened in 1987.

Time: GMT -11 hrs. EST -6 hrs. EDT -7 hrs.

Climate: Wet and tropical, annual rainfall up to 750 cm in mountainous areas.

Entry requirements
Passport: Required by all except US citizens with proof of citizenship. There are regulations concerning unexpired duration of passports.
Visa: Not required by US citizens with proof of adequate funds. Entry permits granted to all others with proof of onward passage and adequate funds for stays of less than 30 days. Quotas regulating aliens admitted to work in each year allow the admission of 1,000 Western Samoans and 400 people from all other countries, not including US or government employees.
Customs: Import of certain items may be subject to regulation or licensing.

Health precautions
Mandatory: Vaccination certificates required for yellow fever if travelling from infected areas.

Air access
Airlines serving: Hawaiian, Polynesian, Samoa Aviation.
International airport: Pago Pago International (PPG), 14 km from town; duty-free shop, restaurant, shops and car hire.

Surface access
Main ports: Pago Pago.

National transport
Road: Most of the 70-80 km road network on Tutuila, 90 per cent of it surfaced. A new road and two bridges are being built in the Fagatogo area of Pago Pago.
Buses: Limited bus service offers transportation to passenger requirements.

Working hours
Government and business: (Mon-Fri) 0730/0830 – 1800/1730; (Sat) 0830 – 1200.
Banking: (Mon-Fri) 0830 – 1700; (Sat) 0800 – 1200.

Telecommunications: The US State Department is to develop Pago Pago as a communications relay centre for its diplomatic and aviation interests in the Pacific.
Telephone: Dialling for American Samoa, IDD access code + 684 followed by subscriber's number (no area code necessary). There were 7,800* telephones in 1987.
Telex: Telex service available.

Banking
Head offices: Economic Development Bank of American Samoa, Amerika Samoa Bank.
American Samoan branches: Bank of Hawaii.

Weights and measures: US system.

Hotels: Pago Pago, Tutuila
Herb and Sia's, P.O. Box: 430 (tel: 6335413).

Pago Pago Rainmaker, P.O. Box: 996 (tel: 6334241; tx: 782511; fax: 6335959).

Car hire: Pago Pago, Tutuila
Avis: (tel: 69991630; tx: 530).

Morris Scanlan: (tel: 6335520).

Pacific Tours & Travel: (tel: 6335224).

Pavitt-U-Drive: (tel: 6991456).

Samoa Tours: (tel: 6335884).

Samoan Holiday: (tel: 6334692).

Taxis: Pago Pago, Tutuila Black Ace Taxi Service (tel: 6335445/5859).

Business advice and information
Development Bank of American Samoa, PO Box 9, Pago Pago, 96799.

Office of Economic Development and Planning, Territorial Planning Commission, Pago Pago, 96799 (tel: 633-5156).

Office of Tourism, Convention Centre, Pago Pago, 96799 (tel: 633-5187; fax: 633-1250).

Andorra

Tom Burns

Elections to vote in Andorra's 28-member council of elders at the end of 1988 reflected a new mood in the extremely ancient micro-state of Andorra. Virtually all the candidates – there are no political parties as such just as there are no trade unions – had the term 'progressive' built into their manifesto. And for the first time there were electoral posters. Such stylistic details alone heralded new times.

More important was the fact that when the second round of voting took place on 17 December, it emerged that Mr Josep Pintat, Andorra's *cap de govern* or prime minister, could no longer count on a comfortable majority of like-minded counsellors or elders to keep him in power. Mr Pintat is considered a representative of the old school of Andorran politics. Those who had made the most of the 'progressive' and 'renewal' catchwords when they were canvassing each and everyone of Andorra's 7,000-odd voters had made impressive gains in the polls. Men such as Mr Oscar Ribes who represent a new approach to the challenges facing Andorra in the 1990s look set to make the running in the new council of elders, an assembly that is also known as the council of the valleys, and which is formed by four representatives from each of the seven valleys that together make up Andorra.

The overriding hurdle that faces Andorra concerns coming to terms with the single market Europe. There is a consensus over what sort of deal the tiny state should make with the European Community. Andorra ought to have a third country status in everything that concerns agriculture and should seek a form of customs union for other sectors. There is lessagreement, however, over the trade-off steps that Andorra should take in order to secure such a deal.

Old school politicians, men closely linked with Andorra's banking, tourism and retailing sectors, would prefer to leave as much as possible, and preferably all, of the state's idiosyncrasies untouched. A newer generation of political leaders advocates, in contrast, a number of fundamental institutional changes. The latter do so out of a sense of expediency, for the EC is unlikely to meet Andorran negotiating positions while the state remains politically rooted in the Middle Ages.

The position of Spain, for example, is an important one when it comes to striking an EC deal. There are some 20,000 Spaniards working in Andorra and they, like all other non-Andorrans, are completely disenfranchised. It is not just that they cannot vote; they have inadequate health care protection and are denied trade union rights. The socialist government of Mr Felipe Gonzalez in Madrid had made it clear that up-to-date social security and labour law provisions in Andorra are a virtual prerequisite for an agreement with the EC.

Madrid's influence over Andorran affairs is considerable. Constitutionally the occupier of the Catholic Bishopric of Seu d'Urgell, a Spanish episcopal see that extends into Andorra, is one of the two co-princes of what is technically called the Principality of Andorra. The other co-prince, under an agreement that has lasted for just over 700 years, is the French head of state, for the latter is held to be the legal successor of the medieval Counts of Foix who, back in the thirteenth century, contested ownership of Andorra with the Seu d'Urgell bishops.

However, the chief leverage that Madrid can exert over Andorra lies with the dependence of the Principality on Spanish tourism. Often likened to a giant shopping emporium, Andorra is a centre for cheap, duty free, manufactured goods and at least two-thirds of the bargain hunters who crowd into its capital city Andorra la Vella are Spaniards, thus making a decisive contribution to the Andorran economy. In theory Madrid could severely curtail the duty free zone advantages that Andorra enjoys by simply enforcing stricter customs control.

KEY FACTS

Official title: The Principality of Andorra

Head of state: Co-Princes Monsignor José Martí Urgell) and François Mitterrand (President of France)

Head of government: President Josep Pintat Solans

Capital: Andorra la Vella

Official Languages: Catalan

Currency: French Franc and Spanish Peseta

Exchange rate: Fr6.31 per $; Peseta 118 per $ (Nov 1989)

Area: 468 sq km

Population: 50,000 (1989)

GDP per capita: $9,834 (1987)

Labour force: 17,604 (1983)

The Spanish Government, in its bid to encourage changes in Andorra as part of the EC negotiations, can also apply pressure on the Principality's pretensions to remain as a privileged financial zone, an offshore centre, albeit wedged between Spain and France. Madrid is particularly concerned about allegedly excessive banking secrecy in Andorra that encourages tax evasion and the laundering of illicit funds. There is little doubt that the Madrid government will voice such concerns in the course of the Principality's negotiations with the EC.

The likelihood is that a change in Andorra's often archaic customs is very much on the agenda. It will come in part because of Spanish, and also of French, pressure and in part, too, because, the 'progressive' elements of the Andorran political establishment in the aftermath of the December 1988 elections, will be seeking to update the Principality's traditional practices so as to better negotiate with the EC. A change in the franchise is one possibility as is a move towards a more proportional voting system. There is also likely to be an overall reform of Andorra's company law in order to encourage more fluid foreign investment. Under present legislation Andorrans must own two-thirds of a company based in the Principality.

The age-old restrictions on foreign equity ownership have contributed to the considerable wealth of the Andorran nationals. Numbering only some 15,000 out of a total resident population of close to 45,000, the Andorran nationals are estimated to have a per capita income in the region of $15,000 a year. The restrictions however look increasingly incompatible with a Europe that is committed to deregulation and the free transfer of capital. They are, in addition, counter-productive in the eyes of some who argue that the Principality now has to encourage investment in order to build up tourism.

Expansion of tourism

The Principality's tourist authorities are tenaciously promoting skiing in winter and hill trekking in summer under a well-funded 'Escape to Andorra' campaign. Tourism is viewed as the natural currency earner for Andorra and as one which will help offset the losses in the retail sector that will inevitably follow with the implementation of customs union with Europe. Part and parcel of such ambitions is Andorra's clear intention to profit from the high profile exposure that the surrounding area of Spain will be receiving in 1992 when Barcelona stages the summer Olympics. Andorra plans to stage a so-called small European nations Olympics that will bring together athletes of other continental micro-states such as San Marino, Liechtenstein and Monaco.

There are plans, also, for attracting some of the high-tech investment that is currently pouring into Barcelona and to its surrounding area of Catalonia. Two new road tunnels are presently being built through the Pyrenees that will considerably improve Andorra's communications with both Toulouse and France and with Barcelona and Spain. The conditions for research and skilled manufacturing investment are potentially there but as yet there has not been anything to suggest that a major transfer of resources to the area is likely to come from France or Spain.

Outlook

It is more probable that under a new political leadership, with a relaxation of arcane restrictions, and in the framework of an understanding with the EC, Andorra will develop during the 1990s along similar lines to the other micro-states which it will be inviting to its tailor-made Olympic programme. Andorra will consolidate its status as a noted financial centre, as a high premium retirement zone and as an above-average tourist area.

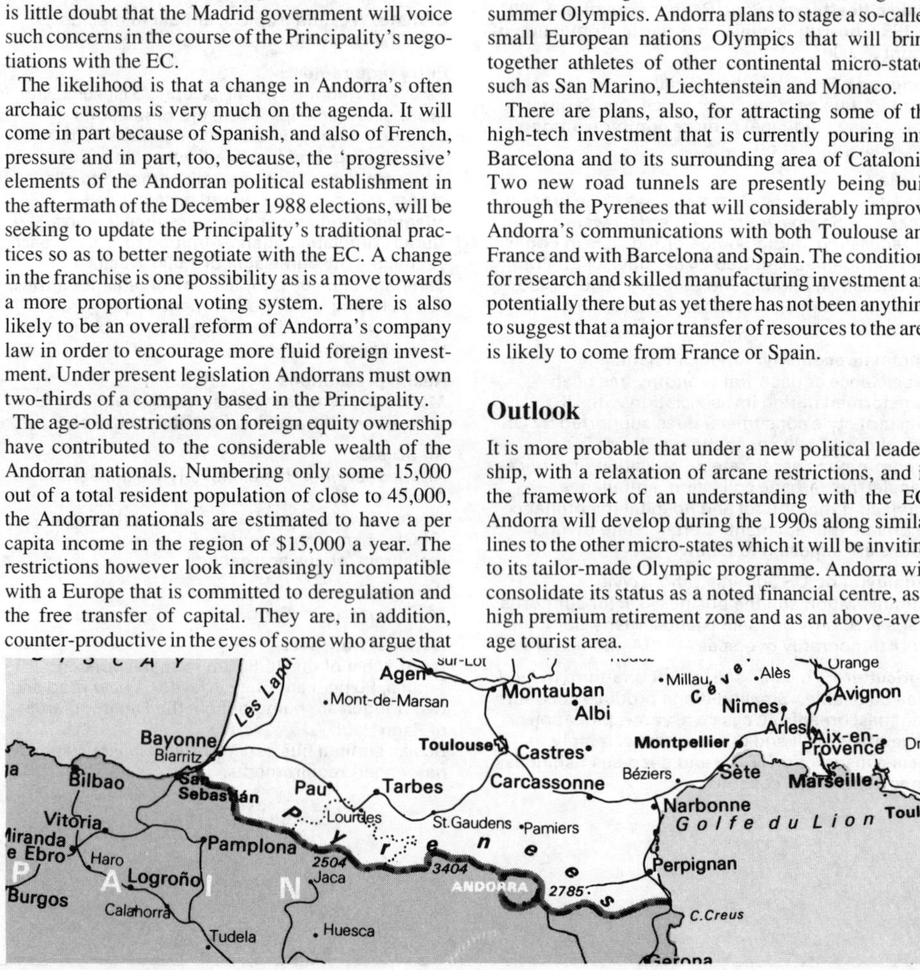

For all the changes that will inexorably alter Andorra's political, social and even business profile, there are some elements in the Principality's make-up that will remain frozen in time. Andorra, founded by Charlemagne, is intensely loyal to its traditions whenever possible. One dearly-held tradition is that Andorrans do not pay income tax. Another is that they pay a special levy to the Bishop of Seu d'Urgell every Christmas consisting of seven hams, 14 capons and 28 cheeses – one ham, two capons and three cheeses from each of the valleys that send representatives to the council of elders.

COUNTRY PROFILE

Political system: Co-principality under joint sovereignty of president of France and Spanish bishop of Urgell. Day-to-day government by an elected head of government and the 28-member General Council of the Valleys.

Main city: Andorra la Vella (population 19,072 in 1987).

Media
Press: Several local weeklies, including *Poble Andorra* with a circulation of 3,000, and *Andorra 7*. *Broadcasting:* The state-owned radio and television stations are not operating at present. However around six TV stations and several radio stations can be received from France and Spain.

Domestic economy: Economic activity based on small-scale agriculture, commerce and tourism.

Employment: Large immigrant workforce, mainly Spanish. Tourism and allied industries major employer.

External trade: Reliance on imports of foodstuffs, raw materials, manufactures and consumer goods from Spain and France. Major entrepôt for numerous European goods owing to favourable excise duties.

Agriculture: Traditional small-scale farming. Principal crops tobacco and potatoes. Also animal husbandry and forestry.

Industry: Small manufacturing sector primarily servicing tourism.

Energy: Near self-sufficiency in hydro-electricity.

Banking: Seven commercial banks with some 34 branches. French and Spanish currencies in use. Strict secrecy laws in force.

Legal system: Based on French and Spanish civil codes.

BUSINESS GUIDE

Time: GMT + 1 (GMT + 2 from late Mar to late Sep).

Climate: Warm summers and moderately cold winters; temperatures range from 0°C to 30°C.

Entry requirements: For nationals of France and Spain only an identity card is required; others require a passport or similar. There are no health requirements or currencyrestrictions.

Air access: Nearest international aiports are Barcelona (Spain), 225 km, and Toulouse (France), 180 km from Andorra.

Surface access
From Spain Rail: Barcelona to Puigcerda, then by bus to La Seu d'Urgell and Andorra. Madrid to Lleida (Lérida), then bus to La Seu d'Urgell and Andorra. *Road:* Barcelona-Andorra via Puigcerda and La Seu d'Urgell; Barcelona-Andorra via Igualada, Calaf, Pons and La Seu d'Urgell. *From France Rail:* Trains to Ax-les-Thermes, Hospitalet or La Tour de Carol (Toulouse-Perpignan lines), then bus to Andorra. *Road:* From Hospitalet or La Tour de Carol to Pas de la Casa, on the frontier, then Port d'Envalira.

Hotels: Around 250 hotels, most with modern facilities.

City transport
Buses: Constant minibus services link all the villages; there are 186 km of roads of which 162 km are paved.

Public holidays
Fixed dates: 1 Jan, 6 Jan, 19 Mar, Good Friday, Easter Monday, 1 May, 24 Jul, 15 Aug, 8 Sep, 8 Dec, 25 & 26 Dec.

Telecommunications
Telephone: Automatic service connects every part of country. Approximately 350 phones per '000 inhabitants. International dialling code: 33 628. *Telex:* Modern telex service. Country code: 590 AND. *Telegram:* Service available main hotels and Spanish and French post offices in Andorra la Vella.

Postal service: Service operated by Spanish and French in capital. Andorran stamps used for international correspondence. Postal service within Andorra free.

Banking
Banc Agricol i Comercial d'Andorra, Mossen Cinto 6, Andorra la Vella (tel: 21-3-33; tx: 201).

Banca Cassany SA, Avinguda Meritxell 39-41, Andorra la Vella

Banc Internacional, Avinguda Meritxell 32, Andorra la Vella (tel: 20-0-37; tx: 206 AND).

Banca Mora SA, Placa Coprinceps 2, Les Escaldes (tel: 20-6-07; tx: 222).

Banca Reig, Avinguda Meritxell, Andorra la Vella (tel: 22-6-18).

Credit Andorra, Avinguda Princep Beullock 19 (tel: 20-3-26; tx: 200 CREDIAND).

La Caixa, Pl Rebés, Andorra la Vella (tel: 20-0-15; tx: 261).

Principal newspapers
Poble Andorra, Avinguda Meritxell 112, Andorra la Vella (tel: 22-5-00).

Other useful addresses
French Post Office, Avinguda Meritxell 57, Andorra la Vella (tel: 20-4-08).

Ministry of Finance, Andorra la Vella.

Ministry of Commerce, Industry and Agriculture, Andorra la Vella.

Ministry of Tourism and Sport, Andorra la Vella.

Pas de la Casa Customs, Franco-Andorran frontier (tel: 55-1-20).

Police (tel: 21-2-22).

Sant Julia de Loria Customs, Spanish-Andorran frontier (tel: 41-0-90).

Servei de Telecomunicacions d'Andorra STA, Avinguda Meritxell 110 (tel: 21-0-21).

Sindicat d'Iniciativa de les Valls d'Andorra, C/Dr Vilanova, Andorra la Vella (tel: 20-2-14).

Spanish Post Office, C/Joan Maragall, Andorra la Vella (tel: 20-2-57).

Hotels: Andorra La Vella
Andorra Center, Doctor Nequi 12 (tel: 24999; tx: 377 AND; fax: 28329).

Andorra Palace, Prat de la Creu (tel: 21072; tx: 208 AND).

Andorra Park, Roureda Guillemó (tel: 20979; tx: 377 AND; fax: 28329).

Avet d'Or, Av Meritxell 31 (tel: 20558).

Bellavista, Av Meritxell 26 (tel: 21288; tx: 305 AND).

Bellpi, Av Santa Coloma 30 (tel: 20651).

Cassany, Av Meritxell 28 (tel: 20636).

Celler d'En Toni, Verge del Pilar 4 (tel: 21252).

Cerqueda (Santa Coloma), c/ Mn Lluis Pujol (tel: 20235).

Consul, Plaça Rebés 5 (tel: 20196).

Conxita, Av Dr Mitjavila 26 (tel: 21298).

Cornella, Av Meritxell 61 (tel: 21480; tx: 341).

Costa, Av Meritxell 44, 3 (tel: 21439).

Davos, c/ Esteve Dolsa s/n (tel: 60615).

Eden Roc, Av Dr Mitjavila 1 (tel: 21000; tx: 54134; fax: 60319).

El Racó d'En Joan, c/ de la Vall 20 (tel: 20811).

El Roure, Av Dr Mitjavila 24 (tel: 25483).

Enclar, Roureda de Sansa 18 (tel: 20310).

Festa Brava, c/ La Llacuna 7 (tel: 20741).

Flora, Antic C/ Major 23 (tel: 21508; tx: 209).

Florida, c/ La Llacuna 11 (tel: 20105; tx: 262).

Freixanet, c/ la Creu Grossa s/n (tel: 20549).

Galàxia, Av Meritxell 7 (tel: 26975).

Garcia, Av Princep benlloch 51 (tel: 20968).

Garden (Santa Coloma), Av d'Enclar 93 (tel: 21751).

Hostal Calones, Antic c/ Major 8 (tel: 21312).

Hostal del Sol, Plaça Guillemó 3 (tel: 23701).

Hotel de l'Isard, Av Meritxell 36 (tel: 20092; tx: 377 AND).

Indalo, Av Dr Mitjavila 27 (tel: 21871).

Internacional, Mossén Tremosa 2 (tel: 21422; tx: 269).

Jaume I, Roureda de Sansa 2 (tel: 20061; tx: 203).

L'Arribada, Av Princep Benlloch 67 (tel: 20867).

La Pedrera, Av Dr Mitjavila 6 (tel: 20014).

Les Arcades, Plaça Vuillermó 5 (tel: 21355).

Les Fonts, c/ de l'Agüeta 7 (tel: 21393).

Mercure Andorra, Av Meritxell 58 (tel: 20773; tx: 208; fax: 29018).

Montserrat, Av Meritxell 68 (tel: 20083).

Nevada, c/ M Cinto Verdaguer 1 (tel: 24361).

Normandia, Av Princep Benlloch 49 (tel: 20125).

Parador Santa Coloma (Santa Coloma), Av d'Enclar 104 (tel: 21804).

Pensió La Rosa, Antic c/ Major 16 bis (tel: 21810).

President, Av Santa Coloma 40 (tel: 22922; tx: And 233 HOPRE).

Principat, Av Princep Benlloch 57 (tel: 20219).

Pyrenees, 20 Av Princep Benlloch 20 (tel: 20508; tx: 209 AND).

Reata (Santa Coloma), Av d'Enclar 78 (tel: 22040).

Residència Albert, Av Dr Mitjavila 16 (tel: 26975).

Residència Benazet, c/ La Llacuna 19 (tel: 20698).

Residència President, c/ Roureda de Sansa, s/n (tel: 22922; tx: 233).

Viena, c/ de la Vall (tel: 29233; tx: 411; fax: 29906).

Angola

Michael Griffin

An almost surreptitious handshake in Gbadolite, Zaïre on 22 June 1989 signalled the beginning of the end of the 14-year-old Angolan civil war which has claimed hundreds of thousands of lives and crippled one of Africa's wealthiest nations. Though the cease-fire was rescinded shortly afterwards, the Gbadolite meeting resulted in a working model of whatever settlement will finally be agreed between the two war-weary sides. Pressure from South Africa's new government will bring Uniao Nacional para Indepencia Total de Angola (UNITA) back to the negotiating table in 1990 when an independent Namibia will have gnawed away at its essential supply routes. The fate of Savimbi remains a significant stumbling block on the track to peace but excessive delay risks the shunting of UNITA into a historic and impotent sideline, for events are moving rapidly within the region.

Although the United States opposed Angola's admission to the International Monetary Fund (IMF), President Bush's sub-Saharan policy is far more fluid than his predecessor's and Pretoria's economic ambitions in Angola's oil and diamond industries have already displaced destabilisation as a priority in policy-making circles. But peace, when it comes, will present as many challenges to the rigid, privileged elite governing Angola as war has done and, in seeking to release the country's economic promise, investors and donors can only unearth the same vexed ghosts that have plagued the nation since the beginning of the First Liberation War in 1961.

The historic handshake in Gbadolite was witnessed by 17 African heads of state at a summit hastily convened by President Mobutu of Zaïre on the eve of his first visit to post-Reagan Washington, and the circumstances leading up to it remain the subject of intense speculation. President Eduardo dos Santos had vowed time and again never to parley directly with Dr Jonas Savimbi, leader of the UNITA and he may not even have been aware of his presence at the meeting until the very last moment. Certainly, Dos Santos carried no brief from Luanda to conclude a unilateral cease-fire – even one as vague as the Gbadolite Declaration which failed to mention Savimbi in any of its 11 causes – without first consulting the politburo, the party or his Cuban allies. Diplomats were full of praise for his audaciousness in clasping Savimbi's hand but, in Luanda, there was talk of an imminent palace coup, master minded by hardline elements of the Movimento Popular de Libertacao de Angola (MPLA), reluctant to cede place or the lucrative perks of an economy on a permanent war footing.

The UNITA leader, for his part, urgently needed to reassert his public posture in the light of the New York Accord of December 1988 which transformed SWAPO from a *nom de guerre* into a mergent, sovereign state and paved the way for the gradual withdrawal of the 50,000 Cuban troops estimated to be on Angolan soil. Their departure and the linked independence of Namibia in November 1989, will effectively reduce the Angolan conflict from a superpower contest to a local, ethnic disturbance, all the more irritating for the impact it will have on business. Events subsequent to Gbadolite served only to cast Savimbi and his movement in the role of sore losers particularly since the most signficant concession – that UNITA was not solely the stooge of Pretoria but an authentic voice of Angolan dissent – had already been ceded in the more moderate language of Luanda's announcement.

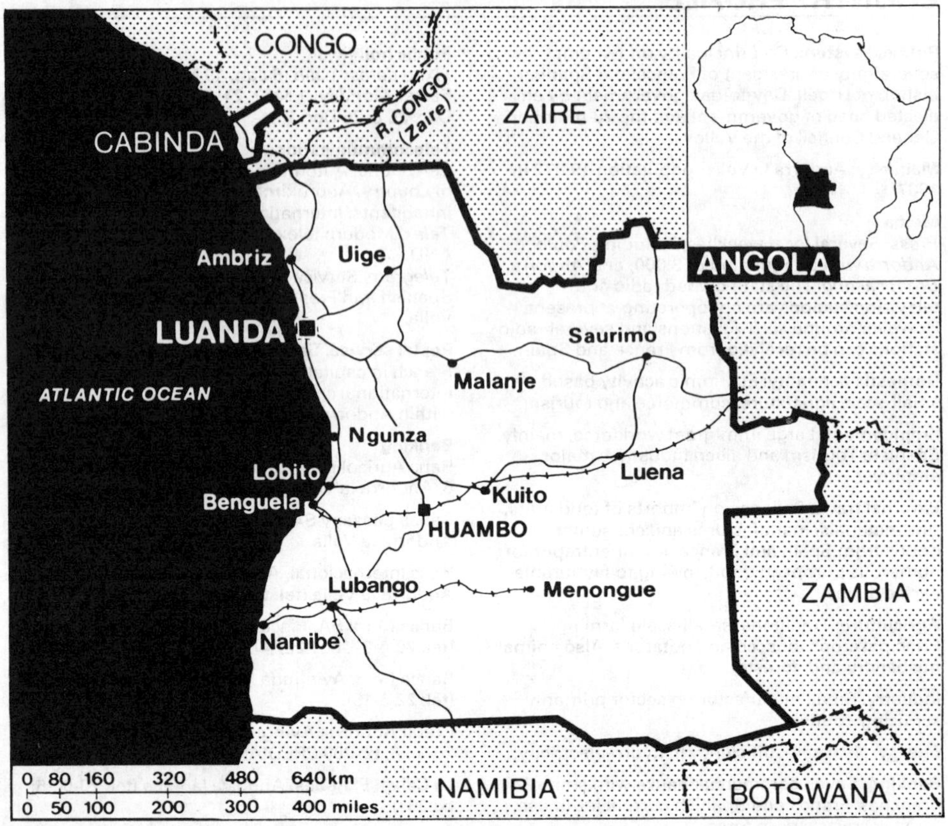

Both men, it seems, may have been royally bamboozled for, when the assembled photographers asked the Angolans to re-enact the gesture for the world's press, they refused and quit the conference hall. The only immediate winner was Mobutu who, after a lifetime spent destabilising Angola, flew to Washington as the self-appointed peacemaker in Africa's thorniest conflict. And yet in a war as protracted as Angola's, even this tentative ice-breaking ceremony must be regarded as setting an important precedent, the first of many small steps towards the adoption of an acceptable peace formula.

In addition to the cease-fire, the Gbadolite meeting provided amnesty for UNITA fighters and their voluntary integration into the Angolan armed forces, an end to US and South African interference and the recognition of the Angolan constitution. This latter clause, which re-affirmed the legitimacy of the MPLA as the sole, political party, failed to satisfy the USA's unrealistic demand for multi-party elections in Angola. But UNITA cadres, it was reported, would also be integrated into government structures, which would substantially dilute the existing power blocs within the 30-35,000-strong MPLA. None of this was remarkably new, having been presented as a proposal earlier in the year. The real breakthrough appeared to be in Savimbi's alleged acceptance of 'special treatment', construed to mean temporary exile in a country of his choice. President Kaunda of Zambia announced that Savimbi had agreed to these conditions but his tidings were premature.

Two months later, the Gbadolite Declaration was in shreds and Savimbi announced that his forces were prepared to resume hostilities. It is questionable whether they had ever ceased. In July 1989, an attack on the power lines cut Luanda's electricity supply for

KEY FACTS

Official title: People's Republic of Angola

Head of state: President José Eduardo dos Santos

Head of government: President José Eduardo dos Santos

Ruling party: Movimento Popular de Libertação de Angola – MPLA-PT

Capital: Luanda

Official Languages: Portuguese (various African languages spoken)

Currency: Kwanza (Kw) = 100 Lwei

Exchange rate: Kw29.9 per $ (Nov 1989)

Area: 1,246,700 sq km

Population: 9.2m (1987)

Trade balance: $118m (1987)

Foreign debt: $1.5bn (1987)

10 days, convoys were attacked in Huila province and a civilian airliner was downed. Some observers blamed the infractions upon poor communications between UNITA and its roving bands of raiders while others accused entities within Angola of seeking to wreck an accord which can only undermine the present concentration of political and economic powers.

Both sides are exhausted but the thought of reconciliation is even more threatening. FAPLA, the armed forces, consumes 60 per cent of Angola's oil proceeds and enjoys first dip of imported and locally produced goods but it controls barely 30 per cent of the national territory and has made no serious attempt to widen the security net beyond the main roads or the urban zones. Despite an impressive war record, this omission has helped to amplify the anxiety caused by UNITA's attacks and also sown the suspicion that the state of emergency is being artificially prolonged in some areas of the country. Certainly the prospect of demobilising some of its 57,000 battle-hardened men in an economy without jobs, food, a viable currency or a considered agricultural policy must fill the government with dread.

Similar fears haunt the MPLA, a party whose revolutionary roots have become entangled with privilege and corruption and which, over the last 14 years, has signally failed to address the burning issues of education, health, employment or social justice which fuelled its successful struggle against the Portuguese. An immense gulf now divides its powerbase in the cities – swollen by refugees – from the rural areas where the remaining farmers are milked of their surplus under a two-tier exchange rate system weighted in favour of the urban consumer. Though it is impossible to gauge popular support for the rebels' cause, the insertion of UNITA cadres within the party structures or even the peacetime, social fabric of Angola threatens to repeat the events of 1975 when a similar coalition deteriorated into the present bloodletting.

The Angolans scrupulously avoided celebrating the cease-fire and it is debatable whether reconciliation is possible given the scale of injuries inflicted upon the population since 1975. According to UNICEF, 100,000 people died from war-related causes in the first half of the decade and there were 10,000 amputations in 1985 – 80 per cent of them non-combatant women and children – due to systematic mining of fields, riverbanks and footpaths. In 1988, SEAS, the government welfare arm, reported that 1.5m people had been displaced or otherwise affected by war and 90 per cent of the population was judged to be on the brink of destitution. And yet appeals for emergency assistance have largely been ignored by donors either through US hostility to the Dos Santos regime or because Angola's oil wealth placed the country beyond the pale of humanitarian concern. In 1989, the government requested $273m in emergency aid, $112m in food and $143m in non-food needs, but appeals in the past have won less than a third of estimated requirements.

Meanwhile, the regional peace initiative agreed in New York and which links the withdrawal of Cuban troops to the holding of free elections in Namibia proceeded apace – though not without incident or delay. The first plane-load of Namibian refugees did not leave Angola until June 1989 under the gaze of the United Nations Transition Assistance Group (UNTAG). Compelled to implement a plan devised a decade earlier and which was ill-prepared for the scope for organised sabotage available to South African paramilitary units in Namibia, UNTAG staff were deeply sceptical of a trouble-free handover of power following the November elections. If, as many observers predict, the new government in Namibia proves unable to guarantee security within its immense territory, UNITA attacks may show signs of slackening but will not be ended.

KEY INDICATORS

	Unit	1983	1984	1985	1986	1987
Population	m	8.34	8.54	8.75	8.98	9.23
Agricultural production	1979-81=100	99.5	99.6	99.3*	100.8	101.8
Food production	1979-81=100	102.0	101.9	102.1	103.4	104.0
Oil production	'000 tonnes	732	840	954	1,161	1,423
Diamond output	'000 carats	1,034	970	900	200	900
Coffee production	'000 tonnes	22	27	25	16*	15*
Exports	$m	1,796	1,970	2,105	1,655	1,327
Crude oil	$m	1,495	1,608	1,905	–	–
Refined oil products	$m	82	90	93	–	–
Imports	$m	816	1,037	1,350	1,015	1,209
Balance of trade	$m	980	933	755	640	118
Exchange rate	Kw per $	29.9	29.9	29.9	29.9	29.9

* estimated figure

Economic revival

A durable peace is still remote but international capital is once more marching into Luanda, buoyed up by the euphoria of recent events. Indeed, one sure motive for attending the Gbadolite meeting was to send potential investors a signal of the government's good faith and its willingness to find common ground with UNITA. This pragmatism – and the golden opportunities that exist in the country's mineral and hydrocarbons sectors – has won influential friends in Portugal, West Germany, France, Italy and Belgium. In spite of US hostility, the IMF board voted on 19 July 1989 to admit Angola to the Fund, paving the way for World Bank investments of over $300m per year in 1990 and, more importantly, bilateral aid that will result from IMF supervision.

Despite its marxist stance and the long-running war, the government has a proven ability to foster a capital intensive private sector free from interference and to maintain workmanlike relations with companies based in the United States and South Africa. Indeed, the government maintains with some justice that its programme of nationalisation in 1975 stemmed less from economic ideology than the very real problem of what to do with the thousands of farms and factories abandoned by the Portuguese. So it is that highly profitable, export industries exist side by side with a moribund domestic economy. 75 per cent of the 14 major food commodities have to be imported, and exports such as coffee have either fallen dramatically or, like sisal, cotton and sugar-cane, disappeared entirely. New investment is lured by the potential of Angola's mineral wealth but Luanda's greatest challenge – and the one that concerns the IMF – is the long-term problem of how to restore production within the domestic economy.

The key to revival is the Saneamento Economic e Financiero (SEF) launched in 1988, a structural adjustment programme devised locally but with IMF priorities in mind. The SEF's main proposals include the cutting of subsidies to parastatals, their closure or eventual privatisation, stimulation of the private sector through greater price flexibility, increased autonomy and the provision of commodities, the revision of foreign investment codes and exchange rate adjustment. Though much of the SEF agenda has yet to be implemented (and is therefore regarded as pure rhetoric), there have been radical improvements at vital levels of economic activity. The virtual legalisation of the *Kandonga* or black market has resulted in the re-opening of private shops with goods priced at the parallel rate: fruit, vegetables and local transport are free of price controls; a network of government-run shops, to be privatised later, has been set up to replace the Portuguese 'bush traders' and provide peasant producers with agricultural inputs, food and consumer items. At a structural level, a peasants' union has been formed to elicit a greater understanding of the needs of people in the rural areas, now regarded as a priority area for economic motivation.

These measures are not extreme enough for the IMF, particularly in the thorny area of the exchange rate. In March, the government outlined a plan for a two-phased, 100 per cent devaluation of the *kwanza*, thereby doubling the current dollar rate from Kw29.92 (unchanged since 1975) as against the parallel rate of Kw2,500-Kw3,000. The IMF is reportedly seeking an adjustment five times as great, as well as the elimination of 'payments in kind' and the '*Bastimento*' or ration-card, systems which ensure a livelihood for factory and government workers but which also fuel the unofficial markets and have converted Angola from a money to a barter economy.

With average monthly earnings ranging from Kw9-13,000, any attempt to mop up excess liquidity through devaluation will dangerously marginalise the 615,000 waged employees and their dependants, most of whom only survive the present hardships by playing middlemen in the *kandonga*. Similarly, efforts to reduce the import bill by ending or cutting back on purchases of subsidised food – up from 156,000 tonnes in 1987 to 220,000 tonnes in 1989 and set to rise alarmingly in the light of a four-year drought in the south – will meet stiff resistance from within the government as well as the urban areas which currently account for 37 per cent of the country's 10m people.

The rush to the cities has traditionally been blamed upon insecurity but an equally compelling reason has been the lack of an effective *kandonga* in the rural areas which have therefore been deprived of everything from clothing and medicines to tools, kerosene and building materials. Regional authorities (recently awarded wider powers including the right to import directly according to their needs) have had some success with local experiments in liberalisation, notably in Lobito and Lubango, two priority areas for reconstruction. But there is no indication that central government is prepared to bite the bullet and extend flexible pricing to the growers of cereal or cash crops, thus providing some incentive for the displaced to return to the soil. In this matter, MPLA policy has followed the pattern set by the Portuguese which was to make the countryside serve the interests of the towns and it will take several years to reverse this trend.

Oil sector confidence

The oil sector, therefore, will continue to dominate for a long time to come and, despite the war, Angola remains one of the most active exploration and development areas in the world. The main deposits are located offshore in the Cabinda enclave or in blocks south of the River Zaïre, thus rendering personnel and infrastructure all but immune to sabotage. New blocks at the mouth of the Kwanza river and in Namibe have also created a stir among geologists and future finds will no doubt swell estimates of proven reserves which currently stand at 1.7bn barrels. The cost of such wealth has been an unhealthy dependence upon the global oil market and, when it collapsed in 1986, Angola's exports slumped from $2bn to $1.3bn, driving debt service as a proportion of exports from 18.7 per cent to 28.6 per cent in the space of a year. With prices once again firm and now enhanced by the prospect of improved national security, the oil industry is enjoying a new burst of confidence.

Despite its hostility to the Luanda government, the USA still consumes 75 per cent of Angolan crude, over half of which is pumped out by the Cabinda Gulf Oil Company, a subsidiary of the Chevron Corporation. Between 1986-90, investment in the sector was forecast to top $3bn and production has risen far more steeply than was originally predicted by Sonangol, the government corporation responsible for development of the sector. Output increased from 129,600 barrels per day (b/d) in 1982, to 276,000 b/d in 1986, to 450,000 b/d in 1989, far above official estimates of 370,000 b/d by 1990 and there are indications that even more may have been extracted due to poor monitoring of tankers.

Late in 1988, a consortium of Western banks agreed to lend the government $120m to exploit and develop the new Takula field in Cabinda and, in July 1989, fresh tenders were invited for blocks 9-13, attracting a large number of participants. The potential of these new, southern blocks remains to be seen.

Attempts to revive the diamond industry have been plagued by UNITA's highly effective campaign of attacks on both the mines and any foreign contractor seeking to exploit them on behalf of Endiama, the state mining authority. Angola's deposits, which include six untouched kimberlite pipes and a generous proportion of large, high-quality stones, make it potentially one of the top three producers in the world but, in addition to the problem of insecurity, Endiama faces huge transportation difficulties and a smuggling tradition which reportedly accounts for over 35 per cent of all mined stones.

Under the Portuguese, output stood at 2.4m carats in the peak year of 1975 but, after a precipitous, post-independence fall in production, it stabilised briefly at 1.48m carats worth $225m in 1980. In 1984, UNITA attacks led to the closure of the Cuango operations but, by 1986, Endiama reported sales of 1m carats worth $180m. With the price of diamonds quadrupling to $160 per carat in 1988, the time seemed ripe for a fresh bid to resuscitate the industry and, in May 1989, De Beers signed a letter of intent to develop three of the kimberlites and to resume sales of Angolan output through the Central Selling Organisation. But UNITA attacks had continued undeterred by the private security firms contracted to defend the mines and, in September, the working at Catoca was reported to have shut down operations.

IMF membership and the rise in oil prices will raise Angola's international credit rating but the USA's refusal to recognise the government will still deter domestic, American banks from participating in the billion dollar reconstruction programmes devised for the Lobito Corridor and the four provinces of the south. Compared with other African oil producers, notably Nigeria, Angola's debt burden is highly conservative, especially in the light of the massive cost of the war. In the years following the oil price collapse, borrowings grew from $2.46bn to $4bn but, of this amount, half was estimated as being due the Soviet Union for military purchases. Repayment became unmanageable in 1987 when there was a flurry of reschedulings and in July 1989, shortly before the IMF announcement, the Club of Paris rescheduled a further batch of unspecified debt for a 10-year period. But any improvement obtained through increased revenues or declining military expenditure is more than likely to be swallowed up by the government's ambitious and costly reconstruction programmes. The total cost of rehabilitating the Benguela Railway – essential if the government is to ' repossess' the war-torn, agriculturally productive interior – has been set at $575m while a similar amount is likely to be spent on the Southern Rehabilitation Programme for Huila, Namibe, Cunene and Cuanda Cubango.

Outlook

The key to Angola's future remains the security situation as it has been since independence in 1975. Despite positive moves in several crucial directions, it is still difficult to predict the outcome of the peace programme given the obfuscation of South African and US policy-makers. Certain questions need to be addressed: to what extent is Namibian independence a unanimous strategy to redesign the map of the Frontline states over the long term and to what extent is it simply an attempt to placate international opinion without jeopardising real, political control of the territory. The intimidation of SWAPO returnees in defiance of the UN in the run-up to the November elections looked suspiciously like the beginning of a systematic destabilisation of the fragile, new nation. But the losses suffered by the 'new look' Nationalist Party in South Africa's elections in September 1989 – mainly to the more moderate Democratic Party – suggest that President de Klerk had underestimated the domestic demand for a revision of Pretoria's hardline approach both in and outside the country. If the latter proves to be the case, then there are positive implications for Angola in the coming year.

South African companies have already shown considerable interest in Angolan diamonds and commercial interests may succeed in shouting louder than their more cautious, military counterparts. With the withdrawal of Cuban troops well under way and Moscow undoubtedly committed to a peaceful resolution of the military stalemate, a continuation of US supplies to UNITA would seem purposeless as well as strategically obsolete, particularly in the light of recent changes in American policy towards RENAMO in Mozambique and the Contras in Nicaragua. So, tentatively, the final part in the jigsaw of peace assembled by US Secretary of State, Chester Crocker – the talks between UNITA and the Angolan Government – can be expected to continue discreetly through 1990 with diplomatic recognition by the USA as the final prize for reconciliation.

COUNTRY PROFILE

Political system: One-party Socialist republic. The leadership of the country centres on the MPLA-PT (Movimento Popular de Libertacao de Angola – Partido de Trabalho). The party congress convenes once every four years to elect the president and 75 committee members. The committee elects an 11-member Political Bureau (which includes the president) as its executive. Gained independence from Portugal on 11 November, 1975.

Political parties: Movimento Popular de Libertação de Angola (MPLA-PT) is the only legal party.

Main cities/towns: Luanda (population 1.3m), Huambo (120,000), Lobito, Lubango, Soyo, Malange.

Language: The official language, and the language for all documentation, is Portuguese. French, Spanish and English are occasionally used in business.

Media
Press: National dailies include *Jornal de Angola* and *Diário da República* (government news sheet). Regional newspapers are published in several towns. A few periodicals are also available (mainly political).
Radio: National radio service with broadcasts in Portuguese, English, French, Spanish and vernacular languages, operated by Radio Nacional de Angola.
Television: Limited TV service operated by Televisao Popular de Angola (TPA).

External trade: Steady growth of oil exports and tight import controls helped reduce current account deficit. Trade balance remains vulnerable to shifts in world demand and price for oil, diamonds and coffee.
Exports: Crude petroleum and derivatives (around 96% of total value in 1989), coffee (2%) and diamonds (1.6%). Also small quantities of sisal, cement, iron-ore, fishmeal, timber and cotton. Main destinations: USA, France, Netherlands, Spain, UK. Also increasing countertrade arrangements with Brazil and USSR.
Imports: Machinery, transport equipment, military equipment, foodstuffs, raw materials, textiles and clothing, medical and chemical goods. Main sources: USA, France, Brazil, Portugal, USSR.

Agriculture: 13.7 per cent of GDP in 1986. Largest sector of the economy, employing over 70 per cent of the workforce. Only 2-3 per cent of land is cultivated. Infrastructural constraints, shortages of inputs and essential commodities have combined with rural insecurity to hamper development of Angola's potential as a food producer. At least 50 per cent of national food requirements are now imported. Main cereal or root crops are cassava, maize (declining), wheat and bananas. Major cash crops are coffee, cotton, sugar cane and sisal but only the first is now produced in exportable quantities and in volumes only six per cent of 1975 output when Angola was also self-sufficient in foodstuffs. Fisheries offer considerable potential chiefly in Namibe and Kwanza Sul provinces. Agreements have been signed with USSR, Italy, Spain and Portugal. Also commercial forestry in Cabinda and Moxico. Livestock farming has been disrupted by banditry in the south and by the current four-year drought.

Industry: Sector is predominantly state-owned, with production centred on food processing, brewing, light manufacturing, electrical goods (e.g. radio production), textiles and construction materials. Also production of steel, motor vehicles, detergents, tobacco, bicycles and chemicals. Activity concentrated in Luanda, Huambo and Lobito. Industrial output has been sluggish owing to shortages of foreign exchange for vital imported inputs and need to divert scarce resources towards defence. Around 60 per cent of total production is accounted for by nationalised industries.

COUNTRY PROFILE

Mining: Rich in mineral reserves, including copper, gemstones, iron-ore, manganese, phosphates, salt and uranium. Diamonds are second largest foreign exchange earner and the Catoca field is the world's third largest. As a result of the war and smuggling, production fell to around 0.7m carats in 1985. Iron-ore mining, primarily in the Huila, Huambo and Bié provinces, severely hampered by security problems.

Hydrocarbons: 19.9 per cent of GDP in 1986. Economic performance largely determined by level of oil production and exports (mainly to USA). Production, which reached 450,000 b/d in 1989, comes mainly from offshore fields in the northern enclave of Cabinda (jointly operated by Cabinda Gulf Oil and the state-owned Sonangol company). Oil reserves estimated at 1.7 bn barrels. Angola is not a member of OPEC. Main domestic refinery in Luanda is operated by the state-owned Petrangol. Liquefied natural gas exports began in 1983 and reached nearly 180,000 tonnes in 1986. Natural gas reserves estimated at over 50bn cubic metres.

Membership of international organisations: African Development Bank, Arab Bank for Economic Development in Africa (recipient), CMEA (observer status), Economic Commission for Africa, GATT (provisional), ILO, IMF, Lomé Convention, OAU, SADCC, UN, WHO.

BUSINESS GUIDE

Time: GMT +1.

Climate: Luanda and northern regions: Oct-Apr, hot and humid, usual temperature range 28-32°C, maximum likely 34°C; Apr-Sep hot but less humid with daytime temperature range 25-30°C and cool evenings. Southern regions: more temperate. Rainfall can be frequent and heavy, particularly in Apr.

Entry requirements
Passport: Required by all.
Visa: Required by all (except certain transit passengers). Exit permit also required. Visas can be obtained in Luanda via the inviting company or government agency.
Currency: Import of local currency up to Kw1,000 is permitted, but export is prohibited. No restrictions on import of foreign currency, declaration of which is compulsory. N.B. Credit cards are not generally accepted.

Health precautions
Mandatory: Cholera vaccination certificate required if arriving from an infected area.
Advisable: Yellow fever vaccination and malaria prophylaxis essential. Cholera, typhoid and polio vaccinations recommended. Water precautions must be taken. N.B. Medical services in Angola are free to foreign nationals.

Representation overseas: Algiers, Belgrade, Brazzaville, Brussels, Cairo, Dar es Salaam, Havana, Kinshasa, Lisbon, London, Lusaka, Moscow, Paris, Rome, Stockholm.

Air access
National airline: TAAG, (Linhas Aéreas de Angola).
Other airlines: Aeroflot, Air Gabon, Air Zaire, Cubana, Ethiopian, Interflug, Sabena, TAP-Air Portugal, UTA, Varig.
International airport: Luanda-4 de Fevereiro (Code: LAD), 4 km from city, restaurant, bank, currency exchange. No taxis.

Surface access
Main ports: Cabinda, Lobito, Luanda, Namibe.
Overland access: Not practicable; borders with adjacent countries have been closed. There are plans to re-open the Benguela railway, but this seems unlikely unless UNITA attacks diminish.

Hotels: Bookings must be made one month in advance and on official requisition to Anghotel. Bookings cannot be made by airline companies or at airport. N.B. Strictly enforced curfew in Luanda from 2400 to 0500.

Electricity supply: 220 V AC, 50 cycles.

Credit cards: Not generally accepted. Hotel Presidente Meridien in Luanda will accept Amex.

City transport
Taxis: Restricted service operated by state company. No service operated from airport to Luanda.

National transport
Air: Regular domestic flights connecting main centres. Passengers on internal flights required to carry official authorisation *(guia de marche)*. Separate helicopter/plane services to Cabinda enclosure.
Road: 8,000 km network of major roads serving centres is being extended, but some stretches have been badly damaged. Full identity papers should be carried at all times.
Buses: Extensive but crowded service. Flat fare in Luanda.
Rail: Service operated by PCFT and CFB. Three classes; second and third class usually fairly basic; no sleeping cars or air-conditioning; some restaurant cars operated. Routes include: Luanda-Malanje (some railcar services; refreshments scheduled on some services; journey times vary, e.g. 7-13 hrs); Benguela-Luau and Namibe-Menongue.

Public holidays
Fixed dates: 1 Jan, 4 Feb ('Beginning of the Armed Struggle'), 1 May, 17 Sep (National Hero's Day), 11 Nov (Independence), 10 Dec (Foundation of MPLA), 25 Dec (Family Day).

Working hours
Government and business: 0830 – 1230, 1430 – 1800. (Exceptions: Sat 0830 – 1230, Sun closed.)
Banking: (Mon-Fri) 0830 – 1230, 1430 – 1530.

Telecommunications

Telephone: Telephone dialling for Angola: IDD access code + 244 followed by area code (2 for Luanda) followed by subscriber's number. Satellite link with Lisbon. International calls bookable through the operator at least six hours in advance.
Telex: Telex facilities available from main hotels. International telex code: AN.
Other: State-operated postal and telegraphic services are reliable but often subject to delays.

Banking
Central bank: Banco Nacional de Angola (BNA).
Main commercial bank: Banco Popular de Angola (BPA).

Trade fairs: Exhibitions and fairs organised by the Angolan Chamber of Commerce. Luanda International Trade Fair (FILOA), held every November, attracts a number of foreign exhibitors.

Representation in capital: Algeria, Belgium, Brazil, Bulgaria, Cape Verde, China PR, Congo, Cuba, Czechoslovakia, Denmark, Egypt, Finland, France, Gabon, German DR, German FR, Ghana, Hungary, Italy, Korea DPR, Netherlands, Nigeria, Norway, Poland, Portugal, Romania, Sao Tomé & Principe, Spain, Sweden, Switzerland, Tanzania, USSR, UK, Vatican, Vietnam, Yugoslavia, Zai\03re, Zambia.

Hotels: Luanda
Panorama, P.O. Box: 6493 (tel: 37841/6).

Presidente Meridien, Avenue 4 de Fevereiro, P.O. Box: 5791 (tel: 372260/4; tx: 3120 hotang).

Other useful addresses
ANGOP (news agency), PO Box 2181, Luanda (tel: 34945).

Associacoa Comercial de Luanda, PO Box 1275, Edificio Palacio de Comercio, le Andar, Luanda (tel: 22453).

Banco Nacional de Angola (BNA), Avenida 4 de Fevereiro 151, Luanda (tel: 39141; tx: 3005).

Campiano (travel agents), Av Commandante, Valodia 8B.

Direccao de Emigracáoe Fronteiras de Angola (visa queries), Defa, Luanda (tel: 30314, 30019; tx: 3127).

Direccao dos Servicos de Estatistica (statistical agency), PO Box 1215, Luanda.

Direccao dos Servicos de Comercio, PO Box 1337, Largo Diogo Cao, Luanda.

Importang (state import agency), Calcada do Municipio 10, PO Box 1003, Luanda (tel: 37994; tx: 3169).

National Planning Committee, PO Box 1205, Luanda (tel: 39052; tx: 3082).

National Tourist Agency, Centro de Informaca e Turismo de Angola, Palacio de Vidro, PO Box 1240, Luanda (tel: 72750).

Radio Nacional de Angola, Caixa Postal 1389, Luanda.

TAAG (Linhas Aéreas de Angola), Rua Luis de Camoes 123, Andar, Luanda (tel: 23523/34880).

Televisao Popular de Angola (TPA), Caixa Postal 2002, Luanda.

Anguilla

Tony Thorndike

Economic development in Anguilla, mainly attributable to a very buoyant tourist trade, has now reached such a scale that labour shortages threaten future growth. Inflationary pressures have increased and concern is growing that the quality of the tourist product may suffer if development is unchecked. But popular confidence in the government was expressed with the return to office of Emile Gumbs' administration in the 1989 general election.

The current economic and political situation in this small British Dependency sharply contrasts with the recent past. Anguilla, named because of its eel-like shape, suffered centuries of neglect as part of the former colonial grouping of St. Kitts–Nevis–Anguilla. But in 1967 the independently spirited islanders broke away, precipitating a crisis which ended in British troops and police being despatched to the rebel territory two years later. The British finally conceded Anguilla's secession in 1980.

Economic boom

Anguilla's history of poverty can largely be traced to its limestone sub-base which greatly limits agricultural possibilities. In addition, rainfall on the relatively flat scrub-covered island is generally poor and unpredictable. The result was heavy emigration (over 4,000 Anguillans live in the United States) and a tradition of seafaring and schooner building which still persists. Salt-gathering was also a traditional industry, but that has nearly died out. Fortunes changed with the gradual exploitation of the many extraordinary beaches and diving sites. At first development was slow, hampered by inadequate infrastructure and bankruptcies but, with the opening of new airport facilities and promotional offices, and the imposition of stricter financial rules on foreign investors, the tourist sector now dominates the economy.

Revenue from tourism reached EC$51.3m (US$19m) in 1987, 25 per cent over 1986, and that for 1988 nearly EC$70m (US$26m), representing 23.4 per cent of GDP. Total visitor arrivals reached 65,050 in 1988, an increase of 30 per cent over the previous year. Sixty six per cent were from the United States, but the fastest growing market was Western Europe. As European visitors tend to stay longer than their North American counterparts, the average length of stay increased from 7.5 bed nights in 1987 to 8.9 in 1988. Many hotels reported high occupancy rates, especially during the high (winter) season. There is also a buoyant day-excursionist market, with visitors from nearby overcrowded St. Maarten/St. Martin, and their number has increased with the introduction of new ferry services. The island community was deeply shocked when two US visitors were raped and killed in early 1989. The assistance of Scotland Yard was sought and the murderer swiftly caught and convicted: the image of peaceful Anguilla was restored.

The tourist sector was largely responsible for the construction boom, and the number of hotel rooms rose by a third to 580 in 1987-88. However, nearly all this increase followed foreign investment and concern continues to be expressed at the declining number of locally-owned facilities, especially guesthouses. The government has pledged to positively encourage local investment; for example, import concessions are no longer available to foreign entrepreneurs. Residential construction was also strong, both for foreign owners attracted by the virtual absence of taxation and liberal residency regulations, and for nationals (and returning islanders) who enjoy very reasonable mortgage facilities.

Fortunately the tourist boom has not led to the neglect of other sectors. Agriculture, mainly vegetables and root crops, continued to do well as a result of good rainfall, while the fishing industry recorded strong growth, boosted by a new fish processing plant. As usual, lobsters were the most profitable catch, the majority being exported to Puerto Rico. By contrast, the financial services sector recorded lower growth than hitherto. Reforms in the early 1980s dispelled the unsavoury reputation it had acquired and, for a period, the record of non-resident deposits was good, with growth reaching 90 per cent annually. The slowdown was not attributable to the signature in 1987 by Britain on Anguilla's behalf of the information exchange agreement permitting access by US federal agencies to private bank accounts if 'laundering' of illegal funds was suspected; rather it is that competition from other Caribbean centres has been intense.

Overall, the economy grew by 8.2 per cent in 1988, compared to 11.7 per cent in 1987. The reduction was traceable to the overheating of the small economy, which sharply increased the rate of labour mobility and cut unemployment to 5 per cent. The labour shortage means that joblessness is virtually voluntary. Increasing inflation – to over 5 per cent in 1988 – has been an unwelcome result.

Government revenue was boosted by the high level of economic activity, rising to EC$25.1m (US$9m) in 1988. But increased civil service salaries resulted in a 29 per cent increase in government expenditure to EC$22m (US$8.1m) and a smaller surplus than hitherto available for capital spending. British aid therefore continued to be necessary. There are few sources of revenue except customs duties (up to 30 per cent), bank licences, land transfer tax (5 per cent) and tourist bed tax. Fortunately European Community assistance has resulted in improved roads, while Cable and Wireless completed a EC$13m (US$4.8m) telecommunications modernisation programme in 1988.

Political stability

Three parties contested the February 27 general election. Chief Minister Emile Gumbs was returned to power, his Anguilla National Alliance (ANA) winning three of the seven seats with support from one Independent. The ANA won four seats in the 1984 poll. The Anguilla United Party (AUP) won two seats and the Anguilla Democratic Party (ADP) one. The AUP was led by Ronald Webster, the mercurial leader of the 1967 rebellion. He had enjoyed virtually unan-

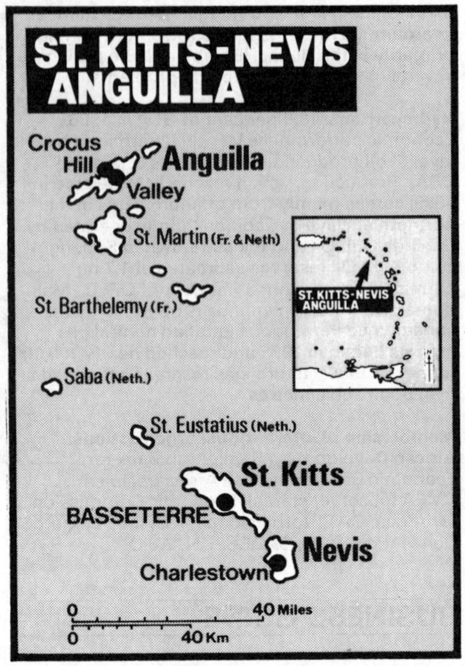

imous support at that time, and for a number of years thereafter, until the electorate tired of his authoritarianism. He was defeated in the 1984 election, his party then being called the Anguillan People's Party (APP). He announced his retirement to concentrate on his extensive real estate interests. However, the appeal of politics proved too much, but he was again defeated.

Little change in direction is expected; in any event, personalities rather than policies divide the various loose party groupings. It is expected that new constitutional provisions will be introduced during 1989, giving greater autonomy.

Labour shortages and inflation will result in a further slowing down of the economy in 1989 and 1990. Construction is expected to decline following the completion of several major hotel projects, giving a breathing space for the tourist industry to consider whether further expansion will be consistent with maintaining the quality of the presently up-market product. Greater consideration will also have to be given to identifying new sources of government revenue if reliance on British aid for capital expenditure, and all the restrictions that it entails, is to be reduced.

KEY FACTS

Official title: Anguilla

Head of state: Queen Elizabeth II

Head of government: Chief Minister Emile Gumbs

Ruling party: Anguilla National Alliance

Capital: The Valley

Official Languages: English

Currency: East Caribbean Dollar (EC$) = 100 cents

Exchange rate: EC$2.69 per $ (Nov 1989)

Area: 96 sq km (Anguilla 91 sq km, Sombrero 5 sq km)

Population: 7,200 (1988)

GDP per capita: $3,562 (1987)

GDP real growth: 8.2% (1988)

Unemployment: 5% (1988)

Inflation: 5% (1988)

Aid flows: $2.3m (1988)

Visitor numbers: 65,050 (1988)

COUNTRY PROFILE

Political system: British Caribbean dependency with executive power resting with an appointed British governor, assisted by an executive council (chief minister, two ex-officio members, and not more than three other ministers). The legislature is the House of Assembly having a five-year term (seven members elected, two appointed, two ex-officio). The governor is responsible for defence and external affairs, but is required to consult the chief minister on matters relating to internal security, the police and civil service.

Main town: The Valley.

Media
Press: One weekly – the *Vantage*. Government publishes *Information Service Bulletin* and *Official Gazette* monthly. Other publications: *The Chronicle, Anguilla Life, What to do in Anguilla.*
Radio: The government-owned Radio Anguilla transmits over 14 hrs/day and the private Caribbean Beacon Radio transmits for 24 hrs/day.
Television: ABC Channel 2 (Anguilla Broadcasting Corporation. A private cable 12-channel/24-hour TV service is also in operation, using US satellite programming.

External trade: Island remains dependent on workers' remittances from abroad and grants-in-aid from the UK. Substantial trade deficit.
Exports: Lobsters, livestock.
Imports: Mainly foodstuffs, construction materials and consumer goods from Europe, USA, Puerto Rico and Trinidad.

Agriculture: Mainly small-scale farming of peas, corn, sweet potatoes, vegetables and tropical fruits. The average cultivated plot is less than half an acre. Traditional livestock-raising (goats, sheep, poultry) is important, with some animals raised for export. Lobster fishing is a major employer and principal earner of foreign exchange. A large part of the catch is exported to the US Virgin Islands and Puerto Rico.

Industry: Apart from tourism, centred almost entirely on fish processing and construction work. Tourism is developing rapidly and has become the major industry.

BUSINESS GUIDE

Time: GMT -4.

Climate: Sub-tropical with a mean annual temperature of 27°C. Hottest from Jul to Oct, coolest from Dec to Feb. Cooling trade winds blow throughout the year. Rain mainly falls between Sep and Dec.

Entry requirements
Passport: Valid passport required by all.
Visa: Generally not required except for a few specific countries. Onward or return tickets must be held.
Currency: No restrictions on import/export of local or foreign currency.

Health precautions: Vaccination certificates for cholera and yellow fever required if arriving from infected areas. Water precautions advisable.

Air access
Airlines serving: Air BVI, American Eagle, Carib Aviation, Coastal Air Transport, LIAT, Windward Islands Airways.
International airport: Wall Blake (AXA), 3 km from The Valley. Daily air service with St Maarten (Netherlands Antilles) and St Thomas (US Virgin Islands). Also regular air services from Antigua, St Kitts and San Juan (Puerto Rico). Air taxi and charter services are available from Air Anguilla, Carib Aviation and Tyden Air.

Surface access
Main ports: Road Bay, Sandy Ground, Blowing Point.
Water access: Daily ferry services operate between St Martin (French Antilles) and Anguilla.

Hotels: Several high-quality hotels and a wide selection of villas and apartments. 10 per cent service charge is usually included in bill. Eight per cent tax is added to room rates.

Car hire: Readily available. National licence needed to obtain temporary local licence.

City transport
Taxis: Most generally used form of transport; readily available and inexpensive.

National transport
Road: Over 90 km of road, of which over half is surfaced. Public roads cover all parts of the island.

Public holidays: 1 Jan, Good Friday, Easter Monday, Labour Day, 30 May (Anguilla Day), Whit Monday, Queen's Birthday, August Monday, August Thursday, Constitution Day, Separation Day, 25 Dec, 26 Dec.

Working hours
Government and business: (Mon-Fri) 0800 – 1200, 1300 – 1600.
Banking: (Mon-Fri) 0800 – 1500; also (Fri) 1500 – 1700.

Telecommunications
Telephone: Dialling code for Anguilla: IDD access code + 1-809-497 followed by subscriber's (four-digit) number.
Telex: Facilities operated by Cable and Wireless. Public booth available in The Valley.

Banking
Commercial banks (head offices): Caribbean Commercial Bank, National Bank of Anguilla, Trade Continental Bank & Trust Co.
Foreign branches: Barclays Bank International.

Electricity supply: 120 V AC, 60 cycles.

Hotels: Anguilla
Anguilla Great House, Rendezvous Bay, P.O. Box: 157 (tel: 6061; tx: 9337 GRHOUSE; fax: 6019).

Bay View Beach Apartments, Blowing Road, P.O. Box: 69 (tel: 6239).

Cap Juluca, Maunday Bay (tel: 6666/6779; tx: 9355).

Cinnamon Reef Beach Club, Little Harbour (tel: 2727; tx: 9307 cinmon).

Cocoloba Plantation, Barnes Bay, P.O. Box: 332 (tel: 6871/2/3/4; tx: 9332; fax: 6332).

Cove Castles, Shoal Bay West, P.O. Box: 248 (tel: 6801).

Coyaba Beach Club, The Cove, West End, P.O. Box: 117 (tel: 6601).

Cul-de-Sac, The Valley, P.O. Box: 173 (tel: 6461).

Easy Corner Villas, South Hill, P.O. Box: 65 (tel: 6433/6541; fax: 6410).

Emerald Estates (Hotel Resort), Sandy Hill (tel: 4323).

Fountain Beach, Shoal Bay (tel: 3491).

Inter Island, Lower South Hill, P.O. Box: 194 (tel: 6259).

La Siena, Meads Bay, P.O. Box: 200 (tel: 6827; fax: 6829).

Malliouhana Beach, Maids Bay, P.O. Box: 173 (tel: 4976111; tx: 9316; fax: 6011).

Massara Resort, Katouche Bay (tel: 3200).

Meads Beach Resort, Meady Bay, P.O. Box: 202 (tel: 6100).

Pelican Cove Beach, Cove Bay (tel: 6501).

Rainbow Reef Villas, Seafeathers Bay, P.O. Box: 130 (tel: 2817).

Rendezvous Bay Hotel, The Valley, P.O. Box: 31 (tel: 6549).

Sea Grape Beach Club, Meads Bay, P.O. Box: 65 (tel: 6433/6541/6990; fax: 6410).

Sea House Apartments, Blowing Cove, P.O. Box: 17 (tel: 6751).

Shoal Bay Resort, Shoal Bay (tel: 2011).

The Mariners, Sandy Ground, P.O. Box: 139 (tel: 4972671, 4972815; tx: 9319; fax: 2901).

Yellow Banana Hotel, Stoney Ground, P.O. Box: 63 (tel: 2626).

Car hire: Avis (Bennie & Sons): Blowing Point (tel: 6221; tx: 9348).

Connors Car Rental: The Valley (tel: 6433).

H & R Car Rental: The Valley (tel: 2656).

Island Car Rental: Airport Road (tel: 2312).

Other useful addresses
Anguilla Chamber of Commerce, PO Box 321, The Valley (tel: 2710; tx: 9327).

Anguilla Hotel and Tourism Association, PO Box 321, The Valley (tel: 2944; tx: 9347).

Cable & Wireless Public Booth, The Valley (tx: 9302).

Director of Tourism, The Valley (tel: 2759; tx: 9313; fax: 2751).

Antigua and Barbuda

Tony Thorndike

Despite a general election which returned the Antigua Labour Party government to power, in effect the preserve of the Bird family, and a continuing economic boom largely sustained by tourism, political and economic prospects are uncertain. The degree and range of corruption up to the highest levels is scandalous while international concern is growing at the increasing debt burden. Political infighting within the government continued after a lull during the election campaign. Other problems included increased labour unrest, partly due to labour shortages and inflation, and overdevelopment of the tourist industry.

The reputation that Antigua and Barbuda has developed for official corruption and suspect accounting led *The New Yorker* to conclude that the twin-island state 'represents the shabbiest aspect of Caribbean politics', and a 'playpen' for politicians who use high office as 'a vehicle for increasing one's personal wealth'. Prime Minister Bird, almost 80, has dominated Antiguan politics for nearly four decades, except for a period in the 1970s when his Antiguan Labour Party was voted out of power, and he has become one of the wealthiest men in the Commonwealth Caribbean. He presides over a dynasty characterised by bitter division, as the two eldest legitimate sons fight for the succession, worth millions of dollars. In the background is the patriarch's 26-year old mistress, Cutie Francis, ever anxious to keep her patron in power to protect her growing business empire. It is indeed a sordid tale, and one which the Antiguan people will have to pay for as the growing number of debtors knock on the door of the Treasury.

Election victory

The result of the general election on 9 March 1989 was a foregone conclusion, such is the domination of the Antigua Labour Party (ALP). It had won all 16 Antiguan seats (one other is for Barbuda) in the 1984 poll; this time it won 15, with 63.8 per cent of the votes cast, losing only one to the United National Democratic Party (UNDP). The only other change was in Barbuda, where Hilbourne Frank of the Barbuda People's Movement defeated the sitting member and his arch-rival Eric Burton of the Barbuda National Party, which is close to the ALP. The turnout was the same as in 1984, a relatively low 61 per cent, reflecting a degree of voter apathy.

The ALP dates from 1944 when its parent, the Antigua Trades and Labour Union (ATLU), was formed by Vere Bird. Since then it has been in power except during 1971-76 when it was ousted by the now defunct Progressive Labour Movement (PLM). The PLM also emerged from a union base, that of the Antigua Labour Union (ALU), an ATLU breakaway. The UNDP was created in 1985 mainly representing middle-class interests. It later attempted to widen its appeal by merging with the remnants of the PLM but to little effect, one problem being ineffectual leadership. Its share of the vote was only 31 per cent. However, the most important struggle is between the two unions, as many a hapless employer bears witness.

The third party is the radical Antigua Caribbean Liberation Movement (ACLM), led by former teacher Tim Hector. In many ways it is the most interesting, not only because of its often incisive analyses of local affairs but also since the party newspaper, *Outlet*, is deservedly the most read. Its exposures of official corruption and other wrong-doing are legendary, even though some have fallen wide of the mark. It has survived numerous attempts to gag it either through libel writs or through sabotage. The ALP government has never lost opportunities to remind its close ally, the United States, of ACLM's links with Cuba. But matters took a gruesome and sinister turn in May 1989 when Hector's wife Arah was found brutally murdered. Although her alleged murderer was a drug addict, it was rumoured that he had been encouraged to kill her by the 'drug barons' whose activities, and alleged official links, were being exposed by *Outlet*. What was offensive to many was the publication of a police picture showing her nude and dismembered body in the ALP party newspaper.

Scandal and corruption

The recent history of corruption includes the highly profitable sale of passports to non-nationals (in 1988 the US government demanded the recall of Antigua's Ambassador for his alleged involvement), laxity in casino operations and maladministration of foreign aid and loans. The latter has reached the stage where donor officials, including those from the World Bank, question the state's suitability for assistance. But two scandals in particular have attracted international attention as they have highlighted the struggle for succession within the ALP, between Deputy Prime Minister Lester Bird and elder brother, Vere Bird Jr.

The first scandal involved the upgrading of facilities at the Vere Bird International Airport, where it was alleged that Vere Bird Jr. embezzled some US$4m out of a US$11m contract. The prime minister was forced to agree to an inquiry which, although clearing his son from criminal wrongdoing, severly criticised his behaviour. Demands by eight cabinet members, including Lester Bird, for Vere Bird Jr. to be dismissed were resisted by the prime minister. A third brother, Ivor, then entered the fray. Managing the family TV station, he forced Lester Bird off the board and over the airwaves denounced him and other ministers who had given evidence against his brother. But Ivor Bird's credibility was badly dented by revelations in *Outlet* that he was acting for a New York property developer involving land which was being offered at a very cheap price that had been compulsorily purchased by the government. Ultimately the ATLU executive stepped in and insisted on an end to the recriminations.

Despite his father's antagonism towards him, Lester Bird reaped the benefit for a period. He had already engineered the support of the ALP youth arm and Vere Bird's power base in the ATLU was badly dented. He also admitted during the election campaign that there was 'some merit' in the corruption charges made by the opposition parties. But there was a cloud over him as well, in the shape of the disposal of the Royal Antiguan Hotel, a large 360-room complex built with the aid of a DM120.5 (worth US$45m at the time, later US$75m) loan from West Germany. *Outlet* published documents that purported to show that the ownership of the hotel site had been transferred to a company, Deep Bay Development, in which Lester Bird and another minister were the only shareholders. He insisted that the shares were held in trust for the government. There followed a bitter argument between Lester and his father over the disposal of the hotel, the prime minister preferring a prominent business family, the Hadeeds, to handle the sale, disregarding the talks that Lester Bird was conducting with the Trump organisation who were possible buyers. Lester Bird won, but only after he threatened to resign unless consensus in cabinet decision-making was restored. Another state-owned hotel, the Halcyon Cove, was subsequently sold to business associates of the Hadeeds.

Other scandals also surfaced. One involved the ownership of electrical generating equipment at the Crabbs power station, the breakdown of which caused numerous blackouts during late 1988. Extravagant claims of sabotage by 'Cuban-trained' engineers were made by the minister concerned, but it became clear that not only was the government responsible for maintenance, which had been lacking, but that the turbines were not new but reconditioned. Further-

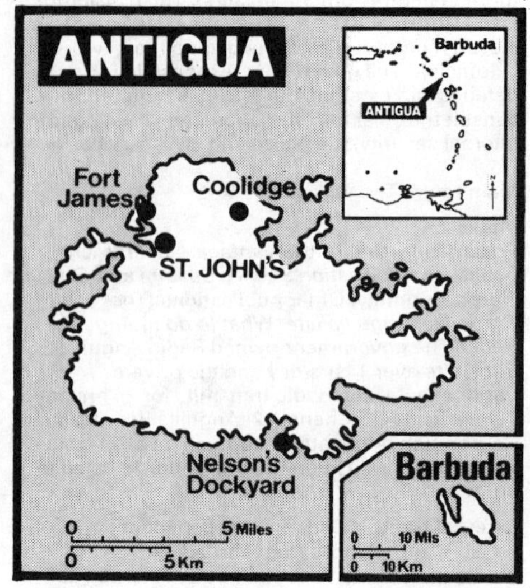

KEY FACTS

Official title: Antigua and Barbuda

Head of state: Queen Elizabeth II

Head of government: Prime Minister Vere C. Bird

Ruling party: Antigua Labour Party (ALP)

Capital: St John's

Official Languages: English

Currency: East Caribbean Dollar (EC$) = 100 cents

Exchange rate: EC$2.69 per $ (Nov 1989)

Area: 280 sq km (Antigua), 161 sq km (Barbuda)

Population: 77,500 (1988)

GDP per capita: $3,399 (1987)

GDP real growth: 6.6% (1988)

Inflation: 1.2% (1987)

Trade balance: –$229.75m (1987)

Visitor numbers: 375,750 (1988)

more, they were still owned by a British company despite US$65m having been spent on the construction and equipping of the plant in 1987. Another scandal surrounded the sale of sand from Barbuda at little more than EC$1 per cubic metre to a US contractor (whose company is indirectly three-quarters owned by three ministers, including Lester Bird), who resold at up to EC$23. The 1,000 or so Barbudans were furious and, when the contractor and another announced a scheme to build a quarantine station on the island to house South American llamas destined for the US pet market, as well as a 'wildlife park', some 300 of them blockaded the port to prevent around 200 llamas from landing. Many of the animals died from the unaccustomed heat and their plight was extensively recorded by the world's press as they languished on a small islet off Antigua, their fate unknown.

KEY INDICATORS

	Unit	1984	1985	1986	1987	1988
Population	'000	75.1	75.6	76.3	80.0	–
GDP	Ec$m	–	464.8	537.2	624.8	–
GDP per capita	Ec$	–	6,148	7,041	7,810	–
Inflation	%	4.2	0.9	3.0	1.2	–
GDP real growth	%	6.5	6.7	7.6	8.7	6.6
Exports	$m	35.2	28.3	30.9	28.5	–
Imports	$m	–150.17	–174.8	–284.15	–258.25	–
Current account	$m	0.6	–23.1	–126.6	–83.3	–
Exchange rate	EC$ per $	2.7	2.7	2.7	2.7	2.7

* estimated figure

Prosperity on credit

There is little doubt that corruption and economic mismanagement have contributed to an uncomfortably large debt burden, although another reason is the extraordinarily high material expectations of most Antiguans, bolstered by high wages. Credit also helped to sustain growth, which in 1988 amounted to 6.6 per cent (1987: 8.7 per cent), especially in the private and tourist construction sectors. Many workers hold more than one job, yet there remains considerable excess labour in the public sector. Skilled and professional manpower is at a particular premium. What aid is granted is in loan form because of the relatively high income per capita.

Direct official debt rose from US$240m to US$250m in 1988, or over US$3,000 per inhabitant; it was a manageable US$63m as late as 1986. In fact the total is much more if public utility and state-owned property debt is taken into account: annual interest on the Royal Antiguan Hotel loan alone is over US$9m. The reality is about US$850m and it will be over US$900m by 1990.

A good deal is owed to commercial banks, some of which have experienced defaulting to the tune of over US$45m. Commercial loans were especially directed to tourist facilities but these are proving to be less profitable than hoped, despite high prices. Fortunately, arrears due to the Caribbean Development Bank were paid off, thus restoring the state's eligibility for concessionary finance. But making matters worse is a continuing deficit on the recurrent account, one reason being the abolition of income tax in 1976 and the near exhaustion of alternative sources of revenue. The 1987 deficit was EC$53m, partly due to inefficient collection and supplementary expenditure; that for 1988 is expected to exceed EC$70m. More than EC$54m was set aside in the 1988 budget for official debt servicing, up from EC$35.9m the previous year. Recurrent expenditure was set at EC$263.5m (US$97.6m), compared to EC$288.5m (US$106.9m) in 1987, which included substantial extra spending. Since high public sector pay rises were awarded there is little doubt that planned expenditure will be considerably exceeded. As before, the deficit was to be met by commercial bank borrowing on the hope that increased revenue from the tourist industry would be sufficient security.

Economic prospects

The completion of a number of large projects, notably Heritage Quay, the first part (US$13m) of the extensive US$35m development of the harbour area in the capital, St. Johns, and the need to repay pressing debts will mean a slowdown in growth. There is also some doubt whether the tourist industry, which contributes 60 per cent of GDP, will be able to continue to expand. There were 176,893 overnight visitors and 198,857 cruise ship passengers in 1988, representing increases of 11.1 per cent and 29.5 per cent respectively over 1987. Visitors from Western Europe sharply increased as a result of more air services, especially from West Germany. Tourist expenditure rose from US$186.7m to US$221.9m. But net earnings were about 30 per cent of these totals, since 70 cents in every tourist dollar leaves the economy because of extensive foreign ownership and to pay for imported food and materials for the sector.

The problem is looming over-capacity as nearly 30 more, and larger, hotels are being built or are at the advanced planning stage. There were about 3,500 rooms available in early 1989 and a report prepared for the Ministry of Tourism warned that unplanned capacity – up to 11,000 additional rooms are envisaged – would lower the occupancy rate from 68 per cent to only 41 per cent. Other problems include increasing prices for often indifferent service, and general harassment of tourists. Labour shortages and disputes, infrastructural deficiencies and pollution will make cleaner and, above all, cheaper destinations such as Mexico all the more attractive.

Labour shortages have also affected other sectors, especially manufacturing. Nonetheless, factory space was all taken up although high wages has led to stagnant exports. The fortunes of the once dominant agricultural industry were helped by better rainfall; together with the fishing industry it contributed 12 per cent to GDP.

Fortunately, despite the debt crisis and other problems, Antigua enjoys the support of the United States and can probably count on Washington to deflect any financial pressure from the international financial institutions. There remain two small US military bases on Antigua, remnants of a once large facility. On the regional front, the general air of prosperity and Vere Bird's unwillingness to share power was behind Antigua's unwillingness to participate in the Eastern Caribbean unity discussions prompted by the prime ministers of St Lucia and St Vincent in May 1987. Instead closer relations were proposed with the immediate neighbours Montserrat and St Kitts-Nevis, a good deal of whose trade has to transit through Antigua.

COUNTRY PROFILE

Political system: Independent state within the Commonwealth with British monarch as head of state, represented by a governor-general. Legislative power rests with the bicameral parliament which consists of a 17-member House of Representatives (16 Antiguan seats and one representing Barbuda) elected every five years by universal adult suffrage, and a 17-member appointed Senate.

Main town: St John's (population 20,000), Codrington (on Barbuda).

Media

Press: Main newspapers are *The Nation, The Worker's Voice, Rappore* and *The Outlet* (only daily).

Broadcasting: One government radio and television service (The Antigua and Barbuda Broadcasting Service) plus one private radio service (ZDK). Also private cable TV (CTV) offering ten channels of US television 24 hrs/day.

External trade: Heavy dependence on imported food and energy. Large trading deficit is only partially offset by re-exports (mostly manufactured goods) and earnings from tourism and capital inflows.

Exports: Machinery and transport equipment (30-35 per cent of total value), textiles (8-10%), manufactured goods (40-44%), chemicals, beverages and tobacco. Main destinations: USA (30-35% of total value), UK, Canada (10-12%), CARICOM (20-24%), Venezuela.

Imports: Petroleum products (50-52 per cent of total value), vehicles and parts, foodstuffs, textiles/clothing, paper, medicinal and pharmaceutical products, machinery and plant. Main sources: USA (48-50% of total value), UK (16%), Venezuela (13%), CARICOM (9-11%), Canada (4-6%).

Agriculture: Contribution to GDP has dropped to eight per cent due to factors such as soil depletion and drought, and only about 2,000 people are employed on the land. Sugar industry was closed in 1985. Small amount of cotton grown. Policy is to encourage self-sufficiency in food. Fishing is a growth area.

Industry: Sector contributes four per cent to GDP. Activity is centred on food processing, galvanised sheet, paints and light industries (mainly assembly of household appliances, vehicles, garments, paper products). Rum distilling and light manufacture are the major export industries. Tourism accounts for 60 per cent of GDP, with annual earnings of over US$187m, and contributes to buoyancy of construction sector.

Mining: Known deposits of high quality barytes, limestone and clay.

COUNTRY PROFILE

Energy: All the country's requirements are currently imported, mainly petroleum from Trinidad and Tobago.

Membership of international organisations: CARICOM, Commonwealth, GATT, UN, UNESCO, FAO, ICAO, ILO, IMF, IMO, OAS, OECS, WHO, World Bank. Associated with the EC (Lomé Convention).

Time: GMT -4.

Climate: Tropical and humid with temperature range from 21°C to 32°C. Little variation throughout year, although driest from Jan to Mar.

Entry requirements
Passport: Required by all except citizens of USA, Canada and UK, if they have proof of identity.
Visa: Required by all except citizens of the Commonwealth, EC countries, Scandinavian countries, Japan, USA, Austria, Switzerland and certain other countries for visits up to six months.
Currency: No restrictions on foreign or local currency entering or leaving the country. Declaration on arrival. Visitors must satisfy immigration that they have sufficient for length of stay.

Health precautions
Mandatory: Yellow fever vaccination certificate if arriving from an infected area.
Advisable: Water from tap is safe to drink.

Air access
National airline: Antigua is a shareholder in LIAT.
Other airlines: Air BVI, Air Canada, Air Martinique, American, British Airways, Eastern, BWIA, Lufthansa, Windward Islands Airways.
Tax: Departure tax is payable on leaving the country.
International airport: V.C. Bird International (ANU), 9 km from St John's; duty-free shop, restaurant, bank, post office, car hire.

Surface access
Main port: St John's, Nelson's Dockyard, Crabbs Slipway.

Hotels: Over 100 hotels and guest houses available. Six per cent government tax on all room rates; 10 per cent service charge is usual; one per cent tax on currency exchange. Book well in advance if arriving during Carnival time (i.e. late Jul, early Aug).

Car hire: National or international licence required to obtain visitor's driving permit. Driving on the left. Maximum speed 88 kph.

City transport
Taxis: Fixed rate system – lists available, but not metered and advisable to negotiate fares in advance.

National transport
Air: Scheduled daily services between Antigua and Barbuda operated by Four Island Air Services and LIAT.
Road: Network connects all major centres. Over 1,000 km of roads, at least 40 per cent of which are all-weather.
Buses: Restricted service.

Public holidays: 1 Jan, Good Friday, Easter Monday, Labour Day, Whit Monday, Carnival (2 days early Aug), 1 Nov (Independence Day), 25 Dec, 26 Dec.

Working hours
Government: (Mon-Fri) 0800 – 1200, 1300 – 1630. N.B. Offices close at 1500 on Fridays.
Banking: (Mon-Thu) 0800 – 1300; (Fri) 0800 – 1300, 1500 – 1700.
Shops: (Mon-Sat) 0800 – 1200, 1300 – 1600/1630. Exception: (Thu) some open mornings only.

Telecommunications
Telephone: Dialling code for Antigua and Barbuda: IDD access code + 1-809 followed by subscriber's (seven-digit) number. Domestic service provided by Antigua Public Utilities Authority. International services through Cable and Wireless Ltd.
Telex: International services provided by Cable and Wireless Ltd, based in St John's, 24 hrs/day.

Banking
Banking centre: St John's. Antigua has a developing offshore banking trade.
Commercial banks (head offices): Antigua Commercial Bank, Bank of Antigua, Fidelity Bank.
Development bank: Antigua and Barbuda Development Bank.
Foreign branches: Bank of Nova Scotia, Barclays Bank International, Canadian Imperial Bank of Commerce, Royal Bank of Canada, Swiss American Bank.

Electricity supply: 230/115 V AC, 60 cycles.

Hotels: Antigua
Anchorage, Dickenson Bay, P.O. Box: 147 (tel: 20267; tx: 2160 anchor).

Hotels: St John's
Antigua Horizons, Long Bay, P.O. Box: 54 (tel: 4632006; tx: 2172 anchofar).

Blue Waters Beach, Soldiers' Bay, P.O. Box: 256 (tel: 20290; tx: 2066; fax: 20293).

Flamingo, Queen Elizabeth Road, P.O. Box: 466 (tel: 21266; tx: 2086).

Halcyon Cove, Dickenson Bay, P.O. Box: 251 (tel: 20256; tx: 2132 halcove; fax: 4620271).

Half Moon Bay, P.O. Box: 144 (tel: 4300/5; tx: 2138; fax: 4306).

Hawksbill Beach, P.O. Box: 108 (tel: 20301/21515; tx: 2189; fax: 21515).

Jolly Beach, Lignumvitae Bay, P.O. Box: 744 (tel: 20068; tx: 2126).

Long Bay, P.O. Box: 442 (tel: 32005; tx: 2172 anchofar).

St James's Club, Marmora Bay, P.O. Box: 63 (tel: 31113; tx: 2088; fax: 31430). @guide first = **Car hire:** St John's @guide first = Antigua Car Rentals: Barrymore Hotel, Fort Road (tel: 4621815).

Avis: Old Parham Road, PO Box 1076 (tel: 4614522; tx: 2145).

Budget: Powell Estates, Coolidge (tel: 4623009).

Dollar: Cross & St Mary's Streets (tel: 4620362).

Ramco Car Rentals: Upper Nevis Street (tel: 4623397).

Business advice and information
Antigua and Barbuda Tourist Office, Corner Long and Thames Street, PO Box 363, St John's (tel: 4620029, 4620480; tx: 2122).

Antigua and Barbuda Manufacturers' Association, PO Box 1158, St John's (tel: 4621536).

Antigua Hotel Association, Long Street, PO Box 454, St John's (tel: 4623702, 4620374; tx: 2172).

Antigua Chamber of Commerce, Cross and Redcliffe Streets, PO Box 774, St John's (tel: 4620743; tx: 2105).

Cable and Wireless Telex Bureau, St Mary's Street, St John's (tx: 2145).

Ministry of Economic Development, Tourism and Energy, Queen Elizabeth Highway, St John's (tel: 4620092/9).

Argentina

Jimmy Burns

Towards the end of May 1989, a number of Argentine cities were engulfed in an unprecedented wave of violence and looting in protest at vast food price increases. In a Latin American country which had traditionally prided itself on its self-sufficiency in agricultural produce and its relatively stable social environment, the riots surfaced as a dramatic reminder of a long-running economic crisis. The riots and the subsequent declaration of a 30-day state of siege were also a sad final chapter on the civilian presidency of Raul Alfonsin, who had been elected in 1983 on a wave of popular support only to see his party defeated by Peronist leader Carlos Menem in the elections of 1989.

Alfonsin's policies discredited

The riots in Cordoba, Rosario and Buenos Aires followed the latest package of 'emergency measures' aimed at arresting a growing crisis of confidence and stability. The measures included a return to the single exchange rate fixed and regulated by the government; export duties of 20 per cent for industrial goods and 30 per cent for agricultural goods; a 4 per cent tax on the first sale of agricultural land; suspension of all state-financed public works; threats of legal action against price 'speculators'; rises in public sector tariffs and a six month suspension of tax relief schemes for industrial promotion. In a nationally televised speech, Mr Alfonsin blamed the crisis on unpatriotic behaviour. But the causes were rather more complex. They stemmed from an unpalatable mix of economic mismanagement, endemic structural economic problems and the absence of a strong political institutional framework capable of ensuring the effectiveness of democracy in Argentina.

Mr Alfonsin had taken charge of the country following the collapse of a military regime which had been badly tainted by an appalling human rights record and a humiliating defeat in the Falklands War. His principal concerns when he came to power in 1983 were the issues of human rights and politics. The economy did not initially top his list of priorities. At the time Alfonsin's decision to bring the generals to trial and restore fundamental human rights such as freedom of expression turned him into something of an international superstar – an example not just to the developing world, but to many Western countries who had not seen such a bold transition from dictatorship to parliamentary rule since the immediate post-war years in West Germany. However, Alfonsin also inherited a weak economy – high inflation and a large foreign debt – which he tackled from the outset with only limited success.

The last months of the Alfonsin presidency in 1989 saw the inevitable results of economic decisions which were taken too late and without sufficient planning. They were testimony too to the govern-

KEY FACTS

Official title: Argentine Republic
Head of state: President Carlos Saúl Menem
Ruling party: Partido Justicialista
Capital: Buenos Aires
Official Languages: Spanish
Currency: Austral (Au) = 1,000 centavos
Exchange rate: Au1,801 per $ (Dec 1989)
Area: 2,766,889 sq km
Population: 31.96m (1987)
GDP per capita: $2,862 (1987)
GDP real growth: –3.1% (1988)
Inflation: 4,427% (1988)
Trade balance: $4,234m (1987)
Foreign debt: $59.6bn (1988)

ment's inability to break the deeply entrenched interests which had survived the period of military rule, namely the armed forces themselves and the trade unions. Once regarded as a man full of good intentions, Alfonsin had by April come to be regarded by the majority of his fellow countrymen as a consummate failure, who had brought a nation dangerously close to ungovernability.

Economic crisis deepens

Over the previous 12 months, Argentina had paid only $100m interest on its $60bn foreign debt. Small wonder that relations between Buenos Aires and the international financial community had rarely been worse. In reality, that most dreaded of Third World demons, a debt moratorium, was in place without it yet being officially declared. Hopes that the International Monetary Fund – which had over the previous six years seen budget deficit targets repeatedly broken – would extend a fresh standby loan of $2bn, and commercial banks loans of $3bn, had come to nothing. The IMF had cut Argentina off in 1988 in the middle of a $1.4bn standby credit. It said it would hold back $800m until Argentine policies improved. They didn't. On the contrary they got steadily worse.

In general the view of the international financial community was one of resignation. Apart from downgrading Argentina's credit ratings and making some provisions for bad debts, the international financial community decided from fairly early on in 1989 that it would await the outcome of the next presidential elections scheduled for May before making any serious moves.

At least part of the explanation for the IMF's tough stand on Argentina, it was suggested at the time, was the continuing quarrel between the Fund on the one side and the World Bank and the United States Treasury on the other about the role of the two multilateral institutions in handling the international debt crisis.

IMF officials, particularly Managing Director Michael Camdessus, were reported to be still piqued by the new leading role that the World Bank had taken in backing an earlier Argentine anti-inflation drive and conceding a $1.2bn loan (even though the World Bank subsequently suspended disbursement of $350m related to reforms in the banking and financial sectors).

With inflation running into triple figures annually from the second year of the Alfonsin administration, and a total since March 1985 of 8,400 per cent, Argentine workers, businesses and farmers all found themselves worse off than at the end of military rule. Since the government introduced its first radical anti-inflation package in June 1985, real wages had fallen by 23 per cent, monthly interest rates had rarely fallen below 25 per cent, and the austral, the new Argentine unit of currency, had depreciated by 150 per cent in just two months. And this was not the end of it, by any means. Falling tax revenues, even by Argentine standards of mass evasion, had reached new limits, as had public spending, which meant an ever increasing budget deficit and losses in the public sector estimated at $8.5m per day.

In the face of such a crisis, the best the government could think of was token readjustments of the economy, such as those described above, and cabinet reshuffles. On 31 March, the day before April Fool's Day, Mr Juan Sourrouille resigned after serving as President Alfonsin's hand-picked economy minister for 1,500 days. He was followed by his team of under-secretaries, debt negotiators and central bank officials. Although Mr Sourrouille was the minister most closely identified with the government's economic mismanagement, his resignation had arguably rather less to do with economics and a great deal more to do with politics.

Presidential campaign

As Argentina prepared itself for the scheduled presidential elections on 14 May, all the opinion polls suggested that the ruling Radical Party was heading towards a crushing defeat. President Alfonsin, even had he wanted to, was unable to stand for reelection because the country's constitution prohibited him from serving two successive terms. The Radical candidate was a poor substitute: a lacklustre provincial governor from Cordoba called Eduardo Angeloz. On the election trail, Mr Angeloz had spent what little energies he had on publicly attacking Mr Sourrouille and his economic team for plunging Argentina into crisis. As the *Financial Times* noted at the time: 'The reality of Mr Sourrouille's resignation is that Argentina's underlying economic crisis was not significantly different last week from two months or two years ago. What has happened is that in the context of a presidential election, economic icebergs have started to move ominously together, symbolised by a rapid depreciation of the austral since February 6. But those icebergs have not suddenly appeared; simply President Alfonsin's government has chosen the easier route to ignoring them. The largest and ugliest iceberg of all, the fiscal deficit was probably well past 10 per cent of gross domestic product in 1988.'

As it turned out, the appointment of Juan Carlos Pugliese, the president of the House of Deputies, as the new economy minister proved little more than a stop-gap measure quite ineffectual in dealing with the deep-rooted crisis of confidence affecting the country. A loyal Radical Party member, Pugliese's appointment had brought him to kamikaze-style prominence, where honour was expected to be found in self-sacrifice. Even worse, to many Argentines, his appointment symbolised the internal corrosion of Radical Party politics – an expedient, rather cynical device simply designed to lull the wavering voters, who were estimated to represent as much as 30 per cent of the electorate.

The strategy failed. The 1989 presidential campaign was in a sense the most democratic ever staged in Argentina. Focussed almost exclusively around the country's two major political parties, the Radical Party and the Peronist Party, the lead-up to polling day was conducted without violence or allegations of fraud. Foreign observers were impressed by the general climate of mutual tolerance. Such a mood suggested that Raul Alfonsin's greatest achievement had to do with the simple fact of becoming the first democratically elected Argentine president in six decades to hand over the baton to another democratically elected president.

The two main candidates for the presidency, Eduardo Angeloz of the Radical Party and Carlos Menem of the Peronist Party, were both lawyers; both had studied at the same university at the same time; both were provincial governors, and both were in their late fifties. By all accounts, they respected each other, but in all other respects their political personalities revealed themselves as different as chalk from cheese.

Mr Angeloz is of Spanish and Swiss ancestry. Lacking the charisma of Mr Alfonsin, he was disliked by the younger members of the Radical Party, the so-called 'Young Turks', who had over the previous six years manoeuvered themselves into positions of influence throughout the state apparatus centred on Buenos Aires. On his campaign trail, Mr Angeloz distanced himself from the populist rhetoric favoured by the 'Turks', concentrating instead on sober assessments of the country's economic problems. Anything but flashy in his appearance or delivery of speech, Mr Angeloz talked much as a European politician would: that change is possible but can be hard to achieve. He said he wanted to lead Argentina towards free markets and balanced budgets. He wanted to stop cosseting industry and give the Argentine central bank more authority over money supply and interest rates. (In the past the central bank had been forced to play second fiddle to the economy ministry and had little control over the inflationary policies adopted by the provincial banks.) Among his boldest projects was the pledge that he would privatise three sacred cows of the state sector: the Atomic Energy Commission, the petroleum monopoly, Yacimientos Petroliferos Fiscales (YCM), and Fabricaciones Militares. The latter was Argentina's huge army-controlled industrial complex – reputedly the country's largest holding company, with interests ranging from chemicals and arms manufacturing to shipbuilding and trains. The aim was to reduce the army's main power base and thus their influence on society. Similarly, Mr Angeloz promised to tackle that other political 'monopoly', the trade unions, which throughout the Alfonsin administration had fuelled the general climate of instability by staging a series of general strikes. Mr Angeloz said he was prepared to put his plans for sweeping new

KEY INDICATORS

	Unit	1984	1985	1986	1987	1988
Population	m	30.10	30.56	31.03	31.50	31.96
GDP (1985 prices)	Au bn	41.5	39.6	41.8	42.6	41.3
GDP per capita	$	2,255	2,274	2,340	2,343	2,862
Inflation (average rate)	%	650	769	190	232	4,427
GDP real growth	%	2.6	–4.5	5.5	1.9	–3.1
Agricultural production	1979-81=100	107.6	105.5	108.7	107.2	–
Wheat	'000 tonnes	13,200	8,700	8,900	10,100	–
Soya beans	'000 tonnes	7,000	6,500	7,100	7,000	–
Industrial production	1980=100	92	84	94	93	–
Manufacturing	1980=100	91	82	92	92	–
Construction	1980=100	47	44	48	55	–
Mining production	1980=100	101	98	98	94	–
Crude petroleum production	(mn					
barrels)	174.7	167.7	157.2	155.7	164.4	
Exports	$m	8,100.0	8,396.0	6,852	6,360	9,134
Meat	$m	404.3	385.7	464.7	599.3	607.2
Wheat	$m	965.8	1,133.2	395.1	351.0	–
Corn	$m	748.2	765.9	654.2*	298.5	381.6
Hides and skins	$m	304.9	288.9	340.7	354.2	367.7
Wool	$m	216.4	196.3	135.2	140.8	184.4
Imports	$m	4,118	3,518	4,406	5,392	4,900
Capital goods	$m	643	648	1,730	2,360	800
Raw materials	$m	3,225	2,514	2,293	2,500	3,820
Balance of trade	$m	3,982	4,878	2,446	968	4,234
Current account	$m	–2,495	–952	–2,859	–4,239	–1,609
Exchange rate (average)	Au per $	0.0677	0.6018	0.943	2.144	8.753

* estimated figure

laws aimed at breaking the labour monopoly and the closed shop to a national referendum over the head of Congress if necessary. Such tough talk was the kind that won Margaret Thatcher several election victories in Britain. But in Argentina, it failed to divert popular attention from the poor record on the economy with which the Radical Party as a whole was identified. For the country's 6m voters dependent on salaries, life under a Radical government had translated itself into a firm disincentive to vote Radical on 14 May.

On the eve of the poll, newspapers updated their statistics to point to the glaring social contradictions confronting Latin America's least populated and potentially most resourceful nation. Monthly inflation at 40 per cent surpassed the previous record of 38 per cent in March 1976, the month in which a military coup led by General Jorge Videla had toppled the government of Isabel Peron. It was estimated that 30 per cent of Argentine households were classifiable as poor, lacking sufficient income to cover basic necessities of clothing, diet and education. The United Nations Children's Fund estimated that 20,000 Argentine children annually died prematurely from diseases directly related to malnutrition. Some 2m people lived in slums around Buenos Aires in conditions familiar to countries lacking a tenth of Argentina's natural resources. In his caretaker capacity, the new economy minister Juan Carlos Pugliese, had carried out three economic packages including price controls, but had failed to stem the rot. Price increases continued to leapfrog throughout the election campaign.

For the Peronists, and for large sectors of the population, Carlos Menem cut the kind of colourful, dynamic figure not seen since the days of the party's founder and chief mentor, General Juan Peron. With long hair and bushy sideburns brushed in the style of one of Argentina's most famous nineteenth century *caudillos* (a Spanish or Latin American military dictator), Menem entered the campaign trail as an unashamedly Third World populist. Of Syrian stock, and brought up in the provinces, he epitomised the self-made man in a country of recent immigration. He was surrounded by much media hype about his passion for fast cars, planes and women. His own marriage to Zulema, a similarly Arab-blooded Argentine who, unlike Menem, had stuck to her Muslim origins, was reported to have been saved from the rocks by the Vatican. Zulema, so the story went, had threatened to abandon her husband half-way through the campaign trail unless he paid more attention to her. It was a demand that Menem could hardly ignore given that the Argentine Constitution discriminates against divorcees. Once reconciled with Menem, Zulema almost outdistanced her husband on the way to the polls. Her hair and body reshaped, by her own admission, in the style of Linda Evans of *Dynasty* fame, Zulema made no secret of her ambition to take up where the late Eva Peron had left off. She declared herself a devout Peronist, and promised to use all her powers to help the dispossessed.

Menem himself was something of a political maverick, underpinned by elements of orthodoxy. He had been a loyal supporter of General Peron, whose regime, spanning much of Argentine's post-war history, had been marred by political violence and corruption. He was imprisoned by the military junta that took power in 1976 and was elected governor of La Rioja when Alfonsin was elected president. His main achievement in office was to have attracted foreign investment to a previously poverty-stricken province, but at the expense of huge tax breaks and by printing money, which distorted the true extent of economic mismanagement. Menem was little known internationally except in neighbouring Paraguay where he conducted an open friendship with General Stroessner. Inside Argentina his friends were as diverse as the Peronist movement itself. They ranged from right-wing trade unionists and military officers to left-wing students and members of the Radical Party enamoured with the idea of a Third World Movement. In order to reach the top of the Peronist hierarchy, Menem had overcome formidable odds. As a first-generation Syrian immigrant, he was a Sunni Muslim who had converted to Roman Catholicism. This is the first time that a Syrian-Lebanese had risen to such heights in Argentina, where the political class had traditionally been dominated by German, Spanish or Italian stock.

General Peron had come to power using his influence within the armed forces. His second wife Isabelita came to power simply because of the Peron name. Menem had chosen the democratic path of internal primaries to impose his leadership on the Peronist Party. His rival in the primaries, former economy minister Antonio Cafiero, presented himself as the leader of a 'renewalist' faction capable of breaking with the party's authoritarian past and taking it into the modern world of social democracy. Cafiero ran an ill-organised primary campaign in which he had failed to portray himself convincingly as a moderate in the Alfonsin mould. By contrast, Menem – both in the primaries and his subsequent race for the presidency – owed much of his appeal to the naive popular nostalgia surrounding Peron, who died in 1974. Menem tapped an Argentine form of ancestor worship. His campaign rhetoric matched the style. He talked boldly of recapturing the Falklands, of staging a productive 'revolution', of reviving Argentina's links with the Third World and Latin American solidarity. He declared himself a champion of the poor who would slay once and for all the dragon of financial speculation by the rich. In this role he represented both the extreme nationalism of Argentina and the anti-Buenos Aires, anti-intellectual vein which Peron had so fervently promoted. As he and Zulema plunged into one crowd after another, and toured the slums, Menem appeared to regenerate the old Peronist slogan: 'Shoes yes, books no'. For large sectors of the population who under Alfonsin had seen the priorities of government aimed squarely at such vague intellectual concepts as cultural freedom and political morality rather than at keeping prices down, such phrases were not without attraction. One of Menem's main campaign promises was a 'salariazo', a massive wage increase for all those whose wage earnings had been badly dented. He also promised a five-year moratorium on the payment of debt interests.

Menem victory for Peronists

On 14 May, Carlos Menem was swept to victory with a convincing slice of the popular vote. Early returns gave him 47.2 per cent of the vote against 37 per cent for the Radical Party. Other parties fared badly, confirming the political dominance of the country's two

COUNTRY PROFILE

Political system: Republic with executive power vested in the president, who is elected by an electoral college for six years. Legislative power is held by the bicameral Congress, consisting of the Chamber of Deputies (254 members elected by universal suffrage for four years) and the Senate (46 members nominated by provincial legislatures for nine years).

Main cities/towns: Buenos Aires, Córdoba, Rosario, Bahía Blanca, Santa Fé, La Plata, Mar del Plata.

Media
Press: Major daily newspapers in Buenos Aires include *La Nación* (circulation 210,648), *La Prensa* (65,000), *Clarín* (480,000), *Crónica* (morning 330,000; evening 190,000), *Ambito Financiero* (business), *El Cronista Comercial* (100,000) and *La Razón* (evening 180,000). *The Buenos Aires Herald* (20,000) is in English and *Le Quotidien* is in French. Many provincial towns have a daily newspaper. Also numerous periodicals and magazines available throughout the country.
Radio: Numerous private and state stations. Over nine million radios – serving at least 90 per cent of population.
Television: Numerous (commercial) privately and state-owned stations. Four channels serve Buenos Aires.

External trade: Argentina has managed to maintain a balance of trade surplus in recent years due to a policy of strict import controls. Heavy foreign debt burden and slump in capital inflows has however kept overall balance of payments in deficit. Argentina adopted an increasingly tough stance on foreign debts in 1987, demanding easier repayment terms and relief from conditions imposed by the International Monetary Fund.
Exports: Major exports (1987): Meat (9.4% of total value), hides and skins (5.6%). Main destinations (1987): EC (27% of total value), Latin America Free Trade Zone ((20.6%), USA (14.1%), USSR (10.1%), Netherlands (9.7%), China (4.2%), Japan (3.5%), Iran (3%), others (7.8%).
Imports: (1985): Machinery and equipment (39% of total value), chemicals and rubber (25%), mineral products (15%), basic metals and manufactures (8%), transport equipment (6-8%), agricultural products (3%). Main sources (1987): EC (31.1% of total value), Latin America Free Trade Zone (29.6%), USA (16.2%), Japan (7.6%), Netherlands (2.1%), USSR (1.6%), China (0.2%), others (11.6%).

Agriculture: Accounts for approximately 15 per cent of GDP and employs about 16 per cent of the workforce. Production is concentrated on the pampas. This occupies a vast area of central Argentina which, owing to fertile soil and favourable climate, is responsible for 85 per cent of the country's agricultural exports. Traditionally, the sector has been dominated by the rearing of livestock and in particular beef for export. However, in recent years there has been a significant shift towards crop production. This has been primarily due to changes in the relative prices offered on the international markets and the introduction of improved production techniques, including increased mechanisation, increased use of fertilisers, better pest control measures and new plant varieties. Cereal production in particular has improved although there have been fluctuations in output due to adverse weather conditions and export price variations. Principal crops grown are wheat, maize, sorghum, soya (world's third largest producer), oil seeds, rice, sugar-cane, citrus fruits, grapes (world's fifth largest wine producer), tea and tobacco. In contrast, livestock output has declined in recent years and exports have fallen dramatically. Major causes of this have been ineffective marketing, the introduction of import restrictions on fresh and frozen meat, especially by the USA and Japan; and increased competition from other leading meat exporters such as the EC, Brazil and Uruguay.

Industry: Manufacturing accounts for approximately 25 per cent of GDP and employs about 20 per cent of the working population, while construction accounts for around four per cent of GDP. The severe slump in industrial output in the early 1980s gave way to a period of limited expansion, mainly due to increased local supplies of agricultural produce and the imposition of strict import controls. Major sectors of production include food, textiles, machinery and transport equipment, consumer durables, industrial chemicals, metal working, engineering, paper, iron and steel and electrical equipment. Traditional labour-intensive and natural resource-based industries, which suffered more than most under the government's open economy policy of the late 1970s and early 1980s, have been among the quickest to recover. In particular, food processing and beverages, which are seen as occupying a pivotal role in the future development of the economy, owing to their use of domestically produced agricultural inputs, high productivity and a good export earnings potential, have demonstrated high levels of growth. Production of vehicles has also recovered to some extent, but in 1985 production in most other sectors, including machinery, equipment, construction materials, textiles and consumer durables, remained depressed. The construction sector has suffered especially from sharp cuts in public capital expenditure programmes.

Mining: Sector remains largely undeveloped, contributing only 0.3 per cent to GDP, in spite of a rich variety of mineral deposits. Iron-ore is the principal mineral extracted, mostly in Río Negro province, but output is only sufficient to supply about half of the requirements of the country's largest blast furnace complex, the remainder being made up from imports. Other minerals extracted in smaller quantities include copper, lead, zinc, tin, gold, silver and uranium. Production of these remains depressed, but a mining promotion law has been passed to develop mining activities considered beneficial to the economy. In particular, copper is being encouraged, as the large domestic market for this is at present met by imports.

Energy: Much emphasis has been placed on the development of energy production as a necessary step towards further economic expansion. Domestic oil production, all of which is derived from onshore fields, accounts for virtually all of domestic consumption, with crude oil imports having ceased in 1983 (some lighter oil products are imported in small amounts). In 1987 however, decline in production and increase in demand again made imports necessary. About 70 per cent of the oil sector is controlled by the state-owned Yacimientos Petrolíferos Fiscales. A major obstacle to growth has been the low returns offered to private oil companies, but a relaxation of government policy is expected to lead to increased development and production from these. Output has fallen slightly from the 1981 peak to 2.428m cu metres in 1985. Oil reserves are estimated at 2.3bn barrels. Several natural gas reserves have been found, although at present output is insufficient to meet demand and imports (mainly from Bolivia) are required. However, production has increased steadily to nearly 524.8bn cu feet per year, and it is expected that there will eventually be an exportable surplus. Natural gas reserves are estimated at 26.41m cu metres. Coal is mined almost exclusively in the Río Turbio area, and a major expansion programme is underway – principally involving improvements to infrastructure. Installed electricity generating capacity is estimated at 12,000 mW, and production has been increasing with the development of several hydroelectric projects, although 45 per cent of output remains thermal. Nuclear capacity is also being developed, and the country's third nuclear plant was scheduled to come on line in 1987, although this has now been delayed due to financial constraints.

Membership of international organisations: CCC, FAO, G-77, GATT, IADB, IAEA, ICAC, ICAO, IDA, IFAD, IFC, IHO, ILO, IMO, IMF, INTELSAT, Inter-American Development Bank, Interpol, IOOC, ISO, ITU, IWC (International Whaling Commission), IWC (International Wheat Commission), Latin American Integration Association, NAM, Organisation of American States, PAHO, SELA, UN, UNESCO, UPU, WFTU, WHO, WMO, World Bank, WTO, WSG.

major groups. In the previous presidential election, the Radicals won 51.8 per cent of the vote and the Peronists 40.2 per cent of the vote. The tables had clearly turned. The results showed a greater margin of superiority for Menem than had been predicted by the polls. They meant that one major element of political uncertainty disappeared overnight. Not only had Menem secured nearly half the popular vote, he had also achieved about 10 per cent more than his party's natural level of support. There had been a strong Peronist showing throughout the country, even in Cordoba and the Federal Capital: regarded as bastions of support for the Radical Party. Menem, moreover, emerged from the election in the uniquely strong political position, for an Argentine leader, of having a majority in both houses of Congress. Prior to the vote the Radicals had hinted that they might try and block the Peronists with their allies in Congress and within the Electoral College, which determines who should be president in the event of a close vote.

On election day the Radicals experienced at first hand a famous Argentine psychological phenomenom, the *voto de castigo* – the highly volatile punishment vote of the working class. One of Eduardo Angeloz's repeated campaign slogans *se puede* (it can be done) had been matched by a Peronist slogan which aptly summed up the frustrations of large sectors of the population: 'If it can be done, do it now, the country needs it'. Quite apart from capturing the vote of the economically disenchanted, Menem counted on the support of the trade union movement (traditionally loyal to the Peronist Party), influential sectors of the armed forces and the Church, and business leaders, some of whom had done well out of the corporate state forged by Peron. Beyond this initial support, however, the longer term consequences of the Menem victory remained uncertain. Indeed, there was no contemporary parallel in Latin America for a leader elected to office with such a limited record and such a vague programme.

Menem found himself being sworn in as president in a situation of extreme crisis. Although Alfonsin had planned to see out his constitutionally entitled term of office until December 1989, he relinquished power on 8 July. By then the state was barely recovering from the psychological shock of the food riots and a further downturn in Argentina's international standing. In June, US regulators told banks to write down their medium-term and long-term loans to Argentina by 20 per cent. The US bank regulatory body told banks that it had classed their Argentine loans as 'value-impaired'. The measure meant that the banks had to set their reserves against 20 per cent of their Argentine credits, while retaining them in the hope that future negotiations might yield some payment from the debtor. The move came more than a year after Argentina had stopped paying interest on its medium-term and long-term foreign bank debt. The arrears were expected to more than double by the end of 1989. Argentina continued to pay interest on trade credits borrowed from banks and on almost all of its foreign government and international agency debt. Desperate to bolster its flagging foreign reserves, Argentina had also borrowed $125m from a group made up of Brazil, Mexico and Venezuela.

Initially, neither the Radicals on the way out nor the Peronists on the way in appeared prepared to shoulder the burden of making hard decisions on the economy. Only when Menem was sworn into office did something approaching a programme begin to emerge. Menem's first cabinet line-up surprised some observers because of its apparent diversity. But on closer examination, it revealed itself as a reflection of the corporatism which lay at the root of the Peronist Party. Ministers were drawn from the trade union movement, from big business and from sectors of the political class most closely identified with the conservative wing of the Argentine Church. A surprise choice was that of Domingo Cavallo, a former governor of the central bank and an economist by training, as the new foreign minister. Jose Bordon, who had been widely tipped for the foreign ministry having been largely responsible for the Peronist Party's foreign affairs platform document, was instead given the post of minister of public works. As governor of the province of Mendoza, Bordon had been involved with the sale of the loss-making vineyard Giol, owned by the provincial government. But perhaps the most astonishing appointment was that of Miguel Roig to the economy ministry. Roig had until recently been a senior executive of the Brazilian-based multinational Bunge and Born, which had massive business interests spanning the continent in sectors such as grain trading, construction and food processing. As a symbol of the private sector, Miguel Roig had not voted for Carlos Menem as president. But Menem decided to offer him the post on the advice of Jorge Triaca, head of the plastic workers' union, who was picked as minister for labour.

Sweeping new politics

In his first days at the helm, Menem surprised many observers by the speed with which he and his senior ministers adopted pragmatic policies which went against the main drift of much of the rhetoric of the election campaign. At the foreign ministry, Domingo Cavallo wasted little time in offering Britain an olive branch over the vexed Falklands issue. He promised to put Argentina's demands for sovereignty over the islands on the back burner, and urged negotiations instead on the resumption of diplomatic and commercial ties. One of the reasons behind the move was that Cavallo believed that a restored relationship with Britain would pave the way for more fruitful trade links between Argentina and the rest of the European Community.

Miguel Roig died of a heart attack on 14 July. But his successor, Nestor Rapanelli, also came from Bunge and Born and promised to carry out a thorough reconstruction of the economy, including sweeping privatisation, budget cuts and early negotiations with the International Monetary Fund. Five days before he died, Roig had announced an emergency economic programme which he claimed would change the face of the country through an initial period of extreme austerity. It included a 600 per cent increase in state-controlled prices on items such as petrol, gas, electricity and telephones; a new officially-fixed exchange rate of 650 australes to one US dollar, and a promise to privatise rapidly all state-owned companies except those related to 'national security'. He believed that the sharp increases in public tariffs could be borne, as unofficial estimates suggested that $5bn – substantially more than last year's trade surplus – was in private circulation in the domestic economy. As part of the package, both private and public sector employees received a one-off additional payment of 8,000 australes, along with an advance payment of a portion of July's salary. Other measures included the suspension of a costly system of tax relief benefits for the promotion of industry and a modification of central bank regulations. The central bank was also to be prevented from using its reserves to finance national and provincial government losses. These measures were aimed at drastically reducing Argentina's fiscal deficit, which had risen in mid-1989 to about 15 per cent of GDP, and at cutting inflation back from its projected monthly level of 200 per cent in July. Few economists in Argentina doubted that such measures were urgently needed. The operating deficits of the 10 largest state-run companies ranged from $1bn in the case of YCM to $22m in the case of Obras Sanitarias Nacional, which deals with sewage disposal. For years the deficits had been financed in a variety of ways, ranging from straight subsidies from the Treasury through to transfers of cash, bonds and creative accounting. According to one official estimate, there were 40,000 judicial cases outstanding against publicly-owned companies, many a result of failure to pay bills on time, or at all.

Nevertheless, having generated huge expectations in the course of its election campaign, the presidency of Carlos Menem was facing an uphill struggle in delivering his programme without major political and social unrest. As with the other plank of the economic programme – the three-month price and salary freeze imposed in July – the future success of the stabilisation was crucially dependent on the support of business and trade unions, neither of which had in the past seemed prepared to put aside their own vested interests. The appointment of Nestor Rapanelli to the economy ministry appeared to symbolise the president's commitment to the essentials of austerity. But quite apart from a public scandal over his links to a fraudulent financial operation in Venezuela, Rapanelli was by late 1989 having to steel himself for dogged opposition to his privatisation plans. The government insisted that privatisation of telecommunications, shipping, gas and electricity, some of the large loss-makers, would not mean unemployment. And yet a major part of the public sector's financial problems could be traced to over-manning. Moreover, privatisation was not unqualified good news for large sectors of the Argentine business community. Many of the large 'price-forming' companies had in recent years come to depend almost entirely on state contracts, which last year cost the Treasury $2bn in over-pricing for goods and services. To add to this, industry was already suffering a sharp recession, with unemployment and lay-offs becoming commonplace. A further promise from President Menem's economic team was a revision of the country's taxation laws which, as one observer reported, had 'more escape holes than Swiss cheese'. The tax reforms promised a universal 10 per cent value added tax, and tighter, simpler rules imposing income tax payments of 20 per cent. To many analysts, this too seemed to be something of a tall order in a country where only 30,000 out of a population of 32 million bothered to pay their taxes.

Part of the government's problem in setting out on its bold programme appeared to be one of credibility. Sweeping economic programmes as drawn up first by Roig and then Rapanelli had been tried during the early years of the Alfonsin administration and had failed. Moreover, Menem's privatisation programme appeared to be underpinned by an essential contradiction: that of a Peronist trying to undermine what General Peron himself had built up. And there was no small irony in the fact that the economic policies initially adopted by Menem were precisely those defended by his opponent Mr Angeloz during the election campaign. Arguably, Peronism is and had always been nothing more or less than pragmatism elevated to the status of a creed. As the *Financial Times* remarked: 'If it suits the greater interest of Argentina, do it. Once it suited Peronism to have in state hands all major industry. Now it does not.'

Drought affects harvest

Not even Peron, however, had the power to dictate the weather, and here too the Menem administration inherited a major problem from the Alfonsin government. In 1989 Argentina was hit by one of the worst droughts in recent years. This played havoc with the country's grain harvest which is so crucial to its overall economic performance. The main problem had been carried over from 1988 when the main annual rainfall had been unusually concentrated in the month of March. Since then a dry spell persisted, with the result that in the province of La Pampa 400,000 hectares were seriously damaged by fires caused by unusual dryness. The local summer season in Christmas 1988 was the dryest on record since 1871 in the province of Buenos Aires. The knock-on effect of these conditions, which were inevitably attributed to structural changes in world climatic conditions as a result of deforestation in Brazil's rain forest and thinning of the ozone layer, left many Argentine farmers on the verge of bankruptcy and the Treasury having to look for funds elsewhere. In 1988, Argentina had managed to find relief from its balance of payments crisis from high international prices for grain as a result of the US drought. However, the severe drought early on in 1989 impeded harvesting in many areas where crops, particularly maize, did not grow properly. The total harvested area fell by almost 9 per cent in comparison with 1987-88, the total sown area of 18.3m hectares was down by 4.3 per cent in comparison with 1987-88 and was the lowest for 20 years. The total harvest was forecast in 1989 to be in the region of just over 27m tonnes or 25 per cent less than the previous season. The maize harvest fell by almost 48 per cent from last season to 4.8m tonnes, sorghum down 40 per cent to 1.9m tonnes, and soyabeans down 21 per cent to 7.8m tonnes. Agricultural products, including meat and related products, account for almost 80 per cent of Argentina's foreign currency earnings so that the drought contributed to a worsening trade deficit. On the basis of earlier projections, it had been hoped that 1988's trade surplus of $3bn would be raised to $4.45bn, but this was revised sharply downwards to about $2.9bn.

Trade union and military power

Clearly doubts remain about the president himself. If he was capable of saying all things to all men in the lead-up to his election, only to reverse half of what he had promised within days of his election, what is there to stop Carlos Menem from changing his mind again? To many observers internationally, he remains an enigma – a curious, somewhat disconcerting product of a continent which remains largely untamed socially and politically. Within Argentina itself, Menem remains in part a hostage to institutions like the trade unions and the armed forces, which gave him their tacit support initially. If Argentina is ever to modernise, trade union reform and an armed forces without the power to dictate policy remain key elements. But in the first months of the Menem administration, there were already indications that the trade unions still regarded 'reform' as a euphemism for confrontation. Although their monopoly of power based on jobs for life was being made increasingly unmanageable in the context of an economic crisis, no trade union leader appeared in a hurry to relinquish his position. General Peron exercised complete control over his trade unionists. Those who fell into line were rewarded with control over pension funds, while those who didn't were eliminated. Mr Alfonsin tried first to introduce new legislation curbing the powers of the unions, and then attempted to buy them off by appointing a trade union official as his minister of labour. Both tactics failed.

On the political front, the biggest problem facing Menem was the military question. One of the first moves by the new defence minister, Italo Luder, was to appoint new chiefs for each of the services in an attempt to insure a hold on the military. Nevertheless, the main concentration of power within the armed forces appeared still to be focussed on the figure of Colonel Mohamed Ali Seineldin, who was linked to a series of attempted rebellions during the government of Alfonsin. The junior and middle ranking officers grouped around Seineldin all belong to a generation that earned their kudos in the so-called 'dirty war' against left-wing guerrillas in the 1970s. They also fought in the Falklands War but returned embittered with the poor leadership of their generals and convinced that they could have won had they been better equipped.

Ever since the restoration of civilian rule, Seineldin and his officers have been demanding the end of the trials for human rights abuses, significant increases in the defence budget and a public statement from the civilian authorities recognising their contribution to society. The officers are a tightly-knit group, with their own internal literature and code of conduct. Seineldin himself is a commando regarded as an impetuous fighter. When Menem was elected, Seineldin was still under detention for his part in the last military rebellion against Alfonsin in December 1988. But it was clear that both Menem and his defence minister were facing strong pressure from within the armed forces to declare an amnesty. The problem for the government, however, was that even if it declared such an amnesty it would not necessarily assure itself of stability within the armed forces. Seineldin's supporters have become almost a separate political class, characterised by a fanatical nationalism which conflicts with the policies adopted by the Menem administration on issues like the Falklands and foreign investment. This nationalism is partly a consequence of the Falklands War. Although the Argentine armed forces have historically been pro-Western, the war pitched Argentine officers like Seineldin against Britain, the United States and most of Western Europe. The officers emerged from the War with their ideological ties with the West firmly broken, even though the new strategic context in which they were supposed to move remained unclear.

Comparisons were being made late in 1989 with Bolivia. There a government managed to confront and overcome hyperinflation through considerable ruthlessness, including using a military-backed state of emergency, breaking strikes wherever necessary, and by using force against miners, traditionally the country's most powerful union. Argentina under Menem was soaring beyond Bolivia's worst inflationary records. And yet the prescription for dealing with the situation had yet to prove a remedy. Not only could Menem not count on the support of the key figure in the army, he was also risking alienating one of the main pillars of his support, the trade unions. Argentina's powerful metal workers union, the Union Obrero Metalurgico (UOM), pledged its moral support for Menem only to follow this up with a demand for a 200 per cent wage increase for its 320,000 members.

BUSINESS GUIDE

Time: GMT -3.

Climate: Ranges from sub-tropical in the north to sub-antarctic in the south. The densely populated central zone (including Buenos Aires) is temperate: summer from Dec to Mar – hot and humid with temperature 26°C to 35°C. From Apr to May – temperature 10°C to 25°C. From Jun to Aug – winter with temperature 5°C to 20°C, nights can be cold with temperatures below freezing point. From Sep to Nov – spring with temperature 12°C to 25°C.

Entry requirements
Passport: Required by all except nationals of neighbouring countries with identity cards.
Visa: Check regulations with embassy/consulate as they change frequently. Generally all business visitors require a visa. Tourist visa usually required by nationals of EC countries, USA, Japan, Australia, Canada, most Commonwealth countries, Eastern bloc countries and certain others. Special requirements for British citizens; check with Argentine interests section at the Brazilian embassy in London.
Currency: No restrictions on import/export of local or foreign currency.

Health precautions
Mandatory: None.
Advisable: Typhoid and polio vaccinations recommended. Malaria prophylaxis advisable for visits to some lowland tropical areas. Water precautions should be taken outside main towns.

Representation overseas: Abidjan, Addis Ababa, Ankara, Asunción, Athens, Baghdad, Bangkok, Beijing, Beirut, Belgrade, Berlin, Berne, Bogotá, Bonn, Brasilia, Brussels, Bucharest, Budapest, Cairo, Canberra, Caracas, Copenhagen, Damascus, Dakar, Dublin, Guatemala City, The Hague, Havana, Helsinki, Islamabad, Jakarta, Jeddah, Kingston, Kinshasa, Lagos, La Paz, Lima, Lisbon, London, Madrid, Managua, Manila, Mexico City, Montevideo, Moscow, Nairobi, New Delhi, Oslo, Ottawa, Panama City, Paris, Port-au-Prince, Port of Spain, Prague, Quito, Rabat, Rome, San José, San Salvador, Santiago, Santo Domingo, Seoul, Sofia, Stockholm, Tegucigalpa, Tehran, Tel Aviv, Tokyo, Warsaw, Washington.

Air access
National airline: Aerolíneas Argentinas.
Other airlines: Aeroflot, Aeroperú, Air France, Alitalia, Austral, Avianca, Canadian Airlines International, Canadian Pacific, Cruzeiro, Cubana, Eastern, Ecuatoriana, Iberia, KLM, LAB, Lade, Lan-Chile, LAP, Lufthansa, Pan Am, Pluna, Quantas, Swissair, Varig, Viasa.
International airports: Buenos Aires-Ezeiza (EZE, Buenos Aires BUE), 50 km from city, duty-free shop, restaurant, bank, car hire. Downtown domestic terminal Jorge Newbery, commonly known as 'Aeroparque' (AEP) caters for certain international flights to neighbouring countries – Brazil, Chile, Paraguay, and Uruguay, duty-free shop, restaurant, bank, car hire. Córdoba-Pajas Blancas (COR), 15 km from city. Jujuy-El Cadillal (JUJ), 40 km from city. Rosario-Fisherton (ROS), 18 km from city.

Surface access
Main ports: Buenos Aires, Ensenada (La Plata), Rosario, Bahía Blanca. There are numerous smaller ports and some specialised terminals (for oil, cereals, raw materials etc).
Water access: Ferry and hydrofoil services on the Río de la Plata link Colonia and Montevideo (Uruguay) with Buenos Aires. Also ferries operate from Paraguay on the Paraná river.
Overland access: Four branches of the Pan-American Highway run from Buenos Aires to the borders of Bolivia, Brazil, Chile and Paraguay. Entry from Uruguay is possible via bridges over the river Uruguay at Puerto Colón, Puerto Unzué and the Salto Grande dam. Most routes surfaced throughout (but particular routes should be checked) – distances make journeys time-consuming (e.g. Santiago-Buenos Aires over 1,400 km). Many bus services available on these routes. Not all are direct but connections can be made. Train or combined train/bus journeys possible from Uruguay, Brazil, Paraguay and Bolivia. Rail link with Chile exists, but there are currently no passenger services.

Hotels: Wide range available, graded from one to five stars. Service charge and tax of up to 24 per cent are usually included in bills.

Car hire: Available in Buenos Aires and most main centres. International driving licence required – which must be validated by Automóvil Club Argentino (ACA) or a driving permit can be obtained from the Directiou General de Transito (Cnel. Roca 5252 – 0700 to 1300).

BUSINESS GUIDE

City transport
Taxis: Generally have yellow roofs; can be hailed or found on ranks; metered within cities; tips not necessary, through generally expected from tourists.
Buses: Major towns have good local services. Within Buenos Aires there is a comprehensive 'pay as you board' service, operating 24 hrs/day; buses often crowded.
Metro: (Known as *Subte*); operated within Buenos Aires from early morning to late at night; tokens needed (can be purchased at booking offices); fixed fare system.

National transport
Air: Extensive services operated by: Aerolíneas Argentinas; Austral (particularly in southern part of country); LADE (serves smaller centres). Demand for services is high and it is advisable to book flights in advance. Monthly unlimited travel tickets available.
Road: Total network around 250,000 km, a quarter of which is paved. Roads linking major centres are in good condition. Off major routes, gravel and dirt roads are well-maintained. Automóvil Club Argentino (ACA) is useful for information and advice on driving conditions.
Buses: Long-distance bus services are operated by a number of companies, mostly centred on Buenos Aires (e.g. routes to Mar del Plata, Córdoba, San Martín de Los Andes, Mendoza).
Rail: A comprehensive rail system links main towns. Long-distance Pullman services, with air-conditioning, sleeping facilities and restaurants, are recommended. It is advisable to book well in advance. Information/booking facilities are good in Buenos Aires. Generally services have improved in the last few years. Unlimited travel tickets are available.
Water: Regular sailings to Rosario and Corrientes via Paraná River. Flota Fluvial operates services on the Plate, Paraná, Paraguay and Uruguay Rivers. Patagonian ports also served but sailings irregular – check locally.

Public holidays: New Year's Day, Maundy Thursday, Good Friday, 1 May (Labour Day), 25 May (Revolution Day), 10 Jun, 20 Jun, 9 Jul (Independence Day), 17 Aug, 12 Oct, 8 Dec, 25 Dec (Christmas).

Working hours: Post offices: (Mon-Fri) 0800 – 2000; (Sat) 0800 –1400. (Open until 2400 for telegrams).
Government and business: (Mon-Fri) 0730 – 1300 (1230 – 1930 in winter).
Banking: (Mon-Fri) 1000 – 1600 (with regional variations).
Shops: (Mon-Sat) 0900 - 1900 (some close at 1200 on Sat).

Telecommunications
Telephone: Dialling code for Argentina: IDD access code + 54, followed by area code (Buenos Aires 1, La Plata 21, Resistencia 722, Córdoba 51, Mar del Plata 23, Rosario 41, Santa Fé 42) followed by subscriber's number. Automatic telephone system throughout country, but local systems tend to be overloaded.
Telex and telegram: Facilities available at main telephone offices in Buenos Aires (24 hrs/day) and other main towns. Telex generally available at larger hotels and telegram facilities at ENTEL offices in Retiro and Once stations (Buenos Aires).

Banking
Banking centre: Buenos Aires.
Central bank: Banco Central de la República Argentina.
Commercial banks (head offices): Banco Agrario, Comercial e Industrial de San Juan, Banco Arfina, Banco Argentino de Comercio Sarmiento, Banco Argentino Inversión, Banco Argentino del Atlántico, Banco Comercial de Dolores, Banco Comercial de la Plata, Banco Comercial del Tandil, Banco Comercial de Norte, Banco Comercial de Tres Arroyos, Banco Continental, Banco Cooperativo de Caseras, Banco Co-operativo del Este Argentino, Banco de Avellaneda, Banco del Buen Ayre, Banco de Catamarca, Banco de Chivilcoy, Banco de Coronel Dorreyo, Banco de Coronel Pringles, Banco de Crédito Argentino, Banco de Crédito Comercial, Banco de Crédito Rural Argentino, Banco de Entre Ríos, Banco de Galicia y Buenos Aires, Banco de Hurlingham, Banco de Italia y Río de la Plata, Banco de la Ciudad de Buenos Aires, Banco de la Edificadora de Olavarria, Banco de la Junza, Banco de la Nación Argentina, Banco de la Provincia Buenos Aires, Banco de la Provincia de Córdoba, Banco de la Provincia de Corrientes, Banco de la Provincia de Jujuy, Banco de la Provincia de la Riója, Banco de la Provincia del Chubut, Banco de la Provincia de Salta, Banco de la Provincia de San Luis, Banco de la Provincia de Santa Cruz, Banco de la Provincia de Tucumán, Banco del Interior y Buenos Aires, Banco del Iquaza, Banco del Norte y Delta Argentino, Banco del Sud, Banco Delta, Banco de Mendoza, Banco de Oeste, Banco de Olaviarra, Banco de Pavana, Banco de Préstamos de la Provincia, Banco de Quilmes, Banco de Río Negro y Neuquén, Banco de San Juan, Banco Edificador, Banco Empresario de Tucumán Coop, Banco Español del Río de la Plata, Banco Ganadero Argentino, Banco Hipotecario Nacional, Banco Hispano Italo Libanés, Banco Internacional, Banco Israelítea Río del de la Plata, Banco Mercantil Argentino, Banco Municipal de Paraná, Banco Municipal de Rosario, Banco Municipal de Tucumán, Banco del Chaco, Banco Nacional de Desarrollo, Banco Popular Argentino, Banco Provincal de Salta, Banco Provincial de Santa Fé, Banco Provincial de Santiago del Estero, Banco Regional, Banco Regional de Cuyo, Banco Regional del Norte Argentino, Banco Regional del Salado, Banco Regional del Sureño, Banco Río de la Plata, Banco Roberts, Banco Rural (Sunchales) Cooperativo, Banco Santafesinó Inversion y Desarrollo, Banco Shaw, Banco Sirio Libanés del Río de la Plata, Banco Social de Córdoba, Banco Supervielle Société Général, Banco Tornquist, Banco Unido del Litoral, Banco Velox, Caja Nacional de Ahorro y Seguro, Nuevo Banco de Azúl, Nuevo Banco Italiano, Nuevo Banco Santiago del Estero, Nuevo Banco Santurce.

Electricity supply: 220 V AC, 50 cycles.

Representation in capital: Albania, Algeria, Australia, Austria, Belgium, Bolivia, Brazil, Bulgaria, Canada, Chile, China PR, Colombia, Costa Rica, Cuba, Czechoslovakia, Denmark, Dominican Republic, Egypt, El Salvador, Finland, France, Gabon, German DR, German FR, Greece, Guatemala, Haiti, Honduras, Hungary, India, Indonesia, Iran, Iraq, Ireland, Israel, Italy, Japan, Republic of Korea, Lebanon, Libya, Mexico, Morocco, Netherlands, Nicaragua, Nigeria, Norway, Pakistan, Panama, Paraguay, Peru, Philippines, Poland, Portugal, Romania, Saudi Arabia, Spain, Sweden, Switzerland, Syria, Thailand, Turkey, UK, USA, USSR, Uruguay, Vatican, Venezuela, Yugoslavia, Zaïre.

Hotels: Buenos Aires
Baven, Avenida Callao 346 (tel: 8041600).

Bisonte, Paraguay Esq Libertad 1057 (tel: 3948041; tx: 23750; fax: 3939086).

Buenos Aires Sheraton, San Martín 1225 (tel: 3116331; tx: 9222; fax: 3116311 EXT 2661).

City, Bolívar 160 (tel: 346481; tx: 21484).

Claridge, Tucumán 535, 1049 (tel: 3227700; tx: 24261).

De Las Américas, Libertad 1020 (tel: 3933432; tx: 23883).

Elevage, Maipú 960, 1006 (tel: 3132082; tx: 17374).

Gran Buenos Aires, Marcelo T. de Alvear 797, 1058 (tel: 3123001/5; tx: 22594).

Lancaster, Avenida Córdoba 405 (tel: 3113021; tx: 22092).

Libertador, Avenida Córdoba y Maipú 1054 (tel: 3922095; tx: 24264).

Panamericano, Carlos Pellegrini 525, 1009 (tel: 3936062; tx: 23000; fax: 3936017).

Plaza, Florida 1005 (tel: 3126001; tx: 22488; fax: 3132912).

Presidente, Cerrito 846, 1010 (tel: 497671; tx: 22269).

Salles, Cerrito 208, 1010 (tel: 350091; tx: 17894).

Savoy, Callao 181, 1022 (tel: 400151/8; tx: 22327).

Hotels: Córdoba
Crillon, Rivadavia 85, 5000 (tel: 46093; tx: 51792).

Gran Dora, Entre Rios 70, 5000 (tel: 42031; tx: 51744).

Hotels: Rosario
Majestic, San Lorenzo 980 (tel: 60027).

Car hire: Buenos Aires
Avis: Suipacha 268, 7th Floor C (tel: 459286; tx: 117928); airport (tel: 6200362).

Budget: Carlos Pellegrini 977 (tel: 3138169; tx: 25043).

Europcar National: Esmeralda 1084 (tel: 3113583; tx: 23954).

Hertz: Esmeralda 985 (tel: 3111885; tx: 25415).

Business advice and information
Banco Central de la República Argentina, Reconquista 266, 1002 Buenos Aires (tel: 40-0161; tx: 012-1137, 012-2210, 012-2802).

Cámara Argentina de Comercio, Aveneida L N Alem 36, Industria y Producción de la República Argentina, Florida 1, 4/28 piso, 1005 Buenos Aires (tel: 331-8051; tx: 18693).

Cámara de Comercio Exterior de la Federación del Comercio e Industria, Avenida Córodoba 1868, Rosario, Santa Fe (concerned with import and export).

Cámara de Comercio Exterior de Bahía Blanca, Alsina 19/21, 8000 Bahía Blanca.

Cámara de Comercio Exterior de la Federación del Comercio e Industria, Avenida Córdoba 1868, Rosario (tel: 23896).

Cámara de Comercio Exterior de Santa Fé, San Martín 2231, 3000 Santa Fé.

Dirección Turismo Municipalidad Buenos Aires, Sarmiento 1551 (tel: 46-1251).

Instituto Nacional de Planificación Económica, Hipolito Yrigoyen 250, 8/28, Buenos Aires.

Ministry of the Economy, Hipólito Yrigoyen 250, 1310 Buenos Aires (tel: 34-6411; tx: 21952).

Principal newspapers
Clárin, Piedras 1743, 1140 Buenos Aires (tel: 270061).

Crónica, Garay 130, 1063 Buenos Aires (tel (3611001).

El Cronista Comercial, Alsina 547, 1087 Buenos Aires (tel: 33-3015).

La Nación, Bouchard 557, 1106 Buenos Aires (tel: 313-1003; tx: 18558).

La Prensa, Avda de Mayo 567, 1319 Buenos Aires (tel: 331-1001).

La Razón, Gral Hornos 690, 1272 Buenos Aires (tel: 26-9051).

The Buenos Aires Herald, Azopardo 455, 1107 Buenos Aires (tel: 34-8477).

Australia

Michael Gawenda

During 1989 a new and ominous phrase entered the lexicon of economic jargon in Australia, and in a very real sense, it summed up – even if it simplified – the key issue that was almost certain to dominate economic debate in the country well into the 1990s. From the average householder who was now paying a record mortgage rate of 17 per cent on his bank home loan, to the key civil servants and other economic decision makers, the phrase 'the foreign debt trap' came to signify everything that was fundamentally wrong with the Australian economy. Put simply, the Australian economy had taken on a mountain of foreign debt in the previous five years and that debt was growing at an ever increasing rate. Australia was living beyond its means, and some economists were starting to suggest that the country was heading down the 'Argentine road' of crushing debt, high inflation, and economic stagnation.

With a general election due some time before May 1990, Bob Hawke's Labor Government was locked in to an economic strategy that had given the country the highest interest rates in the OECD group of countries and as a result, the strong probability of a recession some time in 1990 if not earlier. Treasurer Paul Keating, who in the August 1988 budget, had promised Australians that after three years of falling living standards, he was about to 'bring home the bacon', now sombrely warned Australians that they would have to accept several years of economic pain and restraint in order to bring the foreign debt problem under control. The alternative, he said bluntly was a 'severe economic downturn and a lot more pain'. That downturn might now come whatever restraint was imposed by the government.

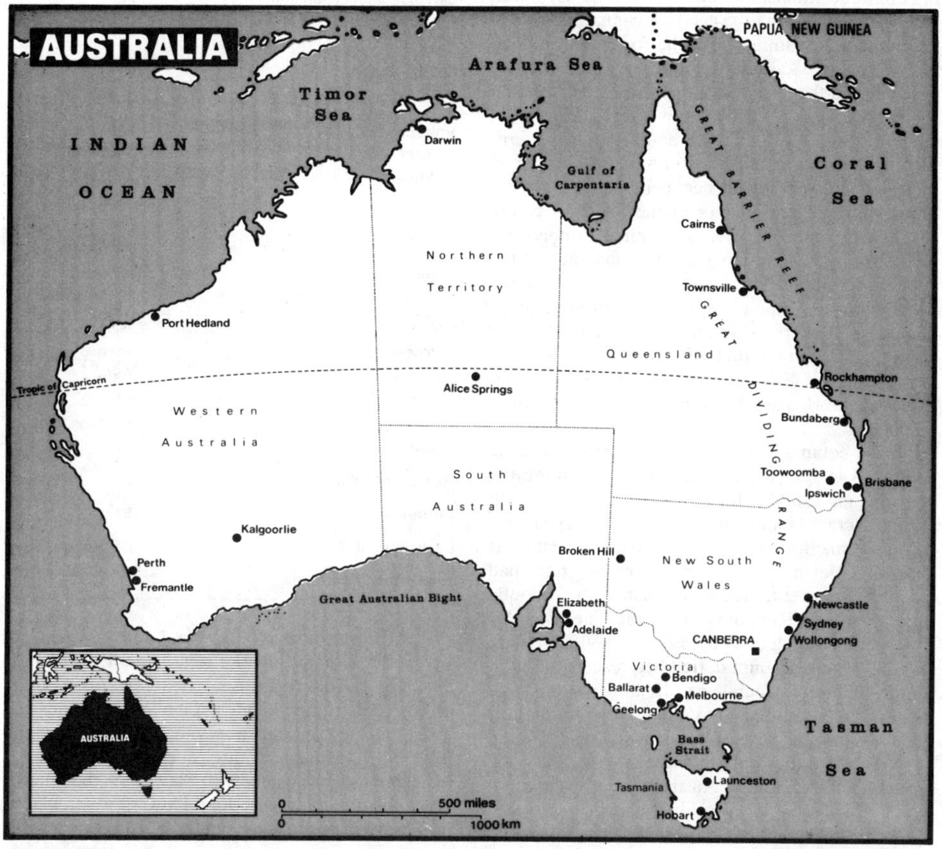

Over-optimistic forecasts

What had gone wrong since that budget in 1988 when Keating talked about the economy in glowing terms and painted a bright future just around the corner? A quick look at those 1988 budget forecasts vividly revealed what had gone wrong. At the heart of the forecasts was Keating's confidence that the current account deficit, which was just over $A14bn in the 1987-88 financial year, would fall to just over $A9bn in 1988-89. Inflation, according the forecasts, would fall from 8 per cent to 4.5 per cent, wage increases under Australia's centralised wage fixing system would be contained to 5 per cent, and economic growth would moderate slightly with GDP growth down from around 3 per cent to 2.5 per cent.

'So the broad picture is now in good shape', Keating said. 'The macro economic settings are right, the current account problem is under control, and we are set for sustainable growth. What now needs to be tackled is micro-economic reform and this government is committed to getting these reforms right too.'

There were some grounds for Keating's optimism. Labor had turned around a $A4bn budget deficit in 1985 to a surplus of $A1.5bn by 1988 and had managed to cut real government spending by 1.5 per cent. In a period of economic growth when unemployment had fallen to the lowest levels since 1980, wage increases had been limited to rises lower than the inflation rate by the Prices and Incomes Accord between the government and the trade union movement. And for a while in the first half of 1988, a combination of high commodity prices and what proved to be a temporary slow-down in imports seemed to suggest that the current account deficit, and hence the country's level of foreign indebtedness, was at last heading downwards.

Within two months of the 1988 budget, it was clear that the budget forecasts were pipe-dreams. GDP growth, rather than moderating, surged, and as a result imports flowed in at unsustainable levels. The loosening of monetary policy by the government in 1987 in the wake of the stock market crash had lowered interest rates and as a result, had fed a housing boom that saw house prices double in the major cities, Sydney and Melbourne, during 1988. Easy money and strong consumer demand led to private investment growth rates of around 6.5 per cent, a level not seen in Australia since the early 1970s. But the Australian economy – despite all the talk since the mid-1980s of industry restructuring so that the reliance on commodity exports would be eased, and the need for import replacement manufacturing development – could not cope with rapid growth without an explosion of imports and a rapidly accelerating current account problem. By the beginning of 1989, the economy was growing at an annual rate of almost 4 per cent and the first six months of the year had seen the current account deficit pass the figure predicted for the whole of the financial year in the 1988 budget forecasts. Even the unemployment figures, which showed that unemployment, at 6 per cent, was lower than at any time since the mid-1970s, was cause for alarm rather than celebration. The economy was clearly over-heated. The out-going governor of the Reserve Bank, Mr Bob Johnston, bluntly summed up the problem: 'In the event, the crucial miscalculation

KEY FACTS

Official title: Commonwealth of Australia

Head of state: Queen Elizabeth II

Head of government: Prime Minister Bob Hawke

Ruling party: Australian Labor Party

Capital: Canberra, ACT

Official Languages: English

Currency: Australian Dollar (A$ = 100 cents)

Exchange rate: A$1.26 per $ (Nov 1989)

Area: 7,682,300 sq km

Population: 16.5m (1988)

GDP per capita: $15,097 (1988)

GDP real growth: 3.7% (1988)

Labour force: 7.6m (1986)

Unemployment: 6% (1989)

Inflation: 7.6% (1988)

Trade balance: –$1,136m (1988)

Foreign debt: $84.3bn (1987)

we all made, including myself, was to underestimate the power of the economy, the pressure of growth, the exhuberant spirits at large and we have gone on for a very long time underestimating how strong the economy is. We have got to have a strategy of winding back spending and in the process of doing that, lowering domestic strains, the pressure on imports, and as well as that, lowering the need for capital inflow.'

Keating, too, admitted that the forecasts had been too optimistic. He said the economy was 'too effervescent' and that as a result, inflation would remain around the 7 per cent mark and foreign debt would continue to grow. But he was emphatic: the government's macro-economic settings were sound and there was no need for further spending cuts. What's more, with a general election on the horizon, Keating said the government was determined to honour a long-term promise to introduce personal income tax cuts of $A5.5bn in July 1989, part of a deal with the trade union movement designed to moderate wage demands. In order to slow the economy, fiscal policy was tightened. In the first three months of 1989, the prime lending rate rose from 15 per cent to 18 per cent, and home mortgage rates went from 13.5 per cent to 15 per cent.

But the boom showed no sign of easing. Keating continued to rely on monetary policy to bring the economy to a 'soft landing' – a slow-down in growth without a crash landing into a recession. Interest rates continued to climb. By July 1989, the prime rate was almost 21 per cent, and home mortgage rates had reached 17 per cent, a record in Australia's economic history. Still the economic news remained bad. By June, net foreign debt had passed the $A100bn mark and the current account deficit had reached $A17bn for the 1988-89 financial year. Keating admitted that interest rates would have to remain high for the foreseeable future and the current account problem would be with Australia well into the 1990s. Enter the foreign debt trap. Out of Australia's current account deficit of some $A17bn for the financial year 1988-89, $A10bn would be interest payments to overseas lenders. The foreign debt trap meant that the Australian economy was stuck in a vicious cycle in which interest payments on total debt ensured high current account deficits. Ominously, despite record interest rates, the Australian dollar moved downwards in 1989, falling some 13 per cent against the US dollar in the first eight months of the year. The depreciation helped exporters, but because foreign debt was written mainly in US dollars, the interest burden on foreign debt grew larger and the currency depreciated.

The government, in a precarious position politically, implemented the promised $A5.5bn income tax cuts which added $A16 a week to average weekly earnings. Most economists agreed that given Australia's economic problems, tax cuts of this magnitude, which would simply add to demand pressures in an overheated economy, were absurd. In fact many, including Melbourne University economics lecturer Dr Neville Norman, argued that what was needed was not a tax cut, but a tax increase in the form of a general consumption – or VAT – tax to dampen demand and encourage savings. Dr Norman was convinced that Australia's low savings ratio was the key to its debt problem.

'We are a developing nation and a developing nation will have debt', he said. 'Australia's foreign debt, unlike the debt of some Latin American countries, is private company debt, borrowings for productive investment. We need that investment. What we need though is higher internal savings to fund that investment. We can't continue to borrow from overseas at the current rate.'

KEY INDICATORS

	Unit	1984	1985	1986	1987	1988
Population	m	15.58	15.79	16.02	16.25	16.53
Gross domestic product (GDP)	A$bn	201.87	225.91	247.17	278.06	314.96
GDP per capita (Constant)	A$	13,816	14,307	14,402	17,111	15,097
GDP real growth	%	7.5	4.9	2.1	4.3	3.7
Inflation	%	4.4	6.4	9.0	8.3	7.6
Wages	1980=100	95.2	100.0	107.8	113.6	121.2
Volume of exports	1985=100	91	100	105	114	114
Wool	1985=100	77	100	99	109	104
Wheat	1985=100	108	100	120	113	58
Iron ore	1985=100	97	100	94	95	112
Beef	1985=100	93	100	116	140	141
Coal	1985=100	86	100	102	114	112
Volume of imports	1985=100	93	100	97	100	119
Exports	A$m	26,417	32,601	33,792	37,797	42,173
Wool	A$m	1,915	2,894	3,062	4,314	5,869
Wheat	A$m	2,454	2,570	2,877	2,006	1,409
Iron ore	A$m	1,611	2,147	1,850	1,674	1,743
Beef	A$m	1,023	1,287	1,585	1,983	1,994
Coal	A$m	3,937	5,061	5,318	5,043	4,612
Imports	A$m	23,560	37,054	39,093	41,816	45,925
Balance of trade	$m	–884	–1,284	–2,071	–480	–1,136
Current account	$	–8,549	–8,717	–9,726	–8,725	–11,218
Exchange rate	A$ per $	1.43	1.59	1.50	1.38	1.28

* estimated figure

Political problems for Labor

There were signs in the midst of this economic crisis that the seven-year-old Hawke Labor government was perhaps running out of steam, though Keating – and even Hawke to a lesser extent – were, in public at least, determined to put a brave face on things and insist that the government was on the right track and determined to make the tough decisions needed to get the economy into balance. In May 1989, the opposition Liberal Party leader John Howard was replaced by long-term rival and former leader Andrew Peacock at a stormy party meeting, basically because there was a widespread perception that the dour and colourless Howard lacked Peacock's public appeal. Immediately, the opinion polls showed that the opposition was ahead of the government and would win any general election quite easily. They showed that the electorate, burdened by record interest rates, no longer was convinced that the Labor government had the answers to Australia's economic problems. Industry Minister Senator John Button admitted that the government had only a 50-50 chance of surviving at the forthcoming election and added this key reason: the restructuring of Australian industry, on which all of Australia's future economic well-being hinged, had not happened as quickly or as thoroughly as the government had hoped. He pointed to the fact that despite three years of strong economic growth, employment in manufacturing had remained static and commodities still accounted for more than 70 per cent of Australia's export earnings.

Worried backbenchers, aware that their constituents deeply resented those 17 per cent mortgage rates, urged Keating and Hawke to set up a mortgage relief scheme, and to introduce a luxury tax on some consumer items in order to rein in imports. At the same time, the Australian Council of Trade Unions was pushing for a 6 per cent general wage increase in 1989 despite the July tax cuts in order to maintain living standards – a promise Hawke and Keating had given the ACTU leadership in 1988 when the economic picture was decidedly rosier. Keating bluntly refused seriously to consider any sort of mortgage relief scheme, nor was he, in the current world trade environment – when Australia was arguing for the removal of export subsidies and quotas on primary products – likely to contemplate a tax on luxury imports which would amount to a form of import duty. While the opposition talked in general terms of the government's economic failings, it too had failed to release any detailed economic programme for reform, hoping that the old adage that governments lose elections rather than oppositions win them would be proved true in Australia at the next election.

In 1985, Keating was a firm advocate for a consumption tax, but was over-ruled by Hawke and the rest of the Labor Party. Privately, he still believed such a tax to be both desirable and inevitable in Australia, but with the opposition opposed to such a tax, Keating bowed to political pressures and ruled out any such move. The government's options therefore, were limited, and with no major fiscal policy changes mooted – the government had also ruled out any substantial further cuts in government spending – it seemed that monetary policy would remain the weapon with which the government would tackle Australia's growing debt problem in the short term. The government's survival would therefore depend on whether the record high interest rates merely slowed the economy and led to a shallow recession, or whether they led to a dramatic collapse and a severe down-turn.

International economic factors

One factor outside the control of the government and which could well decide whether Australia experienced just a slow-down in economic activity – leading to slightly higher unemployment, a moderate squeeze on profits, and a manageable fall in living standards – or a major down-turn, was the future shape of the international economy, especially the US economy in 1990 and beyond. According to the OECD forecasts, United States GDP was expected to grow by around 2 per cent in 1990, down from about 3 per cent in 1989 and 4 per cent in 1988. The total OECD economic group was expected to have economic growth of around 2.5 per cent in 1990, down from about 3 per cent in 1989. All these forecasts represented a reasonably moderate decline in economic activity and if achieved, commodity prices, on which Australia's export health depended, and on the maintenance of which hinged Australia's ability to avoid a recession,

COUNTRY PROFILE

Political system: Australia, a parliamentary democracy, is a federation of six states in which legislative powers are divided between the bicameral Australia Federal Parliament, with a 148-member House of Representatives and a 76-member Senate, and six state parliaments. Under the constitution, the governor-general, who represents the British sovereign, is formally the head of state and chief executive. A cabinet of senior ministers, headed by a prime minister, holds day-to-day executive responsibility and is formed by the party with a majority in the House of Representatives, which is elected to a three-year term by compulsory universal adult suffrage. There are some 900 local government bodies at the city, town, municipal and shire levels.

Political parties: Ruling party: Australian Labor Party (85 seats in the House of Representatives, following July 1987 elections); Liberal Party (45 seats), National Party (18 seats).

Population: Australia's population of 16,249,000 (Jun 1987) is growing at a rate of 1.2 per cent per annum. 63 per cent live in the eight major conurbations on the coast; indeed, much of the interior is practically uninhabitable because of the lack of rain, the absence of cultivable soil and the considerable distances involved in travel. About 20 per cent of the population were born outside the country, and about 1.5 per cent are aboriginals – more than half of them living in the cities.

Main cities/towns: Sydney, NSW (state capital, population 3,431,000 in 1986); Melbourne, Vic (state capital, pop 2,942,000); Brisbane, Qld (state capital, pop 1,171,000); Adelaide, SA (state capital, pop 993,000); Perth, WA (state capital, pop 1,025,000); Newcastle, NSW (pop 420,000*); Canberra, ACT (national capital, pop 265,000*); Wollongong, NSW (pop 240,000*); Hobart, Tas (state capital, pop 175,000*); Gold Coast, Qld (pop 175,000*); Geelong, Vic (pop 142,000*); Darwin, NT (territorial capital, pop 65,000*).

Language: English. Aboriginal dialects are becoming scarce.

Telecommunications: Overseas telephone, telegram, facsimile, phototelegram, telex, leased circuit, audio broadcast and data transmission services are provided by submarine cable, satellite links and short-wave radio.

Media
Press: More than 500 newspapers are published in Australia. Dailies: Main metropolitan dailies include *The Sun-News Pictorial* (Melbourne), *The Age* (Melbourne), *Daily Mirror* (Sydney), *The Telegraph* (Sydney) and *The Sydney Morning Herald* (Sydney). The national daily, *The Australian*, is printed simultaneously in Sydney, Melbourne, Perth and Brisbane. Weeklies: Main weeklies include, *Sunday Telegraph* (NSW), *The Sunday Mail* (Qld) and *Sunday Sun* (Qld). Business: South Pacific editions of international publications, such as *Newsweek* and *Time*, and various international newspapers are widely available.
Broadcasting: The Australian Broadcasting Corporation (ABC) operates a nationwide non-commercial service. *Radio:* 127 stations broadcast ABC programmes, and there are 135 commercial radio stations. *Television:* Of more than 137 TV services operating, some 50 are private commercial stations and the remainder transmit programmes of the Australian Broadcasting Corporation (ABC).

Domestic economy: Australia has traditionally relied on primary production, and is an important producer and exporter of farm products. Considerable investment in export-orientated mining and energy projects has led to a growth in the volume of coal and iron ore exports, but has also contributed to a rapid rise in gross external debt. Australia is the world's largest exporter of alumina, wool, beef and veal, mineral sands, coal, live sheep and refined lead, and among the top suppliers of wheat, sugar, iron ore, bauxite, nickel and other commodities. The services sector accounts for about 73 per cent of employment and about half of GDP. Protective barriers around industry (now being dismantled) and periodic mining have helped maintain a high-wage economy. A traditional current account deficit on the balance of payments is financed largely by a net inflow of private capital. But a 20 per cent fall in the terms of trade over the 1984-86 period resulted in an unsustainable deterioration in Australia's external position and a substantial rise in overseas debt. A combination of wage restraint (real wages have declined 6.5 per cent since mid-1985), unaccommodating monetary policy, firm fiscal policy (a projected budget surplus of A$580m in 1987-88) and a significant depreciation of the currency have contributed to a turnaround in the balance of payments. The current account deficit was expected to become a surplus by mid-1989.

Employment: Since 1967 employment in manufacturing and agriculture has declined, while employment in community services, personal services, finance and other service industries has increased. About 55 per cent of those in employment belong to one of 319 trade unions.

External trade: Export marketing organisations have been established under government statutes to supervise and promote export of main income-earning commodities, and the Export Market Development Grants scheme provides taxable cash grants for development of overseas markets. The tariff system is used to protect the car, textile and footwear manufacturing industries, with some use of import quotas.
Exports: In 1986, exports increased by 4.1 per cent to A$33,706m. Main exports are coal (15.8% of 1986 total, and valued at A$5,400m in 1987), wool (7.1% of 1986 total, A$3,900m in 1987), wheat (8.5%), iron ore (5.7%), petroleum oils and products (4.5%), beef and veal (4.5%), alumina (4.2%), aluminium (2.9%) and gold (2.8%). Main destinations are Japan (25.4% of 1987 total), Far East (23.6%), Europe (17.8%), North America (13%) and New Zealand (5.6%). Merchandise exports in 1988 were US$32,770m.
Imports: In 1986, imports rose by 7.9 per cent to A$35,734m. Main requirements are road vehicles (8.7%), industrial machinery (10.6%), office equipment (6.4%), electrical goods (5.3%), textiles (4.9%) and petroleum oils and products (4.7%). Main sources are USA, Japan, UK, German FR, New Zealand, Taiwan, Singapore, Saudi Arabia and Indonesia. Merchandise imports in 1988 were US$33,876m.

Agriculture: Agriculture accounts for about four per cent of GDP, employs six per cent of the workforce (1987) and contributes about 38 per cent of export earnings. Pastures cover about 40 per cent of the land area and six per cent is devoted to other agricultural activities. Australia is an important producer of wheat, but because of the EC/USA subsidy war and the slump in world grain prices, Australia's wheat production fell from 22m tonnes in 1983-84 to 12m tonnes in 1987-88. Sugar is also an important rural industry which has been greatly affected by low prices and the international markets. Other industrial crops include barley, oats, grain sorghum, maize, rice, tobacco and cotton. Fruit and vegetables such as oranges, apples, pears, bananas, pineapples, grapes, carrots, potatoes, onions and cauliflowers are grown for both the domestic and export markets. Wine, too, is selling well abroad with exports at 21.46 litres in 1986-87.
Livestock-raising, especially of cattle (23.3m head in 1987) and sheep (160m head in 1987), is an important part of the agricultural sector. Many farmers, though, are in severe difficulty because of poor international markets and high borrowing costs. Australia leads the world in wool exports and is a major beef exporter mainly to the USA, Japan, South East Asia and Great Britain. Forestry and fishing play a relatively small role in the economy.

Industry: Manufacturing accounts for 29 per cent of GDP, although it employs only 20 per cent of the workforce. The sector has suffered from poor investment in recent years, despite boosts to exports (because of the low exchange rate). Motor manufacture is a major activity, and other activities include aircraft maintenance and construction and avionics; petrochemicals, fertilisers, plastics, pharmaceuticals, cosmetics and toiletries; electrical and electronic goods; light and heavy engineering; shipbuilding; food processing; iron and steel production; aluminium, copper and lead smelting; zinc, nickel, gold and tin refining; motor vehicle manufacture; oil refining; pulp and paper manufacture; and textile, clothing and footwear manufacture. Some 90 per cent of all business enterprises are small businesses (manufacturing enterprises with less than 100 employees or service enterprises with fewer than 20). Tourism is a major and growing industry, creating up to 400,000 jobs and earning A$3,700m in 1987 with 1.8m tourists visiting.

Mining: Mining accounts for about 11 per cent of GDP and it employs only 130,000. Proven bauxite deposits at Weipa (Queensland) total 6.18bn tonnes (1986), a quarter of world resources. Iron ore mining has recovered somewhat from the slump caused by slack Japanese demand in the early 1980s, and investment is returning. Australia is the world's second largest nickel producer, with deposits of 8.96m tonnes (1986), and copper production was 261,000 tonnes in 1984-5. Lead and zinc deposits total 16.7m tonnes and 24.3m tonnes respectively, and new investment is occurring as production increases. Tin production is stagnant due to slack demand, but gold mines have reopened after the slump of the 1970s, making the country the fifth largest producer, and output 100 tonnes, from 20 tonnes in 1980. Diamonds are attracting investment – Australia is the fourth largest supplier of industrial diamonds. Australia leads the world in production of mineral sands (bauxite, rutile, zircon, ilmenite and monazite), and manganese, tungsten and copper are also important. Reserves of low-cost uranium account for about 20 per cent of proven world reserves, although mining is subject to economic and political restraints. Recoverable coal reserves (36bn tonnes*) are mostly accessible and close to ports. Australia's massive coal exports (now the largest quantity exported anywhere in the world) are destined mainly for the Japanese steel industry.

Hydrocarbons: Crude oil production of about 500,000 b/d, with 90 per cent from the Bass Strait field, provides about 80 per cent of domestic requirements. Natural gas is supplied from various fields to ACT, five state capitals and several towns in NSW and Vic. Exports of LNG to Japan are scheduled for 1989 as the second phase of the North West Shelf gas project, which is to supply 40 per cent of WA energy requirements by 1993.

Energy: Although a net exporter of energy, Australia imports about 15 per cent of its crude oil requirements, mainly from Saudi Arabia, Indonesia and UAE. Each state has its own pattern of electric power development, and there is no national grid as such.

Financial markets: There are six stock exchanges, with the principal two in Sydney and Melbourne. Non-residents may trade freely on the Sydney Futures Exchange.

were likely to remain fairly buoyant in 1990. But there were signs that the OECD forecasts might be over-optimistic. In mid-1989 the US inflation rate was up around the 5 per cent mark and interest rates were rising: inflation, while still relatively low, was on the rise in Japan and West Germany, with the inevitable pressure on interest rates. Were growth to dip dramatically in the USA during 1990, commodity prices would tumble and a recession in Australia would be inevitable.

Corporations in trouble

The government's tight monetary policy had by mid-1989 already caused major problems for two of the country's largest corporations. The Melbourne-based Hooker Corporation, the country's largest builder and land developer, with major interests in the USA came close to going into receivership in June 1989, as high interest rates brought to an end a prolonged property boom and saw Hooker struggling to service its $A1.7bn debt. The company's executive chairman and the man who owned just over 50 per cent of the company, George Herscu, stepped down from his post as the company's big creditors gave it time to sell off $A1bn worth of assets, including its US shopping complexes, in an effort to reduce its debt to manageable proportions. And the Bond Corporation which owned Australia's largest television network, as well as brewing interests in the USA and Britain and gold mines in Chile, saw its share price fall by 75 per cent as investment analysts wondered how Bond would manage to service its debt of almost $2bn in the immediate future. Few doubted that other highly geared Australian companies would also experience difficulties – with some going into liquidation – if interest rates, as expected, remained high well into 1990.

Outlook

Most economic forecasters were agreed that the Australian economy in 1990 was heading for a reasonably sharp slow-down and would remain shackled by trade weaknesses that would boost the nation's soaring foreign debt. They agreed that the slow-down would be accompanied by higher unemployment, higher inflation, and much lower private investment. There was also general agreement that the current account deficit would continue to swell Australia's foreign debt and that it would remain around its mid-1989 level of about 5 per cent of GDP, double what the government estimated was required to bring debt under control. At the same time, economic growth was tipped to fall to around 2 per cent in 1990 and even lower by the end of the year, and inflation was expected to increase slightly to around 7.3 per cent. Most economists agreed that high interest rates, coupled with tight monetary policy, was a crude instrument with which to tackle Australia's economic problems, an instrument that would dampen private investment on which sustained growth was dependent, as well as dampening general demand which the policy was designed to achieve.

What was most worrying was that during a three year period of relatively high commodity prices, Australia's foreign debt problem had not eased, and the micro-economic reforms that Keating had said were so crucially important if the vital structural changes in the Australian economy were to be implemented, did not materialise. While the Prices and Incomes Accord had delivered moderate wage increases in preceding years, it was doubtful whether a deregulated financial system, introduced by Keating, could co-exist with a centralised and regulated wage fixing system in which productivity was not the key measure for any wage increases. Even some leading Labor government ministers, including Finance Minister Senator Peter Walsh, and Treasurer Keating himself, privately agreed that labour market deregulation was necessary in Australia. They also privately, and occasionally publicly, argued for a major privatisation programme that would involve selling off Qantas, the Commonwealth-owned international airline, and Australian Airlines, the domestic carrier, together with the Commonwealth Bank, but the Labor Party was extremely unlikely to agree to such a major departure from party policy. Nor was the trade union movement likely to agree to any deregulation of the labour markets or any major change to work practices in key economic areas.

In June 1989, for instance, an inquiry into Australia's docks and shipping industry recommended major changes which included the halving of the workforce which, the report said, was inefficient, engaged in out-moded work practices, and helped make Australia's coastal shipping industry among the most expensive in the world. After union resistance to the report's recommendations, the government put the whole issue in the 'too-hard' basket indefinitely. But it was just such reforms which would determine Australia's long-term economic health.

The Hawke government could argue that during its seven years in office, it had implemented some major reforms of the Australian economy. It had deregulated the financial system, floated the dollar, and turned around a budget deficit that amounted to more than 4 per cent of GDP in 1985 to a budget surplus of almost $A5bn in 1989. Keating quite rightly pointed out that few Western governments, including the US government, had managed such a dramatic turn-around. He could also point to the fact that Australia's unemployment rate of around 6 per cent compared favourably with the rate in most OECD countries, and this at a time when participation rates were the highest in the country's history.

But the fact remained that even with these achievements and reforms, Australia had not solved its fundamental structural problems and so was not well placed to handle a slump in commodity prices when that slump, inevitably, would come. Australians might no longer be living 'off the sheep's back' as they had done through much of the twentieth century, but primary products and commodities still dominated the country's exports, leaving it exposed to the dramatic and swift effects of the commodity cycle. Australia's economy would remain fragile and vulnerable, until, in the long term, this changed.

COUNTRY PROFILE

Legal system: Judicial power is vested in the High Court. Parliament has in addition created the Federal Court and the Family Court to deal with specialised areas of federal law. All states and the Northern Territory have their own system of courts, with supreme courts and magistrates' courts, which are vested with federal jurisdiction in some areas.

Membership of international organisations: ANZUS, AIOEC, Asian Development Bank, CCC, Colombo Plan, Commonwealth, DAC, ELDO, ESCAP, GATT, FAO, IAEA, IATP, IBA, IBRD, ICAC, ICAO, ICO, IDA, IEA, IFAD, IFC, IHO, ILO, IMF, International Lead & Zinc Study Group, INTELSAT, INTERPOL, IOOC, IPU, IRC, ISO, ITC, ITU, IWC, International Whaling Commission, IWC, OECD, South Pacific Agreement on Regional Trade and Economic Co-operation, South Pacific Commission, South Pacific Forum, UN, UNESCO, UPU, WHO, WIPO, WMO, World Bank, WSE.

Time: GMT + 8 hrs in Western Australia; GMT + 9.5 hrs in the Northern Territory and South Australia; GMT + 10 in New South Wales, Queensland, Tasmania and Victoria; add 1 hr for summer time in all but Queensland, Northern Territory and Western Australia from 30 Oct-18 Mar.

Climate: Ranges from tropical to temperate. Nov-Mar (summer) – tropical in the north, hot, humid and rainy in coastal areas, and warm to hot with mild nights in the south. Apr-Sep (autumn/winter) – warm to hot by day and cool at night in the north, and cool and showery with many sunny days in the south. Jan-Feb generally the hottest months, and average annual temperatures vary from 27°C in the far north to 13°C in the far south. Some 70 per cent of the continent is arid. Extremes of heat and cold occur in the interior.

Entry requirements
Passport: Required by all.
Visa: Visas must be obtained before arrival by all except holders of Australian and New Zealand passports. Short-term visitors' visas may be obtained for stays up to six months.
Currency: There is no limit on currency importation, but only A$5,000 of Australian notes and coins which may be taken out.
Customs: Personal effects are allowed duty-free. Strict quarantine regulations make it inadvisable to carry food, fruit, vegetables, seeds, animals or plants without prior approval. Aircraft cabins are sprayed before disembarkation. Importation of certain items is prohibited, including narcotic and dangerous drugs, firearms and birds. Both import and export of protected wildlife or goods derived therefrom (i.e made from skins, feathers, shell, bone, etc.) is strictly prohibited. There are restrictions or quotas on the commercial importation of equipment, machinery, motor vehicles, footwear, clothing and other goods; preferential duty rates may apply to goods from some countries (New Zealand, Canada, Papua New Guinea, South Pacific Forum countries and certain developing countries).
Income tax: Income tax is levied on income derived directly or indirectly from Australian sources, except where specifically exempted.

Health precautions
Mandatory: Vaccination certificates for yellow fever or cholera if travelling from an infected area.

Representation overseas: Abu Dhabi, Accra, Algiers, Amman, Ankara, Apia, Athens, Baghdad, Bandar Seri Begawan, Beijing, Beirut, Belgrade, Berlin, Berne, Bonn, Brasilia, Brussels, Buenos Aires, Cairo, Colombo, Copenhagen, Damascus, Dar es Salaam, Dhaka, Dublin, The Hague, Hanoi, Harare, Honiara, Hong Kong, Islamabad, Jakarta, Jeddah, Kingston, Kuala Lumpur, Kuwait, Lagos, Lima, Lisbon, London, Lusaka, Madrid, Manama, Manila, Mexico City, Moscow, Nairobi, Nauru, New Delhi, Nicosia, Ontario, Ottawa, Paris, Port Moresby, Port Vila, Pretoria, Rangoon, Rome, Santiago, Seoul, Singapore, Sliema (Malta), Stockholm, Suva, Tarawa, Tehran, Tel Aviv, Tokyo, Tripoli, Vienna, Vientiane, Warsaw, Washington DC, Wellington.

Air access
National airline: Qantas.
Other airlines: Aerolineas Argentinas, Air Calédonie, Air New Zealand, Air Niugini, Air Pacific, Air Zimbabwe, Alitalia, All Nippon Airways, Ansett, Australian, British Airways, CAAC, Canadian, Cathay Pacific, Continental (Air Micronesia), Dragonair, East-West, Garuda, Hawaiian, JAT, JAL, KLM, Lauda Air, Lufthansa, MAS, Norfolk, Olympic, Philippine, Polynesian, Royal Brunei, Singapore, Solomon Islands Air, Thai Airways, United, UTA.
Tax: An airport tax of A$20 is levied on departure.

BUSINESS GUIDE

International airports: Brisbane International (BNE), 6.5 km from Brisbane, Qld, with duty-free shop, bar, buffet, restaurant; Cairns (CNS), 4 km from Cairns Qld, with duty-free shop, hotel reservations; Darwin (DRW), 8 km from Darwin, NT; Kingsford Smith (SYD), 11km from Sydney, NSW, with duty-free shop, bar, buffet, restaurant, bank, post office, shops; Llanherne (HBA), 22km from Hobart, Tas; Perth (PER), 10 km from Perth, WA, with duty-free shop, bar, buffet, restaurant, bank, post office, shops; Townsville (TSV), 5 km from Townsville, Qld; Tullamarine (MEL), 21 km from Melbourne, Vic, with duty-free shop, bar,buffet, restaurant, bank, post office, shops; West Beach (ADL), 8 km from Adelaide, SA, with bar, buffet, restaurant, post office, shops.

Surface access
Main ports: Dampier, Port Headland, Sydney, Newcastle, Port Kembla, Melbourne, Fremantle (Perth), Gladstone, Hay Point (coal), Port Walcott, Westernport, Weipa, Brisbane, Geelong and Port Adelaide.

Hotels: Hotels should be booked well in advance. A 10 per cent tip is optional.

Credit cards: Major international credit cards, such as American Express, Carte Blanche, Diners Club plus Mastercard and Visa affiliates can be used for a wide range of goods and services in city and resort areas, but use may be restricted elsewhere. Some taxis accept credit card payments, but they are not normally accepted at grocery stores and petrol/gas filling stations.

Car hire: Hire cars are widely available. Current overseas licences are recognised, but International Drivers Permits are recommended. The required third-party insurance is normally included in car hire charge. Use of seat belts is compulsory and speed limit in towns is generally 60 km/hr. Driving is on the left-hand side of the road. Trams have the right of way.

City transport
Taxis: Metered taxis operate in all main cities and towns from major hotels, shopping areas and signposted taxi ranks. Radio- controlled taxis are listed in local telephone directories. Tipping is not expected, but a tip of A$0.30 or the balance of the fare rounded up to the nearest dollar is sometimes given.

National transport: Travellers are not permitted to carry fruit, vegetables or plants into the State of Victoria.
Air: Air transport is regularly used and well developed. Regular services linking main centres and over 450 airfields are operated by Ansett Airlines, Australian Airlines, East-West Airlines, Air Queensland and over 25 other operators. Charter aircraft are widely available.
Road: The 865,000 km road network (1986) includes primary routes linking main centres and secondary routes serving urban and other populated areas. Seek advice from the appropriate local automobile association before travelling in remote areas, as roads may be affected by weather conditions.
Buses: Air-conditioned express coach services link main centres, including Tasmania via ferries. Buses and, in some capital cities, trams, provide good services on main town routes, but convenient cross-town transport is not always available. Several sightseeing and tour services operated. Melbourne's public transport system includes a City Loop underground railway, and fares apply to all public transport within appropriate zones.
Rail: Railways, mainly government-owned and operated, provide express inter-urban passenger services, electrified suburban services and long-distance freight services, using a 44,000 km network of tracks. Long-distance passenger trains are air-conditioned, with dining and sleeping facilities. Advance booking is recommended.
Water: There are more than 30 ports. Sea transport is extensively used for internal and international freight shipment. Ports are currently constructing containerised cargo facilities.

Public holidays: In addition to official public holidays observed throughout Australia, extra statutory holidays are observed in individual states and the ACT, some of which are listed below.
Fixed dates: 1 Jan, 26 Jan (Australia Day), 25 Apr (Anzac Day), 5 Jun (Foundation Day, WA), 25-27 Dec.
Variable dates: Hobart Cup Day, Royal Hobart Regatta Day, Labour Day (WA and Tas), Labour Day (Vic), Good Friday, Easter Saturday, Easter Monday, Bank Holiday (Tas) and Easter Tuesday (Vic), Labour Day (Qld) and May Day (NT), Adelaide Cup Day (Adelaide area only), Birthday of HM the Queen (except WA), Alice Springs Show Day (NT), Tennant Creek Show Day (NT), Katherine Show Day (NT), Darwin Royal Show Day (NT), Bank Holiday (NSW) and Picnic Day (NT), Melbourne Show Day (Melbourne only), Labour Day (NSW), Labour Day (SA), Hobart Show, Recreation Day (Northern Tas), Melbourne Cup Day (Melbourne area only), Proclamation Day (SA) and general holiday (ACT and NT).

Working hours: Late night shopping (to 2100) is available in Sydney, Perth and Darwin on Thu and in Melbourne, Brisbane, Hobart and Canberra on Fri.
Government and business: (Mon-Fri) 0900 – 1700.
Banking: (Mon-Thu) 0930 – 1600; (Fri) 0930 – 1700.
Shops: (Mon-Fri) 0900 – 1700; (Sat) 0900 – 1200.

Social customs: Tipping is not generally customary, although it is usual to tip up to 10 per cent for good service in better restaurants.

Telecommunications
Telephone: Dialling code for Australia, IDD access code + 61 followed by area code (8 for Adelaide, 7 for Brisbane, 62 for Canberra, 89 for Darwin, 02 for Hobart, 3 for Melbourne, 9 for Perth, 2 for Sydney, 77 for Townsville) followed by subscriber's number. In 1986, Telecom Australia's 5353 exchanges served over 9m telephones, with 99 per cent of the telephone services connected to automatic exchanges. Developments include mobile telephones (Sydney and Melbourne), a national telefinder service for radio paging, digital data and packet switching networks and fibre optic cable links. There is a submarine cable link to New Zealand, Hawaii and Canada, and this system is being developed to double capacity. Satellite links. Short wave radio is less frequently used these days.
Telex: Telex facilities are available at General Post Offices in Brisbane, Canberra, Melbourne, Sydney, Adelaide, Perth, Hobart and Newcastle. The Overseas Telecommunications Commission operates 24-hr public telex services in Sydney, Melbourne, Brisbane and Canberra.
Fax: The Overseas Telecommunications Commission will accept documents for transmission over the counter or it operates a free courier collection in Sydney, Melbourne, Brisbane, Adelaide and Perth.
Telegram: Cables at urgent rate (2-4 hr delivery), ordinary rate (4-6 hr delivery) or letter rate (24 hr delivery) can be sent via the telephone operator or local post offices.

Banking
Central bank: Reserve Bank of Australia. Specialist banks: Australian Resources Development Bank (Melbourne), Commonwealth Development Bank of Australia (Sydney), Primary Industry Bank of Australia (Sydney). *Head offices in Melbourne:* Australia and New Zealand Banking Group, National Australia Bank, National Commercial Banking Corp of Australia. *Head offices in Sydney:* Australian Bank, Commonwealth Banking Corp, Commonwealth Bank of Australia, Commonwealth Savings Bank of Australia, Martin Corporation, Westpac Banking Corp. *State banks:* Bank of Adelaide, Bank of Queensland, Rural and Industries Bank of Western Australia, State Bank of New South Wales, State Bank of South Australia, State Bank of Victoria. *Australian branches:* Bank of New Zealand, Banque Nationale de Paris. (Most of the big international banks have representatives in Australia).
Australian subsidiaries and representatives: Canadian Imperial Bank of Commerce Australia Ltd, Hongkong and Shanghai Banking Corp, First Chicago Australia Ltd, Grindlays Australia Ltd, Marac Australia Ltd, Pittsburgh National Seldon & Co Ltd, Société Générale Australia Ltd. *Merchant banks:* ABN Australia Latd, AMP Morgan Grenfell, Australian European Finance Corporation, Australian International Finance Corporation, Australian United Corporation, BA Australia Ltd, Barclays Australia Ltd, BT Australia, Capel Court Corp, Chase-NBA Group, Citinational, Commercial Continental, Delfin-BNY Acceptances, Development Finance Corp, Elder's Finance and Investment Co, Euro-Pacific Finance Corp Ltd, Hambro Australia, Hill Samuel Australia, International Pacific Corp Ltd, Lloyds International Ltd, Martin Corp, Partnership Pacific Ltd, Rothwells Ltd, Schroder, Darling and Co Ltd, Standard Chartered Australia Ltd, Trans City Holdings, Tricontinental Holdings, Wardley Australia.

Advertising: Advertising practice is subject to the provisions of the Trade Practices Act. There are some restrictions on alcohol, tobacco, medical products and advertising to children. No advertising for cigarettes is permitted on TV. The Advertising Standards Council interprets and applies the self-imposed codes created by advertisers, advertising agencies and media. In addition to commercial press, TV and radio advertising, outdoor advertising is available in many forms, as well as direct mail facilities and house-to-house distribution of samples and literature.

Trade fairs: 15th International Watch Clock and Jewellery Fair, Feb, Sydney; International Gift Trade Fair, Feb, Sydney; International Enamelling Exhibition, Feb, Adelaide; Personal Computer Show, Mar, Sydney; Australian International Electronic Communications and Information Technology Exhibition, Mar, Sydney; Australian International Office Technology Exhibition, Mar, Sydney; International Motor Show, Mar, Melbourne; The Sydney International Home Show, Mar, Sydney; Decorex 90, Australian International Interior Design Exhibition re: Australian Trade Promotions, Apr, Melbourne; Environment 90 International Exhibition, May, Sydney; Matpak International Materials Handling Exhibition, May, Sydney; Swimming Pool and Spa Association, May, Gold Coast; Australias International Engineering Exhibition, May, Sydney; Australian International Electrical and Electronic Industries Exhibition, May, Melbourne; The Australian Personal Computer Show, May, Melbourne; International Technology Exhibition, May, Sydney; Society of Motion Picture and TV Engineers International Congress and Exhibition, Jun, Sydney; Australian International Office Technology Exhibition, Jun, Melbourne; International Catering Trade Fair, Jun, Sydney; AUSTEXT (Australias International Textile Exhibition, Jul, Melbourne; LABEX, International Laboratory Equipment Exhibition, Jul-Aug, Sydney; Australian International Home Show, Aug, Sydney; German Music Publishers Fair, Sep, Sydney; HOTEX, Australian International Hotel Accommodation and Retail Food Equipment Exhibition, Sep, Melbourne; AIFE, Australian International Food and Drink Exhibition, Sep, Melbourne; AUSPLAS Trade Exhibition of the Plastics Industry, Oct, Melbourne; The Australian International Franchising Exhibition, Oct, Melbourne; AUSPLAS, Australias International

BUSINESS GUIDE

Exhibition and Conference for the Plastics Industry, Oct-Nov, Sydney.

Electricity supply: 220-250 V AC, with 3-pin flat plug fittings and bayonet-type light sockets. Leading hotels also supply 110 V outlets for razors and small appliances.

Weights and measures: Metric system (gradual conversion from imperial system, which is still in use).

Representation in capital: Argentina, Austria, Bangladesh, Belgium, Brazil, Burma, Canada, Chile, China PR, Cyprus, Denmark, Egypt, Fiji, Finland, France, German DR, German FR, Ghana, Greece, Hungary, India, Indonesia, Iran, Iraq, Ireland, Israel, Italy, Japan, Jordan, Kenya, Republic of Korea, Laos, Lebanon, Libya, Malaysia, Malta, Mauritius, Mexico, Netherlands, New Zealand, Nigeria, Norway, Pakistan, Papua New Guinea, Peru, Philippines, Poland, Portugal, Saudi Arabia, Singapore, South Africa, Spain, Sri Lanka, Sweden, Switzerland, Thailand, Turkey, Uganda, USSR, UK, USA, Uruguay, Vatican, Venezuela, Vietnam, Yugoslavia, Zambia.

BUSINESS DIRECTORY

Useful telephone numbers
Emergency Services: 000
VicRail Information: 619 1111 (Melbourne)

Hotels: Adelaide
Adelaide, 62 Brougham Place, North Adelaide, 5006 (tel: 267-3444; tx: 82174; fax: 2390189).

Adelaide Parkroyal, 226 South Terrace, 5000 (tel: 223-4355; tx: 82156; fax: 2320769).

Adelaide Travelodge, 208-223 South Terrace, 5000 (tel: 223-2744; tx: 89232).

Barron Town House, 164 Hindley Street, 5000 (tel: 211-8255; tx: 82941; fax: 2311179).

Gateway Hotel, 147 North Terrace, 5000 (tel: 217-7552; tx: 88325; fax: 2165131).

Grosvenor, 125 North Terrace, 5000 (tel: 213-2961; tx: 82634; fax: 2310765).

Hilton International Adelaide, 233 Victoria Square, 5000 (tel: 217-0711; tx: 87173; fax: 2310158).

Old Adelaide Inn, Corner O'Connell & Gover Streets, North Adelaide, 5006 (tel: 267-5066; tx: 89271; fax: 2672946).

The Meridien Lodge, 21-29 Melbourne Street, North Adelaide, 5006 (tel: 267-3033; tx: 88018; fax: 2390275).

Hotels: Brisbane
Bellevue Brisbane, 103 George Street, 4000 (tel: 221-6044; tx: 41057; fax: 2217474).

Brisbane City Travelodge, Roma Street, 4000, P.O. Box: 575 (tel: 238-222; tx: 141778; fax: 2832288).

Brisbane Parkroyal, Corner Alice & Albert Streets, 4000 (tel: 221-3411; tx: 40186; fax: 2299817).

Gateway, 85-87 North Quay, 4000 (tel: 221-0211; tx: 41335; fax: 2361035).

Gazebo Ramada, 345 Wickham Terrace, 4000 (tel: 831-6177; tx: 41050; fax: 8325919).

Lennon's Brisbane, 66-76 Queen Street, 4000 (tel: 222-3222; tx: 40252; fax: 2219389).

Mayfair Crest International, Corner Ann & Roma Streets, King George Square, 4000, P.O. Box: 934 (tel: 229-9111; tx: 41320; fax: 2299618).

Metropolitan Inn Motel, Leichhardt and Little Edward Streets, 4000 (tel: 831-6000; tx: 42099).

Ridge Motel, Cnr Leichhardt & Henry Streets, 4000, P.O. Box: 283 (tel: 831-5000; tx: 41566; fax: 8322589).

Sheraton Brisbane, 249 Turbot Street, 4000 (tel: 835-3535; tx: 44944; fax: 8354960).

Summit Apartments, Cnr Allenby & Leichhardt Streets, Spring Hill, P.O. Box: 4000 (tel: 839-7000; tx: 44522).

Tower Mill Motor Inn, 239 Wickham Terrace, 4000 (tel: 832-1421; tx: 40382; fax: 8351013).

Hotels: Canberra
Canberra City, 74 Northbourne Avenue, 2600, cnr Northborne Av & Cooyong St,, P.O. Box: 2600 (tel: 49611; tx: 62050).

Canberra International Motor Inn, 242 Northbourne Avenue, Dickson 2602 (tel: 476966; tx: 62154; fax: 487823).

Canberra Parkroyal, 102 Northbourne Avenue, Dickson, 2600 (tel: 491411; tx: 61516; fax: 496878).

Lakeside International, London Circuit, 2601 (tel: 476244; tx: 79062374; fax: 573071).

Rex, Northbourne Avenue, Braddon, 2601, P.O. Box: 517 (tel: 485311; tx: 62363; fax: 488357).

Town House Motor Inn, 60 Marcus Clarke Street, 2601 (tel: 488011; tx: 62057).

Hotels: Darwin
Darwin Travelodge, 122 The Esplanade, 5794, P.O. Box: 3821 (tel: 815388; tx: 85273; fax: 815701).

Diamond Beach Hotel and Casino Resort, Gilruth Ave, 5790, P.O. Box: 3846 (tel: 467666; tx: 85214; fax: 819186).

Four Seasons, Dashwood Crescent, 5790, P.O. Box: 401 (tel: 815333; tx: 85309; fax: 410909).

Sheraton Darwin Hotel, 32 Mitchell Street, 5790, P.O. Box: 1750 (tel: 820-000; tx: 85991; fax: 811765).

Hotels: Hobart
Four Seasons Downtowner, 96 Bathurst Street, 7000 (tel: 346333; tx: 58074; fax: 310898).

Hobart Pacific Motel, Kirby Court, Knocklofty (tel: 346733; tx: 58129; fax: 311197).

Sheraton Hobart Hotel, 1 Davey Street, 7000, P.O. Box: 1601 (tel: 354-535; tx: 58037; fax: 238175).

South Port Town House, 167 Macquarie Street, 7000 (tel: 344422; tx: 58087).

Wrest Point Federal, 410 Sandy Bay Road, 7005 (tel: 250112; tx: 58115; fax: 252424).

Hotels: Melbourne
Australia, 266 Collins Street, 3000 (tel: 63-0401; tx: 30988).

Chateau Melbourne, 131 Lonsdale Street, 3000 (tel: 663-3161; tx: 31483).

City Gardens Melbourne, 335 Abbotsford Street, 3051 (tel: 320-6600; tx: 37742).

Crossley Lodge Motel, 51 Little Bourke Street, 3000 (tel: 662-2500; tx: 31980; fax: 6636556).

Gordon Place Apartments Motel, 24-32 Little Bourke Street, 3000 (tel: 663-5355; tx: 35468).

Hilton Internat Melbourne on the Park, Corner Wellington Parade & Clarendon Streets, 3002, P.O. Box: 103 (tel: 419-3311; tx: 33057; fax: 4195630).

Hyatt on Collins, 123 Collins Street, 3000 (tel: 657-1234; tx: 38796; fax: 6503491).

Marco Polo Inn, Cnr Harker Street & Flemington Road, 3051 (tel: 329-1788; tx: 31093; fax: 329).

Melbourne Airport Travelodge, Link Road, Melbourne Airport, 3045, P.O. Box: 94 (tel: 338-2322; tx: 34263).

Menzies at Rialto, 495 Collins Street, 3000 (tel: 620-9111; tx: 136189; fax: 6141219).

Noah's, Corner Exhibition & Little Bourke Streets, 3000, P.O. Box: 68B (tel: 662-0511; tx: 32779).

Park Plaza International (tel: 650-7488).

President Melbourne, 63 Queens Road, 3004 (tel: 529-4300; tx: 30987; fax: 511042).

Regent, Collins Place, 25 Collins Street, 3000 (tel: 65-30000; tx: 37724; fax: 6504261).

Rockmans Regency, Corner Lonsdale & Exhibition Streets, 3000 (tel: 662-3900; tx: 38890; fax: 6634297).

Royal Parade Travelodge, 441 Royal Parade, Parkville, 3052 (tel: 380-9221; tx: 37080; fax: 3876448).

Sheraton Melbourne, 13-19 Spring Street, 3000 (tel: 650-5000; tx: 31021; fax: 6509622).

Southern Cross, 131 Exhibition Street, 3000 (tel: 653-0221; tx: 30193; fax: 6502119).

St Kilda Road Travelodge, St Kilda Road & Park Street, 3205 (tel: 699-4833; tx: 30869; fax: 6901603).

Town House, 701 Swanston Street, Carlton 3053 (tel: 347-7811; tx: 31271; fax: 3478225).

Victoria, 215 Little Collins Street, 3000 (tel: 653-0441; tx: 31264; fax: 6509678).

Windsor, 103-115 Spring Street, 3000 (tel: 653-0653; tx: 30437; fax: 6545183).

Hotels: Perth
Chateau Perth, Cnr Hay Street & Victoria Ave, 6000 (tel: 325-0461; tx: 92872; fax: 2212448).

Freeway, 55 Mill Point Road, 6151 (tel: 367-7811; tx: 92483).

Hilton International Parmelia, Mill Street, 6000 (tel: 322-3622; tx: 92365; fax: 4810857).

Kings Ambassador, 517 Hay Street, 6000 (tel: 325-6555; tx: 92616; fax: 2211539).

Merlin Perth, Corner Plain Street & Adelaide Terrace, 6000 (tel: 323-0121; tx: 94131).

New Esplanade, 18 The Esplanade, 6000 (tel: 325-2000; tx: 93327).

Orchard Perth, 707 Wellington Street, P.O. Box: 7244 (tel: 327-7000; tx: 95050; fax: 3277017).

Perth Ambassador, 196 Adelaide Terrace, 6000 (tel: 325-1455; tx: 95799; fax: 3256317).

Perth International, 10 Irwin Street, 6000 (tel: 325-0481; tx: 92999; fax: 3232902).

Perth Parkroyal, 54 Terrace Road, 6000 (tel: 325-3811; tx: 92316; fax: 2211564).

Sheraton Perth, 207 Adelaide Terrace (tel: 325-0501; tx: 92938; fax: 3254032).

Townhouse, 778-788 Hay Street, 6000 (tel: 321-9141; tx: 92419; fax: 4812250).

Transit Inn, 37 Pier Street, 6000 (tel: 325-7655; tx: 92739; fax: 3257655).

Hotels: Sydney
Auburn Travelodge, 95 Station Road, Auburn, 2144 (tel: 648-8888; tx: 25095; fax: 6485653).

Best Western Florida Motor Inn, 1 McDonald Street, Potts Point, 2011 (tel: 358-6811; tx: 21128).

Boulevard, 90 William Street, 2011 (tel: 357-2277; tx: 24350; fax: 3563786).

Cambridge Hotel, 212 Riley Street, Surrey Hills 2010 (tel: 212-1111; tx: 23813; fax: 2811981).

Camperdown Travelodge, Cnr Missendon Road & Marsden Street, Camperdown, 2050 (tel: 516-1522; tx: 26689).

Chateau Sydney, 14 Macleay Street, Potts Point, 2011 (tel: 358-2500; tx: 22490).

City Gardens Sydney, Corner City Road & Myrtle Street, Chippendale, P.O. Box: 416,2007 (tel: 690-9100; tx: 74389; fax: 6982007).

BUSINESS DIRECTORY

Gazebo Ramada, 2 Elizabeth Bay Road, Potts Point, 2011, P.O. Box: 1143 (tel: 358-1999; tx: 121569; fax: 3562951).

Hilton International Sydney, 259 Pitt Street, 2001, P.O. Box: 3934 (tel: 266-0610; tx: 25208; fax: 2656065).

Hilton International Sydney Airport, 20 Levey Street, Arncliffe 2205, P.O. Box: 75 (tel: 597-0111; tx: 170795).

Holiday Inn Menzies, 14 Carrington Street, 2000, P.O. Box: 5290 (tel: 20232; tx: 20443; fax: 2903819).

Hyatt Kingsgate, On William Street, Kings Cross, 2011, P.O. Box: 252 (tel: 356-1234; tx: 23114; fax: 3564150).

Hyde Park Plaza, 38 College Street, 2010 (tel: 331-6933; tx: 22450; fax: 3316022).

Inter-Continental Sydney, 117 Macquarie Street, 2000 (tel: 230-0200; tx: 176890; fax: 2512342).

Manhattan, 8 Greenknowe Avenue, Potts Point, 2011, P.O. Box: 791 (tel: 358-1288; tx: 24910; fax: 3573696).

Manly Pacific Parkroyal, 55 North Steyne, Manly 2095 (tel: 977-7666; tx: 73097; fax: 9777822).

North Sydney Travelodge, Blue Street, 2060 (tel: 92-0499; tx: 26644; fax: 9223698).

Northside Gardens, 54 McLaren Street, North Sydney, 2060 (tel: 922-1311; tx: 121173; fax: 9224939).

Old Sydney Parkroyal, 55 George Street, 2000 (tel: 20524; tx: 72279; fax: 2512093).

Park Plaza International (tel: 331-6933).

Parramatta City Motel, Corner Great West Highway & Marsden Street, Parramatta, 2150 (tel: 635-7266; tx: 70261).

Regent of Sydney, 199 George Street, 2000 (tel: 238-0000; tx: 73023; fax: 2512851).

Ryde Travelodge, 1188 Victoria Road, West Ryde, 2114 (tel: 858-5333; tx: 70262).

Sebel Town House, 23 Elizabeth Bay Road, 2011 (tel: 358-3244; tx: 20067; fax: 3571926).

Sheraton Wentworth, 61-101 Phillip Street, Central Business District 2000 (tel: 230-0700; tx: 121227; fax: 2279133).

Southern Cross, Corner Elizabeth & Goulburn Streets, 2000 (tel: 20987; tx: 26324; fax: 2111806).

Top of the Town, 227 Victoria Street, Kings Cross, 2011, P.O. Box: 59 (tel: 33-0911; tx: 123945; fax: 334972).

Wynyard Travelodge, 7-9 York Street, 2000 (tel: 20254; tx: 26690; fax: 2909888).

Car hire: Main offices
Avis, 214 William Street, Kings Cross, Sydney, NSW (tel: 357-2000 tx: 21320).

Budget, Melbourne, Vic (tel: 320-6333; tx: 39088); 93 William Street, Kings Cross, Sydney, NSW (tel: 339-8888).

Europcar (Natcar), Corner of Peel & Queensbury Sts, North Melbourne, Vic 3000 (tel: 329-5000; tx: AA 152056 NATCAR).

Hertz, 97 Franklin Street, Melbourne 3000, (tel: 347-3044; tx: 31230).

Business advice and information
Australian Chamber of Commerce, PO Box E139, Queen Victoria Terrace, ACT 2600 (tel: 732381; tx: 62507; fax: 733646).

Australian Dairy Corp, Dairy Industry House, St Kilda Road, Melbourne, Vic 3004 (tel: 819-4000)

Australian Industrial Development Corporation, 212 Northbourne Avenue, Canberra, ACT 2600 (tel: 479411; tx: 62307).

Australian Trade Development Council, Department of Trade and Resources, Canberra, ACT 2600.

Australian Wheat Board, Ceres House, Lonsdale Street, Melbourne, Vic (tel: 605-1555; tx: 30196).

Australian Wool Corp, Wool House, Royal Parade, Parkville, Vic (tel: 341-9111).

Brisbane Chamber of Commerce, 243 Edward Street, Bisbane, Qld 4000 (tel: 221-1766; tx: 145636; fax: 221-6872).

Chamber of Commerce & Industry South Australia, 136 Greenhill Road, Unley, S.A. (tx: 88370; fax: 272-9622).

Confederation of Australian Industry, PO Box E14, Queen Victoria Terrace, ACT 2600 (tel: 73-2311; tx: 62733; fax: 73-3196).

Department of National Development and Industry, Tasman House, Hobart Place, PO Box 5, Canberra, ACT 2600.

Hobart Chamber of Commerce, 65 Murray Street, Hobart, Tas (tel: 344325; tx: 58296).

International Chamber of Commerce, Commerce House, Brisbane Avenue, Barton, PO Box E118, Queen Victoria Terrace, Canberra, ACT 2600.

International Trade Department Centre, Edgecliff Centre, 203 New South Head Road, Edgecliff, NSW 2027 (tel: 329297).

State Chamber of Commerce and Industry (Victoria), World Trade Centre, Commerce House, Flinders Street and Spencer Street, Melbourne (tel: 611-2233; tx: 31255; fax: 6112266).

New South Wales Government Department of Industrial Development & Decentralisation, GPO Box 4169, Sydney, NSW 2001 (tel: 272741; tx: 20972 AA; Cable: WALESPREM).

Northern Territory Development Corporation, GPO Box 2245, Darwin, NT 5794 (tel: 894211; tx: 86397 DCDEV AA).

Western Australia Chamber of Commerce and Industry, 14 Parliament Place, Perth, WA (tel: 321-5681; tx: 93609).

State Chamber of Commerce and Industry (incl. Sydney Chamber of Commerce), 93 York Street, Sydney NSW 2000 (tel: 299-7888; tx: 27113; fax: 2903278).

Western Australian Development Corporation, 28th Floor, City Mutual Tower, 197 St George's Terrace, Perth, WA 6000 (tel: 322-7933; tx: 95220 WADC AA).

Banking
Central bank
Reserve Bank of Australia, 65 Martin Place, GPO Box 3947, Sydney, NSW 2000 (tel: 230-0277, 234-9333) tx: 20106).

Commercial banks
Australia and New Zealand Banking Group, 55 Collins Street, PO Box 537E, Melbourne, Vic 3000 (tel: 658-2955; tx: 39920).

Australia & New Zealand Savings Bank Ltd, Collins Place, 55 Collins Street, Melbourne, Vic 3000 (tel: 658-2955; tx: 39920).

Barclays Bank Australia Ltd, Barclays House, 25 Bligh Street, Sydney, NSW 2000 (tel: 233-6622; tx: 71237; fax: 2213060).

Commonwealth Bank of Australia, Pitt Street and Martin Place, Sydney, NSW 2000 (tel: 227-7111; tx: 120345/120267; fax: 1912).

Commonwealth Savings Bank of Australia, GPO Box 2719, Pitt Street & Martin Place, Sydney, NSW 2001 (227-7111; tx: 120345, 120267, 24514; fax: 2326573, 2351653).

National Australia Bank, 500 Bourke Street, PO Box 84A, Melbourne, Vic 3000 (tel: 605-3500; tx: 30241).

Natwest Australia Bank Ltd, 39th Level, Qantas International Centre, International Square, George Street, Sydney, NSW 2000 (tel: 250-8500; tx: 177326/7 NWAB HO; fax: 251-2763).

Rural & Industries Bank of Western Australia, 54-58 Barrack Street, Perth, WA (tel: 320-6206; tx: 92417 RIBWA; fax: 3206444).

State Bank of Victoria, PO Box 267D, 385 Bourke Street, Melbourne, Vic 3000 (tel: 604-7000; tx: 32910, 34055; fax: 03 6022150).

State Bank of New South Wales, GPO Box 2628, Sydney, NSW 2001 (tel: 226-8000; tx: 121550; fax: 226).

State Bank of South Australia, 97 King William Street, GPO Box 399, Adelaide, SA 5001 (tel: 210-4411; tx: AA 82082; fax: 2104758, 2123056).

Westpac Banking Corporation, 60 Martin Place, GPO Box 1, Sydney, NSW 2001 (tel: 226-3311; tx: 20122).

Australian branches Bank of New Zealand, 333 George Street, GPO Box 507, Sydney, NSW 2001 (tel: 290-6666; tx: 123240).

Banque Nationale de Paris, 12 Castlereagh Street, PO Box 269, Sydney, NSW 2000 (tel: 232-8733; tx: 20132).

Travel information
Australian Tourist Commission, 324 St Kilda Road, PO Box 73B, Melbourne, Vic 3000 (tel: 267-1233, 690-3900; tx: 31911).

Automobile Association of the Northern Territory (AANT), 79-81 Smith Street, Darwin, NT 5790 (tel: 813837; tx: 85748 AANT).

Australian Capital Territory Tourist Bureau, Canberra Centre, Northbourne Avenue, Canberra City, ACT 2601 (tel: 233-3666).

National Roads and Motorists Association (NRMA), 151 Clarence Street, Sydney, NSW 2000 (tel: 456-4640).

NSW Government Travel Centre, 16 Spring Street, Sydney, NSW 2000 (tel: 231444; tx: 24395).

Northern Territory Government Tourist Bureau, 31 Smith Street, Darwin NT 5750 (tel: 816611/3).

Queensland Government Tourist Bureau, Corner Adelaide and Edward Streets, Brisbane, QLD 4001 (tel: 312211; tx: 40235).

Royal Automobile Club of Queensland (RACQ), 300 St Paul's Tce,Brisbane, QLD 4001 (tel: 253-4444; tx: 41007).

Royal Automobile Club of Tasmania (RACT), Corner Patrick & Murray Streets, Hobart, Tas 7001 (tel: 382200; tx: 58071 RACT).

Royal Automobile Club of Victoria (RACV), 123 Queen Street, Melbourne, Vic 3000 (tel: 607-2211; tx: 30223 RACV).

Royal Automobile Club of Western Australia Inc (RACWA), 228 Adelaide Terrace, Perth, WA 6000 (tel: 325-0551; tx: 93106 RACWA).

South Australian Government Travel Centre, 18 King William Street, Adelaide, SA 5000 (tel: 212-1644).

Tasmanian Government Tourist Bureau, 80 Elizabeth Street, Hobart, Tas 7000 (tel: 300211).

Victoria Tourism Commission, 230 Collins Street, Melbourne, Vic 3000 (tel: 619-9444).

Other useful addresses
World Trade Promotions, 291 Sussex Street, Sydney, NSW 2000 (tel: 02-267 5122). (Trade fairs and exhibitions.)

Austria

W. L. Luetkens

Austria always has been rich in paradox. During 1989 it managed to combine the discovery of a succession of scandals in public life with sterling economic performance. The country was run by a coalition of opposites, the conservative People's Party and the Socialist Party. And quarrelsome though that coalition proved to be, it was a survivor and managed to agree on what may turn out to have been a historic decision to apply for membership in the European Community (EC). Moreover the two parties, by far the largest in the country, undertook to stick by that decision, regardless of the outcome of the election of a new parliament due in the autumn of 1990 at the latest. The significance was that no other combination of parties and certainly no single party was likely to command the two-thirds parliamentary majority which would eventually be required to ratify any deal with the EC.

The application to Brussels was approved in the Austrian parliament by the coalition parties and the smaller Freedom Party (FPOe), but not by the Greens. It was submitted in July 1989. The chief stipulation was that the EC would have to accept explicitly the status of everlasting neutrality adopted by Austria in 1955. That could make for difficulties when serious negotiations with the Community begin, probably not before it has completed its internal market in 1992.

Dr Alois Mock, the Austrian foreign minister, said upon handing in the application that the reservation meant that Austria would steer clear of alliances and would not permit foreign military bases upon its territory. Within the European Commission and among some members of the EC there were fears that Austrian neutrality would be hard to fit into the political co-operation of the Twelve – let alone into a European confederal or federal system. Vienna liked to cite the case of neutral Ireland which does belong to the EC, but that overlooks the fact that, unlike Ireland, Austria abuts on the Soviet sphere of influence.

In August 1989 Moscow expressed concern about the Austrian application. An *aide mémoire* handed over in Vienna, while phrased in friendly terms, did say that, in the EC, Austria would forfeit the scope for putting a policy of neutrality into practice. Officially Vienna hoped that increasingly friendly East-West relations would dispel these reservations.

Free trade agreement

Austria wishes to join the EC largely for economic reasons, the chief of which is that Austrian business will find itself at a disadvantage in the EC once its internal market has been perfected. Under a free trade agreement, Austria already enjoys duty- and quota-free trade in industrial goods with the EC. But only membership can give it a voice in the determination of Community policy and a safeguard against non-tariff barriers, bureaucratic pitfalls and the possible emergence of a fortress mentality in the Community. Dr Franz Vranitzky, the Chancellor, a Socialist, rejected as inadequate a trial balloon flown by Mr Jacques Delors, Chairman of the European Commission, suggesting the alternative of an arrangement between the EC and the European Free Trade Association, of which Austria is a member.

About two-thirds of Austrian foreign trade habitually is done with the EC. Moreover Austria behaved like a *de facto* member of the European Monetary System. The exchange rate of the Schilling was managed to track the Deutsche Mark, the lead currency in the EMS – a policy which had the support of both business and trade unions in spite of the occasional grumble from industrialists. Only under extreme pressure was the policy at all likely to be abandoned.

Besides insisting upon Austrian neutrality, the main conditions Vienna wanted to be met in the negotiations for EC membership were the safeguarding of the highly developed welfare system, assurances that environmental protection would not be watered down, and that Austrian smallhold and mountain peasants would be assured of a living. In addition, all parties agreed that some way must be found to reduce the stream of heavy EC lorries across Austria. Residents along the Brenner route between Italy and West Ger-

KEY FACTS

Official title: Republic of Austria

Head of state: President Dr Kurt Waldheim

Head of government: Chancellor Dr Franz Vranitzky (Socialist Party)

Ruling party: Coalition of Socialist Party and People's Party

Capital: Vienna

Official Languages: German

Currency: Austrian Schilling (S) = 100 Groschen

Exchange rate: S13.1 per $ (Nov 1989)

Area: 83,853 sq km

Population: 7.6m (1988)

GDP per capita: $13,900 (1987)

GDP real growth: 1.0% (1987)

Labour force: 3.37m (1987)

Unemployment: 5.3% (1988)

Inflation: 1.9% (1988)

Trade balance: –S61.6bn (1988)

many suffered greatly from the resulting noise and pollution. The Community had accepted that there was a problem. Vienna wanted to solve it by restricting lorry traffic at night, by encouraging the use of less noisy vehicles, and by transferring traffic from road to rail. The issue led to a conflict with Italian and German lorry owners. The former blockaded the Brenner route for a few days in September 1989.

During the discussion preceding Vienna's application for EC membership it was pointed out that, between the conclusion of the agreement for free industrial trade in 1970 and 1988, Austrian exports to the Community had multiplied by five whereas imports had risen only 3.8 times. The implication was that Austrian industry would be able to cope with increased competition within the EC. There is a rule of thumb that both productivity and labour costs are rather lower than in West Germany, balancing out at least partially. However it is widely accepted that – even without being in the EC – Austrian industry will have to diminish its dependence upon goods of low added value such as steel and forestry products, and to increase expenditure on research and development. Once in the Community some service industries and in particular the financial institutions would face problems if competitors from elsewhere in the EC, and especially from West Germany, decided to move in in force.

West German influence over the Austrian economy is strong. West Germany is Austria's largest trading partner by far, has a strong position in the Austrian markets for cars and other consumer goods, and owns a respectable chunk of Austrian industry. In September 1989 an agreement was reached for MAN of Germany to take majority control of the lorry-building facilities of the Austrian Steyr group. There was a certain amount of public concern when, in 1988, it was found that a German publisher owned stakes in Austrian newspapers with an aggregate market penetration of almost 60 per cent.

Links with West Germany are crucial to the overall performance of the Austrian economy. When German business is flourishing, demand, especially for components, spills over into Austria. When the Germans are pushed to meet export demand, the Austrians, with similar technology, benefit. When things are going well in West Germany, easily the largest source of foreign tourists in Austria, the Austrian tourist industry – a crucial element in Austrian external payments – reaps the benefit.

The Wifo economic research institute forecast that gross domestic product would rise by a real 3 per cent in 1990, following upon an increase of 4 per cent in 1989. Consumer prices would rise by 2.7 per cent in 1989 and 3.3 per cent in 1990 and the unemployment ratio would decline from 4.9 per cent in 1989 to 4.7 per cent in 1990. Wifo attributed the sound growth performance to good export demand and to the stimulus administered to private consumption and investment by the tax reform of 1988.

KEY INDICATORS

	Unit	1984	1985	1986	1987	1988
Population	'000	7,552	7,557	7,559	7,563	7,600
Gross domestic product (GDP)	Schbn	1,278.7	1,354.1	1,432.5	1,487.5	–
GDP per capita	Sch	169,314	179,185	189,509	196,681	–
Inflation	%	5.6	3.2	1.7	1.4	1.9
GDP real growth	%	2.0	2.9	1.7	1.0	–
Unemployment	%	4.5	4.8	5.2	5.6	5.3
Wages (monthly earnings)	1985=100	94.3	100.0	104.5	107.8	–
Share prices	1985=100	52.4	100.0	125.0	106.5	101.0
Interest rate (official disc.)	%	4.5	4.3	4.0	3.0	4.0
Agricultural production	% growth	3.9	–1.4	–1.8	–2.7	0.4
Industrial production	1980=100	104	109	110	111	116
Investment goods	1980=100	101.0	112.8	114.6	115.8	–
Consumer goods	1980=100	106.0	108.8	108.9	108.0	–
Intermediate goods	1980=100	104.0	106.9	106.9	120.4	–
Manufacturing	1980=100	104	109	110	109	115
Exports	S bn	314.5	354.0	342.3	342.4	382.8
Imports	S bn	392.1	431.0	408.1	412.4	444.4
petroleum	Sch bn	40.0	41.6	19.4	18.6	15.1
Foreign exchange	$m	3,586	4,112	5,534	6,772	6,711
Balance of trade	S bn	–77.6	–77	–65.8	–70	–61.6
Current account	$m	–264	–273	86	–429	–658
Exchange rate	S per $	20.0	20.7	15.3	12.6	12.3

* estimated figure

External balance

Usually Austrian external payments show a hefty trade deficit, but a current account almost balanced by income from foreign tourists. Wifo's figures for the current deficit were S4.1bn in 1989, followed by S7.9bn in 1990. As a high-grade debtor, Austria has generally easily covered such amounts by importing capital.

During 1988, foreign visitors spent S120bn in Austria which, after deducting the S74bn spent abroad by Austrians, left a healthy net surplus. The tourist industry stands to gain from increasing complaints of pollution along the Italian Adriatic coast which may deflect holiday makers (especially of Austrian nationality) to the Alps. But like manufacturing industry, the Austrian tourist trade faces the need to restructure in order to increase income from each of the more than 20m Austrian and foreign tourists expected each year. Little suitable space is left for additional recreational facilities, especially ski lifts, and the share of private rooms in total available tourist accommodation is unduly high.

Under the socialist-conservative or 'red-black' coalition of Dr Vranitzky, a banker, econonomic policy followed the widespread international trend towards deregulation and privatisation, though at a distance. The pattern is essentially that of a mixed economy. Small manufacturing businessess, often with sterling export performance, are especially well represented. State ownership remains strong in banking and heavy industry. But, on the West German pattern, state-controlled businesses are under pressure to behave like profit-conscious members of the private sector.

In keeping with the acceptance of market forces, the central bank, in 1989, went a long way towards liberalising cross border capital movements. It signalled that it would continue along this road to fall in with the European Community. Equally, it took a firm stand on the hard currency policy linking the Schilling to the D-Mark.

Social partnership

By and large, Dr Vranitzky managed to ensure the acquiescence of the trade union federation, OeGB, in this economic policy. In spite of occasional militant noises, the unions stuck to their traditional moderation which down the years caused strikes to be almost unknown and wage settlements to be reasonable. The employers, also in spite of occasional militant noises, remained wedded to consensualism too. This co-operative attitude has often been cited as an important reason for Austria's advance from a war-ravaged, largely agricultural state to one where GDP per head matches or exceeds that of the United Kingdom. It is an attitude that also goes far towards explaining why it had proved possible to yoke together the Socialist and People's Parties.

Besides making for industrial peace the consensual system or social partnership as it is called in Austria, can also make for rigidities and backscratching. That created an opening for Dr Jörg Haider, head of the Freedom Party, and the *enfant terrible* of Austrian politics who regularly cocked snooks at the political establishment. As leader of the only effective opposition party, the Greens being disorganised and divided, he was was bound to appeal both to protest voters and to those who felt constricted by the system.

Dr Haider, who was born in 1950, challenged the corporate organisation of Austrian labour and business. He railed against the longstanding rule requiring all businesses of any size to belong to the chambers of commerce organisation, headed by the *Bundeswirtschaftskammer* in Vienna, as representative of their interests. Likewise workers in businesses of any size must belong to Chambers of Labour, though they are rather overshadowed by the trade unions. Inevitably, the chambers system has produced bureaucracies which may be more common but no more popular in Austria than in most other countries.

What really enabled Dr Haider to make hay was the discovery of a number of scandals which caused the resignations of several Socialist ministers and other high-ups in the party. Because of the widespread state ownership of businesses there always is the danger that business misdeamours will also reflect on politicians.

One of the scandals surrounded the loss of the freighter Lucona and six of her crew in 1977 together with a cargo declared as a uranium processing plant, but subsequently suspected of having been no more than scrap iron. In 1988 a warrant was issued for the arrest of one Udo Proksch, owner of a Vienna coffee house and friend of many top Socialists, on suspicion of insurance fraud and of homicide by having the Lucona blown up. He had by then disappeared abroad. A parliamentary committee, reporting in 1989, charged that some of Proksch's friends in high places were involved in a cover up.

Guns for Iran

Another scandal, fraught with some danger to the coalition, led to the arraignment of a number of directors of the state-owned Noricum arms manufacturer on charges of having broken the law by selling guns to Iran, a country then at war. Allegations were widely made in the Austrian press that the deal could not have gone through without the knowledge of several Socialist ministers. The affair prompted Dr Vranitzky to propose the phasing out of exports of armaments. That could severely embarrass several publicly owned and private concerns. On top of all that, several highly placed Socialist party officials resigned when faced with allegations of tax evasion.

The electorate seems to have given the personally popular Dr Vranitzky, head of the Socialist Party (SPOe) since 1988, some credit for working to clean up his party, because the OeVP received a much worse mauling than the Socialists did in elections to several provincial legislatures in 1989. The more fundamental reason was that it was easier for People's Party (OeVP) supporters than for Socialists to drift to the non-socialist FPOe. Moreover, in the popular mind, the scandals appeared to reflect upon politicians generally, rather than upon any one party. Austrians share the cynicism about politicians common in countries with an autocratic past. The success of the FPOe should not be exaggerated. It did receive 29 (previously 16) per cent of the votes in traditionally nationalist Carinthia, but only 16 per cent (9 per cent) in Salzburg, and 16 per cent (6 per cent) in Tyrol. That was enough for Dr Haider, with the support of the local OeVP, to supplant the Socialist head of government in nationalist-minded Carinthia.

There has been unending discussion of what Dr Haider's politics really are. His party used to be an odd mixture of continental-style free marketeering liberals and a right and even extreme right wing which in the nineteenth century favoured union with Germany and to this day looks upon Austria as an essentially German state. Dr Haider sided with the nationalists. But that does not necessarily make him a Nazi, neo or otherwise.

The likely diagnosis is that Dr Haider is a born opportunist and a bright rebel with the gift of the gab. He appeals especially to younger people who chafe under the rigidities of the consensual system, but also to the extreme Right. He certainly did alarm the Slovene minority in Carinthia.

Dr Haider's successes were a step in a gradual erosion of the traditional Austrian party system. At one time, membership in a major party could almost become a way of life. There were rival 'red' and 'black' sports clubs. Party membership often was the route to preferment or, for instance, a council house flat. That explains why roughly 1.5m Austrians in a population of all ages of 7.6m hold party membership cards – a proportion probably unique in any democracy.

Sociological change and the reduction of party animosities as a result of post-war prosperity assisted Dr Vranitzky to shift his party towards the non-ideological political centre. He had to move carefully, but could bank on his personal popularity in the country to muffle the Left.

Rough ride

His coalition partners in the OeVP had a rather rougher ride. Theirs is a Christian Democratic party serving the often contradictory interests of the middle classes, especially the smaller employers; the peasantry; the white collar people; and those blue collar workers whose Christian beliefs have kept them apart from the formerly Marxist SPOe. Shortly after the poor showing of the OeVP in the provincial elections, Dr Mock, whose qualities did not include charisma, was replaced both as party leader and vice-chancellor – but not as foreign minister – by Josef Riegler.

A change of style became apparent at once. The new man (like Dr Vranitzky) said openly that the ultimate decision about membership in the EC must depend on the terms negotiated. That may sound trite, but was more guarded than Dr Mock had been. Riegler also

COUNTRY PROFILE

Political system: Federal democratic republic with nine Länder (states), each with its own state legislature and government. Bicameral parliament consisting of the Nationalrat, whose 183 constituent members are elected every four years by proportional representation, and the Bundesrat, containing 63 members elected for various terms by the state legislatures. Executive power rests with the head of the federal government, who is the Chancellor and usually the leader of the largest party in the Nationalrat. The President is elected by popular vote every six years.

Main cities/towns: Vienna (population 1,515,300), Graz, Linz, Innsbruck, Salzburg, Klagenfurt, St Poelten, Bregenz, Eisenstadt.

Media

Press: National dailies include *Kurier* (circulation 385,543), *Die Presse* (71,732), *Neue Kronen-Zeitung* (978,510), *Der Standard* and *Wiener Zeitung* (25,500, government newspaper), all published in Vienna but read throughout the country. Quality regional papers also exist, e.g. *Salzburger Nachrichten* (65,529). *Weeklies and business:* Weekend editions of newspapers tend to be larger than dailies and contain a large amount of advertising. Several weekly illustrated magazines and technical and trade magazines published. Major business/political magazines include *Profil* (94,000), *Wochenpresse* (52,593, weekly) and *Trend* (80,800, monthly). German magazines are also widely read.

Broadcasting: Radio and TV administered by State-run corporation owned by the Federal Government.

Radio: Approximately 2.7m receivers in use. There are three national programmes, 10 local programmes and an overseas service. In Vienna, Radio Blue Danube broadcasts in English, French and Spanish; and Austrian National Radio (ORF) puts out a news summary in English each morning at 0805 and in French at 0810.

Television: Approximately 2.5m TV sets in use. There are two channels broadcast for an average of 12-15 hours per day. Videotext on a limited scale. Skychannel, Superchannel, and cable increasingly available.

Domestic economy: Gradual shift of economic base since 1950s from agriculture to heavy industry and services. The pace of change has led to growth in industrial output, high living standards and moderate levels of unemployment. Large public sector with government investment in industry now being partly privatised. Economy also characterised by heavy dependence on tourism and trade with FRG. Broad consensus among trade unions, business and government on commitment towards full employment and sustained economic growth. Government economic policy directed towards reducing the budget deficit (4.3 per cent of GDP in 1988) and debt burden and keeping inflation down. Aim is to reduce federal budget deficit to 2.5 per cent of GDP by 1992 by cutting public sector spending which will involve measures such as the removal of subsidies to nationalised industries.

Employment: Moderate levels of unemployment, falling slowly. Wage increases moderated in line with inflation which, with strong productivity gains, have led to a gradual decline in unit labour costs. Industrial relations are very good, with no major disputes.

External trade: Regular trade deficit owing to growth in import demand for both consumer and capital goods. However, income from tourism and net capital inflows largely offset trade deficit. Leading exports include machinery and transport equipment, chemicals (chiefly plastics and pharmaceuticals) and manufactured goods. Major imports are mineral fuels, chemicals, machinery, foodstuffs and consumer durables. From 1988, there was renewed growth in trade with USSR and other CMEA countries. In 1988, 68 per cent of imports came from the EC, FRG providing 45 per cent of total. Of Austrian exports 64 per cent went to EC, 35 per cent to FRG. Other individual major trading partners are Italy, Switzerland, France and USA.

Agriculture: Contributes 4 per cent to GDP and employs about 9 per cent of the labour force. Dominated by small scale farming (50 per cent of farms cover less than 10ha), although trend is towards larger, more mechanised units leading to increased productivity. Farming is concentrated in Upper Austria, the northern part of Lower Austria, Burgenland and Styria. Principal products are milk, beef, veal, pork, sugar-beet, maize, barley and wheat. Although output fluctuates the country remains almost 90 per cent self-sufficient.

Industry: Contributes about 30 per cent to GDP. Nationalised companies administered by holding company Österreichische Industrieholding AG contribute 20 per cent to GDP and produce over 20 per cent of exports. In 1987, however, the government initiated a series of reorganisations and privatisations, aimed at reducing inefficiency and bureaucracy. Traditional industries such as steel and textiles have been declining in recent years, and government industrial policy is to switch from these to newer industries such as electronics, chemicals and vehicles.

Mining: Deposits of various minerals, notably magnesite (world's largest producer), iron, lead and zinc ores, salt, graphite, coal and gypsum. Commercial exploitation restricted by very small number of viable deposits and geological difficulties.

Energy: Despite successful energy conservation programme and increase in local oil and gas production, still heavy dependence on energy imports, especially gas from the USSR. Domestic oil and gas reserves are expected to be depleted by 1990. Near self-sufficiency in electricity, owing to rapid development of hydro-electricity. Nuclear programme has been halted due to powerful environmentalist opposition.

Banking: Two large state-controlled banks, Creditanstalt-Bankverein and Österreichische Länderbank, have large industrial holdings of their own and, as financiers to industry, exert wide influence on Austrian business. Full services also provided by savings banks and co-operative banks, whose role is proportionately much greater than that of similar institutions in the English-speaking world.

Legal system: Civil law system with Roman law origins. Constitution adopted 1920 – repromulgated 1945.

Membership of international organisations: ADB, Council of Europe, DAC, ECE, EFTA, EMA, ESRO (observer), FAO, GATT, IAEA, IDB, IBRD, ICAC, ICAO, IDA, IEA, IFAD, IFC, ILO, International Lead and Zinc Study Group, IMF, IMO, INTELSAT, INTERPOL, ITU, IWC, OECD, UN, UNESCO, UPU, WFTU, WHO, WIPO, WMO, WTO, WSG.

made statements suggesting that he would like to continue the Grand Coalition beyond the next general election, though without excluding the possibility of teaming up with the FPOe. His attitude helped to overcome the crisis between OeVP and SPOe caused by the election of Dr Haider as head of the Carinthian Government with support from the local OeVP.

The small Green Party was hobbled by differences between a hard Left and a *bourgeois* wing concerned mainly with the environment. But no party could be indifferent to green issues. For instance, regulations were planned to discourage the use of disposable plastic containers and to demand a deposit on the sale of batteries to ensure that they are turned in for recycling rather than thrown out with the rubbish.

Greenery played an important part in Austrian external relations. Vienna was relieved when, in 1989, West Germany scrapped plans to build a uranium processing plant at Wackersdorf in Bavaria, less than 130 km from the Austrian border. There was less joy in Vienna when Hungary, with its own green wave, decided to drop a hydro-electric project at Nagymaros on the Danube in which Austrian engineers and banks were deeply involved. A storm caused by allegations concerning the wartime record of President Kurt Waldheim appeared to die down in 1988 after a commission of historians had found that he had not been personally involved in war crimes. But the affair could always blow up again.

When it came to office early in 1987, the coalition had to put to rights the financial troubles which had overtaken much of state-owned industry and a structural budget deficit. Progress was made on both issues. By 1989 Dr Ferdinand Lacina, finance minister, was on the way towards the planned reduction of the budget deficit to 2.5 per cent of GDP by 1992 instead of the 9 per cent threatened for that year before the 'red-black' coalition took over. But the windfalls of unexpectedly strong economic growth and the sale of some state businesses were not going to last for ever.

State-owned industry had been given until 1990 to break even and, by 1989, had made good progress. In this case, too, positive cyclical factors obscured remaining structural weaknesses.

Progress had also been made towards reforming the state pensions system which promised to become increasingly burdensome as the age structure of the population deteriorated; and with a tax reform intended to improve incentives. In both cases a second round of measures was in prospect after the election of a new parliament due no later than autumn 1990.

Outlook

The election was going to show whether Dr Haider's FPOe could continue to eat into the electoral support of the two main parties. In both the SPOe and the OeVP there were people who would rather ally themselves with the FPOe than continue the 'red-black' or Grand Coalition. Dr Vranitzky himself strictly excluded a coalition with the FPOe as long as Dr Haider ran it; Riegler, too, showed a preference for the Grand Coalition. The outcome was going to depend on how big a fright Dr Haider managed to give the big parties – and how they reacted to it.

Regardless of the outcome, Vienna's wish to join the EC looked sure to survive, since the three parties had all backed the initial application. Opinion polls showed a popular majority in favour, important because any deal agreed with the EC will have to be submitted to a referendum. But referendums can be unpredictable. There were unmistakable apprehensions, especially among Socialists in eastern Austria, who were afraid that joining the Community would favour the already more prosperous west. Others were afraid of harder competition or of lessened environmental protection once they were in the EC.

That touches upon a centuries-old debate whether Austria should orient itself towards East or West. Post-war Austria is, undeniably a Western country. But the revolutionary changes in Eastern Europe encouraged those who felt that it should try to be a link between the European blocs.

Since the war, Austria has been remarkable for political and social stability. But as 1990 approached the consensual system of social partnership was seen to be on trial. The Haider phenomenon threatened established political patterns, greatly assisted by the discovery of the scandals of the 1970s and occasional jobbery at local level. Yet scandals or no, anyone visiting the country will find that, on the whole, it is well – if bureaucratically – run. Perhaps that is the greatest Austrian paradox.

BUSINESS GUIDE

Time: GMT + 1 (GMT + 2 from late Mar to late Sep).

Climate: Moderate central European. Wide variations in temperature with altitude, from -4°C (Jan) to 25°C (Jul). Rain possible Apr-Nov, snow Dec-Mar, warm and sunny May-Oct.

Entry requirements
Visa: Visas not required for nationals of most countries for visits of up to three months.
Health: Vaccination certificates required if travelling from infected areas.
Currency: No restrictions on amount of Austrian or foreign currency taken into country. Up to S100,000 in Austrian currency may be taken out and there is no limit on the export of foreign currency. Remaining restrictions, mainly on export of short-term capital by Austrian residents, were in 1989 in process of being liberalised.
Customs: Personal baggage and items for personal use not dutiable.

Air access: Regular flights operated by most international airlines.
National airlines: Austrian Airlines, also Lauda Air and Tyrolean Air.
Main international airport: Schwechat (VIE) 17 km from Vienna.
International airports: Graz (GRZ), Salzburg (SZG), Innsbruck (INN), Klagenfurt (KLU), Linz (LNZ).

Surface access: Good rail and road links with all surrounding countries. Motorists should check advisability of routes, especially in winter, with ÖAMTC or ARBÖ (Austrian Automobile clubs).

Hotels: Generally of a high standard with a large selection available in most cities. Classified from five stars to one star. Rates vary according to category (between about S2,500 and S300 for a single room in Vienna) but are generally cheaper outside the capital. Credit cards accepted: Eurocard and Visa, and less widely, American Express and Diners Club.

Restaurants: Wide choice in Vienna. Service charge of 10.5-15 per cent included in bills. Customary to leave five per cent tip as well.

Car hire: Self-drive and chauffeur-driven services available at railway stations, airports and in major cities. Rates vary with size of car from about S300 to S2,000 per day, plus additional charge per kilometre and fuel is extra. Also available on weekly basis at cheaper rates. VAT at 20 per cent (32 per cent if rental over 21 days) must be paid but insurance is optional. Speed limit 100 kph on most roads, 130 kph on motorways, in built-up areas 50 kph, unless otherwise stipulated.

City transport: Good public transport network covering all of Vienna. Frequent bus, tram, city railway and metro services. It is advisable to purchase tickets in advance at official salespoints or at a tobacconist ('Trafik'). Block tickets are available at S56 for four rides (1989), each to include any transfers within capital's system.
Taxis: Widely available from stands or via radio/telephone services. Fares are metered but in some areas zone charges or set charges for standard trips apply, e.g. from Vienna Airport to the city the sum of S110 is added to the fare to cover the return journey. Tip about 10 per cent.

National transport
Air: Austrian Air Services or Tyrolean Airways operate regular flights between Vienna and Graz, Linz, Klagenfurt, Innsbruck and Salzburg. Also between Innsbruck and Graz and Linz; and Linz and Salzburg.
Road: Over 200,000 km of roads including some 1,400 km of motorways and around 35,000 km of main roads.
Rail: State-owned network of almost 6,000 km, most of which is electrified. Also about 20 private railways covering a total 600 km. Frequent intercity service from Vienna to Salzburg, Innsbruck, Graz and Klagenfurt.
Water: 350 km of navigable waterways. Passenger service between Vienna and the Black Sea and on upper Danube. Austrian Federal Railways operate passenger services on all the larger lakes.

Public holidays
Fixed dates: 1 Jan, 6 Jan (Epiphany), 1 May, 15 Aug (Assumption), 26 Oct (National Day), 1 Nov (All Saints'), 8 Dec (Immaculate Conception), 25 and 26 Dec. Each province has a holiday on the day of its patron saint.
Variable dates: Easter Monday, Ascension, Whit Monday and Corpus Christi.

Working hours
Government and business: (Mon-Fri) 0800 – 1600 (many offices/companies do not work Fri pm).
Banking: (Mon-Fri) 0800 – 1230 and 1330 – 1500 (1730 on Thu, head offices open without break), local variation.
Shops: (Mon-Fri) 0800 – 1800 (outside central Vienna lunch break usually between 1230 and 1500), half day Sat. On every first Sat in the month, many shops are open until 1700 (usually without a lunch break).

Social customs: Exchange pleasantries for a few minutes before getting down to business. Shake hands with everyone in group on meeting and departure. Address business people by their titles. Give hostess flowers or confectionery.

Business language and interpreting/translation: Available through the Interpreters' Institute of Vienna University. Commercial firms can usually find someone to interpret without too much difficulty. The language spoken in Austria is German, but English is widely spoken.

Telecommunications
Telephone: International dialling code: 43 followed by 1 for Vienna. Reduced rates for calls outside working hours.
Telex: Widespread use by Austrian firms. Hotels have service available to residents. Public telex facility, apart from those at trade fairs and conferences, is at Öffentliche Fernschreibstelle in Vienna. Country code 47 A.
Telegram: Available at any post office.

Postal service: No deliveries on Saturday, and only larger post offices open on Saturday mornings (0800 – 1200 hours), with the exception of main station offices in main cities which are open 24 hours a day, seven days a week. Stamps can be purchased at tobacconists and newsagents as well as post offices.

BUSINESS GUIDE

Advertising: Covers all media. TV advertising is popular but restricted facilities and high demand means space must be booked several months in advance. Newspapers (especially weekend editions) and magazines are also popular. Poster sites are available under the auspices of the Municipal Authorities. Advertising in cinemas and by direct mail is also widely used. Information can be obtained from Österreichischer Werberat in Vienna (tel: 65050). Total advertising spend (1987) S8,100m. Main media for adverts newspapers, TV and magazines. Major categories advertised: retail, services and food.

Trade fairs: Most important is Vienna International Fair held every spring and autumn. Several others including biannual industrial fair held in Graz, Young International Fashion Fair every Mar and Sep in Salzburg, annual timber fair in Klagenfurt, tourism and catering in Innsbruck, textiles in Dornbirn, and many others covering a wide range of industries.

Electricity supply: 220 V AC.

Representation in capital: Afghanistan, Algeria, Albania, Argentina, Australia, Belgium, Bolivia, Brazil, Bulgaria, Canada, Chile, China, Colombia, Cuba, Czechoslovakia, Denmark, Ecuador, Egypt, Finland, France, German Democratic Republic, Federal Republic of Germany, Greece, Guatemala, Holy See, Hungary, India, Indonesia, Iran, Iraq, Ireland, Israel, Italy, Ivory Coast, Japan, Korea DPR, Republic of Korea, Kuwait, Lebanon, Libya, Luxembourg, Malaysia, Mexico, Moroccco, Netherlands, New Zealand, Nicaragua, Nigeria, Norway, Oman, Pakistan, Panama, Peru, Philippines, Poland, Portugal, Qatar, Romania, Saudi Arabia, Senegal, South Africa, Spain, Sweden, Switzerland, Thailand, Tunisia, Turkey, UAE, United Kingdom, Uruguay, USA, USSR, Venezuela, Yugoslavia, Zaire.

BUSINESS DIRECTORY

Hotels: Vienna
Ambassador, Neuer Markt 5, A-1010 (tel: 51466; tx: 111906; fax: 513 2999).

Beethoven, 6 Millôeckergasse, A-1060 (tel: 5874482(0); tx: 112574; fax: 587448239).

Bellevue, Althanstrasse 5, A-1091 (tel: 345631(0); tx: 114906; fax: 316799).

Bristol, 1 Kâerntner Ring, A-1015 (tel: 515160; tx: 112474; fax: 51516550).

Clima City, Theresianumgasse 21a, A-1040 (tel: 051996; tx: 131106; fax: 6509604).

Clima Johann Strauss, Favoritenstrasse 12, A-1040 (tel: 5057624, 657624; tx: 132699; fax: 5057628).

Clima Villenhotel, Nussberggasse 2C, A-1190 (tel: 371516; tx: 115670; fax: 371392).

Cottage, Hasenauerstrasse 12, A-1190 (tel: 312571/2; tx: 134146; fax: 31257110).

Hilton, A-1030, Am Stadtpark (tel: 752652; tx: 136799; fax: 75265220).

Hotel am Schubertring, Schubertring 11, A-1010 (tel: 717020; tx: 111569).

Im Palais Schwarzenberg, Schwarzenbergplatz 9, A-1030 (tel: 784515; tx: 136124; fax: 784714).

Imperial, 1 Kâerntner Ring 16, A-1010 (tel: 501100; tx: 112630; fax: 50110410).

Inter-Continental, Johannesgasse 28, A-1037 (tel: 71122; tx: 131235; fax: 7133389).

Kaiserhof, Frankenberggasse 10, A-1040 (tel: 5051701; tx: 136872; fax: 5058875).

Köenig von Ungarn, Schulerstrasse 10, A-1010 (tel: 51584; tx: 116240; fax: 526520/8).

Marriott, Parkring 12a, A-1010 (tel: 515180; tx: 112249; fax: 51518/6736).

Pullman Belvedere, Am Heumarkt 35/37, A-1030 (tel: 7525350; tx: 111822; fax: 752535 844).

Rosen Burgenland, Wilhelm Exner-Gasse 4, A-1090 (tel: 439122; tx: 114293).

Rosen International, Buchfeldgasse 8, A-1080 (tel: 435291; tx: 114293).

Royal, Singerstrasse 3, A-1010 (tel: 524031, 51568; tx: 112870).

SAS Palais, Weihburggasse 32, A-1010 (tel: 515170; tx: 136127; fax: 5122216).

Sacher, Philharmonikerstrasse 4, A-1015 (tel: 51456; tx: 112520; fax: 51457/8/10).

Savoy, Lindengasse 12, A-1070 (tel: 934646; tx: 132491).

Car hire: Vienna (area code 1)

Avis: Weyringergasse 33, A-1040, (tel: 5055839, tx: 112813); new office opening 3rd floor, A-1010 (tel: 5873595; tx: 136507).

Budget: Schwechat Airport (tel: 7770).

Europcar: Kaerntner-Ring 14 (tel: 5054166, 5054200, 5054481; tx 112181).

Hertz: international reservations, Ungargasse 37, A-1030 (tel: 7131596/0; tx: 135232).

Banking
Österreichische Nationalbank (central bank), A-1090 Vienna, Otto Wagner-Platz 3 (tel: 43600; tx: 114778).

Creditanstalt-Bankverein (commercial bank), A-1010 Vienna, Schottengasse 6 (tel: 531310; tx: 133030; fax: 5350203).

Genossenschaftliche Zentralbank AG (Central Institute of the Austrian Raiffeisen Banks), A-1011 Vienna, Herrengasse 1 (tel: 71707-0; tx: 136989).

Girozentrale und Bank der Österreichischen Sparkassen AG (Central Bank of the Austrian Savings Banks), A-1011 Vienna, Schubertring 5 (tel: 711940; tx: 132591).

Österreichische Länderbank AG (commercial bank), A-1011 Vienna, Am Hof 2 (tel: 531240; tx: 115561).

Verband Österreichischer Banken und Bankiers (Association of Austrian Banks and Bankers), A-1013 Vienna, Börsegasse 11 (tel: 5351771; tx: 132824).

Zentralsparkasse und Kommerzialbank Wien, A-1030 Vienna, Vordere Zollamtsstrasse 13 (tel: 71191-0; tx: 133615).

Principal newspapers
Die Presse, 1010 Vienna, Parkring 12A (tel: 51414).

Kurier, 1072 Vienna, Lindengasse 52 (tel: 96210; tx: 132631).

Neue Kronen-Zeitung, 1190 Vienna, Muthgasse 2 (tel: 36010; tx: 114327).

Profil 1010 Vienna, Marc-Aurel-Str 12 (tel: 661670; tx: 136404).

Trend, 1010 Vienna, Marc-Aurel-Str 12 (tel: 567677; tx: 111853).

Wiener Zeitung, 1037 Vienna, Rennweg 12A (tel; 787631; tx: 131805).

Wochenpresse, 1070 Vienna, Lindengasse 52 (tel: 96210; tx: 135869).

Other useful addresses
Austria Presse-Agentur (APA) (Co-operative Agency of the Austrian Newspapers and Broadcasting Co), A-1199 Vienna, Gunoldstrasse 14 (tel: 36050; tx: 114721).

Interpreters' Institute of the Vienna University (tel: 347649 ext 298).

Ministry of Economic Affairs, Stubenring 1, 1010 Vienna (tel: 75000; tx: 111780).

ÖAMTC (Österreichischer Automobil-Motorrad und Touring Club), A-1010 Vienna, Schubertring 3 (tel: 711990; tx: 133907).

Öffentliche Fernschreibstelle (public telex office), A-1010 Vienna, Börseplatz 1.

Österreichische Fremdenverkehrswerbung (Austrian National Tourist Office), A-1040 Vienna, Margaretenstrasse 1 (tel: 58866; tx: 32222306; fax: 5886620).

Österreichischer Rundfunk, Werbeabteilung (Austrian Radio Advertising Dept), A-1136 Vienna, Würzburggasse 30 (tel: 82910 or 8291 ext 4502/4503; tx: 2250).

Österreichisches Statistisches Zentralamt (central statistical office), Hintere Zollamtstrasse 2b, A-1030 Vienna (tel: 711280; tx: 132600).

Vereinigung Österreichischer Industrieller (Association of Austrian Industrialists), A-1030 Vienna, Schwarzenbergplatz 4 (tel: 711350; tx: 131717).

Wiener Börse (Vienna Stock Exchange), A-1011 Vienna, Wipplingerstrasse 34 (tel: 53499; tx: 132447).

Zentralstelle Aussenhandel u. Zoll (Ministry of Economic Affairs), Export/Import Licensing Office, A-1030 Vienna, Landstrasser Hauptstrasse 55-57 (tel: 711020; tx: 131300).

Bahamas

Tony Thorndike

Although the prosperous Bahamian economy continues to grow, prospects are not as positive as they have been in the past few years. One problem is a slower than expected expansion in the critically important tourist and offshore finance sectors, perhaps because of the adverse publicity surrounding the drug trade and crime level. Allegations of Prime Minister Pindling's involvement in drug-related scandals continue to be heard, but he and his government now co-operate fully with the US Government.

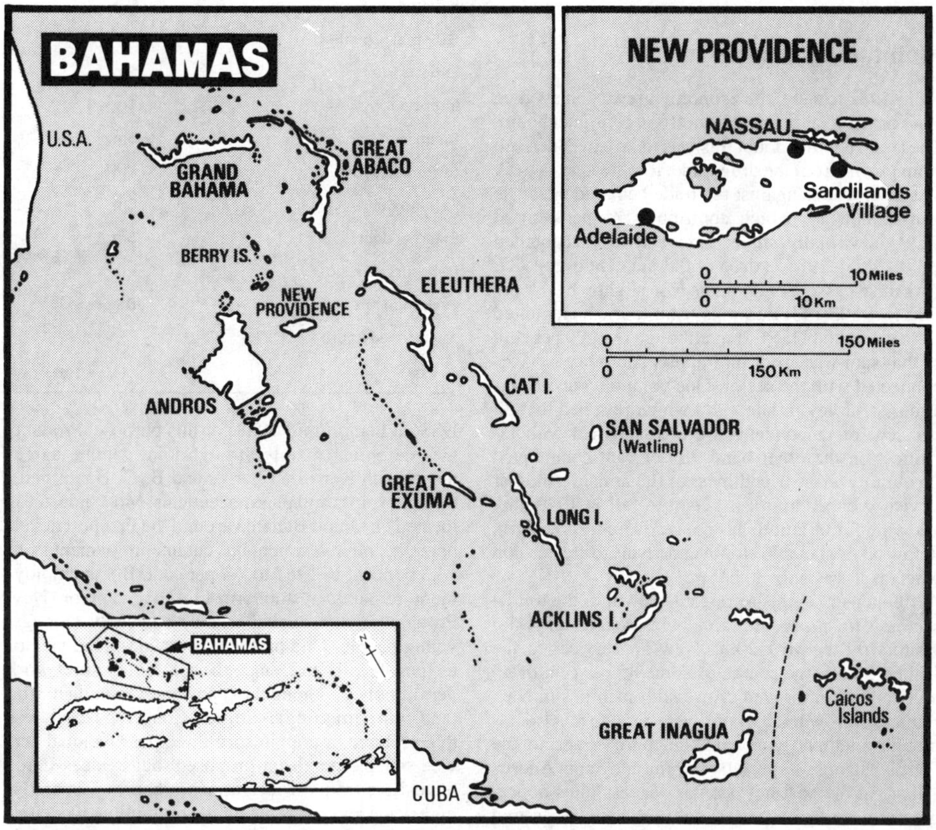

Because the archipelago of 700 islands and over 2,000 uninhabited cays (pronounced 'keys') is at its most northerly point barely 50 miles from the Florida coast, development has inevitably been strongly influenced by the United States. But, as the archipelago stretches in an arch some 700 miles long towards the the edge of the Caribbean Sea, the attractions of the more far-flung of the islands – the Family Islands – for smugglers are many. During the days of Prohibition the wares were liquor; now they are narcotics. The huge sums involved have tempted many Bahamians, and corruption has reached to the highest levels of government. There are an abnormal number of millionaires, many of whom are politicians. An added attraction to drug smugglers is the offshore finance industry. Although tighter regulations have been introduced, the strict secrecy laws have enabled the 'laundering' of illegal funds. The reputation of the industry has suffered as a result. But the government has since 1985-86 taken action against the drug traders, particularly when it became known that an alarming number of young people in the islands had become addicted.

Political longevity

The success of Prime Minister Sir Lynden Pindling and his Progressive Liberal Party (PLP) in the June 1987 elections made him the longest serving leader in the Commonwealth Caribbean, having been in office since 1967. The result, the PLP's fifth victory, proved that the self-styled 'Black Moses' has not lost his appeal, especially among the less prosperous – mostly the 85 per cent black population. The PLP came to power in the historic 1967 election when the white-based United Bahamian Party – the preserve of the merchants and traders, the so-called 'Bay Street Boys' – was defeated. It was a long struggle, for the PLP was formed as early as 1953 to represent the working class. Assaults on it by various right-wing groups have been singularly unsuccessful, the latest being that of the Free National Movement (FNM), formed out of a number of rival factions in 1981.

The PLP won 31 of the 49 seats, the FNM the remainder. The FNM increased its representation by five, but only because the Assembly was expanded by seven extra seats. The very small socialist Vanguard Party attracted minimal support. The FNM failed to make an impression because of its upper class connections and too close identification with the United States and Washington's scarcely concealed anti-Pindling campaign. Kendall Isaacs, its demoralised leader, stepped down in favour of Cecil Wallace-Whitfield.

The election campaign was dominated by the drug crisis. The Columbian Medellín cartel had begun to use the Bahamas as a staging point in the late 1970s and, even now, the US Drug Enforcement Agency (DEA) estimates that some 70 per cent of all cocaine and 50 per cent of marijuana entering the United States does so through the Bahamas. Bribery became rife and Pindling was openly accused in a NBC TV broadcast in 1983 of complicity. He was forced to establish a Royal Commission in which, although he himself was exonerated (but he could not adequately explain why his official salary covered only one-eighth of his spending), several ministers and officials were named. Intense US pressure forced him to dismiss them and to co-operate with the DEA and the US Coastguard to the extent of permitting virtually complete freedom of movement throughout the territory. The result was the capture of huge amounts of narcotics and the arrest of several senior Colombian and other traffickers, notably Carlos Lehder Rivas. But the DEA distrusted Pindling and attempted to trap him until the agency was warned off by powerful members of Congress who felt that Pindling was being made a scapegoat.

Co-operative measures intensified after the election, including the signature of a Mutual Legal Assistance Treaty (M-LAT) providing for collaboration in the investigation of criminal matters. By 1988 the United States Government publicly acknowledged the 'excellent' work done by the Bahamian Government in the war against the drug trade, but still criticised the level of drug-related corruption. Indeed, in May 1989 two members of the US Senate strongly criticised Congressional delay in ratifying M-LAT and the Senate voted by 57 to 40 against a motion seeking to 'decertify' the Bahamas for US aid because of alleged failure to co-operate with the US authorities over anti-drug measures. Pindling's confidence was such that he could brush off the accusations made by Lehder at his trial in Miami that he had personally given Pindling US$400,000.

KEY FACTS

Official title: Commonwealth of The Bahamas

Head of state: Queen Elizabeth II

Head of government: Prime Minister Sir Lynden Pindling

Ruling party: Progressive Liberal Party

Capital: Nassau

Official Languages: English

Currency: Bahamian Dollar (B$) = 100 cents

Exchange rate: B$1.00 per $ (fixed)

Area: 13,864 sq km

Population: 234,000 (1988)

GDP per capita: $11,317 (1988)

GDP real growth: 2.0% (1988)

Inflation: 4.2% (1987)

Trade balance: –$774.3m (1987)

Foreign debt: $233m (1987)

Visitor numbers: 3.2m (1988)

As ever the champion of the poor, Pindling announced a campaign to enable the black majority to obtain more 'economic justice', although no details were given as to how. But he was acutely aware that the economy was not growing as it had done in the recent past.

Maintaining prosperity

From 1981 to 1986 the economy grew at an average 6 per cent a year, boosted by a strong consumer boom. There can be little doubt this was to an extent derived from profits from the drug trade, for it reached a peak just as measures against the trade began to bite. The tourist industry, which contributes 35 per cent of GDP, is virtually static: stopover arrivals totalled 1,474,245 in 1988, a drop of 0.3 per cent over 1987, but cruise passenger landings rose slightly by 4.9 per cent to 1,505,143. Earnings from tourism amounted to an estimated US$1.2bn, although over 70 per cent of that sum was spent on imports and other services connected with the sector. One point of concern was a tourist survey in late 1988 which revealed that 32 per cent of respondents were dissatisfied with facilities; on the other hand 78 per cent commented favourably on the friendliness of Bahamians. Another is visitor expenditure: it began to fall in 1979 and, allowing for inflation, has not recovered. As a result, it now takes about 40 stopover arrivals to create a job whereas it was only 30 before.

Strenuous attempts to restore growth in the tourist sector were made, the aim being a total of 4-5m visitors by the year 2000. New air services to the United States (the source of some 85 per cent of all tourists) and improved cruise ship facilities in Nassau's overcrowded harbour were foremost. Also, although attempts to spread the tourist trade to the Family Islands – especially Exuma and South Andros – have in the past not been as successful as hoped, greater means of access should produce the desired results. Fortunately the sharp increase in ship calls and other visitors to Grand Bahama recorded from the mid-1980s has been sustained.

The manufacturing and agricultural sectors are also being promoted in order both to enhance diversification and to widen employment possibilities. The Inter-American Development Bank approved a US$6m loan in mid-1988 for this purpose. Products include citrus (mainly through joint venture agreements with Kenwood Farms and B. G. Hammond), cement, spirits and pharmaceuticals. Most industry is located on Grand Bahama around the Freeport development, and investment has cut unemployment from 14.2 per cent in 1987 to 7.4 per cent. But unemployment remains a stubborn 11.4 per cent on New Providence, the capital island, especially amongst young people. One problem is a widespread lack of technical skills and, since there is a very strict work permit policy, potential expansion of manufacturing (and with it, training) is limited. Concerned to address this problem, a compulsory three-year national service scheme has been proposed, but it faces fierce opposition from a wide variety of bodies. Generally, high wage levels has meant the promotion of capital-intensive rather than labour-intensive industry. Private sector activity in the construction sector fell from 1987. However, although the number of housing starts fell by 5 per cent, their value increased by nearly 19 per cent, a reflection of still high disposable incomes.

Government capital expenditure has since 1987 risen sharply to maintain economic growth, aimed at improving social services, road and airport rehabilitation and improvements and the upgrading of tourist facilities. As a result, public sector borrowing has been high because of the narrow tax base (and poor records of collection), and local bank liquidity levels were affected. The total debt in 1988 exceeded US$635m, with debt servicing amounting to US$104.6m. Of this, internal debt totalled US$445m. External reserves were put at US$175.4m. Although the social services element was limited, expenditure went some way to alleviate social distress in the poorer districts of New Providence where most of the estimated 60,000-plus Haitian immigrants live – most of them illegally. One problem is the lack of public housing, mainly because of very high land prices.

The 1988 financial record clearly showed substantial government spending. There was an overall deficit of US$81.2m, expenditure of US$518.8m (1987: US$547m) being set against revenue of US$432.6m. Demonstrations by public sector workers protesting insufficient pay increases in November 1988 led to an additional and unbudgeted allocation of US$13m. The trade deficit swelled from US$881.6m to over

KEY INDICATORS

	Unit	1984	1985	1986	1987	1988
Population	'000	222	225	238	242	234
GDP (1988 dollars)	B$m	1,594	2,424	2,487	2,596	2,648
Inflation	%	4.0	4.6	5.4	6.0	4.2
GDP real growth	%	3.0	3.0	2.6	4.4	2.0
Exports fob	$m	261.9	295.8	293.4	273.1	273.6
Imports fob	$m	866.2	1,096.2	1,012.8	1,154.7	1,047.9
Balance of trade	$m	–604.3	–800.4	–719.4	–881.6	–774.3
Current account	$m	–45.5	–148.7	–26.7	–166.0	–132.4
Foreign debt	$m	2271	2163	2523	233	–
External debt	$m	226.7	215.5	252.1	232.7	–
Debt service ratio	%	3.5	4.2	3.3	2.7	–
Exchange rate	B$ per $	1.0	1.0	1.0	1.0	1.0

* estimated figure

COUNTRY PROFILE

Political system: Independent state within the Commonwealth with British monarch as head of state, represented by a governor-general. The governor-general is advised by a cabinet made up of the prime minister (the leader of the main party in the House of Assembly), the attorney-general and at least seven other ministers (chosen on the prime minister's advice). The bicameral legislature consists of a 16-member Senate appointed by the governor-general and a House of Assembly which currently has 49 members (subject to regular reviews, as are boundaries). The parliamentary term is five years.

Main cities/towns: Nassau (on New Providence island), Freeport (on Grand Bahama island).

Media
Press: The main dailies are *The Nassau Guardian, The Tribune* and *Freeport News.*
Broadcasting: The government-owned Bahamas Broadcasting Service operates four commercial radio and one television station as ZNS. Freeport is covered by cable TV. American TV can also be received.

External trade: Regular trade deficit traditionally offset by invisible earnings from tourism, offshore financial business and shipping.
Exports: Petroleum and products (95 per cent of total value in 1985), pharmaceuticals, chemicals etc, rum, aragonite, salt, agricultural products, cosmetics and perfume. Main destinations: USA, Canada, EC and UK.
Imports: Crude oil (65 per cent of total value in 1987), machinery and transport equipment, manufactured goods, chemicals, foodstuffs. Main sources: UK, Norway, Canada, Saudi Arabia, Nigeria, USA.

Agriculture: Although less than three per cent of land area is cultivated, near self-sufficiency has been achieved in poultry, meat, eggs, fruit and vegetables. Recent expansion of export crops such as limes, pineapples, papayas, avocados, cucumbers and mangoes. Also development of commercial fishing, mainly conches, crawfish and crab. Sector contributes five per cent to GDP.

Industry: Sector is small-scale, contributing around 12 per cent to GDP. The largest contributor to the industrial sector the crude oil trans-shipment terminal operated by Burmah Oil; the 350,000 b/d capacity oil refinery run by the Bahamas Oil Refining Co ceased refining in 1984. Re-exports of crude and refined oil (mainly to the USA) are estimated to account for around 16 per cent of GNP. Other activity is centred on the production, for export, of light rum, chemicals and pharmaceuticals. The companies manufacturing these products are largely foreign-owned and located in the Freeport area on Grand Bahama. Other light industries include food processing, confectionery, garments, small boat building and furniture making. The construction industry is also important. Bahamas is main tourist destination in the region and tourism contributes 70 per cent of GDP and employs two-thirds of workforce.

Energy: No commercial energy resources have been discovered, although oil exploration is underway.

Banking: Offshore banking is a major activity. There are some 390 institutions licensed to carry on banking and/or trust business either within or from the Bahamas.

Membership of international organisations: FAO, GATT (de facto), ICAO, ILO, IMF, UN, WHO, Commonwealth, Economic Commission for Latin America, CBI, Caricom, Caribbean Basin Initiative, IBRD, IMO, Inter-American Development Bank, ITU, state associated with the EC (Lomé Convention), UNESCO, VPU, WIPO, WMO.

US$960m. To help balance the books, greater attention had to be given to the offshore banking sector.

Banking reform and fine-tuning

The Bahamian offshore finance industry was for long a leader in the international banking field. In 1983 the IMF calculated that it was the third largest banking centre in the world, after Britain and the United States. But by 1988 it had slipped to eleventh place. Virtually no growth was recorded in that period yet the increase for all other offshore centres averaged 78 per cent. Negative publicity, easy tourist earnings, work permit difficulties and bureaucratic inefficiency were all to blame, as was increased competition from other Caribbean centres, notably the Cayman Islands, which surpassed the Bahamas in registrations in 1985. The sharp hike in licence fees in November 1988 did not help.

The Bahamian industry can be traced as far back as 1906, but the modern foundation is The Banks and Trust Companies Regulation Act of 1965. There are no income, corporation, dividend, inheritance or capital gains taxes to pay. Neither is there exchange control and the Bahamian dollar is on par with the American, and a wide variety of professional skills are at hand. At the end of 1988 there were 391 licensed banks and trust companies (1986: 374), 280 being authorised to deal with the public, about half foreign-owned. Their combined assets exceeded US$160bn. Eurocurrency dealing is a speciality which received a big boost in the 1970s with the need to recycle petro-dollars. The main business is taking deposits from non-Americans and lending them to mainly North American customers. There are also a large number of trusts whose assets are not recordable. One area showing some promise is offshore insurance, but the business is small compared to Barbados and the Cayman Islands. By contrast, ship registration is a growth industry. By September the Bahamas' registry consisted of 810 ships with a total tonnage exceeding 10.8m gross tonnes, compared with 190 ships and 3.5m tonnes in 1985. The offshore industry as a whole employs over 3,000.

In order to develop a better image, there is now limited co-operation with the United States to trace criminal-linked deposits, commensurate with security requirements. One major qualification is that alleged tax offences are excluded. The reason is simple: tax evasion is by definition not an offence as there is no direct taxation in the Bahamas. In addition, an explicit code of conduct by the Association of International Banks and Trust Companies is now in place. Legislative fine-tuning is seen as another way to recapture Bahamian market share. The 1866 Companies Act has been superceded by a streamlined and flexible International Business Companies Act, while trusts will receive further protection from the Domicile of Trust Act, which declares in which jurisdiction settlements could occur. Work also began in 1989 on the $25m Bahamas Financial Centre office complex in Nassau. But only time will tell whether these initiatives will restore confidence in the Bahamian finance industry.

BUSINESS GUIDE

Time: GMT -5 (GMT -4 Apr-Oct).

Climate: Semi-tropical with temperatures ranging from 20°C in winter to 30°C in summer. The rainy season is Jun-Oct.

Entry requirements
Passport: Required by all except nationals of Canada, UK and USA having proof of identity, assured return passage and means of support for periods up to three weeks.
Visa: Not required by nationals of EC countries (except Spain), Commonwealth, certain other European, Latin American and African countries. (N.B. visas must be obtained by visitors from communist countries and Republic of South Africa). Assured onward/return passage is required in all cases.
Currency: No restrictions on import of any currency or on export of coins. Export of Bahamian notes (not exchangeable abroad) up to B$70 is permitted.

Health precautions
Mandatory: Yellow fever vaccination certificate if arriving from an infected area.
Advisable: Typhoid, polio vaccinations. Water precautions.

Representation overseas: London, Washington, Ottawa, Port-au-Prince.

Air access
National airline: Bahamasair.
Other airlines: Aero Coach Aviation, Air Canada, Air Jamaica, Braniff, British Airways, Chalks International, Continental, Delta, Eastern, Midway, Pan Am, Paradise Island Airlines, Piedmont Aviation, Turks & Caicos National, TWA.
International airports: Freeport International (FPO), 5 km from city, duty-free shop, bar, restaurant, buffet, shops, car hire; Nassau International (NAS), 15 km from city, duty-free shop, restaurant, bank, post office, car hire. Paradise Island International, 5 km from city on Paradise Island.

Surface access
Main ports: Freeport, Nassau.

Hotels: Wide variety available. Bills usually include 15 per cent service charge, as well as the six per cent hotel room tax.

Car hire: National licence valid for three months – for longer periods international licence required. Rates vary according to season. Traffic drives on the left.

City transport
Taxis: Metered in New Providence; fixed-rate system – meter starts at B$1.20 and 20c for every one-fifth of a mile; 15 per cent tip is usual.

National transport
Air: Very extensive air charter network (over 50 landing sites throughout islands) – local enquiries should be made for particular requirements.
Road: Main centres well-served by surfaced roads.
Buses: Few conventional buses but 'jitneys' (mini-buses) serve New Providence Island.
Water: Ferry and mail-boat services are operated but for business travellers length and timing of journeys may prove a major drawback.

Public holidays: 1 Jan, Good Friday, Easter Monday, Whit Monday, first Friday in Jun (Labour Day), 10 Jul (Independence), first Mon in Aug (Emancipation Day), 12 Oct (Discovery Day), 25 Dec, 26 Dec.

Working hours
Government and business: (Mon-Fri) 0900 – 1730.
Banking: (Mon-Thu) 0930 – 1500; (Fri) 0930 – 1700.
Shops: (Mon-Sat) 0900 – 1700.

Telecommunications
Telephone: Dialling code for Bahamas: IDD access code + 1-809 followed by area code (Freeport 352, Nassau 32) followed by subscriber's number. Automatic internal service on main islands.
Telex: One public telex exchange in the Telecommunications Office, East Street, Nassau. Efficient telegraph service to all parts of the world.

Postal service: Air Service takes about one week to Europe, surface mail takes about two months.

Banking
Banking centre: Nassau.
Central bank: Central Bank of the Bahamas.
Commercial banks (head offices): American Security Bank International (Nassau), Arawak Trust Co, Artoc Bank & Trust, Bahamas Development Bank, Banca della Svizzera Italiano (Overseas), BankAmerica Trust & Banking Corp (Bahamas), Bankers Trust Co, Bank Leu International, Bank of Boston Trust Co (Bahamas), Bank of Nassau, Bank of Montreal (Bahamas & Caribbean), Bank of Montreal International, Bank of New Providence, Bank of Nova Scotia International, Barclays Bank International, Canadian Imperial Bank of Commerce, Central Bank of the Bahamas, Charterhouse Bank and Trust International, Chase Manhattan, Citibank, Commonwealth Bank, Deltec Banking Corp, Finance Corporation of Bahamas, First Home Banking Centre, Handelsbank N.W. (Overseas), Lloyds Bank International (Bahamas), People's Penny Saving Bank, Republic National Bank of New York (International), Natwest, Royal Bank of Canada International, SFE Banking Corp, Swiss Bank Corp (Overseas), Wardley International Bank.

Advertising: Provided on all forms of media, though no trade journals are published.

Electricity supply: 120 V AC, 60 cycles.

Representation in capital: Austria, Barbados, Belgium, Brazil, Canada, Costa Rica, Cyprus, Dominican Republic, France, The Gambia, German FR, Guyana, Haiti, Israel, Italy, Luxembourg, Mexico, Monaco, Netherlands, Norway, Panama, Peru, Sweden, Switzerland, UK, USA.

Hotels: Nassau/New Providence Island
Cable Beach, Cable Beach, P.O. Box: N-4920 (tel: 327-7342/9; tx: 20281).

Emerald Beach, Cable Beach, P.O. Box: N-7108 (tel: 327-8001; tx: 20116).

Nassau Beach, Cable Beach, P.O. Box: N-7756 (tel: 77-711; tx: 20117; fax: 3277615).

Pilot House, East Bay Street, P.O. Box: N-4941 (tel: 322-8431; tx: 20472).

Royal Bahamian Meridien, P.O. Box: N-10422 (tel: 327-6400; tx: 20317; fax: 6648602).

South Ocean Beach, South West Road, P.O. Box: N-8191 (tel: 326-4385; tx: 20296).

Wyndham Ambassador Beach, West Bay Street, P.O. Box: N-3026 (tel: 327-8231; tx: 38220204).

Business advice and information
Bahamas Agricultural and Industrial Corp, PO Box N-3217, Nassau (tel: 322-3740/3).

Bahamas Chamber of Commerce, PO Box 665, Shirley/Collins Avenue, Nassau (tel: 322-2145; tx: 179).

Bahamas Development Bank, Adderley Building, Bay Street/Rawson Square, PO Box N-3034, Nassau (tel: 322-8721/2).

Bahamas Employers Confederation, PO Box N-166, Nassau (tel: 322-4471).

Bahamas Information Services, Nassau Court, PO Box N-8172.

Central Bank of the Bahamas, PO Box N4864, Nassau (tel: 322-2193; tx: 20115).

Freeport Commercial and Industrial Ltd, PO Box F-2666, Freeport, Grand Bahama (tel: 352-6711).

Hotel Corporation of the Bahamas, PO Box N-9520, Nassau (tel: 327-8395).

Ministry of Tourism, PO Box N-3701, Nassau (tel: 322-7501/4, 322-8634/7).

Bahrain

Nicholas Childs

Among Bahrainis, the wave of optimism which initially greeted the Gulf War cease-fire in August 1988 was undoubtedly dampened to some extent by the lengthy stalemate in the UN-sponsored peace talks between Iraq and Iran, the continued unpredictability of Iranian politics in the post-Khomeini era, and the fact that the anticipated post-war reconstruction boom was slow to materialise. There was still a degree of confidence that the fighting between the Iraqis and Iranians had actually come to an end, that Tehran was embarked on a new course of moderation, and that a strong upswing in Gulf business activity would eventually come. And Bahrain's economy, in particular, with its strong banking sector and what, by regional standards, is a relatively diversified industrial base, still stands to be a major beneficiary of any such rejuvenation. At the same time, significant economic challenges lie ahead, not least the fact that Bahrain faces increasingly stiff competition in the future, both regionally and internationally, in its traditional spheres of expertise.

A changing political scene

On the surface, the tiny island of Bahrain – actually two main islands, and over 30 smaller ones – has appeared extraordinarily stable ever since it regained full independence from Britain in 1971.

Following the 1979 Islamic revolution in Tehran and the start of the Gulf War, Bahrain's staunchly pro-Western, pro-Iraqi policies have caused some friction between the island and Iran. But, to complicate the picture, there is a particular history of ancient entanglements between the two countries.

In 1783, the Arab rulers of Zubarah, ended nearly two centuries of Persian domination of Bahrain. Their descendants still run the island today.

Following the toppling of the Pahlari regime, the new Islamic republic also revived old claims to sovereignty over Bahrain. Throughout the 1980's the Bahraini authorities were vigilant in detecting any Iranian inspired challenges to their authority.

Yet, in spite of this, at the end of the decade the government was stable. The authorities clearly operate a relatively efficient security service. The ruling family, led by the Emir, Sheikh Isa Bin Sulman al-Khalifa, also remains popular. The al-Khalifas have adopted relatively benevolent policies towards their inhabitants, and have remained quite close to the community. And Bahrainis still enjoy what, by regional standards, is a markedly liberal society.

But Bahrainis still have more reason than most to view anxiously the evolution of the government of the new Iranian president, Hashemi Rafsanjani. Since the Gulf War cease-fire, Bahrain has gradually increased its contacts with Iran again. But the process looks likely to be a slow one. Indeed, the Bahrainis themselves hosted the first post-war Gulf Co-Operation Council (GCC) summit in December 1988, at which the member states agreed to take a cautious approach to the issue of relations with Iran. They linked the full normalisation of ties to a lasting settlement in the Gulf War and a commitment by Tehran not to interfere in their internal affairs.

The defence budget increased steadily in the late 1980s. The 1988 figure was BD66.5m ($176.4m) which was actually BD10m ($26.5m) higher than originally, and well ahead of the 1987 total of BD51.5m ($136.6m). And the planned defence budgets for 1989-90 were BD69.4m ($184m) and BD72.9m ($193.4m) respectively.

But in modern times Bahrain, because of its small size, has generally relied on outsiders for protection. Between 1861 and 1971, that protection came from Britain, and links between the two countries remain strong. But now, the ultimate guarantor of Bahraini security is the United States.

Indeed, Bahrain probably has a stronger strategic relationship with Washington than any other Gulf state except Saudi Arabia, and possibly Oman. When the British withdrew their permanent presence from the Gulf in 1971, the US Navy took over the former Royal Navy installations in Bahrain as their main facility in the region, and it provided a lynchpin for American tanker escort operations in the Gulf in 1987-88.

In 1985, the US Army Corps of Engineers also began work on a large new military air base in the south of the island. It was due to be completed in 1989, at a cost of about $90m, and it undoubtedly figures prominently in US contingency plans for reinforcing its presence in the Gulf in an emergency.

At least the closeness of the relationship between Manama and Washington gave the Bahrainis access to pieces of advanced US military technology which were denied to some of their neighbours by the pro-Israeli lobby in the US Congress. In particular, Bahrain bought a squadron of F-16 fighters in 1988 to bolster its fledgling air force.

At the regional level, Bahrain has close cooperation with Saudi Arabia. Indeed, the causeway between Bahrain and Saudi Arabia's Eastern Province, opened in 1986, was probably intended as much as a political signal to Tehran of this fact as it was meant to facilitate economic links between the island and Saudi Arabia. And, on most aspects of regional and international policy, Manama closely shadows Riyadh.

But one area in which Bahrain's policy was actually ahead of Saudi Arabia's was that of relations with the communist bloc. Like Riyadh, Manama had been increasing its contacts with the Soviet Union. Then in April 1988 it also became the latest Gulf Arab state to establish full diplomatic relations with China. Saudi Arabia was now the only GCC state not to have full formal links with Peking. In part, this trend was a sign of a maturing foreign policy throughout the Gulf. At the same time, while these closer ties were unlikely to threaten the basic relationships between the GCC states and the West, they were also partly a recognition that major trading opportunities exist in the Soviet Union and China, particularly now that both countries were embarked on economic restructuring programmes.

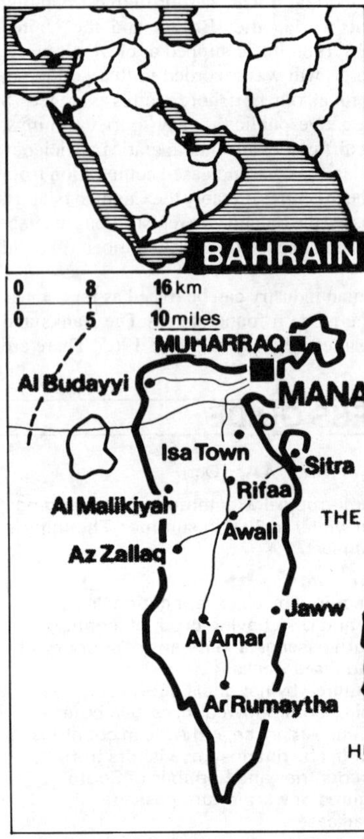

Slow growth and deficits

On the economic front, early forecasts for 1989 predicted annual growth only slightly higher than that achieved in 1988. Moreover, the government's two-year budget for 1989-90, published at the beginning of 1989, appeared to confirm the suspicion that the country's economy still faced a number of medium-term challenges.

First, the authorities predicted an increase in the planned government budget deficit, to BD100m ($256m) in each year. Total government spending for 1989 was projected at BD530m ($1,406m), and for 1990 BD40m ($1,432m).

One problem for the Bahraini authorities has been the continued sluggishness of the oil market. The island's own oil deposits have always been modest: its own oil field only produces about 42,000 barrels a day. But Saudi Arabia supplies about 200,000 b/d to

KEY FACTS

Official title: State of Bahrain

Head of state: Shaikh Isa bin Sulman Al-Khalifa

Head of government: HH Shaikh Khalifa bin Sulman Al-Khalifa

Capital: Manama

Official Languages: Arabic

Currency: Bahraini Dinar (BD) = 1,000 fils

Exchange rate: BD0.38 per $ (Nov 1989)

Area: 676 sq km (35 islands)

Population: 457,600 (1989)

GDP per capita: $7,040* (1987)

GDP real growth: –2.8% (1987)

Inflation: 0.3% (1988)

Oil reserves: 140m* barrels (1988)

Trade balance: –$12.8m (1988)

* estimated figure

feed the Sitra refinery. And oil revenue remains the government's main source of income.

Another difficulty has been that continued high public spending levels will exacerbate Bahrain's mounting trade deficit problem. Of course, ultimately, the hope is that the development spending will allow import substitution by broadening the island's industrial base.

Finance at the crossroads?

The banking sector has long been a mainstay of the Bahraini economy. But, for most of the 1980s, it was depressed by the contraction of local business and the need to provision for substantial bad debts, particularly in Latin America.

In the mid-1970s, Bahrain established itself as a major centre for offshore banking units (OBUs) attracted by the prospects of business in the then oil-rich Gulf. The OBUs make a substantial contribution to the domestic economy. And, after years of contraction, in the late 1980s they were showing some sustained signs of recovery.

Seven of the top eight OBUs showed profits in 1988, compared to just two in the previous year. Indeed, the largest, the Arab Banking Corporation, showed record net profits of $142m. The total assets held by the OBUs were also up for the second successive year, to stand at the end of 1988 at $68.1bn.

The results for the local commercial banks, however, remained mixed. The National Bank of Bahrain was among the leaders to record profits substantially higher than in 1987. But the total assets of the commercial banks declined in 1988 by 4.2 per cent, and overall loans and advances were also down.

It seemed that the financial market would be boosted by the government's need to finance its budget deficit, chiefly by issuing treasury bills. The industrial expansion planned by the government should also provide new lending opportunities. And there was still the promise of substantial reconstruction business, and a general revival of Gulf business activity.

Nevertheless, rationalisation of the banking sector seemed inevitable. Certainly, Bahrain will not be able to take its traditional pre-eminence in the financial field for granted. As the relative attraction of the Gulf has declined with the slump in the oil market, so Bahrain will have to compete more on level terms with other offshore banking centres like Hong Kong. And the post-war era is also likely to be marked by increased regional competition, since Bahrain's neighbours are also developing their own financial systems.

Perhaps the most significant development for Bahrain's economy in 1989 was that, after much delay, its stock exchange finally opened in June. Initially, a total of just 29 local and joint venture companies were registered, with a total capitalisation of approximately $1,600m.

The Bahrainis clearly have ambitions to host the main regional stock trading centre, dealing in shares of GCC and even other foreign companies. Certainly the reservoir of untapped Gulf investment funds is very great. And Bahrain still has the advantage over potential rivals like Kuwait of its financial track record and a pool of resources at hand among its OBUs.

In the light of the opening of the stock exchange, the Bahraini government was also apparently planning a major privatisation programme. That, at least, could provide funds to help alleviate its budget problems, although the authorities are likely to want to retain their stakes in the island's major industrial concerns.

Diversification: a watchword

Even so, the government made plain its ambitious plans to diversify the economy further, and at the same time attract investors from the main industrialised countries – and especially from the United States, Britain and West Germany – to set up new joint ventures.

KEY INDICATORS

	Unit	1984	1985	1986	1987	1988
Population	'000	400	420	410	420	480
Gross domestic product (GDP)	BD m	1,877.0	1,746.0	1,606.3	1,638.1*	–
GDP per capita	BD	4,693	4,156	3,736	3,757*	–
Time deposit rate	%	7.0	6.7	5.6	5.0	5.5
Inflation rate	%	0.3	–2.7	–2.3	–1.0	0.3
Consumer prices	1985=100	102.7	100.0	97.7	96.0	96.3
Industrial production						
Crude petroleum	1985=100	88.5	100.0	89.0	88.4	89.5
Refined petroleum	1985=100	108.3	100.0	132.0	131.0	129.5
Volume of exports						
Refined petroleum	1985=100	111.8	100.0	138.5	137.3	135.5
Aluminium	1985=100	117.2	100.0	143.6	96.7	74.9
Exports	$m	3,092	2,787	2,344	2,384	2,410
Petroleum	$m	2,712.3	2,442.3	1,970.2	1,967.0	1,830.3
Aluminium	$m	223.4	147.6	209.3	162.2	189.9
Imports	$m	3,480	3,107	2,405	2,714*	2,594
Crude petroleum	$m	1,644.7	1,423.7	998.4	1,249.2	1,061.4
Balance of payments	$m	72	100	38	–12.8	–
Current account	$m	205	52	–66	–153	–
Exchange rate	$ per BD	2.66	2.66	2.66	2.66	2.66

* estimated figure

Leading this particular drive was the head of the industrial development centre (IDC) of the ministry of trade and development, Sheikh Homoud Khalifa al-Khalifa. The next aim, he said, was for Bahrain to replace 30 per cent of its imports in the next five years. And he pointed out that the government itself has allocated $2.6bn to industrial projects over the next five years.

IDC officials pointed to a number of incentives for investment in Bahrain. These include a 70 per cent government contribution to the cost of feasibility studies for proposed projects, the possibility of 100 per cent foreign ownership, subsidised energy, the duty-free import of the necessary plant and raw materials, personal and corporate tax exemption, no restrictions on the repatriation of income, and duty-free access to other GCC markets. The priority areas for the government were apparently aluminium and related industries, petrochemicals, engineering, electronics, and healthcare.

A desire to create significant numbers of new jobs was also behind the campaign. Bahrain, like the other Gulf States, faces a mounting threat of unemployment. According to the most recent figures from the country's central statistics bureau, 34 per cent of the population is at present under 14 years of age. Hence, the local labour force is expected to double by the end of the century.

Better industrial performance

As it is, Bahrain's industrial base is already very broad for a country of its size. In the hydrocarbons sector, the Sitra refinery continued to operate close to its 250,000 b/d capacity. But it is one of the oldest refineries in the Gulf, and the Bahrain National Petroleum Company (Bapco), which operates it, plans a major modernisation programme to increase its efficiency, particularly in the area of high-value refined products. The total cost of the Bapco programme is expected to be $640m.

Meanwhile, in 1989 work had started on the $74.5m expansion of the Bahrain National Gas Company's (Banagas') gas liquefication plant. The contractor is the Tokyo-based JGC Corporation. It is intended to increase the capacity of the Banagas plant from 110m cu ft/d to 170 cu ft/d.

Much of Banagas' production is taken up fuelling aluminium Bahrain's (Alba's) smelter. With strong world demand for aluminium, much is clearly expected from Alba in the future. And 1988 was a record year, with total production reaching nearly 183,000 tonnes.

Work was under way in mid-1989 to expand Alba's capacity to 205,000 tonnes a year. But a much more ambitious expansion is also planned for the longer term. This would raise capacity to around 400,000 tonnes a year, at a total cost of over $1,000m.

Part of this new scheme will be an 800mW power station, which will probably be linked eventually to the national grid.

Elsewhere, the island's iron pelletising plant resumed production in November 1988, having been shut down in 1986 when, as the Arab Iron and Steel Company, it ran into serious financial difficulties. It now operates as the Gulf Industrial Investment Company (GIIC), a subsidiary of the Kuwait Petroleum Corporation.

The Arab Shipbuilding and Repair Yard (ASRY), which is owned by the Organisation of Arab Petroleum Exporting Countries, recorded its first operating profit in 1988 since it was set up in 1977. The first four months of 1989 also showed a profit. This was despite the end of the 'tanker war', which provided some lucrative business. However, ASRY abandoned plans to buy a floating drydock to complement its existing 450,000-tonne capacity graving dock. Instead it planned to invite bids for the design of a second grading dock to cater for small vessels, which the company is not equipped to handle efficiently. There appeared to be a gap in the market here following the demise at the beginning of 1988 of the Bahrain Slipway Company, which had specialised in maintaining small craft like tugs and oil rig support ships.

Unfortunately for the Bahrainis, the trend towards greater industrialisation and a growing population threw up two important resource problems in the late 1980s. One was the issue of future power supply and the other that of water. Both these difficulties are commmon throughout the GCC. Unfortunately, tentative plans for a GCC-wide power grid are proceeding very slowly. At least, in Bahrain, this problem will be alleviated slightly when the new ASRY power station is complete.

But the issue of water supplies may be more acute. The authorities had already restricted water consumption to 60m g/d, but underground water supplies were still being depleted at a rapid rate. At least the government was planning two new 15m g/d reverse osmosis plants to try to improve the situation.

Finally, in terms of generating new trade and business, the government planned a new port and 'free zone'. Despite dredging, the new generation of large container ships will not be able to negotiate the Khawr Qulaiyah, the inner channel of Mina Sulman. And international links would be improved if the ambitious plans of the national carrier, Gulf Air – actually jointly owned by Bahrain, Qatar, Abu Dhabi and Oman – came to fruition. Having won finance for an initial order of four new Boeing 767-300 airliners worth $260m, Gulf Air intended to buy a further 18 aircraft, costing about $1bn, with options on another 18. The airline made net profits of BD2.58m ($6.8m) in 1988, and expected to treble this in 1989. But financing its expansion plans might still be a problem.

Outlook

There was still an air of relative optimism in Bahrain at the end of the 1980s, as in the rest of the Gulf. The stability of the government hardly seemed in question. And better regional relationships, especially with Iran, and an improving business climate, were still widely anticipated, albeit that progress on these fronts was likely to be at a steadier pace than was originally expected in August 1988. Economic prospects continued to rely heavily on oil, but industrial expansion seemed to be ahead in all the right areas.

At the start of the new decade, perhaps the biggest danger is that political stability could foster stagnation. The ruling al-Khalifa family are fortunate to be assisted by several able ministers. But the new, harsher competitive atmosphere which is likely to pervade the post-war Gulf will not allow any complacency where development plans are concerned.

COUNTRY PROFILE

Political system: Bahrain is a constitutional state headed by the Amir and administered by a cabinet of ministers formed at the time of independence from Britain in 1971 to succeed the Council of State.

Population: About a quarter of the population of 457,600 (1989) are foreigners, mainly from other Middle Eastern countries and the Indian subcontinent.

Main cities/towns: Manama (capital, population 150,000), Muharraq (65,000), Jidhafs (33,000), East and West Rifa'a (30,000), Isa Town (40,000), Hidd (7,000), Budaiya (2,471), Awali (1,800). Hamad, a new town, was planned to accommodate up to 60,000 people.

Language: The official language is Arabic, but English is widely spoken. Farsi (Persian), Hindi and Urdu also frequently used.

Media
Dailies: Akhbar Al Khaleej (Arabic) (circulation 22,000) and the *Gulf Daily News* (English) (11,500).
Weeklies: Several Arabic weeklies and magazines.
Business: Falcon Publishing Group, and Al-Hilal Publishing and Marketing Group each produce several magazines and directories on business, industry, finance, banking and travel. International newspapers carry regular supplements on Bahrain.
Radio: The state-owned Bahrain Broadcasting transmits two services in Arabic and two in English. Otherwise, it is possible to receive Capital Radio, the Voice of America, BBC World Service and services from Oman, Qatar, Abu Dhabi and the popular, music-oriented Aramco service from Dhahran.
Television: Bahrain Channel 4 (colour) in Arabic broadcasts eight hours daily: Bahrain Channel 55 (colour) in English broadcasts nightly. Other TV stations received in Bahrain include RAK (Ras Al-Khaimah), Dubai 10, Dubai 33, Dubai 41, Kuwait 8, Kuwait 10, Qatar 9, Qatar 11, Qatar 37, Oman 8, Abu Dhabi 5, Abu Dhabi 31, Saudi 6, Saudi 29 and Aramco 3 (English), depending on conditions. There were 108,000 sets* in 1988.

Domestic economy: Bahrain's hydrocarbon reserves are small and crude oil production is declining; it is expected to contribute less than half of GDP by 1998. Agricultural and fishing sectors are modest. The oil slump of 1984-86 severely depressed the economy, necessitating cuts in government spending, and led to the curtailment of major construction projects, but there was some stabilisation in the late 1980s. Foreign banks were attracted during the late 1970s and early l980s in large numbers by favourable offshore banking policies, and new industries (ship repair, aluminium, petroleum services) were also encouraged in an effort to limit dependence on oil revenues. Bahrain benefited from aid and investment from other Gulf states, especially Saudi Arabia, and with the opening of the causeway link between Bahrain and Saudi Arabia, further Saudi investment was made.

Employment: Foreigners comprise about 58 per cent of the workforce, predominantly in the construction sector. The unemployment rate is an estimated 10 per cent. Agricultural workers represent about 25 per cent of the rural population.

External trade
Exports: Main exports are mineral fuels and lubricants (86.8 per cent of 1985 total), manufactures classified by material (9.1%), machinery and transport equipment (2.8%), other manufactured goods (0.5%), and chemicals (0.2%). Main destinations are Saudi Arabia (24.9% of 1985 total), Japan (15.3%), United Arab Emirates (14.7%), China PR (7.9%), USA (7.9%), India (7.3%), Kuwait (3.7%), Hong Kong (3.1%), Taiwan (2.6%), German FR (2.3%) and UK (2.0%).
Imports: Main imports are mineral fuels and lubricants (46.2 per cent of 1985 total), machinery and transport equipment (20.9%), manufactures classified by material (11.3%), other manufactures (6.8%), food and live animals (6.8%), chemicals (5.1%) beverages and tobacco (1.3%), and non-oil crude materials (1.2%). Main sources are USA (20.5% of 1986 total), UK (20.3%), Japan (12.8%), German FR (13.8%), Italy (6.0%), Australia, Netherlands, France, India and United Arab Emirates.

Agriculture: Development of agriculture is limited by labour shortages, lack of water and salinity of the soil, and the sector provides only 1 per cent of GDP. The major crop is alfalfa for animal fodder, amd farmers also produce dates, figs, mangoes, pomegranates, melons, papayas, canteloupes, water turnips, potatoes and tomatoes for the local market. The six-year agricultural plan for 1982-88 boosted agricultural production from 6 per cent to 16 per cent of domestic requirements, and achieved self-sufficiency in egg production, 50 per cent of requirements in milk production, and 75 per cent of vegetable requirements. The plan, and its successor, emphasised drainage to reduce salinity, improvement of the soil and new irrigation and cultivation techniques; experiments with hydroponics are also under way. The land tenure system, under which over 60 per cent of cultivable lands is held on three-year leases, discourages the stability needed for development.

Livestock: Lack of grazing inhibits livestock production. One large dairy has annual milk production of 500,000 litres. Small dairy farmers, responsible for 15 per cent of production, have established a co-operative and constructed a milk pasteurising plant. A poultry processing plant and feed mill instigated by the government boosted poultry production and helped reduce import requirements.

Industry: Bahrain's most prominent non-oil industry is the ALBA aluminium plant, which supplies various downstream manufacturing plants as well as the GARMCO rolling mill (a joint venture between six Gulf states, opened in 1986 and intended to export to the whole region). Export-oriented light and medium industries have been attracted to 'free' industrial zones established at Mina Sulman, Ma'amir, the North Refinery, Abu Gazal and North Sitra, including plastics, paper, steel-wool and wire-mesh producers, marine service industries, block factories, aluminium extrusion and assembly and asphalt plants, cable manufacturing, prefabricated building and furniture. Major petrochemicals activity. Total refinery capacity is estimated at 250,000 b/d.

Hydrocarbons: The oil sector remains the mainstay of the economy, providing nearly 60 per cent of government revenues and 70 per cent of export revenues (1987/88). The country has proven reserves of 140m barrels in its onshore field at Jebel Al Dukhan (1988), but these will dry up by 1996 at current levels of extraction. Output, at 44,000 b/d in 1986, was well down on the 76,600 b/d of 1970 but still higher than the levels of the mid-1980s. Nevertheless, oil export values were estimated to have fallen to $1.97bn in 1987, from $2.6bn in 1984. Consequently, attention is shifting toward importing crude oil from Saudi Arabia, which already provides 83 per cent of oil requirements, and which allows Bahrain a free hand in exploiting the Saudi field which they jointly control. There is at least a 50-year supply ofnatural gas, estimated at 209bn cu metres, and whose exploitation offers opportunities for liquefaction plants as well as supplying fuel for industry. Non-associated gas from the Khuff zone is re-injected into the oil field to maximise petroleum production.

Shipping: Bahrain's ship-repair industry is largely based on ASRY, the ship-repair operation sited off Muharraq, which is jointly owned by the seven OAPEC countries and was purpose-built for servicing VLCCs and ULCCs up to 500,000 dwt. Other services in the area include running repairs, overhauls, casualty and damage repairs on tankers, fishing boats and other vessels and, at Mina Sulman, custom service for offshore oil exploration vessels. ASRY was operating at 90 per cent of capacity in 1987, reduced its deficit in 1987 and stood to gain from the ending of the Gulf War in 1988.

Banking: Bahrain's banking centre is located in Manama. Banking rapidly became a major source of national income as the number of offshore banking units increased rapidly from 1976. The boom in offshore banking was slowed after 1984 by falling oil revenues and by the Saudi Arabian Government's actions to restrain offshore trading in the Riyal. By 1988 several of the offshore companies had pulled out.

Stock exchange: The Bahrain Stock Exchange opened in 1989, with an initial registration of 29 companies, with a total capital of about $1,600m.

Legal system: Traditional Majlis, an open court at which anyone may present a petition to the Amir to deal with as he sees fit, takes place in his palace for several hours each week, and co-exists with other systems of government.

BUSINESS GUIDE

Membership of international organisations: Arab League, FAO, G-77, GATT, Gulf Co-operation Council (GCC), ICAO, ILO, IMF, IMO, INTERPOL, Islamic Conference Organisation, Islamic Development Bank, ITU, NAM, OAPEC, UN, UNESCO, UPU, WHO, World Bank.

Time: GMT +3 hrs.

Climate: Exceptionally hot and humid in summer and mild in winter. Best time to visit: from Oct to Apr. Humidity high Jul-Sep, with temperatures to 44°C (average 35°C).

Entry requirements
Passport: Required by all except certain seamen.
Visa: GCC and British nationals do not require entry visas. For all others, entry visas must be obtained in countries where a Bahraini consulate is available and extended locally. Travellers with onward reservations may obtain a 72-hr (extendable) transit visa on entry, although such visas are not mandatory for workers joining an oil rig by charter or sea fare, who must instead supply a letter from their employer. Journalists must make prior arrangements with the Ministry of Information.
Prohibited entry: Israeli nationals or anyone holding a passport with an Israeli visa.
Currency: Any currency may be freely imported and exported.
Customs: Personal effects duty free. Pornographic and obscene literature and pictures, arms and ammunition, cultured or undrilled pearls are all prohibited.
Income tax: There is no income tax in Bahrain.

Health precautions
Mandatory: Certificate of vaccination against yellow fever and cholera, if travelling from infected area.
Advisable: TAB vaccination.

Representation overseas: Amman, Baghdad, Beirut, Bombay (consulate), Cairo, Geneva (permanent representative), Jeddah, Karachi (consulate), Kuwait, London, New York (UN), Montreal (consulate), Nicosia, Paris, Tehran, Tunis, Washington DC.

Air access
National airline: Gulf Air (jointly with Qatar, Oman and the United Arab Emirates), PO Box 138, Bahrain.
Other airlines: Air India, Air Lanka, Biman Bangladesh, British Airways, Cathay Pacific, Cyprus Airways, Egyptair, Iraqi Airways, KLM, Korean, Kuwait Airways, Lufthansa, MEA, Philippine, PIA, Qantas, Royal Jordanian, Saudia, Syrian Arab, THY Turkish, TWA, UTA, Yemenia.
International airport: Bahrain International, Muharraq (Code: BAH), 6 km from city, with bar restaurant, buffet, bank, shops, hotel reservations. Aiport charge of BD3 payable on leaving the country.

Surface access
Main ports: Mina Sulman (Manama, containers) and Sitra (main oil terminal).
Overland access: The Saudi-Bahrain Causeway has opened up links between Bahrain, Saudi Arabia and Qatar. Bus and taxi services available to cross the causeway. Normal Saudi Arabian visa regulations apply.

Hotels: First-class hotel capacity for 175,000 average-stay visitors. A 12 per cent service charge is usual. Major hotels and most restaurants are licensed. All major credit cards accepted.

Credit cards: All major credit cards are accepted by major hotels and trading establishments in Bahrain.

Car hire: Insurance is compulsory and international driving licences must be validated at the Ministry of Interior Traffic Headquarters (near Isa Town) before use in Bahrain. Car hire firms are listed in the local telephone directory, and it is generally recommended to compare prices.

City transport
Taxis: Taxis (with orange side wings and black-on-yellow number plates) are plentiful and fares are regulated. Agree a fare with the driver before booking. Taking a cab from a hotel taxi rank generally costs double the usual rate, and recommended fares from the airport are displayed outside the arrivals terminal. Fares increase after midnight. Mini-cabs or 'pick-ups' can be hailed from any bus stop, and their standard fare is substantially lower than the taxi fare.

National transport
Road: There are good tarmac roads between centres, and six-lane highways from a ring road by-pass system for Manama and Muharraq.
Buses: The comprehensive public bus service covers urban and rural areas. It is inexpensive, but often unreliable. The bus fleet was reduced in response to low utilisation rates. Fleet of 108 buses undergoing gradual replacement by automatic transmission vehicles and new terminal at Isa Town.
Water: Dhow trips are arranged most weekends to sand bars and nearby islands from the old wharf (Mina Manama) on King Faisal Road. Boat trips to neighbouring islands are frequently arranged on Friday and publicised in the local press.

Public holidays
Fixed dates: 1 Jan, 16 Dec (National Day).
Variable dates: Eid al-Fitr, Eid al-Adha, Al Hijra, Ashura, Prophet's Birthday.

Working hours: Friday is weekly holiday, when government offices, businesses and most shops are closed. Many close early on Thursday. Regular hours subject to change during month of Ramadan. Some banks and businesses close on Saturday.
Business: (Sat-Thu) 0730 – 1200, 1430 – 1800.
Government: (Sat-Thu) 0700 – 1300.
Banking: (Sat-Wed) 0800 – 1200 and 1600 – 1800; (Thu) 0800 – 1100.
Shops: (Sat-Thu) 0800 – 1200, 1530 – 1830; some are open for a few hours on Fri in the Souk.

Social customs: Women should dress conservatively in town – sundresses, shorts and tight jeans are liable to attract unwelcome attention and offend against Islamic codes. Pork should not be eaten in the presence of Muslims. It is polite to avoid eating, drinking or smoking before Muslims during daylight hours in the month of Ramadan (it is illegal to do so in public).

Telecommunications: Excellent cable and satellite services using latest digital exchange technology. Three earth stations, two working to the Intelsat satellite network and the third to the Arab satellite (Arabsat), together with a channel submarine cableto the UAE and Qatar, form Bahrain's International outlets.
Telephone: Dialling code for Bahrain: IDD access code + 973, followed by subscriber's six-digit number (no area code required). There are about 140,000 telephones in use. IDD access code for international calls from Bahrain is 00 followed by the country code etc.
Telex: Well-established telex facilities. Public telex booths at Bahrain International Communications (BATELCO), Main Office, Mercury House, Manama and sub-office in Muharraq. Telex code 490, country code BN.
Telegram: Arabic and English telegrams can be sent.
Other: BUREAUFAX international facsimile transmission service (to 23 countries); IDAS (International Data-base Access Service); Reuters Monitor and Dealing Service; full telephone services for private cars from Mobile Automatic Telephone Service (MATS).

Postal service: Most business mail delivered to private boxes rented from main Post Offices.

Banking: Central Banking Authority is Bahrain Monetary Agency (BMA).
Main banking centre: Diplomatic area, Manama.
Commercial banks: Al Ahli Commercial Bank, Algemene Bank Nederland NV, Arab Bank Ltd, Bahraini Saudi Bank, Bank of Bahrain and Kuwait, Bank Melli Iran, Bank Saderat Iran, Banque du Caire, Banque Paribas, British Bank of the Middle East, Chase Manhattan Bank NA, Citibank NA, Grindlays Bahrain Bank BSC, Habib Bank Ltd, National Bank of Abu Dhabi, National Bank of Bahrain, Rafidain Bank, Standard Chartered Bank, United Bank Ltd.

Trade fairs: As well as the major hotels, exhibitions are held at the Bahrain Exhibition Centre in Manama. A new Bahrain Exhibition Centre is under construction in Manama and due to open in 1991, with more than 7,000 square metres of airconditioned exhibition space, as well as conference facilities. Exhibitions are organised by Arabian Exhibition Management (PO Box 20200, Manama; tel: 250033; fax: 242381).

Electricity supply: 230 V AC everywhere except Awali, which has 110 V; various types of plug fitting, normally three-pin flat.

Weights and measures: Metric system (local units also in use).

Representation in capital: China PR, Denmark, France, India, Iran, Iraq, Japan, Jordan, South Korea, Kuwait, Netherlands, Oman, Pakistan, Saudi Arabia, Sweden, Switzerland, UK, USA.

Hotels: Manama
Al Jazira, Al Khalifa Road, P.O. Box: 5898 (tel: 258810; tx: 8999 jazotl).

Aradous, Wali Al Ahed Avenue, P.O. Box: 5878 (tel: 241011; tx: 8900 aradous).

Bristol, near Shaikh Issa Road, P.O. Box: 832 (tel: 258989; tx: 8504 bristol).

Delmon, P.O. Box: 26 (tel: 234000; tx: 8224; fax: 246236).

Diplomat, Building no.59, Rd no. 1701, P.O. Box: 5243 (tel: 531666; tx: 9555; fax: 530843).

Gulf, Gudaibiya Bay, P.O. Box: 580 (tel: 233000; tx: 8241 gulfotel; fax: 713040).

Hilton, Palace Avenue, Diplomatic Area, P.O. Box: 1090 (tel: 535000; tx: 8288; fax: 533097).

Holiday Inn, Muharraq Causeway, P.O. Box: 5831 (tel: 531122; tx: 9000/1 holinn; fax: 530154).

Middle East, Al Zubara Road, Gudaibiya, P.O. Box: 838 (tel: 254733; tx: 8296; fax: 233910).

Oasis (formerly Grundy), Shk. Mubarak Building, Government Road, P.O. Box: 5908 (tel: 259979; tx: 8970 grundy).

Oriental Palace, Tijjar Road, P.O. Box: 5546 (tel: 233331; tx: 9392 arabtl).

Palace Inn, Juffair, P.O. Box: 1028 (tel: 714261; tx: 9126 pallin).

Ramada, Bani Otbah Avenue, Gudaibiya, P.O. Box: 5750 (tel: 714921; tx: 8855; fax: 742809).

Regency Inter-Continental, King Faisal Road, P.O. Box: 777 (tel: 231777; tx: 9400 regency; fax: (973) 240028).

Royal, Old Palace Road, Gudaibiya, P.O. Box: 154 (tel: 714901; tx: 8777 royal).

Sheraton, 5 Palace Ave, P.O. Box: 30 (tel: 533533; tx: 9440 sherat; fax: 255718).

Tylos, Government Road, P.O. Box: 1086 (tel: 252600; tx: 7482 tylhtl; fax: 252611).

Vendome, Gudaibiya Avenue, Gudaibiya, P.O. Box: 710 (tel: 257777; tx: 8888 vendom).

BUSINESS DIRECTORY

Useful telephone numbers
Emergency services: 999

Directory enquiries: 181

International enquiries: 191

International bookings: 151

Operator: 100

Car hire
Avis: airport (tel: 320242; tx: 8683); Hotel Delmon (tel: 234423; tx: 8683).

Budget: central reservations (tel: 534100; tx: 8867).

Europcar: PO Box 800 (tel: 682999; tx: 8489); airport (tel: 323999); Gulf Hotel (tel: 713999).

Taxis
Airport taxis (tel: 320982).

In Manama (tel: 253349).

Muharraq airport (tel: 322727).

Speedy Motor Services, radio-controlled cabs (tel: 682999; tx 8489).

Business advice and information
Bahrain Chamber of Commerce & Industry, PO Box 248, Manama (tel: 250369, 233913; tx: 8691 ghurfa; fax: 241294).

Bahrain National Gas Co, PO Box 29099, Manama (tel: 756222; tx: 9317 bangas).

Bahrain Petroleum Company BSC, Awali (tel: 754444; tx: 8214).

Bahrain Television Company, PO Box 1001, Manama (tel: 681811).

Central Statistics Organisation, PO Box 5835, Manama (tx: 8835 cso).

Ministry of Commerce & Agriculture, PO Box 5479, Diplomatic Area, Manama (tel: 531531; tx: 9171 tejara).

Ministry of Development & Industry, PO Box 235, King Faisal Highway, Manama (tel: 291511; tx: 8344 tanuya).

Ministry of Finance & National Economy, PO Box 333, Government House, Government Road, Manama (tel: 262400; tx: 8933 maleya).

Ministry of Information, PO Box 253, Ministry of Information Complex, Isa Town (tel: 681811, 683988; tx: 8399 inform).

Banking
Bahrain Monetary Agency (central banking authority), Diplomatic Area, PO Box 27, Manama (tel: 535535; tx: 9144 bahmon bn; fax: 534170).

Arab Bank Ltd, PO Box 395, Yousif Engineer Bldg, Govt Ave, Manama (tel: 255988; tx: 8232 arabnk bn; fax: 231640).

Bahraini Saudi Bank, PO Box 1159, Govt Ave, Manama (tel: 263111; tx: 7010 bsb bn; fax: 263048).

Bank of Bahrain & Kuwait, PO Box 597, Govt Ave, Manama (tel: 253388; tx: 8919 bakubk bn; fax: 275785).

British Bank of the Middle East, PO Box 57, BBME Bldg, Al Khalifa Ave, Manama (tel: 242555; tx: 8230 bbme bn; fax: 256822).

National Bank of Bahrain (BSC), PO Box 106, Govt Ave, Manama (tel: 258800; tx: 8242 nbofx bn; fax: 263876).

United Bank Ltd, PO Box 546, Sh Khalifa Bldg, Govt Ave, Manama (tel: 254032; tx: 8247 pakbnk bn; fax: 234143).

Bangladesh

Kevin Rafferty

Moudud Ahmed, vice president of Bangladesh, has set an ambitious target for his country – he wants it to be among the next wave of newly industrialising countries (NICs). 'The NICs today are Hong Kong, South Korea, Singapore and Taiwan. Indonesia, Malaysia and Thailand are among the countries seeking to join them. We will be in the next wave', he asserts confidently. It is a difficult leap to make from the Bangladesh of the late 1980s to his vision. It is a land that continues to battle against winds and tides, rains, floods and drought alike, and does this with limited natural and human resources. Just to make the task more difficult, the country's inheritance includes politicians who squabble continually over the limited spoils of power.

Physical and historical inheritance

A recent report from the International Monetary Fund gives a good outline of the fragile superstructure of Bangladesh. It says 'It is a small country with few natural resources, but a large population – one now approaching 110m and growing at about 2.5 per cent a year. Poverty is widespread; income per head averages about $180, among the lowest in the world. Infant mortality is high, life expectancy short and basic health care rudimentary. The country is highly vulnerable to droughts, floods and cyclones. Agriculture is the mainstay of Bangaldesh's economy, accounting for about one half of gross domestic product. Much of the small industrial sector is agro-based and the export base is narrow ....'

The land that is today called Bangladesh lies in the flat delta of the great Ganga-Brahmaputra river system. There are just a few hilly wrinkles on the eastern edges of a land that is pancake flat, criss-crossed by giant rivers, roaring tributaries, meandering streams and smaller water courses everywhere. This water, almost everywhere in sight, is both the country's lifeblood and its constant nightmare. It helps to create one of the most fertile soils in the world, but it also makes the building of infrastructure difficult and expensive. The capital city, Dhaka, is surrounded on three sides by water, an asset in older days when rulers sought a defensible stronghold, but a handicap in modern times when communication is vital. The road journey from the capital to the main port of Chittagong less than 320 km away is still punctuated by two ferry crossings. But even the great volume of water does not give Bangladesh assured irrigation for its crops. In 1989, as so often in the past, Bangladesh's economy was hit both by drought and by floods.

The handicaps of the physical inheritance have been exacerbated by history. For more than a hundred years what is now Bangladesh was a territory lying at the forgotten or neglected end of someone else's domains. During the British raj in India, Bangladesh was the relaxed rural half of Bengal. The state of Bengal was split at Indian independence. The western part, including Calcutta and most of its industrial hinterland became West Bengal state of India; the eastern half became East Pakistan in a country ruled 1500 km away first from Karachi, then from the army garrison of Rawalpindi/Islamabad. East Pakistan got some industry, mainly jute mills, but the lion's share of the industrial investment was located in West Pakistan.

At Bangladesh's independence in 1971 per capita income was about $80. Since then Bangladesh has had a constant battle against the weather, the wretched human inheritance, the fragile political system and fickle politicians.

Above all, the country is handicapped by the large number of people crammed into a small area of land. Adding together all its land area, including the rivers and the fields that are regularly inundated with flood waters and therefore uninhabitable, Bangladesh has 143,998 sq km or less land than Great Britain, but its population is almost exactly double that of Britain's. It is true that city states like Hong Kong and Singapore have a greater density of population and have still prospered, but Hong Kong has only 5.8m people and Hong Kong's inhabitants enjoy a head start in terms of literacy, education and even basics like regular food and safe water. A glimpse at statistics put out by the World Bank and Asian Development Bank shows just how rock-bottom poor Bangladesh is. In Asia only war-torn Afghanistan and Cambodia score lower. Bangladesh's literacy rate is about 25 per cent. Barely half the population gets adequate calorie and protein intake every day. Vast numbers of Bangladeshis do not have access to safe water supplies.

KEY FACTS

Official title: People's Republic of Bangladesh

Head of state: President Hussain Mohammad Ershad

Capital: Dhaka

Official Languages: Bangla (English also spoken)

Currency: Taka (Tk = 100 poisha)

Exchange rate: Tk31.0 per $ (Nov 1989)

Area: 143,998 sq km

Population: 104.5m* (1988)

GDP real growth: 2.6% (1988)

GNP per capita: $170 (1988)

Labour force: 29.40m (1986)

Unemployment: 25% (1985)

Inflation: 9.4% (1988)

Trade balance: –$1,444m (1988)

Foreign debt: $9.56bn (1987)

* estimated figure

A glimpse of life in Dhaka or in the rural areas reveals just how difficult it is to make a living and to struggle out of poverty. There are people everywhere, yet the basic means of livelihood and transport might have come from a century or more ago. The country has two main crops, rice for food and jute, the main cash crop. Rice production has slowly climbed to over 16m tonnes, but if this were divided exactly equally among the rapidly growing population it would mean that each Bangladeshi would get just about half a kilo of rice a day. Jute has long faced a declining world market and inexorable competition for its very basic uses which are sacking, carpet backing and hessian. The synthetic substitutes are made by big multinational companies like ICI whose products are not at the mercy of the winds or tides or the pricing policies of politicians as is Bangladesh jute. And each year, of course, there are new mouths to feed, so that the task of economic development is rather like trying to push a stone up a steep mountain.

Economy

All things considered, Bangladesh has been fighting hard and making some slow progress in the face of all the handicaps. The World Bank and the IMF both gave the country a good report card. The IMF reported in 1989 that 'Bangladesh's performance under the [structural adjustment] programme so far has been broadly satisfactory. During the first two years of the arrangement the economy's external position strengthened markedly. And structural reforms have been sustained against difficult odds. However, output and price objectives have not been fully met.' The aim of the IMF adjustment programme was to raise economic growth, create new employment, lower the inflation rate and reduce the fiscal and current account deficits to about 7 per cent of gross domestic product. In addition, the ambition was to increase agricultural and industrial production, improve competitiveness and increase domestic savings.

Among the policy measures adopted were price incentives to stimulate agricultural output; simpler investment regulations and more flexible exchange-rate management to boost industrial output and exports; improvements in tax policy and administration and the management of public sector enterprises; greater reliance on market-oriented policies; and a greater effort to recover bank loans.

The weather played havoc with the best efforts to get on top of the situation. There were floods in August 1987, among the worst that Bangladesh has seen. They caused considerable loss of life, damaged crops and infrastructure and disrupted supplies. Politically inspired labour unrest added to the turmoil. Then in 1988 Bangladesh was hit by even worse floods. More than three-quarters of the land area was inundated; several thousand people lost their lives; crops, infrastructure, transport, communications, electricity, water and sewage and industry were all disrupted. For several weeks, after the main floods had subsided, low lying parts of the capital were best accessible by boat. The budgetary cost of relief and rehabilitation after the 1988 floods was estimated at almost 2 per cent of GDP. As if this was not enough, towards the end of 1988 Bangladesh was struck by a major cyclone. And in early 1989 the failure of the early rains brought drought.

The impact of all this has been shown in economic growth rates, which have been much lower than expected, averaging just under 4 per cent a year. In 1988-89 the best estimates were that growth would be only 2 per cent. Given the steady 2.5 per cent increase in population, this would mean that Bangladeshis on average would be worse off at the end of the year. Inflation rates also remained higher than expected and savings were hit by having to repair the damage.

Yet again jute, for long the mainstay of Bangladesh's export earnings, but a vulnerable crop, was badly hit by the weather. The drought in early 1989 meant that farmers were only able to plant about half of the area expected. Jute production of 4.5m bales had been expected, but looked likely to fall short by 2m. Even with a carry-over of 2m bales from the 1988 harvest Bangladesh could face serious difficulties. It is not a good time for the crop. Jute Minister Myeedul Islam told an International Jute Organisation meeting in March 1989 that the 'industry was passing through a critical phase, threatening its survival.' The ever-in-

KEY INDICATORS

	Unit	1984	1985	1986	1987	1988
Population	m	96.73	98.66	100.62	102.3	104.5
Gross domestic product (GDP)	Tk bn	349.92	416.96	462.01	539.17	594.10
GDP per capita (Constant)	Tk	4,156	4,226	4,592	5,270	5,685*
Inflation	1980=100	10.5	10.7	11.0	9.5	9.4
Industrial production	% change	5.1	1.1	–1.9	9.5	0.4
Agricultural production						
Rice	'000 tonnes	14,279	14,392	23,110	23,120	21,900
Jute	'000 tonnes	931	813	1,216	859	883
Sugar-cane	'000 tonnes	7,056	6,769	6,640	6,896	7,207
Tea	'000 tonnes	42	38	38	38	41
Pulses	'000 tonnes	263	266	191	185	188
Oil seeds	'000 tonnes	263	266	269	–	–
Condiments & spices	'000 tonnes	289	293	295	–	–
Tobacco	'000 tonnes	47	49	46	40	42
Exports	Tk m	23,640	27,997	26,761	33,030	40,967
Jute goods	Tk m	10,484	11,647	8,716	7,260	9,287
Raw jute	Tk m	3,186	3,655	3,423	2,582	2,471
Imports	Tk m	71,673	77,266	82,204	83,781	96,709
Petroleum	Tk m	7,953	9,698	6,434	6,598	7,490
Balance of trade	$m	–1,408	–1,287	–1,421	–1,369	–1,444
Current account	$m	–543.0	–539.5	–614.7	–334.7	–286.9
External debt	$m	5,613	6,536	7,859	9,506	–
Debt service ratio	%	10.6	14.1	17.1	16.5	–
Exchange rate	Tk per $	25.35	28.00	30.41	30.95	31.73

* estimated figure

creasing population meant that farmers were growing rice and wheat first. The best hope for the jute crops could be new high-yielding varieties, which could produce five tonnes per hectare, or three times existing yields. Besides the income to the farmers from the sales of raw jute, more than 60 mills in Bangladesh employ nearly 175,000 workers, a considerable portion of the infant industrial labour force in a mainly rural but overcrowded country.

The economy is propped up in this sickly state by annual commitments of more than $2bn in aid. At their meeting in Paris in April 1989 aid donors pledged $2.3bn in a package comprising almost $1.5bn in project aid, nearly $600m in commodity aid and $180m in food aid. But the Bangladesh economy is in such a poor state that the country cannot find the local taka funds to use in conjunction with the money provided by the foreign donors. A World Bank investigation in early 1989 discovered that a huge sum of $4.6bn in project aid was unused, waiting for local funds.

Even higher aid spending would be needed to begin to make a dent in the flood control measures that are essential for Bangladesh's long-term survival. According to one estimate, a 20-year flood control plan would cost $10bn. Flood damage has been increased by felling of the few forests that remain. 40 years ago forests covered 20 per cent of the country, but now they have dwindled to about 9 per cent. Most Bangladesh rural homes are made of bamboo, timber and mud.

Some of the appalling longer-term prospects were beginning to hit home to the country's planners. Dhaka, the capital and largest city with 5m people, is already groaning under some of its problems. More than 70 per cent of the capital's population are classified as 'poor', and more than 2m people live in slums in shanty towns. The city needs about 50,000 new houses a year to cope with the swelling population,

COUNTRY PROFILE

History since 1945: The area today called Bangladesh was once part of united British India. In 1947 (when the subcontinent became independent) it became East Pakistan. It had a majority of the population of all Pakistan, but a much smaller land area and always had much less political, military and economic clout. The feeling of neglect by rulers more than 1,000 miles away in Karachi and Rawalpindi led to growing tensions between the two parts of Pakistan. General elections in 1970 gave the eastern-based Awami League an absolute majority in the National Assembly and committed to a programme of autonomy. But military rulers and parties in the West didn't like this. After unsuccessful negotiations the military arrested the Awami League leader and began a crackdown. This led to nine-month guerilla struggle which was ended on 16 December 1971 when the Indian Army backed by freedom fighters entered Dhaka. A secular, socialist democratic republic of Bangladesh was declared. The new country has had an unsettled history. Army majors assassinated the first leader Sheikh Mujibur Rahman, in August 1975. General Ziaur Rahman, a Pakistan army major turned freedom fighter, took over, but was himself assassinated in 1981 after rivalry among senior military officers.

Political parties: The ruling Jatiya Party won all but 50 of the 300 parliamentary seats in 1988 elections. But the party's hold was exaggerated because the major opposition parties boycotted the elections, as well as claiming that the results were rigged. The country has a profusion of political parties – about 28 at the last count. There are two main opposition groups. The Awami League claims good roots in the countryside as well as among intellectuals and is headed by Sheikh Hasina Wajed, daughter of the assassinated first leader. The other main party is the Bangladesh Nationalist Party under Begun Khaleda Zia, widow of President Ziaur Rahman. Each of these parties heads a multi-party alliance. Sometimes the opposition manages to run joint campaigns against the government, but both have refused to co-operate with the new system of parliamentary democracy under a strong president.

Population: Bangladesh's population of almost 110m by 1989 is one of the highest densities in the world, at 760 per sq km. Growth is averaging 2.6 per cent per annum. The 1981 census (published 1986) revealed that 47 per cent are under 14, and that 85 per cent live in rural areas. 86 per cent are Muslims and 10m are Hindus, and there are 250,000 Christians and Buddhists. Islam was made the state religion in 1988.

Main cities/towns: Dhaka (population 4.2m), Chittagong (pop 1.67m), Chalna, Khulna, Narayanganj, Sylhet and Comilla.

Language: The official language is Bengali, known locally as Bangla, which is spoken by 98 per cent of the population; however, Urdu and English (especially in business circles) are also used.

Media
Press: Dailies: There are 48 national daily newspapers (seven in English and 43 other daily periodicals). Ittefaq is the most popular Bengali daily. Weeklies: There are some 118 weekly newspapers. In addition there are 28 fortnightly and 23 monthly publications. Business: There are few trade publications.
Radio: National broadcasting services maintain eight radio channels with regional coverage. External services are broadcast in several languages.
Television: Government services maintain one TV channel in Dhaka and one in Chittagong, with colour transmissions. There were 350,000* sets in 1986.

Domestic economy: The economy is mainly agricultural, and remains vulnerable to changes in weather and worldwide commodity markets. In 1987, severe drought was followed by some of the worst flooding of the century, in which up to eight million people were reported homeless. Civil strife, lack of capital, reliance on jute exports and rapid population growth have also hampered economic development over the years. Bangladesh relies heavily on foreign aid to keep the economy going. Aid provides 40 per cent of government revenues, 50 per cent of foreign exchange and 90 per cent of the budget. Another important contributor to earnings is remittances from Bangladeshis working abroad, which have actually been higher than earnings from jute, the main export. In recent years, the government has tried to open up the economy by emphasising private enterprise, encouraging foreign investment and setting up an export processing zone at Chittagong. The last two years have seen a successful start to industrial exports.

External trade: Bangladesh is one of the 32 developing nations designated in 1985 to receive exemption from US competitive trade requirements in accord with the Generalised System of Preferences.
Exports: Main exports are jute manufactures and raw jute (together accounting for 40 per cent of all exports) garments, fish, leather and tea. In the late 1980s Bangladesh also began assembling electronic goods for export. Main destinations are USA (23.6% of 1986 total), Japan (8.0%), UK (5.5%), Italy (5.1%), Belgium (4.2%), Pakistan (4.0%), Iran (3.6%), USSR (2.9%), China (2.8%) and German FR (2.8%).
Imports: Main imports are manufactured goods, machinery and transport equipment, petroleum and petroleum products, chemicals and pharmaceuticals, vegetable oil and fats, inedible crude materials and foodgrains. Main sources are Japan (13.9% of 1986 total), USA (8.5%), Singapore (7.4%), United Arab Emirates (4.7%), German FR (4.4%), UK (3.6%), Canada (3.6%), USSR (3.4%), China (2.9%) and Denmark (2.8%).

Agriculture: Agriculture employs 85 per cent of the labour force and accounts for 50 per cent of GDP and traditionally provides the bulk of exports. The land is very fertile, but agricultural productivity is hampered by fragmentation of holdings, reliance on traditional crops, drought and floods. Total foodgrain production is 17m tonnes. Most available agricultural land is subsistence smallholdings, except for large tea plantations. Rice is the most important crop. Production has reached about 16m tonnes. Other crops are jute (the biggest cash crop), sugar-cane, tea and tobacco with oil seeds, cotton, spices, fruit and vegetables being other minor crops. Livestock raised include cattle, buffaloes, sheep, goats and poultry. Forests occupy 16 per cent of total land area and provide timber and fuel for domestic consumption. Inland and marine fishing have developed steadily in recent years with substantial government assistance, providing an important source of protein and output grows three per cent a year. About 80 per cent of the catch is from inland waters.

Industry: The industrial sector contributes 14 per cent to GDP and accounts for 11 per cent of the labour force. The New Industrial Policy aims to increase the involvement of the private sector, denationalising much of the country's state industry. Main activities are jute and cotton spinning and weaving. Textiles and clothing industries are the fastest growth areas. By 1988 textile factories employed 200,000 people, mostly women. Other products include cigarettes, sugar, paper, flour, edible oils and fish products, soap, skins and hides, tea, spare parts and simple machinery. Tanning is being developed. Heavy industries include several fertiliser factories, cement and clinker grinding factories, an oil refinery, a steel mill and several steel rolling mills. An Export Processing Zone in Chittagong has almost 40 factories operating (1988). The government is proceeding with plans to bridge the river estuary which divides east and west Bangladesh.

Mining: Bangladesh is self-sufficient in natural gas, with proven reserves of 350bn cu metres (enough for 90 years) and production of over 13m cu m a day. There are plans to develop further fields and exploration continues. Oil prospecting has been unsuccessful offshore, but there are hopes of onshore deposits. Current plans include possible development of limestone and coal deposits. Other resources present include peat, white clay and mineral-bearing sands.

Energy: Natural gas is available near the Bakhrabad centre of production in the east, with plans for distribution throughout the country by pipeline. Electricity is generated by a hydroelectric plant at Karnaphuli Reservoir; capacity, at 1,000mW, is very low, but the Government hopes to electrify half the country by 1990. There is little prospect that the planned nuclear power plant will be built, however.

Membership of international organisations: Afro-Asian People's Solidarity Organisation, Asian Development Bank, CCC, Colombo Plan, Commonwealth, EC trade and co-operation agreement, ESCAP, FAO, G-77, GATT, IAEA, IBRD, ICAO, IDA, IFAD, IFC, ILO, IMF, IMO, INTELSAT, INTERPOL, IOC, IRC, Islamic Conference Organisation, Islamic Development Bank, ITU, NAM, OIC, South Asian Association for Regional Co-operation, UN, UNCTAD, UNESCO, UPU, WHO, WMO, World Bank, WTO.

but public and private sectors together are building just 2,000 new homes a year. As it is, water and sewage authorities can meet only 45 per cent of Dhaka's peak water requirements. A United Nations survey said only one in three homes in the capital had access to safe drinking water. And the Dhaka population was expected to grow to 10m by the turn of the century. No wonder that the city planners are beginning to talk of the 'urban nightmare'.

Government optimism

In the circumstances it is hard to see why Vice President Moudud should have any cause for optimism, let alone for seeing Bangladesh pressing hard on the heels of prosperous places like Hong Kong. But Moudud's argument is that the existing NICs are already being priced out of their traditional markets. 'Wages are going up, their cost of living is going up, so the cost of their products is also going up', he argues. 'In addition, certain facilities which they used to enjoy, particularly from the United States, such as the generalised scheme of preferences, have now been withdrawn. Altogether now they will lose business of about $10bn. They are being forced to go upmarket. Bangladesh can offer a place for relocation of industries, and it is already taking place.'

He cites a deal that a Bangladesh concern made with a Korean company to manufacture Puma shoes for Germany. The Korean factory could no longer make them profitably and the company was worried about the order before it decided to come to Bangladesh. 'Now Puma shoes would be manufactured in Bangladesh' says Vice President Moudud. 'The goods will go to Germany, the Koreans will be able to keep their commitment. And we will have 1,900 new jobs and the foreign exchange. All sides will benefit.' The vice president claims that Bangladesh has a head start over other countries in attracting foreign investment because of its low wages, $30 a month for unskilled workers and $40-50 for skilled workers. These figures include medical and other benefits.

But Bangladesh has had a low wage structure ever since independence. Many other countries in other parts of the world have plentiful supplies of workers and low wage rates, and neither foreign nor domestic industries are falling over themselves to pump money in. So what is special about Bangladesh today? The new thing according to Moudud is the dedication to private enterprise. He points out that textile and jute mills have been privatised and so have two out of the six nationalised commercial banks. Bangladesh is encouraging new private banks. As prime minister, a job he did before becoming vice president in 1989, Moudud boasted that 'I have out-Thatchered Thatcher' – claiming to have sold 550 factories to the private sector. Not all of these were fully privatised. In some of them the state only sold part of its holding.

Moudud's personal belief is that no manufacturing industry should be in the public sector. 'We are trying to withdraw.' He pledges that the government will persevere in its new policy, simply because it is not possible for the government to be responsible for everything. 'We can give directions and help them by improving opportunities, but people must earn their own living. Socialism has given them the hope that the state would provide everything. But this is not possible. The state will provide facilities, the support, the encouragement, the guidelines, the policy framework and infrastructural support. But we want to bring self confidence in every individual to use his own talents.'

Moudud believes that the policy is paying off and Bangladesh is beginning to develop a flourishing private sector. Garment factories employ 500,000 people, 80 per cent of them women. By the end of 1989 the country expects to be the world's third largest garment exporter to the USA. The vice president boasts that the country has already moved well beyond garments and into electronics. 'We are even assembling domestic computers and will be exporting Bangladesh-assembled computers to the United States this year. We are assembling video-cassette recorders which are being exported, something that was unthinkable three years ago. We are making radios on a large scale. Our television assemblers are pricing multinationals like Philips out of the market. That's because we have reduced the duty on components to 8 per cent. Previously some [brand name components] attracted duties of 150 per cent. But for television, now you don't need brand-name parts any more. The components are all standardised. You can buy them as chips in Hong Kong godowns, and that's what is happening; and the multinationals are less competitive.'

Moudud promises that the government has speeded up its decision-making process. 'We have set up the Board of Investment. It is a single-roof sanctioning procedure. Decisions will be taken inside 120 days maximum, and in fact we have been taking an average of 65 days.' He admits to impatience – which is why he is pinning his faith on private investment as a catalyst to official aid. He grumbles that aid is too slow. 'We want to see the use of our gas, we want our oil to be explored and discovered. Typically we have to wait years for the damn Official Development Assistance, seven years to negotiate, then the consultant comes and takes three years to tell us what to do and what not to do, and then lining up the finance takes ten years. But that's too long. For Bangladesh ten years is like a hundred years. In ten years, there will be even more unemployed. We have to act now.'

He claims that Bangladesh's poor image in the world outside is exaggeratedly bad.

Political situation

A more powerful deterrent may prove to be the political situation. By 1989 Hussein Mohammed Ershad had survived in power for more than seven years, the longest term of any leader of independent Bangladesh. But it had not been an easy ride and even in 1989 the president faced opposition from more than a score of political parties and groupings who opposed him and claimed that he was running an authoritarian if not dictatorial regime. Just before Ershad's visit to the UK earlier in the year, several political parties, workers' groups, other organisations and individuals took British newspaper advertisements to make their case against the Ershad years. They alleged that: Ershad and his ruling Jatiya Party had come to power only after a referendum, presidential election and two general elections had been rigged; that corruption was higher than ever, particularly through the use of foreign aid; that landlessness and rural poverty had increased from 43 to 77 per cent under his rule; that government spending particularly on the military had risen by 150 per cent, while economic growth had fallen; that law and order had broken down and instead thugs ruled the day under government protection.

The problem for the opposition is that though they agree in opposing Ershad and condemning his rule, they cannot get together for a sustained campaign, let alone to present alternative, or credible policies. There is a general problem that Bangladeshis are prone to argument and squabbling. Indeed it has often been said that if there are two Bengalis there are the makings of at least three political parties. More particularly, the Bangladesh opposition in the late 1980s was split into two main camps, each riven apart by deep-seated historical suspicion and rivalry. The two main groupings centred around the Awami League of Sheikh Hasina Wajed, who had seven other parties and splinter groups around her; and Begum Khaleda Zia and her Bangladesh Nationalist Party, with a similar number of parties around her umbrella. Both women are the inheritors of the political mantles of Bangladesh's first two leaders, both of whom were assassinated. Hasina is the daughter and only surviving family member of Sheikh Mujibur Rahman, 'Banga Bandu, the father of Bangladesh', murdered by army majors in 1975. Khaleda is the widow of General, later President, Ziaur Rahman, who himself was killed in a coup in 1981.

At various times the two groups had managed to combine in opposition to Ershad, but their unity always proved short-lived. Khaleda's BNP had remained steadfast in its boycott of the 1986 election, but the Awami League had contested it, though later its MPs quit parliament. In 1989 fresh accusations erupted between the two women. Hasina told a public rally in April that Zia had been the 'ringleader' of the killers of her father and that when Zia had assumed power, he had protected the assassins. Two days later, Khaleda's special assistant, Ehsanul Kabir, responded by accusing Hasina of being behind the murder of Zia and helping the bloodless coup in 1982 which toppled 'a democratically elected (BNP) government'. The danger of the opposition fighting each other became so great that in May a group of 55 intellectuals called upon the main opposition groups to reconcile their differences and to join with fundamentalist Muslim groups to unite in the 'imperative need ... to fight against autocracy.' The intellectuals made demands for 'all clashes and strife to cease and a truly representative government to emerge through free and fair elections, establishing the rights of the people and democracy.'

In 1989 three former members of Ershad's party, including two ex-ministers, tried to bring some unity out of the opposition by forming yet another political party. Salauddin Qader Chowdhury, Anwar Zahid – both ex-ministers – and former Jatiya party treasurer Morshed Khan had resigned before the March 1988 elections, claiming that these polls would not solve the political crisis in which opposition boycotts and strikes had brought Bangladesh to a halt for long periods. The three men claimed that they wanted to put together a bridging party that would be liberal and democratic, in favour of a free economy, give encouragement to the private sector and yet support social welfare programmes. They were hoping to enlist members of both the Awami League and the BNP as well as former bureaucrats and army members unhappy with the authoritarian drift of Ershad's rule. Typical of Ershad's tough ways, the moment the two ministers had resigned, they were expelled from the ruling party, their passports seized and cases alleging irregular business transactions and nonpayment of income tax were filed against them. All of Chowdhury's office and home phones were cut off and the former minister complained that the president could not stand anyone who had the temerity to go against him.

In July 1989 Ershad pushed through a surprise constitutional amendment which would limit the president to a maximum of two terms of five years each. Assuming that Ershad contests the 1991 election, he would have to retire by October 1996 at the earliest. The opposition scorned the amendment and the offer to hold an election at any time and invite foreign observers, on the grounds that the amendment was simply a ruse to legitimise the previous controversial elections. Most of the opposition parties want to see a return to a parliamentary rather than a presidential system. Though they do not have the capacity to challenge and defeat Ershad's strong-man rule, the opposition parties, especially when they combine, have considerable mischief-making value, at least enough to destroy the dreams of quick progress to NIC status.

BUSINESS GUIDE

Time: GMT + 6 hrs.

Climate: The climate is hot and humid with average annual rainfall of 2,500 mm, mostly falling Apr-Oct and with monsoon rains Jun-Sep. Likely temperature Apr-Sep is 38-40°C; Nov-Mar 21-27°C by day or as low as 10°C by night.

Entry requirements
Passport: Required by all except certain seamen.
Visa: Visas are usually required, except for nationals of EC countries (not Denmark), Australia, the Maldives and certain other countries for short stays. Holders of Taiwanese and Israeli passports may encounter difficulties. Border formalities may be protracted when entering by road from India. Regulations change at short notice; check with local diplomatic centres.
Currency: No more than Tk100 may be imported or exported. Foreign currency must be declared on arrival, and no more than the amount imported may be exported, It is best to retain all foreign exchange receipts. Hotel bills and international air tickets purchased in Bangladesh must be paid in hard currency or with travellers' cheques.
Customs: Personal effects are allowed duty-free provided they are declared on entry. No export licences are required for export of gifts or souvenirs whose total value does not exceed Tk 100. All imports are under licence, and there are restrictions on the importation of luxury goods. Trade with Israel, South Africa and Taiwan is prohibited.

Health precautions
Mandatory: Vaccination certificates are required for typhoid and cholera, and for yellow fever if travelling from an infected area. Health regulations may change, and it is best to make detailed enquiries before travelling.
Advisable: Anti-malarial precautions (although officially eradicated).

Representation overseas: Abu Dhabi, Algiers, Ankara, Baghdad, Bangkok, Beijing, Beirut, Belgrade, Berlin, Bonn, Brasilia, Brussels, Bucharest, Buenos Aires, Cairo, Calcutta, Canberra, Colombo, Dakar, Doha, Geneva, Islamabad, Jakarta, Jeddah, Kabul, Kathmandu, Kuala Lumpur, Kuwait City, London, Moscow, Nairobi, New Delhi, Ottawa, Paris, Rangoon, Rome, Singapore, Stockholm, Tehran, Thimphu, Tokyo, Tripoli, Warsaw, Washington.

Air access
National airline: Biman Bangladesh.
Other airlines: Aeroflot, British Airways, Burma Airways, Dragonair, Druk-Air, Emirates, Gulf Air, Indian, Iraqi Airways, Kuwait Airways, PIA, Royal Nepal, Saudia, Singapore, Thai International.
Tax: Airport tax of Tk 200 is payable on departure from both airports.
International airports: Zia International (DAC), 20 km from Dhaka, with duty-free shop, bank, post office and car hire; Patonga (CGP), 22.5 km from Chittagong.

Surface access: It is possible to travel by road from a number of points in India, including West Bengal, Assam, and Tripura. May be difficult during monsoon seasons, however.
Main ports: Chittagong, Chalna.

Hotels: Hotel bills must be paid in a hard currency or with travellers' cheques. Provincial towns have government rest houses with fairly spartan accommodation, for which booking well in advance is advisable.

Credit cards: Major credit cards, such as American Express, Visa, etc. are accepted in some places.

Car hire: Car hire can be arranged by the Bangladesh Parjatan Corp or is available from the hotels Sheraton, Sonargaon and Purbani in Dhaka and from the Hotel Agrabad in Chittagong. Driving is on the left-hand side of the road.

City transport
Taxis: Taxis are available at main hotels and airports. Negotiate fares before undertaking a journey. A 10 per cent tip is usual. Rickshaws and autorickshaws are available, but are not recommended for use at night. Autorickshaws should be metered, but often are not. Negotiate fares in advance.

National transport
Air: Biman Bangladesh serves main centres. There are regional airports at Jessore, Ishurdi, Saidpur, Sylhet, Cox's Bazar, Comilla and Thakurgaon.
Road: Around 57 per cent of the 9,300 km road network outside municipal areas is metalled. Road conditions are generally moderate; frequently difficult in monsoon season. Numerous ferry crossings make journey times unpredictable.
Rail: There are regular services on the 2,750 km railway network which serves main centres, but they are slow and subject to delay. Six classes of rail travel are available: 'Air-conditioned', 'First' and 'Express' are recommended. The main rail routes are Dhaka-Chittagong (journey time five hrs) and Khulna-Ishurdi-Chilhati (journey time 10 hrs). Rail services are being upgraded.
Water: There are 7,000 km of navigable inland waterways, with regular passenger steamer services to the five main ports: Dhaka, Narayanganj, Changpur, Barisal and Khulna. Passage should be booked well in advance.

Public holidays
Fixed dates: 1 Jan (Bank Holiday), 21 Feb (Martyr Day), 26 Mar (Independence Day), 15 Apr (Bengali New Year), 1 May (May Day), 1 Jul (Bank Holiday), 23 Oct (Durga Puja), 7 Nov (National Revolution Day), 16 Dec (Victory Day), 25 Dec (Christmas Day), 31 Dec (Bank Holiday).
Variable dates: Shab-e Qadr, Eid-ul-Fitr, Eid-ul-Azha, Muharram, Eid-e-Miladunna. Actual dates of Muslim festivals depend on the sighting of the moon.

Working hours
Government and business: (Sat-Thu) 0730 – 1400; (Fri) 0730 – 1230.
Business: (Sun-Thu) 0800/0900 – 1230, 1330 – 1700.
Banking: (Sat-Wed) 0900 – 1300; (Thu) 0900 – 1200.

Social customs: It is offensive to drink, eat or smoke in public or before Muslims during the month of Ramadan. Muslim women should not be photographed unless it is certain that no objection will be made.

Telecommunications
Telephone: Dialling code for Bangladesh, IDD access code + 880 followed by area code (2 for Dhaka, 31 for Chittagong) followed by subscriber's number. In 1987, there were some 160,000* telephones with 90 per cent automatic service, 51 per cent of them in Dhaka and 12 per cent in Chittagong. IDD services can be erratic.
Telex: Telex facilities are available at leading hotels.
Telegram: Normal, fast and express telegram service is available from post and telegraph offices.

Banking
Main centre: Dhaka. *Central bank:* Bangladesh Bank. *Head offices:* Agrani Bank, Bangladesh Krishi Bank, Bangladesh Shilpa Bank, Janata Bank, Pubali Bank, Rupali Bank, Sonali Bank, Uttara Bank. An Islamic Bank was established in 1983. *Bangladesh branches:* Algemene Bank Nederland, American Express International Banking Corp, Arab Bangladesh Bank, Bank of Credit and Commerce, Banque Indozuez, French International Bank, Grindlays Bank, Habib Bank, Standard Chartered Bank, State Bank of India.

Advertising: Advertising is available in the press, on radio and TV, and in cinemas. Outdoor sites in major towns and some direct mail are also available.

Electricity supply: 220/240 V AC, with British-type 2 or 3 round pin plug fittings and bayonet lamp sockets.

Weights and measures: Metric system (local units also in use).

Representation in capital: Afghanistan, Algeria, Australia, Austria, Belgium, Bhutan, Brazil, Bulgaria, Burma, Canada, China PR, Cuba, Czechoslovakia, Denmark, Egypt, France, German DR, German FR, Ghana, Greece, Hungary, Iceland, India, Indonesia, Iran, Iraq, Italy, Japan, Korea PDR, Republic of Korea, Kuwait, Libya, Mauritius, Mongolia, Morocco, Malaysia, Nepal, Netherlands, New Zealand, Nigeria, Norway, Pakistan, Poland, Romania, Saudi Arabia, Somalia, Spain, Sri Lanka, Sweden, Switzerland, Syria, Thailand, Turkey, UAE, UK, USA, USSR, Vatican, Vietnam, Yugoslavia.

Hotels: Chittagong
Agrabad, Agrabad Commercial Area, P.O. Box: 147 (tel: 500111/20; tx: 66237).

Hotels: Dhaka
Dhaka Sheraton, 1 Minto Road 1000, P.O. Box: 504 (tel: 505061; tx: 642401; fax: 412972).

Purbani International, 1 Dilkusha Commercial Area, Motijheel 2 (tel: 254081/6; tx: 642460).

Sonargon, Karwan Bazar, P.O. Box: 3595 (tel: 315001; tx: 642426).

Car hire: Dhaka
Bangladesh Parjatan Corp: 125A Motijheel Commercial Area, Dhaka (tel: 255769).

Business advice and information
Bangladesh Jute Mills Corp, Adamjee Court, 4th Floor, Motijheel Commercial Area (tel: 235171; tx: 65662).

Export Promotions Bureau, 122 Motijheel Commercial Area (tel: 232245/9; tx: 642204).

Federation of Bangladesh Chambers of Commerce & Industry, 60 Motijheel Commercial Area (tel: 282880, 250566).

Metropolitan Chamber of Commerce & Industry, PO Box 2641, 122/124 Motijheel Commercial Area (tel: 245747; tx: 5540).

Planning Commission, G O Hostel, Sher-e-Bangla Nagar, Dhaka.

Banking
Bangladesh Bank (central bank), Motijheel C/A, PO Box 325, Dhaka 1000 (tel: 235000; tx: 65657).

Agrani Bank, 9D Dilkusha C/A, Motijheel, PO Box 531, Dhaka 1000 (tel: 250906; tx: 642757).

Bangladesh Krishi Bank, 83-85 Motijheel C/A, Dhaka 1000 (tel: 240031; tx: 642526).

Janata Bank, 110 Motijheel C/A, PO Box 468, Dhaka 1000 (tel: 236216; tx: 65840).

Rupali Bank ltd, Rupali Bhaban, 34 Dilkusha C/A, PO Box 719, Dhaka 1000 (tel: 256021; tx: 65635).

Sonali Bank, Motijheel CA, PO Box 147, Dhaka 1000 (tel: 252990; tx: 642644).

Travel information
Automobile Association of Bangladesh, 3/B Outer Circular Road, Dhaka, 17 (tel: 243444, 250541).

Bangladesh Parjatan Corp (National Tourist Organisation), Old Airport Building, Tejgaon, Dhaka, 15 (tel: 327855, 325421, 325416; tx: 642206).

Barbados

Canute James

The political scene in the island changed dramatically in early 1989, with the formation of the New Democratic Party to challenge the hegemony of the Barbados Labour Party and the Democratic Labour Party. The NDP, led by Richard Haynes, a former finance minister, emerged after public rows between Haynes and Prime Minister Erskine Sandiford. There were early indications that the emergence of the new party would reduce the incumbent DLP's support, increasing the chances of the BLP regaining office. The economy continued its steady, though unspectacular, growth in 1988, although the agricultural sector continued to decline, while there was increasing dependence on tourism.

Ruling party splits further

The split in the government worsened when Haynes left the DLP, after spending some months as a backbencher. He had earlier resigned as finance minister, claiming Sandiford had made key appointments to the Central Bank, including that of the governor, without Haynes' knowledge. In leaving the ruling party, Haynes took with him three other disaffected backbenchers. This gave the new NDP more sitting members than the three of the opposition BLP. Haynes eventually replaced BLP leader Henry Forde as leader of the parliamentary opposition. The NDP is not offering any policies which are radically different from those pursued by the BLP and the DLP, but members of the older parties suggested that Haynes could garner support from young Barbadians, and could also split the DLP's support in the next elections which are due in May 1991. However, the DLP has played down reports that Sandiford planned an earlier election, with the date determined by the rate of growth of NDP support, and whether this threatened to remove the ruling party's base.

The economy, which grew by an average 2.6 per cent per year between 1983 and 1987, expanded by 3.5 per cent in 1988. The deficit on the current account of the balance of payments moved from $53.5m in 1987 to $500,000 in 1988, and the fiscal deficit, which was equivalent to 7.6 per cent of GDP in 1987 was reduced to 4.2 per cent in 1988. Servicing the foreign debt took almost 16 per cent of what was earned from exports and tourism in 1988, against 18 per cent the previous year. Unemployment remained at 18 per cent, while inflation in 1988 was 4.8 per cent, against 3.4 per cent in 1987.

The offshore financial services sector, comprised of banks, captive insurance companies and foreign sales corporations established by US firms, continued to grow, and in 1988 contributed $20m to the economy. But the prospects for further expansion have been made uncertain by US legislators who have removed exemptions granted to US companies under a bilateral treaty between Barbados and the US Government.

The government started negotiations with the US authorities in early 1989, seeking a clarification of the terms of the treaty under which the exemptions were granted. Barbadian officials said the changes made unilaterally by the US legislators would not destroy the offshore sector, but would make it more difficult to attract new clients.

Tourism spurs growth

Tourism continued to be the basis of the growth of the economy. In 1988 visitor arrivals grew by 7 per cent to reach 451,500. The government then set a target of 5 per cent per year, but officials admitted that achieving this could be difficult because of increasing competition from other resorts. Visitor expenditure in 1988 also improved to $425m, with the real value added in the sector growing by 10.5 per cent. The average length of stay was 6.7 days, up from 6.5.

The United States remained the largest single market, accounting for about a half of the tourists visiting the island. But arrivals from the United States fell 2.5 per cent in 1988, although this was made up by growth of 10.7 per cent in the European market. An expansion in summer charters from Britain led to a 28 per cent growth in the British market in 1988, lifting arrivals to just over 101,000, and helped reduce the seasonality problems for the industry. More hotels are remaining open in the low summer period. This led to an increase in 1988 in the average hotel occupancy level, which reached 68 per cent, although still below the industry's target of 75 per cent. The overall improvement in the industry was reflected in a reduction in the hotel industry's indebtedness to local banks. At the end of 1988 this stood at $63m, $2.2m less than a year earlier.

A new threat emerged for the industry in mid-1989 with the hardening of the US dollar against other major currencies. This threatened to take the steam out of the rapidly growing European market by making a holiday on the island increasingly expensive, as the Barbados dollar is tied to the US dollar. But this did not halt plans for expanding the island's capacity beyond the 14,000 beds in hotels, apartments, guest houses and villas. New hotels are planned to meet increasing volumes of visitors. There was also growth of 28 per cent in cruise visitors in 1988, with the volume reaching 291,000.

Output of the manufacturing sector grew 6.7 per cent in 1988. This was led by 29.7 per cent growth in chemicals, and 24.9 per cent in wooden furniture because of an increase in exports. The performance in the larger industries was more moderate as garment production was up 4.3 per cent, following the 15.7 per cent increase in 1987 when the local market was protected. The slump in electronics which followed the pullout of major US companies in 1986 appeared to have ended with growth of 3 per cent in 1988. But production of processed food, reflecting the state of the agriculture sector, fell 5.6 per cent.

The manufacturing sector was assisted by a growth in exports, indicating that some manufacturers were taking advantage of new regulations within the Caribbean Economic Community which increased opportunities for duty free trade. Despite this, sales to Caribbean Community markets in 1988 were a half of what they were in 1983.

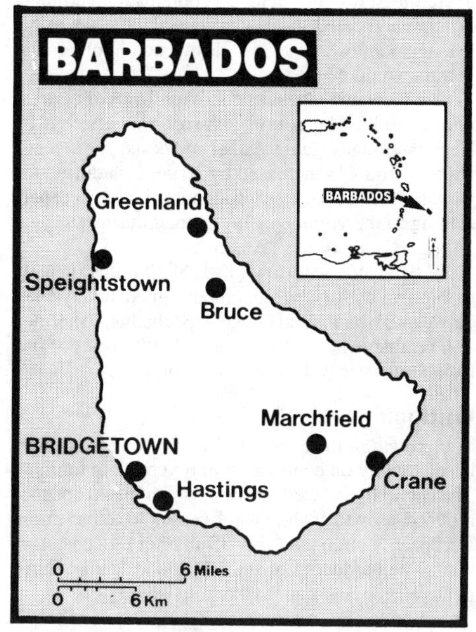

Low sugar output

The decline in agriculture continued in 1988, and was reflected in the performance of the sugar industry in the first half of 1989. The sector's contribution to national output fell from 7.6 per cent in 1981 to 5 per cent in 1988. The government has been unable to get new farmers to make use of about 6,000 acres of arable land which is unused. The sugar industry, the pillar of agriculture, was aiming at a production target of 80,000 tonnes for the 1989 harvest – the same as was produced in 1988. Output, however, reached 68,000 tonnes, which officials described as being 'the lowest since 1931'. The industry was hit by poor weather, with dry spells when rain was needed and heavy rains during the ripening period.

The decline in production created problems for the industry in meeting its commitments to its guaranteed markets, mainly its quotas of about 54,000 tonnes to the European Community, 7,500 tonnes to the United States and 12,000 tonnes for domestic consumption. Officials spoke of having 'a mechanism' for dealing with any shortfalls in meeting these commitments, but were worried about the likely effect of a proposed cut in prices paid by the European Community. The proposal for a 2 per cent reduction in the price paid by the EC could cost Barbados about $3.50 per tonne – a loss which the financially weak industry can ill

KEY FACTS

Official title: Barbados

Head of state: Queen Elizabeth II

Head of government: Prime Minister Lloyd Erskine Sandiford

Ruling party: Democratic Labour Party

Capital: Bridgetown

Official Languages: English

Currency: Barbados Dollar (BD$) = 100 cents

Exchange rate: BD$2.00 per $ (fixed)

Area: 430 sq km

Population: 254,000 (1988)

GDP per capita: $4,233 (1988)

GDP real growth: 3.5% (1988)

Inflation: 4.8% (1988)

Trade balance: –BD$361.0m (1987; merchandise)

Foreign debt: $621m (1987)

Visitor numbers: 451,500 (1988)

afford. Nevertheless, the industry continued to be a major foreign currency earner for the island, bringing in about $35m in 1988. The industry, and the rest of the agriculture sector, was due for changes in late 1989, following recommendations made to the government by a British company which had been contracted to study the sector's problems. Government officials indicated that wide-ranging changes in the management of the sugar industry and in other subsectors would be made.

There were mixed results for other forms of agriculture in 1988. The cotton industry was affected by labour shortages, poor weather and pests. The acreage under cotton was increased by 30 per cent during the year, but one out of every four acres was not reaped, leading to the volume of lint ginned during the year falling by a quarter to 120 tonnes.

The fish catch was down in 1988, but this is likely to improve with a new agreement signed in May 1989 between Barbados and Guyana. Barbadian fishermen will be able to harvest just under 2,000 tonnes of fish from Guyanese waters over six months.

Outlook

The growth in the economy is expected to continue, based mainly on continued improvement in tourism. The economy is faced, however, with the prospect of reduced growth if the island ceases to attract more European visitors and the US market remains stagnant. The hardening of the US dollar against European currencies can make the island less of an attraction for European tourists.

Other sectors are not likely to do better than they have in 1988. Manufacturing will continue to improve, with new attempts at attracting business in data processing through more modern facilities. The growth in this sector could be higher if local manufacturers overcome their reluctance to take advantage of opportunities for freer trading within the Caribbean Community.

There is little indication that agriculture will do better, with the major crop, sugar, suffering from problems of productivity. Despite the confidence of government officials, the industry is faced with losing a part of its valuable EC quota if it cannot fulfil its commitments. This would be a severe setback to an industry which cannot compete at world market prices, and which has a history of weak finances.

The government of Prime Minister Sandiford will weigh the benefits of an early election to the ruling DLP, given the emergence of the NDP led by Richard Haynes, the former finance minister. If the NDP does split support for the ruling party, this will improve the political stock of the BLP, led by Henry Forde.

KEY INDICATORS

	Unit	1984	1985	1986	1987	1988
Population	'000	260	253	254	254	254
GDP	BD$m	2,075	2,433*	2,297	2,499	–
GDP per capita	BD$	8,234	9,616*	9,062	9,850	8,512
Inflation	%	4.6	4.0	1.3	3.4	4.8
GDP real growth	%	3.6	1.1	5.1	2.6	3.5
Agricultural production	1979-81=100	81.3	79.1	87.9	78.0	74.9
Sugar	'000 tonnes	100.5	100.4	98.6	110.0	83.3
Manufacturing sector growth	%	1.9	–9.5	5.1	–6.6	6.8
Stop-over tourist arrivals	'000	368	359	370	422	447*
Exports & re-exports	BD$m	784.4	707.8	552.3	313.7	348.9
Domestic exports	BD$m	583.7	496.5	420.6	214.5	–
Sugar & molasses	BD$m	65.6	57.9	57.0	62.2	–
Rum	BD$m	6.5	8.0	8.8	7.2	–
Chemicals	BD$m	25.9	21.1	18.9	21.9	–
Electrical components	BD$m	335.9	304.1	241.7	51.7	–
Clothing	BD$m	64.8	45.0	35.5	23.8	–
Imports cif	BD$m	1,324.7	1,210.7	1,181.1	1,035.9	1,170.4
Canada	BD$m	75.5	62.3	78.4	81.3	–
CARICOM	BD$m	155.0	180.2	126.0	143.3	–
UK	BD$m	99.3	111.2	127.9	120.5	–
USA	BD$m	633.2	500.0	470.3	333.6	–
Balance of trade	BD$m	–537.3	–513.8	–628.8	–722.2	–361.0
External debt	$m	390	457	6128	621	–
Debt service ratio	%	3.4	5.5	7.4	–	–
Exchange rate	BD$m per $	2.0	2.0	2.0	2.0	2.0

* estimated figure

COUNTRY PROFILE

Political system: Independent state within the Commonwealth, with executive power vested in the British monarch, represented by a governor-general. Legislative power is exercised through the bicameral parliament, comprising a 21-member Senate (appointed) and a 27-member House of Assembly (elected every five years).

Main cities/towns: Bridgetown (population 100,000), Vistius, Speightstown, Holetown.

Media
Press: Two daily newspapers: *Barbados Advocate* and *Daily Nation*; a few weeklies, and monthlies, including *Barbados Business Update*.
Radio: Four national radio services (approximately 19 hrs/day); also cable radio (Rediffusion). *Television:* Government-owned Caribbean Broadcasting Corporation operates CBC-TV colour/commercial station, transmitting between 0545 – 1000 and 1600 – 2300 daily.

External trade: Regular deficit is partially offset by invisible earnings from tourism and capital inflows.

Exports: Electrical/electronic components (61% of total value in 1985), sugar and molasses (12%), clothing (9%), chemicals (4%). Main destinations: USA (68% of total value in 1985), Guyana (13%), Trinidad and Tobago (8%), UK (8%), Canada (3%).

Imports: Machinery (16% of total value in 1985), electrical components (16%), food and beverages (15%), fuels (5%), motor cars (3%). Main sources: USA (41% of total value in 1985), CARICOM countries (15%), UK (9%), Canada (5%).

Agriculture: Around 65 per cent of the total land area is under cultivation, mostly for sugar-cane production. Sugar sector contributes about three per cent to GDP, agricultural sector overall about seven per cent. Production of sugar has, however, declined since 1980 (to a record low of 83,400 tonnes in 1987), owing to labour unrest, widespread cane fires and a contraction in world demand. Emphasis has been placed on diversifying production away from sugar and more towards the farming of sea island cotton, green vegetables and market garden produce. Commercial fishing of flying fish and tarpon has increased dramatically in recent years with the island's 680-boat fleet taking an annual catch of nearly 3,000 tonnes.

Industry: Sector contributes 12 per cent to GDP and accounts for 90 per cent of exports. Production is centred on light manufacturing and assembly of electrical and electronic goods, food processing, clothing, sugar-refining, petro-chemicals and beverages. Most new foreign-owned export-oriented industries are based on the island's nine purpose-built industrial estates, which are largely managed by the Barbados Industrial Development Corporation. After earlier dramatic growth in the electronic/electrical assembly industry, a slump in the computer market led to the closure of several plants including the US-owned companies Intel and Corcom which had dominated the island's industry. Manufacturing performance as a whole has been disappointing. More traditional industries such as sugar refining, textiles and clothing and building materials have suffered the greatest decline. Emphasis is placed on expanding the number of value-added joint venture assembly industries. Tourism is major foreign exchange generator, contributing around 10 per cent to GDP and employing 16 per cent of the workforce. After a long recession in the industry, the number of long-stay arrivals increased by 12 per cent in 1984 to 367,652 and to 369,770 in 1986, a three per cent increase over 1985. Tourist expenditure also rose to BD$647m in 1986, though cruise ship arrivals slumped for the fourth consecutive year. Around 41 per cent of all arrivals came from the USA, 20 per cent from Canada and 20 per cent from CARICOM countries.

Hydrocarbons: Production of oil from the island's single Woodbourne oilfield has risen from 305,000 barrels in 1980 to 697,200 barrels in 1985 (56 per cent of domestic needs) although production was reduced in 1986 due to the fall in oil prices. Exploration currently underway in northern parish of St Andrew and off the island's west coast. Total oil reserves estimated at 3-4m barrels. Natural gas reserves meanwhile total around 392m cu metres, with about 88 per cent of production currently being flared.

Energy: By stepping up oil exploration and reducing the level of domestic consumption, Barbados aims to achieve oil self-sufficiency by 2000. Also plans to exploit gas potential and to expand solar and wind energy programmes.

BUSINESS GUIDE

Membership of international organisations: FAO, GATT, ICAO, IFC, IFAD, ILO, IMF, UN, UNESCO, WHO, WIPO, WMO, Caribbean Association of Industry & Commerce, CARICOM, Commonwealth, Inter-American Development Bank, Organization of American States, state associated with the EC (Lomé Convention), World Bank.

Time: GMT -4 (GMT -5 in summer).

Climate: Generally cool, with temperature around 26°C, except during rainy season (Jul-Nov).

Entry requirements
Passport: Required by all except nationals of Canada and USA who have proof of identity.
Visa: Citizens of Commonwealth, EC countries, USA, Canada and Australia do not require tourist visas.
Currency: No restrictions on import/export of foreign currency. Export of local currency limited to amounts declared on arrival.

Health precautions
Mandatory: Yellow fever vaccination certificate if arriving from an infected area.
Advisable: Typhoid/polio vaccination.

Representation overseas: Brussels, Caracas, London, New York, Ottawa, Port of Spain, Washington, Bangkok.

Air access
National airline: Caribbean Airways.
Other airlines: Air Canada, Air Caribe International, Air Martinique, American, British Airways, BWIA, Cubana, Cruzeiro, Guyana Airways, LAV, Liat, Pan Am.
Tax: Departure tax is payable on leaving.
International airport: Grantley Adams International (BGI), 18 km from Bridgetown; duty-free shop, restaurant, bank, shops, hotel reservations, car hire.

Surface access
Main port: Bridgetown Harbour.

Hotels: Wide range of first-class hotels available. Eight per cent government tax and 10 per cent service charge generally applied.

Car hire: Local drivers permit must be obtained. Available at Sir Grantley Adams Airport on presentation of national driving licence. Traffic drives on the left and is often heavy in the morning and at night. Strict speed limits of 20 mph in Bridgetown and Speightstown and 30 mph elsewhere.

City transport
Taxis: Can be hailed, ordered by telephone or found on ranks; Tourism Board publishes list of standard fares.

National transport
Road: Over 2,000 km of surfaced road. Main roads radiate from Bridgetown.
Buses: Frequent and efficient standard fare service operating throughout island.

Public holidays: 1 Jan, Good Friday, Easter Monday, first Mon in May, Whit Monday, Kadooment (Aug), United Nations Day (first Mon in Oct), 30 Nov (Independence), 25 Dec, 26 Dec.

Working hours
Government and business: (Mon-Fri) 0800 – 1600/1630; (Sat) 0800 – 1200 (except government offices.
Banking: (Mon-Thu) 0000 – 1500. Also (Fri) 0900 – 1300, 1500 – 1700.

Telecommunications
Telephone: Dialling code for Barbados: IDD access code + 1-809 followed by subscriber's (seven-digit) number. Overseas calls can be made direct from most public telephones on the island. Dial 0 for overseas operator.
Telex and telegram: Facilities available at most hotels and at Cable and Wireless (West Indies) offices. Additional international services include facsimile, leased circuit, international database access service.

Banking
Banking centre: Bridgetown. Barbados is a flourishing offshore banking centre.
Central bank: Central Bank of Barbados.
Commercial banks (head offices): Barbados National Bank.
Specialist banks: Barbados Development Bank, Caribbean Development Bank.
Foreign branches: Bank of Nova Scotia, Barclays Bank International, Canadian Imperial Bank of Commerce, Chase Manhattan Bank, Citibank, Royal Bank of Canada, Royal Bank Trust Co.

Electricity supply: 110/115 V AC, 50 cycles.

Representation in capital: Brazil, Canada, China PR, Colombia, Republic of Korea, UK, USA, Venezuela.

BUSINESS DIRECTORY

Hotels: Bridgetown
Hilton, P.O. Box: 510 (tel: 426-0200; tx: 2272; fax: 4368946).

Hotels: Christ Church
Accra Beach (tel: 427-7866; tx: 2723).

Best Western Sandy Beach, Worthing (tel: 435-8000; tx: 2382; fax: 4287564).

Caribbee Beach, Hastings (tel: 436-6232; tx: 2314).

Casuarina Beach Club, St Lawrence Gap (tel: 428-3600; tx: 3862502).

Coconut Court, Hastings (tel: 427-1655; tx: 2281).

Fairholme, Maxwell (tel: 428-9425; tx: 2314).

Ocean View, Hastings (tel: 427-7821).

Regency Cove, Hastings (tel: 427-7924; tx: 212).

Rockley Resort & Beach Club, Worthing, P.O. Box: 35 W (tel: 427-5890; tx: 2319).

Sandy Beach, Hastings (tel: 436-6232; tx: 2314).

Silver Sands (tel: 428-6001; tx: 2365).

South Winds Beach, St Lawrence (tel: 428-7181; tx: 3862427).

Southern Palms Beach Club, St Lawrence (tel: 428-7171; tx: 2332).

Welcome Inn, Maxwell Coast Road (tel: 428-9900; tx: 2226).

Windsor Arms, Hastings (tel: 427-7831; tx: 2641).

Hotels: St James
Barbados Beach Village, Fitts Village (tel: 425-1440; tx: 2514; fax: 4240996).

Buccaneer Bay, Paynes Bay (tel: 432-7981; tx: 2505; fax: 7230).

Colony Club, Porters, P.O. Box: 429 (tel: 422-2335; tx: 2268; fax: 4221726).

Coral Reef Club (tel: 422-2372; tx: 2407; fax: 4221776).

Discovery Bay Beach, Holetown (tel: 432-1301; tx: 2559).

Glitter Bay (tel: 422-4111; tx: 2397; fax: 3940).

Inn on the Beach, Holetown (tel: 432-0385; tx: 2314).

Sandy Lane (tel: 432-1311; tx: 2225; fax: 4322954).

Settlers Beach (tel: 422-3052; tx: 2570).

Tamarind Cove, Paynes Bay, P.O. Box: 429 (tel: 432-1332; tx: 2268; fax: 4221726).

Hotels: St Michael
Paradise Beach Cunard, Freshwater Bay (tel: 424-0888; tx: 2256; fax: 4240889).

Hotels: St Peter
Cobbler's Cove (tel: 422-2291; tx: 2314).

Eastry House (tel: 422-2201; tx: 2203; fax: 2202).

Heywoods (tel: 422-4900; tx: 2519).

Sandridge Beach, Roadview (tel: 422-2361; tx: 2644; fax: 4221965).

Hotels: St Philip
Crane Beach (tel: 423-6220; tx: 2381).

Marriott's Sam Lord's Castle (tel: 423-7350; tx: 82908; fax: 3862234).

Car hire: Bridgetown
Corbins: (tel: 427-7755)

Courtesy: (tel: 429-7970).

Dear's: (tel: 429-9277).

Sunny Isles: (tel: 428-8009).

Sunshine: (tel: 427-1234).

L E Williams: (tel: 427-1043).

Business advice and information
Barbados Board of Tourism, Harbour Road, PO Box 242, Bridgetown (tel: 427-2623/4).

Barbados Chamber of Commerce, Kensington Manor, Fontabelle, St Michael, Bridgetown (tel: 426-0747,426-2056).

Barbados Industrial Development Corp., PO Box 250, Pelican House, Princess Alice Highway, Bridgetown (tel: 427-5350; tx: 2295 bardev; fx: 426-7802).

Barbados Manufacturers Association, Room 206, Plantation Building, Bridgetown (tel: 426-4474, 427-9898).

Caribbean Airways, Lower Bay Street, Bridgetown (tel: 426-9900; tx: 2251).

Caribbean Broadcasting Corporation, PO Box 9000, Bridgetown.

Cable and Wireless (West Indies), Lower Broad Street, Bridgetown.

Central Bank of Barbados, PO Box 1016, Bridgetown (tel: 436-6870; tx: 2251).

Belgium

Susan Traill

The Belgian economy revived with unexpected vigour in 1988-89. The political scene was untypically peaceful after the agreement in summer 1988 to devolve major powers to the regions and to give Brussels its long-promised regional status. But critics accused the government of devolving its problems, especially the chronic budget deficit in which there was little real improvement. The Flemish coalition parties pressed for completion of the federalisation process by summer 1990, a move which would allow for devolution of other divisive issues such as the legalisation of abortion.

In his March 1989 speech to parliament, King Baudouin warned of political fragmentation if central government did not strengthen its policies regarding public finances and overall management of the economy. He also warned against too centralised and remote decision-making structures in the regions. Press editorialists commented that the two largest parties, the Flemish Christian Socialists (CVP) and the French-speaking Socialists (PS), had consolidated their power in the north and south respectively. Both parties made gains in the European Parliament elections of June 1989. The PS became the leading party in the new Brussels executive voted in at the same time.

Devolution and division

Devolution served to mask underlying differences between the main coalition parties, the centre Christian Socialists (CVP and PSC) and left-wing Socialists (PS and SP). Central government was left with a comfortable vacuum regarding its financial responsibilities once 30 per cent of its total budget was transferred to the regions and linguistic communities in 1989. Commentators noted a 'softening of resolve' concerning reforms to cut public spending, a point that was not lost on the IMF in its otherwise congratulatory 1989 report on the Belgian economy. But with one devolution phase not yet fully effective and a further phase in the pipeline, real reductions in government expenditure were virtually impossible.

The ideological differences between the coalition parties began to emerge more clearly by summer 1989. In June, an interministerial study was ordered to try and break the deadlock over the law on state companies, aimed at turning losses eventually to profit as a precursor to privatisation. The Socialists want to retain a greater degree of state control, worker control or monopoly than the Christian Socialists are willing to accept. The reform of public credit institutions to reduce the number from six to two to better face Europe-wide competition from 1992 was agreed in principle, but reservations remained. There was little progress on social reforms, especially that of the costly health insurance system. In contrast, reforms to strengthen Belgian industry for the opening up of the EC internal market in 1992, such as new laws on mergers and transparency of share ownership and transactions during takeover bids, stock exchange reform and tax reform, were adopted without much trouble.

The next stage of devolution is unlikely to be harmonious and could lead to the government's demise if the Flemish push the pace. This final phase will involve the setting up of regional parliaments and consequent reform of the national parliament, and the final delineation of national and regional responsibilities. A major divergence between the CVP and PS appeared in early 1989 over the possible abolition of parity in the national parliament between Flemish and French speakers. Prime Minister Wilfried Martens called for proportional representations which would give the Flemish a permanent majority of seats.

From 1989, the three regions (Flanders, Wallonia and Brussels) became responsible for public works, transport, housing, water, most aspects of environment and energy policy, scientific research and export promotion. The three language communities (Flemish, French-speaking and German-speaking) are responsible for education, health and cultural affairs. Central government remains responsible for overall economic policy, monetary and financial policy, labour law, social security policy (including unemployment pay), technical standards, public procurement, consumer protection and foreign trade policy. The central government only transferred 85 per cent of the funds necessary for new responsibilities in 1989, imposing an immediate budget- cutting exercise on the regions and communities.

The June 1989 elections for the Brussels regional council resulted in a coalition similar to the five-party central government coalition. The PS, election winner with 18 out of 75 seats, set up with the PSC (9 seats), CVP (4 seats), SP (2 seats) and Volksunie (VU | 1 seat), and in addition the Front Democratique des Francophones (FDF | 12 seats). The Flemish parties were guaranteed three out of seven cabinet posts. Country-wide elections for the local administrations took place in October 1988.

In February 1989, six parties (CVP, PS, SP, PSC, VU, FDF and French-speaking Liberals PRL) agreed to limit election campaign spending to BF50m per campaign and BF200,000-500,000 per constituency and candidate in return for government grants totalling BF15.4m. The grants will be divided between the parties on the basis of votes at the previous election, in addition to a small set sum. The Flemish Liberal Party (PVV) and the Greens (Ecolo-Agalev) did not agree with the scheme. The extreme nationalist Flemish party, Vlaamse Blok (VB), was not invited to take part.

KEY FACTS

Official title: The Kingdom of Belgium

Head of state: King Baudouin I

Head of government: Prime Minister Wilfried Martens (CVP)

Ruling party: Coalition of Christelijke Volkspartij (CVP), Parti Socialiste (PS), Parti Social Chrétien (PSC), Socialistische Partij (SP) and Volksunie (VU)

Capital: Brussels

Official Languages: Dutch, French and German

Currency: Belgian Franc (BF) = 100 centimes

Exchange rate: BF39.0 per $ (Nov 1989)

Area: 30,518 sq km

Population: 9.92m (1988)

GDP per capita: $15,345 (1988)

GDP real growth: 4.2% (1988)

Labour force: 4.2m (1987)

Unemployment: 10.9% (1988)

Inflation: 1.2% (1988)

Trade balance: $1,105m (1988)

Foreign debt: BF1,087bn (1988)

Anti-nuclear view prevails

The anti-nuclear influence of the Socialists resulted in Belgium giving the lead within NATO against deployment of short-range nuclear missiles in early 1989. By April 1989, the government called for negotiations on short-range nuclear missiles to start as soon as possible. It declined to take any decision on extending the range of its Lance missiles before 1991-92, when the US-Soviet talks on conventional weapons are due to come to a conclusion. At the NATO summit in June 1989, Belgium declared any increase in defence spending 'unacceptable'. The 1989 and 1990 defence budgets, set at around BF100bn, contained zero growth. In December 1988, the government awarded Italian firm Agusta a BF12bn helicopter order, with 73 per cent of the contract value guaranteed to Belgian firms. In May 1989, French firm Dassault won a BF11bn order to supply anti-radar devices and Sidewinder missiles on condition that 80 per cent of the contract value be placed with Belgian firms.

In June 1989, Belgian Air Force Colonel Guy Binet was convicted of selling NATO secrets to 'a foreign power'. Belgian businessman Gustave Keteleer, on trial in Brussels for fraud in 1989, was said to have handled bonds used to pay for US arms to Syria and Iran, part of the Iran-Contra affair. Belgian-owned Drexel Lambert Bank was also allegedly involved in the deal. Arrest warrants were issued in January 1989 for two Antwerp businessmen accused of supplying false documents and acting as intermediaries for West German chemicals supplied to Libya's alleged chemical weapons plant.

In June 1989, the government withdrew a BF300m loan to China in protest at the killings in Tiananmen Square and offered political asylum to Chinese dissidents. Prime Minister Wilfried Martens expressed Belgium's support of moves towards democracy in Poland during a visit there in March 1989. Closer relations with the Soviet Union were established with the signing of an investment protection agreement in February 1989. Bell-Alcatel won a major contract to supply telecommunications equipment.

Relations with Algeria improved with the signing of a new gas supply agreement in June 1989. Belgium agreed to increase its volume of imports over five years from 3m cubic metres to 4.5m cubic metres a year at a lower price of $2.30 per million BTU. Belgium also agreed to pay BF11bn in compensation for its abrogation of the original agreement. Also in June, the Belgian-Libyan agreement to revive trade led to the release of Dr Jan Cools, who had been held hostage in Lebanon for six months. Belgium was to send 30 paramilitary police to Namibia in September 1989 for six months as part of the United Nations observer force for the elections. Relations with former colony Zaire were in crisis from late 1988 to summer 1989 when Zaire stopped debt repayments and called into question the 1960 independence settlement with Belgium.

KEY INDICATORS

	Unit	1984	1985	1986	1987	1988
Population	'000	9,858	9,859	9,910	9,920	9,920
Gross domestic product (GDP)	BF bn	4,534	4,843	5,115	5,323	5,608
GDP per capita	BF	459,931	491,226	516,145	536,593	565,323
Inflation	%	6.3	4.9	1.3	1.6	1.2
GDP real growth	%	2.3	0.7	2.2	2.4	4.2
Unemployment	%	14.1	13.3	12.3	11.9	10.9
Share prices	1985=100	90	100	144	170	168
Industrial production	1985=100	98.6	100.0	101.0	103.6	108.8
Investment goods	1980=100	101.9	105.9	106.3	104.5	–
Consumer durables	1980=100	99.6	100.6	104.7	106.5	–
Consumer non-durables	1980=100	109.1	111.4	113.2	115.5	–
Intermediate goods	1980=100	100.0	101.0	101.3	102.4	–
Manufacturing	1980=100	102	104	106	108	114
Construction	1980=100	57	54	54	54	60
Exports (FOB)	BF bn	2,992.1	3,167.7	3,070.3	3,100.1	3,381.1
Imports (CIF)	BF bn	3,195.8	3,317.8	3,065.2	3,110.1	3,386.5
Petroleum	%	14.0	11.9	7.1	6.6	–
Balance of trade	$m	–1,218	–472	556	–180	1,105
Current account	$m	–55	669	3,055	2,794	3,379
Foreign exchange	$m	3,608	3,969	4,630	8,363	8,306
Exchange rate	BF per $	57.79	59.36	44.66	37.34	36.77

* estimated figure

Preparing for 1992

In late 1988, the government set up a 1992 committee headed by the Socialist MP Anne-Marie Lizin to co-ordinate policies to help strengthen Belgian industry for the opening up of the EC internal market. The government already had four measures in the pipeline. The first phase of tax reform, voted in in December 1988, reduced and simplified personal taxation as of the 1989 tax year. The top rate was reduced from 70 per cent to 55 per cent and the average cut was 10 per cent. The government introduced a bill to reduce corporation tax as of 1990 from the 43 per cent standard rate to the average effective rate applied after deductions prior to 1990 of around 32 per cent. In the new system, deductions will be limited, tax returns will be simpler and the cost of collection less. The government planned to keep a special tax regime for the 'co-ordination centres' of multinational companies, a successful measure in attracting new investment. The easing of the personal tax burden also reduced the social security burden for companies, since the two percentages are linked. However, social security charges remain among the highest in the EC.

The law on share transparency came into effect in June 1989. Shareholders trading in publicly quoted stock of companies with capital of over BF250m must inform the Banking Commission and the company whenever their holding of voting shares reaches 5 per cent of total voting rights. Companies may decide to lower the threshold to 3 per cent. Belgium's second largest bank, BBL, was the first to do so when launching a new share issue in June 1989. The law increased the supervisory powers of the Banking Commission. The government drafted new anti-trust legislation in summer 1989, setting up a monopolies commission to which all mergers will have to be submitted for approval.

Under the stock exchange reform, banks will be allowed to buy up to 25 per cent of the equity of the 200 brokerage firms from January 1990. This will strengthen the brokers and should bring some of the *hors-Bourse* business back into the exchange. Smaller share and bond operations will have to be conducted on the exchange. The Computer Assisted Trading System (CATS) was introduced in January 1989. In June 1989, the Peterbroeck Committee recommended major reductions in brokerage fees through improved trading techniques to increase competition and boost transactions.

The government also opened up the management of public debt with a view to developing the secondary market in government stock and so increasing public finances. The share of foreign banks underwriting government issues was increased from 5.6 per cent to 9.6 per cent. Banking fees for holding government debt were reduced, while commissions for placing it were increased. With regard to the six public credit institutions – to be amalgamated into two groups – the government proposes to limit its guarantee and so force them to compete with the commercial banks.

Société Générale de Belgique, the holding company raided in 1988 by Italian businessman Carlo De Benedetti and rescued by French financial group Suez, contributed significantly to the 47 per cent increase in stock exchange operations in 1988 to BF395bn. The raid helped hasten share transparency and stock exchange reforms. The company, whose activities represent about 30 per cent of Belgian GDP, reorganised into three industrial divisions and one financial division, shed some peripheral activities and acquired new companies to streamline its business. The space, defence and telecommunications division of loss-making subsidiary ACEC were sold off, but jobs were guaranteed.

Budget musical chairs

The 1988 budget deficit was BF433.9bn, 7.7 per cent of GNP, better than expected but largely on account of the unexpected surge in economic growth. The aim of the 1989 budget, drawn up in March, was to reduce the deficit to BF405.4bn or 6.9 per cent of GNP. But GNP was expected to rise by 5.7 per cent, too high an estimate. BF561.8bn of total government receipts of BF1,563.6bn were transferred to the regions and communities to cover expenditure of BF587.7bn.

The Walloon and Brussels regions drew up balanced budgets of BF100bn and BF30bn respectively. A large part of the Brussels budget went to paying the losses of the urban transport company, one of the heaviest state company loss-makers. The Flemish region and community seemed headed for an uncharacteristic deficit of over BF20bn in 1989 under pressure from the new SP members of the executive coalition. The PVV criticised the executive for failing to adopt a definitive budget for the first half of 1989, claimed that the deficit was higher than stated and warned against following the debt spiral example of central government.

The 1990 central government budget was set at BF850bn with the deficit to be held to BF405bn, but BF60bn in cuts were necessary to achieve this. By July 1989, no ready solution had been found. In 1989, the government hoped to increase revenues by BF35bn through more active management of the reserves of the central bank, a law to cut down rampant tax evasion in the construction industry, reorganisation of the finance ministry and an increase in excise taxes. But debt service charges will increase from 1990, when rescheduling agreements with the banks expire. The 1989 debt service was already BF442bn.

Unexpected boost to the economy

GDP growth doubled in 1988 to 4.2 per cent, investment rose by 13.7 per cent, industrial production by 5.4 per cent, while inflation fell to 1.2 per cent and unemployment fell to 9.6 per cent from 11.1 per cent in 1987. Private consumption increased by 2.5 per cent, while government consumption stagnated. Export growth picked up to 8.9 per cent and that of imports to 8.7 per cent. The balance of payments surplus increased to $3.1bn. In 1989, the pace of economic growth is expected to slow to 3.25 per cent and investment to 8.25 per cent. Industrial production will rise by 4 per cent and the unemployment rate should remain unchanged, as will private and government consumption increases. Inflation, however, is expected to rise to 3 per cent.

The metalworking, machinery and plastics industry federation, Fabrimetal, agreed wage increases of between 1.7 and 3 per cent (to be negotiated by sub-sector) for 1989-90. The government offered a 2 per cent pay increase to civil servants in 1989, but the trade unions pressed for a two-year agreement. Negotiations were complicated by the devolution of budgets and powers.

In Flanders, industrial production rose by 3.1 per cent in 1988 and private investment soared by 23.5 per cent. Unemployment fell, especially the number of long-term unemployed. Economic priorities for 1989 were investment and employment. With this in mind, the executive decided to close the loss-making KS coalmines at end-1992 rather than end-1996 and use the funds set aside for coal for new investment and job creation. After violent clashes between miners and police, a redundancy settlement was reached in June 1989. In contrast, the executive decided to continue subsidising the shipbuilding sector because of the return to the economy in subcontracting, taxes and jobs.

The 1989 priority of the Brussels economic programme was urban renewal, implying continued strong investment in construction. In Wallonia, 1988 industrial production rose by 6.2 per cent and private investment by 27.5 per cent, while unemployment fell, a performance which should be sustained in 1989. The executive's region-aided investment policy therefore proved a success, but the executive decided to tighten aid conditions in 1989.

Investment and orders increase

The metalworking sector saw 1988 orders increase by 7.6 per cent in volume and deliveries by 5.7 per cent. Exports rose by 5.6 per cent and imports by 5.9 per cent. Employment fell by 0.8 per cent. Prices rose by 2.3 per cent and investment increased to BF45bn, mainly in additional production capacity, which was up by 15 per cent (excluding the car sector). Investment is expected to increase by 24 per cent to BF55bn in 1989.

The primary metals processing sector increased deliveries by 14 per cent, with growth led by steel. Main steel company Cockerill-Sambre had a 1988 net profit of BF7bn with BF20bn in cash flow. In 1989, the company planned to repay BF4.2bn of its BF9.2bn debt to the Walloon Region, its 98 per cent shareholder since devolution. The steel giant expected increased profits and production in 1989.

Shipbuilding saw a 20 per cent increase in 1988 orders and a 10 per cent rise in deliveries. In contrast, railway equipment deliveries were down by 40 per cent and orders by 60 per cent. Armament and general machinery orders rose by 6 per cent each, mainly for export. Deliveries of electrical and electronic goods increased by only 1 per cent while orders rose by 12 per cent. Deliveries of telecommunications equipment were down by 10 per cent, largely due to delays in Belgian government contracts, but orders increased by 40 per cent. 1988 car production increased by 3 per cent, turnover by 5 per cent and jobs by 4 per cent.

1988 turnover in the chemicals industry increased by 12.6 per cent to BF834bn, while the volume of production increased by 8 per cent and profitability improved by 7 per cent. Construction had a good year with a 19 per cent increase in the number of homes and 50 per cent rise in offices built. Investment was BF421.8bn, a 15 per cent increase. Growth should continue in 1989, with a 6 per cent rise in the number of homes and 20 per cent increase in offices and investment of BF461bn, a 9.3 per cent increase. Turnover in the diamond industry rose by 28 per cent in 1988 to BF461bn. Exports rose by 25 per cent, imports by 30 per cent, and prices, by 13.5 per cent.

Outlook

The political calm that reigned in 1988-89 is unlikely to continue into 1990 as the debate on the next phase of devolution gets under way. Party differences could reach a critical point over the urgent need to cut spending on loss-making state companies and on social services, including health insurance, as debt service charges place an extra burden on the national budget from 1990. The coalition may have to continue debudgetising items, a practice it is pledged to repeal. The government faces two other apparently less crucial but no less volatile issues, the legalisation of abortion, which has produced a crisis of conscience within the CVP, and the linguistic quarrel over the Fourons district, which although nominally resolved continued to simmer through July 1989.

Much work remains to be done to put the machinery of the 1989 devolution into effect and to clarify the responsibilities of central and regional governments. There is no reliable estimate so far of the cost of state reform and the extra bureaucracy involved, but some account will have to be given before the next phase is agreed. Business federations complained of confusion and additional cost in submitting project applications to both the regional and national levels. Simplification of procedures will be essential if the major reforms, agreed by central government to help Belgian business face stiffer international competition from 1992, are to bear full fruit.

The pace of GDP growth is expected to slow to 2.75 per cent in 1990. The increase in investment will ease to 4.25 per cent. Private consumption will remain relatively strong with an increase of 2.75 per cent, sustained by the growth in real disposable income. Exports and imports are expected to stay buoyant with growth of 6 per cent and 6.25 per cent respectively. The balance of payments surplus will remain at $2.75bn. Inflation is expected to fall to 3.5 per cent. Unemployment, at 9.75 per cent, will remain at roughly the same level as in 1988 and 1989.

COUNTRY PROFILE

Political system: Constitutional and hereditary monarchy, consisting of nine provinces. Legislative power vested in monarch and bicameral parliament (Senate and Chamber). Senate has 184 members, 51 elected by provincial councils, 26 co-opted by the elected senators, 106 directly elected and 1 member by right (Prince Albert). The Chamber of Representatives has 212 members (deputies) directly elected by proportional representation for a period of four years. Both chambers can propose and veto legislation, though the political make-up of the government is determined by the composition of the Chamber of Representatives. The existence of three communities (French, Flemish and German-speaking) and of three regions (the Walloon, Flemish and Brussels regions) is constitutionally recognised and various powers have been devolved to their councils (parliaments) and executives (governments).

Main cities/towns: Brussels (population 1m), Antwerp, Ghent, Charleroi, Bruges, Liège and Mons.

Media
Press: Over 30 dailies in both French and Dutch, including *Le Soir* (circulation 208,833), *La Libre Belgique* (80,813, French), *De Standaard, De Morgen* (51,947) and *Het Laatste Nieuws* (301,306, Dutch). Various magazines and journals covering political and economic affairs. Principal business papers/magazines: *L'Echo de la Bourse* (30,000), *De Financieel Ekonomische Tijd* (24,830), *Tendances/Trends, Intermediaire, Belgian Business* (65,700), *Industries et Sciences, Prospects* (English).
Broadcasting: Over 3m TVs and 4.5m radios in use. 'Radio-Télévision Belge de la Communauté Française' (RTBF) broadcasts in French, and 'Belgische Radio en Televisie' (BRT) in Dutch. There are two commercial TV stations: RTL-TVi (French language) and VTM (Dutch). Several foreign broadcasts received, either direct or via cable. Cable TV covers most of the country. 625 independent radio stations in 1988.

Domestic economy: Characterised by broad industrial base with heavy dependence on external trade. Economic growth picked up after the introduction of tight fiscal/monetary policies in 1982, including the 1982 devaluation of the Belgian franc, and wage and price controls. Further efforts to stimulate exports and reduce unemployment and public spending were constrained by a growing debt burden and low levels of industrial activity.

Employment: Steady decline in employment in traditional industrial and construction sectors, and an increase in service sector. Unemployment rate stabilised owing to employment support measures, rise in service sector vacancies and more work-sharing. Real wages continue to lag behind inflation due to wage increase restrictions and price control. Workforce highly skilled and highly unionised.

External trade: Regular trade deficit reversed from 1983 onwards owing to strong growth in exports. In 1986 the current account showed a surplus for the first time since 1973. Many companies export more than 80 per cent of their production. Major world exporter of carpets, diamonds, copper products, glass, chemicals and motor vehicles. Around 70 per cent of imports consist of producer goods, of which metal products and petroleum predominate. Most trade still conducted with EC countries particularly West Germany, France and the Netherlands.

Agriculture: Although small-scale (contributing less than three per cent to GDP), cultivation is intensive, especially in Flanders. Self-sufficient in sugar, eggs, butter and meat, and exporter of vegetables and horticultural produce. Fishing less important, only for domestic market.

Industry: Large-scale, export-based industrial sector, with manufacturing contributing over 22 per cent to GDP. Main activities include metal working (leading EC producer of non-ferrous metals), chemicals, food processing, diamond cutting, textiles, glass and motor vehicle assembly. Rationalisation of traditional industries together with upturn in export demand led to growth in industrial production after period of rapid decline.

COUNTRY PROFILE

Mining: Only coal, clay and sand is mined on any substantial scale. Coal production from the Kempische Steenkoolmijnen in Limburg has steadily fallen due to declining demand for coke from industry and rapid change-over to nuclear fuel, and the mines will be closed in 1992.

Energy: Around 80 per cent of requirements met by imports. Since introduction of energy conservation programme in 1982 dependence on imported gas and oil has however fallen. Nuclear power accounts for 66 per cent of electricity output.

Banking: Divided into three main groups – commercial banks, public credit institutions and private savings banks. Some 85 commercial banks, of which largest are Générale de Banque, Banque Bruxelles Lambert and Kredietbank. The National Bank of Belgium is responsible for implementing monetary and credit controls, while the Institut Belgo-Luxembourgeois du Change maintains control over foreign exchange transactions.

Legal system: Civil law system influenced by English constitutional theory but largely French.

Membership of international organisations: ADB, BENELUX, BLEU, CCC, Council of Europe, DAC, EC, ECE, ECOSOC, EIB, ELDO, EMS, ESRO, FAO, GATT, IAEA, IBRD, ICAC, ICAO, ICES, ICO, IDA, IDB, IEA, IFAD, IFC, ILO, IMF, IMO, INTELSAT, International Lead and Zinc Study Group, IOOC, IPU, ITC, ITU, NATO, OAS (Observer), OECD, UN, UNESCO, UPU, WEU, WHO, WIPO, WMO, WSG.

BUSINESS GUIDE

Time: GMT + 1 (GMT + 2 from late Mar to late Sep).

Climate: Temperate, with warm weather from May-Sep. Summer temperature averages around 18°C, winter around 1°C. Hottest month Jul, coldest Jan. Rain throughout the year, wettest months Oct-Nov.

Entry requirements: Visas not required by nationals of USA, Japan and most western European countries for period of up to three months.
Currency: No restrictions on foreign or local currency movements. However, regulations are liable to change.
Customs: Personal effects and goods (up to the value of BF25,000 for non-European residents and BF7,500 for European residents) duty free.

Air access: Frequent first class and economy flights linking Brussels with over 90 international cities.
National airline: Belgium World (SABENA).
Main international airport: Brussels (BRU), 12 km from city centre.
International airports: Antwerp (ANR), Ostend (OST), Liège (LGG).

Surface access: Good road and rail access with the main centres of the Netherlands, France, Germany and Luxembourg. Also daily crossings by boat or jetfoil from Ostend or Zeebrugge to UK.
Main ports: Antwerp, Ghent, Zeebrugge, Ostend, Brussels, Liège.

Hotels: Advisable to book hotel or pension in advance either direct or through Belgium Tourist Reservations. No bookings can be made through national tourist office. All tariffs must be displayed by law. Average cost of hotel room with breakfast BF3,500/night. Service charge usually included. Major credit cards accepted.

Restaurants: Wide variety and numerous. Many serve Belgian specialities. Service charge is included in bill. Small tip for good service.

Car hire: Available at airports and in most main towns. Speed limits: urban roads 60 kph, main roads 90 kph. Maximum speed on motorways 120 kph, minimum speed 70 kph. Drive on the right, overtake on the left. Wearing of seat belts compulsory. Usually prohibited for children under 12 to sit in front seats.

City transport
Taxis: Readily available. Standardised fare system, which includes service charge. Chauffeur-driven cars cheaper on long runs.
Buses: Regular flat fare tram and bus service.
Other: Metro: Networks in Brussels and Antwerp.

National transport
Air: Limited internal service.
Road: Extensive road network. Toll-free motorways serve all main towns with the exception of those in the Ardennes. Comprehensive coach service, particularly to rural areas, operated by Société Nationale des Chemins de Fer Belges (SNCB) and Société Nationale des Chemins de Fer Vicinaux (SNCV).
Rail: Belgium said to have densest railway network in world. Service operated by Société Nationale des Chemins de Fer Belges (SNCB). First and second class service serves all main towns. Combined tickets allowing for stop-overs in main towns offer best value. Special airport shuttle service operates from Brussels Central Station and North Station. Express trains (TEE) ensure rapid connections with all French, Dutch and German cities.
Water: Over 1,500 km inland waterways, 50 km per 1,000 sq km. Service operated by Administration des Voies Hydrauliques. Inland canals connect with major French, Dutch and German ports.

Public holidays
Fixed dates: 1 Jan, 1 May, 21 Jul (National Day), 15 Aug (Assumption), 1 Nov (All Saints'), 11 Nov (Armistice Day), 25/26 Dec. Fixed-date holidays that fall on a Sun. are observed on the following Mon.
Variable dates: Easter Monday, Ascension Day, Whit Monday. (Also: 11 Jul – Dutch-Speaking Community Day and 27 Sep – French-Speaking Community Day).

Working hours
Business: (Mon-Fri) 0830 – 1630.
Government: (Mon-Fri) 0900 – 1700.
Banking: (Mon-Fri) 0900 – 1200 (or 1300) and 1400 – 1530 (or 1600).
Shops: (large variations) typically 0900 – 1900.

Social customs: Attention to language sensibility is a must. Two culturally distinct halves – Dutch-speaking Flanders in north, French-speaking Wallonia in south. Brussels speaks mainly French. Polite to be punctual and to shake hands on arrival and departure.

Business language and interpreting/translation: Available from embassies, tourist offices, chambers of commerce and many commercial agencies. German only spoken by a minority. English is widely understood.

Telecommunications
Telephone: Local and international direct dialling (00). Cost of local call BF10. International dialling code: 32 followed by 2 for Brussels. For 'telefax' transmission of documents dial 1214 or 1314 for conference calls.
Telex: Available in most major hotels. Country code 46 B.
Telegram: Service available from telegraph offices (open 24 hrs/day) and direct by telephone.

Postal service: Post offices open 0930 – 1200 and 1400 – 1600 (Mon-Fri), longer in main towns. Mail can be sent 'Poste Restante' to main post offices – passport required when collecting. Average cost letter inland BF13.

Advertising: Complicated by language problems. Information available from Ministry of Economic Affairs. Advertising spend (1985) BF30,817m. Key categories: department stores, shops, automotives and banking.

Trade fairs: Several international exhibitions, of which Brussels International Trade Fair and International Fair of Flanders in Ghent are most important. Information available from Ministry of Economic Affairs Exhibitions Office. Belgium ranks second in Europe for the number of international conventions.

Electricity supply: 220 V AC.

Hotels: Brussels
Arcade Stephanie, Avenue Louise 91-93, B-1050 (tel: 539-0240; tx: 25558; fax: 5380307).

Archimède, 22/24 Rue Archimède, B-1040 (tel: 231-0909; tx: 20420).

Bedford, 135 Rue du Midi, B-1000 (tel: 512-7840; tx: 24059; fax: 5141759).

Belson, 805 Chaussée de Louvain (tel: 735-0000; tx: 64921; fax: 7356043).

Europa, 107 Rue de la Loi, B-1040 (tel: 230-1333; tx: 25121).

Grand Scheers, 132 Boulevard Adolphe Max, B-1000 (tel: 217-7760; tx: 21675; fax: 2175750).

Hilton, 38 Boulevard de Waterloo, B-1000 (tel: 513-8877; tx: 22744; fax: 5137233).

Hyatt Regency, 250 Rue Royale, B-1210 (tel: 217-1234; tx: 61871; fax: 2178444).

Mayfair, 381/383 Avenue Louise, B-1050 (tel: 649-9800; tx: 24821; fax: 6486699).

Métropole, 31 Place de Brouckère, B-1000 (tel: 217-2300; tx: 21234; fax: 2180220).

Palace, Rue Gineste 3, B-1210 (tel: 217-6200; tx: 65604; fax: 2187691).

Park Yser, Avenue de l'Yser 21, B-1040 (tel: 734-4991).

President Nord, 107 Boulevard Adolphe Max, B-1000 (tel: 219-0060; tx: 61417; fax: 2181269).

Queen Anne, 110 Boulevard Emile Jacqmain, B-1000 (tel: 217-1600; tx: 22676).

Ramada, Chaussée de Charleroi 38, B-1060 (tel: 539-3000; tx: 25539; fax: 5389014).

Royal Windsor, 5 Rue Duquesnoy, B-1000 (tel: 511-4215; tx: 62905; fax: 5116004).

Sheraton, 3 Place Rogier, B-1210 (tel: 219-3400; tx: 26887; fax: 2186618).

Car hire: Brussels

Avis: international reservations (tel: 724-06-25; tx: 24051); Rue Americaine 145 (tel: 537-12-80; tx 24051).

Budget: (tel: 376-85-31; tx: 69082).

Europcar: international reservations, Rue St Denis 117, 1190 (tel: 344-76-30; tx: 65687); 235 Avenue Louise (tel: 640-94-00; tx: 20539).

Hertz: international reservations (tel: 735-40-50); Boulevard Lemonnier 8 (tel: 513-28-86; tx: 22140).

Business advice and information
Association of Belgian Contractors (ADEBOM), Avenue Grand-champ 148, B-1150 Brussels (tel: 771-61-08; tx: 21449).

Belgian Association of International Trading Houses (ABNEI), Israelietenstraat 7, B-2000 Antwerp (tel: 232-75-47; tx: 31588).

Belgian Foreign Trade Office (BDBH-OBCE), Boulevard Emile Jacqmain 162, B-1210 Brussels (tel: 219-44-50; tx: 21502 bexpo b).

BUSINESS DIRECTORY

Brussels Business Federation (VEB-UEB), rue Botanique 75, B-1030 Brussels (tel: 219-32-23).

Chambres des Agences-Conseils en Publicité (CACP), 28 Avenue du Barbeau, B-1160 Brussels (tel: 672-23-87).

Commission de la Bourse de Bruxelles (Stock Exchange), Palais de la Bourse, Rue H Maus 2, Brussels (tel: 511-34-06).

Committee of Belgian Consulting Firms, (BUROBEL), Avenue Louise 430, B-1050 Brussels (tel: 648-10-55; tx: 21591 Cebi b).

Fédération des Entreprises de Belgique (FEB-VBO Belgium Business Federation), 4 Rue Ravenstein, B-1000 Brussels (tel: 515-08-11; tx: 26756 febb).

Federation of the Metalworking, Mechanical and Electro-technical Engineering and Plastics Processing Industries (Fabrimetal), 21 rue des Drapiers, B-1050 Brussels (tel: 510-23-11).

Flemish Economic Association (VEV), Brouwersvliet 15, Box 7, B-2000 Antwerp (tel: 231-16-60; tx: 72893 vev b).

INBEL (Belgian Institute for Information and Documentation), Avenue des Arts 3, B-1040 Brussels (tel: 217-11-11).

Institut National de Statistique, 44 Rue de Louvain, Brussels B-1000 (tel: 513-96-50).

Ministry of Economic Affairs, Square de Meeus 23, B-1000 Brussels (tel: 512-66-90; tx: 61406 SEGECO B).

Ministry of Economic Affairs and Planning, 23 square de Meeûs, 1040 Brussels (tel: 511-19-30; tx: 21062).

Ministry of Finance, 12 rue de la Loi, 1000 Brussels (tel: 233-81-11).

Ministry of Foreign Trade, 2 rue des Quatre Bras, 1000 Brussels (tel: 516-83-11).

National Credit Insurance Office (OND-NDD) Square de Meeus 40, B-1000 Brussels (tel: 512-38-00; tx: 21147 OFNADUC B).

Tourism and Information, Brussels (TIB), Hôtel de Ville, Grand Place, B-1000 Brussels (tel: 513-89-40; tx: 65206).

Union of Walloon Enterprises (UWE), rue Capitaine Crespel 42, B-1050 Brussels (tel: 513-45-34).

Banking

Banque Nationale de Belgique (central bank), Boulevard de Berlaimont 5, B-1000 Brussels (tel: 221-21-11; tx: 21105; fax: 2213100).

Association Belge des Banques/Belgische Vereniging der Banken, 36 Rue Ravenstein, Box 5, B-1000 Brussels (represents privately owned banks (tel: 512-58-68).

Association des Caisses d'Epargne Priveés (private savings banks), 34-35 Place de Jamblinne de Meux, B-1040 Brussels (tel: 736-99-20; tx: 63186).

Banque Bruxelles Lambert, Avenue Marnix 24, B-1050 Brussels (tel: 517-21-11; tx: 21421).

Banque Paribas Belgique, WTC - Tour 1, Avenue Emile Jacqmain 162, B-1210 Brussels (tel: 219-30-10).

Caisse Generale d'Epargne et de Retraite/Algemeene Spaar-en Lijfrentekas (state-owned), Rue Fosse aux Loups 48, B-1000 Brussels (tel: 213-61-11 ; tx: 26860).

Generale de Banque, Montagne du Parc 3, B-1000 Brussels (tel: 516-21-11; tx: 21283).

Kredietbank, Arenbergstraat 7, B-1000 Brussels (tel: 517-41-11; tx: 21207).

Principal newspapers

Belgian Business, 42 Ave du Houx, 1170 Brussels (tel: 673-81-70; tx: 23830).

De Financieel Ekonomische Tijd, 5 Browersvliet, Bus 3, 2000 Antwerp (tel: 231-57-56; tx: 32614).

De Morgen, 84 Moutstraat, 9000 Ghent (tel: 21-80-91; tx: 11264).

Het Laatste Niews, 105 Emile Jacqmainlaan, 1000 Brussels (tel: 219-32-90; tx: 21495).

La Libre Belgique, 127 Blvd Emile Jacqmain, 1000 Brussels (tel: 211-27-77; tx: 21550).

L'Echo de la Bourse, 131 Rue de Birmingham, 1070 Brussels (tel: 526-55-11; tx: 23396).

Belize

Trevor Petch

With a general election due by the end of 1989, the pace of political activity had begun to build up towards all-out campaigning by the middle of the year. The governing United Democratic Party (UDP) of Prime Minister Manuel Esquivel emphasised its record on economic policy, taking measures to ensure a visible distribution of benefits in its March budget, while the opposition People's United Party (PUP) cast around for a decisive political issue which would turn the campaign in its favour. All the indications were that the election would be close, with the result determined by a handful of marginal seats where local and personality factors would play as important a role as national questions.

During 1988 it had appeared that the decisive issue might be that of a possible outline agreement to the long-standing border dispute with Guatemala, but by mid-1989 that seemed unlikely. Although a Joint Commission between the two countries met regularly from June 1988, after a year of talks it appeared that the two countries had failed to find sufficient common ground to enable a draft treaty to be agreed. A number of economic and security co-operation questions were resolved, but the main problem separating the two countries – Belize's absolute commitment to the non-negotiability of its sovereignty, and Guatemala's historic ambition for its own sovereign access to the Atlantic through its own territorial waters – apparently proved irreconcilable, at least within the time limit represented by the onset of campaigning for the presidential election in Guatemala in November 1990. No further substantive negotiations are to be expected before 1991. Nevertheless, Guatemala's acceptance of the necessity of direct negotiations with Belize, together with the dialogue which has taken place on secondary matters, indicate that a solution may be possible then. With Guatemalan approval, Belize is set to join the Organisation of American States during 1990.

Agriculture – still a struggle

Belize continued to experience strong export-led growth during 1988, with real GDP increasing by about 7.5 per cent. Although agricultural production was affected by poor weather, international prices for the major exports – sugar, citrus concentrate and bananas – continued to develop favourably. The outlook for the medium term for these products remains good: all three benefit from protected markets or tariff benefits which will continue into the 1990s.

During late 1988 and early 1989, a number of disease problems affected various agricultural sectors, including an outbreak of sigatoka in the banana fields, worries over rust in the canefields and a small infestation of Mediterranean fruit fly. The latter has been a periodic brake on attempts to develop the export of alternative crops such as vegetables and tropical fruit. A combination of production and marketing problems have also hampered the development of exports of other agricultural products such as beef, red kidney beans and cocoa, and the exploitation of new markets, particularly in CARICOM. The fishing industry (primarily lobster, conch, crab and shrimp) continues to be affected by over-exploitation, while the infant shrimp-farming industry, despite many years of effort, has yet to resolve initial production problems.

The UDP had pinned many of its hopes for rapid development on attracting overseas private investment in large projects, many of which have been slow to arrive or have never got off the ground at all. During 1989, however, a major hotel development in Belize City was eventually begun, although under the auspices of the Ramada chain rather than Camino Real as originally intended, and with a significant injection of capital required from the US government agency OPIC (Overseas Private Investment Corporation) and

KEY FACTS

Official title: Belize

Head of state: Queen Elizabeth II

Head of government: Prime Minister George Price

Capital: Belmopan

Official Languages: English (Spanish also spoken)

Currency: Belize Dollar ($BZ) = 100 cents

Exchange rate: $BZ2.00 per $ (fixed)

Area: 22,973 sq km

Population: 174,000 (1988)

GDP per capita: $1,138 (1987)

GDP real growth: 7.6% (1988)

Inflation: –0.3% (1988)

Trade balance: –$60.9m (1988; merchandise)

Foreign debt: $123.9m (1988)

Visitor numbers: 136,000 (1988)

a smaller shareholding from the government itself. Also during 1989, the national Petroleum Corporation of Jamaica (Petrojam) re-opened the Libertad sugar factory in Belize for the production of wet alcohol as part of its own ethanol production scheme. Nevertheless, the long-term viability of this project remained in doubt, primarily because of the problem of securing an adequate cane supply at a price economic both for Petrojam and for the cane farmers.

Despite the doubts over the prestige projects, there has been a significant amount of investment on a smaller scale, and this contributed during 1988 to a significant increase in imports of capital equipment. During 1989, the expansive tendencies of the domestic economy were accentuated by a heavy programme of government investment (with much of the funding from foreign aid sources) in roads, bridges, telecommunications and other projects, including completion of the refurbishment of the international airport and commencement of a new hospital in Belize City. The forthcoming elections appeared to have played some part in the timing of this burst of activity, which followed a tax-cutting budget.

After a crushing defeat in 1984, the PUP has slowly recovered. A right-wing faction having split away in early 1985, by 1989 it appeared the more unified of the two parties. The candidate selection process for the governing UDP was particularly acrimonious and surrounded by a number of accusations of rigged conventions. These developments, reflecting internal tensions within the party between the three separate political strands which united to form an opposition to the PUP in 1973, appeared likely to cost it both votes and some seats. At least one disaffected UDP supporter had already announced that he would run as an independent by July 1989. The PUP was unable to find an issue on which it could seize to press home a decisive political advantage, although it chose as a major focus the Citizenship Bond Scheme by which foreigners investing US$25,000 in an interest-free bond became entitled to Belizean citizenship. The PUP itself had planned such a scheme, but it was implemented by the UDP government, with Hong Kong Chinese as the target market. Both this issue, and the campaign in general, developed into the vicious mud-slinging of all Belizean elections.

Problems to be solved

The new government faced a number of strategic problems, most of them familiar. Despite the favourable external situation and the success of the UDP government in bringing the public sector under control, poor infrastructure, caused partly by the maintenance difficulties inherent from the geography and climate, and an irregular and expensive power supply continued to hamper further development. In the absence of discovery of oil in commercial quantities despite decades of optimistic exploration, a first attempt at development of hydro-electric power announced during 1989 (to be completed using Chinese aid) and the purchase of electricity from Mexico, primarily to supply a new industrial free zone on the northern border, offered some alleviation to at least one of these obstacles. Any reduction in the bill for imported energy would be welcome: any increase in the general standard of living tends to be translated directly into imports of consumer goods, a population of only 174,000 spread over an area twice that of Jamaica representing an inadequate market for efficient import substitution industries. At the same time, while urban unemployment continues to be held in check largely by emigration, there are significant labour shortages in agriculture, which are made up by immigration, legal or otherwise, from the surrounding countries.

Since those who come in are largely of Hispanic descent, and those who leave largely Afro-Belizean, the alteration of the ethnic balance of Belizean society

has created a number of tensions. The UDP government relaxed its initially harsh position regarding immigration, somewhat in response to the needs of the agricultural economy, and it was hoped during 1989 that an influx of UN funds to alleviate the effects of the refugee problem in Central America as a whole would go some way to relieving some of the pressures on Belizean infrastructure and social resources.

A second, increasingly severe problem is the use of Belize by international cocaine traffickers as a transshipment route. These supply lines follow those established earlier in the 1980s for the smuggling of marijuana, domestic production of which has been brought under a fair degree of control by aerial spraying. The shift to cocaine has brought with it a surge in domestic drug abuse and related crime; Belize lacks the resources to deal effectively with either the drug smugglers or the problems which follow in their wake.

A third area of increasing concern is that of the social and physical environment, and in particular the effects upon it of the development of the tourist industry. Belize is becoming a more popular destination both for water-sport enthusiasts and for nature and adventure tourism. One effect of this has been to encourage rampant land speculation in a number of resort areas of the country, which will undoubtedly worsen as investors see profitable opportunities in providing first-class facilities, which are at present lacking. Along with tourism development have come the familiar problems of pollution and ecological damage, with the fear that easy earnings from tourism will exacerbate the problems in generating balanced agricultural and light industrial development. In an attempt to limit the future negative impact of unrestricted tourism, the government introduced an outline policy strategy early in 1989, but the tensions between quick income-earning opportunities and longer-term social and ecological effects will prove difficult to resolve.

KEY INDICATORS

	Unit	1983	1984	1985	1986	1987
Population	'000	160	162	166	170	173
GDP	BZ$m	308.5	337.3	343.9	393.4	393.6*
Inflation	%	5.4	3.9	3.3	0.8	2.0
GDP real growth	%	–2.0	4.5	4.1	3.3	5.0*
Exports	BZ$m	130.3	186.4	180.3	185.3	198.7
Sugar	BZ$m	68.3	65.1	45.9	62.9	54.0
Citrus	BZ$m	13.7	19.6	24.2	22.2	31.0
Exchange rate	BZ$ per $	2.0	2.0	2.0	2.0	2.0
External debt	$m	–	96.5	118.4	121.4	138.4
Debt service ratio	%	–	4.0	11.6	9.5	7.3

* estimated figure

COUNTRY PROFILE

Historical profile: Mayan civilisation flourished for several centuries until approximately 1000 AD (although some continuation to 16th century); highly developed astrological calendars and concept of zero among attainments; probable first settlement date 1638; British sovereignty 1670; convention on border with Guatemala 1859; British colony 1884; universal adult suffrage 1954; Legislative Assembly elected 1954; internal self-government 1964; name change from British Honduras Jun 1973; independence 21 Sep 1981.

Political system: Independent state within the Commonwealth with the British monarch as head of state, represented by a governor-general. The governor-general is advised by the cabinet (led by a prime minister) which holds executive power. The legislature is the National Assembly consisting of the Senate (eight members, five chosen by the prime minister, two by the opposition leader and one by the governor-general) and the House of Representatives (28 members elected by universal adult suffrage for a five-year term).

Main cities/towns: Belize City (population 50,000), Orange Walk, Dangriga, Corozal, Belmopan (capital, 4,000).

Media
Press: Five weeklies: *Amandala, The Beacon, The Belize Times, The People's Pulse* and *The Reporter* – no dailies.
Radio: Radio Belize transmits in Spanish and English. *Television:* Controlled satellite services available on about twelve television stations.

External trade: Balance of payments steadily improved after 1985, but trade deficit has increased owing to greater level of economic activity, particularly in investment and construction.
Exports: Sugar including molasses (34% of total value in 1986), citrus concentrate (34%) bananas, shellfish and clothing. Main destinations: USA, UK, Canada, Mexico, CARICOM countries.
Imports: Machinery and transport equipment (typically 25% of total value), food etc (25%), manufactured goods, mineral fuel, chemicals etc. Main sources: USA (40% in 1985), EC (16%), Canada, Jamaica and Mexico.

Agriculture: Sector forms mainstay of economy, contributing 20 per cent to GDP and accounting for about 65 per cent of foreign exchange earnings. Production of sugar (the main cash crop, accounting for half the total exports) has fluctuated widely owing to recurrence of disease, falling world demand and inclement weather. Growth has switched to banana and citrus production, mainly oranges. Winter vegetables, papayas, mangoes and cocoa are also grown for export, while rice, maize, roots, beans and vegetables are produced for local consumption. Areas of farmland are also given over to illegal growing of marijuana. Dairy farming, livestock and bee keeping are also important. Fishing, mainly for lobsters, conch and shrimp, contributes significantly to the economy. Earnings from the fish exports (largely to the USA) amounted to $BZ12m in 1988, compared with $BZ35m from citrus concentrate. Approximately 40 per cent of total area is considered to have arable potential but only around 10-15 per cent is cultivated; 45 per cent of total area is forest, much of which is commercially exploitable.

Industry: Sector is small-scale with manufacturing contributing around eight per cent to GDP. Production is centred on agricultural processing (sugar and citrus fruits) and clothing for export. Emphasis is on encouraging the establishment of processing and assembly for export. A number of small import-substitution industries exist. Alcohol produced from sugar cane under an agreement with the Jamaican state petroleum company, Petrojam.

Tourism: Tourism, centred on diving and sailing on the coral reef and archaeology and wildlife tours inland, has expanded rapidly since 1984. Earnings in 1988 were approximately US$27.55m. Modernisation of the international airport and extensive construction of new hotel rooms in Belize City is expected to lead to further growth.

Energy: Commercial energy requirements are supplied by petroleum and products, all of which are imported. Despite an extensive oil exploration programme, commercial quantities have so far not been found. An UEP scheme (2.4 mW capacity) was scheduled to be started in 1989.

Membership of international organisations: CARICOM, Commonwealth, Economic Commission for Latin America, FAO, GATT, IDA, IFAD, IFC, ILO, IMF, OAS from 1990, state associated with the EC, UN, UNESCO, WHO, WMO, World Bank.

BUSINESS GUIDE

Time: GMT -6.

Climate: Sub-tropical with temperatures ranging between 10°C and 30°C. Hottest months between Mar-Sep and rainy season Jun-Oct.

Entry requirements
Passport: Required by all. Assured onward passage required.
Visa: Not required by citizens of the Commonwealth, USA, Japan and certain European and Latin American countries. USA visa required if arriving via Miami.
Currency: Declaration on arrival. N.B. Keep a copy of the declaration form because travellers are not permitted to export more than this amount of currency. A maximum of $BZ100 can be exported. Travellers should have at least $BZ75 for each day of intended stay. This figure includes moderate accommodation and food for a day in Belize City.

Health precautions
Mandatory: Yellow fever vaccination certificate if travelling from infected area.
Advisable: Anti-malaria; tetanus, typhoid, polio and rabies vaccinations. Water precautions should be taken.

Representation overseas: Berne, Caracas, Hong Kong, Kingston, London, Mexico City, New York, Panama City, San José, San Salvador, Seoul, Taipei, Tel Aviv, Tokyo, Washington.

Air access
Airlines serving: Aerovias, Continental, Taca International, Tan Sahsa.
International airport: Philip Goldson International (BZE), 16 km from Belize City; duty-free shop, bar, restaurant, buffet, post office, shops, bank.

Surface access
Main port: Belize City.
Overland access: Main routes are from Melchor de Mencos (Guatemala) and Chetumal (Mexico).

Hotels: Approximately 163 hotels and guest houses. Book well in advance. Five per cent government tax and usually 10 per cent service charge.

Car hire: Foreign or international licence acceptable for 30 days. Driving on the right.

City transport
Taxis: Fixed rates apply within Belize City (higher at night). No meters. Tipping is discretionary.

National transport
Air: Maya Airways and Tropic Air operate domestic services to main centres.
Road: Over 1,500 km of surfaced road – but difficulties in rainy weather often reported.
Buses: Service operates within Belize City; long-distance coach services to major centres.
Water: Regular ferry services and small boats ply to offshore cays.

Public holidays: 1 Jan, 9 Mar (Baron Bliss Day), Good Friday, Holy Saturday, Easter Monday, 1 May (Labour Day) 24 May (Commonwealth Day), 11 June (Queen's official birthday) 10 Sep (National Day), 21 Sep (Independence), 12 Oct (Colombus Day), 19 Nov (Garifuna Settlement Day), 25 Dec, 26 Dec.

Working hours
Government: (Mon-Fri) 0800 – 1200, 1300 – 1700; closes 1630 on Fridays.
Banking: (Mon-Fri) 0800 – 1300; also (Fri) 1500 – 1800.

Telecommunications
Telephone: Dialling code for Belize: IDD access code + 501 followed by area code (Belize City 2, Belmopan 8) followed by subscriber's number. Most telephones in Belize City are nearly 100 per cent automatic.
Telex: Modern facilities available at Belize Telecommunications Ltd (BTL) Public Booth, Belize City, Belmopan and San Ignacio.

Banking
Banking centre: Belize City.
Central bank: Central Bank of Belize.
Commercial banks: Atlantic Bank, Bank of Nova Scotia (Scotiabank), Barclays Bank International, The Belize Bank Ltd.

Electricity supply: 110/220 V AC, 60 cycles.

Representation in capital: Colombia, Panama, UK (also Mexico, Honduras and USA in Belize City).

Hotels: Belize City
Bellevue, P.O. Box: 428 (tel: 2-77051; tx: 275 bellevue).

Chateau Caribbean, 6 Marine Parade (tel: 2-72813; tx: 230).

Fort George, 2 Marine Parade, P.O. Box: 321 (tel: 2-77242/3; tx: 220; fax: 3820).

The Villa, 13 Cork Street (tel: 2-45743).

Hotels: Belmopan
Belmopan Convention Hotel, Bliss Parade (tel: 22130).

Circle A Lodge, 37 Halfmoon Avenue (tel: 22296).

Car hire: Belize City Elijah Sutherland: 127 Neal Pen Road (tel: 73582).

Europcar National: International Airport Road, Ladyville (tel: 2117).

Lionel Gorden: 113 Amara Avenue (tel: 72184).

Smith's Auto Rentals: 125 Cemetery Road (tel: 73779).

Business advice and information: Belize Export and Investment Promotion Unit (BEIPU), Carlton House, PO Box 291, 7 Cork Street, Belize City (tel: 44138, 44193; tx: 121 chambers; fax: 30755).

The Belize Bank, 60 Market Square, Belize City (tel: 77132/5; tx: 158 bzebank; fax: 72712).

Belize Tourist Bureau, PO Box 325, Belize City (tel: 77213).

Belize Chamber of Commerce and Industry, PO Box 291, Belize City (tel: 73148; tx: 121 chambers).

Central Bank of Belize, PO Box 852, Treasury Building, Belize City (tel: 77216; tx: 225; fax: 77106).

Benin

Kaye Whiteman

The regime of President Mathieu Kérékou went through a major crisis in January 1989, which it was lucky to survive. The background was economic, coming at the moment of clinching a difficult deal with the IMF. But the strikes and riots of that month also marked a personal crisis of confidence for an increasingly embattled president.

The turn of the year into 1989 marked a critical period for the Peoples' Republic of Benin. Both economically and politically, the regime of President Kérékou deteriorated through the 1980s and in January 1989 both crises converged. Most importantly, the situation of the economy had reached breaking point, and the place where the break came was in the banking system. This had been entirely state controlled since the introduction of the Marxist–Leninist state in 1974 (the year the country changed its name and became a 'peoples' republic'). Confidence in the system was already low, and lowered further by the depleted state of public finances. Revelations of exorbitant loans from the state bank, the Banque Commerciale du Benin (BCB), to government officials led to a run on the bank, the only one in Benin. This triggered a full-scale flight of CFA francs to neighbouring Togo, and a situation in which the BCB could no longer pay depositors, and non-government employers could no longer pay staff. This compounded an already grave situation in which the treasury was already unable to pay government employees.

The main focal point of discontent was the education system. On 9 January 1989, teachers in Cotonou and Porto Novo went on strike in protest at non-payment of salaries, to be immediately joined by university students who had likewise not received grants, and were fearful that the announcement of 10 January that the government's negotiations with the IMF had been 'crowned with success' would mean a cutting back on grants. Fears of retrenchment in the civil service after an IMF deal caused industrial action to spread to civil servants. Discontent was particularly high in Porto Novo, traditional seat of radical politics in the country, where economic problems were also more acute due to a new year tightening up over the border in Nigeria, following a new anti-smuggling decree. Traders in Porto Novo have always done well out of contraband to and from Nigeria. It was in Porto Novo that a violent demonstration broke out on 23 January, in which public buildings were attacked, shops looted, pictures of President Kérékou were burnt, and two policemen were reportedly beaten to death. Following this 100 paratroopers were sent to the town to restore order and troops were ordered to shoot on sight. Numerous arrests were made, especially in Porto Novo, of those alleged to have stirred up discontent, including a former foreign minister, Tiamiou Adjibade.

The president, now fighting for his political life, adopted a carrot and stick approach. While continuing the security clampdown, and threatening to sack striking workers and students who did not return to work, he ensured the payment of some of the arrears owed. Members of the armed forces were in the forefront, as there had been reports of serious discontent. Benin has an army that is 4,350 strong (including a 350-man air force and 200-man navy). There are also 2,000 gendarmes and a people's militia of 1,500–2,000. In the army, it was said to be only the presidential guard of about 1,000 who were totally loyal to the president. Only they have access to the armoury, and only they were regularly paid. The settlement of arrears was carried out with help from an extraordinary subvention to the recurrent budget from France, the first such subvention for many years.

Military discipline was also at stake in the trial before the state security court of four army plotters alleged to have tried to overthrow the government in June 1988. The plotters were alleged to have planned the assassination of President Kérékou, and to have received assistance from the Libyan charge d'affaires, Mr Abousetta, who had been expelled the previous June. The court gave the leader, Captain Amadou Abdourahmane, 20 years detention, and the three others received lesser sentences. A fifth plotter, Mudashiru Durossimmi, was sentenced to life *in absentia* in July 1989. At the same time, the trial of those accused of having been involved in a much more serious attempt to overthrow the government in March 1988, was adjourned *sine die*. These included a former minister of education, Col. Hilaire Badjogoume, and Captain Hountoundji, said to have been the leader of the plot, who was earlier thought to have died in detention. Also accused *in absentia* was Colonel Francois Kouyame, a popular former minister who had escaped from detention in Parakou in December 1988, said to have joined the growing ranks of the Benin opposition in exile, largely based in Paris and Lagos.

Although, at the beginning of April 1989 some 50 people were released in an amnesty, some of whom had been in detention for four years, there was much concern at the conditions of Benin's many political detainees. A student, Luc Togbadja, was said to have died in prison in March, and in April Amnesty International expressed concern that eight people detained without trial since January may have been tortured. In recognition of the concerns of human rights campaigners, in mid-June the president visited the military camp to establish, he said, the number of political detainees.

Financial help from abroad

Meanwhile, President Kérékou's efforts to rehabilitate his position concentrated on keeping the economy afloat, as well as going through an exercise of renewal of political institutions. The January announcement that a deal had been done with the IMF was firmed up in May, with the approval of a $45m adjustment loan from the World Bank, for Benin's economic recovery programme 1989-91. The thrust of this programme was to give 'an enhanced role to the private sector', with trade liberalisation and market deregulation, further development of the agricultural sector, and pruning of the inefficient public sector, a restructuring of the banking sector and of the civil service. The programme also included measures to minimise any negative social impact from the programme. At about the same time the delayed 1989 budget was ratified, proposing expenditure of CfA104.5bn with a deficit of CfA28.7bn, or 6 per cent of GDP, but including draconian proposals for a 50 per cent reduction in allowances paid to civil servants and the military, and early retirement for 750 of them. The budgets of state and party were separated, and extra-budgetary accounts eliminated.

Shortly after, details of the deal with the IMF became known. Approval was given for a structural adjustment facility of $27m for the three years 1989-92. Of the total, $7.8m was to be disbursed during the first year. These were important amounts for a small

KEY FACTS

Official title: People's Republic of Benin	**Area:** 112,622 sq km
Head of state: President Mathieu Kérékou	**Population:** 4.3m (1987)
Ruling party: Parti de Révolution Populaire du Bénin	**GDP per capita:** $351 (1986)
Capital: Porto Novo	**GDP real growth:** –8.0%* (1987)
Official Languages: French (Fon, Yoruba, Bariba widely spoken)	**Inflation:** 5.0% (1986)
Currency: CFA Franc = 100 centimes	**Trade balance:** –$385.1m (1987)
Exchange rate: Franc CFA316 per $ (Nov 1989)	**Foreign debt:** $1,133m (1987)
	* estimated figure

country, and reflected the fact that this was Benin's first IMF programme. At the same time, the adjustment programme involved some very harsh political decisions, especially in the area of cutting back the top-heavy civil service, and deregulating the state sector. Reorganising the collapsed banking sector was a high priority, with particular emphasis on persuading outside capital to re-establish private banks. Among these was a Benin–Nigeria bank, set up as part of Nigeria's contribution to the rehabilitation of the Benin economy. This contribution also involved financial assistance, and provision of oil at concessional prices in an attempt to limit the smuggling of Nigeria's subsidised petrol over the border.

The main support for Benin, however, was from the French. Even before the IMF deal, despite the doubts about Kérékou's position, they had helped the regime to survive, and once Benin was tied up with the IMF, became even more supportive, announcing in May two loans worth a total of $26m to help ease the government's liquidity shortage, and help in the banking sector's restructuring. The new inflow of funds helped the government's situation somewhat, but the payment of salaries was still behind, and there was sporadic evidence of deep discontent in the two big towns, Cotonou and Porto Novo. Apart from providing banking specialists, France attached customs experts to the port of Cotonou to help improve tax collection.

Part of Kérékou's strategy for survival was to go through a new electoral process, with elections on 18 June 1989 for a new single list in the Revolutionary National Assembly (ANR). Official figures stated that 86 per cent of the 2m voters went to the polls, and of those 89.6 per cent supported the list of the ruling Parti de Revolution Populaire (PRPB), who make up the 206 peoples' commissars of the ANR. The fact that the 'no' vote increased from 1.5 per cent in 1984 to 8.7 per cent in 1989 excited some comment, but there was still scepticism as to the authenticity of the figures, in view of the evident high level of discontent in the country. In July, the ANR chose its 15-member permanent committee which, at the beginning of August, unanimously selected Kérékou as president of the republic for another five-year term.

KEY INDICATORS

	Unit	1983	1984	1985	1986	1987
Population	m	3.81	3.93	4.04	4.17	4.30
Gross domestic product (GDP)	CFA bn	439.4	466.2	499.8	507.2	–
Real GDP growth	%	–2.0	0.5	2.4	–3.2	–
GDP per capita	Franc CFA	115,328	118,626	123,713	121,630	–
Agricultural production	1979-81=100	104.6	130.0	136.4	145.8	130.9
Food production	1979-81=100	103.1	125.9	132.2	139.3	126.4
Exports	$m	66.5	166.7	122.9	100.5	123.0*
Imports	$m	294.0	288.1	297.6	314.2	508.1*
Balance of trade	$m	–227.5	–121.4	–174.7	–213.7	–385.1*
Total external debt	$m	714	681	819	942	1,133
Debt service ratio†	%	11.2	15.3	15.5	24.5	15.9
Exchange rate	CFA per $	381.1	437.9	449.3	346.3	300.5

* estimated figure

COUNTRY PROFILE

Political system: Executive power is held by the president, nominated by the Parti de Révolution Populaire du Bénin at a congress where the National Council of the Revolution instituted a National Revolutionary Assembly (196 members elected by universal adult suffrage). The term of institutions has been increased from three to five years. Gained independence in 1960 as the Republic of Dahomey. Renamed Benin 1975.

Political parties: The Parti de Révolution Populaire du Bénin is the only legal political party, basing its activities on Marxism-Leninism.

Population: Most of the population retain animist beliefs, although 15 per cent are Christians and 13 per cent are Muslims.

Main cities/towns: Cotonou (population 420,000), Porto-Novo (150,000, capital), Parakou (68,000), Abomey (57,000), Natitingou (39,000), Ouidah (60,000).

Language: French is the official language and the universal business language; African languages (Yoruba, Bariba and Fon) are widely used in everyday life.

Media
Press: The government information bureau publishes the daily *Ehuzu*, and the fortnightly *Journal Officiel de la République du Benin.* The Chamber of Commerce publishes a weekly bulletin.
Broadcasting: The state-owned Office de Radiodiffusion et de Télévision du Bénin broadcasts radio services in French, English and 18 local languages, and television transmissions for eight hours weekly.

External trade: Large trade deficit only partly offset by foreign remittances and other invisible earnings, notably transit trade and aid transfers. In 1986 the balance of payments weakened further because of low cotton and petrol prices.
Exports: Cocoa, cotton, palm products, coffee. Main destinations: Spain, German FR, France, Portugal, Netherlands.
Imports: Textiles, machinery, foodstuffs, petroleum products. Main sources: France, Austria, Brazil, Netherlands, Italy.

Agriculture: Sector contributes around 43 per cent to GDP (1986). Subsistence farming (mainly collectivised) and animal husbandry form cornerstone of economy, employing 85 per cent of workforce. Principal food crops are yams, cassava, sorghum, beans, millet, maize and rice. There is virtual food self-sufficiency. The chief cash crops, cotton and palm oil, are farmed mainly on large industrial plantations. The most recent statistics suggest that cotton output has been declining. Other cash crops include coffee, sugar, cocoa, karité nuts and tobacco. Livestock farming is important in the north; fishing is confined mainly to inland waters.

Industry: Sector is small-scale, contributing around 14 per cent to GDP (1986) and employing 16 per cent of the workforce. Production is centred on processing primary products (palm oil, fats, sugar, beverages, cotton) for export, and the manufacture of consumer goods and construction materials for home consumption. Cement production by the state-owned companies, SCB and Sonaci, and oil refining are the main heavy industries. Negotiations are in progress with the IMF to reform state enterprises.

Mining: Activity confined to extraction of limestone for local cement industry. Known reserves of phosphate, chromite, uranium, low grade iron-ore, marble and gold.

Hydrocarbons: Production of oil from offshore fields currently running at 450,000 tonnes per annum (9,000 b/d), sufficient to meet domestic demand. Exploration of oil and gas continuing with Swiss support.

Energy: Traditional fuels account for over 70 per cent of total energy consumption. A large part of the electricity is imported from the hydroelectric Akosombo dam in Ghana.

Membership of international organisations: African Development Bank, African States Associated with the EC (Lomé Convention), Arab Bank for Economic Aid in Africa (recipient), Economic Commission for Africa, French Monetary Area, GATT, IDA, IFAD, FAO, ILO, IMF, OAU, UN, West African Economic Community (observer status), WHO, WIPO, World Bank.

Time: GMT +1.

Climate: Equatorial in the south with average daytime temperatures reaching 30-38°C. Main dry season from Jan to Mar. Rainy season from May to Jul and from Sep to Dec. Very humid in coastal areas. The north is tropical with more extreme temperatures, and single dry and rainy seasons. Length of rainy season varies with location but it is generally very wet from Jul to Oct.

Entry requirements
Passport: Required by all except nationals of OAU countries with proof of identity.
Visa: Required by all except nationals of France, German FR, Denmark and OAU countries. Advisable to obtain a multi-entry visa. Visas cannot be obtained from neighbouring Togo or Burkina Faso. Exit visas required if visit exceeds 48 hours, except for French nationals. A return or onward ticket is also required.
Prohibited entry: Nationals of South Africa, Israel, Chad and Morocco.
Currency: No restriction on import of local or foreign currency, but amounts must be declared on arrival. Currency exports are limited to the declared amounts.

Health precautions
Mandatory: Yellow fever vaccination certificate required.
Advisable: Cholera, typhoid and polio vaccinations. Malaria prophylaxis should be taken as risk exists throughout country. Water precautions are necessary, especially outside towns.

Representation overseas: Abidjan, Accra, Algiers, Beijing, Bonn, Brussels, Bucharest, Havana, Kinshasa, Lagos, Moscow, Niamey, Ottawa, Paris, Tripoli, Washington.

Air access
National airline: Transports Aériens du Bénin (TAB). (Benin is also a shareholder in Air Afrique).
Other airlines: Aeroflot, Air Afrique, Air Burkina,

BUSINESS GUIDE

Cameroon, Ghana Airways, Libyan Arab, Nigeria Airways, Sabena, UTA.
International airport: Cotonou-Cadjehoun (Code: COO), 4.8 km from city; car hire, taxis.

Surface access
Main ports: Cotonou.
Overland access: Routes from Burkina Faso, Togo, Nigeria and Niger.

Hotels: Available in all main towns. Accommodation in Cotonou is limited. Advance booking essential. Service charge usually included in bill, otherwise 10 per cent tip.

Credit cards: Access/Mastercard accepted by hotels, etc.

Car hire: Available in Cotonou. Chauffeur-driven services are recommended. Insurance/liability position should be checked. International driving licence required.

City transport
Taxis: Fixed charge within towns, but advisable to negotiate fares in advance. Tipping is optional.

National transport: Travel by foreigners outside Cotonou is restricted and visitors are advised to clear plans with the authorities.
Air: Regular government-run services between Cotonou, Parakou, Natitingou, Kandi and Djougou. Airport tax of 250 CFA on all internal flights.
Road: Good main roads in south connecting towns to Cotonou and Porto Novo. Mainly laterite, but the main coast road, which connects Lagos with Accra, is surfaced, and the road north from Cotonou is surfaced to Savalou. Bus services link towns on these main routes. In northern areas some roads are only passable in dry season.
Rail: Services run by OCBN. Two classes available on railcars; second class rated fairly basic; restaurant and sleeping cars not operated but light refreshements available on some services. Daily services include: Cotonou-Parakou (journey time guide, seven to eight hours); Cotonou-Pobé (four hours); Cotonou-Segboroué (2.5 hours).

Public holidays
Fixed dates: 1 Jan, 16 Jan (Martyr's Day), 1 Apr (Youth Day), 1 May, 26 Oct (Armed Forces Day), 30 Nov (National Day), 23 Dec (Women's Day), 25 Dec, 31 Dec (Production Day).
Variable dates: Eid al-Fitr, Eid al-Adha (Tabaski).

Working hours
Government and business: (Mon-Fri) 0800/0830 – 1230, 1500 – 1830.
Banking: (Mon-Fri) 0800 – 1130, 1500 – 1630.
Shops: (Mon-Sat) 0830 – 1300, 1600 – 1930; (Sun) 0800 – 1200. Shops that open Sun mainly close Mon a.m.

Telecommunications
Telephone: Direct dialling via Paris. Dialling code for Benin: IDD access code + 229 followed by subscriber's number (there are no area codes).
Telex: Facilities available in Cotonou at Post Offices and good hotels.

Banking
Central bank: Banque Centrale des États de l'Afrique de l'Ouest.
Main centre: Cotonou. *Headquarters:* Banque Commerciale du Bénin, Banque Beninoise pour le Développement.

Advertising: In the press, radio, Benin's seven cinemas and on hoardings and trains. Direct mail not effective and all advertising is strictly controlled. No TV advertising.

Electricity supply: Electricity supply 220 V AC 50 cycles.

Representation in capital: Algeria, Bulgaria, China PR, Cuba, Egypt, France, German FR, Ghana, Korea PDR, Libya, Niger, Nigeria, USSR, USA, Zaire.

Hotels: Cotonou
Aledjo PLM Azur, P.O. Box: 2292 (tel: 330561; tx: 5180; fax: 331574).

Sheraton, Boulevard de la Marina, P.O. Box: 1901 (tel: 300100, 301256; tx: 5111; fax: 300155).

de la Plage, Boulevard de France, P.O. Box: 36 (tel: 312560; tx: 5032).

du Port, P.O. Box: 7067 (tel: 314443; tx: 5032).

Hotels: Natitingou
Tata Somba PLM Azur, P.O. Box: 4 (tel: 216590; tx: 5354).

Other useful addresses
Agence Benin-Presse, PO Box 120, Cotonou.

Banque Centrale des États de l'Afrique de l'Ouest, BP 3108, Avenue des Barachois, Dakar, Senegal (tel: 21-16-15, 23-10-42; tx: BCEAO 21815 SG, 21817 SG).

Caisse Centrale de Co-opération Economique (CCCE), PO Box 38, Cotonou (tel: 31-35-80, 31-45-81; tx: 5082).

Chambre de Commerce et d'Industrie du Bénin, PO Box 31, Cotonou (tel: 31-32-99).

Import/Export Alimentation de Benin, PO Box 53, Cotonou.

Institut Nationale de la Statistique et de L'Analyse Economique, PO Box 323, Cotonou (tel: 31-40-81; tx: 314103).

Mission Permanente d'Aide et de Coopération, PO Box 476, Cotonou (administers aid from France).

Société Nationale d'Equipement, PO Box 2042, Cotonou (deals with capital goods).

Société Nationale de Commercialisation et d'Exportation du Benin (Sonaceb), PO Box 933, Cotonou (tel: 31-28-22).

Société Nationale de Commercialisation des Produits Pétroliers (Sonacop), PO Box 245, Cotonou (tel: 31-22-90; tx: 5245).

Société Nationale d'Importation du Bénin, PO Box 2042, Cotonou.

Bermuda

Tony Thorndike

A surprise general election resulted in serious losses for the governing party, mainly because of popular dissatisfaction with Bermuda's economic performance caused in the main by poor tourist receipts. The delicate ethnic balance within the government was also upset.

The election on 9 February 1989 centred largely on economic, immigration and environmental issues, and a majority of the 42,352 electorate who cast their vote expressed dissatisfaction at the record of the United Bermuda Party (UBP) and the aloof leadership of its first black leader, John Swan. But the small population continues to enjoy one of the highest standards of living in the world, with a per capita income in 1989 exceeding $20,000, and a sustained consumer boom has created inflationary problems. Not many Bermudans will support the call for wage restraint made in the March budget, although the slowdown in the tourist trade is beginning to have effect.

Political shock

Just over three years after the last election in October 1985, Premier John Swan went to the country in an attempt to quell backbench unrest. The 31 UBP seats were whittled down to 23, the party's share of the vote falling to an all-time low of 50 per cent. The opposition Progressive Labour Party's representation rose to 15 seats from the seven it won in 1985, winning one seat from the PLP-breakaway National Liberal Party which was left with only one seat. The restoration in the PLP's fortunes was welcomed with relief by its leader Frederick Wade following its poor showing in 1985.

The UBP has been in power since the granting of self-government by Britain in 1968. Its traditional base was the conservative white merchant aristocracy established with British colonialism in the eighteenth century, and latterly with the business community. By responding to social change and needs it not only survived but prospered. John Swan assumed the leadership in 1982 and forced through a series of reforms in the teeth of strong opposition. The traditionalists were eventually forced out of office by Swan via two successive elections in 1983 and 1985. But they, to a man very conservative, regrouped as the 'Concerned Bermudians', and their persistence

KEY FACTS

Official title: Bermuda
Head of state: Queen Elizabeth II
Head of government: Premier John W. Swan
Ruling party: United Bermuda Party
Capital: Hamilton
Official Languages: English
Currency: Bermuda Dollar (BD$) = 100 cents
Exchange rate: BD$1.00 per $ (fixed)
Area: 55.4 sq km

Population: 57,700 (1987)
GDP per capita: $18,000* (1987)
GDP real growth: 4.0% (1987)
Inflation: 5.3% (1988)
Trade balance: –BD$530m (1986)
Foreign debt: $627m (1987)
Visitor numbers: 583,388 (1988)

* estimated figure

was rewarded when most were returned to ministerial office in 1989. Their return was at a cost: seven of the eight defeated UBP members were black, an unforeseen development that seriously damaged the ethnic balance within the party and its claim to national status.

There were a number of reasons for the UBP's poor showing. The downturn in the vital tourist industry was one. Poor publicity was blamed by some, but there is little evidence to support this. More to the point was a lack of consistency in planning, leading to an unacceptable erosion of scarce open areas on the 55 square-mile island. Other issues included inadequate consultation over new development projects, increasing pollution (and widespread disagreement over the suitability and cost of a refuse incinerator) both on land and in water, seemingly uncontrolled car imports and consequent traffic congestion and immigration. Concern at the latter was shared by the workforce. At the senior management level in the financial services sector, expatriate employment reached over 7,000, and their families compete for middle-level positions to the detriment of trained Bermudans. At the manual level, increasing numbers of Azoreans predominate. Neither was AIDS far from the public mind. Bermuda has one of the highest incidences of the condition in the world: at 1.6 cases per 1,000 it compares poorly with the United States's 0.3.

There was also the divisive independence issue. Independence was first proposed in the late 1970s and was recommended by a Royal Commission in 1978. But the prospect attracted little support, especially from within the UBP. However, the issue was revived by Swan in 1982. Twice he tried to persuade fellow Bermudians, in 1982 and 1986, but his proposal for a referendum was rejected. He then tried a new tack: the establishment of a UBP 'Independence Review Committee'. This published an Interim Report in July 1987 which argued the case for sovereignty and made a variety of suggestions to cater for defence needs and foreign representation. The defence question touched sensitive nerves both within and without Bermuda, as there has been a US naval facility on the island since the 1940s. Housing both US and British personnel, it is frequently used by nuclear-armed submarines. Indeed, the British government imposed a veto in August 1988 on a proposed consultation visit by the United Nations Decolonisation Committee because of the sensitivity surrounding the US base.

Following the Interim Report, Swan announced that

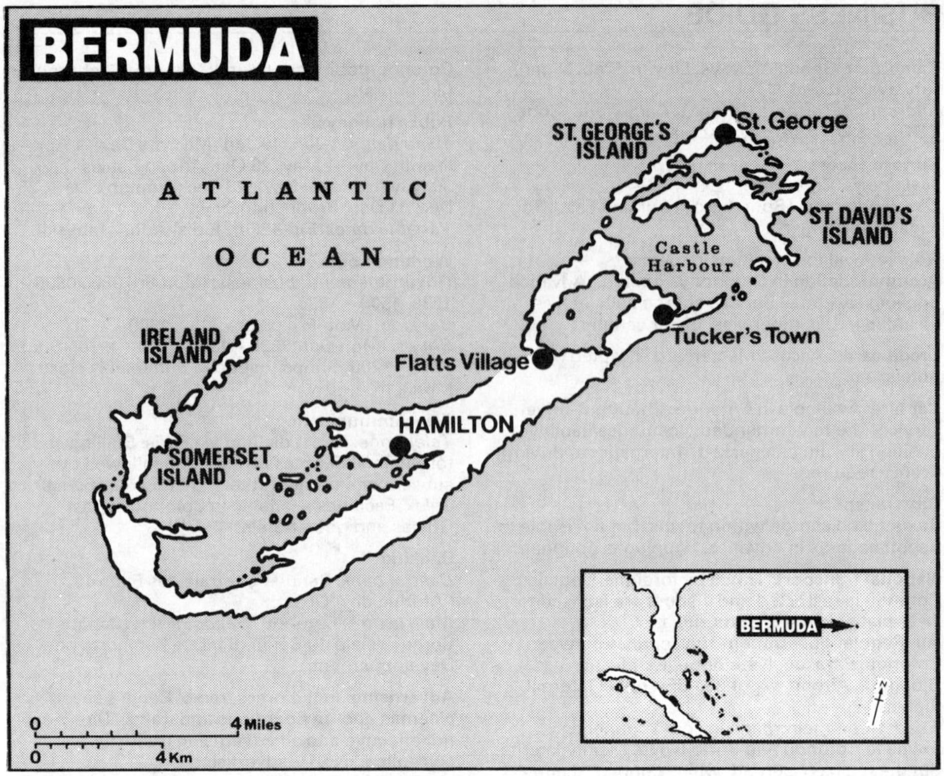

discussions would be sought with Britain, and that one option was for defence and diplomatic representation to remain British responsibilities, possibly in conjunction with the United States and Canada. However, the talks were aborted by British insistence that there was neither any possibility of 'limited independence' as advanced by Swan, nor any prospect of Britain sharing its remaining reserve powers. Britain, said Foreign Office Minister Tim Eggar, would willingly grant full independence (subject to arrangements securing the US base) if it was clear that there was public support for such a move. Embarrassingly, as the talks were proceeding in June 1988, a respected opinion poll was published suggesting 80 per cent of islanders opposed independence, compared to 66 per cent in 1982. It was clear that Swan could not continue his campaign and the independence issue featured relatively low down on the election agenda.

The two opposition parties had naturally made much of Swan's pro-independence posture, loudly proclaiming that he had no national mandate to proceed. But the resurgence of the UBP's major opponent, the Progressive Labour Party (PLP), was not attributable to that. Rather it was a combination of public disquiet and a reform of PLP policy after a highly-damaging internal party split. The PLP was founded in 1964 specifically to represent black working class interests. Its espousal of socialist policies became linked to a strong pro-black position under the leadership of Mrs Lois Browne-Evans; as a result it became associated in the public mind with outbreaks of racial violence in the early 1970s. A change of heart followed and a more centralist platform evolved, just as the UBP moved in the same direction, albeit from the right. But

COUNTRY PROFILE

Historical profile: Said to have been discovered in 1503 (by Spaniards led by Juan de Bermudez); English mariners stayed for some time 1609-10; charter given by James 1 to Virginia Company to include Bermuda in dominion 1612; first permanent settlers arrived shortly afterwards; islands subsequently sold to City of London; became property of Crown 1684; oldest British colony; representative government 1623.

Political system: British Crown colony with broad measure of internal self-government since 1968. The appointed governor is guided on most internal matters by a cabinet appointed from the bicameral legislature comprising a Legislative Council (11 appointed members – five chosen by governor, three recommended by premier, three by opposition leader) and a House of Assembly of 40 elected members (two from each of the 20 constituencies) elected for a five-year term.

Main cities/towns: Hamilton (population 3,500), St George (St George island).

Media
Press: The only daily is *The Royal Gazette.* There are several weekly productions, including *Mid-Ocean News, Bermuda Sun* and *Bermuda Times.*
Broadcasting: Two television and three radio channels operated by the Bermuda Broadcasting Corporation, which are commercial and privately owned. Also one independent radio station, the St George's Broadcasting Company which also operates a micro-wave pay-TV service. Bermuda Cablevision Ltd began an island-wide cable television service in July 1988.

External trade: Large trade deficit is offset by net invisible earnings from tourism and international business.
Exports: Rum, flowers and re-exports of petroleum and pharmaceutical products. Main destinations: Canada (26% of total value in 1987), USA (21%), Sweden (15%), UK (10%).
Imports: Manufactured goods (35-40% of total value), food, beverages, tobacco (25%), chemicals etc, mineral fuels, raw materials. Main sources: USA (55% of total value in 1987), UK (15%), Canada (10%), Japan, France, German FR, Hong Kong.

Agriculture: Less than six per cent of total area is cultivated arable land, most of which is used by tenant farmers for growing fruit, vegetables and flowers. Although self-sufficient in eggs and milk, around 80 per cent of food requirements need to be imported. Small fishing industry.

Industry: Manufacturing and construction combined contribute around 10 per cent to GDP and employ less than five per cent of the workforce. Major activities include ship repair, small boat building and manufacture of paints, perfumes, pharmaceuticals, mineral water extracts and handicraft souvenirs. Emphasis is on encouraging light industry in the Freeport area on Ireland Island North.

Tourism: Tourism is the mainstay of the economy, accounting for over 50 per cent of foreign exchange earnings and employing directly and indirectly 65 per cent of the workforce. Stop-over and cruise ship arrivals numbered respectively 425,020 and 158,368 in 1988.

Fiscal system: The island's liberal tax/exchange regulations, good international communications network has attracted numerous 'captive' insurance and finance companies. Estimated that offshore financial and corporate activities account for 16 per cent of GNP and some 40 per cent of foreign currency earnings.

the socialist tendency remained strong, and when the party was comprehensively defeated in 1985, those favouring a harder leftist line assumed the upper hand and engineered the defeat of the moderates. The moderate minority, which controlled two seats, left to form the National Liberal Party (NLP). The NLP also took away some of the PLP's trade union support, but that was more apparent than real as the trade unions, led by the influential Bermuda Industrial Union, began to show their independence, partly because of labour shortages.

An extra significance of the election result was that PLP support was much more multi-racial than hitherto. Bermudan society is characterised by subtle class and racial differences which are effectively hidden from casual visitors. To the average Bermudan, however, they are very real.

Economic retrenchment

About half of Bermuda's GNP (which reached US$1.38bn in 1987-88) derives from tourism. The industry also accounts for 65 per cent of employment. All records were broken in 1987 with 476,853 stopover visitors and 154,455 cruise ship excursionists. But 1988 saw a decline in the stopover category to 425,020, mostly of US visitors, although the excursionist trade increased slightly to 158,368. The 1988-89 winter season saw a further 11 per cent decline in stopover arrivals and the industry braced itself for a total less than 400,000 for the first time in 15 years. Because of this, the economy grew by less than 0.5 per cent in 1988. To worsen matters, inflation rose to 5.3 per cent and the balance of payments deficit swelled to US$56m.

Not surprisingly, the 1989 budget was restrictive. Although no new taxes were imposed, utility prices were raised. The foreign exchange allowance for overseas travel was also slashed. These measures were followed in May by higher interest rates. But calls for wage restraint fell on deaf ears: not surprisingly, given very high expatriate salaries and lifestyles to match. To make the point, a prolonged industrial dispute by the dock workers resulted in a pay rise of 16.4 per cent. Recurrent expenditure was budgeted at US$274.7m, down from US$279m in 1988, and revenue at US$307.4m, the surplus being for capital projects.

Steady financial centre

Fortunately the well-tuned offshore financial sector performed well, earning US$285m in 1988. Bermuda specialises in corporate rather than offshore finance and, as such, tends to reflect international business climates rather more than do other Caribbean financial centres. The dominant industry is insurance, which employs the majority of the 6,500 workforce in the sector as a whole. Over US$3bn turnover is generated annually, mainly for 'captive' insurance companies established by multinational corporations to handle their corporate risks, and increasingly those of other companies as well, up to 40 per cent of business in some cases.

About 6,300 companies are registered. All are tax free – guaranteed to 2006 – and operate under 'exempt' licences. Such a licence means that the '60-40' rule does not apply, whereby a company must be at least 60 per cent Bermudan owned. Of these, 1,300 are insurance companies (all but 200 being 'captives') and a large number of others are concerned with the licensing of patents, trademarks, software and copyrights. About 650 are shipping concerns. Over 3.5m tonnes of shipping is registered under Bermuda's 'flag of convenience' laws, rendering US$1.5m worth of fees per year. One advantage is that the local shipping register is technically part of the British register, backed up by British law, but with fewer restrictions. The company pre-registration process is vigorous to exclude criminal elements and to maintain the excellent bankruptcy record. Despite very high rents, the availability of technical and professional expertise is probably unrivalled. The '60-40' rule excludes offshore banking, which explains why there are but three banks. The two biggest are the Bank of Bermuda (recently expanded to include a Cayman Island institution) and N.T. Butterfield and Son. The third, the Bermudan Commercial Bank, one-third owned by Barclays, is growing fast. All three specialise in tax-free interbank and international business. Although no minimum deposit is required, new customers are accepted only after exhaustive enquiries.

A major boost to the insurance sector was the ratification in late 1988 of a tax treaty with the United States whereby US-owned 'captives' are exempt from the 4 per cent withholding tax imposed by the US Treasury on gross premiums, and which permits US corporations to tax-deduct the cost of conventions held in Bermuda. Bermuda was thereby put on the same footing as Barbados, its arch rival in the world insurance business. There is no doubt that present economic difficulties are not only temporary but that Bermuda's prosperity will not be challenged.

BUSINESS GUIDE

Time: GMT -4 (GMT -3 from Apr to Oct).

Climate: Sub-tropical with temperatures usually ranging between 16°C during winter (Nov-Mar) and 28°C during summer (Apr-Oct). No marked rainy season. Occasional hurricane-force winds in Sep.

Entry requirements
Passport: Required by all except USA and Canadian nationals holding proof of identification. British Visitor's Passport acceptable for Bermuda but not USA – for journeys via that country 'full' passport necessary. Return/onward ticket required by all travellers.
Visa: Required by nationals of Eastern European countries, Cuba, Argentina, Iran, Iraq, Libya and certain others (unless permanent residents of USA). Stays over three weeks are not automatically allowed without a permit – applications should be made at airport.
Currency: No restriction on import of foreign exchange but export must not exceed that declared on entry. No restriction on import of local currency; export limited to BD$250.

Air access
Airlines serving: Air Canada, American, British Airways, Continental, Delta, Eastern, Pan Am, Piedmont Aviation.
International airport: Kindley Field (BDA), 19 km from city of Hamilton; bar, restaurant, bank, shops, hotel reservations.

Surface access
Main ports: Hamilton, St George. Weekly cruises link Bermuda with New York in summer months.

Hotels: Generally expensive and advisable to book well in advance. Six per cent government tax payable on check-out on room rates. 10 per cent tips are expected except where 10-15 per cent service charge added to bill.

Car hire: Self-drive cars are not available and local licences are not issued for short stays. Numerous mopeds available for hire.

City transport
Taxis: Metered taxis available; also available by hour or day – legal rates applicable; those with blue pennant driven by qualified guides and are available on time basis for sightseeing; 10-15 per cent tip is usual.

National transport
Road: Around 250 km of well-surfaced roads. Traffic drives on left; speed limit is 35 kph. Two-wheeled vehicles are favoured.
Water: A number of harbour points are served by ferries.

Public holidays: 1 Jan, Good Friday, 24 May (Bermuda Day), Queen's Birthday (Jun), Cup Match and Somers Day (Jul/Aug), Labour Day (first Mon in Sep), 11 Nov (Remembrance Day), 25 Dec, 26 Dec.

Working hours
Government and business: (Mon-Fri) 0900 – 1700.
Banking: (Mon-Fri) 0930 – 1500; also 1630 – 1730 Fri.
Shops: (Mon-Sat) 0900 – 1700.

Telecommunications
Telephone: Dialling code for Bermuda: IDD access code + 1-809-2 followed by subscriber's (six-digit) number. There are no area codes. Direct dialling to UK, North America, West Indies, Europe and most other countries worldwide. Service is now digital electronic and there are public phone kiosks throughout the island.
Telex: Modern international facilities available.

Postal service: Advisable to use air mail service which is fast.

Banking
Banking centre: Hamilton. Bermuda has substantial offshore banking activities.
Commercial banks (head offices): Bank of Bermuda, Bank of N T Butterfield & Son, Bermudan Commercial Bank.
Joint stock/venture banks: Bermuda Commercial Bank Ltd, Bermuda Provident Bank, Bermuda Shields Trustee Co Ltd, Chartered Trust Co Ltd, European American Finance (Bermuda) Ltd, Montreal Trust (Bermuda) Ltd, Pierson, Heldring & Pierson (Bermuda) Ltd, Schroders (Bermuda) Ltd, Schroder International Trust Co Ltd. *Trust company:* International Trust Co of Bermuda Ltd.

Advertising: Available on all forms of media.

Electricity supply: 115-230 V AC, 60 cycles.

Business advice and information
Bank of Bermuda, Front Street, Hamilton (tel: 95-4000; tx: 3212).

Bermuda Chamber of Commerce, PO Box HM-655, Front Street, Hamilton (tel: 95-4201; fax: 92-5779).

Department of Tourism, Global House, 43 Church Street, Hamilton (tel: 92-0023).

Government Information Services, 113 Front Street, Hamilton (tel: 92-6470, 92-6384; tx: 3558 news).

Government Statistical Department, 74 Church Street, Hamilton (tel: 95-5151 ext 1613).

Hotels: Hamilton
Bermudiana, P.O. Box: 842 (tel: 95-1211; tx: 3264; fax: 8563).

Elbow Beach, P.O. Box: 455 (tel: 36-3535; tx: 3268; fax: 368043).

Grotto Bay Beach & Tennis Club, 11 Blue Hole Hill, P.O. Box: 1291 (tel: 93-8333; tx: 3476).

Hamiltonian & Island Club, P.O. Box: 1738 (tel: 95-5608; tx: 3493).

Harmony Club, P.O. Box: 299 (tel: 36-3500; tx: 3563; fax: 362624).

Inverurie, P.O. Box: 1189 (tel: 36-1000; tx: 3481).

Princess, P.O. Box: 837 (tel: 95-3000; tx: 3242).

Hotels: St George
Marriott's Castle Harbour, P.O. Box: 841 (tel: 93-2040; tx: 3219; fax: 938288).

Bhutan

Michelle Misquitta

Bhutan has sometimes, misleadingly, been called the Switzerland of Asia. There are indeed plenty of superficial similarities. The two countries are landlocked and small. With 47,000 sq km of area, Bhutan is slightly larger than Switzerland. The majestic Himalayan peaks are higher and mightier and even more awesome than the Alpine peaks and pretty valleys of Switzerland. There the similarities end. Switzerland has been able to use its special strategic position at the centre of Europe to enhance its status and specialness even though surrounded by Austria, France, Italy and West Germany, all of them much bigger, more populous and potentially mightier countries. Indeed, Switzerland has been able to preserve a very special independence. Bhutan has vast giants on its doorstep, but there are only two of them and they are China and India so that its room for political manoeuvre is limited. Moreover, whereas Switzerland has been able to turn its handicaps into assets, so that it has become almost the richest country in the world, Bhutan is trapped by its bleak landlocked position and is virtually the poorest country on earth. Its $150 annual per capita income estimated by the World Bank sits sadly and pathetically beside the $21,330 income of Switzerland.

Relations with India and China

In spite of its disadvantages, Bhutan is slowly beginning to make its way in the world, both politically and economically, though always carefully so as not to offend its giant neighbours. It was not long ago that the West had no name for Bhutan. Locally, the Bhutanese call their country *Druk Yul* – which means 'the Land of the Thunder Dragon'. Bhutan preserved a self-imposed isolation for hundreds of years, through rule by a priestly theocracy and then by an absolute monarch. Even the intrepid explorers of the British East India Company hardly managed to penetrate the ice curtain. Only in the past decade or so did Bhutan itself decide to join the rest of the world.

But it must tread carefully. It has a special relationship with India. Under Article 2 of the 1949 Indo-Bhutan Treaty, Bhutan 'agrees to be guided by the advice' of India in its external relations. Until recently, this meant that India was big brother and mentor to the young Himalayan kingdom. Only in the last few years has Bhutan opened other embassies abroad, and now has them in Dhaka, Geneva, New York and Kuwait besides its embassy in the Indian capital of New Delhi.

In October 1989 Bhutan's Minister of Foreign Affairs Lyonpo Dawa Tsering arrived in Beijing for Sino-Bhutanese border talks. China's Foreign Minister Qian Qichen described the relations between the two countries as 'very good all along' even though they have not yet established diplomatic relations and have yet to solve disagreements about their border. In Bhutan's capital of Thimpu, with a population of about 40,000, only India has a formal diplomatic office. Other countries that have diplomatic relations have their mission chiefs based in Delhi and concurrently accredited to Bhutan. India has objected to the Hindu kingdom of Nepal developing too close a relationship with China, so Bhutan has to be extremely cautious in its approach.

Gradual modernisation

Bhutan has also begun to increase its communications links with the outside world. Its fledgling airline Druk Air now offers direct flights between Paro, the nearest airport to Thimpu, and Calcutta, Delhi, Dhaka, Kathmandu and even Bangkok. The airline uses British Aerospace 146 aircraft.

But bringing Bhutan properly into the twentieth century is a different and more time-consuming matter. Only in 1907 did the theocracy give way to absolute monarchy. Only in the 1950s did the modern economic revolution start with the abolition of serfdom, introduction of paper money and the building of modern roads. Although an east-west road has been built, the difficult terrain with its high mountains and steep valleys means that communications between one valley and the next are difficult. Belief in old customs dies hard, and the 7,500m mountain Gangkar Punsum was recently put out of bounds to foreign tourists after local people complained that they were disturbing the spirit of the still unclimbed peak.

Tourism

Tourists number only about 2,000 a year, and have to pay a tax of between $110 and $250 a day depending on the season in Bhutan. The number is deliberately limited – and visitors are advised to apply three months or more in advance to secure their visas – and the Bhutanese are conscious of the need to preserve their land and its history against foreign depredations.

KEY FACTS

Official title: Kingdom of Bhutan

Head of state: Druk Gyalpo Jigme Singhe Wangchuk

Capital: Thimphu

Official Languages: Dzongkha

Currency: Ngultrum (Nu = 100 chetrums) – par with Indian rupee

Exchange rate: Nu16.7 per $ (Nov 1989)

Area: 47,000 sq km

Population: 1,345,000 (1987)

GDP per capita: $150 (1987)

GDP real growth: 6.7% (1984/85)

Trade balance: –$63.0m* (1987)

Foreign debt: $40.7m (1987)

* estimated figure

KEY INDICATORS

	Unit	1985	1986	1987	1988	1989
Population	m	–	–	–	6.7	–
Agricultural production						
Rice	'000 tonnes	1900	2,000	1,855	2,000	–
Maize	'000 tonnes	78	98	85	100	–
Cassava	'000 tonnes	100	105	112	115	–
Sugar cane	'000 tonnes	190	200	205	210	–
Bananas	'000 tonnes	92	100	106	110	–
ODA Disbursement	$m	13	13	14	–	–
Electricity	m kWh	–	–	–	–	–
Imports (OECD/USSR)	$m	–	–	–	–	–
Exports (OECD/USSR)	$m	–	–	–	–	–
External debt	$m	378	435	594	–	–
Debt service ratio	%	4.4	3.2	1.9	–	–
Trade balance	$m	–	–200	–	–	–
Exchange rate	R per $	–	–	–	–	150

* estimated figure

Some Buddhist temples and shrines have also been closed to foreigners, another reminder that it will be a difficult and unusual struggle to bring this remote kingdom fully into the twentieth century. One of the aims of the current sixth Five-Year Plan is to preserve and promote Bhutan's rich cultural and historical inheritance.

Economy

That plan runs from 1987 to 1992 and involves spending of Ngultrum 8811.2m ($730m). India is providing the lion's share of the funds, 40 per cent, though this is a considerable reduction on the early days of Bhutan's development when New Delhi provided 99 per cent of the funds. The king, Jigme Singhe Wangchuk, is the chairman of the planning commission, testifying to its importance. In the past most attention has been paid to education and health, but now a large chunk of the spending is being devoted to industry – centred on the two main urban areas of Thimpu and Puntsholing on the border with India – to export promotion and diversifying sources of taxation and government revenue.

Bhutan is on the list of the United Nations least developed countries. It is also getting assistance from the World Bank and from the Asian Development Bank. In 1988 the Asian bank approved two loans worth $3.65m on top of a loan of $3.3m the previous year. The money was provided to upgrade industrial estates and to provide foreign currency to Bhutan Development Finance Corporation to help private sector enterprises. Total ADB lending to Bhutan is more than $30m. In addition, the bank is helping the country with policy guidance.

Given the years of backwardness and the poor communications, development will be a Himalayan-high struggle. Most Bhutanese villagers have not even seen a simple thing like a shop. Unlike Switzerland, landlocked Bhutan is not at the centre of a modern developed region like Europe. Both of its neighbours are poor, and transport links in the mountainous region are as poor as anything anywhere in the world. India and China are mutually suspicious, and India has been stingy in permitting development of international transport communications involving India.

Some economists say that the $150 a year per capita income attributed to Bhutan is an understatement in real terms because the typical farm size in this country of 1.5m people is large by poor country standards. But agriculture is basic, consisting of production of small amounts of rice, buckwheat, mountain barley and millet. Buffaloes and yaks are main draught animals, testifying to the tough nature of the land. By any indicators, Bhutan scores poorly in international tables. Only 23 per cent of its children are enrolled in primary school, one of the lowest rates in the world. There is one doctor per 23,310 people. Life expectancy is 48 years. Manufacturing is a tiny element in GDP. Clearly, Bhutan has a long and steep mountain to climb.

COUNTRY PROFILE

Political system: Bhutan is a monarchy governed by the King, a nine-member Royal Advisory Council and a six-member council of ministers. Although the King is the chief executive authority, all laws are enacted and all important issues debated by the National Assembly which meets twice a year. 101 members are elected for three years and the other 49 members include officials, monastic representatives (10 seats), ministers and members of the Royal Advisory Council. This latter body has nine members, one representing the King, two the monasteries, and six regional representatives. Each family has one vote for National Assembly elections.

Political parties: There are no political parties.

Population: The population of 1,298,000 (1987) includes some 10,000 Tibetan refugees. Population growth rate is 2 per cent per annum, despite an infant mortality rate of 15 per cent (UN estimate). The state religion is Mahayana Buddhism. Bhutan enjoys a close relationship with India, which effectively guarantees its national defence.

Main cities/towns: Thimphu, Paro.

Language: The official language is Dzongka, which bears similarities to Tibetan.

Media
Radio: There are several radio stations, including those broadcasting flood information, with some 11,000 radios in use (1981).

Domestic economy: The economy is based almost entirely on agriculture and forestry, and Bhutan is almost self-sufficient in food production. Tourism and postage stamps are important sources of foreign exchange. India and international organisations are supporting Bhutan's series of five-year development plans with aid. Priorities in the current plan include extending the road network, improving education and health facilities, building further hydroelectric power stations, and improving irrigation. The government also seeks to decentralise decision-making.

Employment: Public administration accounts for 10.4 per cent of employment.

External trade: Bhutan is one of the 32 developing nations designated in 1985 to receive exemption from US competitive trade requirements in accord with a law known as the Generalised System of Preferences.
Exports: Main exports are cement, oranges, cardamom, timber, gum, rosin and handicrafts. Main destinations are India (90 per cent of all exports), the Middle East, Singapore and Western Europe.
Imports: Main imports are raw materials, fuel, food and consumer goods, mainly fron India.

Agriculture: Agriculture accounts for 90 per cent of employment (1985) and provides 50 per cent of GDP. About 15 per cent of the land area is fertile lowland arable and 65-70 per cent is forested. Main crops are rice, wheat, maize, mountain barley, millet, potatoes, pulses, tobacco, spices and fruit, especially apples and oranges. Cattle, sheep, goats, buffaloes, yaks, horses, donkeys, mules and poultry are raised. Some timber is exported.

Industry: The industrial sector accounts for 18 per cent of GDP. Manufacturing provides 1.5 per cent of employment (1981) and provides 14 per cent of GDP. Small-scale local industries produce woodwork, weaving, textiles, soap, metals, handicrafts, carpets, matches, food processing and plywood manufacture. A cement works development project is being undertaken with Indian aid and co-operation.

Mining: Mineral resources remain mostly untapped. Minerals present include dolomite.

Energy: There is some hydroelectric power generation (22m KWh in 1981), with projects to develop further use of hydroelectric resources carried out with Indian aid and co-operation.

Membership of international organisations: Asian Development Bank, Colombo Plan, ESCAP, FAO, G-77, IBRD, IDA, IFAD, IMF, NAM, South Asian Association for Regional Cooperation, UN, UNESCO, UPU, WHO, World Bank.

BUSINESS GUIDE

Time: GMT + 6 hrs.

Climate: In Thimphu, likely temperature range Jun-Sep (monsoon season) is 10-20°C; Dec-Feb, the likely maximum is 13°C with temperatures frequently below freezing. In southern areas/Duars Plain, temperatures are generally higher. Annual rainfall of 7,500 mm is sometimes recorded; over 5,000 mm usual.

Entry requirements
Passport: Required by all.
Visa: Visas are required by all, and application at least three months in advance is advisable. Tourists are admitted only on package tours.

Health precautions
Mandatory: Vaccination against cholera and yellow fever if arriving from an infected area.
Advisable: Anti-malarial precautions.

Representation overseas: Dhaka, Geneva, New Delhi, New York, Kuwait.

Air access
National airline: Druk-Air.
International airport: Paro (near Thimphu). Direct flights to India and Thailand only.

Surface access: Overland access is via the road from the Indian frontier (Jaigaon) to Phuntsholing.

Hotels: All hotel bookings are made through the Bhutan Tourism Corp. Private hotels are open only to Bhutan nationals, some Indian nationals and certain business contacts; state hotels are of adequate standard.

Car hire: Certain services are available, and local enquiries are recommended.

City transport
Taxis: Certain services are available, and local enquiries are recommended.

National transport
Road: The road network comprises some 2,000 km of road, largely surfaced. There are about 850 cars and some 1,150 jeeps.
Buses: Bus services are available between main centres. Local enquiries are recommended.

Public holidays
Fixed date: 17 Dec (National Day).
Variable dates: Paro Tsechu (Mar-Apr), Thimphu Tsechu (Sep-Oct).

Telecommunications
Telephone: Dialling for Bhutan is via the International Operator. Telephone service is restricted to main centres. There were about 1,300 lines in 1986.

Weights and measures: Metric system.

Representation in capital: Bangladesh, India.

Hotels: Paro
Olathang (tel: 58/99).

Hotels: Thimphu
Bhutan, P.O. Box: 111 (tel: 2546, 2596).

Business advice and information
State Trading Corp of Bhutan, 52 Trivoli Court, Ballygange Circular Road, Calcutta, 700019, India.

Banking
Royal Monetary Authority (RMA) (central bank), PO Box 154, Thimpu (tel: 2540; tx: 890206).

Bank of Bhutan, PO Box 75, Phuntsholing (tel; 225; tx: 890304).

Travel information
Bhutan Tourism Corp, PO Box 159, Thimphu (tel: 2570, 2854; tx: 312220 in Delhi; cable: BHUTRISM).

Bolivia

Patrick Knight

The new Bolivian Government faces immense challenges, but has several things in its favour.

Although the new government in Bolivia faces the immense challenge of getting the country moving again without provoking a return of the hyperinflation of the early 1980s, it has several things in its favour. One is that the price of tin and most of the wide range of other minerals which Bolivia produces, has risen significantly from the lows of the early and mid-1980s. Following re-organisation, Bolivia's tin production rose by at least 30 per cent in 1989 to 15,000 tonnes and is expected to rise further in future.

Another is that after more than a decade and a half of discussion, $1bn of World Bank finance has now been arranged for the construction of a gas pipeline which will carry 2 – 3m cubic metres a day of natural gas to the Brazilian border, as well as for a major power plant and two petrochemical plants to be built there which will guarantee Bolivia an income of at least $250m a year for the next 20 years. Once this new industrial pole begins to operate, and with plans beginning to take shape for the development of the massive quantities of iron ore, manganese and other minerals which are located on both sides of the border, this part of Bolivia will begin to develop fast. Bolivian gas will replace oil in a large cement plant on the Brazilian side of the border. It may also be used to make iron and steel and it has been proposed that a free zone should be set up in the area.

Improving navigation on the River Paraguay, which would benefit not only Bolivia, but Paraguay, Brazil and Argentina too, is also rising on the agenda. This could give Bolivia an outlet to the Atlantic as well as one to the Pacific through Chile. This is the good news which the strange hybrid government headed by Jaime Paz Zamora, but effectively run by ex-president and strong man Hugo Banzer Suarez, had to welcome it.

The bad news is that the economy is still stagnating – the 3 per cent growth in each of the past two years has been the result of the large inflow of foreign aid and capital, rather than any significant increase in production, while unemployment remains between 20 – 25 per cent. If Brazil starts to buy large amounts of gas from the mid-1990s onwards, Argentina will cease to do so in 1992. This will cut off what is now Bolivia's major source of income, as gas exports earned slightly more than $200m this year, 30 per cent of all export earnings. While the prospects for increased gas exports are promising in the medium term, although the reserves are not in fact so large as to give a cushion for more than another 20 years, there has so far been little sign of Bolivia's oil production increasing significantly. Revenues from the oil industry which now produces less than 20,000 barrels per day, are still the main source of government funding. Oil revenues provide 70 per cent of its income, and although the size of the public sector deficit was reduced dramatically – from more than 50 per cent of GNP at the commencement of the Paz Estenssoro Government to about 3 per cent at the end of it – and the income from taxation was raised to about 16 per cent of GNP, 90 per cent of government revenues still have to be spent on salaries and wages.

It remains to be seen whether the two virtually opposite tendencies within the new Bolivian Government will be able to work together in some sort of harmony; or whether men who until very recently used to be avowed political enemies, one from the left the other from the right, will spend all their time and energy fighting one another. The new president, Paz

Zamora, of the Movimiento Izquierdo Revolucionario, MIR, polled slightly less than 20 per cent of the popular vote for the presidency in the elections of May 1989. He came third, behind the finance minister of the previous administration, Gonzalo Sanchez de Lozada of the Movimiento Revolutionario Nacional, MNR, who got 23.07 per cent, and Hugo Banzer Suarez of the Acción Democratica Nacionalista, ADN, who got 22.74 per cent. In the previous presidential elections of 1985, Banzer had obtained a slightly larger number of votes than Paz Estenssoro. But when Congress came to decide which of the two men should govern the country, a different result prevailed and Dr Paz came to power, with Banzer supporting his policies. This time, Banzer's party, which has four seats less than Lozada's MNR in Congress, decided that it was not going to be kept from power a second time. It therefore made the unlikely alliance with the left wing MIR rather than playing second fiddle again. Paz Zamora, vice-president during the disastrous presidency of Hernan Siles Suazo, when inflation reached 24,000 per cent, has said that Hugo Banzer, responsible for the imprisonment and exile of the new president in the early 1970s, has been completely converted to democracy. Whether that is the case or not, men from Banzer's party now occupy 10 of the country's 17 ministries, including all the most important ones such as finance,

KEY FACTS

Official title: Republic of Bolivia

Head of state: President Jaime Paz Zamora

Ruling party: Movimiento de la Izquierda Revolucionario (MIR) and Acción Democrática Nacionalista (ADN)

Capital: La Paz (administrative); Sucre (legal)

Official Languages: Spanish (Quechua, Aymará, Guarani also spoken)

Currency: Boliviano (B) = 1m pesos

Exchange rate: B2.88 per $ (Nov 1989)

Area: 1,098,581 sq km

Population: 6.99m (1988)

GDP per capita: $724 (1988)

GDP real growth: 2.8% (1988)

Inflation: 16.0% (1988)

Trade balance: $57.4m (1988)

Foreign debt: $5.7bn (1988)

foreign affairs, mining, transport, agriculture and the armed forces. Paz Zamora, a nephew of the outgoing president, says that he remains committed to continuing the main threads of the policies of the previous government. The free exchange rate will be maintained, the new simplified tax system will not be altered and efforts will be continued to maintain the public sector deficit under tight control. Before he said all that, the Bolivian currency fell sharply against the dollar, then it rose again.

The policies of Gonzalo Sanches de Lozado, finance minister of the previous government, reduced the public sector deficit from almost half of the GNP to a very respectable 3 per cent. Apart from cutting inflation to between 10 – 12 per cent last year, his policies have resulted in the Central Bank being able to attract sufficient dollars, both from the narcotics business and from the buoyant contraband trade, to be able to increase its reserves from only $10m in 1985 to $183m last year. This was enough to be able to tolerate the continuing deficit in the balance of trade. The trade deficit was a weak point of the Paz Estenssoro administration, as Bolivia is now importing far more than it is earning from legal exports. Officially, exports total some $700m and goods worth some $600m are coming in. In fact, smuggled goods are estimated to take the total of imports to almost $1bn. The big problem is that although the Lozada medicine halted inflation and restored some degree of stability to the Bolivian economy, it has not yet proved successful in promoting sufficient growth to bring about any halt in the decline of per capita incomes, estimated to have fallen by 12.5 per cent during the four years of previous government. Nor has it lowered the rate of unemployment, which at between 20 – 25 per cent is one of the highest in Latin America. Nevertheless, it would be churlish not to admit that the ground has been prepared for a return to growth. Whether this in fact occurs will depend to some extent on the responsibility of Bolivia's politicians but to a greater extent on events outside Bolivia's control. These include the price of oil and minerals, as well as the earnings of Bolivia's number two export, cocaine, the price of which is estimated to have fallen by a third in the past couple of years.

KEY INDICATORS

	Unit	1984	1985	1986	1987	1988
Population	'000	6,250	6,430	6,550	6,800	6,990
GDP (1985 prices)	B$bn	2,776	2,771	2,691	2,749	2,827
Inflation (average rate)	%	1,281	11,750	276	14.6	16.0
GDP real growth	%	–2.9	–1.8	–2.9	2.2	2.8
Rice production	'000 tonnes	166	173	137	130	–
Maize production	'000 tonnes	489	554	538	543	–
Sugar-cane	'000 tonnes	–	3,159	2,870	2,730	–
Tin production	'000 tonnes	19.9	16.1	10.5	8.1	10.2
Lead production	'000 tonnes	8.4	6.2	3.3	9.2	8.4
Zinc production	'000 tonnes	37.8	38.1	33.3	39.1	39.6
Crude oil production	'000 tonnes	969	917	1,020	–	–
Natural gas production	m cu metres	4,906	4,648	4,556	4,565	4,811
Exports	$m	724.5	623.4	563.8	565.8	601.1
Tin	$m	247.8	186.7	103.9	68.5	76.9
Natural gas	$m	375.7	372.6	328.7	248.6	215.0
Imports	$m	412.3	462.8	596.5	646.3	485.1
Consumer goods	$m	94.9	148.7	–	–	–
Raw materials	$m	203.0	317.9	–	–	–
Capital goods	$m	185.1	290.0	–	–	–
Balance of trade	$m	312.2	160.6	–51.0	–127.6	57.4
Current account	$m	–174.9	–282.4	–384.2	–426.8	–198.0
Debt service ratio	%	49.8	46.8	42.1	43.9	34.8
Exchange rate (average)	B$ per $	0.003	0.44	1.92	2.05	2.35

* estimated figure

Gas pipeline finally financed

The main event of the past few years for Bolivia will probably turn out to have been the signing in late 1988 of the agreement with Brazil for the construction of the long delayed pipeline which will carry gas to the Brazilian border and complementary works. The World Bank has agreed to advance Bolivia $1bn for this project. One-quarter of the money will be utilised for the building of the 650km, 14-inch gas pipeline from near Santa Cruz to the border; another quarter will go on building a 300 mW gas-fuelled power station; and the remaining half will go on building two petrochemical plants, one to make the fertiliser urea, the other to make polyurethane. Under the terms of an agreement with Brazil, that country will undertake to buy between 2 – 3m cubic metres of gas daily for the next 25 years. In addition it will buy some of the electric power, which will allow the expansion of industry in this region. Brazil has also undertaken to purchase 330,000 tonnes of urea annually during the next 25 years, 100,000 tonnes of which must remain in Brazil, the remainder may be exported, and 100,000 tonnes of polyethlene, all of which can be exported if it is not required in Brazil. Gas will replace oil in a large cement plant in the region, while a 1.8m tonnes-a-year sponge iron plant is planned and a steel mill may follow, as there are massive reserves of minerals straddling the border in the region of Puerto Suarez/Corumba.

The result of all this will be annual earnings of at least $250m for Bolivia, while it will also help put the two countries trade in a more balanced position. At the moment, Brazil provides 30 per cent of Bolivia's imports, worth some $260m in 1988. However, Bolivia only sold Brazil goods worth $18m in exchange and a large debt has been accumulated in recent years. In future, Bolivia may buy some of Brazil's debt at whatever discounted price it is offered. It may also buy up some of Argentina's debt in the same way, as a means of reducing its own. The income from the sale of gas, power and petrochemicals will soon be needed urgently: in 1992 the contract to export natural gas to Argentina comes to an end and with more gas of its own than it is able to make use of, Argentina will not renew the contract. Exports of hydrocarbons are still Bolivia's largest export earner, responsible for $214m in 1988.

A combination of improved prices and increased production and exports meant that the contribution of minerals rose sharply in 1988 and again during 1989, with exports earning $196m in 1988. Bolivia was on course to export 14,600 tonnes of tin in 1989, compared with 10,573 tonnes in 1988. This 30 per cent rise in tin exports was mainly due to the fact that the re-organisation of the state-run tin company, Comibol, once responsible for virtually all Bolivia's tin production, began to show results. Following the sacking of no less than 24,000 miners in 1986, Comibol's production sank to less than 5 per cent of all tin mined in 1987. But six mines were re-opened last year and for the first time in two decades, Comibol, which in 1985 was losing $9m a month, was expected to generate a profit of $3m. It produced 40 per cent of the country's total tin production in 1989. It is anticipated that Bolivia's production of tin will increase to 17,000 tonnes by 1993. What actually happens will depend on the price, but this seems to be likely to remain firm, as demand looks like being sustained. The legislation has been changed to permit Comibol to undertake joint-venture projects with other companies from within Bolivia and abroad in future and several projects are likely to come to fruition. Bolivia's gold production is also becoming significant. However, it is difficult to quantify, as it is being largely carried out by freelance miners and much of the gold is smuggled out.

One feature of the Paz Estenssoro Government was the policy of buying back, at an average of about 11 per cent of its face value, a substantial proportion of the country's foreign debt with private foreign banks. So far, some $450m of the $700m total debt to private banks has been bought by the government. However, Bolivia has still close to $4bn of debt with Paris Club countries and other international institutions; although it can be expected that these will be less severe and demanding with Bolivia than with most of its neighbours. Bolivia's heroic efforts to bring inflation under control by the textbook method has won it many friends in the international financial community, anxious to be able to demonstrate that the economy of one Latin American country apart from Chile is working well, and to argue that orthodoxy works.

Control of drugs

Another considerable source of income for Bolivia are the funds which enter the country from the various international organisations and governments, estimated to have totalled $100m last year. Among this is some $55m which comes into Bolivia each year destined for the control of drugs. This somewhat inconvenient fact helps to explain why Bolivia has been able to carry out policies which would have resulted in a revolution in any other country, although it rather tarnishes the country's image and weakens the orthodox case. As in the case of all Latin American countries involved in the drugs business, it is a very complicated and often contradictory issue. It is estimated that 300,000 people are involved in some way or other with the cocaine business, 50,000 of them directly. Some 60,000 hectares is reckoned to be planted to coca in Bolivia at the moment. Although some small areas have been switched to growing other crops in recent years, as a result of government incentives given to farmers, the area planted with coca is reported to have increased by 40 per cent during the Paz Estenssoro years. The export of coca and partly refined cocaine is calculated to contribute between $250 – 300m to the country's reserves each year. The proceeds of the export of cocaine forms the main bulk of the money which is deposited in the central bank each year; no questions are asked about its origins and it permits Bolivia to pay its way. This amount is reported to be considerably less than was the case in the early 1980s, when drugs reckoned to contribute more than $500m to the Bolivian economy. This fall is not due to the efforts of the 60 or so staff in the US embassy reported to be dedicated to controlling the drugs business, rather the opposite. Production in the Andean region has risen so much that the price of cocaine has fallen sharply in the past few years, indicating that all the efforts to control its use and distribution have been in vain. The Bolivian authorities are forced to have a very ambivalent attitude towards drugs, to tolerate, if not encourage the drugs trade. Without it the country's condition would be far more serious than it in fact is. This is one good reason why the formula to halt inflation which was adopted in Bolivia, with its extremely simple economy, its small industrial sector and its almost subsistence economy, would not be successful in any other, more complex country without a massive income from drugs to keep it going. No country, apart perhaps from Peru and Colombia, could hope to attract an amount equivalent to 50 per cent of legitimate exports from illegal exports, if it allowed dollars from those exports to be deposited with no questions asked.

Outlook

Bolivia has still not begun to solve the problem of how to offer employment to the vast numbers of people who are still leaving the countryside and migrating to the cities. This phenomenon is occurring several decades later than it did in most of its neighbours. It is estimated that 60 per cent of the Bolivian population of 7m, still increasing at one of the fastest rates on the continent, 2.9 per cent annually, is now urban, as opposed to 40 per cent 10 years ago. This is having the same serious consequences as elsewhere. It is estimated that 80 per cent of all Bolivians live in extreme poverty, compared with 40 per cent in most Latin American countries. The minimum wage is now about $1.50 per hour, one of the lowest on the continent. 20 per cent of the babies born in Bolivia die in infancy, again the highest proportion in Latin America; while the fact that one in every five Bolivians is illiterate, again a higher proportion than in any other country in the region, is becoming more apparent now that so many live in cities. One major problem is that there is little formal work for the migrant masses in Bolivia's cities. Some sources claim that 50 per cent of the working population is involved in the black economy, which is of as large a scale as the official economy. The only saving grace, if this is true, is that the per capita income is doubled to $1,000 a year. It is estimated that 40 per cent of the country's industrial capacity is now idle. The problem is that with a per capita average income of some $500, only about 500,000 Bolivians have enough purchasing power to buy anything apart from the basics. The market is too small to offer an attractive and profitable opportunity for the manufacture of any but the most basic of goods, which in any case cannot now compete with imported goods at the present relatively low tariff levels. Although the policy of virtually freeing the exchange rate helped to curb inflation, it has also made almost all Bolivian industry unviable.

Finding a niche in the modern world is not proving easy for Bolivia.

COUNTRY PROFILE

Political system: Constitutional republic with executive power vested in the president and his appointed cabinet, and legislative power vested in the bicameral congress, comprising a Senate (27 members) and a Chamber of Deputies (130 members). Bolivia is divided into nine departments, each of which elects three senators. The prefect of each department is appointed by the president.

Main cities/towns: La Paz (population about 1m), Cochabamba, Santa Cruz, Oruro, Potosi, Sucre.

Media
Press: Six principal dailies in La Paz, of which *Presencia* (circulation 90,000), *Hoy* (25,000) and *El Diario* (45,000) circulate nationally. Three dailies in Santa Cruz, two in Cochabamba and one each in Trinidad, Oruro and Tarija.
Radio: Around 150 radio stations broadcasting in Spanish, Aymara, Quechua. The majority are local and commercial. *Television:* Government-owned commercial TV station (Empresa Nacional de Televisín) and university television service broadcasting mainly in Spanish. Also privately-owned TV channels serving La Paz (5 channels), Cochabamba (5), Santa Cruz (6), Oruro (2).

External trade: Regular trade surplus since 1980-84 due to squeeze on imports but deficit registered from 1986-88 due to increased imports. Economic reactivation plan announced in 1987 aims to boost exports. Current account remains in deficit, largely as a result of the rising cost of debt servicing.
Exports: (1988): Natural gas (36.5% of total value), tin (13%), zinc (10%), silver (7%), antimony (3%). Illegal exports of cocaine are estimated to be as high as $2bn per year. Main destinations (1988): Argentina (38% of total value), UK (11%), USA (17%), German FR (6%), France (1%).
Imports: (1988): Raw materials and intermediate goods (35% of total value), capital goods for industrial use (42%), consumer goods (22.5%), transport equipment (17%), construction materials (3%). Main sources (1987): Brazil (19% of total value), USA (24%), Argentina (13%), EC (19%), Japan (9%).

Agriculture: Contributes 22 per cent to GDP and employs around half of the work force. Production of food crops (rice, maize, potatoes and wheat) and meats has not kept up with the increase in population. Consequently, food imports from the main suppliers, USA and Argentina, have risen. The main agricultural exports are soya beans, cane-sugar, coffee and increasingly timber (around 65 per cent of the land area is covered by forests – most of it unexploited). Emphasis is currently being placed on improving soya bean output and developing the export potential of the livestock, and in particular leather exports. Sugar and soya are the main crops for industrial use. A major growth area is the cultivation of coca leaf, used to manufacture the drug cocaine. Periodic droughts and flooding affect the Altiplano region, the main subsistence/food crop farming area. Production of crops for industrial use and export takes place largely around Santa Cruz.

Industry: Sector contributes around 20 per cent to GDP (manufacturing 11 per cent). Production is primarily for the small domestic market and is centred on the processing of minerals (mainly tin, lead and zinc smelting) and agricultural products. Oil refining and cement production are also important activities. Main constraints to development are: shortages of raw materials and capital goods, weak domestic demand, poor infrastructure and low levels of investment.

Mining: Sector's contribution to GDP has declined from 10 per cent in late 1970s to 7 per cent, though it remains an important export earner. Major world tin producer, but output has steadily declined owing to falling world prices and high extraction costs. Total tin reserves estimated at 1.6m tonnes – mainly low quality ore. Also world's leading producer of antimony and major producer of tungsten. Other minerals extracted include gold, silver, zinc, lead, copper and limestone. Large reserves of lithium and potassium as yet unexploited. Private-owned mining enterprises account for the majority of mining output. Recent emphasis on developing local refining and smelting industries.

Hydrocarbons: Proven oil reserves of 158m barrels and natural gas 133bn cu m. Gas sales from the Tita field to Argentina (through the 526 km pipeline to Yacuida) are the main source of foreign exchange. Annual production of gas is around 5bn cu metres, exports around 2.1bn cu metres, excluding liquid gas sales to Brazil. Almost all the oil extracted (mainly from the Rio Grande, Camiri, and more recently the La Pena and Santa Cruz fields) is consumed locally. Output has fallen from a peak of 48,000 b/d in 1973 to under 21,000 b/d. The state oil company, Yacimientos Petrolíferos Fiscales Bolivianos, accounts for over 60 per cent of total crude production.

Membership of international organisations: Andean Group, FAO, G-77, IADBN, IAEA, IATP, ICAO, ICO, IDA, IFAD, IFC, ILO, IMF, INTELSAT, Inter-American Development Bank, Interpol, ISO, ITC, ITU, IWC (Wheat), IWC (Whaling), Latin American Integration Association, NAM, Organisation of American States, PAHO, SELA, UN, UNESCO, UPU, WHO, WMO, World Bank, WTO.

BUSINESS GUIDE

Time: GMT -4.

Climate: Varies with area. Tropical in the lowlands, temperate in the higher valley regions, cooler in the Altiplano areas. Average temperature in La Paz ranges from –2°C at night to 22° C during the day; the rainy season is from Nov-Mar.

Entry requirements
Passport: Required by all.
Visa: Tourist card or visa required by all except citizens of USA and certain Western European countries (not France). Exit stamps needed before departure. Non-tourist visitors are required to obtain a Determined Object Visa. All visitors must register with the immigration authorities.
Currency: No restrictions on import/export of local or foreign currency. Visitors may be required to prove they have equivalent of $500 before being let in.
Customs: Visitors are not required to declare valuable personal effects such as cameras and radios. Business samples are free of duty so long as they are re-exported within 90 days.

Health precautions
Mandatory: Yellow fever vaccination certificate required from visitors arriving from infected areas, and from those travelling to infected Bolivian regions.
Advisable: Typhoid, paratyphoid and polio vaccinations. Malaria risk exists in rural areas – prophylaxis recommended. Owing to altitude of La Paz and other places in the Altiplano, sufferers from heart or lung complaints should seek medical advice before leaving for Bolivia. Water precautions should be taken.

BUSINESS GUIDE

Representation overseas: Argentina, Belgium, Brazil, China, Colombia, Costa Rica, Cuba, Czechoslovakia, Ecuador, Egypt, France, German FR, Hungary, Israel, Italy, Japan, Republic of Korea, Malta, Mexico, Nicaragua, Panama, Paraguay, Peru, Portugal, South Africa, Spain, Switzerland, Taiwan, UK, USA, USSR, Uruguay, Vatican, Venezuela, Yugoslavia.

Air access
National airline: Lloyd Aéro Boliviano (LAB).
Other airlines: Aerolíneas Argentinas, Aeroperú, Cruzeiro, Eastern, Lan-Chile, Lineas Aéreas Paraguayas, Lufthansa.
Tax: An airport tax of US$15 is payable on leaving Bolivia.
International airports: La Paz/El Alto (LPB), 15 km from city, duty-free shop, bar, restaurant, buffet, bank, post office, shops, hotel reservations (restricted hours in some instances); Santa Cruz-Viru-Viru (SRZ), 16 km from city, duty-free shop, restaurant.

Surface access
Main ports: Although Bolivia is landlocked, access to the sea (and in certain cases freeport facilities) has been granted by: Paraguay (via river Paraguay); Brazil (Belem, Santos, Corumba, Port Velho); Argentina (Buenos Aires, Rosario). Also access via Peru (Matarani) and Chile (Arica and Antofagasta).
Overland access: Possible by road from: Peru (Peruvian part of route may be difficult); Argentina (Bolivian roads may not be passable after rain); Chile (road not generally recommended). N.B. travellers who do use overland routes are generally advised to check border-post opening hours. Rail services connect La Paz with Chile (Arica and Calama), and Argentina (Buenos Aires).It is also possible to take the steamer on Lake Titicaca from Puno (Peru) to Guaqui (Bolivia).

Hotels: Wide range available. Advisable to book in advance. Service charge and local tax added. Tipping 10 per cent. Credit cards are only accepted in the international hotels.

Car hire: International licence issued by member of Federación Inter-Americana de Touring y Automovil Clubs; drivers must also buy a *hoja de ruta*, a form which notes your itinerary; drive on right. Hertz has an office in the Sheraton Hotel, La Paz.

City transport
Taxis: Remise, or yellow cabs. Fixed rates applicable per passenger – frequently shared; within La Paz cabs with green flags (*trufis*) ply along fixed routes. Tips not expected.

National transport
Air: Lloyd Aéreo Boliviano (LAB), TAM (army airline) and Línea Aerea Imperial (LAI) operate domestic services to main centres.
Road: Around 40,000 km of road, but less than 5 per cent paved and only 20 per cent classed as all-weather roads. La Paz-Oruro is paved, Oruro-Cochabamba partly paved, Cochabamba-Santa Cruz paved but in poor state. The branch of the Pan-American Highway which connects Argentina with Peru crosses Bolivia from south to north-west. Toll permit system in operation; garage and petrol services sparse outside main centres.
Buses: Services run between and within main towns but can be slow and unreliable; city routes frequently crowded; lorries also used for basic inter-town passenger transport. Advisable to reserve seat in advance.
Rail: North-south line from La Paz to Oruro and Uyuni, with spurs to Cochabamba, Uncia, Potosi, Sucre and Villazón. Eastern system runs from Santa Cruz to Yacuiba and to Corumbá at the Brazilian border. Sleeper and first class services available; advance booking essential. Railcar (*ferrobus*) available on some routes. Expreso del Sur: special train service from La Paz to Buenos Aires, Argentina.

Public holidays: 1 Jan, Carnival (Feb/Mar), Maundy Thursday, Good Friday, Easter Monday, 1 May, Corpus Christi, 6 Aug (Independence), 2 Nov (All Souls'), 25 Dec. Additional holidays are observed in the various departments (e.g. La Paz Day – 16 Jul).

Working hours
Government and business: (Mon-Fri) 0900 – 1200, 1400 – 1800; (Sat) 0900 – 1200 (business only). Public offices: (Mon-Fri) 0800 – 1200, 1430 – 1830. Afternoon hours later in provinces.
Banking: (Mon-Fri) 0830 – 1130, 1430 – 1700.
Shops: (Mon-Fri) 0900 – 1200, 1400 – 1800; (sat) 0900 – 1200.

Telecommunications
Telephone: Efficient automatic telephone service between La Paz, Santa Cruz, Oruro and Cochabamba. Possible to reverse the charges when calling the USA, Brazil and Chile. Dialling code for Bolivia: IDD access code + 591 followed by area code (La Paz 2, Potosi 62, Sucre 64, Tarija 66, Trinidad 46) followed by subscriber's number.

Postal service: Mail is delivered to post office boxes (*casilla*), and to the door. Mail should therefore be addressed to the addressee plus *casilla* number and town.

Banking
Banking centre: La Paz.
Central bank: Banco Central de Bolivia.
Commercial banks (head offices): Banco Agrícola de Bolivia, Banco Boliviano Americano, Banco del Estado, Banco de Santa Cruz de la Sierra, Banco Hipotecario Nacional, Banco Industrial, Banco Mercantil, Banco Minero de Bolivia, Banco Nacional de Bolivia.
Foreign branches: Banco de la Nación Argentina, Banco do Brasil, Banco Popular del Perú, Citibank, First National Bank of Boston.

Electricity supply: 110 V and 200 V in La Paz and 220 V in Cochabamba and most other towns; US-type flat-prong plugs.

Representation in capital: Argentina, Belgium, Brazil, China PR, Colombia, Costa Rica, Czechoslovakia, Cuba, Ecuador, Egypt, France, German FR, Israel, Italy, Japan, Korea DPR, Republic of Korea, Mexico, Panama, Paraguay, Peru, Portugal, Romania, South Africa, Spain, Switzerland, Uruguay, UK, USA, USSR, Vatican, Venezuela, Yugoslavia.

Hotels: Cochabamba
Capitol, Calle Colombia at 25 de Mayo (tel: 24510).

Gran-Ambassador, Calle España, P.O. Box: 264 (tel: 48777; tx: 6240).

Gran-Cochabamba, Plaza Ubaldo Anze, P.O. Box: 500 (tel: 43300; tx: 6314).

Hotels: La Paz
Crillón, Plaza Isabella Católica, P.O. Box: 2350 (tel: 352121/30; tx: 3275).

Eldorado, Avenida Villazón, P.O. Box: 77 (tel: 363355; tx: 3625).

Gloria, Calle Potosí, P.O. Box: 8620 (tel: 370010; tx: 2414).

La Paz, Avenida Camacho 1277, P.O. Box: 797 (tel: 356950; tx: 32413275).

Plaza, Avenida 16 de Julio (tel: 378311; tx: 2674; fax: 343397).

Sheraton, Avenida Arce (tel: 356950; tx: 3275).

Sucre Palace, Avenida 16 de Julio 1636, P.O. Box: 1014 (tel: 36345; tx: 2445).

Hotels: Santa Cruz
Bolivia, P.O. Box: 1958 (tel: 36292).

Gran Cortez, Avenida Cristobal de Mandoza 280, P.O. Box: 626 (tel: 31234; tx: 7616).

Las Palmas, P.O. Box: 915 (tel: 30366).

Los Tajibos (Holiday Inn), Avenida San Martin y 3er Anillo, P.O. Box: 2966 (tel: 30022; tx: 4241; fax: 33915).

Motel Asturias, Modes St 154 (tel: 39611).

Santacruz, Calle Pari no 59, P.O. Box: 398 (tel: 48811, 48843; tx: 7620).

Business advice and information
Agencia Promotora de Inversiones Castedo Laygue, casilla 2355, La Paz (deals with agricultural and industrial development/investment).

Banco Central de Bolivia, PO Box 3118, La Paz (tel: 374151; tx: 3228, 3398).

Bolivian Tourism Institute, Calle Mercado 1328, La Paz (tel: 367463/4; tx: 2534 inbotur).

Cámara Departamental de Industria y Comercio de Santa Cruz, Avda Suárez de Figueroa 127, 3\28 y 4\28, Casilla 180, Santa Cruz (tel: 34555; tx: 4298).

Cámara Departamental de Comercio de Oruro, Pasaje Guachalla, Casilla 148, Oruro (tel: 50606; tx: 2230).

Cámara Departamental de Comercio de Sucre, Casilla 33, Sucre (tel: 21320; tx: 2694).

Cámara Nacional de Comercio, Avenida Mariscal Santa Cruz No 1392, 1st Floor, Edificio Cámara Nacional de Comercio, Casilla 7, La Paz (tel: 324340/350042; tx: 2305 BV; fax: (591)2).

Cámara Nacional de Exportadores, Casilla 20744, Plaza Avaroa esq. B. Salinas, La Paz.

Cámara Nacional de Industrias, Casilla 611, Avenida Mariscal Santa Cruz 1392, 14th Floor, Edificio Cámara Nacional de Comercio, La Paz.

Confederación de Empresarios Privados de Bolivia, Edificio Cámara Nacional de Comercio, 7th Floor, Casila 20439, Avenida Mariscal Santa Cruz 1392, La Paz.

Central Bank, Ayacucho esq, Mercado. Cajón Postal 3118, La Paz (tel: 374151).

Central Post Office, Av Mcal. Santa Cruz esq. Calle Óruró, La Paz.

Corporación Minera de Bolivia, Avenida Mariscal Santa Cruz 1092, Casilla 349, La Paz.

Lloyd Aéreo Boliviano SAM, Casilla 132, Aeropuerto 'Jorge Wilstermann', Cochabamba (tel: 25900; tx: 6290).

Ministry of Economic Planning, Avda Arce 2147, La Paz (tel: 372060; tx: 5321).

Ministry of Industry, Commerce and Tourism, Avda Camacho esq Bueno, Casilla 1372, La Paz (tel: 372044; tx: 3259).

Pubol Advertising Agency, Avenida 6 de Augusto, Casilla, La Paz (tel: 341088).

Principal newspapers
El Diario, Loayza 118, Casilla 5, La Paz. Tel: 356835; tx: 5530

Hoy Avd 6 De Agosto 2170, Casilla 477, La Paz. Tel: 326683; tx: 2613

Presencia, Avda Mariscal Santa Cruz, Casilla 1451, La Paz. Tel: 372344; tx: 2659

Botswana

Gillian Cronje

'As long as women want to wear diamonds, Botswana will be able to supply them,' declares Debswana's chief executive, Louis Nchindo. Debswana, the diamond mining company which is jointly owned by De Beers Consolidated Mines and the Botswana Government, is the world's largest producer of gem stones, the third largest in terms of carats. Output in 1989 was estimated at 16m carats, compared to 15.2m carats in 1988 and 13.2m carats in 1987. Diamonds account for 78 per cent of the country's national income and earned over $1bn in foreign exchange in 1988.

Some experts predict that diamond output will continue to expand for some time to come, but according to Debswana's public statements, production has reached a plateau. Prospecting at an annual cost of $4m has apparently failed to discover commercially viable deposits, but it may not be in De Beers' interest to publicise any major finds at present. Earnings could increase in other ways: a new re-crushing plant, built at the cost of $105m at Debswana's Jawaneng mine, was scheduled to go into operation in April 1990. Using the most modern technology it will recover diamonds missed during the initial processing. Its output is estimated at 2m carats a year.

Mining

Less glamorous than diamonds but consisting of the same carbon as the precious stones is Botswana's coal, in what may be the largest southern African deposits. The government, with West German aid, is expanding the Coal Utilisation Project, which will market and distribute coal and promote it as a substitute for fuel wood. Coal also plays an important role in satisfying the steeply growing demand for electricity. The commissioning at the end of 1989 of a coal-fired 33mW generator at the Morupule power station brought the country's total generating capacity to 180mW and reduced its reliance on the South African grid.

New mining developments include the exploration by a Canadian company for gold under the Kalahari desert, at a depth of almost 2 km below the surface. This expensive venture stems from the belief that the western outcrop of South Africa's Witwatersrand reef runs underneath Botswana. Small-scale gold mining has long been carried out near Francistown. A new mine at Sashe went into production in 1988 with the creation of 200 jobs; another – Map Nora – is expected to generate gross revenues of $6.5m a year.

Botswana RST's copper-nickel mining operation at Selebi-Phikwe, until recently the least hopeful of the country's mineral enterprises, showed increased profits in the year ending June 1989 because of favourable world nickel prices – 'a much needed relief to the company', comments the annual report. As it is one of Botswana's largest employers, RST's improved fortunes must also have come as a relief to the government. At the beginning of 1988, there was talk that Amax, which holds 30 per cent of the shares, was considering pulling out of Botswana.

Agriculture

After mining, beef is the most important foreign exchange earner, bringing in some $80m a year. The European Community (EC) takes about 70 per cent of the meat, with Britain the largest single customer. Although cattle are the traditional source of livelihood for rural dwellers who form the great majority of the population, trends have increasingly favoured big ranchers. Five per cent of ranchers now own half the national herd, and 45 per cent of the population owns no cattle at all. Cattle-keeping is heavily subsidised, but this use of public funds benefits mainly the small elite of privileged ranchers – a telling criticism frequently made by the opposition.

The seven-year drought which came to an end in 1988 reduced the national herd by a third, to 2m head. This increased to 2.3m in 1989, but the pre-drought total is unlikely to be regained before 1992. As a result of the drought Botswana exported only an estimated 14,000 tonnes of beef to the EC, against its quota of 19,000 tonnes. The government set aside some $35m for a two-year drought recovery programme, and the 1988–89 harvest was expected to be a grain bumper crop of 102,000 tonnes – more than twice the amount harvested in the drought years. Subsidies were made available for water drilling schemes, and owners of more than 60 cattle were invited to apply for grants of up to 60 per cent of borehole costs.

Thanks to the diamonds, Botswana's economy has grown at an annual average of 13.8 per cent since 1970. According to Bank of Botswana projections, the national income increased by 8.7 per cent in real terms in 1989 to $1.65bn, with a per capita income of $1,600. But this national prosperity is not widely felt, largely because of the government's conservative fiscal policy. At the end of 1988 foreign reserves exceeded $2bn, covering 30 months' imports; in the first two months of 1989 they rose by a further $107m. At that point President Quett Masire's government came under growing pressure from all sides, even from within his ruling Botswana Democratic Party (BDP), to spend more on job-creating projects and the diversification of the economy, not least because 1989 was an election year.

Botswana's population growth of 3.5 per cent means that more than 20,000 people enter the labour market each year, but there are job vacancies for 6,000 at most. Even the huge new Sua Pan soda ash project will create only 450 permanent jobs, although more labour is required in the initial stages. There is a desperate shortage of skills, but only 20 per cent of new job seekers have had secondary education; half of the new entrants have only been to primary school; the rest have had no schooling at all. Kenneth Koma, leader of the main opposition party, the Botswana National Front (BNF), has been highly critical of what he regards as the government's lack of dynamism in creating a more balanced economy. The BNF wants more money spent on industrial development and training, and measures to ensure a more equal distribution of land.

Industries

Officials in Gaborone admit that some of the existing schemes to encourage local industry have performed below expectation, and they blame the failures on the absence of indigenous business traditions, skill shortages, and on the import of cheap manufactures, mainly from South Africa. The department of industrial affairs administers the Financial Assistance Policy (FAP) which was specifically established to create employment and diversify the economy. The range of incentives offered by FAP consists of a number of five-year grants, in particular the unskilled labour grant, training grants and tax holidays. There is also a local preference scheme to safeguard industries from foreign competition when tendering for government contracts. Tariff protection may also be offered for a period of eight years in terms of the Infant Industry Clause of the Southern African Customs Union (SACU) to which Botswana belongs. (Other SACU members are Lesotho, Swaziland and South Africa; in effect, SACU is run by Pretoria.)

Despite the difficulties surrounding FAP, local industry can also boast of some important successes. For instance Marothodi Ltd, owned 48 per cent by the Botswana Development Corporation (BDC) and 52 per cent by its workers, designs clothes and hand-prints textiles, selling all it can produce. Formed in 1970, BDC is a wholly-owned government organisation which has invested in about 60 enterprises in all sectors of the economy except mining. It was singled out for praise by the World Bank for its sound procedures and strong financial performance. The BDC owes its achievements to a strong and growing economy, a conservative investment and lending strategy, independent management and a highly qualified staff,

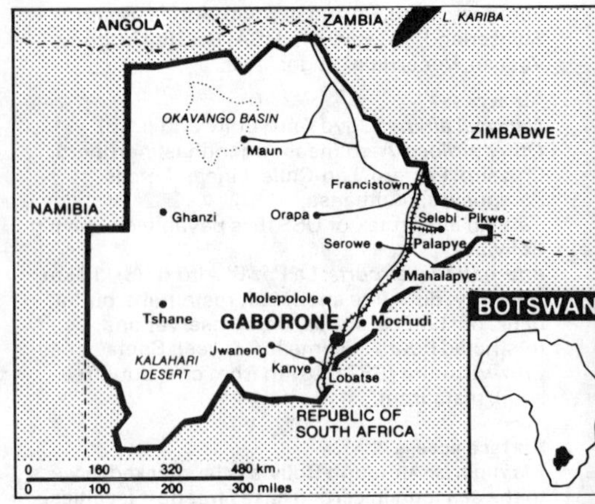

KEY FACTS

Official title: Republic of Botswana
Head of state: President Dr Quett K. J. Masire
Ruling party: Botswana Democratic Party
Capital: Gaborone
Official Languages: English and Setswana
Currency: Pula (P) = 100 thebe
Exchange rate: P1.96 per $ (Nov 1989)
Area: 582,000 sq km

Population: 1.21m (1988)
GDP per capita: $973 (1988)
GDP real growth: –1.3% (1988)
Inflation: 8.4% (1988)
Trade balance: $386m (1988)
Foreign debt: $518m (1987)

KEY INDICATORS

	Unit	1984	1985	1986	1987	1988
Population	m	1.05	1.09	1.13	1.17	1.21
Gross domestic product (GDP)	P m	1,538	1,661	1,893	2,171	2,144
GDP per capita	P	1,465	1,524	1,675	1,856	1,772
Inflation	%	8.6	8.1	10.0	9.8	8.4
GDP real growth	%	2.0	8.0	14.0	14.7	–1.3
Agricultural production	1979-81=100	93.4	91.5	89.0	91.5	99.6
Food production	1979-81=100	93.3	91.4	88.9	91.4	99.6
Crop production	1979-81=100	125.6	128.5	138.8	115.8	141.1
Livestock	1979-81=100	102.3	98.5	96.1	98.3	100.3
Mining production	1985=100	102	100	102	104	116*
Copper/nickel	'000 tonnes	52	52	41	35	47
Coal	'0000 tonnes	392.8	437	499	579	605
Diamonds	m carats	12.9	12.7	13.1	13.2	15.2
Tourist arrivals	'000	56	60	65	62	–
Exports	P m	865.2	1,405.7	1,616.4	2,662.9	2,575.8
Diamonds	P m	622.4	1,066.2	1,198.5	2,249.8	1,976.5
Copper/nickel	P m	68.8	121.2	121.1	118.1	370.2
Meat	P m	62.7	98.3	120.8	103.6	–
Imports	P m	907.4	1,101.0	1,332.5	1,571.2	1,871.8
Balance of trade	P m	–42.2	304.7	283.9	1,091.7	704.0
Total external debt	$m	267.6	343.3	390.4	517.6	–
Debt service ratio	%	3.5	5.2	4.2	3.7	–
Exchange rate	P per $	1.28	1.89	1.87	1.68	1.82

* estimated figure

the 1989 issue of the World Development Report says. The corporation was once criticised in World Bank circles for crowding out the private sector, but it has been instrumental in forming the Botswana Investment Trust which is reported to be working on the privatisation of some of BDC's profitable enterprises. The establishment of Botswana Stock Brokers Ltd with the help of the BDC has given rise to speculation about the possible establishment of a stock exchange in Gaborone.

The BDC has shown consistent profits, except for one year when Air Botswana's operating losses led to a small deficit. But the airline was separated in 1988 from the BDC parent company and established as a separate state-owned corporation with a 1988–89 budget of $9.5m. Since then it has actively expanded the country's transport facilities through lengthening the runways and adding maintenance hangars at the international airport in Gaborone. A second international airport at Kasane in the north is due for completion in 1991; a £5m contract for the work went to a British company. In 1989 Air Botswana started twice-weekly flights to Windhoek, bypassing Johannesburg, with plans for more frequent services after Namibia's independence.

Air Botswana also bought a 46-seater turbo-prop passenger aircraft from a French–Italian consortium. A declaration of intent was signed with Zambia, Zimbabwe, Uganda and Tanzania to establish a joint venture for the operation of wide-bodied aircraft on international routes. This was part of the preparation for boosting the tourist sector, as was the upgrading of the Nata–Maun road which cuts through the Nxai Pan game reserve. Maun is also the starting point for safaris into the Okavango basin and the Chobe game park. But tourism was not the only consideration: high on the agenda were plans to lessen Botswana's dependence on South African communications. Improvements of the road from Maun to Mamuno on the border with Namibia could be seen in this light. Although the long-cherished plans for a trans-Kalahari rail link were ruled out on grounds of cost, Botswana Railways was still considering the feasibility study for a rail link with the Gobabis terminal in Namibia. Such a line would connect Botswana with the Atlantic coast, as well as opening up an otherwise inaccessible but mineral-rich part of the desert.

Botswana Railways has also established handling and storage facilities at both Gaborone and Francistown in the north. At present almost all Botswana's rail traffic goes via South Africa. But with repairs on the Limpopo line between Bulawayo in Zimbabwe and Maputo in Mozambique near completion, Botswana has an alternative direct outlet to Maputo. The Francistown–Maputo line via Bulawayo measures 1,275 km compared to the 1,725 km between Francistown and Durban. This route would even be slightly shorter than that between Gaborone and Durban. As the north of the country is expected to generate growing amounts of traffic, the direct rail line to Maputo will become increasingly cost-effective. These considerations were behind the government's 1989 donation of 145,000 sleepers, produced by Kwena Concrete, a Botswanan company, to help repair the Limpopo line.

The contract for laying the 175 km branch line from Francistown to the $360m Makgadikgadi and Sua Pan soda ash project was won by an Italian firm. Mining was to begin early in 1991 at the rate of 300,000 tonnes of soda ash and 650,000 tonnes of salt a year. South African interests own 52 per cent of the new mining company. Botswana signed loan agreements with three South African banks for half the total cost of the Sua Pan scheme; a further third of project financing was to come from shareholder equity; the rest from Botswanan banks and foreign international institutions.

The bulk of the output will go to the South African AECI company, where it will replace raw materials previously imported from the American soda ash producer, Ansac. There was considerable criticism of the Sua Pan scheme from Botswana's partners in the Southern African Development Co-ordination Conference (SADCC) who considered that it entrenched Pretoria's economic stranglehold over the sub-region. The SADCC was formed to reduce the dependence of its members on the South African economy; Botswana has always been a leading member of the organisation, whose headquarters are in Gaborone. But the SADCC was not the only opponent of the soda ash project. According to AECI, Ansac tried to block the Sua Pan scheme by warning AECI's customers that the project was not viable, and by threatening to dump soda ash on the South African market. Botswana hopes that its black African neighbours will also become customers. For instance, Sua Pan could meet the Zimbabwe Industrial Development Corporation's annual requirement of up to 40,000 tonnes of soda ash, 28,000 tonnes of coarse salt and 11,000 tonnes of fine salt.

Southern Africa

Commercial links between Botswana and Zimbabwe improved after the signing in September 1988 of an amended trade agreement. This answered some of Harare's complaints that its imports from Botswana were really South African goods with very little Botswanan content. Under the amended pact, made-in-Botswana goods enjoy duty-free entry into Zimbabwe only if the product has a minimum local content of 25 per cent. Basically, relations between the two countries are good, and there is increasing defence and security co-operation in the light of South Africa's destabilisation activities in the region. The government in Pretoria regards both countries as hostile. In August 1989 it completed work on a high-voltage electrified fence along 24 km of its border with Botswana to stop guerrillas from crossing into South Africa; a similar barrier was erected along the border with Zimbabwe in 1985.

The government in Gaborone firmly denies Pretoria's frequent claim that 'armed terrorists' are being infiltrated across its borders from Botswana. Such claims are often followed by hot pursuit raids from South Africa, apparently aimed at political refugees but killing indiscriminately. In December 1988 a heavily-armed South African contingent attacked the border village of Ditlharapeng, burned several huts and killed two of the villagers. Later that month – only four days after the signing in New York of the Angola–Namibia agreement which was to bring peace and goodwill to the whole of southern Africa – a 14-year-old boy fell victim to a South African commando raid on Gaborone. Although Pretoria has not come up with concrete evidence regarding its claims that insurgents use Botswanan territory to infiltrate South Africa, Botswana has proof of South Africa's operations. The December attack on Gaborone came on the heels of 10-year gaol sentences meted out by a Gaborone court to two South African commandos captured in an earlier raid, and may have been an act of retaliation. But border violations by South Africa are so frequent that villagers have complained to President Masire about the nightly loss of sleep caused by the intruders. In 1989 the commander of the Botswana Defence Force, Lt-General Mompati Merafhe, found renewed evidence of South African hit squads operating in the country.

During the 1989 election campaign Kenneth Koma, the opposition leader, compared his BNF to the southern African liberation movements, in particular the ANC and SWAPO, except that it did not wage an armed struggle; but he felt that Botswana needed liberating from Masire's colonial-style BDP. His criticisms of the government's conservative fiscal policies, its record on employment, housing and land distribution increased the BNF's following, particularly among urban voters. The government's authoritarian manner in dealing with the student strikes of 1988–89 had lost its support among the younger generation. The BDP was itself divided, with a welfarist section favouring a more equal distribution of Botswana's wealth, but there was little doubt in the run-up to the elections that it would again be returned.

COUNTRY PROFILE

Political system: The legislature is the 38-member National Assembly, 34 of whom are elected for five years by universal adult suffrage. The assembly elects a president who has executive power. He appoints a cabinet responsible to the assembly. The president is an ex-officio member of the assembly. The approval of a House of Chiefs is needed for some measures, but it cannot veto legislation. Gained independence in 1966. Formerly Bechuanaland.

Political parties: The Botswana Democratic Party is the ruling party, with 34 parliamentary seats in the 1989 elections; others include the Botswana People's Party (12 seats) and the Botswana Labour Party (one seat).

Population: Botswana (total population 1,256,000), has a varied ethnic structure in which the largest group are the Bamangwato (40 per cent of the total). 75 per cent of the population live in the eastern 10 per cent of the country.

Main cities/towns: Gaborone (population 120,000), Francistown, Serowe.

Language: English and Setswana are the official languages, the latter being the language of state.

Media
Press: Botswana Guardian, Gazette, Mmegi, in addition to the government owned *Botswana Daily News.*
Radio: National service in Setswana and English operated by Radio Botswana, based in Gaborone.
Television: South African and Bophuthatswana transmissions can be received.

External trade: Balance of payments depends heavily on earnings from diamonds, which have risen considerably since the early 1980s. In 1988 alone the significant surplus in the trade balance was sufficient to offset the persistent deficit in the services account, as well as bolstering the current account to a surplus of P708m. International reserves have been escalating too, and amounted to P4.2bn (US$2.2bn) at the end of 1988, providing an import cover (c.i.f. duty inclusive) of 28 months. Foreign currency reserves in 1989 were about US$3bn.
Exports: Diamonds (73 per cent of total value in 1988), copper/nickel (16%), beef (4%). Main destinations: Europe (about 89% of total value), Southern African Customs Union (about 4%).
Imports: Food (about 16 per cent of total value), machinery and electrical goods (16%), transport equipment (16%) and petroleum products (7%). Main sources: Southern African Customs Union (79% of total value), other African, mainly Zimbabwe (8%).

Agriculture: Sector contributed about three per cent to GDP in 1988 although it employs 80 per cent of the workforce. Production divided between small, traditional farms and around 360 large-scale commercial units. Climate and poor soil suited to extensive ranching, with the result that livestock produce accounts for about 80 per cent of marketed output. In the past, rearing of livestock has been hampered by frequent outbreaks of foot-and-mouth disease – now largely controlled – and more recently by drought. Adverse weather and soil conditions have made expansion of crop production difficult, but output is expected to increase after the 1988 and 1989 rains which broke a six-year drought. Government aims for self-sufficiency in basic foodstuffs, such as maize, millet, beans and sorghum. Land development and irrigation programmes underway.

Industry: Manufacturing sector is underdeveloped, contributing five per cent to GDP in 1987-88. Wide range of consumer products manufactured by relatively small number of enterprises. Also large number of village industries, mainly producing handicrafts. Small domestic market has hampered attempts to stimulate production; therefore, emphasis on expansion of export market for traditional products, such as textiles and processed meat, and import substitution.

Mining: Exploitation of rich mineral reserves, notably diamonds, has provided key to Botswana's rapid economic growth, with sector currently accounting for 44 per cent of GDP in 1988. All diamond mining carried out by Debswana (De Beers Botswana Mining Co). Output continues to increase, reaching 13.9m carats in 1988, and Botswana is now the world's third largest producer of uncut diamonds after Australia and Zaïre. The main diamond mines are at Orapa, Letlhakane and Jwaneng. Copper-nickel is mined at Selebi-Phikwe. Production increased in 1988 to 46,518 tonnes (from 33,194 tonnes in 1987). Soda ash mining is developing significantly. It is also hoped to develop gold, iron-ore, potash, manganese, chromite and uranium reserves. There is a major joint venture with RSA.

Energy: Largest known coal reserves in Africa, though of low quality. So far only exploited by one mine, at Morupule. Output, which is mainly used in power stations, has increased to keep pace with expansion of electricity generating capacity. New coal mines at Serowe and Kgaswe being developed, and export potential also being considered. No known domestic oil reserves. At present, all oil provided via South Africa. Total installed electricity generating capacity (all coal-fired) 223 mW.

Membership of international organisations: African Development Bank, Arab Bank for Economic Development in Africa (recipient), Commonwealth, EC/ACP, Economic Commission for Africa, FAP, GATT, IDA, IFAD, IFC, ILO, IMF, OAU, Southern African Customs Union, Southern African Development Co-ordination Conference, UN, WHO, World Bank.

Time: GMT +2.

Climate: Sub-tropical, with hot summers and dry winters. Temperatures range from about 5°C to 22°C in Jul to 18°C to 33°C in Jan.

Entry requirements
Passport: Required by all.
Visa: Required by all except Commonwealth and most West European countries and USA.
Currency: Export of local currency is restricted (up to P200 may be freely taken out of the country on departure). Botswana is not part of the Rand area.
Customs: Member of Southern African Customs Union, therefore virtually no restrictions on movement of goods from South Africa, Namibia, Lesotho and Swaziland.

Health precautions
Mandatory: Cholera if arriving within six days of travelling from infected area.
Advisable: Anti-malarial prophylaxis for visitors to northern game reserves.

Representation overseas: Brussels, Harare, London, Lusaka, Stockholm, Washington.

Air access
National airline: Air Botswana.
Other airlines: Air Malawi, Air Zimbabwe, British Airways, South African Airways, Zambia Airways, Lesotho Airways, Kenya Airways.
International airports: Francistown (Code: FRW), 7 km from city; Gaborone (Code: GBE), 15 km from city; buffet, car hire.

Surface access
Overland access: Bus services to Zimbabwe and South Africa.

Hotels: First-class hotels available in all main towns. Generally advisable to book in advance – essential at weekends and during public holidays.

Credit cards: American Express, Access/Mastercard, Visa, Diners'.

Car hire: Foreign or international driving licence required. Speed limits: 120 k/h on main roads, 60 k/h in built-up areas. Seat-belts must be worn. Facilities available to hire Avis car in South Africa and deposit it in Botswana, or vice versa. Except by special prior arrangement, hire cars are not generally available for driving from Botswana to Zimbabwe or Zambia.

City transport
Taxis: Available in capital. 10 per cent tip is usual.

National transport
Air: Regular flights connecting Gaborone, Francistown, Maun, Selebi Phikwe operated by Air Botswana, which also provides direct charter flights to other airstrips throughout country.
Road: Over 8,000 km of roads, of which over 1,900 km are tarred and a further 1,000 km gravelled. Most major towns are connected by good roads, and projects are under way to improve many other roads.
Buses: Bus services remain underdeveloped. Services between Gaborone and Francistown, going on to Nata and Maun.
Rail: Main Cape Town (South Africa)-Bulawayo (Zimbabwe) railway runs for over 700 km through Botswana, linking several towns. This section is operated by Botswana Railways.

Public holidays
Fixed dates: 1 Jan, 2-3 Jan, President's Day (July), 30 Sep-1 Oct (Botswana Day), 25 Dec, 26 Dec.
Variable dates: Good Friday, Easter Monday, Early May Bank Holiday, Late May Bank Holiday.

Working hours
Business: 0730 – 1230, 1345 – 1630.
Banking: 0815 – 1245. (Exceptions: Sat 0815 – 1045, Sun closed: there is some variation in hours between summer and winter).
Shops: 0800 – 1730.

Telecommunications
Telephone: Land lines connect the 12 main towns by microwave links. Dialling code for Botswana: IDD access code + 267 followed by six-figure subscriber's number (there are no area codes). Over 12,000 telephones in use.
Telex: Telex facilities available at main post office in Gaborone and at major hotels everywhere. DHL have an office in Gaborone.

Representation in capital: China PR, German FR, Libya, Sweden, USSR, USA. *High commissions*: India, Nigeria, UK, Zambia, Zimbabwe. *Consuls*: Belgium, Denmark, France, Guyana, India, Iraq, Italy, Japan, Kenya, Korea DPR, Korea (Republic), Lesotho, Malawi, Mozambique, Netherlands, Norway, Switzerland, Tanzania.

Hotels: Gaborone
Gaborone Sun, Nyerere Drive N, P.O. Box: P Bag 0016 (tel: 351111; tx: 2433, 2672; fax: 351111).

Oasis, Tlokweng Road, P.O. Box: 30331 (tx: 2497).

President, The Mall, P.O. Box: 200 (tel: 353631; tx: 2434; fax: 351840).

Other useful addresses
Bank of Botswana, PO Box 712, Gaborone (tel: 351911; tx: 2405, 2448, 2827, 2935; fax: 372984).

Botswana Development Corporation Ltd, PO Box 438, Gaborone (tel: 351811; tx: 2251; fax: 373539).

Botswana Enterprise Development Unit (promotes industrial & rural development), PO Box 0014, Gaborone.

Botswana Meat Commission, Private Bag 4, Lobatse (tel: 330321; tx: 2420).

Department of Information and Broadcasting, Private Bag 0060, Gaborone (tel: 352541, 352261; tx: 2394).

Brazil

Patrick Knight

Every economic pointer demonstrated the seriousness of the deterioration in the Brazilian economy during 1989, as the crisis which began in the early 1980s, when the country was first unable to meet its debt servicing commitments, intensified and moved towards its final crisis. Inflation, which touched almost 1,000 per cent in 1988, was on course to be at least twice that in 1989, while the country's economy shrank by an estimated 1 per cent. Industrial production was on course to be 2.5 per cent below that of the previous year and the situation was only made less serious by a very good harvest, particularly of soya and citrus. The public sector deficit was running at the record level of 6 per cent in the middle of the year, as the government failed to get its accounts under any sort of control.

The third of the unorthodox economic packages attempted by the Sarney government, unwilling to take the tough action needed to put its affairs in order, ended in complete failure, with inflation leaping from 6 per cent in May, to 24 per cent in June and close to 30 per cent in July. Even before the latest attempt, it had become completely clear that the unorthodox economic packages attempted by several Latin American countries cause far more problems than they solve and that they serve a political, rather than an economic purpose. All they achieved was to slow the rise in consumer prices for a short time, while when they collapsed – as all of them did – inflation always returned with renewed vigour. The three freezes in Brazil did an ever increasing amount of damage to the public sector companies, which unlike those in the private sector, were unable to protect themselves by finding ways round the freezes. The private sector was able to take advantage of the very high real interest rates paid by a government desperate for funds to meet its obligations, yet not prepared to alter the taxation system to share the burden more justly.

A problem which did not previously affect Brazil seriously, capital flight, has also now become an important issue. The Bank for International Settlements estimated that $32bn has left the country during the past few years.

It is true that this is relatively far less than has fled Argentina, Mexico or Venezuela, but it is a record for Brazil, where investors had previously preferred to leave their money, as it was obtaining an attractive return and seemed safe.

Partly encouraged by the exchange rate freeze in the early part of 1989, many international companies operating in Brazil remitted as much in profits and capital during the first half, $2bn, as they did during the whole of 1988. Many were taking advantage of the short term opportunity given to them by the government and, in fact, brought back the money via the black market to end up with several times more capital than they started out with. Other companies are now taking money out of Brazil and leaving it out.

One of the effects of the failure to get the public finances under control has been that it was not possible for Brazil to get anywhere near meeting the targets set for the economy in agreement with the IMF. Because of the lack of an agreement, none of the $2.8bn due to come to Brazil last year from the IMF itself, from other official agencies, from the private banks and from Japan, was forthcoming. Because of this lack of new funds, and determined to maintain its reserves of foreign exchange above the critical level of $6bn, the Brazilian Government suspended payment of interest and capital during the year in a new moratorium.

Trade surpluses maintained

The only positive thing to be said about the performance of the Brazilian economy is that it is proving possible for the country to maintain its impressive trade surpluses. Exports continued to run at double the rate of exports, producing surpluses in excess of $16bn, which was almost, but not quite, enough to meet all debt commitments. This success in exporting has benefited those industries in Brazil which are able to sell a significant proportion of their production abroad. These include textiles, paper and pulp, steel, footwear and, to some extent, the motor industry, although this industry has clearly failed to meet the challenge of new technology and very few important investments are now being made in the country compared with those in the United States, Europe and several South East Asian countries. Aided by large influxes of capital, the industries able to export have been able to invest heavily in modernising their machinery and equipment. In contrast, those industries only selling in Brazil, as is the case of most public sector companies apart from the Vale do Rio Doce Mining company and to some degree, Petrobras, have been hit hard by the drying up of external capital. Prevented from raising tariffs to adequate levels, public sector companies have not been able to make the investments needed to cope with the increased demand which will come when the Brazilian economy starts to move ahead again.

Although the crisis has continued to deepen and intensify, it is by no means certain that it has yet reached the point where the government has no alternative but to tackle the underlying problems shackling the Brazilian economy. The crisis could easily drag on for several more years yet. Bolivia's economy has been in a profound crisis for many years and only during the government of Siles Suazo did things reach such a point that powerful vested interests were prepared to face sacrifices to allow real reforms to take place. However, although it is often cited as being exemplary, Bolivia is a bad example to take, as it must be remembered that the economy there is fuelled by income from illegal cocaine exports equal to that obtained from legitimate exports.

It is as if Brazil were obtaining a parallel income of about $24bn a year, which would enable a transition to be made relatively painlessly.

The more relevant example is Argentina which has been in crisis since the 1950s, and whose economy for the past 35 years has stagnated and slipped steadily down the league table. It remains to be seen whether the new government led by Carlos Menem will in fact have the power and the will to impose measures which, if they are to succeed, must prejudice the interests of some of the most powerful groups in society. Groups which, like their counterparts in Brazil, have been able to benefit from the financial chaos of recent years and have taken the opportunity to strengthen themselves.

A debate rages in Brazil as to whether the economic difficulties of the past few years have been caused by the excessive weight of the foreign debt, or whether the causes are internal, reflecting the profound weaknesses of many Brazilian institutions and their chronic inability to adjust to new circumstances. Many critics firmly place all the blame for the present difficulties on the fact that in its attempt to meet all its debt obligations Brazil has been forced to transfer so much of its national product abroad that incomes have been depressed here, thus provoking the crisis. Some suggest that the remittances of capital abroad in debt servicing, which have far exceeded any new investments, are having the same effect on the Brazilian economy as did the enforced reparations which were imposed on Germany following the First World War. It is suggested that it was these remittances which provoked the hyperinflation and economic chaos of the 1920s in that country, which helped provoke the world recession and which pushed a desperate Germany to adopt fascism and caused the Second World War. In fact, even now debt servicing commitments form less than 2 per cent of GNP and the share of the national product which is exported is still below 10 per cent, considerably less than that of many other countries, notably in Europe, although considerably higher than has been usual for Brazil.

Brazil's trade surplus is the third largest in the world, and by far the largest of any country in the Third World. Despite the tremendous effort and sacrifice which has been made by Brazil in increasing exports, the debt has continued to grow, while creditor countries have not been prepared to resume their lending. There has been a sharp reduction in average incomes and the gradual process of income re-distribution which had begun during the late 1970s, has been reverted. A greater proportion of wealth is now concentrated in fewer hands than ever before.

Exporting success postponed crisis

Although not sufficient to persuade bankers to start lending again, the success in exporting has postponed the crisis, which has reached its peak in several neighbouring countries, and as a result several extremely damaging phenomena have taken place. During the late 1960s and through the 1970's, the state sector dominated development in Brazil. The steel, electricity generating and transmission, oil, telecommunications and transport industries concentrated the great majority of investments, and the public sector gained

KEY FACTS

Official title: Federative Republic of Brazil

Head of state: President Fernando Collor de Mello

Capital: Brasília

Official Languages: Portuguese

Currency: Cruzado (Cz) = 100 centavos

Exchange rate: Cz5.92 per $ (Nov 1989)

Area: 8,511,965 sq km

Population: 144m (1988)

GDP per capita: $2,449 (1988)

GDP real growth: 0.0% (1988)

Inflation: 683% (1988)

Trade balance: $19,182m (1988)

Foreign debt: $112bn (1988)

great prestige, until it came to dominate the economy. The years of economic crisis have seen the virtual end of the period of state sector dominance which has resulted in what many regard as the excessive strengthening of some parts of the private sector. Although apparently so powerful, the state sector enterprises always depended on outside finance, rather than growing by means of their own earnings, so they were very vulnerable to this support being cut off. Ever since they were established, the state sector companies in Brazil, such as Eletrobras, Petrobras, Telebras and Siderbras – enterprises too large and complex and requiring investments too substantial to be raised by the private sector – have encouraged the intermediate industries, such as the motor industry, textiles, pulp and paper, petrochemical, heavy industry and the rest which are overwhelmingly in private hands. To achieve this, the rates charged to customers for these utilities have invariably been subsidised by the government out of taxes, that is to say by the community as a whole, although where applicable, rates to the private consumer have also frequently been set below cost. To obtain the finance needed for expansion plans, all the state sector companies have been encouraged to borrow, mainly from abroad, and it is these state companies which are responsible for the bulk of the foreign debt. But because of the drying up of foreign capital in recent years, the usual source of new finance has been closed to the companies, while the successive devaluations which have occurred have meant that the weight of their debts has risen substantially.

The state firms have all been particularly hard hit by the various price freezes of the past few years, which have worsened the state of their finances, already overburdened by years of easy living and bad habits. Almost all of the state sector is in deep crisis and, worst of all, much needed investments in new capacity are not being made, so that the Brazilian economy runs the real risk of being strangled in a few years' time through a lack of electricity, steel and communications systems. The telephone system is already chronically congested, while several thousand kilometres of highway are breaking up each year through lack of maintenance. Although it has become fashionable to talk of privatising the state sector companies – and many of them would doubtless benefit from being in more dynamic hands – the fact that the state is running them is not the real problem. The main problem now is that they are not being allowed to charge economic charges for their services at a time when the private sector they were structured to serve is now quite strong enough to stand on its own feet.

Wealth transfer to private sector

The soaring inflation of the past few years, and the insatiable demand for short term finance by a government unable or often unwilling to raise sufficient income for its expenditure out of taxes, has meant that very high real interest rates have been paid to those able and willing to lend. Only the cushioned private sector has been in a position to lend, as despite the various freezes, most firms have managed to keep receipts far enough ahead of expenditure and made profits large enough to have plenty over to invest in government securities. Most companies now get at least a third of their profits from their investments in government stocks, a phenomenon responsible for a massive transfer of wealth from the public to the private sector. Because of the way power is structured in Brazil, no government has been prepared to tackle such problems so far, but finding a formula to reconcile the interests of both the state and the private sector will be a major task of the new administration.

Disastrous Sarney government

The government of José Sarney has now come to an inglorious end and Brazil has a new president. It will be some time before historians tell us whether the economic deterioration which Sarney presided over was in fact the result of his own weak, vaccilating and vain character, his failure to produce a medium term strategy and his inability to confront any group in society. Or whether his miserable performance would have been the fate of any president following the two decades of military rule and that the same economic deterioration would have occurred whoever had taken over in 1985. The misfortunes of Raul Alfonsin and Alain Garcia seem to suggest that it is the moment, rather than the man, although Julio Sanguinetti did a good job in Uruguay under circumstances as difficult as those of any of his neighbours. Perhaps Brazilian society was not ready for the profound reforms which are clearly urgently required. It may be the case that only after the economic situation of the great mass of society has deteriorated sufficiently will it be recognised that a new president must be allowed to make the profound reforms which are needed. Otherwise Brazil can be expected to emulate Argentina and slide gradually down the economic league table, out of the first 10, where it now is, and eventually out of the first 20 – only maintaining a place at all because of its sheer size and the performance of some activities it will prove hard to wreck, such as agriculture, mining and the initial processing of primary products.

One thing which has not yet been recognised in Brazil is that it has lost for decades to come the largely undeserved high regard it enjoyed during the 1970s in the eyes of the international financial and business community. Since the early 1980s Brazil has tried to do the impossible, to persuade the banks and financial institutions in the developed world to start to lend once again, as during the years of great growth, the late 1960s and early 1970s. It has not yet sunk in that for several reasons, the developed countries are now in no mood to aid Brazil, however well it performs. For one thing, there is not now perceived to be any ideological threat to the continent, as was thought to be the case following the success of Fidel Castro in winning power in Cuba. The Soviet Union has now made it clear that its period of imperialist expansionism is over, for the time being at least. The private banks are now able to make far larger profits from lending to consumers in their own countries and have no interest in throwing good money after bad. They will not willingly lend any more while they are not getting back what they have already advanced.

It needs to be recognised that there is no desire among governments in the developed countries to see new rivals emerge from the developing world. The developed countries have had enough trouble adapting to the challenge from Japan and the other Asian countries – they have no wish to facilitate the emergence of yet more competitors. Brazil must now accept that it is on its own, and that is very difficult for a country which has a long history of dependence, relying upon other countries to take its decisions for it, to face up to. The new reality requires the sort of profound reform of the structures and the balance of power achieved by Japan and West Germany, to give the country an agility and a flexibility which is now completely lacking. At the moment, initiatives are usually bogged down by a bureaucracy more interested in minutiae than opening the way for growth and change, which reflects the interests of the deeply entrenched establishment. A reason for pessimism is that this stifling bureaucracy is still not seen as an obstacle to prosperity by the mass of Brazilians, who nurture a great affection for the restrictive practices which surround them. Most Brazilians, isolated from the progress and rising living standards in much of the rest of the world, have no desire to break out of the trap they do not know they are in, if it means making some sacrifices in the short term. To achieve a real revolution in Brazil, which will result in the liberation of the creative forces which are latent in the country, will be almost as hard as it is proving to be in the Soviet Union, a society with a similar bureaucracy and plethora of restrictive practices, which came into existence for the same reason – to benefit the powerful few and keep the masses in check.

KEY INDICATORS

	Unit	1984	1985	1986	1987	1988
Population	m	132.7	135.6	138.5	141.5	144.4
GDP (1985 prices)	Cz bn	1304.9	1413.8	1521.1	1575.9	1571.6
GDP per capita (current $)	$	1,618	1,629	1,972	2020	2449
Inflation	%	210	223	145	230	683
GDP real growth	%	5.1	8.3	7.6	3.6	0
Agricultural production	1979-81=100	113.0	125.3	112.6	127.7	–
Sugar-cane	m tonnes	222.3	247.2	238.5	273.9	–
Oranges	m tonnes	13.4	14.2	13.3	15.0	–
Industrial production	1980=100	91	99	110	111	107
Manufacturing growth rate	%	6.1	8.3	11.3	1.0	–3.4
Mining sector growth rate	%	27.3	11.5	3.7	–0.7	0.4
Crude oil output	m b	173.5	205.8	216.6	216.2	211.6
Crude steel output	'000 tonnes	18,386	20,450	21,770*	–	–
Exports	$m	27,005	25,639	22,349	26,225	33,783
Coffee	$m	2,564	2,369	2,006	1,959	1,998
Soybeans & products	$m	2,566	2,540	1,640	2,325	3,046
Iron-ore	$m	1,605	1,658	1,615	1,615	1,892
Imports	$m	13,916	13,153	14,044	15,052	14,601
Petroleum	$m	7,473	6,203	3,584	4,724	4,099
Balance of trade	$m	13,089	12,486	8,305	11,173	19,182
Current account	$m	42	–273	–5,304	–1,450	–
Exchange rate	new Cz per $	0.00185	0.00620	0.01366	0.03923	0.26238

* estimated figure

Thirst for change

As is the case in the Soviet Union, the Brazilian people are becoming tired of the excuses of their leaders and are thirsting for change. This is the fundamental reason for the major upset which occurred on the Brazilian political scene at the beginning of 1989.

Early in the year it appeared that the contest to be the first president elected by popular suffrage since 1960, would involve only well known names. These would include Leonel Brizola, the once fiery nationalist, whose very mention used to send shivers through the presidents of the multinationals whose operations in Brazil he had threatened to nationalise. Then there was Ulysses Guimaraes, part of the furniture in Brasila, architect of the new constitution. There was ex-vice-president Aureliano Chaves, the leader of the workers' party, Luis Inacio, 'Lula' da Silva, and Paulo Maluf, a strong candidate from the right. But suddenly a new star rose in the firmament in the form of Fernando Collor de Mello, governor of the small state of Alagoas, in the north east. Collor's claim to fame was that on assuming power in his small state, he had sacked several of his predecessors – well-paid political appointees, men known as 'Maharajas' in Brazil, because of the huge size of the salaries they are paid, which have no relation to the work they do, but which are a token of reward for electoral services rendered. Collor was also only 39, little more than half the age of many candidates. He did not look back to the past, as his rivals tended to do and he just repeated what the very unsophisticated electorate wanted to hear, that Brazil needed to change. When in the final round of the presidential elections, in December 1989, he defeated his socialist rival, Luís Inácio Lua da Silva, what was questionable was whether Collor, despite his reformist-sounding rhetoric, a man closely allied to the establishment in Brasilia and to the all-powerful groups which dominate the media, would in fact prove capable of taking on the entrenched groups he claimed as enemies in his election campaign.

Congress versus president

Sarney's years were certainly made particularly difficult by the relationship which came to exist between the unusually weak presidency and a congress anxious to increase and assert its authority after two decades of impotence, when it had been systematically weakened and often humiliated by the military. One of the results of the first of the unorthodox reform plans, the Cruzado Plan, was that it enabled the PMDB party to win an overwhelming majority in the 1986 elections. In the normal Brazilian tradition, politicians from all ideological positions abandoned their previous parties and joined the one which was clearly going to win. The result was that the PMDB, when it could be persuaded that what it was voting for was favourable to the institution of congress, was capable of vetoing any initiative taken by the president. The past few years have seen a deep stalemate between congress and the presidency, one demonstration of which is Brazil's unworkable new constitution. Besides being inordinately long and detailed, this document places a huge number of obstacles in the way of any president congress does not want to aid. The constitution has woven yet another new web of bureaucracy in a country already crippled by it. *De facto*, the country has been largely operating under a parliamentary system for the past three years, further limiting Sarney's freedom.

Need for electoral reform

The stalemate demonstrates the need for a profound electoral overhaul in Brazil, as the present system of proportional representation, by which a candidate is elected by voters throughout the state, rather than in a constituency, means in practice that congressmen are not answerable to anybody for their decisions. The congress to be elected in 1990 will certainly fundamentally revise the new constitution and much of the highly restrictive legislation will probably be consigned to the scrapheap.

Positive results of exporting

Although many criticise the fact that Brazil has been exporting so much during the past few years, in fact the results have been highly positive for those sectors which have been involved. The pulp and paper industry, for example, benefiting from high external prices, has been able to embark on a programme of massive investment which will result in the duplication of its capacity during the next few years, which it would not have been able to undertake otherwise. About $10bn is to be spent during the next five years by an industry which has demonstrated that Brazil is one of the world's most favoured locations for new mills, because of the very favourable growing conditions, as well as the fund of experience which has been accumulated.

The textile industry, which, to guarantee the continuity of production has opted to export about a quarter of its production each year – however attractive the internal market might appear for short periods – has also been able to invest more than $30bn during the past few years, much of it the proceeds of overseas earnings. Output is likely to increase from 1.2m tonnes of textile products to almost 2m tonnes by the mid-1990s; the number of garments made increasing to more than 2bn.

When Brazil set about increasing its steel making capacity to 25m tonnes, it was not envisaged that in 1988 and 1989 40 per cent of all that steel would be exported and that the industry would top the league table of exporters. But because of the slump in internal demand and the strong demand for steel from some parts of the world, notably South East Asia, Brazil's steel exports earned more than $2bn in 1988 and 1989. Despite the impressive exporting performance, in contrast to the private sector exporters, because the industry was forced to charge so little for the steel it sells on the home market and because not all the steel companies were in a position to export, the earnings are not sufficient to finance the next phase of expansion which the industry must make if it is to be in a position to meet the demand which will occur by the year 2000. This is not the case in the aluminium industry, in which the state owned CVRD mining company, Brazil's largest private concern, the Votorantim mining group and several multinationals all participate. Aluminium also achieved record export earnings in 1988 and 1989, again because of the slump in internal demand. In the case of this industry, which does not have major foreign debts, earnings are being utilised to duplicate output, again taking advantage of Brazil's unique situation as a low cost producer. The fact that the private sector is investing so heavily is of great relief to Brazil's heavy industry, which in previous times relied almost entirely on the public sector for its orders.

Another victim of the squeeze on state enterprises, which tend to be treated as political footballs in Brazil, subject to the whims of the president and his advisors, has been the oil monopoly Petrobras. This company plans to double its output of oil by 1995 to 1.5m barrels per day. This is perfectly feasible, given the fact that Brazil's reserves are rising far faster than consumption, boosted by the major finds which are still being made offshore Rio de Janeiro state, off Rio Grande do Norte, and in the Amazon jungle. It is now becoming clear that there are major reserves of hydrocarbons hidden beneath the jungle, even if mainly in the form of gas. Petrobras plans to completely

COUNTRY PROFILE

Political system: Federal republic of 23 states, three federal territories and one federal district. Executive power is exercised by the president (currently serving a six-year term, though there are plans to shorten it to four) aided by the ministers of state. The legislature is the bicameral National Congress consisting of the Federal Senate (72 members, two-thirds directly elected, one-third indirectly; members elected in rotation, for eight years) and the Chamber of Deputies (487 members directly elected from constituencies for period of four years). Constitutional re-organisation in process.

Main cities/towns: Brasília (population about 1.8m), São Paulo (10.8m), Rio de Janeiro (7m), Belo Horizonte, Porto Alegre, Recife, Belém, Salvador, Manaus.

Media
Press: Around 250 daily newspapers with a combined circulation of under 5m. Due to distribution difficulties in such a large country there are no national dailies. However numerous regional papers are published. Main US and European newspapers available in Rio de Janeiro and São Paulo. *Brazil Herald* (circulation 18,000) incorporating *Daily Post* is only local English language daily. Main papers are *O Dia* (207,000), *Jornal do Brasil* (200,000), *O Globo* (350,000) and *Jornal do Comércio* (in Rio de Janeiro, 31,000) and *O Estado de São Paulo* (230,000), *Fôlha de São Paulo* (211,900), *Gazeta Mercantil* (leading business paper, 80,000), *Jornal da Tarde* (120,000), *Diário Popular* (90,000) and *Diário do Comércio e Indústria* (50,000). Numerous periodicals are also available.
Broadcasting: Over 2,000 radio stations (including AM, FM, SW and Tropical Wave) and 177 TV stations. Satellite connection with rest of world.

External trade: Record trade surplus achieved in 1984-85 owing to rapid growth in exports (particularly coffee, soya, orange juice, metallurgical products, footwear, chemicals and oil derivatives) and sharp reduction in oil and capital goods imports. Current account also recorded surplus in 1984 despite heavy debt interest payment burden leading to large services deficit. The trade balance remained in surplus in 1985-86 although the continuing structural deficit on services meant the current account returned to deficit.
Exports: (1986): Coffee (10% of total value), soya (7%), iron ore (7%), steel & products (8-9%), capital goods (8-9%), textiles/footwear (7-8%), transport materials (6-7%), fuel/lubricants (6-7%), chemicals (5-6%). Main destinations (1985): USA (26% of total value), German FR (5%), Netherlands (6%), Italy (4%), France (3%), UK (2-3%), Argentina (2%), Japan (5%), Middle East (5-7%).
Imports: (1985): Mineral fuels and products (49% of total value), machinery and electrical equipment (15%), chemicals (11%), vegetable products (7%), transport materials (4%). Main sources (1985): Middle East (20% of total value), USA (20%), German FR (6%), France (2%), UK (2%), Argentina (3-4%), Japan (4%).

Agriculture: Sector contributes 13 per cent to GDP and accounts for around 40 per cent of exports. Only 20 per cent of arable land is cultivated and large-scale farming is concentrated in the south and south east of the country. Major agricultural exports are coffee (world's largest producer and exporter), sugar cane (world's largest producer and exporter – used domestically as an important source of alternative energy through the production of fuel alcohol), soya (world's second largest producer after USA), orange juice (supplies 85 per cent of world market for orange juice concentrates), tobacco, cocoa, cotton, butter, maize and cattle (around 10 per cent of total world trade). The emphasis on export crops, rather than food stuffs, has led to a heavy dependence on imports, especially of wheat. Domestic production of wheat accounts for less than a third of consumption. Selective channelling of credits towards export orientated crops has also led to serious decline in production of staple foods, such as rice, potatoes, beans, cassava, sorghum and maize. Government plans to address the problem of food supply by irrigating more land for wheat production, improving credit facilities and raising food prices. Fishing (cod, crab, lobsters and shrimps) is an important source of food, and the sector is being developed with exports of fish products becoming important in recent years. Brazil's timber reserves estimated to be third largest in the world. Softwoods (particularly parana pine and eucalyptus) are grown in the southern regions and mainly supply local pulp and paper industry, while hardwoods are felled for export. Most of the country's woods are used primarly as fuel.

Industry: Sector is well developed with manufacturing contributing over 27 per cent to GDP and industrial goods accounting for around 60 per cent of exports. The role of the government in manufacturing is substantial with the state accounting for more than two-thirds of the activity in the sector. Production is concentrated on capital goods, construction materials, chemicals and petrochemicals, smelting (steel and non-ferrous metals), rubber, motor vehicles, sugar and wood processing. Technologically based industries have seen rapid growth, especially the electronics and information processing sectors. However traditional industries such as textile and clothing, beverages and food processing remain important and continue to account for around 50 per cent of total output. Activity is largely centred in the south east, especially around São Paulo. Emphasis has been on encouraging large private companies to diversify into local manufacture of consumer and capital goods for the domestic market. Industrial production slumped during period 1980-83 after almost a decade of sustained growth. Relaxing of government's austerity programme and upturn in export demand led to marked improvement in capital goods and intermediate goods industries during 1984-86. Domestic demand for consumer goods was high in 1985 and substantial growth was also recorded in this sector after several static years but domestic demand was sharply cut back in 1987 and industrial growth weakened rapidly. Brazil is eighth largest steel producer and sixth largest producer of motor vehicles in the world. It is also a major exporter of armaments (including naval vessels) and boasts a diversified electronics and aerospace industry.

Mining: Vast mineral resources, particularly in states of Pará, Minas Gerais and Rondônia. Major world producer and exporter of iron ore and bauxite. Other minerals produced are manganese, coal, zinc, nickel, lead, copper, cadmium, cobalt, chrome, gold and tin. In addition, there are abundant reserves of gypsum, titanium, phosphates, platinum, beryllium, precious and semi-precious stones, thorium and zirconium. Important developments are the opening of the Grande Caraja mine, which is estimated to hold the world's largest iron ore reserves (18bn tonnes), the exploitation of coal resources under the National Coal Programme (Prócarvão) and the expansion of gold production at Nova Lima.

Hydrocarbons: Production of crude oil has more than doubled in last four years to around 600,000 b/d, making Brazil the third largest producer in Latin America. The rapid growth of domestic production and falling oil prices have led to sharp fall in value of crude imports. Largest oil fields are Rio Grande Norte, Sergipe, Bahia, Ceára and Rio de Janeiro (offshore fields account for over 75 per cent of known reserves). Owing to intensified oil exploration and development, particularly in the

modernise its transport system to take account of the fact that oil imports will be gradually reduced in the next few years. It plans to build several thousand km of new pipelines, new refineries, new and mainly small ships for coastal deliveries, as well as continuing with the offshore search programme and the massive expenditure involved in bringing the oil ashore from fields which lie under some of the deepest waters yet tackled. Petrobras had planned to spend $3.7bn on its investment projects during 1989, but because its income was cut by $500m a month due to the price of fuels being held down below production costs in a fruitless effort to beat inflation by this means, investments are having to be reduced by half. The date of self-sufficiency had to be pushed back, and suppliers' plans hurriedly altered.

Agriculture

Much of agriculture did very well during 1989, aided by good growing weather. But one sector in particular was also badly prejudiced by government policy, the sugar and alcohol industry. Because of rapidly increasing demand and a reduction in the price paid to planters, the amount of alcohol produced from sugar cane has not been sufficient to meet demand for the past three years. The situation became critical at the beginning of 1989, when alcohol ran short in some parts of the country and motorists had to leave their cars at home. Ships were ferrying alcohol from one part of the country where the harvest had begun, to the other where it had not, and carrying it back again later. The problem has again been a lack of any clear definition by this government of what the policy should be for the alcohol programme. Although there was plenty of warning, no action was taken to reduce demand, or to increase supply, so that the inevitable finally happened. Brazil may now not only lose the opportunity to increase its export earnings from the present high sugar price, but the world's largest sugar producer may actually have to import some sugar or alcohol this year, something which really encapsulates the incompetence of the Sarney government. The argument rages as to whether Brazil's alcohol programme is or is not viable, and whether or not it should be gradually phased out and the massive amount of land now devoted to cane gradually switched to something else. This would happen naturally if the industry ceased to be subsidised.

Soya producers, although responsible for an all-time record crop in 1989, also had reason to complain last year. Because the exchange rate was frozen for several months – another part of the ill-fated 'Summer Plan' – those producers who had not arranged to sell their product at the beginning of the year, and only a few wise ones did, (the rest bet on a repeat of the 1988 drought in the United States) the price they received fell steadily throughout the year and some farmers were severely prejudiced. If sugar and soya had a mixed year, coffee also had to face a severe shock with the ending of the quota system operated by the International Coffee Organisation and the subsequent slump in prices. Brazil's earnings from coffee were down by some $500m last year and will be down by at least $800m this, although because the tax levied on coffee by the government has been reduced, farmers' earnings will not be so badly affected. The government will not have the funds to buy up excess stocks in future, however, and overall coffee production is likely to fall. In fact, Brazil stands to lose far less by the termination of the Coffee Agreement than countries which depend to a far greater degree on coffee. These days coffee forms only 7 per cent of Brazil's export revenues; it used to form more than half.

Another industry which had a very good year was citrus, with earnings of more than $1bn. Brazil is responsible for 80 per cent of all the orange juice which is traded around the world and it is a com-

COUNTRY PROFILE

offshore Campos Basin, and falling domestic consumption, self sufficiency in oil climbed to over 50 per cent, and is expected to reach 100 per cent by 1993. National petroleum products industry supplies 35 per cent of the domestic market. State owned oil company, Petrobrás, maintains control over exploration and extraction and operates 10 out of 12 oil refineries. Total reserves of crude petroleum estimated at 2.3bn barrels, but new finds may push this up to 4bn barrels. Exploration work in Sergipe, Alagoas, Bahia, upper Amazon basin and Campos basin meanwhile boosted proven natural gas reserves to over 86bn cu metres. Production of gas has risen to 5.4bn cu metres, the bulk of which is used locally as petrochemical raw material and as a fuel for steel smelting. First section of north-east gas pipeline (Nordestão) is being completed, to bring gas from Ceará and Rio Grande do Norte to industrial consumers in north east coastal towns. A complementary 65 km undersea pipeline from offshore fields is under construction. Plans have also been drawn up to build a 3,000 km pipeline from western Amazonia to the industrial south-east.

Energy: Gradual replacement of oil by hydro-electricity as primary energy source. Imports of oil and derivatives down 18 per cent in 1985. Continued expansion of country's vast hydroelectric potential (estimated at 213,000 mW). Hydroelectric plants currently provide some 92 per cent of nation's electricity. First and second stage of Itaipú power station (12,600 mW) completed and first stage of giant Tucurúi dam operational. Plans to build 17,000 mW power station on Xingu river in Amazon region. Brazil's nuclear programme has however been sharply reduced due to financial constraints. Angra nuclear power plant, which was officially opened in 1985, supplies 25 per cent of Rio de Janeiro's electricity. As a result of the government-sponsored fuel alcohol programme, Proácool, nearly 85 per cent of all vehicles sold in 1984 were alcohol-powered. Substitution of alcohol for petroleum was estimated to save Brazil around $2bn per year in foreign exchange. However, declining oil price has meant that alcohol is no longer such an attractive alternative to oil and in March 1987 the government suspended its investment in the programme until at least 1989-90.

Membership of international organisations: CCC, FAO, G-77, GATT, IADB, IAEA, ICAC, ICAO, IDA, IFAD, IFC, IHO, ILO, IMF, IMO, INTELSAT, Inter-American Development Bank, International Coffee Organisation, IPU, IR, ISO, ITU, IWC (Wheat), Organisation of American States, PAHO, SELA, UN, UNESCO, UPU, WHO, WIPO, WMO, World Bank, WTO.

Time: GMT -3 (except western states including towns of Campo Grande, Corumba, Cuiabá, Manaus: GMT -4. Fernando de Noronha Islands: GMT -2. Acre: GMT -5.)

Climate: Mainly tropical and sub-tropical; particularly humid in Amazon region and along the coast. Temperate on the southern coast and on the higher lands. Rainy season in south from Nov to Mar, driest from Jun to Aug. Temperatures in Rio de Janeiro: winter (Jun-Sep) 14-30°C, summer (Dec-Mar) 20-39°C.

Entry requirements
Passport: Required by all. Passports should be valid for six months beyond the date of arrival.
Visa: Required by all except nationals of EC countries, Scandinavian countries, Canada, most South American countries and certain others.
Currency: No restriction on import/export of local currency. Foreign currency import unlimited but amounts must be declared; export of foreign currency allowed up to declared amount. Regulations liable to change at short notice.
International credit cards widely used, though cash advances only paid in local currency.

Health precautions
Mandatory: Yellow fever vaccination certificate if arriving from infected areas.
Advisable: Yellow fever vaccination essential for visits to infected areas within Brazil; these include Mato Grosso, Rondônia and states surrounding the Amazon. Typhoid/paratyphoid and polio vaccinations also recommended. Malaria prophylaxis advisable for visits to Amazon regions. Water precautions should be taken.

Representation overseas: Abidjan, Abu Dhabi, Accra, Algiers, Ankara, Asunción, Athens, Baghdad, Bangkok, Beijing, Beirut, Belgrade, Berne, Bissua, Bogáta, Bonn, Bucharest, Budapest, Buenos Aires, Cairo, Caracas, Copenhagen, Dacca, Dakar, Damascus, Gaborone, Georgetown, Guatemala City, Helsinki, The Hague, Havana, Islamabad, Jakarta, Jeddah, Kingston, Kinshasa, Kuwait City, Lagos, La Paz, Lima, Lisbon, London, Luanda, Lusaka, Managua, Manila, Maputo, Mexico City, Monaco, Montevideo, Moscow, Nairobi, New Delhi, Panama, Paramaribo, Paris, Port-au-Prince, Port of Spain, Prague, Praia, Pretoria, Oslo, Quito, Rabat, San José, San Salvador, Santiago, Santo Domingo, Seoul, Singapore, Sofia, Stockholm, Tegucigalpa, Tehran, Tel Aviv, Tokyo, Tripoli, Tunis, Vatican City, Vienna, Washington, Warssaw, Yaoundé.

Air access
National airline: VARIG (Viação Aérea Rio Grandense). Cruzeiro and Transbrasil also operate international flights.
Other airlines: Aeroperú, Aerolíneas Argentinas, Air France, Alitalia, Avianca, British Airways, Canadian Airlines International, Eastern, Iberia, JAL, KLM, Ladeco, Lan-Chile, LAP, Lloyd Aéro Boliviano, Lufthansa, Pan Am, Pluna, Royal Air Maroc, Suriname Airways, SAS, South African Airways, Swissair, TTAG Angola, TAP, Vasp.
International airports: Belem-Val de Cans (BEL), 12 km from city; Belo Horizonte-Pampulha (BHZ), 9 km from city, restaurant, buffet; Brasilia-International (BSB), 15 km from city, duty-free shop, bar, restaurant, buffet, bank, post office, shops, hotel reservations, car hire; Confins Internacional (CNF), 45 km from city; Manaus-Eduardo Gomes International (MAO), 14 km from city duty-free shop, bar, restaurant, buffet, bank, post office, shops, car hire; Porto Alegre-Salgado Filho (POA), 8 km from city, restaurant, buffet, bank; Rio de Janeiro (RIO), 12 km from city, bar, hotel, taxi, duty-free shop, restaurant; International-Galeão (GIG), 20 km from Rio, bar, restaurant, buffet, bank, post office, shops, hotel reservations, car hire; Santos Dumont (SDU), in city centre (air shuttle to São Paulo every half hour at peak times); Salvador-Dois de Julho (SSA), 35 km from city, restaurant, buffet; São Paulo (SAO); Guarulhos (GRU), 25 km from city, restaurant, bank, post office, shops; São Paulo – Cogonhas (CGH), 14km from city; Viracopos (VCP), 95 km from São Paulo, bar, restaurant, buffet, bank, post office, shops.

Surface access
Main ports: Paranaguá, Recife, Rio de Janeiro, Santos and Vitória. Also numerous smaller ports including Belém, Tubarão, Santana, Maceió, São Sebastião, Madre de Jesus, Areia Branca, Rio Grande.
Overland access: Possible by road from Argentina, Bolivia, Paraguay, Uruguay (eg bus services Curitiba-Asunción, Porto Alegre-Montevideo, Rio de Janeiro-Buenos Aires). Rail links run from Argentina and Uruguay.

Hotels: Graded from one- to five-stars. Wide range available in main towns but sometimes heavily booked (e.g. during Carnival) and advance booking advisable. Listings available from local

modity which is constantly gaining popularity as new markets are found. Last year saw the first shipment of frozen juice to the Soviet Union, delivered to a packing factory built with Brazilian capital and technology. The import of this juice will be paid for with the proceeds of the export of Russian apple juice, also processed at the same plant. Until now, Russian juice has been of too low a quality for export. If this project is successful, and it is almost certain that it will be, this market could have an enormous potential for Brazil. The same is true for Japan, which is to open its doors to imported juice for the first time in 1992. Again Brazil, still with almost unlimited areas of land available for growing citrus, is in a very good position to take advantage of Japan's new policy, basically aimed at aiding the United States.

If there are question marks over coffee and sugar, farmers are now rushing to plant rubber in the state of Sao Paulo, where it has been found that the plant does much better than in its natural habitat, the Amazon. In the south, rubber is not subject to the diseases which makes growing on plantations in the jungle region unviable. Farmers have been attracted by the fact that the Brazilian government pays twice the world price for rubber, in a forlorn attempt to persuade tappers to go into the jungle to collect the sap from the trees.

Amazon future brighter

The fate of the Amazon region has become a major issue in Brazil during the past few months. The one positive result is that it is gradually sinking in that the long-time practice of the haphazard cutting down of the forest by companies taking advantage of tax incentives which allow them to do it for nothing, has been demonstrated to be wholly negative. It is now being realised that timber worth billions of dollars has been going up in smoke every year. Once again, vested political interests have ensured that this system has not been tampered with until now. But Brazil is at last paying some attention to world public opinion, although the criticism must be sustained if the bad habits are to be broken. The lack of manpower to police any of the laws, which in theory should prevent massive clearances, demonstrates – if it needed to be demonstrated – that what Brazil needs is not more laws, but fewer, more realistic ones which can be enforced.

Deterioration of social services

The general bankruptcy of the state has meant that the last few years have seen a serious deterioration in the social services in Brazil, particularly education and health. Most state schools and universities were closed for several months by teachers' strikes during 1989, at a time when Brazil needs more than anything enough trained people to implement new policies and take advantage of new opportunities.

BUSINESS GUIDE

tourist offices. Only five-star hotels are not price controlled. Service charge usually included in bill, if not 10 per cent tip usual.

Credit cards: Amex, Diners, Mastercard and Visa widely accepted.

Car hire: International driving licence required, which must be validated by Automovil Club do Brasil. Traffic is often congested in main cities. Petrol is of poor quality and expensive. Service stations are rare on some roads (e.g. Belo Horizonte-Brasília) and often closed on Sundays.

City transport
Taxis: In Rio do Janeiro several types. So called 'common' taxi (yellow with checkered stripe), radio taxi (white, with a red and yellow stripe). Mostly metered and can be hailed. A 40 per cent surcharge operates between 2300 – 0600, on Sundays and public holidays. Tipping optional.
Buses: Extensive services operate in all main centres. Efficient though crowded. Two types – regular and special (frescões).
Metro: Two line service in Rio de Janeiro. Line one goes from Botafogo Station to Saenz Peña Station (Tijuca): (Mon-Sat) 0600 – 2300. Line two cuts across the city's centre, from Estácio Station to the Maria de Graca Station: (Mon-Sat) 0600 – 2000. Also two line network in São Paulo. Integrated bus/metro tickets available.

National transport: When travelling between cities on public transport, visitors must carry passports, as proof of identity is required.
Air: Regular domestic and charter flights to all main cities. Main form of long distance travel. Air taxis available at most domestic airports. Advance booking not necessary for shuttle flights between Rio de Janeiro and São Paulo (about one hour). Main internal airlines are Varig, Cruzeiro, Transbrasil and Vasp.
Road: All main centres are connected by surfaced highways, particularly good roads in the north. Many of the local roads are however in need of urgent repair. In total around 1.4m km of roads supervised by the Departamento Nacional de Estradas de Rodagem (DNER).
Buses: Frequent inter-city services between main centres. Standards are variable although many routes are now served by modern high quality coaches. Sleeping berths (*leito*) available on some routes.
Rail: State and privately-owned railways operate limited services to most main centres throughout country. Service is generally slower than bus and long-distance travelling can be uncomfortable. Good sleeper services with restaurant cars operate between São Paulo, Rio de Janeiro and Belo Horizonte.
Water: Services on São Francisco river between Juazeiro and Pirapora and up the Amazon to Manaus. Hydrofoil service between Rio de Janeiro and Niteroi.

Public holidays: New Year's Day, 20 Jan (Rio de Janeiro only), Jan (São Paulo only), Carnival (Feb-Mar), Good Friday, Easter Monday, 17 Apr (Tiradentes' Day), 1 May, Corpus Christi, 7 Sep (Independence Day), 12 Oct (Nossa Senhora de Aparecida), 2 Nov (All Souls'), 15 Nov (Proclamation of the Republic), 25 Dec (Christmas), 31 Dec (New Year's Eve). Certain fixed-date holidays may, by Presidential decree, be shifted to the previous Monday.

Working hours
Government and business: (Mon-Fri) 0900 – 1200; 1400 – 1800. Government departments usually 1100 – 1800. N.B. In Rio de Janeiro and São Paulo there is no siesta break; in Brasília there is a three hour siesta, 1200 – 1500.
Banking: (Mon-Fri) 1000 – 1600.
Shops: (Mon-Fri) 0900 – 1900, (Sat) 0900 – 1300. Shopping centres (Mon-Sat) 0900 – 2200.

Telecommunications: A 40 per cent tax is added to cost of all international communications.
Telephone: Long distance calls can be made from state telephone company offices, post offices and hotels. Trunk dialling system linking all main cities. Public telephone booths (*orelhões*) operated by tokens, which can be bought from newsagents, cafés etc. Dialling code for Brazil: IDD access code + 55 followed by area code (Belo Horizonte 31, Brasília 61, Rio de Janeiro 21, São Paulo 11) followed by subscriber's number.
Telex and telegram: Country telex code: BR. Telex messages can be sent from main post offices during business hours and from hotels at nights and weekends. Telegrams can be sent over the phone or at post offices and hotels.

Postal service: Postal charges are high. PO Box (*Caixa Postal*) address and post code number should be used when possible.

Banking
Head offices: Brasília.
Banco do Brasil, Banco Nacional de Credito Cooperativo, Banco Nacional de Habitâçao.
Head offices: Rio de Janeiro.
Asociação dos Bancos do Estado do Rio de Janeiro, Banco Boavista, Banco Bozano Simonsen de Investimento, Banco Bozano Simonsen, Banco Brascan de Investimento, Banco do Estado do Rio de Janeiro, Banco Internacional, Banco Lar Brasilia (Chase Manhattan), Banco Mercantil do Brasil, Banco Nacional, Banco Nacional Brasileiro, Banco Nacional do Desenvolvimento Economico, Banco Novo Rio de Investimentos, Banco Pinto de Magalhães, Banco Sul Brasileiro, Unibanco-Banco de Investimento do Brasil, Unibanco-União do Bancos Brasileiros.
Head offices: São Paulo.
Banco America do Sul, Banco Auxiliar de São Paulo, Banco Bamerindus de São Paulo, Banco Bandeirantes, Banco Bradesco de Investimento, Banco Brasileiro de Descontos, Banco Cidade de São Paulo, Banco de Tokyo (Bank of Tokyo), Banco de Comercio e Industria de São Paulo, Banespa (Bank of the State of São Paulo), Banco do Estado do Paraná, Banco do Grande São Paulo, Banco Europeu para a America Latina, Banco Expansão, Banco F. Barretto, Banco Finasa de Investimento, Banco Francês e Brasileiro, Banco Geral do Comercio, Banco Itaú, Banco Noroeste de Investimento, Banco Noreoste do Estado de São Paulo, Banco Nova America, Banco Piratinga, Banco Real, Banco Real de São Paulo, Banco Regional de Desinvolvimento do Extremo Sul, Banco Republica, Banco Safra, Banco Safra Investimentos, Banco Sogeral, Banco Sudameris Brasil, Banco Sumitomo Brasileiro, Caixa Economica do Estado de São Paulo, Casa Bancario F. Matarazzo, Safra Credito Financiamento e Investimentos, Safra SA Credito Imobiliario, Sindicato dos Bancos nos Estados de São Paulo.

Trade fairs: Numerous local/regional and some international trade fairs held at Anhembi Park in São Paulo. Since 1977, there has been an international exhibition centre (Riocenter) at Jacarepaguá near Rio de Janeiro. Main international events are: International Textile Fair (São Paulo – Feb/Mar), International Fair of Office Equipment (São Paulo – Sep), International Maritime Exhibition (Rio de Janeiro – Oct), International Packaging, Paper and Graphic Art Fair (São Paulo – Oct alternate years).

Electricity supply: 110 V AC, 60 cycles (Rio de Janeiro and São Paulo), 220-240 V AC, 60 cycles (Brasilia).

Representation in capital: Algeria, Angola, Argentina, Australia, Austria, Bangladesh, Belgium, Bolivia, Bulgaria, Cameroon, Canada, Chile, China PR, Colombia, Costa Rica, Cuba, Czechoslovakia, Denmark, Dominican Republic, Ecuador, Egypt, El Salvador, Finland, France, German DR, German FR, Ghana, Greece, Guatemala, Guyana, Haiti, Honduras, Hungary, India, Indonesia, Iran, Iraq, Israel, Italy, Ivory Coast, Japan, Republic of Korea, Kuwait, Lebanon, Libya, Malaysia, Mexico, Morocco, Netherlands, Nicaragua, Nigeria, Norway, Pakistan, Panama, Paraguay, Peru, Philippines, Poland, Portugal, Romania, Saudi Arabia, Senegal, South Africa, Spain, Suriname, Sweden, Switzerland, Syria, Thailand, Togo, Trinidad and Tobago, Turkey, USSR, UK, USA, Uruguay, Vatican, Venezuela, Yuogoslavia, Yemen Arab Republic, Zaïre.

BUSINESS DIRECTORY

Hotels: Brasilia
Carlton, Setor Hoteleiro Sul, Q5 lote G, 70300 (tel: 224-8819; tx: 1981).

Das Americas, Setor Hoteleiro Sul, Q4 lote D, 70300 (tel: 223-4490; tx: 1484).

Eron, Setor Hoteleiro Norte, Q5 lote A, Plano Piloto (tel: 226-2125; tx: 1422).

Nacional, Setor Hoteleiro Sul lote A, 70300 (tel: 226-8180; tx: 1062).

San Maiw, Setor Hoteleiro Sul, Q5 lote C (tel: 226-2211; tx: 3744).

Torre Palace, Setor Hoteleiro Norte, Q4, lote A, 70710 (tel: 226-3360; tx: 1905).

Hotels: Recife
Mar, Rua Br de Souza Leao 451, Boa Viagem (tel: 341-5433; tx: (81)1073).

Miramar, Rua dos Navegantes 363, Boa Viagem (tel: 326-7422; tx: 2139).

Othon Palace, Avenida Boa Viagem 3722, 50000 (tel: 326-7661; tx: 2141).

Recife Palace, Avenida Boa Viagem, 4070 (tel: 325-4044; tx: 4528).

Hotels: Rio de Janeiro
Atlantico Sul, Avenida Sernambetiba 1800, Recreio (tel: 327-8411; tx: 21813).

Caesar Park Ipanema, Avenida Vieira Souto 460, 22420 (tel: 287-3122; tx: 21204; fax: 247-7975).

Copacabana Palace, Avenida Atlantica 1702, 22021 (tel: 255-7070; tx: 22248; fax: 255-7070).

Everest Rio, Rua Prudente de Morais 1117, 22420 (tel: 287-8282; tx: 22254).

Excelsior Copacabana, Avenida Atlantica 1800, 22021 (tel: 257-1950; tx: 21076).

Gloria, Rua do Russell 632 (tel: 205-7272; tx: 23623).

Inter-Continental Rio, Avenida Prefeito de Mendes Morais 222, 22600 (tel: 322-2200; tx: 21790; fax: 3222758).

Lancaster, Avenida Atlantica 1470, 22021 (tel: 541-1887; tx: 37858).

Leme Palace, Avenida Atlantica 656, 22010 (tel: 275-8080; tx: 23265).

Luxor Continental, Rua Gustavo Sampaio 320, 22010 (tel: 275-5252; tx: 21469).

Luxor Copacabana, Avenida Atlantica 2554, 22041 (tel: 257-1940; tx: 23971).

Luxor Regente, Avenida Atlantica 3716, 22070 (tel: 287-4212; tx: 23387).

Marina Palace, Rua Delfim Moreira 630, Leblon Beach (tel: 259-5212; tx: 30224).

Meridien Copacabana, Avenida Atlantica 1020 (tel: 275-1122; tx: 23017; fax: 5416447).

Miramar Palace, Avenida Atlantica 3668 (tel: 247-6070; tx: 21508).

Nacional Rio, Avenida Niemeyer 769, 22450 (tel: 322-1000; tx: 23615).

Ouro Verde, Avenida Atlantica 1456, 22021 (tel: 542-1887; tx: 23848).

Praia Ipanema, Avenida Vieira Souto 706 (tel: 239-9932; tx: 31280).

Rio Othon Palace, Avenida Atlantica 3264, 22070 (tel: 255-8812; tx: 22655).

Rio Palace, Avenida Atlantica 4240, 22070 (tel: 521-3232; tx: 21803; fax: 2271454).

Rio Sheraton, Avenida Niemeyer 121, 22450 (tel: 274-1122; tx: 23485).

Sol Ipanema, Avenida Vieira Souto 320 (tel: 227-0060; tx: 21979).

Trocadero, Avenida Atlantica 2064, 22021 (tel: 257-1834; tx: 22655).

Hotels: São Paulo
Augusta Boulevard, Rua Augusta 843 (tel: 257-7844; tx: 21552).

Augusta Palace, Rua Augusta 467 (tel: 266-1277; tx: 37956).

Bourbon, Avenida Vieira de Carvalho 99 (tel: 223-2244; tx: 32781).

Brasilton Sao Paulo, Rua Martins Fontes 330, CEP 01050 (tel: 258-5811; tx: 25558; fax: 2573137).

Bristol, Rua Martins Fontes 277 (tel: 258-0011; tx: 24734).

Caesar Park, Rua Augusta 1508 (tel: 285-6622; tx: 22539; fax: 2886146).

Comodoro, Avenida Duque de Caxias 525 (tel: 220-1211; tx: 36261).

Eldorado Boulevard, Avenida São Luis 234 (tel: 256-8833; tx: 22490; fax: 2889117).

Eldorado Higienopolis, Rua Marques de Itu 836 (tel: 222-3422; tx: 30546).

Excelsior, Avenida Ipiranga 770 (tel: 222-7377; tx: 26187).

Grande Hotel Ca d'Oro, Rua Augusta 129 (tel: 256-8011; tx: 21765; fax: 2310359).

Hilton, Avenida Ipiranga 165, P.O. Box: 30522 (tel: 256-0033; tx: 21981; fax: 2573137).

Holiday Crowne Plaza, Rua Frei Caneca 1360 (tel: 284-1144; tx: 33096; fax: 2513121).

Maksoud Plaza, Alameda Campinas 150 (tel: 251-2233; tx: 30026; fax: 2514202).

Merak, Avenida Lavandisca 262-270 (tel: 542-6866; tx: 32014).

Metropolitan Plaza, Alameda Campinas 474 (tel: 287-4855; tx: 35667).

Mofarrej Sheraton, Alameda Santos 1437 (tel: 284-5544; tx: 34170; fax: 5544176).

Nikkey Palace, Rua Galvão Bueno 425 (tel: 270-8511; tx: 35187; fax: 2706614).

Normandie, Avenida Ipiranga 1187 (tel: 228-5766; tx: 24519).

Novotel São Paulo Morumbi, Rua Ministro Nelson Hungria 450, CEP 05690 (tel: 533-1211; tx: 25662).

Othon Palace, Rua Libero Badaró 190 (tel: 239-3277; tx: 34082).

Planalto, Avenida Caspar Libero 117 (tel: 227-7311; tx: 21607).

San Michel, Largo do Arouche 200 (tel: 223-4433; tx: 22457).

San Raphael, Largo do Arouche 150 (tel: 800-8000, 220-6633; tx: 22457).

Hotels: São José dos Campos, São Paulo
Novotel São José dos Campos, Avenida Dr Nelson d'Avila No 2.200. CEP 12200 (tel: 21-1011; tx: 3383).

Principal newspapers
Brazil Herald, Rua do Rezende 65, 20.231, Rio de Janeiro. Tel: 2212772

Diário Commércio e Indústria, Rua Alvaro De Varvalho 354, 01.050 Sa\08o Paulo, SP. Tel: 256 5011; tx: 21436

Diário Popular, Rua Major Quedinho 28, 1\28-6\28, 01.050 São Paulo, SP. Tel: 258 2133; tx: 21213

Fûlha De São Paulo, Alameda Barão De Limeira 425, Campos Elísios, 01.202 Sa\08o Paulo, SP. Tel: 874 2222; tx: 22930

Gazeta Mercantil, Rua Major Quedinho 90, 8\28, 01.050 São Paulo SP. Tel: 256 3133, tx 23748

Jornal de Brazil, Av. Brazil 500, São Cristovao, 20.940, Rio de Janeiro, RJ. Tel: 585 4422; tx: 23690

Jornal Do Comércio, Rua do Livramento 189, 20.221 Rio de Janeiro, RJ. Tel: 253 6675, tx: 22165

Jornal de Tarde, Av. Eng. Caetano Álvares 55, 02.550 São Paulo, SP. Tel: 856 2122; tx: 23511

O Di\01a, Rua Riachuelo 359, 20.235 Rio de Janeiro, RJ. Tel: 272 2000; tx: 22595.

O Estado de São Paulo, Av Eng Caetano Álvares 55, 02.550 São Paulo. Tel: 856 2122

O Globo, Pob 1090 Rua Irineu Marinho 35, 20.233, Rio de Janeiro, RJ. Tel: 272 2000; tx: 22595

Other useful addresses
Abril Press News Avency, Avenida Otaviano Alvesde, Lima 4400, CP2372, 01.000 São Paulo, SP (tel: 266-011; tx: 22094).

American Express, Avenida Atlântica 2316, Copacabana, Rio de Janeiro.

Asociação Brasileira de Imprensa, Rua Araujo Porto Augre 71, Castelo.

Banco Central do Brasil, Edifício Sede, 70074 Brasília, DF (tel: 224-1453; tx: 1702).

Brazilian Chamber of Commerce, Avenida Rio Branco, 91, 8\28 S/10/12, 20040 Rio de Janeiro (tel: 221-1901).

Cebitur (Tourist Office), Avenida Brigadeiro Faria Lima 1323, São Paulo.

Central Post Office, Praça do Correio, São Paulo.

Central Post Office, Setor Hoteleiro Sul, Brasília.

Central Post Office, Rua Primeiro do Marco 64, Rio de Janeiro.

Confederacâo Nacional do Comercio, Dutra 4\28 Andar, Quadriga SCS Brasília DF.

Confederacâo Nacional da Industria, Avenida Nilo Pecanha 50, 2000 Rio de Janeiro (tel: 231-9260).

Departmento Nacional de Telecomunicacôes (DENTEL), Via N2, Anexo do Ministério das Comunicacôes, Bloco R, 70.066 Brasília DF (tel: 223-3229; tx 2233229).

Embratur (Empresa Brasileira de Turismo), Praça Mava 7, 11th & 12th Floors, Rio de Janeiro.

Fundação IBGE, Rua Franklin Roosevelt 166, Rio de Janeiro.

Industrial Development Council, Praça Mauá 7, 16\28 Andar, Sala 1602 Rio de Janeiro.

Jornal do Barasil, Avenida Brasil 500, CEP 2000, Rio de Janeiro.

Ministry of Finance, Esplanada dos Ministérios, Bloco P, 5\28, 70.048 Brasília, DF (tel: 223-2729; tx: 1142).

Ministry of Industry and Commerce, Esplanada dos Ministérios, Bloco 6, 70.053 Brasília (tel: 223-7784; tx: 1066).

PETROBRAS (Petroleo Brasileiro), Avenida Republica do Chile, CP809, Rio de Janeiro.

Rêde Ferroviária Federal, SA (Federal Railway Corp), Praça Procópio Ferreira 86, 20.224, Rio de Janeiro RJ (tel: 291-2185; tx: 21372).

Servicos Aéreos Cruzeiro do Sul, SA, Avenue Silvio Noronha 365, 20.000 Rio de Janeiro, RJ (tx: 21765)

Tourist Information Centre, Barata Ribeiro 272, Copacabana, Rio de Janeiro.

VARIG SA, Almte Silvio Noronha 365, 20021, Rio de Janeiro, RJ (tel: 297-5141; tx: 22363).

British Virgin Islands

Tony Thorndike

One of the most tranquil of Caribbean territories hosts both elite tourism and a rapidly-growing offshore financial centre. Both ensure enviable living standards and lifestyles. The political hiccup in 1987 is now well in the past and political stability is once again the norm.

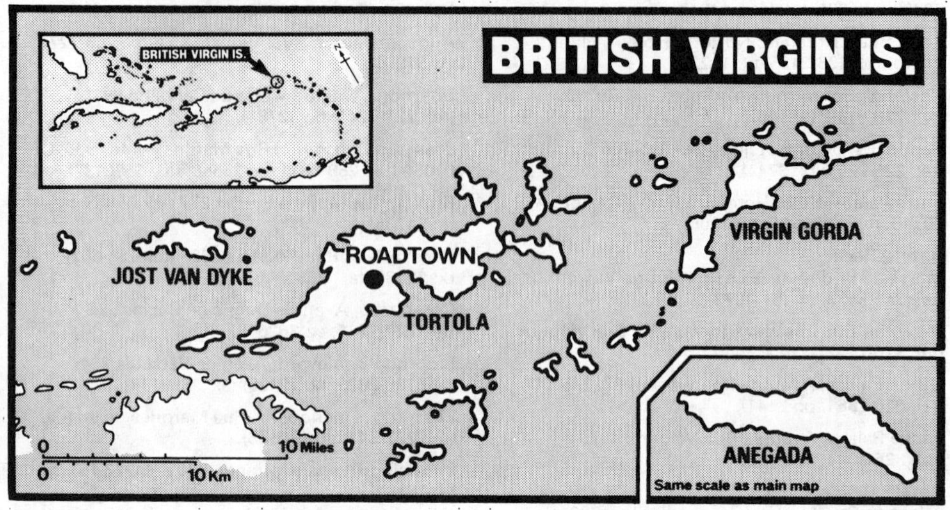

As a tourist poster puts it, 'We are not for everyone'. The 14,500 population is scattered over 16 out of some 36 islands tucked away to the north-east of the bustling US Virgin Islands, and the contrast between the two groups of Virgins (they are each geographical extensions of the other) is considerable. The British Virgin Islands, or BVI for short, has worked hard to protect an image conducive to relaxed lifestyles and international business largely untainted with money laundering and narcotics smuggling, although a number of smugglers have been apprehended. Politically, there is no talk of independence from Britain.

The Caribbean's new financial centre

The growth of the offshore financial services sector has been both rapid and outstanding. Its success stems from stability, the use of the US dollar as the official currency and, crucially, one piece of legislation making offshore company registration and operation flexible and easy, the International Business Companies Ordinance of 1984. 7,135 companies were registered under its auspices by December 1988, twice the figure of a year earlier; a month later the figure exceeded 12,300. The BVI attracted a disproportionate amount of business fleeing Panama which it is not likely to lose. Admittedly the attrition rate is high, but nonetheless, 18,000 registrations are estimated for 1990. But as the bonanza continues, official concern is growing about over-reliance on a single market. An expansion in banking activity is therefore likely and already by end-1988 offshore deposits exceeded $250m. However, the territory is moving with great caution in that area.

Offshore company legislation followed in the wake of the cancellation of double taxation treaties by Britain and the United States. The key is that International Business Companies (IBCs) cannot conduct business in the BVI or own land there, but can engage in any business except insurance. There are no residency qualifications for principals, and lawyers and accountants equipped to act as 'sleeping partners' are available, mostly in the sleepy capital of Roadtown on the most populous island, Tortola. An IBC need only have one shareholder whose identity (or that of any director) need not be disclosed. No audited accounts are necessary, and reporting requirements are skeletal. There is no liability for income tax and fees are remarkably low, as little as $350, although there is a 1 per cent profits tax. Many IBCs are private trusts established by wealthy individuals; others are connected to multinational companies aimed at reducing tax liabilities, often arising out of transfer pricing between subsidiaries. Several IBCs are US mutual funds (unit trusts) for private placement with non-US investors, avoiding the need to register with the US Securities and Exchange Commission and protecting clients against changes in United States tax laws.

But rapid growth has its problems. A major weakness is the weak regulatory environment, and concern is growing about the true nature of some IBCs. Accordingly a joint government-private sector committee was established in 1988 (British government financial advisers were also involved) to make recommendations both to curb abuse and to maximise income. It was also charged to consider areas of expansion. Chief Minister Lavity Stoutt led a delegation to Britain in July 1988 to discuss the sector's future as well as British development aid, and a visit was also made to the Channel Islands to learn from their offshore experience. Offshore insurance legislation is a top priority. Attracted by the BVI's image, insurance business is growing but these companies are registered under the BVI's existing Companies Act. Ship registration is also a possibility. But shortages of trained personnel and limited infrastructure are factors limiting expansion; office space in Roadtown is at a premium despite minimal client contact. There is also the growing problem of immigration. In the past that was largely restricted to manual labour, but the influx of expatriate financial professionals has made clear the need for trained BVI islanders, an issue that the government has not really tackled.

Up-market tourism

Yacht-based tourism is the hallmark of the BVI, although extensive hotel construction, which resulted in 993 rooms being ready for the 1988-89 winter season, threatens the balance. Unfortunately the 1988-89 yachting season saw a decline and a number of chartered yachts left for elsewhere. Hotel under-occupancy resulted in a sharp slowdown in construction. But the more exclusive hotels, such as those which occupy their own islands, like Peter Island, saw little reduction in business. Virgin Gorda benefited from greater hotel development, centred on the highly picturesque 'Baths', a series of unusually-shaped rocks. 174,001 stopover visitors landed in 1988, 8 per cent over 1987, nearly 100,000 of whom were yacht-based. In addition there were some 29,487 excursionists, mainly from nearby St. Thomas, USVI, and another 36,843 were from cruise ships. However, the growth in excursionist traffic – a 28 per cent increase over 1987 – was not entirely welcome as it created congestion and relatively little income.

Gross tourist receipts in 1988 were over $150m, accounting for half of GNP and a third of employment. Although there is growing environmental concern, especially over reef damage, the industry contrasts with that of St. Thomas with its excessive hotel and condominium development.

There is little other economic activity, virtually all commodities and materials having to be imported. One resource, fish, is victim to much poaching by foreign trawlers. Infrastructural improvements, financed largely by assistance from Britain, the European Community and the Caribbean Development Bank is proceeding, with plans for new port facilities in Tortola, better airport facilities and improved water supply and health services. Work on a much delayed $17.8m electricity generating project finally began in 1989. The European Investment Bank withdrew funding when a British tender was accepted in preference to a lower one submitted by a West German contractor; the BVI Social Security Board loaned $6m to fill the gap.

KEY FACTS

Official title: British Virgin Islands

Head of state: Queen Elizabeth II

Head of government: Chief Minister H. Lavity Stoutt

Ruling party: Virgin Islands Party

Capital: Road Town

Official Languages: English

Currency: US Dollar ($) = 100 cents

Area: 153 sq km

Population: 14,500 (1989)

GDP per capita: $9,491 (1987)

GDP real growth: 15.1% (1987)

Inflation: 5.2% (Jan–Aug 1988)

Trade balance: $35.2m (1987)

Foreign debt: $14.2m (1988)

Aid flows: $4m (1987)

Visitor numbers: 203,488 (1988)

The territory's budget has shown improvement since 1986 as a result of buoyant economic conditions. Despite supplementary spending of $4m, mainly to cover a 25 per cent public sector pay increase, and a record expenditure of $30m in 1988, a current account surplus was recorded. The 1989 budget set recurrent expenditure at $27.3m and capital spending at $2.7m. Revenue was forecast at $30.5m and because of the expanding financial sector and better tax collection, this should be exceeded.

Political stability

The tranquillity normally associated with BVI politics was upset in 1987 when the then Chief Minister George Romney had to resign because of an association with one Shaun Murphy, whose Financial Management and Trust company was allegedly used for laundering illegal funds, although Romney himself was cleared of any wrongdoing. The episode showed how easily naive leaders of small islands can be duped. The ensuing election in September 1987 led to the return of veteran politician Stoutt. His Virgin Islands Party won five seats to the United Party's two. Romney, an Independent, was re-elected, no doubt a wiser man.

But the administration was then threatened when Stoutt's deputy, Omar Hodge, was forced to resign amid allegations of bribery and conflict of interest. Despite his insistence that the deal, a $5,000 transaction with Peter Island, was in the course of normal business, a court of inquiry found against him. His case was not bettered when he attempted to prevent the report's publication. He subsequently, and unsuccessfully, sued for wrongful dismissal and then helped form a new political party, the Independent People's Movement. Hodge's replacement was the opposition leader Ralph O'Neal, who was tempted across the floor with a ministerial appointment. O'Neal's replacement was Conrad Maduro, who was subsequently named Chairman of the Public Accounts Committee. The Hodge affair generally illustrates the importance of personalities and the weakness of party politics in the BVI and the broadly partisan style of doing business. It is a political style which so far has worked well.

COUNTRY PROFILE

Historical profile: Discovered by Columbus 1493; Sir Francis Drake visited channel (which now bears his name) 1595; settled by Dutch 1648; English settled 1666; Tortola taken over by English 1672.

Political system: British Caribbean dependency with an appointed governor who is responsible for defence, security and external affairs. The executive council (the governor, attorney-general, chief minister and three other ministers) has responsibility for finance. The Legislative Council is composed of nine elected members plus the attorney-general, the speaker and one member nominated by the governor and is elected for a four-year term.

Main cities/towns: Road Town (population 12,500, on Tortola island).

Media
Press: Newspapers include *The Island Sun* (bi-weekly) and *The Beacon* (weekly).
Broadcasting: Radio ZBVI operates a daily commercial station from Tortola. Cable television.

External trade: Large trade deficit offset by tourist spending, capital inflows and by workers' remittances from overseas.
Exports: Handicrafts, fruit, vegetables, fish, rum and gravel. Main destination: US Virgin Islands.
Imports: Machinery and equipment (35-37% of total value), foodstuffs (18-20%), fuel (13-15%). Main sources: USA, US Virgin Islands, Puerto Rico, UK.

Agriculture: Sector contributes under 10 per cent to GDP. Production centred on livestock farming, fishing, food crops (mainly fruit and vegetables) and sugar cane for rum production. Main areas of activity are Tortola, Virgin Gorda and Jost Van Dyke. Expansion of tourist industry has increased dependence on imported foodstuffs, mainly from US.

Industry: Tourism is the mainstay of the economy, with gross tourist earnings amounting to around US$150m per year. Most important activities are yachting and boat chartering, but new hotels are attracting more stopover arrivals. Other industries include rum-distilling and construction materials.

Banking: Significant capital inflows due to islands' liberal tax and exchange laws and close proximity to the USA, the US Virgin Islands and Puerto Rico.

Time: GMT -4.

Climate: Sub-tropical, with no marked seasonal variation in temperature (generally 24°C to 30°C during the day, 10°C cooler at night). Rainfall generally low although tropical storms may occur from Aug to Nov.

Entry requirements
Passport: Required by all, though birth certificate is sufficient for US and Canadian citizens.
Visa: Not required by nationals of EC and Commonwealth countries, USA, Japan, Scandinavian countries and certain others. A return or onward ticket is required.
Currency: No restriction on import of foreign currency but amounts should be declared. Exports limited to the amounts declared on arrival.

Health precautions
Mandatory: None.
Advisable: Typhoid and polio vaccinations. Water precautions.

Air access
National airline: Air BVI.
Other airlines: American, Eastern, Liat.
International airport: Beef Island (EIS), 15 km from Road Town on Tortola, bar, restaurant, buffet, shops; Virgin Gorda (VIJ) 5 km from Spanish Town on Virgin Gorda.

Hotels: Expensive but wide range available. Bookings should be made well in advance. Seven per cent hotel tax and 10 per cent service charge usually added to bill.

Car hire: Can be hired on Tortola and Virgin Gorda. Temporary licences can easily be obtained from rental agency on production of national driving licence. Traffic drives on the left.

City transport
Taxis: Widely available. Taxi rank in Road Town is opposite central post office and Taxi Assocaiation on Wickhams Cay. Tipping optional. Can be hired on a time basis.

National transport
Air: Regular flights to Puerto Rico, US Virgin Islands and Antigua. Air BVI operates domestic flights linking Tortola, Virgin Gorda and Anegada. Charter services are also available.
Road: The main highway from Beef Island through Road Town to West End is surfaced. There is a bridge connecting Beef Island with Tortola. There is also a surfaced road on the northern ridge from east to west. Roads on Virgin Gorda are in variable conditions.
Water: Various types of boats ply between islands. Regular ferry service between Road Town or West End (Tortola) and Charlotte Amalie (St Thomas, US Virgin Islands).

Public holidays: 1 Jan, Commonwealth Day (Mar), Good Friday, Easter Monday, Whit Monday, Queen's Birthday (Jun), 1 Jul (Territory Day), 6-7-8 Aug (August Festival) 21 Oct (St Ursula's Day), 14 Nov (Prince Charles' Birthday), 25 Dec, 26 Dec.

Working hours
Government and business: (Mon-Fri) 0830/0900 – 1630/1700.
Banking: (Mon-Fri) 0900 – 1400; also (Fri) 1600 – 1730.

Telecommunications
Telephone: Automatic telephones system links all major islands in the group. System is currently being extended to all islands. IDD from the islands has also been introduced. Dialling code for BVI: IDD access code + 1-809-49 followed by subscriber's (five-digit) number. There are no area codes.
Telex: Public facilities available at the Central Telegraph Office, Road Town (on Tortola).

Banking
Banking centre: Tortola.
Commercial banks (head offices): Bank of Nova Scotia, Barclays Bank International, Chase Manhattan Bank, Development Bank of the British Virgin Islands, First Pennsylvania Bank.

Electricity supply: 120/208 V AC, 60 cycles.

Hotels: Tortola
BVI Aquatic, West End (tel: 495-4541).

Long Bay, Road Town, P.O. Box: 433 (tel: 495-4252).

Moorings/Mariner Inn, Road Town, P.O. Box: 139 (tel: 494-2332; tx: 9716).

Prospect Reef Resort, Road Town, P.O. Box: 104 (tel: 494-3311; tx: 7931).

The Castle, Road Town, P.O. Box: 206 (tel: 494-2515).

Car hire: Tortola
Anytime Car Rentals: PO Box 258, Road Town (tel: 42875).

Avis (Nibbs Car Rental): PO Box 3447, Road Town (tel: 43322). tx: 7914).

Budget: Wickhams Cay, PO Box 311, Road Town (tel: 42639; tx: 932).

International Car Rentals: PO Box 117, Road Town (tel: 42516/7)

National Car Rental: PO Box 121, Road Town (tel: 43197; tx: 7967).

Business advice and information
Bank of Nova Scotia, PO Box 434, Road Town, Tortola (tel: 42526; tx: 7951).

Barclays Bank International, PO Box 70, Road Town, Tortola (tel: 42171; tx: 7982).

British Virgin Islands Tourist Board, Joshua Smith Building, PO Box 134, Road Town, Tortola (tel: 43134).

BVI Hotel and Commerce Association, PO Box 376, Wickhams Cay, Road Town, Tortola (tel: 43514/2947; fax: 42226 c/o The Moorings, Tortola).

Cable and Wireless (West Indies), PO Box 440, Road Town, Tortola (tel: 44444).

Chase Manhattan Bank, PO Box 435, Road Town, Tortola (tel: 42662; tx 7935).

Development Bank of the BVI, PO Box 275, Road Town, Tortola (tel: 43737).

First Pennsylvania Bank, PO Box 67, Road Town, Tortola (tel: 42117; tx: 7970).

Virgin Islands Broadcasting Ltd, PO Box 78, Road Town, Tortola (tel: 42250).

ZBVI Radio, PO Box 78, Road Town, Tortola (tel: 42250).

Brunei

Michelle Misquitta

At first sight there is little special about Brunei. It is a micro-state of mostly jungle washed by the South China Sea, which might be anywhere in tropical south-east Asia. But two things give it away and mark this tiny 5,700 sq km strip of territory – actually two tongues of land on Borneo island separated by the Temburong river – as unique. One is flares burning off surplus oil and gas just off the coast; the other is the magnificent 1,788-room palace of the ruler Sultan Sir Hassanal Bolkiah (to give him his short title).

Oil and gas wealth

For the rest, Brunei seems to have gone out of its way to be unostentatious. Jungle is jungle is jungle, and Brunei has its inhabitants who still live in traditional long houses where they reside in extended families or small villages of 60 to 80 people each. In the capital Bandar Seri Begawan there are rows upon rows of rickety wooden stilt houses jutting into the bay as they have done for decades. True enough, the television aerials and electricity wires going into the traditional houses immediately indicate that they are not poor; but there is little in the homes of ordinary Brunei people to suggest that this is one of the richest countries in the world in per capita terms and challenges Japan for the title of richest in Asia. According to the World Bank, Brunei's per capita income in 1987 was $15,390, probably an underestimate.

Its ruler the sultan is a different matter. His fortune and claim to fame as 'the richest man in the world' according to the *Guinness Book of Records*, with assets of the order of $25bn, has gone before him. He does not like the title, claiming that the recorders have appropriated Brunei's oil and gas wealth and attributed it to him. But even the sultan's friends have difficulty drawing the line between the sultan's assets and those of the country. He is, after all, virtually an absolute ruler, and is not really tainted by democracy even though the proclamation of independence of 1984 describes Brunei as a 'democratic and independent Malay Muslim monarchy.' The constitution has been in abeyance for some years. His sparkling palace and his much-heralded fleet of aircraft, fast cars and the motorcycle outriders who attend his journeys speak of an ordinary man who likes to flaunt his wealth.

That wealth is such that according to one of his biographers, 'if the sultan's wealth were used to buy Moët & Chandon champagne, then the bottles, placed end to end, would reach the moon and back with enough left over to circle the earth three times ....'

That wealth, whether the sultan's or the country's, comes almost exclusively from the country's substantial earnings from its oil and gas reserves. Precisely how much it adds up to is difficult to know. Oil output peaked at 261,000 barrels a day (b/d) in 1979 but is being held at 150,000 b/d to help conserve reserves. In addition Brunei sells 5m tonnes of liquefied natural gas each year to Japan. These resources are many times more than enough for Brunei's own use – since the country uses only about 1 per cent of its oil production for domestic energy. After all, Brunei's total population of 240,000 is hardly a medium-size town in US or European terms.

But the gas reserves are estimated to run out in just over 20 years at current extraction rates, and the oil will run out soon after that. Apart from that Brunei's exports consist of some timber, pepper and rubber, none of which makes much of an impression on the trade statistics dominated 96 per cent by oil and gas.

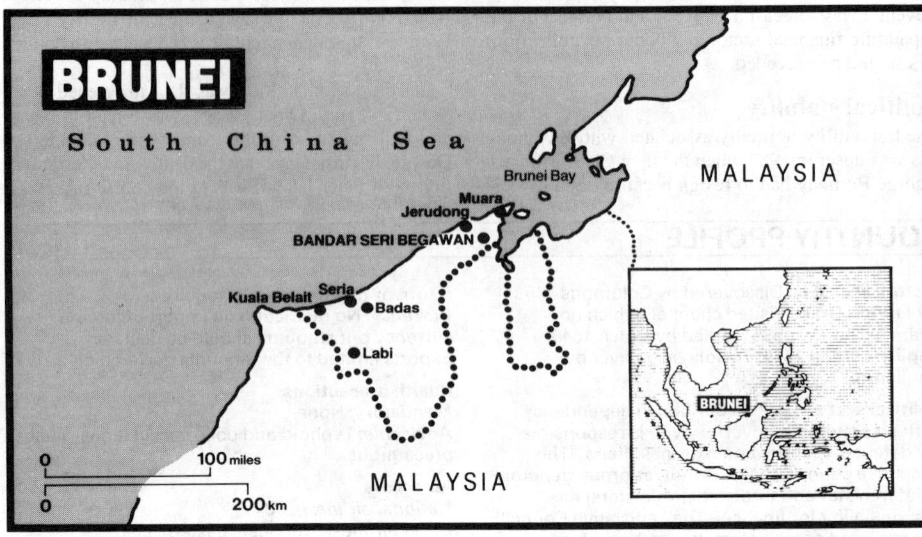

Gradually Brunei has increased its share in the ownership of its oil and gas resources. The oil is produced in co-operation with Shell, hence the expression 'Shellfare state' to describe Brunei. The gas is a three-way partnership between Brunei, Shell and Mitsubishi of Japan. The country has the world's largest gas liquefaction plant, at Lumut. In addition, Brunei is going downstream in the oil industry and thanks to construction of a 10,000 b/d refinery in Seria will be self-sufficient in refined oil products.

Diversification plans

There is a lively argument whether Brunei should worry or not about its heavy dependence on oil and gas. Officially, the country is dedicated according to the 1986-90 development plan to diversify. Civil servants talk of food processing, ceramics factories, cement plant, glass- and optics-making based on the fine silica sand to be found in the country. Some officials will wax eloquent about how the huge reserves give Brunei valuable time to develop its entrepreneurial skills in a vast range of ways.

Hard reality is that it is taking a long time to see the emergence of entrepreneurs. The largest part of the 85,000 strong workforce – a huge 46 per cent – works for the government. Ordinary Brunei citizens have a generous welfare system, cheap food, no income tax. In addition civil servants get a free pension scheme and soft loans to buy cars and houses. So why work too hard? One local observer says 'Brunei has a word for *mañana* (tomorrow) but it doesn't have the same sense of urgency.'

The sultan himself has hardly set a good example in his well-publicised business partnerships. Whatever the truth about them, the best that can be said is that he has been unfortunate in choosing his friends. By 1989 it also became apparent that Qaf, a company billed a few years ago as launching an ambitious scheme for Brunei citizens to share in ownership of local companies, was facing hard times. Qaf (which is an Islamic name for a mythical mountain) was listed on the Singapore stock exchange in 1984 through the reverse takeover of a local group Ben and Company. The prospectus promised 'as part of efforts to promote participation by nationals of Brunei in the equity of companies operating in Brunei,' to offer 14 per cent of the group to 'selected Brunei institutions'. The offer was not made and the *Financial Times* wrote that squabbles within the royal family killed the plan before it could get off the ground. The paper claims that Qaf had fallen out with its royal backers. It may have been Brunei's first corporate baby, but it was orphaned shortly after its birth.

Pessimists ask how long Brunei can go on enjoying all the privileges when the oil and gas run out. There are some entrepreneurs in the country, but they are mostly Chinese who number about 40,000, though not all have full citizenship. The sultan's declaration in 1984 that his country would forever be 'a sovereign democratic and independent Malay Muslim mon-

KEY FACTS

Official title: Sultanate of Brunei (Negara Brunei Darussalam)

Head of state: Sultan and Yang Dipertuan, Sir Muda Hassanal Bolkiah Mu'izzaddin Waddaulah

Capital: Bandar Seri Begawan

Official Languages: Malay (Chinese, English also spoken)

Currency: Brunei Dollar (B$ = 100 cents)

Exchange rate: B$1.96 per $ (Nov 1989)

Area: 5,765 sq km

Population: 300,000* (1988)

GDP per capita: $15,390 (1987)

GDP real growth: 2.2% (1988)

Inflation: 2.0% (1988)

Trade balance: $0.8bn (1988)

Foreign debt: $147m (1987)

* estimated figure

archy' did not exactly encourage the Chinese, many of whom had begun to emigrate. Brunei has to import more than 80 per cent of its food requirements. Agriculture employs only a small portion of the workforce. The local fishing industry supplies only 65 per cent of domestic demand.

On the other hand, if properly invested, the immense reserves of the country – $20-25bn at least – should provide a suitable cushion for Brunei to enjoy the good life for some time into the foreseeable future. In 1989 the head of the Brunei Investment Agency claimed that income from the investments is already yielding more money than the oil. If the agency chooses its friends carefully it should be able to keep its citizens happy.

Pressures for change?

Whether there will be political pressures for change is more difficult to judge. Many Chinese living in Brunei are unhappy, feeling they are disadvantaged but the Chinese are not really players in the game. The Malay tradition has been to accept the prerogatives of royalty, though the government may invite trouble by its heavy-handed way of dealing with opposition. In 1988 the government dissolved the country's main political party, the Brunei National Democratic Party, and arrested and detained its president and secretary-general. The party had been set up with the sultan's approval, but had apparently angered him by asking for an end to the country's 26- year-old emergency and requesting the sultan to step down as prime minister and call elections.

In 1988 the sultan publicly said he saw no reason to move towards British style constitutional monarchy. He claimed that the Brunei people 'showed no signs of wanting to change a system that had worked well since time immemorial'. The ruler said he was in touch with his people through constant travel. The Brunei government is dominated by the royal family since the sultan is also prime minister and minister of defence, Prince Jefri is finance minister and Prince Mohamed is foreign affairs minister.

In 1989 at the Commonwealth summit in Kuala Lumpur the sultan agreed to buy some expensive military toys from Britain. He placed a £250m order for 16 Hawk-100 two-seat single engine fighter-bombers and for three Exocet armed corvettes. Until now the Brunei armed forces have operated only helicopters and its navy is basically a coastal force. But there is no obvious threat to Brunei from abroad. It is a member of the Association of South East Asian Nations. In addition the oil fields which the new weapons are obviously intended to protect, enjoy the help of a batallion of Gurkha infantry.

COUNTRY PROFILE

Political system: Brunei became independent in 1984 with its Sultan as ruler. Previously it had been a protectorate of Britain. The proclamation of independence describes Brunei as a 'democratic and independent Malay Muslim monarchy'. The constitution, which has been partly in abeyance since the early 60s, provides for various appointed councils (religious, legislative etc.) formally to advise the Sultan, who is head of State.

Political parties: The Brunei National Democratic Party was dissolved in January 1988 just three years after it was formed in May 1985. The other major party was the National United Party formed in 1986.

Population: Over half of the 241,400 (mid-1988 estimate) inhabitants live in the immediate district of Brunei/Muara. Most are Malays, with about 18 per cent Chinese. Population growth is fairly high at 3.7 per cent per annum and about 46.1 per cent are aged 20 or under.

Main cities/towns: Bandar Seri Begawan (population 147,300, including suburbs), Kuala Belait, Seria, Tutong, Bangar.

Language: Bahasa Melayu is the official language; English is common in business circles.

Media
Press: An English-language weekly, *Borneo Bulletin*, is published in Kuala Belait. *Pelita*, a Malay-language paper and *Brunei Darussalam* (English Language) are published by the government. Brunei Shell Petroleum publishes *Salam* in English and Malay. Various foreign newspapers circulate, including the *Singapore Straits Times* and several Chinese newspapers published in Sarawak and Singapore.
Radio: The government-operated radio station transmits two services with programming in Malay, Chinese and English. There were 87,000 receivers in 1988.
Television: The government-operated TV station transmits two services with programming in Malay and English in colour. Malaysian Channels RTM 1 and 2 are received. TV runs on the PAL system. There were 58,500 sets in 1988.

Domestic economy: Brunei's economy is based on the considerable reserves of natural gas and oil and economic development has concentrated on developing infrastructure, including redeveloping the international airport (completed) and constructing a deep-water port, a sports stadium, roads, hospitals, schools and numerous government buildings. The National Development Plan for 1986-90 shifts emphasis to agriculture, fishing, timber and financial services. Company taxation is the government's chief source of revenue, and surpluses in recent years have been invested abroad, yielding substantial investment income. The Brunei Investment Agency is responsible for holding and managing the government's reserve fund and external assets. There is a high standard of living, with medical services, education and pensions provided free and subsidies for food and housing. The Brunei Darussalam University was established in 1985.

Employment: According to the 1986 population survey, total working population is about 81,000, of which the government sector is the largest employer of 41,000, followed by Brunei Shell Petroleum Sdn Bhd and Royal Brunei Airlines. About 39,000 are working in community services, followed by 9,000 in construction and 6,000 in mining, quarrying and manufacturing. Unemployment rate had increased from 3.6 per cent in 1981 population census to 6.1 per cent in 1986 population survey.

External trade
Exports: Main exports are crude petroleum, natural gas, and petroleum products. Main destinations are Japan, Singapore, Thailand, USA, Republic of Korea, Philippines and Malaysia (Sarawak).
Imports: Main imports are machinery and transport equipment, manufactured items, food and live animals, other manufactures and chemicals. Main sources are Singapore, Japan, USA, UK, Peninsular Malaysia, Australia, Taiwan, Netherlands and China PR.

Agriculture: Agriculture employs a small proportion of the workforce. Approximately 15 per cent of the total land area of 567,758 hectares is cultivated or under grazing. Agriculture accounts for some one per cent of GDP and 80 per cent of food requirements is imported. Tropical fruit production is entirely a smallholder's activity producing a wide range of varieties. Vegetable production is intensive, concentrated on the fertile alluvial plain close to the urban centres and smallholder rice farming is on the decline. The government has initiated a pilot large-scale rice mechanisation project aimed at increasing rice production and the evaluation is still in progress. Poultry meat production is increasing and Brunei Darussalam is approaching self-sufficiency in eggs. The government has invested in cattle and buffalo production schemes designed to increase domestic supply of beef. Brunei Darussalam is embarking on development programmes to increase agricultural productivity as a first step towards food security in the country. Fish is a principal source of protein, but local fishing industry supplies only 37 per cent (1988 figure) of domestic demand. The Fisheries Department is investing in a trawling industry, with plans for a fishing port with ice-making capacity (a fish landing complex has been completed in Muara) and marketing co-operatives. 70 per cent of the land area is primary and secondary tropical rain forest, and 37 per cent of the land area is Forest Reserve. Logs and sawn timber and timber production are for domestic consumption only.

Industry: The present industrial structure of Brunei Darussalam, long dependent on export of oil and gas, consists mainly of small-scale enterprises. Some examples of these are the saw mills, brick factories, tile factory, coffee production and fibreglass products. Two new additions are an ice-cream and aerated drinks factory and a textile factory producing ready-made garments. Brunei Darussalam is now working towards a diversified economy that would be self-sustaining with a meaningful degree of self-reliance. Its investment policy is open and flexible and it welcomes investors, both local and foreign, in any productive industrial activity. To date, 23 sites in all the four districts have been identified for industrial activities and would be offered to investors at attractive terms.

Hydrocarbons: The mining sector accounts for about 62.5 per cent of GDP in 1987. Oil and natural gas are the main products. Brunei Shell Petroleum Co Sdn Bhd is the sole oil producer, operating five offshore oilfields and two onshore oilfields. Oil production reached a peak at 261,000 b/d in 1979 and was gradually reduced to 150,000 b/d in 1988 in line with the government's conservation policy introduced in 1980. Production will continue at this level into the next century. Brunei's total gas production continues to be well above 800m scf/d, of which some 745-750m scf/d is for export to Japan with whom it has a long-term contract commitment. This constitutes some 155 cargoes a year.

Energy: Three per cent of oil production is used for domestic energy requirements.

Main companies: Brunei Shell Petroleum Co Sdn Bhd, Brunei LNG Sdn Bhd, Brunei Coldgas Sdn Bhd, Brunei Shell Tankers Sdn Bhd, Brunei Shell Marketing Sdn Bhd.

Membership of international organisations: ASEAN, Commonwealth, country in association with the EC, ESCAP, IMO, Interpol, Organisation of the Islamic Conference, UN.

BUSINESS GUIDE

Time: GMT + 8 hrs.

Climate: Tropical, with a high average humidity of 82 per cent and temperatures ranging 23-35°C. Rainfall is variable, from 2,500mm/year on the coast to 7,500mm/year in the interior, with Oct-Mar the wettest months and monsoon rains in Nov-Dec.

Entry requirements
Passport: Required by all.
Visa: UK nationals are exempted from visa requirements for stays of up to 30 days; Thais, Indonesians, Malaysians, Singaporeans, French, Federal Republic Germans, Japanese, Benelux countries, Swiss, Canadian, South Korean and Philippine visitors for holidays of up to 14 days; Malaysians and Singapore citizens require sponsors if staying over 14 days. However, they need visas if they intend to stay longer or take up employment. All others require visas.
Currency: Unlimited import of both foreign and local currency. Export of local currency is restricted to B$1,000, and export of foreign currency is limited to amount imported.
Customs: Personal effects are allowed duty-free.
Income tax: There is no income tax.

Health precautions
Mandatory: Vaccination certificates for yellow fever are required for travellers over one year of age travelling from an infected area.
Advisable: Chest X-ray and blood film examination for malaria are required for the issue and renewal of labour permits as Brunei Darussalam is malaria-free.

Air access
National airline: Royal Brunei.
Other airlines: Cathay Pacific, MAS, Merpati, Philippine, Singapore, Thai Airways International.
International airport: Bandar Seri Begawan International (BWN), 10 km from city, with car hire and taxi service.

Surface access: Road connections betwen Brunei and Sarawak are good. The bitumen road between Miri and Kuala Belait has now been completed. Resurfacing of many roads is in progress.
Main ports: Muara, Bandar Seri Begawan, Kuala Belait, Lumut. Facilities are being expanded in 1986-90.

Hotels: Rooms in major hotels have air conditioning, telephones, TV, bathrooms and showers. A 10 per cent service charge is usual.

Credit cards: American Express, Diners Club, Mastercharge, Visa and other major credit cards are accepted at some hotels.

Car hire: Self-drive and chauffeur-driven car hire is available. An international driving licence is required.

City transport
Taxis: Taxis are available from the international airport to all parts of Brunei, but are otherwise scarce. Fares are by negotiation. Tipping is not usual.

National transport
Road: The 1,448 km road network serves the towns, and is being expanded and modernised. A main highway links Bandar Seri Begawan with Kuala Belait and Seria, with a new road linking Muara and Tutong providing access to western districts. There is no road connecting the Tembutong district but water taxi service is available. There are 101,009 vehicles in Brunei (1986).
Buses: Services operate betwen Bangar Seri Begawan and Kuala Belait and Seria, and also serve rural areas.
Water: River taxi service is available.

Public holidays
Fixed dates: 1 Jan, 23 Feb (National Day), 31 May (Anniversary of the Royal Brunei Armed Forces), 15 Jul (Sultan's Birthday), 25 Dec.
Variable dates: Chinese New Year, Me'raj, First day of Ramadan, Anniversary of the revelation of the Koran, Hari Raya Puasa, Hari Raya Haji, First day of Hajrah, Maulud.

Working hours
Business: (Mon-Fri) 0800 – 1200, 1315 – 1700; (Sat) 0800 – 1200.
Government: (Mon-Thu, Sat) 0745 – 1215, 1330 – 1630.
Banking: (Mon-Fri) 0900 – 1200, 1400 – 1500; (Sat) 0900 – 1100. Some banks open during lunch-hour.
Shops: (Mon-Sat) 0730 – 1900/2200. Post offices: (Mon-Thu, Sat) 0730 – 1600; (Fri) 0830 – 0930.

Social customs: The right hand should be used for offering or receiving anything, from food to money. Muslims do not eat pork or drink alcohol. Refusal of offered refreshment is discourteous.

Telecommunications: The system is one of the most modern among developing countries.
Telephone: Dialling code for Brunei, IDD access code + 673 followed by area code (2 for Bandar Seri Begawan, 3 for Seria and Kuala Belait, 5 for Temburong, 4 for Tutong) followed by subscriber's number. There were approximately 38,000 telephones in 1987 with 100 per cent automatic service. 24-hr overseas telegram and telephone service available at Telecommunications Counter, General Post Office, Bandar Seri Begawan.
Telex: No public system, but facilities are found in major hotels.

Banking

Main centre: Bandar Seri Begawan.

Central bank: Main duties carried out by the Treasury Department.

Head office: International Bank of Brunei Berhad.

Brunei branches: Citibank, Hongkong & Shanghai Banking Corp, Malayan Banking Berhad, Overseas Union Bank, Standard Chartered Bank, United Malayan Banking Corp.

Advertising: Press advertising is only accepted by *Borneo Bulletin*. The TV station accepts advertising in Malay and English.

Electricity supply: 230 V AC, with 3-pin round or 3-pin square plug fittings.

Weights and measures: The metric system is now in full use.

Representation in capital: Australia, France, German FR, Indonesia, Japan, Republic of Korea, Malaysia, Netherlands, Oman, Pakistan, Philippines, Singapore, Thailand, UK, USA.

Useful telephone numbers: Bandar Seri Begawan
Police: 22333

Fire: 22555

Ambulance: 22366

Directory enquiries: 0213

International calls: 000

Hotels: Bandar Seri Begawan
Ang's Jalan Tasek Lama, P.O. Box: 49 (tel: 43553; tx: 2280).

Brunei, 95 Jalan Pemancha, P.O. Box: 50 (tel: 42372-9; tx: 2287).

National Inn, Jalan Tutong, P.O. Box: 1090 (tel: 21128; tx: 2345).

Sheraton-Utama, Jalan Bendahara 2087, P.O. Box: 2203 (tel: 44272; tx: 2306; fax: 21579).

Hotels: Kuala Belait
Sea-View, Lot 3678 Jalan Seria, P.O. Box: 127 (tel: 32651; tx: 3301).

Sentosa, 93 Jalan McKerron, P.O. Box: 252 (tel: 34341).

Car hire
Avis: Unit 2, Block 4, Bangunan Hasbullah Jalan Gadong, Bandar Seri Begawan (tel: 24921, 21421; tx: 2300 BEYNEL); International Airport (tel: 42284).

Haji Morni Ag Ahmad: 2557 Jalan Seria, Kuala Belait (tel: 32941; tx: 3310).

Syarikat Yuran Rent-A-Car: 144A Jalan Pemancha, PO Box 1603, Bandar Seri Begawan (tel: 24054, 29355; tx: 2447; cable: YURAN).

Academic/research institutes
University of Brunei Darussalam, Gadong, Bandar Seri Begawan, Brunei Darussalam (tel: 27001 (PABX0; tx: BU 2729 UNIBRUN).

Business advice and information
Brunei State Chamber of Commerce, PO Box 2246, Bandar Seri Begawan (tel: 28533; tx: 2309; cable: CITYBK).

Chinese Chamber of Commerce, 9 Jalan Pretty, Kuala Belait (tel: 34374).

Economic Development Board, Ministry of Finance Building, Bandar Seri Begawan 2011 (tel: 40243).

Indian Chamber of Commerce, Suite 12,2F Jalan Gadong, Bandar Seri Begawan (tel: 23886; cable: BEEJAY).

Malay Chamber of Commerce, Jalan McArthur, Bandar Seri Begawan (tel: 24766).

Banking
The Hongkong & Shanghai Banking Corporation, cnr Jalan Sultan and Jalan Pemancho, PO Box 59, Bandar Seri Begawan (tel: 42305; tx: 2273).

International Bank of Brunei, Bangunon 1BB, Lot 199, Jalan Roberts, PO Box 2729, Bandar Seri Begawan (tel: 20686; tx; 2320).

Overseas Union Bank Ltd (Singapore), Unit G5, RBA Plaza, Jalan Sultan, Bandar Seri Begawan 2089 (tel: 25477; tx: 2256).

Travel information
Information Bureau, Information Section, Broadcasting and Information Dept, Prime Minister's Office, Bandar Seri Begawan 2041 (tel: 40400/1, 21872/3, 40527; tx: 2614).

Royal Brunei Airlines, PO Box 737, Bandar Seri Begawan (tel: 25931).

Bulgaria

Judy Dempsey

In late 1989 Bulgaria found itself belatedly swept up in the process of democratisation, albeit in a less dramatic way than neighbouring Romania, and less comprehensively than Czechoslovakia or Hungary. On 10 November 1989 the ageing Todor Zhivkov was replaced by Petar Mladenov as Chairman of the Communist Party. By 11 December, Mladenov was calling for constitutional change to abolish the Communist Party's monopoly on power, to strengthen the role of parliament and separate the government and the Communist Party. Elections were planned for the spring of 1990 and a special party congress due to be held on 26 March to enact election laws enabling the creation and legalisation of opposition prties.

As Hungary and Poland had forged ahead with political and economic reforms during 1989 and the rulers of Czechoslovakia and the GDR had, at the year's end, been forced to introduce democratic reforms in the face of popular pressure, Bulgaria in contrast, had adopted a much more cautious attitude towards introducing perestroika and glasnost. This was understandable. Unlike Hungary and Poland, there had not been pressure either from inside or outside the ruling communist party to speed up the process of change. Besides, Mr Todor Zhivkov, the former president and leader of the Bulgarian Communist Party (BCP), had refrained from shaking up the party in order to push through changes, something which the Hungarians, Poles and particularly Mr Gorbachev had undertaken as a means of introducing a new and younger generation into the ranks of the party to lead the revolution of perestroika. The political culture in Bulgaria had always been vastly different from those countries which inherited certain European political or democratic experiences. Traditions of a strong state and a weak, or non-existent civil society in Bulgaria inhibited calls to introduce and promote real change.

Nonetheless, what is taking place in Eastern Europe in general and particularly in the Soviet Union has had, and will continue to exert, a considerable impact in Bulgaria. The burgeoning, for instance, of several independent political groups was in itself an indicator of this change. Mr Todor Zhivkov's failure to adjust to change, or at least prepare the ground for his successor, proved to be of considerable importance.

Whither the road to perestroika?

This was the question asked during a meeting of the BCP's central committee on 13 and 14 December 1988. The plenum had been expected, particularly in the light of the dramatic sackings from the politburo in the early summer. Mr Zhivkov, a cautious and hesitant leader who has looked somewhat askance at the changes taking place in the Soviet Union, made a number of uncompromising changes in the party's leading bodies. In one sweep, he removed Mr Chudomir Aleksandrov from the central committee. Mr Aleksandrov, thought at one time to be an heir apparent to Mr Zhivkov, had been head of economic and cadre policy in the central committee. But his views, particularly those which advocated getting rid of what he once unwisely termed 'dead wood' from the party, lost him his job. Probably his downfall deprived the closet party reformers of an 'opposition' spokesman for more perestroika. Mr Stoyan Mikhailov, the liberal-minded head of the central committee's ideology and cultural department, who had supported the idea of much greater glasnost in the media, was also dropped from the central committee. The official explanation for the political demise of both men simply stated that it was 'for reasons of expediency'. The unofficial, and indeed more plausible cause for their eclipse from the political scene and a view which is widely accepted as the closer version of the truth, was that both men were in a hurry to speed up reforms. In a country where the pace of change and time are inextricably linked to Mr Zhivkov, they had, in retrospect, little chance of surviving politically under the present leadership.

Mr Zhivkov also took the opportunity to demote both Mr Stoyan Markov, a candidate member of the politburo, and a rather ineffectual chairman of the committee on quality, as well as Mr Ognyan Doynov, the chairman of heavy machine-building. Politically, their exit to obscurity cost Mr Zhivkov little political damage. The old faces on the politburo remained intact. But what was of significant importance for the coming months, was the promotions to the politburo itself. First, 60-year-old Mr Dimitar Stoyanov, the then interior minister, was promoted to full politburo rank. His ministerial ranking was taken over by Mr Georgi Tanev, the former minister of transport. He was to play a major role in the escalating dispute between Turkey and Bulgaria in mid-1989 and in temporarily resolving the status of the country's ethnic Turkish minority by issuing them with passports to travel to neighbouring Turkey. That decision actually led to serious economic difficulties for the country. Mr Ivan Panev, head of the party organisation in Sofia, the capital, was also promoted. As head of the country's largest party organisation, and in a system where the party chiefs of the capital cities wield considerable influence, his political future ought not to be ignored.

Thus, 1989 opened up with a government with new ministers, an even tighter politburo which tended even more to reflect the cautious views of Mr Zhivkov and an economy which was in need, if not of urgent attention, then certainly of some structural changes.

The sluggish economy

One of the main problems with the Bulgarian economy is that it has been subjected to frequent tinkering over the last three years of the 1980s. A spate of reforms have been initiated, at least on paper, but in practice, they appeared to be haphazard and thus are taking time to bite. One of the reasons is that Bulgaria

KEY FACTS

Official title: People's Republic of Bulgaria
Head of state: President Petar Mladenov
Ruling party: Bulgarian Communist Party
Capital: Sofia
Official Languages: Bulgarian
Currency: Lev = 100 stotinki
Exchange rate: Leva 0.84 per $ (Dec 1989)
Area: 110,912 sq km

Population: 8.9m (1988)
GDP per capita: $2,750 (NMP, 1987)
GDP real growth: 5.1% (NI, 1987)
Labour force: 4.1m (1987)
Trade balance: –leva 140m (1987)
Foreign debt: $5,075m* (1986)

* estimated figure

lacks a highly skilled workforce as well as a skilled managerial class, two problems which beset all the other countries of Eastern Europe. Also, implementing economic reforms is easier said than done, as Mr Zhivkov, in one of his most critical speeches to date, spelt out.

The occasion was the convening of the central committee on 4 and 5 May 1989 to discuss agriculture. This plenum had been long overdue. It had been scheduled sometime in the autumn of 1988, then later, for the spring of the following year. It is understood that the delay was due to uncertainty about how far to go with introducing a much greater degree of private ownership into agriculture. Again, like the rest of Eastern Europe and the Soviet Union, the whole debate about broadening small and private-scale farming has run up against one problem: the nature of ownership in a communist-type system.

Mr Zhivkov tried to address this crucial question at the plenum on agriculture. First, he acknowledged that agriculture had been neglected over the years. With Bulgaria's rush towards industrialisation and modernisation, a trend common to all the east European countries in the early 1950s, which was in imitation of the Stalinist model, little consideration or attention had been given to agriculture. The one exception was in Hungary. Mr Zhivkov, trying to catch up on lost time, which was leading to poorer food supplies to the cities, unveiled a new programme for agriculture. The key issue, he said, was restoring the status of the villages. This was a clear reference to young people who had deserted the villages in their droves for the towns, leaving the villages to be run by their ageing parents and the small household plots unattended. As an incentive, Mr Zhivkov called for better social services in villages, an improved infrastructure and a better education system. 'A village without a school is a forest without birds' said Mr Zhivkov. And he continued, 'The present situation is catastrophic indeed in view of modern requirements: 84 per cent of the people are with elementary and lower education and only 2.5 per cent have higher and semi-higher education. No organised and purposeful work has been done in recent years to raise the qualifications of the people engaged in rural production.' He then pin-pointed other deficiencies and shortfalls in agriculture. Machinery was obsolescent, people were recruited regardless of their qualifications, land was under-used.

But that was only the aperitif to the analysis. Mr Zhivkov suggested that land could be leased by individuals for up to a period of 50 years. He also suggested opening up the market for food supplies, allowing producers to sell both in private and state markets. Indeed this package of reforms came close to what Mr Gorbachev had been proposing in the Soviet Union. He also bemoaned the organisation of agriculture.

However, in a speech which went beyond revealing the true state of agriculture, and in some ways could be interpreted as a criticism of the package of reforms introduced over the past three years, Mr Zhivkov said that since the July 1987 plenum, 'we failed to achieve the expected results in a number of spheres ... we have not yet carried out significant breakthroughs in most of the key sectors ... we took daring decisions about the attributes of power ... we closed a number of ministries ... all this however failed to provide the revolutionary impetus demanded by the July concept and the country's needs ....' And in an analysis which, ironically, could have been presented by the ousted Mr Aleksandrov, Mr Zhivkov blamed the sorry state of affairs on the preservation of the 'self-interested hierarchy' and 'obsolete forms of organisation'. Interestingly, he also blamed 'corporations, industrial combines and ministries', institutions which he himself had created over the past few years. Mr Zhivkov was now admitting that these new decision-making giants were a mistake.

KEY INDICATORS

	Unit	1984	1985	1986	1987	1988
Population	m	8.96	8.96	8.96	8.97	8.99
Net national income	leva bn	24.6	25.0	26.9	28.2	–
NI per capita	leva	2,703	2,747	3,022	3,142	–
NI real growth	%	4.6	1.8	5.3	5.1	–
Retail trade turnover	% growth	3.8	3.2	3.5	3.9	–
Labour productivity	% growth	2.0	2.1	5.2	5.3	–
Agricultural production	% growth	7.0	–12.3	11.7	–4.3	–
Industrial production	% growth	4.2	3.2	4.0	3.3	–
Cement	'000 tonnes	5,717	5,296	5,721	5,600	–
Steel	'000 tonnes	2,878	2,926	2,965	3,044	–
Exports	leva m	12,987	13,739	13,350	13,816	–
CMEA	%	75.4	75.6	81.7	81.4	–
Non-CMEA Western counties	%	9.1	8.5	6.9	6.8	–
Imports	leva m	12,842	14,067	14,353	13,967	–
CMEA	%	78.5	75.5	79.2	80.5	–
OECD	%	13.8	15.2	15.4	15.1	–
Balance of trade	leva m	145	–328	–1,002	–140	–
Exchange rate	leva per $	1.01	1.03	0.95	0.87	0.85

* estimated figure

This sudden reversal of policies which, between 1986-88, were deemed to show the way forward for Bulgaria's road to perestroika, is due largely to the realisation that tinkering with the system rather than structurally changing it, could not provide the real ingredients for a leaner and more efficiently run economy.

Indeed, economic statistics indicate that the economy is not peforming as well as had been expected. The Five Year Plan for 1986-1990 had set a target annual growth of 6 per cent. However, actual growth rates for 1986 and 1987 were 5.5 per cent and 5.1 per cent respectively, not a poor figure when set against very ambitious targets. In other sectors, the picture has been mixed. Agriculture dropped by 0.7 per cent compared to 1987. The biotechnological and chemical industry, one of the sectors which had been allotted the greatest attention over the past decade, grew by 13.8 per cent, while electronics, another important sector, grew by 9.7 per cent. The consumer goods sector grew by 3.6 per cent while generation of electricity, which had been subjected to repeated droughts during the 1980s, grew a modest 3.6 per cent. However, growth of foreign trade, which has been re-organised several times since 1985, recorded only a 1.1 per cent increase, while exports grew by 4 per cent and imports, subject to more restrictions, fell back to 1.8 per cent.

But the statistics also highlighted some other weaknesses, for example, in the retail goods sector. Although sales of retail goods had grown by 6.1 per cent, the authorities, in their list of 30 items stated that some, such as food, were subjected to shortages. Again, as in previous central committee meetings, Mr Zhivkov stressed the need to press ahead with economic reforms. But with the exception of some changes in foreign trade, the numerous plans advocated by the politburo seem not to have been implemented largely for the reason that many people remain unsure about how they should be implemented.

Glasnost rears its head

As its east European neighbours are fast realising, any attempt to push through economic reforms demands three things: a thorough reappraisal of previous economic policies (as well as mistakes), the promotion of those who are known more for their pragmatic and technocratic (if not liberal) inclinations than their political loyalty, and a more open media to explain the goal of the economic reforms.

In Bulgaria, the leadership had been rather slow about addressing these two main issues. First, the official view has been for some time that the reforms in Bulgaria started back in 1956 during the famous April Thesis, in which Mr Zhivkov consolidated his power. For well over thirty years, he has been at the helm. Thus, any policies which have been introduced over that period of time owe their origin and legitimacy to Mr Zhivkov who, naturally enough, is reluctant to admit that fundamental mistakes in planning have been made. Second, even though Mr Zhivkov has been captain of a ship sailing at fewer knots per hour than some of his east European neighbours, he appears still unwilling to bring younger blood into the central committee and the politburo, possibly on the grounds that they could be perceived as some threat to the status quo.

But during 1989 the perceptible need for change was making itself felt. Undoubtedly, the fact that most Bulgarians have access to Soviet television, where they can see for themselves the immense changes that Mr Gorbachev has introduced, has provided an important yardstick for those prepared to question current policies in Bulgaria. But also, quite remarkably for a country which has known of little opposition in the post-war period, many independent groups have blossomed since early 1989.

The first signs of the flowering of these groups which, during the following year did not represent an opposition to the ruling communist party, in the Polish or Hungarian sense, came in 1988, with the founding of an independent environmental movement. The authorities reacted promptly to this movement, content to ban it rather than tolerate it. But given the changes taking place in Eastern Europe and in East-West relations, the authorities were by mid-1989 slowly coming to terms with the existence of independent groupings. These cover a broad spectrum, from the Independent Association for the Defence of Human Rights, and the Independent Committee on Religious Freedom, to the Muslim Initiative Group. But as long as these groups did not overtly challenge the authorities, they were tolerated. However, when many of them spoke out in defence of the ethnic Turkish minority in Bulgaria, the leadership responded by expelling some of the organisers.

The ethnic Turkish problem is one of those issues which cannot be ignored because it has considerable foreign policy and economic implications for the Bulgarian leadership. Yet, at a time when the Soviet Union is trying desperately to come to terms with its own ethnic minority problems, latent problems which have come to light through the policy of glasnost, Mr Zhivkov has rather surprisingly adopted completely the opposite course.

Essentially, the Bulgarian authorities hold the view that an ethnic Turkish minority does not exist. The one million strong Muslim population, they say, are of Bulgarian origin. They also say that 'Turkish-speaking Bulgarians' were forced to adopt Turkish customs, language and religion during the Ottoman Empire. After the Second World War, the Turkish-speaking Bulgarians were allowed to emigrate to Turkey. However, in the mid-1980s, the Bulgarian authorities decided that the remainder of the ethnic Turkish minority should adopt Bulgarian customs and language. A widespread campaign, which really amounted to a policy of forced assimilation, was implemented. And those who resisted were imprisoned.

The problem was that the issue remained unsolved. And by 1989, with the advent of glasnost, several independent human rights groups took up the cause of the ethnic Turks, demanding, among other things, that they be given their own language schools, newspapers and books. The upshot was that Mr Zhivkov and his interior minister, Mr Georgi Tanov, took two steps to quell the growing unrest among the ethnic Turkish minority. They expelled the leaders of those who led demonstrations and strikes. And in an extraordinary policy, the authorities granted passports to at least 130,000 ethnic Turks to leave the country for Turkey. Some of the Turkish minority claimed that they had been forced to leave. Others said that they were simply taking advantage of the new laws on more liberal travel to leave the country altogether. Whatever the reasons, this mass exodus has created two problems for the authorities.

The first is economic. Bulgaria has had a labour shortage for some time. Many of those that left were factory workers, dentists, doctors and engineers. But the majority were agricultural workers. No sooner had the exodus started, than Mr Georgi Manov, the minister for agriculture, admitted that it had exacerbated the already chronic labour shortage: so much so, that pupils, students and pensioners were conscripted to collect the 1989 harvest. Mr Manev also said that the labour shortfall would have to be made up by working week-ends as well, something which caused resentment among the workforce who made their views known to *Trud*, the trade union daily.

The other repercussion concerns foreign policy. There is no doubt that Bulgaria's international image had improved markedly following the early 1980s when certain countries failed to substantiate allegations that Bulgaria had been behind the attempts to assassinate the Polish-born Pope John Paul II. In addition, the country had tightened up its control over drug-trafficking, something difficult to do, given Bulgaria's position as a transit route from Europe to the Middle East. But these more positive developments were somewhat undermined following the action over the Turkish problem. Indeed, during 1989, the Bulgarian foreign ministry again became concerned about its relations with its neighbours. Not only were its relations with Turkey strained, but it had still to resolve a dispute with neighbouring Romania. The emission of chlorine fumes from a factory in southern Romania concerned the inhabitants of Ruse, an old merchant city on the Danube located in the north of Bulgaria on the border with Romania. The complaints of the inhabitants of Ruse precipitated the birth of the independent environmental movement. In addition, the Soviet Union, whose ambassador in Turkey had offered to mediate in the dispute between Sofia and Ankara, indicated that it too was unhappy with the difficult relationship between a Warsaw Pact country and a NATO country. In late summer of 1989, it appeared that Mr Zhivkov was prepared to tackle the thorny issue of the ethnic Turkish minority by actually allowing them to leave. Their exodus, in effect, would make Bulgaria into a homogeneous, Slav country, something which Mr Zhivkov had worked towards for some time.

Outlook

There are few signposts on the road ahead. A few years ago, before Mr Gorbachev came to power, a road without signs would have been an unthinkable situation for any east European country. Then, the signs were clearly marked. The directions, all coming from Moscow, allowed few turnoffs, few side-roads, few stops along the way. Everything was predictable. And it was along this straight ideologically marked road that Bulgaria unswervingly and loyally drove. But Mr Gorbachev has changed all of that.

COUNTRY PROFILE

Political system: Unicameral 400-member National Assembly elected every five years. The Assembly elects, from among its members, the State Council, which exercises overall guidance over domestic and foreign policy. The Council of Ministers, which is also elected by the Assembly, is the supreme executive and administrative body and is being substantially restructured. Until December 1989 political power was held by the Bulgarian Communist Party, which together with the Bulgarian Agrarian People's Union dominated the Fatherland Front (the largest socio-political organisation). Following calls by the new Chairman of the Party for constitutional change, the abolition of the Communist Party's monopoly of power, the strengthening of parliament and the separation of the party and government, a special party congress will be held on 26 March 1990 to enact election laws allowing for opposition parties.

Main cities/towns: Sofia (population 1.1m), Plovdiv, Varna, Rousse, Bourgas, Stara Zagora, Pleven.

Media
Press: 17 dailies including the Communist Party paper *Rabotnichesko Delo* (circulation 850,000). Numerous scientific and technical journals. Business publications in main European languages include *Bulgarian Foreign Trade* (13,000, bi-monthly), *Economic News of Bulgaria* (15,000, monthly), *Sofia News* and *Bulgaria Today* (25,000).
Radio: Four home service radio programmes and several local stations. Foreign service broadcasts several hours each week. *Television:* Bulgarian Television broadcasts around 90 hours each week.

Domestic economy: Good economic growth during 1980s, led by manufacturing sector. Export revenue is about 50 per cent of the national income (1987). Emphasis on the development of machine building, electronics, chemical industry and biotechnologies. The ninth Five-Year Plan (1986-1990) forecasts further rises in production and improvement of product quality. Rights and responsibilities of the economic organisations have been extended in recent years. Since the beginning of 1987 a reform in economy management has been in progress to decentralise management functions and transform the economic organisations into self-managing units. The branch economic ministries are being replaced by new-style institutions as voluntary economic groupings of self-managing economic organisations.

Employment: Industry is the major employer, (38 per cent of workforce), followed by agriculture (19 per cent). Real wages and labour productivity have increased steadily. Ninth Five Year Plan envisages real per capita incomes rising by 18 per cent.

External trade: In the period 1981-85 foreign trade turnover showed an average annual increase of 10 per cent. Economy is heavily export-orientated, and the total value of exports is equivalent to over 50 per cent of national income. Also significant exports of chemicals, consumer durables and ferrous metals. Imports mainly consist of capital goods, raw materials and fuels. Around 75 per cent of all trade is conducted with CMEA countries, particularly with the USSR. Adverse trade balance with CMEA countries has been notably reduced. The proportion of trade with OECD countries has declined, while that with developing countries has rapidly increased. Percentage of total trade in 1987 with CMEA countries was 81 per cent; with developed Western countries 11 per cent; with developing countries eight per cent.

Agriculture: Sector, which is considered to be among the most efficient in CMEA, accounts for 12.5 per cent of NMP and in 1987 unprocessed agricultural products accounted for 1.9 per cent of total exports and industrial goods of agricultural origin 16 per cent. Rapid growth of co-operative farms and agro-industrial units. Cultivation of private plots is encouraged, helping towards basic food self-sufficiency. Principal crops are wheat, maize, sugar-beet, barley, sunflower seed, tobacco, tomatoes, grapes. Also important is dairy and poultry farming. Long-term development of agriculture based on further concentration and specialisation, mechanisation, improved irrigation, increased grain production and expansion of dairy sector.

Industry: Mainstay of the economy accounting for 60 per cent of NMP. Rapid development of electrical and electronic engineering, biotechnological and chemical industries, metallurgy, and machine building. Also large-scale production of construction materials, vehicles, processed foods, textiles, consumer durables. Largely export based with dependence on material inputs from USSR. Current emphasis on modernisation, reorganisation and expansion of high-tech industries.

Mining: Small deposits of iron, lead, zinc and manganese. Large deposits of copper ore recently discovered in the Sredna Gora mountains. Ambitious programme for the development of coal reserves under way. Exploration for oil and gas concentrated in the north of the country and in the Black Sea.

Energy: Nuclear power is now main source of energy.

Banking: The Bulgarian National Bank (Bulgarska Narodna Banka) is responsible for normal central banking activities. Since 1987, when the banking system was restructured, other institutions have been concerned with foreign trade, industrial finance and commercial banking, including the Bulgarian Foreign Trade Bank, the Bank for Economic Projects, the Economics Bank and the State Savings Bank.

Membership of international organisations: CCC, CMEA, WFTU, FAO, IAEA, ICAO, ILO, IMO, International Lead and Zinc Study Group, International Medical Organisation, International Organisation of Journalists, International Radio and Television Organisation, IPU, ITC, ITU, IWC, UNESCO, UPU, Warsaw Pact, WHO, WIPO, WMO, WTO.

BUSINESS GUIDE

Time: GMT + 2 (GMT + 3 from late Mar to late Sep).

Climate: Summer hot and dry, Apr-Sep average temp 23°C. Cold winters, average temp –1°C, with heavy snow.
Clothing: Mediumweight for most of year, heavy overcoat in winter.

Entry requirements: Visas, which are not obtainable at border, required by nationals of most Western European countries. Prohibited entry to nationals of South Africa.
Currency: No Bulgarian currency can be taken in or out of the country. No restrictions on amounts of foreign currency taken in. It can be exchanged at Balkantourist offices and the Bulgarian National Bank. Advisable to keep all receipts for customs departure.
Customs: Small quantities of spirits, wines and beverages duty-free. Also gifts to the value of 50 Leva. Valuable personal effects should be declared verbally to Customs on entry. No restrictions on goods bought for foreign exchange in Corecom shops, where Eurocheques, travellers' cheques and credit cards can be used.

Air access: Main international airlines operate flights to Sofia.
National airline: Balkan-Bulgarian.
Main international airports: Sofia (SOF), 9 km from city centre; Varna (VAR) and Bourgas (BOJ).

Surface access: Frontier exit/entry points – Yugoslavia (by rail: Dragoman; by road: Kalotina, Zlatarevo, Gjueshevo, Vrashkachuka); Turkey (by road/rail: Svilengrad; by road: Capitan Andreevo); Romania (by rail and road: Rousse, Kardam; by road: Durankulak; by river ferry: Kalafat-Vidin); Greece (by road and rail: Koulata).
Main ports: Varna, Bourgas.

Hotels: Deluxe, first and second class ratings system. Hotels for visitors run by Interhotels – Balkantourist, Balkantourist, Orbita and Cooptourist. Advisable to book in advance. Some hotels provide guests with special coupons which can be used in other restaurants and cafes. Main international credit cards accepted.

Car hire: Service available from Balkantourist, large hotels and at airport. International driving licence recommended. Most car hire accounts transacted in hard currency (between $0.40 and $0.60 per km). Drivers normally given special petrol coupons which can be used throughout the country. Speed limits: out of town 80 kph and 120 kph on motorways, in town 60 kph. Drinking and driving strictly prohibited.

City transport: Efficent and cheap tram and bus service.
Taxis: Available and cheap. Official taxis have meters. A 5-10 per cent tip in local currency is usual.

National transport
Air: Balkan-Bulgarian Airlines operate 8 domestic lines connecting Sofia with all main towns. Shuttle service to Varna and Bourgas. Fares are generally inexpensive.
Road: Good road network.
Rail: Approximately 6,500 km of track connecting all main towns. First-class travel recommended. Necessary to make reservations.

Public holidays
Fixed dates: 1/2 Jan, 3 Mar (Liberation from Turkey), 1/2 May, 24 May (Education Day), 9/10 Sep (National Days), 7 Nov (Russian Revolution).

Working hours
Government and business: (Mon-Fri) 0830 – 1730.
Banking: (Mon-Fri) 0800 – 1200, 1300 – 1700, also Sat 0800 – 1100.
Shops: (Mon-Sat) generally 0800 – 1900.

Social customs: A nod of the head means 'No', a shake of the head means 'Yes'.

Business language and interpreting/translation: Business languages: English, German, French and Russian. Service provided free of charge by foreign trading organisations (must be arranged in advance). Also available on paying basis from main hotels and Balkantourist as well as from the World Trade Centre (INTERPRED). Cyrillic alphabet used.

Telecommunications
Telephone: Direct dialling to main European cities. Public telephones operated by 5 stotinki piece. Credit card facilities also available. For trunk calls dial 121, international calls 0123 and general enquiries 144. IDD access code to Bulgaria: 359 followed by 2 for Sofia.
Telex: Available from most major hotels and central post office (open 24 hours). In many cases international telex cards are not accepted. Payment is usually in hard currency. Country code 67 BG.
Telegram: Available from any post office.

Postal service: Mail for overseas destinations should be clearly marked 'airmail'.

Advertising: All advertising media available. Agency – Bulgarreklama. Some advertising services also supplied by Bulgarian Chamber of Commerce, whose monthly bulletin takes some foreign paid adverts.

Trade fairs: Annual Plovdiv International Fairs in spring (consumer goods and related equipment) and autumn (technical goods). Annual exhibitions in Sofia in spring (usually with a technical theme).

Electricity supply: 220 V AC.

Representation in capital: Afghanistan, Albania, Algeria, Argentina, Austria, Belgium, Brazil, China PR, Colombia, Congo, Cuba, Czechoslovakia, Denmark, Egypt, Ethiopia, Finland, France, German Democratic Republic, Federal Republic of Germany, Ghana, Greece, Hungary, India, Indonesia, Iran, Iraq, Italy, Japan, Kampuchea, Democratic People's Republic of Korea, Kuwait, Laos, Lebanon, Libya, Mexico, Mongolia, Morocco, Mozambique, Netherlands, Nicaragua, Peru, Poland, Portugal, Romania, Spain, Sweden, Switzerland, Syria, Turkey, United Kingdom, USA, USSR, Uruguay, Venezuela, Vietnam, Yemen PDR, Yugoslavia.

BUSINESS DIRECTORY

Hotels: Sofia
Balkan Sheraton, 5 Lenin Square (tel: 87-65-41; tx: 23030; fax: 871038).

Bulgaria, 4 Russki Boulevard (tel: 87-19-77; tx: 22499).

Moskva Park, 25 Nezabravka Street (5 mins from city centre) (tel: 71261; tx: 22411).

Novotel Europa, 131 Boulevard Georgi Dimitrov (tel: 31261; tx: 22051).

Rodina, 4 Russki Pametrick Square (tel: 5331; tx: 22200).

Sofia Grand, Narodno Sobranie De Square (opp National Assembly) (tel: 87-88-21; tx: 23405).

Vitosha New Otani, Boulevard Anton Ivanov 100, Lozenets (5 mins from city ctr) (tel: 62-41-51; tx: 8652297).

Business advice and information
Bulgarian Chamber of Commerce and Industry, 11A Alexander Stambolijski Boulevard, 1040 Sofia (tel: 872631; tx: 22374) publish a monthly newspaper, *Economic News of Bulgaria.*

Banking
Agricultural and Cooperative Bank, G.Dimitrov Blvd 37, Plovdiv 4018 (tel: 231876, 238946; tx: 44324).

Bank for Economic Projects, Legue Street 17, Sofia 1000 (tel: 801737; tx: 23390).

Biochim Commercial Bank, Ivan Vazov St 1, Sofia 1000 (tel: 541366, 874911 (ext 314); tx: 23862).

Bulgarian Foreign Trade Bank, Sofiiska Komuna Street, 2, Sofia (tel: 8551; tx: 22031).

Bulgarian National Bank (Bulgarska Narodna Banka) Sofiiska Komuna Street 2, Sofia (tel: 8551; tx: 22392).

Construction Bank, Kiril i Metodii St 17/19, Sofia 1000 (tel: 83841, 832320; tx: 23887).

Economic Bank, Slavianska St 8, Sofia 1000 (tel: 870741; tx: 23910).

Electronics Commercial Bank, Chapaev St 55, Sofia 1574 (tel: 707447; tx: 23745).

Transport Bank, Shipka St 5, Varna 9000 (tel: 223073, 231865; tx: 77303).

Other useful addresses
Balkantourist, Vitosha Boulevard 1, Sofia (tel: 43331; tx: 22583).

Balkan-Bulgarian Airlines, Narodno Sabranie 12, Sofia (tel: 884433, 884493); at Sofia airport (tel: 71201; tx: 22299, 22480).

Biotechnological and Chemical Industry Association, Triyaditsa St 2, Sofia 1000 (tel: 8601; tx: 22360, 22369).

Bulgarian Association for Tourism and Recreation, Lenin Square 1, Sofia (tel: 84131; tx: 23101, 23553).

Bulgarian Foreign Trade Magazine, Parchevich 42, Sofia 1000.

Bulgarian Industrial Association, 134 Rakovski St, Sofia (tel: 882501; tx: 23523, 23607).

Bulgarreklama (advertising agency), 42 Parchevich Street, Sofia 1040 (tel: 85151; tx: 23781).

Central Co-operative Union, Rakovski Street 99, Sofia (tel: 8441; tx: 22058).

Central Statistics Office, 10, 6-ti Septemvri Street, Sofia (tel: 878711; tx: 22001).

Corecom (foreign exchange shop), 8 Tsar Kaloyan Street, Sofia (tel: 85131; tx: 22176).

Communications Association Gourko St 6, Sofia 1000 (tel: 882095, 889416; tx: 22515; fax: 803871).

Construction and Building Industry Association, Kiril i Metodii St 17, Sofia 1000 (tel: 83841; tx: 22182, 22183).

Electronics Association, Chapaev St 55, Sofia 1113 (tel: 7351; tx: 23745).

Energetics Association, Triyaditsa St 8, Sofia 1000 (tel: 86191; tx: 22707, 22708).

Forestry and Timber Industry Association, Antim I St 17, Sofia 1000 (tel: 86171; tx: 22481, 22815).

Heavy Machinebuilding Association, Slavianska St 8, Sofia 1000 (tel: 870741; tx: 23490, 23499).

Industry for Man Association (consumer goods), Aksakov St 29, Sofia 1000 (tel: 86131; tx: 22387).

Interhotels – Balkantourist, 5 Triaditsa Street, 1000 Sofia (tel: 873901; tx: 22052).

International Fair – Plovdiv, G Dimitrov Boulevard 37 (tel: 553191, 553146, 26129, 26139; tx: 44432).

Interpred (represents foreign firms), 16 Bulgaro – Savetska Drouzhba Boulevard, Sofia (tel: 71464646; tx: 23284).

Medical Industry Association, Bademova Gora St 20-a, Sofia 1404 (tel: 592111; tx: 23098).

Metallurgy and Mineral Resources Association, G Dimitrov Blvd 22, Sofia 1000 (tel: 83851; tx: 22502).

Burkina Faso

Nicolas Kulibaba

The government of Captain Blaise Compaore weathered its second year in power in a political atmosphere of scepticism and malaise. Through an unwieldy mix of revolutionary rhetoric and liberal economic reforms, the Front Populaire regime sought to win back the support of both youthful radicals, most alienated by the bloody ouster of Thomas Sankara, and traditional elites, most displaced by him. But in the byzantine world of Burkinabe politics, Compaore must keep a sentry's eye over the plethora of ideological splinter groups and special interests that have emerged in the post-Sankara era. Though the regime earned high marks from international financial institutions and foreign donors for economic management, it was hardly immune to inflammatory charges of corruption and ideological betrayal from critics at home. In September 1989, a coup attempt resulted in the execution of Major Lingani and Captain Zongo, the second and third-ranking members of the government.

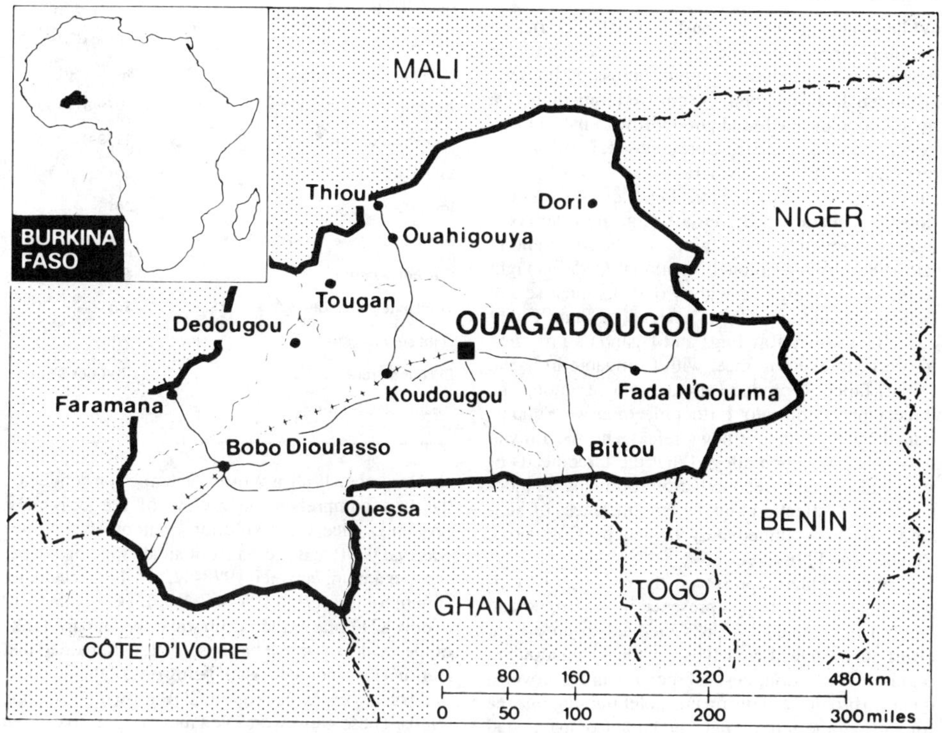

The death of Burkina Faso's charismatic young president, Thomas Sankara, in a 1987 coup d'etat heralded what many Burkinabe consider to be an era of disillusionment. Sankara's successors, a military junta of one-time colleagues in arms, faced a Herculean political labour: how to engineer repudiation of their revolutionary comrade without discrediting the revolution that brought them to power.

The response to Sankara's death eas electrifying, and unusual even by the standards of coup-plagued Africa. A darling of the Third World media, Sandara had risen to become an outspoken critic of the West and radical reformer of the status quo almost immediately upon his ascension to power in 1983. However, in a sudden split with fellow members of the Conseil National de la Revolution (CNR) – Compaore, Jean-Baptiste Boukari Lingani, and Captain Henri Zongo – Sankara was gunned down and ignominiously buried in a shallow grave in October 1987. While Burkina Faso's trade unions and civil servants immediately rallied to support Compaore's Front Populaire and its policy of 'national rectification', the regime has since failed to win little other than grudging tolerance among the population at large.

Alleging that the basic goals of the 1983 revolution would remain intact, Compaore set out to dismantle much of the political structure put in place by Sankara. Among the first victims of the new regime were local cells of the Comite de Defense de la Revolution (CDR), ideological outposts of the CNR whose presence throughout the country had displaced and threatened bureaucratic and ethnic elites. These were replaced by scaled-down Comites Revolutionnaires, ostensibly radical, but unthreatening to the economic elites whose support was so essential to Compaore.

Other political reforms which critics allege have betrayed the revolutionary mandate of government included salary increases for civil servants, a failure to enforce price ceilings on domestically-produced foodstuffs, suppression of mandatory contributions by civil servants for special health and education projects, lowered prices for beer, and abolition of a variety of rural taxes.

While Compaore's reforms went far in buying time for the regime in an atmosphere of scepticism, they also raised serious questions about the ideological underpinnings and future direction of the Front Populaire. The absence of transparency in policy-making by the regime and the emergence of ideological factionalism among its supporters further fuelled an absence of confidence in the public at large.

Factionalism within...

Although the political core of the regime was located in a small group of military units most loyal to Compaore, Defence Minister Lingani and Economic Promotion Minister Zongo, Burkina's leaders allowed a combative cluster of civilian rivals to spring up around them. Most closely associated with the regime were remnants of the Union des Communistes Burkinabe (UCB), a mixed civilian and military group of former Sankara loyalists, headed by former university rector Clement Ouedrago, and the civilian-dominated Union des Luttes Communistes (ULC), led by one-time Housing and Urban Affairs Secretary Moise Traore. Both groups officially fused in 1988 to form a coalition, Pour le Parti (PLP). However personal rivalries within the PLP, nudged along by a directive from Compaore, led to the emergence of yet another coalition in 1989, dubbed the Organisation pour la Democratie Populaire-Mouvement du Travail (ODP-MT).

Compaore's impatience with rivalries among the regime's supporters came to a head in April 1989, when several previously favoured groups were deprived of representation in the reorganised cabinet as the result of their refusal to join the ODP-MT. Suddenly in disfavour were the Groupe des Communistes Burkinabe (GCB) and a dissident core of the ULC. Wishing to leave the gates of political opportunity open to the GCB and others, Compaore charged Clement Ouedraogo with the task of enticing numerous other groups into the fold of the ODP-MT.

Ouedraogo's task was a formidable one, however. Among the numerous organised political associations with whom he had to negotiate were a variety of groups built on the remnants of pre-revolutionary political parties, including the Rassemblement Democratique Africain, the Union Progressive Voltaic, and the Groupe des Democrats et Patriots. The latter distributed tracts which, while claiming loyalty

KEY FACTS

Official title: Burkina Faso

Head of state: Captain Blaise Compaore

Capital: Ouagadougou

Official Languages: French (More, Dioula, Gourmantche, Peul widely spoken)

Currency: CFA Franc = 100 centimes

Exchange rate: Franc CFA316 per $ (Nov 1989)

Area: 274,200 sq km

Population: 8.3m* (1987)

GDP per capita: $216 (1987)

GDP real growth: –3.2% (1987)

Inflation: 4.2% (1987)

Trade balance: –$418m (1986)

Foreign debt: $861m (1987)

* estimated figure

to the Front Populaire, charged ODP-MT constituent groups with being self-interested 'sectarian imposters' and implied that they operated as mere fronts for political patronage.

....and without

While Compaore might boast of small victories in building support at home, his regime enjoyed little new success in building support across its borders. Although the Front Populaire enjoyed something of a honeymoon of newfound goodwill with its southern neighbours, Togo and the Cote d'Ivoire, it had at least one specific failure elsewhere.

In recognition that the Front Populaire's assassination of Sankara had driven a wedge into the community of revolutionary regimes on the continent, in late 1988 Libya's Colonel Muammar Qadafi sought to reconcile Compaore with two of his most severe critics. Summoning Ghana's Head of State Flight Lieutentant Jerry Rawlings and Uganda's President Yoweri Museveni to meet with Compaore in Syrte, Qadafi failed to produce any agreement among the three. Opportunities for further meetings were undermined by a premature news release by the Libyan press agency, which alleged that the three heads of state had agreed to reconciliation following an independent investigation into the circumstances of Sankara's death – immediately denied by Compaore, Rawlings and Museveni.

Further problems presented themselves to Compaore with the continued presence in the Congo and Nigeria of Sankara loyalists pledged to an all-out effort to avenge themselves on the Ouagadougou regime. The willingness of neighbouring governments to provide a platform and potential staging area for anti-regime activity indicated that Compaore had yet to earn diplomatic credibility among even his peers in the region.

Economic headway

Located in the arid and marginally-productive Sahel, Burkina Faso faces the chronic dilemma of cyclical drought and good deficits. Nonetheless, newly available data indicates that in spite of ecological uncertainty and political turmoil, the country may be making much-needed economic headway.

KEY INDICATORS

	Unit	1983	1984	1985	1986	1987
Population	m	7.5	7.7	7.9	8.1	8.3
Gross domestic product (GDP)	CFA bn	380.9	390.5	479.9	507.8	539.1
Real GDP growth	%	0.0	2.3	14.7	11.8	3.2
GDP per capita	Franc CFA	50,789	50,714	60,745	62,691	64,952
Inflation	%	8.3	4.8	6.9	–2.6	4.2
Agricultural production	1979-81=100	106.9	107.4	127.8	146.4	134.7
Food production	1979-81=100	99.0	93.3	91.4	88.9	91.4
Exports	$m	159	175	159	233	–
Imports	$m	472	409	439	650	–
Balance of trade	$m	–315	–234	–280	–418	–
Current account	$m	–60	–3	–66	–124	–
Total external debt	$m	422	433	540	669	861
Debt service ratio	%	5.7	5.7	16.9	8.9	–
Exchange rate	CFA per $	381.1	437.9	449.3	346.3	300.5

* estimated figure

In December 1988, a World Bank mission conducted the first comprehensive review of the Burkinabe economy since the revolution. Among its findings was that GDP had grown from an average of 3.1 per cent per annum between 1978-82, to a 4.6 per cent per annum for the period 1983-87. While per capita GDP of only $217 places the nation squarely amid the ranks of the world's poorest nations, the fact remained that Burkina Faso could boast of being one of the very few in Africa to enjoy per capita GDP growth during the 1980s. These gains were even more impressive given the nation's accelerating population growth rate, which has climbed from 2.7 per cent per annum in the 1970s to 3.5 per cent in the 1980s.

Also salutary were the World Bank's findings regarding financial management under the Sankara and Compaore regime's. Running against the grain of conventional wisdom regarding military regimes on the continent, the World Bank observed high levels of competence among the nation's financial managers, adequacy of accounting and other financial records, committed efforts to avoid deficit financing, and adherence to budgetary commitments.

However valued kudos from multilateral financial institutions might be, Burkina Faso's land-locked economy remained structurally unstable, plagued by heavy reliance on imported foodstuffs, manufactures, and energy, and low levels of technological development.

Troubling, too, are recent indications that the medium-term progress noted by the World Bank might not have taken into account short-term deterioration under the Front Populaire. A resurgent outbreak of violent street-crime in the nation's cities and several incidences of corruption by Front Populaire appointees undermined public confidence in Compaore's stewardship of the public trust. Charges that Sankara's death might have been directly related to a planned crackdown on high-level pilferage of gold from newly-exploited mines in Burkina's northern provinces still circulated widely and lent credence to the charge that Burkina's unpopular regime might indeed represent the abandonment, rather than the claimed 'rectification', of Sankara's nationalist ideal.

COUNTRY PROFILE

Political system: Legislative and executive power rests with the military rulers. Popular participation has been sought through the nationwide creation of Comités Révolutionnaires. Gained independence in 1960. Formerly a province of French West Africa. The administration of Thomas Sankara came to power during a military coup in August 1983, and he was killed in October 1987 in a coup led by Captain Blaise Compaore.

Political parties: There are no legal political parties.

Population: Estimated at 8.3m (1987).

Main cities/towns: Ouagadougou (population 300,000), Bobo-Dioulasso (200,000), Koudougou (28,000).

Language: French is the official language and the universal medium for documentation; African languages include More, Dioula, Gourmantche and Peul.

Media
Press: One national daily circulating in Ouagadougou and main cities/towns. Several periodicals are published – mainly economic and industrial. *Sidwaya* is a government-controlled daily paper.
Radio: Radio Burkina broadcasts in French and 13 African languages.
Television: Télévision Nationale du Burkina provides transmissions seven days a week to Ouagadougou and Bobo Dioulasso, Koudougou and Ouahigouya, in French and African languages.

External trade: Balance of payments depends to some extent on workers' remittances and foreign aid. French aid declined drastically during the Sankara years (from $88m in 1982 to $19m in 1985) but could pick up again under the new government. Foreign debt remained relatively low at $720m at the end of 1987, and the ratio of debt to current account exports was 300.8.
Exports: Raw cotton, live animals, shea-nuts, hides and skins, rubber products, sesame seeds, gold (steep rise in export). Main destinations: France, China, Ghana, Côte d'Ivoire, UK, Japan.
Imports: Transport equipment, petroleum products, cereals, machinery. Main sources: France, Côte d'Ivoire, USA, Ghana, Netherlands.

Agriculture: Sector accounts for 40 per cent of GDP and around 50 per cent of export earnings. Subsistence farming and nomadic stockraising predominate. Principal food crops are sorghum, millet, yams, maize, rice and beans. Large areas are unsuitable for cultivation. The 1986-87 grain harvest of 1.8m tonnes allowed a surplus of 20,000 tonnes to be set aside as a reserve. There are plans to mechanise farming and open up new areas for development. Farming was allocated a third of the 1981-90 development plan budget, and the 1987 cotton harvest of 180,000 tonnes was twice the 1983 level. Cotton is the main cash crop; other are shea-nuts, sesame and sugar cane. Livestock production is concentrated in the north, mainly for export to Côte d'Ivoire (which severely restricted its Burkina beef imports in recent years) and Ghana.

Industry: Sector contributes around 15 per cent to GDP and employs 13 per cent of the workforce. Production is centred on the processing of agricultural commodities (flour milling, sugar refining, manufacture of cotton yarn and textiles) and production of consumer goods including moped/bicycle assembly, footwear and soap manufacture. Foreign investment is minimal and development remains handicapped by the chronic shortages of raw materials and spares.

COUNTRY PROFILE

Mining: Activity confined to extraction of gold-bearing quartz at Poura (reserves estimated at 30,000 kg), marble and antimony. Viable deposits of zinc and silver at Perkoa, and some 13.3m tonnes of manganese deposits at Tambao, as well as known reserves of limestone, bauxite, nickel, phosphates and lead. Exploitation of resources hindered by weak infrastructure.

Energy: Heavy dependence on imported energy, particularly fuel oil. All electrictity is thermally generated. Construction of a 10 mW hydroelectric station at Kompienga was due to be completed by 1988. Also plans to build power stations on the White (Nankambe) and Black (Mouhoun) Volta Rivers and at Pama.

Membership of international organisations: African Development Bank, African States in Association with the EC (Lomé Convention), Arab Bank for Economic Development in Africa (recipient), Economic Commission for Africa, Economic Community of West African States, French Monetary Area, GATT, IDA, IFAD, ILO, IMF, Islamic Conference, Islamic Development Bank, OAU, UN, West African Economic Community, WHO, WIPO, World Bank.

BUSINESS GUIDE

Time: GMT.

Climate: Tropical. Dry season from Nov to Mar, when Harmattan wind blows keping the humidity low. Temperatures in Ouagadougou range from 14°C at night to over 35°C during the day. Main rainy season is from Jun to Oct. Highest rainfall in the south, lowest in the far north where arid desert climate prevails.

Entry requirements
Passport: Required by all.
Visa: Required by all except nationals of France, Belgium, German FR, Italy, Luxembourg, Netherlands and most surrounding West African countries. An onward or return ticket is also required. Exit permits (obtainable from police on Ouagadougou) required before leaving the country.
Prohibited entry: Nationals of South Africa and Israel.
Currency: No restrictions on import/export of foreign currency or local currency.

Health precautions
Mandatory: Yellow fever vaccination certificate.
Advisable: Cholera vaccination is essential; typhoid and polio vaccinations recommended. Malaria prophylaxis should be taken as risk exists throughout the country. Water precautions are also advisable.

Representation overseas: Abidjan, Accra, Algiers, Bamako, Beijing, Bonn, Brussels, Cairo, Copenhagen, Havana, Jeddah, Lagos, Moscow, New York, Ottawa, Paris, Riyadh, Rome, Tripoli, Washington.

Air access
National airline: Air Burkina (Burkina Faso is also a minority shareholder in Air Afrique).
Other airlines: Aeroflot, Air Afrique, Air Algérie, Air Ivoire, Ethiopian, UTA.
International airport: Ouagadougou (Code: OUA), 8 km from city, restaurant, car hire.
National airport: Bobo-Dioulasso (Code: BOY), 6km from city.

Surface access
Road: Most practicable during dry seasons – from Mali (Bamako) and Niger (Niamey), when buses operate on these routes. The road from Ghana is being improved. Also possible from Côte d'Ivoire, Benin and Togo.
Rail: Daily express service from Abidjan (Ivory Coast) to Bobo Dioulasso and Ouagadougou. Sleeping/dining cars.

Hotels: Available in Ouagadougou and Bobo Dioulasso – limited availability elsewhere. Advisable to book in advance. Service is included in bills.

Electricity supply: 220/380 V AC, 50 cycles.

Car hire: National licence plus permit or International driving licence required. Use of chauffeur-driven cars advised.

City transport
Taxis: Unmetered and available in main centres. 10 per cent tip usual.
Buses: Frequent in Ouagadougou.

National transport
Air: Air Burkina serves Ouagadougou, Bobo Dioulasso, and other main centres. Light aircraft can be chartered from Air Burkina. The airline also operates flights to surrounding countries, Mali, Togo, Benin and Côte d'Ivoire.
Road: Condition varies; some roads are only passable in dry season, though international roads are all-weather.
Buses: Services from Ouagadougou to main towns.
Rail: Daily railcar service runs Ouagadougou-Bobo Dioulasso and on to Côte d'Ivoire; two classes; some restaurant cars, sleeping accommodation and air-conditioning. Service can become overloaded.

Public holidays
Fixed dates: 1 Jan, 8 Mar (Women's Day), 1 May, 4 Aug (National Day), 15 Aug (Assumption), 1 Nov (All Saints'), 25 Dec.
Variable dates: Easter Monday, Ascension, Whit Monday, Eid al-Fitr, Eid al-Adha, Ashura.

Working hours
Government and business: (Mon-Fri) 0700/0800 – 1200/1230 and 1500 – 1730/1830. Early closing (at 1700 hrs) Mon and Thu.
Banking: (Mon-Fri) 0800 – 1200. Some banks also open 1500 – 1800.

Telecommunications
Telephone: International telephone facilities available from main hotels. Dialling code for Burkina Faso: IDD access code + 226 followed by subscriber's number (there are no area codes).
Telex: International telex facilities available from main hotels.

Banking
Central bank: Banque Centrale des États de l'Afrique de l'Ouest.
Main centre: Ouagadougou.
Head offices: Banque Internationale du Burkina, Banque Internationale pour le Commerce, l'Industrie et l'Agriculture du Burkina, Banque Nationale de Développement, Banque pour le Financement du Commerce et des Investissements, Union Révolutionnaire de Banque.
Foreign branches: Banco Exterior de España, Banque Internationale pour l'Afrique Occidentale, Banque Nationale de Paris, Caisse Centrale de Coopération Économique.

Representation in capital: Algeria, China PR, Cuba, Egypt, France, German FR, Ghana, Iran, Korea PDR, Rep. of Korea, Libya, Netherlands, Nigeria, Switzerland, USSR, USA.

BUSINESS DIRECTORY

Hotels: Bobo-Dioulasso
Ran, P.O. Box: 50 (tel: 98-18-45/46).

Hotels: Ouagadougou
Independance, Rue de la Resistance du 17 Mai, P.O. Box: 127 (tel: 30-60-60; tx: 5201).

Pullman Silmande, P.O. Box: 4733 (tel: 30-00-96, 30-01-76; tx: 5345).

Ran, Avenue Binger, P.O. Box: 62 (tel: 34-255; tx: 52188).

Other useful addresses
Air Burkina, Avenue Loudun, PO Box 1459, Ouagadougou (tel: 33-61-55; tx: 203).

Autorité des Aménagements des Vallées des Voltas (AVV), PO Box 524, Ouagadougou.

Banque Centrale des États de l'Afrique de l'Ouest (central banking authority for the members of the West African Monetary Union), PO Box 356, Ouagadougou (tel: 36-715; tx: 5205).

Burkina Auto Location (car hire), Hotel Independence (tel: 33-27-20).

Chambre de Commerce, d'Industrie et d'Artisanat du Burkina, PO Box 502, Ouagadougou (tel: 33-61-48; tx: 5268).

Division du Tourisme et de l'Hôtellerie, PO Box 624, Ouagadougou (tel: 33-24-48; tx: 5555).

Faso Tours (national travel agency), PO Box 1318, Ouagadougou (tel: 33-53-50; tx: 3345).

Groupement Coopératif de Ventes Internationales des Produits du Burkina (Cooproduits), PO Box 91, Ouagadougou (tx: 5224).

Groupement des Petits Commerçants, PO Box 952, Ouagadougou.

Institut de la Statistique et de la Démographie, PO Box 374, Ouagadougou (tel: 35-537).

Office National des Commerce Extérieur, PO Box 389, Ouagadougou (tel: 33-62-25; tx: 5258).

Société de Commercialisation, PO Box 531, Ouagadougou (tel: 33-30-07; tx: 5274); PO Box 375, Bobo-Dioulasso (tel: 39-04-23).

Syndicat des Entrepreneurs et Industriels, PO Box 446, Ouagadougou.

Télévision Nationale du Burkina, PO Box 7029, Ouagadougou (tel: 33-68-01; tx: 5231).

West African Economic Community, PO Box 643, Ouagadougou.

Burundi

Diana Hubbard

Following the terrible bloodbath of August 1988, in which as many as 20,000 people were killed when ethnic tensions between the majority Hutu and minority but elite Tutsi rose to boiling point, the country returned to a period of relative peace during the ensuing 12 months. The troubles did result in a temporary freezing of donor aid, as many Western nations expressed scepticism over official versions of the uprising, but relations were gradually restored and 1989 saw a resumption of aid flows designed to support economic restructuring.

In a new government formed shortly after the massacre, President Buyoya appointed a Hutu prime minister for the first time since 1965 but, while this went some way to appeasing the Hutu majority, a few months later Buyoya had to deal with an unsuccessful coup attempt reportedly fomented by his predecessor Jean-Baptiste Bagaza. As the year passed, Buyoya succeeded in rehabilitating his government with Western nations, a point which was underlined in September 1989 when the Belgian Prime Minister Wilfried Martens paid the first official visit since independence of a head of government of the former colonial power and by the announcement that Pope John-Paul II would visit in 1990.

Hutu prime minister

As many of the 63,000-odd refugees who fled into Rwanda during the massacres began to return home, President Buyoya announced the setting up of a 24-member committee of national unity in early October to investigate the conflict. Just two weeks later, on 19 October, Buyoya announced the formation of a new government and named 38-year-old Adrien Sibomana – a Hutu – as prime minister. With a total of 11 Hutu in the 23-member cabinet, the majority ethnic group is far better represented than in the past but still disproportionately since Hutu constitute 85 per cent of the country's 5m population, and strategic portfolios, such as defence and foreign affairs, remain in the hands of Tutsi ministers.

Although ethnic tensions, which had been at their worst in rural areas, diminished noticeably, on 10 March 1989, 10 men – mainly Tutsi, like the president – including five army officers, were arrested and accused of plotting to assassinate Buyoya at a military parade planned for the following day. Concrete facts remained sparse but there were strong suspicions of the involvement of former President Jean-Baptiste Bagaza, overthrown by Buyoya in a bloodless coup in September 1987 – and perhaps more significantly, of backing for Bagaza from Libya. Following the arrests, some 70 Libyans living in Burundi, including several diplomats, were given 48 hours to leave the country. There is some basis for the rumour of a Libyan connection, since Buyoya has proved himself more conservative than his predecessor and has actively sought a rapprochement with Burundi's huge neighbour Zaïre and its Western-oriented leader President Mobutu, and has also made considerable efforts to repair relations with the Catholic Church. Under Bagaza, many foreign missionaries were expelled and the holding of some church services was banned.

Main exports

In addition to the delicate political situation, Buyoya also tried to come to grips with the economy. Like neighbouring Rwanda, Burundi has been hard hit by the depressed world coffee price. Coffee represents around 80 per cent of export revenue and brought in an estimated $90m in 1988. The harvest has remained fairly stable in the late 1980s, totalling 36,300 tonnes in 1987–88 and 33,500 tonnes in 1988–89. Cotton is the only other main export, although a large proportion of the crop is consumed domestically. Volume is modest – 7,515 tonnes in 1988 – and production and transport costs relatively high.

Despite poor prospects for export earnings, the government decided to increase spending in the 1989 budget, mainly in an attempt to appease social pressures. It provided for a 38 per cent increase in current expenditure to BF34.8bn, with large allocations for national education and health. The increase in capital spending, was more restrained and was targeted at BF5.5bn, up from BF4bn in 1988.

KEY FACTS

Official title: Republic of Burundi

Head of state: Major Pierre Buyoya (chairman of the Military Council for National Redemption)

Capital: Bujumbura

Official Languages: Kirundi, French (Kiswahili widely spoken)

Currency: Burundi Franc (BFr) = 100 centimes

Exchange rate: BFr157 per $ (Nov 1989)

Area: 27,834 sq km

Population: 5.15m (1988)

GDP per capita: $232 (1987)

GDP real growth: 2.0% (1987)

Inflation: 4.5% (1988)

Trade balance: –$77.7m (1987)

Foreign debt: $755m (1987)

KEY INDICATORS

	Unit	1984	1985	1986	1987	1988
Population	m	4.98	4.72	4.86	5.00	5.15
Gross domestic product (GDP)	BFr m	120,451	141,347	140,842	143,590	–
Real GDP growth	%	–0.1	11.7	3.8	2.0	–
GDP per capita	BFr	24,187	29,946	28,980	28,718	–
Inflation	%	14.3	3.6	1.9	7.0	4.5
Agricultural production	1979-81=100	100.9	114.6	119.1	121.2	125.8
Food production	1979-81=100	102.0	115.1	120.9	121.1	127.7
Coffee production	'000 tonnes	26.2	32.5	31.3	37.0*	33.0*
Tea production	'000 tonnes	3.4	4.1	3.6	4.0*	5.0*
Cotton (lint) production	'000 tonnes	6.5	7.1	7.9	7.0*	8.0*
Exports	BFr m	12,367	13,533	19,393	11,117	17,981
Coffee	BFr m	10,555	11,511	17,057	7,891	15,505
Imports	BFr m	22,383	22,754	23,195	24,465	28,885
Petroleum	BFr m	3,930	3,896	3,055	3,607	3,665
Balance of trade	BFr m	–10,016	–9,221	–3,802	–13,348	–10,904
Current account	$m	–39.3	–42.2	–37.6	–96.5	–62.1
Public foreign debt	$m	334	415	528	718	–
Total external debt	$m	346	446	552	755	–
Debt service ratio	%	16.8	18.9	20.6	38.5	–
Exchange rate	BFr per $	119.7	120.7	114.17	123.6	140.4

* estimated figure

Late April 1989 saw the successful holding of a major donors' conference in Bujumbura, which brought participants together to examine the 1988–92 Five-Year Development Plan. Total planned investment is around $1bn, with forecast real growth in GDP of 5 per cent a year. The conference seemed to give the green light to major donors, particularly on the funding of structural adjustment, the pace of which had slowed following the 1988 massacres. In July 1989, the IMF announced that it was releasing the final tranche, of SDR8.5m, of its three-year structural adjustment facility announced in August 1986, bringing the total IMF funding since that date to SDR21.6m. Japan had earlier pledged Y2.4bn to support the structural adjustment.

Project funding

Other donors announced project funding during the year, including SDR25.1m in May from the World Bank to promote the private sector in agriculture, improve extension services and the supply of inputs such as fertiliser and tools. As much as 90 per cent of the population is employed in the agricultural sector and a large proportion is still not integrated into the money economy. France has concentrated on power and telecommunications with a FFr40m package to expand the power network and install transformers and a low-voltage network to bring power to several small towns, and a further FFr31.5m to help finance the construction of telephone exchanges, including a new 11,000 line exchange in Bujumbura and a 4,000 line exchange in Ngarara, to the north of the capital.

COUNTRY PROFILE

Political system: The 30-strong Military Committee for National Salvation (MCNR) which took over in a military coup on 3 September 1987 is chaired by the Republic's President, Major Pierre Buyoya. Received independence from Belgium in 1962 as a Kingdom; became a Republic after the first military coup in 1969.

Political parties: The only political party is the Union pour le Progrés National, which has 1.2m members.

Population: The Hutu people comprise 85 per cent of the population, with the Tutsi making up the other 15 per cent.

Main cities/towns: Bujumbura (population 200,000), Gitega (30,000).

Language: Kirundi is the national language, French the administrative language, and Kiswahili is also spoken.

Media
Press: Main newspapers are *Le Renouveau du Burundi* (French daily) and *Ubumwe* (Kirundi weekly), both government controlled.
Radio: Radio service in Kirundi, Swahili, French and English operated by the state broadcasting station Voix de la Révolution.
Television: TV service established in 1984.

External trade: Regular trade deficit owing to depressed world coffee prices. Also large services deficit due to salary transfers and high transport costs, but encouraging signs in improved levels of foreign investment and aid.
Exports: Coffee (65 per cent of total value), tea (4-6%), cotton fabrics (2-4%). Main destinations: German FR (30-35 per cent of total value), Finland (8-10%), Japan (4-6%), Kenya (4-6%), USA, Italy, UK, Netherlands, France, Belgium/Luxembourg.
Imports: Machinery and equipment (16-20 per cent of total value), petroleum products (14-16%), food (9-12%). Main sources: Belgium/Luxembourg (14-16 per cent of total value), Iran (12-15%), France (9-13%), German FR (9-13%).

Agriculture: Mainstay of economy, contributing 53 per cent to GDP and employing about 85 per cent of the total population. Most land under cultivation is devoted to subsistence crops – mainly cassava, bananas, sweet potatoes, pulses, maize and sorghum. Cattle rearing is also an important source of food, as is fishing on Lake Tanganyika. The main cash crop is coffee, more than 90 per cent of which is arabica, and which is being encouraged with higher producer prices. Other cash crops include tea, cotton (increasingly important), palm oil and tobacco. Output of cash and food crops has fluctuated in recent years owing to the effects of drought, soil erosion and price fluctuations.

Industry: Manufacturing sector contributes around 16 per cent to GDP and is based almost entirely in Bujumbura. Current production includes beer, soft drinks, cigarettes,glass, textiles, insecticides, cement, oxygen and coffee processing.

Mining: Gold mining resumed in 1984, producing 1,115 troy ounces. Substantial nickel reserves (up to five per cent of world total) have been found, but low world prices and an inadequate infrastructure mean extraction is not economically viable. Also known deposits of cassiterite and uranium. Phosphates and limestone may be used for cement production.

Energy: Reliance on imported petroleum (from Iran) and electricity (from Zaïre). Wood is the main source of energy, providing 95 per cent of the subsistence sector needs. An 8 mW hydroelectric power plant was built in 1984 on the Mugere river, and an 18 mW dam opened in 1986. Large peat deposits are also being developed, and petroleum has been located under the Ruzizi Plain and under Lake Tanganyika. A pipeline to the Tanzanian coast is under consideration.

Membership of international organisations: African Development Bank, African States Associated with the EC (Lomé Convention), Arab Bank for Economic Development in Africa (recipient), Communauté Economique des Pays des Grands Lacs (CEPGL), Economic Commission for Africa, GATT, IDA, IFAD, FAO, ILO, IMF, OAU, UN, WHO, WIPO, World Bank, WTO, Zone d'Echange Préférentielle (ZEP).

Time: GMT +2.

Climate: Equatorial around Lake Tanganyika (including Bujumbura); hot and humid with frequent winds, temperature 23°C to 30°C. Temperate elsewhere with average temperature 20°C. Rainy season from Oct-May (except brief dry period Dec-Jan), long dry season from Jun-Sep.

Entry requirements
Passport: Required by all.
Visa: Required by all.
Prohibited entry: Nationals of South Africa, except some residents.
Currency: Unlimited import of foreign currency but amounts must be declared. Export of foreign currency limited to amount declared on arrival. Import/export of local currency limited to BFr2,000.

Health precautions
Mandatory: Yellow fever vaccination certificate if arriving from an infected area.
Advisable: Yellow fever and cholera vaccinations are considered essential. Malaria prophylaxis should be taken as risk exists throughout the country. Water precautions are essential except in major hotels. AIDS risk.

Representation overseas: Addis Ababa, Algiers, Beijing, Bonn, Brussels, Bucharest, Bukavu (Zaïre), Cairo, Dar es Salaam, Geneva, Kampala, Kigali, Kinshasa, Moscow, Nairobi, New York, Ottawa, Paris, Riyadh, Tripoli, Washington.

Air access
National airline: Air Burundi.
Other airlines: Aeroflot, Air France, Air Rwanda, Scibe, Air Zaïre, Cameroon, Ethiopian, Kenya Airways, Sabena.
Tax: Departure tax of BFr1,000 is levied.
International airport: Bujumbura (code: BJM), 11 km from city; restaurant, currency exchange. A new terminal was opened in 1986.

Surface access
Main port: Bujumbura (on Lake Tanganyika). Ferries operate from Tanzania (Kigoma) and Zaïre (Kalemi) to Bujumbura. Nearest sea port is Dar-es-Salaam (Tanzania).
Overland access: Good roads from Bukavu (Zaïre) and Kigali (Rwanda) to Bujumbura. Roads from Tanzania are generally in poor condition.

Hotels: Advisable to book in advance. Very little accommodation available outside Bujumbura. Service charge not usually levied; 10 per cent tip is usual.

Electricity supply: 220 V AC.

Car hire: Local firms only. Services unreliable. International driving licence.

City transport
Taxis: Limited availability in Bujumbura.

National transport
Air: Air Burundi operates internal flights linking Bujumbura with Gitega; aircraft can be chartered at Bujumbura.
Road: Mostly unsurfaced and can be difficult in rainy season. Surfaced routes are being extended and local advice should be sought. Very little public transport available.
Water: Local boats may be available on Lake Tanganyika.

Public holidays
Fixed dates: 1 Jan, 1 May, 1 Jul, 15 Aug, 18 Sep, 1 Nov, 25 Dec.
Variable dates: Good Friday, Easter Sunday, Ascension.

Telecommunications
Telephone: Dialling code for Burundi: IDD access code + 257 followed by area code (2 for Bujumbura, 40 for Gitega) followed by subscriber's number.
Telex and telegram: Facilities available from Direction des Télécommunications in Bujumbura.

Banking
Central bank: Banque de la République du Burundi.
Main centre: Bujumbura.

Representation in capital: Belgium, China PR, Cuba, Egypt, France, German FR, Korea DPR, Libya, Romania, Rwanda, Tanzania, USA, USSR, Vatican City, Zaïre.

Hotels: Bujumbura
Burundi Palace, Avenue de l'Uprona, P.O. Box: 225 (tel: 2920).

Club du Lac Tanganyika, Chaussée d'Uvira, P.O. Box: 2784 (tel: 5011).

Novotel, Avenue du Peuple Murundi, P.O. Box: 1015 (tel: 22600; tx: 5097).

Source du Nil, Avenue du Stade, P.O. Box: 2072 (tel: 25222; tx: 5030).

Other useful addresses
Air Burundi (tel: 5684411).

Banque de la République du Burundi, PO Box 705, Bujumbura.

Caisse Centrale de Mobilisation et de Financement (finances development projects etc.), PO Box 8, Bujumbura (tel: 5642; tx: 83).

Chambre de Commerce et de l'Industrie du Burundi, PO Box 313, Bujumbura (tel: 2280).

Cambodia

Jacques Bekaert

In 1989, major international efforts in Jakarta and in Paris failed to bring peace to this land which has been at war since at least 1970. In February 1989 the second JIM (Jakarta Informal Meeting) took place in the Indonesian capital but resulted in even fewer accomplishments than the first one, in July 1988 in Bogor. One of the main results of the first JIM was to link the total withdrawal of Vietnamese forces from Cambodia to the end of foreign military assistance to the anti-Vietnamese resistance and to Phnom Penh. It was considered a victory for Hanoi and its protégé, Hun Sen, but was accepted by all parties involved. Very little happened during JIM II. The best part of a week was spent producing a concensus document which barely masked the basic failure of the conference.

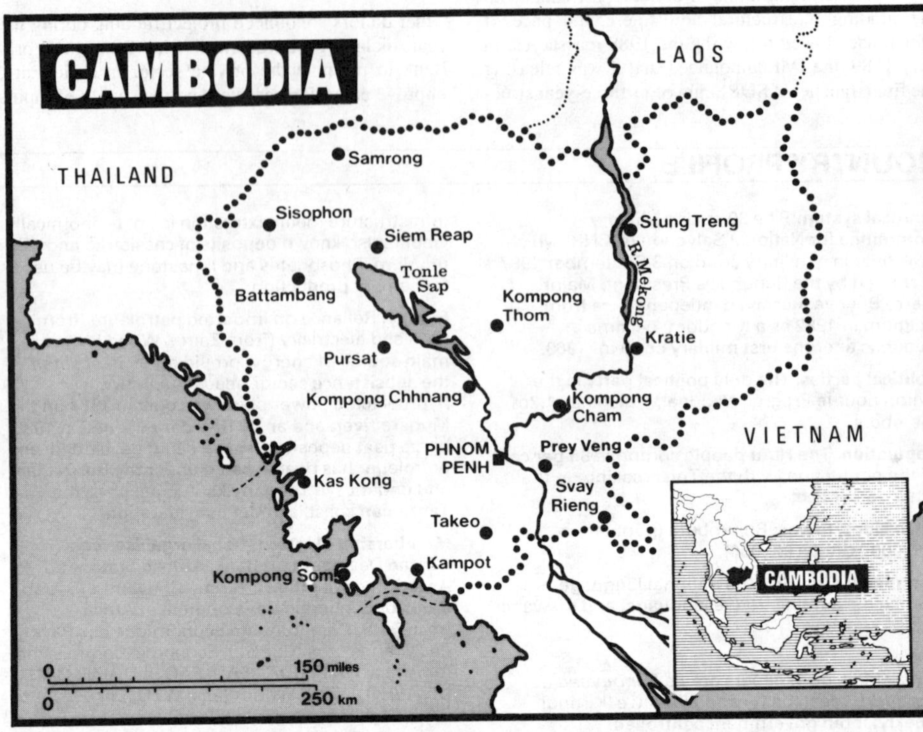

The fifth encounter, in early May also in Jakarta, between Prince Sihanouk and Prime Minister Hun Sen, the two leading Cambodian personalities, seemed more promising. From Phnom Penh, Hun Sen brought a new flag, a new name (State of Cambodia instead of People's Republic of Kampuchea) and a new constitution voted a couple of days before by the National Assembly. Sihanouk was impressed. Hun Sen was confident the prince would eventually abandon the Party of Democratic Kampuchea (Khmer Rouge – PDK) and return to Phnom Penh where he would become head of state, with powers confirmed by the new constitution.

The International Conference on Cambodia, which met in Paris from 29 July 1989 until the end of August, was convened with the mission to find a political settlement that would allow for the sending to Cambodia of an international monitoring commission. Vietnam indeed, on 5 April 1989, had promised that it would withdraw the rest of its 'volunteer' army by the end of September.

A preliminary meeting at La Celle St Cloud between Hun Sen and Sihanouk on 24 July, followed by a session of the four Cambodian factions, only served to underline the extent of disagreements existing between the khmers themselves. For the conference, everybody that mattered came to Paris: the five permanent members of the UN Security Council, Japan, Australia, India, Canada, the six countries of ASEAN, Vietnam and Laos – all represented at the opening of the conference by their foreign ministers. A month later the co-chairmen, France and Indonesia, admitted it was better to suspend the conference. Once again it had been impossible to convince the four factions to be more flexible in order to achieve a durable peace.

In Paris, Phnom Penh and Hanoi adopted a tougher line on the Khmer Rouge: the PDK had to disappear as a political, social and military organisation, they said. Only former members could be accepted in a proposed supreme council which would organise general elections after the Vietnamese withdrawal. The resistance refused Hun Sen's conditions and added the question of the Vietnamese settlers in Cambodia to the list of sensitive subjects to be dealt with in Paris. The problem of 'foreign residents' in Cambodia, objected Vietnam, was an internal matter to be considered by the sole authorities of the country, not by an international conference.

None of the major powers present in the French capital contributed much to the conference. The Paris conference demonstrated, if nothing else, how little Washington, Beijing or Moscow were ready to do: they all talked about peace and waited. Nothing happened.

Since there was no credible international control of the Vietnamese withdrawal, its authenticity was challenged even before it started. By the end of August 1989, Hanoi declared it had only 26,000 troops left inside Cambodia. Western intelligence sources claimed 40-50,000 was closer to the truth. Thai military spokesmen mentioned 80-100,000. As for the resistance, it accused Vietnam of leaving behind some 30,000 regular *bo doi* (infantrymen) disguised as Phnom Penh troops, plus 100,000 'civilians' in arms, former soldiers ready to assist the local militia. So, alas for the people of Cambodia, it was back to the battlefield.

Inside Cambodia

At the beginning of 1989, it looked as if the People's Republic of Kampuchea was ready to emerge from the relative isolation imposed on the Vietnamese-installed regime by its Thai neighbour. Prime Minister Chatchai Choonhavan's new policy of changing the 'battlefield into a market place', brought its first results when, at the end of January, Hun Sen, paid a surprise visit to Bankok where he was treated as an honoured guest.

Trade between Thailand and Cambodia, while not always officially sanctioned, became an important factor in the relations of the two countries. No official figures exist. A lot of goods are smuggled, with the interested blessing of local Cambodian officials, from the Thai province of Trat or from Singapore to Koh Kong, Kompong Som and Kampot. Cars, motorcycles, fabric, medicine, hi-fi and video equipment are among the most popular items, later to be found in the markets of Phnom Penh, a city where private cars became the rule rather than the exception.

In February the council of ministers in Bangkok decided to legalise the importation of logs from Cambodia. Cambodian timber soon arrived legally in Thailand. Many Thai MPs and businessmen made the trip to Koh Kong to capitalise on the new open border policy. Everything became possible – at a price. Khmer Rouge make millions of batht every month in 'transit fees', paid for the privilege of mining for gemstones in Khmer Rouge conrolled areas near Pailin. The money goes to the war effort.

In order to counteract resistance propaganda, which obviously was having some degree of success among Khmers peasants, the Phnom Penh regime launched its own *perestroika.* In early April a party cadres' conference showed the way to major transformation of the country's political and economic structure. The accent was now on private initiative and the constitution was modified several times to allow the legal existence of the private sector. Land was returned to the peasants, the samaki (solidarity) groups abolished.

On 30 April 1989, at the end of a two-day extraordinary session, the National Assembly voted unanimously to change the name of the country and its flag. The state of Cambodia was born. The words 'social-

KEY FACTS

Official title: Cambodia

Head of state: President of the Council of State, Hun Sen

Capital: Phnom Penh

Official Languages: Khmer (French also spoken)

Currency: Riel (R = 100 sen)

Exchange rate: R150 per ($) (Nov 1989)

Area: 181,035 sq km

Population: 7.6m (1987)

Labour force: 4.1m* (1986)

Trade balance: –$200m* (1986)

ism' and 'revolution' were dropped from the constitution which now insists on the importance of private property and the need to guarantee it. Buddhism was declared the state religion. In July, during its 17th session, the National Assembly declared the permanent neutrality of Cambodia. It also instated compulsory military service.

The ruling People's Revolutionary Party of Cambodia (PRPC), while still considered a Marxist-Leninist party by some of its members and brother parties around the world, continues to play the leading role, arousing the suspicion of supporters of the resistance as to the true nature of the state. The party controls every aspect of life in Phnom Penh and even more so in the provinces. But its ideological content became unclear, with some senior officials announcing in April 1989 that 'Marxism-Leninism was a mistake of the past'. Prime Minister Hun Sen himself said, a few weeks before the constitution was changed, that 'people do not like socialism'.

KEY INDICATORS

	Unit	1984	1985	1986	1987	1988
Population	m	7.1	7.3	7.4	7.6	–
Agricultural production						
Rice	'000 tonnes	1,970	2,100	2,000	1,855	2,000
Maize	'000 tonnes	90	92	98	95	100
Cassava	'000 tonnes	95	100	105	112	115
Sugar cane	'000 tonnes	185	190	200	205	210
Bananas	'000 tonnes	84	92	100	106	110
External debt	$m	285	318	435	594	–
Debt service ratio	%	0.4	4.4	3.2	1.9	–

* estimated figure

The economy

Because of the lack of precise regulations and the uncertainty regarding the war, foreign investments in Cambodia were kept at a minimum level. There is one Japanese saw mill in the suburbs of the capital and a couple of hotels and restaurants financed mostly by overseas Cambodians. Thai entrepreneurs as well as Thai International Airways have several projects for tourist development, especially near Angkor Wat, including direct flights from Bangkok to Siem Reap. But the reality of war may interfere with such initiatives. At the end of July 1989, the industry ministry in Phnom Penh issued a list of projects under its authority open to domestic and foreign investments, including pharmaceutical projects, ice and brewery, bicycle tyres and garment factories; a sugar refinery, a plastic plant, food and food seasoning plants.

Because of war and drought, there was some alarm regarding the rice crops of late 1989 and in mid-August the ministry of agriculture predicted some shortages. Some 57,000 hectares less than in 1988 had been planted.

The resistance, even after the Paris conference, received strong reassurances from Beijing that the amount of military support would not drop. Thailand, while more open to the Indo-Chinese countries, was not about to abandon its own special relations with the resistance. From May to September 1989 huge supplies of military hardware were sent by the USSR to the Cambodian army, including a squadron of MIG-21, tanks and artillery. But it is said that corruption is raging inside this badly-paid army and by August brand new Soviet-made military equipment was allegedly already on sale at the Thai border.

The relative fragility of the Cambodian regime, as well as that of the two non-communist resistance groups, raises serious questions about a possible return to power of the Khmer Rouge. Even Vietnamese military sources agree that the Khmer Rouge are better disciplined, better trained and have better commanders than any other Cambodian faction. Chinese Prime Minister Li Peng, in July 1989, warned the resistance against any attempt to overthrow the Phnom Penh regime. Such action, Beijing believes, would prompt Vietnam to send its troops back to Cambodia. The Khmer Rouge themselves see a protracted war, and favour political and psychological warfare over big battles. By the end of 1989 the main question seemed to be: can the Phnom Penh government effectively control the country? In other words, can anyone stop the Khmer Rouge?

COUNTRY PROFILE

Main city: Phnom Penh (population 650,000* in 1985).

Domestic economy: Like most of Indochina, Cambodia's economy is almost completely agricultural. With its small population the country would have good economic potential. It could be a food exporter but dislocation continues with the unabated guerrilla struggle and economic activity is very basic.

External trade
Exports: No figures for exports have been available for some years. A quantity of goods (largely rubber and timber) are regularly bartered for goods from the Soviet Union. Main destinations (on a cash basis) are the USSR (typically 84.1% of total exports), Japan (6.6%), EC (6.2%, of which UK 4.3%), USA (1.2%), Turkey (0.8%), and Spain (0.7%).

Agriculture: Agriculture employs 85 per cent of the workforce, and benefits from good soil and a generally satisfactory climate. Most of the irrigation systems created by forced labour in the 1970s are still in use; crops include rubber (virtually the only export), rice (both hill and lowland types), maize, cassava and fruit and vegetables. Cambodia also produces jute and sawn timber products. Despite crop failures in the early 1980s, an underlying improvement has been evident. Cattle stocks are improving, but fish represents the only animal protein for most people.

Industry: Most of what little industry Cambodia possessed was wrecked during the unrest of the 1970s, under Pol Pot, and particularly by the virtual closure of the towns in the agricultural drive. No industrial production figures have been issued since 1973; past activities have included textiles, paper, plywood, rubber goods (tyres and footwear), food processing, brewing, cement and both light and heavy fuel oils. Development is hampered by the absence of adequate transport and other structural amenities.

Mining: There are deposits of iron ore, copper, manganese, gold and bauxite, but exploitation is largely prevented by the absence of transport facilities. At present phosphates are the only economically viable minerals, and these are mined for use in the local fertiliser industries. Offshore oilfields have been located to the south, but these too remain unexploited.

Membership of international organisations: Asian Development Bank, Colombo Plan, ESCAP, FAO, G-77, GATT, IAEA, IBRD, ICAO, IDA, ILO, IMF, IMO, Interpol, IRC, ITU, NAM, UN, UNESCO, UPU, WFTU, WHO, WMO, World Bank, WTU for CGDK.

Time: GMT + 7 hrs.

Climate: Generally hot and very humid, with a rainy season from Jun to Oct/Nov. Likely temperatures in Phnom Penh are 22-30°C in Nov/Dec and 24-34°C in Apr.

Entry requirements
Passport: Required by all.
Visa: Visas are required by all, and nearly impossible for the individual traveller to obtain. Tourist visas are issued, but tourist must be part of a tour. Application should be made well in advance of proposed visit. Tourism remains very limited.

Health precautions
Mandatory: Vaccination certificates for cholera or yellow fever if travelling from an infected area.
Advisable: Anti-malarial precautions.

Representation overseas: Hanoi, Havana, Moscow, New Delhi, Ulan Bator, Vientiane.

Air access
National airline: Air Kampuchea (one route only, to Vietnam); otherwise, Hang Khong Vietnam.
Other airlines: Aeroflot, Lao Aviation.
International airport: Pochentong (PNH), 8 km from Phnom Penh.

Surface access: Overland access is via Vietnam from Ho Chi Minh City by car, travelling on Route Number One, but more expensive than plane.
Main ports: Kompong Som (formerly Sihanoukville).

National transport
Buses: Some bus or passenger truck services are available. 13,500 km of roads; about 15 per cent surfaced.
Rail: 612 km of track. Rail services operate between Phnom Penh – Aranyaprathet and Phnom Penh – Kompong Som. The Thai border route is closed.

Public holiday: 7 Jan (National Day).

Telecommunications
Telephone: Dialling for Cambodia is via the International Operator. Availability of telephone service is limited. All external services are via Vietnam.

Representation in capital: Bulgaria, Cuba, Czechoslovakia, German DR, Hungary, India, Laos, Mongolia, Poland, USSR, Vietnam.

Hotels: Phnom Penh
Monorom.

Samakhi.

Business advice and information
Chambre de Commerce et de l'Agriculture, Vither Preah Baksei Cham Krong, Phnom Penh.

National Bank of Kampuchea, Phnom Penh (tel: 24363).

Banking
National Bank of Kampuchea (central bank), 26 Song Ngoc Ninh, Phnom-Penh.

Cameroon

Diana Hubbard

Although the depression in the world market for Cameroon's main exports – oil, cocoa and coffee – continued unabated through 1989, major progress was finally made on the process of restructuring the economy. The breakthrough came in September 1988, with the conclusion of a standy-by arrangement and compensatory financing facility from the IMF and, after an unusually long gap, with the June 1989 announcement of a $150m World Bank structural adjustment loan (SAL). This was rapidly followed by a Paris Club meeting of official creditors and an agreement to reschedule a substantial component of the country's external public debt.

The financial year 1989–90, which started on 1 July 1989, saw the introduction of a realistic budget and the ensuing months brought the start of an extensive programme to restructure the vastly inefficient public sector and the very delicate decision to slash producer prices for the main cash crops. The social repercussions were muted and the political scene has remained quiet. President Paul Biya has shown few indications to date of being the kind of tough leader generally needed to carry through such swingeing measures but he may yet confound his critics.

Paul Biya takes over

With a series of steady real growth rates throughout the decade from 1975 to 1985 – averaging as high as 8 per cent a year once oil came on stream in major quantities in 1978 – Cameroon established itself in the early 1980s as the success story of West Africa. A combination of a wide range of agricultural exports and conservative financial management under President Ahmadou Ahidjo had laid the foundations of the economy and offshore oil reserves then added the icing to the cake. When Ahidjo suddenly retired in November 1982, naming Prime Minister Paul Biya as his successor, Biya inherited a strong and healthy economy.

However, there were some noticeable shortcomings in Ahidjo's administration. He had held on to the political reins very tightly and had set up one of the strongest 'police states' in the continent. Cameroon was strictly a one-party state and, even within that party, critics dared not raise their voices. In the economic sphere, Ahidjo was at pains to avoid a repeat performance of neighbouring Nigeria's oil-fuelled spending sprees. The greater part of oil revenues were placed in 'off-budget accounts' (the CHB) and investment in badly needed infrastructure, such as roads, schools and hospitals, was frugal.

With the state coffers brimming over in late 1982, Paul Biya announced that his would be a 'New Deal' government, with greater transparency in its financial transactions and the gradual introduction of democracy, although still within the framework of a single party, which he was later to rename the Cameroon People's Democratic Movement. His first few months in office were almost euphoric but, behind the scenes, Ahmadou Ahidjo, at that time still the party chairman, was beginning to regret his hasty exit. Paul Biya was starting to do things his own way and was not proving to be the handpicked puppet intended by Ahidjo. Cabinet reshuffle followed cabinet reshuffle, as Biya began to doubt the loyalties of several old retainers from previous governments.

In August 1983 an alleged coup plot fomented by Ahidjo was uncovered and in February 1984 the two main suspects were tried and sentenced to death. The third suspect – none other than Ahidjo himself – was by now installed in his retirement villa in the south of France and was sentenced to death *in absentia*. Just two months later, on 6–7 April, Biya's own presidential guard staged a coup attempt in Yaoundé, with the loss of several hundred lives during two to three days of fighting. Biya held on to power, but only just, and the following weeks were marked by arbitrary arrests and summary executions with the true fate of many leading personalities not properly known; Biya's New Deal halo had become a little tarnished.

Oil bonanza

Meanwhile, public spending was steadily rising – mainly on much needed development projects, such as the first ever tarred road to link Yaoundé with the commercial capital Douala. But with the increased spending there also came increased corruption and a gradual emptying of the CHB. In line with Ahidjo's policy, Biya was always at pains to underline the importance of agriculture and – cushioned by expanding oil revenues – he continued to raise producer prices for coffee, cocoa and cotton, even when the world market prices were falling. Diversification was encouraged with, for example, major investment in a brand new rubber scheme in the south of the country near Kribi, where some 15,000 hectares of dense rain forest was cleared to set up the Hevecam plantation. The Semry rice project (set up in the far north of the country in 1980 to provide food and a livelihood for thousands of people living in the arid, almost Sahelian, region which is also the most densely populated of Cameroon's 10 provinces) was expanded with a second and then a third plantation.

Oil production also continued to expand. The first, and still the largest, operator was the French firm Elf Aquitaine, through its local subsidiary Elf Serepca. Elf produces mainly from the Rio del Rey field, close to the border with Nigeria, although it holds offshore permits over a wide area. Production from Rio del Rey started in commercial quantities in 1977, followed by another French operator Total which brought the Moudi field on stream in 1983, and the Shell offshoot Pecten, which operates the Lokele field. In 1986, Cameroon's oil output peaked at a total of 9.2m tonnes, dropping back to 8.8m tonnes in 1987 and around 8.3m tonnes in 1988. Cameroon's geological structures were never considered as interesting oil-wise as those of neighbouring Nigeria, or even Gabon, but there is an added factor which has acted as a major deterrent to exploration and development. Cameroon's Hydrocarbons Code is by far the most unattractive among the African oil producers, with the government enforcing swingeing taxes and taking a major share of output: so much so that of Elf Serepca's 1988 production of 6.16m tonnes, it ended up with just 1.09m tonnes. Indeed, Elf was the only company prepared to keep up any sort of exploration after the oil price crash of 1986, although on only a very modest scale, and in 1988, it did bring the new Biboundi concession, still in Rio del Rey, into production.

Agriculture

On the agricultural front, despite a superficially healthy picture created by factors such as the extremely fertile volcanic soils around Mount Cameroon in the south west, and the fact that, apart from the north and far north, Cameroon has rarely suffered to any great degree from drought, both coffee and cocoa farming suffer from a structural weakness because of

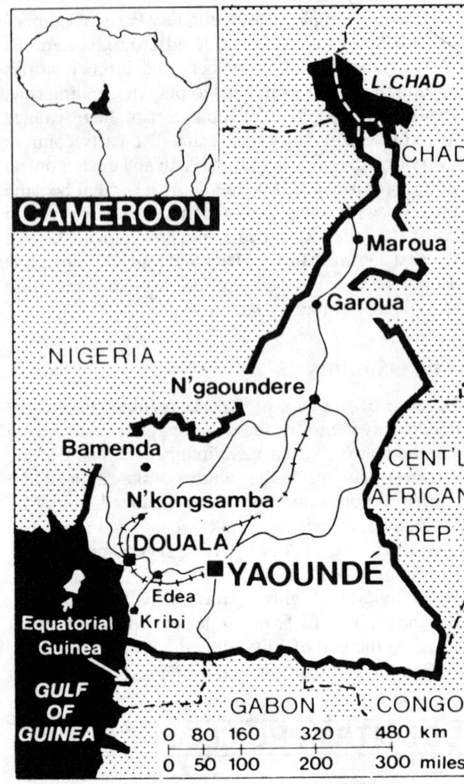

KEY FACTS

Official title: Republic of Cameroon

Head of state: President Paul Biya

Ruling party: Rassemblement Démocratique du Peuple Camerounais (RDPC)

Capital: Yaoundé

Official Languages: French, English (various African languages spoken)

Currency: Franc CFA = 100 centimes

Exchange rate: Franc CFA316 per $ (Nov 1989)

Area: 475,442 sq km

Population: 10.7m (1988)

GDP per capita: $1,185 (1988)

GDP real growth: –2.0% (1987)

Inflation: 8.6% (1988)

Trade balance: $398m* (1988)

Foreign debt: $4.0bn (1987)

* estimated figure

the tiny size of the average smallholding – only about 2.5 hectares – and the fact that young people have left the land and farmers are growing older. The average cocoa farmer is 52. Conscious of these shortcomings, the government launched a programme known as the Exploitations Agricoles de Moyenne Importance (EAMI) in November 1987, designed to attract younger and relatively well-educated people to set up medium-sized farms, using some modern technology, but not over-sophisticated or over expensive. Unfortunately, the deepening economic crisis subsequently diverted the government's attention away from EAMI and, certainly, little was heard of the programme in 1989.

Cameroon's coffee is grown mostly in the west of the country, where many farmers have formed small co-operatives, and roughly comprises 80 per cent robusta and 20 per cent arabica. Total output has tended to fluctuate more widely than that of cocoa because of the bushes' vegetative cycles. Production in 1988 was estimated at 90,000 tonnes, a similar figure to the 91,300 tonnes in 1987. The peak in recent years came in 1986, when 133,700 tonnes of coffee were harvested.

Cocoa, produced mainly in the centre and centre south, has remained more stable, with output in 1988 of 125,000 tonnes, compared with 128,900 tonnes in 1987 and 127,000 tonnes in 1986. The cocoa development board, Sodecao, is to oversee one of the largest projects to be launched during the economic crisis, calling for total investment of $285m, of which the World Bank is lending $103m, and aimed at revitalising the industry and boosting output by around 38,000 tonnes a year by the year 2000.

Cotton has gone from strength to strength during the 1980s, with an all-time record output in 1988–89 of 165,400 tonnes of seed cotton. Expansion of land given over to cotton was promoted by the cotton development board, Sodecoton, and very high yields – an average 581kg/ha were achieved. The future of cotton growing, however, will be carefully examined during the economic restructuring. Sodecoton has operated as a social venture as much as a profit-seeking organisation introducing cotton in the arid north on land which was previously uncultivated, establishing a cash crop, providing employment and integrating the cotton farmers into the money economy. It has also built roads and set up the infrastructure for basic health care and education. In 1989 Sodecoton became one of the state corporations targeted to sign a performance contract with the government and was likely to see its social vocation reined in.

Also due for a performance contract was Cameroon's largest agricultural enterprise and the biggest employer – with some 22,000 people on its payroll – the Cameroon Development Corporation (Camdev). With headquarters at Limbe in the anglophone west of the country, Camdev inherited a vast and scattered area of plantations, most of which had belonged to German settlers and were expropriated during the First and Second World Wars. It supervises production of a wide range of crops, and is a major producer of palm oil and rubber, followed by bananas, tea, and modest quantities of pepper, avocados, coconut and mangoes. Generally a rather sleepy and outmoded organisation, Camdev has recently made a bold step into banana expansion via a joint venture contract with the US producer Del Monte, with plans to boost banana production to 80,000 tonnes by 1991.

A sector identified by the World Bank as under-exploited is Cameroon's vast forest resources. At a roundtable meeting in Yaoundé in April 1989, donors were told that Cameroon has the potential to become Africa's largest producer and exporter of forestry products. Less than half the 17.5m of dense forest is currently exploited and forestry contributes only about 10 per cent of agricultural exports. Vertical integration would also give a boost to local industry.

KEY INDICATORS

	Unit	1984	1985	1986	1987	1988
Population	m	9.87	10.17	10.46	10.82	10.67
Gross domestic product (GDP)	CFA bn	3,195	3,839	4,135	4,005	3,770
GDP per capita	CFA '000	324	375	395	370	353
Inflation	%	11.2	1.3	7.7	6.0	8.6
GDP real growth	%	7.5	8.1	1.6	-2.0	–
Agricultural production	1979-81=100	107.9	109.0	115.1	113.7	119.7
Food production	1975=100	105.5	110.5	114.5	115.4	118.1
Coffee output	'000 tons	130	116	132	96	138
Cocoa output	'000 tonnes	120	118	123	131	130
Industrial output	1975=100	215.8	238.6	245.0*	229.0*	206.3
Oil production	'000 b/d	150.0	187.0	174.3	167.0	168.0
Exports	$m	1,568	1,679	2,090	1,689	1,703*
Oil	%	74.3	64.4	46.8	52.9	–
Imports	$m	1,059	1,164	1,650	1,435	1,305*
Balance of trade	$m	509	515	440	254	398*
Current account	$m	–183	–586	–564	–893	–
Total external debt	$m	2,712	2,917	3,666	4,028	–
Debt service ratio	%	8.7	8.7	12.1	15.8	–
Exchange rate	CFA per $	437.9	449.3	346.3	300.5	297.8

* estimated figure

A tropical forest action plan is to be drawn up and a more rational exploitation policy encouraged.

Structural adjustment

The performance contracts for public corporations and parastatals are a major feature of the $150m World Bank SAL, eventually signed, after protracted negotiations, in June 1989.

The initial breakthrough in the crisis had come in September 1988, when the IMF announced an SDR69.5m 18-month standby arrangement and an SDR92.7m compensatory financing facility to compensate for the shortfall in export earnings from oil, coffee, cocoa and timber in the year ended June 1988. The transition to IMF subservience was not, however, smooth. The first tranche of the standby was drawn down without hitch but, by the first quarter 1989, rumours were circulating of the Cameroon Government's failure to comply with the Fund's performance criteria, particularly financial criteria, with suggestions that some export earnings were still being diverted into bank accounts outside the franc zone.

These tensions also contributed to further delays on the conclusion of the SAL with a quite unexpected nine month gap between the granting of the IMF and then the World Bank loans. The World Bank SAL announcement was swiftly followed by what President Biya had for so long maintained was "unthinkable': a Paris Club rescheduling of Cameroon's external debt with official creditors. Despite its problems, Cameroon still has a relatively high per capita GDP and consequently did not fall into the category of 'debt distressed' countries identified at the Toronto Summit of 1988 as qualifying for exceptional debt rescheduling terms. The Paris Club will, however, bring considerable relief on the debt service due up to the end of 1990.

Another austerity budget was unveiled on 1 July 1989, fixed at CFA600bn for fiscal 1989–90, exactly the same amount as for 1988–89. Allowing for inflation of around 8 per cent, the budget is a reduced one in real terms; however, the debt rescheduling has facilitated a cut in the debt service component, allowing the government to increase projected recurrent spending by just over 13 per cent to CFA425bn from CFA375bn; the capital budget, at CFA172bn is more than 22 per cent down on 1988–89, when it totalled CFA225bn, but the debt service element is down from CFA150bn to CFA97bn.

The year 1990 will be an extremely taxing one for the government, with a whole panoply of World Bank prescribed reforms to introduce, although the complete restructuring is expected to take at least five years. One of the first items on the agenda is the technically insolvent banking sector. Many of Cameroon's 10 commercial banks have been technically bankrupt since 1987, as the government exhausted the CHB and began to withdraw its sizeable domestic bank deposits. At the same time, arrears to contractors and suppliers to the public sector grew even longer and those same contractors and suppliers called on their bankers for overdraft facilities.

The first to fall to the liquidator, on 19 July 1989, was Société Camerounaise de Banque (SCB), in which the French bank Crédit Lyonnais had held a 25 per cent share. In its place, however, Crédit Lyonnais has set up a new bank Société Commerciale de Banque–Crédit Lyonnais Cameroun, in which it has a 65 per cent share with the remaining 35 per cent held by the government. The new bank has taken over whatever good business SCB had left, while the government has set up a new debt recovery company (the Société de Recouvrement des Créances du Cameroun) to try to recover SCB's bad debts. In September 1989 came the liquidation of the much smaller Banque Paribas Cameroun, with the liquidation of the wholly state-owned Cameroon Bank expected to follow. Both the national development bank, Banque Camerounaise de Développment (BCD) and the agricultural development bank Fonader are to be closed down. The remaining commercial banks all need restructuring with a large injection of finance needed to offset the massive proportion of bad debts or non-performing loans which they have accumulated.

Many of the commercial banks and the two development banks were among a list of 75 enterprises studied by a presidential commission set up in June 1987 to evaluate the worst performing companies in the public and parastatal sectors. It looks as if more than 20 of these will be liquidated and those considered potentially viable will be privatised. Just under 40 major enterprises will remain under the aegis of the government, including the utilities, transport companies and agricultural marketing organisations. Many of them, however, will have to sign performance contracts, laying down financial targets. In a first round of such contracts in July 1989, the water board Société Nationale des Eaux du Cameroun (SNEC), the electricity board Société Nationale d'Electricite (Sonel) and the state-owned palm oil producer Société Camerounaise de Palmeraies, all signed such contracts, with the rest to follow in stages up to April 1990.

The most difficult decision, however, and one which was called for by the World Bank for several years previously, was the cut in producer prices for the main cash crops. Despite the overall downward trend in world prices for cocoa, coffee and cotton since 1985, the government had resolved not to cut its officially fixed producer price, in an effort both to stem the rural exodus and to encourage increased production. With the start of the cocoa season on 1 September 1989 came a presidential decree cutting the producer price by a hefty 40 per cent to CFA250 from CFA420 a kilo for high-grade beans: and as the cotton and coffee seasons opened (in November and December) similar cuts were also expected.

Political scene

It remains to be seen whether the farmers accept these vast cuts in income as stoically as the rest of the population has borne austerity measures. Still to be tackled is the problem of the over-staffed and under-employed civil service, which has benefited from a vast range of perks, such as free use of telephones at home, return air tickets once a year to Europe for middle-grade employees upwards, and subsidised and free housing. With no immediate signs of an upturn in commodity prices, Cameroon's period of economic austerity is likely to be protracted. On the political front, the composition of the new cabinet after a government reshuffle in July 1989 was almost less interesting than the fact that two very significant appointments in the security services were made at the same time. Almost unbelievably, the long-serving head of the secret service under Ahidjo, Jean Fochivé, who lost his job shortly after the April 1984 coup attempt, was brought back, along with the former secretary of state for internal security, Denis Ekani: two men with a reputation for extremely tough handling of any potential or actual political opposition.

After the spate of elections in late 1987 (municipal) and April 1988 (legislative and presidential), the political arena was quiet in 1989. The next major event on the agenda is the party congress, held only once every five years, now slated for 1990.

On the external front, 1989 saw an official visit by President Biya to Belgium but perhaps the most significant event was Cameroon's first ever participation in a francophone summit, in Dakar in May 1989, with observer status. Having firmly shunned both the francophone and anglophone communities for many years, this marked a policy shift; and there were indications in the autumn of 1989 that Cameroon would also be seeking to play a role in the British Commonwealth.

COUNTRY PROFILE

Political system: Executive power is vested in the president, who appoints a cabinet. The president is Paul Biya. Legislative power is held by the unicameral National Assembly whose membership was raised to 180 in the April 1988 elections. Both the president and the assembly are elected every five years. In practice, political power is held by the central committee of the Rassemblement Démocratique du Peuple Camerounais/Cameroon People's Democratic Movement (RDPC/CPDM). East Cameroon gained independence in 1960, West Cameroon in 1961. A completely unified state was formed in 1972.

Political parties: The Cameroon People's Democratic Movement (RDPC/CPDM) replaced the Cameroon National Union (CNU) in March 1985. It is the sole legal party. The party congress is held every five years.

Population: Cameroon has a highly diversified population comprising some 200 ethnic groups and 24 linguistic groups. There are about 200,000 Europeans in the country, mainly French speakers. Islam dominates in the north and Christianity in the south of the country. Population in 1988 was estimated to be 10.8m.

Main cities/towns: Douala (population 1m), Yaoundé (national capital, pop 800,000), Edea (80,000), Garoua (100,000), Buea (21,000), Limbe (formerly Victoria, 60,000), Nkongsamba, Bamenda, Bafoussam.

Language: French and English are the official languages.

Media
Press: The official daily paper *Cameroon Tribune*, in French, is also published in English as a weekly. There are several French-language weeklies, including *le Messager*, *le Républicain* and *l'Africain*.
Radio: From Yaoundé, Radiodiffusion Nationale du Cameroun operates government service in French, English and local languages. Stations in all ten provincial centres.
Television: Cameroon Television (CRTV) began broadcasting in December 1985 and today covers 75 per cent of the country.

External trade: Trade account has been in surplus since 1976, with earnings from oil assuming increasing significance. But lower oil export revenues have reduced surplus from \$515m in 1985-86 to \$70m in 1986-87. This considerably increased the current account deficit (\$583m in 1985-86) which was never in surplus because of high profit and dividend transfers.
Exports: Crude petroleum (estimated 48 per cent of total value in 1988), coffee and cocoa (23 per cent), aluminium, wood, cotton, rubber. Main destinations: France (35.6% of 1985 total value), USA (14.5%), Spain (12.1%), Netherlands (11.4%), German FR (9.7%), Italy (6.8%).
Imports: Machinery and transport equipment (typically 30-32% of total value), iron and steel (8-9%), paper and allied products (2-3%), alumina (2-3%), pharmaceuticals (2-3%). Main sources: France (55.4% of total 1985 value), German FR (9.8%), Japan (6.3%), Italy (6.3%), USA (6.2%), UK (5.5%).

Agriculture: Agriculture and forestry provide employment for 60 per cent of working population and account for 33 per cent of foreign exchange earnings. Despite recurrent drought food production has steadily risen, and currently meets virtually all local requirements. Main cash crops are cocoa (world's fifth largest producer), coffee, cotton, tobacco, rubber, palm oil, sugar and bananas. Production of tropical wood for export and for local wood/pulp industry is also important. About 40 per cent of the land area is forested. Millet, cassava, maize and plantains are the main subsistence crops. Livestock-raising and fishing are important sub-sectors. Most agricultural production is in the hands of smallholders, with the exception of rubber and palm, which are run under the plantation system.

Industry: Manufacturing sector contributes around 20 per cent to GDP and is growing in importance. Production is centred on import substitution (radios, soap, tyres, footwear manufacture) and the processing of agricultural commodities (sugar refining, palm oil processing, cotton spinning, tobacco and wood/pulp production) and oil refining (petroleum). Also aluminium smelting at Edéa (using imported bauxite), steel processing and leather tanning. The government is a major participant in the industrial sector, mainly through the Société Nationale d'Investissement. Emphasis on reducing sector's dependence on imports, decentralisation away from Yaoundé and Douala, and increased local participation in small and medium-sized enterprises.

Mining: Bauxite deposits of some 800m tonnes at Adamaoua have been identified but remain unexploited. Also massive deposits of iron-ore at Kribi and potential reserves of uranium, rutile, industrial clays and low-grade nickel and cobalt. Tin is mined on a very small scale.

Hydrocarbons: Offshore oil production (light crude) began in 1978. Estimated current reserves around 40m tonnes and 1987 output approximately 176,000 b/d. Rio del Rey field accounts for 70-75 per cent of total output. Main oil company is Elf Serepca. Although not a member of OPEC, it generally follows OPEC pricing policy. Crude oil exports go mainly to the USA. Cameroon's first refinery (Sonara) at Cap Limboh was opened in 1981. Capacity at the plant is 41,000 b/d, sufficient to meet local needs and provide a surplus for export. There are extensive gas reserves in Kribi and other areas, but plans for a liquefied natural gas plant have been indefinitely postponed.

Energy: The country has a massive hydroelectric potential. In 1986-7 production from the Edéa Dam complex, Lagdo near Garoua, and Song-Loulou generated enough to meet 95 per cent of needs. Total electricity output was 2,600 gigawatt/hour, much of which went to the aluminium industry. Large new hydro station planned for early 1990s.

Membership of international organisations: African Development Bank, African States Associated with the EC (Lomé Convention), Arab Bank for Economic Development in Africa (recipient), CEEAC, Economic Commission for Africa, French Monetary Area, GATT, IDA, IFAD, ILO, IMF, Islamic Conference, Islamic Development Bank, OAU, UDEAC, UN, WHO, WIPO, World Bank.

Time: GMT +1.

Climate: Tropical in the south with fairly constant average temperatures throughout the year, ranging between 18°C at night and 30-32°C during the day. Rainy season mainly between Jun and Oct. The north is generally dry and dusty with extreme ranges in temperature.

Entry requirements
Passport: Required by all. Should extend for at least three months.
Visa: Required by all. Difficult to enter country if passport contains South African visa.
Prohibited entry: Nationals of South Africa.
Currency: Foreign currency import is unlimited, but amount must be declared. Export is limited to declared amount. Local currency import is unlimited, but export is limited to Franc CFA 25,000 if leaving the CFA area.

BUSINESS GUIDE

Health precautions
Mandatory: Yellow fever vaccination certificate required.
Advisable: Cholera, typhoid and polio vaccinations recommended. Malaria prophylaxis should be taken as risk exists throughout the country. Water precautions should be taken outside the main centres.

Representation overseas: Addis Ababa, Algiers, Bangui, Beijing, Bonn, Brazzaville, Brussels, Cairo, Jeddah, Kinshasa, Lagos, Libreville, London, Luanda, Madrid, Monrovia, Moscow, New York, Ottawa, Paris, Rio de Janeiro, Washington.

Air access
National airline: Cameroon Airlines.
Other airlines: Aeroflot, Air Afrique, Air Gabon, Ethiopian Airlines, Lufthansa, Nigeria Airways, Sabena, Swissair, UTA.
International airports: Douala International (Code: DLA), 10 km from city, duty-free shop, bar, restaurant, buffet, bank, post office, shops; Yaoundé (Code: YAO), 4 km from city, restaurant, buffet; Garoua International. Yaoundé is unable to take wide-bodied aircraft.

Surface access
Main ports: Douala, Kribi, Garoua (on River Bénoué).
Overland access: Road access is possible from Nigeria, Chad, Gabon and the Central African Republic. However, these routes are considered rough and may become impassable during rainy seasons.

Hotels: Good hotel accommodation is available in main centres. Advisable to book in advance, and confirmation should be obtained from the selected hotel in writing or by telex. Service charges usually added to bill.

Electricity supply: 220 V AC, 50 cycles; plugs are of the two-pin round type.

Credit cards: Major credit cards are generally accepted.

Car hire: Chauffeur or self driven cars are available in Yaoundé and Douala, but can be expensive. International driving licence required.

City transport
Taxis: Not metered but have minimum fare and fixed prices; long journeys and daily hire (around CFA 15,000) should be negotiated; 10 per cent tip is optional.

National transport
Air: Cameroon Airlines operates domestic service between main centres, including several daily flights between Yaoundé and Douala. Efficient services and best means of transport.
Road: Total 31,800 km. Surfaced roads between main centres. Major routes are from Douala to Limbé, Buea, Bafoussam, Sangmelima, Bamenda, and Yaoundé (new all-weather road). Most other roads are unsurfaced and are often impassable during rainy season.
Buses: No fixed schedules and subject to suspension in rainy season for routes outside main centres. Modern coach services were introduced between Yaoundé and Douala in 1987.
Rail: There are 1,168 km of track. Services from Douala to Yaoundé and N'gaoundéré and from Douala to N'kongsamba and Kumba. Second class cheap but uncomfortable. Sleeping facilities are available on some trains. New express three-hour service between Yaoundé and Douala introduced in 1987, with greatly improved facilities.

Public holidays
Fixed dates: 1 Jan, 11 Feb (Youth Day), 1 May, 20 May (National Day), 15 Aug (Assumption), 25 Dec.
Variable dates: Good Friday, Ascension, Ramadan, Eid al-Fitr, Eid al-Adha.

Working hours
Government and business: (Mon-Fri) 0730 – 1200, 1430 – 1800. North west and south west provinces (Mon-Fri) 0730 – 1430, (Sat) 0730 – 1230.
Banking: (Mon-Fri) 0730 – 1200, 1430 – 1630. (Some banks: (Mon-Fri) 0800 – 1330).
Shops: (Mon-Sat) 0700/0800 – 1230, 1430/1500 – 1830/1900.

Social customs: Hand shaking is customary. Cameras should be used with discretion.

Telecommunications
Telephone: Telephones often available in restaurants, post offices, though difficult to get a line in Douala and Yaoundé. Dialling code for Cameroon: IDD access code + 237 followed by subscriber's number. There are no area codes.
Telex: Facilities available in main centres at central post offices and large hotels. Yaoundé and Douala have excellent international telex facilities at Intelcam offices.
Fax: Available at Intelcom.
Telegram: Available but not officially in use during night hours.

Banking
Central bank: Banque des États de l'Afrique Centrale.
Main centre: Douala.
Commercial banks: International Bank of Africa-Cameroon, Bank of Credit and Commerce Cameroon, Banque Internationale pour l'Afrique Occidentale-Cameroon, Banque Internationale pour le Commerce et l'Industrie du Cameroun, Banque Paribas Cameroun, Cameroon Bank, Méridien Bank, Société Camerounaise de Banque, Société Générale de Banques au Cameroun, Standard Chartered Bank Cameroon.

Advertising: Main advertising agency is Cameroon Publi-Expansion (CPE). Advertising in the press, radio, cinemas, posters, telephone directory. No facilities for direct mail advertising.

Representation in capital: Algeria, Belgium, Brazil, Canada, Central African Republic, Chad, China PR, Congo, Egypt, Equatorial Guinea, France, Gabon, German DR, German FR, Greece, Israel, Italy, Côte d'Ivoire, Rep. of Korea, Liberia, Libya, Morocco, Netherlands, Niger, Nigeria, Poland, Saudi Arabia, Senegal, Spain, Switzerland, UK, USA, USSR, Vatican, Zaïre.

BUSINESS DIRECTORY

Hotels: Douala
Akwa Palace, Avenue de la Liberté 52, P.O. Box: 4007 (tel: 42-26-01; tx: 5322).

Arcade, Rue Ivy (tel: 42-21-08; tx: 6183).

Ibis, Avenue Pierre Lotti, P.O. Box: 12086 (tel: 455800, 425817; tx: 5558; fax: 423605).

Meridien, Avenue des Cocotiers, P.O. Box: 3232 (tel: 42-61-36, 42-12-42; tx: 5822; fax: 423507).

Sawa Novotel Douala, Avenue de Verdun, P.O. Box: 2345 (tel: 42-08-66, 42-14-70; tx: 5532).

Hotels: Yaoundé
Central, P.O. Box: 6 (tel: 22-34-22).

Hotel des Députés, P.O. Box: 24 (tel: 22-46-63; tx: 8341).

Indépendence, P.O. Box: 474 (tel: 22-29-64).

Mont Fébé Sofitel, P.O. Box: 711 (tel: 23-40-02; tx: 8263; fax: 231500).

Terminus, P.O. Box: 212 (tel: 22-00-21).

Car hire: Douala

Hertz: Novotel Sawa, PO Box 5482 (tel: 42-99-18; tx: 5532).

Europcar: PO Box 1265 (tel: 42-86-19; tx: 5908).

Car hire: Yaoundé

Hertz: Hotel Mont Fébé, PO Box 711 (tel: 23-40-02; tx: 8263).

Other useful addresses
Banque des États de l'Afrique Centrale, PO Box 1917, Rue du Docteur Jamot, Yaoundé (tel: 22-25-05; tx: 8343 KN).

Cameroon Airlines, PO Box 4092, Douala (tel: 42-25-25; tx: 5245).

Cameroon Press and Publishing Co, PO Box 1218, Yaoundé (tel: 23-40-12; tx: 8311).

Chambre d'Agriculture, de l'Elevage et des Forêts, PO Box 287, Yaoundé (tel: 22-38-85, 22-28-44); PO Box 400, Douala (tel: 42-52-80; tx: 8243 chagri).

Chambre de Commerce, d'Industrie et des Mines: PO Box 4011, Place du Governement, Douala (tel: 42-28-88, 42-36-90; tx: 5616); PO Box 36, Yaoundé (tel: 22-47-76); PO Box 59, Garoua (tel: 27-11-77); PO Box 211, Limbe (tel: 33-42-30); PO Box 944, Bafoussam (tel: 44-13-44; tx: 7046); PO Box 551, Bamenda (tel: 36-18-20; tx: 5842).

Department of Statistics, PO Box 25, Yaoundé (tel: 22-07-88; tx: 8203).

EC Delegate, PO Box 847, Yaoundé (tel: 22-13-87, 22-21-49; tx: 8298 delefed).

Ministry of Tourism, Yaoundé (tel: 22-44-11; tx: 8318).

Office National de Commercialisation des Produits de Base, PO Box 378, Douala (tel: 42-67-76, 42-50-88; tx: 5260) – sole marketing agency for cocoa, coffee, cotton, groundnuts, palm kernels.

Société Camerounaise de Promotion Yaoundé (activities include commercial and industrial promotion).

Syndicat des Commerçants, Importateurs et Exportateurs du Cameroon (SCIEC), PO Box 562, Douala (tel: 42-03-04).

Syndicat des Industriels du Cameroon, PO Box 1516, Yaoundé (tel: 22-24-68; tx: 8286 syndicam); PO Box 673, Douala (tel: 42-30-58; tx: 5342 syndicam).

Canada

Bernard Simon

Having basked for many years in its virtually unrivalled political and economic stability, Canada has come face to face with a number of serious challenges. The free trade agreement with the USA, implemented in January 1989, is opening new vistas for Canadian businesses, but also exposing them to more intense competition and raising concerns about the attractiveness of Canada to foreign investors. At the same time, the high taxes and massive federal and provincial budget deficits which have funded one of the industrial world's most generous social security systems are casting a shadow over the confidence of both the domestic and foreign business communities in Canada's long-term future.

Renewed tensions between English and French-speaking Canada raised by the debate over Quebec's inclusion in the 1982 constitution, the problems of absorbing a wave of ethnically diverse immigrants and familiar complaints about uneven economic progress in different parts of the country are also putting some strain on Canada's social and political fabric.

Politics

The ruling Progressive Conservative Party, led by Prime Minister Brian Mulroney, comfortably won a second term in office in a general election held in November 1988. With the next poll not due until 1993, the Tories have thus virtually assured themselves of at least nine successive years in office. As of mid-1989, the Tories had 167 seats in the House of Commons, the Liberals 83 and the New Democrats 43.

So far the Tories have had mixed success in putting their own imprint on the country after more than 20 years of almost unbroken Liberal rule. During its first term, the Mulroney government loosened foreign investment rules, made substantial progress in deregulating the energy and transport sectors, negotiated a free trade pact with the USA and set in train the privatisation of several well-known state-held corporations, notably Air Canada, De Havilland of Canada and Canadair. All these were important initiatives, accomplished in most cases in the face of concerted opposition, notably from the other two political parties and organised labour.

But in other areas, the Mulroney government has had difficulty in implementing its agenda. Despite the priority given early in its first term to cutting the yawning federal budget deficit, it has trimmed the shortfall by only about C$4bn to a projected C$30bn in the fiscal year ending 31 March 1990. The business community is especially critical of the fact that the bulk of this modest reduction was achieved through tax increases, rather than spending cuts. Not until the budget tabled in May 1989 did Ottawa make a serious attempt to start pruning Canada's generous – but expensive – social security system. The 1989–90 budget provided for a tax clawback of benefits, such as child allowances and old age pensions, paid to relatively wealthy Canadians.

The Mulroney government chalked up a notable achievement in April 1987 by persuading the premiers of all 10 provinces to agree to terms for Quebec to sign the 1982 Canadian constitution. Known as the Meech Lake Accord, the agreement provided for recognition of Canada's only predominantly French-speaking province as a 'distinct society'. It also gave the provinces significant new powers in the appointment of Supreme Court judges and federal senators. However, the accord will come into force only if it is ratified by all 10 provincial legislatures before mid-1990. Meech Lake was initially hailed as one of the most significant steps towards the reconciliation of English and French Canada since the waning of separatist fervour in Quebec in the early 1980s. But with the notable exception of Quebec, where the agreement is still widely supported, doubts have grown that the wider powers granted to the provinces will seriously weaken Canada's already fragile federal system. The agreement was also increasingly seen as another example of the Conservative government's efforts to win crucial electoral support in Quebec at the expense of English Canada. Although the leaders of both main opposition parties welcomed the accord, influential groups in both parties have urged a review of policy. By mid-1989, eight provinces had ratified the accord. But the winds of provincial politics were not blowing in the federal government's favour. In keeping with their tradition of putting different parties into office at the federal and provincial levels, Canadian voters have thrown out Conservative administrations in several provinces. In New Brunswick, one of the staunchest opponents of the Meech Lake Accord, the Liberals made a clean sweep in a 1987 election. And in Manitoba a minority Tory government, which took office in 1988, shored up its support by insisting that changes be made in the agreement before it will recommend ratification. Prince Edward Island and Newfoundland, now also under Liberal rule, added their voices to the opponents of the agreement. The future of the Meech Lake Accord, and with it relations between Quebec and the rest of Canada, will be influenced by developments within the two opposition parties, particularly the Liberals.

After an unhappy return to politics lasting five years, Liberal leader John Turner announced in spring 1989 that he will step down as soon as the party chooses a successor. Turner's years in office marked a nadir in the Liberals' fortunes. Since its defeat in 1984, the party has been racked by dissent: first over the free trade agreement, then over the Meech Lake Accord and, almost continuously, over Turner's leadership.

KEY FACTS

Official title: Dominion of Canada
Head of state: Queen Elizabeth II
Head of government: Brian Mulroney
Ruling party: Progressive Conservative Party
Capital: Ottawa
Official Languages: English, French
Currency: Canadian Dollar (C$) = 100 cents
Exchange rate: C$1.17 per $ (Nov 1989)

Area: 9,976,139 sq km
Population: 26.1m (1989)
GDP real growth: 4.5%* (1988)
Unemployment: 7.5% (July 1989)
Inflation: 4.0% (1988)
Trade balance: $8,881m (1988)
Foreign debt: $320bn (1989)
* estimated figure

KEY INDICATORS

	Unit	1984	1985	1986	1987	1988
Population	'000	25,130	25,360	25,610	25,650	25,950
GDP	C$m	444.74	478.77	506.48	549.69	598.73
Industry share prices	1985=100	84.5	100	110.3	126.8	–
Consumer prices	1985=100	96.2	100	104.2	108.7	113.1
Industrial production	1985=100	90.8	100	100.4	105.6	111.8
Gold production	1985=100	93.3	100	119.9	136.4	153.4
Volume of exports	1985=100	96.0	100	106.9	112.0	122.5
Wheat	1985=100	128.0	100	73	51	57
Wood	1985=100	–	100	91	241	–
Wood pulp	1985=100	1100	100	109	117	116
Newsprint	1985=100	99.0	100	93	107	–
Volume of imports	1985=100	93.8	100	113.7	123.8	–
Exports	C$m	117.0	124.2	125.3	129.7	144.8
Crude petroleum	C$m	4,391	5,952	3,847	4,816	4,120
Natural gas	C$m	3,586	3,983	2,604	2,457	2,948
Wheat	C$m	3,907	3,207	1,837	2,230	3,751
Wood	C$m	6,923	5,234	4,932	5,751	–
Wood pulp	C$m	3,912	3,388	3,995	5,464	6,323
Newsprint	C$m	3,703	5,692	7,093	6,380	6,190
Imports	C$m	100.7	110.1	118.2	122.7	137.7
Balance of trade	$m	15,967	12,574	7,565	9,063	8,881
Balance of payments	$m	1,995	–1,470	–7,600	–7,060	–8,330
Exchange rate	C$ per $	1.30	1.37	1.39	1.33	1.23

* estimated figure

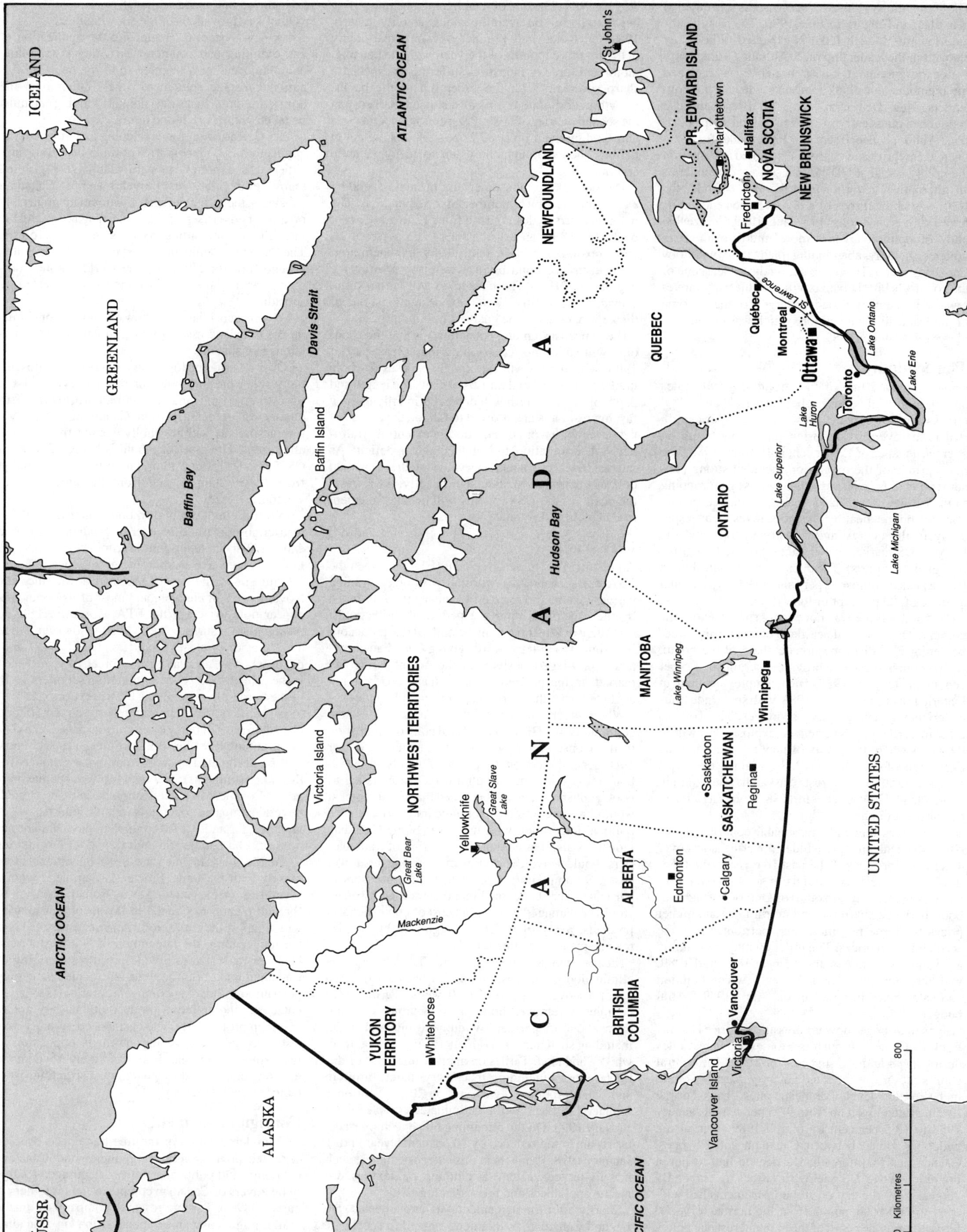
ICELAND
GREENLAND
ATLANTIC OCEAN
Davis Strait
Baffin Bay
Baffin Island
ARCTIC OCEAN
Victoria Island
NORTHWEST TERRITORIES
Great Bear Lake
Great Slave Lake
Yellowknife
Mackenzie
YUKON TERRITORY
Whitehorse
ALASKA
USSR
PACIFIC OCEAN
BRITISH COLUMBIA
Vancouver
Vancouver Island
Victoria
ALBERTA
Edmonton
Calgary
SASKATCHEWAN
Saskatoon
Regina
MANITOBA
Lake Winnipeg
Winnipeg
Hudson Bay
ONTARIO
Lake Superior
Lake Huron
Lake Michigan
Toronto
Lake Erie
Lake Ontario
Ottawa
Montreal
Québec
QUEBEC
St Lawrence
NEWFOUNDLAND
St John's
PR. EDWARD ISLAND
Charlottetown
Halifax
NOVA SCOTIA
NEW BRUNSWICK
Fredricton
UNITED STATES
C A N A D A
0 Kilometres 800

The convention to choose a successor is scheduled to take place in Calgary in June 1990. The party's attitude towards Meech Lake is expected to be a key element in the leadership race. Although Turner supported the agreement, it is opposed by many rank and file members as well as some likely leadership contenders. New Democrat leader Ed Broadbent also announced plans early in 1989 to relinquish his position. Although Broadbent got high marks for intelligence, integrity and charisma, he failed to transform the mildly socialist NDP into a serious contender as an alternative government. Like the Liberals, the NDP was in the throes of a divisive debate about its future direction during 1988 and 1989. A sizeable body of opinion favours more middle-of-the-road policies, notably abandoning pledges to withdraw Canada from NATO and to nationalise at least one of the country's big six banks. Whether the party moves towards the centre will depend largely on the outcome of the leadership convention, which was due to take place in December 1989.

The economy

Canada enjoyed one of the strongest growth rates among industrial countries between 1983 and 1989. Gross domestic product grew by over 3 per cent in real terms every year during that period, with an average advance of 4.3 per cent in 1987 and 1988. The mainsprings of the expansion included strong consumer spending (with personal savings rates dropping to their lowest level in 15 years); an unprecedented housing boom; healthy business investment, especially in the energy and forestry sectors; and the buoyant US market (which takes close to four-fifths of Canada's exports). The unemployment rate dipped from a peak of almost 12 per cent in 1983 to a national average of 7.3 per cent in June 1989.

The good fortune was not evenly spread across the country. The industrial heartland of Ontario and Quebec enjoyed a disproportionate share of the boom, with output in these two provinces surging by 5.5 per cent in real terms in 1988. The unemployment rate in Ontario fell to 5 per cent in 1988, with some industries experiencing severe labour shortages. The Toronto area, in particular, experienced explosive growth to the point where the influx of new residents and the expansion of the business base strained the city's infrastructure. By contrast, Saskatchewan's output contracted by 2.9 per cent in 1988, largely as a result of searing drought.

Inflation was kept under reasonable control, with the rise in the consumer price index hovering at just over 4 per cent for most of the past five years. By 1988, however, capacity constraints in some booming industries (especially in southern Ontario), a tight labour market, a sharp rise in interest rates and higher prices for some raw materials contributed to some acceleration in prices. The inflation rate was running at an annualised rate of about 5 per cent in mid-1989, with the latest push coming from new consumption and sales taxes imposed in the April 1989 federal budget.

Signs of stronger upward pressure on prices led the Bank of Canada to push up domestic interest rates during 1988 and the first half of 1989. The central bank's action was reflected not only in a sharp climb in the absolute level of domestic rates – banks' prime lending rate moved up from 9.75 per cent in January 1988 to 13.5 per cent by March 1989 – but also a widening spread between Canadian and US rates. Canada has for more than a decade had to pay a premium above US yields to attract investors. By August 1989 the gap on some instruments had widened to 400 basis points from the normal 150–200 basis points. One result of the high interest rate policy was a strong advance in the Canadian dollar. The currency rose steadily from 76 US cents in late 1987 to a peak of just over 85 US cents at the beginning of August 1989. The bite of the high interest rate policy began to be felt on overall business activity in early 1989. Gross domestic product grew only marginally in the first five months of the year, with the housing, motor and forestry industries suffering an especially sharp reversal. With the exception of the banks, growth in corporate profits also slowed markedly in the second quarter of 1989. Prospects in Alberta were damaged by the drop in oil prices, while a big cut in fishing quotas brought hardship to many Atlantic communities.

The outlook for consumer and business spending was dampened by a number of tax increases. The manufacturers' sales tax rose from 12 to 13 per cent on 1 June 1989, an extra surcharge on personal income tax took effect on 1 July 1989 and unemployment insurance contributions were to go up at the beginning of 1990. These tax hikes will further strain Canada's competitiveness in export markets, one of the key concerns of the business community.

The concensus among economists is that the economy will slide into recession during at least part of 1990, with the growth rate for the year as a whole unlikely to be higher than 1.5 per cent. The length and depth of the downturn will depend primarily on how fast interest rates are allowed to fall, on the course of key commodity prices and on the extent of Mulroney's determination to cut the budget deficit. As interest rates fall back, the value of the Canadian dollar is expected to slide significantly. As always, the health of the US economy will be a key element in Canada's performance.

Free trade

An historic free trade agreement (FTA) between the USA and Canada, the world's two biggest trading partners, came into force on 1 January 1989. Initiated by the Canadians during a period of protectionist sentiment in Washington in the mid-1980s, the accord is designed as a step towards giving Canadian exporters a privileged position in the neighbouring US market, in the process ensuring that Canada is not excluded from the world's emerging trading blocs.

While widely supported by the Canadian business community, the FTA was the focal point of opposition to the Conservative government in the November 1988 general election. Opponents, mainly in the labour movement and in cultural industries (such as book publishing), were concerned that the agreement would hasten complete economic and perhaps even political integration with the USA. Some Canadian industries also feared that the lowering of trade barriers would leave them vulnerable to bigger, lower cost US competitors. Canadian food processors in particular were concerned that their competitiveness would be damaged by the high cost of domestic farm products, many of which are regulated by supply-management marketing boards.

The most tangible element of the FTA is the gradual elimination of all customs duties between the two countries over a period of 10 years. Tariffs were completely removed on the date of implementation from about 15 per cent of dutiable bilateral trade, including such items as computers, furs, animal feed, whisky and skis. Tariffs on another one-third of dutiable trade – including machinery, paint, furniture and paper products – were cut by 20 per cent on 1 January 1989, and will be eliminated completely by January 1993. On the remaining 50 per cent of trade, tariffs are being reduced by 10 per cent a year until 1 January 1998. Products in this category include such heavily protected items as clothing and textiles, domestic appliances and processed foods.

Shortly after the agreement was implemented, the US and Canadian governments agreed to accelerate the elimination of duties on a number of items. In order to qualify for duty-free treatment, at least half the direct production costs of the final products must be incurred in Canada or the USA.

Other key provisions of the FTA include:

- A new system for reviewing the application of anti-dumping and countervailing duty laws. Many complaints can now be referred to five-member, binational dispute settlement panels rather than the domestic courts. Panel decisions are binding and must be taken within strict time limits.
- A 5–7 year target for developing new rules for the application in both countries of trade-distorting subsidies. US exporters in particular have registered many complaints over the years against Canadian subsidies, including regional development grants, tax breaks for advertisers in Canadian magazines and the rates charged for cutting trees in state-owned forests. The bilateral examination is expected to lay valuable groundwork for efforts by the world trading community as a whole to reduce the barriers created by subsidies.
- US bank subsidiaries are now exempted from limits on the domestic assets of foreign banks with operations in Canada.
- Over a period of four years, Ottawa will substantially relax the threshold for official review of US investments in Canada. Indirect acquisitions (a change in ownership of a Canadian company's foreign parent) will be wholly exempt from review after 1992. The asset threshold is being lifted from C$25m to C$150m. Cultural industries are excluded from these changes, and from the agreement in general.
- With limited exceptions, both countries will bar restrictions on trade in energy products with one another. This includes non-discriminatory access for US consumers to Canadian energy supplies.

Criticism of the accord has died down since the November 1988 election, and most of its opponents appear now to accept the FTA as a *fait accompli*. Antagonism resurfaces periodically, however, when a US subsidiary closes production facilities in Canada, or when a Canadian firm is forced to lay off staff in the face of strong competition from suppliers south of the border. In general, the FTA is expected to lead to the same type of integration between the US and Canadian arms of many companies as presently exists in the automobile industry. Factories in each country will be designated to supply particular products for the entire continent rather than for two separate markets. Movement in this direction has already started. Several companies operating in both countries – Celanese, RTZ and ICI for example – have bought out minority shareholders in their Canadian subsidiaries to enable them to integrate these operations more closely with US units. Procter & Gamble is among several companies moving away from the concept of 'branch plants' in Canada in favour of subsidiaries which supply the entire North American market. Over the longer term, the integration of North American business may lead to a widespread restructuring of Canadian business. The trend is expected to step up pressure on Canadian companies in many sectors to cut costs and improve productivity if they are to maintain profit levels. The business community may also become more vociferous in its opposition to recurring tax increases and is likely to demand quicker action to reduce government spending, especially on social security.

Foreign investment

The implementation of the free trade pact, coupled with receptive government policies and Canada's economic and political stability, encouraged a surge in takeovers of Canadian companies by foreigners in early 1989. A Toronto research company estimated that foreign companies spent $10bn on Canadian takeovers in the first half of 1989, more than three times the total for the whole of 1988.

Acquisitions during 1988 and 1989 included some

COUNTRY PROFILE

Political system: Canada is a federation of 10 self-governing provinces in which the federal government has authority over areas of national interest, while provincial governments have specific authority over local matters, including education, hospitals and public lands (including natural resources). The British sovereign is head of state, represented locally by an appointed governor-general who appoints a prime minister and cabinet. The governor-general is represented in each province by a lieutenant-governor.

Political parties: Ruling party: Progressive Conservative Party (held Alberta, Manitoba, Nova Scotia, Northwest Territories and Saskatchewan in mid-1989). Other parties: Liberal Party (Quebec, Newfoundland, New Brunswick, Prince Edward Island, Ontario); New Democratic Party (Yukon); Social Credit Party (British Columbia).

Language: English and French are the official languages, with French predominating in Quebec (Montreal is the second largest French-speaking city in the world).

Media

Press: Daily newspapers published in metropolitan centres circulate widely throughout the region, and French language dailies are published in seven cities. There is no national newspaper as such, but the *Globe and Mail* based in Toronto is distributed throughout the country. The *Financial Post* (daily, Toronto) and the *Financial Times* (weekly, Toronto) are read by businessmen throughout the country. Magazines: *Macleans* is published weekly and *Chatelaine* and *Saturday Night* are published monthly. In addition, most popular US publications are widely read in Canada.

Broadcasting: Programmes put out by US radio and TV stations are widely available.

Radio: There are 307 AM and 151 FM privately-owned radio stations. The CBC owns and operates 73 AM stations, with 18 affiliates, and 28 FM stations, with 29 affiliates; there are also 25 community-owned stations.

Television: The CBC English and French network consists of 22 owned and operated stations plus 37 privately-owned and affiliated stations: CTV, a privately-owned and financed English network, has 28 affiliated stations; Global, also privately-owned and financed, is a small network serving southern and eastern Ontario. TVA, an independent French network, consists of 10 stations and operates in Quebec and parts of New Brunswick. Alberta, Ontario and Quebec also have educational networks. In addition, there are 700 privately-owned and operated cablevision companies who transmit US and Canadian transmissions, some of which would be otherwise unavailable, to subscribers. The CBC launched a 24-hour TV news channel in August 1989. Pay-television, first introduced in 1983, has become increasingly popular, with over 1m households subscribing to at least one service.

Domestic economy: Strung out across six time zones and with 85 per cent of its population living within 200 miles of the US border, there is a tendency toward regional segregation, with each looking to the USA for markets and raw materials. Canada is the most dependent of the major industrialised countries on foreign direct investment, with over 23.7 per cent of total assets of non-financial industries foreign-controlled in 1986 (37.6% for the oil and gas sector, over 40% for the manufacturing sector and 31% for the mining sector). Over 20 per cent of US direct investment abroad is in Canada. Ontario has the largest manufacturing and agricultural sectors, with the nation's finance and services capital in Toronto. The motor and steel industries are also centred on southern Ontario. Quebec has a strong manufacturing base and significant hydroelectric power resources. Montreal is enjoying a revival after a decade of decline caused by fears of Quebec separatism. Alberta is the most important energy-producing province. British Columbia accounts for 50 per cent of the country's forestry sector and 25 per cent of its fishing industry, with the third largest manufacturing sector. Manitoba's agricultural sector is traditionally important, with modest growth in manufacturing. Saskatchewan's agricultural sector accounts for 25 per cent of national output and its mining sector – notably uranium – has increased. Forestry and fishing have a greater role in Newfoundland and Nova Scotia, but hopes of an offshore oil and gas bonanza have yet to materialise.

Employment: In recent years there has been a rise in employment in the services sector.

External trade: The USA accounts for 74 per cent of exports and 71 per cent of imports (April 1989).

Exports: Main exports are motor vehicles and parts (25.9% of the 1988 total), other manufactures (19.9%), other metals and minerals (13.5%), pulp and paper (10.5%), energy materials (8.6%), lumber and saw mill products (5.9%) and chemicals and fertilisers (5.1%). Main destinations are USA (74.1% of 1988 exports), Japan (6.0%), UK (2.5%) and other EC countries (8.3%).

Imports: Main imports are machinery and equipment (31.7% of 1988 total), motor vehicles and parts (26.1%), industrial materials (18.0%), consumer goods (10.7%), foodstuffs (6.3%), energy materials (1.7%) and construction materials (1.8%). Main sources are USA (69.1% of 1988 total), Japan (6.2%), UK (3.6%) and other EC countries (8.3%).

Agriculture: Agriculture, fishing and trapping account for 2.7 per cent of GDP, and forestry for another 0.7 per cent (1988). Canadian agriculture has become increasingly capital-intensive, and agricultural employment has dropped. Canada is the world's second biggest wheat exporter, with its high-quality spring wheat commanding a premium price on world markets. It is also a sizeable producer of other grains, notably barley, rapeseed and oats. Prairie farmers were hit by severe drought in 1988, and received roughly C$6bn in aid from a variety of government support programmes. About three-quarters of forest products output is exported. The east coast fishery was hard hit in 1989 by quota cuts.

Industry: Manufacturing accounts for 19.7 per cent of GDP and 19 per cent of employment and construction for another 7.1 per cent of GPD (1988). Production of primary metals and transport equipment has grown in recent years, reflecting exceptional growth in the automobile industry. 85 per cent of automobile production is exported to the USA. Motor vehicle output surged by 21 per cent in 1988 to 1.98m units. Several Far East car companies have invested in Canada in recent years. Hyundai of South Korea opened its first North American plant at Bromont, Quebec in 1988. Canada is the world's largest exporter of pulp (28 per cent of world total) and newsprint (37 per cent), with most production located in British Columbia, Ontario and Quebec. The output of the forest products industry grew by 2.5 per cent in 1988, reaching 26m tonnes of wood pulp and 10m tonnes of newsprint. The main manufacturing sectors include petroleum refining, pulp and paper mills, motor vehicle manufacturers, steel, sawmills and planing mills, the dairy products industry, motor vehicle spare parts and accessories, metal stamping and pressing industry, smelting and refining, industrial chemicals, food processing, commercial printing, communications equipment manufacturers, feed industry, plastics fabricating industry, aircraft and aircraft parts manufacturers. Capacity utilisation improved from 1985, reaching 90 per cent at the end of 1988, but some slackening was evident during 1989.

Mining: Mines, quarries and oil wells account for six per cent of GDP (1988). Canada is the world's largest exporter of nickel, zinc, uranium, potash and asbestos, and the second largest producer of molybdenum. Main metals produced include iron ore (39,840,000 tonnes in 1988), pig iron (9,643,000 tonnes), primary steel (14,699,000 tonnes), primary copper (756,000 tonnes), nickel (199,000 tonnes), primary lead (352,000 tonnes), and primary zinc (1,278,000 tonnes). Main minerals produced include asbestos (705,000 tonnes in 1988), gypsum (8,979,000 tonnes), cement (11,927,000 tonnes), lime (2,504,000 tonnes), salt (10,647,000 tonnes), potash (8,190,000 tonnes), gold (130.5m grammes) and uranium (12,325 tonnes).

Hydrocarbons: There are substantial reserves of gas, oil and coal concentrated in western Canada. Reserves of conventional crude oil were estimated at 705.5m cubic metres at the end of 1987. Natural gas reserves were estimated at 2,320bn cubic metres. Oil and gas output rose in volume terms during 1988, but a drop in world prices pushed the value of production down by 12 per cent to C$18bn. After a boom in capital spending between 1986 and 1988, investment by the oil and gas industry slowed in 1989. Several big projects, including the development of oil and gas fields off the Newfoundland and Nova Scotia coasts, were delayed by the drop in oil prices. According to the National Energy Board, Canada's ability to export conventional crude oil will be increasingly limited during the 1990s by diminishing supplies. Coal production has grown rapidly with the development of big new mines in British Columbia. Some 40 per cent of production was exported in 1988, mainly to Japan.

Energy: In addition to natural gas, oil and coal, Canada has significant undeveloped hydroelectric resources. Oil made up 41.2 per cent of Canada's energy needs in 1987, followed by natural gas (29.2%), electricity (15.6%) and coal (14%). Oil's share has steadily declined since the 1970s, standing at 43.5 per cent in 1983. Electricity is provided by utilities usually owned by provincial governments. Hydroelectricity accounts for about 60 per cent of total power generation, but the proportion is likely to rise over the next few years as several big new projects, mainly in Quebec and Manitoba, are commissioned to tap some of Canada's vast water resources. Possible exports of water to the USA from Quebec and Ontario are under consideration.

Membership of international organisations: Canada belongs to all major multilateral organisations, including GATT, UN and its agencies, IMF and World Bank.

of the biggest deals in Canadian corporate history, such as Amoco's $5.2bn purchase of Dome Petroleum, Stone Container's takeover of forest products group Consolidated Bathurst and a $2.8bn bid by Amax for nickel producer Falconbridge. With an unusually high level of foreign ownership in many key Canadian industries (tobacco, electronics, transport equipment, cement and oil and gas are some examples), takeovers and mergers involving non-Canadian companies are a sensitive political issue in Canada. Suspicion is focused especially towards US companies, which make up about four-fifths of total foreign investment in Canada. Investment Canada, the agency which monitors foreign investment, has set down conditions on several recent deals involving employment levels, transfer of technology and other measures to ensure minimum dislocation of the Canadian economy. Nonetheless, the Mulroney government prides itself on the favourable climate it has created for foreign investors since coming to office. Investment Canada has not turned down a single one of the 651 applications submitted since it replaced the more restrictive Foreign Investment Review Agency in 1985. (Six applications were withdrawn after Investment Canada raised objections to them.)

Privatisation

The federal government and several provinces are in the process of privatising some of the state-owned corporations which play a key role in many sectors of the Canadian economy. Enthusiasm for the sell offs, even among those responsible for them, was generally less than wholehearted. Most privatisations were strongly resisted by trade unions and opposition parties, and the search for suitable buyers has often been a controversial and protracted process. A myriad of enterprises on both the federal and provincial levels are likely to remain in public hands for the foreseeable future.

Among federal Crown corporations, the bulk of the sell offs have been in the transport sector. Through two public share issues, the first in October 1988 and the second in July 1989, Ottawa disposed entirely of its 100 per cent stake in Air Canada, the country's biggest airline. VIA Rail, the loss-making passenger train operator, indicated in mid-1989 that it is willing to sell off some of its routes to private enterprise. Talks also began in 1989 on putting some airports under private management. Earlier privatisations included De Havilland Aircraft of Canada, the commuter aircraft maker which was sold to Boeing Co; and Canadair, the executive jet manufacturer sold to Montreal-based Bombardier.

The biggest test of the federal government's commitment to continuing the sell off of assets is Petro-Canada, the giant integrated oil company created by the Trudeau government in the late 1970s and expanded thereafter by the purchase of several foreign-owned energy groups' Canadian assets. Petro-Canada, with assets of $8.6bn, has been viewed for the past decade as a symbol of Canadian efforts to achieve greater domestic control over the energy sector. The Mulroney government has floated the possibility of privatising it, but no firm decision had been taken by the middle of 1989.

Right-wing governments in British Colombia and Saskatchewan were the most active privatisers among the provinces. The Social Credit government in British Colombia sold off such diverse businesses as highway maintenance and nurseries. The Saskatchewan government outlined plans during 1989 to dispose of its interest in Potash Corp. of Saskatchewan, one of the world's leading potash producers, the natural gas division of Saskatchewan Power Corp. and a general insurance business.

BUSINESS GUIDE

Time: Pacific Daylight Time = GMT -8hrs; Eastern Standard Time = GMT -5hrs (daylight savings time in operation from early Apr to late Oct). There is a total of six time zones.

Entry requirements
Passport: Required by all except US citizens permanently resident in USA.
Visa: Visas are not required by citizens of France, German FR, Japan, UK and USA. For stays of longer than three months it is necessary to register with an immigration officer.
Health: No vaccination certificates required.
Currency: There are no restrictions on the import and export of currency.
Customs: Personal effects are allowed duty-free. Certain items, such as plants, meat, cereals, dairy products and live animals are subject to import licensing.
Income tax: Income tax clearance certificates are not required before leaving the country.

Business centres: Toronto (population 3,427,168); Calgary (672,326); Edmonton (785,465); Montreal (2,921,357); Ottawa (819,263); Vancouver (1,380,729).

Hotels: It is advisable to book rooms in advance. Provincial sales tax is applied to hotel bills in some provinces. For visitors travelling by car, high-standard motels are available around all major towns and cities where rates are considerably less than those charged by city-centre hotels. Most large hotels have facilities for small displays or exhibitions, and smaller rooms may be rented as sample rooms. Most hotels impose a substantial surcharge on telephone calls.

Credit cards: Credit cards are widely used.

Car hire: Car hire is widely available. Overseas driving licences may be used for the first three months of a visitor's stay (six months in British Columbia). Driving is on the right-hand side of the road.

City transport
Taxis: Good taxi services operate in all major cities; rates vary between cities.
Buses and underground: Toronto, Montreal, Vancouver and Edmonton have efficient, safe and clean underground systems. Most cities have reliable and extensive bus services.

National transport: Canada has since 1985 implemented a far-reaching deregulation of the tightly protected transport sector, including privatisation of certain companies.
Air: There are frequent and extensive air services connecting all towns and cities of importance with 60 major airports and 700 smaller ones lacking control tower facilities. Recently privatised Air Canada and Canadian Airlines International serve the main routes, and several regional carriers operate. Air travel is the most widely recommended form of travel between major cities.
Road: There are about 845,000 km of roads, about 81 per cent surfaced. Motorways connect some large industrial centres and most large cities have a motorway network.
Buses: Long-distance coach services link all major centres. They are very well air-conditioned, and it is often recommended that travellers keep a sweater handy. Inter-city buses carried more passengers in 1986 than rail and air services combined.
Rail: The rail network comprises 95,670 km of track. The government-owned Canadian National Railway and Canadian Pacific Rail are the two main railway services, but passenger services are operated by VIA Rail, a government agency. Air-conditioning, refreshment facilities and sleeping accommodation are available on long-distance passenger services. Canrail passes give unlimited travel for certain areas and routes.
Water: The St Lawrence Seaway provides deep-water passage from the Atlantic to the Great Lakes, and there are 3,017 km of canals, mainly used for leisure. There are 25 large deep-water ports and about 650 smaller ports and government wharves.

Public holidays: In addition to statutory public holidays observed throughout the country, additional holidays are observed in certain provinces.
Fixed dates: 1 Jan, 1 Jul (Canada Day), 11 Nov (Remembrance Day), 25-26 Dec.
Variable dates: Good Friday, Easter Monday, Victoria Day, Civic Holiday (all provinces except Quebec), Labour Day, Thanksgiving Day.

Working hours: Working hours vary throughout the country. Some small businesses close completely in Jul and Aug and government departments may work variable hours during the summer months.
Government and business: (Mon-Fri) 0900 – 1700.
Banking: (Mon-Thu) 1000 – 1500/1630; (Fri) 1000 – 1800. Some banks in large centres operate much longer hours, including Saturdays.
Shops: There is a five-day working week, but most retail stores in cities open on Sat and a few on Sun as well. Late shopping (to 2100) on Thu or Fri is common in large cities; in suburban shopping centres, supermarkets often stay open until 2100 or 2200 (Mon-Fri).

Telecommunications
Telephone: Dialling code for Canada: IDD access code + 1 followed by area code (604 for Vancouver, 416 for Toronto, 514 for Montreal, 613 for Ottawa) followed by subscriber's number. It is possible to determine the cost of a long-distance call for a set number of minutes by dialling O (for Operator) and asking for rate and route information. Ringing tone is a 2-second tone separated by a 4-second silence; number unobtainable is indicated by a recorded message; number engaged is indicated by rapid pips.
Fax: Fax facilities are available at all major hotels and businesses.

Postal service: Post office opening hours are generally (Mon-Fri) 0800 – 1745. Mail sent to General Delivery, Central Post Office at any Canadian city must be collected *in person* by the addressee, and proof of identity will be required. It should be marked 'hold for 15 days', which is the longest period possible, and should have the return address clearly marked. Stamps can often be purchased at hotels or from vending machines outside post offices or in shoppping precincts. Mail service can be erratic, hence the popularity of courier services, especially for businesses.

Delivery services
Courier: Widely used both within and between cities. A wide choice of carriers is available. Federal Express serves US cities from Canada.

Advertising: Advertising is available in the press, on commercial radio and TV and via direct mail. Cinemas are increasingly used.

Electricity supply: 120-240 V (mostly 120 V) 60 cycles AC, with two-pin flat-prong plug fittings (or three-pin with one round and two flat prongs) and screw-type lamp sockets.

Weights and measures: Metric system (Imperial and US systems also still in use).

Cape Verde

Aidan Meehan

The 10 islands and smaller structures which make up the Cape Verde archipelago are located some 600 km from the coast of Senegal at their nearest point. Indeed, since the African coastline formed the first point of reference for the Cape Verdes, the islands even took their identity – Les Iles de Cap-Vert in French – from the promontory bearing that name above Dakar.

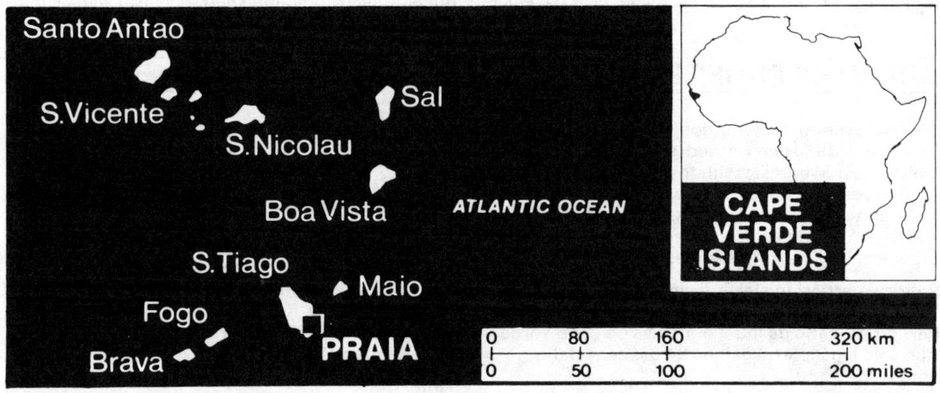

This romantic-sounding nomenclature, however, hides the reality, which is the exact antithesis of the image conjured up by ideas of the Green Cape. The Cape Verde islands are dry; almost unbelievably so. Their barren, volcanic slopes lead down into a transparent sea whose flawless, white sand bed reflects a virtually ceaseless sunshine back into desiccated air. The almost 300,000 Cape Verdians still living in their homeland (twice that many have already fled to the United States, Portugal and the African continent) nonetheless produce a kind of living from these apparently barren slopes; and what they cannot produce themselves is made up from international subsidies and remittances from the other two-thirds of Cape Verdian humanity.

This Cape Verdian version of the human condition did not augur well for a newly independent state born out of Portuguese colonialism in 1975. At the time, one of the country's senior figures was heard to remark that he doubted that 'enterprise Cape Verde' would ever be a workable phenomenon. There were few teachers, even fewer health workers and no organs of elected power at all. The economy, such as existed, was in disarray, dependent on a small fishing industry, irregular exports of a specialist cement base called pozzolana and the landing charges levied on the international airport's best customer, South African Airways, whose fleet was forbidden to overfly independent African countries en route to Europe and North America.

In the 15 years since independence, much has changed: even the climate has been kinder since 1985. Then the maize crop stood at some 2,000 tonnes; two years later it reached 21,000 tonnes even if, by the end of 1989, the result appeared to be more like half that for the current year. On the negative side, however, the tainted manna represented by South African Airways has been withdrawn, as the carrier was forced in 1988 to cease its direct flights to the United States. The result is a drastic cut in income from the international airport on the island of Sal.

Politically, the unwholesome quarrel with Guinea-Bissau after the 1980 coup d'etat in that country has been almost forgotten and relations have returned to normal state-to-state conditions. The central power of the party has also been called into question. The PAICV, heir to the PAIGC formed in the forest guerrilla struggle in Guinea-Bissau, held its Third Party Congress in November 1988, an event which opened up the ranks to younger party militants at the expense of the old guard from the maquis. Wrongly interpreted as a shift to the moderates by some, the true significance of the rearrangement was that it allowed both a discussion of fresh policies and a renewal of the party's ranks. President Aristides Pereira, despite health problems early in 1989, remains the undisputed and popular leader both of the party and the country and has showed remarkable skill in maintaining the continuity of the party's democratic socialist philosophy while encouraging economic and political experiment.

Indeed, socialist Cape Verde appears to be eager to join the ranks of the lucrative off-shore banking centres. A semi-secret mission was sent to Europe and North America in mid-1988 to seek reactions to such a scheme both from businessmen and from aid donors who might have been shocked by the idea. The mission returned with the news that the proposal had been more than well received, leaving the way open for the party congress to approve proposed legislation to turn the idea into reality. At the same time, the congress also voted to widen opportunities for foreign investors in the country, most notably by voting to introduce legislation to end the state's monopoly over virtually all forms of economic life including ownership of land. The national bank is the only institution to see its state-owned status preserved.

Elsewhere, a proposal has even been made to privatise parts of the health service – uniquely efficient in Africa – and, though the proposal was only briefly aired, the fact that it was discussed at all gives an indication of the shift in attitudes.

These events do not flow from what might be called an abandonment of the revolution. On the contrary. The revolution's first objective is to ensure the survival of the Cape Verdian people and their independence as a sovereign state. With few natural resources, the country has depended to a greater or lesser extent since independence on the remittances sent home by the overseas Cape Verdian communities. In recent years, these remittances have covered almost 95 per cent of the commercial deficit which in 1986 amounted to $106m, or over 30 times the value of exports. Clearly this cannot be allowed to go on forever, even if that were to be possible.

Thus, the search has been going on for some time to seek alternative sources of revenue. A ship repair yard, CABNAVE, on the island of Sao Vicente has only been modestly successful in attracting foreign vessels; and light industry, principally in clothing, can only be a small part of the solution. If the islands want investment, then they will have to offer a realistic investment climate.

Tourism

A different kind of climate – the meteorological one – may provide part of the answer. After years of resisting the temptation to allow the growth of a tourist industry, the government has now relented and, spurred on by the decisions of the party congress, is to exploit the natural beauty and limitless sunshine of some of the islands, pricipally Sal which already has its international airport which, with the departure of the South Africans, is sadly under-used. Ten hotels and tourist installations are due to be built on the island, by Swedish, Finnish and West German interests. These will provide jobs and foreign currency. The hope is that, by using the sparsely populated Sal the cultural effect of the tourist invasion will be less drastic and the transport costs of importing all the trimmings required by tourists will be reduced to a minimum by using the existing airport facilities.

These new developments do not, however, signal the abandonment of the popular socialism which has been the Cape Verde Republic's hallmark for a decade and a half. Self help is still very much the order of the day and the extension of social schemes in health and education remain a government priority. The commitment to agriculture, which will always provide employment for the vast majority of islanders, is as strong as ever and the government is still on target to achieve the afforestation programme of 115,000 hectares planted with trees by the end of the century – from 3,000 hectares in 1975. In 1988 alone 8,000 hectares were planted.

The Republic of Cape Verde has distinguished itself

KEY FACTS

Official title: Republic of Cape Verde

Head of state: President Aristides Maria Pereira

Ruling party: Partido Africano da Independência de Cabo Verde (PAICV)

Capital: Praia

Official Languages: Portuguese and Creole (national language)

Currency: Cape Verde Escudo (CV Esc) = 100 centavos

Exchange rate: CV Esc77.2 per $ (Nov 1989)

Area: 4,033 sq km (ten islands and ten islets)

Population: 357,000 (1988)

GNP per capita: $500 (1987)

GNP real growth: 6.0%* (1986)

Trade balance: –$124m* (1988)

Foreign debt: $131.2m (1987)

* estimated figure

in many ways since independence, not least by its mere survival, but perhaps most significantly by its ability to attract and keep foreign aid without sacrificing sovereignty. The country has benefited from its much-vaunted strategic position, but it has also had the advantage of a particularly astute and rational leadership. President Pereira is regarded as the elder statesman of the Portuguese-speaking African countries and his small nation has hosted a number of key meetings, most notably between the Angolan and South African belligerents. The head of state is received as warmly in Washington as he is in the Kremlin and Beijing and, in spite of the country's often restated commitment to socialism and the radical group of African countries, Cape Verde has been able to play the role of go-between in a number of regional and continental conflicts.

The result is a bitterly poor country which survives without reducing itself to beggar status and whose one-party government remains apparently invincibly popular.

COUNTRY PROFILE

Political system: Constitution adopted 5 September 1980 and revised 12 February 1981. Executive power rests with the president, elected for five years by the 86-member People's National Assembly, itself elected for five years by adult suffrage. Formerly a Portuguese overseas province, it gained independence in 1975.

Political parties: In elections to the national Parliament, held in December 1985, the ruling Partido Africano da Independência de Cabo Verde (the sole legal party) gained 94 per cent of the vote.

Population: Besides the 357,000 Cape Verdeans resident on the islands, there are some 300,000 living abroad, many in the United States.

Main cities/towns: Praia, on Santiago Island (population 25,000), is the national capital; Mindelo, on São Vicente (30,000), is the commercial centre.

Language: Portuguese is the official language, while most inhabitants speak Creole, the mother tongue.

Media
Press: *Voz di Povo*, a bi-weekly government publication. The state and the Catholic Church also publish a number of periodicals.
Broadcasting: Programmes in Portuguese and Creole by Emissoria Oficial da Republica da Cabo Verde and Voz da São Vicente. An experimental television service, TVEC, broadcasts in the evening.

External trade: Large trade deficit partly offset by income from ship servicing, workers' remittances from abroad and substantial aid flows.
Exports: Mainly frozen and fresh fish, garments, bananas, livestock produce, salt and entrepôt trade. Main destinations: Italy, Portugal, Spain, Netherlands, France, Angola and Argentina.
Imports: Mainly foodstuffs such as rice, wheat and maize. Also small quantities of fuel, machinery and textiles. Main sources: Portugal (35% of total value), Netherlands (14-15%), Spain (5-6%).

Agriculture: Sector forms backbone of economy, with over 70 per cent of the population engaged in subsistence farming. Approximately 10 per cent of total area is cultivable. Recurrent drought, soil erosion, disease and a weak infrastructure have reduced agricultural production considerably. About 85 per cent of food requirements are currently met by imports, largely in the form of food aid. Main crops include maize, beans, bananas, sweet potatoes, sugar-cane, coffee and ground nuts. Small quantities of vegetables and livestock produce are exported. Fishing (shellfish and tunny) is of growing importance, supporting some 20,000 people, and currently accounts for over 60 per cent of export revenues.

Industry: Little industry except for ship repair and fuelling, fish processing and canning, cement production, flour milling, soft drinks and rum plants, cigar manufacture and garment making.

Mining: Activity largely confined to exploitation of pozzolana (volcanic derivative) – temporarily suspended – and production of salt on Sal by evaporation method.

Energy: Dependence on imported petroleum and products. Hydro-electricity potential.

Membership of international organisations: African Development Bank, African States Associated with the EC (Lomé Convention), Arab Bank for Economic Development in Africa (recipient), Economic Commission for Africa, Economic Community of West African States, GATT, IDA, IFAD, ILO, IMF, IMO, Inter-state Commission to Combat Drought in the Sahel, Non-Aligned Movement, UN, World Bank.

Time: GMT -1.

Climate: Hot with very little rainfall. Temperatures range from around 20°C at night to 32°C during the day. Rain most likely from Aug to Sep.

Entry requirements
Passport: Required by all.
Visa: Required by all except nationals of Cape Verde Islands, Portugal and Guinea-Bissau.
Currency: No restriction on import of foreign or local currency, but amounts must be declared. Export limited to amounts declared on arrival or CV Esc6,000 or equivalent.

Health precautions
Mandatory: Yellow fever and cholera vaccination certificates if arriving from an infected area.
Advisable: Cholera, typhoid and polio vaccinations. Malaria risk exists in some areas. Water precautions should be taken.

Representation overseas: Algiers, Bissau, Conakry, Dakar, The Hague, Lisbon, Luanda, Moscow, Paris, Rome, Washington.

Air access
National airline: Transportes Aéreos de Cabo Verde (TACV).
Other airlines: Aeroflot, Cubana, Gambia Air Shuttle, South African Airways, TAAG-Angola, TAP-Air Portugal, Varig.
International airport: Ilha do Sal-Espargos (Code: SID).

Surface access
Main ports: Mindelo (São Vicente), Praia (Santiago).

Hotels: Accommodation is very limited and restricted to only four islands: Fogo; Sal; Santiago; São Vicente.

Electricity supply: 220 V AC.

City transport
Taxis: Available on main islands.

National transport
Air: Transportes Aéreos de Cabo Verde (TACV) flies to all islands except Brava.
Road: Buses available on main islands.
Water: Boats ply between the islands.

Public holidays
Fixed dates: 1 Jan, 20 Jan (National Hero's Day), 8 Mar (Women's Day), 1 May, 1 Jun (Youth Day), 12 Sep (Independence), 25 Dec.

Working hours
Banking: (Mon-Fri) 0800 – 1230, 1430 – 1800.

Telecommunications
Telephone: Dialling code for Cape Verde Islands: IDD access code + 238 followed by subscriber's number. There are no area codes.
Telex: International telex code: CV.

Banking
Central bank: Banco de Cabo Verde. *Main centre:* Praia. *Foreign branch:* Banco Nacional Ultramarino.

Representation in capital: Brazil, Cuba, France, Portugal, Senegal, USA, USSR, (Diplomatic representation for most countries is in Guinea-Bissau or Senegal.)

Hotels: Cape Verde Islands
5 de Julho, Mindelo, São Vicente (tel: 313750).

Morabeza, Santa Maria, P.O. Box: 33, Sal.

Porto Grande, Mindelo, São Vicente, P.O. Box: 103 (tel: 313113, 313232).

Praia-Mar, Praia, Santiago, P.O. Box: 75 (tel: 614153, 614355; tx: 79).

São Filipe, Fogo.

Other useful addresses
Associacao Commercial e Agricola de Sotavento de Cabo Verde, PO Box 78, Praia, Santiago (tel: 612991).

Associacão Comercial Barlavento, PO Box 62, Mindelo (tel: 313281).

Banco de Cabo Verde, Av. Amilcar Cabral, Cidade da Praia (tel: 613472, 613153, 613352, 613264; tx: 99350).

CABETUR (tourist agency), PO Box 318, Praia, Santiago (tel: 613008, 613075; tx: 6032).

Direcção-Geral das Alfandegas (customs body), PO Box 98, Praia, Santiago (tel: 613835, 613026; tx: 6058).

Direcção-Geral do Comércio (trade body), PO Box 105, Praia, Santiago (tel: 614159; tx 6041).

Direcção-Geral do Turismo, PO Box 211, Praia, Santiago (tel: 614473; tx: 6041).

Direcção-Nacional das Pescas (national fisheries authority), Praia, Santiago.

Emissora Oficial da República de Cabo Verde, PO Box 26, Praia, Santiago (tel: 613729).

EMPA (state trading organisation), PO Box 107, Praia, Santiago (tel: 249; tx: 54 EMPA CV).

Empresa Caboverdeana das Infraestructuras de Pescas (fisheries infrastructure corporation), PO Box 59, Mindelo, São Vicente (tel: 2434; tx: 99384).

Empresa Nacional de Aeropuertos e Seguranca Aerea, Aeroporto Amilcar Cabral, Ilha do Sal (tel: 411394; tx: 4036 ASA).

Empresa Nacional de Combustivels (national combustibles corporation), PO Box 1, Mindelo, São Vicente.

Ministeria da Industria e Energia – Unidade de Promocao Industrial, PO Box 145, Praia, Santiago (tel: 613949, 614444; tx: 6035).

National Statistical Service, AV 12 Septembre, Praia, Santiago.

National Tourist Office, PO Box 105, Praia, Santiago.

TACV Airlines, 11-13 Rua Gerra Mendes, Praia, Santiago.

Cayman Islands

Tony Thorndike

In this the most prosperous of all Caribbean territories, economic expansion is set to continue, driven by exceptionally buoyant tourism and offshore financial sectors and a remarkable property boom. Although there was a general election, political affairs seem of little consequence as the islanders once again set their face against constitutional advance. But there is evidence of a creeping 'colour bar', with harassment of black immigrant workers contrasting with the comparative welcome given to white North Americans.

The results of the 16 November 1988 election reflected continuing boom conditions with the return of the more conservative members of the colonial legislature. While to the visitor the capital – George Town, on the main island of Grand Cayman – seems a little sleepy, to Caymanians it and surrounding areas have changed virtually out of all recognition as development has surged ahead.

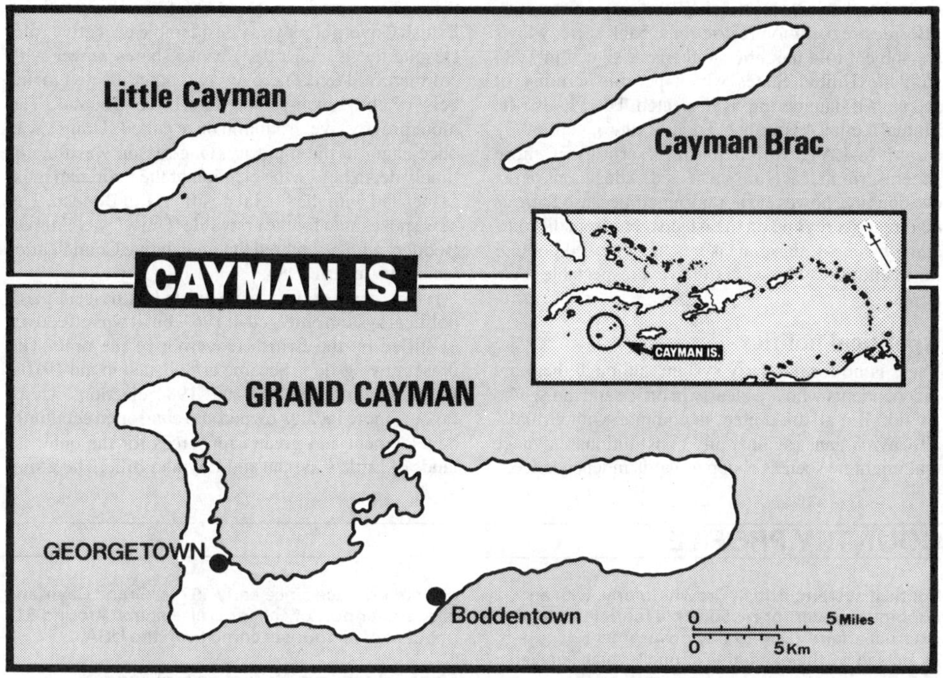

Boom

The total GDP in 1987 was calculated to be CI\$329m (US\$396m, the US dollar being worth only 84 Caymanian cents), giving a per capita income of US\$17,400, the highest in the Caribbean (excluding Bermuda). Overall growth, 10 per cent annually for the past three years, shows no sign of slowdown. Inflation, however, is a concern at over 6 per cent. Some attribute this to labour scarcity and to consequent high wages, but a more realistic explanation is the rapid rise in land prices due to the prolonged property boom. Land sales in 1988 totalled CI\$131m (US\$158m) compared to CI\$93.4m worth in 1987, representing nearly 7,000 transactions. There is also a very heavy reliance on imports and the trade deficit in 1988 reached CI\$160m (US\$193m). Exports were a derisory CI\$1.8m, down from CI\$2.2m in 1987, being mostly second-hand cars for Miami car dealers. All but CI\$8m of the deficit was covered by service sector earnings.

Government revenue (totalling CI\$79.6m in 1988) derives mainly from company and bank fees and licences (CI\$6.2m), stamp duty from land sales (CI\$12.2m) and import duties. Expenditure in 1987 totalled CI\$71.8m, of which CI\$54.8m was for recurrent purposes. Development plans include airport upgrading and improved water desalination plants and better sewerage systems. Public debt stood at US\$24.8m at the end of 1988, but reserves and surplus funds totalled US\$21.6m.

There is little economic activity outside the service sectors and construction. The once-famous turtle industry is now limited, mainly due to strict import regulations by the US Department of Agriculture. The tourist attractions of the Caymans, best epitomised by Grand Cayman's Seven Mile Beach, are well-known. However, expansion in 1988 was limited, there being only a 4.6 per cent increase in stopover visitors from the 1987 figure, 209,044 to 218,709. Shortage of rooms and the ravages of Hurricane Gilbert in September 1988 were the main reasons. The completion in 1990 of the Caymans' largest hotel, with 360 rooms, will ease the problem. Cruise ship passenger visits, however, rose by 16 per cent from 271,748 to 315,585. High prices do not seem to deter tourists: they merely add to the desirable image of exclusivity and high finance.

World financial centre

The Caymanian financial industry has assumed world status. Some 530 banks, 18,000 companies and 360 insurance units, the latter virtually all 'captives', established by multinational companies to handle their insurance needs and increasingly those of others, are registered. Offshore banking is the sector's speciality: in 1984 banking business rivalled that of the Bahamas, but by 1989 it was over 50 per cent higher. Unusually for the Caribbean, less than 30 per cent of banks are US-owned and there is a strong Asian presence.

Although much is brass plate, with only 72 banks having a physical presence, 42 of the 50 largest banks in the Western world have branches in the Caymans. Also, 250 of the banks are authorised trusts in addition to the many thousands of other trusts thought to be operating, providing a wide variety of tax and estate planning in a very confidential manner. Bank establishment is relatively cheap and there are no reserve requirements for those institutions with 'B' licences, whose operations are restricted to foreign business only. Eurocurrency dealing is the major activity. While some business is tax driven, the ease of restrictions provides the main attraction. Conscious of the possibilities for abuse, there is strict vetting of applications and impressive self regulation by the banks themselves. Customers are also screened by banks. Total foreign bank assets were over US\$260bn.

Insurance has become a strong growth area after the passing of the 1979 Insurance Act, and the Caymans are the most popular domicile after Bermuda. Substantial business is expected from Barbados when the latter's tax treaty with the United States (which gives tax exemption on premium income) ends in December 1989. Ship registration is another potential growth area.

The key to this expansion is a mixture of stability and favourable legislation and the misfortunes of other Caribbean centres. Absence from taxation goes back two centuries, but other legislation is more recent, dating from the 1960s, culminating in the Confidential Relationships Preservation Act of 1976. Very considerable investment has been made in electronic communication facilities and the government directly sponsored Cayman Airways to ensure schedules to and from North America convenient to business people. The range of professional expertise available is impressive. The original spurt was provided by some institutions departing the Bahamas when that country became independent in 1973.

KEY FACTS

Official title: Cayman Islands

Head of state: Queen Elizabeth II

Head of government: Governor George P. Lloyd

Capital: George Town (Grand Cayman)

Official Languages: English

Currency: Cayman Islands Dollar (CI\$) = 100 cents

Exchange rate: CI\$0.83 per \$ (Nov 1989)

Area: 260 sq km

Population: 24,900 (1989)

GDP per capita: \$18,239 (1987)

GDP real growth: 10% (1988)

Inflation: 5.2% (1988)

Trade balance: –\$9.7m (1987)

Visitor numbers: 534,294 (1988)

While their move was premature at the time, when the Bahamas offshore industry began to be infiltrated with illegal, mainly drug-related funds, and political uncertainty and corruption spread, the slow stream became a flood. Recently a large amount of business has moved from Panama. But not all applications succeeded because of strict registration requirements.

Besides registration restrictions, bank secrecy laws are subject to a narcotics agreement signed in 1984 with the United States which permits scrutiny of accounts if 'laundering' is suspected. There is also the Mutual Legal Assistance Treaty (1986), commonly called M-LAT, which permits scrutiny if more general criminal acts are suspected. Allegations of tax evasion are, however, not accepted because there is no direct tax regime on the islands. Far from discouraging business, these measures have proved popular for institutions who wish to have a respectable location.

Apolitical politics

There is no formal party system, although there are Independents who are clearly in office and those who are not. But all are united, in common with virtually all Caymanians, against any constitutional advance that might be seen as a step towards independence.

Seven sitting members were returned in the election together with five newcomers. Most of the former were associated with the more conservative group, once called 'Dignity', and Benson Ebanks resumed his position as First Elected Member of the Executive Council: in the restricted Caymanian constitution there is no provision for a chief minister, and the British Governor plays a considerable executive role. Despite the formal title, Ebanks shares power with Norman Bodden. There are two other elected members of the seven-person Executive Council. The more progressive grouping, once called 'Unity', was once again in the minority. One reason was the untimely death before the election of the politician most associated with the liberal position, Jim Bodden. The new spokesman for the erstwhile 'Unity' was Truman Bodden, who re-entered the Legislative Council after losing his seat in 1984.

In fact, party politics have existed. A nascent party political system emerged in 1961, but it was effectively stifled by the British Governor of the time. The present two groups became established in the 1970s, Unity winning the 1980 and 1984 elections. Unity favours more locally-owned and controlled economic development and greater resources for the outer islands of Little Cayman and Cayman Brac. The group argues that Caymanians have become complacent, that Britain does not have much to offer and that constitutional advance (though not necessarily leading to independence) is both feasible and necessary. But these relatively modest policies seem not to attract much support, rather the business of business appears to be more important.

The immigrant question

The boom has brought severe labour shortages and therefore increased immigration. In 1989 it was estimated that of the 24,900 population only 63 per cent were Caymanian, compared to 79 per cent in 1980. But there is subtle colour discrimination. Most Caymanians are mixed race/white, and black immigrants – mainly Jamaican – appear less favoured than whites, most of whom are North American. Indeed, there is increasing reliance on Honduran and other Central American labour on strict project-based permits. Immigration laws are constantly being amended, particularly to prevent the grant of residency status. Residency rules for electoral purposes were changed in 1987 to favour islanders; but they also had the effect of excluding *emigré* Caymanians from what limited political process there is.

COUNTRY PROFILE

Political system: British Crown Colony, with an appointed governor responsible for defence and external affairs. Day-to-day government exercised by the executive council which consists of three official members appointed by the governor and four of the 12 elected members of the Legislative Assembly. Parliamentary term: four years.

Main cities/towns: George Town (on Grand Cayman; other islands are Cayman Brac and Little Cayman).

Media
Press: Only daily newspaper is *The Caymanian Compass*. Monthly magazines – *Financial and Business Journal, Newstar.*
Broadcasting: Consists of the government radio station, Radio Cayman, and education/cultural station ICCI-FM. There is no television service.

External trade: Large deficit usually covered by invisible earnings and capital inflows from tourism and financial services.
Exports: Shrimps, honey, jewellery, second-hand cars. Main destinations: USA, Costa Rica, UK, Canada.
Imports: Foodstuffs, petroleum and derivatives, machinery and transport equipment, tourist related goods. Main sources: USA (70-75 per cent of total value), Japan, UK, Jamaica, Switzerland.

Agriculture: Despite efforts to promote agricultural production, the islands remain heavily dependent upon imported foods (mainly from Florida, Costa Rica and Jamaica). There is very little cultivable soil, except on Cayman Brac. Production is centred on cattle raising, poultry and pig farming, vegetable and tree crops such as mango, citrus, bananas, avocadoes, coconuts. Fishing is a traditional operation, though commercial turtle farming has virtually ceased. There has, however, been a small expansion in lobster and shrimp fishing.

Industry: Sector makes only a very small contribution to the economy, diversification being hampered by factors such as high labour costs and full employment. Activity centred on building materials (concrete blocks and tiles), tourist related industries such as jewellery, printing and food processing.

Tourism: Most important economic sector, contributing 50 per cent to GNP and employing over 30 per cent of workforce. Industry has grown at a steady pace since early 1970s. Grand Cayman is a regular port of call for cruise ships. Around 81 per cent of all tourists come from the USA.

Financial services: No foreign exchange regulations, no direct tax, flexible company law, good communications, political stability and stringent secrecy laws have established the Caymans as one of the world's leading banking and finance centres. Known as the 'Switzerland of the Caribbean', the Caymans house over 17,000 registered companies, 500 offshore banks (most of which are engaged in the Eurocurrency market business) and over 350 captive insurance operators. Licence fees from banks and insurance together with company fees contributed over $14m to government revenue in 1987.

Membership of international organisations: State associated with the EC.

Time: GMT -5.

Climate: Prevailing NE trade winds moderate otherwise hot climate. Average temperatures between 24°C and 29°C. Rainy season from May to Oct but annual rainfall is low.

Entry requirements
Passport: Required by all except citizens of UK and colonies, nationals of USA and Canada (for stays up to six months where proof of identity and assured onward passage held).
Visa: Not required by British, Commonwealth citizens, nationals of EEC countries, Japan, USA, Israel, South Africa and certain other countries. However, visitors should note that US visas are required for travel via the USA.
Currency: No restriction on import of foreign or local currency. Canadian, British and US as well as most other currencies are readily acceptable.

Health precautions
Mandatory: None.

Air access
National airline: Cayman Airways.
Other airlines: Air Jamaica, American Airlines, Northwest.
Tax: CI$6 or US$7.50 payable on departure.
International airport: Owen Roberts International (GCM), 3.5km from the centre of George Town, duty-free shop, bar, restaurant, buffet, money exchange, shops.

Surface access
Main port: George Town.

Hotels: Wide choice available throughout the islands, mainly at the beach. 10-15 per cent service charge is added to bill in lieu of tipping and a government tax of six per cent is added to the room rates.

Car hire: International licence recommended. Local permit obtainable on production of national licence. Traffic drives on the left.

City transport
Taxis: Readily available at hotels. Fares are based on a fixed place-to place tariff. Tipping optional.

National transport
Air: Cayman Airways operates domestic services (Grand Cayman to Little Cayman and Cayman Brac). Executive Air Services offer charter service.
Road: Over 175km of road, mostly surfaced. Speed limits of 50, 40, 30, 25 mph strictly enforced.
Buses: Regular bus service between West Bay and George Town.

Public holidays: 1 Jan, Ash Wednesday, Good Friday, Easter Monday, third Mon in May (Discovery Day), Queen's Birthday (Jun), first Mon in Jul (Constitution Day), Mon after Remembrance Sun (Nov), 25 Dec, 26 Dec.

Working hours
Government and business: 0830 – 1700. (Exceptions: Sat, Sun closed).
Banking: 0900 – 1430. (Exceptions: Fri 0900 – 1300, 1430 – 1630; Sat, Sun closed).

Telecommunications: Modern automatic telephone system, telegraph and telex services operated by Cable and Wireless (West Indies) Ltd, George Town. Dialling code for Cayman Islands: IDD access code + 1-809-94 followed by subscriber's (five-digit) number. There are no area codes.

Electricity supply: 110 V AC.

Hotels: Grand Cayman
Ambassadors Inn, P.O. Box: 1789 (tel: 949-7577; tx: 4432; fax: 9497050).

Beach Club, West Bay Beach, P.O. Box: 903 (tel: 949-2023; tx: 4253).

Brac Reef Beach, Cayman Brac BWI, P.O. Box: 235 (tel: 948 7323).

BUSINESS GUIDE

Caribbean Club, West Bay Beach, P.O. Box: 504 (tel: 947-4099; tx: 4346).

Caribe Inn, North Side, P.O. Box: 1410 (tel: 9477636).

Cayman Kai, North Side, P.O. Box: 1112 (tel: 947-9556; tx: 384953).

Grand Pavilion, P.O. Box: 69 (tel: 947-4666; tx: 4435).

Holiday Inn, Seven Mile Beach, P.O. Box: 904 (tel: 74444; tx: 4438; fax: 74213).

Hyatt Regency, P.O. Box: 1698 (tel: 949-1234; tx: 4334; fax: 8528).

Radisson (opening June 1990), Seven Mile Beach, on the Beach.

Ramada Treasure Island, P.O. Box: 1817 (tel: 97-777).

Spanish Cove, P.O. Box: 1014 (tel: 949-3765; tx: 3094262).

Sunset House, P.O. Box: 479 (tel: 949-7111; tx: 4367).

Car hire: Grand Cayman

Ace Hertz: PO Box 53, (tel: 92280, 92932; tx: 4277).

Budget: Walkers Road (tel: 95605; tx: 309, 4249).

Cico-Avis: PO Box 400 (tel: 92468; tx: 4243).

Dollar Car Rental: PO Box 666, Palm Grove Inn, Airport (tel: 92981; tx: 4431).

Europcar National: PO Box 1105 (tel: 94790).

Holiday Payless Rent-a-Car: The Island Center, Walkers Road, Airport (tel: 97074; tx: 4284).

Business advice and information

Cayman Airways, PO Box 1101, George Town, Grand Cayman (tel: 92311, 98272).

Cayman Islands Department of Tourism, PO Box 67, George Town, Grand Cayman (tel: 97999, 98487).

Government Information Service, Tower Building, Grand Cayman (tel: 97999).

Cayman Islands Chamber of Commerce, PO Box 1000, George Town, Grand Cayman (tel: 94746).

Cable and Wireless (West Indies) Ltd, PO Box 293, George Town (tel: 97800).

Radio Cayman, PO Box 1110, George Town, Grand Cayman (tel: 97799).

Banking

The Bank of Nova Scotia, PO Box 689, Grand Cayman (tel: 97666; tx: 4330).

Barclays Bank International, PO Box 68, Grand Cayman (tel: 97300; tx: 4219).

Canadian Imperial Bank of Commerce, PO Box 695, Grand Cayman (tel: 98666; tx: 4222).

Cayman National Bank and Trust Co, PO Box 1097, Grand Cayman (tel: 94655; tx: 4313).

CIBC Bank and Trust Co (Cayman), PO Box 695 (tel: 98666; tx: 4222 CP).

First Home Banking, PO Box 914, Grand Cayman (tel: 95774; tx: 4265).

The Royal Bank of Canada, PO Box 245, Grand Cayman (tel: 94600; tx: 4244).

Washington International Bank and Trust, PO Box 609, Grand Cayman (tel: 98144; tx: 4214).

Central African Republic

Diana Hubbard

In a tough economic climate, the government of President André Kolingba made slow but steady headway with its programme of economic reform and adjustment. The second tranche of a World Bank structural adjustment loan (SAL) was made available in December 1988 and the third year programme of an IMF Structural Adjustment Facility (SAF) was scheduled for agreement in September 1989. More than one half of the proposed 2,000 redundancies among civil servants were agreed by mid-1988 but the government also had to deal with vociferous cash-strapped students, protesting about delays in grant payments, and contain a potentially damaging toxic waste scandal.

With ex-Emperor Bokassa safely behind bars and serving a life sentence with hard labour, political dissension was more evident externally than at home, particularly on the northern border with Sudan. Firm world diamond prices helped to offset the depressed picture for coffee and cotton and efforts were made to improve the collection of taxes and duties. The road to economic recovery is still a long one but external donors are sufficiently impressed to maintain aid flows in the interim.

Toxic waste scandal

Although the foreign press quickly packed their bags and departed in February 1988 when the long-running trial of Bokassa ended with Kolingba commuting a death sentence to life imprisonment, the Central African Republic (CAR) was back in the limelight just six months later when a series of scandals broke involving the dumping of toxic waste in several African countries. One year and a presidential inquiry later, the real story in CAR was still not clear. Versions range from an allegation that 70,000 tonnes of waste was transported from the port of Pointe-Noire in Congo over a considerable distance to CAR, where it was dumped, to the presidential commision's findings that 'no toxic waste was imported into or buried in the CAR'. Whatever the truth, and most observers now give credence to an account of two small river vessels loaded with waste travelling through Congo and on into CAR, the man at the centre of the scandal, National Assembly President Maurice Metho was voted out of his job in March 1989 and replaced by a long-serving deputy Michel Docko.

Diplomatic links with Israel

The beginning of 1989 also saw a decision by Kolingba to follow the trend set by several fellow conservative francophone states since 1982 and to resume diplomatic relations (broken off in 1973) with Israel. In fact, commercial links with Israel were never really disrupted because of the large number of Israelis involved in the export of diamonds from CAR and, again in common with many other African countries, unofficial links already existed through Israeli security men within the presidency. The action did little, however, to ease already strained relations with CAR's northern neighbour Sudan. With law and order in southern Sudan virtually non-existent and a rapprochement between the government in Khartoum and Libya, feathers were further ruffled in May when the presidential plane carrying Kolingba to his first official visit to Israel was turned back when it started to fly over Sudan. The incident resulted in the official rupture of relations between the two countries and an embarrassed Kolingba had to wait until July to make a rescheduled – but successful – visit to Jerusalem. By that time, the Sadiq el Mahdi government in Sudan had been ousted by a group of army officers who seemed set on more pragmatic government than their predecessors, and on more peaceful relations with CAR.

The Israeli connection could also prove of assistance to Kolingba in the thorny issue of diamond smuggling and tax evasion. Buoyant world diamond prices – up by a massive 27 per cent in 1988 – have maintained CAR's diamonds as its leading export, with revenue of CFA14.1bn in 1988. This figure, however, is for official earnings on estimated exports of around 340,000 carats. It is also estimated that a staggering 300,000 carats were smuggled out of the country. Consequently, aid donors – and France in particular – are now making disbursements dependent on improved performance in tax collection. Towards the end of 1989, it looked as if the lobby to cut export tax on diamonds from 12 per cent to 6 per cent, would win; its argument was that this cut in export tax would boost total tax revenue as smuggling would become far less commercially attractive.

KEY FACTS

Official title: Central African Republic

Head of state: President General André Kolingba

Capital: Bangui

Official Languages: French (Sango widely spoken)

Currency: Franc CFA = 100 centimes

Exchange rate: Franc CFA316 per $ (Nov 1989)

Area: 622,984 sq km

Population: 3.0m (1988)

GNP per capita: $330 (1987)

GNP real growth: 0.4%* (1987)

Inflation: –4.0% (1988)

Trade balance: –$70.6m (1987)

Foreign debt: $435m (1987)

* estimated figure

Cotton and coffee: low world prices

The main aid donors (France, the World Bank and the IMF) came closer during 1989 to harmonising their reform programmes for CAR. A particular point of dissent had been policy on cotton growing. The 1988–89 harvest was exceptionally good and at 11,500 tonnes was more than 50 per cent up on the 1987–88 figure of 7,500 tonnes. Ironically, however, while cotton exports boost the trade balance, the large subsidy which the government makes on current producer prices means that CAR's cotton is not really competitive on the world market. The World Bank, in particular, has advocated a cut in the land surface given over to cotton and diversification into other crops, a policy originally not much favoured by the French.

While there are no moves to date to treat coffee – the country's third export after diamonds and timber – in a similar way, any increases in output (an estimated 12,000 tonnes in 1988–89) were more than wiped out by low world prices.

There is, however, more optimism among donors about long-term prospects for CAR's considerable forest resources. Production has been neglected in the late 1980s, with output of logs down to 154.4m cu metres in 1987 from 244.4m in 1984, although log exports have remained steadier and totalled 52 cu metres in 1987. In 1989, work started on the preparation of a comprehensive forestry policy, with the emphasis on environmental considerations and the long-term protection of resources. Donors would like to see a selective policy, under which the most valuable trees would be identified and then cut. CAR's land-locked position weighs heavily on transport costs for timber and, consequently, value added is to be encouraged via the setting up of veneer manufacturing.

Despite the government's commitments to reform, revenue collection in 1988, at CFA40bn, was still 11 per cent under the target of CFA45bn. Savings on salaries in 1989 through the programme of redundancies should help to check spending for the year and the disbursement of the third year tranche of the SDR15.2m SAF agreed in December 1988 with the IMF was expected in the last quarter, together with the second tranche of the $40m World Bank SAL agreed in June 1988. This would then lead to a new rescheduling of external debt, with the Paris Club on concessional Toronto terms. Relief of around CFA15bn was already obtained from the Paris Club in December 1988. Following on this, French President Mitterrand announced that France would wipe out the debt of the poorest countries, including CAR, taking effect from 1 January 1990, reducing CAR's total external debt to around CFA130bn, representing a debt service ratio of around 35 per cent of export earnings.

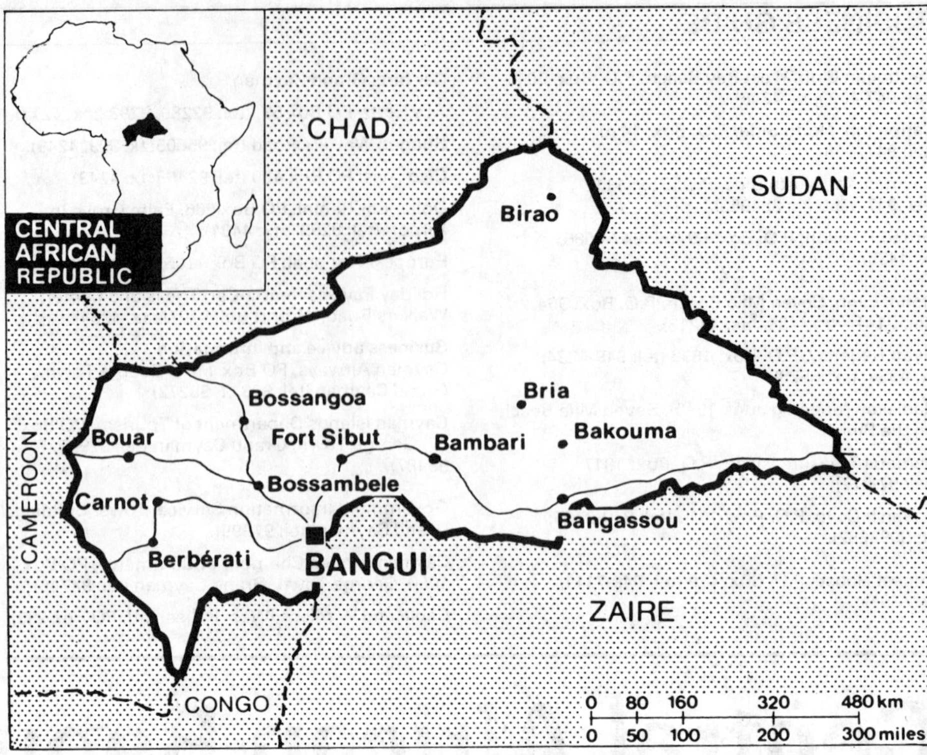

Meanwhile, the government earned praise from donors over the steady pace of privatisation. In 1989, the state insurance company SIRIRI was successfully sold off to a group of private nationals and more sales of viable public sector enterprises are expected. The emphasis in project lending has been very much on energy. The IDA announced a $18m loan in January 1989 for a $75.7m project to increase energy production and improve distribution throughout the country. This includes construction of a new hydro-electric dam at M'Bali to the north west of Bangui, being built by the Chinese and scheduled for completion in October 1990. Smaller projects are planned near Bambari on the M'Boumou river, at M'Baiki in Lobaye, together with an extension of the Carnot plant in Haute Sangha.

COUNTRY PROFILE

Political system: Republic with part military part civilian government. Gained independence in 1960. Formerly one of the four territories included in French Equatorial Africa. A new Constitution was approved by a national referendum in November 1986, creating a one-party state; at the same time General André Kolingba was returned as President.

Population: The Central African Republic has an estimated three million inhabitants, including up to 15,000 Europeans. Half are Christians and almost half are animists; there is a small Muslim minority.

Main cities/towns: Bangui (population about 300,000), Berberate (38,000), Bassangoa (35,000).

Language: French is the official language, although African languages such as Sango are widely used.

Media
Press: The only national daily is *Elesongo*. A few periodicals are also published in Bangui, including *Journal Officiel de la République Centrafricaine*.
Radio: State-owned Radiodiffusion-Télévision Centrafrique broadcasts in Sango and French.

External trade: The trade deficit has been reduced through import cuts, but the long-term debt remains a severe burden. At the end of 1987 the ratio of debt to current account stood at 330.7.
Exports: Diamonds (estimated 32% of total value in 1986), coffee (32%), timber (21%), cotton (11%). Main destinations: Belgium (36.5% of total value in 1985), France (22.1%), Italy (12.2%), Spain (9.6%), German FR (7.4%).
Imports: Consumer goods, food, capital goods, oil. Main sources: France (70.9% total value in 1985), German FR (7.5%), Japan (6.4%), Italy (6.2%) and Netherlands (3.2%).

Agriculture: Mainstay of economy, contributing 37 per cent to GDP and employing 85 per cent of workforce. Mainly subsistence farming and animal husbandry. Production hampered by soil erosion, widespread drought, and under-developed marketing and infrastructure. Significant forestry potential, including over 60 species of commercially viable trees. Main crops are millet, sorghum, cassava, groundnuts, maize and plantains. In 1986 coffee equalled diamonds as principal foreign exchange earner; the harvest of 11,400 tonnes was some 4,000 tonnes below the country's ICO quota. The output of cotton and cotton lint was expected to decline following the IMF-prescribed withdrawal of subsidised producer prices. Tobacco is also grown. Timber is the third largest export earner but continues to decline in volume (down by about a third in 1986, compared to 1985.) Around 12 per cent of the total land is arable; much of the rest is savannah.

Industry: Sector is small-scale and under-developed, with manufacturing contributing around eight per cent and construction around four per cent to GDP. Production is centred on brewing, tanning, food processing. textiles and timber. Also new import substitution industries such as motor cycle and bicycle assembly. Main industrial centre is the Bangui district.

Mining: Diamond mining in west of country is the main mineral activity, providing 50 per cent of GDP. 1987 output was expected to be about the same as 1986 – 360,000 carats (72 per cent gemstones). Production has, however, fluctuated due to smuggling and series of disputes between foreign mining companies and the government. Small quantities of gold are also produced. Other mineral deposits include uranium, limestone, iron-ore, copper and manganese.

COUNTRY PROFILE

Energy: Strong dependence on imported fuels. Possible petroleum reserves on the border with Chad. The bulk of electricity production is generated at the Boali hydro-station.

Membership of international organisations: African Development Bank, African States in Association with the EC (Lomé Convention), Arab Bank for Economic Development in Africa (recipient), Customs and Economic Union of Central Africa, Economic Commission for Africa, French Monetary Area, GATT, IDA, IFAD, ILO, IMF, OAU, UN, WHO, WIPO, World Bank.

BUSINESS GUIDE

Time: GMT +1.

Climate: Hot all year with temperatures up to 36°C. Dry season from Nov to Feb with cooler nights. Rainy season from May to Oct. The north is very hot all year round.

Entry requirements

Passport: Required by all except certain African and French document holders.

Visa: Required by all except nationals of France, German FR, Italy, and certain African countries (including most West African nations). An onward or return ticket is generally required.

Currency: No restrictions on import of foreign currency, but amount must be declared; export up to declared amount allowed. Unlimited import of local currency; export limited to francs CFA 25,000.

Health precautions

Mandatory: Yellow fever vaccination certificate required.

Advisable: Cholera, typhoid and polio vaccinations recommended. Malaria prophylaxis should be taken as risk exists throughout country. Water precautions are necessary – bilharzia risk in some areas. AIDS risk.

Representation overseas: Abidjan, Algiers, Baghdad, Beijing, Belgrade, Berne, Bonn, Brazzaville, Brussels, Bucharest, Cairo, Khartoum, Kinshasa, Lagos, Libreville, N'Djamena, Ottawa, Paris, Port Louis, Rabat, Rome, Tokyo, Vatican City, Washington, Yaoundé.

Air access

National airline: Central African Republic is a shareholder in Air Afrique.

Other airlines: Air Gabon, Cameroon, Lina Congo, UTA.

Tax: Airport tax of CFA2,000 on departure.

International airport: Bangui-M'Poko (Code: BGF), 10 km from city, restaurant, post office, shops, car hire.

Hotels: Good standard hotels are available in Bangui – limited accommodation elsewhere. It is necessary to book well in advance of visit. Where service charge is not included in bill a 10 per cent tip is usual.

Electricity supply: 220/380 V AC, 50 cycles.

Car hire: Self or chauffeur driven cars available. International driving licence required.

City transport

Taxis: Available in Bangui; fares by negotiation.

National transport

Air: Small light aircraft can be chartered from Air Afrique.

Road: Eight main roads run from Bangui to the main towns and those that are surfaced are toll roads. The Trans-African Lagos-Mombasa highway passes through the Central African Republic. Most other roads can become impassable during rainy season. (N.B. Spare parts and petrol stations tend to be sparse outside Bangui.)

Buses: Limited coach service operates between Bangui-Bangassou.

Water: Boats ply from Bangui on the Oubangui River which runs into the River Congo, connecting Bangui to Zaïre and the Congo (including Brazzaville, from where railway runs to Pointe-Noire). Also services from Salo on the Sangha River.

Public holidays

Fixed dates: 1 Jan, 29 Mar (President Boganda's Remembrance), 1 May, 13 Aug (Independence), 15 Aug (Assumption), 1 Nov (All Saints'), 1 Dec (National Day), 25 Dec.

Variable dates: Easter Monday, Ascension (third week in May), Whit Monday, Mother's Day (late May).

Working hours

Government and business: (Mon-Fri) 0700 – 1300, 1430 – 1700, (many offices shut at 1300); (Sat) 0700 – 1200.

Banking: (Mon-Sat) 0700 – 1200.

Telecommunications

Telephone: Dialling code for Central African Republic: IDD access code + 236 followed by area code (61 for Bangui) followed by subscriber's number.

Telex: Telex facilities available at the main post office in Bangui and at good hotels.

Banking

Central bank: Banque des États de l'Afrique Centrale.

Main centre: Bangui. *Head offices:* Banque Centrafricaine d'Investissements, Banque Internationale pour l'Afrique Occidentale, Banque Nationale Centrafricaine de Dépôts, Caisse Centrale de Coopération Économique, Union Bancaire en Afrique Centrale.

Representation in capital: Cameroon, Chad, China PR, Congo, Egypt, France, Gabon, German DR, German FR, Iraq, Côte d'Ivoire, Japan, Korea PDR, Rep. of Korea, Libya, Nigeria, Romania, Senegal, Sudan, USA, Vatican, Yugoslavia, Zaïre.

BUSINESS DIRECTORY

Hotels: Bangui

Minerva, P.O. Box: 308 (tel: 61-02-33).

New Palace, P.O. Box: 108 (tel: 61-45-88).

Novotel, P.O. Box: 1015 (tel: 61-02-79; tx: 5297).

Rock, P.O. Box: 569 (tel: 61-01-55; tx: 5227).

Safari, Boulevard Général de Gaulle, P.O. Box: 134 (tel: 61-42-20).

Sofitel Bangui, Boulevard Général de Gaulle, P.O. Box: 1015 (tel: 61-30-38/48/43; tx: 5340).

St Sylvestrie, P.O. Box: 308 (tel: 61-38-00).

Car hire: Bangui

Tsiros: PO Box 68, Bangui (tel: 61-44-46).

Delsarte Frères: PO Box 122, Bangui Airport (tel: 61-26-61).

Business advice and information

Chambre de Commerce, PO Box 813, Bangui (tel: 61-42-55).

Chambre d'Industrie et d'Artisanat, PO Box 252, Bangui (tel: 61-27-77).

Chambre d'Agriculture, d'Elevage, des Eaux, Forêts, Chasses et de Tourisme, PO Box 850, Bangui (tel: 61-09-33).

Caisse Centrale de Coopération Économique, PO Box 817, Bangui (tel: 613634; tx: 5291).

European Development Fund, PO Box 1298, Bangui (tel: 61-30-53, 61-01-13).

Office Centrafricain de Tourisme (OCATOUR), PO Box 655, Bangui (tel: 61-45-66).

Banking

Banque des États de l'Afrique Centrale (BEAC) (central bank), PO Box 1917, Rue du Docteur Jamot, Yaoundé, Republic of Cameroon (tel: 22-25-05; tx: 8343).

Banque Internationale pour l'Afrique Occidentale (BIAO), PO Box 910, Bangui (tel: 61-17-68, 61-46-33).

Union Bancaire en Afrique Centrale (UBAC), PO Box 59, Bangui (tel: 61-29-90).

Chad

Nicolas Kulibaba

A long-sought-after peace agreement between President Hissène Habré and Colonel Muammar Qadafi, signed in mid-1989, may prove to be a pyrrhic victory for the Chadian head of state. Violent defections of Habré's closest allies early in 1989 left him bereft of trustworthy advisors and presiding over a government dominated by one-time opponents. Political rupture within the government was but a symptom of deep seated inter-ethnic tensions and clan rivalries which it seemed would inevitably intensify. As the spectre of increased hostility with Libya diminished, prospects for confrontation with the Sudan – which was providing refuges for many of Habré's opponents – increased dramatically. With political stability as a prerequisite for rehabilitation of Chad's beleaguered economy, Habré could well be denied the opportunity to profit from the era of good feeling that most assumed would follow in the wake of peace with Libya.

Nearly 10 years of overt hostility between Libya's Colonel Qadafi and the on-off regime of Hissène Habré came to a tentative close in 1989, with the successful negotiation of a peace agreement. In exchange for abandonment of its claims to the mineral-rich Aouzou Strip in northern Chad, Libya obtained not only the repatriation of hundreds of its troops, captured during the climactic battles of 1987, but also the withdrawal of French troops from bases adjacent to the disrupted territory.

Negotiations leading to the agreement involved a game of military chess, in which France gradually withdrew its Legionnaire troops and surveillance equipment from Libya's southern flank, to rearguard air bases in the Central African Republic. However, the deterrent force represented by the French appeared to give way to more covert guidance to N'Djamena from US and Israeli advisors, ever-watchful of Libyan adventurism and the recent dissolution of political stability in the Sudan.

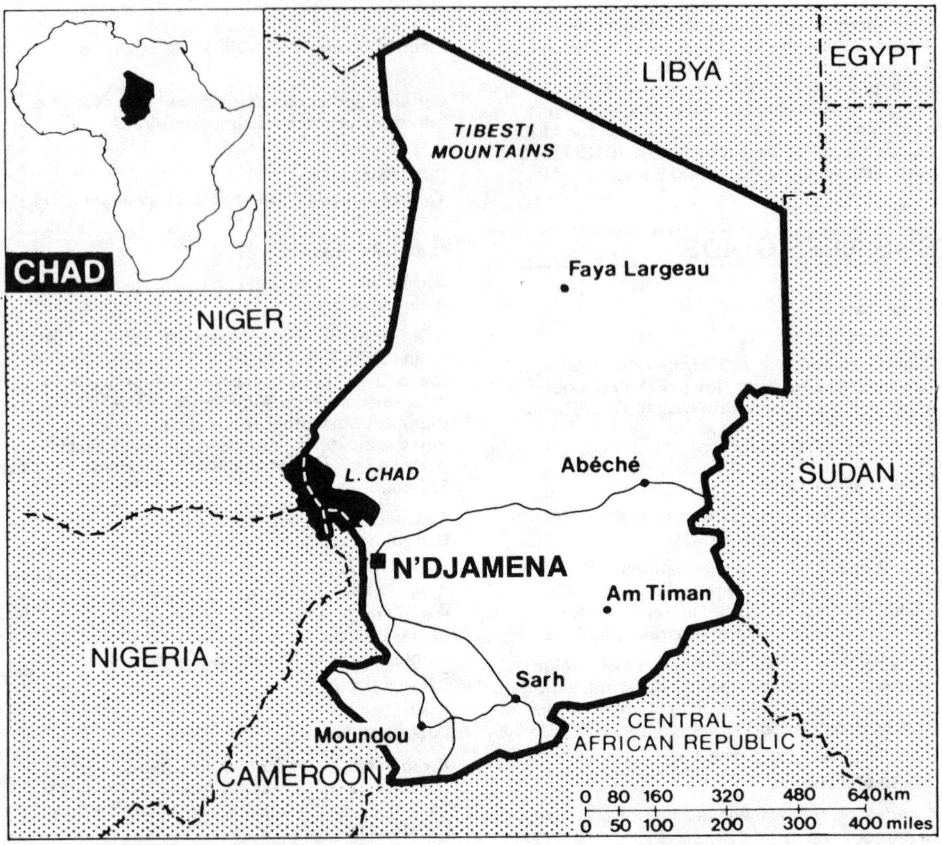

Political rupture

Habré's diplomatic victory in the Aouzou Strip was overshadowed by dramatic events earlier in the year which led to defection by the president's inner circle of advisors and a shift in the regime's centre of gravity.

The first signs of treachery within the presidential palace appeared early in the year, with the arrest of arms traffickers at Ounianga, along the Sudan border. Alleged to be funnelling weapons to anti-Habré forces in Sudan's Darfur province were a number of individuals closely associated with the commander in chief of the *Forces Armées National Tchadiennes* (FANT), Hassan Djamous. Djamous, the leading strategist and field commander in Chad's 1987 victory over Libyan forces at Ouadi Doum and Faya Largeau, was one of Habré's most trusted military aides. Compounding Djamous' problems were the activities of his brother-in-law, Fadoul Siddick. A former lieutenant to Habré's arch-rival, Goukkouni Oueddei, Siddick returned to N'Djamena in late 1987 where he resided with Djamous. In early 1988 Siddick was arrested for acting as a covert agent for the Libyan Government, and died shortly afterwards while in military detention.

On 2 April, amid rumours of an impending coup d'etat, Djamous and several colleagues fled N'Djamena, with loyalist troops in hot pursuit and under orders to avoid armed confrontation anywhere near the capital. Wounded in the escape attempt, Djamous was captured by government troops and died several weeks later in N'Djamena's Hospital of the Martyrs.

Also captured were Ibrahim Mahamat Itno, Habré's former minister for territorial administration and the president's principal coalition builder, and Tahar Guenassou, one-time minister of the interior. Another of Habré's key military advisors and a former commander in chief of the FANT, Idris Deby, eluded government troops, obtaining asylum in the Sudan.

For those with an inside line on clan rivalries within the ethnic Goran coalition at the heart of the regime, the defections appeared to stem from growing disillusionment with Habré's political programme of reconciliation with former opponents. Djamous, Itno, Guenassou and Deby were members of the Zagawa Goran who, along with Habré's Anakaza clan and the non-Goran Hadjerai formed the historical nucleus of the ruling regime.

The death of Idris Miskine in 1984 in unexplained circumstances alienated a number of Habré's Hadjera allies and caused the collapse of the first leg of the coalition tripod. The remnants of troops once loyal to Miskine regrouped in their Guera homeland, where they operated as the opposition *Mouvement de Salut National*. The latest defections not only deprived the president of the last of his original coalition allies, but also made tenuous N'Djamena's control over the Zagawa heartland of Biltine, along the Sudanese border.

However, Habré's formula for political success in recent years was to co-opt opposition groups through

KEY FACTS

Official title: Republic of Chad

Head of state: President Hissène Habré

Ruling party: Union Nationale pour l'Indépendence et de la Révolution (UNIR)

Capital: N'Djaména

Official Languages: French (Arabic and various African languages spoken)

Currency: Franc CFA = 100 centimes

Exchange rate: Franc CFA316 per $ (Nov 1989)

Area: 1,284,000 sq km

Population: 5.4m* (1988)

GDP per capita: $130 (1987)

GNP real growth: 0.5% (1987)

Inflation: 6.5%* (1987)

Trade balance: –$149.5m (1987)

Foreign debt: $318m (1987)

* estimated figure

recruitment of their leaders into the regime's power structure. While this gradually diluted the personal influence of such allies as Djamous and Deby, it also facilitated a coalition of ethnic groups and regional interests necessary to the survival of the regime. Hence, while Habré may have smarted from the loss of his Zawaga allies, he at least defused threats by General Djibril Negue Djogo's *Front Démocratique Tchadien*; Mahama Senoussi Khatir's *Comité d'Action et de Coordination-Conseil Démocratique Revolutionnaire.* Perhaps the most important recruit to the ruling coalition was the pre-eminent leader of the economically critical southern provinces, Wadal Abdelkader Kamougue, leader of the *Mouvement Revolutionnaire du Peuple Tchadien.*

KEY INDICATORS

	Unit	1984	1985	1986	1987	1988
Population	m	4.90	5.02	5.14	5.27	5.40
Inflation	%	20.4	5.0	–13.0	–2.7	12.8
Exports	$m	131.0	89.9	98.6	110.6	140.6
Imports	$m	181.0	240.3	287.9	366.1	419.4
Agricultural production	1979-81 = 100	98.2	102.4	106.1	97.8	100.6
Food production	1979-81 = 100	99.5	102.7	106.7	98.5	101.2
Balance of trade	$m	–50.0	–150.4	–189.3	–255.5	–278.8
Current account	$m	13	–65	–59	–11	–16
Total external debt	$m	156.2	183.3	235.1	317.8	–
Debt service ratio	%	3.0	8.7	3.1	3.8	–
Exchange rate	CFA per $	437.9	449.3	346.3	300.5	297.8

* estimated figure

Diplomatic strains

The shifting landscape of opposition to Habré's government resulted in increased tensions between Chad and several of its neighbours. Most problematic were relations with the Sudan whose regime changed in a 1989 coup d'etat. Because Khartoum's Arab regime was preoccupied with internal dissent and secessionist activity in several of its provinces, its ability and willingness to deal with Chadian dissidents was dubious. During 1989 there were increased movements of Zagawa populations into the Sudan, where they formed a tentative alliance with groups loyal to Adoum Togoi, a former Chadian defence minister who replaced Habré as defence minister following the latter's 1980 split with President Goukkouni Oueddei.

Relations with Khartoum reached their most fevered pitch in May, following the capture of Djamous and unrelated inter-ethnic violence between Sudanese Arabs and Chadian Fours, which N'Djamena alleged left 600 civilians dead. Habré attributed the violence to instigation by Islamic clerics of the Oumma clan, recruited by Tripoli.

A further thorn in the flesh of Habré's regime was the Benin Government of Mathieu Kérékou, which served as a base for the opposition activities of Adoum Togoi, who frequently shuttled between Cotonou and Tripoli. However, Kérékou's support for the dissidents waned somewhat, following Chad's peace agreement with Libya and as the result of domestic political turmoil in Benin.

Economic progress

As in the past, Chad remained heavily dependent upon concessionary aid from the international community. Economic uncertainty stemmed only in part from N'Djamena's preoccupation with civil war. Even in the best of times, the nation's desert ecology and cyclical drought raise questions regarding the viability of Chad's largely agricultural economy.

The prospects of peace and continuity for development programming resulting from *détente* with Libya opened the floodgates of international goodwill. Grant and concessionary assistance from multilateral sources in 1989 alone made available nearly $200m in funds for agriculture, petroleum exploration, and road rehabilitation, while the government of France assured budgetary support of Ffr40m per annum. A new development thrust in the Borkou-Ennedi-Tibesti (BET) corridor was to target a region long neglected as the result of civil war, while petroleum prospecting by major Western interests began in Chari-Baguirmi, Batha, Quaddai, Karem, Sedigi and the BET. Further facilitating these efforts would be the revived *Banque Tchadienne de Développement*, which enjoyed financial support from a variety of donors.

While cotton produced in the Chadian south continued to be the nation's largest export, generating more than 40 per cent of its foreign exchange revenues, new European and North American sponsored programmes to develop production of gum arabic, dates and food grains in the central and northern regions of the country were scheduled to start during 1990.

COUNTRY PROFILE

Political system: Executive and legislative power is held by a 33-member council headed by the president, consisting of 24 ministers and nine secretaries of state. The Conseil National Consultatif has 30 members. Gained independence from France in 1960.

Political parties: The Union Nationale pour l'Indépendence et la Révolution (UNIR), formed in 1984, is the sole party. Since the end of the civil war, several parties – particularly in the south – which used to belong to the umbrella opposition movement, Gouvernement de l'Unité Nationale Tchadienne (GUNT), have joined UNIR. The remnants of GUNT are still based in Libya.

Population: Chad's population is estimated at 5.4m (1988); this would compare with a figure of around four million in 1979. The population is composed largely of Muslims and animists.

Main cities/towns: N'Djaména (capital, formerly Fort Lamy, population 551,700), Sarh (106,500), Moundou (95,800), Abéché (77,200), Faya Largeau (17,600).

Language: French is the official and the business language; Arabic is spoken by most of the population. English is not used.

Media
Press: A few newspapers and periodicals are published, including *Info-Tchad* (a daily news bulletin) and Journal Officiel de la République du Tchad. Also an Arabic weekly paper, *Al-Watan*.
Radio: Government radio station, Radio Diffusion Nationale Tchadienne, broadcasts from N'Djaména, with stations at Moundou, Sarh and Abéché. *Television*: Télé-Chad, transmitting since December 1987 in French and Arabic.

External trade: Widening trade deficit is more than offset by official transfers, mainly development aid from France, German FR, USA and the EC.
Exports: Raw cotton (typically 75 per cent of total value), meat and live animals, textiles, hides and skins, natron, gums and resins. Main destinations: Cameroon, France, Nigeria, CAR, Congo, Sudan.
Imports: Machinery and transport equipment, food, petroleum. Main sources: France, USA, Cameroon, Italy, German FR, Nigeria, Congo.

Agriculture: Sector forms mainstay of economy, contributing 60 per cent of GDP and all exports. 85 per cent of population engaged in farming. Around 20 per cent of land is arable, and most of this is the southern flood plains of the Logone and Chari Rivers. Subsistence farming and livestock predominate in the north. Main food crops are sorghum, millet, dry beans, sesame, potatoes, rice and maize. Government statistics show that cereal production in 1986 was 681,064t. Cash crops include oil seeds – groundnuts and sesame – which totalled 117,600 tonnes in 1986 – sugar cane (239,00 tonnes), tobacco (100,000 tonnes). The 1986 output of cotton, the main export crop, was 89,469 tonnes, according to government sources.

Industry: Sector is small-scale and under-developed, contributing only seven per cent to GDP. Activity is centred in N'Djaména, Moundou and Sarh and is based on agriculture, particularly food processing, textiles, brewing, tobacco processing, leather and construction materials.

Mining: The only minerals extracted in any quantities are soda and rock salt (which are exported mainly to Nigeria) and natron (used in preservation of meat and in tanning). Known deposits of chromium, tungsten, titanium, iron, wolfram, gold, uranium and tin. Other mineral deposits, and possibly petroleum, are thought to lie in the Aouzou Strip, along the Libyan border.

Hydrocarbons: Oil reserves in Kanem region and Doba rift valley, although not of exportable quality. Oil deposits near Lake Chad said to be considerable. Refinery in N'Djaména. Electricity supply provided by two power stations in N'Djaména and plants in Moundou, Sarh and Abéché.

Membership of international organisations: African Development Bank, African States Associated with the EC (Lomé Convention), Arab Bank for Economic Development in Africa (recipient), Economic Commission for Africa, French Monetary Area, IDA, IFAD, IMF, Islamic Conference, Islamic Development Bank, OAU, UN, WIPO, World Bank.

BUSINESS GUIDE

Time: GMT +1.

Climate: Hot and arid in northern desert regions, and wet and tropical in the south. Southern rainy season from May to Oct, central rainy season from Jun to Sep with temperatures ranging from 20°C at night to as high as 40°C during the day. Dry season throughout the rest of the year, lower evening temperatures.

Entry requirements
Passport: Required by all except nationals of certain African countries.
Visa: Required by all except nationals of German FR, Nigeria and certain other African countries. A return or onward ticket is also required. All visitors must register with authorities on arrival. Exit permits must also be obtained before leaving.
Currency: No restriction on import of foreign currency, but amounts must be declared; export limited to declared amount. Unlimited import of local currency which must also be declared.

Health precautions
Mandatory: Yellow fever vaccination certificate required if arriving from an infected area.
Advisable: Yellow fever, cholera, typhoid and polio vaccinations. Malaria prophylaxis recommended as risk exists throughout the country. Water precautions necessary outside the capital.

Representation overseas: Algiers, Baghdad, Bangui, Beijing, Beirut, Bonn, Brazzaville, Brussels, Bucharest, Cairo, Jeddah, Khartoum, Kinshasa, Lagos, Moscow, Paris, Tripoli, Washington, Yaoundé.

Air access
National airline: Air Tchad (Chad is also a minority shareholder in Air Afrique).
Other airlines: Air Afrique, Ethiopian, Sudan Airways, UTA.
Tax: Airport tax of CFA3,000 is levied on departure to destinations outside Africa. Tourist tax is CFA2,000.
International airport: N'Djaména (Code: NDJ), 4 km from city, restaurant.

Surface access
Overland access: Main overland points of entry are by ferry across the Chari River to N'Djaména from Cameroon, or from Maiduguri (Nigeria) via Cameroon. Also, access possible by road from Bangui (Central African Republic) from Maroua (Cameroon), Niger and Sudan. Road conditions are variable and access can be very difficult especially in rainy season.

Hotels: Reservations should be made well in advance of visit. If service charge is not included in bill a 10 per cent tip is usual. Limited availability outside of N'Djaména.

Electricity supply: 220 V AC, 50 cycles.

Car hire: Limited availablity in N'Djaména. French or international driving licence is required.

City transport
Taxis: Available in N'Djaména; set-fare system in operation; 10 per cent tip is usual; can be hired on a time basis or by the day.

National transport
Air: N.B. Government authorisation is required to travel outside N'Djaména. Restricted domestic service operated by Air Tchad. Scheduled services occasionally commandeered by armed forces.
Road: There are surfaced roads around N'Djaména; most other roads are not in good condition and are often impassable during rainy season (Jun-Oct). Permits and four-wheel drive vehicles required for all travel outside N'Djaména.

Public holidays
Fixed dates: 1 Jan, 1 May, 25 May, 11 Aug (Independence), 1 Nov, 28 Nov, 25 Dec.
Variable dates: Easter Sunday, Easter Monday, Eid al-Fitr, Eid al-Adha, Maulid al-Nabi.

Working hours
Government and business: (Mon-Sat) 0700 – 1400, (Fri) 0700 – 1200. Specific times vary within this period.
Banking: (Mon-Sat) 0700 – 1100, (Sun) 0700 – 1000.

Telecommunications
Telephone: Dialling code for Chad: IDD access code + 235 followed by subscriber's six-figure number.
Telex: Telex facilities available at Office des Postes et Télécommunications, N'Djaména and other main post offices in Moundou, Abéché and Sarh.

Banking
Central bank: Banque des Etats de l'Afrique Centrale.
Main centre: N'Djaména. *Head offices:* Banque de Développement du Tchad (BDT), Banque Tchadienne de Crédit et de Dépôts, Banque Internationale pour l'Afrique au Tchad. *Foreign branches:* Banque Centrale des Etats de l'Afrique Equitoriale et du Cameroun, Caisse Centrale de Co-opération Economique.

Representation in capital: Algeria, Central African Republic, China, Egypt, France, German FR, Nigeria, Sudan, USA, USSR, Zaïre.

Hotels: N'Djaména
La Chari, P.O. Box: 118 (tel: 51-33-43).

La Palmeraie (tel: 51-55-16).

La Tchadienne Novotel, P.O. Box: 109 (tel: 33-11/12/14/15; tx: 5308).

Car hire: N'Djaména
COOP Taxi: PO Box 812 (tel: 51-43-16).

Habiba: PO Box 126 (tel: 51-43-72).

SONER: PO Box 248 (tel: 51-31-98; tx: 5285).

Banking
Banque des États de l'Afrique Centrale (BEAC) (central bank), PO Box 1917, Rue du Docteur Jamot, Yaoundé, Republic of Cameroon (tel: 22-25-05; tx: 8343 KN).

Banque Internationale pour l'Afrique au Tchad (BIAT), PO Box 87, N'Djaména (tel: 51-43-14, 51-43-21, 51-56-84; tx: 5228 afrbk).

Banque Tchadienne de Crédit et de Dépôts (BTCD), PO Box 461, N'Djaména (tel: 51-41-90, 51-41-95, 51-42-03; tx: 5212 btcd).

Other useful addresses
Direction du Tourisme, PO Box 86, N'Djaména (tel: 51-50-32, 51-23-03, 51-23-05).

Chambre Consulaire du Tchad, PO Box 458, N'Djaména (tel: 51-52-64).

Chile

Stephen Hays

Towards the end of 1989 Chile's central bank was forecasting an 8.5 per cent rise in GDP over the full year. This represents the most outstanding economic performance in Latin America and Chile's struggling democratic neighbours are bitter that it has been achieved under the aegis of a military dictatorship, while their economies have turned in lacklustre performances. But with Chile poised, in March 1990, to restore the first elected president and legislature since the 1973 coup, the challenge for the new government will be to maintain this momentum in an economy that shows signs of overheating.

From the 5 October presidential plebiscite in 1988 that resulted in defeat for General Pinochet's bid to extend his rule for a further eight years, the process of transition to democracy has accelerated rapidly. In July Chileans voted overwhelmingly to approve a package of constitutional reforms agreed to by the military government and the opposition parties. The voting was compulsory and over 85 per cent of the electorate voted in favour of the reforms. Only 8 per cent voted against. The changes in the constitution included increasing the number of directly elected members to the Senate, thus making it a more genuinely democratic body; abolishing Article 8, which outlawed marxist groups; shortening the next presidential term to four years from eight; balancing the number of civilians and military representatives in the powerful National Security Council, and easing the process for further amendments to the charter. General Pinochet, who agreed to the reforms with great reluctance, played down the significance of the vote and portrayed the result as an endorsement of the constitution designed by the military. In reality, the president's concessions appeared to be a tacit acknowledgement of the success of the 17-party centre-left opposition coalition in securing majority support among Chile's population for their vision of the

KEY FACTS

Official title: Republic of Chile
Head of state: President Patricio Aylwin
Capital: Santiago
Official Languages: Spanish
Currency: Chilean Peso (CH$) = 100 centavos
Exchange rate: CH$272 per $ (Nov 1989)
Area: 756,626 sq km
Population: 12.8m (1988)
GDP per capita: $2,518 (1988)
GDP real growth: 7.4% (1988)
Inflation: 14.7% (1988)
Trade balance: $2,219m (1988)
Foreign debt: $21.7bn (1988)

country's future. After years of internal disputes and failure to agree on common policies, the opposition parties have been welded into an effective alliance under the leadership of Patricio Aylwin – head of the Christian Democratic Party and the coalition's presidential candidate. Aylwin managed to achieve an easy victory in December's election over former finance minister, Hernan Buchi, the presidential candidate for the the right-wing National Renewal and Independent Democratic Union parties.

A government under Aylwin is certain to lead to future friction with the armed forces, due to the continuing military tutelage enshrined in the constitution. A particular source of conflict is the constitutional clause, which would prevent a civilian president removing the military's commanders-in-chief. Patricio Aylwin caused a storm by suggesting that General Pinochet should retire as commander of the army when he hands over power in March. Pinochet is currently allowed to retain his post for another four years and has warned that the incoming government will have 'serious problems' if it attempts to remove him, or bring members of the armed forces to trial for human rights abuses.

Given the likelihood of Aylwin's success in the polls, business leaders have been taking a close look at the opposition's economic programme. Their plan, called 'Growth with Social Equality', seeks to reassure voters that no radical changes are in store for the successful export-oriented policies of the present administration, while at the same time promising a more equitable distribution of wealth. Nevertheless, certain specific proposals in the opposition's programme are alarming businessmen. The opposition has backed industry-wide collective bargaining (even if only to establish minimum wages), which businessmen oppose because of the widely differing conditions between companies in a single industry. After the tough Pinochet years, unionisation will also become easier and workers are promised greater job security and higher pension rights and family bonuses.

The key plank of the opposition's social programme is a 'Solidarity Fund', which will channel more spending into housing, health and education, and will be financed by new taxes. Corporate taxes are expected to rise from their current level of 10 per cent to 15-20 per cent – still moderate by international standards. The plan is careful to condition spending on economic growth and higher revenues. The opposition's aim is for the fund to rise gradually over three to four years to 4 per cent of GDP. As the December elections approached fears of a revival of the right-wing death squads that characterised the early years of the Pinochet regime were fuelled by the murder of a marxist politician, Jacar Neghme. Neghme was the first opposition leader killed in a decade, and headed the moderate wing of the Movement of the Revolutionary, that had pledged support for Aylwin's campaign. The '11th of September Front', which claimed responsibility for the killing in alleged revenge for attacks by left-wing guerrillas on police, said that other deaths could follow.

Economy

Chile experienced the sixth consecutive year of economic growth in 1989 since the recovery from the severe recession of 1982-83. Much of the credit for this expansion in the economy can be given to policies that have encouraged the diversification of exports away from copper – the country's prime foreign currency earner. The trade surplus for 1989 was forecast to be similar to 1988's record level of $1.7bn, but 43 per cent higher than the surplus in 1987.

The continuing strength of the economic boom led to problems with 'overheating'. Rising industrial output was unable to keep pace with demand increasing at a faster rate due to a substantial recovery in the value of real wages and falling unemployment. A surge in both the price index and imports was the result. Overall unemployment stood at 6.4 per cent of the workforce between February-April compared to 8.6 per cent a year earlier and despite a 3.2 per cent growth in the labour force in the intervening period. Some 233,000 new jobs had been created by April, with the most dramatic increases in employment seen in the construction sector (23.7 per cent), mining (19.4 per cent) and industry (15.3 per cent). In July imports exceeded exports by $16m, the first such monthly deficit reported for four years. Over the January-July period imports leapt 40 per cent in comparison to 1988. Meanwhile, the consumer price index reached an annualised rate of 18.4 per cent by August compared to 12.7 per cent for the whole of 1988. A worried government had already applied the brakes. In June the central bank had responded to the upsurge by devaluing the peso and raising interest rates, and these measures had begun to take effect by the third quarter of the year with price rises slowing down. Inflation for the whole of 1989 was estimated at 15 per cent – the upper end of the central bank's initial target.

On the export front, copper remained the leading export with sales $400m higher in the first half of 1989 than 12 months earlier, due to the continuance of a buoyant world market that saw the total value of Chile's copper exports rise by 62 per cent during 1988. Copper continues to fall, however, in overall importance among the country's exports. At one time in 1970, it contributed as much as 87.5 per cent of exports, but between 1984-88 this had slipped to an average of 44.1 per cent.

There have been significant advances in other export-oriented sectors, notably in fruit and forestry, despite the losses incurred by exporters due to the US Food and Drug Administration's ban on imports of Chilean fruit in March 1989. Preliminary figures for the 1988-89 season, ending in June, showed that fruit exports totalled 91.4m boxes, a 7.2 per cent increase over the same period of the previous season, although the financial effect of the FDA ban was dramatic, following the discovery of two grapes laced with cyanide by Pinochet opponents in the Port of Philadelphia. Despite the expansion in volume, central bank figures show that the value of fresh fruit exports was 21.9 per cent or $375m lower in the first five months of 1989, with exporters reporting total losses of between $150-$200m at the end of May. According to the Exporters Association, some 10-15m boxes of fruit were destroyed as a result of the US ban. Pre-ban fruit exports for the season had originally been estimated at around 100m boxes valued at $650m. Growers are not expected to suffer from the FDA's action as the authorities have undertaken to compensate them from a reserve fund established with record copper revenues.

In the forestry sector, the government's promotion of tree planting is paying off in increased acreage and higher export earnings. Exports of forestry products reached $730m in 1988 and the pace of expansion is accelerating into the 1990s. The National Forestry Corporation has estimated that investments totalling over $3.7bn will be required for saw mills, pulp plants and infrastructure over the period 1992-2003 to exploit the volume of timber becoming available and ensure a sustainable yield in the following years. With the massive investments in the cellulose sector expected to raise capacity from 479,200 tonnes in 1988 to 1.35m tonnes in 1992, it is possible that earnings from forestry products could overtake the economic primacy of copper by the end of the century.

KEY INDICATORS

	Unit	1984	1985	1986	1987	1988
Population	m	11.9	12.1	12.3	12.5	12.8
Gross domestic product (1985) bn	pesos 2,515	2,577	2,723	2,879	3,092*	
Inflation (average rate)	%	19.9	30.7	19.5	19.8	14.7
GDP real growth	%	6.3	2.4	5.7	5.7	7.4
Agricultural production	% growth	7.5	5.6	8.8	3.2	5.5
Fishing	% growth	11.5	5.6	10.0	3.5	2.6
Manufacturing production	% growth	9.8	1.2	8.0	5.5	8.7
Construction	% growth	4.2	16.1	1.3	10.6	6.1
Mining production	% growth	4.4	2.2	1.4	0.0	4.2
Copper	1985=100	96	100	103	103	106
Exports	$m	3,651	3,804	4,199	5,224	7,052
Copper	$m	1,587	1,761	1,771	2,101	3,475
Agricultural produce	$m	428	515	683	796	930
Industrial goods	$m	1,260	1,168	1,420	1,824	2,273
Imports	$m	3,288	2,954	3,099	3,994	4,833
Consumer goods	$m	552	505	585	691	831
Food	$m	492	246	169	209	258
Capital goods	$m	597	650	735	1,101	1,370
Intermediate goods	$m	2,097	1,867	1,947	2,395	2,833
Balance of trade	$m	363	850	1,100	1,230	2,219
Current account	$m	–1,781	–1,328	–1,137	–808	–168
Disbursed foreign debt	$bn	19.9	20.3	20.2	21.1	19.0
Exchange rate (average)	pesos per $	98.7	161.1	193.0	219.5	245.1

* estimated figure

Debt and finance

With the election approaching, foreign and domestic investors have been moving quickly to take advantage of Chile's foreign debt conversion programme, before the new government takes power. Although a position on Chile's foreign debt obligations was not included in the Alywin coalition's election policy document, the subject is known to be high on the opposition's planning agenda.

Ricardo Ffrench-Davis, a leading opposition economist, has proposed that debt swaps under the Chapter 19 rules for foreign companies should be limited to productive investments that bring in fresh capital in a ratio of one-to-one with the amount of debt converted. He believes that Chapter 19 has allowed too many simple transfers of assets to foreign investors, who have captured the lion's share of the discount on Chilean debt paper. This amounts to a subsidy for foreign investors that the country can ill afford, he argues. Anticipation by investors of a change in the programme's rules produced a rush to use the mechanism before election time. The central bank approved $732m in swaps from January-July 1989, compared to total swaps of $890m in the whole of 1988. From mid-year, the bank had a further $1bn worth of requests awaiting approval.

Since the debt swap programme's inception in 1985 there have been $2.6bn in Chapter 19 swaps. Another $2.4bn has been swapped under Chapter 18, which allows Chileans to pre-pay their debt. Along with some portfolio swaps, prepayments and buybacks from the central bank and capitalisations, Chile's commercial bank debt has been reduced during the last four years by $7.6bn. The opposition has also been highly critical of so-called 'bargain-basement' or 'fire-sale' prices for companies privatised with the use of the debt swap programme, which increase the 'subsidy for foreign investors'. According to central bank figures, $153m of Chapter 19 investments were for companies being privatised. Purchases of companies under temporary government administration as a consequence of the 1982 economic collapse accounted for another $405m of swaps. This total of $658m represents 25 per cent of all Chapter 19 operations by the middle of 1989. The government counters opposition 'sell-out' charges by noting that if foreigners had not been bidding for the companies to be privatised they would have fetched even lower prices.

The opposition is not proposing to halt the debt swap programme completely, but rather is aiming to pare it down, with transactions restricted to investments that increase exports and provide access to new markets and technology. Ffrench-Davis contends that the medium-term benefits of swaps to the balance of payments is questionable. Capital and profit remittances could well outpace debt interest payments in the future. He points out that while interest payments can be negotiated, it would be difficult for the government to change its treatment of remittances.

The authorities have ruled out debt-bond exchanges, along the lines of the Mexican model, as it says debt buybacks are totally voluntary and therefore do not prejudice the banks against Chile should it one day wish to return to the capital markets. This may happen sooner rather than later, as Chile will have to seek fresh loans to meet a bunching up of repayments in 1991. In that year, $3.5bn on interest and principal repayments fall due – equivalent to 15 per cent of the country's GDP. By the end of June, Chile's total debt was $16.7bn, with much of the previous commercial bank debt having been replaced by credits from multilateral and official lenders.

Mining

Chile's mining sector turned in a strong performance in 1989, due to higher world prices for key exports. By the first half of the year shipments of metals and minerals had earned $2.09bn, some 51.5 per cent of the country's total export revenues and 28.6 per cent up on the sector's foreign sales during the same period of 1988. Exports of copper rose from $1.4bn to $1.86bn, iodine and nitrates from $70m to $73m and iron ore from $49m to $56m. A report from the US embassy in Santiago published in the first half of the year, concluded that Chile's mining boom will continue over the next decade. The report looked at projects in various stages of development, or awaiting final approval, and estimated that copper production would increase from 1988's 1.43m tonnes to 2m tonnes by 1991.

The La Escondida copper mine, expected to become the world's third largest producer after Chile's Chuquicamata and El Teniente mines, is due to come on-line at least four months ahead of schedule in April 1991, because of the fast pace of construction. The mine will produce an annual output of 320,000 tonnes of copper over its estimated 52-year lifespan. The deposit, located some 3,200 metres above sea level in the province of Antofagasta, possesses reserves of 1.9bn tonnes with an approximate grade of 2 per cent copper. Once production begins at La Escondida, the private mining companies estimate that their share of Chile's copper output should increase to 50 per cent by the late 1990s.

Chile's 18,000 kg per year gold output is set to double by 1991 as a result of the development of new and existing deposits, according to the Chilean Copper Commission. Almost $400m is being invested in several mines which, once in full operation, are expected to make Chile the second largest gold producer in Latin America after Brazil and the seventh largest in the non-communist world. An output of 36,000 kg a year would represent 2 per cent of Western gold production. Patricio Jarpa, the Copper Commission's vice president, said Chile's high gold production was expected to earn the country an extra $200m a year. In addition, the investments would more than double silver production to 1.2m kg from the current 520,000 kg, netting a further $200m in export revenues.

The privatised nitrates company, Soquimich, invested $45m in 1989 to increase production, develop new products and boost foreign sales. A year earlier the company had posted profits of $48.5m, some 36 per cent higher than 1987, on sales of $250m. Soquimich is aiming to become the world's largest producer of the specialist fertiliser potassium nitrate, and is already the second largest international producer of iodine. During 1989 the company signed an investment agreement with Ise Chemicals and Mitsubishi of Japan to increase iodine output by 20 per cent to 4,500 tonnes a year. Mitsubishi dropped its plans to purchase a 40 per cent stake in Soquimich after protests that the Japanese iodine producer could thwart the Chilean company's project to further expand its iodine ouput to 8,000 tonnes per year by 1995 – generating a potential annual export revenue of $144m.

The stock market

Foreign investment in Chilean companies continued unabated in 1989. In what was viewed an important vote of confidence in the post-Pinochet administration, Salomen Brothers launched a $60m investment fund that marked the first step towards the internationalisation of the Santiago stock exchange. The Salomen Fund, which went on offer in September and was heavily oversubscribed, plans to invest 75 per cent of its capital in Chilean equity and debt securities in its first six months. Three other smaller funds for direct foreign investment in Chilean equities have also been authorised. Citibank has created two small funds of $4m each and Batterymarch, the Boston-based investment company, also brought in $15m under its Luxembourg-based Latin American equity fund. Compania de Telefonos de Chile (CTC) is planning a sale of some $95m of shares in international equity markets at the start of 1990. The intention of the sale is to reduce the stake of Bond Corporation International (BCIL), the Hong Kong listed company through which Australian entrepreneur Alan Bond controls 52 per cent of CTC following the company's controversial privatisation. Under its first full year of Bond management in 1988, CTC doubled its profits to $63m, up from $31m in 1987. Estimated profits for 1989 were $102m.

COUNTRY PROFILE

Political system: Unitary state governed by military junta since 1973. Executive power held by the president, legislative power exercised by a government junta of four. The 1980 constitution allows for possibility of return to civilian rule by 1989.

Main cities/towns: Santiago (population about 4.9m), Valparaíso, Viña del Mar, Concepción, Antofagasta, Arica, Iquique, Coquimbo, La Serena, Temuco.

Media
Press: Principal national dailies are *El Mercurio* (circulation 120,000), *La Nación* (20,000), *La Segunda, La Tercera* (170,000), *Diario Oficial de la República de Chile (15,000), Las Ultimas Noticias* (150,000), *La Epoca* (50,000). Foreign newspapers and magazines at bookshops and hotels.
Radio: Over 300 radio stations broadcasting on short and medium wave and FM. Most are commercial and in Spanish. *Television:* Four TV channels and numerous stations, mostly commercial. Televisión Nacional de Chile operates 108 stations. Cable TV network available in some areas of Santiago, one of whose four channels is a satellite link with the US Armed Forces Radio and Television Service.

External trade: Export earnings vulnerable to shifts in world copper prices. Small trade surplus registered over 1981-84 as a result of the sharp decline in imports and depreciation of the peso, rising in 1985 and 1986 as non-copper exports increased and imports were restrained. Current account remains heavily in deficit owing to large deficit in invisible trade. Capital account surplus has increased due to positive movement in short-term capital and debt rescheduling.
Exports: Copper (48% of total value in 1988), fruit (10%), basic metals (8%), fishmeal (7%), paper/cellulose (6%), timber (3%). Main destinations: USA (22% of total value in 1987), German FR (19%), Japan (11%), Brazil (7%), UK (6%), Italy (5%).
Imports: Intermediate goods (56% of total value in 1988), capital goods (26%), consumer goods (21%). Main sources: USA (19% of total value in 1987), Japan (10%), Brazil (9%), German FR (8%), Argentina (4%), Venezuela (4%), France (2%).

Agriculture: Although only 11.5 per cent of land is arable, soil is fertile and well irrigated, particularly in the central area and main river valleys. Sector, including fisheries, employs about 16 per cent of workforce and contributes eight per cent to GDP. Dependence on imported foodstuffs reduced slightly by improved wheat and potato harvest (two main food crops). However, food imports, mainly from USA, remain high. Other important crops are oats, barley, rice, beans, lentils, maize and chickpeas. The main industrial crops are sugar-beet, sunflower and rape seed. Important cash/export crops are onions, garlic, apples, table grapes, citrus fruits, avocados, pears, plums and nuts. Livestock farming concentrated in the south. Fishing has grown rapidly over last five years, and Chile now ranks as the most important fish producer in Latin America. Total fish catch (including varieties of hake, mackerel, sardine, anchovies) over 5m tonnes. Forestry also important sector, with 4.5m hectares of natural forest suitable for commercial exploitation. Timber supplies for production of sawn wood and pulp have risen to over 10m cu metres – mainly fast growing radiata pine.

Manufacturing: Sector contributes 21 per cent to GDP and employs 14 per cent of workforce. Activity is largely concentrated in and around Santiago. Financial conglomerates control a substantial section of the denationalised industries, though small firms with less than 10 employees still dominate. Export based industries include petrochemicals, paper, base metals, plastics, rubber and food processing (particularly fish and malted barley). Home market industries include textiles, footwear, cement, motor vehicles, local food, beverage and tobacco processing and machinery. Despite industrial restructuring sector remains generally under-capitalised, vulnerable to import competititon and burdened by debts.

Mining: Sector contributes eight per cent to GDP, is the main export earner and the major centre of foreign investment. Activity concentrated on copper, of which Chile is the world's leading producer and holds around 20 per cent of world's proven reserves. The state-owned copper enterprise, CODELCO, holds 70 per cent of national reserves and administers the four largest mines: Chuquicamata, El Teniente, Andina and El Salvador. Copper will also be extracted from the Escondida mine, the biggest proven deposit in the world, under a project led by BHP Australia, at an investment cost of US$1.2bn. Also mining of molybdenum (a by-product of copper mining), silver, gold (the El Indio mine ranks among the highest grade mines in the world), iron-ore, manganese, lead, natural nitrates, mercury, marble, coal, sulphur and limestone. Plans to develop lithium reserves – estimated that production could soon account for 20 per cent of world output.

Hydrocarbons: Oil is extracted mainly from offshore fields at Straits of Magellan, and from onshore at Tierra del Fuego and the southern mainland. Reserves are estimated at 736m barrels and production is around 140,000 b/d – equivalent to domestic refining capacity. An intensive prospecting and drilling programme is being carried out by the state-owned Empresa Nacional del Petróleo (ENAP), currently the only oil producer in the country. Natural gas production, the bulk of which comes from the Magellanes field, has steadily fallen since the early 1970s to some 6bn cu metres in 1985. Total gas reserves are estimated at 122bn cu metres.

Energy: Around 60 per cent of domestic energy requirements satisfied by petroleum and gas, seven per cent by coal, 12 per cent by hydroelectricity, the balance by wood. Plans to wipe out petroleum imports (currently accounting for 30 per cent of total energy requirements) by 2003. Also greater utilisation of vast hydro-power and coal reserves, particularly in copper smelting.

Membership of international organisations: CCC, CIPEC, ECOSOC, FAO, G-77, GATT, IADB, IAEA, ICAO, IDA, IFAD, IFC, IHO, ILO, IMF, IMO, INTELSAT, Inter-American Development Bank, Interpol, IPU, ITU, Latin American Integration Association, Organisation of American States, PAHO, SELA, UN, UNESCO, UPU, WHO, WIPO, WMO, World Bank, WSG, WTO.

Time: GMT -4 (GMT -3 from Oct to Mar).

Climate: Generally hot and dry in north, Mediterranean in central region (cool nights) and wet and cold in the south. Temperatures in Santiago range from 10°C to 30°C in summer (Dec-Mar) and 0°C to 25°C in winter (Jun-Sep). The rainy season in the Santiago area is from May-Aug.

Entry requirements
Passport: Required by most nationals, with exception of citizens of Argentina, Brazil, Paraguay and Uruguay where national identity card is sufficient.
Visa: Required by all except short-term visitors (up to 90 days) from countries that have diplomatic relations with Chile. All visitors required to complete a tourist card in duplicate, obtainable from Chilean consulates, airline offices and on most aircraft bound for Chile. Check latest regulations as they are liable to change at short notice.
Currency: No restrictions on import/export of foreign or Chilean currency. International credit cards widely accepted. Receipts for money changed on entry should be retained. Entrance tax for tourists from certain countries operating. Difficult to change pesos for foreign currency when leaving country.

Health precautions
Mandatory: None.
Advisable: Typhoid vaccination recommended. Water precautions should be taken. Not advisable to eat uncooked vegetables, or fruit which cannot be peeled.

BUSINESS GUIDE

Representation overseas: Amman, Ankara, Asunción, Athens, Bangkok, Beijing, Beirut, Berne, Bonn, Brasilia, Brussels, Bucharest, Buenos Aires, Cairo, Canberra, Caracas, Copenhagen, Damascus, Guatemala City, The Hague, Jakarta, Lima, Lisbon, London, Madrid, Manila, Monaco, Montevideo, Nairobi, New Delhi, Oslo, Ottawa, Panama City, Paris, Port-au-Prince, Pretoria, Quito, Rabat, Rome, San José, San Salvador, Santo Domingo, Seoul, Singapore, Stockholm, Tegucigalpa, Tel Aviv, Tokyo, Vatican City, Vienna, Washington, Wellington.

Air access
National airlines: Lan-Chile (Línea Aérea Nacional de Chile) and Ladeco (Lineas Aéreas del Cobre).
Other airlines: Aerolíneas Argentinas, Aeroperú, Air France, Alitalia, Avianca, British, Canadian Airlines International, Eastern, Ecuatoriana, Iberia, KLM, LAP, LAB, Lufthansa, Pan Am, Pluna, Swissair, Varig, Viasa.
Tax: Airport tax of US$12.50 or equivalent in CH$ levied on all passengers leaving from Santiago de Chile (Pudahuel Airport). Only exemptions are children up to two years and holders of Uruguayan diplomatic passports.
International airports: Arica-Chacalluta (ARI), 18 km from city; bar, restaurant, buffet, shops, car hire. Santiago-Comodoro Arturo Merino Benítez (often known as 'Pudahuel') (SCL), 16 km from city; bar, restaurant, buffet, bank, post office, shops, hotel reservations, car hire.

Surface access
Main ports: Antofagasta (typically 16-18% of total tonnage), San Vicente (around 25%), Valparaíso (20-22%), San Antonio (20%); also, Arica, Iquique, Puntas Arenas, Coquimbo, Chacabuco, Puerto Montt, Talcahuano.
Overland access: Possible from Argentina (e.g. coach service, via Mendoza), Bolivia (e.g. train routes La Paz-Arica and La Paz-Antofagasta), Peru (via Pan-American Highway, including coach service).

Hotels: Numerous luxury and first-class hotels as well as good hotels in lower price range. The Stars Classification System will soon be in operation. Bookings may be made at the Sernatur information office at Pudahuel Airport. A 16 per cent hotel tax is added to bill. Service charge is usually included, but an extra 5-10 per cent tip is usual.

Car hire: National or international licence accepted. Can be arranged at the airport and in most major towns. A large deposit may be required. All car drivers require Carnet de Passages et Douanes issued by automobile club. Traffic drives on the right.

City transport
Taxis: Cheap and widely available in main towns. Initial charge (*Bajada de Bandera*) displayed on front windscreen. Large blue taxis do not have meters. Radio taxis charge higher fares. Within Santiago: Coloured black and yellow; can be hailed but scarce at rush hour; mostly metered but for long journeys fares should be negotiated in advance; extra charges at night and on Sundays; checking destination beforehand often advised; tipping not usual.
Metro: Efficient metro system in Santiago consisting of two main lines: (San Pablo-Escuela Militar line and Los Héroes-Callejón Ovalle line). Trains run 0700 – 2245.

National transport
Air: Lan-Chile Airlines and Ladeco operate frequent flights linking main centres. Seats should be booked well in advance. Air taxi services also available.
Road: Pan-American Highway runs north-south, 3,100 km. Highway has been extended as Southern Highway as far south as Coihaique but only parts are suitable for cars; 12-14 per cent of total network classed first-class highway. Only possible to reach Punta Arenas by land from RíO Gallegos (Argentina).
Buses: Frequent inner city service. Express coaches linking main centres generally recommended (e.g. Santiago-Arica, typically one departure/day; Santiago-Valparaíso, approx hourly service).
Rail: Fast diesel-electric train service. The main line runs from Santiago to Puerto Montt (includes sleeper service, restaurant cars, air-conditioning, typical total journey time around 18 hrs); Japanese train links Santiago-Concepción (first-class service – journey time around nine hours including bus service from Chillán to Concepción). See local newspapers for all transport schedules. The northern line no longer operates.

Public holidays
Fixed dates: New Year's Day, 1 May, 21 May (Navy Day), 2 Jun (bank holiday), 29 Jun (San Pedro y San Pablo) 15 Aug (Assumption), 11 Sep (Day of Liberation), 18 Sep (Independence), 19 Sep (Armed Forces Day) 12 Oct (Discovery of America) 1 Nov (All Saints'), 8 Dec (Annunciation), 25 Dec (Christmas).
Variable dates: Good Friday, Holy Saturday. N.B. main holiday period Jan to early Mar.

Working hours
Business: (Mon-Fri) 0830 – 1230, 1400 – 1800; (Sat) 0900 – 1300.
Government: (Mon-Fri) 0830 – 1720.
Banking: (Mon-Fri) 0900 – 1400.
Shops: (Mon-Fri) 1000 – 2000; (Sat) 1000 – 1330.

Telecommunications
Telephone: Most of Chile's telephone services supplied by the Compania de Teléfonos de Chile, although small independent companies operate in Arica, Valdivia and Coyhaique. Dialling code for Chile: IDD access code + 56 followed by area code (Concepción 41, Santiago 2, Valparaíso 32) followed by subscriber's number.
Telex: Telegram, telex and facsimile services available in main towns through Telex Chile, Transradio Chilena and ITT Communicaciones Mundiales. Calls may be made collect. The telecommunications system is linked to the rest of the world by satellite, and Chile has an efficient internal and international postal service. Country telex code: CZ, CK, LL, CT.

Banking
Banking centre: Santiago
Central bank: Banco Central de Chile.

Trade fairs: FISA industrial fair held every Oct-Nov in Santiago. International Air and Defence Show, Mar-Apr, Santiago.

Electricity supply: 220 V AC, with two-pin plugs and screw type bulbs.

Representation in capital: Argentina, Australia, Austria, Belgium, Brazil, Canada, China, Colombia, Costa Rica, Denmark, Dominican Republic, Ecuador, Egypt, El Salvador, France, German FR, Greece, Guatemala, Haiti, Honduras, India, Indonesia, Israel, Italy, Japan, Jordan, Republic of Korea, Lebanon, Netherlands, New Zealand, Norway, Panama, Paraguay, Peru, Philippines, Romania, South Africa, Spain, Sweden, Switzerland, Syria, Turkey, UK, USA, Uruguay, Vatican, Venezuela.

Hotels: Santiago
Carrera, Teatinos 180, P.O. Box: 2272 (tel: 6982011; tx: 340222; fax: 721083).

Conquistador, Miguel Cruchaga, 920 Casilla 10315 (tel: 6965599; tx: 240888; fax: 6965599).

Galerías, San Antonio 65, P.O. Box: 3319 (tel: 384011; tx: 340649; fax: 330077).

Holiday Inn Crowne Plaza, Avenida Libertador B. O'Higgins 136 (tel: 381042; tx: 441306; fax: 336015).

Libertador, Avenida Libertador B. O'Higgins 853 (tel: 394213; tx: 645143).

Santa Lucia, Huérfanos 779 (tel: 398201; tx: 242292).

Santa María, Avenida Sta María 2050 (tel: 2326614).

Sheraton San Cristobal, Avenida Santa María 1742, P.O. Box: 16058 (tel: 745000; tx: 440004; fax: 2236656).

Tupahue, San Antonio 477, P.O. Box: 2910 (tel: 383810; tx: 240842; fax: 395240).

Business advice and information
Agencia Informativa Orbe de Chile Ltda, Philips 56, Santiago (tel: 6966431).

Asociación de Radiodifusores de Chile, Pasaje Matte 956, Office 801, Casilla 10476, Santiago (tel: 398755).

Banco Central de Chile, Agustinas 1180, Santiago (tel: 6962281; tx: 40569 CENBC-CL).

Comisión Económica para América Latina y el Caribe (CEPAL)/Economic Commission for Latin America (ECLA), United Nations Building, Avenida Dag Hammarskjold s/n, Casilla 179D, Santiago.

Cámara de Comercio de Santiago de Chile (Chamber of Commerce), Santa Lucía 302, 3rd Floor, Casilla 1297, Santiago (tel: 330962; tx: 240868).

Comisión Chilena del Cobre (Chilean Copper Commission), Agustinas 1161, 4th Floor, Santiago (tel: 726219).

Comisión Chilena de Energía Nuclear (Chilean Nuclear Energy Commission), Amunátegui 95, Santiago (tel: 6990070; tx: 340468 cochen).

Corporación Nacional del Cobre de Chile (CODELCO) (Copper Corporation), Huérfanos 1270, Santiago (tel: 6988801).

Comisión Nacional de Energía (Energy Commission), Teatinos 120, 7th Floor, Santiago (tel: 6981757).

Comité de Inversiones Extranjeras (Foreign Investment Committee), Teatinos 120, 10th Floor, Santiago (tel: 6983246, 6984254).

Compañía de Aceros del Pacífico (Pacific Steel Corporation), Huérfanos 669, 8th Floor, Santiago (tel: 395666; tx: 240280/240288 CAP CL).

Corporación de Fomento de la Producción (CORFO) (Development Corporation), Moneda 921, Santiago (tel: 380521).

Empresa Nacional de Minería (ENAMI) (National Mining Corporation), Mac-Iver 459,2nd Floor, Casilla 100-D, Santiago (tel: 396061).

Empresa Nacional de Petróleo (ENAP) (National Petroleum Corporation), Ahumada 341, Santiago (tel: 381845; tx: 240447).

Instituto de Promoción de Exportaciones (PROCHILE) (Export Promotion Bureau), Avenida Pedro de Valdivia 0193, Santiago (tel: 2317108, 2316283; tx 240836 PROCH CL).

Instituto Textil (Textile Institute), Bandera 566, Office 101-5, Santiago (tel: 6960210).

Líneas Aéreas del Cobre (Ladeco), Huérfanos 1157, Departamento de Ventas, Santiago (tel: 712960).

Línea Aérea Nacional (LAN CHILE) (Chilean National Airline), Agustinas 1197 (Departamento de Reservas y Ventas), Santiago (tel: 723523).

Ministry of Economy, Development and Reconstruction, Teatinos 120, 10th Floor, Santiago (tel: 725522; tx: 240558).

Ministry of Finance, Teatinos 120, 12th Floor, Santiago (tel: 6982051; tx: 241334).

China

Michael B. Yahuda

The Beijing massacre of the pro-democracy demonstrations of 4 June and the subsequent campaign of intimidation and suppression have cast a deep pall over Chinese politics and they have set back the prospects for economic reform and development. Deng Xiaoping and his fellow octogenarians claim that they were confronted by a counter-revolutionary rebellion that was precipitated by the unchecked spread of 'bourgeois liberalisation' (or Western values). The General Secretary of the Communist Party, Zhao Ziyang and several prominent reformers linked with him were dismissed for their alleged role in fomenting rebellion and for not standing up to 'bourgeois liberalisation'. A conservative hardline current is in the ascendant. Superficially, order has been restored, but a new political crisis cannot be long delayed if only because of the advanced age of many of the key leaders. Deng Xiaoping himself was 85 years old in August, 1989.

Meanwhile, the central government is still grappling with the problems of cooling an over heated economy amid inflationary conditions and continued concern about food production. These problems have been exacerbated by the sanctions imposed by Western governments outraged by the events of June and by the reluctance of the international business community to make substantial new commitments to a country in such uncertain conditions.

Origins of the crisis

Deng Xiaoping and his fellow revolutionary veterans profoundly misjudged the character of the student demonstrations that began in mid-April upon the death of the reform-minded Hu Yaobang who had been forced to resign as General Secretary of the Communist Party two years earlier by the same veterans. Deng thought that he was confronted with a new Cultural Revolution. On 25 April, he told a high level Party meeting: 'This is not an ordinary student movement. It is turmoil... What they are doing now is altogether the same stuff as what the rebels did during the Cultural Revolution. All they want is to create chaos under the heavens.'

As leading victims of the 10 years of terror, Deng and his colleagues are obsessed with the dangers of the revival of mob rule. But the students were not like the hysterical Red Guards incited by Mao with hatred for his opponents, who were filled with intolerant revolutionary fervour against those with whom they disagreed, until they turned upon each other, they were manipulated by radical leaders associated with Mao. The student demonstrators of 1989 were a grass roots protest. Their initial demands were moderate and even when the troops were first sent to suppress them in May, they stopped them with pacifist 'people's power'.

Deng had also misjudged the response of the urban population. When his message was conveyed in an authoritative *People's Daily* editorial of 26 April, that for good measure added that there was a counter-revolutionary conspiracy by a 'tiny handful', the students challenged that by demonstrating the following day only to be joined by tens of thousands of ordinary citizens. By early May, they were joined by intellectuals, journalists, Party members of leading organisations and by workers. Hundreds of thousands of people took to the streets and joined in the student demands for the resignations of Deng Xiaoping and Premier Li Peng. Having rejected Party General Secretary Zhao Ziyang's more accommodating approach, Deng had no alternative but to call in the army. Even then, no one expected the overwhelming brutality of the massacre on the night of 3–4 June 1989.

By his action as Chairman of the Military Affairs Commission, and that of his permanent vice-chairman, Yang Shangkun (who also holds the post of president of the People's Republic of China), showed that that position, with its executive command of the armed forces, was more important than the leadership of the Communist Party (nominally his superior). All the attempts at political reform and the development of legally distinct institutions were shown to be hollow. The political reforms that had taken place, such as the relaxation of political controls, had been allowed because they were seen as necessary for the economic reforms and the policies of the Open Door.

Intellectuals spoke out, but they could not change the structures of power. However, these in turn lacked legitimacy and belief in themselves, and had been corrupted. Meanwhile, the economic reforms had run into trouble because of inflation and the overheating of the economy. Retrenchment was introduced in September 1988. Jobs were lost and many urban groups suffered a decline in their standard of living. Instead of seizing the opportunity to establish a reconciliation between the rulers and the ruled through proper dialogue with the demonstrators, Deng and his veteran comrades resorted to brute force.

Current crisis

The hardline regime sought to consolidate its rule by a number of measures that superficially appeared to be restoring order. First, it staked a claim to legitimacy on the grounds that the events of 4 June were a suppression of a counter-revolutionary attack upon the army that had been instigated by conspirators at home and abroad. The conspiracy aimed at nothing less than the overthrow of the government, the replacement of socialism with capitalism and the establishment of a bourgeois republic that would be dependent on a foreign power – presumably the United States. That in Deng's view was what 'bourgeois liberalisation' was about.

Second, that claim provided a basis for instituting a reign of terror against the pro-democracy activists and their supporters. No national figures for the arrests and the executions have been published – perhaps because of the storm of outrage in the West that the executions of the first few days provoked. However, by the end of July, local newspapers had revealed that more than 5,000 and 3,000 people had been arrested in Beijing and Jiangsu Provinces respectively. No figures are available for the other 27 provinces and municipalities.

Third, an ideological campaign of 'political education' was launched among the 200m urban dwellers so that in their workplaces or neighbourhood committees they should 'voluntarily' declare their adherence to the new line. Yuan Mu, the official spokesman of the state council warned the Chinese people against talking to foreigners as they do not have 'labels on their foreheads' disclosing their true identity.

Fourth, under the guise of stamping out corruption, tax evasion and profiteering as evidence of 'bourgeois liberalism', the private and collective businesses that employ about a tenth of the urban workforce had their operations restricted.

Nevertheless, the hardliners have given indications of the political turmoil and uncertainty that lurks not far below the surface. The brief communique of the Central Committee Plenum that was held on 24–26 June 1989 (i.e. three weeks after the massacre, a delay which is significant in itself) made a point of stating that it was Deng and the veterans who took the initiative from the outset. Moreover, although it announced the dismissal of Zhao Ziyang from all his posts, it went on nevertheless to assert that he would be further investigated. By openly stating this, the leadership clearly intended to warn the reformers that even further punishment may await them.

If final authority is vested with men in their eighties, there must be grave doubts about the duration of the current order. Chinese officials up and down the hierarchy will be loath to commit themselves to the new order beyond the minimum necessary lest they find themselves over-exposed when the wheel of political change turns yet again. That cautious immobilism is further strengthened by the indecision concerning Zhao Ziyang. It suggests that political conflict is still continuing at the highest level.

The Central Committee Plenum appointed new leaders without resolving the problem of the new hierarchy of power. The Party secretary of Shanghai, Jiang Zemin, was promoted to General Secretary of the Communist Party and the standing committee of the politburo was enlarged to six with three new members replacing two who had been ousted. All six belong the 'younger' generation with the oldest aged 72 and the youngest 56. However, the only official

KEY FACTS

Official title: People's Republic of China

Head of state: State President, Yang Shangkun

Head of government: Premier of the State Council, Li Peng

Ruling party: Communist Party of China; Chairman of the Military Commission: Deng Xiaoping; General Secretary: Jiang Zemin

Capital: Beijing (Peking)

Official Languages: Putonghua (Mandarin Chinese – Peking dialect)

Currency: Renminbi Yuan (Rmb = 100 fen)

Exchange rate: Rmb3.71 per $ (Nov 1989)

Area: 9,596,961 sq km

Population: 1,100m (1989)

GDP real growth: 11.2% (1988)

GNP per capita: Rmb836* (1988)

Labour force: 528m (1988)

Unemployment: 2.3% (1986)

Inflation: 18.5% (1988)

Trade balance: –$5,500m* (1988)

Foreign debt: $30.2bn (1987)

Visitor numbers: 31m (1988)

* estimated figure

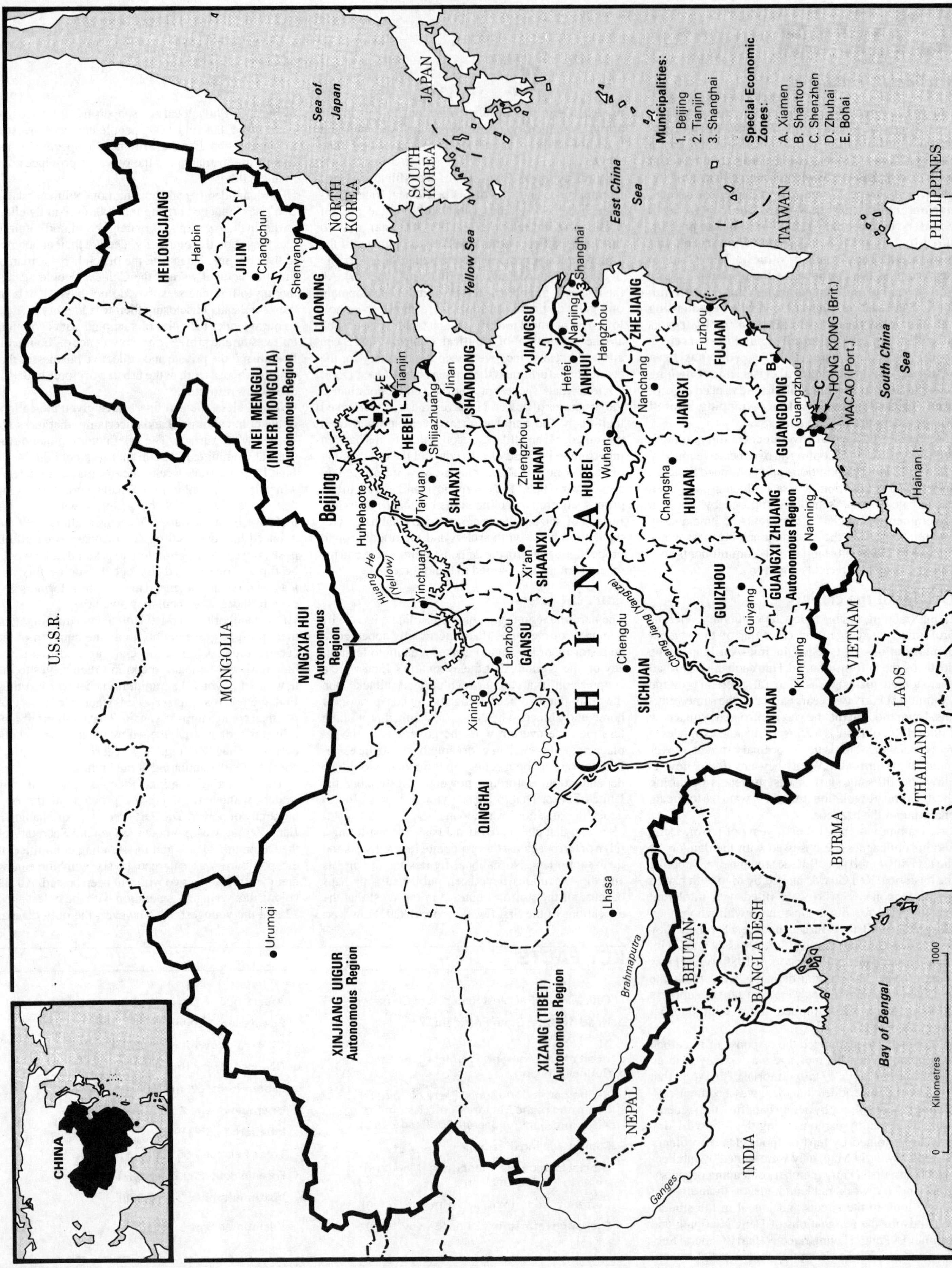
Municipalities:
1. Beijing
2. Tianjin
3. Shanghai
Special Economic Zones:
A. Xiamen
B. Shantou
C. Shenzhen
D. Zhuhai
E. Bohai
CHINA
U.S.S.R.
MONGOLIA
NORTH KOREA
SOUTH KOREA
JAPAN
Sea of Japan
Yellow Sea
East China Sea
South China Sea
TAIWAN
PHILIPPINES
Hainan I.
HONG KONG (Brit.)
MACAO (Port.)
VIETNAM
LAOS
THAILAND
BURMA
BANGLADESH
BHUTAN
NEPAL
INDIA
Bay of Bengal
Ganges
Brahmaputra
Chang Jiang (Yangtze)
Huang He (Yellow)
HEILONGJIANG
Harbin
JILIN
Changchun
LIAONING
Shenyang
NEI MENGGU (INNER MONGOLIA) Autonomous Region
Huhehaote
Beijing
HEBEI
Tianjin
Shijiazhuang
SHANXI
Taiyuan
SHANDONG
Jinan
JIANGSU
Nanjing
Shanghai
ANHUI
Hefei
ZHEJIANG
Hangzhou
FUJIAN
Fuzhou
JIANGXI
Nanchang
GUANGDONG
Guangzhou
HENAN
Zhengzhou
HUBEI
Wuhan
HUNAN
Changsha
GUANGXI ZHUANG Autonomous Region
Nanning
GUIZHOU
Guiyang
YUNNAN
Kunming
SICHUAN
Chengdu
SHAANXI
Xi'an
GANSU
Lanzhou
NINGXIA HUI Autonomous Region
Yinchuan
QINGHAI
Xining
XINJIANG UIGUR Autonomous Region
Urumqi
XIZANG (TIBET) Autonomous Region
Lhasa
Kilometres
0
1000

photograph of this, the highest body of the Party, showed the 82-year-old Yang Shang Kun in central position. The fact that he had no right to be in the photo at all only signalled where the power was located.

The political crisis in China is no longer about succession, but it is about the survival of the communist system itself. A large question mark exists as to how long the regime can outlive its first generation of leaders.

KEY INDICATORS

	Unit	1984	1985	1986	1987	1988
Population	m	1,030	1,040	1,054*	1,069	1,090
GDP	$m	291.4	282.8	271.7	293.4	388.2
GDP (Constant 1980 prices)	Rmb bn	1,283.5	1,624.2	752	823	915
Consumer prices	% increase	8.9	8.8	6.0	7.3	18.5
Agricultural prices	% increase	4.1	8.9	7.2	10.0	–
Industrial output	Rmb bn	701.5	827.77	1,115.70	–	–
Heavy industry	Rmb bn	364.1	429.27	583.30	–	–
Light industry	Rmb bn	337.4	398.47	532.40	–	–
Industrial production	1980=100	141.2*	166.7*	185.2	–	–
Agricultural output	Rmb bn	361.2	380.9	394.7	–	–
Agricultural production	1980=100	151.9	160.1	165.9	–	–
Government consumption of GDP	$m	97.0	115.7	127.3	144.3	–
Exports	$m	26.7	28.2	29.6	39.1	52.5
Imports	$m	26.7	40.8	37.2	38.9	58.0
Balance of trade	$m	447	–12,592	–7,589	240	–5,500
Current account	$m	2,509	–11,417	–7.034	300	–5,370
External debt	$m	12,082	16,722	21,939	30,227	–
Debt service ratio	%	6.5	6.3	7.9	7.1	–
Exchange rate	Rmb	2.32	2.94	3.10	4.0	3.76

* estimated figure

Economic problems deepen

The policies of retrenchment that had been introduced in September 1988 to cool the overheated economy and to reduce inflation worked with mixed success at best before the June crackdown. But the impact of those events amid continuing political uncertainty, the passivity of a demoralised urban workforce and the curtailment of foreign economic relations will have significantly adverse effects upon the economy.

Since 1978, China's economy has grown by more than 10 per cent per annum. Living standards on average have more than tripled. But these ostensibly favourable developments have masked many deep-seated problems. The return to family farming and the greater use of the market mechanism resulted in a rapid expansion of agricultural production until 1984. But since then, the output of grain in particular has stagnated at below the 1984 high. That has given rise to serious concern especially as the population has been increasing by 15m people a year to reach a total of 1,100m in April 1989.

In industry the nub of the problem is that the reforms have undermined the old administrative controls of the centralised system while at the same time effective macro-economic control mechanisms are not in place. Take the question of prices, for example. It is freely acknowledged by all concerned that prices in China are irrational. Some key commodities are still under the strict control of the state planning commission, a broad range of others are subject to varying degrees of central and local control and still others are subject to market fluctuations. Moreover, up to a third of the state's budgetary expenditures are set aside as subsidies to urban dwellers and the loss-making state run enterprises. The attempt at a thoroughgoing price reform in the late spring of 1988 ended in inflation and a run on the shops and the banks. China is in a half-way house between a communal Stalinist economy and a reformed market one: the country suffers the worst of both worlds.

The main objectives of the strategy of retrenchment introduced since September 1988 were: to reduce the rate of economic growth; to cut inflation; to bring investment under control; to bring aggregate demand and supply into equilibrium; to increase farm output and energy supplies. By April 1989 some initial success had been recorded in cutting the scale of fixed investment. The government report claimed that more than 18,000 fixed assets investment projects had been halted or delayed with a saving of 12 per cent of the surplus work load of all projects. There had also been a reduction in institutional purchases. The financial situation was said to be improving with credit being brought under control and savings rising. However, the prime minister warned that the Chinese people must prepare for several years of austerity.

However, other signs suggest that the central government was experiencing severe difficulties in controlling the economy even before the cataclysmic events of June. The austerity programme had hurt the state industries which the government sought to protect. The tighter price controls which were extended to cover commodities that had previously been freed, made it even less attractive for those enterprises producing low, fixed price raw materials to produce more than the minimum necessary to fulfil their obligations. The production of coal, for example, grew by only 2.2 per cent in February 1989. In the course of 1988, while the value of heavy and light industrial output grew by 18.8 and 22.6 per cent respectively, coal output increased by only 4.5 per cent. More generally, within heavy industry, processing industries grew far more rapidly than power and raw materials, further intensifying the problem.

There were problems even in controlling fixed investment. Arising out of the greater economic powers exercised at provincial, local and enterprise levels, a variety of practices have developed to circumvent the policies of the central government. The main one arises out of the two tier price structure that was introduced in 1985. Devised as an incentive to stimulate output based on higher productivity, the scheme allows economic units to sell at market prices output beyond that contracted to the state. The surplus production is usually priced well above the state price – as to be expected in a situation where demand generally exceeds supply. The more important organisations, however, devised arrangements by which they supplied each other's needs at the lowest price, thus ensuring supplies and evading tax simultaneously. The scheme also stimulated widespread corruption in which organisations sold at high prices supplies acquired at the low price.

Another widespread malpractice has had a disastrous effect in agriculture where officials have used funds earmarked for paying farmers for investment purposes (often of a non-productive kind such as guest houses). In the past when farm payments were due, the officials acquired further subventions from the state. But because of the financial squeeze in 1988–89 the extra funds have not been forthcoming and officials have required the farmers to accept deferred payment certificates (i.e. official IOUs). It is reliably reported that in the first half of 1989, the shortfall amounted to half of the 20bn Yuan originally allocated.

Meanwhile, rural industries and construction work has been hit. This has swelled the numbers of unregistered unemployed in both urban and rural areas. The state run industries which account for the bulk of the 100m urban workforce are operating at reduced capacity. Nevertheless workers are still paid bonuses and regular wages and they, together with state officials, are scheduled to receive wage rises in compensation for the continuing inflation. Thus while supply dwindles, while distribution slows and the factories are half idle, consumer demand remains high. This only fuels the inflationary pressure.

The official inflation rate for 1988 was 18.5 per cent, but the December to December rate was 22 per cent. The rate for February 1989 was 27 per cent. But the rate for urban areas was much higher (30–40 per cent). Moreover, the official rate is misleading as it under-represents that on which most disposable income is spent.

Thus the austerity programme has been able to slow down growth without tackling what economists describe as the in-built tendency of the economy to expand. In particular, the government has been unable to impose tight financial discipline or to alter a pricing system that favours processing rather than primary industries. The failure to implement bankruptcy laws or to accept urban unemployment has deepened the existing industrial inefficiencies. Thus it remains to be seen how long the government can press the stop button without having to bow to irresistible pressure to release. The government appears to believe that a further two years of cooling of the economy will allow more balanced growth in a more stable economic environment. Others fear that once the stop button is released, all the problems of overheating will recur even more strongly.

Fiscal problems

The state budget has been running a deficit for several years. The deficit for 1988 was 8,049m Yuan and the planned deficit for 1989 is 7,400m. The minister of finance revealed in his report of March, 1989 that the actual deficit for 1988 – if foreign borrowing (13bn) and domestic bond issues (13.1bn) were excluded from government revenue – was 34.15bn, or more than four times the official figure. The more conservative elements who have predominated since June 1989 regard the continuing budgetary deficits as contributing significantly to inflation.

However, the growth of governmental revenue was consistently below the growth rate of the economy as a whole. According to the minister of finance the proportion of financial revenue to national income

COUNTRY PROFILE

Political system: The political strongman continues to be Deng Xiaoping. Deng owes his position to his own influence and his control of the army (through his position as Chairman, State Military Affairs Commission). But 1987-88 saw an important step in the changing of the guard to a new, younger generation. Zhao Ziyang was made General Secretary of the Party. Li Peng became premier and three other younger leaders in their 50s, 60s and 70s, Yao Yilin, Qiao Shi, Hu Qili, joined the Standing Committee. Yang Shangkun was chosen as the new president in a largely ceremonial office. Jiang Zemim became the Party's general secretary in June 1989. The formal political system is as follows: the National People's Congress (NPC), composed of about 3,000 deputies elected from each region, elects the head of state, the state president, and the national government, the State Council, to administer the country. Choice of the State Council's premier is made on the advice of the Central Committee of the Party. When the NPC is not in session, its decisions are administered by the NPC Standing Committee. Local government is vested in elected people's congresses in 22 provinces, three municipalities (Beijing, Shanghai and Tianjin) and five autonomous regions, Guangxi Zhuang, Nei Mongol, Ningxia Hui, Xinjiang Uygur and Xizang (Tibet), with lesser congresses elected at town and commune level. The highest authority of the Chinese Communist Party is the National Party Congress, whose members are elected by local congresses and which elects an executive central committee. The central committee elects a political bureau (politburo), which is headed by the Party secretary-general.

Main cities/towns: Beijing (Peking, urban population 4.86m in 1986), Shanghai (pop 6.98m), Tianjin (pop 5.38m), Shenyang (pop 4.2m), Wuhan (pop 2.7m), Guangzhou (Canton, pop 2.6m)

Media
Radio: Domestic radio services broadcast in the major languages and dialects, with external services in numerous languages. In 1986 the 167 stations served an estimated 239.4m sets.
Television: TV services are broadcast by Central People's Television Network. The 104 local stations broadcast to 69.7m sets.

Domestic economy: The economy is dominated by more than 100,000 state-owned enterprises, mainly in urban areas, which account for 77 per cent of industrial output, with some 304,000 collective or co-operative enterprises, mainly in rural areas, accounting for 22 per cent. Success with rural reforms which linked household remuneration to output has led to a larger-scale restructuring of the economy away from a rigid centralised planned economy to a mixed economy governed by legislation. The move away from the rigid, centralised system to private enterprise was successful in boosting the economy and a growth rate of 9.3 per cent was reported to the ADB in 1987. But this led to dangerous overheating of the economy. Inflation was officially 7.5 per cent, but in some areas almost 20 per cent. In 1988 there were furious debates about whether essential prices should be allowed. Eventually these were passed but offset by smaller wage increases. A 20-year plan (1981-2000) anticipates a 10-year period of reorganisation, with emphasis on developing agriculture, light industry, tourism, energy and transport, and on raising the standard of living and developing the economic infrastructure. But the 1986-90 plan envisages a slowing of growth, as the country consolidates the achievements of the early 1980s and attempts to rectify the foreign trade deficits caused by capital equipment imports. The government hopes to reduce and eliminate state subsidies, especially for agricultural products, and carry out wage reform to link wages with economic performance. Fujian and Guangdong provinces have had special rights and privileges regarding foreign trade and investment since 1979-80. There are four Special Economic Zones (SEZ), areas intended to attract foreign investment and technology and to minimise the effects of contact with the West; Shenzhen, Zhuhai and Shantou, all in Guangdong province, and Xiamen (Amoy), Fujian province. In addition, 14 ports have been designated Open Coastal Cities and empowered to offer the same incentives and investment conditions as the SEZs, and special areas have been designated Economic and Technical Development Zones in each.

External trade: Rapid opening of the Chinese economy led to growth in leaps and bounds in China's external trade throughout the 1980s. Total trade had gone from very small figures to more than $82bn by 1987. There were worries in China in 1985 when imports rose sharply and created a yawning trade gap but by 1987 the position was improved enough for China to reopen purchase of goods from Japan.
Exports: China's exports rose by 28 per cent in 1987, to $39.5bn.Primary products (food and oil) have declined in significance since the late 1970s, although they still represented over 20 per cent of exports in 1987. Manufactured goods (especially clothing and textiles) have increased sharply with the shift towards small-scale manufacture. Exports rose to 66.4 per cent in 1987. There has been a distinct softening of the official ban on trade with Taiwan. Main destinations are Hong Kong (31.1% of 1986 total), Japan (15.1%), USA (8.4%), UK (4.6%), USSR (3.9%), Singapore (3.8%), Jordan (3.3%), German FR (3.2%), Netherlands (1.5%) and Poland (1.4%).
Imports: Imports rose by 0.7 per cent in 1987 to $43.2bn. Main categories are light industrial and metal products, machinery and transport equipment, chemicals, food, crude materials. Main sources are Japan (28.6% of 1986 total), Hong Kong (12.8%), USA (10.8%), German FR (8.3%), Australia (3.2%), Italy (2.6%), Canada (2.3%), UK (2.3%), France (1.7%), and Brazil (1.6%).

Agriculture: Chinese agriculture employed 75 per cent of the workforce (1986), and has made major advances since 1983 with the shift toward local autonomy and private initiative; rural incomes rose by a real 5.3 per cent in 1987, compared with a 1.7 per cent urban increase. National grain production rose in 1987 to 402.4m tonnes; cotton output was up 18 per cent; rapeseed and other oil bearing crops up another 18 per cent but sugar and jute crops declined. Other important crops include tea (up eight per cent). Numbers of livestock increased. Fisheries developed; aquatic products were up 14 per cent.

Industry: Industrial production has soared in the mid-1980s, due to strong consumer demand and increasing foreign involvement. (It was, however, a cause of overheating and foreign exchange problems). Output rose by 18 per cent in 1985 before slowing to seven per cent in 1986 and rose again by 16.5 per cent in 1987. The greatest increases were by joint venture sector – up 98 per cent. Light industry grew 16.8 per cent and heavy industry developed alongside, up 16 per cent. Basic industries like iron and steel, non-ferrous metal industries and industrial chemical output grew, but there were shortages of raw materials and energy supplies. A third of the urban workforce is now self-employed or working in collectives. There is considerable foreign investment coming from Hong Kong (which accounts for 80 per cent of new funding in the Shenzhen development zone), from Japan and from the USA. Total industrial output by the four special economic zones was worth Yuan 11.2bn in 1987, up 52 per cent. In 1987 China signed new contracts for overseas projects and labour service worth $1.74bn. Heavy industry is concentrated in the north-eastern provinces, coastal areas and in major cities in south and central China, producing pig-iron, steel, rolled steel, timber, cement, plate glass, sulphuric acid, soda ash, caustic soda, chemical fertilisers, pesticides and plastics. Main machinery manufactured includes mining equipment, power generating equipment, machine tools, motor vehicles, tractors, walking tractors, locomotives and steel ships. Light industrial activity occurs throughout the country, and current policy is to modernise light industry in southern China. Main consumer goods produced are bicycles, sewing machines, wrist watches, TVs, radios, cassette recorders, cameras, washing machines and refrigerators. Main light industrial products are cotton yarn, cloth, woollen piece goods, paper products, sugar, beer, cigarettes, chemical pharmaceuticals and detergents. The tourist industry has been expanding. China received 26m visitors in 1987, and earned $1.8bn, up 20 per cent.

Mining: Verified exploitable reserves of coal are estimated at 770bn tonnes, enough to last 400 years at present rates, and coal-mining is being expanded although transport is a major problem. 1987 output was 920m tonnes. Of the coal-mining projects planned for co-operation with foreign firms, the development of the Pingshuo mine reserves at Shanxi (1.5 bn* tonnes) is the most important. There are deposits of practically every extractable mineral in some degree, but exploitation is limited by transport, development costs and sluggish world markets. Copper, bauxite, gold, lead and zinc are receiving top priority. There has been some slowing of the recent boom in foreign petroleum exploration of the continental shelf, estimated to contain 30-50bn barrels of crude oil, but onshore production continues to boom. China is already the world's sixth largest oil producer (134m tonnes in 1987).

Energy: Bottlenecks in energy production have been identified as a major drawback to industrial development, especially in the south-east. Coal accounts for 20 per cent of electricity generation and hydro-electric power for the remaining 20 per cent, but transport problems have made it uneconomic to use coal and/or oil in some regions, and attention is turning to the two planned domestic nuclear power stations in Shanghai and Liaoning, as well as the joint-venture project in Guangdong province. China aims to raise its installed capacity to 570bn kwh by 1990. Power generating units with a total capacity of 8.1m kwh were installed in 1987. Total electricity generated was 496bn kwh.

Membership of international organisations: ADB, CCC, EC trade and co-operation agreement, ESCAP, FAO, IAEA, IBRD, ICAO, IDA, IFAD, IFC, IHO, ILO, IMF, IMO, INTELSAT, ITU, UN, UNESCO, UPU, WFTU, WHO, WIPO, WMO, World Bank.

Time: GMT + 8 hrs.

Climate: Varies considerably from north to south. In the north there are extremes of temperature, ranging from -15°C (Jan) to 40°C (Jul-Aug), with possible dust storms in Apr-May and most rainfall and humidity in Jul-Aug. In the south, the climate is sub-tropical, quite warm until the end of Oct.

Entry requirements
Passport: Required by all.
Visa: Visas are required by all, and are generally arranged for business visitors by a host organisation in China. Tourist visas are also available. Changes in itinerary after arrival may require approval. For travel to all but 257 specified destinations, provinces, travel permits are also required. Regulations have recently been relaxed somewhat, and the list of permit-free areas is growing rapidly.

dropped from 31.9 per cent in 1979 to 19 per cent in 1988. But these figures should not be taken at face value since they pertain only to the revenue sources that fall within the purview of the ministry of finance. State revenues also include the so-called 'extra-budgetary revenues', meaning the earnings of state institutions at all levels. In 1988 these totalled 227bn Yuan, or 88 per cent of the official revenue cited in the budget. Local governments directly controlled only 2.5 per cent of the extra budgetary funds. Some 17 per cent belonged to military and other non-economic institutions. The rest, that is 80 per cent, were controlled by state-owned enterprises and the government departments responsible for them.

There appears to be no mechanism for monitoring, let alone regulating, these funds even though they have been the principal sources for financing what the ministry of finance and state planning commission regard as the excess investment in capital construction responsible for the overheating of the economy in the last few years. The state enterprises and institutions should not be regarded as profit maximising (and therefore cost reducing) entities but rather as bureaucratic organisations concerned to enlarge the scope of their operations and extend their control over resources. Hence they do not respond to macro-economic signals. Instead of seeking ways to reform this system, the conservative leadership has reverted to administrative means. More than 130,000 government inspection teams comprising over 530,000 personnel have been dispatched by the centre to look into tax evasion and other malpractices thought to deprive the government of several billion Yuan of revenue a year. The difficulty is that the government is both player and referee in the game. Tax payments are in essence politically negotiated. As an observer noted, 'what you pay, and to whom you pay it, depends mainly on who you are'.

Meanwhile, by far the largest item of governmental expenditure remains subsidies to loss-making farms and to urban consumers. Together they amount to over 30 per cent of the total. The subsidy is somewhat disguised as that given to loss-making state enterprises is deducted from state revenue. At the same time, the core of government budgeted revenue and expenditure is derived from the taxes and profits of centrally controlled state-owned enterprises. This was the primary reason for introducing the contract management system so as to locate the unit for negotiating tax rates in the individual enterprises rather than in the powerful state bureaucracies within whose systems they operate. But the new system is also open to manipulation. In 1988 the actual profits of locally controlled contracted state enterprises rose by 21.3 per cent, but income and regulatory taxes paid by them fell by 9.5 per cent.

Stagnating agricultural production

A growing concern of the government is the stagnation of grain production since the bumper harvest of 1984, when an output of 404m tonnes was reached. Since then, production has varied between 380m and 400m tonnes. Production in 1988 was 390m tonnes. As the population continues to grow by 15–16m people a year, per capita grain consumption has dropped from 400 kg to 360 kg per annum.

China's leaders are justifiably proud of the fact that with only 7 per cent of the world's arable land, the country is basically able to feed and clothe 20 per cent of the world's population. But the margins are narrow. The initial successes of the early 1980s, when agricultural output increased markedly, were regarded as the result of reforms that allowed the return to family farming after 25 years of collectivisation and that permitted a greater role to the market. But it was not realised that the increased yields in large part were due to better utilisation of the capital construction (drains, irrigation etc.) that had taken place during the years of collectivisation. As a result, investment in agriculture was relatively neglected. In addition to a relative reduction in inputs, the neglect of ecological matters and the expansion of rural housing and industry led to the decline in arable land – roughly in the order of 0.5 per cent a year. Moreover, as industrial price rises reduced the profit margins of peasants they shifted from grain production where a good proportion of the output was fixed at low state prices to the more profitable cash crops.

Belatedly in 1988, the government increased investment in agriculture by 15 per cent. But even so, at 18.1bn Yuan that only accounts for just over 7 per cent of budget expenditure. For 1989 the government decided to increase the compulsory purchase price of grain by 18 per cent and to give preference in the supply of artificial fertiliser to grain producers. However, the increase was more than offset by the increase in the cost of diesel oil, fertiliser, pesticides, farm machinery etc. The government was therefore resorting to administrative rather than economic measures to raise grain production. This would necessarily create distortions and reduce efficiency. The target for 1989 was 410m tonnes. Meanwhile, China had once again to resort to the world market to make good the shortfall in supplies.

The austerity programme hit the countryside hard. The fastest growing sector in industry was the private and collective enterprises in the new townships in the rural areas. The value of their output has grown to exceed the value of agricultural production even though they employ only 15 per cent of the 380m rural labour force. They have been an important source of enhanced income to peasant families. However, with supplies and capital becoming scarcer, these firms have been hard hit; rural employment has been hard hit; social order is being threatened as millions shift around the country in search of work. At one point in the spring of 1989, for example, some 30,000 a day were pouring into Canton's railway station. By the end of March, 2.5m such people were calculated to have descended upon Guangdong Province since the beginning of the year. Measures were taken by the police and security forces to stem the flow at source.

The Politicisation of education

On the eve of the student demonstrations, Deng Xiaoping told a visiting foreign dignitary that however deep the economic problems, those of education were even worse. He then alluded to 10 years of relative neglect of education. Deng's remarks have subsequently been interpreted to mean the neglect of political education.

That alleged problem has now been addressed. In August 1989, it was announced that the university intake of new students in the autumn would be reduced by 30,000. China's premier university, in Beijing, will admit only 800 compared to 2,200 in 1988. Moreover, the first year will be spent in army barracks, where the student freshmen will receive political education in suitably spartan conditions from the army that suppressed the 'counter-revolutionary rebellion'. It was also announced that China would continue to send students abroad, but the 30,000 or so graduates who are notionally eligible would first have to spend a year or two in factories or in the countryside.

This reversion to Maoist practices reflects the peculiar outlook of the hardline leadership in Beijing which believes such measures can revive the spartan idealism of the 1950s.

Yet genuine education is in crisis in China. Ten years of relative neglect of education has resulted in a demoralised and underpaid cohort of teachers. More than 230m people (just under a quarter of the population) are officially described as illiterate. The prevailing attitude is officially said to be that 'it is useless to go to school to study'. There is widespread resistance not only to the nine-year system of compulsory education, but even to universal *elementary* education. Peasant worker families apparently believe that their children can be more profitably employed elsewhere than in school. In a country of 1,100m people, according to official figures there were in 1988 only 113,000 postgraduate students and 2,066,005 undergraduates in institutions of higher learning. A further 1,728,000 adults attended university and college courses. Only 13m youngsters attended secondary schools and 165m were enrolled at primary schools.

These are dismal figures for a country intent upon modernisation. Per capita spending on education ranks China among the lowest in the world. Although the 1989 budget calls for a 10 per cent increase in spending on education after a 15 per cent increase in the previous year, the true priorities of the authorities are revealed by other figures. The consumption spending by institutions during 1987 was more than twice as high as all educational expenditure during the previous year. The fact that China's leaders have chosen this year to reduce student enrolment by two thirds for political reasons reflects a view of priorities that can only damage the prospects for modernisation in the long run.

Foreign economic relations

As a result of the austerity programme, foreign economic activity began to slow down in early 1989. Nevertheless, the overall outlook was good, especially in the eastern and southern coastal regions. In 1988, for example, trade grew by nearly 25 per cent to top $100bn; $8.94bn in foreign capital was actually used (an increase of 16.4 per cent over the previous year); and approval was given to nearly 6,000 enterprises to use foreign investment (the largest number in any year). Moreover, $1.3bn was earned in contracted projects abroad and international tourism brought in $2.2bn.

More broadly, a pattern of variegated foreign economic relations emerged that seemed to promise much despite signs of discontent in some regions. The south-eastern provinces of Guangdong and Fujian were becoming integrated in an external Chinese nexus involving the overseas Chinese of South-east Asia as well as Hong Kong and increasingly Taiwan. Although precise figures are elusive, Hong Kong had become a major conduit for these relations. Between 2–3m of the workforce for Hong Kong factories were now located in southern China. Individuals and companies from Taiwan established joint ventures and invested in Fujian as well as conducting trade to the value of $2.8m. These all tended to be on a relatively small scale. Large-scale investment by Western, Japanese and American firms were also directed mainly to the coastal areas. Liaoning in the north-east being particularly attractive to Japan. Major infrastructure investment was developed through loans from the World Bank, the Asian Development Bank and Japanese ODA loans.

But much of this halted or suspended as a result of political uncertainty in China. Western sanctions and the reluctance of the international business community to invest anew in China hurt the Chinese economy although at the end of 1989 it was difficult to gauge the severity of the set-back. The most visible change was caused by the cut back in tourism. Hotels were down to occupancy rates of 15–25 per cent only. Some of this was likely to pick up as the emotional reaction to the crackdown receded, but it would clearly take many months if not longer, before there would be a return to the 1988 figures.

The Chinese Government has sought to make much of the new opening to the Soviet Union and Eastern Europe. But while the provinces bordering on the Soviet Union (Heilongjiang in Manchuria, Inner Mongolia and Xinjiang) may hope for some local invigoration, the prospects as a whole are modest. The trade agreements for 1989 which have the advantage of not involving hard currency came to just over $6bn for all the socialist countries – or about 5 per cent of China's foreign trade for 1988. It is useful without being decisive.

Foreign trade

China's overall import and export trade value increased by 24.4 per cent in 1988 to reach a total of $102bn. Exports grew by 20.6 per cent and imports by 27.9 per cent. The deficit in the first six months of 1989 grew to $3.82bn, thought to be due to a decline in the growth of exports (6.1 per cent) because of low state procurement prices for export goods and because of the rising costs of production. Imports were cut to basic materials but because of the shortfalls in supply within China, imports continued to grow at previous rates to meet the demand.

The expected decline in Western business interest in China is unlikely to be reflected as greatly in trade with China. The former was bound up to a large extent in exaggerated expectations about the prospects of the Chinese domestic market and about continued market liberalisation. Trade, however, is less risky and China has a good record as a trading country despite its recurrent domestic upheavals. Although China must now adjust to a less welcoming international economic environment in which soft loans will be harder to come by, its trade should be the least affected of its broad range of foreign economic relations.

Foreign policy

Ironically, China has suffered its gravest setback in foreign relations in 20 years, precisely at the point when it had at last assured itself of the peaceful international environment deemed necessary for the domestic goals of modernising the country.

China acted as host to President Bush in February and to President Gorbachev in May of 1989. Neither objected to the visit of the other. For the first time, China was able to enjoy good and normal relations with both superpowers simultaneously. The events of the previous 18 months had shown that the two superpowers could reach agreements on arms control measures and on diffusing regional conflicts between themselves without involving China. Even the Soviet withdrawal from Afghanistan was negotiated and acted upon without any Chinese involvement. This meant that China was no longer of central concern to the two superpowers in the management of their strategic relations. If that in turn meant that China was somewhat marginalised in global affairs, it also yielded tangible benefits in significantly reducing the military threat of the Soviet Union and in proffering China greater opportunities to engage in an omni-directional foreign policy and in asserting a more coherent regional role.

The summit meeting with Gorbachev was overshadowed by the student demonstrations, but it promised to result in substantial benefits to both sides. The Soviet Union was committed to withdrawing its airforce and three-quarters of its ground forces from Mongolia. A further 200,000 troops were to be withdrawn from 500,000 or so deployed to the north of China. If fully carried through, these measures would substantially reduce the Soviet threat to China's heartland for the first time in 20 years. China also hoped through normalising relations to stabilise the Korean peninsula by ending Kim I Sung's capacity to play off the two communist giants for his own purposes. China also hoped to reduce still further the military claims upon the budget and to improve the economic conditions of the three provinces bordering the Soviet Union as well as to upgrade some of its heavy industry facilities and energy and transport supplies. For his part, Gorbachev gained an important foreign policy triumph, cut back on military expenditure, gained access to Chinese consumer goods and hoped through the good offices of China to gain access to the Pacific Basin. He also hoped that China would be instrumental in attracting joint ventures and development in the Soviet far east. Meanwhile, the two reforming communist giants could develop normal relations based on a sober recognition of common and conflicting interests.

The new relationship with the Soviet Union was achieved with the blessing of both Japan (which Premier Li Peng visited in April) and the United States. China was in no danger of losing its preferential position in the western Pacific. Moreover, the new relations with the Soviet Union increased the pressure on Vietnam which was now firmly committed to withdrawing its troops from Cambodia by the end of September. That also opened the door to re-establishing contacts with Vietnam at vice foreign ministerial level for the first time in 10 years.

However, these favourable developments were rocked by the Beijing massacre and its aftermath. Outraged Western governments reflecting the horror felt by their peoples and parliaments imposed sanctions in the form of postponing loans and high level contacts. The United States also stopped sending military-related technology. Any hopes that China's leaders entertained of turning to the Soviet Union to fill the gap were swiftly dashed. In addition to being in dire need of capital itself, the Soviet Union under Gorbachev's leadership sought reform at home and a new partnership with the West. Therefore, it saw no advantage in aligning itself with the hardliners in Beijing.

The West is unlikely to impose sanctions against China for a long period especially once martial law is lifted and the domestic suppression is relaxed. But Western countries are unlikely to restore relations to their previous warmth. Moreover, once the outside world witnessed the ruthlessness with which military force was used against students and civilians at home, the United States was likely to reappraise its policy of supplying China with military-related technology in view of the misgivings of China's neighbours, many of whom are old friends and allies of the United States.

Hong Kong

The events of April–June 1989 had a dramatic impact on Hong Kong. The massacre and its aftermath destroyed any credibility that the People's Republic of China still enjoyed with regard to honouring its future commitments to the territory. But these events also politicised the people of Hong Kong. In May, hundreds of thousands and perhaps even a million people demonstrated their solidarity with the pro-democracy movement in Beijing. Never before had such vast numbers discovered their political voice. They identified themselves as both Chinese and Hong Kongese. The lack of faith in the communists led to demands for the right of abode in Britain that were unlikely to be realised. It also increased the rate of emigration of young professionals. But it also transformed the process of transition to 1997: the rate of democratisation was likely to be accelerated significantly. A Bill of Rights would also probably be promulgated. The transitional period of eight years to the attainment of Chinese sovereignty in 1997 was likely to be marked by more acrimony between the parties concerned.

In the harsher international environment, China's need for Hong Kong as an international gateway is probably greater than before, but at the same time the mood of defensive truculence in Beijing will make it more difficult for the negotiators of the People's Republic of China to take a more conciliatory approach. These are testing times for both China and Hong Kong.

BUSINESS GUIDE

Currency: It is illegal to use foreign currency within China PR. Foreign Exchange Certificates must be obtained from the Bank of China to pay for services in hotels, shops, etc, air and rail fares and places catering solely for foreigners. RMB may be requested, but cannot be taken out of the country although the FEC certificates can (and are valid for six months). Foreign currency must be declared on entry, and it is always best to retain details of this and all exchange transactions.
Customs: Certain items produced in China before 1949, such as embroidery, silks, porcelain, scrolls and objets d'art, may be subject to export restrictions.

Health precautions
Mandatory: Vaccination certificates required for cholera or yellow fever if travelling from an infected area.
Advisable: Anti-malarial precautions. There are occasional outbreaks of cholera in southern China.

Representation overseas: Accra, Addis Ababa, Aden, Algiers, Amman, Ankara, Antananarivo, Apia, Athens, Baghdad, Bamako, Bangkok, Banjul, Bangui, Beirut, Belgrade, Berlin, Berne, Bissau, Bogotá, Bonn, Brasilia, Brazzaville, Brussels, Bucharest, Budapest, Buenos Aires, Bujumbura, Cairo, Canberra, Caracas, Christchurch, Colombo, Conakry, Copenhagen, Cotonou, Dakar, Dhaka, Damascus, Dar es Salaam, Dublin, Freetown, Gaborone, Georgetown, The Hague, Hanoi, Harare, Havana, Helsinki, Islamabad, Kabul, Kampala, Kathmandu, Khartoum, Kigali, Kingston, Kinshasa, Kuala Lumpur, Kuwait, Lagos, Libreville, Lima, Lisbon, Lomé, London, Lusaka, Luxembourg-Ville, Madrid, Manila, Maseru, Mexico, Mogadishu, Monrovia, Moroni, Moscow, Muscat, Niamey, Nairobi, New Delhi, Nicosia, Nouakchott, Ouagadougou, Ottawa, Paramaribo, Paris, Port Louis, Port Moresby, Prague, Pyongyang, Quito, Rabat, Rangoon, Reykjavik, Rome, Sana'a, Santiago, São Tomé, Sofia, St Julian's, Stockholm, Suva, Tehran, Tirana, Tokyo, Tripoli, Tunis, Ulan Bator, Victoria, Vienna, Vientiane, Warsaw, Washington, Wellington, Yaoundé.

Air access: Services are improving, but CAAC flights are still erratic. When flying from Hong Kong via Guangzhou, it will be necessary to stop over and arrange forward flights into the interior (no advance booking for destinations other than Beijing).

National airline: CAAC (Civil Aviation Administration of China).

Other airlines: Aeroflot, Air France, Air Mongol, Alitalia, All Nippon Airways, British Airways, Canadian Airlines International, Cathay Pacific, Chosonminhang Korean Airways, Dragonair, Ethiopian, Finnair, Interflug, Iranair, Iraqi Airways, JAL, JAT, LOT-Polish, Lufthansa, MAS, Northwest, Philippine, PIA, Qantas, SAS, Singapore, Swissair, Tarom, Thai International, United.

Tax: An airport tax of Rmb20 is payable on departure from China.

International airports: Capital (PEK), 26 km from Beijing (Peking), with duty-free shop; Hongqiao (SHA), 12 km from Shanghai, with restaurant, shops; Baiyun (Can), 7 km from Guangzhou (Canton).

BUSINESS GUIDE

Surface access: The Kowloon-Canton Railway Corp (KCR) is the main carrier of passengers to and from China, with express trains serving Kowloon-Guangzhou and an indirect Kowloon-Lowu service. The Trans-Siberian Express operates two weekly services, one USSR train serving Moscow-Beijing via Harbin and one Chinese train via Ulan Bator. Bus services link Guangzhou to the Hong Kong border, and it is possible to hire a taxi for the three-hour journey (about $100). Hovercraft services operate three times a day between Hong Kong and Guangzhou, and also serve Shekou, Shenzhen and Zhuhai. A motorway is under construction between Hong Kong and Guangzhou.
Main ports: Tianjin (containers), Dalian, Guangzhou, Huangpu (containers), Lianyungang, Qingdao, Qinhuangdao, Shanghai (containers).

Hotels: There is a shortage of accommodation in peak seasons, and visitors often complain about low standards, service, the plumbing and the occasional mice. The main hotels in major cities are of a reasonable standard, but most hotels are frugal, often with fixed-time, fixed-menu meals and even cold water only during certain hours in the evening and morning. There are new hotels built with foreign assistance (in Nanjing, Guangzhou, Beijing and Zhanjiang) and international-standard joint-venture hotels (in Tianjin, Hainan, Xiamen, Fujian, Hangzhou and Shenzhen). Charges for government guest houses, which are used to accommodate hotel overflows, are high. Reservations for business visitors are made by their host organisations. Chinese-foreign joint-venture hotels are able to accept bookings from outside China. Hotel reservations for over 70 Chinese cities are being computerised.

Credit cards: Credit cards are all but unknown, although international-standard hotels and restaurants may accept major cards such as American Express, Visa or Mastercharge. Over 40 cities accept Visa and Mastercharge for payment in Friendship stores or for cash advances at some Bank of China branches. Use of cash or travellers' cheques is more usual. Cash withdrawals from banks are possible with major cards, but are not encouraged and frequently cause long delays.

Car hire: Self-drive car hire is not available. Chauffeur-driven cars may be hired at a standard rate on a daily basis. Bicycle hire is available in some towns, but it is advisable to carry proof of identity when riding.

City transport
Taxis: Taxi service is available in all major cities. It can be difficult to find a taxi except at hotels, and it is likely to be time-consuming. Metered taxis may be hailed in the street in Guangzhou and Shenzhen. Taxis hired by the day charge on the basis of distance travelled and waiting time. If possible, it is best to retain a taxi until returning to the hotel, paying the driver a small waiting fee during appointments or meals. Tipping is not practised.
Metro: Beijing has an underground railway.

National transport
Air: CAAC provides regular services between the major cities, with first-class service on some routes. Flights are always overbooked and seats should be confirmed as a matter of course. Independent regional airlines also operate. Tickets not booked through an official guide/travel service should be booked and collected well in advance. Allow plenty of time for inevitable and often prolonged delays in services. Airport announcements are not multilingual.
Road: There are 915,000 km of internal roads, but most are narrow and poorly surfaced, making long-distance travel time-consuming. A super-highway is under construction to link Beijing and Tianjin.
Buses: There are extensive local bus services in main cities, generally inexpensive and crowded. Detailed route maps are available in Beijing, but lack of maps in English makes bus transport difficult in most places.
Rail: The rail network has over 53,000 km of track, with some electrified services, although most locomotives are steam-powered. Rail services operate between main cities. Deluxe rail services, with opulent German-made sleeping cars and private dining coaches, are available. Generally, rail travel is comfortable, but time-consuming because long distances are involved.
Water: Inland waterways and coastal shipping services are an important form of transport.

Public holidays
Fixed dates: 1 Jan, 1 May (Labour Day), 1-2 Oct (National Day).
Variable dates: Chinese New Year (up to three days).

Working hours: Sun is taken as a weekly holiday.
Government and business: (Mon-Sat) 0800 – 1200, 1300 – 1700.

Social customs: Tipping is not practised, although it is courteous to thank hotel and restaurant staff for their services. It is customary to present a business card. Taiwan (Formosa) is considered a province of China, and should not be referred to as a country.

Telecommunications: China plans to spend $22bn in 1986-90 on developing the overstretched telecommunications system. In particular it hopes to quintuple the telephone system to 33m lines by 2000 (reaching 10m by 1990).
Telephone: Telephone, telex and telegraph calls from China frequently arranged by hotels but also possible through post offices; it is often recommended that international calls are paid for with a credit card as this may be cheaper; outgoing international calls are somewhat more expensive than calls from abroad to China. It is often possible to dial directly out of the country from major hotels, etc; although it takes patience, line quality is often excellent. Dialling for China is mostly via the International Operator. IDD code 86, followed by 1 for Beijing, 411 for Dalian, 591 for Fuzhou, 20 for Guangzhou, 513 for Nantong, 532 for Qingdao, 21 for Shanghai, 24, for Shenyang, 755 for Shenzhen, 22 for Tianjin, 577 for Wenzhou, 27 for Wuhan, 592 for Xiamen. Advance booking of telephone calls within China generally recommended; outside main cities service may not be so extensive or fast but telephone service largely automatic. Visitors are often recommended to keep a note of important telephone numbers (e.g. hotel, embassy) as directories may not be readily available. There were some 8m telephones in 1987, of which 5m were in towns and cities.
Telex: Telex facilities are available from Administration of Telecommunications offices in Beijing (automatic service) and Shanghai.
Telegram: Telegrams service is available from Administration of Telecommunications offices in Beijing and Shanghai, or from any telegraph office.

Postal service: Speedpost is now available so that letters can be sent or received within 48 hours from most parts of the country and of the world. There are domestic courier services.

Banking
Central bank: People's Bank of China.
Foreign exchange bank: Bank of China.
Domestic banks: Agricultural Bank of China, Bank of Communications, China International Trust & Investment Corp, China Investment Bank, China & South Sea Bank, China State Bank, Kincheng Banking Corp, Kwangtung Provincial Bank, National Bank of China, National Commercial Bank, National Industrial Bank of China, People's Bank of China, People's Construction Bank of China, Sin Hua Trust, Savings & Commercial Bank, Yien Yieh Commercial Bank.
Chinese branches: Bank of East Asia, The Chartered Bank, Hongkong and Shanghai Banking Corporation, Oversea-Chinese Banking Corporation, Standard Chartered Bank.

Advertising: Advertisements may be placed in the daily press, on TV and on radio. Outdoor advertising may be used in certain urban sites. Contact the Ministry of Foreign Economic Relations and Trade.

Trade fairs: Woodworking Machinery Show (Mar, Beijing), China Textile Machinery Exhibition (Apr, Beijing), 3rd International Agro Expo (Apr, Beijing), International Electronic Production Conference (Apr, Beijing), 5th Wire Asia Conference and Exhibition (Apr, Shanghai), 5th International Exhibition of Construction Equipment and Materials (May, Shanghai), Miconex (May, Beijing), 3rd International Exhibition on Die and Mould Technology and Equipment (May, Shanghai), 14th World Mining Congress and Exhibition (May, Beijing), 6th International Exhibition on Automotive Industry (Jun, Beijing), Energy China (Jun, Beijing), Non Destruction Inspection and Industrial Measurement (Jul, Beijing), China Petroleum and Natural Gas (Nov, Beijing).

Electricity supply: 220 V AC, 50 cycles, with 2- or 3-pin flat plug fittings, with 2-pin round as well in some hotels, and mostly screw-type light bulb fittings.

Weights and measures: Metric system (with Chinese units in use).

Representation in capital: Afghanistan, Albania, Algeria, Argentina, Australia, Austria, Bangladesh, Belgium, Benin, Brazil, Bulgaria, Burkina Faso, Burma, Burundi, Cameroon, Canada, Central African Republic, Chad, Chile, Colombia, Congo, Cuba, Czechoslovakia, Denmark, Egypt, Equatorial Guinea, Ethiopia, Finland, France, Gabon, German DR, German FR, Ghana, Greece, Guinea, Guyana, Hungary, India, Iran, Iraq, Ireland, Italy, Japan, Jordan, Kenya, Korea PDR, Kuwait, Laos, Lebanon, Liberia, Libya, Luxembourg, Madagascar, Malaysia, Mali, Malta, Mauritania, Mexico, Mongolia, Morocco, Nepal, Netherlands, New Zealand, Niger, Nigeria, Norway, Oman, Pakistan, Peru, Philippines, Poland, Portugal, Romania, Rwanda, Senegal, Sierra Leone, Somalia, Spain, Sri Lanka, Sudan, Sweden, Switzerland, Syria, Tanzania, Thailand, Togo, Tunisia, Turkey, Uganda, USSR, UK, USA, Vanuatu, Venezuela, Vietnam, Yemen AR, Yemen PDR, Yugoslavia, Zaïre, Zambia, Zimbabwe.

Hotels: Beijing
Beijing, 33 Dong Chang'an Avenue, Cnr Wangfujing St, (tel: 558331; tx: 22426).

Beijing International Airport, Beijing Capital Airport (tel: 522931, 552515).

Beijing-Toronto (Jinglun), Jianguo Menwai Street (tel: 5002266; tx: 2100112; fax: 5002022).

Beiwei, Bei Wei Lu, Xuan Wu (tel: 338631).

Diaoyutai Guest House, San Li He Lu (tel: 668541).

Fragrant Hills (Xiangshan), Fragrant Hills Park (tel: 285491; tx: 22626).

Great Wall Sheraton, Lian Ma Qiao Dong San Huan Bei Rd, North Donghuan Road No 6 (tel: 5005566; tx: 22002; fax: 5001938).

Holiday Inn Lido, Jichang Road, Jiangtai Road (tel: 5006688; tx: 22618; fax: 5006237).

Hua Du, 8 Xinyuan South Road, Chaoyang District (tel: 5001166; tx: 22028; fax: 5001615).

Jianguo, Jianguomenwai Da Jie (tel: 5002233; tx: 22439; fax: 5002871).

Nationalities (Minzu), Fu Xingmennei Avenue (tel: 658541).

Overseas Chinese (Hua Chiao) Mansion, Wang Fu Jing Main Street, Beikou (tel: 558851).

Peace (Heping Bin Guan), 4 Chin-yu Hu-T'ung (tel: 555131, 558841; fax: 5731).

Qianmen, 175 Xongan Road, Xuanwu District (tel: 338731).

Shangri-La, 29 Zhzhuyuan Road 100081 (tel: 8412211; tx: 222322; fax: 1471/2).

Xi Yuan, Erligo, Western District, Nr Zoo (tel: 868821; tx: 228315).

Xin Qiao, 2 Dongjiaomin Xiang, Chongwenmen (tel: 557731).

Yanjing (Yanjing Fandian), 19 Fuxingmenwai Ave (tel: 8687216; tx: 20028).

You Yi Friendship, Baishiqiao Road, Haidian District (tel: 890621).

Hotels: Guangzhou
Baiyun, Huanshi Dong Road (tel: 67700).

Central, Guoji Lu, Sanlyuanli (tel: 678331; tx: 44664).

China, Liu Hua Lu (tel: 666888; tx: 44888; fax: 677014).

Dong Fang, Liu Hua Road (tel: 669900; tx: 444439).

Garden, 368 Huanshi Road, Dong Lu 510064 (tel: 338989; tx: 44788; fax: 350467, 350706).

Guangzhou, Haizhu Square (tel: 338168, 61556).

Liu Hua Guest House, Huanshi Lu West (tel: 678398; tx: 44298).

Nanhu Resort, Nanhu South Lake, Tonghe, Shahe (tel: 777930, 776367; tx: 44298).

Overseas Chinese (Hua Chiao), Zhan Q'an Road (tel: 663488; tx: 3931).

Renmin Dasha (People's Mansion), 207 Chang Di (tel: 661445).

Revolving Palace, Zumialou, Foshan (tel: 85622; tx: 1374).

White Swan (Baietan), 1 Southern Street, Shamian Island (tel: 886968; tx: 44688; fax: 861188).

Hotels: Nanjing
Jinling, Zin Jie Kou Square (tel: 741999, 742888; tx: 34110; fax: 643396).

Nanjing, 259 Zhong Shan Bei Lu Road (tel: 34121; tx: 34102).

Hotels: Shanghai
Cypress (Longbo), 2419 Honqiao Road (tel: 329388; tx: 33288; fax: 329323).

Dahua Guest House, 914 Yan'an Road, West Shanghai (tel: 523079).

Hengshan, 534 Hengshan Road (tel: 377050; tx: 33009).

Hilton International, 250 Hua Shan Road (tel: 550000; tx: 33612; fax: 553848).

Hua Ting Sheraton, 1200 Cao Xi Bei Lu (tel: 391000; tx: 33589; fax: 550830).

International (Guoji), 170 Nan Jing Road West (tel: 225225, 291010).

Jing'An Guest House, 370 Hua Shan Road (tel: 563050; tx: 7777).

Jinjiang, 59 South Mao Ming Nan Road (tel: 534242; tx: 33011).

Nikko Longbai, 2451 Hongqiao Road (tel: 593636; tx: 30138; fax: 593012).

Overseas Chinese (Hua Chiao), 104 Nan Jing Xi Road, West Shanghai (tel: 226226; tx: 4321).

Peace (Heping), 20 East Nanjiag Road (tel: 211244).

Shanghai Guest House, 505 Bei Lu (Nr Jing'An Si) (tel: 312312; tx: 33002).

Shanghai Mansions, 20 Soochow Road (North) (tel: 246260, 244186; tx: 33007).

Hotels: Shenzhen
Bamboo Garden, Tung Men Bei Road (tel: 222934, 221848; tx: 420390; fax: 234835).

Century Plaza, Kin Chit Road (tel: 220888; tx: 81320; fax: 234060).

East Lake, Shenzhen Reservoir (tel: 22727; tx: 42236).

Honey Lake Country Club (Xiangmihu Dujia, Shennan C Road (tel: 23493).

Jingling Guest House.

Nanhai, Shenkou (tel: 92888; tx: 42879; fax: 92440).

New World (UTC).

Overseas Chinese, 45 Wo Ping Lu (tel: 22811).

Shenzhen Bay, Shahe, c/o 1411 New World Ctr, E Wing, Salisbury Rd, H.K. (tel: 38459, 38464).

Shenzhen International (Yayuan Binguan), 1 Dong Men Bei Lu (tel: 259369, 259440; tx: 420299; fax: 230702).

Shenzhen-Shanghai.

Hotels: Tianjin
Astor, 33 Taier Zhuang (tel: 311688; tx: 23266).

Crystal Palace, You Yi & Binshui Roads (tel: 310567; tx: 23277; fax: 310591).

Hyatt, Jie Fange Road North (tel: 318888; tx: 23350; fax: 310021).

Sheraton Tianjin Hotel, Zi Jin Shan Road, He Xi District (tel: 313388; tx: 23353; fax: 318740).

Hotels: Wu Han, Hubei province
Jianghan, 211 Shengli Street, Hankou (tel: 21253; tx: 40150).

Useful telephone numbers: Beijing
Police: 110.

International calls, English-language: 337431/553536.

Local, long-distance enquiries: 116.

Cable and telex information: 664900.

Taxis: 557671.

Airport-flight enquiries: 552515/555531, ext 382.

Business advice and information
Administration of Telecommunications, 11 Xichanganjie Street, Beijing; Nanjing Road East 30, Shanghai.

Centre for Market and Trade Development, International Trade Research Institute, Ministry of Economic Relations and Foreign Trade, Andingmenwai, Beijing (tx: 22168 MFTPK CN).

China Council for the Promotion of International Trade, Fuxingmenwai Dajie (tel: 662835, 507766 ext 7107; tx: 22315 CCPIT CN; cable: COMTRADE BEIJING).

China International Trust and Investment Corporation, 2 Qianmen Dong Dajie, 14th floor, PO Box 6200, Beijing (tel: 757181, 753600; tx: 22305 CITIC CN; cable: 0207).

China National Chemicals Import & Export Corp, Erligou, Xijiao, Beijing (tel: 890931, 891179; tx: 2243).

China National Instruments Import & Export Corp, Erligou, Beijing (tel: 890931; tx: 22242).

China National Light Industrial Products Import & Export Corp, Donganmen Dajie, Beijing (tel: 555680; tx: 22282).

China National Machinery Import & Export Corp, Erligou, Beijing (tel: 890931; tx: 22242).

China National Metals & Minerals Import & Export Corp, Erligou, Beijing (tel: 890931, 892376; tx: 22241).

China National Offshore Oil Corporation, PO Box 2118, Beijing (tel: 550231; tx: 22186 EQUIP CN).

China National Technical Import Corp, Erligou, Beijing (tel: 890931, 892116; tx: 22244).

China Ocean Shipping Agency, Dong Changan Jie, Beijing (tel: 553424; tx: 22264).

Chinese Export Commodities Fair, Guangzhou Foreign Trade Centre, Guangzhou Customs Administration, Xi Changan Jie, Beijing (tel: 556105, 5579984; Executive Office: 30848,32622).

Customs General Administration of the People's Republic of China, Taipingqiao Dajie, Beijing (tel: 661437, 662564).

International Trade News, Director, Advertising Department, Building 3, Ministry of Foreign Trade Compound, Beijing (tx: 22168 MFTPK CN).

Ministry of Foreign Affairs, Chaoyangmennei Dajie, Beijing (tel: 555831, 552190, 550257, 558801).

Ministry of Foreign Economic Relations and Trade, 2 Dong Changan Jie, Beijing (tel: 553031; Customs, ext 548; tx: 22168 MFTPK CN, 22560 FICPO CN).

Shanghai Advertising Corp, Yuanmingyuan Lu, Shanghai (tel: 21366; cable: ADVERCORP).

State Administration for Industry and Commerce of the People's Republic of China, 10 Sanlihe Dong Lu, Xichengqu, Beijing (tel: 863031; cable: 2092).

State Administration of Import and Export Commodity Inspection, 2 Dong Changan Jie, Beijing, (tel: 553031 ext 433, 554302; cable: 7526).

Banking
People's Bank of China (central bank), San Li He, West City, Beijing (tel: 863907; tx: 22612).

Bank of China, Bank of China Bldg, 410 Fuchengmen Nei Dajie, Beijing (tel: 118311; tx: 22254).

Bank of Communications, 200 Jiang Xi Rd, Shanghai (tel: 213400; tx: 33438).

China & South Sea Bank Ltd, 17 Xi Jiao Min Xiang, Beijing.

Kincheng Banking Corporation, 17 Xi Jiao Min Xiang, Beijing (tx: 73405).

Kwangtung Provincial Bank, 17 Hsi Chiao Min Hsiang, Beijing.

National Commercial Bank Ltd, 17 Xi Jiao Min Xiang, Beijing.

National Industrial Bank of China, 130 Peking Road East, Shanghai.

Yien Yieh Commercial Bank Ltd, 17 Xi Jiao Min Xiang, Beijing.

Travel information
China International Travel Service, Dong Changan Jie, Beijing (tel: 554192, 755374; tx:22350; fax: 5122068).

China Travel Service (HK) Ltd, 77 Queen's Road, Central, Hong Kong (tel: 5 259121, 5 215247; tx: 73344 HKCTS HX).

Visa Office of the Ministry of Foreign Affairs of the People's Republic of China in Hong Kong, NCNA Building, 387 Queen's Road East, Hong Kong (tel: 5 744163/083; tx: 72206 VOPRC HX).

Colombia

Peter Nares

Colombia's 1990 presidential election campaign, when it got under way in mid-1989, aroused scant public interest. The mainstream candidates of the two largest parties, the ruling Liberals and the opposition Conservatives, had little new to offer voters, either in political or economic spheres. They produced no credible blueprints for terminating the 25-year guerrilla war or for countering the rising security threat posed by drug gangs and right-wing paramilitary groups. While the candidates canvassed for votes, political turbulence prevailed in many rural regions as extremists of the right and left fought it out for supremacy. Yet despite the relentless violence, the economy remained relatively vigorous and, to the surprise of many foreign observers, the convulsed Andean republic continued to attract both foreign investment and credit. There were fears, however, that economic activities might slacken in the latter half of 1989 because of the security situation, high interest rates, steep inflation and external factors such as the uncertain future of the International Coffee Agreement (ICA).

President Virgilio Barco

As Liberal President Virgilio Barco neared his last year of office, he still retained the respect of much of the electorate in 1989. An engineer, Barco was voted into office in 1986 when he defeated the Conservative candidate, Alvaro Gómez. On his assumption of the presidency, Barco lowered the political temperature. In contrast with his populist predecessor, Conservative President Belisario Betancur, Barco adopted an aloof, technocrat's approach. Of patrician background, Barco lacks the common touch of the traditional Colombian politician, and initially the public welcomed his new presidential style. In exchange for competent management, voters were prepared to accept four years of drab, if honest government.

Barco, who will step down in August following the 1990 presidential elections, has lived up to the public's expectations insofar as his years in office have been unmarred by either political scandal or major administrative bungling. His personal integrity has never been questioned while, as a moderate reformer, he is given credit for his well-meaning, though not particularly successful, efforts to modernise the Colombian bureaucracy and to raise living standards in low-income regions.

The president, when he assumed office, was confronted with two major challenges. First, on the security front, he had to combat left-wing guerrillas, right-wing extremists, and drug-traffickers, whose violence had reached such proportions that the authorities had virtually lost control of extensive areas of the country. Second, despite rampant insecurity, the president had to maintain economic growth and to reduce unemployment, which had topped 12 per cent when he succeeded Betancur.

In the opinion of most analysts, Barco fared significantly better in the economic field than in the security sphere. As he approached the close of his term, he had relatively little to show for his endeavours to keep the country's leftist guerrillas at bay. The most intransigent of the four main rebel groupings, the pro-Castro ELN National Liberation Army, had rejected all peace overtures, and it continued to attack oil field and goldmining installations in an economic war that has occasioned multi-million dollar losses.

The army was powerless to halt the ELN offensive which was aimed primarily at Colombia's principal

petroleum export outlet: the 500-mile Caño Limón-Coveñas pipeline between the nation's largest Eastern Plains oil field and the Caribbean coast. During the first half of 1989, the pipeline was attacked by insurgents approximately 20 times, and flow down it of foreign-bound crude was paralysed for days at a time. As a consequence, the petroleum sector suffered losses of some $110m in the initial five-and-a-half months of the year.

A second left-wing guerrilla organisation, the Soviet-aligned FARC Revolutionary Armed Forces, clashed frequently with the army in 1988. Then, the following year, its combatants declared a provisional truce after the movement expressed tentative interest in the possibility of opening peace talks with the government. But it remained to be seen whether the FARC would resume fighting or whether it would opt, once and for all, for negotiations.

A third guerrilla force, the radical nationalist M-19, did, however, begin truce talks in earnest with the authorities and it seemed possible, in mid-1989, that its commanders might disband the movement and return to civilian life. The M-19 initiated its truce in response to a peace offer which Barco made to the guerrillas in 1988. Accordingly, the administration could claim that it had made at least some progress in its efforts to restore peace to troubled zones. To this, though, sceptics objected that the M-19 was no longer

KEY FACTS

Official title: Republic of Colombia
Head of state: President Virgilio Barco Vargas
Ruling party: Partido Liberal
Capital: Bogotá
Official Languages: Spanish
Currency: Colombian Peso (Col$) = 100 centavos
Exchange rate: Col$415 per $ (Nov 1989)
Area: 1,138,914 sq km

Population: 30.2m (1988)
GDP per capita: $1,739 (1988)
GDP real growth: 3.7% (1988)
Unemployment: 10.9% (1989)
Inflation: 28.1% (1988)
Trade balance: $648m (1988)
Foreign debt: $17.8bn (1988)

a viable fighting force because of the heavy losses and desertions it had suffered over the years. Rather than face military defeat, it was argued, the debilitated M-19 had agreed as a last resort to open negotiations with the government.

Certainly, in military terms, the M-19's decision to start peace talks was expected to have little bearing on the outcome of the insurgency. With the ELN fighting on and with the position of the FARC still to be defined, prospects for negotiating a lasting settlement of the guerrilla war, then, were as uncertain as ever in mid-1989.

Right-wing extremists

Barco, though he failed to terminate the insurgency, did however meet with some success when finally, in 1989, he cracked down on right-wing terrorists. The latter had been ever-more active in rural zones where their death squads had murdered hundreds of radicals. In several instances, the right-wing extremists had been financed by cocaine barons who had used their narco-dollars to purchase vast ranches. To protect their newly-acquired farms against left-wing rebels, drug traffickers founded private armies which launched murder campaigns against not only marxists but radicals as whole.

For years, the authorities were unable to halt the terrorists. Then, in 1989, the picture changed when the security forces raided a number of farms in the remote Eastern Plains where they discovered camps where right-wing gunmen were being trained. A major drive was launched against the death squads, and, temporarily, there was a marked decline in their activities. Nevertheless, the continuing dangers posed by gunmen were graphically illustrated in May 1989 when terrorists tried to murder the chief of the DAS security police, General Miguel Maza, who had spearheaded the fight against right-wing extremists. The murder attempt failed, and the general escaped unhurt.

The hard-pressed security forces, fighting as they were political terrorists of both the right and left, also had to combat the country's immensely wealthy cocaine gangs. Reportedly between $500m and $1bn from drug trafficking enters Colombia annually, and the leaders of the country's two biggest narcotics rings, the Medellín and Cali cartels, remain at large, though they have been forced underground. The government, in 1988–89, closed down dozens of clandestine cocaine laboratories and seized tons of the alkaloid. But the drug smuggling persisted, and it was unlikely that an early solution to the problem would be found.

Colombia today has one of the world's highest per capita homicide rates. The security situation, in fact, is so precarious that even constantly escorted public figures are no longer safe. For example, in 1988 the former presidential candidate, Alvaro Gómez, was kidnapped in a Bogotá street by M-19 guerrillas who killed his bodyguard. Gómez was later released unharmed. In other incidents the following year, a Liberal presidential pre-candidate, Ernesto Samper, was shot and seriously injured at Bogotá Airport while, at Sasaima, the emerald mining magnate Gilberto Molina and more than a dozen other people were gunned down at a party by terrorists.

Against this background of interminable violence, some foreign analysts express surprise that the economy functions at all – let alone that it continues to expand. True, in many sectors expansion in recent months has been steady rather than dynamic. Notwithstanding, by Latin American criteria, Colombia's economic situation could be decidedly worse.

KEY INDICATORS

	Unit	1984	1985	1986	1987	1988
Population	m	28.1	28.6	29.2	29.7	30.2
GDP (1985 prices)	pesos bn	4,816	4,966	5,255	5,535	5,742
Inflation	%	16.1	24.1	18.9	23.3	28.1
GDP real growth	%	3.4	3.1	5.8	5.3	3.7
Agricultural production	% real gwth	1.8	1.6	3.3	6.1	3.0
Iron-ore production	'000 tonnes	444	439	461	560	570*
Gold production	'000 troy oz	731	1,142	1,279	853	937*
Silver production	'000 troy oz	153	169	149	167	190*
Platinum production	'000 troy oz	10	12	14	22	23*
Crude oil production	'000 b/d	170	175	300	385	375
Industrial production	1980=100	102.9	105.7	112.5	121.1	123.6
Sugar	'000 tonnes	1,177	1,367	1,272	1,293	1,361*
Cement	'000 tonnes	5,276	5,308	5,915	5,898	6,329*
Exports	$m	4,273	3,650	5,331	5,661	5,164
Raw coffee	$m	1,799	1,784	3,046	1,689	1,696
Fuel oil	$m	445	409	408	860	966
Imports	$m	4,027	3,673	3,409	3,793	4,516
Capital goods	$m	1,574	1,281	1,377	1,617	1,195*
Intermediate goods	$m	2,487	2,106	2,048	2,192	1,898*
Consumer goods	$m	431	347	382	457	422*
Balance of trade	$m	246	-23	1,922	1,868	648
Current account	$m	–1,401	–1,809	383	336	–467
Exchange rate (average)	pesos per $	100.8	142.3	194.3	242.6	299.2

* estimated figure

The economy

With an external debt of some $17.3bn in mid-1989, the country retains its reputation as a model debtor. Unlike its neighbours, Colombia has not had to renegotiate its foreign loans and it has an impeccable repayment record.

The international reserves dropped more than $400m to some $3.3bn in the first five months of 1989. But treasury officials were confident they would be able to make good this loss later that year when international bankers were expected to grant the nation a $1.7bn credit package. The administration predicted GDP should augment by 4 per cent or more in 1989, compared to 3.7 per cent in 1988. By the spring of 1989, the unemployment rate had dropped to 10.9 per cent against 12.6 per cent a year earlier.

The financial sector continued to prosper, in no small measure because of its high interest rates. Its gains in 1988 increased nearly fourfold to almost $167m.

On the external front, non-traditional exports were up by roughly a quarter in January-November 1988. As a result, coffee, which has traditionally accounted for a half or more of dollar earnings, generated only a third of the nation's export income despite the fact that in 1988 it brought in its customary $1.6bn or so. The second major export was petroleum, shipments of which were worth nearly $1bn – a fall of almost $370m on the previous year. The decline was attributable to guerrilla attacks on oil installations and to the world price situation.

The textile and leatherware sectors in 1988 increased their exports by $90m to $334m thanks to modernisation programmes, aggressive marketing and expanding overseas demand. Coal exports were up $56m to $307m, primarily because of steadily rising output at the Cerrejón Norte mine, while flower exports grew to around $200m. The chemicals and metalworking industries were among other sectors which significantly augmented their dollar earnings. The 1988 trade surplus of some $648m would have been considerably higher had it not been for constant guerrilla raids which reduced oil export revenue.

Investment

In the foreign investment sphere, approved capital inflow in 1988 amounted to $190m, against $133m the previous year. The positive trend persisted in the first two months of 1989 when foreign investment authorisations totalled $105m. The rising inflow of capital from abroad was traceable in no small part to Renault's acquisition, for nearly $53m, of the state IFI Industrial Institute's stake in the vehicle-assembly group, Sofasa – the Colombian manufacturer of Renault models. Toyota, for its part, was also reportedly interested in buying into Sofasa.

In other sectors as well, there was continued interest from overseas investors. Swiss entrepreneurs, for example, are planning to invest $30m in the building materials company, Pavco, while Shell has purchased Hocol's oil operation in the Huila region. Drummond, of Alabama, meanwhile, is pressing on with its $580m project to develop the northern Loma steam-coal field for export markets.

The government, in its drive to attract foreign capital and technology to the banking sector, suffered a setback in 1988. Its bill to open up the industry to foreign investment provoked opposition in Congress where nationalists objected to clauses that would permit 100 per cent foreign ownership of finance houses in certain circumstances. The bill's opponents pressed for a 49 per cent ceiling on foreign equity in the banking sector. The government was due to reintroduce the bill in 1989, and it was possible a compromise formula would be found. But pending the outcome of events in Congress, it was unlikely the finance sector would draw in much new foreign investment. On the national investment front, a new coal project was unveiled by the Colombian corporation, Prodeco, which proposes to exploit the northern Calenturitas field in a development that could generate $2.5-3bn from exports during the 30-year life of the programme.

The government in 1988-89 continued its efforts to resurrect the stalled $410m Papelcol paper plant at Caloto, Cauca Valley. Construction of the complex has been paralysed by financing problems, and a further $120m at least will be required to complete the plant. Foreign investment is being actively sought for the venture. Another major project, the Cerromatoso ferro-nickel mine, picked up steam in 1988-89 thanks to buoyant world prices, and it earned some $180m from exports in 1988.

On the negative side, however, Barco's management of the economy has been marred by the unusually high level of inflation, which exceeded 13 per cent in the first five months of 1989. The picture was also clouded by the loss in workers' purchasing power. In many cases, annual wage hikes failed to keep abreast of inflation in 1989, and largely as a consequence commercial sales (excluding fuel) dropped some six per cent in the first quarter of 1989; industrial output was down nearly 3.1 per cent.

The exorbitant cost of credit continued to hamper business expansion plans, and industrialists called for the reimposition of interest controls. The latter were lifted in January 1989 and, by mid-year, interest rates had risen by as much as four per cent. The government, reacting to pressure from businessmen, ordered state banks to reduce interest rates in June 1989. By the summer of 1989, business associations, worried by stagnant or falling demand, had started to caution that the economy was heading, if not for recession, at least for a slow-down. The administration dismissed such fears as alarmist, despite the negative statistics in sectors such as the construction industry, whose building activity was down nearly 20 per cent in January–April 1989.

Builders were particularly hard hit by the authorities' decision to mothball the Bogotá Metro rail programme, which had to be indefinitely deferred because of financing difficulties. With external loan servicing commitments accounting for more than 40 per cent of export earnings, government economists warned it might be perilous to augment the foreign debt further by embarking on a relatively low-priority project such as a metro. Nonetheless, it appeared probable that an Italian consortium would in due course construct the rail network, though almost certainly not during the Barco government's term of office.

The economic war

Given the intensity of political violence in Colombia, at first sight it may appear strange that only a limited number of economic sectors have been directly affected to date by the guerrilla war. This, however, is readily explainable: most industrial companies are located in or near cities where guerrillas have been able to make only minor inroads. The rebels operate primarily in rural zones where there are few, if any, industrial concerns. Hence the mainstream economy continues to expand, and the majority of businessmen have been able to all but ignore the insurgency.

But in country regions there is a different story. In northern Colombia dozens of ranches have been virtually abandoned by their proprietors in the face of threats by insurgents. Meat production has slumped in such zones and unemployment has grown because of the minimal level of new investment. But in densely populated central areas, most farming regions remain tranquil because the guerrillas still lack the manpower and backing to subvert such districts. The same is true of most coffee-growing localities where there have been few rebel incursions. Guerrillas aside, coffee planters in any event have other problems to consider. Crop diseases, replanting and adverse weather conditions reduced harvests in 1988-89, and the situation was aggravated in June 1989 when Mild's prices dropped substantially because of the likely collapse of the International Coffee Agreement.

The controlled world coffee market will be thrown open to market forces if the ICA founders, and Colombia – to compensate for resultant over-supply and depressed prices – have to augment its yearly Mild's exports by 3.5m bags or more to sustain its current income level from the crop. Some brokers predict this should be possible because, with 9.5m bags on stockpile, the country can increase exports short term – and by a sizeable margin. But other experts warn that, without the ICA, world prices could slump further. Moreover, they caution, Colombia's creaking port system may be unable to cope long-term with greatly increased coffee shipments; accordingly, annual

COUNTRY PROFILE

Political system: Executive power vested in the president, elected by universal adult suffrage for four years. The president appoints a cabinet (and various regional governors etc). The legislature is the bicameral Congress, which comprises a Senate of 114 members and a House of Representatives of 199 members, all of whom are elected for four-year terms by universal suffrage.

Main cities/towns: Bogotá (population 5-6m, 2,640 metres above sea-level), Cali, Medellín, Barranquilla, Bucaramanga, Manizales, Cartagena.

Media
Press: Main daily newspapers are *El Tiempo* (circulation 200,000), *El Siglo* (65,000), *La República* (20,000) and *El Espectador* (from Bogotá, 215,000), *El Colombiano* (123,707), *El Mundo*, (Medellín, 60,000) and *El País* and *Occidente* (Cali, 50,000). Principal US and Western European dailies at larger news-stands in main cities.
Broadcasting: Numerous stations/channels – largely commercial. The National Institute of Radio and TV (INRAVISION) operates three TV stations, broadcasting in Spanish. Some hotels and an ever-increasing number of residences in Bogotá have cable television.

External trade: Since 1984 Colombia has registered favourable trade balances thanks in no small measure to the successful diversification of its external commerce. Traditionally coffee has accounted for half or more of hard-currency income, but the relative importance of the commodity is declining as earnings from oil, coal, ferronickel and other non-traditional products rise. In 1988, oil, coal, ferronickel, emerald and flower exports together brought in almost as much as coffee. The balance of export income derives from agricultural commodities such as bananas, tobacco and cotton, and from industrial and manufactured items such as cement, plastics, leather goods and books. The debt service burden has grown to over 40 per cent of export income. But the reserves, which exceeded $3bn in the first half of 1989 are relatively stable because of conservative fiscal management and tight monetary controls.
Exports: Coffee (37% of total value), light crude, fuel oil and coal (together 25%), textiles (5-8%), chemical products (4-6%), bananas (5-7%), printed products (2-3%). Main destinations (1987): USA (40% of total value), German FR (12%), Venezuela (4%), Holland (4%), Japan (4%), France (2%), UK (2%).
Imports: (1987): Mechanical and electrical equipment (28% of total value), chemical products (19%), plastics and rubber products (6%), mineral products (3%). Main sources (1987): USA (36% of total value), Japan (9%), German FR (8%), Brazil (4%), Canada (4%), France (4%).

Agriculture: Long the mainstay of the economy, agriculture still accounts for 17 per cent of GDP, but it now ranks second to industry and manufacturing which have overtaken it, if only marginally. Agricultural exports, primarily coffee, earn $2bn or more annually, and in the past, when Mild's prices have been high, have generated over half of dollar income. After Brazil, Colombia is the world's second largest producer of coffee. Other import cash-export crops include bananas, exotic fruits, cut flowers, tobacco, cotton, sugar cane and cocoa. Exports of artificially reared shrimps are rising as the fledgling sector gathers strength. Shipments of fresh and processed tropical fruit are also expected to augment. The illegal export of cocaine and to a lesser extent marijuana continues, and it is estimated that the narcotics trade may earn between $500m and $1bn yearly, despite government control campaigns. Output of some staple food crops, such as rice and wheat, is frequently insufficient to meet rising domestic demand. Meat production has been hit by guerrilla violence in ranching zones. The government has launched emergency development programmes in low-income farming districts as part of its plans to boost agricultural output and to pacify politically turbulent regions. Some five per cent of total area is cultivated arable land, 25 per cent pasture and 50 per cent forest and woodland.

Manufacturing: Contributes 24 per cent to GDP and employs 15 per cent of the workforce. Food processing, beverages and tobacco are the largest industries, followed by chemicals, textiles and clothing, capital goods industries and motor vehicles. Metals, cement, electrical engineering and paper also important. Emphasis on local sourcing, foreign investment and expansion of export capacity. Main industrial centres Medellín, Bogotá and Cali.

Mining: Fastest growing sector in the economy, mining's contribution to GDP has grown from under two per cent to over five per cent and the positive trend should persist in view of rapidly rising coal output, particularly in northern zones near the Caribbean coast. Second largest producer of gold and platinum in Latin America. Also production of emeralds, iron-ore, silver, copper, lime, sulphur, manganese, phosphates and salt. High-quality hard coal reserves, estimated at more than 16bn tonnes. Development centred on El Cerrejón open-cast mine in Guajira. Production at Cerrejón mines is now approaching 15m tonnes a year. Also plans for the large-scale exploitation of major steam coal deposits at La Loma are now being prepared.
Hydrocarbons: Production of crude petroleum has boomed in recent years owing to significant new Eastern Plains discoveries. Proven oil reserves exceed 2bn barrels making the country self-sufficient save for small quantities of gasoline which is still imported because of refining capacity limitations. Mean crude output in 1988 was 370,000 b/d and exports have grown to over 150,000 b/d in 1989. Current refining capacity around 215,000 b/d. Known gas reserves total around 4.1 trillion cu ft, representing 30 years supply at the current rates of extraction. Output, which has risen to over 500m cu ft/day, consumed mainly by local downstream industries and north coast residents.

Energy: Self-sufficient in energy since 1984. Oil gradually being replaced by other energy sources, in particular hydro-power, gas and coal. But growth in demand has been slower than the 10% per annum anticipated, and the emphasis is currently on overhauling existing capacity rather than on new installations.

Membership of international organisations: Andean Group, FAO, G-77, GATT, IADB, IAEA, ICAC, ICAO, ICO, IDA, IFAD, IFC, IHO, ILO, IMF, IMO, INTELSAT, Inter-American Development Bank, Interpol, IRC, ISO, ITU, Latin American Integration Association, NAM, Organisation of American States, PAHO, SELA, UN, UNESCO, UPEB, UPU, WFTU, WHO, WIPO, WMO, World Bank, WSG, WTO.

Mild's earnings could diminish by $100-200m. But even if this situation were to materialise, the rising exports of other sectors would cushion the effect of a drop in coffee earnings.

In the banana-exporting Urabá region, security rather than economic considerations are perturbing planters. Left-wing guerrillas continue to infiltrate banana plantations and foment labour unrest. But despite this, banana exports were up in value in 1988 when they earned over $250m.

Another sector that has been the constant target of guerrillas is the goldmining industry. Rebel raids temporarily paralysed the operations of the country's largest gold-extraction company, Mineros de Antioquia, on a number of occasions in 1988–89. National gold output was valued at $413m in 1988 and, short term, it is unlikely to rise significantly. But the sector's prospects should improve in time with the planned exploitation of the remote Guainia gold field which could enter production in the 1990s.

The sector which has suffered most at the hands of the guerrillas is the oil industry. Its export earnings continue to be cut because of rebel raids on the Caño Limón pipeline. The ELN insurgents, who repeatedly disrupt flow down the line with bomb attacks, are trying to halt the extraction operations of multi-nationals such as Occidental. They are also seeking the nationalisation of the oil sector. Yet despite the ELN's ravages, multinationals continue to sign exploration contracts with the government, which can offer them both attractive drilling terms and promising geological prospects. In 1988, 80 exploration wells were sunk, and crude output averaged some 375,000 barrels a day (b/d). Guerrilla raids permitting, mean production in 1989 was expected to grow to at least 430,000 b/d, and it was hoped crude exports would bring in over $1bn, provided the Caño Limón pipeline could be safeguarded by the army – an improbable eventuality, to judge by the military's poor showing in recent months.

To counter the insurgent threat, defence expenditure has been augmented, and the military are no longer held on a tight rein, as was the case during the Betancur government, which tried to appease the rebels with numerous concessions in an abortive bid to negotiate a truce. Barco, by contrast, has adopted a firmer stance. When, for instance, he finally unveiled his peace plan in 1988, he offered the guerrillas no more than basic amnesty terms and vague promises of reforms. With the exception of the rag-tag M-19, the rebels either reacted cautiously to the president's offer or they rejected it outright.

Probably Barco, the pragmatist, had expected as much. Evidently in his view, hard-line guerrillas such as the ELN will fight on regardless of any truce overture. So, say Barco's advisers, the government has only two courses of action open to it. First, it must strengthen the armed forces to enable them to keep the guerrillas at bay in remote regions where they can occasion minimal economic damage. Second, the authorities must neutralise the rebels' political appeal by pushing through crash development programmes in impoverished regions.

This stick-and-carrot approach produced few results in 1988-89. The army was unable to push back the rebels to isolated regions. Rather, in many areas a military stalemate persisted, with neither side able to gain the ascendancy. Meanwhile, government plans to upgrade public services in turbulent zones, to woo away peasants from the guerrillas, failed in most cases to have their desired effect. In numerous districts, the insurgents remain as active as ever. The government had originally earmarked $1.1bn for emergency development projects in 1986-90. But budgetary curtailments reduced the scope of the programme and delayed its inception.

In brief, yet further fighting, it seems, lies ahead. Significantly most of the prospective candidates in the 1990 presidential elections have been hesitant to formulate clear-cut peace plans. Apparently, like most Colombians, they suspect the insurgency could be an intractable problem that could drag on for ever.

The election campaign

Colombians tend to vote for personalities rather than for policy, and the 1990 election campaign is no exception to this rule. In mid-1989, some half-dozen mainstream aspirants were vying for the candidacies of the ruling Liberals and the opposition Conservatives – the only two parties with sufficient strength to win general elections. Most of the pre-candidates were well-tried establishment figures with few new wares to offer the electorate.

On the Conservative side, former Foreign Minister Rodrigo Lloreda was the favourite for the candidacy and, unlike many of his opponents, he did have a meaningful platform. Lloreda, a wealthy newspaper proprietor, would give priority to export promotion in order to reduce unemployment. He would also boost food output and revamp the educational system to emphasise technical skills. A technocrat, Lloreda is highly regarded by the business establishment. His principal rivals for the candidacy include Gabriel Melo, a caucus lightweight, and Alvaro Leyva, a progressive Conservative who is viewed sympathetically by moderates and not a few radicals.

For the Liberals, there were four main pre-candidates in 1989: Luis Galán, a charismatic leader with a large middle-class following; Hernando Durán, a forceful ex-mayor of Bogotá; Ernesto Samper, an affable, British-style Liberal; and finally the maverick Alberto Santofimio, the Liberal Party baron of his home region, Tolima.

Galán and Durán were generally considered to be the principal contenders in mid-1989. The leader of the former New Liberalism breakaway faction, Galán is a moderate who, given his personal popularity, could sweep to a landslide presidential victory if he were to be selected as candidate. But he has more than his share of enemies among the ranks of his fellow Liberals who have still not forgiven him for splitting the party in the past when he founded his dissident New Liberalism faction. Durán has a deserved reputation as a man of action. A hard-liner in security matters, he has said that, if elected, he will increase military strength to meet the guerrilla threat. Durán has considerable caucus support, and many Liberal barons, who harbour misgivings about Galán, may opt for him. Samper is dismissed by some observers as a bland politician who lacks leadership qualities. At the bottom of the list, Santofimio has too controversial a background to have much chance. He was once associated with a political group which allegedly had links with drug traffickers. The socialist UP Party and far-left candidates are unlikely to make more than a marginal showing in the elections.

Provided the Liberals do not split their ranks, they should carry the polling, if only because they are Colombia's largest party. Also, though Barco's Liberal government has not been unduly popular, it is generally respected and it has not antagonised voters. The Liberals' 1986–90 track record is therefore not a major issue in the polling.

For investors, the election outcome should not be of pressing concern, as little divides the two major parties idealogically. All the main pre-candidates are pro-business moderates and, regardless of which party emerges victorious at the polls, the emphasis is likely to be on continuity in economic affairs. Two warning notes, however, may be in order. First, the Conservatives have for months been calling for renegotiation of the external debt in order to free funds for short-term development. Thus, in the unlikely event of a Conservative victory, debt renegotiation could be a possibility. Second, Galán, should he win his way to the presidency, may not necessarily be as responsive as Barco to foreign investors; he has a nationalist streak. A Galán government, though it would in theory welcome capital from abroad, could in practice be expected to do so cautiously.

Such considerations apart, the investment climate should remain generally positive in Colombia for the forseeable future.

BUSINESS GUIDE

Time: GMT -5.

Climate: Tropical with irregular rainy seasons in coastal areas, temperate on the plateaux and cold in the Andes region. Daily temperature range in Bogotá fairly constant throughout year from 8°C to 22°C; heaviest rain from Mar to May and Oct to mid-Dec.

Entry requirements

Passport: Required by all with few exceptions (e.g. certain nationals of Ecuador and certain tourist visitors from Trinidad and Tobago).

Visa: Required by all except nationals of EEC countries (excluding Greece, Luxembourg, France and Portugal), Austria, Scandinavian countries, Switzerland, Japan, Israel, Liechtenstein, Republic of Korea, Barbados and certain American countries. US citizens require tourist entry cards, obtainable from nearest Colombian consulate. Citizens of certain neighbouring countries also exempted from visa requirement under certain other conditions.

Currency: No foreign exchange restrictions for foreign visitors. For foreign residents and for nationals there are limitations on the import/export of foreign and local currency. N.B. Central Bank will only change a maximum of $250 worth of travellers' cheques per day. Generally travellers' cheques are difficult to change (try Banco de la Republic).

Health precautions

Mandatory: none.

Advisable: Yellow fever vaccination essential for visits to some regions. Malaria risk exists in lowland areas – prophylaxis recommended. Water precautions should be taken outside main towns. Also risk of hepatitis – precautions recommended. Visitors to Bogotá need to acclimatise to altitude.

Representation overseas: Abidjan, Asunción, Beijing, Beirut, Belgrade, Berlin, Berne, Bonn, Brasília, Brussels, Bucharest, Budapest, Buenos Aires, Cairo, Canberra, Caracas, Copenhagen, Guatemala City, The Hague, Georgetown, Kingston, La Paz, Lima, Lisbon, London, Madrid, Managua, Manila, Mexico City, Monaco, Montevideo, Moscow, Nairobi, New Delhi, Oslo, Ottawa, Panama City, Paris, Port-au-Prince, Port of Spain, Prague, Quito, Rome, San José, San Salvador, Santiago, Santo Domingo, Seoul, Sofia, Stockholm, Sydney, Tegucigalpa, Tehran, Tel Aviv, Tokyo, Vatican City, Vienna, Warsaw, Washington.

Air access

National airline: Avianca (Aerovías Nacionales de Colombia).

Other airlines: Aerolíneas Argentinas, Aeroperú, Air France, Air Panamá, British Airways, Copa, Ecuatoriana, Eastern, Iberia, Lacsa, Lufthansa, Ladeco, SAM, Varig, Viasa.

BUSINESS GUIDE

Tax: Airport tax of US$15 payable on departure.
International airports: Barranquilla-Ernesto Cortisso (BAQ), 10 km from city, car hire; Bogotá-El Dorado (BOG), 12 km from city, duty-free shop, bar, restaurant, buffet, bank, post office, shops, hotel reservations, car hire; Cali-Palmaseca (CLO), 19 km from city, restaurant; Cartagena-Crespo (CTG), 2km from city; Medellín-Rionegro (MDE), 15 minutes' flight by scheduled and frequent helicopter service to city.

Surface access
Main ports: Barranquilla, Buenaventura, Cartagena, Santa Marta.
Overland access: Possible from Ecuador and Venezuela by road.

Hotels: Graded from one- to five-stars by the National Tourist Corporation. Tourist tax of 5 per cent on rooms; most hotels charge minimum fee per night for insurance against theft. Advisable to book well in advance. Higher tariff operates from Dec to Apr. Service charge normally added to bill, otherwise 10 per cent tip. A 5 per cent tax is added to all hotel bills. Major credit cards widely accepted in hotels, restaurants and shops.

Credit cards: Amex, Diners', Visa and Mastercharge are widely used.

Car hire: National or international licence acceptable for up to three months, but must be accompanied by an official translation. Traffic drives on the right and during the working day is heavily congested in main towns.

City transport
Taxis: Within Bogotá, usually metered with minimum charge and extra at night, holidays, Sundays and for out of town journeys; can be hailed; tipping not usual. For unmetered taxis fares should be agreed in advance of journey. Shared taxis, *colectivos*, operate within cities and suburbs. Green and white taxis have English-speaking drivers and can be rented by the hour/day at major hotels.

National transport
Air: Frequent and cheap air services between Bogotá and all main centres. Major air companies operating internal flights are Avianca, SAM, ACES, Aeropesca and SATENA. Around 15 smaller companies also operate domestic services. Advisable to book in advance.
Road: Approximately 75,000 km of road network – less than 20 per cent surfaced; highway links Bogotá-Cali; for other journeys local enquiries advisable.
Buses: Usually flagged down, and in most cases no scheduled stops. Within cities buses tend to be crowded; long-distance bus journeys generally for more adventurous travellers and care is needed – Bogotá-Medellin inter-city service, however, is fairly reliable and comfortable.
Rail: Main route is from Bogotá to Santa Marta (twice weekly; typical journey time–25 hrs). The rail service has deteriorated, however, and taking a train is not recommended.
Water: Cargo boats and some hire boats operate on the four major hydrographic systems. Most traffic on Magdalena river.

Public holidays
Fixed dates: 1 Jan, 1 May, 20 Jul (Independence Day), 7 Aug (Battle of Boyacá), 8 Dec (Immaculate Conception), 25 Dec.
Variable dates: Maundy Thursday, Good Friday. If they do not fall on a Mon, the following holidays are observed on the Mon next: Epiphany (6 Jan), St Joseph's (19 Mar), St Peter & St Paul (29 Jun), Assumption Day (15 Aug), Colombus Day (12 Oct), All Saints' Day (1 Nov), Independence of Cartagena (11 Nov), as well as the following normally variable dates – Corpus Christi, Sacred Heart of Jesus/Thanksgiving Day, Ascension.

Working hours
Business: (Mon-Fri) 0800 – 1200, 1400 – 1800.
Banking: (Mon-Fri) 0900 – 1500.
Shops: (Mon-Sat) 0900 – 1900 in the largest cities; 0900 – 1230, 1430 – 1830 elsewhere.

Telecommunications: Major hotels can handle long distance telephone calls, telefax or telex messages.
Telephone: In Bogotá the telecommunications office at Cra. 13 no 26-49, opposite Tequendama Building, is open from 0900 to 2045. Dialling code for Colombia: IDD access code + 57 followed by area code (Bogotá 1, Cali 23, Medellín 4, Cartagena City 53) followed by subscriber's number.
Telex: Country code: CO.

Postal service: Separate post offices for surface and air mail. The former not recommended. Some pharmacies and newsagents in Bogotá have airmail collection.

Banking
Banking centre: Bogotá. *Central bank:* Banco de la República. *Head offices:* Banco Anglo Colombiano, Banco de Bogotá, Banco Cafetero, Banco de Caldas, Banco de Colombia, Banco Colombo Americano, Banco de Colombo Americano, Banco Colpatria, Banco Comercial Antioqueño, Banco del Comercio, Banco Central Hipotecaria, Banco Crédito y Comercio de Colombia, Banco de Crédito, Banco del Estado, Banco Extebandes de Colombia, Banco Ganadero, Banco Grancolombiano, Banco Industrial Colombiano, Banco Internacional, Banco Internacional de Colombia, Banco de Occidente, Banco Popular, Banco Real de Colombia, Banco Santander, Banco Sudameris Colombia, Banco Tequendama, Banco de los Trabajadores, Caja de Crédito Agrario Industrial y Minero, Caja Social de Ahorros, Corporación Financiera Colombiana.
Foreign branches: Banco Francés de Comercio Exterior, Bank of Tokyo, Banque Nationale de Paris, Chase Manhattan Bank, Crédit Suisse Zurich, Deutsch-Südamerikanische Bank, First National State Bank, Kleinwort-Benson, Marine Midland Bank, Swiss Bank Corp.

Trade fairs: Most important is Bogotá's biennial International Industrial Fair in July (next one to held in July 1990). In Medellín there is a textile show each Sep.

Electricity supply: 110/120 V AC; two-pin flat blade plugs.

Representation in capital: Argentina, Austria, Belgium, Bolivia, Brazil, Bulgaria, Canada, Chile, China PR, Costa Rica, Czechoslovakia, Denmark, Dominican Republic, Ecuador, Eygpt, El Salvador, Finland, France, German DR, German FR, Greece, Guatemala, Haiti, Honduras, Hungary, India, Iran, Israel, Italy, Japan, Rep of Korea, Lebanon, Mexico, Netherlands, Nicaragua, Panama, Paraguay, Peru, Poland, Portugal, Romania, Spain, Sweden, Switzerland, UK, USA, USSR, Uruguay, Vatican, Venezuela, Yugoslavia.

Hotels: Bogotá
Bacata, Calle 19 no 5-20, P.O. Box: 16815 (tel: 2432210; tx: 43198).

Belvedere, Transv. 18 no 100-16 (tel: 2577700).

Bogotá Plaza, Calle 100 no 18A-30, P.O. Box: 90827 (tel: 2572200; tx: 45891).

Charleston, Carrera 13 no 85-45 (tel: 2571100; tx: 44355; fax: 2180605).

Continental, Avenida Jimenez no 4-16, P.O. Box: 3627 (tel: 2821100; tx: 43115).

Cosmos 100, Calle 100 no 21A-41 (tel: 2566015; tx: 42127).

Hilton International, Carrera 7 no 32-16, P.O. Box: 27533 (tel: 56020; tx: 44467; fax: 77480).

La Fontana, Avenida 127 no 21-10, P.O. Box: 101281 (tel: 2740200; tx: 45382; fax: 2160449).

Tequendama, Carrera 10 no 26-21, P.O. Box: 240112 (tel: 2861111; tx: 45418).

Hotels: Cali
Aristi, Carrera 9A no 10-04 (tel: 822521; tx: 51154).

Inter-Continental Cali, Avenida Colombia no 2-72 (tel: 823225; tx: 55599; fax: 822567).

Hotels: Cartagena
Capilla Del Mar (on Bocagrande Beach), P.O. Box: 3939 (tel: 651033; tx: 37739; fax: 655145).

Caribe, Bocagrande, Cra I 2-87, P.O. Box: 1330 (tel: 650155; tx: 37811).

Hilton International, El Laguito, P.O. Box: 1774 (tel: 650666; tx: 37645; fax: 650661).

Las Velas, El Laguito, Calle I 1-160, P.O. Box: 18987 (tel: 650000; tx: 37715).

Hotels: Medellín
Inter-Continental Medellín, Variante Las Palmas, P.O. Box: 51855 (tel: 2-660680; tx: 66685; fax: 2661548).

Nutibara, Calle 52A no 50-46, P.O. Box: 1067 (tel: 2319111; tx: 65263).

Veracruz, Carrera 50, no 54-18, P.O. Box: 4266 (tel: 2315511; tx: 65307; fax: 2318881).

Car hire: Bogotá
Avis: Calle 99, No 11-26 (tel: 6104455; tx: 43297).

Budget: Avenida 15 No 107-08 (tel: 2136020, 2134721, 2136383).

Hertz: Hotel Tequendama, res. Carrera 10, No 26-35, PO Box 241372 (tel: 2430309, 2347961), and Hotel Hilton (tel: 2885651 ext 181).

Car hire: Bogotá – El Dorado Airport
Avis: (tel: 2682421; tx: 43297).

Business advice and information
American Express, Carrera 7 No 35-20, Bogotá (tel: 2875782).

Asociación Nacional de Industriales (ANDI), Carrera 13 No 26-45, Bogotá (tel: 2349620).

Banco de la República (central bank), Carrera 7 No 14-78, Bogotá (tel: 2831111; tx: 044559, 044735, 045407; fax: 2866008, 2866109).

Caja de Crédito Agrario ndustrial y Minero, Carrera 8 No 15-43, Bogotá (tel: 2844600).

Cámara de Comercio de Bogotá (international chamber of commerce), Apartado Aéreo 29824, Carrera 9a, No 16-21, Bogotá (tel: 2819900, 2860511; tx: 45574 cacbo).

Carbones de Colombia (CARBOCOL), Carrera 7, No 31-10, Pisos 5-18, Bogotá (tel: 2873100).

Corporación Autónoma Regional del Cauca (CVC), Carrera 3 No 10-10, Cali (tel: 2873100).

Department Administrativo de Aeronáutica Civil (DAAV), Aeropuerto Internacional El Dorado, Bogotá (tel: 2669200).

Department Nacional de Planeación, Calle 26 No 13-19, Botogá (tel: 2824055).

Empresa Colombiana de Petróleos (ECOPETROL), Carrera 13 No 36-24, Bogotá (tel: 2856400).

Empresa Nacional de Telecomunicaciones (TELECOM), Calle 23, No 13-49, Bogotá (tel; 2694077).

Fondo de Promoción de Exportaciones (PROEXPO), Calle 28, No 13A-15 Piso 35-42 (tel: 269077).

Instituto Colombiana de Comercio Exterior (INCOMEX), Edifico Centro Comercio Internacional, Calle 28 No 13A-15 (tel: 2812200).

Instituto de Fomento Industrial (IFI), Calle 16, No 6-66, Pisos 7-15, Bogotá (tel: 2822055).

Comoros

Michael Griffin

On 2 June 1989, the federal assembly approved a package of constitutional amendments which appeared to assure President Ahmed Abdullah a third term of office when his mandate was to expire in October 1990. Included were the restoration of the post of prime minister and the direct election of island governors but the most far-reaching measure was the repeal of Article 16, which limited the president to just two six-year terms. However, in November 1989 he was shot dead by a disgruntled senior army officer, and supreme court president Haribou Chebani took over as president in accordance with the constitution.

Constitutional reform

The government carefully stage-managed the movement for constitutional reform, painting the putative third term as a response to popular demands for continuity and stability. There were lengthy consultations in each of the three islands but delegates were screened and the bid of a few opposition participants to table a degree of multi-party democracy as a quid pro quo were ignored. But before the amendments could become law, they had first to be submitted to a federal referendum and popular feeling was running high, fanned by reports of corruption, cuts in wages and a looming structural adjustment programme.

Ali Bazi Selim, a former minister and the only deputy to oppose the amendments, showed far more consistency than his colleagues in the islands' quasi-legal dissident parties. Said Hassan Said Hachim, co-founder of the Opposition Unie with the former prime minister, Ali Mroudjae, returned to the government fold in September 1988. Abdullah Mouazoir, the head of the Union pour une République Démocratique aux Comores (URDC), endorsed the president in fulsome terms in January 1989. These two reconciliations confirmed Abdullah's oft-quoted aphorism that 'there is no opposition in the Comoros, only rival successors' and both were widely believed to have been angling for the plum job of prime minister, abandoned in 1985.

But the organised opposition poses far less of a threat than the profound discontent that has infected the islands in 1989, including even Anjouan, the president's chief power base. The traditional rivalry between Mutsamudu, the main harbour, and Domoni, the presidential fief ruled by his unpopular son Nassuf, erupted in a spate of anti-government tracts last spring which accused Ahmed Abdou, the president's special advisor, and Mikidache Abdourahmane, the minister of planning, of selling out Anjouan for their own personal gain. A similar strain of animosity has gripped Moheli, the smallest and least influential of the islands.

The importance which the government attaches to such discontent can be gauged from its decision to trade off a pledge for free, gubernatorial elections against an affirmative result in the referendum on the president's future. But the holding of such elections is more likely to fuel the current, secessionist mood than to extinguish it and, rather ominously, the presidential guard, which is led by mercenaries, commenced an unusual, island-wide manoeuvre in Anjouan in July 1989.

The economy

The timing both of the referendum and the presidential elections that were to ensue, was crucial if Abdullah were to prevent them evolving into a popular vote on the structural adjustment programme which was slowly hammered out with the World Bank during 1987–88. Finance Minister Said Ahmed Said Ali, has laboured hard and long to make the contents more palatable but few Comorians can still doubt but that the agreement with Washington spells the end of the good life, in so far as it ever existed. With unusual speed, the federal assembly nodded through a series of SAP-linked reforms, including the abolition of the tobacco monopoly, the reorganisation of the post and telecommunications services and a new system of personal taxation.

But this was only the start. One fifth of all 8,500 government posts were to be axed when the adjustment programme came into force in January 1990 and across-the-board wage cuts of 5 per cent became immediate in July 1989. The most painful measure – whose scale is still not known – was to be a blanket rise in tax on all imported goods from food to cement and consumer items. Regarded as vital if the government is to dent its perennial trade deficit, the rises will have same impact as a currency devaluation which is technically unthinkable since the Comoros is a member of the franc zone.

Ironically, adjustment comes at a time when the terms of trade in vanilla, the islands' chief export, have shown a marked improvement. Export value climbed from an all-time low of $4.25m in 1987 – when buyers opted for lower grade Indonesian pods trading at less than half the agreed price – to $13.5m in 1988. Clove prices continued their decade-long slide but exports by volume rose to nearly 2,000 tonnes – an all-time peak – and demand for the 60 tonne ylang-ylang crop strengthened, pushing sales to $2.5m. Total exports for 1988 amounted to $20.4m compared to $11.6m in the previous season.

The cancellation of the Comoros' debts to France, proposed by President Mitterand in Dakar in May 1989, will also help to ease the government's crushing debt load. In 1987, service on the $202.9m total amounted to $9.2m – more than 80 per cent of exports in what was a remarkably poor year – and the level is expected to rise during the mid-1990s. Rescheduling is an urgent priority with the Arab funds who are waiting for an IMF agreement before resuming their once-generous investments in the islands.

If relations with France have warmed, the Pretoria connection was abruptly shut down. Publicly, the government denies its close military and political links with South Africa despite the evidence of Sun International's $11.4m tourist resort taking shape on

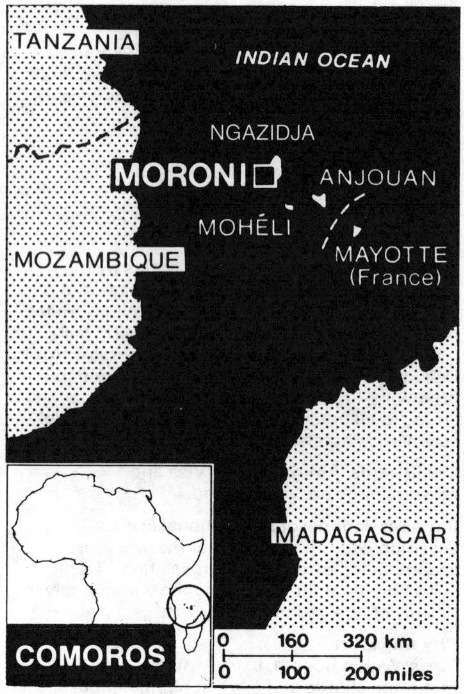

Grand Comoros. In April 1989, the government summarily dismissed Pretoria's commercial attache who subsequently alleged that South African aid was being systematically pillaged by the mercenaries of the presidential guard.

This gesture helped to bridge the gulf that separates the Comoros from the Frontline states such as Tanzania and Mozambique. Both have been deeply concerned by South Africa's growing influence. In April 1989, the minister of foreign affairs of Mozambique elicited a pledge that the Comoros would never be used to destabilise its neighbours and, in Tanzania in July, the president was building yet more diplomatic bridges.

But despite the rupture with Pretoria, the uncrowned king of the Comoros, Bob Denard, remains securely in the saddle and there is little prospect that he will leave his tropical retreat for some years to come.

KEY FACTS

Official title: Federal Islamic Republic of the Comoros

Head of state: President Haribou Chebani

Ruling party: Union Comorienne pour le Progrès

Capital: Moroni

Official Languages: Arabic, French (Comoran, Likala widely spoken)

Currency: Franc CFA = 100 centimes

Exchange rate: Franc CFA316 per $ (Nov 1989)

Area: 2,171 sq km

Population: 424,000 (1987)

GDP per capita: $370 (1987)

GNP real growth: 1.7%* (1987)

Trade balance: –$34.0m (1985)

Foreign debt: $203m (1987)

* estimated figure

COUNTRY PROFILE

Political system: The president is elected by universal adult suffrage for a six-year term although he may rule for an unrestricted number of terms. He nominates the prime minister and a council of ministers of up to nine members. Each island has a governor, appointed by direct elections. The legislature is the 42-member Federal Assembly elected by universal adult suffrage for five years, and is responsible for each island's resources. Gained independence in 1975, except for Mayotte, which remains a French overseas territory.

Political parties: The official political party is the Union Comorienne pour le Progrès, but it is not the sole vehicle of political activity. The Opposition Unie, formed by former ministers, and the Union pour une République Démocratique aux Comores (URDC) both contested the legislative elections in 1987. A host of less organised groups are based in Paris, notably the UNDC, CNPAC and FD. The legality of opposition groups remains discretionary. Although provided for in the Constitution, true multi-candidate elections have not so far taken place. The white mercenaries who overthrew the post-independence government in 1978 still have a commanding influence.

Main cities/towns: Moroni (on Njazidja (Grand Comoro), population 22,000), Mutsamudu (on Anjouan, pop 14,000), Fomboni (on Moheli, pop 7,000).

Language: French is the dominant language, although Arabic and numerous African languages are spoken. English is rarely encountered.

Media
Radio: Radio Comoros broadcasts in Comoran, French, Arabic and Swahili.

External trade: Both trade deficit and current account deficit were reduced in 1986, according to central bank figures, but the first half of 1987 suggested a renewed deterioration in the balance of payments. Foreign aid, mainly from France, accounts for over 40 per cent of GDP. Consistent trade deficit due to erratic pricing for main commodities. Balance of payments support provided regularly by France, also its main aid partner, and by Japan.

Exports: Vanilla (65% of total value), cloves (18.5%), ylang-ylang (12.5%), small amounts of perfume, coffee and copra. Main destinations: USA (69%), France (25%) 1988.

Imports: Fuels (over 20% of total value), rice (16%), vehicles (5%). Main sources: France (55%), Bahrain (oil), Kenya (building materials), Botswana (meat), Brazil (rice).

Agriculture: Although as much as 50 per cent of total land area is cultivated and 80 per cent of the working population employed in the sector, it remains severely under-developed (due to poor soil, adverse weather conditions and inadequate facilites) and over 50 per cent of the country's food requirements have to be imported. The government has instituted projects to attempt to achieve self-sufficiency and 60 per cent of EC aid for 1986-90 was allocated to this end. Major food crops grown are cassava, sweet potatoes, rice and bananas; while main cash crops are ylang-ylang, vanilla, cloves.

Industry: Sector is under-developed, with activity confined to distillation of essences (particularly from ylang-ylang), vanilla processing, soft drinks, plastics and woodwork.

Energy: Heavy dependence on imports for all fuel requirements. Under the 1983-90 Development Plan, over $15m has been allocated for the construction of a hydroelectric plant.

Membership of international organisations: African Development Bank, African States, associated with EC (Lomé Convention), FAO, IDA, IFAD, ILO, IMF, Islamic Conference, OAU, UN, UNESCO, WHO, World Bank.

BUSINESS GUIDE

Time: GMT +3.

Climate: Tropical. Dry season May to Oct with average temperature 24°C. Rainy season from Nov to Apr with temperature 27°C to 35°C. Very hot and humid on coasts, cooler on inner highlands.

Entry requirements

Passport: Required by all.

Visa: Required by all; can be bought on arrival at airport. Onward ticket and exit permit also needed.

Currency: No limit on import/export of local currency. Foreign currency must be declared on arrival and any amounts unused may be taken out on departure. (Check current regulations, as they are liable to change.)

Health precautions

Mandatory: Yellow fever vaccination certificates have been requested from visitors arriving from infected areas. Check current regulations.

Advisable: Typhoid and polio vaccinations recommended. Malaria prophylaxis advisable as risk exists throughout the country. Water precautions should be taken.

Representation overseas: Brussels, Paris.

Air access

National airline: Air Comores.

Other airlines: Air France, Air Madagascar, Air Mauritius, South African Airways.

International airport: Moroni-Hahaya (Code: HAH), 20 km from Moroni, on Grand Comoro; bar.

Surface access

Main ports: Moroni (Njazidja – formerly Grand Comoro) and Fomboni (Nzwani): both have offshore anchorage for larger vessels. Irregular services. Mutsamudu (Mwali) takes medium-sized ships at quay.

Hotels: Advisable to book in advance. Limited first class accommodation available on Njazidja, Nzwani and Mahore, but several high-quality developments in progress.

Car hire: International driving licence required.

City transport
Taxis: Service provided by 'taxi-brousse' – bush taxis – on each island.

National transport

Air: Each island is served by Air Comores. Flights two or four times weekly.

Road: Surfaced roads on Njazidja and Nzwani; other islands' roads can be difficult in rainy season. Mwali has only very basic tracks.

Water: Small boats, which can be hired, ply between islands.

Public holidays

Fixed dates: 1 Jan, 13 May, 6 July.

Variable dates: Start of Ramadan, Eid al-Fitr, Eid al-Adha, Islamic New Year, Prophet's Birthday.

Working hours
Government and business: (Mon-Thu) 0730 – 1200 and 1500 – 1730, (Fri) 0730 – 1100, (Sat) 0730 – 1200.
Banking: (Mon-Thu) 0700 – 1200; (Fri) 0700 – 1130.

Telecommunications
Telephone: Dialling for Comoro Islands: through operator.

Banking
Central bank: Banque Centrale des Comores.

Main centre: Moroni.

Head offices: Banque des Comores, Institut d'Emission des Comores. *Foreign branches:* Banque Française Commerciale, Caisse Centrale de Co-opération Economique.

Electricity supply: 220 V AC.

Hotels: Mayotte
Le Rocher, P.O. Box: 42 (tel: 42).

Hotels: Njazidja (Grand Comoro)
Itsandra, P.O. Box: 3 (tel: 2365).

Le Coelacanthe, P.O. Box: 404 (tel: 2275).

Le Karthala, P.O. Box: 53 (tel: 2000).

Novotel Ylang-Ylang, Moroni, P.O. Box: 404 (tel: 730240; tx: 255).

Car hire: Njazidja (Grand Comoro)
Itstandra Hotel: PO Box 3 (tel: 2365).

Moroni-Omar Cassim & Sons: PO Box 700 (tel: 2382).

Other useful addresses
Air Comores (Société Nationale des Transports Aériens), PO Box 544, Moroni.

Banque Centrale des Comores, Moroni (tel: 73-10-02; tx: 213 KO).

Banque Internationale des Comores, BO 175, Moroni (tel: 73-02-43; tx: 242).

Chambre de Commerce, d'Industries et d'Agriculture, PO Box 763, Moroni (tel: 2116).

Institut d'Emission des Comores, PO Box 405, Moroni.

Office National du Commerce, Moroni.

Radio-Comoros, PO Box 250, Moroni.

Société Hotelière et Touristique des Comores, Itsandra Hotel, Njazidja (tel: 2365).

Société Internationale des Comoros, BP 175 (tel: 73-02-43; tx: 242).

Congo

Susan Traill

Congo's agreement with the IMF expired in April 1988 and was not renewed. The IMF accused the government of overspending. Commercial bankers, who visited in early 1989, reported that Congo had insufficient funds to allow for a new rescheduling agreement. Even so, government and multilateral agencies continued to lend. Oil production rose in 1988 and is expected to increase again in 1989, encouraging creditors to take a longer term view of repayment.

To celebrate his ten years in power, President Denis Sassou-Nguesso promoted himself from colonel to general. In July 1989, the congress of the Parti Congolais du Travail (PCT) was expected to endorse Sassou-Nguesso for a further five-year term. It was also due to debate the 1990-94 economic development plan which focuses on financial austerity and the development of productive resources. The plan was to be adopted in December 1989 at the congress to mark the twentieth anniversary of the founding of the PCT. In August 1988, Sassou-Nguesso gave amnesty to former party officials Claude-Ernest Ndalla and Jean-Pierre Thystere Tchicaya, convicted of involvement in 1986 car bomb incidents on the twenty-fifth anniversary of the armed forces revolution. However, those involved in the 1987 rebellion led by Captain Pierre Anga were not released. Sassou-Nguesso declared in early 1989 that they would be tried and judged.

Congo scored a series of diplomatic successes in 1988-89. The high spot was the peace agreement negotiated between South Africa, Angola and Cuba in Brazzaville in December 1988. In July 1989, following the breakdown in the June 1989 cease-fire between Angola and UNITA, Brazzaville was the venue for a meeting between President Sassou-Nguesso, Angolan President Dos Santos and Gabonese President Omar Bongo to review the situation. The EC–ACP summit to negotiate Lomé IV took place in Brazzaville in March 1989. Also in 1989, the African Petroleum Producers' Association (APPA) chose the Congolese capital for its headquarters and Congolese Mining Minister Aime Emmanuel Yoka as its first president.

Media accusations of illegal arms dealing flared in March 1989 and the French government blocked sales to Congo of 50 Mistral missiles. Various reports claimed they were destined ultimately for South Africa, UNITA, Angola or the Soviet Union. French businessman Thierry Mailler was arrested in Paris but claimed the deal had been sanctioned by the French authorities. In Congo, the hardline M22 faction was said to have arranged the deal and members of the ruling Mbochi clan were said to have leaked the details to the press. Relations with France continued as normal, however, lending credence to the argument that the deal had had official backing.

IMF discontent but others still lend

The IMF claimed that Congo had failed to meet austerity criteria in 1986-88 despite the government's cutting of civil service expenditure, closure of loss-making state companies, reduction of state investment and the introduction of measures to improve food production in order to reduce the country's large food import bill. Congo's situation exemplifies the untenable conditions of so many African countries that have submitted to IMF terms: IMF agreement paves the way for new lending, and total debt and debt service continue to rise. At end-1987, Congo's total debt was $4.4bn. The projected debt service was $895m for 1988 and $802m for 1989, almost equal to total budget expenditure in both years. The government therefore fell further into arrears. By early 1989, total debt was estimated at 150 per cent of GNP. Commercial bankers agreed that the government was in no position to continue repayments. They started to prepare a debt/equity swap covering $300m in commercial debt and due to be launched in September 1989. Manufacturers Hanover Trust and a French

KEY INDICATORS

	Unit	1984	1985	1986	1987	1988
Population	m	1.7	1.7	1.8	1.8	1.9
Gross national product (GNP)	$bn	2.0	1.9	1.6	1.9	1.6*
GNP per capita	$	1,176	1,118	889	1,056	842*
Inflation	%	12.7	6.1	–1.7	11.7	16.5
GDP real growth	%	2.9	1.0	–6.9	–4.2	–0.6
Agricultural Production	1979-81 = 100	102.0	106.9	113.0	114.2	115.9
Food production	1979-81 = 100	102.1	106.8	113.2	114.5	116.1
Timber production (logs)	'000 cu m	233	243.1	289.6	327.4	312.1
Crude petroleum production	'000 b/d	130.0	113.8	117.5	126.0	135.0
Exports	$m	1,268.4	1,144.8	672.6	876.7	816.5
Crude petroleum	$m	1,055.0	890*	840*	429	386
Wood	$m	56.3	50.6	83.7	–	–
Imports	$m	617.6	630.2	512.6	419.9	444.0
Balance of trade	$m	650.8	630.1	160.0	456.8	372.5
Current account	$m	210.2	–161.3	–600.7	–222.7	–327.5
Total external debt	$m	2,032	3,077	3,721	4,636	–
Debt service ratio	%	25.2	32.5	40.4	18.6	–
Exchange rate	CFA per $	437.9	449.3	346.3	300.5	297.8

* estimated figure

KEY FACTS

Official title: People's Republic of the Congo

Head of state: President General Denis Sassou-Nguesso

Ruling party: Parti Congolais du Travail

Capital: Brazzaville

Official Languages: French (Lingala and Kikongo widely spoken)

Currency: Franc CFA = 100 centimes

Exchange rate: Franc CFA316 per $ (Nov 1989)

Area: 342,000 sq km

Population: 1.9m (1988)

GDP real growth: –0.6% (1988)

GNP per capita: $842* (1988)

Inflation: 16.5% (1988)

Trade balance: $372.5m (1988)

Foreign debt: $4.4bn (1987)

* estimated figure

co-ordinating committee drew up a list of Congolese companies for equity investment in the agriculture, forestry and transport sectors.

The 1989 budget was set initially at CFA235.8bn (receipts and expenditure in balance) with CFA46.8bn earmarked for investment and CFA189bn for administration. This compares with a final 1988 budget of CFA204.4bn, with CFA31.7bn for investment and CFA172.6bn for administration. Civil service salaries were cut by 10 per cent in November 1988 and there is a hold on recruitment. Priorities for investment in 1989 were agriculture, infrastructure maintenance and primary health care.

French aid continued to flow in 1989. Credits included CFA500m for early retirement in the public sector, FFr36.25m for urban roads, CFA20bn for the construction of a 900-bed hospital in Brazzaville, FFr9m for cocoa plantation rehabilitation and marketing in the northern Sangha region, FFr62m for phase one of the telecommunications re-equipment programme, FFr38m for railways and ports, and FFr15m for the Boko food project. The European Investment Bank (EIB) announced a Ecu16m loan in January 1989 for the palm oil programme, co-financed with France, ADB and BDEAC. This included construction of an agro-industrial complex for Sanghapalm and a vegetable oil refinery at Brazzaville. West Germany pledged aid of DM29m for 1988-89, including DM17m for transport and irrigation and DM12m for rural technical assistance. The USA agreed to finance 15 village-level food projects in 1989-90. The Soviet Union continued iron ore prospection at Yanga Kubanza, while France continued potassium prospection.

Oil production was reported to be 7.3m tonnes in 1988, up from 6.3m tonnes produced in 1987, but revenues remained low. Elf's 12m tonne Tchibouela field, which came on stream in December 1987, and Agip's 10m tonne Zatchi field, which came on stream in September 1988, were due to reach full production in 1989 with annual output of 1,572,500 tonnes and 310,000 tonnes respectively. Total 1989 oil production was forecast at 7m tonnes but could go higher. Prices were expected to strengthen in the second half of the year. Elf signed a production agreement with the government in May 1988 for the 33m tonne Tchendo field and will invest CFA34bn in its development. Agip conducted exploration and test drilling on three offshore sites in 1989. British Petroleum (BP) agreed in September 1988 to invest $25m over four years in exploration at the Kayes B offshore block.

Total timber production in the first half of 1988 was 188,266 cubic metres with 183,434 cubic metres exported, 27 per cent and 22 per cent up respectively on the first half 1987 figures. Eucalyptus production is increasing, boosted by French aid to Unite d'Afforestation Industrielle de Congo (UAIC). In mid-1988, Royal Dutch Shell announced it was going into joint venture with UAIC to plant 24,000 ha in the next seven years for a total investment of $50-60m. Cocoa production was 2,305 tonnes by July 1988, roughly double the previous year's harvest. Railway company CFCO suffered a drop in turnover to CFA22.9bn in 1988 from CFA25.5bn in 1987, partly due to the fall in manganese shipments from Gabon.

Outlook

Increased oil production in 1988-89 should be sustained in 1990 and the strengthening of world oil prices from mid-1989 should improve export earnings and government receipts. Timber production and exports will also increase in 1989-90. The longer term outlook for oil is promising with the development of Elf's Tchendo field. Amoco's offshore gas find in late 1988 could open the way for diversification of the hydrocarbons sector if reserves prove commercially viable. Shell's commitment to forestry development is particularly encouraging and could set a precedent for diversification by other oil companies.

These positive signs no doubt played a large part in the commercial creditor banks' decision to try a debt/equity swap in 1989. The fruits of such arrangements are usually slow to mature, but whether or not the swap is a success, it demonstrates a significant shift in lending policy to the longer term outlook. Government agencies and commercial banks seem to be more satisfied than the IMF with Congo's attempts at austerity. Therefore, Congo may be able to do without a new IMF agreement. The government will avoid a new agreement for as long as possible to ease internal tensions and in the hope of being able to return to its former policy of balancing influence from the industrialised West with that of the Eastern bloc.

For Western governments, Congo's importance as an intermediary in finding a solution to the southern African crisis is a strong incentive to support the economy. Congo's diplomatic role is likely to increase rather than diminish in the coming years.

COUNTRY PROFILE

Political system: The chairman of the Parti Congolais du Travail is president of the republic. The legislature is the 153-member People's National Assembly elected from a list approved by the Party. The Assembly is responsible to the president, who is in turn responsible to the Party. Became independent in 1960 and proclaimed a socialist republic in 1963.

Political parties: The ruling Parti Congolais du Travail (PCT) is the only legal political party. The national party congress is held every five years. The central committee meets twice a year. There are three official affiliate organisations for PCT trade unionists, women and youth: the Confédération Syndicale Congolaise (CSC), Union Révolutionnaire des Femmes Congolaises (URFC), and Union de la Jeunesse Socialiste Congolaise (UJSC).

Population: Estimated in 1987 at two million. Population growth is estimated at 3.4 per cent. 56 per cent of the population live in urban centres of some kind, the highest urbanisation rate in Africa. Two-thirds live within 510 km of the Congo-Ocean Railway, which links Pointe Noire on the Atlantic coast with Brazzaville, the national capital, in the interior. The Government is anxious to discourage the urban drift at present.

Main cities/towns: Brazzaville (population 600,000), Pointe Noire (350,000), Loubomo, Nkayi, Madingou-Kayes, Kinkala.

Language: French is the dominant language, and the language for all business documentation. Numerous African languages are spoken, including Lingala and Kikonga.

Media
Press: Daily newspapers include the government-owned *Mweti* (four times a week), *Agence Congolaise d'Information* (official bulletin), and *La Semaine Africaine* (Catholic). Several journals and periodicals, including *Etumba* and *Le Stade*.
Radio: Radiodiffusion-Télévision Nationale Congolaise broadcasts programmes in French, Lingala and Kikongo.
Television: Broadcasts 80 hrs/week.

External trade: Regular trade surplus from oil export earnings, but serious problems since 1986 due to the contraction of world oil markets. A mounting current account deficit. Net outflows, imports of services, freight charges and interest payments have caused a sharp growth in foreign debt. Aid and loans mainly from France and World Bank.
Exports: Crude petroleum (90% of total value in 1985), timber (5%). Main destinations: USA (55% of total value in 1985), Spain (17%), France (9%), Italy.
Imports: Machinery and transport equipment, metals and products, food. Main sources: France (typically 50% of total value in 1985), Spain (5%), Japan, Italy, German FR.

Agriculture: Sector has been overshadowed by development of petroleum industry. Although still employing 30 per cent of the workforce, contributes less than 10 per cent to GDP. Less than one per cent of the total land area is under cultivation (mainly in the alluvial Niari Valley). Farming is small-scale with output concentrated on subsistence crops such as plantains, cassava, yams, groundnuts, manioc, potatoes, wheat, maize, beans and paddy rice. Despite a steady growth in food production, the Congo relies heavily on food imports. The government aims to achieve food self-sufficiency by the year 2000. The main cash crops are coffee, cocoa, tobacco and sugar. Attempts to expand production of other cash crops include the rehabilitation of oil palm estates and a major new cocoa project. The main agricultural export is timber, mainly Okoumé logs (of which the Congo is a major world supplier). Around half the total timber output (around 600,000 cu m per year), is used for wood processing. Approximately 60 per cent of the country is covered by woodlands and forests, much of it unsuitable for commercial exploitation. Since the end of the 1970s a major afforestation programme involving fast-growing tropical hardwoods has been under way.

Industry: Manufacturing sector is largely under-developed, contributing less than five per cent to GDP. Construction contributes a further six per cent. Around 100 manufacturing enterprises operating chiefly in the Brazzaville and Pointe-Noire districts and in the Niari Valley. Activity is centred on agro-food and timber processing, textiles and oil refining. There are also a few small-scale import substitution industries (footwear, soft drinks, metal working, chemicals) and a cement plant. The private sector accounts for around 73 per cent of industrial turnover. Emphasis is on joint-venture enterprises, particularly in pulp/paper and light manufacturing.

Mining: Lead, zinc and gold are extracted in small quantities. Deposits of high-grade iron-ore, phosphates, copper, potash and bauxite known to exist.

Hydrocarbons: Petroleum sector is mainstay of economy, providing over 70 per cent of government revenue and 90 per cent of total export earnings. Nine main oilfields in production (seven offshore) with combined reserves of over 750m tonnes. Current rate of production is around 120,000 b/d, 70 per cent by Elf-Congo. The new Tchibouela field is producing at around 30,000 b/d. Congo is not a member of OPEC. Output from the 20,000 b/d capacity oil refinery at Pointe-Noire covers domestic needs. Plans to export liquefied natural gas (reserves 65,000m cu metres) have been abandoned in view of unfavourable market conditions.

Energy: Net exporter of energy. Besides local supplies of oil, fuelwood is the major source of energy. Installed electricity generating capacity is 120 mW. A major project to be completed by 1990 is the construction of a 100 mW hydroelectric plant on the River Lefini.

COUNTRY PROFILE

Membership of international organisations: African Development Bank, African States in Association with the EC (Lomé Convention), Arab Bank for Economic Development in Africa (recipient), Customs and Economic Union of Central Africa, Economic Commission for Africa, French Monetary Area, GATT, IDA, IFAD, ILO, IMF, OAU, UN, WHO, World Bank, WTO.

BUSINESS GUIDE

Time: GMT +1.

Climate: Equatorial or sub-equatorial. Main dry season from Jun-Sep with average temperatures ranging from 15°C at night to 32°C during the day. Rainy season from Oct-May with higher average temperatures and high humidity. Generally hotter and more humid in Congo Basin, drier and cooler in highlands.

Entry requirements
Passport: Required by all.
Visa: Required by all except tourists from German FR and some surrounding French-speaking African countries, who are staying no more than 15 days.
Currency: There is a maximum on amount of local currency that can be exported, currently CFA25,000. Foreign currency must be declared on arrival and export is allowed up to that amount.

Health precautions
Mandatory: Yellow fever vaccination certificate.
Advisable: Cholera, typhoid, polio vaccinations recommended. Malaria prophylaxis should be taken as risk exists throughout the country. Water precautions should be taken. AIDS risk.

Representation overseas: Algiers, Baghdad, Bangui, Beijing, Belgrade, Berlin, Bonn, Brussels, Bucharest, Cairo, Conakry, Dakar, Havana, Kinshasa, Libreville, Luanda, Maputo, Moscow, N'Djaména, Paris, Rome, Tirana, Washington, Yaoundé.

Air access
National airline: Lina Congo operates internally. (Congo is also a minority shareholder in Air Afrique).
Other airlines: Aeroflot, Air Afrique, Air Gabon, Cameroon, Ethiopian, Interflug, TAAG-Angola, TAP-Air Portugal, Sabena, Swissair, UTA.
Tax: Departure tax of CFA200 is levied, for domestic as well as international flights.
International airport: Brazzaville-Maya Maya (Code: BZV), 4 km from city; restaurant. Pointe Noire (Code: PNR), 6 km from city.

Surface access
Main ports: Pointe-Noire, Brazzaville (inland river port). Entry from Zaire possible by ferry from Kinshasa to Brazzaville across the Congo. Cars can be carried. Service operates hourly during the day, 0800 – 1200 and 1400 – 1700.
Overland access: There is a road from Lambaréné, in Gabon, to Loubomo and Brazzaville; this is not surfaced all the way. Entry from Cameroon practicable in dry season only. Surfaced road from the Cabinda enclave of Angola.

Hotels: Good hotels available in Brazzaville, Pointe-Noire and Loubomo; few elsewhere. Advisable to book well in advance. 10 per cent tip is usual.

Electricity supply: 220/230 V AC, 50 cycles.

Car hire: Available from main hotels in Brazzaville and Pointe-Noire. International or national driving licence accepted.

City transport
Taxis: Freely available in Brazzaville, Pointe-Noire and Loubomo; tipping is not usual. Can be hired by the hour or day. Fares should be negotiated in advance of journey.

National transport
Air: Lina Congo operates domestic service to main towns. Brazzaville–Pointe-Noire several times daily; regular flights from Brazzaville to Kindamba, Makoua, Loubomo, Owando, Ouesso, Impfondo, and other main towns.
Road: Main route from Pointe-Noire, through Brazzaville to Ouesso is not uniform in quality; section from Loubomo to Pointe-Noire liable to become impassable in rainy season. 243 km of tarred roads; other routes mainly tracks which can be impassable when wet.
Rail: Congo-Océan (CFCO) operates a daily service from Brazzaville to Point-Noire. Dining cars and sleepers available. The service can be irregular due to derailments, although improvements are being made to tracks. Brazzaville to Pointe-Noire journey-time 11-15 hrs. A mineral line links M'Binda, on the Gabon border, with the Congo-Océan line near Loubomo. A passenger service is operated under the responsibility of CFCO.
Water: Ferries on the Rivers Congo and Oubangui are a principal form of transport.

Public holidays
Fixed dates: 1 Jan, 18 Mar (Death of Marien Ngouabi), 1 May, 15 Aug (National Day), 1 Nov (All Saints'), 25 Dec (Children's Day), 31 Dec (Foundation Day).

Working hours
Business: (Mon-Fri) 0800 – 1200 and 1430 – 1730, (Sat) 0800 – 1200.
Government: (Mon-Fri) 0700 – 1400, (Sat) 0700 – 1200.
Banking: (Mon-Sat) 0620 – 1300.

Shops: (Mon-Sat) 0800 – 1200 and 1500 – 1800/1900. Some shops close on Mon afternoons; a few open Sunday mornings.

Social customs: There are strictly enforced regulations about photographing public buildings and places. Those wishing to use a camera should obtain a permit from Ministry of Tourism.

Telecommunications
Telephone: Dialling code for Congo: IDD access code + 242 followed by subscriber's six-digit number (Brazzaville numbers begin 81, 82 or 83).
Telex: Facilities available at post offices and some hotels.
Telegram: Facilities available at post offices and some hotels.

Banking
Central bank: Banque des États de l'Afrique Centrale.
Main centre: Brazzaville. *Head offices:* Banque Commerciale Congolaise, Union Congolaise de Banques (UCB), Banque Internationale du Congo, Banque Nationale de Développement du Congo.
Foreign branches: Caisse Centrale de Co-opération Économique.

Advertising: Limited possibilities. Press and cinema advertising available. The national radio and TV station accepts advertising through AFRIMEDIA agency only.

Trade fairs: A biennial trade fair is held in years with even-numbered dates in Pointe Noire.

Hotels: Brazzaville
Meridien, Boulevard Marechal Lyautey, P.O. Box: 588 (tel: 83-09-10, 83-09-12; tx: 5368).

Olympic Palace, P.O. Box: 65 (tel: 83-25-02; tx: 5277).

PLM Azur M'Bamou Palace, P.O. Box: 1094 (tel: 81-20-10/12; tx: 5366).

Pullman Cosmos, Avenue Albert 1er, P.O. Box: 1188 (tel: 83-48-46; tx: 5342).

Hotels: Pointe-Noire
Atlantic Palace, Boulevard du Général de Gaulle, P.O. Box: 939 (tel: 94-17-21; tx: 8336).

Novotel, Boulevard du Général de Gaulle, P.O. Box: 790 (tel: 94-12-00; tx: 8240).

PLM Azure Zambia, 54 Avenue N'teta, P.O. Box: 561 (tel: 94-02-72; tx: 8297).

Victory Palace, Rue Bouvandzi, P.O. Box: 124 (tel: 94-07-04; tx: 8284).

Car hire: Brazzaville

Avis: Hotel M'Bamou Palace, PO Box 1135, Brazzaville (tel: 81-21-10, 81-10-40; tx: 5366 KG).

Car hire: Pointe-Noire

Hertz: Novotel, Pointe-Noire (tel: 94-12-00; tx: 8204 KG).

Banking
Banque des États d'Afrique Centrale (BEAC) (central bank), PO Box 1917, Rue du Docteur Jamot, Yaoundé, Republic of Cameroon (tel: 22-25-05; tx: 8343 KN).

Banque Commerciale Congolaise, PO Box 79, Brazzaville (tel: 81-08-80; tx: 5237 bankcgo).

Banque Internationale du Congo, PO Box 33, Brazzaville (tel: 81-03-08; tx: 5339 bidc).

Banque Nationale de Developpement du Congo, PO Box 2085, Brazzaville (tx: 5312 bndc).

Union Congolaise de Banques, PO Box 147, Brazzaville (tel: 81-10-68; tx: 5206 ucobank).

Other useful addresses
Agence Congolaise d'Information (ACI), PO Box 2144, Brazzaville (regulatory body for palg grove planting).

Bureau pour le Développement de la Production Agricole, PO Box 2222, Brazzaville.

Caisse Centrale de Co-opération Economique (CCCE), PO Box 96, Brazzaville (tel: 83-03-20).

Chambre de Commerce, d'Agriculture et d'Industrie de Brazzaville, PO Box 92, Brazzaville (tel: 83-10-89).

Chambre de Commerce, d'Agriculture et d'Industrie de Kouilou-Niari, PO Box 665, Pointe-Noire (tel: 94-04-30, 94-12-80).

Direction Générale du Tourisme, PO Box 456, Brazzaville (tel: 83-09-53; tx: 5325).

Direction de la Statistique, PO Box 2031, Brazzaville (tel: 83-43-24).

Institut de Développement Economique de la République Populaire du Congo, c/o The Presidency, Brazzaville.

Office Congolais des Bois (OCB), PO Box 1229, Pointe Noire (tel: 94-82-48).

Office Congolais de l'Entretien Routier (OCER), PO Box 2073, Brazzaville.

Office National du Commerce (ONC), PO Box 2305, Brazzaville (tel: 83-43-99; tx: 5309).

Société Nationale de Recherche et d'Exploitation Pétrolieres (Hydro Congo), PO Box 2008, Brazzaville (tel: 83-35-60; tx: 5215).

Syndicat des Commercants, Importateurs et Exportateurs de l'Afrique Equatoriale (SYCOMIMPEX), PO Box 84, Brazzaville.

Cook Islands

Kevin Rafferty

(map on page xxiv)

Even in terms of the mini- and micro-states of the South Pacific the Cook Islands is unique. The 15 islands comprising the territory have a land area of a mere 234 sq km and 18,000 inhabitants. But there are as many Cook Islanders, a people of Polynesian Maori stock, living in New Zealand as in the Cooks themselves. So it is perhaps appropriate that the country is not independent but self-governing in free association with 'big brother' New Zealand. In 1989 there was a change of political guard with a new government elected and the Cook Islands Party led by Geoffrey Henry, scion of the Henry family that had led the islands in the early 1980s. The Henry family claimed to be descended from Britain's King Henry VIII.

KEY FACTS

Official title: Cook Islands

Head of state: Queen Elizabeth II

Head of government: Prime Minister Geoffrey Henry

Ruling party: Cook Islands Party

Capital: Rarotonga

Official Languages: English

Currency: New Zealand Dollar (NZ$ = 100 cents)

Exchange rate: NZ$1.70 per $ (Nov 1989)

Area: 234 sq km (15 islands) – Rarotonga (67 sq km); Mangaia (51)

Population: 18,000* (1988)

GDP per capita: $1000* (1986)

Labour force: 4,000* (1988)

Inflation: 10.0% (1986)

Trade balance: –$20m* (1987)

Visitor numbers: 36,000* (1986)

* estimated figure

Relations with New Zealand

In the general election of January 1989 Henry's party won 12 of the 24 seats. But it was able to claim the support of two MPs of the Democratic Tumu party and thus oust the previous coalition led by Dr Pupuke Robati. The election probably also marked the final departure from politics of Sir Tom Davis, now in his mid-70s. Davis had been in power since 1983 until he was ousted by a 1984 cabinet coup of ministers unhappy with his dictatorial ways and particularly for his habit of biting the New Zealand hand that helps to feed the Cook Islands and keep it alive.

'Sir Tom was an embarrassment for us in so far as the relationship with New Zealand was concerned,' complained Foreign Affairs and Police Minister Norman George. 'Our cabinet and the rest of the people have always been very close to New Zealand and Sir Tom's behaviour was an embarrassment to all of us because it created bad feeling.'

Relations became particularly strained when David Lange became prime minister of New Zealand because of his policy of not allowing nuclear weapons or nuclear-powered ships into the area. Davis, a doctor and a surgeon who had previously worked on the US space programme, disagreed violently with this attitude. He accused the New Zealand government of being 'dictatorial'. This brought a written response from Lange that if he was as dictatorial as Davis suggested, then Davis 'might like to take a final step and make the Cook Islands independent.'

New Prime Minister Henry exhibited a very different attitude from his predecessors. He claimed that his government has 'great respect for New Zealand's non-nuclear policy. I think it's impressive and I can say I know many Pacific leaders who approve of the stance taken by the New Zealand government.' On the question of independence, Prime Minister Henry claimed to have 'no hangup' on a subject that, he predicted, would become an issue in due course. 'The New Zealanders are nudging us towards that direction. Possibly our young people may choose that course. I personally feel that our relationship [with New Zealand] is unique and something worth preserving.'

Economy

What any government of the Cooks must try to preserve is the support and aid that it gets from its big neighbour to the south. Without its links to New Zealand the Cooks would be lost, and particularly would not be able to live as well as it does. Per capita income is NZ$1,600, huge for a small group of economically fragile and vulnerable islands.

The Cooks' own economy cannot support even its small population. It has to rely on subsistence farming, fishing, some exports of agriculture and clothing and even sales of postage stamps to produce exports that are generally only 20 per cent of the value of imports. Cyclone Sally in 1987 caused $27m worth of damage and set back the economy by two years. The trade deficit is running at about $30m a year. The government is the main employer in the modern monetary economy with more than 40 per cent of the labour force. In the northern atolls inhabitants subsist on coconuts and fish and do some copra processing; in the south paid labour is more prevalent. Wages are relatively high with a minimum wage of more than NZ$1.7 per hour for workers over 18 years old.

To try to generate more domestic income the Davis government set up an international financial centre on the Cooks but this is making slow progress in the face of competition from elsewhere and the small and unknown nature of the Cooks in the outside world.

The country is able to afford its high standard of living because of income flows from New Zealand. Remittances from Cook Islanders working in New Zealand help to bridge the gap. So does aid worth about NZ$8m a year. Most particularly, tourism is invaluable, tourist earnings alone almost cover the trade deficit. These would not be possible without the good relations and good physical links with New Zealand.

The big brother is after all hundreds of miles distant from the Cooks, and Tonga and New Caledonia are physically closer to parts of New Zealand than the Cooks. Only Western Samoa benefits from good links to New Zealand on quite the same scale and in 1989 the Samoans felt the cold breath of recession from Wellington anxious to get rid of some of its responsibilities. The Cooks are smaller and if they were independent they would be more vulnerable to the storms that can blow up in a cruel world.

COUNTRY PROFILE

Political system: The Cook Islands are self-governing in free association with New Zealand, with a 24-member legislative assembly elected to a five-year term by universal adult suffrage, an elected prime minister and a cabinet of ministers. The upper legislative house of hereditary chiefs, the House of Ariki, has advisory functions only. Local affairs are handled by traditional councils and village committees in the outer islands. More Cook Islanders live in New Zealand than inhabit the islands themselves. The British sovereign is head of state.

Political parties: Cook Islands Party.

Population: Cook Islands population 17,463 (1986 census). The population consists largely of Polynesian Maori stock, with some New Zealanders and a small number of Europeans. The 1970s and 1980s have seen a substantial migration to New Zealand.

Main town: Avurua, Rarotonga (population 9,400* in 1987).

Language: Most Maoris are bilingual, speaking English as well as their own Polynesian dialect.

Media

Press: Cook Islands Broadcasting and Newspaper Corporation publish *Cook Islands News* in English and Maori six days a week (Mon-Sat). *Kia Orana* is a weekly in English and Maori. The *Cook Islands Sun* is a quarterly in English.

Radio: Cook Islands Broadcasting and Newspaper Corporation (CIBNC) broadcasts 17 hrs/day, seven days/week in English and Maori, including bulletins from Australian and New Zealand national stations. Radio Kia Orana Country FM103 operates an FM station broadcasting general programmes 18 hrs/day, seven days/week in English and Maori.

Television: There is a television service planned for 1990.

Domestic economy: The economy is mainly based on tourism and the finance centre. Subsistence agriculture and fishing continue to be important activities, while remittances from migrant workers, aid from New Zealand, sales of postage stamps and export of agricultural produce and clothing continue to have significant roles in the economy. The Cook Islands suffer from an adverse balance of trade, and imports are generally five times the value of exports. In the northern atolls, people subsist mainly on coconuts and fish, with some copra processing; in the southern atolls there has been a shift to paid labour, although many work their own plantations. A significant offshore banking business has been developed. Major projects include expansion of the electricity system and improvements in the harbour and shipping service.

COUNTRY PROFILE

Employment: The government is the main source of paid work, employing 42 per cent of the labour force in 1986. The Public Service Association has the largest membership of worker associations and is relatively influential, but other unions are relatively small (Cook Islands Industrial Union of Waterside Workers and the Airport Workers' Union).

External trade: There is free trade with New Zealand. Other avenues are already well established and are being expanded.
Exports: In 1987, exports rose by 3.3 per cent to US$4m. Main exports are clothing (40%), fresh and canned fruit (35%), copra (10%) and handicrafts. Main destination is New Zealand (over 75%), Tahiti, Japan, UK, Australia and South Korea.
Imports: In 1987, imports rose by 1.3 per cent to US$20m. Main imports are manufactured goods, food and live animals, machinery and transport equipment, mineral fuels, chemicals, beverages and tobacco and raw materials. Main source is New Zealand (usually 75-85 per cent of total).

Agriculture: Rich volcanic soil on southern islands. Subsistence farming and fishing for local consumption: citrus fruit, pawpaws, capsicums, beans, pineapples, aubergines and other vegetables produced for export, 90 per cent exported to New Zealand. Livestock includes pigs, poultry and cattle. Development plans by Atiu island for coffee and arrowroot starch and cattle farming. Cattle farming, coffee and exotic fruits are also being developed on Mauke. The majority of banana exports are from Aitutaki. There is a rapidly expanding pearl farming industry on Maniniki and Penryhn which is now a significant foreign-exchange earner with excellent prospects. The recently over-fished trochus shell (for shell buttons) population has been re-seeded and the results are being closely monitored. There is a limited number of local part-time fishermen who are unable to cater for all internal requirements. Fisheries Agreements were signed in 1987 allowing Taiwanese and Korean vessels to operate within the 200-mile exclusive economic zone for a set fee.

Industry: The tourist industry has now become the country's main source of income. Secondary industries include agricultural exports, clothing manufacture, fruit canning/processing, electronic component assembly and handicrafts. New hotels are under construction.

Legal system: The High Court of the Cook Islands has three divisions: Civil, Criminal and Land. Appeals from all divisions are to the Cook Islands Court of Appeal. Appeals from this court go to the Privy Council in London.

Membership of international organisations: Asian Development Bank, Asia Pacific Development Centre, Development Information Network for Asia and the Pacific, Fertiliser Advisory Development and Information Network for Asia and the Pacific, Economic and Social Commission for Asia and the Pacific (associate member), Food and Agriculture Organisation, Forum Secretariat, International Civil Aviation Organisation, Pacific Islands Development Programme, South Pacific Commission, South Pacific Forum, World Health Organisation.

BUSINESS GUIDE

Time: GMT -10 hrs.

Climate: Damp and tropical, mild from Apr-Nov but Dec-Mar hot and humid, with likelihood of hurricanes. The mean temperature is 23.9°C, and average rainfall over 2,000 mm/year, heaviest on the forested volcanic slopes of the southern islands.

Entry requirements
Passport: Required by all with exception of New Zealand citizens.

Visa: Not required for stays of up to 31 days. Permits can be applied for on arrival for stays of up to four months.

Customs: Import licences are required for goods from countries other than New Zealand. Incoming passengers are permitted to bring in a maximum of 200 cigarettes, one litre of wine and spirits or 4.5 litres of beer.

Air access

National airlines: Air Rarotonga, Cook Islands International (inaugurated Aug 1986).

Other airlines: Air New Zealand, Cook Islandair, Hawaiian, Polynesian.

International airport: Rarotonga (RAR), near Avarua.

Surface access

Main ports: Avatiu (on Rarotonga), Mangaia (for pineapple exports) and Aitutaki.

Car hire: Car, scooter and bicycle hire are available on Rarotonga. International or Commonwealth national driving licence must be presented to obtain local licence. Driving is on the left.

City transport

Taxis: Taxi service is available on Rarotonga.

National transport

Air: Cook Island Air and Air Rarotonga operate inter-island services. Airstrips for small planes on Aitutaki, Penryhn, Rakahanga, Mitiaro, Atiu, Mauke and Mangala.

Road: A surfaced road of 32-33 km runs around Rarotonga coast.
Buses: Bus service is available. Hourly in Rarotonga from 0730 to 1600 Mon-Fri.
Water: Inter-island shipping services are provided by the National Line. There are harbours on Aitutaki, Atiu, Penrhyn and Suwarrow, but local advice should be sought before approaching.

Public holidays
Fixed dates: 1 Jan, 25 Apr, 1 Nov, 25-26 Dec.
Variable dates: Good Friday, Easter Monday, Birthday of HM the Queen, Constitution Day, Gospel Day.

Working hours
Government and business: (Mon-Fri) 0800 – 1600.
Banking: (Mon-Fri) 0900 – 1500.
Shops: (Mon-Fri) 0800 – 1600; (Sat) 0800 – 1200.

Social customs: Tipping and bargaining should be avoided. *Dress:* Brief attire (e.g. bikinis) should not be worn in towns or villages. Nude or topless sunbathing will cause offence.

Telecommunications

Telephone: Dialling code for Cook islands, IDD access code + 682 followed by subscriber's number (no code necessary). Automatic telephone exchange in Rarotonga with 2,300 private and 600 business lines; Aitutaki has a manual exchange with about 160 telephones. International telecommunications via Cable and Wireless and Peacesat satellite links.

Telex and telefax: Telex and Facsimile services available on Rarotonga.

Telegram: Telegram service available on all islands.

Banking: Legislation to enable Cook Islands' development as an offshore financial centre and tax haven was enacted in 1981/82. *Main centre:* Avarua (on Rarotonga). *Head office:* Cook Islands Development Bank. *Cook Islands branches:* European Pacific Banking Corporation, Post Office Savings Bank.

Electricity supply: 230 V AC.

Representation in capital: New Zealand and France.

BUSINESS DIRECTORY

Hotels: Aitutaki, Rarotonga
Aitutaki Resort, P.O. Box: 342 (tel: 20234).

Hotels: Amuri, Aitutaki
Dive Rarotonga Hostel, Rarotonga (Kavera), P.O. Box: 38 (tel: 21873; tx: 62046 RG).

Hotels: Avarua, Rarotonga
Arorangi Lodge, P.O. Box: 51 (tel: 27379,; tx: 62034 RG).

Beach, P.O. Box: 700 (tel: 22461; tx: 26016 COUNSEL RG).

Edgewater, P.O. Box: 121 (tel: 25435; tx: 62059 RG).

Kiiki, P.O. Box: 68 (tel: 21937; tx: 62076).

Lagoon Lodges, P.O. Box: 45 (tel: 22020; tx: 62076 MAEVA RG).

Palm Grove Lodges, P.O. Box: 23 (tel: 20002; tx: 62076 ATTN; fax: 20969).

Puaikura Reef Lodges, P.O. Box: 397 (tel: 23537; tx: 62045 ATTN).

Rarotongan, P.O. Box: 103 (tel: 25800; tx: RG62003; fax: 25799).

Tamure Resort, P.O. Box: 17 (tel: 22415; tx: 62007).

Hotels: Rarotonga
Manuia Beach, P.O. Box: 700 (tel: 22461; tx: RG62060).

Pacific Resort, P.O. Box: 790 (tel: 20427; tx: RG62041).

Car hire
Rarotonga Ace Rent-a-Car: PO Box 257 (tel: 21901).

Avis: PO Box 74 (tel: 22833).

Budget: PO Box 607 (tel: 20888).

Cook Islands General Transport Ltd (tel: 24442).

Haurua Enterprises (tel: 27190).

Odds and Ends (tel: 20942).

Polynesian Rentals (tel: 20888).

Rental Cars (CI) Ltd: PO Box 326 (tel: 24442).

Tipani Rentals: PO Box 72.

TPA Rental Cars (tel: 20611).

Vaima Rentals (tel: 22222).

Business advice and information
Cook Islands Development Bank, PO Box 113, Rarotonga (tel: 2065; tx: 2006).

Ministry of Trade, Labour, Industries and Transport, Head Office, Avarua, PO Box 61 (tel: 28810).

Rarotonga Chamber of Commerce, Avarua, Rarotonga.

Travel information
Air Rarotonga (tel: 22888).

Cook Islandair (tel: 26304).

Cook Islands Tourist Authority, PO Box 14, Rarotonga (tel: 29435; tx: 62054; fax: 21435).

Government Information Office, PO Box 106 (tel: 20725; tx: 62006; fax: 20856).

Costa Rica

Roberto Espíndola

The ruling Partido Liberación Nacional (PLN) was almost torn apart by infighting over the presidential nomination for the 1990 elections, but in the end rallied around the victor, Carlos Manuel Castillo, who will face the candidate of the opposition Partido Unidad Social Cristiana (PUSC), Rafael Angel Calderón. The infighting, however, left three clear factions within the PLN: the one around Castillo and former President Oduber and *Figueres* or 'neoliberal' wing; the 'progressive' or Social Democratic wing led by former Party Secretary Rolando Araya – the loser in the nomination struggle – and former President Monge, and finally the group around President Arias. This reflects Arias' increasing isolation, his domestic policies criticised by both left and right.

A key issue within the presidential campaign became the accusations made in Washington by a former Panamanian diplomat, who claimed that Panama's military leader, General Manuel Antonio Noriega, would have given $0.5m to PUSC in 1986 to fund Calderon's unsuccessful bid for the presidency. PUSC activists replied by pointing out that there had been similar charges about the $100,000 which had found their way into Arias' campaign coffers.

Months away from the presidential election of February 1990, opinion polls were giving an advantage of seven percentage points to Calderon over Castillo, although the PUSC itself had only a three-point advantage over PLN. This perhaps reflected the electorate's greater familiarity with Calderon's name, as he had run for the presidency twice before, in 1982 and 1986. In terms of domestic policies, there is no great difference between the two; in the area of foreign policies – and particularly concerning regional conflicts – Calderon may be expected to adopt a more aggressive stand, whilst Castillo could be expected to act in greater co-ordination with Washington.

Debt strategy

The success of structural adjustments in improving the economic situation encouraged the administration of President Arias to aspire to become the first Latin American country to resolve its foreign debt, running at $4.8bn and the largest per capita indebtedness in the world. President Arias believed the relatively small size of the debt – compared with other countries – and efforts to improve the economy and make internal structural adjustments, would enable the government to take advantage of the Brady Plan. Under this plan, the $1.6bn debt to foreign commercial banks could be bought for a much lower price, as its market value was estimated at 16–20 per cent of this amount.

Finance Minister Fernando Naranjo claimed that the government had fulfilled its obligations to international financial organisations and – after a second deal with the IMF – an accord with the World Bank was in the pipeline. Negotiations with the Paris Club in May, together with the legislative assembly's approval of SAL II and the signing of agreements with the IMF and the AID, were expected to clear the way for bilateral agreements worth $1bn.

In May 1989 the IMF executive board approved a loan agreement for $95m, including a $55m standby facility to run for one year as a support to the balance of payments, while in exchange the government committed itself to limiting its operating deficit for 1990 to 11bn colónes, or 2.75 per cent of GNP. This deal also gave the government $40m to pay interest arrears on the debt to commercial banks – unpaid since 1986. Part of this sum, $14m, is earmarked to begin buying back debt from commercial banks.

Washington agreed that the success the Arias administration had in applying structural adjustments and in its negotiations with creditors and international financial organisations were sufficient to include Costa Rica among the first beneficiaries of the Brady Plan. The application of the Brady Plan will not be easy, as Costa Rica is indebted to some 200 foreign commercial banks – which must be consulted – with most of them only willing to re-sell discounted debt if they are paid in fresh money. The latter requirement will force the government to search for $250m from the IMF or World Bank, or – alternatively – to use its international reserves, estimated at $400m.

As part of the negotiations with foreign banks, central bank president Eduardo Lizano announced early in 1989 that there was little chance of resuming full interest payments on the debt to foreign commercial banks. The government even agitated the ghost of a cessation of payments to foreign creditor banks. In fact, since 1987 monthly payments to creditor banks had already been limited to just 40 per cent of amounts due, causing over $200m in interest arrears to accumulate.

The economy

In June 1989 the government asked the World Bank for an extension of the deadline for the legislative assembly to ratify SAL II, the structural adjustment plan required for a $100m loan. This was the second extension requested within three months by the Arias administration. As part of the agreement, Japan would grant a loan to match the one from the World Bank.

KEY FACTS

Official title: Republic of Costa Rica
Head of state: President Oscar Arias Sánchez
Ruling party: Partido Liberación Nacional
Capital: San José
Official Languages: Spanish
Currency: Colón (C) = 100 céntimos
Exchange rate: C83.0 per $ (Nov 1989)
Area: 51,100 sq km

Population: 2.85m (1988)
GDP per capita: $2,235 (1988)
GDP real growth: 3.0% (1988)
Inflation: 20.8% (1988)
Trade balance: –C6,785m (1988)
Foreign debt: $4.72bn (1987)

SAL II intends to achieve a reform of the economy to make it more competitive abroad, increasing exports and – according to officials – allowing 'access to external resources'. The agreement was signed with the World Bank in December 1988, giving the legislative assembly three months for its ratification. However, parliamentary opposition to the plan forced the government to seek extensions.

The opposition PUSC blocked the deal, claiming that it left local industries unprotected in its effort to increase their efficiency and to orient their production to the external markets. Trade unions and the Left claimed that the changes proposed would be detrimental to workers, increasing unemployment and the price of basic consumer goods; they also argued that the changes would lead many small and medium size businesses to bankruptcy.

The SAL II terms continue the SAL I programme signed with the World Bank in 1985, and aim to streamline the economy by boosting non-traditional exports, increasing domestic savings and reducing protectionism for basic grains. It involves reducing import duties to a maximum 40 per cent on consumer goods and 20 per cent on inputs and capital goods by January 1991, and modernising the system of export incentives to avoid unfair-trade retaliation in foreign markets.

Guaranteed prices paid to basic grain producers will come down to no more than 40 per cent above the world market price and the Consejo Nacional de la Produccion will authorise private grain imports if there is a shortage. This gave rise to concern in the agricultural sector about the risk of foreign competition, with opposition spokesmen pointing to the inability of local farmers to compete with heavily-subsidised foreign producers of corn, beans and rice. It is feared that competition will prove too fierce for small farmers, who will go bankrupt and join the unemployed.

Financial reforms required by SAL II include the modernisation of the state banking system, another area of intense controversy, even within government ranks. Trade union leaders, in general, supported the modernisation of the economy proposed by President Arias, but opposed the opening up of banking to private interests.

The privatisation proposal placed control of the central bank in the hands of the private sector. Proponents argued that it will strengthen the state bank and the whole financial system by increasing foreign investment, but opponents fear that the existence of a nationalised banking system, established in 1948, could be threatened.

War on inflation

President Arias launched in February 1989 a 'war on inflation', with a package of measures to contain inflation, which reached 25.3 per cent the previous year. The main measures were a freeze on public services' rates, rigorous price control, harsher penalties for speculators, together with others aimed at stimulating production, such as more credits, reductions of import and export taxes, and a slowdown of devaluations.

Arias claimed that his administration would meet the targets of its social programmes and cut inflation by half in his last 12 months in office. He announced in late February that the government was to wipe out this year's 8bn colónes fiscal deficit to help halve the inflation rate. A breakdown of the government's austerity measures included a 3.2bn colónes cut in the decentralised, autonomous government agencies, a reduction of 400m colónes in tax exonerations for

KEY INDICATORS

	Unit	1984	1985	1986	1987	1988
Population	m	2.4	2.5	2.7	2.8	2.9
GDP (1985 prices)	C bn	196.5	197.9	208.9	220.2	226.6
GDP per capita	$	1,468	1,527	1,667	1,674	1,686
Inflation (average rate)	%	12.0	15.1	11.8	16.9	20.8
GDP real growth	%	8.0	0.7	5.5	5.4	3.0
Agricultural production	% real gwth	10.1	−5.5	4.8	4.0	5.3
Coffee	'000 tonnes	137	124	122	156	144
Exports	C m	44,818	49,242	62,732	72,714	100,047
Coffee	C m	11,903	15,938	21,942	20,996	24,049
Bananas	C m	11,178	10,493	12,137	14,351	18,787
Sugar	C m	1,576	684	622	1,065	–
Beef	C m	1,939	2,709	3,910	3,924	4,096
Imports	C m	48,705	55,408	64,244	86,788	106,832
Petroleum	C m	3,544	4,431	3,813	5,236	5,231
Balance of trade	C m	−3,887	−6,166	−1,512	−14,074	−6,785
Current account	$m	−251	−291	−161	−386	−268
Tourist arrivals	'000	274	262	259	278	329
Exchange rate (average)	C per $	44.53	50.5	56.0	62.8	75.8

* estimated figure

COUNTRY PROFILE

Political system: Executive power is held by the president elected by popular vote for a four-year term – if a 40 per cent vote for any candidate is not obtained a second ballot is held. The term is not renewable. The president appoints and is assisted by a 22-member cabinet. The legislature is the unicameral 57-member Legislative Assembly elected at the same time as the president, also for a four-year term.

Main cities/towns: San José, Puerto Limón, Alajuela, Cartago, Heredía, Puntarenas, Liberia.

Media
Press: The three major daily newspapers are *La Nación* (circulation 90,000), *La República* (60,000) and *La Prensa Libre* (50,000). Weeklies are also published in certain provincial towns. There are few periodicals and trade papers, including *La Tribuna Económica, Comercialización Internacional, Agroindustria. The Tico Times* (12,000) publishes weekly in English.
Radio: There are over 100 commercial radio stations and several cultural and religious networks. *Television:* There are eight commercial stations, covering 90 per cent of the country. Direct transmission from USA via satellite.

External trade: Dependence on few agricultural exports makes balance of trade highly susceptible to world commodity price shifts. Ability of the country to import is restricted by regular shortages of foreign exchange, due primarily to heavy debt burden.
Exports: Coffee (typically 25% of total value), bananas (19%), cattle and meat (4%), cocoa, sugar, fertilisers and non-traditional products (43%). Main destinations: USA (44% of 1987 total), German FR (14%), Guatemala (4%), El Salvador (3%), Italy (3%), Netherlands (3%), UK (2%), Japan (1%).
Imports: Primary commodities (typically 49% of total value), consumer goods (22%), oil and fuel (4%), machinery and equipment (3-5%). Main sources: USA (37% of 1987 total), Japan (8%), Venezuela (7%), Guatemala (5%), German FR (5%), Mexico (5%).

Agriculture: Most important sector of the economy, contributing over 18 per cent to GDP, employing over 28 per cent of the workforce and accounting for over 60 per cent of total export earnings. Most important cash crops are coffee, bananas and sugar. A considerable amount of meat, mostly beef, is also exported. Production of cash crops has risen in recent years, but increased exports have tended to be offset by falling world prices. Crisis in banana industry since 1985 has led to sharp drop in export earnings from bananas. Staple food crops include rice, maize and beans. Around 8-10 per cent of total area is cultivated arable land, 25 per cent pasture and over 50 per cent woodland and forest. Irrigation is mostly in the Nicoya region.

Industry: Manufacturing sector contributes up to 20% to GDP and is concentrated on export-orientated processing of agricultural products. Main activities are food processing, textiles, chemicals and plastics. Also important are cement and metal industries, petroleum refining and pulp/paper processing. The government has promoted growth in the manufacturing sector with investment incentives and tax holidays, and the Caribbean Basin Initiative has contributed. There are industrial free zones *(Zonas Francas)*, where incentives apply, at Puerto Limón, Puntarenas and Cartago. Privatisation of state-run industries is in progress.

Mining: Discovery of valuable bauxite deposit in Boruca area prompted large-scale investment in an aluminium smelting plant. Small quantities of gold, silver, manganese and mercury are also mined.Commercial deposits of iron-ore may also be present.

Energy: Electricity generation is 80 per cent hydroelectric and 20 per cent thermic. Major sector for investment. Some oil production.

Membership of international organisations:
Central American Common Market, Central American Democratic Community, FAO, G-77, IADB, IAEA, ICAO, ICO, IDA, IFAD, IFC, ILO, IMF, IMO, INTELSAT, Inter-American Development Bank, Interpol, IPU, ITU, IWC, ODECA, Organisation of Central American States, PAHO, SELA, UN, UNESCO, UPEB, UPU, WHO, WMO, World Bank, WTO.

exporters and a 400m colónes slash in wheat subsidies. These cuts, when added to a 3.9bn colónes increase in public service rates, came to 7.9bn colónes, and other adjustments added by President Arias brought the total to 8.5bn colónes. Government departments and state enterprises made savings by suspending new investment projects, cutting back personnel and revamping costly procedures.

The Central Permanente de Trabajadores – gathering most Costa Rican unions – decided that Arias' 'declaration of war' would not be enough to stop inflation and launched its own campaign, including a general strike in April.

Trade deficit

One of the most controversial issues in 1989 was the growing trade deficit which reached $163m in 1988 and was expected to exceed $200m by December, prompted by a fast rise in imports. The growth in imports reflected increased purchases abroad of basic grains and of luxury goods. This is, according to local economists, a consequence of increased liquidity in the economy. The government tried to justify the import spree as showing the use of credits available to local firms in the purchase abroad of capital goods for industry, but such credits were apparently also used to import the semi-elaborated inputs required by the *maquiladoras* (in bond manufacturers).

Meanwhile, exports in 1988 showed an increase of 9.7 per cent over 1987. This, according to government officials, was the direct result of a 29.2 per cent increase in non-traditional exports, as the export of goods classified as traditional dropped 3.6 per cent.

Growth in industrial exports showed mainly the development of textiles for export through the use of *maquiladoras*. There were 26 textile *maquiladoras* in 1984, but they had increased to 145 by 1988. This helped take export earnings from $34m in 1984 to $260m in 1988.

Government critics argued that the *maquiladoras* are of little benefit to the economy as the export earnings generated are lower than the imports of inputs they require. They added that the typical *maquiladora* is owned by a foreign firm, takes advantage of the cheapest labour force available – 85 per cent of the *maquiladoras'* workers are women – and brings little or no external inputs of investment or technology.

Coffee drive

Coffee producers and exporters, as well as the government, reacted to the suspension in early July 1989 of the International Coffee Agreement's export quotas, by launching a drive to conquer the largest possible share of the now unregulated world coffee market. Although the suspension of the export quotas will have an initial effect on the economy, producers and exporters were confident that Costa Rica will benefit from a free coffee market, claiming that the ICA's quotas were unfavourable to them and were 'biased' towards other producing countries.

Despite such optimism, coffee exports were suspended in July to avoid the effects of the sharp price drop which followed the export quotas' suspension. Out of the 1988–89 harvest of 3.6m 46-kilo bags, Costa Rica had by July, already sold abroad 60 per cent, leaving just some 831,260 bags to be exported. Official sources indicated that they expect earnings from coffee exports of $300m in 1988 to remain at that level this year.

The export quotas guaranteed price levels, but also meant reductions in earnings when the ICO cut back the total export quota to respond to falls in prices. Out of a total ICO export quota of 58m 46-kilo bags in 1988, Costa Rica's quota was 2.46 per cent, which initially meant 1.8m 46-kilo bags, but after several cuts that had been reduced to 1.5m bags, with losses in export earnings of over $25m.

Outlook

Industrial unrest will increase as the elections get nearer, with increasing pressure on the government not to alienate voters. Credit conditions will also be affected by the electoral climate, loosening up to stimulate expansion.

It is likely that the government can meet its goal of keeping the annual inflation rate below 12 per cent. Unemployment is expected to get down to 5.4 per cent in 1989, whilst the economy is likely to grow by 3.5 per cent.

Devaluations will continue, with the colón reaching a rate of 88 to the US dollar by year-end and the parallel rate probably reaching some 92 colones. Exports are expected to grow by 11 per cent, fetching $1.4bn, whilst imports are expected to grow by more than the 8 per cent anticipated in official estimates. Public sector savings should reach 6 per cent of GDP this year and in 1990, whilst investment is to be restricted to 6.8 per cent of GDP.

BUSINESS GUIDE

Time: GMT -6.

Climate: Hot and humid in lowland coastal areas; temperate and warm in central highlands (San José). Dry season Dec to May; average temperature in San José 25°C. Rainy season from Jun to Nov; average temperature 25°C.

Entry requirements
Passport: Required by all.
Visa: Required by all except nationals of EC countries, (excluding France, Greece, Ireland), Canada, Japan, Israel, Scandinavian countries, Austria, Switzerland and certain others. Exit permits normally required for stays of more than 90 days. Regulations are liable to change at short notice; check with embassy/consulate.
Currency: No restrictions on import of foreign or local currency. Foreign currency should be changed only in banks and authorised bureaux. Visitors may change excess local currency back to US dollars, but only at main offices of state commercial banks and on production of onward airline ticket and passport.

Health precautions
Mandatory: None.
Advisable: Typhoid, polio vaccination. Malaria risk in some low lying areas – prophylaxis advisable. Water precautions should be taken outside San José.

Representation overseas: Berne, Beirut, Bogotá, Bonn, Brasília, Bridgetown, Brussels, Bucharest, Buenos Aires, Caracas, Guatemala City, The Hague, Kingston, La Paz, Lima, Lisbon, London, Madrid, Managua, Mexico City, Montevideo, Ottawa, Panama City, Paris, Quito, Rabat, Rome, San Salvador, Santiago, Santo Domingo, Seoul, Taipei, Tegucigalpa, Jerusalem, Tokyo, Vatican City, Washington.

Air access
National airline: Lacsa (Líneas Aéreas Costarricenses).
Other airlines: Aeronica, Copa, Eastern, Iberia, KLM, Mexicana, Pan Am, Sahsa, SAM, Taca International, Varig.
International airport: San José-Juan Santamaría International/El Coco (SJO), 18km from city; duty-free shop, bar, restaurant, buffet, bank, post office, shops, hotel reservations, car hire.

Surface access
Main ports: Caldera, Puntarenas (Pacific); Limón/Moir (Caribbean).
Overland access: Possible via Pan-American Highway which runs from Nicaragua, through San José, to Panama. International coach services (operated by Tica Bus and others) run from Managua and Panama City to San José. It is advisable to book well in advance.

Hotels: Advisable to book well in advance. Three per cent tourism tax, 10 per cent sales tax, and 10 per cent service charge will be added to bill.

Car hire: Temporary permit must be obtained from local traffic authorities on production of national licence.

City transport
Taxis: Within San José: can be hailed or ordered by telephone (35-99-66, 21-84-66, 21-25-52); have yellow number plates; checking fixed-rate fare beforehand is generally advised. Tipping is unnecessary.

National transport
Air: SANSA is the main domestic carrier and operates regular flights to provincial towns. A number of smaller airlines provide internal flights as well.
Road: Total network of some 30,000 km of all-weather roads. Main routes are the Pan-American Highway; San José to Caldera; San José to Guapiles; and San José to Puerto Limón. A new road is being built to link Pacific coast towns.
Buses: Regular services in main towns. Long-distance services operate between San José and most major centres.
Rail: Two routes are operated from San José, with regular daily services to Puntarenas and to Puerto Limón. Trains are slow but comfortable.

Public holidays: New Year's Day, 19 Mar (St Joseph's Day), Holy Thursday, Good Friday, Holy Saturday, 11 Apr (Juan Santamaria's Day), 1 May, Corpus Christi, 29 Jun (St Peter and St Paul), 25 Jul (Annexation of Guanacaste), 2 Aug (Virgin of Los Angeles), 15 Aug (Assumption), 15 Sep (Independence Day), 12 Oct (Columbus Day), 8 Dec (Immaculate Conception), 25 Dec, 28-31 Dec (bank holidays - San José only).

Working hours
Government and business: (Mon-Fri) 0800/0830 – 1100/1130, 1300 – 1600/1700/1730; (Sat) 0800 – 1100. Government only: 0800 – 1600 (Mon-Fri).
Banking: (Mon-Fri) 0900 – 1500, 1530 – 1800.
Shops: (Mon-Sat) 0800 – 1200, 1400 – 1800.

Telecommunications
Telephone: Dialling code for Costa Rica: IDD access code + 506 followed by subscriber's number (no area codes needed). Internal automatic telephone system connects main towns.
Telex: Public telex facility is available in San José at Radiográfica Costarricense, Calle 1, Av. 5 (open 0700-2200).
Telegram: Facilities available in San José, Limón, Puntarenas and Golfito.

Banking
Banking centre: San José
Central bank: Banco Central de Costa Rica. *Head offices:* Banco Anglo-Costarricense,Banco de Comercio, Banco de Costa Rica, Banco Crédito Agrícola de Cartago, Banco de la Construcción, Banco Latinamericano (Costa Rica), Banco Lyon, Banco Nacional de Costa Rica, Banco Popular y de Desarrollo Comunal, Banco de San José, Compañía Financiera de Londres (Lloyds), Corporación Franco Americana de Finanzas (Costa Rica), Corporación Internacional de Boston, Financiera del First Pennsylvania.
Foreign branches: Banco Centroamericano de Integración Económica, Banco Lyon, Banco de Santander, Bank of America, Bank of London & South America, Bolivar International Merchant Bank, Citicorp, First National City Bank (Costa Rica).

BUSINESS GUIDE

Electricity supply: 110 V AC, 60 cycles.

Representation in capital: Argentina, Belgium, Brazil, Bulgaria, Canada, Chile, Colombia, Czechoslovakia, Dominican Republic, Ecuador, El Salvador, France, German FR, Guatemala, Honduras, Israel, Italy, Japan, Republic of Korea, Mexico, Netherlands, Nicaragua, Panama, Peru, Poland, Romania, Spain, Switzerland, Taiwan, UK, USA, USSR, Uruguay, Vatican, Venezuela, Yugoslavia.

Hotels: San José
Ambassador, Calle 26 & 28 Avenida Central, P.O. Box: 10186 (tel: 21-81-55; tx: 2315).

Balmoral, P.O. Box: 3344 (tel: 22-50-22; tx: 2254; fax: 211266).

Bougainvillea, Barrio Tournon, P.O. Box: 69-2120 (tel: 33-66-22).

Cariari, Centro Colon, P.O. Box: 737 (tel: 39-00-22; tx: 7509).

Corobici, Autopista General Cañas, P.O. Box: 85480 (tel: 32-81-22; tx: 2700; fax: 315834).

Europa, Calle Central, Avenida 5, P.O. Box: 72 (tel: 22-12-22; tx: 3242; fax: 213976).

Gran Costa Rica, P.O. Box: 527 (tel: 21-40-00; tx: 2131).

Holiday Inn Aurola, San José-Aurola (tel: 33-72-33; tx: 3545; fax: 551036).

Irazú, P.O. Box: 962 (tel: 32-48-11; tx: 2307; fax: 323159).

Royal Dutch, P.O. Box: 4258 (tel: 22-14-14-; tx: 2925).

Sheraton Herradura, Ciadad Cariari, P.O. Box: 7-1880 (tel: 39-00-33; tx: 7512; fax: 39-22-92).

Torremolinos, Calle 40, Avenida 5 Bis, P.O. Box: 2029 (tel: 22-52-66; tx: 2343).

Car hire: San José
Ada Rent-a-Car (General RaC): PO Box 64 (tel: 22-79-29).

Avis: Sabana Norte, next to Nissan dealer, (tel: 32-99-22; tx: 3144).

Budget: (tel: 23-32-84; tx: 3084).

Europcar National: PO Box 1128, 36th St, 7th Ave (tel: 33-44-06, 23-93-09; tx: 3075).

Hertz: PO Box 10030 (tel: 21-18-18; tx: 2566).

Business advice and information
Banco Central de Costa Rica, Apdo 10058, San José (tel: 33-42-33; tx: 2163 bancent).

Cámara de Comercio de Costa Rica, Urbanización Tournón, Apdo 1114, 1000 San José (tel: 21-00-05, 21-02-01; tx: 2646).

Cámara de Industrias de Costa Rica, Apdo 10003, Calle 13-15, Avda 6, San José (tel: 23-24-11; tx: 2474).

Centro de Promoción de Exportaciones e Inversiones (CENPRO), Apdo 5.418, San José (tel: 21-71-66; tx: 2385).

CINDE, Apdo 7.170, 1000 San José (tel: 33-17-11; tx: 3514).

Colegio de Periodistas de Costa Rica, Sabana Este, Apdo 5.416, San José.

Ferrocarriles de Costa Rica SA (FECOSA), Apdo No 1, San José (tel: 26-00-11; tx: 2393).

Instituto Costarricense de Turismo (main office), Calle 5-7, Avenida 4, Apdo 777, San José (tel: 23-17-33); tourist information office, Calle 5, Avenida 0-2 (tel: 22-10-90).

Líneas Aéreas Costarricenses SA (LACSA), Adpo 1.531, San José (tel: 32-35-55; tx: 2188).

Ministry of Economy and Trade, Apdo 10.216, 1000 San José (tel: 22-10-16; tx: 2414).

Ministry of Finance, Apdo 5.016, San José (tel: 22-91-22; tx: 2277).

Ministry of Foreign Trade, La Llacuna 12\28, Avda Central, calle 5, San José (tel: 22-58-55; tx: 2936).

Ministry of National Planning and Economic Policy, Avda 3 y 5, Calle 4, San José (tel: 21-95-24; tx: 2962).

Unión de Cámaras, Apdo 539-1.002, Paseo de Estudiantes, San José (tel: 33-35-55; tx: 3644).

Côte d'Ivoire

Gerald Bourke

The heavy dependence of the Ivorian economy on depressed world cocoa and coffee markets has continued to take its toll, wreaking havoc with public finances and rendering the prospect of a resumption of debt servicing even more remote. Under the circumstances, the crisis of confidence in industry has persisted, with profits and turnover tumbling. President Houphouet-Boigny, 84, has spent months away from home with the result that some – though not all – crucial decisions have been put on the back burner.

Finance and the economy

Deteriorating terms of trade have further undermined the economy. Gross Domestic Product at current market prices is officially estimated to have risen by just 0.2 per cent in 1988 to CFA3,072.7bn. However, allowing for inflation – the so-called 'African consumer price index' rose by 8.2 per cent – the economy shrank by 3.2 per cent in real terms. GDP per capita declined even more sharply, given the estimated 2.2 per cent increase in the population.

Customs figures indicated that Côte d'Ivoire recorded a rare trade deficit during the first eight months of 1988. This was put at CFA1.4bn, compared with a surplus of CFA72.1bn during the same period in 1987. However, by the end of the year official estimates suggested, rather optimistically, that a surplus of CFA344.6bn had been achieved. This was marginally above the CFA334.6bn surplus recorded for 1987. Export revenues were estimated to have fallen by 4 per cent to CFA856bn, while imports were put at CFA511.4bn, 9 per cent down on 1987.

Detailed figures for the first six months of 1988 showed that cocoa bean exports slumped by 54.3 per cent compared to the same period in the previous year to CFA67.5bn, while raw coffee sales fell by almost 20 per cent to CFA45.1bn. Estimates by the Economy and Finance Ministry – again somewhat optimistically – put the current account deficit for 1988 at CFA209.2bn, down from the CFA231.5bn recorded in 1987. Meanwhile, a sharp increase in net outflows from the public and parastatal sectors led to a sharp rise – from CFA67bn to CFA185bn – in the capital account deficit. The overall balance of payments deficit is estimated to have risen from CFA51.5bn to CFA78.2bn.

The sharp drop in export earnings sent the public finances further awry. Caisse de Stabilisation, the state commodity marketing board which was once a major source of budgetary revenues, again incurred substantial operational losses during the 1988-89 (October-September) crop marketing year. This was due to the wide gap between market prices and those paid to local growers. In 1989, the government abandoned the restrictive policy on cocoa exports it had operated during the previous year. Last January, after months of negotiations and market speculation, it confirmed having sold 400,000 tonnes of the 1988-89 crop – estimated at 780,000 tonnes – to Sucres et Denrees (Sucden), a French commodity broker. The government described the deal – then the biggest of its kind ever concluded – as a response to the 'speculative attacks' that had undermined its policy of securing remunerative prices for local growers. Few details were disclosed, although traders believed that given the huge gap between market and domestic prices, the sale was made at a loss.

Indeed, it did little to ease the cash crisis afflicting the farm sector. Earnings remained well below the level needed to provide growers, traders, exporters, bankers and the government with adequate funds. Throughout the 1988-89 season, many farmers across the country's cash crop belt said they had received little or nothing in payment for their produce. Local bankers and shippers, anticipating a continued decline in world prices, were reluctant to make full use

KEY FACTS

Official title: Republic of Côte d'Ivoire

Head of state: President Félix Houphouët-Boigny

Ruling party: Parti Démocratique de la Côte d'Ivoire

Capital: Yamoussoukro (political); Abidjan (economic centre)

Official Languages: French

Currency: Franc CFA = 100 centimes

Exchange rate: Franc CFA316 per $ (Nov 1989)

Area: 322,463 sq km

Population: 11.6m (1988)

GDP per capita: $857 (1988)

GNP real growth: –3.9% (1987)

Inflation: 7.0% (1988)

Trade balance: $815m (1988)

Foreign debt: $13.5bn (1987)

of the credits available from the central bank for the purchase of cocoa and coffee in the interior at the government's guaranteed price level of CFA400 a kilo. Intermediate traders, who traditionally collect crops directly from growers on the strength of funds advanced by shippers, were obliged to limit purchases and issue credit notes. The result, as harvesting proceeded, was an accumulation of stocks in the interior and mounting grower dissatisfaction.

The government continued to put the blame for these problems on 'Western speculators' and became increasingly critical of intermediate buyers for not paying farmers. New incentives to encourage farming co-operatives to play a bigger role in the marketing of both cocoa and coffee were unveiled last July. The move reflected the government's belief that the Caisse de Stabilisation could make considerable savings if co-operatives sold direct to shippers, bypassing intermediate traders. Marketing costs have traditionally accounted for up to 30 per cent of the FOB price guaranteed by the board. In 1987-88, co-operatives marketed only 18 per cent of the cocoa crop and 30 per cent of the coffee. The target for 1989-90 was set at 40 per cent for both commodities.

In mid-1989, the government cleared what remained of the 1988-89 crop, selling a further 150,000 tonnes to Sucden and 280,000 tonnes to Philipp Brothers (Phibro), a US trading house. A month later, the two companies bought 500,000 tonnes of Côte d'Ivoire's 1989-90 crop. The sale confirmed what had been suspected for months: that the government had effectively abandoned the much- maligned International Cocoa Agreement and had decided to sell privately from now on.

As Côte d'Ivoire's cash crisis deepened, President Houphouet-Boigny came under increasing pressure from the country's foreign creditors – particularly the International Monetary Fund and World Bank – to reduce the CFA400 a kilo price officially guaranteed to cocoa growers. In July he finally capitulated, cutting it to CFA250. As a result, the World Bank forecast that Caisse de Stabilisation could post a modest profit in 1989-90, after three seasons in the red.

While cocoa production continued to expand rapidly, there was a correspondingly sharp fall in the output of coffee. This trend, coupled with moves within the International Coffee Organisation (ICO) to take greater account of increased demand for more popular, higher quality arabicas grown mainly in Latin America, at the expense of African robustas, led to a series of cuts in the country's export quota. In October 1988, Côte d'Ivoire's share of the global ICO quota for 1988-89 had been fixed at just over 3.62m (60 kilo) bags. This was equivalent to 6.95 per cent of the total, compared to the 7.35 per cent share it had enjoyed in 1987-88. In January 1989, its quota for the season was reduced to 3.52m bags; three subsequent cuts, in February, March and April brought it down to 3.29m bags. The prospect of an unregulated coffee market, following the collapse of crucial talks between producer and consumer members of the ICO in London last July, looms ever larger. Côte d'Ivoire would lose out heavily in the event of a free-for-all, given the weakening demand worldwide for robustas.

The adverse effect of the marketing board's losses on public finances has been considerable, obliging the government to rely even more heavily on local commercial banks to fund commodity exports and priority recurrent spending. By mid-1989, its domestic arrears had reached CFA400bn – equivalent to 13 per cent of GDP. With the volume of loans outstanding to banks expanding much more rapidly than their capital and deposit bases, they in turn became much more dependent on the central bank and foreign banks to support lending activities.

The government's increased reliance on credit from the domestic banking system was partly due to the fact

that budget revenues were substantially below anticipated levels. The 1989 budget provided for just a 1 per cent increase – to CFA644.43bn – in total spending, with both recurrent and capital outlays slated to rise by the same margin to CFA499.48bn and CFA145bn respectively. Given the constitutional requirement for a balanced budget, the recurrent revenue projection of CFA499.48bn seemed highly optimistic, given that receipts in 1988 – from income, indirect, export and import taxes – were almost 10 per cent below the original forecast.

Capital outlays for agriculture and transport were set to rise by 21 and 5 per cent respectively, while social investment – for health, housing and education – was sharply reduced. Significantly, investment funds from domestic sources – bank borrowing and tax receipts – were projected to rise by 17 per cent to CFA67.5bn, while external sources – multilateral agencies, bilateral creditors and commercial banks – were expected to provide CFA77.5bn, 10 per cent less than anticipated for 1988.

As in previous years, there was no provision in the budget either for the Caisse de Stabilisation's losses or for debt servicing. In 1988, these contributed to an increase in public sector deficit from CFA235bn to an estimated CFA350bn – 11.5 per cent of GDP. How-

KEY INDICATORS

	Unit	1984	1985	1986	1987	1988
Population	m	9.84	10.25	10.69	11.14	11.61
Gross domestic product (GDP)	CFA bn	2,858	3,137	3,244	3,118	2,963
GDP per capita	Franc CFA	290,477	306,049	303,461	279,892	255,211
Inflation	%	4.2	1.8	7.3	0.4	7.0
Agricultural production	1979-81 = 100	105.6	117.4	118.3	124.2	124.0
Food production	1979-81 = 100	118.9	119.8	121.5	127.6	131.1
Exports	$m	2,625	2,761	3,187	3,092	2,354
Imports	$m	1,487	1,410	1,640	1,839	1,539
Balance of trade	$m	1,138	1,351	1,547	1,253	815
Current account	$m	–58.2	63.6	–300.9	–950.5	–1,280.8
Total external debt	$m	8,179	9,837	11,142	13,555	–
Debt service ratio	%	20.9	20.1	21.4	19.6	–
Exchange rate	CFA per $	437.9	449.3	346.3	300.5	297.8

* estimated figure

ever, the depreciation of the dollar – in which some 40 per cent of Côte d'Ivoire's estimated $12bn public debt is denominated – together with the continued suspension of principal and interest payments to official and commercial creditors, brought debt service payments down by over one-third to CFA216.3bn.

Creditors within the London and Paris Clubs have received no payments from the government since early 1987, with the result that by mid-1989 outstanding arrears had reached CFA800bn. During that time, the secondary market value of Côte d'Ivoire's commercial debt declined from around 80 per cent of its face value to less than 15 per cent. The only external debt payments now being made are to the IMF and, more irregularly, the World Bank.

Last summer, tentative agreement was reached between the government, the Fund and the Bank on the broad outline of a new economic and financial reform package. The previous standby agreement with the Fund was suspended in May 1988 when the government failed to meet a deadline for the resumption of interest payments to commercial creditors – the first since the 1987 default. Around the same time, the Bank froze approval of non-project loans.

The major obstacle to an accommodation with the Washington institutions was removed when President Houphouet-Boigny agreed to lower the cocoa producer price. The draft agreement with the Fund, scheduled to run from mid-1989 to the end of 1990, envisaged further increases in import taxes and duties – the largest single source of budgetary revenues – the elimination of ad hoc tax exemptions and a 20 per cent reduction in non-wage public spending. However, President Houphouet-Boigny resisted pressure for a cut in the government's wage bill, which accounts for almost half of total budgetary outlays. The Fund forecast that if the agreed reforms were fully implemented, the real budget deficit would stabilise at around 11 per cent of GDP in 1989.

However, the concessions wrung from the government were not expected to close the country's huge financing gap – debt servicing obligations plus the current account deficit – or facilitate a resumption of payments to official and commercial creditors. According to the Fund, the government would be faced with a financing gap of CFA1,200bn over the duration of the programme, leaving London and Paris Club creditors with the prospect of having to reschedule all principal and interest due up to the end of 1990, as well as all outstanding arrears and previously rescheduled debt.

This would provide relief amounting to some CFA550bn. An additional CFA250bn could be raised from multilateral sources – the Fund, World Bank, the EC – and France. To help meet the residual CFA400bn gap, a donors' conference was being planned for autumn 1989, although it was not expected to yield more than CFA125bn. With no new money likely to be provided by commercial creditors, only substantial debt reduction concessions can resolve what the IMF considers a 'major economic and financial crisis'. Indeed, under the scenario envisaged by the Fund – external viability by the year 2000 – the 18-month stabilisation programme would be followed up by a three-year enhanced financing facility incorporating a comprehensive debt reduction element.

With the public finances in disarray and the economic outlook gloomy, industry has continued to suffer. Value-added in the manufacturing sector is officially estimated to have increased by 2.2 per cent to CFA642.5bn during 1988, following its 2.0 per cent rise in 1987. However, having corrected for price increases, it is believed to have fallen by 1.7 per cent in real terms. In 1987, the real decline in value-added had been 0.6 per cent. Similarly, while there was another nominal rise in industry's pre-tax turnover, real turnover fell again. The worst hit sectors were the vehicle assembly industry, leather goods and footwear, clothing and textiles, chemicals and fertilisers; most of them adversely affected by cheap, often fraudulent, imports. The construction and allied industries, however, continued to buck these trends.

Under the circumstances, there has so far been little response to the government's development strategy, unveiled in March 1988, to carry Ivorian industry into the next century. The strategy – which focuses on the need to modernise the import substitution sector, expand the range of export-oriented industries which process local raw materials and create new, high value-added manufacturing units – identifies 25 areas in which future development efforts should be concentrated. All told, it envisages the overhaul or construction of 224 new factories at a cost of CFA463bn. There is no indication as to where the money might come from; the strategy simply advocates easing

COUNTRY PROFILE

Political system: Executive power rests with the president (elected by universal suffrage for a five-year term) who appoints a council of ministers. The National Assembly is the legislature (175 members elected for five years). The president can ask the Assembly to reconsider a measure or submit it to referendum. The Parti Démocratique de la Côte d'Ivoire is the only officially recognised political party. Côte d'Ivoire gained independence from France on 7 August 1960. Under a recent constitutional amendment the president of the National Assembly is empowered to assume the office of President of the Republic for an interim period of 45-60 days, prior to a presidential election, in the event of a sudden presidential vacancy.

Political parties: The sole registered political party is the Parti Démocratique de la Côte d'Ivoire.

Population: There are more than 60 ethnic groups among the 10.6m inhabitants (1988). Although the urban population comprises some 55 per cent of the total, the overall population density is very low at around 30 per square kilometre.

Main cities/towns: Abidjan (population 2,500,000), Bouaké (600,000), Daloa (100,000), Man (75,000), Korhoga (60,000), Gagnoa (60,000), San Pedro (55,000).

Language: French is the official language, and the medium for all documentation. Numerous African languages are also spoken.

Media
Press: The only daily morning newspaper is *Fraternité Matin* with a circulation of 100,000; in 1987, an official evening paper, *Ivoir' Soir* (circulation 50,000), was launched. Weekly publications include the government weekly, *Fraternité Hebdo* (20,000), *Ivoire Dimanche* (75,000), *Abidjan 7 Jours, Le Griot* and *Le Guido*.
Radio: Broadcasts by Radio Télévision Ivoirienne in French, English and local languages. The commercial service is known as Radio Abidjan-Inter. There are 2.5m receivers.
Television: Transmissions (including colour) for several hours a day with stations in many parts of the country. There are 500,000 sets.

External trade: Regular trade surplus offset by substantial outflows on the invisibles account (mainly repatriation of profits, outward remittance of earnings and foreign debt servicing). In recent years lower export earnings, have helped increase current account deficit.
Exports: Cocoa beans (29% of total value in 1988), coffee (14%), timber (11%), petroleum products (8%). Main destinations: France (16.5% of total value in 1987), USA (11.6%), German FR (10%), Italy (8%), Netherlands (7.1%), Spain, UK, Japan, Mali, Burkina Faso, Belgium/Luxembourg, Portugal, Morocco.
Imports: Machinery and equipment (17% of total value in 1987), food products (15.4%), energy products (11%). Main sources: France (34.3% of total value in 1987), Nigeria (7.2%), Netherlands (5.9%), Italy (5.6%), Japan (4.8%).

Agriculture: Sector forms mainstay of economy, contributing over 30 per cent to GDP and employing two-thirds of the workforce. World's leading producer of cocoa and third largest producer of coffee, which account for about 40 per cent of exports. Production of principal cash crops, which are grown in the southern half of the country, has been badly affected by shifts in world commodity prices. Other export crops include sugar, cotton, rubber, bananas, pineapples, palm oil and coconut palm. Forestry also important, though exports of wood and wood products have steadily fallen; a reafforestation programme is in progress. Diversification has led to improved output of basic food crops (mainly rice, maize, manioc) and the country is now nearer its long-term aim of self-sufficiency. Recent emphasis on co-operative farming, overhauling marketing system and use of selective incentives. Also management training and technical advice programmes. Approximately 12 per cent of total area is cultivated, 25 per cent is pasture, 20 per cent forest.

Industry: Sector contributes 12 per cent to GDP and is based on the processing of domestic primary products and on import substitution. Activity concentrated in Abidjan. Aided by foreign capital, manufacturing expanded rapidly after independence but its rate of growth has slowed in recent years. Current emphasis is on export-orientated industry, particularly textiles, leather, footwear, palm oil, sugar, cocoa and timber industries.

Mining: Of little importance at present, although some diamond mining at Tortiya and some small scale gold mining. Significant deposits of iron-ore are located at Mount Klahoyo. Feasibility studies being carried out on nickel, bauxite and cobalt deposits. There is a plan to exploit gold deposits at Ity.

Hydrocarbons: At the end of 1988 the country's biggest operational field, Espoir, offshore Jacqueville, was shut down. The closure of the only other field, Belier, off Grand Bassam, was planned for 1990. Crude oil production, which is entirely for domestic consumption, has declined from its peak of around 7m barrels a year in 1984. Country has emerged as francophone West Africa's main refinery centre. Exports of refined petroleum products to neighbouring countries have rapidly increased since expansion of Vridi oil refinery (current capacity 80,000 b/d). Abundant deposits of gas in coastal region around Jacqueville.

Energy: Hydro-electric generation provides around 40 per cent of electricity needs.

Membership of international organisations:
African Development Bank, EC (Lomé Convention), Arab Bank for Economic Development in Africa (recipient), Communauté Economique de l'Afrique de l'Ouest, Conseil de l'Entente, Economic Commission for Africa, Economic Community of West African States, FAO, GATT, IDA, IFAD, ILO, IMF, International Cocoa Organisation, International Coffee Organisation, OAU, UN, West African Monetary Union, WHO, WIPO, World Bank.

credit restrictions so as to allow a freer flow of finance to established and prospective entrepreneurs, a more thorough tapping of domestic savings potential, as well as measures to encourage the repatriation of funds held abroad by Ivorians and to attract foreign investment.

Nor has the 1985 overhaul of the government's investment code, which expanded the range of incentives on offer, yielded much in the way of dividends. The incentives include duty free entry of imports required to build and equip new plants, as well as tax exemptions on operational licences, property, profits, capital gains and reinvestments. The tax breaks are most favourable for companies willing to build factories outside Abidjan, which has absorbed 70 per cent of new industrial investments since independence.

Last January, a UN-sponsored industry seminar in Abidjan called into question the commitment of the government, which accounts for more than 50 per cent of capital investments in the country, to private enterprise. It also found that production costs, particularly of raw materials, labour, telecommunications, electricity, fuel and transport, were excessively high. Speakers were especially critical of the high prices of homegrown raw materials – including cocoa, coffee, cotton and rubber – used by the agro-industrial companies which dominate the manufacturing sector. The cost of labour, notably of expatriate managers, was also condemned, and companies were urged, whenever possible, to give preference to the recruitment of Ivorian nationals.

Meanwhile, the World Bank carried out an illuminating survey of the range of taxes applied to industry, and recommended reforms which it said would improve both government revenue and the incentives environment. The Bank was particularly critical of the 'overly complicated' system of rebateable contributions which firms must make to the National Investment Fund; the 'excessive' number of enterprises benefiting from customs duty and value-added tax exemptions; the distorting effect of the wide variation in tariff rates, and the prevalence of underinvoicing. Under the terms of the stabilisation agreement provisionally reached with the IMF in mid-1989, the government undertook to address many of these shortcomings.

Politics

President Houphouet-Boigny, who officially turned 84 in October but is believed to be as much as five years older, spent much of 1989 in Europe, ostensibly seeking moral and material support from Côte d'Ivoire's creditors. While in Paris, discreet but firm pressure was apparently applied by the French government in an effort to persuade him to step down. While acknowledging that he has had bouts of ill-health, he gave no indication whether he intends to run for a seventh consecutive five-year term in the presidential elections scheduled for the last quarter of 1990. In the event of a sudden presidential vacancy, the constitution stipulates that the chairman of the National Assembly would become interim head of state and be responsible for organising elections within two months. The current chairman of the Assembly is Henri Konan Bedie, 56, who, like Houphouet-Boigny, is a member of the Baoule tribe.

During the president's absence, dozens of marches were held throughout the country in support of his stand against Western commodity speculators. These took place, somewhat ironically, against a background of growing discontent in the countryside over the non-payment of cash crop growers. They were organised by leading members of the ruling Parti Démocratique de la Côte d'Ivoire (PDCI), many of whom took advantage of Houphouet-Boigny's stay in Europe to consolidate their power bases. As the economy deteriorated, a number of underground opposition parties sought to capitalise on the growing grassroots disaffection with the PDCI. The most prominent of these is Le Front Populaire Ivoirien (FPI), a loose coalition of socialist groupings whose secretary-general, Laurent Gbagbo, returned home from six years of self-imposed exile in France in September 1988.

The government's flirtation with Pretoria has continued to blossom. In October 1988, Houphouet-Boigny received the (then) South African president, P.W. Botha, in Yamoussoukro. A month later, several business executives, members of the South African Foreign Trade Organisation, visited Côte d'Ivoire. *Fraternité-Matin*, the official daily, continued to publish editorials condemning sanctions and advocating the need for dialogue with Pretoria.

Houphouet-Boigny also played host to a stream of visitors involved in the talks aimed at bringing to an end Angola's civil war and South Africa's rule of Namibia. In particular, he was in the forefront of moves to persuade Angola's president Jose Eduardo Dos Santos, to reach a workable compromise with the Unita rebel movement.

Outlook

President Houphouet-Boigny's decision to finally cut the cocoa producer price, thus paving the way for a fresh injection of funds from multilateral agencies and the rescheduling of official and commercial debts, will help ease the severe pressure on public finances and encourage some diversification of farm output. Nevertheless, given the gloomy outlook for cocoa prices on the world market, due in part to the prospect of continued bumper Ivorian harvests, the economy seems set to remain in the doldrums for several more years.

The crisis of confidence among manufacturers which took hold in the mid-1980s will not be easily healed either. Further industrial lay-offs, company closures and disinvestments seem inevitable. Only a meaningful effort by the government to bring down production costs and stamp out rampant commercial fraud can restore some of the sector's lost competitiveness.

On both the economic and political fronts, the outcome of the party, municipal, parliamentary and presidential elections which are scheduled for the last quarter of 1990 will be crucial.

The fundamental question is whether Houphouet-Boigny will stand again or whether, having dominated Ivorian politics for half a century, he will finally give way to a younger man.

BUSINESS GUIDE

Time: GMT.

Climate: Varies from south to north. In the southern coastal region the climate is tropical and humid. Dry seasons from Dec to Apr and from Aug to Sep. Rainy seasons from May to Jul and from Oct to Nov. Temperatures remain fairly constant throughout the year ranging from about 20°C at night to 30-33°C during the day. In the north the climate is more extreme, but less humid, with a dry season from Nov to Apr and a rainy season from May to Oct.

Entry requirements
Passport: Required by all except particular document holders of certain African countries.
Visa: Required by all except nationals of France, German FR, Italy, Denmark, Sweden and certain neighbouring African countries. A visa is not required for visitors from the following countries for stays not exceeding three months: Andorra, Belgium, Cameroon, Chad, Central African Republic, Congo, Gabon, Mauritius, Monaco, Netherlands, UK and Ireland. A surety bond is required by visitors who do not have a return ticket.
Currency: No limit to the amount of local or foreign currency which can be taken in. A maximum of 25,000 CFA francs per person is allowed to be taken out. Unused travellers' cheques must be declared on exit.

Health precautions
Mandatory: Yellow fever vaccination certificate.
Advisable: Cholera vaccination essential. Hepatitis, typhoid and polio vaccination recommended. Malaria prophylaxis should be taken as risk exists throughout the whole country. Water precautions should be taken though piped water is regarded as safe in Abidjan.

Representation overseas: Accra, Addis Ababa, Algiers, Bangui, Beijing, Berne, Bonn, Brasilia, Brussels, Buenos Aires, Cairo, Conakry, Copenhagen, Beirut, Dakar, Geneva, Kinshasa, Lagos, Libreville, London, Luanda, Madrid, Monrovia, New York, Ottawa, Paris, Rabat, Rome, Tokyo, Tunis, Vatican City, Vienna, Washington, Yaoundé.

Air access
National airline: Air Ivoire operates internally and to Mali and Burkina Faso. Air Afrique, owned by several West and Central African countries, has its headquarters in Côte d'Ivoire and is the main international airline.
Other airlines: Air Burkina, Air Gabon, Air Zaire, Alitalia, British Airways, Cameroon, Egyptair, Ethiopian, Ghana Airways, Iberia, MEA, Nigeria Airways, Royal Air Maroc, Sabena, South African Airways, Swissair, UTA, VARIG.
International airport: Abidjan-Port Bouet (Code: ABJ), 16 km from city; duty-free shop, restaurant, bank, post office, shops, car hire.

Surface access
Main ports: Abidjan, San Pedro, Sassandra. Container and ro-ro facilities.
Overland access: Road: Possible from Accra, Ghana (new road), Burkina Faso, Guinea, Liberia. NB: Regulations regarding vehicle documents should be checked before travelling. Rail: Regular daily services from Burkina Faso, connecting Ouagadougou with Abidjan. Sleeping and restaurant facilities available.

Hotels: Wide range in Abidjan and in the main centres. Tipping usually 10-15 per cent.

Electricity supply: 220 V AC, 50 cycles.

Car hire: Self-drive and chauffeur-driven cars can be hired in Abidjan, Bouaké, Daloa, Gagnoa, Man and Sassandra. Most of the cars are Fiats, Renaults and Peugeots. International driving licence required. Seat belts must be worn in front seats.

BUSINESS GUIDE

City transport
Taxis: Red taxis with meters can be hailed or ordered by telephone in main centres. Two tariffs operating, one from 0600 – 2400, the other 2400 – 0600. The early morning tariff is double that for the day and evening. Bush taxis run to all parts of the country.
Buses: Extensive service, with air conditioned vehicles, operated in Abidjan by state-run Sotra and countrywide by private operators.

National transport
Air: Air Ivoire operates regular flights from Abidjan to Bouaké, Daloa, Man, Korhogo, Sassandra, San Pedro, Berebi, Tabou, Bondoukou, Odiénné and Guiglo.
Road: Extensive network of 6,000 km good roads stretching from south to north, with transverse roads interconnecting. 43,000 km of other roads.
Buses: Coach services operate daily between Abidjan and major cities. Scheduled services include Abidjan-Man, Abidjan-Korhogo and to towns on international route Accra-Abidjan.
Rail: A 1,147 km network from Abidjan to Ouagadougou passing through Agboville, Dimbokro, Bouaké, Katiola and Ferkessédougou. Seven express trains (with first and second class, sleeping and dining cars) run each week in both directions.

Public holidays
Fixed dates: 1 Jan, 1 May, 2 May (Bank Holiday), 15 Aug (Assumption), 1 Nov (All Saints'), 7 Dec (National Day), 25 Dec, 26 Dec.

Variable dates: Good Friday, Easter Monday, Ascension Day, Eid al-Fitr, Mon in late May, Eid al-Adha, Mon in late Aug.

Working hours
Government and business: (Mon-Fri) 0800/0800 – 1200 and 1430 – 1730, (Sat) 0800 – 1200.

Banking: (Mon-Fri) 0800 – 1130 and 1430 – 1630.

Shops: (Mon-Fri) 0800 – 1200 and 1430 – 1830, (Sat) 0800 – 1200 and 1430 – 1730.

Telecommunications
Telephone: STD telephone service to Western Europe and USA. Dialling code for the Ivory Coast: IDD access code + 225 followed by subscriber's number. There are no area codes.

Telex: Telex facilities available at main hotels and Central Post Office.

Banking
Central bank: Banque Centrale des États de l'Afrique de l'Ouest.

Main centre: Abidjan.

Head offices: African Development Bank, Banque Internationale pour le Commerce et l'Industrie de la Côte d'Ivoire, Banque Réal de Côte d'Ivoire, Banque Nationale pour le Développement Agricole, Crédit de la Côte d'Ivoire, Société Générale de Banque en Côte d'Ivoire, Société Ivoirienne de Banque, Banque Internationale pour l'Afrique Occidentale – Côte d'Ivoire.
Branches: American Express International Banking Corp, Banco do Brasil, Bank of Credit & Commerce International, Banque Paribas, Barclays Bank International, B T Overseas Representative, Citibank, Ljubljanska Banka, Ecobank.

Representation in capital: Algeria, Argentina, Austria, Belgium, Benin, Brazil, Burkina Faso, Cameroon, Canada, Central African Republic, Chad, Chile, China PR, Colombia, Congo, Czechoslovakia, Denmark, Egypt, Ethiopia, France, Gabon, German FR, Ghana, Guinea, Hungary, India, Israel, Italy, Japan, Rep of Korea, Lebanon, Liberia, Mali, Mauritania, Morocco, Netherlands, Niger, Nigeria, Norway, Poland, Romania, Rwanda, Senegal, Spain, Sudan, Sweden, Switzerland, Thailand, Tunisia, UK, Uruguay, USA, USSR, Vatican, Zaïre.

BUSINESS DIRECTORY

Hotels: Abidjan
Golf Inter-Continental, 08 Boulevard Langunaire, P.O. Box: 18 (tel: 43-10-44; tx: 26112; fax: 430544).

Grand, 17 Rue Montigny, P.O. Box: 1785 (tel: 33-54-24; tx: 23807).

Hamanieh, Boulevard de Lorraine 11, P.O. Box: 548 (tel: 36-91-55; tx: 43357).

Hilton, Ave Delafosse, P.O. Box: 2185, 01 (tel: 32-83-22; tx: 22636; fax: 332218).

Hotel de France, 19 Blvd de Marseille, 01, P.O. Box: 690 (tel: 32-85-85).

Ibis Marcory, Blvd Giscard d'Estaing, 15, P.O. Box: 594 (tel: 36-92-55; tx: 42130).

Ibis Plateau, 7 Blvd Roume, 04, P.O. Box: 1185 (tel: 32-01-57; tx: 22608).

Ivoire Inter-Continental, Boulevard de la Corniche, Abidjan 08, P.O. Box: 01 (tel: 44-10-45; tx: 23555; fax: 440045).

Massarana, Ave de la Côte d'Ivoire, 04, P.O. Box: 874 (tel: 35-45-65,35-46-56).

Novotel, 10 Ave du Général-de-Gaulle, 01, P.O. Box: 3718 (tel: 32-04-57; tx: 23276; fax: 332636).

Palm Beach, P.O. Box: 2704 (tel: 35-42-16; tx: 42236).

Sebroko-Frantel, Autoroute de l'Ouest, P.O. Box: 1380 (tel: 32-09-33; tx: 23828).

Tiama, P.O. Box: 643, 04 (tel: 32-08-22; tx: 3494).

Hotels: Bouaké
Harmattan, P.O. Box: 1357 (tel: 63-33-95; tx: 2493).

Ran, P.O. Box: 556 (tel: 63-20-16; tx: 69128).

Hotels: Yamoussoukro
President Sofitel, P.O. Box: 1024 (tel: 64-01-81/83; tx: 72104).

Car hire: Abidjan

Abidjanaise de Location Automobile (Budget): 8 Rue du Sénateur Lagarosse (tel: 35-60-11).

Avis: Hotel President Yamoussoukro (tel: 64-01-08; tx: 22526).

Civicars: 165 Boulevard de Marseilles (tel: 35-89-51).

Europcar: 42 Rue du Dr Calmette (tel: 35-11-35, 35-88-06).

Hertz: Agence Centrale Carrefour, Hopital de Treicheville (tel: 35-77-47; tx: 42462).

Locauto: Rue Pierre et Marie Curie (tel: 35-54-95, 35-54-72).

Mattei: Boulevard de la République (tel: 32-30-34, 32-98-87).

Car hire: Abidjan airport

Avis: (tel: 36-72-73; tx: 22526).

Europcar: (tel: 36-79-80).

Hertz: (tel: 35-77-47; tx: 42462).

Banking
Banque Centrale des États de l'Afrique de l'Ouest (central bank), BP 3108, Ave des Barachois, Dakar, Senegal (tel: 21-45-01; tx: 3154).

African Development Bank, 01 PO Box 1387, Abidjan (tel: 21-07-11, 21-15-50; tx: 717498).

Bank of Credit and Commerce International, Jeceda Building, Blvd de la Republique, Abidjan (tel: 21-54-10).

Banque Internationale pour l'Afrique Occidentale, 10 Avenue J. Anoma, PO Box 1274, Abidjan 01 (tel: 21-07-22).

Banque Internationale pour le Commerce et l'Industrie de la Côte d'Ivoire, Avenue Franchet d'Esperey, PO Box 1298, Abidjan 01 (tel: 21-03-79).

Barclays Bank, Alpha 2000 Building, Abidjan (tel: 21-28-04).

Société Générale de Banque en Côte d'Ivoire, 5 Avenue J. Anoma, PO Box 1355, Abidjan 01 (tel: 21-03-33).

Société Ivoirienne de Banque, 34 Boulevard de la République, Abidjan (tel: 21-00-00).

Other useful addresses
Air Afrique, 3 Avenue Barthe, PO Box 3927, Abidjan 01 (tel: 21-22-31; tx: 3785).

Air Ivoire, 2 Avenue du Général-de-Gaulle, PO Box 1027, Abidjan 01 (tel: 21-34-29).

American Express, 14 Boulevard de la République, PO Box 1297, Abidjan (tel: 21-02-11).

Customs Authorities, Boulevard de la République, Abidjan (tel: 21-52-23).

Department of Foreign Trade, Abidjan (tel: 21-26-27).

Fraternité Matin, PO Box 1807, Abidjan 01 (tel: 21-27-27; tx: 3718).

General Surveillance Co, PO Box 795 (tel: 21-12-90, 21-38-04). (responsible for import controls.)

Ministère du Tourisme, PO Box V 184, Abidjan (tel: 33-29-92, 32-49-70; tx: 23438 MINITOUR).

Office National du Tourisme, PO Box 184, Abidjan (tel: 21-00-88, 21-07-33; tx: 23438).

Société Nationale d'Operations Petroliéres de la Côte d'Ivoire, PO Box V 194, Abidjan (tel: 21-40-58; tx: 2135).

Société pour le Développement Minier de la Côte d'Ivoire, PO Box 2816, Abidjan (tel: 21-29-94; tx: 3552).

UTA Airlines, 11 Avenue J Anoma, PO Box 1527, Abidjan 01 (tel: 21-22-31).

Wagon-lits (railway information), Boulevard de Marseilles, Abidjan (tel: 21-20-66, 21-39-10).

Cuba

Roberto Espíndola

In 1989, which started with the first withdrawals of Cuban troops from Angola, followed by the visit to Cuba of Soviet President Mikhail Gorbachev, the political system was shaken and the government embarrassed by the scandal of several senior officers found guilty of corruption and of links with international drug traffickers.

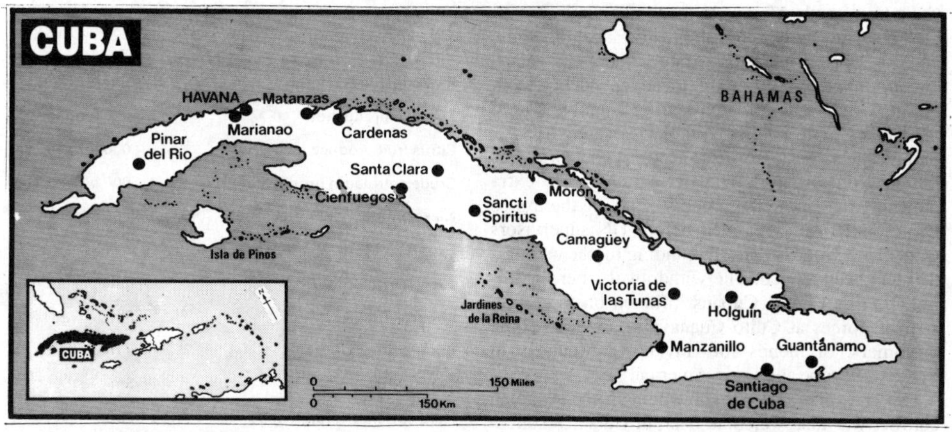

A former commander of the Cuban troops in Ethiopia and in Angola, General Arnaldo Ochoa was executed in July with three associates after being found guilty of having helped drug traffickers to smuggle some six tonnes of cocaine into the United States, as well as on more general grounds of corruption and contraband. The highly decorated General Ochoa, who had received the title of 'hero of the Republic' in recognition of his work in Angola, was indeed a popular hero and his involvement in drug running came as a major shock for ordinary Cubans.

Having returned to Cuba in 1986, Ochoa's expertise and taste for clandestine operations led to his developing contacts with Department Z – later to become Department MC – of the interior ministry; the job of this secret department was to bypass the US embargo of trade with Cuba by paying US residents to take clandestine shipments of essential spare parts, equipment and medicines to the island. Ochoa and his associates at the interior ministry made contact with Colombian drug traffickers and agreed to be paid to allow cocaine shipments to be taken to Cuba and transferred to light aircraft or fast boats destined for the United States. The Cuban officers had no difficulty in making use of the secret facilities of Department MC, keeping the drug shipments away from the border guards or air force control. They were reportedly paid $3.4m, which they put in secret accounts in Panama or kept hidden in Cuba. The Ochoa group was caught after the government received a warning from Panamanian officials, who told Havana that Washington had evidence of Cuban officers' involvement in the drugs trade and was getting ready to expose them. The Cuban government then announced the arrest of Ochoa and his associates before they were denounced abroad.

This scandal was particularly embarrassing for President Castro and his government, as Havana had vehemently denied on several occasions US charges accusing Cuban officials of involvement in the drugs trade, claiming that such charges were 'US propaganda'. Having to accept the involvement of senior and much admired officers with the Medellín cartel not only conceded a major point to Washington, it also eroded the moral tone which had characterised the Cuban political process. On the other hand, there were signs that the population welcomed the government's willingness to punish corruption among top officials.

The execution of Ochoa and three other officers, and the imprisonment of several others, was followed in August by a large-scale purge of senior officials, with generals and ministers arrested, demoted or forced to retire. In the short term, their replacement represented a reassertion of President Castro's personal control over the political system; in the long term it poses questions about the presence of factions preparing for the succession of President Castro.

The political connotations of this scandal were heavily hinted at in charges against Ochoa of being 'hypercritical' and 'a populist'; cryptic references which related to Ochoa's criticisms of the settlement of the Angola conflict – where he thought Cuba could have obtained a better deal – and of what he saw as a lack of jobs on offer for the troops returning from Africa. His 'hypercriticisms' then became 'populist' when he began to argue the need for an opening of the political system and a liberalisation of the economy, basically advocating the case for a Cuban 'perestroika'. His political stand, besides his frequent use of expensive presents to cultivate friendships and alliances, had also made Ochoa the potential leader of a faction within the armed forces and the ministry of the interior.

Although Havana's willingness to purge top officials involved in the drugs trade was praised in Washington – claiming that it vindicates charges previously made by US officials – there are no signs of a likely rapprochement between the two governments. Despite Havana's demonstrated willingness to remove two of the main obstacles to an improvement in relations – its military presence in Angola and the involvement of senior officials in the drugs trade – US official attitudes remain hostile, with military manoeuvres conducted dangerously close to Cuba's airspace and territorial waters, and plans for an exile-controlled television station beaming to the island from Florida. Although there are signs from business and political circles that they would welcome an end to the economic blockade of Cuba, any improvement in relations is likely to take place at a very slow pace under the present US administration.

Human rights remains a contentious issue between Havana and Washington, reawakened in August after the US Government denounced the detention of four human rights activists, including Elizardo Sanches, a figure previously denounced by other dissidents who considered him 'too soft' because of his attempts to seek a compromise with the government. Dissident sources claimed that at least 21 activists had been arrested in 1989; some dissidents close to the Catholic Church remained free and even managed to publish small newsletters of church-related news. The tolerance of the latter is, however, believed to be related to the government's desire not to spoil relations with the church and the likelihood of a papal visit to the island.

No perestroika

The Cuban leadership fiercely defended their refusal to imitate the political reforms proposed by Soviet leader Mikhail Gorbachev, to the point of defending the actions of the Chinese Government in dealing with pro-democracy dissenters. The visit to Havana by Gorbachev in April led to the signing of a treaty of friendship and co-operation, with the Soviet leader explicitly accepting the Cuban leadership's right to follow a path different from the one he has promoted. But Gorbachev also took the opportunity to emphasise the need for greater efficiency in the Cuban economy, in order to make possible 'more efficient and dynamic' economic relations between the two countries.

Relations between the two countries did not improve in the following months, with Havana banning some Soviet and East European publications after the publication of analyses by Soviet economists calling for a revision in the favourable trade terms granted to Cuba. Although their views did not claim to represent Moscow's official opinion, the increasing delays in deliveries of Soviet spares, manufactured goods and oil supplies were interpreted, at least by some Cuban officials, as a form of pressure. Those signs were often confirmed by President Castro's resolute defence of centrally-planned socialism and the role of the state, probably more out of an awareness of the political consequences which a relaxation of state control would have than out of any doctrinaire conviction. Rather than a return of orthodox communists to key positions, there has been a return of the leftist socialist ideologues of the 1960s, some of whom have become the official interpreters of the process of 'rectification of errors and negative tendencies' initiated by President Castro.

KEY FACTS

Official title: Republic of Cuba
Head of state: Dr Fidel Castro Ruiz
Ruling party: Communist Party
Capital: Havana
Official Languages: Spanish
Currency: Peso = 100 centavos
Exchange rate: Peso 0.79 per $ (Nov 1989)
Area: 114,524 sq km
Population: 9.7m* (1986)
GNP per capita: $2,238 (1985, GSP)
GNP real growth: 3.7% (1987, GSP)
Foreign debt: $10.7bn (1987)

* estimated figure

Withdrawal from Angola

An area where Havana and Moscow were in agreement was in the withdrawal of Cuban troops from Angola as part of a deal concluded at the November 1988 US-mediated talks in Geneva between Angola, Cuba and South Africa. At those negotiations a timetable was agreed for the withdrawal of some 52,000 Cuban troops from Angola under UN supervision, linked to the granting of independence from South African rule to neighbouring Namibia and to guarantees of respect of Angolan sovereignty. Although the first stage of the troop withdrawal was timetabled for April, Havana scored some extra points by beginning the process of taking the troops home on 10 January, even before UN supervisors were in place and before UN funding for the withdrawal had been agreed. The withdrawal, after the humiliating defeat the Cubans inflicted on the South African forces at Cuito Cuanavale, provided some diplomatic dividends for Havana, particularly in Western Europe and Latin America.

The end of military involvement in the Angolan conflict was also welcome back in Cuba. Despite a remarkable level of popular support maintained for that commitment, the length of the conflict and the cost in casualties had begun to take its toll, with young people beginning to express doubts. The return of the troops was proceeding smoothly in 1989, without any apparent effect on the level of employment. The Ochoa scandal, however, brought to the surface alleged criticisms about a failure to provide sufficient or adequate jobs for the returning fighters. There were also criticisms from Soviet officials about the number of men and women under arms kept by Havana, claiming that, being protected by the Soviets and in the current state of improved relations between Moscow and Washington, Cuba did not need such large armed forces.

Besides the boost to unemployment which a demobilisation could cause, Havana's reluctance to accept the Soviet guarantees and reduce its armed forces reflects the annoyance of the Cuban leadership at what they see as attempts by the two superpowers to negotiate over the future of Central America and of Cuba, without including Havana in their deliberations.

KEY INDICATORS

	Unit	1982	1983	1984	1985	1986
Population	m	9.8	9.9	10.0	10.1*	9.7*
Global social product	$bn	19.2	20.2	21.6	22.6	–
GSP per capita	$	1969	2052	2182	2238	–
GSP real growth	%	2.5	5.2	7.4	4.8	–
Raw sugar production	m tonnes	8.1	7.5	8.3	8.2	–
Tobacco production	'000 tonnes	45	30	45	46	46
Citrus fruit production	'000 tonnes	550	615	650	732	786
Crude petroleum production	'000 b/d	11	13	15	15	–
Nickel production	'000 tonnes	38	39	38	34	35
Exports (CMEA)	P m	4940	4754	4893	5310	–
Exports (Non CMEA)	P m	760	770	570	672	–
Imports (CMEA)	P m	5537	5403	6058	6640	–
Imports (Non CMEA)	P m	628	815	1149	1247	–
Total CMEA aid	$m	729	744	761	–	–
Hard currency foreign debt	$bn	3.2	3.3	3.4	3.5	–
External debt	$m	5,888	6,689	5,964	7,050	9,359
Debt service ratio	%	9.4	8.5	8.1	7.1	6.4
Exchange rate	P per $	0.8	0.8	0.9	0.9	–

* estimated figure

An indebted economy

The economy remained in serious trouble for the third year running, reflecting mainly the effect of external pressures caused by declining Soviet economic support and a growing indebtedness to the market economies. After a negative growth of -3.7 per cent in 1987, the economic decline slowed down in 1988, but it is doubtful whether or not it began to be reversed, doubts enhanced by the government's reluctance to provide statistical information for that year beyond estimates of a 2 per cent increase on the 1987 output. Despite a particularly good sugar harvest in the 1988-89 season, estimated to have exceeded 8.2m tonnes, the economy remains in trouble, unable to grow or even to maintain existing supply levels to consumers.

The external debt to Western creditors continued to grow in 1987 and 1988, as exports failed to earn sufficient convertible currency to service the existing debt and pay for the imports required by the economy. In 1987 that debt went up to $6.09bn and the following year increased by $305.7m reaching $6.4bn by year end. The 1988 increase was due to the rescheduling of previous debts and to a slight increase in short-term credits from suppliers; the latter went up by $67.5m, taking the total short-term debt to suppliers up to $1.43bn. Yet another factor pushing up the debt in 1988 was a $189m increase in government-guaranteed export credits, which took that form of indebtedness up to $2.63bn. This hike in foreign indebtedness led to a clampdown on imports from market economies in the last quarter of 1988, which managed to prevent any further increase in indebtedness during that quarter; the total debt, however, had gone up to $6.41bn in the first quarter of 1989.

In general, Cuba owes $2.84bn to the Paris Club countries – Western industrialised nations plus Japan – and $2.36bn to commercial banks, plus the $1.43bn owed to suppliers, $46m in multilateral loans and a similar amount in miscellaneous credits. According to Banco Nacional de Cuba (BNC), convertible currency reserves improved by $43.5m in 1988, reaching $97.5m. That amount, though, is still well below the minimum considered safe both by BNC and by foreign creditor banks.

The trade balance in convertible currency registered a $55.3m surplus in 1988, as exports reached $1.11bn and imports totalled $1.05bn; that was the consequence of a $39.3m increase in Cuban sales abroad, whilst imports grew by just $24m. This improvement was reached despite a sharp $101.4m drop in the hard currency earned by re-selling Soviet-supplied oil, caused by the fall in international oil prices and by domestic requirements, boosted as a consequence of new industries coming on stream. The surplus, however, also reveals the increasing clampdown on imported industrial inputs as a consequence of the need to save hard currency.

Although imports to be paid in convertible currency represent just 18.3 per cent of total imports, they are purchases essential to local industry. Their absence is reflected in a decline in the goods available to the local population. This is a process which began in 1987 when the government decided to aim – as a minimum – for supplies to remain at the 1986 level. In 1988 that target proved impossible to attain and by the second half of that year supplies had clearly fallen below 1986 levels, raising the possibility of consumer protests and a proliferation of black market activity.

Excluding purchases of fuels and sugar – required to meet export commitments – the convertible currency imports made in 1988 covered less than 30 per cent of those which were in fact required. A major drain on convertible currency has been the need to buy sugar abroad in order to meet commitments, mainly with the Soviet Union: over the last three years the government has had to spend $89.6m on such sugar purchases. In 1988, the government had to buy 1.1m tonnes of sugar abroad to meet deliveries to the Soviet Union; in 1989, and despite a very good harvest, sugar had to be bought from Thailand and Brazil to meet deliveries for 4.3m tonnes to the Soviet Union. Despite such purchases, the sugar sector improved its export earnings from $222.5m in 1987 up to $383.4m in 1988, partly helped by much improved international prices. In terms of export volume, there was also an improvement, with 2.3m tonnes exported to the market economies in 1988, as compared with 1.5m tonnes in 1987. Sugar has returned to its place as Cuba's most important export, replacing the re-sale of Soviet oil, which declined from $286.3m in 1987 down to $184.9m in 1988.

Other exports which registered improvements in 1988 were nickel, tobacco, citrus, seafood and computer screens, which registered an aggregate improvement of $79.1m. Tourism, by comparison, earned the economy $125m in 1988 – 20 per cent more than in 1987 – and, after the investment of $35m in 1988 and some $70m this year, a substantial increase was expected for 1989.

Trade with other members of Comecon – although still accounting for 87 per cent of Cuba's total trade – declined by 0.9 per cent in 1988, reflecting a $31.6m drop in exports and a $66.5m cut in imports. The situation worsened in 1989, with significant delays in the arrival of Soviet oil and of manufactured goods from other Comecon partners. Some observers blamed such difficulties on a rapid deterioration in relations between Havana and Moscow, despite a record 9bn rubles trade protocol the two governments signed for 1989.

Outlook

The Ochoa scandal and the subsequent purge of senior officials has allowed President Castro to reassert his rule and even to make conciliatory gestures towards Washington. The episode, in the long term, points to the troubles which will surround President Castro's eventual succession and to the internal political stresses which accompany an economy under serious external pressures.

COUNTRY PROFILE

Political system: Executive power rests with the council of ministers. It is appointed by the unicameral legislature, the National Assembly of People's Power, which is indirectly elected for a five-year term and meets twice a year. Elected from and by the National Assembly is the 31-member Council of State, whose president is also the head of state, president of the council of ministers and commander-in-chief of the armed forces. The constitution adopted in 1976 places the Communist Party of Cuba as the leading force in both the state and society.

Main cities/towns: Havana (population 2m), Camagüey, Marianao, Santiago de Cuba, Holguín, Guantánamo, Matanzas, Cienfuegos, Santa Clara.

Media
Press: The weekly edition of the official daily *Granma* is published in French and English. Other major national dailies are: *Juventud Rebelde* (evening) and *Tribuna de La Habana*, and there is also an extensive provincial press.
Broadcasting: Services controlled by the Instituto Cubano de Radiofusión. There are five national radio and two television stations.

External trade: Balance of payments is reliant on special terms of trade with socialist countries and foreign currency earnings from sugar. In recent years CMEA countries have accounted for over 85 per cent of trade. However increasing emphasis is being placed on non-sugar exports to market economies, although ability to expand trade with non-CMEA countries is constrained by growing hard currency debt, protectionist policies in developed countries and US trade embargo on all Cuban goods.
Exports: Sugar (raw, refined & molasses; typically 70-80% of total value), nickel (8%), foodstuffs and beverages (6%), tobacco and products (2%), petroleum products. Main destinations: USSR (typically 70% of total value), eastern Europe (12%), Far East (4%), non-communist countries (14%).
Imports: Machinery and transport equipment and other capital goods (typically 34% of total value), fuels (23%), raw materials and foodstuffs (19%), basic manufactures (14%). Main sources: USSR (typically 66% of total value), eastern Europe (16%), Far East (3%), non-communist countries (13% – France 12%, Spain 13%, UK 11%, Japan 10%, German FR 8%).

Agriculture: Sugar-cane farming remains cornerstone of economy, accounting for around 50-80 per cent of export earnings and contributing around nine per cent to GSP. Ranked as world's largest exporter (6.8m tonnes in 1983-4). Despite massive investment and rapid mechanisation, sugar production has been held back by rust disease, falling world prices and labour shortages. Exports of raw sugar are largely destined for the USSR and other CMEA countries, usually under countertrade arrangements. Other important export crops are citrus fruits, coffee and tobacco. Production of staple foodstuffs (rice, maize, potatoes, yams, arum, meat and dairy produce) has increase dramatically since the 1960s. Estimated that 30 per cent of total area is cultivated, mainly by large state farms and co-operatives. Commercial fishing has increased and now ranks third in exports after sugar and nickel.

Industry: Sector contributes around 40 per cent to GSP and employs some 20 per cent of the workforce. Activity, which is state-controlled, is centred primarily on sugar and food processing, oil refining and the production of construction materials, steel, nickel, paper, chemicals, textiles, non-electrical machinery and consumer durables. Emphasis is on expanding export-oriented industries and reducing dependence on imported capital goods and spares. New projects include the recently completed 5m-tonne capacity cement works at Cienfuegos, a 150m-litre capacity rum distillery at Santa Cruz del Norte, a textile plant at Santa Clara, a nickel refinery at Punta Gorda and the rehabilitation of the chemicals, fertilizer and steel industries. Tourism sector is expanding rapidly, visitors (mostly from Canada) numbering 114,000 in the first half of 1987, and convertible currency earnings from the sector reportedly exceed $100m per year.

Mining: Fourth largest nickel reserves in the world, estimated at 19m tonnes. Production, which is located at Nicaro and Moa, reached a record level of 41,000 tonnes in 1983 but in 1986 was only 35,100 tonnes. Despite USA embargo on Cuban nickel exports and low world prices, sales of semi-processed nickel have risen, particularly to the USSR. Other minerals extracted include chromite (10m tonnes of reserves), cobalt, copper, iron-ore, salt, stone and coal.

Hydrocarbons: Offshore oil production has risen threefold since 1978 to around 20,000 b/d, sufficient to meet eight per cent of local needs. New discoveries made by Petróleos Mexicanos off north coast.

Energy: Heavy dependence on imported oil, mainly from USSR/Venezuela under oil swap agreement. Sufficient oil refining capacity to process nearly all imports. Emphasis on energy conservation, increased by use of bagasse as oil substitute and expansion of electricity capacity. A new 330 mW thermal power station is planned at Matanzas and a nuclear power station at Cienfuegos.

Banking: As well as the central bank, Banco Nacional de Cuba, two other state-controlled banks are: Banco Financiero Internacional, which finances Cuba's foreign trade, and Banco Popular de Ahorros.

Membership of international organisations: FAO, ILO, UN, UNESCO, WHO, GATT, Comecon, Economic Commission for Latin America, International Bank for Economic Co-operation, International Investment Bank.

BUSINESS GUIDE

Time: GMT -5 (GMT -4 from Mar-Oct).

Climate: Tropical with an annual mean temperature of 25°C. The rainy season is from May-Oct and hurricanes are frequent.

Entry requirements
Passport: Required by all with isolated exceptions. Visitors must register with immigration officials the day after arrival.
Visa: There are extensive regulations on visas, and careful prior checking is essential. Most Eastern and a few Western European countries do not require visas. Exit visa required for visitors staying more than 14 days.
Currency: Import/export of local currency prohibited. Foreign currency must be declared on arrival. Visitors should retain white exchange paper each time money is changed. Cuban pesos can then be exchanged for foreign currency on departure.

Health precautions
Mandatory: Yellow fever vaccination certificate if arriving from infected area.
Advisable: Typhoid, polio vaccination. Water precautions should be taken.

Representation overseas: Accra, Addis Ababa, Aden, Algiers, Antananarivo, Athens, Baghdad, Beijing, Beirut, Belgrade, Berlin, Berne, Bissua, Brasília, Brazzaville, Brussels, Bucharest, Budapest, Buenos Aires, Cairo, Colombo, Conakry, Copenhagen, Cotonou, Damascus, Dar es Salaam, Freetown, Georgetown, The Hague, Hanoi, Harare, Helsinki, Kabul, Kampala, Kigali, Kuwait City, Lagos, Lima, Lisbon, London, Luanda, Lusaka, Madrid, Malabo, Managua, Manila, Maputo, Mexico City, Monrovia, Montevideo, Moscow, New Delhi, Nicosia, Oslo, Ottawa, Panama City, Paris, Phnom Penh, Prague, Praia, Quito, Rome, Sofia, Stockholm, Tehran, Tirana, Tokyo, Tripoli, Ulan Bator, Vatican City, Victoria (Seychelles), Vienna, Vientiane, Warsaw, Washington.

Air access
National airline: Cubana (Empresa Cubana de Aviación).
Other airlines: Aeroflot, Air Panama, CSA, Iberia, Interflug, Mexicana.
International airport: Havana-José Marti (HAV), 18km from city.

Surface access
Main ports: Antilla, Cienfuegos, Guayabal, Havana, Mariel, Matanzas, Nuevitas, Santiago de Cuba.

Hotels: Advisable to book in advance. US dollars are accepted. Tipping is officially forbidden though the practice is growing. Most hotels demand to see exchange paper which covers the amount of the bill.

Car hire: A few cars and jeeps are available at some tourist centres and at the airport.

City transport
Taxis: State and private taxi service. Usually ordered through hotel; metered or operated on fixed-rate system – checking beforehand advisable.

National transport
Air: Cubana operates domestic service to main centres. Internal flights generally arranged through Cubatur.
Road: The Central Highway runs for over 1,100 km, virtually from end to end of the island; total network exceeds 30,000 km, at least 40 per cent surfaced although some roads/tracks may not be passable in wet weather.
Buses: Services in towns are generally rated as inexpensive but invariably crowded. Coaches link main centres – although difficulties with bookings etc have been recorded.
Rail: Around 5,000 km of public service track; railcars operate between Havana-Cienfuegos, Cardenas-Jaguey, and also Havana-Santiago de Cuba – some services on this route offer refreshments, air conditioning.

Public holidays: 1 Jan (Liberation Day), 1 May, 25-27 Jul (Revolution Days), second Mon in Oct (Beginning of Wars of Independence).

Working hours
Government and business: (Mon-Fri) 0830 – 1230, 1330 – 1730.
Banking: (Mon-Fri) 0830 – 1200, 1330 – 1500; (Sat) 0830 – 1030.
Shops: (Mon-Sat) 1230 – 1930.

Telecommunications
Telephone: Dialling code for Cuba: IDD access code + 53 followed by area code (Havana 7) followed by subscriber's number. IDD service to Havana only. International calls from Cuba can be subject to long delays.
Telegram: Can be sent from all post offices in Havana, RCA offices in other main centres and major hotels.

Postal service: Unreliable and slow. Try to address letters to PO boxes within Cuba and use air mail for international correspondence.

BUSINESS GUIDE

Advertising: Advertising facilities are limited and state-controlled and generally not available to foreign companies.

Trade fairs: Considerable emphasis on trade fairs and exhibitions. Havana is principal centre. Prominent events are: biennial spring medical fair (1989, 1991...), and the annual autumn International Trade Fair.

Electricity supply: 110-120/220-240 V AC, 60 cycles.

Representation in capital: Afghanistan, Albania, Algeria, Angola, Argentina, Austria, Belgium, Brazil, Bulgaria, Canada, China PR, Congo, Czechoslovakia, Denmark, Ecuador, Egypt, Ethiopia, Finland, France, German DR, German FR, Ghana, Hungary, India, Iran, Iraq, Italy, Japan, Kampuchea, Korea PDR, Laos, Lebanon, Libya, Malta, Mexico, Mongolia, Netherlands, Nicaragua, Nigeria, Panama, Peru, Poland, Portugal, Romania, Spain, Sweden, Switzerland, Syria, Turkey, UK, USSR, Vatican, Venezuela, Vietnam, Yemen PDR, Yugoslavia.

Hotels: Havana
Capri, 21 y N, Vedado (tel: 320511).

Comodoro, 1a y 84, Miramar (tel: 225551).

Libre, L y 23, Vedado (tel: 305011; tx: 511835).

Nacional, 21 y O, Vedado (tel: 78981/2; tx: 511832).

Presidente, Calzada esq. a G, Vedado (tel: 327521).

Riviera, Paseo y Malecón, Vedado (tel: 305012/9; tx: 511832).

Triton, 3a Ave esq a 70, Miramar (tel: 225531).

Car hire: Havanautos at Capri Hotel, Marazul Hotel and airport.

Business advice and information
Banco Nacional de Cuba, Apdo 736, Havana (tel: 311612; tx: 51146).

Cubana Airways, Apdo 6215, Calle 23 No 64, Vedado, Havana (tel: 74911; tx: 511737).

Cubatur (Empresa de Turismo Nacional e Internacional), Calle 23, No 156, Apdo 6560, Vedado, Havana (tel: 324521; tx: 511366).

Cámara de Comercio de la República de Cuba, Apdo 370, Calle 21, No 661, Vedado, Havana (tel: 30356/7/8; tx: 51174).

Export Promotion, Infanta No 16, Vedado, Havana (tel: 709064; tx: 511174).

Cyprus

Michael Jansen

The new president of the Republic, Dr George Vassiliou, who took office in February 1988, set the tone of developments during 1988-89. He did this by launching a peace offensive at the United Nations and in Europe and the USA. At home he reached out to the Turkish Cypriots, estranged from the Greek Cypriots since the 1974 occupation by mainland Turkish troops of more than a third of the island. On both the international and internal planes Vassiliou transformed the image of the Greek Cypriots: by showing some flexibility and a readiness to take the concerns of the Turkish Cypriots into account, he demonstrated that it was the Turkish Cypriot leader, Rauf Denktash, who was inflexible and backward looking, rooted in old wrongs and past fears. This widened the gap between Denktash's ruling party and the two main Turkish Cypriot opposition parties which, increasingly, found Vassiliou's conciliatory message convincing and appealing.

The new United Nations initiative

In August 1988 United Nations Secretary General Perez de Cuellar launched his fourth initiative with the aim of achieving a Cyprus settlement by 1 June 1989. He began by inviting the island's two leaders to meet for the first time over a tête-à-tête lunch in Geneva on 24 August. The two men agreed to three rounds of intensive talks, during which they would discuss 'all aspects' of the Cyprus problem and review progress, on a periodic basis, with de Cuellar. The principle of confidentiality was meant to be observed and the ideas of the two sides were not supposed to be committed to paper so they would not freeze in fixed positions. However, in December Denktash both broke silence and submitted working papers to the Secretary General, forcing the Greek Cypriots to come up with a comprehensive proposal for a 'federal solution' at the end of January 1989. This proposal was drafted by the leaders of all the Greek Cypriot political factions and represented a consensus of the community. The leaders of the two main Turkish Cypriot opposition parties, Oskar Ozgur of the Republican Turkish Party and Mustafa Acinji of the Communal Liberation Party, representing nearly half of the Turkish Cypriot electorate, welcomed the proposal, reasserted their support for a 'federal' rather than a 'confederal' solution and demanded that Denktash honour a commitment to federalism he made to the Turkish Cypriot assembly.

Though the two sides did not reach agreement by 1 June, nor by the end of the month when they met in New York with the Secretary General, they committed themselves to further talks during the summer with the object of producing an 'outline' of a settlement by September. Such deadlines kept the two men talking, though there was little hope that they would change their very different visions of a future Cyprus: Vassiliou is working towards an integrated federal republic consisting of two communal provinces or regions, while Denktash insists that there should be two independent states, with controlled borders, joined in a loose confederation.

People's power

Though the positions of the two leaders seemed irreconcilable, party leaders and individuals on both sides, but more especially on the Turkish Cypriot side, demonstrated that the status quo was not acceptable. In particular, Turkish Cypriots opposed to Denktash's policy of total separation took advantage of increased opportunities for meeting with Greek Cypriots. In October 1988, on the occasion of United Nations Day, the island's peacekeeping force invited ordinary people from both communities to a party held at the Ledra Palace Hotel in the UN buffer zone to celebrate the award of the Nobel Peace Prize to all the UN forces. Although only one hundred Turkish Cypriots attended, against the wishes of their government, as compared with over a thousand Greek Cypriots, the intermingling of the two peoples produced a psychological breakthrough. This was followed by a series of people-to-people events, which boosted the policy of rapprochement being actively pursued by the United Nations and the Vassiliou government. In December 1988 Turkish Cypriot opposition leaders spent several days in the Republic as the guests of Greek Cypriot politicians. At the end of January 1989, the London-based Friends of Cyprus group sponsored two days of meetings between leading members of both communities, again at the Ledra Palace. During these sessions an ambitious plan for intercommunal meetings was drawn up, but it had to be scrapped when Denktash blocked Turkish Cypriot participation. However both external and popular pressure forced the Denktash administration to relax its ban and Turkish Cypriots were permitted to take part in a series of important events: a conference of the International Union of Students across the dividing 'Green Line' in Nicosia in April; two conferences of party leaders, one in Prague in May, one in Nicosia in June; meetings of journalists, sports writers, medical technicians and other specialised groupings. The most striking, encouraging fact emerging from these contacts was that the young people of the two communities, who have grown up since the 1974 separation, reacted in a normal, friendly fashion to one another. For them the past did not count. This potentially easy relationship between the youth was first demonstrated during a party at the Ledra Palace following the International Union of Students conference in April. There the students mingled freely and even participated in folk dancing – together. Commenting on this phenomenon, a Turkish Cypriot newspaper supporting the Denktash view stated, sourly, that it 'was well known in the history of

KEY FACTS

Official title: The Republic of Cyprus

Head of state: President George Vassiliou

Ruling party: none – President Vassiliou is an independent

Capital: Nicosia

Official Languages: Greek and Turkish

Currency: Cyprus Pound (C£) = 100 cents

Exchange rate: C£0.49 per $ (Dec 1989)

Area: 9,251 sq km

Population: 688,000 (1988)

GDP per capita: C£2,873 (1988)

GDP real growth: 6.9% (1988)

Labour force: 261,000 (1987)

Unemployment: 2.8% (1988)

Inflation: 3.5% (1988)

Trade balance: –$1,030m (1988)

Foreign debt: C£461m (1988)

Cyprus that Turkish and Greek Cypriots never danced together'.

The Greek Cypriot Women's Walk Home movement staged its third -and most spectacular - demonstration against the division of the island. On 19 March 1989 some 3,000 women, including 200 from Europe, the USA, Canada and Australia, crossed the 'Green Line' into Turkish army-occupied territory at two places, one a chapel perched on a high hill above the busy Nicosia-Larnaca road, the other a deserted bicommunal village in the Famagusta district. The objective of the women was to show that the status quo was unacceptable, that they wanted to lift the barriers and live side-by-side with the Turkish Cypriots.

Economic developments

In 1988, the first year of the implementation of its customs agreement with the EC, the economy of the Republic matched its 1987 record performance. It sustained a 7 per cent growth rate in the GDP, while keeping inflation at a low 3.4 per cent and reducing the unemployment rate from 3.4 to 2.8 per cent, achieving near full employment. The single factor causing concern was the gradually growing fiscal deficit, which reached C£85m in 1988, or 4.3 per cent of the GDP, as compared with 4.1 per cent in 1987. It was predicted that, during 1989, the fiscal debt would rise to 4.5 per cent and the high 7 per cent growth rate would fall to 4-5 per cent. The low level of unemployment was expected to continue, but inflationary pressures seemed to be building up.

Imports (non-military) reached C£866.8m in 1988 and exports C£330.8m (including C£95.1m in re-exports). Cyprus continued to have a favourable balance of trade with the Arab countries, importing C£41.7m against C£110.5m in exports. The serious trade imbalance was with Western Europe. Although exports to the EC grew in 1988 to C£141.5m, accounting for over 40 per cent of export earnings, this could not compete with an import bill of C£472.5m. This has, however, been more than compensated for by the tourism sector, accounting for 15 per cent of the GDP. Its gross receipts for 1988 were C£360m, a 17.5 per cent increase over 1987. Cyprus's current account balance was a positive C£6.2m. In the first three months of 1989 a further 15 per cent rise in earnings from tourism was recorded. Some 1.3m tourists were expected in 1989.

In its 1989-93 Economic Action Plan the Vassiliou government seeks to redress the imbalance with Europe by diversifying and improving the design and quality of its manufactured goods, which account for 49 per cent of exports, while retaining its present small and medium scale manufacturing units. Cyprus is due to receive C£31.0m (C£7.0m in grants and C£24.0m in loans) of EC development assistance for this purpose. Farmers, who accounted for 12.5 per cent of overall visible export earnings in 1988, are being encouraged to produce high-priced export crops like avocados and kiwis, as well as the staple potatoes and citrus, and to upgrade their vinestock for the production of more competitive wines. On 1 June 1989 the government announced a package of measures to check the uncontrolled tourist development along the coast which threatens the island's environment. These include the withdrawal of tax incentives, the imposition of stricter controls on new buildings and a freeze on permits for new tourism construction for a period of ten months. The objective of the plan is to 'upgrade the tourist industry' through ensuring 'balanced development', taking into account the island's shortage of manpower and limited natural resources. The government also hopes to expand other sectors in order to reverse the brain drain that takes most university graduates overseas. On the public finance front, major changes are proposed, including tax reforms and the introduction in 1990 of VAT

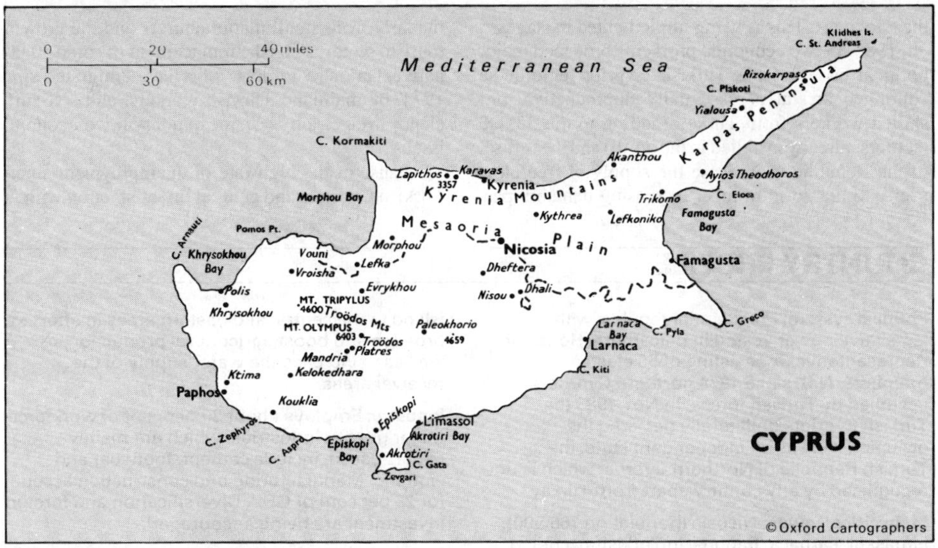

to offset the C£40m lost by tariff reductions under the Customs Union Agreement.

The economy of the north

The Republic's prosperity, achievement of full employment and EC orientation widened the already wide gap with the economy of the breakaway state in the north. The per capita income in the Turkish Cypriot area was one-third of that in the Republic, where per capita income compares favourably with that in newer members of the EC. Although the Turkish Cypriots control 36.4 per cent of the territory of Cyprus (containing 70 per cent of the productive capacity of the island in 1974), the economy – subservient to political and military considerations – is now only one-tenth the size of the economy of the Republic, having less than one-tenth of the Republic's GDP. This despite the fact that the growth rate between 1982-87 was almost equal: 5.3 per cent in the north, 5.9 per cent in the Republic. And as a result of the low per capita income in the north, the income differential between the two regions continued to increase in 1988-89. The Turkish Cypriot economy has two major internal structural problems: an overlarge bureaucracy and a large agricultural sector - the main productive sector - which has low productivity. Furthermore, the economy has been dominated by a few entrepreneurs with close political ties to the Denktash administration and the mainland. These internal weaknesses have been exacerbated by the region's close connection with the economy of the Turkish mainland, which has exported its 80 per cent inflation rate, currency devaluations – from 38 Turkish Lira to the C£ in 1974 to 1,700 to the C£ in 1989 – and inconsistent development record. The ties between the Turkish Cypriot economy and that of the mainland were formalised in protocols signed in 1986 which, in effect, create an economic union between

KEY INDICATORS

	Unit	1984	1985	1986	1987	1988
Population	'000	657.4	665.2	673.1	680.4	688
Gross domestic product (GDP)	C£m	1,336.6	1,480.12	1,598.0	1,778.7	1,976.5
GDP per capita	C£	2,033	2,225	2,361	2,614	2,873
Inflation	%	8.0	5.8	4.3	3.9	3.5
GDP real growth	%	8.8	4.7	3.8	7.0	6.9
Unemployment	%	3.3	3.3	3.7	3.4	2.8
Agricultural production	1980=100	115.3	115.8	114.4	121.8*	–
Crops	1980=100	104.6	101.2	95.7	101.8	–
Livestock	1980=100	135.9	144.2	150.9	160.4	–
Industrial production	1985=100	102	100	101	109	–
Mining	1980=100	85.0	100	91.1	97.2	95.6
Tourist arrivals	'000	665.9	769.7	827.9	948.5	–
Exports	C£m	307.1	266.5	237.8	279.9	357.9
Food/live animals	C£m	83.3	59.2	66.5	70.8	–
Beverages/tobacco	C£m	28.7	24.8	22.3	19.3	–
Machinery/transport	C£m	28.8	31.5	33.0	29.1	–
Fuels/lubricants	C£m	21.2	23.5	15.2	15.5	–
Imports	C£m	796.5	762.3	659.1	711.4	877.7
Food/live animals	C£m	89.8	82.5	72.4	73.9	–
Fuels/lubricants	C£M	144.8	137.0	82.9	87.5	–
Machinery/transport	C£m	209.1	199.4	160.2	174.2	–
Balance of trade	$m	–702.4	–708.2	–693.6	–761.3	–
Current account	$m	–2,070	–153.1	–10.3	96.5	1.0
Exchange rate	$ per C£	1.70	1.64	1.94	2.08	2.14

* estimated figure

the two areas. This is being implemented in stages.

In 1988-89 two economic problems produced major political crises in the Turkish Cypriot region: the continued arrival of essentially unproductive, unskilled workers from Turkey (adding to the 52,000 'settlers' already installed in the north) and the refusal of the Republic to increase the supply of free electricity to the north to meet increasing demand, particularly from small manufacturers and the growing tourism sector. Thus the replacement of some 30,000 native Turkish Cypriots, who have emigrated since 1974, by mainland Turkish workers, and self-sufficiency in electricity became major politico-economic issues.

Because of the high rate of unemployment among Turkish Cypriots and general dissatisfaction with the performance of the economy, some 600 semi-skilled workers were permitted, in 1988-89, to cross daily into the Republic to take up jobs on construction sites along the Larnaca-Aya Napa coast. This infiltration of the 'Green Line' – and the return of the Turkish Cypriot workers with high Greek Cypriot wages in their pockets – also served to break down long-standing barriers of fear and suspicion.

COUNTRY PROFILE

Political system: Presidential republic with legislative power vested in unicameral House of Representatives, consisting of 56 elected members. N.B. since 1974 northern Cyprus occupied by Turkish forces. In Nov 1983 the Turkish Cypriots unilaterally declared the occupied zone as an independent state, the Turkish Republic of Northern Cyprus, which is not recognised by any country apart from Turkey.

Main cities/towns: Nicosia (population 166,900), Limassol, Larnaca, Paphos. In Turkish-occupied sector: Famagusta, Kyrenia, Morphou.

Media
Press: Dailies: There are 17 local dailies with a total circulation of about 86,000 (10 Greek, six Turkish, one English). *Weeklies:* There are 13 weeklies: seven in Greek, five in Turkish and one in English (total circulation of weeklies in Greek and English 35,500). Most have a political bias.
Periodicals: There are 47 periodicals, several of which are business publications and some are in English. *Business: The Cyprus Bulletin* (circulation 28,000) is an official publication on Cyprus affairs published by the Cyprus Press and Information Office in seven languages every fortnight.
Broadcasting: All radio and TV programmes are broadcast by Cyprus Broadcasting Corporation (CyBC). Programmes broadcast in Greek, Turkish and English. Some radio programmes in Arabic and Armenian.

Domestic economy: Considerable levels of economic growth since late 1970s, based on increased earnings from manufactured and agricultural exports and on a rapid expansion of earnings from tourism. The New Development Plan (1989-93) aims at fast growth – five per cent per annum – under conditions of stability, and places emphasis on restructuring, technological upgrading and modernisation. Economy remains export-orientated with exports of goods comprising mainly manufactured products and, to a lesser extent, agricultural products.

Employment: Balanced distribution of the labour force. Low and declining levels of unemployment, and average earnings are rising faster than inflation.

External trade: Widening trade deficit offset by earnings from tourism and capital inflows, largely comprising foreign loans and direct investments. Major exports include clothing, footwear, cement, potatoes, citrus fruits and wine. Europe is major export market, with exports to Europe increasing over 1986-88. Exports to Middle East were rising during 1987-88 but remained lower than their level in the early 1980s owing to increased domestic consumption and higher demand from manufacturing and tourism. Imports were rising during 1987-88. Principal imports are manufactured goods, oil, machinery and transport equipment. UK, West Germany, Italy, Greece and Japan are major suppliers of imported goods. Agreement on full customs union with EC over 15 years came into effect in January 1988.

Agriculture: Sector contributes about eight per cent to GDP and is a significant earner of foreign exchange. Major crops are potatoes, grapes, citrus fruits and barley. Large-scale water development programme under way conveying surplus water from the south western part of the island to the central and eastern areas in effort to broaden and boost agricultural production, as well as to increase the water supply of the receiver areas.

Industry: Employs about 30 per cent of workforce. Major growth industries, which are mainly export-based, include cement, footwear and clothing. Manufacturing and construction account for 26 per cent of GDP. Diversification and foreign investment are being encouraged.

Tourism: Most rapidly expanding sector of the economy. Largest single earner of foreign exchange.

Mining: Commercially mined minerals include iron pyrites, gypsum, asbestos, chrome ore, copper, salt and marble. Production of most of these falling. Efforts to revive output constrained by the high level of production costs in relation to export/selling price.

Energy: Almost total dependence on fuel imports, principally oil for electricity generation. Successful energy conservation programme under way involving exploitation of solar energy.

Legal system: Based on common law with civil law modifications.

Membership of international organisations: AALCC, CCC, Commonwealth, Council of Europe, CSCE, FAO, G-77, GATT, IAEA, IBRD, ICAO, ICO, IDA, IFAD, IFC, ILO, IMF, IMO, INTELSAT, INTERPOL, ITU, Non-Aligned-Movement, OIC (Observer), UN, UNESCO, UNIDO, UPU, WFTU, WHO, WIPO, WMO, WTO. In October 1987 the Republic of Cyprus signed a protocol providing for the second stage of the Cyprus-EC Agreement for Customs Union.

BUSINESS GUIDE

Time: GMT + 2 (GMT + 3 from late Mar to late Sep).

Climate: Mediterranean. Summers hot and dry, although more humid in coastal areas, changeable winters with occasional rain. Temperatures range from 0°C (in mountains) to 40°C. Hottest months Jul and Aug, coldest Jan and Feb.

Entry requirements
Visa: Visa not required by nationals of most European, Commonwealth and Gulf countries, the USA, Japan and a few others. N.B. Legal entry only by Larnaca and Paphos airports and ports of Paphos, Larnaca and Limassol. Occupied Turkish sector can be entered only via Turkish mainland (entry to the Turkish sector is considered illegal by the government of the Republic of Cyprus).
Health: Tap water considered safe to drink.
Currency: Only C£50 in Cyprus bank notes may be taken into or out of the country. There is no limit on foreign currency imports, although amounts in excess of the equivalent of US$1000 should be declared at customs on form DNR. Amounts so declared may be re-exported on departure.
Customs: Personal items duty-free. Unauthorised export of antiquities is prohibited. Permission of Cyprus Museum is required.

Air access: Direct flights from several major international airports.
National airline: Cyprus Airways.
Main international airport: Larnaca (LCA), 5 km south of city.
International airports: Paphos (PFO),11 km from city (72 km from Limassol). Nicosia International airport has been closed to traffic since 1974.

Surface access: By ship from Greece, Syria, Israel, Italy, Lebanon and Egypt.
Main ports: Larnaca and Limassol. Also port at Vasilikos serving industrial complexes in the area. The island's main port at Famagusta has been under Turkish occupation since 1974.

Hotels: Approximately 380 hotels (one to five stars) and hotel apartments (A to C class) plus several tavernas. Visitors should book well in advance especially during peak holiday season (Apr-Oct). Cyprus Tourist Board operates rating system, both for hotels and any other licensed tourist accommodation. Tipping is not obligatory. A 10 per cent service charge and a three per cent entertainment tax are added to the bill. American Express, Diners, Eurocard, Visa and Bank Americard are all accepted. Travellers' cheques must be endorsed.

Restaurants: Wide variety with many specialities available. A 10 per cent service charge and a three per cent tax are normally added to bills. Price lists are required to be displayed.

Car hire: Available in most parts of the island. Rates vary from C#12 to C#24 per day depending on size of car, and are also subject to seasonal variation. Prestige service is also available, but at higher prices. Cheap rates for hire periods of more than one week. Visitors should book cars well in advance June-September. National or international driving licence required. Driving is on the left.

City transport
Taxis: Efficient service throughout the island by metered taxis. Share taxis available from ports and airports. Prices are regulated. Between 2300 and 0600 an additional 15 per cent is charged. Tipping is standard practice.

National transport
Buses: Efficient 'Intra-cud' (inter-city) bus service available. All buses run from central bus depots, connecting towns and villages.

Public holidays
Fixed dates: 1 Jan, 6 Jan (Epiphany), 25 Mar (Greek Independence Day), 1 April (National Day), 1 May, 15 Aug (Assumption), 1 Oct (Independence Day), 28 Oct (OKHI Day), 25 and 26 Dec.
Variable dates: Green Monday, Good Friday, Easter Saturday and Easter Monday, Christmas and late Summer holidays.

Working hours
Business: (Mon-Fri) (winter) 0800 – 1300 and 1430 – 1730, (summer) 0730 – 1300 and 1600 – 1830; (Wed and Sat) half day.
Government: (Sat) 0730 – 1300; (Mon-Fri) (winter) 0730 – 1400, (summer) 0730 – 1330.
Banking: (Mon-Sat) 0830 – 1200 and (specially for tourists) 1530 – 1730.
Shops: (Mon-Fri) (winter) 0800 – 1300 and 1430 – 1730, (summer) 0730 – 1300 and 1600 – 1830; (Wed and Sat) half day.

BUSINESS GUIDE

Social customs: The custom of handshaking is practised. It is considered impolite to refuse drinks offered at a first meeting. Cypriots customarily offer fruit preserves to guests.

Telecommunications
Telephone: Approximately 211,200 direct exchange lines. Service is automatic and direct dialling is available. International dialling code: 357. For Nicosia – prefix +2, Larnaca +41, Limassol +51 and Paphos +61. IDD access code 00.
Telex: Available in major hotels. Country code 605 CY.
Fax: Rapidly increasing use of facsimile service.

Advertising: No code of advertising but CBC operates its own code of practice which restricts cigarette and tobacco advertising. Information can be obtained from the Cyprus Advertising Agents Association at the Cyprus Chamber of Commerce and Industry.

Trade fairs: The Cyprus International Fair held annually for two weeks every summer (end May – mid Jun). Information from Cyprus State Fairs Authority.

Electricity supply: 240 V AC.

Representation in capital: Australia, Bulgaria, China PR, Cuba, Czechoslovakia, Egypt, France, German Democratic Republic, Federal Republic of Germany, Greece, Holy See, India, Iran, Israel, Italy, Lebanon, Libya, Nicaragua, PLO, Poland, Romania, Switzerland, Syria, USA, USSR, United Kingdom, Yemen PDR, Yugoslavia.

BUSINESS DIRECTORY

Hotels: Larnaca
Golden Bay (on beach), Larnaca-Dhekelia Road, P.O. Box: 741 (tel: (41)23444; tx: 4570; fax: (041)23451).

Palm Beach, Larnaca-Dhekelia Road, P.O. Box: 394 (tel: (41)57500; tx: 3822; fax: 20788).

Hotels: Limassol
Amathus Beach, Amathus Avenue, P.O. Box: 513 (tel: (51)21152; tx: 2683; fax: 29343).

Churchill (on beach), 28th October Avenue, P.O. Box: 1626 (tel: (51)24444; tx: 3447; fax: 23494).

Curium Palace, 2 Byron Street, P.O. Box: 48 (tel: (51)63121; tx: 3150; fax: 59293).

Elias Beach, Ancient Amathus District, P.O. Box: 4300 (tel: (51)25000; tx: 5111).

L'Onda Beach, P.O. Box: 2000 (tel: (51)21821; tx: 4342; fax: 20040).

Poseidonia Beach, P.O. Box: 1206 (tel: (51)21000; tx: 2805).

Sheraton (on beach), P.O. Box: 1064 (tel: (51)21100; tx: 5626; fax: (51)24394).

Hotels: Nicosia
Acropole, 17 Heroes Street, P.O. Box: 4084 (tel: 462025).

Asty, 12, Prince Charles Street, Ayios Dometios (tel: 473021; tx: 2618).

Averof, 11, Averof Street (tel: 63447).

Cleopatra, 8, Florina Street, P.O. Box: 1397 (tel: 445254; tx: 2316).

Crown, 12 Phillellinon Street, P.O. Box: 4041 (tel: 465264; tx: 3131).

Europa, 16 Alkaios street, P.O. Box: 5029 (tel: 454537; tx: 4020).

Excelsior, 4 Photios Stravrou, Pitta Street (tel: 442062; tx: 2067).

Hilton, Archbishop Makarios III Avenue, P.O. Box: 2023 (tel: 464040; tx: 2388; fax: 453191).

Kennedy, 70 Ragaena Street (tel: 475131; tx: 2202).

Ledra, Grivas Dhigenis Avenue, P.O. Box: 1390 (tel: 441086; tx: 3355).

Philoxenia, Eylenja Avenue, Eylenja, P.O. Box: 5466 (tel: 499700; tx: 2850).

The Churchill, 1, Achaenes Street, P.O. Box: 4145 (tel: 448858; tx: 3280; fax: 445506).

Car hire: Nicosia

Autohansa: Eleftherias Square (tel: 472352; tx: 2628).

Avis: PO Box 2276, 2 Homer Avenue (tel: 472062, 462507; tx: 2625).

Budget: (tel: 462042; tx: 2802).

Europcar: 32 Archbishop Makarios III Avenue (tel: 444365; tx: 4464).

Hertz: Eleftherias Square (tel: 477783; tx: 2611); Petrolina Petrol Station, Griva Dhigeni Avenue (tel: 353413, 475464).

A. Petsas and Sons Ltd: C Pantelides Avenue (tel: 462650, 474357).

Car hire: Larnaca
Avis: 43 Archbishop Makarios III Avenue (tel: 57132; tx: 2625).

Europcar: 7-8 Joanna Court, Artemidos Avenue (tel: 57442/62; tx: 2416).

Hertz: 33F Arch. Makarios Avenue (tel: 55145).

Car hire: Limassol

Avis: Nicosia-Limassol Road (tel: 24192).

Budget: Nicosia Road (tel: 23672).

Europcar: 4B Blue Sea House, 28 October Avenue (tel: 24025; tx: 4472).

Hertz: Anna Court G3, Nicosia-Limassol Road (tel: 23758).

A. Petsas and Sons Ltd: Sea Breeze Bldg, Old Limassol-Nicosia Road (tel: 23672).

Car hire: Paphos
Avis: Apostolou Pavlou 87, Vasilica Centre No 9 (tel: 43235; tx: 2625).

Budget: Green Ct, Apostle Paul Av (tel: 35522).

Europcar Office 4: 1 Constantinos Court, 77 Poseidonos Avenue (tel: 34149-69).

Hertz: 54A St Paul's Avenue (tel: 33985).

Car hire: Larnaca Airport

Avis: (tel: 57132, 57920).

Europcar: (tel: 57442/62).

Hertz: (tel: 22388).

Chambers of Commerce
Cyprus Chamber of Commerce and Industry (6th Floor), Grivas Dhigenis Ave No 38, PO Box 1455, Nicosia (tel: 449500, 462312; tx: 2077).

Larnaca Chamber of Commerce and Industry, PO Box 287, 12 Gregoriou Afxentiou Street, Skouros Building, 4th Floor, Larnaca (tel: 55051; tx: 3187).

Limassol Chamber of Commerce and Industry, PO Box 347, 25 Spyros Araouzos Street, Berengaria Building, 3rd Floor, Limassol (tel: 62556; tx: 2890).

Nicosia Chamber of Commerce and Industry, 3 Deligeorgi Street (tel: 449500; tx: 2077).

Paphos Chamber of Commerce and Industry, Demetra Court, Grivas Dhigenis Avenue, PO Box 61, Paphos (tel: 35115; tx: 2888).

The Famagusta Chamber of Commerce and Industry, Stylianides Bldg (3rd Floor), 1 Koumandarias Street, PO Box 3124, Limassol (tel: 70165).

Banking
Central Bank of Cyprus, Box 5529, 36 Metochiou Avenue, Nicosia (tel: 445281; tx: 2424).

Arab Bank, 85 Digenis Akritas Ave, Nicosia (tel: 457111; tx: 4825). Branches in Limassol, Larnaca, Paphos, Aya Nappa.

Bank of Cyprus Ltd, Box 1472, 86-90 Phaneromeni Street, Nicosia (tel: 464064; tx: 2451; fax: 464096).

Barclays Bank, Eletheria Square, Nicosia (tel: 448838; tx: 4949).

Co-operative Central Bank Ltd, Box 4537, Gregorius Afxentiou Street, Nicosia (tel: 442921; tx: 2313).

Cyprus Development Bank, PO Box 1415, 50 Archbishop Makarios III Avenue, Alpha House, Nicosia (tel: 457575; tx: 2797).

Cyprus Investment and Securities Corporation, 58-60 Digenis Akritas Avenue, PO Box 597, Nicosia (tx: 4449).

Cyprus Popular Bank Ltd, Box 2032, 39 Archbishop Makarios III Avenue, Nicosia (tel: 450000; tx: 2272).

Hellenic Bank Ltd, PO Box 4747, 92 Digenis Akritas Avenue, Nicosia (tel: 447000; tx: 3311).

Lombard Banking (Cyprus) Ltd, Styl Lenas Square, Nicosia (tel: 474333; tx: 2262).

National Bank of Greece, 36 Arch Makarios III Avenue, Nicosia (tel: 442211; tx: 2445).

Principal newspapers
Cyprus Bulletin, Nicosia (tel: 451001).

Other useful addresses
Central Post Office, Eleftheria Square, Nicosia (tel: 703219).

Cyprus Broadcasting Corporation, PO Box 4824, Nicosia (tel: 422231; tx: 2333).

Cyprus Employers and Industrialists Federation, 30 Grivas Dhigenis Avenue, PO Box 1657, Nicosia (tel: 445102).

Cyprus Hotel Association, PO Box 4772, Nicosia (tel: 445251; tx: 4747).

Cyprus News Agency, PO Box 3677, 97 Ay Omoloyitae Avenue, Nicosia (tel: 458413).

Cyprus State Fairs Authority, PO Box 3551, Makedonitissa, Nicosia (tel: 448918; tx: 3344).

Cyprus Telecommunications Authority, PO Box 4929 (tel: 310271; tx: 3288).

Cyprus Tourism Organisation, Zena Building, 18 Theodotou Street, Box 4535, Nicosia (tel: 443374; tx: 2165).

Department of Statistics and Research (tel: 302349).

MAM (The House of Cyprus Book), PO Box 1722, Nicosia (tel: 72774).

Ministry of Commerce and Industry, G A Araouzos Street, Nicosia (tel: 403441/8; tx: 2283).

Ministry of Finance (tel: 403201; tx: 3399).

Press and Information Office, Nicosia (tel: 402640; tx: 2251).

Czechoslovakia

Judy Dempsey

Czechoslovakia's 'Velvet Revolution' surprised observers by its speed, and was characterized by the Communist Party's failure to appreciate the significance of almost daily developments in November and December. As 'people power' became an intractable force, an almost farcical process of panic decisions reflected the erosion of the Communist Party's grip on events. On 24 November 1989 hard-line Secretary Milos Jakes was replaced by Karel Urbanek. Following that the Party's dominant role was removed from the constitution on 29 November. On 7 December Ladislav Adamec resigned as prime minister, to be replaced by Mr Marion Calfa. Two days later, the unpopular President Gustav Husak (installed at the time of 1968 invasion) announced his impending resignation and a new cabinet with an unprecedented communist minority was installed, leaving only the choice of president in contention.

The two obvious candidates for the presidential post were Mr Alexander Dubcek (prime minister at the time of the Soviet invasion) and the well-known dissident author, Mr Vaclav Havel and de facto leader of the Civic Forum grouping. The decision that the job should go to a Czech rather than a Slovak paved the way for Mr Havel to be sworn in on 29 December. Mr Dubcek was appointed chairman of the Federal Assembly on 27 December.

While Poland and Hungary enthusiastically embraced economic and political reforms throughout 1989 Czechoslovakia, in contrast, adopted a cool attitude towards the changes taking place in the Soviet Union and in neighbouring countries. Reluctance to introduce a more flexible political system and to tackle the economy stemmed largely from the particular style of leadership which had ruled Czechoslovakia since 1968 after the Soviet-led invasion crushed the Prague Spring, the movement which attempted to give socialism 'a human face'. Since Mr Mikhail Gorbachev, the Soviet leader, has all but discarded the 'Brezhnev Doctrine', which essentially gave the Soviet Union the right to interfere in other countries by force if necessary, in order to maintain the status quo, the leadership in Prague tried desperately throughout 1989 to maintain a status quo which owed its legitimacy to the 'old thinking'. But as a younger generation of Czechoslovaks was growing increasingly restless with the slow pace of the reforms in their own country, the leadership was forced to adopt a different stance towards change. If not, the prospects for maintaining the domination of the Communist Party were likely to be undermined even further.

The legacy of 1968

The Prague Spring of 1968 held the key to the level of discontent during 1989 and to the increasingly frantic policies adopted by the Communist Party of Czechoslovakia (CPCz). Because the leadership owed its position to the post-1968 'normalisation' period, any attempt at filling in that particular blank spot in history would almost certainly have precipitated the end of its political life. Witness what had taken place in Hungary during 1989 when those who had persistently referred to the Hungarian 1956 Uprising as a 'counter-revolution' were soon dropped from important political posts as the new leadership in Hungary began openly to address its past.

But in Czechoslovakia – unlike Hungary, where the pressure for change had come largely from within the ruling communist party, or Poland, where it had come from outside the party – any pressure for change from within the CPCz was such that so far, it had not been forceful, coherent or indeed prepared enough to deal with the highly sensitive issue of 1968. In addition, those exerting pressure from outside the party, while increasing all the time, had lacked a solid platform on which to articulate their views. After 22 years of virtual silence, largely due to the means deployed by the regime towards maintaining stability, it was not surprising that the Czechs and Slovaks remained cautious about precipitating change. However, during 1989 the younger generation came to play a much greater and eventually decisive role in heightening the degree of impatience with current policies.

Yet despite the apparent atrophy among the country's leadership and its continuing official standpoint – that those who had participated in the Prague Spring had intended to overthrow the socialist order – both the Soviet Union and some of its east European allies began speaking much more openly about the consequences of the invasion.

Indeed, this was high on the agenda when Mr Aleksander Yakovlev, a prominent member of the central committee of the CPSU, visited Czechoslovakia on 11 November 1988 for a five day visit. During his stay, he gave an interview to *Rude Pravo*, the Czech party daily, in which he criticised current policies in Czechoslovakia. 'If you profess democracy and open information, you should also publish different views, whether you like them or not...the one thing we must not allow is factual inaccuracies...we write about life only formally, as in a protocol, but nowhere can we read about what is interesting everywhere, in your country, in our country, or in the countries of our neighbours.'

The Czechoslovak authorities shrugged off the criticism, which was also directed at the state of the economy. And, as if by coincidence, as an indicator of the growing economic problems facing the leadership, a few weeks before Christmas 1988, the authorities introduced new customs rules which in effect banned the export of a large number of items, ranging from food, textiles and clothes, to household appliances, spare parts and even toilet paper. The reaction from Czechoslovakia's neighbours was immediate, sharp and critical. However, on 22 November, Mr Ivan Kulhanek, a foreign ministry spokesman, justified the new regulations on the grounds that 'Czechoslovak citizens had welcomed the restrictions against the export of selected goods...as a necessary step to protect the domestic market, especially before Christmas....' The 'customs war', as it became known, was much more than introducing restrictions on exports. It was a poignant reminder to Czechoslovak economists about the deteriorating state of the economy.

KEY FACTS

Official title: The Czechoslovak Socialist Republic

Head of state: President Vaclav Havel

Head of government: Prime Minister Marian Calfa

Ruling party: Interim government with minority participation by Communist Party and majority from loose opposition grouping of Civic Forum

Capital: Prague

Official Languages: Czech and Slovak

Currency: Koruna or Crown (Kcs) = 100 heller

Exchange rate: Kcs14.8 per $ (Nov 1989 – commercial); Kcs9.53 per $ (tourist)

Area: 127,896 sq km

Population: 15.61m (1988)

GDP per capita: $6,045 (NMP, 1986)

GDP real growth: 2.0% (NMP, 1987)

Labour force: 7.6m (1987)

Inflation: 2.3% (1986)

Economic Performance

Compared with other east European countries, the economy appears in good shape. The shops are full of consumer goods, despite declining standards in quality. Indeed, Soviet visitors often refer to Prague as the New York of the socialist world. But in this context, comparisons are relative. On a much broader scale, economists in Prague were becoming increasingly concerned about productivity levels, quality and the future direction of the economy. These problems were reflected in the 1989 budget. During one of the budget discussions, the Czechoslovak leadership announced that the 1989 growth rate would be cut from 3.5 per cent to 2.2 per cent. A 3 per cent growth rate was forecast for 1990. It was also estimated that the original planned cumulative Domestic Net Material Product growth of 19 per cent for the whole of the Five Year Plan (begun in 1986) would in fact only reach 13-14 per cent.

The disappointing forecasts for 1989, were, however, coupled with the introduction of new industrial and agricultural procurement prices which took effect from 1 January, 1989. The idea behind the changes was to make prices reflect, or at least come closer to, both domestic and international levels. This was bound to have an effect on consumer prices since enterprises, accustomed to subsidies in the past, would inevitably be forced to increase prices for labour and production. Another spin-off of the price changes was the introduction of one single rate for the Crown, the Czechoslovak unit of currency. Several different rates had existed in the past, which made a uniform accounting system difficult to introduce. In addition, it created a distorted overall view of the performance of the economy, given that prices for some consumer items were charged (in the area of exports, for instance) at different rates from others.

However, even though there had been some administrative and financial changes introduced into the running of the Czechoslovak economy, they did seem to go far enough. This was the message delivered by Mr Ladislav Adamec, the prime minister, in office since October 1988, as he commented on the 1989 State Plan. He had no hesitation in saying that 'the management system, the inadequate industrial structure and the backwardness of the tertiary services sector have reached dangerous dimensions.' Yet despite these criticisms, the prime minister did not envisage higher capital investments for the modernisation of light industry, once the show-piece of Czechoslovakia's economy in the inter-war period. Indeed, output for 1989 was planned to be 0.8 per cent lower than the original Five Year Plan had anticipated. The picture for industry was no less colourful. The growth of gross output in Czechoslovak industry slowed down from 3.2 per cent in 1986 to 2.3 per cent in 1987 and to 2.0 per cent in 1988. It was estimated that 25 per cent of all industrial companies would fall short of their planned targets. And the Federal Statistics Office bluntly stated that 'labour productivity did not correspond to the needs of the economy', a situation which could have significantly negative effects for the consumer economy. The other problem which had to be tackled was the transition from an extensive to an intensive development of the economy, a transition which the authorities admitted had been long overdue. This was confirmed by the Federal Statistics Office which in early 1989 clearly showed that investments for extensive development of the economy had increased by 4.5 per cent over the 1987 level even though the plan had envisaged a 2.4 per cent growth.

Extensive economic development had been part of Eastern Europe's post-war programme. Under the command economy, which was pushed to extremes by Stalin, each country had been exhorted to modernise and industrialise at a pace which had paid little attention either to the existing infrastructure or to the political culture. Such plans had been reined in during the 1960s when more liberal-minded Czechoslovak economists, recognising the potentially irreparable damage which was being inflicted on the country's traditionally successful sectors, such as light industry, tried to divert investments away from extensive to more intensive growth. Those policies, alas, were quickly shelved after 1968. Twenty-two years later, the cost of this policy has been particularly obvious in the kinds of investments allocated to scientific research and technological development. In his important speech to the Federal Assembly on 20 June 1989, Mr Adamec went so far as to say that while 200,000 people were engaged in scientific and technological research and development, 'most industrial sectors are lagging 10 or more years behind the world.' Other economists in Prague say the lag is wider, a phenomenon which is a sad reflection on the country's post-war economy which could boast one of the finest light-industrial sectors in Eastern Europe.

But that speech, of which the original draft was so radical and critical that it had not been presented to parliament, went further than previous analysis of the past two decades. 'People', said Mr Adamec, 'were not pacified by mere words or repeated promises' but were demanding 'guarantees based on securities, truth and credibility' and that 'demands for the speedier rectification of long-term shortcomings are getting louder.' The prime minister went so far as acknowledging that 'the government does not have enough means to rectify the existing state of affairs speedily. Since becoming prime minister, Mr Adamec, in tones which reflected deep pessimism rather than any hint of optimism, said 'the task that we have taken upon ourselves looks bigger and more difficult than it seemed at the beginning...'

KEY INDICATORS

	Unit	1984	1985	1986	1987	1988
Population	m	15.5	15.5	15.6	15.6	15.6
Net material product (NMP)	$bn	80.4	80.1	93.7	–	–
NMP per capita	$	5,187	5,168	6,045	–	–
NMP real growth	%	3.2	2.9	12.0	2.0	–
Investment	% growth	1.5	2.8	2.0	3.7	–
Agricultural production	% growth	4.4	0.5	1.6	0.9	–
Industrial production	% growth	4.0	3.1	2.6	2.3	–
Coal mining	% growth	1.6	–2.1	–1.5	–	–
Metallurgy	% growth	1.7	2.2	2.0	1.9	–
Chemicals	% growth	3.0	3.9	3.4	3.0	–
Rubber	% growth	3.0	3.4	3.1	–	–
Textiles	% growth	2.2	2.9	2.5	–	–
Food	% growth	2.0	1.2	1.6	0.8	–
Imports	Kcs bn	113.7	105.8	103.0	–	–
Mineral fuels	Kcs bn	35.0	31.6	38.2	–	–
Manufactures	Kcs bn	10.0	10.5	11.4	–	–
Machinery and equipment	Kcs bn	36.6	37.7	40.3	–	–
Exports	Kcs bn	114.5	104.9	123.2	–	–
Chemicals	Kcs bn	6.7	6.1	7.3	–	–
Machinery and equipment	Kcs bn	55.4	55.7	66.0	–	–
Balance of trade	Kcs bn	1.2	0.9	37.0	–	–
External debt	$bn	3.3	3.6	3.9	3.7	–
Exchange rate	Kcs per $	6.4	6.1	5.4	5.2	5.5

* estimated figure

Younger, critical voices

Although Mr Gorbachev had repeated that each country was free to choose its own model of socialist development and that the old days of interference in the internal affairs of another country were over, his attempts to introduce perestroika and glasnost in the Soviet Union have nonetheless provided a source of enormous inspiration for thousands of young Czechoslovaks. And it is not only the Soviet Union which they have seen change so rapidly. Neighbouring Hungary had been forging ahead with major political reforms which were deemed vital if the economy was to emerge from its present crisis. Against this background, throughout 1989, the Czechoslovak leadership had been confronted with one of the most remarkable developments since 1968: the open discontentment of the younger generation.

The first signs came on 15 January 1989, the twentieth anniversary of the death of Jan Palach, the young student who committed suicide in protest against the crushing of the Prague Spring. Initially, hundreds of young people gathered in Wenceslas Square in the centre of Prague. By 17 January, the numbers had swollen to 5,000. But neither the authorities nor the police turned a blind eye to these unprecedented gatherings by allowing the commemorations to continue. On the contrary, the police used truncheons, dogs and water cannon to disperse the thousands of people who had gathered peacefully to commemorate the anniversary. The scale of the force shocked not only the international community. Czechs and Slovaks, young and old, and even establishment figures in the artistic community all openly criticised the leadership of Mr Milos Jakes, the dour party leader, for the harsh, tough, reaction by the authorities. Any hope that the leadership had been seriously contemplating major reforms was dissipated over those four days. In addition, the leadership pressed ahead in condemning Mr Vaclav Havel, the banned Czechoslovak playwright, to a prison term on the grounds that he had illegally placed flowers in Wenceslas Square.

The youth, mere babies in 1968, temporarily retreated. But the January events, instead of crushing any critical voices, provided a catalyst for the opposition to galvanise itself. This did not apply wholly to Charter 77, the independent human rights movement, which was itself surprised by the scale of the demonstrations. Instead resentment that change was not taking place in Czechoslovakia spread throughout society. People saw what had been taking place in Hungary and the Soviet Union and they wanted to share in these experiences too. This was the message conveyed repeatedly to the authorities throughout

1989, during which time the Czechoslovak party was bombarded with open letters. For example, more than 1,000 Czechoslovak artists, many of them prominent actors, film and theatre directors, painters, sculptors, art critics and playwrights, all lent their support to, and signed, an appeal requesting the release from prison of those who were arrested in January. But this amounted to no ordinary appeal. For the first time, establishment figures broke their silence, a development which must have caused considerable concern among the leadership. That letter was soon followed by another one, this time from 670 scientists, scholars and researchers which was addressed to the prime minister. Sections of it are worth quoting as a means of indicating the impact the Soviet Union was having on various sections of Czechoslovak society.

Under the impact of Soviet developments, changes that are necessary have begun to be discussed in our country, too, first only in regard to the economy and now also in the field of society's democratisation. Unfortunately, the highly excessive intervention by security police units against rallies meant to commemorate the death of Jan Palach makes us fear that deeds do not match words...the purpose of our profession is to create new intellectual and material values, and criticism is an inalienable part of this process...we understand that the process of restructuring and democratising Czechoslovak society will not be an easy one. We do hope, however, that it will lead to democracy, which opens up room for a pluralism of views. From our work in science, we know that nonconformist views can also be valuable and that they sometimes produce results of great importance and benefit....'

This measured, reasoned letter was followed by several others. The most important one, endorsed by well over 10,000 signatories, was drawn up by ten independent groups. But instead of being addressed to the prime minister of Czechoslovakia, it was also sent to the prime ministers of Bulgaria, Hungary, East Germany, Poland and the Soviet Union , politely asking them all to review their attitude towards the 1968 invasion of Czechoslovakia.

Even before this petition, the reactions from Eastern Europe towards the past and current events in Czechoslovakia had been revealing to monitor. Hungary went the furthest in questioning what took place in 1968 by broadcasting an interview with Mr Alexander Dubcek, the former communist party leader of Czechoslovakia who was ousted from power in 1968. But for broadcasting the two-part television interview with Mr Dubcek, the authorities in Budapest received a sharp rebuke from Prague which accused them of interfering in the internal affairs of another country. The Polish authorities for their part did not ban Vaclav Havel's plays, which were performed in Poland during 1989. But as an indication of the growing differences within Eastern Europe towards glasnost and perestroika, both East Germany and Romania sided with Czechoslovakia by sharply criticising both the opposition in Prague and the pace of the political reforms in neighbouring Hungary. Clearly, the old days of socialist fraternity and solidarity, in which east European communist leaderships once spoke with unanimity, were fast disappearing throughout 1989.

The breakdown of the fraternity

But the breakdown in monolithic thinking went far beyond words. Nowhere was this more noticeable than in the Hungarian decision to postpone work on the controversial Nagymaros-Gabcikovo dam complex.

The dam, which had been designed to divert the River Danube by 180 km, was a mammoth plan mapped out in the early 1950s, the days of giant, prestige projects. At the time, the Hungarian communist party was opposed to the idea, partly on the grounds that it would be uneconomic. However, both the Soviet Union and Czechoslovakia insisted that the dam be constructed. Work started in the early 1980s, after a treaty had been signed between Budapest and Prague. But as ecological costs to the environment became more apparent, it gave rise to a large 'green' movement in Hungary, which was centred on the 'Duna Koer' or Danube Circle, an independent group. In a nutshell, after much debate in parliament and outside, in May 1989 the Hungarian authorities postponed work on the dam. The significance of this decision was twofold. First, it clearly demonstrated the strength of the green movement in Hungary and the realisation by the Hungarian authorities that its popularity could only increase from such a decision. But second, it sparked off a furious row with Czechoslovakia, which had already invested much time and money into building part of the dam at Gabcikovo, Slovakia, just on the Hungarian border.

The authorities in Prague at first accused certain circles in Hungary of trying to discredit former decisions made by the communist party. It also said Hungary had broken an international treaty. But of even more importance, the dispute over the Nagymaros-Gabcikovo dam highlighted the latent differences between both countries towards responding to independent pressure. Interestingly, this time round Czechoslovakia received no support from the Soviet Union which, during 1989, seemed more reconciled to allowing its east European allies to sort out their own disputes themselves.

COUNTRY PROFILE

Political system: Federal state of two republics, Czech and Slovak, both with equal rights. Supreme power formally rests with the bicameral Federal Assembly, consisting of the Chamber of the People, which is elected regionally, and the Chamber of the Nations, elected by the Czech and Slovak National Councils. The President of the Federal Assembly appoints a cabinet, including the Prime Minister. Before the dramatic events in late 1989, political power was held by the Communist Party which dominated the five-party National Front. On 10 December 1990 a new cabinet with a Communist minority was installed. Civic Forum constitutes interim government under President Vaclav Havel. Alexander Dubcek is Chairman of Federal Assembly. Election planned for June 1990 but it remains to be seen whether these will be just parliamentary or whether those for local authorities, Czech and Slovak assemblies will take place first.

Main cities/towns: Prague (population 1.2m), Bratislava, Brno, Ostrava, Kos\06ice, Plzen\06.

Media
Press: Dailies: There are 11 national dailies (six Czech, five Slovak), mostly published in Prague. The party newspaper is *Rude Pravo* (circulation 1,108,000). Also some provincial dailies. *Weeklies:* Numerous consumer magazines. *Business:* About 50 technical journals published monthly, covering all sectors of the economy, also an economics weekly *Hospodárske Noviny* (125,000). Business publications in foreign languages produced by Rapid Advertising Agency and Orbis Agency.
Radio: Approximately 4.5m receivers in use. There are five major national stations and numerous local ones. Programmes are broadcast in Czech and Slovak and there is also a foreign service.
Television: Over 4m receivers in use. Networks serve both Czech and Slovak populations from Prague and Bratislava.

Domestic economy: Centrally planned economy with a bias towards heavy industry. Emphasis being placed on intensive development of the economy, with investment targeted on improving productivity and efficiency, and increasing exports. Current five-year Plan (1986-90) involves a more flexible approach to economic management and the restructuring of the wide industrial base.

Employment: Labour force totals some 7.6m, of which 40 per cent are employed in industry and 14 per cent in agriculture and forestry. Labour shortages have been experienced in some sectors, leading to the employment of foreign workers.

External trade: Trade with the USSR accounts for about 45 per cent of total foreign trade and other CMEA countries account for a further 25 per cent. Much of this takes the form of exchanging exports of machinery for import of energy and raw materials. Major exports are machinery and transport equipment (over 50 per cent of total), consumer goods, chemicals, mineral raw materials and metals. Major imports are capital goods and raw materials.

Agriculture: Nearly 7m ha of agricultural land, of which 70 per cent is arable. Sector dominated by 1,700 co-operative farms and 200 state farms. Growth has been sluggish (less than five per cent per annum since 1980) despite high degree of mechanisation, and the country remains about 90 per cent self-sufficient in food. Main crops are wheat, barley, sugar-beet, rye and potatoes. Also intensive production of livestock and dairy produce.

Industry: Accounts for around 60 per cent of NMP though its share of the labour force is relatively small. The sector is state-owned and is dominated by heavy industry, although there has also been expansion of traditional industries such as textiles, clothing, leather, footwear, glass and wood products. Rapid growth of electronics industry as part of the policy of restructuring the sector towards high-technology products. Projects also under way to reduce the reliance on imports of energy and raw materials.

Mining: Mineral production insufficient to satisfy domestic demand although a wide variey of minerals are mined. These include iron ore, antimony, uranium, lead, zinc, copper and tin; plus a range of non-metallic minerals such as fluorite, graphite and bentonite.

Energy: Although domestic coal production accounts for around 60 per cent of total energy consumption, there is a heavy dependence on imports of oil and gas, mostly from the USSR. Emphasis on developing nuclear energy (four nuclear power stations are currently being built) and hydro-electricity, which in 1985 accounted for 14 per cent of electricity generation.

Banking: All banks are nationalised. Most important are Ceskoslovenska Statni Banka (state bank), Ceskoslovenska Obchodni Banka (commercial bank), Zivnostenska Banka (trade bank), Investicni Banka (investment bank).

Legal system: Civil law system based on Austrian-Hungarian codes, modified by Communist legal theory.

Membership of international organisations: CCC, CMEA, FAO, GATT, IAEA, ICAO, ICO, ILO, IMO, International Lead and Zinc Study Group, IPU, ISO, ITC, ITU, UN, UNESCO, UPU, Warsaw Pact, WFTU, WHO, WIPO, WMO, WSG, WTO.

Outlook

Against this background, it would seem that while the Soviet Union keeps a close eye on its allies, it is simply not prepared to force the east Europeans into implementing perestroika and glasnost.

During 1989, the view from Moscow was such that Mr Gorbachev had more than enough problems in his own garden without trying to attend to other ones. In addition, the last thing the Soviet leader wanted was unrest in the orchards. And if the Czechoslovak leadership had managed to maintain stability, while making half-hearted attempts to introduce changes, then so be it. Except the problem was that the leadership under Mr Jakes, certainly in 1989, was in no position to confirm that all was well in the orchard. Young people wanted change and they had taken to the streets to call for it. In short, stability in Czechoslovakia could no longer be taken for granted.

However, it does seem that the leadership, sooner or later, will have to address the question of 1968. Once that has been achieved, it could breath new life into a country which can still boast the finest democratic traditions in all of Eastern Europe. True, this is what the leadership seemed to fear most in 1989: the renaissance of a political culture which runs contrary to the present style of leadership in Czechoslovakia. Yet without acknowledging and indeed tapping these traditions, the leadership could face more discontent.

In the meantime, the leadership has granted the media greater leeway. It also moved closer to improving its contacts with the Catholic Church and with the Vatican by filling, in July 1989, three important bishoprics. Eleven of the 13 bishoprics had remained vacant for some time due to disagreements between the Vatican and Prague. On this front, there are grounds for optimism. The same cannot be said for the Czechoslovak economy, which is in need of capital investments and modernisation. Only time, coupled with a more flexible leadership, will pull the country out of its malaise. Although the youth are growing more impatient, the Czechoslovak spirit, so accustomed to waiting, will almost certainly be ready to respond eagerly to the changes when they come about.

BUSINESS GUIDE

Time: GMT + 1 (GMT + 2 from late Mar to late Sep).

Climate: Continental, with warm summers and cold winters. Average mean temperature nine degrees. Snow throughout winter, rain likely all year, frequent thunder storms in summer.

Entry requirements: Visas required by all except CMEA nationals. Passports should extend for at least five months from the date of visa issue. Business visas valid for three months. Currency slips issued at frontier must be surrendered on departure. Travel regulations liable to change at short notice.
Health: No vaccination certificates required unless arriving from an infected area.
Currency: Equivalent of (West German) DM30 for each day of visit to be exchanged on arrival, either in cash or vouchers. No restrictions on amount of foreign exchange that may be taken in or out. Local currency may not be taken in or out.
Customs: personal effects and all goods purchased in TUZEX shops are duty-free. Most other goods, excluding glassware, priced at Kc\09s300 or less are duty-free.

Air access: Major international airlines operate regular services.
National airline: Ceskoslovenske Aerolinie.
Main international airport: Ruzyne (PRG) 16 km from Prague.
International airports: Bratislava (BTS), Poprad-Tatry.

Surface access: Crossing points in Austria, GDR, FRG, Hungary, Poland and USSR.
Main ports: Although country is land-locked, it has access to the sea through its two main rivers. The River Elbe connects with Hamburg on the North Sea and River Oder with Szczecin on the Baltic.

Hotels: Bookings best msde in advance through CEDOK and Pragotour or direct with hotel concerned. De luxe, A, B and C categories. 10 per cent tip is usual. Diners Card, Eurocard, Amex, Carte Blanche, Visa, Mastercard and Bank American card accepted at major hotels.

Restaurants: Most hotels have their own dining rooms, and there are numerous restaurants. 10 per cent tip usual.

Car hire: Available through CEDOK, Avis, Pragocar and Rekrea. Minimum hire time usually three hours. Average hire charge $25/day, excluding insurance. 50kph speed limit in built-up areas. Seat belts obligatory. Petrol coupons can be bought at discount in advance from TUZEX.

City transport
Taxis: Good service operating in all main towns. Basic fare Kcs5km. Surcharges for journeys between 22.00 – 06.00. Taxis can be ordered by telephoning 203941-5/202951-5. Aerotaxis can be booked at Ruzyne Airport.
Buses: Reliable flat-fare service. In Prague (1Kcs) tickets can be bought in advance from tabak shops and other shops displaying sign Predprodej Jisdenek.
Metro: Three flat fare lines operate in Prague (lines A, B, C). Service runs from 0500 – 2400. For information on city transport telephone 229252.

National transport
Air: Ceskoslovenske Airolinie operates extensive domestic network. Regular daily flights from Prague to Brno, Bratislava, Ostrava, Presov, Holesov, Kosice Piestany, Bystrica, Karlovy Vary and Poprad.
Road: Several major highways linking Prague with main towns (usually marked with an E). Total length of road around 74,000 km, of which 90 per cent tarred. Buses generally quicker and more comfortable than trains.
Rail: Prague connected to all major towns. Seats should be booked in advance on main routes; tickets available from CEDOK or railway ticket offices. Network covers 13,142 km (24 per cent electrified).
Water: There are 480 km of navigable waterways. Major river ports at Prague, Melnik, Usti and Labem, Decin, Bratislava and Komarno.

Public holidays
Fixed dates: 1 Jan, 1 May, 9 May (National Day of Liberation), 28 Oct (Foundation of Czechoslovakia – 1918), 25 Dec, 26 Dec.
Variable dates: Easter Monday. If an offical holiday falls on a Thursday or Tuesday, the Sunday or Saturday is usually treated as a working day and the Friday or Monday observed as a holiday instead.

Working hours
Government and business: (Mon-Fri) 0800 – 1700
Banking: (Mon-Fri) 0800 – 1200.
Shops: (Mon-Fri) 0830 – 1800 and, in most cases Saturday mornings until 1100.

Social customs: Shake hands on meeting and departing. Punctuality is important.

Business language and interpreting/translation: Can be supplied on request. Most Czechs (not Slovaks) speak some German.

Telecommunications
Telephone: Direct dialling available from major towns. International dialling code: 42 followed by 2 for Prague and 7 for Bratislava. Local calls cost Kcs1. Directory enquiries dial 120 (local), 0135 (international enquiries). For telegram service dial 127.
Telex: Available in most major hotels. Country code: 66C.

Postal service: Most mail travels by air at no extra cost. Postal rate within Prague Kcs1, overseas airmail Kcs6 (up to 10g).

Advertising: Usually conducted through national agencies. Made In (mainly promoting foreign products in Czeckoslovakia), INCHEBA (Bratislava) and Rapid. As well as issuing magazines which carry advertisments, these will also place adverts in other media, and give help with all forms of publicity. Advertising space can be purchased on TV and radio, in cinemas, newspapers, magazines and posters.

Trade fairs: Most important is Brno International Engineering Fair (held over eight days in Sep). Brno Fair Authority organises smaller specialised exhibitions covering a wide range of goods. Major chemical fair (INCHEBA) held annually in Bratislava.

Electricity supply: 220 V AC.

Representation in capital: Afghanistan, Albania, Algeria, Argentina, Austria, Belgium, Bolivia, Brazil, Bulgaria, Burma, Canada, China PR, Colombia, Costa Rica, Cuba, Denmark, Ecuador, Egypt, Finland, France, German Democratic Republic, Federal Republic of Germany, Ghana, Greece, Hungary, India, Indonesia, Iran, Iraq, Japan, Kampuchea, Democratic People's Republic of Korea, Lebanon, Libya, Mexico, Mongolia, Morocco, Netherlands, Nicaragua, Nigeria, Norway, Peru, Poland, Portugal, Romania, Spain, Sudan, Sweden, Switzerland, Syria, Tunisia, Turkey, USA, USSR, United Kingdom, Uruguay, Venezuela, Vietnam, Yemen Arab Republic, People's Democratic Republic of Yemen, Yugoslavia.

Hotels: Prague
Inter-Continental, Namesti Curievych 1, 1101 (tel: 2311812; tx: 122681).

Interhotel Alcron, Stepanska 40, 11123 (off Wenceslas Square) (tel: 245741/9; tx: 121814).

Interhotel Ambassador, Vaclavske Namesti 5, 11124 (tel: 221351; tx: 122237).

Interhotel Esplanade, Washingtonova 19, 11000 (tel: 222552; tx: 121067).

Interhotel Flora, Vinohradska 121, 13000 (tel: 274241; tx: 122238).

Interhotel Jalta, Vaclavske Namesti 45, 11000 (on Wenceslas Square) (tel: 265541/9; tx: 121580).

Interhotel Olympik, Sokolovska tr 138, 18676 (tel: 828541/9; tx: 121436).

Interhotel Olympik 11 – Garni, Namesti Invalidovne 186-76 (tel: 8284519; tx: 011236).

Konopiste Motel, Benesov u Prahy , 25601 (tel: 2748).

Panorama, Miloska 7, 14063 (tel: 416111; tx: 123576).

Parkhotel, 20 Veletrzni 7 (tel: 3807111; tx: 122278).

Pruhonice Club Motel, Pruhonice u Prahy, 25243 (tel: 723241/9; tx: 122611).

Car hire: Prague.

Avis, Budget, Europcar: Stepanska 42 (tel: 2352809, 2352825; tx: 122641); Ruzyne Airport (tel: 367807; tx: 122729).

Europcar: Intercontinental Hotel (tel: 2319595); Interhotel Forum (tel: 410213).

Hertz: Hotel Intercontinental, Namesti Curieovych (tel: 2319595; tx: 121353).

BUSINESS GUIDE

Business advice and information
Chamber of Commerce of Czechoslovakia, Argentinska 38, 17005 Prague 7 (tel: 8736; tx: 121862).

Banking
Ceska Statni Sporitelna (Czech state savings bank), Vaclavske Namesti 42, 11398, Prague 1.

Ceskoslovenska Obchodni Banka AS (commercial bank), 115 20 Prague 1, Namesti Prikope 14 (tel: 2132; tx: 122201).

Slovenska Statni Sporitelna (Slovak state savings bank), Leningradska 24, 80100, Bratislava.

Statni Banka Ceskoslovenska (state bank), Namesti Prikope 28, 11003 Prague 1 (tel: 2112; tx: 12202).

Zivnostenska Banka (trade bank), Namesti Prikope 20, 11380 Prague 1 (tel: 22346/9, 23343/4; tx: 122313).

Principal newspapers
Hospodarske Noviny, Na Fiorenci 3, 112 86 Prague 1 (tel: 2323451; tx: 121184).

Lidová Demokracie, Karlovo Nám 5, 120 78 Prague 2 (tel: 291505; tx: 121403).

Mladá Fronta, Panská 8, 112 22 Prague 1 (tel: 221661; tx: 122468).

Rudé Pravó, Na Porici 30, 112 86 Prague 1 (tel: 249851; tx: 121184).

Other useful addresses
Advokatni Poradna (legal advice bureau), Narodni 32, Prague 1 (tel: 224782-7).

Brno Trade Fairs and Exhibitions, Vystaviste 1, 602100 Brno (tel: 3141111; tx: 62239).

Cechofracht (state forwarding/clearing agent), Namesti Prikope 8, 11183 Prague 1, (tel: 222951/2/3; tx: 122510).

Cedok (travel and hotel corporation), Namesti Prikope 18, 11135 Prague 1 (tel: 224251; tx: 121109).

Central Automotoclub CSSR, Opletalova 29, 11395 Prague 1 (tel: 223544).

Central Post Office, Jindrisska 14, Prague 1 (tel: 264841).

Ceska Statni Pojistovna (insurance/management for foreign affairs), Spalena 16, Prague 1 (tel: 2148111).

Ceskoslovenske Statni Drahy (state railways), Namesti Prikope 33, 11005 Prague 1.

Ceskoslovenska Tiskova Kancelar (news agency), Opletalova 5-7, 11144 Prague 1.

CSA (Czechoslovak Airlines) (external services), Ruzyne Airport, 16015 Prague 6 (tel: 367760, 367814).

Interpreter and translation services: Cedok, Namesti Prikope 18, Prague (tel: 220825). Or Prazska Informacni Sluzba, Letenska 2, Mala Strana, Prague 1 (tel: 534255).

Advokatni Poradna (legal advice bureau), Narodni 32, Prague 1, (tel: 224782-7).

Made In (publicity, advertising agency), Kotorska 16, Prague 4.

Ministry of Finance, Prague.

Ministry of Foreign Trade, Ul Politickych Veznu 20, Prague 1 (tel: 2126).

Rapid (advertising agency), Ulice 28, Rijna 13, Prague 1 (tel: 267941/5; tx: 121142).

Slov-Air (domestic flights), Iranka Airport, Bratislava.

State Library, Prague 1 – State Mesto, Klementinum, Namesti dr Vacka (tel: 267241).

Statisticka Rocenka Ceskoslovenska Socialisticke Republiky (general statistics), Spalena 51, 11302, Prague 1.

Denmark

Hilary Barnes and Xueling Lin

Confronted with rising unemployment, massive foreign debt and a balance of payments deficit running for 25 consecutive years, the affluent Danes are facing a bitter pill of reforms that may not be easy to swallow. The right-of-centre minority coalition government formed in May 1988, after a very indecisive election sparked off by a question of defence policy rather than economics, succeeded in moderating wage increases and curbing some of the Danes' spendthrift ways. Now the question is whether this government will be able to push through its major tax reform, covering cuts in corporate and income tax and reductions in welfare benefits, or whether they will be thwarted by the socialist bloc. If they do not succeed in drumming up a majority, however frail, a new election is inevitable.

Political outlook

Prime Minister Poul Schlüter, head of the centre-right minority coalition that came into power in May 1988, did a surprisingly good job of keeping the fragile coalition together and succeeded in getting the controversial 1989 Finance Act passed, which contained proposals to reduce the wealth tax, trim welfare spending and privatise state assets.

Immediately after the 1988 vote, the issue which sparked off the election, of whether NATO vessels carrying nuclear weapons should be allowed in Danish harbours, faded rapidly. NATO appeared to have forgiven this particular Danish prank and any Danish doubts about Danish membership in NATO have not resurfaced.

The greater difficulty was how to form a sufficiently stable coalition in the 179 seat Parliament (*Folketing*) with eight parties jostling for the reins. The party that stood in the balance was the Radical Liberals who held the decisive 10 seats between the socialist and non-socialist blocs.

Mr Schlüter succeeded in persuading the Radical Liberals onto his side with the offer of two key posts, the economy and the environment ministry, so that together with the Liberals and his own Conservative party, the coalition have 67 seats in the *Folketing*.

Few observers at the time would have betted on this minority coalition lasting more than a year, far less the four-year term theoretically available to it. Surprisingly, they have succeeded in passing the one year benchmark, as well as turning the minds of the electorate from foreign policy to the more pressing issue of putting their domestic economic house in order. But they have not been able to do so without major concessions to both the left and the right.

The socialist bloc, made up of the more moderate Social Democrats and the more extreme Socialist People's Party, dealt a very serious blow to the coalition's plan to reign in government expenditure by refusing to allow cuts in unemployment benefits and pension expenditure. (The coalition had hoped to get government expenditure in real terms back to 1987 levels in 1990.) On the question of tax cuts for 1988-89, the coalition had to reluctantly accept the support of the far-right Progress Party, which was pressing for cuts in the wealth tax, in order to get the 1989 Finance Act passed.

The Progress Party was founded in the 1970s essentially as a protest party against the high levels of taxation, but has since taken on an anti-immigrant turn which has made it a party *non grata* in the *Folketing*. Neither the socialist nor non-socialist parties want to join hands with the Progress Party for fear of being associated with its 'racist' stand, despite the party holding 16 seats in the *Folketing*.

If the May 1988 election is any indicator, the Progress Party will not be able to be kept out in the cold for ever. They were the only party to make a substantial gain in the 1988 election, jumping from nine to 16 seats. Whether this was due to their 'lower taxes, more private initiative' stand or their anti-immigrant propaganda is unclear. But if the tax reform launched by the coalition results in a 1989-90 election, the

KEY FACTS

Official title: The Kingdom of Denmark
Head of state: Queen Margrethe II
Head of government: Prime Minister Poul Schlüter (Conservative)
Ruling party: Coalition of Conservative People's Party, Liberal Party and Radical Liberal Party
Capital: Copenhagen
Official Languages: Danish
Currency: Danish Krone (Kr) = 100 \18re
Exchange rate: Kr6.86 per $ (Dec 1989)

Area: 43,080 sq km
Population: 5.13m (1988)
GDP per capita: $21,060 (1988)
GDP real growth: -0.4% (1988)
Labour force: 2.75m (1987)
Unemployment: 8.7% (1988)
Inflation: 4.5% (1988)
Trade balance: Kr9.2bn (1988)
Foreign debt: $41bn (1988)

Progress Party stands to win even more seats since anti-immigrant feeling has picked up since the 1988 election.

The two small non-socialist parties in the *Folketing*, the Christians and the Centre Democrats, were part of the coalition that existed before the 1988 election but were not included in the post-1988 government. Freed of having to tow a common government line, they have voted from issue to issue. (e.g. The Christians voted unsuccessfully against the government for a law extending shop opening hours for fear this would erode church-going, while they voted with the government on the 1989 Finance Act.)

The major change, however, in the political scene is the change in attitude within the Social Democrats. After being the dominant party in the 1970s, which built up the welfare state, they now accept that something drastic needs to be done. The Social Democrats have put forward their own tax reform package, which includes cuts in income tax for lower income earners, but leaves marginal rates unchanged; increased compulsory pension contributions, and reductions in corporate tax but on a broadened tax base.

The government is very eager to try and compromise with the Social Democrats, because without a broad consensus nothing will be implemented. Mr Schlüter has repeatedly said that he is 'very open to suggestion' concerning the tax reform.

The two possibilities are either that the government will be obliged to compromise so much that the reform will have lost all its bite, or alternatively a new election will be held in 1989-90 where this time the issue will not be foreign policy but which tax package is going to save Denmark.

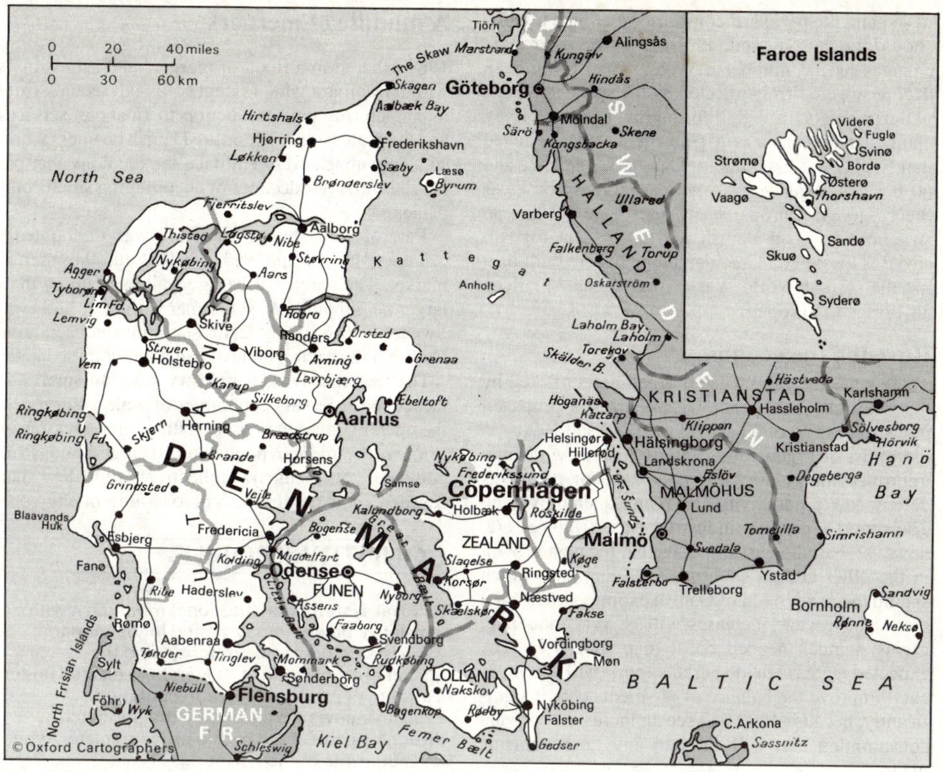

Foreign policy

Denmark's reputation as NATO's most reluctant member has not improved since May 1988 when the country appeared to be on a direct collision course with NATO and faced a possible loss of membership with the alliance.

The rupture in 1988 arose when the centre-left majority in the *Folketing*, against the wishes of the minority coalition government, passed a resolution requiring the government to inform visiting warships that Denmark did not allow nuclear weapons on its territory. This placed Denmark's membership in NATO in serious jeopardy, as NATO warships from the USA and Britain would no longer be able to visit Danish ports as part of NATO military exercises – the British and Americans having a policy of '(we) neither confirm nor deny the presence of nuclear weapons'.

The Danes theoretically had not allowed nuclear weapons on Danish territory since 1956, but they had tacitly accepted the possibility by not asking visiting NATO vessels to disclose what weapons they were carrying.

NATO was not pleased by this sudden desire to be fussy, especially coming from a country that makes one of the lowest contributions in NATO. (Defence spending hovers at only 2.2 per cent of GDP.) The then US Secretary of State George Schultz warned the Danes that if the resolution was implemented, it would have 'serious consequences' for US-Danish defence co-operation. The UK expressed doubts whether the 14,000-strong UK Mobile Force, intended to reinforce Denmark in a crisis, would be able to get to Denmark by sea if the resolution was strictly interpreted.

The issue forced an election in May 1988 which weakened the minority coalition's hand even further, but, ironically, the Social Democrat party, which had prompted the resolution, was unable to form a coalition to take over.

The matter solved itself when Prime Minister Poul Schlüter won over the Radical Liberals, who support the non-socialist bloc on most economic matters but have a pacifist foreign policy, into his Conservative-Liberal coalition government.

The concessions the new government made to the Radical Liberals for switching sides were: to confirm Denmark's 1956 policy; to send notes out to diplomatic missions in Copenhagen reaffirming that Denmark did not permit nuclear weapons on its territory in peace time; and adding a sentence into the letters of clearance given to governments when permission for naval visits is asked for, stating that it is assumed that the visit will comply with the rules laid down by the Danish Government. However, no mention of nuclear weapons is made in the letters of clearance.

In effect Denmark went back to its position of turning a blind eye to the presence of nuclear weapons on board visiting NATO ships.

Danish defence spending has not increased since the May 1988 election, even though the Danish defence ministry admitted that it would need to push up its present defence spending about three-fold to meet its alliance commitments. Denmark has only 17,000 soldiers, with 55,000 reserves and a Home Guard of 75,000. Conscription in Denmark takes the form of a lucky draw, where a young man of 18 can avoid doing military service if he pulls a 'free number'. This is because the quota of conscripts is so low that there used to be an over-supply of eligible young men. Suggestions from some Conservative members of parliament that the number of conscripts should be

KEY INDICATORS

	Unit	1984	1985	1986	1987	1988
Population	'000	5,111	5,116	5,125	5,129	5,130
Gross domestic product (GDP)	Kr bn	565.3	615.1	663.6	692.3	723.6
GDP per capita	Kr '000	110.6	120.2	129.5	135.0	141.1
Inflation	%	6.3	4.7	3.6	4.0	4.5
GDP real growth	%	4.4	4.2	3.1	–0.7	–0.4
Unemployment	%	10.1	9.2	8.1	8.0	8.7
Agriculture	1985=100	99	100	103	101	100
Industrial production	1985=100	96	100	104	101	102
Retail sales (value)	1980=100	146	154	164	165	169
Exports (value)	Kr bn	165.3	179.6	171.8	175.3	187.4
Exports (volume)	1980=100	119	123	126	128	135
Imports (value)	Kr bn	171.8	191.5	184.7	174.0	178.2
Imports (volume)	1980=100	110	118	126	125	123
Balance of trade	Kr bn	–6.5	–11.9	–12.9	1.3	9.2
Current account	$m	–1,637	–2,728	–4,490	–3,047	–1,828
Net external debt	$bn	29.9	41.9	47.0	*40.0	41.0
Exchange rate	Kr per $	10.4	10.6	8.1	6.8	6.7

* estimated figure

raised and the period of conscription of nine months should be lengthened, in light of demographic changes as the number of young men turning 18 declines, were firmly rejected by the socialist bloc.

Surprisingly, a US information agency survey claimed that as many as 70 per cent of Danes believed that NATO was essential for Denmark's defence and 60 per cent considered American military bases as 'necessary' in Europe. The Danes have not had an American base on Danish territory except for two airbases located on Greenland which were built during the second World War as part of the American air-bridge to Europe.

Revising the economy

There was one small piece of good news offered by the 1989 OECD report on Denmark – a projected industrial export of 6 per cent for 1989 and 1990 – in the midst of a generally gloomy picture with little improvement in the current balance of payments deficit, low productivity and rising unemployment. The report's positive industrial export figures are, however, based on assumptions that the high growth in the other OECD countries will continue, which should mean a growth of Danish export markets, and that Danish wage increases will be very moderate, giving a much needed boost to productivity. The expected increase in industrial exports will not, however, improve the balance of payments deficit significantly, because 1990 will see an increase in private consumption that will eat into any improvements made.

The government may console itself with the fact there was negative growth in private consumption in 1987 and 1988, and a zero per cent growth predicted for 1989 – a result of the 'potato diet', a tax reform introduced in 1987 to reduce private consumption by removing or lowering many taxable income deductions. As a result of the fall in domestic demand in 1987-88 the current balance of payments deficit dropped to 1.7 per cent of the GDP in 1988, the lowest level since 1975, but once demand recovers, as is expected in 1990, it will be difficult to keep the deficit from rising again.

Prime Minister Poul Schlüter had cause to feel moderately satisfied after the 1989 spring round of collective wage negotiations, because although he was unable to convince the trade unions to delay their demand for a cut in the working week from 38 to 37 hours, the negotiations did produce a lower wage increase than was feared. The wage increases agreed on for 1989-90 hovered at 4-4.5 per cent, taking the cut in working hours into account, which it is estimated will be a few per cent under the OECD European average. Productivity, however, stayed at a gloomy low, with Denmark near the bottom of the pile amongst the OECD countries. Over the period 1986-88, the Danes had a productivity of minus O.4 per cent against 1.8 per cent for the OECD as a whole. Another cause of worry in Mr Schlüter's bag of troubles is a rise in international interest rates. Net debt appeared to have stabilised at about 40 per cent of GDP in 1988. An economic survey in 1989 from the Danish Economic Ministry projects, however, that given the current Danish economic trends and average interest on debt staying at the 1988 level of about 10.5 per cent, in five years the debt will have risen to 42 per cent of GDP and the current account deficit from 1.7 per cent to 1.9 per cent of GDP.With a 1 per cent increase in the interest rate, the net debt after five years will have risen to 44 per cent of GDP and the current account deficit up to 2.5 per cent of GDP. The only consolation Mr Schlüter can offer himself is that unlike domestic wage increases, there is little he can do to change international interest rates. And given Denmark's record frequency for holding elections, nine in 17 years, he may not be around in five years to bear the burden.

A muddle of mergers

1989 ushered in a flood of mergers and acquisitions within Denmark which swept across all sectors; from manufacturing and production to financial services and the professions. The entire Danish business world suddenly appeared to embrace the ideology that big is beautiful, after decades of encouraging small companies.

Part of this change of heart was undoubtedly prompted by the approach of 1992 and the internal market. The danger of being swallowed up by their larger neighbours, West Germany, the UK and even Sweden removed any last reservations conservative Danish industry might have had about joining hands.

The fear that Danish industry was too small and lacked the necessary economies of scale to survive in the open EC market was not without good reason, as prior to the merger boom the 10 largest industrial companies in Denmark collectively sold less than Sweden's car company, Volvo. Some of the companies that merged found themselves in an awkward position, because although on the EC scale they were still only medium-sized players, in Denmark's small market they had reached a near monopoly situation that had to be referred to the European Commission.

The merger between the pharmaceutical companies Novo and Nordisk Gentofte to form one of the largest insulin producers in the world – together they hold about 65 per cent of the Western European market – was referred to the Commission but was cleared. The acquisition by the charter tour company, Spies Rejser, of its chief rival Tjaereborg – together they dominate the Danish travel market with an 85 per cent market share – was pending in mid-1989, but looked likely to get the Commission's green light.

In the financial services sector, the top three Danish insurance companies, Baltica, Hafnia and Topsikring, pushed forward an aggressive diversification. They acquired stock broking firms, real estate agents, ambulance services and furniture manufacturers in preparation, they claimed, for the liberalisation of the

COUNTRY PROFILE

Political system: Constitutional monarchy, with legislative power jointly vested in the monarch (who has no personal authority) and the unicameral Folketing (179 members elected under a system of proportional representation). Executive power exercised by the monarch through a cabinet led by the prime minister, who is responsible to the Folketing.

Main cities/towns: Copenhagen (population 1.4m), Aarhus, Odense, Esbjerg, Aalborg.

Media
Press: Well-developed press with one of the highest per capita newspaper circulations in the world. Most publications privately owned and tend to have fairly strong political leanings.
Dailies: There are 47 dailies with total circulation of about 1.8m. Most popular include *Aktuelt, Ekstrabladet* (circulation 229,509), *BT* (202,750), *Politiken* (152,215), *Jyllands-Posten, Morgenavisen* (128,601), and *Berlingske Tidende* (128,815). *Business:* There are two business dailies: *Børsen* (43,041) and *Erhvervs-Bladet* (104,386). Most popular business magazines include *Børsens Nyhedsmagasin* (monthly) and *Penge & Privatøkonomi* (monthly). *Weeklies:* Most popular include *Familie Journalen* (360,500), Billed-Bladet, Hjemmet (266,835) and *Alt for Damerne* (118,754).
Broadcasting: Approximately 2m TV licences (which include radio) in use. Programmes broadcast by Radio Denmark. Country's first national advertising-financed television station, TV2, introduced Oct 1988. West German and Swedish programmes can also be received in south of country. Over 1m households are cabled. Non-commercial teletext and commercial videotext services are available. Radio Denmark's Programme 3 broadcasts news in English at 0810 on weekdays, and in German (at 0815) and French (at 0820) only May to Aug.

Domestic economy: Balanced economy with well-developed, export-based manufacturing sector. Also important agricultural sector. Upturn in economic performance from 1982 owing to deflationary policies combined with expansion of exports. This was followed by a period of severe balance of payments problems.

Employment: Around three per cent of labour force employed in agriculture, 28 per cent in industry and nearly 70 per cent in services. Unemployment reduced slightly due to upswing in economic activity and reduction in working week but increased in 1987.

External trade: Regular current account deficit, despite steady export growth. External position weakened by heavy debt burden. Principal export commodities include meat and preparations, dairy produce, fish, chemicals, basic manufactures, machinery and ships. Major imports are intermediate goods, fuels and lubricants and consumer goods. Around 45 per cent of trade turnover is with other EC countries, primarily West Germany, UK, France and Netherlands. Other significant trading partners are Sweden, Norway and USA.

Agriculture: Contributes five per cent to GDP and 21 per cent of exports. Organised into local cooperatives which are united in national federations. Intensive farming concentrated on livestock production, mainly pigmeat, beef, veal, poultry and dairy produce. Also world's third largest fish exporter. In 1987 new legislation was introduced to tackle the urgent problem of nitrate pollution, caused largely by animal slurry. It is now an offence to have non-green fields in winter.

Industry: Highly developed and diversified sector, which is almost wholly under private ownership. Small manufacturing firms predominate. Production, which has increased rapidly in recent years, is heavily based on imported raw materials, components and semi-manufactures. Main industries include food processing, chemicals, metal production, engineering, shipbuilding, paper, textiles, electronics and furniture.

Mining: Few indigenous mineral resources except for salt, sulphur, chalk and clay.

Energy: Considerable reserves of oil and gas which currently meet over 20 per cent of domestic energy requirements. Emphasis on energy conservation and conversion of power stations from imported coal to locally produced gas.

Main companies: Novo Industri, United Breweries, Den Danske Bank and Copenhagen Handelsbank.

Banking: Commercial banking network comprises over 70 regional and local banks, largest of which are Copenhagen Handelsbank and Den Danske Bank. Also over 130 savings banks.

Legal system: Civil law system. Constitution adopted 1953.

Membership of international organisations: ADB, CCC, Council of Europe, DAC, EC, ELDO (Observer), EMS, ESRO, FAO, GATT, IAEA, IBRD, ICAC, ICAO, ICES, ICO, IDA, IDB, IEA, IFAD, IFC, IHO, ILO, International Lead and Zinc Study Group, IMF, IMO, INTELSAT, INTERPOL, IPU, ISO, ITC, ITU, IWC, NATO, Nordic Council, OECD, UN, UNESCO, UPU, WHO, WIPO, WMO, WSG.

financial services sector that is already on its way because of 1992. This movement infected the mortgage credit institutions, Kreditforeningen, Nykredit and BRF, which went all out to acquire real estate dealers and shares in mortgage institutions in other EC countries, primarily in the UK, to secure a steady supply of clients.

The savings banks, which could not raise external equity capital until a law in 1988 was passed to allow them to convert to limited liability companies, are the next most likely candidates to break out of the straitjacket imposed by Danish law, which does not actually allow such a flow in financial services. The Danish central bank has tolerated the changes that have taken place up until 1989, but it is unclear if it will continue to do so.

Environment, taxes and 1992

The two major sore points between Denmark and the EC were, and will probably continue to be, the environment and indirect taxation. Like Holland, Denmark does not believe that the EC goes far enough on environmental issues.

They clashed in 1988 over whether Denmark could ban canned beer and soft drinks and keep its returnable bottle scheme for beverages on the grounds that returnable bottles were more 'environmentally friendly'. Foreign brewers argued that the obligatory returnable bottle scheme prevented them from selling beer in Denmark and protected the local breweries. Surprisingly, the European Court of Justice gave priority to Denmark's environmental argument over the argument that the free trade of beverages was being inhibited.

Denmark scored another green point in 1989 when they passed a law forcing cars sold in Denmark from 1991 to be fitted with American-standard catalytic convertors instead of the less clean European-standard ones decided on by the Commission. For a little while it looked as if the Danes would land up in the Court of Justice again, but the Commission eventually bowed to growing pressure from the European Parliament, and a change to the American-standard catalytic convertors was made.

While Denmark's clashes with the EC over environmental matters will die down as the EC gets greener, the more substantial hurdle of Denmark being grossly out of line with the rest of the EC over indirect taxation will not. If anything, the situation appears to be worsening.

Denmark is fighting very hard to keep its across-the-board 22 per cent VAT (MOMS in Danish) on all goods and services as well as high purchase taxes on many luxury goods and white goods. But this is very difficult given EC pressure to harmonise towards a two-tier system at lower levels. The situation is aggravated by Denmark's sharing a border with the Federal Republic of Germany, which has a two-rate VAT scale of 7 and 14 per cent. Even before border controls were liberalised, border shopping by Danish families picking up cheap goods in West Germany was very popular. Ever since Denmark gave in to EC pressure to raise the non-taxed allowance on goods brought into Denmark, video machines and microwave ovens have been packed into the boot. The Danish state is already losing millions in lost tax revenues – an estimated loss of DKr770m a year – because of cheaper petrol bought in West Germany alone.

BUSINESS GUIDE

Time: GMT + 1 (GMT + 2 from late Mar to late Sep).

Climate: Temperate; temperatures range from about 22 °C to about -7°C. Some rain throughout the year.
Clothing: Medium weight throughout the year, warm clothing for winter months.

Entry requirements: Visas required for visitors from eastern Europe, Arab countries, and South Africa; most others need only passports.
Health: None.
Currency: No restrictions on amount of foreign currency taken in or out of country, or on amount of Danish currency taken in.
Customs: Personal effects duty-free, plus duty-free allowance.

Air access: Regular flights by most major international airlines.
National airline: Scandinavian Airline System (SAS) – jointly owned with Sweden and Norway.
Main international airport: Copenhagen (CPH) at Kastrup, 10 km from capital.
International airports: Aalborg (AAL), Aarhus (Tirstrup) (AAR), Billund (BLL), Esbjerg (EBJ).

Surface access: Frequent ferry services from UK, Germany, Poland, Sweden and Norway. Road and rail access from several European countries via Germany.
Main ports: Copenhagen, Aarhus, Hirtshals, Frederikshavn and Esbjerg.

Hotels: No official rating system. Luxury single bedroom with breakfast costs between Kr1,070-2,000 per night in Copenhagen, elsewhere between Kr580-880. Bills include 15 per cent service charge. All usual credit cards accepted. Advisable to book accommodation in Copenhagen in advance especially during summer.

Restaurants: All bills include 15 per cent service charge and 22 per cent VAT.

Currency: Danish krone floats against all foreign currencies, but the exchange rate with certain other West European currencies is stabilised within the European Monetary System's grid.

Foreign exchange information
Trade policy: Generally no limitations on imports.
Car hire: Available throughout the country at main Danish State Railway (DSB) stations and all airports. Can be booked in advance through stations, international car hire firms and travel agents. Valid driving licence required, which must be carried when driving. Most firms stipulate a minimum age of between 20-25. Speed limits 50 kph in built-up areas, 80 kph out of town and 100kph on motorways. Even for minor speed limit offences, drivers are liable to pay heavy fines on the spot. If payment cannot be made, the car may be detained.

City transport
Taxis: Good service in all major towns. Can be hailed in the street, by telephone or at ranks, and display green 'Fri' sign when available. Costs about Kr12 plus Kr8 to Kr10 per kilometre, and fare includes a tip.
Buses: Frequent efficient services in main cities. Flat-rate fares usual e.g. in Copenhagen Kr7 for a ticket which can be used interchangeably on any bus within one hour of purchase.

National transport
Air: Daily flights operate between Copenhagen and Odense, Billund, Esbjerg, Karup, Skrydstrup, Stauning, Sonderborg, Thisted, Aalborg and Aarhus.
Road: About 70,000 km of roads including 593 km of motorway. Bus routes mainly supplement rail network.
Buses: Good bus service in Copenhagen including night buses until 0230. Country bus network operates where there are no railways.
Rail: Approximately 2,500 km of railways operated by DSB and a few private companies, providing a very efficient service linked to the ferry services.
Water: Frequent ferry services connecting islands of Zealand, Funen and Lolland and Jutland peninsula, operated by DSB. Also over 400 km of inland waterways.

Public holidays
Fixed dates: 1 Jan, 5 Jun, 24-26 Dec, 31 Dec.
Variable dates: Maundy Thursday, Good Friday, Easter Monday, Great Prayer Day, Ascension Day, Whit Monday.

Working hours
Business: (Mon-Fri) 0800 – 1600 or 0830 – 1630.
Government: (Mon-Fri) generally 0900 – 1700.
Banking: (Mon-Fri) 0930 – 1600 (Thu 1800).
Shops: (Mon-Fri) 0800 – 1700 or 0900 – 1730, (Sat) close at 1300 or 1400.

Telecommunications
Telephone: Efficient automatic service. International dialling code: 45. There are no Area Codes.
Telex: Available 24 hours a day at the Central Telegraph Office in Copenhagen. Can also be dictated over the phone on 'Fonotelex' by dialling 0022. Country code 55DK.
Telegram: Can be sent from any post office or by telephone.

Postal service: Post offices open 0900 – 1700 Mon-Fri plus Sat mornings. Post office at Copenhagen Central Station open late Mon-Fri, Sat 0900 – 1600, and from 1000 – 1700 on Sun.

Advertising: Most widely used media are the press and cinemas. Direct mail also popular. There is no advertising on TV or radio, and poster sites are heavily regulated. Total advertising spent (1984) Kr1939m primarily on cars and finance.

Trade fairs: Several specialised international fairs covering most products held throughout the year. Most, such as Scandinavian Fashion Fair, Scandinavian Furniture, the Fisheries Fair and the Chem Tech International Fair, are held at the Bella Centre in Copenhagen. Other major venues are at Herning and Aalborg.

Electricity supply: 220/380 V AC.

Representation in capital: Argentina, Australia, Austria, Belgium, Brazil, Bulgaria, Burkina, Canada, Chile, China PR, Colombia, Costa Rica, Cuba, Czechoslovakia, Egypt, Finland, France, German Democratic Republic, Federal Republic of Germany, Ghana, Greece, Holy See, Hungary, Iceland, India, Indonesia, Iran, Ireland, Israel, Italy, Ivory Coast, Japan, Democratic People's Republic of Korea, Republic of Korea, Lesotho, Libya, Mexico, Morocco, Netherlands, Norway, Pakistan, Poland, Portugal, Romania, Saudi Arabia, Spain, Sweden, Switzerland, Thailand, Turkey, Uganda, Uruguay, USA, USSR, United Kingdom, Venezuela, Yugoslavia.

Hotels: Copenhagen
Bel Air, Loejtegaardsvej 99, DK-2770 (tel: 513033; tx: 31240).

Copenhagen Admiral, Toldbodgade 24, DK-1253 (tel: 118282; tx: 15941; fax: 323607/325542).

Grand, Vesterbrogade 9, DK-1620 (tel: 313600; tx: 15343; fax: 313350).

Hotel d'Angleterre, Kongens Nytorv 34, DK-1050 (tel: 120095; tx: 15877).

Kong Frederick, V Voldgade 25-27, DK-1552 (tel: 125902; tx: 19702; fax: 935901).

Park, 3 Jarmers Plads, DK-1551 (tel: 133000; tx: 15692; fax: 143033).

Plaza Sheraton, Bernstorffsgade 4, DK-1577 (tel: 149262; tx: 15330; fax: 939362).

SAS Globetrotter, Engvej 171, DK-2300 (tel: 551433; tx: 31222; fax: 451558145).

SAS Royal, Hammerichsgade 1, DK-1611 (tel: 141412; tx: 27155; fax: 141421).

Sara Dan, Kastruplundgade 15, DK-2770 (tel: 511400; tx: 31111).

Savoy, Vesterbrogade 34, DK-1620 (tel: 314073; tx: 27215).

Scandinavia, Amager Boulevard 70, DK-2300 (tel: 112324; tx: 31330; fax: 570193).

Sheraton, 6 Vester Søgade, DK-1601 (tel: 143535; tx: 27450; fax: 321223).

Car hire: Copenhagen
Avis: Sjaeliandsbroen 1, 2450 (tel: 31-79-79, 31-50-20; tx: 19221).

Budget: Nyropsgade 6 (tel: 13-39-00; tx: 27414).

Europcar: Gammel Kongevej 70 (tel: 24-66-77; tx: 16027).

Hertz: Ved Vesterport 3 (tel: 12-77-00; tx: 21022).

Inter-Rent: Jernbanegade 6 (tel: 11-62-00; tx: 27440).

Business advice and information
Grosserer Societetet, Børsen (royal exchange), DK-1217 Copenhagen (tel: 91-23-23; tx: 19520).

Banking
Danmarks Nationalbank (central bank), Havnegade 5, DK-1093 Copenhagen (tel: 633-14-14-11; tx: 27051).

Copenhagen Handelsbank AS (commercial bank), Holmens Kanal 2, DK-1091 Copenhagen (tel: 12-86-00; tx: 19177).

Den Danske Bank of 1871 AS (commercial bank), Holmens Kanal 12, DK-1092 Copenhagen (tel: 15-65-00; tx: 27000).

Den Danske Bankforening (bankers' association) Bankernes Hus, Amaliegade 7, DK-1256 Copenhagen (tel: 12-02-00; tx: 16102).

Jyske Bank (Bank of Jutland), Vestergade 8-16, 8600 Silkeborg (tel: (06)82-11-22; tx: 62231).

Privatbanken A/S, Torvegade 2, PO Box 1000, 2300 Copenhagen S (tel: 11-11-11; tx:65872).

Provinsbanken A/S, Kannikegade 4-6, 8100 Arhus C (tel: (06)12-25-22; tx: 64403).

Sparkassen SDS, 8 Kongens Nyttorv, 1050 Copenhagen K (tel: 13-13-39; tx: 15745).

Principal newspapers
Alt for Damerne, Vognmagergade 11, 1148 Copenhagen K (tel: 151925; tx: 16705).

Berlingaske Tidende, Pilestraede 34, 1147 Copenhagen K (tel: 157575; tx: 27094).

Billed Bladet, Vesterbrogade 16, 1506 Copenhagen K (tel: 231611; tx: 27094).

Børsen, Møntergade 19, 1116 Copenhagen K (tel: 320102; tx: 22903).

B.T., Kr Bernikowsgade 6, 1147 Copenhagen K (tel: 141234).

Ekstra Bladet, Rådhuspladsen 37, 1585 Copenhagen V (tel: 118511; tx: 16885).

Erhvervs-Bladet, Vesterbrogade 12, 1620 Copenhagen V (tel; 524001; tx: 19785).

Familie Journalen, Vigerslev Allé 18, 2500 Valby, Copenhagen (303333).

Jyllands Posten, Morgenanvisen, Grøndalsvej 8260 Viby J (tel: 146677; tx: 68747).

Politiken, Politikens Hus, Rødhuspladen 37, 1585 Copenhagen V (tel: 118511; tx: 16885).

Other useful addresses
The Central Telegraph Office, Købmagergade 37, DK-1150 Copenhagen K (tel: 120903).

Copenhagen Airtaxi, Copenhagen Airport Roskilde, DK-4000 Roskilde (tel: 391114; tx: 43234).

Copenhagen Stock Exchange, Nikalaj Plads 6, DK-1067 Copenhagen K (tel: 933366; tx: 128613).

Dansk Arbejdsgiverforening (employers' confederation), Vester Voldgade 113, DK-1503 Copenhagen V (tel: 93400; tx: 16464).

Danmarks Agentforening (association of commercial agents of Denmark), Børsen DK-1217 Copenhagen K (tel: 144941).

Danmarks Statistik, Sejrøgade 11, DK-2100 Copenhagen.

Danmarks Turistråd (tourist board), HC Andersen's Boulevard 22, DK-1553 Copenhagen V (tel: 111325; tx: 27586).

Det Økonomiske Råd (economic council), Kampmannsgade, DK-1604 Copenhagen V (tel: 135128).

Forened Danske Motorejere (FDM) (the Danish motoring organisation) Blegdamsvej 124, DK-2100 Copenhagen \17 (tel: 382112).

Industriraadet (federation of Danish industries), HC Andersen's Boulevard 18, DK-1533 Copenhagen V (tel: 152233; tx: 9112217).

Ministry of Economic Affairs, Slotsholmsgade 12, 1216 Copenhagen K (tel: 923222; tx: 16833).

Ministry of Finance, Christiansborg Slotsplads 1, 1218 Copenhagen K (tel: 923333; tx: 16140).

Ritzaus Bureau 1/S (news agency), Mikkel Bryggersgade 3, DK-1460 Copenhagen K.

Teknisk Forlag AS (technical press) (publishing house) Skelbaekgade 4, DK-1717 Copenhagen V.

Thomson Communications (Scandinavia) AS, Hestemøllestrede 6, Postboks 2181, DK-1017 Copenhagen K.

Djibouti

Gillian Cronje

In June 1989 President Hassan Gouled Aptidon paid his first official visit to France to reaffirm the strong economic and security links between the two countries. There were rumours that cost-cutting in France's military budget would lead to a reduction in the 4,000-strong French naval base in Djibouti. This base forms the mainstay of Djibouti's economy, and Gouled was relieved to hear that it would be maintained at its present strength. He was also assured that French aid would continue. This totalled FFr37.5m in 1989 as part of a 10-year pact which was about to expire. Gouled asked that the agreement be renewed for at least another five years.

Djibouti likes to regard itself as a 'haven of peace' surrounded by turbulent countries. This peace was shattered in March 1989, when the rebellion in Somalia spilled across the border and caused tribal clashes. The Issas are the ethnic majority in Djibouti; in Somalia they form the bulk of the rebellious Somali National Movement. The border with Somalia was officially closed two months later, but the influx of Somali refugees continued. Gouled threatened more stringent measures and said the illegal immigration had forced the unemployed to engage in 'undesirable activities', including political agitation.

At the same time Djibouti suffered a major disaster: the town, which has an average annual rainfall of 140mm, was flooded by 507mm of rain in less than one week. There were dozens of deaths and 200,000 people were made homeless. The magnitude of the misfortune for the country was increased by the fact that three-quarters of its population lives in the capital itself. 70 per cent of the city was seriously damaged, and it was the poorest quarter – Balbala, with its dirt roads and crumbling walls – which suffered most. Donations, in money and in kind, arrived in answer to the government's international appeal. France contributed FFr10m and Japan $100,000 to help rebuild Balbala, but the floods showed that more than repair work was required. The need to rebuild practically the entire city has long been recognised. In fact, an urban rehabilitation plan was launched in 1985; but its implementation was slow because of the expense involved. The first phase alone was costed at FFr120m.

The housing situation was serious even before the floods. In January 1989 there were clashes between the police and Balbala inhabitants which left at least four people dead and over a hundred wounded. The confrontation happened during protests at the forcible 'cleaning-up' of the shanty town through the destruction of offending houses. The 1989 rains might help to galvanise Djibouti's aid donors into providing the necessary funds for urban reconstruction.

KEY FACTS

Official title: Republic of Djibouti

Head of state: President Hassan Gouled Aptidon

Ruling party: Rassemblement Populaire pour le Progrès (RPP)

Capital: Djibouti-ville

Official Languages: French, Arabic (Somali and Saho-Afar widely spoken)

Currency: Djibouti Franc (DFr) = 100 centimes

Exchange rate: DFr175 per $ (Nov 1989)

Area: 23,200 sq km

Population: 500,000* (1988)

GDP per capita: $430* (1987)

Trade balance: –$182m (1987)

Foreign debt: $187m (1988)

* estimated figure

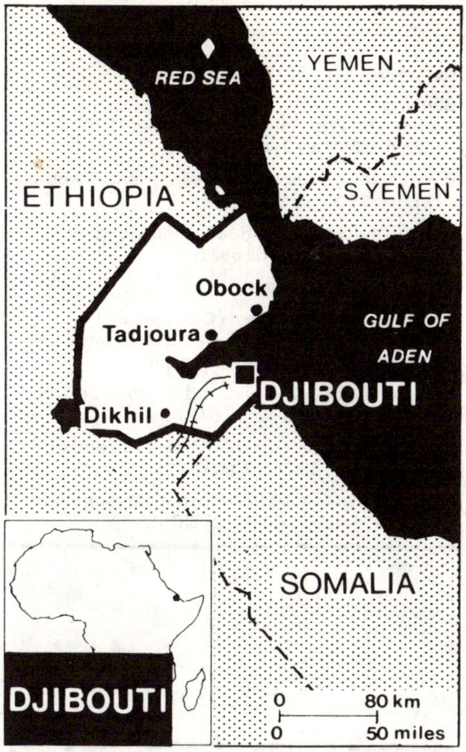

Development schemes

Meanwhile there is hope of improved economic fortunes in two important development schemes involving the energy and transport sectors. Djibouti relies on imported fuel to meet almost 90 per cent of its energy needs, but has one of Africa's most extensive geothermal resources near Lake Assal in the east-central part of the country. In July 1989 an IDA credit of $9.2m was announced in support of a $38m project to develop the first section of the Assal Geothermal Steam Field. The project requires the drilling of four wells to help drive generating units, the construction of a steam-gathering system and transmission line, and the installation of two generators. Djibouti's electricity company will receive management training to gain expertise in running the geothermal development and production. Eventually Djibouti may become an exporter of electric power to its neighbours.

Increasing traffic in Djibouti port, which also serves Ethiopian needs, has led to major improvement work. Phase I of the project, which should be completed in mid-1990, involves the rehabilitation of some of the berths. Phase II provides for the supply of additional container handling equipment, fork-lifts, cranes, trucks and tractors. Land reclamation is being carried out for the expanded facilities. The container yard will be able to offer repair services. When the work is completed, Djibouti can expect a significant growth in its transshipment business. The modernised harbour will also support the free trade zone which is expected to encourage processing.

Phase II is being financed by a $9.4m Kuwaiti loan. Djibouti is on excellent terms with its Arab neighbours and active in the Arab League; but it is also a conscientious member of the Organisation of African Unity. In 1989 it became the first member state (after Nigeria) to act in accordance with a resolution designed to prevent the dumping of toxic substances in Africa: possession of nuclear materials or waste is now punishable with life imprisonment in Djibouti.

COUNTRY PROFILE

Political system: Republic with executive power vested in Council of Ministers, headed by a prime minister, and responsible to the directly-elected president. Government is formed from 65-member Chamber of Deputies (elected for five years by universal suffrage). Gained independence in 1977. Formerly called French Territory of Afars and Issas (TFAI).

Political parties: Rassemblement Populaire pour le Progrès is the only legal political party.

Population: The 1985 population estimate of 300-350,000 has been greatly swelled by the numbers of refugees from the Ogaden and Eritrean wars in Ethiopia and Somalia, and the 1988 total was estimated at anything up to 500,000. Of the Djibouti nationals, two-thirds live in the capital, Djibouti. About half of the national total are Issas, of Somali origin, and most of the remainder are Afar; there are about five per cent Arab and 10 per cent European residents; except for the latter, Islam is universal.

Main cities/towns: Djibouti-ville (population 200,000).

Language: French and Arabic are the main business languages, although English is understood by the larger trading houses.

Media
Press: Principal newspaper is *La Nation de Djibouti* (weekly).
Radio: One state-owned television station, two AM and one FM radio stations. Radio service in French, Afar, Somali and Arabic (21 hrs/day).
Television: TV service five hrs/day.

External trade: Large visible trade deficit ($236m in 1986) partly offset by net income from services and transport, aid flows and concessionary loans.
Exports: Skins and leather, cereals, re-exports. Main destinations: France, Italy, Somalia, Yemen PDR.
Imports: Machinery and transport equipment (typically 12-14% of total value), food, beverages and tobacco (12-14%), basic manufactures (6%), textiles. Main sources: France, Japan, UK, Ethiopia, Bahrain.

Agriculture: Under-developed sector due to poor terrain (mostly desert). Around 95 per cent of food requirements are imported. Over 50 per cent of population are nomads engaged in herding of goats, sheep and camels. Projects under consideration include increasing the amount of arable land by irrigation schemes and development of fishing industry to take 1,200 tonnes a day by 1990.

Industry: Sector accounts for around 15 per cent of GDP and is limited to construction and small-scale concerns such as mineral water bottling, tanning, dairy and animal food plants. Contracts have been awarded for the setting up of some new industries (cement, tiles, paints, meat processing). Foreign investment is being encouraged, although affected by political uncertainties in the region. The port is Djibouti's main economic activity. It has a capacity of 40,000 tonnes a year.

Mining: Surveys have indicated the presence of a few minerals such as copper, gypsum and sulphur, but as yet no minerals are mined commercially.

Energy: Recent discoveries of gas reserves are hoped to make the country self-sufficient in gas and provide a surplus for export. Current geothermal survey has resulted in the construction of a power station.

Membership of international organisations: African Development Bank, African States Associated with EC (Lomé Convention), Arab Fund for Economic and Social Development, Arab League, FAO, G-77, IGGAD, ILO, IMF, Islamic Conference, OAU, UN, WHO.

Time: GMT +3.

Climate: Very hot and arid from Apr to Aug; average temperature 32°C and can reach 45°C. Slightly cooler from Oct to Mar, with occasional light rain.

Entry requirements
Passport: Required by all.
Visa: Required by all except nationals of France on short stay visits. 10-day visas are available on arrival for nationals of EC countries and USA. An onward ticket is also required.
Prohibited entry: Nationals of Israel, South Africa.
Currency: No restrictions on import/export of local or foreign currency.

Health precautions
Mandatory: Yellow fever vaccination certificate if arriving from an infected area.
Advisable: Cholera, typhoid and polio vaccinations. Malaria prophylaxis recommended as risk exists throughout the country. Water precautions should be taken.

Representation overseas: Addis Ababa, Baghdad, Brussels, Cairo, Jeddah, Mogadishu, Nairobi, New York, North Yemen, Paris, Riyadh, Sana'a, Tokyo, Tunis.

Air access
National airline: Air Djibouti.
Other airlines: Aeroflot, Air France, Air Madagascar, Air Tanzania, Alyemda, Ethiopian, Somali, Yemenia.
Tax: Airport tax of DFr2,000 payable on departure from Djibouti.
International airport: Djibouti-d'Ambouli (Code: JIB), 6 km from city. Freight facilities have been upgraded.

Surface access
Main ports: Djibouti (12 berths and a container terminal).
Overland access: Due to political situation, entry from Somalia may be impossible. Surfaced road from Addis Ababa (Ethiopia) which can be difficult, with problems caused by political situation. Rail link with Ethiopia.

Hotels: Available in Djibouti-limited elsewhere. Service charge is normally included. Tipping is not usual.

Electricity supply: 220/380 V AC, 50 cycles.

Credit cards: Generally not accepted, except by airlines and Sheraton Hotel.

Car hire: Available in Djibouti and at the airport. International driving licence recommended. A temporary licence can be obtained on presentation of national licence.

City transport
Taxis: Available in main towns. Tipping is not usual; official tariff but usual for visitors to be charged 50 per cent more; similar increase at night.

National transport
Air: Air Djibouti operates a daily domestic service to Obock and Tadjoura from Djibouti. Dikhil and Ali-Sabieh can be reached by chartered aircraft.
Road: Surfaced roads to Ethiopian border and to Arta and from Djibouti to Tadjoura; most roads are in need of repair. Local advice should be taken as to when to travel by road.
Rail: Certain towns are served on the Djibouti-Addis Ababa track. Passenger transport reduced due to political situation.
Water: Ferry service linking Djibouti with Tadjoura and Obock .

Public holidays
Fixed dates: 1 Jan, 1 May, 27 Jun (Independence), 25 Dec.

Variable dates: Lailat al-Miraj, Eid al-Fitr, Eid al-Adha, First of Muharram, Ashura, Prophet's Birthday.

Working hours
Business: (Sun-Thu) 0730 – 1200, 1530 – 1800; (Sat) 0730 – 1200.
Government: (Sat-Wed) 0700 – 1200, 1530 – 1730; (Thu) 0700 – 1200.
Banking: (Sun-Thu) 0715 – 1145.
Shops: (Sat-Thu) 0800 – 1200, 1600 – 1930.

Telecommunications
Telephone: Dialling code for Djibouti: IDD access code + 253 followed by subscriber's number. There are no area codes. 100 per cent automatic service; outside town of Djibouti there are very few telephones. Radio telephone network with North Yemen and Saudi Arabia. An international radio telephone service connects Djibouti with Europe.
Telex: Available at main post office (Boulevard de la République).

Banking
Central bank: National Bank of Djibouti. *Others:* Bank of Credit & Commerce International, Banque de Djibouti et du Moyen Orient, Banque Indosuez-Mer Rouge, Banque pour le Commerce et l'Industrie, Caisse Centrale de Co-opération Économique, Commercial and Savings Bank of Somalia, Commercial Bank of Ethiopia, Caisse de Développement de Djibouti.

Representation in capital: China PR, Egypt, Ethiopia, France, Iraq, Libya, Oman, Saudi Arabia, Somalia, USSR, USA, Yemen AR, Yemen PDR.

Hotels: Djibouti
Ali-Sahieh, Ave Clémenceau Menelik, Place du 27 Juin (tel: 35-24-25/28-78/30-59).

Menelik, P.O. Box: 1153 (tel: 35-11-77; tx: 5937).

Plein Ciel, Blvd Bonhoure, P.O. Box: 1869 (tel: 35-38-41; tx: 5941).

Sheraton, Plateau du Serpent, P.O. Box: 1924 (tel: 35-04-05; tx: 5912; fax: 355892).

Banking
National Bank of Djibouti (central bank), PO Box 2118 (tel: 35-27-51/2; tx: DJ 5838).

Banque de Djibouti et du Moyen Orient, PO Box 2471 (tel: 35-11-33; tx: 5943).

Banque Indosuez (Mer Rouge), Place Lagarde, PO Box 88 (tel: 35-30-16; tx: 5829).

Other useful addresses
Chambre Internationale de Commerce et d'Industrie de Djibouti, BP 84, Place Lagarde (tel: 35-10-70, 35-08-26, 35-06-73; tx: 5967).

Ministry of Foreign Affairs and Co-operation, PO Box 1863 (tel: 35-33-42; tx: 5720).

Ministry of Trade, Transportation and Tourism, PO Box 24 (tel: 35-25-40; tx: 5871).

Ministry of Industry and Industrial Development, PO Box 175 (tel: 35-03-40; tx: 5871).

Dominica

Tony Thorndike

Dominica's tenth year of independence was celebrated with higher than expected economic growth. Prudent management of the economy by the Freedom government, and generous aid, was largely responsible. However, the crucial banana industry on the jungle-clad island faces problems with the impending single European market.

Although still one of the poorest of the Eastern Caribbean states, Dominica has during the 1980s seen an economic transformation. The Dominica Freedom Party government and, above all, its redoubtable leader Eugenia Charles, is entitled to the credit, having restored Dominica's overseas image as well as its economic prospects.

Steady growth

When the Freedom government came to power in 1980 the island had not only been devastated by a hurricane (and was to suffer another soon after the election), but also by mismanagement. Since then substantial overseas aid has transformed much of the infrastructure, especially the once-appalling road network, inflation has been slashed from 30 to under 2 per cent and unemployment cut from around 20 to below 10 per cent; indeed there were serious labour shortages in the agriculture and construction sectors by 1989. A 4.5 per cent GDP growth was recorded in 1987 and a lower rate was forecast for 1988. But in the event recorded growth reached 5.6 per cent. It was not a surprise as exports were up by over 35 per cent, construction output by 31.2 per cent and tourist arrivals by 18.1 per cent in the first half of the year alone. The construction boom was eloquent testimony of local confidence and the general appeal of the island. It was a fitting conclusion to the three-year Structural Adjustment Plan which commenced in 1986. All the targets agreed with the main donors – the International Monetary Fund, the United States Agency for International Development (AID) and the Caribbean Development Bank (CDB) – were met.

Dominica's credit rating is high and the island is viewed as a model of multinational donor orthodoxy. But there was one concern: external public debt rose to nearly EC$200m (US$74m) at the end of 1988. This has prompted the government to attempt to increase the proportion of grant in aid packages. In fact grant financing has been relatively high, one element being the high esteem in which Prime Minister Charles is held in Washington after her central role in arranging the United States' intervention in Grenada in 1983.

Tight economic management was also in evidence in the 1989-90 budget. Considerable effort was made in 1988-89 to control expenditure and there were virtually no overruns. Although the emphasis since 1982 has been to shift from direct to indirect taxation, the 20 per cent consumption tax was removed from about a dozen basic commodities. Once again a surplus was planned, with expenditure at EC$108.9m (US$40.3m) and revenue at EC$131.5m (US$39.2m). The budget surplus for 1988-89 was EC$21m (US$7.8m) after expenditure of EC$85.7m (US$31.7m). Miss Charles announced that the surplus would go towards the capital programme, planned to be EC$108m (US$40m), up from EC$76.6m from the previous year. The most important public sector capital project is hydro-electricity. If all goes to plan, Dominica will be self-sufficient in electricity by 1990-91.

Future problems

Although construction activity provided the greatest spurt to economic growth in 1988, the good performance of the banana industry, the mainstay of Dominica's exports, was a contributory factor. 1988 production was 72,800 tonnes (1987: 63,874 tonnes), of which 70,400 tonnes were exported, not far short of the 74,000 tonnes target. Gross earnings rose from EC$86m to EC$99.5m. Quality control is far more effective than before with the introduction in 1986-87 of field-packing stations.

But there is considerable uncertainty, for virtually all banana exports go to Britain under the preferential access and price arrangements of the Lomé 111 Convention of the European Community. The convention governs commodity exports from the 64 Africa-Caribbean-Pacific (ACP) countries to the Community, the third version of which established national markets for certain products at protected, i.e. high, prices. Under the Banana Protocol, Windward Island (including Dominica) bananas enjoy such a facility with Britain, but it means that West Germany and the Benelux countries, with neither quotas, tariffs nor special access agreements, are able to purchase much cheaper 'dollar' bananas from Central and South America. The high banana retail price paid by British consumers would not be tolerated over much of the Community and, in any event, the arrangements could not survive the introduction of the unified market in 1992. Given the low productivity of Commonwealth Caribbean banana producers, due to topography and peasant-based land holdings (which is especially the case in Dominica), the prospect of lower prices is very worrying.

Negotiations between the ACP countries and the Commission for a new (Lomé IV) convention – to take effect on 1 January 1990 – commenced in October 1988. So far as bananas are concerned, the most likely outcome appears to be a 20 per cent tariff on 'dollar' bananas and a minimum import price for ACP produce commensurate with Community free market philosophy. Needless to say, Prime Minister Charles has assumed a high profile role to ensure the highest price to protect farmers who have no other source of cash crop income.

The manufacturing sector also experiences problems, in that case the sluggish regional CARICOM market. Despite incentives, investment is unsatisfactory. However, coconut-based oil and soap production increased in 1988 due to intensified marketing. Other products include coffee, cartons, garments and footware. A new box plant to service the banana industry built under the Puerto Rican twin-plant '936' programme opened in 1989 despite representations from the regional boxing organisation, Winera. Miss Charles rejected the argument on the grounds that Winera never invested in Dominica as promised. Dominica qualified for '936' funding by signing a Tax Information Exchange Agreement with the United States which is aimed at apprehending tax evading US nationals and 'laundering' of illegal funds overseas. Citrus production, used for concentrate for the US market, remains static at about 11,000 tonnes per year. Forestry remains relatively unexploited despite large wood reserves, but both the European Community and the United Nations have funded development schemes. A more unusual export is water, to parched Aruba.

Tourism will never be a major industry due to the volcanic base of Dominica and the consequent lack of white sand beaches. Consequently Dominica appeals mainly to nature lovers. Encouragingly, stopover arrivals in 1988 totalled 32,445, a 16 per cent rise over 1987. One reason was the heavily-promoted Reunion 88 programme aimed at enticing the many *emigrés* back to celebrate a decade of independence. But increased hotel capacity was another factor. Cruise ship passengers visiting Dominica sharply declined from 12,080 in 1987 to 4,294.

Conviction politics

The next election is due in 1990 and Miss Charles' aim is for a third victory. Like Mrs Thatcher in Britain, she can best be described as a 'conviction politician'. She founded the Freedom party in the early 1970s as a reaction to the increasingly oppressive legislation and downright poor government of the long-standing Dominica Labour Party (DLP). The 1980 election was the turning point when the DLP was defeated. This followed the resignation by public demand of DLP Prime Minister Patrick John and the naivety of his successor Oliver Seraphine, who allowed his government to be tainted with widespread corruption. There followed three armed attempts to overthrow her government, all implicating John and the small defence force which he had founded when in power. The force leader was eventually hanged and John imprisoned. Military help was given by France from neighbouring Martinique. Relations between Dominica (formerly French) and France are warm and can be readily seen by the number of aid projects, notably the new airport at Canefield near the capital, Roseau.

Freedom again won in the July 1985 election with 15 out of the 21 seats. Since 1980 the DLP has gone through numerous splits and defections, giving Freedom 17 to 4. By 1987 Micheal Douglas emerged as DLP leader, although the party image suffers because of his and his brother Rosie's links with Libya, Cuba and the Soviet bloc. A new third party – the seventh in a decade – the United Workers' Party, emerged in 1988. It is led by Edison James, formerly head of the Banana Marketing Corporation. But neither will seriously challenge Freedom, even if they form an alliance, so entrenched is the Charles administration.

A major event in 1990 will be a referendum on the question of political union incorporating Dominica and the other islands of the Windward group – St Lucia, St Vincent and Grenada. Originally proposed for 1987, the referenda on all four islands follows the initiative of the prime ministers of St Lucia and St Vincent. The outcome is awaited with interest.

KEY FACTS

Official title: Commonwealth of Dominica

Head of government: Prime Minister Eugenia Charles

Ruling party: Dominica Freedom Party

Capital: Roseau

Official Languages: English

Currency: East Caribbean Dollar (EC$) = 100 cents

Exchange rate: EC$2.69 per $ (Nov 1989)

Area: 750 sq km

Population: 81,200 (1987)

GDP per capita: $1,550 (1987)

GDP real growth: 5.6% (1988)

Inflation: 1.7% (1988)

Trade balance: –$23.95m (1988)

Visitor numbers: 36,739 (1988)

COUNTRY PROFILE

Political system: Executive power rests with the president of the republic, who acts on the advice of the cabinet. The president is elected by the unicameral legislature, the House of Assembly, for five years, renewable once. A candidate may be elected unopposed if the prime minister and leader of the opposition agree, otherwise a secret ballot is held. The House of Assembly has five members nominated by the prime minister, four nominated by the leader of the opposition, one ex-officio member and 21 members directly elected by the 21 constituencies. Parliamentary terms run for five years. The prime minister is the leader of the majority in the House of Assembly and the leader of the opposition is appointed by the president as leader of main grouping outside the government.

Main cities/towns: Roseau (population 11,000), Portsmouth.

Media
Press: Three weeklies *The New Chronicle, Voice of the Island* (independent) and the *Official Gazette.*
Radio: Government-operated radio service and two privately-owned religious stations. *Television:* Two local television companies provide 12 channels of cable television.

External trade: Balance of payments deficit is supported by IMF loan and bilateral aid from USA, France and UK.
Exports: Bananas (64% of total value), soap (18%), coconut products, citrus fruit, fruit juices, vegetables, essential oils, water. Main destinations: UK (has traditionally taken about 70 per cent of banana exports), CARICOM countries.
Imports: Traditionally machinery, transport equipment, food, manufactured goods, chemicals. Main sources: UK, USA, CARICOM countries.

Agriculture: Sector is mainstay of economy, contributing around 30 per cent to GDP. Bananas are the main crop, with exports destined mainly for the UK. Production totalled nearly 72,800 tonnes in 1988. Other export crops include coconuts/copra, oranges, grapefruit and limes. Livestock-raising and fishing are under-developed and geared towards domestic consumption. Over 40 per cent of the total area is covered by forests with a number of commercially viable wood species present.

Industry: Sector is small-scale and centred on agricultural processing, mainly coconut oil, copra and soap, canned fruit juices, cigarettes, cigars and rum. Water bottling for export is also important. Light manufacturing industry is expanding. Tourism is an important and growing source of foreign exchange. Development has been limited by the inadequacy of air services and poor infrastructure.

Energy: Around 80 per cent of island's electricity needs is met by hydro power, of which there are plentiful resources. A hydroelectricity expansion programme is in progress.

Membership of international organisations: FAO, GATT, IFAD, ILO, IMF, UN, UNESCO, UPU, WHO, WMO, CARICOM, Commonwealth, Economic Commission for Latin America, Organisation of American States, state associated with the EC, World Bank.

Time: GMT -4.

Climate: Sub-tropical with year-round trade winds moderating the heat. Daytime temperatures range from 24°C to 32°C; coolest from Dec to Mar. Driest from Jan to May. Rainy season from Jun to Nov; rainfall much higher in mountain areas.

Entry requirements
Passport: Required by all except nationals of USA and Canada with proof of identity. Onward ticket also mandatory.
Visa: Generally only required by nationals of communist countries.
Currency: Export of foreign and local currency must not exceed import.

Health precautions
Mandatory: Yellow fever vaccination certificate required if arriving from an infected area.
Advisable: Typhoid, polio vaccination. Water precautions should be taken.

Representation overseas: Kingston, London, Washington, Frankfurt, Port of Spain, Fort-de-France (Martinique, consulate), Bangkok.

Air access
Airlines serving: Air BVI, Air Caribe, Air Guadeloupe, Air Martinique, LIAT.
Tax: EC$15 payable on departure.
Airports: Canefield (DCF), 4 km from Roseau; Melville Hall (DOM), 57 km from Roseau, bar, buffet, post office, shops.

Surface access
Main port: Roseau.

Hotels: Limited availability. Bills include a 10 per cent service charge and 10 per cent tax.

Car hire: Temporary driver's permit required. Obtained on production of national driving licence.

City transport
Taxis: Available at airports and through hotels. Fixed rate system.

National transport
Road: The network is over 750 km – most of which is classified as first class, having been restored with US and European aid after extensive hurricane damage.

BUSINESS GUIDE

Public holidays: 1 Jan, Carnival (2 days in Feb or Mar), Good Friday, Easter Monday, Labour Day, Whit Monday, first Mon in Aug, 3 Nov (Independence), 4 Nov (Community Day of Service), 25 Dec, 26 Dec.

Working hours
Business: (Mon-Fri) 0800 – 1600. (Sat) 0800 – 1600.
Government: (Mon-Fri) 0800 – 1600/1700.
Banking: (Mon-Fri) 0800 – 1200/1300; also (Fri) 1500 – 1700.

Telecommunications: Telex, telegraph and fax services available from Cable and Wireless.
Telephone: Dialling code for Dominica: IDD access code + 1-809-44 followed by subscriber's (five-digit) number. There are no area codes. Fully automatic network.

Banking: Agricultural Industrial and Development Bank (AID), Bank of Nova Scotia, Banque Française Commerciale, Barclays Bank International, Dominica Co-operative Credit Union, National Commercial Bank of Dominica, Royal Bank of Canada.

Foreign missions: China PR, Republic of Taiwan, France, Venezuela.

Electricity supply: 220 V AC, 50 cycles.

Hotels: Roseau
Anchorage, Castle Comfort, P.O. Box: 34 (tel: 82638/9; tx: 8619; fax: 4485680).

Castaways, Mero, P.O. Box: 5 (tel: 96244/5; tx: 8628; fax: 4496246).

Continental Inn, 37 Queen Mary Street (tel: 82214, 82215).

Excelsior, P.O. Box: 413 (tel: 91501/2).

Reigate Hall, Reigate (tel: 448-4031; tx: 8654; fax: 4484034).

Sisserou, Castle Comfort, P.O. Box: 134 (tel: 3111; tx: 8628; fax: 3300).

Springfield Plantation Inn, Archbold Plantations, P.O. Box: 41 (tel: 91401; tx: 8625).

Business advice and information
Division of Tourism (National Development Corporation), PO Box 73, Valley Road, Roseau (tel: 82186, 82351; tx: 8649 tourist).

Division of Industry (National Development Corporation), PO Box 293, Valley Road, Bath Estate, Roseau (tel: 82045; tx: 8642).

Dominica Association of Industry and Commerce, PO Box 85, 18 Cork Street, Roseau (tel: 82874; tx: 8625 DAIC).

Dominica Broadcasting Corporation, Victoria Street, Roseau (tel: 83283).

Dominica Hotel Association, PO Box 270, Roseau (tel: 84436; tx: 8607).

Banking: Bank of Nova Scotia, 28 Hillsborough Street, Roseau (tel: 85800; tx: 8671).

Banque Française Commerciale, Queen Mary Street, Roseau (tel: 84040; tx: 8629 bantil).

Barclays Bank, 2 Old Street, Roseau (tel: 82571; tx: 8618).

Dominica Co-operative Credit Union, Great Marlborough Street, Roseau (tel: 82191).

National Commercial Bank, 64 Hillsborough Street, Roseau (tel: 84401, 84403; tx: 8620).

Royal Bank of Canada, Bay Street, Roseau (tel: 82771; tx: 8637).

Dominican Republic

Canute James

The government of President Joaquín Balaguer came under increasing political pressure amid confusion over whether he would stand for another term in the election due in May 1990. Continuing disagreement within the opposition Partido Revolucionario Dominicano (PRD), and the failure to heal the split in the party, increased the popularity of former president Juan Bosch, leader of the Partido de la Liberación Dominicana (PLD). The economy continued to deteriorate despite improvement in overall growth. The completion of the devaluation of the peso led to higher inflation which was partly responsible for triggering outbreaks of civil unrest and general strikes. But there was no indication what the government would do to ease the problems as it appeared confused about whether to conclude a pact with the International Monetary Fund.

Political violence

The political climate became increasingly violent in the first half of 1989, leading to a two-day general strike in June in which several people were killed in clashes between the military and strikers. The strike was called to protest aspects of the government's economic policy, and to support demands for an increase in the national minimum wage to cushion the impact of devaluation of 50 per cent in 1988, and which was running at about the same level in early 1989 with the unification of exchange rates.

This was preceded by clashes between the military and agricultural workers in February 1989, and further clashes the following month. Anti-US sentiment was displayed in demonstrations marking the anniversary of the US invasion of the country in 1965. Attention to the political and economic problems of the country was temporarily diverted with the trial of former president Salvador Jorge Blanco on charges of corruption. Jorge Blanco, who was president between 1982 and 1986, was accused of allowing purchases of military equipment from companies owned by two of his associates who were also being tried. Later, the former president faced additional charges of assisting drug traffickers to use the country as a transshipment point for drugs going from South America to the United States. He was charged with allowing the construction of illegal airstrips to facilitate the traffickers.

Jorge Blanco fled to the United States in 1987 and was subsequently convicted *in absentia*. But he returned home in November 1988 seeking a retrial and abrogation of his 20-year term in prison. On his return the former president was imprisoned, but was later released under orders of President Balaguer.

It was becoming clear in early 1989 that Juan Bosch could be the winner of the election due in May 1990. President Balaguer said he would not run, but reversed this position when he was chosen as the candidate of the Partido Reformista Social Cristiano (PRSC). But despite this, he still faced opposition from within the party from a faction led by Fernando Alvarez Bogaert, which feels that the deterioration of the economy would mean certain defeat for the incumbent president.

Balaguer uncertain

Balaguer later compounded the confusion over his intentions by rejecting the PRSC's endorsement, saying he did not want to be president after 1990. This brought little joy to the warring factions of the PRD, one led by Jacobo Majluta and the other by Jose Francisco Pena Gomez. The split, which contributed to Balaguer's victory in 1986, has weakened the party and many efforts to reunite the factions failed. Despite talking reconciliation, Majluta insisted on his candidacy in 1990, while Pena Gomez said he had just under 20 per cent of the electorate's support and could form a coalition with other parties.

The growth of the economy slowed in 1988 because of a contraction in government spending as the economic situation worsened. The temporary halt to nickel exports and the loss in foreign earnings during the dispute between the government and Falconbridge Dominicana also contributed to the much reduced rate of growth. The devaluations of the peso during the year saw inflation rising towards 50 per cent against 16 per cent in 1987. The value of exports in 1988 increased 25 per cent to reach $890m, and with imports being held at $1.85bn – marginally higher than in 1987 – through controls on imports, the trade deficit was reduced, with the current account deficit in 1988 being $300m, $30m more than 1987.

The government eventually decided to end the cumbersome multiple exchange rate system, and unified the rate at the start of 1989 at 6.35 pesos to the dollar. This represented a depreciation of 84 per cent against the dollar since 1984. But although the government

KEY FACTS

Official title: Dominican Republic

Head of state: President Joaquín Balaguer

Ruling party: Partido Reformista Social Cristiano

Capital: Santo Domingo

Official Languages: Spanish (English widely spoken)

Currency: Dominican Republic Peso (RD$) = 100 centavos

Exchange rate: RD$6.75 per $ (Nov 1989)

Area: 48,400 sq km

Population: 6.9m (1988)

GDP per capita: RD$2,260 (1988)

GDP real growth: 0.8% (1988)

Inflation: 46.0% (1988)

Trade balance: –$714.5m (1988)

Foreign debt: $4.2bn (1988)

regarded this realignment of the currency as one measure it would have to implement in an agreement with the International Monetary Fund, it did not approach the Fund to conclude the agreement.

The government continued to be concerned at a possible adverse public reaction to IMF austerity measures which would accompany a credit package and a structural adjustment programme. In 1984 riots in which an estimated 100 people were killed followed the attempts of former president Salvador Jorge Blanco to implement an austerity programme agreed with the IMF.

In December 1988 President Balaguer attacked the IMF, blaming the institution for many of the Caribbean region's economic problems. Yet eight months earlier the president said he would be submitting a structural adjustment programme to the IMF, opening the door to new commercial credits and refinancing arrangements for the $4bn foreign debt. This followed a rejection by commercial bank creditors of the government's proposals for refinancing before the conclusion of an IMF programme.

Fear of IMF

However, by February 1989, President Balaguer indicated that he was not keen on any pact with the IMF. He referred to the effects of the riots in 1984, and said the country would no longer make sacrifices to meet its foreign debt obligations. Instead the government would be selective in how it serviced its foreign debt.

Roberto Martinez, the finance minister, said the public foreign debt was $3.4bn, and that servicing this in 1988, with arrears accrued at the end of 1987, would demand $1.2bn. He said this was 165 per cent of the value of the country's exports in 1988.

The government presented a budget for 1989 totalling 6.3bn pesos, with just over half slated for public investment programmes which have always been favoured by President Balaguer. Despite these expenditures, the country continued to be plagued by poor utilities.

Frequent power blackouts have adversely affected tourism and local industry. A $105m loan offered by the World Bank to improve electricity generation could not be drawn down because the government was unable to meet the conditions.

The mining sector was preoccupied with a dispute between Falconbridge Dominicana and the government over the imposition of special duty on exports. The company accounts for about 5 per cent of the world's nickel production. The new duty, along with the unified exchange rate, would have reduced the company's earnings from its exports, and appeared to violate the company's contract which stated that it was not liable for such taxes on exports.

The dispute coincided with an announcement by the parent company, Falconbridge of Canada, that it was purchasing the 17.5 per cent interest in Falconbridge held by Armco, a US steel producer. This was intended to give the Canadian major 85 per cent of the Dominican firm. The government holds 10 per cent, with the rest of the equity held by private shareholders.

Although the dispute led to a halt in shipments, Falconbridge Dominicana continued production. In what appeared to be moves to end the dispute, the company made three shipments in March 1988, but then announced that no more would be made, and that production would be halted until an agreement was reached with the government. This followed publication of the company's figures for 1987 which showed earnings of $13.4m in 1987, against losses of $2m in 1986.

Dispute resolved

The dispute was eventually resolved in May 1988, with an agreement that the company would pay a 33 per cent tax on the first 60 cents of profit per pound

of exports, and 55 per cent above this level. The government was anticipating earnings of $60m from the new measure, although the company lost about 10,000 pound in production.

Despite this, nickel output reached 75m pound for 1988, 5m pound more than 1987. More significantly, higher market prices lifted Falconbridge Dominicana's profits for 1988 to $88.4m.

The progressive reduction in the country's US sugar quota was reversed in 1988 when it was increased by 30 per cent to 176,000 tonnes. This followed appeals by the government and others in the Caribbean against the reduced access to the US market, and the effects on US domestic production of a drought. Sugar output for 1988 of 865,000 tonnes was about 50,000 tonnes higher than 1987 production.

KEY INDICATORS

	Unit	1984	1985	1986	1987	1988
Population	m	6.1	6.2	6.4	6.5	6.9
GDP	RD$m	10,355	13,866	17,501	21,745	31,407
GDP per capita (constant)	RD$m	1,698	2,237	2,448	2,988	2,260
Inflation	%	26.9	37.5	9.7	15.9	46.0*
GDP real growth	%	–0.5	–2.6	3.2	8.1	0.8
Agricultural production	1975=100	–	–	–	–	–
Sugar-cane	'000 tonnes	10,271	8,419	8,208	8,772	–
Coffee	'000 tonnes	144.2	132.2	137.0	134.3	–
Cocoa	'000 tonnes	34.5	34.5	36.2	38.7	–
Tobacco	'000 tonnes	27.9	31.5	26.0	28.7	–
Ferro-nickel production	'000 tonnes	24.3	25.4	24.1	32.5	27.3
Silver production	'000 tons	37.6	49.2	41.0	35.7	45.0
Exports	US$m	868.1	735.2	717.6	711.3	890.1
Sugar/molasses	US$m	300.0	190.1	146.2	145.2	142.1
Cocoa	US$m	75.9	64.8	67.8	72.4	69.7
Coffee	US$m	94.9	57.6	115.9	61.4	66.2
Gold/silver alloy	US$m	131.8	113.6	111.8	120.0	98.1
Imports	US$m	1,257.1	1,293.0	1,245.8	1,591.5	1,608.0
Balance of trade	$m	–389.0	–547.4	–629.6	–880.2	–714.5
Tourist arrivals	'000	640	775	878	1,069	–
External debt	$m	3,092	3,441	3,519	3,695	–
Debt service ratio	%	10.1	12.7	16.3	–	–
Exchange rate	RD$ per $	1.0	3.1	2.9	3.8	6.14

* estimated figure

Expansion continued in the tourism sector, with major hotel building projects underway in the resort centres. In enlarging the capacity of the industry, the complement of 18,000 rooms in 1989 was being increased by about 6,000 more rooms.

The expansion of the country's free zones saw employment in these rising to about 80,000, with more zones planned by the government and private investors. The expansion was assisted by increased interest in the Dominican Republic as a location for the manufacture of footwear.

Outlook

The economic outlook for the country is far from bright. The deterioration will continue until the government finds the political will to tackle the structural imbalances and reopens access to sources of new money. It appears that this will not be possible without an agreement with the International Monetary Fund. The longer the government waits, the more harsh and difficult will be the economic adjustments it will have to make. This, and the fear of adverse reaction in an increasingly volatile society, has prevented the government from concluding an agreement with the IMF.

But it is still having to deal with public anger at the state of the economy without the IMF agreement. The economy will continue being unable to finance its foreign debt obligations from its export earnings, and will need refinancing from commercial bank and Paris Club creditors. Again, an agreement with the IMF would be necessary before the creditors make any new agreements.

Increased earnings from nickel and tourism will bring temporary relief to the economic pressures, but a continuing balance of payments deficit will fuel pressure on the rate of the peso, forcing the government to consider further devaluations. The inflation from this will again increase public anger in a situation in which efforts are being made to hold wages not prices.

The gloomy economic outlook is not helped by uncertainty on the political scene. President Balaguer's handling of the economy, and his on-again, off-again candidacy for the 1990 presidential election, has helped to split the PRSC between his supporters and those of Fernandez Alvarez Bogaert.

The continuing split within the opposition PRD indicates that it will not be able to offer a credible alternative, with indications that the victor could be former president Juan Bosch of the PLD.

COUNTRY PROFILE

Historical profile: Discovered by Christopher Columbus in 1492; his brother (Bartholomeo) founded Santo Domingo; his son (Diego) was first governor of Spanish colony; ceded to France 1795; regained by Spain 1808; independent 1821; ruled by Haiti 1822-44; Dominican Republic 1844.

Political system: Executive power rests with the president, elected for a four-year term by direct popular vote. The legislature is the bicameral National Congress – the 30-member Senate is elected for a four-year term, one member for each province plus one for the *Distrito Nacional*; the House of Representatives (currently 120 members) is elected under similar conditions but on the basis of population (one member/50,000 population).

Main cities/towns: Santo Domingo (population 1.6m), Santiago de los Caballeros, La Vega, San Cristobal, Puerto Plata, San Pedro de Macoris.

Media
Press: Most major daily newspapers are independent and include *Listín Diario*. No technical press.
Broadcasting: Around 150 radio and six television stations, all private/commercial except for the government-owned Radio Televisión Dominicana, which accepts sponsored programmes, but is not fully commercial.

External trade: Heavy trade dependence on imports and increased debt service payments have widened the balance of payments deficit.
Exports: Sugar/molasses (20 per cent of total value in 1986), coffee, gold and silver, ferro-nickel, cocoa (beans), tobacco (leaf). Main destinations: USA (77 per cent of total value in 1985), Switzerland, Venezuela, Netherlands, Puerto Rico, Spain, Japan, Belgium, Canada.
Imports: Petroleum and products (typically 25% of total value), machinery and transport equipment (10-12%), food etc (10-12%), chemicals, pharmaceuticals etc, iron and steel, textiles. Main sources: USA (33% of total value in 1985), Venezuela (26%), Netherlands Antilles, Japan, German FR, Canada, Puerto Rico, Spain, UK, Brazil, Netherlands, Mexico.

Agriculture: Sector employs over half the workforce and accounts for some 70 per cent of export earnings. The principal commercial crop is sugar-cane, production of which has fluctuated due to vagaries of weather, falling export demand and labour shortages. Coffee (grown on smallholdings), cocoa and tobacco are also important cash crops. Rice, vegetables and citrus fruits are grown for home consumption.
Cattle-raising has expanded considerably and commercial fishing is being developed. Estimates of cultivated arable land vary, e.g. 18-25%; pasture 17-30%; woodland/forest 25-40%. Soil is generally fertile and rainfall/water availability is adequate. Hurricane damage occurs.

Industry: Sector contributes around 24 per cent to GDP and is largely state-owned. Activity is centred on sugar-refining (which is the dominant industry), cement production and the processing of foodstuffs, tobacco, beverages and textiles. Other light industries include plastics, rubber, chemicals and paper. Emphasis is on encouraging joint ventures that utilise a high percentage of local materials, expanding facilities at the main industrial free-zones (La Ramona, San Pedro de Macoris, Santiago) and overcoming the serious supply/energy problems. Tourism is main source of foreign exchange earnings ($400m in 1985).

Mining: Gold, silver and ferro-nickel are all mined in significant quantities, though mineral production has declined since the late 1970s. Bauxite production was discontinued in 1983 but resumed in 1987. Mineral exports accounted for 28 per cent of total exports in 1984, compared to 40 per cent in 1979. Gypsum, limestone and marble are mined for the domestic market. Also known deposits of copper, iron, titanium and platinum.

Energy: Heavy dependence on imported oil, most of which is used to generate electricity. Oil refinery at Nigua produces around 30,000 b/d. Small deposits of oil located at Charco Largo.

Membership of international organisations: Economic Commission for Latin America, FAO, GATT, ILO, IMF, Inter-American Development Bank, Organisation of American States, UN, UNESCO, WHO, World Bank.

Time: GMT -4.

Climate: Tropical with temperatures ranging from 27°C during the dry season (Nov-Apr) to 37°C between Jun and Oct when humidity is highest.

Entry requirements
Passport: Required by all except nationals of USA, Canada and certain other countries with proof of identity. New passport required if expiry date is within three months of visa application or if the applicant has visited, or intends to visit, Cuba.
Visa: Not required for short stays for visitors from certain Western European and Latin American countries and Japan.
Currency: Import of foreign currency must be declared and not be exceeded by export. Import/export of local currency prohibited.

Health precautions
Mandatory: Yellow fever and cholera vaccination certificate if arriving from infected area.
Advisable: Typhoid/TAB polio vaccinations. Malaria precautions recommended. Water precautions advised and general health insurance.

Representation overseas: Bogotá, Bonn, Brasília, Buenos Aires, Caracas, Guatemala City, Lima, Madrid, Managua, Mexico City, Monaco, Montevideo, Panama City, Paris, Port-au-Prince, Quito, Rome, San José, San Salvador, Santiago, Seoul, Taipei, Tegucigalpa, Tel Aviv, Tokyo, Vatican City, Washington.

Air access
National airline: Dominicana.
Other airlines: Air Aruba, Air Canada, Air France, ALM, American Avianca, Continental, Copa, Dominair, Hispaniola Airways, Iberia, LAV Pan Am, LIAT, Turks & Caicos National, Viasa.
International airport: Santo Domingo-Las Americas (SDQ), 30 km from city, duty-free shop, bar, restaurant, buffet, bank, post office, shops, hotel reservations, car hire; La Union-Puerto Plata (POP), bank, duty-free shop, restaurant, bar, car hire.

Surface access
Main port: Santo Domingo, Haina.
Overland access: Main route runs from Haiti via Elias Pina.

Hotels: Bills usually include five per cent government tax and 10 per cent service charge. They are considerably more expensive during the winter.

Car hire: National or international licence required. Chauffeur-driven cars can be negotiated with taxi drivers outside main hotels.

City transport
Taxis: Normal means of transport in cities. In Santo Domingo most taxis ply fixed routes, stopping on request, with single tariff for any distance or route. No tip expected. Taxis which travel off these routes may be difficult to find, especially at night.

National transport
Air: Services operated by Servicios Aeros Dominicanos between Santo Domingo, Santiago, Puerto Plata and Barahona. Aeroplanes with pilot can also be hired.
Road: Over 17,000 km of road, around half surfaced – highways link Santo Domingo with Hinguey, Montecristo, Dajabon, San Juan, Elias Pina.
Buses: Service in Santo Domingo generally described as cheap though crowded. Fairly numerous services from Santo Domingo to Puerto Plata, La Romana – journey times vary. Also to Barahona and Samana.
Rail: Besides lines used for transporting sugar there is track between La Vega and Sanchez but no scheduled passenger services are listed.

BUSINESS GUIDE

Public holidays: 1 Jan, 6 Jan (Los Reyes), 21 Jan (Altagracia), 26 Jan (Duarte Day), 27 Feb (Independence), Good Friday, 14 April (Pan-American Day), 1 May, 16 Aug (Restoration Day), 24 Sep (Las Mercedes), 12 October (Colombus Day), 24 October (United Nations Day), 1 Nov (All Saints' Day), 25 Dec.

Working hours
Business: (Mon-Fri) 0800 – 1200, 1400 – 1800.
Government: (Mon-Fri) 0730 – 1430.
Banking: (Mon-Fri) 0830 – 1730.

Telecommunications
Telephone: Dialling code for Dominican Republic: IDD access code + 1-809 followed by subscriber's (seven-digit) number. There are no area codes.
Telex: External and internal (telex and telegram) service operated by the government, ITT-America Cables and Radio Inc and RCA Global Communications Inc.

Banking
Banking centre: Santo Domingo.
Commercial banks: Commercial banks: Banco Español, Banco Antillano, Banco Dominicano del Caribe, Banco de los Trabajadores de la República Dominicana, Banco de Reservas de la República Dominicana, Banco Hipotecario Dominicano, Banco Metropolitano, Banco Nacional de la Construcción, Banco Intercontinental, Banco Popular Dominicano, Banco Dominico Hispano, Banco del Comercio Dominicano, Banco Panamericano, Banco del Progreso Dominicano, Banco Nacional de Crédito, Banco del Exterior, Banregión, Banco Universal, Banco Gerencial y Fiduciario, Banco Mercantil, Banco Cibao.
Foreign branches: Bank of Nova Scotia, Chase Manhattan Bank, Citibank.

Advertising: Widely used in all forms of media except posters. The Santo Domingo Chamber of Commerce produces the weekly *Comercio y Producción.*

Electricity supply: 110 V AC.

Representation in capital: Argentina, Brazil, Chile, Colombia, Costa Rica, Ecuador, El Salvador, France, German FR, Guatemala, Haiti, Honduras, Israel, Italy, Japan, Jordan, Republic of Korea, Mexico, Nicaragua, Panama, Peru, Spain, Taiwan, USA, Uruguay, Vatican City, Venezuela.

Useful telephone numbers
Police: 682-3151. Police (radio patrol): 533-1074. Red Cross: 689-4288. Fire Department: 682-2000. Centro Médico Nacional (hospital): 682-0171.

Hotels: Santo Domingo
Cervantes, Cervantes 30 (tel: 688-2261).

Concorde, Avenida Anacaona, P.O. Box: 1974 (tel: 562-8222; tx: 3460179; fax: 5628938).

Continental, 16 Avenida Máximo Gómez (tel: 689-1151;; tx: 3460425; fax: 6878397).

El Embajador, 65 Avenida Sarasota, P.O. Box: 1340 (tel: 533-2131; tx: 3460528).

Hispaniola, Avenida Independencia, P.O. Box: 2113 (tel: 535-7111; tx: 3460033; fax: 5354050).

Jaragua Resort, Avenida George Washington 367 (tel: 686-2222; tx: 0758; fax: 6860528).

Lina, Avenida Máximo Gómez, P.O. Box: 1915 (tel: 689-5185; tx: 3460278; fax: 6865521).

Naco, 22 Avenida Tiradentes (tel: 562-3100; tx: 3460382).

Napolitano, Avenida George Washington (tel: 687-1131; tx: 3264101).

San Gerónimo, Avenida Independencia 205, P.O. Box: 15 (tel: 533-8181; tx: 44319).

Santo Domingo, Avenida Independencia, P.O. Box: 2112 (tel: 532-1511; tx: 3460125).

Sheraton, Avenida George Washington 365, P.O. Box: 1493 (tel: 685-5151; tx: 3460529).

Business advice and information
Aerolíneas Argo, Avenida 27 de Febrero 409, Santo Domingo (tel: 566-1844; tx: 3460531).

American Chamber of Commerce, PO Box 95-2, Hotel Santo Domingo, Santo Domingo.

Asociación de Industrias de la República Dominicana, Avenida Sarasota 4, Santo Domingo.

Banco Central, Avenida Pedro Henríquez Urena, Santo Domingo (tel: 688-2011; tx: 3460052).

Cámara Oficial de Comercio, Agricultura e Industria del Distrito Nacional, Apdo Postal 815, Arz Noel 52-Altos, Santo Domingo (tel: 682-2688, 682-7206).

Dominicana de Aviacón, Leopoldo Navarre, Edificio San Rafael, PO Box 1415, Santo Domingo (tel: 687-7111).

ITT-America Cables and Radio Inc, Julio Verne 21, Santo Domingo (tel: 682-3115).

RCA Global Communications Inc, Edificio Diez, Calle Conde 203, Santo Domingo (tel: 682-3722).

Secretaría de Estado de Turismo, César Nicholas Penson 59, Santo Domingo (tel: 685-2388; tx: 3460303).

Car hire: Santo Domingo
Avis: Ave Abraham Lincoln (tel: 532-2969; tx: 3460174).

Budget: Ave J F Kennedy Esq, Lope de Vega (tel: 567-0175/6).

City: (tel: 533-9295).

Easter Island

(map on page xxiv)

COUNTRY PROFILE

Political system: Easter Island (also known as Rapanui) is administered as a district of Chile, with governor and locally elected council.

Main town: Hanga Roa.

Language: Spanish is the official language. Polynesian is widely spoken, however. English is not used.

Media
Radio: Day-long radio service provided by Chilean Air Force. *Television:* Closed-circuit private TV system, using video cassettes.

External trade
Exports: Main exports are tunafish, avocados and pineapples to Chile.
Imports: Main imports are food, fuel, construction materials and machinery from Chile.

Agriculture: Although the island is predominantly grassland, pine, eucalyptus and fruit trees have been planted in recent years. Main crops are bananas and pineapples, sweet potatoes, yams, sugar cane, maize, potatoes, tomatoes, castor beans, melons, grapes, avocados. Soil erosion led to drastic reduction in the numbers of livestock; there are now about 1,500 sheep (as compared to 50,000 in 1950), mainly Corriedale and some Merinos. There are wild horses in addition to those used as draught animals and for riding. Poultry raised includes pigeons, quail and ducks. Fishing for lobster, tuna and king fish is an important local source of protein.

Industry: Aside from agriculture, tourism is main industry. Approximately 2,700 visited in 1986.

Time: GMT -6 hrs.

Climate: Subtropical, cooled by constant winds. Average rainfall 1,250mm falling mainly in Jun/Jul; average temperature range 16-27°C.

Entry requirements
Passport: Required by all.
Visa: Required by all. Assured onward passage is necessary.

Air access
Other airlines: Lan-Chile.
International airport: Mataveri International (IPC).

Surface access
Main ports: Hanga Roa.

Car hire: Make local enquiries regarding availabilities of car hire.

National transport: There are no surfaced roads; four-wheel drive vehicles, motor cycles and horses are the main means of transportation. Minibuses are used by tourists.

KEY FACTS

Official title: Isla de Pascua (Easter Island)

Head of state: President of Chile

Head of government: Governor Sergio Rapu Haoa

Capital: Hanga Roa (only inhabited centre)

Official Languages: Spanish (Polynesian also spoken)

Currency: Chilean Peso (CH$ = 100 centavos)

Exchange rate: CH$272 per $ (Nov 1989)

Area: 180 sq km

Population: 2,100 (1988)

Unemployment: 3% (1987)

Visitor numbers: 4,000* (1987)

* estimated figure

Ecuador

Alfredo W. Forti and John M. Zindar

As Ecuador begins its second decade of democracy, initiated with the return of civilian rule in 1979, the new social democratic government of President Rodrigo Borja faces the challenge of solving an economic crisis with severe adjustment policies, while at the same time maintaining social and political stability. Skilfully balancing left-leaning social and foreign policies with conservative economic measures, Borja is held up as a new model of political pragmatism in Latin America. Observers have called Borja's early policies 'supply-side socialism'. This year, the new president's popularity will be tested by legislative elections, the results of which will largely depend upon the economic situation.

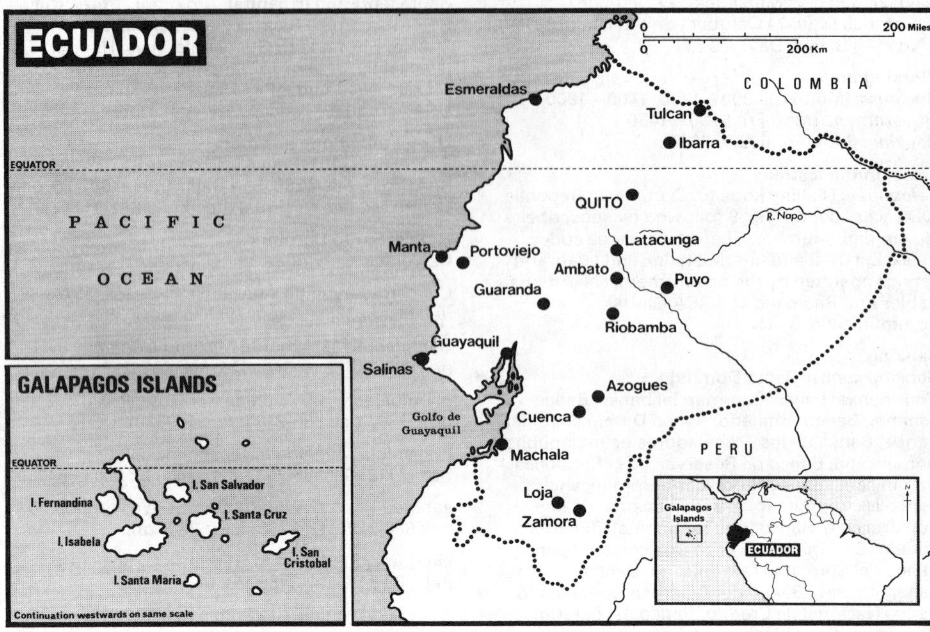

A new government inherits economic chaos

The start of President Rodrigo Borja's social democratic administration in August 1988 marked a sharp departure from the previous conservative government of Leon Febres Cordero. Borja, a 47-year-old law professor and founder of the Democratic Left Party (ID), obtained 47 per cent of the popular vote (52.8 per cent excluding spoiled and blank votes) in run-off elections, beating populist candidate Abdala Bucaram with a margin of about 200,000 votes. The failure of Cordero's conservative and market-oriented policies was in part a result of his 'authoritarian' style that polarised Ecuadoran society. In addition, Cordero faced a number of adverse factors that converged to create what developed into a national crisis: (1) an opposition-dominated congress, (2) a sharp decline in oil prices in 1986, (3) a rebellious military, and (4) the earthquake in March 1987 that disabled the main oil pipeline for six months.

What Borja found when he took office was an inflation-ridden economy, a fiscal deficit that represented 17 per cent of the GDP, a rapidly declining sucre that went from 262 per dollar in January to about 560 in mid-July (a 165 per cent devaluation in one year), negative international monetary reserves ($319.7m) and only $9m of liquid reserves in the Central Bank. As a result of these devastating economic indicators, Ecuador had suspended all payments on its $11bn foreign debt.

Ecuador's population growth of almost 3 per cent requires the creation of approximately 100,000 new jobs per year. Unemployment figures in 1988 showed an historically high 12 per cent and underemployment ran at more than 50 per cent. In areas such as education, housing and health the Cordero administration was unable to fulfil promised goals. All of this created a state of social tension characterised by frequent general strikes, opposition rallies and increased human rights abuses by the government.

The national emergency plan

Within weeks of taking office, the Borja administration adopted an adjustment plan designed to address, in the short term, the three principal economic problems: inflation, the fiscal deficit and external imbalances. The main aspects of the plan included: (1) currency devaluation of 56 per cent followed by weekly mini-devaluations of 2.5 per cent, (2) the taking control of private sector foreign exchange transactions by the Central Bank, (3) tough restrictive monetary policies to reduce the growth of the money supply below the inflation rate, (4) a severe programme of austerity that reduced government spending and reformed tax law, and (5) the continuation of the moratorium on foreign debt service.

The economic emergency plan was characterised by gradual adjustments, as opposed to the 'shock therapy' applied in other Latin American countries. The government was aware of the social cost the plan had, and in order to balance such hardships to the lower income population, a number of 'cushioning' measures were implemented. There was a rise in salaries and a policy contemplating freezing prices for 12 basic products. In addition, the government created a 50bn sucre fund for social development, and a government Agency of Small Business Support with a 4bn sucre budget administrated by the Central Bank.

Borja deals with social instability

Situated between two violence-ridden nations, Peru and Colombia, Ecuador has also experienced unrest in the form of two leftist guerrilla groups. Organised in 1984 and 1986 respectively, Alfaro Vive Carajo (AVC) and Montoneras Patria Libre stepped up their bombings, kidnappings and bank robberies during the last years of Febres Cordero's government. Cordero's response through the police and armed forces led to many reports of human rights violations. Congress created a Special Human Rights Commission that denounced the arbitrary treatment of guerrilla group members and political dissidents by security forces. Many of these accusations against Cordero are still pending. The Supreme Court tried the former president in 1989 over the disappearance of $150,000 of government funds, apparently used to finance foreign training for anti-subversive security units without the knowledge of Congress.

Borja, who was a leading critic of Cordero's response to political violence, used an innovative approach to the problem by engaging in a process of private dialogue with the AVC leadership. This initiative succeeded in convincing the guerrillas to lay down their arms and incorporate themselves into the open political process. Only Venezuela, in the mid-1960s, has accomplished a similar feat. This successful neutralisation of insurgent violence reflects Borja's exceptional political skill in defusing domestic violence and social unrest.

A broken political alliance

Ecuador's Congress consists of a 71-member unicameral legislature. 12 of the members, called National Legislators, are elected for four-year periods. The remaining 59, called Provincial Legislators, are elected for two-year terms. Following the 1988 elections, Borja's ID obtained 29 seats. With his alliance that included the Christian Democratic Popular

KEY FACTS

Official title: Republic of Ecuador

Head of state: President Rodrigo Borja Cevallos

Ruling party: Izquierda Democratica

Capital: Quito

Official Languages: Spanish (Quechua also spoken)

Currency: Sucre (S) = 100 centavos

Exchange rate: S604 per $ (Nov 1989, official); S596 per $ (free)

Area: 270,670 sq km

Population: 10.2m (1988)

GDP per capita: $1,477 (1988)

GDP real growth: 12.8% (1988)

Inflation: 58.3% (1988)

Trade balance: $589m (1988)

Foreign debt: $10.5bn (1988)

Democracy Party (DP, which holds seven seats), the communist Left Broad Front (FADI, two seats) and the leftist Frente Radical Alfarista (FRA, two seats), Borja's party commanded a comfortable parliamentary majority.

By mid-1989, however, differences over economic policy had caused FADI to separate itself from the alliance. The Popular Democracy Party, which also participated in the executive branch, withdrew from the governing alliance with the resignations of Minister of Industry and Commerce Juan Jose Pons, and Housing Board Chairman Juan Pablo Mocagatta. Nevertheless, an agreement was reached between both parties by which DP's Wilfrido Lucero was re-elected as president of the Ecuadoran Congress, and ID's Antonio Rodriguez replaced fellow party member Nicolas Issa in the vice-presidency. Thus, the ID – DP alliance was maintained in Congress.

The prolonged recession

Ecuador has not been able to sustain significant economic growth since 1986. Growth did rebound sharply in 1988 when normal oil production resumed after the March 1987 earthquake-damaged facilities were repaired. Recession returned in 1989 and has continued into 1990. Borja has not been able to promote the revival of growth hoped for because austerity measures have severely dampened domestic demand (down an estimated 2 per cent in 1988 and another 1 per cent in 1989) and government spending, while a contraction in world trade has been especially adverse for Ecuador's commodities. Global prices for virtually all of Ecuador's primary exports have softened in recent years.

Progress has been made in the battle against inflation, which, compared to many of Ecuador's Latin American neighbours, is relatively mild. Inflation has been high enough (around 50 per cent in 1989), however, to discourage investment and feed ever-increasing wage demands and unrest by the unions. The monthly wage floor rose 16 per cent in late 1988. The unions have recently demanded a minimum wage increase of 170 per cent while the government has only offered a 23 per cent rise. Union action is, nevertheless, weakened by severe unemployment (12 per cent) and underemployment (50 per cent). While labour mounted eight general strikes against the Cordero government, Borja's adroit political skills have thus far neutralised militant union action. Gross fixed capital investment is dropping in 1990 for the third consecutive year. New foreign investment, which declined by a sharp 30 per cent in 1988, has yet to recover.

Borja's government will be able to bring inflation down to about 20 per cent by year-end 1990. The tightening of fiscal and monetary policy, completely out of control by the end of Cordero's term, has been one of Borja's primary accomplishments. Cordero doubled the money supply (M2) and drastically increased government spending during his last year in office, in an attempt to improve his party's chances at the 1988 electoral polls. The fiscal deficit, as a percentage of GDP, jumped from 2.3 per cent in 1987 to 17 per cent in 1988. Borja reversed these trends in 1989 by cutting the budget deficit back to 4 per cent of GDP and the growth of the money supply to 30 per cent.

Domestic interest rates have traditionally been kept artificially low to stimulate expansion of local production. Ecuador's economy, however, has been plagued by a severe lack of domestic savings and significant capital flight. Cordero's free market policies did attract the temporary return of some capital. Borja's President of the Monetary Board, Abelardo Pachano, revised the domestic interest rate regime in April 1989. The annual interest rate on most Central Bank loans was boosted from 23 per cent to 36 per cent. Earlier, the maximum allowable spread

KEY INDICATORS

	Unit	1984	1985	1986	1987	1988
Population	m	9.1	9.4	9.7	9.9	10.2
GDP (1985 prices)	sucre bn	1,063.7	1,109.9	1,144.3	1,081.4	1,220.0
GDP per capita	$	1,363	1,268	1,163	960	991
Inflation (average rate)	%	31.3	28.0	23.0	29.5	58.3
GDP real growth	%	4.2	4.3	3.2	-5.2	12.8
Agricultural sector	% real gwth	10.6	10.0	9.9	7.5	3.8
Bananas	'000 tonnes	1,924	1,970	2,316	1,962	–
Cocoa	'000 tonnes	60	131	90	85	–
Coffee	'000 tonnes	90	121	118	118	–
Manufacturing sector	% real gwth	-1.8	0.2	-1.7	0.3	0.7
Petroleum production	'000 b/d	255	275	280	165	300
Exports	$m	2,622	2,905	2,186	2,021	2,203
Crude petroleum	$m	1,678	1,825	912	739	875
Bananas	$m	135	220	263	267	298
Coffee	$m	175	191	299	192	152
Cocoa	$m	96	138	71	83	78
Fish/fish products	$m	165	166	313	419	422
Imports	$m	1,567	1,611	1,631	2,054	1,614
Balance of trade	$m	1,055	1,294	555	-33	589
Current account	$m	-148	149	-613	-1,131	-597
Exchange rate (average)	sucres per $	62.5	69.6	122.8	170.5	301.6

* estimated figure

between private bank rates to borrowers and savers had been reduced from 19 to 15 percentage points. Interest rates are, nevertheless, still negative in real terms as inflation has not yet been sufficiently controlled.

The suffocating debt

Ecuador's external debt totalled an estimated $12.5bn by year-end 1989. Debt service ratios have climbed in recent years to a high of 82 per cent of exports in 1988. The value of Ecuadoran debt on the secondary market dropped to 10 per cent of face value in 1989. Long-term public debt accounts for the vast majority of the external debt (85 per cent). Private commercial banks hold 55 per cent of the debt while official bilateral and multilateral debt accounts for 30 per cent. The government's 1989 budget included interest payments for official creditors only. Ecuador has attempted to reschedule its debt six times since 1982. No new agreement with the IMF or commercial banks has been implemented since 1987. Paris Club (official) debt was refinanced in January 1989.

Cordero, who had been considered a model leader by, and received favourable conditions from, the international financial community (particularly the Reagan Administration), unilaterally suspended interest payments on the debt in January 1987. The March 1987 earthquake, which severely damaged the economic infrastructure, ensured that the moratorium would be extended indefinitely. A small payment was made in late 1987 but the moratorium resumed in early 1988. The new Borja government maintained the moratorium but made another small payment ($15m) in June 1989, with the hope of accelerating chances for a new agreement with creditors.

Interest payment arrearage has grown substantially, to $1.52bn by late 1989. Predictably, relations with creditors have been difficult. In December 1988, US Government banking authorities declared loans to Ecuador value-impaired, forcing banks to further build reserves against such loans, and further degrading Ecuador's ability to attract new credit. Citibank, the United States' largest bank, seized $80m of Ecuadoran Central Bank deposits to cover an outstanding trade loan. This unusual action reflected, and intensified, the acrimonious nature of the external debt dispute. Ecuador retaliated by closing all public accounts and suspending transactions with Citibank. The Inter-American Development Bank mediated a resolution in July which restored financial relations. Ecuador hopes to join soon the 'Group of Eight', the co-ordinating organisation of Latin America's largest debtors.

Ecuador requires close to $1bn in new credit to restore liquidity and to begin to restructure the economy, re-establishing the country's growth and creditworthiness. 'Debt-for-nature swaps', such as the one sponsored by the World Wildlife Fund and Nature Conservancy that purchased $9m in Ecuadoran debt in April 1989, have no impact on the debt burden. Borja is carefully implementing an austerity programme that he hopes will soon qualify Ecuador for substantial debt relief as provided for in the plan devised by US Treasury Secretary Nicholas Brady.

Trends in trade

The significant adjustment of Ecuador's three-tiered exchange rate regime in May 1989 improved the business environment for the export sector and brought the country one step closer to a Brady Plan agreement with the IMF. Exporters are now allowed to exchange foreign currency earnings at the full 'intervention' rate, generally 10 per cent higher than the free market rate. Previously, exporters were given an exchange of 5 per cent below the intervention rate. The measures in May also accelerated the annual depreciation (from 2.5 to 3.0 per cent per week) of the sucre at the intervention rate, now used for virtually all transactions. This has narrowed the differential with the free market rate to 6 per cent.

Foreign exchange reserves have steadily declined since 1985. Lack of foreign exchange is responsible for the interest payment moratorium begun by Cordero in 1987. Despite the interruption in debt service, Ecuador has been unable to strengthen its reserve position. Reserves stood at an estimated $400m (about one quarter's worth of annual imports) at year-end 1988, down almost 50 per cent from 1985. A sharp deterioration in Ecuador's terms of trade, export production shortfalls and a surge in imports accounts for the declining trade surpluses and weak convertible currency earnings. The 1989 trade surplus barely exceeded \$300m, down 75 per cent from 1985. The current account deficit continues to grow (estimated \$500m in 1989) after the significant improvement in 1988 when the earthquake-damaged oil production facilities were repaired.

Oil has dominated the economy of Ecuador since it began exporting the commodity in 1972. Ecuador is one of two Latin American nations belonging to the oil cartel OPEC. Oil production increased 78 per cent – 310,000 barrels per day (b/d) – in 1988 after production was interrupted for six months by earthquake damage the previous year. Production increased to 370,000 b/d in 1989. Oil, which has earned up to 70 per cent of annual foreign exchange income, now only accounts for 40 per cent of export value. Ecuador has considered withdrawing from OPEC due to a recent cut in the country's production quota (230,000 b/d). Declining world prices will minimise the amount of expected oil-generated revenue and future foreign investment. Borja repeatedly states that foreign investment in the exploration and extraction of oil and other primary resources is welcome. The government has recently announced its intention to have the state oil company, CEPE, begin a phased takeover of the principal pipeline and the eastern jungle fields that CEPE now operates jointly with Texaco.

Bananas have been the only recent bright spot in the agricultural export sector. Exports increased in 1988 by 16 per cent to a record \$310m. Ecuador was able to respond to world shortages due to hurricane damage in the Caribbean Basin and expand market shares, particularly in the Middle East. In the medium term, however, banana prices will weaken in the face of oversupply. The recent breakup of the international coffee commodity agreement signals a decline in the earning potential of that crop; and cocoa prices will continue to be depressed. The production of Ecuador's other major export, shrimp, declined sharply in 1989 due to rising production costs, unfavourable ocean currents and new stiff competition from China.

Outlook

The emergency economic plan implemented by the Ecuadoran Government has produced positive results in controlling the fiscal deficit, and rationalising monetary and exchange rate policies. These achievements, a product of contractive austerity measures, have made it difficult for the production segments of the economy to recover. This situation may eventually breed social instability. Nevertheless, the government's conscious attempt to implement a policy of gradual adjustment, in order to minimise the social cost, may be rewarded with both domestic and international support. Ecuador could be next in line to be considered eligible for debt reduction under the Brady Plan. Economic recovery will depend heavily upon the economic and commercial policies of the United States, which usually purchases over 60 per cent of Ecuador's exports.

This year marks the mid-term of Borja's presidency. Congressional elections will take place and the results, as was the case during Cordero's term, will set the tone of the following two years. Present conditions foretell a favourable outcome for ID due to a combination of factors that include the popularity of President Borja and the lack of serious challenges from the opposition.

Ecuador has become a new actor at the regional level. Foreign Minister Diego Cordovez has played a significant role as mediator in regional conflicts such as the Panamanian crisis. The probable inclusion of Ecuador into the Latin American Group of Eight (replacing Panama) will signal the Latin governments' recognition of Ecuador's new role. President Borja's European tour in the autumn of 1989 ended with his personal participation at the Non-Aligned Nations summit in Yugoslavia.

The 1990s present a challenging task for Ecuador's leadership – consolidation of democracy with economic growth. The task is being carried out at a high social cost. Timely support by the international community will determine the future prospects for this promising nation, which thus far represents a haven of peace and relative stability in the region.

COUNTRY PROFILE

Political system: Executive power rests with president, elected by direct vote for a four-year term. The president appoints a cabinet (and provincial governors etc). The legislature is the unicameral Chamber of Representatives (currrently 71 members elected for a two or a four-year term by province on basis of population).

Main cities/towns: Quito (population 1.1m, 2,805 metres above sea-level), Guayaquil (pop over 2m), Cuenca, Ambato, Esmeraldas, Manta.

Media

Press: Principal dailies are *El Comercio* (circulation 130,000), *Hoy* (55,000), *Ultimas Noticias* (in Quito, 90,000) and *El Universo* (174,000), *Expreso* (30,000), *La Razón* (28,000), *El Telégrafo* (in Guayaquil, 35,000). US newspapers also available in Quito and Guayaquil. A number of periodicals and trade magazines are also published.

Radio: Over 470 commercial stations, 10 cultural and 10 religious stations. Broadcasts in numerous languages. *Television:* Several (commercial) stations and a cable network. Broadcasts in Spanish and English.

External trade: Reduction in imports since 1982, though currency devaluation and restrictions allowed for improvement in trade balance despite static export revenue. Current account deficit also diminished due to debt rescheduling, reduced insurance and freight charges from the falling imports, and new exchange regulations, but rose in 1986. Collapse in world oil prices in 1986 led to loss of export earnings and decision in early 1987 to restrict debt servicing.

Exports: 1988: Oil and derivatives (45%), shrimp (18%), bananas (14%), coffee (8%) and cocoa. Main destinations (1987): USA (61.2% of total value), German FR (3.9%), South Korea (2.5%), Japan (2.4%), Panama (2.1%).

Imports: Mineral products, chemicals, industrial machinery and equipment, motor vehicles, construction materials, processed foodstuffs. Main sources (1987): USA (30% of total value), Japan (13.4%), German FR (9%), Brazil (6.2%), Mexico (5%).

Agriculture: Once the mainstay of the economy, agriculture contributed about 15 per cent to GDP in 1986. However, the sector still employs around one-third of the population, most of whom are engaged in subsistence farming. The sector has suffered from low levels of mechanisation and irrigation, and lack of financial incentives. In coastal regions the main crops are bananas (Ecuador is world's leading exporter), cocoa, coffee, oil palms, sugar-cane, cotton, rice and maize, while the Sierra produces legumes, maize, wheat, potatoes, rye and barley. Cattle mainly reared in the highlands; small-scale poultry farming in Manabi Province. Agricultural output has fluctuated in recent years owing to flooding and shifts in world cocoa and banana prices. On average, increases in food production continue to lag behind population growth. Sea fishing and shrimp farming has grown in importance, with shrimp accounting for 5 per cent of total exports in 1985, and Ecuador was the world's chief exporter of shrimps in 1986. Although around half of the total land is covered by forests, production of hardwoods and balsa wood has fallen due to cost and difficulty of removal.

Industry: Despite rapid growth during 1970s, manufacturing sector contributes under 20 per cent to GDP and employs only about 12 per cent of the workforce. Food processing, textiles, household products and pharmaceuticals account for over two-thirds of manufacturing activity. Other important industries are metals, non-metallic minerals, cement, paper and pulp, petrochemicals, glass, rubber and plastics. Output has stagnated in recent years owing to continued shortages of imported raw materials, shortage of skilled labour, depressed home demand, heavy debt burden and a fall in investment levels. Activity is concentrated in Quito and Guayaquil. 'Free zone' offering incentives for manufacture of export products at Esmeraldas.

Mining: Zinc, gold, silver, copper, lead, sulphur, titanium, antimony and manganese mined in small quantities. Recent interest has been in gold, with a joint government and private mining venture in the Nambija region starting in 1985. Emphasis is currently on attracting more domestic and foreign investment.

Hydrocarbons: Petroleum sector is the driving force in the economy, accounts for some 40 per cent of government revenues and nearly 70 per cent of export earnings. Major inland exploration began in the early 1960s, with important discoveries being made at Lago Agrio. Exports began in 1972 after completion of the trans-Andean pipeline linking the oilfields of Oriente Province with refinery at Esmeraldas. Despite recent lack of funds for exploration, development of untapped wells has helped boost crude production to around 300,000 b/d in 1986. Plans to raise total oil output to 345,000 b/d by 1989, despite OPEC quota restrictions. However, severe earthquakes in north-east in March 1987 damaged main pipeline and stopped oil exports for several months. Exploitation of oil resources undertaken by state-owned oil company, CEPE, and by foreign conglomerates such as Texaco. Total recoverable reserves estimated at around 1bn barrels, equivalent to ten years output at present extraction rate. Current emphasis is on developing exploration, and expanding refining capacity to increase production of lighter products. Natural gas reserves in Oriente as well as Gulf of Guayaquil. Production of natural gas remains under 40m cu metres p.a. though output of liquefied petroleum gas has risen since opening of Shushufindi plant.

Energy: In the light of country's dwindling petroleum reserves, government has expanded hydro-electricity generation to over 50 per cent of total generating capacity. Emphasis on exploiting vast hydroelectric potential, with stations being built on Pastaza river (Agoyán plant), and across the Paute gorge. The hydroelectric project at Daule – Peripa will also provide irrigation for 100,000 hectares in the Daule region and Santa Elena peninsula.

Membership of international organisations: ECOSOC, FAO, G-77, IADB, IAEA, ICAO, ICO, IDA, IFAD, IFC, IHO, ILO, IMF, IMO, INTELSAT, Inter-American Development Bank, Interpol, IRC, ITU, Latin American Integration Association, NAM, OPEC, Organisation of American States, PAHO, SELA, UN, UNESCO, UPEB, UPU, WFTU, WHO, WMO, World Bank, WTO.

BUSINESS GUIDE

Time: GMT -5 (except Galapagos Islands: GMT -6).

Climate: Temperate in the Sierra, including Quito, with annual average temperature 20°C which remains fairly constant throughout the year, although daily temperature range is high; driest from Jun to Oct.

Entry requirements
Passport: Required by all except certain nationals of Colombia.
Visa: Required by all visitors who intend to stay longer than 90 days and nationals of Eastern bloc, certain Latin American and other countries for any length of stay. All other visitors require tourist card valid for up to 90 days. These are obtainable on arrival. N.B. If visit (for other than tourist purposes) is likely to exceed 90 days an exit permit with both tax and police clearance is necessary.
Currency: No restrictions on import or export of foreign or local currency. International credit cards generally accepted in Quito and Guayaquil. Travellers' cheques can be difficult to exchange outside main towns. Dollar travellers' cheques are the most easily negotiable.

Health precautions
Mandatory: None.
Advisable: Yellow fever vaccination essential for visits to Pastaza province in the east. Typhoid, polio, tetanus vaccinations recommended. Malaria prophylaxis advisable in some low lying areas (below 1500m). Gamma globulin injection against hepatitis also recommended. Neither milk nor tap water are safe to drink. Bottled mineral water widely available.

Representation overseas: Asunción, Beijing, Belgrade, Berlin, Berne, Bogotá, Bonn, Brasilia, Brussels, Bucharest, Cairo, Caracas, Guatemala City, The Hague, Havana, Kuwait City, La Paz, Lima, Lisbon, London, Madrid, Managua, Mexico City, Monaco, Montevideo, Moscow, Ottawa, Panama City, Paris, Port-au-Prince, Prague, Rome, San José, Santiago, San Salvador, Santo Domingo, Seoul, Stockholm, Tegucigalpa, Tel Aviv, Tokyo, Vatican City, Vienna, Warsaw, Washington.

Air access
National airline: Empresa Ecuatoriana de Aviación (Ecuatoriana), SAETA.
Other airlines: Aerolíneas Argentinas, Aeroperú, Air France, Air Panamá International, Avianca, Eastern, Iberia, KLM, Ladeco, Lufthansa, Varig, Viasa.
International airports: Guayaquil-Simón Bolivar (GYE), 5 km from city centre, duty-free shop, restaurant, buffet, currency exchange, post office, shops, car hire; Quito-Mariscal Sucre (UIO), 8 km from city centre, duty-free shop, bar, restaurant, buffet, bank, post office, shops, car hire.

Surface access
Main ports: Esmeraldas, Guayaquil, Manta, Puerto Bolívar, San Lorenzo.
Overland access: Practicable from Peru and particularly Colombia (via the Pan-American highway). Coach services operate Cali-Quito, Lima-Quito; both at least once a day.

Hotels: Wide range available in Quito and Guayaquil. A government tax of 10 per cent and a service charge of 10 per cent payable on all rates.

Car hire: Major companies operate from Quito and Guayaquil. International licence required. Police checks are common.

City transport
Taxis: Can be hailed or found on ranks; fares by meter; tipping not usual.

National transport
Air: Air transport is well developed. Several internal airlines including ANDES (freight), SAETA, SAN and TAME (military airline operating civilian schedules) operate domestic services to main towns. Air-taxi and charter services are available from Guayaquil and Quito.
Road: Most parts of the country are accessible by surfaced or all-weather roads. Major routes run north-south in the coastal lowlands and the sierra. The Pan-American Highway runs from Tulcan via Ibarra, Quito, Riobamába, Cuenca, Loja to Macará. Good roads link the Sierra to the coastal ports.
Buses: Services link main towns, for example, Quito-Esmeraldas, Quito-Manta, Guayaquil-Manta, together with luxury service Quito-Guayaquil. Reservations should be made for long distance services. Services within cities are often cheap but crowded.
Rail: Routes include Quito-Riobamba, Guayaquil-Bucay, Alausi-Huigra, Sicambe-Cuenca and Ibarra-San Lorenzo. Rail travel generally uncomfortable and unreliable.

Public holidays
Fixed dates: 1 Jan, 1 May, 24 May (Battle of Pichincha), 24 Jul (Bolivar's Day), 10 Aug (Independence), 9 Oct (Independence of Guayaquil), 12 Oct (Discovery of America), 1 Nov (All Saints' – not official), 2 Nov (All Souls'), 3 Nov (Cuenca Day), 6 Dec (Foundation of Quito/Quito only) 25 Dec.
Variable dates: Carnival (Feb), Good Friday.

Working hours
Business: (Mon-Fri) 08.30 – 1230, 1430 – 1830.
Government: (Mon-Fri) 0830 – 1630.
Banking: (Mon-Fri) 0900 – 1330. (N.B. These hours apply to Quito, Guayaquil and certain other large cities; elsewhere local variations include closing on Mon, opening Sat and later in the afternoon.)

Telecommunications
Telephone: All large towns have long distance telephone facilities. Collect calls to USA and Canada. Dialling code for Ecuador: IDD access code + 593 followed by area code (Guayaquil 4, Quito 2) followed by subscriber's number. Overseas calls from Ecuador may be subject to long delays. Local service deteriorated with rapid growth of customers.
Telex and telegram: Facilities available at hotels and from state telecommunications agency, IETEL, in Guayaquil and Quito. Country telex code: ED.

Postal service: Generally unreliable.

Banking
Central bank: Banco Central del Ecuador. *Head offices in Guayaquil:* Banco Continental, Banco de Crédito de Hipotecario, Banco del Progreso, Banco de Guayaquil, Banco del Azuay, Banco de Pacífico, Banco Industrial y Comercial, Banco Sociedad General de Crédito, Banco Territorial, Filanbanco, La Previsora SA. *Head offices in Quito:* Banco Amazonas, Banco Bolivariano, Banco Central del Ecuador, Banco Consolidado, Banco Ecuatorianode Desarrollo, Banco Ecuatoriana de la Vivienda, Banco de Cooperativas del Ecuador, Banco de la Producción, Banco de los Andes, Banco del Pichincha, Banco de Préstamos, Banco Internacional, Banco Nacional de Fomento, Banco Popular del Ecuador, Banco Bolivariano, Banco Consolidado, Caja de Crédito Agrícola-Ganadero.
Foreign branches: (Guayaquil) Banco Holandés Unido, Bank of America, Bank of London and South America, Citibank. (Quito) Banco Holandés Unido, Bank of America, Bank of London and South America, Citibank.

Trade fairs: Annual manufacturers' fair takes place in Guayaquil in October.

Electricity supply: 110 V AC, 60 cycles.

Representation in capital: Argentina, Austria, Belgium, Bolivia, Brazil, Bulgaria, Canada, Chile, China, Colombia, Costa Rica, Cuba, Czechoslovakia, Dominican Republic, Egypt, El Salvador, France, German DR, German FR, Guatemala, Honduras, Hungary, Israel, Italy, Japan, Rep. of Korea, Mexico, Netherlands, Panama, Paraguay, Peru, Poland, Romania, Spain, Sweden, Switzerland, UK, USA, USSR, Uruguay, Vatican, Venezuela, Yugoslavia.

Hotels: Guayaquil
Continental, Chile y 10 de Agosto (tel: 329-270; tx: 3799; fax: 325454).

Grand Guayaquil, Boyaca y 10 de Agosto, P.O. Box: 9282 (tel: 329-690; tx: 43394; fax: 327251).

Oro Verde, 9 de Octubre y García Moreno, P.O. Box: 9636 (tel: 327999; tx: 43744; fax: 329350).

Palace, Chile 214, P.O. Box: 608 (tel: 321-080; tx: 43435).

Ramada, Malecón y Orellana (tel: 312-200).

Unihotel, Clemente Ballén 406, P.O. Box: 563 (tel: 327-100; tx: 043773; fax: 328352).

Hotels: Quito
Alameda Real, Roca 653 y Avenida Amazonas, P.O. Box: 358-A (tel: 562-345; tx: 2224; fax: 560824).

Colón International, P.O. Box: 3103 (tel: 547-300; tx: 2542).

Inter-Continental, Gonzáles Suárez 2500, P.O. Box: 2201 (tel: 230-300; tx: 22223).

Business advice and information
ANDES Airlines, Apartado 3317, Aeropuerto Simón Bolivar, Guayaquil (tx: 3228).

Asociación de Comerciantes e Industriales, Boyacá 1416, Guayaquil.

Banco Central del Ecuador, Av. 10 de Agosto y Briceño, Plaza Bolivar, Quito (tel: 519-384, 571-807; tx: 2359 Bancen ed, 2165 Bancen ed).

Cámara de Comercio de Quito, Apartado 202, Avenida Amazonas y de la República, Quito (tel: 453-011).

Centro de Desarrollo (CENDES), Casilla 2321, Quito (tel: 543-486).

Consejo Nacional de Desarrollo (Development Agency), Avenida 10 de Agosto y Checa, Quito.

Corporación Estatal Petrolera Ecuatoriana, Avenida 6 de Diciembre, Casilla 5007/8, Quito (tel: 528-105, 526-756, 525-420; tx: 022213, 022297).

Dirección Nacional de Turismo (Dituris), Reina Victoria 514 y Roca, Quito (tel: 239-044).

Federación Nacional de Cámaras de Comercio de Ecuador, Avenida Olmedo, Guayaquil.

IETEL (public telex service), Avenida 6 de Diciembre, Quito.

Instituto Nacional de Estadística, 10 de Agost, 229, Quito (tel: 519-022).

Ministry of Finance and Public Credit, Avda 10 de Agosto 1661 y Jorge Washington, Quito (tel: 544-500; tx: 2358).

Ministry of Industry, Trade, Integration and Fisheries, Juan León Mera y Roca, Quito (tel: 524-666; tx: 2166).

Organizaciones Norlops SA (Advertising Agency), Avenida 18 de Septiembre 290, Quito (tel: 529-525).

Publicistas Advertising Agency, 1 de Mayo 812, Guayaquil (tel: 393-300).

TAME Airlines, Avenida Amazonas 1354 y Colón, 6\28,PO Box 2665, Quito.

Car hire: Quito
Budget: Mariscal Sucre Apt (tel: 459-052, 240-763).

Europcar National: Casilla 11019 (tel: 549-848).

Hertz: Santa Maria 517 Amazonas (tel: 540-071).

Car hire: Quito International Airport
Hertz: (tel: 246-381).

Egypt

Anthony Walker

Egypt enters the 1990s facing huge challenges in economic and social policy. If the 1980s were a time when illusions were punctured about Egypt's ability to cope with unrestrained population growth and a sluggish economy, the coming decade is likely to provide an ever sterner test of leadership.

Egypt's $50bn foreign debt puts it in the front rank of Third World debtor nations alongside Brazil, Mexico and Argentina. Its debt-service ratio (obligations on debt as a percentage of export earnings) is one of the highest among all developing countries. Its prospects of attracting new credit are limited. Egypt exhibits all the classic symptoms of a Third World country struggling to keep its head above water. And yet somehow it manages to stay afloat, confounding the direst predictions of imminent economic collapse. Egypt has lived with the problem of foreign debt since the Khedive Ismail put the country into hock to the British and the French in the last century, and there is no reason to believe things will be any different a hundred years from now. What makes the present situation more difficult, however, is the fact that the rate of population increase – a doubling of the present 54m in 25 years is predicted – is putting an intolerable strain on the country's limited resources.

Egypt was a net food exporter in the early 1970s. It now imports more than 50 per cent of its food requirements, and, in spite of an ambitious land reclamation programme, there seems little prospect of it bridging the 'food gap' unless population growth is restrained, and dramatically so. In a predominantly Muslim state, where there is no firm consensus about the need for a crash birth control programme, prospects of real progress towards reducing the rate of population increase are bleak. The crushing burden of too many people dependent on insufficient arable land has presented Egypt with perhaps the greatest challenge in its long civilisation. Few would dispute that the leadership requires enormous reserves of will and political courage to tackle this most critical problem head-on.

Foreign policy

Paradoxically, just at a time when Egypt's economic fortunes are at an increasingly low ebb, its external relationships have perhaps never been better. Its return to the Arab League at the Casablanca summit in May was a triumph for President Hosni Mubarak's quiet perseverance. Egypt's ostracism from Arab forums followed Anwar Sadat's maverick 1979 peace treaty with Israel.

Since coming to power in 1981, Mr Mubarak has edged Egypt back into the Arab mainstream, while not putting in jeopardy the 'cold' peace with Israel. It has been a balancing act accomplished with considerable skill and patience. Egypt's reward was its reception in Casablanca. Even Syria and Libya, the two 'rejectionist' states which had been Cairo's most implacable foes, extended a welcoming hand. Mr Mubarak was able to re-establish in Casablanca the beginning of a fragile dialogue with Libya's mercurial Colonel Qadafi. Egypt and Syria have been warily (in Egypt's case particularly) nudging towards each other. At the same time, Egypt keeps its lines of communications open to Israel, thus demonstrating its uniqueness among the Arabs, as the only Arab state to maintain relations with Tel Aviv. It was a measure of Egypt's self-confidence that directly after Casablanca, Mr Mubarak despatched Mr Boutros Ghali, his Minister of State for Foreign Relations, to Israel. The visit hardly raised a ripple of protest in the Arab world. All this was like money in the bank to Egypt, for it demonstrated to the Americans in particular Egypt's value as a moderating factor in a volatile region, and as a possible bridge between the Palestinians and the Israelis. Egypt is willing and, indeed, anxious to facilitate the beginning of a genuine peace process that would help resolve the Israeli-Palestinian dispute.

Egyptian officials played an absolutely critical role in pushing Mr Yasser Arafat and the Palestine Liberation Organisation towards accepting a formula long demanded by Washington for the beginning of a dialogue. Mr Arafat visits Cairo so often these days that the Egyptian capital has come to resemble a second home. Egypt has also been successful in refurbishing its relations with Gulf states. A visit to Egypt by King Fahd of Saudi Arabia just prior to the Casablanca summit emphasised the normalisation of relations with the Gulf. Egypt's interest in closer ties with Gulf states is not altogether altruistic. The Egyptians hope that increased investment will result, and that the Gulf will start providing additional aid. Gulf businessmen remain wary of Egypt, however. Many have had their fingers burned in joint ventures in Egypt. The corrupt and inefficient Egyptian bureaucracy is legendary.

KEY FACTS

Official title: Arab Republic of Egypt
Head of state: President Hosni Mubarak
Ruling party: National Democratic Party
Capital: Cairo
Official Languages: Arabic
Currency: Egyptian Pound (LE) = 100 piastres
Exchange rate: LE2.59 per $ (Nov 1989)
Area: 1,101,499 sq km

Population: 54m (1989)
GDP per capita: $731* (1987)
GDP real growth: –1.5% (1987)
Inflation: 17.6% (1988)
Trade balance: –$6.8bn (1988)
Foreign debt: $50bn (1989)

* estimated figure

Egypt's other success has been to begin to re-establish a reasonably well-rounded commercial relationship with the Soviet Union. The Soviets are becoming more active in the commercial field in Egypt, in both general trading and in project work. Most important for Egypt, the Soviets are now engaged in modernising antiquated plants they supplied during the 1950s and 1960s period of close economic co-operation. With Egypt's foreign exchange resources squeezed tight, and therefore its ability to import Western technology and equipment much reduced, its need for Eastern bloc assistance on favourable and barter terms is that much greater.

Arab Co-operation Council

Another interesting development on the foreign policy front has been the formation of the Arab Co-operation Council, grouping Egypt, Iraq, Jordan and North Yemen in a new regional constellation whose ostensible aim is economic integration. Agreement to form the ACC was signed in Baghdad in February 1989, but the organisation has made a slow start. Critics have questioned whether it makes sense to group four of the sickest economies – with the possible exception of that of Iraq – in the Arab world, and the suspicion remains that the real purpose of the ACC is political. Some observers have seen the grouping as both a counterweight to the Gulf Co-operation Council, dominated by Saudi Arabia, and as a means of Iraq further seeking to isolate its mortal enemy, Syria.

While the ACC charter speaks of unifying the economies of the four members, this is likely to prove easier said than done. Each of the four countries is burdened by foreign debt, and the degree of real complementarity possible between the respective economies is open to question. Observers have asked whether this might not be another ill-fated attempt by the Arabs to formalise a much-parroted desire for unity – another version of the pan-Arab dream. Mr Mubarak said pointedly of the ACC experiment: 'The wheel started turning and some think it is moving slowly, but we want it that way ... we fear that if it turned faster there would be a setback, then our peoples would lose confidence and our rivals would rejoice.'

Politics

Egypt's internal politics have been marked in the past year by the continuing tug-of-war between the authorities and Muslim extremists. It seems for the moment that the government is winning. Its tough 'carrot and stick' approach appears to be holding the extremists in check, while at the same time allowing the mainstream Islamic trend sufficient latitude to express its views in parliament and in the press. Egypt's Interior Minister, Major General Zaki Badr, a blunt-speaking disciplinarian, has virtually declared war on the militant Islamic tendency, and at any hint of trouble dozens, if not hundreds, of activists are rounded up under Egypt's emergency laws. Some observers believe that Islamic activism has 'peaked', and that the threat to established secular institutions from the religious right is now receding, but that judgment appears premature. Egypt's economic difficulties, and the likelihood that circumstances will get worse before they get better, are likely to continue to provide fertile ground for political activists.

The Muslim Brotherhood appears to have lost some credibility since its good showing in the last People's Assembly election in early 1987. Its performance in the parliament has been indifferent, and its lack of a saleable political programme has been exposed. Mr Mubarak's ruling National Democratic Party continues to dominate national institutions. Meanwhile, the broad left shows no sign of welding itself into an effective political force. The National Progressive Unionist Party, or *Tugammu*, is seeking to make its voice heard in the union movement, but at this stage it does not appear to pose any serious threat to the established order.

KEY INDICATORS

	Unit	1984	1985	1986	1987	1988
Population	m	47.19	48.50	49.61	51.26	51.90
Gross domestic product (GDP)	$m	28,134	30,642	34,689	35,057	37,956
GDP per capita	$	596	632	699	684	731
Inflation	%	17.1	12.1	23.9	19.7	17.6
Agricultural production	1979-81 = 100	117.1	124.0	130.0	135.1	138.1
Food production	1979-81 = 100	122.7	129.6	137.1	144.4	148.5
Exports	LE m	2,198	2,599	2,054	3,046	4,094
Petroleum & products	LE m	1,264	1,765	1,041	1,078	1,277
Cotton	LE m	340	299	308	272	319
Imports	LE m	6,783	6,276	7,247	7,523	14,679
Balance of trade	$m	–5,386	–4,503	–4,011	–4,327	–6,751
Current account	$m	–1,158	–1,454	–1,285	407	–1,190
Tourist arrivals	'000	1,560	1,650	1,680	1,720*	–
Suez Canal dues	LE m	665	654	1,010	1,100*	–
Total external debt	$m	30,513	34,798	37,861	40,264	–
Debt service ratio	%	15.8	17.7	15.9	12.7	–
Exchange rate	$ per LE	1.43	1.20	1.43	1.43	1.43

* estimated figure

Economy

The economy is Egypt's critical battleground, and there is no sign that the authorities are winning the struggle. In fact, the portents are ominous. Egypt's almost insatiable demand for food imports puts ever increasing pressure on its balance of payments. Lack of new credit means the government has to dig even deeper into its limited cash reserves to fund purchases of wheat and other vital commodities such as cooking oil and sugar. Egypt faces a financing gap this year of some $2bn and will require substantial outside assistance to help make up for the deficit. Prospects are not promising. Egypt is already one of the world's largest aid recipients. It receives more than $2bn annually from the USA in civil and military grant aid. Only Israel is more generously endowed by the USA. Egypt is also the recipient of substantial dollops of aid from other Western donors. International lending institutions such as the World Bank are heavily involved in attempts to rescue the Egyptian economy, and yet Egypt somehow reminds one of the drowning man who is not quite able to reach the liferaft that has been cast in his direction.

One of Egypt's main problems is that the era of easy credit is over. Most Western credit agencies have either suspended cover for Egypt, or have imposed very strict criteria for new lending. Western banks – private sector debt to banks in Egypt is estimated at about $3bn – have virtually stopped lending to Egypt. Business is restricted to trade finance for only the most reliable clients. Several foreign banks are trying to assist Egypt to refinance some of its $50bn foreign debt, but are not finding the Egyptian authorities receptive. These schemes have focused particularly on Egypt's $5.7bn military debt to the USA. The repayments schedule for this debt is particularly burdensome, and until the recent increase in US interest rates, the interest on the debt of an average of some 12.1 per cent was much higher than the market rate. Officials within the government have argued that Egypt should follow the example of other countries liable for large military debts to the USA such as Turkey and Israel and utilise cheaper commercial bank finance, but these views have been rejected. Egypt is clearly banking on further Paris Club reschedulings of its official debt, and eventual US forgiveness of the military debt altogether. Bankers report that while letter of credit business remains relatively buoyant for tradeable consumer goods, there is little money available for new project financing. Government-guaranteed export finance has almost dried up. Egypt's economy is in a deep recession. The public sector is in trouble. Starved of new investment, and desperately short of foreign exchange to cover imports of raw materials and spare parts, many public sector factories are operating well below capacity. Loss making public companies such as the Misr Dairy are a continuing burden on the national budget. Government leaders have at last begun talking about the need to privatise loss-making enterprises, but as things stand it is unlikely that the private sector would be interested in taking over government 'dinosaurs' that are grossly overmanned and heavily indebted.

Discussion about the Egyption economy in the past year has focused on prospects for a new agreement with the International Monetary Fund that would enable a second Paris Club rescheduling of some $10-12bn of official debt. Desultory discussion continued for most of 1989, with the IMF seeking Egyptian agreement to push ahead with exchange and interest rate reforms, and also to institute further cuts in subsidies. Egypt has responded sluggishly to these demands. It has increased energy prices, raised taxes on cigarettes, phased out the cheapest (and heavily subsidised) bread loaf and increased interest rates to encourage a higher level of domestic savings in Egyptian pounds. Egypt and the IMF were expected to agree on a new reform programme late in 1989, but judging by Egypt's past record of respecting Fund agreements, there cannot be much optimism that the latest one will stick. A 1987 agreement collapsed within a few months after Egypt failed to meet performance targets on reductions in its budget deficit. Again the deficit is likely to be the sticking point. The IMF wants Egypt to reduce its deficit to about 12 per cent of GDP from the present (1988-89) figure of about 18 per cent. That seems unrealistic, given the huge pressures on the government to increase salaries and wages of a bloated public sector at a time when inflation is estimated to be running at between 25-35 per cent.

The cabinet approved a budget for fiscal 1989-90 that envisages expenditures of LE33.6bn compared with LE28.7bn last year ($=LE2.50 approximately). The budget provided for a narrowing of the deficit from LE7.2bn to LE6.9bn or about 10 per cent of GDP. But this projection appears thoroughly unrealistic. Budget figures should be regarded as indicative at best. Revenues from tax and customs receipts, Suez canal dues and oil sales are expected to reach LE26.7bn compared with LE21.5bn last year. Egypt's other serious worry is unemployment and under-employment. The unemployment rate exceeds 20 per cent or almost 3m out of a total workforce of about 14m. This is a potentially explosive situation in a country where there is no system of unemployment benefits. An added concern for the authorities is that unemployment is particularly high among university graduates. The government has been unable in recent years to fulfil its promise of providing jobs for all graduates. The waiting list is now at least five years, and there seems little prospect of the pool of unemployed graduates being mopped up soon. It is estimated that some 450,000 young people are entering the job market each year, and with the population rate increasing at some 3 per cent annually that staggering figure is unlikely to diminish.

Defence

Egypt's defence sector received a mega-shock in April when Mr Mubarak sacked the long-serving defence minister, Abdel-Halim Abu Ghazala. Mr Ghazala had long been regarded as the second most powerful figure in the government – some might have described him as more powerful than the president himself. He headed a defence empire that operated outside the effective control of the government. It generated its own income, it was engaged in farming and other sideline activities such as contracting, and, of course, it operated a large (by regional standards) arms industry. None of this has changed under the new defence minister, Youssef Abu Taleb, but there is no question now that Mr Mubarak has his own man installed in this highly sensitive post. The reasons for Mr Ghazala's removal are obscure, but it is thought that Mr Mubarak had wanted for quite some time to clip the wings of his strongest potential rival. It was a measure of the president's growing self-confidence that he felt able to do so without fearing a backlash from professional officers who had prospered under Mr Ghazala's aggressive management of the defence sector. General Abu Taleb, a popular and respected former governor of Cairo, has been given the task of curbing defence spending, and perhaps dismantling some of Mr Ghazala's more grandiose schemes. The new defence minister will not have an easy time. Budgetary pressures, and a reduction in earnings from defence industries following the winding down of the Gulf War – Egypt supplied more than $1bn in small arms and ammunition to Iraq during the conflict – are just two of the problems facing General Abu Taleb. There may also be morale problems in the military as falling living standards in the community also affect a hitherto pampered officer class.

Egypt's spending on security amounts to about 25 per cent of budget outlays, according to some estimates. The question is whether the country can continue this high level of spending on the military at a time of relative calm in the Middle East, and at a time when the country is deep in recession. Among expensive projects that General Abu Taleb may be tempted to either scrap or scale down is the plan to assemble a sophisticated American-built tank, the M1A1, at a newly-constructed plant outside Cairo. This project was very much Mr Ghazala's baby and will not necessarily be accorded the same priority under his successor. Another vexed issue in the defence sector is the future of the Arab Organisation of Industrialisation (AOI), which was established in 1974 as a joint venture between Egypt, the United Arab Emirates, Qatar and Saudi Arabia to compete with Israel's military industries. When Egypt signed its peace treaty with Israel, its joint venture partners froze their involvement in the enterprise and with it some $700m in capital investment. Egypt has been trying to persuade its partners to reactivate their involvement and, more important, release frozen funds for much-needed new investment. Negotiations are continuing on a formula that might allow the AOI to receive a new lease of life, but these discussions are proceeding slowly. Egypt's former partners are not anxious, it seems, to resume their involvement under the original terms. Finding a new formula is not proving easy.

Oil and gas

Egypt, which is facing the worrying prospect of being a net oil importer by early next century unless major new finds are made, is trying to speed up the exploration effort. In 1989, some 40 new oil and gas explor-

COUNTRY PROFILE

Political system: Democratic socialist state with executive power vested in president of bicameral National Assembly, who appoints a cabinet of ministers. The legislature is the 458-member People's Assembly.

Population: The population is growing at a rate of 2.8 per cent p.a., creating social pressures especially in the larger conurbations. Population 53.5m (1988 estimate).

Main cities/towns: Cairo (population 14.5m in 1988), Alexandria (5m), Aswan, Luxor, Fayoum, Ismailia, Port Said, Suez, Asyot, Minia. Seven new cities are being built near the existing population centres, to accommodate the rapidly growing population.

Language: The official language is Arabic; French and English are widely spoken, especially in business circles.

Media
Press: Cairo is the major publishing centre in the Middle East. Principal dailies include *Al-Ahram* (circulation 480,000, and 650,000 on Friday), *Al-Akhbar* (750,000), *Al-Gomhouriya* (550,000). Weeklies with largest circulation are *Akhbar al-Yom* (over 1m), *Al-Mussawar, Akher Sa'a*, and the economic weekly, *Al-Ahram Al-Iqtisadi*. Also daily English language paper, *Egyptian Gazette*, and daily French papers, *Le Progrés Egyptien* and *Le Journal d'Egypte*. Main English language commercial weekly is *The Middle East Observer*. Virtually no national trade press.
Radio: National service broadcasting in several foreign languages. Over 12 stations operated by ERTV. About 10m receivers in 1987.
Television: Two channels operated by ERTV broadcast 20 hrs/day. 4.5m sets in 1985.

Domestic economy: Egypt's economy has relied since the 1970s on petroleum sales, remittances from workers abroad, tourism and Suez Canal tolls. Slowing growth rates in the 1980s highlighted the deep structural imbalances in the economy, such as the heavy food subsidies approved by former strongly socialist governments which the current administration is attempting to remove. Foreign investment slowed to a trickle, and the general lack of activity put the banking system under considerable pressure. But in 1987 there were signs of recovery, as oil revenues increased and tourism improved. Inflation remains a serious problem, however.

Employment: The natural increase in the workforce is being swelled by the return of migrant workers to Egypt, as construction projects dry up elsewhere in the Middle East. Unemployment is a major problem, particularly in the cities, and reached nine per cent in 1987.

External trade: The balance of payments relies heavily on earnings from oil, tourism, Suez Canal dues and workers' remittances. The regular trade and current account deficit is only partly offset by rising US aid flows. The current account deficit fell to $1.36bn in 1987 from $2.25bn in 1985.
Exports: Main exports are hydrocarbons (60.1% of 1988 exports) and raw cotton and cotton yarn (18.2%). Other exports include engineering products, oranges, cotton fabrics and potatoes. Main destinations are Italy (30.8% of 1985 total), Japan (10.5%), German FR (9.9%), France (7.9%), Romania (5.2%), USSR (3.3%).
Imports: Main imports are foodstuffs (24.6% of 1986/87 imports), transport goods and machinery (23.8%), chemicals (10.4%), wood and paper products (8.9%), metal products (7-9%) and oils and lubricants (5.5%). Main sources are USA (18.9% of 1985 total), German FR (8.8%), France (8.2%), Italy (7.9%), Japan (6.0%), UK (5.0%).

Agriculture: The sector contributes 16 per cent to GDP and employs around 30 per cent of the workforce (1987), but its importance is declining in relation both to industry and to population growth. Cultivation is concentrated in the Nile and Delta regions and less than five per cent of total land area is cultivated. Smallholders whose operations are generally controlled by co-operatives dominate. Production, which is highly labour-intensive, has been disappointing due to price controls, fragmented land tenure, increased soil salinity and consumer preference for imported foods. Rapid population growth has contributed to dependence on imported food, especially grain. Some 65 per cent of food requirements is imported. Cotton is the main cash crop, accounting for over 11.5 per cent of export earnings and providing the basis for the local spinning and weaving industry. There is also multiple cropping of other cash crops, in particular potatoes, tomatoes, berseem (Egyptian clover) and sugar cane and beet. Major subsistence crops include maize, sorghum, rice, wheat, beans and vegetables. Livestock farming and freshwater fishing are also important. Current Government policy emphasises increased use of machinery and fertilisers, stabilisation of farm prices and canal drainage, but in its efforts to stimulate production the government has abolished all price fixing systems except for cotton and cane sugar. Meanwhile, work continues on major irrigation projects.

Industry: The manufacturing sector contributes around 15 per cent to GDP and 75 per cent of production originates from the relatively inefficient public sector. The government is now stressing private investment, however, and it is looking for a real growth rate of 7.3 per cent per annum in 1988-1993. Agro-allied industries (food processing and textiles in particular) account for 40 per cent of GDP. Emphasis is placed on diversification and import substitution, the development of downstream chemicals and heavy industry, including the Helwan iron and steel complex, the Nag Hammadi aluminium plant and El Dikheila integrated steel works. Other products include cement, fertiliser, sulphuric acid, paper and board, motor vehicles, tyres, glass, pipes, electrical goods and arms and military equipment. The construction sector suffers from shortages of materials. New city of Tenth of Ramadan, between Cairo and Ismailia, is being developed as an industrial centre.

Mining: The sector is small-scale with activity centred on mining of iron ore, phosphate rock, gypsum, limestone, salt, silica, kaolin, clays, barytes and fluorspar. There are reserves of coal, uranium, chromium, tantalum and molybdenum.

ation agreements were signed, and additional acreage was opened up for bidding. One of the factors that encouraged renewed interest in oil and gas prospects in Egypt was the unveiling of a new 'gas clause' that provides incentives for companies to explore for and develop gas deposits. Under the 'take or pay' arrangement, companies will be compensated for gas utilised at a benchmark Mediterranean price for gas, less 15 per cent to cover infrastructure costs. The contracts specify that gas should be paid for either in cash or oil. As long as local oil production exceeds domestic demand, companies can be reasonably certain of getting paid. However, there are few illusions among foreign oil company representatives in Cairo about the likelihood of payment difficulties once Egypt becomes a net importer.

Most interest among oil explorers has focused on the Western Desert, where initial successes indicated prospects of a promising new oil province. But in the past year early optimism has diminished somewhat, although companies are continuing to bid for new acreage and commit substantial resources to the exploration effort. Exploration is also continuing in the Gulf of Suez and new finds are being made, although there seems little prospect of the discovery of big new reservoirs. About 80 per cent of Egypt's 870,000 barrels per day (b/d) production comes from the Gulf of Suez fields, most of which reached maturation several years ago. Egypt is also opening up new acreage in the Sinai, and Amoco has taken a lease of 19,000 sq km. Amoco plans to drill two wells initially, and also to conduct seismic research. Esso drilled a couple of shallow wells in the Sinai in the 1940s, but there has been almost no activity since. Other areas of interest include the Nile Delta and waters offshore. Both are relatively un-explored. Oil company executives report that negotiations with the Egyptian Petroleum Corporation (EGPC) have become progressively tougher over the years. They note that Egypt is squeezing them harder on the cost recovery terms and on share of profit oil. But at the same time they praised EGPC for a more businesslike approach under its new chairman. Delays in negotiating new exploration agreements have been much reduced. EGPC has also been adopting a more realistic position on oil pricing. Previously it was very slow to adjust prices in line with movements in the market.

Agriculture

Increases in agricultural production continue to lag behind the rate of population growth. The yearly increase in agriculture output is hovering around 2.5 per cent compared with a population increase of between 2.7 and 3 per cent. Problems associated with narrowing the food gap have persuaded the authorities to redouble their efforts to open up new agricultural land in an attempt to reduce growing dependence on food imports. Egypt imports more than two-thirds of its wheat and vegetable oil, one-half of its sugar and one-third of its feed-corn. Egypt's land reclamation bureau has identified about 3.4m acres out of 17m surveyed as being suitable for reclamation. This would be in addition to the 7m acres presently in use. Some 2.8m acres (of the 3.4m) would be irrigated with Nile waters. The remaining 600,000 would be watered from stocks of underground water in places like the Western Desert and Sinai. The present target is to reclaim 130,000 acres annually. If this rate is maintained for the next decade, Egypt will have broadly matched in 10 years the 1.32m acres reclaimed since 1952, the year that brought Gamal Abdel Nasser to power. Egypt's land reclamation bureau is presently working on two big schemes: one to the west of the Nile Delta, and the other to the east. Total acreage to be reclaimed in the two schemes is 900,000. Egyptian officials insist that in spite of setbacks over the years in the land reclamation effort, new methods of irrigation and new incentives to farmers will ensure more productive use will be made of reclaimed land.

Outlook

Egypt's outlook is, as always, precarious. It has managed to stagger into the 1990s without serious mishap. Mr Mubarak has managed to maintain order, and at the same time promote limited liberalisation and democratisation. Egypt enjoys an impressive degree of press freedom, and its representative institutions are less constrained than others in the Arab world. But the government's margin for error is being gradually eroded by the continuing deterioration of the economy. Unemployment and inflation have reached the sort of levels that should be flashing amber warning signals at the government. Egypt's stagflationary problems are such that it would seem that a bold new approach is required to jolt the country out of its present economic malaise. The question is whether the leadership has the will or the imagination to do it.

COUNTRY PROFILE

Hydrocarbons: The oil sector forms the mainstay of the economy, contributing over 15 per cent to GDP and accounting for around 55 per cent of export earnings. Proven reserves are sufficient to last over 20 years at current rates of extraction. Production increased to over 900,000 b/d in 1987 and is expected to remain stable at this level. However, domestic consumption is increasing at a faster rate than production and petroleum refining capacity has risen to over 330,000 b/d in an effort to meet domestic requirements and conserve foreign exchange. Activity is controlled by the state-owned Egyptian General Petroleum Corp. Main producing fields are Morgan, Ramadan, July and Ras Ghareb. An extensive network of pipelines runs from Suez to Alexandria. The development of natural gas resources from Abu Maadi, Abu Gharadiq and Abu Qir fields is of increasing importance. Total gas reserves are estimated at up to nine trillion cubic feet, and production is expected to double by 1992, meeting much of Egypt's domestic energy requirement.

Energy: A high priority energy programme is now centred on coal-fired power stations, but also includes the construction of hydro-electric, nuclear and gas-powered electricity generating stations. Efforts to improve energy efficiency are hampered by high level of energy subsidies.

Membership of international organisations: African Development Bank, FAO, GATT, IDA, IFAD, IDA, ILO, IMF, Islamic Development Bank, OAPEC, OAU, UN, UNESCO, WHO, WTO.

Time: GMT +2. (GMT +3 from May to Sept)

Climate: Warm and dry from Nov to Mar, though temperatures can rise to over 38°C, with spells of intense heat in Cairo and Upper Egypt, from Apr to Oct. Rainfall limited to northern coastal area. Hot dusty wind from desert (Khamaseen) blows in Apr-May and Aug-Sep.

Entry requirements
Passport: Required by all except certain Palestinians and some merchant seamen and must be valid for at least three months subsequent to date of visit.
Visa: Required by all except nationals of Arab countries (except Libya) and Malta, for visits exceeding 24 hours. Easily renewed, and transit visas (up to seven days) can be obtained locally from Cairo airport. All visitors must register with authorities within seven days of arrival (those staying in a hotel are automatically registered).
Prohibited entry: Nationals of South Africa, Libya and South Korea.
Currency: Following the return to a partially free market for foreign exchange, a rate of around LE2.2 per $ is used for most transactions. Currency declaration forms must be completed on arrival and retained for presentation to customs authorities and bureaux de change. Import and export of Egyptian currency limited to LE20; export of foreign currency must not exceed the amount imported.

Health precautions
Mandatory: Vaccination certificates against cholera and yellow fever if travelling from infected area.
Advisable: Typhoid and paratyphoid. Avoid drinking tap water. Raw vegetables are inadvisable except in top-class restaurants.

Representation overseas: Abidjan, Abu Dhabi, Accra, Addis Ababa, Amman, Ankara, Antananarivo, Athens, Bangkok, Bangui, Beijing, Belgrade, Berlin, Bern, Bissau, Bogatá, Brasilia, Brazzaville, Brussels, Buenos Aires, Bucharest, Bujumbura, Budapest, Bamako, Canberra, Caracas, Colombo, Conakry, Copenhagen, Cotonou, Dakar, Dar es Salaam, Dhaka, Dublin, Freetown, Georgetown, Guatemala City, The Hague, Hanoi, Harare, Havana, Helsinki, Islamabad, Jakarta, Kampala, Kathmandu, Khartoum, Kigali, Kinshasa, Kuala Lumpur, Lagos, La Paz, Libreville, Lima, Lisbon, Lomé, London, Luanda, Lusaka, Madrid, Manila, Maputo, Mexico City, Mogadishu, Monrovia, Montevideo, Moscow, Muscat, Nairobi, N'Djamena, New Delhi, Niamey, Nicosia, Nouakchott, Oslo, Ottawa, Ouagadougou, Panama City, Paris, Port Louis, Prague, Rangoon, Rome, San Salvador, Santiago, Singapore, Stockholm, Tel Aviv, Tirana, Tokyo, Valletta, Vatican City, Vienna, Vientiane, Warsaw, Washington DC, Wellington, Yaoundé.

Air access
National airline: Egyptair.
Other airlines: Aeroflot, Air Algérie, Air Djibouti, Air France, Air Malta, Air Sinai, Alitalia, Austrian, Alyemda, Balkan, British Airways, China, CSA, Cyprus Airways, El Al, Emirates, Ethiopian, Finnair, Garuda, Gulf Air, Iberia, Interflug, Iraqi, JAL, JAT, Kenya Airways, KLM, Kuwait Airways, LOT, Lufthansa, Malev, MEA, Olympic Airways, PIA, Royal Air Maroc, Royal Jordanian, Sabena, SAS, Saudia, Singapore, Somali, South African Airways, Sudan Airways, Swissair, Tarom, TWA, Tunis Air, Turk Hava Yollari, Yemenia, ZAS Airline of Egypt.
Tax: Airport tax is levied on departure.
International airports: Cairo (Code: CA1), 23 km from city. Alexandria (Code: ALY), 7 km from city. Avoid buying tickets at the last moment as some airline offices close for several hours at a time.

Surface access
Main ports: Alexandria, Port Said (under development), Suez/Port Ibrahim. Car ferry service operates from Genoa/Alexandria, Venice/Pyraeus/Alexandria, Latakia/Alexandria, Jeddah/Suez. Nile steamer service three days/week between Wadi Halfa (Sudan) and Aswan.

Hotels: Wide range available. Bills are quoted in US dollars and may now be settled in Egyptian currency. Necessary to book well in advance. 10 per cent tip usual.

Credit cards: Except airline tickets, the free market exchange rate is used in calculating credit card transactions. Visa, Amex, Diners' Club; Mastercard less frequently accepted.

BUSINESS GUIDE

Car hire: International driving licence and third-party insurance needed. Hire charges should be negotiated in advance. Max speed limit on main roads is 90 kph; heavy fines for infringements. Traffic in Cairo is heavily congested.

City transport
Taxis: Metered and unmetered taxis are readily available, but meters where fitted are often not used. Fares should be agreed in advance. Air-conditioned limousines are available at airports and main hotels. Fares are usually listed in the major hotels. Tipping usually 10 per cent.
Metro: The Cairo metro was built in the mid 1980s and offers a limited network.

National transport
Air: Egyptair operates domestic services from Cairo to Luxor, Aswan, Hurghada, Abu Simbel and Alexandria. Air Sinai operates services to North and South Sinai. If planning to fly south in the winter, book well in advance. Travel to certain areas of Nile Delta is restricted.
Road: 19,000 km surfaced network includes good roads linking Cairo/Alexandria, Cairo/Port Said/Ismailia/Suez/Sinai, Cairo/Fayum/Luxor/Aswan. Speed limit of 90 kph, rising to 100 kph on the Cairo-Alexandria desert road. Substantial fines for speeding.
Buses: Frequent service from Cairo to Alexandria, Port Said, Suez and other major cities. Cheap tram/bus service in Cairo and Alexandria (but crowded).
Rail: 3,500 km of track (1988). Express and through trains from Cairo to Alexandria, Luxor and Aswan. Four classes available; certain routes have air-conditioned sleeping cars and buffet service. Suburban service around Cairo.

Public holidays: When a public holiday falls on a Friday, the following Saturday will be taken as a holiday.
Fixed dates: 25 Apr (Sinai Liberation Day), 1 May, 18 Jun (Evacuation Day), 23 Jul (Revolution Day), 6 Oct (Armed Forces' Day).
Variable dates: Muslim holidays observed include Ramadan Bairam (Eid al-Fitr), Kurban-Bairam (Eid al-Adha), Muslim New Year, Mouled al-Nabi (Prophet's Birthday), Ashura Day; Coptic festivals: Palm Sunday, Sham el-Nessim (Easter).

Working hours
Government and business: 0900 – 1400 in summer (exceptions: Thu 0900 – 1300, Fri closed); 0900 – 1300, 1600 – 1900 in winter (exceptions: Thu 0900 – 1300, Fri closed).
Banking: (Mon-Thu) 0830 – 1230; (Sun) 1000 – 1200.
Shops: 1000 – 2000 in summer; 1000 – 1900 in winter (Mon and Thu 1000 – 2000). Closed Sun.

Telecommunications: A major updating of services is under way, including new exchanges in Cairo and Alexandria.
Telephone: Automatic system in main towns, though service is subject to delays. International calls should be booked well in advance. Dialling code for Egypt: IDD access code + 20 followed by area code (3 for Alexandria, 97 for Aswan, 2 for Cairo) followed by subscriber's number. Note: Egyptian residents/firms are indexed under the initial letters of their first names. Non-Egyptians are indexed under initials of surname.
Telex and telegram: Telex and telegram: Available at Cairo Central Telegraph Office (24 hour service), main hotels and post offices. International telex code: UN.

Postal service: Advisable to use airmail, which to Europe takes around five days. Postal censorship in force and internal postal service highly unreliable.

Trade fairs: Cairo International Book Fair, January; Cairo International Trade Fair, March. Both at the Exhibitions Ground, Nasr City, Cairo. For further information contact GOIEF.

Electricity supply: 220-440 V AC in most areas; in some rural districts 110-380 V AC is still found.

Weights and measures: Metric system (local units also in use).

Representation in capital: Afghanistan, Albania, Algeria, Angola, Argentina, Australia, Austria, Bahrain, Bangladesh, Belgium, Bolivia, Brazil, Bulgaria, Burkina Faso, Burma, Burundi, Cameroon, Canada, Cape Verde, Central African Republic, Chad, Chile, China, Colombia, Congo, Côte d'Ivoire, Cyprus, Czechoslovakia, Denmark, Ecuador, Eire, Ethiopia, Finland, France, Gabon, German FR, German DR, Ghana, Greece, Guatemala, Guinea, Guinea-Bissau, Hungary, India, Indonesia, Iran, Iraq, Israel, Italy, Japan, Jordan, Kampuchea, Kenya, North Korea, Kuwait, Lebanon, Liberia, Malaysia, Mali, Mauritania, Mauritius, Mexico, Mongolia, Morocco, Nepal, Netherlands, Niger, Nigeria, Norway, Oman, Pakistan, Panama, Peru, Philippines, Poland, Portugal, Qatar, Romania, Rwanda, Saudi Arabia, Senegal, Sierra Leone, Singapore, Somalia, Spain, Sri Lanka, Sudan, Sweden Switzerland, Syria, Tanzania, Thailand, Tunisia, Turkey, UAE, Uganda, UK, Uruguay, USA, USSR, Vatican, Venezuela, Vietnam, Yemen AR, Yemen PDR, Yugoslavia, Zaïre, Zambia.

Hotels: Cairo
Atlas Zamalek, 20 Gamet el Dowal el Arabeya (tel: 3464175, 3465782; tx: 93281; fax: 3476958).

Baron Heliopolis, Aruba Street, P.O. Box: 2531 (tel: 66-87-01; tx: 93445; fax: 907077).

Belair, Mokattam Hill, P.O. Box: 2755 (tel: 920192; tx: 93210).

Cairo Inn, 26 Syria Street, Mohandesseen (tel: 3490661; tx: 93957).

Cairo Sheraton, Galaa Square, Giza, P.O. Box: 11 (tel: 3488600, 3488700; tx: 92041, 22382; fax: 3489051).

Continental Savoy, 10 Opera Square (tel: 911030, 911049; tx: 92726).

El Borg, Saray El Gezira Street (tel: 40-61-79; tx: 94148).

El Gezira Sheraton, Gezira, P.O. Box: 264 (tel: 3411333, 3411555; tx: 228123; fax: 3405056).

El Nil, Ahmed Ragab Street, Garden City (tel: 3542800/4/5; tx: 93284).

El Salam Hyatt, Abdel Hamid Badawi Street, Heliopolis, P.O. Box: 5614 (tel: 452155; tx: 92184; fax: 455755).

Green Pyramids Hotel, Pyramids Road, Giza, P.O. Box: 22 (tel: 856786; tx: 93701).

Marriott, Baraya El, Gezira, Zamalek, P.O. Box: 33 (tel: 3408888; tx: 93464; fax: 340-6667).

Mena House Oberoi, Pyramids Road, Giza (tel: 855444; tx: 92316; fax: 853424).

Meridien, Corniche el Nil, P.O. Box: 2288 (tel: 845444; tx: 325).

Meridien Heliopolis, 5 El Ouruba Street (tel: 901819; tx: 20300; fax: 908533).

Movenpick, Cairo Airport, P.O. Box: 2741 (tel: 664977, 679799; tx: 93402; fax: 202667374).

Nile Concorde, El Tahrir Street, Dokki (tel: 701873; tx: 93655).

Nile Hilton, Tahrir Square (tel: 740777, 750666; tx: 92222; fax: 760874).

Novotel, Cairo Airport, P.O. Box: 8 (tel: 661330; tx: 93711, 20953).

President, Taha Hussein Street, Zamalek (tel: 3400652, 3400718; tx: 93655; fax: 3400652).

Ramada Renaissance, Cairo to Alexandria desert road, near Pyramids, Giza, P.O. Box: 70 (tel: 538111; tx: 93595).

Ramses Hilton, 1115 Corniche el Nil (tel: 744400, 758000; tx: 94260; fax: 757152).

Safir, Midan El Misaha (tel: 3482828, 3482424; tx: 20350).

Semiramis Inter-Continental, Corniche El Nil, P.O. Box: 60 (tel: 3557171; tx: 94257).

Shepheard's, 182 El Nil Street, Agouza (tel: 699346; tx: 20693).

Sheraton-Heliopolis, Uruba Street, Horreya, P.O. Box: 2466 (tel: 667700, 665500; tx: 93350; fax: 666108, 907546).

Siag Pyramids Hotel, 59 Marioutya Road, Ahram, P.O. Box: 107 (tel: 857399; tx: 93522).

Sonesta, 4 el Tayaran Street, Nasr City (11 km from airport) (tel: 609444; tx: 92721918; fax: 619980).

Windsor, 19 Alfi Bey Street, P.O. Box: 2045 (tel: 915810, 915277; tx: 93839).

Hotels: Hurghada
Magaweesh (Club Méditerranée) (tel: 914972; tx: 92035).

Sheraton, 48 B Giza Street, Orman Tower Building, Red Sea (tel: 40785, 40779, 40604; tx: 92750; fax: 40604).

Hotels: Ismailia
PLM Azur Etap, Forsan Island, P.O. Box: 77 (tel: 765322; tx: 63038).

Hotels: Luxor
Isis, Khaled Ibn El Walid Street (tel: 82750; tx: 92078).

Luxor, Opposite Karnak Temple (tel: 82400; tx: 24126).

Movenpick Jolie Ville, Crocodile Island (tel: 84855; tx: 23825; fax: 84936).

PLM Azur Etap Luxor, Corniche el Nil (tel: 82011; tx: 92080).

Savoy, Corniche el Nil (tel: 82200; tx: 92160).

Sheraton, P.O. Box: 43 (tel: 84544, 84472; tx: 24437; fax: 84941).

Winter Palace, Corniche el Nil (tel: 82222; tx: 82160).

Business advice and information
Cairo: Cairo International Fair (GOIEF), Exhibitions Ground (tel: 836136; tx: 92600 fair).

Federation of Chambers of Commerce, 4 Midan El Falaki (tel: 3551164; tx: 92645).

General Organisation for Industrialisation, 6 Khalil Agha Street, Garden City (tel: 3546335).

Investment and Free Zone Authority, PO Box 1007 (tel: 906804; tx: 92235).

Ministry of Economy and Foreign Trade, 6 Adly Street (tel: 933489, 908911).

Ministry of Industry, 2 Latin America Street, Garden City (tel: 3550641, 3551855).

Banking
Central Bank of Egypt (central bank), 31 Kasr el Nil, Cairo (tel: 751529; tx: 223866).

Bank of Alexandria, 49 Kasr el Nil, Cairo (tel: 913822; tx: 92069,92738 bnalx).

Banque du Caire, 22 Sharia Adly Pasha, Cairo (tel: 749669; tx: 92022).

Banque Misr, 151 Sharia Mohamed Bey Farid, Cairo (tel: 912711; tx: 92242).

National Bank of Egypt, 24 Sharia Sherif, Cairo (tel: 354470; tx: 92832).

El Salvador

Barbara Annis

In 1989, President José Napoleòn Duarte lay dying and saw the disintegration of most of the things for which he had long worked. El Salvador's notorious far-right National Republican Alliance (ARENA) party wrested the presidency from the beleaguered and discredited Christian Democrats (DC). Surely the story will be recorded as a modern tragedy.

Nine years after Archbishop Oscar Romero of San Salvador was gunned down in his church while serving Mass, President Duarte condemned ARENA party founder and national assemblyman Roberto D'Aubuisson as the intellectual author of the murder. Duarte had sworn during his inauguration in 1984 that he would bring Romero's killers to trial.

Since then, however, Duarte has been unable to legally prove D'Aubuisson's link to the slaying, despite allegations stemming from the arrest in 1989 of the man who admits he drove the getaway car. Duarte came close to his inaugural vow, but narrowly missed his objective. Contained within the symbol of the slain archbishop and Duarte's inability to bring the killers to justice lies the tragedy of the failed president.

Now in power, as liver cancer consumes Duarte, is ARENA party's aptly named Alfredo Christiani – symbol of Salvador's extreme, death-squad-linked, old-right's rebirth as new-right moderates and economic miracle workers. With Christiani at the helm, ARENA successfully shunned its long-time association with the death squads and won the election on a platform of economic reform.

The economy of El Salvador, however, is irrevocably intertwined with the ongoing civil war. Inaugurated 1 June 1989, Christiani calls for the demobilisation of the FMLN rebels as a starting point for getting the country back on the road to economic stability. It is a feat he cannot achieve, having rejected outright the FMLN's tentative bids for peace.

Instead, El Salvador will continue to get by on the approximately $400m yearly it receives from the USA, which constitutes about half of its foreign currency income. That amount has increased from $65m in 1980, but has decreased from almost $700m in 1985. Representatives of 50 governments attended Christiani's inauguration, applauding the democratic transition of power. If Christiani is in fact able to establish his legitimacy, he can also expect additional aid from some of those sources.

Legitimacy in the international community will not be easily won by Christiani. The months since his inauguration are fraught with social polarisation. Even moderate Christian Democrats within and out of the government are far removed from ARENA politics and policies. Political assassinations continue, but are now frequently attributed to factions within ARENA doing each other in as well as to guerrillas and death squad members. Old alliances are being shaken and new loyalties established.

Modern Latin American man

Christiani is far more than ARENA with clean hands and a smile on its face, maintain his supporters. He represents the best hope for the country, they say. There is no denying, however, that Christiani queerly mimics US policy as it has evolved since the end of the Carter administration.

Christiani won the presidency partly by attacking corruption among the ruling Christian Democrats. He attacked the bureaucracy, and he scorned its attempts to implement social and economic justice policies prescribed by the Carter administration and abhorred by El Salvador's elite.

A graduate of Washington's Georgetown University, Christiani even manages to capture certain speech inflections and intonations of Reagan administration free-market, cold war stalwarts such as Jeanne Kirkpatrick.

The Salvadoran president's family fortune was made in coffee, the traditional backbone of the Salvadoran economy. One of Christiani's first acts in office was to announce his intention to privatise El Salvador's coffee exports. A government-run coffee agency was created in 1980 as part of reforms that included nationalising commodity trade and banks. Christiani says he will double coffee production during his first year in office and the benefits will go mostly to the growers. He suggests the establishment of private marketing agencies to parallel government-run agencies.

Nevertheless, El Salvador's coffee harvest in 1988 was the lowest in 30 years. It is a favourite target for guerrillas, disgruntled growers fighting government policies and the forces of nature.

Christiani's task with agriculture and other sectors as yet unaddressed will be to walk a fine line: attempting to maintain his current supporters by opening the economy to foreign and domestic investors, while dismantling State social welfare programmes at a pace that won't bring the whole house down, all the while holding leftist rebels at bay. He must also toss enough economic bones to the military and far-right extremists fast enough to keep them from abusing human rights so badly that the USA is forced to cut off aid.

The insurgents

Knowledgeable observers predict El Salvador will become increasingly polarised during the Christiani administration. They see disenfranchised Christian Democrats, purged from the government, eventually joining with broad-spectrum FMLN guerrillas as the so-called democratic centre shifts inevitably to the right under ARENA.

Former president José Napoleòn Duarte called the Christian Democrats the 'wedge' between the extremes on the right and the left. With that wedge removed, one theory suggests increased numbers of right-left clashes will force greater collusion between the centre and the left, which have more in common with one another than with minority interests represented by ARENA.

In a mid-1989 peace proposal, for instance, the FMLN offered to suspend attacks on economic infrastructure, stop raids on military establishments, halt car bombings and urban insurgency in exchange for maintaining the main reforms of the Christian Demo-

KEY FACTS

Official title: Republic of El Salvador

Head of state: President Alfredo Felix Cristiani Burkard

Ruling party: Partido Alianza Republicana Nacionalista (ARENA)

Capital: San Salvador

Official Languages: Spanish

Currency: Colón (C) = 100 centavos

Exchange rate: C4.99 per $ (Nov 1989)

Area: 21,400 sq km

Population: 5.1m (1988)

GDP per capita: $955 (1988)

GDP real growth: 1.6% (1988)

Inflation: 19.7% (1988)

Trade balance: –$305m* (1987)

Foreign debt: $1.8bn (1988)

* estimated figure

crats – agrarian and bank nationalisation and commodity controls. Christiani says he is open to a dialogue with the rebels, but will not change his economic programmes or consider power sharing arrangements.

The other economy

The World Bank and the International Monetary Fund (IMF) appear to recognise a Salvadoran economy separate from the war economy supported primarily by the USA. Perhaps not surprisingly their prescription for its success is essentially the same as their prescription for a successful economy practically anywhere.

The sister institutions, for example, have proposed El Salvador devalue its currency. The currency had remained relatively stable at $1:C5 between 1986-89. It began to slip in the black market in late 1988 and had fallen to C6.20 to the dollar just before the elections. The multilaterals were proposing a devaluation of between 50-100 per cent.

They argue a devaluation will help El Salvador's exports become more competitive in international markets – a contention considered dubious given their strong commodity base where prices worldwide are low and regulated. All of El Salvador's commodity exports have seen a steady decline in recent years, mostly due to the war.

Christiani has said El Salvador's economic future lies with its manufacturing sector. He has likened the country to Japan. The sector shows little signs currently, however, of fulfilling that promise. It has only limped along, directing its products mostly to El Salvador's small impoverished internal market since the beginning of the war.

The US Chamber of Commerce in El Salvador believes in painting as rosy a scenario as possible. It appears to hold faith that Christiani will be able to significantly turn things around. At a Washington briefing AMCHAM representatives said the Salvadoran election results provide optimism tempered by realisation of serious problems that continue to confront the country.

Admitting that projected growth for 1989 was less than 1 per cent, the business sector laid most of the blame on the poor coffee harvest. Strong winds early in 1989 stripped flowering trees, resulting in a 45 per cent reduction in coffee exports and a $100m drop in exports compared to 1988. That reduction also meant a $25m drop in coffee export tax revenues and a C200m fall in rural family wage income. Further results of the poor coffee harvest are a projected increase in the trade deficit from $420m to about $530m.

One positive note the Salvadoran business sector was able to strike was the high probability that US economic support funds would continue to flow and would give some dollar exchange relief.

In mid-1989 a severe shortage of dollars in the banking system was seriously restricting imports and adversely affecting overall business activity. As a consequence, the Central Bank was forced to adjust its import policy by requiring the extension of letters of credit from 90 to 180 days and to limit the availability of dollars at the official rate to essential goods and services.

The business community also maintained high hopes for the country's ability to exploit its potential to export fresh agricultural products, both traditional and non-traditional, and seafood and handicraft items. Critics find those assertions ridiculous in the face of 40 per cent unemployment and 55,000 people in uniform. A high percentage of illegal Salvadoran refugees to the USA are former soldiers who were unable to find employment after they had completed their terms in the army. What if the war were to end tomorrow?

However, El Salvador's business sector says it is pinning its hopes on Christiani finding a solution to the war. Christiani began his attack on that problem primarily by proposing several reforms to the Salvadoran criminal code. He suggested, among other changes, making it a criminal offence to head or organise a group that spreads doctrines that might

KEY INDICATORS

	Unit	1984	1985	1986	1987	1988
Population	m	4.8	4.8	4.9	5.0	5.1
GDP (1985 prices)	C m	14,053	14,331	14,421	14,825	–
GDP per capita	(current) $	845	1,170	812	942	955
Inflation (average rate)	%	11.6	22.2	31.9	24.9	19.7
GDP real growth	%	1.5	2.0	6.3	2.8	1.6
Agricultural sector	% real gwth	3.3	–1.1	–3.1	2.1	–3.8
Food production	1979-81=100	103.9	99.1	104.1	101.0	–
Exports	C m	1,814	1,698	3,563	2,955	3,044
Coffee	C m	1,131	1,131	2,453	1,758	1,790
Cotton	C m	25.9	76.9	24.5	15.0	–
Imports	C m	2,201	2,165	3,860	4,478	5,034
Petroleum	C m	325.8	333.1	369.0	521.7	405.0
Balance of trade	C m	–387	–467	–297	–1,523	–
Current account	$m	–188.5	–188.7	–17.1	–	–
Exchange rate	C per $	2.5	2.5	5.0	5.0	5.0

* estimated figure

COUNTRY PROFILE

Political system: Republic with executive power vested in the president (who is elected for five years), assisted by the vice-president and council of ministers. Legislative power is held by the unicameral Legislative Assembly, which has 60 members elected for three years by universal suffrage.

Main cities/towns: San Salvador, Santa Ana, San Miguel.

Media
Press: In San Salvador there are four daily newspapers - *Diario de Hoy* (circulation 87,260), *La Prensa Gráfica* (97,312), *Diario Latino* (20,000) and *El Mundo* (58,032). Also provincial newspapers in major towns.
Radio: There are over 70 commercial radio stations, and four commercial television channels.
Television: There are two government-run educational TV channels and four commercial channels.

External trade: Regular balance of trade deficit since 1981 owing to contraction in sales of manufactured goods to CACM, declining cotton exports and falling world coffee prices.
Exports: 1987: Coffee (60% of total value), industrial manufactures (31%), miscellaneous manufactures (7%), sugar (2%), cotton (5%). Main destinations (1987): USA (45% of total value), German FR (17%), Guatemala (12%), Japan (4%).
Imports: Basic manufactures (typically 30% of total value), raw materials (25%), chemical products (16%), foodstuffs (15%), machinery and transport equipment (10%). Main sources: USA (36% of total value), Guatemala (13%), Mexico (9%), Venezuela (7%), Japan (6%), German FR (5%), Costa Rica (4%).

Agriculture: Contributes about 25 per cent to GDP and employs around 45 per cent of the working population. Principal export crop is 'mild' coffee, which accounts for over 60 per cent of export earnings. However, production has been badly affected in recent years by the negative results of the agrarian reform, which is a failure. Nationalisation of the bank and other institutions has also contributed, as have guerrilla activities and weather conditions, but to a lesser extent. Other major crops are cotton, sugar-cane, maize, beans and rice. A programme of agrarian reform has been under way since 1980, and over 20 per cent of the country's farmland has now been redistributed to farm labourers and tenant farmers. Approximately 32 per cent of total area is cultivated arable land, 30 per cent pasture and 10 per cent woodland/forest.

Industry: El Salvador has the most developed industrial sector of any CACM country, with manufacturing accounting for up to 16 per cent of GDP. Major industries include textiles, leather, pharmaceutical drugs and machinery. The development of agro-industries has led to the expansion of food processing and frozen shrimp has become a leading export. Other goods manufactured for the export market include computer components, integrated circuits, automobile parts, lighting fixtures, TV sets and clothing. Production throughout the sector has been badly affected by guerrilla activity, shortages of capital and strikes, which have forced many firms to close down and others to operate at well below capacity.

Mining: Small amounts of gold and silver are mined, and reserves of copper, iron-ore, lead, mercury, sulphur and zinc are also thought to be present.

Energy: Electrical energy is produced by four hydroelectric installations and one geothermal plant. Total installed capacity about 650 mW.

Membership of international organisations: Central American Common Market, FAO, G-77, IADB, IAEA, ICAC, ICAO, ICO, IDA, IFAD, IFC, ILO, IMF, IMO, INTELSAT, Inter-American Development Bank, Interpol, ITU, Organization of American States, IWC (Wheat), ODECA, PAHO, SELA, UNESCO, UNO, UPU, WFTU, WHO, WIPO, WMO, World Bank, WTO.

subvert public order or disseminate propaganda that subverts the public order.

The intention of the so-called reforms appears to have been to try to undercut popular based support for the leftist insurgents. Critics, mostly outside El Salvador, charged the reforms put peaceful and violent demonstrators in the same category and unfairly infringed on freedom of the press. Some critics suggested the mostly undefined reforms left far too much interpretation to the government and were meant to serve as a warning to its opponents.

Washington's view

Much of the US Congress and the Bush administration remain sharply split over El Salvador. For example, a furore was raised when it was learned that Vice President Dan Quayle had visited Roberto D'Aubuisson while on a trip to El Salvador. Quayle defended the visit, saying he had warned D'Aubuisson the United States would not tolerate ongoing human rights abuses. Critics of the visit said Quayle's visit gave tacit recognition of D'Aubuisson's unofficial power to influence the course of events in El Salvador and appeared to some as a signal that the United States now supports a non-reformist policy there.

Congress only appropriated approximately $85m for the Salvadoran military in fiscal year 1990, and withheld $5m in economic assistance pending significant progress in solving a September 1988 massacre of peasants. A report accompanying the appropriation charged human rights conditions in the country had deteriorated in the 1988-89 period.

BUSINESS GUIDE

Time: GMT -6.

Climate: Semi-tropical. Dry season from Nov to Apr; temperature range 15°C to 32°C. Rainy season from May to Oct; average temperature 25°C. Generally temperature depends on altitude and coastal areas are hotter and more humid than upland areas.

Entry requirements
Passport: Required by all.
Visa: Not required by citizens of: Colombia, Costa Rica, Finland, Guatemala, Honduras, Japan, Mexico, UK, and Western European countries (except France).
Currency: Restrictions on import and export of foreign and local currency. US dollars are accepted almost everywhere.

Health precautions
Mandatory: Yellow fever vaccination required if arriving from an infected area.
Advisable: Typhoid, polio, hepatitis, tetanus vaccinations. Malaria prophylaxis should be taken as risk exists throughout the country. High rabies risk. Water precautions are essential.

Representation overseas: Asunción, Bogotá, Bonn Brasilia, Brussels, Buenos Aires, Caracas, Guatemala City, Lima, London, Managua, Mexico City, Madrid, Montevideo, Ottawa, Panama City, Paris, Quito, Rome, San José, Santiago, Santo Domingo, Taipei, Tegucigalpa, Tel Aviv, Tokyo, Vatican City, Washington.

Air access
National airline: TACA International.
Other airlines: Continental, Copa, Emerald, Lacsa, Pan Am, Sahsa.
International airport: San Salvador-El Salvador International (SAL), 44 km from city; car hire, restaurant.

Surface access
Main ports: Acajutla, La Unión/Cutuco, La Libertad (fishing only).
Overland access: Roads run from Guatemala and Honduras. Scheduled bus/coach services have been varied (e.g. Ciudad Guatemala-San Salvador, Tegucigalpa-San Salvador) but for these and any services in operation account should be taken of political circumstances and local enquiries made.

Hotels: Advisable to book in advance; 10-15 per cent tip is usual.

Car hire: Foreign or international licence required.

City transport
Taxis: Can be hailed or ordered by telephone; fixed rate system is not rigidly followed – check before proceeding; no taxis have meters. Tipping is unusual.

National transport
Air: No scheduled internal services but charter flights are available.
Road: Pan-American Highway runs (over 600 km) through country; total network around 16,000 km, approximately 45 per cent all-weather – linking most areas/main centres.
Buses: Good services between major towns; often faster than trains. City buses are crowded but the service is good.
Rail: Lines radiate from San Salvador (to Metapan, Ahuachapan, La Unión/Cutuco, Puerto Acajutla); trains are slow and primitive; no restaurant cars.

Public holidays: 1 Jan (New Year's Day) , 13 Apr (Good Friday), 16 Apr (Easter), 1 May (Labour Day), 10 May (government only), 15 June (Corpus Christi), 22 Jun (government only), 29 Jun and 30 Jun (banks only), 4-6 Aug (San Salvador Festival), 15 Sep (Independence Day), 12 Oct (Discovery of America), 2 Nov (All Souls' Day), 5 Nov (First Call of Independence), 24-25 Dec (Christmas), 31 Dec (Bank Holiday).

Working hours
Government and business: (Mon-Fri) 0800 – 1600; (Sat) 0800 – 1200. N.B. Government offices close at 1600.
Banking: (Mon-Fri, San Salvador) 0900 – 1300, 1345 – 1600; times vary in other cities.
Shops: (Mon-Fri) 0900 – 1200, 1400 – 1800; (Sat) 0800 – 1200.

Telecommunications
Telephone: Dialling code for El Salvador: IDD access code + 503 followed by subscriber's number. No area codes are needed. For telephone information dial: 14 (24-hr service); international enquiries dial: 19.
Telex: Public telex available at Antel, the government telecommunications office, and at embassies.
Telegram: Service available at Antel offices only.

Banking
Central bank: Banco Central de Reserva de El Salvador.
Head offices: Ahorro Préstamos e Inversiones, Ahorro Metropolitanos, Asociación de Ahorro y Préstamo Atlacatl, Asociación de Ahorro y Préstamo la Popular, Banco Agrícola Comercial de El Salvador, Banco Capitalizador, Banco de Comercio de El Salvador, Banco de Crédito Popular, Banco Cuscatlán, Banco de Desarrollo e Inversiones, Banco de Fomento Agropecuario, Banco Financiero, Banco Hipotecario de El Salvador, Banco Internacional de El Salvador, Banco Salvadoreño, Banco Mercantil, La Central de Ahorros, Construccion y Ahorro, CRECE, Crédito Immobiliario, Financiera Nacional de la Vivienda, Financiera Salvadoreña.
Foreign branches: Banco Centro-Americano de Integración Económica, Banco de Santander y Panamá, Bank of America, Bank of London & South America, Citibank.

Advertising: Available in the press, cinemas, commercial TV and radio stations, periodicals and posters.

Electricity supply: 110 V AC, 60 cycles.

Representation in capital: Argentina, Brazil, Chile, Colombia, Costa Rica, Dominican Republic, Ecuador, France, German FR, Guatemala, Honduras, Israel, Italy, Japan, Mexico, Nicaragua, Panama, Paraguay, Peru, Spain, Taiwan, USA, Uruguay, Vatican, Venezuela.

Hotels: Costa del Sol
Izalco Cabana Club (tel: 236764; fax: 503 240363).

Tresoro Beach (tel: 340600; tx: 20037).

Hotels: San Salvador
Alameda, Alameda Roosevelt y 43 Avenida Sur (tel: 239999; tx: 20435).

Camino Real (Westin), Boulevard de los Héroes (tel: 233344; tx: 20130).

El Salvador Sheraton, 89 Avenida Norte, Colonia Escalón (tel: 242222; tx: 20166).

Presidente, Avenida de la Revolución, Colonia San Benito, P.O. Box: 888 (tel: 243044; tx: 20037).

Siesta, P.O. Box: 01-217 (tel: 232500; tx: 30 104).

Terraza, 85a Avenida Sur y Calle Padres Aguilar, Colonia Escalón (tel: 235444; fax: 244674).

Car hire: San Salvador Avis: 43 Avenida Sur 137 (tel: 24-2623, 23-6321; tx: 20518 AVIS SAL).

Budget: 85 Avenida Sur 220 (tel: 23-5677, 24-4420).

Hertz: Calle Los Andes J-16, Colonia Miramonte (tel: 26-8099; tx: 20153).

Car hire: International airport Avis: (tel: 24-6268, 39-9268; tx: 20518 AVIS SAL).

Hertz: (tel: 24-5881, 39-9481).

Other useful addresses
AESA Aerolíneas de El Salvador, SA de CV, Centro Comercial Beethoven sobre Paseo Gen. Escalón, Apdo (06)1830, San Salvador (tel: 24-6166).

ANAES (advertisers), Edificio Banco, Capitalizador, 5to Piso No 512, San Salvador (tel: 21-6979).

ANTEL (telephone co), Calle Rubén Darío y 5a, Ave Sur, San Salvador.

Asociación Salvadoreña de Industriales, Colonia Roma, Apdo Postal (06)48, San Salvador (tx: 20235).

Banco Central de Reserva de El Salvador, la Calle Poniente y 7a Avenida Norte, San Salvador (tel: 22-1222; tx: 20088).

Buró de Convenciones y Visitantes de la Ciudad de San Salvador, Suite 221-222, Hotel Presidente, Avda La Revolución, Col. San Benito, Apdo 2124, San Salvador (tel: 24-0508; tx: 20037).

El Salvador Tourist Board (Instituto Salvadoreño de Turismo), Calle Rubén Darío 619, San Salvador (tel: 22-1208, 22-8000; tx: 20775 ISTU SAL).

Ministry of Economy, Centro de Gobierno, 4a Avda Norte 233, San Salvador.

Ministry of Finance, 3a Avda Norte y 13 Calle Poniente, San Salvador.

Ministry of Foreign Trade, Paseo General Escalón 4122, Apdo 0119, San Salvador (promotes foreign investment, free zones etc and deals with industrial development, joint ventures etc) (tel: 24-3000; tx: 20269).

Press Association of El Salvador, Paseo Gral Escalón 4130, San Salvador (tel: 23-8943).

TACA International Airlines, Edificio Caribe 2nd Floor, San Salvador (tel: 23-2244; tx: 30156).

Equatorial Guinea

Susi Pollak

In 1988, when Equatorial Guinea hoped to negotiate a structural adjustment package with the IMF, President Teodoro Obiang Nguema asked France rather than Spain, the former colonial power, to support its case at the World Bank. 'We feel like orphans, being the only Spanish-speaking country in Equatorial Africa,' he explained. 'All the other countries in the region are French-speaking. That is why we want to become part of the francophone group.' He said this during a private visit to Paris, and nobody asked him why he forgot to mention his closest neighbour, Nigeria, which is larger than all the francophone states of the region combined. At the time, Equatorial Guinea was already a member of the Economic Community of Central African States (CEEAC) and therefore of the franc zone, with the convertible CFA franc replacing the old *ekwele* and France its largest aid donor.

There were good reasons why the existence of Nigeria slipped Obiang's mind: the 1987–88 row with Lagos about his government's links with South Africa was not long behind him. Relations with Nigeria have officially improved since the expulsion from Equatorial Guinea of seven South Africans, but many officials in Lagos remain privately convinced that the South Africans have left behind a spy network and installations. Indeed, Pretoria promised to continue its 'aid' to Equatorial Guinea, despite the expulsion of its personnel.

After the row with Lagos the Obiang government expressed fear of Nigerian military intervention and felt that its 2,000-strong army – beefed up by 600 Moroccan troops – would be unable to withstand an onslaught from Lagos. France therefore undertook to reorganise Equatorial Guinea's security forces; a recent defence agreement provides for the setting up by France of a motorised infantry unit and other military assistance. In 1988 French budgetary aid amounted to CFA780m; total French aid the previous year was CFA8.8bn. French investment has concentrated on the acquisition of large timber plantations on the mainland, near the border with Gabon.

At the end of 1988 the country's foreign debt was still growing, and debt service charges amounted to about 50 per cent of export earnings. Cocoa is the main source of foreign exchange. The 1985–90 $19m cocoa rehabilitation package co-financed by the World Bank and the EDF is beginning to show results, with production up by 15 per cent in 1988, at 10,000 tonnes. This is still only a quarter of the pre-independence output. The $19m went mainly towards helping the small-holders who farm in Rio Muni, the mainland part of Equatorial Guinea. The big plantations on the island of Bioko (formerly Fernando Po) remain untended, although some cocoa is still harvested from the old trees. Soon after Obiang's visit to Paris, the IMF's $16m structural adjustment facility was unblocked for banking reforms and public investment. A five-year aid package worth $58m was agreed with six countries (France, Spain, Italy, Japan, China and the USA). Most of the money was earmarked for the rehabilitation of the cocoa plantations, and the rest was to go towards boosting the fishing sector and improving communications.

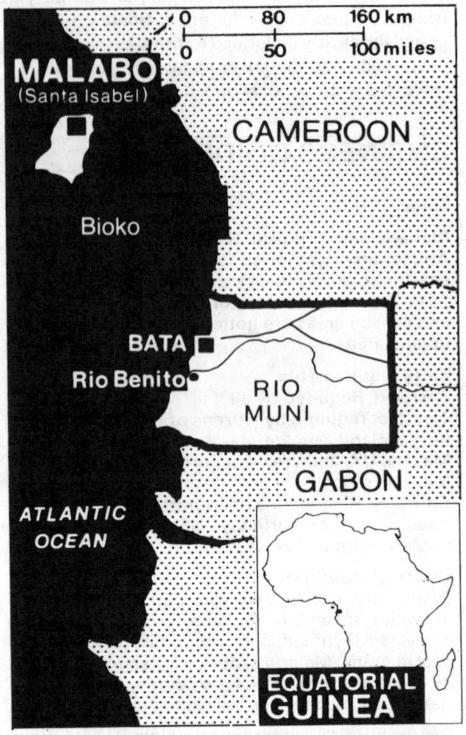

Relations with Spain

Relations with Spain have long been overshadowed by Spanish criticism of the brutal dictatorship run by Obiang and his predecessor. The executions of opponents in Malabo and the suppression of opposition is chronicled only in the Spanish press. But in January 1989 Obiang headed a 50-strong delegation to Madrid, and Spain announced it was writing off a third of Equatorial Guinea's official debt of 5,000m pesetas, with the remainder to be paid off over 14 years with 8 years grace. The visit resulted in a cultural co-operation agreement, including education and health, but there was no sign of the more substantial aid Obiang wanted. Madrid merely promised to contact private interests which might explore the possibility of investment in Equatorial Guinea.

A few months later Obiang tried again to strengthen economic relations with Spain. He sent one of his ministers to Madrid, who during his visit, denied that some 2,000m pesetas worth of Spanish aid had ever reached Malabo, or that Spanish language and culture were being replaced by French. There was talk of a new co-operation agreement between the two countries, but Madrid made it clear that this depended on democratic processes, while Malabo insisted that democratic pluralism was 'unviable in sub-Saharan Africa'.

Presidential elections – the first since independence – took place on 25 June 1989, with Obiang as sole candidate. But people were allowed to express their opposition by putting 'no' on the ballot papers. The borders were sealed off before the elections and no independent reporters were allowed to witness the poll. Obiang won.

KEY FACTS

Official title: Republic of Equatorial Guinea

Head of state: Gen. Teodoro Obiang Nguema Mbasango

Ruling party: Partido Unico Nacional de los Trabajadores (PUNT)

Capital: Malabo

Official Languages: Spanish (Fang widely spoken)

Currency: Franc CFA = 100 centimes

Exchange rate: Franc CFA316 per $ (Nov 1989)

Area: 28,051 sq km

Population: 410,000* (1987)

GDP per capita: $399 (1987)

Foreign debt: $199m (1988)

* estimated figure

COUNTRY PROFILE

Political system: Executive power is vested in the Supreme Military Council (Council of Ministers), headed by the president. A new constitution, known as the Basic Law (adopted in 1982), makes provision for presidential and general elections for the 11-member State Council and a House of Representatives every seven years. Equatorial Guinea gained independence from Spain in 1968.

Political parties: The Partido Unico Nacional de los Trabajadores (PUNT) is the only legal political party.

Population: 410,000 (1987 estimate), mostly Roman Catholic. The mainland province of Rio Muni (26,000 sq km) has 75 per cent of population, 90 per cent of whom belong to the Fang ethnic group. The island province of Bioko (2,000 sq km) consists of Bubis, Fangs and Creoles. The island of Pagalu (17 sq km), 600 km southeast of Bioko, and three rocky outcrops – Elobey Grande, Elobey Chico and Corisco – close to the mainland, are all part of Rio Muni Province.

Main cities/towns: Malabo (population 35,000, on island of Bioko), Bata (Rio Muni).

Language: Fang, Bubi, Spanish, Ibo, Creole.

Media
Press: Spanish paper *Ebano* published irregularly in Malabo, and *Portopoto* (Fang).
Radio: Two radio stations (one commercial) broadcast in Spanish and local African languages.
Television: There is a limited television service.

External trade: Trade balance and current account heavily in deficit. Strong reliance on aid. IMF facilities withdrawn in 1986.
Exports: Cocoa (estimated 70% of total value in 1986), coffee (24%), timber (6%). Main destinations (1984): Netherlands, German FR, Italy, France.
Imports: Foodstuffs (estimated 25% of total value in 1986), petroleum products (25%), transport equipment. Main sources (1984): France, Spain, China PR, Hong Kong, Italy, Sweden.

Agriculture: Mainstay of economy, but food imports needed. Main cash crops: (Bioko) cocoa (up 15 per cent in 1987-88, compared to 1986-87); Rio Muni: coffee, timber. Main food crops are cassava, sweet potatoes, bananas, palm.

COUNTRY PROFILE

Industry: Under-developed sector with activity centred on very small-scale food and timber processing.

Mining: Activity limited to exploitation of alluvial gold. Reserves of copper, iron-ore, uranium, tantalum and manganese; all judged to be uneconomic for exploitation.

Energy: Petroleum reserves have been discovered both on- and offshore, which, when developed, should meet domestic requirements. Offshore gas reserves have also been found. Hispanoil, Elf, Getty, Pecten and CFP-Total have exploration rights. Certain potential fields contested with neighbouring Gabon.

Membership of international organisations: African Development Bank, African States in Association with the EC (Lomé Convention), Arab Bank for Economic Development in Africa (recipient), Central African Customs Union, Economic Commission for Africa, GATT (provisional), IDA, IFAD, IMF, OAU, UN, WHO, World Bank.

Time: GMT +1.

Climate: Equatorial with heavy rainfall for most of the year except for slightly drier period from Dec to Feb. The mainland Rio Muni is drier and cooler than Malabo. Average temperature 26°C throughout the year, and generally very humid.

Entry requirements
Passport: Required by all.
Visa: Required by all.
Currency: Unlimited import/export of foreign currency.

Health precautions
Mandatory: Yellow fever vaccination certificate required if arriving from an infected area.
Advisable: Cholera vaccination considered essential; yellow fever, typhoid, polio vaccinations recommended. Malaria prophylaxis advisable as risk exists throughout the country. Water precautions should be taken.

Representation overseas: Addis Ababa, Beijing, Havana, Lagos, Libreville, Madrid, Moscow, New York, Paris, Rabat, Washington, Yaoundé.

Air access
National airline: Ecuato Guineana De Aviación.
Other airlines: Aeroflot, Cameroon, Iberia, Nigeria Airways, Royal Air Maroc.
International airport: Malabo (Code: SSG), 7 km from city; Bata, 6km from city. Most travellers enter from Douala (Cameroon).

Surface access
Main ports: Malabo, Bata, Luba, Mbini and Kogo.
Overland access: Access by semi-surfaced road from Gabon to Mbini and Bata, though route is not generally recommended.

Hotels: Accommodation is very limited but there are hotels in Malabo and Bata. Essential to arrange hotel before travelling, preferably through local business contacts. Food is rarely available at the Bata Hotel and, in Malabo, air conditioning is available only in some rooms in the Apartotel Impala.

National transport
Air: Ecuatoguineana de Aviación operates a service between Bata and Malabo.
Road: On Bioko a surfaced road links major towns in the north. On mainland Rio Muni a surfaced road links Bata with Mbini and a partly surfaced road links Bata with Ebebiyin (near Gabon border). Other roads are unsurfaced and can be difficult.
Water: Weekly boat service between Malabo and Bata.

Public holidays
Fixed dates: 1 Jan, 5 Mar (Independence Day), 1 May, 25 May, 10 Dec (Human Rights' Day), 25 Dec.

Working hours
Government and business: (Mon-Fri) 0830 – 1500; (Sat) 0830 – 1200, (alternate Sat) 1000 – 1200.
Banking: (Mon-Fri) 0830 – 1300.

Telecommunications
Telephone: Dialling code for Equatorial Guinea: IDD access code + 240 followed by subscriber's number. There are no area codes.
Telex: Limited telex facilities at central post offices in Malabo and Bata.

Banking
Central bank: Banque des États de l'Afrique Centrale, Malabo (headquarters in Yaoundé, Cameroon). *Main centre:* Malabo. Banque Internationale pour l'Afrique Occidentale, Banco de Crédito y Desarrollo (development bank), Banco Exterior de Guinea Ecuatorial (Guinextebank).

Electricity supply: 220 V AC, 50 cycles.

Representation in capital: Cameroon, China PR, Cuba, France, Gabon, Korea PDR, Nigeria, Spain, USA, USSR.

Hotels: Bata
Panafricaine.

Hotels: Malabo
Apartotel Impala (tel: 2492).

Eureka (tel: 3319).

Other useful addresses
Banco de Crédito y Desarrollo (credit and development bank), 1 Avda de la Libertad, PO Box 39, Malabo (tel: 2146).

Banco Exterior de Guinea Ecuatorial, Carretera de Aeropuerto, Malabo (tel: 2001).

Banque des États de l'Afrique Centrale (BEAC), PO Box 1917, Yaoundé, Cameroon (tel: 222505; tx: 8343).

Banque Internationale pour l'Afrique Occidentale, Calle de Argelia No 6, PO Box 686, Malabo (tel: 2367, 2887).

Camara de Comercio Agricola y Forestal de Malabo, PO bxo 51, Malabo (tel: 151, 152).

Comite Sindical de Cacao (cocoa growers' organisation), Bioko.

Direccion General de Correos y Telecomunicaciones, Malabo.

Empresa Estatal de Comercio Interior y Exterior, Malabo.

Empresa Guineano-Espanola de Petroleos (Gepsa), Malabo.

Ethiopia

Andrew Lycett

Political and economic developments in Ethiopia in 1988–89 were dominated by efforts to reach peace – or, at least, some form of accommodation – in the long-running war between the People's Democratic Republic of Ethiopia, centred in Addis Ababa, and its secessionist northern province of Eritrea. In September 1989 substantive talks between representatives of the Ethiopian Government and the leading freedom movement, the Eritrean People's Liberation Front (EPLF) began in Atlanta, Georgia, under the aegis of former US President Jimmy Carter. Later in the same month, Cuba was due to withdraw the remaining 2,000 troops it had stationed in Ethiopia for over a decade. At the height of the Ogaden war between Ethiopia and Somalia in 1978, Cuba had 20,000 troops fighting for marxist Ethiopia.

Russian pressure

Both events indicate that Ethiopia, for so long the standard-bearer of unrepentant Marxism–Leninism, is beginning to experience its own perestroika. The outcome is difficult to predict. But the Soviet Union, which has backed its African ally unreservedly since President Mengistu Haile Mariam's army-based revolution overthrew Emperor Haile Selassie in 1974, has begun to put pressure on it to halt the wars which have crippled its economy ever since. The Soviet Union is reliably reported to have provided arms worth $5–6bn to Ethiopia in the intervening 15 years. Some of this bill has been paid off – but at a crippling price to Ethiopia, as cotton and coffee crops have been bartered in exchange. Now Moscow has warned the Ethiopians that, when its present contracts run out in 1990, it will provide no further weapons for their wars, particularly those in Eritrea and neighbouring Tigray. Understanding the long-established link between events in Eritrea and those in Southern Sudan,

KEY FACTS

Official title: People's Democratic Republic of Ethiopia

Head of state: President Mengistu Haile Mariam

Ruling party: Workers' Party of Ethiopia

Capital: Addis Ababa

Official Languages: Amharic

Currency: Birr = 100 cents

Exchange rate: Birr 2.05 per $ (Nov 1989)

Area: 1,251,282 sq km

Population: 47.9m* (1988)

GDP per capita: $113 (1988)

GDP real growth: 8.0% (1987)

Inflation: 7.0% (1988)

Trade balance: –$1,125m* (1987)

Foreign debt: $2.6bn* (1987)

* estimated figure

the Soviets may be trying to broker a Horn of Africa peace, similar to the one they underwrote for the United States in Southern Africa in 1988, agreeing to help speed a Cuban withdrawal from Angola in return for a South African withdrawal from Namibia.

Addis Ababa has begun to look further afield – to North Korea, China, East Germany and Zimbabwe – to provide the arms it needs, if only to defend itself against potential hostilities from neighbouring Somalia. Relations with Israel, broken in 1973 when Ethiopia joined the OAU boycott at the time of the Arab–Israel war, have recently improved. According to one report, Israel has begun to supply Ethiopia with light and medium weapons and anti-tank artillery.

According to President Mengistu, Ethiopia's army was draining the economy of 50 per cent of its regular budget of Birr 1.6bn ($765m) in 1987-88, largely through its war in Eritrea. However, it was a succession of government defeats on the ground in Eritrea in early 1989 which finally convinced the Ethiopian head of state of the need for an end to hostilities. The crunch came in Tigray rather than Eritrea. In February the Ethiopian army, having been pushed back into the garrison town of Mekele by the Tigray People's Liberation Front (TPLF), was driven out of this important strategic link in its chain of supply from the capital to Eritrea. The Russians reportedly withheld the ammunition needed to hold on to the important town.

In February and March 1989 thousands of young soldiers were reported to have died as the Ethiopian army was repeatedly pushed back by the TPLF. These losses led to the press-ganging of youths who were sent to fight at the front. In April Major-General Mulatu Nagash, commander of the Ethiopian forces in Eritrea, was court-martialled. Five other top-ranking officers were arrested and held responsible for the defeats in the north.

Coup attempt

These arrests were accompanied by widespread changes in the army. Many senior generals were retired and over 40 colonels promoted to the rank of brigadier-general. These moves added to the sense of confusion. As President Mengistu left Ethiopia on 16 May 1989 on an official visit to East Germany, a coup was planned to overthrow his regime. However, the security forces had prior information about the plot. The two main conspirators (chief of staff Major-General Merid Neguisse and airforce commander Major-General Amha Desta) intended to use the airforce, based largely at Debre Zeit, to make it appear that their takeover was a *fait accompli*. However, security forces, under Minister of Internal Affairs Tesfaye Wolde Selassie, arrested Generals Merid and Amha as they were making their final plans, shortly after President Mengistu had departed for East Germany. Loyal army and airforce units were deployed to ensure no support was forthcoming for the plotters. In Addis Ababa 300 troops held out against Mengistu's government, barricading themselves inside the ministry of defence for a day. When Defence Minister Haile Giorgis Habte Mariam tried to remonstrate with them, he was shot. Brigadier-General Kumulatchew Dejene, former deputy commander in Eritrea, was killed trying to seize Addis Ababa airport so as to prevent loyal troops arriving in the capital from other bases.

In Eritrea itself the revolt was unco-ordinated but more effective. It was led by Major-General Demissie Bulto, commander of the 2nd revolutionary army in Eritrea. Under his command rebels took over the Asmara radio station ad started broadcasting in the name of the 'Eritrea Revolutionary Forces and Popular Police Forces'. This proved a mistake because it allowed loyal troops to suggest that the rebels had handed the province over the EPLF. Nevertheless, the message from the radio station was clear: the rebels wanted an end to the war in Eritrea. They called for a transitional government drawn from all opposition forces. After 24 hours loyalist forces retook the radio station. Many senior officers were killed including General Tadesse Tessema, operations chief of the army in Eritrea. Hundreds of officers were arrested. According to one report, virtually the entire staff of the ministry of defence and the airforce command were arrested, as was the majority of the Asmara command.

When Mengistu returned from his travels, he pronounced the rebels to be a disenchanted elite with no 'social base at all'. The severed head of the leader of

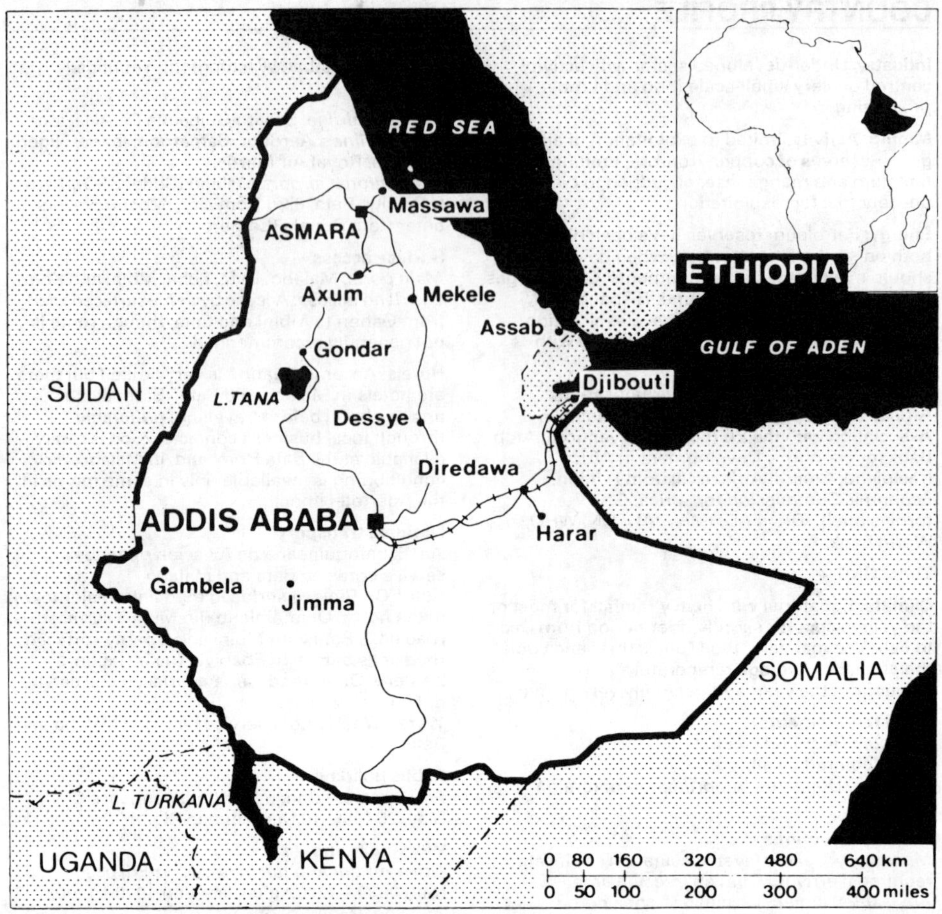

KEY INDICATORS

	Unit	1984	1985	1986	1987	1988
Population	m	41.6	43.4	44.9	46.2	47.9
Gross domestic product (GDP)	Birr m	10,001	9,881	10,823	11,134	11,228
GDP per capita	Birr	240	228	241	241	234
Inflation	%	8.4	19.0	–9.8	–2.4	7.0
GDP real growth	%	–3.5	–20.2	24.2	8.0	–
Agricultural production	1979-81=100	88.5	95.9	105.2	102.5	106.8
Food production	1979-81=100	122.7	129.6	137.5	144.6	148.5
Exports	Birr m	862.7	689.4	941.6	735.2	–
Coffee	Birr m	545.7	432.7	725.6	414.0	–
Hides & skins	Birr m	96.3	103.6	109.9	123.7	–
Imports	Birr m	1,631.2	1,733.9	1,922.8	1,859.9	–
Crude petroleum	Birr m	370.4	289.4	223.9	219.0	–
Balance of trade	$m	–381.5	–507.7	–455.4	–1,125.0*	–
Current account	$m	–130.4	105.9	–327	–	–
Total external debt	$m	1,5434	1,879	2,187	2,590	–
Debt service ratio	%	13.6	19.4	22.9	–	–
Exchange rate	Birr per $	2.07	2.07	2.07	2.07	2.07

* estimated figure

the revolt in Eritrea, Major-General Demissie Bulto, was paraded through the streets of Asmara.

In a report on human rights violations in Ethiopia published in August 1989, Amnesty International alleged that 'several thousand suspected opponents of the government in Ethiopia, including many prisoners of conscience, are currently imprisoned'. It stated that some 180 people were still detained following the abortive May coup. Among them were Fanta Belay, minister of industry and former commander of the airforce, and General Tesfaye Berhanu, commander of the navy. Amnesty added: 'Many are held in harsh conditions and have reportedly been tortured repeatedly over a lengthy period. Methods of torture include beating on the soles of the feet, electric shocks and submersion in water. The prisoners include many people arrested on account of their ethnic or regional origin in areas where there has been recent armed opposition to the government.'

Moves to peace

The coup and its aftermath did, however, serve to speed moves which were already underway to bring about peace in Eritrea. In April 1989 former US President Jimmy Carter, returning from a trip to Ethiopia, had met EPLF secretary-general Issayas Aferworki at Khartoum airport. Aferworki travelled to Washington the following month. The EPLF accepted a formula which seemed to fall short of an outright demand for independence. In June President Mengistu put forward his proposals for peace. Although the US Government has yet officially to put its weight behind an armistice, the EPLF duly met with the Ethiopian Government in Atlanta in September 1989. At the conclusion of these talks Aferworki expressed optimism about the results, but said that civil war in Ethiopia would probably continue for several months. He stated that the Eritreans had 'reason to believe they are getting close to a solution'.

Despite the talks, fighting continued in Eritrea. In September 1989, just after the Atlanta parley had ended, the EPLF and the Ethiopian People's Democratic Movement (EPDM), operating together as the Ethiopian People's Revolutionary Democratic Front (EPRDF), announced they had defeated a government force of 18 brigades, which they estimated to number 40,000 men.

Around the same time the TPLF said that, fighting with the EPDM in the north-eastern province of Wollo, it had killed over 1,000 government soldiers in fierce fighting in late August and early September. Although the government did not confirm any details of the battles, the TPLF claimed the rebels had overrun the whole of north-western Wollo and forced the army to evacuate its garrison of 2,000 soldiers at Lalibela, best known for its twelth-century churches carved out of rock.

As if civil war was not enough, popular morale, just beginning to recover from the droughts of 1984–85 and 1987–88, suffered a further reverse in late December, when some 8,000 people were reported to have died within six weeks in an epidemic centred on Welaita, 320 km south of Addis Ababa.

Development delayed

After a period when the government seemed seriously concerned about improving the economy, development took second place to political and military activity throughout most of 1988–89. The Five-Year Plan presented in September 1988 had proposed spending Birr 6bn ($2.9bn), with defence taking the lion's share, followed (in descending order) by agriculture, industry, mining and transport.

The question troubling the minds of planners in Addis Ababa was how to finance this Plan. Transport is the sector which has received the most attention from outside donors. In April 1989 the World Bank's IDA announced a $72m credit to help Ethiopia modernise and increase its transport capacity. The money was to go to the ministry of transport and communications, the marine transport authority, the Ethiopian Freight Transport Corporation (EFTC), the road transport authority, the maritime and transit services corporation, the Ethiopian customs and Excise administration and the trucking industry. The African Development Bank was also to provide $55m, the European Investment Bank $12.5m and the Ethiopian Government itself $12.5m to bring total financing for the transport sector to $164.6m. Two particular projects will be initiated: the construction of two multipurpose berths at Assab port, including land works and cargo handling equipment, and the rehabilitation of old trucks and the procurement of new trucks for the parastatal EFTC.

Meanwhile, Ethiopian Airlines continues to operate as one of the most profitable airlines in Africa. In the financial year ending June 1989 it made a net profit of $24m on revenues of $240m. Over the next 10 years it plans to spend $2.1bn on new planes and equipment. As the first part of this investment programme, which means purchasing two planes a year by 1990, it officially announced in June 1989 a $387m deal to buy six Boeing 757s powered by Pratt and Whitney engines. With the addition of Mogadishu to its route network in June 1989, Ethiopian Airlines now flies to 29 destinations in Africa. The Seychelles, Botswana and Namibia are to be added soon. In 1988 it carried 583,000 passengers, and expected growth of between 15 and 20 per cent in 1989.

During 1989 Ethiopian Airlines has introduced its new Ag-Cat turbo-engined crop spraying aircraft to the local market. It hopes soon to be able to export the aircraft to neighbouring African countries.

In the short term, though, Ethiopia depends for its economic and physical survival on the successful management of its agriculture, which accounts for 40 per cent of GDP, 80 per cent of employment and 85

COUNTRY PROFILE

Political system: Rule by a president elected for a five-year term. The ruling Workers' Party of Ethiopia at present has an 11-member politburo, with six alternate members, as its main executive arm. The politburo is headed by the party General Secretary, currently President Mengistu Haile Mariam. There is also a central committee, comprising 134 full members and 64 alternates, which is the main forum of political debate. Political opposition has been suppressed. Armed opposition from secessionist elements in Eritrea and Tigre has been increasing. A new Constitution, stressing the Marxist-Leninist character of the state, was approved by 81 per cent of the population in a referendum conducted in February 1987, and adopted when the People's Democratic Republic of Ethiopia was declared in September 1987.

Political parties: The ruling Workers' Party of Ethiopia, established 1984, is the only legal political party.

Population: Estimated at 43.5m (1986). The Ethiopian Coptic Church is influential particularly in the north, but there are many Muslims in the south.

Main cities/towns: Addis Ababa (population around 2m), Asmara (400,000), Dire Dawa (80,000), Harar, Massawa, Assab, Jimma.

Language: Amharic is the official language, but English is widely understood in business circles.

Media
Press: Daily newspapers in the capital include *Addis Zemen* (Amharic), and *Ethiopian Herald* (English). Others including *Hebret*, are published in Asmara. Several weekly and other periodicals, in Amharic and English, are published in Addis Ababa.
Radio: Service by Voice of Revolutionary Ethiopia, chiefly in Amharic, but also in English, Arabic, Somali, French and local languages.
Television: The state-controlled commercial Ethiopian Television broadcasts to most of the country via a new microwave link-up. There are 30,000 sets.

External trade: The balance of trade deficit in 1987 was partly due to donors' suspension of aid because of state control over the economy. The decline is also the result of falling world coffee prices, aggravated by internal fighting and drought. The debt service ratio is still comparatively low at 20 per cent.
Exports: Coffee (77 per cent of total value in 1986), hides and skins (12 per cent), pulses, live animals. Main destinations: USA, German FR, Japan, GDR.
Imports: Machinery, transport equipment, petroleum and petroleum products. In 1988, 1.3m tonnes of food aid were required. Main sources: USSR, Italy, German FR, Japan, UK.

Agriculture: Mainstay of the economy, accounting for about 40 per cent of GDP, 80 per cent of employment and 85 per cent of exports. The chief cash crop is coffee, grown on the plateaux in the south-west. Other cash crops: cotton, sugar. Also exports of hides and skins. The main food crops are maize, sorghum, wheat, barley, millet and teff. Drought, guerrilla activity, soil erosion and transport problems have led to sharp drop in production, and killed off much of the country's livestock. By the end of 1989 there are indications that another massive famine may occur.

Manufacturing: Sector contributes about 16 per cent to GDP and is based on the processing of agricultural raw materials. Principal among these is food processing, but textiles, handicrafts, and leather production are also significant. Growth in the sector has been constrained by lack of raw materials and the war in Eritrea where a lot of industrial activity was concentrated. Other major problems include outdated machinery and techniques, and the need for imports throughout the sector. The Ten Year National Development Plan places great emphasis on import substitution.

Mining: At present activity is limited to small-scale gold mining. An ore-processing plant is being built, untapped reserves of copper, zinc, potash and iron-ore.

Energy: Considerable hydroelectric capacity. Present installed generating capacity is about 1,200 mW, but it is hoped to increase it by another 1,000 mW by the end of the decade. Wood has always been a traditional source of energy, but it is hoped increased availability of electricity will reduce this, as many forests have now become severely depleted. There are prospects of locating petroleum in the Ogaden region, and Soviet as well as Western companies have started prospecting.

Membership of international organisations: African Development Bank, African States in Association with the EC (Lomé Convention), Arab Bank for Economic Development in Africa (recipient), Comecon (observer status), Economic Commission for Africa (headquarters in Addis), IGGAD, ILO, IMF, OAU (headquarters in Addis), UN, WHO, World Bank.

per cent of exports. However, since 1974, agriculture has fallen at an average rate of 0.4 per cent a year. The state farms, which have received the bulk of expenditure, account for only 3 to 4 per cent of total crop production which, according to the FAO, is likely to reach 7.4m tonnes in 1989 – still well short of the needs of a rapidly expanding population.

New thinking on agriculture

Ethiopia's population is currently estimated at some 48m, and this figure is growing at 3 per cent per annum. That means an extra 1m mouths to feed each year, and something approaching 65m people by the year 2000. To feed this number agricultural production needs to double to 14m tonnes by the turn of the millennium.

This stark reality has led to a radical rethinking of the policy of collectivisation and centralised marketing. In early 1988 the government agreed to raise producer prices by an average of 7.7 per cent. Private grain traders, who disappeared with the revolution, were given licences again. Farmers were permitted to sell their produce on the open market rather than exclusively through the Agricultural Marketing Corporation. Improved farming practices, including better use of inputs and research, has also been studied.

These reforms encouraged the World Bank and the European Community to commit $160m to Ethiopian agriculture, initially mainly to buy fertilisers and other priority inputs. This largesse led in turn to pressure for more radical reforms. It was argued that in Ethiopia peasants do not own their land. This results in little attention being paid to land conservation. Production is poor, soil erosion high.

The ministry of agriculture is therefore examining a number of proposals which include the leasing of land for periods of up to 25 years, allowing the hiring of labour, and adopting even more liberal pricing and marketing policies. To encourage this trend the EC announced in April 1989 that it was granting ECU 10.5m to the sectoral import programme for agriculture in Ethiopia. This would help relieve short-term foreign exchange constraints on the purchase of essential agricultural inputs.

If Ethiopia seizes the opportunity to end its wars in Eritrea and other outlying provinces, its economy could within a short time become one of the most dynamic in Africa.

BUSINESS GUIDE

Time: GMT +3.

Climate: Depends on altitude. Coastal and lowland regions very hot and dry throughout the year. On plateau (including Addis Ababa and Asmara) dry season from Oct to May with temperature range from as low as 6°C in Dec to 26°C in Mar (light rain from Feb to Apr). Temperatures can fall sharply at night during the dry season. Rainy season from Jun to Sep with average temperature 21°C.

Entry requirements
Passport: Required by all.
Visa: Required by all except nationals of Kenya. Business visas are valid for 30 days. For longer stays visitors must register with the Immigration Office for an Alien's Registration Card and an exit visa. Transit visas are available at Bole Airport.
Prohibited entry: Nationals of Iraq, Syria and 'white' nationals of South Africa.
Currency: Import/export of local currency restricted to 10 birr. Foreign currency import unlimited but amounts must be declared on arrival and export is allowed up to declared amount.

Health precautions
Mandatory: Yellow fever vaccination certificate required if travelling from an infected area.
Advisable: Yellow fever, cholera, typhoid and polio vaccinations. Malaria prophylaxis recommended for visits outside Addis Ababa or Asmara. Water precautions should be taken.

Representation overseas: Abidjan, Accra, Aden, Ankara, Athens, Beijing, Belgrade, Berlin, Bonn, Brussels, Cairo, Dakar, Djibouti, Havana, Jeddah, Khartoum, Kinshasa, Lagos, Mexico City, Monrovia, Moscow, Nairobi, New Delhi, London, Paris, Rome, Sana'a, Stockholm, Tokyo, Tripoli, Washington.

Air access
National airline: Ethiopian.
Other airlines: Aeroflot, Air Afrique, Air Djibouti, Air Zaire, Alitalia, Alyemda, CAAC, Kenya Airways, Lufthansa, Saudia, Yemenia.
Tax: Airport tax of Birr 8 on departure.
International airports: Addis Ababa-Bole International (Code: ADD), 8 km from city, bar, buffet, restaurant, bank, post office, shops, car hire; Asmara-Yohannes IV (Code: ASM), 9 km from city, restaurant, currency exchange.

Surface access
Main ports: Assab, Massawa. Regular sailings by Ethiopian Shipping Lines.
Overland access: A road links Addis Ababa and Nairobi (Kenya), forming part of the Trans-East African Highway. However, there may be difficulties gaining access through customs at Kenyan/Ethiopian border. In practice the only method of gaining entry to the country is by flying to Addis Ababa. The rail route from Djibouti to Addis Ababa cannot be used by visitors.

Hotels: Available in Addis Ababa and other main centres. Service charge of 10 per cent and tax of 2 per cent added to bills but a small tip is usual in addition to the service charge.

Electricity supply: 220 V AC, 50 cycles.

Credit cards: Only Hilton Hotel accepts credit cards.

Car hire: National driving licence acceptable for up to one month; a temporary licence should then be obtained from the authorities. Availability of cars is low, but they can be obtained in Addis Ababa from National Tour Operations.

City transport
Taxis: Available in main cities; fare by negotiation; tipping is not usual; can be hired on a time basis.

National transport
Air: Special permission is required for travel outside Addis Ababa. Check current regulations with the Immigration Office or Tourist Commission. Ethiopian Airlines operate domestic service to main towns.
Road: All-weather network connects principal towns. The system is undergoing expansion. Driving on right.
Buses: Coach services (liable to suspension) have been scheduled: Addis Ababa-Asmara; Addis Ababa-Gondar.
Rail: Line runs from Addis Ababa to Dire Dawa (and on to Djibouti). However, visitors are not allowed to use this line at present; check current regulations. A new link was planned between Addis Ababa and Assab.

Public holidays: The Ethiopian calendar year is divided into twelve months of 30 days each and one month of five days (six in a leap year).
Fixed dates: 7 Jan (Christmas), 19 Jan (Epiphany), 2 Mar (Battle of Adowa), 6 Apr (Victory Day), 1 May, 11 Sep (New Year's Day; 12 Sep in year before leap year), 27 Sep (Feast of the Finding of the True Cross).
Variable dates: Good Friday, Eid al-Fitr, Eid al-Adha, Prophet's Birthday.

Working hours
Government and business: (Mon-Fri) 0800 – 1200, 1300 – 1600; (Sat) 0800 – 1200.
Banking: (Mon-Fri) 0800 – 1200, 1300 – 1600; (Sat) 0800 – 1200.
Shops: (Mon-Sat) 0800 – 1300, 1500 – 2000. Local variations.

Telecommunications
Telephone: Dialling code for Ethiopia: IDD access code + 251 followed by area code (1 for Addis Ababa, 4 for Asmara). Calls to Europe and neighbouring African countries can be made from Addis Ababa and Asmara.
Telex and telegram: Available in Addis Ababa from the Regional Telecommunications Service offices and from major hotels.

Banking
Central bank: National Bank of Ethiopia. *Banks in Addis Ababa:* Agricultural & Industrial Development Bank, Commercial Bank of Ethiopia, Housing and Savings Bank.

Representation in capital: Algeria, Argentina, Austria, Belgium, Bulgaria, Burundi, Cameroon, Canada, China PR, Cuba, Czechoslovakia, Egypt, Equatorial Guinea, Finland, France, German DR, German FR, Ghana, Greece, Guinea, Hungary, India, Indonesia, Iran, Italy, Ivory Coast, Jamaica, Japan, Kenya, Korea PDR, Rep. of Korea, Liberia, Libya, Malawi, Malaysia, Mexico, Netherlands, Niger, Nigeria, Pakistan, Philippines, Poland, Romania, Rwanda, Saudi Arabia, Senegal, Sierra Leone, Spain, Sudan, Sweden, Switzerland, Tanzania, Thailand, Turkey, Uganda, UK, USA, USSR, Vatican, Venezuela, Vietnam, Yemen AR, Yemen PDR, Yugoslavia, Zaïre, Zambia, Zimbabwe.

Hotels: Addis Ababa
Ethiopia, Unity Square, P.O. Box: 1131 (tel: 447400; tx: 21072).

Ghion, Ras Desta Damtew Ave, P.O. Box: 1643 (tel: 447130/34; tx: 21112).

Hilton, P.O. Box: 1164 (tel: 518400; tx: 21104; fax: 510064).

Ras, Churchill Road, P.O. Box: 1632 (tel: 447060).

Wabe Shebelle, Ras Abebe Anegay Ave, P.O. Box: 3154 (tel: 447187/90; tx: 21416).

Hotels: Asmara
CIAAO Asmara, 62 Avenue Mene (tel: 110122; tx: 21112).

Car hire: Addis Ababa
National Tour Operations: (tel: 444838; tx: 21370).

Banking
National Bank of Ethiopia (central bank), PO Box 5550, Addis Ababa (tel: 447430/9; tx: 21020).

Commercial Bank of Ethiopia, National Theatre Square, PO Box 255, Addis Ababa (tel: 447420, 447429).

Commercial Bank of Ethiopia, Adwa Avenue, PO Box 219, Asmara (tel: 111844).

Other useful addresses
Agricultural and Industrial Development Bank, PO Box 1900, Addis Ababa (tx: 21173).

Addis Ababa Chamber of Commerce, PO Box 2458, Addis Ababa (tel: 158055; tx: 21231).

Asmara Chamber of Commerce, PO Box 856, Asmara (tel: 11388; tx: 110814).

Board of Telecommunications of Ethiopia, PO Box 1047, Addis Ababa.

Campaign and Central Planning Supreme Council (NRC-CPSC), PO Box 1013, Addis Ababa.

BUSINESS DIRECTORY

Central Statistical Office, PO Box 1143, Addis Ababa (tel: 113010).

Economic Commission for Africa, PO Box 3001, Addis Ababa.

Ethiopia-Djibouti Railway Co, PO Box 1051, Addis Ababa (tel: 47250; tx: 21414).

Ethiopian Airlines, PO Box 1755, Addis Ababa (tel: 21012).

Ethiopian Chamber of Commerce, PO Box 517, Addis Ababa (tel: 448240, 444005; tx: 21213).

Ethiopian Food Corporation, Mortgage Building, Higher 21, Kebele 04, Addis Ababa (tel: 158522; tx: 21292).

Ethiopian Import-Export Corporation, PO Box 2313, Addis Ababa (tel: 152400; tx: 21099).

Ethiopia News Agency (ENA), PO Box 530, Patriots Street, Addis Ababa (tel: 120014; tx: 21068).

Ethiopian Television, PO Box 5554, Addis Ababa.

Ethiopian Tourism Commission, PO Box 2183, Addis Ababa (tel: 158055; tx: 21231).

Ethiopian Transport Construction Authority, PO Box 1770, Addis Ababa.

Ministry of Foreign Trade, PO Box 1769, Addis Ababa (tel: 448200).

Falkland Islands

COUNTRY PROFILE

Political system: The Falkland Islands are a dependent territory of the United Kingdom, executive authority being vested in the British monarch and exercised by the governor on her behalf. The governor is advised by the executive council, over which he presides, composed of three of the elected members of the Legislative Council, elected by the body to sit on the executive council for a period of twelve months, and two ex-officio members, the chief executive and the financial secretary. The Legislative Council is composed of eight members elected by universal adult suffrage and two ex-officio members the chief executive and the financial secretary. The Legislative Cojncil is expowered to pass laws for the government of the islands, subject to the approval of the British monarch acting through her secretary of state for foreign affairs. General elections are held every four years. Defence and foreign affairs are the responsibility of the British Government.

Media
Press: Two newspapers. There is one monthly, *Penguin News*, and the government publishes *The Government News* and *The Gazette*.
Broadcasting: One radio station, operated jointly by Falkland Islands Broadcasting Service and British Forces Broadcasting Service provides for 24 hours/day listening; includes taped imported programmes, BBC World and National Services.

External trade
Exports: Wool, sheepskins/hides have been virtually the only exports. Main destination: UK.
Imports: Food, machinery and transport equipment, timber, manufactured goods. Main source: UK.

Agriculture: Soil is generally poor – peat over clay (peat is used as fuel). Virtually all available land has been used for sheep ranching although small areas of arable are cultivated (e.g. for potatoes, hay crops; vegetable crops have been grown by individual households; good crops/specimens recorded where care/protection provided – up to present time there has been no commercial growing). A hydroponic garden facility constructed in Port Stanley is now yielding good quality crops of salad vegetables for local and fishing fleet consumption.

Livestock: Sheep stock are a Corriedale/Polwarth mixture with small admixture of certain other breeds, e.g. Romney; average clip over 3.55 kg; a fine wool is produced, suitable for garments, knitting, tweeds, etc.; wool and sheepskins have been virtually the only export; up to 2m kg of wool/year is exported to the UK market. A smallscale wool processing and knitting plant at Fox Bay has been established since 1985, providing for consumption of raw wool and producing high-quality yarns and garments. Productivity is not high – average ewe only produces, in her lifetime, just over two lambs to reach maturity. The 1987-88 farming returns detail the following numbers of animals: sheep (704,612), cattle (5,969), horses (1,837), pigs (17), poultry (2,362) and dogs (915). Mutton is the principal source of protein and is supplemented during winter by beef.

Industry: Small industrial units, serving both on- and off-shore commitments. Hand-knitted local garments produced for sale to visitors. Fishing: The Falkland Islands have historically had no fishing industry. However, as a result of the declaration of the Interim Conservation and Management Zone by the British Government in 1986, the Falklands entered into managing and policing a fish reserve and generating significant revenues through the annual award of fishing licences. Revenues from the fishing sector amounted to 30m in 1989, representing 75 per cent of the Islands' annual budget.

Time: GMT -4 (GMT -3 from Sep to Apr).

Climate: Temperatures range from minus 6°C to 21°C with occasional minimum of minus 10°C and maximum 25°C. Rainfall around 700 mm/yr. Strong to gale-force winds are frequent during spring and early summer.

Entry requirements
Passport: Required by all.
Visa: Not required by citizens of UK and Colonies, nationals of certain other countries. Prior checking is advisable.

Health precautions
Mandatory: None.
Advisable: Typhoid, polio vaccinations, yellow fever.

Air access
International airport: Mount Pleasant, 48km from Port Stanley (PSY).

Surface access
Main port: Port Stanley.

National transport
Air: The Falkland Islands Government Air Service (FIGAS) operates three 'Islands' land-based aircraft to the majority of settlements according to bookings and weather conditions.
Road: At present there is a limited amount of surfaced road – mainly around Port Stanley and Mount Pleasant. Gravel roads are a common feature, and new gravel roads are projected on both East and West Falkland.

KEY FACTS

Official title: Falkland Islands

Head of state: Queen Elizabeth II

Head of government: Governor William Hugh Fullerton

Capital: Port Stanley

Official Languages: English

Currency: Falkland Pound (£) = 100 pence

Exchange rate: £0.63 per $ (Nov 1989)

Area: 12,173 sq km (24,346 with East and West Falkland and adjacent islands)

Population: 1,919 (1986) (plus garrison of 2,000 soldiers)

BUSINESS GUIDE

Telecommunications: Direct satellite telephone, telefax and telex links now in operation (operated by Cable & Wireless). A number of IDD subscribers have been accessed to the network. Under an agreement signed September 1988 Cable & Wireless will install a new telephone system throughout the Islands. Dialling code for Falkland Islands: IDD access code + 500 followed by four-digit subscriber's number. There are no area codes.

Business advice and information
Falkland Islands Association, 2 Greycoat Place, London SWP 1SD (tel: 01-222-0028).

Falklands Islands Company, Port Stanley (tel: 300; tx: 2418; fax: 2603).

Falkland Islands Development Corporation, Port Stanley (tel: 2211; tx: 2428; fax: 2606).

Falkland Islands Government Office, Falkland House, 14 Broadway, London SW1H OBH (tel: (1) 222-2542; tx: 8950476 figlon; fax: (1) 222-2375).

Falkland Islands Government Secretariat (Chief Executive), Port Stanley (tel: 2618; tx: 2423; fax: 2212); Financial Secretary (tel: 2617; tx: 2423; fax: 2212).

Governor's Office, Government House, Port Stanley (tel: 2201; tx: 2414; fax: 2201).

Standard Chartered Bank, Box 166, Ross Road, Port Stanley (tel: 352; tx: 2422).

Hotels: Port Stanley
Emma's.

Malvina House.

Upland Goose Hotel (tel: 455; fax: 2206).

Federated States of Micronesia

T. R. Hughes

(map on page xxiv)

President Bush signed legislation in July 1989 elevating diplomatic relations with the Federated States of Micronesia (FSM), as well as with the Marshall Islands, to full embassy status with an exchange of resident ambassadors. It was a welcome uplift to the new Pacific state's national pride. The United States freed the islands from the Japanese in the Second World War and, for 40 years, administered them under United Nations trusteeship.

Like the Marshalls, the FSM opened a new chapter in its history in November 1986 by becoming an independent nation under a compact of free association with the USA, enjoying self-government and conducting its own foreign affairs but leaving its defence in American hands. Under the compact, the United States will provide $1.3bn of economic and service aid over 15 years. The basic link of free association remains indefinitely but can be ended at any time while defence and aid are subject to renegotiation at the end of that period.

The islands of the FSM, sitting lush green and gem-like in vast expanses of the Western Pacific ocean, were devastated as US troops wrested them from the Japanese in island-by-island battles. The Micronesian economy was completely destroyed. Although the United States committed itself, as the administering authority, to rebuild the islands economically and politically, Americans admit that the old Garden of Eden vision of low-slung coral atolls and aquamarine waters rich with fish tended to blind them to the reality that Micronesia – the name means 'little islands' – was one of the most backward of American outposts. There was little or no industry, scarce employment, poor roads, erratic electricity and water supplies. Nature was partly to blame: the islands are resource-poor and remote.

But some Micronesians, including the FSM's former president Yosiwo Nakayama, complained that the USA allowed their islands to become recipients of welfare handouts so as to tie them to Washington which sees Micronesia as a Pacific defence line.

Economic hurdle

The current presidential incumbent, John Haglelgam, who at 41 is from the modern generation of political leaders, recognises the need for aggressive efforts by the islanders to change the economy from a mainly agricultural and fishing subsistence base to a more productive, export-oriented one. The government is the biggest source of jobs for the 86,000 population.

Economists compare the American stewardship with the period when the islands were under the pre-war Japanese mandate. Foreign investment, together with trade and tax policies which encouraged Micronesian exports, resulted in the Japanese Government phasing out subsidies by 1932. Micronesian exports rose through the late 1930s although most of the benefits flowed to the Japanese rather than to the Micronesians.

The reverse occurred during the US trusteeship: exports sagged while US subsidies increased. Now that the FSM is in the hands of the Micronesians and since the compact gives them favoured access to the US market for certain locally-produced items, the islanders hope they can exploit the lesson of history – and this time take a share of the profits in any foreign business partnership.

Even though there is a natural tendency by the local firms to look first to the USA mainland because of their long political and economic affiliation with the USA, the response in the first three years of the compact has been comparatively muted. However, the competition is keen and representatives from Japan, China, Australia, New Zealand have been more aggressive and more successful than US firms although the area will never be a major Western-style market for products and services. The Japanese have long been active in giving development aid and seeking trade.

Two of the smallest populated states, Kosrae and Yap, are the most energetic in seeking foreign investment. Kosrae was the first to finish its new airport inaugurated in May 1988. It is working with US, Chinese and Nauruan business partners in a manufacturing venture and hotel project. Yap is linked to another foreign group to establish a small tuna fishing fleet, a cold storage plant and a small canning plant.

New capital

Construction work is proceeding apace on rapidly-growing Pohnpei where a new capital is being built 10 km south-west of Kolonia in the Palikir valley (a Korean firm is involved in this project). A resort development is also planned with foreign help. And Truk, which has been dogged by financial problems, is planning an industrial complex on the island of Dublon when enough power and water supplies are available.

Although all the four states are blessed with international airports and harbours with adequate facilities for commercial traffic, the scattered islands pose a headache for anyone trying to improve the physical infrastructure. Transportation services are expensive and communications difficult. And there are also inter-state rivalries to be considered.

Internationally, the FSM received a shot in the arm with its admission as a full member of the South Pacific Forum in 1987. This acceptance of sovereignty by its Pacific neighbours was seen as a big step towards seeking full membership of the Asian Development Bank and access to its development aid. But some Western countries, Britain among them, are not happy over the legalistic interpretation of FSM's claim to be a fully independent nation. The termination of the strategic trust status by the UN Trusteeship Council has still to be ratified by the Security Council which has ultimate responsibility.

Political strains within the complicated federal structure could pose a threat to economic development as states jostle with each other for national attention and slow down decisions on pressing issues.

Within congress, old rivalries also rise to the surface. Trukese members, with nearly half the congressional seats, are quick to complain if they fail to get what they see as their fair share, while other state delegations complain of Trukese domination of the congress. An outburst by a Trukese senator complaining that Pohnpeians had taken most of the executive branch jobs led to a two-week boycott by three Pohnpeian legislators. The Speaker declared the session one of the most trying ever faced. In effect, President Haglelgam, who comes from Yap, has the daunting task of preserving unity among nine Micronesian and Polynesian groups to make his policies work.

KEY FACTS

Official title: Federated States of Micronesia

Head of state: President John Haglelgam

Head of government: President John Haglelgam

Capital: Kolonia (on Pohnpei)

Official Languages: English

Currency: US Dollar ($ = 100 cents)

Area: 607 volcanic islands and coral atolls

Population: 86,000* – Truk: 39,000; Pohnpei: 24,000; Yap: 10,000; Kosrae: 7,000

Labour force: 20,000* (1988)

Visitor numbers: 5,000* p.a.

* estimated figure

COUNTRY PROFILE

Political system: The Federated States of Micronesia, a former Japanese colony and a part of the Trust Territory of the Pacific Islands, administered by the US under a UN mandate since 1947, was formally linked with the US in Nov 1986 by the Compact of Free Association. The Compact gives the islands independence, but maintains defence.

Population: 86,000 (1987 estimate).

Main town: Kolonia (on Pohnpei), Moen (Truk), Colonia (Yap), Tofol (Kosrae) but Kosrae State Government is based at Lelu.

Language: English is the official language, but several Micronesian languages and Japanese are widely spoken.

Media

Radio: Radio broadcasts transmitted at least 18 hrs/day in English and the main local dialects.

Television: TV coverage fairly widespread, with some cable relay.

Domestic economy: Under the Compact of Free Association the US Government is to supply $1.5bn over a 15-year term, of which 30 per cent is for infrastructural development. Typhoon damage to Truk in November 1987 caused a set-back and aid from its neighbours was vital to

COUNTRY PROFILE

the state. Yap has targetted foreign investment as its best hope in developing the economy. Feasibility studies are in progress along with development of the tourism industry.

External trade
Exports: Main exports are coconut oil, copra and fish. Some beef, fruit and vegetables have been exported. Main destinations are US, Japan and Guam.
Imports: Main imports are food and beverages, machinery, construction materials, textiles and fuel.

Agriculture: Subsistence farming of coconuts, taro, breadfruit, bananas, yams, sweet potatoes and vegetables, with pig and poultry raising, is the main occupation. Commercial farming practised where climate and soils are favourable. Fishing is mainstay of economy, supplying primary source of protein and export revenues, and received 32 per cent of all government investment in 1985-1989, compared with 12 per cent for agriculture and forestry.

Industry: Small-scale industries exist, including handicrafts, fish processing, bottling, copra processing, bakeries and boat building. The sector received 20 per cent of government investment 1985-1989. The Pohnpei Agricultural and Trade School has begun to turn a profit on its small coconut-products plant, which makes 'Oil of Ponape' toiletries. Tourism has potential, but is still relatively undeveloped.

Mining: Small mineral deposits exist, but exploitation is hampered by a shortage of land to accommodate displaced population and doubts about economic viability.

Membership of international organisations: Association of Pacific Island Legislatures, ESCAP (associate member), South Pacific Commission, South Pacific Forum.

Taxation: The US Tax Code now applies in territory.

Time: GMT +12 hrs on Yap, Pohnpei and Truk.

Climate: Warm and humid, temperatures 23-30°C, humidity around 80 per cent. Rainfall variable, minimum 250 mm/year, and can occur in downpours. Hurricanes are possible.

Entry requirements
Passport: Required by all except US citizens with proof of citizenship.
Visa: Not required by US citizens with proof of adequate funds. Entry permits granted to all others with proof of onward passage and adequate funds for stays of less than 30 days. Special regulations may apply to some non-tourist destinations.
Health: Precautions against tetanus, typhoid and paratyphoid are advisable. Leprosy has been endemic for generations, but since 1960 the latest upsurge has reached epidemic proportions, with an attack rate of 15 per 1,000 population, largely in Pohnpei state which includes islands of Kapingamarangi and Pingelap.

Air access
Other airlines: Air Micronesia.
International airports: Pohnpei (PN); Truk (TKK); Yap (YAP).

Hotels: There are small hotels on several of the islands.

National transport
Road: Outside administrative areas, the road network, which is being extended, frequently consists of tracks not passable to ordinary vehicles. A surfacing programme is in progress, however.
Water: Frequency of inter-island services governed by weather conditions, but trips to inhabited islands take place at least every 30-45 days.
Taxis: Cheap and readily available in most centres.

Public holidays
Fixed dates: 18 Feb (President's Day), 10 May (Constitution Day), 12 July (Micronesian Day).

Working hours
Government and business: (Mon-Fri) 0800 – 1700.

Telecommunications
Telephone: Dialling code for Federated States of Micronesia, IDD access code + 691 followed by area code (Kosrae 370, Pohnpei/Panape 320, Truk 330, Yap 350). There are telephone exchanges on Pohnpei, Truk and Yap.

Banking: Bank of Guam, Bank of Hawaii.

Hotels: Kolonia, Pohnpei
Blue Rose Inn, Kolonia 96941, P.O. Box: 162 (tel: 340).

Cliff Rainbow, Kolonia 96941, P.O. Box: 96 (tel: 415).

Hifumi Inn, Kolonia 96941, P.O. Box: 811 (tel: 382).

Nan Madol, Kolonia 96941 (tel: 525).

Pohnpei, Kolonia 96941, P.O. Box: 430 (tel: 330).

South Park, Kolonia 96941, P.O. Box: 829 (tel: 255).

Village, Kolonia 96941, P.O. Box: 339 (tel: 3202797; tx: 6803).

Hotels: Kolonia, Yap
Esa, Yap 96943, P.O. Box: 141 (tel: 435).

Rai View, Kolonia 96943, P.O. Box: 130 (tel: 218).

Hotels: Moen, Truk
Christopher Inn, Moen 96942, P.O. Box: 37 (tel: 652).

Truk Continental, Moen 96942 (tel: 727).

Car hire: Rental cars are available in the major administrative centres, either through Hertz, Avis, National or Thrifty or from local firms.

Other useful addresses
Kosrae State Government, Lelu, Kosrae, FSM 96944.

Office of the Governor, Yap State Government, Colonia, Yap, FSM 96943.

Pohnpei State Government, Kolonia, Pohnpei, FSM 96941.

Truk State Government, Moen, Truk State, FSM 96942.

Fiji

Kevin Rafferty

(map on page xxiv)

The tiny Pacific island country of Fiji, a minnow in world terms but a large fish among the micro-states of the vast ocean, continued to tear itself apart over the repercussions of the series of coups d'etat in 1987. In 1989 two significant things happened. Dr Timoci Bavadra, the Fijian prime minister and Labour Party leader ousted in the military coup, died of cancer in November at the age of 55. Earlier in the year the government had declared that native Fijians now outnumbered Indians by a slender margin of 3,000 in a population of more than 700,000. If the figures are to be believed, it was the first time since 1946 that the Fijians have been the largest racial group in what they consider to be their own country. However, many non-biased commentators doubt whether it is a cause for any rejoicing, except by the narrow-minded Fijian nationalists running the country.

Two years after the military coup abolished allegiance to the British Queen, took Fiji out of the Commonwealth and established a republic in the 332 islands of Fiji, it is hard to miss royal nostalgia. Superficially at least all the trappings of loyalty are still there: the portrait of Queen Elizabeth II is everywhere, on banknotes, in the High Court, in the prime minister's office, on top of the gate of the official residence of President Ratu Sir Penaia Ganilau. The Fiji flag still features the British union flag in its top left-hand corner. Public holidays celebrate the Queen's birthday and that of Prince Charles, the heir to the British throne.

The links to the British Crown go back to 1874 when the Fijian chiefs handed over the country to the protection of Queen Victoria. Fijian people who had celebrated cannibalism enthusiastically were converted to Christianity by Wesleyan Christian Methodists. Fiji's religion became Christianity – apart from the Indians whom the British imported to work as indentured labour in the sugar plantations who retained their Hindu or Muslim faiths.

KEY FACTS

Official title: Republic of Fiji

Head of state: President Ratu Sir Penaia Ganilau

Head of government: Prime Minister Ratu Sir Kamisese Mara

Capital: Suva (on Viti Levu)

Official Languages: English, Fijian, Hindi

Currency: Fijian Dollar (F$ = 100 cents)

Exchange rate: F$1.50 per $ (Nov 1989)

Area: 18,333 sq km (about 332 islands, 110 inhabited)

Population: 740,000* (1988)

GDP per capita: F$1,335 (1988)

GDP real growth: –2.5% (1988)

Labour force: 247,000 (1988)

Unemployment: 10% (1986)

Inflation: 11.7%* (1988)

Trade balance: –$54.3m (1988)

Foreign debt: $446.2m* (1987)

Visitor numbers: 208,155 (1988)

* estimated figure

Power shift

Two years after the coup, the nominal rulers of Fiji also remain the same. President Ganilau is the same man who used to be called the governor general when Fiji still recognised the Queen as its head of state. The prime minister in 1989 was still Ratu Sir Kamisese Mara, who had taken Fiji to independence in 1970 and ruled for most of the period since, except for a short interval in the 1970s and Bavadra's brief period in power.

But real power has shifted from the government leaders to an alliance of the army – represented in cabinet by home minister and army chief Major General Sitiveni Rabuka (the same man who led the coup, but promoted three ranks) – the council of Fijian chiefs and the Methodist Church. Throughout 1989 a battle raged over exactly what kind of new constitution the country should have. The prime aim would have to ensure 'Fiji for the Fijians'. A final draft of a new constitution offered indigenous Fijians a built-in majority in parliament to ensure their domination. In addition, the president and the prime minister will have to be ethnic figures. But not all the Fijians were happy with what was offered. Some leading army officers resented a clause in the draft constitution preventing an army leader from simultaneously holding ministerial office. A report from a group of army officers damned Bavadra and his friends for daring to continue to criticise the government. It declared 'their adamant refusal to accept reality and the objectives of the coup, coupled with their selfish and unrealistic ideologies about the destinies of the nation, has compelled us to believe that, in the best interests of national peace, prosperity and stability, they should be neutralised.' The officers said they didn't care about international reaction, adding 'such action on our part would naturally result in further criticism and condemnation, both local and international, being levelled against us, but then why should we care? This is our country.'

The power of the Methodist Church and particularly of its fundamentalist wing was also seen. God and the teachings of Jesus Christ are enshrined in the constitution – or at least the stern Methodist interpretation of them. In mid-1989 the government relaxed some of the stricter provisions of its controversial Sunday observance decree. Under the new decree all commercial activities are banned, apart from those considered in the public interest. Organised sport, public entertainment and any form of assembly in a public place are also forbidden. However, agricultural work, sugar-cane milling, public transport during specified hours, the baking and selling of bread, marriages, funerals and family picnics will be permitted. Before the relaxation no form of work was allowed on Sunday. One farmer was fined in 1989 for picking a lettuce from his garden. Some of the more fervent Christian fundamentalists are still pressing for stricter observance and are prepared to use intimidation as a means of getting their way.

Prime Minister Mara, aged 69 in 1989, had several times said he would retire in December after two years heading the interim government. But the president asked him to postpone his departure at least until elections could be held. Mara agreed on condition that Rabuka also made a choice. The two men do not get on well together and the prime minister has openly said he felt his authority had been undermined by Rabuka's twin roles as cabinet minister in charge of security and also army commander. He gave the

KEY INDICATORS

	Unit	1984	1985	1986	1987	1988
Population	'000	690	700	700	700	710
Gross domestic product (GDP)	F$m	779	984	1,056	972	948
GDP per capita	F$	1,129	1,406	1,509	1,389	1,335
Inflation rate	%	5.4	4.4	1.8	5.7	11.7
Volume of exports						
Sugar	1985=100	93.4	100.0	79.6	103.9	99.4
Coconut oil	1985=100	144.3	100.0	134.3	61.8	41.2
Exports	F$m	280.1	264.0	312.6	406.0	518.1
Sugar	F$m	116.6	118.3	141.6	196.9	209.7
Coconut oil	F$m	18.47	7.64	3.91	3.02	3.42
Re-exports	F$m	82.3	80.7	70.4	74.6	83.2
Imports	F$m	487.0	508.0	493.6	465.6	658.7
Petroleum	F$m	107.1	115.5	82.1	75.6	88.6
Balance of trade	$m	–162.8	–174.8	–122.4	–26.8	–54.3
Current account	$m	–26.9	–12.6	4.5	–4.6	30.1
Debt service ratio	%	9.7	10.6	10.3	12.1	–
External debt	$m	413.4	443.7	440.9	446.2	–
Exchange rate	F$ per $	1.08	1.15	1.13	1.24	1.43

* estimated figure

COUNTRY PROFILE

Political system: Until 1987 Fiji recognised the British sovereign as head of state, but then in a series of coups Lieutenant Colonel Sitiveni Rabuka overthrew the recently elected mixed Fijian and Indian government of Timoci Bavandra of the Labour Party and then declared the country to be a republic. The colonel, later promoted to brigadier and head of the armed forces, objected to the fact that native Fijians had lost control of their own country to immigrant Indians (who form 48.7 per cent of the population against 46 per cent Fijians). Former governor-general Sir Penaia Ganilau was sworn in as president and head of state, Ratu Sir Kamisese Mara again became prime minister. Meanwhile the Fijian Council of Chiefs and the military and other interests were wrestling with a new constitution which would formalise political control of the country in the hands of the Fijians. Prior to the 1987 coups, a carefully balanced constitution gave a House of Representatives 22 Fijians, 22 Indians and eight general electors (for Chinese, Europeans and others).

Population: Of the 715,375 inhabitants registered in 1986, only 329,305 were ethnic Fijians, while 348,704 were of Indian origin, 5,000 were Chinese, and 4,000 were Europeans.

Main cities/towns: Suva (on Viti Levu, population 69,665 in 1986), Lautoka (on Viti Levu, pop 28,728), Nadi (on Viti Levu, pop 7,709), Ba (on Viti Levu, pop 6,515), Nansori (on Vita Levu, pop 5,242, Labasa (on Vanua Levu, pop 4,917), Savusavu (on Vanua Levu, pop 2,179), Sigatoka (on Viti Levu, pop 2,097), Levuka (on Ovalau, pop 1,106).

Language: English, the official language, is widely used in business circles, but Fijian, Hindi and Cantonese are also commonly spoken.

Media
Press: One English daily newspaper, the *Fiji Times* published in Suva. A number of other papers also published in Fijian and Hindi. *Fiji Post* is a bi-weekly newspaper. *Island Business* is a monthly trade magazine; others include *Construction and Commercial Trade News*. *Fiji Beach Press* is a monthly tourism newspaper. *Fiji Magie* is also a monthly tourism magazine.
Radio: Fiji Broadcasting Commission, a government funded body, broadcasts a fully commercial service on three wavebands; Fiji 1 in English and Fijian, Fiji 2 in English and Hindi, and Fiji 3 in English. FM96, a commercial radio station, broadcasts 24 hours in English. *Television:* The government has undertaken to set up a TV service for Fiji, Western Samoa and Tonga via satellite from a ground station to be built in Fiji.

Domestic economy: The impact of political turmoil was quickly felt on Fiji's economy. Real growth of gross domestic product actually slumped by 6.3 per cent as business was badly affected and tourists were scared away. Fiji has a small, open economy, highly dependent on sugar as its main export and tourist earnings to offset the normal large deficit on the trade account. Until the military intervened, the country had been expecting a good year, especially since sugar output and prices had risen. Before the political turmoil, tourist arrivals were running at about 214,000 a year and earnings were almost F$145m. In the second quarter of 1987 (when the troubles started) tourist arrivals dropped to 40,000 against 57,000 in the same quarter of the previous year. Sugar production was damaged by the severe drought. There was a knock-on impact on business, through a decline in confidence which led to the increase in capital flight. The Fiji dollar was devalued by a total of 33 per cent in 1987. In 1986 the US dollar was worth between 1.10 and 1.15 Fijian dollar, but by October 1987 the US currency was worth F$1.50. However the situation appeared to have stabilised in 1988. To encourage more foreign investment, the government announced the establishment of tax-free zones and tax-free factories. By the end of July 1989 there were 44 garment factories operating in the TFZs as well as TFFs.

Employment: Labour force expanding at 2-3 per cent per annum. Unemployment estimated at 11 per annum in 1987, but is hard to measure because of large subsistence sector. High industrial and civil service wages, protected by strong trade unions, impede absorption of excess labour in these sectors. Real wages of 21,000 civil service employees rose sharply in the early 1980s, while 85,000 paid employees' wages rose five per cent; and self-employed sugar growers' incomes halved (some 120,000 are self-employed).

military man an ultimatum, asking him to choose between the army and government. The 41-year-old Rabuka said he would give up the cabinet post. However, there he put a sting in the tail. He added that 'should the need arise', he would step down as head of the army to become prime minister.

Ethnic tensions

Tension has continued between the Fijians and the Indians, not only because the Indians are mostly non-Christian and therefore do not agree with the Methodist ideas of how Sunday should be celebrated. One Sunday in the middle of October several Indian temples and mosques in Lautoka, Fiji's second city, were firebombed by Fijian fundamentalist Christians. Among those arrested was a 16-year-old schoolgirl who said she was a member of the Methodist Youth League.

Shortly after this, India decided to recall permanently its Ambassador T. Srinivasan and downgrade its Suva embassy to a consulate. The Fiji government added fuel to the flames when it instructed Srinivasan to leave Fiji within three days, expressing its 'displeasure with the actions and statements of Mr Srinivasan over the past two years'. New Delhi reacted describing the expulsion as 'unnecessary' and 'thereby further embittering bilateral ties.'

The Fijian nationalists may rejoice about being more numerous than the Indians, but the figures had damaging implications for the country as a whole, and particularly for the economy. At the last reliable registration in 1985, the figures were 330,141 ethnic Fijians and 347,445 Indians (Chinese and Europeans together accounted for 10,000). If the Fijians have overtaken the Indians in the population, it implies a bigger brain drain even than had been thought. It remains a possibility that the figures have been falsified to provide the nationalists with a morale booster.

What is certain is that the Indians have been trying to get out and that this is sapping Fiji's economic strength. There has been a drain of much of the professional talent that Indians supplied. Nervous people have been fleeing mainly to Australia and New Zealand. In early 1989, the Fiji Medical Association reported that it had lost 150 of the country's 280 doctors; the Law Society said that 110 of 150 solicitors had left; the Institute of Accountants lost 150 out of 400 members; the Public Service Commission began a recruiting drive to replace 4,000 out of 18,000 staff who had quit; and the Fiji Teachers Union said that there had been 800 resignations since the coup. The government acknowledged the brain drain in allocating an extra F$5m for the Public Service because of 'a serious loss of skills and manpower'. Volunteer doctors from Thailand and China had to be recruited through the United Nations Development Programme to make good the shortage.

For all this haemorrhaging of vital talent, the economy has picked up since the coup. A major earner, tourism, is returning to normal, at least in terms of numbers. The tourist board hopes to set a new record of more than 300,000 arrivals by 1990. However, hotels have complained that the new tourists are mainly cheaper groups and back-packers, and that there has been a decline in richer individual visitors more likely to spend heavily.

The country's foreign exchange reserves have recovered from a low of F$131m in 1987 and were at reasonably healthy levels for a developing country, providing several months' cover for imports. This was thanks, in part, to injections of aid from countries that saw an opportunity in the coup and rejoiced in the cooler relations between Fiji and its closest neighbours and important trading partners, Australia and New Zealand. France was quick to offer help, and Malaysia, a country with its own racial problems, offered assistance in setting up a television network.

This progress offers the nationalists no cause to rejoice that they can manage and prosper without the Indians. As a report from the International Monetary Fund experts noted: 'The native Melanesians are involved mainly in subsistence agriculture and public sector employment. The people of Indian descent are engaged in sugar cultivation and trading activities.' The Indian people have provided the dynamism to an economy that has plenty of potential. They are also the main tax-payers to a budget that has long suffered from constraints.

There are plenty of flashpoints still to come. One of the main ones is the large-scaled expiry in 1991 of 30-year sugar-cane leases granted to Indians. Sugar is Fiji's main cash crop and export earner bringing 40 per cent of exports. A former Fiji High Court Judge Kishor Govind, now living in exile, put forward a bleak picture which – he said – was the choice of the Indians. 'They can risk displacement now with the hope of the regime collapsing in the short term, or wait for their economic ruin when the regime is established and their leases expire.'

There is no lack of bitterness and pettiness in the regime. At the behest of the army commander the government banned foreign journalists from reporting the funeral of ousted Prime Minster Bavadra, saying that the local – carefully controlled – press should be allowed full flower.

However much the Fiji nationalists may wish to shut off the rest of the world and go their own way, Fiji is highly vulnerable. Its import bill is up to 50 per cent higher than exports, themselves highly dependent on sugar. To fill the gap it needs tourism and aid, both decided by foreigners.

COUNTRY PROFILE

External trade
Exports: In 1987, exports totalled $409m. Main exports are raw sugar (55.7% of 1987 total), unrefined gold (15.1%), fish (6.2%), molasses (3.2%), coconut oil (0.9%) and timber. Main destinations UK (41.4% of 1987 total), Australia (20.4%), Malaysia (9%), USA (5.3%) New Zealand (5%), Japan (3.5%), Canada (1.6%).
Imports: In 1987, imports totalled $466m. Main imports machinery and transport equipment (19.4%), foodstuffs (18.5%), manufactured goods (22.3%), petroleum products (16.2%), raw materials (11%). Main sources Australia (28.6% of 1987 total), New Zealand (16.9%), Japan (12.1%), Singapore (11%), UK (4.7%), USA (5.2%), Taiwan (3.8%), China (2.5%) and Hong Kong (2.0%).

Agriculture: Historically, 85 per cent of land is granted to Fijian clans (*Matagali*) and by law cannot be sold, which has led to under-employment of some land. Soil is generally fertile and easily worked. Sugar normally accounts for half agricultural output, but has declined both in quality and in quantity. Refined sugar production fell 20 per cent to 401,000* tonnes in 1987. However sugar earned a record $147m in 1987 and expects to do the same in 1988. The Fiji Sugar Corporation aims to diversify into ethanol and to encourage other uses for spare land, especially rice (50 per cent of which is imported). Other crops include rice, bananas, pulses, tobacco, citrus, sorghum, cocoa, ginger and broom corn. The government has invested $10m in a five-year project to develop 2,100 hectares of Sigatoka Valley (Viti Levu) for agricultural production to reduce food imports. Beef and dairy cattle, pig, goat and poultry rearing meet the bulk of local demand for fresh meat, milk and eggs. Fiji received further Australian funds for Yalavou beef-raising project on Viti Levu. Forestry is becoming a major development sector. In 1987 $4.7m worth of timber was exported. In addition to natural rain forest, large new plantations of pine and hardwood were established in the late 1970s. Exports of pine chips to Japan and sawn pine to Australia began well in 1987. The country hopes to produce 1.8m cu m by 2000, and to quadruple income to $F 30m. Fishing for local consumption include skipjack, yellowfin and commercial species, and prawns and oysters are raised in fish farms. Bêche-de-mer, shark fins, trochus, mother-of-pearl and turtle shells are sold.

Industry: Manufacturing industry (excluding sugar milling) accounts for 13 per cent of GDP and sugar accounts for one-third of industrial output. Sugar cane is crushed at four local mills and exported as raw sugar and molasses. Copra milling, which produces coconut oil and oil cake for export, is dominated by two large under-used factories at Suva, but a small-scale rural copra mill intended to cut growers' energy and transport costs is under construction at Savusavu on Vanua Levu to produce coconut oil for a local soap factory. Tourism, the biggest single foreign exchange earner, employs some 16,000 people. A plan for new hotel projects worth $168m to develop further the tourist industry is impeded by lack of local equity to add to foreign investment. Visitors dropped as the political climate was uncertain. Total number of visitors was around 190,000 in 1987. Timber milling and wood-related industries are expected to expand now that 1970s pine plantations are ready for harvest, and British Petroleum is investing $42m in a joint venture with the Fiji Pine Commission to mill pine and export wood chips to Japan. There are two breweries, a flour mill and a steel rolling mill.

Mining: Gold has been mined for export at Vatukoula (on Viti Levu) for some years, and has benefited from the sharp rise in world prices. Gold production rose 53.1 per cent to 2,856 kg in 1986. Total exports in 1987 were worth $60m, but accessible ore is expected to be exhausted within 10 years. Large low-grade copper deposits have been discovered at Namosi near Suva. Oil exploration is actively underway.

Energy: In 1979-84 Fiji carried out a major hydroelectric scheme at Monasavu on Viti Levu, with enough output to cover demand until 1992. Work on another two mini-hydroelectric schemes is expected to start 1988. The projects in Vanua Levu and BA Viti Levu are expected to supply electricity to 70 villages and other processing industries.

Main companies: Fiji Sugar Corp, Fiji Pacific Fishing Co, Fiji Motibhai & Co, Fiji Ika Corp, Punja & Sons Ltd, Lees Trading Company Ltd, Air Pacific Ltd, Fiji Post and Telecommunications Ltd, Fiji Electricity Authority and Fijian Holding Ltd.

Membership of international organisations: Asian Development Bank, Colombo Plan, Economic and Social Commission for Asia and the Pacific, FAO, G-77, GATT, IBRD, ICAO, IDA, IFAD, IFC, ILO, IMF, IMO, INTELSATT, Interpol, ISO, ITU, South Pacific Agreement on Regional Trade and Economic Co-operation (SPARECA), South Pacific Bureau for Economic Co-operation (now known as Forum Secretariat), South Pacific Commission (participating member), South Pacific Forum, UN, UNESCO, UPU, WFTU, WHO, WIPO, WMO, World Bank. Associated with the EC under the Lomé Convention.

Time: GMT + 12 hrs.

Climate: Hot and damp, tempered by cool winds from May-Oct. Maximum temperature during summer (Dec-Apr) 32°C, when hurricanes and cyclonic storms sometimes occur, rarely falls below 18°C during the rest of the year.

BUSINESS GUIDE

Entry requirements
Passport: Required by all except certain document holders and certain seamen.
Visa: Visitors must produce proof of funds for maintenance and assured onward passage. Temporary visitors' visas valid for 30 days are issued on arrival, may be extended for up to six months.
Currency: No restriction on importation of currency. Foreign currency notes to a value equivalent to amount brought in and F$100 in Fijian currency notes may be taken out. Transfers of funds to accounts outside Fiji subject to Exchange Control legislation.
Customs: Personal effects allowed duty-free. Strict animal and plant quarantine regulations; fruit or plant material should not be brought in. Many agricultural and manufactured items subject to import embargoes and licensing, and the list is subject to alteration. Details available from the Ministry of Commerce and Industry in Suva.

Health precautions
Mandatory: Vaccination certificates for smallpox and yellow fever if travelling from an infected area and for cholera when travelling from Bangladesh, India, Kiribati, Nauru, Pakistan or Tuvalu.

Representation overseas: Auckland, Brussels, Canberra, London, Los Angeles, New Delhi, New York, Sydney, Tokyo, Vancouver, Washington DC, Wellington.

Air access
National airline: Air Pacific.
Other airlines: Air Calédonie, Airline of Marshall Islands, Air New Zealand, Air Pacific, Canadian Airlines International, Fiji Air, Polynesian Airlines, Qantas, Solomon Islands Air, Turtle Airways.
Tax: F$10 payable on departure.
International airports: Nadi International (NAN), 200 km from Suva; duty-free shop, restaurant, buffet, bank, post office, car hire; Nausori (SUV), 19 km from Suva with milk bar.

Surface access
Main ports: Labasa, Lautoka, Levuka, Savasava, Suva.

Hotels: Numerous tourist hotels of all types, frequently in scenic locations around the islands. Advisable to book rooms well in advance.

Credit cards: Most major credit cards accepted at hotels, restaurants, shops and rental car agencies, tours, cruises and travel agencies. American Express, Diners Club, Visa, JCB and Master Card have representatives in Suva.

Car hire: Chauffeur-driven and self-drive car hire available. Current overseas or international licence acceptable for six months. Driving on the left-hand side of the road, speed limits 50 km/hr in towns and villages, 80 km/hr on highways.

City transport
Taxis: Metered taxis available in main centres. It is advisable to negotiate fares for long journeys in advance.

National transport
Air: Fiji Air operates several daily flights between Nadi International and Nausori Airport and most other domestic services. Air Pacific operates main routes – between Nadi, Suva and Labasa. Sunflower Airlines, Air Coral Coast and Turtle Island Airways operate on parts of Viti Levu and are available for charter. Helicopters can be chartered from Pacific Crown Aviation, Suva.
Road: A 3,300 km road network, about one-third of which is tar-sealed. On Viti Levu, a 500 km coastal highway links main centres. A trans-insular road on Vanua Levu connects Labasa with Savu Savu. Short-term visitors advised to use air services.
Buses: Air-conditioned buses operate daily between Suva, Nadi and Lautoka.
Water: Small inter-island vessels operate from Suva and Lautoka. A regular ferry service connects Suva and Labasa, Ovalau and Koro Island.

Public holidays
Fixed dates: 1 Jan, 25 Dec and 26 Dec.
Variable dates: Good Friday to Easter Monday, Aug Bank Holiday, Fiji Day, Diwali, Prophet's Birthday. Dates of Muslim festivals vary, as they depend on lunar calendar.

Working hours: Shops closed on Sun.
Business: (Mon-Fri) 0800 – 1300, 1400 – 1600.
Government: (Mon-Thu) 0800 – 1300, 1400 – 1630; (Fri) to 1600.
Banking: (Mon-Thu) 0930 – 1500; (Fri) 0930 – 1600.
Shops: (Mon-Fri) 0800 – 1800; (Sat) 0800 – 1300.

Social customs: Tipping is not customary. Clothing may be casual, but should be modest: swimsuits are acceptable on beaches and around hotel pools, but should not be worn in public.

Telecommunications
Telephone: Dialling code for Fiji, IDD access code + 679 followed by subscriber's number (no area code required). Fiji is connected to the international system via Commonwealth Pacific Cable and to neighbouring Pacific island territories via Suva Post Office Radio. All main centres linked by telephone, with 51,216 subscribers, 34 telephone exchanges and 90 rural radio telephone stations in 1984.
Telex: International telex facilities available at Fiji International Telecommunications Ltd (FINTEL), Victoria Parade, Suva, or at major hotels.
Fax: Facilities are available.
Telegram: Ordinary and deferred overseas telegram facilities available at all telegraph offices.

Banking
Fijian banks: National Bank of Fiji, Reserve Bank of Fiji.
Fijian branches: ANZ Banking Corp, Bank of Baroda, Bank of New Zealand, Fiji Development Bank, Westpac Banking Corp.

Advertising: Advertising available in local press, radio and cinema. Few sites suitable for outdoor advertising.

Electricity supply: 240/415 V AC, with flat three-pin plug fittings and bayonet-type light fittings. Wiring rules of the Standard of Associations of Australia in use. Larger hotels have 110 V conversion units for electric shavers.

Weights and measures: Metric system (replacing imperial system).

Representation in capital: Australia, China PR, France, India, Israel, Japan, Republic of Korea, Malaysia, Marshall Islands, New Zealand, Papua New Guinea, Tuvalu, UK, USA.

Hotels: Suva, Viti Levu
Grand Pacific, Government Buildings, Victoria Parade, P.O. Box: 2086 (tel: 23011; tx: 2321).

Outrigger Motel, 147 Queen Elizabeth Drive, P.O. Box: 912 (tel: 25428; tx: 2139).

President, Queen's Road, P.O. Box: 1351 (tel: 361033; tx: 2630).

Southern Cross, 63 Gordon Street, P.O. Box: 1076 (tel: 31423; tx: 2162).

Sunset Apartment, Corner Gordon and Murray Street, P.O. Box: 485 (tel: 23021/22; tx: 2141).

Suva Courtesy Inn, Malcolm and Gordon Streets, P.O. Box: 112 (tel: 312300; tx: 2272; fax: 301300).

Suva Peninsula, Cnr McGregor Road and Pender Street, P.O. Box: 888 (tel: 313711; tx: 2196).

Suva Travelodge, Victoria Parade, P.O. Box: 1357 (tel: 314600; tx: 2159 FJ; fax: 314379).

Tradewinds Hotel, Queen's Road, Lami, P.O. Box: 3377 (tel: 362854; tx: 2152).

Car hire: Nadi, Viti Levu
Avis (airport): PO Box 9088 (tel: 72233; tx: 5187).

Budget (airport): PO Box 9164 (tel: 72735; tx: 2483).

Dominion: (tel: 72262).

Hertz: PO Box 9367 (tel: 72771; tx: 2415).

Khan's Rental Cars: PO Box 3 (tel: 70088).

United Touring Fiji: PO Box 9172, Nadi Airport (tel: 72411; tx: 5171 FJ).

UTC: (tel: 72463).

Car hire: Suva, Viti Levu
Avis: (tel: 313833; tx: 5187).

Budget: (tel: 49299).

Dominion: (tel: 25977).

Hertz: (tel: 23206; tx: 2415).

UTC: (tel: 25616).

Academic/research institutes
University of the South Pacific: Laucala Campus, PO Box 1168, Suva (tel: 313900; tx: FJ 2276);

Alapua Campus, PO Box 890, Alapua, Apia, Western Samoa (tel: 21671; tx: 291 USP SX). Founded 1970. Various institutes including Institute for Research, Extension and Training in Agriculture, Institute of Marine Resources, Institute of Pacific Studies and Institute of Social and Administrative Studies (supplies region with short-term courses in subjects such as auditing, financial management and social welfare).

Business advice and information
Bureau of Statistics, PO Box 2221 Suva (tel: 315144).

Commonwealth Development Corporation, 371 Victoria Parade, Suva (tel: 25607).

Department of Information, PO Box 2225, Government Buildings, Suva (tel: 211250/1).

Development Bank of Fiji, PO Box 104, Suva (tel: 25661; tx: 2578).

Fiji Trade and Investment Board, PO Box 2303, Government Buildings, Suva (tel: 315988, 313295; tx: 2355).

Ministry of Primary Industries, PO Box 358, Rodwell Road, Suva (tel: 311233).

National Bank of Fiji, 107 Victoria Parade, PO Box 1166, Suva (tel: 311999; tx: 2135).

National Marketing Authority of Fiji, PO Box 5085, Raiwaqa, Suva (tel: 385888; tx: 2413).

Reserve Bank of Fiji, PO Box 1220, 4th Floor, Dominion House, Thomson Street, Suva (tel: 313611; tx:2164 FJ).

Suva Chamber of Commerce, 7th Floor, Honson Building, Thomson Street, PO Box 337, Suva (tel: 26183).

Banking
Reserve Bank of Fiji (central bank), PO Box 1220, 4th Floor, Dominion House, Thomson Street, Suva (tel: 313611; tx: 2164).

Australia & New Zealand Banking Group Ltd, 4th Floor, Civic House, Town Hall Rd, PO Box 179, Suva (tel: 314000; tx: 2194).

Fiji Development Bank, PO Box 104, Suva (tel: 25661; tx: 2279).

Merchant Bank of Fiji Ltd, Suva.

Westpac Banking Corporation (Australia), Town Hall Road, PO Box 238, Suva (tel: 311666; tx: 2133).

Travel information
Fiji Visitors Bureau, Thomson Street, PO Box 92, Suva (tel: 22867).

Finland

Enrique Tessieri

After a long period of regional geopolitical stability, 1992 in the West and perestroika in the East were imposing demanding economic as well as political challenges on neutral countries like Finland. Relations with the USSR continued to hold a special place, even if Finnish–Soviet trade levels had dropped by a third and glasnost was attempting to teach the USSR to be more politically tolerant of its neighbours. Domestically, an eventual tightening up of economic policy was in the offing if Finland wanted to blow away the winds of recession which loomed over the economy.

Finland was feeling the winds of change from outside and within. Although still politically premature, some optimistic observers believed that once-sensitive political topics such as relations with the USSR, consensus and even EC membership would be more openly debated by Finns in coming years. This condition naturally hinged on how far Europe could mend its ways and whether Soviet leader Mikhail Gorbachev's policies would galvanise or falter. On the other side of the coin, the government would be forced to take some decisive steps to curb inflation in the 1990 budget. Inflation almost doubled in 1987-89 and was expected to reach 6.7 per cent by the end of 1989.

Economic performance

The biggest challenge to the government as the parliamentary elections of 1991 approached was how to curb mounting inflation and contain the deteriorating current account balance while sustaining a moderate level of economic growth. Some analysts felt that the government was obliged to implement an economic package if it wanted to 'cool down' the accelerated pace of the Finnish economy. During the 1980s, Finland's annual GDP rate grew on an average by 4 per cent and reached a staggering 5.2 per cent in 1988. This recent leap in GDP growth was fuelled mainly by investments in the real estate sector and was unwelcome news, because all forecasts were supposed to point towards smaller growth in 1988.

The second biggest challenge to the country's economic policy makers was how to curtail the deficit in the current account balance. The deficit for 1988 reached FIM12.5bn and was expected to drop to FIM16.5bn by the end of 1989. If the government did not take measures to eradicate the deficit in the 1990 budget, analysts believed that the current account deficit might drop a further FIM2bn by the end of 1990.

During the 1980s, Finland had been taking decisive steps to open up its domestic market. In 1986-87, the Bank of Finland lifted all restrictions on the financial market. This measure led to an unprecedented level of private indebtedness that was responsible for fuelling domestic consumption, demand, prices and inflation. Domestic consumer debt as a percentage of disposable income rose from 80.8 per cent in 1987 to 95.7 per cent in 1988 while saving ratios for the same group declined from 1.8 per cent in 1987 to -2.0 per cent in 1988.

Finland also went ahead in October 1988 with steps to deregulate its markets as well as to do away with price controls. This measure put strains on the Finnish economy, and did not help contain inflation since many of Finland's small markets continued to be dominated by monopolies and price cartels. This was especially true of the real estate sector, where housing prices in Helsinki went up an incredible 40.6 per cent in 1988. Inflation was mainly responsible for turning

KEY FACTS

Official title: Republic of Finland

Head of state: President Mauno Koivisto

Head of government: Premier Harri Holkeri (National Coalition Party)

Ruling party: Coalition of National Coalition Party Social Democratic Party, Swedish People's Party and Rural Party

Capital: Helsinki

Official Languages: Finnish and Swedish

Currency: Markka or Finnmark (FIM) = 100 penniä

Exchange rate: FIM4.16 per $ (Dec 1989)

Area: 338,144 sq km

Population: 4.95m (1988)

GDP per capita: $18,163 (1987)

GDP real growth: 5.2% (1988)

Labour force: 2.55m (1988)

Unemployment: 4.5% (1988)

Inflation: 5.1% (1988)

Trade balance: FIM6.1bn (1988)

Foreign debt: FIM64.6bn (1988)

Helsinki into Europe's second most expensive city after Oslo.

In a bid to get a grip on inflation, the government reached a stabilisation agreement with the labour sector in autumn 1988. The accord called for an extra-moderate wage quarterly increase of 1.5 per cent, or 6 per cent total for 1989. Wage increases for 1989, however, were expected to reach 8 per cent. Furthermore, the consumer price index, also stipulated in the stabilisation agreement to be 4 per cent, was forecast to reach 5 per cent by the end of 1989. In March 1989, the Bank of Finland revalued the Finnish markka by 4 per cent in a bid to keep interest rates high and thereby curb domestic consumption.

Even if the economic picture looked gloomy, ministry of finance officials were confident that there was still enough time to implement measures before the economy skidded off its tracks. Hopeful of a 'soft landing' rather than 'overkill,' measures to improve the general economic situation were expected to decrease GDP annual growth to 3.3 per cent in 1989 and for the following five years, until 1994, to an average 2.5 per cent.

KEY INDICATORS

	Unit	1984	1985	1986	1987	1988
Population	'000	4,882	4,908	4,918	4,925	4,950
Gross domestic product (GDP)	FIM bn	307.9	336.8	360.3	393.6	–
GDP per capita	FIM	63,068	68,623	73,261	79,919	–
Inflation	%	7.1	5.8	2.9	4.1	5.1
GDP real growth	%	3.3	3.5	2.3	3.8	5.2
Unemployment	%	5.2	5.0	5.4	5.1	4.5
Agricultural production	1979-81=100	117.5	114.4	113.6	102.4	102.6
Manufacturing production	1980=100	112	116	118	123	128
Investment goods	1980=100	128	137	142	150	–
Consumer goods	1980=100	106	109	109	112	–
Intermediate goods	1980=100	113	115	116	122	–
Exports	FIM bn	80.9	84.0	82.7	87.5	91.7
USSR	FIM bn	15.4	18.1	16.7	13.5	–
Sweden	FIM bn	10.0	11.1	12.2	13.1	–
UK	FIM bn	9.7	9.1	8.6	10.0	–
German FR	FIM bn	7.8	7.8	8.0	9.6	–
Imports	FIM bn	74.7	81.4	77.6	86.5	85.6
USSR	FIM bn	17.3	17.1	11.8	12.5	–
German FR	FIM bn	10.4	12.1	13.1	15.1	–
Sweden	FIM bn	9.2	9.6	10.5	11.2	–
UK	FIM bn	5.8	5.8	5.0	6.2	–
Balance of trade	FIMbn	6.3	2.6	5.1	1.0	6.1
Current account	$m	–10	–766	–731	–1,949	–2,988
Exchange rate	FIM per $	6.0	6.2	5.1	4.4	4.2

* estimated figure

Politics

The most significant political event to hit Finland was Mr Kalevi Sorsa's resignation as foreign minister. Mr Sorsa, a veteran Social Democratic Party (SDP) politican, was named the new speaker of parliament. Although some political observers felt that SDP Chairman Mr Pertti Paasio's appointment to the helm of the foreign ministry was a bit of muscle-flexing on his part, others believed that Mr Sorsa's new posting was to groom him for the next presidential elections of 1993. Of all the favourites for the presidency, Mr Sorsa was the only one to have been named speaker of parliament, a politically important post which ranks second in line after the president.

It was not clear whether President Mauno Koivisto was willing to run for a third term or not. However, Centre Party chairman Mr Paavo Vaeyrynen, who came in second place in the presidential elections of February 1988, was keenly eyeing the presidency.

After having dominated the Finnish political scene for almost 50 years, the Centre Party was left out in the cold after the parliamentary elections of March 1987. The historical coalition cabinet, made up of the SDP, the Conservative Party (Kokoomus), Rural Party and Swedish Folk Party, was expected to complete its mandate although co-existence had been at times difficult. Before 1987, Finnish politics kept Kokoomus in opposition for 21 years.

The municipal elections of October 1988 were the most uneventful political event. In a campaign that lacked hard issues, the election results did not force the SDP or Kokoomus to form a new cabinet even if the Centre Party was proclaimed the victor of the municipal elections. What caused more debate was voter turnout, which reached an all-time low of 69.9 per cent. In the 1984 municipal elections voter turnout was 74 per cent and in 1980, 78.1 per cent.

The most tragic news of 1989 was the suicide of Nokia's chairman and CEO, Mr Kari Kairamo. The rapid international growth of Nokia, Finland's largest privately-owned company, was due to Mr Kairamo. Under his wing, the company's net sales grew from FIM2bn in 1975 to an impressive FIM22bn in 1988. Many questions remained unanswered about the motives behind the death of Mr Kairamo, a man with great visions for the future who openly encouraged closer economic and political integration between Finland and Western Europe.

Foreign policy

After over four decades of relative geopolitical calm, Finland's successful and even prosperous foreign policy was slowly being opened up for national debate. 1992 and perestroika were forcing Finland to embark once again on the onerous task of reassessing its place within a fast-changing Europe where neutrality and domestic consensus will be likely to lose ground. If Finland's foreign policy has allowed it to co-exist with the USSR, consensus has ensured domestic peace at a high price because it has encouraged self-censorship and killed debate to a great degree.

Although the single most important issue on the foreign policy agenda continued to be how far to go in integration with Western Europe, Finland took a decisive political tilt towards the West in May 1989 when it became a member of the Council of Europe. Before its membership, Finland was the only Western European country – apart from the principalities of Monaco and Andorra – to stay out of the Council, which Helsinki perceived to be too vocal about human rights violations in the Soviet bloc. Another foreign policy victory for Finland was its election to the UN Security Council in October 1988.

History has taught Finns to become geopolitical pragmatists, well aware of the virtues and vices of having a common border with the USSR. But Finnish politicians are generally business oriented and therefore see closer co-operation with the EC as paramount if the country wants to secure its vital economic links.

How much Finland's foreign policy would be obliged to change in the 1990s was an area of interesting debate. The answer to this question naturally hinges on how far Europe can mend its ways and if Soviet leader Gorbachev's reforms will be able to overhaul an economically inefficient and totalitarian USSR. If, on the one hand, Finland is being obliged to integrate into an economic scheme such as 1992, it is also, on the other hand, slowly attempting to find its cultural and historical roots with other Finno-Ugric tribes of the East like the Estonians and Ingrians, who are today a part of the Soviet Union.

For the time being, Helsinki still considered it too premature to jump on any integration bandwagons, hoping that EFTA would be able to bridge the gap between the EC and Finland after 31 December 1992. There was also a lot more debate in the local press on 1992 and its effect on Finland's foreign policy. This polarised public opinion into two camps: those who feel that 1992 is an essentially positive development for Finland, and those that feel that it is basically negative.

Finnish–Soviet trade

Even if increased trade with Western Europe has offset dwindling trade levels with the Soviet Union, both countries have yet to find a suitable substitute for oil to counter imbalances. Even so, the Finns are not too enthusiastic about raising trade levels with the Soviets for the moment because Western Europe is presently employing most of Finland's industrial capacity. In the past, the high level of Finnish–Soviet countertrade was kept up by high international oil prices. Ninety per cent of all exports from the Soviet Union come in the form of energy; around 60% of this being oil.

After international oil prices began their nose-dive at the end of 1985, Finnish–Soviet countertrade dropped by roughly a third in the last years of the decade and was characterised by heavy trade imbalances in Finland's favour. Finland's biggest gripe was that this trade imbalance in the clearing house account – that exceeds the credit limit of R300m – has been interest-free for the Soviets. Faced with this financial anomaly, both Finland and the USSR signed in September 1988 new bilateral finance agreements to come into force by 1990. Trade balances in the clearing house scheme would be reviewed by both parties on a quarterly basis and if they exceeded the new credit limit of R200m, the difference would be paid in convertible currency that had been subject to interest rates.

To weight lopsided trade imbalances heavily in Finland's favour, in autumn 1988 Finnish Export Credit turned the most recent deficit above the credit limit into a R300m loan, on which the Soviets would have to pay hard cash and Libor interest rates if the trade imbalance did not shift in the USSR's favour during the duration of the loan. A similar three-year

R287m loan was also granted by Finland in 1987 to the USSR when the sum well exceeded the credit limit. The first instalment of this loan was expected to be paid in either hard cash or goods in 1989 and the third and final one in 1991.

The hard cash element in Finnish–Soviet bilateral trade was a monumental step in the history of trade between both countries. These new agreements force both parties to take special care not to create trade imbalances while allowing neither country much room to increase trade dramatically so long as oil prices do not reach pre-1986 levels.

Energy, and in particular oil, is one way of turning Finnish–Soviet barter trade into hard cash. Thus Finland purchases Soviet oil and trades part of it off for hard currency on international spot markets.

Oil continued to play a key role in trade between both countries. Since 1987, Suomen Petrooli (SP), a subsidiary of the Russian oil giant Soyuznefteexport, came into the Finnish oil trading picture. Before SP's entry, state-held Neste, Finland's largest company, had a monopoly not only on Soviet oil but on all oil imports to Finland.

If Neste traded off Soviet oil for hard cash on international spot markets, SP's role has been to import 'expensive' Soviet oil to help write off trade deficits in the clearing house account. Already in July 1988, SP had traded a 2m tonne consignment of oil and, later on in 1988, a further 3m tonnes to East Germany and Poland. The cost of the deal was speculated to have been around FIM2.5bn and helped lower the trade deficit by some R150m.

There is a great deal of secrecy in Finnish–Soviet trade when it comes to volumes and prices; thus the exact price at which Finland purchases its oil, gas or electricity from the USSR is not known. Even so, Finnish–Soviet pricing schemes are regulated by intricate pricing formulas. Their purpose is to protect bilateral trade from dramatic price fluctuations. In 1986, when oil prices plunged, Soviet oil prices to Finland lagged a month behind Rotterdam and North Sea Brent prices.

Energy and Trade

1988 and 1989 were characterised by important trade agreements with the Soviet Union. After three years of negotiation, Finnish state utility company Imatran Voima (IVO) signed a new 10-year (1990-2000) electricity import agreement with the Russian company Machinoexport. Under the accord, Finland would continue importing some 600mW of electricity from its eastern neighbour until 1993, when the Soviets pledged to boost electricity imports by a further 300mW. The total cost of the 10-year deal was placed at FM5.5bn.

Apart from the USSR, Finland also imported 0.9TWh of electricity from Sweden in 1988. The agreement with Sweden was due to end in 1993 and negotiations on 1994-95 quotas were under way. These negotiations would be complicated because of Sweden's plans to begin part of its total nuclear phase-out at the beginning of 1995.

The new deregulation law that came into force in October
1988 created some interesting changes in the Finnish energy market. For instance, the ministry of trade and industry gave the green light to Teollisuuden Voimansiirto (TVS) to import some 300mW of electricity from the Soviet Union in the 1990s. TVS is a recently-founded firm owned jointly by three private utility companies and Nokia. Before TVS, IVO had a monopoly over electricity imports to Finland.

Although the new deregulation law was certain to have an impact on the electricity sector, it was expected to have no effect on the national oil industry market. The main obstacle in the way of companies other than Neste and SP importing oil from the USSR, as well as from Western markets, has been the paramount importance of oil in Finnish–Soviet trade. Some observers, however, believed that this might change, especially after 1992, when EFTA countries such as Finland will have to increase trade reciprocity with the EC. Some also feared that bilateral trade with the USSR might be severely undermined if Soviet monopolies were obliged to become profitable and oil trade were liberalised.

Gas also dominated a great part of trade news. In June 1989, both Neste and the Russian Soyuzgasexport tentatively agreed to increase gas supplies to Finland to around 4bn cubic metres by the mid-1990s and even up to 5bn cubic metres by the end of the decade. Gas consumption in Finland reached some 1.6bn cubic metres in 1988 but was expected to reach 2.0bn cubic metres by the end of 1989. The final agreement was due to be signed by both countries in October 1989, when Soviet leader Gorbachev was expected in Helsinki for an official visit.

This deal with Soyuzgasexport was done in the expectation of a major gas agreement which was yet to be signed between Sweden and the USSR. If the Swedes and the Soviets could agree on price and volumes during 1989, gas should be flowing through Finland via a Gulf of Bothnia submarine pipeline to Stockholm by 1993-94.

In order to meet new gas supply demands in the 1990s Neste planned to invest some FIM2bn that would include, among other things, a new 500km-long pipeline that would run from the town of Imatra, right next to the Finnish–Soviet border, through southern Finland to the south-western coastal towns of Turku and Uusikaupunki, where the submarine pipeline would run across the Gulf of Bothnia to Gaevle/Forsmark and southward to Stockholm. The cost of the submarine pipeline would be just under FIM1bn, split equally between Neste and Swedish state-held utility Vattenfall.

In face of increasing electricity consumption demands forecast to jump from 58.7TWh in 1988 to 77TWh by the year 2000, the paper and pulp industry actively promoted the commissioning of a fifth nuclear plant. Although the coalition government has declared a temporary nuclear moratorium, many observers were confident that Finland would order its fifth nuclear plant – most likely a Soviet 1000mW VVER-type reactor – after the parliamentary elections of 1991. Finland was on the verge of ordering a fifth nuclear plant just before the Chernobyl catastrophe.

At the end of 1988, the question of Soviet free-trade zones also came up. According to local press reports, the Soviet politburo has given the go-ahead for a first-ever created free-trade zone within the USSR that would be located by the south-eastern part of the Finnish–Soviet border. The Soviets believe that incentives like tax breaks, cheap labour and the proximity of Leningrad make the free-trade zone attractive to Finnish high-tech firms. So far, however, no great interest was expressed by Finnish companies because the zone lacks adequate infrastructure to house sophisticated industries. The USSR was also toying with the idea of creating other free-trade zones in the Kola Peninsula, the Baltic States, as well as declaring Estonia's capital city of Tallinn, just across the Gulf of Finland from Helsinki, a free-port zone.

COUNTRY PROFILE

Political system: Republic with legislative power held by unicameral parliament and the president. Parliament consists of 200 members elected by system of proportional representation for four years. President now directly elected for six-year term, with final decision being taken by a 301-member electoral college (itself elected by direct popular vote) if no candidate wins an overall majority. President appoints a council of state, headed by a prime minister, which is responsible to parliament, for general administration of country.

Main cities/towns: Helsinki (population 486,100), Tampere, Turku, Lahti, Espoo and Vantaa.

Media
Press: Tends to be regional and influenced by political parties. *Dailies:* Over 90 dailies, of which 12 are published in Swedish. Most popular are *Helsingin Sanomat* (Helsinki, circulation 453,597), *Uusi Suomi* (Helsinki, 154,794), *Ilta-Sanomat* (Helsinki, 208,655), *Aamulehti* (Tampere, 141,896) and *Turun Sanomat* (Turku, 131,731). *Weeklies:* Most popular are general interest magazines, such as *Apu* (276,660), *Valitut Palat* (Reader's Digest, monthly, 330,662) and *Seura* (284,060), illustrated news magazines, such as *Suomen Kuvalehti* (110,116), and women's magazines such as *Eeva, Anna* (157,340), *Me naiset* (108,482) and *Kotiliesi* (190,773). *Business:* Several covering most aspects of trade and industry, including *Kauppalehti* (daily newspaper, 80,274), *Fakta* and *Talouselämä*.
Broadcasting: Under general control of Finnish Broadcasting Company 'Oy Yleisradio Ab', which is state-controlled.
Radio: Over 2.5m radios in use. Three domestic programmes transmitted plus a foreign service broadcasting to Europe, Africa, Middle East and Far East. Private local radio started in 1985; there were 34 stations broadcasting in June 1988.
Television: Approximately 1.9m TVs in use, of which over 345,000 also received cable TV in Feb 1988. Finnish Broadcasting Company transmit on two channels, which also carry programmes produced by the commercial TV company 'MTV Oy'. There were over 162 cable stations in June 1988, the biggest being Helsinki Televisio (HTV). Several satellite channels are available.

Domestic economy: Market-orientated economy with heavy dependence on foreign trade. Expansion of industrial exports and high levels of investment allowed for sustained levels of economic growth, GDP growing at an average of three per cent p.a. between 1980 and 1988. Living standards have steadily increased.

Employment: Around 58 per cent of labour force employed in service industries, 30 per cent in industry and construction and 11 per cent in farming and forestry. Unemployment rate held at 6 per cent, though vast regional variations. Moderate incomes policy in operation.

External trade: Falling oil import costs and growth in exports (particularly to OECD countries) reversed regular trade deficit and offset foreign debt burden. Forestry products account for 41.6 per cent of exports, metal and engineering products account for around 39.3 per cent of exports. Also exports of raw materials and production necessities (54.4%), consumer goods (22.8%), investment goods (18.9%), chemicals (9%), fuels (3.2%), crude oil (4.6%) and food, textiles and clothing. Principal imports include energy products (oil, gas and coal), chemicals, raw materials, intermediate and consumer goods, transport equipment and machinery. Around 64.6 per cent of exports go to Western Europe and 16.7 per cent of total trade is conducted bilaterally with the USSR. Other major trading partners include FRG, Sweden, UK, Japan, France and the USA.

COUNTRY PROFILE

Agriculture: Farming and forestry contribute over eight per cent to GDP. Forestry is main supplier to local chemical, pulp and paper industries. Although farm holdings are small and productivity remains low, Finland is 85 per cent self-supporting in food grains, dairy products and root crops.

Industry: Highly developed export-based sector centred largely on the forest products industry. Recent expansion of the metal and engineering industries (machines, transport vehicles and electro-technical products), chemicals, plastics, textiles and clothing. Also important are food processing, rubber and shipbuilding. State-owned companies account for less than 16 per cent of total industrial output. Current industrial policy based on diversification/restructuring and improvements in export price competitiveness.

Mining: Around a dozen ore mines in operation producing mainly copper, iron, zinc, iron-pyrites, limestone, chromium, cobalt and nickel. Deposits are generally small. Prospecting activity is being intensified to curb imports and refining technology is a mjor focus of development work.

Energy: Domestic energy resources – mainly hydro and to a much lesser extent peat and waste wood – make up 30 per cent of total consumption. Expansion of nuclear power has reduced dependence on imported coal and oil. Industry however is still heavily reliant on imports of oil principally from the USSR. Increasing amounts of Soviet oil are also imported for re-export to the world market. Continuing efforts to conserve energy and increase domestic energy production.

Main companies: Kansallis-Osake-Pankki (bank), Union Bank of Finland, NESTE Corporation, Nokia Corporation, Oy Wärtsilä, Rauma-Repola, Pohjola Insurance, Kone Corporation, Valmet, Outokumpu.

Banking: System headed by Suomen Pankki (the Central Bank of Finland), which is responsible for monetary policy and foreign credit. Several commercial banks including the Union Bank of Finland and Kansallis-Osake-Pankki. Around 370 co-operative banks, around 260 savings banks, five mortgage banks and the state-owned Postipankki (Postal Bank). Main foreign banks: Citibank, Midland Montagu, PKbanken of Sweden and Indosuez.

Legal system: Civil law system based on Swedish law. Constitution adopted 1919.

Membership of international organisations: ABD, ASDB, ASDBs, CCC, Council of Europe, DAC, EFTA, FAO, GATT, IAEA, IBRD, ICAC, ICAO, ICES, ICO, IDA, IDB, IFAD, IFC, IHO, ILO, International Lead and Zinc Study Group, IMF, IMO, INTERPOL, IPU, ITU, IWC, Nordic Council, OECD, UN, UNESCO, UPU, WHO, WIPO, WMO, WSG. Free trade agreement with EC and co-operation agreement with CMEA.

BUSINESS GUIDE

Time: GMT + 2 (GMT + 3 from late Mar to late Sep).

Climate: Temperate, with warm summers and cold winters. Summer temperatures average around 17°C, winter around -1°C although can reach as low as -30°C in Feb. Heavy snow from Dec to Apr.
Clothing: Lightweight in summer. Medium weight and heavy topcoat in winter – fur hat recommended.

Entry requirements: Visas not required for nationals of most European countries, and of several other countries.
Health: Polio vaccination advised for those under 17.
Currency: No restrictions on import of currency, but amount of foreign currency must be declared and not be exceeded by amount taken out. Up to FIM10,000 may be taken out.
Customs: Personal effects duty-free, plus duty-free allowance.

Air access: Regular daily flights by most major international airlines.
National airline: Finnair.
Main international airport: Helsinki-Vantaa (HEL), 19 km from capital.
International airports: Jyväskylä (JYV), Kemi (KEM), Kokkola (KOK), Oulu (OUL), Rovaniemi (RVN), Tampere (TMP), Turku (TKU) and Vaasa (VAA).

Surface access: Daily ferry services from Sweden, twice weekly from FRG and Poland. Reservations should be made in advance as these tend to be heavily booked, especially during summer and weekends.
Main ports: Helsinki, Kotka, Hamina, Mariehamn, Vaasa, Turku, Pori, Sköldvik, Rauma and Oulu.

Hotels: Classified into three price categories. Generally of a high standard. Rates vary depending on location, facilities and season. In Helsinki a single room costs from FIM150 to FIM650 per night, including breakfast and 15 per cent service charge. Accommodation should be booked well in advance especially during summer. If accommodation is unobtainable, a place may be found through 'Hotellikeskus' (accommodation clearing-house) at the Central Railway Station in Helsinki. All major international credit cards are accepted.

Currency: The markka is related to a trade-weighted basket of currencies – US dollar (8.1%), British pound (13.4%), Swedish krona (19.1%), German mark (19.3%), others (40.1%).

Foreign exchange information
Account restrictions: Three types of account are authorised for non-residents: (1) convertible-from which funds may be freely transferred abroad, (2) restricted-for payment between Finland and countries with currencies that are not freely convertible into US dollars, (3) capital-for transactions that require the permission of the Bank of Finland.

Car hire: Available in most major towns. Rates vary from about FIM100 to FIM250 per day, plus FIM1.00 to FIM2.70 per km, depending on type of car. Rates include oil, maintenance and insurance. Minimum age limit (usually between 19-23) and at least one year's driving experience required. Speed limits, 50 kph in built-up areas, 80-100 kph on normal roads and 120 kph on motorways. Wearing of seat belts is compulsory. The use of headlights at all times is obligatory.

City transport
Taxis: Available from ranks, by telephone or can be hailed in the street. Identifiable by 'Taksi' sign on roof. Basic fare about FIM10 plus FIM3 per km. Extra charges at night and weekends. Tipping is not expected.
Buses: Good services throughout country. Flat fare of FIM6.50 in Helsinki. Also 10-journey ticket obtainable for FIM45.
Trams: Service in Helsinki.

National transport
Air: One of the densest and cheapest internal networks in Europe. Finnair provides connections between Helsinki and Ivalo, Joensuu, Jyväskylä, Kajaani, Kemi, Kittilä, Kokkola, Kuopio, Kuusamo, Lappeenranta, Mariehamn, Mikkeli, Oulu, Pietarsaari, Pori, Rovaniemi, Savonlinna, Tampere, Turku, Vaasa and Varkaus.
Road: Approximately 75,000 km of roads, of which about 11,000 km are main roads and about 200 km are motorways. Efficient coach services cover the entire country, and are main form of transport in Lapland.
Rail: Network of over 6,000 km (including over 1,000 km electrified), operated by state railway company 'Valtionrautatiet'. Relatively inexpensive and there are several passes available allowing travel over a set period. Seat reservation is obligatory on special express trains.
Water: Important method of transport, owing to large number of lakes (187,888), which cover 31,500 sq km.

Public holidays
Fixed dates: 1 Jan, 1 May, 6 Dec, 24, 25 and 26 Dec.
Variable dates: Epiphany, Good Friday, Easter Monday, Ascension Day, Whit Monday, Midsummer and All Saints' day.

Working hours
Government and business: (Mon-Fri) 0800 – 1600, close 1515 during summer.
Banking: (Mon-Fri) 0915 – 1615.
Shops: (Mon-Fri) 0900 – 1700 or later, (Sat) 0900 – 1600, close 1400 during summer.

Social customs: When invited into a Finnish home for the first time it is usual to give flowers. Skål is customary – guests do not drink until their health is proposed by their host. Punctuality is important. Business meetings are sometimes held in saunas.

Business language and interpreting/translation: Available through tourist offices or through classified section of telephone directory under Käännöstoimistoja ja kielenkääntäjiä.

Telecommunications
Telephone: Direct dialling to most towns and most European countries. Local calls from telephone boxes require FIM1 or FIM5 coins. International dialling code: 358 followed by 0 for Helsinki. Outgoing international code 990, followed by country code.
Telex: service available to most parts of the world. Public facility at Central Post Office in Helsinki. Country code 57 FF.
Telegram: Can be sent from post offices or hotel desks.

Postal service: Stamps available from Post Offices book/paper shops, kiosks, stations and hotels. Mail may be sent 'Poste restante' c/o Main Post Office in Helsinki (or other city).

Advertising: Total advertising spend (1984) FIM6179m. Key categories include automotive, food, entertainment, transport and banking. All usual media are available to advertisers. Advertisements can be placed on TV through the commercial TV company 'MTV Oy' or via cable TV. Press advertisers are advised to consider whether their target audience is Finnish or Swedish-speaking. Radio advertising is limited to private local stations. Cinema films and posters are also widely used.

Trade fairs: Most important are Helsinki International Fair (consumer goods) and Helsinki International Technical Fair, which are held alternately every autumn at Helsinki International Fair Centre, at Pasila about 6 km from capital. All exhibitions held at the Centre are organised by Finnish Fair Corporation, 'Osuuskunta Suomen Messut'.

Electricity supply: 220 V AC.

Representation in capital: Argentina, Austria, Belgium, Brazil, Bulgaria, Canada, China PR, Colombia, Cuba, Czechoslovakia, Denmark, Egypt, France, German Democratic Republic, Federal Republic of Germany, Greece, Holy See, Hungary, India, Indonesia, Iran, Iraq, Israel, Italy, Japan, Democratic People's Republic of Korea, Republic of Korea, Mexico, Netherlands, Norway, Peru, Poland, Portugal, Romania, South Africa, Spain, Sweden, Switzerland, Turkey, USA, USSR, United Kingdom, Venezuela, Yugoslavia.

BUSINESS DIRECTORY

Hotels: Helsinki
Aurora, Helsinginkatu 50 (tel: 717400; tx: 125643).

Helsinki, Hallituskatu 12 (tel: 171401; tx: 121022; fax: 176041).

Hesperia, Mannerheimintie 50 (tel: 43101; tx: 122117; fax: 4310898).

Inter-Continental, Mannerheimintie 46-48 (tel: 40551; tx: 122159).

Kalastajatorppa, Kalastajatorpantie 1 (tel: 48511; tx: 121571; fax: 4581668).

Klaus Kurki, Bulevardi 2 (tel: 602322; tx: 121670; fax: 608538).

Marski, Mannerheimintie 10 (tel: 641717; tx: 121240).

Olympia, Läntinen Brahenkatu 2 (tel: 750801; tx: 122101; fax: 750801).

Palace, Eteläranta 10 (tel: 171114; tx: 121570; fax: 654786).

Ramada Presidentti, Eteläinen Rautatiekatu 4 (tel: 6911; tx: 121953; fax: 6947886).

Rantasipi Airport (4km airport), Takamäantie 4, 01511 Vantaa, P.O. Box: 53 (tel: 826822; tx: 121812).

Seurahuone, Kaivokatu 12 (tel: 170441; tx: 122234; fax: 160441).

Tapiola Garden (in Espoo), Tapiontori (tel: 461711; tx: 122304; fax: 462332).

Torni, Yrjönkatu 26 (tel: 131313; tx: 125153; fax: 131361).

Vaakuna, Asema-aukio 2 (tel: 171811; tx: 121381; fax: 171811).

Car hire: Helsinki
Avis: Fredrikinkatu 67 (tel: 41155; tx: 122255).

Budget: (tel: 6945300; tx: 121230).

Europcar: Hitsaajankatu 7C, 00810 (tel: 7556133; tx: 122561); airport, Vantaa (tel: 826677).

Hertz: Oy Polarpoint AB Itälahdenkatu 20 (tel: 675211; tx: 123229).

Banking
Suomen Pankki (The Bank of Finland) (central bank), Snellmaninaukio, POB 160, 00171 Helsinki (tel: 1831; tx: 121224).

Kansallis-Osake-Pankki, Aleksanterinkatu 42, 00100 Helsinki 10 (tel: 1631; tx: 124412).

Osuuspankkien Keskuspankki Oy (Central Union of the Co-operative Banks), Arkadiankatu 23, 00100 Helsinki 10 (tel: 4041; tx: 124714).

Pankkien Neuvottelukunta (Joint Delegation of the Finnish Banks), Eteläranta 10, 00130 Helsinki (tel: 629712).

Postipankki (State-owned), Unioninkatu 20, 00007 Helsinki 7 (tel: 164-3932; tx: 123816).

Skopbank (Central Bank of the Finnish Savings Banks), Mikonkatu 4, 00100 Helsinki 10 (tel: 13341; tx: 122284).

STS Bank (merged with Peruspankki), Paasivuorenkatu 3, SF-00530, Helsinki (tel: 73181; tx: 126196).

Suomen Pankkiyhdistys.r.y. (Bankers' Association), Kansakoulukatu 1A, PO Box 1007, 00101 Helsinki (tel: 6948422).

Suomen Säästöpankkiliitto (Savings Bank Association), Pohjoisesplanadi 35A, 00101 Helsinki 10 (tel: 13341; tx: 125768).

Union Bank of Finland (incorporating Bank of Helsinki), Aleksanterinkatu 30, 00100 Helsinki 10 (tel: 1651; tx: 124407).

Principal newspapers
Helsingin Sanomat, Ludviginkatu 6-8, PO Box 975 (tel: 1221; tx: 124897).

Ilta-Sanomat, Korkeavuorenkatu 34, PO Box 375, 00101 Helsinki (tel: 1221; tx: 124897).

Uusi Suomi, PO Box 139, 00101 Helsinki (tel: 53031; tx: 124898).

Other useful addresses
Central Chamber of Commerce, Fabianinkatu 14A, 00100 Helsinki 10 (tel: 650133).

Central Statistical Office, Annankatu 44, 00100 Helsinki (tel: 17341; tx: 122656).

Confederation of Finnish Industries, Eteläranta 10, SF 00130, Helsinki 13 (tel: 661665).

Finland Travel Bureau Ltd, Mail Department, PB319, 00101 Helsinki 10 (poste restante service).

Finnish Foreign Trade Association, Arkadiankatu 4-6B, 00100 Helsinki 10 (tel: 69591; tx: 121696).

Helsinki Stock Exchange, Fabianinkatu 14, 00100 Helsinki 10 (tel: 624161; tx: 123460)

Liiketyönantajain Keskusliitto (Confederation ofCommerce Employers), Eteläranta 10, 00130 Helsinki 13 (tel: 19281; tx: 122167).

Main Post Office, Mannerheimintie 11, 00100 Helsinki 10.

Matkailun Edistämiskeskus (tourist board), Asemapäällikönkatu 12B, 00521 Helsinki (tel: 144511; tx: 122690).

Meilahti Hospital, Haartmanink 3 (tel: (9)0-4711).

Ministry of Finance, Snellmaninkatu 1A, 00170 Helsinki (tel: 1601; tx: 123241).

Ministry of Trade & Industry, Aleksanterinkatu 10, 00170 Helsinki (tel: 17361; tx: 125472).

NESTE (largest industrial corporation), Keilaniemi, 02150 Espoo (tel: 358-0-4501; tx: 124641).

Oy Suomen Tietotoimisto (news agency), Lönnrotinkatu 5, 00120 Helsinki 12 (tel: 646224; tx: 124534).

Suomen Työnantajain Keskusliitto (Finnish Employers' Confederation) Eleläranta 10, Helsinki 13 (tel: 17281; tx: 124635).

Tullihallitus (Board of Customs), Erottajankatu 2, 00120 Helsinki (tel: 6141).

Ulkomaankaupan Agenttiliitto (Finnish Foreign Trade Agents' Federation) Mannerheimintie 42A 00260 Helsinki 26 (tel: 446768).

France

Julia Crawford

France's minority Socialist administration emerged from its first year of government much stronger than many had expected, with both President François Mitterrand and Prime Minister Michel Rocard still riding high in the public opinion polls. Aided by continuing divisions among the right-wing opposition, Rocard's moderate negotiating style appeared well suited to the demands of minority government, and the electorate seemed to be developing a taste for consensus politics. Unless moves to reform the opposition succeeded in putting an end to the disarray of the right and centre parties, the government looked likely to remain in power at least until the next general election in 1993.

Political situation

When Mitterrand appointed Rocard as prime minister after the general elections of June 1988, many observers did not expect the new premier to last. Some even said Mitterrand had appointed Rocard, who had nurtured presidential ambitions, out of spite. For while Mitterrand was re-elected with almost as big a majority as General de Gaulle himself (54 per cent in the second round presidential run-off against Jacques Chirac of the Gaullist RPR), the subsequent parliamentary elections failed to produce a majority for the Socialists. They won 277 seats in the 577-seat parliament, only five more than the centre-right UDF – RPR opposition. It was only thanks to the surprisingly good performance of the Communists, who won 27 seats, that the left as a whole ended up with a majority in the national assembly. But after Mitterrand's campaign commitments to moderation, the socialists could hardly form a coalition with the Communists.

Given the Socialists' parliamentary situation and Mitterrand's campaign commitments to moderation, the president in fact had little alternative but to appoint Rocard to the premiership – and, having made the appointment, to keep him there. Rocard was the

KEY FACTS

Official title: The French Republic

Head of state: President François Mitterrand

Head of government: Prime Minister Michel Rocard

Ruling party: Parti Socialiste

Capital: Paris

Official Languages: French

Currency: Franc (Fr) = 100 centimes

Exchange rate: Fr6.04 per $ (Dec 1989)

Area: 543,965 sq km

Population: 55.9m (1988)

GDP per capita: $16,859 (1988)

GDP real growth: 3.7% (1988)

Labour force: 24.0m (1987)

Unemployment: 10.1% (1988)

Inflation: 3.1% (1988)

Trade balance: –$8.1bn (1988)

Foreign debt: $40bn (1986)

only obvious candidate who combined the moderate Socialist outlook and the stature to carry off the political balancing act required. But despite the government's assertions that it sought an 'opening to the centre', no formal coalition agreement was reached, nor were any prominent members of the moderate opposition included in the government. Although Rocard did include some non-Socialists in the cabinet, he was left leading what was still essentially a minority government.

True, the government could only be brought down by an alliance between the Communists and the conservative parties. The Communists were unlikely to agree to such an alliance, as its logical consequence would be a return to power of the right-wing parties. But still Rocard looked to have a hard task to hold the government together and push its legislative programme through. In the event, the prime minister's skills seemed well suited to his task. He was also helped by the continuing disarray of the opposition, and the fact that the tide of French politics seems to be veering away from its traditional extremism. The electorate has since 1986 systematically rejected strong majority government: first by voting in a right-wing prime minister, Jacques Chirac, who was forced to 'co-habit' with Socialist President Mitterrand from 1986 to 1988, and then by failing to give the Socialists an absolute majority. This has tended to make French politics look less presidential and more parliamentary. Another factor in the trend towards consensus has been the increasing intrusion of Europe into national politics, with all the mainstream parties now committed to European integration.

Rocard's first major test was restoration of peace between the warring European and Melanesian communities in the French Pacific territory of New Caledonia and the negotiation of an accord for the island's future, subsequently approved in a national referendum. The more enduring demonstration of his aplomb was the fact that the government managed to sail through the first year of its legislative programme with remarkable ease. However, the going was not always easy. The record level of abstentions in the November 1988 referendum on New Caledonia (63 per cent) was seen as a setback for the prime minister. In late 1988 and early 1989 the government faced a series of strikes from discontented employees, ranging from nurses, prison workers and transport workers to public sector workers in Corsica. As the government's popularity began to slide and the Socialists did badly in a couple of by-elections, the opposition launched a parliamentary vote of no confidence in an attempt to bring the government down.

But the opposition, still rent by internal strife, was unable to capitalise on the situation. As the government's combination of moderation and determination gradually put an end to the strikes, not even an embarrassing series of financial scandals could cause serious electoral damage: the March 1989 municipal elections saw a respectable performance by the Socialists despite an abstention rate of 30 per cent, the highest in this type of election since 1945.

The municipal elections, where local alliances play a strong role, marked the continuing decline in relations between the Socialists and Communists. A national pact between the two parties, although finally concluded as in 1983, was only reached after several months of talks and a considerable degree of acrimony. The elections brought gains for the Socialist

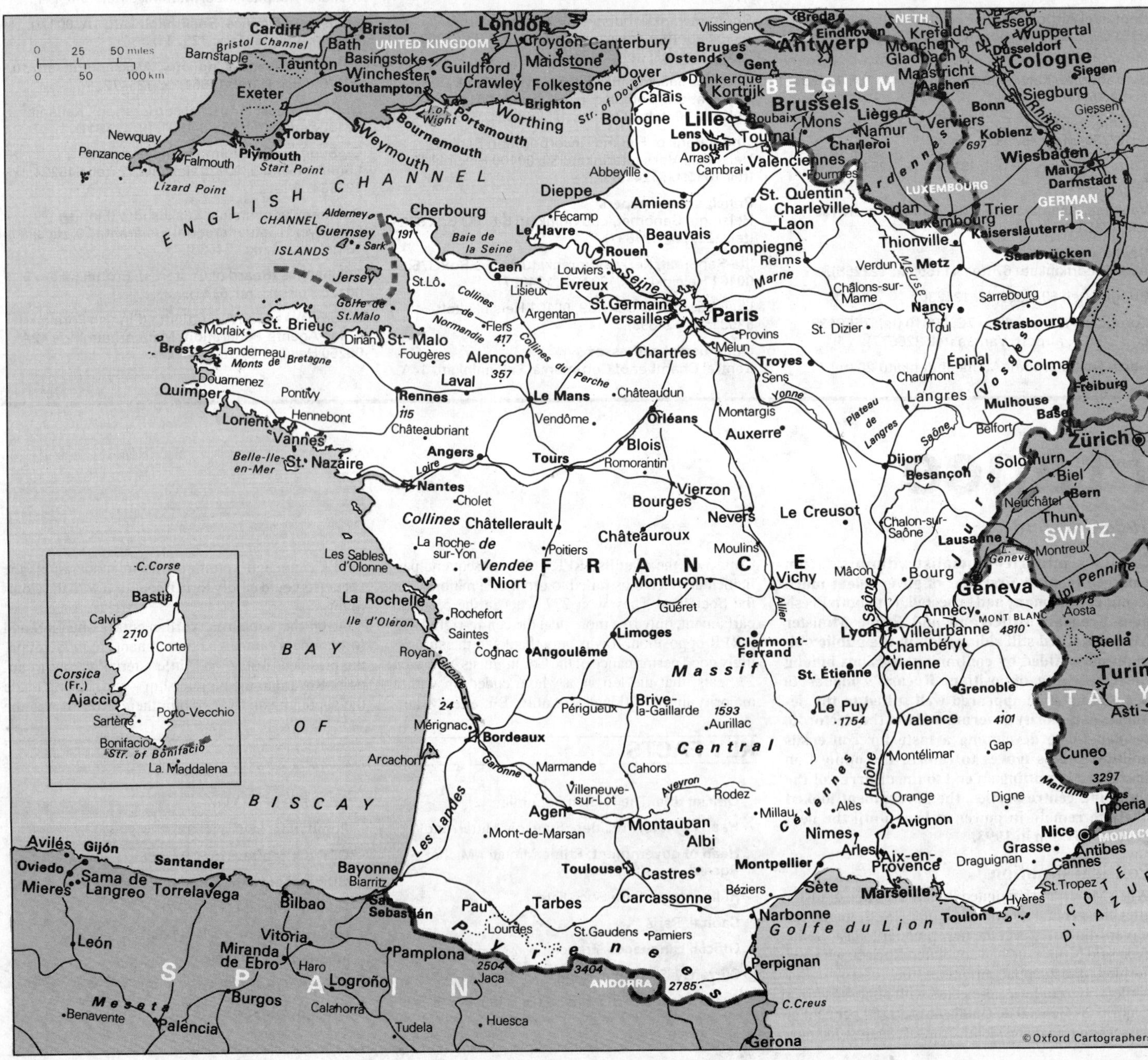

Party notably in the catholic west. Its recapture of Nantes, Brest and Redon seemed to confirm its ascendancy in this part of the country where, traditionally, it has been weak. The Communist vote continued its decline of recent years, although it held up in certain areas.

But the real surprises of the municipal elections were the strong performance of the environmentalist greens and the rise of several young, reform-minded members of the opposition RPR and UDF parties. The greens gained over 10 per cent of the first-round vote in many towns, allowing them to maintain their candidate in the second round, and up to 15 per cent in some places. Green leader Antoine Waechter stressed that his party had become a national force, but refused Socialist overtures to conclude any sort of formal alliance.

Contrary to expectations, the centre-right UDF party generally lost out in the municipal elections, with Socialist gains often coming at its expense. In Lyon, the former UDF mayor was roundly ousted by a rising star of the RPR, Michel Noir – despite the presence on the UDF list of former prime minister Raymond Barre. And although the RPR also suffered some setbacks, these were largely compensated by its successes in Lyon and Paris, where RPR leader Jacques Chirac carried the majority of arrondissements in the first round.

This time both the UDF and RPR were adamant they would not form second-round alliances with Jean-Marie Le Pen's ultra-right wing Front National. The FN, while it did not match Le Pen's 14 per cent score in the first round of the presidential elections, nevertheless held its ground with 10 per cent of the vote or more in some parts of the country.

The rise of a new, young generation of UDF – RPR mayors – notably Noir in Lyon – seemed to reflect a desire for change among right-wing voters. They had delivered a vote of confidence for younger men who wanted to unify and reform the opposition, emphasising more democratic, humanitarian policies. The opposition had still not learned the single most important lesson of the presidential election: that it needed to heal its internal rifts if it were to present a credible government alternative to the Socialists. Internal bickering and power-mongering had if anything increased, especially since the centrist CDS party broke ranks from the UDF to assert its separate parliamentary identity after the general elections of 1988.

The breakaway of the CDS, one of the five parties to form the UDF, came as a blow to former president and UDF leader Valéry Giscard d'Estaing, whose strategy was to use the UDF as his power base. After Chirac's defeat in 1988, Giscard hoped to negotiate with Mitterrand from a position of strength with a view to forming a Socialist – UDF coalition. Without the CDS he lost much of his bargaining power. The CDS breakaway also upset another presidential aspirant, François Léotard, head of the Parti Republicain, the biggest party in the UDF with some 70 seats. Léotard believed that the UDF should merge with the RPR and become a major opposition grouping. Rivals for what remained of the UDF, Giscard opted for a more conciliatory centrist position, Léotard for an uncompromising shift to the right.

When, in the wake of the municipal elections, Noir led a cross-party initiative to reform the right, it had a considerable degree of public support. For two brief weeks 12 'reformers' from the RPR, UDF and CDS looked poised to launch their own campaign for the European elections. It seemed they might succeed in mobilising a unified campaign of young conservatives which could include the centrists – but only at the expense of dumping the traditional opposition leaders for a new breed of younger men. Inevitably the old guard kicked back, and the reformers fell back into line.

KEY INDICATORS

	Unit	1984	1985	1986	1987	1988
Population	m	55.0	55.2	55.4	55.6	55.9
Gross domestic product (GDP)	Fr bn	4,361.9	4,700.1	5,052.5	5,301.3	5,654.5
GDP per capita	Fr	79,307	85,147	91,200	95,347	101,154
Inflation	%	7.5	5.7	2.5	3.3	3.1
GDP real growth	%	1.3	1.9	2.3	1.9	3.7
Unemployment	'000	2,340.3	2,458.4	2,516.6	2,621.7	2,562.9
Labour costs	1985=100	94.3	100.0	104.5	109.4	112.8
Share prices	1985=100	85.6	100.0	153.3	177.6	162.1
Retail sales	1980=100	146	160	170	181	–
Interest rates	%	9.5	9.5	9.5	9.5	9.5
Agricultural production	1979-81=100	109.5	106.7	101.7	104.9	103.7
Industrial production	1985=100	99	101	102	104	108
Manufacturing	1980=100	97	98	98	100	105
Energy	1980=100	122	129	135	140	144
Construction	1980=100	84	83	87	92	98
Exports	$m	92,214	95,927	117,990	139,663	160,638
German FR	$m	13,668	14,405	19,011	–	–
Italy	$m	10,139	10,467	13,933	–	–
UK	$m	7,368	7,875	10,391	–	–
USA	$m	7,546	8,258	8,738	–	–
CMEA	$m	3,083	3,050*	2,716	–	–
OPEC	$m*9,079	7,060*	–	–	–	
Imports	$m	96,865	101,203	120,343	148,912	168,727
EEC	$m	52,087	54,900*	71,640	–	–
CMEA	$m	3,856	3,450*	4,080	–	–
OPEC	$m	11,911	12,060*	–	–	–
USA*$m	7,974	8,200*	9,000	–	–	
Balance of Trade	$bn	–4.7	–5.3	–2.4	–9.2	–8.1
Current account	$m	–876	–35	3,055	–4,077	–3,547
Foreign exchange	$bn	19.1	24.3	28.4	29.6	22.4
Exchange rate	Fr per $	8.7	8.9	6.9	6.0	6.0

* estimated figure

In May, the UDF and RPR finally announced that they would run a joint list for the European elections, headed by former president Valéry Giscard d'Estaing of the UDF. The RPR said it had chosen to 'sacrifice' the leadership of the electoral list to the UDF in the interests of unity, although the choice of Giscard was obviously anathema to the centrists and reformers. CDS leader Pierre Mehaignerie then announced that the centrists would run a separate campaign for the European elections. The candidate chosen to head the CDS list was Simone Veil, leader of the liberal grouping in the European Parliament, former president of the Strasbourg assembly and former minister under Giscard. Although not actually a member of the CDS, Veil had lent her support to the reform movement; she had a solid track record in European politics and a good image with the public. Noir's 'Band of 12' said they would not join anybody's list, although they lent their presence to the Veil campaign in the run-up to the vote.

Mehaignerie's aim was to capture 15 per cent of the vote in the 18 June poll, but in the event the centrist vote fell seriously short. Veil's list polled only 8.4 per cent, in fifth place behind the UDF – RPR (28.9 per cent), the Socialists (23.6 per cent), the ultra-right FN (11.7 per cent) and the greens (10.6 per cent). The Communists scored only 7.7 per cent. Again abstentions were at a record high of almost 50 per cent – perhaps not surprising after 10 electoral Sundays in just over a year. The relatively low score of the Socialist list was seen above all as a setback to the presidential ambitions of its leader Laurent Fabius, former prime minister and now leader of the national assembly.

Giscard's strong comeback in the first real test since his 1981 presidential defeat reaffirmed his desire to lead a reunified centre-right opposition. In the wake of his victory, he proposed that the UDF and RPR should merge, but this was promptly rejected by the RPR. Indeed, the scale of Giscard's success seemed to less than please other presidential aspirants in the opposition who thought he could be safely shunted off to Strasbourg. Chirac, who appeared to be recovering some of his morale after his defeat in the 1988 presidential election, was quick to stress his RPR party's role in the election success. And Léotard launched an attack on both Chirac and Giscard, saying that the candidature of either for the next presidential elections would spell disaster for the opposition. 'Their ability to govern the country is unquestionable, but what's in doubt is the ability of the French to want them to govern,' he told the press.

Léotard had lent belated support to Giscard's candidature in the European elections in the hope of capturing the UDF presidency from his rival. But, when Giscard showed no sign of relinquishing it, Léotard began efforts to build an alternative power base: a bid to replace Jean-Claude Gaudin as president of the UDF parliamentary grouping, and overtures to the centrists. Describing the UDF as an 'obsolete structure', he proposed 'a new deal' to the CDS, which nevertheless remained sceptical.

Indeed, as the 'Band of 12' initiated a new call for opposition reunification within a decentralised structure, all the parties concerned seemed to have jumped on the reformist bandwagon – in varying degrees, according to their aspirations. In mid-1989 it remained to be seen whether the opposition could really put its house in order in time for its 'estates general' scheduled for 1990. This grand assembly of RPR,

UDF and CDS aimed to draw up a programme for government and reach an agreement on primaries for the next presidential election.

The Socialists also faced a possibly momentous congress in March 1990, with party General Secretary Pierre Mauroy calling for a wide-ranging policy debate and overhaul of the party's structures. This became more pressing after the disappointing performance of the party in the European elections. Although all the main party factions were now agreed on the need for economic vigilance, Rocard could be under pressure to give his policies a more solidly left-wing flavour. Many in the party felt that, with economic growth of 3.7 per cent in 1988, austerity could be relaxed a bit to give more emphasis to the rights of employees and the underprivileged. Most of the motions drawn up before the 1989 summer recess also opposed working arrangements with the centre and right-wing parties – and even the Rocard supporters seemed to have stopped talking about 'opening to the centre'.

The main factions within the party had, however, gradually been converging on policy – all had now had a spell in power and had generally had the edges rounded off their Socialist idealism of the early 1980s. The congress seemed more likely to provide a stage for personal power struggles between a number of presidential aspirants, notably Laurent Fabius (leader of the national assembly) and Lionel Jospin (education minister and former party secretary). Mitterrand, at over 70, will not run again in the next presidential election in 1995. Fabius, who counts his supporters mainly among the parliamentary party, might look for an opportunity to upset the strained accord between the party majority of Mitterrandists and Mauroyists. Jospin and Mauroy were also rumoured to be trying to form a new majority within the party, with the agreement of the Rocardians. A good deal of manoeuvring was likely both before and during the congress, although the unifying force of the president seemed likely to prevent open war.

Foreign policy

Foreign and defence policy has always been the privileged domain of the president. This inevitably led to a certain degree of tension between Mitterrand and Chirac during their period of co-habitation (1986-88), which disappeared with the appointment of Rocard. In 1989 France continued to play a significant role on the European and world stage. In January it hosted an international conference in Paris which called for more urgent steps to ban chemical weapons, and in March it was largely instrumental in convening an international conference in The Hague to call for more effective environmental protection. In July Mitterrand hosted the G7 summit in Paris, timed to coincide with celebrations for the bicentenary of the French Revolution.

But the main axis of French foreign policy remains moves towards European integration – even if this means some domestic economic headaches, especially in the field of taxation. On 1 July 1989 France assumed the six-month presidency of the EC from Spain. Mitterrand's top priorities for his European presidency were: economic and monetary union, progress on citizens' and workers rights, protection of the environment and fiscal harmonisation. Less pressing but still important items on his daunting agenda were renewal of the Lomé Convention and the Community's budget for 1990.

France has been perhaps the most vocal country in favour of European monetary union. At the Madrid summit in 1989, Mitterrand said he would convene an intergovernmental conference immediately after 1 July 1990, the date by which all capital movements within the Community were due to be freed. The conference was expected to lead to monetary union through revision of the EC's founding Treaty of Rome. Mitterrand said that if some member states – notably Britain – maintained their opposition to monetary union, he would not rule out an agreement 'signed by less than 12 states'.

Another area where Mitterrand was likely to clash with British Prime Minister Margaret Thatcher was the area of workers' rights. Paris wanted the EC to draw up a charter of the main principles governing protection of European employees, along with a set of operational provisions. It seemed likely that, to pacify the British, the charter would contain only outline rules on, for example, employee consultation, continuous training and encouragement of employee mobility. It would be left to the European Commission to draw up the operational provisions, which would then be discussed on a 'case-by-case' basis. Mitterrand was likely, however, to face less opposition on the setting up of a European environment agency, as agreed in The Hague. This new body, which was intended to play only an informative and advisory role, could be in place by late 1990 or early 1991.

Moderate policies

Against the background of a recovering economy, the government pursued its commitment to pragmatic, market-oriented policies aimed at job creation through non-inflationary growth. This included being tough with striking workers who were dissatisfied with wage restraint at a time of economic boom. Faced with a spate of unrest in the run-up to the municipal elections, Rocard's method was to give ground only where he felt it justified, on a 'case-by-case' basis. And if proof were needed of the government's new-found pragmatism (no more socialist experiments as in 1981), its labour law reforms were a case in point. Despite calls from the left to reinstate redundancy laws scrapped by the previous right-wing administration, Employment Minister Jean-Pierre Soisson did not do so. He left the final word in laying off workers with the companies concerned, not with the government. But he also offered some comfort to the unions by strengthening works councils' rights to information on employment levels and beefing up training incentives.

In the field of industrial policy, too, the government's watchword was continuity. True to Mitterrand's electoral commitments, the government said it would neither reverse the privatisations implemented by the previous government nor introduce new privatisations. In the 1989 review of heads of state companies, most of the existing company chairmen were reappointed. Industry Minister Roger Fauroux emphasised that industry should be run by industrialists, not politicians.

The government nevertheless moved to substantiate its commitments to make welfare, housing, labour and education national policy priorities. Its first move was to introduce a minimum guaranteed income, counterbalanced and partly financed by a new wealth tax. It also amended the previous government's immigration laws with a view to ensuring that the country's tangled residence permit and naturalisation laws be applied more humanely. In the face of considerable right-wing opposition, the government also pushed through a bill to curb the steep rise of rents in Paris and give more protection to tenants.

The government also began moves radically to overhaul and upgrade the national education system. Inevitably, when Education Minister Lionel Jospin launched negotiations with teachers' unions at the beginning of 1989, the teachers were only interested in talking about one thing – money. Jospin originally planned to spend an extra FFr10bn over five years to employ more teachers and improve their salaries, but the total was driven up to FFr11.6bn for 1989-94, plus an extra FFr6.2bn over the following five years. The increase in state spending would be partly compensated by a reduction in the defence budget. Considerable uncertainty remained over the reform element of the government's package, however. The national assembly passed a general 'orientation law' on education, but concrete changes in school curricula and timetables, for example, depended on the detailed application laws due to follow. While cynics said the Jospin reforms were doomed to failure like their predecessors in the face of vested interests, some of his proposals suggested that he might be made of sterner stuff.

Another highly controversial reform was the establishment of a new broadcasting authority, the Conseil Supérieur de l'Audiovisuel (CSA), to replace the much criticized CNCL. The CSA, responsible for ensuring 'pluralism' in radio and television programmes, was given wider powers than its predecessor, including power to impose sanctions on offenders. At the same time, the Bill introduced tighter limits on television advertising breaks from July 1989. Indeed, the government's concern to improve the quality of French broadcasting continued to arouse controversy. Other moves included the creation of a joint presidency for the two remaining state TV channels, A2 and FR3, whose joint share of the viewing audience had slumped to 33 per cent compared with 61 per cent for the national private channels.

The government was also expected to face growing hostility in autumn 1989 to another draft Bill on broadcasting. This aimed to impose strict quotas on 'French' and 'European' content for prime time viewing, and limits on financial links between producers and broadcasters. Private channels La Cinq and M6, both in financial trouble, said the proposals would spell the end of private TV in France, while TF1 said they would make profitability impossible. On the other hand, producers' representatives were pressing for even tougher controls. The main motivation behind the proposals was to stem the rising tide of foreign films and soap operas (particularly American) reaching French screens. The industry is already regulated by a tangle of rules and quotas, but these have tended to be flouted, especially by La Cinq.

Healthy economic growth

Despite gloomy predictions in the wake of the 1987 stock market crash, the French economy performed better in 1988 than it had done for a decade (GDP up 3.7 per cent). A slight slowdown was expected for 1989 and 1990, but growth seemed likely to continue at a healthy level of about 3 per cent, fuelled by strong investment and exports. Wage and price increases remained moderate while job creation was stronger than expected.

Nevertheless, the unemployment level remained high at around 10 per cent and would perhaps rise further owing to demographic factors. The other black spot in an otherwise rosy picture was the persistence of the foreign trade deficit.

The strongest contributor to growth was the dynamism of business investment, which rose by 9.1 per cent in 1988 after only 4.6 per cent the previous year. In industry, the figure was even higher, at around 12 per cent. This trend had not yet made good the decline in industrial investment which lasted for a decade after the first oil crisis, but nevertheless pointed in the right direction. Corporate investment was expected to remain strong in 1989 and 1990, although growing at a slightly slower rate than 1988.

Another encouraging factor was that companies started to invest to increase production capacity, whereas since 1984 fixed investment had been primarily geared to productivity gains. Company profit margins improved to 29.9 per cent in 1988, matching their best levels of the early 1970s. This was due primarily to wage moderation, with hourly wage rates up 3.4 per cent. However, the rise in corporate profit margins was forecast to be curtailed in 1990 with the expected increase in competitive pressures.

Exports, which surged by 7 per cent in 1988 after only 2.7 per cent in 1987, were also expected to remain buoyant. A slackening of the international environment was likely to be offset by two positive developments. First, France's loss of market share was likely to be compensated in 1990 by its improved competitiveness – the result of several years' restraint in unit labour costs – and by faster-growing productive capacities. And secondly, exports of services were expected to grow at a more sustained pace.

However, imports have been growing even faster than exports (8.6 per cent in 1988), owing largely to the boom in corporate investment. This caused the trade deficit to widen to FFr33bn, after FFr31bn in 1987 and a small surplus in 1986. Despite an improvement in early 1989, the government was predicting an overall deficit of FFr35bn for 1989. Most worrying was the deficit on manufactured goods which plunged to FFr35bn in 1988 from FFr5bn in 1987. The agro-food and service sectors earned surpluses, however. The government expected the deterioration of the trade account in manufactured goods to slow in 1989 and cease in 1990, owing to the stabilisation of French export market shares and the relative moderation of domestic demand. Private consumption growth (2.6 per cent in 1988) was expected to slow down slightly in 1989 and 1990, reflecting the growth of real disposable income. Inflation was expected to remain modest at about 2.5 per cent for 1990.

One of the main bonuses of 1988's strong growth was that job creation was higher than expected. Prime Minister Michel Rocard, when he took power in mid-1988, was predicting a deterioration in the unemployment situation, partly because training and job creation schemes set up by the previous government would run out of budget resources and partly because of demographic factors. France is one of the few European countries where the active population is still growing strongly. After average annual job losses of about 53,000 in 1984-87, the trend started to reverse in 1988. Job losses in industry slowed down while 200,000 jobs were created elsewhere, mostly in the service sector. From a peak of 11 per cent in 1987, the unemployment rate fell to just over 10 per cent. The government expected job creation, while remaining strong, to decelerate slightly in 1989 and 1990. The OECD was predicting that the unemployment rate would creep up again to 10.7 per cent in 1990.

One of the government's main aims was to continue reducing the budget deficit, but it was faced with pressures on both the revenue and expenditure side. In spring 1989, Rocard said he aimed to bring the deficit down from FFr115bn in 1987 to FFr100bn in 1989 and FFr90bn in 1990, within the framework of national policy priorities and preparation for the EC single market. But as well as the balancing act required to finance its reforms, particularly of the education system, the government also faced the shrinkage of two major revenue sources: value added tax and the withholding tax on savings. France has among the highest VAT rates in the EC and they will have to come down to avert the danger of a diversion of trade. It has the highest withholding taxes on savings, and these will also have to fall to avoid a flight of capital.

State spending was expected to rise significantly in 1990, owing to the delayed effects of 1988's public sector pay agreement, an expected FFr20bn rise in the cost of the public debt and the effects of government reforms. Priority ministries such as education, research and justice were to be given generous credits, but the 'non-priority' ministries seemed unlikely to be satisfied with their meagre deal. For example, Interior Minister Pierre Joxe was pressing for more money in the face of police pay demands and Industry Minister Roger Fauroux wanted FFr15bn to inject

COUNTRY PROFILE

Political system: Republic with bicameral parliament consisting of the Senate (currently 319 members indirectly elected for nine years) and the Assembly (currently 577 members including 17 overseas representatives elected for five years). System of government based on the Constitution of the Fifth Republic. The president, elected every seven years, appoints the prime minister and other members of the government. Since 1982 much administrative and financial power traditionally held by the state has been devolved to the 22 regions and 95 departments of metropolitan France.

Main cities/towns: Paris (population 9.3m), Marseille, Nice, Bordeaux, Lyon, Nantes, Rouen, Saint-Etienne, Toulouse, Strasbourg, Le Havre, Lille, Grenoble, Toulon, Valenciennes, Lens, Nancy, Tours, Bethune, Clermont-Ferrand.

Media
Press: Large number of daily, weekly and monthly newspapers and magazines. Some regional newspapers have a larger readership than national titles. *Dailies:* Major national newspapers include *Le Monde* (circulation 362,443), *Le Figaro* (433,496), *France-Soir* (410,679), *Le Matin* and *Le Parisien Libéré* (339,271). Most popular regionals include *Ouest-France, La Voix du Nord* (377,219), *Sud-Ouest* (356,989) and *Le Progrès* (353,608). *Weeklies:* Newspapers include *France Dimanche* (721,000) and *Journal du Dimanche.* Also a large number of general and special interest magazines. *Business:* Large number covering all aspects of business including *Les Echos* (daily, 72,992), *La Tribune de l'Economie* (daily, 75,000), *L'Expansion* (200,565), *Le Nouvel Economiste* (117,090) and *La Vie Française* (125,000). Press information available from Agence France Presse and various directories such as L'Annuaire de la Presse and Tarif Media.
Broadcasting: All French broadcasting overseen by independent regulatory commission, the CSA. Actual broadcasting done by Télédiffusion de France.
Radio: Almost 50m receivers in use. Programmes produced by state-owned Radio France and local private stations. Commercial stations operating outside France, such as Radio Luxembourg, Radio Monte Carlo and Europe No 1, also received.
Television: About 20m receivers in use. There are 6 channels. Antenne 2 and FR3 will remain state-owned, while Télévision Française 1, Canal Plus, La Cinq and M6 are privately owned. Canal Plus is a pay TV channel and has more than 1m subscribers. The state-owned channels accept advertising, but only between programmes. Development of satellite and cable TV under way with some 56m homes cabled by the end of 1986 and the launch of a French DBS satellite in the same year.

Domestic economy: Mixed economy with large agricultural, industrial and service sectors. High levels of growth during 1970s halted by onset of world recession. Current economic policy essentially deflationary, aimed at reducing external debt, trade deficit and public expenditure. Major industrial restructuring and modernisation programme now almost complete, with emphasis on production for the home market and expansion of new export capacity. Privatisation of nationalised companies got under way in 1987 but was slowed down by international stock market crisis in October 1987. The Socialist government which took power in early 1988, promised neither to continue privatisation nor to re-nationalise. It has said, however, that it will seek new ways to raise capital for state-owned companies.

Employment: Virtual stagnation of economy combined with intensified industrial restructuring and rationalisation has led to high levels of unemployment although upward trend showed signs of reversing in 1988/early 1989. Minimum wages are revised periodically to keep pace with inflation, while the government may decide on an extra boost (recent policy is a token few per cent points extra), real wage increases continue to lag behind inflation. Of total workforce around eight per cent employed in agriculture, 35 per cent in industry and the remainder in services.

External trade: After a small surplus in 1986, the trade balance plunged into deficit and has continued to deteriorate, owing mainly to the deficit on manufactured goods. High corporate investment has contributed to a level of imports higher than exports, but the government expects the deterioration to halt in 1990. Major exports include metals, chemicals, industrial equipment, consumer goods and agricultural products. Over half of total exports destined for EC, particularly West Germany and Italy. Principal imports are oil, machinery and equipment, chemicals, iron and steel and foodstuffs. Major non-oil suppliers include West Germany, Italy, Benelux, UK and USA.

Agriculture: Major European food producer, with self-sufficiency in dairy produce, and substantial exporter of livestock produce, wine and fruit and vegetable. Despite steady increase in production and exports, sector has been developing in an unbalanced manner owing to higher feed prices and price distortion under EC's Common Agricultural Policy. Co-operatives account for between 30-50 per cent of agricultural output.

Industry: Broad industrial base incorporating large, capital-intensive, state-owned sector and small and medium-sized manufacturing enterprises. Industrial policy aimed at reconquest of the domestic market, promotion of 'new technology' sector and internationalisation of state-owned companies.

Mining: Significant producer of iron ore, bauxite, potash and coal. Also seventh largest uranium producer in the world. In effort to reduce dependence on imported minerals, exploration for lead, zinc, barium and tungsten has been stepped up. Mining sector contributes less than 2 per cent to GDP.

Energy: Heavy dependence on imported energy, oil accounting for around 20 per cent of total import bill. Nuclear power targeted to form the cornerstone of future energy policy but nuclear programme has been cut back since 1981. Also development of hydro-electricity and solar energy potential. Domestic production of gas falling, though consumption steadily increasing.

Main companies: Elf Aquitaine, Air Liquide, L'Oréal, BSN, Groupe Française des Pétroles, Thomson, Bull, CGE, LVMH, Cap Gemini.

Banking: System divided into commercial banks and finance companies, co-operative institutions and specialised credit institutions. Financial deregulation is now largely complete.

Membership of international organisations: ADB, CCC, Council of Europe, DAC, EC, EIB, ELDO, EMA, EMS, ESRD, FAO, GATT, IAEA, IATP, IBRD, ICAC, ICAO, ICES, ICO, IDA, IDB, IFAD, IFC, IHO, ILO, IMF, IMO, INTELSAT, International Lead and Zinc Study Group, INTERPOL, IOOC, IPU, IRC, ISO, ITC, ITU, IWC, NATO, OAS (Observer), OECD, South Pacific Commission, UN, UNESCO, UPU, WEU, WFTU, WHO, WIPO, WMO, WSG, WTO.

into state-owned companies – compared with FFr4bn wanted by the finance ministry!

By mid-1989 the government already seemed to have relaxed some of its original targets. Whereas in spring 1989 it had set the goal of a 1.5 per cent cut in civil service employment levels (a cut of 15-18,000 jobs), it later revised this to an increase of 3,000 in civil service jobs. The reasons may have been partly economic – healthy fiscal receipts and upwardly revised economic forecasts allowed it more leeway – but they were also political. Faced with continuing pressure from public sector unions, the government decided not to give them more reason to complain, especially as the administration was determined to hold out on pay rises.

Business environment

The approach of the single European market seemed to have a profound impact on French business attitudes and strategies. In the late 1980s a large number of private and state-owned companies embarked on ambitious international expansion plans to extend their presence in Europe and North America through joint ventures, partnerships and foreign takeovers. Statistics from Hambros Bank put the number of takeovers at 156 in 1988 compared with only 80 in 1987. And this trend continued in 1989, with France consistently among the top five acquiring nations in European mergers and acquisitions activity. The size of takeover bids also continued to break records.

Spectacular deals concluded in 1989 included state-owned aluminium group Péchiney's FFr7.5bn acquisition of American Can; food group BSN's FFr16.5bn purchase of Nabisco Europe; and the FFr23bn operation launched by Compagnie Financière de Suez to win control of insurer Victoire. Suez's bid followed Victoire's acquisition of German insurance company Colonia, which was itself indicative of French companies' new boldness: Colonia was considerably bigger than Victoire in revenue terms. State-owned oil group Elf-Aquitaine announced it would acquire American chemicals company Pennwalt Corp. for $1bn, after earlier acquiring 25 per cent of Britain's Enterprise Oil. Elf's mergers and acquisitions expenditure rose to 50 per cent of total investment in 1988 from 20 per cent in 1987 and only 9 per cent in 1986.

Large groups like Thomson, the state-owned defence and electronics group, or Compagnie Générale d'Electricité, the privatised telecommunications and heavy engineering group, also continued their ambitious international expansion strategies involving acquisitions. CGE, for example, announced a merger of its Alsthom engineering operations with the power division of Britain's General Electric Company (GEC). Another high-profile Franco-British tie-up consolidated in 1989 was that between luxury products group LVMH and Britain's Guinness. These moves were aimed at giving French groups the necessary critical size to compete in global markets.

One of the question marks overhanging the development of French industry was the financing of state-owned companies' international development, given that the government had promised not to introduce further privatisations. Most of the state-owned groups turned a healthy profit in 1988, but carried a level of long-term debt way above most of their international competitors. Industry Minister Roger Fauroux's stated aim was to allow state-owned groups to continue their expansion without getting further into debt. But he looked likely to have to fight for funds, given the government's overall budget constraints. One possible solution being mooted was the creation of a holding company which would be authorised to raise funds on the capital markets to strengthen state-owned companies' capital and reserves. The 1989 bourse flotation of part of Péchiney's new subsidiary Péchiney International could also point to a possible solution for other companies.

Fauroux, himself a former industrialist, stressed that industry should be run by industrialists, not politicians. He believed that the government, as state shareholder, should provide financial backing to help nationalised companies in their strategic development, but that they should be treated as 'normal companies' operating in a free-market environment.

State-owned car-maker Renault was a case in point. The government argued to the EC Commission in Brussels that its decision to grant Renault a FFr12bn debt write-off was the final stage of the company's 'normalisation' process. Although Renault continued to be a source of discord between Paris and Brussels, the government said the auto maker would receive no more government subsidies and would be treated as an 'ordinary company'.

French attitudes to foreign investment also changed, with EC companies (especially Italian) increasingly entering the capital of French firms. And with 1993 drawing closer, even attitudes to the Japanese – the long-standing *bête noir* of French car manufacturers – seemed to have softened. For example, after long-running resistance from the industry and the authorities, Subaru's plans to build a small car plant near Angers, in north-west France, looked likely to get a green light from the government. Moreover, Subaru was holding co-operation talks with Renault which would allow it to satisfy local content rules. Renault President Raymond Lévy was quoted as saying his group's future did not lie in 'splendid isolation' but in 'co-operation and alliances'.

Bourse reform

While increased takeover activity had begun to worry many French companies, a series of embarrassing financial scandals served as a catalyst for much-needed reform of the French bourse. The government, deeply embarrassed by scandals surrounding state-owned Péchiney's acquisition of American Can and the 1988 stock market raid (allegedly government-backed) on privatised bank Société Générale, moved quickly to introduce new safeguards. The bourse watchdog body *Commission des Opérations de*

BUSINESS GUIDE

Time: GMT + 1 (GMT + 2 from late Mar to late Sep).

Climate: Temperate throughout most of the country. Mediterranean in the south-east. Jul and Aug are the hottest months (average temperatures 21-28°C), Jan and Feb are the coldest (average temperatures 2-12°C). Driest month is Mar, wettest Aug.
Clothing: Lightweight for summer, medium weight and topcoat for winter.

Entry requirements: Visa required by all except EC, Swiss and US nationals.
Health: Vaccination certificates not usually required. EC nationals covered for medical treatment.
Currency: No limit on amount of French or foreign currency that may be taken in.
Customs: Personal effects duty-free.

Air access: Regular flights by all major international airlines.
National airline: Air France.
Main international airport: Orly (PAR-ORY), 14 km south of Paris, Charles-de-Gaulle (PAR-CDG), 30 km north-east of Paris.
International airports: Bordeaux (BOD), Lille (LIL), Lyon (LYS), Marseille (MRS), Nice (NCE), Toulouse (TLS).

Surface access: Good rail, road and sea connections with all surrounding countries.
Main ports: Marseille, Boulogne, Nice, Calais, Dieppe, Dunkirk, Cherbourg, Le Havre, Rouen.

Hotels: Classified into deluxe and one to four star. Reservations (either direct or through centralised booking offices) should be made in advance during holiday season. Single rooms are rare and rates are usually quoted for a double room. All major credit cards accepted. Tip usually 12-15 per cent of bill if no service charge added.

Restaurants: Very wide variety and numerous. It is advisable to book in advance, particularly for Sunday lunch. Tip 12-15 per cent if no service charge included.

Currency: The French franc floats in relation to all other currencies, but is stabilised within the European Monetary System against other EMS currencies.

Foreign exchange information
Account restrictions: No restrictions on non-resident accounts held by foreign companies.
Trade policy: Few restrictions on imports from GATT, OECD and EC member countries. Products from non-member countries may not be imported unless they are a general exception, or import licences or prior authorisation are obtained.

Car hire: All major international hire companies have offices in Paris and other main towns. Drive on the right; *priorité à droite* applies in some built-up areas – cars coming out of side turning on the right have priority. Speed limits: 130 kph on toll motorways, 110 kph on dual carriageways, 90 kph on other roads and 60 kph in towns. Note that these limits are reduced on wet roads. Wearing of seat belts compulsory in front seats.

City transport
Taxis: Only available from ranks (*stations de taxi*). Standard fare of around Fr8 plus charge per km. Tipping usually 10-15 per cent.
Buses: Extensive flat fare service.
Metro: Comprehensive network in Paris and Lyon. Flat fare. N.B. same tickets may be used on both buses and metro.

National transport
Air: Major cities are linked by Air France, Air Inter and UTA. Frequent internal air services to many regional airports are operated by Air Inter the national domestic airline, Air-Alpes, Air Alsace, Air Vendée, Air Littoral, Rousseau Aviation and Touraine Air Transport. Certain services operate only during summer.
Road: Good local bus service. Also some long-distance coach services operated by Société Nationale des Chemins de Fer Français (SNCF). Approximately 1.5m km of roads, including 5,000 km of motorways, most of which are toll roads (*autoroutes à péage).*
Buses: In Paris same tickets may be used on buses and metro; a *carnet* of 10 tickets is cheaper. Few long-distance services.
Rail: SNCF operates nationwide network to over 6,000 destinations. Approximately 38,000 km of track of which one-third electrified.
Water: Extensive inland navigable waterways.

Public holidays
Fixed dates: 1 Jan, 1 May, 8 May (Victory in Europe Day), 14 July (Bastille Day, *Fête Nationale*), 15 August (Assumption), 1 Nov (All Saints'), 11 Nov (Armistice Day) and 25 Dec.
Variable dates: Easter Monday, Ascension Day, Whit Monday.

Working hours
Business: (Mon-Fri) 0900 – 1200 and 1400 – 1800.
Government: (Mon-Fri) 0830 – 1800 (staggered nine-hour day with two-hour lunch break).

Bourse (COB), was given wider powers to investigate abuses and harder-hitting powers to impose sanctions. Finance Minister Pierre Bérégovoy said the new law aimed to give the COB 'wide institutional, financial and legal autonomy'. But doubt still remained about the COB's independence, as well as its will and resources to take effective action against a seeming epidemic of insider dealing scandals.

On introducing the Bill, the government also announced it would include the obligation for an investor exceeding one-third of a listed company's capital to launch a public bid. That clause, however, was dropped from the final text, with jurisdiction to be left to the COB. The law also included a clause whereby shares in companies held by their own subsidiaries will cease to carry voting rights from 1 July 1991. This will end the practice, long used by French companies, of protecting themselves against possible hostile takeover bids by farming out up to 10 per cent of their equity to safe hands.

Meanwhile, Paris's stated ambition to become the leading continental financial centre in post-1992 Europe started to be taken seriously even outside France. The breathtaking pace of reform undertaken by governments and market authorities in the 1980s has gone a long way to raising Paris's international stature. While financial reform has allowed the launch of sophisticated new financial products and markets, heavy investment by the bourse authorities has revolutionised the bourse's previously antiquated trading practices.

The socialists Jacques Delors and Pierre Bérégovoy started the modernisation of the bourse in the early 1980s, and the Chirac government continued in the same direction. Continuous trading, introduced in June 1986, was progressively extended, first to French shares on the main market, then to shares on the cash and second market and finally, in 1989, to foreign shares. Also from March 1988 continuous trading was progressively introduced on the bond market. After the enormous success of the financial futures market, MATIF, started in 1986, a traded share options market, MONEP, opened in September 1988. New classes of options were launched on the MONEP in 1989, while the MATIF introduced its first non-domestic futures contract (on Deutschemark interest rates).

At the beginning of 1988, French stockbrokers were allowed to expand their capital base and the scope of their activities. Many of the French brokerage houses were acquired by banks and investment houses, both French and foreign. By mid-1989, fewer than 10 remained independent. The brokers' monopoly on bourse trading is to be lifted from 1992. Future reform plans include the scrapping of the country's five regional bourses and the integration (by 1991) of regional stocks into a single national trading system; and the removal of stocks from the main market if they do not meet a certain minimum level of trading. Two important reforms introduced in 1989 were designed to bring Paris further into line with London: the freeing of brokers' commission rates and the opening up of market-making.

The market-making reforms were expected to remove some of the price advantage that London has in trading large blocks of French shares. Indeed, one of the main motivations behind the modernisation drive was the fear that Paris would lose out more and more to London. Twenty-four major French stocks are traded in London, with more of these changing hands in the City than in Paris on some days. Bourse authorities said that while they were happy about the internationalisation of French stocks, they wanted to ensure that the largest slice of this trade took place on the domestic market.

A new system ('COCA') linking brokers' computerised order books and the central market's continuous trading system was due to be installed in late 1989. This was expected to speed up back-office work and guarantee smoother order flows. March 1990 was expected to see the full launch of a massive FFr400m project ('Relit') to link the bourse, member firms and financial institutions nationwide. Started in 1986, it aimed to speed up settlement and delivery, cutting the time between negotiation and delivery to five days, compared with the present 10-20.

By mid-1989 the bourse had recovered to well above the levels preceding the 1987 share slump. Paris had been one of the markets hit hardest by the crash, with the index falling to as little as 253 in January 1988 compared with its March 1987 peak of 460. By the end of 1988 the index had climbed back to 415.3 and in mid-August 1989 it stood at over 500.

BUSINESS GUIDE

Banking: (Mon-Fri) 0900 – 1200 and 1400 – 1600. Some banks close on Mondays and all close early on the day before a bank holiday.
Shops: (Mon-Sat) variable hours.

Social customs: Handshaking is usual on meeting or parting with personal acquaintances. Many people take their holidays in August.

Business language and interpreting/translation: Business managers generally prefer to speak French. Translation expensive. Details available from chamber of commerce and consular offices.

Telecommunications
Telephone: System has been simplified with all subscribers now having an eight figure number, and there are now only two regions – Paris and the provinces. Most public telephones are operated by coins though a telephone card system is being introduced. Local calls cost Fr1 for three minutes. International dialling code: 33, followed by, for Paris 1 and the new eight figure number, for the provinces simply the eight figure number.
Telex: Facilities available in central post offices of most major towns, and two public telex offices in Paris. Country code 42F.

Postal service: Letters can be sent for collection c/o Poste Restante, Poste Centrale in the town to be visited.

Advertising: Numerous advertising agencies operating throughout the country. Television advertising is strictly controlled, by Régie Franc\04aise de Publicitié, and is costly. Limited advertising on radio. All other media widely used – newspapers and magazines dominate. Total advertising spend (1984) Fr21,020m. Key categories include food, leisure, entertainment, services, retailing, furniture and transport.

Trade fairs: Over 50 major international exhibitions held every year, mostly in Paris. Information on exhibitions and other promotional events from the Conference and Incentive Department of the French National Tourist Office.

Electricity supply: 220 V AC.

Representation in capital: Afghanistan, Albania, Algeria, Angola, Argentina, Australia, Austria, Bahrain, Bangladesh, Belgium, Benin, Bolivia, Brazil, Bulgaria, Burkina Faso, Burma, Burundi, Cameroon, Canada, Central African Republic, Chad, Chile, China PR, Colombia, Comoros, Congo, Costa Rica, Cuba, Cyprus, Czechoslovakia, Denmark, Dijoubti, Dominican Republic, Ecuador, Egypt, El Salvador, Equatorial Guinea, Ethiopia, Finland, Gabon, German Democratic Republic, Federal Republic of Germany, Ghana, Greece, Guatemala, Guinea, Haiti, Holy See, Honduras, Hungary, Iceland, India, Indonesia, Iran, Iraq, Ireland, Italy, Ivory Coast, Japan, Jordan, Kenya, Republic of Korea, Kuwait, Laos, Lebanon, Liberia, Libya, Luxembourg, Madagascar, Malaysia, Mali, Malta, Mauritania, Mauritius, Mexico, Monaco, Mongolia, Morocco, Mozambique, Nepal, Netherlands, New Zealand, Nicaragua, Niger, Nigeria, Norway, Oman, Pakistan, Panama, Paraguay, Peru, Philippines, Poland, Portugal, Qatar, Romania, Rwanda, San Marino, Saudi Arabia, Senegal, Seychelles, Singapore, Somalia, South Africa, Spain, Sri Lanka, Sudan, Sweden, Switzerland, Syria, Tanzania, Thailand, Togo, Tunisia, Turkey, Uganda, USSR, United Arab Emirates, United Kingdom, USA, Uruguay, Venezuela, Vietnam, Yemen Arab Republic, People's Democratic Republic of Yemen, Yugoslavia, Zaïre, Zambia, Zimbabwe.

BUSINESS DIRECTORY

Hotels: Paris
Ambassador Concorde, 16, Boulevard Haussmann (tel: 42469263; tx: 650912; fax: 40220874).

Baudin (near Gare du Nord), 10, Rue Pierre Semand (tel: 42813711; tx: 660836).

Burgundy, 8 Rue Duphot (tel: 42603412; tx: 213380).

Campaville Gare du Nord, 26 Rue de l'Aqueduc (tel: 42392626; tx: 216200).

Elysées Bassano, 24 Rue de Bassano (tel: 47204903; tx: 611559).

Excelsior Opéra (near opera), 5 Rue Lafayette (tel: 48749930; tx: 283312).

Hilton, 18 Avenue de Suffren (tel: 42739200; tx: 200955; fax: 47836266).

Holiday Inn, 10 Place de la République (tel: 43554434; tx: 210651; fax: 47003234).

Inter-Continental, 3 Rue de Castiglione (tel: 42603780; tx: 220114; fax: 42611403).

Keppler (near Arc de Triomphe), 12 Rue Keppler (tel: 47206505; tx: 620440; fax: 47230229).

L'Horset Astor, 11, Rue d'Astorg (tel: 42665656; tx: 642737; fax: 42651837).

La Pérouse, 40 Rue la Pérouse (tel: 45008347; tx: 613420).

Le Bristol République, 112 Faubourg St Honoré (tel: 42669145; tx: 280961; fax: 42666868).

Le Colbert (near Notre Dame), 7 Rue de l'Hôtel Colbert (tel: 43258565; tx: 260690).

Le Grand Inter-Continental, 2 rue Scribe (tel: 42681213; tx: 220875; fax: 42661215).

London Palace, 32 Boulevard des Italiens (tel: 48245464; tx: 642360).

Mercure Paris Etoile, 27 Avenue des Ternes (tel: 47664918; tx: 650679; fax: 47637791).

Meridién Paris Etoile, 81 Boulevard Gouvion St Cyr (tel: 47581230; tx: 290952).

Ritz, 15 Place Vendôme (tel: 42603830; tx: 220262; fax: 42602371).

Saxe Résidence, 9 Villa de Saxe (tel: 47839828; tx: 270139).

Car hire: Paris
Avis: 5 Rue Bixio (tel: 45503231; tx: 201143).

Budget: (tel: 43875555; tx: 281386).

Europcar: 3 Avenue du Centre (Les Quadrants) (tel: 30438282; tx: 696783).

Hertz: 120 Boulevard Magenta (tel: 42853203; tx: 650957).

Chambers of Commerce
International Chamber of Commerce, 38 Cours Albert-1er, Paris (tel: 42618597).

Chambre de Commerce et d'Industrie de Bordeaux, 12 Place de la Bourse, 33076 Bordeaux (tel: 909128; tx: 541048).

Chambre de Commerce et d'Industrie de Boulogne, Quai Gambetta, 62204 Boulogne-sur-Mer (tel: 339292; tx: 110013).

Chambre de Commerce et d'Industrie de Grenoble, 6 Boulevard Gambetta, 38028 Grenoble (tel: 472036; tx: 320824).

Chambre de Commerce et d'Industrie de Lyon, Palais du Commerce, 20 Rue de la Bourse, 69289 Lyon (tel: 381010; tx: 310828).

Chambre de Commerce et d'Industrie de Marseille, Palais de la Bourse, 13222 Marseille (tel: 919151; tx: 410091).

Chambre de Commerce et d'Industrie de Nice, 20 Boulevard Carabacel 06007 Nice (tel: 559155; tx: 460041).

Chambre de Commerce et d'Industrie d'Orléans et du Loiret, 23 Place du Martroi, 45044 Orléans (tel: 620840; tx: 760912).

Chambre de Commerce et d'Industrie Paris, 27 Avenue de Friedland, 75382 Paris (tel: 5619900; tx: 650100).

Chambre de Commerce et d'Industrie de Rouen, 34 Rue Bouquet, 76000 Rouen (tel: 35984728; tx: 770036).

Chambre de Commerce et d'Industrie de Toulouse, 2 Rue d'Alsace-Lorraine, 31002 Toulouse (tel: 252100; tx: 531877).

Banking
Banque de France (central bank), 39 Rue Croix des Petits Champs, 75001 Paris (tel: 42615672 tx: 220932).

Association Française de Banques, 18 Rue la Fayette, 75009 Paris (tel: 42469259).

Banque Française du Commerce Extérieur, Boulevard Haussmann 21, 75009 Paris (tel: 42474747; tx: 660370).

Banque Indosuez, Boulevard Haussamann 96, 75008 Paris (tel: 45612020; tx: 650409).

Banque Nationale de Paris SA, 16 Boulevard des Italiens, 75009 Paris (tel: 42444546 tx: 280605).

Banque Paribas, Rue d'Antin 3, 75002 Paris (tel: 42981234; tx: 210041).

Caisse Centrale des Banques Populaires, Montmartre 115, 75002 Paris (tel: 40390000; tx: 210993).

Compagnie Financière de Crédit Industriel et Commercial (CIC Group), Rue de la Victoire 66, 75009 Paris (tel: 42808080; tx: 290692).

Crédit Agricole, Boulevard Pasteur 91-93, 75015 Paris (tel: 43235202; tx: 250971).

Crédit Foncier de France, SA, 19 Rue des Capucines, 75050 Paris (tel: 42448000; tx: 230079).

Crédit Lyonnais SA, Boulevard des Italiens 19, 75002 Paris (tel: 42957000 tx: 615310)

Société Générale, Boulevard Haussmann 29, 75009 Paris (tel: 42982000; tx: 290842).

Principal newspapers
France Dimanche, 6 Rue Ancelle, 92525 Neuilly-sur-Seine Cedex (tel: 47384312; tx: 611462).

France Soir, Rue de Bercy, 75112 Paris (tel: 45082800).

La Tribune de l'Economie, 12 Rue Béranger, 75003 Paris (tel: 48049999; tx: 230735).

La Vie Française, 2 Rue Béranger, 75003 Paris (tel: 48049999; tx: 670092).

La Voix du Nord, 8 Place du Général de Gaulle, 78167 Lille.

Le Figaro, 25 Ave Martignon, 75398 Paris Cedex 08 (tel: 42568080; tx: 211112).

Le Monde, 5 Rue des Haliens, 75427 Paris Cedex 09 (tel: 42479727; tx: 650572).

Le Nouvel Economiste, 22 Rue de la Tremoille, 75008 Paris (tel: 47230105; tx: 613572).

Le Parisien Libéré, 25 Ave Michelet, 93400 Saint Ouen (tel: 42528800; tx: 660041).

Le Progrès, 93 Chemain de Saint-Priest, 69680 Chassieu (tel: 78908118).

Les Echos, 67 Ave de Champs Elysées, 75381 Paris Cedex 08 (tel: 45621968; tx: 290275).

L'Expansion, 67 Ave de Wagram, 75017 Paris (tel: 47631211; tx: 650242).

Sud Ouest, 8 Rue de Cheverus, BP 521, 33000 Bordeaux (tel: 56909272; tx: 570670).

Other useful addresses
Agence France Presse (news agency), 11-15 Place de la Bourse, 75002 Paris.

ANIT (public information service) 8, Avenue de l'Opéra, 75001 Paris (tel: 1/2603738).

La Bourse de Paris (Stock Exchange), Palais de la Bourse, 4 Place de la Bourse, 75080 Paris.

Caisse Centrale de Co-opération Economique (CCCE), 233 Boulevard Saint-Germain, Paris (tel: 45503220; tx: 20750).

Centre Française du Commerce Extérieur, 10 Avenue d'léna, 75116 Paris (tel: 7236123).

DATAR (government investment agency), 1 Avenue Charles Floquet, 75007 Paris (tel: 7836120; tx: 200970).

Direction Générale des Impôt, Centre des Non-Résidents, 9 Rue d'Uzés, 75094 Paris.

Direction du Tourisme, 17 Rue de l'Ingénieur Robert Keller, 75740 Paris.

Institut National de la Statistique et des Etudes Economiques, 18 Boulevard Adolphe Pinard, 75675 Paris.

Ministère de l'Industrie, des P & T et du Tourisme, Service d'Étude des Stratégies et des Statistiques Industrielles, 83-85 Blvd du Montparnasse, 75270 Paris Cédex 06 (tel: 45564108).

Ministére des Finances et du Budget, 139 Rue de Bercy, 75572 Paris Cédex 12 (tel: 40040404).

Ministry of the Economy, Finance and the Budget, 246 Boulevard St Germain, 75700 Paris (tel: 42603300; tx: 217068).

Ministry of Foreign Trade, Paris.

Ministry of Industry and Territorial Development, 101 Rue de Grenelle, 75700 Paris 9tel: 45563636).

Service de la Répression des Fraudes et du Contrôle de la Qualité, 44 Boulevard de Grenelle, 75732 Paris (tel: 2806722).

French Guiana

Greg Chamberlain

'Guiana is not the El Dorado some hope to find,' remarked prefect Jean-Pierre Lacroix in October 1989 after a year of rising crime and unusual social unrest in France's biggest remaining colony. For centuries the target of dreamers and crooks from Europe who ripped off French Government subsidies for development schemes that invariably came to naught, Guiana is now the object of desire of the poor beyond its borders. Forty per cent of its 94,000 people are now recent immigrants (most of them illegal) – from Haiti, Suriname, Brazil and Guyana. Almost all have come in search of one of the continent's highest standards of living, fashioned by a French Government intent on making it a 'showcase' of French technology and culture in Latin America, as Paris officials are fond of putting it.

More than 10,000 refugees from the festering civil war in neighbouring Suriname are still camped on the banks of the Maroni river, which forms the frontier. Officially they are waiting to go home, but many would prefer to have French citizenship and stay. A new cease-fire and peace treaty in mid-1989 between Suriname's civilian government and rebel leader Ronnie Brunswijk soon broke down when the Surinamese army balked at forgiving the rebels who, it charged, were being accommodated and assisted by French authorities. Fighting erupted again and repatriation was halted. France spent some $7m in 1988 to feed and house the refugees, who were partly blamed along with other immigrants for a 60 per cent upsurge in violence and burglaries in 1988-89. That, accompanied by sharply increased drug abuse, produced two nationwide one-day protest shutdowns and the biggest demonstration in the colony's history in September 1989. Both unions and business demanded more police and stricter immigration control. But the main trade union, the far leftist Guianese Workers' Union (UTG), denounced as provocative the doubling of the police force with 100 French riot police. Guianese are also cool to the 3,000 troops France stations in Guiana.

The other side of the immigration coin was shown by the simultaneous 11-day strike of 800 workers, nearly half of them Brazilian labourers, over low pay and poor conditions on the three major development projects under way in the country. These are the expansion of the highly-successful European space base at Kourou, the building of a $400m dam at Petit Saut (which is to provide the colony with all its energy for 30 years from 1992) and a 60 km diversion of the country's main road, from Cayenne to St Laurent. The space base is the dynamo of the economy and Kourou is the country's fastest growing town. But France's socialist government is sensitive to its tendency to become a racist white ghetto and launched the Phèdre Plan in June 1989 to ensure the town's wealth was shared by the whole country. Kourou, exempt from local taxes, however remained the modern symbol of Guiana's legendary corruption, attracting shady local businessmen who also bankroll the Guianese Socialist Party (PSG) which has dominated the colony's political life for the past decade. The frequent strikes and general malaise of 1988-89 seemed to infect the PSG itself and its potentates – Elie Castor, the mildly pro-independence deputy in the French parliament and president of Guiana's General Council, and his rival, Georges Othily, president of the Regional Council. Hardliners led by Cayenne mayor and PSG leader Gérard Holder expelled Othily from the party in June 1989 for his refusal to go along with the PSG's

highly partisan distribution of state funds and patronage. But Othily got his revenge three months later when party dissidents enabled him to capture the colony's single seat in the French senate, defeating incumbent Raymond Tarcy, the official PSG candidate. The dissidents were purged, but the damage to the party was done and Othily's leadership was reinforced. Holder stepped down as secretary-general in November 1989 and was replaced by one of his aides, Antoine Karam.

The right remained divided, with young St Laurent mayor Léon Bertrand, the other of Guiana's two deputies, maintaining his break with the local Gaullist RPR party. The other main right-wing group, the UDF, got a new leader in Elie Chow Chine, replacing the veteran Claude Ho A Chuck, who has formed his own party.

All Guiana's politicians, however, are united in their broad suspicion of France, despite the massive subsidies, minimum wage and social welfare which are only dreams on the rest of the continent. The French cash has nevertheless reduced the once-vocal pro-independence movement to nothing, despite occasional cries that France is 'pillaging the wealth' of Guiana. Many fear a single Europe in 1992 will ride roughshod over local problems, apart from ending the comfortable network of local monopolies and corruption.

Guiana is also vulnerable because its highly artificial economy produces virtually nothing. Schemes to roll back the Amazonian rain forest (or 'Green Hell' as

KEY FACTS

Official title: French Guiana

Head of state: President François Mitterand

Head of government: Commissioner (prefect) Jean-Pierre Lacroix

Capital: Cayenne

Official Languages: French (Creole widely spoken)

Currency: French Franc (Fr) = 100 centimes

Exchange rate: Fr6.31 per $ (Nov 1989)

Area: 91,000 sq km

Population: 93,540* (1989)

Inflation: 4% (1987)

Trade balance: –Fr2,048m (1987)

Foreign debt: $1,200m (1988)

* estimated figure

some call it) which blankets 82 per cent of the country and open it up to farming and cattle-raising have come to little because of high costs, poor transport and planning, and the difficult climate. Only 0.14 per cent of the land is farmed and even this area is shrinking. Nearly everything is imported, most of it from France. Exports cover a mere 14 per cent of imports. Only in October 1989 was the colony's first cement factory being planned. However, exports of fish and shrimps (62 per cent of all exports and sold to France and the United States) continued to grow steadily, though shrimp catches were down in 1989 after two good years. The other major exports, timber and rice, also showed solid results. Gold production rose by nearly two-thirds in 1987, the latest figures available, but smuggling undoubtedly masked a higher production.

COUNTRY PROFILE

Historical profile: First reported European sighting 1496; first French settlement 1604; numerous changes in control between French, English, Dutch, Brazilian, Portuguese – and border disputes 1654-1915; overseas department status 1946.

Political system: Overseas department of France with additional status as a region of France, represented by two members of the French National Assembly and one member of the Senate. Local administration is through a directly-elected departmental council *(Conseil Général)* of 19 members, and an indirectly-elected 31-member regional council *(Conseil Regional).*

Main towns: Cayenne (population 45,000), Kourou (13,000) (mainly new town around Space Centre), St Laurent-du-Maroni (10,000).

Media
Press: Daily papers are *La Presse de Guyane* (circulation 16,000) and *Guyane-Matin. France-Guyane* (4,000) published four times a week. US and metropolitan French papers are available. Several periodicals but no trade press.
Broadcasting: Radio-Télévision Française d'Outre-mer (RFO) (TDF) broadcast in French. There are several independent radio stations. There are approximately 15,000 TV sets in use.

External trade: Heavily dependent on France for financial aid. Balance of trade deficit mainly due to high imports of food and fuels, and undeveloped export potential.
Exports: Shrimps (57% of total value in 1987), timber (8%), gold (7%), rice (6%), fish (4.4%), rum. Main destinations: France (38% of total value in 1987), French Antilles (25%), USA (22%), Japan (11%).
Imports: Food (22% of total value in 1987), machinery and transport equipment (21%), petroleum products (10%). Main sources: France (62% of total value in 1987), Trinidad and Tobago (9%), Japan (3%), USA (4%), German FR (3%).

Agriculture: Only 0.14 per cent of total land area is cultivated and production is dominated by crops for domestic consumption such as rice, maize and bananas, while sugar-cane is grown for rum production. Small number of cattle farms have also been established. Vast timber resources (covering 82 per cent of country) but poor infrastructure means these have not been fully exploited. Fishing, especially for shrimps, also growing in importance.

Industry: Manufacturing virtually non-existent except for small factories and a few saw-mills. Limited industrial activity around European Space Agency's rocket-launching site at Kourou.

Mining: Only mineral exploited commercially at present is gold – with output of 514 kg in 1987. Substantial deposits of bauxite and kaolin have been found but extraction is not economically viable in the foreseeable future.

Energy: Relies to a large extent on petroleum imports. Domestic electricity consumption was 237 kWh in 1987.

Membership of international organisations: Franc zone, WFTU.

Time: GMT -3.

Climate: Tropical. Generally hot and humid with heavy rain. Dry season from Aug to Dec with average temperature 28°C. Rainy season from Dec to Jul with temperature range 22°C to 32°C.

Entry requirements
Passport: Required by all except nationals of France, certain EC and certain francophone countries holding national identity card.
Visa: Required by all except nationals of EC countries and Switzerland. N.B. Entry regulations liable to change at short notice – prior checking with embassy/consulate is essential.
Currency: No restriction on import of foreign currency but amounts should be declared, and amount taken out must not exceed that taken in. Check latest regulations with embassy/consulate.

Health precautions
Mandatory: Yellow fever and cholera vaccination certificates required if arriving from infected areas or if staying for more than 14 days.
Advisable: Yellow fever, typhoid, polio vaccinations. Malaria prophylaxis advisable outside Cayenne. Water precautions should be taken, though tap water in Cayenne is safe.

Air access
Airlines serving: Air France, Cruzeiro, Surinam Airways, Minerve.
International airport: Cayenne-Rochambeau (CAY), 17 km from city; duty-free shop, bar, post office, shops, hotel reservations, car hire.

Surface access
Main ports: Dégrad des Cannes, Kourou, Le Larivot, Saint Laurent du Maroni.
Overland access: It is normally possible to reach Cayenne from Suriname by bus along the coast road but rebel activity in eastern Suriname has cut off this route. There is no road access from Brazil.

Hotels: Good standard accommodation available in Cayenne, Kourou and St Laurent. Advisable to book in advance. Rates normally include service and taxes, if not 10 per cent tip usual.

Car hire: Available in Cayenne and at the airport. International licence accepted.

City transport
Taxis: Available in main towns; fares by negotiation.

National transport
Air: Air Guyane serves main centres. (Bookings can be made through Air France.)
Road: Cayenne district is served by a good road system. The only major road runs from Cayenne, via Kourou, to St Laurent on Suriname border.
Buses: Scheduled services on Cayenne to St Laurent route.
Water: Major form of travel. Motor boat serves coastal towns. River boats link interior centres with coast.

Public holidays: 1 Jan, Easter Monday, 1 May, 4 May (Ascension Day), 8 May (Armistice Day), 10 June (Slavery Day) 14 Jul (Bastille Day), 15 Aug (Assumption), 1 Nov (All Saints'), 11 Nov (Armistice Day), 25 Dec (Christmas Day).

Working hours
Government and business: (Mon-Fri) 0730/0800 – 1300, 1430/1500 – 1800/1830 (closed Wed and Fri afternoons).
Banking: (Mon-Fri) 0745 – 1200, 1445 – 1645.

Telecommunications
Telephone: Dialling code for French Guiana: IDD access code + 594 followed by subscriber's number. No area code needed.
Telex: International service available in Cayenne and Kourou, through hotels and the Chamber of Commerce (Cayenne only).
Telegram: Service available at major hotels and at TSF radio station in Cayenne. Telegrams are relayed via Suriname or Martinique.

Banking: Banque Nationale de Paris Guyane (commercial bank), Banque Française Commerciale Antilles-Guyane, Crédit Populaire Guyanais, Crédit Martiniquais.

Electricity supply: 220 V AC, 50 cycles.

Hotels: Cayenne
Beauregard, Route de Remire (tel: 35-41-00; tx: 910310).

Le Polygone, Carrefour du Larivot RN 1 (tel: 30-54-00; tx: 910329).

Novotel, Chemin Hilaire, Route de Montabo (tel: 30-38-88; tx: 910581).

PLM Montabo, Route de Montabo (tel: 312575; tx: 910555).

Hotels: Kourou
Les Relais de Guyane, Avenue des Roches, BP 814, 97310 Kourou (tel: 32-00-66; tx: 91-05-14).

Hotels: Saint-Laurent
Star, Rue Thiers, 97320 (tel: 34-10-84).

Car hire: Cayenne Avis: 91 Rue Lt Becker (tel: 31-25-27; tx: 910590).

Budget: Shell Station (tel: 31-15-40; tx: 910391).

Europcar National: PO Box 84897339 (tel: 30-16-23, 30-18-27; tx: 910562).

Hertz: Route de la Madeleine (tel: 30-61-71).

Car hire: International airport Avis: (tel: 31-15-53).

Business advice and information
Agence Régionale de Developpement du Tourisme et des Loisirs de la Guyane, 12, Rue Lalouette – BP: 801 – 97338 Cayenne (tel: 30-09-00; tx: 910364).

Air Guyane, Rochambeau, Cayenne (tel: 31-72-00).

Banque de la Guyane, Place Victor-Schoelcher 2, Cayenne (tel: 31-15-15; tx: 522).

Chambre de Commerce et d'Industrie de la Guyane, PO Box 49, Place de Grenoble, Cayenne (tel:30-30-00; tx: 910537 ccicay).

Jeune Chambre Economique de Cayenne, 2 bis rue Docteur Saint-Rose, BP 1094, Cayenne.

Radio-Télévision Française d'Outre-mer (RFO), Rue de Dr Devèze, BP336, 97305 Cayenne (tel: 31-15-00; tx: 910526).

French Polynesia

David Robie

(map on page xxiv)

As for millions of Europeans, the theme Europe 1992 is on the mind of thousands of Pacific Islanders living in France's three territories in the South Pacific. But the 'no barriers' plan is causing anxiety and resentment – particularly among the Tahitians. Many see it in terms of an 'invasion threat' from European Common Market nationals and business people, further eroding their prospects for sovereignty. In mid-1989 several initiatives by the Tahitian coalition government, political parties and citizens groups renewed pressure on France over the future of metropolitan rule and nuclear tests in Polynesia.

KEY FACTS

Official title: French Polynesia (Polynésie Française)

Head of state: President of France

Head of government: President Alexandre Leontieff

Ruling party: Coalition led by Te Tiarama

Capital: Papeete (on Tahiti)

Official Languages: French, Tahitian

Currency: CFP Franc (CFP Fr = 100 centimes)

Exchange rate: CFP Fr113 per $ (Nov 1989)

Area: 3,600 sq km (35 islands and 83 atolls)

Population: 188,814 (1988)

GDP real growth: 0%* (1987)

Labour force: 65,200 (1988)

Inflation: 2.6% (1988)

Trade balance: –CFP Fr79,376m (1988)

Foreign debt: $397m (1987)

Visitor numbers: 135,387 (1988)

* estimated figure

KEY INDICATORS

	Unit	1984	1985	1986	1987	1988
Inflation	%	10.7	7.8	–0.7	2.0	2.6
Copra production	tonnes	7,389	13,442	13,707	14,980	11,298
Exports	CFP Fr m	5,084	6,564	5,112	9,094	8,111
Imports	CFP Fr m	85,622	88,864	92,667	90,587	87,487
French imports	CFP Fr m	39,994	41,316	46,430	47,299	46,691
French military spending	CFP Fr m	19,794	21,599	22,733	22,541	23,552
French civilian spending	CFP Fr m	23,921	26,268	28,754	29,454	35,633
French investment	CFP Fr m	3,721	1,500*	–	–	–
External debt	$m	167	223	282	397	–
Debt service ratio	%	12.0	4.9	5.7	2.0	–
Exchange rate	CFP Fr per $	160	164	132	105	109
Visitor numbers	–	101,595	122,086	161,200	142,800	135,387

* estimated figure

Among the developments was an unprecedented special session of the territorial assembly called to debate a proposed popular vote over nuclear tests. Tahitian President Alexandre Leontieff ordered the parliamentary session following pleas by two pro-independence parties and a week-long hunger strike by protesters just before Bastille Day.

Other major issues considered were a controversial $300m five-year economic and social development plan for French Polynesia initiated by French Prime Minister Michel Rocard and Overseas Territories Minister Louis le Pensec, and the future of Tahiti's relationship with Europe starting in 1993.

Ironically, the pressure on France has come just when Rocard's socialist government had prepared the most crucial diplomatic initiative in the South Pacific since nuclear tests began at Moruroa and Fangataufa atolls 23 years ago. Officials in many of the regional governments regard Rocard, who visited both French-ruled territories and independent Pacific states, as the most positive influence in French–Pacific relations in more than two decades. They believe his efforts will go a long way towards establishing a more popular French presence in the region. The officials regard France as having vital technical expertise in areas such as deep-sea mining and other development assistance that will be beneficial to the South Pacific. French cultural influence is also valued.

Some Pacific officials, however, are more suspicious about the apparently enlightened new attitude shown by Paris. One factor which has been troubling some Melanesians is the growing French influence in post-coup Fiji. Critics also believe France is trying to buy influence in the region to blunt South Pacific criticism over nuclear testing and to isolate some of the independence factions in New Caledonia and French Polynesia. Former New Zealand Prime Minister David Lange made a harsh criticism of 'cheque-book diplomacy'.

Fears about Europe 1992

While Paris has been making overtures to independent South Pacific states, however, islanders in New Caledonia and Tahiti have been disappointed in the lack of French support for their case over Europe 1992. Tahitian voters boycotted the 1989 European parliamentary elections: 90 per cent of the 110,000 registered voters refused to cast a ballot. And the biggest boycott was in the ghettos of the airport suburb of Faaa, the fiefdom of pro-independence and anti-nuclear mayor Oscar Temaru, where a mere 2 per cent of voters took part.

The implications for Tahiti of a fully integrated European Common Market only became apparent to the 41-seat territorial assembly during May 1989. Assembly deputies realised then that under the so-called Project Europe 1992 plan, or 'Europe Without Frontiers', the 320m citizens of the 12 nations forming the European Community (EC) would soon be allowed to live, work and set up businesses in the French overseas territories without restriction.

'Having already seen their islands, especially Tahiti, being invaded during the past 25 years by 30,000 metropolitan French people,' said Tahiti historian and author Dr Bengt Danielsson, 'the assemblymen were, of course, outraged by this very real threat to become totally submerged by another and much bigger invasion wave. An equally distasteful prospect, which also shocked the assemblymen was the freedom big European business firms would soon have to invest and open branch offices in their islands and thus complete the French political domination with a multinational economic colonisation.'

Leontieff's government insisted that Tahiti should have the right to take part in any direct negotiations between France and the EC dealing with the movement of community citizens between Europe and Tahiti – and to give its view on any decisions.

An internal assembly report indicates that the right of access of French citizens is also at stake. This has long been a controversial issue in Tahiti with a series of Tahitian governments trying unsuccessfully to curb the right of free access enjoyed by French citizens, even those with criminal records. The outmoded French immigration laws currently applying to Tahiti date as far back as 1939. During 1989 these laws were challenged over a deportation order against two EC foreigners, but they have yet to be tested over French immigration.

In spite of the 1984 autonomy statute – claimed at the time to rival the Cook Islands' relationship with New Zealand – the French Government still retains control over justice, national defence, money, much of external trade (such as with New Zealand) and international airline rights as well as immigration.

In this year of the French revolution's bicentenary, noted Health and Environment Minister Jacqui Drollet, leader of the pro-independence party Ia Mana Te Nunaa, it was time for France to honour its commitment to one of the constitutional principles of the Fifth Republic: 'Faithful to its traditional mission, France will lead the peoples for which it has taken responsibility towards the freedom to administer themselves and to democratically manage their own affairs.'

COUNTRY PROFILE

Political system: French Polynesia is a French overseas territory whose high commissioner, appointed by the French government, directs all local state services, including defence, foreign affairs, immigration, civil law, justice, information, foreign trade, international communications, higher education and scientific research. A French law of Sep 1984 conferred conduct of certain local affairs (local budget, health services, primary education, culture, social welfare, public works, agriculture and sports) to council of 10 ministers whose president is elected by 41-member Territorial Assembly, itself elected by universal adult suffrage. Territory represented in French Parliament by two deputies and a senator.

Political parties: Coalition of Te Tiarama, Here Ai'a la Mana Te Nunaa, Te Ea No Maohi Nui, Amuitahira'a No Oteania, Ai'a Api and Te Atira A Porinetia; opposition: Tahoeraa Huiraatira, Tavini Huiraatira.

Population: The population of 188,814 (1988) is growing at some four per cent a year, and is gradually converging on Tahiti – population 131,309 (1988) – as conditions become more difficult.

Main cities/towns: Papeete (on Tahiti, population 23,555), Faaa (on Tahiti, pop 2,408), Fare (on Huahine, island pop 4,479), Uturoa (on Raiatea, island pop 8,560).

Language: French and Tahitian are the official languages; in tourism and business, English is often spoken as well.

Media
Press: Two daily newspapers in French: *La Dépêche de Tahiti* and *Les nouvelles de Tahiti* and two weeklies: *Tahiti Sun Press* (in English) and *Tahiti Rama*.
Radio: In addition to the government-operated service, there are 10 private radio stations.
Television: The government operates colour TV service. There are two channels.

Domestic economy: Aside from tourism and black pearls, the economy is based on smallholdings growing fruit and vegetables, and plantations providing copra and coconut oil for export. The five-year plan for 1984-88 promoted tourism, but also emphasised development of the islands' 3,998,651-sq km economic zone through high seas, coastal and lagoon fisheries, pearl farming and aquaculture. Deep-sea resources (mainly tuna) are fished exclusively by Asian fleets under licence. Development of the remote archipelagos (Marquesas, Australs, Tuamotu-Gambiers) has commenced with construction of more airstrips. Maintenance and servicing of French nuclear testing centre, Centre d'Experimentation du Pacifique (CEP) has had dramatic impact on the islands' economy.

Employment: The establishment of CEP has caused a shift of workers away from agriculture and fishing towards tertiary activities.

External trade
Exports: Typically, the main exports are black pearls, coconut oil and its derivatives, copra, beer, vanilla, fish and shells of various kinds. Main destinations are France (typically, 50-60% of exports), USA (12-18%), Italy (10-12%), New Caledonia (6-10%) and Japan (5-6%).
Imports: Main imports are machinery and equipment (including transport equipment), food, mineral fuels, textiles and raw materials. Main sources are France (52% of 1987 total), other EC countries (13%), USA (10%), Japan (5%), Australia (4%), New Zealand and Singapore (3%).

Agriculture: Agriculture accounts for only about four per cent of GDP, but it contributes to practically all exports and its development is a central plank of government policy. Primary products: copra (for coconut oil processing), vanilla, mother-of-pearl shells, taro and cultured pearls. Weather permitting, local production supplies over 60 per cent of overall demand for some vegetables (lettuce, tomatoes, green beans, turnips, carrots, potatoes and onions). Fruit produced for export (mangoes and pineapples), for fruit juice factories (pineapples and grapefruit) and for local market (including bananas, oranges, pineapples, limes, mangoes, avocados, pawpaws. Local production supplies about 28 per cent of demand for dairy products (1986) and 83-87 per cent of demand for pork. All food imports were quite sharply reduced in 1986, as domestic production resumed full capacity. Although 70 per cent of the islands' land area is covered in forest, conditions limit exploitation to random felling, and almost all timber is imported. Plantations recently developed are intended to yield productive forest zone of 11,250 hectares of Caribbean pine within 25 years, and 30 hectares of wood for local cabinet-making is planted per year. Pearl farming is established in lagoons (although low prices have depressed the business since 1985), and green mussel, prawn, live-bait and fresh-water shrimp aquaculture under development.

Industry: Tourism is main industry, accounting for 20 per cent of GDP. In 1987, the islands had 142,800 visitors. The oil mill, Huilerie de Tahiti, purchases all copra produced and processes it into coconut oil and meal (for animal feed), soap-making and monoi (scented coconut oil). Other industries include breweries, soft drinks and fruit juice factories and power station. Several small concerns produce textiles and handicrafts. Solar energy is much used as a power source.

Legal system: For legal matters, consult a lawyer familiar with French law.

Membership of international organisations: South Pacific Commission. Associated with EC.

BUSINESS GUIDE

Time: GMT -10 hrs.

Climate: Two seasons: warm and moist (Dec-Feb) average temperature 27°C; cool and dry (Mar-Nov), average temperature 21°C. Rainfall varies, depending on relief of island and exposure to prevailing winds, but heaviest Nov-Mar. Most islands are mountainous (volcanic) and ringed with coral reefs; the Tuamotu and Gambier groups are mainly low-lying atolls.

Entry requirements: Entry requirements are liable to change, so check with a French consulate or an airline before travelling.
Passport: Required by all.
Visa: No visa required for a stay of up to 14 days for citizens of most countries, but an onward or return ticket is necessary.
Customs: Personal effects allowed duty-free. All baggage coming in from Fiji and Samoa except for hand luggage, is fumigated. Travellers should carry clothing and toilet articles for an overnight stay in their hand luggage and have their hotel collect other baggage from the airport after fumigation. Import of some household pets, dangerous goods and drugs prohibited.
Income tax: No income tax for individuals, and no sales tax. There is a business licence tax.

Health precautions
Mandatory: Vaccination certificates for yellow fever if travelling from an infected area.

Representation overseas: The Tahiti Tourist Promotion Board (Office du Développement du Tourisme) maintains offices in Aukland, Hong Kong, Los Angeles, Paris, Santiago, Sydney, Tokyo. In Europe information available through Maisons de la France offices. There is also OPATTI (Office de Promotion et d'Animation Touristiques de Tahiti et ses Iles).

Air access
National airline: Air Tahiti (domestic flights only; international flights by UTA).
Other airlines: Air Calédonie, Air France, Air New Zealand, Air Tahiti, Continental, Cotam (French military), Hawaiian, Lan-Chile, Minerve, Qantas, UTA.
International airport: Faaa International (PPT), 6 km from Papeete; restaurant, bank and car hire.

Surface access: Over 90 per cent of foreign visitors arrive by air.
Main ports: Papeete (harbour installations can accommodate ships up to 35,000 tons).

Hotels: Over 3,000 tourist rooms and bungalows; all tourist hotels have restaurants.

Credit cards: American Express, Diners Club, MasterCard and Visa accepted throughout Tahiti.

Car hire: Numerous car hire establishments; rates include insurance. Drivers must hold a licence valid for at least one year and must be at least 21 years old. Driving on the right-hand side of the road.

City transport
Taxis: Fares are controlled and should be displayed in each cab. In Tahiti, they double between 2300 and 0500. Information on fares available at the Tahiti Tourist Promotion Board at the airport and in downtown Papeete.

National transport
Air: Over 25 airfields in addition to Faaa International, and some 45 small aircraft for local travel. Inter-island baggage limit is 10kg. Air Tahiti operates scheduled flights to Moorea, Huahine, Raiatea, Bora-Bora, Maupiti, Rangiroa, Manihi, Takapoto, Tubuai, Nuku-Hiva (Marquesas), Ua Huka, Hiva Oa, Ua Pou, Anaa, Makemo, Hao, Rurutu and Mangareva (Gambiers), and several other atolls. Air Moorea operates daily services between Tahiti and Moorea. Both airlines also offer air taxi services, charters, circle island flights and transportation to other islands. Other air operators include Tahiti Conquest Airlines (TCA), Tahiti Helicopters and Pacific Helicopter Services.
Road: There are approximately 200 km of road on Tahiti, including a circular 120 km asphalt road around the main part of the island, and 100 km of road on Moorea.
Buses: 'Le Truck' runs an unscheduled transportation service between Papeete and outlying districts, leaving approximately every half hour for nearby areas and daily for distant points.
Water: There is a scheduled boat service between Papeete and Moorea.

Public holidays
Fixed dates: 1 Jan, 5 Mar, 1 May, 14 Jul (up to 10 days), 8 Sep, 1 Nov, 11 Nov, 25 Dec.
Variable dates: Good Friday, Easter Monday, Ascension Day, Whit Sunday and Monday, Armistice Day.

Working hours
Government and business: (Mon-Fri) 0730/0800 – 1700.
Banking: (Mon-Fri) 0745 – 1530; (Sat) 0745 – 1130 (some banks only).
Shops: (Mon-Fri) 0730 – 1130, 1400 – 1700; (Sat) 0730 – 1130.

Social customs: Tipping is not customary, and is contrary to traditional Tahitian hospitality.

Telecommunications
Telephone: Dialling code for French Polynesia, IDD access code + 689 (no area code required) followed by subscriber's number. Tahiti has an

BUSINESS GUIDE

automatic telephone network, and the manual inter-island services are gradually being adapted to automatic. There were 17,600 telephones in 1987.
Telex and telefax: Service available at main Post Office in Papeete, at the Chamber of Commerce (within certain hours) and at some hotels. Most hotels have a telefax.
Telegram: Cable service available at post offices seven days/week, but within restricted hours.

Banking
Banque de Polynésie, Banque de Tahiti, Banque Indosuez, Paribas, Socredo.

Advertising: Advertising available in local newspapers. Correspondence and trade literature should be in French.

Electricity supply: 220 V AC, 60 cycles (check with hotel before using appliances).

Weights and measures: Metric system.

Representation in capital: Austria, Belgium, Chile, Denmark, Finland, Germany FR, Italy, Rep. of Korea, Monaco, Norway, Sweden.

Hotels: Tahiti
Climat de France, Punaauia (tel: 426621; tx: 339).

Ibis, Ave du Prince Hinoi, Papeete, P.O. Box: 4545 (tel: 423277; tx: 319).

Matavai, Papeete, P.O. Box: 32 (tel: 426767; tx: 222).

Pacific, Boulevard Pomare, P.O. Box: 111 (tel: 437282).

Royal Papeete, Papeete, P.O. Box: 919 (tel: 20129; tx: 384).

Royal Tahitien, Pirae in Rue Temarii, P.O. Box: 5001 (tel: 428113).

Sofitel Maeva Beach, Papeete, P.O. Box: 6008 (tel: 428042; tx: 214 FP; fax: 426107).

Tahara'a Hyatt Regency, Point Venus, Papeete, P.O. Box: 1015 (tel: 441122; tx: 225).

Tahiti, Papeete, P.O. Box: 416 (tel: 429550; tx: HOTAHITI 406 FP).

Tahiti Park Royal Beachcomber, Papeete, P.O. Box: 6014 (tel: 4251; tx: 276; fax: 436106).

Car hire: Papeete (on Tahiti)
André: (tel: 429404).

Avis: PO Box 1683, Rue Charles Vienot (tel: 429649; tx: 298).

Hertz: PO Box 1617 (tel: 420471; tx: 255 FP).

Pacificar: Rue des Remparts, Pont de l'Est (tel: 424304, 424364).

Robert: (tel: 429720).

Taxis: Papeete (on Tahiti) Airport (tel: 36520).

Station du Jasmin (tel: 420894).

Station du Marché (tel: 420292).

Station Vaima (tel: 429835).

Business advice and information
Caisse Centrale de Co-operation Economique, BP 578, Papeete, Tahiti.

Chambre de Commerce et d'Industrie de la Polynésie Française, PO Box 118, Rue Cassiau, Papeete, Tahiti (tel: 420344; tx: 274).

Institut Territorial de la Statistique, BP 395, Papeete, Tahiti.

Service des Affaires Economique, BP 82, Papeete, Tahiti.

Service de Tourisme (Papeete Tahiti), Fare Manihini, Blvd Pomare, BP 4527, Papeete (tel: 429330; tx: 254).

Syndicat des Importateurs et des Négociants, PO Box 1607, Papeete, Tahiti.

Syndicat d'Initiative de la Polynésie Française, BP 326, Papeete.

Travel information
Air Moorea, BP 6019, Papeete (tel: 424834; tx: 314).

Air Tahiti, Blvd Pomare, BP 314, Papeete (tel: 422333).

OPATTI (Office de Promotion et d'Animation Touristique de Tahiti et ses Iles), Fare Manihini, Blvd Pomare, BP 65, Papeete (tel: 429626; tx: 254).

Tahiti Tourist Promotion Board (Office du Développement du Tourisme), PO Box 65, Papeete, Tahiti (tel: 429626; tx: 254 OFFTOUR FP; fx: 436619).

Gabon

Susan Traill

Despite Gabon's achievements in cutting public spending in 1988, negotiations with Paris Club donors for the much-needed 1989 rescheduling agreement dragged on account of oil revenue problems. The IMF loan agreement which expired in December 1988 was not renewed. However, financial aid continued to flow. Austerity also continued, but the political scene remained calm. The inauguration of the Rabi-Kounga oilfield and new oil finds produced a 'Rabi boom', with oil companies flocking into Gabon, and renewed confidence in the long-term strength of the economy.

Although public sector workers went on strike against pay cuts in December 1988, there was no opposition to President Omar Bongo's policies within the ruling Parti Democratique Gabonais (PDG). Relations with France were calm and there was no need for the traditional playing off of French versus US donor influence. In May 1989, President Bongo took the unprecedented step of inviting Abbe Paul Mba-Abbesole, leader of the dissident French-based Mouvement pour le Redressement National (MORENA) to Libreville to discuss MORENA's proposals for a multi-party democracy in Gabon. MORENA wants to rewrite the 1961 constitution to include the multi-party system as a permanent feature. A first step towards democracy would be to introduce competing candidates for the PDG in the constituencies at general elections. MORENA recommends a maximum of four parties. MORENA also wants a substantial reduction in the bureaucracy, including a cut in the number of ministries from 36 to nine.

This last point is probably of most interest to President Bongo as it fits in well with the austerity programme. PDG members, especially those who could lose their cabinet status, were angry at his meeting with Mba-Abessole, as were many MORENA members. MORENA is split between the followers of Mba-Abessole and those of Max Anicet Koumba Mbandinga. The latter accused Mba-Abbesole of selling out to the regime. MORENA disclosure of government plans to set up a nuclear waste or uranium processing or, as the government claimed, uranium storage company led to the cancellation of this programme in August 1988.

The government expelled 500 illegal immigrants, mostly political dissidents from Equatorial Guinea, in summer 1988. The government has computerised residency permit data to help reduce illegal immigration. But of the estimated 110,000 expatriates at mid-1988, only 37,000 had papers in order.

Gabon signed a countertrade agreement with inflation-ridden Brazil in September 1988, under which Gabon will exchange oil and timber for railway rolling stock, helicopters and aeroplanes, household goods and foodstuffs.

Reforms generate aid

The government issued a decree cutting public sector wages by 15 per cent in mid-1988. However, President Bongo's insistence that state companies should apply the reductions fully was met with a series of strikes in December 1988. The Air Gabon strike was only resolved when management agreed to a lesser pay cut.

In January 1989, the government raised taxes on tobacco to generate extra income of CFA3bn a year. The 'solidarity' tax of 3 per cent on the majority of wages was maintained but the levy was increased to 4.5 per cent for salaries above CFA1m a month. Car and motorcycle tax was incorporated in the price of petrol. In April 1989, price controls and quantitative restrictions were lifted on a wide range of imported goods. Price controls were maintained on bread, sugar, rice, locally-brewed beer, mineral water, soap,

KEY FACTS

Official title: Gabonese Republic

Head of state: President Omar Bongo

Ruling party: Parti Démocratique Gabonais

Capital: Libreville

Official Languages: French (Fang widely spoken)

Currency: Franc CFA = 100 centimes

Exchange rate: Franc CFA316 per $ (Nov 1989)

Area: 267,667 sq km

Population: 1.09m (1988)

GDP per capita: $2,988 (1988)

GDP real growth: –12.3%* (1988)

Inflation: 6% (1988)

Trade balance: $404.3m (1988)

Foreign debt: $2.1bn (1987)

* estimated figure

cooking oil and petrol. Quota restrictions were retained for sugar, soap, cement and oil.

The EC lent $2.6m for the audit of 18 state companies in 1988 to provide the basis for decisions to restructure or privatise. In February 1989, the government decided to offer some of its 51 per cent stake in Societe Nationale de Bois Gabonaise (SNBG) to private investors, but to maintain the company's timber marketing and buying monopoly on the grounds of quality and the importance of timber to the economy. SNBG reduced its 1988 losses to CFA300m from CFA3.2bn in 1987, partly by cutting the number of staff from 691 to 500. Air Gabon, 80 per cent state owned, cut its 1988 losses to CFA4.5bn from CFA7.4bn in 1987. The company plans to become profitable again by 1990 through further staff reductions and other austerity measures.

The main problems under discussion between the government and Paris Club donors in the first half of 1989 were low oil prices, the unfavourable French franc - US dollar exchange rate, and the increasing cost of the debt service. The debt service is expected to rise to CFA173bn in 1989 (from CFA64bn in 1988) and again to around CFA200bn by 1991-92. Gabon pressed for a more growth-oriented programme in negotiations with the IMF in December 1988, but it seems unlikely that the IMF will agree.

In May 1989, the World Bank and African Development Bank (ADB) backed a $100m three-year programme for road maintenance. Italy had agreed CFA6bn for roads and CFA3bn for health in January 1989. UNDP granted $1.3m for technical assistance to strengthen the mines ministry and help implement prospection programmes. Total structural adjustment aid in 1988 included $50m from the World Bank, UA45m from ADB, CFA16bn from France, a CFA2.5bn grant from Canada and $2.6m from the EC.

Austerity and boom

GDP declined by 3.6 per cent in nominal terms to CFA983bn in 1988. Inflation was 6 per cent. Oil revenues were down by 14.6 per cent to CFA227.5bn. Total exports were probably below the 1987 level of CFA386.6bn. Government oil receipts, at CFA75.7bn, were 5 per cent below the budget estimate. The government was also only able to effectuate CFA12.1bn of the CFA38.6bn budget-cut target, but the budget deficit was reduced by 26 per cent to CFA78.1bn.

1989 budget expenditure was set at CFA358bn, about the same level as final expenditure in 1988. Total revenue was set at CFA271bn, CFA184bn from taxes, CFA65bn from oil and CFA22bn in other non-fiscal receipts, giving a budget deficit of CFA87bn. CFA96bn was earmarked for investment spending, CFA17bn for restructuring and privatisation of state companies and the rest for projects.

Oil output was 7.9m tonnes in 1988, roughly the same as in 1987, with exports valued at CFA280bn. Gabon's OPEC quota was raised to 166,000 barrels a day (b/d) in November 1988 and again to 175,000 b/d in spring 1989, equivalent to 8.75m tonnes a year. The government entered negotiations with OPEC for a substantial further increase. The commissioning of the Rabi-Kounga onshore oilfield in January 1989 greatly increased Gabon's production capacity. When the new field is up to full capacity in 1990, it will produce about 6m tonnes a year. The offshore Obando field operated by LAMSO and British Gas (which bought out Tenneco's Gabonese interests in late 1988) reached full capacity in 1988 with production of 2.2m tonnes. Obando and the traditional oilfields, where production is falling, are expected to pump some 6m tonnes in 1990, which, with Rabi-Kounga output, would far exceed Gabon's OPEC quota. 1989 production is expected to be within quota. Total reserves are estimated at 100m tonnes.

The opening of Rabi-Kounga, operated by Shell and Elf, fuelled new investment by an increasing number of oil companies attracted by the lucrative $5/b production cost of the onshore site. By mid-1989, 39 oil companies were established in Gabon. Exxon decided to set up its regional headquarters for West and Central Africa in Libreville. LASMO will invest $70m in offshore and onshore exploration in 1989-93. Elf made one offshore find and one onshore find in late 1988.

Timber production stabilised at 1.35m cubic metres in 1988 and export receipts were CFA50bn. Manganese production was 2.4m tonnes in 1988 and is expected to be 2.6m tonnes in 1989, 1.2m tonnes of which will be exported via Congo. The rest will be shipped through the new port of Owendo, officially opened in January 1989. Manganese reserves are about 200m tonnes and Gabon has 35-40 per cent of the world market. Uranium producer Comuf announced in late 1988 that it could increase production to full capacity of 1,500 tpa within 6-12 months if world market conditions improved. But export re-

KEY INDICATORS

	Unit	1984	1985	1986	1987	1988
Population	m	0.95	0.98	1.02	1.06	1.09
Gross domestic product (GDP)	CFA bn	1,535.8	1,645.8	1,173.4	1,035.0	970.0
GDP per capita	CFA Fr '000	1,616	1,679	1,150	976	890
Real GDP growth	%	–	16.1	–2.1	–12.9	–12.3
Inflation	%	5.9	7.4	6.3	–1.0	–
Agricultural production	1979-81=100	117.1	124.1	129.9	135.1	138.1
Food production	1979-81=100	89.9	96.6	106.8	103.7	108.5
Exports	$m	2,017.8	1,951.4	1,074.2	1,286.4	1,195.6
Imports	$m	733.2	854.7	979.1	731.8	791.2
Balance of trade	$m	1,284.6	1,096.7	95.1	554.5	404.3
Current account	$m	112.7	–162.5	–1,057.6	–449.1	–615.6
Total external debt	$m	909	1,113	1,608	2,071	–
Debt service ratio	%	11.1	10.6	13.9	5.1	–
Exchange rate	CFA per $	437.9	449.3	346.3	300.5	297.8

* estimated figure

ceipts suffer from the unfavourable French franc – US dollar exchange rate as well as low prices. Important phosphate reserves of at least 50m tonnes were discovered in central Gabon at end-1988, but commercialisation will be difficult on account of transport problems. Cocoa production more than doubled in 1987-88 to 2,305 tonnes by July 1988, but the government stated that the Cocoa Fund, which gives producers a guaranteed price, would be exhausted by 1989-90 without a revival in world prices.

Outlook

Further Paris Club rescheduling seems likely given Gabon's good behaviour in cutting public spending and the losses of state companies such as Air Gabon and SNBG. With new oil production and discoveries, governments, multilateral agencies and commercial banks will continue to lend. The setting up of new oil company operations, such as Exxon's regional headquarters, and hot competition for the award of new exploration and production licences could bring a welcome boost to government receipts from the sector.

With this support, the government programme to reduce the debt service from 1994 stands a good chance of success. However, estimates from the government's independent financial advisors of 7 per cent annual growth in 1989-92 seem over-optimistic. Austerity could give rise to greater unrest as the cuts begin to have their full effect. It is likely that the government will take steps to soften the impact as much as possible without alarming the IMF. The government will have to take major decisions on the fate of state companies in 1989-90. The private sector will be put to the test on whether or not it is willing to buy into government companies. Oil companies, such as Shell, which have diversified worldwide operations to include other natural resources, may be persuaded to invest in export products such as timber.

COUNTRY PROFILE

Political system: Executive power rests with the president (elected by universal adult suffrage for seven years) who appoints and leads a council of ministers, including a prime minister. The legislature is the National Assembly (111 elected members). Gained independence in 1960.

Political parties: The Parti Démocratique Gabonais is the only legal political party. The party Congress sets policy, but in addition there is a 27-member political bureau and a 253-member central committee. The political bureau can override the council of ministers by issuing decrees. The central committee operates in an advisory capacity. There are four specialised organisations within the party, the École des Cadres du Parti, Union des Jeunes du PDG (UJPDG), Union des Femmes du PDG (UFPDG), and the Fédération des Syndicats Gabonais (FESYG).

Population: The 1.3m inhabitants (1988) comprise some 40 different ethnic groups, of which the Fangs are the largest (40 per cent of the total) and the Bapounous (20 per cent) are also highly significant. Nearly half of the population live in the urban centres. There are some 25,000 Europeans, mainly of French nationality.

Main cities/towns: Libreville (population 400,000), Port-Gentil (200,000), Lambaréné (50,000), Franceville (200,000).

Language: French is the official language and the medium for all documentation; African languages such as Fang are widely spoken.

Media
Press: Main daily is the *Gabon-Matin*; others include *L'Union*. Several periodicals are published (mainly economic and government).
Radio: La Voix de la Rénovation broadcasts 18 hrs/day. Also a provincial network. Broadcasts in French and African languages. Africa No.1 and Radio II also broadcast 18 hours a day.
Television: National service operated by Radiodiffusion-Télévision Gabonaise. Programmes are transmitted by satellite to other African countries.

External trade: A sharp fall in oil earnings since 1986, and higher import costs, have resulted in a negative trade balance, reversing earlier surpluses. The current account deficit has increased six-fold between 1985 and 1986.
Exports: Petroleum (83 per cent of total value in 1985), timber, manganese, uranium. Main destinations: France (43% of total value in 1986), USA (17%), Spain, Brazil.
Imports: Machinery and transport equipment (typically 30-32% of total value), food and agricultural products (13-14%), manufactured goods (10-12%), metal products (10-12%). Main sources: France (56% of total value in 1986), USA (3%).

Agriculture: Sector contributes only four per cent to GDP and suffers from urban drift, underinvestment and weak rural infrastructure. Less than one per cent of total area under cultivation. Around 65 per cent of food requirements are imported. The Fifth Development Plan (1984-90) aims to increase production of traditional crops (cassava, maize, groundnuts, vegetables) and export-orientated crops (cocoa, coffee, sugar-cane, palm oil and rubber). Thick forests cover 76 per cent of the country and exports of timber, mainly okoumé wood, have risen in recent years. Particularly suitable for making plywood, okoumé accounts for three-quarters of total wood production, and 30 per cent is processed locally into plywood. Fishing sector offers considerable potential, but remains underdeveloped.

Industry: Manufacturing sector is relatively small, contributing around five per cent to GDP. Activities include wood processing, foodstuff industries, chemicals, ship repair, metal working, textiles and clinker crushing. Industrial expansion is hampered by a narrow domestic market, shortages of skilled labour and weak infrastructure. Construction industry contributes around eight per cent to GDP.

Mining: Activity is concentrated on extraction and export of manganese ore (reserves 200m tonnes) and uranium (reserves 35,000 tonnes). Both are crudely refined before export, the manganese as a 51 per cent concentrate and the uranium as 74 per cent pure 'yellow cake'. Manganese goes mainly to Europe, but also to USA and the Far East. Uranium goes mainly to France (about 10 per cent of France's requirements), the rest to Belgium and Japan. Large deposits of iron ore, barytes (used in paint-making) and niobium await the completion of the Trans-Gabon railway to make them accessible.

Hydrocarbons

Energy: In 1986 electricity production was 897.1m kWh, with 195.7m kWh thermally produced, the balance of 701.4m kWh by hydro-electric plants. The Tchimbélé Dam controls the flow of water to the Kinguété plant, which supplies the Libreville area with electricity, producing 337m kWh in 1986. In addition, the Tchimbélé plant produced 219m kWh in that year.

Membership of international organisations: African Development Bank, African States in Association with the EC (Lomé Convention), Central African Customs Union, Economic Commission for Africa, French Monetary Area, GATT, IAEA, IDA, IFAD, ILO, IMF, International Cocoa Organisation, Islamic Conference, Movement of Non-Aligned Countries, OAU, OPEC, UDEAC, UN, WHO, WIPO, World Bank.

BUSINESS GUIDE

Time: GMT +1.

Climate: Equatorial with an annual mean temperature of 28°C and high levels of humidity. The rainy seasons are between Oct and mid-Dec, and between mid-Jan and May. Dry season from Jun to Sep.

Entry requirements
Passport: Required by all except nationals of certain African countries with national ID card.
Visa: Required by all except certain West African countries.
Currency: No restriction on import of foreign and local currency but amounts must be declared. Export of local or foreign convertible currency limited to franc CFA200,000 in bank notes.

Health precautions
Mandatory: Yellow fever vaccination certificate required if arriving from an infected area.
Advisable: Yellow fever, cholera, typhoid and polio vaccinations. Malaria prophylaxis recommended as risk exists throughout the country. Water precautions should be taken.

Representation overseas: Abidjan, Abu Dhabi, Algiers, Bangui, Beijing, Belgrade, Bonn, Brasilia, Brazzaville, Brussels, Bucharest, Buenos Aires, Cairo, Caracas, Dakar, Doha, Jeddah, Kinshasa, Lagos, Lomé, London, Madrid, Malabo, Manila, Moscow, Nouakchott, Ottawa, Paris, Pyongyang, Rabat, Rome, Seoul, Tehran, Tokyo, Washington, Yaoundé.

Air access
National airline: Air Gabon
Other airlines: Air Afrique, Air Zaire, Cameroon, Ecuato Guineana de Aviacion, Lina Congo, Nigeria Airways, Royal Air Maroc, Swissair, UTA.
International airport: Libreville-Léon M'Ba (Code: LBV), 12 km from city; restaurant, currency exchange; Port Gentil (POG), 6 km from city.

Surface access
Main ports: Port Gentil, Owendo.
Overland access: Main roads connect Gabon with Cameroon, Congo and Equatorial Guinea. These are semi-surfaced but generally are in good condition and well maintained.

Hotels: Available in Libreville, Port Gentil, Lambarene and other main centres.Advisable to book in advance. Service charge usually included in bill, if not a tip of 10-15 per cent is usual. Most hotels accept major credit cards.

Electricity supply: 220-30 V AC, 50 cycles.

Credit cards: Can be used in some hotels and restaurants in Libreville, but generally not widely accepted.

Car hire: Available in main towns, at airports and through hotels. International driving licence required. Charges are high.

City transport
Taxis: Unmetered 'collective' and private taxis available in main towns; tipping is not usual; rates vary according to the time of day.

National transport
Air: Air Gabon and Air Affaires Gabon operate scheduled and charter flights to all main centres.
Road: Except for routes Libreville to Ndende, Booué and Bitam, roads can be difficult in rainy season. Under 10 per cent of total network (7,100 km) is asphalted.

BUSINESS GUIDE

Buses: Regular coach and minibus services link Libreville with Lambaréné, Oyem, Mouila and Bitam. Some services are subject to rainy season conditions.
Rail: Regular services operate on the Transgabon railway linking Libreville with Booué, via Ndjolé and Franceville. Two classes; generally new rolling stock; air-conditioned railcars for some services but no refreshment or sleeping accommodation scheduled.
Water: A ferry service (taking two hours) has operated since August 1986 between Libreville and Port Gentil.

Public holidays: Additional holidays may be announced at short notice.
Fixed dates: 1 Jan, 12 Mar ('Anniversaire de la Renovation'), 1 May, 17 Aug (Independence), 1 Nov (All Saints'), 25 Dec.
Variable dates: Easter Monday, Eid al-Fitr, Whit Monday, Eid al-Adha.

Working hours
Government and business: (Mon-Fri) 0730 –1200, 1430 – 1800; (Sat) 0800 – 1300.
Banking: (Mon-Fri) 0730 – 1130, 1430 – 1630.
Shops: (Mon-Sat) 0800/0830 – 1200/1230, 1430/1500 – 1830/1900. (Some shops closed on Mon.)

Telecommunications
Telephone: Network comprises an international automatic exchange and 14 domestic automatic exchanges. Dialling code for Gabon: IDD access code + 241 followed by subscriber's number. There are no area codes.
Telex: Facilities available at Main Post Office in Libreville and at major hotels.

Banking
Central bank: Banque des États de l'Afrique Centrale.
Main centre: Libreville.
Commercial banks: Banque Internationale du Gabon, Banque Internationale pour le Commerce et l'Industrie du Gabon, Banque de Paris et des Pays-Bas Gabon, Banque Gabonaise de Développement, Union Gabonaise de Banque, Bank of Credit & Commerce International, Barclays Bank, Caisse Centrale de Coopération Économique, Citibank, Interbanque Libreville.

Advertising: Extremely limited facilities. The only active agency is Havas-Gabon.

Representation in capital: Belgium, Cameroon, Central African Republic, China PR, Congo, Egypt, Equatorial Guinea, France, German FR, Italy, Ivory Coast, Japan, Korea PDR, Rep. of Korea, Mauritania, Morocco, Nigeria, Philippines, Romania, Senegal, Spain. Togo, UK, USA, USSR, Venezuela, Yugoslavia, Zaïre.

Hotels: Libreville
Dowé (Novotel), Boulevard de l'Indépendance, P.O. Box: 56 (tel: 764750; tx: 5451).

Gamba, P.O. Box: 10074 (tel: 732736; tx: 5373).

Monts de Cristal, P.O. Box: 554 (tel: 762520; tx: 572460).

Okoumé Palace Inter-Continental, Boulevard de Nice, P.O. Box: 2254 (tel: 732619, 732185; tx: 5271; fax: 731629).

Rapontchombo (Novotel), P.O. Box: 48 (tel: 764742; tx: 5350).

Sheraton, P.O. Box: 4064 (tel: 742929; tx: 5432; fax: 766161).

Other useful addresses
ADL (Aeroport de Libreville), PO Box 363 (tel: 73-61-28; tx: 5459).

Air Affaires Gabon, PO Box 629, Franceville (tel: 73-20-10; tx: 5360).

Air Gabon, PO Box 2206, Libreville (tel: 76-08-25; tx: 5213).

Banque des États de l'Afrique Centrale, PO Box 1917, Rue du Docteur Jamot, Yaoundé, Republic of Cameroun (tel: 22-25-05; tx: 8343 KN).

Chambre de Commerce, d'Agriculture, d'Industrie et des Mines du Gabon, PO Box 2234, Libreville (tel: 72-20-64, 72-07-53; tx: 5554).

Compagnie Minière de l'Ogoué (Comilog), PO Box 578, Libreville (tel: 72-24-74).

European Development Fund, PO Box 321, Libreville (tel: 73-22-50).

Ministerie du Tourisme, PO Box 4031 (tel: 76-42-29).

Société Gabonaise de Financement et d'Expansion, PO Box 2151, Libreville.

Société Gabonaise de Raffinage, PO Box 530, Libreville (tel: 75-23-65; tx: 8217).

Société Equatoriale de Travaux Petrolières Maritimes, PO Box 493, Libreville (tel: 75-35-09; tx: 8225).

The Gambia

Kaye Whiteman

The period 1988–89 was politically quiet in the Gambia, at least on the surface, with President Dawda Jawara taken up with the chairmanship of the Economic Community of West African States (ECOWAS), the West African regional grouping. Below the surface, however, there were continued rumblings over the negative impact on public opinion of the continuing harshness of the economic recovery programme, and over charges of corruption against officials.

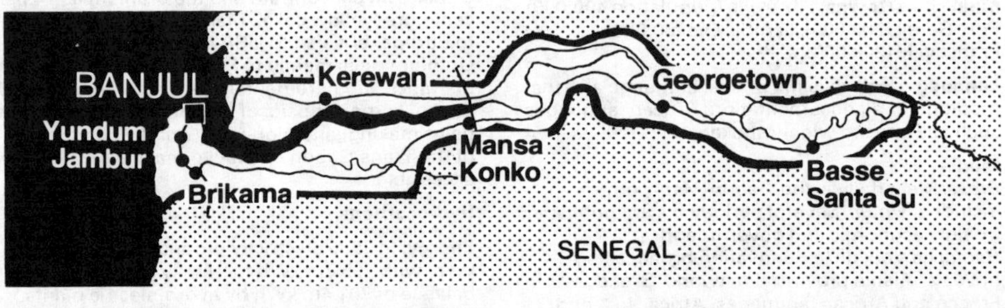

Chairman of ECOWAS

At the organisation's summit in Lomé in June 1988, President Dauda Jawara was made chairman of ECOWAS, an organisation to which the Gambia has always attached considerable importance, because of the need for a counterweight to offset too close a relationship with Senegal in the Senegambian Confederation. In this capacity, Banjul was host to a number of ECOWAS-related meetings. These included notably a meeting on indebtedness in November 1988 (it has been calculated that the total debt of ECOWAS' 16 member states is in the neighbourhood of $60bn, including notably both the Nigerian and the Ivoirian debt), and the budget session of the ECOWAS council of ministers in December 1988.

Unfortunately, it proved impossible for the Gambia to host the 1989 summit in Banjul, as a culmination of the Jawara chairmanship, despite an earlier intention, as the infrastructure in the Gambian capital was still not sufficiently developed, in spite of substantial hotel accommodation. So the summit went to Ouagadougou, and President Jawara did not continue a second year in the chair as some had predicted. Banjul has made a new commitment, however, to host the 1990 summit, which will also mark the fifteenth anniversary of the group.

Also, faced with a major crisis between two member states (Senegal and Mauritania), following riots in April 1989, Jawara was not able to use the ECOWAS chair to bring about any effective mediation, although the presidents of the two feuding countries did meet for a handshake at the ECOWAS summit in June. There were some limited cases of attacks on Moorish Mauritanians in the Gambia, and looting of shops, but on the whole the country was seen as a haven for Mauritanian refugees from Senegal.

A major development was the effective collapse of the eight-year old Senegambian Confederation, long considered to have been blocked on the economic and monetary front. Ironically, in November 1988, ob-

KEY FACTS

Official title: Republic of The Gambia
Head of state: President Sir Dawda Kairaba Jawara
Ruling party: People's Progressive Party
Capital: Banjul
Official Languages: English (Mandinka, Wolof, Fula spoken)
Currency: Dalasi (D) = 100 bututs
Exchange rate: D7.62 per $ (Nov 1989)
Area: 11,295 sq km
Population: 810,000* (1988)
GDP per capita: $290 (1987)
Inflation: 11.7% (1988)
Trade balance: –$31.0m (1988)
Foreign debt: $319m (1987)

servers felt that after years of stagnation, the economic side of the confederation seemed to have made some progress, with the announcement from Senegal that a free trade area would be set up, involving free circulation of goods originating inside the zone, although it was far from being the customs union envisaged in the original Senegambia Treaty. This slight progress made the abrupt Senegalese decision in August 1989 to 'freeze' the confederation and withdraw its troops from the Gambia, which had been there since they helped put down the 1981 putsch, the more surprising. It was seen as the measure of the changed political climate in Dakar, above all since the row with Mauritania.

Both the latter part of 1988 and the first half of 1989, the mid-term between elections, were quiet politically. The main interest was centred on two meetings, first in Banjul and then in the UK, between President Jawara and Sheriff Dibba, leader of the main opposition party, the National Convention Party (NCP), which holds five seats in the country's parliament, the only opposition party to have seats there. President Jawara, disclosing at the end of the year that there had been this contact, said that these meetings marked a new departure, and a 'change for the better'. Mr Dibba described them as being 'in the interest of national unity'. There was some speculation that the contacts might presage a coalition government or even a merger of the two parties, but there was no immediate follow-up. Nor did the president meet the leaders of the other two smaller opposition parties, the Gambian People's Party and the People's Democratic Organisation for Independence and Socialism.

Considerable interest was aroused by a libel trial that lasted for several months in which Sanna Manneh, journalist with the newspaper *the Torch* was accused of defaming several ministers, whom he had accused of corruption. In April 1989, the judge found him guilty on one count, but acquitted him on two others. And on the count on which he was found guilty, the judge ruled that there had been no intention to defame, and cautioned the journalist before discharging him. This was seen in Banjul as a virtual acquittal. No move was made against the ministers involved, however. One of them, Saihou Sabally, was promoted from agriculture to finance, after the resignation because of illness of the incumbent minister Sheriff Sisay in February 1989. (Sisay, who had been a founder member of the People's Progressive Party (PPP) with Jawara in the late 1950s, died shortly after). Concern at corruption among officials, however, led to at least 12 officials including three parastatal managing directors' being investigated or sent on leave. A presidential inquiry was set up to look into allegations of corrupt enrichment against Works and Telecommunications Minister Momodou Cham.

The major preoccupation of the Gambian Government continued to be the state of the economy. In November 1988, the International Monetary Fund (IMF) approved a three-year Enhanced Structural Adjustment Facility (ESAF) at about the same time as a similar facility was offered to Senegal. It involved 20.52m SDRs, of which 6.84m were available straight away. ESAFs are usually made available to countries which have met with IMF approval over a period of years. The Gambia introduced its economic recovery programme in 1985, and it was subject to conditionalities on usual IMF policy lines, such as elimination of subsidies. The ESAF approval cleared the way for a number of other international donors, such as the USA and the UK, to make new funds available.

At the same time, the World Bank made available a new $23m structural adjustment loan (with other donors the total amount available was $34m), to assist in the liberalisation of the marketing of groundnuts during the 1989–90 season, which would open the market to all private traders, without a fixed producer price, in which traders would compete for maximum purchase.

KEY INDICATORS

	Unit	1984	1985	1986	1987	1988
Population	'000	720	740	770	790	810
Gross domestic product (GDP)	$m	205	170	191	229	–
GDP per capita	$	285	229	248	290	–
Inflation	%	22.1	18.3	56.6	23.5	11.7
Agricultural production	1979-81=100	106.6	104.8	107.3	109.6	111.8
Exports	$m	90.77	62.86	64.03	65.45	74.11
Imports	$m	99.48	74.91	83.48	95.03	105.16
Balance of trade	$m	–8.71	–12.06	–19.44	–29.58	–31.05
Total external debt	$m	230.1	245.7	270.7	318.9	–
Debt service ratio	%	6.8	2.1	9.9	12.9	–
Exchange rate	Dalasi per $	3.58	3.89	6.94	7.07	6.71

COUNTRY PROFILE

Political system: Executive power rests with the president, who is elected every five years. Legislative power is vested in the 49-member House of Representatives. Gambia gained independence from the UK in 1965. The Federation of Senegambia, established in 1982, has created the framework for special links with Senegal.

Political parties: The ruling party is the People's Progressive Party (PPP),

Population: The 788,000 population (1988 estimate) includes an expatriate (largely European) contingent of some 50,000.

Main cities/towns: Banjul (capital) (population 45,000, plus more than 110,000 in the neighbouring conurbations around Kombo St Mary). Other towns are Brikama, Gunjur, Soma, Farafenni, Bansany, Georgetown, Basse.

Language: English is the official and commercial language; some French is spoken.

Media
Press: Main newspapers are *The Gambia Weekly* (est. 1989), a weekly government publication, and *The Gambia Times*, published fortnightly by the ruling party. Other publications are *The Nation, Gambia Onwards, The Torch, Toiler* and *The Gambian.*
Radio: Radio Gambia, a government non-commercial station, operates 19 hours daily in English and Gambian languages. The commercial station, Radio Syd, also broadcasts in English, French and African languages, and provides tourist information in German and Swedish.
Television: Senegalese transmissions are received.

External trade: Heavy reliance on bilateral and multilateral aid to offset regular balance of payments deficits. Groundnuts and tourism are main sources of foreign exchange.
Exports: Groundnuts (about 50 per cent of total value). Main destinations: Switzerland, Italy, UK, France, Belgium.
Imports: Food, live animals, beverages, basic manufactures, machinery and transport equipment, mineral fuels. Main sources: UK, France, German FR, Netherlands, Japan, USA, China PR.

Agriculture: Agriculture provides half of GDP and employs 80 per cent of the workforce. Groundnuts provide around half of all export earnings. Production of food crops (rice, maize, millet, sorghum, cassava) is insufficient to meet local needs, but receives a great deal of official encouragement. Small-scale fruit and cotton farming is important and some livestock are exported to neighbouring countries for breeding. Fishing has also increased in importance with the annual catch rising to over 11,000 tonnes. Illegal fishing by foreign trawler fleets remains a problem. Current emphasis is on increasing cotton, livestock, rice and fish production, and irrigating swamp areas around the River Gambia.

Industry: Manufacturing sector is small-scale and under-developed. Main activities (most of which are centred around Banjul, particularly in the Kanifing Industrial Estate) include groundnut and fish processing, brewing, footwear, perfume, cement and brick production. The government has recently set up an industrial estate outside Banjul, and aims to attract foreign investors interested in light manufacturing. Tourism has made strong growth to become the second biggest foreign currency earner. Tourist arrivals in 1986-87 were 10 per cent up on the 78,465 arrivals in 1985-86. There were 112,381 tourist arrivals in the 1988-89 season.

Mining: Known deposits of kaolin, tin, ilmenite, zircon and rutile, but these are unexploited.

Energy: Total dependence on imported petroleum. Fuelwood is the main source of domestic energy. Hydroelectric resources currently being developed on the River Gambia.

Membership of international organisations: African Development Bank, African Groundnut Council, African States in Association with the EC, Arab Bank for Economic Development in Africa (recipient), Commonwealth, Economic Commission for Africa, Economic Community of West African States, GATT, IDA, IFAD, IFC, IMF, Islamic Conference, Permanent Inter-State Committee on Drought Control in the Sahel, OAU, OMVG, UN, WHO, World Bank.

Time: GMT.

Climate: Sub-tropical with distinct dry and rainy seasons. Dry season from Nov to May with average temperatures around 21-27°C. The dry Harmattan wind keeps the humidity low. Rainy season from Jun to Oct with high humidity and average temperatures around 26-32°C.

Entry requirements
Passport: Required by all.
Visa: Required by all except nationals of Commonwealth countries, ECOWAS member countries, and some European countries.

BUSINESS GUIDE

Currency: No restrictions on import of foreign or local currency, although a declaration form must be completed on arrival. Export allowed up to amount declared on import, plus D75 in local currency and the equivalent of D250 in other currency. Thriving black market in hard currency is best avoided.

Health precautions
Mandatory: Yellow fever vaccination certificate and cholera vaccination certificate required by all.
Advisable: Cholera, Typhiod and polio vaccinations. Malaria prophylaxis should be taken as risk exists throughout the country. Water precautions are necessary.

Representation overseas: Brussels, Dakar, Freetown, Jeddah, Lagos, London, Washington.

Air access
National airline: Gambia Airways does not operate any aircraft – it acts as an agent.
Other airlines: Air Gabon, Air Senegal, British Airways, Ghana Airways, Nigeria Airways. Gambia Air Shuttle operates two aircraft.
Tax: Airport tax of about $12 payable on departure.
International airport: Banjul-Yundum International (Code: BJL), 29 km from city; bar, bank, post office, shop.

Surface access
Main ports: Banjul.
Overland access: Road access to Banjul possible from Senegal (Dakar), by the Trans-Gambia Highway and the South Bank Trunk Road, or more directly, via the car ferry from Barra to Banjul. Gambia Public Transport Corp provides a daily luxury coach service from Senegal across country to Banjul.

Hotels: Around 20 hotels (3,000 beds) in Banjul and on the coast. Book well in advance, especially if arriving during tourist season (Nov-May). Many Gambian hotels are geared to package holidays. 10 per cent tip is usual.

Electricity supply: 240 V AC, 50 cycles.

Credit cards: Access, Visa accepted by certain hotels.

Car hire: National or international driving licence accepted. Car hire facilities are somewhat limited and local enquiries (e.g. to Ministry of Information & Tourism) are advised.

City transport
Taxis: Fares fixed by government regulations and tend to be expensive. Tariff is displayed by taxi drivers. However it is advisable to agree fare in advance.

National transport
Road: 3,000 km of roads of which 450 km are paved. Highways run along each bank of the River Gambia; the Trans-Gambia highway runs north to south, crossing the river at Farafenni-Mansa Konko (car ferry); other inland roads may become impassable in the rainy season.
Buses: The Gambia Public Transport Corporation (GPTC) operates cheap and reliable services linking Banjul with the coastal hotel area and other main centres. There are several commercial bus services such as Arudalaye and Transgambia services.
Water: An hourly car ferry service operates between Banjul and Barra (journey time – 20-30 minutes).

Public holidays
Fixed dates: 1 Jan, 1 Feb (Confederation Day), 18 Feb (Independence), 1 May (Labour Day), 15 Aug (Assumption), 25-26 Dec.
Variable dates: Good Friday, Easter Monday, Eid al-Fitr, Eid al-Kabir (Eid al-Adha), Ashura, Maulid al-Nabi.

Working hours
Government and business: (Mon-Thu) 0800/0900 – 1600/1700; (Fri) 0800 – 1230.
Banking: (Mon-Thu) 0800 – 1300; (Fri) 0800 – 1100.
Shops: (Mon-Thu) 0800/0900 – 1200, 1400 – 1700; (Fri) 0800/0900 – 1200, 1500 – 1700; Sat mornings only.

Telecommunications
Telephone: Dialling code for The Gambia: IDD access code + 220 followed by subscriber's number. There are no area codes. 24-hour service on calls to UK.
Telex: Facilities are available at the Post Office, Marina Road, Banjul. 24-hr service at the Gamtel Co office in Banjul.

Banking: *Central bank:* Central Bank of the Gambia.
Others: Banque Internationale pour le Commerce et l'Industrie du Sénégal, Gambia Commercial and Development Bank, Standard Chartered Bank of Gambia.

Representation in capital: China PR, Iraq, Korea PDR, Mauritania, Nigeria, Senegal, Sierra Leone, USSR, UK, USA.

Hotels: Banjul
Atlantic, Marina Parade, P.O. Box: 296 (tel: 28601; tx: 2250; fax: 27861).

Carlton, Independence Drive, P.O. Box: 639 (tel: 27258).

Wadner Beach, Mile 2 (tel: 8119; tx: 2219).

Other useful addresses
Banque Internationale pour le Commerce et l'Industrie du Sénégal, Wellington and Picton Street, Banjul (tel: 28145).

Central Bank of the Gambia, 1-2 Buckle Street, Banjul (tel: 28103; tx: 2218).

Central Statistics Office, Central Bank Building, Buckle Street, Banjul (tel: 28105).

Commercial and Development Bank, Leman and Buckle Street, Banjul (tel: 28865).

German Democratic Republic

Michael Simmons

In the space of just a few weeks in the autumn of 1989, the GDR changed from being one of the most stable, predictable and conservative members of the Warsaw Pact/CMEA alliance to being one of the most intractably democratic countries on the new European landscape. Communism as a controlling force was dispensed with and new market-oriented forces, and West German investment intentions, took over in the economy. The old elite, shown up as hugely corrupt and fallible, were swept aside and a fundamentally new leadership, almost unknown before the political earthquake took place, set about achieving a special relationship with the Federal Republic.

Clearly, it will be years before 'the new order' settles down, politically and economically, and after the initial heavy spadework by New Forum as an opposition, a series of multifarious bodies began to emerge – at the end of 1989 – determined that all vestiges of the corrupt past should be rejected. For a country which once had the strongest Communist Party outside the Soviet Union, the break with the communists was necessarily traumatic. The basic unevenness of the relationship with West Germany is likely to persist for several years.

The last phase of Erich Honecker's tenure as leader of the Socialist Unity Party (SED) of East Germany began in the wake of his hugely sucessful visit to West Germany. He was suddenly confronted with the realities of East Germany's faltering economy, with internecine differences within the Warsaw Pact which led to some reformists baying for his blood and with his own advanced age. He was born in 1912 to working-class parents when Kaiser Wilhelm was still fairly secure on his throne. In the Stalinist 1930s, he spent a brief but formative period in Moscow absorbing thinking he never managed to shake off. During the Nazi period he was arrested and was imprisoned for the duration of the war. Such conditionings deeply influenced many of his speeches on German and European issues. With his departure, frustrations that had undoubtedly built up among middle management and its equivalent in the civil service bureacracy of the GDR – notably among men and women in their forties, very much the post-war generation – since Mikhail Gorbachev became Soviet leader in early 1985, meant that fundamental change in political and economic management had to come fairly soon.

KEY FACTS

Official title: The German Democratic Republic

Head of state: President Manfred Gerlach

Head of government: Prime Minister Hans Modrow

Ruling party: Government of National Responsibility (until March 1990 elections)

Official Languages: German

Currency: DDR Mark = 100 pfennig

Exchange rate: mark 1.77 per $ (Dec 1989)

Area: 108,333 sq km

Population: 16.67m (1988)

GDP per capita: mark 16,105 (NI, 1988)

GDP real growth: 8.2% (NI, 1986)

Labour force: 8.6m* (1988)

Trade balance: mark 3.0bn (1988)

* estimated figure

Honecker's warning

An important moment of truth – as opposed to propaganda and window-dressing – came with the meeting of the party central committee in East Berlin in early December 1988. Until that moment, the usual stance of the East Germans to the Gorbachev–reformist way of thinking had been one of qualified defiance and satisfaction with the status quo. The GDR, it was constantly reiterated, had the most stable and, arguably, the most successful economy in the entire CMEA area. It did not matter that there was a special and hugely preferential trading arrangement with West Germany; it did matter that growth rates were consistently on an upward curve.

At this meeting, Honecker showed that he was far from satisfied with the country's economic situation. Not only were important macro-economic plan targets not being met, he declared, but investment planning, farm production and energy consumption were all showing trends which were a cause for deep concern. A forthright dressing down of this nature is a rare event in the GDR calendar. In farming, the party leader warned that there had been shortfalls in grain and sugar beet production as well as 'indefensible losses' in other areas. In capital investment, there had been intolerable delays in completing key projects – a development which could only have a knock-on effect on other areas of the economy. So far as energy was concerned – and this has long been a sensitive area – his warning was that far too much was being consumed.

There was a perceptible slowing down in real economic growth rates during 1988 but, as if in response to the admonitions and exhortations of the party leader at the preceding year's end, there was some recovery during the first half of 1989. Provisional statistics published, with the usual prompt efficiency, at the beginning of July 1989, showed that national income in the year's first six months was 4 per cent higher than in the comparable six months of the previous year. Industrial output, according to these figures, was 4.4 per cent higher than in the first half of 1988 and 'net production and productivity' increased by 6 per cent. Costs were cut by 1 per cent, which in global terms must be a significant figure. How these cuts were achieved was not made clear.

In the construction industry, according to the statisticians, there was a 4.3 per cent improvement while productivity went up by only 4.8 per cent (which is considerably lower than the growth just noted in industry). However, official pleasure was recorded at the 15 per cent increase – to well over 31,000 – in the number of office and personal computers produced, alongside roughly comparable increases in related computer and robotic equipment.

In agriculture, which had been another hugely sensitive area at Honecker's December talk-in, it was noted that state purchases of fatstock, milk and eggs were even higher than expected. But, it was also noted, somewhat darkly, that 'considerable efforts' were now being put into the further irrigation of productive land. Drought in the Schwerin area was a problem in mid-1989.

Trade became an increasingly vexed question towards the end of 1989. The wholesale shift to economic reform in Czechoslovakia, Hungary and Poland as well as the Soviet Union – meant that bilateral commercial relations with most CMEA partners were more difficult than hitherto and that multilateral relations, as thrashed out interminably in the committee rooms of CMEA itself, were almost impossible. Not only were there arguments about quality and standards of goods exchanged – arguments with far-reaching implications – but there were also increasingly open disputes about pricing and availability of specific items. With West Germany, however, the gradually improving political climate was accompanied

KEY INDICATORS

	Unit	1984	1985	1986	1987	1988
Population	m	16.6	16.6	16.7	16.7	16.7
GDP (1985 prices)	mark bn	722.5	748.6	770.0	789.4	810.9
National income (1980 prices)	mark bn	222.1	232.8	252.2	261.2	268.4
NI per capita (1985 prices)	mark	13,791	14,532	15,172	15,695	16,105
NI real growth	1970=100	189	199	207	215	221
Labour productivity	1980=100	127	138	150	160	171
Gross industrial production	1970=100	204	213	221	228	235
Oil, gas and coal	1970=100	209	218	221	229	233
Plastics	1970=100	272	278	273	269	297
Chemicals	1970=100	275	283	284	292	307
Metallurgy	1970=100	190	196	204	210	209
Machine tools	1970=100	317	344	364	387	404
Electronics	1970=100	513	616	702	782	872
Engineering	1970=100	288	289	315	349	369
Precision mechanics	1970=100	401	442	457	530	520
Construction	1970=100	158	162	167	170	176
Exports	mark bn	90.4	93.5	91.5	89.9	90.2
Machinery and equipment	%	47.0	46.6	46.1	48.0	47.6
Fuels, minerals and metals	%	17.5	20.0	21.0	16.7	15.1
Industrial consumer goods	%	15.1	14.1	15.0	16.0	16.4
Imports	mark bn	83.5	86.7	90.5	86.6	87.2
Machinery and equipment	%	26.0	26.8	28.0	34.1	37.0
Fuels, minerals and metals	%	41.6	42.5	40.3	38.0	33.5
Industrial consumer goods	%	5.7	6.2	6.6	5.7	5.7
Balance of trade	mark bn	6.9	6.8	1.0	3.3	3.0
Exchange rate	mark per $	2.70	2.30	2.18	1.80	1.76

* estimated figure

by gradually improving trade relations. The unique and massively preferential tariff arrangement that exists between the two Germanies – and in many sectors no tariff at all – continued to be of special importance to the GDR. This meant that in the first four months of 1989, for example, both countries' statisticians were able to point to big increases in bilateral turnover. On West German figures, East German exports to West Germany were up by 5 cent in the first four months of the year, while imports from West Germany to the GDR were up by all of 13 per cent.

In the course of the 1986-90 Five-Year Plan, the faltering of the economy led among other things to a lowering of sights in terms of targets. The average planned annual increase in the national income, set during the year 1985, had been 4.5 per cent, but once the plan period had reached the half-way point, it was realised that these figures were unrealistic – a realisation which came about partly as a result of more difficult trading terms with CMEA partners. A consequence of this realisation was that by 1989 the expected growth rate had fallen to 4 per cent. The adjustments, taken together, meant that pretty certainly the Five-Year Plan targets as a whole would not be met. The authorities' way of presenting this unpalatable and out-of-the-ordinary development were expected to be fascinating to behold.

Reform factors

History will look back on the second half of the 1980s in Europe as the time when reform was constantly at the top of the agenda. In East Germany everybody acknowledged this to be the case, but those in authority – and especially those closest in thinking to Honecker himself – stuck to their guns and refused to follow the Gorbachev path in this direction. They preferred instead to congratulate the Soviet leadership politely on the very necessary changes it had been bringing to the Soviet Union itself and to concentrate, in the case of the GDR, on one of the other Gorbachev maxims – that each of the socialist countries should be allowed to follow its own path. The collapse of the Honecker regime was evidence that, having tried too hard, its fall would be a very heavy one.

Thus, in April 1989, an important commentary in the party newspaper, *Neues Deutschland*, underlined that the GDR would continue for the foreseeable future to hold firm to the principle of the planned economy. (The Gorbachev aim is to reduce the role of 'the Plan' to the minimum.) *Neues Deutschland* added that 'both sides of the dialectic inter-relationship', central planning and management on the one hand and greater autonomy for the producers on the other, were in fact being improved 'with the goal of showing to even greater advantage in our country the merits, ideals and values of socialism'. Indeed, the newspaper article went further, and pointed out that this 'system' was being further developed, to include 'the principle of self-generation of resources'. This particular principle, it was stated, would direct the interest of the producers to increased peak performances in goods and technology and to reducing costs. The short-term aim now was to bring more plant managers round to this way of thinking so that more disposable national income would be generated. 'The only concern,' said *Neues Deutschland*, 'is the benefit of this for our economic and social policy.' This harnessing together of 'economic' and 'social' policy was the mainstay of Honecker's thinking ever since 1971.

After the local elections, held throughout the country shortly after the *Neues Deutschland* article appeared, Honecker came forward at the meeting of the State Council (of which he was chairman) to 'assess' the results as an indication of faith in the tried and trusted policies aimed at the welfare of the people, peace and socialism. His effusive thanks took in a mention of the 'comprehensive dialogue' on basic issues which was characteristic of 'socialist democracy' as practised in the GDR and which provided 'a sound foundation' for mastering the tasks to come. However ponderous such thinking might sound to some, it was, nevertheless, an important element in Honecker's Germany. While 'reform' went ahead by leaps and bounds, however faltering, in Hungary, Poland and the Soviet Union itself, the note sounded in East Germany remained one of caution. In June 1989 Joachim Herrmann, in charge of propaganda at the SED central committee, publicly expressed deep anxiety which he said was affecting the people of the GDR at the 'developments' down the road in Hungary. Even in 1988, party officials in East Berlin castigated fellow East Germans for grumbling about GDR conditions when in fact they didn't know how lucky they were. 'In Hungary,' the argument ran, 'they talk and talk, but they don't produce. In Poland and the Soviet Union, there is all sorts of dissent. Why should the GDR go in that direction...?' Symptomatically, Mr Herrmann was one of the very first Politburo members to be sacked – on the same day as Mr Honecker himself.

One area where 'reformed' thinking was making headway was in the environment. Statements from the leadership on short-term economic developments were regularly laced with promises to do more about the environment and to combat pollution. Cleaner tap-water had become a significant preoccupation. Concrete moves, in the form of new equipment, some of it imported, to cut down on sulphur dioxide emissions – responsible for acid rain – were being taken at several of the country's bigger, and dirtier, manufacturing plants. In addition, partly as a result of some dissident activity, a more open and critical public discussion on environmental questions was being encouraged, though with the caveat that reducing pollution is an extremely costly business and money is not plentiful.

Gorbymania

It was the visit by Mikhail Gorbachev to the Federal Republic in June 1989 that had an unsettling effect in East Germany. Almost anyone in the country who wanted to – and that meant almost everybody – could turn an electronic knob during that visit and watch every recorded moment of it on television from his or her armchair. The contrast between the popular wel-

COUNTRY PROFILE

Political system: In 1989, a socialist state in which the supreme organ of state power is the Volkskammer (People's Chamber), a unicameral body of 500 members. Executive branch of government is the Ministerrat (Council of Ministers) whose chairman is appointed by the Chamber. Before the dramatic events in late 1989, five parties, mass organisations and many non-party individuals formed the National Front, led by the Socialist Unity Party (SED). Parliamentary elections planned for March 1990.

Main cities/towns: Berlin (population 1.2m), Dresden, Karl-Marx-Stadt, Leipzig, Magdeburg, Erfurt, Potsdam, Rostock, Halle, Gera, Schwerin, Cottbus.

Media
Press: 129 provincial dailies in the 14 districts with combined circulation of 9.3 million. Leading dailies *Neues Deutschland* (Berlin, circulation 1,092,811), *Berliner Zeitung* (Berlin, 203,653), *Junge Welt* (Berlin, 1,380,477), *Sächsische Zeitung* (Dresden, 560,800), *Das Volk* (Erfurt, 401,236), *Leipziger Volkszeitung* (485,000), *Ostsee-Zeitung* (Rostock, 292,469). Also more than 30 weeklies and 542 journals/periodicals.
Radio: Local, national and external radio service. Radio Berlin International broadcasts in 11 languages on medium and short-wave.
Television: Two major TV channels transmitting over 176.8 hours programmes per week. West German TV programmes also received.

Domestic economy: Closely integrated with that of the USSR and other CMEA countries. Virtually all economic sectors state-owned. Management of economy is highly centralised, pattern of growth determined by five-year plan. Emphasis in recent years on full exploitation of modern technology, particularly in export-orientated industry. Economic growth averaging 4.3 per cent p.a. between 1978-84, and about 4.0 per cent p.a. between 1985-88.

Employment: Around 37.4 per cent workforce engaged in industry, 10.8 per cent in agriculture and 10.3 per cent in trade. Labour shortages in some sectors.

External trade: Foreign trade plays a very important role in the economy of the GDR, with more than 50 per cent of the national income being redistributed through foreign trade channels. Machinery and equipment are the most important export items of the German Democratic Republic. These include electrical engineering equipment, machine tools, textile machinery, printing machinery, electronic equipment, and machinery for the food processing and chemical industries. With regard to imports, more than 60 per cent of the overall requirements of raw materials are imported. Various foodstuffs and consumer goods are also imported, as well as a great amount of machinery and equipment. The CMEA member countries account for around two-thirds of the GDR's foreign trade turnover, followed by the developed market-economy countries and the developing countries.

Agriculture: Approximately 6.2 million ha of agricultural land, of which 40 per cent given to cereals. Economic importance of sector declined, despite increased specialisation and higher productivity. Main crops: barley, wheat, rye, potatoes, sugar-beet. Animal husbandry also significant.

Industry: Mainstay of economy, contributing 72 per cent of GDP. Most important activities: engineering, vehicle assembly, construction, electronics, chemicals, metallurgy, machine tools, food processing. Industrial enterprises organised into over 126 centrally managed combines and another 85 combines belonging to industries managed at county level.

Mining: General lack of mineral resources, though self-sufficiency in lignite, rock-salt, potash and fluorspar. Tin and copper mining also significant.

Energy: Most domestic oil requirements supplied by pipeline from USSR. More than 80 per cent electrical energy supplied by locally mined lignite. Also rapid development of nuclear energy (to supply 15 per cent of needs by 2000) and natural gas. Successful programme of energy conservation undertaken over last three years.

Banking: State-controlled system headed by Staatsbank der Deutschen Demokratischen Republik. Visitors may open a foreign exchange account with Staatsbank.

Legal system: Civil law system modified by Communist legal theory.

Membership of international organisations: CMEA, IAEA, ICES, ILO, IMO, IPU, ITU, UN, UNESCO, UPU, Warsaw Pact, WFTU, WHO, WIPO, WMO, WTO.

come for the Gorbachevs from ordinary West Germans and the reserve that was publicly expressed by many East Germans would not have escaped the average viewer. In spite of this, the East German media tended to play down the visit – concentrating more on what the Soviet leader had to say about the GDR's 'contribution to peace' and on general principles underlying the visit than on the substance. In doing so they managed to accentuate the growing isolation of the country (and a small handful of others) in their continuing rejection of the Gorbachev road.

For worried members of the East German establishment this was, of course, a disconcerting development. The concern of the latter group was deepened by the conservative reaction of the leadership (including Mr Egon Krenz, who took over briefly after Mr Honecker's departure) as it spoke up in support of the Chinese Army and the way in which it had suppressed the 'counter-revolutionary' students in Tienanmen Square in the centre of Beijing in early June 1989.

But it was the Gorbachev visit to East Berlin for the 40th anniversary celebrations of the founding of the state which tipped the balance of popular discontent. Even while the Soviet leader was still in the country, the first demonstrations were forming in East Berlin, Dresden and Leipzig. The steady flow, which became a flood, of East Germans wanting to leave their country in order to settle in the west added force to the protestors' arguments, which grew more and more forceful each day. By the end of the birthday month (October), it was clear that an era in German history had ended, and that President Gorbachev, in his storm-tossed summit with President Bush, had once again pulled off something very extraordinary in international politics – though his precise role in the GDR will not be unfolded for a long time.

The Stalin factor

The lingering ghost of Stalin haunted the East German leadership in the late 1980s. Honecker's career would have been without foundation if Stalin had not been Soviet leader in his early days and if he had not backed Honecker's predecessor, Walter Ulbricht, in the first crucial days after the GDR state was founded in 1949. So it was perhaps inevitable as the pro-Gorbachev establishment in Moscow set about with a will dismantling the Stalin myth, that the East Germans went in the opposite direction. Ulbricht, according to the party newspaper in August 1988, should now be applauded as 'a revolutionary, internationalist, patriot and statesman'. Ulbricht's widow, Lotte, was fêted in the same tone on her 85th birthday. Nothing was said of Ulbricht's differences with Moscow – which were largely responsible for his abrupt departure from office in 1971 – and little was said of his period as national leader from 1949 on.

Stalin was also a major talking point in the Humboldt University of East Berlin. Here a group of progressive historians, working with pro-glasnost Soviet colleagues, uncovered all sorts of less than palatable details about how several hundred German communists were executed, on Stalin's orders, in the years before the Second World War. It was not an issue on which Honecker seemed willing to pronounce publicly, though he did ban the Soviet information magazine, *Sputnik*, at about the time the revelations were being made. The offending magazine included an article on the 1939 Stalin–Hitler non-aggression treaty and clearly did not show the communists in a very favourable light. The East German leadership defended the banning of the magazine, saying that it contained nothing that would strengthen German–Soviet friendship, and instead was offering 'distorted accounts of history.'

Economy

The national economy was showing signs of faltering in the last years of the 1980s, but the faltering was by no means all the responsibility of the GDR itself. External terms of trade, prevailing prices for essential imports, hiccoughs in the implementation of bilateral arrangements, a static work-force – all contributed. So, in the thinking of closet reformists, did the failure to reform management, a process which, once set in motion, would allegedly galvanise management to new heights of achievement and inculcate in them new objectives to be reached in the national interest.

Meanwhile, at every available turn, and especially when announcing economic results, the leadership did not hesitate to exhort the population to work still harder, to keep to stipulated delivery dates, and to sustain and improve upon the quality and the range of goods on offer. The 'velvet glove' element in these exhortations concealed consumer dissatisfaction with the range, quality and availability of goods in the shops and also impatience, at factory procurement level, with the flow of semi-manufactures and sometimes raw materials.

A series of indices not so frequently encountered

BUSINESS GUIDE

Time: GMT + 1 (GMT + 2 from late Mar to late Sep).

Climate: Temperate; warm summers, cold winters with temperature range from -3°C to 24°C.
Clothing: Lightweight for summer, medium weight and heavy topcoat for winter.

Entry requirements: Entry visa must be obtained before travelling unless traveller holds hotel vouchers. Entry possible from West Berlin at any time, in which case, permit can be obtained through any Interhotel. Transit visas issued for visit to East Berlin. Entry forms completed at customs point must be returned on departure. Visitors requested to register with authorities within 24 hours of arrival.
Health: Inoculation certificates may be requested from Asian and African visitors.
Currency: Minimum of M25 for each day of visit must be exchanged on entry. No local currency may be taken in or out. Evidence of spent currency may be requested on departure.
Customs: Gifts and personal effects duty-free.

Air access: Direct flights from some West European countries including Austria, Belgium, Cyprus, Denmark, Finland, Greece, Italy, Malta, Netherlands, Sweden and Turkey. During Leipzig trade fairs direct flights between Leipzig and Denmark, Belgium, FRG, France, Greece, Italy, Netherlands, Austria and Switzerland. Regular transit between Schönefeld and Berlin (West).
National airline: INTERFLUG (operates 54 foreign routes).
Main international airport: Berlin (SXF) - Schönefield, 25 km from city.
International airports: Dresden (DRS), Erfurt (ERF), Leipzig (LEI).

Surface access: Approved crossing points: Sassnitz (from Sweden), Warnemünde (from Denmark), Helmsted/Marienborn (from Hannover), Rudolphstein/Hirschberg (from Munich), Zinwald/Georgenfeld (from Czechoslovakia), Frankfurt/Oder (from Poland), West Berlin.
Main ports: Rostock, Stralsund, Wismar.

Hotels: INTERHOTEL – the leading hotel chain of the GDR, comprising 32 hotels in 13 towns. Bookings, inclusive of visa application, at all Interhotels directly or through TRAVEL SERVICE (Unter den Linden, 23, Berlin, DDR-1080 (tel: 20923224; tx: 114143; fax: 2294095). TRAVEL SERVICE also organises single of group journeys for tourists. Payment in foreign currency, acceptance of Diners Club, American Express, Visa, Japan Credit Bank, Eurocard or Eurocheque.

Car hire: International driving licence required. Speed limits 100 kph motorways, 80 kph main roads, 50 kph built-up areas.

City transport: Cheap and reliable service in major towns.
Taxis: Metered fare, average M0.8 per km. Additional standing charge M0.5 per trip.
Buses: Flat busfare M0.2/trip. Leipzig and Berlin operate electric train service. Fare varies M0.2 – M0.7/trip. Metro in Berlin.

National transport
Road: Approx 47,300 km, of which 1,850 are motorways and 11,250 main roads. Secondary roads poorly surfaced. Petrol stations infrequent. Recovery of vehicles expensive.
Rail: Reliable and speedy network connecting Berlin with all main towns. Total length of track 45,000 km. 1st and 2nd class service available. Supplements for express trains, seat reservations and sleeping cars.
Water: Around 2,500 km of navigable inland waterways. Inland ports at Dresden, Riesa, Meipen, Torgau, Potsdam, Wittenberg.

Public holidays
Fixed dates: 1 Jan, 1 May, 7 Oct, 25-26 Dec.
Variable dates: Good Friday, Whit Monday.

Working hours
Business: (Mon-Fri) 0800 – 1600.
Banking: (Mon-Fri) 0800 – 1600 and (Sat) 0830 – 1200.
Shops: (Mon-Fri) 1000 – 1900.

Business language and interpreting/translation: Business visitors will need to speak German. Translation available from Intertext agency, Interhotel group and International Trade Centre.

Telecommunications
Telephone: Efficient local service. Coin-operated booths in most towns. Cheap rates for International direct-dialled calls. International dialling code: 37 followed by 2 for Berlin and 41 for Leipzig.
Telex: Public service available in main towns and main hotels. Local charge 10pfg/minute. Dialling code 69 DD.
Telegram: Service operated by post office. Standard rate 10pfg/word (local), 15pfg/word (Europe). Minimum 10 words.

Postal service: Letters should be at least 9cm by 14cm. Stamps available from hotels, slot machines and post office.

Trade fairs: International Leipzig fairs every Mar and Sep. Latter concentrates on engineering and electrical goods, former on machinery, chemicals, motor vehicles and printing. Consumer goods covered at both events.

Electricity supply: 220 V AC.

Representation in capital: Afghanistan, Albania, Algeria, Angola, Argentina, Australia, Austria, Bangladesh, Belgium, Brazil, Bulgaria, China PR, Colombia, Congo, Cuba, Czechoslovakia, Denmark, Ecuador, Egypt, Ethiopia, Finland, France, Federal Republic of Germany, Ghana, Greece, Guinea, Guinea-Bissau, Hungary, India, Indonesia, Iran, Iraq, Italy, Japan, Kampuchea, Democratic People's Republic of Korea, Kuwait, Laos, Lebanon, Libya, Mali, Mexico, Mozambique, Mongolia, Morocco, Mozambique, Netherlands, Nicaragua, Nigeria, Norway, Pakistan, Panama, Peru, Philippines, Poland, Portugal, Romania, Somalia, Spain, Sudan, Sweden, Switzerland, Syria, Tunisia, Turkey, United Kingdom, Uruguay, USA, USSR, Venezuela, Vietnam, Yemen Arab

elsewhere in the planned economies were used by the exhorters. Thus, emphasis was given to the proportion of goods coming out of this or that factory bearing the special 'quality' mark; to the number of rejected goods (presumably of appalling quality and unsellable) returned to the producer; to the measurable savings achieved on the production line in economies on raw materials and the weight of components; to the number of new and 'restructured' work places that had been created through rationalising the work force and the jobs that they do, and so on.

There was also a continued reiteration that more savings could and should be achieved in the area of scrap – and official statistics are now routinely given for the amounts of scrap metal, waste paper and empty bottles that have been collected, and, by implication, put back into the economy. At a rough calculation, economies on the production line were expected by the planners to generate approximately a 20 per cent increase in the national income. At the same time, as the quality and quantity of some consumer goods (children's toys, for instance) was visibly falling off, emphasis was given to the efforts to produce more such goods wherever possible. In addition, new emphasis was being given to the increasing role being played by the 'private' service sector. Practitioners in this area can only operate officially with an authorised licence, but in areas like baking, hairdressing, tailoring and shoe repairing, applicants were coming forward in steadily increasing numbers. Car servicing, watch and umbrella repairing were also newly burgeoning areas.

Computers are an area where the GDR had been trying particularly hard. Aspirations which may have been entertained by Bulgaria earlier in the 1980s to be the main supplier to all CMEA in this sophisticated area seemed by the end of the decade to have been taken over, with some efficiency, by the GDR. Reports of quantities produced of computers, and computer equipment, are not always specific, but are usually qualified with the view that output is 'satisfactory.' There were reports in mid-1989 that a significant boost in this area was coming from notionally 'illegal' imports of equipment from either or both Japan and West Germany. These imports were said to be breaking the Cocom embargo, which is supposed to eliminate sales of 'strategic' goods, or goods that could be used for strategic purposes, to countries which are members of the Warsaw Pact. Despite the Honecker strictures that some capital investment projects were running behind schedule while others were getting out of hand in terms of expense, several dozen projects a year 'of great importance to the national economy' were in service in early 1989. Here, too, pride of place was given to projects which will improve and increase the flow of computers and ancillary equipment. Others singled out for mention as important will produce distributors, metal-cutting equipment and beer. Around 40 per cent of East Germany's foreign trade is with the Soviet Union, and the stated aim, since the signing of intensive bilateral agreements in 1988, is for closer economic integration. There is pressure – originating in Moscow – on, and therefore from, the GDR leadership for closer adherence to agreed delivery dates and for closer relations between individual GDR enterprise and combine managers with their partners in the reformed Soviet Union. Here complications arise when a trader used to commercial transactions on behalf of the amorphous state comes into direct contact with the newly profit-motivated partner. By the same token, the serious debates about the future, if any, of the CMEA trading organisation focused on how to reconcile the reformed economies with the unreformed. While the East Germans' own economic viability was made that much easier because of their own special relationship with the West Germans, they were nevertheless watching with some concern how other CMEA countries were succeeding in Brussels in getting their own exclusive bilateral trading agreements with the EC. Usually they have been the recognisably reformed countries – notably Hungary – and they have been the most critical of CMEA's effectiveness. The East Germans kept a low profile in this debate, but the Romanian veto, on, for instance, the question of industrial standards, will not have gone unheard. The discrepancies in economic policies among CMEA's member countries, with Hungary at one end of the scale and Romania at the other, are a reason for continued scepticism in East Berlin on the prospects for wholesale change in the CMEA corporate way of doing.

Aloneness

East Germany's 'aloneness' in this and other areas had, by the end of the 1980s, become the undercurrent in all public debates, whether in the volkskammer (parliament) or in international fora where the country is listened to with respect. The departure of Erich Honecker as the personification of this way of thinking had been overdue in the minds, if not the spoken words, of many SED party members for a number of years. It remains to be seen what role these people will have in shaping the country's uncertain future.

BUSINESS DIRECTORY

Republic, People's Democratic Republic of Yemen, Yugoslavia, Zaïre, Zimbabwe.

Hotels: Berlin
Domhotel (opening 1990), Platz der Akademie (tel: 2204509; tx: 114141).

Grand Hotel, Friedrichstrasse 158-164 (tel: 20920; tx: 115197; fax: 2294095).

Interhotel Berolina, Karl-Marx-Allee 31 (tel: 2109541; tx: 114331).

Interhotel Metropol, Friedrichstrasse 150/153 (tel: 22040; tx: 114141).

Interhotel Stadt Berlin, Alexanderplatz (tel: 2190; tx: 114111/13).

Interhotel Unter den Linden, Unter den Linden 14 (tel: 2200311; tx: 112109).

Palast, Karl-Liebknecht-Strasse (tel: 2410; tx: 115050).

Car hire: Berlin
Avis, Europcar and Hertz: Hotel Metropol, Friedrichstrasse 150-153, DDR-1080 (tel: 2204695/6; tx: 114144).

Budget: Budapesterstr 18a (tel: 2611357).

Europcar: Berlin Airport, Schönefeld (tel: 6722418).

Business advice and information
Chamber of Foreign Trade (Kammer für Aussenhandel-KFA), 1100 Berlin, Schönholzerstrasse 10-11 (tel: 48220; tx: 114840; fax: 4828408).

International Trade Centre (Internationales Handelszentrum GmbH), 1086 Berlin, Friedrichstrasse (tel: 20480, 2060; tx: 114381).

Banking
Staatsbank der Deutschen Demokratischen Republik (state bank), 1086 Berlin, Charlottenstrasse 33-33A (tx: 114671).

Bank für Landwirtschaft und Nahrungsgüterwirtschaft der DDR, 1080 Berlin, Clara-Zetkin-Strasse 37.

Deutsche Aussenhandelsbank AG, 1080 Berlin, Unter den Linden 24-30 (tel: 22870; tx: 0114411).

Deutsche Handelsbank AG, 1080 Behrenstrasse 22, Berlin (tel: 2202911; tx: 114665).

Principal newspapers
Berliner Zeitung, 1026 Berlin, Karl-Liebknecht Str 29 (tel: 2440; tx: 114854).

Junge Welt, 1026 Berlin, Karl-Liebknecht Str 29 (tel: 2440; tx: 114857).

Leipziger-Volkszeitung, 7010 Leipzig, Peterssteinweg 19 (tel: 715432; tx: 51495).

Neues Deutschland, 1017 Berlin, Franz-Mehring-Platz 1 (tel: 5850; tx: 112051).

Ost-See Zeitung, 2500 Rostock, Richard Wagner Str 1A (tel: 3650; tx: 31317).

Säschische Zeitung, 8010 Dresden, Julian-Grimau-Allee (tel: 4864240; tx: 2251).

Other useful addresses
Allgemeiner Deutscher Nachrichtendienst (news agency), 1026 Berlin, Mollstrasse 1 (tel: 2350; tx: 114601).

Berliner Absatz Organisation (information on companies trading in GDR), Hardenbergstrasse 16-18, 1000 Berlin 12 (West Berlin).

Berolina Travel Ltd, 22 Conduit Street, London W1R 9TD (tel: 01 629 1664).

Central Bureau for International Licensing Trade of the GDR, 1020 Berlin, Schicklerstrasse 5-7.

Deutsche Auslands-und Rückversicherungs AG (state insurance agency), 1020 Berlin, Inselstrasse 16 (tel: 2700171; tx: 114402).

East European Statistics Service, 10 Boulevard St-Lazare, 1030 Brussels, Belgium (tel: 022184349; tx: 21108).

Flight bookings: Alexanderplatz 5, Berlin 1026 (tel: 2109181; tx: 114617); Rampische Strasse 2, PSF 328, Dresden 8012 (tel: 4956013; tx: 26351); Hermann-John-Strasse 6, Erfurt 5020 (tel: 51351; tx: 61423 IFTOE dd); Sachsenplatz 1, Leipzig 7010 (tel: 286246; tx: 51349).

Interflug, Gesellschaft für Internationalen Flugverkehr GmbH, 1189 Berlin-Schönefeld, Zentralflughafen (tel: 6720; tx: 112891/3).

Interhotel (for hotel bookings), 1035 Berlin, Simplonstrasse 52-56.

Interwerbung (advertising agency) GmbH, 1157 Berlin, Hermann Dunckerstrasse 89 (tel: 5090981; tx: 112106 inter dd).

International Leipzig Trade Fair, 7010 Leipzig, Leipziger Messeamt, markt 11/15, Postfach 720 (tel: 71810; tx: 0 512 294).

Intertext, Fremdsprachendienst der DDR (translation service), 108 Berlin, Friedrichstrasse 169/170 (tel: 2335005, 2337283).

Ministry of the Chemical Industry, 1086 Berlin, Leipziger Str 5-7 (tel: 232; tx; 1152311).

Ministry of County-Controlled and Food Industries, 1086 Berlin, Leipziger Str 5-7 (tel: 232; tx: 1152311).

Ministry of Finance, 1086 Berlin, Leipziger Str 5-7 (tel: 232; tx: 1152311).

Ministry of Foreign Trade (Ministerium für Aussenhandel), 1080 Berlin, Unter den Linden 44-60 (tel: 2330; tx: 114437).

Ministry of the Glass and Ceramics Industry, 1086 Berlin, Leipziger Str 5-7 (tel: 232; tx: 1152311).

Reisebüro der Deutschen Demokratischen Republik (tourism), 102 Berlin, Alexanderplatz 5.

Statistical Pocket Book of the German Democratic Republic, State Central Administration for Statistics, Hans Beimler Strasse 70/72, 102 Berlin.

German Federal Republic

John Davies

The West Germans were struggling om 1989 to keep pace with the problems – and opportunities – thrown up by the dramatic changes in Eastern Europe. After coping somehow with a flood of East German refugees, the Bonn republic embarked on a drive to spur both economic prosperity and political reform across the divide. With the frontier, including the ugly Berlin Wall, becoming more open, the question of German reunification, with all its complex and emotional overtones, became a more real issue. Throughout Eastern Europe, the West Germans were anxious to encourage moves towards market economies, smoothing the way to foreign business investment and trade, and were in the forefront of efforts to round up financial help for Poland and Hungary. They wanted the European Community to open doors eventually to Eastern Europe, while keeping up the EC's own momentum towards becoming a more closely knit and barrier-free body. All this took place at a time of economic buoyancy but political disarray within West Germany, which was shaping up for a series of state polls followed by a hard-fought federal election.

One thing West Germans value is a quiet life. Since the war, they have created a society of unparalleled affluence, a tidy country that believes in hard work yet goes about its affairs in an atmosphere of unhurried orderliness. In their comfortable and well managed existence, the West Germans have been thoroughly jolted by the rush of events in Eastern Europe. As the communist regime in neighbouring East Germany tottered before the storm of popular dissent towards the end of 1989, the West Germans were both overjoyed and concerned. On the one hand, they welcomed political freedoms in the East, but at the same time there were worries that turmoil might get out of hand and that the future offered not just breathtaking new perspectives but also incalculable political and economic risks.

As the crisis built up over months, the Federal Republic found itself welcoming a seemingly never ending stream of trains and buses packed with East German refugees seeking a new life in the West. Town halls and other public buildings all over the country were thrown open to shelter refugees in rows of bunk beds. Refugees were also taken into private homes, often by relatives already in the West but in some cases by total strangers. Many West Germans – some remembering their own or their families' earlier migration from Eastern Europe – donated clothes and toys for refugee families.

The Bonn government at first tried to stem the flow of refugees by temporarily closing its office in East Berlin and its embassies in Prague and Budapest, which were being beseiged by East Germans intent on going to the West. But there was no stopping them. At one stage, the West German embassy in Prague was jammed with thousands of refugees, with many more crowded in the streets outside, all waiting for papers entitling them to go to the West. As it became more on the defensive, East Germany ran special trains to shuttle its disaffected citizens from Prague to West Germany, which had no option but to take them. Some of the first refugees left their cars behind in Hungary to flee across the border on foot, but as the exodus grew and as border controls relaxed, many East Germans drove to the West in long lines of old fashioned Trabant cars.

Ironically, the flow of refugees showed no signs of slackening until the beleaguered East German regime ended travel curbs and simply threw its borders open for its citizens. By then about 200,000 East Germans had already made their way to West Germany. At that point, the West Germans found themselves under a different kind of pressure from the East. Border crossings, especially in Berlin, were swamped with day trippers, gaping wide-eyed at life in the West and at the shops packed with glittering consumer goods. Confronted with this new throng, the West German authorities began to have second thoughts about their policy of handing out DM100 to East German newcomers. But they had to think about how to bring

KEY FACTS

Official title: The Federal Republic of Germany

Head of state: President Richard von Weizsäcker

Head of government: Chancellor Helmut Kohl (Christian Democrat)

Ruling party: Coalition of the Christian Democratic Union, the Christian Social Union and the Free Democratic Party

Capital: Bonn

Official Languages: German

Currency: Deutsche Mark (DM) = 100 pfennig

Exchange rate: DM1.77 per $ (Dec 1989)

Area: 248,708 sq km

Population: 61.2m (1988)

GNP per capita: $19,261 (1988)

GNP real growth: 3.6% (1988)

Labour force: 26.2m (1988)

Unemployment: 8.7% (1988)

Inflation: 1.1% (1988)

Trade balance: DM127.9bn (1988)

KEY INDICATORS

	Unit	1984	1985	1986	1987	1988
Population	'000	61,181	61,015	61,080	61,170	61,200
Gross national product (GNP)	DM bn	1,769.9	1,844.4	1,945.2	2,017.8	2,121.8
GNP per capita	DM	28,929	30,229	31,847	32,987	34,670
Inflation	%	2.4	2.2	–0.2	0.2	1.1
GNP real growth	%	3.3	2.0	2.5	1.7	3.6
Unemployment	%	9.1	9.3	9.0	8.9	8.7
Wages	1985=100	96.3	100.0	103.5	107.6	112.1
Share prices	% rise	12.6	33.0	35.2	–7.9	–16.5
Interest rate (official disc.)	%	4.5	4.0	3.5	2.5	3.5
Industrial production	1985=100	96	100	102	102	106
Investment goods	1985=100	91	100	104	105	108
Consumer goods	1985=100	99	100	102	103	106
Road vehicles	1985=100	89.4	100.0	104.8	107.4	–
Mechanical engineering	1985=100	94.2	100.0	105.5	102.9	–
Construction	1985=100	108.9	100.0	106.4	106.0	–
Electrical engineering	1985=100	90.1	100.0	104.9	105.2	–
Retail sales growth	1980=100	111	114	117	120	122
Private consumption	DM bn	1,003.6	1,038.4	1,068.6	1,112.8	1,156.8
Government consumption	DM bn	350.2	365.6	382.7	397.0	411.5
Exports	DM bn	488.2	537.2	526.4	527.4	567.7
France	%	12.6	11.9	11.8	12.1	–
Netherlands	%	8.6	8.6	8.6	8.7	–
UK	%	8.4	8.6	8.5	8.8	–
Italy	%	7.8	7.8	8.1	8.7	–
USA	%	9.6	10.3	10.5	9.5	–
Imports	DM bn	434.3	463.8	413.7	409.6	439.8
Netherlands	%	12.2	12.6	11.6	11.0	–
France	%	10.6	10.6	11.4	11.6	–
Italy	%	7.9	8.0	9.2	9.6	–
UK	%	7.7	8.0	7.2	7.1	–
Balance of trade	DM bn	54.0	73.4	112.6	117.8	127.9
Current account	$bn	9.75	16.98	39.85	45.63	48.64
Exchange rate	DM per $	2.8	2.9	2.2	1.8	1.8

* estimated figure

currency stability to East Germany, whose Ost-Mark was rapidly losing value. The West Germans also had to waive their stringent car pollution regulations for the army of Trabants chugging around, spewing out exhaust fumes.

Money for homebuilding

The refugee influx was generally seen as an economic boost for West Germany, despite the headaches. In time, the newcomers from East Germany and the continuing flow of ethnic Germans from other parts of Eastern Europe would add an extra one percentage point to the country's economic growth rate, according to some economists. Many new arrivals from East Germany were skilled workers who readily found

jobs, relieving pressure in an already tightening labour market. They found jobs much faster than many of the ethnic Germans arriving from Poland and the Soviet Union, some of whom were not fluent in the German language.

One of the biggest problems, however, was housing. The Bonn government in November launched a programme to pump more cash into public-supported homebuilding and to provide more tax concessions for private building. The government was concerned to prevent friction, as some West Germans were already grumbling that newcomers were getting favourable treatment for housing as well as competing for jobs.

With the ousting of Erich Honecker as East German communist leader and then the opening of the Wall, the Bonn government redoubled its efforts to stem the tide of refugees by helping to make life more tolerable in East Germany. It offered substantial financial aid to the East provided political reforms went ahead. This would be the latest of many handouts which Bonn has given the East. For years the communists exploited Bonn's charity to extract large sums of money whether to subsidise East-West roads, to help clean up the environment or to buy the freedom of detainees. Now, government and business in West Germany were preparing to take a more direct hand in modernising and revitalising the East German economy, including such basics as its antiquated and hopelessly inadequate telephone system. The task would involve huge investment but could yield lucrative rewards.

An end to division?

For decades the West Germans had resigned themselves to the division of Germany, but now the situation seemed to change abruptly. With the opening of the Wall after 28 years, there were emotional scenes. Chancellor Helmut Kohl interrupted his long-awaited visit to Poland to fly to Berlin, where he told a rally that Germany, East and West, was one nation. Reunification is envisaged in West Germany's basic law or constitution, which was meant to apply to all Germans, including those in the East. With East Germany likely to become ever more dependent on the Bonn republic, West German politicians are likely to voice support more loudly for some kind of reunification. But as politicians have regained control of events, there have been words of caution all round – not only from Britain and the Soviet Union, but also in Bonn and East Berlin. Chancellor Kohl stressed that such a step would require the support of East Germany and of the wartime victors. The implication was that reunification – which would shift the European balance of power, with far-reaching and incalculable consequences – would need time and careful planning.

But Chancellor Kohl dismissed fears that West Germany might be lured into a more neutralist position in order to bring about reunification. West Germany was firmly anchored in the European Community, he stressed. Since the changes in East Berlin, Bonn has reinvigorated its two-pronged policy of bolstering reform in Eastern Europe and simultaneously backing closer co-operation within the EC, including moves for monetary union. Its call for a special EC summit in November to consider Eastern Europe was consistent with this approach of progressing on both fronts and, possibly, of interlocking them. Bonn agreed to give wide-ranging aid to Poland, spelt out during Mr Kohl's Warsaw visit, and pressed other EC nations to join in providing urgent practical aid to both Poland and Hungary.

In this way, Bonn was pursuing Germany's traditional interest in building economic links with Eastern Europe while consolidating its large share of developed West European markets. As a result, it may well be that a future market-oriented Eastern Europe will be drawn more closely into the orbit of an EC in which Bonn, through its sheer economic strength, is a dominant force.

COUNTRY PROFILE

Political system: Federal Republic with power shared by a federal government based in Bonn and 10 states (Länder) plus West Berlin, each with its own constitution, parliament and government. Under constitution (Grundgesetz) federal legislative power vested in a directly elected lower house (Bundestag) and a federal council representing state governments (Bundesrat). Federal executive power rests with the government, led by the federal chancellor. The president, who is elected by specially convened federal convention every five years, is constitutional head of state with little influence on government. The ten Länder are: Baden Württemberg, Bavaria, Bremen, Hamburg, Hesse, Lower Saxony, North Rhine-Westphalia, Rhineland-Palatinate, Saarland and Schleswig-Holstein.

Main cities/towns: Bonn (population 290,000), West Berlin (1.9m), Hamburg (1.6m), Munich (1.3m), Cologne (1.0m), Essen, Frankfurt, Dortmund, Düsseldorf, Stuttgart, Duisburg, Bremen, Hanover.

Media

Press: Extensive range of publications. *Dailies:* Roughly 600 newspapers published regionally. Highest circulation is *Bild Zeitung* (4.3m). Also influential are *Frankfurter Allgemeine Zeitung* (circulation 360,000), *Süddeutsche Zeitung* (378,420) and *Die Welt* (225,000). Weeklies: Most influential include *Der Spiegel* (957,068), *Die Zeit* (468,000) and *Stern* (1,478,926). Also TV and radio magazines and various other general/special interest publications. *Business:* Numerous trade publications, plus journals published by the Chambers of Commerce and Trade Unions, including *Börsenzeitung and Handelsblatt* (daily, 112,000), and *Wirtschaftswoche* (weekly, 118,882). *Broadcasting:* Private radio and TV interests have become more firmly established in recent years. The long-established system of publicly controlled radio and TV stations and networks has been forced to yield ground. More than 5m of West Germany's 26m households have plugged into the cable TV network (and through it to satellite TV).

Domestic economy: Increased investment and strong export sales have spurred economic growth. Consumer spending remains high and inflation is relatively low. The wide-ranging industrial sector has been transformed technologically to remain competitive in world markets, where it generally has a reputation for advanced products, high quality and reliable delivery.

Employment: A shortage of skilled workers has developed even though unemployment has remained around 2m. Influx of ethnic Germans has put pressure on the labour market. Relatively high wage costs, compared with other countries, always a concern to employers and a spur to more capital-intensive production techniques. Trend to shorter hours accompanied by efforts to operate more flexible working time in factories and other businesses. But unions resistant to pressure for Sunday work in highly capital-intensive plants.

External trade: One of the world's main trading nations, West Germany favours open markets. In international negotiations it is normally a strong advocate of trade liberalisation. Many West German industries sell more than half their goods abroad. Trade and current account surpluses have been running at record or near-record levels in recent years.

Agriculture: Number of farms down drastically to 700,000 from 1.6m in 1949. Many are small-scale and part-time family farms. While some farms have grown in size, average farm is smaller than in Britain and France. Under EC pressure to cut surplus production and subsidies, Bonn offered financial inducements to ease way for farmers to reduce output and phase out excessive crop land. Output predominantly wheat, barley, sugar beet, pigmeat and dairy products.

Industry: Leading European producer of motor vehicles and accessories, industrial plant, machine tools, electrical goods, scientific instruments, chemicals, pharmaceuticals and consumer goods. Traditional industries (steel, shipbuilding) have contracted because of foreign competition and weaker demand. Strenuous efforts made to modernise industry through use of electronics and more flexible production techniques.

Mining: Relatively few natural resources other than large supplies of black and brown coal, so largely dependent on imports. Mining of coal has fallen steadily owing to drop in demand from local steel and electricity industries.

Energy: Energy-saving, a top priority at the time of oil crises in 1970s, still intently pursued. Despite lower crude oil prices, use of oil down in absolute terms and as proportion of primary energy consumption. With vast reserves of (relatively expensive) coal, West Germany has a long-term arrangement under which a special levy on top of the price of electricity subsidises the use of coal by generating companies. This has helped smooth the way for coal industry to scale down output and phase out jobs, especially since lower steel production has meant reduced demand for coal. Nuclear power accounts for a third of electricity output, but there is strong opposition to nuclear plants.

Banking: Sophisticated system underpinning country's economic strength. Three main categories: central bank (Bundesbank) basically concerned with monetary stability; multi-purpose banks, including commercial, co-operative and (publicly owned) regional Landesbanks and savings banks; and specialist banks, including mortgage banks and instalment credit houses. Many banks have important shareholdings in industrial companies and bankers sit on the supervisory boards of many companies. Biggest national banks are: Deutsche Bank, Dresdner, Commerzbank, Westdeutsche Landesbank, Bayerische Vereinsbank. Around 260 foreign banks operate.

Leading industrial companies: Daimler-Benz (1988 turnover DM 73.5bn), Siemens (DM 59.4bn), VW (DM59.2bn), Veba (DM 44.4bn), BASF (DM 43.9bn), Hoechst (DM 41.0bn), Bayer (40.5bn), RWE (DM 38.9bn) (in year to 30 June 1989), Thyssen (DM 29.2bn), Bosch (DM 27.7bn).

Legal system: Civil law sysytem. Basic Law (constitution) adopted 1949. Federal constitutional court reviews legislative acts.

Membership of international organisations: ADB, CCC, Council of Europe, DAC, EC, EIB, ELDO, EMS, ESRO, FAO, GATT, IAEA, IBRD, ICAC, ICAO, ICES, ICO, IDA, IDB, IFAD, IEA, IFC, IHO, ILO, IMF, IMO, INTELSAT, International Lead and Zinc Study Group, INTERPOL, IPU, ITC, ITU, NATO, OAS (Observer), OECD, UN, UNESCO, UPU, WEU, WHO, WIPO, WMO, WSG, WTO.

Arms control and Mr Gorbachev

The sweeping changes under way in Eastern Europe strengthened pressures in West Germany for more progress on international arms control. They also added to the astonishing popularity of Mr Mikhail Gorbachev, the Soviet president, whose visit to Bonn and Stuttgart in June 1989 was an emotional affair. While Moscow's calls for disarmament clearly stemmed from its economic plight, many West Germans were anxious to seize the opportunities for arms agreements. The West German foreign minister, Mr Hans-Dietrich Genscher, was particularly keen to goad the superpowers to negotiate about removing short-range nuclear weapons from Europe. Chancellor Kohl himself took up this case early in 1989 in the face of misgivings among his own more conservative colleagues and opposition within NATO, especially from Britain. The eventual NATO compromise was open to various interpretations, but Mr Genscher chose to present it as a victory, clearing the path towards future agreement with the Russians on both conventional forces and short-range nuclear arms. The NATO compromise at least cleared the way for a warm welcome for President George Bush in Bonn in May 1989.

President Bush and his secretary of state, Mr James Baker, have been realistic enough to recognise the pivotal position of West Germany as Eastern Europe gropes towards reform and modernisation. They have at times shown a clear willingness to concede to the West Germans an influential role in determining Western reaction to events in the East. Earlier in 1989, however, disarmament controversies and growing debate about German reunification seemed to lend a sharper edge to Bonn's relations with the USA and with the other Western allies. Strong feeling developed, for instance, against low-flying aircraft exercises, and curbs were eventually agreed. Social Democrats and more left-wing activists also demanded that the rights of allied forces be curbed in West Germany. Nearly 45 years after the end of the war, it was unreasonable, they argued, that allied troops should have more rights on German soil than NATO troops had in other host countries. Critics also voiced astonishment that, so long after the war, the West Germans still had no automatic right to fly in the belt along the border with East Germany or through the air corridors to Berlin.

Political uncertainty

The turmoil in East Germany and the arrival of so many refugees added new elements of uncertainty to West German politics. But the crisis, raising eventual prospects of German reunification, gave Chancellor Kohl a chance to take new initiatives and to rebuild support for this government ahead of the crucial elections in 1990. Mr Kohl was certainly in need of a political boost. Despite a buoyant economy, there had been deepening disillusionment with the Bonn government – a coalition of Mr Kohl's free-market but welfare-conscious Christian Democratic Union (CDU), its Bavarian sister party, the Christian Social Union (CSU), and the small liberal-minded Free Democratic Party (FDP).

'Red-green' co-operation

The federal government's problems, if anything, grew steadily worse while the economy picked up. Even though international criticism of West Germany's earlier lagging economic performance abated, domestic disenchantment with the Bonn coalition grew. A heavy blow came with the West Berlin election result in January 1989. The city's governing Christian Democrats under Mr Eberhard Diepgen suffered an unexpected defeat. In their place a 'red-green' coalition was patched together by the Social Democrats under their moderate local leader, Mr Walter Momper, and the 'Alternative Liste' (AL), the city's militant equivalent of West Germany's environmental party, the Greens. A further setback came in March 1989 when the Christian Democrats lost control of the city administration in Frankfurt and were replaced by a 'red-green' coalition. Then in June 1989 the Christian Democrats and their Bavarian partners lost ground heavily in the nationwide election for West Germany's representation in the European parliament in Strasbourg. The CDU/CSU notched up just 37.8 per cent of the vote, only a whisker ahead of the Social Democrats and down from 45.9 per cent in the previous Euro-poll in 1984. In all three of these contests, radical right-wing elements caused a stir by winning seats. In particular the Republicans headed by Mr Franz Schönhuber showed they were a force to be reckoned with by gaining 7.5 per cent of the votes in West Berlin and going on to pick up 7.1 per cent of the votes nationwide in the Euro-poll. The Republicans garnered support from voters dissatisfied with all major established parties.

'Red-green' co-operation and the rapid rise of the Republicans threatened to upset all political calculations in Bonn. Fears were aroused of political fragmentation and of shifting party alliances. If the Republicans succeeded in winning seats in the Bundestag election set for December 1990, it was possible that neither the existing coalition nor even a 'red-green' coalition might muster a majority. This spectre encouraged speculation that the junior coalition partner, the FDP, might switch support to the Social Democrats or even to a 'red-green' coalition.

Rabbits from a hat

The Christian Democrats under Chancellor Helmut Kohl came under strong pressure to halt the drift of disillusioned supporters to the Republicans, but were divided over how to do this. Moderate members, keen on social welfare and a liberal humanitarian approach to immigrants and foreign policy, warned against any shift to the right. Paradoxically, the CDU's deeper troubles seemed to strengthen Mr Kohl's hand, initially at least. In their disarray in the months after the Euro-election, the Christian Democrats seemed to have no clear alternative to him, even though many felt he lacked leadership and vision. Mr Kohl, who had always made up for a shortage of charisma with pragmatism and dogged staying-power, tried to rally his party with calls for unity. He had shown before that he could produce rabbits from a hat in time to impress voters and hoped that skilful handling of events could do the trick again.

In a tactical move in August 1989, Mr Kohl proposed to drop Mr Heiner Geissler as party general-secretary and replace him with the younger and widely respected Mr Volker Rühe, until then known as a foreign affairs expert. Mr Geissler, party chief for a record 12 years and at times a cabinet minister, had earned the enmity of the right wing. He was accused of dragging the CDU too far left of centre in his bid to attract new voters, notably young people and women. Mr Kohl took a considerable political risk by moving against Mr Geissler, intensifying a struggle for position and influence in the party ahead of its congress in Bremen in September 1989. But it was a typical attempt by Mr Kohl to regain the political initiative through a swift change in personnel, while professing there was no change in policy. He made a similar move with a cabinet reshuffle in April 1989. He unceremoniously dropped the hapless Mr Rupert Scholz after just 11 months as defence minister, used the opportunity to shift Mr Gerhard Stoltenberg from finance to defence, and bound the CSU more tightly to him by bringing in its new leader, Mr Theo Waigel, as finance minister.

But the CDU and its Bavarian counterpart had a formidable task ahead – irrespective of whether Mr Kohl soldiered on as chancellor or whether he fell victim to one of the deadly intrigues for which the CDU has a reputation. Ahead of the Bundestag election, polls loomed in four states and all threatened to be hard fought. In Lower Saxony a coalition of Christian Democrats and the FDP was defending a majority of just one seat. The situation was wobbly for Mr Ernst Albrecht, the CDU state premier, because one of his party's state parliamentarians was openly flirting with the Republicans. For this reason, there was speculation that a poll might come even sooner than expected. The Christian Democrats were also struggling to mount a credible challenge to the Social Democratic governments in the industrial and coal-mining states of North Rhine/Westphalia and Saarland. The CSU, meanwhile, had to do something to stem threatened defections to the Republicans in the Bavarian state poll.

Touching a raw nerve

The Social Democrats gained hope from the Christian Democrats' troubles but presented a less-than-resolute image to voters. The SPD put off nominating its own chancellor-candidate until after the state elections. This meant it would be uncertain for many months whether the party's standard bearer might be Mr Oskar Lafontaine, the Saarland premier, or perhaps Mr Hans-Jochen Vogel, the SPD's federal parliamentary leader. The party continued to profess reluctance to join forces with the Greens, aware that close association risked alienating voters. But at the same time leading SPD members held a series of discussions with prominent Greens, exploring common ground. Other SPD members conspicuously commended certain aspects of the Free Democrats' policies, while Free Democrats in turn spoke about the party's 'bridging function' in helping to form stable governments with moderate policies.

The SPD's problems were compounded by the realisation that it, too, had lost voters to the Republicans in the Euro-election. While the Republicans were branded by their more virulent critics as neo-Nazis, the new party in fact condemned Nazism. Mr Schönhuber, a former TV broadcaster in Bavaria, spoke bluntly about issues which many politicians touched with extreme caution. In particular he opposed the influx of immigrants from Eastern Europe and asylum-seekers from all over the world, including Sri Lanka, Lebanon and Turkey. He also criticised West Germany's membership of the European Community as a threat to jobs and an obstacle to German reunification. The Republicans touched a raw nerve in West German politics. They drew support from various dissatisfied segments of the population. In parts of rural Bavaria, they won over voters unhappy that wealthy Germans were buying up Bavarian houses for weekend or holiday use, pricing local families out of the market.

The Republicans' Euro-poll success prompted them to move their headquarters from Munich to Bonn in readiness for their campaign to enter the Bundestag. Their Bavarian origins and their strength in that state posed an awkward problem for the CSU, which had a fragmented leadership since the death of Mr Franz Josef Strauss in 1988. Mr Waigel, the CSU's president and its leader in Bonn, took a strong line on a number of sensitive issues, including German reunification, as his party attempted to staunch the loss of support to the Republicans.

Surge in investment

While domestic political wrangling and troubled relations with Eastern Europe cast long shadows over Bonn, the country's economy showed brighter trends – only to be taken for granted by West Germans and their main trading partners. After moderate though hesitant growth for some years, West Germany's gross national product increased by 3.4 per cent in real terms in 1988. This was almost double the 1.8 per cent rate of expansion in 1987. The momentum carried on into 1989.

The driving forces were an upsurge in investment, which grew by 8.4 per cent in 1988, and strong demand for exports, particularly to other EC countries. With orders still running high, many companies were at or close to the limit of their capacity utilisation. Output of all kinds grew strongly: machine tools, cars, heavy trucks, chemicals. The strong rise in corporate profits helped to encourage further investment. In many cases, expansion and modernisation were seen as necessary to take advantage of the more open EC market which the 12 member countries are trying to create by the end of 1992, with competitive conditions being made more uniform and with internal barriers of all sorts being dismantled.

One effect of the stronger economic growth was to ease pressure on the federal government budget. But the Bonn coalition remained embarrassed by handling of the budget. After all, it came into office in 1982 with a promise to put government finances on a sound footing and made great play of its success in doing so in the ensuing years. Now it had raised doubts among voters over whether its grip on budget finances was as strong as it claimed. As it happened, the budget shortfall to be made up by new borrowing was some DM35.3bn, compared with an earlier official forecast of DM40bn. The borrowing requirement for 1989 was expected to fall to well below DM30bn. However, finances were helped by a DM10bn payment to the government from the Bundesbank, reflecting profits in 1988 from foreign currency movements. By contrast, the central bank had been able to hand over only a fairly paltry DM240m out of its 1987 profits. With the buoyant economy producing higher tax proceeds than expected, the government's finances appeared well capable of coping in 1990 with concessions due under the third stage of the federal government's sweeping tax reform.

Embarrassment of riches

The more buoyant economy in 1988-89 took the wind out of the sails of foreign critics who for years had been urging West Germany to step up its economic performance to help stimulate other countries. But the critics pointed out that the West Germans were benefiting to a large extent from sales abroad, especially in Europe. True enough: the West Germans again built up an embarrassment of riches in their international payments. The trade surplus, long attacked as excessive by the USA, rose to a new record of DM128bn in 1988, compared with DM117.7bn the year before. Taking account of insurance, spending by tourists, service payments and transfers, the current account showed a near-record surplus of DM84.9bn. The previous year the surplus had slipped to DM80.8bn from the earlier record of DM85.0bn in 1986. With export earnings to the USA down, the West Germans more than made up for the slack through a hefty trade surplus with their European partners. More than 54 per cent of West Germany's export income came from other EC countries. It is true that just over half its imports also came from the EC, but they nevertheless lagged DM80bn behind what it creamed from these countries in export earnings.

As economic activity picked up, inflation became more of a worry, though it remained far less than in most other countries. After being barely perceptible for a couple of years, inflation reached 1.2 per cent in 1988 and then gathered pace to an annual rate of around 3 per cent by mid-1989. As interest rates rose in Britain and the USA and as the Deutsche-Mark came under pressure on foreign exchanges, the Bundesbank brought in more restrictive measures. It signalled a tighter policy by announcing a lower monetary growth target of 5 per cent for 1989. It engineered higher money market rates through such devices as its securities repurchase agreements with banks and progressively lifted its official discount and Lombard rates. By mid-1989 these benchmark rates were up to levels not seen since 1982. The Bundesbank was concerned above all to narrow the gap with interest rates abroad and so discourage capital outflows. By this means it hoped to ease pressure for a decline in the value of the D-Mark against the dollar, fearing that such a devaluation would make imports dearer and fuel inflation.

Interestingly, the flow of long-term capital from West Germany showed a big increase in 1988. This came as little surprise, partly because of the higher interest rates abroad and bouts of D-Mark weakness. It also indicated that West German investors, shy of controls, were sending money abroad to avoid the controversial withholding tax on interest earnings *(Quellensteuer)*. Plans for this tax came under heavy fire from the Bundesbank and commercial bankers, but the federal government persisted with introducing it from January 1989. Then, recognising the unpopularity of the tax, the government back-pedalled as soon as Mr Waigel replaced Mr Stoltenberg as finance minister in the April cabinet reshuffle. Mr Waigel's first major move was to abandon the tax as from July 1989. The whole exercise exasperated bankers, who spent heavily to set up arrangements to deduct the tax from their customers and then to undo those arrangements.

Computerised stock trading

The government showed more reluctance, however, to drop the stock exchange turnover tax (*Börsenumsatzsteuer*), which has also been strongly criticised. As a result of this tax, some stock exchange business has been diverted from West Germany to other financial centres, notably London. The volume of stock market business was in any case slack after the crash of 1987 and then as a consequence of relative D-Mark weakness and higher interest rates. The share market, however, steadily climbed out of its post-crash trough during 1988. Share prices and the volume of business showed bouts of liveliness as 1989 progressed.

Stock exchange authorities continued to press on with plans for a West German options and futures exchange, aiming to begin business at the beginning of 1990. Legislation came into effect in August 1989 to clear away archaic legalistic obstacles, including the possibility that such trading might have come under gambling laws that inhibited recognition of debts. In preparation for the 1990 start-up, the West Germans have co-operated with the Swiss Soffex exchange and have kept a wary eye on the Liffe exchange in London, which jumped the gun by introducing trading in a West German bond futures contract in 1988.

The big banks, which dominate the West German stock exchanges, took a decision in July 1989 which is expected to have far-reaching consequences. They resolved that a computer exchange should be set up, initially to trade outside the hours of business for floor trading but eventually for parallel trading. The computer exchange would be dominated by large-scale business in major shares for large clients, while floor trading would continue to be the domain of trading on behalf of personal investors and in second-ranking stocks. The decision was scarcely revolutionary when compared with computer dominance in trading abroad, but it sent ripples through the West German financial community. It meant that the banks had come down against extending the hours of floor trading, instead opting for full-scale computer operations alongside the computerised options and futures exchange. It raised further doubts about whether all seven of the regional stock exchanges outside Frankfurt could survive for much longer – a touchy issue in a country that values decentralisation. It also posed a threat to the livelihood of brokers who have traditionally operated on the floor of exchanges.

Banking on insurance

With this decision the banks showed they were determined to press on with the efforts made in recent years to improve West Germany's standing as an international financial centre. Simultaneously the banks – and insurance companies – have been building up links and operations abroad to take account of the wider opportunities and increased competition in a more open EC. In one move, the private bank of Sal Oppenheim disclosed in July 1989 that it would merge its insurance operations, including Colonia, with non-French insurance operations of the Victoire group of France. Oppenheim would retain a substantial minority stake in the Franco-German insurance group, which would be one of Europe's largest insurers. The 200-year-old Oppenheim bank was also to change its legal form and increase its capital base, evidently bringing in considerable funds arising from the insurance deal. Oppenheim saw such a deal as opening better prospects for both its insurance and banking operations to develop their interests in Europe and further afield. The deal, however, touched off a fierce struggle between Victoire's major shareholders in France. Cie Financière de Suez, already pressing to extend its influence into insurance, made a complex bid to establish control over Compagnie Industrielle and therefore over Victoire and the German operations.

West German banks have futher intensified their already growing involvement in insurance. They have been spurred along by the trend for affluent West Germans to invest about a third of their new savings in life assurance, encouraged by tax advantages that do not apply to savings put in traditional bank accounts. Deutsche Bank, the country's biggest bank, which has been in the forefront of this drive, received the formal go-ahead in 1989 to set up its own life assurance company. In keeping with its reputation for forcing the pace with innovation, it announced plans to offer a more transparent type of life assurance policy, giving policy holders a clearer picture of what is happening to their payments. The big Munich-based Allianz insurance group, which was deeply offended by Deutsche's thrust into life assurance, reached a co-operation agreement in March 1989 with Dresdner, the country's second largest bank. Under this pact, to apply in some regions only, life assurance and other forms of insurance would be offered to Dresdner customers. The third of West Germany's 'big three' commercial banks, Commerzbank, announced a complex arrangement in May 1989 to co-operate with DBV, an institution dealing in life assurance, health insurance and other forms of insurance.

With EC financial markets becoming even more open and competitive, further big changes were shaping up in the publicly owned banking sector and in the co-operative banking movement of West Germany. Like their commercial rivals, these banks are coming under stronger pressure to offer individuals and businesses a wide range of services, both domestic and foreign. A report commissioned by the savings banks from McKinsey management consultants was expected to accelerate the trend towards mergers among the many small banks owned by local community authorities. More controversially it was also bound to intensify long-running debate about the need for a shake-up of the big regional Landesbanks. On a similar track, the association of co-operative banks pressed ahead in 1989 with plans for a more rational and up-to-date organisational structure. This would involve local mergers to cut the number of independent co-operative banks by half in the next ten years, as well as the merging of five big central co-operative banks into the DG Bank, which already has big national and international business. Apart from their organisational set-up, the publicly owned and co-operative banks faced the difficult question of how to raise more funds to finance expansion of business. Deutsche Bank, for its part, simply tapped its shareholders for well over DM1bn in one of its periodic rights issues in February 1989.

The power of big banks

The perennial question of the power of West German banks surfaced again during 1989. The Free Democrats in Bonn – and notably their president, Count Otto Lambsdorff – renewed a call for legislation to place a low ceiling, possibly 15 per cent, on bank shareholdings in industrial companies and to restrict the number of their seats on companies' supervisory boards, with a possible ban on seats in competing companies. Count Lambsdorff also criticised the traditional practice of investors giving banks proxy control of their voting rights. This made a nonsense of shareholder meetings where votes were overwhelmingly controlled by the big banks, he argued. The commercial banks, which took the latest political debate very seriously, pointed out that they had already disposed of substantial industrial holdings and given up some supervisory board seats. They also drew attention to the advantages of their direct involvement in industry. Deutsche Bank, for instance, stepped in to rescue the trading house of Klockner and Co in October 1988, when it fell victim to heavy losses from oil futures trading, and then arranged its takeover in 1989 by Viag, the newly privatised and diversifying industrial and aluminium group (which is co-operating closely with the Bayernwerk electricity utility). The FDP's campaign led to steps for the issue of banking power to be investigated by a special working party set up by the coalition parties in Bonn.

As usual, Deutsche Bank was the main target in the renewed controversy over the power of banks – and specifically its 28 per cent stake in Daimler-Benz. Daimler now was being transformed into a wide-ranging high-technology group, involving not just cars and trucks but also aerospace and electronics. It was only to be expected, therefore, that some concern would be voiced about West Germany's biggest bank being the key shareholder in the country's biggest industrial group. The long-running plan for Daimler to take over the aerospace and defence company of Messerschmitt-Bölkow-Blohm (MBB), continued its tortuous path in 1988-89. As expected, the Cartel Office in West Berlin rejected the merger in April 1989 and the plan then went to the economics minister, Mr Helmut Haussmann, for his decision. The issue thus went full circle, as the Bonn government had been a prime mover in the whole exercise. It wanted to find a way to end its subsidies of the Airbus airliner project by finding a wealthy partner for MBB, which is West Germany's major stakeholder in the international Airbus consortium. But the Daimler/MBB merger plan has been anything but straightforward. Daimler's wily chief executive, Mr Edzard Reuter, drove a hard bargain all along. There was much debate about the monopolistic aspects of the deal, as Daimler had already swallowed the MTU aero and marine engine group and the resuscitated AEG electrical and electronic company. The Monopolies Commission in August 1989 advised Mr Haussmann to approve the Daimler/MBB deal, though only under certain conditions. These would involve the sale of some engine or defence equipment operations to reduce monopolistic concentration, and preferably the disposal of Deutsche's stake in Daimler. Even that qualified approval went too far for the Monopolies Commission's own chief, Professor Ulrich Immenga, who found himself in a dissenting minority on this issue. To underline his opposition to a Daimler/MBB merger he resigned when the commission released its report.

Realignment in industry

Daimler-Benz, however, is just one of the more spectacular cases of industrial realignment going on in Europe. The pace is being forced by a combination of factors – application of new technologies, the high cost of research and development, the need for wider markets, more powerful competition and greater risk. All these factors are spurring along EC efforts to create a single European market free of internal obstacles by the end of 1992. And as such a market becomes more of a reality, the pressures for corporate realignment have become even more intense.

In electronics and telecommunications, for instance, moves for realignment have continued unabated. Siemens, the West German standard bearer in this sector, joined forces with Britain's General Electric Company (GEC) in 1988-89 to pursue a hostile takeover bid for Plessey of the UK. Siemens, which is defending a strong position at home, also streamlined its top management structure in 1989 and pressed on with efforts to expand in the USA. It worked out agreement with IBM for a virtual takeover of Rolm, the office communications operation which IBM had wrestled with for some years. With pressures and risks high in electronics, Nixdorf found itself in an unenviable position in 1988-89. Once a stock market favourite, it reported a heavy loss for 1988, its marketing chief and chief executive departed, and it was struggling to maintain its independence. At SEL, the communications company where control has shifted from ITT of the USA to Alcatel of France, Mr Gerhard Zeidler took over the chief executive's job from Mr Helmut Löhr at a difficult time in March 1989. The West German Bundespost, meanwhile, introduced its more liberalised telecoms regime when its new business-like structure came into operation from July 1989.

With outsiders anxious to secure places in the EC, the Hanover-based Hanomag found itself wooed by Komatsu, the Japanese construction equipment company. Hanomag, once part of the ill-fated Esch empire, stood to gain expanded production and more jobs as part of Komatsu's plan to buy a large stake. Komatsu hoped to use the long-established German company to benefit from an upsurge in building in the EC. With the Japanese boosting car production in the UK, Robert Bosch decided in 1989 to build a £100m plant in South Wales. This was in line with its strategy of meeting expanded foreign sales out of manufacturing abroad rather than through high-cost exports from West Germany. Linde, meanwhile, widened its fork-lift truck operations by taking over Lansing Bagnall of the UK.

Fending off predators

In retailing, big groups struggled to press ahead with cross-border plans. Mr Erivan Haub's family-owned Tengelmann group, through its A&P unit in the USA, made an unsuccessful bid to take over the Gateway chain of food stores in Britain in 1989, but then moved to join forces with the successful Isosceles consortium in the hope of becoming an active partner in running the UK group. The Asko retail group built up by the controversial and somewhat secretive Mr Helmut Wagner announced plans to link up with Dutch, French and UK chains, although immediately upset them over a stake it bought in the Dutch group.

While many companies in Europe and the USA warily watched for takeover predators, more West German companies resorted to the traditional defence by placing a limit on the voting rights of any single shareholder. But this tactic, long adopted to preserve the independence of prestigious German companies, came under attack from shareholder champions such as Count Lambsdorff. Businessmen were also alarmed by a demonstration that the tactic might not necessarily fend off raiders. To general surprise, the energy conglomerate Veba snapped up a 46 per cent holding in Feldmühle-Nobel, part of the old industrial empire of the Flick family. Although Veba did not acquire voting rights to match its holding, it enjoyed the support of Feldmühle's management and was considered likely to play a decisive role at the company.

Veba, which has considerable electricity interests, caused a stir in 1989 with the disclosure that it was negotiating further long-term arrangements to reprocess spent nuclear fuel in France. This disclosure was followed by the official *coup de grâce* for the controversial nuclear reprocessing plant planned for Wackersdorf in Bavaria. West Germany's decision to abandon the plant was taken not just because of the history of violent protest at the Wackersdorf site. Above all, spent nuclear fuel could be reprocessed more cheaply, as well as with less fuss, at La Hague in France and Sellafield in Britain. The Bonn government signed framework reprocessing agreements with France and Britain, while German electricity utilities worked out details with the reprocessing operators.

Further sweeping changes are shaping up in the West German electricity sector, spurred along by EC efforts to remove national subsidies and open the way to Community-wide competition. During 1989 West Germans closely scrutinised their long-term arrangement or *Jahrhundertvertrag* under which electricity users pay a surcharge to subsidise high-priced local coal used in power stations. With elections looming in the coal states of North Rhine–Westphalia and Saarland, Chancellor Kohl was anxious to limit futher damaging cutbacks in coalmining. But a consensus was emerging that the long-term deal needed drastic revision after 1995. It was generally assumed that West Germany's EC partners, particularly France, would sanction further coal subsidies only if the French could step up cross-border electricity sales. Industrial users in West Germany were keen to buy electricity from France, which produces three-quarters of its power relatively cheaply from nuclear plants.

The nuclear image

German utilities were coming under added pressure to cut electricity prices and many indicated this would be possible, now they did not face the extra costs of Wackersdorf. Interestingly, some utilities were eyeing prospects for diversification more intently. One diversifier, RWE, pressed ahead with its re-organisation after taking over the German oil refining and petrol distribution network of Texaco of the USA in 1988.

Nuclear power plants, which produce just a third of West Germany's electricity, remained deeply unpopular. Their image was not helped by scandals over alleged illicit transport and bribes in the nuclear industry. Nor was the nuclear image improved by the controversial decision to abandon the high temperature reactor at Hamm and continued legal wrangling over whether to carry on with the fast breeder reactor at Kalkar. Social Democrats and especially the Greens continued to argue in 1989 for nuclear plants to be shut or phased out and increasing attention was focused on the cost and complexities of dismantling unwanted nuclear plants.

In preparation for the next federal election, the Social Democrats came up with a scheme to levy environmental taxes on various forms of energy. This plan was being hammered out ahead of the party congress in Bremen in December 1989 and highlighted the importance which all parties attached to environmental issues.

Wrangling – and terrorism

With the Germans inclined to wrangle interminably over taxes, environmental levies seemed likely to be a long-running saga ahead of the federal poll due in

December 1990. The country also appeared heading for a dust-up between trade unions and employers in 1990. The IG Megall engineering union headed by the calculating Mr Franz Steinkühler signalled a tough campaign for pay rises and shorter hours. Overall, West Germany entered 1990 in a fairly grim mood – anxious about the consequences of turmoil in East Germany and about lingering terrorism at home. The Red Army Faction sent a tremor through West German society with its assassination of Mr Alfred Herrhausen, head of Deutsche Bank, on November 30. By blowing up his armoured car near Frankfurt, the terrorists removed a towering figure in German business – a cool, forceful but reflective man close to Chancellor Kohl. He was killed just days after his bank finalised plans to take over Morgan Grenfell, the UK merchant bank. But his vision encompassed not just Western Europe but was focusing more on the challenges emerging in East Germany and throughout Eastern Europe. The murder of such a figure – a man determined to speak out despite the clear physical risk he ran – was likely to make even the most courageous German businessmen think twice about taking a high public profile while terrorism remained a threat. All this set a sober and disquieting stage for West Germany's domestic political struggles in 1990, with its state and federal elections.

BUSINESS GUIDE

Time: GMT + 1 (GMT + 2 from late Mar to late Sep).

Climate: Temperate; warm summers, fairly cold winters. Warmest month July. Rain throughout year.
Clothing: Medium weight for winter, lightweight for summer.

Entry requirements: Visa not required for nationals of most European countries and several others. N.B. regulations differ slightly for travel to West Berlin. Travel agent or German consulate should be consulted prior to departure.
Health: Vaccination certificates not usually required, unless arriving from infected area. EC nationals covered for medical treatment.
Currency: No restrictions on the movement of local or foreign currency. Imports of money for investment subject German regulations.
Customs: Personal effects duty-free.

Air access: Regular flights by all major international airlines.
National airline: Lufthansa.
Main international airport: Cologne-Bonn (CGN) 28 km from Bonn, 18 km from Cologne, Berlin-Tegel (BER) 8 km from city, Düsseldorf (DUS) 13 km from city, Frankfurt (FRA) 10 km, Hamburg (HAM) 11 km, Hanover (HAJ) 11 km from city, Munich (MUC) 10 km, Stuttgart (STR) 14 km.

Surface access: Good rail, road and sea connections with all surrounding countries. Access to West Berlin by rail or autobahn from Hamburg, Hanover, Munich and Nuremberg.
Main ports: Bremen, Bremerhaven, Hamburg, Kiel, Wilhelmshaven.

Hotels: No offical rating system. Single room with breakfast costs up to DM300, including VAT and 10-15 per cent service charge. Advisable to book well in advance, especially when trade fairs are being held. All major credit cards accepted.

Restaurants: Wide range of types and prices. Do not wait to be seated by waiter; sit at empty table and signal to waiter or waitress. French and English spoken widely in restaurants. Bill usually includes service charge and VAT. Extra tip optional.

Currency: Deutsche mark floats against all foreign currencies, but is included in European Monetary System float. No multiple exchange rates in Germany.

Foreign exchange information
Account restrictions: No restrictions on deposits of non-residents. *Trade policy:* Few import restrictions. No import duties on goods from EC states. Licences only required for military equipment, drugs and agricultural products.

Car hire: Widely available. Daily rates DM55-155 plus DM0.39-1.33 per km, depending on size of car. Weekly rate for medium-size car about DM860 with unlimited km (special weekend rates available). Speed limits: built up areas 50kph, normal roads 100kph, autobahns 'recommended' top speed of 130 kph. Information from automobile clubs such as Allgemeiner Deutscher Automobil Club eV (ADAC), Automobil Club von Deutschland eV (AvD) and Deutscher Touring Automobil Club eV. Wearing of seat belts compulsory.

City transport: Buses, trams, metro and electric railway services in many towns. Good taxi service in all main cities.

National transport
Air: Frequent services linking West Berlin, Hanover, Cologne/Bonn, Düsseldorf, Frankfurt, Hamburg, Bremen, Munich, Nuremberg and Stuttgart. Early-morning flights provide direct links between many of these centres.
Road: Good nationwide coach services operated by the German Federal Railways and others. Almost 500,000 km of road, including 6,500 km motorways (autobahns).
Rail: Reliable nationwide service operated by German Federal Railways (Deutsche Bundesbahn). Major cities linked by EuroCity trains (EC), and Inter-city trains (IC Zug) – first and second class. Advisable to book in advance. Over 32,000 km of track, more than one-third electrified.
Water: Seaports on the Baltic and North Sea coasts linked to inland waterways and railways. Over 500 km of navigable inland waterways, which are used extensively.

Public holidays
Fixed dates: 1 Jan, 1 May, 17 Jun (Day of Unity), 24-26 Dec, Christian festivals such as Epiphany and Assumption are observed in some regions. Many businesses, especially in manufacturing industry, close down for a month in July/August.
Variable dates: Good Friday, Easter Monday, Ascension Day, Whit Monday. Other festivals such as Corpus Christi and Day of Prayer and Repentance, are celebrated in some regions.

Working hours
Business: (Mon-Fr) usually 0800 – 1730.
Government: (Mon-Fri) usually 0800 – 1700 (close 1500 in Bonn on Fri).
Banking: (Mon-Fri) various hours between 0830 and 1530 (most open until 1800 on Thu).
Shops: (Mon-Fr) usually 0800 or 0900 – 1830 (close 1400 on Sat, 1800 on first Sat of month). Some shops may stay open Thu evening under recent agreement.

Social customs: Shake hands with host. No longer important to address people by their title, though academic titles 'Herr/Frau Doktor' and 'Herr/Frau Professor' should always be used. However, people in small towns more title conscious than those in cities. Do not try to pay bill if invited to a restaurant during business hours. If dining at a German's home it is considered impolite to arrive late, gift of flowers a social 'must', do not drink until the host has his or her glass.

Telecommunications
Telephone: International dialling code: 49 followed by 228 for Bonn and 69 for Frankfurt. Local calls from public call boxes cost 20 Pfennig, long-distance calls DM1. Cheaper rate for long-distance calls between 1800 and 0800.
Telex: Widely used. Available in main hotels, at fairs and exhibitions and at Pressehaus 1 (Bonn) and Fernmeldeamt 1 (West Berlin). Country code 41D.
Telegram: Available through post offices.

Postal service: Air mail outside Europe pays variable surcharges according to weight and distance. Stamps available from hotels, slot machines and post offices. Use of post code (Postleitzahl) obligatory.

Advertising: Ethical controls operate throughout country and comparative advertising is banned. All media are widely used. Detailed information available from directories such as Leitfaden durch Presse und Werbung. Major advertising media: newspapers, consumer magazines and TV.

Trade fairs: Host to several of the world's most important international trade fairs, especially in such fields as textiles, foodstuffs, machinery, printing, books, toys, hardware and household goods. Information on events published by Ausstellungs-und Messe-Ausschuss der Deutschen Wirtschaft, in Cologne, which lists in its calendar about 140 fairs and exhibitions of supraregional or international significance. The major fair cities are West Berlin, Cologne, Düsseldorf, Essen, Frankfurt, Hamburg, Hanover, Munich, Nuremberg and Stuttgart. .

Electricity supply: 220-250 V AC.

Chambers of Commerce
Berlin – Industrie– und Handelskammer zu Berlin, D-1000 Berlin 12, Hardenbergstrasse 16-18 (tel: 31801; tx: 183663; fax: 3180278; teletex: 308606).

Bonn – Industrie– und Handelskammer Bonn, D-5300 Bonn, Bonner Talweg 17 (tel: 22840; tx: 8869306).

Cologne – Industrie– und Handelskammer zu Köln, D-5000 Cologne 1, Unter Sachsenhausen 10-26 (tel: 16400; tx: 8881400).

Düsseldorf – Industrie– und Handelskammer zu Düsseldorf, D-4000 Düsseldorf 1, Ernst–Schneider–Platz 1 (tel: 35571; tx: 852815).

Frankfurt – Industrie– und Handelskammer Frankfurt /Main, D-6000 Frankfurt/Main 1, Börsenplatz (tel: 21970; tx: 411255).

Hanover – Industrie– und Handelskammer zu Hanover–Hildesheim, D-3000 Hanover 1, Schiffgraben 49 (tel: 31070; tx: 922769).

Munich – Industrie– und Handelskammer zu München und Oberbayern, D-8000 Munich 2, Max-Joseph-Strasse 2 (tel: 5116; tx: 523678).

Münster – Industrie– und Handelskammer zu Münster, D-4400 Münster, Sentmaringer Weg 61 (tel: 7071; tx: 892817).

Nuremberg – Industrie– und Handelskammer zu Nürnberg, 106, Hauptmarkt 25-27 (tel: 13350; tx: 622114).

Stuttgart – Industrie– und Handelskammer Mittlerer Neckar, Sitz Stuttgart, D-7000 Stuttgart 1, Jaegerstrasse 30 (tel: 2005-0; tx: 722031).

Deutscher Industrie– und Handelstag (DIHT) (Federation of German Chambers of Commerce), Adenauerallee 148, D-5300 Bonn (tel: 1040).

Banking
Deutsche Bundesbank (central bank), D-6000 Frankfurt/Main, 50, Wilhelm-Epstein Strasse 14 (tel: 1581; tx: 414431).

Bayerische Hypotheken und Wechselbank AG, Theatinerstrasse 11, D-8000 Munich 2 (tel: 23661; tx: 528650).

Bayerische Landesbank Girozentrale, Brienner Strasse 20, D-8000 Munich 2 (tel: 217101; tx: 5286270).

Bayerische Vereinsbank AG, Kardinal-Faulhaber-Strasse 1, D-8000 Munich 2 (tel: 21321; tx: 529921).

Bundesverband Deutscher Banken (Federal Association of German Banks), D-5000 Cologne 1, Mohrenstrasse 35-41, Postfach 100246 (tel: 16631; tx: 8882730).

Bundesverband der Deutschen Volksbanken und Raiffeisenbanken, D-5300 Bonn 1, Heussallee 5, Postfach 120440 (tel: 5090; tx: 886779).

Commerzbank AG, Postfach 100505, Neue Mainzer Strasse 32/36, D-6000 Frankfurt/Main (tel: 13620; tx: 4152530).

Deutsche Bank AG, Taunusanlage 12, D-6000 Frankfurt/Main 1 (tel: 71500; tx: 417300).

Deutscher Sparkassen – und Giroverband, D-5300 Bonn 1, Simrockstrasse 4, Postfach 1429.

DG Bank (Deutsche Genossenschaftbank), Am Platz der Republik, D-6000 Frankfurt/Main 1 (tel: 744701; tx: 412291).

Dresdner Bank AG, Jürgen-Ponto-Platz 1, D-6000 Frankfurt/Main (tel: 2630; tx: 415240).

Hessische Landesbank Girozentrale, Junghofstrasse 18-26, D-6000 Frankfurt/Main (tel: 13201; tx: 4152910).

Norddeutsche Landesbank Girozentrale, Georgsplatz 1, D-3000 Hanover 1 (tel: 1030; tx: 921620).

Westdeutsche Landesbank Girozentrale, Herzog Str 15, D-4000 Düsseldorf (tel: 82601; tx: 8588216).

Bank für Gemeinwirtschaft, Box 110222, Theaterplatz 2, D-6000 Frankfurt/Main 1 (tel: 2580; tx: 4122154).

Principal newspapers
B Handelsblatt, 4000 Düsseldorf 1, Kasernenstr 67, Postfach 1102 (tel: 83880; tx: 17221391).

Der Spiegel, 2000 Hamburg 11, Brandstwiete 191 Ost-West Strasse, Postfach 110420 (tx; 2162477).

Die Welt, 5300 Bonn Godesberger Allee 9 (tel: 3041; tx: 885714).

Die Zeit, 2000 Hamburg 1, Postfach 106820, Speersort 1, Pressehaus (tel: 3280-0; tx: 886433).

Frankfurter Allgemeine Zeitung, 600 Frankfurt a.M., Hellerhof Str 2-4, Postfach 100808 (tel; 75910; tx: 41223).

Stern, Gruner und Jahr AG, 2000 Hamburg 36, Postfach 302040 (tel: 41181; tx: 2195213).

Süddeutsche Zeitung, 8000 Munich 2, Sendlingerstr 80, Postfach 202220 (tel: 21830; tx: 523426).

Wirtschaftswoche, 4000 Düsseldorf 1, Kasernenstr 67, Postfach 3734 (tel: 8388-0; tx; 8582048).

Hotels: Berlin
Ambassador, Bayreuther Strasse 42/43 (tel: 219020; tx: 184259; fax: 21902380).

Berlin, Kurfurstenstrasse 62 (tel: 269291; tx: 184332).

Bristol Kempinski, Kurfurstendamm 27 (tel: 884340; tx: 183553; fax: 8836075).

Inter-Continental, Budapester Strasse 2 (tel: 26020; tx: 182894).

Hotels: Bonn
Bergischer Hof, Münsterplatz 23 (tel: 633441/42).

Bristol, Poppelsdorfer Allee/Prinz Albert Strasse (tel: 26980; tx: 8869661; fax: 2698222).

Continental, Am Hauptbahnhof 1 (tel: 635360/63535).

Etap, Königshof, Adenauerallee 9-11 (tel: 26010; tx: 886565).

Hotel am Tulpenfeld, Heussallee 2-10 (tel: 219081; tx: 0886328).

Steigenberger, Am Bundeskanzlerplatz (tel: 20191; tx: 886363).

Sternhotel, Markt 8 (tel: 654455; tx: 886508).

Hotels: Bremen
Crest, August-Bebel-Allee 4 (tel: 23870; tx: 244560).

Park, Im Burgerpark (tel: 34080; tx: 244343; fax: 3408602).

Plaza, Hillmannplatz 20 (tel: 17670; tx: 246868).

Hotels: Cologne
Ambassador, Barbarossaplatz 4a (tel: 235181; tx: 8881780).

Dom, Domkloster 2a (tel: 20240; tx: 8882919; fax: 2024260).

Excelsior Ernst, Domplatz (tel: 2701; tx: 8882645; fax: 135150).

Inter-Continental, Helenen Strasse 14 (tel: 2280; tx: 8882162; fax: 2281301).

Hotels: Dortmund
Drees, Hohe Strasse 107 (tel: 12290; tx: 822490; fax: 1299555).

Novotel Dortmund-West, Brennaborstrasse 2 (tel: 65485; tx: 8227007; fax: 650944).

Parkhotel Witterkindshof, Westfalendamm 270 (tel: 596081; tx: 822216).

Westfallen, Strobelallee 41 (tel: 1204245; tx: 822413).

Hotels: Dusseldorf
Breidenbacher Hof, Heinrich-Heine-Allee 36 (tel: 13030; tx: 8582630; fax: 1303830).

Hilton, 20 Georg-Glock-Strasse (tel: 437703; tx: 8584376; fax: 4377650).

Holiday Inn, Konigsallee Graf Adolf Platz (tel: 38730; tx: 8586359; fax: 3873390).

Inter-Continental, Karl Arnold Platz 5 (tel: 45530; tx: 8584601; fax: 4553110).

Nikko, Immermannstrasse 41 (tel: 8340; tx: 8582080; fax: 161216).

Ramada, Am Seestern 16 (tel: 591047; tx: 8585575).

Hotels: Essen
Bredeney, Theodor-Althoff=Strasse 5 (tel: 714081; tx: 857597).

Sheraton, Huyssenallee 55 (tel: 20951; tx: 8571266; fax: 231173).

Hotels: Frankfurt
Arabella, Lyoner Strasse 44/48 (tel: 66330; tx: 416760; fax: 6633666).

Continental, Baseler Strasse 56 (tel: 230341; tx: 412502; fax: 232914).

Frankfurt Marriott, Hamburger Allee 2/10 (tel: 6000; tx: 412573; fax: 79552432).

Hessicher Hof, Friedrich-Ebert-Anlage 40 (tel: 75400; tx: 411776; fax: 7540954).

Inter-Continental, Wilhelm-Leuschner-Strasse 43 (tel: 26050; tx: 413639; fax: 252467).

Sheraton, Frankfurt Airport (tel: 69770; tx: 8414189294; fax: 69772209).

Hotels: Hamburg
Atlantic Kempinski, An der Alster 72/79 (tel: 28880; tx: 2163297; fax: 247129).

Hamburg Marriott, ABC Strasse 52 (tel: 35050; tx: 2165871; fax: 4035051777).

Holiday Inn, Leopoldstrasse 200 (tel: 228060; tx: 2165287; fax: 2208704).

Inter-Continental, Fontenay 10 (tel: 414150; tx: 211099).

Plaza, Marseiller Strasse 2 (tel: 35020; tx: 214400).

Ramada Renaissance, Grosse Bleichen (tel: 349180; tx: 17403141; fax: 34918431).

Vier Jahreszeiten, Neuer Jungfernstieg 9/14 (tel: 34940; tx: 211629; fax: 3494602).

Hotels: Munich
Hilton, Am Tucherkpark 7 (tel: 48040; tx: 529437; fax: 48044804).

Holiday Inn, Leopoldstrasse 200 (tel: 5218645; tx: 780020; fax: 78002672).

Königshof, Karlsplatz 25 (tel: 551360; tx: 523616).

Sheraton, Arabellastrasse 6 (tel: 924011; tx: 522391).

Vier Jahreszeiten Kempinski, Mazimilianstrasse 17 (tel: 230390; tx: 523859; fax: 23039693).

Westpark, Garmischer Strasse 2 (tel: 51960; tx: 523680; fax: 5196649).

Hotels: Stuttgart
International, Plieninger Strasse 100 (tel: 72021; tx: 7255763; fax: 7202210).

Schlossgarten, Schillerstrasse 23 (tel: 20260; tx: 722936; fax: 2026888).

Schlossgarten, Schillerstrasse 23 (tel: 20260; tx: 722936; fax: 2026888).

Steinberger Graf Zeppelin, Arnulf-Klett-Platz 7 (tel: 299881; tx: 722418).

Other useful addresses
Allgemeiner Deutscher Automobil Club (ADAC) eV, Am Westpark 8, D-8000 Munich 70 (tel: 7676/0).

Arbeitskreis Deutscher Marktforschungsinstitut (ADM) (association of German market research institutes), Burgschmiedstrasse 2, D-8500 Nuremberg (tel: 3951).

Arbeitsgemeinschaft der deutschen Exporteurvereine (federation of German exporters), Gotenstrasse 21, D-2000 Hamburg 1 (tel: 2801337).

Aussenhandelsvereinigung des deutschen Einzelhandels (AVE) (foreign trade association of the German retail trade), Weyerstrasse 2, D-5000 Cologne 1 (tel: 217617).

Ausstellungs- und Messe-Ausschuss der Deutschen Wirtschaft, Lindenstrasse 8, D-5000 Cologne 1.

Automobil Club von Deutschland (AvD) eV, Lyoner Strasse 16, D-6000 Frankfurt/Main 71 (tel: 6606/0).

Berlin Economic Development Corporation (BEDC), Budapester Strasse 1, D-1000 Berlin 30 (tel: 2636/1; tx: 184467).

Bund Deutscher Unternehmensberater (BDU) (association of German business consultants), Gotenstrasse 161, D-5300 Bonn (tel: 379001).

Bund Deutscher Verkaufsförderer und Verkaufstrainer (BDVT) (association of German sales promoters and sales trainers), Mühlenfeld 69, D-4005 Meerbusch 2 (tel: 4121).

Bundesamt für gewerbliche Wirtschaft (federal office for trade and industry), Frankfurter Strasse 29-31, D-6236 Eschborn 1 (tel: 4041).

Bundesanstalt für Arbeit (federal labour office), Feuerbachstrasse 42, D-6000 Frankfurt/Main (tel: 71111).

Bundesstelle für Aussenhandelsinformation (BfA) (German foreign trade information office), Blaubach 13, D-5000 Cologne (tel: 20571).

BUSINESS DIRECTORY

Bundesverband des Deutschen Gross– und Aussenhandels (federation of German wholesale and foreign trade), Kaiser-Friedrich-Strasse 13, D-5300 Bonn 1 (tel: 26004/0).

Bundesverband der Deutschen Industrie (BDI) (federation of German industries), Gustav-Heinemann-Ufer 84-88, D-5000 Cologne 51 (tel: 3708/0).

Bundesverband der Dolmetscher und Uübersetzer eV (interpreters), Rüdigerstr 79a, D-5300 Bonn 2 (tel: 345000).

Bundersverband deutscher Markt– und Sozialforscher (BVM) (association of German market researchers), Schlossstrasse 4, D-6050 Offenbach (tel: 8059266).

Bundesvereinigung der Deutschen Arbeitgeberverbände (BDA) (federation of German employers' associations), Gustav-Heinemann-Ufer 72, D-5000 Cologne 51 (tel: 37950).

Bundesvereinigung der Fachverbände des deutschen Handwerks (federation of German crafts), Johanniter-Strasse 1, D-5300 Bonn (tel: 5451).

Bundesvereinigung deutscher Einkaufsverbände (BEV) (federation of German purchasing associations), Neumarkt 14, D-5000 Cologne (tel: 219456).

Centre for Economic Promotion, Landesgewerbeamt Baden – Württemberg (contact: Dr Reuss), Postfach 831, D-7000 Stuttgart 1 (tel: 1232222; tx: 723931).

Centralvereinigung Deutscher Handelsvertreter-und Handelsmakler-Verbände (CDH) (federation of German trade agents and brokers' associations), Geleniusstrasse 1, D-5000 Cologne 41 (tel: 514043).

Deutsche Presse-Agentur GmbH (dpa) (news agency), D-2000 Hamburg 13, Mittelweg 38.

Deutsche Zentrale für Tourismus (tourist Board), D-6000 Frankfurt/Main 1, Beethovenstrasse 69 (tx: 4189).

Deutscher Gewerkschaftsbund (DGB) (federation of German trade unions), Hans-Böckler-Strasse 39, D-4000 Düsseldorf 30 (tel: 43010).

Deutscher Kommunikationsverband (BDW) (association of German advertising consultants), Adenauerallee 209, D-5300 Bonn (tel: 211047).

Deutscher Public Relations Gesellschaft (DPRG) (German society for public relations), Flandrische Strasse 4, D-5000 Cologne 1 (tel: 212499).

Deutscher Touring Automobil Club eV, Amalienburgstrasse 23, D-8000 Munich 60 (tel: 8111048).

Deutsches Institut für Wirtschaftforschung (DIW) (economic research institute), Königin-Luise-Strasst 5, D-1000 Berlin (tel: 829910).

Gesellschaft für Konsum-, Markt– und Absatzforschung (GfK) (market research institute), Burgschmietstrasse 2, D-8500 Nuremberg (tel: 3951).

Gesellschaft Public Relations Agenturen (GPRA) (association of PR agents), Sierichstrasse 43, D-2000 Hamburg (tel: 2701936).

Hauptgemeinschaft des Deutschen Einzenhandels (HDE) (central association of German retail trade), Sachsenring 89, D-5000 Cologne 1 (tel: 322091).

Ifo-Institut für Wirtschaftforschung (economic research institute), Poschinger Strasse 5, D-8000 Munich (tel: 92241).

Infratest (market research institute), Landsberger Strasse 338, D-8000 Munich (tel: 56001).

Institut der Deutschen Wirtschaft (economic research institute), Gustav-Heinemann-Ufer 84, D-5300 Bonn (tel: 37041).

Institut für Wirtschaftforschung (HWWA) (economic research institute), Neuer Jungfernstieg 21, D-2000 Hamburg 36 (tel: 35621).

Kontakstelle zur Einfuhr aus Entwicklungsländern (central office for imports from developing countries), Kaiser-Friedrich-Strasse 13, D-5300 Bonn (tel: 218051).

Kooperationsbüro des rationalisierungs-Kuratoriums (German industrial standardisation board), Düsseldorfer Strasse 40, D-6236 Eschborn (tel: 4951).

Ministry of Economic Co-operation, 5300 Bonn 1, Karl-Marx-Str 4-6 (tel: 5351; tx: 8869452).

Ministry of Economics, 5300 Bonn 1, Villemombler Str 76 (tel: 6151; tx: 886747).

Ministry of Finance, 5300 Bonn 1, Graurheindorfer Str 108 (tel: 6821; tx: 886645).

Statisches Bundesamt (federal statistical office), Postfach 5528, D-6200 Wiesbaden (tel: 751).

Verband beratender Ingenieure (VBI) (association of consulting engineers), Zweigstrasse 37, D-4300 Essen (tel: 792044).

Verband der Fertigwarenimporteure (VFI) (association of finished good importers), Gotenstrasse 21, D-2000 Hamburg (tel: 2801337).

Verband unabhängig beratender Ingenieurfirmen (association of independent consulting engineers), Winston-Churchill-Strasse 1, D-5300 Bonn (tel: 217064).

Verlag W Kohlhammer (publisher of federal statistics), Philip-Reis-Strasse 3, D-6500 Mainz 42 (tel: 59094).

Verlag moderne Industrie (publisher of 'Who's Who in Marketing and Communications'), Justus-von-Liebig-Strasse 1, D-8910 Landsberg (tel: 1251).

Zentralausschuss der deutschen Werbewirtschaft (ZAW) (central committee of the German advertising industry), Villichgasse 17, D-5300 Bonn (tel: 351025).

Ghana

Adotey Bing

The Economic Recovery Programme (ERP) continued in 1989 with its focus on liberalisation and openness. It was still hitting urban workers in particular. The external balance continued to be covered, but local investment was still not high enough to guarantee sustained growth. Meanwhile 110 district assemblies had been set up and commenced deliberating. They would soon be making their own demands (political and economic) on the system. How these various expectations would be met constituted the primary question for 1990.

Economic background

Immediately after independence, Ghana, led by Kwame Nkrumah, encouraged foreign capital to set up an industrial base. This option was abandoned as having failed after about three years. Nkrumah then pursued a state-owned and managed industrialisation strategy. Major policy instruments were import controls, and credit management. Some important industrial enterprises were thus established, and manufacture's share of GDP went from 10 per cent in 1960 to 14 per cent in 1970. A number of enterprises were inefficiently run and some of the managers were corrupt. But unstable and plummeting cocoa prices, coupled with the West's general hostility towards Nkrumah's Pan-Africanism and socialism placed additional obstacles before the programme. His overthrow in 1966 ushered in a period of liberal economic policies increasingly badly managed, first by the military National Liberation Council (NLC) that overthrew him, and later by the government of Kofi Busia. The National Redemption Council which overthrew Busia in 1972 began by reversing his 90 per cent devaluation of the cedi. It followed this with nationalistic and populist measures such as 'Operation Feed Yourself' and 'Feed Our Industries'. However, it not only mismanaged the import licence system, but was responsible for phenomenal increases in money supply and government borrowing. The first resulted in scarcity in essential commodities and the second in hyperinflation.

KEY FACTS

Official title: Republic of Ghana

Head of state: Flight Lt Jerry John Rawlings

Capital: Accra

Official Languages: English (Twi, Fante, Ga, Ewe, Dagbani, Adanme widely spoken)

Currency: Cedi (C) = 100 pesewas

Exchange rate: C291 per $ (Nov 1989)

Area: 239,460 sq km

Population: 14.1m* (1988)

GDP per capita: $624 (1987)

GDP real growth: 6.0%* (1988)

Inflation: 31.3% (1988)

Trade balance: –$124.7m (1987)

Foreign debt: $3.1bn (1987)

* estimated figure

Thus for a decade after 1970, much of it under the NRC which later metamorphosed into the Supreme Military Council (SMC), economic policy produced unenviable results. Real GDP stagnated, export volumes declined and industrial capacity utilisation plummeted to around 20 per cent. The neglected state-owned enterprises made losses, and high inflation coupled with negative real interest rates reduced the domestic savings rate to 3 per cent and investment to 2 per cent of GDP. Real wages fell by 80 per cent and the central government deficit rose to 15 per cent of GDP. The resulting internal price distortions provided greater returns to traders and renters than to producers.

Between 1978 and December 1981 when the Provisional National Defence Council (PNDC) led by Jerry Rawlings came to power, there were rapid and chaotic changes in regimes, reflecting the underlying political crisis brought on by over a decade of economic and social decline.

Economic Recovery Programme

On coming to power the PNDC adopted an IMF-supported Economic Recovery Programme (ERP), phase II of which came to an end in 1988. Its main objectives were stabilisation and liberalisation. Its first target was excess liquidity which it tackled by demonetising all C50 notes, at the time the highest denomination note. Then from April 1983 to May 1986, the cedi was successively devalued from C2.75 to C90 to the dollar. In September of that year, a flexible exchange rate policy was adopted obviating the need for further 'official devaluations'. The exchange rate is now determined at a weekly auction. Price controls were removed in 1984 and a policy of remunerative agricultural producer prices adopted in particular to encourage production of cocoa and cash crops. This has taken the cocoa price from C360 per (30kg) bag to C5,500 per bag in June 1988. Exporters can now retain 20 per cent of the foreign exchange they earn. Exceptions are the cocoa marketing organisation (Cocobod) which retains only 10 per cent and the mining sector where Ashanti Goldfields Corporation (AGC) can retain 45 per cent and the State Gold Mining Corporation (SGMC), 35 per cent.

To encourage foreign private investors a new investment code was adopted. In June 1988 32 state-owned enterprises, out of a total of 235 (181 of which are majority owned), were listed for privatisation. Strenuous efforts have gone into rehabilitating the neglected, and by 1982 very inefficient, transport and communications infrastructure. Lastly, in a bid to eliminate the central government deficit, public expenditure has been cut back and subsidies reduced or removed. Redeployment from the public sector has taken an estimated 45,000 people off government or parastatal pay roles.

In response real GDP growth has averaged 6 per cent since 1984. Cocoa and timber export volumes increased. Overall total export earnings have held up, but the stubbornly negative trade balance has necessitated continued inflows of capital. Inflation however, having fallen in 1985, rose to double the 1982 level in 1988. The public deficit of 6.3 per cent of GDP in 1982 was turned into a surplus in 1986.

International multilateral agencies were quick to applaud the ERP, and Ghana has received around $2.5bn in external support. In November 1987 the IMF approved a total of SDR375.3m over three years, followed by an Extended Fund Facility (EFF) worth SDR245.4m and a SDR129.9m Structural Adjustment Facility, also to be spread over three years. On the domestic scene, the approach improved the position of cash crop farmers and some local businessmen. But resistance continued to come from industrial and office workers who in many respects bore the brunt of the policies.

Current policies and programmes

The basic objectives of liberalisation, promotion of the export sector and reliance on private investors for growth, continued. The ERP has now become a rolling programme. The current phase is attempting to make the financial sector a more efficient mobiliser of investible surpluses. A $100m loan from the IDA was agreed in 1988 for this purpose. But there is also promise of further rehabilitation and stimulation of

KEY INDICATORS

	Unit	1984	1985	1986	1987	1988
Population	m	12.39	12.72	13.05	13.39	14.13*
Gross domestic product (GDP)	C m	270,561	343,048	511,373	746,000	–
GDP per capita	C	21,837	26,969	39,186	55,713	–
Inflation	%	39.7	10.3	24.6	39.8	31.3
GDP real growth	%	2.6	5.1	5.2	4.8	6.0
Agricultural production	1979-81=100	129.1	123.7	130.1	132.1	141.0
Food	1979-81=100	130.3	124.8	131.2	133.3	142.3
Cocoa production	'000 tonnes	175	205	226	188	290
Gold production	'000 oz	287	310	287	298	327
Diamond production	'000 carats	346	632	558	436	225
Manganese	'000 tonnes	287	349	333	295	262
Exports	$m	566.0	632.4	773.4	826.8	–
Imports	$m	533.0	668.7	712.5	951.5	–
Balance of trade	$m	32.9	–36.3	60.9	–124.7	–
Current account	$m	–38.8	–134.2	–43.0	–96.9	–
Total external debt	$m	1,900	2,175	2,656	3,124	–
Debt service ratio	%	16.4	15.6	16.7	–	–
Exchange rate	C per $	36.0	54.4	89.2	153.7	202.4

* estimated figure

the export sector, along with institutional changes and more liberalisation. At the end of 1988 import licences were abolished. Special attention is being given to gold which is now considered a more reliable source of export earnings than Ghana's traditional export commodity, cocoa. Nevertheless, although cocoa export earnings were down in 1988, cocoa output is to be increased, and is expected to reach 300,000 tonnes in 1995. But by 1991 it is planned that it will contribute only 40 per cent of export earnings. Other commodities being developed are cotton, bottles, packaged pineapples, limes, mangoes, avocados, shellfish and beer.

The private sector, especially foreign investment, is being encouraged. At a meeting with potential foreign investors, Secretary for Finance and Economic Planning Kwesi Botchwey said the private sector would be expected to provide 50 per cent of investment by 1990. This now seems very optimistic. The 1989 budget contained important changes in company tax regulations. A three-tier corporate tax system was introduced. The highest, 55 per cent, was for banking, insurance, commerce and printing. The lowest is now 45 per cent, to be paid by manufacturing, farming, and the export sector. The middle band of 50 per cent will be paid by all other branches of the economy. Meanwhile a stock exchange is being mooted, but no firm plans seem to have been made. Some effort has been put into developing a rural banking system and by July 1987, 119 such banks had been established with total deposits of C2.2bn. About 40 per cent of their lending goes to agriculture.

Recent performance

Real GDP growth for 1988 was estimated at 6 per cent. Latest figures for Gross Fixed Investment place it at 10.7 per cent of GDP in 1987, well up on the 3.5 per cent of GDP recorded in 1982. Inflation was officially reported to be 25 per cent in 1988 (unofficially 33 per cent). Consumer prices rose 40 per cent in 1987. Figures for 1988 (up to September only) indicated a growth in money supply of only 7 per cent, reflecting the government's growing determination to bring inflation down to 5 per cent by 1990. The impact of this policy on the productive sector in particular remains to be seen. During 1989 a tight money policy was continued, with difficult results for manufacturing in particular. Despite the fact that all banks are under instructions to lend at least 20 per cent of their portfolios to agriculture, many did not do so. The continuing weakening of the cedi as determined through the auction tended to keep internal prices moving up. The exchange rate in April 1989 was C263 to the dollar compared with an average of C200 to the dollar for 1988.

Although the minimum wage was increased to C170 per day in 1989, the increase was much less than past rates of inflation and since it has no legal backing is not likely to be adhered to by all concerned. Thus its impact is likely to be minimal. Food prices showed a tendency to rise under the ERP's liberalisation policy, reaching an estimated 33 per cent in 1988. Real interest rates have been negative for a number of years and are in part responsible for the low level of domestic savings (10.7 per cent of GDP in 1987). On the other hand a high nominal interest rate discourages producers and encourages speculators. In 1989 official controls over the maximum lending and minimum deposit rates were abolished. In response Ghana Commercial Bank reduced its interest rate on loans to a standard 28 per cent. Exceptions were agriculture (23 per cent), export trade (24 per cent) and manufacturing (25 per cent).

The government budget surplus of C38,592m in 1988 was the largest of recent times, comparing favourably with the 1987 surplus of C30,463m. The surplus was sufficient to meet the requirements of development expenditure as well as lending to, and investment in, public utilities. Latest figures showed government spending falling as a share of GDP: by 1987 it was down to 10.6 per cent from the abnormally high 20.3 per cent of 1983. Since 1985 the government has been able to keep the current account balance positive. Latest figures on the structure of government revenue showed that since 1982 more of it had been coming from trade and less from tax on domestic production and incomes. Tax receipts from international trade were 40 per cent of revenue, and those from incomes were down a third to 21 per cent, while taxes from domestic production amounted to 24 per cent. The proposed cutback of 12,000 jobs from the public sector will reduce the level of expenditure further, as will those on education. Students will now be awarded loans which will be repayable on employment. But the government will only be saving money if the correct level of interest charges is chosen, and students on graduation find employment at pay levels sufficient to service their debts. On the other hand the decentralisation needed to allow the district assemblies to function properly might increase government spending during the early 1990s.

External trade and payments

Latest figures showed a C1.96bn ($11.2m) trade surplus in 1987. Non-traditional exports such as handicrafts and fish, as well as agricultural products such as pineapples, contributed an increasing – if still small – share (5 per cent in first half of 1988) to export earnings. Traditionally, exports are dominated by cocoa which in 1987 generated 59 per cent of earnings. In 1988 a fall in world cocoa prices led to a loss of an estimated $100m. In a bid to increase export sales of worked wood a timber exhibition was held in London in July 1988.

While the trade balance since 1982 has fluctuated, one year positive and the next negative, the balance on current account has remained positive. Between 1982 and 1987, OECD and OPEC countries disbursed an estimated $1,559m. This annual average of $260m has ensured that in each year the capital account balance has been positive. Yet it is $290m short of a 1985 estimate of what was necessary annually.

A consultative club meeting, the fifth, took place in March 1989, leading to pledges of $900m. According to the World Bank, $2.2bn was owed as publicly guaranteed debt (disbursed and outstanding) in 1987. Of this, the amount owed to multilateral sources was 36 per cent, down from 38 per cent in 1982. In November 1988 Ghana's exposure to multilateral agencies increased when it gained access to SDR368m of the IMF's new Enhanced Structural Adjustment Facility (ESAF). The agreement will replace the three-year Extended Fund Facility and Structural Fund Facility arranged in November 1987. The terms of the new deal are an improvement over the 1987 ones. Larger amounts can be borrowed for longer periods (10 rather than three years), and at lower interest rates (0.5 compared with 7 per cent). By 1987 external arrears had been reduced to $90m from $600m in 1983.

In March 1988 the government permitted the opening of foreign exchange bureaux, allowed to buy and sell foreign currency. Within three months of authorisation there were over 40 such bureaux. By early 1989 there were over 100, appearing in cities outside Accra. These will help to channel some of the large sums of foreign exchange entering the country informally into official or semi-official channels.

Agriculture

Food production in 1988 was much improved over 1987. Maize output was 600,000 tonnes, up 32 per cent on the previous year. 1988 rice output was 50 per cent up at 95,000 tonnes, and cassava was 13 per cent up at 3.3m tonnes. Altogether agriculture accounted for over 50 per cent of GDP in 1987, a drop from the 57 per cent of 1982. The 'agricultural export sector' which was only 3.3 per cent of GDP in the early 1980s, returned 20.6 per cent of GDP by 1987, mainly because of the policy of flexible exchange rates. Industrial crops such as cotton, rubber, sugar cane, tobacco and kenaf have not done so well in recent years. The exception is oil palm. Indeed the scale of output of the oil palm sub-sector led to plans in 1988 to rehabilitate the Bonsa Tyre factory, the country's tyre producing facility.

Forestry accounted for 4.6 per cent of GDP in 1987. In the early phase of the ERP forestry was considered to have great potential and considerable investments were made to boost output. But log production in 1987 was 951,000 cubic metres, just 6 per cent up on the previous year. Lumber production, at 284,000 cubic metres, did somewhat better at 22 per cent above the 1986 level though in value terms earnings increased by 63 per cent between 1986 and 1987. Nevertheless, the industry is beset with problems. Preliminary results of a government investigation revealed that each year some $49m in unpaid taxes was owed by timber. Nine major companies were found to have not paid their taxes on profits totalling $127.6m between 1986 and 1988. As a result a forest products inspection bureau was set up to supervise sawmills and to develop the wood work industry.

Mining

Mining accounted for 1.8 per cent of GDP in 1987. Figures for the first four months of 1988 showed that gold output was 116,000 fine ounces, well up on the 1987 figures. Other buoyant sub-sectors were manganese (76,000 tonnes compared with 253,000 tonnes for the whole of 1987), and bauxite (95,000 tonnes compared with only 195,000 in 1987). Considerable emphasis is being placed on this sector. There are an estimated 1.8bn ounces in gold reserves. Possible production levels could reach 2.7m ounces compared with the current 400,000 ounces. New shafts are being sunk by AGC and the SGMC is engaged in rehabilitating its plant and equipment. Plans are that these two alone will be producing around 500,000 ounces by 1990. AGC reported profits of C9.13bn ($46m) for 1988, 17 per cent up on 1987. The company produced 305,000 ounces in 1988 and was expected to produce 348,000 ounces in 1989. Official earnings from diamonds increased after a decision to pay local prospectors at the foreign exchange bureaux rate. In December 1988 alone, 10,000 carats were bought by the Diamond Marketing Corporation. Before the change some 4,100 carats were being bought in a whole year.

Industry

By 1987 industry accounted for 15.9 per cent of GDP compared with 6.2 per cent in 1982, while manufacturing contributed 9.9 per cent of GDP in 1987 compared with 3.6 per cent in 1982. Industry, which galloped along at 12.7 per cent in 1985, settled down to a canter at 6 per cent in 1986. Latest figures showed that industrial capacity increased by 11 per cent in 1987 and by 10.3 per cent in 1988, reaching 32 per cent by 1988. This is still below the ERP target however. Local producers are beginning to call for protection from liberalised trade gales now buffeting their enterprises. Garments factories seem to have been particularly badly hit. A report of a survey in 1988 said that nearly all garments factories operating in the country three to four years before had ceased to do so, because of competition brought on by liberalisation. To date the government has indicated that it will not heed local producers' calls to protect them.

Political scene

The district assembly elections resulted in a new layer of representation: some 110 districts averaging between 60 to 70 persons each. No financial or literacy hurdles were placed in the path of prospective candidates. They did however have to be vetted for suitability. Some 400 women were elected representing some 10 per cent of elected candidates. The number of teachers that stood and were successful was high. There is no indication of the extent to which the 'old politicians' gained election to the assemblies.

Rural areas tended to have a high turnout while the towns had considerably lower ones. For example, Upper East and other rural areas reported turnouts of around 80 per cent. By contrast Accra reported 30 per cent. Those elected will comprise two-thirds of the district assemblies; the government will appoint the other one-third.

The assemblies have local responsibility for health, education, maintenance of public infrastructure and education. They are empowered to raise local revenues by levying taxes. A number of branches of the government have been decentralised to accommodate the districts and their functions. These new departments will function as departments or wings of the assemblies and do what the assemblies decide. The assemblies are expected to each have a minimum of 17 district administrative staff, who they will presumably employ and pay. An important area of decentralisation has been the legal system. Public tribunals have been extended to the district and community levels.

Decisions of the district assemblies within their legislative jurisdiction have to be deposited with the secretariat of the PNDC immediately after they have been passed. If within 21 days of this happening, no objection has been raised to them, they automatically become law. The chair of the assembly will be the district secretary, appointed by the PNDC. At the moment nothing is publicly known about the future development of the assemblies, although it seems almost certain that there will have to be, sooner or later, moves towards regional assemblies.

The assemblies are likely to channel energies away from national and towards local politics. They will provide intensified scope for local lobbying and with it increased opportunities for the dispensation of political favours. Harnessing local energies could lead to more local projects. Local businessmen are likely to grow in wealth and influence. The assemblies seem almost certain to institutionalise the altered balance of power between rural and urban peoples.

Internal relations

The PNDC celebrated the tenth anniversary on 4 June 1989 of the first coming to power of Flt-Lt Rawlings with the rallying call 'Freedom, justice and accountability'. In a plea to the middle classes Rawlings in his address called for reconciliation and forgiveness between the PNDC and those who felt they had been wronged by the events of 4 June 1979. This did not prevent the chairman and secretary of the Ghana Bar Association being arrested and detained for a few weeks in June. Although no announcement was made concerning the arrests, they seem to have been connected with the fact that the Bar Association was about to hold a meeting to mark the anniversary of the abduction and murder of three high court judges in 1982. The person responsible, at the time a member of the PNDC, was found guilty and executed. The murder of the judges, as well as an army officer, had been used at the time to try to remove the PNDC from power. Clearly the government's nerves were still too raw.

Meanwhile relations with workers have also been strained. Two disputes have stood out: one at the Ghana Broadcasting Corporation which led to 32 workers being sacked, and the other at Ghana Textiles and Printing (GTP). 1,000 workers were sacked by the government and made to reapply for their jobs.

The release in December 1988 of Kwame Karikari, former head of the Ghana Broadcasting Corporation and chairman of the New Democratic Movement (NDM) who had been detained without charge since July 1987, signalled the beginning of the end of the open confrontation between the government and the left. In May 1989 the remaining two members of the NDM, Akoto Ampaw and Kwesi-Adu Amankwah – political officer of the Trade Union Congress – were released after two years in detention also without charge.

Foreign relations

The visit by Nigerian head of state Ibrahim Babangida signalled improved relations between Ghana and Nigeria. Following the apparent rapprochement with Togo in July 1988, relations with Togo have worsened. An estimated 124 Ghanaians were expelled from there towards the end of the year souring relations. After 15 years the Ghana-Côte d'Ivoire border re-demarcation commission has arrived at an agreement on the definition of the 400-mile border between the two countries. Special arrangements have been arrived at with Britain concerning the conditions that will apply in the event that UK investments estimated at £150m are expropriated. The essence of the agreement will be to provide legal protection for British companies in the event of expropriation.

COUNTRY PROFILE

Political system: Executive and legislative power is vested in the Provisional National Defence Council, and 'Committees for the Defence of the Revolution' handle local matters. Ghana gained independence from the UK in 1957.

Political parties: There are no active political parties. The PNDC's political wing has been dissolved, and its activities have been taken over by the PNDC secretariat.

Population: Ghana's population of around 13.7m in 1987 is growing at an annual rate of 2.7 per cent. The Akan people comprises 44 per cent of the population, the Mole-Dagbani 16%, the Ewe 13%, the Ga-Adangbe 8%, the Guan 4% and the Gurma 4%.

Main cities/towns: Accra (population 1,450,000), Kumasi (600,000), Tamale (265,000), Sekondi-Takoradi (260,000), Tema (220,000).

Language: The principal African languages spoken are Twi, Ewe, Fante, Ga and Dagbani.

Media
Press: Main national dailies are the *People's Daily Graphic, The Pioneer* and *The Ghanaian Times*; main weeklies include the *Weekly Spectator, Mirror, Echo* and *Business Weekly*. A number of periodicals in English and African languages are published.
Radio: Ghana Broadcasting Corporation runs three radio stations. They include a radio service in English, Hausa and other African languages and an external radio service in English, French and Hausa.
Television: There is one state-run TV channel.

External trade: After 1985 the current account improved because of rising export earnings from cocoa, gold and timber.
Exports: Cocoa (54% of total value in 1986), gold (15%), wood (4%), manganese, diamonds. Main destinations: USA (21% of total value in 1986), UK (16%), USSR (11%).
Imports: Machinery and transport equipment (typically 28-35 per cent of total value), basic manufactures, petroleum and products, chemicals, food, beverages and tobacco, miscellaneous manufactures. Main sources: UK (21% of total value in 1986), German FR (11%), Nigeria.

Agriculture: Main sector of the economy, accounting for 50 per cent of GDP and employing around half of the work-force, mainly smallholders. Cocoa is traditionally the most important crop. Output fell in the 1970s, due to fluctuating export demand, low producer prices, poor weather conditions, smuggling and a shift towards food crop cultivation. In 1983-84 the harvest was down to 160,000 tonnes, but recovered and reached 230,000 tonnes in 1986-87. Other cash crops for export include bananas, kola nuts, limes, coffee, copra, shea nuts and palm kernels. Timber is also exported, and export levels have been rising again since 1982, reaching 329,000 cu metres in 1986. Crops grown for the local agro-industry include rubber, sugar, cotton, oil palms and kenaf. Subsistence farming of food crops (cassava, plantains, rice, maize, sorghum, millet, yams) has been affected by prolonged drought and shortages of fertilisers, but there has been a recovery, particularly in maize and rice production. The annual domestic fish catch averages over 200,000 tonnes, satisfying over 75 per cent of domestic demand. Fish farms have been set up in the northern regions in an effort to achieve total fish self-sufficiency. Estimated 43 per cent of total area is potential arable land (estimates of cultivated area as low as 12 per cent of total area have appeared); around 60 per cent is woodland/forest; (estimated 25 per cent rain forest).

Industry: Sector has suffered relative decline over past 10 years and currently contributes about 10 per cent to GDP, as opposed to 20 per cent in the early 1970s. Apart from traditional agro-allied industries such as wood/cocoa processing and brewing, Ghana has a number of medium-sized modern enterprises, including an aluminium smelter vehicle assembly plants, textiles, cement manufacture, oil refining and glass making. Most of these were established during the 1960s, usually with state control or participation, but a new aluminium rolling mill has been established recently. Virtually all industries were affected in 1981-84 by a shortage of imported raw materials and parts, recurrent power cuts and rising costs, but a new industrial policy has scored significant successes in reversing the trend. Current policy centres on maximising capacity utilisation.

Mining: Second largest sector of economy and major foreign exchange earner. Gold is the principal mineral export, accounting for around 15 per cent of total exports. Proven reserves estimated at 57m kg. Ashanti Goldfields Corp (55 per cent owned by the government and 45 per cent by the Lonrho Group) accounts for 85 per cent of total output. The other major producer is the State Gold Mines Corp. Total production of gold (288,000 oz in 1986) is scheduled to rise to over 1m oz by 1990. Diamonds (mostly industrial and mined by the Ghana Consolidated Diamonds Ltd in the Birim Basin), and manganese-ore (mined at Nsuta by the National Manganese Corp) each account for 1-2 per cent of export earnings. Ghana has vast bauxite reserves near Kibi, but heavy transport/extraction costs have so far held back commercial development of local bauxite for use by the Valco aluminium smelter.

Hydrocarbons: Oil was first brought on-stream from the Saltpond Field in 1978. Production is currently 3,000-5,000 b/d. Exploration for more oil is underway in the offshore area of the Tano. A refinery is located at Tema with a capacity/throughput of 1.25m tonnes/year.

COUNTRY PROFILE

Energy: Heavy dependence on imported oil, primarily from Nigeria. Virtually all the country's electricity requirements are met by the 912 mW Akosombo hydro-power plant and the 160 mW Kpong plant. Supplies were affected in 1983-85 by prolonged drought and technical problems, but have greatly improved.

Membership of international organisations: African Development Bank, African States in Association with the EC, Arab Bank for Economic Development in Africa (recipient), Commonwealth, Economic Commission for Africa, Economic Community of West African States, FAO, GATT, IDA, IFAD, ILO, IMF, International Cocoa Organisation, OAU, UN, WHO, WIPO, World Bank.

BUSINESS GUIDE

Time: GMT.

Climate: Tropical. Hottest months in Accra Mar-Apr, coolest Aug. Rainy season Apr-Oct. Cooler and less humid in the north.

Entry requirements

Passport: Required by all.

Visa: Required by all except nationals of Economic Community of West African States (ECOWAS). Multiple entry visas can only be obtained after arrival.

Prohibited entry: Holders of South African passports.

Currency: Unlimited import of foreign currency. Visitors are given Exchange Control Form T5 on arrival, on which declaration of foreign currency and all transactions made during length of stay must be recorded. Export of foreign currency limited to amount declared on arrival. Import of local currency prohibited; export limited to C20. It is advisable to check latest currency regulations prior to visit.

Health precautions

Mandatory: Yellow fever vaccination certificate.

Advisable: Yellow fever and cholera vaccinations considered essential, typhoid and polio vaccinations recommended. Malaria prophylaxis should be taken as risk exists throughout the country. Water precautions are essential; bilharzia risk is present.

Representation overseas: Abidjan, Addis Ababa, Algiers, Beijing, Belgrade, Berlin, Berne, Bonn, Brasilia, Brussels, Cairo, Canberra, Conakry, Copenhagen, Cotonou, The Hague, Harare, Havana, Jeddah, Lagos, Lomé, London, Luanda, Monrovia, Moscow, New Delhi, Ouagadougou, Ottawa, Paris, Prague, Rome, Tehran, Tokyo, Tripoli, Tunis, Washington.

Air access

National airline: Ghana Airways.

Other airlines: Aeroflot, Air Afrique, Balkan, British Airways, Egyptair, Ethiopian, KLM, Libyan Arab, Nigeria Airways, Swissair.

International airport: Accra-Kotoka International (Code: ACC), 10km from city; duty-free shop, bar, buffet, restaurant, bank, post office, taxis.

Surface access

Main ports: Tema, Takoradi.

Overland access: Coastal road runs from Lagos (Nigeria) through Cotonou (Benin) and Lomé (Togo) to Accra. The condition of this road is variable. A generally good road links Abidjan (Ivory Coast) with Kumasi. Borders with Togo, Côte d'Ivoire and Bukina Faso were re-opened in 1984. But border guards have been known to refuse entry even though all borders are officially open.

Hotels: Available in Accra, Kumasi, Takoradi and other regional capitals. Advisable to book in advance. 10 per cent government tax and 10 per cent service charge are usual.

Electricity supply: 220 V AC, 50 cycles.

Car hire: Not widely available, and expensive. International driving licence recommended; this must be endorsed by Police Licensing Officer if stay exceeds 90 days. Traffic drives on the right.

City transport

Taxis: Available in Accra, best to hire daily. Tipping is not usual; fare by negotiation; current rates are often posted in hotels.

Buses: Extensive services in Accra run by city authority and private operators.

National transport

Air: Ghana Airways operates between main centres including Accra, Kumasi, Takoradi, Sunyani and Tamale.

Road: Over 30,000 km of road. All-weather roads connect main towns, though many are in a state of disrepair. Main routes are Accra-Tema, Accra-Takoradi, Accra-Kumasi, Accra-Koforidua, Accra-Ho.

Buses: State run services connect major centres. Subject to delays and cancellation. Not recommended for business users.

Rail: Network links the three main cities with daily services: Accra, Kumasi and Takoradi. Two classes; air-conditioning and restaurant cars not available; sleeping accommodation available on some services.

Public holidays

Fixed dates: New Year's Day, 6 Mar (Independence Day), 1 May (Labour Day), 5 Jun, 1 Jul, 25 Dec, 26 Dec, 31 Dec (Revolution Day).

Variable dates: Good Friday, Holy Saturday, Easter Monday.

Working hours

Government and business: (Mon-Fri) 0800 – 1230, 1330 – 1700.

Banking: (Mon-Thu) 0830 – 1400; (Fri) 0830 – 1500. Some banks in Accra open Sat morning.

Shops: 0800 – 1200, 1400 – 1730. (Exceptions: Wed, Sat 0800 – 1300.)

Telecommunications

Telephone: Dialling code for Ghana: IDD access code + 233 followed by area code (21 for Accra) followed by subscriber's number. Dialling from Ghana: through operator. 85 per cent automatic service. Service cannot be relied upon.

Telex and telegram: Rates subject to review. International facilities available in Accra.

Banking

Central bank: Bank of Ghana.

Main centre: Accra.

Head offices: Agricultural Development Bank, Bank for Housing and Construction, Barclays Bank of Ghana, Ghana Commercial Bank, Ghana Co-operative Bank, Merchant Bank (Ghana), National Investment Bank, National Savings & Credit Bank, Premier Bank, Social Security Bank, Standard Chartered Bank Ghana.

Representation in capital: Algeria, Australia, Benin, Brazil, Bulgaria, Burkina Faso, Canada, China PR, Cuba, Czechoslovakia, Denmark, Egypt, Ethiopia, France, German DR, German FR, Guinea, Hungary, India, Italy, Ivory Coast, Japan, Korea PDR, Rep. of Korea, Lebanon, Liberia, Libya, Mali, Netherlands, Niger, Nigeria, Pakistan, Poland, Romania, Saudi Arabia, Spain, Switzerland, Togo, UK, USA, USSR, Vatican, Yugoslavia, Zaïre.

BUSINESS DIRECTORY

Hotels: Accra

Ambassador, Independence Ave, P.O. Box: 3044 (tel: 664646; tx: 2113).

Continental, Liberation Road, P.O. Box: 5252 (tel: 775361; tx: 2113).

Star, Fourth Circular Road, P.O. Box: 652 (tel: 77728; tx: 2113).

Hotels: Kumasi

Catering Rest House, P.O. Box: 3179 (tel: 3656; tx: 2113).

City, Rain Tree Road, P.O. Box: 1980 (tel: 3293; tx: 2113).

Hotels: Takoradi

Atlantic, Axim Road, P.O. Box: 273 (tel: 33019; tx: 2113).

Hotels: Tema

Meridian, P.O. Box: 330 (tel: 2878; tx: 2113).

Car hire: Accra
Avis: PO Box 7914 (tel: 228799, 228760; tx: 2300, 2134).

Other useful addresses
Accra: Bank of Ghana, PO Box 2674 (tel: 666902, 666921; tx: 2052).

Ghana Industrial Holding Corporation, PO Box 2784.

Ghana Investment Centre, PO Box 193.

Ghana National Chamber of Commerce, PO Box 2325 (tel: 662427, 662210).

Ghana National Procurement Agency, Ministries Post Office, Private Mail Bag, Accra.

Ghana National Trading Corporation, PO Box 67, Ghana House, Accra (tel: 664871).

Ghana Tourist Board, State House, PO Box 46 (tel: 665461, 666533).

Ministry of Trade, Industries and Tourism, PO Box M47 (tel: 663327).

Statistical Service Board, PO Box 1098 (tel: 666512).

Gibraltar

Tom Burns

One image of Gibraltar is a hopeful one that illustrates the future of the Rock as a sound financial centre. The best example of this rosy picture in 1989 came with the opening of yet another bank on the crowded crown colony. The new institution was the property of the Royal Bank of Scotland and Banco de Santander; it was a joint venture between a British and a Spanish bank. RBS and Santander had decided that Gibraltar was an integral part of their strategy to develop in unison a pan-European banking network.

Another image is much less sanguine. Despite the co-operation between Spanish and British-based financial and business institutions, the status of Gibraltar remains clouded by a profound tension. A British crown colony, claimed by Spain and occupied by a population that is as small as it is determined to run its own affairs, Gibraltar continues to be an irritant in Hispano-British relations.

The best illustration of the problems arising from this tension came last year when Spanish customs officers who were in hot pursuit of alleged Gibraltarian smugglers landed their coastguard vessel on the isthmus that links Gibraltar to Spain, drew their weapons to arrest the suspects and were promptly arrested by the Rock's constabulary. The alleged smugglers were allowed to go free and the Spanish customs officers were committed to trial (they jumped bail and returned to Spain). Furious about what they perceived to be an abject lack of co-operation against the supposed smugglers by the Gibraltarian authorities, police on the Spanish side of the Rock's border adopted retaliation tactics and insisted on thorough customs checks of all vehicles passing through the frontier. The long queues that inevitably built up were reminders of the 'siege' years when Spain unilaterally sealed the border with the British colony.

KEY FACTS

Official title: The City of Gibraltar

Head of state: Queen Elizabeth II (represented by Governor)

Head of government: Chief Minister Joe Bossano

Capital: Gibraltar

Official Languages: English

Currency: Gibraltar Pound (G£) = 100 pence (sterling)

Exchange rate: G£0.63 per $ (Dec 1989)

Area: 5.8 sq km

Population: 30,077 (1988)

GNP per capita: G£4,481 (1987)

Inflation: 4.4% (1988)

Trade balance: –G£98.7m (1988)

Potential for offshore finance

Gibraltar is clearly an offshore financial centre of considerable potential. It offers sophisticated, confidential and secretive services, with considerable tax breaks for individuals and corporations, it is in the sterling area and in the European time zone. In addition, it has high net worth offshore clients on its own doorstep, by virtue of the large expatriate community living along the Costa del Sol in neighbouring Spain. And yet there will continue to be a question mark hanging over that potential while mistrust and disagreements govern its judicial and sovereign status.

Mr Joe Bossano, now in his third year as chief minister of Gibraltar, has turned in a creditable performance in his efforts to boost the Rock's self-sufficiency and its belief in its own future. After long years in opposition to Sir Joshua Hassan, Mr Bossano was elected in a landslide poll to provide just that. The veteran Sir Joshua represented Gibraltar's conservative old guard and its colonial links with Britain. Mr Bossano, who made his political mark as a tough trade union leader, stood for self-assertion. Inevitably he was caricatured as a Gibraltarian version of Malta's erstwhile dynamic leader Mr Dom Mintoff.

The analogies were however misplaced. Mr Bossano has shown himself to be more than adept at instilling business confidence in the Rock and has shown far greater diplomatic skills than he was originally credited with. He has, moreover, embarked on ambitious plans to reclaim land so as to ease the acute shortage of office space and of housing on the Rock, and his team has worked hard to improve social services. Most important of all, the Bossano government has worked closely with business leaders on the Rock to create self-regulatory controls aimed at ensuring financial transparency. Gibraltar's image as an off-shore centre was badly tarnished by the Barlow-Clowes scandal.

The Gibraltarian leader's article of faith is that the crown colony can go it alone, negotiating a third country status with the EC and acting as an independent and efficient centre for multiple services on the southern tip of Europe. He passionately believes that Gibraltar has an identity all of its own. Astutely, he has worked towards forging relations not with Madrid and London – for he mistrusts both power centres – but with the authorities of the immediate Spanish hinterland.

If Gibraltar is booming thanks to its own curious status and facilities, so are the areas on the Spanish side that encircle the Rock. Investment is a two-way process and, with more of it entering Spain from Gibraltar, the balance is tipped in favour of the surrounding hinterland. Increasingly, Gibraltarians are living on the Spanish side of the border, where housing conditions, communications, leisure facilities and business opportunities are in many ways far more enticing than on the Rock itself. Mr Bossano makes a point also of cultivating the expatriate community on the Costa del Sol and encouraging confidence in the Rock's financial services.

Dispute over airport use continues

Madrid and London, whom Mr Bossano tends to ignore, at least publicly, do however cast a long shadow over the chief minister's record. This is best illustrated by the continuing dispute over the use of Gibraltar airport. An agreement between Britain and Spain foresees the dual use of the airport, which is situated on the isthmus linking Spain to the Rock proper. This isthmus area is in fact no-man's land according to the 1713 Treaty of Utrecht, by which Spain, at the end of the War of the Spanish Succession, ceded the Rock's citadel in perpetuity to the British crown. The present impasse is the result of Mr Bossano's adamant refusal to allow any Spanish involvement in the running of the airport. In this the chief minister is staunchly supported by the Gibraltarian man in the street, but the consequence is that flight links between Gibraltar and the rest of the world are severely restricted (there are no flights to Spanish destinations) and overdue plans to enlarge and improve the landing strip have been shelved.

Spain and Britain have an ongoing negotiating process to discuss the 'whole future of Gibraltar' that is governed by the so-called Brussels Agreement of 1984 which paved the way for the re-opening of the Spain-Gibraltar border. Spain, which will not renounce its sovereignty claims, has tabled compromise formulas such as condominium, by which responsibility over the Rock is shared by the British and the Spanish crowns or by a straight leaseback of the Rock to Britain. Spain, which claims that it will continue to respect Gibraltar's special identity, its institutions and its planned third country or associated status with the EC, moreover accepts Britain's insistence, as built into the Brussels Agreement, that there will be no change in Gibraltar's status without the consent of the Gibraltarian people themselves.

Mr Bossano, however, refuses to be a party to the Brussels negotiating process. The core of the election campaign that swept him to power was a blanket rejection of the Brussels Agreement, and he has not wavered from his stand. London, which believes in the words of former Foreign Secretary Sir Geoffrey

KEY INDICATORS

	Unit	1984	1985	1986	1987	1988
Population	'000	28.3	28.8	29.2	29.7	30.1
Gross national product (GNP)	G£m	89.1	96.7	109.4	121.3	135.1
GNP per capita	G£	3,414	3,676	4,110	4,481	–
Inflation	%	7.7	3.2	4.4	4.4	4.4
Imports	G£'000	66,098	113,163	111,654	140,962	144,800
UK	G£'000	23,335	33,451	36,307	49,386	–
Japan	G£'000	3,261	5,872	6,599	9,883	–
Spain	G£'000	2,577	7,361	11,226	16,282	–
Netherlands	G£'000	2,187	2,966	3,634	4,593	–
Exports	G£'000	25,074	48,348	44,321	51,731	46,100
UK	G£'000	3,224	5,993	8,420	10,024	–
Morocco	G£'000	1,307	1,798	2,877	1,366	–

* estimated figure

Howe, that Gibraltar's long term interests are best served through close co-operation with Madrid, has done what it can reasonably do to undermine Mr Bossano's obstinacy. One milestone last year in the changing British attitude to its colony was the decision to almost completely withdraw its troop presence from the Rock. The gentle, and the not so gentle prods that London delivers to Mr Bossano in the hope of steering him towards Spain have not however achieved any results so far.

While the underlying dispute over Gibraltar continues, there will always be caveats about its future. Instances of confidence that mark new bank openings, especially if they are Hispano-British joint ventures, will be undermined by others of complete distrust, as when security personnel clash and ensuing border queues starkly highlight Gibraltar's economic fragility.

COUNTRY PROFILE

Political system: Gibraltar is governed by House of Assembly, modelled on British parliamentary lines, responsible for defined domestic matters. The British Government, through the Governor, remains directly responsible for security, defence and foreign affairs.

Media
Press: Only daily newspaper is *Gibraltar Chronicle* (circulation 2,700). Weeklies: *Panorama* (2,500), *Vox* (which publishes a Spanish section, 1,000), *Gibraltar Post, The Democrat* (1,000) and *The People* (also carries a Spanish section, 1,800).
Broadcasting: Over 7,000 combined radio and TV licences issued. Programmes broadcast by Gibraltar Broadcasting Corporation (GBC) and include advertising. GBC-Radio broadcasts in English and Spanish, the British Forces Broadcasting Service in English only, GBC-TV in English only.

Domestic economy: The economy is primarily dependent on service industries. Its main base is centred on providing services to tourism, the commercial port, the offshore financial services, banking and the Ministry of Defence. Since the full opening of the land frontier with Spain in 1985, the number of tourists visiting Gibraltar has rapidly increased to over 3.7m in 1988. Tourism expenditure in 1988 was estimated at some 36m. The closure of the Naval Dockyard in 1985 saw the formation of a commercial ship-repair yard – Gibraltar Ship-repair Ltd. The port also provides an important source of income with over 2,700 ships calling there in 1988. The Gibraltar Government has also invited the setting up of light industries there, by making available a package of incentives and other benefits to successful companies. Gibraltar's popularity as an offshore finance centre has also increased during the past year. Gibraltar became an associate member of the EC when the United Kingdom joined in 1972, and is included for all matters, except VAT, the Customs Union and the Common Agricultural Policy.

Employment: The level of employment in 1988 (April) stood at 12,995 with the Gibraltarian workforce totalling some 8,800. Employment in the financial sector continued to increase from 940 in October 1987 to 1,158 in April 1988. Unemployment is relatively low and stood at just 4 per cent of total workforce at the end of January 1989.

External trade: Regular trade deficit largely offset by invisible earnings. Heavy dependence on imports of food, consumer goods, raw materials, fuels and equipment from UK. Re-exports mainly petroleum products, tobacco, manufactured goods and wine to UK, Spain and and Morocco.

Legal system: Based on English law. New constitution adopted 1969.

Time: GMT + 1 (GMT + 2 from late Mar to late Sep).

Climate: Temperate with hot and dry summers, and fairly mild rainy winters; temperatures vary between about 10°C and 29°C.
Clothing: Lightweight for summer, medium weight and light raincoat for winter.

Entry requirements: As for United Kingdom.

Air access: Direct flights from London and Manchester operated by GB Airways and Air Europe Ltd. GB Airways also operate direct flights to Tangier and Tetuan.
National airline: GB Airways.
International airport: Gibraltar (GIB) is served by North Front 2.5 km from town centre.

Surface access: Regular catamaran services from Tangier, and road access from Málaga through La Línea frontier in Spain.

Hotels: No offical rating system. reservations should be made in advance especially during summer. Some hotels reduce rates between Nov and Mar.

Car hire: Available through local car hire firms and travel agents. UK International driving licence and evidence of insurance required. The British Automobile clubs AA and RAC have agents in Gibraltar.

National transport: Bus and taxis services available. Total of about 45 km of roads. No railway network.

Public holidays
Fixed dates: 1 Jan, 25 Dec, 26 Dec.
Variable dates: Commonwealth Day, Good Friday, Easter Monday, May Day, Spring Bank Holiday, Queen's Birthday, Late Summer Bank Holiday.

Working hours
Business: (Mon-Fri) 0900 – 1300 and 1430 – 1800. (Sat) 0900 – 1300.
Banking: (Mon-Thur) 0900 – 1530, (Fri) 0900 – 1530 and 1630 – 1800.
Shops: (Mon-Fri) 0900 – 1300 and 1500 – 1900, (Sat) 0900 – 1300.

Business language and interpreting/translation: The official language is English but Spanish is sometimes used when dealing with Spanish business.

Telecommunications
Telephone: Rates for calls from public telephone boxes dependent on time of day. International direct dialling facilities available to over 75 countries. International dialling code: 350. There are no area codes.
Telex: Two public telex call offices. Country code 405 GK. Information from Gibraltar Telecommunications International, Main Street.
Telegram: Available 24 hours a day from Gibraltar Telecommunications International.

Banking: Domestic and offshore services available. There are some 14 banks operating in Gibraltar, and around eight offshore banks.

Electricity supply: 240 V AC.

Hotels
Both Worlds Aparthotel, Sandy Bay (tel: 76191; tx: 2236).

Bristol, 10 Cathedral Square (tel: 76800; tx: 2253).

Caleta Palace, Sir Herbert Miles Road, Catalan Bay (tel: 76501; tx: 2345; fax: 71050).

Continental, Engineer Lane (tel: 76900; tx: 2303; fax: 75366).

Gibraltar Beach, Sandy Bay (tel: 76191; tx: 2236).

Holiday Inn, 2, Governor's Parade (tel: 70500; tx: 2242; fax: 70243).

Montarik, Bedlam Court, Main Street (tel: 77065/90; tx: 2226).

Ocean Heights, Montague Place (tel: 75548; tx: 2371).

Queen's, Boyd Street (tel: 74000; tx: 2269).

Rock, Europa Road (tel: 73000; tx: 2238; fax: 73513).

Business advice and information
Gibraltar Information Bureau, Arundel Great Court, 179 Strand, London WC2R 1EH (tel: 836-0777; tx: 266303; fax: 2406612).

The Gibraltar Chamber of Commerce, PO Box 29, Suite 11, Don House, 38 Main Street (tel: 78376; tx: 2151; fax: 78403).

Banking
Algemene Bank Gibraltar Ltd, PO Box 100, 2/6 Main Street (tel: 79220; tx: 2234).

Banco de Bilbao (Gibraltar) Ltd, 260/262 Main Street (tel: 77796, 77871, 77896; tx: 2153).

Banco Central Sa, 198/200 Main Street (tel: 73625, 73650, 73675; tx: 2154).

Banco Español de Credito, 114 Main Street.

Banco Hispano Americano, Don House, 30/38 Main Street (tel: 74199; tx: 2152).

Bank of Credit & Commerce Gibraltar Ltd, 20/22 Main Street (tel: 73328, 77305; tx: 2337).

Banque Indosuez, 206/210 Main Street (tel: 75090; tx: 2216).

Barclays Bank PLC, 84/90 Main Street (tel: 78565; tx: 2231).

Gibraltar & Iberian Bank Ltd, PO Box 408, Giberia House, 21 Cannon Lane, (tel: 73800, 74170; tx: 2364; fax: 79310).

Gibraltar Trust Bank (Credit Suisse), Neptune House (tel: 78399; tx: 2144).

Hambros Bank Ltd, 32 Line Wall Road (tel: 74850; tx: 2251).

Hong Kong Bank & Trust Co Ltd, Hadfield House, 3 Library Street (tel: 79500; tx: 2363).

Jyske Bank (Gibraltar) Ltd, PO Box 143, 76 Main Street (tel: 78325, 72782; tx: 2215).

Lloyds Bank PLC, 323 Main Street (tel: 77373; tx: 2108).

National Westminster Bank, 57 Line Wall Road (tel: 77737; tx: 2114).

Republic National Bank of New York (Gibraltar) Ltd, Neptune House, Marina Bay, PO Box 5578 (tel: 79374; tx: 2291).

Royal Bank of Scotland (Gibraltar) Ltd, 1 Corral Road (tel: 73200; tx: 2124).

United Bank of Gibraltar Ltd, Giberia House, 21 Cannon Lane (tel: 74170, 73800; tx: 2366).

Principal newspapers
Gibraltar Chronicle, 2 Library Gardens (tel: 78589; tx: 2205).

Panorama, 93-95 Irish Town (tel: 79797).

The Democrat, PO Box 241, 1211 Bell Lane (tel: 74261).

The People, PO Box 193 (tel: 72867).

Vox, 52 Devil's Tower Road (tel: 77414).

Other useful addresses
Department of Trade & Industry, The Haven, 23 John Mackintosh Square (tel: 79533; fax: 71406).

Economic Planning and Statistics Office, Cathedral Square (tel: 75515; tx: 2223).

Gibraltar Government Tourism Agency Ltd, Cathedral Square (tel: 76400; tx: 2223; fax: 79980).

Greece

Kerin Hope

Greece appeared set for a period of political uncertainty after the defeat of Prime Minister Andreas Papandreou's socialists in a June 1989 general election that failed to produce a clear-cut result. Another election loomed as soon as an unprecedented coalition of conservatives and communists finished investigating a series of financial scandals held responsible for the socialists' downfall after almost eight years in power. Meanwhile, soaring public sector deficits threatened a debt crisis, making it clear that a new government's first task would be to impose a package of austerity measures.

The Koskotas scandal, as it was known, came to overshadow every other issue in Greece from the autumn of 1988, even the prime minister's open love affair with a 34-year-old air stewardess. Months afterwards, however, it was still not clear whether the illegal dealings of a 35-year-old banker, George Koskotas, would have come to light had it not been for Mr Papandreou's sudden illness in August. The 69-year-old prime minister was rushed to London for open-heart surgery where, before the operation, he declared his intention of divorcing his American wife of more than 30 years and marrying the stewardess, Ms Dimitra Liani, who was constantly at his bedside at Harefield Hospital.

During Mr Papandreou's two-month absence, the Bank of Greece, the country's central bank, launched an enquiry into irregularities at the Bank of Crete, the lynchpin of the Koskotas business empire. It also included a media group, Grammi, which controlled three daily newspapers, six magazines and a commercial radio station, as well as Greece's most popular soccer team, Olympiakos of Piraeus. The result was the uncovering of the biggest financial scandal in Greek history.

Mr Koskotas rose to prominence when he bought the bank in 1984 while employed as a middle-ranking executive at its Athens headquarters. A Greek brought up in the United States, he claimed to have transferred $100m from abroad to invest in Greece. But after he fled the country in November and was arrested in Massachusetts, he was accused of embezzling more than $200m from the bank. Fighting extradition, Mr Koskotas claimed in tape recordings sent to Greece from his jail cell that Mr Papandreou had blackmailed him into providing funds needed by his Panhellenic Socialist Movement (PASOK) to fight the upcoming election. Millions of dollars belonging to state corporations were deposited in the Bank of Crete and the interest skimmed off in order to fill the party coffers and pay off socialist officials, among them several cabinet ministers, according to Mr Koskotas.

Mr Papandreou denied any wrongdoing, but half-a-dozen PASOK appointees were jailed during a preliminary legal investigation on charges of accepting bribes and illegally diverting state funds. At the same time, more financial scandals were revealed, including allegations of kickbacks and illegal arms exports in deals arranged by officials at the state-owned Hellenic Arms Industry. The minister of defence procurements, Mr Stathis Yiotas, resigned in December after submitting an eight-page letter filled with evidence of corruption. He charged that the government bought 40 French-made Mirage 2000 jets at twice the price paid by Switzerland for the same aircraft, and then decided they should be equipped with French air-to-air Magic missiles that cost 10 times as much as the American equivalent, the Sidewinder.

KEY FACTS

Official title: The Hellenic Republic
Head of state: President Christos Sartzetakis
Head of government: Prime Minister Xenophon Zolotas
Capital: Athens
Official Languages: Greek
Currency: Drachma (Dr) = 100 leptae
Exchange rate: Dr161 per $ (Dec 1989)
Area: 131,957 sq km

Population: 10m (1988)
GDP per capita: $5,242 (1988)
GDP real growth: 1.0% (1988)
Labour force: 3.9m (1987)
Unemployment: 7.6% (1987)
Inflation: 13.5% (1988)
Trade balance: –$6,702m (1988)
Foreign debt: $17.7m (1988)

In what was called the 'maize scandal', Greece offered no defence when taken to court in Luxembourg by the European Community shortly before the election. The case concerned forgery of documents by a Greek state company, ITCO, in order to pass off a consignment of Yugoslav maize sold to Belgium in 1986 as Greek and so avoid paying duty on imported produce sold to an EC member. No Greek official turned up at the hearing and Community officials noted that the Greek government had so far failed to take any action against the officials involved.

Two days after the election, US police arrested the staff of the state-owned Mortgage Bank's representative office in New York and shut down its branch offices in other American cities. They were investigating charges that the bank had helped Greek Americans to evade taxes by sending up to $700m to Greece for deposit in high-earning foreign currency accounts. The bank's overseas offices were intended to provide consultative services for investment in Greece, but the publicity given to the arrests, coming on the heels of the Koskotas scandal, seemed likely to deter any would-be investor.

The scandals, however, appeared to have little effect on Greece's capacity to raise loans abroad, although the political furore they caused, in combination with the prime minister's frail health, succeeded in paralysing economic policy-making for months before the election. During the latter half of 1988 there had been little cause for concern. The current account deficit for the year totalled just $1.01bn, only a little over the forecasts, while annual inflation came down from 15.6 to 13.8 per cent, compared with a hoped-for 12 per cent. The economy grew by 3.5 per cent, the best showing ever under the socialist government.

Tourism, still the major foreign exchange earner, did reasonably well in 1988, despite a terrorist attack in July aboard a Greek island cruise ship, the City of Poros, in which nine people were killed. A collision in October in Piraeus harbour between an Italian freighter and a Greek cruise ship carrying several hundred British schoolchildren could have been a

KEY INDICATORS

	Unit	1984	1985	1986	1987	1988
Population	m	9.90	9.93	9.97	9.99	10.01
Gross domestic product (GDP)	Dr bn	3,806	4,618	5,480	6,256	7,446
GDP per capita	$	3,411	3,365	3,926	4,625	5,242
GDP real growth	%	2.7	3.1	0.8	–0.1	1.0
Inflation	%	18.4	19.3	23.0	16.4	13.5
Share prices	1980=100	60	67	89	–	–
Unemployment	'000	71	89	110	118	–
Wages	1985=100	83.4	100.0	112.7	123.5	146.3
Agricultural production	1979-81=100	104.78	113.02	105.56	98.79	108.20
Manufacturing production	1985=100	97.5	100.0	99.3	97.3	102.3
Investment goods	1980=100	81.8	81.2	79.3	76.3	–
Intermediate goods	1980=100	87.4	97.4	101.5	89.2	–
Consumer goods	1980=100	108.0	111.4	111.5	110.1	–
Exports	$m	4,394	4,293	4,513	5,612	5,933
Exports (volume)	1982=100	138.7	138.5	161.9	–	–
Imports	$m	8,624	9,346	8,936	11,112	12,005
German FR	%	18.2	16.9	24.0	–	–
Italy	%	10.5	9.3	13.1*	–	–
France	%	7.5	6.4	9.1*	–	–
Balance of trade	$m	–4,230	–5,053	–4,423	–5,500	–6,702
Current account	$m	–2,132	–3,276	–1,676	–1,223	–958
Exchange rate	Dr per $	112.7	138.2	140.0	135.4	141.9

* estimated figure

major disaster, but thanks to skilled rescue efforts, only one child was drowned. Tourist arrivals rose by 2.8 per cent to 8.3m, while income reached $2.8bn, up from $2.2bn in 1987.

Shipping earnings also improved as both the tanker and freight markets maintained their recovery, totalling $1.38bn, up from $1.2bn in 1987. And in a remarkable departure from the traditional view among Greek shipowners that investment at home should be avoided, a group of London-based owners organised the Greece Fund, based in Britain and underwritten by foreign merchant banks. The fund, which is quoted on the London stock exchange, was set up by the Schroder merchant bank and is managed in co-operation with a subsidiary of Ergobank, a privately-owned Athens bank. It imported Dr3.5bn ($20m) to acquire shares quoted on the Athens exchange.

Long-promised stock market reforms were finally legislated in the autumn of 1988, but were slow to take effect. The new law abolished the 60-year-old monopoly on dealing held by the exchange's 32 brokers, allowing banks and brokerage companies to hold seats, and provided for setting up an over-the-counter market in secondary issues, as well as a computerised transaction system. Stiff penalties were mandated for insider trading. But there were few immediate changes: only two licences for brokerage firms had been issued six months after the law passed. No new issues were listed after four companies raised their capital by a total of Dr4.7bn late in 1988.

Volume averaged around Dr250m ($1.4m) a day for 1988 but slumped to around Dr100m ($570,000) in the first quarter of the year. The average price to earnings ratio rose from 7 to 8 last year, still among the lowest in Europe. Prices picked up considerably on the eve of the election, only to slump again when it failed to produce a clear result. Officials at the exchange say it remains an uphill task to persuade the highly secretive, family oriented Greek business community that it might be worthwhile to float capital on the market. Only 120 issues are listed, all of them Greek firms, and only one-quarter of the top 200 companies are quoted on the exchange. Even without the prevailing political difficulties, it would take more than a year to make the new legislation effective. That leaves very little time for the market to develop before 1992, when the major European financial players are expected to move in.

Reforms in the banking system continued as the Bank of Greece dismantled more of its old-fashioned regulations. Interest rates were to be fully freed from the end of 1989, and procedures were under way to permit commercial banks to handle forward dealing in foreign exchange, a necessary prerequisite if Greece is eventually to join the European Monetary System. At present only spot transactions in foreign currencies are allowed, while the Bank of Greece holds exclusive rights for forward dealing. But the slow pace of building an infrastructure for a securities market based on both private and institutional capital is likely to leave Greece less than adequately prepared for the barrier-free market. More legislation is needed, bankers point out, to modernise underwriting and syndication regulations in order to attract a broader investor base and create a real secondary market.

The pace of mergers and acquisitions of Greek firms by foreign investors increased as more companies were put on the market. The trend reflected a tendency among older Greek businessmen, who took over family-owned companies in the 1960s, to opt for wealthy retirement rather than face the problems of remaining competitive after 1992. Government investment incentives and lower operating costs made Greek firms attractive to foreign buyers. Most activity focused on food and beverages, consumer goods, tourism and light manufacturing. Recent buyers include Jacob Suchard of Switzerland, which extended its holdings in Greece to a 25 per cent interest in the Ion chocolate firm and bought out the Bingo confectionary group. The old-established Metaxa brandy concern has been sold to Grand Metropolitan, while the Cypriot-owned 3E group, which holds the Greek franchise of Coca-Cola, acquired the wine and spirit makers Caviros and Botrys as well as the Papoutsanis soap and oil company. Another wine-maker, Boutaris, went into a joint venture with Britain's United Distillers in order to promote their products through each other's marketing networks. In another sector where there is growing foreign interest, the French insurance group Union des Assurances de Paris bought into the Greek firm Interamerican.

The goverment, however, had little success in its efforts to auction off the 'ailing industries' – debt-ridden firms which were taken over by the state in an effort to revive output and avoid large-scale unemployment. In addition, the major state projects made little progress. Harbour works got underway for the $600m Soviet-Greek alumina project in central Greece, but there were delays over machinery purchases in Western Europe. A major find of natural gas was made at Epanomi outside Salonica, but construction details of a billion dollar pipeline to carry Soviet natural gas from the Bulgarian border to Athens remained unresolved.

The future of the North Aegean Petroleum Company, which has pumped seabed oil off Thassos since 1974, was uncertain. The oilfields are expected to be exhausted in the next two years, but no plans to resume exploration were announced. The state-owned Public Petroleum Corporation, which in 1987 had tried to take over the consortium, said it was not interested in acquiring a majority holding owned by Denison Mines of Canada, which was seeking to offload its 68 per cent share in NAPC.

The Athens metro expansion, the Rion-Antirion bridge, the Acheloos river diversion and Spata airport projects were effectively frozen by the political impasse, although possibilities for financing them through the European Investment Bank were discussed. The metro and airport plans could, however, be quickly revived as both are crucial to Athens' bid to stage the 1996 Olympics.

While the political uncertainty seemed unlikely to affect the flourishing private sector, the swift rate of increase in the budget deficit prompted both the IMF and the European Commission to issue strong warnings in reports released in mid-1989 that austerity measures should quickly be enforced. At that point, the budget appeared headed for a Dr1,900bn ($10.8bn) deficit, more than 30 per cent higher than the socialists' forecast of Dr1,400bn ($8bn). The higher deficit would represent around 20 per cent of GDP. The 1989 budget had called for a 20 per cent increase in revenues but, in the first five months of the year, the rise reached only 6 per cent. In addition, government spending was expected to soar in the second half of the year as public sector salary costs rose, partly due to thousands of additions to the state payroll during the run-up to the election. At the same time, Greece's foreign currency reserves were estimated to have fallen by 30 per cent in the first six months of the year. However, the country's net financial benefits from the EC, as one of the Community's poorest members, totalled Dr338bn ($1.93bn) for 1989, with an additional Dr700bn ($4bn) due over the next five years under the Integrated Mediterranean programmes for development of southern Europe's more backward areas. Both the IMF and EC forecast growth of around 3 per cent and inflation of around 14.5 per cent, but stressed that stabilisation measures and a freeze on government spending were urgently needed. But there seemed little chance the warnings would be heeded: with another election campaign due in the autumn, neither the conservatives nor the communists would be willing to risk losing votes through applying unpopular economic measures.

The conservative-communist coalition emerged after two weeks of energetic political horse-trading following the inconclusive election result. The conservative New Democracy party, headed by Constantine Mitsotakis, finished well ahead of PASOK with 44.25 per cent to 39.15 per cent of the vote. The left coalition, an alliance of Greece's old-fashioned pro-Moscow communist party and left wing progressives, including former Eurocommunists, came in a distant third with 13.12 per cent.

The ND total would have normally been enough to secure a comfortable working majority in the 300-member parliament under Greece's reinforced proportional voting system. But the socialists revised the electoral law early in the year, following normal government practice in an election year, so that it became next-to-impossible for one party to secure an overall parliamentary majority. ND took 145 seats, Pasok 125 and the Left Coalition 28, with others taking two seats. But the socialists' calculation that the communists would be willing to join them in a coalition misfired. Given a mandate by President Christos Sartzetakis, ND and PASOK each failed to form a government. When the communists received their mandate, for the first time ever, Mr Mitsotakis, 69, made a shrewd but statesmanlike move and stepped aside so that a coalition could be formed.

A popular former naval officer, Tzannis Tzannetakis, 62, who had proved an effective manager of the ND election campaign, was sworn in as premier, and the Left Coalition were handed two sensitive cabinet posts: the justice ministry and the interior ministry, which is responsible for overseeing parliamentary elections. The coalition set the seal on a political reconciliation between right and left after 40 years of division. It dated back to the 1946-49 civil war in which communist insurgents were defeated by right-wing nationalists with the help first of Britain, then the United States.

It was agreed that the coalition government's mandate would be 'catharsis', or 'cleansing', a word borrowed from ancient Greek tragedy, and extend only to supervising investigation of the major scandals: the Koskotas affair, the arms purchases, the maize sale and allegations of widespread telephone tapping of officials and prominent personalities of all political persuasions.

Parliament swiftly voted to pursue enquiries into the possible involvement of Mr Papandreou and several former socialist cabinet ministers – Deputy Prime Minister Agamemnon Koutsogiorgas, Industry Minister George Petsos, National Economy Minister Panayotis Roumeliotis and Finance Minister Dimitris Tsovolas – in the banking scandal. Most politicians predicted that elections would most likely be held again in November, under the same electoral system which had given the communists a sizeable boost by making several concessions towards a simple proportional system.

But even if one party secured a majority or formed a working coalition, the prospect of yet another election still loomed. Parliament had to elect a new president by spring 1990 and, under the constitution, the successful candidate needed to capture a total of 180 votes for election. If he failed – which seemed not unlikely, even though the presidency is now a largely ceremonial post – parliament would be dissolved and new elections held. The new chamber could then elect a head of state with a simple majority.

Mr Papandreou's own political future also seemed in doubt after he came down with a severe attack of pneumonia immediately after the election while taking his turn to try and form a government. He made a quick recovery and shortly afterwards married Ms Liani, but a large question mark hung over the extent of his role in the financial scandals and his unwillingness to investigate them while in power. To most analysts, however, it seemed the socialist era was over. Debate centred on whether a younger leader could prevent a split in PASOK if Mr Papandreou retired, and on which ND leadership aspirant could carry the day. The well-liked former labour minister, Giorgos Gennimatas, 48, was seen as the top contender to head PASOK, while former Athens mayor Miltiades Evert, 46, was mentioned as the conservatives' rising star.

Two pressing foreign policy issues involving the United States will confront the next government. One is the future of the four American bases in Greece, under threat of closure by mid-1990 if no new agreement is reached. The old pact expired on 20 December 1988 and under its terms Greece gave a 17-month notice period for the bases to shut down. More than a dozen rounds of negotiations brought no apparent progress, but the socialists were clearly playing for time as the elections drew near. The Americans made it clear they would be happy with a renewal of the previous five-year agreement. The socialist government declared that even if an new agreement was reached, the US air force support base at Athens airport would have to go, and the accord approved in a referendum before it could take effect. The conservatives are in favour of retaining the bases: their presence ensures that Greece receives upwards of $300m a year in US military credits as a form of rent. The communists would prefer they were shut down.

At the same time, a decision was still pending over

the extradition of a Palestinian suspected of placing a bomb aboard a PanAm airliner over Hawaii, killing one passenger and injuring several others. The supreme court, Greece's highest judicial authority, ruled in favour of extradition early in 1989, but the final decision was to be taken by the justice minister. The justice minister in the interim government, Nikos Konstantopoulos, who happened to be the defendant's lawyer, said no decision would be taken until after the autumn election. The United States indicated that if the defendant was not handed over, it would reconsider the 7:10 ratio in military aid to Greece and Turkey, which the Greeks consider a deterrent to what they see as Turkish expansionism in the Aegean. There was even a suggestion that Greece might be declared unsafe for American tourists, a serious economic threat since the tourist industry has not recovered fully from a similar 'travel advisory' issued by the US State Department in 1985.

Relations with Turkey took a turn for the worse over alleged mischief-making by Ankara among the ethnic Turkish minority in Thrace during the election period. Earlier, Mr Papandreou and his Turkish counterpart, Mr Turgut Ozal, had reasserted their commitment to a reconciliation effort known as 'the spirit of Davos', after the Swiss ski resort where it was launched in 1988. But the hoped-for improvement in bilateral trade failed to materialise and political suspicions among senior officials on both sides proved hard to overcome. Although ND backs better bilateral relations, it seemed unlikely that any other Greek politician would seek to emulate Mr Papandreou's personal rapprochement with Mr Ozal.

COUNTRY PROFILE

Political system: Parliamentary democracy, with supreme executive power vested in the president who is elected for five years by the unicameral parliament. Parliament consists of 300 deputies elected for four years by universal adult suffrage. President appoints prime minister and, upon his recommendation, a cabinet to form the government.

Main cities/towns: Athens/Piraeus (population 3m), Salonika, Patras, Canea, Heraklion (Crete), Volos, Larissa, Kavalla.

Media
Dailies: Of the 20 national dailies, 16 are published in Athens and four in Salonika. Most popular are *Eleftheros Typos* (circulation 85,000), *Apoghevmatini* (130,000), *Ethnos* (150,500), *Avriani* (115,000), *Ta Nea* (155,000) and *Vradyni* (71,914), all published in Athens in the afternoon. Popular morning newspaper *Kathimerini*. Also several regional dailies with generally low circulation although *Macedonia* (47,989) and *Thessaloniki* (36,040) are popular. *Athens News* (10,000) is published daily in English.
Weeklies: Several general and special interest magazines published.
Business: Include financial papers such as *Naftemboriki, Kerdos, Express* and *Economicos Tachidromos*.
Broadcasting: Approximately 3.5m radios and 3m TV sets in use. State-supervised companies, Elliniki Radiophonia Tileorassi – Hellenic radio and Television (ERT 1) and ERT 2, operate from Athens. Local government-operated radio stations in Athens and Salonika. There are many private radio stations in Athens. Most popular are: 9.84 FM, TOP FM92, 4, ANTENA 97, 1 FM. There are two TV stations in Thessaloniki, ERT3 and TV100. Satellite TV also available: Europe and CNN are satellite channels. Cable TV available in a few months' time.
Advertising: All usual media are used. All television broadcasts are commercial, and two of the four radio programmes accept advertising. Posters are widely used, under control of site owner. Cinema firms are also popular, and direct mail is becoming increasingly important. Newspaper and magazine advertising is constrained by relatively low circulation figures. Total advertising spending (1988) Dr39,301m. Key categories include food, retail, cigarettes, alcohol, clothing and electrical appliances.

Domestic economy: Mixed economy with heavy dependence on agriculture, tourism and shipping. Economic growth has been undermined by large debt burden, high inflation and low levels of investment. Stabilisation programme introduced in October 1985 appeared to be halting deterioration in economy.

Employment: Around 45 per cent of labour force employed in services, 25 per cent in industry and 25 per cent in agriculture. Steady 2.5 per cent annual growth of labour force has kept unemployment rate at over eight per cent despite expansion of public sector employment. Wage indexation system in force.

External trade: Growing current account deficit owing to increased debt payments, falling invisible earnings (principally from shipping and expatriate workers' remittances). Visible exports, which picked up in 1986, consist mainly of fruit and vegetables, textiles, petroleum products, tobacco, clothing and metals. Principal imports include raw materials, fuels and lubricants, chemicals, machinery, and transport equipment, foodstuffs, basic manufactures and manufactured consumer goods. Since entry into the EC in 1981 trade with Europe has increased, particularly with W Germany, Italy, France, UK and the Netherlands.

Agriculture: Important sector of economy contributing around 17 per cent to GDP and accounting for over 20 per cent of exports. Farm holdings are small and fragmented. Sector also handicapped by weak infrastructure, low levels of technology and generally poor soil. However, with the exception of meat, dairy products and animal feeds, Greece is self-sufficient in foodstuffs. Main crops include wheat, barley, maize, fruit, vegetables, oil seeds, tobacco, cotton and sugar beet.

Industry: Sector forms mainstay of economy with manufacturing contributing about 20 per cent to GDP and accounting for nearly 50 per cent of exports. Small family-owned companies dominate, most of which are situated in the Athens area. Production has been sluggish and relatively few industries remain competitive on a European scale. Major industries include textiles, clothing and footwear, chemicals, food processing, metals manufacture and cement.

Mining: Relative wealth of natural resources including large deposits of bauxite, lignite, magnesite, ferrochrome, ferronickel, lead, zinc, uranium and manganese. Mining activity is however small scale and sector contributes less than 3 percent to GDP.

Energy: Around 55 per cent of primary energy requirements met by imported oil. Current emphasis on developing indigenous oil resources (largely from the Prinos oil field) and expanding hydroelectric and geothermal usage. Also to increase use of lignite for electricity generation and introduce peat, wind and solar power.

Banking: System headed by the Bank of Greece, which acts as cashier and banker for the Greek state. Around 14 indigenous commercial banks including the National Bank of Greece, Credit Bank, Ionian and Popular Bank. Also some 17 foreign banks, three investment banks, two mortgage banks and two specialised credit institutions.

Legal system: New constitution promulgated 1975 and amended in 1986.

Membership of international organisations: CCC, EC, EIB (associate), EMA, FAO, GATT, IAEA, IBRD, ICAO, IDA, IFAD, IFC, IHO, ILO, IMF, IMO, INTELSAT, INTERPOL, IOOC, ITU, IWC, NATO, OECD, UN, UNESCO, UPU, WHO, WIPO, WMO, WSG, WTO.

Time: GMT + 2 (GMT + 3 from late Mar to late Sep).

Climate: Hot and dry summers, from Apr to Oct, with average temperatures 24°C – 32°C. Mild winters in south but cooler in north with temperatures falling as low as 0°C.
Clothing: Lightweight for summer, mediumweight with light topcoat for winter.

Entry requirements: Visas not required by nationals of most West European countries, and some others. N.B. regulations likely to change at short notice; Greek consulate or travel agent should be consulted.
Health: Vaccination certificates may be required if travelling from infected area.
Currency: Up to Dr100,000 in Greek currency may be taken into the country and up to Dr20,000 taken out, in denominations of up to Dr5,000 banknotes. No restrictions on movement of foreign currency.
Customs: Personal effects duty free.

Air access: Regular direct flights by all major international airlines.
National airline: Olympic Airways.
Main international airport: Athens (ATH). The East Airport is for international and charter flights. The West Airport (Hellinikon) is only for the Olympic Airways flights, both domestic and international.
International airports: Alexandroupolis (AXD), Corfu (CFU), Heraklion (HER), Ioannina (IOA), Kos (KGS), Mykonos (JMK), Paros (PAS), Rhodes (RHO), Salonika (SKG), Skiathos (JSI), Thira (JTR).

Surface access: Frequent passenger ferry services from Italy to Piraeus, and car ferry service between Ancona (Italy), Brindisi, to Igoumenitsa and Patras. Road and rail routes from most parts of Europe via Italy, Yugoslavia or Bulgaria.
Main ports: Heraklion, Igoumenitsa, Patras, Piraeus, Rafina, Salonika, Volos.

Hotels: Numerous hotels in all main towns. Classified as de luxe, A,B,C,D and E. Single de luxe room without breakfast costs about Dr10,000 per night, including 15 per cent service charge but excluding tax. A small tip will also be expected. Advisable to make reservations well in advance, especially between May and September. All major credit cards are accepted. N.B. drinking water is not always purified outside main cities.

Currency: The drachma floats, with rates set by the Bank of Greece and the commercial banks. The drachma is not included in the European Monetary System, but is part of the European Currency Unit (ECU) basket and floats against the ECU.

Car hire: All major car hire companies have offices in Athens and some other main towns. Rates vary depending on size of car, length of hire and season (in summer a medium size saloon car costs about Dr25,000 per week). International driving licences recognised, but British, Belgian, Austrian and West German full licences also accepted. International insurance Green Card is valid, provided Greece is mentioned. Wearing of seat belts is compulsory.

BUSINESS GUIDE

City transport
Taxis: Reliable service in Athens, costing: flag – Dr25 and Dr34 per km (minimum charge Dr200). Extra charges for luggage, waiting time, journeys outside Athens/Piraeus and journeys after midnight: 01.00 – 05.00 double charge Dr58 per km; 05.00 – 06.00 additional Dr30.
Buses: Flat rate of Dr30 within city limits.

National transport
Air: As well as the international airports, there are a further 25 other airports all connected by regular services operated by Olympic Airways.
Road: Approximately 40,000 km of road, of which about 9,000 km are classed as main roads.
Rail: Over 2,500 km of track operated by Hellenic Railways Organisation Ltd, with services to most main towns.
Water: About 80 km of navigable inland waterways, plus several regular ferry services along the coast and connecting the various islands.

Public holidays
Fixed dates: 1 Jan, 6 Jan (Epiphany), 25 Mar (Independence Day), 1 May, 15 Aug (Assumption), 28 Oct (OCHI Day), 25 and 26 Dec. August is traditionally the month when summer holidays are taken.
Variable dates: Clean Monday, Orthodox Good Friday, Orthodox Easter Monday and Orthodox Whit Monday.

Working hours
Business: (Mon-Fri) generally 0800 – 1400 and 1700 – 2000, tend to close earlier during summer and on Mon and Wed afternoons.
Government: (Mon-Fri) usually 0800 – 1500.
Banking: (Mon-Fri) 0800 – 1400.
Shops: (Mon, Wed and Sat) 0800 – 1500, (Tue, Thu and Fri) 0800 – 1400 and 1730 – 2030. Winter working hours (Oct-Apr): (Mon) 1300 – 1900, (Tue, Wed, Thu, Fri) 10.00 – 19.00, (Sat) 09.30 – 15.00.

Business language and interpreting/translation: Can often be provided on a daily basis by hotel or travel agent. List of agencies usually available from commercial departments of embassies.

Telecommunications
Telephone: Efficient, relatively cheap system, with automatic dialling throughout most of country. International dialling code: 30 followed by 1 for Athens and 661 for Corfu.
Telex: Available at main post offices and at larger hotels. Country code 601 GR.
Fax: Available at larger hotels and main post offices.
Telegram: Can be sent from post offices or by telephone. Services (domestic and international) operated by Greek Telecommunications Organisation (OTE).

Postal service: All overseas mail automatically sent air mail. Internal service can be subject to delays.

Banking: State banks make up 80 per cent, remainder foreign and small private banks. Liberalisation of banking system was initiated in 1987 involving rationalisation of role of central bank (Bank of Greece), dismantling compulsory reserve requirements and regulated interest rates, and the development of an active capital market. In addition to the commercial banks, the investment banks are Hellenic Industrial Development Bank (ETBA), National Investment Bank for Industrial Development and Investment Bank. The Agricultural Bank of Greece, National Mortgage Bank of Greece, National Housing Bank, Postal Savings Bank and Deposits and Loans Fund are specialised credit institutions.

Trade fairs: Most important is the Salonika International Trade Fair, held annually in September. Other more specialised events take place in Athens and Piraeus, as well as Salonika.

Electricity supply: 220 V AC.

Representation in capital: Albania, Algeria, Argentina, Australia, Austria, Bangladesh, Belgium, Bolivia, Brazil, Bulgaria, Cameroon, Canada, Chile, China PR, Colombia, Costa Rica, Cuba, Cyprus, Czechoslovakia, Denmark, Dominican Republic, Ecuador, Egypt, Ethiopia, Finland, France, Gabon, German Democratic Republic, Federal Republic of Germany, Ghana, Haiti, Holy See, Honduras, Hong Kong, Hungary, Iceland, India, Indonesia, Iran, Iraq, Ireland, Israel, Italy, Japan, Jordan, Republic of Korea, Kuwait, Lebanon, Liberia, Libya, Luxembourg, Madagascar, Malaysia, Malta, Mexico, Morocco, Netherlands, New Zealand, Nicaragua, Norway, Pakistan, Panama, Paraguay, Peru, Philippines, Poland, Portugal, Romania, Saudi Arabia, South Africa, Spain, Sweden, Switzerland, Syria, Taiwan, Thailand, Togo, Tunisia, Turkey, United Kingdom, Uruguay, USA, USSR, Vatican City, Venezuela, Yemen Arab Republic, Yugoslavia, Zaïre.

BUSINESS DIRECTORY

Hotels: Athens
Astir Palace, Vasilissis Sophias Street/Panepistimiou Street (tel: 3643112/9; tx: 222380; fax: 20788).

Athenaeum Inter-Continental, 89-93 Syngrou Avenue (tel: 9023666, 9220007; tx: 221553; fax: 9217653).

Chandris, 385 Syngrou Avenue (tel: 9414824; tx: 218112).

Dorian Inn, 17 Piraeus Street (tel: 5239782; tx: 214779).

Grand Bretagne, Constitution Square (tel: 3230251; tx: 219615; fax: 3228034).

Hilton, 46 Vassilissis Sofias Avenue (tel: 7220201; tx: 215808; fax: 7213110).

Ledra Marriott, 115 Syngrou Avenue (tel: 9347711; tx: 223465; fax: 9358603).

NJV Meridien, Constitution Square (tel: 3255301; tx: 210568/69).

Plaza, 78 Acharnon/Katrivanou Street, P.O. Box: 104 (tel: 8225111; tx: 214686).

Stanley, 1 Odysseos Street, Karaiskaki Square, P.O. Box: 107 (tel: 5240142; tx: 216550; fax: 5244611).

Titania, 52 Panepistimiou Avenue, P.O. Box: 142 (tel: 3609611; tx: 214673).

Chambers of Commerce
Athens Chamber of Commerce and Industry, 7 Akadimias Street, Athens 10671 (tel: 3602411/19; tx: 215707).

Heraklion-Crete Chamber of Commerce and Industry, PO Box 154, Heraklion, Crete (tel: 282420;tx: 282292).

Larisa Chamber of Commerce and Industry, P O Box 108, 44 Papkyriazi Street, Larisa (tel: 222219; tx: 295183).

Patras Chamber of Commerce and Industry, 58 Michalakopoulou Street, Patras (tel: 277679 or 275879; tx: 312179).

Piraeus Chamber of Commerce and Industry, 1 Plateia Loudovikou, Piraeus 18531 (tel: 4177241 or 4179480; tx: 212970).

Thessaloniki Chamber of Commerce and Industry, 29 Tsimiski Street, Thessaloniki (tel: 275341; tx: 412115).

Banking
Bank of Greece (central bank), Odos El Venizelou 21, Box 105, Athens 10250 (tel: 3201111; tx: 215102).

Agricultural Bank of Greece, Odos Panepistimiou 23, 10564 Athens (tel: 3230521; tx: 222160).

Algemene Bank Nederland NV, 3 Pararigopoulou Street (tel: 3243973/7).

Bank of Crete, Odos Voukourestiou 22 and Valaoritou, Athens 10671 (tel: 3606511; tx: 218633).

Commercial Bank of Greece, Box 16, Odos Sophocleous 11, Athens 10235 (tel: 3210911; tx: 216545).

Credit Bank, 40 Stadiou Street, GR-102 52 Athens (tel: 3245111; tx: 218691).

Ergobank, Panepistimiou 36, Athens 10679 (tel: 3601011; tx: 218826 erba GR; fx: 3645786).

General Hellenic Bank, SA, 9 Panepistimiou Street, 102 29 Athens (tel: 3241283, 3246451; tx: 215702, 218628).

Hellenic Industrial Development Bank SA (ETBA), Panepistimiou 18, Athens 10672 (tx: 215203 etva GR; fx: 3621023).

Investment Bank SA, 1 Korai Street, 10564 Athens (tel: 3230214/6; tx: 214239).

Ionian and Popular Bank of Greece, El Venizelos St 18/20, 102 43 Athens (tel: 3225501, 3254501, 3230055/9; tx: 215269 IPAT GR, 215969 IPAT GR).

National Bank of Greece SA, Odos Aeolou 86, Athens 10232 (tel: 3210411; tx: 214931/8).

National Investment Bank for Industrial Development SA, 14 Amalias Avenue, 10236 Athens (tel: 3242651/9; tx: 6113; fax: 3296211).

National Mortgage Bank of Greece, PO Box 667, Odos El Venizelou 40, Athens (tel: 7799301; tx: 215026).

Other useful addresses
Athenagence (ANA) (news agency), Odos Pindarou 5, Athens 10671 (tel: 3639816; tx: 215300).

Centre for Planning and Economic Research (KEPE), 22 Ippokratous Street, Athens 10680 (tel: 3627321/5).

EC Press and Information Office, 2 Vassilissis Sophias Avenue, Athens 10674 (tel: 743982/3/4).

Ellinikos Organismos Tourismou (tourist organisation), Odos Amerikis 2, Athens 10564 (tel: 32231111/9; tx: 215832).

Federation of Greek Industries, Odos Xenophontos 5, Athens 10557 (tel: 3237325; tx: 216247).

Greek Radio Television SA (ERT), 432 Messoghien, Ag Paraskevi, Attica (tel: 6595970).

Hellenic Organisation for the Promotion of Exports (HOPE), 1 Mitropoleos Street, 10557, Athens (tel: 3247011/16; tx: 220201).

Ministry of Agriculture, Odos Aharnon Street 2-6, Athens 10432 (tel: 3291206; tx: 215308).

Ministry of Commerce, Kanningos Square, Athens 10432 (tel: 3616251; tx: 215282).

Ministry of Finance, Odos Karageorgi Servias 10, Athens 10184 (tel: 3224071; tx: 216373).

Ministry of National Economy, Syntagma Square, Athens 10563 (tel: 3230911; tx: 221086).

Ministry of Environment and Town Planning, Odos Amaliados Street 17, Ampelokipoi, Athens 11523 (tel: 6431461; tx: 216374).

Ministry of Industry, Energy and Technology, Odos Zalokosta 10, Athens (tel: 3630911; tx: 216325).

National Statistical Service of Greece, 14-16 Lycourgou Street, Athens 10552 (tel: 3248511/18; tx: 216734).

Greenland and Faroe Islands

Hilary Barnes and Xueling Lin

Burdened by very heavy debts and large losses in their main fishing industries, Greenland and the Faroe Islands were struggling in 1989 to rein in expenditures. Both territories are part of the Kingdom of Denmark and receive large subsidies from the Danish State, although maintaining their own home rule legislatures. Taxes are collected by the home rule authorities and are considerably lower than in the rest of Denmark – a situation that may not be able to continue given the serious economic situation.

Greenland

Greenland was a colony, closed to trade with outsiders, until 1953 when it became an integral part of Denmark. Then a programme of rapid development was introduced at great cost, and has since been continued by the Greenland home rule government which came into existence in 1979.

The large expenditure by the home rule government has, however, created a debt burden which the Greenlanders, an Inuit (Eskimo) race, are now having to face. With loans in the region of DKr3bn taken out in Denmark and outside, the Greenland home rule government came under very serious criticism for uncontrolled expenditure and poor administration from an advisory committee set up by the Danish prime minister's office.

The debt was broken down into: loans for the purchase of houses with the Danish mortgage credit institutions; loans for ships and cutters from the Danish Ship Credit Fund; the Greenland government-owned company Proeks' drawing of credit with the Danish bank, Handelsbank; and foreign loans. Both the Danish mortgage credit institutions and the Danish Ship Credit Fund have put a brake on any further loans.

The heavy losses run up by Proeks, which is the Greenland government's production and sales company for fishing and is totally dominant in this main industry, has also taken a toll on the government's credibility.

Proeks' problems started in 1986 when they began to put large quantities of shrimp into storage in the hope of better prices. The plan backfired when shrimp prices went down and Proeks was landed with a bill for about DKr50m in storage charges. There appears to be a glimmer of hope at the end of the tunnel though; the advisory committee expected Proeks' 1988 loss of DKr378m to drop to DKr190m for 1989 because of more stable shrimp prices and increasing cod fishing.

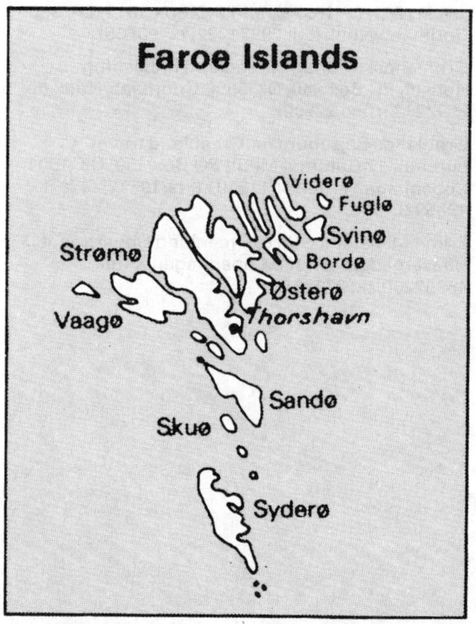

Greenland's economy has a relatively narrow base – the export sector made up of the fishing industry, an internal production sector which is completely dominated by building and construction and the service industry, in particular trade, transport and public service production.

The public sector traditionally plays a very large role in Greenland's economy. It employs almost two-thirds of all wage earners and its total purchase of goods and services – for consumption and investment – made up 75 per cent of the GNP in 1986. Given the present strains on the economy, this trend is getting a much needed review with more private housing being suggested and a change in the monopoly position of government companies.

As part of the Kingdom of Denmark, Greenland is part of NATO, but it opted out of EC membership in 1985. The EC has an agreement with Greenland allowing fishing in Greenland waters by EC fishing vessels in exchange for access to EC markets for Greenland fish products.

Ever since World War 2, Greenland has had two US airbases at Thule and Soendre Stroemfjord. The United States decided in 1989 to give up the base at Soendre Stroemfjord at the end of 1991, which caused some anxiety, the base being used as an airport for most commercial flights between Greenland and Denmark. From 1992 the Danish and Greenland governments, which have approached Nato for help with the running costs, will have to decide whether the base should be closed down.

About one-fifth of Greenland's population of about 55,000 are expatriate Danes and they make up the dominant 'foreign' group in the territory. Many of the expatriate Danes are employed in the Greenland public sector, particularly in health services which are still administered by the metropolitan government.

The home rule government would like to continue taking over areas of administration from the Danish State – as they have been ever since 1979 when they obtained home rule – but there is the problem of whether they have sufficiently qualified Greenlanders to fill the posts when the expatriate Danes leave. The home rule government admits that this is a serious problem and has put off taking over the health sector for another five years.

Faroe Islands

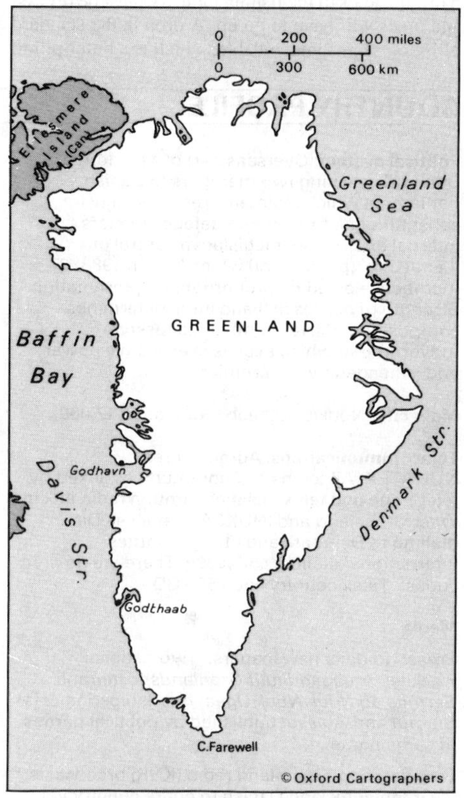

The Faroe islanders have had to face up to a large bill in the form of a DKr6.7bn foreign debt after several years of a booming economy spurred on by large investments from the home rule government.

The islanders obtained home rule in 1949, but their economy has still remained very dependent on the Danish State – in the form of direct transfers which made up more than 14 per cent of disposable national income and state guarantees for loans.

The 1989 report by a special Danish committee on the islands painted a very grim picture, stating that estimates for net debt were considerably higher than earlier thought, which they blamed on deficient statistics in fundamental areas such as freight, shipping, tourism and wages.

Fishing, the 47,000 islanders' main industry, has been doing badly with losses nearly doubling in 1989 to about DKr150m. The reason for this dangerous downturn appears to be over-fishing combined with over-investment.

The catch per fishing day has been dropping by an estimated 5 per cent every year since 1983. Fishing in the Barents Sea has also been substantially reduced because of negotiations with the USSR and Norway.

KEY FACTS

Official title: Greenland

Head of government: Premier Jonathan Motzfeldt

Ruling party: Coalition of Siumut Party and Inúit Ataqatigiit

Capital: Nuuk (Godthaab)

Official Languages: Greenlandic and Danish

Currency: Danish Krone (Kr) = 100 \18re

Exchange rate: Kr6.86 per $ (Dec 1989)

Area: 2,175,600 sq km

Population: 54,524 (1988)

Labour force: 23,062 (1987)

Trade balance: –Kr1,103m (1987)

All three parties have agreed to cut down on catches; the Faroese cod and haddock quota being cut by more than half.

The committee did not mince words about what it thinks the home rule government ought to do – they have recommended that all state projects come to a halt, subsidies to the fishing industry must be revised and taxes will have to go up. A drop in the standard of living seems unavoidable, which is a bitter pill to swallow since the islanders have had one of the highest per capita incomes in the world and are used to six per cent wage increases in the private sector.

Outlook

Politically, the possibility of carrying out at least some of the committee's suggestions appears reasonably good. After several breakdowns in the parliament (Lagting) in negotiations over how to implement the economic belt-tightening, a coalition led by Jogvan Sundstin of the Folkeflokken was formed in May 1989.

The Folkeflokken, a right-centre liberal party, is in principle in favour of independence from Denmark, but in practice cannot afford to strive for it given the islanders' economic plight.

COUNTRY PROFILE

Political system: Overseas part of Kingdom of Denmark, electing two members to Danish parliament which maintains responsibility for constitutional, foreign and defence matters. Internal affairs under legislative control of 'Landsting' (parliament) which has (in 1988) 27 members elected by proportional representation. Strength of parties in 'Landsting' determines composition of five-member 'Landsstyre' (government) which exercises executive power, and is headed by the premier.

Main city: Nuuk/Godthaab (population 12,000).

Telecommunications: Administered by NUNATEK. All towns and communities linked by telephone and telex facilities through radio link in West Greenland and INUKSAT satellite. Direct dialling to Denmark and other countries. International dialling code: 299. There are no area codes. Telex country code 503 GD.

Media

Press: No daily newspapers. Two national weeklies *Atuagagdliutit/Grønlandsposten* and *Sermitsiaq Nuuk/Nuuk Ugeavis*. Also periodicals *Siumut* and *Atassut* published by political parties of same name.

Broadcasting: Greenland radio (KNR) broadcasts in Greenlandic and Danish to entire country via INUKSAT. TV programmes (mostly Danish in origin) transmitted by KNR-TV, which also produces some programmes of its own. Each local community has one or more private TV stations which are allocated a 15-minute broadcast daily on KNR-TV (30 minutes on Sunday). Approximately 25,000 radios and 10,000 TVs in use.

Advertising: As for Denmark.

Domestic economy: Small economy centred on fishing and livestock farming. Heavy dependence on subsidies from Denmark, largely in the form of block grants. Comprehensive modernisation programme and expansion of fish exports allowed for steady increase in living standards.

External trade: Exports consist mainly of fish and fish products, zinc and chrome ores, lead and skins. Major markets include Denmark, Japan, W Germany, Netherlands, Norway, Sweden and the USA. Imports, which include foodstuffs, fuels, chemicals, consumer goods and capital goods, mainly come via Denmark.

Agriculture: Largely confined to sheep farming in the south. Also small scale reindeer farming. Production of lamb and reindeer meat is mainly for domestic consumption. Fishing, sealing and hunting is mainstay of economy. Principal products include cod, shrimp, salmon and seal.

Industry: Mainly fish processing and packaging. Also some tanning and leatherworking in the south.

Mining: Commercial mining of lead and zinc by Den Sorte Engel at Marmorilik. Known reserves of coal, uranium, iron ore and possibly oil.

Energy: Almost total energy dependence on Denmark.

Legal system: As for Denmark.

Membership of international organisations: As for Denmark, except EC.

BUSINESS GUIDE

Time: GMT-3 (-2 summer). Exceptions: Scoresbysund GMT-1 (GMT summer); Eastgreenland, Mesters Vig GMT (GMT summer); Qaanaaq & Thule GMT-4 (GMT-4 summer).

Climate: Arctic; temperatures at Nuuk/Godthaab vary between about -12°C and 11°C.

Entry requirements: As for Denmark except that diplomats need a visa.

Air access: Regular flights from Copenhagen to Kangerlussuaq/Søndre Strømfjord (SFJ) and, during summer, Narsarisuaq (UAK) operated by Grønlandsfly. Also weekly connection between Reykjavik (Iceland) and Nuuk/Godthaab, operated by Grønlandsfly.

National airline: Grønlandsfly A/S.

Surface access

Main port: Nuuk/Godthaab; ports (and heliports) at all towns.

National transport: Grønlandsfly operate daily flights Kangerlussuaq-Nuuk-Narsarisuaq, and twice a week to East Greenland. Other towns are linked by helicopter. Greenland Trade operate two passenger liners on the west coast. Also other smaller vessels in other areas.

Public holidays

Fixed dates: 1 Jan, 1 May, 21 Jun, 3 Jul, 24, 25 and 26 Dec and 31 Dec.

Variable dates: Maundy Thursday, Good Friday, Easter Monday, 'Store Bededag', Ascension Day/General Prayer Day and Whit Monday.

Working hours: As for Denmark.

Banking: Two local banking establishments: Gro\18nlandsbanken A/S and NUNA - Bank A/S. Monetary policy and administration headed by Danish National Bank.

Chambers of Commerce: There is no chamber in Greenland. Commercial information may be obtained from: KNI (Greenland Trade Department), Strandgade 100, Postboks 100, DK-1004, Copenhagen K, Denmark.

BUSINESS DIRECTORY

Principal newspapers

Atuagagdliutit/Grønlandsposten, PO Box 39, 3900 Nuuk (tel: 347070).

Sermitsiaq, PO Box 150, 3900 Nuuk.

Other useful addresses: All in Denmark except for Godthaab/Nuuk.

Arktisk Institut, 10 L.E. Bruunsvej, DK-2920 Charlottenlund (tel: 451640543)

Atuagagdliutit/Grønlandsposten (advertising), PO Box 39, DK-3900 Godthaab/Nuuk.

Commission for Scientific Research in Greenland, 10 Øster Voldgade, DK-1350 Copenhagen K (tel: 451113666; tx: 27125).

Greenland Home Rule Government Denmark Office, Sjaeleboderne 2, 1122 Copenhagen K (tel: (01)134224; tx: 15804; fax: 451322024).

Greenland Trade Shipping Department, Grønlandshavnen, DK-9220 Aalborg Ost (tel: 458157677; tx: 69688).

Grønlandsfly A/S (Greenlandair), PO Box 192, 12 Gammel Mønt, DK-1004 Copenhagen K (tel: 451112241; tx: 16973); and P O Box 1012, DK-3900 Godthaab/Nuuk (tel: 29921222; tx: 90665).

Grønlandske Forlag (Greenland Publishing House), PO Box 840, DK-3900 Godthaab/Nuuk (tel: 29922122; fax: 22500).

Grønlands Rejsebureau (Greenland travel bureau), 12 Gammel Mønt, PO Box 130, DK-1004 Copenhagen K (tel: 451131011; tx:19512; fax: -1 138592).

Prime Minister's Office, Greenland Department, 3 Hausergade, DK-1128 Copenhagen K (tel: 45932200; tx: 27125).

Grenada

Tony Thorndike

Political controversy has once again become Grenada's hallmark as the New National government, painfully put together in the aftermath of the US-led 'intervention' of October 1983 split asunder as Prime Minister Herbert Blaize dismissed a prominent minister, denounced his own party and formed another. Contenders for power grew with the prospect of a general election, in an atmosphere of deepening animosity and recrimination. Fortunately, the once-sluggish economy improved, but a government liquidity crisis caused by poorly thought-out taxation policies added to the problems of the 'Spice Island'.

The high hopes for political and economic stability that followed the trauma of 1983 have mostly been dashed. Expectations that economic largesse would follow the American troops, who sealed the fate of the marxist regime which had killed its leader and many more besides, were never met, although more than US$100m of US aid was received over the next three years. The mendicant mentality of most Grenadians was one ingredient, but the ultimate problem is political: poor and uninspired leadership with little self-confidence and sense of national pride. As a result of numerous defections, and a major challenge to his premiership, Prime Minister Blaize eventually announced his plan to form his own party in opposition to his own government, an extraordinary development by any standard. The risk of a return to power of Sir Eric Gairy is growing, although probably within an alliance. Gairy's eccentricities, corruption and authoritarianism helped prompt the 1979 coup which ushered in the revolutionary government.

Political disintegration

The general election was constitutionally due at the end of 1989 (the term of the present Parliament ends on 28 December), but the constitution permits its calling for up to 90 days after the limit. Prime Minister Blaize exploited this provision, however, the potential outcome of the poll was thrown into further confusion by his death in December 1989. The once dominant New National Party (NNP) which he helped to put together for the previous election was progressively reduced to only 3 members by defections. With Blaize's own defection and his formation of The National Party (TNP) in August 1989, it is no longer the governing party. The NNP which won 15 of the 16 seats in December 1984 has been reduced to 3 House of Assembly members and, with the formation of the National Party (NP) by Blaize at the end of August 1989, is no longer the governing party. Up to then, six defectors, plus the one elected member of Gairy's Grenada United Labour Party (GULP), became members of the National Democratic Congress, formed in July 1987. The slide continued when Blaize lost the leadership of the NNP to Dr Keith Mitchell, the dynamic Minister for Public Utilities and Works. An ambitious politician with a finger in many pies, he was adept at the art of patronage and many ordinary Grenadians had reason to be grateful to him. Mitchell called his faction of the NNP the 'New Thrust' and never lost an opportunity thereafter to criticise Blaize, but the prime minister regained the initiative by dismissing Mitchell and another minister in July; another resigned in sympathy. Even with new leadership by the new Prime Minister Ben Jones, the TNP seems doomed, particularly since there is little possibility of reconciliation between the TNP and NNP.

Sensing Blaize's numerical vulnerability in the House of Assembly, Mitchell soon afterwards tabled a vote of no-confidence but the NDC did not support it, pleading technical difficulties. Later the NDC tabled its own, but Blaize side-stepped the twin threats by proroguing parliament just two days before the critical sitting. Although constitutional, Blaize's action was regarded by most Grenadians, many regional observers and Blaize's erstwhile supporters in the regional right-wing Caribbean Democratic Union, as a denial of the democracy which many hoped had been restored after the revolutionary experience of 1979-83. Blaize backed up his offensive by announcing an investigation into the financial management of five government departments and four statutory bodies, all of which were associated with the dismissed ministers.

Formed out of a number of political groupings in the aftermath of the US-led military action, the NNP was artificial from the start. It was the result of a desperate search for unity by regional leaders, after experience of the People's Revolutionary Government (PRG) led by the New Jewel Movement. Indeed, the schisms that developed were inevitable. The senior faction was the old Grenada National Party (GNP). The GNP was created in 1956 by Blaize to combat Gairy, whose oratory as a trade union leader and championship of the working class made him a major force. Not surprisingly, the GNP represented the conservative middle classes and business community. The other two major constituents were a group formed in exile during PRG rule spearheaded by university lecturers Keith Mitchell and Francis Alexis, and another formed in 1984 by George Brizan. Despite his age (he is in his 70s) and infirmity, Blaize insisted on being leader and successfully garnered US support.

The NNP swept to power in 1984 with considerable financial aid from such other right-wing parties as the US Republicans and the West German Christian Democrats, the fear being either a return of GULP (which captured 36 per cent of the vote) or victory by the left. But the latter was shattered after the deep split that had developed in the NJM which led to a vicious struggle for power and, eventually, death. The NJM leadership implicated in the killing of PRG Prime Minister Maurice Bishop were arrested and 14 of the 17 were convicted of murder and sentenced to hang. More fortunate were those who supported Bishop and who survived. They created the Maurice Bishop Patriotic Movement but it commanded little electoral support. Neither is it likely that it will grow to any significant extent, at least not until the long appeal process involving the convicted 17 is exhausted. Although there is a good deal of public indifference towards the trial, whereby substantiated reports of brutality at the prison (by, for example, Amnesty International) are shrugged off, the drama serves as a reminder of revolutionary excess.

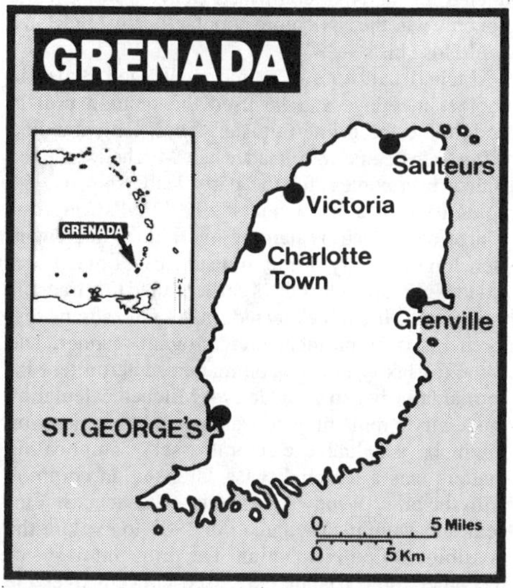

Blaize's authoritarianism and over-enthusiasm to root out any sign of revolutionary resurgence and to support the United States in all respects (even to the extent of voting against sanctions against South Africa in the United Nations, the only Third World member to do so) soon led to dissatisfaction within NNP ranks. But he was able to play Brizan, Alexis and Mitchell off against each other until 1986 when two ministers resigned and a rapprochement between Alexis and Brizan was reached. It was not long before they also resigned, the issue being massive (and politically selective) redundancies in the public sector. After a period of time they helped form the National Democratic Congress (NDC). The sole GULP member, Marcel Peters, joined them: he had been disowned by Gairy because he disobeyed Gairy's orders to boycott parliament. Three other former ministers followed in succession. The NNP tried to minimise the impact of the NDC by engaging in blatant patronage, in which Mitchell took a leading role. Matters went further: a party commission in 1987 recommended that all key civil service appointments be filled by NNP supporters and that the party vet all employees on government projects. This was partially put in hand with several 'reliable' retired public service workers being persuaded to return to fill posts left by those 'retrenched'.

KEY FACTS

Official title: Grenada
Head of state: Queen Elizabeth II
Head of government: Prime Minister Herbert A. Blaize
Ruling party: National Party
Capital: St George's
Official Languages: English
Currency: East Caribbean Dollar (EC$) = 100 cents
Exchange rate: EC$2.69 per $ (Nov 1989)

Area: 345 sq km
Population: 103,400 (1987)
GDP per capita: $1,346 (1987)
GDP real growth: 5.8% (1988)
Inflation: 5.8% (1988)
Trade balance: –$59.6m (1988)
Foreign debt: $48.7m (1988)
Visitor numbers: 210,632 (1988)

At the NDC convention in November 1988 Nicholas Brathwaite was elected leader. A respected educationalist, Brathwaite had headed the interim government established after the US military action pending the 1984 election. He was also nominated to fight the Carriacou seat held by Blaize. An opinion poll conducted for the NNP confirmed Blaize's worst fears: not only was the NNP more popular locally but Blaize could lose his seat.

Mitchell had for some time challenged Blaize in cabinet meetings and he used the opinion poll to undermine the prime minister. Their long-standing differences came to a head when Mitchell defeated Blaize in a contest for the party leadership by 190 votes to 155 at the January 1989 NNP Congress. Blaize responded by naming his faithful lieutenant, Ben Jones, deputy prime minister. But a much heralded NNP rally a few days earlier failed to attract the thousands Mitchell expected: many Grenadians disliked his overconfident and arrogant manner, and some his laxity in financial matters. Intermittent informal talks began with Gairy – Mitchell calculating that Gairy's remaining appeal to the older generation might be a valuable electoral asset. Complicating matters was a regional twist. Grenada, in common with the other Windward Islands – St Lucia, St Vincent and Dominica – agreed in 1987 to explore the possibility of political union. The prime ministers of the other three islands expressed concern over the political confusion in Grenada and in June 1989 sought a meeting with Blaize to discuss his electoral intentions, knowing his lukewarm attitude to the initiative. As St Lucia's John Compton remarked, 'general elections in Grenada is not a matter for purely national concern'. Referenda on the question of union are planned for all four islands in 1990.

Blaize also lost some overseas support because of his anti-socialist paranoia. The oppressive Emergency Powers Act was specifically aimed at 'subversives' and he and Jones made it clear that the US-trained Special Service Unit, a paramilitary branch of the police force, would be used whenever necessary. He was true to his word, but the problem was that force was used against legitimate opposition. A list of banned persons was issued naming prominent journalists and several opposition leaders and activists from other islands. In March 1989 a long list of banned books was published, including anti-apartheid material, to widespread international criticism. But the government permitted a ceremony of remembrance at the site where Bishop and others were killed by the revolutionary army on 19 October 1983, although care was taken that no names were recorded.

As for other parties, Gairy dismissed his apparently chosen successor Raymond Fletcher and had himself re-elected GULP leader. Fletcher went to launch a new party, the Grenada People's Movement, with Fennis Augustine, former PRG High Commissioner in London. Former IMF official Davidson Budhoo announced the formation of another party in August 1989, the People's Party for Growth and Accountability. Budhoo attempted to draw together some of the anti-Blaize parties but in the event only the Maurice Bishop Patriotic Movement appeared to respond. The election, when it comes, will be closely contested (although the NNP and GULP may agree not to oppose each other) and Grenadians will not be the only people anxious to know the result.

Economic growth

Fortunately for the beleaguered government, the economy continued to grow: by 6 per cent in 1987 and 5.8 per cent in 1988. Greater tourist income and an upturn in private investment (especially construction) were primary reasons. The banks enjoyed high liquidity levels, partly due to conservative risk policies, but in spite of heavy government borrowing because of chronically poor revenue records. Deposits increased considerably despite lower real interest rates, but these also spurred demand for consumer and construction credit. As a result the overall level of debt (US$84.5m, of which US$48.7m was external) became a matter of concern; debt servicing took up over 25 per cent of recurrent expenditure in both the 1988 and 1989 budgets. The trade deficit also grew (by 5.1 per cent) in 1988, with exports worth US$32.5m against imports of US$92.1m.

Tourist stopover arrivals increased by 7.6 per cent to 61,775 and cruise ship passengers by 6.4 per cent to 135,980, producing overall earnings of US$28.2m. Although welcome, these increases do not approach those recorded elsewhere in the region, mainly because of high prices and poor promotion. About a quarter of arrivals are returning *emigré* Grenadians who spend relatively little. Mitchell announced ambitious plans to boost the cruise ship trade with new quays capable of handling up to six ships at once, a prospect which had a mixed reception.

Manufacturing fared better: output increased by over 10 per cent in 1988, but traditional agriculture experienced little growth, only 2.1 per cent. Banana output grew by 12.8 per cent to 9,344 tonnes (9,132 tonnes exported) and that of nutmeg and mace by 2.9 per cent and 5.9 per cent to 2,879 tonnes and 332 tonnes respectively. Concern was expressed by the Grenada Cooperative Nutmeg Association when Indonesia, the only other nutmeg producer, and which has a price agreement with Grenada, cut prices. An Association delegation was swiftly despatched to Jakarta. The cocoa industry did not fare well, production decreasing by 21.8 per cent to 1,402 tonnes. But hopes for a revival were boosted by Can$5m worth of rehabilitation aid from Canada, and the reorganisation of the industry.

Government liquidity crisis

The government's fiscal problems began with the sudden introduction of value added tax (VAT) in the 1986 budget. Income tax and 17 other taxes were abolished, but the much needed simplification created fiscal disaster as revenue plummeted due to poor collection methods and public resentment. The gap was first partially filled by a business levy of 2.5 per cent on all sales which was even less popular, and then by change in VAT incidence. In the 1989 budget Blaize announced that the tax would be collected from importers and manufacturers at source, and not at the retail point. Furthermore, a 10 per cent sales tax would be imposed on all locally-produced goods. But the prospects of better collection remain slim. In 1989 the Eastern Caribbean Bank recorded that since 1986, revenue collected has been only 56 per cent of annual targets. Property tax collection was 78 per cent, the business levy 50 per cent and VAT a measly 42 per cent. The effect on the public services was an unofficial pay freeze. A strike and a demonstration in mid-June 1989 expressed widespread anger, making it unlikely that Blaize could delay negotiations much longer.

With such a record, budgetary forecasts are somewhat fanciful. Nonetheless, recurrent expenditure in 1989 was estimated to be EC$159.1m (US$59m) compared to EC$147.9m in 1987. The revenue estimate in 1988 was EC141.1m but less than EC$120m was collected. EC$162m (US$60m) revenue was sought in 1989. Capital spending was reduced from EC$102.5m to EC$88.5m, two-thirds financed by borrowing, including EC$24.8m by debentures and EC$8.7m from the National Insurance scheme. Given the structural nature of Grenada's economic problems, there is some speculation as to whether a new administration would fare much better.

COUNTRY PROFILE

Political system: Independent state within the Commonwealth with the British monarch as head of state represented by the governor-general. Executive power is vested in the cabinet, led by the prime minister, which is responsible to the bicameral parliament (13-member Senate and 15-member House of Representatives).

Main cities/towns: St George's (population 110,000).

Media
Press: Weeklies include *The Grenadian Voice, Grenada Guardian, The Informer, The National, The Tribune* and *Indies Times.*
Broadcasting: Radio Grenada (government-owned) broadcasts daily. A television station opened in 1986 now covers the St George's, St David's and St John's areas.

External trade: Widening trade deficit (due mainly to falling world prices for its major agricultural exports) is only partly offset by invisible earnings from tourism, workers' remittances from abroad and aid flows.
Exports: Nutmeg (35% of total value in 1986), cocoa (15%), bananas (13%), mace (7%), clothing. Main destinations: UK (30-35% of total value), Trinidad and Tobago (20-25%), German FR (8-10%), USA, Canada, Netherlands.
Imports: Foodstuffs (25-30% of total value), petroleum (15-17%), manufactures, machinery and transport equipment. Main sources: UK (25-30% of total value), Trinidad and Tobago (25-30%), Hong Kong, Japan, USA, Finland.

Agriculture: Sector forms the mainstay of economy, contributing 20 per cent to GDP (1987) and accounting for 65 per cent of domestic exports. Activity centres on the traditional farming of nutmeg and mace (world's second largest producer), cocoa and bananas. Earnings from these commodities (except nutmeg) have declined in recent years due to falling export demand and crop damage. Sugar-cane, lime, and a wide variety of fruit and vegetables are grown for local consumption and industrial use. Livestock production is also important for local consumption, though meat and milk still have to be imported. Around 45 per cent of total area is cultivated arable land, 5 per cent pasture, 10-13 per cent woodland/forest. Agricultural development policy is geared towards the rehabilitation of the cocoa industry, the promotion of new export crops, greater provision of fertilisers and other inputs and privatisation of state farms.

Industry: Activity centres on the processing of agricultural products and their derivatives, e.g. sugar, rum, coconut oil, lime juice, honey. Furniture, handicrafts and garments are also manufactured for export to the CARICOM market. New industries include a $1.5m soft drinks factory at Tempe, spice packaging and fish processing.

Tourism: Hotel and restaurant sector contributes 21.3 per cent to GDP and is a major source of foreign exchange. The number of tourists staying on the island (mainly from USA and UK) increased with the opening of the Point Salines International Airport. During 1988 arrivals from the USA numbered 12,640 and from the UK 7,139.

Membership of international organisations: CARICOM, Commonwealth, Economic Commission for Latin America, FAO, GATT (de facto), IFAD, ILO, IMF, Organisation of American States, Organisation of Eastern Caribbean states, state associated with the EC (Lomé Convention), UN, WHO, World Bank.

BUSINESS GUIDE

Time: GMT -4.

Climate: Tropical marine with an annual mean temperature of 28°C. Rain occurs mainly between Jun and Dec. Driest from Feb to May.

Entry requirements

Passport: Required by most citizens. An onward ticket is essential.

Visa: Required by all except nationals of Commonwealth, Scandinavian and EC countries, USA and certain other countries.

Currency: Import of local currency unlimited subject to declaration. Export limited to amount declared on arrival.

Health precautions

Mandatory: Yellow fever and cholera vaccination certificates required if arriving from an infected area.

Advisable: Typhoid, polio vaccination.

Representation overseas: Brussels, Caracas, London, New York, Ottawa, Port of Spain, Washington.

Air access

Main airlines: ALM, Air Antilles, British Airways, BWIA, Canada 3000, LIAT (in which Grenada is a shareholder).

Tax: Airport service charge of EC$25 on departure.

International airport: Point Salines (GND), 9 km from St George's; duty-free shop, restaurant, shops, hotel reservations (restricted hours in some instances).

Surface access: The following shipping lines call at Grenada: Harrison, Bernuth, Atlantic, Royal Netherlands, Geest, Royal Caribbean Cruise Line, Chandris American Line, Cunard. Regular boat services from St Vincent, Martinique and Trinidad. Twice-weekly ferry service to Carriacou island (Grenadian Grenadines). Other shipping lines: Seaburn Cruise Line, Clipper Cruise Line, Carnival Cruise Line, Regency Cruises.

Main port: St George's.

Hotels: Normally necessary to book in advance. Except for town hotels, most are located near beaches. Eight per cent sales tax on food and beverages and 10 per cent service charge added to bill.

Car hire: International licence required or local permit obtained (valid national licence must be presented to local Traffic Department). Traffic drives on the left.

City transport
Taxis: Widely available. Fares are regulated.

National transport
Road: Total network over 800 km, much of it all-weather.
Buses: Public and private services run from the market place in St George's to all parts of the island. Cheap but often slow and few run during the late afternoons, evenings and Sundays.

Public holidays: 1 Jan, 7 Feb, Easter, Labour Day, Whit Monday, Corpus Christi (Jun), first Mon and Tue in Aug (Emancipation Day), 25 Oct (Thanksgiving Day), 25 Dec, 26 Dec.

Working hours
Government and business: (Mon-Fri) 0800 – 1145, 1300 – 1600.
Banking: (Mon-Fri) 0800 – 1200; also Fri 1500 – 1800.
Shops: (Mon-Sat) 0800 – 1145, 1300 – 1600. Exception: closed Sat 1300 – 1600.)

Telecommunications

Telephone: Dialling code for Grenada: IDD access code + 1-809 followed by area code (Carriacou 443, St George's 440) followed by subscriber's number. 100 per cent automatic service. Major extensions of system completed – to 6,300 lines with new exchanges.

Telex: Modern international services operated by Cable and Wireless Ltd, St George's: open 0700 – 1900 (Mon-Sat) and 1600 – 1800 (Sun).

Banking: Bank of Nova Scotia, Barclays Bank International, Grenada Bank of Commerce, Grenada Co-operative Bank, Grenada Development Bank, National Commercial Bank of Grenada.

Electricity supply: 220/240 V AC, 50 cycles.

Representation in capital: Belgium, Canada, UK, USA, Venezuela.

BUSINESS DIRECTORY

Business advice and information

Grenada Chamber of Industry and Commerce, PO Box 129, DECO Building, Mt Gay, St George's (tel: 4485).

Grenada Hotel Association, PO Box 440, Mt Gay, St George's (tel: 1590).

Grenada Industrial Development Corporation, Frequente Industrial Estate, St David's (tel: (444)1035, 1040).

Grenada Manufacturers' Association, Grande Anse, PO Box 213, St David's (tel: (444)4637).

Grenada Tourist Board, PO Box 293, The Carnage, St George's (tel: 2279, 3371, 2001; tx: 3422).

Banking

Bank of Nova Scotia, Halifax Street, St George's (tel: 3274; tx: 3452).

Barclays Bank, Halifax Street, St George's (tel: 3232; tx: 3421).

Grenada Bank of Commerce, Halifax Street, St George's (tel: 3521).

Grenada Co-operative Bank, Church Street, St George's (tel: 2111).

Grenada Development Bank, Halifax Street, St George's (tel: 2382).

National Commercial Bank of Grenada, Halifax Street, St George's (tel: 3566).

Hotels: Grand Anse Beach, St George's

Cinnamon Hill, P.O. Box: 292 (tel: 444-4301/2; tx: 3425 grnhtl).

Hibiscus Inn, Grand Anse Beach, P.O. Box: 279 (tel: 444-4233; tx: 3425 grnhtl).

Ramada Renaissance, P.O. Box: 441 (tel: 444-4371; tx: 3468; fax: 4444800).

Spice Island Inn, P.O. Box: 6 (tel: 444-4258; tx: 3425 grnhtl).

Hotels: Lance aux Epines, St George's

Calabash, P.O. Box: 382 (tel: 440-4234; tx: 3425 grnhtl).

Horse Shoe Beach, P.O. Box: 174 (tel: 440-4410; tx: 3422 grnhtl).

Secret Harbour, P.O. Box: 11 (tel: 444-4548/9; tx: 3425 grnhtl).

Hotels: St George's

Balisier, Richmond Hill (tel: 440-2346; tx: 3203425).

St James, Grand Etang Road (tel: 440-2041).

Car hire

St George's Arnold's Limousine Service: Grenville Street, St George's (tel: 1781).

Avis/Spice Island Rentals: Lagoon Road (tel: 3936).

David's Car Rentals: Archibald Avenue (tel: 2399, 3038).

Juliana Aird & Co: St John's Street (tel: 2504).

Car hire

St David's Y and R Car Rentals, L'Anse aux Épines (tel: (444)4448).

Guadeloupe and Martinique

Nicolas Kulibaba

Relative calm and conciliatory political gestures by the metropole shifted the focus of public attention away from secessionist activity in 1989 and onto efforts to stem deterioration of the islands' economy. Rising unemployment, declining agricultural productivity and demographic change remain barriers to economic and political autonomy in these under-developed Caribbean outposts of France, highlighting the role of the welfare state.

French colonies until 1946, Guadeloupe and Martinique were granted semi-autonomous legislative home rule in 1983, under the decentralisation programme of socialist President François Mitterrand. While under the administrative guidance of *préfets* appointed by the central government, locally-elected *conseils généraux* are legally empowered to formulate departmental budgets, to impose local taxation, make development grants and to elect representatives to regional policy-making bodies, the *conseils régionaux*. Home rule, however, stops short of security and defence matters. The presence of more than 1,200 gendarmes and police, who report directly to central authorities, has been the object of much controversy, particularly regarding their sometimes heavy-handed role in suppressing militant successionist activity.

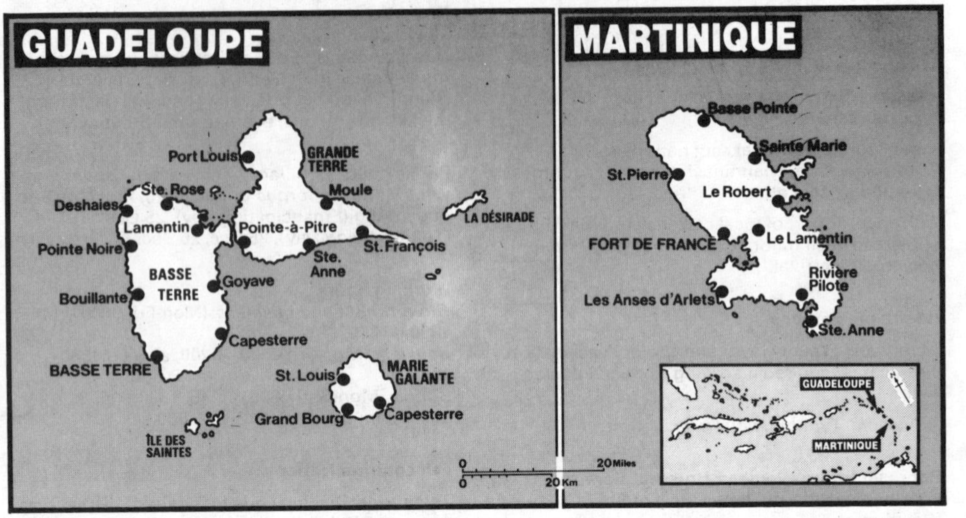

The colonial legacy

The bonds of Guadeloupe and Martinique to 'La France hexagonale' are those of dependency, stemming from the islands' former status as colonial suppliers to the metropolitan markets, the effects of which are deeply gouged even into the modern institutions of political and social life. Both departements rely heavily on economic subsidies to their outmoded agricultural infrastructure and inadequate social services. Like many of their Antillean neighbours, the islands are historically oriented around the production of cane sugar, rum, tropical fruits and vegetables, markets for which are increasingly competitive and subject to supply glut. Efforts to diversify the island economies through the expansion of tourism and artisan industry have registered limited gains. However, rapid growth of the service economy has also added to the existing sensitivity of Guadeloupe and Martinique to fickle European and North American markets and to competition from Caribbean neighbours.

Efforts by central government to stabilise the economies of the *départements outre-mer* have been stymied in recent years by accelerated social and demographic change. The descendants of white, colonial-era planters and administrators, known locally as *békés*, until recently constituted a mere 6 per cent of the islands' population. Historical privilege, however, resulted in control by this minority of nearly 80 per cent of island lands. A wave of new migration from mainland France, which began in the 1970s, has since brought about a major shift in the size of the departements' white population of civil administrators, technocrats and entrepreneurs. At the same time, economic frustration, high unemployment and the absence of formal barriers to immigration between the islands and the metropole have resulted in the annual migration of between 15,000 and 20,000 persons each year. As a result of such mass movement in the 1980s, more than one-third of the native population of Guadeloupe and Martinique currently resides in Europe or elsewhere in France's overseas domain.

In spite of chronically high unemployment, the relative affluence of Martinique and Guadeloupe compared with other islands in the region has attracted an estimated 40,000 clandestine immigrants from Haiti and Dominica in recent years. Their desperation and willingness to work for subminimum wages has undercut employment by local workers, increasing social tensions and resulting in incidents of violent retribution.

Radicalism and change

Long a hotbed of separatist activity, Guadeloupe passed through 1989 without a revisitation of the violent outbursts that had characterised much of the past decade. The return of calm was due, at least in part, to the continued incarceration of the leaders of the Alliance Revolutionnaire Caraibe (ARC), whose campaign of terrorist bombings spread from Guadeloupe to Martinique beginning in 1983.

The mainstream of separatist activity continues to be represented by the Union Populaire pour la Libération de la Guadeloupe(UPLG), the leading independence movement on the archipelago. In April 1989 it again threatened the island with disruption to protest the arrest of seven of its members and the surveillance of others by security forces. The UPLG militants were accused of complicity in a series of bombings which occurred in Guadeloupe in January 1988. Responding to the arrests, the UPLG called for a general strike and erected barricades on main roads leading to the cities of Pointe-a-Pitre, Basse Terre and Saint Louis. Following a brief, if violent, confrontation between the protestors and the police, Justice Ministry officials released a number of detainees within several days of their arrest, leading UPLG founder Dr Claude Makouke to declare a victory for the separatist movement.

The government's move toward appeasement, however, fell within a more conciliatory approach to the islands and recognition that the political centre in France's overseas departements has moved considerably leftward in recent years. Among the individuals arrested by the authorities in April was Jean Barfleur, a physician whose UPLG cell in St Louis has captured six municipal council seats in March elections, drawing more than 40 per cent of all votes cast.

How far the Mitterrand government might be willing

KEY FACTS

Official title: Guadeloupe
Head of state: President of France
Head of government: Préfet Jean-Paul Proust
Capital: Basse-Terre
Official Languages: French
Currency: French Franc (Fr) = 100 centimes
Exchange rate: Fr6.31 per $ (Nov 1989)
Area: 1,705 sq km
Population: 339,700 (1989)
GDP per capita: $3,600 (1985)
Inflation: 3.5 % (1987)
Trade balance: –Fr5.7bn (1987)
Foreign debt: $67m (1987)
Visitor numbers: 441,175 (1986)

KEY FACTS

Official title: Martinique
Head of state: President of France
Head of government: Préfet Jean-Claude Roure
Capital: Fort-de-France
Official Languages: French
Currency: French Franc (Fr) = 100 centimes
Exchange rate: Fr6.31 per $ (Nov 1989)
Area: 1,100 sq km
Population: 337,000 (1989)
GDP per capita: $4,761 (1985)
Inflation: 3.6% (1987)
Trade balance: –Fr5.5bn (1987)
Foreign debt: $59m (1987)
Visitor numbers: 562,000 (1987)

to go in defusing separatist frustrations became evident in June 1989. France's National Assembly agreed to consider a proposal, sponsored by Guadeloupe's Communist Party deputy, which would grant an amnesty to 13 Guadeloupe activists accused of acts 'intended to diminish political and social order and the authority of the Republic'. Rumoured to enjoy the support of President Mitterrand, the amnesty would free separatists arrested for activities prior to 14 July 1988 – virtually all who remained behind bars at the time of the proposed legislative action. The most controversial of these is Luc Reinette, the mastermind of the ARC's campaign of terror which claimed several lives during a three-year period earlier in the decade.

In anticipation of a debate over the proposed amnesty, judicial authorities in Paris postponed the trial of Reinette and two of his accomplices – Henri Bernard and Henri Amedien – until later in the year. This, in turn, prompted the three to declare an indefinite hunger strike which was abandoned after it failed to generate popular support in Guadeloupe.

While Mitterrand's first term in office coincided with a period of unease in the overseas departements, the president's second term has heralded an era of improved goodwill. His sensitivity to the issue of racism in France and his support for a wide array of

COUNTRY PROFILE — GUADELOUPE

Historical profile: Carib name was 'Karukera' (island of beautiful water); first European visitor, Columbus 1493; settled by French 1635; overseas department status 1946.

Political system: Overseas department of France with additional status as a region of France, represented in the French National Assembly by three deputies and in the Senate by two senators. Local administration is through a directly elected Assembly of 41 members and a 42-member departmental council (*conseil général*), also directly elected.

Main cities/towns: Pointe-á-Pitre (population 76,500, straddles islands of Grand Terre and Basse Terre (mountainous)), Basse-Terre (pop 24,800, on the island of Basse Terre, capital of Guadeloupe and dependencies of Marie-Galante, Les Saintes, Désirade, St Barthélemy and St Martin), Grand Bourg (on Marie-Galante).

Media

Press: Main daily is *France Antilles*.

Broadcasting: Three radio and one television (France Région 3) stations operating.

External trade: Massive trade deficit is only partially offset by earnings from tourism and aid flows from France.

Exports: Bananas, sugar, rum. Main destinations: France (71% of total value), Martinique (19%), French Guiana (3%), Réunion (3%), Anguilla (0.5%).

Imports: Foodstuffs, consumer goods, clothing, machinery and transport equipment, petroleum etc. Main sources: France (65% of total value), Italy (4%), German FR (4%), USA (3%), Martinique (3%), Japan (3%).

Agriculture: Main sector of the economy, though country relies heavily on imported food from France. Activity is based primarily on sugar-cane and its by-products, both for local refining and export. Many of the large sugar plantations have, however, been turned over to the cultivation of bananas and pineapples. Offshore fishing is a traditional source of food. The main catch includes lobsters, crab and octopus. Estimated 25 per cent of total area cultivated arable land, 10 per cent pasture, small area (perhaps five per cent) with arable potential; 15 per cent woodland/forest (includes national park of around 3,000 ha).

Industry: Sector contributes around 12 per cent to GNP and employs 15 per cent of the workforce. Main activities include sugar refining (only four factories now operating), rum distilling, food processing, cement and bricks manufacture, box and mail/wire making, mineral water bottling and ship repair. There is an industrial freeport at Jarry. Tourism: Rapid development of sector during 1970s with number of tourists arrivals rising to over 440,000.

Membership of international organisations: Franc zone.

Time: GMT -4.

Climate: Sub-tropical with annual mean temperature of 27°C. Levels of humidity and rainfall highest around Basse-Terre.

Entry requirements
Passport: Required by all except nationals of France, most other EC countries (including UK) and certain other countries with proof of identity for short visits.
Visa: Required by all except nationals of EC and most other western European countries.
Currency: No restrictions on import/export of foreign currency.

Health precautions
Mandatory: Yellow fever vaccination certificate if travelling from an infected area.
Advisable:

Air access

National airline: Air Guadaloupe.

Other airlines: Air France, Air Canada, Air Nova, American, LAV, LIAT.

International airport: Pointe-à-Pitre-Le Raizet (PTP), 3 km from city; duty-free shop, restaurant, buffet, bank, post office, shops, hotel reservations, car hire.

Surface access
Main port: Pointe-à-Pitre.

Hotels: If service charge is not added, 15 per cent tip is usual.

Car hire: International licence required.

City transport

Taxis: Plentiful but generally regarded as expensive.

National transport

Air: Air Guadeloupe, Air St Barthélemy and Liat operate frequent services to all the dependent islands from Point-à-Pitre.

Road: Total network around 3,000 km – including about 500 km of national highway; secondary roads can be tortuous.

Buses: Inexpensive minibus services on most routes. Organised bus tours available.

Public holidays: 1 Jan, Carnival (Shrove Tuesday and Ash Wednesday), Good Friday, Easter Monday, 1 May, Ascension, Whit Monday, 14 Jul (National Day), 21 Jul (Schoëlcher Day), 15 Aug (Assumption), 1 Nov (All Saints'), 11 Nov (Armistice Day), 25 Dec.

Working hours

Government: (Mon-Fri) 0730 – 1300, 1500 – 1630.

Banking: (Mon-Fri) 0800 – 1200, 1430 – 1600.

Shops: (Mon-Sat) 0800 – 1230, 1500 – 1800.

Telecommunications

Telephone: Dialling code for Guadeloupe: IDD access code + 590 followed by subscriber's six-digit number. 100 per cent automatic service; public telephones available.

Telex: No public facilities, but Chamber of Commerce at Pointe-á-Pitre will allow business visitors to use their telex.

Banking

Banking centre: Pointe-à-Pitre.

Commercial banks: Caisse Régionale de Crédit Agricole Mutuel de la Guadeloupe.

Foreign branches: Banque des Antilles Françaises, Banque Française Commerciale, Banque Nationale de Paris, Banque Régionale d'Escompte et de Dépôts, Caisse Centrale de Coopération Economique, Crédit Martiniquais, Société Générale de Banque aux Antilles.

Electricity supply: 220/380 V AC, 50 and 60 cycles.

Hotels: Gosier
Arawak, P.O. Box: 396 (tel: 84-12-74; tx: 919733).

Auberge de la Vieille Tour (tel: 84-23-23; tx: 919751).

Callinago, Pointe de la Verdure (tel: 84-25-25; tx: 919778).

Ecotel (tel: 84-20-20; tx: 919828).

La Creole Beach Resort (tel: 84-15-00; tx: 919836).

Marissol Bas du Fort (tel: 90-84-44; tx: 919855).

Novotel Fleur d'Epée, Bas du Fort, 97190 (tel: 90-81-49; tx: 919847).

Salako (tel: 84-14-10; tx: 919835).

Village Soleil, Marina Bas du Fort (tel: 90-85-76; tx: 919094).

Hotels: St François
Méridien, P.O. Box: 37 (tel: 88-51-00; tx: 919774).

Car hire
Avis: Hotel Marissol, Gosier (tel: 90-84-44).

Budget: Novotel Hotel, Gosier (tel: 90-81-49; tx: 919847).

Europcar National: 3 rue Becquerel, Z.I. de Jarry (tel: 26-60-64; tx: 919964).

Hertz: Route de la Gabarre, Pointe-à-Pitre (tel: 82-88-44; tx: 029745).

Car hire: International airport
Avis: (tel: 82-33-47; tx: 919798).

Budget: (tel: 82-95-58).

Europcar National: (tel: 82-50-51).

Hertz: (tel: 82-00-14).

Business advice and information
Air Guadeloupe, 97110 Abymes (tel: 82-28-35; tx: 919008).

Chambre de Commerce et d'Industrie de Basse Terre, PO Box 17, 6 rue Victor Hugues, 97100 Basse Terre (tel: 81-16-56; tx: 919781).

Chambre de Commerce et d'Industrie de Pointe-à-Pitre, Assainissement BP 64, 97152 Pointe-à-Pitre (tel: 90-08-08; tx: 919780).

Direction de la Promotion Touristique, Préfecture de la Guadeloupe, rue Lardenoy, 97100 Basse Terre (tel: 81-15-60; tx: 919717).

France-Antilles, 1 rue Paul Lacave, Pointe-à-Pitre (tel: 90-25-25; tx: 919728).

Office Départemental du Tourisme, 5 Square de la Banque, PO Box 1099, 97181 Pointe-à-Pitre (tel: 82-09-30; tx: 919715).

Office du Tourisme, 23 rue Delgrés, Pointe-á-Pitre.

budgetary supports for the ailing island economies have, at least, bought him the indulgences of patience and good faith among Guadeloupe and Martinique's leaders. Encouraging a more active role for the departements' economic and social councils has also been welcomed by their constituent trade, labour and management groups, who are being asked to take greater responsibility for the programming of budgetary resources.

Economic rescue

While the French government may be short on a precise strategy for turning around the island economies, it has earned high marks for its concerted efforts. The principal challenge facing both Martinique and Guadeloupe is structural in nature: how to increase productivity in the midst of profound social and economic change.

The foundation of the local economy remains agriculture. However, cane and sugar production have been in steady decline for the past seven years due to the combined influences of outmoded production and refining technology, uneven and inequitable land distribution and the absence of market incentives. Erosion of the sugar economy has, in turn, triggered decline in the island rum industry. More alarming – and less predictable – have been recent declines in other agricultural commodities. Efforts to diversify agricultural production in the 1970s led to the development of farming of garden vegetables and fruits for the European and North American market. However, recent years have seen output of certain crops drop sharply, by as much as 40 per cent for citrus and 33 per cent for garden vegetables in the last three years alone.

Compounding the islands' problems has been a drop-off in tourism, once heralded as the most promising of new economic avenues. Revenue from foreign (non-French) tourism declined between 17 and 20 per cent for Guadeloupe and Martinique during each of the last two years. This decline is attributed to the relatively poor performance of the US dollar against the franc, underlining the fickle nature of tourism as a source of foreign exchange earnings. Sluggish tourism arrivals, in turn, have resulted in the cancellation of daily flights between the islands and North American hubs. Minor compensation for a decline in North American arrivals has come from the increase in tourism from the French mainland, with the number of these visitor nights increasing by 30 per cent between 1986 and 1988. In a campaign to counteract these trends, the central government has accorded extensive grants and credits to the Antillean departements, including a 1989 tourism promotional budget of FFr9.9m. Additional budgetary support for development in the islands has been drawn from such diverse sources as the Fonds Europeen de Developpement Regional (FFr78m in 1988-89) and the Fonds Social Europeen (FFr62m).

A particular target of government assistance are programmes aimed at lowering unemployment (23 per cent in 1989) in both departements. A centrepiece of government strategy is *Service Militaire Adapté*, which replaces mandatory military service with professional training for under-educated and unemployed youth. Incentives for employment generation in industry are provided through tax and capital equipment credits that can be directly tied to the creation of jobs. Parliamentary debate, however, has heard charges that the socialist government's economic remedy for the overseas departements may be too scattershot, requiring a more precise targetting of development funding. Citing the recommendations of Guadeloupe and Martinique Economic and Social Councils, the emphasis on future development funding appears to be shifting toward weaning the islands away from their dependence on external commerce and toward greater self-sufficiency.

Chronic trade deficits (FFr5bn for Guadeloupe in 1987 and FFr4.6bn for Martinique) not only underline the fact of structural dependency, but also hint at the immensity of the economic challenge facing the islands in the years ahead.

COUNTRY PROFILE — MARTINIQUE

Historical profile: Name said to be either from Martinica (given in honour of St Martin) or corruption of Madinina (native name); discovered by Columbus in 1493; settled by French 1635 – in face of indigenous Indian hostility; Marie-Josephe Rose Tascher de la Pagerie – Napoleon's Empress Josephine born Les Trois Ilets 1763; St Pierre destroyed by eruption of Mount Pelee 1902; overseas department status 1946.

Political system: Overseas department of France with additional status as a region of France, represented in the French National Assembly by three deputies and in the Senate by two senators. Local administration is through a directly elected Assembly of 41 members and a departmental council (*conseil général*) of 36 members, also directly elected.

Main cities/towns: Fort-de-France (population 110,000), Lamentin, Sainte Marie, Schoelcher.

Media
Press: France Antilles is the major daily.
Broadcasting: Three radio and one television (France Région 3) stations operating.

External trade: Large trade deficit is only partly offset by invisible earnings from tourism and workers' remittances from abroad.
Exports: Bananas (55-69% of total value), petroleum products, rum, sugar, pineapples. Main destinations: France (65% of total value), Guadeloupe (24%), German FR (6%), French Guiana (3%), USA (0.2%).
Imports: Foodstuffs, machinery and transport equipment, crude oil, consumer goods. Main sources: France (65% of total value), UK (6%), Italy (4%), German FR (3%), Japan (3%), USA (2%).

Agriculture: Mainstay of economy, employing 13 per cent of the workforce and accounting for 70-80 per cent of exports. Around 37 per cent of total land area is cultivated, 25 per cent is forest and 19 per cent savannah. The majority of farms are privately run by small-holders. Activity is centred on the production of sugar-cane, pineapples and bananas, mainly for industrial processing and exports. Crops such as sweet potatoes, yams, manioc, beans, cabbages and tomatoes are grown primarily for domestic consumption. Small quantities of aubergines, avocado pears, flowers and limes are exported. Virtually all the island's meat requirements are met by imports. Inshore fishing (cod, lobster, crayfish, crab, clams) is undertaken all year round.

Industry: Sector contributes only 10 per cent to GDP and employs under 15 per cent of the workforce. Major industries include a 500,000-tonne capacity oil refinery, a cement works, rum distilling, sugar refining, fruit canning, soft drinks manufacture and a polyethelyne plant. Five industrial zones have been set up and tax exemptions introduced to encourage light industrial development.

Tourism: Major sector of the economy with tourists coming mainly from France and the USA. The total number of arrivals has steadily increased, reaching over 562,000 in 1987.

Membership of international organisations: Franc zone.

BUSINESS GUIDE

Time: GMT -4.

Climate: Sub-tropical with an annual mean temperature of 26°C. Rain heaviest in the north. Rainy season from Jul-Nov.

Entry requirements
Passport: Required by all except nationals of France, most other EC countries (including UK), and certain other countries with proof of identity for short visits.
Visa: Required by all except nationals of EC and most other western European countries and US citizens for a maximum of three months.
Currency: No restrictions on import/export of foreign currency.

Health precautions
Mandatory: Yellow fever vaccination certificate if travelling from an infected area.
Advisable: Typhoid, polio vaccination. Water precautions.

Air access
National airlines: Air France, Air Martinique.
Other airlines: Aeromaritime, Air Canada, Air Nova, American, LAV, LIAT, Minerve.
International airport: Fort-de-France-Lamentin (FDF), 14km from city; restaurant, car hire, shops.

Surface access
Main port: Fort-de-France.

Hotels: Five per cent room tax. If service charge is not added, 15 per cent tip is usual.

Car hire: International licence required, supported by valid foreign licence for periods after 20 days.

City transport
Taxis: Extra charges at night: plentiful but generally regarded as expensive; cheaper shared taxis are available.

National transport
Air: Charter flights and helicopter trips provided by Air Martinique.
Road: Well developed network of more than 1,500 km, much of it surfaced.
Buses: Limited services (shared taxis compete).
Water: Regular ferry services Fort-de-France to Pointe de Bout and Anse Mitan.

Public holidays: 1 Jan, Carnival (Shrove Tuesday and Ash Wednesday) Good Friday, Easter Monday, 1 May, Ascension, Whit Monday, 14 Jul (National Day), 21 Jul (Schoëlcher Day), 15 Aug (Assumption), 1 Nov (All Saints'), 11 Nov (Armistice Day), 24 Dec, 25 Dec.

Working hours
Business: (Mon-Fri) 0800 – 1200, 1430 – 1700 (Sat) 0800 – 1200.
Government: (Mon-Fri) 0730 – 1300, also (Mon-Fri) 1500 – 1630.
Banking: (Mon-Fri) 0800 – 1200, 1430 – 1700.

Telecommunications
Telephone: Dialling code for Martinique: IDD access code + 596 followed by subscriber's six-digit number. 100 per cent automatic service.
Telex: No public telex facilities, though facilities available at hotels.

BUSINESS GUIDE

Banking
Commercial banks: Banque des Antilles Françaises, Banque Française Commerciale, Banque Nationale de Paris, Caisse Centrale de Coopération Economique, Crédit Agricole Mutuel, Crédit Martiniquais, Société Générale de Banque aux Antilles.

Advertising: No trade press, though newspapers, television and radio take advertising. Small tax on posters.

Electricity supply: 220/380 V AC, 50 cycles.

Car hire: Airpor
Avis: (tel: 51-26-86).

Budget: (tel: 51-22-88).

Europcar National: (tel: 51-51-51).

Hertz: (tel: 51-28-22).

Car hire: Fort-de France

Avis: 4 rue Ernest Deproge (tel: 70-11-60; tx: 912376).

Budget: 12 Rue Felix Eboue, BP 1006 (tel: 63-69-00, 70-06-10).

Hertz: 24 rue Ernest Deproge, PO Box 1007 (tel: 60-64-64; tx: 912000).

Business advice and information
Agence pour le Développement Economique de la Martinique, 26 rue Lamartine, 97207 Fort-de-France (tel: 73-45-81; tx: 912633 chamco).

Agence pour le Développement du Tourisme de la Martinique, Anse Goureaud (tel: 61-61-77).

Air Canada, Aeroport du Lamentin (tel: 51-29-81).

Air France, 26 rue Ernest Deproge 97200 Fort-de-France (tel: 63-75-52; tx: 912478).

Air Martinique, Aeroport du Lamentin (tel: 51-09-90; tx: 912385).

American Airlines, Aeroport du Lamentin (tel: 51-12-29).

Chambre de Commerce et d'Industrie de la Martinique, rue Ernest Deproge 50, 97205 Fort-de-France (tel: 63-00-08; tx: 912633).

Chambre des Métiers, Morne Tartenson, Fort-de-France (tel: 72-32-22).

Office du Tourisme, Pavillon du Tourisme, Boulevard Alfassa, Fort-de-France (tel: 63-79-60; tx: 912678).

Banking
Banque des Antilles Françaises, 34 rue Lamartine, Fort-de-France (tel: 73-93-44; tx: 912636).

Banque Française Commerciale, 6-10 rue Ernest Deproge, Fort-de-France (tel: 63-82-57).

Banque National de Paris, Avenue des Caraibes, Fort-de-France (tel: 63-66-67).

Caisse Centrale de Coopération Economique, 12 Boulevard Général de Gaulle, Jardin Desclieux, 97206 Fort-de-France (tel: 73-31-02; tx: 912313).

Crédit Agricole Mutuel, rue Ernest Deproge, Fort-de-France (tel: 73-17-06).

Crédit Martiniquais, rue de la Liberte, Fort-de-France (tel: 59-93-00).

Société Générale de Banques, rue de la Liberté, Fort-de-France (tel: 71-69-83).

Guam

T. R. Hughes

(map on page xxiv)

Guam's voters, riding on the wave of an economic boom, decided against any change in the political spectrum in the November 1988 elections. They wasted no time in relaunching their crusade for a commonwealth status once the Bush administration and the US Congress had settled into office. With low inflation and decreasing unemployment, Guamanians may be justified in thinking they have exported all their problems to Washington. An unincorporated territory, the 549 sq km island is geographically the westernmost part of the United States and an important link in the US defence chain. It is home to some 126,430 people of which 23,600 are US military personnel and their dependents. The US military presence is a substantial pillar of the local economy.

Reflecting the general prosperity, unemployment fell to an all-time low of three per cent during the third quarter of 1987 and the government is attacking the housing problem, a pressing social sore, by building several hundred new units. Economists say the gap in living standards between Guam and the rest of the USA continues to close.

The US air force base and naval commands jointly contribute around $400m as their share of the federal government's annual spending of some $680m, more than half of which goes on local salaries. The military occupies the bulk of the land held by the federal authorities amounting to about a third of the total land ownership.

In the private sector, tourism is the big revenue earner. Visitors to Guam totalled 483,900 in 1987, up 18 per cent over the previous year, and the momentum seems to be continuing. Nearly 80 per cent of tourists are from Japan spurred by the strength of the yen against the US dollar. Tourism, which took off in 1967 when Pan Am inaugurated direct flights between Guam and Tokyo, provides about 13,000 jobs (roughly a third of private sector employment) with a payroll of some $150m. But economists estimate the visitor industry may be pumping as much as $400m, and probably more, every year into the economy.

Construction boom

Tourism has also helped to generate a construction boom which is trying hard to catch up with the demand for new hotels, shops, restaurants and tourist-oriented businesses. Guam's *Annual Economic Review* says that in 1987 tourism 'somewhat reached a saturation point awaiting additional hotel rooms' and government policy makers have kept the problem at the top of their agenda. They confidently expected the current feverish bout of building work will raise the number of rooms by 3,183 to a total of 7,047 by 1991. Policy makers are also giving priority to tourist-enhanced measures such as improving roads, lighting, sewage and water supplies.

The light manufacturing sector is small but growing with activities in boat building, printing, textiles, furniture and varied products for the local market. Aquaculture also shows promise with its shrimps and prawns possibly replacing some of the $7m fish imported every year and also as an export. There are 11 private pond farms covering more than 100 acres. But the offshore fish business is running into severe price competition from Micronesia.

The manufacturing and commercial sectors rely heavily on imported commodities, mainly from the USA and Japan, and the island has a big adverse trade gap. In Pacific terms, however, it has an enviable physical infrastructure, reinforcing its aim to become a major transshipment point for cargo to the USA and Micronesia. The international airport is among the most modern in the Pacific and can handle jumbo jets. Guam claims, with justification, to have the finest deep water harbour between Hawaii and the Philippines with more than 2,000 feet of docks and 21 acres of container yards. There is a high standard of international communications. Guam sees itself as a possible replacement for Hong Kong when the Chinese take over the British colony in 1997. And, with Palau and the Northern Marianas, it could be an alternative home for the American naval and air presence in Subic Bay and Clarke Field in the Philippines if these bases were abandoned.

Voting pattern

Politically, Guamanians stuck to a familiar voting pattern in 1988. In a split ticket result, they returned the Republican non-voting member of the US Congress, Ben Blaz, to Washington and elected 13 Democrats and eight Republican senators to the local unicameral legislature – the same margin won by the democrats two years' earlier. The governorship, not due for election until 1990, is under republican control. Blaz, a former US Marine general, faced an

KEY FACTS

Official title: Territory of Guam
Head of state: President of the USA
Head of government: Governor Joseph F. Ada
Capital: Agana
Official Languages: English and Chamorro
Currency: US Dollar ($ = 100 cents)
Area: 549 sq km
Population: 126,430 (1988)
Labour force: 45,190 (1987)
Visitor numbers: 483,954 (1987)

unexpectedly tough fight to beat Ben Pagelinan by 16,696 votes to 13,477. Pagelinan is a former chief of staff in the scandal-racked administration of ex-Governor Richard Bordallo and made much during the campaign of Blaz allegedly missing important votes in congressional committees. Blaz cannot vote on the floor of the House but he can vote in committee and is a *de facto* ambassador for his territory. At local level, supporters of the feminist vote were heartened by the trend to give women an increasing share of legislative seats. Four women were elected in 1986 and this time the number rose to seven. A popular win was by Madeleine Bordallo, a democrat like her husband. It was the second bit of good news for the Bordallos within weeks. Only a month before, the ex-Governor had won a major court battle to overturn a nine-year gaol sentence for bribing businessmen (he had been sentenced to 30 years but the sentence was reduced). He escaped the charges because the US appeals court in San Francisco ruled that such bribery was only a crime if it involved state officials and Bordallo was only a territorial one.

With the elections out of the way, Guamanians renewed their efforts to alter the territory's status to that of a commonwealth. Reintroducing the Guam Commonwealth Bill on the first day of the new Congress, Blaz effectively gave the green light to a new flurry of activity as Guam politicians voiced increasing frustration at Washington's slow pace in considering reform. At heart is a controversy over the proposal to recognise the special rights of the indigenous Chamorro population in the decision-making process. The principle is not entirely at variance with the US constitution which pledges to look after the interests of the original inhabitants. Chamorros make up 42 per cent, the biggest ethnic group, of Guam's multiracial population (Blaz was the first Chamorro to win a general's stars) but the special rights concept, if at the expense of other groups such as the Filipinos and the US mainlanders, may not find favour in the US Congress when it considers the relevant legislation.

At least Guamanians could not complain at receiving an unsympathetic hearing from Bush's new Interior Secretary Manuel Lujan, who probably knows far more about Guam than any of his predecessors. But Lujan himself unwittingly provoked a backlash from the territory in mid-1989 by attempting to have the United States stop reporting to the UN Decolonisation Committee (the Committee of 24), as it has done voluntarily for years and in much the same way as reporting on Puerto Rico was dropped when it became a commonwealth. Guam's Governor Joseph F. Ada (R), who heads the territory's panel on self-determination, professed himself shocked and offered direct testimony to the committee. Blaz promptly put himself on a collision course with Ada by supporting Lujan. The gloomy consensus was that the expected new pressure by the Committee of 24 would not do Guam's case any good in or out of the US Congress.

COUNTRY PROFILE

Political system: Guam is a virtually self-governing unincorporated territory of the US represented by an elected non-voting delegate to the US House of Representatives. Local executive power rests with a governor elected to a four-year term who heads a 15-member cabinet of departmental directors. The 21-member legislature is elected to a two-year term by popular vote.

Main town: Agana.

Media
Broadcasting: Several commercial radio services, 24-hr/day broadcasting. Commercial and public service TV, including colour transmissions and cable service.

Domestic economy: Tourism is now the leading area of the local economy, contributing over $400m in 1987. Some agriculture, with staple foods grown on smallholdings. Imports generally 10 times the value of export income. Some 346 US firms have established themselves as foreign sales corporations (FSCs) in Guam, with extensive assistance from the USA. Small industries have grown with US encouragement; military bases provide additional employment.

Employment: About 40 per cent of local employment is with either US or territorial government. Labour force totals 45,190 (1987).

External trade: Guamanian exports enter a number of countries, including Australia, Japan and US, free of duty.
Exports: Main exports are re-exports of iron and steel scrap. Main destinations are USA and Micronesia.
Imports: Main imports are fuel, machinery and transport equipment, food and manufactured items. Main sources are US and Japan (about 80 per cent of total), Hong Kong, Taiwan and the Philippines.

Agriculture: Most agricultural activity is part-time market gardening on smallholdings. Main crops are fruit, vegetables, maize, sweet potatoes, cassava, bananas, breadfruit, coconuts and sugar cane. Livestock includes pigs, cattle and poultry. Fishing is important source of protein.

Industry: Tourism is the leading industry, providing 18 per cent of GDP and generating a third of private-sector employment. 483,954 people per annum visited the island (1987). Further development of infrastructure for tourist areas is planned. About 80 per cent of tourists arrive from Japan. Industry employs about 3.3 per cent of labour force, including oil refining, rock/cement production, printing, garment manufacture, watch manufacture and food processing. About one-third of land area devoted to military use; US military personnel and dependents comprise about one-fifth of population.

Membership of international organisations: Association of Pacific Island Legislatures, South Pacific Commission, Worldwide Chamber of Commerce.

BUSINESS GUIDE

Time: GMT + 10 hrs.

Climate: Warm and humid, temperatures 24-30°C, Dec-May generally cooler and drier. Rainfall, up to 300 mm/month and averages 2000mm p.a., heaviest Jul-Sep. There are occasional tropical storms.

Entry requirements
Passport: Required by all except US citizens with proof of citizenship departing from US points.
Customs: Guam is a duty-free port.
Income tax: The US income tax system prevails, but money due is paid to the government of Guam in lieu of local income tax.

Health precautions
Advisable: Precautions against tetanus, typhoid and paratyphoid advisable.

Air access
Other airlines: All Nippon Airways, Air Micronesia, Hawaiian Air, JAL, Northwest.
International airport: Guam International (GUM); duty-free shop, bar, restaurant, first class lounge, buffet, currency exchange, hotel reservations, shops, car hire.

Surface access: 12 steamship lines serve Guam.
Main ports: Apra Harbour. Island Shipping Lines, established in 1985, is developing services to the Federated States of Micronesia.

National transport
Road: Over 600 km of surfaced roads.

Public holidays
Fixed dates: 1 Jan (New Year's Day), 16 Jan (Martin Luther King Day), 4 Jul (Independence Day), 21 Jul (Liberation Day), 10 Nov (Veterans Day), 8 Dec (Lady of Camarin Day), 25 Dec (Christmas Day).
Variable dates: President's Day (3rd Mon in Feb), Guam Discovery Day (1st Mon in Mar), Good Friday, Memorial Day (last Mon in May), Labor Day (1st Mon in Sept), Columbus Day (2nd Mon in Oct), Thanksgiving Day (4th Thurs in Nov).

Working hours
Banking: (Mon-Thu) 1000 – 1500, (Fri) 1000 – 1800.

Telecommunications
Telephone: Dialling code for Guam, IDD access code + 671 followed by subscriber's number (no area codes necessary). Mainly automatic telephone service, over 4,214 business lines in 1987.
Telex and telefax: 24-hr telex service is available at good hotels. Facsimile service also operational.
Fax: Service now available.
Telegram: Telegram service available.

Banking

Head offices: Bank of Guam, Guam Savings and Loan Association.

Guam branches: Allied Bank, American Savings and Loan Association, Bancorp Finance of Hawaii-Guam, Bank of Hawaii, Bank of Orient, California First Bank, California Overseas Bank, Citibank, Citicorp Banking Corporation, First Commercial Bank, First Hawaiian Bank, First Savings and Loan, Hongkong Bank, Hongkong Bank of California, Hongkong and Shanghai Banking Corp, Metropolitan Bank and Trust, Oceanic Bank.

Weights and measures: US system.

Representation in capital: Japan, Rep of Korea, Philippines, Nauru.

Hotels: Guam
Guam Dai-ichi, Agana 96911 (on Tumon Beach), P.O. Box: 3310 (tel: 6465881; tx: 7006188; fax: 6466729).

Guam Reef, 1317 Pale San Vitores Rd, Tumon, P.O. Box: 8258 (tel: 6466881; tx: 6329; fax: 671-646-5200).

Hilton International Guam, Caller Service No 7350-V, Tamuning 96910 (tel: 6461835; tx: 6234; fax: 6646038).

Car hire: Tamuning
Avis: (tel: 6461613; tx: 6482); airport (tel: 6467676).

Budget: airport (tel: 6460366/67).

Hertz: PO Box 7855 (tel: 6465875; tx: 6357).

National: PO Box 8638, Tamuning (tel: 6467972; tx: 6390).

Toyota Rent-A-Car: PO Box 6428 (tel: 6461871/2; tx: 96911).

Business advice and information
Guam Economic Development Authority, Suite 911, ITC Building, 590 South Marine Drive, Tanuning, Guam 96911 (tel: 6494141/2/3; fax: 6494146).

Travel information
Guam Visitors Bureau, PO Box 3520, Agana.

Guatemala

Barbara Annis

In 1990, Guatemala is once again braced for a presidential election. But how joyless this election compared to the hope-filled democratic celebration of 1985, which saw victory by the first civilian government in a decade. Led by once immensely popular Christian Democrat (DC) Vinicio Cerezo, that government is now racked from the effects of corruption, personal and systemic; three failed coups; ongoing insurgency and belligerence from the military and private sectors, and an economy beyond its control. Cerezo and the DC fell into disgrace.

Only the military, which indisputably continues to hold the country in its iron grip – itself fractionalised and leaderless – prevents the country's total plunge into anarchy. What promises are made for what spoils, as more than 20 political parties, most of them right of centre, jockey for leadership, and attempts at coalitions are made?

The DC is the only party to hold primaries. All other deals are worked out behind the scenes by power brokers far from the scrutiny and influence of the Guatemalan voter. Currently running on the DC slate, with a runoff scheduled for August 1989, are former Foreign Minister Alfonso Cabrera Hidalgo, Development Minister Rene de Leon Schlotter and Raquel Blandon de Cerezo – estranged from her husband Vinicio.

The president's premature designation of his boyhood chum Cabrera as his successor, as he shortly afterward retired to the seclusion of his farm in the isolated Guatemalan countryside, was the kiss of death for the ruling faction of the DC. Cabrera has long been charged with a panoply of infractions – alleged to have committed all manner of corrupt acts – and is associated with Guatemala's burgeoning drug trade.

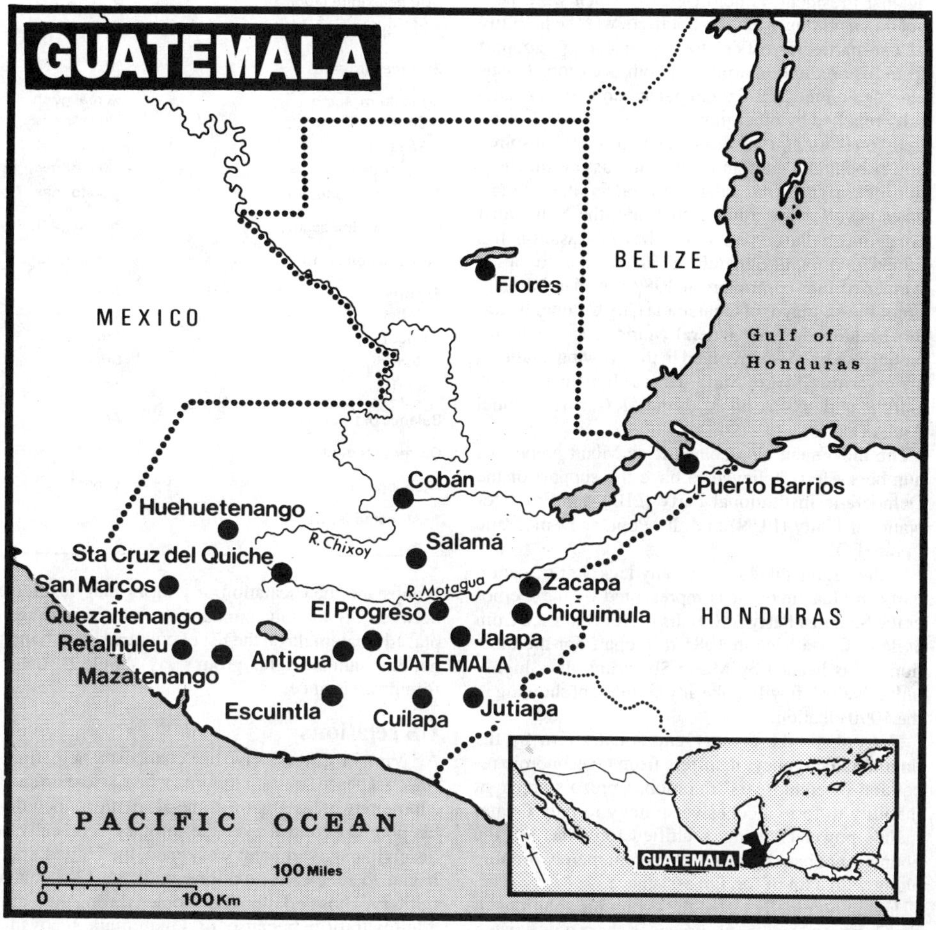

Political parties

Traditionally Guatemala's leading parties of the extreme right are the National Liberation Movement (MLN) and Authentic Nationalist Central (CAN), while the key party of the centre-right has been the National Union of the Center (UCN). These parties sometimes work within the framework of government, but they are also strongly associated with international paramilitary operations. The traditional right made such a poor showing in the 1985 elections (just 3 per cent of the vote nationwide) that members are desperately trying to create new parties with cleaned up images.

The last year has seen the development of a spate of new right and moderate-right political groupings, apparently disassociated with the extra legal activities of the old far-right. The Plan for National Advancement Party (PAN), for instance, led by Guatemala City Mayor Alvaro Arzu, took an early lead in urban public opinion polls. He has a youthful, energetic image and managed to avoid being discredited despite having held public office – a rare event in Guatemala. He also has the backing of important business sectors.

General Hector Mario Lopez Fuentes was the candidate for the hard-right MLN until he fell out with its long-time leader Mario Sandoval Alarcon. The MLN's main link is with the agricultural oligarchy. Lopez has since formed his own party, the National Unification Movement (MUN), which reportedly receives funding from El Salvador's ruling ARENA party. Sandoval's former running mate, Lional Sisniega Otero, also now leads his own party, the Party of Anti-Communist Unification (PUA).

The CAN was organised around former President General Carlos Manuel Arena Osorio (1970-74), and is sometimes allied with the MLN. Its leader, Juan Carlos Simons, however, resigned in 1985 to form the 'official' new right movement, which he claims is a mass movement – not just political – against corrupt and authoritarian government and which calls for a free society and free economy.

Also in the race but among the more moderate elements is Fernando Andrade, ex-foreign minister and so-called architect of the country's transition from military to civilian government. He is the candidate of the dominant wing of the Revolutionary Party (PR) and hopes to gather votes from disenchanted former Christian Democrats.

The rather tired UCN party has once again launched newspaper publisher Jorge Carpio Nicolle, who ran second in the 1985 election. His party received an injection of new life from Jorge Serrano Elias, who gave up his own presidential candidacy to run as vice president with Carpio. Serrano is favored by US conservatives and is a protestant evangelist, as is born-again General Efrain Rios Montt, still another candidate.

Rios Montt directed the bloody war against Guatemala's insurgents as Chief of State before he was ousted in a coup by other military officers. He too is running on a law-and-order ticket.

KEY FACTS

Official title: Republic of Guatemala
Head of state: President Vinicio Cerezo
Ruling party: Partido Democracia Cristiana Guatemalteca
Capital: Guatemala City
Official Languages: Spanish (Quiché, Cakchiquel, Mam, Kekchi also spoken)
Currency: Quetzal (Q) = 100 centavos
Exchange rate: Q2.85 per $

Area: 108,890 sq km
Population: 8.7m (1988)
GDP per capita: $1,502 (1988)
GDP real growth: 3.5% (1988)
Inflation: 10.9% (1988)
Trade balance: –$396m (1987)
Foreign debt: $2.8bn (1988)

Military sorting things out

The number of current or former military candidates running for presidential office reflects the numerous factions in their ranks. A primary reason for the failure of the military-led coup attempts in 1987-88 against President Vinicio Cerezo, which were both put down without a single shot fired, was the inability of one particular officer to muster enough support from his brethren in arms. On both occasions troops marched on the presidential palace only to be peacefully repelled by other units.

In one attempt to consolidate military-inspired power behind a civilian façade, and assure the election of a rightist candidate, several military candidates have formed a loose coalition, the Nationalist Guatemalan Party (PANAC). Mario Castejon has joined forces with General Anibal Guevara, so-called winner of the controversial 1982 presidential elections, the ex-mayor of Guatemala City Colonel Abundio Maldonado and General Mario Lopez Fuentes among others. Also involved in the growing coalition are ex-army Chief of Staff General Benedicto Lucas Garcia and ex-President General Carlos Manuel Arana Osorio.

The movement also counts Rios Montt among its numbers. He is believed to have the support of the Democratic Institutional Party (PID), the Front for National Unity (FUN) and the Popular Democratic Force (FDP).

To the degree there remains any legal left-of-centre party in Guatemala, it is represented by the Democratic Socialist Party (PSD). Its leaders returned from exile in Costa Rica in 1985 to prepare for the elections. It is headed by Mario Solorzano. It is highly unlikely the left will make any significant showing in the 1990 election.

Not only is the country emotionally torn by the ongoing insurgency, it suffers from an economic recession of eight years' duration. Cerezo's apparent inability to form a cohesive strategy to rally Guatemala's power elite into a unified force has left the country, not bankrupt like others, but merely limping along with a lot of foreign assistance.

Having nominally tested the economic solutions of the Christian Democrats, however, there now exists a widespread perception among the Guatemalan electorate that only a get-tough attitude from the presidential palace will rescue the country from its economic doldrums and social disarray. Only the political right is believed to have the answers. Unfortunately, this perception translates to a likely return to the repression of the past, with all of its accompanying abuses.

KEY INDICATORS

	Unit	1984	1985	1986	1987	1988
Population	m	7.7	8.0	8.2	8.4	8.7
Gross domestic product (1988$)	$m	–	12,196	12,214	12,596	13,042
GDP per capita (1988 $)	$	–	1,515	1,490	1,500	1,502
Inflation (average rate)	%	3.4	18.6	36.9	12.3	10.9
GDP real growth	%	0.5	–0.6	0.1	3.1	3.5
Agricultural sector	% real gwth	1.6	0.4	–0.8	3.6	3.0
Coffee	'000 tonnes	140	162	159	159	–
Seed cotton	'000 tonnes	–	172	130	90	–
Sugar-cane	'000 tonnes	6,410	7,000	7,000	6,900	–
Petroleum production	'000 tonnes	237	184	289	–	–
Manufacturing sector	% real gwth	0.5	–0.8	0.7	1.7	2.5
Construction sector	% real gwth	–28.4	–8.4	3.1	10.5	18.3
Exports	$m	1,132	1,060	1,044	978	–
Coffee	$m	365.5	450.8	502.4	355.1	377
Sugar	$m	71.3	43.8	51.7	51.4	64.9
Cotton	$m	75.8	71.6	24.3	16.2	34.3
Imports	$m	1,278.6	1,174.9	959.5	1,479.4	1,557.0
Balance of trade	$m	–142	–121	143	–396	–
Current account	$m	–377.4	–246.3	–17.6	–442.5	–
Exchange rate	Q per $	1.00	1.00	1.88	2.50	2.62

* estimated figure

Human rights

Guatemalan President Vinicio Cerezo and Costa Rican President Oscar Arias together took a strong position on regional peace. They dared to defy US policy, steadfastly refusing to support the contras, and pushed instead for an indigenous solution to the various conflicts broiling in Central America. Arias was awarded a Nobel Peace Prize for his efforts and Cerezo was able to divert some international attention from the quickly deteriorating human rights situation in his own country.

For example, as part of the so-called Esquipulas Peace Accord, worked out between Central American leaders, a National Dialogue was called in Guatemala to discuss solutions to its armed conflict. Three prominent members of the Unitary Representation of the Guatemalan Opposition participating in the Dialogue were then quickly forced to flee the country after numerous death threats and a bombing attack.

This incident, along with increased involvement by the military in the administration of the national police; human rights abuses directly attributable to the security and defence forces; unexplained massacres in the countryside; intimidation of human rights groups and the escalation of joint military exercises between US and Guatemalan forces have once again placed Guatemala on the list of countries which international human rights groups say should be denied foreign assistance.

US relations

As Vinicio Cerezo saw his chance to help find a regional peace solution diminish, he also experienced a loosening of his grip on domestic matters, including his nominal control over the military. Ironically, as his civilian government weakened, the United States began to renew its relations with the Guatemalan military. Those relations had cooled during the Carter administration because of Guatemala's abysmal human rights record. The last two years, however, have re-established the United States as the number one arms supplier to Guatemala and have seen an increase in US military assistance and increased cooperation between US Army teams, the National Guard and the Guatemalan military in training and other kinds of exercises.

For instance, the Washington Office On Latin America, a human rights monitoring group, reports that 45 members of the 13th Army Corps of Engineers in Fort Knox, Kentucky arrived in Guatemala on 26 April 1989 to assist in highway construction in the department of Solola. It is recognised as an area of armed conflict. The project is presented as civilian marketing assistance but has potential military use.

In March 1989, the Guatemalan Army rented two Chinook helicopters from the United States to transport over 1,500 people that the army said had voluntarily come out of hiding in the mountains of northern Quiche province.

National Guard units from Kentucky, Georgia, Iowa and Oklahoma have been engaged in medical exercises in the province of Chimaltenango – they are in uniform and are armed.

In mid-May a Blackhawk helicopter with US crewmen aboard was forced to land near the Guatemalan city airport; five of the crew were reported to be injured.

Forty troops from Fort Bragg conducted operations with elite Kaibil troops from the Guatemalan army in the Peten province of Guatemala, reportedly in parachute training and survival.

Economy boosted by foreign aid

The Guatemalan Government has managed to hold the currency firm at Q2.70 to the dollar since early 1988, despite persistent rumours of imminent devaluation, and to hold inflation at a steady 12 per cent. There is considerable pressure, however, particularly from the agrarian sector, to devaluate, and criticism from the business and other sectors that the economy is propped up by foreign assistance rather than growing on its own.

The Bank of Guatemala reported a positive balance of dollars for the first four months of 1989 compared to 1988. Nevertheless, income for the same period actually decreased $5.4m to $585.3m. Dollar outflow was restricted as a result of the Central Bank's efforts, which included higher interest rates.

The economy was bolstered by new loans from the multilateral lending institutions, including $70m in Economic Support Funds from USAID, $40m in a stand-by loan from the IMF and $80m from the World Bank. Total foreign assistance in 1988 equalled $252.6m. 62 per cent came from the USA, the rest from Germany and Belgium. Germany also pledged a cash infusion of $76.7m for the fiscal year 1988-89.

The multilateral lending institutions are taking the lead in pushing Guatemala's depressed manufacturing sector toward a more competitive stance in the international markets. The sector has largely focused on internal consumption and import substitution. Industry generates approximately 16 per cent of the economy.

The Central Bank optimistically predicts the economy will grow at a rate of 4.5 per cent compared to 3.5 per cent in 1988 – the first positive growth to be registered in the country in eight years. Those projections may have to be revised, however, depending on the effects of the failure of members of the International Coffee Organisation (ICO) to reach price agreements. Coffee continues to generate a hefty percentage of Guatemalan economic revenues. Coffee futures in general have dropped, but Guatemala grows a mild, relatively rare type of coffee in high demand worldwide.

The Central Bank also reports an expansion of 3.0 per cent in commerce and 3.2 per cent in agriculture. Banking and construction have boomed. Guatemala recorded 18.4 per cent growth in construction and 5 per cent in banking in 1988. Manufacturing showed more moderate growth of 2.5 per cent. Nevertheless, population growth continues at 3 per cent and wealth is unevenly distributed, benefiting only approximately 14 per cent of the population.

Tourism soared in the first eight months of 1988, drawing approximately 170,000 visitors, a 15.5 per cent increase over the same period of the previous year. El Salvador is the greatest source of Guatemalan tourism. An estimated 220,000 tourists visited Guatemala in 1988. Approximately 55 per cent were from El Salvador. They generated income of approximately $77.8m.

The National Coffee Association (ANACAFE) reported it had exported $9.6m in coffee for the first half of the 1988/89 season. $73.1m worth of coffee was shipped to members of the ICO. An additional $22.9m worth was shipped to non-members.

Non-traditional exports in Guatemala reached $215.6m in 1988, an increase of 20 per cent over 1987. That number fell short of expectations, however. The Non-traditional Export Guild had projected a 34 per cent growth rate over 1987. That rate was influenced by the devaluation of the Quetzal. In 1988 non-traditional exports, led by clothing, accounted for 17.7 per cent of all exports. Traditional products – coffee, bananas, sugar and cotton – accounted for 62.7 per cent.

Cardamom, for years a big non-traditional revenue producer, fell on hard times beginning in 1984. Oversupply worldwide and less demand have seen the price plummet from $626 per hundredweight bag in 1984 to $121 per hundredweight in 1988. Guatemala exported $100m worth of the aromatic spice in 1984 and just $34.8m worth in 1988. Simultaneously, production increased dramatically. Guatemala's cardamom production increased from 160,000 100-pound bags in 1984, to 304,400 100-pound bags in 1988.

COUNTRY PROFILE

Political system: Executive power is held by the president, assisted by an appointed cabinet, while legislative power is vested in the 100-member unicameral National Congress. A new constitution came into force in 1986, following the elections in December 1985, to replace the 1966 constitution suspended in 1982.

Main cities/towns: Guatemala City (population estimated to be two million), Quezaltenango, Mazatenango, Antigua.

Media
Press: There are several daily newspapers including *El Gráfico* (circulation 60,000), *Prensa Libre* (68,500) and *La Hora* (evening 20,000). Also various periodicals and trade papers.
Broadcasting: The government operates five radio stations. There are also six educational (with some broadcasts in English) and over 100 commercial stations. One government and three commercial TV stations operate from Guatemala City.

External trade: Guatemala's ability to trade has been limited in recent years due to a severe shortage of foreign exchange caused by reduced export earnings and a heavy debt burden.
Exports: 1985: Coffee (43% of total value), cotton (7%), bananas (7%), cardamom (6%), sugar (4%), fresh meat, vegetables. Main destinations (1986 estimated): USA (49% of total value), El Salvador (12-15%), German FR (5%), Japan (5-8%), Honduras (4%), Costa Rica, Netherlands, Italy, Nicaragua.
Imports: 1985: Machinery and transport equipment (19% of total value), basic manufactures (17%), chemicals etc (23%), mineral fuels etc (23%), food products (7%). Main sources (1986 estimated): USA (70% of total value), El Salvador (4%), Venezuela (7%), Mexico (8%), Costa Rica (5%), Japan (8%), German FR (9%), UK, Honduras, Italy.

Agriculture: Contributes more than 25 per cent to GDP and employs about 55 per cent of the total workforce. Production is mainly export-oriented, the major cash crops being coffee (the largest single earner of foreign exchange), sugar-cane, cotton, cardomom and bananas. Falling world commodity prices prompted the government into a policy of diversification, leading to increased exports of tobacco, fruit and vegetables and meat. Adverse weather conditions in recent years have caused severe damage to many plantations and the threat of guerrilla activity has also led to reduced output. Maize is the main food crop, although rice and wheat are also grown. Commercial fishing is under-developed, though small amounts of shrimps are caught for export. Approximately 15 per cent of total area is cultivated arable land, 10 per cent pasture and 55 per cent forest. Long-term rural development plan introduced in 1986 to help peasants to buy land and to provide them with technical assistance.

Industry: One of the most developed industrial sectors in Central America, with the manufacturing sector contributing up to 16 per cent to GDP and employing up to 15 per cent of the workforce. Manufacturing is based predominantly on agricultural inputs, with major firms involved in food processing, rubber, textiles, paper and pharmaceuticals. Other important industries are oil refining, cement, metals, electrical goods assembly, plastics, chemicals, fertilisers and cigarettes. Most of the value-added industries are export-based, so growth in the sector has been severely curtailed by the collapse of the Central American market. Social unrest, high energy costs, shortages of imported materials and a slump in private and public investment have severely hampered industrial production, with many firms closing down operations completely.

Mining: Iron, lead, zinc, antimony, tungsten, cadmium and silver are mined on a small scale, but in general output has fallen owing to lack of investment and rising fuel costs. Nickel-ore mining at Chalac-el-Estor ceased in 1982.

Energy: Several petroleum deposits have been located and proven reserves total more than 20m barrels, although actual reserves may be up to 100m barrels. Development has been slow due to the reluctance of foreign companies to invest in the area owing to political unrest and unfavourable terms. Oil production is small, averaging around 3,000 b/d from 10 fields. Guatemala refines oil imported primarily from Venezuela. Plans are underway to increase hydroelectric capacity, to meet at least two-thirds of domestic demand by the end of the century.

Membership of international organisations: CCC, Central American Common Market, FAO, G-77, IADB, IAEA, ICAC, ICAO, ICO, IDA, IFAD, IFC, IHO, ILO, IMF, IMO, INTELSAT, Inter-American Development Bank, Interpol, IRC, ISO, ITU, IWC, ODECA, Organisation of American States, Organisation of Central American States, PAHO, SELA, UN, UNESCO, UPEB, UPU, WFTU, WHO, WMO, World Bank.

Time: GMT -6.

Climate: Coastal regions hot and humid all year. Highland regions (including Guatemala City) temperate. Dry season from Nov to May with temperature range 23°C to 30°C. Rainy season from Jun to Oct with average temperature 22°C.

Entry requirements
Passport: Required by all.
Visa: Required by all except nationals of EC countries (excluding Greece, Irish Republic), Scandinavian countries, Japan, Austria, Switzerland and certain other countries.
Currency: No restrictions on import/export of foreign currency.

Health precautions
Mandatory: Yellow fever vaccination certificate required if arriving from an infected area.
Advisable: Typhoid, polio vaccinations. Malaria risk exists in some areas – prophylaxis recommended. Water precautions should be taken. Hepatitis innoculation recommended.

Representation overseas: Bogotá, Bonn, Brasilia, Brussels, Buenos Aires, Cairo, Caracas, Lima, Managua, Mexico City, Monaco, Montevideo, Ottawa, Paris, Quito, Rome, San José, San Salvador, Santiago, Santo Domingo, Seoul, Taipei, Tegucigalpa, Tel Aviv, Tokyo, Vatican City, Vienna, Washington.

Air access
National airline: Aviateca (Empresa Guatemalteca de Aviación).
Other airlines: Aeronica, Continental, Copa, Eastern, Iberia, KLM, Lacsa, Mexicana, Pan Am, Sahsa, SAM, Taca International, Tropic Air.
Tax: Tourist tax of Q20 payable on departure.
International airport: Guatemala City-Aurora (GUA), 7 km from city; duty-free shop, bar, buffet, restaurant, bank, hotel reservations, post office, shops, vaccination centre, car hire.

Surface access
Main ports: Champerico, Puerto Barrios, San José, Santo Tomás de Castilla and the Quetzal Port.
Overland access: Pan-American Highway runs through country (Mexico-El Salvador); there are other roads from El Salvador, Honduras and Mexico and there is a route via Melcour de Mencos from Belize – plans for any journey should be made in the light of current political, and in certain cases, road conditions. It is possible to use scheduled bus or train services but some of these have been subject to suspension. Bus services from San Salvador, Tegucigalpa, Managua, San José and Panama.

Hotels: Most charge 10 per cent room tax.

Car hire: Any valid licence usually acceptable.

City transport
Taxis: Good service in Guatemala City. Fares by negotiation but there are set rates for journeys from airport to certain destinations; tipping is discretionary (5-10 per cent).

National transport
Air: Aviateca operates domestic service to major centres. Schedules liable to change at short notice.
Road: Total network around 18,000 km, up to 20 per cent surfaced; using unpaved roads can be difficult due to effect of weather (wet or dry) on volcanic ash. Paved roads are of fair quality.
Buses: Numerous services within Guatemala City - said to be (outside usual rush hours) less crowded than some cities. Bus services cover major towns.
Rail: Regular services operate from San José (on Pacific coast) to Puerto Barrios (on Atlantic coast) via Guatemala City. There are also routes to the Mexican border and San Salvador. Service is not very reliable.

Public holidays: 1 Jan, Holy Wednesday, Holy Thursday, Good Friday, 1 May, 30 Jun, 15 Aug, 15 Sep, 20 Oct, 1 Nov, 24 Dec, 25 Dec, 31 Dec.

Working hours
Government and business: Generally (Mon-Fri) 0800 – 1800, but many variations by individual offices.
Banking: Generally (Mon-Fri) 0830 – 1430, but many variations by individual banks.

BUSINESS GUIDE

Shops: (Mon-Fri) 0800 – 1800.

Telecommunications
Telephone: Dialling code for Guatemala: IDD access code + 502 followed by area code (Guatemala City 2) followed by subscriber's number.
Telex and telegram: Service operated through Guatel offices. Local telegram service available at post offices.

Banking
Central bank: Banco de Guatemala.
Head offices: Banco Agrícola Mercantil, Banco del Agro, Banco del Café, Banco del Ejército, Banco de Occidente, Banco de los Trabajadores, Banco Granai & Townson, Banco Industrial, Banco Inmobiliario, Banco Internacional, Banco Metropolitano, Banco Nacional de Desarrollo Agrícola, Banco Nacional de la Vivienda, Corporación Financiera Nacional, Crédito Hipotecario Nacional de Guatemala, Financiera Guatemalteca, Financiera Industrial y Agropecuaria.
Foreign branches: Banco Centroamericano de Integración Económica, Bank of America, Lloyds Bank International.

Advertising: Available on TV and radio stations, press, cinemas, posters.

Electricity supply: 110 V AC, 60 cycles.

Representation in capital: Argentina, Austria, Belgium, Bolivia, Brazil, Canada, Chile, Colombia, Costa Rica, Dominican Republic, Ecuador, Egypt, El Salvador, France, German FR, Honduras, Israel, Italy, Japan, Republic of Korea, Mexico, Nicaragua, Paraguay, Peru, Portugal, South Africa, Sweden, Switzerland, Taiwan, USA, Uruguay, Vatican, Venezuela.

Hotels: Antigua
Ramada Inn, 9a Calle y Salida a Ciudad Vieja, P.O. Box: 286 (tel: 011015; tx: 6351).

Hotels: Guatemala City
Camino Real Westin, Avda. Reforma Calle 14 (tel: 680211; tx: 5940; fax: 374313).

Conquistador Sheraton, Via 5, 4-68, Zona 4 (tel: 64691; tx: 5340; fax: 324533).

Cortijo Reforma, Avda Reforma 2/18, Zona 9 (tel: 66712/16; tx: 5203).

El Dorado, 7a Avda 15-45, P.O. Box: 2814 (tel: 317777/82; tx: 5415).

Maya Excelsior, 7a Avda 12/46, Zona 1 (tel: 82765).

Ritz Continental, 6a Avda 'A' 10/13, Zona 1 (tel: 81671/5; tx: 5424).

Car hire: Guatemala City
Budget: Avenida Reforma 15-00, Zona 9 (tel: 367669, 322591; tx: 6059).

Europcar National: 14 Calle 1142, Zona 10 (tel: 680175, 683057).

Hertz: 7a Avenida 14-76, Zona 9 (tel: 312421; tx: 5056).

Car hire: International airport
Avis: (tel: 310017).

Europcar National: (tel: 318365).

Hertz: (tel: 311711).

Business advice and information
Asociación Guatemalteca de Agentes de Viajes (AGAV) (Guatemalan Association of Travel Agents), 6a Avda 8-41, Zona 9, Apdo 2735, Guatemala City 9tel: 310320; tx: 5127).

AVIATECA-Aerolineas de Guatemala, Avenida Hincapié, Aeropuerto 'La Aurora', Zona 13, Guatemala City (tel: 318222; tx: 5960).

Banco de Guatemala, PO Box 365, 7a Avenida 22-01, Zona 1 Guatemala City (tel: 534053; tx: 5231, 5461).

Cámara de Comercio de Guatemala, 10A Calle 3-80, Zona 1, Guatemala City (tel: 82681/2/3; tx: 5478).

Cámara de Industria de Guatemala, Ruta 6, 9-21, Zona 4, Apdo 214, Guatemala City (tel: 61311/14/18; tx: 5402).

Centro Nacional de Promoción de las Exportaciones, 6A Avenida Torre Profesional 5\28, Zona 14, Apdo 1237, Guatemala City.

Comité Coordinador de Asociaciones Agrícolas, Comerciales Industriales y Financieras (CACIF), Edif Cámara de Industria de Guatemala, Ruta 6, No 9-21, Zona 4, Guatemala City 9tel: 310651; tx: 6133).

Dirección General de Radiodifusión y Télévisión Nacional, 5a Avenida Zona 1, Guatemala City (tel: 25045).

Inforpress Centroamericana 9a Calle 'A' 3-56, Zona 1, Guatemala City (tel: 29432).

INGUAT (Instituto Guatemalteco de Turismo), 7A Avenida 1-17, Centro Cívico, Guatemala City (tel: 311333; tx: 5532).

Ministry of Economy, Palacio Nacional, Guatemala City.

Ministry of Finance, Palacio Nacional, Guatemala City.

Guinea

Susi Pollak

On the occasion of Guinea's thirtieth independence anniversary, General Lansana Conté called for patience and tolerance: 'We have to wait at least five more years before we see the end of the tunnel,' the President said. In the meantime 'we must continue to suffer'. Conté and his colleagues freely admit to being disappointed at the results of the first stage of economic reconstruction, drawn up in consultation with the World Bank. They adopted the IMF's free market philosophy and acted in accordance with it. In the first phase of the Structural Adjustment Programme (SAP I) which was launched in 1985, 30,000 out of 70,000 public servants were made redundant; 41 public enterprises put on the market for privatisation; 16 earmarked for liquidation.

The aim of SAP I was to free the economy from state controls and to establish the framework for a market-oriented system. Phase II was launched with a $65m IDA credit in 1988 and is to end in 1993. It aims at improved finance management and the continued restructuring of the civil service and public enterprise sector – in other words, more of the same medicine. After three years of austerity, incomes were down in real terms while the cost of essential commodities continued to rise. In 1987 and 1988 there were riots by students and the unemployed.

When he saw the IMF's deputy director for Africa at the end of 1988, Conté pleaded for a more flexible application of SAP II because of the danger of social unrest. Minister for Planning and International Co-operation Eduard E. Benjamin, said: 'We are very concerned about the human aspect of the retrenchment programme.' Benjamin himself is a former World Bank employee. In March 1989 the World Bank responded with an IDA credit of $9m to help Guinea design social policies. This loan covers a pilot scheme which is supposed to help reduce the suffering of low-income groups by improving their access to employment opportunities and basic social services.

Agriculture was supposed to be the centrepiece of SAP I, according to the World Bank, but the sector's growth was held back by inadequate research and extension facilities. To make up for this, the IDA granted Guinea a $18.4m agricultural loan to be used during the 5-year period of SAP II on research and extension work. This should allow 200,000 rural families to increase their income by producing more food, helping also to fulfil the government's aim of achieving food self-sufficiency by 1993. Cash crops like grapefruit and bananas are being officially encouraged. Exports of coffee, a relatively new crop, reached 9,000 tonnes in 1988.

KEY FACTS

Official title: Revolutionary Republic of Guinea

Head of state: President Colonel Lansana Conté

Ruling party: Military Committee for National Redress (CMRN)

Capital: Conakry

Official Languages: French

Currency: Guinean Franc (GFr)

Exchange rate: GFr298 per $ (Nov 1989)

Area: 245,957 sq km

Population: 6.5m (1987)

GDP per capita: $320 (1987)

GDP real growth: 5.9%* (1987)

Trade balance: $116m (1987)

Foreign debt: $1.8bn (1988)

* estimated figure

Mineral wealth

The real centrepiece of Guinea's economy is its mineral wealth: bauxite and diamonds. Bauxite accounts for 60 per cent of foreign exchange earnings, and the 1987–88 aluminium boom on the world market, with prices climbing almost 60 per cent, was reflected in increased government revenues. Frigua, currently the only alumina producer in Africa, is being modernised with the help of an ECU48m credit from the EIB. The government owns 49 per cent of the company, which should become a highly profitable venture, if aluminium's fashionable reputation as an environment-friendly metal is any indication. Another new bauxite development is a joint Soviet–Guinea scheme for a new mine in the Boke region; this will have an annual capacity of 10m tonnes of ore.

Diamond production also continues to look up, with a steep rise in output from 49,000 carats in 1987 to 200,000 carats worth $52m in 1988. The stones are mined by the Areador Mine, a joint venture between Bridge Oil of Australia and the Guinean Government, each of which holds 50 per cent of the shares. In November 1988 the Australian parent company announced it had sold a 188.77 carat gem diamond from Guinea for 'what is understood to be the highest price per carat ever paid for a rough, uncut diamond': A$10.2m ($8.62m). A few months later Areador extended its mining over an additional 405 sq km.

With these prospects Guinea's economic outlook should be rosy, but the total disbursed foreign debt, which stood at $1.421bn at the end of 1986, requires heavy servicing. Projections for 1987–92 show amounts from $50m to $165m a year. Of these payments, interest charges represent just under 25 per cent, which implies that Guinea's capital repayments are more substantial than is usual in Africa. The trouble is that balance-of-payments figures and other statistics concerning expenditure and revenue are unreliable. A $14m IDA credit seeks to remedy this position by financing about half of a four-year scheme designed to strengthen government planning and economic management. The project includes civil service training, commercial bank supervision, the preparation of balance-of-payment data and of economic legislation. Being so closely monitored by foreign consultants who report to the World Bank may not be confidence-building, but it may be essential in the case of Guinea.

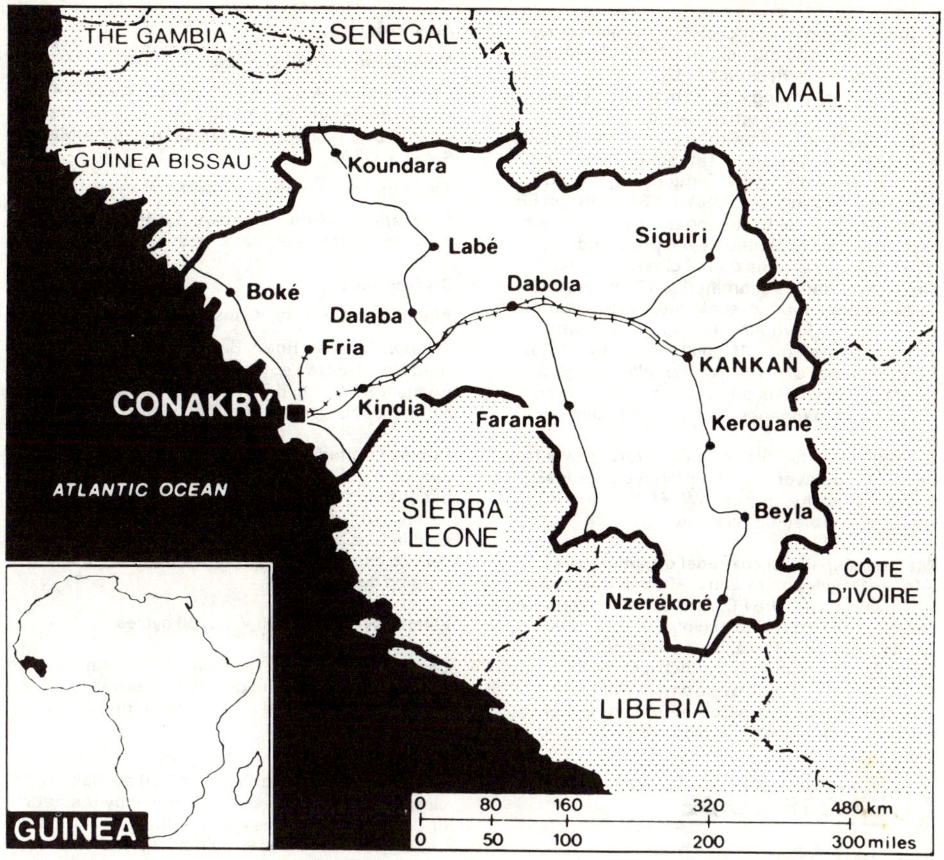

Conté himself rails publicly against the shortcomings of his administration, particularly at the 'generalised corruption, incessant embezzlement, laxity in implementing budget estimates, and the malfunctioning of our administrative system – a situation that has finally paralysed all the recovery programmes that have been launched'. These failures he ascribed to 'a wrong interpretation of the government's message and to the very nature of the Guinean mentality'. It was therefore necessary, he said, to draw up a new constitution to promote a new Guinean mentality. The draft would be prepared by 'Guineans chosen for their competence and wisdom'. He set no time-table on the change to the new system, and observers consider that the present army rule will continue for two or three years, provided that the army disunity on which he himself commented does not interfere with the government's ability to function as an entity before the changes can be effected.

COUNTRY PROFILE

Political system: Political power rests with the Military Committee for National Redress (CMRN). Formerly a French colony. Gained independence on 2 October 1958.

Population: Most of the estimated 6.5m inhabitants (1987) are Muslims. There is a small number of Roman Catholics.

Main cities/towns: Conakry (population 600,000), Kankan (60,000), Siguiri, Kindia, Boké.

Language: French is the official language, but African languages are in daily use. English is seldom used.

Media
Press: One weekly newspaper, *Horoya*, which is the organ of the CMRN. Also two official publications: *Journal Official de Guinée* and *Le Travailleur de Guinée.*
Broadcasting: Television and radio programmes in French, English, Arabic, Portuguese and local languages. Services operated by Radiodiffusion-Télévision Guinéenne (RTG) from Conakry and carry advertising.

External trade: Trade surplus generated by bauxite and alumina exports, but revenues from this source projected to decline. Although diamond exports are rising, the balance of payments situation is precarious, despite large volumes of concessionary assistance from the World Bank, IMF and foreign aid donors, including the USSR. Guinea does not belong to the West African Monetary Union (franc zone).
Exports: Dominated by bauxite, alumina and diamonds. Agricultural products (coffee, fruit, bananas, pineapples) make up remainder. Main destinations: USA, USSR, France, Spain, German FR.
Imports: Mainly foodstuffs, machinery and transport equipment, petroleum products, textiles and construction materials. Government is currently discouraging imports. Main sources: France, USA, USSR, Brazil, Trinidad and Tobago.

Agriculture: Around 80 per cent of population engaged in subsistence farming. Only three per cent of land is cultivated arable land, though considerable potential for development. Main cash crops are bananas, groundnuts, oil palm, cotton, pineapples, citrus fruits and coffee. Main subsistence crops are rice, cassava, maize and vegetables. Output has stagnated due to transport problems, low levels of mechanisation, poor marketing and lack of vital inputs. A large-scale road improvement programme under way is specifically designed to aid farmers. Emphasis on achieving self-sufficiency in foodstuffs, particularly cereals, largely through agro-pastoral farms and co-operatives. Also incentives to farmers and provision of fertilisers. Agricultural aid-financed projects especially important. Also small livestock, fishing and forestry sector.

Industry: Manufacturing sector provides up to 25 per cent of GDP. Except for aluminium smelting it is small-scale and designed to meet local requirements. Aluminium smelting from locally-mined bauxite is being modernised with French aid. The other main industries, which are textiles, food processing and plywood, are handicapped by supply bottlenecks and shortages of skilled labour. The new investment code and economic liberalisation are expected to attract more foreign capital.

COUNTRY PROFILE

Mining: Most dynamic sector of the economy, accounting for 25 per cent of GNP and almost all export earnings. Reserves of bauxite put at over 500m tonnes, among largest high-grade reserves in the world, but output of bauxite from three main sites (Boké, Fria-Kimbo and Kindia-Debele) has slumped to around 11m tonnes. Bauxite is processed locally by the Friguia company and exported to Cameroon, USA, USSR, German FR, Spain and France. Up to 2bn tonnes of iron ore reserves to be exploited by the proposed Mifergui-Nimba mine on the Liberian border. Diamond reserves estimated at 400m carats (93 per cent gem quality), exploited by joint venture companies at Kerouane and Macenta. Annual production well over 100,000 carats (36,600 carats in first quarter of 1988). Also small-scale gold mining. Proven deposits of uranium, titanium, copper and manganese not yet exploited.

Energy: Considerable potential hydro-power from several large rivers as yet untapped potential. Current capacity only 80 mW. Exploration for offshore petroleum underway.

Membership of international organisations: African Development Bank, African States in Association with the EC, Arab Bank for Economic Development in Africa (recipient), Economic Commission for Africa, Economic Community of West African States, IDA, IFAD, ILO, IMF, Islamic Conference, Islamic Development Bank, OAU, Permanent Inter-State Committee on Drought Control in the Sahel, UN, WHO, WIPO, World Bank.

BUSINESS GUIDE

Time: GMT.

Climate: Tropical and humid. South: rainy season Jun-Oct, rainfall particularly heavy in Conakry, average temperature range from 22°C to 30°C. Dry season from Nov to Apr, likely temperature range 24°C to 35°C. North: generally cooler and drier.

Entry requirements

Passport: Required by all.
Visa: Required by all except nationals of Guinea and Mali. Before departure, visitors must have their return or onward tickets and visa certified by the Department de la Régulation Aérienne et Maritime.

Prohibited entry: 'White' nationals of South Africa (transit through airport only).

Currency: No restrictions on foreign currency and travellers' cheques. Exact amount should be declared on entry and declaration form produced on departure. The import and export of local currency is strictly forbidden.

Health precautions
Mandatory: Yellow fever vaccination certificate.
Advisable: Cholera, typhoid and polio vaccinations. Malaria prophylaxis essential as risk exists throughout the country. Water precautions should be taken.

Representation overseas: Abidjan, Accra, Addis Ababa, Bamako, Banjul, Beijing, Belgrade, Berlin, Bissau, Bonn, Brazzaville, Brussels, Bucharest, Cairo, Dakar, Dar es Salaam, Freetown, Havana, Jeddah, Kinshasa, Lagos, Maputo, Monrovia, Moscow, Nairobi, Ottawa, Rabat, Paris, Rome, Tokyo, Tripoli, Washington.

Air access

National airline: Air Guinée.

Other airlines: Aeroflot, Air Afrique, Air Ivoire, Air Zaire, Ghana Airways, KLM, Linhas Aereas da Guinée-Bissau, Nigeria Airways, Royal Air Maroc, Sabena, UTA.

Tax: Airport tax (variable) levied on departure.

International airport: Conakry (Code: CKY), 13 km from city, restaurant, bank, post office.

Surface access

Main ports: Conakry, Kamsar.

Overland access: Road: Best route is the coastal road from Sierra Leone (Freetown) to Conakry. Roads from Ganta (Liberia) to N'zérékoré and from Mali (to Kankan and Siguiri) can be difficult.

Hotels: Limited first-class accommodation available in Conakry and Kankan; good hotels are expensive. Advisable to book well in advance, and to have written confirmation. Hotel bills may be paid in foreign currency or by credit cards. A service charge is usually included in the bill. Tipping optional.

Electricity supply: 220 V AC, 50 cycles.

Car hire: International and national driving licence required. Driving outside city limits with chauffeur and special authorisation only.

City transport

Taxis: Available in Conakry, limited availability elsewhere; can be hired from hotels by the hour or day. Standard fares apply within towns, but for longer journeys fares should be agreed in advance. Tipping is optional.

National transport

Air: Air Guinée operates regular domestic service between Conakry, Boké, Kankan, Kissidougou, Labé, Macenta, N'zérékoré, Siguiri.

Road: A few main roads are surfaced, e.g. from Conakry north to Kindia and Kissidougou, and parts of the road east to Freetown in Sierra Leone. Most roads are laterite and become impassable during the rainy season (May-Oct). Road improvement under way.

Buses: Coach services Conakry-Kindia-Gaoual and Dabola-N'zérékoré.

Rail: Narrow-guage railway from Conakry to Kindia and Kankan, which is in poor condition.

Public holidays

Fixed dates: New Year's Day, 1 May (Labour Day), 17 May, 15 Jul, 28 Sep, 2 Oct, 1 Nov (All Saints'), 22 Nov, 25 Dec.

Variable dates: Easter Monday, Ascension, Tabaski. Other Muslim holidays observed.

Working hours

Government and business: (Mon-Thu) 0800 – 1500; (Fri) 0800 – 1300; (Sat) 0800 – 1500.

Banking: (Mon-Sat) 0800 – 1300.

Telecommunications

Telephone: Dialling code for Guinea: IDD access code + 224 followed by subscriber's number. There are no area codes.

Telex: Facilities available at major hotels in Conakry. International telex code: GE.

Banking

Central bank: Banque Centrale de la République de Guinée.

Main centre: Conakry. *Head offices:* Banque Internationale pour le Commerce et l'Industrie de la Guinée, Banque Guinéenne du Commerce Extérieur, Société Général de Banques en Guinée, Banque Internationale pour l'Afrique en Guinée, Islamic Bank of Guinea.

Representation in capital: Algeria, Benin, Bulgaria, Cape Verde, China PR, Congo, Cuba, Czechoslovakia, Egypt, France, German DR, German FR, Ghana, Guinea-Bissau, Hungary, India, Iraq, Italy, Japan, Korea DPR, Lebanon, Liberia, Libya, Mali, Morocco, Nigeria, Poland, Romania, Saudi Arabia, Senegal, Sierra Leone, Switzerland, Syria, Tanzania, USSR, USA, Vietnam, Yugoslavia, Zaïre.

BUSINESS DIRECTORY

Hotels: Conakry
G'Bessia, Conakry Airport, P.O. Box: 743 (tel: 61145).

Hotel de l'Indépendence (Novotel), P.O. Box: 287 (tel: 445021; tx: 2112).

Hotel de l'Unité, P.O. Box: 683 (tel: 444711; tx: 3139).

Car hire: Conakry
Avis: Grand Hôtel de L'Indépendence, PO Box 287 Conakry (tel: 44-47-47, 44-15-15; tx: 22268).

Europcar: BP 587 (tel: 44-15-29; tx: 22283).

Other useful addresses
Conakry: Agence Guinéenne de Presse, PO Box 191.

Air Afrique, PO Box 77 (tel: 44-47-70/72).

Air Guinée, 12 Av de la République, PO Box 12 (tel: 44-29-81 for reservations, 46-15-37 for airport, 44-46-14 for office).

Banque Centrale de la République de Guinée, 12 Boulevard du Commerce, PO Box 692 (tel: 44-17-34/35, 44-18-76, 44-26-34, 44-31-63; tx: 22225 GE).

Banque Guinéenne du Commerce Extérieur, PO Box 143 (tel: 44-26-31; tx: 635).

Chambre de Commerce, PO Box 545.

Chambre Economique de Guinée, PO Box 609.

Comité d'État pour la Co-operation avec l'Europe Occidentale.

ENTRAT (state forwarding firm), PO Box 315.

Entreprise Nationale Import-Export (Importex), PO Box 152 (tel: 44-28-13, 44-28-09; tx: 625).

Office National des Hydrocarbures (Onah).

Radio-Télévision Guinéenne (RTG) PO Box 391.

Statistical Office, Bureau du Premier Ministre (tel: 44-21-48).

Tourist Office, PO Box 617, Square des Martyrs.

Guinea Bissau

Aidan Meehan

A tiny wedge of Portuguese-speaking territory covering 36,000 sq km, Guinea-Bissau is surrounded by the francophone countries of Senegal and the Republic of Guinea. Its 948,000 people are among the poorest in the world, as well as among the least healthy. As much as a third of the land surface of the state can disappear when the rivers are in flood and the tide is high; when the rains fail, however, and the tide rises anyway, the paddy fields of the country's staple crop, rice, can become salinated, resulting in the total loss of the harvest and the virtual sterilisation of the soil.

The Portuguese came to this part of Africa over five centuries ago and at one time held a series of forts and trading stations – principally for the export of slaves – all along the coast and beyond the Bight of Biafra. Incompetence and weakness, however, robbed them of their possessions to the profit of the marauding French and English, and left them with the remnants nobody else wanted. Thus, Guinea-Bissau remained Portuguese largely by default, a sort of consolation prize for the plucky Portuguese who had shown the way along the coast.

That was during the nineteenth century, but little had changed by the mid-1960s when one of the more interesting guerrilla struggles against colonial rule in Africa was mounted by a movement calling itself a party: The African Party for the Liberation of Guinea-Bissau and Cape Verde (PAIGC), led by an agronomist named Amilcar Cabral. Though he never lived to see the struggle victorious (he fell under a Portuguese-hired assassin's gun in January 1973), Cabral inspired the PAIGC to take a new, humanistic road to revolutionary warfare, while judiciously enlisting the help of Western news teams whom he brought into the forests in his international campaign against the Portuguese.

The result was independence on 24 September 1973, recognised by the post-coup Lisbon regime a year later, and a Guinea-Bissau outwardly committed to non-alignment, democratic socialism, economic flexibility and African unity – particularly where the Cape Verde Islands were concerned, since they had supplied many of the PAIGC's fighters and leaders and were still engaged in their own liberation struggle against the Portuguese.

The Guinea-Bissau economy, however, was in a shambles. The only city, Bissau itself, had been kept afloat by the false economy generated by the occupying Portuguese forces which had resulted, for example, in the construction of what was then West Africa's most sophisticated brewery. The countryside was either suffering from the after effects of the guerrilla war, or was about to be gripped by one of the most severe droughts in recorded history. The new state was in no position to follow Cabral's advice and 'give concrete benefits' to the people who had supported the struggle. Instead, the economy went into accelerating decline and, despite intensive aid particularly from the Scandinavian countries, the social benefits of health, education and other services which the party had promised (and often delivered) throughout the war became impossible to distribute. The government itself was in the hands of Amilcar Cabral's half-brother, Luiz, who was no match for his sibling either in leadership or common sense, and who consistently alienated his comrades, arousing suspicion rather than solidarity.

The crisis was translated in November 1980 into a coup d'etat against the Luiz Cabral régime which, at the time, was described by the victors – led by the

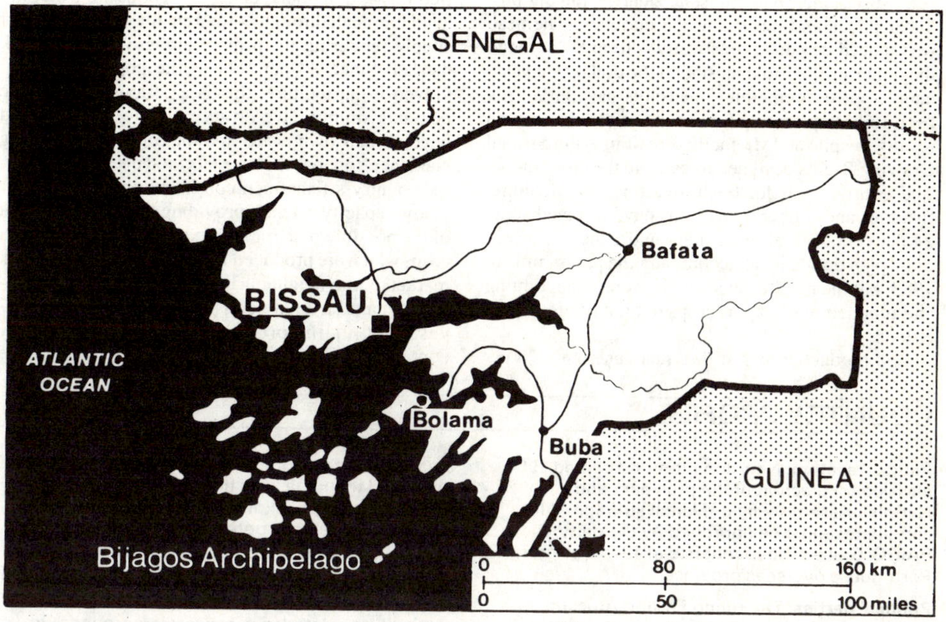

current President Joao Bernardo (Nino) Vieira – as being run by a 'Cape Verdian clique'. Minor attacks on Cape Verdians were reported in the capital and formal links with the party in the islands were broken (by the islanders). Nino Vieira's heroism during the war and his impeccable non-Cape Verdian credentials won him and his new 'Council of the Revolution' widespread popular acceptance, even if enthusiastic support was often lacking.

He has proved himself since to be a shrewd and deft politician, surviving three serious attempts at counter-coups and he restored internal democracy to the country while maintaining the party in power and even increasing its popularity. Links with the Cape Verdes are again amicable and the country is once again accepted as an equal member of the 'Lusophone' (Portuguese-speaking) African community consisting of Guinea-Bissau, Cape Verde, Angola, Mozambique, and Sao-Tomé and Príncipe. Since the coup attempt in 1985 political life has stabilised and, though a half-hearted opposition group exists in Lisbon, the Vieira government seems relatively secure for the moment.

The original sins which brought Nino to power have nonetheless resolutely failed to disappear. In spite of a genuine attempt to spread the coverage of health care, the country is so afflicted by tropical diseases and hygiene problems that life expectancy was estimated to be only 42 years in 1987, while 45 per cent of children die before the age of five and infant mortality remains one of the highest in the world with 200 deaths per thousand live births. Only 5 per cent of the population officially have access to clean water and over 90 per cent are infected with the malaria parasite.

The picture is hardly more encouraging in education. While there has been an increase in the number of graduates from universities and training institutes since independence (there were only five such graduates in 1973), illiteracy remains high at around 70 per cent of the total population and 95 per cent of adults. Only 20 per cent of secondary-school-age children actually attend such institutions while 62 per cent of their younger compatriots are able to receive primary education. The quality of the instruction they receive, however, and the materials they have to work with are another matter. Much greater investment – political as well as economic – will have to be made in both these spheres before any real improvement can be expected.

Economy

Such spending, if it does come, is unlikely to be pulled from the national purse. Guinea-Bissau, as one of the International Monetary Fund's 10 poorest nations on

KEY FACTS

Official title: Republic of Guinea-Bissau

Head of state: President Brigadier Gen. João Bernardo Vieira

Ruling party: Partido Africano da Independência da Guiné e Cabo Verde (PAIGC)

Capital: Bissau

Official Languages: Portuguese (Creole and Balante widely spoken)

Currency: Guinea Peso (GP) = 100 centavos

Exchange rate: GP650 per $ (Nov 1989)

Area: 36,125 sq km

Population: 922,000* (1987)

GNP per capita: $170 (1987)

Inflation: 30%+ (1986)

Trade balance: –$30m (1987)

Foreign debt: $387m (1987)

earth, had a per capita GNP of only $170 in 1987. In the context of the economic liberalisation ushered in by the 1980 coup d'etat, the government began to implement an IMF financial and economic stabilisation programme in 1984. Its aim was to free up the market by reducing state interference, imposing monetary discipline, reducing state salaries, raising producer prices and devaluing the currency. The peso was indeed devalued drastically – from 84 pesos to the dollar in 1983 to over 650 to the dollar five years later – but the remaining measures have either not been fully implemented or have not had the desired results, despite an IMF facility totalling 6.8m Special Drawing Rights designed to back up the programme. The priority areas for fresh investment (agriculture, fisheries and timber extraction) have received some 24 per cent of total investment under the new economic plan drawn up to accompany the programme in 1983, but the results, once again, were disappointing and production of export crops did not rise to expectations.

Food production has still to recover from a 70 per cent fall in output which took place during the colonial period between the 1950s and 1970. Though increases are being maintained, they are not sufficient to reduce the country's dependence on imported food. A mistake (perhaps an inevitable one) made by the government immediately post-independence was carried forward: namely, consumer goods are either not available at all through normal channels or are in short supply rendering them more attractive to the black market. Since 83 per cent of the population is involved in agriculture, this is a serious lacuna. The peasants simply have no incentive to produce food beyond their own needs and the tiny urban working class therefore has to import its food needs. Similarly, the country's fishermen consistently return with far below capacity catches, providing less than 5 per cent of the possible potential for export. Much of the cash crops which are produced finds its way on to the black market or into neighbouring countries where the supply of manufactured goods is better. Meanwhile, investment in import-consuming urban industry continues to be a waste of resources, since these industrial units are incapable of producing virtually anything the peasants could ever hope to afford to buy.

The solution, or at least part of it, would be for foreign lenders and donors to supply money for the purchase of manufactured goods to be placed on the peasant market, either by an efficient state sector or by a network of private traders. In this way the peasants could be encouraged to grow both food and cash crops and contribute to the re-equilibrium of the hopeless imbalance in trade. For this to happen, however, would require an entirely new attitude not only on the part of the authorities of Guinea-Bissau but also those of the major international financial institutions.

In 1990, this small, often heroic country continues to live just above the limit of survival, and dependent on outside altruism. Even the hopes generated by thoughts of oil wealth beneath the surrounding seas have dissolved as the richer beds in the disputed zones were awarded to Guinea-Bissau's neighbours, leaving only the pickings to the heirs of Amilcar Cabral's African revolution.

COUNTRY PROFILE

Political system: New constitution adopted 1984 allowed for election of the president and a National Assembly by universal suffrage. All candidates must be approved by the party, the PAIGC. Gained independence in 1974. Formerly a Portuguese overseas province.

Political parties: The ruling Partido Africano da Independência da Guinée Cabo Verde is the only legal political party.

Population: Guinea-Bissau's population, estimated in 1987 at 924,000, is among the poorest in Africa, with a per capita GDP of less than $200 p.a. Some 40 per cent are Muslims, but animism is also widespread.

Main cities/towns: Bissau, Bafata, Gabu, Cantchungo.

Language: The official language is Portuguese, but Creole is widely used in daily business. Very little English is used and all correspondence and literature should be in Portuguese and French.

Media
Press: The government publishes *Nô Pintcha* three times per week. No English-language services.
Broadcasting:
Radio: Radiodifusao Nacional da República da Guiné-Bissau broadcasts services in Portuguese.
Television: There is no television service.

External trade: Large trade deficit and heavy dependence on foreign aid and credits.
Exports: Cashew nuts (53 per cent of total value in 1986), fish products (11%), palm nuts (10%), timber (10%), groundnuts (7%), cotton (5%). Main destinations: Portugal, India, France, Spain, Senegal.
Imports: Food, fuel, transport equipment, machinery and equipment. Main sources: Portugal, USSR, Senegal, France, Netherlands.

Agriculture: Employs over 80 per cent of population; receives top priority under development plan. Only eight per cent of total area is cultivated; inland areas are largely savannah, coastal areas are forest and mangrove swamps. Sector dominated by subsistence farming and animal husbandry. Chronic food shortages, despite emphasis on food self-sufficiency and cooperative farming. Main food crop paddy rice (currently up to 100,000 tonnes a year); other food crops millet, sorghum, plantains, root crops, some maize. In 1986 cashew nuts overtook groundnuts as the main export crop; other cash crops include palm kernels, coconuts, tobacco, sugar. Fishing sector has grown in importance since late 1970s. Joint ventures encouraged. Exports of fish and shellfish expected to increase as the country's enormous fish resources are exploited. Forestry also offers considerable export potential.

Industry: Manufacturing sector is small-scale (under 200 industrial units), contributing less than five per cent to GDP. Production is agriculture-related; e.g. processing groundnuts, fish processing, rice dehusking, sugar refining. Also large brewery plant, small Citroën assembly plant, brick making and textiles. A drive to modernise transport facilities.

Mining: There are some 200m tonnes of bauxite reserves in the region of Boé, but exploration costs are too high to justify extraction. Also known deposits of phosphate (in Cacheu and Oio), gold and possibly diamonds.

Energy: Dependence on imported oil, though offshore exploration underway. Hydroelectric potential remains largely untapped.

Membership of international organisations: African Development Bank, African States Associated with the EC (Lomé Convention), Arab Bank for Economic Development in Africa (recipient), Economic Commission for Africa, Economic Community of West African States, GATT, IDA, IFAD, ILO, IMF, Islamic Conference, Islamic Development Bank, OAU, UN, WHO, WIPO, World Bank.

Time: GMT.

Climate: Tropical with rainy season from mid-May to Nov and dry season from Dec to Apr. Average temeratures range from 20 to 38°C in Apr-May, and from 15 to 33°C in Dec-Jan. High humidity from Jul to Sep.

Entry requirements
Passport: Required by all.
Visa: Required by all except nationals of all ECOWAS countries. An exit visa, obtainable from the central police station in Bissau, is necessary before leaving. Visa extensions also available from central police station.
Currency: Import/export of local currency prohibited. Foreign currency import unlimited but amounts should be declared, export allowed up to declared amount. Travellers' cheques can be cashed in all banks but personal cheques can only be cashed in the National Bank in Bissau.

Health precautions
Mandatory: Yellow fever vaccination certificate.
Advisable: Cholera, typhoid and polio vaccinations. Malaria prophylaxis should be taken as risk exists throughout country. Water precautions should be taken.

Representation overseas: Algiers, Berlin, Brussels, Conakry, Dakar, Havana, Lisbon, Moscow, Stockholm, Washington.

Air access
National airline: Transportes Aereos de Guiné Bissau.
Other airlines: Aeroflot, Air Sénégal, Cubana, Europe Aero Service, Gambia Air Shuttle, TAP-Air Portugal.
International airport: Bissalanca (Code: BXO), 11 km from Bissau.

Surface access
Main ports: Bissau.
Overland access: The road from Senegal is not generally recommended.

Hotels: Accommodation is very limited and difficult to obtain at short notice. Reservations should be made well in advance, preferably through business contacts. Hotel tariffs are liable to change at short notice, therefore confirmation of booking is recommended. Hotel 24 Septembro has air-conditioned rooms.

City transport
Taxis: Available in Bissau and serve all main towns. Long distance taxis leave from the market square in Bissau.

National transport
Air: Transportes Aereos de Guiné Bissau serves around 60 small airports and airfields.
Road: Total network over 3,250 km, of which about a third is all-weather. A road improvement programme is in progress.
Water: Boats serve most towns on the coast and up-river. Tickets available from the Guin\emar Office.

Hotels: Bissau
24 Septembro, P.O. Box: Bissau.

Bissau Sheraton (opening winter 1988), Avenida 14 de Novembro.

Ronda, Amilcar Cabral, Bissau.

Tamar, Ave 12 de Septembro, Rua 3 Bissau.

Other useful addresses
Bissau: Associacão Commercial e Industrial e Agricola da Guiné-Bissau.

Banco Nacional da Guiné-Bissau, PO Box 38 (tel: 215433; tx: 969249).

Empresa Nacional de Comercio Geral, PO Box 5 (tel: 212925).

Empresa Nacional de Pesquisas e Exploracao Petroliferas e Mineiras (Petrominas), 58 rua Eduardo Mondlane (tel: 212279).

Guyana

Tony Thorndike

A massive devaluation, a condition of agreement with the International Monetary Fund, triggered widespread strikes and reduced many to near penury. Although Guyana is by all standards bankrupt, the government conceded nothing to those demanding political reform and prospects remain very poor for what was once one of the richest countries in the Commonwealth Caribbean.

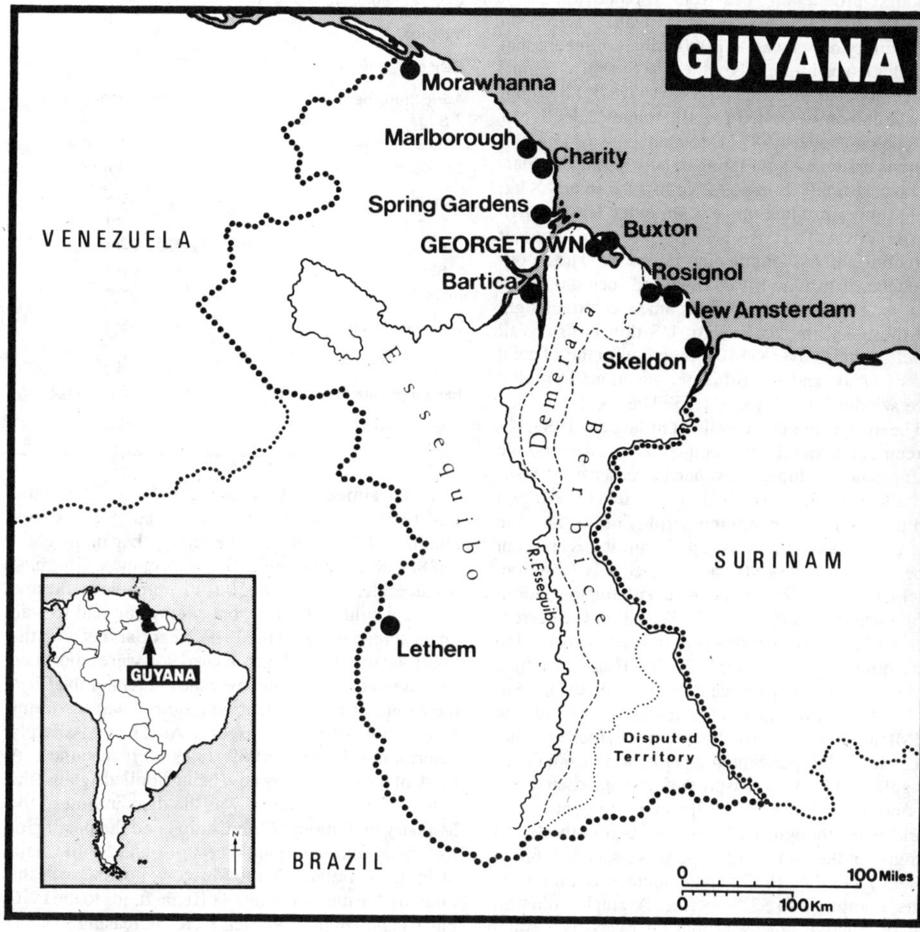

The IMF agreement, the result of tortuous negotiations, brought much needed financial relief. But while on the international front Guyana has moved from militant pro-socialist positions and has inaugurated a relatively liberal foreign investment code, internally the 'paramouncy' of the ruling People's National Congress remains little changed in practice. President Hoyte's position is by no means secure and, given the enormous economic pressures facing the country, it is unlikely that the government will be able to ignore demands for internationally-observed elections in 1990, as it has done in the past.

State socialism and paramouncy

The decline of the economy can be fairly laid at the door of one man: Forbes Burnham, founder of the People's National Congress (PNC) and prime minister from independence in 1966 to his death in 1985. Once regarded as a supporter of foreign investment and an ally of the USA, Burnham became a militant espouser of Third World anti-imperialism. Under his authoritarian leadership the economy was extensively socialised and huge debts accumulated.

But socialism as such was not the problem. Nationalisation of most of the economy (85 per cent) was in any event inevitable given the colonial domination of the economy by Bookers (sugar and retail) and Demerara Bauxite and Reynolds Metal (bauxite). The tragedy was that what emerged was a parody of socialism where nepotism, corruption and human rights abuse became the norm. The extent of corruption in all aspects of economic life and overstaffing of state enterprises was a scandal, especially as an added ingredient was racial discrimination. Ethnic division between the Afro-Guyanese (40 per cent) and the majority Indo-Guyanese (60 per cent) is deep and was exploited for political ends, a phenomenon that is by no means dead. The PNC is broadly associated with the Afro-Guyanese and has had to resort to corrupt elections and intimidation to retain power; on the other hand the People's Progressive Party (PPP) is predominantly Indo-Guyanese. Furthermore, the PNC can counter those who criticise its brand of state socialism by pointing to the marxist-leninist foundation of the PPP and the orthodox pro-Moscow marxist leadership of Dr Cheddi Jagan. It should, however, be noted that many Indo-Guyanese support the PPP less out of ideology than out of ethnic solidarity. The only other credible party is the Working People's Alliance, a leftist nationalist multi-ethnic group with strong support from intellectuals, distrusted by both the dominant parties.

To ensure the PNC's dominance, Burnham announced the 'paramouncy' of the party in 1974, virtually rendering all organs of government and parastatal organisations arms of the party. Although Burnham's successor President Desmond Hoyte has since denied this, for example saying in mid-1989 that there was now a 'clear distinction' between party and government, it is clear that the PNC remains the political bedrock of the long-suffering country. In a nutshell, therefore, there seems little alternative to ethnic-orientated socialism. The result is deep national disunity, capital flight, an extensive parallel economy and the emigration of the best young people. The scale of emigration generally has been such that the population is nowhere near the projected 900,000; it is more likely to be in the region of 700,000.

Decline and privation

It was some time after Burnham's death before Hoyte could make tentative steps to reform the economy since there were many in the government wedded to Burnham's ideas. But a fall in GDP of over 20 per cent in real terms between 1977 and 1987 and a 30 per cent fall in per capita income in the same period made change inevitable. Exports fell 55 per cent in value, imports rose by 66 per cent and production in all sectors fell back. The overall total of external debt was nearly US$2bn in 1989, including US$1.2m in debt service arrears but excluding accumulated interest. As a result Guyana has the highest debt per capita ratio in the Western hemisphere with a debt to exports ratio in 1988 of over 700 per cent. The strikes in 1989 could mean a ratio of approaching 1,000 per cent by 1990 unless production greatly improves! Reserves in 1988 were a negative US$700m (1987: US$647m).

Economic performance in 1988 was, as in previous years, dismal. The reasons included bad weather, reduced world prices (especially for sugar), strikes and input shortages. Brave forecasts by Finance Minister Carl Greenidge in his 1987 budget speech were dashed as production of bauxite, sugar and rice, the

KEY FACTS

Official title: Co-operative Republic of Guyana	**Area:** 214,970 sq km
Head of state: President Desmond Hoyte	**Population:** 720,000 (1988)
Ruling party: People's National Congress	**GDP per capita:** $995 (1988)
Capital: Georgetown	**GDP real growth:** –3.7% (1988)
Official Languages: English	**Inflation:** 40% (1988)
Currency: Guyana Dollar (G$) = 100 cents	**Trade balance:** $20.6m (1988)
Exchange rate: G$30 per $ (Nov 1989)	**Foreign debt:** $1.3bn (1988)

three major export earners, fell. Production of all grades of bauxite fell to under 1.3m tonnes; that of calcined bauxite fell by 6 per cent, metal grade fell 17 per cent and calcined by 31 per cent: overall earnings declined by 8 per cent to US$80m. Sugar export earnings were down to US$75m, a 21 per cent fall, although production rose from 156,000 tonnes in 1987 to 167,550 tonnes. Since Guyana's European Community quota was 167,000 tonnes, it meant that sugar had to be imported for the first time in a century to meet local needs. Rice production was hit by fungi disease and only reached 132,100 tonnes, a decline of 8.3 per cent over 1987. Declared gold production (there is extensive smuggling to help sustain the parallel market) fell from 662.2 kg in 1987 to 584.8 kg, while timber production, at 4.4m cubic feet, was 15 per cent down. The crisis became ever more apparent with constant and lengthening power cuts, as a consequence of turbine breakdown and fuel shortages. Very long queues for petrol became a common sight and the black market rate for US dollars on 'Wall Street' crept to over G$40, compared to the official rate of G$10 and the officially sanctioned parallel 'free window' rate of G$21.75. The local currency had been substantially devalued in January 1987.

KEY INDICATORS

	Unit	1984	1985	1986	1987	1988
Population	'000	800	790	750	750	720
GDP (1985 prices)	G$m	1,961	1,964	1,919	1,922	1,851
GDP per capita (1985 prices)	G$	2,451	2,486	2,559	2,563	2,571
Inflation	%	25.2	15.1	7.9	28.7	40.0
GDP real growth	%	1.1	0.2	–2.3	0.1	–3.7
Agricultural sector	% real gwth	3.4	0.0	3.7	–4.9	–9.0
Sugar	'000 tonnes	245	246	249	224	168
Rice	'000 tonnes	179	156	183	146	130
Bauxite production	m tonnes	1.33	1.63	1.47	1.36	1.0
Exports	G$m	808	875	979	2,367	–
Sugar	G$m	271	282	356	906	709
Bauxite	G$m	359	422	351	842	821
Rice	G$m	80	57	57.2	155	139
Imports	G$m	747	958	942	–	–
Balance of trade	G$m	61	–83	37	–	–
Current account	$m	–96.4	–96.6	–	–	–
Exchange rate (average)	G$ per $	3.83	4.25	4.27	9.76	10.0

* estimated figure

Greenidge's March 1989 budget reflected economic performance. Although revenue increased due to better collection methods, the deficit soared to G$259m and the combined operating surplus of all the state enterprises was only G$492m. Central government expenditure for 1989-90 was expected to be G$8.8bn, of which US$5.15bn will be recurrent. But the overall deficit was calculated to reach G$2.23bn. The current account balance of payments deficit was expected to rise from US$86.7m to US$138.7m. Faced with this, the bombshell – a 70 per cent devaluation to G$33 to the US dollar – was inevitable. It was also a condition of IMF agreement. Virtually overnight, food and fuel prices leapt 200 per cent and petrol went from G$13 per gallon to G$39. Transport prices were such that it was no longer economical for many to continue to work, even though a 20 per cent across-the-board increase in the minimum wage was conceded, but at G$29.97 per day, the US equivalent was a mere 90 cents, compared to $2.50 before. A graphic illustration of relative cost was that an overseas airmail stamp cost nearly half a day's wages. Interest rates reached 45 per cent. Public reaction was immediate. Strikes broke out and those in the crucial export sectors lasted about six weeks, leading to losses estimated at US$130m. But the government stood firm and serious, taking advantage of division within the trade union movement.

IMF agreement

Recourse to foreign aid donors and sources of concessionary finance was made almost impossible because of US$180m worth of arrears owed to the IMF, World Bank and the Caribbean Development Bank. The first steps to reform began in 1987 with the privatisation (the more ideologically, compatible term is 'disaggregation') of some small enterprises and a plan for a future stock exchange. A new foreign investment code in which Hoyte insisted there 'were no sacred cows' was trumpeted. One important result was a joint venture agreement between the state-owned bauxite company, Guymine, and Reynolds to develop a huge US$25m bauxite facility in the interior, the planned output of which would be larger than the total production of all other facilities as a whole.

Formal negotiations with the IMF commenced in April 1988, the aim being a financing agreement which would enable a bridging loan from commercial and bilateral sources to be used to pay the arrears due to the Fund and other multilateral agencies. Once paid, the Fund's disbursement would then be used to pay the bridging loan and to support an Economic Recovery Programme (ERP) agreed with the Fund.

Talks aimed at a securing a US$1.7bn (later US$1.93bn) supplementary financial package (including $1.2bn debt rescheduling) began in early 1988, Canada chairing a Guyana Support Group. It became clear that US$550m in immediate aid and credits would be needed, but everything had to wait on an agreement with the IMF. Negotiations with the Fund and the World Bank in late 1988 were adjourned because of disagreement over the content of the ERP, the extent of the devaluation required and the terms of a public sector pay freeze. As services sharply deteriorated in early 1989, talks were resumed. A mark of the pressure by now being applied against the beleaguered government was the dismantling of the Ministry of National Mobilisation and the posting of its strictly marxist head, Vice-President Ranji Chandisingh, as Ambassador to Moscow; not only was the ministry the main conduit of official funds to the PNC but Chandisingh's politics were increasingly unattractive. The only good news was the extension to Guyana of Caribbean Basin Initiative eligibility by the United States.

Agreement in principle was finally reached after some hard bargaining in Washington in February 1989. The USA also agreed to resume wheat supplies on concessionary terms; not a moment too soon as food queues were threatening public order. The ERP was to encompass structural adjustment policies, the reduction of the parallel market, infrastructural rehabilitation and the restoration of national creditworthiness. But the long negotiations fostered rumours of devaluation. Huge withdrawals of funds from the banks and demand for credit fuelled inflation, while the 'Wall Street' exchange rate went to over G$60. Hoyte attempted to devise a Social Impact Amelioration Programme to relieve the most vulnerable in the society, but little could be done for lack of finance.

The IMF agreement was in the form of a 'memorandum of policy' pending monitoring of the economy to the end of 1989 when it was expected that the normal letter of intent would be signed. Published at the end of April, there was an immediate problem: ERP production targets included within it were unattainable from the start, mainly because of the strikes. There seemed only one tangible result, the sharp reduction in parallel market activities (the black market rate fell back to around G$42) as the very high interest rates began to bite.

Central to the memorandum was an Enhanced Structural Adjustment Facility of US$181m, payable as soon as the arrears to the Fund were cleared, and a standby facility of US$80m. External financing from other sources eventually met the US$1.93bn target, covering arrears, project finance and balance of payments support. It was agreed that the US$180m bridging loan would be paid in November; of this US$130m would be repaid to the support group, the remaining US$50m being loaned. Lines of credit were opened by Britain and Canada. As for debt rescheduling, negotiations opened with the Paris Club in May. The terms agreed were generous, Guyana being the first country outside Africa to be granted 'exceptional debt relief'. All long-term debt was to be repayable over 20 years with a 10-year grace period; short- term debt over 10 years, with a five-year grace period. There was no debt cancellation or interest rate reduction: however, in subsequent negotiations with individual creditors, Britain was among those which hinted that some of Guyana's obligations might be written off. In the Caribbean, debts to Barbados and CARICOM were also rescheduled. With all this settled, Hoyte travelled to the USA in June to request further assistance and to drum up private investment. But criticisms remained: that of PNC political domination.

Power games

Hoyte's accession to the premiership in August 1985 was by no means assured, his main contender being Hamilton Green, now prime minister. Green continues to enjoy support amongst the business community and exerts a powerful influence on proteges within ministries. By contrast, Hoyte tends to work with, and rely on, technocrats whose loyalty to him appears to be tenuous. Then there is the circle around Viola Burnham, Burnham's widow, consisting of loyalists working to preserve Burnham's socialist programme. While the latter have been effectively isolated, the Green faction has strengthened and Hoyte's majority in the cabinet is a slender one. Economic reconstruction is therefore not only critical for Guyana but for Hoyte as well.

Hoyte called an election to help consolidate his position in the party soon after his accession, and in this he largely succeeded, most of the 42 PNC members returned being ready to accept him. But so far as the opposition parties were concerned, since the PNC is in control of the electoral process, seats were not so much won as allocated. Eight were given to the PPP and one each to the WPA and the United Force, a one-time important PNC ally. Corruption of the democratic process was not new. Indeed, the PNC itself was a product of betrayal. Burnham was a co-founder of the PPP with Jagan to promote national independence, but a leadership struggle resulted in Burnham forming the PNC. Conscious of British and American antagonism towards Jagan, he helped engineer a system of proportional representation which led to electoral victory in 1964. Admittedly, victory also depended on the support of the right-wing United Force, but in the rigged 1969 poll that group was dispensed with. It was therefore predictable that under his rule human rights were severely curtailed, culminating in the draconian National Security Act of June 1985. Successive reports by such bodies as Amnesty International and Americas Watch told a sorry tale.

Admittedly the atmosphere of fear that permeated the country during the latter days of Burnham's rule has eased, but the 1988 report of the Guyana Human Rights Association singled out police brutality and restricted access to information as continuing problems. Hoyte's problem has been to ensure that badly needed reforms do not fundamentally affect the position of the PNC. For instance, an Integrity Commission was established to identify corruption and to recommend action, yet political corruption was excluded from its terms of reference. Then there was the saga of PNC relations with the trade union movement. Although the General Agricultural Workers' Union (GAWU) is allied to the PPP, the PNC has in the past been able to rely on the support of most unions in the Trades Union Congress (TUC), especially the Guyana Mine Workers' Union (GMWU) representing Afro-Guyanese bauxite workers. But a portent of problems ahead was the dissassociation by this union from the PNC in 1984.

The trade union crisis developed in September 1988 at the TUC conference. Six unions, including GAWU and GMWU, were already outside the PNC camp and they were joined by a seventh, the Guyana Public Service Union (GPSU). They protested against the strident lobbying by PNC activists to ensure that none of their nominees would be elected to TUC posts. The seven walked out of the conference and formed the Federation of Independent Trade Unions in Guyana (FITUG). GPSU leader George Daniels was elected president and was promptly sacked from his post in the civil service. A 'day of protest' was called but there was little response except in the bauxite industry: the government responded by refusing to recognise the two dissident unions involved, which led to a 15-day strike. The anti was then increased with the budget. FITUG was at the forefront of protest, demanding that the entire budget be withdrawn. The government refused negotiation. Strains appeared in the leadership, two of the seven members wanting to prolong the strike, as indeed did many of the rank and file bauxite workers. The TUC, for its part, while criticising the budget, did not support the strikes at all. FITUG was further weakened by the decision of the GPSU to return to the TUC fold after a new leadership had been elected. The inter-union squabble went further as the sugar unions in FITUG accused each other of poaching members. The PPP joined the fray, concerned that its influence over the sugar workers was eroding. It directed its anger in particular at the Working People's Alliance, which appeared to have made effective inroads into the trade union world, especially the sugar sector.

With such divisions in the opposition hiding behind thin veils of unity, it is not surprising that the PNC can remain in power. Nonetheless, the fact remains that the PNC's political legitimacy is questionable and the country remains ethnically divided. For Guyana to pull itself out of the appalling economic mess that 25 years of PNC rule has bequeathed, a clear commitment to corruption-free democracy and national unity is sorely needed.

COUNTRY PROFILE

Political system: Republic with executive power vested in the president, who appoints and leads a cabinet including a prime minister and four vice presidents. Legislative power is held by the unicameral National Assembly, consisting of 12 regional representatives and 53 members elected for five years by universal suffrage. The leader of the party with the largest number of votes becomes the president.

Main cities/towns: Georgetown (population 185,000), New Amsterdam (pop 20,000), Linden (pop 30,000).

Media
Press: There is one daily newspaper – the *Guyana Chronicle* (circulation 60,000) – which is state-owned – and a new independent weekly, *Stabroek News*. Several other weeklies and periodicals are occasionally available, including *New Nation* (26,000), *Mirror* (20,000), *Catholic Standard* (10,000) and *Sunday Chronicle* (100,000) when newsprint is available. There is no trade press.
Radio: Guyana Broadcasting Corporation operates two radio stations (Radio Rorima and Voice of Guyana) on behalf of the government. *Television:* Two local TV services rebroadcasting TV received by satellite from the USA.

External trade: Persistent trade deficits (due to low world prices for major exports) and high debt service burden have had a serious effect on the balance of payments position. Founder member of Caribbean Common Market (CARICOM) but trade with other member countries has fallen dramatically due to Guyana's failure to repay debts to the organisation.
Exports: 1987: Bauxite (58% of total value), sugar (27%), rice (6%), shrimps (5%). Main destinations: UK (22-27% of total value), USA (15-20%), EC excluding UK (20-30%), Canada (6-8%), Trinidad and Tobago (6-8%), Venezuela, Japan, South Korea.
Imports: Fuels and lubricants (up to 40% of total value), machinery and transport equipment (12%), consumer goods (12%), chemicals (7%). Main sources: Trinidad and Tobago (25-35% of total value), USA (18-25%), UK (10%), Venezuela, Brazil.

Agriculture: Contributes 26 per cent to GDP. Cultivation of cash crops is confined to the alluvial coastal plain. Main cash crops are sugar (which accounts for nearly 30 per cent of total export earnings), rice and coconuts. However, sugar production has declined in recent years due to adverse weather conditions and strikes. Rice production has also been hit by the weather and has been damaged by disease. Guyana is self-sufficient in sugar, rice, vegetables, fish, meat and fruit, and increased government investment in the sector is expected to improve production of many other products. Emphasis is being placed on the cultivation of oil palms, soya and corn, and on the development of dairy farming. Valuable timber resources, but less than 10 per cent of this is exploitable. Fishing is also a valuable source of income and Guyana is the area's largest exporter of shrimps.

Industry: Processing of sugar, bauxite, rice, coconuts, timber, gold and diamonds accounts for about 75 per cent of total industrial output. Remainder of sector consists of small-scale manufacturing units, mostly producing consumer goods, and account for almost five per cent of GDP. Expansion of sector has been hampered by a lack of domestic energy supplies, inefficiency and a lack of technical and managerial personnel. A serious shortage of foreign exchange has also caused the closure of many firms relying on imported inputs. The government is, however, trying to expand the country's industrial base with a policy of diversification, and a limited encouragement of foreign investors to work with the predominant state sector.

Mining: Contributes about five per cent to GDP and remains of great importance to the economy in spite of declining output. Bauxite is the country's most important mineral, accounting for 35-45 per cent of total export earnings. Production slumped in the early 1980s, prompting the government to seek outside help for the future management of the industry. However, results since 1985 have been encouraging. There are plans to rehabilitate the alumina plant which ceased production in 1982. The only other minerals mined commercially are gold and diamonds. Earnings from these, which were severely limited by widespread smuggling, increased following reform of foreign exchange regulations.

Energy: Heavy dependence on imported oil. Considerable hydroelectric potential upon which the future development of the economy depends. Hydroelectric projects, with a combined capacity of almost 4000 mW are planned. The International Development Association has also financed a project to evaluate Guyana's petroleum reserves. Initial results have proved promising.

Membership of international organisations: CARICOM, CCC, CBD, FAO, G-77, GATT, IADB, IBA, ICAO, ICJ, IDA, IFAD, IFC, ILO, IMF, IMO, Inter-American Development Bank, Interpol, IRC, ISO, ITU, NAM, OAS, PAHO, SELA, UN, UNESCO, UPU, WFTU, WHO, WMO, World Bank.

BUSINESS GUIDE

Time: GMT -3.

Climate: Tropical with high humidity and rainfall. Temperature nearly constant throughout the year at 27°C to 28°C. Two dry seasons, from Feb to Mar and Aug to Oct. Two rainy seasons, from Nov to Jan and Apr to Jul.

Entry requirements
Passport: Required by all.
Visa: Required by all except nationals of CARICOM countries and Suriname. Regulations alter frequently; check with embassy/consulate.
Currency: All currency (and jewellery) should be declared. Declaration forms must be kept and produced on departure to enable re-export of foreign currency. Up to G$40 may be taken in or out of country in local currency.
Tax: Departure tax of G$50 payable following visits of more than 48 hours.

Health precautions
Mandatory: Yellow fever and cholera vaccination certificates required if arriving from an infected area.
Advisable: Yellow fever, typhoid and polio vaccinations. Malaria prophylaxis recommended. Drinking water precautions should be taken at all times.

Representation overseas: Beijing, Brasília, Brussels, Caracas, Havana, London, Lusaka, Moscow, New Delhi, Ottawa, Tehran, Washington.

Air access
National airline: Guyana Airways.
Other airlines: BWIA, Cruzeiro, Cubana, LIAT, Surinam Airways.
International airport: Georgetown-Timehri (GEO), 28 km from city; bar.

BUSINESS GUIDE

Surface access
Main ports: Georgetown, New Amsterdam.
Overland access: Coastal road runs from Suriname border to Georgetown, via a ferry across the Berbice river at New Amsterdam. Entry from Brazil is possible at Lethem but a flight will have to be taken from there to Georgetown. There are no road connections to Venezuela.

Hotels: Available in Georgetown, Linden and New Amsterdam. Advisable to book in advance. 10 per cent tip is usual.

Car hire: Available in Georgetown but must be booked well in advance. Petrol shortage can cause problems. National or international driving licence accepted.

City transport
Taxis: Widely available in major towns and can be found on ranks. Standard fares for city journeys. Fares for longer trips should be negotiated in advance. 10 per cent tip is usual.

National transport
Air: Only practicable method of travelling to the interior of the country. Guyana Airways operate domestic network, but permits are needed from the Ministry of the Interior for non-nationals. Early booking essential. Charter facilities available at Georgetown.
Road: Outside Georgetown roads are limited. A coastal road links Georgetown, Rossignol, New Amsterdam and the Suriname border. Another coast road runs west from Georgetown, via the Demerara river, to Parika. There is a new road inland to Linden.
Buses: Service in Georgetown is good but crowded. Coaches link Georgetown with towns on the coastal roads, and with Timehri and Linden.
Water: Services link coastal towns and several inland places via rivers. Services include New Amsterdam-Ituni, Georgetown-Bartica, Rosignol-New Amsterdam.

Public holidays: 1 Jan (New Years' Day), 23 Feb (Republic Day), Phagwah, 13 Apr (Good Friday), 16 Apr (Easter Monday), 27 Apr (Id al-Fitr, end of Ramadan), 1 May (Labour Day), 5 May (Indian Heritage Day), 2 Jul (Caribbean Day), 4 July (Id al Adha, feast of the Sacrifice) 6 Aug (Freedom Day), 2 Oct (Eid al-Adha, Deepavali, Prophet's Birthday), 25-26 Dec (Christmas). When a public holiday falls on a Sun, the Mon following is taken as the holiday.

Working hours
Government and business: (Mon-Thu) 0800 – 1230, 1300 – 1630; (Fri) 0800 – 1200, 1300 – 1530.
Banking: (Mon-Thu) 0800 – 1230; (Fri) 0800 – 1230, 1500 – 1700; (Sat) 0800 – 1100.
Shops: (Mon-Fri) 0800 – 1130, 1300 – 1600; (Sat) 0800 – 1130.

Telecommunications
Telephone: Dialling for Guyana: IDD access code + 592 followed by area code (Georgetown 2, New Amsterdam 3, Linden 4) followed by subscriber's number.
Telex: These services are operated by Guyintel from their offices at Bank of Guyana Building in Georgetown. Some hotels also offer telex facilities.
Fax: Fax facilities available at Guyana Telecommunication Corp, Bank of Guyana Building.

Banking
Central bank: Bank of Guyana.
Head offices: Guyana Bank for Trade and Industry, Guyana National Co-operative Bank, National Bank of Industry and Commerce, The Republic Bank.
Foreign branches: Bank of Baroda, Bank of Nova Scotia (Scotiabank).

Advertising: Available on the two radio stations and in the press, cinemas, posters and on two TV transmission stations.

Electricity supply: Electricity supply is not standardised; Georgetown generally 110 V AC, but some supplies are 220 V AC, 50 cycles. Elsewhere supply is 110 V AC at either 50 or 60 cycles.

Representation in capital: Brazil, Canada, China PR, Colombia, Cuba, German DR, German FR, India, Korea PDR, Libya, Suriname, Trinidad and Tobago, UK, USA, USSR, Venezuela, Yugoslavia.

BUSINESS DIRECTORY

Hotels: Georgetown
Guyana Pegasus, Seawall Road, P.O. Box: 101147 (tel: 52856; tx: 2234; fax: 260532).

Park, 38 Main Street (tel: 54912; tx: 2297).

Tower, 74/5 Main Street, P.O. Box: 101010 (tel: 72011; tx: 2257).

Business advice and information

Bank of Guyana, 1 Church Street and Avenue of the Republic, PO Box 1003, Georgetown (tel: 63261/7, 63251/9, 71531/5; tx: 2267).

Berbice Chamber of Commerce and Development Association, New Amsterdam.

CARICOM (Caribbean Community and Common Market), Bank of Guyana Building, PO Box 10827, Georgetown.

Consultative Association of Guyanese Industry, East Street, PO Box 10730, Georgetown.

Georgetown Chamber of Commerce, 156 Waterloo Street, Cummingsburg, Georgetown (tel: 56451).

Guyana Airways Corporation, 32 Main Street, PO Box 10223, Georgetown (tel: 67201; tx: 2242).

Guyana Broadcasting Corporation, PO Box 10760, Georgetown (tel: 69231).

Guyana Manufacturers' Association, Church Street, Georgetown.

Guyana National Co-operative Bank, 1 Lombard Street, Georgetown (tel: 57810).

Guyana National Trading Corporation Ltd, 45-47 Water Street, Georgetown (tel: 66191).

Guyana Overland Tours, PO Box 10173, 6 Avenue of the Republic, Robbstown, Georgetown (tel: 69876).

Ministry of Finance, Main and Urquhart Sts, Georgetown (tel: 67241).

Ministry of Internal Trade and Consumer Protection, 95 Carmichael St, Georgetown (tel: 62505; tx: 2288).

National Bank of Industry and Commerce, 38-40 Water Street, Georgetown (tel: 63231).

New Guyana Marketing Corp, Robb Street, Georgetown.

Upper Corentyne Chamber of Commerce and Development Association, Corriverton.

Haiti

Greg Chamberlain

President Prosper Avril survived his first year in power with the tacit support of Washington and for want of any viable alternative. But slow progress towards promised elections and growing lawlessness and uncertainty eroded business confidence and dealt a heavy blow to the already-blighted economy. Many of the mostly foreign-owned light assembly re-export plants, which provide tens of thousands of jobs around the capital, planned to close. Continued massive smuggling, which has cut prices for consumers since the collapse of the 29-year Duvalier family dictatorship in 1986, also boosted the black market for the US dollar, which had reached 40 per cent above the official rate by September 1989. Local production was badly hit both by the fall in prices and by the extra cost of imports, to be paid for in dollars.

The government is still losing much customs revenue because of the contraband, aggravating the effects of the suspension of aid by the United States and others in protest against the army's role in bloodily destroying national elections in November 1987. But this was made up for by improved and broadened tax collection. The regime was still hard put, however, to pay the restive army, government employees and the monthly national oil bill and had nothing left over for investment. Foreign experts calculated that real per capita income had sunk to the level of 1960.

In mid-1989, however, the United States restored a few million of the $70m in direct aid it had cut off, and the World Bank, IMF and others began offering help again. But the glimmer of hope was dimmed by an IMF-required austerity package in September 1989, featuring a 30 per cent import surcharge on all but the most essential items, whose prices at once soared as merchants sought compensation. These measures and other new taxes were immediately denounced by both rich and poor. The charitable mood of the creditors of Latin America's poorest country illustrated General Avril's success in persuading them he still had things on target for a new attempt to establish democracy and that he was seriously cracking down on cocaine smuggling at the behest of the United States. Haitians were not so convinced however.

The spectacular scenes of army democracy and Avril's extensive talks with opposition leaders in the weeks after he came to power in September 1989 soon faded. Avril weeded out the more reformist of the junior officers who had made him president, but he allowed their young leader, Sergeant Joseph Hebreux, to continue at his side, as at least a symbol of the movement. The general's struggle to control the army led him to various concessions, notably the reappointment of many of the officers purged by the junior officers' uprising and the omission of clauses restricting army rights and powers when he partially restored the suspended constitution in March 1989. He was also forced to free a group of leftist soldiers, led by Sergeant Patrick Bauchard, whom he had jailed after an alleged attempted coup in October 1988.

Factionalism and Avril's self-defensive favouritism towards the 1,100-man presidential guard brought things to a head in late March 1989, however, when he attempted a purge a few days after a visit from a top US State Department official who demanded that he sack drug lords in the military.

The Leopards anti-subversion unit attacked Avril's mountain villa at dawn and drove him in handcuffs to the airport and, it was hoped, exile. But Hebreux and the reformists rescued him and the commanders of the army's three main units were removed. Washington also helped by refusing to deal with army chief General Herard Abraham, whom the rebels installed as president for a few hours. But the revolt continued for a week as the Leopards were joined by the Dessalines Battalion, the army's second biggest unit which had seen its notorious commander and leading drug trafficker, Colonel Jean-Claude Paul, first dismissed by Avril and then mysteriously killed by poison in November 1988. After a bizarre 21-hour heavy artillery battle (mostly using blanks) between the presidential guard and the Battalion, in which some 40 people were killed, the revolt ended.

Avril, strongly backed by the United States as the 'last best hope for democracy', claimed exiled Duvalierists and drug barons were behind the rebels, but the conflict was largely inspired by Avril's cosseting of the presidential guard. He dissolved the two rebel units, dispersed their soldiers to the provinces and said he would cut the army from 7,500 to 6,000 men. Resentment continued however. The troops' anger at the exiling of the rebel leaders to the United States was increased by being far from the pleasures and comforts of the capital. Divisions also persisted within the presidential guard. Avril made his hardline old ally, Acedius St Louis, a general and moved him up from information minister to interior and defence (February 1989) and then to be, in addition, chief of military intelligence (June 1989). Another close friend, General Gerard Lacrete, recently military attaché in Washington, was appointed number three in the army hierarchy, after Avril and Abraham.

Although for the most part the army rank-and-file are the least likely sector to be interested in civilian rule, since it would cut off the unusual economic

KEY FACTS

Official title: Republic of Haiti

Head of state: President Brig. Gen. Prosper Avril

Capital: Port-au-Prince

Official Languages: French and Creole

Currency: Gourde = 100 centimes

Exchange rate: Gourde 5 per $ (Nov 1989)

Area: 27,749 sq km

Population: 6.3m (1988)

GDP per capita: $319* (1988)

GDP real growth: –0.2%* (1988)

Inflation: 4.0% (1988)

Trade balance: –$127.8m (1988)

Foreign debt: $836.5m (1988)

* estimated figure

opportunities which accompany holding power, Avril was obliged by foreign pressure to pay lip service to a democratic future. In February 1989, he managed to gather much of the opposition together for a week-long parley on the subject. After some initial hostility, the opposition accepted his new electoral supervisory council, which travelled the country and consulted a wide range of politicians to build a base for its authority to organise local and national elections in eight stages from April 1990 and culminating in a two-round presidential vote in October and November the same year. Rumours persisted, however, that Avril had little intention of stepping down. They were fed by his remarks that the presidential chair was 'too hot' for a civilian and that 'responsible politicians' had to 'educate' the masses before any elections. He rejected demands for a coalition government to oversee a transition to democracy, though gave a few minor posts to low-level opposition figures. He also denounced 'professional agitators', whom he accused of plotting 'the destruction of the country'. A leaked government internal memo said violence might be needed to crush a peasant uprising supposedly being planned by leftists and priests in central Haiti.

Avril's failure to control increasing nationwide lawlessness reinforced the idea that he might try to stay on as a 'saviour'. Uniformed soldiers joined former Duvalierist thugs and agents, still holding sway in rural areas, to stage armed robberies and to beat, arrest or kill members of supposed 'communist' grassroots organisations, especially in the northwest and the Artibonite Valley. Though he appealed for the gunmen to hand over their weapons, praised human rights, promised to end 'unmentionable practices' of the past and announced closure of the Fort Dimanche political prison, Avril did little to investigate or punish the violence and other Duvalierist killings and atrocities since the Duvaliers fled. In December 1988 he allowed the Duvalierist former Port-au-Prince mayor Franck Romain, whose men had carried out the massacre of a church congregation three months earlier, to slip out of his embassy refuge and into exile in the neighbouring Dominican Republic, to the fury of the opposition. In September 1989 he reduced the 1986 life imprisonment sentence for murder on former secret police chief Luc Desyr to 30 years.

The conservative hierarchy of the Catholic Church, dominated by the papal nuncio, Father Paolo Romeo, meanwhile cracked down on militant priests and on the church's radio station, *Radio Soleil*, which spearheaded much of the revolt against the Duvaliers. The fiery leader of the 'liberationists', Father Jean-Bertrand Aristide, was finally brought down (after the failure of official attempts to transfer him abroad) when his order, the Salesians, was persuaded to expel him in December 1988 for 'incitement to violence and advocacy of class struggle'. Though he remained immensely popular and continued to call for revolution and rail against 'imperialists in cassocks', he was deprived of the authority of a pulpit and withdrew to run an orphanage. A few months later, the hierarchy attacked *Radio Soleil*. After forcing out its radical Belgian director, Father Hugo Triest, and replacing him with a Haitian aide of the unpopular archbishop of Port-au-Prince assisted by an ex-accountant to the Duvalier family, the bishops sacked most of the station's journalists in July 1989 for criticising the church. Programmes airing listeners' complaints about the regime were dropped.

The 'democratic sector' politicians remained divided and largely ineffective in 1989. Concrete action was led by the left, headed by the main trade union federation, CATH. Its various calls for general strikes were only partly successful, but they helped

KEY INDICATORS

	Unit	1984	1985	1986	1987	1988
Population	m	5.18	5.27	5.36	5.44	5.52
GDP	$m	1,840	1,845	1,856	1,844*	1,998*
GDP per capita	$	341	342	309	300*	319*
Inflation	%	6.5	10.6	3.3	−11.4*	4.0
GDP real growth	%	0.3	0.3	1.0	−0.2	−0.2
Agricultural production	1979-81=100	105.9	108.5*	109.8*	107.1	100.8
Food production	1979-81=100	106.8	110.0	115.5	110.3	103.2
Coffee production	'000 tonnes	37.3	36.9	37.8	30.0	31.0*
Sugar production	'000 tonnes	5,958	5,727	3,150*	3,000*	3,000*
Exports	Gourdes m	893.1	871.2	930.5	1,100.7	1,001.0
Coffee	%	237.2	226.4	279.4	124.4	187.7
Imports fob	$m	337.9	344.7	303.2	311.2	283.9
Balance of trade	$m	−123.3	−121.7	−112.5	−101.1	−127.8
Current account	$m	−103.0	−94.7	−45.2	−31.2	−53.0
External debt	$m	656	704	698	804	–
Debt service ratio	%	4.3	4.7	4.6	5.1	–
Exchange rate	gourdes/$	5.0	5.0	5.0	5.0	5.0

* estimated figure

COUNTRY PROFILE

Historical profile: First European sighting – Columbus 1492; settled by Spanish; island of Hispaniola divided between France and Spain 1697; slave rebellion 1791; slavery abolished 1793; François Toussaint (Toussaint Louverture) drove back Spanish, English, French forces – he was captured by French and died in French gaol but struggle continued under his generals; independence declared 1804 – first black republic in world – by Jean-Jacques Dessalines; various monarchical periods until 1859; occupied by USA 1915-34.

Political system: Under the constitution approved by referendum in March 1987, executive power is held by an elected president and cabinet of ministers. The bicameral National Assembly consists of a 77-member Chamber of Deputies and a 27-member Senate, both directly elected. However, the Assembly was dissolved and the constitution suspended after a military coup in June 1988. The constitution was largely restored in March 1989 and new elections promised at an unspecified date.

Main cities/towns: Port-au-Prince (population 1m), Cap Haitien, Gonaives, Les Cayes, Jérémie, Port-de-Paix, Jacmel, St Marc.

Media

Press: Dailies: *Le Matin, Le Nouvelliste, L'Union* (official). Weeklies: *Haiti Observateur, Haiti Progrés, Haiti en Marche, Le Petit Samedi Soir, Journal du Commerce.*

Radio: One government-owned station plus 13 commercial stations. *Television:* Government-owned television station and independent pay-cable station with four channels.

External trade: Large capital inflows (mainly aid from the USA and France) and workers' remittances only partly offset perennial trade deficit.

Exports: Manufactured goods (65% of total value), coffee (20%), cocoa, essential oils, mangoes (4%), sugar. Main destinations: USA/Puerto Rico (77% of total value), Italy, France, German FR, Belgium, Canada, Netherlands.

Imports: Machinery and transport equipment (20% of total value), basic manufactures (20%), food and live animals (25%), petroleum etc (10%-12%), chemicals (12%), animal/vegetable oils (12%). Main sources: USA/Puerto Rico (65% of total value), Japan, Canada, Netherlands, France.

Agriculture: Largest sector of economy, employing 70 per cent of the workforce but contributing only 30 per cent to GDP and 40 per cent of export earnings. Main cash crops are coffee, which accounts for around 30 per cent of export earnings, mangoes and sugar. Only 10 per cent of cultivation is carried out on large plantations and most cash crop farming is undertaken on rented plots by numerous smallholders. Subsistence farming and animal husbandry dominate, particularly cultivation of maize, rice, sorghum, beans, fruit and vegetables. Production is largely outside the cash economy. Recurrent drought, absence of irrigation, low producer prices and a weak infrastructure have kept production levels down and necessitated the import of more foodstuffs, particularly cereals. Estimated 47 per cent of total area cultivated land and 20 per cent pasture.

Industry: Sector contributes around 15 per cent to GDP and is concentrated in Port-au-Prince. Besides the traditional food processing, construction and textile industries, there are many US-owned light manufacturing and assembly plants. These offshore 'cheap labour' industries (mainly assembly of sports goods, toys, electrical components) have significantly contributed to foreign exchange earnings, but the sector's output fell during the 1980s because of political instability and under-developed infrastructure.

Mining: The mining and export of bauxite ceased in 1983 with the closure of Reynolds mine at Miragoane. Known, but commercially unviable, deposits of copper, silver, gold, marble, lignite and natural asphalt.

Energy: Heavy dependence on imported fuels. Electricity supply restricted to main towns.

Membership of international organisations: FAO, GATT, Economic Commission for Latin America, ILO, IMF, Inter-American Development Bank, Organisation of American States, UN, UNESCO, WHO, World Bank.

remind Avril of the limits to his power. By early 1989, several left-wing groups had split and had also broken with the exiled Stalinist figure Ben Dupuy, owner of the widely-read weekly, *Haiti-Progres*. An impressive new politician, Chavannes Jean-Baptiste, leader of the Papaye Peasant Movement in central Haiti, appeared on the scene, however, as grassroots organisations steadily grew despite repression. In 'the Republic of Port-au-Prince', as the politically isolated capital is sometimes dubbed, conservative ex-finance minister and former presidential candidate Marc Bazin, along with social democrat Serge Gilles, emerged as future political heavyweights. The two men formed a 'National Alliance for Democracy and Progress' in August 1989 in preparation for elections. At the same time, another group, with less support, including former Duvalierist Hubert de Ronceray, mulatto businessman Louis Dejoie and ex-agriculture minister Gerard Philippe-Auguste, formed a loose conservative coalition, as did a score of small leftist parties. Populist Christian Democratic leader Sylvio Claude, whose party was also split, and the Unified Communist Party remained outside the alliances.

Many doubted, however, that credible elections could be held amid the daily violence and intimidation, any more than they could in 1987. Avril himself remarked that given the country's extreme poverty, the elections might be free but they would not be honest. The torture and murder soon afterwards of three supporters of ex-president Leslie Manigat by regime thugs destroyed the remaining credibility of Avril, who a week later snubbed a new US ambassador who arrived with fighting talk about human rights. Poverty meanwhile continued to generate corruption and scandal. The president of the national bank, Ernest Ricot, was sacked in July 1989 in connection with a foreign currency and sugar trading racket. The government also came under fire for its concession of a large tract of North Eastern Haiti to a Dominican Republic company to set up a duty free zone. It denied it was negotiating with a US firm, Lund International, to lease the barren Ile de la Tortue and part of the nearby mainland to foreigners for 200 years as a virtually independent state. Despite protests, thousands of tonnes of US toxic waste remained on the shore near the town of Gonaives after being dumped there under a lucrative deal in 1988 with powerful Duvalierists. Avril announced plans for a $15m national housing fund to be financed by the private sector, tax breaks for those setting up social projects in poor areas and the leasing of 'all available' state lands to 'needy families' for nine years. But little action was subsequently seen. All the best land is already in the hands of the wealthy and the redistributed plots would not be allowed to be sold or passed on to descendants. The arrest and severe beating in November 1989 of three leftist leaders, including Evans Paul and trade unionist Jean-Auguste Mesyeux, for supposedly plotting to kill Avril and blow up the presidential palace added to pessimism.

Haiti's English-speaking Caribbean neighbours made a new attempt to prod the regime towards democracy by sending a four-man delegation of foreign ministers to the country in August 1989 at the invitation of Avril, who had stressed that isolated Haiti wanted to take part in regional activities. The mission offered technical help to stage elections. But the regional body, Caricom, at its summit a month earlier, once again brushed aside Haiti's long-standing membership application for fear its size and poverty would swamp the organisation.

According to Professor Leslie Manigat, whom the army overthrew in 1988 after five months and who has now resumed his teaching career in exile, Haiti is 'in danger of dying'. The country was becoming 'ungovernable', he said, bringing foreign intervention nearer. On a starker level, experts estimate that the country will be a virtual desert in 20 years time if nothing continues to be done about chronic deforestation and soil erosion. Local wood provides three-quarters of Haiti's energy supply.

BUSINESS GUIDE

Time: GMT -5.

Climate: Year round temperature in Port-au-Prince varies only slightly from 24°C to 27°C. The rainy season is from May-Nov.

Entry requirements
Passport: Required by all except nationals of USA and Canada with proof of identity.
Visa: Not required by nationals of UK and colonies, German FR, Canada, USA, Austria (subject to length of stay and/or onward passage being assured).
Currency: No local or foreign currency restrictions on import or export. Travellers' cheques and credit cards widely accepted. Difficult in banks and almost impossible in hotels and shops to exchange foreign currency other than US dollars.

Health precautions
Mandatory: Yellow fever vaccination required if arriving from an infected area.
Advisable: Yellow fever, typhoid, polio vaccinations. Malaria risk in low lying areas – prophylaxis essential. Water precautions should be taken.

Representation overseas: Bogotá, Bonn, Brasília, Brussels, Buenos Aires, Caracas, Dakar, Lima, London, Madrid, Mexico City, Nassau, Ottawa, Panama City, Paris, Rome, Santiago, Santo Domingo, Seoul, Tel Aviv, Tokyo, Vatican City, Washington.

Air access
Airlines serving: Air Canada, Air France, Air Jamaica, ALM, American Eastern, Haiti Trans Air, Pan Am, Surinam Airways.
International airport: Port-au-Prince-Mais Gaté (PAP), 15 km from city; duty-free shop, bar, buffet, bank, car hire.

Surface access
Main ports: Port-au-Prince, Cap Haitien, Gonaives.
Overland access: Possible from Dominican Republic – although sometimes bureaucratic delays.

Hotels: Five per cent tax and 10 per cent service charge usually added to bill.

Car hire: Cars can be hired in Port-au-Prince, at the airport and in Petionville. International licence required. Hire cars have registration numbers beginning with 'L'.

City transport
Taxis: Shared taxis (*publiques*) identified by red ribbon in window and registration number beginning 'P'. Large out of town taxis (*camionettes*) also available. Tipping not usual.

National transport
Air: Occasional internal flights operated by the government, using military aircraft. Otherwise other towns can be reached from Port-au-Prince by charter flights.
Road: Total network is around 4,000 km – not all passable/practicable in wet weather. Surfaced roads are from Port-au-Prince to Cap Haitien, Port-au-Prince to Jacmel and Port-au-Prince to Les Cayes.
Buses: Unscheduled services operate from Port-au-Prince to Les Cayes, Jacmel, Jérémie, Hinche, Port de Paix and Cap Haitien.

Public holidays: 1 Jan, 2 Jan (Heroes of Independence), Mardi Gras (Shrove Tuesday), Good Friday, 14 Apr *(Pan American Day), 1 May, 18 May (Flag Day), Ascension *, 15 Aug *(Assumption), 17 Oct *(Death of Dessalines), 24 Oct *(UN Day), 1 Nov *(All Saints', half day), 2 Nov *(All Souls'), 18 Nov (Army Day), 5 Dec (Discovery of Haiti), 25 Dec. Dates marked * are not official holidays.

Working hours
Business: 0800 – 1200, 1330 – 1700. (Exceptions May-Sep 0700 – 1200, 1330 – 1600; Sat-Sun, closed all year).
Government: 0800 – 1400.
Banking: (Mon-Fri) 0900 – 1300.

Telecommunications: International telephone and telex facilities available at main hotels and at offices of TELECO. Dialling code for Haiti: IDD access code + 509 followed by area code (Port-au-Prince 1) followed by subscriber's number.

Banking
Banking centre: Port-au-Prince.
Central bank: Banque Nationale de la République d'Haiti.
Commercial banks: Banque National de Crédit, Banque Industrielle et Commerciale d'Haiti, Banque de l'Union Haitienne, Banque Populaire Haitienne.
Foreign branches: Bank of Nova Scotia, Banque Nationale de Paris, First National City Bank of New York, First National Bank of Boston, Royal Bank of Canada.

Electricity supply: 110-220 V AC, 60 cycles.

Representation in capital: Argentina, Brazil, Canada, Chile, Colombia, Costa Rica, Dominican Republic, Ecuador, El Salvador, France, German FR, Honduras, Israel, Italy, Japan, Liberia, Mexico, Panama, Peru, Spain, Taiwan, UK, USA, Vatican City, Venezuela.

Hotels: Port-au-Prince
Castelhaiti, St Gerard, P.O. Box: 446 (tel: 20624; tx: 2030424).

Christopher, P.O. Box: 962 (tel: 21452; tx: 2030469).

El Rancho, route El Rancho, P.O. Box: 71 (tel: 72080/4; tx: 2030245).

Oloffsson, Avenue Christophe, P.O. Box: 1720 (tel: 20139; tx: 2030471).

Plaza Holiday Inn, rue Capois 10, P.O. Box: 1429 (tel: 23722/20821; tx: 2030356; fax: 20822).

Royal Haitian, Martissant 1 (tel: 40458; tx: 3420194).

Villa Creole (tel: 71570, 71571; tx: 3490319).

Business advice and information
Banque Nationale de la Républic d'Haiti, Angle Rue Ferou et Rue Américaine, Port-au-Prince (tel: 24142; tx: 2020026).

Chambre de Commerce et de l'Industrie d'Haiti, PO Box 982, Port-au-Prince (tel: 20281).

Commissariat à la Promotion Nationale et à l'Administration Publique, Palais des Ministères, Port-au-Prince (tel: 24148/2; 21027/2; tx: 0083).

Haitian International Business Center, 444 Brickell Avenue, Brickell Suite 650, Miami, Florida 33131, USA (tel: 374.8300; tx: 51936).

Office National du Tourisme d'Haiti, Avenue Marie Jeanne, Port-au-Prince (tel: 21720).

Office National pour la Promotion des Investissements, PO Box 2223, Port-au-Prince (tel: 22476, 23346; tx: 2030390).

Hawaii

T. R. Hughes

(map on page xxiv)

Although Hawaii is one of America's smallest states in both land area and population, its robust economy invites envious glances from the US mainland. The state exceeded the mainland national growth consistently over the past four years. Set in the Central Pacific 3,400 km south-west of San Francisco and so far away as to be almost a political nonentity, the 50th state of the USA has a service-based economy closely integrated with the mainland. But the markedly different lifestyle and culture give it a chance to play an important role as America's bridge to Asia and the Pacific islands.

Worrying to Hawaii residents in the current economic climate is the increasing influence of the Japanese who, while welcome as high-spending tourists, evoke anxiety as investors in real estate and business. Tourism and foreign investment leaped from 1985 onwards. A study by a Los Angeles accounting firm estimated, that $3.3bn out of a total of $12.7bn real estate in the United States bought by Japanese in 1987 was located in Hawaii.

While Japanese money has flowed beneficially into construction and other commercial ventures as well as into research and academic areas, helping to push up the growth rate in state revenues from an average of 6.8 per cent in 1983–86 to 11.2 per cent in 1987, there is apprehension over the growing Asian involvement – an issue in the November 1988 elections. The inflow of Japanese capital was blamed for forcing up the cost of real estate, hitting both ends of the population spectrum. Young people had difficulty finding affordable homes while pensioners worried about property taxes. Hawaiians certainly found some of the bigger speculative land deals unpalatable. Lurid allegations of racketeering, fraud, conspiracy and breach of contract emerged in a court hearing in 1989 involving the sale of a one-time sugar plantation to a Japanese firm. In another incident, a Japanese national bought an island in Kaneohe Bay only to be deported for falsifying documents to get into the United States.

Senator Rick Reed introduced a bill in the state legislature in 1989 that would prohibit non-residents owning residential land. This 'highlights the strong undercurrent of feeling among Hawaiians about the apparent loss of their homeland to foreigners' *(Pacific Islands Monthly)*. On the other hand some Japanese hotel owners have shown themselves model citizens, supporting local conservationists and voluntarily providing more affordable housing than demanded of developers.

Economic outlook

Although the economy was slowing in 1989 after outpacing the mainland for several years, there was confidence in the business community that the recent spate of investments and new projects soon to come on stream would keep things running strongly into 1990. The Bank of Hawaii predicted a 'fairly strong outlook' for 1989. But it forecast that the state's 10 per cent nominal economic growth of 1987 and 1988 would be cut back in 1989 to about 8 per cent to match the decline in the USA nationally. And there were also likely to be physical constraints in the state reining back the flourishing construction industry – there is a labour shortage – and a flattening out of the eastbound, mainly Japanese, tourist trade. Hawaii, however, is not likely to fall below the national rate of growth.

Tourism, which took off in the 1970s, is the mainspring behind a changing economic pattern. Federal defence spending, a big source of income and jobs in the immediate post-Second World War decades, declined from 13.7 per cent of the Gross State Product (GSP) in 1983 to 9.5 per cent in 1988 and was thought to have slipped behind the rapidly-rising and partly tourist-driven construction industry in 1989 in its economic importance. Agriculture's share was also down.

While the population rose by 6 per cent in the five years up to 1988, job opportunities went up twice as fast and the unemployment rate is one of the lowest in the nation at 3.2 per cent of the total labour force of 530,000 (compared to the national average of 5.6 per cent). Inflation, however, was up on the national average with the consumer price index at 5.5 per cent in 1988.

The Bank of Hawaii forecast a GSP of $23.4bn for 1989, up from $21.7bn in 1988 and $19.7bn in 1987. The tourist industry contributed a projected $8.5bn (but probably higher) in 1989 with some 6.3m visitors against $7.8bn in 1988 with 6.1m arrivals.

Federal government spending was put at $4.7bn in 1989 of which defence accounted for $2.2bn compared with $4.6bn in 1988 of which defence accounted for $2.1bn. The Bank forecasts completed construction work at $2.6bn in 1989, manufacturing industry at $2.3bn and agriculture at $260bn.

Hawaii's public infrastructure, however, is not keeping pace with the boom. Local prosperity has resulted in a big demand for cars and trucks, and traffic congestion, especially in Honolulu, is a problem. Apart from roads, there is need for better water supplies, more schools and educational institutions and public spaces.

While the number of westbound visitors, mainly those from the American mainland and Europe, was fairly static at about 4.2m over the past four years, the eastbound tourist trade jumped 25 per cent. Spending by the Asian and Japanese doubled from $1.6bn in 1986 to more than $3bn in 1988 thanks partly to the fall of the dollar against the yen and other foreign currencies. But Hawaii also benefited from various pluses: its own promotional efforts, a rise in Asian incomes with new emphasis on leisure and travel, and Tokyo's nudging of its own citizens to travel abroad as a means of redressing trade imbalances.

The enhanced purchasing power of the visitors was dramatic. The average Japanese tourist spent $400 a day, possibly more, in 1988 during a 5.8-day stay against $110 a day by the westbound visitor. But the average American stayed twice as long.

Hoteliers responded to the increased flow of visitors by putting up prices and running the risk of hurting Hawaii's image as a traditional tourist mass market – and also exposing the state to sharp competition from other tropical resorts such as South America and the Caribbean.

Tourism's success boosted the construction industry with a demand for new and renovated hotels and resort development. The industry was expected to double its output in the four years to the end of 1989, the most rapid expansion on record. Barring a change in the nation's fortunes, the industry seemed to be heading for new peaks. A shortage of skilled workers meant importing some from the US mainland.

Defence

Defence has long been an important component in the economy supporting more than 100,000 military personnel and their families and directly providing jobs for more than 20,000 Hawaiian civilians. These high-skill positions, worth an average of $30,000 a year, are half as much again as the average state wage. The defence sector grew substantially in the first half of the 1980s under President Reagan's build-up of national defence forces. Growth has slowed in recent years as the US Congress has come to grips with massive budget deficits, and with increased emphasis on labour-saving technology, the sector is unlikely to grow much in the 1990s.

Hawaii's diversified manufacturing industry lags behind the economy as it fights off low-cost imports of food and other products. The industry is too fragmented to take advantage of mass-production methods and labour costs are high.

Agriculture is Hawaii's fourth biggest primary industry after tourism, defence and manufacturing. Sugar production, a predominant agricultural activity for more than a century, is one of the most efficient in the world. Despite this, it remains vitally dependent on government subsidies. Pineapple production has shrunk considerably during the past 20 years. Like sugar, pineapple producers find it difficult to compete with cheaper imports.

For trade and other relations with Pacific states, Hawaii is a major entrepot, financial and service centre. It would like to expand to take in Pacific rim nations but lacks the marketing facilities and infrastructure of US mainland cities.

1989 election

True to political form, Hawaii voted for the Democratic candidate, Michael Dukakis, in the 1988 presidential election giving him, in fact, a bigger share of the vote – 56 per cent – than any of the mainland states. The election also saw the convincing return of Senator Spark M. Matsunaga, 72-year-old Japanese American Democrat, and Congressman Dan Akaka, whose district includes Honolulu's suburbs and outer islands.

The vote did not go all one way. Congresswoman

KEY FACTS

Official title: State of Hawaii	**Population:** 1,098,000 (1988)
Head of state: President of the USA	**GNP per capita:** $19,381 (1988)
Head of government: Governor John Waihee	**GNP real growth:** 3% (1987–88)
Capital: Honolulu, on Oahu	**Labour force:** 516,000 (civilian, 1988)
Official Languages: English and Hawaiian	**Unemployment:** 3.2% (civilian, 1988)
Currency: US Dollar ($ = 100 cents)	**Inflation:** 5.5% (1988)
Area: 16,641 sq km (8 major islands)	**Visitor numbers:** 6.1m (1988)

Patricia Saiki was re-elected from the staunchly-democratic urban Oahu district where, in 1986, she had defeated a democrat of Samoan ancestry in a bitter fight for an open seat to become the first Republican ever elected from Hawaii to the US House of Representatives.

In another key Republican victory, Frank F. Fasi was re-elected Mayor of Honolulu, America's biggest Pacific city. Unlike most US cities, a Honolulu city election is almost akin to a nation-wide contest since the electorate encompasses Oahu, the island on which 75 per cent of Hawaiians live. After many years as a maverick Democrat, Fasi scored as a Republican in 1984. In 1988, he seized on the rocketing real estate prices and, in a controversial campaign proposal, called for the state legislature to put a ceiling on foreign investment. Fasi's victory meant that he was well-placed to run for governor in 1990 when Governor John Waihee (Democrat) is expected to seek re-election. If Fasi decides not to stand, Republicans may well turn to Ms Saiki. Either way the Republicans will have a tough fight on their hands as Hawaii has unfailingly returned a Democrat to the gubernatorial mansion since achieving statehood in 1959.

COUNTRY PROFILE

Political system: The Hawaii State legislature is bicameral, with a 25-member Senate elected to a four-year term and a 51-member House of Representatives elected to a two-year term. Each of the major islands is a county. Oahu is known as the city and county of Honolulu; Molokai and Lanai are included in the county of Maui; Niihau is included in the county of Kauai; the island of Hawaii is known as Hawaii county; Kahoolawe is not inhabited. In addition to its two US senators, Hawaii elects two members of the US House of Representatives.

Political parties: Democratic Party (one seat in the US House of Representatives, following 1988 elections); Republican Party (one). The Democrats hold two seats in the US Senate.

Population: Hawaii has a very mixed racial structure. The original population was of Polynesian origin, but in later centuries Chinese, Japanese, Filipinos, Portuguese and Americans blended in and inter-married. Less than half the population are aged under 30. The capital, Honolulu, is a typically prosperous, modern high-rise American city, and tourism exerts a dominant influence on the state.

Language: English and Hawaiian are the official languages, English is universally spoken in business and political circles.

Media
Press: The main dailies read statewide are the afternoon *Honolulu Star-Bulletin* (circulation 97,950) and the morning *Honolulu Advertiser* (circulation 93,088). Other dailies include the *Hawaii Tribune-Herald* (Hilo), *Maui News West Hawaii Today* (Kailua-Kona) and one Japanese-English and one Chinese-English newspaper (Honolulu). *The Garden Island* is published five days a week on Kauai. Weeklies: weekly suburban newspaper. *Pacific Business News*, a weekly newspaper, and *Hawaii Business*, a monthly magazine, specialise in business news. *Honolulu Magazine* is a general interest monthly.
Radio: There are 47 radio stations: 27 AM, 18 commercial FM and two non-commercial FM, including public radio. *Television:* There are 16 commercial VHF or UHF TV stations and two non-commercial stations, including public TV. All carry news and other programmes of all four national US networks (three commercial and public TV). Cable TV had 265,029 subscribers, 195,438 on Oahu, in 1987.

Domestic economy: According to the US Bureau of Economic Analysis, Hawaii, per capita personal income in Hawaii is about 103 per cent that of total US. Gross State Product was estimated in 1988 at $21.3bn, and the labour force at 516,000. The economy depends on US federal expenditure, tourism, agriculture and construction. Land is at a premium, with 38 per cent government-owned and 24 per cent in the hands of the six largest private owners who are sometimes reluctant to sell their holdings. US military personnel and their dependents comprise about 11.4 per cent of the state's population. The rate of civilian unemployment in 1988 was less than the US national average at 3.2 per cent, with 25.3 per cent employed in services, including tourist-related business and industries. Agriculture accounts for 2.2 per cent of employment (1987), and sugar and pineapples dominate the export economy; sugar exports in 1987 were $354m, while pineapple receipts totalled $251m. Nowadays there is pressure to diversify into other crops and livestock. Papaya, coffee, bananas, passion fruit, oranges and guavas are grown, and macadamia nut production is expanding rapidly. Flowers and horticultural products earned $56m in 1987. Tourism receipts rose as visitors totalled 5,799,800 in 1987, but manufacturing remained a minor contributor to the economy as international commodity and textile markets became tougher.

External trade: Honolulu is one of the US foreign trade zones where imported goods may be stored, sold wholesale, exhibited and otherwise manipulated or manufactured without being subject to customs duty until withdrawn from the zone into US customs territory.

BUSINESS GUIDE

Time: GMT -10 hrs, EST -5 hrs, EDT -6 hrs.

Climate: Temperatures average 24°C year round. The winter months are slightly cooler; it is warmest Jun-Sep. Rainfall is variable, but occurs mostly Nov-Mar; average annual rainfall in Honolulu is 63.5cm.

Air access
National airlines: Air America, Air New Zealand, Air Tungaru, Aloha, Aloha Islandair, American, American Trans Air, Canadian, China, Continental, Delta, Garuda, Hawaiian, JAL, Korean Air, MAS, Northwest, Pan Am, Philippine, Qantas, Singapore, TWA, United.
International airport: Honolulu International (HNL), 10 km from Waikiki; duty-free shop, car, restaurant, buffet, bank, hotel reservations, post office, shops, car hire; General Lyman (ITO), 3 km from Hilo on Hawaii; Kahului (OGG), 8 km from Kahului on Maui; Keahole (KOA), 13 km from Kailau-Kona on Hawaii.

Surface access
Main ports: Hilo, Honolulu, Kahului, Nawiliwili, Kawaihae. Considerable automation and containerisation. The Foreign Trade Zone port at Honolulu has extensive entrep\04t facilities.

Hotels: Hotels are numerous in Waikiki, a section of Honolulu, and all the main hotel chains represented. Accommodation is of a generally high standard. Book well in advance if travelling from Christmas-Mar or Jun-Aug.

Credit cards: All major cards are accepted.

Car hire: US, Canadian or international driving licence required.

City transport
Taxis: It is technically illegal for taxis to cruise for business or pick up passengers on the street.

National transport
Air: Inter-city and inter-island services are operated by Air Molokai-Tropic Airlines, Aloha Airlines, Aloha Islandair, Hawaiian Airlines, Panorama Air, United Airlines. There are also air taxi services.
Road: The 6,400 km state road network includes freeways, highways and streets.
Buses: Hawaii and Oahu are well served by buses, which are considered inexpensive and convenient.

Public holidays: In addition to US holidays, the following holidays are observed: 26 Mar (Kuhio Day), Good Friday, 11 Jun (Kamehameha Day), Admission Day (3rd Fri in Aug).

Working hours
Government and business: (Mon-Fri) 0745 – 1700.
Banking: (Mon-Thu) 0830 – 1500; (Fri) 0830 – 1800.
Shops: (Mon-Sat) 0900 – 2100 (not all shops, but many); (Sun) 0830 – 1800.

Telecommunications
Telephone: Dialling code for Hawaii, IDD access code + 1 followed by area code (808) followed by subscriber's number. Telephone rates are higher than in the continental US.
Telex: Telex facilities are available through ITT World Communications Inc.

Banking
Head offices: Bank of Hawaii, Bank of Honolulu NA, Bank of Maui NA, Central Pacific Bank, City Bank of Honolulu, First Hawaiian Bank, First Interstate Bank, Hawaii National Bank, Liberty Bank, First Nationwide Bank (being merged).
Hawaii branches: Bank of Tokyo.

Useful telephone numbers
Emergency Services: 911

Operator assistance: 01

Hotel access code: 8

Directory (local calls): 1411

Long distance calls: (area code) + 555.1212

International calls: 0 (for Operator).

Car hire: Honolulu
Avis: 417 Lele Street (tel: 8345524; cable: AVIS CARS HAWAII).

Budget:P.O.Box 15188, Honolulu H1, 96830 (tel: 922-7221)

Dollar Rent-A-Car: 1600 Kapiolani Blvd (tel: 926-4200)

National: 1778 Ala Moana (tel:922-3331). 3049 Ualena Street (tel: 833-3355).

Taxis: Honolulu
Aloha State Taxi (tel: 8473566).

Charley's (tel: 5311333).

SIDA (tel: 8360011).

Business advice and information
Administrator, Foreign-Trade Zone No. 9, Pier 2, 521 Ala Moana, Honolulu 96813 (tel: 5485435).

Business Development and Marketing Division, Department of Business and Economic Development, 250 King Street, PO Box 2359, Honolulu 96804 (tel: 5484621).

Chamber of Commerce of Hawaii, 735 Bishop Street, Honolulu 96813 (tel: 5314111).

City and County of Honolulu Department of Information and Complaints, City Hall, Honolulu 96813 (tel: 5234381).

State of Hawaii Department of Commerce and Consumer Affairs, PO Box 3469, Honolulu 96801.

State of Hawaii Department of Business and Economic Development, Kamamalu Building, 250 South King Street, Honolulu, PO Box 2359, Honolulu 96804 (tx: 7430250).

BUSINESS GUIDE

Travel information
Hawaii Visitors Bureau, 2270 Kalakaua Avenue, Suite 801, Honolulu 96815 (PO Box 8527, Honolulu 96830) (tel: 9231811; fx: 9228991).

Honolulu Department of Transport, Mass Transit Lines, 725 Kapiolani Boulevard, Honolulu (tel: 5311611).

BUSINESS DIRECTORY

Hotels: Honolulu
Ala Wai Terrace, 1547 Ala Wai Boulevard, 96815-1099 (tel: 949-7384; tx: 634178).

Best Western Waikiki Plaza, 2045 Kalakaua Avenue, 96815 (tel: 955-6363; tx: 7238445; fax: 9493413).

Continental Surf, 2426 Kuhio Avenue, 96815 (tel: 922-2755; tx: 7238130; fax: 9242483).

Coral Beach Resort, 2310 Kuhio Avenue, 96815 (tel: 922-3368; tx: 50149).

Coral Seas, 250 Lewars Street, 96815-1981 (tel: 923-3881; tx: 634178).

Halekulani, 2199 Kalia Road, Waikiki Beach, 96815 (tel: 923-2311; tx: 8382; fax: 926-8004).

Hawaiian King, 417 Nohonani Street, 96815 (tel: 922-3894; tx: 7430546).

Hawaiian Monarch, 444 Niu Street, 96815 (tel: 949-3911; tx: 634479).

Hawaiian Regent, 2552 Kalakaua Avenue, 96815 (tel: 9226611; tx: 8700; fax: 922-2852).

Hilton Hawaiian Village, 2005 Kalia Road, 96815 (tel: 9494321; tx: 8380).

Holiday Inn Honolulu Airport, 3401 Nimitz Highway, 96819 (tel: 836-0661).

Holiday Inn Waikiki Beach, 2570 Kalakaua Avenue, Hi 96815 (tel: 9222511; tx: 634226; fax: 9233656).

Hyatt Regency Waikiki, 2424 Kalakaua Avenue, Hi 96815 (tel: 9231234; tx: 1027238278; fax: 9237839).

Inn On The Park, 1920 Ala Moana Boulevard, 96815 (tel: 946-8355; tx: 323170).

Island Colony, 445 Seaside Avenue, 96815 (tel: 923-2345; tx: 634479).

Kahala Hilton, 50000 Kahala Avenue, 96816-5498 (tel: 734-2211; tx: 397148; fax: 7372478).

Monte Vista Condo, 320 Liliuokalani Avenue, 96815 (tel: 923-4766; tx: 181662).

New Otani Kaimana Beach, 2863 Kalakaua Avenue, 96815 (tel: 923-1555; tx: 07430470).

Outrigger Hobron, 340 Hobron Lane, 96815-1039 (tel: 942-7777; tx: 634178).

Outrigger Surf, 2280 Kuhio Avenue, 96815-2699 (tel: 922-5777; tx: 634178).

Outrigger Waikiki, 2335 Kalakaua Avenue, 96815 (tel: 923-0711; tx: 634178).

Pacific Marina Inn, 2628 Wai Wai Loop, 96819 (tel: 836-1131).

Pacific Monarch, 142 Ulunui Avenue, 96815 (tel: 923-9805; tx: 64135).

Prince Kuhio, 2500 Kuhio Avenue, 96815-3696 (tel: 922-0811; tx: 634178).

Princess Kaiulani, 120 Kaiulani Avenue, 96815 (tel: 922-5811; tx: 7430073; fax: 9239912).

Reef, 2169 Kalia Road, 96815-1989 (tel: 923-3111; tx: 634178).

Royal Hawaiian, 2259 Kalakaua Avenue, 96815 (tel: 923-7311; tx: 7431240; fax: 924-7098).

Royal Islander Club, 2164 Kalia Road, Waikiki Beach, 96815 (tel: 922-1961; tx: 634178).

Seaside Surf, 440 Seaside Avenue, 96815-2627 (tel: 922-2383; tx: 634178).

Sheraton Waikiki, 2255 Kalakaua Avenue, Hi 96815 (tel: 9224422; tx: 7430115; fax: 923-8785).

Waikiki Beach Tower, 2470 Kalakaua Avenue, 96815 (tel: 926-6400; tx: 634479).

Waikiki Gateway, 2070 Kalakaua Avenue, 96815 (tel: 955-3741; tx: 634479).

Waikiki Holiday, 450 Lewers Street, 96815 (tel: 923-0245).

Waikiki Marina, 1956 Ala Moana Boulevard, 96815 (tel: 367-6070; fax: 9490996).

Waikiki Pacific Isle, 1850 Ala Moana Boulevard, 96815 (tel: 955-1576).

Waikiki Sand Villa, 2375 Ala Wai Boulevard, 96815 (tel: 922-4744; tx: 8213).

Waikiki Sunset, 229 Paoakalana Avenue, 96815 (tel: 922-0511; tx: 634479).

Waikiki Surf, 2200 Kuhio Avenue, 96815-2899 (tel: 923-7671; tx: 634178).

Waikiki Surfside, 2452 Kalakaua Avenue, 96815 (tel: 923-0266; tx: 634479).

Waikiki Village, 240 Lewers Road, 96815-1984 (tel: 923-3881; tx: 634178).

Waikikian On The Beach, 1811 Ala Moana Boulevard, 96815 (tel: 949-5331; tx: 634479).

Hotels: Kahului (Maui)
Maui Hukilau, 96732 (tel: 877-3311).

Maui Palms, Kaahumanu Avenue, 96732 (tel: 877-0071; tx: 7238582; fax: 5330472).

Hotels: Kihei (Maui)
Haleakala Shores Condos, 2619 South Kihei Road, 96753 (tel: 879-1218; tx: 324562).

Kamaole Beach Royale, 2385 South Kihei Road, P.O. Box: 370, 96753 (tel: 879-3131; tx: 0416).

Kauhale Makai Village By The Sea, 938 South Kihei Road, 96753 (tel: 879-8888).

Kihei Bay Surf Condos, 775 South Kihei Road, 96753 (tel: 244-9551).

Maui Beach, 612-1993 South Kihei Road, 96753 (tel: 879-5455).

Maui Lu Resort, 575 South Keihei Road, 96753 (tel: 879-5881; tx: 634479).

Maui Sunset, 1032 South Kihei Road, 96753 (tel: 879-0674; tx: 677333).

Sugar Beach Resort Condos, 2439 South Kihei Road, Flat 205a, 96753 (tel: 879-7765; tx: 8413; fax: 8797825).

Surf & Sand, 2980 South Kihei Road, 96753 (tel: 879-7744; fax: 0145).

Hotels: Lahaina (Maui)
Hale Ono Loa, 3823 Honoapiilani Road, 96761 (tel: 669-6362).

Kaanapali Beach, 2525 Kaanapali Parkway, 96761-1987 (tel: 661-0011; tx: 405184; fax: 6675978).

Kaanapali Plantation Resort, 150 Puukolii Road, 96761 (tel: 661-4446).

Kahana Villa, 4242 Lower Honoapiilani Highway, P.O. Box: 327, 96761 (tel: 669-5613; tx: 634135).

Kahana Village Condos, 4531 Honoapiilani Road, 96761 (tel: 669-5111; tx: 7430416).

Maui Islander, 660 Wainee Street, 96761 (tel: 667-9766; tx: 5106000517; fax: 6613733).

Paki Maui, 3615 L Honoapiilani Highway, 96761 (tel: 669-8235; tx: 634479).

Sands of Kahana, 4299 Honoapiilani Highway, 96761 (tel: 669-0400; tx: 634479).

Hotels: Maui
Hyatt Regency Maui, 200 Nohea Kai Drive, Lahaina, Hi 96761 (tel: 6611234; tx: 7086676030).

International Colony Club, 2750 Kalapu Drive, 96761 (tel: 661-4070).

Kaanapali Royal, Kaanapali Parkway & 2560 Kekaa Drive 96761 (tx: 8413).

Kaanapali Shores, 3445 Honoapilani Highway, 96761 (tel: 667-2211; tx: 634479).

Mahana, 110 Kaanapala Shores Place, 96761 (tel: 661-8751; tx: 634479).

Maui Eldorado, 2661 Kekaa Drive, 96761 (tel: 661-0021; tx: 7238830).

Maui Kaanapala Villas, 2805 Honoapiilani Highway, 96761 (tel: 667-7791; tx: 634479).

Maui Marriott, 100 Nohea Kai Drive, 96761 (tel: 667-1200; tx: 395114).

Papakea Beach Resort, 3543 Honoapiilani Highway, 96761 (tel: 669-4848; tx: 6695497; fax: 6614683).

Sheraton Maui, 2605 Kaanapali Parkway, 96761 (tel: 6610031; tx: 7431988; fax: 661-0458).

The Whaler On Kaanapali Beach, 2481 Kaanapali Parkway, 96761 (tel: 661-4861; tx: 7051394; fax: 6618315).

Hotels: Napili Beach (Maui)
Coconut Inn, Maui (tel: 669-5712; tx: 7430416).

Napili Bay Resort, 33 Hui Road, 96761 (tel: 669-6044).

Napili Kai Beach Club, 5900 Honoapiilani Road, Lahaina, 96761 (tel: 669-6271).

Napili Point, 5295 Honoapiilani Highway, 96761 (tel: 669-9222; tx: 634479).

Napili Shores, 5315 Honoapiilani Highway, 96761 (tel: 669-8061; tx: 634135).

Napili Surf & Puamala Condos, 50 Napili Place, 96761 (tel: 669-8002).

Hotels: Wailea (Maui)
Maui Inter-Continental Wailea, 3700 & Wailea Alanui, P.O. Box: 779, 96753 (tel: 879-1922; tx: 397306; fax: 8797658).

Stouffer Wailea Beach Resort, 3550 Wailea Alanui Drive, 96753 (tel: 879-4900; tx: 501025; fax: 8796128).

Wailea Elua Condos, 3600 Alanui Drive, 96753 (tel: 879-1595; tx: 67733).

Wailea Villas, 3750 Wailea Alanui, 96753 (tel: 879-1595; fax: 8743554).

Honduras

John Burstein

Honduras remains an island of political calm compared to its neighbours both to the east and south, despite its endemic poverty, which is considered to be among the worst in the hemisphere. Civilian government has persisted for eight years and has shown agility in its response to political crises in the recent past, including acts of political extremism from both the left and the right. Yet the government remains under the commanding influence of the military. Treated preferentially for strategic reasons by the United States, Honduras has withstood a region-wide economic collapse by receiving huge inflows of US aid. In exchange it has offered a base for both US and Nicaraguan contra troops.

Economic austerity measures, long overdue and a subject of great popular opposition, are likely to be phased in both before and following the inauguration of a new president and congress in January 1990.

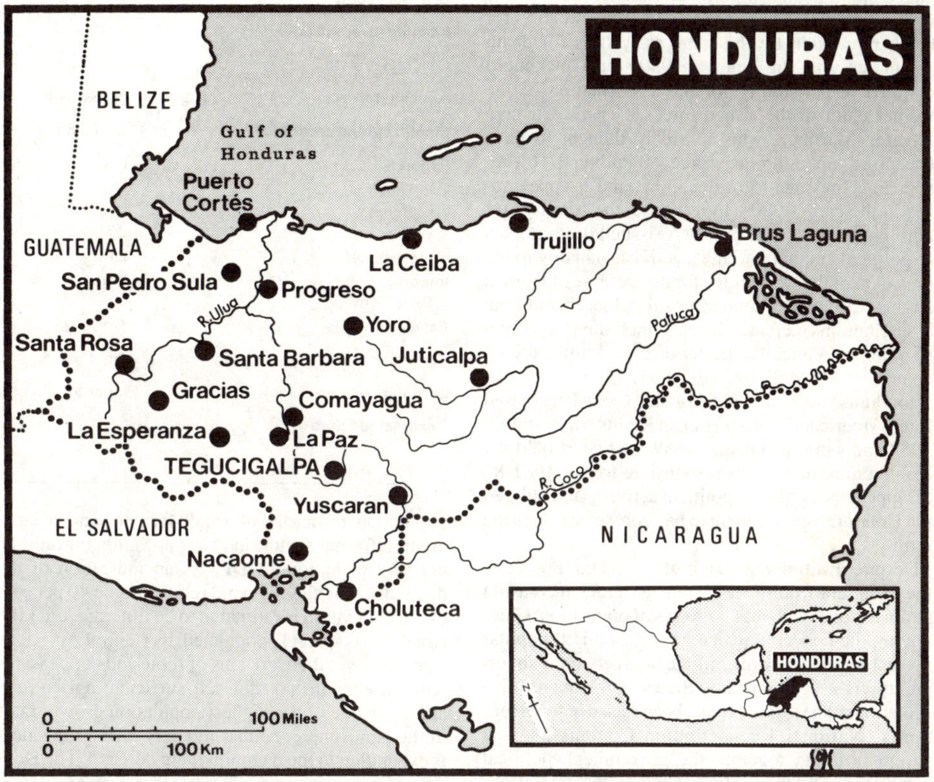

Political renewal

Honduran general elections on 26 November 1989 brought to power the country's third civilian government since the period of nearly uninterrupted military rule from 1963 to 1980. The new president was Rafael Callejas, the candidate of the Partido Nacional. Recent history suggests that the nation's vying political interests are still actively struggling to construct viable democratic institutions.

In 1980, elections were held for a National Constitutional Convention, followed by general elections and the inauguration in 1981 of Liberal party candidate Roberto Suazo Cordova. At the end of his term in 1985, President Suazo created a constitutional crisis by making a last-ditch effort to remain in the presidency or, failing that, to impose his own candidate on his Liberal party. Spurred by public demonstrations against Suazo's *continuista* manoeuvres, however, the chief of the armed forces, with the support of the Catholic Church and the US embassy, brokered a novel polling format which permitted various presidential candidates to run for the same party.

As a result of this unusual single-round voting system, José Azcona Hoyo assumed the presidency in 1986 with only 27 per cent of the vote. Azcona, a liberal, was obliged to cede important government ministries to the major opposing party in a National Unity Pact (PUN) negotiated with top-polling National party candidate Rafael Leonardo Callejas. These ingenious political accommodations have underlined the importance of a civilian head of state for maintaining close relations with the United States, Honduras' major military ally and economic support.

Four political parties are active at the national level, but the Liberal and the National parties – both dating from the turn of the century – are the only ones with mass bases, and the only ones seriously vying for the presidency and control of congress. The general elections will determine the presidency and three designates (vice-presidents), the 132-member national congress and 286 municipal offices.

Both major parties are notable for being rather loose confederations of the factions that make them up. However, the National Party – historically linked to the armed forces and considered to have a more traditional *caudillista* organisation – has increasingly coalesced around the presidential candidacy of Callejas. He is a 47-year-old businessman and economist of conservative bent, who held a ministerial post under the military regime.

Meanwhile, the Liberal party bosses held a hotly contested primary in which it was feared that a left-wing faction, led by Jorge Arturo Reina, might leave. But all united behind Carlos Roberto Flores Facusse, a member of congress and publisher of the newspaper *La Tribuna*. In fact, there appears to be little difference on issues of the economy and conduct of foreign affairs between Flores and Callejas.

The Innovation and Unity party (PINU) and the Christian Democrats – both of social democratic ideology – will also field candidates but, together, these newer parties have drawn less than 1 per cent of the vote in recent elections. Nevertheless, both provide a platform for centre-left political currents, and Christian Democratic presidential candidate (and member of congress) Efrain Diaz Arriviaga has been an effective internationally-recognised dissident voice.

Parties on the far-left of the Honduran political spectrum include the Communist party of Honduras and the Honduran Patriotic Front. Left-wing activists, however, have largely eschewed electoral politics in favour of organising the labour, peasant and student sectors, with increasing success in recent years. For example, there were over 300 work stoppages during 1988 and land occupations are common.

Though considered a step toward democratic institution-building, the 1989 electoral campaign has been marred by persistent rumours of electoral fraud perpetrated by both the Liberals and Nationals through the illegal registry of over 100,000 foreign residents – Nicaraguan refugees, in their majority.

Honduras' praetorian-style armed forces – still in part responsive to the populist ideology of their 1972-80 government – has also expressed concern for the smooth functioning of the electoral process. In June 1989, a declaration from the Superior Council of the Armed Forces (COSUFFAA), the military 'parliament' of 60 high-ranking officers, suggested General Humberto Regalado Hernandez would seek a new five-year term as military chief of staff, in part to

KEY FACTS

Official title: Republic of Honduras

Head of state: President Rafael Leonardo Callejas

Ruling party: Partido Nacional

Capital: Tegucigalpa

Official Languages: Spanish

Currency: Lempira (L) = 100 centavos

Exchange rate: L2.0 per $ (Nov 1989)

Area: 112,088 sq km

Population: 4.8m (1988)

GDP per capita: $851 (1987)

GDP real growth: 3.0% (1988)

Inflation: 4.6% (1988)

Trade balance: –$23.7m (1988)

Foreign debt: $3.3bn (1988)

facilitate a smooth change of government. This would require a significant modification of the Honduran constitution, which has prohibited re-election of the head of the armed forces.

A miniscule guerrilla movement centred around the Cinchonero People's Liberation Movement has engaged sporadically in political violence through the 1980s and this, in turn, has served as justification for harsh repression by the state, as well as for death squad activity. In January 1989, the Cinchonero claimed responsibility for the assassination of the former chief of the armed forces, General Gustavo Alvarez Martinez, who is widely thought to have organised a paramilitary force, 'Batallion 3-16', responsible for some 150 kidnappings and disappearances in the early 1980s. In response, the Alliance for Anti-communist Action (AAA) issued death threats to political and human rights activists, notably to the president of the Committee for the Defense of Human Rights (CODEH), Ramon Custodio. Indeed, political assassinations increased in 1989, including the killing of Salomon Vallecillo, leader of a left-leaning tobacco union.

Honduras was humbled before the Costa Rica-based Inter-American Court of Human Rights when, in July 1988 and again in January 1989, that court held the Honduran armed forces responsible for the 1981-82 disappearances of four political activists and obliged the Honduran government to pay compensation to the families.

Despite institutionalisation of its electoral process, the Honduran political system is under increasing strain to cope with popular forms of opposition. Chief among the concerns which have galvanised popular protest are: land reform, military and economic sovereignty vis-a-vis the United States, and human rights violations. These issues have been raised before Honduras' National Reconciliation Commission – a forum bringing together the government, political opposition, and the Catholic Church, that was established as a result of the 1987 Central American peace plan – has had some success in bringing public attention to acts of political intimidation and human rights violations.

Public discontent reached boiling point in April 1988 when a demonstration of several thousand got out of hand, leading to the burning of a section of the US embassy and the death of at least two Hondurans, prompting the government to briefly impose a state of emergency.

This violent street confrontation was sparked by a highly unusual *cause célèbre*. On April 5, Juan Ramon Matta Ballesteros, a Honduran sought by the United States on charges of drug trafficking, was abducted from his Tegucigalpa home by Honduran *Cobra* special forces and put on a plane to the United States. Not only is extradition of nationals not permitted by the Honduran constitution, but Matta had acquired both a popular following and the support of various rightist and other student and union organisations by spreading some of his personal wealth – estimated at over $1bn – amongst groups and individuals throughout the country. The popular outbreak of violence was widely perceived as an expression of generalised frustration with US infringement on Honduran sovereignty.

At the time of Matta's arrest, public accusations – never receiving investigation – suggested that at least five high-level military officers were involved in drug trafficking and that Honduras, with some 200 airstrips scattered throughout the country, and 350 miles of underpopulated Caribbean coastline, had become a major transit point for drug smugglers.

Economic situation

The primary challenge of the Honduran economy lies in creating jobs and increasing production with a

KEY INDICATORS

	Unit	1984	1985	1986	1987	1988
Population	m	4.2	4.4	4.5	4.7	4.8
GDP	L m	6,462	6,977	7,500	8,109	8,799
GDP per capita	L	1,538	1,586	1,667	1,740	1,833
Inflation (average rate)	%	4.7	3.4	4.4	2.4	4.6
GDP real growth	%	2.8	3.2	3.1	4.9	4.0
Agricultural sector	% real gwth	1.3	2.9	2.4	6.7	2.5
Manufacturing section	% real gwth	8.4	–2.2	2.7	8.4	4.9
Exports	Lm	1451	1529	1709	1616	–
Bananas	Lm	465	547	514	644	–
Coffee	Lm	338	370	644	400	–
Wood	Lm	70	68	65	70	–
Frozen beef	Lm	42	36	40	45	–
Imports	Lm	1621	1607	1584	1627	–
Petroleum	Lm	359	317	194	241	–
Balance of trade	$m	–147.8	–89.6	17.2	–49.5	–23.7
Current account	$m	–374.3	–292.6	–255.2	–306.9	–321.8
Exchange rate (average)	L per $	2.0	2.0	2.0	2.0	2.0

* estimated figure

population primarily of small farmers and a large urban informal sector. Industry accounts for only 15 per cent of national GDP. As an indication of the difficulty in achieving rapid reform, some 40 per cent of the population is estimated to be illiterate. Luckily, Honduras is sparcely populated, by Central American standards, yet there are tens of thousands of peasants without enough land for self-sufficient agriculture, and, with one of the highest population growth rates in the hemisphere, neither industry nor agribusiness is expanding to meet employment demand. The Honduran finance ministry estimates that there are 30,000 additional people seeking to join the workforce each year, only 4,000 of whom find jobs.

During the 1970s the Honduran economy grew at a 5-6 per cent annual rate due to domestic industrialisation spurred by foreign investment and participation in regional trade (though Honduras formally withdrew from the Central American Common Market after the four-day 'Soccer War' with El Salvador in 1969). With the petroleum cartel and the regionally-destabilising Sandinista revolution in Nicaragua, both in 1979, Honduras' growth rate declined dramatically, and by 1983 became negative. Through massive infusions of US aid – which totalled approximately $300m in 1988 – Honduras has managed to ward off major economic recession. But compounding budget deficits and negative trade balances have kept foreign exchange supply precarious and led to arrears in Honduras' debt service, which has doubled since 1981.

Honduras' export economy has long been founded on bananas, coffee and beef, sold to the United States. Other exports of note are wood, shrimp, lobster and sugar. Spurred in part by the duty reductions of the now permanently-established Caribbean Basin Initiative (CBI) of the US government, Honduras has struggled with some success to compete in both the US and extra-hemispheric markets with non-traditional products (e.g. fruits, vegetables and flowers) as well as with goods from free-zone assembly plants and light industry. Coffee prices tumbled in 1989 with the collapse of the quota system of the London-based International Coffee Organisation, creating a negative shock to the Honduran economy.

But the graver crisis for Honduras has been in the area of international finance. Unable through mid-1989 to reach agreement with the IMF and its international creditors on structural readjustment measures, Honduras fell into arrears on debt servicing – becoming one of only nine countries in the world on the World Bank's blacklist – and watched all balance of payments support dry up. A $25m World Bank loan tranche expected in December 1988 was suspended.

Remarkably, even the US Agency for International Development (AID) froze first-semester 1989 payments due to the Honduran government's refusal to implement liberalising reform. Because of its strategic importance for US foreign policy – both as a base for US military forces in the region as well as for the US-supported Nicaraguan counterrevolutionary troops – Honduras had received massive US economic and military aid irrespective of its economic policy.

Devaluation of the lempira has been a long-standing precondition for international financial support. It is the only Central American currency to maintain, officially, its pre-1980s exchange rate against the US dollar (two lempiras to the dollar) and, thus, has come to symbolise Honduran economic strength and even sovereignty. Economists estimate, however, that a free-floating lempira would stabilise at approximately five lempiras to the dollar.

Even the business community, represented by its umbrella organisation, the Honduran Council of Private Enterprise (COHEP), suggested in 1989 that, rather than devalue, the government – including the armed forces – should accept austerity measures. But Azcona, on this point, has said, 'This is a well-organized country...and organized groups will not allow mass dismissals.'

A joint IMF/World Bank mission – echoing earlier US AID recommendations – proposed in March 1989 a 50 per cent devaluation of the lempira through official recognition of the parallel exchange rate, as well as implementation of income tax reform, a salary freeze, significant reduction of the budget deficit, and the selling off of state enterprises.

While business and financial leaders, based primarily in the northern city of San Pedro Sula, have criticised the Azcona government for insufficiently executing the full gamut of neo-liberal reforms urged by US and multilateral advisers, Honduras' well-organised popular sector has attacked the government for its failure to enforce land reform and for generally falling beholden to foreign capitalist interests. Land takeovers increased markedly during the last years of the Azcona government, and the National Agrarian Institute (INA) is a prime focus of criticism by the

National Confederation of Farm Workers (CNTC), the National Association of Honduran Peasants (ANAH) and other peasant groups for the slow pace at which Honduras' 26-year-old agrarian reform is being executed.

Nevertheless, in 1988 Honduras realised 3 per cent growth in GDP (down from 4 per cent in 1987), against population growth of some 3.4 per cent. Independent economists suggest that inflation in 1988 rose as high as 9 per cent, although official figures register 4 per cent, against 2.5 per cent in 1987.

Saddled with a total external debt of over $3.5bn, Honduras is asked to pay nearly $1m a day in its servicing. Thus fully 31 per cent of the 1988 federal budget was set aside for domestic and foreign debt service.

An economic recovery plan in 1989-90 is certain to include cuts in government spending and selling off of some state enterprises owned by the National Investment Corporation (CONADI). In the past year, parastatal companies, such as lumber mills, have already been sold – frequently bought by US businesses which had acquired greatly discounted Honduran paper on the secondary market. But electrical power, water, telephones and forests – placed in government hands during the populist military regime of the 1970s – are likely to continue under the management of government agencies.

Relations with USA

Honduras and the United States are entwined by considerable mutual necessity. Honduras is dependent on US aid dollars to maintain its economy, and the United States relies on Honduras to carry out its chief policy initiative in the western hemisphere: opposition to the leftist Sandinista government of Nicaragua.

The growing military co-ordination between the USA and Honduras is one of the central facts of life in Honduras today. There are now airstrips around the country capable of receiving C-130 cargo planes, a radar station, fuel and munitions depots and new roads and medical facilities. A permanent force of some 1-1,500 US troops is located at the US base of operations, the Soto Cano (formerly, Palmerola) air base, near the town of Comayagua. Nearly constant military manoeuvres and construction projects by the US military, including reserves and national guard, since 1983 have sent a strong message to Nicaragua of US resolve to change the Sandinista government.

With increasing tensions between the United States and Panama, rumours have persisted that the US Southern Command may relocate some of its military and intelligence-gathering installations from the Canal Zone to Soto Cano air base in the 1990s.

Despite considerable domestic opposition from human rights and other popular organisations, the Honduran government has proceeded to negotiate with the United States a 'Third Protocol', significantly modifying a 1954 accord on military co-operation between the two countries. A second protocol has already authorised joint military exercises. The third protocol would give the United States armed forces permission to build permanent structures in Honduras and open further US rights to send ships and aircraft into the country. In addition, a series of agreements on drug trafficking contemplate a new radar system in Honduras, the right being given to the US authorities to board Honduran ships on the high seas and joint narcotics operations.

Honduras also has provided the crucial base from which some 6-8,000 US-supported Nicaraguan Resistance rebels, or contras, have operated during their eight-year war against the Nicaraguan government. The contras have caused the dislocation of some 2,000 Hondurans and they have been the focus of critical attention of pro-sovereignty groups throughout the country. However, since the contras entered into a ceasefire with the Sandinista-controlled government in Managua in 1988 and the Bush administration at least temporarily foreclosed further use of the proxy force, the contra presence, and that of some 100,000 additional Nicaraguan refugees along Honduras' southern border, have turned into irritants in the US-Honduran relationship. Washington acknowledges its responsibility to help Honduras with the influx of foreigners but so far has done no more than maintain the anti-government Nicaraguans in Honduran camps.

The US embassy, sometimes irreverently referred to as the 'shadow government' because of its influence on government policy, will manage an aid package in 1989 of some $207m. Development assistance accounted for $59m; military aid totalled $61.m; and economic support funds (ESF) – given for balance of payments, and frequently classed as a form of security assistance – $87m.

International relations

Honduras has paid a considerable price in its international relations by facilitating the struggle of the Nicaraguan contras. The Central American peace plan, signed by Azcona and the other four presidents of the region in Guatemala on 7 August 1987, included a commitment that no state would allow its territory to be used for the overthrow of another state. Nicaragua has repeatedly denounced Honduras on this point, called for UN Security Council investigation, and even taken the issue to the World Court at the Hague. It did appear in early 1989 that Azcona would join the other Central American presidents in accepting a UN peacekeeping force to facilitate disarming the contras. But by mid-year Tegucigalpa had dropped all co-operative action with its neighbour to the south.

The Honduran military has quietly provided support for the contras in various ways. Its most general form of co-operation has been to cordon off the areas in which the contras function from the press or other civilians. The army's southern battalion has also been known to function in tandem with the contras, principally by holding contra prisoners and arms. The Aguacate army base has also served as a contra medical facility. Presented with the mutiny by various contra leaders against their commander Enrique Bermudez in April 1988, the Honduran military arrested and deported to the United States the rebel leaders, in what was a rare glimpse of the interrelationship between the US, Honduran and contra military commands.

Honduras thus contains a sort of tripwire. Military strategists have long speculated that, should Washington ever embark on direct military intervention in

COUNTRY PROFILE

Political system: Executive power is vested in the president, elected for a four-year term and assisted by a Cabinet. Some legislative functions are vested in the 134-member National Congress.

Main cities/towns: Tegucigalpa (population 620,000), San Pedro Sula, La Ceiba, Puerto Cortés, Tela.

Media
Press: Several newspapers are available, including *El Heraldo* (circulation 25,000) and *La Tribuna* (50,000) in Tegucigalpa; *La Prensa* (40,000) and *El Tiempo* (30,000) in San Pedro Sula. There are also several periodicals and a few trade papers.
Radio: There are over 250 commercial radio stations and one official station run by the government. *Television:* Five or six channels operate from Tegucigalpa and San Pedro Sula.

External trade: Regular balance of payments deficit is partly offset by financial aid flows from USA. Export growth remains constrained by low world commodity prices and dependence on foreign capital.
Exports: 1988: Bananas (40% of total value), coffee (22%), wood (4%), sugar (2%), meat (2%), lead and zinc, shellfish (8%), silver, soap, cotton, tobacco. Main destinations (1988): USA (51% of total value), German FR (11%), Japan (7%), Italy (6%), Guatemala (1%), Panama, Costa Rica.
Imports: 1988: Machinery and transport equipment (23% of total value), basic manufactures (28%), chemicals (22%), mineral fuels, lubricants etc (12%), food and live animals (10%). Main sources (1988): USA (38% of total value), Guatemala (5%), Japan (6%), Mexico (5%), Brazil (2%), German FR (4%), Trinidad and Tobago, Costa Rica (4%).

Agriculture: Principal sector of the economy, contributing 22 per cent to GDP and accounting for 65 per cent of exports. Bananas (grown on the northern lowland) are the main export crop, followed by coffee (grown along the Guatemalan and Nicaraguan border areas). Low world prices and export quotas have reduced both banana and coffee export earnings. Other important agricultural exports include shellfish, frozen meat, wood, cotton and tobacco. The main food crops are maize, rice, sorghum and beans. Production of these staples has steadily risen, though food imports are still necessary to meet domestic demand. Emphasis in recent years has been on land reform and the cultivation of new crops such as cocoa, allspice, cardamom, melons and citrus fruits. Around 10 per cent of total land area is arable, 25 per cent pasture and 40 per cent woodland and forest.

Industry: Manufacturing sector contributes 17 per cent to GDP and employs 13 per cent of the workforce. Manufacturing, which is concentrated around San Pedro Sula, is small-scale and based primarily on the home market. Main products include sugar, concrete, cement, furniture, beer, mineral water, cigarettes, paper products, plywood, textiles, clothing, plastics, oils and fats, soap, household chemicals and processed foods. Manufacturing sector continues to suffer from insufficient investment, lack of spare parts and weak infrastructure.

Mining: Small amounts of gold, silver, lead and zinc are extracted for export. Large reserves of tin, iron, copper and coal.

Energy: Around 25 per cent of energy requirements are imported. Emphasis on offshore oil exploration and development of hydro-power potential. The new 292mW El Cajón hydroelectric plant, and a second hydroelectric dam at Río Lindo/Yojoa, are meeting all the country's electricity needs as well as exporting power to Nicaragua, Costa Rica, El Salvador and Guatemala.

Membership of international organisations: Central American Common Market (note: participation has been affected by dispute with another member, El Salvador), FAO, G-77, IADB, ICAO, ICO, IDA, IFAD, IFC, ILO, IMF, IMO, INTELSAT, Inter-American Development Bank, Interpol, ISO ITU, Organisation of American States, PAHO, SELA, UN, UNESCO, UPEB, UPU, WFTU, WHO, WMO, World Bank.

Nicaragua, it would almost certainly restrict itself to air support and that, if the contras were not considered sufficient as ground troops, Honduran forces might be brought in. Thus any border incident is viewed with critical attention.

On three occasions Sandinista incursions have led to requests for US help from the Honduran military. In March 1988 the Pentagon sent some 3,200 US troops from the 82nd Airborne Division to Honduras for an 'emergency deployment' exercise, responding to a Sandinista incursion some nine miles into Honduras which was reported more extensively from Washington than from the region. The Honduran air force – the strongest in Central America – made a light bombing raid but otherwise the Honduran military did not react. The incident passed leaving embittered Honduran-Nicaraguan relations and considerable criticism within Honduras to the effect that one could be better informed on the state of Honduran security in Washington than in Tegucigalpa.

Honduras' relations with neighboring El Salvador are also tense. The two countries have long disputed access to the Gulf of Fonseca and six pockets of land totalling some 420 square kilometres along their mutual border. The International Court of Justice is expected to rule on this dispute in 1990.

Outlook

Honduras is unlikely to undergo any significant change in its economic structure in the foreseeable future, given that both mass parties advocate similar, neo-conservative economic programmes. However, the difficulty of the Azcona government to impose austerity measures in the face of popular and selective elite opposition suggests a period of mounting social tension. Indeed, most Latin American countries feel the threat of possible popular protest such as rocked Venezuela in 1989.

Foreign grants and loans will continue to be the primary avenue for economic development in the next years, given the low levels of domestic investment. Access to that foreign capital is expected to remain at approximately present levels, especially if demands for improved public administration can be met. These levels may gradually decline, however, if the United States ends its proxy support for the military overthrow of the Nicaraguan regime, thus reducing the importance of Honduras as the strategically-placed US ally in the region. (Alternatively, Honduras might continue to be a front-line state key to US interests if the conflict in neighbouring El Salvador reaches a critical point.)

Chief among the concerns of the Honduran military is control of the country's borders. Refugees along the Nicaraguan and Salvadoran borders, numbering as many as 100,000, will continue to raise major national issues concerning responsibility for their control and sustenance. The Honduran armed forces is facing the problem of controlling and overseeing the demobilisation of contra troops, already receiving dwindling support from the United States. Finally, an unsatisfactory decision regarding the pockets of land claimed by both Honduras and El Salvador by the World Court might rekindle rancour such as preceded the 1969 'Soccer War' between those two countries.

Growing sentiments of anti-Americanism and frustration at reduced standards of living, coupled with restricted government resources devoted to development, will likely find an expanding avenue of release through union, human rights and other mass organisation activities. While few politicians believe there is significant popular support for violent radical movements in present-day Honduras, an increase over the last year in the arrest and even assassination of suspected revolutionaries suggests that the Honduran military will augment its close vigilence over civil society, while leaving the government in civilian hands.

BUSINESS GUIDE

Time: GMT -6.

Climate: Hot, humid and rainy all year in coastal regions. More temperate at high altitudes (including Tegucigalpa which is almost 1,000 m above sea-level). Dry season from Dec to May with temperature range 18°C to 29°C. Rainy season from Jun-Nov with average temperature 18°C to 29°C.

Entry requirements
Passport: Required by all except nationals of most countries (not Cuba) with tourist card issued by a Honduras consulate.
Visa: Required by all except nationals of UK, most Western European countries (excluding France, Austria and Portugal), Canada, certain Latin American countries and Japan. Regulations vary frequently; check with embassy/consulate.
Currency: Some restrictions on import/export of most foreign currency. Amounts of US$ must be declared on arrival, and export allowed up to declared amount. Regulations vary; check with embassy/consulate.

Health precautions
Mandatory: Yellow fever vaccination certificate if arriving from an infected area.
Advisable: Typhoid, polio vaccination. Malaria risk, especially in rural areas – prophylaxis recommended. Water precautions essential.

Representation overseas: Bogotá, Bonn, Brasília, Brussels, Buenos Aires, Caracas, Guatemala City, Lima, London, Managua, Mexico City, Madrid, Monaco, Montevideo, Ottawa, Panama City, Paris, Quito, Rome, San José, San Salvador, Santiago, Santo Domingo, Tokyo, Vatican City, Washington.

Air access
National airlines: TAN (Transportes Aéreos Nacionales) and SAHSA (Servicios Aéreos de Honduras).
Other airlines: TACA, (Iberia and KLM to Guatemala City, then connecting flight), Continental, Eastern.
International airport: Tegucigalpa-Toncontín (TGU), 7 km from city; duty-free shop, bar, restaurant, bank, post office, vaccination centre, shops, car hire.

Surface access
Main ports: Ampala, La Ceiba, Puerto Cortés, Roatan, San Lorenzo, Trujillo/Castilla, Tela.
Overland access: It is possible to reach Tegucigalpa via the Pan-American Highway from Goascorán (on the border with El Salvador) and from El Espino and Guasaule (on the border with Nicaragua). Bus services are run from most Central American countries. Entry from Guatemala is possible via the Western Highway.

Hotels: Bills subject to five per cent sales tax. It is advisable to book in advance.

Car hire: Foreign or international licence required.

City transport
Taxis: Can be hailed, ordered by telephone or found on ranks; also possible to hail and share taxi; fares by negotiation (sometimes a flat rate). Tipping is not usual.

National transport
Air: 190 useable airfields, of which two with long runways. SAHSA, ANHSA, LACSA, ISLEN\05AS and SETCO are main domestic carriers, operating numerous flights between Tegucigalpa, San Pedro Sula, Roatan, La Ceiba, Trujillo and Tela. To reach more remote areas using other (eg Aéreos Servicios) services, local enquiries should be made.
Road: Network of 10,468 km, concentrated along coast (roughly San Pedro Sula to Trujillo) and the area between San Pedro Sula and Tegucigalpa and the Guatemalan border. Condition of main roads is good.
Buses: Frequent services San Pedro Sula to Tegucigalpa; also linking with Juticalpa, Danlí, Choluteca. In general services and extent of network between main centres are favourably reported.
Rail: Along northern coast belt only; limited passenger service; restaurant cars not available.
Water: Irregular service along the Atlantic coast between Puerto Cortés, La Ceiba and Bay Islands.

Public holidays: 1 Jan, Holy Thursday, Good Friday, 14 Apr (Day of Americas), 1 May (Labour Day), 15 Sep (Independence), 3 Oct (Birth of General Morazan), 12 Oct (Discovery of America), 21 Oct (Armed Forces' Day), 25 Dec.

Working hours
Government and business: (Mon-Fri) 0800 – 1200, 1330/1400 – 1700; (Sat) 0800 – 1100.
Banking: (Mon-Fri) 0900 – 1500.

Telecommunications
Telephone: Dialling code for Honduras: IDD access code + 504 followed by subscriber's (six-digit) number. No area codes needed.
Telex: Empresa Hondureña de Telecommunicaciones (HONDUTEL) operates service in main centres.

Banking
Main centre: Tegucigalpa.
Central Bank: Banco Central de Honduras. *Head offices:* Banco Atlántida, Banco Capitalizadora Hondureña, Banco Continental, Banco de Comercio, Banco de Honduras, Banco Hondureño del Café (development bank), Banco de Occidente, Banco Mercantil, Banco Nacional de Desarrollo Agrícola (development bank), Banco Sogerín, Banffaa.

Representation in capital: Argentina, Brazil, Chile, Colombia, Costa Rica, Dominican Republic, Ecuador, El Salvador, France, German FR, Guatemala, Israel, Italy, Japan, Mexico, Niaragua, Panama, Peru, Spain, Taiwan, UK, USA, Uruguay, Vatican, Venezuela.

Hotels: Tegucigalpa
Alameda, Boulevard Suyapa, P.O. Box: 940 (tel: 326920; tx: 1240).

El Prado, Avenida Cervantes (tel: 220127; tx: 1444).

Honduras Maya, Colonia Palmira (tel: 323191; tx: 1145; fax: 327629).

La Ronda, Barrio la Ronda, P.O. Box: 849 (tel: 378151; tx: 1255).

Plaza, P.O. Box: 175 (tel: 372111; tx: 1233).

Sheraton, P.O. Box: 896 (tel: 323439; tx: 1373).

Business advice and information
Associación Nacional de Industriales, Boulevard los Proceres, 4a Avenida, Colonia Lara, Tegucigalpa.

Banco Central de Honduras, PO Box C-58, 1a Calle, Tegucigalpa DC (tel: 222270/10; tx: 1121).

Cámara de Comercio e Industrias de Cortés, 17a Avda, 10a y 12a Calle, Apdo 14, San Pedro Sula (tel: 530761).

Cámara de Comercio e Industrias de Tegucigalpa, Apdo 17-C, Boulevard Centroamérica, Tegucigalpa (tel: 328210).

Consejo Hondureño de la Empresa Privada, Barrio la Plozuela, 5th Floor, Edificio San Miguel, Tegucigalpa.

División Estudios Económicos, Banco Atlántida, Apdo 57-C, Boulevard Centroamérica, Tegucigalpa.

Hong Kong

Kevin Rafferty

Few people in Hong Kong have ever paid much attention to politics or to demonstrating about political matters. They have always been too busy with the business of making money. Until 1989 a gathering of 50 to 100 people would be considered massive for a political debate. All that changed in May and June 1989, when, not once but three times, about 1m people, almost 20 per cent of the colony's total population, spilled onto the streets in a huge outpouring of anger, mourning and grief. They turned the centre of Hong Kong island into a sea of protesting people.

The reason for this unprecedented surge of popular feeling was the action of Beijing's gerontocratic leaders in cracking down on people demonstrating in China for greater democracy. The Chinese of Hong Kong came out in their hundreds of thousands when martial law was declared in Beijing. When their television screens showed unarmed civilian blood being shed in Tiananmen Square, the Hong Kong people again showed their feelings in mammoth marches. On one of the occasions, the number eight typhoon signal was up, normally a sign for the whole of Hong Kong to close down and wait in barricaded homes for the storm to pass. Even the typhoon did not deter the protesters who waved their banners and sang songs of sympathy with China's pro-democracy campaign. After the bloody crackdown, the very British-looking cenotaph on a small green patch in the central banking district was turned into a shrine for the dead, draped with black flags and covered with wreaths. A new version of the Chinese flag, black with white stars (instead of the red with yellow stars) fluttered in mourning in front of the memorial. The whole episode was a stark reminder that Hong Kong is a borrowed place living on borrowed time. It is a British colony that in mid-1997 will be handed over to China. The crackdown ordered by Deng Xiaoping and the old communist guard suddenly raised huge questions as to whether communist rulers of China could be trusted about keeping their promises that Hong Kong would be able to follow its own capitalist lifestyle under China's sovereignty. But the military action also raised immediately pressing questions for the economic and political stability of the Chinese colony.

Economic situation

Hong Kong is a unique place. It is one of the smallest territories in the world, a mere 1071 sq kilometres (or about 404 sq miles) comprising the island of Hong Kong, a small slice of the Chinese mainland and 235 other islands. It is an unpromising place for any kind of settlement, not least because of sharply rising granite hills that limit the areas open for human settlement and mean that there is very little arable land. When Britain first secured a foothold in 1841 by taking possesion of Hong Kong island, Lord Palmerston, the foreign secretary of the time, expostulated about the island's bleakness and promptly sacked Captain Elliott RN who had secured the possession from the ailing Chinese Ching (Qing) emperor. Equally unpromising is the fact that most of Hong Kong's 5.8m population have flooded into the tiny territory as refugees from upheavals in China. Most of them came without any assets of their own. Yet by the late 1980s Hong Kong had already jumped ahead of the lower ranking European countries in international league tables of per capita income. With total GNP of HK$492bn forecast for 1989 or HK$85,640 per capita, this worked out at US$10,979 per head. Hong Kong had almost caught up with the United Kingdom whose GNP per capita was estimated at a little over $11,000. With recent Hong Kong economic growth rates of more than 10 per cent a year in real terms – a figure that slipped to a mere (by Hong Kong standards) 7 per cent in 1988 – the prospect seemed that Hong Kong could take on, and beat, the rest of the world in a very short time.

In general, economic achievements were more than matched by Hong Kong's standing in the world of international trade and finance. With exports of goods and services worth almost $75bn in 1988 and total trade worth $145bn, Hong Kong came in the top 12 trading countries in the world. Its exports were 60 per cent higher than those of its giant neighbour China and five times those of India. In per capita terms, Hong Kong exports goods worth about $12,750 per person a year, about five times the exports of the average Briton, six times those of the average Japanese, and more than ten times those of the aver-

KEY FACTS

Official title: British Dependent Territory of Hong Kong

Head of state: Queen Elizabeth II

Head of government: Governor, Sir David Wilson

Capital: Victoria, on Hong Kong Island

Official Languages: Chinese, English

Currency: Hong Kong Dollar (HK$ = 100 cents)

Exchange rate: HK$7.79 per $ (Nov 1989)

Area: 1,067.7 sq km (including 236 islands)

Population: 5.68m (1988)

GDP per capita: HK$8,260 (1987)

GNP real growth: 6.0% (1988)

Labour force: 2.7m (1988)

Unemployment: 2.2% (1986); plus 1.8% under-employed

Inflation: 7.4% (1988)

Trade balance: –$5,729m (1988)

Foreign debt: $8.4bn (1987)

Visitor numbers: 4.6m (1987)

KEY INDICATORS

	Unit	1984	1985	1986	1987	1988
Population	'000	5,397	5,456	5,532	5,613	5,681
Gross domestic product (GDP)	HK$m	249,630	263,579	291,897	360,240	–
GDP per capita	HK$m	6,450	6,230	7,030	8,260	–
Consumer prices	% increase	–	1.4	2.3	5.5	7.4
Exports	HK$m	221,441	235,152	276,530	378,034	493,070
Imports	HK$m	223,370	231,420	275,955	377,948	498,780
Balance of trade	HK$m	–1,929	3733	575	87	–5,729
External Debt	$m	6,915	7,319	8,214	8,437	–
Debt service ratio	%	17.3	15.8	17.8	17.1	–
Exchange rate	HK$ per $	7.80	7.80	7.80	7.80	7.80

* estimated figure

age US citizen. Hong Kong is the world's largest exporter of clothing, the leader in watch production, churning out 270m pieces a year. It is the largest exporter of toys and leads the world in some less well-known products including exports of candles and of artificial plastic firewood. Backing the export machine was the world's largest container port, a supportive banking and financial system and a government that prides itself on not interfering or trying to guide or direct business. A leading economist boasts that foreign companies placing orders with Hong Kong manufacturers can be sure of getting their products delivered at the right place at the right price and at the right time. 'Everything moves like clockwork in Hong Kong. When goods are finished they are usually packed neatly under polythene and labelled according to the store's own requirements, so that they can actually be wheeled on racks out of the delivery van and straight through the shop's doors for immediate sale.' He contrasts this with the experiences of other developing countries where – he claims – manufacturers could not be so certain of meeting orders or delivery dates because they had a less supportive government machine. 'I've known countries where a manufacturer has accepted an order, and then found that he couldn't complete in time because the customs were holding up supplies of zippers that had to go into the finished garment, or because there was no regular sailing to get the consignment on a ship to the importing port on time.'

It is the boast in Hong Kong that a company can be registered in the morning, in business by the afternoon, and making profits fairly soon thereafter. The entrepreneurial spirit is fostered by two other factors: one is the thirst of the largely refugee population to make money; the other is the very determination of the government to keep out of the hair of businessmen. 'We don't tell people where or how they can make money,' says a government official. 'That's up to them. Government officials can easily get it wrong, and then manufacturers expect help out of public funds.' The government also assists by having virtually the lowest taxation rates in the world. Personal income tax was reduced in the 1989 budget by financial secretary Sir Piers Jacobs by 0.5 per cent to 15 per cent, company tax rates to 16.5 per cent. Even with its low tax rates, the government has been running record budget surpluses boosted by high economic growth rates. The low tax rates mean that businessmen do not find it worthwhile to evade. The surplus in 1988 topped HK$14bn, the highest in a run of record surplus years. Total government reserves in the Exchange Fund are a closely guarded secret, but one way and another the Hong Kong government probably has US$40bn stashed away in accumulated surpluses and earnings from them.

Textiles and clothing

The most notable feature of Hong Kong's manufacturing sector, still the largest contributor accounting for 22 per cent of GNP, is the large number of small enterprises. Many of the colony's 50,000 manufacturing enterprises employ 10 people or fewer. Textiles and clothing industries still take pride of place, accounting for almost 40 per cent of the workforce involved in manufacturing and more than 35 per cent of Hong Kong's domestic exports. The territory has moved worlds away from the textile scene in poor third world countries that are using the industry as a first painful stepping-stone on the way to industrialisation. Hong Kong textile and garment factories are still to be found in unprepossessing buildings in huggermugger Kowloon, sometimes perilously close to the flight path of approaching aircraft about to land at Kai Tak. But inside the building, the making of patterns and cutting of garments are computer-aided to standards that would rival any factory in the West. Garments produced by Hong Kong makers are near the top end of the world market. Indeed some of them sell well in leading international department stores throughout the world including Harrods of London and Saks of New York's Fifth Avenue.

The path to higher and higher quality and fashion is one that Hong Kong was forced to take. It began several decades ago when leading Western countries began applying quota and other restrictions against budding potential rivals from developing countries. The obvious way to get round the restrictions was to improve the quality, so that the price tag – and the profits – on the foreign sales would keep on rising even if there was a limit on the number of garments sold. Western manufacturers themselves were the real losers from their protectionist policies. Quota restrictions gave Hong Kong a secure foothold in the international market that new and upcoming countries were denied; at the same time the restrictions forced Hong Kong to grow up and compete directly on quality with developed countries. Today Hong Kong's textile industry is one of the most modern in the world, able to compete anywhere. In addition, the tiny territory has begun to attract its own fashion designers who make up-market clothes under their own labels. The Institute of Textiles and Clothing at the Hong Kong Polytechnic has also helped to boost knowledge and skills in the fashion business. Some college lecturers boast that local design 'is as good as Japan's and much better than that in the United States.' One local manufacturer grumbled that Europe could no longer keep pace with the territory's move upmarket: he cited a West German chain store that wanted less fashionable goods than Hong Kong was turning out and had gone to South Korea instead.

Electronics

After textiles and clothing, the next biggest sector of manufacturing is electronics, accounting for 18 per cent of domestic exports and 12 per cent of the industrial workforce. Unlike textiles, dominated by migrants who fled from Shanghai after the communist revolution, the electronics industry shows a blend of local and overseas involvement. Government propaganda cites an impressive list of high-quality, high-technology products made in Hong Kong, mentioning 'cassette recorders, hi-fi systems, television sets, wire and cordless telephones with built-in memories and automatic dialling functions, telecommunications equipment, microcomputers, computer memory systems, Winchester and floppy disk drives, read/write magnetic heads, computer printers, switching power supplies, computer-aided design and tested equipment, multi-layer printed circuit boards, electronic modules, liquid crystal displays, quartz crystals and semiconductor devices including integrated circuit wafers (e.g. microprocessors, RAM/ROM memory and logic chips)' Put like this it sounds a highly sophisticated industry, but Hong Kong electronics still has a long way to go. Companies are typically small, so that research and development, the key to the real high-tech game, is tiny in Hong Kong. An official admits that 'We haven't got the research and development because we haven't got defence, aerospace, and other capital-intensive industries.'

Manufacturers themselves are content with a less than high-tech role. As one of them puts it 'as manufacturers we would develop whatever the market will accept rather than have anybody tell us what to do.' Another manufacturer grumbles that in real high-tech development, 'you need a lot of highly paid scientists who spend a lot of time sitting around not contributing to immediate earnings. This is very much *not* Hong Kong's way.'

Plastics and toys

The next biggest industry is plastics with about 10 per cent each of exports and manufacturing and employment. It may seem an unlikely winning industry but plastics sustained the early manufacturing export drive and helps promote new exports, especially toys. Injection and extrusion machines are made locally and much of their production goes to high quality toys. Hong Kong is not the biggest toymaker in the world, but is the world's leading exporter. Several thousand toy factories are strung out from Kowloon to the New Territories – in the mid-1980s they began to spill over into China itself. Most of the factories are small, each with a tiny number of workers.

Indeed in plastics and in toys Hong Kong makers found they got out of their depth when they tried to be too ambitious. Playmates, a big local company that obtained its own stock market listing, devised a chatterbox talking doll called 'Cricket', a successful amalgamation of Hong Kong's traditional toy-making with prowess in electronics. The doll was a huge success and became the seventh most popular toy in the USA. But 70 per cent of the company's sales rested on this single job. Playmates couldn't afford to stand still, so it first devised a companion for Cricket, a boy doll called Corky. Then it invested much money in an even more sophisticated talking model that it hoped to sell for US$100 a doll. This fell flat and had to be abandoned. By 1988 Playmates had hit back with a new product, Tennage Mutant Ninja Turtle. In spite of its name, the doll, based on a US comic book story, was a soaraway success and came second to a Japanese video game system in the great Christmas toy battle in the USA in 1988.

Another problem which Hong Kong manufacturers have encountered is their dependence on the US market and on a small number of companies. In the late 1980s a number of US companies fell on hard times and filed for bankruptcy, an event which spread gloom into Hong Kong just at the most booming economic time.

Watches and clocks

Watches and clocks are another important area for Hong Kong manufacturing, also cashing in on the colony's ability to manufacture quickly and cheaply in response to market shifts. Hong Kong is the biggest producer in volume terms and second in value to Switzerland, having surpassed Japan in 1987. Like textiles, watches have also been subject to protectionist quota restrictions. France tried to penalise Hong Kong in this way. Some manufacturers have tried to get round the protection by setting up plants outside the colony. One local electronics manufacturer took advantage of tax breaks and special depreciation allowances to set up in Ireland. But it found that there was also a price to pay. Notably, labour costs in Ireland were 2.5 times higher than in Hong Kong and workers were less co-operative, anxious not to do overtime when it would push them into higher tax brackets. In Hong Kong manufacturers can count on a labour force that has traditionally been helpful if not docile. Low tax rates mean that Hong Kong workers have been quite happy to work all hours possible – though this flexibility was being called in question in the late 1980s as a tight labour market pushed wages up and meant that in certain areas workers were simply not available. This situation also helped to propel the large scale move of factories across the border into China.

Manufacturing in China

By 1989 the move to China had become so pronounced that there were more Chinese working for Hong Kong manufacturers than there were in the whole of Hong Kong's manufacturing sector. Altogether, manufacturing inside Hong Kong employed just over 900,000 people. Estimates of the number of Chinese working on contracts for Hong Kong employers varied from as few as 2m to a figure of 4m cited by one middle ranking Chinese official. The move to China had started off in the mid-1980s when ever price-conscious Hong Kong manufacturers, especially in the toy and the garment industries, found that they could get part of their production done more cheaply across the border in China. It became common to see lorries travelling between Hong Kong and southern China's Guangdong province with – for example – loads of staring empty eyes of dolls that

COUNTRY PROFILE

Political system: Hong Kong will remain a British crown colony until midnight on 30 June/1 July 1997 when China will take over sovereignty of the 1,071 sq km territory. China has promised that Hong Kong will be a special administrative region and will be able to maintain its own autonomous capitalist system for at least 50 years. The government is headed by a Governor, advised by an Executive Council. The law-making body is the Legislative Council. Both bodies consist of a small number of ex-officio members and a larger number of non-officials. Until 1985 all non-official council members were chosen by the Governor, but in that year 24 of the 57 (including the Governor as President) seats on the Legislative Council were chosen by indirect elections.

Population: The population of 5.7m (1988) has grown at some 1.9 per cent per annum in the last decade, due partly to migration from China PR and Indochina; this is now slowing. 98 per cent are of Chinese descent, and there are 67,900 non-Hong Kong Commonwealth residents (mainly British) as well as 104,300 others, a quarter of whom are Filipinos. 1.26m live on Hong Kong island, 2.2m on Kowloon and the remainder in the New Territories and other islands.

Language: English is universally understood in business and commerce, but most of the population speak Chinese (mainly Cantonese).

Media
Press: There are a total of 67 paid newspapers and around 614 periodicals published. *Dailies:* The 4 English-language dailies are the *South China Morning Post* (circulation 175,000), the *Hong Kong Standard* (circulation 20,000), the *Asian Wall Street Journal* (circulation 25,500) and the *International Herald Tribune*. There are 44 Chinese language dailies. *Weeklies: Asian Week, Far East Economic Review*, and *Hong Kong*, a weekly newspaper published in co-operation with the Hong Kong Tourist Association, lists events and items of interest to visitors. *Business:* There are numerous trade and technical publications in English and Chinese.
Broadcasting: Five companies provide broadcasting services, one of them government-operated. *Radio:* There are 10 different radio services in English and Chinese, three of them commercial. *Television:* there are two commercial TV stations, each with one English and one Cantonese channel. In 1988 some 96 per cent of all homes had at least one TV.

Domestic economy: Following two consecutive years of double-digit economic growth, the overall growth rate of the Hong Kong economy was more moderate in 1988. GDP in 1988 had a growth rate of 7.4 per cent. Yet aside from its natural harbour, Hong Kong has virtually no natural resources. The economy depends on exports of manufactured goods, entrepot trade, tourism and financial and banking activity. The territory has good communications, a consistent free trade economic policy, freeport status, low tax, excellent shipping and cargo-handling facilities (the busiest container port in the world), lack of exchange controls and an efficient banking and insurance sector. Manufacturing remains the most important contributor to GDP – 22 per cent but wholesale and retail trades are almost as important; financial services and real estate for 18 per cent and community and personal services for 15 per cent. Hong Kong is one of the most important financial centres in the world, with a flourishing offshore banking sector. There has been considerable investment in transportation and housing infrastructure. Tourism is also an important contributor to the economy and in 1988 visitor arrivals jumped by more than 24 per cent to 5.6m. Hotels enjoyed a very high 92 per cent occupancy rate. The last few years have seen re-emergence of a vigorous re-export trade, with Hong Kong acting as southern China's most important port.

External trade
Exports: Total value of exports rose moderately in 1988 by 11 per cent to HK$218bn. Main exports are clothing (31% of the 1988 total), plastics, toys, dolls, jewellery (10%), photographic apparatus, optical goods, watches and clocks (9.3%), textiles (7.1%), electrical machinery, household appliances (6%) and telecommunications, sound recording and reproducing apparatus (5.1%). Main destinations were the USA (33.5% of the 1988 total), China (17.5%), German FR (7.4%), UK (7%), and Japan (5.3%). Of these, only Chinese sales improved significantly.
Re-exports: Almost as important are the country's entrepôt services to China and the United States, which rose by nearly 58 per cent in 1987 and 52 per cent in 1988 to total HK$94.9bn and HK$49.5bn respectively. Typically, 39 per cent is in the form of raw materials and 43 per cent in consumer goods.
Imports: Total value of imports in 1988 rose by 32 per cent to HK $498.8bn. Main imports are raw materials and semi-manufactures (43% of the 1988 total), consumer goods (33%), capital goods (16%), foodstuffs (6%), and fuels (2%). Main sources of imports in 1988 were China (31%), Japan (19%), Taiwan (9%), USA (8.3%), Republic of Korea (5.3%), Singapore (3.7%) and the UK (2.6%).

Agriculture: The land area is mountainous, with fertile soils when they are watered, but arid in winter. Agriculture land, including 600 ha of orchards, accounts for nine per cent of the total land area. Agriculture and fishing account for 0.4 per cent of GDP, providing 34 per cent of fresh vegetables consumed, 38 per cent of live poultry, 18 per cent of live pigs, 15 per cent of fresh water fish and 74 per cent of marine fish. Main crops include sweet potato, yams, taro, sugar cane, white cabbage, flowering cabbage, lettuce, chinese kale, radishes and watercress. The offshore fleet has 4,900 vessels and employs about 24,000 fishermen; pond and marine fish farming in the New Territories accounted for three per cent of the production of 287,900 tonnes in 1988.

Industry: Although increasingly eroded by the growth in service industries, manufacturing provides some 22 per cent of GDP and employs 31.5 per cent of the workforce. There are about 50,000 manufacturing establishments of which most are small and medium-sized, producing electronic and electrical goods, toys, plastics and aircraft components in addition to such traditional products as textiles, clothing and clocks and watches. Hong Kong is still a world leader in toy manufacture, with 1,941 toy factories employing 35,049 workers in 1988 and producing HK$10bn of exports; more than 40 per cent of this goes to US companies. There is also a substantial heavy industrial sector, supplying steel and steel products, ship construction and repair, and chemical manufacture. The Hong Kong Industrial Estates Corporation is responsible for providing new land for industrial development and has two industrial estates in the New Territories, one at Tai Po and the other at Yuen Long. Construction activities remained buoyant throughout the territory during 1988 despite the 1997 issue. Tourism is an important source of foreign revenue, attracting some 5.6m visitors in 1988 (22 per cent Japanese, 16 per cent American, 14 per cent other Asian countries) and employing about 120,000 people directly as well as 500,000 indirectly. At least 12 per cent of tourists also visit China PR.

Mining: Mining accounts for 0.1 per cent of GDP (1987), producing mainly feldspar (22,853 tonnes) and kaolin (92,504 tonnes).

Main companies: Swire Pacific, Dairy Farm International, Jardine Matheson, Cathay Pacific Airways, China Light & Power, Hutchison Whampoa, First Pacific International, New World Development, Wharf Holdings, Hong Kong Electric Holdings, Hang Lung Development, MRTC-Mass Transit Railway, HK Telecom, Shui On Group, Winsor Industrial Corporation, Cheung Kong (Holdings), Henderson Land Development, Sime Darby Hong Kong, Burwill International, The Kowloon Motor Bus Company, KH-TVB, Li & Fung.

Financial markets: A unified stock exchange, The Stock Exchange of Hong Kong, opened in 1986, incorporating the Hong Kong, Far East, Kam Ngan and Kowloon stock exchanges. There is also a commodity exchange for futures contracts in gold, sugar and soya beans, and a Hang Seng index futures. Banking deregulation has been accompanied by a tightening of supervision.

Membership of international organisations: Asian Development Bank, ESCAP and Commonwealth (associate member), GATT, IMO, Interpol, Multifiber Arrangement, WMO.

would be placed in the final product assembled in Hong Kong. As labour supply in Hong Kong tightened, manufacturers began to increase the amount of work done in China. A leading toy maker claimed that labour laws in Hong Kong meant that he just could not get the work done in time, whereas China's much laxer system meant that jobs could be completed on time and according to budget. The existence of the Special Economic Zone of Shenzhen immediately across the Shum Chum (Shenzhen) river from Hong Kong was a natural focus encouraging the overspill of Hong Kong industry into China. More powerful was the economic liberalisation in China and the growing contacts between Hong Kong and Guangdong.

A number of manufacturers preferred to bypass the showpiece of Shenzhen. They found that it was easier to strike out into Guandgong province proper and made a deal with a local headman who could supply land and labour at prices up to a fifth of those in Hong Kong. Shenzhen also obtained a growing reputation for greed and for the interposition of time-consuming and expensive politics. By late 1988 and early 1989 (before the Chinese crackdown in Tienanmen Square) some Hong Kong manufacturers including Playmates had decided that they should shift their whole production into China. Cheaper labour was one attraction. The other was that China had plentiful supplies of labour when workers in Hong Kong were becoming difficult to find at any price. By early 1989 Hong Kong's unemployment rate had fallen to a minuscule 1.2 per cent. Such a low unemployment rate meant that there were more jobs than workers. Employers estimated that there was a shortage of 100,000 workers, concentrated in construction, hotels and tourism, retail and manufacturing. By early 1989 there was a chorus of pleas from employers to import workers. There were even suggestions that Vietnamese boat pepole might be allowed from their camps to work in shops or factories – though their inability to speak Cantonese would handicap them even if the hostility of the local population did not. In reality if every potential worker could do every potential job on offer, there would not be a labour shortage. Hong Kong in the 1980s had its fair share of odd job men and women who could be seen on almost every urban street corner, underemployed shoeshine boys and itinerant hawkers. But clearly such people were unwilling to do heavy duty construction-site work and unable to fit in in the more sophisticated Westernised society of hotels.

Hong Kong's great achievement is that its population has travelled so far so quickly. Penniless refugees and sons and daughters of duck farmers in rural Guangdong only a generation ago, Hong Kong Chinese were doing polished jobs in manufacturing and finance that only ten years ago would have been regarded as a Western preserve.

Finance and banking

Equally impressive has been the growth of Hong Kong's non-manufacturing service sector including finance, property, hotels and tourism. As a financial centre, Hong Kong in the 1970s and early 1980s used to boast that it was the third largest centre in the world after New York and London. The emptiness of the boast was revealed by the mid-1980s when Tokyo really came into its own. By the middle of the decade, the Tokyo stock market had surpassed New York in terms of total market capitalisation, and by 1989 it was worth about 40 per cent of the capitalisation of the major world markets. Japanese banks had also powered through to occupy at least nine of the top ten places in world rankings of commercial banks. Japan's mammoth trade and current-account surpluses also revealed the strength and weight of Japanese money in international capital markets. Tokyo, not puny Hong Kong, was the third leg of the great international financial triangle. After all, even at its best levels, the volatile Hong Kong stock market accounted for about 1 per cent of the total world capitalisation.

Nevertheless Hong Kong still has great strengths that Tokyo lacks, and these have drawn large numbers of banks and financial concerns to set up offices in Hong Kong. Most important of all, Hong Kong is an international centre whereas Tokyo is trapped by its very insular Japanese inheritance. Thanks to the widespread use of the English language and British laws, Hong Kong fits in more easily with international ways. Its highly mobile people and their readiness to learn and experience new things make them much more adaptable than Japanese. In consequence Hong Kong probably has hundreds of financial institutions including 160 licenced banks, several hundred stockbrokers, international accountants, lawyers, insurance companies, fund managers. Some of them have been attracted by the more international ambience of Hong Kong, others by a good regional location, still others by the prospect of working with China.

Hong Kong's pre-eminence was not achieved smoothly. Indeed in the 1980s scandals first hit banking and then the stockmarket. The collapse of the mysterious Carrian group in the early 1980s showed that in the boom days banks had a happy-go-lucky attitude to some of their lending, doling out large sums without proper consideration or security. Then seven local banks collapsed or had to be bailed out by the government. Some of them had been imprudent; others had managers or directors who were putting their hand in the till. The collapses provoked the government into stiffening its legislation and improving the powers of the banking commission. Previously in the words of one banking official 'the job of the banking officials was to see that the sums added up properly not to delve into whether directors or managers were guilty of fraud, or to check that they weren't lending to their own families.'

No sooner had the bank scandals died down then trouble hit the stockmarket. In 1986 the Hong Kong market had gone high-tech. A new unified stock exchange opened for business in place of the previous four exchanges. A new fully-computerised trading floor became the scene of business instead of the constantly shifting rugby scrum of activity that marked the four independent exchanges. In New York in early October 1987 Ronald Li Fook-shiu, then chairman of the new stock exchange boasted that 'we are now a stock exchange of international standing and therefore we are trying to make our stocks international.' The boast was exposed a matter of weeks later and in the aftermath of Wall Street's crash of 'Black Monday', 19 October.

Hong Kong was the only major stockmarket in the world to close its doors. It closed not just for a single day but for four days as the authorities tried to work out a rescue package. The fear was not just for the market itself but for the futures exchange. A slump in the Hang Seng Index would have presented traders on the futures exchange with liabilities of more than HK$3bn – whereas the guarantee fund for the exchange had a puny HK$22m in its coffers. When the market reopened it plummeted a record 33 per cent in one day. A roast pig was brought on to the stock exchange, almost a gesture of despair, traditionally used by Chinese when other appeals to the gods have failed. But a complete collapse was avoided as by then a lifeboat fund had been set up to keep the futures traders afloat.

International disapproval of Hong Kong's closure of the stock exchange triggered soul-searching and a clean-up of the market. Former chairman Ronald Li and several other leading lights of the stockmarket were visited by officials of Hong Kong's Independent Commission Against Corruption (ICAC) and subsequent charged with bribery and other offences. A special committee under former Lloyds of London chief executive Ian Hay Davison investigated the way that the Hong Kong market was run. It concluded damningly in June 1988 that 'an inside group treated the exchange as a private club rather than a public utility for the general benefit of members, investors, and issuers. Its executive staff was ineffective, lacking adequate knowledge and experience to cope with the evolving and expanding securities industry, and insufficiently independent of the governing committee.' In another tough phrase, the committee claimed about the securities industry in Hong Kong 'this place is absolutely stuffed with vested interests and they can't be allowed free play.' The government did not escape blame. The Hay Davison committee was quick to point out that the authorities had effectively gone along with the prevailing sentiment that part of the colony's charm was the very freedom from controls. The securities commission supposed to police the stock exchange had throughout been 'passive and reactive'. One reason was that the commission was starved of funds. Most of the Hay Davison reform programme was accepted, though the influential insiders who preferred Hong Kong as a freewheeling place fought hard to water down the more biting reform ideas.

The clean-up started in 1988 when the government vetted the list of candidates for the new stock exchange. Even so, several hundred mainly Cantonese brokers managed to signal that they did not like the new ideas. Only one non-Chinese was elected, Alan Smith, managing director of Jardine Fleming, and one representative of a big international broking house. In addition, the four brokers chosen for individual seats had all been members of the previous much maligned council. The *South China Morning Post* reported in its banner headline 'Local brokers claim victory' then in slightly smaller type 'International firms shut out of line-up for new stock exchange.' Clearly the election was a victory for small-town Hong Kong. The government imposed a new independent chairman of the stock exchange, Sir Quo-Wei Lee, widely respected chairman of the Hang Seng bank. It hired a new chief executive, the 36-year-old Shanghai-born Francis Yuen. Yuen took a considerable salary cut when he left Citicorp, Scrimgeour Vickers in Hong Kong for his new HK$2.5m a year job at the stock exchange. A new tough securities law was passed in 1989 setting up a watchdog commission with sharp teeth to guard the industry. Its main author was Robert Owen, a former British government official, later chairman of Lloyds Merchant Bank, who had been head-hunted for the job of chairman of the Securities and Futures Commission, as the new watchdog was called. Originally Owen wanted far-reaching powers allowing his investigators to march through business premises without a search warrant, to take away the right of silence from people questioned and to use any statement obtained as evidence in criminal proceedings. After howls of disapproval from the local financial community, which protested that the commission would have more powers than the ICAC, Owen compromised and the proposed law was softened. Regulators will need a magistrate's warrant to enter premises, but have powers to prevent documents from 'disappearing' while the warrant is being served. The new law removes the right to silence of a person questioned but information thus provided cannot be used in criminal proceedings.

Later another law was introduced to curb insider trading. The Hay Davison committee had pointed out that ten families controlled companies representing 54 per cent of the stock market's total capitalisation. In a free-wheeling place like Hong Kong, this left the market prone to insider dealing. But once again the laissez-faire financial establishment won the day. Insider dealing became an offence, but a civil one, not a criminal one.

Approach of 1997

Events in China in June 1989 changed many views. True to form the ever volatile stock market plummeted, by 11 per cent when martial law was declared in Beijing, then by about 25 per cent after the People's Liberation Army was let loose on the people in the very heart of the Chinese empire. Forecasts of Hong Kong's growth for 1989, originally a much slower 6 per cent, were revised sharply downwards. Manufacturers who had rejoiced over their new-found ties across the border in China began to wonder about being too dependent on the big brother. Reports from Guangdong province that its exports would be down by at least 25 per cent sent shudders through Hong Kong. The colony's dependence was such that by the end of 1988 more than 90 per cent of Hong Kong companies making garments, toys and electronic goods had at least some of their production done across the border in Guangdong. Equally worrying was that Hong Kong's trade was greatly linked with that of China. The mid- to late-1980s had seen a resurgence of the traditional entrepot role that Hong Kong had played decades before. By 1989 Hong Kong's own domestic exports formed only 44 per cent of its total exports. The rest was made up of re-exports of goods where Hong Kong's superior trading port and customs facilities encouraged other countries, but mainly China, to ship through Hong Kong.

The likelihood of a decline in trade and the question marks over China could not have come at a worse time. China's own economic problems would probably have led to a slowdown in the booming re-export of Chinese goods. But the political crisis raised new questions on both sides: whether Beijing might try to supervise its southern province and curb its capitalist boom; whether Hong Kong industrialists would think it prudent to put all their diversification into China.

The crisis was doubly awkward because in 1989 some Hong Kong electronics manufacturers were charged with 'dumping' by the European Commission and punitive duties were placed on their goods. It was seen as a new get-tough policy from Brussels. Critics, including the *Financial Times* trade editor, commented that the new measures seemed to be at the behest of European manufacturers rather than demanded by fair trade. Hong Kong industrialists protested that they could not practise dumping, that is selling goods below their production cost, because the small local market meant they would go bust. But the Hong Kong case was not helped by the Commission's secretiveness over the figures on which it based its findings.

The greater worry for Hong Kong was the alarm and despondency caused by the Chinese action and the realisation that rulers in Beijing who had turned their troops on their own people would be rulers of Hong Kong after 1997. Already there had been an alarming flight of middle-class managers, the intelligent, educated, articulate, entrepreneurial and professional classes, who had contributed largely to making Hong Kong the fast moving successful international city that it had become by 1989. Even before the Chinese crackdown the flight of emigrants was running at about 50,000 a year. Some US multinationals in Hong Kong reported they were losing about 25 per cent of their managers. The Hong Kong Management Association had started a scheme to try to lure back departing Chinese executives.

Another unusual sign of unease was a sharp rise in dishonesty by employees cheating their bosses so they could make enough money to emigrate. Abuses were getting so serious that some companies were taking out 'fidelity' insurance to guard against being cheated by their own workers. The crackdown simply exacerbated all of these fears. When Singapore announced that it was prepared to take 25,000 Hong Kong emigrants in the 1990s, its consulate was swamped and had to close. Canada, the most popular place for the Hong Kong *emigrés* ran out of application forms for residency permits. Australia and the USA were the two next most popular destinations.

When China's turmoil had subsided, some of the conservative old Chinese managers won back their confidence. Tse-Kai Ann, head of the Winsor Textile group, went to Beijing for discussions on the Basic Law, the new constitution that will govern Hong Kong after 1997 and declared that 3 and 4 June had changed nothing. Other manufacturers also warmed to the new communist party secretary-general Jiang Zemin's promise that 'well water should not interfere with the river water.' One prominent local magnate declared that the number of people killed in China had not been very great and that if Hong Kong behaved itself, China would not interfere. This was the view of the rich, but it was not shared either by those pressing for democracy or by the nervous middle classes.

Unlike other British colonies, Hong Kong is not being given independence, but has only been promised autonomy within China. Communist leaders have spoken out quickly to deter any notions of independence. Such a state would be difficult for any community to feel comfortable with. But for very colonial Hong Kong it poses even more problems than normal. Hong Kong is traditionally run on paternalistic lines. Only as recently as the 1980s did Chinese achieve a majority on the Executive Council, equivalent to a cabinet. Even today, there are large numbers of British civil servant officials at the top of the civil service, and more importantly, at the head of the police. The Executive Council is chosen entirely by the governor and the majority of the Legislative Council (Legco) is chosen by him. The elected members of Legco, 26 out of 57, are all indirectly chosen. No wonder Hong Kong Chinese feel vulnerable faced with hardline Chinese who will be their rulers after 1997.

BUSINESS GUIDE

Time: GMT + 8 hrs.

Climate: Subtropical and monsoonal. Winter (Dec-Feb) is cool and dry, with an average temperature of 15°C and humidity often over 83%.

Entry requirements
Passport: Required by all.
Visa: Nationals of most countries do not require visas for short visits. Regulations regarding entry into Hong Kong are extensive, and travellers are advised to obtain up-to-date information before any journey. Travellers entering for employment or study must obtain an appropriate visa in advance.
Health: Vaccination certificates are not mandatory, but requirements may vary from country to country.
Currency: There are no restrictions on the amount of foreign currency which a traveller may take into Hong Kong. Currency is easily changed, either at banks, hotels or money changers. There are small variations in the rates of exchange offered by these outlets. Local currency is accepted in Macau.
Customs: Personal effects are allowed duty-free.
Income tax: Tax is payable on income or profits arising in or derived from the territory.

Representation overseas: Amsterdam, Athens, Barcelona, Beijing, Brussels, Chicago, Dallas, Frankfurt, Geneva, Hamburg, London, Los Angeles, Manchester, Miami, Milan, New York, Ontario, Osaka, Panama City, Paris, Rome, San Francisco, Shanghai, Singapore, Stockholm, Stuttgart, Sydney, Tokyo, Toronto, Vancouver, Vienna, Washington DC, Zurich.

Air access
National airline: Cathay Pacific.
Other airlines: Air France, Air India, Air Lanka, Air New Zealand, Air Niugini, Alitalia, All Nippon Airways, British Airways, CAAC, Canadian Airlines International, Cathay Pacific, China, Dragonair, Garuda, Gulf Air, JAL, Japan Asia Airways, KLM, Korean Air, Lauda Air, Lufthansa, MAS, Northwest, Philippine, Qantas, Royal Brunei, Royal Nepal, Singapore, South African Airways, Swissair, Thai International, United.
Tax: A charge of HK$100 is made on departure.
International airport: Hong Kong International, Kai Tak (HKG), 7.5 km from Hong Kong in Kowloon, with duty-free shop, bar, restaurant, left-luggage, bank, hotel reservations, post office, shops, Cable & Wireless, police station. A new airport is being considered by the government.

Surface access: The Kowloon-Canton Railway Corp (KCR) is the main carrier of passengers to and from China, with express trains serving Kowloon-Guangzhou and Kowloon-Lowu (32.5 km). Bus services link Guangzhou to the Hong Kong border. Hovercraft services operate four times a day to/from Guangzhou and several times daily to/from Zhuhai. There are frequent daily services to and from Macao by passenger ferry (journey time 2.5-3 hrs), hydrofoil (journey time 75 minutes), jetcats (journey time 70 minutes), high speed ferry (journey time 90 minutes) and jetfoil (journey time 55 minutes).
Main port: Hong Kong.

Hotels: Confirmed hotel rooms and onward flight reservations should be booked well in advance, especially for the peak travel seasons of Apr-May and Oct-Nov. A 10 per cent service charge and 5 per cent tax is added to hotel bills.

Credit cards: Major international credit cards, such as American Express, Diners Club, Visa, Carte Blanche, JCB, Mastercharge and Air Travel are widely accepted, although cash prices may be lower. American Express card holders have automatic access to the London No. 1 Club's half dozen affiliates in Hong Kong, including the Clearwater Bay Golf and Country Club.

Car hire: International driving licence may be used for up to 12 months. In many cases (e.g. for UK licence holders), a Hong Kong licence can be obtained fairly simply. Parking difficulties and traffic congestion should be taken into account when planning to drive in Hong Kong. Chauffeur-driven and self-drive car hire is available.

City transport
Taxis: Metered taxis are widely available, and carry up to five passengers. Green and silver cabs in the New Territories, red and silver in town. Cross Harbour Tunnel tolls are payable by the passenger when the return journey takes this route.
Metro: The Mass Transit Railway (MTR) operates from 0600 – 0100, with 37 stations over a 38 km route. MTR does not sell return tickets, and tickets are only valid for 90 minutes from the time of issue, although tourist tickets are available. Eating, drinking or smoking in stations or trains is prohibited. Passengers are not allowed to carry bulky luggage. The Kowloon-Canton Railway is committed to constructing a 34 km light rail transit system to serve Tuen Mun and Yuen Long in the western New Territories, the first phase of which is now open.

BUSINESS GUIDE

Trams: A flat fare system operates on Hong Kong Island's five-line tram system, which carries about 360,000 passengers/day using 163 cars. The trams are crowded at rush hour, but afford good views of Hong Kong at other times.

National transport: In 1988, an average of 7.5m passengers were served each day by the public transport system.
Road: The government maintains over 1,435 km of surfaced road. Traffic density is high, about 246 vehicles/km. Current development projects include two privately-run cross-harbour tunnels (a third is under consideration). The dilapidated Walled City area is to be demolished to create parking space.
Buses: Bus services are inexpensive and convenient, carrying about 52 per cent of passenger traffic. There are two main private bus companies, China Motor Bus (CMB) and Kowloon Motor Bus (KMB), and private minibus services. Over 3.8m passengers/day are carried by public buses and over 1.5m/day by minibuses.
Water: There are extensive ferry, hovercraft, hydrofoil and coastal services.

Public holidays
Fixed dates: 1 Jan, 25-26 Dec.
Variable dates: Lunar New Year's Celebrations, Day after Ching Ming Festival, Easter Holiday, Day after Tuen Ng Festival, Birthday of HM the Queen, Mon following Birthday of HM the Queen, Sat preceding last Mon in August, Liberation Day, Day following Chinese mid-autumn festival, Day following Chung Yeung Festival.

Working hours: Shopping and services may be restricted during Chinese New Year, but less so at Christmas time.
Business: (Mon-Fri) 0900 – 1700; (Sat) 0900 – 1230.
Government: (Mon-Fri) 0900 – 1300, 1400 – 1700; (Sat) 0900 – 1230.
Banking: (Mon-Fri) 0900 – 1630; (Sat) 0900 – 1230.
Shops: Central District 1000 – 1800; Causeway Bay and Wanchai 1000 – 2130; Tsimshatsui East 1000 – 1930; Tsimshatsui, Yaumatei and Mong Kok 1000 – 2100. Most department stores and shops often open Sun. Some Japanese stores close one day/week, and street markets operate all day and into the night.

Social customs: When shopping, bargaining is often acceptable, but generally not where prices are clearly displayed, as in the larger department stores. Policemen who speak English have a red flash under their shoulder numbers.

Telecommunications: International telecommunications services are provided by Cable and Wireless (HK) Ltd. The Hong Kong Telephone Co Ltd, operating under government franchise, provides telephone and other telecom facilities in Hong Kong.
Telephone: Dialling code for Hong Kong, IDD access code + 852 followed by area code (5 for Hong Kong, 3 for Kowloon, 5 for outlying islands, 0 for New Territories) followed by subscriber's number. There were 2.2m telephone lines and 2.9m exchange lines in 1988, with fully automatic service and 52 per cent of the junction routes between exchanges served by fibre optic cable. Since 1984 a mobile radio telephone system has been in operation. Unlimited free local calls for domestic subscribers.
Telex: Public telex facilities are available 24 hrs at Cable & Wireless offices on Exchange Square. The Resto-Telex service enables an incoming telex to be collected by the addressee, provided full address and telephone in Hong Kong are given in the message. There were over 27,000 telex subscribers in 1988.
Telegram: In addition to placing telegrams at a public telegram counter, telegrams may be telephoned or telexed.
Other: Hong Kong Telephone provides a wide range of services, including direct-dial radio-paging, ship-to-shore telephone, computerised directory enquiries, telephone conferencing and a Viewdata information service. Cable and Wireless supplies leased circuits, Bureaufax, IDAS (International Database Access Service) and international facsimile services. A coast station provides telecomms services to and from ships at sea, as well as an air-to-ground long-range telephone system for aircraft operators.

Banking
Regulatory Authority: Commissioner of Banking.

Advertising: Advertising is available on commercial radio and TV, in cinemas and on poster sites, with some direct mail.

Electricity supply: 200 V AC, 50 cycles, with no uniformity in plug design. Most hotels have shaver adaptors.

Weights and measures: Metric system (Imperial system and local units also in use).

Representatives in Hong Kong: Australia, Austria, Bangladesh, Barbados, Belgium, Belize, Bhutan, Bolivia, Brazil, Burma, Canada, Chile, China PR, Colombia, Costa Rica, Cuba, Cyprus, Denmark, Dominican Republic, Ecuador, Egypt, El Salvador, Finland, France, Gabon, Gambia, German FR, Greece, Guatemala, Guinea, Honduras, Iceland, India, Indonesia, Israel, Italy, Jamaica, Japan, Jordan, Republic of Korea, Liberia, Malaysia, Malta, Mauritius, Mexico, Monaco, Mozambique, Nauru, Netherlands, New Zealand, Nicaragua, Nigeria, Norway, Pakistan, Panama, Paraguay, Peru, Philippines, Portugal, St Lucia, Senegal, Seychelles, Singapore, South Africa, Spain, Sri Lanka, Sweden, Switzerland, Thailand, Trinidad & Tobago, Tuvalu, USA, Uruguay, Venezuela.

Hotels: Hong Kong
Caravelle Hotel, 84-86 Morrison Hill Road, Happy Valley (tel: 5-754455; tx: 65793).

Excelsior, 281 Gloucester Road, Causeway Bay, P.O. Box: 30611 (tel: 5-767365; tx: 74550; fax: 7956459).

Furama Inter-Continental, 1 Connaught Road, Central (tel: 5-255111; tx: 73081; fax: 297405).

Harbour, 116-122 Gloucester Road (tel: 5-748211; tx: 73947 HABOH HX; fax: 722185).

Hilton, 2 Queens Road, Central, P.O. Box: 42 (tel: 5-233111; tx: 73355; fax: 8452590).

Lee Gardens, Hysan Avenue, Causeway Bay (tel: 5-8953311; tx: 75601).

Mandarin Oriental, 5 Connaught Road, Central, P.O. Box: 2623 (tel: 5-220111; tx: 73653; fax: 297978).

Marriott Hotel, 15 Queen's Road Central (tel: 5-8108366; tx: 66899 MARTT HX; fax: 8450737).

Park Lane Radisson, 310 Gloucester Road (tel: 5-8903355; tx: 75343; fax: 5-767853).

Victoria, Shun Tak Centre, Connaught Road, Central (tel: 5-407228; tx: 86608; fax: 476912).

Hotels: Kowloon
Ambassador, 4 Middle Road, at Nathan Road (tel: 3-666321; tx: 43840 AMHOC HX; fax: 690663).

Carlton, 4 miles, Tai Po Road, Tsim Sha Tsui,, P.O. Box: 96161 (tel: 3-866222; tx: 50155 CARHO HX).

Fortuna, 351-361 Nathan Road (tel: 3-851011; tx: 44897; fax: 7800611).

Grand, 14 Carnarvon Road, Tsim Sha Tsui (tel: 3-669331; tx: 44838 GRAND HX; fax: 7237840).

Harbour View (Holiday Inn), 70 Mody Road, East Tsimshatsui, P.O. Box: 98468 (tel: 3-7215161; tx: 38670 HIHV HX; fax: 3-695672).

Holiday Inn Golden Mile, 50 Nathan Road, Tsimshatsui, P.O. Box: 95555 (tel: 3-693111; tx: 56332 HOLIN HX; fax: 3-698016).

Hyatt Regency Hongkong, 67 Nathan Road (tel: 3-111234; tx: 43127 HYATT HX; fax: 7398701).

Imperial, 30/34 Nathan Road (tel: 3-662201; tx: 55893 IMPHO HXY; fax: 3112360).

Kowloon, 19-21 Nathan Road (tel: 3-698698; tx: 47604; fax: 698698).

Miramar, 130 Nathan Road (tel: 3-681111; tx: 44661 MIRHOHX; fax: 691788).

New World, 22 Salisbury Road, Tsimshatsui (tel: 3-694111; tx: 35860 NWHTL HX; fax: 699387).

Park, 61-65 Chatham Road South (tel: 3-661371; tx: 45740 PARKH HX; fax: 7397259).

Prince, Harbour City, Canton Road (tel: 3-7237788; tx: 50950 PRN HXY; fax: 7215545).

Regal Airport Hotel, Sa Po Road, P.O. Box: 89561 (tel: 3-7180333; tx: 40950 HOMRA HX; fax: 7992503).

Regal Meridien, 71 Mody Road, Tsimshatsui East (tel: 3-7221818; tx: 40955 HOMPRO HX; fax: 7236413).

Shamrock, 223 Nathan Road (tel: 7352271; tx: 50561; fax: 7367354).

Shangri-La, 64 Mody Road, East Tsimshatsui (tel: 3-7212111; tx: 36718 SHLA HX; fax: 7238686).

Sheraton Hong Kong, 20 Nathan Road (tel: 3-691111; tx: 45813 HKSHR HX; fax: 7398707).

The Hong Kong, 3 Canton Road (tel: 3-676011; tx: 43838 HONHO HX; fax: 7234850).

The Marco Polo Hong Kong, Harbour City, 13 Canton Road (tel: 3-7215111; tx: 40077 MPHK HX; fax: 7217049).

The Peninsula, Salisbury Road (tel: 3-666251; tx: 43821 PEN HX; fax: 7224170).

The Regent, Salisbury Road, Kowloon (tel: 3-7211211; tx: 37134 REG HX; fax: 7394546).

The Royal Garden, 69 Mody Road, East Tsimshatsui, P.O. Box: 95366 (tel: 3-7215215; tx: 39539 RGHTL HX; fax: 699976).

Business advice and information
American Chamber of Commerce, Swire House 10/F Central, Hong Kong (tel: 5-260165; tx: 83664 AMCC HX).

Chinese General Chamber of Commerce, 24-25 Connaught Road Central, Chinese General Chamber of Commerce Building 7/F, Hong Kong (tel: 5-256385; cable: CHICHACOM).

Chinese Manufacturers' Association of Hong Kong, 64-66 Connaught Road Central, CMA Building, 3rd and 4th Floor, Hong Kong (tel: 5-456166; tx: 63526 MFT HX).

Federation of Hong Kong Industries, 4/F Hankow Centre, 5-15 Hankow Road, Kowloon (tel: 3-7230818).

Hong Kong Export Credit Insurance Corp, South Seas Centre, Tower 1, 2/F, 75 Mody Road, Tsimshatsui East, Kowloon (tel: 3-7233883; tx: 56200 HKXC HX).

Hong Kong Exporters Association, Room 825, Star House, 3 Salisbury Road, Tsimshatsui, Kowloon (tel:3-699851).

Hong Kong General Chamber of Commerce, 22/F United Centre, 95 Queensway, Wanchai, Hong Kong (tel:5-299229).

Hong Kong Industrial Estates Corp, 107 Estate Centre Building, 19 Dai Cheong Street, Tai Po Industrial Estate, Tai Po, New Territories (tel: 0-6531183).

Hungary

Judy Dempsey

If there was one country in Eastern Europe which had gone the furthest in introducing political and economic reforms during 1989, it was Hungary. On the face of it, there was little or no pressure from outside the ruling Hungarian Socialist Worker's Party (HSWP) to introduce such far-reaching changes. After all, unlike Poland, there was and is no broadly based united opposition such as the Polish Solidarity movement. Indeed, the plethora of small groups which sprung up in Hungary during 1989 provided few real substantial challenges to the authorities. Nevertheless, the impulse for reform came from within the communist party itself, and while it was possibly spurred on and encouraged by developments in the Soviet Union, one underlying reason drove the HSWP forward. That was the urgent need to address the country's serious economic problems.

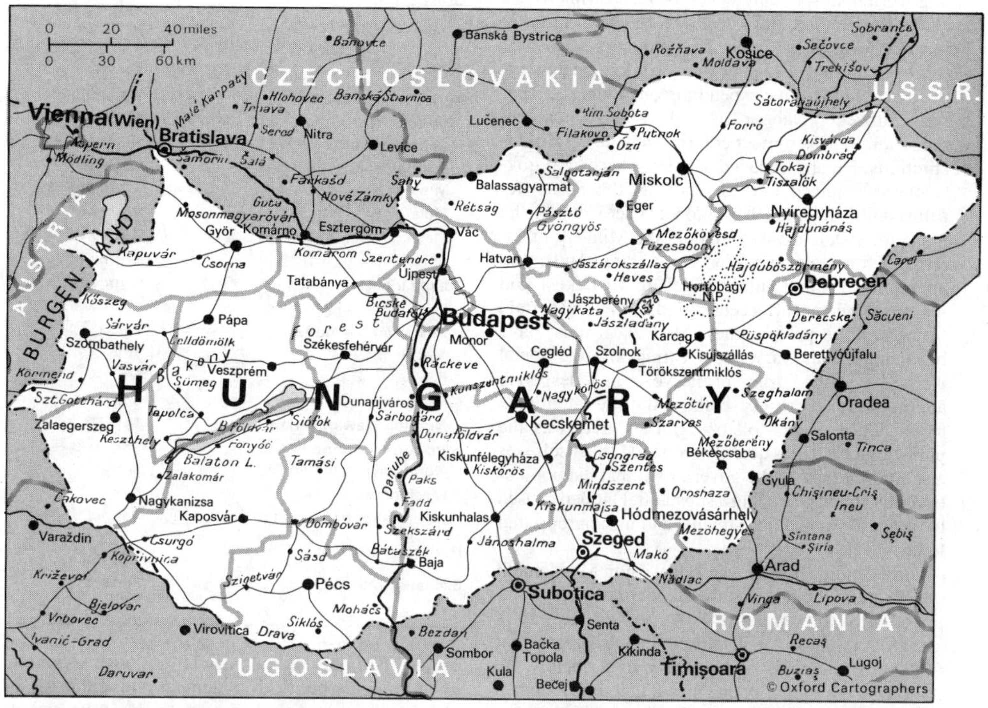

The economic reforms which had been introduced throughout the 1970s and 1980s had run their course. They now lacked the institutional framework in which they could be successfully implemented. In short, the authorities realised that without opening up the political structures, without introducing public accountability, without giving Hungarians a genuine stake in the political system, mechanisms vital for pushing ahead with radical economic reforms, the country was destined to permanent economic decline. But reversing such a decline also entailed a major dilution of the power of the HSWP, as well as the prospects of a a coalition government in 1990, two issues which were to preoccupy the party throughout 1989.

Party handover of power

In order to push ahead with the reforms, it appeared that the authorities, particularly officials such as Mr Rezso Nyers, the veteran father of the economic reforms and a prominent member of the communist party, and Mr Imre Pozsgay, the secretary of state and politburo member, decided in late 1988 and early 1989 to embark on a radical, but untested, road to reform. In practice, it involved the gradual transition from one-party rule to political pluralism. This programme was rife with uncertainties and unpredictables. For, in one-party systems, there is neither a mechanism nor a blueprint whereby the communist party can voluntarily relinquish its leading role. Nor is there any mechanism whereby power can be transferred to the government and parliament – institutions which, since the communist takeovers after the Second World War, were but rubber-stamp bodies, answerable and accountable only to the ruling communist party not to society. In short, during 1989 the Hungarian authorities, like their Polish counterparts, attempted to invest much greater authority and independence in parliament and in doing so, both countries aimed at separating the party from the state.

Signs that this was on the agenda came on 24 November 1988 when Hungary's national assembly or parliament voted in the 40-year-old Mr Miklos Nemeth as prime minister. Little had been expected from Mr Nemeth, a Harvard-trained economist and a relatively unknown quantity. But his speech that day unveiled plans which up till nine months later he had managed to carry out. The government's first task, he said, was to create a modern constitutional state. This involved breaking down the official bureaucracy, creating a dialogue with society and giving the government freedom to make economic and political decisions. His policies were given much greater weight in a speech by Mr Pozsgay who spelt out the government's programme for 1989. This envisaged a reform of the judiciary, the writing of a new constitution, in which the leading role of the party would be excluded, the legalisation of political parties, a new press law and a new electoral law. What Mr Pozsgay in effect proposed was a new concept of reform. Previous reforms, he argued 'remained within the purview of an enlightened absolutist power, which placed it within a protective and paternalistic political concept.' This had failed, he continued, because people were treated as 'political minors, an anonymous mass of subjects in the care of the state.' Both Mr Pozsgay's and Mr Nemeth's programme were, in effect, geared towards the rebirth of political life in Hungary and the renaissance of a civil society.

As a confirmation of his intentions, Mr Nemeth brought into the government Mr Rezso Nyers to head the economics ministry. Mr Nyers had already founded a new political movement, the New March Front in late 1988, one of many which had blossomed almost out of nowhere. His independent views, plus his economic skills and political acumem, served to give the new government a degree of respectability and credibility.

The 66-year-old Mr Nyers did not waste much time in explaining what had to be done. He argued that a coalition government was important if the economic reforms were to continue. He advocated an expansion of private property, the selling-off of state industry, the end of uneconomic, wasteful projects and the reduction of the country's staggeringly high foreign debt, some of the difficult problems which the government would have to deal with throughout 1989.

The economy

But as the debate over political reform forged ahead, sometimes at a frenetic pace which led to exciting developments during 1989, the government remained saddled with a sluggish economy. As it approached the 1990s, the problem of the Hungarian economy was that it was prone not only to inflation – running in 1989 at around 15 per cent – but the authorities also had to focus on reducing the external and domestic debt while at the same time cutting back on public expenditure at home.

By March 1989, Hungary's external hard currency debt had grown to \$14.5bn. Interest payments amounted to \$1.3bn for 1989 alone. In addition, the domestic debt amounted to Ft480bn (\$9.4bn) – 67.9 per cent of Hungary's gross domestic product for 1988. That did not include the Hungarian National

KEY FACTS

Official title: Hungarian Republic

Head of state: Presidential Council of three until elections 25 March 1990

Head of government: Prime Minister Miklós Németh

Capital: Budapest

Official Languages: Hungarian (Magyar)

Currency: Forint (Ft) = 100 fillérs

Exchange rate: Ft63.5 per \$ (Dec 1989)

Area: 93,033 sq km

Population: 10.6m (1988)

GDP per capita: Ft133,200 (1988)

GDP real growth: 0.4% (1988)

Labour force: 4.8m (1988)

Inflation: 15.7% (1988)

Bank's foreign obligations.

The government planned an all-out attack. First, in order to gain the last instalment of a $350m stand-by credit from the International Monetary Fund, Mr Laszlo Bekesi, who was promoted to head the finance ministry in May 1989, said cuts of Ft30bn – Ft40bn in government spending would be made. This would bring the budget deficit down to the originally planned level of Ft19.5bn. This reduction was to be achieved through a 25 per cent cutback in the government's administrative expenditure, including a 20 per cent cutback in labour, as well as a 10 per cent reduction in government contributions to certain institutions. He also reduced subsidies to several government-financed bodies, such as the Scientific Information Institute, the National Peace Council, the Hungarian Defence Association, the Ministry of Internal Affairs and the Workers' Guard. And in an attempt to forestall public criticism, Mr Bekesi said the cuts would not affect education, health and certain social welfare programmes. Mr Bekesi must have been more than wary about eating into the social welfare system. Earlier in the year, in a desperate attempt to raise extra cash, the authorities attempted to introduce a road-toll tax. But so great was the public outcry, it was quickly dropped. That itself was a salutary lesson to the government that the population was already carrying a substantial burden caused by years of wasted investments and misappropriated loans.

Furthermore, neither Mr Bekesi nor Mr Nemeth could ignore the fact that the standard of living continued to fall throughout 1989. At the November 1988 session, it had been disclosed in parliament that 660,000 Hungarians were already living below the poverty level. In reality that meant that their incomes were below Ft3,000 a month. That, combined with a 1 per cent drop in consumption, a 2 per cent reduction in real income, a 6 per cent reduction in real wages and an inflation rate officially set at 15 per cent but reckoned to be running closer to 20 per cent, was enough to breed a certain amount of discontent in the population which inevitably engendered a certain degree of caution in the government.

The authorities were therefore forced to look further afield if they wanted to revive the economy, but without eating into the living standards of the population. And in this context, the government gained much-needed popularity at the expense of Mr Karoly Grosz, the increasingly beleaguered party leader.

The point at issue was the future of the Nagymaros-Gabcikovo dam. For the Hungarian authorities, the dam represented a running sore and a drain on their resources. It had been clear that few officials wanted the project to go ahead, but the political and ideological circumstances prevailing in Eastern Europe in the pre-Gorbachev years had forced the party leadership to defend what often amounted to the indefensible. The dam, first conceived in the early 1950s, and questioned even at that time by the Hungarian authorities, involved re-routing the river Danube 180 km along the Hungarian-Slovak border. One dam would be built at Gabcikovo in Slovakia, the second at Nagymaros in Hungary. A treaty between Czechoslovakia and Hungary was signed in 1977 and work finally began in the early 1980s. But as work proceeded, it gave birth to an independent environmental movement in Hungary called the 'Duna Koer' or Danube Circle. Through public petitions and independent studies, which proved that the dam would do irreparable damage to the environment, besides possibly destroying a natural water-filtering system which could have adversely affected the water supply to Budapest, the issue was finally but hotly debated in parliament.

Curiously, Mr Grosz, apparently oblivious to the changes taking place in the political climate in Hungary, continued in early 1989 to support this vast dam project. Not only did taking such a stance seriously undermine his political judgement and authority, it also handed a political victory to the government which was trying to assert its authority at the expense of a divided party. During a stormy debate in parliament between 10 and 12 May 1989, Mr Nemeth said all work on the dam would be postponed until the government made yet another, but this time independent, assessment of the project.

The government's statement gave the environmental lobby an enormous fillip. It also strengthened the government's standing in society because it showed that it had been prepared, finally, to listen to independent views and respond to public pressure. It was the first taste of a political victory for the opposition. It was also a test-case for the government since it demanded a certain amount of political courage to admit to its East European allies that public pressure precipitated a change of heart. This did not go down well in neighbouring Czechoslovakia which immediately accused Budapest of reneging on a treaty and demanded compensation. But even if Hungary would have to pay compensation to both Czechoslovakia (in the region of $500m) and Austria (which had helped finance construction of the dam at Nagymaros), the postponement of the project had relieved the government of a significant economic burden and granted it a degree of popularity.

Halting this uneconomic project, however, was just one of the moves made by the government to tackle the more obvious problems of the economy. Another, and more novel, if not radical idea, was the proposal to sell off sections of its state-owned industry. This had been mooted by Mr Tamas Beck, the trade minister, when he visited West Germany in February 1989. It was later followed up in an extensive report compiled by a special economic committee in the HSWP, headed by Professor Ivan Berend, the liberal economist and head of the Academy of Sciences. Among other radical measures, the committee suggested turning state enterprises into stock companies through the selling of shares. This would involve commercial banks, insurance companies, enterprises and foreigners. The idea was to raise capital for cash-strapped industries, some of which were profitable. But it also implied a major shift in ideological direction. What the committee argued for was 'a change from state ownership to new common ownership forms and the sale of state-owned enterprises to foreigners and private individuals.'

The proposal represented such a radical shift in economic thinking because in the economic order of the one-party system, it had been the state which had been the owner since property was invested in the state, in this case the ruling communist party. Yet, during 1989, the Soviet Union, Poland and Hungary had been grappling with the question of how to transfer state ownership to the individual, or to foreigners. This meant redefining the whole question of property relations, something which had not been on any economic agenda since the communist takeover. In the case of Hungary, the authorities were preparing to sell off assets which had belonged to the state. By mid-1989 the attempts had borne some fruit. For instance, 49 per cent of Tungsram, the worldwide Hungarian lighting company, was sold off in May 1989 to a consortium of Austrian-led banks. The deal meant that not only would the Hungarian authorities receive much-needed cash, but also that the Austrian banking consortium would have a direct say in the running of the company, from production up to management levels. Such a transfer of control would have been unthinkable even a year before. It was even planned to list Tungsram on the Vienna Stock Exchange in the early 1990s, subject to certain conditions, such as bringing Tungsram's accounting system into line with Western-style accounting. The Tungsram experiment, which is almost certainly not expected to be the only such 'sell-off' will be viewed with interest by Western firms. But as the economists and the government moved further towards the market economy, history was being re-written out on the streets. The party's omnipotence was being challenged.

Burying the past

De-mythologising the party happened slowly and with caution. *Nepszabadsag*, the party daily, in a marvellous article, conveyed the sense of change

KEY INDICATORS

	Unit	1984	1985	1986	1987	1988
Population	m	10.7	10.6	10.7	10.6	10.6
Gross domestic product (GDP)	Ft bn	979	1,034	1,089	1,226	1,411
GDP per capita	Ft	91,800	97,100	102,400	115,500	133,200
GDP real growth	%	2.7	-0.3	1.5	4.1	0.4
Inflation	%	8.3	7.0	5.3	8.6	15.7
Agricultural production	1980=100	109.6	103.5	106.0	103.9	108.4
Crops	1980=100	107.9	102.1	105.9	100.0	107.5
Animal products	1980=100	111.2	105.0	106.2	107.8	109.4
Industrial production	% growth	2.7	0.7	1.9	3.8	0.0
Mining	% growth	–0.6	2.9	–0.1	–0.4	–3.7
Metallurgy	% growth	1.3	0.1	2.8	0.9	4.3
Machinery	% growth	2.2	3.4	3.7	4.7	0.0
Chemicals	% growth	4.0	1.0	1.4	6.1	1.3
Light industry	% growth	3.4	-0.9	1.1	2.9	0.2
Food processing	% growth	3.1	-1.8	1.3	3.2	–2.5
Hard currency exports	$m	4,898	4,490	4,489	5,013	5,835
Machinery	$m	519	594	612	612	731
Fuels	$m	646	363	289	323	225
Foodstuffs	$m	1,433	1,283	1,173	1,229	1,503
Non-food raw materials	$m	1,702	1,658	1,715	2,017	2,472
Hard currency imports	$m	4,302	4,317	4,935	5,388	5,320
Non-food raw materials	$m	2,404	2,372	2,746	3,079	3,322
Machinery	$m	453	525	661	784	730
Hard currency balance of trade	$m	596	173	–446	–375	515
Hard currency debt	$bn	8.8	11.8	15.1	17.7	17.4
Exchange rate	Ft per $	48.0	50.4	45.8	47.0	50.4

taking place in Hungary.

When new owners moved into the office building that used to belong to Ganz-Mavag, the former flagship of Hungarian industry, the new occupants, who are banking and manufacturing specialists, discovered something that they had not detected during their previous visits and surveys. Their old reflexes might, perhaps, have been the reason for the blindness. On the top of the high-rise building, on each of the four sides, was a five-pointed red star. The embarrassing news spread quickly. A letter was sent to the district party committee asking for advice about how to proceed. The party committee, however, had in the past at best dealt only with the installation of stars and never with their removal. In its perplexity, it in turn passed on the request for advice to the local city council, claiming that the issue did not fall under its jurisdiction. The local council argued, however, that the matter was not a city skyline problem but a political one. The request for the removal of the stars then progressed through all the possible floors of the city council's building and finally, as it should be, was sent back to the district party committee. There, the matter was pondered over once more; and a compromise decision was finally reached under the aegis of pluralism and tolerance. If the red stars were really disturbing the new owners in their work, they could be removed. The stars, that is. But not all at once. After all, comrades, dismantling should be done with dignity and moderation.

And so it was, on 16 June 1989 that Mr Imre Nagy, the Hungarian prime minister during the 1956 Uprising who was executed in 1958, was re-buried in Budapest with dignity and full honours.

The event unleashed an enormous collective outburst of emotion from a nation which had not been allowed to speak openly about 1956. For thirty-three years, the Uprising had been called a counter-revolution. During that time, those who had participated in the Uprising, those who had been imprisoned, those who had been executed, had been relegated to the status of non-persons. Whether it was the green light from the Soviet leadership which itself was quickly trying to fill in the blank spots of history, or a decision made by the Hungarian authorities finally to acknowledge one of the great tragedies of Hungarian history, is open to conjecture. But by burying the past with dignity, the Hungarian authorities were preparing for the future.

It had been Mr Pozsgay and Mr Nemeth who argued the case for the reappraisal of 1956. Not just because it was time to come to terms with the past. But because no future government (or indeed communist party) could hope to gain any support in any future free election, if it were not prepared to articulate an issue which ran deep in the national consciousness of the society. On a political note, the re-burial of Imre Nagy led to the downfall of many top communists, including Mr Janos Berecz, who doggedly defended, until it became indefensible, his thesis that 1956 was a counter-revolution. Mr Berecz, who had been once tipped as a possible successor to Mr Janos Kadar, was dropped from the politburo in a reshuffle in July 1989. Others also fell from grace, including the Hungarian ambassador to Moscow, Mr Sandor Rajnai who was recalled for the ignominious role he played in 1956. The chapter closed quietly on 6 July 1989, when Mr Kadar, the party leader from 1956 to 1988 who, despite his unsavoury political role after Soviet tanks crushed the Uprising, helped to forge some national consensus in extraordinarily difficult circumstances, died a disillusioned, lonely man.

However, this burial of the past, instead of bolstering the morale of the HSWP, actually weakened it even further. This was due to the dilatory nature of Mr Grosz who continued to lack a feel for what was taking place in Hungary. Even as the government was preparing to re-bury Imre Nagy, Mr Grosz was reputed to have insisted that the party should not be represented at the ceremony, a miscalculation if ever there was one. Moreover, people such as Messrs Nyers, Pozsgay and Nemeth realised that as the government was preparing for free elections in 1990 and the party was preparing for a congress in October 1989, the party was becoming ill-equipped to deal with the changing political circumstances in Hungary. In short, unless the party adopted a new programme and image, it would fail miserably in any future elections.

Signs of deep disunity in the party were more than apparent throughout 1989. Mr Pozsgay had at one time contemplated splitting the party, and possibly creating a new one. But this was deemed to be impracticable and, besides, would have even further prolonged the confusion and bitterness in the party. The root of the disunity was the growing realisation, especially for the more conservative members, that the old days of one-party rule were coming to an end. And if the party aimed at lEast towards surviving the elections and possibly agreeing to a coalition with the score of opposition groups, then it would require a tighter, more coherent and attractive programme.

This then explains the background to one of the most interesting reshuffles in the party leadership for many years. On 23 June the 113-strong central committee met. By all accounts, it was a meeting designed for 'damage limitation' in the run-up to the congress and the 1990 elections. The power of Mr Grosz was considerably weakened in contrast to that of Mr Nyers who was made chairman of the party. In addition, a new Party Presidium was elected, consisting of four politburo members. They included 59-year-old Mr Grosz, who remained as party leader, but with reduced powers, Mr Nemeth, Mr Nyers and Mr Pozsgay, who was also nominated as the HSWP's candidate for president. Apart from the newly-created Presidium, the central committee also elected a twenty-one member Political Executive Committee which replaced the much smaller politburo. That consisted of a careful balance between reformers and conservatives.

The aim of the reshuffles, as indicated above, was to prepare the ground for the party congress and the elections. It also meant that the party, in theory, would be entering a fresh round of talks with the opposition groups in a somewhat better position, since both Mr Nyers and Mr Pozsgay carried far more weight than Mr Grosz, who had been expected to relinquish the party leadership in October.

Outlook

But as Hungary edges closer towards political pluralism, the path along which it treads remains fraught with uncertainties. For one thing, the economy remains in a sorry state. It requires urgent attention but improvements and tangible results will not be achieved overnight. Politically, too, Hungary goes into the unknown. But unlike previous attempts at reform, the Soviet Union appears to have given the green light to Hungary. Indeed, the removal of Soviet troops testifies to the changing relationship between the Soviet Union and its East European allies as much as it does to the changing nature of East-West relations. In addition, Hungary has some advantages as it looks for signposts along the route. Unlike Poland and needless to say, the Soviet Union, Hungary is a small country of 10 million people. It is also homogeneous and hard-working. These assets look likely to be of increasing importance as the 1990s begin with a year which could be difficult but exciting. During this time, the four stars on the building may have been removed, an act which will have gone beyond mere symbolism.

COUNTRY PROFILE

Political system: Socialist republic with unicameral national assembly (Országgyülés), which is elected by universal suffrage every five years. The Presidential Council of three carries out the functions of head of state. The Council of Ministers has executive power. The leading political force is the Hungarian Socialist Party, adopting in 1989 a social democratic platform. The Party's plans for a popular election in January 1990 were rejected by voters. Multiparty elections are planned for 25 March 1990 after which the President will be selected.

Main cities/towns: Budapest (population 2.115m), Miskolc (pop 0.208m), Debrecen (pop 0.220m), Szeged (pop 0.189m), Pécs (pop 0.183m), Györ, Nyiregyháza (pop 0.119m), Székesfehérvár (pop 0.114m), Kecskemét (pop 0.106m).

Media

Dailies: Five national dailies, including *Népszabadság* (circulation 670,000), Magyar Hirlap (89,000), *Népszava* (276,000) and *Magyar Nemzet* (132,000). The MTI Hungarian News Agency publishes a twin daily, *Daily News-Neueste Nachrichten* in English and German.
Weeklies: Hèti Világgazdaság (economics, 141,000). German language weeklies: *Budapester Rundschau, Neue Zeitung* (4,500).
Periodicals: Other foreign language periodicals: *Hungarian Observer, Hungarian Exporter, Hungarian Economy, Hungarian Travel Magazine, Hungarian Digest (50,000), Hungarian Book Review, Hungarian Business Herald* (4,000), *New Hungarian Quarterly* (5,400), *Hungarian Trade Journal* (18,000). Western magazines can be bought in large hotels and at some news-stalls.
Radio: Radio service operated by Magyar Rádió. English, German and Russian language news broadcasts during summer at 1200 hrs.
Television: TV service operated by Magyar Televizió. Two channels broadcasting around 70 hrs/week.
Advertising: Usually conducted through national agencies. Magyar Hirdetö, Hungexpo and Magyar Média. Advertising space can be purchased also directly on TV and Radio, in cinemas, newspapers, magazines and posters.
News Agency: Magyar Távirati Iroda (MTI), (Hungarian news agency), 1016 Budapest, Fém utca 5-7 (tel: 175-67-22; tx: 224371 mtih).

Domestic economy: Socialist market economy with large industrial base and heavy dependence on foreign trade. Also important agriculture and tourist sectors. Despite endemic labour shortages and few natural resources high levels of economic growth achieved during 1970s. Pace of growth since slowed due to debt burden, higher inflation and reduction in industrial activity. Heavy government deficits. Seventh Five Year Plan 1986-1990 emphasises faster economic growth, a cutback in consumer spending, greater efficiency, production restructuring, lower inflation and, in the latter years of the period, a palpable rise in living standards. Value Added Tax was introduced in September 1987 – the first instance in any socialist country.

Employment: Approximately 37.4 per cent labour force employed in industry and construction, 18.4 per cent in agriculture, 10.9 per cent in commerce and 8.3 per cent in transport and communications. Rapid growth in tertiary sector employment. Wages linked to enterprise profits.

COUNTRY PROFILE

External trade: The majority of foreign trade is a state monopoly, though recently there has been a considerable rise in the number of joint companies which have the right to engage in foreign trade on their own.
Exports: Mainly raw materials, semi-finished products, agricultural produce, machinery and other capital goods. Principal trade partners: USSR (27.6%), FRG (11%), Austria (5.3%), GDR (5.3%), Czechoslovakia (5.4%), Poland (3.3%), Italy (4.2%), Yugoslavia (2.8%), Switzerland and Liechtenstein (2.0%) and USA (3.0%).
Imports: Mainly raw materials, oil, basic manufactures, machinery, chemicals. Principal trade partners: USSR (25%), FRG (12.9%), Austria (7.2%), GDR (6.4%), Czechoslovakia (5.1%), Poland (4.1%), Italy (3.2%), Yugoslavia (2.9%), Switzerland and Liechtenstein (2.6%) and USA (2.2%).

Agriculture: Sector contributes around 18.2 per cent to GDP and accounts for over 20 per cent of export earnings. High productivity. Farming is largely socialised, with co-operatives the dominant form of production. Agriculture and food industry produce on average 50 per cent more food than is consumed domestically. Price support system and subsidies on farming inputs operating. Principal crops include cereals, oil seeds, vegetables, sugar-beet, grapes and potatoes. Also important livestock sector.

Industry: Accounts for 36 per cent of GDP and is major determinant of economic growth. Activity largely confined to state and co-operative sector, which is primarily export based. Major industries include metallurgy, machinery and equipment, building materials, chemicals, food processing, textiles and engineering. Under CMEA joint co-operation and specialisation scheme emphasis on expansion of commercial vehicles, chemicals, pharmaceuticals, aluminium and electronics machinery. In the past four decades co-operation with COMECON CMEA countries determined – though since 1968 to a decreasing extent – the main features of the Hungarian economy. In the 1980s there was a gradual decrease in the proportion of turnover with CMEA member countries.

Mining: Major European producer of bauxite. Also small-scale producer of coal, lignite, manganese ore, oil and natural gas. Few other natural resources.

Energy: Around 50 per cent dependent upon imported oil and gas from USSR and to lesser extent Middle East. Electricity also imported via a grid from Vinnitsa in Soviet Ukraine and exchange with Austria. Steady reduction of imported oil in energy use and increased use of domestic coal.

Banking: Most developed system in Eastern Europe, with National Bank playing important role in economic and financial management of economy. However, under reform of banking system in 1987, various National Bank functions devolved to five commercial banks – Hungarian Credit Bank, National Commercial and Credit Bank, Budapest Bank, Hungarian Foreign Trade Bank and General Banking and Trust Company. Foreign trade and currency transactions conducted by Hungarian Foreign Trade Bank. State Development Bank provides capital funds for main investment projects. Also issues the majority of bonds. National Savings Bank, with extensive branch network, offers banking for the general public, and also foreigners needing foreign exchange accounts. Central-European International Bank is an internationally active off-shore bank owned by the National Bank and six foreign banks. Citibank Budapest is a joint venture between Citibank and Central Bank of Exchange and Credit, a subsidiary of the National Bank. General Enterprise Bank was established in 1985 to take over enterprise fund of National Bank.

Legal system: Based on communist legal theory, with both civil and common law elements, with constitutionality being an important goal since 1988.

Membership of international organisations: CMEA, Danube Commission, FAO, GATT, IAEA, IBRD, ICAC, IDA, ILO, IMF, IMO, IPU, ISO, ITC, ITU, UN, UNESCO, UNCTAD, UPU, Warsaw Pact, WFTU, WHO, WIPO, WMO.

BUSINESS GUIDE

Time: GMT + 1 (GMT + 2 from late Mar to late Sep).

Climate: Continental; average temp between one degree C and three degrees C in winter, and around 20°C in summer. Coldest month Jan, hottest month Jul. Snow and frost in winter. Wettest month May, driest Sep.
Clothing: Medium to heavyweight and heavy topcoat for winter. Lightweight for summer. Raincoat needed in spring and autumn.

Entry requirements: Visa required by all but nationals of CMEA countries, Austria, Finland, Malta, Nicaragua and Sweden. Tourist visas available at frontier (not at railway border stations) and Ferihegy Airport. Visitors must register with authorities within 48hrs of arrival (normally hotels handle registration details). Transit visas valid for 48hrs.
Currency: All foreign currency must be declared on entry if requested. No restrictions on amount of foreign currency taken in or out. Illegal to bring in or take out more than Ft400 per person. Unified exchange rate in operation, which is adjusted weekly. Exchange facilities available at main banks, railway stations (24hrs), hotels and tourist offices.
Customs: Personal effects and goods bought at Intourist shops duty-free. Some restrictions on the goods that can be taken into or out of the country.

Air access: Major international airlines operate direct and indirect flights to Budapest.
National airline: MALÉV Hungarian.
International airport: Budapest (BUD), Ferihegy airport-Terminal II (MALÉV flights) 22 km and Terminal I (all other flights) 16 km from city centre.

Surface access: Good international road and rail connections; access from Vienna, Graz, Belgrade and Ljubljana.

Hotels: Majority of large hotels operated by the three big hotel chains, HungarHotels, Pannonia and Danubius. Reservations should be made in advance directly or through IBUSZ. Amex, Diners, Eurocard, Visa and Master Charge credit cards accepted in larger hotels, restaurants and shops. Tipping usually 10-15 per cent.

Restaurants: Many types and categories. There are restaurants, inns, csárdas and hotels. Food and drink relatively cheap. Tipping 10-15 percent of the bill.

Car hire: Available from IBUSZ travel company, from hotels and at Ferihegy airport. International driving licence and insurance certificate needed. Speed limit in built up areas 60 kph, on main roads 100 kph and motorways 120 kph. Drinking and driving strictly prohibited. Seat belts compulsory.

City transport
Taxis: Besides the state-owned taxi services (Fötaxi, Volántaxi) there are also private taxis operating. Available from ranks, by telephone or can be hailed in the street. Fötaxi (tel: 122-22-22); Volántaxi (tel: 166-66-66); Budataxi (tel: 129-40-00); City taxi (tel: 122-88-55); Lux Taxi (tel: 153-25-32). Basic fare about Ft10 plus Ft11-12 per km. Tipping usually 10-15 per cent.
Buses: Blue tickets used, and the fare is uniformly Ft6.00. Intercity service operates from Engels Ter.
Metro: Identified by large 'M'. Ft.5.00 yellow tickets can be used. Three main lines: the Millennium (yellow line 1), East-West (red line 2), North-South (blue line 3). Tickets on buses, metro etc self-cancelled on mechanical punch machines.
Trams: Yellow tickets used, and the fare is uniformly Ft5.00.
Main railway stations: Déli Pu (Southern RW Terminal), Krisztina krt 37/a, Budapest 1.

Keleti Pu (Eastern RW Terminal), Baross Tér (tel: 142-91-50).

Nyugati Pu (Western RW Terminal), Lenin Körút 111 (tel: 122-78-60).

National transport
Air: No internal air service.
Road: Extensive road network. Country-wide bus service operated by Volán companies. Main terminal at Engels Tér (tel: 172-966).
Rail: Express service operates from Eastern (Keleti) and Western (Nyugati) railway stations. Tickets for most destinations can be booked in advance from central office of MÁV (tel:228-049).
Water: Around 1,700 km of navigable waterways. Hydrofoil service between Budapest and Vienna daily in summer season.

Public holidays
Fixed dates: 1 Jan (New Year's Day), 15 Mar (National Day), 4 Apr (Liberation Day), 1 May, 20 Aug (Constitution and St Stephen's Day), 25-26 Dec (Christmas Holiday). If holidays are close to a weekend, dates may be altered.
Variable dates: Easter Monday.

Working hours: Business, government, banking (Mon-Fri) 0800 – 1630; shops (Mon, Tue, Wed & Fri) 1000 – 1800, (Thu) 1000 – 2000.

Social customs: Shake hands on arriving and departing. Business cards used extensively.

Business language and interpreting/translation: Provided by hotels, travel agencies and OFFI National Translation Bureau (tel: 112-96-16) and private interpreters and translators.

Telecommunications
Telephone: Local 3 min calls can be made from public booths with a 2-forint coin. For operator service dial 117-00-00, 157-33-33. Domestic long distance dial 06, international automatic long distance dial 00. For international information service dial 118-69-77. International dialling code: 36 followed by 1 for Budapest. Since Sep 1989, Budapest numbers are seven-digit, with 1 preceding the previous six-digit number.
Telex: Available from main hotels and at central post office (Városház utca 18, tel: 118-77-28). Country code 61 H.
Telegram: Available at post offices and by dialling 02.

Postal service: Dial 117-22-00 for information on postal matters (English, German, French, Russian language service).

Trade fairs: International trade fairs held in Budapest in late May (technical and capital goods) and late September (consumer goods). Each year there are also a number of specialised exhibitions. Information available from Hungexpo.

BUSINESS GUIDE

Electricity supply: 220 V AC, 50 cycles.

Representation in capital: Afghanistan, Albania, Algeria, Argentina, Australia, Austria, Belgium, Bolivia, Brazil, Bulgaria, Canada, China PR, Colombia, Cuba, Czechoslovakia, Denmark, Ecuador, Egypt, Finland, France, German Democratic Republic, Federal Republic of Germany, Greece, India, Indonesia, Iran, Iraq, Israel, Italy, Japan, Kampuchea, Democratic People's Republic of Korea, Kuwait, Libya, Mexico, Mongolia, Netherlands, Norway, Peru, Poland, Portugal, Romania, Spain, Sweden, Switzerland, Syria, Turkey, United Kingdom, USA, USSR, Uruguay, Venezuela, Vietnam, People's Democratic Republic of Yemen, Yugoslavia.

Hotels: Budapest
Atrium Hyatt, Roosevelt tér 2, H-1051 (tel: 138-30-00; tx: 225485; fax: 188659).

Béke, Lenin körút 97, H-1067 (tel: 132-33-00; tx: 225495).

Buda Penta, Krisztina körút 41-43, H-1013 (tel: 156-63-33; tx: 225495; fax: 556964).

Duna Intercontinental, Apáczai Csere János utca 4, H-1054 (tel: 117-51-22; tx: 225277; fax: 184973).

Flamenco, Tass Vezer utca 7, H-1113 (tel: 161-22-50; tx: 224647; fax: 658007).

Fórum, Apáczai Csere János utca 12-14, H-1368 (tel: 117-80-88; tx: 224178; fax: 179808).

Gellért, Gellért tér 1, H-1111 (tel: 185-22-00; tx: 224363).

Grand Margitsziget, Margitsziget (Margaret Island) (tel: 111-10-00; tx: 226282).

Hilton, Hess András tér 1-3, H-1014 (tel: 175-10-00; tx: 225984; fax: 75100320).

Novotel, Alkotás utca 63-67, H-1444 (tel: 186-95-88; tx: 225496; fax: 665636).

Royal, Lenin körút 47-49, H-1073 (tel: 153-31-33; tx: 224463; fax: 421122).

Thermal, Margitsziget, H-1138 (tel: 111-10-00; tx: 225463, 226681).

Business advice and information
Hungarian Chamber of Commerce, PO Box 106, H-1389 Budapest, Kossuth Lajos tér 6-8 (tel: 153-33-33; tx: 224745).

Ministry of Trade, Honvéd utca 13/15, Budapest V (tel: 153-00-00; tx: 225578).

Principal newspapers
Heti Világgazdaság, 1133 Budapest Vág u 13 (tel: 140-87-76; tx: 22676).

Hungarian Business Herald, 1075 Budapest, Tanács Krt 7.1.3 (tel: 141-72-70).

Hungarian Digest, 1073 Budapest, Lenin Krt 9-11 (tel: 181-15-80).

Hungarian Trade Journal, 1073 Budapest, Lenin Krt 9-11 (tel: 122-12-85; tx: 226207).

Hungarian Travel Magazine, 1088 Budapest, Múzeum u 11 (tel: 136-13-84).

Magyar Hirlap, 1393 Budapest, PO Box 305 (tel: 122-24-00; tx: 224268).

Magyar Nemzet, 1073 Budapest, Lenin Krt 9-11 (tel: 122-24-00; tx: 224269).

Népszabadsag, 1960 Budapest, Blaha Lutzatér 3 (tel: 133-61-30; tx: 225551).

Népszava, 1964 Budapest, Rákóczi u 54 (tel: 122-41-01).

Neue Zeitung, 1391 Budapest, Nagymezö u 49.

New Hungarian Quarterly, 1088 Budapest Rákóczi u 17 (tel: 138-42-14).

Other useful addresses
Állami Biztosító (State Insurance Company), H-1813 Budapest, U\03llöi utca 1 (tel: 117-85-66; tx: 224857).

Federation of Scientific and Technical Societies (MTESZ), Budapest V, Kossuth Lajos tér 6-8 (tel: 153-33-33).

Hungária Biztosító (Hungária Insurance Company), H-1115 Budapest, Bánk bán utca 17 (tel: 182-07-50; tx: 223104).

Hungarian Automobile Club, Budapest XIV, Francia utca 38 (tel: 169-18-31).

Ministry of Finance, 1051 Budapest József Nádor tér 2/4 (tel: 118-20-66).

Ministry of Industry, 1024 Budapest, Mártirok u. 85 (tel: 132-65-70; tx: 225376).

National Planning Office, Roosevelt tér 7/8, Budapest V (tel: 112-34-80, 111-02-00; tx: 224993).

Statistics
Hungarian Central Statistical Office, PO Box 51, Központi Statiszikai Hivatal, 1024 Budapest, Keleti Károly utca 5-7 (tel: 115-28-50; tx: 224308).

Marketing in Hungary, PO Box 106, Budapest V, Kossuth Lajos tér 6-8 (tel: 131-41-55; tx: 224745).

Tourist information
IBUSZ – Hungarian Travel Agency (main Budapest office), VII, Tanács körút 3/c (tel: 142-31-40; tx: 225650).

Banking
Magyar Nemzeti Bank (National Bank of Hungary), H-1054 Budapest, Szabadság tér 8-9 (tel: 153-36-00, 132-33-20; tx: 225755).

Iceland

Jón Hákon Magnússon

The year 1988 marked a downturn in Iceland's economy when one of the most favourable periods in its history had passed and Gross Domestic Product (GDP) decreased for the first time since 1983. GDP was estimated to have fallen 1.4 per cent and was predicted to fall by an additional 1.7 per cent in 1989. The exceptional growth of the years 1984 to 1987, 6.3 per cent on average, came to an abrupt halt in 1988 and the decline looked likely to continue through 1989, with very little economic and political relief in sight.

Political situation

This northernmost island nation of the North Atlantic community was being run at mid-1989 by the most unpopular three-party left-of-centre coalition government since Iceland won its independence from Denmark in 1944. The government, which seemed to be completely unable to cope with the deteriorating economic scenario, was supported by only 23.6 per cent of voters with 76.4 per cent against it, according to polls taken at mid year. The government, led by Mr Steingrímur Hermannsson, prime minister and leader of the Progressive Party, a centrist party with a rural power base, was formed late in 1988, after a three-party coalition government led by Mr Thorsteinn Palsson, leader of the right-of-centre Independence Party, collapsed when the Progressives and the Social Democrats broke away in a dramatic manner during a live TV news broadcast. The other two parties in the Hermannsson government were the Social Democratic Party, led by Mr Jon Baldvin Hannibalsson, foreign minister, and the communist-influenced and anti-NATO People's Alliance led by Dr Olafur Ragnar Grimsson, finance minister.

This unpopular trio of political parties did not have a clear majority in parliament. It could nevertheless sustain a vote of no confidence in the joint chamber of *Althing* (parliament), but it had to rely on support from members of the miniature Citizens' Party, when it tried to pass legislation through both houses of *Althing*. Attempts were made to bring the Citizens' Party into the coalition, without much luck but by mid-1989, since the five-member parliamentary party demanded three ministerial posts for itself, equal to the portfolios of the other three. The polls showed that support for this splinter group, which broke away from the Independence Party during the last elections, had all but evaporated, or in fact dropped from 10.6 per cent support to 0.5 per cent. Its colourful former leader and creator Mr Albert Gudmundsson had left the party to become Iceland's ambassador to France.

Support for the Independence Party in 1989 polls was more than 50 per cent, but nobody can safely predict whether the party will reap the benefit during

KEY FACTS

Official title: Republic of Iceland

Head of state: President Vigdís Finnbogadótti\01r

Head of government: Premier Steingrimur Hermannsson (Progressive Party)

Ruling party: Coalition of Progressive Party, Social Democrats, People's Alliance, Regional Equality Party, Citizens' Party

Capital: Reykjavík

Official Languages: Icelandic

Currency: Icelandic Krona (IKr) = 100 aurar

Exchange rate: IKr62.2 per $ (Dec 1989)

Area: 102,846 sq km

Population: 251,743 (1988)

GDP per capita: $22,976 (1988)

GNP real growth: –1.9% (1988)

Labour force: 127,000 (1988)

Unemployment: 0.7% (1988)

Inflation: 23.5% (1988)

Trade balance: –IKr7,322m (1988)

Foreign debt: $2,427m (1988)

the next elections, which are due in 1991. The governing coalition could easily break up any time before 1991, but yet again nobody is able to predict an early election in this country, where the political climate can be as hot as the famous lava flows from Iceland's numerous volcanos.

Economic downturn

Among the results of the drastic economic downswing is increasing unemployment, which has been almost non-existent in Iceland since the Second World War. Rising unemployment is creating unrest among employees in the private sector. The civil servants are more secure, since the government has not made any serious efforts to cut its ever increasing public sector. Overtime was cut in 1989 to save jobs and surveys show that there is no longer an excess demand for labour as was the case until late 1988. Import of foreign workers has stopped, but Icelanders are now in increasing numbers seeking job opportunities in other countries, mainly in Sweden.

The main task of Mr Hermannsson's left-wing government was to improve the profitability of industry and fishing. The coalition initially promised measures to increase job security, reduce inflation and interest rates, defend the living standards of the lowest paid and cut the trade deficit. However, the results by mid-1989 were mostly the opposite. Inflation from the beginning to the end of 1989 was expected to be 21 per cent and the budget deficit 4-5bn Icelandic krona. The 1989 budget called for increased tax revenues and a more complex tax system than before. The government said it was better to eliminate the budget deficit by increasing taxes than to continue borrowing which in turn increases the rate of inflation.

The country's economic problems are mostly of domestic origin, since external conditions are still favourable, even though fish prices have deteriorated in some markets. In Iceland, the GDP per capita is among the highest in the world. The 1988 GDP amounted to US$5,863m, which is approximately US$23,470 per capita in a nation of 250,000 people. The National Economic Institute (NEI), predicted real earnings per capita to be 6-7 per cent lower in 1989 than 1988. National income in 1989 was anticipated to contract more than the GDP or by around 3 per cent. National expenditure was predicted to fall by just over 3 per cent from 1988. This was mainly due to 3.5 per cent contraction in private consumption and a 3.3 per cent fall in investment expenditure. Public consumption was expected to increase by 1 per cent in 1989.

No increase was foreseen by NEI in the 1989 deficit on the current account, which was expected to run close to 3.5 per cent of GDP, a similar level to that of 1987 and 1988. Contracting purchasing power and weakening of the krona was expected to cut imports and leave a small surplus according to NEI. The services account was also expected to be in the red, excluding hefty interest and repayments on foreign borrowing which looked likely to leave the current account negative to the tune of US$200m.

The economic situation is bleak at best and pointed to a more than 3 per cent cumulative contraction during 1988 and 1989, as a result of a fall in export earnings, unrealistic wage increases and uncontrolled national expenditure. Export production was predicted to increase by only 0.1 per cent in 1989 and the main reason behind relatively poor export prospects was that export production in the marine sector was expected to fall by 4 per cent compared to 1988. This in turn was due to cutbacks in fishing quotas for marine life conservation reasons. Increase in aluminium production was forecast to increase by 5.5 per cent in 1989 and ferrosilicon exports were also increasing. Fish farming production was expected to increase by 21 per cent, but this new field in Iceland is experiencing serious financial problems plus bleak markets and falling prices for farmed salmon in the world market place.

Iceland has developed a comprehensive fisheries management policy to manage the fish stock in the unpolluted North-Atlantic ocean around the 100,000 sq km island. All major fish species are regulated by quotas. The total catch is set on the basis of biological estimates of the status of the stocks and forecasts of the future development, with regard to economic factors. The total catch in 1988 was 1,775,000 tonnes, which is mostly exported to the USA, Eastern and Western Europe and Japan.

The development of the terms of trade of 1989 was still uncertain at mid-year and implied no significant changes in terms of trade. After a stable exchange rate for more than three years, the Icelandic krona was devalued three times in 1988: by 6 per cent in February, 10 per cent in May and finally by 3 per cent in September. The depreciation of the krona continued in 1989. At the beginning of the year it was devalued by 5 per cent and again by 2.5 per cent in February and a month later the central bank used its authority to change the rate of the poor krona by another 2.3 per cent. The krona continued to slide against the US dollar and the total devaluation for the whole year was expected to be over 20 per cent.

Financial markets

After fundamental changes and a steady pace of liberalisation, some reregulation occurred in Iceland's financial markets. As a result of various government measures, the financial market had by mid-1989 become increasingly regulated. The regulations were still very moderate compared with those prevailing a few years back, but they were of concern to the business community. Interest rates are relatively high in Iceland and in spite of a turnaround in the economy from boom to recession, interest rates have shown a tendency to remain much too high for all production industry. The fluctuation of inflation from one period to another has caused large fluctuations in nominal interest rates. The substantial increase in lending and

KEY INDICATORS

	Unit	1984	1985	1986	1987	1988
Population	'000	238	242	244	247	249
Gross national product (GNP)	IKr bn	83.2	113.6	151.9	201.3	246.0
GDP per capita	$	11,028	11,311	15,184	21,059	22,976
Inflation	%	29.1	32.5	21.3	20.6	23.5
GNP real growth	%	2.9	3.1	7.2	8.7	–1.9
Unemployment	%	1.2	0.9	0.7	0.5	0.7
Agricultural production	1979-81=100	103.41	105.66	99.47	97.35	105.37
Total fish catch	1985=100	96.0	100.0	102.0	99.3	107.2
Manufacturing output	1980=100	111.9	113.1	116.0	122.9	–
Exports	IKr m	23,557	38,826	45,093	53,053	61,674
Fish	IKrm	9,989	15,330	19,557	24,190	–
Aluminium & alloys	IKrm	3,636	3,417	4,125	5,081	–
Fishmeal	IKrm	1,547	1,879	2,671	2,222	–
Imports	IKr m	26,744	37,600	45,910	61,237	68,996
German FR	%	12.6	14.0	15.2	–	–
USSR	%	9.1	8.2	6.5	–	–
Denmark	%	9.2	10.0	11.0	–	–
Balance of trade	IKr m	–3,187	1,226	–817	–8,184	–7,322
Current account	$m	–131.3	–115.1	17.2	–191.0	–227.9
Foreign exchange	$m	128	206	310	311	291
Exchange rate	IKr per $	31.7	41.5	41.0	38.7	43.0

* estimated figure

the imbalance between the trend of saving and lending can chiefly be explained by the huge government borrowing from the banks as well as by increase in loans to finance build-up of inventories of the fishing industry.

As of 1 January 1990, commercial banks in Iceland decreased from seven to four, when four banks were due to merge into one – the Bank of Iceland, which became the country's second largest after the National Bank of Iceland. Three private banks agreed in June 1989 to purchase the government majority holding in the Fisheries Bank Ltd (Utvegsbanki Islands hf.) The takeover involved the Industrial Bank (Idnadarbanki Islands hf.), the Bank of Commerce (Verzlunarbanki Islands hf.) and the Union Bank (Althydubanki Islands). The first two commanded together about 14 per cent of the market and the Union Bank another 3 per cent. The new private superbank will be responsible for 29 per cent of bank business in the country which still keeps foreign banks out. The three takeover banks were due to split three ways the treasury's 76.8 per cent holding in the Fisheries Bank, which was earmarked for privatisation by the government. The National Bank of Iceland, which is a state bank, is responsible for 47 per cent of all banking in this island nation. There is still a possibility for another takeover in the decreasing banking community. The National Bank has been negotiating with the Co-operative Society of Iceland (Sambad) to purchase the small and weak Co-operative Bank, as a part of streamlining measures and sales of assets by the cash-strapped Samband Co-op. The government has also indicated that it would like the numerous and often miniature savings and loans banks in Iceland to merge into a single savings bank company.

Energy resources

Iceland's unique position in Europe in terms of energy is primarily related to the fact that the country possesses numerous untapped natural geothermal and hydro power resources. Discussions here centre around those renewable resources that are everywhere in demand owing to the fact that they offer the possibility of extremely economical power production. This is bound to attract the attention of West European countries as regards the economical production of energy. Iceland is seeking partners to harness this energy to power intensive industries like aluminium production.

Relations with the EC

Discussion in Iceland about the EC and EFTA is rapidly gaining momentum. In Iceland there has been a growing awareness about the change in Europe, particularly the development of the EC into a single market by 1992, which will have important political consequences for the country. The Icelandic response to developments in Europe will to a large extent be similar to that of other EFTA countries. Icelanders ask themselves how they can best guard their interests *vis-à-vis* the EC and, no less important, how they can make Icelandic companies at least as competitive as EC companies in order to take advantage of the opportunities provided by the developments in view. At present, Iceland is not considering membership of the EC. Adaptation to the developments through EFTA is the main element in the present Icelandic strategy. It is obvious that a very small country like Iceland will benefit from co-operation with other EFTA countries in matters that are already – or will be – handled jointly by the EC and EFTA.

COUNTRY PROFILE

Political system: Parliamentary republic with legislative power held jointly by the directly elected president and the Althing (63-member directly elected unicameral parliament). Executive power vested in the government, formally appointed by the president, consisting of the prime minister and other ministers but subject to the rule of parliamentarism.

Main cities/towns: Reykjavík (population 93,425), Kopavogur, Akureyri, Hafnarfjordur and Keflávik.

Media
Dailies: There are five dailies published in Reykjavík, which generally represent the views of the main political parties. Most popular are *Morgunbladid* (Conservative, circulation 50,300), *DV* (Independent, 39,000) and *Tíminn* (Progressive, 15,000).
Periodicals: Several monthly and bi-monthly magazines, plus the popular illustrated weekly *Vikan* (14,500). Main business magazine is *Frjáls Verzlun* (8,500). Magazine in English: *News from Iceland.*
Radio: Over 75,000 radio sets in use. Four main radio stations broadcast nationally: Icelandic State Broadcasting Service (RUV) Channel One and Two plus Radio Bylgjan/Radio Stjarnan. Channel One broadcasts on medium and long wave as well as FM.
Television: Approximately 67,000 TV sets in use. Programmes broadcast by Icelandic State Broadcasting Service – Television (RUV) and Channel Two (Stod 2).
Advertising: Newspaper, radio and television advertising are the most effective. The quality of locally produced television advertising material is very high by international standards. Information can be obtained from Samband íslenskra augly\01singastofa (Association of Icelandic Advertising Companies), Hateigsvegur 3, 105 Reykjavík (tel: 2-95-88).

Domestic economy: Open market economy based largely on the fishing industry. Pattern of economic growth thus heavily determined by size of fish catch and world prices of fish products. Record fish catches in 1985-86 and resulting increase in value of fish exports contributed to a trade surplus in 1986, the first since 1978. The year 1988 marked a turn in the economy, after one of the most favourable periods in Iceland's economic history had passed. In 1988 GDP is estimated to have fallen 1.4 per cent and 1.7 per cent in 1989.

Employment: Around 13 per cent of the labour force is engaged in fishing and related industries. Despite economic fluctuations unemployment rate remains low, even if a considerable increase was noticed in 1989. Despite substantial wage increases in early 1988 as well as in 1989, real earnings fell by approximately 2.5 per cent from 1987 to 1988 and a further seven per cent fall was predicted in 1989.

External trade: In 1988 export production declined by one per cent, despite record fish catches. After sharp price rises in 1987, exports of fish products fell substantially in 1988. Prices for aluminium and ferro-silicon, on the other hand, reached almost record heights in 1988. In 1989 export production was predicted to increase by 0.1 per cent; relatively poor prospects due to expected fall of four per cent in marine production. On the other hand a 5.5 per cent increase is expected in aluminium production as well as 21 per cent increase in the production of fish farming operators. Around 70 per cent of exports comprise fish products, mainly white fish, herring, capelin, lobsters and shrimps. Aluminium, ferro-silicon and woollen products account for a further 28 per cent. Heavy import dependence on energy products, raw materials, machinery and tools and intermediate goods for aluminium smelting and manufacturing. Over 50 per cent of trade turnover conducted with EC, of which the UK, West Germany, France, Denmark and the Netherlands are the principal traders. Other main trading partners are the USA, Japan, the USSR and Sweden.

Agriculture: Scarcity of arable land, but good grazing allows for self-sufficiency in meat (mostly lamb), milk, poultry, eggs, cheese and butter. Sector is small-scale, heavily subsidised and organised into co-operatives. Fishing replaced farming early in this century as dominant sector of economy. Fishing industry (including processing) accounts for 17 per cent of GDP and is single most important export earner. Large modernised trawler fleet supplies over 110 freezing plants, which produce white fish fillets, frozen shimps, capelin, scampi, scallops, fish oil and fish meal. The Icelandic Freezing Plants Corporation and Iceland Seafood Ltd are the leading fish exporters. There is rapid growth of inland and off-shore fish farming.

Industry: This sector contributes around 19 per cent to GDP and is centred on fish and food processing, aluminium smelting, ferro-silicon alloys, diatomite production and light manufacturing. A feasibility study on a new aluminium smelter gave promising prospects and a decision on this subject is expected in the near future.

Energy: Rapid development of cheap hydroelectric and geothermal power has led to over 50 per cent self-sufficiency in energy requirements and reduced dependence on imported petroleum fuels, mainly from USSR. Around 81 per cent of homes have geothermal heating. Fishing industry however remains dependent on imported oil.

Banking: In addition to the Central Bank there are four commercial banks operating: the National Bank and Agricultural Bank (both state-owned) and the privately-owned Islandsbanki and Co-operative Bank. There are also over 40 savings banks.

Legal system: Civil law based on Danish law. Constitution adopted in 1874, revised in 1918 and 1944.

Membership of international organisations: CCC, Council of Europe, EC, EFTA, FAO, GATT, IAEA, IBRD, ICAO, ICES, IDA, IFC, IHO, ILO, IMO, INTELSAT, INTERPOL, IPU, ITU, IWC, NATO, Nordic Council, OECD, UN, UNESCO, UPU, WHO, WMO WSG.

BUSINESS GUIDE

Time: GMT.

Climate: Temperate, with mild but stormy winters and cool summers. Rainy in the south. Average temperatures vary between about -1˚C and 12˚C.
Clothing: Mediumweight throughout year, topcoat and raincoat for winter.

Entry requirements: Visas not required by nationals of most West European countries and many others.
Health: Vaccination certificates not required.
Currency: No limits.
Customs: Personal effects duty free, plus duty-free allowance. Tax-free shopping.

Air access: Direct flights from a number of European and North American cities including Amsterdam, Chicago, Copenhagen, Frankfurt, Glasgow, London, Luxembourg, Milan, New York, Oslo, Paris, Zurich.
National airlines: Icelandair (Flugleidir) and Arnarflug (Eagle Air).
Airlines serving: SAS and Lufthansa.

BUSINESS GUIDE

Tax: Departure tax payable.
International airport: Keflávík (KEF) 47 km from Reykjavík, good inexpensive bus service to Reykjavík.

Surface access: Ferry services to Seydistiördur (in the east) from several North European ports during summer.
Main ports: Reykjavík, Akureyri, Hafnarfjördur, Seydisfjördur and Vestmannaeyjar.

Hotels: Most towns have hotels and guest houses available. Between June and September university hostels and boarding schools are also used as hotels. Some hostels and many farms provide bed and breakfast service. No official rating system in operation. Information on accommodation, which should always be booked in advance, available from Tourist Information Centre, Ingólfsstraeti 5, Reykjavík. All usual credit cards are accepted. Tipping is not customary.

Car hire: Available in Reykjavík and several other towns. Rates vary depending on type of car. Minimum age 17 years, and an international driving licence usually required. Advance reservations are necessary between June and August. Self-drive cars not recommended as a method of national transport as road surfaces tend to be poor.

City transport
Taxis: Used extensively. Usually summoned by telephone, but can be hailed in street. Several makes of car used, mostly displaying roof-top 'Taxi' signs.
Buses: Regular services covering centre and suburbs of Reykjavík. Standard fare for any length of journey, even if it involves more than one bus route.

National transport
Air: Icelandair plus several other local services operate regular flights connecting all main towns. Light aircraft readily available for charter and sight-seeing. Ruggedness of terrain and low cost make air travel very popular.
Road: Approximately 8,500 km of roads. Main highways (approximately one quarter of total) are hard-surfaced, and the rest are gravel-surfaced. Main roads usually follow coastline. Regular coach services link even the remote inland areas.
Rail: No railways.
Water: Regular cargo coastal services linking all major ports.

Public holidays
Fixed dates: 1 Jan, 1 May, 17 Jun (National Day – Birthday of Jon Sigurdsson), 25 and 26 Dec, (half days on 24 and 31 Dec).
Variable dates: Maundy Thursday, Good Friday, Easter Monday, First Day of Summer, Ascension Day, Whit Monday and first Mon in August is Bank Holiday.

Working hours
Government and business: (Mon-Fri) usually 0900 – 1700.
Banking: (Mon-Fri) 0915 – 1600, plus 1700 – 1800 on Thu.
Shops: (Mon-Thu) 0900 – 1800, (Fri) 0900 – 2000. Most also open Sat morning except June-August. Kiosks remain open until 23.30 or even later.

Social customs: Handshaking is customary on arrival and departure.

Business language and interpreting/translation: English is used in international business. List of authorised interpreters and translators available from Iceland Tourist Board.

Telecommunications
Telephone: Used a great deal, especially for business purposes. Direct dialling available. For international operator dial 09. International dialling code: 354 followed by 1 for Reykjavík.
Telex: Available in larger hotels and at main telephone headquarters, Austurvöll Square, Reykjavík. Country code: 501 IS.
Fax: Available at most hotels and in most companies.
Telegram: May be sent by phone (dial 06) or from telegraph office in Reykjavík.

Electricity supply: 220 V AC.

Representation in capital: China PR, Czechoslovakia, Denmark, Finland, France, Democratic Republic of Germany, Federal Republic of Germany, Norway, Poland, Sweden, United Kingdom, USA, USSR.

BUSINESS DIRECTORY

Hotels: Keflavík
Flughótel, Hafnarqata 57 (tel: (92) 15222).

Keflavík, Vatnsnesvequr 12 (tel: (92) 14377; fax: (92) 15590).

Kristina, Holtsqata 47-49 (tel: (92) 14444).

Hotels: Reykjavík
City, Ranargata 4a (tel: 18650; tx: 3141).

Esja, Sudurlandsbraut 2 (tel: 82200; tx: 2130; fax: 82130).

Holiday Inn, Sigtúni 38 (tel: 689000; tx: 3135; fax: 680675).

Holt, Bergstadastraeti 37 (tel: 25700; tx: 2233; fax: 623025).

Lind, Raudanarstiqun 18 (tel: 623350; tx: 623150).

Loftleidir, Reykjavik Airport (tel: 22322; tx: 3121; fax: 690391).

Odinsve, Odinstorg (tel: 25640; tx: 3030).

Saga, Hagatorg (tel: 29900; tx: 2255; fax: 623980).

Car hire: Reykjavík
Hertz: international reservations (tel: 29577; tx: 2183); Reykjavík Airport (tel: 689996; tx: 2183).

Inter-Rent: international reservations (tel: 621715, 623515); Skeifan 9, 108 Reykjavík (tel: 686915; tx: 2337; fax: 688663).

There are various other car hire companies in Reykjavík and major towns.

Chambers of Commerce
Chamber of Commerce: Ads Verslunarrinnar, Kringlan 7, 103 Reykjavík (tel: 83088; tx: 2316; fax: 68-65-64).

Banking
Sedlabanki Íslands (Central Bank of Iceland), PO Box 160, Kalkofnsvegur 1, 150 Reykjavík (tel: 699600; tx: 2020; fax: 621802).

Althydubankinn hf (The People's Bank Ltd), Laugavegur 31, 101 Reykjavík (tel: 621188; tx: 4925; fax: 27950).

Bunadarbanki Íslands (Agricultural Bank of Iceland), Austurstraeti 5, 101 Reykjavík (tel: 25600; tx: 2383; fax: 621340).

Idnardarbanki Íslands (Industrial Bank of Iceland), Laekjargötu 12, 101 Reykjavík (tel: 691800; tx: 3003; fax: 623703).

Íslandsbanki, Austurstraeti 19, 101 Reykjavík (tel: 17060; tx: 2047; fax: 623099).

Landsbanki Íslands (National Bank of Iceland), Austurstraeti 11, PO Box 170, 101 Reykjavík (tel: 606600; tx: 2030; fax: 29882).

Samvinnubanki Íslands (The Co-operative Bank), Bankastraeti 7, 101 Reykjavík (tel: 20700; tx: 3142; fax: 28314).

Verzlunarbanki Íslands (Iceland Bank of Commerce), Bankastraeti 5, 101 Reykjavík (tel: 27200; tx: 3027; fax: 19813).

Principal newspapers
DV, Thverholt 11, Reykjavík (tel: 27022; tx: 3079).

Frjáls Verzlun, Ármúla 18, PO Box 1193, 108 Reykjavík (tel: 82300).

Morgunbladid, Adalstrtaei 6, PO Box 1555, Raykjavík (tel: 691100; tx: 2127).

Timinn, Lynghálsi 9, Box 370, Reykjavík (tel: 686300).

Vikan, Háaleitisbrand 1, Reykjavík (tel: 83122).

Other useful addresses
Agricultural Production Board, Baendahöllin, Hagatong, 107 Reykjavík (tel: 19200).

Association of Icelandic Importers, Exporters & Wholesale Merchants, House of Commerce, Kringlan 7, 103 Reykjavík; (tel: 83088; tx: 2316; fax: 686564).

BSI Travel (buses), Umferdarmidstödinni v/Hringbraut, 101 Reykjavík (tel: 22300).

Confederation of Icelandic Employers (Vinnuveitendasamband Íslands), Gardastraeti 41, 101 Reykjavík (tel: 25455; fax: 28421).

Customs Department (head office), Tryggvagata 19, Reykjavík (tel: 600300; tx: 2128; fax: 25826).

Export Council of Iceland, Lágmúli 5, Box 8796, 128 Reykjavík (tel: 688777; tx: 2085; fax: 689197).

Federation of Icelandic Co-operative Societies (Samband of Iceland), Import Division, v/Holtavegur, 104 Reykjavík (tel: 681266; tx: 2260; fax: 680290).

Federation of Icelandic Industries, Box 1407, Hallveigarstigur 1, 101 Reykjavík (tel: 27577; tx: 2085; fax: 25380).

The Icelandic Management Association, Ananaust 15, 121 Reykjavík (tel: 621066; tx: 3000; fax: 29323).

Iceland Tourist Board, Laugavegur 3, 101 Reykjavík (tel: 27488; tx: 2248; fax: 624749).

Keflávík Airport Authority, The Leifur Eiríksson Passenger Terminal Keflávík Airport, 235 Keflávík (tel: 50600; fax: 50610).

Ministry of Commerce, Arnarhvoli 8, Lindargötu, 150 Reykjavík (tel: 609070, 609000; tx: 2092; fax: 621289).

Ministry of Finance, Arnarhvoli, 150 Reykjavík (tel: 25000; tx: 2092).

Ministry for Foreign Affairs, Hverfisgata 115, 105 Reykjavík (tel: 623000; tx: 2225; fax: 622386).

Ministry of Industry, Arnarhvoli, 150 Reykjavík (tel: 25000).

Retailers' Association of Iceland, Húsi Verslunarinnar, Kringlan 7, 103 Reykjavík (tel: 687811; fax: 685569).

The Society of Icelandic Advertising Avencies (SIA), Háteigsvegur 3, 105 Reykjavík (tel: 29588; fax: 623419 – attn: SIA 29588).

Tourist Information Centre, Ingólfsstraeti 5, 101 Reykjavík (tel: 623045; tx: 3169).

India

Kevin Rafferty

In late November 1989 the Indian electorate voted for an end to the Nehru-Gandhi dynasty that has ruled the country for all but two of the 42 years since independence from Britain. Whether it is a temporary or a permanent rebuff to Rajiv Gandhi is difficult to assess, not least because the new Prime Minister Vishwanath Pratap Singh is at the head of a minority government that has plenty of splits and splinters and rivalries of its own.

The verdict of the electorate against the ruling Congress (I) party of Gandhi was clear enough: it was only 192 seats in the 543-seat *Lok Sabha* (House of the People) compared to 415 before the election. But the people's judgement was not one to bring comfort to any party or to any politician who has tasted power. In the north of the country, where Congress had its stronghold, Rajiv Gandhi and his men lost heavily, most notably to V.P. Singh's Janata Dal and its allies. But in the south, Congress won seats handsomely against allies of Singh who had previously formed the state governments there. In consequence the Janata Dal, the second biggest party with 143 seats, was able to form the government only through the help outside the cabinet of the Hindu Nationalist Bharatiya Janata Party, 88 seats, and the marxist-led Left Front, with 52 seats. It seemed an uneasy alliance, in which the BJP would stand out against some of the key things that Singh wanted to do, particularly in repairing communal harmony.

As an immediate illustration and example of the slippery nature of Indian politics, even inside the Janata Dal (actually a coalition of smaller parties), V.P. Singh supporters had to fight hard to ensure the choice of their man as leader of the new MPs. Singh had been the spearhead of the opposition, who had given life and focus to its campaign and had fought tirelessly against growing corruption of the Rajiv Gandhi years in power. It is fair to say that without Singh the opposition would have sunk without trace before the election. But he got no thanks from Chandra Shekhar, one of his leading colleagues in the Janata Dal, who went out of his way to try to stop Singh being chosen as parliamentary leader and thus potential prime minister. The anti-Singh campaign was supported by some millionaire industrialists viciously opposed to Singh and by Congress onlookers. The magazine *India Today* said it was in 'a replay of Indian politics at its filthiest and grimiest. ... money, muscle, treachery, self-interest, blind ambition ... were playing havoc with the people's electoral verdict and holding to ransom the entire political process.' Singh's eventual election was cleverly managed leaving Chandra Shekar dumbfounded but fuming.

Singh showed an encouraging independence in his first days in office. He appointed India's first Muslim home minister, an indication that communal harmony was high on his list of priorities. And he quickly went to pray at the Golden Temple at Amritsar, the Sikhs' holiest shrine, a first step in an unexpected initiative to heal the wounds in Punjab. But the new prime minister has many problems to surmount, some the product of his own party's weakness, others part of the political inheritance.

India itself is a country that has a civilisation stretching back for several thousand years. It was already historic when Europe was primitive and neither Australia nor America had been 'discovered' by modern man. India is by far the most powerful of the seven or eight (depending on whether Burma (Myanma) is counted) countries in the region. In 1987 it lent a helping hand to tiny neighbouring Sri Lanka by send-

KEY FACTS

Official title: Union of India

Head of state: President Ramaswamy Venkataraman

Head of government: Prime Minister Vishwanath Pratap Singh

Ruling party: National Front (coalition)

Capital: New Delhi

Official Languages: Hindi (English, Urdu and 20 other languages widely used)

Currency: Rupee (Rs = 100 paisa)

Exchange rate: Rs16.7 per $ (Nov 1989)

Area: 3,287,590 sq km

Population: 796.6m (1988)

GNP per capita: $330 (1988)

GNP real growth: 1.7%* (1988)

Labour force: 296m* (1988)

Inflation: 9.4%* (1988)

Trade balance: –$7.9bn (1988)

Foreign debt: $46.3bn* (1987)

* estimated figure

KEY INDICATORS

	Unit	1984	1985	1986	1987	1988
Population	m	736.0	750.9	766.14	781.4	796.6
Gross domestic product (GDP)	Rs bn	2,131.8	2,150.2	2,598.8*	3,304.6	–
GDP per capita	Rs	2,855.0	2,863.0	3,446.7	–	–
Wholesale prices	% Change	8.5	5.8	5.4	6.0	8.6
Consumer prices	% Change	8.2	5.6	8.7	8.8	9.4
Industrial production	% Change	9.3	9.1	6.7	11.2	7.3
Cement	m tonnes	29.57	31.89	34.71	–	–
Finished steel	m tonnes	6.56	7.84	7.75	–	–
Jute manufacture	m tonnes	1.15	1.29	1.44	–	–
Sugar	m tonnes	6.12	6.50	7.04	–	–
Tea	m kg	645	657	717	–	–
Paper products	m tonnes	1.36	1.46	1.58	–	–
Agricultural production						
Rice paddy	m tonnes	87.55	96.31	90.8	84.5	102.0
Wheat	m tonnes	44.07	46.89	47.05	44.3	45.1
Pulses	m tonnes	11.96	12.96	13.3	12.0	11.2
Sugar-cane	m tonnes	170.32	171.17	170.65	182.5	196.72
Cotton(lint)	m tonnes	01.45	01.46	0.2	0.2	0.2
Jute & other fibres	m tonnes	1.40	2.29	1.6	1.2	1.3
Potatoes	m tonnes	12.57	10.70	10.4	12.7	14.1
Mineral production						
Iron ore	m tonnes	42.0	42.3	44.2	52.0	–
Bauxite	m tons	2.08	2.34	2.66	2.78	2.78
Copper	'000 tonnes	47.0	48.0	49.2	54.1	51.5
Manganese ore	m tonnes	1.13	1.26	1.28	1.28	1.28
Dolomite	m tonnes	2.3	2.2	2.1	–	–
Crude petroleum	'000 b/d	561.1	606.4	625.0	627.6	655.8
Coal	m tonnes	147.50	153.45	163.307	179.8	195.0
Exports	$m	9,769	9,461	10,460	12,888	14,824
Imports	$m	15,424	17,295	17,728	19,772	22,760
Balance of trade	$m	–5,655	–7,834	–7,268	–6,884	–7,936
External debt	$m	31,272	36,625	41,311	46,370	–
Debt service ratio	%	8.3	11.6	15.2	16.9	–
Current account	$m	–2,703	–5,207	–4,833	–4,146	–4,937
Exchange rate	Rs per $	11.36	12.37	12.61	12.96	13.92

* estimated figure

ing its troops to the teardrop-shaped island off its southern-eastern coast. But by 1989 India was discovering the hard way that neither physical size nor military might were solutions to intractable racial and religious conflicts.

A precarious unity

Any discussion about India should be prefaced with caveats. It is a truism to say that India is not so much a country, more a subcontinent. India is a poor country, according to tables produced by the World Bank or other development agencies. As stated in the latest World Development Report, per capita income of its 800m people was $300 in 1987, putting India on a par with Rwanda and Sierra Leone, slightly ahead of China, Somalia and Togo (each $290), but below Pakistan ($350), Haiti ($360), not to speak of Mauritania ($440) and 75 other developing countries. India is also backward. Consumption of commercial energy is often used as an indicator of economic progress. In this, India scores much worse than China. The 208 kg of oil equivalent that Indians use every year is less than half China's 525 kg and only a fraction of the typical average of 6,575 kg of energy consumed in the rich industrial countries. India is basically a rural country with more than 75 per cent of its population living in villages and largely dependent on agriculture for its livelihood.

But it is also easy to produce generalisations attesting to the strength and vitality of India. Total gross national product is still in the world's top 50, thanks to the 800m population, though India's GNP is about 40 per cent of the total GNP of the United Kingdom, a small island with less than 8 per cent of India's population. Yet in terms of industrial output, India comes in the top 15 in the world. The truth is that India is very much a split, mixed-up country. It has an industrial economy, which is trying hard to compete with fully industrialised countries like the USA and Japan; and it also has a traditional rural economy that has seen some benefits of modern agriculture but remains handicapped by the large population, a lack of capital and the size of landholdings.

In Bombay, Calcutta and other big cities and towns

in India it is easy to marvel at, yet be horrified by, the manifold contrariness and contradictions as they clash together. You can stand at some spots, for example at the Taj Mahal Hotel by the seafront in Bombay or Chowringee in Calcutta, and in the space of five minutes decide that India is one of the most modern, trendy countries in the world and that it is immured in poverty and backwardness. The two hotels are favourite meeting-places for rich businessmen coming from India and abroad, and for young men and women about town seeking a meal or amusement.

These two modern palaces have more or less kept pace with developments in the rest of the world. The hotels are also popular for wedding receptions, attended by people glitteringly dressed up as only Indians in their finery know how. Stand in the cool, never-failing air-conditioning, listen to the gentle splashing of fountains, smell the sweet smell of the well-dressed, watch the neat subservient attendance from doormen bowing as they usher you in to the hurrying scurrying waiters in the coffee shop and the nearly obsequious servants in the plush restaurants punctuating their conversation with 'Sirs' and 'Madams'. You might imagine that you were living in the lotus land of the rich.

Yet the picture of twentieth-century rich India presented in the hotels is not merely a caricature of the country in 1989. There is a modern India that has sprung up since independence. Its main members are businessmen and managers in the industrial sector, senior government servants and army officers, doctors, engineers, lawyers and teachers. Some of them earn salaries that are a pittance by the standards of their counterparts in the West, though their qualifications are often as good. All of them strive to keep their families in a style that would be recognised as middle class, almost anywhere in the world. They would like to have a car, though in India they may have to make do with a motorbicycle, or more likely a motor-scooter. In Delhi it is not uncommon to see a husband driving, his wife sitting sidesaddle behind and holding on to a small child, with another child standing in front on the running board, his father's hands around him clutching the handlebars. Such middle-class families inevitably have a television, a Western-style dining table and chairs, telephone, refrigerator, probably a washing machine and possibly even a video recorder along with a set of the latest Hindi and Western movies.

Similar scenes are common in Bombay. The imposing Gateway of India, built to commemorate the visit of King George V, is a place to meet and feel the cool breezes from the sea, to go for a boat trip to the Elephanta Caves, to sit in the park under Shivaji's statue and eat ice-cream or popcorn. For the poor street performers it is also an opportunity to earn pennies by performing tricks in which young raggedy children ape small monkeys with their dextrous ability, and vice versa. The poor of Bombay do not actually sleep on the open pavement in front of the Gateway. The police keep the area clear, perhaps out of respect for the sensitivities of the hotel guests who would not want their sea view spoiled by any squalid sights. But just around the corner, by an old banyan tree where a saffron-robed, face-smeared *sadhu* sits, is a whole colony of pavement dwellers. They have patched up birds' nest homes, not mere sheets – for this after all is Bombay, rich commercial capital of India. But you can catch their smell as you come and go from the Taj Hotel.

Other cities show similar colonies of rich and colonies of poor. But Mohandas Karamchand Gandhi (often known as the Mahatma or Great Soul) said that really to appreciate India, to know its soul, a person must go to its villages. In terms of statistics, this is still certainly true. About 75 per cent of Indians still live in villages. There are more than half-a-million of these and they differ greatly one from another over the vastness of the great subcontinent that is India. In Punjab state, in the north-west, even though it has been troubled and torn apart by violence between Sikhs and Hindus, you can commonly see tractors, brick houses, numerous workshops by the roadside and forests of television aerials decorating most of the homes. There is an atmosphere of prosperity and the middle class has extended its tentacles right into the heart of rural India. But in the country's Hindi-speaking heartland of Bihar and great wedges of Uttar Pradesh, Madhya Pradesh, farmers are only slowly changing their practices and mostly have to wait for the annual monsoon rains which will decide how bountiful – or otherwise – their crops will be.

India is so vast and passes through so many geographical and physical regions that it is difficult to generalise safely about it. In the far north, India is in the shadow of the great Himalayan mountain range. Only a matter of miles further south, the land falls steeply to the flat alluvial basin of the Indo-Gangetic plain. The influence of the mountains is still felt, in the cool winters and in the downpouring of monsoon rains as the clouds fall over the mountains. The waters turn the thirsty land into fertile, fruitful crop-bearing soil. Further south still, India passes into the tropics so that one local wit commented that Madras had three seasons only, hot, hotter and hottest.

The unity that is India has also been built out of fragments of religion, caste and regionalism. Hinduism dominates, claiming probably 600m followers out of the 800m population. The subcontinent of India was divided into two countries by the British in 1947, India and Pakistan, to provide a homeland for the Muslims of the subcontinent. But today's India has almost 100m Muslims, the same number as in Pakistan itself.

It also has about 20m Christians, almost the same as the whole population of Ireland, Denmark, Norway and Sweden together. Christians are in fact more numerous than Sikhs, whose claims to have the Punjab state all to themselves has thrown northern India into turmoil. Besides these religions there are several million Buddhists, a philosophy that originated in India. There are also smaller religious groups such as Jains, an Indian group that does not believe in killing anything, even insects. More scrupulous Jain members wear face masks for fear of accidentally swallowing an insect. Parsees, or Zorastrians, originally came from Iran fleeing from the advance of Islam; today numbers of them have settled in Bombay and other commercial centres where they are important in business. The Tatas are among influential powerful Parsee families. In addition to these religions there are small numbers of Jews.

Religion is not the only mark that distinguishes and divides communities. Hindus themselves are not united, but are divided according to the way they treat other religions and – more importantly – by caste distinctions. The political scene has been marked by a resurgence of Hindu nationalism. Since the religion started in India and effectively, apart from pockets of Indians abroad, is exclusively Indian, Hinduism can become extremely nationalist. Caste can be not only divisive but is frequently the most powerful deterrent to economic progress. In some parts of the country, untouchables are simply not counted as part of the village, even though untouchability has been forbidden by law. In other parts of India, not only is caste vitally important, but subdivisions of the caste count. In some places one group of Brahmins, the priestly caste, may fiercely contend that they are superior to other sub-groups of Brahmins.

Besides the divisive and frequently damaging impact of religion, there are vast regional differences between parts of India. A major division is between the Hindi-speaking north and the south of the country, whose Dravidian inhabitants have been in India for centuries longer than the Aryans of the north. Many parts of the south have steadfastly resisted attempts to impose Hindi as the national language. Such people would rather speak English, and their resistance has helped to keep English as a nationally important language, not only for the worlds of business and government. The stronghold of Hindi is Uttar Pradesh, India's largest state with about 120m people, Bihar, Madhya Pradesh, Haryana, Rajasthan, all in the north of the country. But the Hindi-speaking belt does not extend as far north-west as Punjab or to the north-east where Bengali holds sway. Language and regional struggles can sometimes be small and local as well as big and national. Two neighbouring states such as Maharashtra, which includes Bombay city, and Gujerat have sometimes shown intense rivalry. That is the reason for their separate existence – since when India won its independence the two were part of the same Bombay state. First Prime Minister Jawaharlal Nehru split it on linguistic lines, creating Marathi-speaking Maharashtra and Gujerati-speaking Gujerat. Other states have also shown deep divisions. Punjab state used to cover large areas of what is now Pakistan and was split at independence (just as Bengal was split with half going to Pakistan). And after independence, the Indian Punjab originally included what later became Haryana state; it was split in two as a political sop to Sikhs. But the Sikhs have not been satisfied and the more extreme of them are seeking an independent state of their own. Other areas of India that have not hit the international headlines yet have been torn by regional squabbles. For many years Telegu-speaking inhabitants of Andhra Pradesh in the south wanted their own state; so far their demands have been resisted.

Where religion and regionalism come together, as in Punjab, violence can be extreme with many lives lost. But even the big modern cities of Delhi, Bombay and Calcutta have resounded to violence, mainly of a religious kind. Hindu against Sikh, Hindu against Muslim, or as in 1989, violence in the wake of calls for strikes by the Hindu nationalist political Shiv Sena group. All in all, by 1989 the unity of India looked increasingly precarious, challenged or under pressure all over the country.

Political situation

All this variety means that political groupings have equally to be all-embracing and able to capture something of each part of the flavour of the country if they are to succeed in forming an India-wide government. For years of India's history this was not a problem. The Congress Party had taken India to independence from the British. Under its leader and India's first prime minister, Nehru, an all-embracing party ruled the country with a share of the jobs going to northerners, southerners, Muslims, Sikhs and even Anglo-Indians. In Nehru's years the Congress Party was sometimes called a 'banyan tree under which many sheltered, but nothing grew.' The demands for unity were vitally important, but soon even a man of Nehru's stature was tested by regionalism and factionalism. Bombay state was broken up. More damaging, the Congress developed a system of powerful local bosses. When Nehru died, a former schoolmaster Lal Bahadur Shastri took over but lived for less than two years. Nehru's daughter, Mrs Indira Gandhi, a 48-year-old widow, was appointed because the old guard of surviving party bosses believed she would be an attractive new face as far as the electorate was concerned, but also malleable and easily manipulated by them. They were right about the first point, and totally wrong about the second. The new leader was preparing to split the hallowed Congress in order to stamp her own dominance over the old bosses.

But by the end of her reign Indira Gandhi's vote-getting magic spell had worn out. She never really trusted anyone apart from herself and her son Sanjay. She

COUNTRY PROFILE

Political system: India is a sovereign democratic republic within the Commonwealth consisting of 25 states and seven union territories. The head of state is the President but real political power is exercised by the Prime Minister who commands the support of the Lok Sabha (the House of the People), the lower house of India's bicameral parliament. The President is elected for six years (renewable) by an electoral college of central and state parliamentary representatives. The Rajya Sabha (the Council of States), the upper house of parliament, consists of eight members nominated by the President and the rest chosen for six-year terms by state legislatures. The Lok Sabha is directly elected by universal adult suffrage for five-year terms, though it can be dissolved earlier by the President at the request of the Prime Minister. Each of India's states has its own governor, elected state assembly with Chief Minister and Council of Ministers.

Main cities/towns: Delhi (population 5.7m in 1986); Calcutta, capital of West Bengal state (pop 12.4m including Greater Calcutta); Bombay, capital of Maharashtra state (pop 8.2m including Greater Bombay); Madras, capital of Tamil Nadu state (pop 4.2m); Hyderabad, capital of Andhra Pradesh state (pop 2.5m including Secunderabad); Bangalore, Karnataka state (pop 2.9m); Ahmedabad, Gujarat state (pop 2.5m); Kanpur, Uttar Pradesh state (pop 1.6m); Pune, Maharashtra state (pop 1.6m). There are 596,000 villages in India.

Media
Press: Dailies: there is no national newspaper, but numerous local dailies are published in several languages, including some 245 in Hindi, 64 in English, 46 in Malayalam, 81 in Marathi, 35 in Gujarati, 26 in Bengali, 43 in Tamil, 78 in Urdu and many others. *The Indian Express* (circulation 522,370) is the largest, being published in 10 centres. *The Hindustan Times* is the leading Delhi newspaper and *The Times of India* is the main Bombay daily. Business: The three daily financial newspapers published in English are the *Economic Times* (Delhi, Bombay and Calcutta editions), *Financial Express* (Delhi, Madras and Bombay) and *Business Standard of Calcutta*. Business periodicals include *Capital* (Calcutta), *Commerce, Business India, Business World* and *Update* (Bombay). There are trade publications covering plastics, engineering, the motor trade, paints, chemicals and textiles.
Radio: There were 30m radio sets in 1987. All India radio planned to start 24 hour broadcasting and increase TV viewing time from nine to 17 hours.
Television: There were 14m TV sets in 1987.

Domestic economy: It is difficult to talk of a single economy for India. The country has more than half a million villages where agriculture, heavily dependent on the seasons, is the mainstay. But it also has a modern industrial sector which ranks India the 15th industrial power in the world. India has developed its own nuclear industry, has a considerable arms making capability and is constructing its own space satellites. Nevertheless progress of Indian industry and the economy generally has been slower than in many other developing countries. At India's independence in 1947 socialism was the vogue word and the country devised heavy duty industrial plans in which the state controlled the economy. Overall growth was about 3.5 per cent a year in the 25-year period from independence to 1977. In the latter years of Mrs Indira Gandhi's rule and the coming to power of her son Rajiv Gandhi, there was a change of direction. In the Seventh Five Year Plan (1986-90) new emphasis was laid on industrial liberalisation and modernisation, on the need to develop an export culture instead of relying on import substitution, plans to generate employment, develop agriculture in low production areas, improve production of basic goods and reduce poverty. Rajiv Gandhi has also welcomed new foreign investment. There are good reasons for the change in emphasis: India's budget deficit and foreign debts have been growing alarmingly. Foreign debts have risen sharply and debt service ratios had crossed 25 per cent of annual foreign exchange earnings. Many government corporations were losing money. But in the agricultural sector slow progress was being made. Thanks to the spread of irrigation India had become self-sufficient in food production. Only when the crippling drought of 1987 followed lean years was the considerable buffer grain stock eaten away so that India had to import small quantities of foodgrains. Although the economic growth rate fell to around 1.2 per cent in 1987, the positive factor was that both the economy and industry managed positive growth rates unlike previous major drought years. Population growth is the country's biggest long-term problem; even with a low birth rate of 1.8 per cent per annum, total population will reach 994 by 2,000, swamping much of the country's growth.

Employment: Employment figures only cover the official sector, with 24m employed in 1984, but another 250m* are engaged in locally-based pursuits. There were 30.1m registered unemployed in 1986. There are large numbers of skilled and semi-skilled workers overseas in the Middle East.

External trade
Exports: Total exports were worth $11.7m in 1987, up 15.3 per cent. Main exports are handicrafts, pearls and precious stones, textiles and clothing, engineering goods, tea, coffee, cashew kernels, iron ore, leather and non-footwear leather manufactures, mineral fuels and lubricants and spices. Main destinations are USA, USSR, Japan, UK and German FR. There are six special export processing zones at Gujarat, Bombay, Calcutta, Madras, New Dehli and Kerala, where foreign companies benefit from simplified customs clearance and tax incentives.
Imports: Total imports were worth $16.2m in 1987, up 2.2 per cent. Main imports are petroleum oil and lubricants, transport equipment, non-electrical machinery and appliances, fertilisers and fertiliser materials, iron and steel and non-metallic mineral manufactures. Main sources are USA, Japan, USSR, German FR and UK.

Agriculture: Agriculture accounts for nearly 35 per cent of GDP and employs 68 per cent of the population (1987), although the Government wants the latter figure to drop to 50 per cent by 2000. Use of improved seed varieties, irrigation and fertilisers has made India self-sufficient in most grains, with an embarrassing stockpile of some varieties; the booming population, however, allows no room for complacency, and in 1987 a severe drought in many areas added to the pressure on producers, with an estimated 7-10 per cent decline from total grain output of 144.1m tonnes in 1986-87. Land for extending cultivation is limited, but there is scope for productivity improvements in other crops. Main crops are wheat, rice, pulses, tea, sugar-cane, cotton, coffee, oilseeds, tobacco, rubber and potatoes. Dairy farming has made India self-sufficient in milk powder and butter. The Government plans to introduce improved techniques to the southern states, and to reform pricing policy.

Industry: Industry contributes 28 per cent of GDP. Overall industry grew 11 per cent in 1987. Manufacturing industry has traditionally been dominated by the huge state-controlled heavy industries favoured by successive governments in the 1960s and 1970s. In the 1980s, however, these have appeared cumbersome and have been losing ground to smaller enterprises, with the steady removal of the special protection formerly provided by the state. Small companies are moving into high-technology products including computers (up by 70 per cent in 1985-86) as well as the traditional light engineering and textiles, but are still hampered by shortages of imported capital goods. The entire system of industrial licensing has been revised to accord with the spirit of freer competition, and many companies are now turning to the private capital markets for expansion funding. Main heavy industries include steel, chemicals, cement and heavy engineering where major development is underway; further developments in petrochemical and fertilisers will result from the recent arrival of oil and gas from offshore fields, which is being piped north-eastward toward New Delhi.

Mining: Mining accounts for less than two per cent of GDP (1987). India is well endowed with mineral resources, mainly coal, iron ore, manganese, bauxite and chromite, but they are not exploited to their full potential. Other minerals present include lead, zinc, tin, silver, mercury and cobalt. There was massive investment in the coal sector in 1986-90, mostly involving large-scale mechanisation of the mines; at present India still has to import coals capable of steel-making. Hydrocarbons: Proven oil reserves are estimated at 500m tonnes, sufficient for 20 years at current rates of exploitation. Over two-thirds of oil production comes from the offshore Bombay High field. Natural gas production doubled in the five years from 1980, with 4.32bn cu metres produced in 1985 from associated gas from oil fields. Besides new pipe systems feeding new industrial centres in Maharashtra and Madhya Pradesh, there are liquefication plants under construction.

Energy: There are five nuclear reactors now in operation, providing three per cent of electricity generating capacity, with plans for two more in Rajasthan and Karnataka. More efficient stoves, solar cookers and biogas plants are being developed to address the problem of energy shortages, and other renewable sources are being explored, including wind and solar generators.

Main companies: Indian Oil Corporation, Food Corporation of India, Tata Group, Steel Authority of India, Birla Group, Oil and Natural Gas Commission, State Trading Corporation, Minerals and Metals Trading, Hindustan Petroleum Corporation, Bharat Petroleum Corporation, Bharat Heavy Electricals, Modi Group, Madras Refineries, Tata Iron & Steel Company (Tisco), RGP Enterprises, Telco-Tata Engineering, JK Singhania, Thapar, Air India, Shriram Group, Western Coalfields, Hindustan Lever, Associated Cement Companies, Central Coalfields, Bangurs.

Financial markets: There are 16 stockmarkets in India. Major financial centres are: Bombay, Calcutta, Delhi, Madras, Ahmedabad, Hyderabad, Bangalore, Madhya Pradesh, Cochin, Pune, Uttar Pradesh, Ludhiana and Gauhati.

Membership of international organisations: AIOEC, ANRPC, Asian Development Bank, CCC, Colombo Plan, Commonwealth, EC trade and co-operation agreement, ESCAP, FAO, G-77, GATT, IAEA, IBRD, ICAC, ICAO, ICO, IDA, IFAD, IFC, IHO, ILO, IMF, International Lead & Zinc Study Group, IMO, INTELSAT, Interpol, IPU, IRC, ITU, IWC, NAM, South Asian Association for Regional Co-operation, UN, UPU, WFTU, WHO, WIPO, WMO, World Bank, WSG, WTO.

was forever making and breaking politicians because she feared that anyone who stayed long enough might develop their own ambitions and therefore should be removed. But she did not really change the system of powerful local bosses, merely shuffled and reshuffled the personalities. She used the slogans of socialism to try to win votes without really ever being a socialist or understanding the economy. In the end she was ousted not by the ballot but by the bullet – from her Sikh bodyguard in late 1984. They were aggrieved at the army operation to quell Sikh dissidents in their stronghold of the Golden Temple at Amritsar. Ironically, it was the Gandhi family who for short-term political reasons had originally built up and given a platform to the Sikh extremist leader Jarnal Singh Bhindranwale killed in the army's assault on the temple.

Indira Gandhi's elder son, Rajiv, inherited his mother's mantle reluctantly. He had only entered politics in his late thirties. Originally he had been an Indian Airlines pilot. When his younger brother Sanjay died in a typical daredevil aircraft stunt, Mrs Gandhi felt she needed another family member close at hand, so he gave up flying to devote himself to politics. Rajiv did not have politics in his blood, but the ruling Congress (I) – the I stands for Indira – could not find any other attractive leader. Initially the image of the modern young man paid off. Rajiv Gandhi had several advantages. He was a fresh face, an instant asset compared to the rather tired old cronies of his mother, some of whose reputations were tainted by constant rumour of corruption. He could be billed 'Mr Clean'. He talked the language of the twentieth century and he had twenty-first century ambitions about technology and the computer age. As a former airline pilot he had at least handled the implements of the twentieth century. He surrounded himself with young men like himself, technocrats impatient for change, men who wanted to go places.

But very quickly, Rajiv Gandhi's clean modern image began to be tarnished. He cashed in on his initial popularity, called an election and won a sweeping victory. However, soon, one by one, he quarrelled with his old friends, the so-called 'Doon School Mafia' since many of them had been at school with him. Rajiv Gandhi clearly exhibited the same suspicious worries and mistrust that his mother had shown. Several former friends quit or were kicked out of his cabinet, accusing the prime minister of wanting to be the only player in a one-man band. Gandhi had to fall back on the old bosses of Congress, much maligned even during his mother's rule. The Gandhi government became smeared by accusations of corruption. It kept its grip at the centre because of its huge 1984 election victory, but by-elections showed a strong trend against the government, and state elections in many parts of India revealed an anti-Gandhi bandwagon. By 1989, with an election due at the end of the year, Gandhi seemed in real trouble. He resorted to some desperate measures to try to hold the high ground.

In domestic affairs he could still use the powers of the central government to weaken the opposition. In April 1989 he did so to bring down the opposition-run government in Karnataka state, whose capital is Bangalore. The state had been ruled for more than six years by an opposition Janata Dal government, led for much of the time by Ramakrishna Hegde. Hegde was an attractive figure and one of the most popular opposition leaders. But he had been plagued by accusations of corruption, though he denied them all and much of the evidence was thin. The central government was doing its best to stir up troubles for Hegde and playing on personality differences in the opposition camp to smear reputations.

Hegde's successor, Somappa Rayappa Bommai, was rather colourless and had a difficult time. One Indian commentator described him as 'an uncaring captain of a ship full of holes and with sails torn, being tossed around on stormy seas.' Bommai's impossible task was to balance the various interest groups in his own party. Late in 1988, 28 government members broke away. The state government and its allies still held 126 of the 224 assembly seats. Then in 1989 a futher 19 members wrote to the Karnataka governor that they had withdrawn their support from the government. The governor, a former Congress (I) Central Government Minister, P. Venkatasubbiah, advised the president of India to dissolve the assembly and place the state under central rule. Most experts regarded the action as unconstitutional, since the constitution says that a government's strength can only be tested on the floor of the House. The Janata Dal government had not been defeated in the House and the chief minister had summoned the assembly for six days after the central government bundled the state government out of power. By this time seven of the 19 dissidents had written to the governor withdrawing what they had said in their first letter. But all this was to no avail. The state governor advised the president and action was taken in the best interests of Rajiv Gandhi's Congress Party.

Gandhi clearly did not want to give the budding opposition to him a chance to settle. Smashing it at state level in a major stronghold could help ruin the chances of the fragile group of parties and personalities staying together harmoniously for the central government election. The nature of the Gandhi government actions had been made clearer earlier in Bihar state. There the Congress-run state government was arguing publicly and one set of Congressmen was refusing to serve with another. But instead of placing the state under central rule, Gandhi sent several ministers to Bihar to crack heads and ensure that party domination was maintained. Similar action and reshuffling of patronage was seen in Rajasthan and Madhya Pradesh when Congress government squabbling got too much. A former Supreme Court judge, V. R. Krishna Iyer, described the action in Karnataka in this way: 'It is central puppetry and the governor was acting like a marionette.'

Village government reforms

Another Gandhi initiative was to propose radical reform of India's village government system. This was widely seen as an attempt to curry favour with the electorate. He proposed a constitutional amendment making it obligatory to hold elections to village councils and other district institutions every five years. The plan was also to give local elected bodies a great measure of authority over the financing and execution of projects within their areas. Decentralisation had long been a main plank of the opposition party, so Gandhi had stolen their thunder. Opposition parties had to protest that they were not against decentralisation, but were opposed to the way that the prime minister was planning to do it. They accused Gandhi of trying to undermine the powers of state government where opposition parties were in the majority and of trying to funnel funds to his own friends.

The Gandhi proposals, claimed the prime minister, would put an end to 'the vice-like grip' that 'power brokers' held over the Indian Government system. The so-called 'scheduled castes' or untouchables were to have seats reserved for them in accordance with their share of the population. In addition, 30 per cent of the seats in the new village council system would be reserved for women. But not everyone was impressed. Bunker Roy who runs an impressive village-based centre in dusty Rajasthan said the proposals had 'given the rudderless ruling party a catchy new slogan for the election year. Power to the People'. Yet nothing could be further from the truth. The more appropriate slogan would be Power to the Powerful ... The Bill simply institutionalises and strengthens the same vested interests at whose hands Panchayati Raj or grassroots democracy has taken a licking.'

Defence and foreign policy

Gandhi also studiously cultivated a pro-Indian tough-guy image in defence and foreign policy, but it was not always as successful as he hoped. At 7.17 am on 22 May 1989 India became the sixth country in the world to develop and successfully test an intermediate range ballistic missile (IRBM – missiles with a range of 500-5,500 km). The missile named *Agni* (fire) was third time lucky since two previous attempts had failed. It carried a dummy warhead of tungsten just over 1,000 km before splashing down in the Bay of Bengal. The weapon is built to carry a one-tonne weapon payload to a distance of up to 2,500 km. Given that India has already tested its nuclear skills, proved with the Pokhara atomic explosion in 1974, *Agni* puts India into the big league. Only the USA, Soviet Union, France, China and Israel have developed IRBM technology. India's argument was that such missiles are already available to China and to other Asian countries such as Saudi Arabia and therefore it is unfair of other countries to complain.

The director of the defence research and development laboratory in Hyderabad, Avul Pakir Jainulabuddin Abdul Kalam, claimed that the launch of *Agni* 'gives us the confidence that we are capable of designing any kind of missile. We are now self-sufficient both in design and missile technology'. He added 'It is a technological strength. Strength respects strength. Weaklings are not honoured, so we should be strong.' Asked whether the weapon was capable of carrying nuclear warheads he replied 'it can carry conventional warheads. It can also carry flowers to offer as a symbol of peace. It can also carry nuclear warheads. In India we are not making nuclear warheads. We provide the carrier and the policy-makers decide how to use it.'

The magazine *India Today* commented that the rocket launch 'has also conveniently dovetailed with a new phase of diplomatic aggressiveness on the part of New Delhi.' It quoted a high level bureaucrat in the ministry of external affairs as saying that south Asia regional co-operation is important 'but we've got to assert ourselves. Our area of concern extends from Afghanistan to the Indian Ocean.' India's declared defence budget is more than Rs132bn ($8.25bn) a year. Its armed forces are the fourth largest in the world. Besides its ability to develop IRBMs, the country is also working on its own major battle-tank, light combat aircraft, a multi-barrelled rocket system, remotely piloted vehicles, advanced electronic warfare systems and radars. Also on the agenda is strengthening India's bluewater fleet, so that the navy will have a three-aircraft-carrier force, more submarines, some of them nuclear-powered, to patrol the Indian ocean. India's military experts complain that even though the navy is much more powerful than Pakistan's, it might be hard pressed to defeat it in battle and could be vulnerable to Pakistani missile boats.

But Prime Minister Gandhi found in 1989 that a powerful military is not a sure guarantee of success. In 1988 the president of the neighbouring Maldive Islands, under siege from a coup attempt, called for help from Delhi. In a short space of time, Indian troops were airlifted in and had mopped up the rebel force, a highly satisfactory clean use of military muscle.

Gandhi was less successful in two of his other tough-guy games. In March 1989, India closed most border trade-and-transit crossing posts with its land-locked neighbour Nepal. The move was seen as a calculated attempt by Delhi to stop Kathmandu's dalliance with China and bring the world's only Hindu monarchy back into India's orbit. Nepal had had the temerity to buy some military equipment from China. In addition, construction contracts that Indian firms regarded as their own were also given to China. For months deadlock continued with Nepal suffering shortages of key supplies but still defiant. Nepal threatened to take the issue to the United Nations. Delhi protested that it had not broken any laws and had allowed treaties which expired in March 1988 to run on for 12 months before imposing what amounts to an economic blockade on Nepal.

Inside India the move against Nepal stirred relatively little feeling. But Rajiv Gandhi's decision to commit troops to Sri Lanka was proving more difficult. The two countries had agreed on the Indian force under a 1987 peace-keeping accord. But in 1989 new Sri Lankan President Ranasinghe Premadasa demanded that all the Indian forces should leave his country by the end of July or otherwise their presence would be a 'slur on (Sri Lanka's) self- respect.' Delhi pointedly refused. In a letter to Premadasa, Gandhi said that if Sri Lanka's deadline remained and bilateral discussions did not take place, 'India would decide the details of the Indian Peacekeeping Force withdrawal unilaterally.' Sri Lanka's foreign minister said that if the Indian soldiers were still operating in the north-east of the country after the deadline, they would 'constitute a foreign army occupying Sri Lanka by force.' Colombo's view is that the Indian forces are in the island at Sri Lanka's request and Gandhi had said that he would withdraw them 'instantaneously' if asked to do so.

India's reluctance to move out was inspired by mistrust of the Sri Lankan Government's promises and fears for the Tamil minority in Sri Lanka. If anything happened to the Tamils, then there would be problems for Gandhi at home in India triggered off by protests in India's own Tamil Nadu state peopled by Tamils. Premadasa's government was anxious to get the Indians out because violence and murders by extremist Singhalese were increasing. It was a vicious circle with violence begetting violence. India was worried about bloodshed if its troops should withdraw. But their very presence was increasing the violent mood of the Singhalese and undermining Premadasa's ability to govern at all. Delhi withdrew a token 600 of the remaining 45,000 troops as a face-saving attempt at reaching an agreement. But then in August 1989 came reports that Indian soldiers had participated in massacres that led to the deaths of at least 51 Sri Lankans, popularly described as India's My Lei. At My Lei in Vietnam, American troops had gone on the rampage and massacred Vietnamese villagers; the allegation in Sri Lanka was that the Indians had done it in cold blood after forcing their victims to lie on the ground. Later Gandhi and Sri Lanka did agree on the withdrawal of the Indian forces by the end of the year.

Only in its dealings with Pakistan, the next strongest country on the subcontinent, with which India has fought three wars, did Gandhi have some success. He went to Islamabad in July, the first bilateral visit by an Indian premier in 30 years. Officials on both sides described the visit as 'an important building block'. But despite smiles Gandhi and his Pakistani counterpart Benazir Bhutto did not solve the problems of mutual claims to the Siachen Glacier, described as 'the world's highest battlefield' where the military of both sides regularly forget that the two countries are not at war.

Domestic affairs

As if problems abroad were not enough, at home Gandhi and his colleagues faced new charges of corruption. In late July 1989 these reunited a previously splintered opposition. The occasion for the unity was a report from the government's own comptroller and auditor-general concerning India's $1.3bn purchase of Bofors field guns from Sweden in 1986. Allegations claiming there were irregularities and pay-offs in the choice of purchase of the guns have been claimed before in a heated Indian press campaign: but parliament had cleared Gandhi. Now new life was given to the allegations: the government's own official declared that there were irregularities in selecting the guns and that the prime minister's own directive on evaluation methods had been flouted. The army, the report noted, on six occasions between 1982 and 1985 expressed its own preference for a French gun rather than the Bofors. The comptroller and auditor-general's report said there was no provision in the contract with Bofors for deducting commissions paid since it had been agreed that middle-men would not be involved in the deal. His suggestion was that there had indeed been kickbacks. At the time of the deal, Prime Minister Gandhi was also defence minister. The prime minister himself declined to attend parliament to answer questions that were raised. The response of Gandhi's own Congress Party members was to abuse the auditor-general. One former minister was heard to declare that the official should be 'kicked in the bottom'. But the effect on the opposition was all the more remarkable. Members chanted 'Rajiv Gandhi is a thief, he is a thief.' The speaker had to adjourn the session. The opposition followed up in an unprecedented show of solidarity declaring that they would all resign *en masse*. More than 70 members of the lower house of parliament handed in their resignations and opposition members of the upper house walked out.

It remains to be seen whether the calculated move to turn parliament into a Congress club would be an act of suicide or would be appreciated by the electorate as an act of stately disdain. In the battle for the minds of the electorate Gandhi could count on All India Radio, responsible for running the country's radio and television, which now reaches every corner of the country. In April 1989 the network came under the control of Professor K. K. Tewari, a former English literature teacher, who became minister of state for information and broadcasting and quickly acquired a nickname of 'Kalashnikov mouth' or 'the minster of insults'. The magazine *Sunday* normally pro-Gandhi, complained that 'at Mand House (the headquarters of All India Radio) words like objectivity and professionalism only arouse derisive laughter.' Tewari was unashamed. 'Credibility', he laughed dismissively, 'a concern of the cocktail circuit.' Congress leaders claimed that the Indian press had campaigned against Gandhi. All India Radio tilted the balance with nightly screening of Rajiv Gandhi's and other cabinet stalwarts' activities and ignored opposition leaders except to ridicule them.

The unity of the opposition is a major question. The opposition parties range from right-wing Hindu nationalists to the Communist Party of India (marxists). In the past it proved possible for the central bloc of opposition to get together and agree for a short time, but not for long before petty personal squabbling fragmented the temporary unity. Even the possession of power between 1975 and 1977 was not sufficient to keep the non-Congress parties united. Throughout the years of Gandhi, mother and son, factional and personal squabbles kept breaking out to reveal the threadbare, much darned, unity. Throughout 1988 and 1989 the most attractive opposition leader was Vishwanath Pratap Singh (later the new prime minister), for a short while, the 'Mr Clean' finance minister in Rajiv Gandhi's government before he resigned, accusing Gandhi of running a one-man show. He became the leader of the opposition National Front embracing a host of parties from left to right.

Singh, from Uttar Pradesh, India's most populous state, was personally an attractive man. His political attractiveness increased when he won a landslide by-election victory over Congress. But Singh's personality was lacklustre and he was short of leadership dynamism. He took an age to answer questions in a very serious long-winded tone, not a display of the character required to cope with the devious and ambitious colleagues around him. Minister Tewari dismissed Singh as 'the Pamella Bordes of Indian politics who will get into bed with anyone.' Even within the central core of parties whose unity was essential if the multi-party opposition was to have a chance of forming a government, there was continuous squabbling. Former president of the Janata party, Chandra Shekar, was quarrelling openly with his own colleague Ramakrishna Hegde and seemed determined to unseat Singh, belittling and doing his best to undermine the National Front leader. Another supposed stalwart of the opposition, Subramanium Swamy, came out in public with allegations of corruption against his colleagues.

Economic performance

With the election in view, talk of renewed and refreshed socialism was in the air. But behind the scenes, Indian officials and even politicians were preparing for more economic liberalisation. Some hard-liners who still staunchly believe that the government must control the 'commanding heights of the economy' were resisting change with all their might. But the hard economic logic of an economy under strain was pushing everyone towards liberalisation.

India's recent economic performance has shown a steady improvement. The overall rate of growth has been much higher than the so-called 3.5 per cent a year that some Indian economists dubbed 'the Hindu rate of growth'. Production of essential foodgrains, mainly rice and wheat, has improved to about 160m tonnes a year. More important, improved assured irrigation means that India is no longer as vulnerable to the failure of the annual monsoon rains. Of course total grain production falls when the monsoon fails, but the experience of the last five years of the 1980s demonstrated that India could survive even two successive bad monsoons and poor harvests without having to import grains. These are considerable achievements that have surprised India's critics. But the economy remains weak and vulnerable.

Per capita income is still low because the economic growth has been eaten away by the inexorably rising population. Both the government budget deficit and overseas debts are large and causes for concern. The trade deficit for the 12 months to the end of March 1989 rose to Rs74bn, up by almost Rs8bn. This was in spite of a 29 per cent growth in exports. Thanks to the rapid depreciation of the Indian currency the deficit in dollar terms only increased by 0.2 per cent and in terms of special drawing rights it actually fell by 0.5 per cent. India's exports in 1988-90 were Rs202.81bn, still below $15bn, a tiny fraction of the exports of much smaller but booming countries like Korea, Taiwan and Hong Kong.

India's current account has also suffered. The deficit in the financial year 1988-89 rose to more than $6bn, much higher than government forecasts. The World Bank predicted that the deficit in the 1989-90 fiscal year would remain at the same sort of level. The government had forecast a $4.8bn deficit for 1989-90. All this helped to produce a sharp rise in India's total foreign indebtedness. The magazine *India Today* claimed that in the last four years of the 1980s, foreign debt had risen by 65 per cent. Total debt was estimated by mid-1989 at about $55bn, making India the fourth largest indebted developing country in the world (after regional Mexico and Argentina). The government figures for debt are much lower, partly because the government excludes money borrowed for non-resident Indians (about Rs120bn), short-term loans and defence loans. Government spokesmen explain the higher debt figures by pointing to a surge of money borrowed by Indian Airlines and Air-India for aircraft and by the Oil and Natural Gas Commission for imported parts, loans which should pay for themselves either in foreign currency earnings or by limiting spending on imports. The need to raise oil production is urgent. Government estimates said demand for petroleum products is rising at 7 to 8 per cent a year. Thus, though domestic production of oil has risen to 34.5m tonnes a year (and will rise to 47-50m tonnes by the mid-1990s), net imports have risen sharply from 16m tonnes in 1986-87 and were expected to double to 32m tonnes by 1990.

In terms of total GNP, says the government, the total debt is still low, about 2 per cent. But another typical touchstone of the burden that debt imposes is the repayments that must be made every year. As a percentage of foreign exchange earnings, these have risen steadily from about 8 per cent in 1980-81 to close to 30 per cent in 1988-89. Though the troubled Latin American countries have had to cope with much higher repayment figures, prudent economists say that a country should begin to get worried when it has to repay 25 per cent or more of its export earnings in interest payments.

The other great weakness of the Indian economy is the concentration of resources in large public sector undertakings, many of which are either losing money or showing a poor return on their investments. Private sector companies have been hemmed in with restrictions. When they reach a certain size, tiny by international standards, they fall under the Monopolies and Restrictive Trade Practices Act (MRTP) meaning that they have to get special permissions to expand. The very biggest companies in India are in the public sector. The Steel Authority of India for example has a turnover of Rs26bn, almost four times that of Tata Iron and Steel Company, the only private sector steel concern and one of the flagships of the big Tata industrial group.

Altogether there are more than 220 public sector enterprises. The top 15 dominate the top 25 companies in India in terms of sales. The public sector as a whole has suffered from having to take over what are called 'sick units', meaning unprofitable companies, since the government has preferred to take over companies rather than let them go bust. Public sector concerns have also been subject to political interference both in the appointment of chairmen and directors and in day-to-day management. The best that can be said is that in recent years, the public sector companies have done better, recording a combined total profit before taxes of Rs71bn in 1988. Economists contend this is still a poor performance given the monopoly position which the public sector enjoys in oil and natural gas, aviation and several other areas. They question is whether the public sector should be involved, for example, in making watches and whether privatised industry subject to true competition would not be impelled to do even better.

Equally depressing have been delays in implementing government-funded projects. According to the government's own report, cumulative delays in 303 projects costing more than Rs200bn amounted to more than 515 years by September 1989. These included investments in steel, coal, fertilisers, power, railways, oil and civil aviation, all in the public sector. The appalling fact is that the official figures are almost certainly under-estimates. A World Bank study of 133 projects finished by 1987 found an average cost overrun of 82 per cent and delays adding 71 per cent to the time. A 4m-tonne expansion of the Bokara steel plant was 12 years late, adding an enormous cost when a committee calculated that a one-year delay in a 1m-tonne steel plant cost Rs5bn in lost revenues.

Officials claim that things are improving, that schedules are being followed more closely and accountability being introduced. Not before time. But thinking is also changing. Signs of this came in the comments of Abid Hussein, senior member of the planning commission, former commerce secretary and confidant of Rajiv Gandhi. In 1988 Hussein was chairman of a high-level government committee whose report called for radical deregulation of the country's capital markets. It also suggested partial privatisation of huge public sector companies like Air-India, Indian Airlines, Bharat Heavy Electrical and Hindustan Machine Tools. The committee suggested that 25 per cent of these companies should be floated and Hussein added 'If I'd had my way the proposals would have been more radical.' The probability was that he would have suggested 40 to 49 per cent public floatation, but his proposals were watered down by more cautious people on the committee, which included India's chief economic adviser, the controller of capital issues and the deputy governor of the Reserve Bank of India. The committee even avoided the use of the word 'privatisation', a sign that socialism is a slogan that still carries weight with the politicians.

Hussein himself questions many of the principles under which India's economy has been run. He asks 'Do the *babus* of bureaucracy believe they know better than the market?' and claims 'Having the State run the commanding heights of the economy sometimes proved good for the commanders, but not for the people.' Hussein contends that India is set on the path of liberalising, deregulating and privatising the economy, and there can be no turning back, no matter what the political colour of the government.

Hussein pays tribute to India's diversified industrial progress but says that it was a costly process. 'India no doubt produced everything in the industrial sphere, but these goods were costly, not competitive, not available in the desired quantities or of sufficient quality. India found that its desire to re-invent the wheel was too costly, too time-consuming and too much of an inefficient way of learning what others had already learnt and done.' He claims that the financial crunch has begun to bite and it is no longer possible for the state simply to dole out money to its own public sector industries – simply because the money is no longer there. He puts the new ideas: 'You must bring competition. There must be a threat to the existence of firms, which will make them perform. And that will only be possible when you have competition. Competition also means that you must curtail the role of the state.' He also calls for managerial changes and granting of more autonomy to individual public sector companies, henceforth in competition with the private sector. On top of this there must be tighter financial control, he says, since 'When I was young I was told that the path to hell was paved with good intentions. Well, I say the path to inefficiency in the public sector is paved with easy access to soft money from the treasury.'

The economic adviser claims that whatever their political colour politicians will have to move along these lines. They will be pushed from two separate directions. The growing middleclass – which Hussein estimates to number 130m – wants continuing economic improvement and a better standard of life. These middle-class Indians, says Hussein, 'cannot be ridiculed or condemned as a parasitical class. They are ordinary people seeking improvements in their lives not members of a luxury or idle class. And their votes will count.'

A left-wing or populist government might be impelled to do something for the rural areas, but such a government has to face the hard financial facts of life. 'If you want to do something for rural India, you have to spend money, and this money has to come from the state. And the state is in a state of near-bankruptcy,' says Hussein pointing out the uncomfortable facts of life. 'It can only find money by diverting some of its other expenditures – and this will once again compel the government to move out of other areas. You certainly cannot handle the Indian economy today by more state control, by less freedom and less entrepreneurship. Liberalisation rests on a few basic principles. You must have faith in your own people that they are not traitors, but patriots. You must also have faith that the market is a better judge of economic matters than the *babus* sitting in the secretariat. You must have confidence that more competition is better than less competition, and that more competition helps to destroy more monopolies than less competition. Any government will say we must improve the lot of the people faster. Can you show me any formula under which by reducing growth you can improve the lot of the poor? If you can, I shall be your slave for life.'

In 1989 India published its eighth Five-Year Plan. This would involve investments worth Rs3,500bn in the public sector, probably to be matched by private sector investment of the same figure or perhaps higher. There was disagreement because Prime Minister Gandhi wanted higher growth than the 6 per cent set as a target by the planning commission. During the seventh plan – 1985-1990, India's economic growth rate was an average 5 per cent a year, still a long way below the double-digit growth rates recorded by South Korea, Singapore, Hong Kong and Taiwan on their journeys away from impoverished developing country status.

One way in which India can help provide resources for faster economic development is by attracting greater foreign investment. Throughout the last years of the 1980s, there had been increasing reports of foreign companies rushing to get involved in India, and India beginning to offer a warm welcome to them. In 1973 Gandhi's mother passed an Act restricting foreign equity in Indian companies to 40 per cent. The Act remains, but the authorities have shown willingness to take advantage of some of the more relaxed provisions, especially for high-tech enterprises (where up to 74 per cent is allowed) and for exporting companies (up to 100 per cent). And in the first quarter of 1989, Delhi approved projects worth another $91m. Among the newcomers was PepsiCo, involved in three plants in Punjab. Foreign funding's share of total private sector expenditure is still small, a mere 4 per cent, compared to 83 per cent in Singapore and 38.5 per cent in Thailand. Most of the investment was provided by US concerns.

In mid-1989 the government announced plans to set up a board for foreign investment to encourage the inflow of funds to the country. However, it seemed likely to be an advisory body which would neither make policy nor lead to changes in the 1973 Act. However, as evidence of its search for new trade ties, in September a Chinese delegation led by Zheng Tuobin, minister of foreign trade and economic relations, visited Delhi, the first high-level trade talks with China for three decades. Trade is only $160m a year between the two most populous countries in the world. The trade meeting followed a visit to Beijing earlier in 1989 by Prime Minister Gandhi. To critics the moves were somewhat perverse given China's crushing of the pro-democracy movement and India's criticism of Nepal for its ties with China.

Washington did not improve its popularity in India in 1989. Carla Hills, the US trade representative, listed India as one of its prime targets under the 'Super 301' clause of the new trade act. India's foreign investment regime and its closed insurance market were among six 'unfair trade practices' alleged by Washington. The USA told India, Japan and Brazil that unless there were talks on these practices, the USA might take retaliatory action. There was no sign that India was prepared to give way to this threatening behaviour. Finance Minister S. B. Chavan said India's nationalised insurance business would not be opened to foreign companies or privatised. The government had begun to liberalise foreign investment, but no outside power could dictate the degree or pace of change. Commerce Minister Dinesh Singh was also asked what India planned to do about the US complaint. He replied bluntly 'Nothing'.

BUSINESS GUIDE

Time: GMT + 5.5 hrs.

Climate: Mainly tropical, but varies greatly from extreme heat in the tropics of the south and the desert of the northwest, to the extreme cold in the northern Himalayas. Nov-Mar is bright and dry in the south, but cold in the north. Bombay is hot and humid; Delhi is dry. Apr-Jun is hot and dry in the south, and more temperate and cool in the north. There are heavy monsoons in the south west in Jun and in the south east in Oct-Nov.

Entry requirements
Passport: Required by all except certain seamen.
Visa: Required by nationals of all countries except Bhutan and Nepal. Additional visas may be required for re-entry by visitors to some neighbouring countries, such as Nepal, Sri Lanka, Afghanistan, Burma and Bangladesh. Visa regulations are lengthy and liable to change, so detailed enquiries should be made well in advance.
Currency: Indian currency may not be imported or exported. There is no restriction on the amount of foreign currency imported, but amounts over $1,000 must be declared, and the amount taken out may not exceed the amount taken in. All money must be exchanged through authorised banks and money-changers. Foreign exchange receipts should be retained.
Customs: Personal effects are allowed duty-free, although high-value items may require a written undertaking to re-export them. Imports from South Africa are prohibited.
Income tax: Visitors whose stay exceeds 90 days must present an Income Tax Clearance and Exemption Certificate issued by an Income Tax Officer on departure.

Health precautions
Mandatory: Vaccination certificates for yellow fever if travelling from an infected area.
Advisable: Vaccination for cholera and anti-malarial precautions. Avoid unboiled drinking water. Salt loss should be replaced by taking extra salt with meals.

Representation overseas: Abu Dhabi, Accra, Addis Ababa, Aden, Algiers, Amman, Ankara, Antananarivo, Athens, Baghdad, Bangkok, Beijing, Beirut, Belgrade, Berlin, Berne, Bogota, Bonn, Brasilia, Brussels, Bucharest, Budapest, Buenos Aires, Cairo, Canberra, Caracas, Colombo, Conakry, Copenhagen, Dakar, Dhaka, Damascus, Dar es Salaam, Doha, Dublin, The Hague, Hanoi, Harare, Havana, Helsinki, Islamabad, Jakarta, Jeddah, Kabul, Kampala, Kathmandu, Khartoum, Kingston, Kinshasa, Kuala Lumpur, Kuwait, Lagos, Lilongwe, Lima, Lisbon, London, Lusaka, Madrid, Malé, Manama, Manila, Maputo, Mexico City, Mogadishu, Moscow, Muscat, Nairobi, Oslo, Ottawa, Panama, Paramaribo, Paris, Phnom Penh, Port Louis, Port of Spain, Prague, Rabat, Rangoon, Rome, Sana'a, Seoul, Singapore, Sofia, Stockholm, Suva, Tehran, Thimphu, Tokyo, Tripoli, Ulan Bator, Victoria (Seychelles), Vienna, Vientiane, Warsaw, Washington, Wellington.

Air access
National airline: Air India.
Other airlines: Aeroflot, Air Canada, Air France, Air Lanka, Air Mauritius, Alitalia, Alyemda, Ariana Afghan, Biman Bangladesh, British Airways, Cathay Pacific, CSA, Druk-Air, Egyptair, Emirates, Ethiopian Gulf Air, Iberia, Indian, Iranair, Iraqi Airways, JAL, JAT, Kenya Airways, KLM, Kuwait Airways, LOT-Polish, Lufthansa, MAS, Pan Am, PIA, Qantas, Royal Jordanian, Royal Nepal, Sabena, Saudia, Singapore, Swissair, Syrian Arab, Thai International, THY-Turkish, Vayudoot, Yemenia, Zambia Airways.
International airports: Baratpur (VNS), 22.5 km from Varanasi; Sahar International (BOM), 29 km from Bombay, with duty-free shop, bar, restaurant, currency exchange, post office; Dum Dum (CCU), 17 km from Calcutta, with duty-free shop, bar, restaurant, buffet, bank, hotel reservations, post office shops; Meenambakkam (MAA), 16 km from Madras, with duty-free shop, restaurant, buffet, bank, hotel reservations, post office, shops; Palam (DEL), 14 km from Delhi, with duty-free shop, restaurant, buffet, bank, post office; Patna (PAT), 8 km from Patna. Airport tax of 100 rupees is levied on departure (50 rupees to Sri Lanka, Pakistan, Bangladesh, Bhutan, Burma, Nepal, Afghanistan).

Surface access: Overland access is possible through Pakistan, Nepal and Bangladesh, but status and opening hours of border crossings should be checked. There is a daily train service between Lahore and Amritsar. Rail connections exist between India and Bangladesh, although rail travel is difficult.
Main ports: Bombay, Calcutta, Cochin, Kandla, Madras, Mangalore, Marmagao, Vishakhapatnam. There are plans to extend containerisation of cargo ports (currently 10 per cent) to 50 per cent by 1990.

Hotels: International-standard accommodation is widely available. Hotel bills must be paid in foreign exchange or in rupees proved to have been purchased in India with foreign exchange, which can be a lengthy procedure. Hotels in main cities are usually heavily booked, and it is advisable to book well in advance. It may be necessary to carry bedding when visiting up-country hotels and rest houses.

Credit cards: Major credit cards, such as American Express, Diners Club, Visa, etc. are accepted by larger hotels, travel agencies and airline offices, as well as some larger stores. The Central Card is issued by the Bank of India, and is widely accepted.

Car hire: Self-drive hire cars are available in Bombay, and chauffeur-driven car hire is available in main cities.

City transport
Taxis: Local taxis of varying standards are usually available. In main cities, metered taxis may not always show current rates, and fares should be negotiated in advance. Tipping is officially discouraged, but is in practice discretionary.

National transport
Air: Indian Airlines operate over 100 flights daily, with domestic services linking all main cities.
Road: There are 1,772,200 km of road, including 833,000 km of surfaced roads and 31,710 km of national highways connecting main cities.
Buses: A number of long-distance express services operate; local services are not generally recommended for visitors.
Rail: The Indian Railways cover over 64,000 km and are the main form of domestic transport. Rail connections are available between all major towns and cities, with air-conditioned coaches and sleeper accommodation available on some routes. A much-needed modernisation programme is in progress; it is hoped to electrify 6,500 km of track by 1990, in addition to the 5,500 km already completed.
Water: There are coastal shipping and ferry services.

Public holidays: The numerous public holidays observed in India tend to vary from state to state, and are usually declared at the beginning of the year in which they fall.
Fixed dates: 1 Jan, 26 Jan (Republic Day), 1 May (May Day), 15 Aug (Independence Day), 2 Oct (Gandhi's Birthday).

Working hours
Business: (Mon-Fri) 1000 – 1700 in Delhi and Madras; 0930 – 1700 in Calcutta; 1000 – 1730 in Bombay.
Government: (Mon-Fri) 0900 – 1730 in Delhi, Calcutta and Madras; 0930 – 1630 in Bombay.
Banking: (Mon-Fri) 1000 – 1400, (Sat) 1000 – 1200 in Delhi, Calcutta and Madras; (Mon-Fri) 1100 – 1500, (Sat) 1100 – 1300 in Bombay.
Shops: (Mon-Sat) 0930 – 1930 in Delhi; 1000 – 1830 in Calcutta; 1000 – 1830 in Bombay; 0900 – 1930 in Madras.

Social customs: Cows are sacred to Hindus, and many Hindus are vegetarian. Sikhs and Parsees do not smoke tobacco. Muslims do not eat pig's flesh in any form, and orthodox Muslims do not drink alcoholic beverages.

Telecommunications
Telephone: Dialling code for India, IDD access code + 91 followed by area code (812 for Bangalore, 22 for Bombay, 33 for Calcutta, 11 for New Delhi, 44 for Madras) followed by subscriber's number. There were about 3.3m telephones in 1986, with 86 per cent automatic service and a satellite relay system, but connections are often difficult. The government aims to be installing 1.5m new lines a year by 1990, however, using digital exchange technology.
Telex: International telex facilities are available 24 hrs to subscribers, at major hotels and from public telegraph/telex offices in main cities.

Banking: Indian banking has traditionally been strongly directed from the centre; even private sector banks (excluding foreign banks) are required to lend to national priority projects. 1985-6 brought a relaxation and improved profitability, however.
Central bank: Reserve Bank of India.
Public sector banks (A): State Bank of India and its subsidiaries: State Bank of India, HQ Bombay; State Bank of Bikaner and Jaipur; State Bank of Hyderabad; State Bank of Indore, HQ Indore; State Bank of Mysore, HQ Bangalore; State Bank of Patiala, HQ Patiala; State Bank of Saurashtra, HQ Bhavnagar; State Bank of Travancore, HQ Trivandrum.
Public sector banks (B): Allahabad Bank, HQ Calcutta; Andhra Bank, HQ Hyderabad; Bank of Baroda, HQ Bombay; Bank of India, HQ Bombay; Bank of Maharashtra, HQ Pune; Canara Bank, HQ Bangalore; Central Bank of India, HQ Bombay; Corporation Bank, HQ Mangalore; Dena Bank, HQ Bombay; Indian Bank, HQ Madras; Indian Overseas Bank, HQ Madras; New Bank of India, HQ New Delhi; Oriental Bank of Commerce, HQ New Delhi; Punjab and Sind Bank, HQ New Delhi; Punjab National Bank, New Delhi; Syndicate Bank, HQ Manipal; UCO Bank, HQ Calcutta; Union Bank of India, HQ Bombay; United Bank of India, HQ Bombay; United Commercial Bank, HQ Calcutta; Vijaya Bank, HQ Bangalore.

Electricity supply: 220 V AC, 50 cycles (sometimes DC) for domestic use, with plug fittings having two or three round pins and bayonet-type lamp sockets; and 440 V AC, 50 cycles 3-phase for industrial use.

Weights and measures: Metric system.

Representation in capital: Afghanistan, Algeria, Argentina, Australia, Austria, Bangladesh, Belgium, Bhutan, Brazil, Bulgaria, Burma, Canada, Chile, China PR, Colombia, Cuba, Cyprus, Czechoslovakia, Denmark, Egypt, Ethiopia, Finland, France, German DR, German FR, Ghana, Greece, Guyana, Hungary, Indonesia, Iran, Iraq, Ireland, Italy, Japan, Jordan, Kampuchea, Kenya, Korea, PDR, Republic of Korea, Kuwait, Laos, Lebanon, Libya, Malaysia, Maldives, Mauritius, Mexico, Mongolia, Morocco, Nepal, Netherlands, Nigeria, Norway, Oman, Pakistan, Panama, Peru, Philippines, Poland, Portugal, Qatar, Romania, Saudi Arabia, Singapore, Somalia, Spain, Sri Lanka, Sudan, Sweden, Switzerland, Syria, Tanzania, Thailand, Trinidad & Tobago, Tunisia, Turkey, Uganda, UAE, UK, USA, USSR, Vatican, Venezuela, Vietnam, Yemen AR, Yemen PDR, Yugoslavia, Zaïre, Zambia.

BUSINESS DIRECTORY

Hotels: Bombay
Airport Plaza, 70c Nehru Road, Vile Parle East, 400099 (tel: 6123390; tx: 1171365).

Ambassador, Churchgate Extension, 400020 (tel: 2041131; tx: 112918).

Astoria, 4 Jamshedji Tata Road, 400020 (tel: 221514; tx: 113428).

Centaur, Bombay Airport 400099 (tel: 6126660; tx: 1171171).

Central Park, Shivsagar Estate, Worli, 400018 (tel: 4925100; tx: 01173662).

Holiday Inn Bombay, Balraj Sahani Marg, Juhu Beach, 400049 (tel: 6204444; tx: 01171266).

Horizon, 37 Juhu Beach, 400049 (tel: 6148234; tx: 1171218).

Leela Kempinski, Bombay International Airport, 490059 (tel: 6363636; tx: 72204, 72339; fax: 6360606).

Nataraj, 135 Netaji Subhash Road, 400020 (tel: 2044161; tx: 112302).

Oberoi, Marine Drive, Nariman Point 400021 (tel: 2025757; tx: 4153/4; fax: 2043282).

President, 90 Cuffe Parade, Colaba, 400005 (tel: 4950808; tx: 114135; fax: 4951201).

Ritz, 5 Jamshedji Tata Road,Churchgate 400020 (tel: 220141; tx: 112520).

Sun-N-Sand, 39 Juhu Beach, 4000049 (tel: 6201811; tx: 1171282).

Taj Mahal, Apollo Bunder, 400039 (tel: 2023366; tx: 112442; fax: 2872711).

Welcomgroup Searock Sheraton, Land's End, Bandra, 400050 (tel: 6425454; tx: 71230140; fax: 6408046).

West End, 45 New Marine Lines, 400020 (tel: 299121; tx: 2892).

Hotels: Calcutta
Airport Ashok, Calcutta Airport 700052 (tel: 569111; tx: 212771).

Great Eastern, 1-3 Old Court House Street, 700069 (tel: 232331; tx: 217571).

Hindustan International, 235/1 Acharya Jagadish Bose Road, 20 (tel: 442394; tx: 7164).

Lytton, 14 Sudder Street, 700016 (tel: 291872/73/75; tx: 3562).

New Kenilworth, 1 Little Russell Street, 700071 (tel: 448394; tx: 3395).

Oberoi Grand, 15 Jawaharlal Nehru Road, 700013 (tel: 292323; tx: 7248; fax: 291217).

Park, 17 Park Street, 700016 (tel: 248301; tx: 217159).

Ritz Continental, 12 J. Nehru Road, 700013 (tel: 232391; tx: 217148).

Rutt-Deen, 21-B Loudon Street (off Park Street), 70016 (tel: 443884, 446911; tx: 212266).

Hotels: Madras
Ambassador Pallava, 53 Montieth Road, Egmore Madras, 600008 (tel: 868584; tx: 7453).

Chola Sheraton, 10 Cathedral Road, 600086 (tel: 473347; tx: 417200; fax: 478779).

Connemara, Binny Road, 600002 (tel: 860123; tx: 418197; fax: 860193).

Imperial, 14 Whannels Road, Egmore, 600008 (tel: 566176).

Ramada, 1/24 G.S.T. Road, Neenambakkam, 600027 (tel: 434238; tx: 416729).

Savera, 69 Dr Radhakrishnan Road, 600004 (tel: 474700; tx: 6896).

Taj Coromandel, 17 Numgambakkam High Road, 600034 (tel: 474849; tx: 417194; fax: 470070).

Welcomgroup Sheraton Park, 132 TTK Road, 600018 (tel: 452525; tx: 0416868; fax: 455913).

Hotels: New Delhi
Akbar, Chanakyapuri, 110021 (tel: 673251; tx: 312863).

Ambassador, Sujan Singh Park, 110003 (tel: 690391; tx: 0313277).

Ashok, 50-B Chanakyapuri, 110021 (tel: 600121; tx: 3165207).

Centaur, Delhi Airport, Gurgaon Road, 110037 (tel: 5452223; tx: 0315477).

Claridges, 12 Aurangzeb Road, 110011 (tel: 3010211; tx: 65526).

Hyatt Regency Delhi, Bhikaiji Cama Place, Ring Road, 110066 (tel: 609911; tx: 316152; fax: 678833).

Janpath, Janpath Road, 110001 (tel: 3320070; tx: 3161546).

Maurya Sheraton, Diplomatic Enclave, 110021 (tel: 3010101; tx: 61447; fax: 3010908).

Meridien, 1 Windsor Place, Janpath, 110001 (tel: 383960; tx: 65312; fax: 384220).

Oberoi Maidens, 7 Sham Nath Marg, 110054 (tel: 2525464; tx: 66303).

Oberoi New Delhi, Dr Zakir Hussain Marg, 110003 (tel: 363030; tx: 74019; fax: 360484).

Qutab, Off Sri Aurobindo Marg, 110016 (tel: 660060; tx: 3162537).

Taj Mahal, 1 Mansingh Road, 110011 (tel: 3016162; tx: 3166874; fax: 3017299).

Taj Palace Intercontinental, 2 Sardar Patel Marg, Diplomatic Enclave, 110021 (tel: 3010404; tx: 62756; fax: 3011252).

Vasant Continental, Basant Lok, Vasant Vihar, 110057 (tel: 678800; tx: 72386; fax: 678899).

Vikram, Ring Road, 110024 (tel: 625639; tx: 2879).

Hotels: Pune, Maharashtra
Blue Diamond, 11 Koregaon Road, 411001 (tel: 663775; tx: 145369).

Business advice and information
Andhra Chamber of Commerce, 2272/2 Angappa Nick Street, Madras.

Associated Chambers of Commerce & Trade of India, 2nd Floor, Allahabad Bank Building, 17 Parliament Street, New Delhi 110001 (tel: 310704; tx: 2537).

Confederation of Engineering Industry, Lodi Estate, New Delhi 110003.

Bengal Chamber of Commerce & Industry, PO Box 280, Royal Exchange, 6 Netaji Subhas Road, Calcutta 700001.

Bengal National Chamber of Commerce, P11, Mission Row Extension, Calcutta.

Bharat Chamber of Commerce, 8 Old Court House Street, Calcutta.

Bombay Chamber of Commerce & Industry, PO Box 473, Mackinnon Mackenzie Building, Ballard Estate, Bombay 400038 (tel: 264681; tx: 0112671).

Calicut and Malibar Chamber of Commerce, Chamber House, Cherooty Road, Calicut 673032.

Cocanada Chamber of Commerce, Commercial Road, Kakinada, Andhra Pradesh.

Cochin Chamber of Commerce and Industry, Aerodrome Road, Cochin 3, Kerala.

Coimbatore Chamber of Commerce, Race Course Road, Coimbatore, Tamil Nadu.

Federation of Indian Chambers of Commerce & Industry, Federation House, Tansen Marg, Delhi 110001 (tel: 344124, 324992; tx: 2546).

Federation of Karnataka Chambers of Commerce and Industry, Kempegowda Road, Bangalore City.

Hindustan Chamber of Commerce, Kondy Chetty Street, Madras.

Indian Chamber of Commerce, Indian Exchange Place, Calcutta.

Indian Merchants' Chamber, Veer Nariman Road, Bombay.

Madras Chamber of Commerce and Industry, Dare House Annexe, 41 Kasturi Ranga Road, Alwarpet, Madras 600018 (tel: 451452; tx: 41536).

Merchants' Chamber of Uttar Pradesh, 15/57 Civil Lines, Kanpur.

Oriental Chamber of Commerce, 6 Clive Row, Calcutta.

Punjab Harayana and Delhi (PHD) Chamber of Commerce and Industry, PHD House, 412 Siri Institutional Area, New Delhi 110016.

Silk and Rayon Export Promotioon Council, Resham Bhavan 78, Veer Nariman Road, Bombay 400020 (tel: 294792; tx: 0113703).

Southern India Chamber of Commerce and Industry, Indian Chamber Buildings, Esplanade, Madras.

State Trading Corp of India, Chandralok, 36 Janpath, Delhi 110001.

Trade Development Authority, PO Box 767, Bank of Baroda Building, Parliament Street, Delhi 110001.

Trade Fair Authority of India, Pragati Maidan, Lal Bahadur Shastri Marg, Delhi 110001.

Travancore Chamber of Commerce, Alleppey, Kerala.

Tuticorin Chamber of Commerce and Industry, South Beach Road, Tuticorin, Tamil Nadu.

Upper India Chamber of Commerce, PO Box 63, Kanpur.

Banking: The Reserve Bank of India (central bank), Central Office, Shahid Bhagat Singh Rd, New Central Office Building, Bombay 400 023 (tel: 2861602; tx: 114222).

Bank of Baroda, 3 Walchand Hirachand Marg, Ballard Pier, PO Box 10046, Bombay 400 038 (tel: 260341; tx: 116345).

Bank of India, Express Towers, Nariman Point, PO Box 234, Bombay 400 021 (tel: 2023020; tx: 112281).

Canara Bank, 112 Jayachamarajendra Rd, PO Box 6648, Bangalore 560 002 (tel: 76851; tx: 845205).

Central Bank of India, Chandermukhi, Nariman Point, Bombay 400 021 (tel: 2026428; tx: 112909).

Indian Bank, 31 Rajaji Salai, PO Box 1866, Madras 600 001 (tel: 514151; tx: 41307).

Industrial Development Bank of India, Nariman Bhavan, 227 Vinay K Shah Marg, PO Box 10020, Nariman Point, Bombay 400 021 (tel: 2027012; tx: 0112193; fax: 2025865).

Punjab National Bank, 7 Bhikaiji Cama Place, Africa Ave, PO Box 274, New Delhi 110 066 (tel: 602303; tx: 3161906, 72193).

State Bank of India, Madame Cama Rd, PO Box 10121, Bombay 400 021 (tel: 2022426; tx: 112995).

Syndicate Bank, PO Boz 1, Manipal 576 119 (tel: 8261; tx: 82242).

UCO Bank, 10 Biplabi Trailokya Maharaj Sarani (Brabourne Rd), PO Box 2455, Calcutta 700 001 (tel: 260120; tx: 214323).

Travel information
Government of India Tourist Office, 88 Janpath, Delhi 110001 (tel: 32005; cable: INDTOUR).

Indonesia

John Murray-Brown

1989 was a mixed year for Indonesia. After the success of its economic reforms, the country's military-backed government still had to be convinced of the need for broader political change, wary perhaps of the tumultuous recent happenings in China. On the home front, the year was characterised by further strengthening of the economy, simmering student unrest, and the brief reappearance of the Islamic extremism which not so long ago might have caused a few jitters for the country's devoutly secular leadership.

In 1989 Indonesia was more than usually preoccupied with the intricacies of palace politics. Though never easy to detect, the military were again said to be disenchanted with President Suharto, particularly over the increasingly embarrassing business exploits of his sons, Bambang and Tommy. The military was widely seen as the sponsor of the discussion of political succession in 1989, an issue prompted by the publication in April of President Suharto's highly controversial autobiography somewhat immodestly titled 'My Thoughts Sayings and Deeds', in which the president hinted he may not stand for re-election in 1993 for what would be his sixth term in office. If it kept its real feelings well hidden, most commentators were agreed that the military is firmly set against a sixth term for President Suharto, and it spent most of the year ensuring he got the message.

Internationally things were also moving. Indonesia emerged from a period of extended diplomatic hibernation to play a role more commensurate with its size and population. President Suharto met with the newly-installed US President George Bush in Washington in June, and was scheduled to go to Moscow for the first time later in the year, a further illustration of Indonesia's staunch commitment to a non-aligned position in foreign policy. In February, Indonesia and China agreed to re-establish formal diplomatic links, 20 years after Beijing was accused of supporting the failed leftist coup of 1965. Indonesia also concluded a year-long amnesty for its own ethnic Chinese left stranded in Indonesia when the two countries first broke off relations. A government census in 1986 had estimated the number of so-called alien Chinese – those who still have not registered as Indonesian citizens and carry Chinese or Taiwanese papers, or in some cases no papers at all – at 250,000. Today there are perhaps 15,000. It now remains to be seen just how much this latest historic rapprochement between Beijing and Jakarta has been overshadowed by the subsequent bloody events in Tiananmen Square.

KEY FACTS

Official title: Republic of Indonesia
Head of state: President Suharto
Ruling party: GOLKAR
Capital: Jakarta, on Java
Official Languages: Bahasa Indonesia
Currency: Rupiah (Rp = 100 sen)
Exchange rate: Rp1,797 per $ (Nov 1989)
Area: 1,919,443 sq km (13,700 islands)
Population: 175.2m (1988)
GDP per capita: $450* (1987)
GDP real growth: 2.1%* (1987)
Labour force: 62.5m* (1988)
Unemployment: 8.9%* (1985)
Inflation: 8.0%* (1988)
Trade balance: $5,726m (1988)
Foreign debt: $52.6bn (1987)
Visitor numbers: 1,050,014* (1987)

* estimated figure

On a more positive note, Indonesia was again the diplomatic ringmaster for a series of key meetings between Kampuchea's warring factions in an effort to end the 10 year-old Indochina conflict. After hosting the so-called JIM, or Jakarta Informal Meeting, Mr Ali Alatas, the country's foreign minister, was chosen to co-chair the International Peace Conference on Kampuchea in Paris in July. There was some diplomatic movement on Indonesia's own vexed issue of East Timor, which continues to cause Indonesian officials considerable headaches. Pope John Paul II was due to visit this former Portuguese colony in October. A delegation from the Portuguese parliament was also discussing a possible trip to this devoutly Catholic island which Indonesia annexed in 1970. In 1988 Indonesia officially declared the province open. It had been closed to outsiders, except for very limited official trips by diplomats and journalists. 1989 also marked a turning point for Indonesia's other trouble spot, Irian Jaya. In November the province, the eastern half of the island of New Guinea, completed 20 years since the controversial Act of Free Choice of 1969 where the government took a sample of pro-Indonesian opinion from among tribal chiefs.

Closer to home, on the political front Golkar, the government party, surfaced from its National Congress in the previous October with the military firmly back with the party whiphand. In many countries it might have caused a political sensation. In just six days Golkar transformed its top leadership, blooded more than 20 new members into the Central Committee and appointed a leader whom few at the outset

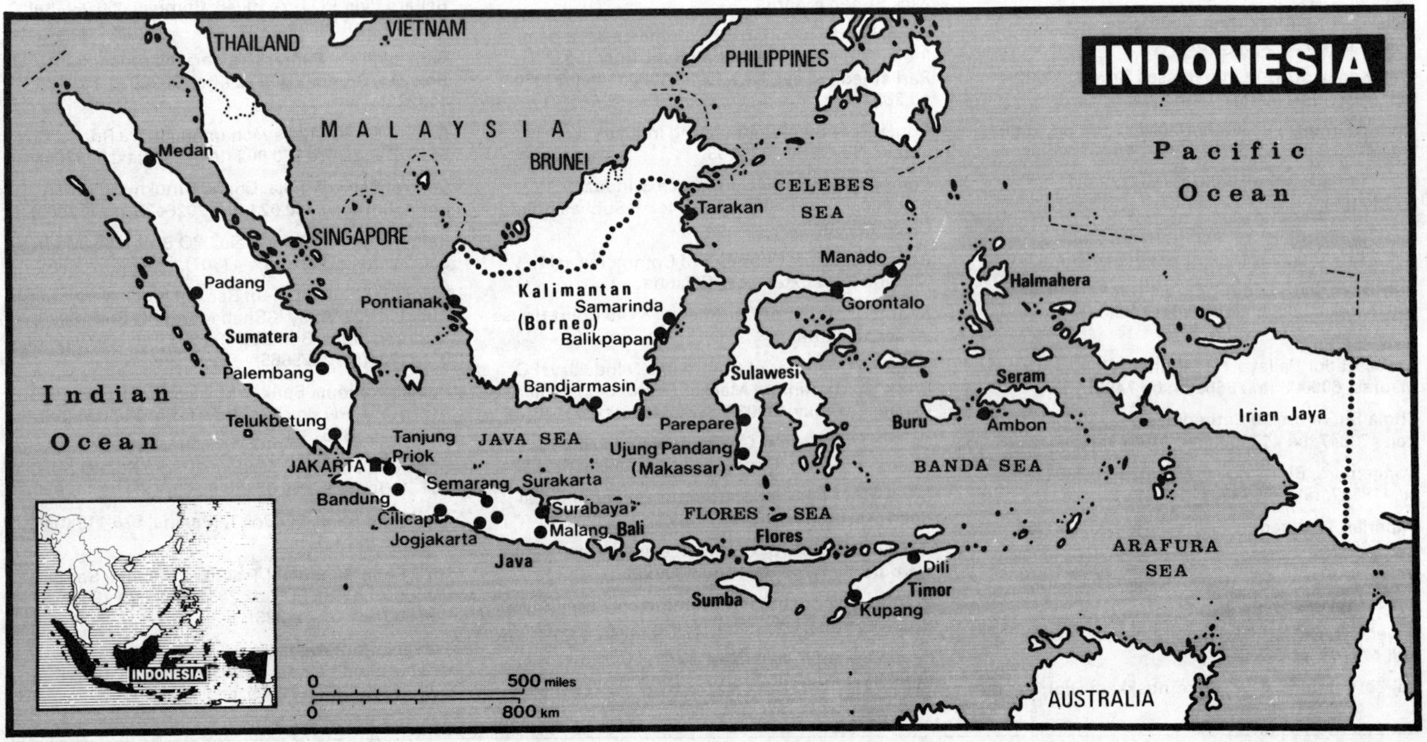

considered even an also-ran. But this was a typically Indonesian contest where the craving for consensus overrode any need for real debate. Nonetheless, to regain control of the leadership of Golkar represented an important victory for the military, an institution which has played the lead role in Indonesia's politics ever since General Suharto seized power in the mid-1960s after the coup attempt.

At the last election in 1987, Golkar won more than 70 per cent of the vote, a performance achieved largely without the strong arm support of the security forces. What is more, Mr Sudharmono, the previous Golkar chief of the party, became increasingly independent, largely at the military's expense. General Wahono, the new leader, is clearly President Suharto's own choice: loyal, with no independent power base. He is also acceptable to the military.

The need to appease the military had become all too apparent. Pertamina, the state oil company, for so long led by a soldier, was to have its first civilian president-director. In August, President Suharto chose Mr Faisal Abda'oe, the former finance director, in a move which more than anything underlined Pertamina's improved housekeeping since it almost bankrupted the country with debts of $10bn in 1974. If some saw that as a step in the right direction, even more significant was the dismantling of Kopkamtib, the military's murky but hugely powerful state intelligence apparatus set up by President Suharto after the coup. Kopkamtib was disbanded and replaced by Bakorstanas, another curious political acronym. Some defence experts say the new agency has few of the wide-ranging and often ill-defined powers of political control enjoyed by Kopkamtib. There were others who felt the change may have been more symbolic than real. Either way it was taken as a welcome sign of Indonesia's growing political maturity and more open government.

Another indication was the role of the non-government lobby groups which began to dictate much of public debate in 1989. All sorts of equity issues were raised – consumer protection, industrial pollution and the problem of land rights in a traditional setting. Walhi, a local green pressure group, dared to sue the government – or more accurately the ministries of Forestry and Environment and the Investment Board – under the new Environmental Impact Law in a case involving a private timber and rayon project in Sumatra which locals claimed was causing pollution and deforestation. In July the case had twice been to court and twice adjourned, apparently because the government has still to select its defence council.

After oil, forestry is Indonesia's most important industry with export earnings reaching $2bn in 1988. However, Indonesia is now losing its forests at the rate of 1m hectares every year; more than any other country apart from Brazil. And many Indonesians are alarmed at the extravagance of the logging industry – a motley group of military and Chinese business interests, who are seen to make huge profits, avoid tax and do little to replant the forest. There was particular concern voiced over a pulp timber project involving the US company Scott Paper in the distant province of Irian Jaya, which ecologists claimed not only damaged the environment but also threatened to displace a local tribal community. Exactly what was the military's role in all this was never very clear. In the student unrest at the start of the year it was suggested the military was actually encouraging the protests. Certainly in parliament it has been the military faction which has provided much of the debate on fundamental issues; for example the relationship between the executive and the legislature.

As in 1988, discussion of the economy provided the real focus for debate. If the statistics were encouraging, what was also noticeable was the renewed popular backing for the government's handling of the economy, reflected in the emergence of an Indonesian

KEY INDICATORS

	Unit	1984	1985	1986	1987	1988
Population	m	159.94	163.39	168.66	171.61	175.22
Gross domestic product (GDP)	$bn	84,857	85,288	74,710	68,752	80,603
GDP per capita	$m	560	520	500	450	–
Wholesale prices	% increase	13.0	4.4	8.5	17.7	7.8
Consumer prices	% increase	10.5	4.7	5.8	9.3	8.0
Crude oil production	'000 b/d	1,181.5	1,256.8	1,158.1	1,177.5	–
Volume of exports	1985=100	90	100	108	104	–
Crude petroleum	1985=100	135	100	117	115	–
Refined petroleum	1985=100	278	100	185	327	–
Rubber	1985=100	101	100	92	93	–
Exports	$m	21,902	18,590	14,824	17,135	19,465
Crude petroleum	$m	10,214	7,217	4,438	4,575	–
Petroleum products	$m	1,456	832	729	989	–
Rubber	$m	952	715	674	1,011	–
Imports	$m	13,882	10,259	10,718	12,512	13,492
Balance of trade	$m	5,707	5,822	2,458	4,674	5,726
External debt	$m	31,952	35,999	43,039	52,581	–
Debt service ratio	%	14.6	19.8	27.3	27.9	–
Current account	$m	–1,856	–1,923	–3,911	–2,098	–1,189
Exchange rate	Rp per $	1,025.9	1,110.6	1,282.6	1,643.8	1,685.7

* estimated figure

export lobby, the revived fortunes of the country's tiny stock exchange and restored public confidence in the currency, the Rupiah. The deregulation policies of the government's Western-trained technocrats has very quickly become a rallying point for groups of every political colour who want reform and political change.

One of the main objects of reform has been to improve business efficiency and boost non-oil exports at a time of uncertain trading conditions. Indonesia, like other Asian countries, is deeply worried by the threat of both the European single market in 1992 and the US-Canadian trade pact. In particular, Jakarta is concerned at what is seen as the increased protectionism of US trade policy under its 1988 Omnibus Trade and Competitiveness Act. Indonesia has made strides to reduce its own trade barriers, slashing tariffs and replacing import and trade monopolies in key areas like plastics and on many steel products. The government is also committed to put in place new laws protecting intellectual property rights – a major bone of contention, particularly with US trade officials.

In 1987 Indonesia introduced amendments to its copyright law, increasing fines and sentences for offenders much in line with the USA. In August 1989 a bilateral copyright accord signed with the USA came into effect covering computer software in addition to audio and audiovisual tapes – which made it more comprehensive than a similar deal struck with the European Community (EC). Revision of the trademark law was mooted, while a new patent which is understood to include the sensitive area of pharmaceuticals was making its way through parliament. However, the reforms also have a clear political impact. In some ways they represent the most serious challenge to the maze of vested interests which have dominated the economy and underpinned President Suharto's New Order regime. Some observers would argue the current moves should be seen as perhaps the last hope to revive an economy long stifled by sluggish growth, chronic unemployment and, above all, endemic corruption.

It was always going to be an uphill struggle to change the economic ways of such a deeply conservative society. But once again the perseverance of cabinet technocrats like Finance Minister Johannes Sumarlin was well rewarded. After improvements in 1988, the non-oil account, with manufacturing showing encouraging gains, again recorded increases; foreign investment approvals rose to record levels and the country won hearty support from its main aid consortium, the Inter-Governmental Group on Indonesia (IGGI) which meets in early June every year in the Hague. According to official figures, non-oil exports rose to $12,048m in fiscal 1988-89, accounting for 62 per cent of total exports. Investment figures for the first six months of calendar 1989 reached $2.8bn, which compares with $4.4bn for the whole of the previous year. The head of the Investment Board said 75 per cent of all investment, both foreign and domestic, was currently directed at export industries in traditional sectors like textiles and in new areas like shoes and food processing. What is more, officials claim as a result of better approval procedures around 80 per cent of total investment commitments are now realised, a marked improvement. Also interesting was the trend in the source of foreign investment, with Taiwan, South Korea and Hong Kong leading the way as semi-skilled manufacturers relocated to avoid increasing domestic production costs.

Oil and gas also continued to make a vital contribution to the external account. Indonesia successfully negotiated the extension of a number of major oil production contracts, after conceding some ground on tax treatment. More significantly, Indonesia and Japan seemed poised to conclude negotiations for the expansion of Indonesia's gas facilities with liquefied natural gas (lng) set to overtake oil as the single largest export commodity. After months of ritual courtship, talks got under way in Jakarta in June with the Japanese buyers demanding revision of the pricing formula and changes in the offtake agreement. By August Pertamina had conceded price changes, but in return agreement was expected on the construction and financing of three new lng production trains at Bontang in East Kalimantan and Arun in North Sumatra. The main question concerned the timetable for the development of the resources – whether to sell now or hold out for better prices. With one eye on the potential budget receipts, the government clearly felt it had little choice.

In many ways the government has also had little choice but to push on with reform, the main thrust being directed at banking, the financial markets and trade and investment sectors. Officials described the banking reforms as the most comprehensive restructuring of the industry since the mid-1960s. Under the new regulations, foreign banks are allowed to operate outside Jakarta – a move aimed not so much at the retail sector but at facilitating trade finance, at a time when many industries are switching into export lines in response to government incentives. The regulation is restricted to six leading provincial cities and will allow the 11 existing foreign banks to set up branch offices and new banks to create local joint ventures. The new rules opened the way for seven new Japanese banks to form local partnerships, challenging the monopoly of Bank of Tokyo in trade finance for locally-sited Japanese investments. Perhaps of greater impact were changes in the capital adequacy requirements of banks. The government appeared determined to instill competition efficiency, and above all financial prudence. It also wants to cut down on inter-group lending – the practice where big businesses establish banks to act as their own private lending arms, often without adequate internal controls.

The stock exchange also received a badly needed fillip in 1989. The government eliminated the tax advantage bank deposits previously enjoyed over stock dividends. It also relaxed trading regulations, lifting the limits on daily price swings, previously set at 4 per cent. In December 1988 the exchange experienced a genuine bull run for the first time in its 11-year history, with the index pushing through 400 compared with a level of 70 in early 1988. In 1989 the trend continued, if at less of a breakneck pace. Much of the activity has been stimulated by new rules allowing foreigners to buy stock. In the present market this rule only applies to eight of the 25 companies listed. The remaining stocks are foreign joint ventures who use the market to meet the government's local equity requirement without having to take on board a local partner. However, in practice many foreign brokers have been able to book stock through local nominees, an anomaly the government will be forced to address. Foreign interest in the market was reflected in the launch of two new funds, one by Indo-Suez, the French bank, the other by Jardine Fleming. This was followed by listings by Bakrie and Brothers, the first Indonesian domestic company to float stock. Astra, a large local Chinese group involved in car assembly and agribusiness, floated a bond issue last year.

The reforms also had their effect on the government's fiscal account. In January, in his new budget for 1989-90, President Suharto was able to announce a mildly expansionary spending plan, with civil servants' salaries up 25 per cent – the first increase in four years. Growth in the budget was to be financed by increases in tax collection and further contributions from the country's main aid donors. Capital and development spending was set to increase 48 per cent in a bid to help relieve employment pressures with 9m new job seekers entering the labour market over the coming five years. Total planned spending, both capital and routine, was raised 26 per cent to Rp36,570bn. Aid contributions in the form of grants or soft loans tied to the purchase of goods from the donor country was expected to rise a massive 58 per cent to Rp11,330bn, accounting for 31 per cent of total revenues.

According to Sumarlin, the main reason for the increased spending projections was the marked improvement in the implementation of foreign aid projects. The price of oil, still Indonesia's largest export earner, was decisive also. Indeed in May the government announced an increase in its budget estimates for average oil prices from $14 to $15.50 a barrel, still

COUNTRY PROFILE

Political system: Executive power rests with the president, who is elected to a five-year term by the People's Consultative Assembly, which meets to decide general policy and for presidential elections. Half its 1,000 members are appointees representing regional assemblies, political parties and other groups. The House of Representatives contributes the other 500 members of the assembly with 100 appointed members and 400 directly elected by citizens who are married or over age 17. The cabinet is appointed by and responsible to the president. In 1987 President Suharto was re-elected for a fourth term of office.

Political parties: In elections held in April 1987, the government alliance Golkar won 299 seats in the House of Representatives. The Development Unity Party won 61 seats and the Indonesian Democratic Party obtained 40 seats.

Population: Population is 175.6m (1988). Although 95 per cent of the population are of Malay origin, there are some 300 minorities. Population growth is around 2.1 per cent, and over 74 per cent of the inhabitants live in rural areas. Indonesia has the world's largest Muslim population, although on Bali the Hindu faith predominates.

Main cities/towns: Jakarta, Surabaya, Bandung, Semarang and Yogyakarta on Java; Denpasar on Bali; Medan and Palembang on Sumatra; Ujung Pandang (Makassar) on Sulawesi (Celebes); Banjarmasin, Balikpapan, Samarinda and Pontianak on Kalimantan (Borneo); Jayapura on Irian Jaya (New Guinea).

Language: Bahasa Indonesia is the official language; English is common in business circles, and many older Indonesians speak Dutch.

Media
Press: English-language dailies which circulate widely in West Java are *The Indonesia Times, Indonesia Observer* and *The Jakarta Post*; Indonesian-language dailies include *Kompas, Berita Buana, Merdeka, Suara Pembaruan and Suara Karya*. Indonesian weeklies include *Tempo* and *Fokus*. Business: *Indonesian Commercial Newsletter* is a bi-monthly, bilingual business publication.
Radio: Radio Republik Indonesia (RRI) is the government-owned and operated network, with external services in several languages (including English, French, German and Japanese). There are many private radio stations with high listenership. *Television:* TVRI, the government-owned and operated TV network, broadcasts about seven hrs/day, with some colour programming. Advertising is banned.

Domestic economy: Agro-procurement and supply and agro-industrial companies are important, but oil and gas dominate the economy, supplying 40 per cent of export revenues in 1988. The oil price slump of 1986-87, however, reduced its share of government revenue from 66 per cent in 1984-85 to around 40 per cent in 1986-87. Much of the manufacturing sector is in foreign hands, with a thriving locally owned construction sector, and the Government is simplifying regulations to stimulate further investment. Repelita IV (the five-year plan from April 1984) aimed to raise living standards, achieve a more equitable distribution of wealth, develop agriculture and rural areas and improve infrastructure and industrialisation. With the need to develop non-oil exports, there was emphasis on export-orientated and labour-intensive industries. Building a motor industry was an important part of the five-year plan, but the industry suffered from too many firms, all importing parts for assembly. The government has resorted to tight monetary control to prevent capital flight and reduce inflationary pressures. Dependence on imported petroleum products has been sharply reduced since 1985, with the expansion of domestic processing capacity. Since 1966, Indonesia has received regular inflows of foreign aid from various international organisations and from Japan, the US and the Netherlands. According to some sources, Indonesia is poised to encounter a 'spiral of deregulation, investment, exports and growth', in which GDP could expand by five per cent a year 1989-1993. A fifth five-year plan, Repelita V, was launched in April 1989. It set targets for investment, manufacturing output, export growth and employment criteria. The latter is of key importance in the context of a population growth of two per cent annually and the glaring income disparities among the population. GDP was estimated to have reached 5.4 per cent in 1988, up from 3.5 per cent a year earlier. The non-oil sector, led by manufacturing and mining, grew by nearly six per cent in 1988, and this growth is expected to continue.

Employment: Labour force was 67.5m strong in 1987, growing at the rate of 1.8m per year. One-third or more of the labour force is out of work or underemployed.

External trade
Exports: Total exports stood at $16.78bn in 1987. Main exports are oil and oil products ($5,166m in 1986-87), liquefied natural gas ($2,804m), wood and wooden products ($1,418m), vegetable products, coffee ($826m), palm oil ($103m), textiles ($682m), artificial resins and plastics and basic metals and metal products including tin ($180m). Main destinations are Japan (44.9% of 1986 exports), USA (19.6%), Singapore (8.4%), Netherlands (3.1%), Republic of Korea (2.4%), Hong Kong (2.3%), Australia and German FR (2.3%).
Imports: Total imports were $12.41bn in 1987. Main imports are machinery and electrical equipment ($1,907m in 1986-87), chemical products ($1,728m), mineral products ($133m), base metals ($922m), vehicles, aircraft and ships ($888m), food and beverages ($704m), clothing and textiles ($412m). Main sources are Japan (29% of 1986 imports), USA (14%), Singapore (9.1%), German FR (6.7%), Saudi Arabia, Australia (3.9%) and UK (3.8%).

Agriculture: Agriculture accounts for about 25 per cent of GDP (1988), and smallholdings alone employ about 55 per cent of the labour force. It is predicted that agriculture's contribution to GDP will decline to just over 20 per cent in 1993. After two years of modest growth, due mainly to long droughts, the sector's output increased by around 3.2 per cent in 1988, compared with 2.4 per cent in 1987. Agricultural products make up 25 per cent of non-oil export earnings, with forestry making up another 25 per cent, and provide the basis for other industries. After planting more high-yield varieties, investing in irrigation systems, doubling the use of fertilisers and trebling the use of pesticides, Indonesia has achieved self-sufficiency in rice, its major crop, with 28.3m tonnes produced from lowland paddies in 1988; there are now surplus rice stocks of about 1m tonnes. Cassava, maize, sugar, sweet potatoes, bananas and many other fruits and vegetables are grown for local consumption. Large estates, now undergoing rehabilitation, produce coffee, tea, rubber, coconuts and oil-palm nuts, mostly for export. Indonesia is the world's second largest producer of rubber (the second largest agricultural export commodity), and is seeking self-sufficiency in sugar. Forest products have been the third most important export earner, and Indonesia has some of the world's largest remaining reserves of tropical hardwoods (estimated at 191m acres in 1987). Recent legislation has aimed to reduce the rate of felling and to ban the export of logs, and has increased the proportion used locally in timber-processing. Fishing is also an important local industry. Output was 2,500 tonnes in 1986.

well below what Indonesian crudes were fetching at the time. World Bank Vice President Mr Attila Karaosmanoglu said Indonesia's need for special donor assistance could well end if oil prices continued their upward trend. The technocrats take a somewhat different view, arguing for assistance in order to maintain the pace of economic reform. Indonesia faces a considerable debt burden while internal debate about the direction of economic and industrial policy is far from resolved. Since the cancellation of projects worth some $10bn back in 1984, the technical ministries, for whom commercial patronage is often a lever of political power, have been urging more capital spending, on projects where there is often little strategic or economic programme for financing them. Some sceptics now worry that a revival in oil prices could even be the deathknell of a reform programme which in its first stages in 1986 was a direct response to the oil price collapse.

For the technocrats the need to contain public spending and so limit its foreign debt exposure remains a central plank in the programme. Total debt remains a heady $50bn according to the World Bank, while over 30 per cent of Indonesia's debt is denominated in Japanese Yen, a currency which has dramatically appreciated since the Plaza Accord of 1985. According to the Bank debt service, both capital and interest repayments now absorb some 36 per cent of the country's export revenues.

In June the government won a significant concession from Japan, the country's largest donor and most important trade and investment partner, under a Yen 50bn loan. Under the terms of the new loan Indonesia has the option to draw down the amount in dollars which will afford the government a valuable hedge against adverse currency movements. The borrowing also helps to top up Indonesia's undrawn commercial standby credits which stand at around $2.4bn – a policy which provides a cushion against falls in the oil price. Central Bank Governor Adrianus Mooy also noted that the terms – both pricing and maturity – were appreciably better than on previous deals, again reflecting the growing confidence of the international business community in Indonesia's economy. However, pressure on the capital account remains. In May it was revealed the government was preparing a major rescue operation for the country's largest private sector steel works – Cold Rolling Mill Indonesia (CRMI). A controversial project from the start, this $825m steel plant – a joint venture between Krakatau, the state steel corporation, and companies owned by Mr Liem Sioe Liong, Indonesia's richest industrialist and a close business colleague of President Suharto – was understood to face serious problems in meeting interest payments on its $620m debt. With that sort of gearing many bankers questioned whether the project was ever feasible. There was talk of a possible $350m cash injection raising serious questions about the role of government in what was deemed a private sector project, where the original financing was provided without sovereign guarantees. The troubles at CRMI also provided a sharp reminder of the place of the so-called private sector in Indonesia – the politically connected business groups who have long prospered as a result of government largesse. Many commentators recalled the financial collapse at Indocement, another Liem concern, where the government agreed in 1985 to take a $320m, 35 per cent equity stake. In a curious way the CRMI crisis, however painful, may provide useful ammunition for the reformists at a time when they are trying to block a number of industrial and infrastructure projects where the economics are less than proven. There was even some suggestion that the story of CRMI's troubles was deliberately leaked to the press, in order to strengthen the reformers' hand in the debate.

The battle, however, is far from over. A new petrochemical deal with Shell in which Bimantara, a company owned by President Suharto's second son, is taking an equity stake looks to some like a rerun of CRMI. The Suharto connection is seen by many as tantamount to a government guarantee. If the economic feasibility is perhaps sounder, there is nonetheless the feeling that Indonesia should be concentrating on labour intensive projects rather than trying to pick winners with large-scale import substitution industries.

Perhaps the single greatest threat to Indonesia's hard earned economic and political stability is the growing spectre of unemployment. Many Indonesian economists feel it unfair to compare their country with either Malaysia or Thailand, which have a far more manageable geographic size and smaller population. Indonesia may be Asia's largest oil exporter and the only member of the Organisation of Petroleum Exporting Countries (OPEC), but it has a population of more than 170m with a per capita income of less than $500. It was back in late 1988 that the government was given the grimmest reminder to date of the enormity of the problem when six young school leavers were crushed to death in a stampede for jobs outside a provincial labour exchange.

COUNTRY PROFILE

Industry: Manufacturing industries are relatively small, accounting for 14.4 per cent of GDP (1986). The share of manufacturing has been increasing in recent years, accounting for more than 15 per cent of GDP (1988) compared with 12 per cent in 1982. The value of manufactured exports exceeded the value of oil and gas exports in 1988 and manufacturing is likely to be the leading growth sector for the next few years. Development of certain intermediate and heavy industries has been encouraged. Main products are textiles, petrochemicals, cement, fertilisers, basic metals, motor cycles, chemicals, paper and household goods, mainly for domestic consumption. Consumer electronics are produced for export as well as for the rather flat domestic market. There has been a recent proliferation of plywood factories in response to official pressure to switch from exporting raw logs, and Indonesia is now the world's largest export supplier. In 1984 Toyota and Mitsubishi Motors opened engine factories, making a total of 9,000 commercial vehicle engines p.a. Honda is to become involved in passenger vehicles, although the country's car industry is operating at only 60 per cent of capacity. Indonesia is the world's largest exporter of LNG (liquefied natural gas), produced from enlarged plants in Kalimantan and Sumatra and sold mainly to Japan under long-term, contract. LNG facilities are being expanded. Growth of tourism is being strongly encouraged, with construction of small and medium-sized hotels and a tourist complex at Nusa Dua on south Bali. Tourist arrivals in 1987 were up 27 per cent to 1,050,014; 8% came from Singapore, 11% from Australia, 12% from Japan and 6% from the USA.

Mining: Mining and quarrying accounted for 11.1 per cent of GDP (1986), but its share has fallen continuously in recent years as production has generally dropped in response to depressed world prices. Indonesia is a major producer of tin (245,000 tonnes in concentrate in 1986-87) and nickel ores (1,681,000 tonnes in concentrate), and has reserves of bauxite (1986-87 production: 686,910,000 tonnes), copper (254,720 tonnes), manganese, gold, silver and iron sand. Coal is mined in Sumatra, with estimated reserves of 21bn tonnes. Oil and gas receipts represent 55 per cent of government revenue and 68 per cent of total export earnings. The state-owned oil company Pertamina has sole rights to exploration, development, production, refining and distribution of oil and gas. There are huge reserves of oil, about 8.7bn barrels, and oil exploration continues to increase. Most oil is produced by foreign companies. Natural gas is supplied from two very large fields at Arun in North Sumatra and Badak in East Kalimantan, though very large offshore discoveries have been made around the Natuna Islands in the South China Sea; nearly all is liquefied. Two million tonnes of LPG a year are sold to the Republic of Korea. The role of oil and gas peaked in the early 1980s when it contributed over four-fifths of total exports. Although oil and gas earnings are still significant, its contribution to GDP is declining and is expected to fall from 19.8 per cent to 16.3 per cent over the next five years.

Energy: Indonesia is almost totally dependent on oil and natural gas for its energy supply. Coal and hydro-electric power contribute 13 per cent of energy consumed, with projects planned to develop coal-fired and hydro electric power generation in order to preserve oil for export. Oil contributes 65 per cent and natural gas another 22 per cent (1986-87). The Government aims by 1988-89, however, to generate 44.7 per cent of electricity from coal, hydro electric and geothermal plants.

Main companies: Pertamina, PT Perkebunan XII, Astra International, Garuda Indonesian Airlines, PT Perkebunan XXI/XXII, PT Wijaya Karya, PT Unilever Indonesia, Perum Perhutani, Perusahaan Aspal Negara, PT Budidharma Daja Manunggal, PT United Tractor, PT Teijin Indonesia Fibre Corp, PT Tjipta Niaga, PT Pantja Niaga, Marison NV, PT British American Tobacco, Pabrik Gula Rejo Agung Baru, PT Semen Cibinong, PT Goodyear Indonesia, PT Bayer Indonesia, PT Waskita Karya, PT Multi Bintang Indonesia, PT Batu Sepatu, PT Sucaco (Supreme Cable Manufacturing Corp), PT Barata Metalworks & Engineering, Biro Asri, PT Philips Ralin Electronics, PT Industri Sondany, PT Dos Ni Roha, PT Sumber Matraman.

Membership of international organisations: ANRPC, ASEAN, Asian Development Bank, Association of Tin Producing Countries, CCC, CIPEC, Colombo Plan, EC trade and co-operation agreement, ESCAP, FAO, GATT, IAEA, IBA, IBRD, ICAO, IDA, IFAD, IFC, IHO, ILO, IMF, Islamic Conference Organisation, Islamic Development Bank, NAM, OIC, OPEC, UN, UNESCO, UPU, WFTU, WHO, WIPO, WMO, World Bank, WTO.

BUSINESS GUIDE

Time: GMT + 7 hrs in West Zone (Java, Sumatra, Bali); GMT + 8 hrs in Central Zone (Kalimantan, Sulawesi, Timor); GMT + 9 hrs in East Zone (Molucca, Irian Jaya).

Climate: Tropical, uniformly hot and humid with monsoon rains from Oct-Apr and dry season from May-Sep.

Entry requirements

Passport: Required by all except for holders of certain documents issued by the UN or Indonesia and certain seamen.

Visa: Visas are usually required for all except UK, US, Canadian, Australian, Scandinavian and ASEAN passport holders. Regulations are subject to 'interpretation'; if in doubt, obtain a visa. Special permission from the Immigration Department must be obtained for travel to Irian Jaya or Timor.

Prohibited entry: Documents, including passports, issued by Israel are not acceptable.
Currency: The maximum amount of currency which may be imported or exported is Rp50,000 per person. Exchange rates for foreign currency are generally the same at both banks and money changers. Major currencies or travellers' cheques may be exchanged at most banks, except in the provinces.
Customs: Personal effects are generally allowed entry, with electrical goods subject to customs duty. Importation of arms and ammunition, pornography, advertising material printed in Bahasa Indonesia, certain pharmaceutical preparations and cosmetics, built-up TVs and built-up vehicles is prohibited. Films and videos require a permit from the censorship authorities, which can take up to a month. Some other products are subject to testing or registration before import is allowed.
Income tax: New income tax laws were introduced in 1984 to widen the base of taxation throughout the country and sales taxes were applied in 1985.

Health precautions
Mandatory: Vaccination certificates for yellow fever or cholera if travelling from infected areas.
Advisable: Cholera vaccination and anti-malarial precautions. All water should be boiled before drinking.

Representation overseas: Abu Dhabi, Addis Ababa, Algiers, Ankara, Antananarivo, Baghdad, Bangkok, Belgrade, Berne, Bonn, Brasilia, Brussels, Bucharest, Budapest, Buenos Aires, Cairo, Canberra, Caracas, Colombo, Copenhagen, Dakar, Damascus, Dar es Salaam, Dhaka, The Hague, Hanoi, Helsinki, Islamabad, Jeddah, Kabul, Kuala Lumpur, Kuwait, Lagos, London, Madrid, Manila, Mexico City, Moscow, New Delhi, Ottawa, Paramaribo, Paris, Prague, Rangoon, Port Moresby, Pyongyang, Rome, Seoul, Singapore, Sofia, Stockholm, Tehran, Tokyo, Tunis, Vatican City, Vienna, Vientiane, Warsaw, Washington, Wellington.

Air access
National airline: Garuda Indonesian Airlines.
Other airlines: Air Micronesia, Bouraq Indonesia, Cathay Pacific, China, CSA, JAL, KLM, Lufthansa, Philippine Airways, PT Mandala, MAS, Merpati, Nusantara, Qantas, Royal Brunei, SABENA, Saudia, SIA, Singapore, Swissair, Thai International, UTA.
Tax: Airport tax of RP9,000 is levied on international departures (RP1,200 on domestic flights).
International airports: Cengkareng International (CGK), 20 km from Jakarta, on Java; Halim (formerly the main international airport), 10 km from Jakarta; Ngurah Rai (DPS), 13 km from Denpasar, on Bali; duty-free shop, restaurant, bar, buffet, hotel reservations, post office, shops; Polonia (MES), 2 km from Medan, on Sumatra.

Surface access
Main ports: Tanjung Priok, Jakarta (containers, being modernised); Tanjung Perak, Surabaya; Belawan, on Sumatra (undergoing modernisation).

Hotels: International-standard hotels have air conditioning and often business centres, where translation and secretarial services are normally available. A 20 per cent tax and service charge is normally added to the bill, so tipping with small change is usual.

Credit cards: American Express and Visa cards are accepted at international-standard hotels, but are not accepted for domestic air fares. International car hire companies, such as Avis and Hertz, will also accept credit cards.

Car hire: Car hire, mostly chauffeur-driven, is available in major towns and cities. Except for international car hire operators which accept credit cards, full payment for car hire is made up-front.

City transport
Taxis: Taxis can be obtained at hotels, airports and railway stations. Standard fares are usual from Bali Airport to major hotels. Metered taxis, which are preferable, are available in Jakarta, but it may be necessary to insist on the use of the meter. Fares are very reasonable. Taxis may also be hired by the hour, which is less expensive for longer journeys. In Jakarta, it can be difficult to hail taxis, so engage one at the hotel and retain it until returning. A 10 per cent tip is usual.

National transport
Air: Garuda operates extensive domestic services, including daily services between Jakarta, Surabaya and Medan. Other routes are also served by Bouraq Indonesia Airlines, PT Mandala Airlines and Merpati Nusantara.
Road: Extensive road network includes over 186,000 km of road, 25 per cent of which is surfaced. A 525 km highway links key areas in Jambi and South Sumatra. Although roads are reasonable, distances are great and, except to Bandung, most visitors travel by air outside Jakarta. Secondary roads are frequently impassable in the rainy season.
Buses: Express coach services link the main cities. Local bus services are inexpensive, but their use is complicated, they are often crowded, and service may be interrupted in the rainy season. For short distances, three-wheelers called *bajajs* are available.
Rail: The rail network, limited to Java, Sumatra and Madura, comprises 8,600 km of track, mostly on Java. Java and parts of Sumatra have air-conditioned express rail services with sleeping and dining cars only between major cities. There are several trains daily from Jakarta to Bandung and Surabaya. Ordinary services can be slow, with many stops.
Water: There are extensive scheduled and non-scheduled inter-island sailings.

Public holidays
Fixed dates: 1 Jan, 17 Aug (Independence Day), 25 Dec.
Variable dates: Good Friday, Ascension of Mohammad, Ascension of Jesus Christ, Waisyak Day (Jun), Eid al-Fitr, Eid al-Adha, Islamic New Year, Birthday of Mohammad. Actual dates of Muslim festivals may vary, as they depend on the sighting of the moon.

Working hours
Business: (Mon-Fri) 0800/0900 – 1600/1700; (Sat) 0800/0830 – 1230/1400.
Government: (Mon-Thu) 0800 – 1500; (Fri) 0800 – 1100; (Sat) 0800 – 1400.
Banking: (Mon-Fri) 0800/0830 – 1600/1700; (Sat) 0800/0830 – 1300.
Shops: 0900/1000–2100/2200 (some close at 1730).

Social customs: Indonesia is predominantly Muslim, and alcohol is not considered essential to social intercourse. Use only the right hand for giving and receiving, especially food, and do not start to consume food or drink until invited by the host to do so. *Pribumi* is used to describe anything indigenous or native to Indonesia, and occurs in commercial or business contexts with reference to local participation, local capital investment or local loans.

Telecommunications
Telephone: Dialling code for Indonesia, IDD access code + 62 followed by area code (22 for Bandung, 21 for Jakarta, 61 for Medan, 31 for Surabaya) followed by subscriber's number. International telephone calls from Jakarta are via communications satellite at all times. Busy lines can make certain calls subject to long delays, especially to Singapore. It is generally cheaper to make a collect (reversed charge) telephone call. A helpful telephone directory is published for Jakarta, where some seven-figure numbers have been introduced. (Numbers starting 48 are now 489, numbers starting 54 are now 548, and numbers starting 81 are now 819.)
Telex: Good international telex facilities are available at major hotels and chief telegraphic offices in major cities. A 24-hr public telex facility is available at the Directorate General for Post and Telecommunications in Jakarta.
Telegram: International cable delivery to and from Indonesia is good, but within the country deliveries are not so reliable.

Postal service: The postal system is generally good, but international parcels and packages are subject to long delays at Customs. Internal mail between main centres is quick and reliable via the express postal service *(pos kilat)*. Mail to outer islands can be subject to long delays. There are specialised delivery and courier services, but these are generally slow.

Banking
Main centre: Jakarta.
Central bank: Bank Indonesia.
State-owned banks: Bank Bumi Daya; Bank Dagang Negara, Bank Ekspor Impor Indonesia, Bank Negara Indonesia 1946, Bank Pembangunan Indonesia, Bank Rakyat Indonesia, Bank Tabungan Negara.
Leading banks with foreign exchange dealings: Bank Bali, PT Bank Buana Indonesia, Bank Central Asia, Bank Duta Ekonomi, Bank Wiaga, Bank Pacific, Bank Umum Nasional, Overseas Express Bank, Pan Indonesian Bank.
Other head offices: Bank Arta Pusara, Bank Amerta, Bank Bumi Arta Indonesia, Bank Dagang Nasional Indonesia, Bank Danamon Indonesia, Bank Indonesia Raya, Bank N.I.S.P., Bank Perdania, Bank Susila Bakti, Sejahtera Bank Umum, South East Asia Bank, United City Bank.
Indonesian branches: Algemene Bank Nederland, American Express International Banking Corp, Bangkok Bank, Bank of America, Bank of Tokyo, PT Bank Perdania (joint venture), Chartered Bank, Chase Manhattan Bank, Citibank, European Asian Bank, First National Bank of Chicago, Hongkong & Shanghai Banking Corp, Standard Bank.

Advertising: Advertising is available in newspapers, where it is limited by government regulations on size of papers, in a wide range of magazines, in cinemas and on public and private commercial radio. Outdoor advertising is widely available on bus exteriors, hoardings, bus stop shelters and so forth.

Trade fairs: A number of international trade fairs and exhibitions take place in Jakarta, notably on medical, technical and scientific themes.

Electricity supply: 110 V AC, 50 cycles, with 220 V AC, 50 cycles increasingly used.

Weights and measures: Metric system.

Representation in capital: Afghanistan, Algeria, Argentina, Australia, Austria, Bangladesh, Belgium, Brazil, Brunei, Bulgaria, Burma, Canada, Chile, Czechoslovakia, Denmark, Egypt, Finland, France, German DR, German FR, Greece, Hungary, India, Iran, Iraq, Italy, Japan, Korea PDR, Republic of Korea, Malaysia, Mexico, Morocco, Netherlands, New Zealand, Nigeria, Norway, Pakistan, Papua New Guinea, Philippines, Poland, Romania, Saudi Arabia, Singapore, Spain, Sri Lanka, Sweden, Switzerland, Syria, Thailand, Tunisia, Turkey, UK, USA, USSR, Vatican, Venezuela, Vietnam, Yemen Arab Republic, Yugoslavia.

BUSINESS DIRECTORY

Hotels: Bandung, Java
Istana, Jalan Lembong 21-44 (tel: 433025; tx: 28409; fax: 51725).

Kumala Panghegar, Jalan Asia Afrika 140, P.O. Box: 506 (tel: 52141; tx: 28623; fax: 431583).

Panghegar, Jalan Merdeka 2, P.O. Box: 506 (tel: 57584/7; tx: 28276; fax: 431583).

Panorama Panghegar, Km 17, Lembang, P.O. Box: 507 (tel: 1006030; tx: 28276; fax: 431583).

Patrajasa, Jalan Ir. H Juanda 132, 40132 (tel: 82590, 81664).

Hotels: Denpasar, Bali
Bali Beach, Sanur, 80001, P.O. Box: 275 (tel: 88511; tx: 35133; fax: 87917).

Bali Hyatt Sanur, Sanur, P.O. Box: 392 (tel: 88271; tx: 35127; fax: 87693).

Bali Oberoi, Kuta Legian Beach, Jalan Kayu Aya, P.O. Box: 351 (tel: 51061; tx: 35125; fax: 52791).

Club Bualu, P.O. Box: 6 (tel: 71310; tx: 35231).

Kuta Palace, Jalan Pura Bagus Teruna, Legian Beach, P.O. Box: 244 (tel: 51433/6; tx: 35234).

Melia Bali Sol, Nusa Dua Tuban, P.O. Box: 1048 (tel: 71510; tx: 35327; fax: 71360).

Nusa Dua Beach, P.O. Box: 1028 (tel: 71210; tx: 35206; fax: 71229).

Pertamina Cottages, Kuta Beach, P.O. Box: 121 (tel: 51161; tx: 35131).

Putri Bali, Nusa Dua, P.O. Box: 01 (tel: 71210; tx: 35247; fax: 71139).

Sanur Beach, Sanur, P.O. Box: 279 (tel: 8011; tx: 35135; fax: 8010).

Tandjung Sari, Sanur, P.O. Box: 25 (tel: 8441; tx: 35157 TANSARI).

Hotels: Jakarta, Java
Borobudur Inter-Continental, Jalan Lapangan Banteng Selatan, P.O. Box: 329 (tel: 370108; tx: 44156 BDOIHC IA; fax: 359741).

Horison, Jalan Pantai Indah, Taman Impian, Jaya-Ancol, P.O. Box: 3340 (tel: 680008; tx: 42824; fax: 684044).

Hyatt Aryaduta Jakarta, Jalan Prapatan 44-46, P.O. Box: 3287 (tel: 376008; tx: 46220; fax: 349836).

Indonesia, Jalan MH Thamrin, P.O. Box: 54 (tel: 323942; tx: 61233 HIPA JKT; fax: 321508).

Jakarta Hilton International, Jin Jend Gatot Subroto, Senayan, P.O. Box: 3315 (tel: 583051; tx: 46673,46698 HILTON; fax: 583091).

Jayakarta Tower Hotel, Jl Hayam Wuruk 126,, P.O. Box: 803 (tel: 6294408; tx: 41113 JAHOTEL; fax: 6295000).

Kartika Chandra, Jalan Gatot Subroto, P.O. Box: 185 (tel: 511008,; tx: 46470; fax: 5204238).

Kartika Plaza, 10 Jalan MH Thamrin, P.O. Box: 2081 (tel: 321008; tx: 61575 BOLKAR IA; fax: 322547).

Mandarin Oriental, Jalan MH Thamrin, P.O. Box: 3392 (tel: 321307; tx: 61755 MANDA JKT; fax: 324669).

Marco Polo, 19 Jalan Cik Ditiro (tel: 325409; tx: 46607; fax: 3107138).

President, Jalan MH Thamrin 59, 10350 (tel: 320508; tx: 61401 PREHO; fax: 333631).

Putri Duyang Cottage, Taman Impian Jaya Ancol (tel: 680108, 680611; tx: 42843 TIJA JKT).

Sabang Metropolitan, 11 Jalan HA Salin, P.O. Box: 2725 (tel: 354031; tx: 44555 SABANG; fax: 371083).

Sahid Jaya, 86 Jalan Jendral Sudirman, P.O. Box: 41 (tel: 587031; tx: 46331 JKT; fax: 583168).

Sari Pacific, Jalan MH Thamrin 6, P.O. Box: 3138 (tel: 323707; tx: 44514; fax: 323650).

Wisata International, Jalan M.H. Thamrin, P.O. Box: 2457 (tel: 320408; tx: 61787; fax: 324597).

Hotels: Medan, Sumatra
Danau Toba International, 17 Jalan Iman Bonjol (tel: 327000; tx: 51167).

Pardede International, 14 Jalan Ir. Juanda (tel: 323866; tx: 51438).

Polonia, 14 Jalan Jend Sudirman 14-18 (tel: 325300; tx: 51376).

Tiara Medan, Jalan Cut Mutiah, P.O. Box: 328 (tel: 516000; tx: 51721; fax: 510176).

Hotels: Surabaya
Garden Palace, Jl Yos Sudarsoll Jalan Pemuda (tel: 479251; tx: 34184; fax: 516111, 516138).

Hyatt Bumi Surabaya, Jalan Jend. Basuki Rakhmat 124-128, P.O. Box: 60271 (tel: 511234; tx: 34316; fax: 470508).

Mirama, Jalan Raya Darmo 68-76, P.O. Box: 232 (tel: 69501; tx: 34154).

Simpang, Jl Pemuda 1-3 (tel: 42150; tx: 31607).

Useful telephone numbers: Jakarta

Police: 510110

Ambulance: 119

Fire: 113

Directory (local): 108

Directory (other Indonesian): 106

International information: 102

International operator: 101

Domestic connections: 100

Car hire: Jakarta
Avis: Jalan Diponegoro 25 (tel: 332900, 349206; tx: 44243); Soekarno-Hatta Airport (tel: 5507088; tx 44243).

PT Bluebird Taxi: (tel: 353556).

Hertz: Jalan Maluku, 13 Jakarta Pusat (tel: 336942; tx: 45755).

Business advice and information
American Chamber of Commerce in Indonesia, 8F Citibank Building, 55 Jalan HM Thamrin, Jakarta (tel: 5780800; tx: 44368).

ASEAN Secretariat, Ji Sisigamangaraja 70A, Jakarta (tel: 716451; tx:47213).

Business Advisory Services, Kuningan Plaza Building, Jalan Rasuna Said Kav C-11-14, Jakarta (tel: 5177295).

Central Bureau of Statistics, Jalan Dr Sutomo 18, Jakarta (tel: 372808).

Co-ordinating Board for Capital Investment (Badan Ko-ordinasi Penanaman Modal), Jalan Gatot Subroto 6, Jakarta (tel: 511648, 512008).

Department of Agriculture, Jalan Harsono RM 2, Jakarta (tel: 782131).

Department of Communications, Jalan Merdeka Barat 8, Jakarta (tel: 366322).

Department of Finance, Jalan Lapangan Banteng Timur 2-4, Jakarta (tel: 349895).

Department of Industry, Jalan Gatot Subroto Kav 52-53, Jakarta (tel: 511738).

Department of Information, 9 Jalan Medan Merdeka Balat, Jakarta (tel: 374392).

Department of Mining and Energy, Jalan Merdeka Selatan 18, Jakarta (tel: 305440, 360232).

Department of Trade, Jalan Ridwan Rais 5, Jakarta (tel: 366318).

Directorate General of Posts and Telecommunications, Kebonsirih 37, Jakarta.

Indonesia-British Association (IBA), Lippo Life Building, 2nd Floor, Jalan HR Rasuna Said Kav B-10, Jakarta 12910.

Indonesian Chamber of Commerce and Industry (KADIN), Jalan Medan Merdeka Timur 11, Jakarta (tel: 377459, 367906).

Indonesian Exporters Association (GPEI), Jalan Kramat Raya 4-6, Jakarta (tel: 346892, 350099).

Indonesian Importers Association (GINSI), PO Box 2744/DDK, Wisma Nusantara, Jalan Maja Pahit 1, Jakarta (tel: 360643, 367269).

Jakarta Chamber of Commerce and Industry, Kadin Jaya Graha, Jalan I H Juanda 38, Jakarta (tel: 365609, 370943).

National Development Planning Board (BAPPENAS), Jalan Taman Suropati 2, Jakarta (tel: 336207, 348990).

Pertamina (State oil and gas company), Jalan Merdeka Timur, Jakarta (tel: 3031; tx: 44152, 44302).

Banking
Bank Indonesia (central bank), Jalan M H Thamrin 2, Jakarta (tel: 372408; tx: 44164).

Bank Bumi Daya, Jalan Imam Bonjol 61, PO Box 106, Jakarta 10002 (tel: 333721; tx: 61117).

Bank Dagang Negara, Jalan M H Thamrin 5, PO Box 338/JKT, Jakarta (tel: 321707; tx: 61628).

Bank Ekspor Impor Indonesia, Jalan Lapangan Setasiun 1, PO Box 32, Jakarta Kota (tel: 673122; tx: 42702).

Bank Negara Indonesia 1946, Jalan Lada 1,PO BOx 1946 KB/JAK, Jakarta Kota 11110 (6901217; tx: 42674).

Bank Rakyat Indonesia, Jalan Jenderal Sudirman Kav 43-43, Jakarta (tel: 586111; tx: 44728).

Travel information
Directorate General of Tourism, Jalan Kramat Raya 81, PO Box 409, Jakarta (tel: 359001; tx: 45625; cable: INTO).

Ikatan Motor Indonesia (IMI), Gedung KONI, Pusat Senayan, Kotakpos 609, Jakarta (tel: 591102).

Courier services
Birotika Semesta PT, Jalan H R Rasuna Said Setiabudi Building, Jakarta (tel: 517989).

City Link Indo PT, Jalan Thamrin Wisata Hotel, Jakarta (tel: 333432).

DHL, Wisma Metroploitan II, Jalan Jend Sudirman, Jakarta (tel: 5781616).

Inter Pacific Citra PT, Jalan Panglima Polim Raya 15, Jakarta (tel: 735613).

LTH International, Jalan Tanah Abang Timur 16A, jakarta (tel: 371108).

OCS, 11th Flr, Skyline Building, Jalan M H Thamrim, Jakarta (tel: 326704).

Pronto, Prince Building, 1, 4th Flr, Jalan Jend Sudirman, Jakarta (tel: 588059, 586567, 583959).

TNT-Skypak, 20th Flr, Summitmas Tower, Jalan Jend Sudirman, Jakarta (tel: 5781157/8).

Interpreter and translation services
Commercial Advisory Foundation in Indonesia (CAFI), Jalan Probolinggo 5, Jakarta (tel: 344485).

Henry Suryopranoto SH Associates, Jalan Hayam Wuruk, Harco Building D34-35, Jakarta (tel: 656021).

Iran

Anthony Hyman

Ayatollah Khomeini's death in June 1989 produced a wave of emotion among Iranians, and had a major impact on Iran's political system. It was an orderly transition through national elections, a referendum and new appointments, but beneath the surface a bitter power struggle went on between rival politicians and factions. Many observers expected a more moderate and pragmatic approach to international relations to emerge before long. Though the cease-fire with Iraq held, moving beyond a fragile truce to a full peace settlement looked problematic.

The appointment of President Ali Khamenei as Iran's spiritual leader of the nation, by a vote of the Assembly of Experts, took place within hours of Khomeini's death. There was no way, though, in which he or any other clergyman could fill the unique place of Khomeini, or inherit his unique moral authority. A presidential election in August was a landslide won by the shrewd, pragmatic and popular ex-Speaker of the *Majlis* (Iran's Parliament), Hojatoleslam Ali Akbar Hashemi Rafsanjani. His victory at the polls, with a claimed 94.5 per cent majority of votes cast, was hardly surprising, coming after scores of independent candidates for the post of president had been barred. Rafsanjani took over as a full executive president, rather than a ceremonial one. Iran's new Constitution established a framework for a new leadership structure, giving a greatly enhanced role to the president, and abolishing the post of prime minister. The proposed changes were put to a referendum on the same day as the presidential elections.

Pragmatic politics

Rafsanjani's first moves on becoming president confirmed expectations that he was trying to purge the most outspoken radical elements in the government, by dropping the ministers for the interior and intelligence. A radical majority in the *Majlis* had urged him to retain Ali Akbar Mohtashemi as interior minister, but he was among those dropped in August 1989 when Rafsanjani proposed a new cabinet of 22 members, with 12 new members and a marked technocratic bias. In place of Mohtashemi, the cleric Abdullah Nouri was named as interior minister. Mohtashemi, regarded as one of the hardliners among Iran's anti-Western radicals, had been Iranian ambassador to Damascus from 1982-86. He remained intimately linked to Hizbollah and other extremist groups operating in the Lebanon responsible for hijackings and other terrorist acts. Another hardliner ousted was Mohammad Reyshahri, ex-intelligence minister. Mir Hossein Mousavi lost his post as prime minister when the reorganisation of government took place.

Various explanations were advanced as to why Ahmad Khomeini, the politically ambitious son (and private secretary) of Ayatollah Khomeini, had not challenged Rafsanjani for the post of president. It was widely believed in Tehran circles that Ahmad Khomeini would continue trading on his father's reputation, and soon enter the *Majlis* or even take up the post of Speaker. However, the young Khomeini lacked an independent power-base.

A dark horse for the future could be Ayatollah Montazeri, Khomeini's former designated heir. Though rebuked and publicly humiliated by Khomeini shortly before his death, Montazeri retains public esteem and respect for his forthright criticisms of the government. As one of the most senior clerics, and a man of principle, he is unlikely to sit back quietly in the holy city of Qum if he disagrees with policies.

As far as political freedoms were concerned, there was optimism that Khomeini's death and the end of the war would usher in a more liberal era. The Freedom Movement, Iran's sole legally-recognised opposition party led by the Islamic Republic's first prime minister, Mehdi Bazargan, was hopeful that constitutional liberties would be respected.

As Iran's president, the 54-year-old Rafsanjani showed determination to implement plans for rapid economic growth and improvement of Iran's foreign relations with the West. He pledged to follow 'the path of Islam and the Quran', also aiming to reduce the state role in the economy.

Plain speaking about the country's economic crisis and the harmful proliferation of bureaucracy came from many official sources. 'Jobs seem to get scarcer by the day, despite the fact that the nation has embarked on a broad reconstruction programme in

KEY FACTS

Official title: Islamic Republic of Iran

Head of state: President Hashemi Rafsanjani; Leader of the Islamic Revolution, Hojatoleslam Sayed Ali Khamenei

Capital: Tehran

Official Languages: Farsi

Currency: Rial (Rl); 10 rials = 1 toman

Exchange rate: R72.2 per $ (Nov 1989)

Area: 1,648,000 sq km

Population: 52.7m (1988)

GDP real growth: –0.2%* (1987)

GNP per capita: $3,408.2 (1988)

Inflation: 15%* (1987)

Oil reserves: 92.9bn barrels (1988)

Trade balance: $144m (1988)

Foreign debt: $4.7bn* (1987)

* estimated figure

which hundreds of towns and cities are to be rebuilt', stated Morteza Alviri, head of the parliamentary budget and planning committee in August 1989.

The alarming scale of drug addiction was faced up to with harsher laws and an officially-led campaign against drugs initiated in January 1989. At least one million Iranians, mainly young men, are addicted to heroin or opium. Some 200,000 addicts had registered by the end of July 1989 for compulsory treatment over six months, with rehabilitation in labour camps newly set up in each province. 700 drug traffickers were reported hanged in Iran in the first six months of 1989.

The basic questions of sovereignty over the Shatt al-Arab waterway and freedom of navigation in Gulf waters for Iraq were behind the failure to sign a peace settlement between Iraq and Iran, more than a year after the cease-fire. Ayatollah Khamenei accused Iraq in July 1989 of blocking progress in stalled peace negotiations by refusing to withdraw its forces behind international borders. Tehran claimed that Iraq was in occupation of 1,150 sq miles of its territory. In spite of rearming and giving large budget allocations to defence industries, both governments gave signs of confidence that fighting would not resume.

Iran demobilised its *baseej* (volunteer militia) force in 1989, also reducing the length of military service. However, the continued holding of some 100,000 prisoners of war was one of the humanitarian issues which was ignored. 70,000 Iraqi POWs were held by Iran, though some of them did not wish to be returned to the mercies of the Saddam Hussein regime, because in captivity they had 'followed the line of Imam Khomeini'.

Within Iran's military establishment, there continued much infighting and rivalry between the regular armed forces and the *Pasdaran* (Revolutionary Guards Corps). They formed in effect parallel armies, airforces and even navies, causing intense competition and wastage of resources.

KEY INDICATORS

	Unit	1984	1985	1986	1987	1988
Population	'000	45,960	47,820	49,760	51,250	52,740
Gross national product (GNP)	$m	169,197	164,502	164,500*	177,000*	179,750*
GNP per capita	$	3,681.4	3,440.0	3,305.9	3,453.6	3,408.2
Crude oil production	'000 b/d	2,032.4	2,192.3	2,037.1	2,297.6	2,305.4
Exports	$m	12,422	13,328	8,322	11,685*	9,084*
Petroleum	$m	12,255	13,115	7,183*	10,515*	8,170*
Imports	$m	15,370	11,635	10,521	9,570	8,940*
Balance of trade	$m	–2,948	1,693	–2,199	2,115*	144
Exchange rate	Rl per $	90.03	91.05	78.76	71.46	68.68

* estimated figure

A shattered economy

In Tehran's de-luxe hotels they still have up on the walls the brash slogan written in the first flush of the Revolution, 'Death to the USA'. In countless murals and peeling graffiti on Tehran walls, there are threats to the USA, USSR and the 'lesser Satans', Britain and France. The USA is routinely attacked as public enemy number one, yet these anti-imperialist slogans from ten years back have lost much of their political potency. Many Iranians do firmly believe that Washington is somehow orchestrating the array of problems faced by the country. The public mood is ambivalent though – a blend of distrust and fatal attraction. US dollars are the preferred foreign currency to hoard. Many Iranian professionals still dream of going off to live in California, joining their relations and friends.

Iran's shattered economy badly needs outside help to repair essential plant for its oil, refining, petrochemicals, ports and infrastructure, besides expansion of the electricity industry on which industrial production depends. Severe shortages of electricity were causing regular six-hour cutoffs of heating and light in Tehran and other cities in 1989. On cold winter nights, by the dim light of candles in freezing rooms, the war still seemed on.

The positive side is Iran's debt position, in striking contrast to that of Iraq. Even through the war years, Iran managed to reduce its foreign debts by strict housekeeping, paying back $8bn to foreign banks. With virtually no medium-term debt, and only an estimated $4.8bn in short-term debt, Iran's finances and economy look at first sight in remarkably good shape. But appearances can be deceptive.

Large-scale credits are judged vital. Only with foreign help can reconstruction go forward at the desired pace. There are current disputes over Iran's frozen assets in the USA, and with Britain and other EEC states over Iranian responsibility for terrorist acts and hostage-taking in Lebanon, besides the death *fatwa* (edict) of Khomeini against British writer Salman Rushdie. These and the failure to achieve a peace settlement between Iraq and Iran make the risk factors too great for foreign governments to consider stepping in to guarantee large credits. Imports of machinery, spare parts and other scarce foreign items are dependent on the regime's system of import licences and permits, viewed by many as little better than licensed corruption. Minor trade liberalisation was introduced from the end of 1988 for private sector imports of pharmaceutical products and foodstuffs. Private traders willing to provide their own foreign exchange were in principle permitted to import.

The mood within the influential *bazaari* class was far from easy to assess. They were behind the success of the Islamic Revolution, and many undoubtedly are still loyal to its ideals. However, there are many alienated one-time supporters of the Islamic Revolution. Bitterness against the *mullahs* is more natural because high-ranking clerics have such a high profile in the *Majlis*, and a near-monopoly of senior posts in the government. Thus the *mullahs* get the blame for the regime's failures on the economic front. The huge inequalities of wealth in Iran were also a focus for sharp criticism. The growth of a new, overprivileged merchant class was attacked in a hard-hitting report by a radical *Majlis* group, which declared that the war profiteers were the real winners of the eight year-old war. *Kayhan* newspaper published a survey in January 1989 which blamed 'pseudo-religious foundations and corrupt economic centres within the system' as the 'source of poverty in Iran'. Such criticisms of corruption were widely credited by Iranians. A long trial of the chairman of a bogus charity foundation (Nabovat) ended, also in January, with the chairman Musavi Qumi being sentenced to death for embezzlement of over $80m of public funds. Many more so-called charity organisations were ordered to close down in 1989.

Oil saves the economy

The regime's stated economic goal of moving away from acute dependence upon oil is still remote from achievement. In 1989, just as under the late Shah, well over 90 per cent of foreign exchange earnings were dependent on oil exports. However, oil production increased from 2.3m to 3m barrels a day (b/d) by April 1989. This jump marked the first time that level had been reached by Iran since the war started in 1980.

Iran's oil engineers surprised foreign experts by the speed with which oil production was increased during the first half of 1989, and in particular with their ability to repair war damage to installations. Even the Abadan refinery was brought back into partial operation by May 1989. The Oil Ministry planned to meet the shortfall in refining capacity by building three new refineries at Bandar Abbas, Bandar Taheri and Arak, with a total capacity of some 420,000 b/d. Oil exploration also achieved some breakthroughs, with announcements in 1989 of large and medium size oilfields. However, just 20 oil exploration rigs were in operation in the first months of 1989, compared to 60-plus rigs before the Revolution, when Iran for a time was producing oil at a rate of 5m-6m b/d.

The most worrying long-term economic impact of the war will be the effect of serious damage to installations and oil wells, besides neglect and mismanagement on the country's oilfields. Many officials admit that Iran will sooner or later have to face up to the urgency of arranging foreign credits. Rafsanjani, then *Majlis* Speaker, bluntly told a West German trade delegation visiting Tehran at the end of 1988 that the German share in Iran's reconstruction would depend on large-scale trade credits. Demands for colossal war reparations by Iran from Iraq as a condition for peace – with fixed capital losses alone placed at $25bn – were quietly shelved after the success of Iraqi military pushes in 1988.

Foreign trade patterns

The USSR is set to become one of Iran's most important partners in trade and development in the 1990s. The visit by Rafsanjani to the USSR, only weeks after Khomeini's death, sealed agreements which should greatly expand Soviet-Iranian economic links. The Soviet Union has going for it the basic fact that it shares a long border with Iran, and that Iran seeks to diversify its sources of supply and trading routes.

Iran's natural gas exports to the USSR, broken off in 1979 after a price dispute, were set to resume. They will depend on expansion of Iran's gas pipeline infrastructure. There were technical obstacles to be solved with the IGAT-2 pipeline, and reorientation of Iran's IGAT-1 pipeline. The cut-off of Iran's gas exports to Soviet Central Asia in 1979 was followed by a boom in Iran's own domestic market for gas. In the shared region of the Caspian Sea, bilateral co-operation in oil and gas exploration and development was going ahead.

This leaning to the USSR did not prevent Iran's opening to the West. Prime Minister Hussein Musavi, accompanied by a high-powered delegation, made his first ever official trip to Europe in January 1989, visiting Rome and Warsaw. Italy's business with Iran increased to a point where by 1986 it briefly replaced Britain as Iran's third largest trading partner. West Germany, Japan and Britain were the main partners, with a two-way trade worth $1.6bn in 1988. In spite of a series of diplomatic crises, bilateral trade with Britain was expected to remain relatively high. The UK exported £248m worth of goods, importing £140m from Iran in 1988.

Besides the USSR, Comecon states of Eastern Europe were pushing hard bilateral links with Iran. Romania, the GDR (East Germany), Hungary, Bulgaria, Yugoslavia, Czechoslovakia and Poland all established a substantial presence as technical partners with Iran. Each of these states found useful markets for its products and expertise in heavy industry, chemicals and agriculture, mainly in exchange for oil shipments. Even here, the best Western technology is preferred, where available. Iran had reportedly completed negotiating a contract in summer 1989 to equip its Air Force with the joint Yugoslav-Romanian *Orao* (Eagle) fighterbomber, which is powered by Rolls Royce engines and is heavily dependent on British components. The Comecon bloc states also include North Korea, one of the principal suppliers of armaments to Iran through the war years, and naturally well-considered in Tehran. South Korean firms are also very active in Tehran, along with the Japanese. Representatives of companies from Far Eastern states appear confident they will win prize reconstruction contracts. Iran's imports from Socialist bloc countries were estimated at around 15 per cent in 1988 – three times the pre-Revolution total. This was, admittedly, a shrunken market, where barter deals swapping Iran's oil for Eastern Europe's industrial products and technical help were convenient.

Revival of agriculture

Agriculture was one of the stated priorities in rebuilding Iran's war-shattered economy. However, the allocation of scarce resources between rival ministries did not favour agriculture. The great bulk of the 1989 budget was devoted to oil, petrochemicals and heavy industry, together with reconstruction of the cities devastated by the war. Agriculture is bound to gain, though, from reversion to peacetime production of factories turned over to armaments. More tractors and spare parts, together with fertiliser, will gradually become available for Iran's farmers. In addition, the engineers of the Jehad-i-Sazandegi (or 'Reconstruction Crusade') organisation can be expected to devote resources to their original purpose, instead of the war effort.

Iran's staple crop is wheat. Current wheat production, officially stated at 6.5m tonnes, satisfies 70 per cent of domestic needs. But with a projected population of 65m by 1995, increasing by some 6,000 each day, Iran will be importing up to 5m tonnes of wheat annually, unless yields are greatly improved. Total imported foodstuffs cost annually $2bn, or a full third of state revenues in 1988. At a seminar on dry farming in January 1989, Iran's deputy-minister for agriculture, Javed Angeji, urged the necessity of exploiting the best scientific methods in order to improve crop yields . Significantly, allocation of state funds for agronomists are on the up; the budget for Iran's cereal research units from Nauroz (21 March 1989) was set at Rls 950m ($13.6m at the official rate).

Iran has had a success story in the production of fruit and vegetables, as well as processing and canning for domestic and limited export sales. These have steadily increased through the 1980s. There are, however, familiar problems of distribution and lack of processing plants. In the cities, fresh fruits are relatively costly, and virtually beyond the reach of ordinary working people, whose wages have lagged far behind inflation. Iran's livestock production faced big difficulties from 1984 because of a shortage of foreign exchange to purchase imported fodder. To improve stock, thoroughbred cattle were being imported from Western Europe, and more help has been given to Iran's nomads in finding pastures for their herds. However, a growing volume of meat (200,000 tonnes in 1985-86) has had to be imported each year to satisfy domestic demand. Private enterprise is largely responsible for the thriving specialised sector of chicken and egg production. Some poultry farms are big and use the most modern methods, especially ones close to the capital. The government is now actively encouraging small farmers to supplement their incomes by raising chickens. Poultry farmers have steadily expanded production since the Revolution, but not kept pace with demand.

Distortions of a war economy

The all-out war effort showed Iranian capabilities and ingenuity in producing a wide range of armaments. But this was at great cost to Iran's manufacturing industry, languishing since the early 1980s. Only the oil industry, vital for export earnings, along with chemicals and petrochemicals, escaped the squeeze on new investment. The current emphasis is on petrochemicals, but the exigencies of war budgets resulted in relatively small units being planned, which critics maintain will prove uneconomic. Economic difficulties affected the Iranian public in recurrent shortages and long queues for poor quality subsidised food – when available. The real value of wages and salaries was eroded by high inflation. The Iranian rial was being traded in the street at 15 to 20 times less than the official rate for foreign currency.

During the war, many of the country's most modern, best equipped industrial plants were turned over to munitions. This was especially true of the factories in the Tehran region for assembling motor vehicles. The process of re-conversion will be complex and is likely to be long. The priority must be to carry out repairs to Iran's oil facilities and infrastructure in general. In Tehran, Tabriz, Isfahan and other industrial centres, factories were operating at levels of around 30 per cent capacity. The causes were basic ones, needing time as well as foreign exchange to resolve; acute shortages of raw materials, lack of electricity and parts for machinery, and shortages of trained managers and skilled personnel. Labour relations in Iran's factories tended to be poor, with high absenteeism of the workforce. Many of the country's large factories are heavily over manned, turning in large annual

COUNTRY PROFILE

Political system: Iran is an Islamic republic which took its present form after the removal in February 1979 of Shah Reza Pahlavi. Executive power rests with a Cabinet, which answers to the parliament *(Majlis)*, but in practice decisions are vetted by the clerical Council of Guardians. It is through this channel that Hojatoleslam Sayed Ali Khamenei has exercised personal authority, since the death of Ayatollah Khomeini.

Political parties: During *Majlis* elections held in May 1985, the Islamic Republican Party won practically all of the 270 seats. Elections in April 1988 produced gains for the radical faction within the government. In 1987 the Islamic Republican Party was dissolved.

Population: Iran's population is growing at the very high rate of 3.6 per cent p.a., with a marked shift toward the towns in recent years. The 1986 census result of 48m was unexpectedly high. Tehran's population has doubled since 1979 to eight million. In 1988 population was estimated to be 49.3m.

Main cities/towns: Tehran (population estimated 8m), Isfahan, Mashhad, Tabriz, Shiraz , Ahwaz, Abadan.

Language: Farsi; some English, French and German.

Media
Broadcasting: Voice and Vision of the Islamic Republic of Iran runs the radio and two TV stations.

Domestic economy: Oil revenues constitute 48 per cent of all government income and contribute about 15 per cent to GDP (1986-87). Owing to the international oil glut and the war with Iraq, oil revenues have not met budgeted targets, despite a stabilisation of prices in 1986-87, and imports have risen dramatically. Money in circulation quintupled in the six years following the revolution of 1980, causing severe inflation, and the accumulated government deficit represents a major stumbling-block in the future. In 1986 *Majlis* tried unsuccessfully to restrain government spending, and its only real prospect seemed to lie in a confirmation of the peace with Iraq. In mid-1988, Bandar Abbas was still the only non-oil port to be fully operational, because of war damage and other disruption.

Employment: Although no official figures are given, unemployment is estimated at three million (15 per cent of the workforce). Of those employed, 48% work in agriculture, 18% in manufacturing industries, 9% in commerce, 7% in construction and 12% in service industries. Some two million men are in the armed forces.

External trade: Shortages of foreign currency have produced a boom in countertrade, which obscures the extent of foreign trade; Iran is also understandably reluctant to declare the full extent of its oil traffic in the Gulf, but oil revenues are thought to have totalled some $10bn in 1987.
Exports: Main exports are crude oil (89.4% of total in 1984), petroleum products (8.3%); non-oil items represented only 2.3%. Main destinations are Japan (16.5% of 1985 exports, and rising), Italy (10.0%), Turkey (9.1%), Singapore (7.3%), Syria (6.8%), France, German FR, Netherlands and UK (0.5%).
Imports: Main imports are machinery (15.2% of 1983/84 total) iron and steel (12.6%), electrical goods (8.4%), grain (5.5%), vehicles and spare parts (5.0%), cotton (4.4%), plastics and raw materials (3.0%), meat (2.8%), fertiliser (2.1%), animal fats (1.9%), pharmaceuticals (1.9%, textiles (1.8%) and chemicals (1.8%). Main sources are German FR (16.3% of requirements in 1985), Japan (13.4%), UK (6.7%), Italy (6.0%), Singapore (4.0%), Syria, Turkey and Sweden.

Industry: Manufacturing has been badly affected by foreign currency shortages and by war disruption. Activities have included textiles, petrochemicals and refining, food and beverage processing, furniture, printing, plastic and metal manufactures, leather/leather goods, bricks, cement, glass, automobile assembly, fertilisers, chemicals and electrical goods. Although there is a strong revival in company formation barely a third is in the productive sector, with services proving the most popular area of expansion.

Minerals and mining: Development of copper, lead, zinc and coal is under way.

Hydrocarbons: Reserves of natural gas are estimated at 13 trillion cu m (1987). Oil production in 1986-87 was 2.2m b/d*, 25 per cent less than in 1983-84, and air and naval attacks in 1986-87 reduced exports. Total exports are, however, estimated at 1.5m b/d.

Banking: Banking is being progressively brought under Islamic law. Many foreign banks were nationalised in 1979.

Membership of international organisations: CCC, Economic Co-operation Organisation, FAO, G-77, IAEA, IDA, IFC, ILO, IMO, INTELSAT, International Chamber of Commerce, IPU, Islamic Conference Organisation, OPEC, UN, UNESCO, UNIDO, WHO, World Bank.

losses. Only the takeover by state agencies of many large private businesses soon after the 1979 Revolution prevented their closing down for good. However, the future of state enterprises as well as private businesses promises to be much brighter if reconstruction begins in earnest.

Perhaps the key problem for the future is tackling unemployment. Iran's rapidly growing population was placed at 51.7m in the national census, published at the end of 1987. A *Majlis* deputy claimed that this census itself revealed that 3.8m were unemployed. That total would certainly be higher if it took into account not only those without jobs, but also those in seasonal employment, often earning very low wages. The war mobilisation resulted in millions of men serving in the regular armed forces, with millions more joining the *baseej* for short periods. Of one million estimated war-dead, Iran suffered the most with hundreds of thousands more permanently disabled.

The government is aware of the urgency of job creation, but in the short term no reconstruction programme, however ambitious, can hope to find employment for all the jobless – few of whom have special skills or vocational training. There is a general lack of willingness to take up the lowest paid, dirty or tough jobs, especially those at brick kilns and on farms throughout Iran. Over half a million Afghans, mostly arrived as refugees in Iran since 1978, are currently employed in the lowest-paid sectors of industry. Iran's jobless, especially the young, with typically high expectations of lifestyle in the cities, form a large part of the 'deprived class', or *mostazafin*. It was in their name the Revolution was made, and for whom Khomeini claimed to take power in 1979. Yet very few of the mostazafin have yet seen the promised fruits of their Revolution. Their already low living standards actually declined during the 1980s, due to negative growth of the economy. Blame for this used to be given to the so-called 'imposed war' with Iraq, but there is, of course, no longer this excuse for failing to deliver. Apart from Iran's 3.5 per cent annual growth in population, one reason for the problem stems from the regime's failure to reverse the drift away from the land to the cities. The pace of urbanisation actually increased in the 1980s. Far from making Iran self-sufficient in food production – another of the Revolution's objectives – there are shortfalls in most food items. Unless large numbers of uprooted peasants are attracted back to the villages, helping to revive Iran's crisis-ridden agriculture, only mass industrialisation and training of a workforce can provide jobs for all.

President Khamenei inaugurated 'Mines Week' in May 1989, intended to highlight the importance of mining for the economy. Mines and Metals Minister Reza Ayatollahi stated that 73,500 workers were employed in the mining sector. Most of these enterprises are small in scale, but large ones are in operation or nearing completion, notably an iron ore complex at Gol-Gohar, costing $700m, due to be completed in 1990. It is intended to supply the Mobarakeh steel mill under construction near Isfahan.

Iran hopes to step up greatly its mineral exports in the 1990s. Most of the existing schemes for production of non-ferrous metals are working well below capacity, and some have proved to be white elephants. Many projects languished or were put aside for lack of foreign exchange, as is true of the steel industry and much else. However, a number of new state ventures were begun recently, including a project for gold processing, close to Isfahan. Chrome ore, sulphur, copper, cast iron and other minerals at present account for only a small part of non-oil exports. Carpets, caviar, pistachio nuts, dried and fresh fruits, saffron and other agricultural produce are more important. Iran's world famous handmade carpets undoubtedly have huge potential, if production costs can be checked. The export trade in carpets began to recover from 1985, after setbacks caused partly by state obstruction. With exports (mainly to EC countries) of Rls28,217m in the Iranian year 1365 (March 1986-87), Iran's carpets and rugs could contribute as much as Rls100,000m a year with better organisation and marketing, according to official trade projections (which tend to be over-optimistic).

Private investment in Iranian industry has been notable for its scarcity in the 1980s. Trading has provided easy profits without the risks of investment in manufacturing. But given political certainty, firm guarantees for private entrepreneurs, and an unshackling of state controls, the prospect for growth in the shrunken private sector of industry looks good to many observers in Tehran. There is huge unsatisfied demand existing in this domestic market of 52 million people for a wide range of consumer goods.

BUSINESS GUIDE

Time: GMT +3.5 hrs.

Climate: Largely desert, with mountainous regions in the western province of Azarbaijan. Tehran temperatures range from over 38°C to -14°C, with a mean in Jan of 3°C and Jul of 29°C. The Gulf and Caspian areas can be extremely hot and humid. Short rainy season in early winter and spring.

Entry requirements: Travel regulations can change at short notice and should be checked before travelling.
Passport: Required by all except certain seamen.
Visa: Required by all, and must be obtained in advance.
Prohibited entry: Nationals of Jordan and Israel are prohibited entry. There are extensive regulations regarding nationality, transit visas, etc.
Currency: Important and export of local currency is restricted to IR2,000. Money must be changed only through the banking system.

Weights and measures: Metric system.

Health precautions: Details of health requirements should be checked before travelling.
Mandatory: Cholera and yellow fever certificates likely to be required if travelling from infected area.
Advisable: Typhoid, paratyphoid and cholera inoculations and anti-malaria precautions advisable.

Air access
National airline: Iranair.
Other airlines: Aeroflot, Air France, Air India, Alitalia, Austrian, Lufthansa, PIA, Swissair, Syrian Arab, THY Turkish.
International airports: Mehrabad International (Code: THR), 11 km from Tehran, with duty-free shop, buffet, restaurant, bank, post office, shops; Shiraz (Code: SYZ), 15 km from city, with buffet, currency exchange, post office, shops.

Surface access: There are roads from Turkey, the USSR, Afghanistan and Pakistan, though the routes are not always passable. Routes from Iraq were closed, although inevitably the situation would change if the hostilities ceased permanently. There are rail lines from Turkey and the USSR (via Djulfa).
Main ports: Bandar Abbas (containers) and Bandar Anzelli. Bandar Khomeini (containers) and Bandar Bushehr are out of use owing to damage and their position in the war zone. Khorramshahr, the main port, has been totally destroyed in the war with Iraq. Both terminals and most berths on the Kharg Island oil terminal were damaged by bombing in 1985 and 1986, although they were partly restored in 1987.

Hotels: The situation and status of hotels should be carefully checked. The use of the name of an international management chain does not imply a current connection with the chain, but may indicate only a previous link or the usual name by which the hotel is known.

Electricity supply: 220 V AC, 50 cycles.

Car hire: International driving licence (along with two photographs) required.

City transport
Taxis: Taxis are metered and frequently shared. Those hired by telephone are more expensive. Tipping is not expected.

National transport
Road: Surfaced roads serve main centres, and total network comprises 140,000 km, of which 500 km are motorways (1986). Condition of secondary roads may vary.
Buses: Scheduled long-distance coach services may vary, but have included Tehran-Anzali, Khorramshahr-Bandar Khomeini, and Tehran-Zahedan.
Rail: Rail service on 5,000 km network may vary, but three classes of service have operated in the past, with sleeping accommodation, air conditioning and restaurant service available. Expansion to 6,500 km is planned by 1989.

Public holidays
Fixed dates: 20 Mar (New Year – 1 week), 1 Apr (Republic Day) 5 Jun (Revolution Day), 9 Jun (Birthday of 12th Imam), 14 Jul (Martyrdom of Imam Ali).
Variable dates: Eid al-Fitr; Eid al-Adha; Ashoora. Exact dates of Islamic religious festivals are determined by sightings of the moon.

Working hours
Government and business: 0700/0800 – 1300, 1600/1700 – 1900 (closed Thu, Fri and some afternoons).
Banking: 0800 – 1400; (Thu) 0800 – 1200 (closed Fri).

Social customs: Muslim law is strictly adhered to. Always refer to the stretch of water south of Iran as the Persian Gulf or the Gulf – never the Arabian Gulf. Avoid contentious subjects such as Iraq and Israel. It is useful for the visitor to be aware of the differences between the different communities comprising Iran.

Telecommunications
Telephone: Dialling code for Iran, IDD access code + 98 followed by area code (631 for Abadan, 31 for Isfahan, 21 for Tehran). Approximately 1.05m telephones in total (1982), of which 40 per cent are in Tehran, 7 per cent in Korassan and 7 per cent in Isfahan.
Telex and telegram: There are telex and telegraph facilities at the Central Telegraph Office in Tehran.

Banking
Central bank: Bank Markazi Jomhouri Islami Iran.
Main centre: Tehran.
Head offices: Agricultural Co-operative Bank of Iran, Agricultural Development Bank of Iran, Azarbayjan Development Bank, Bank Bimeh Iran, Bank of Industry and Mines, Bank Markazi Iran, Bank Maskan, Bank Mellat, Bank Melli Iran, Bank Omran, Bank Pars, Bank Saderat Iran, Bank Sepah, Bank Tejarat, Commercial Bank of Iran, Construction Bank of Iran, Foreign Trade Bank of Iran, International Bank to Iran, Iran Arab Bank, Iran Islamic Bank, Khazar Development Bank, Khuzestan Development Bank.
Branches: Algemene Bank Nederland, Banca Commerciale Italiana, Banque Nationale de Paris, Dresdner Bank, Grindlays Bank, Lloyds Bank, Kredietbank, Société Générale, Bank of Tokyo.

Trade fairs: Tehran International Trade Fair, held annually in September. Details from Iran International Fair & Exhibition Corp.

Business advice and information
Bank Markazi Jomhouri Islami Iran, Ferdowsi Ave, Tehran, PO Box 11365/8531, 11365/8551 (tel: 310101/9, 310231/9, 311321/9; tx: 212503, 212359, 213119).

Central Ministry of Information, Abasabad Avenue, Tehran.

Chamber of Commerce, Industries & Mines, 254 Taleghani Avenue, Tehran (tel: 836031/9; tx: 212398, 213392).

Export Promotion Centre of Iran, PO Box 11-48, Tajrish, Tehran (tel: 21911; tx: 212896).

International Institute for Commercial Relations Development of Continents, No 7 Third Floor, Bazar Abbassi, Ferdowsi Square, PO Box 11 365-3963, Tehran (tel: 820697; tx: 213636).

Iran International Fair & Exhibition Corp, PO Box 98, 22 Tadjrish, Tehran.

Planning and Budget Organisation (Sazman-e Barnameh va Budje), Meidan-e Baharestan, Tehran.

Tehran Stock Exchange, Taghinia Bldg, 521 South Saadi Ave, Tehran 11447 (tel: 311149).

Travel information
Iran National Tourist Organisation (INFO), Boulevard Elisabeth II, Tehran (tel: 621291/8).

Hotels: Isfahan (area code 31)
Isfahan, Motohary Avenue (city centre), P.O. Box: 16 (tel: 35491).

Hotels: Shiraz
Cyprus (Homa), Sadri Avenue (city centre), P.O. Box: 528 (tel: 28000/12; tx: 332231).

Hotels: Tehran (area code 21)
Asia, 40 Mellat Ave., Serahe Ekbatan, Mellat Ave (tel: 318551).

Atlantic, Taleqani Ave (tel: 890286/8).

Azadi (Hyatt), Evin-Modaress Expressway (tel: 297021/9).

Azadi Grand, Chamran Expressway, Evin Crossing (tel: 298071/298073; tx: 214302).

Beinulmelal (International), Sayyed Khandan Bridge, Dr. Shariati Ave (tel: 840085/9).

Boulevard, 278 opp. Zhaleh Arcade, Keshavarz Blvd (tel: 650459).

Bozorg-e-Tehran (Grand Hotel), close to Motaharri Ave., Vali-e-Asr Ave (tel: 624383).

Buloor (Belair), Sepahbod Qarani Ave (tel: 829881/2).

Bustan, Vali-e-Asr Ave., near Zardusht Blvd (tel: 895661/5).

Caspian, 428 Ayatullah Taleqani Ave (tel: 834066/9).

Commodore, Taleghani Avenue (tel: 642703; tx: 212630).

Crystal, Lalezar Now Ave (tel: 393796).

Diamond, 7th-Tir Sq. Dr. Mufatteh Ave (tel: 822907).

Ekbatab, Taleqani Junction, Vali-e-Asr. Ave (tel: 301664).

En Ghelab, 50 Taleghani Avenue (city centre) (tel: 667251/5; tx: 212599).

Esteqlal, after Jam-e-Jam, Vali-e-Asr Ave (tel: 290011/5).

Evin, Evin Ave (tel: 291030).

Fars, begin of Naser Khosrow St., Imam Khomeini Sq (tel: 303-77).

Ferdowsi, Ferdowsi Sq (tel: 824559).

Firooz, Amir Kabir Ave (tel: 313508).

Gilan, Southern Ferdowsi Ave (tel: 302282).

Gilan Now, Ferdowsi Ave (tel: 316206).

Hafez, Bank Alley, Ferdowsi Ave. (tel: 679063).

Homa (formerly Sheraton), 50 Bizhan Street, Vali-e-Asr, P.O. Box: 11-1961 (tel: 683021/8; tx: 212798).

Independence (formerly Hilton) (tel: 290031/9; tx: 212363).

Inter-Continental Tehran, Dr Fatemi Avenue (tel: 656021/9; tx: 212300 intercon).

Jahan, Mellat Ave (tel: 665596/7).

Keshavarz, 42 Keshavarz Blvd (tel: 899939).

Khayyam, Amir Kabir Ave (tel: 313757).

Kian, Zardosht St, Vali-e-Asr Ave (tel: 650235).

Lalezar, Lalezar Now Ave (tel: 316542).

Marmar, Sepahbod Qarani Ave (tel: 830083/7; tx: 212475).

Mehr, Amir Kabir Ave (tel: 310133).

Miamei, Vali-e-Asr Ave (tel: 620200).

Mina, Dr. Mufateh Ave (tel: 823255).

Naderi, Jamhourie-Islami Ave (tel: 678610).

New Naderi, Goharshad Alley, Naderi Ave (tel: 671356).

Pamchal S, Dr Mufateh (Mobarezan) Ave (tel: 820825).

Parastoo, Shahrokh Alley, Jamhourie Islami Ave (tel: 672422).

Park, Vahdate Islami Ave (Hafez) (tel: 672020/9; tx: 212549).

Pars, 40 Sepand St., Nejatollahi Ave.

Parsian, Dr. Beheshti Ave (tel: 622708).

Pirooz S, Vali-e-Asr Ave., (tel: 533584).

Quds, Taleqani Ave (tel: 6409310).

Sanjar, Dr. Ali Shariati Ave (tel: 751445/7).

Shahrzad, begin of Ekbatab St., Sa'adi Ave (tel: 302217).

Sherafat, opp. Darul-Funoon School, Naser Khosrow Ave (tel: 304857).

Shiraz, opp. Cinema Odeon, N. Sa'adi Ave (tel: 301018).

Tehran Kings, Mosadeq Ave. Valiasr Square (tel: 8983171; tx: 212675 kings).

Victoria, Vali-e-Asr Ave (tel: 890051/5).

Iraq

Janet Sutherland

A key date in the development of Iraq was February 1987, when President Saddam Hussain indicated a radical shift in government policy towards the private and the state sectors, called for greater industrial self-sufficiency, and re-organised the economic ministries. A series of new measures and decisions which emerged nearly every month in the following two-and-a-half years showed that the government had acted seriously in carrying through its intention of economic reform. A change of policy towards encouraging private enterprise and towards the running of state enterprises on more autonomous and business-oriented lines was implemented with considerable vigour. It was confidently expected that all the trends which come into what might be called Iraq's 'new economic policy' would continue and be carried further over the next five to ten years.

Previous economic situation

Two factors in particular characterised the situation prevailing before February 1987. One was that the Ba'ath Party which governs Iraq is nominally socialist, and this meant that economic authority was concentrated in the hands of government-appointed organisations. The other factor was that before the war with Iran, Iraq had huge resources for investment and development and a large surplus. Oil revenues rose from \$144m in 1953 to \$575m in 1972 (when the Ba'ath Party was in power) to \$12,180m in 1979. These were state revenues, so the government had great latitude in financing projects without having to enforce strict disciplines of assessment or operation.

By 1987 Iraq had been running a huge current account deficit for six years. At the time of the end of the war in August 1988, Iraq's international debt was over \$60bn, of which over \$25bn was due to trading transactions with non-Arab Western and Eastern countries (in distinction from financial and commercial aid from other Gulf states). More recent and better estimates of Iraq's debt put it at around \$40bn to the non-Arab world; the debt to its Arab benefactors (principally Saudi Arabia and Kuwait) is also estimated at \$40bn, but this is not expected to be repaid and Iraq regards contributions from the Gulf states as grants rather than loans. It was clear that the task of post-war reconstruction and development in the government's plan could only be carried out by obtaining very substantial credits for a further four or five years.

It was in this situation that Saddam Hussain realised the vital importance of economic and commercial efficiency. In June 1987 he made a speech against uneconomic activity by the state and announced that uncompetitive state enterprises could go to the wall.

The term 'socialist' had been applicable to the Iraqi state and economy only in certain qualified ways. Iraq had been predominantly a centralised command economy before the new economic policy. But although the state had been responsible for development projects, most of the work on state projects had been carried out by private firms, foreign and Iraqi. Thus private business had reaped profits from state contracting.

In the case of projects requiring high technology or complex design and construction organisation, international firms were usually involved. But local firms have also benefited by the expansion of state economic activity.

A calculation by Dr Farouk-Sluglett showed this relationship: between 1972 and 1982 outputs and inputs in the private sector and in the state sector rose at approximately the same rate. Output in the private sector rose by 213 per cent and in the public sector by 203 per cent, and inputs rose by 220 per cent and 205 per cent respectively.

In 1980 about 80 per cent of the inputs and outputs of medium-sized firms were contributed by private firms. By 1982, the private sector contributed 93.8 per cent of total non-oil GDP. But government organisations accounted for about 90 per cent of all imported goods in 1987 and the immediately preceding years, and overall accounted for much the greater part of the gross national product.

The trade budget for 1988 assigned 80 per cent of imports to the state (which included basic foods to a value of about $3bn); 6 per cent of imports were by the mixed sector and 14 per cent by the private sector. In the 1989 foreign trade plan the private sector share was 26 per cent, the so-called socialist sector share 35 per cent and the agricultural sector share 25 per cent.

Economic policy objectives

The keynote of the new economic policy can be stated simply: to make the Iraqi economy more efficient. This apparently simple proposition did, however, involve some major policy shifts.

Iraq has a strong underlying economic position based on the wealth of its resources – in oil and minerals, in its agricultural potential and in its developing industrial base and infrastructure – but also very significantly in the richness of its human resources, since the population contains a relatively high proportion for countries in the region, of qualified administrators, scientists and technocrats and of skilled workers. The enormous debt incurred during the war presents a serious funding constraint for the next few years, but the cost of servicing it should be seen in proportion: the ratio of debt service to GDP is unlikely to exceed about 5 per cent in any future year, and it is expected to begin to fall in the early 1990s and be completely eliminated by the later 1990s. (Debt service may amount to more than a quarter of the value of exports in 1990 to 1992).

Iraq has two over-riding strategic objectives in economic policy for the early 1990s: to increase local production; and to save foreign currency.

In pursuit of these ends in the public sector, the government has drastically pruned the bureaucracy. The number of officials and the extent of red tape are being cut. Efforts are being made to make management of state enterprises more technocratic, professional and commercially oriented. Managers now have more autonomy. A new licence is allowed in the hiring and firing of staff.

State enterprises are now supposed to be able to justify themselves commercially and their directors are expected to make a profit. To put further pressure on improving their performance, private-sector entrepreneurs are now able to set up companies in competition with public-sector establishments. Other competitive agents are being introduced: in banking,

KEY FACTS

Official title: Republic of Iraq
Head of state: President Saddam Hussain
Ruling party: Arab Ba'ath Socialist Party
Capital: Baghdad
Official Languages: Arabic
Currency: Dinar (ID) = 1,000 fils
Exchange rate: ID0.31 per $ (Nov 1989)
Area: 434,924 sq km

Population: 17.2m (1988)
GDP per capita: $2,943* (1987)
GDP real growth: 1.7%* (1987)
Inflation: 25%* (1987)
Oil reserves: 100bn* barrels (1988)
Trade balance: –$879m (1988)
Foreign debt: $15.8bn (1987)

* estimated figure

KEY INDICATORS

	Unit	1984	1985	1986	1987	1988
Population	'000	1,508	1,558	1,611	1,633	1,725
GDP real growth	%	-3	-1	–5	–	–
Crude petroleum production	1985=100	84.5	100.0	117.7	143.5	180.5
Volume of exports						
Crude petroleum	1985=100	84.5	100.0	117.7	143.5	180.5
Exports	$m	9,245	10,342	7,526	9,319	9,304
Imports	$m	9,866	10,511	8,687	7,355	10,183
Balance of trade	$m	–621	–169	–1,161	1,964	–879
Exchange rate	$ per ID	3.22	3.22	3.22	3.22	3.22

* estimated figure

the state-owned Rashid Bank was established in November 1988 to provide competition for the state-owned Rafidain Bank. New investment in the state sector is to be assessed in terms of economic viability, and new public sector enterprises will not be established without feasibility studies.

A more political means of improving economic performance is the opening of outlets for criticism of government and official action. In particular, the new national assembly, inaugurated in April 1989, has powers to examine and criticise both the conduct of state ministries and the operation of state organisations, and top officials are now accountable to the assembly. It can be assumed that a major point of this 'glasnost' is to check public-sector inefficiencies. In a speech in July 1987 Saddam Hussain praised the glasnost policy of the Soviet Union and said that he wanted Iraqi society to be more open; and in November 1988 Saddam Hussain announced that pluralism had been adopted as a policy; this will be given recognition in the new constitution being drawn up by the national assembly. However, its significance is likely to bear more on matters of economic and administrative management than on matters of real politics.

Another major policy programme in respect of public sector industry and services is that enterprises are being sold to the private and to the mixed sectors. In the fields of agriculture and related industries, of tourism and of light industry in general, this disengagement from state ownership and operation is moving towards being comprehensive, and it is extensive for medium-sized firms in manufacturing and services.

Many state businesses and more state farms are being sold outright or else leased to private business. Others are being re-established as mixed sector companies. An example is the Northern Area Beverages and Minerals Drinks Company: more than two-thirds of its shares were offered to the public, and the remainder are held by the Industrial Bank, the State Establishment for Beverages and Food Canning, and the National Company for Food Industries.

Meanwhile, the government's objective in the private sector is to foster expansion and introduce a climate more encouraging to enterprise. To this purpose, regulations are being reduced, and in some cases simply not applied. There is a general liberalisation in economic affairs.

The liberalisation policy is, however, patchy and has met with a setback in the area of prices. A decision was taken in August 1988 to allow market prices to operate and to lift price controls on most goods produced by the private sector. However, following too-high price rises on farmers' produce, the government reinstated the central pricing committee in April 1989, and in June 1989 announced a price freeze on most state produced consumer goods and services. The government food subsidy for 1989 was $750m.

The severe restriction of imports by the government, the extreme overvaluation of the Iraqi dinar, and the weakness of competition in all areas affecting the supply of consumer goods, made for a situation by mid-1989 in which the abolition of all restrictions on the free market would lead to unacceptable levels of inflation or else require an unacceptable curb on consumer demand. The rate of inflation was already high and probably in excess of 30 per cent.

The government knows that the medium- and long-term answer to the dilemma is to stimulate competition and private enterprise production. There is no doubt now that the government regards the private sector as having a vitally important and growing role in the development if Iraq. A crucial concern is how to attract investment into it.

A battery of incentives for private investors was introduced in 1988 and 1989. Among these were notable initiatives to bring in capital from other Arab states, in particular by granting large tax concessions to nationals of Arab countries investing in Iraq.

This encouragement of investment money did not extend to the non-Arab world. Iraq's political history and present stance excludes Western companies or individuals from owning any share in an Iraqi business.

What Iraq is seeking from the OECD countries and non-Arab sources is credit. But its critical shortage of foreign currency would be eased if it could mobilise some of the funds held by its own nationals. The value of private capital held abroad by Iraqis has been estimated at a phenomenal $60bn by some analysts. Iraqi businesses are now able to import any goods paid for by foreign currency funds without any questions asked.

There are also funds available held by Iraqis at home which could make a significant contribution to enlarging private sector investment. To give an illustration: Iraqi private contractors in the construction sector had profits of 35m Iraqi dinars in 1975, which rose to ID156 in 1976, and increased many times more through the construction boom of 1980.

An inhibiting factor on the take-off of its private sector is the political situation in Iraq. This is likely to make the native response to the liberalising of the economy more hesitant and less enterprising than would be wished. A simple example of the difficulties faced under the current dispensation is that no Iraqi national can leave the country without the permission of the authorities.

The new economic policy is marked by the discovery of the profit motive, of competition and of market forces. As there was no 'brains trust' or theoretician associated with the policy, it is assumed that the initiative for introducing it came from Saddam Hussain.

The privatisation aspect of the policy is limited and has only one component. It has been aptly called a policy of commercialisation.

In fact, it is in line with what is happening in other Gulf countries. The state will own and operate the infrastructure – for example, ports, power supply and water supply. The state will continue to own and operate the oil industry and associated oil-based industries, such as petrochemicals. The state will also own and operate most heavy industry when the capital outlay is beyond the will of the private sector or the return on investment is too delayed. The state will also be involved in industries started for reasons of national self-sufficiency, sometimes in participation with the private sector. The private sector, on the other hand, will increasingly take charge of most other industrial and service activities and be encouraged to do so by government and to start new activities.

Iraq's policy may be seen as more singular in two respects: in not being accommodating to any element of Western capital in ownership of joint ventures; and in going for a more ambitious and far-reaching degree of industrial self-sufficiency.

Immediate economic priorities

In general terms the country's top priority is hard currency-raising projects. Specifically this means that Iraq will:

– concentrate on oil and gas facilities as the main revenue earners;

– expand its downstream oil-related industries;

– continue to push ahead with the development of the infrastructure, which is necessary for overall economic development;

– take measures to diversify its economy and establish a stronger industrial base so that it is not so proportionally dependent on oil. Priority will be given to projects which pay off particularly in import substitution: improving agricultural production is significant in this context;

– one further over-riding priority: military expenditure and the requirements of defence must be met and will take a significant share both of the trade budget and of domestic industry.

Iraq's recoverable oil reserves were estimated at 100bn barrels in 1988 and in 1989 the probable figure was revised to 160bn. This is comfortably the second largest in the world, after Saudi Arabia, and about 13 per cent of the global total.

Iraq's OPEC production quota from June 1989 was 2.784m barrels a day. It adhered to its quota in 1988 (2.64m) and was expected to continue to do so. If, however, the OPEC agreement of June 1989 broke down, Iraq would be likely to increase its production. Production capacity is increasing and is set to reach 4.5m b/d in 1991. Export capacity is increasing still more notably: at the beginning of 1990 it was over 5m b/d and this was expected to reach 7m b/d by 1991-92.

Oil exports in 1988 were 1.95m b/d, producing an estimated revenue of $13bn. Exports in 1989 were higher consequent on the cessation of warfare. Iraq's total revenue is of course critically linked to crude oil prices.

During 1989 work proceeded on restoring oil terminals in the Gulf following war damage, and tanker exports recommenced. By late 1989 tanker export capacity at Gulf terminals was 2m b/d. Supertanker exports at the rebuilt Mini al-Bakr offshore terminal (served from Fao) resumed on 1 June 1989. Work also proceeded on major pipeline schemes: the two most important were a strategic line from the south to Haditha in northern Iraq due for completion in 1990, and the IPSA 2 pipeline with a 1.15m b/d capacity to a new terminal on the Red Sea completed in the later part of 1989. Total pipeline export capacity is now 3.1m b/d. Current projects for pipeline and tanker outlets provide for an exceptional strategic flexibility in control over the grade of oil transferred and the place of export.

Proven natural gas reserves are over 1m cubic metres. Most of this is flared, but there are plans to more fully exploit it. Gas-gathering networks are in place in the north and south and the second stage of the trans-Iraq dry gas pipeline was in the course of construction in 1989. The first phase was completed at the end of 1988, running from al-Nasiriya in the south to Baghdad. The second phase takes the pipeline north to Turkey where it could eventually supply gas to Europe. Liquid gas plants and gas treatment plants are scheduled. The oil minister predicted that Iraq would become a major exporter of LPG during 1990.

In mid-1989 spending on oil and gas projects throughout Iraq was running at $5.5bn.

Downstream in the oil industry the refinery at Basra started operating again in 1989 with a 140,000 b/d capacity. Iraqi refining capacity was then 600,000 b/d according to the oil minister in April 1989. The Basra petrochemicals complex (PC1) was built shortly before the war at $1,000m and mothballed. Production was restarted in 1989.

A key refinery known as 'the central refinery' was being built during 1989 near Al-Mussaib, 60 km south west of Baghdad. This was one component of a massive industrial complex planned at Al-Mussaib. Most importantly the government was pushing ahead with one of the largest petrochemical complexes in the Middle East, known as PC2. UK's Bechtel was technical consultant for both projects. The refinery and PC2 together were estimated to cost about $4,000m. Completion of the first phase of PC2 is expected in mid-1991, and the second phase is scheduled for completion two years later.

Major projects to develop the infrastructure were in progress in telecommunications, water and sewerage, roads, railways, airports and ports.

Also of great importance were construction projects in support of agriculture: an extensive dam-building programme, and irrigation, drainage and land reclamation schemes.

The building programme for schools, universities and hospitals was also considerable. Welfare services retained a significant degree of priority.

The rebuilding of towns and installations in the south of Iraq damaged by shelling was a priority for 1989. A $3.1bn reconstruction programme for Basra was undertaken in February 1989 and, astonishingly, completed in June 1989. The complete rebuilding of Fao began in July 1989 and was largely completed by the end of the year.

One estimate was that Iraq needed to spend $30bn to repair war damage to its infrastructure.

Iraq is planning to invest billions of dollars to broaden its industrial base and reduce dependence on crude oil. It had planned and embarked on a vast industrial expansion programme in the late 1970s when oil prices were high. Over $50,000m worth of projects were listed in the 1981 Five-Year Plan. Following the end of the Gulf War in the summer of 1988, several of these were revived by the government for immediate action.

The largest of them were in heavy industry. These included expansion of the iron and steel works at Khor al-Zubair near Basra (most heavy industry is located in the south), and the construction of a second iron and steel complex, bringing steel capacity to 2m tonnes a year. Iraq intends to become self-sufficient in many aspects of iron and steel production. Other projects were in fertiliser plants. A large nitrogenous fertiliser complex was opened in late 1989 at Baiji. Iraq has huge phosphate reserves. A planned threefold increase in fertiliser output will bring it to 4.5m tonnes. It was expected to achieve self-sufficiency in all types of fertiliser during 1990.

Other major projects on hand in mid-1989 were a 200,000 t/y aluminium smelter, a tyre factory to produce 1m tyres a year; and an industrial gas plant. Contracts were awarded for all these projects in late 1988 and 1989.

Plans were also under way to establish a car assembly operation and negotiations were progressing with General Motors.

Other state-initiated industrial developments given the go-ahead in 1988 and 1989 included a pharmaceutical factory, joint Arab ventures for float glass and detergents, and expansion in production of sulphur and cement. Exports of cement in 1989 were due to be 5.5m tonnes, double the amount in 1988; total cement capacity was 20m t/y.

Iraq was expected to continue its very high level of military spending (involving substantial imports) and to extend its already impressively diverse capacities in production of military equipment. It is manufacturing a range of weapons, from small arms and munitions to several complete missile systems, and its own AWACs (airborne early warning system). Areas of Iraqi production strength include electronics. Most remarkable of all was the decision to set up an aerospace industry for local assembly of an advanced jet trainer and for depot-level maintenance of its entire fleet of about 500 advanced combat aircraft. Costs of Iraq's rearmament programme were estimated at about $5bn annually.

Political factors

The cease-fire in the Iran–Iraq war took effect on 20 August 1988 but subsequently peace talks, held in Geneva under the auspices of the United Nations, were deadlocked. Iraq required that in any final peace treaty its sovereignty over the entire Shatt al-Arab waterway should be recognised, and restored to the position before 1975 when it was compromised by a treaty with Iran which Iraq considers invalid. Iran wanted the international boundary to run along the deepest line of the Shatt, which Iraq understandably regarded as a recipe for conflict.

The Shatt al-Arab is the link between the port of Basra and the Gulf. Much of the development of Iraq is being concentrated in the Basra region. Work on the years-long task of clearing the waterway of the debris of war has been held up while the UN talks were stalled and in mid-1989 Iran and Iraq were still in dispute.

Iraq attaches over-riding importance to ensuring full access to the Gulf. Its other important ports are Umm Qasr and Zubair, which are linked by the Khor al-Zubair waterway, whose outlet to the Gulf is through the Khor Abdullah on the borders of Kuwait and Iraq. In 1989 Iraq revealed plans to deepen and widen the Khor Abdullah.

Neither Iraq nor Iran was thought likely to initiate a war again in the early 1990s. But if negotiations made no progress, it looked likely to be an uneasy peace. Both countries continued to build up their military capabilities and acquire more advanced weapons. Without a peace settlement, the possibility of renewed war occurring in the course of time was a spectre to haunt the view of the future.

In this situation Iraq did not undertake a large scale demobilisation. In mid-1989 there were still about 1m men in the armed forces.

In its international position, Iraq is now on the more moderate side of the regional lines, and this was given further recognition in 1988 when the Arab Co-operation Council (ACC) was formed, an alliance linking Iraq with Egypt, Jordan and North Yemen, all moderates of the Arab world. However, Saddam Hussain's dominance and Iraq's relative power looked likely to make other Arab states concerned about how Iraq would exercise its regional role, though Iraq will no doubt try to maintain good relations with other Arab states, except for its hostile view of Syria.

The most prominent political problem the regime had to deal with in the period after the end of the Gulf War was the opposition of elements among the Iraqi

COUNTRY PROFILE

Political system: According to the Provisional Constitution of 1970, the nine-member Revolutionary Command Council (RCC) is the supreme political body. Its chairman is President Saddam Hussein. There is a National Assembly comprising 250 delegates representing an electorate of seven million. The Arab Ba'ath Socialist Party has the majority of seats. There is no real opposition party, but members of the National Assembly are able to examine the actions of government officials. A new constitution is in the process of being drafted to replace the 1970 constitution.

Main cities/towns: Baghdad (population 3.8m), Basra (1.5m), Mosul (1.2m), Kirkuk (800,000), Sulaimaniya (700,000).

Language: Arabic is the official language; French and English are spoken in business.

Media
Press: Arabic dailies include *Al-Thawra* (circulation 250,000), *Al-Jamhouryah* (50,000), *Al-Iraq* (30,000) and *Reyadeh* (40,000). Weeklies include *Waee Al Omal* (25,000) and *Alif Baa* (150,000). The daily *Baghdad Observer* (English) has a circulation of 12,000.
Radio: Domestic service in Arabic, Kurdish and other minority languages; external service in numerous languages.
Television: Regional services transmit six/seven hours per day. There were one million sets in 1988.

Domestic economy: State-controlled economy with a growing private sector, with main revenues from oil and gas and much smaller amounts from agriculture and manufacturing. Private sector consists mainly of food processing, textiles, tourism and trade. The war with Iran resulted in the huge external debt. Since the 1988 ceasefire the government has pressed ahead urgently with reconstruction and development and the economy has been expanding.

Employment: There is a labour shortage which is eased by the employment of foreign labour.

External trade: Iraq publishes no official trade statistics and countertrade also complicates the picture.
Exports: Crude oil represents about 98 per cent of all exports. Oil revenue in 1988 was $13.16bn. Non-oil revenues in 1988 were $70m. Principal destinations are Brazil (15.8%* of 1986 total), Japan (11.6%), Spain (10.2%), Turkey (9.5%), Yugoslavia (9.5%), Italy (8.8%), France (7.3%) and USA (5.6%).
Imports: The public sector accounted for nearly 80 per cent of imports in 1988, and rather less in the 1989 trade plan. Main imports are machinery, spare parts and raw materials for the industrial sector, and also vehicles. About a quarter of imports are for basic foodstuffs (mostly from the USA, and the rest mainly from Turkey). The other main category of imports is manufactured goods: the quotas for these were reduced in the 1988 and 1989 trade plans. Main sources of imports in 1988 (in order of value) were as follows: USA (mainly foodstuffs), Turkey, West Germany, UK, France, Japan, Brazil, Italy, USSR.

Agriculture: Agriculture employs about 30 per cent of the workforce and accounts for less than 10 per cent of GNP. Irrigation systems are badly in need of repair, and salination has affected arable land areas. But the problem is being alleviated by a series of major dams, of which the Bekme project, agreed in 1986, is the largest with a planned capacity of 33bn cu metres. There are also development loans for small farmers, funded by the Arab Fund for Economic and Social Development. Main products are dates (of which Iraq is one of the world's major exporters), barley and wheat.

Industry: The industrial sector consists of heavy industry (all state-owned enterprises), extensive military production and light industry which is increasingly in private ownership. Products of the state sector include iron, steel, cement, bitumen, pharmaceuticals and fertilisers. Oil refining and petrochemicals are important and in process of great expansion. Growing attention is being given to processing and manufacturing, especially in areas of import substitution.

Minerals and mining: Huge resources of phosphates. Sulphur is important and a new processing plant has boosted sulphur exports. Other minerals include iron ore, lead, copper and gypsum.

Hydrocarbons: Iraq has the world's largest proven oil reserves (100bn barrels) and with Iran is the second largest OPEC producer after Saudi Arabia, with a quota of 2,783,000 b/d. Natural gas reserves estimated at 29 trillion cu ft, and a network for its fuller use is being established.

Membership of international organisations: Arab League, FAO, G-77, IAEA, ICAO, IDA, IFAD, IFC, ILO, IMF, IMO, INTELSAT, INTERPOL, Islamic Development Bank, NAM, OAPEC, OIC, OPEC, UN, UNESCO, UPU, WFTU, WHO, WIPO, WMO, World Bank, WSG, WTO.

Kurds. The Kurdish population comprises about 20 per cent of Iraq's total. The ruthlessness of the means used to suppress Kurdish rebellion and to try to destroy the capacity of Kurds to cause trouble to Baghdad in future brought well-publicised condemnation in Western countries. Whatever the diplomatic cost to Iraq may have been, the cost in terms of loss of financial credit was serious: it made it politically impossible for Western countries to go further in extending government-supported lines of credit to Iraq.

An international outcry occurred following an assault against some Kurdish locations on 25 August 1988, when it was alleged that chemical weapons had been used. An extensive and forcible transportation and relocation of large numbers of Kurds from some of their home areas took place throughout 1989.

Outlook

At the end of the 1980s the Iraqi government was optimistic about the economy on account of the following factors:

– higher oil revenue;
– benefits from its economic reform programme;
– larger private sector participation, and the attraction of funds from Arab investors into local enterprises;
– its ability to obtain project-related credit facilities, and also alternative forms of off-balance-sheet financing for trade and project requirements;
– exploitation of the technological experience gained in the military industrial sector.

The burden of international debt is likely to be reduced and eliminated in the medium term. The prospect appeared set for a year-by-year improvement in performance in both the argicultural and industrial sectors and for an across-the-board expansion in economic activity.

Iraq's GDP growth has been estimated at 10 per cent in 1988 and it is predicted that it will be sustained in the range of 5-6 per cent annually between 1990 and 1996.

Iraq was exceptional among national governments in the late 1980s in that it was making strenuous efforts to promote a higher birthrate and at least double its population as quickly as possible. The population growth rate was at least 2.8 per cent per year and was thought to have perhaps risen to over 3 per cent. This seems disconcerting policy.

Iraq's oil revenues were projected to be between $13bn and $19bn annually during the period 1989-1992.

Non-oil revenues were estimated at $70m for 1988 and were expected to be over $100m for 1989. Though export-oriented industries are being encouraged, the increase in non-oil exports was likely to be less significant in the balance of trade than the increase in import substitution through the development of local production and technical competence.

In construction and new projects for tendering, local resources will be used wherever possible; this entails giving preference to local constructors and consultants and requiring building materials to be obtained locally. International contractors may be required to bring in local contractors in joint arrangements. The government will avoid turnkey contracts except in the few cases where there is no reliable alternative.

Government import spending was expected to run at about $9bn a year between 1989 and 1992. In the year to January 1989, nearly 95,000 import permits were issued to the private sector for goods worth a total of $4.2bn. In order of value, the main items were: vehicles and spare parts ($1.1bn); textiles and clothing ($560m); foodstuffs ($440m); electrical items ($410m); household goods ($270m); construction materials ($260m). As these were financed by private foreign exchange, they represented a large mobilisation of offshore currency funds which at a previous time in the Ba'ath Party mind did not legitimately exist.

The current account deficit in 1989 was estimated at about $3.6bn. Some analysts expected this to be reduced to $1.5bn in 1992.

The war with Iran cost Iraq about $150bn, made up of lost revenues from oil, lost production, military expenditure and physical destruction. This left it with a massive debt. The big economic question about Iraq is how it will pay its debt and finance its reconstruction and development.

In 1989, the interest payments and repayment on Iraq's debt to OECD and non-Arab countries would have amounted to about $5bn without rescheduling. Much of the $30-40bn owed to non-Arab countries was due for payment within the next two years, and some payments were in arrears, so rescheduling and new credit are vital to Iraq's programme. During 1989, rescheduling arrangements were made with most of Iraq's creditor countries. Servicing its debt presents particular difficulties between 1990 and 1992 and Iraq is looking for medium- and long-term financing.

New lines of credit are needed if Iraq's revival programme is not to be slowed drastically. These have not been readily forthcoming from Western countries in adequate amounts. The UK Government provided £340m for 1989, and the United States $1,150 in agricultural credits. France, Korea,Brazil, Germany, India, the Soviet Union and Turkey all increased their credit lines. Japan, which has fostered good relations with Iraq can offer substantial long term credit to be repaid in oil.

Iraq will still be short of foreign exchange to meet its programme over the next few years. Some of the financing will be done by countertrade. Foreign trade and contracts will go firstly to those who can provide credit or other means of financing. With that proviso, there is great opportunity for foreign business to benefit by Iraq's willingness to be open to developing trade links. English is widely spoken among an increasingly able management class.

A further burden is defence spending which will remain high while there is no peace agreement with Iran and, moreover, Iraq's imports of military equipment mostly require cash payment.

1990 to 1992 will be the most crucial and difficult period for Iraq's financing. The government expects to become a cash client again in 1995, hoping by then to have repaid its international commercial debt. But financing its ambitious programme might be expected to give Iraq problems for the best part of the 1990s. Some analysts argue that it cannot afford to postpone its adjustment and reconstruction programme. Some economists also question the efficiency of pursuing the extent of industrial self-sufficiency to which Iraq becomes committed.

However, Iraq has been rightly described as the first candidate in the region to join the club of NICs (newly industrialised countries): its resource base in terms of physical endowment and human potential, if combined with effective management and political stability are expected to bring this advancement within the decade.

But it seems that the domestic political situation in the 1990s will affect the rewards reaped as a result of the economic policies of commercialisation and liberalisation. The proper benefits from private sector enterprise and a market economy will not be realised without a considerable measure of liberalisation in the political sphere. There was no serious indication as the 1980s closed that the Iraqi leadership had the will for this or would risk a weakening of its grip over the nation's affairs.

Finally, looking to the 1990s, it seems incongruous that Iraq, a country on the way to becoming a mature economy, should publish no satisfactory statistics and no trade figures at all.

BUSINESS GUIDE

Time: GMT +3 hrs; (GMT +4 Apr-Sep).

Climate: Excessively hot sub-tropical period with no rainfall from May to Sep (38-49°C). Dry and pleasantly warm from Oct to Apr (20-25°C), with occasional heavy rain.

Entry requirements

Passport: Required by all.

Visa: Required by all, except Egyptian, Jordanian, Sudanese and Yugoslav nationals. Visitors must be sponsored and visas must be obtained in advance. Anyone remaining more than 15 days must register with the residence department and may subsequently require an exit visa to depart. Within five days of arrival visitors must attend a hospital or health centre to make an appointment for an AIDS test. On departing a certificate or proof of a forthcoming appointment is required.

Prohibited entry: Nationals of Israel and holders of passports with evidence of travel in Israel or of Jewish relationship are denied entry.

Currency: ID25 may be brought into the country. There is no limit to the amount of foreign currency which may be imported, but it must be declared. Only ID5 per person may be exported.

Customs: Personal effects duty free (typewriters are not considered personal effects). Gold acquired locally may not be exported. Motor vehicles admitted only with valid certificate of insurance. Customs duty is imposed on a motor vehicle after two months (extendable to one year). Imports controlled by licence and certain 'non-essential' goods are prohibited.

Health precautions

Mandatory: Certificates of vaccination against cholera and yellow fever required if travelling from infected areas. See the above note on AIDS testing.

Advisable: TAB vaccination and anti-malaria precautions.

Representation overseas: Abu Dhabi, Aden, Algiers, Amman, Ankara, Athens, Bangkok, Bangui, Banjul, Beijing, Beirut, Belgrade, Berlin, Bonn, Brasilia, Brazzaville, Brussels, Bucharest, Budapest, Buenos Aires, Canberra, Caracas, Colombo, Conakry, Copenhagen, Dakar, Dhaka, Dar es Salaam, Djibouti, Doha, Georgetown, The Hague, Hanoi, Havana, Helsinki, Islamabad, Jakarta, Jeddah, Kabul, Kampala, Kuala Lumpur, Kuwait City, Lagos, Lisbon, London, Lusaka, Madrid, Malé, Manama, Mexico City, Mogadishu, Moscow, Muscat, Nairobi, New Delhi, Ottawa, Paris, Port of Spain, Prague, Rabat, Rome, Sana'a, Sofia, Stockholm, Tokyo, Tripoli, Tunis, Vienna, Warsaw, Washington DC.

Air access: Air tickets issued by TWA are not accepted by Iraqi Airways, and sometimes luggage tagged with TWA tickets will not be forwarded. There are stringent security checks on Iraqi Airways on carry-on luggage. Iraqi Airways may be reluctant to endorse its tickets over to other international carriers.

National airline: Iraqi Airways.

BUSINESS GUIDE

Other airlines: Aeroflot, Air France, Air India, Alitalia, Austrian, Balkan, British Airways, CAAC, Egyptair, Gulf Air, JAT, Kuwait Airways, LOT, Lufthansa, MALEV, PIA, Royal Jordanian, Saudia, Sudan Airways, Swissair, Tarom, THY Turkish, Yemenia.
Tax: Airport tax of ID2 (non-Arab countries) or 500 fils (Arab countries) to be paid on departure.
International airport: Baghdad International (Code: SDA), 20 km from city.

Surface access
Main ports: Khor al-Zubair and Umm Qasr are both operating again, although access is limited to vessels of 2.5-metre draught. Cargo is also moved via Kuwait and containers through Aqaba (Jordan). Cargo is also shipped through Mersin (Turkey).
Overland access: Roads lead in from Jordan (Amman), Kuwait, Saudi Arabia via Kuwait and Turkey (Diyarbakir, Ankara). Roads exist to link Iran and Iraq (Baghdad - Tehran, Basrah - Abadan), but the border area should be avoided. The Orient Express travels to and from Baghdad via Turkey (Istanbul).

Hotels: Hotels should be booked well in advance. A 10 per cent service charge is usually added to the bill. Major hotels may require payment of bills in hard currency, although some embassies and companies are able to arrange payment in local currency. First-class hotels accept major credit cards. Hotels are heavily booked during the Baghdad International Fair (two weeks in early November).

Electricity supply: 220 V AC, 50 cycles with all types of continental two- and three-pin plug fittings in use.

Credit cards: American Express is one of the few cards to be fairly widely accepted, and is recognised by major hotels and airlines. Otherwise, avoid over-reliance on cards.

Car hire: International driving licence required and third-party insurance is compulsory. Driving is on the right-hand side of the road, and international traffic signs are used on major roads. Fuel is available at filling stations in all towns and at about 50 km intervals on major roads. Speed limits 50-100 kph.

City transport
Taxis: Taxis may be found in major cities and at hotels. There is a standard fare system, but normally the fare charged is exactly twice that shown on the meter. Nevertheless, fares should be clearly agreed in advance. Tipping is not expected.

National transport
Air: There is a domestic airport at Mosul and commercial airports at Basra and Kirkuk. Iraqi Airways operates a daily flight to Basra.
Road: Tarmac roads link main centres.
Rail: Rail network includes some three-class services with sleeping accommodation, restaurant cars and air conditioning. Rail links between most major centres, including Baghdad-Mosul, Baghdad-Arbil and Baghdad-Basra.

Public holidays
Fixed dates: 1 Jan, 6 Jan (Army Day), 8 Feb (14th Ramadan Revolution Anniversary), 21 Mar (Now Ruz), 1 May, 14 Jul (Republic Day), 17 Jul (1968 Revolution Anniversary).
Variable dates: Eid al-Fitr, Eid al-Adha, Al Hijra, Ashura, Prophet's Birthday.

Working hours: Weekly closing day is Fri.
Government: 0800 – 1200 (Thu to 1100, closed Fri) in summer; 0830 – 1430 (Thu to 1330, closed Fri) in winter.
Banking: (Sat-Wed) 0800 – 1200 (Thu to 1100, closed Fri) in summer; 0900 – 1300 (Thu to 1200, closed Fri) in winter. Certain Rafidain Bank branches operate after hours.
Shops: 0800/0830 – 1300, 1700 – 1900 (closed Fri).

Social customs: Always refer to the stretch of water south of Iraq as the Arabian Gulf or the Gulf – never the Persian Gulf. Visiting cards are regularly exchanged, often printed in Arabic and English. It is unwise to discuss religion or politics, and desirable to have an informed view on contemporary issues (such as Israel) in case such subjects arise.

Telecommunications
Telephone: Dialling code for Iraq: IDD access code + 964 followed by area code (1 for Baghdad, 40 for Babal Zubair) followed by subscriber's number. Direct international dialling.
Telex: Available at major hotels, with 20 per cent surcharge added to usual telex tariff, and at main post offices.
Telegram: From the telegraph office in Baghdad (Rashid Street).

Postal service: Air mail is more reliable than surface mail.

Banking
Central bank: Central Bank of Iraq.
Main centre: Baghdad.
Head offices: Agricultural Co-operative Bank, Industrial Bank of Iraq, Rafidain Bank, Rashid Bank.

Trade fairs: Baghdad International Fair usually takes place for two weeks in early November each year.

Weights and measures: Metric system.

BUSINESS DIRECTORY

Useful telephone numbers
Police: 104
Directory enquiries: 102
Operator: 5372191
Trunk calls: 103
International operator: 105
Fire: 115
Ambulance: 122
Speaking clock: 106
Emergency hospital: 7195191

Hotels: Baghdad (area code 1)
Al Mansour Melia, Haifa Street, P.O. Box: 6153 (tel: 5370041; tx: 212768).

Al Rasheed Oberoi, P.O. Box: 8070 (tel: 8871300; tx: 213204 rashid).

Al Sadeer Novotel Baghdad, Andalus Square (city centre), P.O. Box: 836 (tel: 7188166; tx: 21343974, 213443).

Babylon Oberoi, P.O. Box: 5273 (tel: 7761964; tx: 213530/31 oberoi).

Baghdad, Saadoun Street (tel: 8877041/4; tx: 212555).

Dar Al Salem, Saadoun Street, P.O. Box: 2182 (tel: 93733, 90506, 96151; tx: 2101 darsalam).

Ishtar Sheraton, P.O. Box: 510 (tel: 8889500; tx: 213146 sherat).

Khayam, Saadoun Street, Alwiyah (tel: 7181971; tx: 212429).

Palestine Meridien, Street 47, Mahallat 102, P.O. Box: 3100 (tel: 8875641; tx: 213151).

Palm Beach, Abou Nawas Street (tel: 8880010/9; tx: 2914 palm).

Hotels: Basra
Basra Sheraton, Shatt El Arab Corniche, P.O. Box: 760 (tel: 219310; tx: 207078 sbsh).

Hotels: Mosul
Ninevah Oberoi, Zakho Road, P.O. Box: 1014 (tel: 811933; tx: 802332).

Car hire: Baghdad
Baghdad Cars: Al Assaf Building, Masbah Square, Baghdad (tel: 97721, 90628).

Sumer Cars: Alwiyah, Saadoun Street, Baghdad (tel: 97721, 90628)

Taxis: Baghdad Adil (tel: 8888187).

Adnan, Maskar al-Rasheed Street (tel: 7191236).

Akhwah, Alwiyah (tel: 7191991).

Embassy, Masbah (tel: 7199388).

Hindiyah, Arasat al-Hindiyah (tel: 7198947).

Mansour (tel: 5414874).

Yasmin, Hindiyah Market (tel: 7197255).

Business advice and information
Baghdad Chamber of Commerce, Mustansir Street, Baghdad (tel: 8886111).

Federation of Iraqi Chambers of Commerce, Baghdad Chamber of Commerce Building, Mustansir Street, Baghdad (tel: 8886111, 8880091/4).

Federation of Iraqi Industries, Federation of Iraqi Industries Building, Khulloni Square, Baghdad.

Iraqi Fairs Administration, Al Mansour Street, PO Box 6188, Baghdad (tel: 5412252; tx: 212641 maaridh).

Ministry of Oil, Baghdad (tel: 519686,5510031; tx: 212216 mipetrol)

Ministry of Trade, Karradat Miriam, Khullani Square, Baghdad (tel: 8876222; tx: 212206).

State Company for Fairs and Trade Services, Mansour, Baghdad (tel: 5417022; tx: 213341 fair).

Banking
Central Bank of Iraq (central bank), PO Box 64, Rashid Street, Baghdad (tel: 8871101; tx: 212203, 212558, 212174).

Rafidain Bank, New Banks' St, PO Box 11360, Massarif, Baghdad (tel: 458001/30, 8870521/30; tx: 2211, 2538A).

Rashid Bank, Tourism Building, Haifa Street, Baghdad (tel: 5388491).

Travel information
British Airways, Sadoun Street, Baghdad (tel: 888-3346).

Iraqi Airways, Saadoun Street, Baghdad (tel: 8870051, 8871175; tx: 28906).

National Tourist Bureau, Baghdad (tel: 7180593, 8878192, 8870237).

Ireland

Paul Tansey

The spectre of political instability came back to haunt Ireland early in the summer of 1989. A snap general election was called for 15 June by the *taoiseach* (prime minister) Mr Charles Haughey. Mr Haughey had presided over a minority Fianna Fail administration for the previous two and a quarter years. Through a mixture of good luck and good management, he had put the Irish economy back on a sounder financial footing. Virtue, however, was not to be rewarded.

Fearing that his precarious parliamentary position was under increasing threat from the parties of the opposition and attempting to capitalise on a brighter economic outlook, Mr Haughey gambled on a quick shuffle of the electoral cards. His objective was to secure an overall majority which would have provided him with a mandate to continue his programme of fiscal correction. But the opposition parties held all the aces in the election campaign. The cuts in government spending engineered by Mr Haughey, particularly in the health sphere, proved to be politically expensive. The electorate took their revenge for the cuts they had suffered by denying Mr Haughey his majority government for the fifth time.

The result of the election was indeterminate. No party, or coalition of like-minded parties, secured an overall majority. In the Irish *Dail* (parliament) of 166 deputies, 83 seats constitutes a majority as one seat is held by the *Ceann Conhairle* (Speaker). Mr Haughey's Fianna Fail party returned to the Dail with 77 seats, a loss of four on the seats they had won at the 1987 election.

Two of the centre-right parties, Fine Gael and the Progressive Democrats, contested the election on the basis of an agreed policy pact. While Fine Gael managed to increase the number of seats it won to 55, a gain of four over its 1987 result, the Progressive Democrats saw their support collapse. Only six of their candidates were returned to the Dail, a long way short of the 14 seats they won when their electoral star was in the ascendant during 1987.

If the parties of the centre-right fared badly, the parties of the left were the real winners in the election. Outright opposition to expenditure cuts aided their cause with the electorate. The Labour Party won three seats, taking its total to 15, while The Workers' Party, broadly Eurocommunist in tone, increased its parliamentary representation from four to seven.

This political arithmetic is of the utmost importance. As a result of the election, Fianna Fail finds itself further away from an overall majority than ever. The centre-right parties of the pact also lost ground, and are not within shouting distance of forming a government. The parties of the left, whilst increasing their representation, made it clear that they were not interested in participating in government. Even on the surface, this is a recipe for political instability.

But things are worse than they seem. Since March 1987, Mr Haughey's minority administration managed to govern with the tacit support of other centrist parties. In essence, both Fine Gael and the Progressive Democrats were prepared to support Fianna Fail, so long as it was seen to be tackling the country's deep-rooted economic ills. This unspoken consensus had served the Irish economy well since 1987. That fragile consensus was, however, shattered by Mr Haughey's precipitate dash to the country and by the antagonisms generated in the bitterly fought election campaign that followed.

The fact that Mr Haughey's party not only failed in its objective of gaining an overall majority, but also lost seats in the election, further weakens his position as the head of a continuing minority administration. As a result, the political landscape now appears more unstable than at any time since the early 1980s, when elections came and went with the seasons.

Irish general elections 1987 and 1989

Party	Number of Seats 1987	1989
Fianna Fail	81	77
Fine Gael	51	55
The Labour Party	12	15
The Workers' Party	4	7
Progressive Democrats	14	6
Independents/Others*	4	6
Total parliament**	166	166

* Including the Speaker
** 83 seats provides a majority government.

KEY FACTS

Official title: Ireland
Head of state: President Dr Patrick Hillery
Head of government: Prime Minister Charles Haughey
Ruling party: Fianna Fáil/Progressive Democrat coalition
Capital: Dublin
Official Languages: Irish and English
Currency: Irish Pound or Punt (Ir£) = 100 pence
Exchange rate: Ir£0.67 per $ (Dec 1989)
Area: 70,283 sq km
Population: 3.54m (1988)
GDP per capita: Ir£5,826 (1988)
GDP real growth: 3.0% (1988)
Labour force: 1.3m (1987)
Unemployment: 18.6% (1988)
Inflation: 2.1% (1988)
Trade balance: Ir£2,088m (1988)
Foreign debt: Ir£9.7bn (1987)

KEY INDICATORS

	Unit	1984	1985	1986	1987	1988
Population*	'000	3,597	3,644	3,537	3,535*	3,540
Gross domestic product (GDP)	Ir£bn	16.4	17.7	18.7	20.0	21.3
GDP per capita	Ir£	4,501	4,917	4,580	6,914	5,826
Inflation	%	8.6	5.5	3.8	3.2	2.1
GDP real growth	%	4.2	1.6	–0.4	4.1	3.0
Unemployment	%	16.4	17.7	18.2	19.0	18.6
Interest rate (official disc.)	%	14.0	10.25	13.25	9.25	8.00
Industrial production	1980=100	124	127	131	146	162
Manufacturing	1980=100	125	131	134	150	169
Investment goods	1980=100	186.8	190.9	–	–	–
Consumer goods	1980=100	102.0	104.7	–	–	–
Intermediate goods	1980=100	114.8	114.4	–	–	–
Exports	Ir£m	8,897	9,743	9,374	10,724	12,301
UK	%	34.4	33.0	34.0	–	–
German FR	%	10.1	10.0	10.8	–	–
Imports	Ir£m	8,913	9,428	8,621	9,155	10,213
Petroleum	Ir£m	977.1	945.0	554.2	511.7	387.5
Balance of trade	Ir£m	–16	315	753	1,569	2,088
Current account	$m	–1,038	–688	–677	375	651
Foreign exchange	$m	2,352	2,940	3,236	4,796	5,087
Exchange rate	$ per Ir£	1.09	1.07	1.34	1.49	1.53

* estimated figure

Until the 1980s, Ireland enjoyed a solid reputation for electoral stability. Fianna Fail, republican but populist, was the natural party of government, its long spans of government being interrupted by occasional periods of coalition, where Fine Gael were always the principal partners.

This tradition of stability has been shattered by the experience of the 1980s. Already in the course of the decade there have been five general elections and none has produced a single-party majority government. The possibility of further elections in the near future cannot be discounted, given the delicate balance of political representation.

The electorate in Ireland has become more volatile. The explanation of this increased volatility lies in the country's economic performance. In particular, it lies in the political impossibility of any Irish government addressing the Irish debt issue without sacrificing political popularity.

Public finance crisis

During the 1970s and through the first half of the 1980s, successive Irish governments borrowed as if there was no tomorrow. Ireland's national debt climbed to the point where it reached Ir£24.5bn by the end of 1987. That's a lot of debt for a country with a population of only 3.5 million people to service it.

Of the total Irish national debt outstanding at end-1987, some Ir£9.7bn was denominated in foreign currencies, making the burden of debt service costs heavier for the domestic economy.

Debt on this scale generates its own momentum. So large do the annual service costs become that, in the absence of major domestic expenditure cuts, the level of the debt expands of its own volition. Interest payments on Ireland's national debt now amount to more than Ir£2bn a year, equivalent to more than 10 per cent of annual Irish Gross National Product.

Clearly, living on borrowed money to this extent means living on borrowed time. At some point, borrowing no longer becomes possible.

Ireland's ratio of national debt to Gross National Product rose very significantly during the late 1970s and early 1980s to stand at 93 per cent by 1982. Attempts by the Fine Gael-Labour Party Coalition to rein back public sector borrowing in the mid-1980s met with only limited success. As a result, the debt/GNP ratio climbed inexorably, reaching 130 per cent by 1986. At this level, the Irish economy was the most highly-geared in Europe. The Fianna Fail minority government took office in March 1987, pledging itself to stabilise the debt/GNP ratio as a matter of urgency. The national debt rose to 133 per cent of GNP in 1987 and stabilised at that level during 1988. Stabilising the debt, even at this very high level, required extensive cuts in public sector borrowing. In turn, this manifested itself in draconian public expenditure cuts.

In the three budgets it introduced since 1987, the minority Fianna Fail administration cut 10 per cent off the volume of current public spending on goods and services. Additional cuts were made in state capital expenditure. While Mr Haughey's zeal with the scalpel earned the approval of the financial markets, it did not win him friends amongst the electorate. As health cuts followed education cuts in an unending procession, Fianna Fail's traditional support amongst the working class began to weaken.

Yet, the public finance crisis remains contained rather than corrected despite the expenditure cuts of the past three budgets. The Public Sector Borrowing Requirement (PSBR) set for 1989 at 6.5 per cent of Gross National Product is still pitched at a level which sent Mr Denis Healey scurrying into the arms of the International Monetary Fund when he was Chancellor of the Exchequer in 1976.

This is the core of the impasse in the Irish political economy at present. The public finance crisis is so deeply embedded that it will require remedial action for years to come. Yet those political parties sufficiently brave to take on the task of fiscal correction find that it involves heavy electoral costs.

Growth in domestic economy

In its bid to double Ireland's take from the European Community's structural funds, the minority Fianna Fail government produced a five-year national development plan in the spring of 1989 embracing an Ir£9bn national investment drive for the five years through 1993, with some Ir£3.5bn being supplied from Brussels. In the course of the plan, the government committed itself to reducing Exchequer borrowing to 3 per cent of GNP by 1993. This would allow, it argued, the national debt to be reduced to 120 per cent of GNP by 1993.

This places in context the scale of the financial problem facing any Irish government over the next decade. Tight expenditure controls and additional cuts are needed to reduce public borrowing still further. The introduction of expenditure cutbacks inevitably entails political unpopularity on a scale which makes the chances of re-election virtually impossible. Yet, even where a government is prepared to take its courage in its hands and seeks to do the decent thing, the prospective rewards are meagre. Reducing national debt to the point where it still stands at one-and-one-fifth times' national income after five painful years is a target that will hardly appeal to the aspiring Cabinet minister.

Politics in Ireland has thus become a negative sum game. If the electorate don't get you, the international bankers will. The extraordinary feature of Mr Haughey's minority administration, spanning 1987 to 1989, was the extent to which domestically induced deflation co-existed peacefully with a high rate of national economic growth. Despite severe deflationary cuts in public spending, the Irish economy has experienced growth over the past three years on a scale not seen since the 1970s.

Real GNP growth over the three years 1987 through 1989 has averaged more than 3 per cent a year. The composition of growth has changed over the period. During 1987, economic expansion was export-led. While exports remained buoyant during 1988 and 1989, domestic demand began to recover sharply by the latter year. Real domestic personal consumption growth has averaged 2 per cent a year over the past three years. Gross investment, which had been negative for most of the 1980s, turned around in 1988 and moved sharply ahead during 1989.

The improvement in the investment tone reflected much more settled conditions in the financial economy. Consumer price inflation fell to a thirty-year low in 1988, when prices rose by just 2.1 per cent on average. While inflation picked up somewhat in the early months of 1989, prices were still rising at an annual rate of less than 3.5 per cent by mid-year.

As a founder-member of the European Monetary System, Ireland's exchange rate closely tracks that of the Deutsche Mark. This exchange rate stability over a prolonged period of time has allowed Irish interest rates to fall steadily towards European levels. By mid-1989, the Irish Central Bank's base lending rate stood at 9 per cent, some five percentage points below the corresponding rate in Britain. In effect, Ireland's membership of the EMS has allowed Irish interest rates to decouple from trends in British interest rates.

Labour market problems continue

The improvements effected at a macro-economic level, particularly in the financial segment of the economy, have been slow to trickle through to real economic variables. In particular, the gains in industrial and agricultural output in recent years have not translated into improved conditions in the labour market.

Ireland has a rapidly-growing labour force and few new jobs with which to accommodate new labour market entrants. As a result, unemployment has remained very high. Ireland's unemployment rate, at around 17 per cent of the labour force, is second only to Spain within the European Community. Moreover, the lack of jobs at home has sent many young Irish men and women in search of employment abroad.

Emigration, which had ceased altogether during the 1970s, has recommenced on a significant scale. From 1981 to the end of the decade, more than 170,000 people left Ireland, equivalent to 5 per cent of the 1981 population. The rate at which young people are leaving the country is accelerating.

COUNTRY PROFILE

Political system: Republic, with legislative power vested in bicameral National Parliament (Oireachtas) consisting of the Senate (Seanad Eireann), which has 60 members; and the House of Representatives (Dáil Eireann), with 166 members elected for five years by single transferable vote system. Executive power is exercised by the Cabinet, led by the Taoiseach (Prime Minister) who is appointed by the President on the recommendation of the Dáil. President is directly elected for seven years.

Main cities/towns: Dublin (population 1.02m) Cork, Limerick, Galway and Waterford.

Media
Dailies: Seven national dailies: *Irish Independent* (circulation 174,788), *Evening Press* (129,695), *Evening Herald* (132,314), *Irish Press* (86,655) and *The Irish Times* (88,739), all published in Dublin; and *Cork Examiner* (63,560) and *Cork Evening Echo* (36,250), published in Cork. All British dailies and Sunday papers are available.
Weeklies: Four national Sunday papers. Most popular is *Sunday Press* (266,019); others are *Sunday World* (366,806), *Sunday Independent* (222,351) and *Sunday Tribune*, all published in Dublin. Also several provincial weekly papers and a number of special and general interest magazines.
Business: Small number of business publications, mostly produced in Dublin, including *Business and Finance* (weekly, 11,270), *Irish Business* (monthly, 9,750) and *Success* (monthly). Most UK trade publications also available.
Broadcasting: Approximately 900,000 TVs in use. Controlled and operated by Radio Telefis Eireann (RTE), which supplies two commercial radio and two commercial TV channels. RTE also finances Radio na Gaeltachta, based in Connemara, which broadcasts radio programmes in Irish. Broadcasts from UK can also be received in many parts of the country. One third of homes are cabled although the system can only carry six channels. Three satellite channels have been allocated but none have been licensed yet.
Advertising: Total advertising spending (1984) Ir£79.6m. Key categories include automotive, building societies confectionery, cosmetics and food. Restrictions applied to advertising expenditure and form and content of advertisements. Promotion of tobacco products is banned completely, and that of alcohol strictly controlled. Commercials on TV and radio are limited to 10 per cent of transmission time. Newspapers, magazines, posters and films are also widely used. Information available from The Advertising Standards Authority in Dublin.

Domestic economy: Mixed economy with large, export-based agricultural sector and largely foreign-owned manufacturing sector. Tourism and construction also important. Economic growth rate fallen since 1979, though significant improvement recorded since 1984. Programme for National Recovery (1987-1990) aims at reducing unemployment to 15 per cent, boosting exports and reducing budget deficit. 1987 austerity budget involved severe expenditure cuts aimed to reduce budget deficit to 6.9 per cent of GNP and public-sector borrowing requirement to 12 per cent of GNP.

Employment: Around half labour force employed in private and public services, some 16 percent in agriculture and the rest in manufacturing and construction. Despite growth in service sector employment and rapid rise in productivity, unemployment continues to increase. Situation exacerbated by steadily growing labour force and emphasis on developing capital-intensive industry. Real earnings have increased slowly (2.3 per cent in 1985) compared with labour productivity (up 17 per cent in 1984).

External trade: Strong rise in exports and lower oil imports costs helped turn regular trade deficit into small surplus, though current account deficit remains large owing to high level of profit repatriation and heavy debt burden. Exports consist mainly of live cattle, meat, dairy produce, textiles, manufactures, machinery and transport equipment and more recently chemicals and electronics. Imports, which in total have risen slower than exports, consist mainly of petroleum and petroleum products, foodstuffs, chemicals, manufactured goods and components, machinery and transport equipment. The UK remains the principal trading partner, but since accession to the EC in 1973 trade with German FR, France and Italy has increased significantly.

Agriculture: Sector accounts for 12 per cent GDP and around 23 per cent of export earnings. Production is centred on livestock farming, accounting for around 87 per cent of output, principally cattle and beef for export and other meats and dairy produce for local consumption. Sector receives substantial development aid and price subsidy under the EC's Common Agricultural Policy. Main crops include barley, sugar beet, potatoes, wheat and turnips.

Industry: Sector accounts for 36 per cent of GDP and over 65 per cent of exports. Indigenous manufacturing base small. Traditional industries such as food and beverages, textiles, paper, non-metallic minerals and machinery dominate, though there has been rapid growth in new export-orientated chemicals and electronic engineering industries. Most of the new capital and skill-intensive industries are subsidiaries of large US and European multinationals and are heavily reliant on imported primary and intermediate inputs.

Mining: Europe's largest zinc and lead deposits located at Navan, County Meath. Production has declined since 1970s. Also reserves of copper, pyrites, gypsum, barytes, limestone, coal and marble.

Energy: Around 70 per cent energy requirements are imported. Dependence on imported petroleum rapidly reduced due to exploitation of domestic gas and peat reserves. Also energy conservation programmed in operation. Exploration for oil under way off the west and south coasts.

Banking: Monetary policy and foreign exchange dealings supervised by the Central Bank of Ireland, a government agency. Four major commercial banking groups: Bank of Ireland, Allied Irish Bank, Ulster Bank and Northern Bank. Also nine Irish merchant banks, 15 industrial banks, 16 foreign banks, four trustee savings banks and a post office savings bank.

Legal system: Based on English common law with substantial modifications. Constitution adopted in 1937.

Membership of international organisations: CCC, Council of Europe, EC, EMS, ESRO (Observer), FAO, GATT, IAEA, IBRD, ICAO, ICES, IDA, IEA, IFAD, IFC, ILO, IMF, IMO, INTELSAT, INTERPOL, IPU, ISO, ITC, ITU, IWC, OECD, UN, UNESCO, UPU, WHO, WIPO, WMO, WSG.

Time: GMT (GMT + 1 from late Mar to late Oct).

Climate: Temperate, with warm summers and mild winters. Rain throughout year. Average temperatures range from about 0°C to 20°C.
Clothing: Mediumweights and raincoat throughout the year.

Entry requirements: Visas not required by EC passport holders, but may be required by others.
Health: Vaccination certificates not compulsory, but can be useful in event of temporary restrictions.
Currency: No limit on amount of currency taken in. Up to Ir£150 in Irish currency and Ir£1,200 in foreign currency may be taken out.
Customs: Personal effects duty-free, plus duty-free allowance.

Air access: Regular flights by most international airlines.
National airlines: Aer Lingus, Ryanair.
Main international airport: Dublin (DUB), 10 km from city centre. Airport bus Ir£2.50.
International airports: Shannon (SNN) and Cork (ORK).

Surface access: Several passenger and car ferry services from UK ports, such as Liverpool, Holyhead, Fishguard and Stranraer (via Northern Ireland). Also several road and rail crossing points from Northern Ireland.
Main ports: Dublin, Dun Laoghaire, Waterford, Cork, Limerick, Sligo, Rosslare, Galway and Dundalk.

Hotels: Classified into four categories: A (de luxe) A(Class 1), B(Class II) and B(Class III). Single de luxe room, including breakfast and 18 per cent VAT, in Dublin costs between about Ir#35 and Ir#75, depending on season and facilities offered. Prices are appreciably cheaper outside capital. Tipping: 10 per cent is customary. All usual credit cards are accepted.

Currency
Exchange rate: Ireland is a member of the European Monetary System. The Irish pound floats within the EMS grid and against other currencies.

Foreign exchange information
Account restrictions: None on non-resident accounts.
Trade policy: No import restrictions, except for armaments for which import licences are required.

Car hire: Available in all main towns, but heavy demand during tourist season. National or international driving licence required. Speed limits 30 mph (48 kph) in built-up areas and 60 mph (96 kph) on main roads. Driving is on the left.

City transport: Good bus and taxi service available in major towns. Destinations in Dublin written in Gaelic. Special bus/rail tickets can be purchased for multiple journeys. Metered taxis can be hailed in the street or called by telephone.

National transport
Air: Daily services between Dublin and Shannon, and Dublin and Cork operated by Aer Lingus.
Road: Approximately 90,000 km of roads, including 16,000 km of main roads. Bus services throughout country, including express coaches connecting main towns.
Rail: Almost 2,000 km of track operated by the Irish Transport Company (CIE). Express services between Dublin and principal cities.
Water: Over 400 km of navigable inland waterways.

Public holidays
Fixed dates: 1 Jan, 17 Mar (St Patrick's Day), 25 and 26 Dec.
Variable dates: Good Friday, Easter Monday, Whit Monday, August Bank Holiday, last Monday in Oct.

Working hours
Banking: (Mon-Fri) 1000 – 1230 and 1330 – 1500, open until 1700 in Dublin on Thu.
Shops and business: (Mon-Fri) 0900 or 0930 – 1730. Supermarkets normally stay open until 21.00 on Fridays.

Telecommunications
Telephone: Direct dialling available throughout country. International Dialling code: 353 followed by 1 for Dublin, 21 for Cork, 91 for Galway, 61 for Limerick, 51 for Waterford and 53 for Wexford.
Telex: Public telex office at Telecom Eireann, Marlborough St, Dublin. Service also available at main post offices and hotels. Country code: 500 EI.

BUSINESS GUIDE

Trade fairs: Several international events held in Dublin. Tend to be held at irregular times and vary from year to year.

Electricity supply: 220 V AC.

Representation in capital: Argentina, Australia, Austria, Belgium, Canada, China PR, Denmark, Egypt, Finland, France, Federal Republic of Germany, Greece, Holy See, India, Iran, Italy, Japan, Republic of Korea, Netherlands, Nigeria, Norway, Portugal, Spain, Sweden, Switzerland, Turkey, USSR, United Kingdom, USA.

Hotels: Dublin
Berkeley Court, Lansdowne Road, Ballsbridge, 4 (tel: 601711; tx: 30554).

Blooms, Anglesea Street, 2 (tel: 715622; tx: 31688; fax: 715997).

Burlington, Upper Leeson Street, 4 (tel: 605222; tx: 93815; fax: 605064).

Buswell's (facing Parliament), Molesworth Street, 2 (tel: 764013; tx: 90622; fax: 764013).

Clarence, Wellington Quay, 2 (tel: 776178; tx: 90710).

Gresham, O'Connell Street (tel: 746881; tx: 32473; fax: 787175).

International, Dublin Airport (tel: 379211; tx: 24612; fax: 425874).

Jury's, Ballsbridge, 4 (tel: 605000; tx: 93723; fax: 605540).

Mont Clare, 14 Clare Street (tel: 616799; tx: 91471; fax: 615663).

Powers Royal (near Trinity College), 47 Kildare Street (tel: 605244; tx: 24288).

Royal Dublin, 40 Upper O'Connell Street (tel: 733666; tx: 4288).

Shelbourne, 27 St Stephen's Green (tel: 766471; tx: 93653; fax: 616006).

Westbury, Grafton Street (tel: 791122; tx: 91091; fax: 797078).

Car hire: Dublin
Avis: 1 Hanover Street East (tel: 776971/6; tx: 33263).

Budget: (tel: 9034668/4678; tx: 53914).

Europcar: Baggot Street Bridge, (tel: 681777; tx: 93784).

Hertz: 12 Upper O'Connell Street (tel: 788322).

Chambers of Commerce
Cork Chamber of Commerce, Carrabeg House, Summerhill, Cork (tel: 509044; tx: 26159).

Dublin Chamber of Commerce, 7 Clare Street, Dublin 2 (tel: 764291; tx: 90716).

Dundalk Chamber of Commerce, 14 The Crescent, Dundalk (tel: 34033).

Limerick Chamber of Commerce, O'Connell Street, Limerick (tel: 45180).

Waterford Chamber of Commerce, Georges Street, Waterford (tel: 72639).

Banking
Central Bank of Ireland, PO Box 559, Dame Street, Dublin 2 (tel: 716666; tx: 31041

Allied Irish Bank Ltd, Bankcentre, Ballsbridge, Dublin 4 (tel: 600311; tx: 93768).

Allied Irish Investment Bank PLC, Bankcentre, Ballsbridge, Dublin 4 (tel: 604733; tx: 93917).

Bank of Ireland, Lower Baggot Street, Dublin 2 (tel: 615933; tx: 93836).

The Institute of Bankers in Ireland (banking association), Nassau House, Nassau Street, Dublin 2 (tel: 793311).

Investment Bank of Ireland Ltd, 26 Fitzwilliam Place, Dublin 2 (tel: 616433; tx: 93811).

National Irish Bank, 7/8 Wilton Terrace, Dublin 2 (tel: 785066).

Ulster Bank, PO Box 145, 33 College Green, Dublin 2 (tel: 777623; tx: 25166).

Ulster Investment Bank Ltd, 2 Hume Street, Dublin 2 (tel: 613444; tx: 93980).

Principal newspapers
Business and Finance, 50 Fitzwilliam Square, Dublin 2 (tel: 764587).

Cork Evening Echo, 95 Patrick Street, Cork (tel: 963300; tx: 6014).

Cork Examiner, 95 Patrick Street, Cork (tel: 963300; tx: 6014).

Evening Herald, Independent House, 90 Middle Abbey Street, Dublin 1 (tel: 731333).

Evening Press, Tara House, Tara Street, Dublin 2 (tel: 713333).

Irish Business, 128 Lower Baggot Street, Dublin 2 (tel: 764587).

Irish Independent, Independent House, 90 Middle Abbey Street, Dublin 1 (tel: 73166).

Irish Press, Tara House, Tara Street, Dublin 2 (tel: 713333; tx: 93752).

Sunday World, Newspaper House, 18 Rathfarham Road, Terenure, Dublin 6 (tel: 901980; tx: 24886).

The Irish Times, 11-15 D'Olier Street, Dublin 2 (tel: 792022; tx: 93639).

The Sunday Independent, Independent House, 90 Middle Abbey Street, Dublin (tel: 731666; tx: 33472).

The Sunday Press, Tara House, Tara Street, Dublin 2 (tel: 713333; tx: 25353).

Other useful addresses
Advertising Standards Authority, IPC House, Shelbourne Road, Dublin 4.

Aer Lingus, Teoranta, Dublin Airport (tel: 370011).

British Airways Office, 60 Dawson Street, Dublin 2 tel: 610666).

CIE (public transport information office), 59 Upper O'Connell Street, Dublin 1 (tel: 787777).

Confederation of Irish Industry, Confederation House, Kildare Street, Dublin 2 (tel: 779801; tx: 24711).

Department of Finance and Department of the Public Service, Government Bldgs, Upper Merrion Street, Dublin 2 (tel: 767571; tx: 30357).

Department of Foreign Affairs, 80 St Stephen's Green, Dublin 2 (tel: 780822; tx: 25300).

Department of Industry and Commerce, Kildare Street, Dublin 2 (tel: 614444; tx: 93478).

Department of Tourism and Transport, Kildare Street, Dublin 2 (tel; 789522; tx: 93478).

Federated Union of Employers, Baggot Bridge House, 84-86 Lower Baggot Street, Dublin 2.

Irish Tourist Board, Baggot Street Bridge, Dublin 2 (tel: 765871; tx: 93755).

Provincial Newspapers Association of Ireland, 24 Dame Street, Dublin 2.

The Stock Exchange, 24-28 Anglesea Street, Dublin 2.

Tourist Information Office, 14 Upper O'Connell Street, Dublin (tel: 747733).

Israel

Andrew Whitley

Israel's second 'government of national unity', born out of the near-deadlock of the November 1988 general elections, proved as incapable as its predecessor of resolving the fundamental questions facing the Jewish state. But its divisions accurately reflected those in the society as a whole, making clear that there are no easy answers.

Under the pressure of the two-year-old Palestinian uprising, or 'intifada', Prime Minister Yitzhak Shamir proposed holding elections in the occupied West Bank and Gaza Strip, to negotiate an interim solution for the region. To some extent this was no more than a tactical ploy, designed to take some of the growing international pressure off Israel. It became a major bone of contention, however, when the Israeli right first blocked Egyptian President Hosni Mubarak's attempt in September 1989 to bring Israelis and Palestinians together for talks and then a

KEY FACTS

Official title: State of Israel

Head of state: President Chaim Herzog

Head of government: Prime Minister Yitzhak Shamir

Ruling party: Coalition led by Likud

Capital: Jerusalem

Official Languages: Hebrew and Arabic

Currency: New Shekel (NIS) = 100 agorot

Exchange rate: NIS1.99 per $ (Nov 1989)

Area: 20,770 sq km

Population: 4.5m (1988)

GDP per capita: $7,677 (1988)

GDP real growth: 1.8% (1988)

Labour force: 1.6m (1988)

Unemployment: 6.4% (1988)

Inflation: 16.3% (1988)

Trade balance: –$3,144m (1988)

Foreign debt: $23.9bn (1988)

Visitor numbers: 1.38m (1987)

US effort to break the stalemate. With the Labour party, the junior partner in the government, openly advocating support for the Mubarak initiative, a full-fledged crisis which threatened to force fresh elections was defused only when Labour realised that the country's electoral arithmetic was working against the centre-left party.

Finance Minister Shimon Peres, the Labour leader, began the long-postponed process of restructuring the economy, to introduce greater market forces and reduce the preponderant role of the state. Although little of substance was achieved by mid-1989, he has shown courage in challenging long-accepted dogmas central to his own party's political wellbeing. In part this is a recognition of changing times as, on virtually all fronts, from the kibbutz to the Histadrut labour federation, the socialist ideals on which Israel was founded have begun to crumble.

Peres' task was complicated by the recession which spread through the economy as 1989 progressed. With unemployment rising to a near record 9.5 per cent, the finance minister's scope for deep cuts in the public sector – widely regarded as a prerequisite for the restoration of growth – was curbed. Inflation has proven to be a continuing headache, inching back up to an annual level of about 20 per cent, well above the average rate in the OECD economies, Israel's principal trading partners. The one beneficial aspect of the slowdown is that it has helped reduce a trade deficit which, in 1988, had swollen to alarming proportions.

US financial aid totalling \$3bn a year – out of net annual transfers to Israel from all sources of approximately \$4.5bn – continues to be an essential prop, stabilising an economy which would otherwise suffer endemic balance of payments crises. Heavy defence expenditure, internal and external debt servicing and an overblown public sector remain the principal problem areas. The foreign exchange crises of the past appeared, however, to have eased in 1989.

At a time when the Jewish state is being pressed on all sides to negotiate over the Palestine Liberation Organisation's (PLO) proposed 'two-state' solution to the Palestinian question, no less critical than its economic aid remains Washington's staunch political backing. Although vague over what political formula it now favours, following King Hussein of Jordan's declared abandonment of responsibility for the occupied West Bank and Gaza Strip in 1988, the USA is still adamantly opposed to the creation of a new state in the Middle East. The Bush administration has, as predicted, proven less fulsome in its support for Israel than that of Ronald Reagan. During its first year in office the USA moved cautiously, judging – probably accurately – that the time was not yet ripe to throw its full weight into the fray.

One development of potentially far reaching importance for Israel arose out of Washington's decision of mid-1989 to begin restricting the number of Soviet-Jewish émigrés permitted to reside in the USA. In the late 1980s fewer than 10 per cent of Soviet Jews leaving their homeland have opted for Israel, their nominal destination on visas issued in Moscow. The resulting loss of fresh blood has demoralised a state whose raison d'etre is the ingathering of the exiles, but which is encountering a drying well of migrants. Many of those newcomers who are arriving, however, are religious, changing the character of the state in the process.

The prospect of a new wave of Soviet migrants, leaving in greater numbers than for two decades, both excites and alarms Jerusalem. A \$2bn programme was announced, to house and settle an expected 100,000 arrivals, with much of the funds slated to be raised from world Jewry. The USA was asked in September to provide \$400m in loan guarantees, but balked at the prospect of some of the cash being used to extend settlements in the occupied West Bank. Indeed, given the decrepit state of its immigrant absorbtion machine, Israel's capacity to handle these new immigrants will be, in its own way, as much a challenge for the government as dealing with the more dramatic, life-and-death events being played out in the occupied territories.

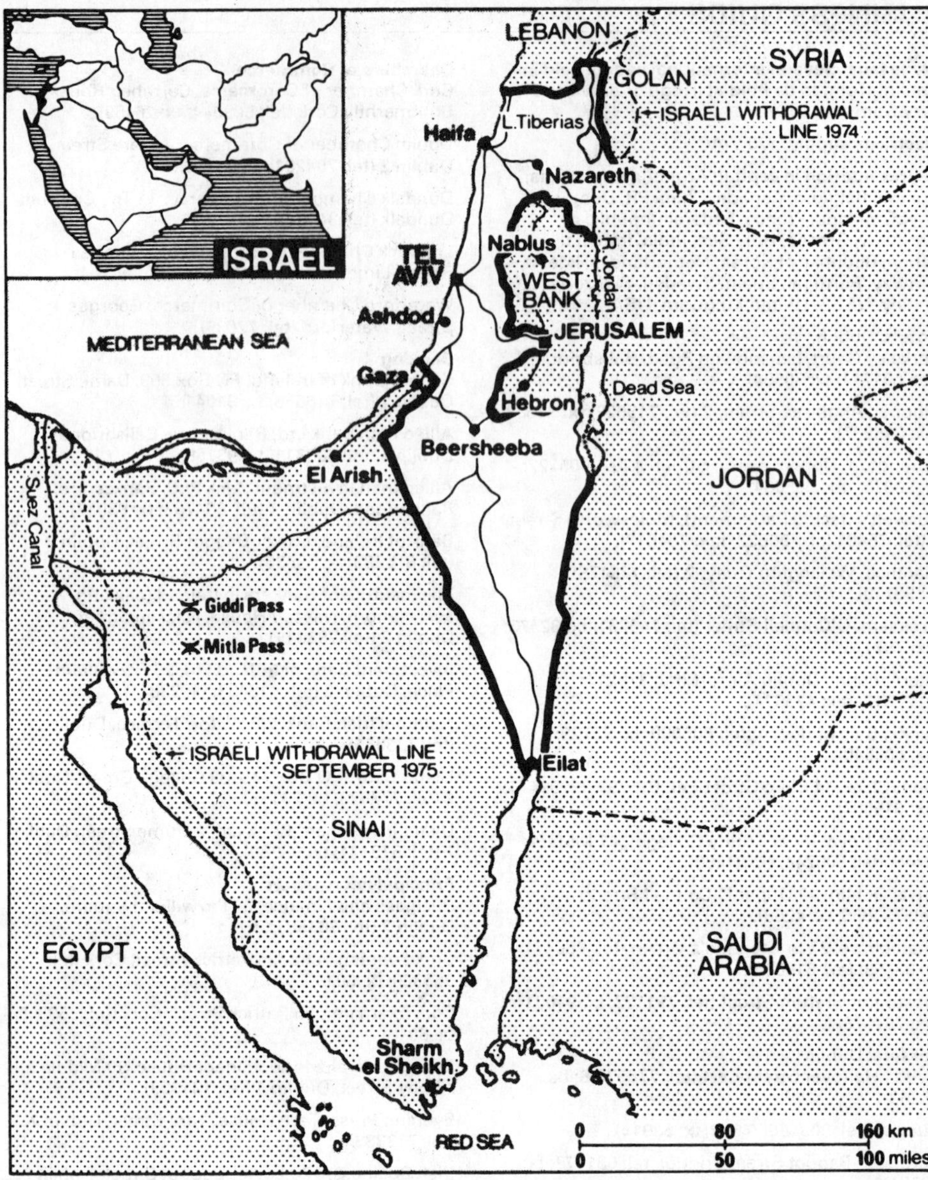

Here, the death roll has risen to well over 700 on the Palestinian side, with at least 40 Israeli fatalities as well. Many thousands of others have been injured as the uprising settles into an increasingly ugly war of attrition. Although the level of frustration among both Israelis and Palestinians over the failure to achieve a breakthrough is high, neither shows signs of giving way, feeling that the situation, for all its costs, can be maintained almost indefinitely.

The November 1988 elections produced a narrow majority for the right-wing bloc of parties led by Shamir's Likud, lending credence to the 73-year-old prime minister's campaign claim to represent a new centre in Israeli politics. Labour suffered a modest setback at the polls, marking the fourth general election in succession that the party, which long regarded itself as the natural party of government, had failed to achieve a victory. A surprise development was the emergence on to the national political stage of a group of ultra-Orthodox religious parties, whose demands – notably over the claimed primacy of the Orthodox stream of Judaism and the stricter observation of religious law – strained relations with the Jewish diaspora and complicated the normally protracted process of forming a government. Municipal elections in March 1989 confirmed the national trend.

For weeks after the 1988 elections, Labour agonised over whether it should serve under Shamir in a grand coalition or go into opposition. A strong doveish faction campaigned hard for the latter; but, in the end, the counsels of Peres and Yitzhak Rabin – anxious to retain his post as defence minister – prevailed. Rabin, himself a former prime minister, has consolidated his reputation as the most popular of Labour's ministers and could re-emerge as the party leader if Peres is deposed. Both the Labour and Likud leaders are troubled by large dissident minorities among their own supporters, requiring skilful political footwork to remain in charge.

New government

The new government, a recast version of the one which had ruled Israel since late 1984, finally took office in late December. The two major differences were that, this time, there would be no rotation of the premiership between the two major party leaders, while several minor parties on both the left and the right were dropped. Although these act as pressure groups on the government, they are incapable on their own of bringing it down. Entitled to a full four-year term, it was evident, however, that the ideological conflicts between Labour and Likud would soon reassert themselves, condemning Israelis to another period of political brinkmanship, and complicating external initiatives to bring a lasting peace to the region.

Despite their fundamental differences of political outlook, notably over the future of the occupied territories, the marriage of interests between Shamir and Peres has been surprisingly durable. And it could remain so, provided that two developments do not simultaneously occur: the Arabs come up with proposals that a majority of Israelis find tolerable, and the US Government chooses to exert the undoubted muscle at its disposal.

An urgent priority for the new government was to contain the damage wrought by the landmark decision of the Reagan administration in its dying days to establish a formal dialogue with the PLO. At a stroke the most friendly US administration Israel had ever known breached the international quarantine in which the Jewish state had been able to maintain the Palestinian body for over a decade.

Attempted cross-border raids from Lebanon and rocket attacks on northern Israeli settlements by dissident Palestinian groups and their Lebanese allies were used, unsuccessfully, as arguments to persuade Washington that Yasser Arafat, the PLO leader, was insincere in his declared renunciation of terrorism. The Bush administration was not, however, ready to retie its own hands in the knots left by former Secretary of State Henry Kissinger a decade earlier. Nor, subsequently, did Shamir have any more success in persuading other major Western countries such as France, Spain and the UK not to enhance the PLO's international prestige by holding high-level meetings with top Palestinian officials, including Arafat.

In May 1989, after a meeting with President Francois Mitterand of France in which he was urged to declare where he stood on the 1964 Palestinian Charter (which called for Israel's destruction and its replacement by a bi-national state) Arafat went so far as to declare the charter 'null and void'. Later, though, considerable doubts were raised as to the extent to which this declaration represented the genuine views of the PLO leadership as a whole.

The pressure was nevertheless on Israel to come up with a credible response to the string of political successes enjoyed by the PLO, including the recognition by over a hundred countries of the Palestinian 'state' declared in the occupied West Bank and Gaza Strip, headed by a retitled President Arafat. Rabin, pursuing a carrot-and-stick policy in his approach to the uprising, announced his own plan for local elections in the occupied territories; not held in the West Bank since 1976, when the Israeli authorities effectively annulled the outcome, and never since 1948 in Gaza. With certain modifications, the plan was later taken up by Shamir, and ratified by the full cabinet on 14 May, shortly before the Israeli leader was to travel to the USA, for his first face-to-face dealings with the Bush administration.

Peace proposals

Shamir's four-point peace plan comprised elections to choose local Palestinians with whom negotiations over the future of the territories could be held, a reaffirmation of the 1978 Camp David principles, a formal end to the state of war between Israel and most Arab states, and an international effort to resettle over 1m Palestinians who are still classified as refugees. The latter idea marked the revival of an earlier, unsuccessful effort to divert international attention away from calls for a political conference on the Middle East – something the Israeli right has always resisted.

Faced with Israeli hostility towards a full-blown international conference, little progress over such an event was made during 1989. A European Community summit in Madrid did call for a conference at

KEY INDICATORS

	Unit	1984	1985	1986	1987	1988
Population	'000	4,160	4,240	4,300	4,370	4,430
Gross domestic product (GDP)	NIS bn	7.6	28.1	43.9	55.6	67.0
GDP per capita	NIS	1,827	6,627	10,209	12,723	15,124
Industrial product prices	1985=100	27	100	145	172	202
Consumer prices	1985=100	25	100	148	178	207
Industrial production	1985=100	96.8	100.0	103.4	108.5	105.0
Volume of exports	1985=100	92	100	111	123	124
Volume of imports	1985=100	96	100	117	131	131
Exports	$m	–	6,189	6,602	7,675	9,064
Imports	$m	–	8,758	9,041	9,613	12,898
Non-military goods	NIS bn	2.45	9.81	9.56	9.37	–
Balance of trade	$m	–	–2,569	–2,439	–1,938	–3,834
Current account	$m	–	–1,318	1,152	1,551	–997
Exchange rate	NIS per $	–	0.29	1.18	1.49	1.59

* estimated figure

which the PLO would be represented in its own right, and not, as in past formulas, as part of a joint delegation with Jordan. And an Arab summit in Casablanca called for an international conference based on UN Security Council resolutions 242 and 338 – an implicit, if not explicit, recognition of Israel's right to exist. But with the USA lukewarm at best about the idea and the Soviet Union distracted by its own problems there seemed little prospect of its realisation in the near future.

Securing his own party's backing for a plan which said that Israel was prepared to discuss the final status of the territories it captured in 1967 was not easy for Shamir. A strong challenge was mounted by his two leading party opponents, the hawkish Ariel Sharon and David Levy. Eventually beaten off, this internal opposition further restricted Shamir's room for manoeuvre.

Not that the longtime Mossad man could be accused of equivocating in public. In a speech to the Likud on 18 May 1989 he said: 'We shall not give the Arabs one inch of our land, even if we have to negotiate for 10 years ... We won't give them a thing.' Later in the year, when the US and Egyptian Governments were trying to prod him into showing greater flexibility, Shamir made clear that even at the risk of a serious rupture with Washington his strategy was the same as it had always had been: to wait for the Arabs to tire of beating against the Israeli rock, and then to sue for peace on Israel's terms. Abroad, the Shamir plan won a qualified endorsement in Washington, but was damned with faint praise by other Western nations and was rejected by the PLO as a patent ploy to circumvent itself.

As Israeli tardiness persisted throughout the summer of 1989 into the autumn, so the level of frustration in Washington at the behaviour of its regional ally also began to rise. In a landmark speech to American Jews in May 1989, Secretary of State James Baker told Israel bluntly that it should 'lay aside, once and for all, the unrealistic vision of a Greater Israel'. At the same time he spoke of the need to fulfill the Palestinians' 'full political rights' – a significant evolution from the Camp David language of a decade earlier, which had referred only to their 'legitimate rights'. In September 1989, he called upon Jerusalem to make up its mind whether or not it stood behind the Shamir proposal, hinting that he was reaching the end of his tether.

Against this background, the prime minister was due to pay a second official visit to Washington in November 1989, expected to prove even more difficult than his first, six months earlier. Indeed, surveying the diplomatic horizons in late 1989 an Israeli official would have little cause for satisfaction. The most optimistic assessment could be that at least with Egypt and the Soviet Union, two problem countries of the past, there had been no backsliding. By any standards, though, political relations with Western Europe, with which Israel has important economic ties, deteriorated.

Egypt, for its own, inter-Arab motives, remained engaged in trying to achieve a breakthrough in the so-called peace process. To that end Butrous Ghali, its veteran minister of state for foreign affairs and the most senior Egyptian official since President Anwar Sadat in 1977, visited Israel in June. The message he had to deliver, however, that the PLO had to be given a central role in the efforts to resolve the Arab–Israeli dispute, was coldly received.

While Israel may have failed significantly to reverse its own diplomatic setbacks at the hands of the Palestinians, the PLO, in turn, became visibly frustrated during the first half of 1989 at the lack of progress in its dialogue with the US Government, held in Tunis at a senior official level. These frustrations were vented through some sharp criticism of the USA for its pro-Israel stance, particularly its backing for the elections plan. As the USA marked time in the Middle East dispute, diplomats from other nations, notably Egypt, Sweden and the Soviet Union, attempted to fill the gap, while denying any formal mediating role.

President Mubarak's initiative, welcomed by the USA, came in the form of a 10-point proposal. Described as being aimed at facilitating the holding of elections in the occupied territories, the carefully drafted plan rested firmly on the principle of exchanging 'land for peace'. Although this point is anathema to the Israeli right, the fact that – remarkably for an Arab state – it made no reference to Palestinian self-determination, nor to the PLO, enabled it to break new diplomatic ground and put Israel on the spot.

After mid-1988, an increasing number of unofficial contacts were held between Israelis and Palestinians associated with the PLO, both in the region and further afield. In early 1989, for instance, Abu Iyad, Arafat's deputy and a known hardliner in the PLO, used a Jerusalem peace group's conference to make an unprecedented direct appeal to Israelis, through a videotaped address, to accept direct talks leading towards a redivision of the former British Mandate territory of Palestine. Later, however, responding to criticism from the increasingly hardline underground leadership of the uprising, the PLO ruled out any further talks by individual Palestinians with Israelis, officials or private citizens. Among Israelis, although opinion polls indicated a shift in public attitudes in favour of talking to the PLO – even among some natural Likud supporters – Shamir's party has not budged an inch on the subject, and shows no likelihood of a change of heart in the near future.

For its part, the PLO did not totally reject the idea of elections in the occupied territories. It subtly modified its initial insistence that these could only be held once Israel had withdrawn from the region to leave open the possibility of a poll held under Israeli auspices, provided international observers were present and they formed part of a scenario leading towards a permanent settlement of the dispute.

Intifada persists

Defying repeated official forecasts – designed as much to lower morale among the 1.5m Palestinians involved as to reassure the Israeli public – that the conflagration was on the wane, the intifada persisted well into 1989. If anything, the violence worsened. It also changed in character, with a rising number of killings by Palestinians of suspected collaborators, along with attacks on Jewish civilians across the old, pre-1967 Green Line border. In the worst of these incidents, a Palestinian forced a bus travelling on the main Jerusalem to Tel Aviv highway over a cliff, killing 15 people. The kidnapping and murder of two off-duty soldiers and the stabbing of three elderly people waiting at a Jerusalem bus stop also aroused widespread revulsion among Israelis.

Taking the offensive, the Israeli army launched repeated sweeps of Palestinian villages, arresting suspected activists who could then be held without trial for up to a year, under relaxed detention procedures granted by Defence Minister Rabin. As of April 1989, army figures revealed that over 20,000 Palestinians had been detained, of whom fewer than 800 had been convicted of any offence.

Increasingly using unconventional methods of fighting the intifada, the army and Shin Bet secret police meanwhile extended their long-standing use of informers to form para-military groups of their own among Palestinian collaborators. Undercover Israeli units working with Palestinians were also responsible for a string of unexplained, cold-blooded killings of activists, despite official denials of the existence of Latin American-style 'death squads'.

Economic sanctions applied include tax raids, seizing the goods of those who either refuse to pay taxes or fall behind, and the issuance of special identity cards to all Gazan adults for work in Israel. The identity card scheme was the latest in a string of measures ultimately aimed at reasserting the authorities' control through the computerisation of all information on Gazans. Extension of the scheme to the West Bank was held up because of the much greater difficulty in policing the West Bank's many crossing points into Israel, but may still be enforced.

In June 1989, in a sober assessment of the task before him, Rabin told the Knesset (Israel's parliament) that the intifada was 'perhaps the most complex confrontation in Israel's history. Israel faced a protracted confrontation with most of the Palestinian population, he said, in which there were no magic cures. Nonetheless, he was determined to stand firm against Palestinian and international 'pressure tactics'.

COUNTRY PROFILE

Political system: Parliamentary system; executive power rests with the government headed by the prime minister. (Presidential role is largely ceremonial). Parliament (Knesset) is unicameral, elected (universal suffrage over age 18) by proportional representation.

Main cities/towns: Jerusalem (population 468,000 in 1987), Tel Aviv-Jaffa (322,000), Haifa (223,000), Bnei Brak (141,000), Holon (141,000), Bat-Yam (132,000), Petach-Tikva (131,000), Rishon LeZion (116,000), Ramat Gan (115,000), Beersheba (115,000).

Language: The official languages are Hebrew and Arabic; English is widely understood.

Media
Dailies: There are 21 daily newspapers, of which the best known are *Davar* (circulation 39,000), *Ha'aretz* (55,000), *Jerusalem Post* (in English) (60,000), *Yediot Aharonoth* (300,000), *Ma'ariv* and *Hadashot*.
Business: Israel Economist (monthly – in English) and *Globs* (weekly) are leading periodicals.
Radio: Comprehensive service in many languages from state-run Voice of Israel. Arguy Radio has a wide listenership in Hebrew. There were some 1,300,000 sets in 1988.
Television: Colour service in Hebrew and Arabic, without advertising, transmitted by Israel Broadcasting Authority. Educational service by Instructional Television Trust. Second channel was running test transmissions in 1989. In 1988, there were some 990,000 sets in use. Programmes can be received from neighbouring countries.
Advertising: Besides radio and press, there is cinema screen advertising, as well as limited poster and increasing illuminated sign facilities.

Domestic economy: Up to 1972 the economy's rate of growth was impressive, averaging 10 per cent per annum, despite persistent inflationary pressures and weak balance of payments. In 1973 a slowdown in growth ensued, due to high oil prices and mounting balance of payments deficits, the massive burden of defence spending (23 per cent of GDP in 1983) and the system of indexed compensation of wage-earners and asset-owners. In the early 1980s the economy entered a period of hyperinflation which peaked at nearly 500 per cent p.a. in 1984. However, in mid-1985 the coalition government achieved considerable success in reducing wage and price increases. Inflation has subsequently been kept under 20 per cent, standing in mid-1988 at 16 per cent. The state participates in all industries and services except agriculture but has promised to reduce its role. In addition to the private sector, there is a significant co-operative sector, involved in industry (Koor Industries – 12 per cent of manufacturing output), banking (Bank Hapoalim, biggest bank), agriculture and construction. There is a considerable element of joint venture between firms in different sectors.

Employment: In 1987 the public service employed 27 per cent of the labour force; other services, 28%; industry, 23%; construction, 7.7%; agriculture, 5.8%; transport, 6.1%. Unemployment is moderate at around six per cent. Real incomes, which fell steeply in 1984-85 as a result of austerity measures, subsequently recovered. The people are well educated – 65 per cent have at least eight years of schooling and 14 per cent are university-trained, with a bias to science and engineering. 1.1m employees out of the total 1.5m labour force (1988) are unionised.

External trade: Israel has a free trade agreement with the USA, allowing unrestricted, tariff-free movement of products and services by the mid 1990s.
Exports: Main exports are machinery and electronics (26.7 per cent of 1987 total), diamonds (22.7%), chemicals (11.4%), textiles and clothing (6.5%), food and drink (4.8%), minerals (2.8%), citrus fruit (2.3%) and rubber and plastics (2.4%). Main destinations are USA (33.9% of 1986 total), UK (7.2%), German FR (5.2%), France (4.4%), Netherlands (4.3%), Italy (3.5%), Belgium/Luxembourg (3.8%), Japan (4.5%) and Hong Kong (3.8%).
Imports: Main import requirements are fuels and lubricants (10.4 per cent of 1987 total), capital goods (18.9%), uncut diamonds (17.2%), other manufacturing requirements (45.6%), foodstuffs (3.2%), and consumer durable goods.

Agriculture: Contributes 7.5 per cent of GDP. Israel is largely self-sufficient in food – importing some cereals, beet sugar and animal feeds. Farms are relatively small but mostly part of larger co-operatives – kibbutzim (of which there were 269 by 1986) or moshavim (co-operative smallholder villages – 458 by 1986) – sharing machinery etc. The major crops are fruits (30% of total production in 1987), vegetables (14%) and livestock (42%). Production of crops with high added-value has risen during the 1980s, e.g. melons, bananas, avocados, celery, tomatoes, flowers, groundnuts. Of the total cultivated land area (4,400 sq km), over half is under irrigation. A major desalination plant is under construction at Ashdod, near Tel Aviv.

Industry: The contribution of manufacturing industry (with mining) to GDP has been broadly constant at 23 per cent. The traditional industries are textiles, clothing and leather goods (nine per cent of the total in 1987), foods (11 per cent), diamond polishing and trading, while the growth sectors are the capital intensive, science-based industries, such as aircraft (executive jets and fighters), electronics, biotechnology, agricultural technology (22 per cent), chemicals and mining (17 per cent). A wide range of goods are made or assembled for the domestic market, including cars, commercial vehicles, electrical goods, paper and paper products. Other important economic sectors are construction, transport and tourism (1.1m visitors in 1988 – 65 per cent from Europe; 20 per cent from USA and Canada) – worth some $1.4bn p.a.

Minerals and mining: Most activity in Negev Desert and Dead Sea area. Main resources: potash, bromine, magnesium and phosphates. Israel is world's largest exporter of bromine (60,000 tonnes p.a.) and bromine derivatives, and produces one sixth of the world output. Potash is extracted from the Dead Sea at Sodom (2.2m tonnes p.a.). Phosphates are mined at Oron (2m tonnes p.a.).

Hydrocarbons: After withdrawal from Sinai in 1982, there are as yet no significant indigenous sources of energy. Practically all oil is imported (7.3m tonnes p.a.) from Egypt, Mexico, Norway and spot markets. Oil exploration – onshore and, more recently, in the Mediterranean (with drilling over 350 wells) – has not produced any major finds. There are oil refineries at Haifa and Ashdod. Some natural gas is piped from the Negev Desert to industrial centres.

Energy: Deposits of oil shale and lignite have been discovered, and an experimental steam boiler, fuelled by oil shale, was built to supply chemical plants near Arad. In further efforts to reduce dependence on oil, greater use is being made of coal (imported) to generate electricity. Plans for nuclear power are under debate. Solar power is widely used for domestic hot water heating, but not for the commercial generation of electricity.

Stock market: Banks act as brokers. Total value of all shares quoted on the stock market was $6.3bn at the beginning of 1985, while it had been $17.5 bn before the price crash of October 1983.

Membership of international organisations: CCC, FAO, GATT, IAEA, ICAC, ICAO, IDA, IDB, IFAD, IFC, ILO, IMF, IMO, IOOC, INTELSAT, INTERPOL, IPU, ITU, IWC, OAS, UN, UNESCO, UPU, WHO, WIPO, WMO, WSG, WTO, World Bank.

For Israelis perhaps the only morale-boosting moment during 1989 came in late July when a special forces unit carried out a daring raid into enemy territory – southern Lebanon – to capture a leading Shi'a Muslim cleric, Sheikh Abdul Karim Obeid. Regarded as a spiritual leader of the Hizbullah armed faction in Lebanon, Sheikh Obeid was believed responsible for the kidnapping of two Israeli soldiers, and his own abduction was clearly intended to force a swap which might also have involved some or all of the Western hostages being held in Lebanon.

As it turned out, the plan backfired badly, earning Israel unusually strong public criticism from the USA and other Western nations, sparking off a flurry of top-level international activity to defuse a threat to kill the Western hostages in revenge, and failing to produce the swap Israel was seeking. From Israel's point of view, the best that could be said for the failed gamble was that at least it had another card to play in the protracted, behind-the-scenes negotiations which had preceded the military action and were likely to resume soon thereafter.

The resignation of Major-General Avram Mitzna as head of Israel's Central Command in May 1989, and his replacement by Major-General Yitzhak Mordechai, a tough soldier previously in charge of Southern Command – responsible for the Gaza Strip – symbolized the weariness within much of the army itself over its thankless job in trying to put down the uprising. As a liberal-minded kibbutz member, Major-General Mitzna found himself frequently at odds with the 70,000, mostly militant, Jewish settlers of the West Bank, who would often take the law into their own hands with reprisal attacks on their Palestinian neighbours.

Although feelings run high among the settlers that they are not properly protected by the army, few of those implicated in the shooting deaths of Palestinians ever feel the full force of the law. As of mid-1989, out of 26 confirmed cases, only 3 settlers had been brought to trial and these only on lesser charges. Indeed, civil rights lawyers agree that a serious erosion of Israel's much-vaunted rule of law has been one of the more serious consequences of the uprising, as seen in the very different treatment meted out to Jews and Palestinians for similar offences.

Intifada's economic cost

The intifada's economic cost has been equally heavy for Israel. Difficult to calculate exactly, government economists estimated that the uprising knocked at least one percentage point off projected gross national product growth during 1988, and was likely to repeat that figure during 1989. A more gloomy assessment came from Gad Ya'acobi, a senior labour minister, who put the cost at NIS1.5bn ($750m) in 1989 and forecast that growth would be reduced by 3–4 per cent.

The defence budget, declining in recent years as a percentage of GDP, was one area which was increased to meet the additional costs imposed by the intifada. Having fallen to 8.6 per cent of GDP in 1988 – the lowest figure in two decades – down from 13.2 per cent, a series of supplements were approved during 1989. Part of this sum was earmarked by the defence industries sector, whose lean times showed no sign of easing.

The Central Bank noted in June 1989 that the intifada had resulted in the loss of $650m in potential export earnings the previous year. Worst hit, predictably, were exports to the occupied territories themselves – Israel's largest 'foreign' market – which dropped by 40 per cent. As belts tightened among Palestinians, much of this business may have been lost forever.

Although regular strike days punctuate the uprising calendar, preventing day labourers from travelling to Israel to work, one apparent paradox of the intifada mad explicable by the worsening economic conditions within the West Bank and Gaza Strip, was that the total number of Palestinian workers involved in the Israeli economy rose significantly during 1987–89, rather than declined as might have been expected. At a time when Jewish unemployment – always a sensitive issue in Israel – has risen to levels which would have been intolerable for any previous government, the participation of over 115,000 Palestinians from the occupied territories in the workforce, some 17 per cent of the total, has complicated race relations still further.

While 1988 was one of the poorest for the Israeli economy, hopes that 1989 would be significantly better were unfulfilled. The government set a modest target of 2 per cent for GDP growth, but 1.0–1.5 per cent appeared more likely, marking the second year running that the GDP had declined in real terms. In stark contrast to the crisis years of 1983 and 1984, the foreign exchange reserves were so healthy during 1989 – partly due to the $3bn annual US grant aid – that the Central Bank was able to repay its foreign debts rather than rolling them over. By June 1989 the foreign debt was down to $24.1bn, compared with $25.2bn a year earlier.

On the other hand, the government's domestic deficit widened much faster than was expected during 1989, attributed largely to lower tax receipts resulting from the domestic recession and cuts in import duties on goods originating in the USA and European Community. Against a treasury forecast of a deficit for the full year of NIS1.7bn, in the first half alone the revenue gap reached NIS2.4bn.

In preaching the need for efficiency at a time when unemployment was soaring, Peres could not be faulted by critics of the Israeli Government's management of the economy. His courage was also demonstrated when he grasped one of the thorniest of nettles: the restructuring of the ownership of the banking sector required after a bank shares scandal provoked the Tel Aviv stock market collapse of 1983.

Fighting tooth-and-nail to resist divesting their control, the original shareholders of the big four banks, including the Labour-affiliated Bank Hapoalim, managed initially to block Peres' plans. These centred on the equalisation of the values of all bank shares, robbing their founders of their voting powers. In the long run, though, the banks' reorganisation appeared inevitable, in the course of which foreign investment would be welcomed by the government.

Another cartel-breaking plan by Labour Minister Moshe Shahal, determined to introduce greater efficiency and competition to the fuel distribution business, also ran into rough water. Faced with the political need to sell off the state-owned Paz oil company – kingpin of the government's much-derided privatisation programme – Shahal was forced to abandon much of his original plans.

Economic outlook

After a difficult two years, the prospects for the economy in 1990 are marginally better, based on the expectation that the corporate shake-out affecting much of Israeli industry has passed its worst and that the banking sector will also rebound from its losses of the recent past. Two encouraging developments for businessmen during the first half of 1989 were a decline in real wages, following several successive years of substantial increases, and two devaluations of the shekel, totalling 15 per cent. Corporate tax was also cut modestly by Peres, from 45 to 43 per cent, together with a promise to reduce it further, to 40 per cent, by 1992; while Israel's traditionally high interest rates staged a welcome decline, engineered by the treasury.

The benefits of these measures, however, were slow to appear, doing little to alleviate the deepening recession in which industry found itself. According to Dov Lautman, president of the manufacturers' association, industrial output fell by 2.5 per cent in the first five months of the year, after declining by 3.5 per cent in 1988. Investment in business also slumped in 1988 to 5 per cent of output, its lowest in decades.

Agriculture has similarly been in the doldrums, as the trade statistics disclose, largely because of bad weather. Total exports came to $578m in 1988, down by 5.8 per cent on the previous year; but this concealed an even bigger decline in the important citrus sector. Forecasts for 1989 were for a 20 per cent recovery in the export of fresh produce, led by avocadoes and tomatoes. Dealing with the exceptionally heavy debt problems of the kibbutzim and moshavim (farm co-operatives) has proven to be a major headache for the banks and the treasury, one that will take many years to resolve fully.

The most celebrated corporate saga of recent times, one which carried far-reaching implications for the rest of Israeli industry, was the six-month-long battle at Koor, the trades union-owned conglomerate responsible for over 10 per cent of industrial output, to avoid compulsory liquidation at the hands of its overseas creditors. Koor's problems of overmanning, heavy debt, sluggish orders and politically-inspired investment decisions reflected those of Hevrat Ha'Ovdim, its parent company and the Histadrut labour federation's economic arm, as whole.

In the end, Koor was able to avoid bankruptcy, thanks in part to the government's assistance, but at the cost of shedding many of its most important components. Aware of the challenge to its own preeminent place in Israeli national life, involving in one way or another some 90 per cent of the adult population, observers looked to the outcome of the organisation's November 1989 elections for clues as to its future behaviour with unusual interest.

BUSINESS GUIDE

Time: GMT +2 hrs; (GMT +3 Apr-Sep).

Climate: Summer (Apr-Oct): temperature range 23°C to 31°C; humidity 70-75 per cent. Winter (Nov-Mar): temperatures 15°C to 20°C. Rainfall: Nov-Mar in periodic downpours.

Entry requirements
Passport: Required by all except certain document holders and seamen.
Visa: Not required by nationals of UK, France, Japan and numerous other countries. Visas on arrival for nationals of USA and certain other countries.
Health: No mandatory vaccinations.
Currency: No limit to the amount of foreign or Israeli currency taken in by visitors. Up to $100 worth of Israeli currency may be converted to other currencies on departure without proof of foreign exchange transactions during stay. There are temporary restrictions for foreign citizens of Palestinian origin. Most foreign exchange transactions are carried out by banks but licensed money-changers operate openly in Arab East Jerusalem, offering a better rate.

Representation overseas: Abidjan, Accra, Ankara, Asuncion, Athens, Bangkok, Berne, Bogota, Bombay, Bonn, Brasilia, Brussels, Bucharest, Budapest, Buenos Aires, Cairo, Canberra, Caracas, Colombo, Copenhagen, Guatemala City, The Hague, Helsinki, Hong Kong, Kathmandu, Kingston, Kinshasa, La Paz, Lilongwe, Lima, Lisbon, Lomé, London, Madrid, Manila, Mbabane, Mexico City, Monrovia, Montevideo, Nairobi, Nicosia, Oslo, Ottawa, Panama City, Paris, Port-au-Prince, Pretoria, Quito, Rangoon, Rome,

BUSINESS GUIDE

San José, San Salvador, Santiago, Santo Domingo, Singapore City, Stockholm, Suva, Tegucigalpa, Tokyo, Vienna, Warsaw, Washington, Wellington, Yaoundé.

Air access
National airline: El Al Israel.
Other airlines: Adria Airways, Air France, Air Sinai, Alitalia, Austrian, British Airways, Cyprus Airways, Iberia, KLM, LOT, Lufthansa, MALÉV, Olympic Airways, Pan Am, Sabena, SAS, South African Airways, Swissair, Tarom, Tower Air, TWA.
Tax: Airport tax is payable on departure from all airports ($11, or shekel equivalent, from Ben Gurion).
International airport: Tel Aviv-Ben Gurion International (Code: TLV), 19 km from city; duty-free shop, bar, buffet, restaurant, hotel reservations, post office, shops, car hire.

Surface access: Normally practicable only by sea. But tourists from Jordan can cross into the West Bank, and thus Israel as well, after obtaining a 'bridge pass' from the Interior Ministry in Amman. Sailings between Larnaca (Cyprus) and Haifa three times a week. By road there are regular buses from Cairo, via Rafa (Gaza Strip) and by taxi from Amman via the Allenby Bridge (permit needed from Jordanian Interior Ministry).
Main ports: Haifa, Ashdod, and (on the Red Sea) Eilat.

Hotels: Plenty of hotels in business and tourist centres. Service charge of 15 per cent usually added to bill. Settlement of bills in foreign currency will avoid payment of local taxes. Many hotels quote prices in US dollars.

Electricity supply: 220 V AC, 50 cycles.

Car hire: National or international licence required. Seat belts are compulsory for drivers and front-seat passengers.

Taxis: Taxis are metered but expensive. For inter-city travel (including Saturdays), share taxis or 'sherut' run between central points in main cities and are not expensive. Some sherut companies, including Arieh and Aviv, will accept advance bookings.

National transport
Air: Arkia operate daily services from Tel Aviv to Jerusalem, Haifa, Eilat, etc.
Road: 5,400 km paved roads in 1988. Main roads are good. Maximum speed 90 kph. 590,000 private cars in 1988.
Buses: Connect all centres of population but can be crowded. Do not operate from sunset on Friday to sunset on Saturday. (See also 'taxis'.)
Rail: Services between Tel Aviv and Haifa (hourly). Seats can be reserved. No services Fri evening or Sat. 572 km of main line (1988).

Public holidays
Variable dates: Purim (Mar), Passover – Pessah (Apr – seven half-days), Independence Day (Apr-May), Pentecost – Shavuot (May-Jun), Tisha B'ab (Jul), New Year – Rosh Hashanah (Sep), Day of Atonement – Yom Kippur (Sep-Oct), Tabernacles – Sukkot (Oct – seven half-days), Simhat Torah (Oct).

Working hours: Offices and shops closed Fri pm and Sat. Most government offices closed all day Fri.
Business: (Jun-Oct) 0730 – 1430; (Nov-May) 0800 – 1300, 1500 – 1800.
Government: (Jun-Oct) 0730 – 1430; (Nov-May) 0730 – 1300, 1345 – 1600.
Banking: 0830 – 1230, 1600 – 1700 (closed Wed, Fri pm, Sat).
Shops: (Jun-Oct) 0800 – 1300, 1600 – 1800; (Nov-May) 0800 – 1300, 1600 – 1900.

Social customs: People are hospitable and informal and culturally diverse. Jewish traditions and customs are generally adhered to, for example kosher eating and Sabbath (Saturday) observance.

Telecommunications
Telephone: Dialling code for Israel: IDD access code + 972, followed by area code and subscriber's number. The country is divided into 11 zones, each with its own area code, e.g. Tel Aviv 3, Jerusalem 2, Haifa 4. Automatic dialling between main towns, but lines can be overloaded during business hours. Telephone directory in English as well as Hebrew. Public pay phones are widely available. 1.95m* telephones in use in 1988.
Telex: Available in main hotels and post offices.

Banking
Central bank: Bank of Israel, Jerusalem.
Main centre: Tel Aviv.

Trade fairs: The main centres for trade fairs and exhibitions are the Israel Convention Centre and the Hilton Hotel in Tel Aviv.

Weights and measures: Metric system, but area is usually measured in *dunam* (=1,000 sq metres).

Representation in country
Embassies in Tel Aviv: Argentina, Australia, Austria, Belgium, Bolivia, Brazil, Burma, Canada, Chile, Colombia, Côte d'Ivoire, Denmark, Dominican Republic, Ecuador, Egypt, El Salvador, Finland, France, German FR, Greece, Guatemala, Haiti, Honduras, Italy, Jamaica, Japan, Liberia, Mexico, Netherlands, Norway, Panama, Peru, Philippines, Romania, South Africa, Sweden, Switzerland, Turkey, UK, Uruguay, USA, Venezuela, Zaïre. *Embassies in Jerusalem:* Costa Rica, El Salvador.

BUSINESS DIRECTORY

Business advice and information

Bank of Israel, PO Box 780, Kiryat Ben-Gurion, Jerusalem 91007 (tel: 552211; tx: 25213/6; fax: 528805).

Central Bureau of Statistics, Hakirya, Romema, Jerusalem (tel: 211211).

Federation of Israeli Chambers of Commerce, PO Box 501, 84 Hahashmonaim Street, Tel Aviv (tel: 288224; tx: 33484).

Government Tourist Information Centres: 7 Mendele Street, Tel Aviv (tel: 223266); 18 Herzl Street, Haifa (tel: 666521); 60 King George Avenue, Jerusalem (tel: 241281).

Israel-British Chamber of Commerce, PO Box 3540, Tel Aviv (tel: 259732; tx: 32111 AMLTIBCC).

Israel Convention Bureau (ISCOB), Israel Tourism Administration, PO Box 1018, Jerusalem.

Israel Trade Fairs Centre/Israel Convention Centre, PO Box 21075, 61210 Tel Aviv (tel: 422422; tx: 361490).

Manufacturers' Association of Israel, Industry House, PO Box 50022, 29 Hamered Street, Tel Aviv (tel: 650121; tx: 342651).
Ministry of Industry and Trade, Foreign Trade Division, Palace Buildings, PO Box 299, 32 Agron Boulevard, Jerusalem (tel: 210111).

Hotels: Jerusalem

American Colony, Nablus Road, P.O. Box: 19215 (tel: 282421; tx: 25362).

Ariel, 31 Hebron Road (tel: 719222; tx: 26243).

Caesar, 208 Jaffa Road (tel: 382156; tx: 26595).

Capitol, Salah A'Din Street (tel: 282561; tx: 25399).

Central, 6 Pines Street, P.O. Box: 1351 (tel: 223111; tx: 26157).

Commodore, Mount of Olives, P.O. Box: 19715 (tel: 284845; tx: 25343).

Diplomat, 6 Etzion Street, P.O. Box: 2494 (tel: 777777; tx: 25335).

Eilon Tower, 34 Ben Yehuda Street (tel: 233281; tx: 26472).

Eyal, 21 Shamai Street (tel: 234161; tx: 26150).

Gloria, Jaffa Gate, P.O. Box: 14070 (tel: 2824311/2; tx: 25329).

Hilton, Givat Ram (tel: 536151; tx: 26155).

Holyland, Bayit Vegan, P.O. Box: 2461 (tel: 630201; tx: 25336; fax: 630205).

Hyatt Regency, 32 Lehi Street, 97665 (tel: 821333; tx: 26174; fax: 815947).

Inter-Continental, Mount of Olives, P.O. Box: 19585 (tel: 282551; tx: 25285; fax: 285384).

Jerusalem Tower, 23 Hillel Street, P.O. Box: 2656 (tel: 222161; tx: 25345).

King David, 23 King David Street (tel: 221111; tx: 25228; fax: 232-303).

King Solomon, 32 King David Street (tel: 241433; tx: 26434).

Kings, 60 King George Street, P.O. Box: 7581 (tel: 247133; tx: 25227; fax: 249830).

Knesset Tower, 4 Wolfson Street (tel: 531111; tx: 26592).

Laromme, 3 Jabotinsky Street (tel: 697777; tx: 26379; fax: 697268).

Lawrence, 18 Salah A'Din Street (tel: 282585; tx: 25379).

Moriah, 39 Keren Hayesod Street (tel: 232232; tx: 25383; fax: 232411).

Mount Scopus, Sheikh Jarrah Quarter (tel: 828192; tx: 26590).

Orgil, 18 Hillel Street (tel: 22461/3).

Panorama, Hill of Gethsemane, P.O. Box: 19768 (tel: 284886).

Pilgrims Palace, Sultan Suleiman Street, Damascus Gate (tel: 284831; tx: 26268).

Plaza Sheraton, 47 King George Street, P.O. Box: 7686 (tel: 228133; tx: 26160; fax: 231667).

President, 3 Ahad Ha'am Street (tel: 631273; tx: 25272).

Ram, 234 Jaffa Road, P.O. Box: 13116 (tel: 535231; tx: 26161).

Ramada, 6 Wolfson Street, P.O. Box: 3369 (tel: 528111; tx: 25252).

Ritz, 8 Ibm Khaldoun Street, P.O. Box: 19130 (tel: 284853/4).

Saint George International, Salah Eddin Street (tel: 282571/5; tx: 26425).

Shalom Jerusalem, Beyit Vegan, P.O. Box: 16060 (tel: 423111; tx: 25341).

Sonesta, 2 Wolfson Street (tel: 528221; tx: 26411; fax: 3240047).

Tirat Bat-Sheva, 42 King George Street, P.O. Box: 7780 (tel: 232121; tx: 28493).

Victoria, 8 Al Masoudi Street, P.O. Box: 19066 (tel: 286220; tx: 26268).

Windmill, 3 Mendele Street (tel: 663111; tx: 26536).

Zohar, 47 Lieb Jaffa Street, P.O. Box: 8261 (tel: 717557; tx: 26409).

Travel information

Automobile and Touring Club of Israel (MEMSI), 19 Petach Tikva, PO Box 36144, Tel Aviv 66183 (tel: 622961; tx: 341118 bxtv).

Ministry of Tourism, PO Box 1018, 91000 Jerusalem (tel: 237311; tx: 26115 tour); International Conventions Division, PO Box 1018, 91009 Jerusalem (tel: 237343; tx: 25218).

Italy

David Lane

The lengthy crisis that followed the collapse of Ciriaco De Mita's five-party coalition in summer 1989 seemed to confirm that Italy's economy runs smoothly in the absence of government; business as usual in spite of the semi-paralysis which afflicts the state machine when governments fall. But with deadlines approaching in 1990 and 1992, industry and finance were increasingly concerned that crucial issues would not be tackled in time. Headed by veteran Giulio Andreotti, holding the prime ministership for the sixth time, the government which succeeded De Mita's faced an over-heating economy, rising inflation and an increasing trade deficit. And the long-standing problem of the public sector borrowing requirement urgently needed curing. Elderly former Bank of Italy governor Guido Carli was appointed to the treasury and the difficult task of finding and administering the economic medicine. Yet even as Andreotti was being accorded his vote of confidence, doubts were being cast over the prospects of the government's survival beyond 1990.

Departure of De Mita

When the world's leaders gathered in Paris in July 1989 for the bicentenary celebrations of the French Revolution, Italy's two representatives were labelled an odd couple. So they seemed, with Prime Minister Ciriaco De Mita, limping on as Rome's lame duck leader, and Foreign Minister Giulio Andreotti, the man named to succeed him, apparently taking care to avoid each other.

The trip to France was the culmination of a bad period for De Mita. He started 1989 as both Italian prime minister and secretary of the Christian Democrat Party, Italy's largest political force. But the party congress in February deprived him of leadership, and declarations made at their congress in May by Socialist Party leaders, partners in the five-party coalition he was heading, left him no alternative to resigning the prime ministership.

One of the factors contributing to De Mita's ire was awareness that he had been lured into an undignified charade of trying to form a new government. Only after seven weeks of 'crisis', a week before the Paris journey, did he acknowledge that his attempts had failed. It was then that President Francesco Cossiga asked veteran Christian Democrat Andreotti to take on the task. De Mita was thus required to act a role in the French capital from which most vestiges of authority had been stripped. Signs of bitterness were understandable.

The end of the forty-eighth government since the Italian Republic was established after the Second World War was flagged by Socialist deputy secretary Claudio Martelli when he told his party's congress that even the driver gets off when the end of the line is reached. But though the message from the Socialists, second largest group in the coalition after the Christian Democrats, seemed clear, his party insisted that De Mita was its choice for a further term in the prime minister's offices at Palazzo Chigi.

With the benefit of hindsight, De Mita would have done best to have tendered his resignation in February. In appointing Arnaldo Forlani as party secretary, the Christian Democrats' congress showed that De Mita was being pushed to the sidelines. But having been deprived of the considerable power he had wielded for seven years as party boss, he decided to hang on as prime minister.

De Mita's downfall could be scored a success for the Socialists, and particularly their leader Bettino Craxi. By insisting that De Mita take prime ministerial responsibilty when Giovanni Goria's government fell in March 1988, the Socialists ensnared the Christian Democrat leader in his party's own rules. These say that the party secretaryship and the prime ministership are incompatible appointments. Once trapped at Palazzo Chigi, baited by the reasoning that a strong government headed by a party leader was needed, De Mita was then easy prey to factions within his own party which considered that he had been pulling the strings of power for too long.

In thinking that he could manage to stay in both saddles De Mita made a serious misjudgement. His own faction was too weak and other Christian Democrat factions had not been sufficiently rewarded during De Mita's secretaryship to give support when the struggle at the Rome congress took place. Party bosses call the shots in Italy. So once ousted from the Christian Democrats' top job, De Mita's days as prime minister were numbered.

European parliamentary elections

Only an exceptionally good result in the elections for the European Parliament in June 1989 could have prolonged De Mita's stay at Palazzo Chigi. This did not happen. Though the Christian Democrats almost maintained the share of the vote won in 1984's European elections (32.9 per cent against 33.0 per cent), the result was a decline compared to the 34.3 per cent share won in 1987's parliamentary elections.

If June 1989's elections were a delusion for De Mita and the Christian Democrats, they were little better for Craxi's party. The Socialists hoped that the outcome of the elections held during the 'crisis' caused by De Mita's resignation would reinforce their hand. There were expectations that Craxi's party would win a significantly larger share of the vote to take it closer to the Communist Party's second ranking. But the ballot box was a disappointment; the Socialist steamroller appeared to have lost momentum. From 11.2 per cent of the vote won in 1984's European elections, it leapt to 14.3 per cent in the parliamentary elections held in 1987. The gain this time was modest. Against hopes of around 17 per cent, Craxi's party moved only one half a point ahead to 14.8 per cent.

Elections which produced some surprises among Italy's minor political forces – with the Republican, Liberal and Social democrat lay parties of the coalition and the neo-fascist Movimento Sociale Italiano losing ground while the 'Greens' made strides – gave consolation to the Communist Party.

For five years the Communists had been a party in retreat. In 1984's European elections they achieved their best-ever result by taking 33.3 per cent of the vote to become Italy's biggest political party, a success which was partly attributable to a sympathy vote for party secretary Enrico Berlinguer who died from a stroke while campaigning. But the party's fortunes then declined and in 1987 it won a 26.6 per cent share, meaning that its advantage over the Socialists had been cut from 22.1 to 12.3 per cent. The question in 1989 was whether Italy's Communists could avoid being eclipsed like their French counterparts. With international attention focused on the rejection of communism in Hungary and Poland and the suppression of liberty in China, the risk seemed real.

By improving its share of the vote achieved in 1987 by one point, to 27.6 per cent, the Communist Party under its new secretary Achille Occhetto showed that it was not a spent force in Italian politics. The party has long been independent from Moscow, and Occhetto has emphasised a political line which appears closer to Socialism and Social Democracy than to marxist-leninist absolutism. In the elections for the European Parliament, Italy's Communist Party signalled to the Socialists that scope was limited for further gains on the left.

The Socialists' failure to achieve a quantum leap in June 1989 was widely thought to rule out the possibility of a left-wing coalition government, the party needing greater weight against the Communists for such a partnership to be feasible. Until that equilibrium is reached Craxi's party seems tied to alliances with the Christian Democrats, a role in which it has shown itself to be astute at a spoiling game.

A 'historic compromise' coalition joining the conservative clerical Christian Democrats to the Communists remains improbable. And as long as the Christian Democrats continue to hang on tenaciously to the governing role which they have held since the republic was established over forty years ago, then Forlani like De Mita before him will be forced to come to terms with the Socialists. In destroying De Mita, whose strength appeared to rival his own, Craxi demonstrated ruthlessness coupled to manipulatory skills in placing the Christian Democrat leader at the mercy of enemies in his own party. Machiavelli would have approved.

The mediaeval Florentine would probably also have raised his cap to Andreotti, who continued to amaze observers with his unsinkability. Seventy years old, with four decades in government and often linked to murky episodes and dark figures, Andreotti is the arch

KEY FACTS

Official title: The Italian Republic

Head of state: President Francesco Cossiga

Head of government: Prime Minister Giulio Andreotti

Ruling party: Coalition of Christian Democratic, Socialist, Social Democratic, Republican and Liberal Parties

Capital: Rome

Official Languages: Italian (and German in Alto Adige)

Currency: Lira

Exchange rate: Lire 1,304 per $ (Dec 1989)

Area: 301,277 sq km

Population: 57.4m (1988)

GDP per capita: $14,430 (1988)

GDP real growth: 3.9% (1988)

Labour force: 22m (1987)

Unemployment: 12.0% (1988)

Inflation: 5.0% (1988)

Trade balance: –lire 12,634bn (1988)

Foreign debt: $60bn (1987)

survivor. But in forming his sixth government, Andreotti could draw strength from the fact that his powers of survival would be spared the ambushes of 'snipers' from within the coalition. One of the most significant measures introduced by the De Mita administration was the abolition of the secret vote on all financial and economic legislation. Since October 1988 Italian parliamentarians vote in secret on matters of conscience and constitution only.

Economic situation

Several major issues remained unresolved while the politicians were engaged in deciding who should form the government to succeed De Mita and under what conditions it should operate. Foremost among the problems facing successive administrations during the 1980s was the public sector borrowing requirement (PSBR). Virtually no major gathering of businessmen, bankers or academics ignored the PSBR. Annual meetings, symposia, round tables and the like dedicated countless hours to debating the problem and how it should be solved.

In 1988 the deficit of Italy's state sector amounted to L124,600bn, equivalent to 11.6 per cent of gross domestic product (the same percentage of GDP as in 1987). With state expenditure continuing to exceed income, a milestone of indebtedness was passed in autumn 1988 when overall public sector debt clicked past the million billion lire mark. At year-end 1988, borrowing by the state and public bodies totalled L1,037,200bn (equivalent to 96.1 per cent of GDP) and put another milestone, public debt equal to GDP, within easy reach.

The leading voice expressing concern over the PSBR belonged to the Bank of Italy's governor, Carlo Azeglio Ciampi. When he addressed the Chamber of Deputies' budget, treasury and planning commission at the beginning of 1989 he noted that Italy's PSBR was proportionately five times greater than the average of the other six principal industrialised countries. 'This puts a brake on growth, accentuates instability, threatens inflation and distances Italy from Europe. The absolute priority of this problem above all others is now recognised by everyone,' said Ciampi.

A few months later, giving his address to the central

bank's annual meting at the end of May, Ciampi returned to the theme. He drew attention to the valuable opportunity which had been missed to straighten public sector accounts while the general economic situation was favourable.

Sergio Pininfarina, chairman of the industrialists' confederation Confindustria, spoke at its annual meeting about political instability and indecisiveness over public sector accounts. 'Criticisms by the Bank of Italy's governor, calls from authoritative international bodies and the suggestions of expert committees working for the government have not been enough to give strength and continuity of action,' he said. Pininfarina noted that measures adopted for reducing the public sector deficit had been based mainly on new revenues, often of a once-off kind, and that little had been done to check the growth of spending.

But piecemeal economic measures and the saga of budgeting over recent years do not encourage expectations of improvement. In September 1987 the PSBR for 1988 was set at L109,500bn, a figure lifted to L114,450bn in May 1988 to take account of 1987's outturn and of measures planned or already introduced. In September 1988 the PSBR forecast was raised to L118,000bn, the eventual result being L6,600bn higher.

When the Finance Act set L117,350bn as 1989's PSBR, many observers considered that this aim was unachievable. Indeed in February 1989 Ciampi drew the politicians' attention to the continuing expansion of public sector expenditure and to the inadequacy of measures to keep the PSBR within the planned figure. Early in 1989 the International Monetary Fund said that it expected Italy's PSBR for the year to exceed L130,000bn.

KEY INDICATORS

	Unit	1984	1985	1986	1987	1988
Population	'000	56,980	57,080	57,236	57,400	57,440
Gross domestic product (GDP)	lire bn	727,230	812,750	896,320	979,680	1,078,860
GDP per capita	$	7,214	7,457	10,505	13,168	18,782
GDP real growth	%	5.4	2.9	2.6	3.0	3.9
Consumer prices	1985=100	91.6	100.0	105.9	110.9	116.5
Share prices	1985=100	59.8	100.0	232.8	224.6	185.1
Interest rate (official disc.)	%	16.5	15.0	12.0	12.0	12.5
Money supply	lire bn	506,159	561,748	–	–	–
Wages (hourly earnings)	1985=100	90.0	100.0	104.8	111.6	118.4
Unemployment	%	10.0	10.3	11.1	11.9	12.0
Industrial production	1980=100	95	97	99	103	108
Manufacturing	1980=100	95	96	99	103	108
Exports	lire bn	129,015	149,698	145,320	144,764	167,430
Machinery/equipment	%	36.1	36.2	39.8	–	–
Textiles/clothing	%	18.0	18.1	12.9	–	–
Chemicals	%	8.0	8.2	11.6	–	–
Imports	lire bn	149,470	167,097	148,300	161,736	180,064
Crude petroleum	%	15.6	15.2	8.4	7.1	5.1
Machinery/equipment	%	20.8	21.0	28.4	–	–
Foodstuffs	%	11.9	12.0	9.5	–	–
Chemicals	%	10.5	10.1	15.8	–	–
Trade balance	lire bn	–20,365	–17,399	–2,980	–16,972	–12,634
Current account	$m	–2,501	–3,546	2,912	–1,663	–5,466
Foreign exchange	$m	20,795	15,595	19,987	30,214	34,715
Exchange rate	lire per $	1,757.0	1,909.4	1,490.8	1,296.1	1,301.6

* estimated figure

Notwithstanding the large deficit to be financed, the Italian authorities encountered no major obstacles to bridging the gap between revenues and expenditure in 1988 and 1989. It was a matter of price. With government paper offering attractive yields, savers were willing to purchase treasury notes, bonds and certificates. Central bank figures show that government securities gave substantial real returns. Against an average rise of 5.0 per cent in the consumer price index in 1988 (4.4 per cent in the previous year), the average interest paid on BOT short-term treasury notes was 11.1 per cent. Savers were offered an average of 10.5 per cent on medium-term fixed rate BTP bonds in 1988.

Italian families had provided much of the finance needed to cover the state sector's deficit. Of total family sector financial assets worth L1,421,800bn at year-end 1988, government securities accounted for 32.1 per cent. Families were showing increasing sophistication in their willingness to switch to savings alternatives offering highest returns. That the banks' share of savings fell sharply during the 1980s was largely attributable to uncompetitive interest rates.

In a financial system where savings instruments like bank deposits, the stock market and mutual funds are unattractive or viewed with suspicion, government paper encounters fewer problems in drawing families' funds. Financing the PSBR was helped by the absence of competition. But it seemed that this relatively easy situation could be changing.

Two factors clouded the horizon. First, the Italian propensity to save was declining. Corrected for inflationary erosion, gross private sector saving averaged 21.6 per cent of national income over the period 1981 to 1988. This was almost one point lower than the average during the 1970s. And the authorities expected the fall to continue, with increasing consumption being stimulated by changing family structure and an ageing population. The effects of this on financing the public sector deficit looked likely to be felt gradually over a long timescale.

Second, the full liberalisation of capital flows within the European Community offered new alternatives. Observers were asking to what extent this threat would materialise in July 1990. Some were convinced that this would be the moment of truth for those who manage Italy's debt-burdened public sector accounts.

Invisible trade

What outflows would result from the liberalisation of capital movements? Since the July 1990 deadline was set, this question occupied the minds of those who managed the Italian economy. They probably drew some confort from recent experiences. Flows into Italy exceeded outflows by nearly L19,000bn in 1988. Italy enjoyed surpluses in direct investment, portfolio investment and loans. Indeed capital flows compensated Italy's worsening current account position in both invisible and merchandise trade.

Invisibles, showing a deficit of L4,286bn in 1988 compared to L650bn in 1987, drew comment from the central bank's governor. 'The deficit position in services is worsening. The surplus earned from tourism diminished by 16 per cent last year and will probably fall further. It is now less than the net outflow due to investment income,' Ciampi said in May 1989.

In tourism Italy faced a battle on two fronts. On one side earnings were static while, on the other, Italians were travelling more widely and more frequently. And with foreign exchange restrictions no longer in force they were spending increasing amounts abroad. In 1988 foreign visitors spent L16,144bn in Italy, about the same as in the previous year. At the same time Italian residents increased their spending abroad by 33 per cent to about L8,000bn. In real terms tourism outflows increased by 29 per cent while revenues slipped by 2.7 per cent.

There was a prophetic note in the Bank of Italy's annual report published at the end of May 1989. 'In the presence of strongly expanding international demand, the real drop in revenues seems due more to competitivity factors other than price; in particular to the scarce improvement in the overall quality of services supplied by the Italian hotel industry, to promotional policies and to environmental care.'

Less than two months later vast quantities of algae, whose presence was blamed on pollution in the Po Valley, invaded resorts in the upper Adriatic, turning the water into an uninviting gelatine. When the news of the algae reached Germany, Austria and Britain, hotels on the Adriatic Riviera from Jesolo to Rimini and Riccione were inundated with cancellations. Spain, Portugal and Greece made gains at Italy's expense in 1988; they profited further as the algae invasion led to more desertions from Italian resorts in 1989.

Merchandise trade

But notwithstanding the negative trend in invisibles, Italy's economic managers were more concerned about the country's merchandise trade in 1988 and 1989. The outturn for 1988 was far from satisfactory with cif-fob customs data revealing a deficit of L12,875bn, against L11,143bn deficit in 1987, in spite of a substantially better energy position. Heavily dependent on imported energy which covered 82 per cent of requirements, Italy enjoyed the benefits of soft oil prices and the weak dollar in 1988. At L15,247bn, its energy bill was the lowest of the decade and more than L3,000bn less than in 1987. But firmer oil prices and a stronger dollar in 1989 contributed to a merchandise deficit which in the first five months had already exceeded 1988s figure.

However, the improvement in Italy's energy trade figures in 1988 was almost offset by a worsening of the deficit in raw materials. Even more alarming for the authorities was the reduction in Italy's surplus on manufactured goods. This surplus fell in 1988, for the fourth successive year. At L19,190bn it was over L2,000bn less than in 1987. Though exports were six per cent higher, Italy lost ground in world markets. This loss of share was attributable mainly to the sectoral composition of global trade, the international cycle being driven by demand for advanced technology capital goods.

'Italian industry remains fundamentally specialised

in traditional sectors,' said Ciampi. Indeed shoes, clothing and textiles continued to be firmly in the vanguard of the country's export efforts. Design flair married to flexible and efficient production has given Italian manufacturers a strong worldwide position. But 1988s figures showed that exports were starting to flag. Increasing competition from low labour cost countries and 1988's weak dollar damaged Italy's footwear, clothing and textiles trade.

Certainly the figures were still excellent in 1988, with L19,002bn surplus from L31,088bn exports and L12,086bn imports. But exports of footwear, clothing and textiles registered an increase of only 2.8 per cent compared to a rise of 12.1 per cent in imports, causing a 2.4 per cent fall in surplus.

And while the surplus in engineering products, another healthy item in Italy's trading accounts, slipped from L14,812bn in 1987 to L13,689bn in 1988, the deficit sectors continued to be headaches for economic managers in Rome. The Italian appetite, coupled to the inability of the country's farmers and food processing companies to produce sufficient quantities of the right products meant that food and agriculture were major red ink items in the merchandise account. In 1988 Italy had deficits of L9,510bn in agricultural products and L8,230bn in food products.

Chemical and pharmaceuticals registered a similar level of deficit, L8,683bn. 'The large deficit in the chemicals sector is due mainly to pharmaceuticals and plastics materials, secondary chemicals in which Italy suffers from a serious technological gap,' explained the Ministry of Foreign Trade in Spring 1989.

Indeed technology was an issue on which attention was increasingly focused. Some observers believed that Italy was slipping behind. Confindustria's deputy chairman Luigi Abete considered that there were important limitations to the process of modernisation which had revolutionised much of Italian manufacturing during the 1980s. 'Modernisation has concentrated on processes and neglected product innovation. We have become good at making automated production systems but have lagged in research and development,' he said.

Abete echoed the concern of Confindustria's chairman Pininfarina, who considered that effort to reduce the research and development gap and state support for innovation was essential. 'It is fundamental for the competitiveness of our industrial system in the international context,' he said. For the industrialists' association competitiveness became a major issue during 1988 and 1989. It identified costs as a crucial factor behind the increasing difficulties faced by Italian producers in selling their goods in world markets.

According to Pininfarina, costs incurred by Italian firms had risen more rapidly in recent years than prices in international markets. Attributing these rising costs largely to the wagebill, in 1989 Confindustria opposed increases in social security contributions and took steps towards modifying the *scala mobile* wage indexation mechanism.

Faced with internally generated inflation, Italian industry was not given the relief of lira devaluation to maintain export competitiveness during 1988. From the realignment agreed at the beginning of 1987 the lira's parity in the European Monetary System remained unchanged. Indeed in May 1989 Ciampi said that action by the central bank to strengthen the lira had been aimed at combating inflation. 'This action has a cost: if prolonged because of inadequate budget policy, it will aggravate the merchandise trade deficit,' he said. Nevertheless, inflation returned to the centre of public attention, as monthly calculations of the consumer price index revealed an upward trend which by mid-1989 had taken inflation to over seven per cent, a level last seen three years before.

'The annual rate of increase in the cost of living approaches seven per cent, more than two points higher than 1988's minimum which was recorded in October,' said Ciampi at the end of May 1989. Expressing the central bank's concern, Ciampi identified the reasons as persistent demand intensity, the increase in international prices of raw materials, the strengthening of the dollar and increases in Italy's indirect taxes.

A price index which was again moving upwards and an inflation differential with trading partners which threatened to widen were adverse signs. Like Ciampi, Pininfarina at Confindustria was also worried. 'The upturn in inflation in the opening months of the year is certainly linked to international factors. But a significant part is caused by increases in social security contributions and taxes which have affected prices and increased firms' costs,' asserted Pininfarina.

Public services

Italy's industrialists were, moreover, deeply concerned about what they saw as hidden costs which competitors in other countries did not incur. Pininfarina described public services and infrastructure as an extremely critical point in the Italian economy. He drew a sharp contrast between the strides made by industry and improving efficiency over recent years and a reverse process of decline in infrastructure and public services.

Italy's poor public services whether post, telephone, air, rail, health or the leisurely-paced bureaucracy represented a cost to users. Businessmen complained about a postal service which often required weeks to carry letters from one side of a city to another. The state-controlled SIP telephone system won notoriety for lengthy delays in connecting new subscribers and the high frequency of poor quality lines and wrong or interrupted connections. Trying to speak or send data from Rome to Milan or Turin, or vice versa, is a lottery; some win, but there are many losers.

Travel, whether air or rail provided various hazards to discomfort the passenger. Few would believe the assertion of former transport minister Giorgio Santus that Italians were glad to fly with state airline Alitalia. There was widespread feeling that the airline had exploited its monopoly position and shown scant regard to users. Winter 1988-89 was a particularly difficult time for air travellers, chequerboard strikes creating havoc with timetables which were already badly disrupted by weather conditions. Regular passengers from Po Valley airports like Milan's Linate find it incomprehensible that Italy's airports are shut down by fog when foreign airports stay open and foreign airlines continue operating in similar conditions.

Strikes also affected railways frequently in 1988 and 1989. Politics and mismanagement played a part in permitting productivity about one-half of that in other European railways, and accounts in which nearly two-thirds of costs were covered by state subsidies in 1988 against less than 25 per cent at the start of the 1970s. The appointment of a special commissioner to replace a board of directors dismissed and under investigation for corruption was the low point of the sorry condition of Italian railways in 1989.

It was unlikely that Pope John Paul II was thinking of the railways or the capital's inadequate underground or surface transport system when he drew parallels between Rome and the Third World, though irate passengers can often be heard to make such comparisons. But he could well have been thinking about the degradation not only in Rome's ugly, dispiriting high-density suburbs but also in parts of its historic centre. And if the Pope were to visit the capital's hospitals unannounced, he might glimpse scenes recalling the times of Florence Nightingale. Notwithstanding the massive funding Italy's health service leaves much to be desired in many areas.

At the root is the fact that the heavily-politicised Italian state machine spends much and spends badly. 'Experience tells us that on one side implementation is considerably inferior to plans, with the public sector finding increasing difficulty in spending funds available for investment, while on the other side the weaknesses of public services management multiply, making them costly and inefficient,' complained Pinifarina.

Italy's industrialists believe that the chronic condition of the public sector will not improve unless new rules of efficiency are introduced. Until then business and the individual will pay the price for poor communications, a snail's pace bureaucracy and the other inadequacies which characterise the country's public sector.

The banking system

The banking system has frequently been pilloried for its poor service. Pininfarina noted that Italian banks needed a more competitive environment. 'Our banking system must face up to international competition. It must raise its efficiency, grow in size, boost its capitalisation and increase its international presence,' he said.

Pininfarina has an ally in Piero Barucci, chairman of Associazione Bancaria Italiana (ABI, the Italian banking association). In addition to being ABI's chairman, Barucci is also chairman of Monte dei Paschi di Siena, the world's oldest bank, so he has had experience of being at banking's front line. Barucci spoke bluntly when he addressed ABI's annual meeting in June 1989. 'Our system is less efficient than others which have been able to work freely on international markets, which operate in currencies having significant markets beyond their borders and which enjoy ownership structures and legislative frameworks that have encouraged reorganisation and a gradual restructuring process,' he said.

ABI was increasingly worried that the banking system would be unprepared for the single market. Barucci called on parliament to create the right conditions for concentration within the system. He wanted less political interference and freedom for the banks to take their own decisions. 'Our capacity to become European banks must recognise how matters stand. We must not delude ourselves that time can compensate for our shortcomings. Operating in a protected market has given us the appearance of strength but made us weaker,' declared Barucci.

Certainly without the refreshing breezes of competition Italian banks have lacked the stimulus to tackle the staff problem. Nearly everyone was agreed that there were too many bank employees, yet their numbers continued to rise, reaching 314,000 at the start of 1989. Paying the salary bill accounted for over 70 per cent of total operating costs.

Gianni Zandano, chairman of Istituto Bancario San Paulo di Torino identified staff costs as a critical area: ')One can assume that the fundamental economic condition which has allowed the Italian system to support staff costs higher than other countries is higher earnings from financial intermediation on the domestic market,' he suggested. Customers pay for the banks' inefficiency through wider interest spreads.

Some banks have taken steps to stem staff growth. Others will have to do so. Greater use of automation can help win improvements in productivity, and Italian banks will have to exploit new technology and different organisational structures if they want to benefit from the liberalisation of branch networks which the Bank of Italy has promised. In May 1989 Ciampi said that the authorities were preparing a new procedure for approving branches. Those who believed that the most recent of the central bank's four-yearly branch plans, announced in 1986, would be the last were proved correct. The tight restrictions on branch growth were to be relaxed.

Ciampi said that expansion plans must be coherent with banks' assets and economic performance, and

with an orderly geographical development linked to technical and organisational capacity. Well run banks with strong asset bases would have the chance to expand. But even for these liberalisation represented a delicate moment, with a risk that too many branches would be established
giving rise to an increase in fixed costs. 'The new approach will mean a different type of network. Bank branches must in general become slimmer,' said Ciampi, calling for efforts to restrain fixed costs and obtain more efficient use of staff and technology.

To the outsider concentration seemed to offer enormous possibilities for raising efficiency through reorganisation and economies of scale. With about 1,100 different credit institutions, the Italian banking system appeared ripe for mergers, takeovers and rationalisation. Indeed a step was taken in this direction in February 1989 when the Cassa di Risparmio di Roma, the capital's savings bank, announced the purchase of a controlling stake in Banco di Santo Spirito.

Some senior bankers considered that this operation was the starting point of a process which would radically change the shape of Italian banking. This would please the Bank of Italy which has been giving encouragement to mergers for several years. There was no lack of scepticism, however, particularly about the prospects of mergers between large banks.

Jobs are lost when banks merge and difficulties arise over staff choice, not least for the top posts. This has been an important deterrent to mergers among state-controlled institutions, which account for about two-thirds of the banking system. Politicians have a profound interest as board appointments are part of a deeply-rooted spoils system and must take account of political power balances. Real mergers mean real job losses, not just tellers, computer operators and legal advisers but also in the boardroom. And fewer directorships mean fewer possibilities of rewarding the political faithful.

An area where mergers and rationalisation would receive a special welcome from the authorities was the *mezzogiorno*, the Italian south. Small banks abound in the *mezzogiorno* and nowhere is their presence more evident than in Sicily. ABI's yearbook reveals that many different banks have head offices on the island. In 1989 the province of Palermo was home territory for 20 banks and Catania for 17. In the mafia heartland of Trapani in Western Sicily and the Caltanissetta and Agrigento provinces in the island's centre and south savers had a choice of 51 local banks for their deposits.

Many are single-branch *casse rurali ed artigiane* (rural and tradesmens' savings banks), though Sicily is also well provided with *banche popolari* (popular co-operative banks) and *banche di credito ordinario* (joint stock corporation banks). But whatever their juridical form, the banks are small. And small means vulnerable.

In April 1989 Ciampi told the parliamentary commission on the mafia that the capacity of criminal organisations to infiltrate the financial system was greatest in those areas where there were weak technical characteristics, organisational shortcomings and inadequate internal controls. Small southern banks were soft points for dirty money to enter the system. With Italy the birthplace of the mafia and its Neapoli-

COUNTRY PROFILE

Political system: Democratic republic with bicameral parliament consisting of Chamber of Deputies (630 members) and the Senate (315 elected members and seven life senators). Both houses elected for five years by proportional representation. President elected by electoral college comprising both houses plus 58 regional representatives. Executive power exercised by Council of Ministers. The President of the Council (Prime Minister) is appointed by the Head of State.

Main cities/towns: Rome (population 3m), Bari, Bologna, Genoa, Florence, Milan, Naples, Turin, Venice.

Media
Press: No national press, and only three or four newspapers have a circulation above 250,000. Therefore magazines important. Much of press controlled directly or indirectly by major industrial groups.
Dailies: Influential dailies include *La Stampa* (Turin, circulation 433,366), *Corriere della Sera* (Milan, 468,072), *La Repubblica* (Rome, 486,000), *Il Giornale* (Milan, 176,912) and the financial daily *Il Sole/24 Ore* (Milan, 170,624).
Weeklies: Very popular. Ten general interest weeklies with circulation over 50,000. *Panorama* and *L'Espresso* sell about 300,000, and *Famiglia Cristiana* (Roman Catholic) about 1.2m copies per week. *Europeo* is another widely-read weekly. Large number of other general and special interest magazines.
Business: Many specialised trade magazines, mostly published in Milan. Major economic/business ones are *Espansione* and *Tempo Economico* (both monthly), and *Il Mondo* and *Mondo Economico* (weekly).
Broadcasting: Main channels controlled by Radiotelevisione Italiana (RAI-TV). Vast number of private stations.
Radio: Approximately 18m receivers in use, served by three national RAI networks and over 4,000 private stations.
Television: Approximately 18.7m TV households. Three state channels plus more than 450 private commercial stations, three of which (Canale 5, Italia Uno and Rete Quattro) have national audience. Some stations from other countries can be received. Some local independents group together to form national networks. Two per cent of households subscribe to state-run teletext service.
Advertising: Complex system of taxation in operation. Mostly conducted through agents. All usual media available. All three state-owned TV networks and two of the three radio networks carry advertising, as do all the private stations. The press, cinema, posters and direct mail are also widely used. Total advertising spend (1984) 5,058bn lire, mainly on food, cosmetics and detergents, automobiles and drinks. TV main medium for advertising.

Domestic economy: Large industrial-based economy with heavy dependence on imported energy. Service sector, particularly tourism, also significant. Economic performance constrained by large and growing public sector deficit, high inflation and regional disparities in economic wealth. Standard of living considerably higher in the north than the south.

Employment: Labour force accounts for 36 per cent of total population. Manufacturing employs about 30 per cent of labour force and agriculture about 12 per cent. Unemployment has almost doubled in last five years to over 12 per cent. Complex wage indexing system ('scala mobile') in operation. Unit labour costs have slowed since 1984 due to strong productivity gains.

External trade: Exports to EC were 57 per cent of total in 1988. Improvement in balance of trade largely dependent on reduction of fuel imports, which currently account for nearly 30 per cent of total import bill. Drop in earnings from tourism and other invisibles recently offset by growth in exports of machinery and transport equipment, textiles and chemicals. Deterioration in 1987 trade balance attributable to intensified foreign competition for exporters, mainly from South East Asia. Also steady increases in imports of raw materials and chemicals. Import of energy products in 1988 was Lire 19,112bn from Lire 180,059bn (10.6 per cent) total. Within the EC, main trading partners are German FR and France. Also increase in trade with USA and CMEA.

Agriculture: Sector contributes around seven per cent to GDP and accounts for 18 per cent of exports. Production is divided between the larger, more fertile, mechanised farms of the north (specialising in meat, dairy, cereal production and new crops like soya bean) and the smaller, labour-intensive farms of the south, which produce mainly wine, olive oil, fruit and vegetables. Italy produces crops of Mediterranean products which fail to satisfy consumers in northern EC markets and which are already in surplus.

Industry: Major industrial sectors include machinery and transport equipment, iron and steel, chemicals, food processing and textiles. Heavy industry is found in Lombardy, Piedmont and Liguria. Problems in the sector include strong and growing competition in traditional products from developing countries, and the small share of the export market taken by high-tech products. Improvements in efficiency are sought in public sector industry and co-operation with the private sector is being encouraged.

Mining: Relatively poor resources, although large quantities of iron ore and pyrites, mercury, lead, zinc, bauxite and marble. Only large sulphur deposit in Europe is in Sicily but is not economically viable. Bauxite mined mainly in Abruzzi, Campania and Apulia, though output has contracted due to fall in demand from aluminium industry. However, output of lead, zinc and particularly copper have all increased.

Energy: Heavy reliance on energy imports (80 per cent of needs). However, plan under way to develop indigenous resources, as well as use of coal and natural gas. To this end, gas is imported under long-term contracts from USSR and Algeria as well as Holland, and a programme to convert power stations from oil to coal is under way. However oil is still a major source of energy. Electricity needs filled by buying extra power generated by France's nuclear programme. Net installed electrical capacity 47,616 mW, of which 72 per cent is thermal. Nuclear energy programme reviewed and narrowed following referenda on nuclear issues in November 1987.

Banking: There are 1,100 separate banking institutions, of which the most common are the single or two-branch rural or tradesmen's banks (more than 700). Banking system is regulated by the Bank of Italy which also regulates monetary and credit policy. Approximately 65 per cent of banking activity is in the hands of public sector banks. Largest is Banca Nazionale del Lavoro and Cassa di Risparmio delle Provincie Lombarde (Cariplo) is the world's largest savings bank. 35 foreign banks operating include Barclays, Citibank and Deutsche Bank. Partial privatisation of leading merchant bank Mediobanca has been undertaken. Deregulation in progress. At year-end 1988 there were 37 foreign banks with branch operations.

Legal system: Based on civil law system with ecclesiastical law influence. Constitution came into effect in 1948.

Membership of international organisations: ADB, ASSIMER, CCC, Council of Europe, DAC, EC, ECOWAS, EIB, ELDO, EMS, ESRO, FAO, GATT, IAEA, IBRD, ICAC, ICAO, ICO, IDA, IDB, IEA, IFAD, IFC, IHO, ILO, International Lead and Zinc Study Group, IMF, IMO, INTELSAT, INTERPOL, IOOC, IPU, IRC, ITC, ITU, NATO, OAS (observer), OECD, UN, UNESCO, UPU, WEU, WHO, WIPO, WMO, WSG.

tan and Calabrian cousins the *camorra* and *ndrangheta*, the authorities were conscious of the need to show resolute action. The Bank of Italy said that the Basle Declaration coincided with its thinking and it gave indications for concrete application to the banking system. ABI gave full support and Barucci emphasised the need for coherent and effective action.

Certainly the Bank of Italy's *Vigilanza* supervisory service had been keeping a close eye on bank operations in the *mezzogiorno*. Of 571 inspections undertaken during the three years 1986 to 1988, 100 were in the three high-crime regions Sicily, Calabria and Campania. Specific interventions affected 105 Italian banks over this period, with 48 banks in these three large *mezzogiorno* regions being called to account for anomalous situations. While specific interventions involved 9.5 per cent of banks nationally, the figure was 30 per cent in Sicily, 21 per cent in Calabria and 15 per cent in Campania.

North-South divide

It was indicative of official interest in the *mezzogiorno's* financial system that 1989's Bank of Italy report gave it a special chapter. Apart from organised crime, other major problems exist and a wide gap separates the south from the centre and north of the country.

Statistics revealed that southern per capita GDP was only 54.9 per cent of the centre and north, and the ratio of per capita family consumption was only 65.9 per cent. In the late 1980s the industrial productivity gap widened and added value per worker in the south was 20 per cent lower than in the centre and north. Production efficiency was lower and business was less profitable. Measures like gross operating margin and return on equity showed that southern firms lagged behind those operating in Italy's centre and north.

Corporate mortality was higher in the *mezzogiorno* and default statistics were worse. Figures at the end of September 1988 showed an average ratio of default to loans of 12.5 per cent in southern banks, compared to 6.5 per cent in central and northern banks. In 1979 the gap was only 1.4 per cent, so it widened considerably during the 1980s. The higher risk faced by the banking system was paid for partly by interest rate differentials. Firms and families in the south earned slightly less on their deposits than their central and northern counterparts. But they paid significantly higher rates when they borrowed, the Bank of Italy's figures revealing a 2.0 per cent difference.

Notwithstanding the resources directed towards the south, it remained at the fringe. Politicians expended vast numbers of words on the *mezzogiorno* problem but did not find the solutions to crush organised crime, reduce the large unemployment gap which separates north from south, bring southern wealth levels closer to the north or narrow the cultural and social divide which splits the country. Sceptics believe that an impoverished south serves the interests of certain political groups, so an absence of will in making improvements is inevitable.

Some industry was attracted to the south, drawn by capital grants and soft loans. But it generally preferred to stay as close as possible to the northern edge of the area eligible for incentives. Ascoli Piceno province and the Abruzzo coastal strip in the central Adriatic have flourishing industrial areas, and the provinces immediately south of Rome, Latina and Frosinone, have also attracted large numbers of well known foreign and northern Italian manufacturing concerns. But few ventured further south. Merchant bankers hunting for acquisitions targets readily admit that buying mandates usually exclude the south.

Preference for the north is understandable. From Milan, Bologna, Brescia, and the other cities and town in and near the Po Valley the mafia and camorra are distant problems. Businesses located in the north are much nearer to markets, and an industrial outlook is well rooted in the labour force.

Industrial growth

Visitors to the thriving cities in Italy's north and centre see the country's successful side which has attracted the attention and compliments of international business and the press. Outsiders are frequently surprised by the contrast between the wealthy north and the impoverished south and between inefficiency in the public sector and the dynamism and results achieved by private sector industry. Partly it is a matter of style, the exploitation of Italian flair for presentation. But the results of the 1980s in industrial renewal are evident.

The outturn for 1988 showed that this was a particularly good year. Total real added value in industry increased by 5.3 per cent, compared to a 3.6 per cent increase between 1986 and 1987, and the 1986/1985 improvement of 2.1 per cent. In manufacturing alone (excluding construction and energy), real added value in 1988 was 5.8 per cent higher than in the previous year. The Bank of Italy attributed the sustained growth in industrial activity to booming demand for investment goods (up by 8.5 per cent). Demand for consumer goods was less strong, though the production of consumer durables rose by 7.4 per cent.

By 1989 Italian industry had enjoyed an exceptional six years' growth; periods of growth during the 1970s did not exceed three years. Moreover, Confindustria believed that the growth phase of the economic cycle would continue, albeit at a slower pace. In summer 1989 Confindustria's research centre published forecasts of GDP growth of between 2 and 3 per cent in 1990 and 1991.

Statistics for capacity utilisation reflected the sharp upward movement of demand and production. Central bank figures showed that utilisation in 1988 was 95.4 per cent, moving above the level reached in 1980. About one-quarter of industry had reached saturation and, notwithstanding investment undertaken over the previous five years, the gap between production and potential narrowed significantly. However, structural changes brought about by new technology, plant modernisation and greater labour flexibility relaxed the capacity constraint.

The continuation of the boom seemed to have ended the labour shake-out. After years of falling payrolls, employment in manufacturing industry rose by 1.1 per cent in 1988. But growth in opportunities was not homogeneous. While the number of jobs increased by 1.8 per cent in the centre and north, the increase in the *mezzogiorno* was a modest 0.2 per cent. Corrected unemployment figures showed that 21.7 per cent of the south's active population was out-of-work in 1988, compared to 8.7 per cent in the centre and north.

Indeed many industrialists in the north, and a good number in the centre, admitted that labour shortages, particularly of skilled workers, was a problem. Computer scientists and electronics engineers had no difficulties in finding employment while, at shopfloor level, experienced engineering mechanics or craftsmen who could maintain high technology automated manufacturing systems were in short supply. Pininfarina at Confindustria noted that the labour market was no longer a market of manpower but a market of ever-increasing professionalism. The industrialists were dissatisfied with the education provided by the state and called for reforms to match education and training to the requirements of the world of work.

In March 1989, while still prime minister, Ciriaco De Mita inaugurated the modernised Fiat plant at Cassino mid-way between Rome and Naples. The event was widely covered by television and the press, and Italians were given a chance to see the types of change which are radically modifying work in manufacturing industry. Automation rules at the plant, allowing flexibility, accurate matching of production to market demand and higher quality. And investment in machinery has increased productivity. When Cassino employed its peak workforce of 10,000 about 1,200 automobiles left the line every day. When it reaches regime in 1990, Cassino is expected to have a daily output of 1,800 vehicles from 7,000 workers.

Cassino produces the Fiat 'Tipo', voted 'car-of-the-year 1989'. It is one of the corporation's principal Italian plants which together achieved an output increase of 9.8 per cent in 1988, to build about 2 million Fiats, Lancias, Autobianchis, Alfa Romeos and Ferraris. Automobile sales generated around 60 per cent of consolidated turnover at Italy's largest private sector group (L25,500bn from 1988's total of L44,300bn).

Success of the 'Tipo' in a market segment where it would challenge Volkswagen's 'Golf' looked likely to be crucial to the Agnelli family's Turin-based corporation for several years. The Group was hoping for another 'Uno', which continued to be its best-selling model with 375,000 sold in Italy alone in 1988. A home market share of 59.9 per cent and improved sales abroad allowed Fiat Auto to win top ranking in European car sales in 1988 with a 14.9 per cent share.

Results in 1988 confirmed the substantial improvements made by Fiat during the decade. Post-tax profits reached L3,030bn against L2,370bn on L38,440bn consolidated turnover in 1987. Self-financing amounted to L5,560bn (L4,670bn in 1987). The Group showed net financial availability of L2,350bn and net assets of L13,550bn at year-end 1988 (L180bn and L11,410bn respectively for 1987). Even difficult sectors like trucks, tractors and earth-moving equipment looked better. Fiat was generally considered to be a picture of corporate health. Even the traumatic departure of Vittorio Ghidella, the group's auto wizard, at the end of 1988 seemed to be quickly forgotten.

Olivetti is another well-known Italian name in Europe, and increasingly on a global scale. Like Fiat, the Ivrea-based information technology and office automation leader recorded a good increase in turnover in 1988, up 14.0 per cent on 1987 to L8,400bn. But the bottom line was disappointing, with post-tax profits lower for the second successive year. Net income of L356bn was the same as five years earlier in 1984.

In his annual report to Olivetti's shareholders, chairman Carlo De Benedetti referred to remarkable changes in the competitive scenario, an increasingly prudent market, contained growth, a shortage of strategic components, a fiercely competitive climate and strong pressure on prices and operating margins. The tough conditions facing Olivetti was a factor in the reorganisation of the group in September 1988 which followed June's appointment of a new managing director.

Another important manufacturing company which reported lower profits in 1988 was Merloni, the 'white goods' maker. The bottom line of the company's accounts showed a L16bn net result on L1,059bn turnover, compared to L18bn on L598bn in 1987. The large increase in sales was due to the acquisition of Indesit which with its existing Ariston brand gave Merloni a 10 per cent share and fourth ranking in the European market. Merloni has leadership in its domestic Italian market for refrigerators, freezers, washing machines and dishwashers.

Merloni Elettrodomestici has gone for growth in a sector where Italy has long enjoyed a prominent European position. In five years the company tripled production volumes and turnover and doubled its European market share. Chairman Vittorio Merloni noted that in 1984 it was a medium-sized family business. He considered that the company had now achieved a critical mass to allow it to compete with global leaders. The company aims for greater efficiency and productivity and believes that it has the flexibility and speed of response to market changes to prepare for the future.

The challenge of internationalisation faced by Merloni is also recognised by manufacturers in another sector in which Italy enjoys international standing: designer clothing. Gruppo Finanziario Tessile (GFT) is a leader among the leaders. The Turin-based family firm produces mens' and womens' clothing for Giorgio Armani, Christian Dior, Ungaro and Valentino.

GFT is the maker for some of the greatest of the most glittering names which currently illuminate world fashion. Producing designer clothing is a big money-spinner as GFT's figures show. Consolidated turnover at the group controlled by the Rivetti family rose by 20.4 per cent in 1988 to reach L1,190bn and net profits increased by 5.5 per cent to L40bn. GFT chairman Marco Rivetti considered, however, that Italy's clothing sector lacked the experience to deal with the challenges of the 1990s.

At a moment when the 'Made in Italy' label is losing some of its competitive advantage, the country's clothing industry is at a junction. Either it can attempt to strengthen its exporting or it can internationalise. GFT has opted for the second direction and Rivetti claimed that its US and Mexican production facilities allowed the group to enjoy strong transatlantic sales growth even against a weak dollar.

Rivetti's group is typical of the newish and little-known family businesses on which Italian manufacturing success was based during the 1980s. Another family concern, albeit with a more recognised name, is Ferruzzi Group. With aggregate turnover exceeding L40,000bn this has become Italy's second largest private sector group. Under Raul Gardini's chairmanship, Ferruzzi Group has grown from being an agri-industry operation to an industrial and financing conglomerate.

Gardini's group maintained its controversial reputation in 1988 and 1989. Originally nicknamed the contadino (peasant), the yacht-loving Gardini was now often called the bucaniere (pirate). Certainly his acquisition of control over Montedison and the subsequent corporate card-shuffling at the beginning of 1988 revealed buccaneering abilities. These were not appreciated by US investors when Montedison made a public offer for its Ausimont subsidiary in January 1989.

Privatisation

Closer to home, Gardini won some more enemies by stating his aims for Enimont, the chemicals joint venture between Montedison and state-owned Enichem which was agreed in December 1988. The group's insistence on the need for a special tax break to reduce tax liabilities on gains arising from asset transfer in the constitution of Enimont was satisfied by a Decree Law. This attracted critical attention from Brussels and adverse comment from Italians who considered tax gifts inappropriate to a business which made record profits in 1988.

When Gardini made clear his intention of acquiring full control over Enimont, a joint venture established with both major partners holding 40 per cent stakes, press and politicians examined the agreement closely and admitted that Gardini had manoeuvred smartly. Montedison kept its best activities out of Enimont and retained the option to confer them later and gain control.

In the run-up to the quotation of Enimont shares on the Milan stock market, 20 per cent of the company's share capital being offered to the public, there was speculation that investors might be shy of buying shares in an operation which looks set, willy-nilly, to fall under the total control of Gardini's Ferruzzi Group. What was originally seen as partial privatisation looked increasingly like a straight change of ownership from state to Ferruzzi Group.

With Montedison's tax break and the structuring of the Enimont deal, there appeared to be little financial benefit for the state in the operation. Possibilities of raising funds through asset disposals were, however, increasingly discussed in 1988 and 1989 as a means of reducing the public debt. One of the obstacles was the sentiment of the stock market which seemed little disposed to absorbing large volumes of new equities, discouraging new quotations of state sector companies.

Yet for the first time in 20 years IRI, the state holding corporation whose subsidiaries include national airline Alitalia, the SIP telephone network, RAI radio and television, the *Autostrade* toll highways, three major banks and leading companies in electronics and aerospace, was able to report a profit; L600bn on turnover of L59,250bn against L375bn loss on L53,000bn turnover in 1987.

Performance by the ENI state hydrocarbons holding corporation was even better. Its net profits reached L1,316bn in 1988, almost twice the L708bn result achieved in 1987, and Chairman Franco Reviglio was forecasting improvement to L1,600bn in 1989. Since 1985, when it returned to profit after four years of loss-making, ENI has not received state funding.

Though profits make state-owned companies more attractive to investors, they also reinforce some political arguments against privatisation. Opposition to reduction of the Italian state's ownership of industry is already considerable, not only from the left but also from the conservative Christian Democrats. Positioning party place-men at the top of companies belonging to IRI, ENI and the smaller EFIM holding corporation is part of political jousting.

Board appointments are rewards in the spoils system, so privatisation is not widely welcomed in party committee rooms or in Rome's Montecitorio parliament building. Indeed Reviglio's declaration regarding the strengthening of ENI's interests in the media, where it already controls the Milanese daily newspaper *Il Giorno* and the news agency *Agenzia Giornalistica Italia*, reflect aims of empire building rather than scaling-down. Reviglio, a Socialist appointee, considers that privatisation should be partial, a way of raising money without losing control.

With the chairmen's appointments expiring at the end of 1989, the scene was set for a lively political struggle over nominations of successors to Reviglio and IRI's chairman, Christian Democrat Romano Prodi. As attacks by new party secretary Forlani on De Mita's choice for top jobs at IRI's RAI radio and television subsidiary showed, even within parties struggles between factions to reward their followers can be bitter. The power and weight of IRI and ENI are such that appointments to the state holding corporations are matters of strategic importance.

'Privatisation and agreements between public and private sector industry are often blocked by heavy political interference, not based on industrial logic but for reasons of tactics and power,' said Pininfarina. With Andreotti, variously known as Beelzebub or *la volpe* (the fox), sitting in the prime minister's office at Palazzo Chigi, it was difficult to imagine that industrial considerations would start to triumph over politics.

BUSINESS GUIDE

Time: GMT + 1 (GMT + 2 from 27 Mar to 24 Sep).

Climate: Temperate in north, Mediterranean in south. Temperatures range from about 4°C to 30°C. Winter is cold and dry on Alps, mild on Italian Riviera, cold and damp in Po valley, mild in southern Italy and Sicily. Summer hot and dry along coast, pleasantly cool in Alps and Appenines. Rainfall not heavy.
Clothing: Mediumweight and light topcoat or rainwear for winter. Lightweight for summer.

Entry requirements: Visa not required by nationals of most European countries. Visitors not staying in a hotel must register with police within three days of arrival. Travel conditions liable to change at short notice due to labour disputes. Advisable to consult travel agent or Italian Consul prior to departure.
Health: Vaccination certificates not required. EC nationals are covered for medical treatment; an E111 form is needed, however.
Currency: No limit on amount of foreign currency brought in, but this must be declared and not exceeded by amount taken out. Up to 600,000 lire in Italian currency allowed in and out of country, but check with bank before departure.
Customs: Personal effects duty-free. Inspections on main international train routes.

Air access: Regular flights by all major international airlines.
National airline: Alitalia.
International airports: Pisa (PSA) for Florence, Turin (TRN), Venice (VCE), Bologna (BLQ), Naples (NAP), Genoa (GOA), Palermo (PMO).
Main international airports: Fiumicino (FCO) 32 km from Rome, Linate (LIN) 7 km from Milan, Malpensa (MSP) 46 km from Milan.

Surface access: By road from France, Switzerland, Austria and Yugoslavia. However, several passes closed during winter. In addition to the Riviera coastal motorway, access from France and Switzerland maintained via St Bernard, Mont Blanc and Frejus tunnels. Also several rail and sea routes.
Main ports: Genoa, Trieste, Augusta, Taranto, Leghorn, Savona, Ancona, Bari, Brindisi, Civitavecchia, Venice, La Spezia, Naples, Palermo (Sicily), and Cagliari and Porto Torres (Sardinia).

Hotels: Classified into five star categories. Rates are fixed with Provincial Tourist Board, and vary according to class, season, services available and locality. Visitors are required by law to obtain an official receipt when staying in a hotel, and hotels are required to report the presence of all foreign guests to the authorities. All major credit cards are accepted in the large, top-category hotels in major cities but not widely elsewhere.

Restaurants: Wide range of standards and types. Service charge of 12-15 per cent usually included but customary to leave tip. Many restaurants do not accept credit cards.

Currency
Exchange rate: Italian lira floats freely against all foreign currencies, but is included in European Monetary System and floats in common with currencies in certain other west European countries. Within EMS the lira is allowed a wider band of fluctuation.

Foreign exchange information
Account restrictions: Under some conditions non-residents can have accounts in foreign currencies. Non-residents may also have 'foreign accounts' in lire.
Trade policy: Few restrictions on imports.

Leading state holding companies: Ente Nazionale Idrocarburi (ENI) – oil, chemicals, textiles, machinery; Istituto per la Ricostruzione Industriale (IRI) – banks, engineering, Alitalia, RAI-TV, SIP (telephones); Ente Partecipazioni e Finanziamento Industria Manifatturiera (Efim) – mainly engineering and metalworking.
State holding companies: Ente Nazionale Idrocarburi (ENI) – oil, chemicals, textiles, machinery; Istituto per la Ricostruzione Industriale (IRI) – banks, steel, engineering, Alitalia, RAI-TV,

BUSINESS GUIDE

SIP (telephones); Ente Partecipazioni e Finanziamento Industria Manifatturiera (EFIM) – mainly engineering and metalworking.

Car hire: Self-drive cars available at between 40,000 and 150,000 lire per day depending on the engine size of the car, plus 600 – 1,300 lire per kilometre. Special weekly tariffs available from about 275,000 to 1,300,000 lire (including charge per kilometre) plus 22 per cent VAT. Official translation of driving licence required. Air and rail travellers can get special deals including car hire. Drive on the right. Maximum speed 50 kph in towns; out-of-town speed limits depend on engine capacity of car and type of road. Seat belts are obligatory by law.

City transport
Taxis: Available in all towns and tourist resorts, usually in ranks at railway stations, or can be called by phone. Fares vary considerably, and unmetered cabs should be avoided. A small tip (8-10 per cent) is usual; some drivers add this to the fare when carrying foreigners.
Buses and metro: All major cities have bus service with one standard fare (700 lire in Rome). Day and monthly tickets also available. Metro (metropolitana) in Rome and Milan with standard single fare as for buses. In Milan tickets last for 70 minutes and can be used on both metro lines and all bus routes. A daily ticket usable for all Rome services is available.

National transport
Air: Alitalia and Aero Transporti Italiani (ATI) operate services connecting Rome to most major towns. Alisarda operate services connecting Rome, Milan and Turin with Sardinia.
Road: Total length of roads approximately 300,000 km, including a toll motorway system (autostrade) which connects most cities. Extensive bus services, operated by several companies, link all major towns. Fares are relatively cheap.
Rail: Efficient and fairly cheap network (over 20,000 km) operated by Italian State Railways (Ferrovie dello Stato), and several small private companies. Advisable to book 'express' tickets in advance.
Water: Ferryboat and hydrofoil services linking the mainland with Sicily, Sardinia and the smaller islands are operated by several lines including the State Railways.

Public holidays: Official list of holidays is not generally published until late in the previous year. Main holiday season is mid-July to mid-September.
Fixed dates: 1 Jan, 6 Jan (Epiphany), 25 Apr (Liberation Day), 1 May, 15 Aug (Ferragosto/Assumption Day), 1 Nov (All Saints'), 8 Dec (Immaculate Conception), 25 and 26 Dec.
Variable dates: Easter Monday. Local public holidays are held on the feast day of each town's patron saint. Businesses often close early on 14 Aug, Christmas Eve and New Year's Eve.

Working hours
Business: (Mon-Fri) 0900 – 1300 and 1400 – 1800.
Government: (Mon-Sat) 0800 – 1400 (senior staff may return 1730 – 2000).
Banking: (Mon-Fri) 0830 – 1330 and also for one hour in afternoon.
Shops: (Mon-Sat) 0900 – 1300 and 1530 – 1930. Hours differ from town to town and in summer. All shops have one half-day closing each week.

Business language and interpreting/translation: Usually available in major towns. Can sometimes be provided at short notice by hotels or travel agents.

Telecommunications
Telephone: Efficient and cheap service. Public telephones use 100 and 200 lire coins or token ('Gettone'), or magnetic cards, available from post offices, tobacconists and some news-stands. International dialling code: 39 followed by 6 for Rome, 2 for Milan, 11 for Turin, 81 for Naples, 41 for Venice and 55 for Florence.
Telex: Available at major post offices. Country code 43.
Telegram: Can be sent from any post office or by telephone. ITALCABLE operates overseas service.

Postal service: Tends to be subject to delays. To ensure quick delivery it is best to use express and/or registered mail service. Stamps sold at post offices and tobacconists. CAI Post courier service from a few main post offices gives rapid delivery in main European cities.

Trade fairs: Over 300 international, national and specialist fairs held every year, including International Book Fair for Children and Youth, Apr, Bologna; Milano Vendemoda Fashion Show, Mar and Oct, Milan; International Exhibition of Footwear Fashions, Mar, Milan; International Household Goods, Crystalware, Ceramics, Silverware and Gift Exhibition, Sep, Milan; International Furniture Fair, Sep, Milan.

Electricity supply: 220 V AC.

Representation in capital: Afghanistan, Albania, Algeria, Angola, Argentina, Australia, Austria, Bangladesh, Belgium, Bolivia, Brazil, Bulgaria, Burma, Cameroon, Canada, Chile, China PR, Colombia, Congo, Costa Rica, Cuba, Cyprus, Czechoslovakia, Denmark, Dominican Republic, Ecuador, Egypt, El Salvador, Ethiopia, Finland, France, Gabon, German Democratic Republic, Federal Republic of Germany, Ghana, Greece, Guatemala, Guinea, Haiti, Holy See, Honduras, Hungary, India, Indonesia, Iran, Iraq, Ireland, Israel, Ivory Coast, Japan, Jordan, Kenya, Republic of Korea, Kuwait, Lebanon, Lesotho, Liberia, Libya, Luxembourg, Madagascar, Malaysia, Malta, Mexico, Monaco, Morocco, Netherlands, New Zealand, Nicaragua, Nigeria, Norway, Oman, Pakistan, Panama, Paraguay, Peru, Philippines, Poland, Portugal, Romania, San Marino, Saudi Arabia, Senegal, Somalia, South Africa, Spain, Sri Lanka, Sudan, Sweden, Switzerland, Syria, Tanzania, Thailand, Tunisia, Turkey, USSR, United Arab Emirates, United Kingdom, USA, Uruguay, Venezuela, Vietnam, Yemen Arab Republic, Yugoslavia, Zaïre, Zambia.

Hotels: Milan
Elcelsior Gallia, Piazza Duca d'Aosta 9, 20124 (tel: 6277; tx: 311160; fax: 656306).

Galileo, Corso Europa 9, 20122 (tel: 7743; tx: 322095).

Hilton, Via Galvani 12, 20124 (tel: 69831; tx: 330433; fax: 6071904).

Michelangelo, Via Scarlatti 33, 20124 (tel: 6755; tx: 340330; fax: 6694232).

Palace, Piazza della Repubblica 20, 20124 (tel: 6336; tx: 311026).

Hotels: Rome
Airport Palace, Viale Romagnoli 165 (tel: 611469; tx: 5692341).

American Palace Eur, Via Laurentina 554 (tel: 5911551).

Bolivar, Via della Cordonata 6 (tel: 6791614; tx: 621399).

Canada, Via Vincenza 58 (tel: 4975385; tx: 613037).

Colosseum, Via Sforza 10 (tel: 4751228; tx: 611151).

Diplomatic, Via Vittoria Colonna 28 (tel: 6542084/6561734; tx: 610506).

Esperia, Via Nazionale 22 (tel: 4744245; tx: 614635).

Giulio Cesare (near Vatican City), Via degli Scipioni 287 (tel: 353308; tx: 613010).

Grand Flora, Via Vittorio Veneto 191 (tel: 497821; tx: 680494).

Grand Ritz, Via Chelini 41 (tel: 803751; tx: 610570; fax: 872916).

Hotel de la Ville, Via Sistina 69 (tel: 6733; tx: 620836).

Mediterraneo, 15 Via Cavour (tel: 464051; tx: 610556).

Metropole, Via Principe Amedeo 3 (tel: 4774; tx: 611061).

Minerva, Piazza della Minerva 69 (tel: 686551; tx: 68327).

Oxford, Via Boncompagni 89 (tel: 4756852; tx: 614387; fax: 4756855).

Plaza Grand, Via del Corso 126 (tel: 672101; tx: 624669).

Regina Carlton, Via Veneto 72 (tel: 476851; tx: 620863; fax: 476851).

Savoy, Via Ludovisi 15 (tel: 4744141; tx: 611339; fax: 4746812).

Sheraton, Viale del Pattinaggio 1 (near Eur) (tel: 5453; tx: 626073; fax: 5423281).

Hotels: Turin
Concord, Via Lagrange 47, 10123 (tel: 5576756; tx: 221323).

Grand Sitea, Via Carlo Alberto 35, 10123 (tel: 5570171; tx: 220229; fax: 548090).

Jolly Ambasciatori, Corso Vittorio Emanuele 104, 10121 (tel: 5752; tx: 221296).

Jolly Ligure, Piazza C. Felice 85, 10123 (tel: 55641; tx: 220167).

Palace, Via Sacchi 8, 10128 (tel: 515511; tx: 221411; fax: 5612187).

Car hire: Rome
Avis: Via Tiburtina (tel: 4124214, 4125429; tx: 610372).

Budget: Via Boncompagni 14/C (tel: 4820966/7; tx: 625143).

Europcar: Via Lombardia 7 (tel: 465802; tx: 612358).

Hertz: Ciampino A/P (tel: 7240095, 7240616; tx: 610379).

Chambers of Commerce
American Chamber of Commerce in Italy, 12 Via Agnello, 20121 Milan (tel: 807955).

British Chamber of Commerce for Italy, Via Agnello 8, 20121 Milan (tel: 876981, 877798; fax: 28100262).

Camera di Commercio, Industria, Artigianato e Agricoltura, Piazza de Pietra, Palazzo Proprio, 00100 Rome (tel: 6783280).

Unione Italiana delle Camere di Commercio Industria (Association of Chambers of Commerce in Italy), Artigianato e Agricoltura, Piazza Sallustio 21, 00187 Rome (tel: 479961).

Banking
Banca d'Italia (central bank), Via Nazionale 91, Rome (tel: 47921; tx: 610021).

Banca Commerciale Italiana, Piazza della Scala 6, 20121 Milan (tel: 88501; tx: 310080).

Banca di Napoli, Via Toledo 177-188, 80132 Naples (tel: 7911111; tx: 710570).

Banca Nazionale del Lavoro, Via Veneto 119, 00187 Rome (tel: 47021; tx: 610116).

Banca Nazionale dell'Agricoltura SPA, Via Salaria 231, 00199, Rome (tel: 85881; tx: 625330; fax: 85883406).

Banco di Roma, Via U Tupini 180, 00144 Rome (tel: 54451; tx: 616184).

BUSINESS DIRECTORY

Cassa di Risparmio delle Provincie Lombarde, Via Monte di Pietá 8, 20100 Milan (tel: 88661; tx: 313010).

Cassa di Risparmio di Roma, 320 Via del Corso, 00186 Rome (tel: 67071; tx: 613541; fax: 67073783).

Cassio di Risparmio di Torino, 31 Via XX Settembre, 10121 Torino (tel: 57661; tx: 220350/1, 221225, 212278, 220215; fax: 638203).

Credito Italiano, Piazza Cordusio, 20123 Milan (tel: 88621; tx: 312401).

Istituto Bancario San Paolo di Torino, Piazza San Carlo 156, 10121 Turin (tel: 5551; tx: 212040).

Monte dei Paschi di Siena, Piazza Salimbeni, Siena (tel: 294111; tx: 570080 PASDIR 1, 572347 PASDIR 1).

Nuovo Banco Ambrosiano, Piazza Paolo Ferrari 10, 20121 Milan (tel: 85941; tx: 335687).

Principal newspapers

Corriere Della Sera, Via Solferino 28, 20121 Milan (tel: 6339; tx: 31001).

Il Giornale, Via Gaetano Negri 4, 20123 Milan (tel: 85661; tx: 333279).

Il Sole/24 Ore, Via Paolo Lomazzo 52, 20154 Milan (tel: 31031; tx: 331325).

La Republica, Piazza Ind Penza 11B, 00185 Rome (tel: 49821; tx: 620660).

La Stampa and Stampa Sera, Via Marenco 32, 10126 Turin (tel: 65681; tx: 221121).

Other useful addresses

Agenzia Nazionale Stampa Associata (news agency), Via della Dataria 94, 00187 Rome (tel: 6786161).

Borsa Valori di Milano, Piazza Degli Affari, 20100 Milan (tel: 8534).

Commissione Nazionale per le Società e la Borsa (commission for companies and the stock exchange), Milan (tel: 877841).

Confederazione Generale dell'Industria Italiana (general confederation of Italian industry), Viale dell'Astronomia 30, 00144 Rome (tel: 59031).

Confederazione Generale Italiana del Commercio (general confederation of Italian commerce), Piazza G.C. Belli 2, Rome (tel: 588783, 580192).

Ente Nazionale Idrocarburi (ENI), Piazzalo E. Mattei, 00144 Rome (tel: 59001; tx: 610082-86).

Ente Nazionale Italiano per il Turismo (tourist board), Via Marghera 2, 00185 Rome (tel: 49711; tx: 621314, 612318).

Istituto per la Ricostruzione Industriale (IRI) Via Vittorio Veneto 85, 00187 Rome (tel: 47271).

Ente Partecipazioni e Finanziamento Industria Manifatturiera (EFIM), Via XXIV Maggio, 43-45, 00187 Rome (tel: 47101).

Istituto Nazionale per il Commercio Estero (Italian government agency for promotion of foreign trade), 21 Via Liszt, 00100 Rome (tel: 59921).

Ministry of Agriculture and Forests, Via XX Settembre, 20, 00187 Rome (tel: 4665; tx: 610148).

Ministry of Budget and Economic Planning, Via XX Settembre 97, 00187 Rome (tel: 47611; tx: 626432).

Ministry of Finance, Viale America, EUR, 00144 Rome (tel: 5997; tx: 614460).

Ministry of Foreign Trade, Viale America 341, EUR, 00144 Rome (tel: 5993; tx: 610083).

Ministry of Industry, Via Vittorio Veneto 33, 00187 Rome (tel: 4705; tx: 622550).

Ministry of Tourism and Performing Arts, Via della Ferratella in Laterano 51, 00184 Rome (tel: 77321; tx: 616400).

Office of the President, Palazzo del Quirinale, 00187Rome (tel: 4699; tx: 611440).

Prime Minister's office, Palazzo Chigi, Piazza Colonna 370, 00100 Rome (tel: 6779; tx: 613199).

Statistics

Istituto Centrale di Statistica, Via Cesare Balbo 16, 00184 Rome (tel: 4673).

Directories

Annuario Politecnico Italiano: Annuario delle Industrie Italiane. Incorporates Annuario Politecnico Italiano srl, Via Silvio Pellico 12, 20121 Milan.

Guida Monaci: Annuario Generale Italiano. Published annually. Guida Monaci SpA, Via F. Crispi 10, 00187, Rome.

Jamaica

Canute James

The general election held on 9 February 1989 saw a change of government which had been predicted by public opinion polls. The conservative Jamaica Labour Party (JLP) administration, led by Edward Seaga, the prime minister, which had been in office since October 1980, was defeated by the social democrat People's National Party (PNP), led by Michael Manley. The PNP took 45 of the 60 seats in the lower house. The Labour party's defeat was heavy despite the government's achievement in getting the economy to grow after a decade of stagnation. But the performance of the economy was adversely affected by a hurricane which hit the island in mid-September, 1988.

PNP wins election

The PNP, which was heavily defeated in the 1980 general election, scored a resounding victory over the JLP on 9 February 1989. The Labour party based its platform for election for a third consecutive term on the state of the economy, which grew by 2.5 per cent in 1986 and by 5.5 per cent in 1987. Seaga also said that his administration had managed to bring a degree of stability, not only to the management of the island's economy, but also to Jamaican life in general.

For his part, Manley, who had been prime minister between 1972 and 1980, argued that although the economy had grown, the standard of living of most Jamaicans was below the level of 1980. The PNP claimed that the structural adjustment of the economy, which Seaga said was behind the economic growth, was responsible for the neglect of social services such as public health and education. Manley promised that if elected, he would rebuild the social services.

Fears that the election would be violent subsided when both Seaga and Manley signed a compact urging party candidates and their supporters to refrain from violence during the campaign. In the event, the death toll in the official campaign reached 13 as party factions clashed at rallies and on election day. Seaga complained that the election was 'the most fraudulent' he had witnessed, but did not repeat this charge when he later conceded defeat.

During the campaign there were charges and counter charges about the government's handling of relief for victims of Hurricane Gilbert, which scored a direct hit on the island on 12 September 1988. Hundreds of thousands of houses were damaged and the agriculture sector severely affected. The PNP charged that the JLP was using the relief funds to shore up its electoral fortunes – a charge which Seaga and the JLP denied.

The election result, which gave the PNP 46 of the 60 seats in the lower house, suggested that Seaga had lost support because of the deterioration in the social services. Voters were also swayed by Manley's repeated undertaking not to pursue the radical socialist policies of the 1970s, indicating that the PNP would be more moderate. The election also brought to an end six years of one-party parliament. Only JLP members had occupied the house since the PNP refused to contest the 1983 election after a row between Seaga and Manley over the state of the electoral register.

In taking office, Manley said he would pursue a policy of continuity, maintaining those of Seaga's programmes and policies which had been successful and changing or dismantling those which were not. Central among these was ensuring amicable relations with the United States, which had been strained under the PNP administration in the 1970s, but which were close under Seaga's administration. It was soon clear that even the more conservative elements in Washington had accepted that Manley would be more

KEY FACTS

Official title: Jamaica

Head of state: Queen Elizabeth II

Head of government: Prime Minister Michael Manley

Ruling party: People's National Party

Capital: Kingston

Official Languages: English

Currency: Jamaican Dollar (J$) = 100 cents

Exchange rate: J$6.34 per $ (Nov 1989)

Area: 10,989 sq km

Population: 2.4m (1988)

GDP per capita: $1,387.5 (1988)

GDP real growth: 1.9% (1988)

Inflation: 8.8% (1988)

Trade balance: –$395m (1988)

Foreign debt: $3.9bn (1988)

moderate. The PNP had got rid of its hard-left faction in the early 1980s, and the social democrat faction had regained control of the party.

Storm damages economy

The economic policy which Manley continued was based on a structural adjustment programme agreed with the International Monetary Fund. The new government promised continuation of the fiscal stringency which was central to the programme, and also tight control on wage increases and maintenance of the twice-weekly central bank hard currency auctions which are used to set the parity of the Jamaican dollar.

But the problems for the new administration were immediate. The effects of the hurricane were pervasive, and the economy, which was growing at an annual rate of 5 per cent before the storm, was depressed. The actual rate of growth for 1988 was less than 2 per cent, and the need for increased imports in the aftermath of the storm put a strain on the island's limited foreign currency. There was also a reduction in export earnings, and the trade and balance of payments deficits increased significantly. Additional government expenditure between September 1988 and the election six months later also led to a wider than forecast fiscal deficit.

These problems put in danger the island's standby credit agreement with the IMF which had been agreed in August 1988, and which allowed drawdowns totalling $118m. The economy failed to meet the performance criteria for the quarters ending December 1988 and March 1989. The government and the IMF subsequently agreed on a new programme, which gave the economy access to $63m which was not drawn down from the previous package.

Seymour Mullings, the finance minister, said the new agreement had been reached without any request for a devaluation of the Jamaican dollar. But the need to bring the fiscal deficit under control led the government to increase prices on a range of imported basic foods by removing the subsidies which had kept prices down. In presenting the budget for the 1989-90 fiscal year, Mullings also announced new taxes on motor vehicles, airline tickets and a mandatory education levy to close a financing gap in the government's projected expenditure.

Although the balance of payments deficit widened in 1988, the current account of the balance of payments improved. This was achieved mainly because of an increase in the new inflows of hurricane-related reinsurance payments, totalling $300.9m. The international reserves improved by $143.3m. The rate of inflation in 1988 was 8.8 per cent, against 8.4 per cent in 1987. Unemployment fell from 21 per cent in 1987 to 18.7 per cent in 1988.

Agricultural output was depressed in 1988 because of the storm in September of that year. Domestic agriculture and bananas for export were most affected, as overall output for the sector fell by 4.4 per cent. The sugar crop of 1988 was not affected as the harvest was completed three months before the hurricane. Sugar production for the year was 220,000 tonnes, 34,000 tonnes more than 1987. Output in the 1989 harvest was affected, however, as fields were flooded for long periods by the heavy rain which accompanied the hurricane, and 30,000 acres of cane were damaged.

The banana industry bore the brunt of the damage. The island's export target of 46,000 tonnes for 1988 was not achieved, and pre-hurricane shipments reached only 28,000 tonnes. The storm destroyed about 20,000 acres of fruit, leading to a halt in exports. Exports were resumed in May 1989, amid indications that the island would be able to ship about 30,000 tonnes for the year.

Bauxite expands

A shutdown of bauxite mines and refineries for a fortnight following the hurricane contributed to a 7.5 per cent decline in ore production in 1988. Output was 7.2m tonnes of bauxite, while alumina production of 1.5m tonnes in 1988 represented a decline of 6.1 per cent.

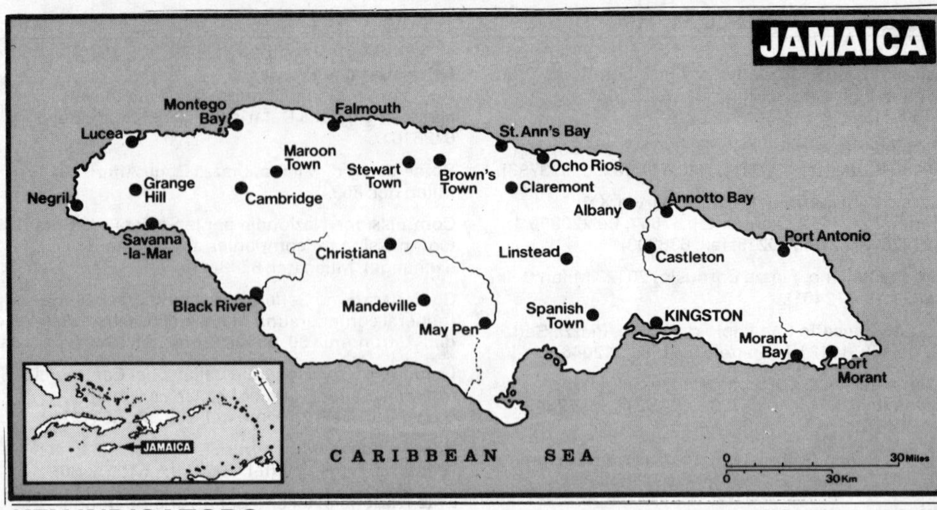

KEY INDICATORS

	Unit	1984	1985	1986	1987	1988
Population	m	2.3	2.3	2.3	2.4	2.4
GDP	J$m	9,355	11,151	13,310	15,717	24,537
GDP real growth	%	–0.9	–4.7	1.9	5.2	1.9
Exports	J$m	2,785	3,043	3,267	3,561	4,145
Alumina	J$m	1,274	1,157	1,155	1,096	1,334
Bauxite	J$m	591	471	534	627	579
Sugar	J$m	172	253	239	338	430
Balance of trade	$m	–334.7	–435.6	–247.9	–356.7	–395.0
Exchange rate	J$ per $	3.94	5.56	5.50	5.50	5.49

After several months of negotiations, the government and the Aluminium Company of America reached agreement on joint ownership of the company's refinery in central Jamaica. The government bought half of the plant, and its rated capacity is being increased from 800,000 tonnes per year to 1m tonnes. There was also an agreement with Alcan on a reduction in the controversial bauxite production levy, and this led to an agreement by the company to rehabilitate its two refineries on the island and run them at a combined rated capacity of 1.1m tonnes per year.

In March 1989 the government and Kaiser concluded arrangements for the reopening of the 1.2m tonnes per year Alpart refinery, the island's largest, which had been closed for three-and-a-half years. These expansions contributed to an 8 per cent increase in raw ore production in the first quarter of 1989.

The new tax regime was accepted by all the companies operating in the island. It was based on a 50 per cent reduction in the levy rate to 3 per cent of realised market price for aluminium ingots sold in major markets in North America, Europe and Asia. But the companies have undertaken to pay income taxes on their Jamaican operations at a rate of 33.3 per cent.

The government also sought to increase its markets for bauxite and alumina. Efforts were made to renew a contract with the Soviet Union, which bought 1m tonnes of ore per year for seven years. The agreement expired in 1989, and the government has indicated that it wants to supply more.

There were also indications that by late 1989 Jamaica and Venezuela would agree on a refining and smelting project, with Jamaican alumina feeding a smelter built in Venezuela.

The supply contract between the previous Jamaican administration and Marc Rich, the commodity trader, was being reviewed by the new government, to determine if it was 'in the best interests of the country', according to Hugh Small, the mining minister. There was no indication what would follow the review, although the government played down reports that it could be involved in another venture with a Rich subsidiary to supply bauxite to a refinery in the US Virgin Islands.

The Jamaican government was unable, however, to make use of a new market which industry spokesmen had earlier said offered great potential. In June 1989 government officials said the island would not be able to supply China with alumina as had been agreed in principle in 1988. The China National Import/Export Corporation had initially wanted 350,000 tonnes of alumina, but the state-owned Bauxite and Alumina Trading Company said in late 1988 that it could supply only 100,000 tonnes over three years because it had to meet its contractual obligations.

However, in June 1989, the government said it would be unable to supply any alumina to China because there was none of the material available. This was partly due to the production shortfall of 30,000 tonnes suffered by the industry when refineries were closed briefly after the hurricane.

Marginal growth in tourism

The effects of the international stock market collapse in the last quarter of 1987, and damage caused by Hurricane Gilbert in September 1988, contributed to a decline in tourist arrivals in 1988.

Although cruise ship arrivals increased 25.9 per cent in the year, stopovers declined by 12.2 per cent to 648,800 visitors, leading to a 1.7 per cent fall in arrivals. The storm damaged several hotels in the main resort areas, but even after they were repaired the sector suffered from other disruptions to telecommunications, electricity and water supplies. The decline was reflected in a fall in the economy's earnings from tourism. Gross visitor expenditure for 1988 fell 11.8 per cent to $525m.

The manufacturing sector also fell victim to the September 1988 storm. Output was running at 5 per cent higher than 1987 up to that month, but was reduced to a negligible level by the end of the year. The worst hit sub-sectors were those in food processing and other agricultural-based operations.

There was, however, expansion in non-traditional manufactured exports during 1988, which grew by 8.4 per cent to reach a value of $223m. This was led by the garment and textiles sub-sector, but the stimulus for non-traditional exports continued to be duty free access to the US market for a wide range of products under Washington's Caribbean Basin Initiative.

There continued, however, to be uncertainty over the future of the fledgling ethanol industry, one of the main non-traditional exports. Amid indications that US legislators would relax barriers to the importation of ethanol from Jamaica on a preferential basis, the new government said it was reviewing all ethanol production projects because of concern about the soundness of the investments and the cost-effectiveness of the product.

The new government announced in April 1989 that one programme inherited from Seaga which it would pursue was the divestment of state-owned hotels. Negotiations were being concluded with local and foreign companies and investors for the sale of 10 resort hotels, with six sale agreements likely to be concluded before the end of 1989, according to Finance Minister Seymour Mullings.

Earlier efforts at divestment saw the sale of a half of the island's largest commercial bank, the only cement producer and a slice of the government's stake in the island's telecommunications company. The response to the offer of the hotels had previously been unenthusiastic. Government officials said the tendency in the industry was for leasing properties, rather than outright purchase. The resorts were acquired by the government in the 1970s when government guaranteed loans were converted to equity after owners threw up their hands when there was a slump in the island's tourism.

But following an apparent change of attitude in the market, there was new interest in purchasing the resorts which contain about 3,000 rooms. Government officials say they expect sales to be based on room prices of between $25,000 and $50,000 each, suggesting that the government could earn as much as $150m from the deals. In its pitch to encourage investors, the government has pointed to steady growth in the island's tourism, with the volume of stayover visitors growing by 75 per cent in the past eight years, and projected to grow another 30 per cent by 1992.

Outlook

Although agriculture will take several months to recover fully from the effects of the hurricane, the economy is expected to resume the levels of growth of 1987. Mullings has predicted growth of 4 per cent for 1989. This can be achieved with expansion in bauxite mining and refining, which will come with the increase in capacity of some plants and the reopening of the island's largest refinery.

Tourism will continue to expand, though exogenous factors, such as the hardening of the US dollar (to which the Jamaican dollar is tied) against other currencies, could reduce the volume of visitors from the European market.

The new government has promised increased incentives for the island's manufacturing sector, and these should provide a basis for the expansion of manufacturing for both the domestic and export markets.

The government is likely, however, to meet resistance from trade unions over some aspects of its economic programmes, particularly a 10 per cent ceiling on wage increases, at a time when inflation is projected to be marginally lower.

Politically, with the next general election due in 1994, the JLP will provide vigorous opposition to the PNP's policies. The local government elections, due in mid-1989, have been postponed for a year, but they should give an indication of the changing fortunes of the two parties. Seaga will also provide answers to questions about whether he will stay on as JLP leader after the defeat in the 1989 general election.

COUNTRY PROFILE

Political system: Independent state within Commonwealth with the British monarch as head of state, represented by a governor-general. He is appointed by the Crown on the advice of the prime minister and must have no affiliation with any political party. Executive power rests with the cabinet – made up of the prime minister and at least 11 ministers. The governor-general appoints a leader of the opposition from members who do not support the government. The legislature is bicameral: the 21-member Senate (13 chosen on the advice of the prime minister, eight on that of the leader of the opposition) and the House of Representatives (60 members elected for a five-year term by universal adult suffrage).

Main cities/towns: Kingston (population 626,000), Montego Bay, Mandeville, May Pen, Spanish Town, Ocho Rios, Port Antonio, Port Morant, Savanna La Mar.

Media
Press: The three major dailies are *The Daily Gleaner*, *The Star* and *The Jamaica Record*. Also numerous periodicals.
Broadcasting: Radio broadcasting operated by the publicly owned Jamaica Broadcasting Corporation (JBC) and the commercial Radio Jamaica Ltd. JBC also operates a television service for about 108 hours per week. A public broadcasting service is in the process of being established for radio and television.

External trade: Heavy reliance on commodity exports and dependence on energy imports continues to cause severe balance of payments difficulties.
Exports: Alumina/bauxite (47% of total value), sugar (7%), bananas (3%), citrus, cocoa, coffee, pimento, rum and gypsum. Re-exports including chemicals, machinery, transport equipment and miscellaneous manufactures represent 16 per cent of total value.
Imports: Fuel (18% of total value), food (8%), transport and equipment (5%), other machinery and equipment (16%). Main sources: USA (49% of total value in 1987), Venezuela (12%-15%), UK (7%), Netherlands Antilles (11-13%), Japan, Canada (4-6%).

Agriculture: Sector contributes around 8.9 per cent to GDP, but employs over 25 per cent of the workforce. The main crop is sugar-cane, which accounts for seven per cent of export earnings (1987) and is an important supplier to the local beverage industry. The sugar industry has, however, suffered enormous losses. Production of the other main export crop, bananas, has yet to fully recover from hurricane damage suffered in 1988; 7,100 acres of bananas for export and 12,000 acres for local consumption were destroyed. In 1992 the target of 150,000 tonnes of fruit per year for the UK should be met. Other important cash crops include coffee, cocoa, pimento (the world's largest exporter), ginger, tobacco, pineapples and coconuts. Rice, maize, various vegetables and citrus fruits are largely grown for local consumption. Commercial fishing and livestock-rearing are important sources of food. Agricultural development is a major government priority and is to be centralised under the Agricultural Development Corporation (ADC) and the Rural Agricultural Development Authority (RADA).

Industry: Sector contributes 22 per cent to GDP (1987) and employs 12 per cent of the workforce. Agro-allied industries dominate, particularly food processing, textiles, sugar refining, paper products, cigarettes and beverages. Other important manufacturing industries (most of which are foreign-owned and heavily dependent on imported materials and components) include chemicals, machinery and tools, glass, cement, metal products. The main capital-intensive industries are petroleum refining at Kingston (capacity 35,000 b/d) and the partial refining of bauxite for export. Emphasis is on expanding facilities at the Kingston and Montego Bay free zones (where an international digital telecommunication facility is to be set up), encouraging foreign investment in export-based manufacturing and exploiting new opportunities offered under the Caribbean Basin Initiative (CBI). Specific area of export targeted by government structural adjustment programme is non-traditional goods; substantial increase in this area over 1983-87 was due mainly to growth in garment sub-sector.

Tourism: Sector has steadily expanded and is now the major earner of foreign exchange. The number of tourist arrivals reached 1m in 1987, compared to 500,000 in 1979. Around 70 per cent of tourist visitors are from the USA, though arrivals from Canada, the UK, German FR and Italy have steadily risen. Emphasis is on hotel construction and refurbishment and privatisation of state-owned tourist facilities.

Mining: Mainstay of economy, generating around 60 per cent of export earnings and important source of government revenue. Activity centred on extraction of bauxite (world's third largest producer) and alumina refining. Known reserves of bauxite around 2bn tonnes, though most is of relatively low quality. Major bauxite mining companies are Kaiser, Alcan and Alcoa. Bauxite production around 7m tonnes in 1986 and 1987 (compared with 12m tonnes in 1980-81), and bauxite exports increased by $23.3m in 1987. Alumina expensive to produce in Jamaica (compared with other producers such as Brazil, Australia, Guinea) due to fuel costs, depreciating currency and bauxite levy, and refineries tend to close and reopen according to market conditions. In an attempt to protect the bauxite industry against world commodity price shifts, the government has sought countertrade agreements with countries such as USA, Colombia, Yugoslavia and the USSR. Other minerals exploited include gypsum, marble, silica and clays.

Energy: Almost all the country's energy needs are imported. Oil is supplied at concessionary rates by Mexico and Venezuela under the San José agreement. The major consumer is the alumina industry, where energy accounts for up to 40 per cent of production costs. The industry is currently being encouraged to use coal rather than oil as the prime energy source. The potential for hydroelectric schemes is small and most electricity is generated in oil-fired power stations. The government plans to convert much of the oil-fired power system to peat, of which the island has vast reserves.

Membership of international organisations: Caribbean Development Bank, CARICOM, Commonwealth, Economic Commission for Latin America, FAO, GATT, ILO, IMF, Inter-American Development Bank, Organisation of American States, state associated with EC (Lomé Convention), UN, WHO, World Bank.

Time: GMT -5.

Climate: Tropical and with a mean annual temperature of 27°C. Months between May and Oct are usually the wettest and most humid.

Entry requirements
Passport: Required by all except citizens of US and Canada entering from their own countries who requqire only proof of identity for stays under six months.
Visa: Not required by some Commonwealth countries, certain European countries and Japan (all subject to certain conditions – prior checking advisable).
Currency: It is an offence to bring in or take out Jamaican currency and all foreign currency must be declared on arrival. No restriction on amount of foreign currency that can be brought in. There are heavy penalties for currency offences.

Health precautions
Mandatory: Yellow fever and cholera vaccination certificates required if arriving from infected areas, especially African and South American countries.
Advisable: Typhoid, polio vaccinations.

Representation overseas: Addis Ababa, Amman, Amsterdam, Beirut, Belize City, Bogota\o1, Bonn, Bridgetown, Brussels, Buenos Aires, Caracas, Castries, Georgetown, Guatemala City, Hamilton, London, Manila, Mexico City, Moscow, Muscat, Nassau, New York, Panama City, Paris, Port-au-Prince, Port of Spain, Ottawa, Rome, San José, Santo Domingo, Tegucigalpa, Washington.

Air access
National airline: Air Jamaica.
Other airlines: Aeroflot, Air Canada, ALM, American, British Airways, BWIA, Cayman Airways, Continental, COPA, Cubana, Northwest.
Tax: J$80 payable on departure.
International airports: Kingston – Norman Manley International (KIN), 17 km from city; duty-free shop, bar, restaurant, buffet, bank, post office, car-hire; Montego Bay – Donald Sangster International (MBJ), 4km from Montego Bay; duty-free shop, bar, restaurant, bank, post office, shops, hotel reservations, car hire (restricted hours in some instances).

Surface access
Main ports: Kingston, Montego Bay, Port Antonio.

Hotels: Around 145 hotels and guest houses available, all graded and mostly primarily for holiday-makers. 10 per cent tip is usual.

Car hire: Widely available. International or national licence accepted; traffic drives on the left.

City transport
Taxis: All have red PPV plates. Advisable to negotiate fares in advance. Taxis in Kingston no longer use meters. 10 per cent tip usual.

National transport
Air: Trans-Jamaican Airlines links Kingston and Mandeville, Montego Bay, Ocho Rios, Port Antonio; also Negril to Montego Bay.
Road: Extensive network of surfaced, all-weather road (over 4,800 km) plus some 11,200 km of secondary roads, half of which is suitable for light motor traffic.
Buses: Those in towns generally rated cheap but crowded; in country often considered slow and sometimes dangerous; minibus services are also available; there are regular, scheduled services over longer distances (e.g. Kingston-Montego Bay; journey time varies, to some extent dependent on route).
Rail: 292 km of track. Several daily trains between Kingston and Montego Bay. First class advisable. Line between Kingston and Port Antonio in disrepair.

Public holidays: 1 Jan, Ash Wednesday, Good Friday, Easter Monday, 23 May (Labour Day), first Mon in Aug (Independence), third Mon in Oct (National Heroes' Day), 25 Dec, 26 Dec.

Working hours
Government and business: 0830 – 1600/1700. (Exceptions: some offices open Sat; Sun closed.)
Banking: 0900 – 1400. (Exceptions: Fri 0900 – 1200, 1430 – 1700; Sat, Sun closed).
Shops: (Mon-Sat) 0900 – 1600/1700. (Exception: Wed and Thu 0900 – 1200/1300 Downtown and Uptown, respectively).

Telecommunications
Telephone: Dialling code for Jamaica: IDD access code + 1-809 followed by subscriber's seven-digit number. No area codes needed. 100 per cent automatic service. Most telephones in Kingston.
Telex: Wide range of international services offered by Jamaican International Telecommunications Ltd (JAMINTEL).

Banking
Banking centre: Kingston.
Central bank: Bank of Jamaica.
Commercial banks: Bank of Commerce Jamaica, Bank of Credit and Commerce, Bank of Nova Scotia Jamaica, Century National Bank, Citibank, Eagle Commercial Bank, First Jamaica National Bank, Jamaica Citizens' Bank, Mutual Security Bank Jamaica, National Commercial Bank of Jamaica, Workers' Savings and Loan Bank.

Advertising: Limited local publication of trade journals but wide use of advertising in the press and on television, radio, cinema and posters.

Electricity supply: 110/220 V AC, 50 cycles.

Representation in capital: Argentina, Australia, Belgium, Brazil, Canada, China PR, Colombia, Costa Rica, Dominican Republic, France, German FR, Haiti, India, Israel, Japan, Korea PDR, Republic of Korea, Mexico, Netherlands, Nigeria, Panama, Peru, Spain, Trinidad and Tobago, UK, USA, USSR, Venezuela.

Hotels: Kingston
Four Seasons Hotel, 18 Ruthven Road, Kingston 10 (tel: 926 8805).

Mayfair, West Kings House Close, P.O. Box: 163 (tel: 926-1610; tx: 529087).

Medallion Hall Resort Hotel, 53 Hope Road, Kingston 6 (tel: 927 5721).

Oceana, King Street/Ocean Boulevard, P.O. Box: 986 (tel: 922-0920; tx: 2230).

Pegasus, 81 Knutsford Boulevard, P.O. Box: 333 (tel: 63690; tx: 2178; fax: 94062).

Terra Nova, 17 Waterloo Road (tel: 926-2211; tx: 2384).

The Courtleigh Hotel, 31 Trafalgar Road, Kingston 10 (tel: 9268174-8).

Wyndham New Kingston, 77 Knutsford Boulevard, P.O. Box: 112 (tel: 926-5430; tx: 2152).

Hotels: Montego Bay
Casa Montego, P.O. Box: 161 (tel: 952-4150; tx: 5321).

Doctor's Cave Beach, P.O. Box: 94 (tel: 952-4355; tx: 5328).

Half Moon Club, P.O. Box: 80 (tel: 953-2211; tx: 5326; fax: 953-2731).

Holiday Inn, P.O. Box: 480 (tel: 953-2485; tx: 5327).

Jack Tar Village Montego Bay, P.O. Box: 144 (tel: 952-4340; tx: 5321; fax: 9526633).

Montego, Federal Avenue (tel: 952 3286).

Roundhill, P.O. Box: 64 (tel: 952-5150; tx: 5348).

Sandals Royal Caribbean, P.O. Box: 167 (tel: 952-5510; tx: 3815333; fax: 9520812).

Wyndham Rose Hall Beach Hotel & Country, Rose Hall (tel: 953 2650).

Hotels: Ocho Rios
Club Caribbean,, Runaway Bay, St. Ann (tel: 973 3507).

Couples, Ocho Rios, St Mary, P.O. Box: 330 (tel: 974 4271; fax: 974 4439).

Jamaica Jamaica, Main Street, Runaway Bay (tel: 973 2436).

Sans Souci Hotel, Club and Spa, P.O. Box: 103 (tel: 974 2353-4; tx: 7496).

Shaw Park Beach, Ocho Rios, St Ann (tel: 974 2552-4).

Turtle Beach Towers, Ocho rIOS, sT. aNN (tel: 974 2801-5/974 2646).

Hotels: Portland
Trident Villas and Hotel, Anchovy, Port Antonio (tel: 993 2602/2470).

Car hire: Kingston
Avis: 3 Oxford Road (tel: 926-1560; tx: 2283).

Bargin Rent-a-Car: 10 Merrick Avenue (Tel: 926-8021/8237)

Europcar National: 19-21 Carlton Crescent (tel: 929-9190/9196; tx: 3540).

Hertz: Pegasus Hotel (tel: 926-3690).

Business advice and information
Jamaica Tourist Board, Dominica Drive, Kingston (tel: 929-9200).

Jamaica Chamber of Commerce, PO Box 172, East Parade, Kingston (tel: 922-0150,922-2081).

Jamaica Economic Development Agency (JAMPRO), 35 Trafalgar Road, Kingston 10 (tel: 929-9450, 929-9461; tx: 49650).

Jamaica Manufacturers Association, 85a Duke Street, Kingston (tel: 922-8880/2).

Jamaica Information Service, Kingston (tel: 926-3740, 926-3590; tx: 2393; fax: 926-6715).

Manpower Division, Ministry of Labour and Employment, 1f North Street, Kingston (tel: 922-9500).

Kingston Free Zone, Lot 27, Shannon Drive, Kingston 15 (tel: 923-5274/8).

The Daily Gleaner, 7 North Street, PO Box 40, Kingston (tel: 922-3400).

Banking
Bank of Jamaica (central bank), PO Box 621, Nethersole Place, Kingston (tel: 922-0750; tx: 2165, 2173, 2167, 2144).

Bank of Nova Scotia Jamaica, Duke Street/Port Royal Street, Kingston (tel: 922-1000; tx: 2297).

CIBC Jamaica Ltd, 121 Harbour Street, Kingston (tel: 922-6120; tx: 2169).

Citibank, 63 Knutsford Boulevard, Kingston (tel: 926-3270; tx: 2115).

Jamaica Citizens' Bank, 4 King Street, Kingston (tel: 922-5850; tx: 2129).

Mutual Security Bank, 30-36 Knutsford Boulevard, Kingston (tel: 929-8950; tx: 2126).

National Commercial Bank of Jamaica, King Street, Kingston (tel: 922-6250; tx: 2139).

Workers' Savings and Loan Bank, 134 Tower Street, Kingston (tel: 922-0196; tx: 2226).

Japan

David Watts

Japan entered a new era in 1989 not only with the inauguration of a new imperial era but also with questions being asked about the political status quo and the norms and values of public life. Just as the monarch finally distanced the ruling house from the Second World War so the political world underwent the start of what promises to be a transformation.

In the space of a year Japan's ruling Liberal Democratic Party (LDP) has seen all the cosy assumptions of 34 years of unbroken rule swept aside in a period of dizzying political change which saw three men holding the office of prime minister. Perhaps nowhere else in the world, with the exception of the pre-Gorbachev Soviet Union, had a country's political make-up seemed so impervious to change through public pressure or reform of the system from within. Yet within the space of a few months the LDP was transformed from something approaching a machine seemingly destined to rule Japan indefinitely to a tired and corrupt-looking group chronically unable to match with political skill the country's superb technological and industrial capabilities.

Despite the shock of electoral defeat and widespread public disapproval the men who controlled the levers of power within the party were still exhibiting a failure to understand the need for real rather than cosmetic reform of the political system which, until 1989, served Japan so well.

The first signs of the impending tornado were barely visible on the horizon when Noboru Takeshita took office as prime minister for what appeared likely to be a period in the political doldrums under the tutelage of a man not known either for the originality of ideas or personal dynamism, particularly in contrast to his predecessor Yasuhiro Nakasone. He was portrayed as a man more comfortable with behind-the-scenes negotiated compromise than with the direct almost brusque style of his predecessor and many Japanese felt it was time for such a man. For the party he was the one who would negotiate badly-needed tax reform through the system after the failure of Mr Nakasone's more direct methods caused outrage in some quarters. Mr Takeshita was seen as a man with the right contacts from his days as party secretary-general to bargain for a bill that seemed to be popular with no-one but those concerned with tax revenues and the need, not only to raise more taxes, but also to make the taxpayers' burden fall more evenly across the board. Until the new proposals were put forward the tax burden had fallen unfairly upon the salaryman, the ordinary office worker whose salary can be predicted with certainty years ahead and for whom there are none of the loopholes available to the self-employed or to the politicians.

But even as the delicate manoeuvring was under way the affair that was to transform the face of Japanese politics was already threatening to torpedo all Mr Takeshita's best-laid plans – the Recruit share scandal. Opposition determination to block what was widely seen as an unpopular tax and to force the government to hold a full-scale debate on the scandal, which had been bubbling through the political world for months, precipitated a final 26-hour parliamentary session to clear the bill. Opposition members outdid each other in attempting to be the slowest to place their ballots with the speaker. But Mr Takeshita eventually won what he saw as his most important political objective and the public was given a new tax which the government would eventually have to start softening.

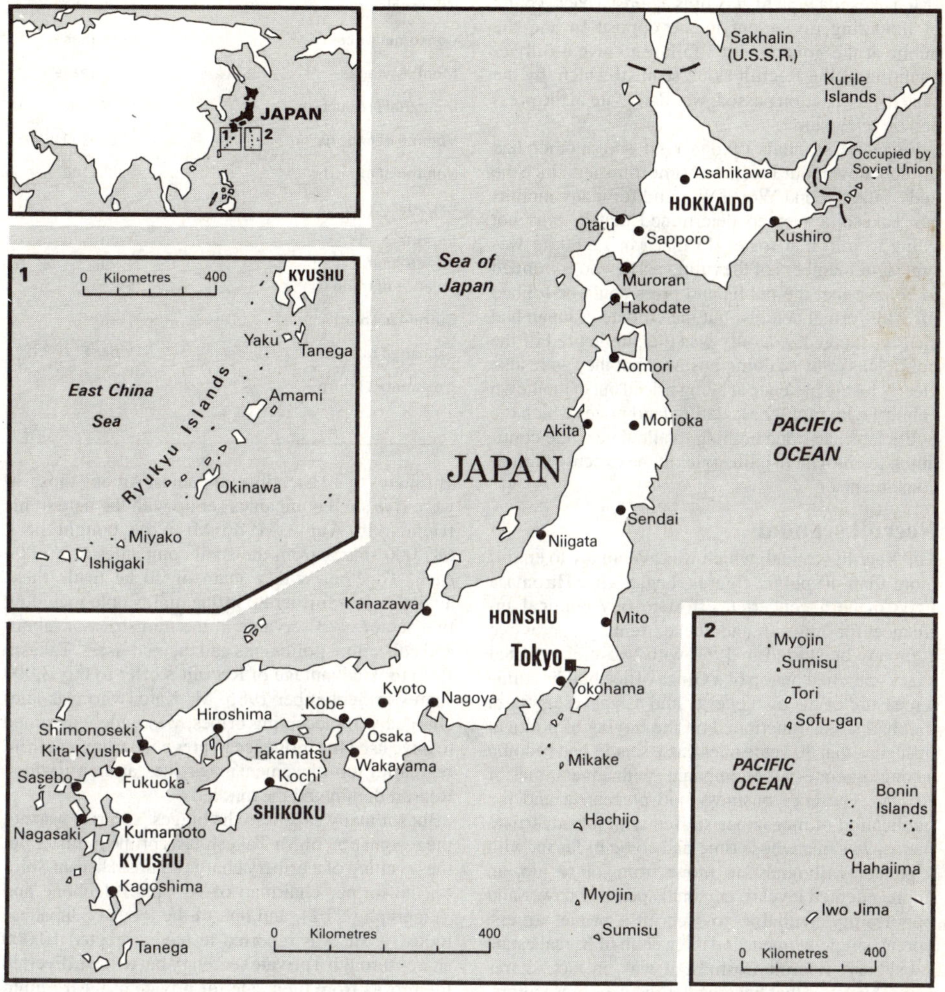

For the first time Japanese consumers were faced with a VAT-type 3 per cent sales tax on even the most modest of purchases – a considerable contributory factor in the growing disaffection with the government over succeeding months. The balancing factor in the government's package was the reduction of tax payable but the amount turned out to be so modest for most people that Mr Takeshita scored no political points despite engineering the most far-reaching overhaul of the tax system in 40 years.

Under the new system taxes on such luxury goods as liquor, cars, electrical appliances, luggage and golf equipment fell from about 20 per cent to a uniform 3 per cent. Imports of luxury items from abroad seemed likely to benefit from increased sales and the Economic Planning Agency estimated that tax cuts would

KEY FACTS

Official title: Japan
Head of state: Emperor Heisei
Head of government: Prime Minister Toshiki Kaifu
Ruling party: Liberal Democratic Party
Capital: Tokyo
Official Languages: Japanese
Currency: Yen (Y)
Exchange rate: Yen 144 per $ (Nov 1989)
Area: 377,728 sq km (3,900 small islands)
Population: 121.8m (1987)
GNP per capita: $16,136 (1987)
GNP real growth: 3.8% (1988)
Labour force: 59.1m (1988)
Inflation: 0.7% (1988)
Trade balance: $94.99bn (1988)
Foreign debt: $145bn* (1987)
Visitor numbers: 2.35m (1988)

* estimated figure

total $46bn during 1989-90. Overall Japanese taxpayers should be paying some $21bn less annually. But for the average salary earner that is scant comfort – with virtually no allowances to make against income he and his wife can only look forward to paying more for daily necessities.

Mr Takeshita won his tax bills against a background of mounting discontent in the opposition and the public at the government's failure to give a full accounting of the Recruit share scandal which, by the time the legislation passed, was daily fare in the press and on television.

Financial scandals of one kind and another had brought down four Japanese prime ministers since the end of the Second World War and for many months Mr Takeshita seemed determined that he was not going to join that roster. In that Mr Takeshita was typical of members of the ruling party who continued to believe that the public and press could be fobbed off with formal denials; but the climate in Japan had clearly changed: not only was the public tired of the politicians' way of doing business but they were also clearly losing patience at being priced out of their own country with rapidly escalating land prices which the politicians had done nothing to alleviate while continuing to enjoy lavish lifestyles at the expense of their constituents.

KEY INDICATORS

	Unit	1984	1985	1986	1987	1988
Population	'000	120,235	121,000	121,490	121,800*	–
Gross domestic product (GNP)	Y bn	298,453	317,252	331,254	345,292	366,366
Wholesale prices	% increase	–0.3	–1.1	–9.1	–3.7	–1.0
Consumer prices	% increase	2.3	2.0	0.6	0.1	0.7
Monthly wages	% increase	3.4	3.2	2.8	2.1	3.7
Industrial production	% increase	9.4	3.7	–0.3	3.4	9.4
Volume of exports	1985=100	95.3	100	99.4	99.8	104.2
Volume of imports	1985=100	100	100	110.5	120.5	140.5
Exports	Y bn	40,325	41,959	35,291	33,316	33,928
Imports	Y bn	32,320	31,076	21,551	21,739	24,007
Petroleum	Y bn	10,804	9678	4064	3,970	3,307
Balance of trade	$bn	44.26	55.99	92.82	96.46	94.99
Current account	$bn	35.00	49.17	85.83	87.00	79.59
Exchange rate	Y per $	237.52	238.54	168.52	144.64	128.15

* estimated figure

Recruit scandal

The Recruit scandal, which was eventually to engulf more than 40 public figures, began after Hiromasa Ezoe made a concerted effort to buy political influence for himself and the extremely aggressive company he started in 1960 with two prefabricated shacks perched on top of a Ginza office building. It is a measure of his own energy and a Japanese system that had so far functioned on the buying of political influence that 30 years later those shacks had become a conglomerate of 27 companies with annual sales of £1.5bn. The core business, job placement and the publication of magazines stuffed with job advertisements, was one whose time had come in Japan, with a greater willingness to move from job to job, an unprecedented level of personal spending power and opportunity available to Recruit's prime targets, young, single women, and the wealth of its real estate subsidiary, Recruit Cosmos. It was, in fact, shares from this latter that became the focus of prosecutors, even though the transfer of shares between private individuals was not illegal and had been a means of transferring money to politicians since large-scale cash donations became illegal in the 1970s.

The affair began humbly enough and would almost certainly never have reached national prominence were it not for the unaccustomed persistence of a district journalist with the *Asahi Shimbun* newspaper in Kawasaki who investigated the purchase of Recruit Cosmos shares by Hideki Komatsu, the deputy mayor of Kawasaki. Mr Komatsu helped Recruit obtain planning permission for a new office building and was one of the original 76 beneficiaries of Recruit largesse.

During the first stage of the scheme, in late 1984 and 1985, the Recruit boss Hiromasa Ezoe made available large numbers of Recruit Cosmos shares to the 76. To make his blandishments even more attractive he offered low-interest loans to those who agreed to buy the shares. The first tranche of these was released for about Y1,200 or £5 a share. Those who sold when the Recruit was first listed on the Tokyo market stood to make profits of £10.60 a share. Three politicians were named in connection with this first phase of the affair and the president of the *Nihon Keizai* newspaper, Japan's leading financial daily, Koh Morita, resigned after admitting that he was one of this initial group.

The second, more aggressive, phase of Recruit's attack came in 1985 when Recruit Cosmos increased its capitalisation allowing Mr Ezoe to sell more than 1m shares to 30 friendly companies. Among those 30 were five shell companies set up with the help of his friends. In August 1986 Mr Ezoe bought back 687,000 shares from the shell companies at Y3,000 each. Together with a team of 20 he made these available to a further 80 influential people including two senior civil servants in the ministries of labour and education, politicians and their advisers. Takashi Kato took advantage of Recruit's offer to buy 3,000 shares in September 1986. Mr Kato, who was later charged with accepting bribes, was widely assumed to have used his influence to prevent parliament from regulating job-placement magazines, an area in which Recruit dominated the market.

But for many Japanese the biggest surprise was not the resignation of Mr Takeshita as prime minister but the levelling of a bribery charge against Hisashi Shinto, the former chairman of Nippon Telephone and Telegraph (NTT), and one of the icons of Japanese industry. He was reported to have accepted 10,000 shares through a private secretary but to have diverted the profits from their sale for private use. Mr Shinto was alleged to have helped Recruit break into the telecommunications market through leasing NTT circuits and helped them get cheap access to a US supercomputer by giving them one which had been bought from the United States as part of the government's measures to alleviate trade friction.

The first major casualty of the affair in December 1988 was Minister of Finance Kiichi Miyazawa who resigned after repeatedly changing his explanation of an aide's purchase of 10,000 shares. There is some evidence that Mr Takeshita had allowed the campaign against the finance minister to go forward as a way of distracting attention from his own involvement. As a tactic it was extremely short-lived and failed to achieve its purpose – each time the prime minister attempted to buy time he only managed to increase public frustration at not getting the whole story.

Pressure on Mr Takeshita increased after Saburo Tsukamoto, leader of the opposition Democratic Socialist Party (DSP), announced that he would step down as party leader after his connection with the affair became known within the party. Even though his involvement was modest by the standards of ruling party members, with only 5,000 of the shares to his name, he elected to go because of the image of the DSP at election-time. His resignation cleared the way for the DSP to attack the government on the Recruit affair. But it was not until April 1989, that Mr Takeshita decided to try and talk his way out of trouble by agreeing to appear before a parliamentary committee. By that time he had already lost control of any credible defensive position he might have put up after the revelation that his support groups received Recruit funds. When it came his explanation was vague, added little to the public's knowledge and, arguably, did more harm than good. Less than two weeks later the Recruit affair claimed its biggest victim when Mr Takeshita told the cabinet of his intention to resign.

The turning point came with a statement from Takashi Ishihara, chairman of the Japan Association of Corporate Executives, that a new leader was needed to break the political stalemate. It was the first sign of several which indicated that the business community was willing to go public with its concerns about the political situation and its effect on Japan's economic interests. 'The Recruit question has caused a grave crisis for the nation's parliamentary democracy.' Mr Takeshita told a television audience in announcing his decision to go. 'As the chief executive of the government and president of the ruling party, I deeply apologise.' Asked about a possible successor he said: 'It is not the place of one resigning to express such opinions.'

For once Mr Takeshita spoke not merely from ingrained modesty but reflected the unusual reality within the ruling party: there was no clear successor ready to step in. In normal circumstances the LDP's tradition of back-room bargaining would have already produced a new leader through a system which, undemocratic though it is, has always managed to achieve 'smooth' transitions of power, at least in public. Mr Takeshita's resignation was not immediate: he would leave office only on the choice of his successor.

The search for that successor made obvious the lack of suitable candidates and underlined the LDP's horsetrading style of internal politics. In fact, there was an excellent candidate available but it was a measure of the entrenched nature of the party leadership and its inability to recognise the gravity of the situation that he was not given a chance to make good on his offer of party reform. That man was Masayoshi Ito the chairman of the ruling party's executive council and a man so unlike his colleagues that he refuses to rub shoulders with the country's business elite, has never held any of the customary politicians' fund-raising parties and rents a modest two-storey house with a tin roof. He had the temerity to suggest to Mr Takeshita that politicians be barred from holding stocks. After fighting shy for health reasons Mr Ito eventually demanded a programme of reforms – reasonable in the circumstances – which the party hierarchy could not accept. His suggestions may have seemed unworldly to those in the LDP but they began to look downright essential to the general public if Japanese at home and abroad were to be able to speak unapologetically about their politicians in future.

Though Mr Takeshita might seem a big enough fish to have been caught by the scandal, there was at least one other of greater influence and import within the party who escaped the net. For a long time it looked as though the former prime minister, Yasuhiro Nakasone, would have to give an accounting before the Diet of his Recruit connections and for some time the opposition held up deliberations with that in mind. Mr Nakasone admitted that three of his employees bought 29,000 shares in September 1986 and sold them again at a profit of just under £300,000. He claimed the money was used to pay for his office administration, though he had no knowledge of the transaction. Although he initially claimed ignorance of NTT's purchase of a third supercomputer, which eventually found its way to Recruit, some library television video tape reminded him of the discussions he had on that very subject with the then President Reagan.

A close associate of Mr Nakasone and his chief cabinet secretary, Takao Fujinami, was also named in Recruit investigations. His secretary reportedly accepted the transfer of 12,000 stocks and Mr Fujinami himself received almost £100,000. But if the old guard of the LDP felt the formula of sacrificing an unpopular prime minister would quickly restore their fortunes, as it had done in the past, they were profoundly wrong. In many ways Mr Takeshita's apologetic departure was merely the first of a series of misfortunes.

Former foreign minister Sosuke Uno was put forward as prime minister. The party clearly saw him as the classic stop-gap until better times and one, moreover, from Mr Nakasone's own faction. Already saddled with those shortcomings in the eyes of the opposition Mr Uno said on accepting the post: 'I am not confident I can handle the job when the party is facing such a difficult time. But now that the party is in an unprecedented crisis, I will commit myself for the party.' Not only opposition party members were angry: 'I am outraged by this nonsense of giving the country's highest office, which only one person can assume, to a sort of managing director of the Nakasone company. I am totally amazed,' said former prime minister Takeo Fukuda. Young hawks in the LDP were unhappy and called for the resignation of senior officials to 'take responsibility' for the situation.

Even before he took office Mr Uno's candidacy was seen as weak and controversial though his mandate was to restore confidence. At best he might be able to restore unity within the party. So it took only the slightest whiff of what in past years would have been no scandal at all to upset the already frail LDP boat.

Until 1989 the maintenance of a mistress by a politician had been if not actually expected as a sign of a man's wealth and power then tolerated. But much to the chagrin of the LDP leadership the rules suddenly appeared to change and the first victim became Mr Uno. The new prime minister's nemesis proved to be a lady who described herself as a geisha and told how he had paid her for sex over a long period. The story surfaced in a weekly magazine and might not have developed beyond that were it not for the foreign press reporting it at some length in Washington and London. Changing attitudes in Japan, among journalists and the public, helped fuel the chase. Journalists seemed to see Mr Uno as fair game and women certainly felt it was time for a change in the rules. Despite the fact that it was clearly not in Japan's interest to have yet another change on the bridge so soon, with the annual summit of industrial nations in Paris only weeks away, the press went after the prime

COUNTRY PROFILE

Political system: Japan's former head of state, the octagenarian Emperor Hirohito, was the world's longest serving monarch with 63 years on the chrysanthemum throne. Until Japan's defeat in the Second World War, Hirohito was regarded as a living god. He died in January 1989. Real power is exercised by politicians and bureaucrats since Japan is a parliamentary democracy with a bicameral legislature known as the Diet. The 511-member house of representatives elected for four-year term and by election by universal suffrage has predominance over the 252-member house of councillors (six-year terms) because of their power over money bills. Since 1955 the prime minister has been the head of the ruling Liberal Democratic Party, a conservative grouping. The LDP president is chosen by the party for two-year terms. In 1986 the LDP recovered its absolute majority in the Diet. The 1989 Recruit scandal shook Japan's establishment. A former minister, two senior civil servants and several top businessmen were indicted on bribery charges arising from their associations with the Recruit company. Mr Noburu Takeshita, then Prime Minister, was forced to resign. The new Prime Minister, Mr Sousuke Uno, was then cited as having had an affair with a prostitute and the Liberal Democratic Party suffered another fall in popularity.

Political parties: Ruling party: Liberal Democratic Party (300 seats in the lower house of the Diet, following Jul 1986 elections); Japan Socialist Party (85 seats), Komeito ('Clean Government Party') (56), Democratic Socialist Party (26), Communist Party (26), New Liberal Club (6), Social Democratic Federation (4). Japanese political style is based on consensus rather than direct confrontation. When the opposition party boycotted the Diet in 1987 the government withdrew controversial tax measures rather than steamroller them through. Factions within the ruling party are as important as the opposition.

Population: Japan's population of 122.09m (1988) is growing at only 0.6 per cent per annum. 80 per cent live on the northern island of Honshu, and density in the major cities (Tokyo, Osaka, Nagoya) is extreme. Buddhist and Shinto religions continue to play a significant role in cultural, philosophical, and even business and political spheres.

Main cities/towns: Tokyo, on Honshu (population in 1987); Yokohama, on Honshu (pop 3.0m); Osaka, on Honshu (pop 2.5m); Kobe, on Honshu (pop 1.4m); Nagoya, on Honshu (pop 2.1m); Kyoto, on Honshu (pop 1.0m); Kita-Kyushu, on Honshu (pop 1.1m); Fukuoka, on Honshu (pop 1.1m); Sapporo, on Hokkaido (pop 1.5m); Hiroshima, on Honshu (pop 1.0m).

Language: Japanese is the official language. Pupils are taught English in school for seven years but this involves a formal grammatical knowledge rather than spoken English. It is hard to move in Japan without some knowledge of Japanese.

Media

Press: Japan has the highest newspaper circulation and readership in the world, and 70m newspapers are published daily. The major national dailies are the *Yomiuri* (circulation 8.8m), the *Asahi* (7.6m), the *Mainichi* (4.2m), the *Chunichi* (2.8m) and the *Sankei* (2m). The four English-language dailies have a readership of about 150,000: the *Japan Times*, the *Mainichi Daily News*, the *Daily Yomiuri* and the *Asahi Evening News*. News magazines in English include *Tokyo Business*, *Business Tokyo* and the more general *Tokyo Journal* which includes an excellent monthly guide to what's on in the capital. Business: The *Nihon Keizai Shimbun* (circulation 2.8m) is the leading financial daily, and there are three major industrial dailies: the *Nikkan Kogyo* (circulation 545,000), the *Nihon Kogyo* (409,000) and the *Nikkei Sangyo* (203,000). The *Nihon Keizai Shimbun* publishes a weekly edition in English, the *Japan Economic Journal*. *Toyo Keizai*, the main economic monthly, also publishes an English version, the *Oriental Economist*. *Business Tokyo* (monthly) gives a useful summary of business developments in Japan. A trade press of 7,000-8,000 publications covers most aspects of business and the economy. In all there are some 150 broadcasting companies, many broadcasting locally only. *Radio:* There are numerous radio broadcasting services, both government and commercial. The English language Far East Network, the radio station of the US forces, is available 24 hrs on both medium wave and FM. Another English language staion, KYTO, is available. *Television:* Of the numerous TV networks, the major national ones are the Nippon Television Network (NTV), the Tokyo Broadcasting System (TBS), Fuji Television and TV Asahi. Nippon Hoso Kyokai (NHK) is the non-commercial public network. Certain Tokyo hotels have closed-circuit English language TV. Some TV programmes including news are available in English through a bilingual converter.

Domestic economy: Japan's economy is second only to the US in absolute terms with total GNP of Yen 344,888bn (US$2,705bn). By 1987 Japanese were richer than Americans with per capita income almost $20,000, thanks to the rising value of the yen against the falling dollar. In the late 1980s Japan has emerged at the world's biggest creditor nation. Its net assets may rise to $1 trillion before the turn of the century, making Japan effectively the world's banker. Japan's economic strength comes despite a shortage of natural resources. Economic growth rates during the 1950s and 1960s were the highest in the world based on high levels of savings and investment, rapid productivity growth and remarkable social consensus. Japanese were prepared to forego immediate reward for long-term benefit. As a result, in large sections of world manufacturing, notably electronics, Japanese producers have no rivals. Manufacturing was the mainstay of the economy, improving quality and price. Japan suffered badly in the 1970s because of the oil price crises, but by the end of the decade oil usage per unit of output was lower than any other industrial country and the economic machine was rolling again. Japan has continually upgraded its economy and shifted from heavy industry with high energy requirements to high technology, high value-added industries such as semi-conductors, industrial robots and computers. In the late 1980s heavy Japanese industries, like shipbuilding and steel, were under pressure from more efficient Korean producers. In the mid 1980s the rapid surge in the value of the yen following the Plaza Agreement of September 1985 posted a high challenge for Japanese companies. From Yen 240 against the US dollar the currency eventually settled down in the 130-125 yen range by mid-1988. Japanese companies pared costs and accepted lower prices so that in mid-1988 Japan was still showing a trade surplus of $90-95bn and a current account surplus of over $80bn. Japan has been much criticised for its soaring trade surpluses but trade has traditionally been a much smaller percentage of national product (peaking at about 13 per cent) than in Western European countries. Growth picked up to 3.5-4.0 per cent range in 1987 and 1988, high for an industrialised country. But Japan's highly efficient manufacturing sector is offset by a labyrinthine distribution system and by a pampered and protected farm sector who form the backbone of support for the ruling LDP. Even so, Japan is the world's largest importer of agricultural products and 60 per cent of its food is imported. Rising Japanese government debt, with debt service taking 20 per cent off central expenditure, has been of concern and in 1988 the government finally passed measures for imposing a general three per cent sales/expenditure tax from 1989.

Employment: Japan has one of the lowest rates of unemployment in the industrial world. It peaked briefly at three per cent but had fallen to 2.7 per cent by 1989 thanks to higher growth rates. Japan's lifetime employment system means that workers will not be laid off or sacked when the economy turns down, so figures are not strictly comparable. There has been a sharp change in distribution of workers. The labour force has risen to about 60m. In 1960, 30 per cent were in primary industry, but this has fallen to 8.8 per cent. Secondary industry employs 35 per cent and tertiary industry 57 per cent. Manufacturing is the most important single employer followed by wholesale and retail trade. Female workers have risen in number (now 36 per cent of the total) as more wives continue to work till their first child. About 28 per cent of workers belong to trade unions, typically a single union in a company.

External trade: Japan's 8000 trading companies, which handle most exports and imports, are dominated by a few large general trading concerns which handle about 50 per cent of trade. Japan continues to run a considerable trade surplus inspite of the appreciating yen. This reached a record $96.5bn in 1987 and is expected to fall to about $80bn in 1988.
Exports: In 1987 exports were worth $224.4bn, and rose to $259.5bn in 1988. Main exports are motor vehicles (about 28 per cent), non-electric machinery (25%), other equipment (21%), metals and metal products (10%). Foodstuffs are a tiny part of exports, less than one per cent. In many areas Japan dominates world markets. In 1986 it exported more than 14m 33mm cameras, almost 28m video cassette recorders, 50m electronic calculators and 170m watches. Destinations for exports are USA (about 35%), South Korea and West Germany (5% each), China (5% in 1986).
Imports: Japan's import profile is unique among industrialised countries. It is the largest food importer in the world, taking more than $20bn of food imports inspite of restrictions on foreign sales of farm products in Japan. Fuel imports are $40bn a year, of which crude oil is half. Japan's imports of manufactured goods are lower than any other industrialised nation, worth about $45bn or only 36 per cent of total imports in 1986 (against 77 per cent in the USA and 68 per cent in West Germany). Import volume rose by 9.3 per cent in 1987. The USA is the biggest source of imports with nearly 25 per cent of the total. China, South Korea, West Germany and the UK each supply about 3-4 per cent.

Agriculture: Agriculture, fishing and forestry account for about 2.7 per cent of national income and 7.2 per cent of employment (1986). Most in farm employment supplement their incomes with non-farm employment. About 72 per cent of land area is mountainous, and almost 70 per cent is covered in forest. Only about 15 per cent of the land area is available for agriculture and stock-rearing, and this is constantly in demand for residential use. Agriculture is highly intensive, making considerable use of technology and capital investment. In 1986, Japan was 70 per cent self-sufficient in food, but only provided about 30 per cent of its cereals and fodder needs. Wet cultivation of rice is the main activity (virtually self-sufficient), with a trend toward production of beef, citrus fruits and tobacco. Wheat, barley, soya beans, potatoes, sweet potatoes, vegetables, fruit, tea and silkworms are also produced. Most forest areas are mountainous, but 37 per cent of forested area contains plantations of Japanese cedar (*sugi*), Japanese cypress (*hinoki*) and Japanese red pine (*akamatsu*). The forest industry supplies the domestic market with timber and wood products. 60 per cent of timber requirements are imported. Fishing contributes substantially to domestic food supply and export earnings. Japanese fishermen catch pollack, pilchards, cod, salmon, mackerel and other fish throughout the north and central Pacific, and are second only to the USSR in whaling. Aquaculture is well developed, and there are inshore fisheries for squid, clams, crustaceans, shallow-water fish and dolphins.

Industry: Manufacturing employs 25 per cent of the workforce and has made strong headway since the early 1980s. The sector has benefitted from innovative technology and considerable financial backing from the government, and leads the world in automated production processes; these in turn, however, are helping to reduce the industrial workforce, which is being increasingly redeployed in the tertiary sector. In the mid-1980s exports were the engine of economic growth, but by 1986 and 1987 the outcry against trade surplus led to softening exports and more domestic pump priming. Particularly affected were computers, video recorders and other electronics, some of which lost money due to glutted demand and slow product development, and also textiles, a traditional activity where production has shifted from natural to synthetic fibres. Automobile production, too, shrank after the record 12.27m units of 1985. The reorientation towards domestic demand has also compensated for the effect of substantial yen appreciation on the export sector. A sustained expansion of private consumption, a major expansion in public investment together with an easing of monetary policy that pushed interest rates to historically low levels, also played a role. Domestic demand is likely to be the main impetus of economic expansion.

Mining: Mining accounts for 0.1 per cent of employment. There are few exploitable mineral resources aside from coal, whose annual production is about 17m tonnes, leaving Japan as the world's largest importer of coal. Molybdenum, manganese, zinc, copper and iron are mined on a small scale. Japan is self-sufficient in sulphur and limestone.

Energy: Over 80 per cent of all energy needs are met by imported fuel, and 60 per cent of all energy needs by imported crude oil. Energy conservation programmes have resulted in reduced oil consumption. Combined with efforts to develop alternative energy sources, this has led to a drop in oil imports in recent years.

Financial markets: In the late 1980s the Tokyo stock market overtook New York in terms of capitalisation to become the world's largest stock market. The Osaka market was also bigger in terms of capitalisation than all except the very big markets of Tokyo, London and New York. There are also stock markets in Kyoto and Nagoya. Inspite of its own peculiar inward-looking ways Japan was increasingly becoming part of the world round-the-clock 24-hour trading system (with New York and London). The very weight of surplus capital meant that foreign investment banks were rushing to set up in Tokyo, even though few of them were profitable and it was hard to get a toe-hold in an exclusive and clubbish Japanese system. Six foreign firms were initially allowed membership of the Tokyo stock exchange in 1986.

Membership of international organisations: Asian Development Bank, ASPAC, CCC, Colombo Plan, DAC, ESCAP, FAO, GATT, IAEA, IBRD, ICAC, ICAO, ICO, IDA, IDB, IEA, IFAD, IHO, ILO, IFC, IMF, IMO, INTELSAT, International Lead & Zinc Study Group, Interpol, IPU, IRC, ISO, ITC, ITU, IWC, OECD, UN, UNESCO, UPU, WFTU, WHO, WIPO, WMO, World Bank, WSG.

minister with all the gusto of their counterparts in the West.

But Mr Uno survived to lead the Japanese delegation to the summit and it took a severe reversal in the upper house elections in July 1989 finally to unseat him. The price the party had to pay was stiff: losing its majority in the upper house for the first time in its history. It was also the source of initial early embarrassment for the new prime minister, Toshiki Kaifu, who found himself rejected by the upper house and the leader of the Japan Socialist Party, Miss Takako Doi, being approved as the country's new leader: another historic first. That was quickly reversed by the constitutionally more powerful lower house but the point was forcibly made: for the first time in more than 40 years a candidate of the ruling party was rejected by the upper house. It was the latest in a series of victories for Miss Doi and the 'madonna strategy' through which her party stressed women candidates and issues at a time when the LDP was weak on both. Miss Doi described the situation thus: 'Today is the beginning of the end for the LDP's monopoly control of politics. This, I believe, is certain.'

But Mr Kaifu, a former education minister, made few concessions to the new realism in choosing his cabinet. Certainly there were two women in the new line up, as directors of the economic planning and environment agencies, but beyond that the mix was very much as usual except for the choice of a new foreign minister, Taro Nakayama, with no previous diplomatic experience.

The duration of Mr Kaifu's tenure was unclear. He could not be seen as anything other than a temporary holder of the office and his connections with Mr Takeshita suggested that the struggle was likely to break out again long before the LDP dared to go to the country in a general election.

Death of the emperor

In the midst of all this political trauma the country had an equally momentous change to accommodate in the death of Emperor Hirohito. It is true that there was ample time to prepare for the transition: the emperor lay on his sickbed for many weeks while the country went into a state of semi-mourning. Commercial activity and entertainment were toned down as the country waited for the inevitable. The world's longest-reigning monarch passed away on 7 January 1989 having spanned Japan's brightest and darkest hours as the leader who took her into war and guided her through the American occupation into an era of peace and prosperity the like of which no other defeated industrial country has enjoyed.

But if the emperor's lingering presence had caused embarrassment and inconvenience for government, populace and business his funeral turned into one of Japan's most important diplomatic events for many years and attracted more foreign dignitaries than the record-breaking 119 country representatives who attended the funeral of President Tito in 1980. It is a measure of Japan's stature that despite the continued underlying tension with neighbouring countries left by the war the funeral roster recorded representatives from 163 countries, including 55 heads of state, 14 members of royal families, 20 vice-presidents and 11 prime ministers who were part of the 10,000 mourners. In fact, such was the crush of bilateral meetings that the original purpose of the gathering was in danger of being lost amid the more immediate pressures of international diplomacy.

The funeral prompted reassessments of the Hirohito era and brought two key historical revelations which cast further light on the late emperor's attitude to the war. The first was that he had tried three times to abdicate between 1945 and 1952 and the second was that President Truman gave a secret order that Hideki Tojo, Japan's wartime prime minister, be persuaded to take responsibility for the war and thus spare the emperor prosecution. Aristides Lazarus, a former Marine Corps lieutenant, said he had kept the mission secret until after the emperor's death but had promised Tojo that his role would eventually be revealed.

For many Japanese the passing of Emperor Hirohito brought hope that the country could put its war role behind it and march into a new future. At his first press conference the new emperor, Emperor Heisei, while not willing to comment substantially on the war role of his father, made it clear that he believed freedom of speech to be the very basis of democracy. His remark followed strong indications that he sees his role as being much closer to the people than his father. He began to use more ordinary forms of speech in contrast to the imperial language used by the late emperor. Other indications of the new style were the reduction of the guard maintained around the emperor, their increased use of civilian clothes, the insistence that imperial progress on the streets of Tokyo should no longer take precedence over ambulances and other emergency vehicles and an imperial willingness to obey traffic lights. It remains to be seen how much Emperor Heisei will be able to break free of his 'minders' in the imperial household agency. There were indications at the press conference not only that the questions had been notified in advance by the journalists but also that the emperor had memorised some of his answers.

The economy

But it will clearly take more than the country's changing image to brush away the trade problems which beset the country no less severely than they did under the *ancien régime*. The economic year began well enough with the March 1989 annual trade figures showing that Japan's current account surplus had shrunk for the first time in seven years in 1988, falling 8.7 per cent to $79.59bn from $87.00bn in 1987. The trade surplus also declined 1.7 per cent during the year to $94.99bn from $96.46bn. It was the first drop in the annual trade surplus in six years, according to the ministry of finance, but ministry officials were uncertain whether it would prove to be a trend that would be maintained during the succeeding year.

An important element in the reduction in the current account surplus was the record 8.40m Japanese who travelled abroad during the year and spent $15.76bn in the process.

The caution of the ministry of finance proved to be well-founded because the trade surplus for March 1989 widened by 1.3 per cent over the previous year to $95.27bn, exceeding the government's forecast by some $14bn. This did not augur well for the achievement of the government's original target of reducing the trade surplus for the 1989-90 fiscal year to $81bn. The ministry of finance's forecast of an $88bn surplus seemed likely to be closer to the mark.

Perhaps the most long-drawn-out and ultimately damaging of the disputes during 1989 may prove to be that over the development of Japan's next-generation jet fighter-bomber, the FSX, a twin-engined re-working of the General Dynamics F16 Fighting Falcon. Already an extremely touchy subject when the Bush administration came to power, the president inherited an agreement under which the aircraft would be co-produced by the two countries. That agreement was strongly opposed by many sectors in Japan and was only concluded after considerable pressure from the US side.

The Bush multi-agency approach to negotiations with Japan, however, opened the way for increased congressional oversight over the agreement and strong pressure from within that body for the United States to scrap it on the grounds that technology transfer would be one-way. The notion of America seeking to renege on an agreement that Japan had only reluctantly acceded to was met with predictable dismay and an increasing lack of faith in the United States.

American congressional concerns centred on the fact that United States' military technology might be used to build up Japan's nascent civil aircraft industry and whatever Japan might learn from the United States would not be reciprocated in the transfer to the United States of Japan's technical know-how in the fields of radar, radar-absorbent paint and carbon-fibre technology, fields in which the United States believes Japan may have an advantage. Long before the United States pushed through the agreement on the FSX there were indications that Japanese industry was making plans to go ahead on its own and during the long-drawn-out days of wrangling in Washington contracts were signed in Japan between the government and Mitsubishi for the aircraft. Development funding has been budgeted for some time and Japanese design teams are already in place.

Whatever the merits of the agreement or the quality of the hybrid aircraft it will produce, the affair has immeasurably strengthened the arguments of those who insist that Japan should no longer be so dependent on the United States for either technological or defence assistance. What may appear to be a victory to congressional leaders may turn out to be truly Pyrrhic.

As if the FSX stand-off were not sufficient for one year's difficulties the USA also announced early on that Japan would probably be one of the countries targeted under the so-called Super 301 section of the Omnibus Trade and Competitiveness Act under which countries that persistently give unfair treatment to American exports can be penalised with tariffs of up to 100 per cent on selected imports from that country. Japan launched a massive lobbying effort in Washington to try and head off its citing under the act, sending such luminaries as the former foreign minister Saburo Okita, the then agriculture minister Tsutomu Hata, and Minister of International Trade and Industry, Hiroshi Mitsuzuka. When the decision came to name Japan in May 1989, the anger in Tokyo was palpable and the feeling strong that it was time to stop humouring the United States. The then prime minister, Mr Takeshita, termed the decision 'regrettable' and the American Ambassador, Michael Armacost was called in to receive a protest. Though the focus of the complaint was narrow (supercomputers, satellites and forest products were named) the Japanese Government was perhaps more stung at being grouped with such lesser industrial luminaries as India and Brazil. Mr Mitsuzuka stated that Super 301 violated the Uruguay Gatt round and Japan would not negotiate outside the Gatt framework. The Americans hoped, however, that Japan would eventually start the 18 months of negotiations provided for under the legislation. At the end of that period the targeted country has three years to remove the barriers to trade before sanctions are enacted.

The shock in Japan at being named as one of the culprit countries served to strengthen the feeling that the USA was hitting at Japan through weakness. It strengthened, too, the ever-present feeling that American culture and life-style is fundamentally weaker and less valid than its Japanese counterpart. In a more sinister vein it reawakened feelings that Western trade policies are motivated by racism. In a new book *A Japan that Can Say No*, co-authored by the chairman of Sony, Akio Morita, this message is spelled out, though not clearly in translation.

Japan invests abroad

For Americans concerned about the impact of Japanese industry and investment the biggest blow of the year came from Mr Morita himself when he announced the purchase of Columbia Pictures for $3.4bn in early October 1988. For many, a Japanese purchase of the White House could hardly have created more impact, such was the feeling that Japanese money was buying a piece of the American soul. More than half the respondents in a poll run by *Newsweek magazine* felt that Japanese economic power might now be a greater threat than Mr Gorbachev's military power.

The purchase was just the latest in a series of stakes and outright purchases in American companies which totalled $16.5bn in 1988. A further $13bn worth of real estate was bought by Japanese interests. The leading Japanese holdings in US companies are now: the $2.6bn outright purchase of Firestone Tire and Rubber Company by Bridgestone; Dai-Ichi Kangyo Bank's 60 per cent stake in CIT Group, the financial concern, worth $1.3bn; Nippon Mining's purchase of Gould Inc., the computer and metal foil maker for $1.1bn; Paloma Industries owns Pace Industries, the $850m manufacturer of water heaters and air-conditioning equipment; Fujisawa Pharmaceutical bought Lymphomed Inc., the drug manufacturer for $798m; an 18 per cent stake in Paine Webber Group Inc. was bought by Yasuda Mutual Life for $300m; Canon Inc. bought 16.7 per cent of Steven Jobs's computer firm Next Inc. for $100m while Sony bought CBS records for $2bn and Mitsukoshi has a 13 per cent stake in Tiffany and Co. worth $93m.

Among the principal purchases of real estate the most visible was the Seibu/Saison group's buying of the Inter-Continental Hotel chain for $2.2bn and Daiichi Seimei's $670m-worth of space in the Citicorp headquarters in New York.

Though not directly affecting foreign interests for the present, the merger of two Japanese banks to make the world's second largest was a strong signal not only of Japanese financial clout but also of the new world of deregulation at home and stronger competition abroad that is now the reality for Japanese bankers. When it finally takes place in April 1990 the merger of the Mitsui and Taiyo Kobe banks will form a single unit with assets of $348bn and capital of $9.02bn.

The near-panic that Japanese investment is causing in the United States is to some extent reflected in Europe though the indications are that Europe will not put up massive walls against Japanese products come 1992. Contrasting with that gloom-tinged view is Britain which has gone all out to win investment of whatever type and technological level. Britain is already the leading host of Japanese investment in Europe and seems set fair to take even more as 1992 approaches. Indeed, a report by Nomura Research Institute says that Japanese companies could turn the UK trade deficit into a surplus within six years and account for a tenth of British manufacturing output by the year 2000.

Following the Nissan plant in Sunderland, now in full production, 1989 saw a string of announcements by Japanese firms on further investment in the British motor industry which will see Japanese car makers responsible for one third of British car output by the mid-1990s at about 500,000 units a year. The key announcement came in July when Honda announced it would take a 20 per cent stake in Rover and build a UK plant at its engine factory site at Swindon. The Swindon site will produce a new model while at Longbridge the company will make 40,000 Honda Concertos a year. The news followed the revelation by Toyota that it would spend £840m on a 200,000 cars-a-year plant at Burnaston near Derby and open an engine plant at Shotton on Merseyside.

Britain's place in the European Community's electronics line-up was confirmed with the announcement by Fujitsu that it will build the Community's first integrated microchip factory worth £100m at Newton Aycliffe in County Durham.

BUSINESS GUIDE

Time: GMT + 9 hrs.

Climate: Climates range from sub-tropical in the south to cooler temperate in the north. On the main island, temperatures range from 35°C maximum in summer (Tokyo is known for its humid, hot summers) to -1°C in winter. The coldest month is Jan and the hottest is Aug. Rainfall is heaviest in Jun and Aug/Sep. South and central Japan is subject to typhoons in late summer and early autumn.

Entry requirements
Passport: Required by all.
Visa: Visas are not required by citizens of certain countries (e.g. German FR, UK) for short stays; certain other nationals (e.g. US) may be exempt from visa fees or require proof of onward passage. Special regulations govern diplomatic, official or transit visitors, or those taking up employment.
Health: Vaccination certificates are not required unless travelling from an infected zone.
Currency: There is no restriction on importation of currency into Japan, but no more than Y5,000,000 may be taken out except by visitors reconverting foreign currency previously imported.
Customs: Personal effects are allowed duty-free. Visitors may purchase souvenir items (pearls, cameras, transistor radios) free of sales tax at designated shops, but they must be taken out of the country within six months.
Income tax: Except for entertainers, temporary visitors are not liable to pay income tax.

Representation overseas: Abidjan, Abu Dhabi, Acra, Addis Ababa, Algiers, Amman, Ankara, Antananarivo, Asunción, Athens, Baghdad, Bangkok, Bangui, Beijing, Beirut, Belgrade, Berlin, Berne, Bogotá, Bonn, Brasilia, Brussels, Bucharest, Budapest, Buenos Aires, Cairo, Canberra, Caracas, Colombo, Conakry, Copenhagen, Dakar, Damascus, Dar es Salaam, Dhaka, Doha, Dublin, Guatemala City, The Hague, Hanoi, Harare, Havana, Helsinki, Islamabad, Jakarta, Jeddah, Kabul, Kathmandu, Khartoum, Kingston, Kinshasa, Kuala Lumpur, Kuwait, Lagos, La Paz, Libreville, Lima, Lisbon, London, Lusaka, Luxembourg-Ville, Madrid, Managua, Manila, Mexico City, Monrovia, Montevideo, Moscow, Nairobi, New Delhi, Oslo, Ottawa, Panama, Paris, Port-au-Prince, Port Moresby, Port of Spain, Prague, Quito, Rabat, Rangoon, Reykjavik, Rome, Sana'a, San José, San Salvador, Santiago, Santo Domingo, Seoul, Singapore, Sofia, Stockholm, Tehran, Tegucigalpa, Tel Aviv, Tripoli, Tunis, Ulan Bator, Vatican City, Vienna, Vientiane, Warsaw, Washington DC, Wellington.

Air access
National airline: Japan Air Lines (JAL).
Other airlines: Aeroflot, Air France, Air India, Air Lanka, Air New Zealand, Air Nippon, Air Pacific, Alitalia, All Nippon Airways, American, Austrian, British Airways, CAAC, Canadian Airlines International, Cathay Pacific, China, Continental, CP Air, Delta Airlines, Dragonair, Egyptair, Finnair, Garuda, Iberia, Iranair, Iraqi Airways, Japan Air System, Japan Asia Airways, KLM, Korean Air, Lufthansa, MAS, Northwest, PIA, Philippine, Qantas, Sabena, Scandinavian, Singapore, Southwest, Swissair, Thai International, Thy Turkish, United, UTA, Varig, Virgin Atlantic.
Tax: Airport tax of Y2,000, on departures from Narita only.
International airports: Fukuoka, Itazuke (FUK), 10 km from Fukuoka; Kagoshima (KOJ), 29 km from Kagoshima; Nagoya, Komaki (NGO), 18 km from Nagoya; Kumamoto (KMJ), 8 km from Kumamoto; Okinawa (OKA), 3 km from Naha; Osaka International (OSA), 20 km from Osaka, with airport bus service to Osaka (journey time 30 mins), Kobe (journey time 40 minutes) and Kyoto (journey time 60 minutes); Sapporo, Chitose (CTS , SPK), 40 km from Sapporo; Tokyo International, Narita (NRT, TYO), 65 km from Tokyo; airport limousine service to Tokyo City Air Terminal (TCAT), and directly to most major hotels. China Airlines flights from Taipei, Taiwan, to Tokyo arrive at Haneda (HND), the former international airport, which now serves largely as a domestic airport, 19 km from city centre.

Surface access
Main ports: Yokohama, Kobe, Osaka, Fukuoka, Nagoya, Tokyo and Nagasaki.

Hotels: Hotels should be booked well in advance. Service charges and taxes are added to the bill, and tipping is not customary. In addition to Western-style hotels, there are traditional Japanese-style inns (*ryokan*) in Tokyo and Osaka.

Credit cards: American Express, Visa, Master Charge, Eurocard and Diners Club (not Access) are widely accepted in major hotels and many restaurants catering to foreigners.

Car hire: An international driving licence is required. Driving is on the left-hand side of the road. Chauffeur-driven cars are often recommended for visitors without command of Japanese and knowledge of the area, as traffic and navigation can be difficult. Symbolic road signs have the expected international meanings, but few direction signs are written in Roman alphabet. A red triangle with white script means 'stop', while a white triangle with a red border and blue script means 'proceed slowly'.

City transport
Taxis: Metered taxis can be easily hired in large cities at hotel entrances or by flagging them down in the street, but don't try to open or close the driver-controlled passenger door). Minimum taxi fares (daytime flagfall Y460–480) increase late at night, and an additional time charge is levied for traffic jams. Few taxi drivers understand foreign languages or read Romanised lettering, so it is advisable to have your destination, including the name of a nearby landmark, written down in Japanese, along with the telephone number if possible. Hotels can often help with this. A map showing the location of the destination is also helpful. Tipping is not customary.
Metro: Efficient underground railway services operate in Tokyo, Yokohama, Osaka, Kyoto, Kobe, Nagoya, Sapporo and Fukuoka, with station names displayed in Roman as well as Japanese lettering. Visitors are often advised to take a minimum fare ticket and go to the Fare Adjustment Window at their destination with sufficient coinage to pay any excess fare. The Tokyo rush hour can be a daunting prospect and is not recommended for the frail or faint-hearted.

National transport
Air: Most domestic flights from Tokyo to Osaka and other Japanese cities are from the former international airport, Haneda (HND), 19 km from Tokyo. Extensive air services provided by JAL, All Nippon Airways, Toa Domestic Airlines, Nihon Kinkyori Airways and others link all main cities and provincial towns.
Road: Extensive road-building programmes are underway, and road transport is the main form of domestic transport. In 1986, the network contained 1.2m km of road. There are good motorways linking Tokyo, Osaka, Kobe, Hiroshima, Yamaguchi, Shimonoseki, Moji, Fukuoka, Kumamoto and Morioka. Tolls are payable on certain roads. Long-distance travel by road is not recommended, as they are very crowded outside main cities.
Buses: An extensive network of frequent coach service link main centres via express motorways, but visitors are advised against coach travel in view of language difficulties and the complexity and number of routes available.
Rail: The 27,000 km rail network constitutes the major means of mass transportation. Japanese National Railways was privatised in Apr 1987 and changed its name to Japan Railways. There are frequent services on the Tokaido Line between Tokyo and Hakata via Nagoya, Kyoto and Osaka, some express (*shinkansen*) and some making more stops, with air-conditioning, restaurant/buffet service and telephone service for calling Tokyo, Yokohama, Nagoya, Kyoto and Osaka, Kobe, Hiroshima and Fukuoka, en route. Seats should be booked well in advance. Other railways with frequent services and many expresses serve the rest of the country. 'Green cars' are equivalent to first class; two pieces of ordinary luggage may be carried free, but there are restrictions on size and weight. Long-distance one-way tickets generally do not permit stop-overs, and ticket refunds are not made after the time of the planned journey. Foreign tourists can make considerable savings by buying a Japan Railpass before arrival in Japan (not available inside the country).

Public holidays: 1 Jan, 15 Jan (Coming of Age Day), 11 Feb (National Foundation Day), 20-21 Mar (Spring Equinox), 29 Apr (Greenery Day), 3 May (Constitution Day), 5 May (Children's Day), 15 Sep (Respect for the Aged Day), 23-24 Sep (Autumn Equinox), 10 Oct (Physical Education Day), 3 Nov (Culture Day), 23 Nov (Labour Day), 23 Dec (Emperor's Birthday). Little work is done on 2-3 Jan. Avoid visits during Golden Week (Apr-May) and the Obon festive season (late Jul-third week in Aug).

Working hours
Business: (Mon-Fri) 0900 – 1700; (Sat) 0900 – 1200 (most companies now close on Saturdays).
Government: (Mon-Fri) 1000 – 1700; (Sat) 1000 – 1200.
Banking: (Mon-Fri) 0900 – 1500. (Sat) banking discontinued.
Shops: 1000 – 1800 (many closed on Wed or Thu).

Social customs: The custom of giving small gifts to business and personal acquaintances is more common than elsewhere. The exchange of business cards (*meishi*) is almost mandatory, and it is good manners to examine received cards carefully and present cards rather formally. Major hotels offer a quick printing service for cards with a Japanese translation on one side. Hotels in large cities can usually arrange interpreter, translation and secretarial services. Shoes are removed before entering Japanese-style houses and restaurants. Restaurants are the usual venue for private social entertaining. Instead of using Mr, Mrs or Miss in polite conversation, the suffix *san* is added to the surname (i.e. *Suzuki-san* instead of Mr Suzuki).

Business language and interpreting/translation: Japan Guide Association, Shin Kokusai Building, 4-1 Marunouchi 3-chome, Chiyoda-ku, Tokyo (tel: 213 2706). Many hotels will often help.

Telecommunications
Telephone: Dialling code for Japan, IDD access code + 81 followed by area code (92 for Fukuoka, 75 for Kyoto, 958 for Nagasaki, 52 for Nagoya, 6 for Osaka, 11 for Sapporo, 3 for Tokyo, 45 for Yokohama) followed by subscriber's number. Hotels generally have adequate English language telephone directories for business addresses and numbers, including foreign trade associations, chambers of commerce and embassies. International calls cannot usually be dialled from public phones, but green phones at Narita airport and in some large hotels are being converted. Green and yellow public phones take larger-value coins. 24-hr/day services available at Tokyo Central Post Office (near the main station) except on Sun and holidays. An English-language information service, the travel-phone, is available at local rates in Tokyo or Kyoto, free of charge elsewhere. Telephone service is automatic, with over 64m telephones (1987): 11% in Tokyo, 3.41% in Osaka, 2.3% in Yokohama and 2.2% in Nagoya.

BUSINESS GUIDE

Telex and telefax: Telex and facsimile facilities are available at good hotels and at Kokusai Denshin Denwa Co Ltd (KDD) offices in Tokyo, Osaka, Yokohama and Nagoya.
Telegram: International cables service is available at good hotels, from the offices of KDD and Nippon Denshin Denwa Kosha, and from larger post offices in main centres.
Other: KDD provides marine radio, international leased circuit, international facsimile telegram, international phototelegram, maritime satellite communications, international datel, international computer access and automex services.

Banking: Japanese banking draws a strict distinction between city (commercial) and investment banking, and allows no overlapping activities. Considerable changes are under way, as Japan accedes to foreign pressure for deregulation.

Main Centres: Tokyo, Osaka and Nagoya.

Central Bank: Bank of Japan.

Advertising: Advertising is available in the press, on commercial radio and TV and in cinemas. Outdoor advertising is widely available.

Electricity supply: 100 V AC, 60 cycles in west Japan (Osaka) and 100 V AC, 50 cycles in east Japan (Tokyo), with flat two-pin plug fittings and screw-type lamp bulbs.

Weights and measures: Metric system (strictly enforced).

Representation in capital: Algeria, Argentina, Australia, Austria, Bangladesh, Belgium, Bolivia, Brazil, Bulgaria, Burma, Canada, Central African Republic, Chile, China PR, Colombia, Costa Rica, Cuba, Czechoslovakia, Denmark, Dominican Republic, Ecuador, Egypt, El Salvador, Ethiopia, Fiji, Finland, France, Gabon, German DR, German FR, Ghana, Greece, Guatemala, Guinea, Haiti, Honduras, Hungary, India, Indonesia, Iran, Iraq, Ireland, Israel, Italy, Ivory Coast, Jordan, Kenya, Republic of Korea, Kuwait, Laos, Lebanon, Liberia, Libya, Madagascar, Malaysia, Mexico, Mongolia, Morocco, Nepal, Netherlands, New Zealand, Nicaragua, Nigeria, Norway, Pakistan, Panama, Papua New Guinea, Paraguay, Peru, Philippines, Poland, Portugal, Qatar, Romania, Rwanda, Saudi Arabia, Senegal, Singapore, Spain, Sri Lanka, Sudan, Sweden, Switzerland, Syria, Tanzania, Thailand, Tunisia, Turkey, Uganda, USSR, UAE, UK, USA, Uruguay, Vatican, Venezuela, Yemen AR, Yugoslavia, Zaïre, Zambia, Zimbabwe.

BUSINESS DIRECTORY

Hotels: Kyoto
Ana Kyoto, Nijo Castle Maz, Horikawa-Dori, Nakagyo-Ku, 604 (tel: 2311155; tx: 5423181; fax: 2315333).

Fujita Kyoto Nishizume, Nijo-Ohashi, Nakagyo-ku, 604 (tel: 2221511; tx: 5422571).

Holiday Inn, 36 Nishihiraki-cho, Takano. Sakyo-ku, 606 (tel: 7213131; tx: 22251; fax: 7816178).

International Kyoto, 284 Nijyo-Aburanokoji, Nakagyo-ku (tel: 2221111; tx: 5422158 INTLHO J).

Kyoto, Kawara-machi-Oike, Nakagyo-ku, 604 (tel: 2115111; tx: 05422126 KYOTEL J; fax: 2114209).

Kyoto Grand, Shiokoji-Horikawa, Shimogyo-ku, 600 (tel: 3412311; tx: 5422551 KYOGRA J; fax: 3413073).

Kyoto Park, 644-2 Sanjusangendo Mawari-machi, Higashiyama-ku, 605 (tel: 5253111; tx: 5422777).

Kyoto Prince, 43 Matsubara-cho, Shimogamo, 606 (tel: 7814141; tx: 5422661; fax: 7814150).

Kyoto Royal, Kawaramachi Sanjo, Nakagyo-ku, 604 (tel: 2231234; tx: 05422888).

Kyoto Station, Shiokoji, Higashi notoin-dori, Shimogyo-ku, 600 (tel: 3617151; tx: 5422456 STHOTL J).

Kyoto Tokyu, Horikawa-Gojyo, Sagaru 580, Shimojyo-ku 600 (tel: 3412411; tx: 5422459; fax: 3412488).

Kyoto Tower, Karasuma-Shijo, Shimogyo-ku, 600 (Kyoto Railway Station) (tel: 3613211; tx: 05423423).

Miyako, Sanjo Keage, Higashiyama-ku, 605 (tel: 7717111; tx: 5422132 MIYAKO J; fax: 7512490).

Mount Hiei, Ipponsugi Hieizan, Sakyo-ku, 606 (tel: 7012111; tx: 05422360 HIEZAN).

New Kyoto, Horikawa, Maruta-machi, Kamigyo-ku, 602 (tel: 8012111; tx: 5422326).

New Miyako, Nichi-Kujoincho, Minami-ku 601 (tel: 6617111; tx: 5423211 NEWMYK J; fax: 6617135).

Westin Kyoto, Takara-ga-ike Prince, Sakyo-ku, Kyoto-Shi, 606 (tel: 7121111; tx: 05422611; fax: 7127677).

Hotels: Narita, Tokyo International Airport
Holiday Inn Tobu Narita, 320-1 Tokko, 28601 (tel: 321234; tx: 3762133; fax: 310617).

Narita International, 650-35 Tomisato-machi, Nanae, Inba-Gun, 28602 (tel: 931234; tx: 3762178 NIHNRT J; fax: 934834).

Narita Prince, 560 Tokko, Narita-shi, 28601 (tel: 331111; tx: 03762147).

Narita View, 700 Kosuge, 28601 (tel: 321111; tx: 3762123 NVHNRT J; fax: 321078).

Nikko Narita, 500 Tokko, 28601 (tel: 320032; tx: 3762165 JHSNRT; fax: 323993).

Hotels: Osaka
Ana Sheraton Osaka, 1-3-1 Dojimahama, Kita-ku, 530 (tel: 3471112; tx: 5236884 ANAEOS J; fax: 3489208).

Hanshin, 2-3-30 Umeda, Kita-ku, 530 (tel: 3441661; tx: 05234269).

Holiday Inn Nankai-Osaka, 28-1 Kyuzaemonoho, Minami-ku (tel: 2138281; tx: 05222939; fax: 2138640).

International Osaka, 58 Hashizume-cho, Uchihon-machi, Higashi-ku, 540 (tel: 9412661; tx: 5293415).

Miyako Hotel Osaka, 6-1-55 Uehommachi, Tennoji-ku, 543 (tel: 7731111; tx: 5277555; fax: 7733322).

New Hankyu, 1-8-1 Shibata, Kita-ku, 530 (tel: 3725101; tx: 05233830).

Nikko Osaka, 3-3 Nishi Shinsaibashi, 1 Chome, Chuo-Ku 542 (tel: 2441111; tx: 5227575 JHSOSA; fax: 2452432).

Osaka Airport, 3/555 Nishi-machi, Hotarugaiki (tel: 8554621; tx: 5286125 APTHTL J).

Osaka Dai-chi, 1-9-20 Umeda, Kita-ku, 530 (tel: 3414411; tx: 05234423; fax: 3414930).

Osaka Grand, 2-3-18 Nakanoshima, Kita-ku,530 (tel: 2021212; tx: 05222301; fax: 062275054).

Osaka Hilton International, 8-8, Umeda 1-chome, Kita-ku, 530 (tel: 3477111; tx: 5242201; fax: 3477001).

Osaka Terminal, 3-1-1 Umeda Kita-ku, 530 (tel: 3441235; tx: 05233738 OSATER J).

Osaka Tokyu, 7-20 Chaya-machi, Kita-ku, 530 (tel: 3732411; tx: 5236751; fax: 3760343).

Plaza, 2-2-49 Oyodo-Minami, 2-chome, Oyodo-ku, 531 (tel: 4531111; tx: 5245557 PLAOSA J; fax: 4540169).

Royal, 3-68 Nakanoshima, 5-chome, Kita-ku, 530 (tel: 4481121; tx: J 63350; fax: 4446570).

Sun Route Minamisenri, 1-20-30, Tsukumodai, Suita City (tel: 8341144; tx: 5287102; fax: 8341159).

Sun Route Sakai, 1-1 Nishi I-cho, Shorinji-cho, Sakai City (tel: 320303).

Toyo, 3-16-19 Toyosaki, Oyodo-ku, 531 (tel: 3728181; tx: 05233886; fax: 3728101).

Hotels: Tokyo
ANA Hotel, 12-33 Akasaka, 1-chome, Minato-ku, 107 (tel: 5051111; tx: 2424625; fax: 5051150).

Akasaka Prince, 1-2 Kioi-cho, Chiyoda-ku, 102 (tel: 2341111; tx: 2324048 AKAPRH J; fax: 2625163).

Akasaka Tokyu, 14-3,2-chome, Nagata-cho, Chiyoda-ku, 100 (tel: 5802311; tx: 2224310).

Capitol Tokyu, 10-3 Nagato-cho, 2-chome, Chiyoda-ku,100 (tel: 5814511; tx: 24290; fax: 5815822).

Century Hyatt Tokyo, 2-7-2 Nishi-Shinjuku, Shinjuku-ku, 160 (tel: 3490111; tx: 29411; fax: 3445575).

Fairmont Tokyo, 2-1-17 Kudan Minami, Chiyoda-ku, 102 (tel: 2621151; tx: 2322883; fax: 2642476).

Ginza Dai-Ichi, 8-13-1 Ginza, Chuo-ku, 104 (tel: 5425311; tx: 2523714 ITIHLC J; fax: 5423030).

Ginza Marunouchi, 4-1-12 Tsukiji, Chuo-ku, 104 (tel: 5435431; tx: 02522214; fax: 5436006).

Ginza Nikko, 8-4-21 Ginza, Chuo-ku, 104 (tel: 5714911; tx: 2522812; fax: 5718379).

Ginza Tokyu, 15-9 5-chome, Ginza, Chuo-ku, 104 (tel: 5412411; tx: 2522601; fax: 5416622).

Grand Palace, 1-1 Iidabashi 1-chome, Chiyoda-ku, 102 (tel: 2641111; tx: 2322981; fax: 2304985).

Haneda Tokyu, 2-8-6 Haneda-Kuko, Ohta-ku, 144 (tel: 7470311; tx: 2466560; fax: 7470366).

Hill-Top, 1 1-1-chome, Kanda Surugadai, Chiyoda-ku, 101 (tel: 2932311; tx: 2226712; fax: 2334567).

Holiday Inn Yaesu, 1-13-7 Hachobori, Chuo-ku, 104 (tel: 5536161; tx: 02523748; fax: 5536140).

Hotel Takanawa, 2-1-17 Takanawa Minato-ku, 108 (tel: 4439251; tx: 2422553 HTLTAK J; fax: 4439025).

Imperial, 1-1-1 Uchisaiwai-Cho, Chiyoda-ku, 100 (tel: 5041111; tx: 2222346 IMPHO; fax: 5819146).

Kayu Kaikan, 8-1 Sanban-cho, Chiyoda-ku, 102 (tel: 2301111; tx: 02323318; fax: 2302529).

Keio Plaza Inter-Continental, 2-2-1 Nishi Shinjuku, Shinjuku-ku, 160 (tel: 3440111; tx: 26874 J; fax: 3440247).

Kokusai Kanko, 1-8-3 Marunouchi, Chiypoda-ku, 100 (tel: 2153281).

Metropolitan, 6-1 Nishi Ikebukuro, 1-chome, Toshima-ku, 171 (tel: 9801111; tx: 2722787 J; fax: 9805600).

Miyako Tokyo, 1-50 Shiroganedai, 1-chome, Minato-ku 108 (tel: 4473111; tx: 2423111; fax: 4473133).

New Otani, 4-1 Kioi-cho, Chiyoda-ku, 102 (tel: 2651111; tx: 24719; fax: 2212619).

Okura, 10-4 Toranomon, 2-chome, Minato-ku, 105 (tel: 5820111; tx: J 22790; fax: 5823707).

Pacific Meridien, 3-13-3 Takanawa, Minato-ku, 108 (tel: 4456711; tx: 22861 J; fax: 4455733, 4455137).

Palace, 1-1 Marunouchi, 1-chome, Chiyoda-ku, 100 (tel: 2115211; tx: 02222580 PALACE J; fax: 2116987).

Sheraton Grande Tokyo Bay, 1-Maihama, Urayasu, Chiba 272-01 (tel: 555555; fax: 555577).

Shiba Park, 5-10 Shiba Koen, Minato-ku, 105 (tel: 4334141; tx: 2422917; fax: 4336327).

Shimbashi Dai-Ichi, 1-2-6 Shimbashi, Minato-ku, 105 (tel: 5014411; tx: 2222233; fax: 5952634).

Shinagawa Prince, 4-10-30, Takanawa, Minato-ku 108 (tel: 4401111; tx: 2425178; fax: 4417092).

Shinjuku Prince, 1-30-1 Kabuki-cho, Shinjuku-ku 160 (tel: 2051111; tx: 2324733; fax: 2051952).

Sunshine City Prince, 3-1-5 Higashi-Ikebukuro, Toshima-ku, 170 (tel: 9881111; tx: 02723749 SUNPRH J; fax: 9830115).

Takanawa Prince, 3-13-1 Takanawa Minato-ku, 108 (tel: 4421111; tx: 2423232; fax: 4460849).

Tokyo Hilton International, 6-2 Nishi-Shinjuku, 6-chome, Shinjuku-ku, 160 (tel: 3445111; tx: 2324515; fax: 3426094).

Tokyo Marunouchi, 6-3 Marunouchi, Chiyoda-ku, 100 (tel: 2152151; tx: 02224655; fax: 2158036).

Tokyo Prince, 3-1 Shibakoen 3-chome, Minato-ku, 105 (tel: 4321111; tx: 2422488).

Tokyo Station Hotel, 1-9-1 Marunouchi, Chiyoda-ku, 100 (tel: 2312511).

Hotels: Yokohama
Holiday Inn Yokohama, 77 Yamashita-cho, Naka-ku, 231 (tel: 6813311; tx: 03822758; fax: 6815082).

New Grand, 10 Yamashita-cho, Naka-ku, 231 (tel: 6811841; tx: 3823411 HNGYOK).

The Yokohama, 6-1 Yamashita-cho, Naka-ku, 231 (tel: 6621321; tx: 03822061; fax: 6623536).

Yokohama Prince, 13-1-3 Isogo-ku, 235 (tel: 7532211).

Yokohama Tokyu, 1-1-12 Minami-saiwai, 1 Chome, Nish-ku, 220 (tel: 3111682; tx: 3822264; fax: 3111084).

Useful telephone numbers
Overseas calls
Tokyo to SE Asia: 211-4211
Tokyo Operator: 0051
Tourist information
Tokyo: 502-1461
Kyoto: 371-5649

Car hire: Tokyo
Avis: 2nd Floor Land K NO3, Akasaka Building, 3-2 Akasaka, 2 chome, Minato-Ku (tel: 583-0911; fax: 5836850).

Hertz: Airport Terminal, Haneda, Ota-ku (tel: 747-8785)

Hertz: Narita Airport, Chuo-To IF, Tyokyaku Terminal Bldg, Narita-Shi, Chiba 286-01 (tel: 476-328897).

ACU Rent-a-Car: 13-2, Hyakunin-cho 1-chome, Shinjuku-ku (tel: 364-2211).

Imperial Hire Cars: (tel: 264-7441).

Isuzu Rent-a-Car: 3-16, Takaido Higashi, Suginami-ku (tel: 334-2411).

Japaren Co: 12-7, Shinjuku 2-chome, Shinjuku-ku (tel: 352-7635).

Kokusai Hire Cars: (tel: 242-5931).

Mazda Rent-a-Car: 1-2-12, Takadanobada, Shinjuku-ku (tel: 209-0620).

Mitsubishi Rent-a-Car: 1-11-8, Kyobashi, Chuo-ku (tel: 563-5271).

Nippon Rent-A-Car: 5-5 Kamiyana-Cho, Shibuya-Ku (tel: 468-7126).

Nissan Rent-a-Car: 5-7, Azabu-dai 1-chome, Minato-ku (tel: 586-2301).

Tokyo Nissan Rent-a-Car: 52-2, Jingumae 5-chome, Shibayu-ku (tel: 407-4431).

Toyota Rent-a-Lease Tokyo: 1 Sanban-cho, Chiyoda-ku (tel: 264-2834).

Car hire: Kyoto
Japaren Co.: 14-1, Kitanouchi-cho, Nishi-kujo, Minami-ku (tel: 681-7956).

Kyoto Hotel Hire Cars: (tel: 211-1818).

Nippon Rent-a-Car: Kyoto Reservation Center 42-13 Kitanouchi-cho, Nishi-9-Jo, Minami-Ku, Kyoto-Shi Kyoto 601 (tel: 075-671-0919).

Nissan Rent-a-Car: 14, Higashi-Sanno-cho, Higashi-kujo, Minami-ku (tel: 661-2161).

Toyota Rent-a-Lease: 30, Shimizu-cho, Kisshoin, Minami-ku (tel: 661-2080).

Car hire: Osaka
Hertz: Osaka Airport (tel: 841-8175).

Japaren Co.: 1-81, Nishisekiya-cho, Naniwa-ku (tel: 632-4881).

Mitsubishi Rent-a-Car: 1-5-23, Minami-horie, Nishi-ku (tel: 538-2428).

Nissan Rent-a-Car: 19-22, Chaya-machi, Kita-ku (tel: 372-0281).

Nippon Rent-a-Car: 1-1-3, Shibata-cho, Kita-ku (tel: 373-2652).

Toyota Rent-a-Car: 3-74, Kawaraya-cho, Minami-ku (tel: 372-5480).

Business advice and information
American Chamber of Commerce, Tosho Building, 2-2 Marunouchi 3-chome, Chiyoda-ku, Tokyo (tel: 211-5861).

Association for the Promotion of International Trade, Nihon Building, 6-2 Otemachi 2-chome, Chiyoda-ku, Tokyo (tel: 245-1561).

British Chamber of Commerce, 1 Kowa Building, 3rd Floor, 11-41 Akasaka 1-chome, Minato-ku, Tokyo 107 (tel: 505-1734).

Council of All-Japan Exporters Association, Kikai Shinko Kaikan Building, 5-8 Shibakaen 3-chome, Minato-ku, Tokyo.

Federation of Economic Organisations (Keidanren), 9-4 Otemachi 1-chome, Chiyoda-ku, Tokyo (tel: 279-1411).

Fukuoka Chamber of Commerce and Industry, 9-28 Hakata-Ekimae 2-chome, Hakata-ku, Fukuoka City, Fukuoka (tel: 441-1111).

International Chamber of Commerce, Japanese National Committee, Tosho Building (311), 3-2-2, Marunouchi, Chiyoda-ku, Tokyo (tel: 213-8585).

Japan Chamber of Commerce and Industry, Tosho Building 2-2 Marunouchi 3-chome, Chiyoda-ku, Tokyo 100 (tel: 283-7824, 283-7601).

Japan Commercial Arbitration Association, Tosho Building, 2-2 Marunouchi 3-chome, Chiyoda-ku, Tokyo (tel: 214-0641).

Japan Committee for Economic Development, Kogo Club Building, 4-6 Marunouchi 1-chome, Chiyoda-ku, Tokyo (tel: 211-1271).

Japan Foreign Trade Council, Sekei Boeki Centre Building, 4-1 Hamamatsu-cho 2-chome, Minato-ku, Tokyo (tel: 435-5950).

Japan Productivity Centre, 1-1 Shibuya 3-chome, Shibuya-ku, Tokyo (tel: 409-1111).

Japan External Trade Organisation (JETRO), 2-5 Toranomon 2-chome, Minato-ku, Tokyo (tel: 582-5511; tx: 24378).

Kobe Chamber of Commerce and Industry, 1-14 Hamabedori 5-chome, Fukiai-ku, Kobe City, Hyogo (tel: 251-1001).

Nagoya Chamber of Commerce and Industry, 10-19 Sakae 2-chome, Naka-ku, Nagoya City, Aichi (tel: 221-7211).

Osaka Chamber of Commerce and Industry, 58-7 Uchiho-machi Hashizume-cho, Higashi-ku, Osaka (tel: 266-6200, 944-6412).

Tokyo Chamber of Commerce and Industry, 2-2 Marunouchi 3-chome, Chiyoda-ku, Tokyo (tel: 283-7500/7610; tx: 222-4920).

Tokyo International Trade Fair Commission, 7-24 Harumi 4-chome, Chuo-ku, Tokyo (tel: 531-3371).

World Trade Centre of Japan, 4-1 2-chome Hamamatsu-cho, Minato-ku, Tokyo (tel: 435-5651).

Yokohama Chamber of Commerce and Industry, 2 Yamashita-cho, Naka-ku, Yokohama City, Kanagawa (tel: 671-7411).

Banking
Bank of Japan (Nippon Ginko) (central bank), 2-1-1 Hongoku-cho, Nihonbashi, Chuo-ku, Tokyo 103 (tel: 279-1111; tx: 22763).

Bank of Tokyo, 3-2 Hongoku-cho 1-chome, Nihonbashi-Hongkucho, Chuo-ku, Tokyo 103 (tel: 245-1111; tx: 22220).

Dai-Ichi Kangyo Bank, 1-5 Uchisaiwai-cho 1-chome, Chiyoda-ku, Tokyo 100 (tel: 596-1111; tx: 22315).

Daiwa Bank, 56 Bingomachi 2-chome, Higashi-ku, Osaka 541 (tel: 271-1221; tx: 63977).

Fuji Bank, 5-5 Otemachi 1-chome, Chiyoda-ku, Tokyo 100 (tel: 216-2211; tx: 22367).

Hokkaido Takushoku Bank, 3-7 Nishi 3-chome, Chuo-ku, Sapporo 060 (tel: 271-2111; tx: 932533).

Industrial Bank of Japan, 3-3 Marunouchi 1-chome, Chiyoda-ku, Tokyo 100 (tel: 214-1111; tx: 22325, 23219).

Kyowa Bank, 1-2 Otemachi 1-chome, Chiyoda-ku, Tokyo 100 (tel: 287-2111; tx: 24275 J).

Long Term Credit Bank of Japan, 2-4 Otemachi 1-chome, Chiyoda-ku, Tokyo (tel: 211-5111; tx: 24308).

Mitsubishi Bank, 7-1 Marunouchi 2-chome, Chiyoda-ku, Tokyo 100 (tel: 240-1111; tx: 22358).

Mitsui Bank, 1-2 Yarakucho 1-chome, Chiyoda-ku, Tokyo 100 (tel: 501-111; tx: J 22378, J 22559, J 22643, J 22644; cable: MITSUIBANK TOKYO).

Nippon Credit Bank, 13-10 Kudan-kita 1-chome, Chiyoda-Ku, Tokyo 102 (tel: 263-1111; tx: 26921, 28788 ncbtok).

Norinchukin Bank (central co-operative bank for agriculture, forestry and fisheries), 8-3 Ohtemachi 1-chome, Chiyoda-ku, Tokyo 100 (tel: 279-0111; tx: 23918).

Saitama Bank, 4-1 Tokiwa, 7-chome, Urawa City, Saitamo Prefecture 226 (tel: 24-2411; tx: 22811).

Sanwa Bank, 4-10 Fushimimachi, Higashi-ku, Osaka 541 (tel: 202-2281; tx: 63234).

Sumitomo Bank, 22 Kitahama 5-chome, Higashi-ku, Osaka 541 (tel: 227-2111; tx: 63266).

Taiyo Kobe Bank, 56 Naniwa-cho, Chuo-ku, Kobe 650 (tel: 331-8101; tx: 78823).

Tokai Bank, 21-24 Nishiki 3-chome, Nakaku, Nagoya 460 (tel: 211-1111; tx: 59612).

Travel information
Japan Automobile Federation, Shiba-Koen, 3-5-8 Minato-ku, Tokyo 105 (tel: 436-2811).

Japan National Tourist Organisation, 10-1 Yurakucho 2-chome, Chiyoda-ku, Tokyo (tel: 216-1901).

Tourist Information Centre, 1-6-6 Yurakucho 1-chome, Chiyoda-Ku, Tokyo 100 (tel: 502-1461); Kyoto Tower Building, Higashi-Shiokojicho, Shimogyo-ku, Kyoto 600 (tel: 075-371-5649).

Jordan

Pam Dougherty

It was in 1989 that Jordanians awoke from their dream that economic comfort and political stability could be taken for granted. The country had for some time, successfully managed to create an aura of moderation and comfort that distinguished it from its neighbours. Then, suddenly it found itself facing a debt crisis and austerity measures that showed up the of the economy and forced King Hussein to turn his attention to internal rather than external affairs.

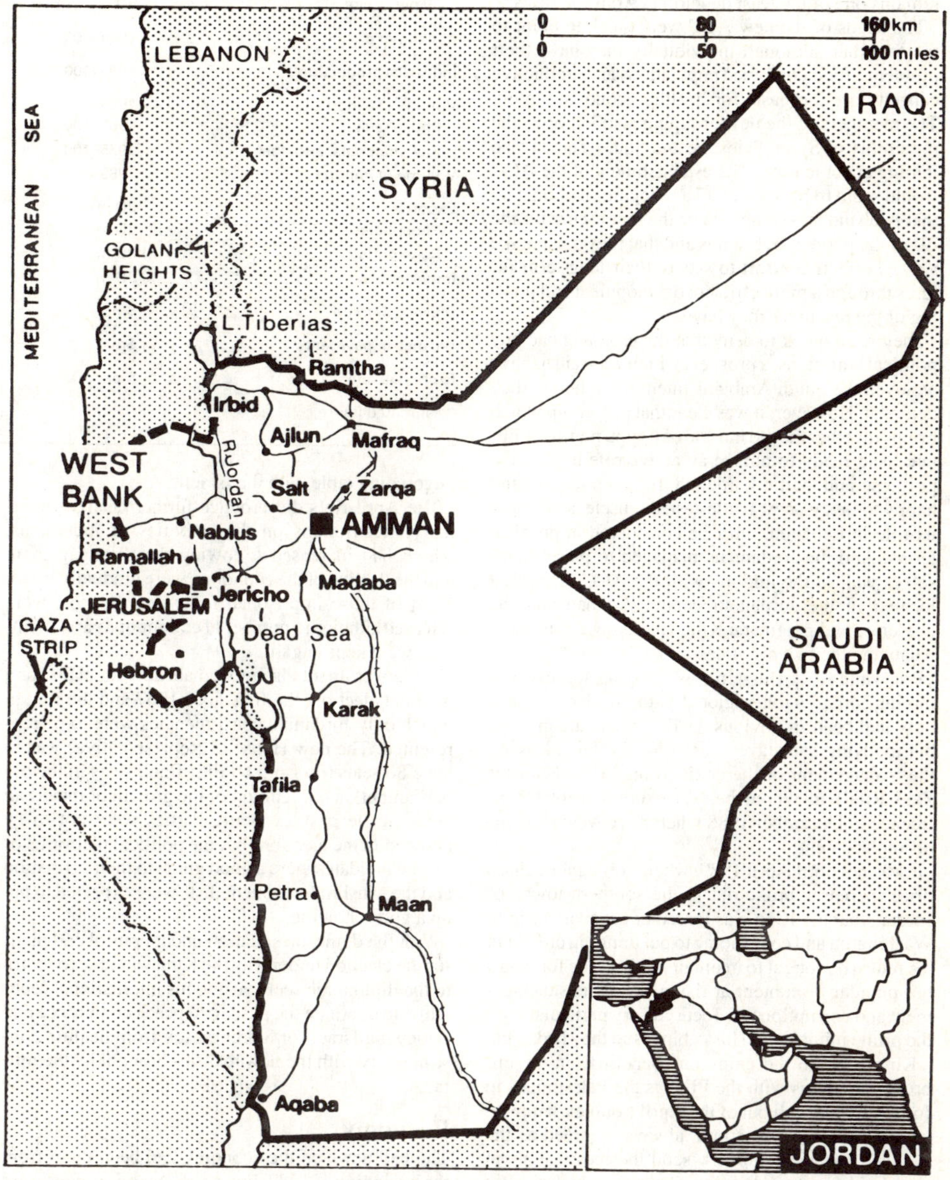

Political situation

In late 1988 and early 1989 King Hussein's attention was very much on his role in regional affairs. In the face of the Palestinian 'intifada' he had finally given up all claims to the West Bank and had severed all Jordan's legal and administrative ties to its people. However, the strong support of Palestinians in the Israeli-occupied territories for the Palestinian Liberation Organisation (PLO) and its Chairman Yasser Arafat threatened to leave King Hussein and his country on the margins of the growing diplomatic effort to achieve a peace settlement with Israel.

King Hussein then set out to rebuild his relationship with Arafat and the PLO, drawing strength as a supporter of the PLO rather than as its competitor and to build his role as a key regional moderator.

In November 1988 it was Jordan that was the location for meetings that brought Iraq, Egypt, the PLO and Jordan together to forge a united moderate position on the Palestinian peace moves. All four accepted the Egyptian solution to the Arab-Israeli conflict of a Palestinian-Jordanian confederation following an Israeli withdrawal negotiated through a Security Council sponsored conference. Jordan was back in the middle of things once more and Arafat and King Hussein were reconciled after a strained 12 months following the November 1987 Amman summit when the Palestinians felt they had very much been put to one side.

In December 1988 it was King Hussein who pushed hard to have the United Nations Assembly meeting in Geneva when United States Secretary of State George Schultz refused Arafat a visa, and who gave the Palestinian leader warm support following the Palestine National Council (PNC) meeting which saw the declaration of the Palestinian state and the acceptance of Israel. Since that time Jordan has not moved from its position of support for the PLO and relations have improved steadily. The move has served King Hussein well at home, helping to allay the fears of his own Palestinian population that the August 1988 disengagement would gradually undermine their security in Jordan.

The King continued to pursue his role as an international statesman into 1989, engaging in a heavy round of travels to promote support for the peace conference and to bring rival Arab leaders together. The disengagement had helped to free Jordan from some of the heavy American pressure to pursue the so-called Jordan option in which the Jordanians were expected to go to negotiations with the Palestinians as their junior partners or perhaps not at all. It had also helped to clear the air between Jordanians and Palestinians as Jordan could no longer be accused of harbouring claims to the Occupied Territories itself.

Instead, the Jordanians set out to carve a new niche, not just as firm supporters of the PLO but as the only Arab state with the qualifications to act as a moderator between the states which would ultimately be involved in an international peace conference – Syria, Lebanon, Egypt and the PLO – but which were sharply divided and between the sworn enemies Syria and Iraq whose rivalry was threatening to destabilise the region further.

At the same time King Hussein was promoting the establishment of a new economic grouping which would, in essence, bring together in the Arab Cooperation Council (ACC) states not covered by the

KEY FACTS

Official title: Hashemite Kingdom of Jordan
Head of state: King Hussein Ibn Talal
Head of government: Prime Minister Mudar Badran
Capital: Amman
Official Languages: Arabic
Currency: Dinar (JD) = 1,000 fils
Exchange rate: JD0.67 per $ (Nov 1989)

Area: 95,400 sq km (including the occupied West Bank)
Population: 2.9m (1988 – East Bank only)
GDP per capita: $1,552 (1988)
GDP real growth: –3.5% (1988)
Inflation: 14% (1988)
Trade balance: –$1,705m (1988)
Foreign debt: $8.1bn (1988)

KEY INDICATORS

	Unit	1984	1985	1986	1987	1988
Population (East Bank)	'000	2,595.1	2,693.7	2,796.1	2,806.2	–
Gross domestic product (GDP)	JD m	1,498.4	1,605.9	1,639.9	1,686.3	–
GDP per capita	JD	577	596	586	581	–
Consumer prices	1985=100	97.1	100.0	100.0	99.7	102.9
Industrial production	1985=100	97.8	100.0	101.4	110.8	101.8
Volume of exports	1985=100	96.0	100.0	102.6	121.9	136.6
Phosphates	1985=100	101.8	100.0	112.2	120.2	126.1
Raw materials, excl. fuel	1985=100	85.0	100.0	122.5	130.5	–
Volume of imports	1985=100	97.7	100.0	105.1	109.0	115.1
Exports	JD m	290.66	310.88	256.03	315.7	381.5
Phosphates	JD m	69.61	66.08	64.80	61.0	–
Re-exports	JD m	29.60	55.54	30.41	66.94	56.71
Imports	JD m	1,071.4	1,074.4	850.20	916.4	1,022.5
Balance of trade	$m	–1,720.7	–1,988.3	–1,426.2	–1,661.4	–
Balance of payments	$m	–264.2	–261.0	–40.0	101.4	–
Exchange rate	$ per JD	2.60	2.54	2.86	2.95	2.69

* estimated figure

two existing Arab blocs, the Gulf Co-operation Council and the Arab Maghreb Union. At first sight the founding members (Jordan, Iraq, Egypt and North Yemen) seem an unpromising group. Jordan, Egypt and Iraq, for all its oil, are all heavily in debt while North Yemen, though in a healthier financial state, remains one of the least developed Arab countries.

The heads of the new ACC were quick to point to the fact that, although individually they have their problems, their combined populations and productive capacity are the basis for healthy economic union. The members of the new grouping also left open the possibility of Syria, Palestine and South Yemen also joining at some time. The establishment of the ACC would seem to be most of all a recognition of the fact that Arab aid is no longer to be the magic cure for the ills of the poorer Arab states and that they must make a more effective effort to secure their long-term futures through a more efficient development and pooling of the resources they have.

They were quick to deny that the grouping had any political ambitions, a prospect which was said to have annoyed the Saudi Arabians intensely although they denied it. However, it was clear that any group which brings together the old muscle of Egypt, now rehabilitated and ready to pursue an active role in regional politics, and battle-hardened Iraq, confident and happy to flex its new battle-tested muscle, as witness its role in Lebanon, must be considered in political terms. And the Jordanians managed to place themselves well in the middle, with the ACC secretariat located in the Jordanian capital Amman and the moderate and pragmatic King Hussein well suited to balancing the two big states.

But while King Hussein was enjoying success promoting Jordan's role regionally, the Jordanian economy had been in obvious decline since late in 1988 but the King had stayed out of local affairs, leaving the management of the crisis to his Prime Minister Zaid Rifa'i in whom he had expressed total confidence in December 1988 when there were calls for his replacement.

It was not until April 1989, when riots against sharp price increases broke out in the southern towns of Jordan, that the King was forced to cut short a trip to Washington and come home to put things in order. On his return he sought to maintain his backing for Rifa'i but popular discontent at rising prices, mismanagement and corruption was focused very particularly on the prime minister and his cabinet and they had to go.

King Hussein did enjoy one benefit from the improved relations with the PLO as the Palestinians in Jordan stayed well out of the April troubles, happy to let the people of the south, always regarded as the throne's strongest support, send the message about popular discontent. Rifa'i was replaced by Zaid Bin Shaker, a relative of the King and former commander of the armed forces who could be relied on for total support if the situation showed signs of getting out of hand.

Bin Shaker's appointment did not meet with much enthusiasm but the new cabinet was accepted as competent and honest and riots gave way to a grumbling discontent as the government officials considered responsible for the economic troubles went unpunished, and to continuing uncertainty as prices continued to rise steadily.

In the first weeks after his return the King kept a fairly low profile, confining himself to a television address, and to receiving a series of delegations from the tribes of the south to declare their loyalty. Gradually, however, he began to go out to the provincial towns, returning to an old habit of travelling in an open-topped car and this time finding more warmth in popular declarations of loyalty.

The King also went on the offensive, blaming outside forces for Jordan's economic plight, referring mainly to pressures exerted to force Jordan to the negotiating table with the Israelis.

The April riots did produce further changes internally. Restrictions on the activities of journalists, which had increased following the disengagement and the decline of the economy, were gradually lifted. Most of those imprisoned following the riots were released although a number of communist party members were kept in gaol.

The government also moved ahead with plans for a national election, the first since 1967 and the first in which only Jordanian constituencies were to be represented. The new House of Representatives was to have 80 seats (up from 60 East Bank seats in the old parliament), a recognition of the growth of the urban areas in the past 22 years. Parties were not to be allowed in the elections but the political platforms of most candidates were a good guide to their affiliation and the Muslim fundamentalists were expected to put up a good showing.

With the domestic situation under control and plans for the election under way the King was able to return to the diplomatic scene, using his international reputation to continue the push for a peace conference and criticising Israel for holding back from dealing constructively with the new realities brought by the intifada.

Economy

The bad news on the economy had been there from the beginning of 1989. Indeed, the warning signs were clear from October 1988 when declining foreign exchange reserves put the dinar under pressure and the Central Bank announced a managed float which quickly turned into an ungraceful slump.

November brought the announcement of an austerity package which included a 12-month ban on luxury imports and a sharp hike in import duties. This was followed by a tight 1989 budget which held spending to around the 1988 level in dinar terms but in real spending value meant a cut.

Jordanians reached the end of the year in shock but thought the worst was over until the news emerged in January that the government was asking for a roll over of repayment on one international loan and had pulled a new dual currency loan of $150m from the market due to a poor response and the realisation that it could no longer meet its existing commitments. Since then the government has been engaged in a battle against a fundamental loss of confidence in the economy that has manifested itself most clearly in the steady decline of the dinar.

By early February 1989 growing awareness in the money markets of the seriousness of the government's position led to a new panic in the money markets and the dinar began to plummet. When the dollar reached 750-800 fils, at a time when the Central Bank rate was set at 570 fils, the government stepped in to close the country's 80 moneychangers and the Central Bank began to set a fixed rate for the major currencies.

The move held the dinar steady and eased the fears of the general public but marked the beginning of a severe foreign exchange shortage as expatriate Jordanians held back from remitting money until the currency settled and Jordanians themselves began chasing dollars in anticipation of further devaluations. The problem was exacerbated by the fact that, although the Central Bank was setting a firm rate, it was unable to intervene to underpin it.

By March the government announced what everyone knew was inevitable: it would have to reschedule the foreign debt and discussions had begun with the International Monetary Fund (IMF) on a programme to restructure the economy away from its dependence on foreign aid and remittances and towards export-oriented local production.

Jordan went into the procedures at least with the advantage of a good previous record on debt repayment and political support from its foreign creditors. Preliminary talks with the IMF went smoothly and in late March the prime minister announced that a package had been agreed and Jordan expected to receive a stand-by credit of around $120m from the IMF and a further $150m structural readjustment loan from the World Bank over the following 18 months.

In a move that was typical of the government's failure to face reality or to communicate with the people, Rifa'i took great pains to announce that the IMF had not asked for either a further devaluation or a reduction on government subsidies on basic foodstuffs and went so far as to say that the government would never accept any measures which would harm the population.

Within days the King had left for a trip that was to include visits to Paris, London and Washington and the government had announced sweeping rises in the price of fuel related products, car registration fees, utilities, alcohol and cigarettes. The rises were part of an overall package designed to raise revenues by JD37m and cut spending by JD17m as a first step towards cutting the budget deficit, by then estimated at JD200-300m. The overall aim was to cut the deficit by 4.5 per cent of GDP in 1989 and a further 3 per cent in 1990.

COUNTRY PROFILE

Political system: Constitutional monarchy with an elected House of Representatives with 80 seats and an appointed Senate. Last full elections were held in 1967 when the House of Representatives still contained seats for both West and East Banks. Between 1978 and 1984 the House of Representatives was replaced by a National Consultative Council appointed by the King. Parliament was revived in 1984 but the House of Representatives was dissolved in July 1988 prior to the disengagement from the West Bank. Elections for a new parliament representing only the East Bank were held in November 1989.

Population: In 1989 population was approximately three million, of whom some 500,000 were officially designated as refugees from Palestine. Population growth is very high at around 3.8 per cent.

Main cities/towns: Amman (population estimated 830,000 in 1988) Salt, Irbid (149,000), Karak, Madaba, Zarqa (291,000), Aqaba.

Language: The official language is Arabic; English is widely spoken, especially in business circles.

Media
Dailies: National newspapers are *Sawt al al Shaab* (circulation 30,000), *Al-Ra'i* (80,000), *Al-Dustour* (60,000) and *Jordan Times* (15,000) (in English with a weekly French language supplement).
Weeklies: There are several in Arabic, including *Akhbar al-Usbu* (circulation 100,000), *Amman al-Masa*, and *Al-Hawadith* and one in English, *Jerusalem Star* (15,000), which also has a small French language supplement.
Radio: National service, in Arabic and English, with advertising. Estimated 550,000 receivers.
Television: Two channels with news and progammes in English and French as well as Arabic. National evening service, with advertising. Estimated 380,000 sets in 1987.

Domestic economy: After the occupation of the West Bank in 1967, massive public investment and sage economic planning and management were initiated, achieving high levels of growth. Performance fell away in the early 1980s as slackening oil prices curbed the appetite of Arab neighbours for Jordan's exports but a strong recovery seemed to be on the way in 1987-88 with GDP growth in 1987. The recovery was not sustained; 1988 saw a decline of 3.5 per cent in GDP and the emergence of a severe debt problem as a drop in expatriate remittances and Arab aid coincided with a growth in debt repayments to $937m. Jordan is now involved in rescheduling its foreign debt and faces a sustained period of austerity and economic restructuring.

Employment: Around 12 per cent of the settled population are engaged in agriculture. The general level of education and training traditionally is high and an estimated 400,000 Jordanians are employed abroad, remitting around $700m in 1988.

External trade
Exports: Main exports: chemicals, including paints, medicaments, detergents, soap and fertilisers (29% of total in 1988), phosphates (24%), potash (20.5%), manufactured goods (11%) and food and live animals (9.5%). Main destinations: Iraq (19% of 1988 total), India (17%), Saudi Arabia (10%), Egypt (5.25%), China (4%), Pakistan (4%), Italy (4%), Kuwait (3.2%), Indonesia (3.2%), Syrian Arab Republic (2.8%) and Japan (2%).
Imports: Main imports: machinery and transport equipment (22.5% of total in 1988), manufactured goods (17.5%), food and live animals (16.9%), mineral fuels including oil (15.3%), chemicals (10%) and raw materials (3.5%). Main sources: USA (12% of total in 1988), Iraq (11%), Saudi Arabia (7%), German FR (6%), UK (6%), Japan (5%), France (4%), Turkey (3%).

Agriculture: Jordan became a net importer of foodstuffs when it no longer had access to its principal growing areas on the West Bank of the River Jordan. Even in a good year, wheat production covers only 25 per cent of requirements, and 92 per cent of the cultivable land is still totally dependent on rainfall. Predominantly smallholdings, but being modernised with seed and fertiliser improvement, irrigation projects and growing mechanisation. Main crops are wheat and barley (north east of country); fruit and vegetables (under irrigation in Jordan Valley), olives, lentils, tobacco. Private irrigation projects are now under way in the southern and eastern desert areas. Animal production, mainly of sheep, goats and poultry is receiving more attention and a number of projects to raise output and improve animal health are underway.

Industry: Major industries are food processing, clothing manufacture, super phosphate fertiliser, petroleum products (crude is piped to Zarqa from Saudi Arabia), manufacture of cement, and tourism. Government has encouraged private industry to diversify and to spread throughout the country. Industrial zones have been established at Sahab (south of Amman), Irbid, Salt and Zarqa, and Aqaba has been further developed as an industrial centre. Other industrial products include sheet glass, furniture, paints, plastics, cigarettes, shoes and leather products, batteries, clothing, confectionery, cosmetics, iron, steel and aluminium products, pharmaceuticals, paper and cardboard.

Minerals and mining: Phosphate is mined at Al-Hasa and Wadi al-Abiad by Jordan Phosphate Mines. Output of phosphate rock is estimated at 6.8m tonnes in 1987, with major development under way at Shidiya. Potash is extracted from the Dead Sea by Arab Potash. Annual output was around 1.2m tonnes by 1988 and was to rise to 1.4m tonnes by 1990.

Energy: Heavy dependence on imported crude petroleum. Oil was discovered in 1982 at Azraq, 70 km east of Amman and the search for commercial amounts continues. Large deposits of oil shale have been found at Lajjun and could in the long term be used to generate electricity. Gas deposits were discovered at Al Risha in north-east Jordan during 1987, and gas turbines have now been connected to the national grid.

Stock exchange: The Amman Stock Exchange opened in 1978 and expanded rapidly. It was constrained in 1985-86 by the recessionary effect of the Gulf War on its export markets, but has made a strong recovery since 1987.

Membership of international organisations: Arab League, CCC, FAO, G-77, IAEA, ICAO, IDA, IFAD, IFC, ILO, IMF, IMO, INTELSAT, INTERPOL, IPU, Islamic Conference Organisation, Islamic Development Bank, ITU, NAM, UN, UNESCO, UPU, WFTU, WHO, WIPO, WMO, World Bank, WTO.

Time: GMT +2 hrs (GMT +3 Apr-Oct).

Climate: Climate and temperature vary with altitude. May-Sep, hot and dry but cool at night except in the Jordan Valley which is well below sea level and low-lying areas such as Aqaba on the Red Sea coast, where temperatures are generally higher. Nov-Mar, mild or warm and sometimes wet, cold at higher altitudes.

Entry requirements
Passport: Required by all (with a few exceptions, mainly for neighbouring states and service personnel).
Visa: Required by all except nationals of certain Middle East states and certain document holders. Nationals of India, Iran, Pakistan, Bangladesh, Sri Lanka and Turkey require approval from Ministry of Interior Affairs, as well as a visa.

The 30 per cent rise in the price of fuel hit hardest and within 48 hours the people of the southern town of Ma'an, where most of the population are involved in the transport business or have low salaried jobs in the army or police force, were on the streets.

Despite popular anger at the measures the new government of Zaid Bin Shaker had no choice but to stick to them if it was to receive IMF support and go ahead with rescheduling.

The new economic team in the cabinet settled into their jobs and began to give a clearer picture of the state of the economy. Jordanians were stunned with the relevation, first that the national debt was $6.5bn and then the news soon after that it was actually $8.1bn, of which $7.2bn was disbursed.

The current account deficit in the balance of payments would reach $1,213.6m in 1989 and $1,278.9m in 1990 making rescheduling unavoidable particularly as expatriate remittances had continued to decline, reaching only $800m in 1988 and expected to drop to $600m in 1990 compared with the 1984 peak of $1.2bn. There was also no question that Arab aid would cover the gap. Only Saudi Arabia had fulfilled its commitment made under the Baghdad agreement of 1978 which promised Jordan $1.2bn annually for 10 years to support it in its role as a front line state.

The government in fact sought to place the blame for its acute financial position on the failure of the Arab states to pay but it was clear that its own overspending and failure to expand the productive base of the economy must be seen as the root cause. The minister of planning now spoke of the need to move from heavy investment in infrastructure to concentrate on more effective use of existing resources and on programmes which would have the largest impact on employment and on the provision of basic needs.

Having explained the seriousness of the situation the government seemed to have little more to say on the economy and it was not until Muhammad Said Nabulsi, who had served as governor of the Central Bank from 1972-85, returned to the post that any moves were made to tackle the by then acute foreign exchange situation which had forced many businessmen to turn to the thriving black market for dollars to cover their import needs.

Nabulsi's first move was to return to a basket of currencies to set the dinar rate, the system that had prevailed until the October 1988 float. The new rate was a little more realistic but did not produce any new currency flows and it was only in mid-July that the Central Bank was able to provide any dollars for the banking system. Its first intervention was to provide $10m to be used exclusively for students abroad, pilgrims, those travelling abroad for medical treatment and for foreign workers who were allowed to remit $100 per month to their home country.

The move helped to bring the black market rate for the dollar down from 820-830 fils to around 730 fils but it rose again sharply the following week on strong rumours of a pending devaluation. The Central Bank then intervened again this time to provide $11m to fund imports of basic foodstuffs, raw materials for industry and agricultural products but the amount was only a small contribution towards the estimated $50m per month needed to fund basic requirements.

Despite government efforts to assure the public that the dinar would be stabilised and that the debt problem was under control, Jordanians, both businessmen and the general public, remained nervous and demoralised. Endless rumours about devaluation, food shortages and price rises continued to spread. As the Eid Al Adha approached rumours of price rises for food caused panic buying among the public and hoarding by merchants until finally Bin Shaker went on a highly publicised visit to the ministry of supply warehouses to show that good stocks of food were available. He then announced a crackdown on merchants found hoarding goods or inflating prices and more than 100 traders were taken before the military courts.

While the government was trying to ease the fears of the public the minister of finance led a team to both the London and Paris clubs to negotiate a rescheduling of debt and the IMF came through with a SDR 16.7m loan for short-term needs and a further SDR 60m to be disbursed over 18 months. Some 43 per cent of Jordan's debt is owed to Paris Club members and it was seeking to reschedule $656m for 1989 and $622m for 1990.

Jordan came away from the Paris talks with an agreement that allowed for a ten year rescheduling with a five year grace period and repayments in semi-annual instalments to begin in September 1994 or 1995 and continue until 1999 or 2000. Jordan also asked for the 'goodwill clause' to be applied allowing for a further rescheduling of 18 months in 1991 and the minister of finance acknowledged that he expected that altogether three reschedulings would be necessary during Jordan's five year economic restructuring programme. The country was left with $370m to be repaid to the Paris and London Clubs over the two years and with the support of the IMF and World Bank loans and foreign aid of around $400m annually the minister of finance was confident it could meet its commitments.

The smooth progress of the Paris Club rescheduling was a boost to confidence but the government had little time to enjoy its achievements as further pressure on the dinar, which it attributed to heavy dumping of dinars on the West Bank, Jordanian and Gulf markets by the Israeli Government, showed its continuing vulnerability.

On 31 July 1989 the Central Bank introduced a two-tier system for the dinar which meant floating it for all but a limited range of strategic commodities for which the bank was to provide currency at a concessionary rate. The move actually formalised the existing situation and was seen in part as a move to test the real market rate in the months leading up to an IMF appraisal mission in November when the government rate and market rates were likely to be brought closer together.

Throughout 1989 the continuing decline in the dinar led to a steady rise in inflation which following two years of zero growth in 1986 and 1987, jumped to 14 per cent in 1988 and was expected to reach a minimum of 20 per cent in 1990 with the July floating likely to push that figure even higher. The government also looked likely to come into conflict with the goals of the IMF-inspired restructuring package as it battled to protect low income groups from the worst of the economic crisis. It was already maintaining heavy subsidies on basic foodstuffs and the provision of cheap currency for these and other strategic items allowed for under the two-tier system for the dinar were estimated to add JD8 – 10m to the subsidy bill. The government was also facing an urgent need to meet the expanding demand for jobs. In 1988 there was a decline in GDP of 3.5 per cent, the maximum expected for 1989 was a growth of 1 per cent with a rise to 3.2 per cent the target for 1990, all in the face of a population growth rate of 3.8 per cent. In the latter half of 1989 the long lines of young Jordanians waiting at the United States, Canadian and Australian embassies in the hope of visas was a potent symbol of the growing problem of unemployment.

Outlook

While it is clear that 1990-95 will be difficult, the Jordanian economy does have certain positive elements. The government spending since 1974 has produced a solid infrastructure which with careful maintenance should carry the country through the restructuring period. There are also signs that local entrepreneurs are beginning to pick up on opportunities in manufacturing for export which the cheap dinar has opened up and are looking with increased confidence to the European and US markets. Agricultural exports have also benefited from the dinar's decline, with Jordan's fruit and vegetables regaining their place in the Gulf markets and exports to Europe expected to treble during the 1989-90 winter season. Tourism is also expanding, again helped by the cheap dinar and growing co-operation with Egypt and Syria in the tourism field should add further potential.

The challenge facing Jordan now is to ensure that the expansion in these areas reaches a sufficient level to meet the goal of putting the country back on its feet and freeing it from dependence on uncertain and politically weighted foreign aid.

BUSINESS GUIDE

Prohibited entry: Nationals or visa-holders from Israel.

Representation overseas: Abu Dhabi, Algiers, Ankara, Athens, Baghdad, Beijing, Beirut, Belgrade, Berne, Bonn, Brussels, Bucharest, Cairo, Canberra, Damascus, Doha, Dubai, Geneva, Islamabad, Jeddah, Khartoum, Kuwait City, London, Madrid, Manama, Moscow, Muscat, New Delhi, New York, Ottawa, Paris, Rabat, Rome, Sana'a, Santiago, Santo Domingo, Tokyo, Tripoli, Tunis, Washington.

Air access
National airline: Royal Jordanian.
Other airlines: Aeroflot, Air Canada, Air France, Alitalia, British Airways, Egyptair, Emirates Air, Gulf Air, Iberia, Interflug, Iraqi Airways, JAT, KLM, Kuwait Airways, Libyan Arab, Lufthansa, MAS, MEA, Olympic Airways, PIA, Saudia, Swissair, Tarom, Tunis Air, THY Turkish, Yemenia.
Tax: Airport tax of JD10 for foreigners and JD25 for Jordanians on departure.
International airport: Amman-Queen Alia International (Code: AMM), 40 km south of city (35 minutes from main downtown hotels, etc); Aqaba International (AQJ), 9 km north of Aqaba.

Surface access: Overland access possible by road from Saudi Arabia and other neighbouring countries.
Main ports: Jordan has 26 km of coastline on the Red Sea around the port of Aqaba, which has a container and ro-ro terminal.

Hotels: There is a good selection of hotels in Amman, as well as the tourist centres of Aqaba and Petra. A service charge of 10-15 per cent is usually added to the bill plus a government tax of 10 per cent on all services at four- and five-star hotels and restaurants. Major credit cards are accepted.

Electricity supply: 220 V AC, 50 cycles. Industrial supply: 220-380 V.

Credit cards: Major credit cards are accepted at hotels and restaurants.

Car hire: National or international driving licence required. Driver must be at least 25 years old and not over 60. Speed limit is 60 kph. Insurance compulsory.

City transport
Taxis: Metered taxis are readily available in Amman and other cities. Can be hired for the journey, or the day for an agreed sum. Since there are no street numbers and few street names, destinations are generally described in relation to landmarks. Tipping approximately 10 per cent.

National transport
Air: Royal Jordanian operate regular flights between Amman and Aqaba. Arab Wings offer a charter service.
Road: Some 7,000 km of surfaced road (1988). Much upgrading in progress. A network of main roads extends from Amman to Aqaba, to all main towns and ultimately to Saudi Arabia, Baghdad and Damascus.
Rail: Network of over 600 km of track, joining in the north with Syria's system. No passenger services.

Public holidays: Friday is the official weekly holiday.
Fixed dates: 1 Jan, 1 May (Labour Day), 25 May (Independence Day), 10 Jun (Arab Revolt and Army Day), 11 Aug (King Hussein's Accession), 14 Nov (King Hussein's Birthday), 25 Dec.
Variable dates: Lailat al-Miraj, Orthodox Easter Sunday, Eid al-Fitr, Eid al-Adha, Al-Hijra, Prophet's Birthday.

Working hours: Seasonal variations, but within the following bands:
Business: (Sat-Thu) 0800/0830 – 1300/1330, 1500/1530 –1830/1930.
Government: (Sat-Thu) 0800/0830 – 1400/1430.
Banking: (Sat-Thu) 0830 – 1430, cashiers closing 1230, 1530 – 1730.
Shops: (Sat-Thu) 0900 – 1300, 1500 – 1900.

Social customs: People are hospitable and courteous, and business and professional people are used to Western visitors. It is not polite to refuse offers of hospitality, for example of coffee or cold drinks during meetings. Care should be taken to respect customs, particularly during Ramadan not to take refreshments or smoke in the company of Muslims.

Telecommunications: The telecommunications system in urban areas is well developed and only around two-thirds of available telephone lines are in use. Plans for expansion and for the improvement of the rural network are now limited by the government austerity programme which has led to a 60 per cent cut in the Telecommunications budget for the next three years.
Telephone: Some 59,000 telephones (1988). Connections with main urban centres generally good, though delays may be experienced in business hours. Direct dialling to most countries. Dialling code for Jordan: IDD access code + 962, followed by the area code (Amman 6, Aqaba 3, Irbid 2, Zarqa 9), followed by subscriber's number. Since 1985 all Amman numbers have six figures. Five-figure numbers beginning with 2,3,4 or 5 were prefaced with a 6, and those beginning with 9, with an 8.
Telex: Over 1,400 telex subscribers in Amman and public telex facilities are available in the main hotels. 24-hour facilities in city centres.
Fax: Increasingly used.
Telegram: Good inland and overseas services.

Banking: Jordan is moving adventurously into new financial services.
Central bank: Central Bank of Jordan.
Main centre: Amman.
Head offices: Arab Bank, Arab Jordan Investment Bank, Arab Land Bank, Bank of Jordan, Cairo Amman Bank, Cities and Villages Development Bank, Housing Bank, Industrial Development Bank, Jordan Co-operative Bank, Jordan Gulf Bank, Jordan Islamic Bank for Finance and Investment, Jordan Kuwait Bank, Jordan National Bank, Petra Bank, Syrian Jordanian Bank.
Foreign branches: Arab Finance Corporation, Arab National Bank, Bank Al Mashrek SAL, British Bank of the Middle East, Citibank, Grindlays Bank, Manufacturers Hanover Trust Company, Osterreichische Länderbank, Rafidain Bank, Société Générale.

BUSINESS GUIDE

Weights and measures: Metric system. Land is measured in dunums (1,000 sq metres).

Representation in capital: Algeria, Australia, Austria, Bahrain, Belgium, Brazil, Canada, Chad, Chile, China PR, Cyprus, Czechoslovakia, Denmark, Egypt, Finland, France, German DR, German FR, Greece, Hungary, India, Indonesia, Iraq, Italy, Japan, Korea DPR, Republic of Korea, Kuwait, Lebanon, Morocco, Netherlands, Oman, Pakistan, Philippines, Poland, Portugal, Qatar, Romania, Saudi Arabia, Spain, Sudan, Sweden, Switzerland, Syria, Tunisia, Turkey, UK, United Arab Emirates, USA, USSR, Vatican, Yemen AR, Yugoslavia.

Hotels: Amman
Alia Gateway, Queen Alia International Airport (tel: (08) 51000; tx: 23361).

Ambassador, P.O. Box: 925390 (tel: 665161/2/3; tx: 21628).

Amra, P.O. Box: 950555 (tel: 815071/80; tx: 23217, 22012; fax: 814072).

Commodore (tel: 665186; tx: 21760).

Inter-Continental, Jabal Amman, P.O. Box: 35014 (tel: 641361; tx: 21207, 21267; fax: 635217).

Jerusalem International, University Street, P.O. Box: 926265 (tel: 665121/9; tx: 22330; fax: 689328).

Marriott, P.O. Box: 926333 (tel: 660100; tx: 21145; fax: 670100).

Middle East, P.O. Box: 19224 (tel: 667160, 667169; tx: 21159).

Philadelphia International, University Road, Jabal (tel: 663100; tx: 21859 PHIL HO JO; fax: 665160).

Plaza, P.O. Box: 950629 (tel: 674111; tx: 23266 PLAZA JO; fax: 674261).

Regency Palace, Queen Alya Street, P.O. Box: 927000 (tel: 660000/15; tx: 22244/5; fax: 660013).

Tyche, P.O. Box: 3190 (tel: 661114/5; tx: 21729).

Hotels: Aqaba
Al Cazar, P.O. Box: 392 (tel: 314131; tx: 62242; fax: 314133).

Al Dhabet, Al Corniche Street, P.O. Box: 1407 (tel: 316480).

Al Jaber, Al Corniche Street, P.O. Box: 579 (tel: 316999).

Al Jameel, Downtown, P.O. Box: 75 (tel: 314118).

Al Manara, Downtown, P.O. Box: 151 (tel: 314366).

Al Nahr Al Khaled, Awqaf Bldg, P.O. Box: 387 (tel: 312456).

Al Nouman, Downtown, P.O. Box: 1009 (tel: 315142).

Al Shati, Downtown, P.O. Box: 702 (tel: 315108).

Al Yamamah, Downtown, P.O. Box: 283 (tel: 314621).

Aqaba, Palace Street, P.O. Box: 1060 (tel: 31516519; tx: 62308).

Aquamarina 1, Al Corniche Street (tel: 316250; tx: 62249; fax: 314271).

Aquamarina 2 (tel: 315165; tx: 63208).

Coral Beach, Al Corniche Street, Aqaba (tel: 313521; tx: 62227; fax: 313614).

Holiday International, Al Corniche Street, P.O. Box: 215 (tel: 312426; tx: 62263).

Miramar, Al Corniche Street, P.O. Box: 60 (tel: 314339; tx: 62275).

Nairoukh, Downtown, P.O. Box: 908 (tel: 312984).

Red Sea, Downtown, P.O. Box: 65 (tel: 312156).

Zahrat Al Urdon, Downtown, P.O. Box: 113 (tel: 314377).

Hotels: Petra
Petra Forum, Wadi Mousa, P.O. Box: 30 (tel: 634200, 83246/7; tx: 64001).

Petra Resthouse, Wadi Mousa, P.O. Box: 5 (tel: 83011; tx: 21607).

Car hire: Amman
Avis: PO Box 6925 (tel: 603500, 670498; tx: 23678); at Amman Airport (tel: (08)51071; tx: 23678).

Budget: central reservations (tel: 604231/9; tx: 24197); Sports City Junction (tel: 630433, 673312; tx: 32220, 24197).

Europcar: Amman Marriott Hotel, PO Box 2020 (tel: 639197, 642401, 660100; tx: 21410).

Business advice and information
Amman Chamber of Commerce, PO Box 287, Amman (tel: 666151; tx: 21543; fax: 666155).

Amman Chamber of Industry, PO Box 1800, Amman (tel: 641684, 642649; tx: 22079; fax: 647852).

Central Bank of Jordan, PO Box 37, Amman (tel: 630301/9; tx: 21250 Bankzi Jo; 21476, 21766 Bakazi Jo).

Civil Aviation Authority, PO Box 7547, Amman (tel: 892282; tx: 21325, 23398; fax: 891653).

Federation of Jordanian Chambers of Commerce, PO Box 7029, Amman (tel: 665492; tx: 24103).

Ministry of Agriculture, PO Box 2099, Amman (tel: 639391; tx: 24176 agri).

Ministry of Public Works and Housing, PO Box 9052 (tel: 668481).

Ministry of Energy and Mineral Resources, PO Box 140027, Amman (tel: 815615/20, 817615/19; tx: 21259 jeasak; fax: 818336).

Ministry of Finance and Customs, PO Box 85, Amman (tel: 636321; tx: 21424).

Ministry of Trade, Industry and Supply, PO Box 2019, Amman (tel: 663191; tx: 21163).

Ministry of Information, PO Box 1854, Amman (tel: 641467; tx: 21523 mnif).

Ministry of Labour and Social Development, PO Box 8160, Amman (tel: 667161).

Ministry of Planning, PO Box 555 (tel: 644466).

Travel information
Ministry of Tourism, PO Box 224, Amman (tel: 642311, 634114; tx: 21741 touris).

Ministry of Transport, PO Box 35214, Amman (tel: 641461; tx: 21541 mot).

Royal Jordanian Airlines, PO Box 302, Amman (tel: 672872 (head office); tx: 21501).

Kenya

Robert Shaw

Kenya's economic fortunes in 1989 were variable with its two major foreign exchange earners, tourism and coffee, suffering respectively from bad publicity and the demise of the ICO's International Coffee Agreement. The year also witnessed the continuing consolidation of the power of the president and the ruling party, the Kenya African National Union (KANU), and a tamer parliament. The country's vice-president for just over one year, Josphat Karanja, resigned in May 1989 and was replaced by the rather low profile minister of finance, Professor George Saitoti.

Macro-economic outlook

As in the previous year the overall statistics were not very encouraging although the country's economy continued to be relatively resilient and active particularly in the private sector. There were fears that the economy may be entering a gentle recession.

Economic growth in 1988 was officially given at around 5 per cent, but pertinent observers believed the figure was nearer 4 per cent. Provisional economic growth estimates for 1989 vary with some pessimists saying the final figure could be as low as 2.5 – 3 per cent. Whatever the final figure, it is likely to be below the country's population growth rate of just under 4 per cent per annum.

The new 1989-93 development plan, with its theme 'Participation for Progress' contains some ambitious and currently unrealistic growth and fiscal targets. It projects that in order to provide employment for the two million entering the labour market during that period, economic growth will have to be in the region of 5.4 per cent and gross investment around 20 per cent of GDP. It was accompanied by the economic survey which portrayed a rather rosy picture of the economy.

KEY FACTS

Official title: Republic of Kenya

Head of state: President Daniel arap Moi

Ruling party: Kenya African National Union

Capital: Nairobi

Official Languages: Kiswahili, English (Kikuyu, Luo, Kikamba, Kiluhya widely spoken)

Currency: Kenya shilling (Ksh) = 100 cents

Exchange rate: Ksh21.5 per $ (Nov 1989)

Area: 582,646 sq km

Population: 23.9m* (1988)

GDP per capita: $360 (1988)

GDP real growth: 6.1% (1988)

Inflation: 8.2% (1988)

Trade balance: –$785m (1988)

Foreign debt: $5.9bn (1987)

* estimated figure

Two of the main macro problems beleaguering the economy were the large balance of payments deficits and the increasing public sector deficits. In both 1987 and 1988 the balance of payments deficits on the country's current accounts were in the region of $500m and the figure for 1989 was not likely to be much less. The public sector deficit continues to hover around 7 – 8 per cent of GDP.

Kenya has had balance of payments deficits in most years since the oil price hikes of the 1970s. Despite the reduced dependence on coffee as the country's primary export over the years and the more prominent role played by tourism, tea and more recently horticultural exports the gap continues to widen.

The saving grace is the large inflow of concessionary aid. According to the ministry of finance, Kenya was expected to receive over $900m in the 1989-90 financial year from aid donors in the form of grants and loans. The most significant donor countries appear to be the USA, Japan, Denmark, West Germany and France. The leading multilateral donors are the World Bank's IDA, the European Community (EC) and the World Bank itself. The significant debt remissions by West Germany, France and the USA have also helped.

An IMF enhanced structural adjustment facility to the tune of $300m was eventually finalised in May 1988, although there were sticking points particularly over the government budget deficit which lengthened the negotiations. This facility was badly needed in order to take the pressure off the country's debt service ratio, as a percentage of exports, which reached an uncomfortable 40 per cent.

The persistently large government budget deficits are likely to become an increasingly vexed issue with donors and in particular with the IMF, as well as with the liquidity strapped private sector. Its roots lie in the government's failure to grapple with its balance of payments problems and the many deficits of its 150 or so parastatal bodies, and in particular the National Cereals and Produce Board (NCPB). Extra expenses to do with the hosting of the All Africa Games in 1987, the 1988 election and the 1989 census have also taken their toll.

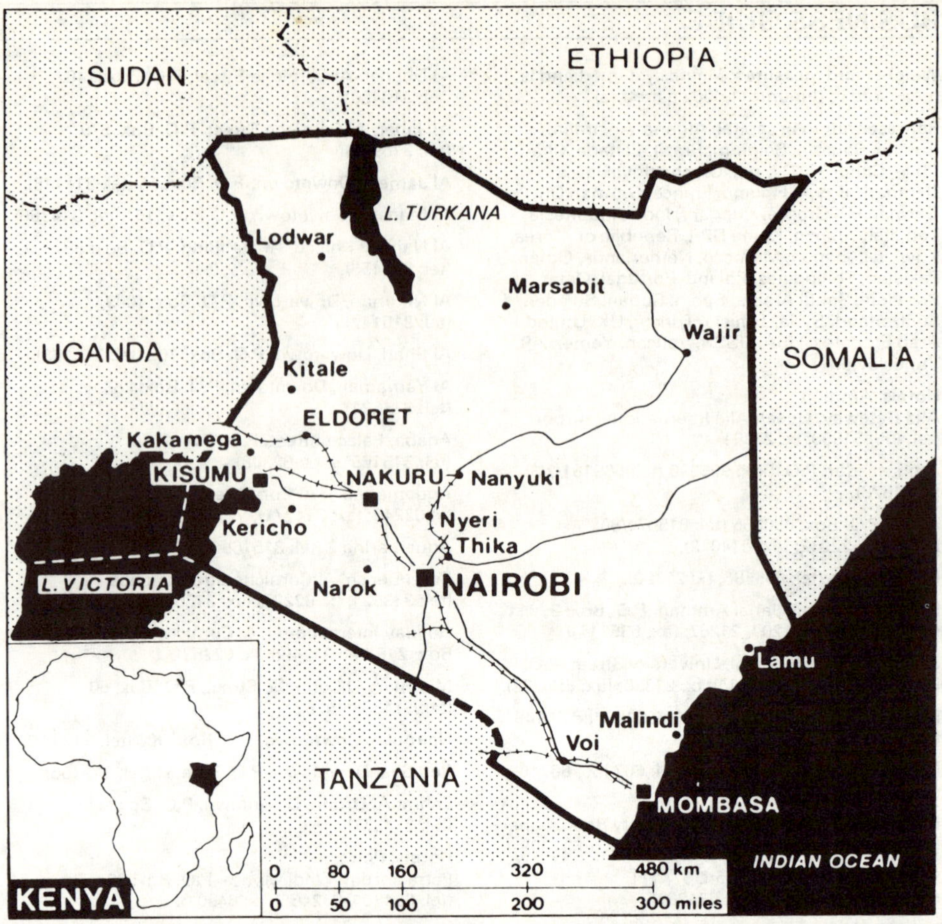

Major foreign exchange earners

Tourism has now overtaken coffee as the country's major foreign exchange earner. It earned the country around $300m in 1988, and the number of visitors was up by 6 per cent to 700,000 over the previous year. While its performance continued to be good there were fears that the spurt of growth witnessed in the second half of the 1980s was slowing down and may well have peaked. The reasons are mainly twofold. The highly competitive beach package holiday business is becoming increasingly saturated. Secondly the spate of bad international publicity received from the rampant poaching in 1988, and the attacks on and murders of several tourists in the country's game parks may well put a damper on safari bookings for the forseeable future.

The demise of the ICO's International Coffee Agreement in early July 1989 brought a series of mixed fortunes to the country's coffee industry. Kenya produces the unwashed Arabica type of coffee of which the world supply and demand equation is less disproportionate. It will have little problem selling off its carryover stock of over 1M bags due to the old quota system. But even though much of Kenyan coffee sells at a premium over other coffees the price has been pushed down sharply by prevailing market trends. Kenya will probably earn the same amount of foreign exchange from coffee in 1989 as it did in the previous two years or so by selling more at lower prices. But Kenyan coffee is also an expensive one to produce, requiring generous amounts of fertilisers and fungicides. When the lower prices start to filter through to the coffee farmer, particularly the smallholder where the economies of scale are at their worst, the production of coffee could become increasingly uneconomic. The official policy is for growth in the coffee industry both in acreage and production, although there is growing pressure on the government to revise this policy and allow some farmers to do away with coffee in favour of more renumerative horticultural crops.

Kenya's third largest foreign exchange earner, tea, has fared better. There is no quota system, despite pressure on Kenya by established producer countries such as Sri Lanka and India to accept one. Kenya has therefore been able to sell all it produces. Kenyan teas generally command good prices due to its special blending quality. Last year prices picked up due to improved demand from the industry's largest export market, Britain, and because of heavy buying by another significant consumer, Pakistan. All in all the country's tea industry is in better shape than the coffee industry, although in the longer term any over supply situation could result in problems particularly to do with weak prices.

The big success story of the late 1980s was the meteoric rise of canned and fresh horticultural exports. Kenya has been a major exporter of canned pineapples for a number of years. Kenya began to export a range of out of season and exotic vegetables and fruits such as French beans, passion fruit, avocados and cherry tomatoes and a variety of flowers including carnations, roses and alstroemerias. In 1988 Kenya exported over 60,000 tonnes of fresh horticultural produce which was roughly double the 1984 figure. The total value of all horticultural exports is now in the region of $100m per annum. This is still quite a way behind the top three foreign exchange earners, but it is conceivable that it could be in their league before 1995 considering the various constraints the others are facing.

Agriculture and industry

Generous rainfall has ensured that the country remains self-sufficient in its staple product: maize. However, there continue to be payment delays to farmers particularly by the NCPB. The NCPB remains the overall buyer and seller of maize and several other foodstuffs, and has a chequered history of enormous losses and graft. It is estimated that its trading losses in recent years account for nearly 20 per cent of the public sector deficit. There has been an attempt to clean up its operations, although there is little evidence of a move towards increased private trade in the commodities it is responsible for, despite World Bank and EC prodding. In 1989 there were periodic shortages of price and distribution regulation food products such as wheatflour, bread and sugar.

Despite bureaucratic constraints, particularly to do with price controls, the industrial sector clocked up quite good growth rates in 1988-89. This led to renewed hope that export-led industrial growth could have a significant role to play in the future.

In the 1970s the country's manufacturing sector grew under the umbrella of blanket protection, and was frequently accused of being insufficient due to limited competition and the small domestic market. During the course of the 1980s the country's industrial base has broadened and grown. Quantitive import restrictions on items that can be produced locally were increasingly replaced by tariff differentials. However there were, and some cases still are, major impediments particularly to do with price controls on most essential items, import licence problems for essential raw materials and the stifling web of regulatory bureaucracy. Price controls on cement and many steel related products were the major cause of the problems facing the former and shortages of such essential items as nails in the latter.

Price controls were lifted on a number of items in

1988 and the import licencing system has undergone continuous refinement and rationalisation to the extent that most essential imports are automatically licensable. These and other changes such as the tax changes in favour of manufacturing exports have come about with a lot of prodding and tempting by such bodies as the World Bank and IMF. For example, the World Bank industrial sector credit for $110m released in late 1988 was conditional on certain changes and reforms taking place. There were also moves to pave the way for export processing zones although it is too early to say what sort of tangible response there will be.

The moves to attract more investment, especially foreign, in export-orientated manufacturing projects stem from two main motives and concerns. First, there is hope that Kenya can attract some of the relocating of manufacturing activities taking place from high labour cost areas to lower labour cost areas. Secondly, there is serious concern by the government and donors alike at the continuing dearth of foreign investment.

The 1989-90 budget

The budget given by the vice-president and minister of finance had two main strands to it. First there were a number of significant proposed reforms and tax changes to do with replacing the sales tax with VAT and making it applicable to a number of services as well as manufacturing, and a 5 per cent presumptive tax on many farm products. Whether VAT will be expanded to encompass other sectors of the economy such as the retail trade remains to be seen. The tax on farm products was interesting because of the belief that farmers get off very lightly particularly in the direct tax network. Both were part of an increasing trend to rely more on indirect taxes. Both corporation and income tax were reduced, but this was offset by the reduction in taxable benefits. Three other areas that were affected were the tightening up of the banking regulations through the Banking Act; the proposed Capital Markets Authority to revitalise the country's rather moribund stock exchange; and the new 'reclassification of traded goods and commodities to the new Harmonised System'. The latter was not very well prepared and resulted in a number of anomolies.

The second strand was the increasingly precarious tightrope being walked by the country as regards its deficits. The budget proposed an increase of around 20 per cent in government expenditure. Yet it was hoped in official circles that concessionary inflows of aid would help fill the gap both on the government deficit and the balance of payments front.

Longer-term problems

If the macro short-term problems particularly to do with foreign exchange and public sector deficits were causing increasing concern, little comfort could be obtained from the longer-term problems thrown up by the country's rapid population growth of just under 4 per cent per annum.

Kenya's population growth is one of the highest in the world. By the end of this century its population is projected to increase from its present estimated level of 23 – 24m to 37 – 38m. Already over half of the country's population is under 14 years of age. Less than one fifth of the country is good cultiveable land and 80 per cent of the population densities are as high as 400 persons per square km.

But the pressure put on the country's land is only one facet of the population problem. Considerable strain is being put on basic infrastructural facilities and services such as healthcare and education, the provision of water particularly in the mushrooming shanty towns, and housing.

KEY INDICATORS

	Unit	1984	1985	1986	1987	1988
Population	m	19.5	20.3	21.2	22.9	23.9
Gross domestic product (GDP)	KSh m	87,781	99,866	116,864	131,216	152,679
GDP per capita	KSh	4,501	4,919	5,512	5,730	6,388
Inflation	%	10.2	29.9	4.0	5.2	8.2
GDP real growth	%	2.1	4.1	7.0	5.8	6.1
Coffee output	'000 tonnes	111	130	116	109	125
Tea output	'000 tonnes	116	147	143	156	164
Maize output	'000 tonnes	1,422	2,650	2,898	2,250	2,685
Wheat output	'000 tonnes	144	250	257	220	240
Industrial production	1985=100	95.6	100.0	105.6	–	–
Exports	KSh m	15,538	16,047	19,737	15,790	19,019
Petroleum	KSh m	2,852	2,385	2,157	1,965	2,316
Coffee	KSh m	4,072	4,613	7,771	3,892	4,905
Tea	KSh m	3,790	3,828	3,456	3,264	3,645
Imports	KSh m	21,944	24,023	26,758	28,904	35,072
Balance of trade	$m	–314	–327	–284	–714	–785
Current account	$m	–126	–113	–38	–498	–455
Tourist arrivals	'000	396	410	422	437	–
Total external debt	$m	3,720	4,390	4,934	5,950	–
Debt service ratio	%	21.6	25.5	22.7	28.8	–
Exchange rate	KSh per $	14.4	16.4	16.2	16.5	17.7

* estimated figure

Another aspect is the increasing number of people, school leavers in particular, looking for jobs. It is estimated that around 400,000 school leavers come on to the job market each year although no more than 30,000 – 40,000 formal jobs are added to the formal sector annually. The labour pool is in the region of 8m persons although there are only 1.5m formal paying jobs. Even in 1988 when the rate of economic growth was regarded as good the growth of total wage employment was only 3.7 per cent, and hence below the rate of growth in the labour force.

The implications of Kenya's population growth are exceedingly serious and are likely to be the root cause of increasing social, economic and political turmoil.

Broad political background

The late 1980s have seen the continuing consolidation of the power of President Moi, who celebrated 10 years in office in October 1988, and of the sole ruling party: the Kenya African National Union (KANU).

The March 1988 elections resulted in several of the country's more outspoken, and controversial, politicians such as Martin Shikuku and Charles Rubia not being returned to parliament – often amid allegations of electoral malpractices. This has resulted in a tamer, and some would argue more pliable, parliament.

The 1988 elections also gave the president the opportunity to choose a cabinet very much in his mould thus dispensing with the last vestiges of inherited alliances or obligations. The main victim of this was Mwai Kibaki. He lost his job as the country's vice-president which he had held since President Moi acceded to the presidency. Mr Kibaki chose not to stand again for the KANU vice-presidency in the party elections later in the same year thus confirming his eclipse. He was replaced by a relative newcomer to the political scene, Josphat Karanja. While being a Kikuyu, Mr Karanja was regarded as a political lightweight. It can be further argued that the party elections resulted in party posts going to people more in line with the president.

The year 1988 also witnessed the enhancement of the president's actual constitutional power when parliament passed the Constitution (Amendment) Act in a matter of days. The new act gave the president the right to dismiss High Court and Appeal Court judges and members of the Public Service Commission without consulting a tribunal.

The mantle of opposition, or criticism, was increasingly taken up by bodies or institutions outside the formal political arena, such as the church and the Press although it was increasingly evident throughout 1988 that the government was getting frequently irritated by the former and intolerant of sections of the latter.

1989 political scene

The 1989 political scene witnessed a continuation and consolidation of the trends already outlined. This involved the various shifts in fortunes within the political hierarchy with particular reference to the Kikuyu tribe including the downfall of one prominent Kikuyu minister and the change in the vice-president; the increasingly pervasive power of the ruling party, KANU; a strident official offensive against sources of irritation whether they were the Press or church leaders.

One other increasingly prominent and disturbing issue was the scale of corruption and mismanagement that was evident or alleged in many sectors of Kenyan society and the economy, and particularly in the public sector.

The political eclipse of Kenneth Matiba in late 1988 and early 1989 can be seen as the final stage in the dismantling of former vice-president Kibaki's power base and a further stage in President Moi's rearrangement of Kikuyu alliances. Mr Matiba was one of the more prominent Kikuyu politicians left. He failed to regain his party post at the KANU elections which he claimed were rigged. He resigned his cabinet post soon afterwards and withdrew completely from the political arena.

While President Moi had no overt role in the well-orchestrated campaign against former vice-president Karanja he did not defend him either. Mr Karanja was subjected to a whispering campaign and then finally to a vote of no confidence in parliament in May 1989. He resigned from his government, parliamentary and party positions soon afterwards and as is usual in these situations he was expelled from KANU. One of the most interesting and extraordinary aspects of the affair was the high degree of Kikuyu involvement in his ouster, thus exposing the rather divided Kikuyu ranks.

He was replaced by the Masai/Kikuyu Professor George Saitoti whose low-key and less Kikuyu-rooted background President Moi obviously feels more comfortable with. Professor Saitoti first came to prominence in 1983 when he was nominated an MP and appointed minister of finance – a post that he has held since.

President Moi has come out of the political events of 1988-89 very strongly, although not surprisingly there are rumblings, particularly among some of the Kikuyu who also form a fair rump of the middle class. While his current position looks unassailable there is the danger that any major economic upset could threaten that position.

One example of his confident position was the release of the seven remaining detainees in June 1989, and his expressed wish that he wanted a new era of tolerance and understanding. The president may also have had an eye on lowering the temperature and ruffled feathers created by the Karanja affair and its aftermath of purges within the party ranks.

The rising role of KANU in the country's affairs continued to be a prominent trend during 1989. The local party hierarchies have become more vociferous and active in day-to-day affairs sometimes to the extent that they appear to be competing with either the local MPs or the local administration or both. Another trend was the increasing role of KANU youthwingers. The *Standard newspaper* printed a report in September stating that the Kisumu KANU branch executive had overruled Kisumu town council's decision to stop KANU youthwingers from collecting revenue from vehicles operating at the local municipal bus stop. In April, two KANU youthwingers interrupted a sermon being given at St Thomas's church, Kerugoya by the Anglican Bishop David Gitari and attempted to grab his microphone.

The deteriorating state-church relations have become a vexed and complicated issue with no end in sight. Indeed, with a fairly docile parliament and little evidence of a resurrection of the subversive organisation 'Mwakenya', various church leaders have become tantamount to an informal opposition. While there does not appear to be any overall concerted attack on the government they have made critical, or conscience-pricking, comments on a number of topical issues ranging from the government's handling of demonstrating medical students to alleged corruption. In September the Bishop of the Church of the Province of Kenya of Maseno South Diocese Dr Henry Okullu took the government, and the police in particular, to task over their action against demonstrating medical students. When he in turn was criticised by three cabinet ministers from his area he replied that their criticism was 'full of irrelevancies' and that 'it is too simplistic to dismiss bishops in this country with a warning that they would not be tolerated or with threats to take them to court'. Dr Okullu's latter comment neatly sums up the official dilemma. While government officials have become bolder in attacking the outspoken bishops and other clergy, they have failed to silence them and in a number of cases have attracted scathing rejoinders which in turn has attracted additional publicity.

The government was more successful in controlling the local press. In April the Attorney-General banned the weekly publication *Financial Review* on the grounds that certain aspects of its coverage of the wheat crisis were incorrect and inflammatory. In particular its references to the government's action being tantmount to 'nationalising' the millers' operations irked the authorities. However the *Financial Review*, with its rather outspoken and populist stance, had raised official hackles for some time and so it was inevitable that something would happen sooner or later. Another blow to the Kenyan Press was in June 1989 when parliament, through a motion of censure moved by the deputy speaker, barred the country's largest selling daily newspaper the *Daily Nation* from covering proceedings in the house indefinitely. The end result was that the Kenyan Press became more nervous and self-censoring.

The increasing and alarming allegations of corruption, graft and mismanagement were taken up by the local press, mainly because many of the instances reported landed up in the courts. In July 1989 six senior customs officials and an official from the ministry of commerce were charged with offences related to the illegal importation of motor vehicles. In the same month several people were charged on 18 counts of conspiring to defraud the NCPB of Ksh51.6m. These were but two examples within a few days of each other.

International relations

There were two cross currents in Kenya's image abroad and in its international relations. Compared to recent years Kenya received less adverse publicity on the human rights front, largely due to the fact that its record was a better one. As a result, strained relations with the Scandinavian countries, for example, have improved. At the same time Kenya's image, and President Moi's in particular, was helped by such events as his mediating and hosting role in the Nairobi talks between Mozambican church leaders and Renamo representatives. President Moi co-chaired the London conference on the ozone layer with Margaret Thatcher. President Moi is increasingly being looked upon as one of Africa's leading statesmen.

The cross current is as a result of the continuing array of adverse international press publicity received from the attacks on and murders of several tourists in the country's game parks, and the murder of the famous lion conservationist George Adamson.

Relations with neighbours

Relations with Tanzania continue to improve, although there has been marginal progress only on greater economic co-operation between the two countries, particularly in the area of tourism. But the picture is not so happy with either Uganda, Somalia, or Sudan. Kenya's relations with Uganda have been chequered ever since President Museveni came to power. This is largely attributed to personal differences between the two presidents. As a result a relatively minor event, the dropping of two bombs on the Kenyan border town of Lokichogio, resulted in the flaring up of a major diplomatic incident. Kenya blamed Uganda for the bombing, although Uganda denied any involvement in it. The turbulence in Somalia has resulted in several border incidents as Somali government forces have pursued and fought with rebel forces in the south. This has often spilled over into Kenyan border towns resulting in killings. Relations with the Sudanese Government remain wary due to suspicions that covert help for the SPLA rebels is being received via Kenya and with official sanction.

COUNTRY PROFILE

Political system: Executive power held by the elected president; legislative power is vested in the 188-member National Assembly, of which 176 members are elected. Kenya gained independence in 1963.

Political parties: The Kenya African National Union is the only legal political party.

Population: Kenya's population of 22.7m (1988 estimate) is growing at an annual rate of 3.8 per cent a year – enough to cancel out all of the country's current economic growth and sufficient to impose a mounting strain on urban services. There are some 80,000 Asians, 40,000 Arabs and 40,000 Europeans in the country. A national census was to be held in 1989.

Main cities/towns: Nairobi (population 1,100,000, some 1,650 metres above sea-level), Mombasa (chief port, population 400,000), Nakuru, Kericho, Thika, Kisumu, Eldoret, Machakos, Nyeri, Kitale.

Language: English and Kiswahili are the official languages, and English is universally used in business. Elsewhere, Kiswahili is the norm.

Media
Press: Principal dailies: *Daily Nation* (circulation 180,000), *The Standard*, *Taifa Leo* (Kiswahili, 80,000 copies), *Kenya Leo*, (Kiswahili) *Kenya Times*. Also national weeklies, including *Weekly Review*, and a number of general and specialised periodicals, including the business paper *East African Report on Trade and Industry*, and *The Kenya Farmer*.
Radio: Voice of Kenya operates extensive national services in Kiswahili, English, Hindi and numerous African languages. There are about 5m receivers (1989).
Television: Voice of Kenya Television operates on four channels for 50 hrs/week. Commercial service in Kiswahili and English. 290,000 sets (1989).

External trade: Kenya's economy has made moderate progress, aided by a domestic austerity programme. But population growth threatens to overtake economic advance. Large current account deficits have been partly funded by foreign borrowing. Tourism is a major foreign exchange earner.
Exports: Coffee (26.6 per cent of total value in 1988), tea (20.2%), petroleum products, fruit and vegetables, cement, sisal. Main destinations: UK (19.9% of total value in 1988), German FR (12.1%), Uganda (8.8%), USA (4.8%), Netherlands (5%).
Imports: Crude petroleum (11.9 per cent of total value in 1988), industrial supplies, machinery and capital equipment, food and live animals, transport equipment. Main sources: United Arab Emirates (11.3% of total value in 1988), UK (18.9%), Japan (12.2%), German FR (9.5%), USA (4.9%).

Agriculture: Sector remains keystone of the economy, contributing 29 per cent to GDP (1988) and providing a livelihood for around 80 per cent of the population. Primary and processed agricultural products also account for over 50 per cent of export earnings. Subsistence farming comprises more than half of output. Principal food crops are maize, sorghum. cassava, beans and fruit. Livestock farming was badly hit by drought in 1984-85. A National Food Programme, started in late 1986, aims at self-sufficiency in basic foods by 1990. Food production (especially maize) is now able to keep up with the rapid population growth with surplus exports in 1987-88. Inadequate storage facilities, little irrigation, recurrent drought and lack of incentives and a severe shortage of arable land have restricted growth. The principal cash crops are

COUNTRY PROFILE

coffee (mainly arabica grown by smallholders), tea, sugar, cotton, pyrethrum, sisal, tobacco, pineapples and wattle. Horticultural produce is increasingly important. Coffee and tea, which together accounted for 60 per cent of export earnings in 1988, have fluctuated over the past few years due to drought and shifts in world commodity prices. Deforestation is a major problem, but forestry and fishing are currently being developed with concessional aid. A Permanent Presidential Commission on Soil Conservation and Afforestation has been created to correct the situation.

Manufacturing: Manufacturing sector contributes 13 per cent to GDP. Principal industries include food and tobacco processing, textiles, glass, chemicals, engineering, tyre manufacture, vehicle assembly and construction materials. Foreign investment, particularly from UK, Japan and USA, plays a significant role. Rapid expansion of import substitution industries during 1970s under protection of high tariff barriers. Current emphasis on developing joint-venture export-orientated industries and encouraging greater utilisation of local raw materials and other inputs. Large sections of industry continue to operate well below full capacity as a result of import controls, rising costs and marketing difficulties. Former power shortages have been controlled (see 'Energy'). Tourist industry is now the largest foreign exchange earner. Arrivals reached 694,000 in 1988, and the government is aiming for 1m by 1990.

Mining: Activity is centred on extraction of soda ash and fluorspar for export. Other minerals include silver, lead, gold, limestone, salt and chromite, but these are not yet commercially viable.

Energy: Around 75 per cent of domestic energy requirements are imported. However, government efforts to reduce oil imports is succeeding. New energy sources being developed include ethanol from molasses and geothermal power. Hydro-electric power currently supplies around 12 per cent of energy needs, and by 1988 four power stations were onstream on the Tana River. Development of electricity sector is a priority in current economic planning; another dam was planned at Turkwel Gorge. Exploratory oil wells in the Rift Valley and north-eastern provinces have produced positive results. Kenya has its own 4.2m tonne capacity oil refinery at Mombasa, which is a major foreign exchange earner.

Membership of international organisations: African Development Bank, African States Associated with the EC (Lomé Convention), Arab Bank for Economic Development in Africa (recipient), Economic Commission for Africa, Commonwealth, GATT, IFAD, IDA, FAO, ILO, IMF, OAU, UN, WHO, WIPO, World Bank.

BUSINESS GUIDE

Time: GMT +3.

Climate: Tropical and hot in coastal regions, temperate, with four seasons, on the plateau and in highlands. Long rains from Mar-Jun and short rains Oct-Nov. Temperatures in Nairobi range from 11°C to 25°C in Jul and 13°C to 30°C in Feb.

Entry requirements

Passport: Required by all.

Visa: Required by all except nationals of Denmark, German FR, Finland, Ireland, Norway, Spain, Sweden, Turkey, Uruguay and most Commonwealth countries; India, Sri Lanka, Nigeria and Australia require visas. Check current regulations as they are liable to change at short notice. Anyone who has spent more than three of the previous twelve months in South Africa, regardless of nationality, will need a visa.

Currency: Import/export of local currency prohibited. Unlimited foreign currency import, but amount must be declared on arrival. Export limited to amount declared.

Health precautions

Mandatory: Yellow fever vaccination certificate if arriving from an infected area.

Advisable: Yellow fever vaccination (and cholera vaccination for visits to coastal areas), typhoid, tetanus, hepatitis and polio vaccinations. Malaria prophylaxis necessary for coastal and other lower altitude regions. Water precautions should be taken – bilharzia is present. AIDS risk.

Representation overseas: Abu Dhabi, Addis Ababa, Beijing, Bonn, Brussels, Cairo, Canberra, Dar es Salaam, Geneva (UN) Harare, Islamabad, Kampala, Kigali, Khartoum, Kinshasa, Lagos, London, Lusaka, Mogadishu, Moscow, New Delhi, New York, Ottawa, Paris, Riyadh, Rome, Stockholm, Tehran, The Hague, Tokyo, Washington.

Air access

National airline: Kenya Airways.

Other airlines: Aeroflot, Air Afrique, Air Botswana, Air France, Air India, Air Madagascar, Air Malawi, Air Mauritius, Air Tanzania, Air Zaire, Air Zimbabwe, Alitalia, British Airways, Cameroon, Egyptair, El Al, Ethiopian, Gulf Air, KLM, Lufthansa, Luxair, Olympic Airways, PIA, Pan Am, Royal Swazi National Airways, Sabena, SAS, Saudia, Somali, Swissair, Uganda, Zambia Airways.

Tax: Airport tax of $10 or its equivalent in convertible currency is payable on departure.

International airports: Mombasa-Moi International (Code: MBA), 13 km from city, duty-free shop, bar, restaurant, buffet, bank, post office, shops, car hire; Nairobi-Jomo Kenyatta International (Code: NBO), 17 km from city, duty-free shop, bar, restaurant, buffet, bank, post office, shops, car hire.

Surface access

Main ports: Mombasa.

Overland access: Road: Routes exist from Uganda, Ethiopia, Sudan (under improvement), Tanzania. Entry by these routes can be difficult. Regulations and conditions should be checked with Kenyan authorities before travelling. Main railway line between Mombasa and the Uganda border.

Hotels: Wide range available in main centres. Advisable to book well in advance during peak season (Nov-Apr). Rating system in operation. 10 per cent tip usual.

Electricity supply: 240 V AC, 50 cycles.

Credit cards: Major cards accepted widely.

Car hire: Can be hired from travel operators and hotels in Nairobi, Mombasa and Malindi. International driving licence accepted, or national driving licence endorsed at a Kenyan police station or at the Registrar of Motor Vehicles in Nairobi (valid for 90 days). Traffic drives on the left.

City transport

Taxis: Available in most major towns. The majority of licenced taxis are metered and often shared, with fares according to time and distance. However fares for long trips should be agreed in advance. Taxis cannot be hailed in the street. A 10 per cent tip is usual. The taxis of companies that run fleets, such as Kenatco, are the most reliable. Some of the older yellow band taxis are unmetered and fares must be negotiated.

Buses: Good and fairly cheap services operate in Nairobi and Mombassa. Also the Matatu – minibuses and vans; these are unregulated and can be overcrowded – not generally recommended for visitors.

National transport

Air: Kenya Airways operates an extensive and regular service linking Mombasa, Malindi, Kisumu, and other major centres, from Nairobi. Local light aircraft companies, such as Sunbird, fly regular services to smaller air fields, such as Lamu, a tourist attraction on the coast. Charter flights are also available to game reserves, such as Masai Mara, and main centres.

Road: Nearly all main towns are connected by 6,700 km of good surfaced roads. In rural areas some of the 48,000 km unsurfaced roads can be difficult in wet weather.

Buses: Several major bus companies operate fast coach services on all major routes, centred on Nairobi. The long distance Peugeot taxi service operates between towns. Cars can be shared; not generally recommended.

Rail: 2,645 km of track. Modern, first and second class service from Nairobi to Mombasa (journey time: approx 14 hours), to Kisumu (track goes on to Uganda), to Nyeri and Nanyuki. Trains often run late, but are fairly comfortable. Advisable to book sleeping compartments in advance.

Public holidays

Fixed dates: 1 Jan, 1 May, 1 Jun (Madaraka Day), 10 Oct, 20 Oct (Kenyatta Day), 12 Dec (Independence/Jamhuri Day), 25 Dec, 26 Dec.

Variable dates: Good Friday, Easter Monday, Eid al-Fitr.

Working hours

Government and business: (Mon-Fri) 0800 – 1300, 1400 – 1700; (Sat) 0830 – 1200/1230. Mombasa offices normally open and close half an hour earlier.

Banking: (Mon-Fri) 0900 – 1400. Open first and last Sat in each month 0900 – 1100.

Telecommunications: Kenya Post and Telecommunications Corporation provides international direct dialling and subscriber trunk dialling to all major towns, and also provides telex, facsimile, data communication and related services.

Telephone: Dialling code for Kenya: IDD access code + 254 followed area code (2 for Nairobi, 11 for Mombasa, 37 for Nakuru) followed by subscriber's number.

Telex: Public telex facilities available from central post offices and main hotels.

Banking

Main centre: Nairobi.

Central bank: Central Bank of Kenya.

Head offices: Barclays Bank of Kenya, Co-operative Bank of Kenya, Commercial Bank of Africa, First Chicago Kenya, Grindlays Bank International (Kenya), Kenya Commercial Bank, National Bank of Kenya, Standard Chartered Bank.

BUSINESS GUIDE

Foreign branches: Algemene Bank Nederland, Bank of Baroda, Bank of India, Bank of Credit & Commerce International, Bank of Oman, Banque Indosuez, Barclays Bank International, Citibank, Habib Bank, Ljubljanska Bank, Lombard Banking Corp, Ottoman Bank. *Development banks:* Development Finance Corp of Kenya, Agricultural Development Corp, Industrial Development Bank, East African Development Bank.

Advertising: Press, cinema, commercial radio and TV available. Outdoor: poster sites at railway stations and illuminated signs. Restrictions: code of ethics applied to pharmaceutical and medicine advertising.

Trade fairs: Nairobi International Show (organised by the Agricultural Society of Kenya, though all sectors of business and industry take part) held every Sep-Oct on a permanent showground. Other trade exhibitions take place at the Kenyatta International Conference Centre in Nairobi.

Representation in capital: Algeria, Argentina, Australia, Austria, Bangladesh, Belgium, Brazil, Burundi, Canada, Chile, China PR, Colombia, Costa Rica, Cyprus, Czechoslovakia, Denmark, Egypt, Ethiopia, Finland, France, German FR, Ghana, Greece, Guinea, Hungary, India, Indonesia, Ireland, Iran, Iraq, Israel, Italy, Japan, Rep. of Korea, Kuwait, Lesotho, Liberia, Malawi, Mexico, Morocco, Nepal, Netherlands, Nigeria, Norway, Oman, Pakistan, Philippines, Poland, Portugal, Romania, Rwanda, Saudi Arabia, Somalia, Spain, Sri Lanka, Sudan, Swaziland, Sweden, Switzerland, Tanzania, Thailand, Turkey, Uganda, UK, USA, USSR, Vatican, Yemen AR, Yugoslavia, Zaïre, Zambia, Zimbabwe.

BUSINESS DIRECTORY

Hotels: Mombasa
Bahari Beach, P.O. Box: 81443 (tel: 471603).

Inter-Continental, Shanzu Beach, P.O. Box: 83492 (tel: 485811; tx: 21153).

Leisure Lodge, P.O. Box: 84383 (tel: 2011/2; tx: 21184).

Manor, Nyerere Ave, P.O. Box: 84851 (tel: 314643).

Mombasa Beach, P.O. Box: 90414 (tel: 471861; tx: 21186).

New Outrigger, P.O. Box: 82345 (tel: 20822/3; tx: 21368; fax: 315801).

Nyali Beach, P.O. Box: 90581 (tel: 471551; tx: 21246).

Oceanic, Mbuyuni Road, P.O. Box: 90371 (tel: 311191; tx: 21298).

Reef, Mwamba Drive, P.O. Box: 82234 (tel: 471771; tx: 21199).

Robinson Baobab, Ukunda, Diani Beach, P.O. Box: 32 (tel: 20268; tx: 21132).

Serena Beach, (17km north of Mombasa), P.O. Box: 90352 (tel: 485721; tx: 21220; fax: 485453).

Shelly Beach, P.O. Box: 96030 (tel: 451001/2/3/4; tx: 21287; fax: 451349).

Trade Winds, Ukunda, P.O. Box: 8 (tel: 2621/2016; tx: 21139).

Two Fishes, Ukunda, on Diani Beach, P.O. Box: 23 (tel: 2720/2037; tx: 21162).

Whitesands, P.O. Box: 90173 (tel: 485926/9; tx: 21175).

Hotels: Nairobi
Ambassadeur, Moi Avenue, P.O. Box: 30399 (tel: 336803; tx: 22223).

Excelsior, Kenyatta Avenue, P.O. Box: 49584 (tel: 26481/2/3/4; tx: 23147).

Fairview, Bishop's Road, P.O. Box: 40842 (tel: 723211; tx: 25584; fax: 721320).

Hilton, Mama Ngina Street & Moi Avenue, P.O. Box: 30624 (tel: 334000; tx: 22252; fax: 339462).

Inter-Continental Nairobi, City Hall Way, P.O. Box: 30353 (tel: 335550/69; tx: 22631).

Jacaranda, P.O. Box: 14287 (tel: 742272/3/4/5/6; tx: 23295).

Milimani, P.O. Box: 30715 (tel: 720760/3; tx: 22613).

Nairobi Serena, Kenyatta Avenue, P.O. Box: 46302 (tel: 725111; tx: 22377; fax: 725184).

New Stanley, Kenyatta Avenue, P.O. Box: 30680 (tel: 333233; tx: 22223).

Norfolk, Harry Thuku Road, P.O. Box: 40064 (tel: 335422; tx: 22559; fax: 336742).

Panafric, Kenyatta Avenue, P.O. Box: 30486 (tel: 720822; tx: 22454).

Sixeighty, Kenyatta Avenue, P.O. Box: 43436 (tel: 332680; tx: 22513).

Hotels: Nanyuki
Mount Kenya Safari Club, P.O. Box: 35 (tel: 22993; tx: 28020; fax: 22494).

Hotels: Tsavo National Park
Kilaguni Lodge, Mtito-Andei, P.O. Box: 30471 (tel: (02)336858; tx: 22033).

Salt Lick Lodge (Hilton), Taita Hills, Bura, via Voi (tel: Mwatate 44; tx: 21088 c/o Nairobi Hi; fax: 339462).

Car hire: Nairobi
Avis: PO Box 49795 (tel: 822186; tx: 22383); or: Kenyatta Avenue (tel: 336794).

Europcar: PO Box 40433 (tel: 334722/5; tx: 22274).

Hertz: PO Box 42196 (tel: 331960; tx: 22228).

Car hire: Nairobi Airport
Avis: (tel: 822186).

Hertz: (tel: 822339).

Car hire: Mombasa
Avis: PO Box 84868 (tx: 22383).

Europcar: Moi Avenue (tel: 314394/5; tx: 21108).

Hertz: Moi Avenue, PO Box 84782 (tel: 316333, 316235; tx: 21286).

Car hire: Mombasa airport
Avis: (tel: 433211; tx: 22383).

Other useful addresses
Nairobi: Agricultural Development Corporation, PO Box 30367 (tel: 338530).

Central Bank of Kenya, Haile Selassie Avenue, PO Box 60000 (tel: 26431; tx: 22721).

Central Police Station, University Way (tel: 22222).

Development Finance Company of Kenya, Finance House, Loita Street, PO Box 30483 (tel: 340401).

East African Report on Trade & Industry, PO Box 30339.

General Post Office, Kenyatta Avenue.

Industrial & Commercial Development Corporation, Uchumi House, Nkrumah Avenue, PO Box 45519 (tel: 29213; tx: 22429).

Investment Advisory and Promotion Centre, National Bank Building, Harambee Avenue, PO Box 55704 (tel: 21401/4; tx: 25460).

Kenya Airways, Airways Terminal, Koinage Street (tel: 29291; tx: 22771).

Kenya Association of Manufacturers, Hughes Building, Kenyatta Avenue, PO Box 30225 (tel: 21602, 21452).

Kenya Association of Tour Operators (for information on conference facilities throughout Kenya), PO Box 48461 (tel: 27005).

Kenya External Trade Authority, PO Box 43137 (tel: 26016).

Kenya Industrial Estates Ltd, PO Box 18282 (tel: 555511).

Kenya Railways, PO Box 30121 (tel: 21211).

Kenya Tea Development Authority, Commonwealth House, Moi Avenue (tel: 21441; tx: 22645).

Kenyatta International Conference Centre, PO Box 30746 (tel: 332383; tx: 22035).

Marketing Society of Kenya, PO Box 44635 (tel: 60034).

Ministry of Commerce & Industry, Co-operative House, Haile Selassie Avenue, PO Box 30430 (tel: 340010).

Ministry of Finance, Treasury House, Harambee Avenue, PO Box 30007 (tel: 338111).

Ministry of Information and Broadcasting, Hakati Road, PO Box 30025 (tel: 23201, 25242, 26051).

Ministry of Planning and National Development, PO Box 3007 (tel: 338111).

Ministry of Tourism & Wildlife, Utalii House, Uhuru Highway, PO Box 30027 (tel: 331030).

National Chamber of Commerce & Industry, Ufanisi House, Haile Selassie Avenue, PO Box 47024 (tel: 20866).

Voice of Kenya, PO Box 30456 (tel: 334567; tx: 22244).

Kiribati

Kevin Rafferty

(map on page xxiv)

In 1989 the unique island country of Kiribati celebrated 10 years of independence from Britain, still one of the poorest states in the world, but immensely proud and determined to make its own way independent of the big powers. It was proving a difficult task. The respected President Ieremaia Tabai declared fiercely that 'handouts is no way .... We must aim to be able to meet our basic needs and to reduce our dependence on the goodwill of others. A very important element of this is to try to live within our own means.' But with per capita income of below $300 a year, it is a hard struggle with no easy help on the horizon.

'Unique' is an overworked word, but is the only word to describe the 33 islands that make up the Republic of Kiribati. Together their land area is only 726 sq km, but they are scattered widely over the Pacific Ocean on both sides of the equator and straddle the international date line. The Gilbert group accounts for almost all the population of 66,000, and is a 700 km-long chain of 16 islands; the Phoenix group is 1,600 km to the east and uninhabited; the Line group is 3,000 km east and only the three northern islands are inhabited. Their land mass is tiny but with the 200-mile economic zone round them, the islands sit in more than 3m sq km of sea. Most of this land is no more than a few feet above water.

All the islands apart from the limestone Banaba are coral atolls with shallow topsoil and a low water absorption making it virtually impossible to grow most crops. Independence for Kiribati coincided with the exhaustion of phosphate deposits in Banaba, resulting in an 80 per cent fall in exports and a reduction by half of government revenues and real gross domestic product. This also threw the islands back on the resources of the sea for their livelihood. Economic prospects are thus limited and the islands' future is highly vulnerable. If the so-called 'greenhouse effect' comes about as feared and the world's oceans rise by a few feet, then the whole of Kiribati could be wiped off the earth.

The vast ocean is the hope for the future, but is also a handicap to a poor people. Tabai himself points out that communications are poor, 'for instance to go to Kiritimati (Christmas Island) from the capital island Tarawa, one has to travel first to the Marshall islands or Fiji, then to Hawaii and back to the eastern part of the country to reach Kiritimati. To go from Tarawa to Kiritimati our people require US visas.' It is also difficult, as the president admits, for a small poor country to police the vast ocean to stop bigger powers poaching the valuable fish resources.

Development ambitions

But Kiribati and Tabai in particular have great ambitions. In spite of his reluctance to take aid Tabai accepted a $2m grant from Japan in 1988 to improve its fisheries industry. The aim is to double the cold storage capacity on Betio, part of the Tarawa atoll. This will improve fish quality and allow the country to get better prices from the quality- conscious Japanese. Tabai is also ambitious for developments on Kiritimati, the biggest island. He is lobbying both in Japan and for support in other Pacific islands for Japan to build its space centre there, including space shuttle retrieving pad, space university, hotel accommodation and other creature comforts. The scheme would also help to stem the drift of population to the main Tarawa atoll which is suffering from overcrowding.

But Tabai has a difficult balancing act to perform.

KEY FACTS

Official title: Republic of Kiribati

Head of state: Beretitenti (president) Ieremia Tabai

Capital: Bairiki (on Tarawa)

Official Languages: I-Kiribati, English

Currency: Australian Dollar (A$ = 100 cents)

Exchange rate: A$1.26 per $ (Nov 1989)

Area: 810 sq km (33 islands and atolls)

Population: 70,000 (1988)

GDP per capita: A$499 (1987)

Labour force: 31,500 (1986)

Trade balance: –A$14.4m* (1987)

* estimated figure

KEY INDICATORS

	Unit	1983	1984	1985	1986	1987
Population	'000	62.1	63.4	63.8	65.5	67.2
Gross domestic product (GDP)	A$m	26.8	29.7	30.4	31.9	33.5
GDP per capita	A$	432	468	476	487	499
Consumer prices (Tarawa)	1975=100	199.6	210.4	219.8	234.3	–
Exports	A$m	4.0	12.5	6.1	2.5	3.1
Copra	A$m	2.2	7.0	4.7	0.5	1.4
Imports	A$m	19.6	20.9	21.6	21.6	–
Balance of trade	A$m	–14.1	–7.4	–10.9	–12.8	–14.4*
Current account	A$m	+5.7	+10.4	6.3	5.7*	5.0*
Exchange rate	$ per A$	1.14	1.23	1.49	1.43	–

* estimated figure

He must be worried about the potential greenhouse effect and is pressing for banning of harmful gases causing it which may destroy his country. But if he is too alarmist, then Japanese and other foreign investors may be scared away. I-Kiribati (Kiribati islanders) reacted angrily to some Australian reports that the greenhouse effect would wipe out five countries including Kiribati: they claimed that the reports themselves were doing more damage by making foreign investors think the country had no future.

Economy

A recent report by the International Monetary Fund on the Kiribati economy makes laconic reading. Exports, the report says, account for less than 10 per cent of GDP and are almost entirely of copra and fish. Imports on the other hand are 50 per cent of GDP, 'reflecting the narrow resource base and the limited opportunities for substitution by domestic production.' Essential food, beverages, consumer goods, petroleum products, machinery and transport equipment, all have to be imported. The country depends on services and transfers, including payments from fishing licences and remittances from I-Kiribati working abroad plus the hated aid. Kiribati also suffers at the same time from high and rising unemployment along with an acute shortage of skilled manpower that can only be met by expatriate labour. The IMF records, 'preliminary national accounts data suggests that real GDP growth has been negligible during this decade [the 1980s] and, on a per capita basis, GDP remains far below that in the days of phosphate mining. Central government expenditure data and import statistics confirm the absence of any firm upward trend in activity.'

Clearly the future depends on Tabai's public relations ability to persuade foreign investors that Kiribati is a great place to be. Foreign investment will attract supporting activities and provide income and education to take the country forward towards his boasts and dreams.

COUNTRY PROFILE

Political system: Kiribati is a democratic republic within the Commonwealth. Formerly the UK Dependent Territory of the Gilbert Islands, Kiribati gained independence in 1979. The House of Assembly (Maneaba ni Maungatabu) consists of 35 representatives elected by universal adult suffrage to a four-year term and one nominee representing Banaba (most of whose inhabitants now live on the island of Rabi in Fiji). The Attorney General is an *ex officio* member (if not already an elected member). Executive power is exercised by a cabinet appointed by the elected Berentitenti (president).

Political parties: There are no political parties.

Population: The 1985 census gave a population of 63,883, of whom about 24,000 live on Tarawa. 85 per cent of the country is Christian.

COUNTRY PROFILE

Language: English is used for official communications; otherwise, I-Kiribati is the norm.

Media
Broadcasting: The government-operated radio service broadcasts seven days per week. Advertising is available.

Domestic economy: Since the cessation of the phosphate industry in 1979, the economy has been based on export of copra and fish (which have not replaced phosphate revenues), aid grants from the UK (until 1985) and international loans for modest development programmes which account for 40 per cent of GNP. Little development of tourism owing to the islands' isolation, high air fares and lack of facilities. Most islanders fish locally, grow their own food and rear pigs and chickens. Living standards are simple. In the early 1980s the UK agreed to establish a trust fund to provide compensation for the population of the former phosphate island of Banaba (Ocean Island), now living on Rati in the Fiji group.

External trade: Remittances from migrant workers are the main sources of income, now that the phosphate rock has been exhausted. Trade figures tend to be erratic.
Exports: Exports were valued at $3m in 1987. Main exports are copra, fish, shark fins. Main destinations are German FR, Marshall Islands, USA, American Samoa, Fiji.
Imports: Imports were valued at $20m in 1987. Main imports are machinery and transport equipment, food, mineral fuels, manufactured goods, miscellaneous manufactures, chemicals, beverages and tobacco. Main sources are Australia, Japan, New Zealand, Fiji, UK, China PR and Hong Kong.

Agriculture: Agricultural development is limited by poor quality of the soil. Commercial or government owned copra plantations on Fanning and Washington islands, but peasant smallholdings are more usual. Most copra exported to Europe by the Copra Co-operative Society. Other crops include breadfruit, pandanus, bananas, pawpaw and a few root crops. Islands also raise pigs and poultry, but fish is an important food resource. Programmes to upgrade subsistence fisheries to small commercial enterprises, and deep-sea fishing carried out by Japanese, Soviet and South Korean fleets under licence.

Industry: Small-scale manufacturing industries include clothing, furniture and handicrafts. Plans to extract salt from seawater using solar power on Kiritimati (Christmas Island, site of the UK atomic test programme). Plans were drafted for the building of a space centre on Kiritimati Island by a Japanese interest at an estimated investment of $8bn.

Mining: Phosphate mining on Banaba ceased in 1979. In 1985, Kiribati, along with Tuvalu and the Cook Islands, signed a tentative agreement to undertake joint exploration for ocean-floor minerals with Japan. The US Geological Survey has already found undersea platinum and copper deposits off the coasts of Kiribati and Vanuatu, and manganese deposits are known to exist around the Line group of islands.

Energy: Annual electricity production and consumption is about 6.5m kWh.

Main companies: Abamakoro Trading, Te Mantari (government fishery company).

Membership of international organisations: Asian Development Bank, Associated with the EEC under the Lomé Convention, Commonwealth, ESCAP, GATT, ICAO, IMF, South Pacific Bureau for Economic Co-operation, South Pacific Commission, SPF, WHO. Membership of the IMF and World Bank signed in 1986.

BUSINESS GUIDE

Time: GMT + 12 hrs.

Climate: Temperatures range from 26-32°C. The westerly gale (Oct-Mar) can be unpleasant.

Entry requirements
Passport: Required by all.
Visa: Nationals of Commonwealth countries, the UK and certain other countries do not require visas, but will be issued with a visitor's permit on arrival.
Currency: There is no limit to the amount of travellers' cheques or currency which may be imported.
Customs: Personal effects are allowed duty-free.

Health precautions
Mandatory: Vaccination certificate for yellow fever if travelling from an infected zone.

Air access
National airline: Air Tungaru.
Other airlines: Air Marshall Islands.
International airport: Bonriki International (TRW) on Tarawa; shop.

Surface access: Cargo services operate from Australia, New Zealand, Japan and the US. Large ships anchor offshore and are off-loaded by barge. Government ship operates Fiji-Kiribati. Further development planned for Betio harbour, which already has container facilities. Remoteness of the group restricts number of large vessels which call.
Main ports: Betio (on Tarawa).

Hotels: In addition to the islands' five hotels, there are rudimentary rest-houses. Travellers' cheques are not accepted.

Credit cards: Credit cards are not accepted.

Car hire: Vehicles may be hired from Tarawa Motors.

City transport
Taxis: It is possible to hire taxis.

National transport
Air: Air Tungaru operates a regular service to all outer islands in the Gilbert group; other services may operate on intermittent basis.
Road: There are 30 km of asphalt road on Tarawa. Construction of a causeway linking Bairiki and Betio will be complete by 1988. The majority of vehicles are motorcycles able to travel beyond the confines of the limited road network.
Buses: There are buses on South Tarawa.
Water: Small boats ply between the islands.

Public holidays
Fixed dates: 1 Jan, 12 and 13 Jul (may vary), 25 Dec.
Variable dates: Good Friday, Easter Monday, first Mon in Aug.

Working hours
Government: (Mon-Fri) 0800 – 1230, 1330 – 1615.

Social customs: Tipping is not expected. In official correspondence I-Kiribati adopt the Western convention of signing their names with initials and surname, but it is customary (and more polite) to address people by their first name. In the I-Kiribati language the letters 'ti' are pronounced 's' ('Kiribati' is pronounced 'Kiribass').

Telecommunications
Telephone: Satellite telecommunication system introduced in 1985. Telephone service is available throughout urban Tarawa. Dialling code for Kiribati, IDD access code + 686, followed by subscriber's number. Radio telephone links available to most outer islands.
Telex: Country code: 771. Automatic facilities.
Telegram: Available from the Telecommunications Department (0800-1230 and 1330-1615).

Banking: The National Loans Board, established in 1974 is being converted to a development bank for financing cottage industry.
Commercial Bank: Bank of Kiribati (associate of Westpac Banking Corporation, Sydney), with offices at Bairiki, Betio and Bikenibeu.

Advertising: The newspaper accepts advertising and there is one cinema.

Electricity supply: 240 V AC, 50 cycles.

Weights and measures: Metric system (imperial units also used).

BUSINESS DIRECTORY

Hotels: Kiribati
Captain Cook, Christmas Island.

Kiribati, Betio.

Otintai, Bikenbeu, Tarawa, P.O. Box: 270 (tel: 28084; tx: 77032).

Robert Louis Stevenson, Abemama.

Seaman's Hotel, Betio.

Car hire
Avis: PO Box 36, Bairi Ki Tarawa (tel: 219090).

Tarawa Motors: (tel: 21090).

Business advice and information
Bank of Kiribati, PO Box 66, Bairiki, Tarawa.

Broadcasting and Publications Authority, PO Box 78, Bairiki, Tarawa.

Ministry of Finance, PO Box 67, Bairiki, Tarawa.

Ministry of Trade, Industry and Labour, PO Box 69, Bairiki, Tarawa.

National Loans Board, PO Box 33, Bairiki, Tarawa (tel: 345).

Travel information
Tourism Division, Ministry of Natural Resources, Bairiki, Tarawa.

High commissions in capital: Australia, UK.

Korea DPR

Aidan Foster-Carter

As the 1990s opened, one of the modern world's most extraordinary attempts at social engineering remained in place; a monument – in many senses – of stability, not to say stagnation, amidst the dynamism of north-east Asia. Kim Il Sung, 78 years old in 1990, is the world's longest serving political leader. Brought to power by Soviet forces in 1945, he went on – despite the debacle of the 1950-53 Korea War – both to cement his personal rule and to distance himself from any great power sponsorship. The first half of his long reign saw rapid industrialisation, and North Korea economically outpaced the South until at least the late 1960s. Thereafter the rot set in, however, as economic reforms were eschewed in favour of ever tighter political control and a personality cult of unprecedented proportions – culminating in the 1980s with the designation of his son Kim Jong Il as his chosen successor. But what was clear in 1988-89 was that this extraordinary regime had little capacity to cope with a rapidly changing environment, notably the huge propaganda success of the Seoul Olympics and the subsequent burgeoning of South Korean economic and other links with Pyongyang's increasingly impatient Soviet and Chinese allies.

Politics and society

On the surface, North Korean politics continued to present itself as a uniquely bizarre spectacle: a kind of stage performance in semi-permanent session, overflowing with boundless loyalty to the 'Great Leader' Kim II Sung, and the 'Dear Leader', Kim Jong II. Thus the year 1988-89 saw two jamborees, big even by Pyongyang standards: the fortieth anniversary of the Korea DPR in September 1988, and the World Festival of Youth and Students in July 1989.

Yet behind the facade of *yuilsasang* or 'ideological monochromaticity', there must have been political debate over a host of pressing problems. Personnel changes provided one key. December 1988 saw the replacement of Prime Minister Li Gun Mo by Yon Hyong Muk. Both were reputedly technocrats, but it was thought that Yon might be more committed to high technology and better foreign economic relations. Although Li's resignation after only two years was officially attributed to ill-health, he had pursued a full programme of activities. Seoul sources speculated that the convening of the fourteenth and fifteenth Workers' Party of Korea Central Committee plenums only 10 days apart in December 1988 indicated a fierce debate over economic policy. Given the state of the economy this was plausible.

It seemed that Kim II Sung himself was semi-retired (he had ayatollah status, so to speak), while day-to-day policy was in the hands of his son. Perhaps this explained what looked like policy paralysis all around, and the puzzling failure to pursue a consistent or rational line on almost anything, be it the economy, or relations with South Korea or Japan.

While the North Korean political elite contained would-be reformers, it was impossible to identify clear factions. Although on past form real change might not be expected until Kim II Sung dies (and he is not yet 80, and enjoys good health), it should not be ruled out that North Korea's growing crisis might require and call forth drastic initiatives even before then. A key question concerned the loyalty of a populace still largely quarantined from information about the outside world, particularly South Korea. Early 1989 saw a trickle of North Korean students in Eastern Europe defecting to Seoul. This could grow.

1989 also saw the publication of the first-ever detailed study of human rights in North Korea. This claimed a *gulag* population of at least 100,000, as well as a remarkable classification system into three quasi-estates : a 'loyal' core of some 5m who live reasonably well; 4 m 'hostiles', mostly doing rural labour; and fully 10m 'waverers' in between. All moral judgements aside, the long-run stability of such a totalitarian edifice in the 1990s must be doubtful. The question is thus not whether North Korea will change, but when and how.

The economy

North Korea in l988-89 showed no signs of tackling, or even admitting, its increasingly serious economic problems. Despite an initially rapid industrialisation – helped by the North having the lion's share of the peninsula's mineral resources – Pyongyang's economy remained largely stagnant for 20 years. Every single plan since the 1961-70 'seven year plan' was *de facto* extended, while for both 1985 and 1986 no plan was announced at all. The third seven year plan (1987-93) incorporated physical targets originally designated as 'long range goals for the 1980s'. Some of these were scaled down as well as extended in time (e.g. steel from 15m to 10m tonnes). The fact that little was heard in 1989 of these formerly much-touted targets suggested problems. North Korea has published no systematic economic data since the mid-1960s.

One sign of economic disarray was the dizzying rate at which economic ministries were reorganised and ministers reshuffled in the late 1980s. Thus Hong Song Nam was moved from planning chief to deputy premier in February 1988 – only to return to his old job in June of the same year. In November 1988, a separate ministry was created for joint venture indus-

KEY FACTS

Official title: Democratic People's Republic of Korea

Head of state: President Kim II Sung

Head of government: Prime Minister Yon Hyong Muk

Ruling party: Workers' Party of Korea

Capital: Pyongyang

Official Languages: Korean

Currency: Won (W = 100 chon)

Exchange rate: W0.96 per $ (Nov 1989)

Area: 122,762 sq km

Population: 21.9m (1988)

GDP real growth: 2–3%* (1989)

Labour force: 7m (1988)

Trade balance: –$530m (1986)

Foreign debt: $1,246m (1987)

* estimated figure

tries. Despite having a joint venture law since 1984, North Korea has attracted almost nobody except the much-milked 'patriotic capitalists' of Chongryun, the pro-Pyongyang organisation of Koreans in Japan.

Although there was limited economic reform on the model of East Germany's *Kombinats* (horizontal linkage of enterprises, within a framework of strict central planning), Stakhanovite exhortation remained the norm in North Korea. In the absence of sorely needed new technology, increased output could only come if people worked even harder. Thus 1988-89 saw not one but two '200-day speed battles': the first leading up to the Korea DPR's fortieth birthday celebrations in September 1988, and the second following on up to Kim Il Sung's birthday (a public holiday) in April 1989. In May, however, the weary populace were rewarded by an unprecedented special bonus of one month's extra wages, around $50 for most workers. There were some suggestions that Yon Hyong Muk, the new premier, favoured such material incentives – hitherto anathema in a country where everything should be done out of loyalty and love for the 'Great Leader'.

The annual budget debate in April 1989 offered few clues. Revenue and expenditure for 1988 were almost in balance at just under Won 32bn (c.$14.5bn), 5 per cent up on 1987. As in previous years, accomplishments announced (almost all as percentage increases, rather than real numbers) for 1988 bore little discernible relation to what the finance minister – Yun Gi Jong, a rare woman in North Korea's political elite – had promulgated in advance the year before. Significantly expenditure on, or investment in, transport rose 20 per cent in 1988; a sector notorious for bottlenecks. Scientific research also received an extra 35 per cent in 1988. For 1989, light industry was to receive 13 per cent more investment. Consumer goods were a real headache, and light industry minister Kim Pok Sin, also a woman, in a keynote speech actually admitted they were 'insufficient'.

It remained to be seen how much longer North Korea could go on over-working and under-rewarding its people. It seemed that serious economic reform must be on the agenda for the 1990s.

Economic relations

Little change was evident in North Korea's economic links, which remained minuscule by the standards of the region – Pyongyang's annual trade volume is under $4bn; Seoul's exceeds $100bn). In particular, no progress was made in settling the ancient saga of Pyongyang's debts. While not large by the standards of the late 1980s – perhaps $3bn – some of these went back to the early 1970s, and North Korea persistently and inexplicably welshed on every attempt at rescheduling. A new proposal in 1988 by a British intermediary for partial forgiveness succeeded only in splitting the creditor banks; no money came forth from Pyongyang. As a result North Korea continued to be wholly cut off from desperately needed advanced technology. At the end of the 1980s it still could not make a silicon chip.

In the absence of substantial Western links – even with neighbouring Japan, trade totals less than $0.5bn – the Soviet Union resumed its traditional role as North Korea's major aid donor and trade partner. Soviet sources were as forthcoming about this as Pyongyang – which liked to attribute everything to *juche* , or self-reliance – was silent about it. Moscow also regularly complained about poor quality, late delivery, and other Korean shortcomings. In the era of *perestroika*, this could not be a happy relationship. Although still perforce committed to Pyongyang for political reasons, the USSR – which was owed at least $1bn by North Korea – demanded an economic return, usually in the form of a share of the output of Soviet-aided Korean factories. Even so, the fact that the USSR's trade surplus with Pyongyang for the first half of 1988 alone was over $550m suggested that Moscow would continue to pay heavily. Perhaps this was the price for Pyongyang's enforced tolerance of the growing Soviet friendship with South Korea.

North–South relations

1988-89 saw some fitful progress in relations between the two Korean states. After several years of negotiations, Pyongyang eventually boycotted the Seoul Olympics by insisting on the unacceptable notion of co-hosting as a condition of the participation. Only six other minor nations joined the boycott, however.

This did not prevent what looked like a breakthrough for dialogue around the turn of the year. In a series of talks about talks at the border village of Panmunjom, both sides seemed ready to make concessions in order to set up a larger-scale meeting of MPs. From Pyongyang's viewpoint, it doubtless helped that Seoul's national assembly boasted an opposition majority. Possibly to its chagrin, nonetheless, even erstwhile firebrands like opposition leader Kim Dae Jung hewed a carefully supra-partisan line in dealings with the North.

Early in 1989, the North also proposed a joint Korean team for the 1990 Asian Games in Beijing. The South seemed enthusiastic, and useful talks were held. Above all, in the wake of South Korean President Roh

COUNTRY PROFILE

Political system: The executive Central People's Committee, consisting of a president, (the head of state), a prime minister and about 15 others, is elected by the Supreme People's Assembly, which consists of about 650 elected members elected in turn by universal suffrage. The Central Committee appoints and directs an Administration Council, with local government vested in nine provincial and four municipal elected people's assemblies. Government at all levels is dominated by the Korean Workers' Party.

Political parties: The Worker's Party of Korea (approximately three million members) is the ruling party. Other 'front' parties are the Korean Social Democratic Party and the Chondoist Chungu Party.

Population: North Korea has a highly homogeneous population, which may be growing at about two per cent p.a. In 1989 the total probably surpassed 20m.

Main city: Pyongyang (population two million in 1986); also Hamhung, Chongjin, Kaesong, Nampo, Sinuiju, Wonsan.

Media
Press: Main dailies are *Rodong Shinmun, Minju Choson, and Pyongyang Shinmun.* The Foreign Trade Publishing House publishes a monthly journal, *Foreign Trade of the DPRK*, which includes listings of specialised corporations, giving telegraphic and telex addresses.
Radio: National and locally produced programmes widely disseminated (e.g. factory, outdoor loudspeakers); external services in several languages.
Television: There are two stations (one in colour), plus a third at weekends.

Domestic economy: The economy, traditionally agricultural, shifted in the 1960s toward industry; in 1986, government figures claimed that industry accounted for 80 per cent of GDP. A series of economic development plans have been implemented, with first priority on increasing heavy industrial output and energy production, and on achieving self-sufficiency in food and industrial raw materials. Current emphasis favours more development of light industry and production of consumer goods. Foreign investment in industry, construction, technology and tourism is now encouraged. The third seven-year plan launched in 1987 called for annual growth of 8 per cent in GNP, an ambitious target. Korea has current foreign debt totalling over $3bn. A rescheduling agreement was signed in 1987 to freeze the country's bank assets if North Korea refused to service its bank loans of $1.4bn.

External trade: Over 60 per cent of all foreign trade is conducted with other socialist countries in the region, or with CMEA members. The USSR accounts for at least one-third, and China one-sixth.
Exports: Main exports are machinery and equipment, non-ferrous metals, rolled steel, magnesia clinker, cement, coal, textiles, ceramics and marine products.
Imports: Main imports are machinery and equipment, mineral fuel, natural rubber, edible oil, food and coking coal.

Agriculture: Agriculture, mostly practised on large-scale collective and state farms, accounts for 40 per cent of employment. About 15-20 per cent of total land area is cultivated arable. There has been considerable investment in agriculture, which has achieved a relatively high degree of mechanisation, but DPR Korea still imports wheat and flour. Fertiliser use has been estimated at two tonnes/ha pa of rice, with six to seven tractors in use per 100 ha. Main crops are rice, maize and potatoes. Other crops include wheat, barley, rape, sugar, millet, sorghum, pulses, sweet potatoes, vegetables, tobacco and silkworm cocoons. Pig-raising and poultry make an important contribution, and other livestock are raised, including cattle, sheep, goats, horses and rabbits. Inshore and offshore fishing, with catches of 2.39m tonnes in 1985, add substantially to GDP. Timber is exploited for domestic use, with about 60 per cent of removed volume thought to be industrial wood.

Industry: Industry accounts for 45-47 per cent of employment. Major manufacturing activities have been diversified to include production of steel,iron, non-ferrous metals, machinery and equipment, fertilisers, plastics and cement. Light industrial products include silk, cotton and rayon textiles, chemicals, processed food, machine tools, hardware and machinery. Development projects in western and eastern industrial zones have included a vinalon factory in Sunchon with productive capacity of 100,000 tonnes pa, a Potash fertiliser complex in Sariwon, a coal mining complex in Anju, steel complexes in Nampo and Chongjin and synthetic rubber plants in Hamhung and Namhung.

Mining: DPR Korea is well endowed with mineral resources, including coal (anthracite and lignite), refractory clays, phosphates, sulphur and graphite, and ores of iron, magnesium, tungsten, copper, lead, zinc, silver, gold, magnesite and nickel. There are no confirmed oil resources, and little natural gas. Non-ferrous metals are an important foreign exchange earner, with 70 per cent of zinc, lead and copper production in the Hamhung district. There have been unconfirmed reports of offshore oil finds.

Energy: Hydroelectric power generation is well-developed, supplying 20-25 per cent of total primary energy requirements. There has been considerable investment in developing exploitation of coal resources.

Membership of international organisations: FAO, G-77, IAEA, ICAO, IPU, ITU, NAM, UN (observer status), UNESCO, UNCTAD, UNIDO, UPU, WFTU, WHO, WIPO, WMO.

Tae Woo's six-point plan of July 1988 for a new approach to North Korea, there came the first signs of economic relations. Small-scale trade began, although Pyongyang steadfastly denied it. And Chung Ju Yung, founder of the massive Hyundai conglomerate, visited his northern birthplace in January-February, returning with wide-ranging proposals for co-operation in industrial development and tourism.

By mid-1989, however, all this had run into the sand, at least temporarily, because of a series of unauthorised visits to Pyongyang by Southern dissidents: the veteran radical priest Moon Ik Hwan at Easter, a female student in July to attend the world youth festival, and a maverick opposition MP whose 1988 visit was uncovered in 1989. North Korea in April suspended all dialogue for three months in protest at Moon's arrest on his return home. In July it was the South's turn to postpone talks, in anger at these unauthorised visits, seen by Seoul as undermining serious government-to-government negotiation, inasmuch as they indicated that Pyongyang deep down still had not accepted the legitimacy of the South Korean regime.

Despite this temporary hiatus, serious and sustained inter-Korean dialogue is bound to come eventually. But North Korea may not be ready for it while Kim Il Sung still lives – in a sense, it goes against all he has ever stood for – while South Korea for its part can afford to wait, as its dynamism continues while the North stagnates. Even so, at a time when some of the world's more intractable disputes (Afghanistan, Angola, Kampuchea) were moving closer to resolution, the continued fragility of Korea's 36-year-old armistice (there is still no peace treaty) looked ever more anomalous.

Foreign relations

1988-89 witnessed North Korea's most serious foreign policy reverses ever, as one after another its communist allies began to break Pyongyang's long and hitherto effective quarantine of the South. The Seoul Olympics were the watershed, and a triumph for Seoul's 'nordpolitik' – only Cuba and a handful of others heeded Pyongyang's call for a boycott. Subsequently, South Korea exchanged trade offices with the USSR, China, Poland, Yugoslavia and Bulgaria, and established full diplomatic relations with Hungary, an event which sent Pyongyang into paroxysms of wrath – it perforce remained tightlipped about the others.

While neither the USSR nor China were likely to go the whole way with Seoul, clearly both had decided that they needed better relations – especially economic – with the Korea which really counts. Sino-South Korean trade passed $3bn in 1988, six times China's trade with the North; while Gorbachev in his Krasnoyarsk speech made specific reference to South Korea. Joint ventures in Siberia were in the offing. Behind all this lay the warming in Sino-Soviet relations which had greatly reduced North Korea's previous capacity, which it had exploited to the hilt, to play off its two major allies against one another.

If old friends thus proved fickle in 1988-89, North Korea did not acquire any new ones. Japanese overtures for better relations were in effect rebuffed. Four rounds of low-level talks (not negotiations) took place with US diplomats in Beijing, with no breakthrough. Diplomatic relations were established with Colombia and Morocco.

North Korea's biggest public relations bash of the year (or indeed ever) crystallised the country's anomalies. In July 1989, Pyongyang hosted the thirteenth World Festival of Youth and Students: a mainly communist event, which North Korea had fondly decided should upstage the Seoul Olympics. Some 20,000 young visitors came, from 180 countries. An estimated $4.5 bn had been spent on a massive building programme, mostly of sports facilities (certainly no discos), including a 150,000 seat stadium – bigger than any in Seoul. Sadly, the foreigners were unimpressed. Even a Soviet newspaper described the extravagance as 'immoral'. What the Pyongyang press regularly denounces as 'pagan infiltration' was also predictably in evidence. A Danish group tried to unfurl a banner which read 'Human rights in North Korea, too!' during Kim Il Sung's opening speech. Delegates from over 50 countries organised an unofficial rally to support Chinese students' struggle for democracy – an issue anathema to Pyongyang. And an alfresco disco had hundreds of foreigners dancing to Michael Jackson, while thousands of their North Korean contemporaries looked on unmoving. As the Czech disc jockey unkindly but truthfully put it, they couldn't dance because they didn't have permission. It was hard to imagine that such a society would survive the 1990s.

BUSINESS GUIDE

Time: GMT + 9 hrs.

Climate: Winters are cold, with temperatures ranging from -3°C to -6°C in Jan, but falling as low as -20°C. Summers are warm and humid, an average temperature in Aug of 25°C. Most rainfall is from Jun-Sep.

Entry requirements

Passport: Required by all.
Visa: Required by all with a few national exceptions. Application for visas should be made well in advance. It is impossible to visit DPR Korea except by official invitation, or by joining group tours from certain countries.
Customs: All import-export transactions are carried out through state agencies and subject to controls.

Health precautions

Advisable: Vaccination against cholera, typhoid and polio.

Representation overseas: Accra, Addis Ababa, Aden, Algiers, Amman, Antananarivo, Bamako, Banjul, Bangkok, Bangui, Beijing, Beirut, Belgrade, Berlin, Berne, Bissau, Brazzaville, Bucharest, Budapest, Bujumbura, Cairo, Conakry, Copenhagen, Cotonou, Dakar, Damascus, Dar es Salaam, Dhaka, Freetown, Geneva, Georgetown, Hanoi, Harare, Havana, Helsinki, Islamabad, Jakarta, Kabul, Kampala, Kathmandu, Khartoum, Kigali, Kingston, Kinshasa, Kuala Lumpur, Kuwait, Lagos, Libreville, Lima, Lisbon, Lomé, Luanda, Lusaka, Malabo, Managua, Maputo, Maseru, Mogadishu, Moscow, New Delhi, Niame, Ouagadougou, Oslo, Paris, Prague, Rome, Sana'a, São Tomé, Singapore, Sofia, Stockholm, Tehran, Tirana, Tripoli, Tunis, Ulan Bator, Valletta, Vienna, Vientiane, Warsaw.

Air access

National airline: Chosonminhang (Civil Aviation Administration of the Democratic People's Republic of Korea).

Other airlines: Aeroflot, CAAC.

International airport: Sunan (FNJ), 24 km from Pyongyang (tx: 5471).

Surface access: Rail services operate to/from Beijing and Moscow.

Main ports: Chongjin, Haeju, Hungnam, Najin, Nampo, Wonsan.

Hotels: In Pyongyang, hotels available are the Koryo, the Changgwangsan, Potonggang and Pyongyang. Booking is normally through government agencies.

City transport

Metro: There is an underground railway in Pyongyang.

National transport: It can be difficult to reach many areas of the interior, although the system is developing. Railways are slow, and inter-city buses non-existent.

Air: Chosonminhang operates domestic services.

Road: The road network (75,112 km) includes motorways between Pyongyang and Wonsan and Pyongyang and Nampo.

Rail: The rail network is put at 8,533 km, 88 per cent of which is said to be electrified (measured by volume of freight hauled) with two classes of accommodation.

Water: Rivers, canals and sea transport provide important internal links.

Public holidays

Fixed date: 16 Feb, 15 Apr, 1 May, 15 Aug, 9-10 Sep (National Day), 10 Oct (Foundation Day of Workers' Party), 27 Dec (Constitution Day).

Telecommunications

Telephone: Dialling for DPR Korea is via the International Operator.

Banking

Central bank: Korean Central Bank. *State banks:* Foreign Trade Bank of the DPRK, Korea Daesong Bank.

Weights and measures: Metric system.

Representation in capital: Albania, Algeria, Bulgaria, China PR, Cuba, Czechoslovakia, Egypt, German DR, Hungary, India, Indonesia, Iran, Libya, Mongolia, Pakistan, Poland, Romania, Syria, USSR, Vietnam, Yugoslavia.

BUSINESS DIRECTORY

Hotels: Pyongyang
Changgwangsan.

Potonggang.

Pyongyang.

Business advice and information
Committee for the Promotion of International Trade of the Democratic People's Republic of Korea, Central District, Pyongyang.

Foreign Trade Publishing House, Pyongyang District, Pyongyang.

Korean Committee for Solidarity with World People, 8-120 Yonggwang Street, Central District, Pyongyang.

Korean General Merchandise Export and Import Corp, Central District, Pyongyang.

Banking
Korean Central Bank (central bank), Nammundong, Central District, Pyongyang.

Foreign Trade Bank of the DPRK, Jungsongdong, Central District, Pyongyang (tel: 34531; tx: 5460).

Korea Daesong Bank, Chungsong-dong, Grand Study House Street, Central District, Pyongyang (tel: 37138; tx: 5479).

Travel information
Korean International Tourist Bureau (Ryohengsa), Central District, Pyongyang (tx: 5499 okkp).

Republic of Korea

Kevin Rafferty

By the common consent of economists, South Korea was the great success story of the 1970s and 1980s. 'Miracle' is only slightly too strong a word to describe a country that shot up from per capita income of less than $100 in 1962 to more than $4,000 in 1989. And along with this came a growing industrial muscle already making a major impact on world markets. But now, in the 1990s, the country faces stiffer tests. Economically, it will have to compete on the same playing field as the industrial giants of Japan, the USA, and Europe, a challenge that has its own difficult and sometimes delicate political ramifications. Politically, Korea's move from militarily dominated authoritarian rule towards democracy is not proving easy. The fact that South Korea's northern border is with communist North Korea imposes its own unique perils. The border is today perhaps the most tense frontier between communism and non-communism in the world. It is in fact not a border between two sovereign states, but merely a truceline between two halves of the same country that are technically still at war.

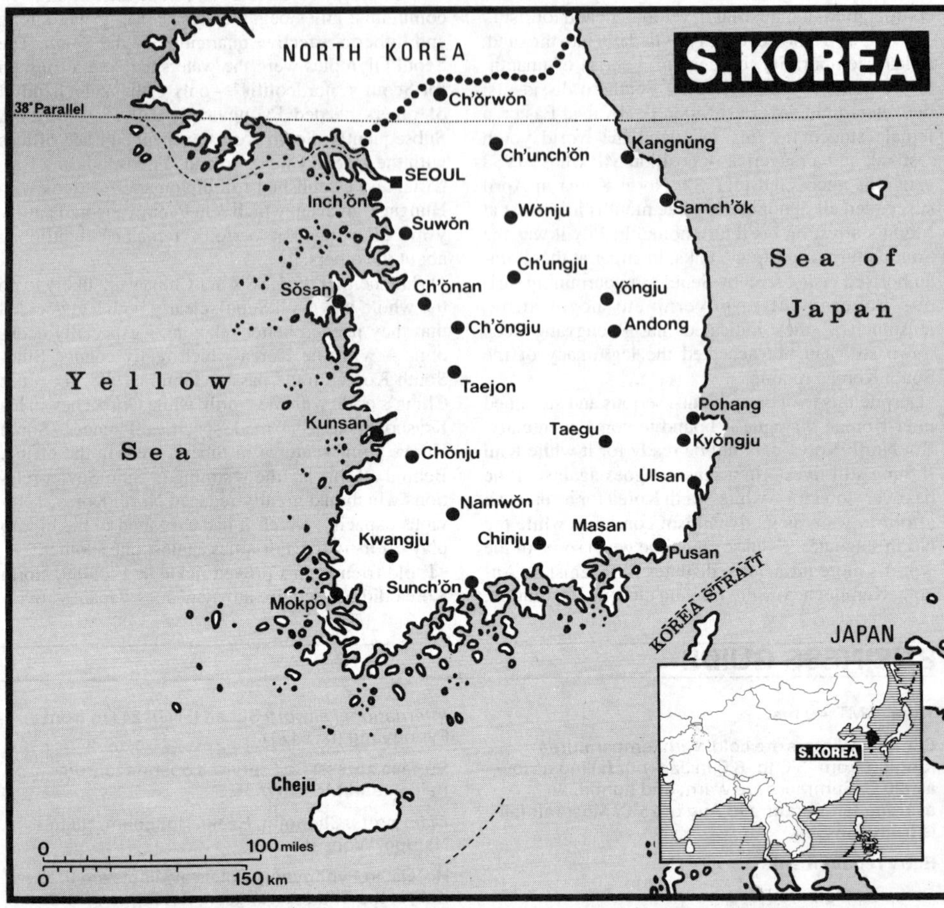

Sustaining the miracle

President Roh Tae Woo, former general and friend and military classmate of previous president Chun Doo Hwan was elected in late 1987, the first president to assume power peacefully and democratically. He won only a minority of the votes, but was chosen ahead of the split opposition candidates. In office, he has sometimes reacted with the heavy-booted reactions that might be expected of a former military man. He has also been helped by the continuing squabbling lack of unity among the opposition leaders.

One of the best ways of observing the great progress that Korea has made is to take a trip through downtown Seoul to the Itaewon district. It offers a ready meeting place for old and new, traditional and modern, Korean and foreign. Big smooth-limbed American soldiers, many of them black, in camouflage combat fatigues mingle with old prune-faced Korean women dressed in *chimachagori* costume; young Koreans in jeans rub shoulders with sleek and loud foreigners just off an aircraft; wiseguy stall holders accost wide-eyed tourists. On the busy road, it often seems one long traffic jam including US army lorries from the nearby military base, modern black business saloon cars, crammed workaday taxis, laden buses and also bicycles ridden by Korean one-man restaurants on the move, able to stop, dismount and offer traditional delicacies in an instant. They are in competition with big international brightly lit shops, showing the logos of Kentucky Fried Chicken, Wendy's and Dunkin' Donuts. But the fast food bicycles still have their favoured traditional customers. Just off the main road, the land slopes steeply down to hundreds of hugger-mugger Korean homes. There, small children and old men sit by the street and play traditional Korean games, local versions of hopscotch and chess, unmoved by the constant hustle-bustle above them. Itaewon is a marketplace for Korea and the world.

On either side of the main road commerce flourishes: shop after shop is packed and piled with consumer goodies. When Hong Kong and Singapore have given up, Korean tailors still offer 24-hour made-to-measure service for anything from shirts and suits to uniforms and tuxedos. Smiling military models are a feature of the shop windows. Silk ties, leather belts and braces, cufflinks and all other accessories are available in shops next door. Fur coats, a Korean speciality, are down the road. There are enough shoes to keep the whole of Europe or North America running, jogging and walking for the rest of the century, shod in brand-name sneakers – you name it, Korea makes it – or smart leather dress shoes, custom-made overnight if you wish.

For entertainment, there are televisions, hi-fis and electronic goods galore, all locally made and bearing Korean brand-names like Samsung or Lucky-Goldstar. Other shop windows show so many beds that the sight of them makes you feel sleepy, and you can feel at home either on casual modern style resting divans or traditional Korean hard couches. Besides all this, there is a veritable cornucopia of Korean souvenirs from giant wooden chests to dresses, ducks and dolls; many bagshops selling everything from delicate Korean eelskin briefcases and handbags to voluminous holdalls, capacious enough to carry the load of a covetous tourist.

As if that is not enough, the pavement itself is filled with open-air stalls selling tee-shirts, tea-cosy woollen hats, socks and stockings in plain and multi-coloured shades; clothing of all shapes and sizes, to fit all shapes and sizes; toy tanks and trucks that trundle on and over every obstacle; guns, from crude plastic waterpistols to sophisticated child terrorist miniature Kalashnikovs; cuddly toys and teddy bears that stare appealingly with brown glassy eyes from a hundred different places at once. Nothing is static. You cannot

KEY FACTS

Official title: Republic of Korea

Head of state: President Roh Tae Woo

Head of government: Prime Minister Lee Hyun-Jae

Ruling party: Democratic Justice Party

Capital: Seoul

Official Languages: Korean (Japanese and English widely spoken)

Currency: Won

Exchange rate: Won 671 per $ (Nov 1989)

Area: 99,091 sq km

Population: 41.97m (1988)

GDP per capita: $4,082 (1988)

GDP real growth: 12% (1988)

Unemployment: 3.1% (1987)

Inflation: 7.1% (1988)

Trade balance: $11.45bn (1988)

Foreign debt: $40.459bn (1987)

Visitor numbers: 1,917,000 (1987)

move on the pavement without being hustled by some trader asking you to step into his shop or look at his goods. There is clearly a lot of determination and dedication behind Korea's 'miracle.'

It is hard to think of any place in Asia which has moved faster or is more stimulating than Korea. By 1989 flights in and out of the country were packed. There were plenty of foreigners, buyers, businessmen, lawyers, bankers, many of them newly awakened to Korea's potential by Seoul's staging of the 1988 Olympics, crowding in in search of opportunities. But Koreans also, thanks to a government directive allowing everyone to travel abroad, are cramming aircraft to look at the outside world. Previously there were restrictions on Koreans under the age of 40 going outside the country.

South Korea's task today of consolidating and extending the economic miracle would be difficult enough without any of the political ramifications. Until 1988 the country faced relatively plain sailing. To its credit, Korea stands out as the only example of a heavily indebted country that has not merely met debt repayments, but has been able to repay the principal. A few years ago, its debts almost reached $50bn, making Korea at the peak about the third most heavily indebted country in the world. By 1989, thanks to a succession of trade and current account surpluses, total debts were expected to fall to below $20bn. Korea's progress is evident enough not only in Seoul but in foreign countries where Korean exports of goods from ships and cars to televisions and video cassette recorders have already made their mark. In some areas, the Koreans are even challenging their close neighbours the Japanese, though this reputation also brings its own share of problems with foreign governments anxious that Korea should not become 'a second Japan.'

KEY INDICATORS

	Unit	1984	1985	1986	1987	1988
Population	'000	40,510	41,800	41,180	41,570	41,970
Gross domestic product (GDP)	Won bn	72,644	80,847	93,371	108,034	125,309
GDP per capita	Won m	1,793,236	1,934,139	2,267,387	2,598,845	2,985,680
Wholesale prices	% increase	0.7	0.9	–1.5	0.5	2.7
Consumer prices	% increase	2.3	2.5	2.8	3.0	7.1
Wages	% increase	8.3	9.9	9.2	11.6	19.7
Industrial production	% increase	15.1	4.4	20.6	18.2	13.7
Manufacturing employment	% increase	3.2	4.2	9.0	15.6	5.6
Volume of exports	1985=100	93	100	112	139	156
Volume of imports	1985=100	94	100	108	131	149
Exports	Won bn	23,571	26,347	30,600	38,892	44,398
Imports	Won bn	24,688	27,089	27,840	33,742	37,898
Crude petroleum	Won bn	4,651	4,848	2,949	3,045	2,698
Balance of trade	$m	–1,036	–19	4,206	7,659	11,445
Current account	$m	–1,372	–887	4,617	9,854	14,161
External debt	$m	42,098	47,159	46,728	40,459	–
Debt service ratio	%	13.6	15.3	16.9	21.9	–
Exchange rate	Won per $	805.98	870.02	881.45	822.57	731.47

* estimated figure

Labour relations

Proofs of big differences between Korea and Japan and of Korea's special challenges came throughout 1989. The economy was disrupted by grumbling labour unrest and strikes, basically for higher wages, but sometimes expressing a wider unhappiness about working conditions. At times, the labour turmoil was bruising and even bloody, and in some cases government riot troops were called in to quell the unrest. One prominent foreign banker in Seoul described the labour problems as 'a cancer that could kill the economic miracle.' Soon-hak Lee, managing director and chief financial officer, international, of the Samsung group, regarded it as quite natural for workers to want their share in the rewards of prosperity having shared the effort. He expressed optimism that workers and management would get together and settle their differences amicably. Not everyone was equally optimistic. 'Korea is not Japan' was the point made time after time by the more pessimistic observers. In Japan workers would be more likely to do as they were told in the interests of communal harmony. 'These Koreans are much tougher, more individualistic people' was the view of one British banker, 'not for nothing are the Koreans called the Irish of Asia.'

In the first half of 1989 there were more than a thousand labour disturbances, compared with a total of 276 in the whole of 1986. As one example of the heightened feeling, Daewoo Shipbuilding and Heavy Machinery, the country's second largest ship building company and part of an industry that was only a few years ago the pride of South Korea's industrial progress and prowess, stood close to bankruptcy. The government and the ailing company negotiated a $600m bail out package. Then the union representing Daewoo's 12,000 workers demanded a pay rise of a whopping 55 per cent. The management offered 24 per cent. This was refused and a strike was called. Feelings ran high and hostile so that 10,000 riot police were mobilised and Daewoo management removed explosive materials from the shipyard. Eventually the workers accepted the deal, but only by a narrow 3 - 2 margin.

Another feature of the labour unrest was a strong touch of xenophobia. Foreign banks were particular targets. In April three Citicorp managers, two of them Americans, were held hostage for 27 hours by union activists at the bank's Diners Club offices. The British Barclays Bank was also a target for the unionists. A floor-to-ceiling poster in Barclays Seoul office showed a group of Korean workers driving away the Barclays eagle. Korean workers successfully unionised the local offices of two US-based multinational companies, IBM and Motorola, the first time that either company had recognised independent unions anywhere in the world. But this was not achieved without violence. At Motorola, the electronics company, some workers poured paraffin over themselves and threatened to immolate themselves to deter police from breaking the strike. A match was in fact struck in the scuffle and some workers badly burned.

Few companies escaped a strike or the implications of one. In May 1989, for example, there were more than 170 companies on strike. Not all were big enough to be noticed by the foreign press since they were mostly medium-sized concerns, frequently suppliers of components to the big conglomerates. But the big *chaebol* (as the giant industrial and trading conglomerates are known) were also a target for unrest. Indeed the situation was complicated by the government's fear that radical students and their supporters were trying to infiltrate the big companies as part of a wider political campaign. The government sent 14,000 riot police in military gear to Ulsan after an appeal by Hyundai to end the 'civil strife' in its shipyard. Roh naturally regarded Ulsan, accounting for 10 per cent of Korea's GNP, as strategically vital. But in modern Korea, it was not always a black and white case of right and wrong. In this case, as Labour Minister Chang Yung Chul admitted, 'the Hyundai management used terror, so it made the situation worse. The general manager hired some people and invaded the union offices.' The minister was right when he said 'greater maturity is needed by both sides.'

The labour strains were worsened because government policy seemed to be to buy off trouble by putting money into everyone's pockets, an easy-going policy that diced with inflation and disaster. It led to high wage rewards, threatening exports that were already under pressure from the rapid appreciation of the won against the US dollar.

By the middle of the year the government was worried and it showed it. In June the president said that the economy had lost $4.5bn so far because of labour unrest, including $1bn in exports. 'It is of great concern to us that the competitiveness of our economy has structurally deteriorated,' he said, appealing to workers not to make 'excessive wage demands' and to help in restoring competitiveness. 'Now that wages have soared, workers are asked to help raise the competitiveness of their enterprises. Employers must exert their sense of resoluteness and self-sacrifice to overcome the adversity facing the nation's economy.' Roh tried to be even-handed when he added for the benefit of the workers 'the incumbent government does not suppress your demands or side with your employers. Neither does it demand unilateral sacrifices by workers as before.'

Fiscal measures

Later in the month the government imposed an economic package of wage and price control as public criticism grew about worsening inflation fuelled by wage increases that were estimated at an average 20 per cent. Future pay deals were to be limited to single figures and an independent national committee with representatives from unions, management, politicians and economists was set up to mediate in pay disputes. Charges for transport, public utilities and other public sector industries were to be curbed. In an attempt to encourage production rather than speculation, the package also offered tax incentives to steer big companies away from the stock market and real estate that had benefited from speculative fever. Small companies, especially exporters, were promised $750m in special support schemes. Added to this was the effect of the appreciating won, an additional factor pinching exports and profits of companies. South Korea's currency was trading at about 660 won against the US dollar, an appreciation of more than 22 per cent since the middle of 1986. Korean officials stated 'Our currency has been revalued enough and its exchange rate against the US dollar should be held as the 660 won level, at least in the near future.' The US treasury differed, claiming that Korea had manipulated its currency. In Washington's view, the appreciation of the won was 'significant', but not significant enough when 'judged against the magnitude of Korea's external surpluses in the last three years, and the much greater appreciation of Korea's competitors in the same period.' Economists at the London stockbrokers UBS Phillips and Drew disagreed with Washington, pointing to Korea's greater internal price pressures than its competitors – meaning that the appreciation of the currency in real effective terms had been much greater.

Trade

The toll was there to see in the trade figures. According to the trade ministry the surplus in the first seven months of 1989 had slipped to a mere $166m from $3.79bn in the same period of 1988. Nevertheless exports in 1989 will be about $65bn. Exports continued to rise, but by less than 10 per cent, while imports grew sharply by more than 20 per cent. The big *chaebol* dominating the country's exports, all reported exports and profitability sharply down. Economic growth for 1989 was still expected to be in the 7 per cent range though this was lower than the 8 per cent original target.

But by August President Roh was again sounding the alarm bells, warning that the economy faced 'a red light signal' because of a 'drastic fall in the export growth rate.' He continued: 'It is imperative to revitalise our exports in order to revive the resilience of our economic growth. In our economy, exports have played the role of a diesel locomotive pulling a long train.' He promised that the government would keep the exchange rate stable. Roh noted that because of the high wage awards, averaging more than 60 per cent from 1987 to 1989, some Korean companies had even closed part or all of their local operations and moved production to China or south east Asia where costs were cheaper.

There are problems aplenty for the years ahead, even for a South Korea that retains its competitive vigour. In 1989 Seoul narrowly escaped being singled out as an unfair trader under the 'super 301' section of the new US trade law. Its escape was a tribute to the market opening measures of the Roh government, and also to some intense lobbying. Six separate groups, including the deputy prime minister, the minister of trade and other parliamentarians went to Washington and other key US cities claiming that Korea was moving towards a more 'responsive and responsible' trade relationship with America. The Koreans pointed out that being listed as a super 301 country could upset the whole regional political process and impair relations with America. US diplomats were aware of Korea's importance in the Pacific balance of power, especially when relations with China had suffered through the Tiananmen Square massacre. Anti-Americanism had already been sparked off in Seoul by demands that Korea open its market to US beef and cigarettes. It was noticeable at the 1988 Seoul Olympics that Americans were unpopular and Soviet athletes were cheered. A South Korea that was growing up and feeling its way in the world resented being pushed around. The Americans damaged their cause by little things such as insisting on scheduling leading Olympic events for the benefit of their own prime time television. That was enough to anger the super-sensitive Koreans.

By 1989 there had been genuine efforts on the trade front with the USA. Seoul was also able to report a falling surplus. In 1987 Seoul's surplus with the USA was almost $10bn, but it fell to $8.5bn in 1988 and to

COUNTRY PROFILE

Political system: Korea's sixth republic started in February 1988 with the inauguration of President Roh Tae Woo. Under the new constitution, the president is elected for a five-year term by direct popular vote and there is a National Assembly whose 299 members serve four-year terms. The assembly has independent powers of investigation and cannot be dissolved by the president. The president has veto power, but the assembly can override by a two-thirds vote. The president appoints government officials including judges, ministers, mayors of five special cities (Seoul, Pusan, Taegu, Inchon and Kwangju) and the governors of nine provinces. Promises of local elections have been made but dates not set.

Political parties: Of the National Assembly's 299 members chosen in April 1988, 224 were elected from local districts while 75 were awarded to each party on the basis of the number of seats it won. Members of Roh's Democratic Justice Party won 125 seats; Kim Dae Jung's Party for Peace and Democracy won 71 seats; Kim Young Sam's Reunification Democratic Party won 59 seats; Kim Jong Pil's New Democratic Republican Party won 35 seats. The four parties derive most of their support from the home provinces of their leaders.

Population: The population of 42.08m (1988) is growing at only 1.21 per cent p.a., compared with 2.9 per cent in 1961. Nearly 60 per cent are under 25 years of age, and 25 per cent live in Seoul. There is a pronounced shift away from agriculture and toward manufacturing and services.

Main cities/towns: Seoul (population 9,991,089 in 1987), Pusan (pop 3,515,000), Taegu (pop 2,030,000), Inchon (pop 1,387,000), Kwangju (pop 906,000), Taejon (pop 866,000), Chonju (pop 426,000), Masan (pop 449,000), Ulsan (551,000), Pohang (pop 251,000), Yeosu (172,000).

Language: The Korean language is a member of the Altaic family with origins in Mongolia. 60 per cent of the vocabulary is borrowed from Chinese. It has its own phonetic character system, Hangul. English is found to a limited extent in business circles.

Media

Press: There are 32 daily newspapers, including six national dailies with a circulation of more than 600,000 each and two English dailies - the *Korea Herald* and the *Korea Times*.

Radio: Educational, religious, commercial and public external broadcasts on four channels in 12 languages. *Television:* There are over 20 TV stations run by the Korea Broadcasting System (KBS) and Munwha Broadcasting Corporation (MBC). The American Forces Korea Network (the US military system) provides both radio and TV broadcasts. KBS3 is an educational station.

Domestic economy: Export-oriented development contributed to better than seven per cent average growth over the past 25 years. Government policy-makers favoured companies with proven track records of earning foreign exchange, giving rise to the octopus-like family-run conglomerates, or chaebol, such as Hyundai, Samsung, Daewoo and Lucky-Goldstar. Although the government has recently emphasised nurturing small and medium industry, the top ten conglomerates still accounted for almost half of Korea's exports in 1987. Profit margins are thin and companies are heavily leveraged. The government is campaigning for companies to issue more equity to decrease their debts and help diversify the ownership of capital. In the past 17 years investment has focussed on heavy industry – shipbuilding, construction, chemicals, machinery, auto production and electronics. In 1988, 58 per cent of facility investments were to be placed in heavy and chemical manufacturing, compared to 12 per cent for light industry and 29 per cent for non-manufacturing. Under the revised current five-year plan (1987-91) per capita GNP hit $3,629 in 1988 and is expected to rise to $5,100 in 1991. The economy grew by more than 12 per cent in 1986 and 1987. In 1988 real growth in GDP was 11.28 per cent. South Korea spends almost two per cent of its GNP on research and development and hopes to boost that to three per cent by the 1990s. Future growth industries are computers, semi-conductors, biotechnology, aerospace and advanced metallurgy.

Employment: Rapid expansion of the economy has enabled maintenance of high levels of employment, and the rate of unemployment has fallen in each of the past five years. In 1987 it was 3.1 per cent. Total workforce is 17.3 million (1988).

External trade: Until 1986 Korea ran considerable deficits on its trade and current accounts, covering the shortfall by hidden borrowing overseas, both from soft loan agencies like the World Bank and from commercial sources. By the mid-1980s Korea's total foreign indebtedness peaked at $47bn, making the country the fourth most heavily indebted in the world. Since then Korea's heavy investment has paid off and the country turned in huge surpluses on both the trade and current account. In 1987 both accounts were in credit by more than $10bn, leading the US, the major export market, to call for tariff cuts. Total debts fell below $40bn. Heavy tariffs on manufactured goods and a bevy of consumption taxes aided the governments export-oriented development. The current averate tariff rate is 18.1 per cent, which the government has proposed to trim to seven per cent by 1993. The government has begun to eliminate special tax incentives for foreign investors. A campaign to diversify markets and import sources as well as deepening Korean production of parts for export items has helped to trim both the surplus with the United States and the deficit with Japan.

Exports: Total exports were worth $59.5bn in 1988. Main exports are machinery and transport equipment (35.8% of 1987 exports), miscellaneous manufactures (32.5%), manufactures classified by materials (21.5%), food (4.4%), chemicals and pharmaceuticals (2.8%), mineral fuels (1.6%). Main destinations are USA (38.7% of 1987 total), Japan (17.8%), Hong Kong (4.7%), German FR (4.2%), UK (3.2%), Canada (3.1%) and Saudi Arabia (2.2%).

Imports: Total imports were worth $49bn in 1988. Main imports are machinery and transport equipment (33.7% of 1987 total), manufactures classified by material (15.2%), mineral fuels (14.7%), other raw materials (14.4%), chemicals and pharmaceuticals (11.2%), other manufactures (5.2%), food (3.9%). Main sources are Japan (33.3% of 1987 total), USA (21.3%), German FR (4.4%), Australia (3.1%) and Malaysia (2.6%).

Agriculture: Agriculture, which employs 20.8 per cent of the workforce and contributes 12.3 per cent of GDP (1987), is subsidised by the government through import bans and heavy tariffs on consumer foods. Main crops are rice, sweet potatoes, barley, potatoes and a wide range of fruit and vegetables, grown mostly on smallholdings. Ginseng and tobacco, (which are controlled by government monopoly), noodles and pears are among the raw food exports from Korea. Livestock production, especially pigs, cattle and poultry, is increasing, but despite considerable efforts, the country depends on imports for animal feed grains. Forestry is expanding, providing plywood, newsprint and other products for export. Much of Korea's lowland forest has been cleared to provide fuel and arable land.

about $5bn in 1989. Fred Bergsten, head of the Institute for International Economics and a former US treasury official under President Carter, told the Senate that Washington should not push for further won appreciation unless the Korean current account surplus stopped falling.

In spite of the progress Korea remains a country that is facing all the problems of adolescence as regards both market opening and becoming a good member of the international trading community. Sometimes it deliberately wants to show its independence – and in Korea's case that can be demonstrated aggressively or sulkily. By mid-1989 Korea was on the verge of abandoning its right as a developing country under the terms of the General Agreement on Tariffs and Trade (GATT) to impose import restrictions if it ran into balance of payment problems. But as a concession for joining the world of adult international trading partners, Korea was seeking special treatment. The government asked for a 10-year grace period for phasing out restrictions – during which the country would be exempt from any new trade challenges under GATT provisions. This seemed intended to prevent the USA from imposing a future super 301 clause.

Earlier in the year Korea had lifted some restrictions on its agricultural imports, but not enough to keep the Americans happy. Not only the Americans have complained about Korea. A much more vigorous European Commission has been watching Korean companies and has several times imposed anti-dumping duties, claiming that Korean concerns are selling goods below their cost of production. In 1989 the Commission imposed anti-dumping duties on Hyundai Merchant Marine, a shipping line, on its freight-rates between Europe and Australia. This was the first time that the Brussels bureaucrats had ever levied duties on a service industry. They forced the Korean line to stop sailing on the route between Europe and Australia, and provoked anger both from Seoul and from European shippers.

Financial markets

Korea is also moving to open up its financial markets. In late 1988 Seoul told the International Monetary Fund that it was deregulating interest rates and preparing to make the won a freely convertible currency, in effect marking Korea's graduation from developing to developed country. When the won will actually be freely convertible is another matter. Bankers in Seoul think that it will be another two to three years at least.

Foreign investors have also expressed a great deal of interest in the Korean stock market. In terms of listed companies, Korea's market is about twelfth biggest in the world. There are more than 500 listed companies and it has been one of the fastest growing in the world. The composite index rose by 93 per cent in 1987 and by 73 per cent in 1988 when market capitalisation rose to about $100bn. This put Seoul ahead in terms of crude market capitalisation of both Hong Kong and Singapore. But both those island territories have stock markets that are wide open to foreigners. Seoul offers only a limited and indirect access, so that foreigners own a mere 2 per cent of the Korean equities. The government attitude to further foreign involvement in the securities market has shown a typical ambivalence: there have been plenty of promises of liberalisation, but actual opening up has been ultra-cautious. Way back in 1981 the government had first published a liberalisation plan that would supposedly open the market to foreign investors by the end of the decade. Under the latest liberalisation scheme, it will be 1992 before the market is open, and that will be a partial one. Korean economists admit that the government is anxious to avoid a rush of foreign funds to Korea. According to Nan Sang Woo, former official and now an economist with the government-backed think-tank the Korea Development Institute, the country 'no longer welcomes massive inflows of foreign capital because we are trying to reduce our balance of payments surplus.' Another powerful factor is that Korean securities firms are too weak and poorly capitalised to compete with big international brokerage houses.

Facing economic problems

Other economic problems surfaced in the painful new world of growing up. Stock market prices have soared, but so too have land prices in Seoul, making the dream of a home of their own an elusive one for many of the young aspiring middle classes who work in the capital. With growth has come inequality of income, which could become a hot political issue.

Optimists who know South Korea believe that if it is competition that is required, the tough Koreans will find a way. Han Seung-Soo, minister of trade and industry made a sly dig at the Japanese in 1989 when he said that Korean investment is for productive purposes rather than for buying stock or government securities. 'And you won't find any multimillion-dollar Picassos hanging on the walls of Korean corporate boardrooms,' the minister said, referring to expensive art purchases by Japanese companies with too much spare cash. Some Korean industries are suffering. The shipbuilding industry, only a few years ago the high-flying star, is decidedly sick. However, economists are optimistic that higher technology manufacturers turning to electronic goods, cars and steel should be able to flourish in spite of the appreciating won, providing that labour problems do not disrupt production. Lee of the Samsung group points out that Korea has already caught up with Japan in some industrial products, notably television manufacturing. In the crucial semiconductor area, his own company is in the process of developing the four mega-D-RAM (dynamic random access memory chips) that will then – he says – be only six months behind the big Japanese companies.

In July 1989 the Korean group Goldstar made a deal with Hitachi, the Japanese electronics giant. Under the agreement, Hitachi will show Goldstar how to make one megabit D-RAMs. It is the first time that a Japanese semiconductor maker has transferred technology to a Korean concern.

Korean companies have also been anxiously looking at investment abroad. In the first half of 1989 investment in overseas projects more than doubled compared to the previous year. Korean companies were looking all over the world, at North America, south east Asia, Europe and at Latin America. Three Korean companies have already announced plans for electronic plants in the UK as part of their protective measures to get round restrictions that may be imposed by a 'fortress Europe' when the European Community becomes a unified market after 1992.

Political situation

One major problem that could interrupt the Korean economic miracle is that life is not merely an economic game. Politics is a new sport for Koreans and so far they have not got the hang of it as a settled democratic game. Roh Tae Woo became president in 1988, in the December 1987 election he had won only 36 per cent of the vote. In the national assembly elections in April the following year, Roh's Democratic Justice Party won only 125 out of 299 seats, the first time that the ruling party had ever been defeated. If the opposition had been able to get together, let alone agree on a policy or political programme, they might have been able to oust the president. A senior government official said accurately 'We have three opposition leaders, all called Kim and all lacking credibility, all obsolete, all part of the discredited past.' The fragmented opposition within the national assembly is only one aspect of the anti-Roh forces. Running street battles between students and workers on the one hand and riot police on the other became a common feature of the Seoul scene in 1988 and 1989. Firebombs and rocks from the students and teargas from the police were the common weapons.

During the first four months of 1989, according to Korean police, more than 300,000 firebombs were thrown, four times higher than the number in 1987, which was reckoned to be a turbulent year. Korean democracy is still a fragile flower. Roh evidently still faces pressure from his former military colleagues who continue to be influential members inside the ruling elite. But middle-class Koreans are also fed up with violence. In May 1989 students took five policemen hostage in the library of Dongui University in Pusan. When hundreds of police stormed the library under a blanket of teargas, the students retreated, but only after dousing their stronghold with paint-thinner. They then tossed a firebomb into the library. Six policemen were killed in the inferno of flames and smoke. President Roh went on national television to warn that unless the violence stopped, he would use his emergency powers, possibly including martial law.

The delicate state of South Korean democracy is made the more complicated because of the border with North Korea. Increasingly, the demands of student radicals have included unification with North Korea and opposition to American forces stationed in the country. A non-political Korean university teacher who comes from an influential family understands both the history of the Korean conflict and some of the aspirations of the students. He says 'Among students there is a growing feeling against military confrontation, and a realisation that whoever gains, Koreans will suffer most.' Anti-American feelings have been fed by small stupidities and shortsightedness of the Americans and by the prickliness of the radical Koreans. Another factor fanning the flames is the feeling among radicals that Washington backed the discredited former ruler Chun and reduced Korea to a pawn in the great power struggle.

The disgraced Chun himself posed a problem, particularly because Roh is not only his successor, but also a former colleague and friend. Chun had run into seclusion in a Buddhist monastery, but there were continuous pressures that he should be brought to trial for corruption during his period in power, or should appear before the opposition-dominated national assembly to account for his actions. 'It's a pity that Korea isn't Japan,' laughed a foreign banker grimly, 'if it were Japan there would be a simple solution: Chun would know that he should go and commit *seppuku* (suicide) to atone for his actions.'

Radical demands that Washington should move its troops out of Korea found some sympathetic ears in the USA though for different reasons. At the moment there are about 46,000 American troops in Korea, nominally under the United Nations command. Inside America itself the pressure comes on budgetary and other grounds. Several Democratic senators tried to introduce legislation calling for a cut in the numbers of troops. New president George Bush's defence secretary Dick Cheney gave a standard response in mid-1989 that the 'Pentagon has no current plans' to reduce forces in South Korea. But the chairman of the joint chiefs of staff Admiral William Crowe said 'I must tell you as a military man, that with the kind of fiscal constraints that we are having put on us, everything in our inventory is open to looking at and certainly being re-examined every year.' One proposal would have the Americans withdraw from their 750-acre base close to Itaewon in the centre of Seoul, and thus reduce numbers, certainly reduce spending and reduce the visibility of American troops in the Korean capital. But Koreans already accuse Washington of dragging its feet over an agreed handover of the 18-hole golf course on the base. This was expected to go ahead in 1990 but demonstrates the clumsiness of Washington's actions even when it is trying to be helpful.

The troops are there to protect the southern half of the peninsula from the communist North. And the sensitive question of North Korea also figures increasingly in politics in 1989. In the last few years South Korea has expanded its ties both economic and diplomatic with communist countries. But the communist arch-enemy across the 38th parallel is a very different matter. Hardliners inside the government have tried to make contact with the North a capital offence. They have failed in this. But they have managed to persuade Roh to pursue a hard line against South Koreans who have visited the North illegally. The visitors include the Reverend Moon Ik-Hwan, a Protestant Minister, Lim Su-Kyong, a student and Father Moon Kyu-Hyong, a Catholic priest who had gone to help the student. All of them were arrested on their return south. Police even arrested other Catholic priests who tried to reach the border village of Panmunjom to escort the student Lim across to South Korea. The priests were stopped by a police roadblock, taken out of their cars and taken away.

Anywhere else in the world, such measures would have seemed unduly harsh. But this was just an indication of how tough the Koreans are and how hardball the politics. Roh and his military backers went further and tried to snare the opposition leaders on the North Korean hook. They were obviously encouraged by public feeling that radical violence had gone too far.

The government accused opposition leader Kim Dae-Jung of having illicit contacts with leader Kim Il Sung of North Korea. The charges were that he had written to the North Korean leader and that a member of his party had illegally visited the North in 1988. It also alleged that Kim and another opposition leader Kim Yong-Sam had helped the Reverend Moon in his trip to North Korea and that Kim Dae-Jung had provided expenses for that trip. Government officials charged that the Reverend Moon was part of Pyongyang's plan to 'form a unification front with South Korean radical intellectuals, religious people, dissidents, politicians and students, split national opinion and create social confusion, with the eventual goal of creating favourable conditions to communise the entire peninsula.'

COUNTRY PROFILE

Industry: Manufacturing industries, expanding and diversifying rapidly, employ 24.7 per cent of the workforce and contribute 30 per cent of national income (1987). Main products include fertilisers and other industrial chemicals, ships, steel, motor vehicles, petrochemicals, rubber, synthetic and natural textiles, garments, footwear, processed foods and a wide range of electric and electronic equipment. South Korea now supplies 70 per cent of Japan's annual two million tonne steel imports; a second steel mill at Kwangyang will increase national capacity by 40 per cent when it comes on stream in 1988. Korea is now one of the world's largest producers of electronic goods, with output rising 45 per cent in 1986. Heavy construction work overseas, especially in the Middle East, is a speciality, although Korean construction companies are suffering from recent cutbacks in the Middle East construction market. Tourism is an important and growing industry, and will recieve a major boost from the 1988 Olympics.

Mining: Small quantities of coal, graphite, gold and silver and ores of tungsten, iron, lead, zinc and copper are mined. South Korea's lack of viable natural resources makes it unusually susceptible to commodity price developments. Coal production rose 1.3 per cent in 1987 to a record 25.6m tonnes.

Energy: Imported energy supplies 80 per cent of demand, with 62 per cent being derived from oil, all of which is imported. Nuclear power has been introduced, and oil's share of power generation dropped in 1987 as nuclear power's share rose from 6.3 per cent to 36 per cent with the construction of two new plants. Another 14 power plants are to be built by 1996, of which 11 will run on coal. Projects are under negotiation for LNG-and LPG-receiving stations, and for a 60,000 b/d oil refinery. Hydroelectric power is also important, and intended to supply 15 per cent of power generation by 1991.

Financial markets: Seoul Stock Market. The deregulating financial climate has stimulated wide international interest since 1986, although direct foreign share ownership is still barred. As one of the few opportunities for speculative investment in South Korea, the stock market has continued to soar. It was one of the few bourses not to plummet in October 1987.

Membership of international organisations: African Development Fund, Asian Development Bank, ASPAC, CCC, Colombo Plan, EC trade and co-operation agreement, ESCAP, FAO, G-77, GATT, IAEA, IBRD, ICAC, ICAO, ICCAT, ICMMP, IDA, IDE, IFAD, IFC, IFDF, IHO, IMF, IMO, INTELSAT, International Whaling Commission, Interpol, IOC, IPC, IPU, IRC, ISO, ITU, IWC, UNCTAD, UNDP, UNESCO, UNIDO, UN Special Fund, UPU, WACL, WHO, WIPO, WMO, World Bank, WTO.

Time: GMT + 9 hrs; EDT + 14 hours. April to October summer time adds one hour.

Climate: Winters dry and very cold, temperatures well below 0°C from Dec to Feb; summers hot and humid, with monsoon rains, tropical storms and occasional typhoons from Jun to Sep.

Entry requirements: Regulations liable to change at short notice and should be checked before departure.
Passport: Required by all except certain UN personnel and seamen.
Visa: Not always necessary for non-business visits, provided onward tickets are held. For stays longer than 90 days, visas must be obtained in advance, evidence of local means of support and a letter of invitation from a local business may be required. Visas must be used within three months of date issued.
Currency: Import/export of Won is limited to W 2,000,000. Local currency is not convertible outside Korea. Foreign currency is not restricted, but must be declared on arrival and departure. Exchange receipts should be retained.
Customs: Personal effects admitted free of duty. High-value items (cameras, watches, etc.) should be recorded in passport to ensure re-export. Visitors may be liable for import duty and tax on items lost or stolen while in the country. Apply for customs clearance prior to importing an automobile. A certificate from the Cultural Properties Preservation Bureau is necessary for exporting antiques. Permission for trade imports or exports must be obtained from the Ministry of Commerce and Industry or from authorised foreign exchange banks, and certain items may be restricted or prohibited (such as ginseng and cuttlefish).
Income tax: Visitors staying longer than 183 days are liable for tax (36-55 per cent with no allowances) on assessed earned income and allowances arising both in Korea and overseas; a tax clearance certificate is required on departure.

Health precautions
Advisable: Immunisation is not mandatory, but vaccination certificate for cholera advisable. Hepatitis is endemic.

Representation overseas: Abidjan, Abu Dhabi, Accra, Addis Ababa, Amman, Ankara, Asunción, Athens, Bandar Seri Begawan, Bangkok, Bangui, Barbados, Beirut, Berne, Bogotá, Bonn, Brasilia, Bridgetown, Brussels, Buenos Aires, Canberra, Caracas, Colombo, Copenhagen, Dakar, Dhaka, Doha, Fasso, Freetown, Guatemala City, The Hague, Helsinki, Islamabad, Jakarta, Jeddah, Kampala, Kathmandu, Khartoum, Kingston, Kinshasa, Kuala Lumpur, Kuwait, Lagos, La Paz, Libreville, Lima, Lisbon, London, Madrid, Manama, Manila, Mbaban, Mexico City, Monrovia, Montevideo, Muscat, Nairobi, New Delhi, Niamey, Nouakchott, Ontario, Ouagadougou, Oslo, Ottawa, Panama City, Paramaribo, Paris, Port-au-Prince, Port Louis, Port Moresby, Port of Spain, Quito, Rabat, Rangoon, Rome, Ruwi, San José, Santo Domingo, Santiago, Singapore, Stockholm, Suva, Taipei, Tehran, Tokyo, Tripoli, Tunis, Vatican City, Vienna, Washington DC, Wellington, Yaoundé, Zomba.

Air access
National airline: Korean Air.
Other airlines: Air France, Asiana Airways, British Airways, Cathay Pacific, Delta, JAL, Japan Air System, KLM, Lufthansa, MAS, Northwest, Saudia, Singapore, Swissair, Thai International, United.
Tax: An airport tax of Won 5,000 is levied on departure.
International airports: Kimpo International (SEL), 17 km from Seoul, duty-free shop, restaurant, bank, hotel reservations, car hire, shops; Kim Hae International (PUS), 26 km from Pusan; Cheju International (CJU), on Cheju Island. A new international airport is under construction at Chongju (due for completion 1992).

Surface access: Daily ferry service between Pusan and Shiminoseki, Japan. Overland access is not practicable.
Main ports: Pusan, Inchon, Masam, Ulsan, Mokpo, Kunsan, Yosu.

Hotels: Book hotels well in advance, especially in spring and autumn. Western-style hotels include a 10 per cent service charge in the bill. Tipping is not usual, although it is on the increase in Western-style hotels.

Credit cards: Main hotels accept major credit cards, but check when booking which ones are accepted for settlement of hotel bills.

Car hire: International driving licences are acceptable, but chauffeur-driven car hire is recommended in the main cities.

City transport
Taxis: Registered taxis carry meters and are clearly marked on the roof. Taxis are plentiful, and available at ranks, by telephone or can be hailed. Call taxis are more comfortable than small taxis, but more expensive. Standard fares for shorter distances, sharing is common and tipping is unusual. In Seoul and Pusan, light brown-coloured taxis are rated better, but are more expensive. A 20 per cent surcharge applies between 0000 and 0400. Taxi drivers suspend use of meters for journeys outside town, so negotiate the fare for such trips in advance. It is advisable to carry written instructions in Korean if possible. There is still a curfew in coastal areas from 2400–0400, so taxis are more difficult to hire as midnight approaches.
Metro: Seoul and Pusan are both developing underground metropolitan transport systems. A subway (Line 1) runs from the Seoul railway station to Chongnyangni, NE of the city, while Line 2 is a 48.8 km circle, Line 3 runs south-east

BUSINESS GUIDE

across the Hangang river and Line 4 runs south-west. Signs are in English and Korean.

National transport: Facilities were expanded and improved in readiness for the 1988 Olympic Games in Seoul.
Air: Korean Air operate daily services between Seoul and Pusan, (50 minutes), Taegu, Cheju, Ulsan and Kwangju, with less frequent services to other centres. Expect to be searched for firearms when embarking on internal flights.
Road: First-class highways link major centres. A 142 km expressway network between Seoul and Taejon is being expanded to four lanes and the whole network is under continual improvement. Roads are poor and unmetalled off the major routes.
Buses: Several lines operate express intercity services along motorways (2,535 buses in 1987). City buses, though cheap and convenient, are crowded. Purple buses have few seats and cost W140 or W150. Green 'seat buses' are more comfortable and air conditione and cost W350. Payment is by token, and there are English-language signs only for city-centre buses.
Rail: All main towns are linked by railway; notices and booking service at main stations often in English. Many trains have sleeping and dining cars. Considerable upgrading is under way.

Public holidays
Fixed dates: 1, 2 and 3 Jan, 1 Mar (Independence Movement Day), 5 Apr (Arbor Day), 5 May (Children's Day), 6 Jun (Memorial Day), 17 Jul (Constitution Day), 15 Aug (Liberation Day), 1 Oct (Armed Forces Day), 3 Oct (National Foundation Day), 9 Oct (Hangul Korean Alphabet Day), 25 Dec.
Variable dates: Folklore Day, Buddha's Birthday, Chusok (Moon Festival) Day. The three days of Chinese New Year are also widely observed by commerce and industry, though not official holidays.

Working hours: Sunday is the weekly closing day. Service establishments open as late as 2200, over weekends and on public holidays.
Business: (Mon-Fri) 0900 – 1200, 1300 – 1800; (Sat) 0900 – 1300.
Government: (Mon-Fri) 0900 – 1200, 1300 – 1800; (Sat) 0900 – 1300. Nov-Feb: 0900–1700.
Banking: (Mon-Fri) 0930 – 1630; (Sat) 0930 – 1300.

Social customs: Korean surnames precede given names, and given names are never used alone, except by intimates. The family names 'Kim', 'Lee' and 'Park' cover more than half the population, which can be confusing. Business associates are normally addressed by title (e.g. Director Kim or Manager Lee, etc). Business entertaining usually takes place in restaurants and wives do not participate. Business visitors should carry a good supply of business cards, which are exchanged on introduction. Hotels can provide bilingual business cards overnight. Use the right hand when giving or receiving. Outdoor shoes should never be worn inside a house. There are certain areas, particularly near the demilitarised zone, where entry and photography are forbidden.

Telecommunications
Telephone: Dialling code for Republic of Korea, IDD access code + 82 followed by area code (Seoul 02; Pusan 51; Incheon 32; Taegu 53) followed by subscriber's number.
Telex: 24-hr public telex office at KTI (Korea National Telegram), 1 Choongmo-ro, Chung-ku, Seoul. Telex facilities are also available at major hotels.
Telegram: Telegram service from post offices (0900 – 1730), with 24 hrs emergency service.

Postal service: Korean authorities sometimes exercise their right of censorship. Overseas mail should be sent via air mail.

Banking: Banking is controlled by the Bank of Korea and the Ministry of Finance.
Main centre: Seoul.
Special banks: Citizens National Bank, Export-Import Bank of Korea, Korea Credit Guarantee Fund, Korea Development Bank, Korea Exchange Bank, Korea Housing Bank, Korea Long-Term Credit Bank, National Agricultural Co-operative Federation, National Federation of Fisheries, National Livestock Co-opeartives Federation, Small and Medium Industry Bank.

Advertising: Commercial advertising may be placed with the daily press, with radio and TV stations and with cinemas. Advertising standards are defined and controlled by the Korea Broadcasting Ethics Committee and the Korea Newspaper Ethics Committee. The Korea Broadcasting Advertising Corporation (KOBACO) grants recognition to advertising agencies for TV and radio.

Trade fairs: The new Korea Exhibition Centre (KOEX) in Seoul opened in 1988, forming part of the newly-built World Trade Centre. 171,144 square metres floor space. Telephone 553-7907/8 or telex K24594 (fax: 557-5784) for enquiries.

Electricity supply: 110 and 220 V AC, 60 cycles, with two-pin square plug fittings and screw-type lamp fittings (gradually switching to 220 V AC).

Weights and measures: Metric system used in commerce; local system also in use, especially relating to land and buildings.

Representation in capital: Argentina, Australia, Austria, Bangladesh, Belgium, Bolivia, Brunei, Brazil, Canada, Chile, Republic of China, Colombia, Costa Rica, Denmark, Dominican Republic, Ecuador, Finland, France, Gabon, German FR, Guatemala, Haiti, India, Indonesia, Iran, Italy, Japan, Liberia, Libya, Malaysia, Mexico, Netherlands, New Zealand, Norway, Oman, Pakistan, Panama, Paraguay, Peru, Philippines, Saudi Arabia, Spain, Sri Lanka, Sweden, Switzerland, Thailand, Turkey, UAE, UK, USA, Uruguay, Vatican City, Venezuela.

Hotels: Seoul
Ambassador, 186-54 2 ka Changchung-dong, Chung-ku (tel: 275-1101; tx: 23269).

Garden, 169-1 Dowha-dong, Mapo-ku (tel: 717-9441; tx: 24742; fax: 7159441).

Hamilton, 119-25, Itaewon-dong, Yongsan-ku (tel: 794-0171/9; tx: 24491; fax: 7950457).

Hilton International, 395, 5-ka, Namdaemun Road, Chung-ku 100, P.O. Box: 7692 (tel: 753-7788; tx: 26695; fax: 7542510).

Hyatt Regency, 747-7 Hannam-dong, Yongsan-ku (tel: 797-1234; tx: 24136; fax: 7986953).

King Sejong, 61-3, 2-ka, Chungmu-ro, Chung-ku 100, P.O. Box: 2583 (tel: 776-1811; tx: 27265; fax: 7554906).

Koreana, 61,1-ka, Taepyong-ro, Chung-ku 100 (tel: 730-9911/20; tx: 26241; fax: 7340665).

Lotte, 1 Sagong-Dong, 100, P.O. Box: 3500 (tel: 771-10; tx: 23533/4/5; fax: 7523758, 7555771).

Palace, 63-1 Banpo-dong, Kangnam-ku, 134 (tel: 532-0101/4100/5000; tx: 22657).

Plaza, 23, 2-ka, Taipyung-ro, Chung-ku 100 (tel: 771-22; tx: 26215/24424; fax: 7558897).

President, 188-3 Ulchi-ro, 1-ka, Chung-ku (tel: 753-3131; tx: 27521; fax: 7527417).

Ramada Olympia, 108-2 Pyong Chang-dong, Chongro-ku (tel: 353-5121/7; tx: 23171; fax: 3538118).

Royal, 6, 1-ka, Myong-dong, Chung-ku (tel: 756-1112; tx: 27239; fax: 7561119).

Sheraton Walker Hill, 21 Kwangjang-dong, Sungdong-ku, P.O. Box: 714 (tel: 453-0121; tx: 28517; fax: 4526867).

Shilla, 202, 2-ka, Jangchung-dong, Chung-ku, P.O. Box: 7000 (tel: 233-3131; tx: 24160, 28937; fax: 2335073).

Westin Chosun, 87 Sokong-dong, Chung-ku, P.O. Box: 3706 (tel: 771-05; tx: 24256; fax: 7521443).

Business advice and information:
Correspondence to government departments and foreign organisations can be addressed simply to the organisation name, followed by 'Seoul'.

American Chamber of Commerce in Korea, 307 Chosun Hotel, 87-1 Sokong-dong, Chung-ku, Seoul (tel: 7523061; tx: 242560; cable: AMCHAMBER).

Association of Foreign Trading Agents of Korea (AFTAK), 45-20 Yoido-dong, Yungdung po-ku, Seoul (tel: 7822206).

Export-Import Bank of Korea, PO Box YPO 641, Seoul (tel: 7841021).

Federation of Korean Industries, 28-1 Uoido-dong, Yungdung po-gu, Seoul (tel: 7830821).

Japanese Chamber of Commerce, Taepyongro 1-ka, Chung-gu, Seoul Press Center Building (tel: 7396962).

Korea Broadcasting Committee, 75, 1-ga Taepong-ro, Chung-ku, Seoul (tel: 7352640).

Korea Chamber of Commerce and Industry (KCCI), 45 Namdaemumro 4-ka, Chung-ku, Seoul (tel: 7590959).

Korean Broadcasting Association, 25, 1-st Taepong-ro, Seoul (tel: 7357117).

Korean Exhibition Centre, 65 Samsung-dong, Gangnam-ku, Seoul.

Korean Trade Promotion Corporation (KOTRA), CPO Box 1621 10-1, 2-ka Hoehyun-dong, Chung-ku, Seoul (tel: 7534180/9; tx: 23659, 27326).

Korean Traders' Association (KTA), World Trade Centre, Korea Building, 10-1, 2-ka Hoehyun-dong, Chung-ku, Seoul (tx: 24265).

Ministry of Finance (MOF), 1st Chungang-dong, Kwachonshi, Kyonggido, Seoul (tel: 5037171).

Ministry of Finance, Foreign Investment Promotion Division, Istchumgang-dong, Kwachomshi, Kyonggido (tel: 5039276).

Ministry of Trade and Industry (MTI), 77-6 Sejong-ro, Chongro-ku, Seoul.

Banking
Bank of Korea (central bank), 110, 3-ka, Namdaemun-no, Chung-ku, Seoul 100 (tel: 7594114; tx: 24711).

Bank of Seoul, 10-1 Namdaemun-no, 2-ka, Chung-ku, Seoul 100 (tel: 7716000; tx: 23311).

Cho Hung Bank Ltd, 14, 1-ka, Namdaemun-no, Chung-ku, Seoul 100 (tel: 7332000; tx: 23321).

Commercial Bank of Korea Ltd, 111-1, 2-ka, Namdaemun-no, Chung-ku, seoul 100 (tel: 7713000; tx: 24611).

Hanil Bank, 130, 2-ka, Namdaemun-no, chung-ku, Seoul (tel: 7712000; tx: 23823).

Korea Development Bank, 10-2, Kwanchul-dong, Chongno-ku, Seoul 110 (tel: 7332121; tx: 27463).

Korea Exchange Bank, 181, 2-ka, Ulchiro, Chung-ku, Seoul 100 (tel: 77146; tx: 23141).

Korea First Bank, 100 Kongpyung-dong, Chongno-ku, CPOB 2242, Seoul 100-160 (tel: 7330070; tx: 23685).

Travel information
Korea Automobile Association, 1, PO Box 2008, Seoul (tel:7855051).

Korean National Tourism Corporation, 10 Ta-dong, Chung-gu, Seoul (tel: 7576030; tx: 28555; fax: 757 5997).

Kuwait

David Bradshaw

Kuwaitis could be forgiven for feeling a touch of self-satisfaction. Their small, vulnerable state had come through the most testing few years since independence. The threat posed by eight years of Gulf War fought virtually on the country's borders – and the terrorism which accompanied it – receded with the cease-fire of August 1988. And the economy, after several years of depression, was enjoying a steady, if unspectacular, upturn based on rising oil prices and the slow recovery of battered business confidence. The bonanza some Kuwaitis expected to follow a Gulf War peace had not happened, but prospects for the 1990s looked promising. Kuwaiti planners said the next decade would be 'rational' and free of the excesses of the boom years of the 1970s and early 1980s.

Building confidence

Economic depression was the result of a disastrous combination of the oil price crash, the collapse of the unofficial Souk al-Manakh stock market in 1982 and the damage done to business confidence by the Gulf War. GDP shrank by 30 per cent in the period 1981-86 from KD6.98bn ($24bn) to KD4.98bn ($16bn). Recession bottomed out in 1985-86 and recovery began in 1987, based on rising oil income. GDP rose to KD5.4bn in 1987 and to KD5.60bn in 1988. GDP in 1989 was expected to grow 5 per cent to KD5.88bn.

Oil provides the lion's share of revenues. Low oil prices have induced Kuwait to exceed its OPEC production quota. During 1988-89 Kuwait's quota stood at just under 1m barrels a day (b/d), but independent analysts believe actual production averaged some 1.5m b/d. Early 1989 production was again way above quota; that of May averaged 1.9m b/d against an OPEC quota of 1.09m b/d. In June 1989 Kuwait, along with the United Arab Emirates, refused to be bound by OPEC quotas, though the oil minister, Sheikh Ali Khalifah al Sabah, promised to peg production at 1.35m b/d.

Assuming a $15 barrel average, fiscal 1989-90 revenues from crude oil and refined products were projected at KD1.95bn ($6.72bn) compared with KD1.8bn ($6.2bn) in fiscal 1988-89. But they seemed likely to be higher than that because of strengthening oil prices. Non-oil income was projected to rise to KD288m ($998m) from KD265m because of increased customs and transit revenues.

1989 imports were planned to rise to $5.75bn and exports to $8.75bn. During the period of declining oil revenues, imports were carefully trimmed to ensure that the balance of payments remained in credit: as exports fell from $12bn in 1984 to $8bn in 1987, so imports fell from $6.5bn to $4.7bn. Japan is the main source of imports; its cars, machinery and electronics account for around 20 per cent of the total. The United States comes second with 12 per cent, West Germany third and Britain fourth. Kuwait's credit rating is high. Japan is Kuwait's biggest customer, followed by Italy and Holland, where Kuwait owns oil refineries and chains of retail outlets.

Kuwait has run a series of nominal budget deficits since the oil price crash. The projected deficits for 1987-88 and 1988-89 were in practice halved by higher than projected oil income. Government revenues for 1989-90 were forecast at KD2.23bn ($7.72bn) – up from KD2.05bn ($7.1bn) in 1988-89 – and spending at KD3.55bn ($12.28bn), up marginally from KD3.4bn. The spending figure includes the statutory allocation of 10 per cent of revenues – KD223m ($771m) – to the Fund for Future Generations. This gives a nominal deficit of KD1.32bn ($4.56bn), but this will be cut, as in the previous two years, if oil prices rise above $15 as they were confidently expected to do. The government said the 1989-90 deficit would be financed by drawing down the General Reserve Fund, as have previous deficits barring that of 1987-88 which was covered by an issue of treasury bonds, a device which was also partly intended to help to stimulate the Kuwait stock exchange.

Kuwaiti deficits are, however, nothing like as bad as they look since Kuwait's income from overseas investments is excluded from the calculations. If investment income was included, then every budget in recent years would have shown a hefty surplus. Investment income has meant that the current account surplus has never fallen below $3.75bn. In 1986, when oil revenues were at their lowest, a 57 per cent surge in investment income to $7.5bn gave a current account surplus of $5.3bn – 25 per cent higher than the year before.

Austerity to come?

Nevertheless, the government is beginning to think hard about trimming some of the expensive social benefits which have made Kuwaiti citizens among the best cared for in the world. The generous Kuwaiti welfare state provides free education, health care, almost free energy and subsidised housing. Critics say it has created a culture of dependency. Now there is talk of taxation, and about charging fees for health and education, which take up 35 per cent of the budget.

That may still be in the future. For the moment, according to Finance Minister Jassim al-Kharafi, outlining spending priorities in the 1989-90 budget, the government would limit expenditure 'without prejudicing social services'. Expenditure on security would grow. He said KD560m had been allocated to construction and KD200m to real estate appropriations, both of which are intended to inject life into the sluggish domestic economy.

Kuwait's 'overseas' economy is anything but sluggish. Kuwait has over $100bn of overseas investments, mostly in the industrialised world. Investment revenues are thought to be running at over $6bn a year, roughly equivalent to oil revenues. The investments are held in two funds. The General Reserve Fund (which also includes investments in Kuwait and substantial loans to Iraq) is worth an estimated $30bn. It has been drawn down to fund recent budget deficits and will fund the $5.5bn of arms purchases Kuwait plans to make over the next few years. The Fund for Future Generations was set up in 1976 and is now worth at least $70bn. Fed annually with 10 per cent of revenues, neither capital nor interest can be touched until the year 2001. Most investments are in equities, but property is becoming increasingly important. One-third of its investments are in the United States, the rest are spread across the world by the Kuwait Investment Authority (KIA) acting through the largely autonomous London-based Kuwait Investment Office (KIO). While the overseas investments of other Gulf Arab states are heavily influenced by political decisions, those of Kuwait are rarely motivated by anything other than profitability. In February 1989 KIA managing director, Fahed Mohammed al-Rashed, said the agency was monitoring economic reforms in the Soviet Union and China with a view to increasing investment there. Kuwaiti institutions have already made modest loans to the Soviet Union; a Soviet delegation visited Kuwait in early 1989 to discuss further loans.

Investment power play

Control of KIO and investment policy rests between the Oil Minister Sheikh Ali, the Finance Minister Jassim al-Kharafi, and the governor of the central bank. All three sit on the board of KIO, which was

KEY FACTS

Official title: State of Kuwait

Head of state: Shaikh Jaber al Ahmad al Sabah

Capital: Kuwait City

Official Languages: Arabic

Currency: Dinar (KD) = 1,000 fils

Exchange rate: KD0.30 per $ (Nov 1989)

Area: 17,818 sq km (including neutral zone)

Population: 1.96m (1988)

GNP per capita: $13,160 (1988)

Inflation: 1.5% (1988)

Oil reserves: 94.5bn barrels (1988)

Trade balance: $1,907m (1988)

Foreign debt: $7,966m (1987)

specifically created in 1982 to oversee investment activities. Mr al-Kharafi believes that Kuwait should not be investing in oil, with which it is already well endowed. The KIO attracted international attention through its big investment in British Petroleum and Spanish criticism of KIO's high-profile raids on the Spanish stock market. Recent investments in Spanish companies have aroused nationalistic feelings among Spaniards who claimed Kuwait is trying to gain control of strategic sectors of the Spanish economy. More serious was KIO's $3bn purchase of 21.6 per cent of shares in BP, despite British Government warnings, which aroused British fears that Kuwait would use its stake to influence the world oil market. Despite Kuwaiti denials, the British Monopolies and Mergers Commission ordered Kuwait to cut its stake in BP to 9.9 per cent. In January 1989 BP shareholders backed a plan to buy back the excess shares. Though Kuwait made a handsome profit, the affair angered many Kuwaitis, and there were veiled (but not serious) threats to pull out some of the $20bn invested in Britain by KIO. These developments showed the vulnerability of overseas investment and the political damage they can do. They were seized upon by those Kuwaitis who feel that excess revenues should not be invested overseas but should be spent on development projects inside Kuwait. A three-man commission appointed by the Emir in 1988 is understood to have recommended a cautious low-profile investment policy, of the kind favoured by Mr al-Kharafi, over the more commercial approach apparently favoured by Sheikh Ali.

KEY INDICATORS

	Unit	1984	1985	1986	1987	1988
Population	m	1.64	1.71	1.79	1.87*	1.96
Gross domestic product (GDP)	KD m	6,381	5,959	4,998	5,600	–
GDP per capita	KD	3,898	3,481	2,792	2,943	2,857
Discount rate	%	6.0	6.0	6.0	6.0	7.5
Consumer prices	1985=100	98.5	100.0	101.0	101.6	103.1
Wholesale prices	1985=100	101.3	100.0	100.6	103.9	108.7
Industrial production						
Crude petroleum	1985=100	109.6	100.0	132.8	125.8	131.8
Crude petroleum	'000 b/d	1,163.0	936.3	1,237.7	971.6	1,396.5
Volume of exports						
Petroleum	1985=100	128.3	100.0	–	–	–
Crude petroleum	1985=100	138.6	100.0	87.4	82.8	86.7
Refined petroleum	1985=100	87.6	100.0	–	–	–
Exports	KD m	3,632.4	3,151.9	2,104.3	2,330.0	1,997.7
Crude petroleum	KD m	1,921.0	1,693.8	970.7	681.4	524.6
Petroleum	KD m	3,179.6	2,952.4	2,229.3	1,940.0	1,783.2
Imports	KD m	2,041.7	1,784.7	1,673.5	1,476.0	1,492.3
Balance of trade	$m	5,604	4,712	1,944	3,545	1,907
Current account	$m	6,374	4,815	5,379	4,676	4,713
Exchange rate	$ per KD	3.38	3.33	3.43	3.59	3.58

* estimated figure

Downstream success

Kuwait has been pumping oil far longer than most of its Gulf Arab neighbours, and oil remains the mainstay of the economy. Its oil reserves of more than 100bn barrels are the third largest in the world, after Saudi Arabia and the Soviet Union, and are large enough to last for two centuries at current rates of extraction. Oil and refined products account for 90 per cent of exports and oil-related products, such as sulphur and fertilisers, make up much of the rest.

Kuwait has gone much further than other Arab oil producers in diversifying into upstream and downstream operations at home and abroad, to insulate itself from the fluctuations of the crude oil market and to establish captive markets for its petroleum products; in the words of one analyst 'to achieve both maximum returns and flexibility in the market.' Crude oil now accounts for less than half of oil exports. In under 10 years, the state-owned Kuwait Petroleum Company (KPC) has established itself as a major force in world petroleum prospecting, production, refining, transport, wholesaling and retailing. Its strategic aim is to achieve maximum flexibility through the vertical and horizontal integration of its oil industry, from oilfield to petrol station. At home, refining capacity has been increased. An expensive eight-year programme to expand the output of domestic refineries was completed in February 1989 when the expanded Mina Abdullah refinery came on stream, following similar modernisation work at Shuaiba and Mina al-Ahmadi. Kuwait's refined products capacity is now over 700,000 b/d.

Abroad, in the early 1980s KPC acquired oilfield leases in the USA, the Gulf of Mexico, the North Sea and the South China Sea. Since 1983, KPC subsidiary Kuwait Petroleum International (KPI), has built up a European refining and retailing network marketing own-brand Q8 products, which now account for over 20 per cent of the Danish market and smaller shares of the markets in Holland, Belgium, Britain, Italy, Sweden and Luxembourg. It owns 4,800 petrol stations in these seven countries. KPI is now planning to fill gaps in its European network, and expand in the United States and the Far East. Following a visit to Kuwait of the Thai energy minister in March 1989, KPC companies will explore for oil in Thailand and introduce Q8 petrol stations there. KPC plans to use Thailand as a springboard into Burma, Cambodia, Laos and Vietnam. Another KPC subsidiary, Petrochemical Industries Company (PIC), has moved into overseas petrochemicals, buying a 25 per cent stake in West Germany's Hoechst, and has set up joint ventures in Turkey, Tunisia and China for fertiliser production using Kuwaiti ammonia. So far, Kuwait has not spent heavily on massive domestic industrial projects, but PIC is going ahead with a plan to build a polypropylene plant at Shuaiba. This may become the basis of a domestic plastics industry. Despite its worldwide expansion, KPC profits have been disappointing. Some of its foreign acquisitions have been loss-makers.

Creating confidence

The government continues its efforts to rebuild the financial system hit by the 1982 collapse of the unofficial stock market, the Souk al-Manakh. Its collapse after casino-like trading based on post-dated cheques, wiped $90bn off the inflated value of local shares leaving the six local banks with some $15bn of unsecured loans. The web of indebtedness and non-performing loans paralysed the local economy and came close to crippling Kuwait's financial system. The government intervened in 1986 with an $8bn bail-out programme which has had some success in salvaging Kuwait's international reputation with a minimum of economic dislocation at home. By May 1989 83 per cent of the debt had been settled, but critics say the government has been too generous to debtors. By pumping in low interest or interest free deposits to the five worst afflicted banks, the government has ensured their survival.

The government has tried, but so far failed, to regenerate activity in the near-moribund official stock market. It has bought stock (it now owns half the shares) and issued treasury bonds. But stock market activity has failed to take off. The imposing new stock market, built of expensive black marble, has a deserted air. Despite a brief flurry of activity after the Gulf War cease-fire, the market moved barely 1 per cent in 1988. The authorities are now planning further measures to breathe life into the exchange. There are plans to list selected foreign companies and in May 1989 the government said banks and investment companies would be allowed to issue unit trusts. Further deregulation of the stock market was planned during 1989.

Stimulating the economy

The stock market depression was symptomatic of a general sluggishness of the private sector. Office rents are down; occupancy rates in 5-star hotels in 1988 averaged only 42 per cent. A 1970s-style boom, which many Kuwaitis gambled would follow the Gulf War cease-fire has not occurred. Neither Iraq nor Iran has enough foreign exchange to finance reconstruction on the massive scale once envisaged. Some Kuwaiti companies built up supplies of machinery and building materials in anticipation of the non-existent boom. They are being forced to liquidate stocks at a loss.

The best news for the local economy was the announcement in January 1989 that the government is definitely going ahead with a $13bn project to build a new town at Subiya, across the bay from Kuwait, and a $1bn causeway to reach it. The scheme will bring a much-needed boost to all sectors of the local economy over the next decade. There were also plans – considered unviable by some – to develop irrigated agriculture in the north.

One problem faced by local businessmen is the small size of the local market. Hopes that the Gulf Co-operation Council might stimulate business have not been fulfilled.

The debate over Kuwait's long-term future continued. Should Kuwait place most reliance on foreign investments and their revenues, or should it spend its oil surpluses building up domestic industry, even if it might not be so profitable? The arguments are more pressing since suitable (i.e. white collar) employment is being found for only half the 2500 graduates each year. Two groups of specialists, one Kuwaiti, the other from the Massachusetts Institute of Technology, have been working on the problem since 1985. Their strategy is to provide the management jobs Kuwaitis prefer by building upon the state's existing international financial operation. One part of the plan is to employ Kuwaitis as managers in industries and businesses owned or part-owned by Kuwait. This is already happening on a modest scale. The Kuwait Hotel Company manages Kuwaiti-owned hotels in Kuwait, Morocco, Tunisia and Egypt; the aim is to build it up into a multinational hotel company based in Kuwait. The Kuwait Real Estate Consortium manages Kuwaiti property holdings in the Arab world. Other parts of the plan are for Kuwait to become a major financial services centre, directing regional oil wealth to investments around the world, offering expertise in commodity trade finance, Islamic banking and shipping. This would require extensive reform of the rules and regulations of the financial sector. Such reforms are being incorporated into the 1990-95 Five-Year Plan now being drawn up.

The government is planning to reduce its central role in the economy, in order both to reactivate the private sector by involving it in areas previously closed to it and also to reduce the government's financial burden. It is encouraging the private sector to invest more in construction projects previously handled by the state, such as housing. It wants banks to get involved in the marketing end of the oil business, and it is thinking of privatising certain state-owned companies.

Kuwaitisation

Many Kuwaitis bemoan the fact that they are a minority in their own state, though without the huge army of immigrant labour modern Kuwait could never have been built. According to the 1985 census, Kuwaiti nationals made up only 40 per cent of the total 1.7m population and only 17 per cent of the labour force. The expatriate population was mainly Arab (63 per cent) but its Asian component (35 per cent) was growing faster. The government is aiming for a majority of Kuwaitis in the total population by 2000, and for the 'Kuwaitisation' of as much of the workforce as possible. This will not be easy. Kuwaitis expect higher wages than foreigners. They are also reluctant to do menial jobs in construction and manufacturing, preferring office work in the public sector, where over nine out of 10 are employed; a law passed in the early 1960s guarantees every citizen a job in government. White-collar jobs presently held by skilled foreigners are increasingly being taken over by educated, young Kuwaitis. The foreign workforce may decline naturally with the end of the main phase of construction, but large numbers of immigrant manual and domestic workers and skilled professionals will almost certainly remain.

Diwaniya democracy

Kuwait is one of the most sophisticated and politically active societies in the Arab world. It was also one of the most liberal; alone among its Gulf neighbours it briefly experimented with a limited form of democracy, giving some 50,000 Kuwaiti males the right to vote for a national assembly. The assembly quickly built a reputation for vocal criticism of government policy. In 1986, when the emirate's fears over the Gulf War and internal subversion were at their height, the assembly was suspended and press censorship imposed. Kuwait's liberal reputation suffered. But while many Kuwaitis want the assembly back, others blame it for making ill-judged and divisive attacks on the widely respected ruling family at a time of crisis. There are no definite signs that the assembly will be restored (although in early 1990 it was being mooted), or that censorship will be lifted.

With the assembly gone, *diwaniyas* – traditional males-only gatherings – have become the most important channel for political and social grievances to be aired. *Diwaniyas* are held at all levels of society. In theory they are open to anyone. Government officials, including ministers, both hold their own *diwaniyas* and attend those of others. They allow the rulers to test the political temperature and try out new ideas, but they are a limited form of popular democracy.

COUNTRY PROFILE

Political system: Executive power rests with the Emir, who appoints a prime minister and then other ministers on his recommendations; the 50-member National Assembly was dissolved in July 1986.

Population: Some 40 per cent of the estimated 1.84m residents (1987) are Kuwaitis, with the remainder coming from other Arab states (63 per cent), and South and East Asia (35 per cent). The Kuwaiti population is growing at around 4.7 per cent per annum. There is a small European community.

Main cities/towns: Kuwait City (capital, population 167,768), Hawalli (944,439), Ahmadi (305,065), Jahra (280,032), Salmiya, Abraq, Kheetan, Farawaniya. (Figures from 1985 census.)

Language: The official language is Arabic. English is widely spoken, especially in business circles.

Media
Press: Five Arabic dailies; two English-language dailies; weekly and monthly magazines in both languages; international newspapers available.
Radio: Service in Arabic, Farsi, English, Urdu.
Television: Channel 1 (Arabic language) 1700 – 2400; Channel 2 (English language), 1900 – 2300; times change during month of Ramadan.

Domestic economy: Mainly dependent on oil (1987 production: 996,000 b/d) and natural gas (206bn cubic feet p.a.). Proven oil reserves estimated at about 92bn barrels, sufficient to last 100 years at present rates of extraction. Slower rate of growth in the period since 1980 than in the boom which lasted up to 1977, despite efforts of the state-owned Kuwait Petroleum Corporation to diversify away from total dependence on crude oil exports by investing in downstream hydrocarbons activities both at home (exploration, production, pipelines, refineries and petrochemical engineering) and abroad (petrol stations, refineries and storage and distribution depots). The economy had begun to recover from the 1982 collapse of the unofficial Souk al-Manakh stock exchange, but was badly affected by the fall in oil prices in 1985-86 and in 1987-88 by the intensification of the war in the Gulf which severely damaged its oil terminals.

Employment: 1985 census figures show that expatriates meet 81 per cent of manpower needs and Kuwaitis only 19 per cent (expatriates of 120 nationalities form 60 per cent of the population).

External trade
Exports: Oil and oil products make up over 85 per cent of exports. Most of the remainder normally represents re-exports, fertilisers and certain raw industrial materials. Main destinations are Japan (21.5% of the 1986 total), Italy (15.1%), Netherlands (11.5%), USA (4.9%), Denmark (4.7%) and German FR (1.3%).
Imports: Main imports are machinery and transport equipment (43.6% of the 1984 total), manufactures classified by material (22.1%), other manufactures (14.5%) and food and live animals (12.2%). Main sources are Japan (29.7% of 1986 total), USA (13.5%), German FR (11.3%), UK (10.6%), Italy (8.2%), and France (7.1%).

Agriculture: Negligible contribution to GDP (0.5 per cent in 1988). Intensive techniques, such as irrigation and hydroponics, are needed to overcome lack of water, soil deficiencies, climatic conditions, pests (red spider). But the sector receives government encouragement, including financial incentives, in an effort to reduce import bills. A major water recycling programme is under construction, although it will not be complete until 2000.

Industry: Main industries are oil and natural gas, fertilisers, construction and building equipment. Nationalised oil industry is controlled by Kuwait Petroleum Corporation (KPC), which also has subsidiaries in other sectors and which produced 107m barrels of refined petroleum products in 1985 as well as 91,000 b/d of liquified petroleum gas. Industrial areas are located in Shuaiba, Mina Abdulla (both south Kuwait) and Shuwaikh (including the port). Petrochemical Industries Company (PIC), a subsidiary of KPC, is the leading industrial organisation, and is involved in the production of fertilisers and other chemicals (salt, chlorine, caustic soda, chlorine gas). Other light industry includes food processing, textiles, furniture, paper, mineral and metallic products, cement, sulphur processing, detergents and construction materials. Efforts to foster growth of non-oil industries have been hindered by small size of domestic market and lack of natural resources other than hydrocarbons.

Stock exchange: After the introduction of new rules in the wake of the 1982 crash, investment slowly resumed. A government-backed programme was introduced to deal with the problem of non-performing loans after 1986.

Membership of international organisations: Arab League, FAO, G-77, GATT, GCC, IAEA, ICAO, IDA, IFAD, IFC, ILO, IMF, IMO, INTELSAT, INTERPOL, IPU, Islamic Conference Organisation, Islamic Development Bank, ITU, NAM, OAPEC, OPEC, UN, UNESCO, UPU, WFTU, WHO, WMO, World Bank, WTO.

Time: GMT +3 hrs.

Climate: Hot, arid summers (May to Oct), with average maximum shade temperatures 38C, and more comfortable winters (Nov to Mar), with some rainfall and cold winds.

Entry requirements
Passport: Required by all (with isolated exceptions).
Visa: Visas or entry permits required for all except nationals of GCC countries; some nationals require an NOC (no objection certificate) issued by the Ministry of the Interior. Most travellers will require sponsorship to obtain a visa, which is not issued at entry point to those arriving without one. In recent years, visa requirements have periodically changed at short notice. Travellers should check with embassies and consulates on the latest position, preferably before departure.
Prohibited entry: Anyone holding an Israeli passport or visa, and anyone with an Israeli customs stamp in his/her passport.
Currency: No restrictions on currency imports or exports.
Customs: Household and personal effects, tobacco and perfume may be imported without restriction; 4 per cent duty may apply on some items; International Car Certificate (Triptyque) required for cars imported for personal use for three-month period, otherwise guarantor or guarantee required.
Prohibited imports: Alcoholic beverages and any materials used for making them (i.e. hops, malt extracts, wine kits); pornographic and/or politically subversive materials; pork products in any form; all goods of Israeli or South African origin. Prohibition of alcoholic beverages strictly enforced.

Health precautions
Mandatory: International vaccination certificates for cholera and yellow fever if travelling from infected areas.
Advisable: TAB vaccination.

Representation overseas: Abu Dhabi, Aden, Algiers, Amman, Ankara, Athens, Baghdad, Beijing, Beirut, Belgrade, Bonn, Brasilia, Cairo, Caracas, Copenhagen, Dakar, Damascus, Dhaka, Doha, Islamabad, Jeddah, Khartoum, Kuala Lumpur, London, Madrid, Manama, Mogadishu, Moscow, Muscat, Nairobi, New Delhi, Ottawa, Paris, Rabat, Rome, Sana'a, Sliema (Malta), Tehran, Tokyo, Tripoli, Tunis, Vienna, Washington DC.

Cease-fire brings new problems

Kuwaitis were understandably relieved when the Gulf War cease-fire came into effect in August 1988. The 1989 independence celebrations were the first for eight years not to be tarnished by fears that the conflict might spill over into Kuwait itself. The war had had a potentially destabilising effect on both state and society. The export of Iran's Shi'ite revolution was particularly feared since Shi'ites – both nationals and expatriates – make up a substantial portion of the population. The Gulf War presented the small and highly vulnerable emirate with a dilemma that was impossible to solve: how to support Arab Iraq without attracting opprobrium from Iran. Only with difficulty could it manage the balancing act; shrewd Kuwaiti diplomacy brought the emirate through the conflict relatively unscathed.

In retaliation for Kuwait's financial and logistical support for Iraq (which was allowed to import goods and weapons through Kuwait) Iran launched a campaign of intimidation against Kuwait, attacked its shipping and fired missiles at its territory. In 1986-87, when Iranian intimidation was at its worst, Kuwait skilfully brought in both superpowers to protect its oil shipments. It was this that eventually led to the presence of a massive Western fleet in the Gulf. At the time, Kuwaiti critics of the policy feared it would drag Kuwait deeper into the conflict; with hindsight the policy was a great success. Eleven Kuwaiti tankers were put under the American flag in 1987; they began to return to Kuwaiti registration in early 1989.

The end of the fighting brought new problems: how to rebuild cordial links wth Iran and adjust to the presence of an assertive and heavily armed Iraq determined to play a dominant role in the Gulf region (and elsewhere in the Arab world). Kuwait's relations with Iran improved steadily; full diplomatic relations were re-established in September 1988. Iranian external attacks, and it appears subversion too, have come to an end. Iran welcomed Kuwait's condemnation of British author Salman Rushdie's book, *The Satanic Verses*. But progress towards a normalisation of relations was interrupted by the hijack and illegal detention by Iran of two small Kuwaiti vessels in the northern Gulf in March and May 1989.

Ironically it was Iraq, whose war effort Kuwait had supported with generous loans, that was again seen as the greater threat to Kuwait. Soon after the cease-fire, Iraq revived its old border dispute with Kuwait, and put pressure on Kuwait to hand over, or lease, two islands, Bubiyan and Warbah, which Iraq said it needed to secure access to the Khor Abdallah channel and the port of Umm Qasr. There was at least one illegal Iraqi military incursion into Kuwait. The question dominated a visit to Baghdad by the Kuwaiti crown prince in February 1989. Iraqi President Saddam Hussein is said to have offered an equal area of mainland Iraqi territory in exchange for the islands. Kuwait, alive to the fact that Iraq has never renounced a historical claim to Kuwait, refused. The construction of a 30 km causeway across Kuwait bay to Subiya where a new town of 100,000 inhabitants is to be built, will underline Kuwaiti sovereignty over this oil-rich but underpopulated area. Despite the border row, Iraq has apparently agreed to supply large quantities of fresh water to Kuwait for domestic, industrial and agricultural use.

Kuwait pressed ahead with defence spending. In 1987 it announced a rearmament programme costing $5.5bn over 10 years. On the shopping list were fighters, minesweepers, tanks and helicopters. It is buying $1.9bn worth of F-18 jets and Maverick missiles from the United States, despite initial opposition from congress, and has put in an order for 200 M1A1 battle tanks, anti-tank missiles and multiple rocket systems. Pro-Western by nature, Kuwait is non-aligned by necessity and good sense. Its policy of keeping good relations with both superpowers paid off during the darkest days of the Gulf War. Kuwait has recently bought some of its arms from the Soviet Union, partly to show the US congress (which turned down its request for Stinger missiles in 1987) that if arms requests continue to be refused or offered only on stiff conditions, then Kuwait will go elsewhere.

BUSINESS GUIDE

Air access
National airline: Kuwait Airways (KU), PO Box 394, Safat, Kuwait (tel: 710788; tx: 23036).
Other airlines: Aeroflot, Air France, Air India, Air Lanka, Alyemda, Austrian, Balkan, Bangladesh Biman, British Airways, Cyprus Airways, Czechoslovak, Egyptair, Gulf Air, Iberia, Iraqi Airways, Japan, JAT, KLM, Lufthansa, Libyan Arab, MALÉV, Middle East, Olympic Airways, Pakistan, Philippine, Royal Air Maroc, Royal Jordanian, Saudia, Sudan Airways, Swissair, Syrian Arab, Tarom, THY Turkish, Tunis Air, Yemenia.
International airport: Kuwait International (Code: KWI), 16 km from city, with hotel accommodation for transit passengers, duty-free shop, buffet, restaurant, bank, hotel reservations, post office, shops; no airport tax. A second runway is under construction, but plans for a new airport have been shelved.

Surface access
Main ports: Mina Abdulla (oil); Mina Ahmadi (mainly oil); Shuweikh (container, ro-ro, slipway); Shuaiba (container, ro-ro, lo-lo).
Overland access: Excellent roads from Saudi Arabian and Iraqi borders.

Hotels: Most visitors are business travellers. Seven five-star hotels have swimming pools and exercise/gymnasium facilities. Although there is no shortage of accommodation, it is still advisable to book hotels well in advance. Small tip for porters is customary. 15 per cent service is usually charged.

Credit cards: Major credit cards (American Express, Diners' Club, Visa and Mastercard or Access) accepted at all hotels and many restaurants and shops.

Car hire: Locally approved/inspected international driving licence (valid for duration of entry permit) and insurance with Gulf Insurance Company or Kuwait Insurance Company are essential. Driving is on the right-hand side of the road.

City transport
Taxis: Both private and shared taxis, which are orange and operate on set routes, are available. Taxis are more expensive from hotel ranks. There is a standard taxi fare in Kuwait City and drivers do not expect a tip. If hiring a taxi for a day or half-day agree the fare in advance. Call taxis are reliable and widely used.

National transport
Road: Network of 3,600 km of good paved roads and expressways link all towns. Considerable expansion of the urban network. There were 543,615 vehicles in 1986.
Buses: Nationwide service operated by Kuwait Transport Company, generally rated good and inexpensive.

Public holidays
Fixed dates: New Year's Day, 25 Feb (National Day).
Variable dates: Lailat al-Mi'raj, Lailat Bara'a, Beginning of Ramadan, Lailat al-Qadar, Eid al-Fitr, Beginning of Hajj, Waqfa, Eid al-Adha, Hijra New Year's Day, Prophet's Birthday.

Working hours: Fri is weekly holiday, when government offices, embassies and banks are closed. Office hours liable to change during month of Ramadan.
Business: 0830 – 1230, 1630 – 2000.
Government: Winter: (Sat-Wed) 0730 – 1330, (Thu) 0730 – 1130; summer: (Sat-Wed) 0700 – 1300, (Thu) 0700 – 1100; Ramadan: (Sat-Thu) 0900 – 1300.
Banking: 0800 – 1200; (Ramadan) 0900 – 1230.
Shops: 0830 – 1230, 1630 – 2100.

Social customs: Strong Islamic influence. Visitors should take care to respect religious customs. Avoid using the term 'Mohammedan' and eating, drinking or smoking in front of Muslims in daylight hours during Ramadan. Never refuse refreshment when first offered. Alcohol is forbidden officially. Be careful when taking photographs.

Telecommunications
Telephone: Dialling code for Kuwait: IDD access code + 965, followed by subscriber's number (no area code). In 1986 there were 235,188 telephone subscribers. There is a mobile cellular system in operation. Kuwait has a ground station to link with the Arabsat telecommunications system. IDD access code for international calls from Kuwait is 00 followed by the country code etc.
Telex: Available at good hotels (surcharge may be added). Telex code 496, country code KT.
Fax: Services available.
Telegram: Facilities available 24 hours at Ministy of Post and Telegraph Main Offices, Abdulla Al Salem Square.

Postal service: PO Box numbers are widely used as few areas have house-to-house deliveries. There are efficient private carrier services. Avoid surface mail for international post. Parcels from abroad are inspected by customs officials on arrival. ZIP codes are now used.

Banking
Banking centre: Kuwait City.
Central bank: Central Bank of Kuwait.
Head offices: Al Ahli Bank of Kuwait KSC, Bank of Kuwait & the Middle East KSC, Burgan Bank, Commercial Bank of Kuwait SAK, Gulf Bank KSC, National Bank of Kuwait, Industrial Bank of Kuwait.
Foreign branches: Bank of Bahrain & Kuwait.

Investment companies: Al-Ahli Investment Co KSC, Arab European Financial Management Co, Arab Trust Co KSC, Kuwait Fund for Arab Economic Development, Financial Group of Kuwait, Kuwait Foreign Trading Contracting & Investment Co SAK, Kuwait International Investment Co SAK, Kuwait Investment Co SAK, Kuwait Real Estate Bank KSC.

Trade fairs: Contact the Information Centre, PO Box 26626, Safat.

Electricity supply: 240 V AC; plug fittings normally three-pin flat type (British).

Weights and measures: Metric system (local units also in use).

Representation in capital: Afghanistan, Algeria, Austria, Bahrain, Bangladesh, Belgium, Brazil, Bulgaria, Canada, China PR, Cuba, Czechoslovakia, Denmark, Egypt, Finland, France, Gabon, German DR, German FR, Greece, Hungary, India, Indonesia, Iran, Iraq, Italy, Japan, Jordan, Rep of Korea, Lebanon, Libya, Malaysia, Mauritania, Morocco, Netherlands, Niger, Nigeria, Norway, Oman, Pakistan, Philippines, Poland, Qatar, Romania, Saudi Arabia, Senegal, Somalia,

BUSINESS DIRECTORY

Spain, Sri Lanka, Sudan, Sweden, Switzerland, Syria, Thailand, Tunisia, Turkey, UAE, UK, USA, USSR, Venezuela, Yemen AR, Yemen PDR, Yugoslavia.

Hotels: Kuwait City
Bristol, Fahad al-Salem Street, P.O. Box: 3531 (tel: 439281/4; tx: 22061).

Carlton, Safat, P.O. Box: 3492 (tel: 2423171; tx: 22064).

Carlton Tower, Fahad al-Salem Street, Safat (tel: 2452740; tx: 44837 cartow).

Hala House, Safat, P.O. Box: 823 (tel: 2553150; tx: 22989 halatel; fax: 2552208).

Holiday Inn, Farwaniya (near airport), P.O. Box: 18544 (tel: 732100; tx: 46460 holinn; fax: 4732020).

Kuwait International, Bnaid Al Gar, 13060 Safat (tel: 25330000; tx: 22039; fax: 2563797).

Marriott, Shuwaikh Port, Safat, P.O. Box: 24285 (tel: 835344; tx: 23596).

Meridien, Fahd al Salem Street, Safat, P.O. Box: 26302 (tel: 2455550; tx: 44458).

Messilah Beach, Al-Bidaa Road, Safat (on the beach) (tel: 5624111; tx: 22215 memotel).

Oasis, Mubarak al Kabir Street, Safat (tel: 2465489; tx: 46121).

Phoenicia, Fahd al-Salem Street, Safat, P.O. Box: 4666 (tel: 421051; tx: 2062 phoenicia kt; fax: 2424402).

Plaza, Safat, P.O. Box: 4450 (tel: 2438800; tx: 30247).

Pullman, Safat, P.O. Box: 21192 (tel: 5634200/4790; tx: 4858).

Regency Palace, Salmiyah (suburbs), P.O. Box: 1139 (tel: 5728000; tx: 46082; fax: 5723109).

SAS, Safat (on the beach), P.O. Box: 26199 (tel: 5657000; tx: 44306; fax: 5652999).

Sahara, al-Sour Street, P.O. Box: 20461 (tel: 2424121; tx: 22102; fax: 2424132).

Sheraton, Fahd al-Salem Street, Safat, P.O. Box: 5902 (tel: 2422055; tx: 22016, 22434; fax: 2448032).

Spring Continental, Salmiyah, P.O. Box: 4187 (tel: 5742655; tx: 2175).

Business advice and information
Industrial Bank of Kuwait, PO Box 3146, Safat (tel: 2457661; tx: 22469).

Kuwait Chamber of Commerce and Industry, Ali Al Salem Street, PO Box 775, 13008 Safat, Kuwait (tel: 2433854/5/6; tx: 22198).

Kuwait Foreign Trading, Contracting & Investment Co, PO Box 5665, Kuwait 13057 (tel: 2449031; tx: 22021).

Kuwait International Fair Co, PO Box 656, Safat (tel: 2458560/1/2/3/4/5; tx: 22540 infair).

Kuwait National Industries Co, PO Box 417, Safat (tel: 815466, 812455; tx: 22165).

Kuwait Petroleum Corp, PO Box 26565, Safat (tel: 2455455; tx: 44874).

Ministry of Commerce and Industry, PO Box 2944, 13030 Safat (tel: 2463600; tx: 22682).

Ministry of Information, PO Box 193, Hilali Street, Murbarak al Kabir Street, Safat (tel: 2415301).

Ministry of Planning Central Statistical Office, PO Box 26188, 13122 Safat (tel: 2428200; tx: 22468 takhtet).

Banking
Central Bank of Kuwait (central bank), PO Box 526, Abdulla Al-Salem Street, Safat 13006 (tel: 2449200; tx: 22173).

Alahli Bank of Kuwait KSC, PO Box 1387 Safat-13014, Mubarak Al Kabir St (tel: 2411101/2; tx: 22067 ahlibank kt; fax: 2424557).

Bank of Bahrain and Kuwait BSC, PO Box 24396 Safat-13104, Commercial Area 9, Ahmed Al Jaber St (tel: 2417140; tx: 23320 kt; fax: 2440937).

Bank of Kuwait & the Middle East, PO Box 71 Safat-13001, Mubarak Al Kabir St (tel: 2459771; tx: 22045 kt; fax: 2461430).

Burgan Bank SAK, PO Box 5389 Safat-13054, Ahmed Al Jaber St (tel: 2439000; tx: 22767 burgank; fax: 2461148).

Commercial Bank of Kuwait SAK, PO Box 2861 Safat-13029, Mubarak Al Kabir St (tel: 2411001; tx: 22004 cbk kt; fax: 2450150).

Gulf Bank KSC, PO Box 3200 Safat-13032, Raed Centre next to Awadi Tower (tel: 2449501; tx: 22001 gulfbank kt).

National Bank of Kuwait SAK, PO Box 95 Safat-13001, Ali Awadi Tower, Ahmed Al Jaber St (tel: 2422011; tx: 22451 natbank kt; fax: 2464156).

Travel information

Gulf Automobile Association, PO Box 827, Safat (tel: 2423864, 2438640).

Laos

Richard Vokes

Despite severe economic problems resulting from a second successive year of drought, Laos' reform-minded leadership is now firmly in control and steering the country through its difficult transition to a more market oriented and open economy. Following the publication of a liberal Foreign Investment Law in July 1988, foreign investment has slowly begun to flow in. Much of this has come from Thailand and has been made possible by the dramatic improvement in relations between Laos and Thailand since Chatichai Choonhaven became Thai prime minister following Thailand's general election in July 1988. Domestically, there is now only limited resistance to the government's 'new thinking' or *jin tanakhan mai*, Laos' own version of *perestroika*, although the delay in promulgating the new constitution provides evidence of some continued differences among the leadership, especially over the direction of future political reforms.

Restructuring foreign relations

The surprise replacement of General Prem Tinsulanond by General Chatichai Choonhaven as the new Thai prime minister in July 1988 has had a profound impact on Lao-Thai relations. Almost immediately, Chatichai launched an initiative on Indo-China aimed at transforming the region from 'a battlefield into a market place', linked to Thailand's rapidly growing economy. Such a move was made possible by the adoption of a positive attitude towards foreign investment by both Vietnam and Laos. The Thais have seen Laos very much as the gateway to this new Indo-China 'market place', since – despite the often strained relations between the two governments since the communists came to power in Laos – extensive trade, both legal and illegal, has continued. Laos, in particular, remains heavily dependent on trade with Thailand and, in spite of efforts over the last 14 years to develop closer ties with Vietnam, much of Laos' import and export trade passes through Thailand. Thus, in spite of the border conflict that had led to a serious outbreak of fighting between the two countries at the end of 1987, Chatichai visited Vientiane for talks with Lao Prime Minister and Party General Secretary Kaysone Poumviharn in November 1988, designed to lay the foundations for improved relations. This was the first visit by a Thai prime minister for a decade. Shortly after the visit, the Thais cut by more than half the list of so-called 'strategic goods' banned from export to Laos and agreed to open more transit points for trade along the border. The underlying importance of economic factors to this new relationship was underscored by the subsequent exchange of high level trade missions in early 1989. The Thai minister of commerce led a 70-member mission to Laos in January, while Lao Minister of Trade, Phao Bounnaphon, led a delegation to Bangkok in April.

Laos also received high level missions from West Germany and France in line with its attempts to improve ties with the West and attract increased aid and foreign investment. Within the socialist bloc,

KEY FACTS

Official title: Lao People's Democratic Republic

Head of state: President Phoumi Vongvichit

Ruling party: Lao People's Revolutionary Party; General Secretary: Kaysone Poumviharn

Capital: Vientiane

Official Languages: Lao (Vietnamese and French also spoken)

Currency: New Kip (= 100 at)

Exchange rate: new kip 580 per $ (Nov 1989)

Area: 236,800 sq km

Population: 3.90m (1988) (Official figures: UN estimate 4.4m)

GDP real growth: 8%* (1984)

GNP per capita: $141 (1988)

Labour force: 1.85m (1988)

Trade balance: –$130m* (1988)

Foreign debt: $740m (1987)

* estimated figure

both political and economic relations with China continued to improve following the resumption of full diplomatic ties in December 1987.

Socialism with foreign investment

As part of the shift to a more market oriented economy, the Lao People's Revolutionary Party (LPRP) is now also committed to a more open economy supported by foreign investment. In July 1988 the 'Law on Foreign Investment in the Lao People's Democratic Republic' was published. This allows foreign participation in all major sectors of the economy and in the development of infrastructure, either through wholly-owned enterprises or joint ventures with state, collective or private enterprises. In the case of joint ventures, foreign capital must constitute a minimum of 30 per cent of the total. The law offers a range of tax and profit incentives to such enterprises as well as guarantees on profit repatriation and against nationalisation, although contracts will be limited to 15 years in the first instance. Priority is to be given to exports and the production of import substitutes, and such enterprises will receive more generous incentives. The main attractions for foreign investors lie in the country's relatively untapped natural resource and primary production base and its tourist potential.

Already by early 1989 over 100 foreign firms had sent representatives to Laos to explore investment opportunities. Most of these were from Thailand. Joint ventures with Thai companies have already been agreed in the fields of timber and wood products, trade, textiles and in the production of *tuk-tuks*, the motorised 3-wheel taxis common in Bangkok. A number of agreements have also been concluded with Thai, French and Australian firms for mineral and gemstone mining. Significantly, in April 1989, the Thai Military Bank became the first foreign bank to open an office in Vientiane since 1975. Although its activities will be geared primarily towards supporting the investment and trade activities of Thai companies in Laos, it forms part of the programme of restructuring of Laos' weak and underdeveloped banking system that is necessary to support the economic reforms. Both the World Bank and the Asian Development Bank are assisting this programme.

Despite the initial burst of interest shown by foreign investors, the growth of foreign investment in the short-term is constrained by the country's extremely poor and underdeveloped infrastructure, a poorly trained workforce and an inefficient bureaucracy. The other key constraint is the lack of a constitution and hence legal framework to govern the operations of foreign investors.

Short-term economic setbacks

Whatever the likely long-term impact of the major changes in economic policy over the last two years, the economy in 1989 was seriously affected by a second successive year of drought which affected the 1988-89 crop season. This time it was the most fertile provinces in the south that were most affected, with paddy production there being cut by half. Overall, national production in 1988 fell to 1.03m tonnes, a fall of 15 per cent compared to 1987, which itself was 20 per cent below the record harvest of 1986. The government has estimated a shortfall equivalent to 225,000 tonnes of milled rice for 1989, more than twice the shortfall in 1988, and is again seeking food aid. Special emphasis is again being given in 1989 to the production of secondary food crops and dry season production, although these will have only a marginal impact on total food availability.

The drought was also one of the factors affecting the production and export to Thailand of electricity, the country's key hard currency export. In August 1988, the level of the Nam Ngum Dam was at its lowest for 16 years, forcing a further cutback in production and exports in 1988 on top of that experienced in 1987. The other principal factor affecting the level of water in the dam was the now serious problem of deforestation. Concern over deforestation was also one of the factors behind the government's introduction of a ban on log exports in 1988. However, the primary purpose of the ban was to try and raise the value added from the forestry sector through the production and export of processed wood rather than raw logs. This has proven difficult given the low quality and high price of processed wood products in Laos and has led to a further significant cut in export earnings. In early

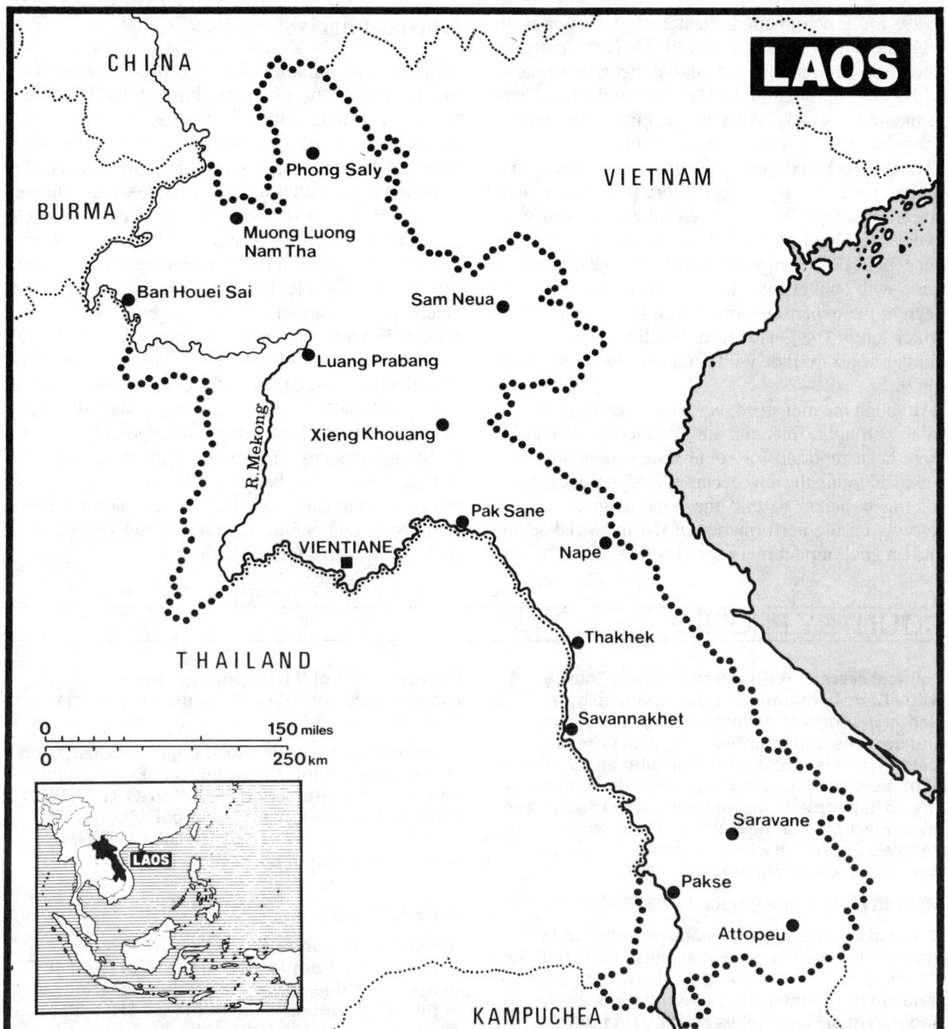

KEY INDICATORS

	Unit	1984	1985	1986	1987	1988
Population	'000	4,320	3,580	3,680	3,780	–
GDP ($m current)	$m	887	835	1,008	654	504
GNP per capita	$m	205	233	274	173	–
Exports						
Electricity	$m	25	26	30	12	11
Tin	$m	1	1	3	2	2
Timber & wood products	$m	5	11	8	33	30
Coffee	$m	9	9	9	9	7
Gypsum	$m	1	1	2	2	1
Exports	$m	44.0	54	55	62	58
Imports	$m	162	193	186	216	188
Balance of trade	$m	–118	–139	–131	–154	–130
Current account	$m	–83	–92	–89	–116	–91
External debt	$m	468	548	599	740	–
Debt service ratio	%	3.4	4.0	2.3	2.0	–
Exchange rate	kip per $	35	35	35	350	–

* estimated figure

1989, the government indicated its willingness to relax the ban on exports of logs for Thai firms entering into joint production and marketing arrangements with Laos, although in the longer term the Lao Government is clearly keen to encourage the establishment of joint ventures within Laos.

The cutbacks in export earnings from electricity and timber have led to a new effort to increase hard currency exports. This has been quite successful, with exports to Thailand, excluding electricity, up by 125 per cent in 1988 compared with 1987. The increase in trade with convertible currency countries has partly been at the expense of trade with the socialist bloc, which until 1987 had been steadily increasing. A similar boost in trade with Thailand was expected in 1989.

Although the industrial sector has been less affected by the drought, financial and production difficulties have been reported for some state economic enterprises adopting the new decentralised managerial and pricing policies, so that the overall impact of the reforms on the performance of the industrial sector and on government revenues is still uncertain.

Glasnost arrives

With the 'new thinking' and the economic restructuring this implies now well established, the LPRP has begun to initiate political changes. In June 1988, elections were held for district-level people's committees; the first elections to take place since the founding of the LPDR in 1975. These were followed in October and November by provincial and municipal elections, while in March 1989 the first national elections, for seats in the Supreme People's Assembly, the country's parliament, were held. In all these elections, the candidates have to be vetted by the Laotian Front for National Construction, the principal mass organisation of the LPRP. They are therefore effectively managed and offer only a limited choice among officially backed candidates. They do, however, provide the leadership with a possible way of removing officials still imbued with the 'old thinking', and have also been seen as an attempt by the government to encourage the view of Laos as a more stable and democratic country, thereby making it more attractive to foreign investors.

Outlook

In spite of the successes of 1988-89, further major improvements in ties with Western nations will depend on the outcome of the attempts in mid-1989 to find a solution to the Cambodia problem. With Vietnam promising to withdraw all her troops from Cambodia by the end of September, with or without an agreement, there is pressure on all parties to reach a solution. The shift towards greater trade with convertible currency countries will continue as economic liberalisation proceeds. This can be expected to put some strain on Laos' relations with the USSR and other Eastern bloc countries, which have already been undermined by alleged inefficiencies in the use of aid by Laos. However, old suspicions and concerns about Thailand's economic domination of Laos are likely to resurface as trade with, and foreign investment from, Thailand increases. This, coupled with the serious domestic economic problems, may help to strengthen resistance to the reform programme, but this is unlikely to be sufficient to seriously weaken the position of the present leadership.

COUNTRY PROFILE

Political system: A Supreme People's Council, with about 45 members, was appointed by a National Congress of People's Representatives and governs under a president, who is head of state. There is also a prime minister and a cabinet of ministers and councillors. Local government is vested in people's committees at town and village level. Elections for the National Congress of People's Representatives (Supreme People's Assembly) were held in March 1989.

Main city: Vientiane (population 377,400 in 1987).

Domestic economy: The economy is primarily agricultural, mainly at a subsistence level, but bombing, defoliation and guerilla warfare have reduced the country's agricultural productivity and potential. Laos relies on funding from the Soviet bloc, Sweden and Vietnam. Foreign aid and loans totalled $100m of which $40m are from the Western bloc.

External trade
Exports: Exports totalled an approximate US$59m in 1987. Main exports are electricity, timber (including cork), coffee, foodstuffs and tin concentrates, and increasingly, chemical products. Main destinations are USSR (typically 40 per cent of exports), Malaysia, China PR, Japan, Singapore, Thailand and UK.
Imports: Imports totalled an approximate US$217m in 1987. Main imports are rice and other food, mineral fuels, machinery and transport equipment. Main sources are USSR, Thailand, Japan, Singapore and France.

Agriculture: Agriculture employs 73 per cent of the workforce, and produces 55 per cent of GDP (1987). Rice, the main crop, is cultivated in irrigated lowland paddies and on drier hill farms. Other crops include maize, sweet potatoes, cassava, pulses, soybeans, groundnuts, fruit, vegetables, sugar-cane, coffee, tobacco and cotton. Livestock raised includes cattle, buffaloes, pigs, goats and poultry. Freshwater fish provide an important source of protein.

Industry: Industry provides seven per cent of GDP. Production of tin concentrates is the main industrial activity. Small-scale manufacturing industries produce beer, cigarettes, detergents, rubber footwear, bricks, plywood, matches, salt, animal feed, veterinary products, handicrafts, alcoholic beverages and soft drinks.

Mining: Tin is currently exploited. Other minerals present include lead, potash, high grade iron ore, zinc and various precious metals. Coal and petroleum are of potential value.

Energy: Some of the output from the hydroelectric dam at Nam Nhum is exported to Thailand.

Membership of international organisations: Asian Development Bank, Colombo Plan, EC trade and co-operation agreement, ESCAP, FAO, G-77, IBRD, ICAO, IDA, IFAD, ILO, IMF, Interpol, IPU, IRC, ITU, NAM, UN, UNCTAD, UNESCO, UPU, WFTU, WHO, WMO, World Bank, WTO.

Time: GMT + 7 hrs.

Climate: Warm and sultry by day, with temperatures from 14-28°C in Nov-Mar to 38-40°C in Apr, but cooler in the mountains, especially at night. Monsoon rains fall from May-Oct with as much as 250mm rainfall/month.

Entry requirements
Passport: Required by all.
Visa: Required by all. The current attitude to casual visitors is unclear. Access is difficult, but is getting easier.

Health precautions
Mandatory: Vaccination certificates for cholera if travelling from an infected area.
Advisable: Anti-malarial precautions.

Representation overseas: Bangkok, Canberra, Hanoi, Havana, London, Moscow, New Delhi, Phnom Penh, Tokyo, Washington.

Air access
National airline: Lao Aviation.
Other airlines: Aeroflot, Hang Khong Vietnam, Thai Airways.
Tax: Airport tax of $1 on departure.
International airport: Vattay (VTE), 4 km from Vientiane.

Surface access: Entry via the rail line running from Bangkok is subject to special permission.

Hotels: At times advance bookings are not accepted.

Car hire: Arrangements are generally made through hotels.

City transport
Taxis: Taxis are available in Vientiane, but often operate along certain routes in manner of buses. Individual hire may require negotiation. Tipping is discouraged.

National transport: Travel outside Vientiane is restricted for foreigners and is generally not allowed without an official escort.
Road: There are adequate roads between main centres. About 21,000 km (1,300 km surfaced).
Buses: There are bus services between main centres.
Water: River transport is important, especially on the Mekong River.

Working hours
Government and business: (Mon-Fri) 0730/0800 – 1130, 1400 – 1700; (Sat) 0800 – 1200.
Banking: (Mon-Fri) 0830 – 1130, 1400 – 1530.

Telecommunications
Telephone: Dialling for Laos is via the International Operator. There were about 5,000 telephones in 1987.

Banking:
Central bank: Banque d'État de la RDP Lao.
State commercial bank: Banque du Commerce Extérieur Lao.
A restructuring of the banking system was undertaken in 1989.

Weights and measures: Metric system (local units also in use).

Representation in capital: Australia, Bulgaria, Burma, China PR, Cuba, Czechoslovakia, Egypt, France, German DR, German FR, Hungary, India, Indonesia, Japan, Kampuchea, Korea PDR, Malaysia, Mongolia, Philippines, Poland, Sweden, Thailand, UK, USA, USSR, Vietnam.

Hotels: Vientiane
Anou, 99 Rue due Bon Constellation, 379 Samsenthai Road.

Imperial, Samsenthai Road (tel: 2881).

Inter, 1 Chou Anou Road.

International, Anon Street (tel: 2408).

Lane Xang, Quai Fa Ngum (tel: 2672).

Peace, 67A Luang Prabang Road.

Sayas, Chou Anon Road (tel: 2514).

Business advice and information
Department of Foreign Trade, Ministry of Industry and Commerce, Vientiane.

Lao Import-Export Company, 43-47 Lanexang Road, Vientiane.

State Committee for the Plan, Vientiane (tel: 2400).

Banking
Banque d'Etat de la RDP Lao (central bank), Rue Yonnet, BP 19, Vientiane (tel: 2000; tx: 4304).

Banque du Commerce Extérieur Lao, BP 19, Rue Pangkham, Vientiane (tel: 2646; tx: 4301).

Lebanon

Michael Jansen

After more than a year of delays, parliamentarians elected a new president in November 1989. But only 17 days after Syrian-backed moderate René Muawed was elected, he was assassinated and succeeded almost immediately by another pro-Syrian, Elias Hrawi.

The failure of the Lebanese parliament and political leadership to elect a new president during 1988/89 had brought the country close to cantonisation. During over 14 years of civil war the president had been the symbol of Lebanese unity and the focus of a genuine popular desire to reconcile and re-establish the Lebanese state. In the absence of a president in the palace at Baabda basic state structures partitioned themselves and the fragments of the Lebanese state were held together only by a tenuous economic interdependence.

The crisis over presidency

After quelling the fighting between pro- and anti-Iranian Shi'a factions in Beirut's southern suburbs in May 1988, Syria worked hard through July and the first half of August with US mediators, shuttling between Beirut and Damascus, to achieve a consensus candidate among the half dozen proposed. Christian political leaders made it clear that they would veto three of these: former President (1970-76) Suleiman Frangie, army commander Michael Aoun and Raymond Edde, a nationally respected figure living in exile in Paris since 1976. The candidate provisionally agreed on was Michel Daher, a lack-lustre Maronite deputy prepared to accept political reform and the continuation of an important role for Syria in Lebanese affairs. This choice reflected the external and internal pressures exerted on the political factions rather than the will of the people who preferred Edde. Then on 16 August, two days before parliament was due to meet, Frangie declared himself a candidate and Damascus switched its support from Daher to Frangie, a longstanding ally and close personal friend of Syrian President Hafex al-Assad. Syria's Muslim and Druze allies in West Beirut were compelled to back Frangie, though they abominated the man as much as mainly Maronite East Beirut. The leaders of the Phalange Party of sitting President Amin Gemayel and the radical 'Lebanese Forces' militia believed Frangie's objective in standing for the presidency was, if elected, to take revenge for the murder of his son and his family by the Phalangist militia then led by Samir Geagea who later became commander of the 'Lebanese Forces'. Ordinary Lebanese objected to Frangie because they considered him responsible for creating the climate of corruption and instability which led to civil war in 1975.

Insistence on Frangie, for whatever motive, was a major Syrian miscalculation. On 18 August only 38 deputies turned up at the villa situated near the museum crossing on the 'Green Line', 13 short of a quorum. While the Syrians saw that deputies living in West Beirut attended, the 'Lebanese Forces' physically prevented those living in East Beirut from going. Failure to elect a president in August created an atmosphere of uncertainty which made it impossible to reach a consensus by 22 September when the speaker of parliament, Hussein al-Husseini, scheduled a last-chance poll. Though the Syrians switched their support back to Daher, and the Americans backed them up, the Christian deputies could not be persuaded to attend the session. A last-minute reconciliation between Aoun and Geagea strengthened the Maronite resolve to boycott Daher. A few minutes

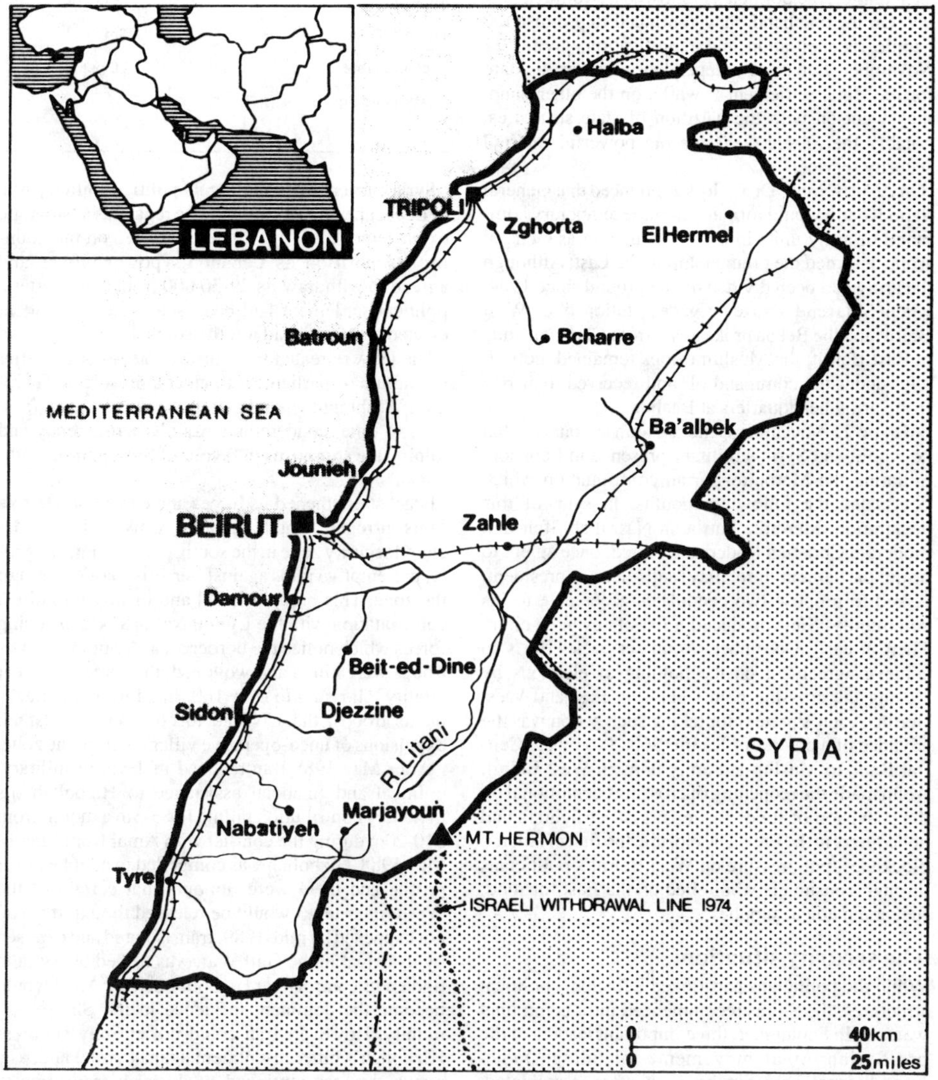

before midnight on 22 September Gemayel appointed Aoun interim prime minister of a six-man military cabinet.

The Muslims promptly condemned the military cabinet and reaffirmed their support for the caretaker government formed by Dr Selim al-Hoss after the assassination of Prime Minister Rachid Karami in 1987. Partition of the two rival cabinets followed: the three Muslim officers – a Sunni, a Shia and a Druze – appointed by Aoun refused to serve and all the Christians based in East Beirut withdrew from Dr al-Hoss' government and handed over their portfolios to Aoun's appointees.

This partitionist situation was reinforced on 18 October when parliament, on the expiry of the term of Speaker al-Husseini, failed to meet and replace him. This time only 22 deputies were present, 17 short of a quorum, the Christians refusing to attend on the pretext that the venue was 'unsafe'. The deputies who met extended al-Husseini's tenure, though the Chris-

KEY FACTS

Official title: Republic of Lebanon

Head of state: President Elias Hrawi

Head of government: Prime Minister Dr Selim El-Hoss

Capital: Beirut

Official Languages: Arabic

Currency: Lebanese Pound (L£) = 100 piastres

Exchange rate: L£413 per $ (Nov 1989)

Area: 10,452 sq km

Population: 2.8m (1988)

Inflation: 700% (1986–87)

Trade balance: –$1,540m* (1986)

* estimated figure

tians argued that the constitution requires that the office should pass to the oldest surviving deputy and insisted that Khazem al-Khalil, a Shi'a like al-Husseini and thus qualified, should succeed. In effect, Lebanon had two executive prime ministers and two parliaments with two speakers.

Thereafter, the politicians followed a two-track policy. On the one hand, they continued to promise the people reconciliation and reunification, giving 22 November, Lebanese independence day, as the date for the 'final settlement'; while, on the other hand, they went on with the partition of state structures, notably the *Sûreté Générale* the powerful internal security service.

On 22 November Dr al-Hoss announced that General Sami Khatib would formally assume temporary command of army units in the West as long as General Aoun assumed the premiership in the East. Although the army had been divided on the ground since 1984, when Shia and Druze brigades stationed in West Beirut and the Bekaa broke away from Christian units in East Beirut, the Muslim troops remained, notionally, under the command of, and received their pay from, army headquarters at Baabda.

Also on 22 November General Aoun announced that he would end Syria's military presence in Lebanon, initiating the Maronite–Syrian confrontation which dominated the following months. In spite of this declaration, Maronite Patriarch Nasrullah Sfeir and moderate Christian leaders attempted, once again, to bring parliament into play and elect a president. Asked to propose two candidates acceptable to his community, Sfeir produced a list of seven or eight. Then in early December, Aoun set back efforts to achieve agreement on the holding of elections by closing the crossing points between East and West Beirut. Aoun's pretext for taking such action was the advance of a credit of LL34m ($60,000) by the Central Bank to Muslim-led army units in West Beirut. Aoun made the reopening of crossings conditional on the provision of funds for repairs of jet fighters and helicopters in the East. The closure, the first of Aoun's blockades, put pressure on West Beirut's politicians through consumers who relied on the East for food, fuel and medical supplies.

The security situation

While the politicians partitioned Lebanon's political structures, the Shi'as recommenced intra-communal warfare. In September, three important leaders of the pro-Syrian Amal movement, including Daoud Daoud, Amal's chief in the south, were assassinated, allegedly by the pro-Iranian Hizbollah. On 22 November clashes between the two factions erupted in the southern suburbs after a failed attempt on the lives of three Hizbollah officials. The Syrians suppressed the fighting but not factional resentment and on 31 December 1988 Amal and Hizbollah renewed their battle on two fronts, in the southern suburbs and in the 'Iklim al-Tuffah', the 'Region of the Apples', adjacent to Israel's security zone in the south. The fighting raged until, on 30 January 1989, Damascus and Tehran compelled their clients to cease fire and agree to a settlement which left Amal in control of the south while Hizbollah was given the freedom to organise and carry out political activity in the southern suburbs. Further fighting broke out there once again in June until the Syrian and Iranian foreign ministers imposed a cease-fire.

External intervention

The 1988-89 crisis brought four new external players – Iraq, the Arab League, France and the USSR – into the Lebanese game. These additional players complicated the already complicated situation being managed, or mismanaged, by the countries with longstanding involvement – Syria followed by the USA, Israel and Iran.

KEY INDICATORS

	Unit	1982	1983	1984	1985	1986
Population	'000	2,640*	2,640*	2,640*	2,670*	2,710*
Exports	L£m	3,450*	3,129*	3,788*	7,910*	19,185
Imports	L£m	14,895*	15,350*	18,687	33,491*	78,275
Trade balance	$m	–2,415*	–2,698*	–2,149*	–1,558*	–1,540*
Exchange rate	L£ per $	4.74	4.53	6.51	16.42	38.37

* estimated figure

Syria carried on defending its politico-military role in Lebanon which it continued to regard as a 'strategic extension of Syria'. Damascus insisted on maintaining its position as Lebanon's power-broker and agreed to withdraw its 29-30,000 troops only after a political settlement had been achieved, a president elected and Israel had left the south.

The USA retreated from the diplomatic scene after the failure of parliament to elect a president. Thereafter Washington merely let its views be known from time to time, condemning Aoun's belligerency and joining the consensus in favour of Syrian troop withdrawals.

Israel strengthened its forces and expanded the role of its surrogate 'South Lebanon Army' in its self-declared security zone in the south, conducting, allegedly, pre-emptive raids against 'terrorist bases' north of the zone. This brought Israel and its ally into direct confrontation with the United Nations peacekeeping forces which police the border area. A number of UN troops were killed and wounded, the force's 11-year fatality toll rising to over 160. In addition to attacks on its troops, the UN had to contend with Israeli expulsions of unco-operative villagers from the zone.

After May 1988 Iran reduced its level of military, political and financial assistance to Hizbollah, its monetary contribution falling to $3-5m a month from $10-20m during the conflict with Amal from March-May 1988. Hizbollah was compelled to adopt a lower profile and there were rumours that certain of the Western hostages would be released though this had not happened by mid-1989. Iran, isolated and exposed by its defeat in the Gulf War, was forced by Syria to tighten its control over Hizbollah so that Amal would become the dominant influence in the Shi'a community. However, Iran could not entirely suppress Hizbollah's bitter resentment against Amal and, as a result, the two continued to skirmish from time to time. For Syria, Amal's ascendancy was essential because Damascus was unwilling to accept the consequences of unchecked Hizbollah raids into northern Israel and the group's campaign to establish a Shi'a 'Islamic State' or canton in Lebanon.

Iraq began open intervention on the side of the anti-Syrian Maronite militants almost as soon as the ceasefire was secured in the Gulf War in August 1988. The Iraqis supplied arms to both the 'Lebanese Forces' and General Aoun, after his elevation to the premiership of the Maronite enclave. Among the weapons supplied were T-54 tanks, multi-barrelled rocket launchers, armoured personnel carriers, heavy artillery and mortars and, reportedly, half a dozen Soviet-designed Frog-7 surface-to-surface missiles which, the Maronites boasted, could hit Damascus. Iraq announced ending these supplies only on 1 July 1989 – once the Maronites had a large stockpile.

Iraq and Libya, which usually backs Syria, were the only two Arab states to recognise the Aoun government. And Iraq gave Aoun crucial diplomatic support for his campaign to 'liberate' Lebanon from Syria: at the Arab summit meeting in May 1989 Iraq was able to convince the Arab leaders that Syrian withdrawal was an urgent necessity. Iraq's motive for intervening in Lebanese affairs was to punish Syria for its support for Iran in the Gulf War.

The second new player was the Arab League, which had given its cover to the original Syrian force which had intervened in Lebanon in 1976, then with the object of restraining the Muslim coalition on the brink of defeating the Maronite militias. In 1988-89 the League set up two committees, its 'Sub-committee on Lebanon' to draw up a political settlement plan and a commission composed of the kings of Morocco and Saudi Arabia and the president of Algeria to implement League decisions by December 1989.

The third outside power to inject itself, once again, into the Lebanese fray was France, the 'mother' country of the Maronites who deny that they are Arabs and resist absorption into the Arab world. France, the former colonial power and author of the undemocratic constitution which gave the Maronites political primacy, as expected, attempted to intervene on the Maronite side, failed and retreated.

The fourth power to intervene was the Soviet Union which put pressure on both Iraq and Syria, in July 1989, to end their conflict by proxy in Lebanon and to co-operate with the Arab League mediation effort.

Hopeful developments

For the war-weary Lebanese, early 1989 offered the hope that Lebanon's problems might be resolved. In January Aoun had eased his blockade at the 'Green Line' crossing points, allowing some supplies of foodstuffs and fuel to cross into West Beirut. The Amal–Hizbollah truce imposed by Syria and Iran at the end of January had eased tension in West Beirut and the South. From 30 January to 1 February the Arab League sub-committee, chaired by the Kuwaiti foreign minister, held separate meetings in Tunis with the two rival prime ministers to try to find a way out of the impasse over the election of a president. At this time the Druze leader, Walid Jumblatt, announced that he would permit the return to the Druze-held Chouf mountains southeast of Beirut of some 300,000 Maronite refugees. Aoun welcomed Jumblatt's offer and he set up a committee to make arrangements for the return of the refugees. Meanwhile the tactical alliance between Aoun and Geagea was crumbling, weakening the solid anti-Syrian Maronite front. Clashes between army soldiers and 'Lebanese Front' militiamen were sparked off by an incident on the ski slopes at the Faraya resort, northeast of Beirut. These clashes reflected the tension between Aoun, who was prepared to compromise in order to secure Lebanese unity, and the 'Lebanese Forces' which favoured partition. On 15 February, in 24 hours of fierce fighting which left 70 people dead in East Beirut, Aoun's army took over 'Lebanese Force' positions in and around the city and in the mountains to the north. During the fighting the West Beirut army commander, General Sami Khatib, offered to send in reinforcements to help Aoun, but was brusquely rebuffed. Aoun's military victory was quickly followed up by a political conclave at Bkirke, the seat of the Maronite Patriarch, which put pressure on the 'Lebanese Forces' militiamen to take to their barracks and cede to Aoun's cabinet their administrative role in the enclave. On 24 February the 'Forces' also handed over control of Beirut port, from which the milita had earned some $250,000 a month in revenues.

These developments prompted progress on the negotiating front. In the third week of February the Arab League sub-committee met with the leaders of the main religious communities, three Christians and three Muslims. The sub-committee discussed with the clerics elections for a president and parliamentary speaker, the appointment of a new prime minster and security arrangements, including the creation of a six nation Arab deterrent force and demilitarisation of the greater Beirut area. The clerics reached agreement on these issues and on reforms involving the redistribution of political power among the communities. To many Lebanese Aoun then appeared as a credible presidential candidate who, if elected, would act on the League's reform package.

Blockades and bombardments

March began badly: on the fifth Jumblatt withdrew his offer to allow Maronites to return to their homes in the Chouf, claiming that he had proof of 'Lebanese Forces' involvement in the murder of a close aide on 9 February. Then on 8 March Aoun made a bid to assert his authority, as the 'legitimate' prime minister of Lebanon, outside the Maronite enclave, which comprises about 30 per cent of the country. On that day a patrol boat belonging to Aoun's 'navy' diverted a ship going to the illegal port at Jiyye, south of Beirut, which is run by the Druze militia. Having ordered the revival of the Lebanese military's 'air-sea joint operations room', Aoun had decided to use it to institute a coastal blockade against Jiyye and the two illegal ports at Khalde and Ouzai operated by the Shi'a Amal movement. He also ordered his naval patrol boats to carry out search and customs operations against ships bound for the legal Muslim ports of Tyre and Sidon, where Aoun's 'navy' was backed up by the Israelis, and for Tripoli in the north.

The Druze retaliated by attacking Maronite forces at Souk al-Gharb in the mountains above the international airport, which Aoun closed down on 12 March by threatening to shell it. Early on the morning of 14 March, Druze gunners lobbed a few shells into Beirut port, causing no casualties. Aoun and the 'Lebanese Forces', released from their barracks, responded with a massive artillery barrage on West Beirut at precisely the time people were going to work and children to school: 46 were killed, over 100 wounded, among them many children. That bombardment, which earned 14 March the name of 'Black Tuesday', destroyed Muslim and Druze confidence in, and support for, Aoun who reiterated his intention of 'liberating' Lebanon from the Syrians, even if he had to destroy Beirut to do so. The Muslim militias, backed by Syria, replied by bombarding East Beirut and instituting a land blockade of the Maronite enclave and a sea blockade of its ports. Intense exchanges of fire went on for eight weeks, re-establishing the Muslim–Christian confrontation lines drawn at the begining of the civil war in 1975. As a result of forceful and continuous pressure exerted by the Arab League backed up by the USA, France and the USSR, the exchanges wound down in May though sporadic shelling continued thereafter. During the lulls in the shelling, Lebanese enterprise reasserted itself and citizens of both sides of the city flooded across the 'Green Line' to go to jobs on the other side, to trade or, simply, to visit friends and family. Not even the assassination on 16 May by car bomb of Shaikh Hassan al-Khaled, the moderate Sunni Muslim Mufti of the Republic, who had worked hard for reconciliation, daunted the will of ordinary Lebanese to retain a practical and profitable but notional unity.

Although, at the end of May, it blocked the adoption by an extraordinary Arab Summit, held at Casablanca, of a comprehensive settlement plan, Syria subsequently found it had to co-operate with a new League committee made up of the kings of Morocco and Saudi Arabia and the president of Algeria, a combination Damascus could not oppose. This committee acted energetically on the imposition of a ceasefire, the lifting of blockades, the election of a president, the redistribution of political power and the withdrawal of foreign forces. League and Soviet pressure secured Iraq's decision on 1 July to stop arming the Maronites, a decision which could have ended the Iraqi-Syrian confrontation in Lebanon but did not. The Syrians promptly demanded the removal of the Frog-7 missiles Iraq had supposedly supplied to the Maronites (though there was some doubt that the missiles had indeed been delivered) and began their own naval search operation against all ships plying to ports in the enclave, the first open involvement by Syrian forces in the conflict. The Maronites refused to yield the missiles and to co-operate with the Arab League proposals designed to break the deadlock on the presidential election issue. This produced further artillery exchanges and sharpened the confrontation between the Maronite leadership and Damascus.

The economic situation

Lebanon maintained some sort of national unity through its economic infrastructure and the interdependence of the two halves of Beirut. The Lebanese had to co-operate in running their public utilities like water and electricity as the installations which provide these essential services are on both sides of the dividing line. Aoun's initial blockade of the West in December caused shortages there of food, fuel and medicines, all imported goods, and the East was severely starved of supplies of everything after the Muslim–Syrian alliance instituted its land and sea blockade in March. Food and fuel supplies fell very low and electricity was cut to one hour a day in the East while the West began to develop self-sufficiency by importing goods through Syria or Muslim-controlled ports in the north and south. Thus even the economic bond between the two sides began to weaken.

The political crisis disrupted very promising economic developments. Following the fall of the Lebanese lira from a rate of LL5 to the dollar in 1984 to LL500 to the dollar at the end of 1987, the exchange rate began to improve in the first quarter of 1988. Much of this improvement was due to the growth of exports of manufactured goods which had doubled since 1984 to reach a value of $700m in 1988, the bulk of the goods being produced in East Beirut. The last two months for which figures were available in 1988 showed a substantial improvement from month to month: from LL8,814m ($19.8m) in September to LL10,588 ($23.7m) in October. The blockades and bombardments made an end to these exports. The first six weeks of exchanges cost $250m in damaged buildings and manufacturing plant. During March-May 1989 more than 300 people were killed, 1,000 injured and 100,000 people left, including an estimated 4,000 professionals. At the end of May losses were estimated at $1.2bn, including exports, and one Lebanese economist estimated that the cost of rebuilding the country would be $25-30bn.

Banking continued to thrive, deposits remaining at a level of $44bn (US$3bn). Commercial banks survived on their offshore activities, though the Intra-Investment Bank and its European-based subsidiaries were under investigation for their financial practices and the Lebanese Central Bank refused to underwrite Intra's external operations. The Central Bank, under its governor Edmund Naim, who had remained steadfastly neutral during the years of crisis and sustained the administrative structures on both sides of the line, came under extreme political pressure from General Aoun in December 1988, though this was thereafter relaxed. The General learnt that he could not afford to weaken the Bank because its operations kept the economy and the governmental structures afloat: in 1974 government expenditure amounted to only 14.9 per cent of the GDP while in 1988 this had increased to 60.1 per cent. From being a free-enterprise, laissez-faire economy, Lebanon had developed a state-run economy, dependent on the Central Bank, which printed and dispensed money, and on its sole remaining source of revenue, the ports, over which the two sides fought so desperately.

Outlook

Not only had life for the ordinary Lebanese become more dangerous, it had also become more expensive, forcing those who could afford to leave to emigrate. The overall consumer price index continued to rise until it topped a rate of over 700 per cent in 1986-87: a basket of essential goods which cost LL60 in 1985 had risen to LL741 in 1988.

Though the prospects for Lebanon, political and economic, were worse than ever, the Lebanese who remained in Lebanon continued to soldier on, and to hope for an improvement, because they had no choice.

COUNTRY PROFILE

Political system: The president is elected for six years and is not eligible for immediate re-election. The constitution allows him executive power assisted by a prime minister and council of ministers, chosen by the president and answerable to but not necessarily from, the 99-member National Assembly. (Political parties have been allocated on the basis of religion to correspond to the communities as shown in a 1932 census which may now not be an accurate reflection of the proportions.)

Main cities/towns: Beirut (population estimated 1.2m), Tripoli.

Language: Arabic is the official language, but English and French are widely spoken in business circles.

Media
Press: Many newspapers in Arabic and, under normal circumstances, also a few French and English.
Radio: Some 1.5m radio receivers in use. Several stations, most community-based.
Television: Some 450,000 TV sets. There are three TV stations, some commercial.

Domestic economy: Banking and commerce have traditionally been the largest contributors to GDP, but have been badly affected by the civil war in recent years. Under dire conditions agriculture and industry have been sustained, but reconstruction has made huge demands on resources. Inflation became especially severe in 1985-1987, leading to a sharp depreciation of the currency.

Agriculture: 30 per cent of total area is cultivated arable or utilised forest. Main crops: wheat, barley, maize, vegetables, potatoes, fruit, olives, tobacco.

Industry: Mainly food processing, textiles, cement, chemicals, oil refining.

COUNTRY PROFILE

Membership of international organisations: Arab League, CCC, FAO, G-77, IAEA, IBRD, ICAO, IDA, IFAD, IFC, ILO, IMO, IMF, INTELSAT, International Wheat Council, INTERPOL, IPU, Islamic Conference Organisation, Islamic Development Bank, ITU, NAM, UN, UNESCO, UPU, WFTU, WHO, WMO, WSG, WTO.

Time: GMT +2 hrs; (GMT +3 Jun-Oct).

Climate: Summer: temperature range 20°C to 30°C, Beirut average 27°C; the coastal region is humid. Winter: coastal region range 10°C to 16°C; colder inland. Snow is usual on mountains. Most rain falls between Nov and Mar.

Entry requirements
Passport: Required by all except certain nationals of neighbouring states.
Visa: Required by all with isolated exceptions.
Prohibited entry: Israeli nationals and travellers holding evidence of visits to Israel.

Health precautions
Mandatory: Vaccination certificates for yellow fever if travelling from an infected area.

Representation overseas: Abidjan, Abu Dhabi, Accra, Algiers, Amman, Ankara, Athens, Baghdad, Beijing, Belgrade, Berlin, Berne, Bogota, Bonn, Brasilia, Brussels, Bucharest, Buenos Aires, Canberra, Caracas, Conakry, Dakar, Doha, Freetown, Havana, Islamabad, Jeddah, Khartoum, Kuwait City, Lagos, London, Madrid, Mexico City, Monrovia, Montevideo, Moscow, New Delhi, Nicosia, Ottawa, Paris, Prague, Rabat, Rome, Sana'a, Stockholm, Tehran, Tokyo, Tripoli, Tunis, Vatican City, Vienna, Washington.

Air access
National airline: Middle East Airlines (MEA).
Other airlines: (services subject to confirmation) Aeroflot, Air Afrique, Balkan, CSA, Iberia, Nigerian Airways, Tunis Air.
International airport: Beirut International (Code: BEY), 16 km from city, with duty-free shop.

Surface access: The main road and rail link to/from Syria is subject to suspension.
Main ports: Beirut, Tripoli, Saida (Sidon), Jounieh.

Hotels: A 10-12 per cent service charge is usually added to the bill, with additional tipping extra.

Electricity supply: 110V or 220V AC, 50 cycles. Supply is subject to fluctuations and blackouts.

National transport
Road: Over 7,000 km, including two international motorways (total length 570 km). Security situation critical throughout the country.
Rail: One line along coast, one east from Beirut. Services subject to suspension.

Public holidays
Fixed dates: 1 Jan, 9 Feb (St Maron), 1 May, 15 Aug (Assumption), 22 Nov (Independence), 25 Dec.
Variable dates: Good Friday, Easter Monday, Orthodox Good Friday, Eid al-Fitr, Eid al-Adha, First of Muharram, Ashura, Prophet's Birthday.

Working hours
Government and business: (Mon-Thu) 0800/0830 – 1230/1300/1400; (Fri) 0800/0830 – 1230/1300. (Winter working hours are longer than summer.)
Banking: (Mon-Fri) 0830 – 1230; (Sat) 0830 – 1200.

Social customs: The life of the people has been totally disrupted by prolonged civil war.

Telecommunications
Telephone: Dialling code for Lebanon: IDD access code + 961, followed by area code (1 for Grand Beirut), followed by subscriber's number. 320,000 telephones in 1979, but for current practical purposes this figure is somewhat academic.

Banking: In recent years many banks have transferred their operations or closed down.
Central bank: Banque du Liban.
Main centre: Beirut.

Weights and measures: Metric system (local units may also be used).

Representation in capital: *mainly in East Beirut* Algeria, Argentina, Australia, Austria, Bahrain, Bangladesh, Belgium, Bolivia, Brazil, Bulgaria, Chad, Chile, China PR, Colombia, Cuba, Czechoslovakia, Finland, France, German DR, German FR, Greece, Haiti, Hungary, India, Iran, Iraq, Ireland, Italy, Japan, Jordan, Korea PDR, Republic of Korea, Kuwait, Liberia, Libya, Mexico, Morocco, Nigeria, Norway, Oman, Pakistan, Paraguay, Poland, Portugal, Qatar, Romania, Saudi Arabia, Spain, Sweden, Switzerland, Tunisia, Turkey, UAE, UK, USA, USSR, Uruguay, Vatican, Venezuela, Yemen AR, Yemen PDR, Yugoslavia.

Hotels: Beirut
Alexandre, Adib Ishak Street, P.O. Box: 11-6861 (tel: 325739; tx: 43422; fax: 338454).

Bristol, Mme Curie Street, Verdun, P.O. Box: 11-1493 (tel: 351400; tx: 20710).

Carlton, Avenue General de Gaulle/Chouran Street, P.O. Box: 3420 (tel: 800240; tx: 20396).

Cavalier, Hamra Street, P.O. Box: 11-9452 (tel: 807830; tx: 22051; fax: 863445).

Coral Beach, Jenah Avenue Chiyah, P.O. Box: 5370 (tel: 317200-2; tx: 20510).

Pavillon, Hamra Street, P.O. Box: 8242 (tel: 350160, 352301; tx: 22562).

Royal Garden, Emile Edde Street, P.O. Box: 113-5872 (tel: 856951/363801; tx: 22565).

Summerland, Jenah, P.O. Box: 11-7791 (tel: 304830; tx: 22622).

Business advice and information
Assocation of Lebanese Industrialists, PO Box 1520, Chamber of Commerce and Industry Bldg, Justinian St, Beirut.

Banque du Liban, PO Box 5544, Masraf Loubnane Street, Beirut (tel: 341230; tx: 20744, 21033).

Beirut Chamber of Commerce and Industry, PO Box 11801, Beirut (tel: 353390; tx: 22269).

Board for Foreign Economic Relations, PO Box 11-5344, Beirut (tel: 483391/5; tx: 20898).

Middle East Airlines, PO Box 206, Beirut International Airport (tx: 20820).

Ministry of Tourism, Beirut.

National Council of Tourism in Lebanon (CNTL), PO Box 11-5344, Central Bank St, Beirut (tel: 340940; tx: 20898).

Trans Mediterranean Airways, PO Box 11-3018, Beirut International Airport (tx: 20637).

Lesotho

Anthony Robinson

This land-locked mountainous kingdom surrounded by South Africa is poised for the biggest transformation in its history as work begins on the $2bn Lesotho Highlands Water Scheme, the biggest civil engineering project currently under way in Africa. The project, which will not be completed until the second decade of the twenty-first century, will supply water to the mines, factories and cities of South Africa's industrial heartland. But it will also make Lesotho self-sufficient in electricity, provide a steady income and open up the country for tourism, irrigated farming and light industry. In the meantime the country remains under the control of a military government three years after Chief Leabua Jonathan was deposed by a military coup reinforced by a South African economic blockade.

The economic blockade which South Africa mounted at the main road and rail crossing points from the Republic into Lesotho in January 1986 underlined the inevitably limited sovereignty of a small country which depends on its big neighbour for its transport links, power supply, food and most other consumer goods and derives the bulk of its income from the remittances of up to 150,000 migrant workers employed across the border. But the blockade, followed in short order by the military coup which

KEY FACTS

Official title: Kingdom of Lesotho
Head of state: King Moshoeshoe II
Capital: Maseru
Official Languages: Sesotho, English
Currency: Maloti (M) = 100 lisente
Exchange rate: M2.65 per $ (Nov 1989)
Area: 30,355 sq km
Population: 1.7m (1988)
GDP per capita: $234 (1987)
GDP real growth: –5.8% (1986)
Inflation: 12.2% (1987)
Trade balance: –$409m (1987)
Foreign debt: $241m (1987)

brought General Metsing Lekhanya to power at the head of a military junta, also marked the beginning of a new and more realistic relationship with South Africa without which Pretoria would not have been prepared to give the green light for the long-mooted Highlands water scheme.

But by 1989 it was no longer clear that the military government was able to offer the kind of stable political environment usually required by investors asked to lock up funds in long-term projects with long-term payoffs. In July the Attorney General's office announced that it was setting up a judicial enquiry into the death of a 20-year-old student just before Christmas 1988. Reports circulating in the capital, Maseru, indicated that General Lekhanya was involved in the fatal shooting incident at Maseru's agricultural college. His bodyguard, an army sergeant, subsequently confessed to having fired the fatal shot. But the Attorney General subsequently found a serious conflict between the statement of the bodyguard and evidence given by a key witness.

In July Charles Mofeli, leader of the opposition United Democratic Party, invited Mrs Margaret Thatcher and President Quett Masire of Botswana to intervene diplomatically to help bring peace to the kingdom. He claimed that Lesotho had reached a constitutional crisis which could only be resolved through an all-party conference called to organise a return to civilian rule.

The rumours surrounding the shooting incident reflected growing popular disenchantment with a military government which was greeted by cheering crowds after the coup which toppled Chief Jonathan. But initial hopes that military rule would soon give way to the restoration of constitutional civilian government failed to materialise, despite significant developments such as the return of prominent political exiles like Ntsu Mokhehle, the 72-year-old leader of the once banned Basutoland Congress Party, and members of the Lesotho Liberation Army which waged a sporadic guerrilla war against Chief Jonathan.

Shortly after the coup in 1986 General Lekhanya announced that he was suspending political activity to allow for the 'healing of wounds inflicted on the people through party political rivalries'. The military junta also returned more authority, at least on paper, to King Moshoeshoe II, who could be the rallying point for a future civilian regime. Meanwhile the military government has also been accused of corruption on a wide scale by political opponents, like Majara Molapo a former top civil servant and relative of Chief Jonathan. In July 1989 Mr Malapo publicly announced his readiness to testify about alleged large-scale embezzlement, drug dealing and other offences involving members of the junta and associates.

Highland Water Scheme progress

Despite the growing political uncertainty, however, preliminary work on the access roads and basic infrastructure for the Highlands Water Scheme is well under way and most of the vital funding appears to be in place. The aim of the scheme is to divert up to 70 cubic metres per second of water northward through a series of tunnels to join up with the Vaal river. At present that water flows uselessly southward to the sea. To reverse the flow a 170 metre high concrete dam will be built at Katse so that the water can flow backwards along a series of tunnels and hydro-power stations.

The southern access road to the construction sites high in the mountains has been completed together with the road bridge just downstream of the main Katse dam site on the Malibamatso river. The main northern access road, being built by LTA of South Africa, is due for completion by November 1990. By that time work should be ready to proceed on the main dams and the tunnels. These will take water northward to the Ash river and beyond to the Vaal Dam south of Johannesburg. From there it will feed into the water system of Africa's most powerful industrial-urban complex.

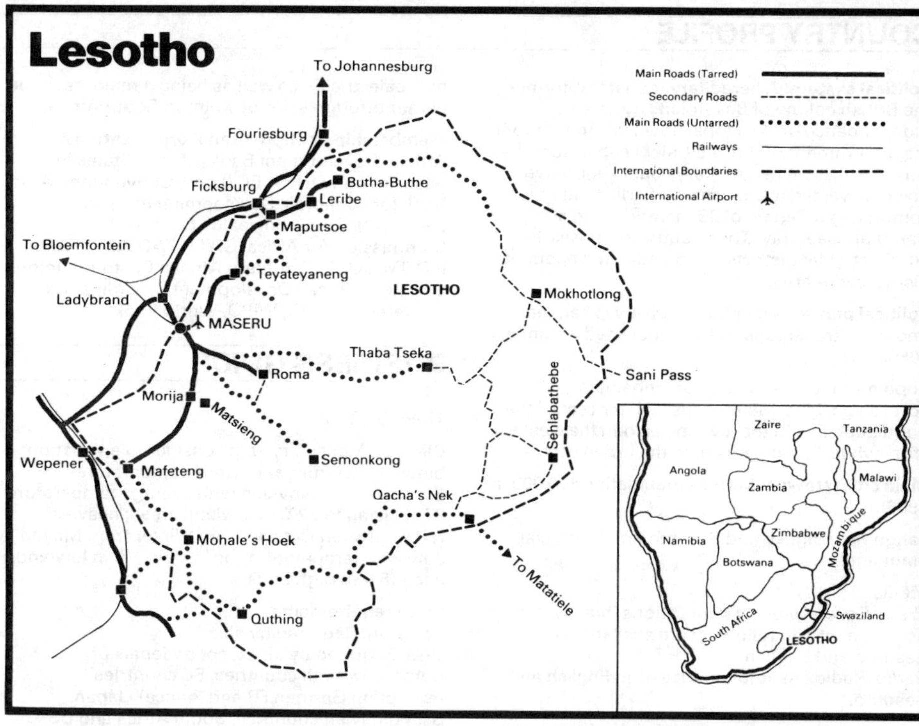

The first deliveries of water should start flowing in 1995. Soon afterwards power should also start to flow from the first stage of the linked hydro-power scheme with an initial 70 MW being delivered from the Muela hydro-power station 45 km north of the Katse dam. By the end of phase two in 2008 this will rise to a maximum of 100 MW after the construction of two more dams at Mohale and Mashai and two more transfer tunnels to take water 75 km north of Katse.

Up to 7,000 workers are expected to be employed at the peak of the construction period, but the long-term benefits are expected to be greater. The new roads in hitherto inaccessible mountain areas are expected to have a useful tourist spin-off, especially if water sports, fishing and other activities can be developed on the new lakes. Water will also be available for irrigation while cheap hydro-power should attract light industry to help absorb some of the surplus population which currently finds employment in South Africa.

Last year Basuto miners working in the South African gold mines, many of them only a few hours from the Lesotho border in the Orange Free State, earned R347m, 21 per cent more than in 1987. The value of remittances sent home to feed and clothe family and relatives increased by 30 per cent. Although the water scheme promises a new, albeit modest prosperity for many Basotho, the country as a whole will still remain dependent for the foreseeable future on its close economic ties with South Africa.

KEY INDICATORS

	Unit	1983	1984	1985	1986	1987
Population	m	1.44	1.47	1.53	1.56	1.62
Gross domestic product (GDP)	Maloti m	394.8	465.5	549.6	654.9	759.7
GDP per capita	Maloti	274	317	359	409	469
Inflation	%	16.9	11.4	14.8	17.2	12.2
Agricultural production	1979-81=100	89.1	91.5	101.0	93.6	96.5
Food production	1979-81=100	86.9	88.9	99.1	91.2	94.3
Exports	$m	31	28	22	25	47
Imports	$m	483	433	314	342	455
Balance of trade	$m	–451	–405	–292	–316	–409
Current account	$m	–10	2	0	–17	–67
Total external debt	$m	135.1	134.7	167.5	187.4	241.0
Debt service ratio	%	4.6	5.3	6.5	4.4	4.4
Exchange rate	Maloti per $	1.1	1.4	2.2	2.3	2.0

* estimated figure

COUNTRY PROFILE

Political system: A hereditary monarchy, formerly the British colony of Basutoland; granted independence on 4 October 1966. In January 1986 the government of the late Chief Leabua Jonathan was overthrown in a military coup. Legislative power is vested in a bicameral parliament comprising a Senate of 33 members and a National Assembly. The country comprises 10 administrative districts, each under an appointed district secretary.

Political parties: All political activity is banned under the transitional rules established in January 1986.

Population: Lesotho has 1.6m inhabitants, according to the 1986 census. 70 per cent of the population live in the lowlands along the western strip, while 16 per cent are in the urban centres.

Main cities/towns: Maseru (population 106,000 in 1986).

Language: English and Sesotho are the official languages.

Media
Press: Several weekly publications (mainly government and religious) are available in Sesotho and English.
Radio: Radio Lesotho broadcasts in English and Sesotho.

External trade: Large trade deficit is offset by positive balance on services, mainly customs receipts, workers remittances and offical aid transfers. Member of the Southern African Customs Union, administered by South Africa.
Exports: Diamonds (negligible after the closure of the diamond mine – only 3.4 per cent of total value in 1986), manufactured goods (37-52%), wool and mohair (35-48% in 1983-86). Main destinations: South Africa (85-90% of total value, including wool and mohair auctioned at South African seaports), EC (5-10%), other Europe (negligible since closure of the diamond mine).
Imports: Manufactured goods (typically 45-60 per cent of total value), foodstuffs (25%), machinery and transport equipment (15-20%). Main sources: Southern African Customs Union (97% of total value), EC (2%).

Agriculture: Contributes 19 per cent to GDP and is the main source of employment. Subsistence farming dominates – mainly animal husbandary and the cultivation of maize. Other food crops include wheat, Sorghum and beans. Crop yields and livestock numbers have fallen since the 1970s due to drought, disease and mismanagement. Cash/export crop farming is being encouraged. The main agricultural exports are wool and mohair. Approximately 13 per cent of total area is potential/cultivated arable land, and 60-80 per cent is pasture.

Manufacturing: Accounts for over nine per cent of GDP and employs around 12,000 persons. Most firms are small and are in joint ventures with the Lesotho National Development Corporation. Production is largely for export (clothing, footwear, textiles) or import substitution (food processing, bricks). Pharmaceuticals and leather/hide processing under development. Tourism is being energetically developed.

Mining: Some diamonds are still mined by digging co-operatives in Lemphane and Liqhobong. Also known deposits of peat, lead, iron-ore and uranium.

Energy: Lesotho at present relies on South Africa for all its power requirements. However, the ambitious Highlands Water Scheme should eventually produce a considerable amount of hydroelectricity, as well as being a major revenue earner through sales of water to South Africa.

Membership of international organisations: African Development Bank, African States in Association with the EC (Lomé Convention), Arab Bank for Economic Development in Africa (recipient), Commonwealth, Economic Commission for Africa, GATT, FAO, IDA, IFAD, ILO, IMF, OAU, Southern African Customs Union, Southern African Development Co-ordination Conference, WHO, WIPO, World Bank.

BUSINESS GUIDE

Time: GMT +2.

Climate: Mostly dry with considerable variation between summer and winter temperatures. Summer from Nov-Jan with average temperature 27°C, rising to 32°C in lowland areas; heaviest rains between Oct and Apr. Winter from May to July with temperatures as low as -7°C in lowlands and -18°C in highlands.

Entry requirements
Passport: Required by all.
Visa: Required by all except nationals of Commonwealth countries, EC countries (excepting German FR and France), Japan, Scandinavian countries, South Africa and USA. Not required by tourists staying less than 30 days. Where applicable, visa is granted at place of entry in Lesotho.
Currency: No restriction on the amount of currency visitors may bring in or take out. South African rand are in general circulation.

Health precautions
Mandatory: Yellow fever and cholera vaccination for visitors from infected areas.

Representation overseas: Beijing, Bonn, Brussels, Copenhagen, London, Maputo, Nairobi, New York, Ottawa, Rome, Washington.

Air access
National airline: Lesotho Airways (operate Maseru to Johannesburg, Durban, Manzini, Gaborone, Harare and Seychelles, as well as charters with Boeing 707).
Other airlines: Air Botswana, Royal Swazi.
Tax: Airport tax of LSM 10 on departure.
International airport: Maseru-Moshoeshoe I (Code: MSU).

Surface access
Overland access: Good, tarred roads from South Africa into south and west of the country. A 3 km rail link connects Maseru with the South African rail system at Marseilles, which is on the Bloemfontein/Natal main line.

Hotels: Bills subject to tax of 12 per cent. Tip of 10 per cent is usual.

National transport
Air: Lesotho Airways operates services linking Maseru with all main centres. Regular flights include Maseru-Qacha's Nek and Maseru-Mokhotlong. Charter also available: Highland Air, PO Box 921 (tel: 324669; tx: 4234).
Road: Surfaced roads link Maseru with Butha Buthe in the north and with Mohale's Hoek in the south. Other roads can be rough, especially in mountain areas. Improvements to road system are being undertaken.
Buses: Coach services have been operated Maseru-Welkom; Maseru-Qacha's Nek.

Public holidays
Fixed dates: 1 Jan, 20 Jan (Army Day), 12 Mar (Moshoeshoe Day), 21 Mar (National Tree Planting Day), 2 May (King's Birthday), 4 Oct (Independence), 25 Dec, 26 Dec.
Variable dates: Good Friday, Easter Monday, Ascension, first Mon in Jul (Family Day), first Mon in Oct (National Sports' Day).

Working hours
Government: (Mon-Fri) 0800 – 1245, 1400 – 1630.
Banking: (Mon-Fri) 0900 – 1300; (Sat) 0900 – 1100.
Shops and business: (Mon-Fri) 0800 – 1700; (Sat) 0800 – 1300.

Telecommunications
Telephone: Dialling code for Lesotho: IDD access code + 266, followed by subscriber's number (no area codes).
Telex: Telex and telegram: Facilities are available in Maseru and other main centres.

Banking
Central bank: Central Bank of Lesotho.
Commercial banks: Lesotho National Bank, Barclays Bank International, Standard Bank.

Representation in capital: China PR, German FR, Korea PDR, UK, USA, USSR.

Hotels: Maseru
Casino, 12 Orpen Road, Private Bag A89 (tel: 22434; tx: 333).

Lancer's Inn, Kingsway/Pioneer Streets, P.O. Box: A84 (tel: 322114).

Lesotho Sun, Hilton Hill Road, Private Bag A68 (tel: 313111; tx: 4372; fax: 322872).

Victoria, Kingsway Street, P.O. Box: 212 (tel: 322002; tx: 4350).

Car hire: Maseru Avis: PO Box 294, Maseru (tel: 314325; tx: 334).

Other useful addresses
Central Bank of Lesotho, PO Box 1184, Maseru 100 (tel: 324281; tx: 4367).

Lesotho Airways Corporation (reservations), PO Box 861, Maseru (tel: 312453, 324513; tx: 4547, 4389).

Lesotho Chamber of Commerce & Industry, PO Box 79, Maseru (tel: 323482).

Lesotho National Development Bank, PO Box 999, Maseru (tel: 314333; tx: 4366).

Lesotho National Development Corporation, Private Bag A96, Maseru (tel: 312012; tx: 4341).

Lesotho Tourist Office, PO Box 1378, Maseru (tel: 322896; tx: 341).

Livestock Marketing Corp, PO Box 800, Maseru (tel: 322064) (sole marketing concern for all livestock and products, including mohair).

Ministry of Information and Broadcasting, PO Box 36, Maseru 100 (tel: 23561; tx: 340).

Ministry of Planning and Economic Affairs, Box 630, Maseru (tel: 323811).

Ministry of Trade and Industry, PO Box 747, Maseru (tel: 322138).

Ministry of Transport and Communications, PO Box 413, Maseru (tel: 323691).

Ministry of Works, PO Box 20, Maseru (tel: 323761).

Radio Lesotho, PO Box 552, Maseru (tel: 323561).

Statistics Bureau, PO Box 455, Maseru.

Trade Promotion Unit, Ministry of Trade and Industry, PO Box 747, Maseru (tel: 323414).

Liberia

Kaye Whiteman

As he approached his tenth anniversary in power, all President Samuel Doe seemed capable of was political survival. The condition of the Liberian economy looked increasingly sombre: even a special group of American financial experts proved unable to resolve Liberia's problems and left at the end of 1988.

The plots of 1988 had their aftermath in the closing months of 1988. The 10 coup plotters of March, including Gabriel William Kpolleh, leader of the banned Liberia Unification Party, were sentenced to 10 years' imprisonment for plotting to assassinate the president. During the trial it was alleged that the plot leader was the late Joe Robert Kaipaye, who was reported to have leapt to his death from the sixth floor of the executive mansion in Monrovia, in March 1988, while being interrogated by President Doe himself. The defendants claimed the trial was biased and corrupt, and defence lawyers accused the judge of incompetence, claiming the jury had been bribed by officials of the ministry of justice. The conduct of the trial embarrassed the government as prosecution witnesses were often seen to be arguing among themselves.

Controversy also continued about the alleged attempted invasion of July 1988 in which a former close colleague of Doe's, General Nicholas Podier, was killed. On 15 November 1988, President Doe ordered the pardon of all those connected with the 'invasion', including two Americans, James Bush and Curtis Williams, who had been detained and charged with treason. Others pardoned included three former army members of the Peoples' Redemption Council, which ruled Liberia from 1980 to 1986, Larry Borteh, Jerry Friday and Christian Moore. Podier was said to have been killed in gunfire on the Ivoirian border. However, Amnesty International earlier claimed that unofficial sources said that the Americans had been on a business trip from Cote D'Ivoire and had not been involved in any invasion, and that Podier was executed 'extra-judicially' by security forces after having been arrested at Robertsfield International Airport. There was also a campaign in the USA on behalf of the two Americans, involving the congressional black caucus and Andrew Young, the mayor of Atlanta. The pardon, which came shortly after the news of the election of President Bush to the White House, was also seen as a response to new threats of suspension of US aid.

Relations with the USA

The question of deteriorating relations with the USA was prominent in the first part of 1989. At the end of 1988, the team of 17 operational experts, known as Opex, sent under a $6.5m contract as part of the US aid programme to assist in the formulation of government policy and in the improvement of financial procedures over a two-year period, left the country. Their final operational report outlined some of the reasons for their dissatisfaction, particularly their powerlessness to cope with the malpractices which they found at all levels of the ministries in which they worked. They concluded that the Liberian Government was incapable of managing its deficits, of undergoing a meaningful budget process, of restraining expenditure or of setting and adhering to priorities for investment. Around 40 per cent of government spending was out of control of the team, which was not allowed access to the accounts of various public companies, notably the Liberian Petroleum Refining Company (LPRC). Salaries were continuing to be paid to non-existent employees, and ministers continued to grant substantial tax exemptions to friends. There also appeared to be no limit to the president's purchase of non-essentials, such as a presidential boat and helicopter, and two bullet-proof Mercedes cars.

The USA appeared to decide that its credibility would be damaged by continuing the mission. The withdrawal of the team led to a worsening of relations as arrears on debts in all directions were allowed to mount up. In March 1989 there was a threat to suspend aid completely from May, only deferred by the payment of $7m arrears in April, following an unusual presidential appeal to Liberians to raise funds by private donation. Opposition to aid to Liberia in congress was fuelled by critics of the Doe regime's human rights record, fed with material from increasingly active opposition groups in the USA. Aid had, in any case, been falling: in 1987 it was $36m, in 1988 $31m, and in $1989 was slated at $22m. In May, limited funds totalling $9.5m were made available for certain development projects, but the rumblings of dissatisfaction continued. For example, Liberia's special status under the Generalised System of Preferences (GSP) was being increasingly threatened by action from human rights and trade union campaigners.

In these circumstances, it was scarcely surprising that Liberia's relations with both the International Monetary Fund and the World Bank continued to be on ice. Total arrears to both institutions was over $550m in March 1989, and was still growing. Liberia itself, facing consistently falling export earnings and declining growth, looked increasingly incapable of paying any debts, and thus set on an irreversible downward economic spiral. In spite of this situation, at the beginning of the year, President Doe announced a 40 per cent increase in civil service salaries over a three-year period, although this did not figure in draft budget proposals. In July, it was announced that the cumbersome five dollar coins ('Doe dollars') would be replaced with banknotes by mid-August. At the same time, the governor of the Bank of Liberia, Thomas Hanson, stressed that the changeover did not mean that the country would be breaking its official link, at parity, with the US dollar. Thus the unofficial rate for Liberian dollars (at least twice the official rate), seemed set to continue.

In July, the economy received a further body blow in the shape of the decision to close down the Liberian–Swedish–American Minerals Company (LAMCO) by the end of 1989. Production of iron ore was stopped immediately, and all of the company's 15,000 employees were told they would be out of work by the end of the year. Although the output at LAMCO's mine, situated at Yekepa in Nimba County near the northern border, had been declining for some years, it still represented the single most important foreign investment in the country. It included the railway from Yekepa to the coast, the port at Buchanan, and the mineral treatment facilities there, all of which were similarly served notice of closure. The main hope in this potentially disastrous situation was the way that had been cleared for the establishment of a new Liberia–Guinea joint venture, probably to be called Nimco, which would exploit the rich deposits on the Guinean side of the frontier, and build a spur to link up with the Liberian railway. LAMCO could then be submerged with the new company, even if the Swedish partners persisted in their apparent wish to pull out.

KEY FACTS

Official title: Republic of Liberia

Head of state: President Samuel K. Doe

Ruling party: National Democratic Party of Liberia

Capital: Monrovia

Official Languages: English (Golla, Kpelle, Kru widely spoken)

Currency: Liberian Dollar (L$) = 100 cents

Exchange rate: L$1.00 per $

Area: 111,369 sq km

Population: 2.5m (1988)

GNP per capita: L$450 (1987)

GNP real growth: –1.0%* (1987)

Inflation: 5.0% (1987)

Trade balance: $63m (1987)

Foreign debt: $1.6bn (1987)

* estimated figure

Political changes

The political situation in Liberia continued to cause concern into 1989. Although the student unrest of the latter part of 1989 calmed down, there were new signs of activity on the part of the opposition parties. Following a warning from the electoral commission (ECOM) that the leaders of the three main parties faced arrest if they met together, the president indicated that such a meeting could take place, while warning that the government would not tolerate any meetings that could provoke chaos and disorder in Liberia. The party leaders – Jackson Doe of the Liberia Action Party (LAP), Bacchus Matthews of the United Peoples Party (UPP), and Ed Kesselly of the United Party (UP) – planned to discuss the failure of economic policy, and human rights issues. President Doe went on to answer the opposition charges of corruption against the government by saying in a TV broadcast at the end of June 1989 that all three opposition leaders had at various times been to him to ask for money or favours.

Meanwhile, President Doe sacked his finance minister yet again. David Farhat, appointed early in 1988, was accused in June 1989 of making a commitment which could 'adversely affect the country', and replaced with Emmanuel Shaw. Also in June, Gray Allison, the defence minister, a fellow Krahn tribesman and formerly close to Doe, was sacked and put on trial in a bizarre ritual murder case. Allison also claimed in court that he had been accused of plotting to overthrow the government. Also in June, Sylvester Moses, a former security boss disgraced two years before, re-emerged as national security minister following the sudden death of the incumbent, Patrick Minikon.

KEY INDICATORS

	Unit	1983	1984	1985	1986	1987
Population	m	2.0	2.1	2.2	2.2	2.3
Gross domestic product (GNP)	L$m	969	937	1,031	1,012	1,063
GNP per capita	L$	500	470	480	460	450
Inflation	%	2.8	1.2	–0.6	3.6	5.0
GNP real growth	%	–4.7	–3.5	4.7	–2.6	–1.0
Agricultural production	1979-81=100	106.9	117.3	115.8	117.3	120.5
Food production	1979-81=100	111.5	114.2	116.8	117.4	121.0
Exports	L$m	427.6	452.1	435.6	408.4	382.2
Rubber	L$m	73.1	91.3	77.1	80.7	89.4
Iron-ore	L$m	267.4	279.0	279.4	248.4	217.7
Diamonds	L$m*17.2	10.9	4.7	6.5	11.0	
Imports	L$m	411.6	363.2	284.4	234.8	307.6
Balance of trade	L$m	46	121	167	149	63
Current account	L$m	–104	–2	56	15	–118
Total external debt	$m	997	1,064	1,229	1,405	1,618
Debt service ratio	%	6.5	4.5	3.9	6.9	–
Exchange rate	L$ per $	1.0	1.0	1.0	1.0	1.0

* estimated figure

COUNTRY PROFILE

Political system: Executive power rests with President, legislative power with the 26-member Senate and the 64-member House of Representatives. Multi-party constitution approved by referendum in July 1984. Established as independent state in 1847.

Political parties: The ruling National Democratic Party of Liberia (NDPL) is opposed by three legal opposition parties (the Liberian Action Party, the Unity Party and the Liberian Unification Party), which in March 1986 formed a 'Grand Alliance'. There are also a number of illegal opposition parties.

Population: The results of the 1984 census showed a population of 'approximately' 2.2m. 1988 estimate is 2.5m.

Main cities/towns: Monrovia (population 250,000, 400,000 including suburbs), Buchanan, Totota, Greenville, Robertsport.

Language: English is both the official and business language.

Media
Press: Five newspapers, the *New Liberian, Daily Star, Standard, Mirror, Observer.*
Radio: Public radio broadcasts in English, French, and African languages. Also two private Christian radio stations. 'Voice of America' also transmits from Monrovia. 175,000 sets in 1987.
Television: Two commercial stations operating between five and ten hours daily. 7,000 sets in 1987.

External trade: Increased outflow of foreign company remittances, debt repayments and falling exports have kept current account in deficit. Balance of payments support provided by USA and German FR in form of concessional aid. The IMF withdrew its loan facility in 1986, and the World Bank suspended disbursement in 1987.
Exports: Iron-ore (61.7 per cent of total value in 1985), rubber (20%), diamonds (2.4%), coffee (4%), timber (5%). Main destinations: German FR (typically 30%), USA (18%), Italy (18%).
Imports: Machinery and transport equipment (27.4 per cent of total value in 1985), other manufactures (23%), foodstuffs (23%), fuels and lubricants (17.5%). Main sources: USA (typically 25% of total), German FR (12.5%), Netherlands (10%), Japan (8%).

Agriculture: Most important sector of economy, contributing 18.5 per cent to GDP and employing 80 per cent of the workforce. Principal cash crop is rubber, which provides 20 per cent of exports. Although mostly grown in foreign-owned plantations, smallholders are responsible for over half the total acreage planted. Coffee, cocoa and timber are also grown for export, but, as with rubber, earnings have been reduced due to falling world prices. Palm oil is produced mostly for the domestic market. Main food crops are rice, cassava and vegetables. The government is attempting to improve production of these to reduce the need for imports of rice which have become necessary to meet domestic demand. Timber is an important export earner with total fellings around 800,000 cu m a year. Commercial ocean fishing is a growing activity, particularly shrimps.

Manufacturing: Sector is relatively under-developed, contributing less than 5.1 per cent to GDP, with very few firms employing more than 50 people. Activity is mainly confined to textiles, food processing, wood products, cement and chemicals. Growth in the sector has been constrained by the small size of the domestic market, the need to import practically all raw materials, and financial problems. However, the government has managed to attract some foreign investment for the setting up of new companies aimed at exploiting local resources. There is also an industrial free zone near Monrovia for the duty-free assembly and manufacture of goods destined for re-export. Liberia is famous for its 'flag of convenience' in the merchant shipping world.

Mining: Most important sector of the external economy, accounting for two-thirds of visible exports, though employing only five per cent of workforce. Liberia is one of the world's major producers of iron-ore (mainly extracted from mines at Mount Nimba and Mano River) although earnings have slumped due to falling world demand. Production estimated at 23m tonnes a year. Major producers are LAMCO, NIOC and the Bong Mining Company. Diamonds and gold are also mined in small quantities, and there are known deposits of bauxite, manganese, columbite, uranium, tantalite, copper, tin, lead and zinc.

Energy: All petroleum requirements are imported. Exploration for indigenous oil reserves undertaken by Amoco off Grand Cape, Mount County. Total installed generating capacity is currently over 300 mW, most of which is diesel-generated. Plans to build a 414 mW hydroelectric project on St Paul River.

Membership of international organisations: African Development Bank, Arab Bank of Economic Development in Africa (recipient), Economic Commission for Africa, Economic Community of West African States, ILO, IDA, IFAD, IAEA, IMF, Mano River Union, OAU, UN, WHO, WIPO, World Bank.

BUSINESS GUIDE

Time: GMT.

Climate: Hot and tropical with high levels of humidity (85-90 per cent) and little temperature variation throughout the year. Average temperatures range between 20-22°C at night and 28-32°C during the day. Wet season lasts from May to Oct with especially heavy rain Jun-Jul.

Entry requirements

Passport: Required by all.

Visa: Required by all except nationals of most West African countries. A return or onward ticket is required when applying for a visa.

Prohibited entry: White nationals of South Africa.

Currency: No restriction on import and export of local or foreign currency.

Health precautions

Mandatory: Yellow fever vaccination certificate required.

Advisable: Cholera vaccination is essential. Typhoid and polio vaccination recommended. Malaria prophylaxis should be taken as risk exists throughout the country. Drinking water should be boiled and filtered.

Representation overseas: Abidjan, Accra, Addis Ababa, Beijing, Beirut, Berlin, Berne, Bonn, Brussels, Bucharest, Cairo, Conakry, Freetown, The Hague, Kinshasa, Lagos, London, Moscow, Nairobi, Paris, Port-au-Prince, Rome, Stockholm, Tokyo, Vatican City, Washington, Yaoundé.

Air access

National airline: Air Liberia.

Other airlines: Aeroflot, Air Afrique, British Airways, Ethiopian,Ghana Airways, KLM, MEA, Nigeria Airways, Sabena, Swissair, Zambia Airways.

International airport: Monrovia-Robertsfield International (Code: ROB), 60 km from city; duty-free shop, bar, restaurant, buffet, post office, shops.

Surface access

Main ports: Monrovia, Buchanan, Greenville, Harper, Robertsport. Monrovia to be expanded with ro-ro and container facilities.

Overland access: Road: Routes from Ivory Coast, Guinea and Sierra Leone are not generally recommended during rainy season.

Hotels: The airport hotel and other major hotels in Monrovia should be booked in advance. Rates are expensive and tipping is optional.

Car hire: Chauffeur-driven or self-drive cars available in Monrovia. International driving licence or national driving licence with permit (valid for up to 30 days) accepted. Traffic drives on the right. Self-drive cars are not generally recommended.

City transport

Taxis: Zoning system in operation. Negotiate fares in advance for long-distance journeys. Tipping is not usual.

National transport

Air: Air Liberia operates domestic service between Monrovia and main towns. Air taxi companies charter planes between Monrovia and airfields throughout the country.

Road: Network of 10,000 km covers most areas, although many roads are untarred. Main highways are from Monrovia to Sanniquellie (with a branch from Ganta to Harper) and from Monrovia to Buchanan.

Rail: There are no passenger railways, but the LAMCO mining company accepts passengers on the mineral line between Buchanan and Nimba county.

Sea: Freight/passenger services between Monrovia and Buchanan.

Public holidays

Fixed dates: 1 Jan, 11 Feb, 15 Mar, 12 Apr, 14 May, 25 May, 26 Jul, 24 Aug, 24 Oct, 29 Nov, 25 Dec.

Variable dates: Decoration Day (2nd Wed in Mar), Good Friday, Fast and Prayer Day (2nd Fri in Apr), Thanksgiving (1st Thu in Nov).

Working hours

Government and business: (Mon-Fri) 0800 – 1200, 1300 – 1600.

Banking: (Mon-Thu) 0800 – 1200; (Fri) 0800 – 1400.

Shops: (Mon-Sat) 0800 – 1900.

Telecommunications

Telephone: Dialling code for Liberia: IDD access code + 231 followed by subscriber's number (no area codes needed). 100 per cent automatic service; but very limited outside Monrovia.

Banking

Central bank: National Bank of Liberia.

Head offices: Agricultural & Cooperative Development Bank, International Trust Co of Liberia, Liberian Bank for Development & Investment, Liberian Trading & Development Bank, Bank of International Credit and Commerce.

Foreign branches: Citibank, Meridien (Liberia).

Representation in capital: Algeria, Cameroon, China PR, Cuba, Egypt, Ethiopia, France, German FR, Ghana, Guinea, Haiti, Israel, Italy, Ivory Coast, Japan, Rep. of Korea, Lebanon, Morocco, Nigeria, Poland, Romania, Sierra Leone, Spain, Sweden, Switzerland, UK, USA, USSR, Vatican, Zaïre.

Hotels: Monrovia

Africa, P.O. Box: 1522 (tel: 23951/224519; tx: 44223; fax: 223732).

Ducor Palace Inter-Continental, Broad Street, P.O. Box: 86 (tel: 224200; tx: 44268).

Car hire: Monrovia

International Automobile Co: (tel: 222486).

Mensah Travel Bureau: PO Box 86 (tel: 222807).

Yes Taxi Co: PO Box 49 (tel: 222970).

Other useful addresses

Agricultural and Co-operative Development Bank (ACDB), PO Box 3585, Monrovia (tel: 224385; tx: 4504).

Air Liberia, PO Box 2076, Monrova (tel: 222144/226404; tx: 4298).

Commonwealth Development Corporation, PO Box 2015, Monrovia.

ELTV (Liberian television system), PO Box 594, Monrovia.

Liberia Chamber of Commerce, PO Box 92, Monrovia (tel: 222218).

Liberian Bank for Development and Investment (LBDI), PO Box 547, Monrovia (tel: 261140; tx: 4345).

Liberian Broadcasting System, PO Box 594, Monrovia.

Liberian Industrial Free Zone Authority, PO Box 9047, Monrovia.

Liberian News Agency (LINA), Ministry of Information, PO Box 9021, Capitol Hill, Monrovia (tel: 222229; tx: 4249).

Liberia Produce Marketing Corporation, PO Box 662, Monrovia.

Ministry of Commerce, Industry and Transportation, PO Box 9014, Monrovia.

National Bank of Liberia, Broad Street, PO Box 2048, Monrovia (tel: 222580; tx: 4215).

National Enterprises Corporation, PO Box 518, Monrovia.

National Investment Commission, PO Box 9043, Monrovia (tx: 4560).

National Ports Authority, PO Box 14, Monrovia.

Statistics Bureau, PO Box 9016, Monrovia (tel: 222622; tx: 4374).

Libya

Andrew Lycett

On 1 September 1989 Libyan leader Muammar Qadafi celebrated two decades in power. A testament to his remarkable political durability, the anniversary also marked a new stage in Qadafi's steady progress towards re-establishing his regime – both internally and in its relations with the outside world – since it was nearly toppled in the aftermath of the American raids on Libya in April 1986.

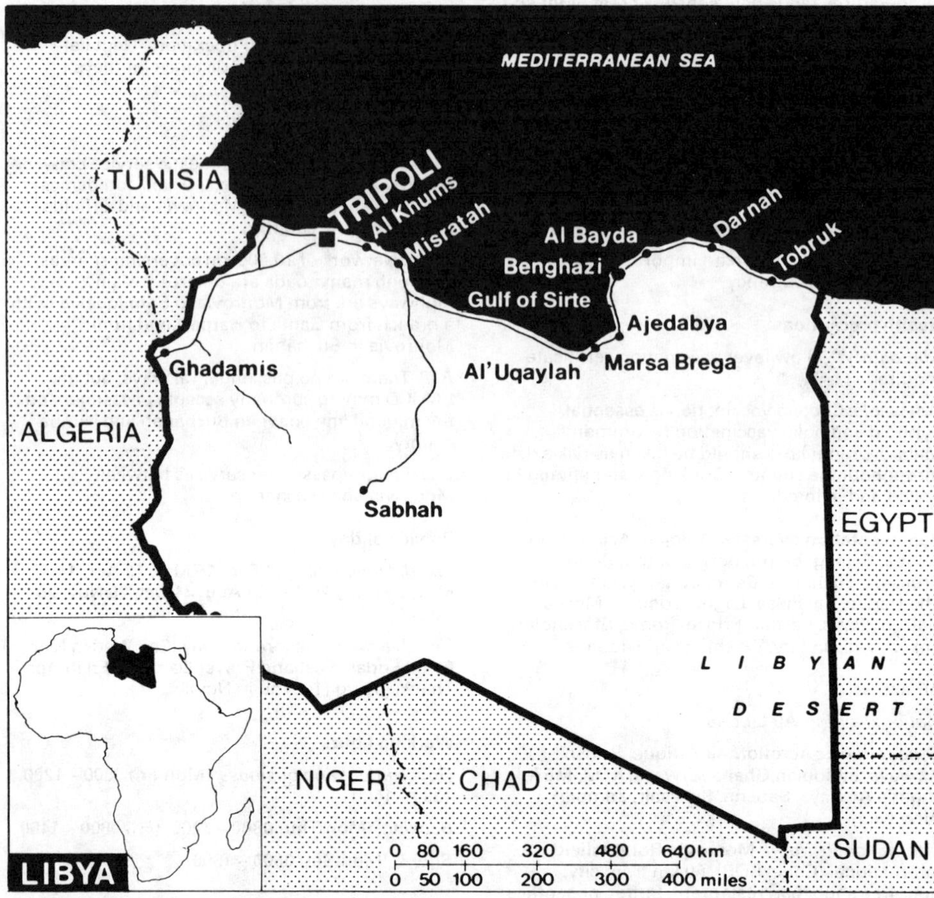

In anticipation of the actual anniversary Qadafi spent an estimated $500m smartening up Tripoli and other Libyan cities for his guests. Libyan workers were reportedly paid LD5 per square metre ($16 at the official exchange rate) to clean, repair and paint rundown buildings. A new complex of 20 villas for VIPs was completed at the last minute at Hadabra Khadra, on the outskirts of Tripoli. The South Korean company Samsung Construction rushed through a $7m contract to build a new town square in Qadafi's home town of Sirte.

Emergence from the wilderness

A distinguished array of Arab leaders came to pay their respects and to mark Qadafi's re-emergence from the global political wilderness to which he had assigned himself in the mid-1980s. Paying his first visit to Libya, King Hasan of Morocco – until recently an implacable opponent of the Libyan leader – sailed into Tripoli harbour in his personal yacht the *Marrakesh* to a noisy reception from sirens and foghorns. Also present were Algeria's President Chadli, Tunisia's President Ben Ali, PLO Chairman Yasser Arafat, President Asad of Syria and, from further afield, President Daniel Ortega of Nicaragua.

Another important visitor was Yusuf Wali, deputy prime minister of Egypt, coming in place of President Mubarak. Perhaps the most significant development in Libya's foreign policy over the year has been the gradual improvement of its relations with neighbouring Egypt. This followed Mubarak's meeting with Qadafi at the Arab summit in Casablanca in May 1989. Full diplomatic relations (broken in 1979 after Egypt signed its peace treaty with Israel) were resumed in June, and subsequent months have seen a gradual improvement in the cold war between Cairo and Tripoli. President Mubarak talked positively on Egyptian television about his meeting with the Libyan leader. 'We discussed various issues,' he said, 'and feel he [Qadafi] is willing to continue the dialogue.'

At the end of May General Yusuf Sabri Abu Taleb, Egypt's newly appointed defence minister, visited the Libyan border, ostensibly to prepare for its opening. At the same time Qadafi's powerful cousin, Ahmed Qadafadam, who has been described as the head of the Libya security service, turned up in the Egyptian capital to resume the talks which had been started in Casablanca. The border between the two countries was subsequently reopened and daily air flights between their two capitals resumed. Tripoli agreed to pay $12m compensation to thousands of Egyptian workers who had been expelled from Libya in 1985.

These moves were mirrored on Libya's western border by improved relations with Tunisia. The rapprochement between the two countries had been improving since diplomatic relations between the two countries were restored in December 1987. In April 1987 a political and economic co-operation pact was signed and the following month Qadafi himself visited Tunis. Over the past year the Libyan leader's improved links with Tunisia have been extended to take in both Algeria and Morocco. An Arab Maghreb Union comprising Libya, Tunisia, Algeria, Morocco and Mauritania was established. A summit meeting of the Union was to have taken place in Tripoli at Qadafi's twentieth anniversary celebrations, but Mauritania's President Ould Taya was not able to attend because of a domestic crisis.

Chad Peace

The good news in Tripoli on 1 September was the announcement in Algiers the previous day that the foreign ministers of Libya and Chad had signed an agreement to end the 16-year-old conflict between their two countries over the status of the mineral-rich Aouzou strip. In a compromise that was partly the work of the Organisation of African Unity, partly an example of the way that the Maghreb Union was working, Libya and Chad agreed to end their dispute. Libya was to withdraw its forces from Aouzou. The issue of sovereignty was to be submitted to the International Court of Justice in The Hague if the two parties could not reach an agreement between themselves within a year. The Libyan and Chadian foreign ministers also agreed to exchange prisoners of war and to halt media attacks on each other.

So appears to have ended a conflict which, in the years following the downturn in oil prices in the early 1980s, has crippled Libya both politically and econ-

KEY FACTS

Official title: Socialist People's Libyan Arab Great Jamahiriya

Head of state: Col. Muammar Qadafi (revolutionary leader); Jadalla Azzuz Talhi (head of state)

Capital: Tripoli

Official Languages: Arabic (Berber, French, English spoken)

Currency: Libyan Dinar (LD) = 1,000 dirhams

Exchange rate: LD0.30 per $ (Nov 1989)

Area: 1,761,132 sq km

Population: 4.3m (1988)

GDP per capita: $5,500 (1987)

Trade balance: $433m (1986)

Foreign debt: $2.3bn (1987)

omically. Since Libya first took over the Aouzou strip from Chad in 1973 (claiming it had been unlawfully ceded to Chad as part of a colonial treaty between Italy and France), Qadafi's intervention in its southern neighbour has cost thousands of Libyan lives and billions of dollars. When Chad forces retook the Aouzou strip in August 1987, and pushed 160 km inside Libya the following month, Qadafi realised his people no longer had the stomach to take the battle to Chad.

For almost two years he jockeyed for position in long-drawn-out talks with Chad, usually mediated by the Organisation of African Unity. In July 1989 Qadafi and Chad's President Hissen Habre met in Mali to discuss terms. Libya was believed to have 1,500 prisoners of war in Chad, where they were being moulded into an anti-Qadafi strike force by the opposition National Front for the Salvation of Libya. Their future was not immediately certain.

Another interesting but controversial visitor to Qadafi's anniversary celebrations was Gianni De Michelis, Italy's new foreign minister. In flying to Tripoli, he became the first foreign minister of a European Community country to meet Qadafi since the EC imposed sanctions on Libya in 1986. Earlier in June Giovanni Prandini, Italian minister of merchant marine, was the first EC minister to visit Tripoli. Signor Prandini met Qadafi's right-hand man, Major Abdul Salem Jalloud, as well as the Foreign Liaison Secretary Jadalla Azzuz Talhi. He indicated that the two countries were moving to an amicable settlement of the claims arising from Italian colonial rule which had bedevilled relations for some time. Joint fishing ventures were discussed. Prandini was followed to Tripoli shortly afterwards by Belgium's Foreign Trade Minister Robert Urbain, who was chasing $90m worth of business for Belgian companies. Libyan–Turkish relations also took a turn for the better following the conclusion of another long-running dispute over payments. Turkish names immediately featured strongly in the lists of companies winning sizeable construction contracts.

KEY INDICATORS

	Unit	1983	1984	1985	1986	1987
Population	m	3.47	3.62	3.79	3.93	4.08
Gross domestic product (GDP)	LD m	8,531	7,574	7,203	6,473	–
GDP per capita	$	8,550	7,240	6,610	5,550	5,500
Inflation	%	11.6	8.0	8.5	–	–
Agricultural production	1979-81=100	122	118	140	109	121
Food production	1979-81=100	123	118	140	109	121
Crude oil production	'000 b/d	1,140	985	1,024	1,308	973
Oil revenues	$m	11,080	11,132	10,658	–	–
Exports	$m	13,185	11,680	10,890	6,157	–
Crude oil	1980=100	60.0	60.0	58.9	–	–
Imports	$m	12,726	11,884	7,999	5,724	–
Balance of trade	$m	459	–204	2,891	433	–
Current account	$m	–1,643	–1,524	2,084	–54	–
Exchange rate	LD per $	0.30	0.30	0.30	0.31	0.27

* estimated figure

Reconciliation

Reconciliation seemed the order of the day. Qadafi demonstrated his desire to end another long-running dispute when in early September he sent Libya's first ever delegation to attend the conference of the Inter-parliamentary Union in London. The presence of the Libyans at the meeting was attacked by British politicians in advance. They thought that at a time when Britain and Libya still have no diplomatic relations following the shooting of WPC Fletcher in St James's Square, London in April 1984, one of Qadafi's minions would use the conference to undermine the idea of parliamentary democracy. But the reality was rather different. Although Qadafi's Green Book is scornful of parliaments – Representation is Fraud is one of his favourite aphorisms – the leader of the Libyan delegation Abdulhamid Amaar was clearly delighted that his country, and its system of general people's congresses, had been accepted into the Union. He responded by issuing an unequivocal condemnation of the murder of the wife of a British soldier by the IRA in West Germany, though his further denial that Libya had supplied weapons to the IRA was greeted with scepticism by Western intelligence experts.

In the longer run, however, Qadafi was more interested in patching up his continuing quarrel with the United States. In December 1988 the outgoing President Reagan raised the stakes once more in his extraordinary battle (mainly of words) with the Libyan leader. He revealed that the Libyans were building a chemical weapons plant at Rabta, some 56 km south-east of Tripoli, close to the town of Shariyan. Speaking at the time of the Pan Am Flight 103 air disaster at Lockerbie, which the USA tried hard to blame the Libyans for, Reagan threatened to bomb the site. The USA promised to provide evidence about the site at a 142-nation chemical weapons conference in Paris in January 1989. Despite showing a number of charts and photographs, and pointing to firms in West Germany, Japan, Switzerland and Italy as culprits, the charge failed to stick. The Libyans do not not deny, however, that Rabta is an industrial site, used for the manufacture of pharmaceuticals.

Under Reagan's successor, President George Bush, relations between Libya and the USA became less volatile. Qadafi appeared to want improved links because he felt the impasse prevented him from developing the oil industry on which his country's economy depends. He had a trump card in his hand. Five US oil companies – Conoco, Marathon, Amerada-Hess, Occidental and W.R. Grace – hold rights to an estimated $2bn worth of oil reserves in Libya. They were forced to withdraw from Libya following Washington's sanctions on Qadafi's regime introduced in 1986. They subsequently reached an agreement with the Libyan National Oil Corporation (LNOC) that it should run their operations and sell their crude entitlements on their behalf for three years – a period which ran out on 30 June 1989. As this cut-off date approached the USA allowed its oil companies to return to Libya through foreign subsidiaries, fearing that Libya might nationalise the assets if it opposed this move. But the USA maintained its ban on the export of US technology to Libya, and American citizens were still not permitted to travel to Libya. In a message to congress on 19 July 1989, President Bush stated he was renewing his ban on trade with Libya for another six months. He added that Libya's politics and actions 'continue to pose an unusual and extraordinary threat to the national security and foreign policy of the US'.

Oil moves

The US oil companies are keen to return to a country which has proved profitable in the past. But, following a meeting between them and the LNOC in Vienna in May, Qadafi made clear they could expect no special favours. He put forward a simple solution: if the chief executive officers of these companies used their power in the US oil lobby, all US sanctions on trade with Libya could be speedily removed and normal relations between the two countries resumed. Again US authorities remain unconvinced that Qadafi is sincere in his desire for reconciliation. In Belgrade in September 1989 the Libyan leader told the Non-Aligned Conference: 'The dollar really should be worthless. It is the devil itself.'

Curiously, an opening between the USA and Libya came in August 1989 when Washington agreed to co-operate in a $6m programme to eradicate a plague of screw worm flies which was affecting North African countries, including Libya. The programme was co-ordinated by the Rome-based International Fund for Agricultural Development.

The main reason for Qadafi's feelers to the outside world was his desire to solve his country's continuing economic problems. Although oil prices improved during 1988–89, around 15 per cent of Libya's oil output of 1.1m barrels a day (b/d) was still committed to barter deals and debt repayments. In the second quarter of 1989, Turkey and the Soviet Union took 30,000 b/d, Italy 20,000, Greece 15,000, Yugoslavia 12,000 and other countries an estimated 50,000.

Much of the hard currency generated by Libyan oil sales continued to go to pay for infrastructural projects, notably the mammoth Great Man-made River (GMR), which involves bringing fresh water to Mediterranean coastal towns from aquifers in Libya's southern desert. Qadafi took a personal interest in the progress of the GMR. It has first priority on resources and foreign exchange. Nevertheless, during 1989 even its contractors, headed by Dong Ah of South Korea, have experienced delays in payments. According to one report, salary payments to many civil servants were up to three months late. According to the Bank of International Settlements in May 1989, Libya's overseas bank deposits fell by 44 per cent in 1988 to $3,429m, largely the result of heavy spending on infrastructural projects and of a consumer boom which followed the opening of the border with Tunisia.

Characteristically, Qadafi has not been deterred from investing in his favourite projects. At the end of August 1989 Dong Ah announced it had been awarded the hotly-contested contract, said to be worth $5.3bn, to build the second phase of the GMR, connecting Tripoli to the southern aquifers. In granting the contract to the South Korean firm (one of several which won construction contracts worth $141m in Libya in the first six months of 1989), Qadafi upset the ambitions of the French construction company Bouygues, which was firmly backed by its own government.

The Libyan leader was not held back either from developing the next phase of Libya's oil industry by his continuing dispute with the US oil companies. In September the LNOC was due to hold talks with two powerful European oil companies (Italy's Agip and West Germany's Veba) about developing new concessions, largely in the Sirte and Ghadames basins, where the American companies would love to be working. Veba is already working on the upgrading of the Amal oil and gas plant, south of Marsa Brega. The new concessions are part of what is known as the third round of exploration and production sharing agreements (EPSA 3). Unlike the first two rounds, EPSA 3 has provisions which allow the contractor to set aside part of the cost of production for rapid cost recovery. Over the previous 18 months LNOC had already signed EPSA 3 agreements with Royal Dutch/Shell, Italy's Societa Energie Montedison, International Petroleum Corporation and Husky Oil (both of Canada), Yugoslavia's INA-Naftaplan, Austria's OMV Exploration and Production, and Brazil's Petrobras Internacional (Braspetro).

One of the bonuses gained from Libya's better relations with its neighbours was the setting up of the Libyan–Tunisian Exploration Company to prospect in the disputed Gulf of Gabes between the two countries. The company held its first meeting in Djerba in August under its new Libyan director-general Tariq Mustafa Hasan. It is responsible for surveying, exploring, developing and producing from the 3,000 square kilometre 7 November oil field, between Libya's Bourri field and Tunisia's Ashtart field. Agip, operator of the Bourri field, is tipped to operate the joint field.

Political problems

On the domestic political front, Qadafi hopes that the new agreement with Chad will continue to cool what was a powerful anti-war lobby. However, the return of large numbers of soldiers from the southern front has added to problems of unemployment in the cities. According to opposition sources there were long-running clashes between Libyan security forces and anti-Qadafi protesters at Tripoli and Benghazi universities in December 1988. On 14 January 1989, Major Misbah Abdul Hafez, head of security at Egdabia, near Benghazi, was seriously injured and his driver killed, following an attack by armed attackers. This led to a clamp-down on dissidents, particularly on a new breed of Libyan-style Islamic fundamentalists. Again, according to opposition sources, eight prisoners were executed in April, including Sheikh Salah, imam of the Abu Mishmasha mosque in Tripoli. In August Amnesty International issued a report listing 40 prisoners of conscience – mainly Islamic fundamentalists – who were arrested without trial following clashes with Qadafi's revolutionary committees between January and May 1989.

In response to general popular concern about the role of the army in Libyan society, Qadafi repeatedly stated his intention to abolish the armed forces and replace them with 'those who could never fail, the Libyan masses'. In August 1989, shortly before his twentieth anniversary celebrations, he announced that the Armed Forces General Command had been replaced by a General Committee on Defence. The post of armed forces commander, headed by Field Marshal Abu Bakr Younis, one of the original Revolutionary Command Council, who seized power with Qadafi in September 1969, was abolished and replaced by a secretary to the general committee.

Despite clear moves away from confrontation, Libya's economy and domestic political structure are still fragile. Future directions in foreign policy are hard to predict. Despite the recent improvement in relations with Egypt, for example, Qadafi continues to predict the overthrow of his neighbour, President Hosni Mubarak. The end to the war in Chad, coupled with reviving oil fortunes, should make the 1990s a happier decade for Libyans than the 1980s. But many contradictions in the fabric of society have yet to be worked out.

COUNTRY PROFILE

Political system: Political power is vested in the 1,112-member General People's Congress (GPC). Executive power is exercised by the General People's Committee, which comprises 20 representatives from local departments. The head of state is elected by the GPC. Independence gained from Italy in 1951.

Political parties: There are no political parties.

Population: The population, estimated at 4.3m in 1988, is growing at the exceptionally rapid rate of 4.5 per cent per annum. Two-thirds of the population live in coastal regions, with half of these residing in Tripoli itself.

Main cities/towns: Tripoli (population 1,200,000), Benghazi (750,000), Misurata (360,000), Zawia (300,000), Homs-Kussabat (200,000).

Language: Arabic is the official language, but French, Berber and English are also spoken.

Media
Press: Newspapers and periodicals published by Jamahiriya News Agency, Press Service and Trade Unions. The main newspapers are *Arraid* and *El Balag*.
Radio: Socialist People's Libyan Arab Jamahiriya Broadcasting Corporation broadcasts in Arabic and English from Tripoli and Benghazi.

External trade: Traditionally strong balance of payments position eroded since 1981 by falling oil earnings, worsening greatly since 1986. Regular current account deficit financed from drawdowns in international reserves, which have fallen dramatically. An austerity budget introduced in 1987. Countertrade is used increasingly in foreign trade.
Exports: Crude oil currently accounts for around 95 per cent and refined petroleum products around five per cent. Main destinations: Italy (29.3% of 1985 total), German FR (17.7%), Spain (9.2%), France (6%), Turkey (5%).
Imports: Machinery and transport equipment (typically 35-40 per cent of total value), manufactured goods (33-37%), foodstuffs (12-15%), chemicals (4-5%). Main sources: Italy (typically 25-30% of total value), German FR (12-13%), UK (6-7%), Japan (5-6%), France (4-5%), Republic of Korea (3-4%). There are payment difficulties for all Libyan imports.

Agriculture: Despite intensive promotion, agriculture only accounts for two per cent of GDP, while employing 17 per cent of labour-force. Cultivable land (five per cent of total land area) is concentrated around oases and northern coastal regions, especially near Benghazi and Tripoli. Climatic conditions and irrigation problems limit output and about 40 per cent of food requirements are imported. Large-scale 'GMMR' (Great Man-Made River) project to increase arable area by 180,000 ha by 1990/95, using water conveyed from reserves under the Sahara via a 1900-km pipeline, completed its first phase on schedule in 1987. Main crops: barley (staple diet for most of the population), dates, castor beans, tomatoes, almonds, pototoes, wheat, oats, olives, citrus fruits, groundnuts. Libya is now self-sufficient in fruit and vegetables, dairy and poultry, but imports a third of its meat requirements.

Industry: Non-oil sector, including construction, accounts for 17 per cent of GDP (1987) and employs nearly 30 per cent of workforce. Traditional small-scale agro-allied industries given way to growth of import substitution industries (such as food processing and building materials' manufacture) and heavy industries (such as petro-chemicals, iron and steel, concrete pipes and vehicle assembly). But major metal smelting projects have been impeded by low incomes and depressed world prices. Other constraints to development are insufficient trained Libyan manpower and small domestic market. Virtually all industry is state-owned.

Mining: Major mineral deposits are iron-ore (which will be used to supply the new steel complex at Misurata), potassium, magnesium, sulphur, gypsum and phosphate. Also potential uranium deposits. Commercial exploitation of minerals restricted by high development costs.

Hydrocarbons: Crude oil production is mainstay of economy, accounting for 40 per cent of GDP and 95 per cent of export earnings. Second most important oil producer in Africa after Nigeria, with proven reserves of 20 bn barrels (most of which are high quality light crudes), sufficient to last 56 years at current level of extraction. Most of the principal oilfields are in the Sirte Basin in north-central Libya, and are operated by the state-owned National Oil Corporation (NOC) in partnership with international corporations, but many US-owned partners have withdrawn since 1986. Production and export (domestic demand absorbs less than 10 per cent of total production) have fluctuated in line with world demand/price shifts and allocation of OPEC quotas, and fell by 25 per cent in 1987 to an average 972,000 b/d. Emphasis is currently on offshore exploration (Bouri field), improving recovery techniques and diversifying into marketing/export of refined oil. At present total refining capacity, located chiefly in Zawia, Brega, Misurata and Ras Lanuf, stands at 360,000 b/d. Also part ownership of refining plants in Yugoslavia and Italy. Extensive natural gas reserves (estimated at 500bn cu metres, with up to 300 bn cu metres in the recently discovered Bouri offshore field); extraction is insignificant and net output has steadily declined to under 3.5bn cu metres per year, of which a third is exported. Natural gas is liquefied in Mersa el Brega.

Energy: Oil self-sufficiency since 1960s. Rapid expansion of electricity capacity with two new 720-mW plants being constructed at Zuwara and Zuweitina. Vast majority of electricity is from oil-fired power plants, but a nuclear station is planned for the Gulf of Sirte.

Membership of international organisations: African Development Bank, Arab Common Market, Arab Fund for Economic & Social Development, Arab League, IAEA, IDA, IFAD, IFC, ILO, IMF, Islamic Conference, Islamic Development Bank, OAPEC, OAU, OPEC, UN, WHO, WIPO, World Bank.

BUSINESS GUIDE

Time: GMT +1 (GMT +2 Apr-Sep).

Climate: Coastal areas enjoy temperate Mediterranean climate. Summer (May-Sep) temperatures up to 40°C, while winter (Nov-Apr) temperatures reach 25-32°C during daytime but can fall to 4-5°C at night. Otherwise, 90 per cent desert, with very hot days (to 50°C), cold nights and sandstorms likely in May-Jun. Low rainfall, mainly Oct-Mar in highlands and semi-desert areas.

Entry requirements

Passport: Required by all.

Visa: Required by all except nationals of Algeria, Malta and Mauritania, who require No Objection Certificates. Arabic translation of passport holder's particulars is necessary. All visitors must register with authorities on arrival – this is usually taken care of by the hotelier.

Prohibited entry: Nationals of Israel and South Africa, and passport holders with Israeli, South African or Egyptian visas.

Currency: No restrictions on import/export of foreign currency. Local currency prohibited. All currency must be declared. Credit cards not normally accepted.

Customs: N.B. Alcohol confiscated on arrival. Imports of alcohol, pork and certain other items prohibited.

Health precautions

Mandatory: Yellow fever and cholera vaccination certificate required if travelling from infected areas.

Advisable: Cholera, typhoid and polio vaccinations. Malaria prophylaxis recommended for visits to certain areas. Avoid drinking tap water.

Representation overseas: Abu Dhabi, Accra, Addis Ababa, Aden, Algiers, Amman, Ankara, Antananarivo, Athens, Baghdad, Bamako, Bangui, Banjul, Beijing, Brasilia, Brazzaville, Brussels, Bucharest, Bujumbura, Budapest, Buenos Aires, Canberra, Caracas, Colombo, Conakry, Cotonou, Copenhagen, Dhaka, Damascus, Djibouti, Doha, Gaborone, Georgetown, Harare. Havana, Islamabad, Kuala Lumpur, Kuwait City, Lisbon, Lomé, Madrid, Malé, Manama, Managua, Manila, Mogadishu, Moscow, New Delhi, Nicosia, Nouakchott, Ouagadougou, Panama City, Paris, Port Louis, Prague, Pyongyang, Rabat, Rome, Sana'a, Seoul, Sliema (Malta), Sofia, Stockholm, Tehran, Tokyo, Tunis, Victoria (Seychelles), Vienna, Warsaw, Washington DC, Yaoundé.

Air access

National airline: Libyan Arab.

Other airlines: Aeroflot, Air Algérie, Air Malta, Alitalia, Austrian, Balkan, Bangladesh Biman, CSA, Egyptair, Interflug, JAT, Korean, KLM, Kuwait Airways, LOT, Lufthansa, Malev, MEA, Olympic Airways, PIA, Royal Air Maroc, Royal Jordanian, Saudia, Sudan Airways, Swissair, Syrian Arab, Tarom, THY Turkish, Tunis Air.

International airports: Benina International (Code: BEN), 29 km from Benghazi; Sebha (Code: SEB), 11 km from town; Tripoli International (Code: TIP), 34 km from city.

Surface access

Main ports: Benghazi, Misurata, Port Brega, Tripoli. Port development at Sirte has been postponed, but a new terminal is being built for the Bouri gas field.

Overland access: Via Tunisia and Egypt by main coast road (1820 km long); through to Algeria (via Sebha and Ghat) and to Niger and Chad (from Sebha).

Hotels: Limited availability and usually expensive.

Electricity supply: 110 or 220 V AC.

City transport

Taxis: Yellow government taxis are generally cheaper than private taxis. Advisable to negotiate fare in advance.

Buses: State run services operate in Benghazi and Tripoli. Can be unreliable and over-crowded.

National transport

Air: Travel between western and eastern provinces generally by air. Libyan Arab Airlines operates domestic service between main centres.

Road: Surfaced road system comprising over 32,000 km linking all main centres. Main roads include the 1,800 km national coast road from the Tunisian to the Egyptian border, Sebha-Ghat, Tripoli-Sebha, Agedabia-Kufra, Sebha-Chad and Niger borders.

Buses: Regular services between main centres, but not practicable to use bus services in towns.

Rail: No commercial railway operating, though there are plans to build network from Tripoli to Ras Ajedi.

Public holidays: A new calendar system was introduced in 1986, in which Gregorian month names were replaced by Libyan equivalents. Consult embassies for further details.

Fixed dates: 3 Jan, 12 Jan, 28 Mar, 11 Jun, 1 Sep, 7 Oct.

Variable dates: Birth of the Prophet, Eid al-Fitr and Eid al-Adha.

Working hours

Government and business: 0700 – 1400 summer, 0730 – 1430 winter. (Exception: Fri closed.)

Banking: 0730 – 0800, 1230 – 1300 and about 1500 – 1700 Mon, Wed and Sat; Fri closed; hours may vary between winter and summer and be reduced during Ramadan.

Social customs: Photography of public buildings is not permitted, though there may not be any notices about it.

Telecommunications

Telephone: Dialling code for Libya: IDD access code + 218, followed by area code (21 for Tripoli, 61 for Benghazi) and subscriber's number. Public telephones available in post offices.

Telex: Available at good hotels.

Banking

Central bank: Central Bank of Libya.

Main centre: Tripoli.

Banks in Tripoli: Agricultural Bank, Aruba Bank, Jamahiriya Bank, Libyan Arab Foreign Bank, Masraf Al Gumhouria, National Commercial Bank, Sahara Bank, Umma Bank.

Banks in Benghazi: Wahda Bank.

Trade fairs: International trade fair held in Tripoli every month.

Representation in capital: Afghanistan, Algeria, Argentina, Australia, Austria, Bangladesh, Belgium, Benin, Brazil, Bulgaria, Burundi, Chad, China PR, Cuba, Czechoslovakia, Denmark, Ethiopia, France, German DR, German FR, Ghana, Greece, Guinea, Hungary, India, Iran, Iraq, Italy, Japan, Jordan, South Korea, Kuwait, Lebanon, Malaysia, Mali, Malta, Mauritania, Morocco, Netherlands, Nicaragua, Niger, Nigeria, Pakistan, Philippines, Poland, Qatar, Romania, Rwanda, Saudi Arabia, Senegal, Spain, Sweden, Switzerland, Syria, Togo, Tunisia, Turkey, Uganda, UAE, USA, USSR, Venezuela, Vietnam, Yemen AR, Yemen PDR, Yugoslavia, Zaïre.

Hotels: Benghazi

Atlas, Shara Latioush, P.O. Box: 31 (tel: 92303).

Kasr El-Jazeera, Shara Ali Uraiet, (city centre), P.O. Box: 285 (tel: 96001/110).

Hotels: Tripoli

Al Kabir, Sharah El Fatah, P.O. Box: 275 (tel: 45940/59; tx: 20780).

Libya Palace, Shara Sidi Ibn-El Imam, P.O. Box: 727 (tel: 31181/91).

Omar El-Khayyam, P.O. Box: 2148 (tel: 95100; tx: 40015).

Banking

Bankin'Ny Indostria, 74 Rue du 26 Juin 1960, Antananarivo 101 (tel: 216-82, 238-64).

Bankin'Ny Tantgaha Mpamokatra, PO Box 183, Antananarivo 101 (tel: 202-51, 236-41).

Banky Fampandrosoana Ny Varotra, Antaninarenina, Antananarivo 101 (tel: 206-91, 264-24).

Other useful addresses

Central Bank of Libya, PO Box 1103, Tripoli (tel: 33591; tx: 20196).

Chamber of Commerce, Trade, Industry & Agriculture for the Eastern Province, PO Box 208-1286, Benghazi.

Chamber of Commerce & Industry for the Western Province, PO Box 2321, Tripoli (tel: 33755; tx: 20181).

General Board of Tourism and Fairs, PO Box 891, Haiti Street, Tripoli (tel: 40840/1).

General National Organisation for Industrialisation, PO Box 4388, Tripoli (tel: 44680/34995; tx: 200990).

Kufrah & Serir Authority, Council of Agricultural Development, Benghazi.

Libya Insurance Co, PO Box 2438, Tripoli.

Libyan Arab Airlines, PO Box 2555, Tripoli (tel: 36021).

National Oil Corporation, PO Box 2655, Tripoli (tel: 46180; tx: 61508).

National Trade Union Federation, PO Box 734, Tripoli.

Tripoli International Fair, PO Box 891.

Liechtenstein

John Wicks

The tiny Principality of Liechtenstein, long one of the world's richest countries, continued to grow even more prosperous in 1989. With gross domestic product estimated at well over $36,000 per capita, it far outranked even its wealthy Swiss neighbour. As in the case of Switzerland, further expansion was subject to a number of limiting factors – first and foremost, a chronic shortage of space and manpower. But the weakening of the Swiss franc, which also serves as Liechtenstein's currency, took at least some of the pressure off the almost totally export-oriented national economy.

Record exports

The best gauge of economic activity for the highly-industrialised Principality is the development of its exports. These reached a new record in 1988. On the former statistical basis, which excludes trade with Switzerland – Liechtenstein's partner in a Customs and Currency Union – foreign sales rose by some 10.2 per cent to SwFr1,371.4m. On the new, expanded basis, which includes trade with Switzerland, exports were up 5.2 per cent to SwFr1,875.7m. Although there are no exact figures for overall economic growth, government circles in Vaduz say that gross domestic product was probably up to about SwFr1,700m in 1988, compared with some SwFr1,500m in 1987. This would seem to be calculated on a different basis from the exports, which are higher than this sum, an apparent discrepancy which can be attributed to the complicated nature of the close Swiss-Liechtenstein links across the open border.

Whatever the case, the Liechtenstein Chamber of Industry and Commerce reports 'considerably better results' for the industrial sector in 1988, with high rates of capacity utilisation and virtually full employment. A survey carried out by the Private Trust Bank Corporation (VPBank) shows that manufacturers viewed the profits situation as 'good to very good' and were optimistic as to their chances in 1989. A subsequent survey by the Chamber indicated that members accounting for 93 per cent of the total labour force termed prospects for 1989 business as good.

According to VPBank, the Principality's service sector also did well in 1988. This is reflected not least in the 13.6 per cent rise to SwFr13,400m in the combined balance-sheet totals of the three banks – VPBank itself, Bank in Liechtenstein and the Liechtensteinische Landesbank. Further growth has been recorded by the banks for the first half of 1989, not only in classical balance-sheet business but also in the field of fee-based financial services.

As far as can be ascertained, the offshore activities for which the country is so well known appear to have flourished further in the past year or two. It may well be that this has been helped along not least by Swiss moves to tighten up on the identification of bank clients and possibly also the virtual demise of Panama as an offshore centre. At the same time, however, the Liechtenstein authorities have in the recent past been trying hard to counter abuses. It is categorically denied in Vaduz that large-scale money laundering takes place via Liechtenstein; this would certainly appear to be true as far as actual bank-note transactions are concerned. Furthermore, the country's Banking Association says it is to promote anti-insider rules similar to those in Switzerland.

In respect of tourism, this remains centred on day-trippers and coach parties. The number of overnight occupancies has been largely unchanged or slightly on the decline for several years now and for 1988 will probably still be well below the record levels of the mid-1970s. There is still something of a shortage of quality accommodation, apart from which the still relatively strong Swiss franc makes the Principality an expensive place to stay.

Limits to growth

As yet, the economy continues to grow. The country's Department for Economic Affairs reports more building activity in 1988 than for any year since the record-breaking 1973, just before the oil crisis. This construction boom took place even though the government booked delays in public-authority projects. Over one half of the volume in terms of actual cubic footage was accounted for by trade and industry. Simultaneously, the labour force increased by a further 5.5 per cent to over 19,000.

This expansion cannot continue unabated. Back in 1980, only less than 13 per cent of the country's territory was classified as 'potential building zones'. In effect, today's reserves are substantially smaller– and very much dearer. Apart from this, it is hard to see how there can be any long-term expansion in the workforce. Liechtenstein already has foreign residents accounting for over one-third of the population; political considerations forbid any increase in this share. Hitherto, the slack has been made up by border-crossing employees from Austria and Switzerland. These now make up no less than a third of the labour force. With internal unemployment at barely 0.1 per cent, Liechtenstein is now faced with shrinking labour reserves in the adjacent Austrian State of Vorarlberg and most definitely in eastern Switzerland. Major companies like Hilti foresee most of their future expansion abroad.

These shortages of land and manpower are, however, still being offset at least in part by the constant upgrading of facilities in the Principality itself, resulting in higher added value from largely unchanged installations.

Political stability

General elections were held in early 1989, following a row between the Coalition partners Patriotic Union (VU) and Progressive Citizens' Party (FBP) in connection with the State Court of Justice. This resulted in a continuation of the Coalition, which has now been in existence since 1938, the VU politician Hans Brunhart remaining Head of Government. Liechtenstein politics are based on three pillars, with Parliament acting as a catalyst between the two sovereigns – the Prince and the people, both of whom have interventionary rights in the political process. On the death in 1990 of Prince Franz Josef II, who had been on the throne since 1938, Prince Hans Adam succeeded him as head of state.

With regard to economic and international affairs in particular, Liechtenstein's fate is tied closely to that of Switzerland due to the Customs and Currency Union. The Liechtensteiners take pains to stress their sovereignty, however, and there is talk of setting up a Swiss Embassy in Vaduz to match the Liechtenstein Embassy in Berne. Liechtenstein is already a member in its own right of various international bodies and has a separate codicil as a member of EFTA and the Swiss-EC Free Trade Agreement.

Prince Hans Adam and members of the government would like to consider United Nations membership at some future date, even though some years ago a majority of Swiss voters turned down a similar move for their country. It is, however, by no means sure that a candidature would find majority support in Liechtenstein either. Theoretically, Liechtenstein could make advances to the European Community. But here, much more than in the case of the UNO, it is hard to see how the little Principality could go it alone without Switzerland – and membership for the Swiss would be fraught with problems and is not much more than a talking point.

KEY FACTS

Official title: The Principality of Liechtenstein

Head of state: Prince Hans Adam

Head of government: Hans Brunhart

Ruling party: Patriotic Union

Capital: Vaduz

Official Languages: German

Currency: Swiss Franc (SwFr) = 100 centimes/rappen

Exchange rate: SwFr1.60 per $ (Dec 1989)

Area: 160 sq km

Population: 28,181 (1988)

GDP per capita: $35,600 (1987)

Labour force: 19,200 (1988)

Unemployment: 0.1% (1988)

Inflation: 1.9% (1988)

Trade balance: SwFr745.2m (1988)

COUNTRY PROFILE

Political system: Union of hereditary monarchy and popular sovereignty. Constitution provides for a unicameral parliament (Landtag), which elects a five-member government, which is thereafter officially nominated by the head of state. The number of seats in parliament was increased from 15 to 25 at the last general election, held in 1989. Elections are held every four years on the basis of proportional representation. Voting rights for women on national issues only achieved in 1984.

Main villages: Vaduz (population 4,890), Schaan, Triesen, Balzers, Eschen-Nendeln. .

Media
Press: Two main dailies: *Liechtensteiner Vaterland* (Vaduz, circulation 8,000) and *Liechtensteiner Volksblatt* (Schaan, 8,000).
Broadcasting: No local commercial radio. Swiss, German and Austrian TV and radio broadcasts received (with commercials).
Advertising: No restrictions on newspaper advertising.

Domestic economy: Highly industrialised, export-based economy with well-developed sector. Ranks as one of the wealthiest countries in the world. Strong economic dependence on Swiss economy through customs and currency union. Tourism sector is important earner of foreign exchange.

Employment: About half the workforce employed in secondary sector. Low levels of unemployment and heavy dependence on foreign workers (mostly Austrian and Swiss).

External trade: Rapid growth in exports of machinery, equipment, metals, ceramics and chemicals. Exports and re-exports mainly to other European countries, including 16.2 per cent (1988) to Switzerland (under customs union). Imports mainly from EC and EFTA countries, consist largely of raw materials and foodstuffs. Financial services and tourism make significant contribution to balance of payments.

Agriculture: Small-scale, employing only about 2 per cent of workforce, compared with 40 per cent in 1930s. Activity concentrated on animal husbandry and farming of fodder cereals.

Industry: Owing to lack of raw materials and small domestic market, sector is export-based and centred on specialised and high-tech production. Manufacturing centred on machine building, precision engineering and metal working industries – combined employ over 60 per cent of industrial workforce. Also traditional industries such as chemicals (mainly pharmaceuticals), textiles, ceramics and food processing.

Energy: 92 per cent dependent on imported energy (1988).

Banking: International offshore financial centre. Largest single supplier of fiduciary funds in Europe. Three main banks in operation: Liechtensteinische Landesbank, Bank in Liechtenstein AG, Verwaltungs und Privat-Bank AG. Close association with Swiss banking system and strict secrecy laws in operation.

Legal system: Largely based on Austrian and Swiss law.

Membership of international organisations: Council of Europe, EFTA, IAEA, INTELSAT, Interpol, ITU, UNCTAD, UNICEF, UNIDO, UPU, WIPO.

Time: GMT + 1 (GMT + 2 from late Mar to late Sep).

Climate: Varies with altitude, generally mild and often windy. Average summer temperature 17°C. Average winter temperature 1°C.
Clothing: Mediumweight throughout year, topcoat for winter.

KEY INDICATORS

	Unit	1984	1985	1986	1987	1988
Population	'000	26.7	27.1	27.3	27.7	28.1
Gross domestic product (GDP)	$m	425.5	489.8	777.8	987.0	1,000*
Inflation	%	2.9	3.4	0.8	1.4	1.9
Unemployment	%	0.4	0.3	0.1	0.1	0.1
Exports	SwFr m	1,060.3	1,192.0	1,239.1	1,244.5	1,371.4
Chemicals	SwFr m	83.0	74.0	71.1	80.1	99.1
Metals	SwFr m	242.0	252.0	248.6	259.6	284.4
Machinery/equipment	SwFr m	500.8	538.6	576.1	601.9	644.8
Imports	SwFr m	434.9	436.5	477.5	509.1	626.2
Chemicals	SwFr m	21.7	22.1	23.3	28.9	37.5
Metals	SwFr m	60.9	64.0	95.0	100.1	117.2
Machinery/equipment	SwFr m	146.4	145.1	153.1	164.0	208.0
Balance of trade	SwFr m	625.4	755.5	761.6	735.4	745.2
Exchange rate	SwFr per $	2.4	2.5	1.8	1.5	1.5

* estimated figure

BUSINESS GUIDE

Entry requirements: Open frontier with Switzerland. On the Austrian border the same entry requirements apply as for Switzerland. Nationals of West Germany, Italy, Greece, Austria, Belgium, France, Luxembourg, Netherlands, Portugal and Spain do not require visas.
Currency: No currency restrictions (currency union with Switzerland). N.B: travellers' cheques must be endorsed by companies, not private persons.
Customs: Restricted amounts of alcoholic beverages, tobacco and gifts (up to value of SwFr100) may be imported duty free.

Air access: Nearest international airport is Kloten, Zurich (ZRH). Approx 130 kilometres from Vaduz.

Surface access: Good road access from Switzerland and to a lesser extent Austria; local trains (Austrian Federal Railways) do not serve Vaduz – station only at Nendeln (halts at Schaan and Schaanwald).
Motorway connections: Balzers, Vaduz, Schaan, Bendern, Ruggell.

Hotels: Around 50 hotels and inns supplying some 1,400 beds. Tips included in hotel and restaurant bills.

Currency
Exchange rate: Responsibility of the Swiss National Bank. The Swiss franc has been floating since 1973, with occasional interventions by the central bank during unstable market conditions.

Foreign exchange information
Account restrictions: By agreement with the Swiss National Bank, accounts may be opened only when the identity of the depositor is ascertained and certain other conditions have been met.
Trade policy: There are no exchange controls over imports or exports. The only quantity restrictions on imports apply to agricultural produce.

Car hire: Service offered by Hertz AG and Beck Taxi in Vaduz. Driver must have held a valid driving licence for at least one year and be over 20 years of age. Speed limit 50 kph in city, 80 kph outside.

City transport: Regular and inexpensive post bus service. Easily obtainable taxi service. Tipping not customary.

National transport: All villages can be reached by postal bus service. Restricted rail network, with stations at Schaan and Nendeln. Nearest main stations at Buchs and Sargans in St Gallen, Switzerland and Feldkirch in Austria.

Public holidays
Fixed dates: 1 Jan, 6 Jan (Epiphany), 2 Feb (Candlemas), 19 Mar (St Joseph), 1 May, 15 Aug (Assumption/National Day), 8 Sep (Nativity of Our Lady), 1 Nov (All Saints'), 8 Dec (Immaculate Conception), 25/26 Dec.
Variable dates: Shrove Tuesday, Good Friday, Easter Monday, Ascension Day, Whit Monday, Corpus Christi.

Working hours
Business: (Mon-Fri) 0800 – 1730.
Government: (Mon-Fri) 0800 – 1600.
Banking: (Mon-Fri) 0800 – 1200, 1330 – 1630.
Shops: (Mon-Fri) 0800 – 1200, 1400 – 1830, (Sat) 0800 – 1600.

Telecommunications: Highly efficient telex and telephone service. International dialling code: 41 75. There are no area codes. Telex code 45 FL. N.B. Swiss stamps cannot be used on letters posted in Liechtenstein.

Business advice and information
Liechtensteinische Industrie-und Handelskammer, PO Box 232, Josef Rheinberger-Strasse 11, FL-9490 Vaduz (tel:22744).

Banking
Bank in Liechtenstein AG, Herrengasse 12, FL-9490 Vaduz (tel: 51122; tx: 889222).

Liechtensteinische Landesbank (National Bank of Liechtenstein), Staedtle 44, FL-9490 Vaduz (tel: 68811; tx: 889400).

Verwaltungs-und Privatbank AG, Aeulestrasse 6, FL-9490 Vaduz (tel: 56655; tx: 889200).

Principal newspapers
Liechtensteiner Volksblatt, 9494 Schaan (tel: 24242).

Liechtenstein-Vaterland, Schloss Str 13, 9490 Vaduz (tx: 889442).

Other useful addresses
Amt für Volkswirtschaft der Fürstlichen Regierung (government economic department), FL-9490 Vaduz (tel: 66111).

Liechtenstein National Tourist Office, Postfach 139, FL-9490 Vaduz (tel: 21443; tx: 889488 Into fl; fax: 20806).

Hotels: Vaduz
Engel and Ratskeller, Staedtle 13 (tel: 21057).

Mühle, Landstrasse 120, 9490 Vaduz (tel: 24141).

Park Sonnenhof, Mareestrasse 29 (tel: 21192; tx: 889328; fax: 00053).

Real, Städtle 21, 9490 Vaduz (tel: 22222; tx: 889484).

Schlössle, Schloss Strasse 68 (tel: 25621; tx: 889311).

Luxembourg

Susan Traill

Luxembourg basked in the glow of the general Western economic revival in 1988 and reaped the harvest of its judiciously planned tax cuts and other incentives with major increases in industrial production and foreign investment. The restructuring of the steel industry was successfully completed, enabling the government to start reducing its financial contribution to the country's largest single employer. However, EC proposals for tax harmonisation hung like a sword of Damocles over the financial centre, Luxembourg's main source of earnings.

Luxembourg held general elections on 18 June 1989, at the same time as the elections for the European Parliament. Despite the Christian Democrat-Socialist coalition's success in managing the economy, each party lost three seats in the Chamber of Deputies. The opposition Liberal Party also lost three seats. The Christian Democrats, with 22 seats, and Socialists, with 18 seats, remain in government and the Liberals, with 11 seats, in opposition. The Communist Party has one seat. The election success story was the Green Party, with four seats, and the Five-Sixths Party, also with four seats, which is pressing for the five-sixths of final salary accorded to public sector employees as their retirement pension to be extended to all pensioners.

The composition of Luxembourg's representation in the European Parliament remains unchanged: the Christian Democrats have three seats, the Socialists two and the Liberals one seat. In February 1989, Luxembourg requested an injunction from the European Court of Justice against a European Parliament resolution, the probable effect of which would be to divert business from Luxembourg and Strasbourg to the European Parliament centre in Brussels.

Luxembourg boosted its efforts to increase trade and attract investment by opening an embassy in Spain and a Board of Economic Development in California in 1988. Trade missions visited the UK and Turkey in 1988 and Japan, South Korea and Scandinavia in 1989. LF567m was earmarked in the 1989 budget for development aid to be channelled through a newly created development aid fund.

1992 tax spectre

EC partners accused Luxembourg of being *anti-communautaire* for its opposition to Commission proposals for tax harmonisation, the cornerstone of the Single Market due to come into effect in 1992-93. Prime Minister Jacques Santer stressed that uniform tax rates were unnecessary to ensure fair competition in the Single Market, since prices are largely composed of non-tax elements. He claimed that uniform tax rates would lead to capital flight, especially of the strong currencies, which in turn would destabilise the EMS and lead to a general increase in the cost of industrial credit and a consequent fall in investment. He suggested that the EC adopt a policy of differing tax rates, like those of the US states, but with gradual moves towards harmonisation.

EC tax harmonisation would hit deposit account operations in Luxembourg, since these are currently tax-free. Short-term bonds placed in Luxembourg francs would also be affected. Dividends from foreign shares will be covered by a later EC directive. Eurobonds would continue to be tax-exempt. The government also complained that increased taxes would push up consumer prices by 4.6 – 7.5 per cent, undermining its success in bringing down inflation and increasing industrial competitiveness.

Industrial competitiveness was strengthened in 1989 by the reduction of corporation tax to 35 per cent and by other tax breaks. Government investment in roads, airport expansion and the railways has helped improve trade prospects. Ten new foreign industrial companies set up in Luxembourg in 1988 creating 500 new jobs. Sixteen new banks opened in 1988, reflecting the continued strength of the financial centre. In April 1988, Luxembourg law was brought in line with an EC directive allowing community-wide marketing of investment fund services from October 1989. Sixty-nine new investment funds were formed in Luxembourg in 1988, bringing the total to 516.

The government announced a revision of existing laws on insider trading and money laundering in 1989 to bring these into line with EC norms. Insider trading will be made a criminal offence and foreign authorities will be given easier access to information on suspects. Those convicted of money laundering will face the same penalties as drug dealers. More inspectors will be recruited by the Institut Monétaire Luxembourgeois and the Commissariat à la Bourse will be given broader supervisory powers. But the government insisted it will retain the banking secrecy law which has helped boost private banking business in Luxembourg.

The government's broadcasting satellite 'ASTRA' was successfully launched in December 1988. By March 1989, 75 per cent of its capacity had been let out and half of the TV channels were already in operation. ASTRA has a 20-year frequency guaranteed by international agreement. In 1989, the government offered tax concessions to audiovisual production companies producing programmes in Luxembourg. By March 1989, five of the 10 projects submitted had been approved and represented investment of LF100m. Several companies were considering building studios in Luxembourg.

Revenues from these operations are designed to help offset the loss of tax receipts from the transfer abroad of most of Compagnie Luxembourgeoise de Télédiffusion's (CLT) activities. Another new revenue-earning measure is the setting up of the Luxembourg shipping register. The Belgian government decided in late 1988 to place the country's entire merchant shipping fleet under the Luxembourg flag from 1990 in order to maintain employment. The reduction in tax and social security costs would roughly have ship overheads of BF40m a year.

Unexpected economic boom

Real GDP expanded by 5.2 per cent in 1988, almost

double 1987's 2.7 per cent growth. Inflation increased to 1.4 per cent, against 0.6 per cent in 1987. Industrial production soared far above expectations by 11 per cent from stagnation in 1987, while unemployment fell to 1.4 per cent in 1988 from 1.6 per cent in 1987. Private consumption rose by 3.2 per cent, roughly the same as in 1987, while the pace of government spending slowed to 2.5 per cent. Gross fixed capital formation rose by 3 per cent as against the 5.2 per cent increase in 1987. Total exports picked up, rising by 7.9 per cent, while the increase in total imports slowed to 6.9 per cent. The trade deficit was reduced to LF28bn in 1988 from LF32.3bn the previous year.

Private sector wages increased by 7-8 per cent in nominal terms in 1988 while public sector wages were set to increase by 3.7 per cent over two years (1988-89). However, company competitiveness was maintained by tax cuts. The January 1988 law giving workers further protection against redundancy was supplemented by a measure in the 1989 budget allowing companies to set up a tax-deductible fund to cover redundancy costs in the event of liquidation. The government also reduced personal taxation in the 1989 budget, increased allocations for the minimum wage by 9.5 per cent, for social housing by 10 per cent and for vocational training provided by the Special Employment Fund by 32 per cent, and stepped up measures for childcare, pensioners and emergency health care.

The government made provision to reduce the long-term debt, largely composed of EC loans to the steel

KEY FACTS

Official title: The Grand Duchy of Luxembourg

Head of state: Grand-Duke Jean of Luxembourg

Head of government: Prime Minister Jacques Santer (PCS)

Ruling party: Coalition of Parti Chrétien-Social (PCS) and Parti Ouvrier Socialiste Luxembourgeois (POSL)

Capital: Luxembourg-Ville

Official Languages: Letzeburgish, French, German

Currency: Luxembourg Franc (LF = Belgian Franc) = 100 centimes

Exchange rate: LF37.0 per $ (Dec 1989)

Area: 2,586 sq km

Population: 377,100 (1989)

GDP per capita: $18,486 (1987)

GDP real growth: 5.2% (1988)

Labour force: 174,900 (1988)

Unemployment: 1.4% (1988)

Inflation: 1.6% (1988)

Foreign debt: $334m (1988)

industry, to LF10bn in 1989 from LF13bn in 1988. This combined with tax cuts and increased social spending enabled the government to trim its accumulated budget surplus to LF4.4bn at end-1989 from LF7.4bn a year earlier.

Steel in profit

Steel producer Arbed's parent company posted a 1988 net profit of LF2.3bn against losses of almost the same amount in 1987. The company expects to double this profit in 1989. Consolidated net profit for Arbed's 150 subsidiaries in 11 countries was $319m in 1988 on turnover of $5.1bn. Reserves at end-1988 were $2.05bn. All subsidiaries were in profit. Arbed has successfully diversified its activities away from total dependence on steel. Its new investment company, Novar, will co-ordinate non-steel operations. Arbed set up a second investment company in 1988, Investar, which it owns 50:50 with state-owned investment company SNCI. Novar and Investar will take part in Arbed's LF50bn 1989-93 investment plan. LF13bn of this is earmarked for steel, including R&D, modernisation of installations and product improvement. In March 1989, Arbed shareholders approved a LF4.2bn capital increase to fund investment. Significantly, the Luxembourg government will not take part and its equity in and liability for Arbed will thus be reduced.

Steel capacity utilisation was 95 per cent in 1988 and production was 3 hours-a-tonne for laminates, making Arbed one of the world's most efficient producers. The company has split its main production between its Luxembourg works (long products) and its Belgian subsidiary Sidmar (flat products). In June 1989, Arbed announced its intention to build a $70m 17,500 tpa steelcord and hose wire factory in the USA which will come on stream in 1992. Luxembourg steel production was 3.66m tonnes in 1988, up from 3.3m tonnes in 1987.

The construction industry showed an 8 per cent increase in the volume of operations in the first 10 months of 1988 due to increased investment in the services sector, public sector and in housing. Demand for home ownership was boosted when the government lowered mortgage rates in early 1988. Although they were raised again later in the year, the 1989 budget included government grants and interest subsidies for home buyers. Construction industry order books, including exports, were strong in early 1989. Not surprisingly, quarrying and production of construction materials showed strong growth in both output and turnover in 1988. Metal goods proved to be the strongest growth sector in 1988 with a 55 per cent increase in output and 58 per cent growth in turnover. Conversely, the machinery industry had the poorest performance with a 19 per cent fall in output and 13 per cent drop in turnover. The food industries and chemicals stagnated, while plastics processing and rubber, office machinery and transport equipment did well.

In the banking sector, total balance sheets reached LF9.9bn at end-1988 ($254bn), an increase of 11.2 per cent. Ninety per cent of the total was transacted in foreign exchange. Loans increased by 10 per cent to reach LF599bn, but the banks are carefully provisioned. Employment in the sector reached 13,737 at end-1988, a 7.8 per cent increase. Sixteen new banks opened in 1988, bringing the total to 143. There was a 7.2 per cent increase in the volume of new issues on the Luxembourg stock exchange in 1988, with two-thirds of the transactions in fixed rate loans. Among bond issues, the main increase was in US dollar, sterling and Canadian dollar bonds, while the number of yen-denominated bonds declined.

Outlook

The 1989 tax cuts and investment incentives make Luxembourg an attractive centre for European operations, and investors responded quickly to the benefits offered. The government will continue to promote investment in the audiovisual sector in 1990, and investment in other new industry through Investar will get under way. A report on automation of the stock market was due to be delivered to the government in summer 1989. It is likely that automation will go ahead to bring the stock exchange in line with other EC financial centres. The government will continue to fight EC tax harmonisation and to maintain Luxembourg's status as a centre for EC institutions.

Although the steel industry is expected to suffer a downturn in demand in 1989-90, it is unlikely the government will intervene, since the restructuring of Arbed should allow the company to cope with any short-term problems. The government's enviable budget surplus will continue to be used for the redeployment of steelworkers, for paying off the long-term debt and promoting investment in general. The government will undoubtedly come under pressure to further improve pensions following the success of the Five-Sixths Party in the June 1989 elections.

The OECD predicted slower GDP growth rates in 1989 (3.5 per cent) and 1990 (2.75 per cent). Industrial production increases are expected to fall to 4.5 per cent in 1989 and 2.5 per cent in 1990. Unemployment should remain low at 1.25 per cent and 1.5 per cent respectively, while inflation will pick up, rising to 2.75 per cent in 1989 and 2.5 per cent in 1990. Growth in private consumption should remain steady at 3.5 per cent and 3.25 per cent respectively, and that of government spending will ease to 1.5 per cent a year. The pace of investment was expected to pick up to 4 per cent in 1989 before slowing to 3.25 per cent in 1990. Exports and imports of goods and services were expected to be roughly in balance, with growth of around 5.5 per cent in 1989 and 4.5 per cent in 1990.

KEY INDICATORS

	Unit	1984	1985	1986	1987	1988
Population	'000	366	366	367	370	372
Gross domestic product (GDP)	$bn	3.83	4.04	5.65	6.84	–
GDP per capita	$	10,464	11,038	15,395	18,486	–
Inflation	%	5.6	4.1	0.3	–0.2	1.6
GDP real growth	%	5.7	3.9	3.4	3.7	5.2
Unemployment	'000	2.7	2.6	2.3	2.7	1.4
Share prices	1985=100	67.4	100.0	151.5	137.9	137.5
Industrial production	1985=100	93.7	100.0	102.7	103.3	115.3
Manufacturing	1980=100	113.9	122.1	125.1	125.3	–
Steel	1985=100	91.8	100.0	95.2	87.7	98.9
Construction	1980=100	81.4	73.8	80.3	86.2	–
Exchange rate	LF per $	57.8	59.4	44.7	37.3	36.8

* estimated figure

COUNTRY PROFILE

Political system: Hereditary and constitutional monarchy, with legislative power exercised by a unicameral Chamber of Deputies, which is elected for a five-year term. Executive power vested in the Grand Duke but normally exercised by the Council of Ministers, led by the head of government.

Main cities/towns: Luxembourg-ville (population 80,000), Esch-sur-Alzette, Differdange, Dudelange, Pétange, Ettelbrück, Diekirch, Wiltz.

Media

Press: Luxemburger Wort (La Voix du Luxembourg, circulation 82,306), Tageblatt, Journal (25,992) and *d'Zeitung* are printed daily in German and French. The French *Le Républicain Lorrain* (24,000) which has an extensive Luxembourg section, is widely read. Numerous journals and magazines including *Handelsblad* (2,100), *d'Revue, Télécran, L'Echo de l'Industrie, Letzeburger Land* (6,500) and *Luxemburg News Digest.*

Broadcasting: Radio-Télé Luxembourg (RTL) operates four radio and three TV stations, broadcasting nationally and in France, Belgium and West Germany. Approximately 255,000 radios and 150,000 TVs in use.

Advertising: Commercial advertising on radio and TV and posters.

Domestic economy: Highly developed but small economy based on banking, other financial services, steel and chemical industries. Tourism and manufacturing also important. Heavy dependence on external trade and immigrant labour.

Employment: Industry, trade and tourism remain major employers, although banking and other financial services account for a growing proportion of total employment (7.6 per cent in 1987). Relative to other Western European countries rate of unemployment remains low, averaging 1-2 per cent during 1980s. Around 45 per cent of the workforce are immigrants.

External trade: Regular trade deficit offset by invisible earnings from financial services and other services, e.g. transport and communications. Heavy reliance on trade with Belgium, West Germany, France, Netherlands and the USA. Main merchandise exports include steel (accounts for 40 per cent), plastics, rubber, machinery and textiles. Imports dominated by energy products, machinery, metals and chemicals.

Agriculture: Sector contributes around three per cent to GDP, with farming concentrated on barley, oats and potatoes in the north, and wheat, fruit and grapes in the south. Food and wine accounted for five per cent of exports in 1986.

Industry: Despite successful diversification programme, sector dominated by the ARBED steel company, which is the country's largest single employer and exporter. After experiencing period of rapid decline, brought on partly by EC's restructuring programme, government initiated major overhaul of steel industry. Other significant industries include chemicals, textiles, rubber, plastics processing, metalworking and vehicle spares manufacture. Overall capacity utilisation increased from 70 per cent in 1982 to 75 per cent at end-1986, with greatest rise in non-metallic mineral products, steel and rubber.

Mining: Deposits of iron ore largely depleted. As late as 1970, iron ore production amounted to 5m tonnes – production ended in Dec 1981.

Energy: Heavy dependence on imported energy, mainly from Belgium and German FR.

Banking: Large banking sector incorporating some 142 banks. Activity orientated towards wholesale banking services, with a large concentration of West German and Scandinavian banks serving corporate customers in Europe, but private banking has rapidly increased.

Legal system: French-based, fiscal law German-based.

Membership of international organisations: Benelux, BLEU, CCC, Council of Europe, EC, EIB, EMS, FAO, GATT, IAEA, IBRD, ICAO, IDA, IEA, IFAD, IFC, ILO, IMF, INTELSAT, INTERPOL, IOOC, IPU, ITU, NATO, OECD, UN, UNESCO, UPU, WEU, WHO, WIPO, WMO.

Time: GMT + 1 (GMT + 2 from late Mar to late Sep).

Climate: Temperate; mild winters with cool summers. Rain fairly frequent except during the months Jul to Sep.
Clothing: Mediumweight throughout year. Raincoat useful.

Entry requirements: Visas not required for nationals of USA, Japan, most Western European countries and many others.
Currency: No restrictions on movement of local or foreign currency. N.B. LF not widely accepted in other countries.
Customs: Personal effects and goods up to specified value duty-free.

Air access: Regular flights by all major airlines.
National airline: Luxair.
International airport: Findel (LUX), 7 km east of capital.

Surface access: Rail connections with Brussels, Frankfurt, Amsterdam, Basle and Paris. Good road links with Brussels, Trier, Paris, Frankfurt and Saarbrücken.

Hotels: A one to five star rating system partially in operation. Bills include service charge. Tipping optional.

Currency
Exchange rate: Value of Luxembourg Franc equals that of Belgian Franc, and is not quoted separately.

Foreign exchange information: Exchange and currency regulations applicable in Belgium apply to Luxembourg.

Credit cards: All main credit cards accepted.

Car hire: Readily available from airport and hotels.

City transport: Regular flat-fare bus service operates within greater Luxembourg-Ville. Metered taxi service with minimum charge. Tipping usually 15 per cent.

National transport: Around 5,200 km of roads and 270 km of railway track. State-run bus and railway services link capital with most main towns. A canal service operates on the Moselle.

Public holidays: If a holiday falls on a Sat or Sun, the Mon following is a holiday as well.
Fixed dates: 1 Jan, 1 May, 23 Jun (National Day), 15 Aug (Assumption), 1 Nov (All Saints'), 25/26 Dec (unofficial: 2 Nov All Souls').
Variable dates: Easter Monday, Ascension Day, Whit Monday (unofficial: Shrove Monday and Luxembourg Fair Day – in Luxembourg-Ville only).

Working hours
Government and business: (Mon-Fri) 0800 – 1800, lunch 1200 – 1400.
Banking: (Mon-Fri) 0900 – 1630.
Shops: Large variations: typically 0900 – 2000, closed Mon morning.

Social customs: Polite to be punctual for appointments and to shake hands on arrival and departure.

Telecommunications
Telephone and fax: Local and international direct dialling. International dialling code: 352. There are no area codes.
Telex: Available at main hotels and Postes Luxembourg-Ville. Country code 402 LU.

Trade fairs: Luxembourg International Trade Fair held every spring and autumn, covering a variety of goods.

Electricity supply: 220 V AC.

Representation in capital: Austria, Belgium, Bulgaria, France, Federal Republic of Germany, Greece, Ireland, Italy, Japan, Netherlands, Portugal, Spain, Switzerland, Turkey, USSR, United Kingdom, USA, Zaïre.

Hotels: Luxembourg-Ville
Aerogolf-Sheraton, Route de Trèves (tel: 34571; tx: 2662; fax: 35234217).

Carlton, 9 Rue de Strasbourg (tel: 484802; tx: 2328).

Cravat, 29 Boulevard Roosevelt (tel: 221975; tx: 2846; fax: 26711).

Dauphin, 42 Avenue de la Gare (tel: 488282).

Holiday Inn, Centre European, Luxembourg-Kirchberg (tel: 437761; tx: 2751).

Inter-Continental, 4 Rue Jean-Engling 1466 (tel: 43781; tx: 3754; fax: 436095).

International, 20-22 Place de la Gare (tel: 485911; tx: 2761).

Le Royal, 12 Boulevard Royal (tel: 41616; tx: 2979; fax: 25948).

President, 32 Place de la Gare (tel: 486161; tx: 1510).

Rix, 20 Boulevard Royal (tel: 27545; tx: 1234).

Senator, rue J. Juncker and rue Wiedal (tel: 492351; tx: 1439).

Car hire: Luxembourg-Ville
Avis: 2 Place de la Gare (tel: 489595; tx: 2298).

Budget: (tel: 433412; tx: 1845).

Hertz: 25 Avenue de la Liberté (tel: 485485; tx: 2720).

Inter-Rent: 88 Route de Thionville (tel: 487684, 400707215; tx: 3488).

Business advice and information
Chambre de Commerce du Grand-Duché de Luxembourg, PO Box 1503, 7 Rue Alcide de Gasperi, L-1615 Luxembourg-Kirchberg (tel: 435853; tx: 60174 chcom; fax: 438326).

Ministère de l'Economie et des Classes Moyennes, 19-21 Boulevard Royal, L-2449 Luxembourg (tel: 4794230, 4794228, 4794231; tx: 3464 ecolu; fax: 460448).

Banking
Association des Banques et Banquiers (Banking Association) Luxembourg BP 13, L-2010 (tel: 29501; tx: 1701 abbl).

Badische Kommunale Landesbank International SA, 7 Boulevard Joseph II, L-1840 Luxembourg (tel: 4535451; tx: 1793).

Banque Nationale de Paris SA, 22-24 Boulevard Royal, L-2952 Luxembourg-Ville (tel: 47641).

Banque Continentale du Luxembourg SA, 2 Boulevard Emmanuel Servais, L-2535 Luxembourg (tel: 474491; tx: 2301).

Banque Générale du Luxembourg SA, Avenue Monterey 27, L-2951 Luxembourg (tel: 47991; tx: 3401).

Banque Internationale à Luxembourg SA, 2 Boulevard Royal, L-2953 Luxembourg (tel: 47911; tx: 3626 bil lu).

Banque de Luxembourg SA, 80 Place de la Gare, L-1616 Luxembourg (tel: 499241; tx: 3425; fax: 494820).

Caisse d'Epargne de l'Etat, 1 Place de Metz, L-2954 Luxembourg (tel: 40151; tx: 3417; fax: 27687).

Deutsche Bank Luxembourg SA, 25 Blvd Royal, BP 586, 2449 Luxembourg-Ville (tel: 468181; tx: 2748).

Institut Monétaire Luxembourgeois (banking supervisory authority), 63 Avenue de la Liberté, L-1931 Luxembourg (tel: 478888; tx: 2766).

Kredietbank SA Luxembourgeoise, 43 Boulevard Royal, L-2449 Luxembourg (tel: 47971; tx: 3418 kblux lu).

Principal newspapers
D'Letzeburger Land, 62 Rue de Strasbourg, BP 2083, 1020 Luxembourg-Ville (tel: 485757).

Handelsblad, 23 Allée Scheffer, 2520 Luxembourg-Ville (tel: 473125).

Le Républicain Lorrain, 17B Rue des Bains, 1212 Luxembourg-Ville (tel: 461711; tx: 1717).

Luxemburgerwort (La Voix du Luxembourt), 2 Rue Christophe Plantin, BP 1908, 2988 Luxembourg-Ville (tel: 49931; tx: 3471).

Tageblatt, 44 Rue du Canal, BP 147, 4050 Esch-sur-Aizette (tel: 547131; tx: 3478).

Other useful addresses
Bourse de Luxembourg SA (stock exchange), 11 Avenue de la Porte-Neuve, L-2227 Luxembourg (tel: 4779361; tx: 2559).

Confédération du Commerce Luxembourg, 23 Allée Scheffer, L-2520 Luxembourg (tel: 473125).

Fédération des Industriels Luxembourgeois, 7 Rue Alcide de Gasperi, L-1615 Luxembourg (tel: 435366).

Groupement des Industries Sidérurgiques Luxembourgeoises (Federation of Iron and Steel Industries in Luxembourg), 3 Rue Goethe, BP 1704, 1637 Luxembourg-Ville (tel: 480001; tx: 2784).

Luxair, Luxembourg Airport, 2987 Luxembourg-Ville (tx: 2372).

Ministry of the Economy and Middle Classes, 19-21 Blvd Royal, 2449 Luxembourg-Ville (tel: 4794244; tx: 3464).

Ministry of Finance, 3 Rue de la Congrégation, 1352 Luxembourg-Ville (tel: 4781; tx: 2790).

Ministry of Foreign Affairs, Foreign Trade and Co-operation, 5 Rue Notre Dame 2240 Luxembourg-Ville (tel: 4781; tx: 3405).

Office National du Tourisme, 77 Rue d'Anvers, Box 1001, L-1010 Luxembourg (tel: 400808; tx: 2715).

Press and Information Service of the Government, 10 Boulevard Roosevelt, L-2450 Luxembourg (tel: 478327; tx: SIPLU 1727; fax: 470285).

Statec (Service Central de la Statistique et des Etudes Economiques), BP 304, 19-21 Boulevard Royal, L-2013 Luxembourg (tel: 4794271).

Syndicat d'Initiative et de Tourisme Ville de Luxembourg, Place d'Armes, L-2011 Luxembourg (tel: 22809).

Macão

Michelle Misquitta

The tiny 17.4 sq km Portuguese-ruled enclave of Macão weathered the political turbulence in China in 1989 better than its neighbour Hong Kong. Indeed, some political commentators say the local Chinese and Macanese are gambling – that word is used advisedly – that when the two territories are returned to China's sovereignty, Hong Kong in 1997 and Macão two years later, it will be Macão that will prosper at Hong Kong's expense. If so, it may presage a return to the days before the British colonialists arrived off the China coast when tiny Macão provided the European eyes and ears to what was happening inside the great forbidden empire of China.

KEY FACTS

Official title: City of the Name of God, Macão, There is None More Loyal

Head of state: Governor Carlos Melancia

Capital: Macão

Official Languages: Portuguese (Cantonese widely spoken)

Currency: Pataca (= 100 avos)

Exchange rate: Pataca 8.03 per $ (Nov 1989)

Area: 17.4 sq km (Macão City, Taipa and Culoane islands)

Population: 443,500 (1988)

GDP per capita: $5,242 (1988)

GDP real growth: 7.5%* (1988)

Labour force: 190,000* (1988)

Inflation: 7.8% (1988)

Trade balance: –$0.28m* (1988)

Foreign debt: $1,405m (1987)

Visitor numbers: 5.54m (1988)

* estimated figure

Reaction to Tiananmen Square

People in Macão were upset by the bloodshed in Beijing's Tiananmen Square when Chinese troops cracked down on pro-democracy demonstrators in June. But in the Portuguese enclave there was no massive outpouring of grief as was witnessed in Hong Kong. The local Chinese kept their feelings to themselves. Cynics said that this was either because Beijing has long maintained a careful grip on Macão that has not been possible in the rumbustious larger British colony, or because Portugal has already taken care of its loyal local Chinese by assuring them of full Portuguese passports, something that Britain has failed to do for its Hong Kong Chinese.

Whatever the reasons, whereas Chinese leaders were fulminating against the outspoken support of the pro-democracy demonstrators in Hong Kong, they had warm words for Macão. No less a person than Premier Li Peng in August smilingly expressed satisfaction with Macao and its colonial master. The friendly relations between China and Portugal were 'unaffected by the counter-revolutionary rebellion', Li declared in saying goodbye to Lisbon's outgoing ambassador in Beijing, Octavio Valerio. The ambassador reciprocated the friendly feelings, declaring 'In the course of settling the Macao issue and implementing the joint declaration, Portugal has all along felt the sincerity of the Chinese side.' These were warm sentiments that were not heard in Hong Kong about the Chinese, nor in Beijing about Hong Kong.

Attitude of Hong Kong

In bigger, more prosperous Hong Kong, there has been a tendency to look down on Macao as the poor relation, almost a country cousin. Hong Kong people have also seen the Portuguese colony as a place to go and do things they were not permitted in the more tightly controlled British colony. Gambling is the obvious example. Hong Kong forbids gambling, except on the stock exchange, the horse race-track, and the Mark Six lottery. That does not stop large and small sums being fluttered away each night in majong games. But the irrepressible gambling instincts of the Chinese can be fully indulged in the open and licensed casinos of Macao. Every weekend, the hydrofoils and ferries plying between the British and the Portuguese colonies are full of gamblers heading to Macao in search of their fortune.

But it is easy to see why Hong Kong on the whole has scorned Macao. For a start the Portuguese colony is much smaller, with fewer than 450,000 people, about 8.5 per cent of the population in Hong Kong. Macão is also poorer, with per capita income of about $5,700, 60 per cent of that in Hong Kong. The local economy is much less diversified, being dependent

COUNTRY PROFILE

Political system: Macão is a Chinese territory under Portuguese administration. The President of Portugal is responsible for foreign relations and international agreements, and appoints the governor after local consultation. He delegates certain executive functions to under-secretaries, who are appointed by the Portuguese President on his nomination, and presides over the 17-member legislative assembly; this is composed of five gubernatorial appointees, six elected members and six members representing the business community, all serving three-year terms. Under the terms of an agreement reached between Portugal and China PR in Apr 1987, Macao is to revert to Chinese ownership on 20 Dec 1999. It will then become the 'Special Administrative Region of Macao, China', and will be administered for 50 years on a 'one country, two systems' basis.

Political parties: There are no organised political parties, except for the Association for the Defence of Interests of Macao, which holds one seat in the Legislative Assembly, and the Macau Democratic Centre.

Population: The vast majority of the population are of Chinese descent. The 1981 census results indicate a total resident population of 241,729. Of these, 177,691 have Chinese nationality and 49,007 Portuguese nationality. The estimated population on 31 Dec 1988 was 434,300.

Language: Portuguese is the official language, but Cantonese is most commonly spoken. English is also used in business and tourist circles.

Media
Press: There are five Portuguese newspapers, the daily *Gazeta Macaense*, and *Jornal de Macau*, the twice-weekly *O Clarim*, and the weekly *Comercio de Macau*, *Tribuna de Macau* and *Expresso do Oriente*. English language press includes the monthly *Macau Travel Talk* and the biennial *Macau Image*. There are several Chinese newspapers, including *O Mun*, *Si Man*, *Wa Kio*, *Tai Chung* and *Sing Pou*.
Broadcasting: Official services (TDM) broadcast in Portuguese and Cantonese, with some commercial programming. Villa Verde Radio also broadcasts commercial programmes. Hong Kong commercial radio is also available. TV: Local television station (TDM) broadcasts 1800 – 2300. Transmissions are received from Hong Kong.

Domestic economy: Macão has few natural resources. Its main economic links are with Hong Kong. Mainland refugees provide cheap labour for flourishing industries, and it is a trading and tourist centre. Gambling alone accounts for 32 per cent of all government revenue (1988). China PR has established its second largest Special Economic Zone in Zhuhai, which has powerful business links with Mac\05ao. A Sino-Macanese joint venture is being set up to sell computer-related products in China. Macão pursues an independent programme of infrastructural development, with considerable recent investment in roads and telecommunications. New banking laws intended to attract more foreign banks and allow a full range of offshore banking services have been enacted to ensure participation in financing the development of southern China. The government offers tax and other incentives for investment in tourism and hotels, electronics manufacturing industry, fishing industry and property development.

External trade
Exports: Total exports grew by seven per cent in 1988 with value of 12bn MOP. Re-exports to China accounted for 5.2 per cent and those to Hong Kong for 55.6 per cent of 1987 total. Main domestic exports are textiles and garments (74% of 1988 total), toys (10%), electronics, artificial flowers, leather goods, porcelain, optical products, furniture, fish and shellfish. Main destinations are USA (34.1%), Hong Kong (14%), German FR (12.1%), France (8.9%), UK (7.7%), China PR (3.8%), Australia and Japan.
Imports: Total imports were worth 10bn MOP in 1988. Main imports are foodstuffs, tobacco, machinery, fabrics, fuel, lubricants, iron and steel, raw hides and skins, motor vehicles, paper and paperboard, ceramic products. Main sources are Hong Kong (34%), China PR (20.3%), Japan (10.5%), EC (7%), Taiwan (5%) and USA (4.5%).

Agriculture: Agriculture and fishing accounted for about six per cent of employment in 1981. About nine per cent of the land area is cultivated. Soils are generally meagre and there is little agricultural production. Fish, prawns and other sea foods are trawled for local consumption and export. Most other foods and most of the territory's water are imported.

for 70 per cent of its exports on textiles, and for 40 per cent of the government budget on the gambling casinos. Not surprisingly Hong Kong Chinese have boasted that Macão has come to success on their coat-tails – since 80 per cent of the visitors who provide the gambling revenue come from the British colony.

There are other reasons for the condescending attitude. Macão is a smaller, shabbier place than its bigger colonial neighbour. A Catholic priest who has long lived in the place says 'Macão works very well for 100,000 people; unfortunately, there are nearly 0.5m living there.' Macão is much less developed in every sense. The territory is still run very much like a colonial possession. The governor's post and all the top jobs are held by Portuguese, who are 2,000 strong with a smattering of the next ranks of jobs occupied by Macanese, the local name for Eurasians, who are closely identified with the Portuguese. Portugal for years kept the civil service closed to local Chinese, and only in 1988 started a scheme offering Chinese Portuguese language training and a chance of government jobs. There are few local Chinese lawyers.

The difficulties have been compounded because Portuguese was the official language of the colony, not a language which has great value in the international market place. In 1989 Macão decided to give Chinese the same official status as Portuguese in government operations. All laws, orders and official forms are to be printed in both languages.

Economic development is still having trouble getting beyond the casino and textile stage. Even the local Portuguese officials admit that the Macão textile companies are very much the overspill product of Hong Kong pushed to Macão in the early days of the Multi-Fibre Agreement (MFA) when Hong Kong had used up its quotas and Macão had some to spare. New industries like electronics, toys, watches which have flourished in Hong Kong have not taken root in Macão. Some years ago the Portuguese administration tried to set up an international financial centre, but it never really got off the ground and certainly not as a challenge to Hong Kong.

In 1989 Macão decided to set up a new body, the Monetary Authority of Macão, intended to strengthen the government's direct powers over the financial markets. It will be modelled on the Monetary Authority of Singapore, headed by a co-ordinating council with Secretary for Economic Affairs Dr Antonio Simoes as its first chairman. Its predecessor, the Instituto Emissor de Macão, set up in 1980 as a quasi private sector central banking body, will be closed.

Chinese influence

The appearance of a tight Portuguese grip on the administration is misleading. Beijing has a role inside Macão that it has never enjoyed in Hong Kong. The Nam Kwong Trading Company used to be Beijing's main official body until the Xinhua News Agency opened an office in Macão. But senior Chinese businessmen who run the chambers of commerce, trade unions and residents' associations have always consulted China before taking any major decisions. The legislative council was dominated by the pro-China businessmen and an alliance of Macanese and Portuguese. The deal was that the Portuguese would have the majority, but would accept Beijing's influence over the territory.

This complaisant alliance was stirred in 1984 when Alexandra Ho, a liberal, Chinese-born Catholic educated in Europe won a seat to the assembly. He is scornful of what he calls 'the old forces' whom he describes as 'obstacles to the development of Macão. They never propose new ideas themselves and always ask what Beijing thinks before they adopt any opinion. They are Beijing's loudspeakers.' But he acknowledges that Macão has to 'live with Beijing because we are not as strong as Hong Kong.' Ho is hoping for social and economic changes in Macão and gained further support in the 1988 legislative elections and in district elections in early 1989.

The government is also pushing ahead with development plans which will see an international airport for Macão by the year 1993, a move intended to increase its independence from Hong Kong. The Macão Jockey Club also announced plans to hold horse-race meetings throughout the year, a plan which could be seen as directed against Hong Kong which has only winter racing.

Senior local Chinese can sit back and participate fully in plans to develop Macão knowing very well that they have their Portuguese passports providing them with a happy exit to anywhere in the European Community countries if things go wrong. Admittedly, with much smaller numbers of local Chinese, Lisbon has taken a much more relaxed attitude and offered full passports to people born in Macão before 1980 and their children. This means 130,000 people will qualify. In addition, more than 4,000 public servants, including 2,200 police and security forces, are being guaranteed jobs in Lisbon, even though the security forces mainly speak Chinese, not Portuguese.

Hong Kong Chinese are cynical about whether Macão will ever be able to take off economically on its own. They point to the closeness to Beijing, 'When they announced plans to make Chinese an official language, Beijing offered its help because China knew there weren't enough qualified local people in Macão. That would never happen in Hong Kong', says a senior Hong Kong official scornfully.

COUNTRY PROFILE

Industry: Manufacturing industry (1987) accounts for 59.5 per cent of employment and provides over 90 per cent of all exports. Textile and garment industries represent 74 per cent of exports. Other main products include fabrics, clothing, furniture, optics, rubber, footwear, enamel and porcelain goods, wines, firecrackers, electronic equipment and plastics (toys). Tourism is an important industry, with 82 per cent of all visitors residents of Hong Kong (1988). Many other tourists visit Macão as part of a trip to Hong Kong or increasingly, as part of a trip to China PR.

Energy: In 1988, more than 90 per cent of all electricity was domestically produced from thermal power plant, and the remainder was supplied by China.

Legal system: The judiciary is regulated by Portuguese law.

Membership of international organisations: EC trade and co-operation agreement, Multifiber Agreement.

Time: GMT + 8 hrs.

Climate: Subtropical and monsoonal. Winter (Nov-Apr) is cool and dry, with an average temperature of 14-23°C. Summer (May-Sep) is hot, humid and rainy, with an average temperature of 27°C; Oct is somewhat less humid. Average annual rainfall ranges from 1,000-2,000mm, with monsoon rains from Apr until Oct.

Entry requirements
Passport: Required by all.
Visa: Transit visas (20 days) or visiting visas (90 days) may be obtained on arrival in Macao. Short stays for US, Canadian, Australian and most West European nationals are often exempted altogether. Nationals of countries without diplomatic relations with Portugal must obtain visas before arrival. Passport and visa regulations are liable to change at short notice.
Currency: There are no restrictions on the amount of currency that may be imported or exported.
Customs: Personal effects are allowed duty-free. Macão is a free port and no import duties are levied, although some imports are subject to a special tax. Registration is required for all imports and an import licence for goods subject to consumption tax, such as beverages, coffee, rice, salt, sugar, wheat, matches, tobacco, bricks, cement and mineral oils, gases, vehicles. There are no export duties on articles purchased in Macão.

Health precautions
Mandatory: Vaccination certificates for cholera and/or yellow fever if travelling from an infected area, or if there is a local outbreak of it.

Representation overseas: Contact Portuguese embassies.

Air access: There are no flights to/from Macão and air access is via Hong Kong.

Surface access: Macão is connected to mainland China by a short causeway, which is being improved. However, most visitors enter via Hong Kong. There are frequent daily services to and from Hong Kong by passenger ferry (journey time 2.5-3 hrs), hydrofoil (journey time 65-75 minutes), jetcats (journey time 70 minutes), hovercraft (journey time 45 minutes) and jetfoil (journey time 55 minutes). It is advisable to book in advance.
Main port: City of Macão.

Hotels: The majority of the small hotels available are air-conditioned. It is advisable to book hotels well in advance, especially on weekends. A 10 per cent service charge and five per cent tax are added to the bill. It is customary to leave a small tip.

Credit cards: Most major credit cards are widely accepted.

Car hire: Car hire is not available, but it is possible to hire mini-mokes. Bicycles can be hired from a number of shops in Avenida Dom Joã IV, near São Francisco Gardens.

City transport
Taxis: Taxis are inexpensive and readily available. There are some 586 licenced, metered taxis (all painted black with cream-coloured tops).
Other: It is possible to hire three-wheeled, two-passenger rickshaws, although they are forbidden to cross the bridge and unsuitable for negotiating hills. It is best to negotiate the fare before undertaking the journey.

National transport
Road: The road network consists of 45 km of roads. Taipa Island is connected with Macão by a bridge and with Coloane by a causeway.
Buses: Bus services operate 0700 – 2400, with services between the ferry pier and the city centre and to/from Taipa and Coloane Island.
Water: Small ferries serve Coloane Island

Public holidays: Numerous festivals are celebrated, including the pre-Lent procession of Our Lord of Pasos; in Apr, the A Ma Festival; in May, the feast of Tam Kong on Coloane Island and the procession of our Lady of Fatima in Macão; and in Nov, the Macão Grand Prix.
Fixed dates: 1 Jan, 25 Apr (Liberty Day), 1 May (Worker's Day), 10 Jun (Camões or National Day), 24 Jun (Municipal Day), 1 Oct (China National Day), 5 Oct (Proclamation of the Republic), 2 Nov (All Souls' Day), 1 Dec (Restoration of Portuguese Independence 1640), 8 Dec (Feast of the

BUSINESS GUIDE

Immaculate Conception), 24-25 Dec.
Variable dates: Lunar New Year's Celebration, Good Friday, Easter Monday, Ching Ming Festival, Tuen Ng (Dragon Boat) Festival, Chung Yeung Festival.

Working hours
Business: (Mon-Fri) 0900 – 1300, 1430 – 1730; (Sat) 0900 – 1300.
Banking: (Mon-Fri) 0930 – 1700; (Sat) 0930 – 1300.
Shops: (Mon-Sat) 1000 – 1900.

Telecommunications: International links to the UK, Portugal and Japan are provided by a satellite earth station.
Telephone: Dialling code for Macão, IDD access code + 853 followed by subscriber's number (no area code required). Service is 100 per cent automatic. International calls can be made from main post offices, but not always from hotel rooms. There are 50,000* (1987) telephone subscribers. International direct-dialling telephone service is provided by Cable and Wireless, which owns 75 per cent of Companhia Telecommunicacoes de Macão (Telemac), the operating company. Two new digital exchanges have been commissioned.
Telex: Telex facilities are available.

Banking
Monetary authority: Instituto Emissor de Macău.
Head Offices: Banco Commercial de Macău, Banco de Cantáo, Banco Hang Sang, Banco do Oriente, Banco Seng Heng, Banco Tai Fung, Banco Weng Hang, Luso International Banking, Bank of China, Macău.
Portuguese banks: Banco Nacional Utramarino, Banco Portugues do Atlantico, Banco Pinto & Sotto Mayor, Banco Totta & Acores.
Foreign branches: Banco do Brasil, Bank of Canton, Bank of Credit and Commercial International, Banque Indosuez, Banque Nationale de Paris, Chartered Bank, Citibank, European Asian Bank, Hong Kong & Shanghai Banking Corp, Overseas Trust Bank.
Offshore banks: Banco Fonseca & Burney, Banco Pinto Sotto Mayor, Totta Acores. Following banking reforms in 1985, the number of banks operating in Macão is likely to increase.

Electricity supply: 220 V AC, 50 Hz in new buildings and 110 V AC for most domestic supply, with various types of plug fittings.

Weights and measures: Metric system.

Car hire
Macau Mokes Ltd: Dept of Tourism Office, I/F Macau Ferry Terminal Building (tel: 78851; tx: 88251).

Business advice and information
Macão Business Centre, PO Box 138, Edificio Ribeiro, Loga D,

Macão Commercial Association (Association Commerciale de Macão), 1st Floor, Largo do Senado 20 (tel: 572042).

Macão Export Promotions Department, 1-3 Rua Pedro José Lobo, International Building (tel: 78221; tx: 88413).

Macão Exporters Association, 10th Floor, Rooms 1001-1002, Tai Fung Building, Avenida Almeida Ribeiro 32 (tel: 75859; tx: 88216).

Macão Industrial Association, PO Box 70, Travessa da Praia Grande No. 56(tel: 574125).

Macão Manufacturers Association, Travessa da Praia Grande No. 56, 1Q (tel: 74215).

Macão Statistics Department, PO Box 3022, Ground Floor, Rua Inacio Baptista, 4D-6, Seaview Garden (tel: 550935; fax: 561889).

Banking
Banco Comercial de Macau SARL, Rua da Praia Grande No 22, PO Box 545 (tel: 569622: tx: 88224 BCM-OM; fax: 580967).

Banco de Cantao SARL, Nos 2F-G Avenida de Almeida Ribeiro, PO Box 165 (tel: 568821; tx: 88294 MABOC OM; fax: 570386).

Banco Weng Hang SARL, 21 Avenida de Almeida Ribeiro (tel: 71633; tx: 88225 BWH OM; fax: 576527).

Luso International Banking Ltd, Luso Bank Building, Avenida da Amizade, No 11, PO Box 817 (tel: 78977; tx: 88220 LIBLM OM; fax: 578517).

Tai Fung Bank Ltd, Tai Fung Bank Building, 32 Avenida Almeida Ribeiro (tel: 84899; tx: OM 88212 TAIFUNG; fax: 570737).

Travel information
Macão Department of Tourism, 1 Travessa do Paiva (tel: 77218, 75156; tx: 88338).

Hotels: Coloane Island
Pousada de Coloane, Praia de Cheoc Van (tel: 28143, 28144).

Hotels: Macao
Bela Vista, Rua Commendador Kou Ho Neng (tel: 573821; tx: 88734).

Central, 26-28 Avendia Almeida Ribeiro (tel: 77700, 77730; fax: 332275).

Estoril, Avenida Sidonio Pais, P.O. Box: 141 (tel: 572081/2/3; tx: 83525; fax: 71215).

Lisboa, Avenida da Amizade, P.O. Box: 85 (tel: 7666; tx: 203 HOTEL OM).

London, 4 Praca Ponte e Horta (tel: 83388).

Mandarin Oriental, Avenida da Amizade, P.O. Box: 3016 (tel: 567888; tx: 88588 MACEX OM; fax: 594589).

Matsuya, 5 Estrada de S. Francisco (tel: 575466).

Mondial, Rua de Antonio Basta (tel: 76503; tx: 6503; fax: 568080).

Pousada de Sao Tiago, Fortaleza De Sao Tiago Da Barra (tel: 78111; tx: 88376 TIAGO).

Presidente, Avenida da Amizade (tel: 553888; tx: 88440 HPM OM; fax: 552735).

Royal, Estrada da Vitoria No 2-4 (tel: 552222; tx: 88494; fax: 563008).

Sintra, Avenida da Amizade, P.O. Box: 698 (tel: 85111; tx: 88324).

Hotels: Taipa Island
Hyatt Regency Macau, Taipa Island, P.O. Box: 3008 (tel: 27000; tx: 88512 HYMAC OM; fax: 561361).

Madagascar

Adrian Hewitt

Madagascar now ranks as the twelfth poorest country in the world according to the latest World Bank indicators in the 1989 World Development Report. The country's Gross Domestic Product is $2.07bn and per capita income only $210 per year. The stock of external indebtedness, which rose rapidly in the early 1980s, has now far outstripped Madagascar's annual output, with public external debt outstanding now approaching $3.5bn. This is an impossible burden for Madagascar's already poor and predominantly rice-farming population to bear.

Nor has the world economic climate been kind to Madagascar in recent years. Despite an increasingly rigorous programme of recovery and reform, during the 1980s industrial output has fallen on average by 2 per cent per annum, and the service sector by 0.5 per cent. Agricultural output alone has risen during the 1980s, by an average rate of 2 per cent per annum, but this is still below the rate of population growth. Madagascar continues to need to import rice with scarce foreign exchange, with food aid credits or as donations. The collapse of world coffee prices in mid-1989 dealt yet another blow to Madagascar's economic recovery, for despite diversification policies, coffee remains by far the leading export earner.

Democracy alive and fairly well

Yet remarkably, among this economic gloom, a robust form of Malagasy democracy continues to survive: it faced some of its sternest tests in 1989 with the presidential contest and the legislative elections, and while on the surface the reins of power have not been transferred, popular sentiments are increasingly being heeded and are now instrumental in promoting changes in government policy. For a while in early 1989, the opposition were within an inch of winning

KEY FACTS

Official title: Democratic Republic of Madagascar
Head of state: President Didier Ratsiraka
Capital: Antananarivo
Official Languages: Malagasy, French
Currency: Franc Malgache (FMG) = 100 centimes
Exchange rate: FMG1,437 per $ (Nov 1989)
Area: 587,041 sq km

Population: 10.9m (1987)
GDP per capita: $190 (1986)
GNP real growth: 1.0%* (1987)
Inflation: 15.0%* (1987)
Trade balance: –$16.5m (1986)
Foreign debt: $3.4bn (1987)

* estimated figure

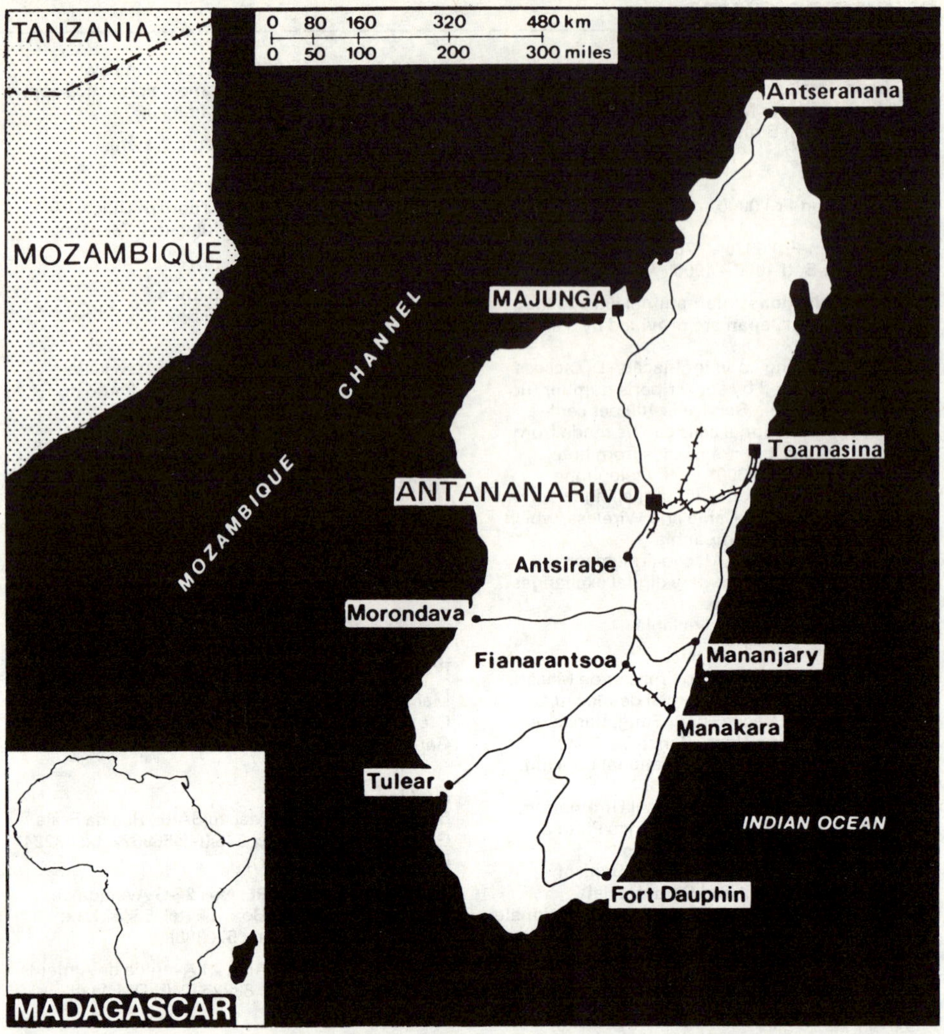

power through the popular vote – which would have made Madagascar more akin to its island neighbour Mauritius than to the mainland African countries – but the parties in opposition to President Ratsiraka and the AREMA (Vanguard of the Malagasy Revolution – the government party) failed, at least this time, to sink their differences and petty rivalries and proved incapable of forming a strong and unified coalition.

Such lively political dissent forms a solid part of Madagascar's tradition. In 1947 a nationalist rebellion against colonial rule was brutally suppressed by the French. The first post-independence government of the Social Democrat Philibert Tsiranana was overthrown in 1972 as a result of strikes and student demonstrations. Attempts by the sitting president, Admiral Didier Ratsiraka, to abolish the opposition and create a one-party state in the early 1980s were fiercely resisted. Far from disbanding or coalescing, new Malagasy political parties are now being formed and the opposition press is again emerging in sprightly form from a period of censorship.

Malagasy presidents are elected for a fixed seven-year term and President Ratsiraka's mandate was due to expire towards the end of 1989, after the parliamentary and local elections had been contested. However, in December 1988, Ratsiraka deftly unsettled the opposition by getting a constitutional amendment passed in the National People's Assembly (which his party dominates) curtailing this, his second seven-year term, and advancing presidential elections to the head of the queue. The opposition were unable or unwilling to agree on a single candidate who stood some chance of defeating the sitting president in the four months electioneering which remained.

Over 4.7m people voted in the presidential election of 12 March 1989. They had a choice of four candidates and, according to the official poll results, Didier Ratsiraka won an absolute majority of the votes cast of almost 63 per cent. The strongest opposition candidate was Manandafy Rakotonirina, with almost one-fifth of the popular vote – a second-best achievement reflecting the fact that opposition parties and factions had failed to coalesce behind a single candidate.

Didier Ratsiraka, 52 years old, was formally declared as re-elected for another seven-year term by the High Constitutional Court on 7 April 1989. Hundreds of appeals against the result alleging electoral fraud were dismissed by the High Court, but were sustained in the case of 43 out of 12,177 polling stations. In these cases the vote was cancelled. The opposition organised a rally to protest against the election results soon afterwards with Manandafy as key speaker. However, electricity supplies to the stadium where the rally was held were disrupted. Later, six people died in rioting and street demonstrations in the aftermath of the presidential elections. An uneasy calm was achieved by the time of the presidential inauguration on 20 April – which was timed to precede a papal visit on 28 April. (Madagascar is about 25 per cent Catholic.)

Both Manandafy and the centre right (VONJY) candidate, Jerome Razanabahiny, turned in creditable performances. Moreover, the incumbent failed to win a majority in Antananarivo or in most of the main provincial centres – Mahajanga, Toliary and Antfiranana – where voting was closely controlled. The sitting president got 70 per cent in Toamasina, his home town, and just squeaked ahead with 52 per cent in Fianarantsoa. The president and his government were particularly disturbed by Manandafy's strong showing. He was, after all, the same man who, as a left-wing professor and student leader, had unseated the government of Philibert Tsiranana in 1972. Furthermore, while all three leading candidates were arguing in favour of better relations with the West and for a constructive attitude toward economic reform and structural adjustment, his was the policy prescription closest to that of the IMF and the World Bank. The former radical academic thereby nearly outflanked the president, though he too had achieved a shift during his last four years in office. The MFM campaigned under the popular slogan of 'farewell to the little red book' *(Veloma ry boky mena)*, President Ratsiraka'a now-defunct *Charter of the Malagasy Socialist Revolution.* It was thus no surprise when Didier Ratsiraka dismissed Manandafy from the Supreme Council of the Revolution (CSR) for having dared to oppose him.

Apart from the president himself – who had expected a larger vote, having achieved 80 per cent at the previous presidential elections – the main casualties of the election were the far left and the orthodox (Moscow-leaning) left leaders. Monja Jaona, a Maoist peasant leader in the 1970s and now 79 years old, received only a paltry 3 per cent of the vote. His political credibility has now collapsed, possibly taking with it that of his party, Monima-K. Seven years ago he was the only candidate to oppose Didier Ratsiraka at the last presidential elections. Then he gained a respectable share of the vote (nearly 20 per cent) and spent some time in gaol after calling a general strike in protest against ballot rigging. The Reverend Richard Andriamanjato, the AKFM (Congress) Party leader since 1958, also lost out. Having sided with President Ratsiraka and having decided not to present himself or his party's candidature in the elections, he was identified with the government and with its failed policies. The AFKM effectively disowned him and he resigned immediately after the elections, subsequently to form a new party – AKFM Fanavazana (Renewal) – to campaign for scientific socialism (marxism-leninism). The AKFM's opposition was also undermined by the very rapid shift in policies on the part of Moscow, especially those concerning Madagascar and socialist states in Africa, which left local communist supporters ill-prepared to establish effective new alliances and party platforms. Rev. Andriamanjato's new party, however, did remarkably well at the legislative elections which followed.

In a conciliatory inauguration speech on 20 April 1989, President Ratsiraka undertook to heed the views of all the people, and 'to reconcile the demands of those who had not voted for him with the aspiration of the majority'. Four months later new blood was brought into the cabinet. In the August reshuffle, the ministry of finance and economic affairs was split

1989 Presidential election: official results

	'000	per cent
Votes cast:		
Admiral Didier Ratsiraka (AREMA)	2,891	62.7
Manandafy Rakotonirina (MFM-MFT)	891	19.3
Jerome Razanabahiny Marojama (VONJY)	688	14.9
Monja Jaona	140	3.0
Total votes cast:	**4,611**	
Spoiled or cancelled	109	
Total attendance:	**4,720**	
Enrolled voters	5,824	
Abstentions	1,104	19.0

1989 Legislative elections:

	Seats	% of Vote
Opposition		
MFM/MFT	7	11.0
VONJY	4	9.7
AKFM-Renewal	3	4.2
Monima K	1	3.0
Government		
AREMA	120	66.8
AKFM (Kongresy)	2	5.3

KEY INDICATORS

	Unit	1983	1984	1985	1986	1987
Population	m	9.40	9.73	9.98	10.55	10.89
Gross domestic product (GDP)	FMG bn	1,220.9	1,369.1	1,553.4	1,806.9	2,210.0
GDP per capita	FMG '000	129.9	140.7	155.7	171.3	202.9
Inflation	%	19.4	9.9	10.5	14.5	15.0
GNP real growth	%	0.5	0.3	2.4	–0.4	1.0
Agricultural production	1979-81=100	109.0	110.8	112.6	116.4	116.9
Food production	1979-81=100	109.7	111.2	113.0	116.6	118.0
Coffee production	'000 tonnes	81	81	82	9	10
Exports	FMG bn	113.4	192.3	181.6	212.7	354.0
Coffee	$m	111.3	142.0	95.9	–	–
Cloves	$m	46.2	36.0	36.9	–	–
Imports	FMG bn	166.8	213.5	265.9	233.9	–
Balance of trade	FMG bn	–53.4	–21.2	–84.3	–11.2	–
Total external debt	$m	–	2,173	2,464	3,012	3,377
Exchange rate	FMG per $	430.5	576.6	662.5	676.3	1,069.2

* estimated figure

into two and technocrats Leon Rajaobelina and Jean Robiarivony appointed to ministerial positions for finance and economy respectively. New ministers were appointed to industry, transport and social welfare, although the sitting prime minister, Colonel Victor Ramahatra, was retained.

Parliamentary elections

The second test for democracy came at the parliamentary elections of 28 May 1989. It would have been constitutionally possible for the opposition to improve on its showing in the presidential elections and emerge with a majority, forcing a period of co-habitation with the Ratsiraka-led government and transforming the National People's Assembly from a rubber-stamp parliament to a powerful legislating and policy-review body. However, not only did the opposition's coalition efforts again fail, but also insofar as the Democratic Alliance across the parties was briefly sustained, it took the almost fatal decision on 30 April to boycott the legislative elections as a protest against the electoral fraud which had occurred previously. Only two weeks before these elections, the opposition decided to reverse its boycott decision and to stand – but as individual parties. The change in policy was justified on the grounds that the threat of boycott had forced modifications to the electoral law: candidates' delegates were to be allowed to monitor the vote in polling stations and a stiffer punishment for fraud was introduced. However, despite its popular appeal, the opposition's vacillations and its incapacity to form a permanent pact between similarly oriented parties was to tell at the 28 May elections.

The government's AREMA party in fact did better than expected, and even better than in earlier legislative elections. With a total of 66.8 per cent of the vote, it actually did better than its leader, Ratsiraka, in the presidential elections. AREMA increased its seats in the national assembly from 117 to 120, and with its close allies the (original) AKFM holding on to two seats, this left the real opposition with only 15 seats.

The turnout, at nearly 75 per cent of the 5.7m registered voters, was also high but was considerably lower in the 19 urban constituencies, including Antananarivo, where the election was more closely monitored. MFM leader Manandafy Rakotonirana claimed that the opposition parties lost 40 to 50 seats as a result

COUNTRY PROFILE

Political system: Executive power is vested in the president, assisted by the Supreme Revolutionary Council. Legislative power is vested in the 137-member National People's Assembly (elections every five years). Madagascar gained independence in 1960.

Political parties: AREMA (Avant-Garde de la Révolution Malgache), AKFM Fanavauzana, AKFM-KDRSM (Independent Congress Party alliance), MONIMA (Mouvement National pour l'Indépendence de Madagascar), MFM (Militants pour un Pouvoir des Petits), UDECMA (Christian democratic union), and VONJY (a faction of the old social democratic party, overthrown in the 1976 military coup).

Population: Madagascar's population of around 10.5m (1987 estimate) is a broad composite of Malayo-Polynesian, Arab and African ethnic groups. The Merinas represent about 20 per cent of the total, while the Betsimisaraka on the east coast account for 10 per cent. There are some 20,000 Europeans, mainly French nationals.

Main cities/towns: Antananarivo (population 1,100,000), Toamasina (57,000), Mahajanga (52,000), Fianarantsoa (49,000), Antsiranana (43,000), Toliary (36,000), Antsirabe (31,000).

Language: French is the usual business language and the medium for all documentation; Malagasy is the other official language. Very little English is spoken.

Media
Press: Principal daily newspapers include *Madagascar-Matin*, *Midi-Madagascar* and *La Croix*. All in Malagasy and French and *Imongo-Vaovao* in Malagasy only. Periodicals include *Revue de l'Océan Indien* and *Lakroan'imadagasikara* (Catholic).
Radio: State monopoly. Radio-Télévision Malagasy (RTM) and Radio Madagasikara broadcast in Malagasy and French.

External trade: Regular trade and current account deficits, largely financed by aid flows and external borrowing.
Exports: Coffee (27.4 per cent of total value in 1985), cloves (20.6%), vanilla (14.8%). Main destinations: France, USA, Japan, German FR.
Imports: Crude petroleum (32.9 per cent of total value in 1985) is the largest single item. Main sources: Saudi Arabia, France, Japan, USA.

Agriculture: Contributes 41 per cent to GDP and employs about 80 per cent of the working population. Main cash/export crops are coffee (82,000 tonnes in 1986), cotton, cloves and vanilla, production of which has fluctuated due to drought and cyclone. Main food crops are rice (basic staple food), maize, bananas and sweet potatoes. The livestock sector is dominant in the west and south of the country. Fishing is small-scale, though it offers considerable potential. Forest preservation and the creation of national parks are receiving large-scale international support.

Industry: Contributes around 15 per cent to GDP. Dominated by food processing and the manufacture of textiles for the domestic market. The first automobile factory opened in 1987 with Iraqi and Libyan help. Major sectors include rice milling, sugar refining, distilling, oil-seed crushing, meat, fruit and vegetable canning, processing of cashew nuts, fruit juices, milk products and jams, cigarettes, soap and rope manufacturing, cotton spinning and brewing. Major capital-intensive industries are oil refining (capacity 800,000 tonnes/year), fertiliser and cement production.

Mining: Sizeable deposits of chrome ore, mica, graphite, gold, bauxite, uranium, iron-ore and nickel. However, as yet only chrome, mica and graphite have been exploited to any great extent, and export earnings from these are limited due to lack of demand. Small quantities of semi-precious stones (garnets and amethysts) are also mined for export.

Energy: Reliance on oil imports, some of which is refined for export. Onshore and offshore oil reserves were discovered in 1980, leading in 1985 to claims of deposits of up to 200m barrels in the coastal waters near Morondava and at Sakhara. Besides fuel wood, hydro-power is the main domestic energy source. But the country's second dam, at Ankorahotra, has been suspended pending evaluation of the oil fields.

Membership of international organisations: African Development Bank, African States in Association with the EC (Lomé Convention), Arab Bank for Economic Development in Africa (recipient), Economic Commission for Africa, GATT, IAEA, IDA, IFAD, IFC, ILO, IMF, Indian Ocean Commission, OAU, UN, WHO, World Bank.

of electoral fraud and that AREMA in fact won only two seats in the capital. But the opposition clearly lost the momentum that was building in their favour in the first quarter of 1989. When on 24 July an armed commando force, claiming to speak for the opposition, seized the national radio station and announced that the Ratsiraka regime had been overthrown, it took only two hours for the government information minister, Simon Pierre, to assert on the radio that the government remained in full control.

Economic austerity continues

While the various forms which structural adjustment was to take remained a major debating point in the electoral period, the reality is that Madagascar's government, and the nation, face a continuing period of austerity for the next few years, as the World Bank and the International Monetary Fund (IMF), representing creditors' interests, insist on further economic reforms. In 1989, Madagascar signed a three-year Enhanced Structural Adjustment Facility (ESAF) loan, worth \$97m, with the IMF, aimed at restoring the balance of payments and recreating a market-based economy. It also projects a recovery of growth to 4.5 per cent. Already a French-owned private bank has been established (all banks were nationalised from 1975), a new tax code more attractive to foreign investors introduced and state intervention in agricultural production and marketing drastically reduced. The 1987 devaluations were so severe (in a bid to stimulate export earnings at any cost) that imported goods are now beyond the reach of most Malagasy people; moreover the debt service has now increased to over 100 per cent of exports (57 per cent after rescheduling): more debt write-offs will be needed before external debt becomes even remotely manageable.

It would be encouraging to see Madagascar as a success story of adjustment (it has been taking the IMF medicine since 1980) but the figures do not prove it yet. Economic growth, currently at 1.8 per cent, is still well below population growth. With the collapse of coffee markets, it is not clear where the projected 4.5 per cent growth is going to come from. Mining (ilnenite/titanium as well as the Malagasy staples of chrome, graphite and mica) could contribute more to output and exports in future. The service sector (tourism and eventually financial services like offshore banking and data processing) could eventually contribute to foreign exchange earnings and assist economic diversification. But this cannot happen in the short term: there is a long struggle ahead to revive the productive economy, restore infrastructure, education and health services, and remotivate a people frustrated as much with domestic bureaucracy as with foreign interference. Perhaps the opposition were smart to keep their collective powder dry until recovery appears more clearly on those Indian Ocean horizons.

BUSINESS GUIDE

Time: GMT +3.

Climate: Tropical, cooler in highlands. In Antananarivo hottest month Dec (15-28°C), coldest Jul (9-19°C). Wettest month (Jan). Rainy season Nov to Mar.

Entry requirements
Passport: Required by all.
Visa: Required by all, except nationals of Lesotho and Malawi.
Currency: Import/export of local currency limited to FMG5,000. No limit on import of foreign currency, but amounts must be declared on arrival, and export allowed up to declared amount. Reconverting Malagasy francs into other currency is virtually impossible.

Health precautions
Mandatory: Yellow fever and cholera vaccination certificates required if arriving from an infected area.
Advisable: Cholera, typhoid and polio vaccinations recommended. Malaria risk exists throughout the country and prophylaxis is necessary. Water precautions should be taken.

Representation overseas: Algiers, Beijing, Bonn, Brussels, Dar es Salaam, Geneva, Moscow, Paris, Port Louis, Rome, Tokyo, Washington.

Air access
National airline: Air Madagascar (one third owned by Air France).
Other airlines: Aeroflot, Air France, Air Mauritius.
Tax: Airport tax of FMG1,500 is levied on departure.
International airport: Ivato-Antananarivo (Code: TNR), 17 km from city; restaurant, currency exchange.

Surface access
Main ports: Antsiranana (Diégo Suarez), Mahajanga (Majunga), Toliary (Tuléar), Toamasina (main port, Tamatave). Few scheduled sea passages.

Hotels: Good hotels are available in Antananarivo, Toamasina, Nosy Be, Ste Marie, and Taolagnaro. A service charge is added to bills at some hotels. Discretionary tipping is usual.

Electricity supply: 110 or 220 V AC, 50 cycles; also 380 V AC, 50 cycles.

Car hire: Available in main centres. International driving licence required.

City transport
Taxis: Flat fare system for short journeys in most towns, otherwise by negotiation; tipping is not usual.

National transport
Air: Air Madagascar operates domestic service to numerous locations (the country has over 100 airfields). Air travel is the most used and generally recommended form of transport. Air Madagascar sell an 'Air Touristic Pass' which gives one month's unlimited travel within Madagascar.
Road: Generally poor and in need of repair. Three main roads leave Antananrivo – (RN4) to Mahajanga, (RN2) to Toamasina and the plateau route south to Fianarantsoa. (RN7).
Rail: Two classes; light refreshements may be available; air-conditioning available on first class trains. Routes include: Antananarivo-Toamasina; Moramanga-Lake Alaotra; Antananarivo-Antsirabe;Fianarantsoa-Manakara. Daily services operate on most routes.

Public holidays
Fixed dates: 1 Jan, 29 Mar (Commemoration (1947) Day), 1 May, 26 Jun (Independence), 15 Aug (Assumption), 1 Nov (All Saints'), 25 Dec, 30 Dec (National Day).
Variable dates: Good Friday, Easter Monday, Ascension, Whit Monday.

Working hours
Government and business: (Mon-Fri) 0800 – 1200, 1400 – 1800.
Banking: (Mon-Fri) 0800 – 1100, 1400 – 1600 (closed in afternoon before a public holiday).
Shops: (Mon-Sat) 0800 – 1200, 1400 – 1800; many grocers open Sun morning.

Telecommunications
Telephone: Dialling code for Madagascar: IDD access code + 261 followed by area code (2 for Antananarivo, 8 for Diego Suarez, 7 for Fianarantsoa, 5 for Tamatave) followed by subscriber's number.
Telex and telegram: Facilities available at the main post office in Antananarivo and the Colbert and Hilton Hotels.

Banking
Central bank: Banque Centrale de la République Malgache.
Head offices: Bankin'Ny Indostria/Banque Nationale pour le Développement Industriel, Bankin'Ny Tantgaha Mpamokatra/Banque Nationale pour le Développement Rural, Banky Fampandrosoana Ny Varotra/Banque Nationale pour le Commerce.

Representation in capital: Algeria, Bulgaria, China PR, Cuba, Egypt, France, German DR, German FR, India, Indonesia, Italy, Japan, Korea PDR, Libya, Romania, Switzerland, UK, USA, USSR, Vatican, Vietnam, Yugoslavia.

Hotels: Antananarivo
Colbert, rue Carayon, P.O. Box: 341 (tel: 20202; tx: 22248 colberhotel).

France, 23 avenue de l'Indépendence, P.O. Box: 607 (tel: 20293; tx: 22322).

Hilton, Rue Pierre Stibbe, Anosy,, P.O. Box: 959 (tel: 26060; tx: 22261 hittels; fax: 26051).

Select, Avenue du 26 Juin, P.O. Box: 589 (tel: 21001).

Car hire: Antananarivo
Air Route Service: (tel: 444-53, 229-98).

Aventours: (tel: 317-61).

Madagascar Airtours: Hilton (tel: 241-92).

Ralrariseta: (tel: 257-70).

Ste Auto Express: (tel: 327-82).

Voyages Bourdon: (tel: 296-96).

Other useful addresses
Antananarivo: Agence Nationale d'Information 'Taratra' (ANTA), 3 Rue du RP Callet, PO Box 386 (tel: 211-71; tx: 22395).

Air Madagascar (tel: 222-22).

Banque Centrale de la République Malgache, Avenue de la Révolution Socialiste Malgache, PO Box 550 (tel: 217-51; tx: 22329).

Bureau de Développement et de Promotions Industriels, PO Box 31 .

Direction du Tourisme de Madagascar, Tsimbazaza, PO Box 610 (tel: 262-98).

Fédération des Chambres de Commerce, d'Industrie et d'Agriculture de Madagascar, PO Box 166 (tel: 215-67).

Institut National de la Statistique et de la Recherche Economique, PO Box 485 (tel: 216-13).

Office Militaire National pour les Industries Stratégiques, 21 Lalana Razanakombana (monitors major industrial projects).

Réseau National des Chemins de Fer, PO Box 259, Soarano (tel: 205-21; tx: 22233).

Société Nationale de Commerce, PO Box 3187 (controls all imports and exports) (tx: 222-58).

Malawi

Thomas Pennant

After two years of slow growth, economic activity in Malawi picked up in 1988 with growth at 3.6 per cent. The recovery was particularly marked in the 'modern' sector following liberalisation of import and foreign exchange controls. Export and import tonnage rose 42 per cent (partly maize imports for over half a million Mozambican refugees) and there were substantial increases in industrial production and private sector investment. For 1989 real GDP growth of 4.5 per cent was forecast – much the same level as achieved in 1983–85, but still uncomfortably close to the population growth rate of 3.7 per cent.

Much of the improvement stemmed from a better financial position, both internally with a reduced budget deficit and externally with more favourable terms for debt outstanding (partial write-offs, longer repayment periods and concessionary interest rates) and increased capital inflows. External debt with Paris and London Club creditors was rescheduled in April 1988 bringing debt relief of K121m. Increased capital inflows were produced by an enhanced Structural Adjustment Facility from the IMF of SDR55.8m in July 1988, following an IMF Stand-by Arrangement providing SDR13m in March 1988. An Industry and Trade Policy Adjustment Credit was arranged with the World Bank, funds being pledged by Japan, the EEC, the ADB, USAID and others as well as the World Bank itself. Covering the two and a half year period to March 1991 this allowed higher import levels facilitating the growth of manufacturing output and investment. There were also substantially increased foreign exchange inflows from donors providing for Mozambican refugees and from migrant workers in South Africa.

Agricultural output increases

Industrial production also expanded because an increase in agricultural output produced more inputs for the processing activities that made up the bulk of the manufacturing sector. Output of tobacco, which accounted for 64 per cent of exports, grew as a 24 per cent increase in Burley output more than offset a decline in production of Flue-, Fire- and Sun/Air-cured tobaccos. Burley prices were favourable because of increased demand for low tar cigarettes in the EEC and for American-style cigarettes in Japan and Taiwan and as a result of drought in some producer countries, particularly the USA. The quantity and quality of Flue-cured tobacco were adversely affected by the poor early-season rains. The fall in output of Fire- and Sun/Air-cured tobaccos occurred as smallholders switched to Burley tobacco and maize. Producer prices for these smallholder tobaccos were raised for the 1988–89 season but this appeared to have done no more than halt the decline in production. Burley production in 1989 was expected to be about 30 per cent up on 1988, with a buoyant market able to absorb this increase.

The quantity of tea produced in Malawi did not increase much in the 1980s but the industry used the limited land available to produce higher quality teas (and higher prices), with substantial replanting and improvement of processing facilities. In 1988 export receipts increased at much the same rate as the overall inflation rate (31.4 per cent) with a 10.4 per cent increase in output. Output in 1989 was expected to be higher (tea did not suffer from the heavy late rains which affected other crops in some parts of the country). With prices firm early in 1989 tea was expected to increase its share of exports from 10.6 to 13.5 per cent.

The other major export crop, sugar, benefitted in 1988 from increased EEC and US import quotas. With increased output expected in 1989 the additional benefit from higher world market prices should be greater, enough at least to maintain sugar's share of total exports around 10 per cent. It seemed likely that the dependence of exports on tea, tobacco and sugar (84.3 per cent of exports in 1988) would increase still further in 1989. Domestic demand from both consumers and industry for maize, groundnuts, pulses, cotton and rice accounted for a higher proportion of output than formerly when there was often a substantial exportable surplus. (Exports of manufactured goods from agro-based industries were expected to rise significantly). Output of most of these increased in 1988, though the picture was less clear than before 1987 when ADMARC lost its purchasing monopoly for smallholder crops other than tobacco and cotton. Rice output should increase strongly from 1990–93 with the completion of irrigation projects. Output estimates for 1987, 1988 and 1989 (forecasts) are given in Table I.

Economic developments

The rate of inflation peaked in the first half of 1988, reaching (year on year) 37.6 per cent in January. Major factors included the 15 per cent devaluation of the Kwacha in January 1988, food price rises following a poor harvest in 1987 and additional demand from Mozambican refugees. Also important were sharp rises in indirect taxation, rents and public utility prices in the 1988 budget. Most of these factors worked their way quickly into higher prices and the inflation rate fell in the second half of 1988. Although not available at the time of writing the mid-1989 year-on-year figure should be below 20 per cent.

The 1988 budget was one of a series of austerity budgets imposed by the IMF and World Bank which reduced public expenditure as a share of GDP from nearly one-third in the early 1980s to barely one quarter in 1988–89. Parastatal bodies were a target for management overhaul and control. By 1988 most had been restored to profitability. Several were associated with major current or prospective capital programmes, including electricity generation and distribution, the second Lilongwe Water Supply project and the replacement of Air Malawi's ageing fleet.

Especially signifiant was the restructuring and return to profitability of the Press Group conglomerate, formerly in the ownership (he would prefer the term trusteeship) of President Banda himself. In two decades this had become involved in every part of the modern sector – agriculture, especially tobacco, manufacturing and publishing, financial services and

commercial activities (where it had partially filled the vacuum left by the displacement of Asian entrepreneurs from most of the country in the 1970s). By the early 1980s Press had grown out of control and was making heavy losses. Restructuring involved transferring the assets and liabilities of Press Holdings from President Banda to Press Group (and ultimately Press Trust, a registered charity 'for the benefit of the people of Malawi'). The assets were then transferred to Press Corporation which was given the task of improving management and profitability so as to repay the loans which the government made to Press Group to cover liabilities of K127m in 1983. An early task was the unscrambling of the joint investment portfolio of Press with the parastatals Malawi Development Corporation and ADMARC. Further changes included the combination of the group's supermarket and hardware chains and substantial investment in basic repair and maintenance of fabric. With improved management and tight financial controls the

KEY FACTS

Official title: Republic of Malawi
Head of state: President Ngwazi Dr Hastings Kamuzu Banda
Ruling party: Malawi Congress Party (MCP)
Capital: Lilongwe
Official Languages: English, Chichewa
Currency: Kwacha (K) = 100 tambala
Exchange rate: K2.75 per $ (Nov 1989)

Area: 118,484 sq km
Population: 8.4m (1989)
GDP per capita: $182 (1987)
GDP real growth: 3.6% (1988)
Inflation: 33.9% (1988)
Trade balance: $139.4m (1987)
Foreign debt: $1.4bn (1987)

group's annual profits increased steadily from 1984 to 1988 when they reached K55.7m with a turnover of K470m. Its subsidiaries in agriculture, distribution, manufacturing and financial services each produced a profit. By 1988 the government loans to Press Group had been reduced to K28m. As the world's largest single grower of tobacco Press should benefit from the high profits on Burley tobacco anticipated for 1989.

External transport costs, long a major constraint, threatened to cripple the economy after 1984 when Renamo first struck effectively at the railway lines to the Mozambican coast. As Table 2 shows, traffic on the Beira and Nacala routes was negligible between 1985 and 1988. The South African route to which most trade was transferred had costs 30 per cent higher than the Dar es Salaam route and twice the level of the Beira and Nacala routes. External transport costs amounted to the equivalent of over 50 per cent of exports. Substantial investment (and foreign aid commitments) were made in renovating the Nacala Line and in establishing the 'Northern Corridor' to Mbeya on the Tazara railway, due for completion in 1992. In the event of a settlement in Mozambique the Nacala Line could quickly be made usable, though not immediately at full capacity because of remaining sub-standard track. The Beira line had deteriorated so badly that it would require years of renovation. But even without it the Dar es Salaam and Nacala routes offered prospects of cost reductions equivalent to a substantial improvement in the terms of trade. Official assumptions were that the Dar es Salaam link will help sustain a growth rate of over 4 per cent to the mid-1990s and that the Nacala Link would permit a higher target.

For most of the period since his return to Malawi in 1958 the succession to Dr Banda was a major political question. Chosen by the youthful leaders of the Malawi Congress Party to add dignity and age-appeal to their independence movement he was expected to retire gracefully from his anticipated figurehead role within a few years. Aged 91 in 1989 he had outlived a number of would-be successors and remained in effective overall control, though there were signs of failing, not least when he lost track in his rambling speeches. His appearances in 1989 were shorter and his former sustained attention to administrative detail lapsed except for brief, apparently arbitrary bouts of intervention when he appeared sometimes to be refighting old battles and old enemies.

Possible political scenarios included a period of something like a regency with power transmitted through his Official Hostess, Cecilia Kadzamira, and her uncle John Tembo, the leading contender to the succession. Despite intermittent attempts to widen their constituency in the Central and Southern Regions (they were strongest in the Central Region) and in core institutions, these two had a narrow power base, deriving mainly from President Banda himself, and antipathy towards them could produce a consensus among influentials for any plausible alternative, including the army commander General Khanga.

From the mid-1980s there was a revival of the political tension present in crises in the mid-1960s, when President Banda fell out with leading members of his independence cabinet and the mid-1970s, after an attempted coup. With economic conditions difficult and salaries declining in real terms corruption became more common, leading to widespread accusations and some arrests.

1989 saw the recurrence of an old theme in Malawian politics: accusations against, dismissal and sometimes detention of northerners in the civil service and education. The Northern Region has only about 13 per cent of Malawi's population, but northerners were well-placed to take advantage of the opportunities for higher education which occurred in the 1960s – mainly because of the higher educational standards of mission schools in the North. After the early intake to the university (and subsequently the civil service) was dominated by northerners, a quota system – one third per 'region of origin' (effectively an ethnic designation) – was introduced in the late 1960s. Changes to this system appeared to set in train an anti-northerner campaign in 1989. A new system of university entry quotas by district rather than region was promoted by politicians from districts whose school-leavers did not attain their share of university places. Though, at least initially, this system was to work within the existing regional quotas it provoked a student protest in January 1989 and there followed dismissals and detentions of staff and students. Early in February 1989 President Banda accused northern civil servants of regionalism in the promotion of the 'Northern Corridor'. (Relations with Tanzania were tense for many years before the re-establishment of diplomatic relations in 1985 because dissidents, mainly from the North, found political asylum there.)

Subsequently a number of northern civil servants were dismissed and some detained. A small number of unexplained deaths were reported to have occurred in custody.

Foreign relations in 1988–89 saw Malawi playing its long-established role of sitting on the fence between South Africa and the Front Line States. Following a state visit in July 1988 from President Chissano of Mozambique, President Botha of South Africa made a short visit in September to hold talks with President Banda. Another visitor, in March 1989, was Mrs Thatcher on her tour of southern Africa. None of these visits had any significant immediate concrete outcome.

Outlook

In 1990 Malawi's per capita GNP may regain the level attained in 1980. For the 1990s to show a signficant improvement on this performance a Mozambican settlement permitting the re-opening of the least-cost external transport routes and the repatriation of refugees seems essential. Also crucial are tobacco and tea prices and the continuation of foreign aid flows. Higher rates of economic growth are needed to provide the means for changing energy sources and agricultural methods to cope with the 3.7 per cent population growth rate revealed by the 1987 census. Whether this rate can be projected into the future – and much besides – depends on the impact of Malawi's reportedly high incidence of AIDS.

KEY INDICATORS

	Unit	1984	1985	1986	1987	1988
Population	m	6.8	7.1	7.3	7.5	7.8
Gross domestic product (GDP)	K m	1,708	1,950	2,191	2,766	3,699
GDP per capita	K	251	275	300	369	474
Inflation	%	20.0	10.4	14.0	25.2	33.9
GDP real growth	%	4.4	4.4	1.1	3.0	3.6
Agricultural production	1979-81=100	108	108	106	104	108
Food production	1979-81=100	103	102	102	101	108
Crop production	1979-81=100	108	109	106	104	108
Livestock production	1979-81=100	109	105	117	112	117
Industrial production	1985=100	99.3	100	101.7	98.0	104.2
Exports	K m	440.7	422.0	462.2	615.1	760.6
Tobacco	K m	255.7	185.5	244.4	373.7	482.9
Tea	K m	112.9	91.4	68.4	61.0	79.6
Sugar	K m	28.9	44.4	39.9	63.4	73.2
Groundnuts	K m	1.1	6.0	15.5	13.2	24.0
Imports	K m	228.9	303.7	286.8	392.4	621.2
Petroleum	K m	45.4	49.6	49.00	72.8	104.1
Balance of trade	Km	211.8	118.3	175.4	222.7	139.4
Current account	$m	0.0	–73.8	–17.8	–9.9	–
Total external debt	$m	885	1,022	1,132	1,363	–
Debt service ratio	%	20.8	29.2	40.5	23.3	–
Exchange rate	K per $	1.4	1.7	1.9	2.2	2.6

* estimated figure

COUNTRY PROFILE

Political system: One party state. Executive power rests with the president (Dr Hastings Kamuzu Banda was appointed Life President in 1971). Parliament consists of the president and the legislature, the National Assembly – 112 members of which are elected for five years, and up to 15 members appointed by the president, as are members of his cabinet. Gained independence in 1964. Formerly known as Nyasaland.

Political parties: The ruling Malawi Congress Party is the only legal political party.

Population: 8.3m (1989 estimate).

Main cities/towns: Blantyre (population 332,000), Lilongwe (capital, 234,000), Zomba (43,000), Mzuzu (44,000).

Language: English is the usual medium for business, while Chichewa is normally spoken in everyday life. Chichewa, Tumbuka, Lomwe and Yao are the main African languages.

Media
Press: One daily (Mon-Fri) newspaper, *The Daily Times*, and one weekly (Sat), *The Malawi News*; plus a limited number of periodicals.
Radio: Domestic radio service broadcasts in English and Chichewa by Malawi Broadcasting Corporation.
Television: No television service.

External trade: Improvements in the trade balance have been held back largely through high transport costs. Sabotage of Malawi's traditional rail links by anti-government rebels in Mozambique has created both export and import difficulties. The debt ratio (total debt outstanding plus use of IMF credit over exports of goods and services) at the end of 1987 was 379.7.
Exports: Tobacco (63 per cent of total value in 1988), tea (10%), sugar (10%), coffee, cotton, groundnuts, pulses. Main destinations: UK (23% of total value in 1987), German FR (9%), USA (11%), South Africa (11%), Netherlands (7%), PTA countries (3%).
Imports: Industrial material, basic and auxiliary (38% of total value in 1987), machinery and equipment (19%), consumer goods (10%), transport (9%), plant, building and construction material (5%). Main sources: South Africa (35% of total value in 1987), UK (20%), PTA countries (10%), Japan (6%), German FR (5%).

Agriculture: Most important single sector of economy, contributing over 38 per cent to GDP in 1985, and providing the livelihood of over 85 per cent of the population. Two sectors: smallholders, growing mainly food crops such as maize and groundnuts; and estate farmers, growing export crops. Tobacco, sugar and tea generated 83 per cent of export earnings in 1988. In the 1980s the marketing of smallholder produce has been liberalised in line with World Bank policy. ADMARC has lost its purchasing monopoly except for smallholder cotton and tobacco. Despite efforts to reverse the trend the smallholder sector has produced a smaller share of exports.

Manufacturing: Sector contributes about 19 per cent to GDP in 1985, and has taken the brunt of economic recession but liberalisation of import and foreign exchange controls led to a recovery in 1988. In previous years expansion has been mostly in the processing of agricultural products, although output of consumer goods and construction materials also increased. The major current constraints on growth are the country's relatively limited resource base, small domestic market and difficulties in importing raw materials and intermediate goods, reduced since 1987 mainly to the benefit of South Africa. The current development plan concentrates on expanding industry to utilise available resources as effectively as possible with an emphasis on small-scale industry. Industrial products include textiles, cigarettes, beer, sugar, ethanol, pharmaceuticals and cement. Tourism, currently limited, is to be developed with foreign involvement.

Mining: 39,000 tonnes of coal was mined in 1988 and there is a prospect of covering existing domestic demand. More conveniently located coalfields at Mwabvi and Lengwe in the southern region are being considered for commercial exploitation. Increased coal production in the late 1980s has reduced coal imports by two-thirds since the mid 1980s. Lime is extracted, and there are known reserves of bauxite, asbestos, graphite, uranium and phosphates.

Energy: All liquid fuels are imported, except for a small amount of ethanol produced from sugar cane. Imported liquid fuels are blended 4:1 with ethanol. Approximately 90 per cent of energy requirements are met by wood burning. A five-year tree planting programme has had to be implemented to replace dwindling resources with shortages becoming acute in the southern region. Net installed electricity generating capacity about 169 mW, with 86 per cent deriving from hydro-electric sources. An additional HEP plant is planned to double output by 1999.

Membership of international organisations: African Development Bank, African States Associated with the EC (Lomé Convention), Non-Aligned Movement, Commonwealth, Economic Commission for Africa, GATT, IDA, IFAD, IFC, ILO, IMF, OAU, Preferential Trade Area, Southern African Development Co-ordination Conference, UN, WHO, World Bank.

Time: GMT +2.

Climate: Tropical, cooler in highlands. Hottest months Oct-Nov, rainy season Nov-Apr. 'Chiperoni' wind can be chilly during Jul-Aug.

Entry requirements: N.B. Strict rules apply concerning clothing and appearance. Male visitors with long hair or shabbily dressed may be refused entry.
Passport: Required by all.
Visa: Required by all except Commonwealth nationals, nationals of certain European countries (including German FR), USA, South Africa.
Currency: Import/export of local currency is allowed up to K20. Foreign currency import unlimited but amounts should be declared. Export allowed up to $15. Currency declaration forms issued on arrival should be kept throughout stay.

Health precautions
Mandatory: Yellow fever vaccination certificate required if arriving from an infected area.
Advisable: Cholera, typhoid and polio vaccinations. Anti-malaria (choice of prophylactic should be checked). N.B. Although tap water is safe to drink in Lilongwe, Blantyre, Limbe and Zomba, other water should be treated with caution. AIDS risk.

Representation overseas: Addis Ababa, Bonn, Brussels, Dar es Salaam, Harare, London, Lusaka, Maputo, Nairobi, New York, Ottawa, Pretoria, Washington.

Air access
National airline: Air Malawi.
Other airlines: Air Tanzania, Air Zimbabwe, British Airways, Ethiopian, Kenya Airways, KLM, LAM, South African Airways, UTA, Zambia Airways.
Tax: Passenger service charge of $10 payable in dollars on departure from Malawi.
International airport: Lilongwe-Kamuzu International (Code: LLW), 28 km from city; bar, restaurant, buffet, bank, post office, shops, car hire.

Surface access
Road: Two main routes from Zambia. Also roads from Zimbabwe, via Lusaka, and from Tanzania.
Rail: Links with Beira (Mozambique) but services not operating in 1987; link with Nacala (Mozambique) with erratic services.

Hotels: Good hotels available in all main commercial centres. However, space can be limited so reservations should be made well in advance and a booking confirmation obtained. Service charge of 10 per cent and a 10 per cent tax are added to bills, and a small tip is occasionally expected.

Credit cards: Only accepted in major hotels, not in shops.

Car hire: Available in main cities. Demand is very heavy so cars should be booked in advance. Self-drive cars are hired at a daily rate, which includes first 40 km. Chauffeurs also charged at a daily rate plus overtime after 1600 and at lunch-time. General speed limit of 80 kph, and 48 kph in urban areas.

City transport
Taxis: In short supply and cannot be hailed. Available by telephone in Blantyre (e.g. tel: 636402). Usually tip of about 30t is expected.

National transport
Air: Air Malawi flies regular services linking Lilongwe, Blantyre, Mzuzu, and Karonga. Also operates charter services to several locations. Twice daily services between Blantyre and Lilongwe (three times Sat; once Sun). This is very heavily used – seats should be booked well in advance.
Road: Over 11,500 km of roads linking main centres. A quarter are main roads and a further quarter secondary roads, most of which are all-weather routes. Other roads become impassable during rainy season. Useful bus network covers most of the country. Luxury coaches operate Blantyre-Zomba-Lilongwe.
Rail: Regular services from Blantyre to Lilongwe but very much slower than buses; connection with ferries at Chipoka on Lake Malawi for services to other lake towns; three classes; first class limited availability; restaurant, sleeping cars.

Public holidays
Fixed dates: 1 Jan, 3 Mar (Martyr's Day), 14 May (Kamuzu Day), 6 Jul (Republic Day, 17 Oct (Mother's Day), 21 Dec (Tree Planting Day), 25 Dec, 26 Dec.
Variable dates: Good Friday, Easter Monday.

Working hours
Government and business: (Mon-Fri) 0730 – 1700, with one hour lunch break.
Banking: (Mon-Fri) 0800 – 1300.
Shops: (Mon-Fri) 0730 or 0800 – 1600 or 1700; (Sat) 0730 or 0800 – 1200.

Telecommunications
Telephone: Dialling code for Malawi: prefix + 265 followed by the subscriber's number. There are no area codes.
Telex and telefax: Public telex and fax facilities in Blantyre and Zomba. Telex country code: MI.

Banking
Central bank: Reserve Bank of Malawi.
Main centres: Blantyre, and Lilongwe.
Head offices: Commercial Bank of Malawi, National Bank of Malawi, Investment and Development Bank of Malawi.

Electricity supply: 230-240 V AC. Three-pin UK plugs the norm in modern buildings.

Representation in capital: Egypt, France, German FR, India, Israel, Republic of Korea, Mozambique, South Africa, Taiwan, UK, USA, Zambia, Zimbabwe.

Hotels: Blantyre
Mount Soche, Glyn Jones Road, P.O. Box: 284 (tel: 620588; tx: 44618; fax: 620124).

Ryalls, Hanover Avenue, P.O. Box: 21 (tel: 620955; tx: 4481; fax: 631680).

Hotels: Lilongwe
Capital, P.O. Box: 30018 (tel: 730444; tx: 4892; fax: 734365).

Lilongwe, Kamuzu Procession Road, P.O. Box: 284 (tel: 721866; tx: 4325).

Lingadzi Inn, P.O. Box: 30139 (tel: 720644/5; tx: 4892).

Other useful addresses
Agricultural Development & Marketing Corp (ADMARC), PO Box 5052, Limbe (tel: 640500, 640044, 640713; tx: 4176).

Air Malawi, PO Box 84, Blantyre (tel: 620177; tx: 4245).

Chamber of Commerce & Industry of Malawi, Henderson Street, PO Box 258, Blantyre (tel: 620670).

Commercial Bank of Malawi, Victoria Avenue, PO Box 111, Blantyre (tel: 633144).

Department of Tourism, Delamere House, PO Box 402, Blantyre (tel: 620300; tx: 4146).

Malaysia

Kevin Rafferty

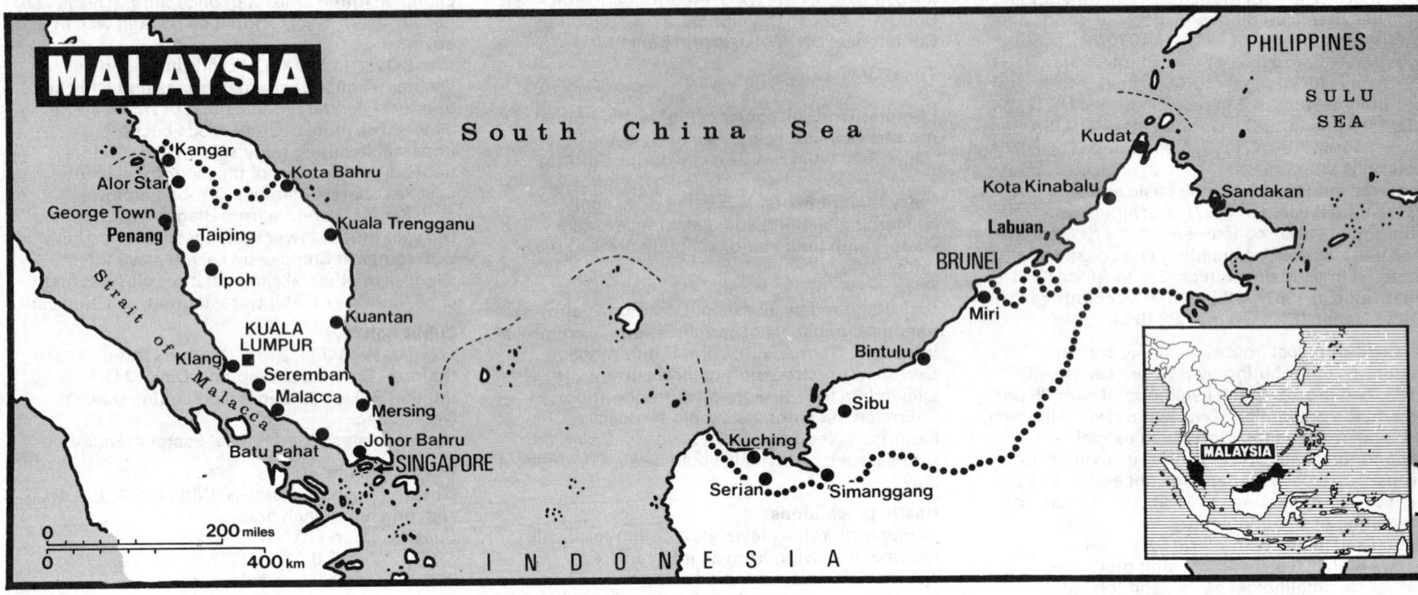

Datuk Seri Dr Mahathir Mohamad, Malaysia's prime minister, shrugged off heart bypass surgery which brought fears for his life, let alone his political future, in early 1989. He was the smiling host of almost 50 heads of government at the Commonwealth summit in October 1989. It was a good platform for the 64-year-old Mahathir to present himself as an internationalist and world-ranking statesman who will lead his country through general elections in 1990 and possibly to the verge of the twenty-first century.

Mahathir's chances and Malaysia's standing have been helped by an economy that continues to boom, so that the country is already on the verge of 'newly industrialising country' (NIC) status, a term which the prime minister himself dislikes. Because of its dependence on commodity exports like oil and natural gas, timber, rubber, tin, palm oil and cocoa, the country remains more vulnerable than most developing countries to world recession. But in its inheritance and balance Malaysia has perhaps the best economic prospects of any Asian economy. It is a country, not a mere city state as are Hong Kong and Singapore, two of the other NICs. It has an enviable balance between land, natural resources and population size, though Mahathir has complained that the 16m population is too small to support his ambitious industrialisation plans. Malaysia's rich natural resources earned it the reputation of being a veritable 'Garden of Eden' on earth. Unlike resource-rich neighbouring Indonesia, which has more than 130m people, Malaysia has room to breathe and expand safely. It also has a good foundation with a per capita income of about $2000 a year, one of the wealthier developing countries.

Political background

Underneath this seemingly continually rosy picture of prosperity and progress, there are major worries and headaches for Malaysia. And chief among them are political concerns. These notably concern Mahathir's hard driving, sometimes prophetically inspired ambitions and whether he is taking the country and its people too quickly along a path that will smash the fragile and delicate racial relationships between Malays, Chinese and Indians.

There can be no doubt about the prime minister's superb, opportunistic brilliance in the dirty game of politics. When the door was opened for him to achieve the prime minister's job, no-one was sure whether he was the right man. That event happened way back in the mid-1970s when Tun Abdul Razak died suddenly in office. Abdul Razak's successor as prime minister Hussain Onn (today Tun Hussain) claimed that he faced a difficult choice in appointing his own successor as deputy president of United Malays National Organisation (UMNO – the main political party for Malays). The choice was between Mahathir, a man who had only recently been re-admitted to the UMNO fold after several years in the wilderness as a radical; Ghafar Baba, a senior politician-businessman but a man of limited educational background; and Tengku Razaleigh Hamzah, a young, rapidly-rising star. Hussain plumped for Mahathir. Mahathir was then ready to step into the prime minister's shoes when Hussain retired, citing ill-health as the reason for stepping down from the hurly-burly of politics and a job which he never really relished.

After Mahathir became prime minister there were still some friends and political allies who were unsure about him. They claimed that two younger UMNO ministers, Datuk Musa Hitam and Razaleigh, enjoyed much greater political backing and warmth of support. The rivalry between the two younger men was intense, with Musa narrowly becoming Mahathir's deputy in a contest among UMNO delegates. Mahathir and Musa were billed as the ideal political team, nicknamed the 2Ms or M1 and M2. Mahathir was assumed to supply the hard-driving vision pushing Malaysia forward, while Musa offered his consummate political skills and made up for the grassroots popular support that Mahathir – it was assumed – lacked. It seemed natural to believe that Musa would even more naturally step into Mahathir's shoes as prime minister – just as Mahathir had stepped into Hussain Onn's shoes. If Musa could continue to stay ahead of his rival Razaleigh, his seemed a certain and popular succession to the top job.

KEY FACTS

Official title: Federation of Malaysia

Head of state: Yang di-Pertuan Agong Sultan Azlan Muhibbuddin Shah Ibni Al-Marhum Sultan Yussuf Izzuddin Ghafarullahu-lahu Shah

Head of government: Datuk Seri Dr Mahathir Mohamad

Ruling party: Barisan Nasional (National Front) coalition

Capital: Kuala Lumpur

Official Languages: Bahasa Malaysia

Currency: Ringgit or Malaysian Dollar (M$ = 100 sen)

Exchange rate: M$2.69 per $ (Nov 1989)

Area: 330,434 sq km

Population: 16.92m* (1988)

GDP per capita: $2,046* (1988)

GDP real growth: 7.4%* (1988)

Unemployment: 8.0%* (1986)

Inflation: 2.5% (1988)

Trade balance: $5,559m (1988)

Foreign debt: $21.675bn (1987)

Visitor numbers: 2.7m* (1985)

* estimated figure

KEY INDICATORS

	Unit	1984	1985	1986	1987	1988
Population	'000	15,190	15,680	16,110	16,530	16,920
Gross domestic product (GDP)	M$m	79,550	77,547	71,144	80,609	90,700
GDP per capita	M$	5,237	4,946	4,416	4,877	5,361
Discount rate	%	5.06	4.13	3.89	3.20	4.12
Consumer prices	% increase	3.5	0.4	0.6	0.8	2.6
Industrial production	% increase	16.0	–3.1	10.3	8.4	14.1
Volume of exports	1985=100	95	100	109	113	188
Rubber	1985=100	106	100	101	108	107
Palm oil	1985=100	92	100	134	126	129
Tin	1985=100	69	100	70	86	85
Petroleum	19850=100	98	100	112	108	124
Exports	M$	38,647	38,017	35,721	45,138	55,334
Rubber	M$	3,672	2,872	3,183	3,915*	5,256
Palm oil	M$	4,547	3,963	3,019	3,263	4,536
Tin	M$	1,162	1,648	652	839	911
Logs and timber	M$	3,982	3,908	4,178	6,055	5,854
Petroleum	M$	8,737	8,698	5,408	6,310	6,128
Imports	M$	32,926	30,438	27,921	31,934	43,411
Balance of trade	$m	2,981	3,577	3,245	5,835	5,559
Current account	$m	–1,671	–613	–122	2,561	1,802
External debt (long term)	$m	15,991	18,150	19,999	21,675	–
Debt service ratio	%	8.5	24.6	13.4	14.3	–
Exchange rate	M$ per $	2.34	2.48	2.58	2.52	2.62

* estimated figure

Power struggle continues

Musa came to grief because he had underestimated Mahathir's political support and the prime minister's clever manipulation of the powerstrings along with a ruthless determination to stay on top. The first steps in a power struggle that has continued until today came in February 1986 when Musa resigned from his posts as deputy president of UMNO and deputy prime minister, claiming that the prime minister was cutting him out of business and had said he could not trust him.

Some political commentators saw this as a ploy by Musa to ditch Mahathir. His expectation was that the UMNO rank and file would rally round to his support and it would be the prime minister who would have to resign. If this was the intention, it failed. Musa miscalculated and forgot that Mahathir's time in power had given him the opportunity to win some supporters and buy others over simply because of the power and perks at his disposal. After party pleadings, Musa stayed on in the UMNO deputy president's job but left the cabinet and government. In the next party triennial elections in 1987 he even teamed up with his hated rival Razaleigh. Razaleigh challenged Mahathir for the UMNO presidency, while Musa fought Ghafar Baba for the deputy president's job.

In a bitterly contested election Mahathir scraped home by 43 votes of 1,479 cast. Ghafar Baba defeated Musa by an even narrower margin – 40 votes – in an election in which 41 papers were put in the ballot box unmarked. It was claimed that the votes defeating Musa were those of Razaleigh supporters who had not forgiven Musa for his earlier rivalry and were not prepared to go along with an obvious alliance of convenience like this. The battle did not end with the party poll and spilled over into the courtroom. A judge declared that because of illegal registration the elections were invalid. That did not stop Mahathir who acted decisively and ruthlessly. He started a new political party Pertubuhan Kebangsaan Melayu Bersatu (Baru) or New United Malays National Organisation (*baru* is the Malay word for 'new').

By 1989 the defeated Musa was re-admitted to the New UMNO fold. He claimed that Mahathir had recently softened his attitude to his opponents. But external observers saw a mixture of humiliation yet political opportunism in Musa's return to the UMNO fold. The former deputy prime minister was prepared to admit defeat, but saw a new chance of power since Mahathir was obviously ailing. His return was announced when Mahathir was convalescing from his operation. When it came to the mid-year cabinet reshuffle, Mahathir had not forgiven Musa sufficiently to reintroduce him to his government. Razaleigh continued his bitter opposition and in 1989 set up his own party called Semangkat '46 (Spirit of '46) – referring to the founding date of the old UMNO to campaign for independence from Britain. Malaysia's first prime minister, Tunku Abdul Rahman and other opposition leaders were among the gathering of the new party's inaugural congress in October. 'We have to destroy the present rule and bring back democracy and freedom to Malaysia,' declared Razaleigh.

One basic problem for Razaleigh is that he cannot afford to go it alone and has had to support his party with an alliance of other groups which are on the surface unhappy bedfellows. His principal ally is the Islamic fundamentalist Parti Islam se Malaysia (PAS). With PAS and the small Berjasa parties, Semangkat '46 has formed what is known as Angkatan Perpaduan Ummah (Muslim Unity Front). For election purposes he is also seeking an alliance with the Chinese dominated but multi-racial Democratic Action Party. At the inaugural congress Razaleigh said the non-Malays should not fear the Muslim Unity Front: 'We shall uphold justice for all groups. Justice is not only for the *bumiputras* (Malays and other indigenous peoples) but also for the non-*bumiputras*'. He added that Islam would be the Front's 'guide in ridding society and government of evil', but non-Muslims would not be forced to become Muslims. However, the Democratic Action Party is still clearly unhappy. Its Secretary General Lim Kit Siang threatened not to enter an alliance with the Front because of PAS's demands that Malaysia should be an Islamic state.

Ethnic and geographical variety

Evidence of by-elections in 1989 indicated that the tide was running in favour of the government coalition, which won five out of seven by-elections in the year to September 1989. Like many things in Malaysia, politics is complicated by the great ethnic, racial and economic variety of the country. Though there are only 16m people, these are divided into 11 states in peninsular Malaysia, plus the two states of Sarawak and Sabah on the north coast of the huge Borneo island. (A tiny enclave on the northern coast of Borneo forms the country of Brunei, while the larger southern part of the island belongs to Indonesia, and is called the Kalimantan province of that country.) Nine of the peninsular Malaysian states have their own hereditary rulers called sultans who together elect the country's Agong (king) for a five-year term. Thus Malaysia is in the peculiar position of having an elected but still hereditary ruler and having nine separate rulers (plus the governors of the four remaining states). But all the rulers are constitutional heads subject to the political rulers. In 1989 Azlan Muhibuddin Shah, Sultan of Perak, was chosen as Agong. He was an internationally distinguished judge, former lord president of Malaysia's Supreme Court, and a big change from his more colourful predecessor the Sultan of Johore who had a reputation as a maverick. At one time in the early- to mid-1980s it looked as if there might be a clash between the monarch and the politicians and Sultan Azlan exercised a cooling influence. In 1988 the Agong, Tuanku Mahmood Iskander, sacked the lord president, Tun Salleh Abas, in a move reputed to have been instigated by Mahathir to bring the judges into line with the government views.

There is great geographical variety between the peninsula and the rest of the country. The bulk of the population is concentrated on the peninsula but the greater part of the land is in east Malaysia, much more rugged land, comprising valuable forest and jungle. In Sarawak in particular there is valuable mangrove forest and both Borneo states are good sources of tropical hardwoods.

The geographical variety is quite matched by racial richness. A slim majority of the country's population consists of Malays and other indigenous people – the *bumiputras*, a word which literally means 'sons of the soil'. Besides Malays, the *bumiputras* include the Ibans of Sarawak and the Kadazans of Sabah. Malaysia also has a large minority of Chinese who today form about 35 per cent of the total population. Indians are the next largest racial grouping, about 9 per cent of the total. The Chinese and Indians are mostly concentrated in the cities and towns like the capital, Kuala Lumpur, Penang, Ipoh and Malacca. Some of the urban areas have a majority Chinese population, and the small Penang state has more Chinese than Malays. On the other hand the Northern and Eastern states of the peninsula, Kelantan, Trengganu, Perlis and Kedah are a Malay heartland where traditional occupations have been farming and fishing: here there are few racial minorities.

Each of the main races has tended to concentrate and congregate in its own racial political party. The government coalition, called the Barisan Nasional (National Front), has long been a large grouping of many parties. For years the practice was that if a new party appeared and gained predominance in any one state, as happened particularly in Sabah and Sarawak where the UMNO writ never ran strongly, and also in Chinese-predominant Penang, that party would be offered a place within the government grand coalition. In this way the government has balanced harmoniously racially but UMNO and the Malays stayed on top. In the early days of Malaya and then – after

1963 – Malaysia, the Malaysian Chinese Association (MCA) was such an important part of the coalition that its leader was finance minister of the country. But since the mid-1970s UMNO has felt strong enough to keep all the important or sensitive portfolios for itself.

Security measures

Prime Minister Mahathir has not been afraid to use the powers at his disposal to stay on top. In 1987 he feared that racial unrest was bubbling up again, so had more than 100 people arrested under the draconian Internal Security Act allowing for detention without trial for renewable periods of up to two years. Those arrested included not merely alleged racial trouble-makers. The opposition claimed that UMNO supporters had been among those stirring up trouble, but hardly any UMNO men were arrested. Instead, those hauled in included a broad swathe of leaders of pressure groups, social and Christian workers, plus a number of opposition MPs including the DAP's Lim and some of his colleagues. In the 1986 general elections the DAP had done so well that it not only overtook the MCA but emerged as the largest opposition party with 24 seats in the 177-member national assembly. Mahathir's sweeping action brought expressions of alarm from many Malaysians including former prime ministers who accused him of trying to run a dictatorship.

But far from bending, the prime minister increased his grip. At his behest the Malaysian parliament passed amendments to the ISA removing the power of judicial review over detention orders in ISA cases. The Act was introduced as far back as 1960 to give the government broad powers to deal with a communist insurgency. Already by the 1980s the insurgents were a diminished force of ageing, mainly Chinese guerrillas in the jungles bordering Thailand and fighting a forlorn battle for the communist cause. On 2 December 1989 the Communist Party of Malaysia (CPM) led by its legendary Secretary-General Chin Peng abandoned its 41-year struggle to overthrow the Malaysian Government. In agreements signed with the Malaysian Government and the Thai military in the southern Thailand town of Haadyai, Chin agreed to disband his 1,100-strong armed CPM units and to destroy all arms and ammunition within 75 days. The deal gave even more force to critics who argue that the ISA is no longer necessary. Mahathir, however, has argued that only the executive can make decisions about national security and must be allowed to act swiftly to avert catastrophes untrammelled by – unelected – judges.

Politics all over Malaysia has much of the rough and tumble about it. It is true at national level and also in the states. In Sabah in 1989 former Deputy Chief Minister Mark Koding founded his own new party. In neighbouring Sarawak rising political awareness among the Ibans, the largest single racial group, was posing problems for a government so far dominated by Malays and Melanaus who are only 20 per cent and 5 per cent respectively of the Sarawak population. The Parti Bansa Dayak Sarawak is in the 13-party national coalition government but was kicked out of the Sarawak government in 1987.

When the old UMNO was declared illegal in February 1988 the party's assets were held in escrow until the courts could decide which of the contesting parties was entitled to them. This caused immense financial difficulty. But new UMNO has been helped by regaining control of United Engineers, a company which has been favoured with a number of government contracts especially for road building. One of them was worth $1.3bn and caused controversy, not least because the company's lack of experience in building roads and closeness to the political party confirmed suspicions that such contracts offered a principal means of generating cash for the UMNO war-chest. In August 1989 United Engineers made a ringgit 146.6m ($55m) takeover of Cement Industries of Malaysia. Critics complain that United Engineers is not the only example of the party's propensity to run its businesses as cash-raising fiefdoms.

The MCA attempts at business fell on stonier ground. Its Multi-Purpose Holdings Investment arm did badly in the mid-1980s, and in 1989 was subject of a takeover bid by Hume Industries led by Qwek Leng Chan. But the bid was withdrawn.

Booming economy

But the best news for Mahathir and UMNO is the booming economy. The next general elections are not due to be held till 1991, but the expectation in 1989 was that the prime minister would cash in on the healthy economy and call the poll in 1990. Bank Negara, the central bank, raised its projection for growth of gross domestic product in 1989 to 8 per cent. Most other economists were forecasting growth of more than 8 per cent thanks to higher than expected domestic demand. The more prudent and cautious Finance Minister Datuk PadukaDaim Zainuddin in his late October budget in 1989 put the figure at 7.6 per cent, down from 8.7 per cent in 1988. He also forecast a further slackening of growth to 6.5 per cent in 1990. The high growth rates have helped fuel inflation, which rose from 2.5 per cent in 1988 to about 4 per cent in 1989. The inflation is partly due to the devaluation of the Malaysian ringgit which in 1989 was worth about 25 per cent less than its level of 1986.

The country is blessed with a vast array of natural commodities, principally oil and gas, rubber, tin, palm oil, tropical hardwoods, but supported by newer crops like cocoa. The biggest export earner is oil and natural gas, fetching about $2.7bn in 1989, a rise of 20 per cent over the previous year. The wealth of the country's natural resources is shown by the fact that oil and gas accounts for about 11 per cent of total exports worth more than $25bn in 1989. This is a much lower share than most oil-exporting countries and gives Malaysia depth and strength. With its vast array of natural resources, the country is protected against weaknesses in individual earnings. In 1989 rubber prices reached a two-year low and export revenues dropped 18 per cent to about $1.6bn. Palm oil earnings remained about the same, worth more than $1.5bn. Malaysia accounts for two-thirds of the world's palm oil and has been involved in a war of words with the US soyabean producers, who have claimed that tropical oils, notably palm oil, have a high cholesterol content, a charge denied by Malaysia. Officials in Kuala Lumpur claimed that the question was a 'trade war disguised as a health issue'. But food producers like MacDonalds hamburgers and Procter and Gamble dropped palm oil to avoid adverse publicity. Malaysia claims that there is plenty of scope for all edible oils and is looking to the Middle East for expansion. In 1989 exports were boosted by heavy Indian buying of oils. Tin exports were boosted by higher prices and reached about $500m in export earnings. The metal was helped by a producer cartel, whose co-operation reduced the tin stockpile from 120,000 tonnes to 25,000 or six weeks' consumption and pushed the price up from $7,200 a tonne at the end of 1988 to a peak of $10,000 in 1989 (though by early 1990 the price had slipped below $7,000 a tonne). Brazil has now overtaken Malaysia as the biggest producer and accounts for 25 per cent of world supply to Malaysia's 20 per cent.

Calls for conservation

But the export earner challenging oil for predominance has drawn Malaysia into domestic and international controversy. This export is timber, earning about $2.5bn a year. International environmentalists have warned that if logging continues at the present rate then Malaysia's tropical forests will be depleted soon after the turn of the century. The cry for conservation has been taken up inside the country at national and local level. Locally the Penans, a Sarawak nomadic tribe, in 1989 got together to stop the logging. They claimed that destruction of the forests had eaten away their homes and hunting grounds. In their home state they formed human blockades to prevent loggers getting to the sites. The tribes were backed by the World Rainforest Movement, an international group of non-governemnt organisations, which petitioned the United Nations General Assembly to respect the rights of the jungle dwellers and halt the destruction of the forests. The UN Food and Agriculture Organisation has also expressed concern about destruction of the forests.

But Prime Minister Mahathir did not care much for this criticism. He went to the Belgrade conference of the Non-Aligned Movement in September 1989 and claimed defiantly that jungle preservation was part of a Western plot to keep cheaper tropical timber out of world markets. He described the Penans as 'unfortunate people' exploited by 'crusading' environmentalists who wanted the tribesmen to be kept as ill-fed, disease-ridden 'museum pieces'. Deputy Prime Minister Ghafar Baba also said there was an international campaign to dissuade Japan from buying Malaysian timber. Nevertheless, the government announced new taxes on timber exports to take effect from March 1990. The reason given was so that timber is available for local processing. Economists said that privately the government was concerned that if it tried to take on the environmentalists head-on it could jeopardise World Bank and other loans to clear new land for oil palm, rubber and cocoa. The ministry of primary industries has begun to advocate tree crop cultivation as part of a programme of 'sustainable forest management.'

Push to industrialisation

Another primary ambition of Mahathir from the early days of his premiership was to begin a 'big push' to make Malaysia an industrial power. The Malaysians, he said back in 1981, no longer wished to be 'drawers of water and hewers of wood.' Mahathir put his personal stamp on a programme of heavy industrialisation based primarily on steel, cars and cement through the state-owned Heavy Industries Corporation of Malaysia (HICOM). The company spent more than ringgit 3bn setting up an iron and steel complex in Trengganu, a cement plant in Kedah, plus the Proton Saga car plant in co-operation with the Mitsubishi group of Japan. In 1989 the motor manufacturer turned in a maiden pre-tax profit though only ringgit 32m ($11.8m), thanks to higher domestic growth boosting demand for cars. The plant just outside Kuala Lumpur began commercial production in 1985 when Malaysia was in recession and domestic car sales had fallen. The car company hopes to boost production to 100,000 units a year from 60,000 in the middle of 1989 and to stay in profit through export sales. Unfortunately its high hopes of selling in the United States, the world's biggest car market, backfired in early 1989 with Proton accusing its US partner of not fulfilling the terms of a November 1987 agreement. Now the Malaysian automobile maker is pinning its hopes on being able to sell more than 10,000 cars a year in the UK.

The prime minister's industrial ambitions have suffered setbacks and left HICOM with heavy debts, but the country has moved a considerable distance from two decades ago when it was mainly a producer of primary commodities. Malaysia has become the

COUNTRY PROFILE

Political system: The Federation of Malaya, comprising the 11 states of Peninsular Malaysia, gained independence in 1957. It changed its name to Malaysia in 1963, when Singapore, Sabah and Sarawak joined the federation. Singapore seceded in 1965. Malaysia is a constitutional monarchy whose king, the Yang di-Pertuan Agong, is elected every five years by a conference of nine hereditary state rulers. The Yang di-Pertuan Agong appoints a council of ministers from the members of the federal parliament, which consists of a 177-member house of representatives (Dewan Rakyat) elected every five years and a senate (Dewan Negara) with two elected members from each state and 32 appointed by the king. Each state has its own constitution and legislative assembly. Malacca, Penang, Sabah and Sarawak are each headed by a governor appointed to a four-year term by the king. The capital, Kuala Lumpur, was designated a Federal Territory in 1974.

Political parties: There are some 24 registered parties, including regional organisations in the various states. Elections held in August 1986 resulted in 148 seats for the Barisan Nasional (National Front). This is a grand coalition of the three big racially based parties, the United Malays National Organisation, the Malaysia Chinese Association, the Malaysian Indian Congress plus a string of smaller parties from the peninsula and the two Borneo states. UMNO, now New UMNO as a result of the in-fighting, dominates the coalition and its president is the country's Prime Minister. The leading opposition party, the Democratic Action Party gained 24 seats though some of its members were arrested and imprisoned in 1987 under the Internal Security Act.

Population: About 82 per cent of the 17.02m population (1988) live in Peninsular Malaysia, with 10 per cent in Sarawak and 8 per cent in Sabah. In Sabah the principal ethnic group is the Kadazans; in Sarawak the Ibans, Bidayuhs and Melanaus predominate. In Peninsular Malaysia itself, the main ethnic group is Malay and there is a substantial minority of Chinese. Population growth is very high at 2.8 per cent p.a.

Main cities/towns: Kuala Lumpur (population 977,100*), Georgetown (capital of Penang, pop 250,000*), Ipoh (capital of Perak, pop 300,000*), Johore Bharu (capital of Johore, pop 250,000*), Malacca (capital of Malacca, 90,000*), Kota Kinabalu (capital of Sabah, 120,000*), Sandakan (pop 120,000*), Kuching (capital of Sarawak, pop 150,000*), Miri (pop 90,000*).

Language: Bahasa Malaysia is the national language, but English is common in commerce and industry. Chinese dialects (Cantonese, Mandarin, Hokkien) are widely used, and Tamil and Punjabi arise among the eight per cent Indian element of the population.

Media
Press: The government controls or owns most of the 42 daily and periodical publications. All newspapers have to be licensed annually. Under a 1984 Act all foreign news must be channelled through the state news agency Bernama. 19 daily newspapers: eight Chinese, five English, four Malay and two Tamil. Circulation is high and sales of newspapers are increasing. Main newspapers are *The New Straits* Times (English, 190,000 copies) *The Star* (English, 160,000 copies), *Business Times, Utusan Malaysia* (Malay), *Nanyang Siang Pau* (Chinese) and *Tamil Nesan* (Tamil).
Radio: Many government-owned stations with separate programmes broadcast in Bahasa Malaysia, Chinese, English and Tamil.
Television: Two TV stations are government-owned and one independent station, TV3. 1.48m sets in 1987.

Domestic economy: Malaysia is abundantly endowed with natural resources – rubber, palm oil, tin, timber, pepper, oil and gas – and has been a stable and fast-growing economy, with wealth from basic resources channelled into diversification through a rapidly expanding industrial sector. The government's 20-year programme (1971-90), known as the New Economic Policy (NEP), is intended to restructure ownership and control of the corporate sector to give Malaysians 70 per cent of equity capital (38 per cent in 1971) and to redistribute the nation's wealth along ethnic lines favouring the Bumiputras (Malays plus indigenous peoples). Bumiputras in 1971 lagged behind foreigners and Chinese and had only 2.8 per cent of the corporate sector. By the mid 1980s the foreign share had come down below 30 per cent while Bumiputras had achieved an 18 per cent share but this was still below target. Development plans originally emphasised import substitution for consumer goods, but have changed since 1981 to development of export-orientated and heavy industries. National growth suddenly collapsed in 1985, when growth rates fell. A combination of plummetting world commodity prices, slower international growth and overambitious Malaysian government plans brought low growth, strained budgets and rising international debts. Relief came in 1987 when growth picked up but Malaysia still faces constraints. Current policy aims to promote private sector growth, continued spending restraint and measures to reduce the current account deficit. The maximum personal income tax rate has been cut from 55 per cent to 40 per cent, and the government is to introduce tax and other incentives which favour large families, aiming for a fourfold increase in population by 2000 in order to increase domestic demand. Plans to privatise the national airline, telecommunications, railways and the container port have been announced.

Employment: Migration to industrial centres, especially Kuala Lumpur, has led to an imbalance in labour supply. The overall unemployment rate is nine per cent (1986) and rising, but plantations face a shortage of labour. Overtime is limited to 32 hours a month to try to create more jobs.

External trade
Exports: The export economy is largely dependent on raw commodities and has suffered since 1981 from low world market prices. Main exports are crude oil (13.9% of 1987 foreign sales),manufactured goods and appliances (44.8%), timber including saw logs (13.9%), chemical products (9.5%), rubber (8.7%), palm oil (8.2%), petroleum products (3.9%) and tin (3.0% in 1984, but only 1.9% in 1986). Main destinations in 1987 were Japan (22.7%), USA (19.5%), Singapore (18.2%), Republic of Korea (5.3%), Netherlands (3.5%), German FR (3.4%), Thailand and UK (3.2% each).
Imports: Imports rose sharply by 14.6 per cent in 1987. The major sectors were intermediate goods which rose 16.7 per cent to 48.3 per cent of total imports, followed by consumer goods to 22.4 per cent and investment goods formed 28.1 per cent of the total. The main suppliers were Japan (21.7%), followed by the USA (18.7%), Asian countries (20.8%), EC (13.4%), Australia (4.1%), Taiwan (3.8%), China (2.9%) and West Asia (2.2%).

Agriculture: Forests cover 27 per cent of the land area, and 22 per cent is cultivated. In 1987, agriculture employed 32 per cent of the labour force and contributed 21.5 per cent to GDP. Malaysia is the world's largest producer of natural rubber, accounting for 34.8 per cent of world production. Most rubber production (97 per cent) is in Peninsular Malaysia, and smallholders account for 60 per cent of output. Since 1970, Malaysia has encouraged diversification into palm oil production, and now accounts for 59.5 per cent of world output. Programmes to replant rubber and palm acreage with better-yielding varieties, and to plant new acreage, are under way. Rice is the main subsistence crop and the government is trying to encourage higher yields. Other main crops are coconuts, cocoa, pepper, sugar cane, pineapples, tobacco, vegetables, sago, tapioca, coffee, tea, maize and groundnuts. Sugar cane and tea are grown as plantation crops; for the rest, smallholders account for most cultivation. Timber logging was cut back by an average 4 per cent per annum in 1983-86, reflecting the conservation policy adopted in response to reductions in total forest area, but Malaysia remains one of the world's largest exporters of tropical hardwood.

Industry: The Malaysian Industrial Development Authority (MIDA) is responsible for promoting and co-ordinating industrial development. MIDA was commissioned to produce a master plan for industry, aiming to improve exports and productivity in the face of declining revenues from commodities. Manufacturing contributed 23.9 per cent of GDP in 1987, becoming the largest contributor to the economy for the first time. Prime Minister Mahathir has encouraged development of heavy industries based on the country's natural resources. But some of the projects had to be cut back because of shortages of funds. The first Malaysian car plant started production in Jul 1985, and steel and copper smelting projects were under consideration in 1987. One of the biggest growing industries has been semiconductors for export. Other industries include food processing (tobacco, cocoa, estate-type agricultural products, beverages, etc.), garments and textiles, rubber products, chemicals, cement and metal products. Despite austerity measures, plans for four cement factories, an iron ore smelter, an aluminium smelter, an iron sheet factory and a light machinery factory, as well as expansion of existing private sector steel-making, are moving forward. The Heavy Industries Corporation of Malaysia (HICOM) is encouraging foreign investment in these projects, preferably on a joint venture basis. Construction has been badly hit by government austerity measures since 1984, and high lending rates have depressed housing demand; the sector shrank by 14 per cent in 1986 and 5.4 per cent in 1987.

Minerals and mining: Malaysia is the world's largest producer of tin, mined mainly in the states of Perak and Selangor; other resources mined include iron ore, bauxite and copper. Weak tin prices, declining reserves, rising production costs and International Tin Council export quotas have depressed tin production in recent years.

Hydrocarbons: Malaysia is self-sufficient in oil, mostly from the offshore fields. The state-owned National Oil Corporation of Malaysia (Petronas) controls the country's 2,500-4,000* mb oil reserves and 79 trillion cu ft natural gas reserves. LNG equipment is still required, and gas production projects in Trengganu and Sabah, to be linked by pipeline to western states of Peninsular Malaysia, are going ahead.

Energy: The government plans to use oil and gas to attract petrochemical and other heavy industries. Development of the as yet unquantified coal reserves and hydro-power potential is intended to diversify sources of energy.

Main companies: Petronas, Sime Darby, Malaysian International Shipping, Malaysian Airlines System, Esso Malaysia, Shell Refining, UMV Corporation, Magnum Corporation, Malaysian Tobacco, Rothmans (M), Consolidated Plantations, Multi-Purpose Holdings (MPH), Federal Flour Mills, Tan Chong Holdings, Malaysian Mining Corporation, Chemical

world's third largest manufacturer of integrated circuits, after the USA and Japan. It is also the world's largest exporter of semiconductors. And in 1990 it is likely to overtake Japan as the top exporter of room air-conditioners. It is also doing well in producing radios and television sets. The country has been helped by its good infrastructure, widespread knowledge of English, plentiful land and lower labour costs than either Japan or its near-neighbour Singapore. Companies like Matsushita Electric Industrial Co and Hitachi of Japan are shifting production to Malaysia and a Hitachi spokesman said that 'the Malaysian (air-conditioner) factory will eventually become one of the company's major factories and, we hope, leading world producer.' Manufacturing accounts for 24.4 per cent of GDP and almost 50 per cent of exports.

Foreign investment

International confidence in the Malaysian economy has been reflected in various ways. Foreign investment has sharply increased. Approvals in the first seven months of 1989 topped $1.7bn, almost up to the level of the whole of 1988. Companies from Japan and Taiwan have had the chance to get into Malaysia. In addition Hong Kong companies have been showing increased interest. The ministry of economic affairs in Taipei has forecast that Taiwanese companies', investments in Malaysia in 1989 could reach $1bn.

However, in 1989 voices were also heard complaining that Malaysia had relaxed its rules and given away too much. Robert Kuok, a millionaire industrialist complained 'for every ringgit of foreign investment, a few more ringgit will flow out and it would be ironic if foreign investments were given preferential treatment.' Datuk Kamil Saleh, director general of the Malaysian Institute of Economic Research, added: 'We may have gone overboard in providing incentives to foreign investors.' Other economists demanded that foreigners should no longer be allowed to purchase local property, a 1987 concession.

Malaysia in 1989 again returned to the Eurodollar bond market, another indication of its improving creditworthiness and confidence. An offering of seven-year 9 per cent bonds for $200m was snapped up with Japanese investors in the lead. Malaysia's debt-service in the mid-1980s had risen to about 25 per cent causing concern about an economy vulnerable to fluctuations in commodity prices. The fact that about a third of the debt is in yen also caused worry inside and outside the country. But the resurgence in economic growth has removed some of the worries and by the end of 1988 total external debt was about $17bn with a debt-service ratio sharply reduced to 13.5 per cent. The government has been replacing foreign with domestic debt and by the end of 1989 total external debts may be down to $15bn, from a peak of almost $21bn in 1986.

But increasing economic growth may still bring problems. The trade account has usually been in healthy surplus except for a few years in the early 1980s when all commodity prices were depressed. In 1988 the surplus was almost $5bn, though this dropped to about $3bn the following year. But the current account was a different matter thanks to the deficit on services. The boom may result in the current account returning to deficit in 1989 and 1990.

Privatisation plans

In August 1989 Mohammed Sheriff bin Mohammed Kassim, director general of the government's Economic Planning Unit said the government had drawn up a 'privatisation masterplan' under which 246 out of 434 state-owned companies would be sold off over the next five years. Of these, 69 companies valued at $1.3bn were targeted for being privatised in the next two years. The companies earmarked for sale come from a wide variety of industries, including agriculture, manufacturing, transport, medical services, utilities and hotels, reflecting the wide involvement of government in the economy. Telecommunications and power are marked for privatisation but not posts. There is still scepticism as to whether the programme will be achieved especially since it has taken five years to sell 22 entities including the national airline and shipping companies, regional utilities and highway construction projects. There have also been accusations of government favouring its own friends in the sell-offs.

Finance and banking

Sometimes prickly, Malaysia has always tried to shrug off its colonial past. In the old days Malaysia was the commodity producing country, – with most of the plantations and industries owned by British colonialists – shipping its produce for sale and export through Singapore. The foreigners have now gone, encouraged by Mahathir's efforts to persuade them to sell to Malaysians. Right until the present day Malaysia and Singapore have been tied closely together, especially through common quotation of shares on each other's stock exchanges. In fact in 1989 the 329 counters on the Singapore exchange included 182 Malaysian companies (accounting for 37 per cent of the Singapore exchange's market capitalisation of $62.2bn). In his budget statement at the end of October 1989, Finance Minister Daim declared the time had come for Malaysia to come out from Singapore's shadow. He ordered that Malaysian companies on the Singapore exchange must delist immediately. The Kuala Lumpur stock exchange he declared had become vulnerable to developments in Singapore. Since 1987 new Malaysian companies being floated on the stock market had been permitted to be quoted in Kuala Lumpur only, not Singapore. Daim declared that the separation 'will make the Kuala Lumpur stock exchange one of the leading stock exchanges in the region.'

But the move reflected the tension in Malaysia between its declared wish to have a free market and yet to control it. Securities and companies laws do not forbid Malaysian companies from being quoted abroad and some of them have been encouraged to float their shares in London. One reason why Kuala Lumpur has lost business to Singapore is better organisation and financial sophistication in the island republic's market. Another is that Singapore brokers charge a 0.5 per cent commission, whereas Malaysian stock brokers charge 1 per cent. The Malaysian brokers were among Daim's warmest supporters in announcing the move.

In another nationalistic move in 1989 the finance minister declared that foreign banks would have to incorporate locally in Malaysia, an action which caused considerable unease. But Daim declared that the foreigners would be allowed to keep their 100 per cent ownership so that 'in practice, therefore, the status quo is maintained.' He claimed the law was intended to remove the advantage that foreign banks had acquired in 1981 when the government imposed minimum capital adequacy ratios on the domestic banks. Foreign bankers disagree that they have an advantage. There are 16 foreign incorporated banks out of 39 commercial banks. Most are single branch operations but the four largest, the Hongkong and Shanghai Banking Corporation, Standard Chartered, Citibank and Oversea-Chinese Banking Corporation, have substantial Malaysian operations. Bank Negara estimates that the foreign banks account for about 25 per cent of all loans.

The banking and financial sector has had more than its share of troubles in recent years. In 1989 Bank Bumiputra, the country's second largest bank, again reported heavy loan losses of ringgit 1.2bn and its parent Petronas, the state oil company, had to pump in ringgit 982m after the bank had written off 90 per cent of its ringgit 676bn capital because of the losses. The losses were said to stem largely from loans made before 1985 when the present management took over.

Nearly a NIC?

By the start of the 1990s Malaysia's high per capita income has encouraged international economists to speak of it as 'nearly a NIC'. Prime Minister Mahathir in July 1989 stated flatly that he did not want the NIC label. 'Malaysia doesn't intend to become a 'NIC'' he said. 'If we become a NIC we will face problems.' He claimed that the expression was developed in the wealthiest countries as a pretext to 'put the brake on these Johnny-come-lately nations.' Once categorised as a NIC, Mahathir warned, the country is stripped of its duty-free trading privileges, its exports subjected to quotas and it is forced to increase the value of its currency. In addition, he claimed, the human rights record of a NIC is scrutinised, 'and invariably found wanting.' Moreover, exploitation of a NIC labour force was often 'discovered and exposed.' He added 'The chances are that these countries will languish in their NIC status and never become fully developed.'

Outlook

The burning question for 1990 is whether Malaysia's so-called New Economic Policy will be continued after the first 20 years of its supposed 20-year programme. The policy was based on the work of the young radical Mahathir who had claimed that the Malays were economically backward in their own country. On account of his radicalism not only was he thrown out of UMNO but the book in which he propounded his thesis was banned (and did not become unbanned until he became prime minister). The policy in effect offered a racial discrimination in favour of the Malays. It has had a partial success. Share-ownership by Malays which was the main evidence of the economic backwardness of the group has risen from 2.3 per cent in 1970 to more than 20 per cent today, though that is still short of the 30 per cent target. A professional Malay middle class has grown up, though that community is still short of entrepreneurs. The other achievement of the policy has been to reduce the foreign ownership of corporate assets from 62 per cent to 26 per cent, well below the 30 per cent envisaged in the original NEP.

The price to pay has been disgruntlement among the Chinese businessmen who are Malaysia's liveliest entrepreneurs. More important, Chinese as a whole feel discriminated against, large numbers of them to the point of feeling themselves second-class citizens. They have not taken to the streets in protest but instead have been leaving the country or sending their children abroad for education, claiming that Malaysia is too small and racial a place for them really to prosper. Chinese are poorly represented among Malaysia's officials, troops and police. Indians fare worst of all and still have only 1.5 per cent of the country's corporate assets. The very poor Malays have also been left out. Mahathir appointed a 150-strong committee of wise men to ponder the issue of the future of the NEP. The chances are that the policy will be extended beyond its 20-year life though with a slightly lighter grip.

Whether this will satisfy the non-Malays remains to be seen. In Malaysia it is all too easy to see the country as a Garden of Eden. But once outside, Malaysia is only one of a number of places competing for attention and investment, pleasant enough. And with only 16m people and a regime that is always increasingly hemmed in and torn by Malay and Islamic demands, the attractions have to be put into context. Malaysia has a long way to go.

COUNTRY PROFILE

Company of Malaysia, Harrisons Malaysian Plantations, BP Malaysia, Malaya United Industries, Genting Associated, Nalin Industries, Associated Pan Malaysian Cement, Perlis Plantations, Amalgamated Steel Mills, Palmco Holdings.

Financial markets: The Kuala Lumpur Stock Exchange is one of the world's most important commodity trading centres, and is linked to the Singaporean exchange.

Membership of international organisations: ANRPC, ASEAN, Asian Development Bank, Association of Tin Producing Countries, CCC, Colombo Plan, Commonwealth, Economic and Social Commission for Asia and the Pacific, EC trade and co-operation agreement, FAO, G-77, GATT, IAEA, IBRD, ICAO, IDA, IFC, ILO, IMF, INTELSAT, Interpol, IPU, IRC, Islamic Conference, Islamic Development Bank, ITC, ITU, NAM, OIC, UN, UNESCO, UPU, WHO, WMO, World Bank, WTO.

BUSINESS GUIDE

Time: GMT + 8 hrs.

Climate: Tropical with high temperatures, high humidity and rainfall year round. Average daily temperatures 21-32°C, except in the hills, where they are 18-24°C. The year is divided into two monsoon seasons: north-east monsoon prevails Oct-Feb, and south-west monsoon from mid-May to Sep.

Entry requirements
Passport: Required by all.
Visa: Visit passes are required by all business travellers and are normally issued to temporary visitors on entry. Visit passes issued for entry into Peninsular Malaysia are not valid for entry into Sabah and Sarawak. Visa rules are complex. Commonwealth citizens (except Indian nationals), British protected persons and citizens of the United States, Japan, Republic of Korea and most Western European nations do not require a visa for visits of up to three months. Citizens of ASEAN countries do not need a visa for visits of up to one month, while for most Eastern European countries, the visa-free period is one week. Visas (when required) and permission for permanent residence or employment must be obtained before arrival.
Prohibited entry: Citizens of Israel or South Africa are not permitted to enter or transit Malaysia.
Currency: Any amount of foreign currency may be imported (except Indian and Indonesian), but visitors may not export more currency than they brought in. It is advisable to declare large amounts of foreign currency on arrival to avoid difficulties on departure.
Customs: Personal goods and a limited amount of tobacco and alcoholic beverages may be imported duty-free. Motor vehicles registered abroad may be imported free of duty for up to three months, after which customs duty and sales tax must be paid. Import of dangerous drugs, pornography, daggers and flick knives, broadcast receivers (with some exceptions) and goods from Israel and South Africa are prohibited. Raw materials and capital machinery for industries complying with the New Economic Policy (NEP) can often be imported duty-free. Special import duties are levied on liquor, tobacco, petroleum products and other items manufactured locally, and specific licensing is imposed on certain items planned for future manufacture.
Income tax: Malaysia has double taxation agreements with 24 countries.

Health precautions
Mandatory: Valid certificate of vaccination against yellow fever if travelling from infected areas.
Advisable: Precautions against malaria; vaccination against cholera, typhoid, hepatitis and polio is also recommended. Tap water is boiled by many people before drinking, though it is generally regarded as safe.

Representation overseas: Abu Dhabi, Addis Ababa, Agdal-Rabat, Ankara, Baghdad, Bamako, Bandar Sri Begawan, Bangkok, Beijing, Belgrade, Berne, Bonn, Brasilia, Brussels, Bucharest, Cairo, Canberra, Cologne, Colombo, Dhaka, Geneva, The Hague, Hanoi, Hong Kong, Islamabad, Jakarta, Jeddah, Kuwait City, Lagos, London, Los Angeles, Manila, Milan, Moscow, Muscat, Nairobi, New Delhi, New York, Ottawa, Paris, Port Moresby, Rangoon, Rome, Sao Paulo, Seoul, Singapore, Stockholm, Suva, Sydney, Tehran, Tokyo, Toronto, Tripoli, Vienna, Vientiane, Warsaw, Washington DC, Wellington.

Air access
National airline: MAS (Malaysian Airline System).
Other airlines: Aeroflot, Air-India, Air Lanka, Air Mauritius, Biman Bangladesh, British Airways, Cathay Pacific, China Airways, CSA, Garuda, JAL, JAT, KLM, Korean Air, Lufthansa, Philippine, PIA, Qantas, Royal Brunei, Royal Jordanian, Saudia, Singapore, Thai Airways, THY-Turkish, UTA.
Tax: An airport tax of M$15 is levied on international departures (M$5 to Singapore).
International airports: Bayan Lepas International (PEN), 17.5 km from Penang, duty-free shop, bar, restaurant, currency exchange, hotel reservations, shops, car hire; Subang International (KUL), 22.5 km from Kuala Lumpur, duty-free shop, bar, restaurant, buffet, bank, post office, shops, car hire; Kota Kinabulu (BKI), 6 km from city; Kuantan (KUA), 16 km from city; Kuching (KCH), 11 km from city.

Surface access: The state of Johore is linked to Singapore by a causeway, with rail and coach services to Kuala Lumpur and Malacca (and on to Thailand). Train services to Thailand three times a week.
Main ports: In Peninsular Malaysia, Dungun, Malacca, Pasir Gudang (a free port), Penang, Port Dickson, Port Kelang and Telok Anson; in Sabah, Kota Kinabalu, Kudat, Labuan, Lahad Datu, Sempoma and Tawau; in Sarawak, Kuching, Bintulu (under development) and Pending Point (under development). Feri Malaysia provides regular services between Peninsular Malaysia, Singapore and Sabah and Sarawak.

Hotels: A 10 per cent service charge is added to hotel bills.

Credit cards: All major cards are accepted in urban centres and hotels.

Car hire: Car hire in all main cities. Driving is on the left-hand side of the road, and use of seat belts in front seats is obligatory. International driving licences are required in Malaysia. Chauffeur-driven cars are available.

City transport
Taxis: In Kuala Lumpur, taxis are metered – 20 per cent additional charge for air-conditioning – hired from taxistands, hailed or ordered by telephone. Higher rates apply 2400 – 0600 (50 per cent extra). Can be hired on a time basis; sharing for long-distance travel is common. Tipping is optional. For taxi service from Subang International airport to destinations in Kuala Lumpur, coupons should be purchased from the coupon booth at the airport and given to the driver. Fare to central KL was around M$16 in 1989. Comfort Radio Taxi Service (tel: 733-0507, 24-hours).

National transport
Air: Over 20 domestic airports. MAS operates extensive network services to main centres and, particularly in Sabah and Sarawak, smaller towns. Since 1986, domestic air services have been progressively deregulated, and MAS is due for partial privatisation.
Road: 78 per cent of the 43,818 km road network in Peninsular Malaysia is paved, and an expansion and improvement programme is under way. Motorways are being built between Kelantan and Trengganu, and between Kedah and Johore Bahru. Monsoon season driving can be difficult. Sabah and Sarawak have a small network of roads centred around Kuching and Sibu. The Ayer Keroh-Pagoh highway now open, reducing travelling time between Kuala Lumpur and Pagoh to two and a quarter hours.
Buses: Over 1,000 routes. Regular coach services operate between main cities. Suburban bus services in Kuala Lumpur; in Sabah and Sarawak rural services are provided by four-wheel-drive vehicles.
Rail: Day and night services link major cities in Peninsular Malaysia; daily service from Kuala Lumpur to Kota Bharu and Tumpat with 1st class carriages. The 2,092 km of single-gauge track is due for modernisation. Sabah's railway links Kota Kinabula and Tenom. There is no railway in Sarawak.
Water: The Straits Steamship Company operates a passenger service between Port Kelang and Sabah and Sarawak every nine to ten days. Frequent ferry services between Penang and Butterworth. There is also a new bridge linking Penang with the mainland.

Public holidays: Malaysia's multi-ethnic and multi-religious population celebrates a variety of holidays – federal, Muslim, Christian, Buddhist and others. In addition to federal holidays, each state has 3-4 additional holidays, one of which is the birthday of its ruler. City Day (1 Feb) is a holiday in the Federal Territory of Kuala Lumpur.
Fixed dates: 1 Jan, 1 May, 31 Aug (National Day), 25 Dec.
Variable dates: Chinese New Year (two days), Good Friday, Vesak Day, Official Birthday of Yang di-Pertuan Agong, Hari Raya Puasa (Eid al-Fitr), Hari Raya Haji (Eid al-Adha), First Day of Muharram, Deepavali, Prophet's Birthday. Actual dates of Muslim festivals may vary, as they depend on sightings of the moon.

Working hours: As well as Friday, the Muslim weekly holiday on Thursday afternoon is observed in the states of Johor, Kedah, Kelantan, Perlis and Terengganu. Other states have a Saturday/Sunday weekend.
Business: (Mon-Fri) 0800/0830 – 1615/1630; (Sat) 0800/0830 – 1230/1245; some variation in Sabah and Sarawak.
Government: (Mon-Fri) 0800 – 1615; (Sat) 0800 – 1245.
Banking: (Mon-Fri) 1000 – 1500, (Sat) 0930 – 1130 in Peninsular Malaysia; (Mon-Fri) 0800 – 1200, 1400 – 1500; (Sat) 0900 – 1100 in Sabah; (Mon-Fri) 1000 – 1500; (Sat) 0930 – 1130 in Sarawak.
Shops: Usually 1000 – 2200 (department stores and supermarkets), 0900 – 1900 (shops) in Peninsular Malaysia; (Mon-Sat) 0800 – 1830 in Sabah; (Mon-Fri) 0900 – 1800, (Sat) 0900 – 1300 in Sarawak.

Social customs: Malaysians place great importance on the correct use of titles. 'Tunku' or 'Tengku' indicates hereditary royalty; 'Tun' denotes membership of a high order of chivalry. 'Tan Sri' and 'Datuk' (or 'Datuk Seri' or 'Dato') indicate knighthood. 'Tuan' or 'Encik' is the equivalent of 'Mr', 'Puan' for 'Mrs', 'Cik' for 'Miss'. Use right hand only for receiving anything (food, drink, money, etc.) and for eating. Refusal of offered refreshment is considered discourteous. Muslims do not eat pork. It is customary to bargain when shopping, except in department stores.

Telecommunications
Telephone: Dialling code for Malaysia, IDD access code + 60 followed by area code (Kuala Lumpur 3; Penang 4; Ipoh 5; Kota Kinabalu 88; Kuching 82; Malacca 6) followed by subscriber's number.

BUSINESS GUIDE

Former six-digit numbers in the Federal Territory of Kuala Lumpur are being progressively replaced with seven-digit numbers. 1.1m subscribers in 1987.
Telex: Public telex facilities available 24 hrs at Telegraph Office, Jalan Raja Chulan, Kuala Lumpur, and frequently at good hotels. 7,980 telex subscribers in 1983. Fax is increasingly widely available.
Telegram: Ordinary, urgent, letter and greetings telegram service available from any telegraph office.

Banking: *Central bank:* Bank Negara Malaysia. *Islamic bank:* Bank Islam Malaysia. *Commercial banks:* Ban Hin Lee Bank, Bank Bumiputra Malaysia Berhad, BankBuruhuh (Malaysia), Bank of Commerce, Bank Utama (Malaysia), Development and Commercial Bank, Hock Hua Bank, Hock Hua Bank (Sabah), Kong Ming Bank, Kwong Lee Bank, Kwong Yik Bank, Malayan Banking, Malaysian French Bank, Oriental Bank, The Pacific Bank, Perwira Habib Bank Malaysia, Public Bank, Sabah Bank, Southern Banking, United Asian Bank, United Malayan Banking Corporation, Wat Tat Bank. *Merchant banks:* Amanah-Chase Merchant Bank, Arab-Malaysian Development Bank, Aseambankers Malaysia, Asian International Merchant Bankers, Asiavest Merchant Bankers (M), Bumiputra Merchant Bankers, D and C Nomura Merchant Bankers, Malaysian International Merchant Bankers, Permata Chartered Merchant Bank, Pertanian Baring Sanwa, Rakyat First Merchant Bankers, UDA Merchant Bankers. *Foreign commercial banks:* Algemene Bank Nederland, Bangkok Bank, Bank of America, Bank of Canton, Bank of Nova Scotia, Bank of Tokyo, The Chartered Bank, Chase Manhattan Bank, Chung Khiaw Bank, Citibank, European Asian Bank, Hongkong and Shanghai Banking Corporation, Lee Wah Bank, Oversea-Chinese Banking Corporation, Overseas Union Bank.

Advertising: Well-established advertising agencies with arrangements with multinational advertising firms operate locally. Most periodicals accept advertising, but regulations can restrict content and presentation. The Commercial Division of the Ministry of Information is responsible for radio and TV advertisements. 420 cinemas with advertising limited to consumer goods. Outdoor advertising and posters widely used; direct mail is available, but multi-lingualism restricts its use.

Trade fairs: Malaysian International Trade Fair of Industrial Development Technology, Machinery and Equipment, and Defence Asia Exhibition and Conference, and several other international fairs are held in Kuala Lumpur each year.

Electricity supply: 220-240 V AC, with three-pin square or round plug fittings.

Weights and measures: Metric system (gradual conversion from Imperial system in progress, with local units also in use).

Representation in capital: Argentina, Australia, Austria, Bangladesh, Belgium, Bolivia, Brazil, Brunei, Burma, Canada, China PR, Czechoslovakia, Denmark, Egypt, Finland, France, German DR, German FR, Iceland, India, Indonesia, Iran, Iraq, Ireland, Italy, Japan, Korea DPR, Republic of Korea, Kuwait, Libya, Netherlands, New Zealand, Oman, Pakistan, Philippines, Poland, Romania, Saudi Arabia, Singapore, Spain, Sri Lanka, Sweden, Switzerland, Thailand, Turkey, UK, USA, USSR, Vietnam, Yugoslavia.

Hotels: Johore Bahru (area code 7)
Holiday Inn, 200 Jalan Dato Sulaiman, Century Garden, 80990 (tel: 323800; tx: 60970; fax: 318884).

Merlin Inn, 10 Jalan Bukit Meldrum, 80300 (tel: 227400, 228581; tx: 60028).

Hotels: Kota Bharu, Kelantan (area code 9)
Perdana, Jalan Mahmud, 15720, P.O. Box: 222 (tel: 785000; tx: 53143; fax: 747621).

Hotels: Kota Kinabalu, Sabah (area code 88)
Capital, 23 Jalan Haji Saman, P.O. Box: 11223 (tel: 231999; tx: 80699).

Hyatt, Jalan Datuk Salleh Sulong, Locked Bag no. 47.88994 (tel: 221234; tx: 80036; fax: 225972).

Shangri-La, 75 Bandaran Berjaya, 88819, P.O. Box: 11718 (tel: 212800; tx: 80001; fax: 212078).

Tanjung Aru Beach, Locked Bag 174, 88999 (tel: 587111; tx: 80751; fax: 217155).

Hotels: Kuala Lumpur & environs
Equatorial, Jalan Sultan Ismail, 50250 (tel: 2617777; tx: 30263; fax: 2617920).

Federal, 35 Jalan Bukit Bintang, 55100 (tel: 2489166; tx: 30429; fax: 2438381).

Fortuna, 87 Jalan Berangan, 0511 (tel: 2419111; tx: 31379; fax: 032418237).

Furama, Komplek Selangor, Jalan Sultan, 50000 (tel: 2301777; tx: 32972).

Grand Pacific, 52/56 Jalan Tun Ismail/Jalan Ipoh, 50480 (tel: 2982177, 4422177; tx: 30886).

Hilton, Jalan Sultan Ismail, 50250, P.O. Box: 10577 (tel: 2422222; tx: 30495; fax: 2438069).

Holiday Inn On-The-Park, Jalan Pinang, 50732, P.O. Box: 10983 (tel: 2481066; tx: 30239; fax: 2435930).

Holiday Inn, City Centre, Jalan Raja Laut, 50350, P.O. Box: 50750 (tel: 2939233; tx: 28130; fax: 2939634).

Malaya, 162 Jalan Hang Lekir, Cecil 50720, P.O. Box: 10644 (tel: 2327722; tx: 30695).

Mandarin, 2/8 Jalan Sultan, 50000 (tel: 2304533; tx: 32143).

Merlin, 2 Jalan Sultan Ismail, 50250 (tel: 2480033; tx: 30487; fax: 2426917).

Merlin Subang, Jalan SS 12/1, Subang Jaya, 47500 (tel: 7335211; tx: 36203).

Ming Court, Jalan Ampang, 50450, P.O. Box: 11295 (tel: 2619066, 2618888; tx: 32621; fax: 2612393).

Pan Pacific, Jalan Putra, 50746, P.O. Box: 11468 (tel: 4425555; tx: 33706; fax: 4417326).

Park Avenue, 16 Jalan Imbi, 55100 (tel: 2428333; tx: 21117 pritel; fax: 2439479).

Park Royal, Jalan Sultan Ismail, 50250 (tel: 2425588; tx: 30486; fax: 2415524).

Petaling Jaya Hilton, Jalan Sultan, Petaling Jaya, 46200 (tel: 2422122, 2422222; tx: 30495).

Plaza, Jalan Raja Laut, 50350, P.O. Box: 12141 (tel: 2982255; tx: 30987; fax: 2920959).

Shangri-La, 11 Jalan Sultan Ismail, 50250 (tel: 2322388; tx: 30021; fax: 2301514).

South East Asia, 69 Jalan Haji Hussein, 50786, P.O. Box: 12681 (tel: 2926077; tx: 31210).

Hotels: Kuala Terengganu (area code 9)
Motel Desa, Bukit Pak Apil 20300 (tel: 623033, 623438; tx: 51477, 32346).

Pantai Primula, Jalan Persinggahan, P.O. Box: 43, 20904 (tel: 622100; tx: 51403).

Hotels: Kuantan (area code 9)
Hyatt, Telok Cempedak, 25050 (tel: 25211; tx: 50252; fax: 507577).

Merlin Inn Resort, Telok Cempedak, 25050, P.O. Box: 46 (tel: 522388; tx: 50285).

Ramada Beach (tel: 587544).

Hotels: Kuching, Sarawak
Aurora, McDougall Road, P.O. Box: 260 (tel: 240281; tx: 70249).

Borneo, Jalan Tabuan, P.O. Box: 1265 (tel: 244121/4; tx: 70389).

Holiday Inn, Jalan Tunku Abdul Rahman, P.O. Box: 2362 (tel: 423111; tx: 70086; fax: 426169).

Kuching Hilton, Jalan Tunku Abdul Rahman (tel: 248200; tx: 70828).

Hotels: Langkawi Island, Kedah
Langkawi Island Resort, Pantai Datuk Syed Omar (tel: 788209; tx: 42044).

Hotels: Malacca
Palace, 201 Jalan Munshi Abdullah, 75100 (tel: 225115, 225355; tx: 25436).

Ramada Renaissance, Jalan Bendahara, P.O. Box: 105, 75720 (tel: 248888; tx: 62992; fax: 249269).

The Inn Melaka, Bangunan Woon Hoe Kan, Jalan Munshi Abdullah, 75100 (tel: 240777; tx: 62995).

Hotels: Penang
Ambassador, 55 Penang Road, P.O. Box: 77 (tel: 24101/7).

Casuarina Beach, Batu Ferringhi (tel: 811711; tx: 40137).

Ferringhi Beach, 12.5 km Jalan Batu Ferringhi, 11100 Batu (tel: 805999; tx: 40634; fax: 805100).

Golden Sands, 87 Batu Ferringhi Beach, 11100, P.O. Box: 222 (tel: 811911; tx: 40627; fax: 811880).

Holiday Inn, Batu Ferringhi Beach, 11100, P.O. Box: 2362 (tel: 811611; tx: 40952; fax: 811389).

Ming Court Inn Penang, 202A MacAlister Road (tel: 368588; tx: 40092).

Orchard, Tanjong Bungah (tel: 803333; tx: MA 40310; fax: 803303).

Palm Beach, 105A Batu Ferringhi, 11100 (tel: 811621; tx: 40404).

Penang Mutiara Beach Resort, Jalan Teluk Bahang, 11050 (tel: 804336; tx: 40822; fax: 803470).

Rasa Sayang, Batu Ferringhi Beach, 11100, P.O. Box: 735 (tel: 811811; tx: 40065; fax: 811984).

Shangri-La Inn, Magazine Road, Georgetown 10300 (tel: 622622; tx: 40878; fax: 626526).

The Merlin Penang, 3 Jalan Larut, 10050 (tel: 376166; tx: 40393; fax: 376615).

Useful telephone numbers
Kuala Lumpur
Emergency: 999
Operator (trunk call enquiries): 102
Directory :103
International service: 108
Telegrams: 104
Tourist Police, KL: 243-5522

Car hire: Johor Bahru (area code 7)
Sintat Rent-a-Car: Holiday Inn, Century Garden (tel: 332313, 323800; tx: 60031 EXTOUR MA).

Car hire: Kuala Lumpur (area code 3)
Avis: 40 Jalan Sultan Ismail, 50250 (tel: 2417144; tx: 30002); airport branch (tel: 7463950).

Budget: (tel: 2611122).

Hertz: Kuala Lumpur Hilton (tel: 2433433; tx: 30956).

National Car Rental: Lot 8, Mid-Level Floor, Hotel Equatorial, Jalan Sultan Ismail (tel: 2480522; tx: 30596 MA); airport branch (tel: 7462025).

Car hire: Kuantan
Sintat Rent-a-Car: Hotel Merlin Kuantan, Telok Chempedak (tel: 528400); airport (tel: 581 226).

Car hire: Penang (area code 4)
Avis: Airport Branch (tel: 839633; tx: 30002).

Hertz: (tel: 830208).

BUSINESS DIRECTORY

National Car Rental: No. 24, Jalan Sultan Ahmad Shah (tel: 629404).

Sintat Rent-a-Car: airport branch (tel: 830958).

Academic/research institutes
University of Agriculture, 43400 UPM, Serdang, Selangor (tel: 356101/10, 355425/8). Founded 1971.

University of Malaya, Pantai Valley 59100, Kuala Lumpur (tel: 560022; tx: MA 37453). Founded 1962.

National University of Malaysia, 43600 Bangi, Selangor (tel: 03-35001; tx: MA 31496). Founded 1970.

Northern University of Malaysia, Bandar Darulaman, 06000 Jitra, Kedah (tel: Jitra (04)772066; tx: MA 42052). Founded 1984.

University of Science, Minden, 11800 Penang (tel: 883822 (PABX); tx: MA 40254 USMLIB). Founded 1969.

Business advice and information
Advertising Standards Authority of Malaysia, c/o Coopers and Lybrand, Hong Kong Bank Building, Leboh Pasar, Kuala Lumpur.

Associated Chinese Chamber of Commerce and Industry, Chinese Assembly Hall, 1 Jalan Maharajalela, 50150, Kuala Lumpur (tel: 232-0473; tx:32995).

Associated Indian Chamber of Commerce, Wisma UOA, 36 Jalan Ampang, PO Box 12564, Kuala Lumpur (tel: 238-7917; tx: 31354).

Malay Chamber of Commerce, 17A Floor, Plaza Pekeliling, Jalan Tun Razak, 50400 Kuala Lumpur (tel: 292-8522).

Bank Negara Malaysia (central bank), Jalan Kuching, PO Box 10922, 57200, Kuala Lumpur (tel: 298-8044; tx: 3201).

Federation of Malaysian Manufacturers, 17th Floor, Wisma Sime Darby, PO Box 12194, 50770, Kuala Lumpur (tel: 293-1244; tx: 32437).

Malaysian Export Trade Centre, Ministry of Trade and Industry, Wisma PKNS, Jalan Raja Laut, 50350 (tel: 292-8122; tx: 33721).

Malaysian Industrial Development Authority (MIDA), 3rd-6th Floor, Wisma Damansara, Jalan Semantan, PO Box 10618, 50720 Kuala Lumpur (tel: 254-3633; tx: 30752).

Malaysian International Chamber of Commerce and Industry, 8th Floor, Wisma Damansara, Jalan Semantan, PO Box 10192, 50706 Kuala Lumpur (tel: 254-2117, 254-1690; tx: 32120).

Ministry of Trade and Industry, International Trade Division, Block 10, Jalan Duta, 50622 Kuala Lumpur (tel: 2540033, 2548044).

National Chamber of Commerce and Industry of Malaysia, Plaza Pekeliling, Jalan Tun Razak, Kuala Lumpur (tel: 298987; tx: 33642).

Perbadanan Nasional Berhad (PERNAS), 16th Floor, Menara Tun Razak, Jalan Raja Laut, Kuala Lumpur (tel: 242-5022).

Sabah Chamber of Commerce, Bangunan Central, Jalan Sagunting, 88000 Kota Kinabulu (tel: 088-54913).

Sarawak Economic Development Corporation, 1st Floor, Bangunan Yayasa Sarawak, Jalan Masjid, PO Box 400 Kuching (tx: 70063).

United Chamber of Sarawak, Chamber of Commerce c/o Ernst and Whinney, Room 301, 3rd Floor, Wisma Bukit Mata, Jalan Tuanku Abdul Rahman, 93100 Kuching, Sarawak.

Banking
Bank Negara Malaysia (central bank), Jalan Kuching, PO Box 10922, 50929 Kuala Lumpur (tel: 2988044; tx: 30201).

Bank Bumiputra Malaysia Bhd, Menara Bumiputra, Jalan Melaka, 50100 Kuala Lumpur (tel: 2988011; tx: 30445).

Development and Commercial Bank Bhd, Wisma On-Tai, 161B, Jalan Ampang, 50450 Kuala Lumpur (tel: 2619077; tx: 31032).

Malayan Banking Bhd, 100 Jalan Tun Perak, 50050 Kuala Lumpur (tel: 2308833; tx: 30438).

Public Bank Bhd, Bangunan Public Bank, 6 Jalan Sultan Sulaiman, 50000 Kuala Lumpur (tel: 2309055; tx: 30584).

United Malayan Banking Corpn Bhd, Jalan Sultan Sulaiman, PO Box 12006, 50935 Kuala Lumpur (tel: 2309866; tx: 30484).

Travel information
Automobile Association of Malaysia, 30 Djalan Datuk Sulaiman, Taman Tun Dr Ismail, Kuala Lumpur (tel: 578138).

MAS information/reservations – KL Airport (tel:746-3000).

Railway Station (tel: 274-7435). Tourist Development Corporation Malaysia (TDC), Putra World Trade Centre, Jalan Tun Ismail, 50480 Kuala Lumpur (tel: 293-5188; tx: 30093).

Maldives

Nalin Wijesekera

Less than an hour by air to the west of Sri Lanka lies an esoteric necklace of 1190 coral islands (202 inhabited) that form the Indian Ocean atoll of the Maldivian Republic. The 181,500 populace of the archipelago are racially mixed, some displaying prominent Aryan features, while others Dravidian, Arabic and even Negroid. The capital Malé embraces 46,000 inhabitants and the nation has remained an independent State – except for a short period of Portuguese rule for 15 years in the mid-seventeenth century. Between 1887 and 1955 the archipelago remained a British protectorate and regained full sovereign status in 1965.

The first settlers in the Maldives could well have been southern Indian fishermen moving to Sri Lanka and losing their way and drifting south. However, cultural and linguistic evidence suggests that modern Maldivians emerged from Aryan stock, which emigrated from Sri Lanka during the great migration of fifth century BC. In fact, Dhivehi, the Maldivian language, although belonging to the Indo-Iranian group, still has strong links with Sinhalese. Because of their geographical location, the Maldive Islands have been always involved in trade, and for centuries Africans and Arabs came to the Maldives for their cowries. This tiny white shell served as money in Indian Ocean countries, and was made into ornaments and used in barter trading in virtually every continent during the twelfth century.

In his quest for economic emancipation, President Maumoon Abdul Gayoom, winner of the International United Nations Environmental Programme Award – the gentle-strong man of the Maldives who is on his third five-year term in the driving seat – has made freedom and independence more meaningful to his people by launching several employment-generating development projects, and has given education a priority niche in his national planning. He has converted the former airbase, Gan island, into an industrial and commercial complex under the Addu Development Authority. The commissioning and further development of the Hulhule International Airport has brought the Maldives closer to the international community and has greatly helped in developing her vast tourist potential. The launching of the Inter-Atoll Transport Project, the Fish Wealth Exploitation Project and the taking over of its only Fish Canning Factory and vesting with the Maldives Fisheries Corporation, and above all the setting up of the Maldives Monitoring Authority and the Bank of Maldives, are some of the far-reaching measures Gayoom has taken to bring the Republic closer to a modern State.

Tourism is a major industry in the Maldives, attracting over 150,000 visitors annually, 86 per cent of tourists originating from West Germany, Italy, France and Sweden. The industry saw a 19 per cent increase in earnings in 1988, while contributing some 22 per cent to the GDP. Recently, the Maldives urged Gulf States to invest in its growing tourist industry. Foreign

KEY FACTS

Official title: Republic of Maldives

Head of state: President Maumoon Abdul Gayoom

Capital: Malé, on Malé Island

Official Languages: Dhivehi

Currency: Rufiyaa (MRf = 100 laari)

Exchange rate: MRf9.32 per $ (Nov 1989)

Area: 298 sq km (1,190 tiny atoll islands) – Malé: 1.5 sq km

Population: 200,000 (1988)

GDP per capita: $405* (1987)

GDP real growth: 8.7%* (1988)

Inflation: 4% (1986)

Trade balance: –$64.5m (1988)

Foreign debt: $72.8m (1987)

Visitor numbers: 114,554 (1986)

* estimated figure

Minister Fathulla Jameel said, that apart from aid, there was virtually no Gulf investment in the islands, although tourism was growing at 15 to 20 per cent annually. Jameel had three days of talks with officials of the Abu Dhabi Fund for Arab Economic Development, which with two other Arab agencies is financing the rebuilding of the Maldives International Airport. Fund officials indicated that they would discuss a feasibility plan for a US$13m hotel in the Maldivian capital of Malé. Said Jameel, 'we have 60 of our islands earmarked exclusively for tourists and our concept is to have one or more hotel on each island,' and added that investment incentives have been greatly enhanced and regulations relaxed.

The Maldivian economy has moved confidently, recording an 8 per cent GDP growth and a 5.6 per cent rise in per capita income through 1988. A new action programme for 1988-90 is now being implemented, and new strategies are being employed to develop the fisheries sector which accounts for 54 per cent of employment, 40 per cent of GDP and in 1988 foreign exchange earnings from fish and fish product exports increased by 49 per cent. The main lines of imports are rice, wheat, sugar, sundry food items, tobacco and beverages, manufactured goods, petroleum products and intermediate and capital goods, while the main export is fish and fish products and, to a smaller degree, garments.

According to President Gayoom, efforts have been made by the government to improve 'social services and to construct new schools in Malé and in other islands to give more opportunities and openings for the education of children'. He said, 'a regional hospital and health centres were built in 1988 that have upgraded health services throughout the country'. Under the efforts to improve communications, the president noted the installation of 'a modern digital telephone exchange in Malé' and added that a project to make telephone services available to other islands is now being implemented. Shipping had a good year in 1989-90 and a profit of US$3.2m has been estimated. The president said that he was satisfied with the improvement in foreign exchange earnings, which had made the balance of payments situation favourable. Imports and exports increased in 1988, while import duties were reduced on more than 500 items, he added, while there was a surplus Budget in 1988. President Gayoom, outlining his policy for 1989-90, said it would be the same as his policy for the previous 10 years. There would be no change in the government's foreign policy, which had earned the respect of the international community.

Like many Third World countries, actual and manmade contingencies have shaken the Maldivian socio-economic and political fabric. A gradual rise in the average sea level is threatening to wipe out the archipelago within the next 30 years, and drinking water could well dry up. Unusually strong tidal waves lash the airport and in the 1988 monsoon left a trail of destruction on several islands, costing some US$4m. The archipelago with its low-lying coral islands stretching nearly 750 km across the Indian Ocean, is nervously watching the rise in sea levels, according to Environmental Affairs Director, Hussein Shihab. 'Some people can talk about it very lightly, but we could get washed away from the face of the earth in the next 30 years. It is a real problem'. He said an estimated rise of 20 to 30cm in the next 20 to 40 years could be 'catastrophic for a majority of the islands which are currently no more than a metre above sea level'. The problem received international attention when President Gayoom told Commonwealth Heads of State, that the 'fearsome environmental trend' of rising sea levels was potentially fatal for his country. 'It will be the death of a nation', Gayoom told the Vancouver meeting, adding, 'we don't know what to do about it. We did not contribute to environmental degradation, and we should not be left to suffer its consequences. We have neither the technology nor the resources to cope with the impending threat'.

On the other side of the coin, in November 1988 up to 400 Tamil-speaking mercenaries landed by boat in the Maldives, stormed the Presidential Palace and tried to seize control of Malé. Political watchers see the Machiavellian hand of former President Ibrahim Nasir – who was forced to resign in 1978 and is now living in Singapore heading an import/export firm – behind the attempted 'coup d'état'. According to President Gayoom, the coup was masterminded by Sri Lankan Tamils hoping to use the archipelago as a launching pad for their separatist struggle in Sri Lanka. Uma Maheswaram, the head of one of Lanka's northern terror groups, the People's Liberation Organisation of Tamil Eelam (PLOTE), had been the main leader in planning the seaborne mercenary invasion of the Maldives, according to Gayoom. If the coup had succeeded, PLOTE was to be given use of one of the 1190 Maldivian islands, as a training base for guerrillas and as a port for smuggling arms into Sri Lanka. In response to a call for help by President Gayoom, India despatched 1600 paratroopers and naval ships, but the mercenaries held nearly 2000 Maldivian troops as hostages and threatened to slaughter them unless the government surrendered. However, the Indian forces overcame the ill-equipped guerrillas, and the coup was effectively crushed. Consequently, thousands of Sri Lankan workers in the Maldives have been returning home, after the decision by the Malé government not to extend their work visas after the end of May 1989.

Apart from this recent abortive 'coup d'état' the Maldives is politically stable, economically growing and socially thrusting towards achieving greater literacy for her people, while President Gayoom has been endeavouring quite successfully to make his country's presence felt internationally by being active and vocal in forums such as the Non-aligned Movement, the Commonwealth and the South Asian Association for Regional Co-operation (SAARC).

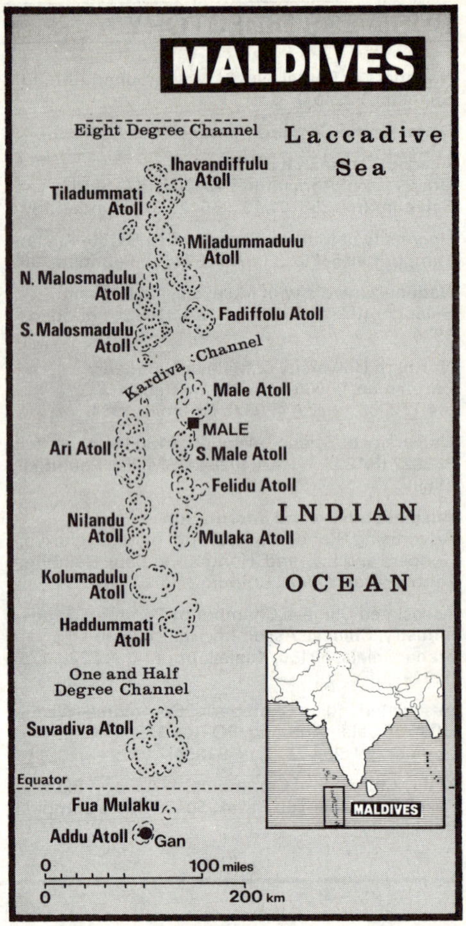

KEY INDICATORS

	Unit	1984	1985	1986	1987	1988
Population	'000	175	177	181	190	–
Gross domestic product (GDP)	MRf m	537	596	651.9	709.7	771.6
GDP per capita	MRf	3,069	3,367	3,602	3,735	–
Exports	$m	17.6	23.0	24.5	30.8	40.1
Dry skipjack	$'000	679	1,306	932	–	–
Fresh skipjack	$'000	6,656	8,757	6,250	–	–
Dry salted fish	$'000	1,048	3,133	2,259	–	–
Imports cif	$m	53.3	52.7	63.8	86.6	104.6
Balance of trade	$m	–35.7	–29.7	–39.3	–55.8	–64.5
Current account	$m	–17.0	–4.5	–8.8	–0.3	–
Tourism (bed/nights)	1985=100	75	100	98	121	158
Fish catch	1985=100	88.9	100	95.8	91.9	124.7
External debt	$m	83.7	84.6	68.7	72.8	–
Debt service ratio	%	16.9	8.5	14.4	–	–
Exchange rate	MRf per $	7.05	7.10	7.15	9.22	8.78

* estimated figure

COUNTRY PROFILE

Political system: The Maldives is a former sultanate. Executive power is held by the president who is nominated by the legislature and directly elected for a renewable five-year term, and who appoints a cabinet. The legislative Citizens' Majlis is made up of 40 members elected by universal suffrage to a five-year term (two from each of the 19 administrative atolls and two from Malé) and eight presidential appointees. The president has the sole power to summon the Citizens' Special Majlis (consisting of cabinet members and elected members) which can alter the constitution by majority vote. Local authority is vested in atoll chiefs and island headmen appointed by the president.

Media
Broadcasting: There were 19,146 radios and 3,211 TV sets in 1985.

Domestic economy: The main economic activities are shipping, fishing and tourism. The population growth rate is high, leading to shortages of land and housing. Development of the tourist industry has required investment in infrastructure and modern passenger boats. Government policy has been to set up resorts on islands not inhabited by Maldivians, preserving a desert island atmosphere. On all islands, water supply is a critical problem. A major programme began in 1984 to improve sanitary conditions and provide drinking water for Malé, including development of Mosque wells and construction of public and private collection tanks, a sewage system and a desalination plant, funded in part by loans from the Saudi Fund for Development. Other plans include construction of housing and land reclamation.

External trade: The Maldives is one of the 32 developing nations designated in 1985 to receive exemption from US competitive trade requirements in accord with the Generalised System of Preferences.
Exports: Main exports are garments (typically 34.6% of 1986 total sales), fresh skipjack tuna (39.8%), dry salted fish (5.9%) and dry skipjack (3.9%). Main destinations are Japan, Sri Lanka, Thailand, Singapore and other Asian countries.
Imports: Main imports are consumer goods (typically, 57% of total purchases), petroleum products (34.5%), intermediate and capital goods (8.9%). Main sources are Sri Lanka, Japan, UK, German FR and Thailand.

Agriculture: Agriculture, mainly fishing, accounts for 30.5 per cent of GDP (1986). Fishing employs 40 per cent of the labour force, providing the main source of protein and export earnings. Traditional craft (*dhoni*) are being replaced by mechanised boats. The soil is shallow and highly alkaline, with poor water-retaining properties. Main crops grown for domestic consumption include coconuts, breadfruit, millet, sweet potatoes, cassava, taro, maize, sorghum, fruit and vegetables.

Industry: Manufacturing accounts for 4.8 per cent of GDP (1986). The Addu Development Authority promotes industrial development on Gan Island, where the government operates a former Japanese tuna cannery and has built factories for packing and storing dried fish. There are two garment factories at Gan and another on Thulusdhoo in Malé Atoll. Other small enterprises in Malé include a PVC pipe manufacturing and a bottling plant. Maldives National Ship Management Ltd (MNSML) operates a fleet of 40 vessels worldwide, carrying 90 per cent of imports (which represented four per cent of cargo carried in 1982). Construction accounts for around seven per cent of GDP. Many modern buildings are under construction in Malé, mostly with imported materials and foreign contractors for large projects. Tourism has become an important sector, accounting for 20 per cent of GDP in 1986, with 85 per cent of tourists arriving from W. Europe, mainly German FR, Italy, France and Japan.

Membership of international organisations: Asian Development Bank, Colombo Plan, ESCAP, FAO, G-77, GATT, IBED, ICAO, IMF, ITU, NAM, OIC, OIE, SAARC, UN, UNESCO, UPU, WHO, World Bank, Commonwealth.

BUSINESS GUIDE

Time: GMT + 5 hrs.

Climate: There is year-round sunshine, with temperatures ranging 26-31°C. Most rainfall occurs during the south-west monsoon season from May-Nov.

Entry requirements
Passport: Required by all.
Visa: Visas are issued on arrival for visits of up to 30 days.
Currency: Local currency may not be imported or exported.
Customs: Personal effects are allowed duty-free. Alcohol, pork, pharmaceuticals, and goods at variance with Islamic culture are prohibited or subject to restrictions. Pornographic material may not be imported.

Health precautions
Mandatory: Vaccination certificates for cholera or yellow fever if travelling from an infected zone.
Advisable: Anti-malarial precautions, outside Malé.

Representation overseas: Colombo, Frankfurt, Karachi, Tokyo, New York.

Air access
Airlines serving: Air Lanka, Emirates, Indian Airlines, PIA, Singapore Airlines. Charter flights are often available from German FR, Switzerland and Italy.
International airport: Malé International, on Hulule Island (MLE), with connecting sea transport to main islands.

Surface access: There are no deep-water ports. Cargo is off-loaded into lighters.

Hotels: There are five hotels in Malé, numerous guest houses and 55 island resorts, mostly in Malé Atoll, but in more distant atolls as well.

Credit cards: American Express is generally accepted; Visa is accepted in some resorts and hotels. American Express has an office in Malé.

Car hire: Due to the small size of the islands, there is little need for transportation. It is possible to walk to all places within the small area of Malé.

City transport
Taxis: There is a taxi rank outside the Government Hospital in Malé. Taxis cannot be hailed in the street.

National transport
Air: Air Maldives Airways operates regular domestic flights. Government charter planes are also available.
Water: Resort islands have regular ferry services. Local boats can be hired but rates are negotiable. Charter vessels are available.

Public holidays
Fixed dates: 1 Jan, 18 Apr, 17 May, 23 Jul, 26 Jul (Independence Day), 14 Aug, 12 Oct, 25 Oct, 11 Nov (Republic Day).
Variable dates: Major Muslim holidays and others. Actual dates of Muslim festivals depend on the sighting of the moon.

Working hours: Starting times may be later during the month of Ramadan.
Government and business: (Sat-Thu) 0730 – 1330.
Banking: (Sun-Thu) 0900 – 1300.

Social customs: Alcohol may not be imported, but it may be purchased.

Telecommunications
Telephone: Dialling code for Maldives, IDD access code + 960 followed by 32 for Hulule Airport and Malé Island, 34 for North Malé Atoll and South Malé Atoll. There were 2,900 telephones in 1987.
Telex: Public telex facilities are available at the Cable and Wireless office in Malé.

Banking: *Central bank:* Maldives Monetary Authority.
Other banks: Bank of Ceylon, Habib Bank, State Bank of India, Bank of Credit and Commerce International (BCCI), Bank of Maldives.

Weights and measures: Metric and imperial systems.

Representation in capital: India, Pakistan, PLO, Sri Lanka.

BUSINESS DIRECTORY

Hotels: Malé
Bi Ya Doo Island Resort, c/o Prabalaji Ent H Maagala, 2 Amir Ahmed Magu Male Biyadhoo (tel: 3516; tx: 77003).

Fiha Lhohi Tourist Resort, Beach Villa (South Male Atoll), 2001 (tel: 342903; tx: 66065; fax: 343803).

Meerufen Fushi Island Resort, c/o Kum S, Ahamadee Bazaar, Malé (tel: 3157; tx: 77002).

Nakachchafushi Tourist Resort, Male Atoll n. (tel: 3846/72665; tx: 66024; fax: 3848).

Villi Varu Island Resort, Viligilvaru Island (tel: 3598; tx: 77003; fax: 3742).

Car hire
Taxi Rank: Malé (tel: 3132, 2122).

Business advice and information
Department of Tourism, 2/F Ghazi Building, Orchid Magu, Henveiru, Malé (tel: 3224, 3226; tx: 66019).

Maldives Monetary Authority, Majeedee Bldg, Marine Drive 20-02 (tel: 2290; tx: 66099).

Maldives National Ship Management Ltd, Ship Plaza, Orchid Magu, Malé (tx: 77038).

Maldives Transport & Contracting Co, MSL Building, Orchid Magu, Malé (tx: 66001/2/3).

State Trading Organisation, Haveeree Higun, Malé (tx: 66006).

Travel information
Tourist Information Unit, Marine Drive, Malé (tel: 5528).

Other useful addresses
Cable and Wireless, 19 Medhuziyaraiy Magu, Malé.

French Honorary Consul (tel: 3760).

German FR Honorary Consul (tel: 2669).

US Consular Office (tel: 2581).

Mali

Susi Pollak

With an economy which basically relies on agriculture, Mali still suffers from the after-effects of recent disasters: locusts and drought. Despite predictions that 1989 would produce a bumper harvest, President Moussa Traoré's ruling Democratic Union of Malian People (UDPM) was looking increasingly to the mining sector for revenues. The financial situation deteriorated seriously after the IMF-approved structural adjustment programme was 'temporarily' suspended in 1987, and although a new IMF package was agreed, the government – fearful of its social and political impact – was slow to implement its more drastic provisions.

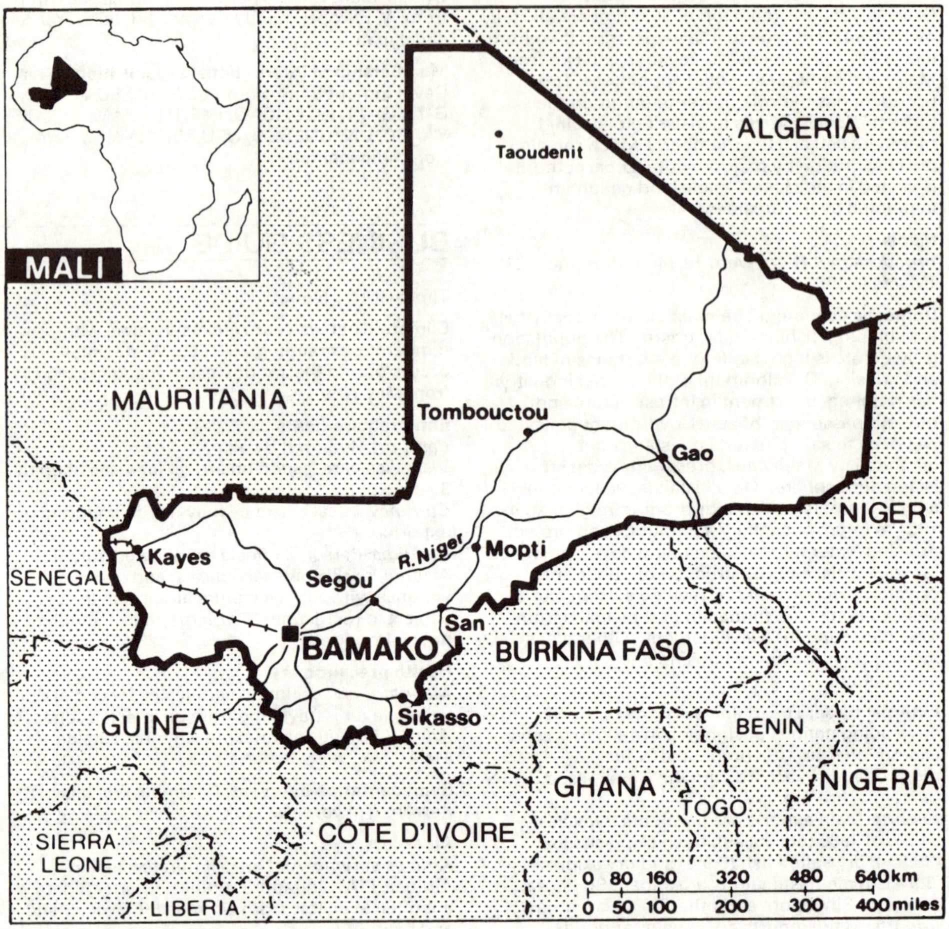

It will take the economy some time to recover from the locust years. Cotton is the principal foreign exchange earner, but the 1987–88 harvest was only 199,000 tonnes, 1.3 per cent down on the previous year. Other cash crops fared even worse: groundnuts, down by 12 per cent; paddy rice, down by 18 per cent. Food crops fared worst of all: the grain harvest was down by 40 per cent compared to 1986–87.

Mineral wealth

Not surprisingly, government prefers to emphasise the non-agricultural sector. According to one ministerial statement, Mali has enormous gold reserves, and exports between 500 and 700 kg a year of the metal, valued at CFA2.9bn. (The uncertainty about the quantity can be ascribed to the existence of large-scale cross-border smuggling. Mali exports gold mined in Guinea and Sierra Leone.) The gold mine at Kalana, near the border with Guinea, which is Mali's first large-scale producer, is estimated to have reserves of 22 tonnes of gold in a rich ore body of 17 grammes of gold per tonne. The mine is operated with Soviet assistance and produced 450 kg of fine gold in 1987 and around 500 kg in 1988. Once construction of the Selingue–Kalana power line is completed, the mine's output should more than double.

A second gold mine is being established at Loulo, near the border with Senegal, where a feasibility study estimated deposits at seven tonnes exploitable over eight years. The mining contract went to Somilo, a joint venture between Mali (51 per cent) and the French Geological and Mining Research Bureau (BRGM). A further venture was announced in 1989 at Syama, in southern Mali, where alluvial gold will be mined by Somisy, a joint venture between the Malian Government (35 per cent) and Australia's BHP-Utah International Inc. (65 per cent). Production was scheduled for 1990.

Phosphates are mined in the Tilemsi region at the rate of 10,000 tonnes a year; the government expects this to increase to 30,000 tonnes. The discovery of diamonds has led to a score of prospecting operations in the Kenieba region, which also has large deposits of bauxite. Marble at Bafoulat is about to be exploited, and there are significant manganese reserves at Angongo, according to official spokesmen.

IMF package

In August 1988 Mali agreed to a new IMF package consisting of a $16.4m standby credit and a three-year structural adjustment facility of $41.8m. This was made conditional on the government's implementation of liberal reforms along the usual IMF lines – basically, the dismantling of public enterprises and support for the private sector. A previous attempt at persuading Mali to adopt such measures came to grief over the government's lukewarm attitude towards the IMF prescription; the 1988-91 structural adjustment looks like running into similar difficulties. An IMF/World Bank delegation which visited Bamako in February 1989 said it was disappointed about the slow pace of progress. Sominex – the state monopoly which handles essential imports and exports – was trimmed but not privatised. The Mali Development Bank, which finances the parastatals, remained unreconstructed.

The government blamed the lack of private sector interest for delays in implementing the IMF plan, but it was clearly worried about the social and political consequences of privatisation. Previous adjustment attempts have caused a dangerous degree of unemployment, and in 1987 Mali appealed to the International Labour Organisation (ILO) for help in alleviating the hardship caused by these public service redundancies.

The government was particularly anxious about some 8,000 unemployed young graduates of technical and higher education. The ILO came up with a placement scheme and a programme which emphasised the 'self-creation of small units of production' – i.e. self-employment. But by the end of 1988 only four such units involving 50 young people were under way, and a total of only 90 people had been helped to find salaried employment. This line of action was therefore abandoned, and the ILO is now helping Mali to formulate a global employment policy involving all groups, starting with the collection of adequate statistical information. This quickly revealed a 34 per cent unemployment rate in the district of Bamako

KEY FACTS

Official title: Republic of Mali

Head of state: President Gen. Moussa Traoré

Ruling party: Union Démocratique du Peuple Malien

Capital: Bamako

Official Languages: French

Currency: Franc CFA = 100 centimes

Exchange rate: Franc CFA316 per $ (Nov 1989)

Area: 1,240,192 sq km

Population: 8.7m (1987)

GNP per capita: $222 (1987)

GNP real growth: 2.0% (1986/87)

Trade balance: –$79.5m (1987)

Foreign debt: $2.0bn (1987)

(79,000 jobless in a population of 230,000), a figure which might prove useful to the government in its attempts to ward off World Bank demands for further redundancies.

Fortunately for Mali, the World Bank is not its only source of aid. In October 1988 it became the first country to benefit from a new debt-relief package from the Paris Club of Western creditor nations: $70m of official debts were rescheduled at lower interest rates. When the IMF facility was announced in August 1988, France added CFA8bn (about $24m) worth of aid in support of the adjustment.

In May 1989 the French Government granted a CFA6.25bn loan for infrastructural work and the rehabilitation of agro-industries. The Islamic Development Bank lent $5m for an oil storage depot in Bamako, and the Canadians CFA4.8bn for various agricultural projects. Even the World Bank remained not entirely aloof. In April 1989 its affiliate, the International Development Association (IDA), lent Mali $33m for an overhaul of its hydroelectric sector, to meet the growing demand for electricity. (At present only 1 in 25 Malians has access to electric power.) A further IDA credit of $26m for the improvement of the educational system was announced in June 1989.

In his address to the 1989 meeting of the UDPM, Traoré said little about the economy but stressed Mali's foreign policy achievements. As his 1988–89 term as chairman of the Organisation of African Unity (OAU) was drawing to a close, he took some credit for the settlement of the Chad/Libya dispute. He also noted developments in the Western Sahara and the deterioration in relations between Senegal and Mauritania – situations which are particularly important to Bamako. Mali has long been active in seeking some understanding between the Saharan liberation movement, POLISARIO, and the Moroccan Government. With Mauritania and Senegal it has close historical and geographic links: Mali, he promised, will make 'the necessary sacrifices to help overcome the after-effects of events no one wanted'.

KEY INDICATORS

	Unit	1983	1984	1985	1986	1987
Population	m	7.74	7.97	8.21	8.44	8.68
Gross national product	$m	1,076	1,051	1,039	1,543	1,931
GNP per capita	$m	139	132	127	183	222
Agricultural production	1979-81=100	118.4	106.8	114.3	123.1	117.2
Exports	$m	166.8	192.0	180.5	191.7	–
Exports	CFA bn	62.90	89.40	79.10	71.20	78.10
Cotton	CFA bn	25.95	39.40	32.80	24.00	31.90
Groundnuts	CFA bn	0.65	0.50	0.50	0.50	0.50
Imports	CFA bn	134.60	160.90	210.80	171.70	148.30
Balance of trade	$m	–74.1	–65.8	–152.4	–141.5	–79.5
Total external debt	$m	1,010	1,287	1,483	1,742	2,016
Debt service ratio	%	1.3	1.9	3.7	2.3	1.7
Exchange rate	CFA per $	381.1	437.9	449.3	346.3	300.5

* estimated figure

COUNTRY PROFILE

Political system: Legislative power is vested in the 82-member National Assembly which is chosen every three years by the UDPM. Mali gained independence from France in June 1960 as part of the Federation of Mali. This was dissolved a few weeks later when Senegal broke away. The Republic of Mali was established in September 1960. The first government was overthrown in 1968 by a military coup, led by Moussa Traoré, now President.

Political parties: The Union Démocratique du Peuple Malien (UDPM) is the only legal political party.

Population: Mali's population of 8.9m (1987 estimate) includes a significant proportion of nomadic tribes. There is a continuing drift toward the Senegal and Niger rivers.

Main cities/towns: Bamako (population 750,000) is the main business centre, Gao, Mopti, Tombouctou, Kayes, Sikasso.

Language: French is the official language and the universal business language; very little English is spoken. Arabic, Bambara, Fulani, Senoufo and Dogon are also spoken.

Media
Press: One daily newspaper, *L'Essor*, published by the UDPM. Several periodicals are also published, mainly by the government.
Radio: The state controlled Radio-Télévision Malienne broadcasts in French, English and local languages. No commercial service.
Television: National colour television network inaugurated 1983.

External trade: Significant amount of foreign trade goes unrecorded. Falling export earnings and heavy dependence on imported energy and manufactured goods led to increasing trade deficits in the mid-1980s. Large current account deficit is partly funded by workers' remittances and official transfers. Strong reliance on overseas aid, particularly from France. GDP declined in 1983-85, then grew again in 1986, with the World Bank providing technical assistance to improve economic management.
Exports: Cotton and cotton products (typically 45-55 per cent of total value), live animals 13-15%), groundnuts (13-14%), fish (2-3%). Main destinations: Belgium (typically 18-20% of total value), France (18-20%), German FR (15-16%), China (7-9%), Ivory Coast (7-9%).
Imports: Petroleum products (typically 18-19 per cent of total value), chemicals and pharmaceuticals (25-27%), machinery and transport equipment (25-27%), foodstuffs (18-20%), construction materials (5-10%). Main sources: France (typically 30-33% of total value), Ivory Coast (23-25%), China PR (10-15%), German FR (4-8%), USA (4-6%), Senegal (4-6%).

Agriculture: Mainstay of the economy, employing 85 per cent of the workforce (largely at subsistence levels) and accounting for virtually all export earnings. Livestock rearing particularly important, contributing about 20 per cent to GDP. Principal food crops are millet, sorghum, paddy rice, maize and groundnuts. The main export crops are cotton, groundnuts, cereals, fresh fruit and vegetables. Fish and live animals also significant export earners. Regular food deficit due to recurrent drought, crop smuggling and inefficient marketing and distribution system. Approximately 20 per cent of the total land area along the Niger river is suitable for cultivation. Government policy aimed at developing production of specific crops, irrigation and liberalisation of internal food markets.

Industry: Sector is small-scale, contributing around six per cent to GDP and employing under eight per cent of the workforce. Manufacturing is concerned mainly with agricultural processing for domestic consumption and export. Other industries include soft drinks, textiles, soaps, plastics, cigarettes, cement, bricks, agricultural tools and equipment. Activity is concentrated in Bamako. Around 90 per cent of production is accounted for by state enterprises.

Mining: Increasing gold mining (500-700 kg p.a. in the late 1980s) and phosphates (10,000 tonnes per annum). Small quantities of salt, limestone and uranium are also mined. Known deposits of bauxite, manganese, iron and tin. Prospecting for lithium, diamonds, copper and petroleum under way. Country's landlocked position and weak infrastructure are major constraints to development of sector.

Energy: Reliance on imported petroleum, though sufficent electricity produced to meet local demand. Bamako fed with hydro-power from the new dam at Selingué. Most other towns rely on diesel generators.

Membership of international organisations: African Development Bank, African Groundnut Council, African States Associated with the EC (Lomé Convention), Arab Bank for Economic Development in Africa (recipient), CILSS, Economic Commission for Africa, Economic Community of West African States, FAO, French Monetary Area, GATT, IAEA, IDA, IFAD, ILO, IMF, Islamic Conference, Islamic Development Bank, Niger Basin Authority, OMVS (Senegal River Development Authority), West African Economic Community, World Bank.

Time: GMT.

Climate: Considerable variation between south, central and northern areas, rain being rare and sporadic in the far north, Sahara region. Bamako: rainy season Jun-Sep/Oct, humidity reaching 80 per cent, likely temperature range 20°C to 36°C; warm dry season Nov-Feb, hot dry season Feb-May with average temperature 35°C.

Entry requirements
Passport: Required by all.
Visa: Required by all except nationals of France and certain African countries. Exit visas and proof of intended departure required. All visitors must register with police on arrival.
Currency: No restrictions on import/export of local currency. Unlimited import of foreign currency although amounts over the equivalent of franc CFA 50,000 must be declared. Restrictions on exports of foreign currency. Credit cards are not widely accepted.

Health precautions
Mandatory: Yellow fever vaccination certificate.
Advisable: Cholera vaccination considered esential, typhoid, tetanus and polio vaccinations recommended. Malaria prophylaxis should be taken as risk exists throughout country. Water precautions must be taken.

Representation overseas: Abidjan, Accra, Algiers, Beijing, Belgrade, Berlin, Bonn, Brussels, Cairo, Conakry, Dakar, Jeddah, Moscow, Ottawa, Paris, Tripoli, Washington.

BUSINESS GUIDE

Air access
Other airlines: Aeroflot, Air Afrique, Air Algérie, Air Ivoire, Air Burkina, Ethiopian, Gambia Air Shuttle, Royal Air Maroc, Sabena, UTA.
International airport: Bamako (Code: BKO), 15 km from city. Airport tax of CFA 2,000-5,000, payable on departure, depending on destination.

Surface access
Main ports: River ports of Bamako, Mopti, Tombouctou and Gao on the Niger.
Overland access: Good road from Niger (Niamey); condition of routes from Côte d'Ivoire and Burkina Faso tends to be variable; those from Senegal and Algeria are not generally recommended. A regular twice weekly rail service from Senegal (Dakar) to Bamako (with sleeping and restaurant cars and facility for conveying vehicles). Journey takes up to 29 hours.

Hotels: Few good hotels available and these can be expensive. It is advisable to book in advance.

Credit cards: Only accepted in main hotels.

Car hire: International driving licence recommended. Hired cars are usually Renaults or Peugeots.

City transport
Taxis: Cheap and widely available but not metered. Official standard fare system in Bamako. Tipping not usual.

National transport
Air: No scheduled services between Bamako and other towns. Charter of light aircraft available from STA.
Road: Main roads run from Sikasso and Bougouni in the south to Bamako, and from Bamako to Mopti and on to Gao via a new tarred road. Condition of roads tends to be variable and secondary roads can be difficult.
Buses: Cheap but generally uncomfortable. Services from Bamako to all main towns.
Rail: Main routes: Bamako-Koulikoro (59 km) leaves Bamako daily at 1100 and 1700; Bamako-Kayes (494 km) leaves Bamako daily at 0945. Two classes, sleeping and restaurant cars available; some air-conditioned cars.
Water: Three river steamers operate up and down the river Niger betwen Aug and late Dec, linking Koulikoro, Mopti, Tombouctou and Gao. Four classes available. First class cabins must be booked in advance through SMERT.

Public holidays
Fixed dates: 1 Jan, 20 Jan (Army Day), 1 May, 25 May (Africa Day), 22 Sep (Independence), 19 Nov (Army Day), 25 Dec.
Variable dates: Easter Monday, Eid al-Fitr, Eid al-Adha, Prophet's Birthday. Travel is difficult during Ramadan.

Working hours
Government and business: (Mon-Thu, Sat) 0730 – 1430, (Fri) 0730 – 1230.
Banking: (Mon-Thu) 0800 – 1430, (Fri-Sat) 0800 – 1230.

Telecommunications
Telephone: Direct telephone links with Paris, Moscow and Cairo. Dialling code for Mali: prefix + 223 followed by area code (22 for Bamako, 27 for Kati, 26 for Koulikoro, 32 for Segou) followed by subscriber's number. Internal service is unreliable.
Telex: Telex facilities available from main hotels and Central Telex Office in Bamako.

Banking
Central bank: Banque Centrale du Mali.
Main centre: Bamako.
Head offices: Bank of Africa, Banque de Développement de Mali, Banque Libyo-Malienne, Banque Malienne de Crédit et de Dépôts, Banque Nationale pour le Développement de l'Agriculture (BNDA).
Foreign branches: Banque de l'Afrique Occidentale, Banque Internationale pour l'Afrique Occidentale, Caisse Centrale de Coopération Économique.

Electricity supply: 220 V AC, 50 cycles.

Representation in capital: Algeria, Burkina Faso, China PR, Cuba, Egypt, France, German DR, German FR, Ghana, Guinea, Korea PDR, Libya, Malaysia, Mauritania, Morocco, Nigeria, Poland, Romania, Saudi Arabia, USA, USSR, Yugoslavia. Palestine Liberian Organisation is represented.

Hotels: Bamako
Grand, Avenue de la Liberté, P.O. Box: 104 (tel: 22-24-81; tx: 578).

L'Amitié (Sofitel), P.O. Box: 1720 (tel: 22-43-21/95/03; tx: 2433 amihotel; fax: 224385).

Le Majestic, Rue de la Nation (tel: 229-20).

Le Motel, P.O. Box: 911 (tel: 236-22).

Les Hirondelles, P.O. Box: 1026 (tel: 22-44-34, 22-44-35; tx: 455).

Hotels: Mopti
Relais Kanaga (Sofitel), P.O. Box: 224 (tel: 22-43-21; tx: 433).

Other useful addresses
Banque Centrale du Mali, PO Box 206, Bamako (tel: 237-56; tx: 574).

Banque Internationale pour l'Afrique Occidentale, Avenue Mohamed V, PO Box 15, Bamako (tel: 256-01, 256-02).

Banque Malienne de Crédit et de Dépots, PO Box 45, Bamako (tel: 253-36).

Chambre de Commerce et d'Industrie du Mali, PO Box 46, Bamako (tel: 22-50-36).

Chambre de Commerce de Kayes, PO Box 81, Kayes.

Direction National du Plan et de la Statistique, Koulouba, Bamako (tel: 22-27-53).

SMERT (Tourist Agency), PO Box 222, Place de la République, Bamako (tel: 22-59-42; tx: 316, 992).

Malta

A special correspondent

At end 1989, the Nationalist Party government which came to power following the May 1987 elections was just past the midpoint of its five year term. The nationalists had been in opposition for the previous sixteen years – although they obtained the majority of the vote in the elections of December 1981, they were left with a minority of the seats in parliament. Constitutional amendments introduced in January 1987 ensured that the party with the absolute majority of votes in future general elections, would be guaranteed the majority of seats.

Under the leadership of Prime Minister Fenech Adami, the nationalist government liberalised economic policies and tried to dilute the *dirigiste* approaches of its predecessor in office, the socialist Labour Party, led by Dr Karmenu Mifsud Bonnici. During 1989, import controls were relaxed further and the government concentrated on the attraction of private sector investment, which it regards as the lynchpin of economic development.

New direct investment is prospected

In the summer of 1988, an industrial development law was passed through parliament, providing tax incentives and other subsidies meant to stimulate private direct investment, especially foreign. The new package was designed to provide larger scope for higher technology industries, in the hope of reducing dependence on clothing firms, which remain extremely vulnerable to international cost competition. Then, in the autumn of 1988, the government passed a law to regulate and promote the establishment of offshore businesses, banks, insurances and private trusts. The law provides for special tax treatment covering offshore companies and lays down criteria regarding confidentiality and the prevention of abuses. An International Business Authority was set up to supervise offshore companies, which the government clearly hoped would become a growth area of the economy.

Meanwhile, major investments to modernise the infrastructure were programmed. The internal and external telephone system is being updated, the construction of a new air terminal was started, and a big power station is being built in the south of the island,

KEY FACTS

Official title: The Republic of Malta
Head of state: President Vincent Tabone
Head of government: Prime Minister Edward Fenech Adami
Ruling party: Nationalist Party (PN)
Capital: Valletta
Official Languages: Maltese and English
Currency: Maltese Lira (Lm) = 100 cents
Exchange rate: Lm0.34 per $ (Dec 1989)

Area: 316 sq km
Population: 350,000 (1988)
GDP per capita: $5,168 (1988)
GDP real growth: 6.2% (1987)
Labour force: 129,300 (1989)
Unemployment: 3.9% (1989)
Inflation: 0.9% (1988)
Trade balance: –$643.5m (1988)
Foreign debt: $294.6m (1987)

at the fishing village of Marsaxlokk. The latter project created heated controversy, as the socialist opposition continued to object to the proposed siting of the power station on environmental grounds. Possibly as a result of this reaction, progress on the power station does not appear to have been as swift as anticipated.

Largely as a result of the liberalisation policies, imports surged forward, growing by 13.8 per cent in 1988 to reach Lm447.3m. Export growth was not too far behind in percentage terms, but from a much smaller base. In 1988, it stood at 13.1 per cent, with total exports reaching Lm235.9m. The widening of the visible trade gap continued in 1989 at roughly the same pace as in the previous year. However, the current account in the balance of payments remains in marginal surplus, due to earnings from services and other invisibles.

While it gave greater importance to the role of the business sector, the government also sought to reduce the weight of the public sector in employment and general economic activity. Soon after it came to power, it claimed that the previous socialist administration had employed many more public sector staff than was necessary, for vote catching purposes. Initial efforts to fire these 'excess' workers were given up following massive public protests organised by the socialists. An Auxiliary Workers' Training Scheme was set up to absorb redundant government employees who had been 'recently' engaged, as well as other unemployed. This procedure meant that on the whole, following an initial decline, the number of workers in the public sector continued to rise well into 1989, from 55,436 in May 1987, to 55,863 in spring 1989 (or 43.2 per cent of the labour force).

The government claimed repeatedly during 1989 that its economic policy was giving results. The total number of gainfully employed had reached 124,296 in early spring 1989, an increase of 3.1 per cent over May 1987 levels. Industrial jobs in early spring 1989 were hardly more than 200 above the level of January 1988, while the number of new jobs created by the tourist sector in the same period stood at some 870.

Tourism remains a mainstay of the economy. As has been usual with all administrations, the aim of diversifying tourist sources has been repeatedly proclaimed. Still, the bulk of tourist arrivals is from the UK. In 1988, 61 per cent of all tourists were British. Gross income from tourism continued to increase steadily, up to 10.5 per cent in 1988 from the previous year, to Lm124.9m. In 1989, gross tourist income trends followed the same growth pattern.

Malta will apply to join the EC

Economic liberalisation policies were twinned with a pro-Western tilt in the island's foreign policy. While maintaining Malta's stance of neutrality and non-alignment, which was written into the constitution in January 1987, the nationalist administration multiplied its contacts with Western European countries, as well as the USA, with which the previous administration had never enjoyed better than cool relations. Technical co-operation with the UK was stepped up during the first half of 1989, following a visit by British Foreign Office Minister Lynda Chalker. However, a long-scheduled visit by Prime Minister Fenech Adami to Bonn continued to be postponed past mid-1989.

The government's pro-Western policies were complicated by its stated determination to maintain the close links established by the socialists when in government with Arab countries, especially Libya. Economic co-operation in trade and joint ventures with Libya were carried over by the Fenech Adami administration, which claimed there was no reason to downgrade economic links that were mutually beneficial to both Malta and Libya. On the issue of Palestine, a Palestinian ambassador was accredited in Valletta, although the government just stopped short of fully recognising the existence of a Palestinian 'state'.

Relations with Eastern European countries, especially the USSR, were also maintained, not least by way of the trade agreements that had been signed by the previous administration. Under a protocol signed in 1986, the Soviet side was to purchase up to \$40m of Maltese goods a year, up to 1990. Light timber carriers were to be built in Malta for Soviet shipping lines. Soviet orders, however, remained well below these levels and a new protocol based on more realistic premises was being discussed during 1989.

A central thrust of the government's foreign policy was to apply to join the European Community during 1990. Meanwhile, existing commercial relations with the Community were normalised following the signing of a renewal to the association agreement between the two sides, which came into effect on 1 April 1989. The association agreement had been allowed to lapse in the early 1980s by the socialist administration, due to disagreement with the Community on the amount of financial aid proposed for Malta. Subsequently, both sides continued to extend the provisions of the old agreement on an autonomous basis. This became quite an anomalous solution to the diplomatic impasse, especially as in the meantime the Community acquired new member states, which had to be fitted somehow into the framework of the association agreement. Under the renewed association, the preferential system by which Maltese industrial exports benefit from full tariff reduction at Community borders, subject to rules of origin, has been confirmed. Malta was also granted some additional tariff reductions on a number of horticultural exports.

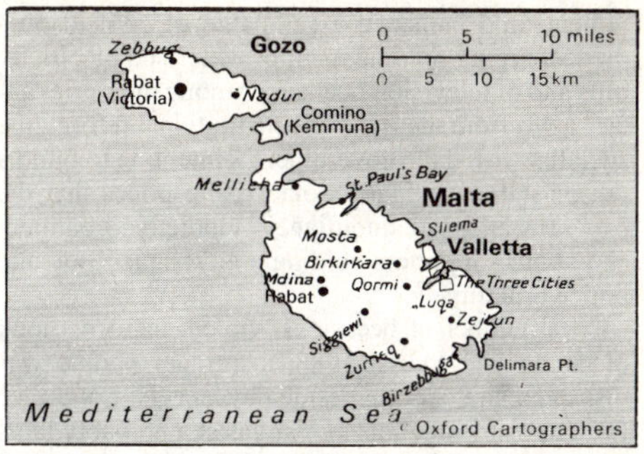

In 1989 too, a third financial protocol between Malta and the Community came into effect. This provided Malta with grants of Ecu 12.5m, soft loans of Ecu 23m, and Ecu 2.5m of risk capital, for a five year period. The Community also agreed to conduct periodic political consultations or 'dialogues' with the Maltese government. The issue of Maltese membership of the EC meanwhile became more topical. The Maltese government repeatedly stated that membership should not pose too many difficulties to the Community, given the island's small size. It has also indicated that it would need a lengthy transition period to enable the Maltese economy to cope with the demands of EC membership.

National reconciliation turns sour

On the internal political front, the government was less than successful in its intention to promote national reconciliation, thereby reducing the deep political polarisation that spread during the late 1970s and early 1980s. If anything, political polarisation may have become more entrenched in Maltese habits. Soon after it took office, the government agreed to the establishment of a parliamentary select committee to propose changes to the constitution, including a widening of the figurehead functions of the president of the republic. These changes were described in almost identical terms in the electoral manifestoes of both nationalists and socialists. They would have made the president the mediator on issues which generated political antagonism, like appointments to the state broadcasting system and the electoral commission.

For most of 1988, after having submitted a preliminary report, the select committee failed to meet. In early 1989, it again started its discussions; they led to stalemate. The real difficulty related to the method of selection of the president under the new proposed system, and to the choice of the first president. The socialists pressed for their former leader and prime minister, the veteran politician Dom Mintoff, to be nominated as president. Such a choice was clearly unacceptable to many of the government's supporters.

Prime Minister Fenech Adami broke the deadlock by moving to nominate and elect a new president under the existing system, for which a simple parliamentary majority is needed. In March 1989, Foreign Minister Vincent Tabone resigned his post and was elected by parliament the following month as president for a five year term. The select committee on constitutional reform collapsed.

Opposition spokesmen denounced the presidential appointment and announced a social boycott of the new president. They also brought to a head a number of ongoing complaints against the government's

KEY INDICATORS

	Unit	1984	1985	1986	1987	1988
Population	'000	332.0	335.0	343.3	345.6	347.0
Gross domestic product (GDP)	Lm m	461.1	472.5	502.0	533.1	591.8
GDP per capita	Lm	1,389	1,410.4	1,462.3	1,542.5	1,705.4
Inflation	%	–0.5	–0.2	2.0	0.5	0.9
GDP real growth	%	0.9	2.6	4.0	6.2	–
Agricultural production	1979-81=100	106	113	107	102	111
Tourist arrivals	'000	479.7	517.9	574.2	745.9	–
Exports (value)	\$m	410.8	421.6	521.6	631.3	–
Exports (volume)	1985=100	97.7	100.0	104.4	105.4	–
Imports (value)	\$m	639.2	672.1	786.5	1,024.6	–
Imports (volume)	1985=100	92.0	100.0	102.1	114.6	–
Balance of trade	\$m	–228.4	–250.5	–264.9	–393.4	–
Current account	\$m	11.3	–22.8	–0.3	6.4	–
External debt	\$m	206.1	209.6	–	294.6	–
Exchange rate	Lm per \$	0.46	0.47	0.39	0.35	0.33

* estimated figure

policies and launched a campaign of civil disobedience, urging non-payment of water and electricity bills and of television licences. Among their grievances, apart from the issue of constitutional reform, the socialists listed the government's intention to build a power station at Marsaxlokk, rising prices that the official statistics camouflaged, vindictive treatment of Labour supporters holding a government job and police brutality.

On a number of occasions, violent incidents took place. In late March 1988, a military parade was disrupted and later that same day, special riot police fired massive amounts of tear gas at a probably inoffensive crowd. In early June, riot police again used tear gas to protect the prime minister and his deputy, when they attended a wedding in a tough working class district; shots were fired, luckily with no fatalities. Later, the government reaffirmed its determination to preserve law and order.

Notwithstanding this effervescence, confrontation was avoided on different occasions. The government came to an agreement with the worker-managed Malta Dry Docks and with Marsa Shipbuilding Yard on a long-term programme of financing and other support to help the ship repair and shipbuilding sectors. Employing over 6,000 workers, they had amassed huge deficits over the years. Negotiations were complicated by claims from the General Workers' Union, a strong ally of the Labour Party, for wage increases.

However, this agreement was bound to have a ripple effect on public sector wages. In 1989, a commission on reform in the public service had to make recommendations that, necessarily, would bring in their wake wage increases to thousands of civil servants. An agreement signed in June 1989 with the teachers' union, where the government has strong political support, and by which teachers were to obtain professional status coupled to prospective wage rises, reflected the emerging pressures for generalised wage increases in the public sector.

Economic growth remains fragile

Since 1987, the deficit in the government budget has widened. This was due in part to the non-arrival of aid funds that were promised to Malta under a financial agreement with Italy, signed in 1986. The Italian parliament failed to ratify the protocol in due time. The first tranche of funds only arrived in December 1988. In the meantime, the government launched long-term loans for Lm40m in 1988 alone, to bridge over the deficits. As the Italian aid funds continue to come on stream, the budget deficits should decrease. Yet, by most standards, due to the overprudent debt policies followed in the 1970s and early 1980s, the government is still underborrowed and enjoys ample margins for further borrowing.

Though there is a structural deficit in the government's finances, current revenues from customs duties and income tax have been rising due to the liberalisation policies and the rise in employment since 1986. On the expenditures side, the government has been careful not to introduce any sizeable increases in social welfare spending, though its electoral manifesto had seemed to promise doing so.

On the whole, the state of the economy will probably determine the political mood of the country during the second half of the Fenech Adami administration. The government declared itself politically and economically confident as to future prospects, because of increasing private sector employment. Business representatives were also expressing renewed confidence by mid-1989. Meanwhile, due to invisible earnings, the current account of the balance of payments was likely to stay positive, despite burgeoning imports.

Even so, there remain some causes for concern. The ongoing export growth remained centred on a big electronics factory owned by a French multinational, set up in the early 1980s. Most new investment announced was in the clothing sector and relatively small-scale. By mid-1989, little data was available about the island's success or otherwise in attracting offshore business. Despite the improvement in the business and economic climate, the foundations for further economic growth were still quite fragile. It will remain so until significant productive investment comes on stream.

COUNTRY PROFILE

Political system: Republic with legislative power vested in unicameral House of Representatives (currently 69 members elected for five years by universal adult suffrage). House elects president (for five years) who then appoints prime minister and cabinet, which holds executive power. The two main parties are the (socialist) Malta Labour Party and the Christian democratic Nationalist Party.

Main cities/towns: Valletta (population 9,400), Sliema, Birkirkara and Qormi.

Language: Maltese and English are the official languages. English spoken by about 80 per cent of the population.

Media
Dailies: *In-Nazzjon Taghna* (circulation 20,000) and *l-Orizzont* (20,000) published in Maltese. *The Times* in English.
Weeklies: Most widely read are *it-Torca* (30,000), *Il-Helsien* (10,000) and *il-Mument* (22,000) in Maltese and *Weekend Chronicle, The Sunday Times* (29,000) and *The Democrat* (12,000) in English.
Business: Few with small circulations. Include *Commercial Courier* (1,500), The Lloyd, and *Industry Today.*
Broadcasting: Supervised by Malta Broadcasting Authority. Approximately 128,000 TV and 20,000 radio licences in use. Xandir Malta (part of TeleMalta Corp.) broadcasts two cable radio services, a wireless service (Radio Malta) and a TV service (Television Malta).
Advertising: Advertisements can be placed on radio and TV through Xandir Malta in Gwardamangia. Television Malta is totally commercial. Newspapers, periodicals, cinemas and poster sites (leased by government) also available.

Domestic economy: Light industry and tourism form mainstay of economy. Also important are ship repairs and agriculture. Economic growth sluggish owing to drop in exports, reduction in tourist revenues between 1980-84 and contraction in manufacturing output. However, tourism again increasing, with arrivals rising by 11 per cent in 1986, 30 per cent in 1987 and five per cent in 1988.

External trade: Regular trade deficit offset by invisible earnings, mainly from tourism and overseas investment. Principal exports include manufactured articles, machinery and transport equipment and agricultural produce. Imports consist mainly of foodstuffs, petroleum products, basic manufactures, chemicals, machinery and transport equipment. Italy, German FR and the UK account for around 60 per cent total trade turnover.

Agriculture: Moderate role in economy. Limited area of land available for agriculture. Potatoes and onions are largest vegetable crops, grapes is largest fruit crop, and flower cultivation flourishing.

Industry: Manufactures include textiles, clothing, synthetic fibres, footwear, wines and beer, furniture, electronic goods, automobile components, measuring/controlling equipment and tobacco products.

Banking: Monetary policy and foreign exchange control exercised by Central Bank of Malta. Three major commercial banks, The Mid-Med Bank, Bank of Valletta and Lombard Bank.

Legal system: Public law based on English common law. Constitution adopted 1964, extensively amended 1974 and 1987.

Membership of international organisations: CCC, Commonwealth, Council of Europe, Eurocontrol, FAO, G77, GATT, IBRD, ICAO, IFAD, ILO, IMF, IMO, INTERPOL, ITU, IWC, NAM, UN, UNESCO, UNICEF, UPU, WHO, WIPO, WMO.

Time: GMT + 1 (GMT + 2 from late Mar to late Sep).

Climate: Mediterranean, with hot summers and warm winters. Temperatures range from about 29°C down to about 10°C. Jan and Feb are coldest months, Jul and Aug the hottest. Best time to visit – Feb-May. Aug and Sep tend to be hot and humid, but usually sea breezes in evening.
Clothing: European clothing suitable for winter, spring and autumn; tropical weight for summer.

Entry requirements: Visa not required by nationals of W European, most African and some other countries. N.B. regulations liable to change at short notice.
Health: Vaccination certificates not required unless travelling from an infected area. Water supplies are filtered, and tap water is safe to drink.
Currency: Up to Lm50 in Maltese currency may be taken into the country and up to Lm25 taken out. No limit on foreign currency taken in, but amount should be declared on entry, then up to this amount may be taken out.
Customs: Personal effects duty-free, plus duty-free allowance.

Air access: Regular flights from most major international airports in Europe and North Africa.
National airline: Air Malta.
Main international airport: Malta (MLA) at Luqa, 8 km from Valletta.

Surface access: Regular car ferry services from Sicily and Italian mainland.
Main port: Valletta; new port at Marsaxlokk.

Hotels: Classified as de luxe, 1A, 1B, 2 and 3. Advisable to book in advance especially during tourist season. All hotel staff English-speaking and many are multi-lingual. Tipping usually 10 per cent if service charge not included in bill.

Restaurants: Numerous, ranging from de luxe to nightclub. Cost of meals ranges from Lm1 for one-course meal with trimmings in pub atmosphere to Lm10 per head in luxury hotel restaurant, including wine. Tipping: 10 per cent of bill if no service charge is made.

Currency
Exchange rate: Value of Maltese lira pegged to calculated value of a basket of certain selected currencies. Composition of basket periodically reviewed, largely to reflect external trade of Malta in these currencies.

Foreign exchange information
Bank accounts: Non-resident accounts may be opened by those non-resident in Malta and also by certain residents not of Maltese origin. Permission to open such accounts can be obtained from authorised dealers, and they can be used where Exchange Control permission would otherwise be required. Foreign investors

BUSINESS GUIDE

require Exchange Control permission to borrow from local sources.
Trade policy: Procurement Division of the Department of Trade is sole importer for a declining range of commodities. Since 1987 more liberal import policies were introduced and import licence requirements were progressively relaxed.

Credit cards: All major credit cards are accepted.

Car hire: Self-drive cars available at daily, weekly and monthly rates (eg about Lm5 per day with unlimited mileage and fully comprehensive insurance). National or international driving licence, which must be endorsed at Lascaris Wharf, required. Speed limits 40 kph in built-up areas, 64 kph elsewhere. Driving is on the left.

City transport: Metered taxis available but meters not often used, therefore advisable to establish price before journey.

National transport
Road: Regular bus services from Valletta to most towns and villages on Malta. Also limited service on Gozo.
Water: Gozo Channel Company operates several daily trips between Malta and Gozo.

Public holidays
Fixed dates: 1 Jan, 10 Feb (St Paul's Shipwreck), 19 Mar (St Joseph), 31 Mar (Freedom Day), 1 May, 7 Jun (Sette Guigno), 29 Jun (St Peter and St Paul), 15 Aug (Assumption), 8 Sep (The Nativity), 21 Sep (Independence), 8 Dec (Immaculate Conception), 13 Dec (Republic Day) and 25 Dec.
Variable dates: Good Friday.

Working hours
Business: (Mon-Fri) 0830 – 1245 and 1430 – 1730.
Government: (Mon-Fri) 0745 – 1230 and 1315 – 1715, (from Jun to Sep) 0730 – 1330 only.
Banking: (Mon-Fri) 0830 – 1230, (Fri) 1700 – 1900, (Sat) 0830 – 1200.
Shops: (Mon-Sat) 0900 – 1300 and 1530 – 1900.

Telecommunications
Telephone: International dialling code: 356. There are no area codes. Direct dialling from Malta to some European countries; others can be reached through Overseas Telephone Exchange – dial 994 (8994 from Gozo). In 1986 telephone numbers previously with 5 digits and starting with 2 or 3 had an extra 2 or 3 added at the beginning.
Telex: Available at first-class hotels and at TeleMalta Corporation offices in Valletta and St Julian's. Country code 406 MW.
Telegram: Can be sent from TeleMalta, in St Julian's, open 24 hours a day (tel: 224131); and Luqa Airport open 0700 – 1830 (tel: 225861).

Trade fairs: International Fair held annually during first two weeks of July at Naxxar. Specialised fairs are also held.

Electricity supply: 240 V AC.

Representation in capital: Australia, Bulgaria, China PR, Czechoslovakia, Egypt, France, Federal Republic of Germany, Holy See, Israel, Italy, Democratic People's Republic of Korea, Libya, Palestine, Spain, Tunisia, USSR, United Kingdom, USA.

Hotels: Valletta
British, 267 St Ursula Street (tel: 226019; tx: 1281).

Cumberland, 111 St John Street (tel: 227732).

Osborne, 50 South Street (tel: 623656/7; tx: 1440).

Phoenicia, The Mall, Floriana (tel: 225241; tx: 1240).

Car hire: Valletta
Avis: 50 Msida Seafront (tel: 225986; tx: 656).

Europcar: 38 Villambrosa Street, Hamrun (tel: 605038, 228745; tx: 989).

Hertz: 66 Gzira Road (tel: 514637; tx: 1255).

Business advice and information
The Malta Chamber of Commerce, Exchange Buildings, Republic Street, Valletta (tel: 624183 or 627233).

Banking
Central Bank of Malta, Castille Place, Valletta (tel: 607480; tx: 1262 citdel; fax: 232403).

Bank of Valletta Ltd, 58 Zachary Street, Valletta (tel: 623261; tx: 1235; fax: 230894).

Lombard Bank Ltd, Lombard House, 67 Republic Street, Valletta (tel: 232631; tx: 1379 lombar).

Mid-Med Bank Ltd, 233 Republic Street, Valletta (tel: 625281; tx: 1370 midval; fax: 230406).

Principal newspapers
Commercial Courier, The Chamber of Commerce, Exchange Bldgs, Republic Street, Valletta (tel: 627233).

Il-Helsien, Marsa Press, Industrial Estate Marsa (tel: 235312).

Il Mumet, Stamperija Indipendenza, 28 Our Lady of Sorrows Street, Pietá (tel: 623641).

In-Nazzjon Taghna, Stamperija Indipendenza, 28 Our Lady of Sorrows Street, Pietá (tel: 623641; tx: 1941).

It-Tórca, Workers' Memorial Bldg, South Street, Valletta.

L'Orizzont, Union Press Publications, Workers' Memorial Building, South Street, Valletta (tel: 624451; tx: 1307).

The Democrat, Stamperija Indipendenza, 28 Our Lady of Sorrows Street, Pietá (tel: 623641; tx: 1941).

The Sunday Times, Strickland House, 341 St Paul Street, Valletta (tel: 224032; tx: 1341).

Other useful addresses
COPE (Confederation of Private Enterprise), Exchange Buildings, Republic Street, Valletta (tel: 230595).

Department of Industry, 90 West Street, Valletta (tel: 229898; tx: 1111).

Department of Trade, Lascaris, Valletta (tel: 224411; tx: 1106).

Employees' Association, 35/1 South Street, Valletta (tel: 227585).

Federation of Industries, Development House, St. Anne Street, Floriana (tel: 234428; tx: 883).

Hotel and Restaurants' Association, 7 Frederick Street, Valletta (tel: 227253).

Malta Development Corporation, House of Katalunya, Marsamxetto Road, Valletta (tel: 221431, 223688, 224218; tx: 1275 devcor; fax: 606407).

Malta Development Corporation, Export Promotion Division, Middle Sea House, St Publius Street, Floriana (tel: 235251; tx: 2966 cpi).

Ministry of Finance, St Calcedonius Sq, Floriana (tel: 220437; tx: 1512).

Ministry of Industry, Auberge d'Aragon, Independence Square, Valletta (tel: 229898; tx: 1111).

Ministry of Tourism, Harper Lane, Floriana (tel: 224444; tx: 1105).

National Tourism Organisation, Harper Lane, Floriana (tel: 224444, 223595, 232010; tx: 1105 holiday).

Marshall Islands

T. R. Hughes

(map on page xxiv)

Three years after signing a Compact of Free Association with the United States, the Republic of the Marshall Islands made a fresh effort in 1989 to promote trade and encourage local and overseas investment. Although it has been self-governing since 1979, the territory, which includes the atomic-test atolls of Bikini and Eniwetak, embarked on a new stage of its development in October 1986 with the ending of the UN trusteeship agreement under which it had been administered by the USA since the end of the Second World War.

It is now a sovereign state responsible for its own foreign policy. But the USA still looks after its defence and provides, as part of the compact, $600m of economic and service aid over a 15-year period. While the free association relationship continues indefinitely, the economic aid and defence aspect of the compact (like that of the neighbouring Federated States of Micronesia) are subject to negotiation at the end of the 15 years. President Bush gave a boost to the republic's national ego in July 1989 when he elevated diplomatic relations to full embassy status with an exchange of resident ambassadors.

KEY FACTS

Official title: Republic of the Marshall Islands
Head of state: President Amata Kabua
Head of government: President Amata Kabua
Capital: Majuro (on Majuro atoll)
Official Languages: English, Marshallese
Currency: US Dollar ($ = 100 cents)
Area: 34 atolls and 870 islets
Population: 43,000* (1988)

* estimated figure

Economic picture

At first glance the economic picture is not promising. With only 175m sq km of land on the islands and atolls dotted around 1.8m sq km of sea, the Marshalls is mainly a subsistence copra-based economy and not well-endowed at that. Coconut oil and copra amount to about 90 per cent of exports currently around $2.5m a year. There is a chronic imbalance of trade with imports far outstripping exports and the outlook is not eased by copra prices seemingly on a constantly downward path. 80 per cent of imports originate in the USA with most of the rest coming from Japan.

The government provides the lion's share of the jobs. One asset for any would-be investor is the republic's favoured access to the US market for certain locally-produced goods. Against that is the problem of raising a skilled workforce. There is no actual shortage of labour. Like many Third World countries, more than half of the resident population of 43,000 is under the age of 15 and the birth rate, at 4.2 per cent, is one of the highest in the world. But the unemployment rate is a severe 25 per cent because of lack of a strong private sector and adequate training facilities.

President Amata Kabua's government gives high priority to developing infrastructure and taking other steps to encourage investment by local and overseas firms in fisheries, tourism and light manufacture. The Majuro airport can accommodate Boeing 727-type aircraft. The container freight yard and a new dock may attract shippers to use the capital as a transshipment point. There is a modern international communications system and a plentiful supply of power.

Apart from the USA, investment over the past few years has come from Japan, China, Australia and Nauru. One attraction for cash-rich overseas businessmen is that they can buy a Marshall Island passport for $250,000 giving them in principle the same right as the Marshallese to settle in the US mainland. The US State Department, however, stipulates that naturalised citizens of the Marshalls must have lived in the islands for five years.

Not unnaturally the republic is looking hard at the one thing it has plenty of – the sea. In addition to fish, it has registered upwards of 1m tonnes of shipping under the Marshalls' flag, with the ships being operated in some cases by Marshallese crews. The government has also commissioned a feasibility study to look into the possibility of extracting and processing seabed minerals within its economic zone after receiving an encouraging report on its underseas wealth from the East–West Center in Honolulu.

Trade mission

In early 1989, a government mission explored new trade openings in Fiji, Australia and New Zealand and returned home impressed by the way tax incentives had attracted new business in Fiji. The trip helped to show how the pattern of commerce in the Pacific basin is gradually changing with more North–South contact as well as East–West. The republic wants to upgrade its own airline at a cost of $40m to operate between Hawaii and Fiji. This, together with other new air and sea routes (the Pacific Forum Line is extending its sailings from Kiribati to include the Marshalls) will bring the islands of the northern and southern hemisphere closer together.

The Marshalls was admitted to the South Pacific Forum as a full member in 1987. This recognition of its independent status was seen as an important step towards joining the Asian Development Bank which, in turn, would unlock substantial development funds. But some Western nations are not happy over the legalistic interpretation of 'independence'. While the UN Trusteeship Council has voted to end the strategic trust, no reference has yet been made to the Security Council which is ultimately responsible.

Whether the new links with the southern island states will be at the expense of those with the USA remains to be seen. In addition to the compact, the USA is still an important source of funding. Separate trust funds have been set up for the people of Bikini and others affected by the US nuclear testing programmes several decades ago.

The long-term US lease of Kwajalein atoll, with some 3,000 United States servicemen, supplements the economy, and low-skill level employment, especially from the nearby, overcrowded Ebeye. The poor living conditions in Ebeye and US rent payments for the missile range were a constant focus of dispute between the Marshallese and Washington in pre-compact days.

Despite some minor problems, by most accounts the relationship with the USA under the compact has gone smoothly particularly since the court settlement of outstanding atomic testing claims by the Bikinians. The Marshallese are a single-culture people, unlike the Federated States of Micronesia, and have successfully blended their traditional leadership with its modern government system and avoid their neighbour's political factionalism.

COUNTRY PROFILE

Political system: The Marshall Islands, a former Japanese colony and part of the Trust Territory of the Pacific Islands, administered by the US under a UN mandate from 1947 to 1986. The Compact of Free Association approved by the US House of Representatives in July 1985 and implemented by the USA in Oct 1986 gives the islands independence, but maintains their defence ties to the USA. This has not dissuaded the islands from debating the idea of joining a Nuclear-free Pacific zone organisation, however.

Main town: Majuro (on Majuro atoll, population 12,000 1987), Ebeye (8,000).

Media
Broadcasting: Radio broadcasts transmitted at least 18 hrs/day in English and the main local dialects. TV coverage fairly widespread, some cable relay.

Employment: Efforts to redevelop coconut production held back by lack of labour. Out-islanders, potential copra harvesters, have migrated to central atolls.

External trade
Exports: Main exports are coconut oil, copra and fish. Some beef, fruit and vegetables have been exported; postage stamps also earn foreign exchange. Main destinations are US, Japan and Guam; relations with Republic of China (Taiwan) are also being developed.
Imports: Main imports are food and beverages, machinery, construction materials, textiles and fuel. Main sources are US and Japan.

Agriculture: Subsistence farming of coconuts, taro, breadfruit, bananas, yams, sweet potatoes and vegetables, with pig and poultry raising, is main occupation. Large areas of potentially arable land not cultivated. The Government is undertaking a five-year programme of replanting 25,000 of 34,000 acres of coconut palms; many are unproductive trees from Japanese (World War II) or German (19th century) settlements. In 1985, US government agreed to pay $42m to clean up radioactive contamination on the Bikini atoll, involving disposal of 30.5 cm of topsoil and resettlement programme. Fishing is mainstay of economy, supplying principal source of protein and export revenues.

Industry: Small-scale industries include handicrafts, fish processing, bottling, copra processing, bakeries and boat building. Majuro atoll's coconut oil plant hopes to expand to refine the oil and make non-dairy creams, soaps and other coconut products. Postage stamps are an important source of revenue.

Mining: Small mineral deposits exist, but exploitation hampered by a shortage of land to accommodate displaced population and doubts about economic viability.

Membership of international organisations: ESCAP (associate), SPF.

Time: GMT + 12 hrs.

Climate: Warm and humid, temperatures 23-30C, humidity around 80 per cent. Rainfall variable, minimum 250 mm/year, can occur in downpours. Hurricanes are possible.

Entry requirements
Passport: Required by all except US citizens with proof of citizenship.
Visa: Not required by US citizens with proof of adequate funds. Entry permits granted to all others with proof of onward passage and adequate funds for stays of less than 30 days. Special regulations may apply to some non-tourist destinations.

Health precautions
Advisable: Precautions against tetanus, typhoid and paratyphoid.

Air access
Other airlines: Air Micronesia, Airline of the Marshall Islands.
International airport: Majuro International (MAJ).

Surface access
Main ports: Majuro (under renovation).

Hotels: There are small hotels.

National transport
Air: Regular scheduled flights between 10 atolls, by Airline of the Marshall Islands.
Road: Outside administrative areas, the road network, which is being extended, may consist of tracks not passable to ordinary vehicles.
Water: Frequency of inter-island services governed by weather conditions; trips to inhabited islands at least every 30-45 days.

Working hours
Government and business: (Mon-Fri) 0800 – 1630.

Telecommunications: Telecommunications system has international satellite link.
Telephone: Dialling code for Marshall Islands, IDD access code + 692 followed by area code (Ebeye 3, Majuro 29).

Banking: New measures being implemented to regulate offshore banking, after the suspension of licences, amid allegations of questionable practices.
Marshall Islands branches: Bank of Guam, Bank of Hawaii.

Business advice and information
Marshall Islands Chamber of Commerce, Majuro.

Tourism Office, Ministry of Resources and Development, PO Box 306, Majuro, Marshall Islands 96960. Tel: 3206.

Hotels: Marshall Islands
Ajidrik, Majuro 96960 (tel: 318).

Eastern Gateway, PO Box 106, 96960 (tel: 337).

Majuro, Majuro 96960, P.O. Box: 185 (tel: 324).

Mauritania

Susi Pollak

Of all the drought-prone countries of the Sahel, Mauritania has been the most affected by desert encroachment. Three-quarters of the land area is desert. Only a minute part of the country is suitable for agriculture, and almost all of this is in the south, in the Senegal River valley, home of the Toucouleur and Soninke people who are ethnically related to the population of Senegal. The rest of the country is arid but suitable in parts for raising livestock on which the livelihood and lifestyle of the Moorish nomads have traditionally been based. In the great drought of recent years more than one third of their 1m cattle died, as did 20 per cent of their 6m sheep and goats, and they were forced into the urban areas where they found it difficult to obtain suitable employment. Nouakchott, built to accommodate 30,000, now has a population of 500,000. Some of the nomadic groups moved towards the lusher south.

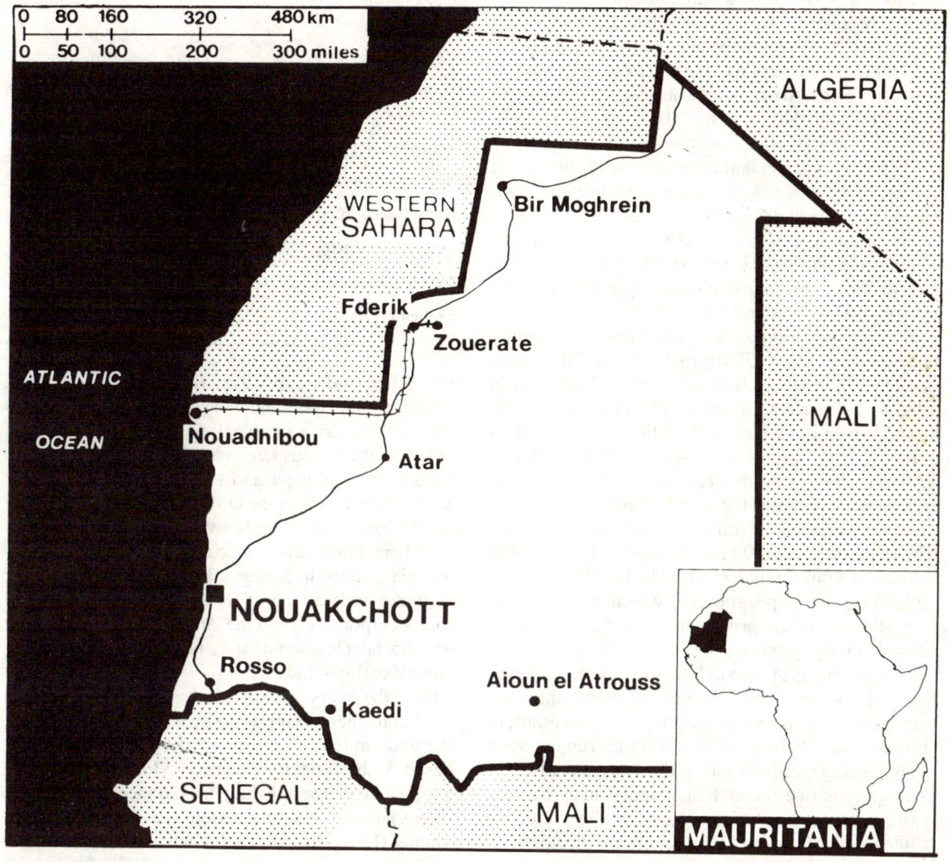

This is the background against which the tragic racial confrontation of 1989 took place – events which had consequences with which Mauritania and its neighbours will have to live for a long time. The crisis came to a head in April 1989 with bloody clashes between the Moors, who control both the army and the government of Colonel Ould Sid Ahmed Taya, and the southern Toucouleur and Soninke.

The Moors consist of two groups: those who are light-skinned and of Arab-Berber descent, and their ex-slaves, the dark-skinned Haratines. All Moors identify themselves with the Arab-speaking people of North Africa. The Toucouleur and Soninke groups are black and ethnically linked to the population of Senegal. The initial attacks on the black southerners are said to have been carried out by the dark-skinned Haratines, but the immediate cause is largely irrelevant.

Tensions between the two groups go back a long time, but they intensified as more irrigable land became available with the completion of the Diama and Manantali dams by the Senegal River Development Organisation (OMVS) which is jointly owned by Mauritania, Mali and Senegal. The area concerned is in the domain of the traditional settled peasantry – the Toucouleur and Soninke. There were rumours that the 'white' Arab-Berber Moors were acquiring the titles to large portions of that land, often in the name of their Haratine allies. The black Mauritanians feared expropriation, that they, the lawful property owners, would be turned into agricultural labourers.

Relations with Senegal

The clashes in Mauritania quickly produced a Senegalese counter-reaction: Mauritanians living in Dakar were attacked and killed, their shops looted, and thousands of them 'repatriated', in many cases forcibly. But the proportion of those affected in Mauritania was much greater. Apart from the 50,000–70,000 Senegalese nationals expelled, some 20,000 black Mauritanians were forced into exile. Many of them had identity papers to show that they, their parents before them and even their grandparents were citizens of Mauritania, but the authorities in Nouakchott described them as people of Senegalese origin; they were stripped of their property and forced into exile.

The dispute between Nouakchott and Dakar produced land claims on both sides. The border was agreed by two previous leaders, Presidents Heydallah of Mauritania and Senghor of Senegal, as running through 'the middle of the Senegal River'; this was necessary for the joint OMVS development treaties to be ratified. Now Mauritania laid claim to St Louis, the Senegalese town from which Mauritania was administered under French colonial rule. The government in Dakar asserted that certain areas on the north bank of the river were rightfully part of Senegal, under a 1933 French colonial decree. Border clashes ensued. In July 1989 Mauritania claimed that Senegal had sent 'armed groups' to attack Mauritanian citizens on their own soil. Intensive Senegalese troop movements were reported, and Mauritania announced that it would 'vigorously oppose any move to attack its independence, sovereignty and territorial integrity'. Until then the source of Nouakchott's military supplies had been France, which also supplies Dakar. But Mauritania, suspecting that the French favoured their Senegalese allies, concluded a military pact with Spain, including the supplies of training and weapons.

Apart from the huge refugee problems created by the forced population removals, the economic dislocation which resulted was considerable. Senegal has long been Mauritania's closest trading partner. With the exception of iron and fish, Mauritania's main exports which are shipped through Nouakchott and Nouadhibou, the established trading routes run overland into Senegal. The April clashes resulted in an immediate shortage of goods. Already in January, when Dakar tried to bring pressure on Nouakchott by blocking their mutual border, long queues of lorries carrying sugar and other foodstuffs, textiles and spare parts, all destined for Mauritania's markets, built up on the southern bank of the Senegal River.

KEY FACTS

Official title: Islamic Republic of Mauritania
Head of state: President Colonel Maaouiya Ould Sid'Ahmed Taya
Capital: Nouakchott
Official Languages: Arabic, French
Currency: Ouguiya (UM) = 5 khoums
Exchange rate: UM86.7 per $ (Nov 1989)

Area: 1,030,700 sq km
Population: 2.0m (1988)
GNP per capita: $503 (1987)
GNP real growth: 0.5% (1986/87)
Inflation: 8.2% (1987)
Trade balance: –$146m (1986)
Foreign debt: $2.0bn (1987)

Maghreb Arab Union

The racial crisis coincided with the formal establishment of the Maghreb Arab Union (MAU), an important economic grouping comprising Algeria, Libya, Morocco, Tunisia and Mauritania. It is possible that this new alliance created the feeling among Mauritania's pro-Arab rulers that they could now afford to disregard the anger which the persecution of the southerners was bound to cause in Senegal and other black African countries. This may turn out to be a miscalculation, as the MAU states – particularly Algeria and Libya – were highly embarrassed by the behaviour of the authorities in Nouakchott, and certain MAU policies which are to Mauritania's advantage could still be reconsidered.

One of the first MAU decisions was to establish a Maghreb High Council for the Exploitation of Iron and Steel, which planned to set up a company to work iron ore reserves around Nouakchott.

Until the disruption of 1989, Mauritania's economy registered successes. GDP growth averaged 4 per cent in 1985-87. Inflation declined from 13.5 per cent in 1985 to 6.3 per cent in 1988. The government followed World Bank recommendations in many of its policies. Total foreign debt at the end of 1988 was $2.04bn, and in 1989 the Paris Club rescheduled an unspecified amount of these debts. The IMF approved a three-year Enhanced Structural Adjustment Facility (ESAF) worth SDR50.85m. It supercedes the 1986 Structural Adjustment Facility. The ESAF is tied to a 1989–91 reform programme which includes the strengthening of the private sector and restructuring of the banking system.

Fishing remains the second-largest source of foreign earnings. Because of the danger of over-fishing by foreign trawlers in north-west Africa, the government entered into a fishery conservation agreement with neighbouring maritime states. The government is encouraging marine research and signed an agreement with Algeria to this effect: Mauritania's Higher Institute for Fish Farming in Nouadhibou will undertake joint projects with the Algerian National Institute for Maritime Sciences and Territorial Development. Another agreement allowed Ghana's Mankoadze Fisheries to fish in Mauritanian waters. Ironically, the flight of the black Mauritanians and Senegalese from Nouakchott seriously damaged the industry: it included some 1,200 local fishermen who left behind idle boats, and the price of fish rose 500 per cent. Nouakchott's boat builders and mechanics were also among those driven into exile.

Despite the serious tensions with Senegal, Mauritania's energy minister and his Senegalese colleague attended a meeting of the OMVS to discuss a schedule for the extension of a high-voltage power line from the Manantali hydroelectric dam in Mali to Mauritania, via Tobene in Senegal. French aid is paying for management training at Sonelec, the national electricity corporation, and the Arab Fund for Economic and Social Development is helping to finance the electrification scheme.

Over the years, Mauritania has received generous Arab and other foreign assistance. By the end of 1988, the total amount of aid granted by the Abu Dhabi-based Arab Monetary Fund was $116m, including a $8.6m loan that year to help the balance of payments. There has also been help from the European Community (EC) – in 1989, a 7m ecu subvention for the import programme – bringing total EC funding to 95m ecus in the current cycle. Assistance arrived in many forms; for instance the French paid FF20m for Mauritania's share in running the joint francophone airline, Air Afrique. Both France and West Germany have written off a significant proportion of Nouakchott's debts.

KEY INDICATORS

	Unit	1983	1984	1985	1986	1987
Population	m	1.67	1.72	1.76	1.81	1.86
Gross domestic product (GDP)	Ouguiya bn	43.21	46.38	55.10	62.70	69.17
GDP per capita	Ouguiya '000	25.80	26.96	31.31	34.64	37.19
Inflation	%	1.0	7.0	13.5	7.8	8.2
Agricultural production	1979-81=100	94.2	94.6	98.6	108.6	107.7
Food production	1979-81=100	94.2	94.6	98.6	108.6	107.7
Exports	$m	354.7	330.3	402.6	449.0	473.0
Imports	$m	626.8	527.5	631.9	581.3	619.0
Balance of trade	$m	−272.1	−197.2	−229.3	−132.3	−146.0
Current account	$m	−213.0	−110.7	−116.1	−85.3	−73.0
Total external debt	$m	1,289	1,334	1,527	1,822	2,035
Debt service ratio	%	10.4	13.0	19.3	17.3	18.2
Exchange rate	Ouguiya/$	54.81	63.80	77.08	74.38	73.88

Outlook

Mauritania's fortunes over the next few years will depend on the government's willingness and ability to retrace its steps on the nationality question. Many black citizens who have been expelled are anxious to return to their farms and jobs in Mauritania, willing to accept that a 'mistake' has occurred in their particular case. Many of the exiles are professionals, civil servants, even members of the armed forces – people who are needed in a country which is as short of skills as Mauritania. There is nobody to replace even the farmers, since few Moors possess agricultural experience, and the courses which were hurriedly set up to train farm operatives are reported to have attracted few promising candidates.

There is no shortage of would-be mediators. Offers to help settle the dispute have come from, among others, Yassir Arafat, the PLO leader, and Zaïre's President Mobutu. But the best chance of reconciliation must lie in the enlightened self-interest of the Mauritanian and Senegalese leaders, with the encouragement of the organisations to which both belong: the OMVS, the Communaute Economique de l'Afrique de l'Ouest (CEAO), and the Organisation of African Unity.

COUNTRY PROFILE

Political system: Power rests with the president who heads the 24-member Military Committee for National Salvation (CMSN). Independence gained in 1960. Formerly a French colony.

Political parties: There are no political parties.

Population: The population of some 2m (1988 estimate) comprises a majority of Arabised Moors. The rest are ethnically linked with the peoples of Senegal and Mali.

Main cities/towns: Nouakchott (population 450,000), Nouadhibou (formerly Port-Etienne, population 45,000), Zouerate (35,000).

Language: Arabic is the official language, while French is usual in business circles. Hassaniya, Pulaar and Wolof are the major languages in everyday use. English is rarely spoken.

Media
Press: Only daily newspaper is *Chaab,* which is printed in Arabic and French.
Broadcasting: Office Mauritanien de Radiodiffusion et de Télévision (OMRT) broadcasts in Arabic and French.

External trade: At the end of 1987 the debt ratio was 347.5. Strong reliance on aid flows and earnings from fish and iron ore. Current account remains in deficit due to large net services payments, but creditors have agreed on long term rescheduling, as the government has introduced structural adjustment programmes.
Exports: Fish and fish products (typically 57-60 per cent of total value), iron ore (40-50%). Main destinations: Japan (10-20% of total value), France (10-20%), Spain (10-15%), Italy, Belguim, UK, Greece, German FR.
Imports: Foodstuffs (30-40 per cent of total value), petroleum products (24-30%), transport equipment (8-12%), consumer goods (8-12%). Main sources: France (typically 18-25% of total value), Spain (15-20%), Senegal (8-10%).

Agriculture: Sector accounts for 12 per cent of GDP. Less than three per cent of total area is cultivated arable; 10 per cent is pasture. Around 80 per cent of population engaged in subsistence farming and nomadic cattle and sheep herding. Recurrent drought and disease wiped out almost one-third of livestock since 1981. Production of foodstuffs restricted to irrigated land in south along the north bank of the Senegal River. In 1987-88, 35 per cent of food requirements met locally; rest supplied by international aid agencies and imported. Main food crops include millet, sorghum, rice, maize, potatoes and dates. New irrigation schemes in the south are expected to step up food production. Mauritanian coastal waters are among the richest in the world and joint-venture fishing is now the most important foreign exchange earner. Main catch is deep sea species and shellfish – particularly shrimps for the Japanese market.

Manufacturing: Sector contributes 12 per cent to GDP. Most important activities are fish freezing and processing and treatment of locally-mined iron-ore. Also various small import substitution industries (brewing, footwear, dairy processing etc), an oil refinery in Nouadhibou at Cansado, newly restarted by the Algerians, and sugar refinery in Nouakchott.

Mining: Extraction of high-grade iron ore from Kedia by the state-owned SNIM, the largest single employer of salaried workers. But reserves at Kedia are projected to last only until 1994. A new mine, Guelb-el-Rhein, 20 miles north of the mainly worked out Kedia mines has been

COUNTRY PROFILE

developed with World Bank assistance, and has overtaken exports from Kedia. Production and export of iron ore picked up in 1984-85 due to temporary strengthening of world markets, but dropped again in 1986-87. Other mineral resources include copper (mining to be revived at Akjoujt); gold (extraction near Akjoujt was due to start in 1988); phosphates (deposits at Bofal), and uranium.

Energy: Primary energy resources consist of fuelwood and small scale thermal electricity (generating capacity 110 mW). All petroleum requirements are imported. Prospecting for oil is underway, mainly in the coastal regions.

Membership of international organisations: African Development Bank, African States associated with the EC (Lomé Convention), Arab Common Market, Arab Fund for Economic and Social Development, Arab League, CEAO, ECOWAS, FAO, GATT, IDA, IFAD, IFC, ILO, IMF, Islamic Conference, Islamic Development Bank, OAU, OMVS (Senegal River Development Authority), UN, WHO, WIPO, World Bank.

Time: GMT.

Climate: Hot and dry. Hottest month in Noukachott – Sep (24-34C); Coldest, Dec (12-29C); Wettest month Aug.

Entry requirements
Passport: Required by all.
Visa: Required by all except nationals of Arab League countries and certain African countries. UK citizens must get their visas from the Consul Generale office in Paris (89 Rue du Cherche Midi XVI).
Currency: Unlimited foreign currency import but amount must be declared on arrival. Unexchanged foreign currency may be exported. Declaration forms must be produced on departure. Import and export of local currency strictly forbidden. Controls are constantly subject to modification.

Health precautions
Mandatory: Yellow fever vaccination certificate required by all, unless arriving from a non-infected area and staying for less than 14 days.
Advisable: Yellow fever, cholera, typhoid, polio vaccinations. Malaria prophylaxis should be taken as risk exists throughout most of country. Water precautions are advisable.

Representation overseas: Abidjan, Abu Dhabi, Algiers, Baghdad, Banjul, Beijing, Bonn, Brussels, Bucharest, Dakar, Damascus, Doha, Jeddah, Kinshasa, Kuwait City, Lagos, Libreville, Moscow, Paris, Rabat, Tehran, Tripoli, Tunis, Washington.

Air access
National airline: Air Mauritanie (Mauritania is also a minority shareholder in Air Afrique).
Other airlines: Aeroflot, Air Afrique, Air Algérie, Air Senegal, Iberia, Royal Air Maroc, UTA.
International airports: Nouadhibou (Code: NDB), 4 km from city; Nouakchott (Code: NKC), 4 km from city.

Surface access
Main ports: Nouadhibou, Nouakchott. The deep-water port near Nouakchott was completed two years ago and traffic should reach 600,000 tonnes for 1989. Nouadhibou remains the largest port.
Overland access: Surfaced road from Dakar (Senegal)to Nouakchott, via ferry over River Senegal at Rosso. Also paved from Mali to Nouakchott.

Hotels: Accommodation is limited and visitors should book well in advance. Service charge is normally included in bill, otherwise 10 per cent tip is usual.

Electricity supply: 127/220 V AC, 50 cycles. Plugs and sockets mostly two-pin (round).

Credit cards: Only accepted in main hotels.

Car hire: International or national driving licence required. Four-wheel drive with chauffeur recommended, although expensive.

City transport
Taxis: Taxis are numerous in main towns. Fares are standardised but not metered. Small tip is usual.

National transport
Air: Air Mauritanie provides a daily service between Nouakchott, Nouadhibou and Zouerate, while flying bi-weekly to other towns in Mauritania such as Kaedi.
Road: Surfaced roads link Nouakchott with Néma in the southeast and Rosso with Akjoujt, via Nouakchott.
Rail: Track inland from coast (Nouakchott-Zouerate) mainly for freight but some passenger services (single-class) scheduled; motor vehicles sometimes carried.

Public holidays
Fixed dates: 1 Jan, 1 May, 25 May (African Liberation Day), 28 Nov (National Day).
Variable dates: Korité (Eid al-Fitr), Tabaski (Eid al-Adha), Maouloud (Prophet's Birthday).

Telecommunications
Telephone: International telephone facilities available in Nouakchott and Nouadhibou. Dialling code for Mauritania: IDD access code + 222 followed by subscriber's number. There are no area codes.
Telex: Available in Nouakchott and Nouadhibou.

Representation in capital: Algeria, China PR, Egypt, France, Gabon, German FR, Rep. of Korea, Korea PDR, Libya, Nigeria, Saudi Arabia, Senegal, Spain, USA, USSR, Vietnam, Zaïre.

Hotels: Nouakchott
El Amane (tel: 5212298).

Marhaba, P.O. Box: 135 (tel: 52094).

Novotel, nr Ambassade de France, P.O. Box: 1366 (tel: 53526; tx: 866; fax: JU-831).

Sabah, P.O. Box: 452 (tel: 51564; tx: 821).

Mauritius

W. G. Huff

Mauritius is best known to the rest of the world as a holiday island with beautiful beaches. Tourism is of course one of the three sectors responsible for the country's remarkable economic success in recent years, the other two being sugar and the textile industry. But this success is beginning to meet serious limitations. The GDP growth rate which was 8.7 per cent in 1986–87 declined to 5 per cent in 1987–88 and 4 per cent in 1988–89. This trend is likely to be reversed; Finance Minister Vishnu Lutchmeenaraidoo predicts an average annual GDP growth rate of 6 per cent in the 1990s. This requires adapting existing economic activities to changing conditions, as well as expanding in new directions such as the provision of offshore banking and related facilities.

Tourism

Statistically this sector has been highly successful, and a government plan envisages a continuing boom, with 315,000 tourists in 1992 and 400,000 by the end of the century. Tourist figures increased from 118,000 in 1982 to 245,000 in 1988. These arrival figures do not tell the whole story. The growth in the number of tourist/nights has been even faster: the visitors stay longer – an average of about 9.5 days in 1976, compared to 11.6 days in 1986 – which means they occupy more hotel rooms and require more services than in the past. Gross foreign exchange earnings from tourism in 1988 were RM2.5bn ($175m); the 1989 figure was estimated at RM2.9bn. The sector employs about 9,000 people directly, and many more indirectly. There are some 60 hotels, and several more are being built.

Whether such rapid expansion is desirable is another question: pollution has become a serious threat. At the beginning of 1989, after an extensive World Bank investigation of the problem, Mauritius was lent $93m for a five-year environmental protection programme which involves the purification of hotel waste waters and sewage before they are discharged into the sea. The number of visitors may have to be scaled down and profits maintained by concentrating to an even greater extent on the luxury end of the tourist trade. It has been government policy for some time to ban charter flights which specialise in package tours. Instead, Air Mauritius operates schedules

KEY FACTS

Official title: Mauritius

Head of state: Queen Elizabeth II

Head of government: Anerood Jugnauth

Ruling party: Coalition led by Mouvement Socialiste Mauricien

Capital: Port Louis

Official Languages: English, French (Creole widely spoken)

Currency: Mauritian Rupee (MR) = 100 cents

Exchange rate: MR15.2 per $ (Nov 1989)

Area: 1,865 sq km

Population: 1.08m (1988)

GDP per capita: $1,714 (1987)

GDP real growth: 7.4%* (1988)

Inflation: 9.1% (1988)

Trade balance: –$146.8 (1988)

Foreign debt: $775m (1987)

* estimated figure

aimed at the upmarket jet set which can afford luxury holidays, and the ministry of tourism's marketing drive pays increasing attention to business visitors with expense accounts. There is some opposition to this policy from investors in the hotel and entertainment sector who favour charter flights to ferry in more and more visitors, to utilise and pay for the new facilities which the government has encouraged them to build.

Mauritius's basic limitation is its small size, a factor which was reflected in the 1989 budget when Lutchmeenaraidoo announced tax concessions for construction companies building high-rise accommodation as an essential space-saving measure. This is a constraint which affects not only the tourist sector but all aspects of life, as does the programme to clean up the environment. For instance the siting of pollution-creating factories – such as the textile industry's dye-works – is to be restricted to a single industrial park with strict environmental rules. Legislation was introduced on the use and disposal of dangerous chemicals, including pesticides and fertilisers which are applied much more profusely in the sugar plantations than would be considered safe in Europe.

Sugar and textiles

Another of Mauritius's basic constraints is the availability of labour. The population has stabilised at just over 1m and manpower shortages are starting to develop. This is felt particularly in the traditional sector of the economy: sugar growing, which was once the island's economic mainstay. Even now 45 per cent of Mauritius's total area – 87 per cent of all arable land – is under sugar cane. About half of the sugar acreage belongs to large plantations; the rest is owned by some 30,000 farmers who work plots of one or two hectares. There were fears in 1989 that the agricultural labour shortage might affect the sugar harvest, particularly since some of the cane was damaged by cyclones. The 1988 harvest of 635,000 tonnes was unlikely to be equalled.

Most of the sugar is exported to the EEC, which buys it at above world market prices under the Lomé treaty. Mauritius's quota is 500,000 tonnes. A further 12,000 tonnes go to the United States, while Britain is the largest single customer. Production has held up in volume terms, but its share of GDP declined from 12 per cent in 1980 to 8 per cent in 1989. The main engine of growth in recent years has been the industial sector to which the plantations have lost so much labour, particularly the textile factories in the export processing zone (EPZ).

The success of the EPZ was built on cheap labour – cheaper, at one stage, than almost anywhere else in the world. But with unemployment at only 3 per cent in 1989, this condition has ceased to exist. With average wage rises of 10 per cent a year in 1987–89 the need for a smaller but skilled workforce and mechanised production became clear. The 1989 budget provided a $13m loan fund to allow the textile industry to invest in modernisation, and to establish a centre for textile technology and a scheme whereby employers can offset double the cost of approved training programmes against tax.

Although the labour scarcity has been an important factor in slowing overall economic growth, it is not the only problem. The country's textiles now face tougher competition from other low-cost areas. To expand, Mauritius must increase the value rather than the quantity of its exports through producing more sophisticated goods. Fashion garments are already replacing the mass produced items. There is substantial vertical integration, with large local spinning and dyeing operations based on new technology. Two weaving mills and two denim factories are being established. To reduce the reliance on textiles there is increasing industrial diversification, with successful new enterprises in electronics, leatherware, offset printing, jewellery and diamond cutting.

Financial services

Towards the end of 1989 Mauritius launched Africa's first offshore banking centre – for bona fide business, the Bank of Mauritius stressed. But the secrecy provisions in its banking law, the concessionary 5 per cent corporate tax rate and the free remittability of profits are designed to attract international interest, and one of the first licences issued went to Britain's Barclays Bank. Mauritius sees a special role for this new venture within the PTA (Preferential Trade Area for Southern and Eastern Africa) and the same applies to its flag-of-convenience plan which was at an advanced stage at the end of 1989.

Until then, only a handful of ships sailed under the Mauritian flag, most of them under the control of two large, old-established firms: Roger and Company and Ireland Blyth, both of which opposed the scheme, particularly the steep increase in registration fees which would follow. Once again, the open register offered anonymity and low taxation: foreign owners and even their shareholders would be exempt from income tax, and only 30 per cent of the crew on these foreign-owned vessels would have to be Mauritian.

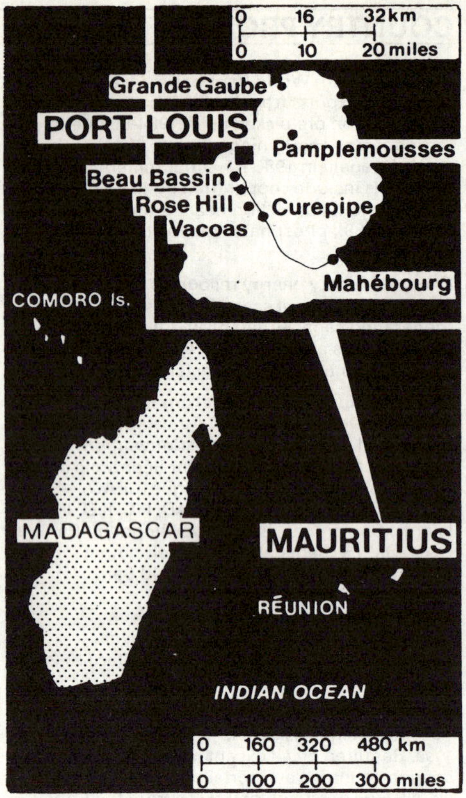

Opponents of Prime Minister Anerood Jugnauth's government fear that both the flag of convenience scheme and the offshore bank are bound to increase contacts with South Africa and provide Pretoria with yet another handy tool for sanctions busting. Lutchmeenaraidoo did not deny that South African funds might be attracted by the offshore scheme, since Mauritius was "open to all banks and institutions which are creditworthy'. Officially the government seeks to reduce contacts with Pretoria to a minimum; it claims that South African investments amount to less than 1 per cent of foreign investment in the country. But a number of scandals involving links with South Africa have put a question mark over the extent of the contacts; they have also led to serious domestic political splits. For instance Sir Gaetan Duval, leader of the pro-South African lobby in Mauritius, resigned in 1988 from the ruling Labour Party coalition in which he served as deputy prime minister and minister of tourism. His difference with Sir Anerood was reported to be over South African versus Asian investment in the hotel sector; Jugnauth preferred the Asian money and snubbed the South African interests. This was seen locally as a South African "defeat'.

In 1989 there was a controversy over the issue of a Mauritian passport to a South African tycoon, and as a result the immigration laws were tightened. Before that foreign investors were often granted the privilege of citizenship, but this practice is to cease – a fact which is also of considerable concern to investors from Hong Kong, who increasingly regard Mauritius as a safe haven from the Chinese.

There are no deep foreign policy differences between the ruling Labour Party coalition and the leftwing Mauritius Militant Movement opposition. The most serious divisions now tend to occur between mutually hostile Hindu sub-groups. There were two recent attempts to assassinate Jugnauth, both at Hindu religious ceremonies. The Hindus form 52 per cent of the population and dominate the government, but Jugnauth insists that he is first and foremost a Mauritian prime minister, not a Hindu one, and he has not been seriously challenged on this point, even by his political opponents.

KEY INDICATORS

	Unit	1984	1985	1986	1987	1988
Population	m	1.02	1.03	1.05	1.07	1.08
Gross domestic product (GDP)	MR bn	14.4	16.6	19.6	22.7	–
Inflation	%	7.3	6.7	1.6	0.6	9.1
GDP per capita	MR	14,118	16,117	18,666	21,215	–
GDP real growth	%	4.7	7.0	8.4	5.9	7.4*
Agricultural production	1979-81=100	103.0	112.0	119.5	120.8	111.9
Sugar production	'000 tonnes	576	646	707	694	631
Exports	MR m	5,180	6,729	9,062	11,362	13,410
Sugar	MR m	2,584.1	2,956.0	3,642.0	4,310.0	–
Imports	MR m	6,494.0	8,119.5	9,199.0	12,765.4	16,846.8
Balance of trade	$m	–41.4	–26.0	58.0	–8.1	–146.8
Current account	$m	–56.1	–29.8	94.2	72.6	–47.1
Tourist arrivals	'000	140	149	165	180	245
Total external debt	$m	547	618	663	775	–
Debt service ratio	%	14.8	11.6	7.3	6.2	–
Exchange rate	MR per $	13.8	15.4	13.5	12.9	13.4

COUNTRY PROFILE

Political system: The appointed governor-general exercises the executive power vested in the constitution by the British monarch. The legislature is the unicameral Legislative Assembly – 62 places directly elected, eight places for the most successful losers overall, the speaker and, if not already elected, the attorney general. The member most likely to command majority support is chosen by the governor-general as prime minister and, on his advice, the governor-general appoints up to 20 members to the cabinet. Mauritius gained independence within the Commonwealth in 1968. In September 1987 it was announced that it was to be a republic, remaining within the Commonwealth.

Political parties: The ruling coalition comprises Mouvement Socialiste Mauricien (MSM), Mauritius Labour Party (MLP) and Organisation du Peuple Rodriguais. Mouvement Militant Mauricien (MMM) and Parti Mauricien Social Démocrate (PMSD) form the opposition.

Population: The population of 1,031,913 (1989 estimate) is a heterogeneous mixture of Indian (52 per cent of the total), French, African, Chinese and British elements.

Main cities/towns: Port Louis (population 141,419), Curepipe, Vacoas.

Language: English and French are the official languages, and are widely understood in business circles; Creole and Hindi are the predominant languages in everyday life.

Media
Press: Several daily newspapers including *Le Mauricien* and *L'Express* and several weekly newspapers and periodicals.
Broadcasting: Mauritius Broadcasting Corporation operates one TV and two radio channels. Broadcasts in English, French, Creole, Hindi, Tamil, Telegu, Marathi, Urdu, Chinese, Gujrati and Bhojpuri.

External trade: Historically, there were current account deficits, but in the late 1980s this improved dramatically after successful implementation of World Bank and IMF structural adjustment programmes. Tourism is an important foreign exchange earner.
Exports: Sugar (34 per cent of total value), clothing (62%), tea (1%). Main destinations: UK (36% of total value), France (22%), USA (13%), German FR (9%).
Imports: Manufactured goods (35 per cent of total value), capital goods (30%), food (11%), chemicals (6%), petroleum products (5%). Main sources: France (12% of total value), USA (12%), South Africa (9%), Japan (7%), UK (7%), German FR (5%), China (4%).

Agriculture: Contributes 13 per cent to GDP and is dominated by sugar-cane cultivation. Approximately 90 per cent of all cultivated land is devoted to the crop, and the sugar industry as a whole employs about 15 per cent of the workforce and accounts for 12 per cent of GDP. Cultivation is undertaken on large plantations and by smallholders. The other export crop is tea, grown by tenant farmers. Around 75 per cent of tea production (green leaf) is controlled by the Tea Development Authority, whose commercial activities are now privately run. Tobacco is grown for the home market. Main food crops include potatoes and other vegetables, output of which is sufficient to meet domestic demand. Much of the island's meat and fish requirements are imported.

Manufacturing: Contributes 25 per cent to GDP and is primarily export-orientated. Activity is co-ordinated by the Export Processing Zone (EPZ). Production of textiles, and in particular knitwear, has become the most significant EPZ industry, and Mauritius is now the world's third largest exporter of pure new wool products. Other industries include the manufacture of watches and clocks, fishing 'flies', spectacle frames and leather goods.

Energy: Approximately 50 per cent of primary energy requirements are imported. Local electricity supply is largely thermal-generated, though rapid development of electricity generation from bagasse.

Banking: In 1989, tax and duty incentives were introduced for foreign banks licensed to engage in offshore banking, with the aim of establishing Mauritius as an offshore banking centre.

Membership of international organisations: African Development Bank, African States Associated with the EC (Lomé Convention), Arab Bank for Economic Development in Africa (recipient), Commonwealth, Economic Commission for Africa, FAO, GATT, G-77, IDA IFAD, IFC, ILO, IMF, Indian Ocean Commission, OAU, UN, WHO, WIPO, World Bank.

Time: GMT +4.

Climate: Maritime – tropical in summer (Nov to Apr), a sub-tropical for rest of year. High humidity especially in inland areas. Summer temperature average 25C to 30C with maximum 35C in Feb. Highest rainfall occurs in summer (and cyclones are likely). From May to Nov, drier and warm with temperatures 19C to 27C. A lowest rainfall from Sep to Nov. Inland areas generally five degrees C lower than coast with higher rainfall.

Entry requirements
Passport: Required by all.
Visa: Required by all except nationals of Commonwealth and EC countries, USA, Scandinavian countries, Japan and certain others. An onward ticket must also be held.
Currency: Import/export of local currency limited to 700 Rps. Any amount of foreign currency may be imported, subject to declaration, and any unused balance may be exported.

Health precautions
Mandatory: Yellow fever and cholera vaccination certificate may be required if arriving from infected areas.
Advisable: Typhoid and polio vaccinations. Water precautions should be taken.

Representation overseas: Brussels, Cairo, Canberra, Islamabad, London, New Delhi, Paris, Washington.

Air access
National airline: Air Mauritius.
Other airlines: Air France, Air India, Air Madagascar, Air Zimbabwe, British Airways, Cathay Pacific, Singapore, South African Airways, Zambia Airways.
International airport: Plaisance (Code: MRU) 3 km from Mahébourg, 45 km from Port Louis; buffet, currency exchange, post office, shops, car hire, banks, duty-free shop.

Surface access
Main ports: Port Louis.

Hotels: Wide choice available but relatively expensive. 10 per cent tax surcharge added to bills (10 per cent tip is usual except where service charge is included). Business visitors staying longer than a day or two generally prefer beach hotels.

Electricity supply: 220 V AC, 50 cycles.

Car hire: Widely available. International or foreign licence accepted; traffic drives on left.

City transport
Taxis: Operate in all towns, villages and resorts. Generally unmetered so it is advisable to agree fare before start of journey. Tipping is not usual.

National transport
Air: Air Mauritius operates inter-island service between Mauritius and Rodrigues and Réunion.
Road: An extensive network throughout the island with 90 per cent surfaced; occasionally congested. A dual highway links Port Louis and Curepipe. Good bus services cover main island.

Public holidays
Fixed dates: 1-2 Jan (New Year), 12 Mar (Independence), 1 May (Labour Day), 1 Nov (All Saints'), 25 Dec.
Variable dates: Maha Shivarathri, Cavadee, Chinese Spring Festival, Ganesh Chatturthi, Eid al-Fitr, Assumption Day, Divali.

Working hours
Business: (Mon-Fri) 0830 – 1630, (Sat) 0900 – 1200.
Government: (Mon-Fri) 0900 – 1600.
Banking: (Mon-Fri) 1000 – 1400. (Sat) 0930 – 1130.

Telecommunications
Telephone: Dialling code for Mauritius: IDD access code + 230 followed by subscriber's number (no area codes required). Automatic system in most areas.
Telex and telegram: Facilities available at Overseas Telecommunications Services Office in Cassis and Port Louis.

Banking
Central bank: Bank of Mauritius.
Head offices: Development Bank of Mauritius, Indian Ocean International Bank, Mauritius Commercial Bank, The State Commercial Bank, Mauritius Co-operative Central Bank.
Foreign branches: Bank of Baroda, Bank of Credit & Commerce International, Barclays Bank, Banque Nationale de Paris, Caisse Centrale de Coopération Économique, Habib Bank, Habib Bank Zurich, Hongkong and Shanghai Bank, Union International Bank.

Representation in capital: Australia, China PR, Egypt, France, India, Republic of Korea, Madagascar, Pakistan, UK, USA, USSR.

Hotels: Inland
Continental, 256 Royal Road, Curepipe (tel: 62036; tx: 4389).

Hotels: Near Port Louis
La Pirogue Wolmar, Flic En Flac (tel: 538441; tx: 4255; fax: 538449).

Le Pearle Beach, Flic en Flac (tel: 538428/9; tx: 4471).

Other useful addresses
Air Mauritius, Rogers House, 5 President John F Kennedy Street, PO Box 441, Port Louis (tel: 087700; tx: 4415).

Bank of Mauritius, PO Box 29, Sir William Newton Street, Port Louis (tel: 004164; tx: 4253).

Chamber of Agriculture, Plantation House, Duke of Edinburgh Avenue, Port Louis (tel: 20271, 20894; tx: 4214 augmaur).

Mauritius Broadcasting Corporation, Broadcasting House, Louis Pasteur Street, Forest Side (tel: 65001; tx: 4230).

Mauritius Chamber of Commerce & Industry, 3 Royal Street, Port Louis (tel: 083301; tx: 4277 chacom).

Mauritius Employers' Federation, Cerne House, Chaussée, Port Louis (tel: 21599, 24298).

Mauritius Export Development and Investment Authority, 2nd Floor, Jamalacs Building, Old Council Street, Port Louis (tel: 087750; tx: 4597 media).

Mauritius Export Processing Zones Association (MEPZA), Sir William Newton Street, Port Louis (tel: 21853; tx: 4612 mepza).

Mauritius Government Tourist Office, Emmanuel Anquetil Building, Port Louis (tel: 011703).

Mexico

George Philip

The first year of any new Mexican presidency is always a period of change. Presidents seek to break from the most unpopular policies of their predecessors and to create a new image of their own. Carlos Salinas has been no exception, although some of his political innovations have masked an underlying continuity in economic management.

This can be seen from Salinas' first cabinet. The economics ministries were all handed out to relatively young, respected but orthodox technocrats – Pedro Aspe, Jaime Serra and Ernesto Zedillo were given the planning, trade and finance ministries respectively. Aspe has a PhD in Economics from MIT, Serra has a PhD in Economics from Yale and Zedillo has both an MSc and PhD in Economics from Yale; when one adds in Salinas' own PhD and two Masters degrees from Harvard, the academic credentials of the economics team are impressive indeed. Less expected was the appointment to some key political positions of experienced operators with a reputation for toughness; Gutierrez Barrios at *Gobernacion*, Manuel Bartlett at Education and Arsenio Farrell at the Labour Ministry. Although Manuel Camacho Solis was made *Regente* of the federal district, Salinas put far fewer of his intimates (with their reputation for liberalism) in political positions than in economic ones. Those whom he did put in the cabinet tended to be in relatively minor positions.

However, observers who believed that this indicated that Salinas would simply pursue a hard-line at home have been proved wrong. What has indeed been surprising has been the moral toughness with which Salinas sought to restore some of the lost legitimacy of the Mexican system. Corrupt trade union leaders, drug mafiosi and crooked businessmen have all felt the sting of presidential power. Salinas himself undoubtedly believes strongly in the rule of law and in the importance of personal honesty. However, his moves against lawbreakers can also be seen as part of a strategy to rehabilitate the PRI in the eyes of the Mexican middle class and (no less important) the US Government. There has been a greater, though by no means total, willingness to accept electoral challenge and Mexican politics is certainly more open and more interesting than at any time since the 1940s. Salinas now faces the same dilemma as Gorbachev and other leaders who have tried to liberalise from above; how do you carry through reforms while remaining in control?

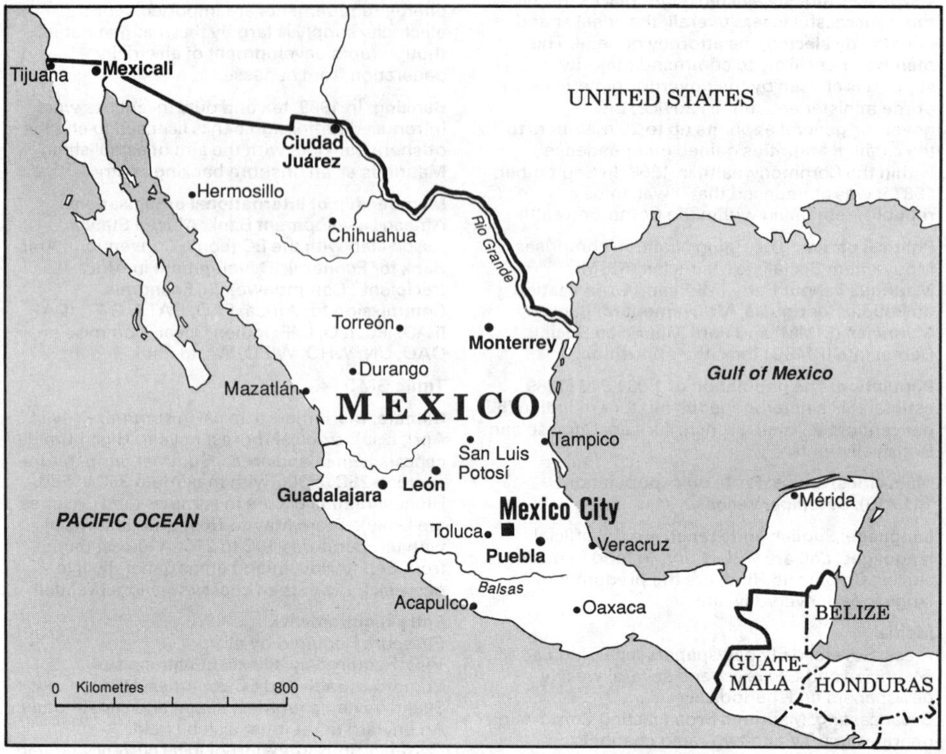

The economy

Mexico has also been seen as a test case of whether a heavily-indebted Latin American economy can recover economically without some kind of rupture with the international financial system. Mexico's external debt is crippling and its internal debt is scarcely less of a problem. Although the non-financial part of the Mexican budget is in surplus, there are real difficulties in raising the loans needed to service existing lending; real interest rates have not fallen below 30 per cent for several years. Obviously rates at this level are a major constraint on economic recovery.

Inflation has been reduced from its high point of around 130 per cent reached in late 1987; it fell in 1988 and a rate of around 20 per cent was expected in 1989. However, efforts to contain inflation by keeping the value of the peso relatively high have led to a slow deterioration in the current balance of payments; if this forces heavy devaluation on the authorities, then the battle against inflation will have been set back. In the longer term, the solution must be to reduce the size of the public sector; in the short-term, the answer is to win the confidence of investors and creditors so as to avoid too sharp an economic adjustment. It is just possible, though by no means certain, that the Mexican authorities will be able to achieve this.

One important factor here has been Salinas' ability to maintain a set of wage controls with the often reluctant consent of the Mexican trade unions. In return there has been some restraint in official prices and at least rhetorical support from business. Salinas is not especially popular with Mexico's labour leaders and has had to manoeuvre carefully in order to avoid antagonising them in any outright manner. However, agreement was reached in June to extend wage and price controls until the spring of 1990; the peso is to be devalued at the rate of one peso a day and administered prices will be kept in check.

The trade balance was somewhat strengthened in the early part of 1989 by an increase in oil prices over expected levels. There has been a subtle but undeniable shift in international expectations about the long-term price of oil. Many now expect oil prices to rise during the 1990s because the current low prices are once again encouraging increases in consumption. From January to May, revenue from oil exports was 40 per cent greater than budgeted. However, the effect of this unexpected gain has been offset in fiscal terms by unexpectedly high domestic interest rates; it is now expected that some 11 per cent of Mexican GDP will be made up by payment of interest on the internal debt.

An important agreement in principle was reached between Mexico and the banks on 24 July. While this still did not amount to a breakthrough, it followed a number of positive developments. Mexico returned to the international bond market in June, raising $100m.

KEY FACTS

Official title: United Mexican States

Head of state: President Carlos Salinas de Gortari

Ruling party: Partido Revolucionario Institucional (PRI)

Capital: Mexico City

Official Languages: Spanish

Currency: Peso (M$) = 100 centavos

Exchange rate: M$2,621 per $ (Nov 1989)

Area: 1,958,201 sq km

Population: 82.7m (1988)

GDP per capita: $2,588 (1988)

GDP real growth: 1.1%* (1988)

Inflation: 114% (1988)

Trade balance: $1.6bn (1988)

Foreign debt: $102bn (1988)

* estimated figure

This could be used to buy back significantly more than $100m of debt on the secondary market and so reduce the total interest cost of Mexico's debt service. More serious amounts of money have been lent to Mexico by the IDB, the World Bank and the IMF. Japan has also agreed to pay in advance for supplies of Mexican crude oil. Yet all of these developments, while welcome, were no substitute for the full-scale renegotiation with the banks along the lines of the Brady Plan.

Such a renegotiation looked imminent by the middle of the year. What was on offer was a general discount of around one-third of the value of total outstanding debt to the banks; this would reduce interest payments by some $3bn a year. In return the banks would receive better guarantees that the remaining two-thirds of the debt would be fully serviced. Moreover, it is likely that the Mexican government's current opposition to debt-equity swaps will prove to be a bargaining position that will eventually be relaxed. Certainly the progress of privatisation, which has been relatively slow so far this year, hinges to some extent on the relaxation of the prohibition of debt-equity swaps.

It was expected that the Mexican authorities would respond to a debt agreement by continuing with other economic reforms which have to some extent been awaiting progress on the main issue. An ambitious programme of trade liberalisation has already taken place under de la Madrid, and the Salinas Government has moved, albeit cautiously, to reduce restrictions on foreign investment. In May a new package of measures was announced. These provided that foreign investments of up to $100m would be approved automatically subject to six conditions being met. The investment would have to be financed externally, located outside the main urban areas of the country and should use adequate technology; it should generate permanent employment for Mexicans, should not lead to a net drain on foreign exchange and should not be larger than $100m. It remains to be seen whether the foreign investors will come.

KEY INDICATORS

	Unit	1984	1985	1986	1987	1988
Population	m	76.8	78.5	79.6	81.2	82.7
GDP (1985 prices)	pesos bn	46,144	47,392	45,655	46,340*	46,850*
GDP per capita (1985 prices)	pesos '000	600	604	574	571*	566*
Inflation (average rate)	%	65.5	57.8	86.2	132.0	114
GDP real growth	%	3.7	2.7	–3.7	1.5	1.1
Agricultural sector	gwth rate %	2.7	3.8	–2.7	1.8	–1.6
Manufacturing sector	gwth rate %	5.0	6.0	–5.7	2.6	2.2
Exports	$bn	24.2	21.7	16.0	20.7	20.7
Petroleum	%	67.5	66.6	38.6	40.7	31.9
Coffee	%	1.9	2.4	5.1	2.4	2.3
Shrimps	%	1.7	1.6	2.3	2.3	1.8
Imports	$bn	11.3	13.2	11.4	12.2	18.9
Balance of trade	$bn	12.9	8.5	4.6	8.4	1.6
Current account	$bn	4.2	1.1	–1.7	4.0	–2.9
Exchange rate (average)	pesos per $	167.8	256.9	611.8	1,378.2	2,273.1

* estimated figure

Political developments

The results of the 1988 presidential elections came as a real shock to the PRI. They had officially forecast that they would get 20 million votes; in the end they managed 9 million. In the federal district, where the electorate is articulate and elections difficult to rig, Carlos Salinas was badly beaten by Cuautehmoc Cardenas. In the second city, Guadalajara, seven of the eight congressional seats were won by the right-wing opposition party, the PAN. What is obvious from detailed analysis of the results is that the regime lost the support of the urban middle class. Many public employees voted for the system – which after all provided them with their jobs. However, the 'educational' middle class (students, teachers, academics) voted overwhelmingly for Cuautehmoc; much of the private sector middle class voted for the PAN. The PRI did retain some support among the urban poor (where it had a plurality but not a majority), and retained its majority support among the peasants. What this means is that, although the PRI just managed to cling onto power, the most dynamic and articulate members of Mexican society lined up alongside one of the opposition parties.

The task of winning back middle class support for the system is a particularly tricky one. What has alienated the middle class more than anything else (apart from economic failure) has been the continuing protectiveness shown by the system to some of its sacred cows. At least in this respect things have changed under Salinas, who has embarked on a major slaughter of these creatures. What remains to be seen is how the vacuum will be filled.

The first dramatic action undertaken by the regime was the arrest of Joaquin Hernandez Galicia ('La Quina') in January 1989. The oilworkers' union leader was plainly an over-mighty baron with whom the incoming regime somehow had to deal. However, the dramatic circumstances of La Quina's fall – the use of troops to arrest him, the charges of murder preferred against him – dominated the headlines in Mexico for many days. The left-wing opposition was put in the difficult position of appearing to support La Quina. Yet in the many debates and discussions about the matter, one point of view commanded conspicuously little support – virtually nobody seriously doubted that La Quina was guilty of serious crimes.

A second major trade union leader fell a few months later, in circumstances which were less dramatic but perhaps more significant. This was Carlos Jongitud who ceased to be 'moral leader' of the teachers' union in April. Mexican trade unions have a great deal of power over their individual members, due to prevailing closed shop arrangements. The main Mexican teachers' union was in the hands of a corrupt and unpleasant bureaucracy which could nevertheless be relied on to ensure that teachers' pay remained very low. The argument was surely put to Salinas that, at a time of economic austerity, it was better to tolerate an unsatisfactory union structure than to risk the budgetary programme. However, Salinas made a very different decision.

In February Salinas had described one of the key objectives of his administration as being the improvement of educational standards in Mexico. In Mexico, after all, the standards provided in state education are poor or even abysmal. Despite the fact that educational provision was one of the key promises of the revolution, most Mexicans are barely literate, and the gap between Mexico and the successful NIC-type developing countries has been growing. The problems have been poor teacher training and worse pay; a teacher in the provinces cannot easily afford to buy a newspaper, let alone a book. Bureaucratic and corrupt trade union organisation did not help; the union 'sold' places (and therefore teaching positions) to virtually anybody with the money and the desire to teach. On the other hand, the political resources available to such a union bureaucracy were obviously considerable. To mention only one of these, the union nominated 14 congressmen who stood for the PRI and were elected in 1988; 14 may not sound a large number out of a total of 500 congressmen, but the government has a majority over the opposition of just 20.

In February, the teachers' union congress confirmed Jongitud's position as 'moral leader' (i.e. Godfather). However, in April dissident leaders called a strike against a low pay offer and also demanded the democratisation of the union; after a strike lasting for several weeks, the government effectively conceded. The pay offer was increased to 25 per cent (which is above the expected level of inflation) and major reforms were made to the trade union.

In both of these cases (as also in the case of the musicians' leader, who was also forced to resign) the government took care to prepare an alternative leadership before removing the existing leaders. The new teachers' union leader, Elba Esther Gordillo, is a formidable political figure in her own right whom Jongitud had kept out because he feared her potential independence. Following the arrest of La Quina, moreover, the Mexican authorities recovered several hundred million dollars held, apparently illegally, in foreign bank accounts.

Despite these evident victories, Salinas' relations with organised labour continue to be difficult. Fidel Velazquez, now aged 89, continues to be the most important union leader in Mexico. Fidel does not especially like Salinas (who was by no means his first choice as president) and has made his general unhappiness clear; however he does not really have anywhere else to go and has been reduced to sulking in his tent. It will meanwhile be hard for other union leaders to emerge until the question of the succession to Fidel is decided; Fidel's own preference as heir apparent is Arturo Romo.

The government's insistence on higher ethical standards was not restricted to union leaders. In February some 150 people were arrested following investigations into stock market fraud; this alleged fraud is said to have lain behind the crash of 1987. Among those arrested was Mr Legorreta, a leading member of one of the most important financial families in Mexico. Further dramatic arrests occurred in May when a leading drug baron, Felix Gallardo, was arrested after spending years successfully on the run. Following subsequent investigations, fully one-third of the Sinaloa police force was relieved of its duties for complicity in the drug trade. More remarkable still was the arrest of the former head of the Mexican security police for the murder of the investigative journalist, Manuel Buendia, in 1984. The allegation was that Zorrilla Perez, the arrested man, arranged for the murder of Buendia to prevent the latter publishing revelations about the links between himself and Felix Gallardo, whom he had been protecting.

Salinas was, however, embarrassed when his choice of police chief for the federal district, Garcia Paniagua, appointed as his deputy Nazar Haro who – as it turned out – was wanted in the United States on a charge of car theft. Nazar Haro was dismissed and Garcia Paniagua sent on a period of sick leave which has proved very extended indeed.

It is likely that these efforts at moral persuasion will indeed convince some of the middle class that Salinas, if not necessarily the system as a whole, is worth supporting. Political reform, however, is probably even more crucial and certainly more difficult. By claiming little more than 50 per cent of the vote last July (a claim which many critics of the regime found exaggerated), the system implicitly admitted that it had been defeated in a number of key states and that it represented only a minority (perhaps a plurality) of several sectors of Mexican society. The opposition could no longer be written off as trivial and unimportant; ways of dealing with it had to be found.

COUNTRY PROFILE

Political system: Federal republic of 31 states and a federal district. Executive power rests with the president (elected by universal adult suffrage for a six-year term, whose mandate is not renewable at any time). The cabinet (normally 19 members) is appointed by the president. Legislative power is vested in bicameral National Congress, consisting of a Senate (two members for each of 31 states and two members for the federal district, serving a two-year term) and a Chamber of Deputies (500 members, of whom 300 are directly elected and 200 distributed on a proportional basis among parties winning fewer than 60 seats directly; three-year terms). Members of neither chamber are eligible for immediate re-election. Each state has its own constitution and is administered by a governor (elected for six years) and an elected Chamber of Deputies.

Main cities/towns: Mexico City (population about 19m, 2,240 metres above sea-level), Guadalajara, Monterrey, Ciudad Juárez, Toluca, Puebla, León, Tijuana, Mexicali, San Luis Potosí, Chihuahua, Torreón.

Media

Press: Around 18 national dailies, of which most influential are: *Excélsior, El Día, La Jornada, Uno Más Uno* (circulation 75,500), *El Universal* (181,375) and *The News* (in English).

Radio: Over 800 commercial and 45 cultural radio stations. Radio VIP in Mexico City broadcasts in English. *Television:* Extensive TV network with 118 commercial stations and eight cultural stations, largest of which are Televisa SA and Instituto Mexicano de Televisión (IMEVISION).

External trade: Balance of payments has swung from deficit to surplus owing to reduction in imports and expansion of non-oil exports. However, government's ability to meet foreign debt commitments depends on maintenance of current account surplus. Recently the trade surplus has suffered from falling oil export revenues following the sharp drop in oil prices in 1986.

Exports: 1985: Crude oil and products (60% of total value), machinery and transport equipment (7%), chemicals (3%), food etc, coffee (2%), minerals, iron and steel, textiles. Main destinations (1985): USA (62% of total value), EC excluding Spain (16%), Japan (7%), Spain (6%), Brazil (1%).

Imports: Machinery and transport equipment (typically 40-45% of total value), agricultural products (13%, of which grains and oil seeds 8-9%), chemicals (10-12%), iron and steel (6-7%), petrochemicals (4-6%), petroleum products (3-5%), processed food, drink, tobacco (3-5%). Main sources (1985): USA (68% of total value), EC (11%), Japan (5%), FRG (4%).

Agriculture: Although employing around 30 per cent of the labourforce, agriculture and livestock contribute only nine per cent to GDP. Farming is small-scale and frequently inefficient. About 50 per cent of total cultivatable land (estimated 19m hectares) held by *ejidos*, rural communities that farm small individual/collective lots. Few large commercial farms except in export-orientated vegetable-producing regions of the north-west. Principal food crops are maize (over 50 per cent of harvested area and 60 per cent of total grain production), sorghum, wheat, rice, barley, potatoes, soybeans and dry beans. Production of basic foodstuffs severely affected by drought, insufficient irrigation (less than 30 per cent of cultivated land is irrigated) and under-developed marketing and infrastructural back-up. Supplies of food grain in particular have not kept up with demands imposed by rapid population growth. Imports of basic foodstuffs amounted to 7.5m tonnes in 1984. Principal export crops are coffee, cotton, fresh fruit, sugar, tobacco and tomatoes. Exports of cattle and frozen fish also important. The value of agricultural exports has declined, though coffee and cotton exports have improved in recent years. Fish production and exports (shrimps, tuna, mackerel, bass, perch, bonito, shark, oysters) have also risen. Mexico possesses vast resources of soft and hard woods, but deforestation is becoming a problem in some areas.

Industry: Sector is well developed, with manufacturing and construction contributing 29 per cent to GDP. Changes introduced since 1985 have liberalised Mexico's trade by more than 80 per cent of total tariffs previously applied. Though the public sector is important, out of 1,155 state-owned companies in 1973, the government has sold 756 as part of its privatisation policy. Most public sector firms are in 'strategic' industries such as base metal manufacture, construction materials, paper and products, chemicals and petro-chemicals, textiles and apparel, food processing, beverages, motor vehicles, electro-domestic appliances, machinery and equipment. The *maquiladoras* (in-bond manufacturers) make, assemble or process components and raw materials brought in 'in-bond' from the USA, and then re-export them duty-free. Apart from tourism, which expanded by 35 per cent from 1984-88, this is the only sector of the economy that expanded significantly since 1982 when the recession began in Mexico. In 1988 there were over 1,000 plants, which together were the country's second largest foreign exchange earner, after oil. Most maquiladoras are owned by US companies. Foreign investment was boosted by the introduction of debt equity swaps in 1986.

Tourism: Tourism is being developed as a major foreign exchange earner. The government stepped up its promotional budgets after the 1985 earthquake, which seriously damaged the tourist resort area of Ixtapa where one hotel had to be closed for three months, and reduced revenues from tourism for two years.

Mining: Leading producer of many minerals including silver, celestine and sodium sulphate, mercury, antimony, cadmium, zinc, lead, copper, manganese, graphite, feldspar and barite. Sector contributes four per cent to GDP, and precious metals account for around 45 per cent of output value. State-owned mines account for 42 per cent of total production. Large new deposits of uranium, gold and silver discovered in Sonora and Ameca respectively. Estimated that only 15 per cent of potential mining areas have been explored to date. Reserves of hard coal estimated at 5.5bn tonnes. Slight recovery in mining activity since 1983 despite depressed world markets.

Hydrocarbons: Fourth largest oil producer in world with 70.9bn barrels of proven petroleum and gas reserves. Potential reserves, perhaps optimistically, estimated at 250bn barrels. (Figures for proven reserves were revised downward in both 1985 and 1986). Exports of petroleum and products, mainly to US, account for 60-70 per cent of total exports and remain a major determinant of overall economic performance. Most oilfields lie in or near Gulf of Mexico. The largest producing field is in the Gulf of Campeche, which produces, on average, 65 per cent of total oil output. PEMEX, the state oil agency, is country's single largest employer and largest single industrial investor. Oil production and exports have been sluggish in recent years (output 2.8m b/d in 1986), owing to falling world demand, insufficient port capacity and heavy debt burden suffered by PEMEX. However, the drop in crude oil revenues partly offset by increased exports of refined products and petro-chemicals and through a marked reduction in domestic consumption of oil products. Expansion and upgrading of refining industry (Mexico has eleventh largest refining capacity in the world) has also improved country's export potential. Crude oil processing from PEMEX's seven large and two small refineries currently 1.4m b/d. Mexico is not a member of OPEC, though in recent years has broadly followed the cartel's export and pricing policies. Despite discovery of new gas reserves and utilisation of previously flared gas from offshore wells, production and exports of gas have fallen due to cuts in US demand.

Energy: Oil and gas account for some 90 per cent of primary energy requirements, remainder made up by hydroelectricity and geothermal power. Plans to construct three nuclear plants subjected to delay, although the Jagrena Verde plant has already been charged with radioactive materials.

Banking: All banks were nationalised in 1982. Banco de México, the central bank, is the sole bank of issue. It acts as a reserve bank and a clearing house for the other banks. Nacional Financiera is the state investment bank and is the principal channel for foreign credits. It will guarantee credits to the government and industrial projects of national significance. As well as promoting economic development, it is involved in the development of the stock market. Banco Nacional de Comercio Exterior also finances large-scale import and export business. The largest of some 20 commercial banks are Banco de Comercio (Bancomer), Banco Nacional de México (Banamex), Banco Serfín, Multibanco Comermex, Banco Internacional, Banco Mexicano Somex, all of which have numerous branches in Mexico.

Membership of international organisations: FAO, G-77, GATT, Group of Six, IADB, IAEA, ICAC, ICAO, ICO, IDA, IFAD, IFC, ILO, IMF, IMO, INTELSAT, Inter-American Development Bank, International Lead and Zinc Study Group, Interpol, IRC, ISO, ITU, IWC, Latin American Integration Association, Organisation of American States, PAHO, SELA, UN, UNESCO, UPU, WHO, WIPO, WMO, World Bank, WSG, WTO.

Salinas began by announcing that there would be a period of debate about political reform. Since then there have been a number of discussions about possible amendments to the Mexican constitution and political system. Two changes are in fact likely to occur. The first is that the few remaining restrictions on the activities of the Mexican Church are likely to be lifted. In the past, Mexican Church and state were bitter enemies; despite (or perhaps because of) this, however, the Church has remained extremely strong in broad social terms, though less so in narrowly political ones. The vast majority of Mexicans still define themselves as orthodox Catholics; few are very much influenced by the Church hierarchy, at least in terms of their political behaviour. However, the government still sees the Church as a potential political resource; taking a conciliatory attitude towards the Church is also one of the few ways in which the system can somewhat increase its popularity without spending money. There is the additional satisfaction that some of the left-wing opposition is broadly pro-Church while some remain hostile; the PRI has rarely given up an opportunity to split the opposition.

The second likely change is that some form of proportional representation is likely to be introduced into the Senate. The last political reform, offered in 1986, gave the opposition a guarantee of 200 seats in the 500-seat Chamber of Deputies. In the event, the opposition won a total of 240; the government's 20-seat majority is very narrow and allows interest groups, which informally nominate congressmen, considerable influence over the government. It also means that the government no longer has the votes to amend the Mexican Constitution without help from at least one opposition party; a two-thirds majority is required for this to happen. It is likely that the government will want a more 'majoritarian' system in the Chamber; in return it is likely to offer guaranteed opposition representation in the Senate. Senators have a six-year term of office while Deputies have only a three-year term; it follows that Senators have slightly more independence from the government. This greater independence, and longer term of office, could provide the basis for the kind of Congressional investigative committee work which Mexico presently does not have. This might be expected to put a slight restriction on the constitutional power of the Mexican president; this may seem little enough, but there is at present virtually no constitutional restraint on presidential power in Mexico.

Formal change, however, is merely one thing. Far more important is the willingness of the government to respect adverse election results. A clutch of municipal election results in December 1988 provided a mixed picture; the opposition won the important municipality of San Luis Potosi (former stronghold of Carlos Jongitud), but government candidates won controversial victories in Guadelajara. The picture remained somewhat mixed following the more important elections of July 1989.

The most important elections occurred in two states. In Baja California expectations were heightened by the fact that the opposition won the state in July 1988. Salinas responded by ordering the replacement of the existing governor and requiring the government machine to pull out all of the stops to win the governorship elections. A huge spending plan was announced for the state, despite the general fiscal restrictions in force, and ministers regularly visited the state and expressed their desire to help. The conservative opposition party, PAN, itself responded by selecting a popular local mayor to run for office. After the elections were held, the PRI admitted, for the first time in its history, that the PAN had won the majority of votes in the state. The state governorship and state legislature are now controlled by the PAN. It will be interesting to see how the situation develops, but most immediate assessments were that an opposition victory in just one of the 32 states of Mexico presented the system with more of an opportunity than a threat.

Elections in Michoacan provided a far more difficult problem. In July 1988 Cuautehmoc Cardenas triumphed overwhelmingly in Michoacan (which is his home state), despite a rough campaign which featured a few assassinations of left-wing activists. It was obvious that the PRI would face a serious challenge in future state elections; as in Baja California, the state governor was sacked and a new man installed. In the first few months of 1989, however, it began to look as though the left-wing opposition coalition might be weakening as splits – assiduously fostered by the system – took their toll. The left opposition was always a somewhat uneasy coalition between people who belonged on the left of the PRI (perhaps associated with the PPS or the PARM), and who had become frustrated by the abandonment of the system of any pretence of revolutionary nationalism, and those who were genuine marxists and therefore natural opponents of the system. Some of the less committed Cardenistas, perhaps wooed by the offer of patronage, began returning to the system. However, Cardenas himself remained steadfastly intransigent, with the support of the Mexican Socialist Party and the 'outside left'. Elections for the local assembly in Michoacan in July gave him the chance to put the pressure back on the system. Cardenas remained hugely popular in his own state, but it was not at all clear that the government would allow him to win. After the votes were counted, the PRI claimed to have won 12 seats against six won by the Cardenistas; after proportional representation was taken into account, this gave the PRI the minimum support necessary to command a majority. The Cardenistas believe, however, that sharp practices (such as invalidating opposition votes on technicalities) prevented them from winning by a similar margin to that claimed by the PRI. Independent observers tend to agree with the Cardenistas.

What these results seem to show is that the system is now willing to accept defeat at the hands of electoral opponents considered broadly compatible; they are not ready to lose to a mobilising left-wing opposition which could not be relied upon to keep its challenge within what the government would regard as acceptable limits. What is not clear is whether this is a balance that can be maintained indefinitely. The Mexican system guarantees that important local and state elections are held at regular intervals; either the PRI must accept an occasional loss, or the bandwagon of protest against electoral fraud will pick up steam once more.

Mexican society

Social change is inevitably far slower than political change, but there are signs of changes taking place here as well. President Salinas has, for one thing, appointed two women as members of his cabinet; they are not in especially important ministries, but at least they (unlike a former minister of tourism) were promoted on merit. It may or may not be a mark of distinction that the first PRI candidate in Mexican history to lose an election for governor was also a woman. Mexico also has an environment minister, Patricio Chirinos, who is close to the president. While the Mexican government did decide to go ahead with the controversial nuclear power development at Laguna Grande, there are signs that it is giving a higher priority to environmental considerations. Given the magnitude of the problems faced by Mexico City (bad air, lead poisoning, inadequate sewage disposal, an impending shortage of water, etc), the authorities may not in fact have a great deal of choice in this respect at least.

The government changed the school year for children in Mexico City; they were given an extra month's holiday in January and made to study in July. January is commonly believed to be the worst month for pollution in the federal capital, and many mothers took advantage of the situation to take their children out of the city during that month. There are also some signs that the government's commitment to decentralisation, rhetoric more than reality for the past decade or more, is being taken more seriously. Some government departments are being moved to provincial capitals, although there is a great deal more that needs to be done.

It is also the case that the momentum of economic growth is gradually beginning to move away from Mexico City. Mexico is continuing to urbanise, but the most rapid areas of growth are to be found in the far north of the country and also, surprisingly, in the south east. In 1940 there were five cities in Mexico with a population of more than 100,000; in 1960 there were 31, and in 1985 there were 64. Provincial politics and society is no longer rural or backwoods. One reason why the nature of provincial politics is increasingly important is that the federal district's share of GDP, which peaked around 1970, is now tending slowly to fall. As the balance of the economy moves further away from the oil revenue/public spending complex towards tourism, manufacturing and certain kinds of agriculture, one would expect this trend to continue. To give only one example, the *maquiladora* (in bond manufacturers) exporters have developed very fast over recent years with the workforce tripling during the 1980s.

The birth rate is also falling very slowly, but it did not begin its decline until the second half of the 1970s. What this means is that the number of Mexicans entering the labour force has presently reached a peak, but will decline only slowly during the coming years. The rate of population growth will still be high for some time to come but, in another 20 years, the worst of Mexico's demographic explosion will be over. If the authorities can manage to keep Mexican society stable for another generation, the country's social problems should then be somewhat easier to resolve.

There are also signs that Mexican politics is becoming less ideological, in the old left-right sense. Mexico's non-aligned foreign policy is less controversial than in the past. Mexico is no longer really supporting the Sandinistas in Nicaragua and the United States is no longer really trying to overthrow them. The Mexican left, like that of many other countries, now professes a healthy scepticism about any form of state communism. It is noteworthy that Fidel Castro attended the inauguration of Carlos Salinas, which the Cardenistas boycotted on the grounds of alleged electoral fraud, while Cardenas himself made his first major foreign trip after the elections not to Cuba or any communist country but rather to the USA. Meanwhile the Mexican government's new-found effectiveness on the drugs issue has helped put additional pressure on Cuba and Panama and won support in Washington.

Similarly the Church/anti-clerical division is now much less important; the Church no longer defines itself as being actively anti-communist and many priests supported the presidential candidacy of Cuautehmoc Cardenas. Twenty years ago the main division within Mexican political society was between Catholicism and business interests on one side, and the marxist left on the other. This is no longer true. The key division now is much more between those who wish to retain the services of a powerful, centralised state in order to push for social reform and those who see the over-mighty state as part of the problem and wish to bring it within the bounds of legality and fiscal rectitude. It is one of the ironies of the situation that much (though not all) of the opposition prefers the first of these two options and most (though again not all) of the government prefers the second.

BUSINESS GUIDE

Time: Generally GMT -6 (Baja California Sur, Nayarit, Sinaloa, Sonora: GMT -7; Baja California Norte: GMT -8). (Clocks put forward one hour in summer).

Climate: Varies with altitude. Tropical southern region and coastlands are hot and wet, while the highlands of the central plateau are temperate. Temperature in Mexico City ranges from 5°C to 25°C with occasional sharp frosts in winter (Dec-Feb).

Entry requirements
Passport: Not required by citizens of Canada and USA with acceptable proof of identity – tourist cards essential. British citizens need only their passports (tourist cards are handed out on the aeroplane before landing). N.B. Travellers cannot leave country without tourist card.
Visa: All travellers not carrying tourist cards need a visa. Business travellers must also possess business cards/permits which state nature of business and firms/agents to be visited. Renewal of visa can take up to two weeks.
Currency: No restrictions on import/export of foreign or Mexican currency.

Health precautions
Mandatory: Yellow fever vaccination certificate required if arriving from an infected area.
Advisable: Typhoid, paratyphoid and poliomyletis vaccinations. Malaria risk exists in some rural areas – prophylaxis recommended. Water precautions should be taken; bottled water is generally available. Avoid raw salads and vegetables.

Representation overseas: Addis Ababa, Algiers, Ankara, Asunción, Athens, Bangkok, Beijing, Belgrade, Belize City, Berlin, Berne, Bogotá, Bonn, Brasília, Brussels, Bucharest, Budapest, Buenos Aires, Cairo, Canberra, Caracas, Copenhagen, Dakar, Guatemala City, The Hague, Havana, Helsinki, Hong Kong, Jakarta, Jeddah, Kingston, Lagos, La Paz, Lima, Lisbon, London, Madrid, Managua, Manila, Montevideo, Moscow, Nairobi, New Delhi, Oslo, Ottawa, Panama City, Paris, Port-au-Prince, Port of Spain, Prague, Quito, Rome, San José, San Salvador, Santo Domingo, Seoul, Sofia, Stockholm, Tegucigalpa, Tel Aviv, Tokyo, Vienna, Warsaw, Washington.

Air access
National airlines: Aeromexico, Mexicana.
Other airlines: Aeroflot, Aerolíneas Argentinas, Aeromexico, Aeronica, Aeroperú, Air France, Air Panamá, Alaska, American, Avensa, Avianca, Aviateca, Continental, Cubana, Delta, Ecuatoriana, Iberia, JAL, KLM, Lacsa, Laredo Air, Lufthansa, Northwest, Pan Am, Taca International, TAN-Transportes Aéreos Nacionales, Tropic Air, United, Varig.
International airports: Acapulco-Gen Juan N Alvarez (ACA), 26 km from city, restaurant, car hire, bank, post office; Cancún (CUN), 5 km from town; Cozumel (CZM), 20 km from town; Guadalajara-Miguel Hidalgo (GDL), 20 km from city, bar restaurant, buffet, bank, post office, shops (restricted hours in some instances); La Paz-Gen Manuel Márquez de León (LAP), 9 km from city, car hire; Mazatlán-Gen Rafael Buelna (MZT), 3 km from city, car hire; Mérida International (MID), 8 km from city, restaurant; Mexico City-Benito Juárez (MEX), 13 km from city, duty-free shop, bar, restaurant, buffet, bank, post office, shops, hotel reservations, car hire; Monterrey-Gen Mariano Escobedo (MTY), 24 km from city, restaurant; Puerto Vallarta-Presidente Gustavo Díaz Ordaz (PVR), 7 km from city, duty-free shop, bar, restaurant, post office, shops, car hire.

Surface access
Main ports: (Gulf of Mexico coast) Coatzacoalcos, Tampico, Veracruz; (Pacific coast) Acapulco, Ensenada, Mazatlán, Manzanillo, Salina Cruz.
Overland access: Possible through numerous points on border with US and, particularly, via the Pan-American Highway, from Guatemala. Various bus and train routes can often be chosen to suit individual needs.

Hotels: Six classified types of hotels, with maximum rates set by government. Advisable to book in advance. 15 per cent value-added tax and service are added to bills. 10 per cent tip usual.

Car hire: Widely available, with or without driver, but often expensive. Foreign licence acceptable.

City transport
Taxis: Usually fitted with meters, but often not used. Agree fare in advance. No tip necessary. Fixed route taxis identified by lime-green colour, rank taxis by coral and those with no fixed route by yellow. Around 17 fixed routes. Number of fingers held out of taxi window indicates number of seats left. Special tourist taxis, *Turismo*, have English-speaking drivers.
Metro: Mexico City Metro, with seven lines, is excellent, but often crowded. Opens 0500 (Mon-Sat), 0700 (Sun). Maps not displayed at all stations, but can be obtained at Insurgentes station on Pink Line.

National transport
Air: Frequent flights to main centres operated by Aeromexico, Mexicana and a number of new domestic airlines.
Road: Total network over 200,000 km, of which nearly one-third surfaced. Note: detailed documentary advice readily available on driving hazards (e.g. animals at night) and procedures (e.g. on narrow bridges) etc.
Buses: Three kinds, first and second class and local. Advisable to book seats in advance in Baja California. City bus service is cheap, but generally crowded. Buses with odd numbers run north-south, even east-west. Peribus service circulates Mexico City.
Rail: Special first class, regular first class and second class service. Routes include Guadalajara-Mexico City; Monterrey-Mexico City; Mexico City-Veracruz-Tapachula; Ciudad Juárez-Chihuahua; services slower than bus. Sleeper services between Mexico City-Guadalajara, Monterrey, Veracruz, Ciudad Juárez, Chihuahua, Mérida. Advisable to book well in advance.

Public holidays: 1 Jan, 5 Feb (Constitution Day), 21 Mar (Birthday of Juárez), Maundy Thursday, Good Friday, Easter Monday, 1 May, 5 May (Battle of Puebla), 16 Sep (Independence), 12 Oct (National Day), 1 Nov (All Saints'), 2 Nov (All Souls'), 20 Nov (Anniversary of the Revolution), 12 Dec (Our Lady of Guadalupe), 25 Dec.

Working hours
Government and business: (Mon-Fri) 0930 – 1330, 1430 – 1900 (considerable regional variation).
Banking: (Mon-Fri) 0900 – 1330; certain banking services available in large cities: 1600 – 1800.
Shops: (Mon, Tue, Thu, Fri) 1000 – 1900; (Wed, Sat) 1100 – 2000.

Telecommunications
Telephone: Automatic telephone system in Mexico City. Heavy tax on all foreign long distance calls. Dialling code for Mexico: IDD access code + 52 followed by area code (Acapulco 748, Chihuahua 14, Mazatlán 678, Mexico City 5, Monterrey 83) followed by subscriber's number. For dialling from Mexico, IDD access code: 00.
Telex: Public telex and telegram facilities available in Mexico City, Acapulco, Guadalajara, Mérida, Monterrey, Tampico, Veracruz, Chihuahua and Puebla. Country telex code: ME.

Banking
Central bank: Banco de México.
Main centre: Mexico City.
Commercial banks (head offices): Arrendadora e Inversionista Latina, Banamex, Banca Confia, Banca Cremi, Banca Promex, Banca Serfín, Banco Aboumrad, Banco BCH, Banco Capitalizador de Ahorros, Banco Capitalizador de Puebla, Banco Continental, Banco del Atlántico, Banco de Oriente, Banco de Pequeño Comercio del Distrito Federal, Banco Internacional, Bancomer, Banco Mercantil del Norte, Banco Mexicano Somex, Banco Nacional de México, Banco Nacional de Obras y Servicios Publicos, Banco Popular de Edificación y Ahorros, Bancreser, Credito Mexicano, Financiera Industrial Mexicana, Financiera Industrial Mexicana, Financiera Multibanco Comermex, Multibanco Comermex, Multibanco Mercantil de Mexico.

*Development banks (head offices):*Banco Nacional de Credito Rural, Banco Nacional de Comercio Exterior, Nacional Financiera.
Foreign branches: American Express International Banking Corporation, Banca Catalana, Banca Commerciale Italiana, Banco de Bilbao, Banque Française de Commerce Extérieur, Banque Nationale de Paris, Banque Paribas, Citibank, Grindlays Bank, Maryland National Bank.

Trade fairs: Agromexico, Feb-Mar, Mexico City.

Electricity supply: 110-117 V AC, 60 cycles. Two-pin flat plugs (as in USA) are used.

Representation in capital: Albania, Algeria, Argentina, Australia, Austria, Belgium, Bolivia, Brasil, Bulgaria, Canada, China, Colombia, Costa Rica, Cuba, Cyprus, Czechoslovakia, Denmark, Dominican Republic, Ecuador, Egypt, El Salvador, Ethiopia, Finland, France, German DR, German FR, Greece, Guatemala, Haiti, Honduras, Hungary, India, Indonesia, Iran, Iraq, Israel, Italy, Ivory Coast, Jamaica, Japan, Rep of Korea, Lebanon, Netherlands, New Zealand, Nicaragua, Norway, Pakistan, Panama, Paraquay, Peru, Philippines, Poland, Portugal, Romania, Spain, Suriname, Sweden, Switzerland, Thailand, Turkey, USSR, UK, USA, Uruguay, Venezuela, Vietnam, Yugoslavia.

Hotels: Guadalajara
Aranzazú, Avenida Revolución 110 (tel: 133232; tx: 681796; fax: 145045).

Camino Real, Avenida Vallarta 5005 (tel: 478000; tx: 681839; fax: 476781).

Castel Plaza del Sol, P.O. Box: 31-444 (tel: 210515; tx: 681603).

El Tapatio, Carretera Chapala, P.O. Box: 2953 (tel: 356050; tx: 681619).

Fénix, Avenida Corona 160, P.O. Box: 1-1151 (tel: 145714; tx: 682885).

Holiday Inn, Avenida López Mateos Sur 2500, P.O. Box: 31214 (tel: 315566; tx: 684004).

Hyatt Regency, Avenida Lopez Mateos Sur y Moctezuma (tel: 227778; tx: 684072; fax: 229877).

Laffayette, Avenida de la Paz 2055 (tel: 301112/520252; tx: 684060; fax: 416263).

Posada Guadalajara, Avenida Presidente, López Mateos 1280 (tel: 212022; tx: 681626).

de Mendoza, Venustiano Carranza 16 (tel: 134646; tx: 682683).

Hotels: Mexico City
Ambassador, Calle Humboldt 38 (tel: 5180110; tx: 1776276).

Aristos Mexico City, Paseo de la Reforma 276 (tel: 2110112; tx: 1771067).

Bamer, Avenida Juárez 42 (tel: 5219060; tx: 1773863).

Camino Real, Mariano Escobedo 700 (tel: 2032121; tx: 1773001; fax: 2506897).

Casablanca, La Fragua 7 (tel: 5663211).

De Cortes, Avenida Hidalgo 85 (tel: 5850322, 5182181; tx: 1772770).

BUSINESS DIRECTORY

El Presidente Chapultepec, Campos Elyseos 218 (tel: 2507700; tx: 1776392; fax: 2509130).

Galería Plaza, Hamburgo 195 (tel: 2110014; tx: 1771808; fax: 5289583).

Geneve Quality Inn, Londres 130 (tel: 2110071).

Holiday Inn Airport, Blvd Puerto Aéreo 502 (tel: 7624088; tx: 1762260).

Holiday Inn Crowne Plaza, 80 Paseo de la Reforma, Columbus Circle (tel: 5667777; tx: 1772029).

Krystal, Avenida Liverpool 155 (tel: 5333500, 2110092; tx: 1761053).

Metropol, Luis Moya 39 (tel: 5108660; tx: 1773655).

Nikko Mexico, 204 Col. Polanco (tel: 2034020; tx: 1763523; fax: 2555586).

Sheraton María Isabel Towers, Paseo de la Reforma 325 (tel: 2073933; tx: 1773936; fax: 5288501).

Hotels: Monterrey
Ambassador, Hidalgo & Emilio Carranza, P.O. Box: 1733 (tel: 409390; tx: 885913; fax: 451984).

Ancira Inter-Continental, P.O. Box: 697 (tel: 432060/9; tx: 382872; fax: 445226).

Holiday Inn, Avenida Universidad 101 (tel: 522400; tx: 382049).

Holiday Inn Crowne Plaza, Avenida Constitucion Ote 300, P.O. Box: 1619 (tel: 449300; tx: 382009).

Jolet, Padre Mier 201, P.O. Box: 10 (tel: 405500; tx: 038700).

Rio, Padre Mier 194 Pte., P.O. Box: 35 (tel: 449040; tx: 383310).

The Inn, Avenue Topo Chico, 64000, P.O. Box: 249 (tel: 522270; tx: 382429).

Car hire: Mexico City
Avis: Dr Velasco 146 (tel: 83-5-781044; tx: 1762483).

Budget: Mexico reservation centre: Atenas No 40 (tel: 5666800; tx: 1776367).

Dollar: Avenida Chapultepec 322 (tel: 5143891, 5257438).

Europcar National: Marsella, Col Juárez 48 (tel: 5330375; tx: 1773984, 1761323).

Hertz: For reservations (tel: 5666902, 5355983; tx: 1776515).

International Airport
Avis: (tel: 7620099, 7623688; tx: 1773073).
Dollar: (tel: 5714422).

Europcar National: (tel: 5718710, 7842241, 5718889).

Hertz: (tel: 7847547, 7847400, 6728372).

Business advice and information
Asociación Nacional de Importadores y Exportadores de la República Mexicana, Monterrey 130, Col Roma-Cuauhtémoc, 06700 Mexico, DF (tel: 5649379).

Asociación Nacional para el Fomento de las Exportaciones Mexicanas, Ed de las Instituciones 7\28, Ocampo 250 Apartado 2674, Monterrey NL (tx: 0383050).

Asociación de Personal Técnico para Conferencias Internacionales AC, Universidad 1855-502, Mexico 20, DF (tel: 5500170).

Banco de México, Avenida 5 de Mayo 2, Apdo 98 bis, 06050 Mexico, DF (tel: 5180500; tx: 176156; fax: 5129990).

Cámara Nacional de Comercio de la Ciudad de México, Paseo de la Reforma 42, Apartado 32005, Mexico, DF (tel: 5922677; tx: 1777262).

Confederación de Cámaras Industriales de los Estados Unidos Mexicanos, Manuel María Contreras 133, 8\18, Cuauhtémoc, 06597, Mexico DF (tel: 5469053; tx: 11773789).

Confederación de Cámaras Nacionales de Comercio, Servicios y Turismo (CONCANACO), Balderas 144, 2\28 y 3\28 Apartado 113 bis, Centro Cuauhtémoc, 06079 Mexico DF (tel: 5114391).

Dirección General de Telecomunicaciones, Lázaro Cárdenas 567 11\18 Piso Ala Norte, Col Navarate, 03020 Mexico, DF (tel: 5194049).

Instituto de Intérpretes y Traductores SA, Río Rhin 40, Mexico 5, DF (tel: 5667722).

Instituto Mexicano del Petróleo, Avenida Eje Central Lázaro Cárdenas 152, Apartado 14-805, Mexico 14, DF (tel: 5676600).

International Telegraph Office, Balderas 14-18 Colón, Mexico DF.

Nacional Financiers, B Avila Camacho 1, Mexico, DF (tel: 3955266).

PEMEX (Petróleos Mexicanos) Avenida Marina Nacional 319, Mexico 17, DF (tel: 2502611).

Public Telex Office, Vallejo y Norte 45, Mexico 2, DF/San Bartolo Naucáplan, Mexico 16, DF.

Secretariat of State for Commerce and Industrial Development, Alfonso Reyes 30, México, DF (tel: 2861823; tx: 1775718).

Secretariat of State for Energy, Mines and Federal Industry, Insurgentes Sur 552, 3\28, 06769 México, DF (tel: 5649789; tx: 1775690).

Secretariat of State for Finance and Public Credit, Palacio Nacional, 1 Patio Mariano, 06066 México, DF (tel: 5182060; tx: 1776397).

Secretariat of State for Tourism, Avenida Pdte Masaryk 172, 3\28, 11587, Mexico, DF (tel: 2508206; tx: 1777566).

Monaco

Patrick Middleton

'I see myself as the managing director of a company.' This has been a repeated theme over the years in the public utterances of Prince Rainier of Monaco. It recurred in July 1989 during the long interview he gave to the Côte d'Azur daily *Nice-Matin* on the fortieth anniversary of his accession to the throne. A close reading of his replies revealed that a justifiable satisfaction over the past was combined with no little anxiety over the future.

Four decades of achievement

The Principality which Rainier took over in 1949 was not in good shape. Its traditional upmarket tourism had been destroyed by the war and was recovering very slowly, while the Société des Bains de Mer (SBM), set up in the last century to manage the country's gaming interests, was under foreign control, with Aristotle Onassis a dominant influence. By 1966 Rainier and his advisers had won control of the company for the Monégasque state.

Today the SBM is a highly effective operator in the leisure market, and employs 10 per cent of local workers. It owns hotels, restaurants, sports facilities, the opera and, of course, the revamped casino. Gaming, however, now contributes only around 4 per cent to the Principality's national income, less than what is earned from the issue of commemorative postage stamps. Tourism, of course, remains important, but the emphasis is shifting to the convention business, with a major new conference centre now under construction.

Two other achievements of the Prince's reign are especially notable. Although it still remains slightly smaller than New York's Central Park, over the past thirty years Monaco's total area has been increased by one-fifth (220,000 square metres) by the reclamation of land from the sea. There are projects for further development, including the construction of one or more artificial offshore islands. Rainier has also sought to diversify his tiny country's economy. Some of the new land (notably in the Fontvieille district) houses *L'industrie à l'étage* (high-rise workshops). Monaco's industrial activities range from canning anchovies to making surgical lighting appliances. Industry represents about 25 per cent of national income, and employs one in four of the workforce.

An uncertain future

And what of the financial sector? Monaco is widely known as a tax haven. Foreigners taking up residence (other than French citizens) pay no personal taxes; the

KEY FACTS

Official title: The Principality of Monaco

Head of state: Prince Rainier III

Head of government: Minister of State: Jean Ausseil

Ruling party: National and Democratic Union (UND)

Capital: Monaco-Ville

Official Languages: French and Monégasque

Currency: French Franc (Fr) = 100 centimes

Exchange rate: Fr6.31 per $ (Nov 1989)

Area: 1.8 sq km

Population: 33,559 (1988)

Visitor numbers: over 1m (1987)

fiscal regime applied to companies is also notably benign. Large amounts of capital pass through the country's almost three dozen banks (employing 5 per cent of local workers), while there is a flourishing sector of offshore management and trust operations. Financial services are an obtrusive presence on the streets of Monaco, and yet behind a reassuring facade of plate-glass and gleaming brass there is an increasing sense of unease.

Pressures on the country's special status come from two directions. Firstly, as 1993 approaches the authorities in Brussels are looking with increasing disfavour at its 'loophole' function. For example, at present VAT payments in France from Monégasque sources are returned to the government of the Principality. This is an important source of income which is almost certain to disappear with the emergence of an EC-wide VAT pool. The shortfall is likely to be made up by some form of personal taxation levied on residents. The French, for their own good reasons, would like to safeguard Monaco's status as far as possible. In his *Nice-Matin* interview Rainier pointedly declared that 'France guarantees our existence', and added that his big neighbour would help the Monégasques 'face up to 1993'. Whether France can do this effectively may depend on a complex intergovernmental bargaining process over such other 'loopholes' as the Channel Islands and Madeira.

Secondly, there is an increasing pressure from national fiscal authorities who are trying to discover exactly who is doing what in Monaco. Moreover, within Europe these authorities are beginning to concert their activity. Paradoxically, even the French taxman, ignoring wider issues of policy, has gone on the attack, and has recently withdrawn the unofficial privilege which allowed Monégasque residents to pay no tax on second homes in France. As a result, some of those affected have chosen to leave the Principality and transform their second homes into *résidences principales*.

A movement out of Monaco is, in fact, becoming more and more noticeable. Until recently, many highly paid foreign executives, especially in the financial services sector, have lived in the Principality, benefiting from tax-free status but almost always in rented accommodation. At the beginning of 1989 the company which owns and lets literally thousands of apartments, seemingly disquieted by the way things look like going, hiked rents by anything up to 100 per cent. Calculating that they would be better off paying French income tax, some of these tenants, too, have crossed permanently into France. The exodus (in which some harassed tax-exiles may join – the Australians have already shown the way) can only weaken the Monégasque economy.

As Rainier III begins his fifth decade as ruler, Monaco is facing the first stages of a major crisis. Much of what has made it attractive to outsiders seems endangered. Some desperate solutions have been canvassed, such as inviting in wealthy Hong Kong Chinese ... but they were far from keen. As they might well have put it, the Principality seems set to live through 'interesting times'.

COUNTRY PROFILE

Political system: Founded on principles of an hereditary and constitutional monarchy. Principality governed under authority of the Prince, a Minister of State and a unicameral National Council (18 members directly elected every five years).

Media
Press: Monaco editions of French newspapers *Nice Matin* and *L'Echo de la Côte d'Azur.* Other French dailies widely read. Official weekly magazine, *Journal de Monaco*, published by Ministry of State. Two monthly magazines: *Gazette de Monaco – Côte d'Azur* (circulation 10,000) and *Monaco Actualité.*
Broadcasting: About 20,000 radios and 25,000 TVs in use. Radio Monte Carlo (RMC), part owned by French government, is important source of government income. Broadcasts in French and Italian. Also overseas broadcasts in 12 languages. Evangelical programmes broadcast in numerous foreign languages by Trans World Radio. Tele Monte Carlo broadcasts 1800 to 2400 daily.

Domestic economy: Small, open and diversified economy based on tourism, banking and insurance but with a significant industrial sector. State income derived from taxes, postage stamps and public monopolies.

External trade: Heavily dependent upon imports from France. Member of the French Franc Zone, with customs and currency union with France.

Industry: Around 700 firms employing 6,000 people, and accounting for 26 per cent of national turnover. Main products are cosmetics, healthcare, pharmaceuticals, plastics, electrical goods, electronics, textiles and food. Also important are construction and public works.

Services: Important sector accounting for 43 per cent of national turnover and employing around 1,200 people. Services sector includes banking, insurance and consultancy.

Banking: 37 foreign banks have branches in Monaco. System is controlled by French bankers' regulations.

Legal system: New constitution adopted 1962.

Membership of international organisations: IAEA, ICAO, IHO, INTELSAT, Interpol, IPU, ITU, UN (Observer), UNESCO, UPU, WHO, WIPO.

Time: GMT + 1 (GMT + 2 from late Mar to late Sep).

Climate: Mediterranean with mild winters and warm summers. Hottest months Jul and Aug, with average temperatures 25°C.
Clothing: Lightweight for summer, mediumweight and light topcoat for winter.

Entry requirements: Visa not required for nationals of most Western European countries. No restrictions on imports of foreign exchange.

Air access: Nearest international airport Nice (NCE), France, 22 km. Heliport facilities in Monte Carlo (MCM). Helicopter link with Nice Airport.

Surface access: Good road and rail connections with France.

Hotels: Classified into one- to four-star and predominantly, four-star/luxury categories. Advisable to book in advance during summer.

Credit cards: All credit cards accepted.

National transport: Around 50 km of roads and 1.6 km of railways (operated by Société Nationale des Chemins de Fer Français).

Public holidays
Fixed dates: 1 Jan, 27 Jan, 1 May, 15 Aug, 1 Nov, 19 Nov, 8 Dec, 25 Dec.
Variable dates: Easter Monday, Fête-Dieu, Ascension Day, Whit Monday.

Telecommunications: International dialling code: 33 93. There are no area codes.

Business advice and information
Direction du Commerce, de l'Industrie et de la Propriété Industrielle, Centre administratif, 8 Rue Louis Notari, 98000 Monaco (tel: 93-15-80-00; fax: 93303974).

Jeune Chambre Economique de Monaco, 2 Avenue Prince Héréditaire Albert, Monaco (tel: 93-15-04-22).

Banking
ABC Banque Internationale de Monaco, Sporting d'Hiver, Place du Casino, PO Box 147, MC98003 Monaco Cedex (tel: 93-50-06-08; tx: 469163).

Banque Centrale Monégasque de Crédit à Long et Moyen Terme, 15 Avenue d'Ostende (tel: 93-30-92-95; tx: 469623).

Banque Industrielle de Monaco SAM, 8 Boulevard des Moulins, BP115 (tel: 93-30-73-11; tx: 469551).

Banque de Placements et de Crédit SAM, PO Box 189, 2 Ave de Grande Bretagne, MC98007 (tel: 50-50-75; tx: 469955).

Compagnie Monegasque de Banque SAM, 23 Ave de la Costa, BP 167 , MC 98003 Monaco Cedex (tel: 50-40-50; tx: 479269).

Crédit Foncier de Monaco SAM, 11 Blvd Albert 1er, MC 98012 (tel: 15-45-00; tx: 469738).

Société de Banques et d'Investissements SAM, 26 Boulevard d'Italie (tel: 93-50-40-40; tx: 469802).

Principal newspapers
Gazette Monaco, Côte d'Azur, 25 Blvd Albert 1er, MC 98000 (tel: 25-20-36).

Journal de Monaco, Monaco.

Monaco Actualité, 5 Rue de l'Industrie, MC98000 (tel: 30-41-79).

Other useful addresses
Conseil Economique, 8 Rue Louis Notari, Monte Carlo (tel: 30-20-82).

Direction du Tourisme et des Congrès, 2a Boulevard des Moulins (tel: 93-50-60-88).

Fédération Patronale Monégasque (employers' federation), 14 Avenue de Grande Bretagne (tel: 93-30-10-19).

Société des Bains de Mer (controls entertainment facilities), Place du Casino, 98000 Monaco (tel: 93-30-99-31).

Union des Commerçants, 14 Avenue de Grande Bretagne (tel: 93-30-85-81).

Hotels: Monte Carlo
Abela, 23 Avenue des Papalins (tel: 92059000; tx: 469162).

Alexandra, 35 Boulevard Princesse Charlotte (tel: 506313; tx: 489286).

Balmoral, 12 Avenue de la Costa (tel: 506237; tx: 479436).

Beach Plaza, 22 Avenue Princesse Grace (tel: 309880; tx: 479617; fax: 502314).

Hermitage, Square Beaumarchais (tel: 506731; tx: 479432).

Hotel de Paris, Place du Casino (tel: 508080; tx: 469925; fax: 255917).

Loews, Avenue des Spélugues (ocean-front) (tel: 506500; tx: 479435; fax: 300157).

Metropole, 4 Avenue de la Madone (tel: 93151515; tx: 489836).

Mirabeau, 1-3 Avenue Princesse Grace (tel: 254545; tx: 479413).

Monte Carlo Beach Hotel, Route du Beach (tel: 93782140; tx: 479432).

Mongolia

Alan J. K. Sanders

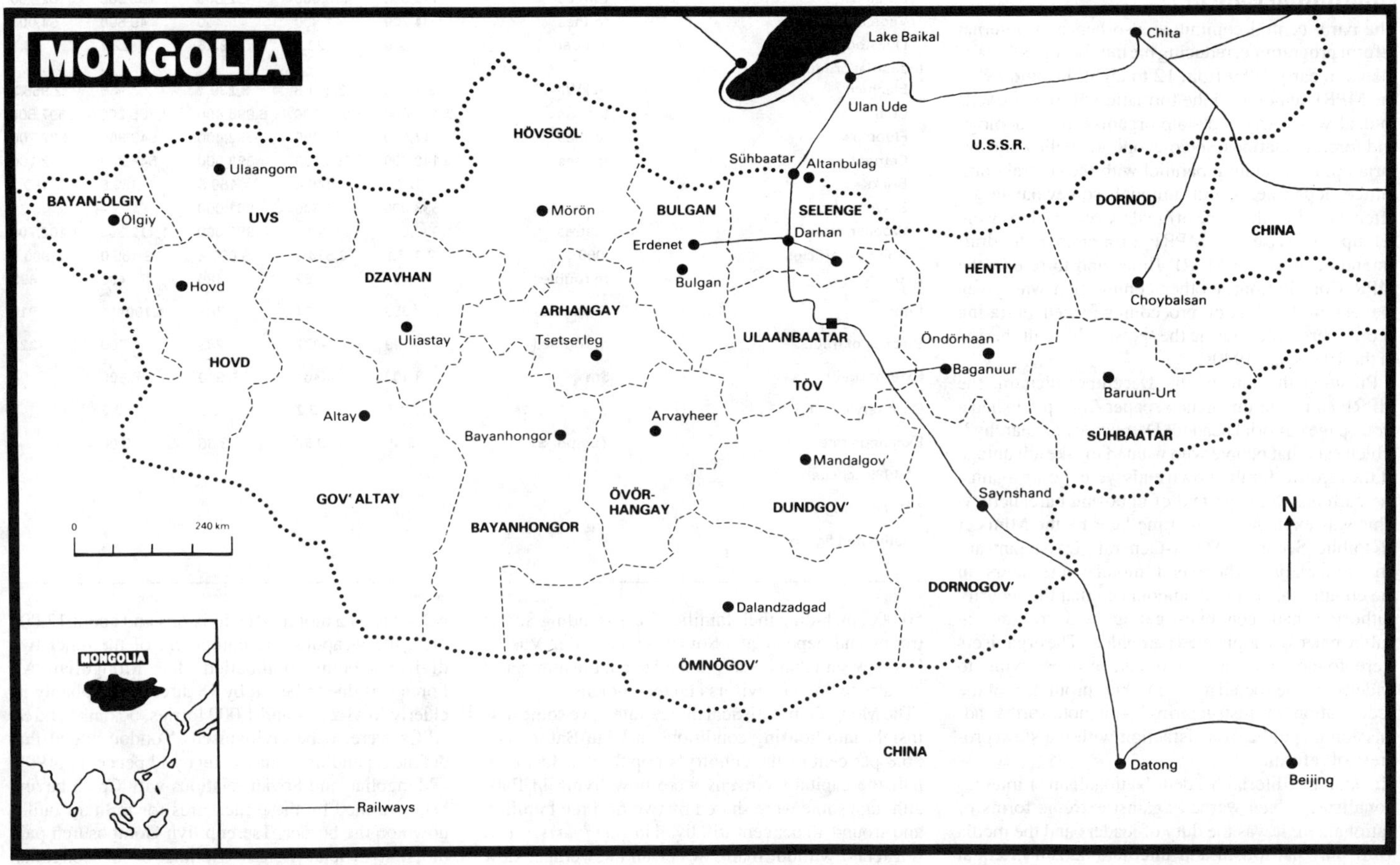

Increasingly aware that the country's programme of economic restructuring was not making headway, the Mongolian party leaders sought to eliminate bureaucratic management methods, overcome inertia and develop popular initiative by applying a more positive policy of glasnost and democratisation, including modest political reform and a fuller and more nationalist appraisal of Mongolia's history. For the first time since he was removed from power in 1984, Yumjaagiyn Tsedenbal was mentioned by name in official documents condemning the 'years of stagnation' under his leadership and accused of continuing some of the stalinist practices of his predecessor, the dictator Marshal Horloogiyn Choybalsan. Mongolia remained firmly committed to its alliance with the USSR, while gradually developing relations with China and seeking to broaden its political and economic contacts with Western and Third World countries and seeking wider membership of international agencies.

A political turning-point

The December 1988 plenary meeting of the central committee of the ruling Mongolian People's Revolutionary Party (MPRP) was a milestone along the country's road of economic and political reform. Despite the restructuring of government bodies which had been taking place since the end of 1987, it had become clear that bureaucracy and lack of drive were stifling economic progress. The plenum approved party politburo proposals calling for the removal of 'blank spots' from the post-revolutionary history of the Mongolian People's Republic (MPR) and emphasising that Tsedenbal had done his best to obscure his links with Choybalsan but had continued to run the country by the same 'administrative command' methods and had used illegal methods to remove political opponents. 'The truth is,' party General Secretary Jambyn Batmtnh told the plenum, 'that our successes and achievements as well as our main shortcomings and omissions are in many ways linked with the activities of Tsedenbal.'

The politburo's proposals also said that until recently Mongolian people's ideas and initiatives 'connected with their feelings of national pride and progressive national traditions' had simply been condemned out of hand as 'nationalism'. Tsedenbal had been unwilling to risk offending the Soviets by disputing their negative view of the historical role of Genghis Khan and the Mongol Empire. The party now called on everyone to 'overcome any instances of indifference to the national cultural heritage and its careful transmission to the next generation'.

Lunar New Year's Day, a Buddhist festival called Tsagaan Sar by the Mongols, was restored as a public holiday in February 1989. The newly established Cultural Foundation began collecting donations for the restoration of the Aryabal (Megdzedjanraysegu) temple in Ulan Bator. A public debate began over the large statue of Stalin, which still dominated one of

KEY FACTS

Official title: Mongolian People's Republic

Head of state: Chairman of the Presidium of the People's Great Hural, Jambyn Batmönh

Head of government: Chairman of the Council of Ministers, Dumaagiyn Sodnom

Ruling party: Mongolian People's Revolutionary Party

Capital: Ulan Bator (Ulaanbaatar)

Official Languages: Mongolian (Russian and Kazakh also spoken)

Currency: Tögrög or Tugrik (Tug = 100 möngö)

Exchange rate: Tug 3.36 per $ (Nov 1989)

Area: 1,565,000 sq km

Population: 2,043,400 (1989)

GNP per capita: $2,140 (1986)

Labour force: 429,000 (1987)

Trade balance: –Roubles 222.7m (excluding credits, 1988)

Foreign debt: $6,500m (1987)

KEY INDICATORS

	Unit	1984	1985	1986	1987	1988
Population	'000	1,820.4*	1,914.7*	1,965.6*	2,071.2*	2,043.4**
Agricultural production						
Grain	tonnes	586,200	889,400	869,400	689,500	814,300
Potatoes	tonnes	122,900	115,000	132,800	137,900	102,700
Vegetables	tonnes	34,200	41,300	46,400	45,600	54,000
Livestock	m head	22.6	22.5	22.6	22.7	23.1
Industrial production						
Electricity	m kWh	2,167.3	2,811.8	3,239.2	2,854.5	2,966.3
Coal	tonnes	5,431,600	6,515,900	6,996,400	7,765,200	8,597,500
Fluorspar	tonnes	747,000	786,700	592,400	543,900	522,700
Cement	tonnes	140,700	150,500	393,500	541,300	502,100
Bricks	m	127.1	146.0	169.8	169.8	180.2
Sawn timber	cu m	583,000	620,600	561,000	–	–
Woollen cloth	metres	980,000	987,800	996,000	1,112,300	1,102,700
Knitwear articles	'000	2,155.9	2,574.0	3,092.3	3,165.0	3,900.0
Exports	m roubles	454	462	480	400	493
Imports	m roubles	653	734	763	1,100***	716
Balance of trade	m roubles	–199	–272	–283	–700	–222
External debt	$m	3,402	4,404	5,510	6,500	–
Debt service ratio	%	4.5	2.2	2.9	2.2	–
Exchange rate	Tug per $	3.45	3.50	3.36	3.00	–

** 1989 census

*** including credit

* estimated figure

Ulan Bator's main streets, while Stalin Street was renamed Natsagdorj Street, in honour of the writer who fell victim to the 'personality cult' in the 1930s. The newspapers published previously suppressed details about the lives of Mongolian revolutionary leaders murdered by Choybalsan.

Continuing reform

The party central committee also began an internal reform programme, reducing the number of its departments in early 1989 from 12 to eight. In June 1989, the MPRP announced the formation of four councils to deal with party and state organisations, ideology and foreign relations, socio-economic policy and agrarian policy, running parallel with the central committee departments and although somewhat larger often including the same officials. Commissions were set up to revise the MPRP Programme, to draft amendments to the MPRP Rules and to re-edit the MPR Constitution; another commission was given the task of deciding on procedures for rehabilitating people persecuted during the 'personality cult' period in the 1930s and 1940s.

Pursuing the line of the December plenum, the MPRP and government newspaper *Unen* published a front-page editorial headed 'Democracy or anarchy?' which said that people who wanted to take advantage of the reforms for their own ends were 'going against the current' on the pretext of opposing bureaucracy. This was explained some time later by the Minister of Public Security, Major-General Agvaanjantsangiyn Jamsranjav, who to end 'months of rumours' in the country, as he put it, announced that the security authorities had been investigating the distribution in public places of duplicated pamphlets. The organisers were found to be not 'anti-state groups trying to undermine the socialist system' but 'promoters of the acceleration of restructuring' – a noteworthy admission of public dissatisfaction with the slow progress of reform.

In another editorial headed 'Nationalism or internationalism?' *Unen* warned against extreme forms of nationalism. It was the duty of leaders and the media to display patriotism and internationalism in equal measure, the editorial said ('internationalism' is the touchstone of loyalty to the USSR). No one could deny that it was thanks to aid from the Soviet Union in particular that the country had reached its present level of development, it added. The usual laudatory press features about aid from the Soviet 'elder brother' had gradually been giving way over the past year to Moscow's own sober self-evaluations and even some Mongolian grumbling about Soviet economic inefficiency. Meanwhile, Sonomyn Luvsangombo was removed from candidate membership of the MPRP politburo and from a deputy premiership following his sacking from the chairmanship of the state construction committee in a welter of criticism about building standards and construction investment policy.

Census results

The results of the national census held in January 1989 indicated that the country's population totalled 2,043,400; the number of citizens of the MPR was 1,998,000. The rate of natural population increase was put at 2.8 per cent per annum, but with the proportion of the population of working age rising by 3.4 per cent there was a prospect of unemployment growing from 27,480 to some 60,000 by 1990. Employment schemes mooted included wider introduction of two- and three-shift working in industry, adoption of a five-day working week and creation of small enterprises in rural areas, as well as the development of service industries and co-operatives.

The census results gave the number of foreign – mostly Soviet – specialists residing in the MPR as 50,000, including their families, but excluding Soviet troops and dependants. Soviet Ambassador Vasiliy Sitnikov said that it was planned to cut the number of expatriate Soviet advisers and technicians.

The Mongolian statistical directorate gave some new insight into housing conditions in Ulan Bator, with 26.8 per cent of the country's population. Just over half the capital's citizens were now living in flats, although some were shared by two or three families, and around 30 per cent still lived in *gers (yurts*, round felt tents) without mains water and sewerage. Low standards were largely to blame for the high infant mortality figures disclosed after decades of concealment, ranging from various Mongolian sources between 64-65 and 70-75 promili (deaths per thousand births). Poor living conditions were responsible for 50 per cent of infant deaths, genetic and ecological conditions for 30 per cent and poor medical services for 20 per cent.

Foreign affairs

The second stage of the partial Soviet troop withdrawal from Mongolia began in May 1989. Altogether 75 per cent of the Soviet troops remaining in Mongolia after the first stage of withdrawal were due to leave in 1989-90 – some 50,000 men, 1,100 armoured personnel carriers (APC), 820 artillery guns and missile launchers, 190 aircraft and 130 helicopters.

Mongolian Defence Minister Colonel-General Jamsrangiyn Yondon announced at the beginning of March that the Mongolian armed forces would be halved to two motor-rifle divisions and about 13,000 men, the weapons and equipment of the other two divisions being mothballed. The Mongolian Air Force was due to be cut by 18 aircraft – probably its elderly MiG-21s – and 1,000 lorries, 60 tanks and 30 APCs were to be civilianised. Yondon added that defence spending would be cut by 11 per cent in 1990.

Mongolian and Soviet relations with China having been warmed by these measures for military build-down on the border, Tserenpiliyn Gombosuren paid the first official visit to Beijing by a Mongolian foreign minister in the 40 years since the establishment of diplomatic relations between the two countries. Three agreements were signed: on setting up an inter-governmental joint commission on economic, trade, scientific and technical co-operation, on exchanges of visits by citizens of Mongolia and China, and on re-opening the Mongolian consulate-general in Hthhot, capital of China's Inner Mongolia Autonomous Region.

Deputy Foreign Minister Humbagyn Oldzvoy stated at mid-year that Mongolia was restructuring its foreign policy and admitted that in the past the country's seat in the UN had been used largely for propaganda purposes. Mongolia joined the Group of 77 in June, and was negotiating relations with the EC and membership of the Asian Development Bank and the International Civil Aviation Organisation. The country was moving towards greater co-operation with the UN, developing nations and Western countries, he said.

COUNTRY PROFILE

Political system: Mongolia, a socialist state, is generally regarded as a Soviet satellite. Executive power is exercised by the Council of Ministers appointed by the People's Great Hural, which also elects an eight-member Presidium to administer the country between meetings of the Hural. The chairman of the Presidium is head of state. The communist Mongolian People's Revolutionary Party (membership: 88,150 in 1986) dominates politics at all levels. Local government is vested in three city and 18 provincial hurals.

Political parties: The Mongolian People's Revolutionary Party is the only official political party. The Mongolian Revolutionary Youth League prepares young people for Party membership.

Main cities/towns: Ulan Bator (population 548,400 in 1989), Darhan (85,800 in 1989) and Erdenet (56,100 in 1989).

COUNTRY PROFILE

Language: Mongolian is the official language; Kazakh is also spoken in western Mongolia; Russian is the principal foreign language.

Media
Press: Principal dailies serving the Party and the government, Ünen (trade unions), and *Dzaluuchuudyn Ünen* (Revolutionary Youth League). *Novosti Mongolii* is published weekly in Russian.
Radio: Services are broadcast in Mongolian and also Russian, Chinese, English, French and Kasakh. There were 226,200 radios in 1988.
Television: There were 122,800 sets in 1988, with a potential audience of 45 per cent of population, mainly in the Ulan Bator and Darhan areas, for Mongolian programmes; Soviet television is, however, received by satellite ground stations all over the country.

Domestic economy: The domestic economy is traditionally based on pastoral activities, although mining is the biggest foreign exchange earner. Mongolia has developed 255 stock-raising co-operative farms and 52 grain and fodder growing state farms. Industries are mainly agriculture-based. Aid from the USSR makes up Mongolia's trade deficit. The current development plan emphasises intensification of agricultural production emphasising capital construction and increasing mining and electricity production.

External trade
Exports: Main exports are: raw materials of plant or animal origin, minerals and metals and industrial consumer goods. Main destinations are CMEA members, other socialist countries and a few capitalist countries.
Imports: Main imports are machines and equipment, fuel, minerals and metals, industrial consumer goods, foodstuffs and related raw materials, chemicals and fertilisers, rubber and construction materials. Main sources are CMEA members, other socialist countries and capitalist countries.

Agriculture: A programme for 1986-90 envisages more consumption of basic foodstuffs, and appears to indicate a loosening of controls on private-sector smallholdings. Less than one per cent of total land area is arable. Cereal production is increasing, predominantly of wheat, barley and oats. Other crops include potatoes, vegetables and fodder crops, particularly hay. For most crops, the state sector accounts for 80-85 per cent of output; for fodder crops, the co-operative sector accounts for 51 per cent of output. Raising of cattle, sheep, horses, goats, camels and yaks yields not only meat and milk, but also goat and camel wool, goat hair, cattle and horse hides and camel, sheep and goat skins. Pig, poultry and rabbit farming are being developed. Marmot, squirrel and wolf pelts are collected. Timber is logged from mountain forests, which cover about 10 per cent of the land area.

Industry: Industrial activity is centred on Ulan Bator and other main cities and is based mainly on agricultural products and mining. Products include bricks, cement, lime, sawn timber, scoured wool, felt, felt boots, woollen fabric, leather footwear, soap, flour, garments, matches, bakery goods, confectionery, meat products, beer and vodka.

Mining: The Erdenet copper-molybdenum complex, an open-pit mining and concentrating development 340 km from Ulan Bator, accounts for over 40 per cent of exports. There are fluorspar mines at Bor-öndör, a gold mine at Ih-altat and a non-ferrous metals mine at Yeröö in Selenge province. Coal production has risen in recent years to 8.5m tonnes. Other minerals present include tungsten, tin, lead and graphite.

Energy: Total generating capacity reached 877 MW in 1988, when 2,966.3m KWh was generated.

Membership of international organisations: CEMA, ESCAP, FAD, IAEA, ILO, International Bank for Economic Co-operation, International Investment Bank, IPU, ITU, UN, UNESCO, UNIDO, UPU, WFTU, WHO, WIPO, WMO.

BUSINESS GUIDE

Time: GMT + 8 hrs.

Climate: Summers are warm and wettish, and winters extremely cold. In Ulan Bator, winter temperatures range from -3.8°C to -47°C, with an average of -26°C in Jan; in summer from 0-38.6°C, with an average of 17°C in Jul. Relative humidity ranges from 65 per cent (Jul-Aug) to 75 per cent (Nov-Feb). Rainfall is low, with an average of 233 mm/year in Ulan Bator (two-thirds of which falls Jun-Aug) and 116-344 mm/year elsewhere. On average, 250 cloudless days are recorded p.a.

Entry requirements
Passport: Required by all.
Visa: Visas are required by all. Applications for visas, along with visas for transit countries, should be made well in advance of planned visit.
Health: No vaccination certificates are required.
Currency: Importation and exportation of Mongolian currency is forbidden.
Customs: Importation of 'hostile' publications and pornography, and export of valuable antiques is strictly prohibited. Customs regulations are enforced by strict examinations.

Representation overseas: Beijing, Belgrade, Berlin, Bucharest, Budapest, Cairo, Geneva (UN), Hanoi, Havana, Kabul, London, Moscow, New Delhi, New York (UN), Paris, Prague, Pyongyang, Sofia, Tokyo, Vientiane, Warsaw.

Air access
National airline: Air Mongol (Mongolian Civil Air Transport or MIAT).
Other airlines: Aeroflot.
International airport: Buyant-Uhaa (ULN), 15 km from Ulan Bator.

Surface access: Ulan Bator is served by the Trans-Mongolian Railway connecting Moscow and Beijing.

Hotels: There are hotels suitable for foreign visitors in Ulan Bator, but in the countryside facilities are basic.

Credit cards: Credit cards are not accepted.

Car hire: Car hire is not available.

City transport
Taxis: Taxis may be hired, but language is a barrier.

National transport: Travel for foreigners is normally arranged in advance, and outlying places are often reached by plane. Some towns (e.g. Darhan) and industrial/mining areas are also served by rail routes (two classes); trains on international routes have sleeping and restaurant cars.
Air: Regular air services provide the best means of long-distance internal travel.
Road: There are 46,700 km of roads and tracks. Except around the cities, roads are mostly unpaved or consist of rough cross-country routes.
Buses: Bus services are available, but roads can be rough and language a barrier.
Rail: Rail services are adequate for north-south travel. In addition to the cities served by the Trans-Mongolian Railway (Sühbaatar, Darhan, Ulan Bator and Saynshand), there are branch lines to various industrial centres and mining towns, including Erdenet, Baganuur and Bor-öndör. Total network in 1986 was 1,750 km.

Public holidays
Fixed dates: 1 Jan, 8 Mar (Women's Day), 1-2 May (Labour Days), 11-12 Jul (Revolution Anniversary), 7-8 Nov (Soviet Revolution Anniversary).
Variable dates: Lunar New Year (Tsaagaan Sar).

Working hours
Government and business: (Mon-Fri) 0900 – 1800; (Sat) 0900 – 1500.
Banking: (Mon-Fri) 0930 – 1230; (Sat) 0930 – 1130.
Shops: (Mon-Fri) 0800 – 1700 (some food shops stay open later); Sat half-day.

Social customs: Requests for permission before taking photographs are generally appreciated.

Telecommunications
Telephone: Dialling for Mongolia is via the International Operator. There were 56,000 telephones in 1988, with 252 exchanges. The main centres have automatic exchanges.
Telex: International telex service is available to non-Communist Bloc countries.

Banking: State Bank of the Mongolian People's Republic.

Weights and measures: Metric system.

Representation in capital: Afghanistan, Bulgaria, China PR, Cuba, Czechoslovakia, German DR, Hungary, India, Japan, Korea PDR, Poland, Romania, USSR, UK, Vietnam, Yugoslavia.

Hotels: Ulan Bator
Altay.

Bayangol (tel: 26770; tx: 232).

Dzaluuchuud.

Ulaanbaatar (tel: 20230; tx: 232).

Business advice and information
Chamber of Commerce and Industry of the Mongolian People's Republic, Nayramdlyn Gudamj 24, Ulan Bator.

Travel information
Juulchin, Ulan Bator (tel: 22884; tx: 232).

Montserrat

Tony Thorndike

The ferocity with which Hurricane Hugo hit Montserrat between 17 and 19 September 1989 resulted in an estimated 95 per cent of all buildings either being destroyed or extensively damaged; the overall cost was estimated at some US$125-150m. The disaster sharply reversed the record of one of the Caribbean's economic success stories and will be bound to delay the planned referendum to decide whether the British Dependency should proceed to independence. Earlier, political tension surrounded widescale criminal investigations by Britain's Scotland Yard into allegations of fraud and the 'laundering' of illegal funds in the nascent offshore finance industry.

Despite the damage, it is likely that Montserrat will recover reasonably quickly. Despite its small size and limited population, the island has an enviable record of economic growth and resilience. But a close look at the figures reveals a high degree of dependency on remittances and aid, and the opponents of independence – probably in the majority – are well aware of Montserrat's economic vulnerability, let alone its exposure to the elements.

Nature's forceful reminder

Although there was adequate warning of Hurricane Hugo, general complacency resulted in little preparation. Winds of up to 150 mph wrought havoc and even after a week, 90 per cent of the 11,500 population were without running water and 60 per cent without sanitation. Although the small airport reopened within a relatively short time, relief efforts were hampered by the loss of the recently rebuilt deepwater jetty. Britain immediately allocated £1m in emergency aid and a frigate, HMS *Alacrity*, lived up to its name being despatched to the rescue. Other assistance came from Canada, Barbados and Jamaica.

Montserrat was the worst hit (other territories severely affected included Antigua, Guadeloupe, Dominica, St Kitts-Nevis, the British Virgin Islands, the US Virgin Islands and Puerto Rico) and preliminary estimates suggested that full restoration would take at least a year. Nonetheless, in November it was announced that most of the hotels would be open, at least in part, to cater for the important winter tourist trade.

Prosperity undermined

Up to the time of Hurricane Hugo there had been a clear growth in prosperity. That much was clear by the tone and content of the March 1989 budget announced by the Premier and Minister of Finance, John Osborne. GDP growth was 12.4 per cent in 1988 (1987: 10.8), pushing annual per capita income well over US$4,000. The main contributors were increased tourist receipts and construction activity. Tourist stopover arrivals increased slightly to 17,876 (1987: 17,029) and the number of cruise ship excursionists rose 12 per cent to some 12,500. By themselves these figures are modest, but earnings were not at EC$29.2m (US$10.8m). Tourists to this mountainous and volcanic island are not normally of the beach-seeking type, but rather long-staying high spenders wishing solitude. A considerable number of retired people, especially Canadians, have made their home in Montserrat. But the more traditional Caribbean-bound tourist will be attracted by the big project planned for Little Bay, the only non-black-sand accessible beach, consisting of a large hotel, marina, 400 villas and a casino. Some of the stopovers are students attending the American University of the Caribbean, an offshore medical school which is in decline due to adverse publicity in the US media and poor enrolment. A more unusual category of visitor consists of musicians attracted to the world-standard Air Studios.

Income tax thresholds were raised and the only new tax was one on borrowing by non-Montserratians. Recurrent expenditure was put at EC$32.1m (US$11.9m) and revenue at EC$32.6m. Capital expenditure was to be EC$22.3m, 10 per cent of which was to be locally financed. Nine donors were to be involved: in June 1989 it was announced that Britain and the European Community would finance the first stage of the rebuilding of the airport, worth EC$77m (US$28.5m), regarded as essential for the tourist industry; and that other donors would finance harbour schemes and the Little Bay tourist project. A widening trade gap was reported, doubtless a consequence of progress. Imports rose to EC$72.5m (US$26.9m) against exports of only EC$6.2m (US$2.3m), the increase almost entirely due to raw materials and capital equipment. Part of this gap was to be met by remittance income, for half the population has migrated to Britain and North America since 1959. This

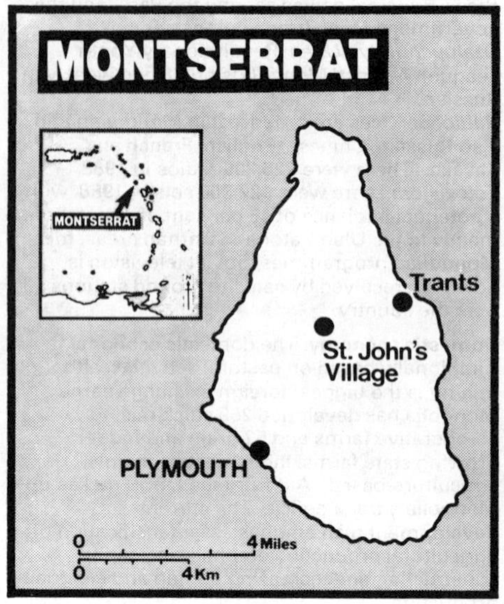

KEY FACTS

Official title: Montserrat
Head of state: Queen Elizabeth II
Head of government: Chief Minister John Osborne
Ruling party: People's Liberation Movement
Capital: Plymouth
Official Languages: English
Currency: East Caribbean Dollar (EC$) = 100 cents
Exchange rate: EC$2.69 per $ (Nov 1989)
Area: 102 sq km
Population: 12,500 (1987)
GDP per capita: $3,997 (1987)
GDP real growth: 12.4% (1988)
Inflation: 1.6% (1988)
Trade balance: –$16.5m (1988)
Visitor numbers: 42,109 (1988)

COUNTRY PROFILE

Political system: British Caribbean dependency with an appointed governor with responsibilities for defence, security and external affairs. Other aspects are dealt with by an executive council (the governor, attorney-general, financial secretary, chief minister, three other ministers). The Legislative Council (the attorney-general, the financial secretary, two nominated members and seven elected members) is presided over by a speaker and serves a five-year term. One elected member is returned from each district by universal adult suffrage.

Main town: Plymouth (population 3,500).

Media
Press: The Montserrat Times and *The Montserrat Reporter* are both weeklies.
Radio: Radio services are transmitted by Radio Montserrat (government-owned); Radio Antilles (private commercial station) broadcasts in English and French.
Television: Two cable television services. External broadcasts are also received (e.g. from Antigua, Puerto Rico, St Kitts, USA).

External trade: Large trade deficit, caused by the need for imports of most basic commodities and a very narrow export base, is offset by earnings from tourism, remittances from Montserratians living abroad and British aid.
Exports: Electrical goods and light fittings (68% of total value), electrical components (21%), polythene bags, garments, ornamental plants, hot peppers, tomatoes, cattle/meat, cotton and cotton goods. Main destinations: CARICOM, USA.
Imports: Food and beverages, machinery and transport equipment, manufactured goods. Main sources: USA, Caribbean, UK, Canada.

Agriculture: Sector has declined in importance in recent years and now contributes five per cent to GDP, employs 10 per cent of labour force and provides 16 per cent of total exports. Montserrat is almost self-sufficient in fruit, ground provisions and vegetables though canned and preserved foodstuffs are imported. Approximately 25 per cent of land area is cultivated, the main crops being cotton, potatoes, tomatoes and carrots. The government is also encouraging the planting of new crops such as onions, pineapples, and avocados. Very unpredictable rainfall has been a major constraint to growth in this sector but this now should be improved with the installation of new irrigation systems. Agro-based industries such as the processing and packaging of tropical fruits and herbal teas for export are being encouraged.

income has kept public external debt down to only US$2m. Despite high consumer demand and virtually full employment, inflation was low at 3.6 per cent. Over 60 workers were imported from Dominica in 1987 and more arrived later, but this was not publicised. After all, Montserrat is no different from other small economies in exaggerating unemployment and discounting success to ensure continued aid.

Other sectors of the economy contribute little. Agriculture recorded modest growth and accounted for 5 per cent of GDP. Decline has been long-term, and a fall off in demand from neighbouring Antigua did nothing for the morale of the handful of full-time farmers. But the value of food imports has been pegged at about EC$10m for a number of years and domestic demand is rising. Manufacturing production, mainly electronic assembly, declined by 1.3 per cent in 1988 due to labour shortages, and a computer data-entry concern closed, adding to the empty premises on the Plymouth industrial estate. The biggest company, W W Electronics, nonetheless employed over 500.

The independence question

Premier John Osborne, a successful businessman and boat builder, has for long made clear his support for independence. At first personal pique was the spur, Osborne being angry at not being able to participate fully in the affairs of the regional organisation, CARICOM. Then there were differences with Britain over aid, exclusion from full British citizenship and a refusal by Britain to fund emergency treatment for a child. But it is by no means clear that the islanders will support him. Not only are most conservative by nature, but many feel that their new-found prosperity would be put at risk.

Independence was not an issue in the August 1987 election when Osborne's People's Liberation Movement (PLM) won four of the seven seats and defeated a divided opposition. The campaign centred on plans for a casino and Osborne's handling of a confrontation with the church over the issue. The PLM first came to power in 1978 when it comprehensively defeated the incumbent Progressive Democratic Party (PDP) government headed by Austin Bramble, taking all seven seats. But the PDP regained ground in 1983, although Osborne's party was victorious with four seats. However, the National Democratic Party (NDP) then emerged, led by businessman Bertram Osborne (no relation). Although the NDP and the PDP talked about an alliance, there was no agreement as to who would be the leader. The result was that the NDP won three seats, the PDP one. As Bramble lost his seat, Eustace Dyer assumed the PDP leadership. The opposition was to be further divided by the resignation in August 1989 from the NDP of one of its two representatives, David Brandt, who alleged he was marginalised by Bertram Osborne and excluded from virtually all policy-making meetings.

Normally few differences separate Montserratian parties (the socialist United National Front faded away after 1984), but the independence question certainly does. The PLM is supported on the issue by Dyer, but the NDP resolutely opposes sovereignty. Ironically, the PDP under Bramble was against independence when the issue first surfaced in 1983. Boosted by this support, Osborne announced that an independence referendum will be held in 1990. Even a cursory reading of the highly partisan Montserratian press prior to the disaster gave the impression that there was no other issue. Although the British Government repeated its policy of granting independence whenever there was popular support for it, inevitably British officials were drawn, unwillingly, into the controversy. Principal Crown Counsel Ray Astin was one victim, being forced to resign after only nine months following allegations that he made disparaging remarks about the island and its lawyers. Perhaps the offshore financial fraud had some influence in this incident.

Offshore fraud

A financial centre based on banks was developed in the early 1980s, and 190 were registered by 1988. Only 'A' licensed banks and trust companies can conduct local business, 'B' licencees being restricted to offshore. In March 1989 the issue of new 'B' licences was suspended as investigations by British police began into suspected fraud and 'laundering' of illegal funds. The licences of three banks were revoked and 62 were required to show why their licences should not suffer a similiar fate. All of them were warned that they had defaulted on fee payments, failed to submit audited accounts and failed to notify change in local agents. The FBI also became involved and several US citizens were arrested.

The opposition was quick to charge the government with negligence and demanded the full text of the investigation, pointing out that Montserrat would have to pay the costs of the investigation, estimated at US$200,000. There is no doubt that the scandal constitutes a major embarrassment for Osborne's government which may affect the outcome of the referendum.

COUNTRY PROFILE

Industry: Manufacturing contributes about six per cent to GDP and accounts for 10 per cent of employment, but is responsible for up to 98 per cent of exports. Tends to be limited to small-scale industries producing light consumer goods such as electronic components, plastic bags, leather goods and various items made from locally-grown cotton. Construction has become the largest sector of the economy - mainly due to the demand for retirement homes. Tourism contributes about 22 per cent to GDP and is seen by the government as playing a vital role in the development of the economy. To this end, a large amount of foreign development aid, mostly from the UK, is used on projects such as the improvement of airport and dock facilities, increasing the availability of hotel beds and building of a hotel/marina complex. Some revenue and much publicity is also generated by Air Studios, whose high quality recording facilities are used by many internationally famous recording artists. Income is also derived from people who have retired to the island, from the sale of postage stamps and from an off-shore medical school. The development of a regional computer services facility was planned for 1988.

Membership of international organisations:
CARICOM, Economic Commission for Latin America (associate), state associated with the EC, OECS.

Time: GMT -4.

Climate: Tropical with mean temperature of 30°C and low humidity due to trade winds. Little variation throughout the year. Wettest months are between Sep and Nov, driest from Feb to Jun.

Entry requirements
Passport: Required by all except nationals of Canada and USA with proof of identity.
Visa: Generally not required by nationals of Commonwealth countries, USA, the EC, Scandinavian countries and certain others. An onward or return ticket should be held.
Currency: No restrictions on import/export of foreign currency, but declaration required on arrival.

Health precautions
Mandatory: Yellow fever vaccination certificate required if arriving from an infected area.
Advisable: Typhoid and polio vaccinations. Water precautions.

Air access
Airlines serving: Air Caribe International, LIAT, Windward Islands Airways.
International airport: Blackburne (MNI), 14 km from Plymouth; bar, restaurant, buffet, shops. Access from Antigua, St Kitts, St Marten.

Surface access
Main port: Plymouth

Hotels: Generally small and accommodation should be booked in advance. Seven per cent room tax is added to bill and 10 per cent service charge is usual.

Car hire: Available at airport, in Plymouth or through hotels. Temporary licence can be obtained on production of national licence. Traffic drives on the left.

City transport
Taxis: Readily available. Legal fixed-rate system.

National transport
Road: Total network over 200 km, over 85% all-weather, linking main centres.
Buses: Buses are privately-owned and readily available. Occasional minibus services operate throughout island.

Public holidays: 1 Jan, 17 Mar (St Patrick's Day), Good Friday, Easter Monday, Labour Day (first Mon in May), Whit Monday, Queen's birthday, August Monday, Liberation Day (Nov), 25 Dec, 26 Dec, 31 Dec (Festival Day).

Working hours
Government and business: (Mon-Fri) 0800 – 1200, 1300 – 1600. Some businesses close early (Wed) 0800 – 1200, and some open (Sat) 0800 – 1230. Shops close at 1230 on Wed.
Banking: (Mon-Thurs) 0800-1300, (Fridays only) 0800 – 1200, 1500 – 1700.

Telecommunications
Telephone: Dialling for Montserrat: IDD access code + 1.809.491 followed by subscriber's (four-digit) number. 100 per cent automatic.
Telex: International facilities operated by Cable and Wireless (West Indies) from Plymouth, where public facilities are available.

Banking
Commercial banks: Bank of Montserrat, Barclays Bank International, East Caribbean International Bank Ltd, First American Bank, Guardian International Bank Ltd, Royal Bank of Canada.

Electricity supply: 230 V AC, 60 cycles; also 400 V, 3-phase.

Business advice and information
Barclays Bank, PO Box 131, Plymouth (tel: 2501; tx: 5718; fax: 4913801).

Montserrat Chamber of Commerce and Industry, PO Box 384, Plymouth (tel: 3640; tx: 5743; fax: 4660).

Montserrat Information Unit, Office of the Chief Minister, PO Box 292, Plymouth (tel: 491-2444; tx: 5720 GORTMNI).

Montserrat Tourist Board, PO Box 7, Plymouth (tel: 2230; tx: 5720).

Hotels: Montserrat island
Belham Valley, Old Towne, Plymouth, P.O. Box: 52 (tel: 5553).

Coconut Hill, Plymouth, P.O. Box: 337 (tel: 2144).

Flora Fountain, Plymouth, P.O. Box: 373 (tel: 3444; tx: 5737).

Morocco

David Bradshaw

1988-89 were better years for the troubled Moroccan economy. Lower oil prices and good harvests helped cut the balance of payments deficit while the weakness of the dollar eased the debt repayment burden. Tight controls on government spending and continuing fiscal reform helped to reduce the budget deficit. But despite five years of austerity and reform the economy remained in a precarious state. Particularly worrying was its continued vulnerability to factors beyond the government's control: the price of oil, the level of remittances, variations in rainfall which affect agricultural output. This vulnerability showed up in the first quarter of 1989.

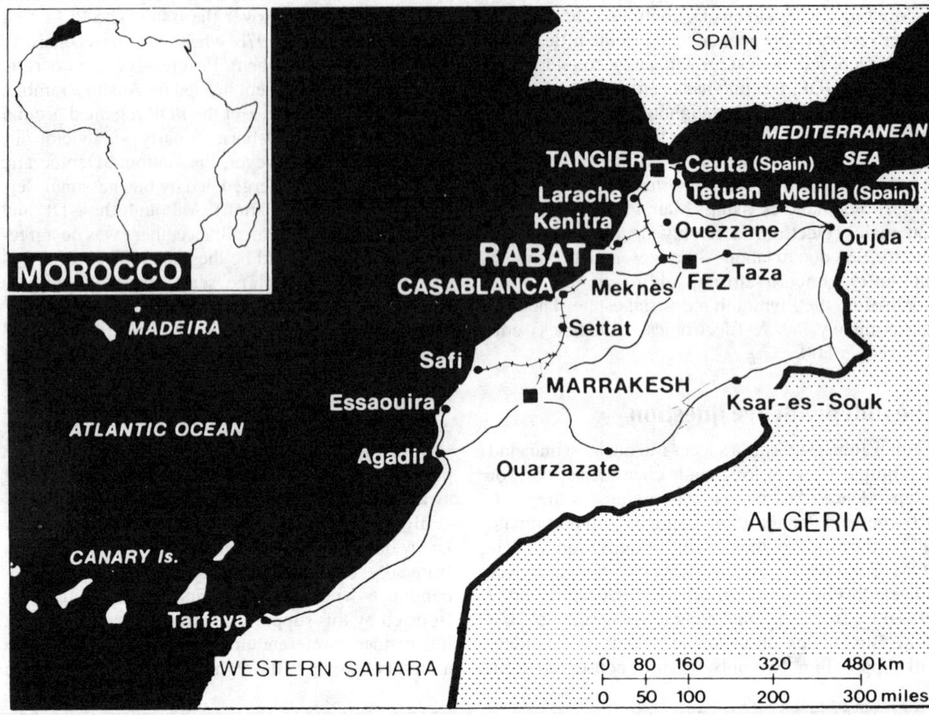

The economy has been in trouble since the early 1980s when it was hit by a series of catastrophic harvests, a depressed world market for its main export, phosphates, the rising cost of oil imports and an expensive war in the Western Sahara. As interest rates and the value of the dollar rose, Morocco was unable to meet repayments on the foreign debt it had accumulated, largely to finance the expansion of the phosphate industry. Since 1983, Morocco has found itself in a debt crisis, forced into debt-rescheduling and an austerity programme prescribed by the IMF.

Emerging from recession

The recent turnaround in the economy can be attributed partly to wide-ranging financial and economic reforms imposed under IMF supervision; and partly to the fortunate coincidence of better rainfall and hence better harvests, and the fall of interest rates, the dollar and the price of oil – all factors beyond the government's control. A good deal of Moroccan macro-economic policy is aimed at reducing the influence of these 'uncontrollable' factors: extending irrigation to reduce the impact of drought and developing domestic energy resources to cut the cost of oil imports.

The government's seriousness about reforming the economy has won it the confidence of the IMF and helped it secure generous rescheduling terms for its foreign debt. While the debt is now some $20bn, – around 120 per cent of GNP – the debt service ratio has fallen from 45 per cent in 1982 to 23 per cent in 1987. In May 1988 Morocco completed a 16-month IMF standby agreement. In August 1988 a new standby loan of SDR220m was approved to run up to December 1989, despite Morocco failing to meet IMF targets for cutting the budget deficit and reducing food subsidies.

In October 1988 the debt was rescheduled at a meeting of Paris Club state creditors to ensure that no more than one-third of export earnings would be allocated to debt servicing.

Austerity and liberalisation

The twin levers of economic policy are a radical squeeze on public spending and a programme of liberalisation. The government's strategy, supported by the IMF and the World Bank, is to enhance the role of the private sector and the free market, streamline the public sector, privatise state land and selected state industries, liberalise trade and implement a development programme aimed at tackling the problems of a weak infrastructure while keeping within strict budget limits. Much progress has been made, but much remains to be done.

Trade has been liberalised so that 90 per cent of imports no longer need prior authorisation. Government spending has been reduced and state revenues increased so that the budget deficit was cut from 12 per cent of GDP in 1982 to 6 per cent in 1987. Tight control of imports and the promotion of exports, together with higher revenues from tourism and remittances, have reduced the balance of payments deficit almost to nothing.

Tax reform is one of the government's most important – and most controversial – financial adjustments. The state needs to widen the narrow tax base to harness more domestic resources for investment. In 1988 tax receipts rose by 20 per cent thanks to higher returns from the value-added tax (VAT) and corporate taxes introduced in 1987. In 1989 profits on construction work became taxable for the first time. There are also proposals to cut back on certain tax holidays. In January 1988 the ministry of finance halved to five years the tax holiday allowed to investors. The powerful business community was outraged. It argues that raising its tax burden works against government efforts to boost exports. The government says five years is still generous and many businesses which benefit have no export potential. The government is now reconsidering the tax-free status granted until the year 2000 to farmers in 1984 to alleviate the effects of drought. A new income tax was adopted by parliament in June 1989, and was due to come into force in 1990. It simplifies the system by rolling the old income tax and a number of other taxes into one.

Privatisation lags behind

Privatisation is another cornerstone of the strategy of liberalisation that has a long way to go. The government says it wants to sell off the 0.4m hectares of farmland held by the state and the reform and sale of many of the 400 state-owned enterprises is a priority of the 1988-93 Five-Year Plan. In 1987, Morocco got a $240m loan from the World Bank to help with the task, but it is taking time to implement. One factor hindering progress is the fear that newly privatised firms will shed labour, deepening an already serious unemployment problem. The state monopoly on fish and food exports has been successfully broken up and turned over to private hands. But it is clearly going to take time to overhaul other state companies to make them attractive enough to find private buyers. The government has drawn up a list of state enterprises it thinks would benefit from privatisation; mining companies and agricultural marketing agencies are among them. But 'strategic' businesses such as the giant

KEY FACTS

Official title: Kingdom of Morocco	**Area**: 711,000 sq km
Head of state: King Hasan II	**Population**: 23.3m (1987)
Capital: Rabat	**GDP per capita**: $720 (1987)
Official Languages: Arabic	**GDP real growth**: 8% (1988)
Currency: Moroccan Dirham (DH) = 100 francs (centimes)	**Inflation**: 2.3% (1988)
	Trade balance: –$752m (1988)
Exchange rate: DH8.35 per $ (Nov 1989)	**Foreign debt**: $17bn (1988)

phosphate monopoly, Office Cherifien de Phosphates (OCP), – Morocco's biggest corporation – the railways, the national airline and the oil and electricity industries are certain to remain in state hands. A privatisation bill went before parliament in mid-1989.

Subsidies on staple foods are a major item of government expenditure which has been reduced under pressure from the IMF and World Bank, although both have expressed impatience at the slowness of the pace. The IMF wanted Morocco to eliminate all subsidies by 1990. Morocco argues that to move too fast risks social unrest. Subsidies on basic goods (flour, sugar and cooking oil) were actually increased in 1989 by over one-third to $156m, possibly in response to the rioting over austerity that shook several Algerian cities in October 1988. The authorities have not forgotten that government attempts to raise prices sparked riots in 1981 and 1984.

At end-1988 the World Bank gave another $200m loan to help with the Structural Adjustment Programme.

Morocco continues to suffer from a shortage of investment. State investment has been cut back as part of the austerity programme; the state says it expects private investors to take over. In 1987 private money accounted for over three-quarters of all industrial investment – some $560m – which went mainly to the textile and leather and agro-industry sectors. Foreign investors have been put off by too much red tape and profit repatriation restrictions. In 1987 foreign investment totalled a pitiful $110m and showed little sign of picking up despite new investment laws. In early 1989 King Hasan ordered the government to cut the red tape that discouraged new investors. A similar pronouncement made some years earlier did nothing to streamline the elephantine bureaucracy.

Morocco relies heavily on outside funding for investment and to finance economic reform. In 1988 the European Community boosted its aid by 60 per cent to $388m over five years; Spain offered aid of $1.1bn over the same period; France provided $132m. The World Bank, the IMF and Arab and African development funds are all putting money into Morocco. Finance Minister Berrada said the financial support was 'an expression of faith in Moroccan stability.'

KEY INDICATORS

	Unit	1984	1985	1986	1987	1988
Population	m	22.9	23.6	23.1	23.3	–
Gross domestic product (GDP)	DH m	107,000	140,000	134,330	140,800	–
GDP per capita	DH	4,000	5,205	6,089	6,046	–
Inflation	%	12.3	7.8	8.7	2.8	2.3
GDP real growth	%	2.4	4.1	5.7	1.0	5.5
Cereal harvest	'000 tonnes	3,707	4,800	7,833	4,347	8,018
Sugar beet crop	'000 tonnes	2,525	2,600	2,625	2,750	2,770
Manufacturing production	1980=100	106.8	107.5	120.0	121.0	–
Mining production	1980=100	115.4	115.4	112.0	110.4	–
Phosphate rock production	'000 tonnes	21,517	20,367	21,427	20,954	24,783
Lead ore	'000 tonnes	144	–	105	105	–
Anthracite	'000 tonnes	838	–	775	634	–
Cement	'000 tonnes	3,496	–	3,742	3,860	–
Exports	DH m	19,111	21,740	22,292	23,390	29,569
Phosphates	DH m	4,619	4,816	3,939	3,080	3,422
Citrus fruits	DH m	1,058	1,493	1,725	1,454	1,610
Imports	DH m	34,397	38,675	34,604	35,271	39,133
Crude petroleum	DH m	8,394	9,931	4,587	5,332	4,380
Balance of trade	$m	–1,408	–1,368	–1,067	–1,069	–752
Current account	$m	–989	–891	–212	175	467
Total external debt	$m	13,969	16,271	18,846	20,706	–
Debt service ratio	%	19.0	24.4	29.0	23.4	–
Tourism receipts	DH m	4,220	6,420	6,730	7,400	–
Exchange rate	DH per $	8.8	10.1	9.1	8.4	8.2

* estimated figure

1988: a very good year

GDP grew an encouraging 8 per cent in 1988 (after the disappointing 1 per cent of 1987) to MD152bn ($18.42bn) thanks largely to an excellent harvest – agricultural output grew by around 10 per cent. At 7.9m tonnes the cereal crop was an all-time high. Manufacturing output grew by just over 4 per cent. One word of caution at this point: official figures grossly underestimate the size of the Moroccan economy. Much economic activity is not formally recorded. The black economy may be as much as one-third the size of the official one.

The 1988 trade deficit was cut to $1.04bn from $1.37bn in 1987 on exports of $3.22bn and imports of $4.26bn. Exports of manufactured goods, fertilisers and phosphoric acid rose while the cost of imports of oil and food fell, thanks to the lower oil price and good harvest. The coverage of imports by exports was almost 75 per cent in 1988 compared with under 48 per cent in 1982. Thanks to a surplus on invisible exchanges of $1.22bn, the result of buoyant tourist receipts and remittances from migrant workers, a tiny current account surplus of $180m was achieved for the second year running. The ratio of budget deficit to GDP was cut to 4.5 per cent, meeting the target set by the IMF.

The central bank predicted that growth in 1989 would be only half that of 1988 because of firming oil prices, rising interest rates and the fact that the 1989 harvest, though very good, was slightly poorer than that of 1988. 1989 cereals output was 7.3m tonnes, the third best on record. The trade deficit was planned at $1bn, but the early signs in 1989 were that it would be considerably worse than that. The first-quarter trade deficit of $560m was 77 per cent bigger than that of 1988 because of a 50 per cent bigger oil bill and a 60 per cent slump in sales of phosphoric acid after India reneged on its purchase contract.

Water for agriculture

Boosting agricultural output is central to the government's strategy. Agriculture employs 42 per cent of the workforce, but its contribution to GDP varies greatly from year to year; 20 per cent in 1986, 15 per cent in 1987. The quality of the harvest has a marked effect on overall economic performance. The series of drought-hit harvests in the late 1970s and early 1980s helped put the economy into recession, while good harvests in four out of five years since 1983 helped propel the economy forward. The 1988 harvest was particularly good; output was 20 per cent up on 1987. That of 1989 was expected to be almost as good as that of 1988. Farming suffers from severe structural problems: small size of holdings, low levels of mechanisation and fertiliser use; a shortage of irrigated land. With the assistance of the World Bank, the African Development Bank and other development funds, the government is trying to modernise Moroccan farming. Agriculture has top priority in the 1988-92 Five-Year Plan, which aims to make Morocco self sufficient. Great emphasis is being put on irrigation. Morocco plans to build one dam a year until the end of the century; the aim is to irrigate 2.4m hectares. In July 1989 contracts were signed with the Soviet Union to build the Mjara dam north east of Rabat. The dam will be Africa's second largest after Aswan, will provide drinking water to Rabat and other northern cities, irrigate 100,000 ha and double electricity production, helping to cut dependence on oil imports. The $1bn project, to be part-financed by Arab funds, will begin in late 1989. Completion of the first phase will be in 1997. 35 per cent of the Soviet contract will be paid with farm products and citrus, the rest through concessionary loans provided by Spain.

The service sector contributes some 35 per cent to GDP. Manufacturing industry, which is reasonably diversified, contributes around 18 per cent to GDP, but its importance is growing. Government policy is to encourage export-oriented industry. Investment in manufacturing in 1988 was up 23 per cent on 1987, almost two-thirds of it going to small and medium enterprises. Over three-quarters of the funds came from private domestic investors, 20 per cent from overseas (France leading the way with over one-third of foreign investment, Spain next with 9 per cent) and only 2 per cent from the state, whose policy is to let the private sector fund industrial growth. The most popular sector with investors was textiles and leather, which took almost 40 per cent of the total; chemicals and related industries took 26 per cent and agribusiness and food processing 18 per cent. The textiles and leather sector continues to boom; 1988 exports were worth $730m.

The state invested heavily in the phosphate industry during the 1980s in order to capitalise on its possession of the largest phosphate reserves in the world. This has meant, on one hand, expanding the production capacity of raw phosphate rock and, on the other, moving downstream into the production of derivatives like phosphoric acid and fertilisers, which are not only more valuable than raw phosphates but are also less vulnerable to fluctuations in world demand. Unfortunately for Morocco, the expansion took place alongside a steady decline of the world phosphate market. The world market for raw phosphates has been stagnant for years; rock exports rose in 1988 but their value fell. Exports of phosphoric acid and fertilisers continue to grow, in value as well as quantity. But Moroccan planners face criticism that their investment in phosphates has not been justified by results and robbed other more promising industries of scarce resources.

Tourism earned Morocco more than $1bn in 1988 making it the second biggest foreign currency earner after remittances from workers abroad, which brought in $1.37bn in 1988. Tourism is one of the fastest growing industries, and is expected to overtake remittances in the 1990s as jobs in Europe begin to dry up. Tourism creates 50,000 jobs directly and another 150,000 indirectly, but there is much room for expansion. Morocco has only 120,000 beds in officially classified hotels. The 1988-92 Plan calls for 40,000 new beds. The industry is an important focus of domestic and foreign investment. Twelve new hotels opened in 1988. Most of the 2.4m tourists in 1988 went on inland tours to Morocco's ancient cities, but Morocco sees its future as a package holiday destination for Europeans and Americans. Morocco aims for 20m tourists a year within 20 years.

Rapid population growth presents Morocco with a major economic and political challenge. The population is growing at a rate of 2.6 per cent per year. 60 per cent of its people are under the age of 25. Half still live below the poverty line. Unemployment is running at at least 20 per cent. Jobs cannot be found for more than half the quarter million young people coming on to the jobs market each year. More investment in job creation is needed if Morocco is to meet the ever-growing demands for more schools, jobs and homes. For several years Morocco has been treading a tightrope between the demands of the IMF and the needs of its poor majority. Social peace may remain at risk as long as so many people have so little, and wealth is so unevenly spread. One fortunate indicator for the Moroccan authorities is that inflation has been limited in recent years to between 2-3 per cent, and looked like staying at this level in 1989.

Curbing extremism

King Hasan has occupied the throne for almost 30 years. At times his grasp on power has seemed distinctly shaky; he has survived coup attempts in 1971 and 1972, a left-wing uprising in 1973 and several assassination attempts as well. But in 1989 he looked as secure as ever.

Morocco likes to describe itself as a constitutional monarchy, but the king has far greater powers than his peers in Britain or Spain. Political parties sponsored by the king dominate parliament after the 1984 election which the oppositon claimed was rigged. The king appoints ministers and they are answerable to him. The small number of legal opposition parties, from far left to extreme right, have little support among the population. They are also not allowed to challenge many aspects of government policy, such as foreign affairs, which is the exclusive preserve of the king and a coterie of close advisers. Moroccan democracy is rather limited.

A danger is that Moroccans, denied a real voice in politics, will turn to the illegal opposition and extremism. This has not yet happened. But the existence of the danger was indicated by disturbances in 1989 among students at the universities of Meknes, Fez, Oujda and Tetouan. The highly efficient intelligence and security services have managed so far to outmanoeuvre the minority of extreme left-wingers and Muslim fundamentalists; the attraction of the latter is reduced by widespread respect for the king's spiritual role as 'commander of the faithful'. But opposition to the regime could well grow if there is no significant improvement in the well-being of the mass of the population, half of which lives below the official poverty line, or if ongoing economic and financial reforms cause standards of living to drop (through the withdrawal of food subsidies, for example). Left-wing parties have become more vociferous in their attacks on government social and economic policy saying that long years of austerity have increased poverty and unemployment, which is put at 1m. They also criticise the growing gap between the wealthy elite and the mass of the poplation. Low wages sparked a number of strikes, notably at the big coal mine at Jerada in the north east, where strikers stayed away from work for two months to February 1989.

The government has faced more criticism over the detention of political prisoners. In March 1989 the Moroccan Association of Human Rights, which was established with the reluctant consent of the authorities at the end of 1988, called for a general amnesty for political prisoners and political exiles. Three political detainees, gaoled after the food riots of January 1984, staged a hunger strike in support of a demand for 'political prisoner' status that attracted a good deal of bad publicity. In May 1989 King Hasan released 228 detainees in an end of Ramadan amnesty. Around 250 others are estimated to be still held.

King Hasan tried to weld the nation behind the drive to pay for the vast Great Hasan II Mosque being built by the Atlantic coast in Casablanca. The mosque, which extends on stilts over the Atlantic breakers, will be one of the biggest in the world, with room for 80,000 worshippers. Its 150-metre high minaret is equipped with a laser beam aimed at Mecca. Moroccans were asked to donate whatever they could afford to the scheme, whose total cost is estimated by some at $500m. Some Moroccans complained of undue pressure from official collectors; others questioned the expenditure of such a sum by a financially pressed nation.

Building Maghreb unity

Relations with socialist neighbour and long-time rival Algeria underwent a spectacular improvement since diplomatic relations were restored in May 1988 after a 12-year rift over the Western Sahara conflict. In February 1989 Algerian President Chadli Bendjedid made the first visit of an Algerian president to Morocco for 14 years. In March King Hasan said parliament would finally ratify the border treaty drawn up with Algeria in 1972. Non-ratification had been a long-standing Algerian grievance. There remained some concern, however, that as long as the Western Sahara conflict defied a solution, it contained the potential to damage the Morocco–Algeria rapprochement.

The Morocco–Algeria rapprochement allowed them and three other Maghreb nations – Tunisia, Mauritania and Libya – to forge ahead with plans to set up a regional common market. The idea of Maghreb unity first appeared when Morocco, Algeria and Tunisia won independence but it failed to take concrete shape because of political quarrels. The Arab Maghreb Union (AMU) was inaugurated in Marrakesh on 17 February 1989, with King Hasan as the first incumbent of the presidency, which will rotate on an annual basis. The AMU was intended to develop economic relations between the member states and enhance the region's negotiating power *vis à vis* other world blocs, such as post-1992 Europe. Members recognised that integrating their very different economic systems would not be easy.

Continued political harmony between Morocco and Algeria was the key to the AMU's success or failure. Both states hoped to reap economic benefits from extending bilateral links under the AMU umbrella, possibly by exchanging Moroccan food and agro-industrial expertise for Algerian energy and heavy industrial products. The two countries planned a number of ambitious economic projects, chief of which was a scheme to build a pipeline to carry natural gas from Algeria to Spain by way of Morocco and the Straits of Gibraltar. Such developments would give both states a concrete interest in burying past political differences.

King Hasan recovered the respected and influential position in Arab politics which was compromised in 1986 by his controversial meeting with the then Israeli Prime Minister Shimon Peres. The king has long been a leading member of the moderate camp in Arab politics. He helped reinforce the moderates in May 1989 by hosting in Casablanca the Arab summit that saw the triumphant return of Egypt to the Arab League and gave Arab backing for the PLO's new moderate posture. An Arab League committee of King Hasan, President Chadli and King Fahd of Saudi Arabia was given a six-month mandate by the summit to bring peace to Lebanon. It was an impossible task. The committee admitted defeat after only two months.

Morocco continued to cultivate close links with Europe, which it believes are at least as vital to Morocco's future as those with the Arab world and Africa. In 1986 Morocco unsuccessfully applied to join the European Community. France remains Morocco's closest European ally, but King Hasan gave British Prime Minister Margaret Thatcher a particularly warm welcome in March 1989. Germany made special efforts in 1988-89 to improve its commercial links. Normally excellent relations with Spain were tarnished by Spanish backing for a UN call for Morocco to hold direct talks with the Polisario Front. The friendship with the United States was slightly clouded by Washington's unhappiness over Libya's presence in the AMU, but its aid, some $120m per year, continued unabated. Though staunchly pro-Western, Morocco kept up good relations with the Eastern bloc, as was reflected by the agreement with the Soviet Union to build the Mjara dam.

Morocco once ritually cut diplomatic ties with states which recognised the self-styled Saharan Arab Democratic Republic (SADR) proclaimed by the Polisario Front. This policy has changed. Diplomatic ties were restored in 1988 with Yugoslavia, one of 71 states to recognise the SADR. King Hasan expressed a wish to rejoin the Organisation of African Unity, which Morocco left after it admitted the SADR as a member in 1984. But this may have to await a settlement in Western Sahara.

Western Sahara

In early 1989 many Maghreb-watchers believed that the Western Sahara conflict was edging towards a solution. They were disappointed, though their optimism had been well founded. Polisario's principal backer is Algeria and its fighters operate from bases in south-west Algeria. In 1988 Algeria and Morocco put many of their political differences to one side for the sake of efforts to build regional economic unity. And when, in February 1989, the five-nation Arab Maghreb Union was formally inaugurated, a settlement in Western Sahara seemed only a matter of time. But while peace remained on the horizon it refused stubbornly to get any closer. Possibly as a result of Algerian pressure, Polisario virtually abandoned military action by the start of 1989. But the expectation that Algeria would push Polisario into a political settlement against its will was not fulfilled.

The former Spanish colony of Western Sahara was annexed by King Hasan in 1976, on the grounds that it was part of 'historic' Morocco hived off by European colonists. This is contested by the Polisario Front which argues that the territory's inhabitants, the Sahrawis, are a distinct people with a right to decide their own future. For Polisario, Western Sahara is a 'decolonisation' issue. Polisario's view is accepted by the UN and the OAU; its self-declared state, the Saharan Arab Democratic Republic, is recognised by 71, mainly Third World, countries and is a full member of the OAU. But Morocco has retained control of the territory by force and because no major power, in Europe or elsewhere, has much interest in creating a scantily populated desert state of doubtful viability in north west Africa. Determined to hold on to the territory, Morocco long resisted pressure for a negotiated settlement. It agreed in 1981 to the principle of a referendum among the Sahrawis but for years did nothing to help arrange one.

In August 1988 came a breakthrough. After 13 years of bitter and at times bloody fighting, Morocco and Polisario accepted, with reservations, a United Nations peace plan for a cease-fire and a referendum among Sahrawis on the territory's future. Each side seemed convinced that the vote would be settled in its favour; but each vowed to accept an unfavourable result. Agreement was reached on the question to be put ('Do you want independence or to be part of Morocco?') and to whom (the 74,600 Sahrawis listed in the 1974 Spanish colonial census). But the two sides could not agree over the conditions under which the referendum was to be held. Polisario argued that the vote could not be 'free and fair' without the withdrawal of Moroccan troops (around 100,000 strong), administrators and several thousand settlers and the reinstatement of Spanish law. Morocco offered merely to restrict its troops to barracks.

Polisario also insisted on direct talks with Morocco to solve outstanding problems and build mutual trust. Direct talks had long been urged by the UN and the OAU, but had been rejected by Morocco which argued that Polisario fighters were merely Algerian mercenaries. But in January 1989 King Hasan finally met a Polisario delegation at his palace in Marrakesh. The two sides viewed the talks in very different ways, however. Polisario saw them as 'negotiations'; Morocco as a dialogue between the king and a group of 'wayward subjects'.

The breakdown of peace moves quashed strong but unconfirmed reports that the two sides had agreed to revive an old compromise settlement giving Western Sahara autonomy within a federated Moroccan state. In February 1989 Polisario leaders firmly rejected any 'third way'.

The conflict thus seemed set to grind on, though at a lower intensity than before. Polisario has not seriously challenged Moroccan control over Western Sahara since the army built the network of defensive barriers which, when a sixth wall was completed in early 1987 enclosed 80 per cent of its area. The barriers steadily marginalised the threat from Polisario, allowing Morocco to get on with an investment and construction programme designed to win the hearts and minds of the Sahrawis and bind the territory irreversibly to Morocco. There was little fighting in the territory after the summer of 1988. Polisario did, however, launch a biggish attack in September 1988 against the defensive wall near Umm Dreiga, leaving dead at least 50 Moroccan troops (and an unknown number of its own). The attack was an embarrassment for Algerian President Chadli Bendjedid; it came just as the Algeria-Morocco rapprochement was making real progress. The attack was probably sanctioned by the hardline faction in the Algerian leadership seeking to undermine Mr Chadli and his reforms. At the end of 1988 the Algerian president strengthened his position and purged the hardliners; since then, Polisario seems to have been under restraint. Algeria's relations with Polisario were complicated by the treaty of the AMU. This stipulates among other things, that no member state may tolerate on its territory any activity or organisation threatening another member state.

The strains of peacemaking after 14 years of war and the uncertainty over Algerian support set up deep tensions within Polisario. At its May 1989 congress, a number of senior officials were removed from their posts. One of them, Omar Hadrani, defected to Morocco in August 1989.

In June 1989 UN Secretary General Perez de Cuellar made a regional tour to try to give the stalled peace process new momentum and declared that he had been encouraged by the will of all the parties to pursue peace. Outsiders saw little progress.

COUNTRY PROFILE

Political system: Constitutional monarchy. Executive power rests with the prime minister and cabinet, who are appointed by the King. Wide powers are conferred on the King who has the right to dismiss the legislative assembly. The legislature is the unicameral Chamber of Representatives with 306 members elected for six years, two-thirds by universal adult suffrage and one-third by an electoral college. Morocco gained independence in 1956.

Political parties: The ruling coalition comprises Mouvement Populaire (MP), Rassemblement National des Indépendants, Parti des Indépendants Démocrates and Union Constitutionnelle; others include Parti de l'Istiglal, Parti du Progrès et du Socialisme and Union Socialiste des Forces Populaires.

Population: Morocco's population of 23.3m (1987 estimate) is 45 per cent urban, and 55 per cent are aged 20 years or younger. Practically the whole indigenous population is Muslim, although there is a small Jewish group in Casablanca and other major cities.

Main cities/towns: Casablanca (population 2.6m), Rabat (capital, 591,000), Marrakesh (1.39m), Meknès (686,000), Salé (483,000), Tangier (492,000), Oujda (869,000), Fez (906,000), Agadir (672,000).

Language: Arabic is the official language; French is the dominant business language in the centre and south, while Spanish is often spoken in the north. Berber is the most widely used in everyday life. English is seldom spoken.

Media
Press: Principal newspapers include *Le Matin*, *Maroc Soir*, *L'Opinion* (French) and *Al Alam* and *Al Ittihadal Ishtiraki* (Arabic). Numerous periodicals are available in French and Arabic, including the business weekly *La Vie Economique*.
Radio: Extensive national and external services in Arabic, French, English, Spanish, Berber.
Television: Over 60 hrs/week, with advertising, in Arabic, French. Nationwide network is basically state controlled, but a second, private, channel was scheduled for launch in March 1989.

External trade: Although exports steadily increased during the 1980s, the trade deficit remains large but is offset by invisible earnings from tourism and especially by workers' remittances from abroad. Current account deficit fell from $891m in 1985 to $212m in 1986 and was expected to balance in 1988.
Exports: Phosphates (typically 24-26 per cent of total value), phosphoric acid (14-18%), citrus fruit (4-6%), clothing (4-6%), fertilisers (3-5%), fish (shell, canned, fresh) (6-9%), carpets (2-3%), tomatoes, canned vegetables. Main destinations: France (typically 22-26% of total value), Spain (9-10%), German FR (6-9%), India (6-8%), Italy (5-8%), Netherlands, Belgium/Luxembourg, UK, Turkey, Japan.
Imports: Machinery and transport equipment (typically 18-22 per cent of total value), crude oil (15.6% in 1986), cereals (typically 5-7%), other foodstuffs (3-5%), chemicals (5-7%), iron and steel (3-4%). Main sources: France (typically 18-20% of total value), Saudi Arabia (13-15%), USA (11-12%), Spain (7-10%), Iraq (7-8%), German FR (4-6%), USSR, Japan, Kuwait.

Agriculture: Sector contributes 15-20 per cent of GDP, employs over 50 per cent of the workforce and accounts for some 25 per cent of exports. Around 70 per cent of the 7.9m ha of arable land is cultivated, mainly by peasant smallholders. Principal crops are wheat, barley, maize (grown in the rain-fed areas) and citrus fruit, beans, chick peas and other pulses, tomatoes (mainly for export), potatoes, olives and oilseeds. Important agro-allied crops are sugar-cane, sugar-beet production being developed to cut down on sugar imports and cotton.Livestock productivity and crop yields have remained low and regular food imports (particularly grain) are necessary to meet domestic food requirements, but in 1986-87 poor rains reduced the cereal crop from 7.7m to 4.2m tonnes. The 1987-88 crop was affected by locust swarms. Agricultural development has also been hampered by the small size of the majority of holdings and restricted access to the EC market. However many, long-term government projects are underway, including irrigation schemes, development of new techniques, and financial incentives to farmers. The fishing sector offers considerable potential, though export markets for the main catch, sardines, are restricted by strong competition from Spain, Portugal and France. Modernisation of the fleet and fishing ports is underway to further the potential of the rich fishing grounds. The local fishing fleet comprises around 2,100 vessels, of which 85 per cent are coastal boats. Around half of the 539,000 tonnes catch (1986) is landed at Agadir.

Industry: Sector accounts for 17 per cent of GDP and is increasing output by 3-4 per cent p.a. Most activity is concentrated in Casablanca. The main industry is the processing of phosphates into phosphoric acid and fertilisers (five plants currently operating and a sixth due for completion around 1990). Followed by oil refining (two refineries at Sidi Kacem and Mohammedia – combined capacity 74,000 b/d), cement, food processing, textiles, paper and timber, metals, rubber, plastics and vehicle assembly. Industrial development has switched in recent years away from import substitution and towards encouraging the manufacture of goods for export, support for small and medium-sized producers, devolution of spending powers to local authorities, and investment in other areas of the country away from Casablanca. Several of the largest state-owned companies are due for privatisation in 1988-89. Total industrial investment rose by some eight per cent in 1987, of which the private sector contributed 72 per cent.

Mining: World's third largest producer (15 per cent of world total) and largest exporter of phosphate rock with estimated reserves of 58bn tonnes. Production of phosphate rock has stagnated since 1981 (down by 0.9 per cent in 1986) and the industry has been geared more towards local processing (mainly fertilisers, phosphoric acid), and sales of phosphates and their derivatives account for around 45-50 per cent of visible export earnings. Responsibility for their extraction, marketing and export is vested in the parastatal Office Cherifien des Phosphates (OCP). Development of reserves at Meskala, Sidi, Hajjaj and Ben Guerir has boosted capacity to 32m tonnes, but production is currently around 21m tonnes (1986). Although phosphates account for 92 per cent of mineral production smaller quantities of other minerals are produced including 500,000 tonnes of anthracite.

COUNTRY PROFILE

Energy: Heavy reliance on imported oil, which accounts for over 80 per cent of total energy requirements. Domestic crude oil production is very low compared with imports, and various exploration projects are under way in the Atlas mountains, the south and east of the country, and offshore. Big shale oil deposits at Timadhit and Tarfaya (estimated at 200bn tonnes) are currently uneconomic. Proven recoverable reserves of gas are estimated at 4bn cu metres. The main producing field is at Essaouira with an annual output of around 60m cu metres. Electricity production depends heavily on imported oil, with hydro output falling due to the drought. Development of coal fired power stations is underway with a 1,300 mW plant now completed at Mohammedia, and a similar station planned at Jorf Lasfar.

Membership of international organisations: African Development Bank, Arab Bank for Economic Development in Africa (subscriber), Arab Fund for Economic and Social Development, Arab League, GATT, IAEA, IBRD, IDA, IFAD, ILO, IMF, Islamic Conference, Islamic Development Bank, UN, WHO, WIPO, World Bank.

Time: GMT.

Climate: Varies with area – Mediterranean on the coast, hotter and drier inland, Saharan in the south. Summer from May to Oct; dry and hot with tempratures 23-28°C on the coast, 30 to 45°C inland. Winter from Nov to Apr; light rain on coast with average temperature 15 to 21°C, dry inland with temperature 20 to 30°C.

Entry requirements
Passport: Required by all.
Visa: Not required for periods of up to 90 days by nationals of France, German FR, Japan, UK, USA, Denmark, Norway, Sweden, Switzerland, Finland, Australia and several other countries.
Prohibited entry: Nationals of Israel, South Africa.
Currency: Import/export of local currency is prohibited. Unlimited import of foreign currency, although amounts over the equivalent of DH15,000 should be declared. Export of foreign currency allowed up to amount declared on arrival. Up to one half of the Moroccan currency purchased by a visitor during his stay may generally be re-exchanged to foreign currency on production of the relevant receipts.

Health precautions
Mandatory: Vaccinations against yellow fever, if coming from an infected area.
Advisable: Typhoid, cholera, tetanus and polio vaccinations recommended. Anti-malaria precautions should be taken.

Representation overseas: Abidjan, Abu Dhabi, Amman, Ankara, Baghdad, Bamako, Beijing, Beirut, Belgrade, Berne, Bonn, Brasilia, Brussels, Buenos Aires, Conakry, Copenhagen, Dakar, Damascus, Doha, The Hague, Islamabad, Jeddah, Kinshasa, Khartoum, Kuwait City, Lagos, Libreville, Lisbon, London, Madrid, Monrovia, Moscow, Nairobi, New Delhi, Nouakchott, Ottawa, Paris, Rome, Tokyo, Tripoli, Tunis, Washington, Yaoundé.

Air access
National airline: Royal Air Maroc.
Other airlines: Aeroflot, Air Afrique, Air Algerie, Air France, Air Mauritania, Alitalia, Balkan, GB Air, Iberia, Iraqi Airways, KLM, Kuwait Airways, Libyan Arab, Lufthansa, Royal Jordanian, Sabena, Saudia, Swissair, TAP Air Portugal, Tarom, Tunis Air.
International airports: Agadir-Inezgane (Code: AGA), 8 km from city, bar, buffet, bank, car hire; Casablanca-Mohammed V (Code: CMN), 30 km from city, bar, buffet, restaurant, bank, post office, car hire, Fez-Sais (Code: FEZ), 12 km from city; Rabat-Salé (Code: RBA), 10 km from city, buffet, restaurant, bank, post office, shops, car hire; Tangier-Boukhalef Souahel (Code: TNG), 15 km from city, restaurant, bank, car hire; Marrakesh (Code: RAK), 6km from city.

Surface access
Main ports: Agadir, Casablanca (major freight port), Jorf Lasfar, Kenitra, Mohammedia, Nador, Safi, Tangier (main passanger port). Regular car ferry and hydrofoil services connect Spain, France and Gibraltar with Tangier and the Spanish administered ports of Ceuta and Melilla.
Overland access: Road access is possible from Algeria via Oujda. The international rail link is also via Oujda.

Hotels: Inexpensive and widely available. Two main types: graded hotels (which are given a one-to five-star rating by the Tourist Board) and small (and usually old) unlisted hotels. Advisable to book in advance. A service charge and local tax is normally added to bill.

Electricity supply: 110-127 V or 220 V AC, 50 cycles.

Credit cards: Major credit cards widely accepted.

Car hire: Widely available but expensive. Major hire companies operate from Agadir, Casablanca and Tangier. National or international driving licences are accepted.

City transport
Taxis: Grands taxis, seating up to six persons, operate along specific routes. Petit taxis are metered and operate within cities. Fares vary considerably and taxi meters rarely work. For fares outside of towns fares should be agreed in advance. 10 per cent tip is usual.
Buses: Agadir, Casablanca, Tangier and other main towns have good bus services. Tickets can be bought in advance of journeys.

National transport
Air: Royal Air Maroc operates domestic services to main centres.
Road: Over 29,000 km of surfaced roads – links between main centres generally good; some of the 30,000 km country roads need care/local knowledge. High Atlas mountains may be impassable in winter. Improvements in progress.
Buses: Frequent, cheap service between towns; long-distance services include: Tangier-Oujda; Fez-Marrakesh; Agadir-Casablanca; Tangier-Casablanca. Advisable to book in advance.
Rail: Limited (1,900 km) but fairly efficient network operated by ONCF. Fares are cheap. Three classes available. Air-conditioning, air-conditioned sleeping cars – couchettes, restaurant cars available; supplements may be payable. Routes include: Oujda-Fez-Rabat-Casablanca; Casablanca-Marrakesh-Agadir; Tangier-Fez-Casablanca.

Public holidays
Fixed dates: 3 Mar (Fête du Trone), 1 May, 23 May (Fête Nationale), 14 Aug (Oued-el-Dahab), 6 Nov, 18 Nov (Independence).
Variable dates: Eid al-Fitr, Eid al-Adha, Hijra, Prophet's Birthday.

Working hours
Government and business: Government: Winter (Mon-Fri) 0800 – 1200 and 1430 – 1800. Summer (Mon-Sat) 0800 – 1600. Ramadan (Mon-Sat) 0900 – 1500. Business: Winter (Mon-Fri) 0800 – 1200 and 1400 – 1800/2000. Summer hours vary, some work (Mon-Fri) 0800 – 1500/1600, others revert to winter hours. Ramadan (Mon-Fri) 0900 –1500/1600.
Banking: Winter: (Mon-Fri) 0800 – 1130 and 1400 – 1830. Summer (Mon-Fri) 0800 – 1600. Ramadan (Mon-Fri) 0900 – 1530.

Telecommunications
Telephone: International calls in most cities can be made direct. Elsewhere calls can be made at Central Post Office (PTT). Dialling for Morocco: IDD access code + 212 followed by area code (8 for Agadir, 7 for Raba, 4 for Marrakesh) followed by subscriber's number). No area code is needed for Casablanca. A major expansion of services was planned for 1989-90.
Telex: Available at good hotels.
Fax: Available at good hotels.
Telegram: Services at main post offices.

Postal service: Postal system is unreliable.

Banking
Central bank: Bank al Maghrib.
Main centre: Casablanca. *Head offices:* Algemene Bank Marokko, Arab Bank (Maroc), Banque Centrale Populaire, Banque Commerciale du Maroc, Banque Marocaine du Commerce Extérieur, Banque Marocaine pour l'Afrique et l'Orient, Banque Marocaine pour le Commerce et l'Industrie, Banque Nationale pour le Développement Economique, Caisse Nationale du Crédit Agricole, Citibank (Maghreb), Crédit du Maroc, Crédit Immobilier et Hôtelier, Crédit Populaire du Maroc, Société de Banque et de Crédit, Société Générale Marocaine de Banques, Société Marocaine de Dépôt et Crédit, Union Bancaria Hispano Maroqui, Union Marocaine de Banques, Wafabank.

Trade fairs: The main centre is Casablanca. A number of fairs of international importance are held every year, including Salon International du Textile et du Cuir (SATEC), Salon du IMME, Salon du Livre (books), Salon du BTP et Artisanat.

Representation in capital: Argentina, Austria, Belgium, Brazil, Bulgaria, Canada, Central African Republic, Chile, China PR, Czechoslovakia, Denmark, Equatorial Guinea, France, Gabon, German DR, German FR, Greece, Guinea, Hungary, India, Iraq, Italy, Côte d'Ivoire, Japan, Mauritania, Netherlands, Nigeria, Norway, Oman, Pakistan, Philippines, Poland, Portugal, Qatar, Romania, Saudi Arabia, Senegal, Spain, Sudan, Sweden, Switzerland, Tunisia, Turkey, UAE, UK, USA, USSR, Yemen AR, Yugoslavia, Zaïre.

Hotels: Casablanca
Holiday Inn, Rond Point Hassan II, Crowne Plaza (tel: 26-87-13, 22-35-27).

Hyatt Regency, Place Mohammed V (tel: 26-41-67; tx: 22628; fax: 269645).

Les Almohades, Ave Moulay Hassan er (tel: 22-05-05; tx: 24927).

Royal Mansour, 27 Ave des FAR (tel: 31-30-11; tx: 22912).

Safir, 160 Ave des FAR (tel: 31-12-12; tx: 24896/24897).

Sheraton, 100 Avenue des Far (tel: 31-78-78).

Tarik, 41 Boulevard de la Corniche (tel: 36-70-73; tx: 22828).

Hotels: Rabat
Ex-Sofitel Rabat, Place Sidi-Makhlouf (tel: 310-91; tx: 32086; fax: 22155).

Hyatt Regency, Aviation Souissi, P.O. Box: 450 (tel: 721-51, 712-34; tx: 31913; fax: 72492).

Rabat Chellah, 2 Rue d'Ifni (tel: 640-52; tx: 31615).

Banking
Bank al-Maghrib, 277 Avenue Mohammed V, PO Box 445, Rabat (tel: 205-31; tx: 31006, 31909).

Banque Marocaine du Commerce Extérieur, 140 Avenue Hasan II, PO Box 13.421, Casablanca 01 (tel: 27-20-49, 27-30-90; tx: 21708, 21805 bmce).

Other useful addresses
British Chamber of Commerce for Morocco, 291 Blvd Mohammed V, Casablanca (tel: 30-37-60).

Chambre de Commerce et d'Industrie de la Wilaya de Grand Casablanca, 98 Blvd Mohammed V, PO Box 423, Casablanca (tel: 26-43-27; tx: 24630; fax: 26-84-36).

Mozambique

Jan Marsh

The Economic Recovery Programme (PRE) adopted by the Frelimo government at the start of 1987 in agreement with the International Monetary Fund (IMF) and the World Bank has succeeded in stemming the economic haemorrhage caused by the South-African backed insurgents of the Mozambique National Resistance (MNF) although it is unlikely that productive capacity can be regained as long as the war continues. Cautious optimism surrounded tentative diplomatic steps towards a cease-fire during 1989. Realism was the keynote of the Frelimo Party's Fifth Congress, held in July 1989, which introduced important political, social and economic changes, further modifying Mozambique's 'Marxist-Leninist' policies and aimed at liberalising state control, facilitating private enterprise and encouraging international investment and aid, without abandoning the socialist principles of central planning, free health care and free education.

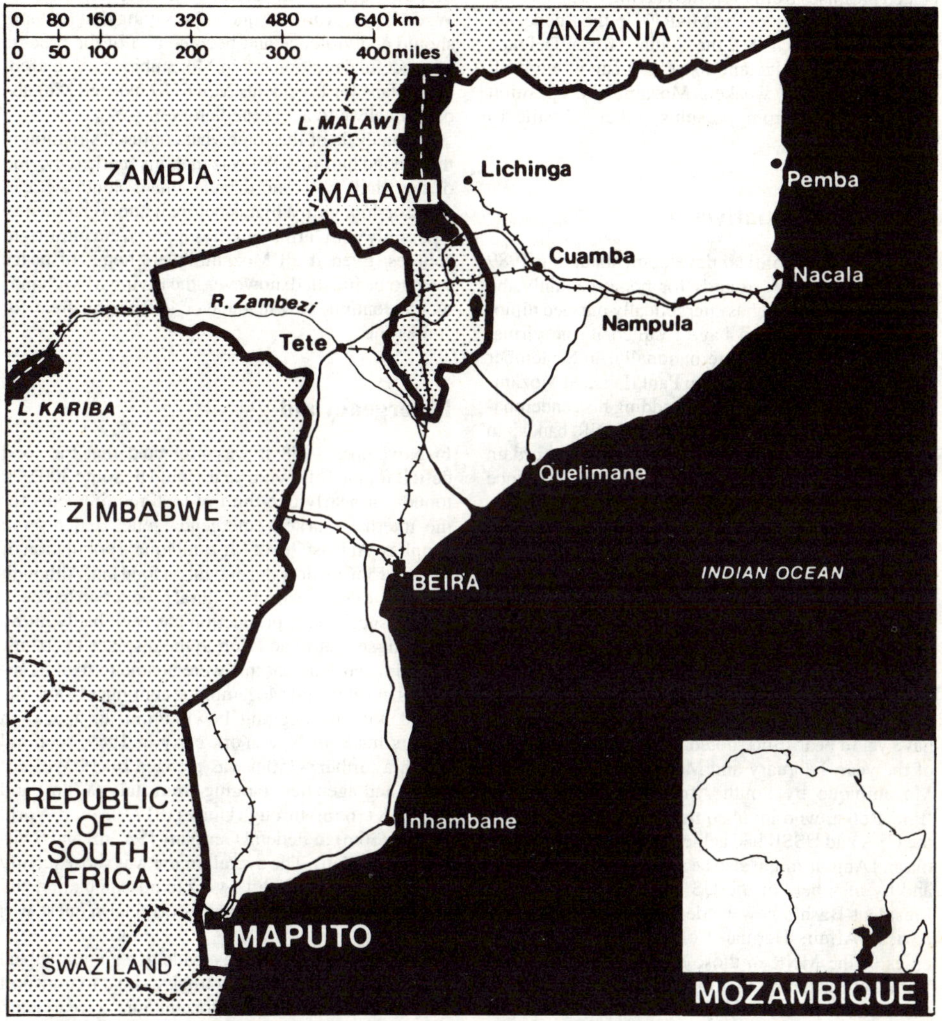

Security situation

'When two elephants fight, it's the grass that suffers' is an African saying cited by Jorge Rebelo, Frelimo secretary for ideological work when recalling the comment of Julius Nyerere, former president of Tanzania, that 'even when elephants make love, the grass still suffers'. The allusion was to the changing superpower relations of 1988-89 and their impact on Southern Africa, in respect of the conflicts caused by South African destabilisation of neighbouring territories and in particular to destructive attacks in Mozambique on economic and civilian targets carried out by the anti-government armed groups of the MNR, who style themselves Renamo and are known by a variety of names in Mozambique, of which the most polite is simply 'the bandits'.

These attacks continued throughout 1989, despite the efforts of the Mozambica Armed Forces (FPLM) who nevertheless succeeded in clearing MNR out of some areas, capturing a number of bases and keeping the Beira corridor generally open to traffic. However, the MNR mounted attacks on provincial centres such as Vila Nova (on the Malawi border), Moamba and Massinga (Inhambane province), Manjacaze (Gaza), as well as numerous attacks close to the capital Maputo. Attacks on the power lines built to carry electricity from the giant Cahora Bassa scheme on the Zambezi river south to the industrial areas of South Africa also continued: between April and December 1988 a total of 674 pylons were destroyed by MNR activity amd other reports estimated that one third of the total of 4,000 pylons were down. Acts of sabotage on other powerlines were sporadically reported, and the MNR continued to threaten transport networks, and to menace rural villages.

The result, as President Joaquim Chissano described in his report on the security situation to the Fifth Congress, was that as many as 1.7m people have been driven from their homes inside Mozambique, and a further 700,000 displaced people have fled to neighbouring states for safety; some are now returning but all remain vulnerable and in need of emergency aid. Conservative estimates are that 100,000 people have been killed since the beginning of 1987, and that half a million more have died of famine and disease directly occasioned by the MNR destruction of food and shelter. Over 1,000 villages have been razed, destroying over 800 rural health posts and over 2,500 village schools. In addition, around 1,000 shops or trading posts, 3,500 trucks and tractors, 44 factories and numerous locomotives and rail trucks have also been destroyed. The human and economic cost to the country is extreme. The Population Crisis Committee in Washington puts Mozambique at the top (or bottom) of its index of human suffering, and infant mortality is now running at 35 per cent.

By the second half of 1989, the extent of MNR activity appeared to be diminishing slightly, though still widespread. At the end of 1988, the detachment of Tanzanian troops which for two years assisted the FPLM in Zambezia province returned home; the Zimbabwean forces helping to secure the road, rail and oil pipeline between the Mozambique port of Beira and Zimbabwe remained in place and there is no

KEY FACTS

Official title: People's Republic of Mozambique

Head of state: President Joaquim Chissano

Ruling party: Partido Frente de Libertação de Moçambique (Frelimo)

Capital: Maputo

Official Languages: Portuguese

Currency: Metical (MT) = 100 centavos

Exchange rate: MT817 per $ (Nov 1989)

Area: 799,380 sq km

Population: 14.7m (1988)

GDP real growth: 3.1% (1987)

GNP per capita: $95 (1988)

Trade balance: –$549m* (1987)

Foreign debt: $3.2bn (1988)

* estimated figure

immediate likelihood of their withdrawal. Military guards are still necessary on the railway, and armed convoys on most major roads.

The climate of coexistence sought by Mozambique in signing the Nkomati agreement with South Africa in 1984 has therefore not been attained, despite repeated promises from Pretoria. Having taken over the training and supply of the MNR from the Rhodesian security forces in 1980, the South African Defence Force appears reluctant to relinquish its control, which successfully weakens Mozambique's position and prevents it from pursuing policies hostile to Pretoria's interests.

Diplomatic initiatives

However, some political developments during 1989 gave Mozambique grounds for hope; certainly the Frelimo government has energetically pursued diplomatic activities which have resulted in the virtual isolation of the MNR internationally. In September 1988, for instance, Pope John Paul II visited Mozambique on his tour of the region, adding his condemnation of the terror inflicted by the MNR bands; in exchange, the Catholic church was allowed to ordain Mozambican priests, a process that had ceased since independence although in general the Catholic hierarchy maintains good relations with the government. The same month Maureen Reagan led a US presidential mission to inspect the traumatic physical and mental effects of the fighting on Mozambique's orphaned children.

Also in September 1988, Chissano had a meeting in Songo (Tete province) with South African President P. W. Botha, who stated that South Africa would curtail its support for the MNR. 'These expectations have yet to bear fruit', noted Mozambique at the end of the year. February and March 1989 saw visits to Mozambique by South African Foreign Minister 'Pik' Botha, who invoked the speculation that what the USA and USSR had achieved in relation to Namibia and Angola might also be applied to Mozambique, and by members of the US administration. In May President Bush's new Under Secretary of State for African Affairs Herman Cohen compared the atrocities of the MNR to those of the Khmer Rouge in Kampuchea, stating that the USA had evidence that assistance was continuing to reach the MNR from South Africa. In July 1989 Cohen visited Maputo after Johannesburg, claiming to have told South Africa to withdraw its support and offering to play a role in a negotiated settlement.

In the nationwide discussions of the proposed agenda for Frelimo's Fifth Congress, some delegates argued for direct talks with the MNR, if this was the only way to stop depredations. Early in 1989 Mozambique began to circulate a position paper to members of the UN Security Council: published in July, this offered a dialogue, if the MNR renounced violence, and guaranteed the right of all to participate in the political process. As the MNR has little policy beyond violence, these conditions more nearly approach American demands for a Western-style 'free society', and may also be connected with Frelimo's revision of its ideological position, of which more below.

In the Central Committee report to the Fifth Congress, Chissano called for a 'national consensus to normalise Mozambican society', generally interpreted to mean the reintegration of MNR personnel. An existing amnesty policy had already reached some 3,000 persons.

Other factors at play are splits and divisions within the MNR: in the bush, in South Africa, in the United States and in Portugal, its main external base. Following the death of spokesperson Evo Fernandes in Portugal in 1988 and a number of defections, the movement is incoherent and has also spawned a splinter group named Unamo, led by the former second-in-command to Renamo leader Afonso Dhlakama. In July 1989 a proposed meeting between Mozambican church leaders and the MNR, planned for Nairobi, failed to materialise when no MNR leadership appeared. The same month saw a visit to Maputo by the South African National Party leader F. W. de Klerk, who was quoted as saying that Renamo should end violence and become part of the process of peaceful development in Mozambique. Also afoot were plans for Presidents Moi of Kenya and Mugabe of Zimbabwe to act as mediators.

One analysis of the somewhat confused situation in the second half of 1989 was that this marked the end of the MNR's second decade, under South African sponsorship, with a hiatus before it attains autonomy – possibly with 'arms length' support from ultra-right sources. Even if all Mozambique's hopes of peace were to be fulfilled, however, the roving MNR bands would continue to wreak havoc in the rural areas for some time.

Emergency aid

In November 1988 it was projected that the food deficit for the following year would be some 300,000 tonnes, or nearly one-third of total requirements for the internal market, emergency relief and reserve supplies. At least 7m of Mozambique's population of around 15m are now dependent on food aid. Although the drastic decline in agricultural production has been halted, subsistence production has been badly hit by the war, so that in addition to the displaced persons, an estimated 3m peasants were unable to feed themselves in 1989, and Mozambique continues to tour the world with its begging bowl. From an appeal to donors made in New York early in 1989, food aid worth a further $350m was pledged by 19 Western states and agencies, ranging from Italy ($78m) and the USA ($65m) through Unicef ($17.9m) and World Vision ($6m) to Federal Germany ($2.3m) and Switzerland ($1.5m). The overall figure was subsequently revised down to $250m owing to miscalculation and double entries. Mozambique's original request was for $329m.

However, despite pledges made to a similar appeal in April 1988, not one new truck or boat to transport supplies had arrived in Mozambique a year later. There are severe difficulties in getting food to certain parts of the country, and in communication: for example, serious starvation in the northern district of Memba (Nampula province) at the start of 1989 was not reported by the local authorities until the church informed the national emergency services. There have also been, for the first time, official reports of theft and diversion of emergency supplies on to the black market – an index not only of the acute situation but also of the administration's loss of control.

In mid-1989 the acting co-ordinator of the Emergency and Disasters Committee (DPCCN) admitted thefts in the ports and railways, and within the DPCCN itself, where several officials were caught red-handed. Up to now, Mozambique has been renowned for the lack of venality in public and commercial life – problems encountered by foreign businesses, for example, are not solved by bribery – so this development is worrying. It is a source of anxiety to the numerous aid agencies now operating in Mozambique (over 50 at the last count) who are also impatient with the incompetent local administration. In response the Mozambican Government is anxious over the tendency of relief agencies to act independently, compete with each other, and carve out little empires. Moreover, wages paid by international agencies distort the local economy: a driver employed by a UN agency, for example, may be earning more than a senior civil servant. If not a poisoned chalice, relief aid may thus cause new problems while solving earlier ones.

Economic situation

While Mozambique is not exceptionally rich in natural resources, it has development potential with regard to agricultural and fishing production, mineral exploitation and local manufacturing, little of which has as yet been tapped, together with geographically useful transport routes and harbours. Post-independence aspirations, already hard hit by natural disasters such as drought, have however been obliterated by the effects of war, reducing the economy to an invalid state. As a result the government was obliged to modify its ideological position and apply to the IMF.

The economic measures taken in exchange for IMF support – of which the third and last tranche was drawn in February 1989, followed by an additional loan of $5m in April – have borne fruit in respect of the aim of arresting the drastic decline and eventually regaining the production levels of 1981. The Economic Recovery Programme (ERP) has achieved small growth rates: agricultural production up slightly in 1987 and 1988, and manufacturing increases in textiles, tyres, hand tools and the assembly of East German radios (for export to Bulgaria). But overall growth for 1987 and 1988 was under target at about 4 per cent each year, and similarly modest results are expected for 1989-90. It should be noted, however, that reliable economic statistics are hard to gather; in much of the country accurate record-keeping is impossible.

The position was reviewed by the IMF and World Bank in March 1988, when continuing reductions were approved in central control of economic affairs, allowing market prices and credit to determine consumption. A steady devaluation of the currency was implemented, in regular monthly reductions: as a result, the official exchange rate dropped from 40 meticais (MT) to the dollar at the beginning of 1987, to MT756 to the dollar in August 1989. Further gradual reductions were to follow.

In mid-1989 Finance Minister Abdul Magid said that the economic indicators planned and approved by the IMF would be 'more or less' achieved in 1989. Inflation was dropping, from 160 per cent in 1987, through 50 per cent in 1988, to an estimated 30 per cent in 1989. Price controls were lifted, reducing the number of commodities with fixed prices from 44 to 28; the phasing-out of price regulation was envisaged. In 1986-88 the number of products subject to state distribution (rationing) fell from 43 to 30 and was expected to be down to 15 by 1990.

The steep price rises bore most heavily on the urban poor, who benefited from controls and rationing, although this has not led to popular unrest. The dire economic situation was blamed on the war, rather than the government. However, there was growing criticism of the army's inability to defeat the MNR, and of the diversion of food and consumer goods on to the black market, and there was still risk of descent into a barter economy.

With additional imports through IMF foreign currency, more goods were available on the market, at a price. But the unofficial markets – known as *dumba nengues* (roughly 'watch out!') – proliferated in the urban areas, selling not only locally-produced foodstuffs but also stolen and diverted goods; recognising that black market trading cannot be eradicated, the government prefers to tolerate the unofficial markets on the grounds that illicit activities can at least be monitored there.

Policy in 1989 was to keep Mozambique's external debt below $5.0bn, with the government endeavouring to persuade creditors to cancel or convert or reschedule loans; in May 1989 Federal Germany cancelled a debt of around $83m, as a sign of its confidence in the PRE, with the result that the foreign debt in mid-89 stood at $4.2bn. Portugal agreed to transform part of its debt into capital for Portuguese companies operating in Mozambique. The United States, in an agreement signed in August 1988, was providing aid funds earmarked for the private sector.

Also in 1989, Mozambique signed a preferential trade agreement with South Africa, reducing import charges on Mozambican exports of fish, cashews, citrus, tyres and cigarettes. This agreement did not bring Mozambique into the Southern African Customs Union. Some 60 South African enterprises were operating in Mozambique, and other major companies were discussing investment: AECI in chemical production, Anglo-American in coal extraction at Moatize, SAPPI in forestry and paper production. In October 1988 the South African trade mission reopened in Maputo, and discussions began on re-starting recruitment of migrant workers from Mozambique to the Witwatersrand mines, which used to be a major source of foreign earnings.

Productivity and earnings

For agricultural producers, prices increased by about 50 per cent in 1988-89, although the incentive value of these measures remains small as long as the war imperils marketing. The most successful production has come from the 'green zone' co-operatives around Maputo, though this makes little contribution to the large quantities of maize and rice which Mozambique has traditonally consumed. Under the economic directives announced at the Fifth Congress, priority was given to supporting peasant families and co-operatives with technical equipment and inputs such as seeds and fertiliser. State farms (formerly mainly plantations owned by large-scale colonial enterprises) were to be oriented towards production for export, when conditions allow. The directives, however, did not establish targets, recognising that these would not be fulfilled, and it remained doubtful whether any change in policy would actually increase food production.

One of Mozambique's major problems is that, following the nationalisation of so many sectors and enterprises at independence, either for political or practical reasons (many former owners and managers left the country, in some cases sabotaging factories) an over large proportion of the labour force outside agriculture is employed by the state. Under the IMF agreement, the public sector workforce was reduced by 30,700. This was critically received in Maputo, where the supply of surplus or unemployed labour has been swollen by an influx from war-threatened rural districts. Earlier governmental promises of 'the right to work' have a hollow ring: policy was in 1989 to redirect such labour into small-scale agriculture, but this can lead to conflict over land in 'green zone' areas around the city. The state-sponsored trade union movement calls for 'sensitivity' in reducing public employment – an implicit warning of unrest. A longstanding problem is that of inefficiency and low expectations: in too many cases workers are paid regardless of performance, owing to breakdowns, lack of supplies and general defeatism. Under the PRE guidelines, pay was to be more closely related to performance and productivity.

Fiscal policy was also revised, to increase state revenue from taxation and reduce economic subsidies. Banking procedures were reformed and it became somewhat easier for enterprises to obtain foreign exchange for imports.

Transport remained a key to foreign currency earnings. The Beira to Zimbabwe rail link remained open, enabling the new managing director of the Beira Corridor Group in Harare to claim that no freight had been lost through any security-related incident for two years. But Beira was handling only half its 30,000 tonne annual capacity. And the Beira–Moatize rail line (for coal exports) was non-operational. Further north, the 600 km Nacala to Malawi rail link remained virtually out of action owing to attacks, despite a new container terminal at Nacala port. The upgrading of the Limpopo rail line (Maputo to Zimbabwe) began in November 1988 with external funding from Canada, Federal Germany and Britain.

In June 1988 South Africa agreed at a meeting with Portugal and Mozambique to pay for electric power from Cahora Bassa, and ironically also promised to provide logistical support to help the Mozambican army guard the powerlines, which are regularly sabotaged by South African-supplied MNR gangs.

Social policies

Since Frelimo came to power on Mozambique's independence in 1975, when over 90 per cent of the nation was illiterate and most people went without medical services, social policy has been based on free access to health care and education. In the early years progress was made in both fields, reducing illiteracy to around 70 per cent and providing basic health care even in remote regions. With the war, however, all this was reversed.

Eighty health workers have been killed or kidnapped by MNR forces. Less than one fifth of the country's 120 districts had a resident doctor. Overall, in 1989 there were 327 doctors in Mozambique, less than 40 per cent of whom were Mozambican. The PRE increased prices for prescribed drugs and introduced fees for hospital consultations, but this revenue amounted to a minimal 3 per cent of health service running costs. There was pressure for the re-privatisation of medicine (and a *de facto* diversion of supplies on to the private market by some health workers). In July 1989 the minister of health reiterated Frelimo's commitment to free health care, but promised legislation allowing companies, churches and non-profit agencies to fund auxiliary services in co-operation with the state.

A similar position existed with regard to education, since the state system was reaching only 40 per cent of primary-age children. In the rural areas schools have been destroyed, and in the urban areas the teacher–pupil ratio is 1:62. Private schools and tuition, hitherto illegal, were to be permitted, in an attempt to meet the demand for instruction without increasing state spending.

On the legal front, the post-independence restructuring of the judicial system was completed in March 1989. At the same time, the revolutionary military tribunal, established to deal harshly with 'counter-revolutionary' and treasonable activities, was abolished, as was the practice of flogging offenders, which had been earlier re-introduced in response to popular demand. Both measures were the subject of human rights criticism from external agencies such as Amnesty International.

Political developments

The assassination of Mozambique's first president Samora Machel in October 1986 (in an air crash on South African soil now generally agreed to have been caused by a false directional beacon) did not occasion any major political changes in the country: the Frelimo leadership has worked as a team since their guerrilla days in the early 1970s and exhibited no signs of personal ambition or rivalry. The difficulties of the late 1980s, as one catastrophe piled on another, did therefore not lead to serious divisions or splits, but the 'old guard' (still mostly in their 50s) may be feeling the strain of unremitting problems – of which Chissano's prostate operation, at a relatively young age, may be a symptom.

New blood and new ideas were certainly among the objectives of the party's Fifth Congress, preparations for which occupied the first half of 1989 and which, despite the existence of an elected People's Assembly, is in fact the chief organ of democracy in Mozambique. Draft theses were discussed throughout the country to sound out public opinion, and the 700 delegates elected as representatives to congress were encouraged to continue the tradition of open debate and critical analysis, so that the leaders could keep in touch with dissent. In the run-up to the Fifth Congress, there was virtual unanimity on the party's failure to end the war, prevent social inequalities and guarantee public integrity and justice under the law.

Some new names were among the leadership elected in July 1989, although few were removed except by death or retirement. The central committee had 160 members (121 at the last congress in 1983) and the politburo had 12 (11 in 1983, and 10 since the death of Machel). Outgoing politburo members were Oscar Monteiro, minister of state administration, transferred to the party secretariat, and veteran Sebastiao Mabote, ex-chief of staff. Newcomers include Foreign Minister Pascoal Mocumbi and the political commissar of the armed forces.

These and other changes – largely reshuffles of the existing pack – were interpreted by the white minority in Mozambique as evidence that Chissano favoured black Mozambicans more than did his predecessor, and may fuel demands for a tribal or regional element in appointments; there was, for instance, some pressure for the whole central committee to be geographically representative. Frelimo remained opposed to such thinking, which was probably more of a means of expressing discontent than a serious challenge to the party, but could not be ignored.

How significant were the ideological changes announced at the Fifth Congress, when Frelimo was redefined as the 'vanguard party of the Mozambican people' rather than as (since 1983) the vanguard of 'the worker-peasant alliance'? All references to Marxism–Leninism and class struggle were dropped in favour of a broader, more nationalistic appeal to unity and social justice.

To some extent this was an alteration in rhetoric only. Without much of a state apparatus and without the means to collectivise agriculture, the notion of Mozambique developing on Marxist–Leninist lines was always somewhat absurd; indeed Frelimo's leaders maintained that the advanced industrial nations, whatever their political colour, with their strong public services, planning controls, compulsory education and state benefits were more closed to socialism than undeveloped peasant societies like Mozambique. The changes in ideology thus reflected, in Chissano's words, 'a real framework, not an ideal one'. They were designed to increase party membership and narrow the gap between party and people. But they also contained a global dimension, for by abandoning the former commitment to 'anti-imperialism' and 'proletarian internationalism', Frelimo was signalling its desire to loosen Mozambique's alignment with the Eastern bloc, perceiving the reality that the West was necessary both for economic recovery and also for any kind of victory over the forces of destabilisation. It remained speculative whether the rhetorical changes were part of the bargain for Western (or IMF) support.

Developments in 1989 indicated how far Mozambique was prepared, or obliged to go to secure this support, but there were no guarantees that Western pressure would succeed in persuading South Africa to call off the MNR. By the time it does so, Mozambique may be too far down the capitalist road to turn back.

COUNTRY PROFILE

Political system: The constitution (adopted 1975, revised 1978 and 1986) states that power belongs to the workers and peasants, united and led by Frelimo. The president (who is also president of the party) holds executive power and appoints and dismisses the Council of Ministers, which is presided over by the prime minister – a new post created in July 1986. President Joaquim Chissano took office in late 1986. The current prime minister is Mario Machunga. The council of ministers is responsible to the People's Assembly (the 250-member People's Assembly is the legislature): the Assembly's functions between sittings are carried out by a permanent commission which it elects.

Political parties: The ruling Partido Frente de Libertação de Moçambique (Frelimo) is the only legal party.

Population: Mozambique's population of 14.7m (1988 estimate) includes some 20,000 Europeans and Asians.

Main cities/towns: Maputo (population around 1m), Beira (main railhead town, population 145,000), Quelimane, Nampula, Nacala, Tete, Angoche, Lichinga, Pemba, Inhambane, Xai-Xai, Chimoio, Mocuba.

Language: Portuguese is the official language and is spoken by most social groups, regardless of ethnic origin. African languages in use include Ronga, Shangaan and Muchope. Portuguese is normal in business circles, but English is also used to some extent.

Media
Press: Principal daily newspapers are *Notícias* (government) and *Diário de Moçambique.* Weeklies include *Domingo* (published Sunday) and *Tempo* (magazine). Several periodicals are also available.
Radio: Radio Moçambique broadcasts a commercial service on three channels from Maputo, in Portuguese, English and local languages.
Television: An experimental television service, TVE, was started in 1981 and is transmitted four days a week.

External trade: Continuing balance of trade deficit caused by falling export earnings and heavy reliance on food and oil imports. Balance of payments deficit widened in recent years due to heavy debt burden and declining revenues from transit earnings and foreign remittances. By 1987 the country was virtually bankrupt.
Exports: 1986 – prawns 60.8 per cent (by value), cashew nuts 13.6%, sugar 6.4%, petroleum products 3.2%, cement 0.24%, cotton 0.8%, tea 0.8%, timber 0.8%, others 13.36%. Main destinations: 1986 – OECD countries (9.75% of value in 1986) mainly to Japan, USA, Spain and Portugal; Eastern Europe (12%, of which 7.7% to GDR).
Imports: 1986 – food 27 per cent (by value), raw materials 20%, machinery and equipment 16%, consumer goods 15%, spare parts 12%, crude oil etc 8.8%. Main sources: 1986 – OECD countries (60% by value) mainly from USA, Italy, France and Portugal; Eastern Europe 12%, chiefly USSR; South Africa 10.3%.

Agriculture: Mainstay of economy, employing up to 85 per cent of the workforce – mainly engaged in subsistence farming. Main cash crops are cashew nuts, tea, sugar, sisal, cotton, copra, oil seeds and some citrus fruits. Maize is the main subsistence crop, but cassava, millet, sorghum, groundnuts, beans and rice are also grown. State farms account for only about 15 per cent of total output but produce most of the cash crops, especially rice and cotton, while 80 per cent of production comes from peasant farms. The remainder derives from a few large private farms and co-operatives. A prolonged period of drought, which worsened in 1986-7, has added to the problems of a lack of skilled manpower, the inheritance of Portuguese colonialism, locally poor soil conditions and, above all, attacks by rebels supported by South Africa. However, the government is attempting to improve productivity by establishing new private farms, offering higher producer prices, dividing up some of the more uneconomic state farms and improving infrastructure. Fishing is of increasing importance and prawns have become the major export. Also important forestry sector, with exploitation of hardwoods from Zambezia, Sofala, Nampula, Manica and Nyasa provinces.

Industry: Predominantly state-owned sector, contributing 26 per cent to GDP in 1986 but suffering severely from general economic contraction. Manufacturing concentrated in food processing (maize, and wheat flour, cashews, sugar, salt); beverages; tyres and hand tools. Other products include textiles, soap, batteries, radios and bicycles. Industrial output rose by 17 per cent in 1987 thanks to availability of foreign exchange under the PRE, which has also prioritised light industry at the expense of heavy industry; production of railway wagons and reinforcing bars, for example, is planned to remain below the levels achieved in 1981.

Mining: Considerable mineral resources including iron, graphite, fluorites, mica, lime clays, tin, gold, nickel and bauxite. Mining currently limited to coal (of which proven reserves around 10bn tonnes), pegmatites, tantalum, copper, marble and semi-precious stones. Production is restricted due to transportation and security problems.

Energy: Developing sector of economy with vast hydroelectric resources only now beginning to be utilised. Mozambique has over 15 major rivers running west-east. The setting up of a national grid, using power generated at Cabora Bassa (shared with South Africa), Lichinga, Cuamba and Chicamba Real is a major requirement of the government's plan to boost industrial production. Large reserves of gas and limited oil reserves have been discovered but have not as yet been exploited.

BUSINESS GUIDE

Time: GMT +2.

Climate: Dry season Apr-Sep, Maputo average temperature range from around 20°C at night to 35°C during the day with considerable variations brought about by wind changes. Rainy season Oct-Mar, likely average daily temperature in Maputo 27°C to 29°C, with maximum of 38°C and high humidity. Inland areas are a few degrees cooler than the coast and rainfall increases with altitude.

Entry requirements
Passport: Required by all.
Visa: Required by all. Applications should be made well in advance. Can be obtained at Mozambican diplomatic missions but usually must be obtained by someone in Maputo (three weeks in advance). Visas are automatically issued for official and business visitors. Tourists only permitted as part of package tours.
Prohibited entry: Foreign nationals who had the status of 'National of Mozambique' before 25 Jun 1975, except those with documentation proving renunciation of this status within 90 days of that date.
Currency: Import/export of local currency prohibited. Unlimited import of foreign currency but amounts must be declared on arrival and equivalent of $25 must be changed in local currency. Export of foreign currency limited to amount declared on arrival.

Health precautions
Advisable: Typhoid and polio vaccinations. Malaria prophylaxis essential as risk exists throughout the country. Water precautions are advisable, especially in rural areas where bilharzia is present.

Air access
National airline: Linhas Aéreas de Moçambique (LAM).
Other airlines: Aeroflot, Citi Air, Lesotho Airways, Royal Swazi National Airways, South African Airways, TAAG – Linhas Aéreas de Angola, TAP – Air Portugal.
Tax: Airport tax of $10 on international departures.
International airports: Beira (Code: BEW), 13 km from city, buffet, restaurant, bank, post office, shops; Maputo-Mavalane (Code: MPM), 8 km from city, bar, restaurant, buffet, bank, post office. Visitors should arrange to be met at the airport on arrival.

Surface access
Main ports: Beira, Maputo, Nacala, Quelimane.
Overland access: Note: The political situation in Mozambique can make travel outside of Maputo inadvisable. Road access possible from all neighbouring countries except Tanzania (new road under construction); good paved roads from South Africa and Zimbabwe. Rail tracks run from Malawi to Nacala, Zimbabwe to Beira and South Africa to Maputo. Passenger services run on the routes from South Africa and Malawi to the Mozambique border, where connections can be made to the country's rail system; freight only from Zimbabwe.

Hotels: Reasonable accommodation is available in Maputo. Bills must be paid in hard currency or by Amex credit card. Tipping is discouraged.

Electricity supply: 220 V AC, 50 cycles.

Car hire: Available with or without a driver from Interfranca, the state-owned hire company, or the National Tourist Office. Only hard currency will be accepted. International licence required.

City transport
Taxis: Only sporadically available in Maputo, and rarely outside. Taxis are metered but for long journeys fares should be negotiated. 10 per cent tip is usual.

National transport: Advisable to check travel regulations. Permission is generally needed to travel outside Maputo.
Air: LAM operates domestic service to main towns; flights often full. Local and charter flights can also be obtained from TTA. Little night time flying.
Road: Good roads connect Maputo to Beira, and Tete in the north; surface of other roads varies. Improvements are being made.
Buses: Services covering most of the country, restricted by security risks and vehicle failure.
Rail: There is a railway network, though no link between Maputo and Beira, but services are liable to disruption due to the security situation.

Public holidays
Fixed dates: 1 Jan, 3 Feb (Heroes' Day), 7 Apr (Women's Day), 1 May, 25 Jun (Independence), 7 Sep (Victory Day), 25 Sep (Armed Forces' Day), 25 Dec.

Working hours
Government and business: (Mon-Fri) 0800 – 1200, 1400 – 1700; (Sat) 0800 – 1200.
Banking: (Mon-Fri) 0800 – 1100; (Sat) 0815 – 0915.

Telecommunications
Telephone: Dialling code for Mozambique: IDD access code + 258 followed by subscriber's number. Good service internally and to other countries by CTT.
Telex: Reliable services available in Maputo and Beira from CTT, and some hotels.

BUSINESS GUIDE

Banking
Central bank: Banco de Moçambique.

Trade fairs: The Agricultural, Commercial and Industrial Fair (FACIM) is held in Maputo annually around Aug/Sep.

Other useful addresses
Aeroporto de Maputo (tel: 732141, 732149).

Agência de Informação de Moçambique (AIM), PO Box 896, Maputo (tel: 30795; tx: 6430).

Agência Nacional de Frete e Navegacao (ANFRENA) (main national shipping agency), Rua do Bagamoyo 366, PO Box 1430, Maputo.

Agência Nacional de Viagens (ANAVIA) (government travel agency), Avda 25 de Setembro 1747, PO Box 965, Maputo.

Banco de Moçambique, Avda. 25 de Setembro 1695, PO Box 423, Maputo (tel: 28151; tx: 6-244, 6-355, 6-357, 6-391).

Banco Standard Totta de Moçambique, Praca 25 de Junho, PO Box 2086, Maputo (tel: 28107, 24597, 22971, 29041; tx: 6223 derby).

Direcção Nacional Portos e Caminhos de Ferro (railways), PO Box 276, Maputo (tel: 24228, 27173; tx: 6208 cfms).

Embassy of the People's Republic of Mozambique, 159 New Bond Street, London W1 (tel: 01-493-5604; tx: 263481 Mozem).

Empresa Nacional de Minas, PO Box 1152, Maputo (tel: 23833; tx: 6413).

Empresa Nacional de Turismo, PO Box 614, Avenue 25 de Setembro 1211, Maputo (tel: 25011; tx: 6552).

Empresa Nacional Petroleos de Moçambique (Petromoc), PO Box 417, Maputo (tel: 27197; tx: 6382).

FACIM, PO Box 1761, Maputo (tel: 23713; tx: 6-411). (Annual international trade fair).

Hidroelectrica de Cabora Bassa (HCB) (operators of Cabora Bassa power complex), PO Box 4120, Maputo.

Imprensa Nacional de Moçambique (publishes statistical bulletins, census information etc), PO Box 275, Maputo.

Interfranca Car Hire, Rua Timor Leste 106 (tel: 25199; tx: 6403).

Linhas Aéreas de Moçambique (LAM), PO Box 2060, Aeroporto, Maputo (tel: 732141, 731688, 732038; tx: 6386).

Ministry of Industry and Energy, Avda 25 de Setembro, 1184/92 3 andar, Maputo (tel: 31029; tx: 6292).

Ministry of Trade, Direccao Nacional de Operacoes Comerciais, PO Box 2930, Maputo (tel: 26091; tx: 6374).

Mozambique Chamber of Commerce, PO Box 1836, Maputo (tel: 741970; tx: 6498).

Mozambique Information Office, 7a Caledonian Road, London N1 (tel: 01-278-8691; tx: 23391).

Office for Foreign Investment Promotion (GPIE), Avda Ahmed Sekou Touré 21, 6 andar, PO Box 1101, Maputo (tel: 742713; tx: 6153).

Radio Moçambique, PO Box 2000, Maputo.

Transportes e Trabalhos Aéreos (TTA) (air taxi services, etc), PO Box 2060, Aeroporto de Mavalane, Maputo.

Hotels: Beira
Dom Carlos, Largo do Farol, P.O. Box: 139 (tel: 711158; tx: 7333).

Hotels: Maputo
Andalucia, Avenida Patrice Lumumba, P.O. Box: 1135 (tel: 23051; tx: 6426).

Cardoso, Avenida Maritires de Muda 707, P.O. Box: 35 (tel: 741071; tx: 6327).

Polana, Avenida Julius Nyerere 1380 (tel: 741001; tx: 6278).

Tivoli, P.O. Box: 340 (tel: 22005; tx: 6297).

Myanma

Nicholas Nugent

Burma, still recovering from the traumatic events of 1988, moved cautiously along the road towards greater democracy. But the military rulers who seized power in September 1988 in an attempt to restore order – they called themselves the State Law and Order Restoration Council – ensured that the moves towards democracy were slow, causing doubts as to whether they really intended to hold elections as promised.

These doubts were strengthened when they placed under house arrest two leaders of the political party which, by virtue of its extensive campaigning, presents the greatest challenge to 27 years of military dominance, the National League for Democracy (NLD). Those detained were party chairman Tin Oo and secretary-general Aung San Suu Kyi, the daughter of Burma's independence hero. Aung San Suu Kyi's British-born husband was effectively detained with her, together with their two sons, causing the British Government to protest at the detention of British nationals.

Party politics

The clampdown on the NLD was precipitated by the repeated defiance by its leaders of the martial law regulations which ban gatherings of more than four people. The authorities moved against Aung San Suu Kyi after she threatened to lead a column of her supporters to the martyrs' memorial on martyrs' day, 20 July, the day when her father and other heroes of Burma's independence struggle are commemorated. In the event she called off the march, but the authorities nonetheless detained her and many other supporters of the NLD. They claimed the party had been infiltrated by communists, though it was privately admitted that its real sin was to have persistently criticised the army.

The NLD is one of more than 200 political parties to have registered with the election commission since the one party monopoly on power was lifted on the eve of the 1988 military takeover. It is the most active and, if the popularity of its charismatic leader – who has the advantage of her father's name and the reverence in which he is held – is anything to go by, it is well placed to win an election. Its leaders say the party will not contest elections unless they are sure they will be fair.

A lengthy process of registering voters aged 18 upwards was taking place in preparation for elections taking place by May 1990. The people will elect a constituent assembly whose job will be to rewrite Burma's one-party socialist-oriented constitution, perhaps preparing the way for a real transfer of power. The clampdown which has arrested NLD members and restrained the movements of its leaders casts doubts on the military regime's preparedness to allow power to pass from military hands.

On seizing power, the military leadership made the return to normality its key prerequisite for fulfilling the previous short-lived regime's pledge to multi-party elections. It stressed that normality meant the restoration of transport and food distribution networks as well as the return of law and order. Just how abnormal life had become was clear from the 10pm to 4am curfew as well as the fact that all schools were closed for a full year, colleges and universities for rather longer, meaning that Burma's 7m young people lost a year of their education. It had not escaped the military's attention that those leading the 1988 democracy movement were all of college age or younger; the reopening of schools and colleges in stages was handled with considerable apprehension.

With the setting up of various student organisations – all of which called for a lifting of the ban on students' unions – it was clear that students would continue to exert pressure for change. One group, the All-Burma Students Democratic Front, set them-

KEY FACTS

Official title: Republic of Myanma

Head of state: Chairman of the State Law and Order Restoration Council: General Saw Maung

Capital: Rangoon (Yangon)

Official Languages: Burmese (with English used in business)

Currency: Kyat (Kt = 100 pyas)

Exchange rate: Kt6.82 per $ (Nov 1989)

Area: 676,552 sq km

Population: 38.3m* (1988)

GDP per capita: Kt1,696 (1987)

GDP real growth: 3.7% (1987)

Labour force: 15.8m* (1988)

Unemployment: 10.0%* (1987)

Inflation: 16.1% (1988)

Trade balance: –$486.6m (1988)

Foreign debt: $4,348m (1987)

* estimated figure

selves well outside the law by uniting those many students, probably more than 10,000, who had fled to camps of the rebel insurgent armies near Burma's border with Thailand to continue their struggle for democracy with arms.

Economic collapse

Meanwhile, with uncertainty continuing over Burma's political direction, its economy had all but collapsed. The lifting in late 1987 of the ban on private internal trading in rice served only to drive up the free market price. Mid-1988 prices were even higher than during the disturbances a year earlier, though statistically there should have been enough rice to go round and some left for export sales. Farmers at first rejected government purchases because they could secure higher prices on the free market. This provoked the authorities to step in and make demand purchases to secure the rice they needed to feed the large army and civil service. The government also released lower price rice on to the market at a subsidised price in an attempt to drive the price down.

Very little rice was left for the government to sell abroad: export earnings of rice by the country which had once been a major rice exporter were once again expected to be negligible, below 200,000 tonnes. A relaxation under which private traders were allowed to export rice and pulses proved ineffective as the cumbersome official machinery for supervising any export deal as well as the unattractive rate of exchange of the kyat to the dollar – around one seventh of the black market rate – removed any incentive for private business people to export rice.

For similar reasons, a new investment law, which dramatically liberalised an economy as closed for potential foreign investors as for Burmese business people, has done little to attract investment. Only in logging and fisheries, where the government has sold concessions to foreign interests, is there evidence of the supposedly more open-door policy. Oil prospectors, for example, whose skills are badly needed to arrest declining production, have still not been allowed in on their terms. Burma imported oil in 1988 for the first time in many years, though this was said to be for the sole purpose of keeping aid projects running. Apart from the political factors which have caused many foreign governments to cut aid to Burma and to ostracise its military rulers, the incentives needed to attract foreign business are still lacking.

The rising prices of rice and other commodities was the reason given for a substantial increase in government salaries in early 1988, which further emphasised the gulf between state employees and others. It is reported that, to pay the higher wages, the government has resorted to printing money, bringing the amount in circulation back up to the level preceding the October 1987 demonetisation of higher value notes, a key trigger of the subsequent pro-democracy movement.

Industry at a standstill

The main source of export earnings was timber, helped by the opening of cross border trade with Thailand and by Thailand's own ban on cutting down trees. And, surprisingly, Burma's annual gems auction in March 1989 produced record sales. But these earnings hardly compensated for the near collapse of the tourist traffic. To eke out its hard currency earnings, the government announced that foreign residents of Rangoon – essentially diplomats and aid personnel – would henceforth have to pay for petrol and foreign telephone calls in hard currency. There were persistent rumours that the military rulers were planning to sell land Burma owns in foreign capitals such as Tokyo, as part of Burma's overseas diplomatic representation, to gain badly needed foreign currency.

What export earnings there were, though, were insufficient to pay for raw materials for Burma's state controlled industry. Without foreign currency, industry ground almost to a halt. Cigarette factories went on to short time working and everything was in shorter supply even than is usual in Burma, where shortages have long been a way of life. Planning and Finance Minister Colonel Abel, who was overseeing Burma's transition to a more capitalist-oriented economy after the acknowledged failure of the Burmese road to socialism (the ideology of the former regime), found few foreign concerns ready to do business with the military leadership. Instead, he concentrated his energy on trying to renegotiate Burma's foreign debt, on which it defaulted for the first time in 1988.

Who rules Burma?

The decision to change the name of the country to Myanma, after the earlier dropping of the appellation 'Socialist', was seen as an example of superficial change disguising little substantive change. The military leadership claimed it wanted to get away from the association of the country with its majority race, the Burmans. They hoped that the use of the country's Burmese name would appeal to the Karen, Kachin, Kayah, Chin, Mon, Shan and other minority races who make up the country. They took the opportunity to change the names of a number of cities too, so that Rangoon, for example, becomes Yangon.

There remains something of a mystery as to who precisely is in charge in Burma. The nominal leader of the State Law and Order Restoration Council is General Saw Maung, the army leader whose seizure of power in September 1988 restored a degree of normality. But it is popularly believed that the country's ruler for 26 years, General Ne Win, continues to direct operations from behind the scenes. Though rarely seen in public, his picture has started to reappear in the official *Working People's Daily* newspaper. Otherwise, the man most often mentioned as a wielder of power is the former head of the military police, Brigadier General Khin Nyunt. In the manifestly military regime that Burma remains, the generals commanding the country's military regions also wield considerable power, the more so since the setting up of military courts empowered to impose sentences (including the death sentence) after summary trial for violations of martial law. But the military rulers preserve their effective anonymity by having very few dealings with outsiders. Most Western diplomats are in any case officially boycotting the ruling council, and demands by political parties for a dialogue with the leadership have not been granted.

The party that for most of Ne Win's long tenure was his instrument of rule, the Burma Socialist Programme Party (BSPP), has found a new existence as the National Unity Party (NUP), and claims to have taken a large proportion of the BSPP's 2.5m membership with it. However, the NUP no longer has a role in government which, following the dissolution of the Pyithu Hluttaw, or parliament, is more firmly in the hands of the army than ever before.

KEY INDICATORS

	Unit	1984	1985	1986	1987	1988
Population	'000	36,392	37,150	37,310	39,140	–
Gross domestic product (GDP)	K m	53,597	55,989	59,028	66,369	71,059
GDP per capita	K	1,473	1,507	1,582	1,696	–
Inflation	%	4.8	6.8	9.3	24.5	16.1
Volume of exports	1985=100	117	100	115	86	53
Rice	1985=100	157	100	134	98	14
Teak	1985=100	109	100	112	82	83
Pulses	1985=100	51	100	83	69	29
Exports	K m	3,171.0	2,803.5	2,201.1	1,755.9	1,051.5
Rice	K m	1,172.2	659.1	654.3	376.4	70.9
Teak	K m	976.2	1065.3	1052.9	649.3	504.0
Pulses	K m	204.7	280.9	188.5	126.8	70.5
Imports cif	K m	2,016	2,402	2,245	2,508	1,538
Balance of trade	$m	1,155	–401.5	–43.9	–752.1	–486.6
Current account	$m	–258.3	–205.5	–157	–208	–
External debt	$m	2,333	3,089	3,791	4,348	–
Debt service ratio	%	40.9	53.1	69.1	59.3	–
Exchange rate	K per $	8.38	7.47	7.33	6.65	6.39

* estimated figure

COUNTRY PROFILE

Political system: A military government took power in September 1988 ending months of unrest. It has taken over the function of the former ruling party, the Burma Socialist Programme Party, and the parliament or Pyithu Hluttaw. The new government is committed to holding multi-party elections by May 1990.

Main cities/towns: Rangoon (population 3m), Mandalay (pop 532,895), Moulmein (pop 219,991), Bassein (pop 335,000) and Akyab (pop 143,000).

Media
Press: In Rangoon, there is one newspaper, the *Working People's Daily*, published in English and Burmese.
Radio: Burma Broadcasting Service broadcasts in Burmese and also in English and other local languages. Some 250,000 radios (1987).
Television: The first station opened end-1980.

Domestic economy: The economy is centralised and based mainly on agriculture and forestry, with an important black market sector. In addition to state-owned enterprises, there is a thriving co-operative sector. The private sector, which includes traders, farmers, small-scale industries and transport enterprises, has no direct access to foreign exchange. Burma has accepted multilateral and bilateral foreign aid, with significant donations from Japan and German FR, and had to seek least developed country status in 1986 in order to help to reschedule external debt repayments of around 60 per cent of export earnings. The target is to transform the economic structure into an agro-based industrial economy.

External trade: Most import demands and exports originate with trade corporations and state boards, with most trade handled by the Myanma Export Import Corp using services of the Burma Five Star Shipping Corp. The picture of external trade presented by official statistics does not reflect the considerable amount of smuggling and black market trading.
Exports: Main exports are agricultural products including rice (typically 42% of total exports, but down to 27% in 1987), forest products like teak and other hardwood (39%, and rising), minerals and gems (7.7%). Main destinations are Singapore (12.4% of 1985 total), China (8.0%), Thailand (6.7%), Japan (6.4%), Malaysia (4.9%), Hong Kong (4.6%), USSR (3.0%), USA (2.6%), UK (2.3%) and German FR (2.0%).
Imports: Main imports are capital goods (57%), intermediate goods (30.9%) and consumer goods (11.9%). Main sources are Japan (31.6% of 1985 total), German FR (12.8%), Singapore (9.6%), UK (4.3%), Czechoslovakia (3.9%), USSR (3.4%), Netherlands (2.9%), Republic of Korea (2.3%), Korea DRP (2.1%) and Malaysia (2.0%).

Agriculture: Agriculture accounts for 27.9 per cent of GDP and provides 70 per cent of employment. About 15 per cent of the total land area is cultivated. Burma is self-sufficient in rice, which is the main foreign exchange earner, but poor harvests in 1985, 1986 and 1987 sharply reduced its earnings. Main crops, are paddy, sugar-cane, groundnuts, sesame, wheat, maize seeds, jute, cotton, beans, wheat and vegetables. Oil palm and rubber plantations are replacing forest in some areas. Cattle, pigs, buffaloes, sheep, goats and poultry are raised for domestic consumption, with elephants used as draught animals. Some 50 per cent of the land area is forest, containing 75 per cent* of world teak resources. Heavy logging is carried out, mostly of export hardwoods, bamboo and fuel, but the country's vast timber potential is still largely untapped. Rivers are well endowed with fish, but marine fish resources are only beginning to be exploited.

Industry: Manufacturing and processing provides 8.7 per cent of employment but contributes 10 per cent of GDP. Small enterprises predominate. Manufacturing industries include food processing (sugar, tobacco, palm oil, rice), cement, cigarettes, aluminium products, paper and nails, steel, cotton yarn, textiles, soap, pharmaceuticals and fertilizers. Textile and jute production is being expanded. New ventures include rubber tyre manufacture and methanol. Small-scale assembly plants produce tractors, trucks, small taxis, pumps, domestic electrical appliances, TV sets and transistor radios. Barges and rivercraft, bricks and tiles, light engineering products, fertilisers, pharmaceuticals, plywood, paints, footwear, toilet goods, cement, textiles, petroleum products, glass bottles and electrical goods are produced in Rangoon. Cottage industries in Mandalay produce copper and brass ware, spinning, weaving and wood carving. A joint venture for producing and installing machinery has been undertaken with a German FR firm, but foreign investment is generally very slow. Tourism, too, is undeveloped and not particularly encouraged.

Mining: Mining employs 7.6 per cent of the workforce and is 1.5 per cent of GDP, but represents the fastest growing sector of the economy. Minerals mined include ores of zinc, tin, and copper, silver, gypsum, limestone and gems such as rubies, sapphires and jade. Low-quality coal is mined. Oil production and natural gas production consumed locally, and exploration is in progress for more.

Hydrocarbons: Burma's established deposits of crude oil are becoming exhausted, and production fell by 12 per cent between 1979-80 and 1983-84. New onshore deposits have been found at Kyontani, and offshore gas deposits south of Rangoon have been located by a Japanese company. Burma is unlikely to become an exporter of fuels, since all local deposits (including the rather poor-quality coal) are urgently needed by the domestic industries. The country is doing its best to develop its considerable hydro-electric generating potential, but is severely limited by the cost factor.

Membership of international organisations: Asian Development Bank, Colombo Plan, ESCAP, FAO, G-77, GATT, IAEA, IBRD, ICAO, IDA, IFC, IHO, ILO, IMF, IMO, Interpol, IRC, ITU, UN, UNESCO, UPU, WHO, WMO, World Bank.

BUSINESS GUIDE

Time: GMT + 6.5 hrs.

Climate: There are three seasons: the rainy season May/Jun-Oct with high humidity and monsoon rains; the hot season Feb/Mar-May with likely temperatures of 37°C (coastal and delta areas) to 40°C (central region); and the cool, dry season Nov-Feb with temperatures of 16°C (central region) to 21°C (coastal and delta areas). Average rainfall p.a. varies from 5,000mm (northern hills and coastal areas) to 2,500mm (delta areas) to 750 mm (central region).

Entry requirements
Passport: Required by all.
Visa: Visas and confirmed onward passage are required by all, with visas generally limited to seven days. No visas are issued on arrival but business visa extensions can be granted in Rangoon. Application should be made well in advance for visitors requiring more than seven days' stay. Not all parts of the country are open to visitors, and some are subject to ongoing guerrilla or bandit activity.
Prohibited entry: Entry is refused to South African, South Korean or Taiwanese passport holders.
Currency: There is no limit on the amount of foreign currency which may be imported, but it must be declared on entry, including American Express cards. Burmese currency may not be exported or imported. Exchange controls are strictly enforced, and it is best to retain all foreign exchange receipts.
The Myanma Foreign Trade Bank handles all foreign exchange transactions.
Customs: Personal effects are allowed duty-free, but customs regulations are restrictive, and it is best to travel light. All jewellery must be declared on entry.
Income tax: Tax clearance certificates for visitors whose stay exceeds six months are required on departure.

Health precautions
Mandatory: Vaccination certificates for cholera, and for plague, yellow fever, typhus or typhoid. Enquiries should be made before travelling.
Advisable: Water should be boiled and filtered before drinking. Anti-malarial precautions. Vaccinations for Japanese encephalitis if visiting between Jun and Oct.

Representation overseas: Bangkok, Beijing, Belgrade, Bonn, Cairo, Canberra, Colombo, Dhaka, Hanoi, Islamabad, Jakarta, Kuala Lumpur, London, Manila, Moscow, New Delhi, Paris, Patan (Nepal), Prague, Rome, Seoul, Tel Aviv, Tokyo, Vientiane, Washington.

Air access
National airline: Burma Airways.
Other airlines: Aeroflot, Bangladesh Biman, CAAC, Thai International.
Tax: There is an airport departure tax of K15.
International airport: Mingaladon (RGN), 19 km from Rangoon, with duty-free shop, bar, restaurant, buffet, bank, post office and hotel reservations.

Surface access: Overland access is not practicable. The border with Thailand is closed for ordinary travel.
Main ports: Rangoon, Pegu, Bassein, Akyab.

Hotels: Accommodation has been described as 'decrepit almost to the extent of charm'. There is no official rating system, but at least the more recent hotels are basic. It is advisable to book hotels well in advance, especially for Nov-Mar. Main Rangoon hotels can arrange interpreter or translation services if necessary. Some Rangoon hotels require payment in foreign currency. A 10 per cent service charge is usually added to the bill, and a 5-10 per cent tip is usual.

Credit cards: American Express credit cards are accepted at some Rangoon hotels.

Car hire: International driving licence required, but cars are generally hired with driver.

City transport
Taxis: Within Rangoon the government operates blue taxis with standard fares; otherwise fares by negotiation. Taxis can be shared or hired on time basis. Three-wheeled taxis and bicycle rickshaws are also available.
Rail: Rangoon has a circular rail system.

National transport: All arrangements for internal travel must be made well in advance via Tourist Burma. Travel outside Rangoon is difficult to arrange except for tourists.
Air: Burma Airways operates services to other main cities. There are some 43 airstrips in use.
Road: There are over 27,000 km of surfaced roads (1979), but road transport is difficult.
Buses: Government co-operatives operate long-distance coach services; local services are not generally recommended.
Rail: Burma Railways operates services to main centres, with sleeping accommodation available on the recommended Rangoon – Mandalay express service. The Mandalay – Lashio – Myitkyina service is also rated good. Restaurant or buffet car facilities are generally not available, and second class travel is not recommended.
Water: There is extensive river and coastal traffic, although the standard of passenger service varies.

Public holidays
Fixed dates: 4 Jan (Independence Day), 12 Feb (Union Day), 2 Mar (Peasants' Day), 27 Mar (Resistance Day), 1 Apr (Bank Holiday), 1 May (World Workers' Day), 19 Jul (Martyrs' Day), 1 Oct (Bank Holiday), 25 Dec.
Variable dates: Full Moon of Tabaung, Thingyan (2-3 days, Apr), Burmese New Year, Full Moon of Kason, Full Moon of Waso, Full Moon of Thadingyut, Full Moon of Tazaundgaing, National Day.

Working hours
Government: (Mon-Fri) 0930 – 1630.
Banking: (Mon-Fri) 1000 – 1430.
Shops: (Mon-Sat) 0900 – 1630; black market to 2000; vegetable market 0500 – 1200.

Social customs: Shoes and socks must be removed before entering any religious building. Revealing or sloppy clothing is not advisable. The title 'U' (pronounced 'oo') is the equivalent of 'Mr' in English.

Telecommunications
Telephone: Dialling code for Burma, IDD access code + 95 followed by area code (2 for Mandalay, 1 for Rangoon) followed by subscriber's number.
Telex: Telex facilities are available to foreigners at main hotels.
Telegram: Telegram service is available from the Posts and Telecommunications Corp in Rangoon.

Banking
State Bank: Union of Burma Bank. *State bank branches:* Myanma Agricultural Bank, Myanma Economic Bank, Myanma Foreign Trade Bank, People's Bank, United Commercial Bank. *Burma branches:* State Bank of India.

Advertising: Advertising is available in the press, in cinemas and on poster sites, which are subject to government control. Direct mail advertising is available.

Trade fairs: There is an annual Gem Emporium.

Electricity supply: 230 V AC, 50 cycles single-phase, using plugs with 2 round pins for lighting, 5 or 15 amp plugs with 3 round pins for power and bayonet-type light sockets; and 400 V AC, 50 cycles 3-phase for industrial use.

Weights and measures: Imperial system (metric system and local units also in use).

Representation in capital: Australia, Bangladesh, China PR, Czechoslovakia, Egypt, France, German DR, German FR, India, Indonesia, Israel, Italy, Japan, Korea PDR, Republic of Korea, Malaysia, Nepal, Pakistan, Philippines, Sri Lanka, Thailand, UK, USA, USSR, Vietnam, Yugoslavia.

BUSINESS DIRECTORY

Hotels: Rangoon
Inya Lake, Kaba Aye Pagoda Road, P.O. Box: 1045 (tel: 62866; tx: 21502).

Kandawgyi, Kanyeiktha Road, P.O. Box: 1467 (tel: 82255, 82327, 83925).

Strand, 92 Strand Road (tel: 81533; tx: 21220).

Thamada, 5 Signal Pagoda Road, P.O. Box: 311 (tel: 71499; tx: 21339).

Business advice and information
Burma Five Star Line Corp, Theinbyu Road, Rangoon.

Burma Ports Corp, 10 Pansoden Street, Rangoon (tel: 83122; tx: 21208, 21219; cable: PORTRUST).

Central Statistical Organisation, New Secretariat, Rangoon (tel: 15555 ext 22).

Myanma Export-Import Corporation, 622-624 Merchant Street, Rangoon (tel: 80266).

Myanma Foreign Trade Bank, Box 203, Theingyazy C Block, Schwedagon Pagoda Road, Rangoon (tel: 84911; tx: 21300).

Posts and Telecommunication Corp, 125 Phayres Street, Rangoon.

Socialist Economic Planning Committee, C/o Ministry of Planning and Finance, Rangoon.

Banking
Union of Burma Bank (central bank), 24-26 Sule Pagoda Rd, Rangoon (tel: 85300; tx: 21213).

Myanma Agricultural Bank, 1-7 Latha St, Rangoon (tel: 73018).

Myanma Economic Bank, 564 Merchant St, Rangoon (tel: 81819; tx: 21213).

Myanma Foreign Trade Bank, 80-86 Maha Bandoola Garden St, PO Box 203, Rangoon (tel: 84911; tx: 21300).

Travel information
Burma Airways, Rangoon (tel: 72911, 84566).

Tourist Burma, 77-91 Sule Pagoda Road, PO Box 559, Rangoon (tel: 78376, 75328, 80321; tx: 21330).

Namibia

Anthony Robinson

Namibia, often described as Africa's last colony, was in 1989 well on its way to independence under the terms of UN resolution 435. This called for UN-supervised elections to a constituent assembly which would draw up a constitution for the former German colony which was ruled by Pretoria from 1915. The South West Africa Peoples Organisation (Swapo), which fought a guerrilla war from bases in Angola for decades, emerged as the largest single party in the new Namibia. But geography and economics will oblige a post-independence government to maintain good relations with a South Africa which is itself in the midst of considerable political and social change.

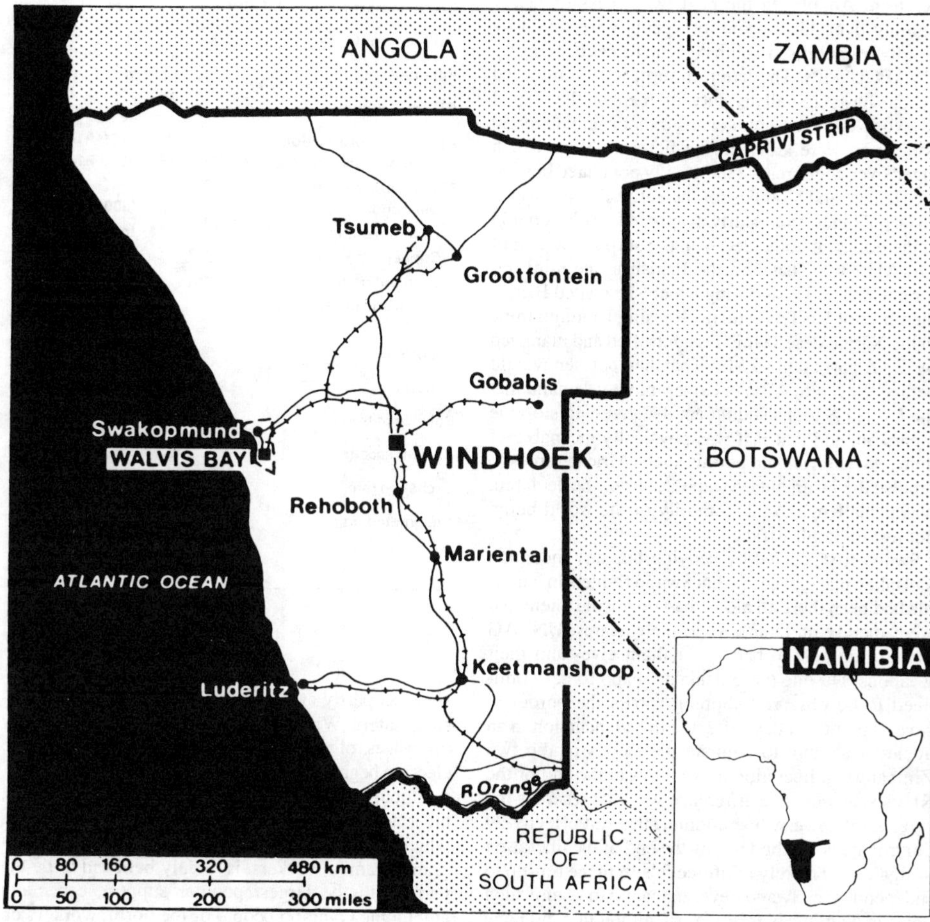

'An irreversible process has been set in motion that will bring Namibia to independence through free and fair elections,' UN Secretary-General, Javier Perez de Cuellar, told a cheering crowd when he arrived at Windhoek airport on 18 July 1989 to see for himself how the independence process was being implemented on the ground. In words addressed mainly to Pretoria's Administrator General Louis Pienaar, he insisted that 'it is imperative that resolution 435 be implemented fully both in letter and spirit'. After pledging that the UN would discharge its responsibility with complete impartiality he added: 'the world organisation expects no less from the South African authorities in this unique international undertaking'. In his diplomatic way Mr de Cuellar was underlining the delicacy of the task facing both the UN and South Africa as they worked together to try and manage the transition to independence.

Although South Africa, through its Administrator General, and the UN through the Secretary-General's special representative, Maarti Ahtisaari, were formally jointly responsible for managing the transition, the original UN independence plan was substantially altered in practice by the major change in superpower relations since the advent of glasnost and perestroika in the Soviet Union. The emergence of a Soviet government interested in regaining international respectability and lessening the burden on its own resources by seeking negotiated solutions to regional problems around the globe is the key which unlocked the Namibian independence process. It meant, in effect, that the path to independence was being co-guaranteed by Moscow and Washington as well as Pretoria and the UN.

What put resolution 435 back on track, after a decade of procrastination, was the agreement signed by Angola, Cuba and South Africa in Washington on 22 December 1988 to end the war in Angola through the withdrawal of all South African forces and the phased repatriation of 50,000-plus Cuban troops and Soviet bloc advisers. For years Pretoria, backed by Washington, insisted that it would not grant independence to Namibia while 50,000 well-armed Cuban troops were present in Angola, posing a potential security threat to South Africa with their sophisticated Soviet-supplied aircraft. A vital part of the New York accords was the setting up, at Pretoria's insistence, of a special Joint Commission to monitor all aspects of the agreement. Both Moscow and Washington were to be represented on the commission which also included representatives from the three signatory powers.

It was this agreement which brought the superpowers in as *de facto* co-guarantors of Namibian independence and it was thanks to a series of joint commission meetings at Mount Etjo in Namibia itself, in Luanda, Havana and Cape Town that the independence process survived its traumatic beginning and subsequent difficulties.

Implementation of Resolution 435

That independence would not be a trouble-free process became clear on April Fool's Day 1988. That day, 1 April, was the agreed date for the start of implementation of Resolution 435. At that point the bulk of the 4,650-strong UN Transition Assistance Group (UNTAG) backed up by 500 policemen (subsequently raised to 1,000) were supposed to have been in position along the border and at various other monitoring points around the country. At the same time all South African forces were supposed to have been withdrawn to barracks leaving policing in the hands of the South West Africa Police (SWAPOL), only lightly armed with side weapons. All Swapo forces meanwhile were supposed to have been withdrawn beyond the 16th parallel over 150 km from the Namibian border under the supervision of a special UN monitoring group.

Reality was very different. Thanks mainly to the opposition of Zimbabwe and other African states to cuts in the original plan to send 7,500 UN troops to Namibia the task of equipping and transporting UNTAG fell heavily behind schedule. Only a handful of UNTAG forces, mainly a British army signals unit, Australian engineers and the advance guard of other contingents from Switzerland, Denmark, Poland, Pakistan and others were in place by 1 April at former

KEY FACTS

Official title: Namibia
Capital: Windhoek
Official Languages: English, Afrikaans
Currency: South African Rand (R) = 100 cents
Exchange rate: R2.65 per $ (Nov 1989)
Area: 823,629 sq km

Population: 1.70m (1987)
GDP per capita: $1,014 (1986)
GDP real growth: 1%* (1986)
Inflation: 16% (1986)
Trade balance: $65m* (1986)
* estimated figure

KEY INDICATORS

	Unit	1983	1984	1985	1986	1987
Population	m	1.5	1.6	1.6	1.7	1.7
Gross domestic product (GDP)	R m	1,751	1,953	2,512	2,929	3,130
GDP per capita	R	1,592	1,627	2,093	2,333*	–
Inflation	%	11.9	9.2	11.2	16.0	–
GDP real growth	%	–5.4	–0.7	0.2	1.0*	2.9
Agricultural production	1979-81=100	92.4	92.8	97.3	99.3	121.6
Food production	1979-81=100	93.6	94.3	98.9	100.9	123.9
Mineral production						
Diamonds	'000 carats	963	930	910	1,010	1,020
Uranium oxide	tonnes	3,713	3,692	3,385	3,800*	3,700*
Copper	'000 tonnes	54	49	48	50	38
Lead (refined)	'000 tonnes	35	29	39	40	41
Zinc (concentrate)	'000 tonnes	56	57	57	60	70
Gold	kg	296	196	194	184	172
Exports	R m	925	1,083	719	795*	–
Imports	R m	977	1,094	562	955*	–
Balance of trade	R m	–52	–12	153	150*	–
Current account	$m	144	146	153	150*	–
Exchange rate	R per $	1.1	1.4	2.2	2.3	2.0

* estimated figure

South African army barracks close to the border like Grootfontein and Ondangwa. There were few, if any, UNTAG forces, actually on the border in the early hours of 1 April when a heavily armed SWAPOL armoured car patrol made the first contact with groups of armed Swapo guerrillas infiltrating across the border from Angola. In the next two weeks of heavy fighting over 330 Swapo insurgents were killed by rapidly re-mobilised South African forces backing up SWAPOL units. The latter, in turn, were reinforced by the hasty re-call to duty of men of the feared Koevoet (crowbar) counter-insurgency unit. Over 30 policemen were killed in the fierce fighting which raged over wide areas of the thickly populated border, the fiercest and bloodiest of the entire 23-year war.

Swapo's incursion nearly scuppered resolution 435 at the outset. It coincided with the surprise arrival of Mrs Margaret Thatcher, the British prime minister, for a whirlwind visit during which she visited British members of UNTAG and the Rossing Uranium mine in the Namib desert which is controlled and managed by RTZ corporation. Mrs Thatcher put her weight behind the South African Government's demand that the United Nations should act decisively to stop the incursion, and that if it was physically incapable of doing so that it should permit South African forces to repulse the insurgents. Swapo's action, she declared, was a challenge to the authority of the world body, not just to South Africa.

Swapo first reacted by denying that any incursion had taken place, arguing that Swapo fighters in Namibia had been there all along, had no hostile intent and were merely on their way to their nearest UNTAG monitoring post to hand over themselves and their weapons. Despite these denials Swapo forces continued to be observed slipping across the border at several points for days after the initial incursion in an apparent attempt to emulate the tactics of the two Zimbabwean liberation movements at the end of the Rhodesian war. The difference was that whereas the rival Zimbabwean liberation armies did effectively control areas of Rhodesia by the end of the liberation struggle, and merely reinforced them in the run-up to independence, Swapo never managed to secure such areas in Namibia before 1 April, although it did have numbers of political commissars with hidden weapons in place.

Swapo's illegal incursion, and its bloody sequel, raised the level of mutual suspicion among the main political groups inside Namibia and gravely damaged Swapo's external image. But it did not de-rail the independence process. This resumed after the Mount Etjo conference of 9 April which confirmed the political will of the superpowers and the major actors – Angola, Cuba and South Africa: to work out a regional settlement of which Namibian independence was the cornerstone. Despite a few weeks slippage in the original programme, the 1 November 1989 election date remained in force. Swapo forces were allowed free passage back to the border and were then taken under UN supervision to bases beyond the 16th parallel, over 150 km from the border. All the local citizen forces, commandos and ethnic forces were demobilised by 27 May and their bases closed by 23 June. On 1 July Mr de Cuellar announced that the number of South African troops had been reduced to 1,443 men confined to the two garrisons of Oshivelo and Grootfontein where they were monitored by UNTAG.

Less than a month after the first of an estimated 41,000 refugees started to return home, a plane full of Swapo leaders and officials arrived at Windhoek airport to an emotional homecoming on 19 June. They were led by Hage Geingob, former director of the UN-funded Namibia Institute, Hidipo Hamutenya, the press and information secretary and Theo Ben-Gurirab, Swapo's veteran diplomatic chief. Sam Nujoma, Swapo's president, was conspicuous by his absence, partly at least because of fears for his personal safety. With Swapo leaders back in the country the process of voter registration and organising for the election began in earnest.

Although Namibia boasts over 40 political parties, most are tiny and electorally insignificant. Swapo, which began life as the political organisation of migrant Ovambo workers, is widely believed to be the party with the largest popular support, especially among the Ovambo people of the north, workers on the diamond and uranium mines and among urban black Namibians in Katatura and other townships. Its main rival, the Democratic Turnhalle Alliance (DTA) is a multi-racial coalition of parties, including several which took part in the ineffective 'transitional government of national unity' created by Pretoria in June 1985. Its most prominent leaders include Dirk Mudge, the veteran leader of the mainly white Republican party, Moses Katjiuonga of the South West Africa National Union (SWANU) and Andreas Shipanga, leader of the breakaway Swapo-democrats. Mr Shipanga is one of the fiercest critics of the autocratic style and corruption of a Swapo leadership which had him jailed in Tanzania and has been accused by the families of Swapo detainees of torturing dissidents.

Both major parties enjoy substantial foreign funding. The main goal of Swapo was to achieve the two-thirds majority at the polls which would enable it to write its own constitution. Its leaders denied that it intended to create a one party state. The main aim of the DTA was to ensure that Swapo did not achieve such a majority and was obliged to sit around the negotiating table after 1 November to hammer out a compromise, multi-party constitution.

Economic boom

Meanwhile the run-up to the elections was accompanied by a short-lived economic boom, especially in Windhoek where the scramble for property and shortage of office space for diplomats, UN personnel, returning politicians and others led to a sharp rise in rents and a construction boom. Investors, including South African-based insurance companies, hurried to erect new office blocks and other accomodation. Higher mineral prices and the second year of good rains also boosted mining revenue, tax revenue and farm incomes while Anglo American and De Beers were among major mining houses to announce new investments in diamonds and gold mining.

At the same time, however, the withdrawal of 50,000 free-spending South African troops from the overpopulated, agricultural north and a further sharp decline in budgetary aid from Pretoria has placed a major question mark over the future economic health of Namibia. Over the last 20 years in particular Pretoria invested heavily in an excellent road network, a nationwide electricity grid, linked to the Calueque hydro-power system in Southern Angola and the South African grid, and other infrastructure. It also subsidised the rail network which has already been handed over to Namibia.

Building up a first-world quality infrastructure, however, has pushed the country into debt and requires expensive long-term maintenance to keep in order. Given the fact that over 90 per cent of the country's imports come from South Africa and that Pretoria will keep control of the Walvis Bay enclave, and hence the region's main deep sea port, it is clear that, in economic terms at least, a politically independent Namibia will have no more economic independence from South Africa than the land-locked mountain kingdom of Lesotho.

How Pretoria exercises its economic and financial power will depend on the policies pursued by a post-independence government. But even prior to the elections, Swapo economics and financial spokesmen were downplaying the party's nationalisation and other socialist policies and emphasising their belief in a mixed economy. The date for independence was unknown at the time of writing. But Namibia is expected to become an independent member of the United Nations, and probably also to join the British Commonwealth, some time in 1990 – barring dramatic surprises in the meantime.

COUNTRY PROFILE

Historical profile: Became German territory 1884 (except Walvis Bay which was British). 1920: mandated to South Africa. 1946: UN refused to sanction South Africa's attempted annexation. 1949: white population given representation in South African parliament. 1950: International Court of Justice (ICJ) ruled South Africa cannot change status of territory without UN consent. 1958: SWAPO (South West African People's Organisation) formed from anti-contract labour movement. 1959: SWANU (South West Africa National Union) formed. 1966: South Africa introduced apartheid laws; UN terminated mandate; SWAPO declared liberation struggle. 1967: UN Council for South West Africa established. 1968: UN renamed territory 'Namibia'. 1971: ICJ ruled South African presence in Namibia illegal, confirmed UN jurisdiction. 1973: UN recognised SWAPO as 'sole authentic representative of Namibian people' and established UN Commission for Namibia. 1975: South Africa organised 'internal' leaders and tribal chiefs into Turnhalle Conference which failed to gain international recognition. 1976: Security Council (SC) unanimously condemned South Africa's occupation and called for UN-supervised elections; Western Contact Group formed to promote independence. Target date set for independence 31 December 1978. 1977: South Africa appointed Administrator-General. 1978: Contact Group's plan for UN-supervised elections incorporated into SC Resolution 435; South African-backed 'internal' elections held in defiance of UN plan; Democratic Turnhalle Alliance (DTA) 82 per cent of 326,000 votes cast. SWAPO boycotted the elections. 1981: Peace conference in Geneva failed. 1983: DTA-led 'internal' government resigned. A multiparty conference of the most important political parties, excluding SWAPO, is called to break political deadlock. 1985: South Africa established Transitional Government (TG) a black majority government consisting of members of six different parties instructing it to draw up a constitution. 1988: South Africa turned down the TG's constitutional draft; independence and peace plan agreed by South Africa, Angola, Cuba, USSR and USA. 1989: Transitional Government resigned. Administrator General of South Africa took over power to govern country towards independence. Monitored by Mr Martti Ahtisaam, Special Representative of the Secretary General of the UN. UN peace plan for Namibia implemented to hold free and fair elections.

Population: The population of 1.252m (1988 estimate) is about 93 per cent black.

Main cities/towns: Windhoek (population 115,000), Swakopmund (15,500), Karibib, Omaruru, Okahandja, Otjiwarongo, Grootfontein, Ondangua, Gobabis, Rehoboth, Mariental, Keetmanshoop, Karasburg, Lüdemitz.

Language: English and Afrikaans are the official languages, and are predominantly used for business purposes. German also spoken. There are numerous African languages and Khoisan in daily use.

Media
Press: Daily newspapers include *Windhoek Advertiser, Allgemaine Zeitung, Die Republikein, Namibian, Times of Namibia, Namibia Nachrichten*. Weeklies are *Windhoek Observer* and *Sunday Republikein*. Bi-weekly is *Namibian Times*.
Radio: Local services in English, Afrikaans, German and several African languages broadcast by South West African Broadcasting Corporation (SWABC). (Namibia is also covered by SWAPO transmissions from neighbouring countries.)
Television: One local TV station under control of SWABC broadcasts on one channel in Afrikaans and English.

External trade: Involuntary member of Rand Monetary Area and Southern African Customs Union (SACU) with all tariffs, trade and exchange controls administered by South Africa. Traditional trade surplus, though fluctuates with changes in world mineral prices.
Exports: Minerals total 83 per cent of exports. Uranium (34% of total value), diamonds (31%), other minerals (21%), karakul pelts, meat and cattle (8%), fish (5%).
Imports: South Africa supplies 80 per cent of imports, covering practically all goods. The UK and German FR also supply foodstuffs and capital goods.

Agriculture: Second most important sector of economy, accounting for about six per cent of GDP. The commercial sector, almost entirely white-owned and situated in the central and southern parts of the country, produces 80 per cent of total agricultural yield. It centres on livestock, mainly cattle and karakul. In 1980 a record number of 3m pelts earned R52m, but in 1981-1982 prices reached record lows as a result of small demand and over-supply. Since 1983 prices started to pick up and the average price paid for pelts in 1987 was R50. Drought in 1980-84 has reduced cattle stock by 50 per cent. The reduction of available land through over-grazing is making serious inroads. There is also a subsistence sector of mainly black farmers, situated largely in the communal areas. Subsistence farming involves 80 per cent of the population and accounts for only two per cent of GDP.

Industry: Small manufacturing sector, contributing only about five per cent to GDP and employing 10 per cent of workforce. Major industries include fish processing, meat canning, production of mining equipment, tyre remoulding, printing and assembly of imported goods. Investment in industry is being encouraged by First National Development Corporation of South West Africa, but growth is hampered by lack of resources, limited market and availability of South African goods.

Mining: Backbone of economy, contributing 36 per cent to GDP and accounting for about 83 per cent of total export earnings (1986). Largest producer of gem quality diamonds. Has world's largest open pit uranium mine (Rossing – owned by Rio Tinto Zinc) and some of world's largest deposits of tin and lithium. There have been cut-backs in the mining of copper and other base metals because of Third World competition on international markets. Other minerals are also present, including caesium and rubidium. There are also gem stones such as aquamarine, amethyst, beryl, garnet and topaz. Extraction is dominated by foreign multinationals (mostly South African, UK or USA).

Energy: Dependent on South Africa for oil imports, although several surveys are underway seeking offshore oilfields. Large gas field has been discovered and exploitation is being considered. Installed electricity generating capacity of the South West Africa Water and Electricity Corporation (SWAWELC) is about 500 mW, much of which is hydro-electric.

Time: GMT +2.

Climate: Sub-tropical and dry. Hottest months Jan-Feb (20-29°C); coldest Jun-Jul (6-18°C).

Entry requirements
Passport: Required by all.
Currency: Import of local currency limited to R200 and export to R6,000. Foreign currency must be declared on arrival, and export is allowed up to declared amount.

Health precautions
Mandatory: Yellow fever vaccination certificate required if arriving from an infected area.
Advisable: Cholera, typhoid, polio vaccinations. Malaria risk exists in some areas. Water precautions should be taken.

Air access
National airline: Namib Air.
Other airlines: Air Botswana, South African Airways, Zambia Airways.
International airport: Windhoek Airport (Code: WDH), 40 km from city; bar, restaurant, buffet, currency exchange, post office, shops, car hire. N.B. Only direct overseas service to Frankfurt.

Surface access
Main ports: Lüderitz, Walvis Bay (South African).
Road: Tarred road from South Africa via Windhoek to northern border with Angola. The road from Botswana is not generally recommended. Roads total 41,762 km, of which 4,382 km tarred.
Rail: Main line from South African border connecting all main centres.

Hotels: Classified from one to five stars. Accommodation is limited and tends to be expensive, so should always be booked well in advance. Bungalow accommodation at nature reserves.

Electricity supply: 220 V AC.

Car hire: Available in Windhoek city centre and at airport. Although roads between major towns are of reasonable standard, the distances involved can be prohibitive. International driving licence required.

City transport
Taxis: Available in main towns; 10 per cent tip is usual.

National transport
Air: Namib Air operates frequent service between Windhoek, Walvis Bay, Swakopmund, Keetmanshoop, Grootfontein, Lüderitz, and other towns.
Road: Network of over 41,000 km, of which some 4,400 km is tarred, and is being expanded; mainly in south where proportion of tarred roads is high.
Rail: Almost 2,400 km of track. Network connects Windhoek with Lüderitz, Walvis Bay, Grootfontein, Gobabis and South Africa. Two classes and restaurant cars available on route running South Africa (de Aar)-Windhoek.

Public holidays
Fixed dates: 1 Jan, 1 May, 3 Oct, 10 Dec, 25 Dec, 26 Dec.
Variable dates: Good Friday, Easter Monday, Ascension.

Telecommunications
Telephone: Dialling code for Namibia: IDD access code + 264, followed by 61 for Windhoek; dialling code for Walvis Bay: IDD access code + 27 followed by 642 (through South Africa). Telephone services are modern and efficient.
Telex: Telex facilities at good hotels. International access code for Namibia is 50-908.

Hotels: Windhoek
Kalahari Sands, 129 Kaiser Street, P.O. Box: 2254 (tel: 36900; tx: 3174; fax: 36900).

South West Safari, Republic Road, P.O. Box: 3900 (tel: 38560; tx: 438; fax: 35652).

Thuringerhof, Kaiser Street, P.O. Box: 112 (tel: 226031; tx: 3227).

Other useful addresses

Bank of South West Africa/Namibia Ltd, Swabank Centre, Bülow Street, PO Box 1, Windhoek 9000 (tel: 37300; tx: 908898).

Chamber of Commerce and Industries, Kaiser Street, PO Box 191, Windhoek (tel: 222000; fax: 33690).

First National Development Corporation of SWA, Private Bag 13252, Windhoek (tel; 306911; tx: 870).

Nauru

(map on page xxiv)

COUNTRY PROFILE

Political system: Nauru became an independent republic in 1969, having been administered by Australia under a trusteeship agreement with the UN. The 18-member parliament elects the president from among its members. President Hammer deRoburt survived two attempts to remove him in Dec 1986/Jan 1987, and was returned to power in a general election. He has been president for most of the period since 1969.

Political parties: Until 1987, all parliament members were independents. The Nauru Party (formerly the only party, formed in 1976) had effectively closed down its activities, but in March 1987 the opposition leader, Kennan Adeang, revived it as the Democratic Party of Nauru, and claimed that eight members of parliament had joined it.

Population: The indigenous population is of Micronesian descent, but a considerable cultural influence derives from Australia.

Language: Micronesian and English are the official languages.

Media
Broadcasting: Limited radio service which broadcasts local news, etc.

Domestic economy: The Republic has spent many years in preparation for the exhaustion of phosphate reserves, investing in overseas property and updating its communications and infrastructural facilities. It operates an airline and a shipping line and supplies cheap housing and free education. It is looking at the possibility of creating an offshore banking system, based on its absence of income tax. Nauru is seeking compensation from the British, Australian and New Zealand governments for the alleged failure to honour agreements about the restoration of the island and the redeployment of its inhabitants.

Employment: Overseas workers comprise nearly 40 per cent of island's total population. Phosphate industry and government service provide the bulk of employment.

External trade: Nauru does not publish import and trade statistics.
Exports: Phosphates are sole export, mainly to Australia (about 50%), New Zealand (about 35%), Japan (3.5%).
Imports: Virtually all necessities imported, including water, mainly from Australia and New Zealand.

Agriculture: Arable land confined to site 100-300 metres between the beach and a cliff, where coconut palms and pandanus are grown. Some vegetables grown, some pigs and poultry kept; coastal canoes supply fish for local consumption.

Industry: Phosphate processing is the only industry.

Mining: Nauru's phosphate reserves represent the highest-grade phosphate ore in the world. In 1979 reserves estimated at about 38m tonnes; at current rates of extraction, about 2m tonnes p.a. (depending on weather and other conditions), they will last until 1995.

Main companies: Nauru Phosphate Corp.

Membership of international organisations: Commonwealth (special member not attending Heads of Government meetings), ESCAP, ICAO, INTERPOL, South Pacific Agreement on Regional Trade and Economic Co-operation, South Pacific Commission (participating member), SPF, UPU.

BUSINESS GUIDE

Time: GMT + 12 hrs.

Climate: Tropical, tempered by sea breezes, with variable rainfall. Temperatures range from 24°C to 34°C in the shade. Monsoon season from Nov-Feb.

Entry requirements
Passport: Required by all.
Visa: Required by all except passengers in transit.
Currency: No limit to the amount of currency, either Australian or foreign, which may be taken into Nauru.
Customs: Importation of firearms, ammunition and animals (including pets) prohibited.
Income tax: No income tax.

Health precautions
Mandatory: Yellow fever vaccination, if arriving from an infected area.

Representation overseas: Agana, Apia, Athens, Auckland, Hong Kong, Honolulu, London, Melbourne, New Delhi, Oakland, Rarotonga, Saipan. Suva, Sydney, Taipei, Tokyo, Tonga.

Surface access: Nauru has its own shipping company, Nauru Pacific Line. Main links are with Australia, New Zealand and Japan.

Car hire: Car hire can be arranged locally.

National transport
Road: A main road (19.3km) circles the island, and all residential areas are linked by surfaced roads.

Public holidays
Fixed dates: 1 Jan, 31 Jan, 17 May (Constitution Day), 26 Oct (Angam Day), 25 Dec.
Variable dates: Easter.

Working hours
Business: (Mon-Fri) 0800 – 1200, 1330 – 1645.
Banking: (Mon-Thu) 1000 – 1500; (Fri) 1000 – 1700.

Social customs: Tipping is not customary.

Telecommunications
Telephone: Dialling code for Nauru, IDD access code + 674 followed by subscriber's number (no area code). Telephone service automatic, with approximately 1,800 telephones in 1986 (300 of which were business lines).
Telex: Telex service is available.

Banking: Bank of Nauru.

Electricity supply: 240 V AC.

Weights and measures: Metric system.

Representation in capital: Australia (for other countries, contact the appropriate representative in Suva, Fiji).

Hotels: Nauru
Meneng (tel: 3210).
Od-N-Aiwo (tx: 33093).

Business advice and information
Nauru Finance Corporation, PO Box 306 (tel: 3390; tx: 33081). Responsible for promoting economic diversification, including offshore banking.

Secretary for Island Development and Industry (tel: 3320; tx: 33081 GOBNARU).

Travel information

KEY FACTS

Official title: Republic of Nauru
Head of state: President Bernard Dowiyogo
Capital: Yaren District
Official Languages: English, Nauruan
Currency: Australian Dollar (A$ = 100 cents)
Exchange rate: A$1.26 per $ (Nov 1989)
Area: 22 sq km
Population: 8,000* (1988)
Labour force: 1,850 (1988)
Trade balance: –$60m* (1987)

* estimated figure

Nepal

Nicholas Nugent

The dispute between Nepal and its large neighbour India, putting the former under virtual trade siege, was the dominant concern of Nepal during 1989. The dispute arose in March 1989 when India refused to renew trade and transit treaties with the land-locked mountain kingdom. India was unhappy about the quantity of goods filtering across the 800 km long border between the two nations. However, it was widely believed that political factors played their part too. India was unhappy about a large arms deal Nepal signed with its northern neighbour, China, whose commitment to peace along its southern border has never entirely been accepted by India. There was also resentment in India that Nepal had introduced a work permit requirement for Indians working in Nepal.

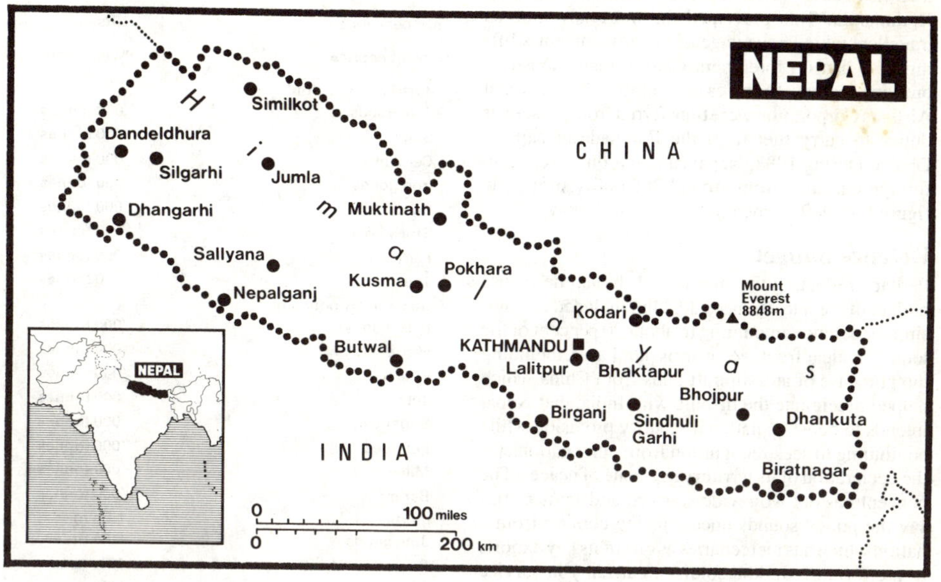

India wanted to combine arrangements for trade and for the transit of third country imports and exports across Indian territory in a single treaty. Nepal refused arguing that the right of a land-locked state to transit goods through neighbouring states was guaranteed by international law and should not be subject to a treaty which covered other matters too. Trade, by contrast, was a bilateral affair which should be dealt with by treaty, said Nepal. With each side holding firmly to its position, India closed 13 of the 15 border crossing points, leaving only those at Raxaul and Jobhani open. The immediate result was a shortage of goods in Nepal, especially in the Kathmandu valley. Fresh vegetables and fuel supplies were worst affected, with senior officials and ministers taking to bicycles as petrol was rationed. Most buses were off the roads, there were long queues at filling stations and the government closed schools and universities to conserve fuel. Some drivers preferred to sell their fuel ration on the black market at many times the regular price. Medicines were also in short supply.

Dispute strains relations

The government claimed that the shortage of domestic fuel for cooking was exerting additional pressure on Nepal's fast declining forests. This was already the subject of ecological concern because of the alleged connection between deforestation and the erosion of the mountain slopes in Nepal, the silting up of major rivers of the northern subcontinent, and severe annual flooding in Bangladesh. The charge that its action had worsened deforestation and erosion rates angered India further. In any case, there seemed little urgency in Delhi to resolve the bilateral differences. Privately, senior Indian officials admitted that India considered there was a need to 'teach Nepal a lesson'. India claimed that a treaty between it and Nepal forbids Nepal's purchase of arms from any other nation, an interpretation of the 1950 treaty with which the government in Kathmandu disagrees.

After a while, though, the main lesson that Nepal seemed to have learnt was of the need to diversify its sources of imports to lessen an unhealthy over-dependence on India. Other neighbours, like Bangladesh, Pakistan and China, all gave support and helped substitute the shortages resulting from India's partial land blockade. But Nepal's Finance Minister B. B. Pradhan, returning from a visit to China, conceded that Nepal's scope for diversifying its import and export trade away from India was limited by the long distances involved in bringing petrol and other goods through the Tibetan region of China, the only other country with which Nepal has a land border. The strains between India and Nepal brought relations between their two governments to a new low point, and threatened the peace and security of the whole region.

However, some Nepalis blamed their government for its handling of the dispute, and for allowing Nepal to become so heavily dependent on India in the first place. At least 75 people were arrested during a student protest march in Kathmandu, during which police fired into the air to disperse the crowd. Though it was believed at the time that the students' union were taking advantage of the trade dispute to advance their campaign for educational and social reforms, there were indications that, in taking their protests on to the streets, they had the support of the two main banned political parties, the Nepali Congress and the Communist Party. (Under the country's non-party panchayat system, political parties are not allowed to exist. The Nepali Congress, though, which has led the campaign for democratic reform, claims an active membership of 50,000 and the support of many more of Nepal's 18m people.)

Some even blamed King Birendra, who effectively heads the government, for provoking the dispute. He was said to have authorised the purchase of arms from China in the clear knowledge that it would arouse India's anger. By the time the National Panchayat, or parliament, adjourned its session in mid-August 1989 – it is not due to meet again till June 1990 – there was still no sign that India would accept repeated calls for talks to try to resolve the dispute. Nepal was learning to live under its straitened economic circumstances.

Economic problems

The slow-down in trade inevitably sapped Nepal's fragile economy. Economic growth was expected to be only about one quarter of the 5.6 per cent target for the year. Worse hit was industry with disruption in the availability of imported raw materials for, among others, the export-oriented textile industry which employs 20,000 workers, as well as of fuel to run the factories. Following the expiry of the trade treaty, India banned the export of most goods to Nepal and imposed heavy duties of most imports from Nepal, effectively making them uncompetitive and therefore unexportable. Before the dispute, around 40 per cent of Nepal's trade was with India.

The situation worsened after Nepal decoupled its rupee from the Indian currency, which had previously circulated freely in Nepal. Henceforth the government regulated all foreign currency which it needed to finance costly alternative imports. Further angry words were exchanged between the two countries after India closed the Budge Budge dock facility at Calcutta port in August, preventing special supplies of oil from Singapore being off-loaded there. India said the closure was for routine maintenance, but Nepal believed India was angered by the fact that

KEY FACTS

Official title: Kingdom of Nepal

Head of state: King Birendra Bir Bikram Shah Dev

Head of government: Prime Minister Marich Man Singh Shrestha

Capital: Kathmandu

Official Languages: Nepali (English also spoken)

Currency: Rupee (NRs = 100 paisa)

Exchange rate: NRs23.9 per $ (Nov 1989)

Area: 147,181 sq km

Population: 18.3m (1988)

GDP per capita: $159 (1988)

GDP real growth: 4.5% (1988)

Inflation: 9.0% (1988)

Trade balance: –$96.1m (1988)

Foreign debt: $947m (1987)

Visitor numbers: 266,000 (1988)

transit trade to Nepal through the port had actually increased since India's own ban on the export of most goods to Nepal.

Another victim of the trade war was Nepal's tourist business. Charter operators cancelled flights rather than risk being unable to refuel aircraft on arrival in Kathmandu. This prevented the more adventurous travellers taking advantage of the government's lifting of a ban on foreigners visiting many sensitive outlying regions. In any case, aircraft of Royal Nepal Airlines Corporation were transferred from passenger duties to carry fuel from the Bangladeshi capital, Dhaka. During 1988, Nepal earned around $6.4m in foreign currency from around 266,000 visitors: the figure for 1989 seemed likely to be well down.

Defence budget

Perhaps reflecting the strains with India, the record budget of Nepalese rupees 20.24bn included a provision for defence amounting to about 20 per cent of the total. It is clear from recent arms purchases, including the purchase of anti-aircraft guns from China which helped precipitate the dispute with India, that Nepal intends to boost its national security provision while continuing to seek recognition from other nations for the declaration of its territory as a zone of peace. (The concept has not won wide support, and some critics say the policy sounds unconvincing coming from a nation which has mercenaries as one of its key exports in the form of Gurkha soldiers, currently in service with armies and police forces in Britain, India, Brunei and Singapore.)

A large part of the remainder of the budget was committed for rural development in an attempt to raise living standards in a country which ranks near the bottom of the international league in per capita GNP terms. Lesser amounts were allocated for agricultural development including irrigation, transport and communications, electricity and power expansion and education and health. To help meet commitments, airport embarkation tax was sharply increased. The finance minister told parliament he would be seeking additional foreign loans. Pakistan was one of the first to offer help in the wake of Nepal's dispute with India, offering commodity loans to the value of US$1m.

Even before the dispute with India, concerns about Nepal's rate of deforestation were recognised. The 1988 Bangladesh flood, the country's most extensive ever, led to major discussions, involving the World Bank among others, as to measures to prevent a recurrence. In Nepal, the concern was more domestic: denuded mountain sides are both unsightly and economically worrying since erosion makes cultivation more difficult. A forestry master plan was implemented, with help from the Asian Development Bank. It provided for a massive programme of reafforestation as well as education to enhance public awareness of the problems created by cutting down forests. The idea was to prevent the level of forestation falling below the current 37 per cent of land area. As much as 240 hectares of forest was being destroyed each day during the shortage of imported fuel, according to a senior forestry department official.

Arrests

Former Foreign Minister Rishikesh Shah, a long-time critic of the non-party panchayat system and of alleged human rights abuses, was jailed on his return from India for claiming in a seminar at a Delhi university that the Indo-Nepal dispute was directly linked to the absence of human rights in Nepal. He said police routinely tortured critics of the government and that more than 300 people were at that time being held in Nepali prisons for political offences. He said they included journalists, trade unionists and teachers.

A former Forestry Minister, Hem Bahadur Malla, was meanwhile convicted of abusing his authority by allowing timber to be sold at a discounted price to a furniture manufacturer. His conviction followed an investigation by the commission for the prevention of the abuse of authority.

KEY INDICATORS

	Unit	1984	1985	1986	1987	1988
Population	'000	16,110	16,630	17,131	17,557	18,300
Gross domestic product (GDP)	NRs m	39,390	41,417	50,428	57,828	67,835
GDP per capita	NRs	2,299	2,510	2,981	3,294	3,707
Consumer prices	% increase	2.8	8.1	19.0	10.8	9.0
Industrial production						
Jute goods	'000 tonnes	21.3	19.2	16.3	18.2	15.8
Sugar	'000 tonnes	17.5	11.0	15.1	24.5	33.5
Cement	'000 tonnes	39.2	31.5	96.0	151.6	185.4
Iron goods	'000 tonnes	12.2	15.3	28.3	34.5	–
Soap	'000 tonnes	5.6	6.5	9.1	11.4	–
Straw board	'000 tonnes	0.4	0.9	0.9	0.4	–
Fertilizer	'000 tonnes	1.6	1.4	2.0	0.8	–
Tea	'000 tonnes	0.8	1.0	1.0	1.0	–
Agricultural production						
Rice (paddy)	'000 tonnes	2,757	2,709	2,804	2,372	2,900
Maize	'000 tonnes	761	820	874	868	902
Wheat	'000 tonnes	634	534	598	686	744
Jute	'000 tonnes	25	33	61	23	26
Sugar-cane	'000 tonnes	509	408	558	586	816
Potatoes	'000 tonnes	383	420	357	391	566
Millet	'000 tonnes	115	124	138	138	150
Barley	'000 tonnes	22	24	23	25	24
Exports	NRs m	1,703.9	2,740.6	3,078.0	2,991.4	4,433
Jute goods	NRs m	212	260	167	164	–
Raw jute and cuttings	NRs m	38	44	2	15	–
Rice	NRs m	76	250	93	14	–
Ghee	NRs m	46	39	47	47	–
Timber	NRs m	8	26	–	–	–
Imports	NRs m	6,521	7,874	9,286	11,851	15,095
Balance of trade	$m	–282.8	–293.9	–350.2	–477.2	–96.1
Current account	$m	–121.6	–119.2	–123.3	–280.2	–47.2
External debt	$m	473	596	747	947	–
Debt service ratio	%	3.9	5.3	9.1	9.7	–
Exchange rate	NRs per $	16.46	18.25	21.23	21.82	23.29

* estimated figure

COUNTRY PROFILE

Political system: Nepal is ruled by a hereditary monarchy advised by councils *(panchayat)* under a system of non-party democracy. Village and town councils elected by universal adult suffrage send representatives to district councils, which send representatives to zonal councils. The unicameral legislature, the Rashtriya Panchayat (National Council), consists of 140 members, of whom 112 are elected for a five-year term, and 28 are royal nominees from different walks of life. The prime minister, who heads the council of ministers, is elected by the members of the Rashtriya Panchayat and then appointed by the king.

Political parties: There are no legal political parties.

Population: The population is a mixture of Indo-Caucasian and Tibeto-Mongoloid tribesmen, but also includes a number of Tibetan refugees. Hinduism and Tibetan Buddhism are the dominant religions; Nepal is the only officially Hindu Kingdom in the world.

Main cities/towns: Kathmandu, Pokhara, Patan, Bhaktapur, Biratnagar, Birgunj, Bhairahawa, Dharan, Nepalguraj, Dhangadi.

Language: Nepalese is the official language, but English is sometimes spoken in urban centres.

Media

Broadcasting: Radio programmes are broadcast in Nepali and English. TV transmissions in English and Nepali started in 1985 in the Kathmandu valley. Recently TV programmes are also transmitted in Pokhara and Biratnagar.

Domestic economy: The economy is based on small subsistence farming villages and is slow to develop, inhibited by lack of resources, lack of power supplies and inadequate internal and external road links. Tourism is an important source of foreign exchange. The current Five Year Plan (1985-1990) emphasises raising living standards to a basic minimum including necessities of food, clothing, housing, education, health and security, and increasing employment through development of the agricultural sector, water resources, industry, export trade and tourism. Nepal relies heavily on foreign aid to finance its trade deficit and development plans. Only about 40 per cent of outstanding public debt has actually been distributed, indicating a limited ability to implement those development projects in hand.

External trade: Nepal is one of the 32 developing nations designated in 1985 to receive exemption from US competitive trade requirements in accord with the Generalised System of Preferences.

COUNTRY PROFILE

There is a 20 per cent rebate on duty payable on imports from Bangladesh, Bhutan, Pakistan, Sri Lanka and Maldives. In 1984 the government abandoned its policy of discouraging non-essential imports, with the intention of increasing import duty revenues.
Exports: Main exports are readymade garments (26% of 1986 total), garments and woollen carpets (12.1%), hides and skins (11.5%), jute products (5.4%), raw jute, ghee (3%), timber (2%), leather goods, spices and handicrafts. Main destinations are German FR (7.2% of 1986 total), USA (26%), India (14%), UK (3.5%), Japan, China PR, Republic of Korea, Singapore, Hong Kong and Canada.
Imports: Main imports are manufactured goods (29.5% of 1986 total), machines and transportation equipment (22.9%), chemicals and drugs (12.5%), mineral fuels and lubricants (11.3%), food and live animals (10.4%) and raw materials. Main sources are India (42.5% of 1986 total), Japan (16.1%), German FR (2.9%), USA (1.8%), UK (1.0%), Australia, New Zealand, Canada and other Asian countries.

Agriculture: Agriculture accounted for 62 per cent of GDP (1987), providing most foreign exchange earnings and 90 per cent of employment. Cultivated areas occupy 16 per cent of total land area, with pasture occupying another 12-14 per cent. Production is extremely vulnerable to adverse weather conditions, with little irrigation practised, and severe soil erosion is becoming a problem. Deforested plains and lower hilltops are terraced for rice production. Development priority is placed on irrigation, increased use of fertilisers and improved seeds. Maize, wheat, millet, barley, sugar-cane, jute, oil seeds, tobacco, fruit, vegetables and medicininal herbs are grown in higher valleys. Cattle, buffaloes, goats, sheep, pigs, yaks and poultry are raised. River fish are an important source of protein. Forests occupy 25-33 per cent of total land area, but will mostly disappear within 15-25 years at present rates of land clearance and exploitation; consequently, afforestation programmes are being implemented and a public awareness programme is under way.

Industry: Industry, mainly small-scale cottage enterprises, accounts for 12 per cent of GDP and employs less than two per cent of the workforce. Jute and sugar processing are the main sectors, with recent growth of import substitution industries such as cotton and synthetic textiles, cement and paper. Other products include leather goods, footwear, pharmaceuticals and agricultural tools. The government offers various incentives to foreign investment in industry, and has established industrial estates in Balaju, Patan Dharan, Hetaunda and other towns. Tourism directly or indirectly employs over six per cent of the labour force, and attracts some 223,000 people a year.

Mining: Few minerals are commercially exploited. Quartz, lignite, copper, cobalt and iron ore are present. The government has invited foreign companies to search for oil.

Energy: Timber resources account for over 75 per cent of energy used. It is estimated that there is potential for hydroelectric power generation sufficient to supply most of the Indian subcontinent, but existing capacity is about 200 mW, and expansion is almost prohibitively expensive but is proceeding gradually. The Asian Development Bank is helping establish a national electricity grid, with hydroelectric power plants under construction at Arun Third 11 and Marsynagdi (due for completion in 1989).

Financial markets: A securities exchange has been set up.

Membership of international organisations: Asian Development Bank, CCC, Colombo Plan, ESCAP, FAO, G-77, IBRD, ICAO, IDA, IFAD, IFC, ILO, IMF, IMO, INTERPOL, IPU, IRC, ITU, NAM, South Asian Association for Regional Co-operation, UN, UN Security Council 1988-89, UNESCO, WHO, WMO, World Bank, WTO.

Time: GMT + 5 hrs 45 mins.

Climate: The narrow lowland belt known as the terai has a warm climate with monsoon rains in summer. The high mountains are permanently snow-covered above 16,000 ft, while the middle mountain area has a temperate climate.

Entry requirements
Passport: Required by all except nationals of India.
Visa: Visas are required by all except nationals of India. Tourist visas for certain areas are issued on arrival for visits of seven days. Applications for other visas should be made well in advance.
Currency: There are no limits on the amount of foreign currency which may be imported or exported, although only nationals of Nepal and India may import Indian currency. Up to 10 per cent of Nepalese currency purchased in the country may be exported. It is best to retain all foreign exchange receipts. There may be limits on the amount which can be re-exchanged.
Customs: Personal effects are allowed duty-free. Importation of certain goods may be prohibited, such as beef, animal fat, arms, explosives and wireless radio transmitters. Cameras and films only in reasonable qualities, for personal use. Except for trade with India, which is possible only through authorised dealers, all imports and exports are under licence.

Health precautions
Mandatory: Vaccination certificates for yellow fever if travelling from an infected area.
Advisable: Vaccination against typhoid and paratyphoid and anti-malarial precautions in the Terai. Health regulations should be checked before departure.

Representation overseas: Bangkok, Beijing, Bonn (FRG), Cairo, Dhaka, Islamabad, Jeddah, London, Moscow, Nairobi, New Delhi, New York, Paris, Rangoon, Tokyo, Washington.

Air access
National airline: Royal Nepal (RNAC).
Other airlines: Biman Bangladesh, Dragonair, Druk-Air, Indian, Lufthansa, PIA, Singapore, Thai International.
International airport: Tribhuvan (KTM), 6 km from Kathmandu.

Surface access: Since the start of the trade dispute there are only two access routes from India, with rail service between Raxaul - Birganj, linking to Kathmandu coach services, and rail access also via Jogbani. Since Mar 1985 it has been permitted to enter Nepal by road from China (Tibet). Nepal can also be approached from Kakarbhita in the east.
Main ports: Nepal is landlocked, and most cargo enters via the Port of Calcutta.

Credit cards: Major credit cards, such as American Express, Visa and Mastercard, are accepted by some hotels.

Car hire: Chauffeur-driven car hire only is available.

City transport
Taxis: Metered taxis can be hailed in Kathmandu. Private taxis available through hotels may cost more.

National transport
Air: Royal Nepal Airlines operate domestic services to about 40 airfields, with some curtailment of services likely during monsoon season to certain mountain regions.
Road: The present road network of over 6,000 km is being extended. Kathmandu, Pokhara and Biratnagar are linked by surfaced road. Transport is difficult outside main centres. However, the completion of the Mahendra Highway in 1988 should make West Nepal accessible throughout the year.
Buses: Bus services, which operate within Kathmandu and to centres served by roads may be crowded.
Rail: The only line serves Jaynagar to Janakpur and Bizalpura.

Working hours
Business: (Sun-Fri) 0900 – 1800.
Government: (Sun–Thur) 1000 – 1700, 1000 – 1500 (Fri); to 1600 Nov-Feb.
Banking: (Sun-Thur) 1000 – 1700, (1000 – 1330 (Fri); to 1600 Nov-Feb.

Telecommunications
Telephone: Dialling code for Nepal, IDD access code + 977 followed by: 1 for Kathmandu, Bhaktapur and Patan, 71 for Bhairawa, 21 for Biratnagar, 51 for Birgunj, 91 for Dhangarhi, 41 for Janakpur, 81 for Nepalgung and 61 for Phokhara.

Banking: The banking sector has recently been opened to foreign banks.
Central bank: Nepal Rastra Bank.
Main centre: Kathmandu.
Head offices: Nepal Bank, Rashtriya Banijya Bank, Nepal Arab Bank, Banque Indo-Suez, Nepal Grindlays Bank.
Representative offices: Standard Chartered Bank, City Bank.

Weights and measures: Metric system (local units also in use).

Representation in capital: Australia, Bangladesh, Burma, China PR, Egypt, France, German DR, German FR, India, Israel, Italy, Japan, Korea PDR, Republic of Korea, Pakistan, Thailand, UK, USA, USSR.

Hotels: Kathmandu
De L'Annapurna, Durbar Marg, P.O. Box: 140 (tel: 221711, 221531; tx: 2205 AAPU NP).

Everest Sheraton, New Baneswor, P.O. Box: 659 (tel: 220567; tx: 2260 HOTEVS NP).

Malla, Lekhtnath Marg, P.O. Box: 787 (tel: 410320; tx: 2238 NP).

Na'rayani, Pulchowk, Hotel Narayani Road, P.O. Box: 1357 (tel: 521711; tx: 2262 NARAYNI NP; fax: 521219).

Shangri-La, Lazimpat, P.O. Box: 655 (tel: 412999, 410108; tx: 2276 HOSANG NP).

Shanker, Lazimpat, P.O. Box: 350 (tel: 410151/2; tx: 2230 NP).

Soaltee Oberoi, Tahachal, P.O. Box: 97 (tel: 211211; tx: 2203 NP; fax: 220191).

Star, Thamel Tuladhar, P.O. Box: 1812 (tel: 11004; tx: 2314 NP).

Woodlands, Durbar Marg, P.O. Box: 760 (tel: 220623, 222683; tx: 2282 HOTWOOD NP).

Yak and Yeti, Durbar Marg, P.O. Box: 1016 (tel: 413999; tx: 2237 YKNYTI NP; fax: 227782).

Business advice and information
Central Bureau of Statistics, PO Box 1284, Singh Durbar, Kathmandu (tel: 13422).

National Planning Commission, PO Box 1284, Singha Dubar, Kathmandu (tel: 215000).

National Trading Ltd, PO Box 128, Teku, Kathmandu (tel: 21962; tx: 2211).

Nepal Chamber of Commerce, PO Box 198, 2 Nepal Bank Building, Dharmapath, Kathmandu (tel: 212005; tx: 2349).

Nepal Economic & Commerce Research Centre, PO Box 285, 7/358 Kohity Bahal, Kathmandu (tel: 215336).

Nepal Industrial Development Corp, NIDC Building, PO Box 10, Durba Marga, Kathmandu (tel: 411221, 411225).

The Netherlands

Ronald van de Krol

The sudden emergence of the environment as a pressing political and economic issue took the Netherlands by surprise in 1988-89. An environment-related squabble within the centre – right coalition brought down the government of Prime Minister Ruud Lubbers in May 1989, marking the first time that a European government had fallen over a 'green' issue. The crisis precipitated a premature general election in September 1989, eight months before national elections had been due to be held. Although the government fell on an environmental issue, it had long become clear that all Dutch political parties, even the coalition partners, were in favour of launching an ambitious and expensive national environment plan to clean up the worst of the country's pollution problem – they simply could not agree on how to pay for it. Equally obvious was the fact that the added cost of pollution controls to business and consumers, while hefty, would do little to dent the buoyancy of the Dutch economy.

Government collapse

The immediate and most obvious cause of the government's fall was an internal dispute over how to pay for the national environment plan. The junior coalition party, the Liberals (VVD), refused to go along with plans to raise extra revenue by scrapping tax breaks for commuters and by levying extra duties on diesel petrol. Mr Lubbers' Christian Democrats (CDA), for their part, refused to give in to VVD pressure, leading to the collapse of the cabinet in early May 1989.

However, in 'green' politics as in every other kind of politics, appearances can deceive. While the environmental dispute was indeed the main factor in the government's unravelling, it was also a smoke screen for other, more deep-seated rivalries and disputes within the cabinet. Nearing the end of two successful four-year terms in office with the larger CDA, the VVD clearly felt the need to raise its own electoral profile by setting itself apart from the CDA. Its decision to make a major issue of fuel taxes and commuters' tax-breaks was designed to appeal to its natural constituency – well-off home-owners in the suburbs of Amsterdam, the Hague, Rotterdam and Utrecht.

But the VVD had other reasons to be annoyed with the CDA. Mr Lubbers, after seven years at the helm and with his personal popularity undiminished, was becoming too heavy-handed for the VVD's liking, acting more like a powerful British prime minister than a traditional Dutch one, who is generally expected to play the role of 'first among equals' in the cabinet rather than that of final arbiter of government policy.

Tension between the VVD and CDA had been building up for much of 1988 and 1989, and the two partners frequently found themselves at loggerheads on a variety of issues, including spending on education and the long-running and complicated saga of the Netherlands' seeming inability to mass-produce a new tamper-free EC passport. So when the final break came, the news was not a great surprise.

As the main player in the cabinet crisis, the VVD went into the September elections under a bit of a cloud. On the one hand, it looked likely to meet disgruntlement among voters for its role in precipitating a cabinet crisis which, to the outside world at least, seemed unnecessary and eminently avoidable. On the other hand, its desire to govern once more with Mr Lubbers and his CDA party met with a cool reception from the Christian Democrats.

The final electoral outcome was not expected to become clear until late 1989, as Dutch general elections are usually followed by a long period of horse-trading among the victorious parties: in the months following an election, the parties which have proved themselves strong enough to form a coalition make sure that their policy objectives are clearly spelt out in detail in a written coalition agreement.

But by the summer of 1989, it appeared that one possible outcome of the election might be a coalition between Mr Lubbers' CDA and the Labour Party (PvdA). While the CDA had been instrumental in reining in state spending in two successive coalitions with the VVD since 1982, a large and vocal section of the party had always expressed doubts about these cuts and had hankered for the days when the party was not associated with the closing-down of old peoples' homes and hospitals.

For this reason, a resurrection of the CDA – PvdA coalition that had been in power before 1982 did not seem at all that far-fetched, though it did raise the question of how both Mr Lubbers and the equally popular PvdA leader Wim Kok could be fitted into one cabinet – and who would be prime minister if they could agree to join forces at all.

One foregone conclusion, however, was that the new cabinet would not contain Finance Minister Onno Ruding, who had made clear beforehand that he wanted to return to private life. As the architect of Mr Lubbers' programme of spending cuts, it was Mr Ruding who managed to maintain fiscal discipline when individual members of the cabinet sought to make special spending pleas for their own departments. As the seven-year-old CDA – VVD coalition drew to an end, it was Mr Ruding, more often than not, who managed to combat the sense of 'austerity fatigue' within the coalition and to reject the calls for increased social spending from both within and outside the government.

Economic policy

The withdrawal of Mr Ruding from Dutch politics made it difficult in mid-1989 to forecast the future of fiscal and economic policy, especially as it usually takes until months after a general election before this portion of the coalition agreement is finally hammered out. But there is certainty about the past: the Dutch economy in 1988-89 put in its strongest performance of the 1980s, crowning the attempts by two successive Lubbers-led cabinets to pull the country out of the recession of the early 1980s. However, despite undisputed success in restoring profits in the private sector, Dutch state finances were still none too rosy in 1988-89, and high unemployment continued to defy the prescriptions of government policy.

When Mr Lubbers, a former businessman and a former economic affairs minister, took office as prime minister in 1982, he found an economy reeling from recession: companies were going bankrupt at an alarming rate; unemployment, which was to peak at more than 17 per cent in 1984, was rising steadily; and interest rates were high, discouraging fresh investment.

Recovery during the first four years of the Lubbers administration was basically export-led, as Dutch firms took advantage of the dollar's strength to export to the USA and to snatch away market share from US companies on other markets. At a later stage, higher foreign sales were followed up by a surge in Dutch corporate investment, helping to spread the recovery to other sectors of the economy.

The growth of exports, of course, had more to do with the recovery in the health of world trade and the movement of exchange rates than with the wisdom of Mr Lubbers' own policies. Yet Mr Lubbers did make a significant contribution in his own right: his pruning of welfare-state spending, while not radical by the standards of less prosperous and perhaps less socially-minded economies, jolted the country out of lethargy, and made plain that cradle-to-grave care by the state could not be financed forever from natural gas revenues, particularly as gas prices were plummeting just as rapidly as the oil prices to which they were linked.

In his own way, Mr Lubbers inspired a mini-revolution: the profit motive became socially acceptable again, and he managed to sever the automatic link that had existed before between rises in private-sector pay and increases in social security and unemployment benefits.

To be sure, there has been no 'Lubbers revolution' to match that of Mrs Margaret Thatcher in Britain. Social spending has been cut, but the foundations of the welfare state are firmly in place, and there has been no serious questioning of the role of the state in caring for the individual. As the OECD put it in its survey of the Dutch economy: Cutbacks in the income-support system have been aimed more at reducing the cost of the programmes than the programmes themselves.'

During Mr Lubbers' second term in office, he moved to cut taxes, bringing down the top rate of value-added tax to 18 per cent from 20 per cent and reducing corporate tax to 35 per cent from 42 per cent. Starting

KEY FACTS

Official title: The Kingdom of the Netherlands
Head of state: Queen Beatrix
Head of government: (caretaker) Prime Minister Ruud Lubbers (Christian Democrat)
Capital: Amsterdam (The Hague is the seat of government)
Official Languages: Dutch
Currency: Guilder (fl) = 100 cents
Exchange rate: fl 1.99 per $ (Dec 1989)
Area: 41,473 sq km
Population: 14.8m (1988)
GDP per capita: fl 30,571 (1988)
GDP real growth: 3.0% (1988)
Labour force: 5.8m (1987)
Unemployment: 7.0% (1988)
Inflation: 0.7% (1988)
Trade balance: $8,171m (1988)

in 1990, the top marginal rate of income tax for individuals was scheduled to come down to 60 per cent from 72 per cent previously. Though Dutch tax-payers still face one of the world's stiffest regimes of tax and social security charges, tax cuts have helped bring the Netherlands' fiscal system more into line with that of its neighbours – a crucial prerequisite for remaining competitive in post-1992 Europe.

Successes and challenges

Going into the elections, Mr Lubbers had one strong selling point – the undisputed strength of the Dutch economy. Export volume rose by 8 per cent in 1988, outpacing a 7 per cent increase in imports and boosting the country's trade surplus to 7.6bn guilders. Gross domestic product expanded by 3 per cent in real terms, more than double the previous year's rise of 1.3 per cent.

In 1988, the Netherlands also managed to extend its recent record for putting in the best inflation performance of the OECD, producing an inflation figure of just 0.7 per cent for the year. Although the Nederlandsche Bank, the Dutch central bank, cautioned repeatedly throughout 1989 that inflationary tendencies were on the rise and must be nipped in the bud, most economists predicted that the Dutch inflation rate in 1990 would be 1 per cent or slightly higher.

Corporate investment, a mainstay of the economy since the mid-1980s, continued to show strong growth. In 1988 alone, industry's investment in fixed assets jumped 10.8 per cent, the highest rate seen since the heyday of post-war reconstruction. With government spending about flat in 1988 and consumer spending down slightly on 1987, corporate investment made the single biggest contribution to the 3 per cent expansion of GDP. All in all, the resurgence of Dutch corporate investment since the recession of the early 1980s was remarkable. Between 1983 and 1988, investment surged by an average annual rate of 6.6 per cent; by contrast, in the recession of 1979-1982 brought on by the second oil crisis, investment had dropped by an average of 4.5 per cent per year.

For all these successes, the economy continued to face problems on two fronts in 1988-89: joblessness, particularly among the low-skilled, long-term unemployed; and state finances.

The Netherlands' unemployment problem is a complicated one. On the one hand, its record of job creation in the latter part of the 1980s was unparalleled in Europe. Employment (as measured in persons rather than person-years) rose by 2 per cent a year in the four years to 1988, twice the EC average. Yet, on the other hand, unemployment itself remained stubbornly high. The explanation for this dichotomy was that the new jobs tended to go to better-trained, younger workers. Older workers with low skills, as well as former 'guest' workers from Turkey and Morocco, benefited little from the economic upturn. At the same time, the chances of finding a job for the long-term unemployed dwindled. In 1984, only 7 per cent of the unemployed had been out of work for four years or more, but by 1988, this percentage had increased to 18 per cent.

Progress had been made, but analysis of the situation was made more difficult by the introduction in January 1989 of a new method of gathering jobless statistics. Under the old method, which produced a 1988 unemployment rate of 13.7 per cent, the social affairs ministry calculated the number of unemployed according to the number of people registered at official labour exchanges. This system came under heavy criticism because the records of the labour exchanges were thought to be out-dated, causing the unemployment problem to be overstated. The new method, which would have produced a 1988 figure of less than 10 per cent, is based on random surveys of households conducted by the Central Statistical Office. Though this system is widely considered to be more accurate, it too does not take account of 'hidden unemployment' among housewives and others, and may understate the problem.

Whatever the true figure, it was clear that some improvement had been achieved, though the overall problem remains intractable. Reschooling and retraining are increasingly being emphasised in an effort to bolster skills. At the same time, various groups in the Netherlands have begun to study the success of Sweden – an equally generous welfare state – in combatting unemployment. Already, the Dutch are exploring the Swedish model of allowing proposed tripartite councils made up of local employers, unions and local authorities to have great leeway in furnishing jobs. But other features of the Swedish system – strictly limiting both access to the dole and an unemployed person's freedom to turn down job offers – would seem less likely to win acceptance in the Netherlands.

In the OECD's analysis, part of the reason why structural
unemployment in the Netherlands is so intractable is that Dutch minimum wage levels are too high and that the range of Dutch wages is too narrow.

In 1987, for example, the Dutch minimum wage (which is not one single figure but varies according to the age of the employee) was equivalent to about 65 per cent of the average wage earned by all Dutch workers. Though this percentage has fallen steadily since a high of 80 per cent in the late 1970s – bringing about, in effect, a relative decline in the minimum wage – there is still not enough of a gap between minimum wages and average wages, and this in turn reduces an employer's incentive to hire someone who has been out of work for years and who may not have the immediate skills needed.

Another problem, in the OECD's view, is that after-tax minimum wages are not much higher than the net minimum income which a potential employee receives while he or she is still on social welfare. 'For married persons on assistance, entering employment near the minimum wage is unlikely to lead to a significant increase in total disposable income,' it said.

Proposals for lowering minimum wages are traditionally controversial in the Netherlands, making this weapon difficult to employ in the fight against unemployment. Instead, the government has tried to approach the problem from the other direction by reducing social-security contributions paid by employers on behalf of lower-paid workers; in this way, by cutting the cost of labour, the government hopes to stimulate the hiring of people at or near the minimum wage. Similarly, by making hiring and firing more flexible, and by allowing employers to take on employees on limited-duration labour contracts for up to one year, the government has tried to make the inflexible Dutch labour market more supple.

As far as state finances are concerned, the government has arguably been making more significant headway in this troublesome area than on the labour market. Total government spending as a percentage of national income fell 2.6 points to 44.6 per cent in 1988. This may still be high by the standards of most other Western countries, but it at least marked the resumption of a downward trend. In 1986 and 1987, for example, government spending as a percentage of national income had shown small increases, marring the good record built up since 1983.

The buoyancy of the economy has helped boost the government's tax receipts. More importantly, higher tax receipts have, in turn, helped compensate for a decline in revenue from natural gas sales that has plagued the Netherlands since the mid-1980s. In 1988, for example, gas revenue fell by a further 3.7bn guilders to 7.3bn guilders, meaning that the Netherlands' gas revenue has fallen by a full two-thirds from their 1985 levels.

All in all, by mid-1989 state finances appeared to be on course, even though the cost of servicing state debt continued to grow. The target Mr Lubbers set for himself in 1986 – reducing the budget deficit to 5.25 per cent of national income in 1990 – looked very likely to be met, whatever the composition of the new

KEY INDICATORS

	Unit	1984	1985	1986	1987	1988
Population	'000	14,362	14,419	14,567	14,660	14,760
Gross domestic product (GDP)	fl bn	400.25	418.2	428.61	431.22	451.23
GDP per capita	fl	27,868	29,003	29,423	29,415	30,571
Gross fixed capital formation	fl bn	74.31	80.25	86.18	86.69	96.67
GDP (1980 prices)	fl bn	337.46	344.23	385.47	390.10	–
Consumer prices	1985=100	97.8	100.0	100.1	99.4	100.1
Share prices	1980=100	77.3	100.0	128.7	129.2	119.7
Interest rate (official disc.)	%	5.00	5.00	4.50	3.75	4.50
Money supply	fl bn	144.36	159.76	167.94	174.53	198.42
Wages (hourly earnings)	1985=100	95	100	102	103	104
Unemployment	%	14.1	12.9	12.0	11.5	7.0
Agricultural production	1979-81=100	109.90	107.91	120.04	115.85	115.35
Industrial production	1985=100	96	100	100	101	101
Manufacturing	1980=100	105	107	111	112	–
Exports	fl m	209,589	224,845	195,064	184,412	204,521
Imports	fl m	199,310	214,839	184,396	184,108	196,882
Trade balance	$m	5,656	5,462	7,168	5,160	8,171
Current account	$m	6,571	4,027	3,620	3,540	5,310
Foreign exchange	$m	9,237	10,782	11,191	16,003	16,075
Exchange rate	fl per $	3.21	3.32	2.45	2.03	1.98

* estimated figure

COUNTRY PROFILE

Political system: A constitutional and hereditary monarchy with legislative power vested in the bicameral parliament (Staten Generaal). The Eerste Kamer (First Chamber) is composed of 75 members and is elected by members of the Provinciale Staten. The Tweede Kamer (Second Chamber) consists of 150 members and is elected directly every four years on the basis of proportional representation. The Second Chamber proposes and amends bills, which can only be approved or rejected by the First Chamber. Executive power is exercised by the Council of Ministers, led by the Prime Minister which is responsible to the Staten Generaal. Ministers may be drawn from outside Parliament. MPs becoming ministers give up their seats to party colleagues on official 'list'.

Main cities/towns: Amsterdam (population 679,140), Rotterdam, The Hague, Eindhoven, Haarlem, Groningen, Utrecht.

Media
Press: Over 50 regional daily newspapers with total circulation of 4m. Five national morning and two national evening papers, including *De Telegraaf* (circulation 706,000), *Algemeen Dagblad* (417,000), *De Volkskrant* (68,500), *NRC-Handelsblad* (215,100) and *Trouw* (122,600). Several weekly magazines, most of which published by United Dutch Publishers (NDU). Eight major financial/business magazines including *Intermagazine* (54,150), and *Elseviers Weekblad* (weekly, 120,000), and *Bedrijfsdocumentaire* (80,165) and *Management Team* (monthly, 110,000). Foreign newspapers and magazines available at kiosks, bookstalls and hotels.
Broadcasting: TV and radio network coordinated by Nederlandse Omroep Stichting (NOS). Programmes broadcast by about 30 organisations, which are allotted transmission time according to the number of subscribers. TV programmes broadcast in original language – often English – with Dutch sub-titles. Overseas broadcasting in Dutch undertaken by Radio Nederland Wereldomroep. About 85 per cent of homes converted to cable TV. Satellite channels available. 30 per cent of homes subscribe to teletext.
Advertising: Total advertising spend (1986) fl 7,128bn. Accepted in all forms of media, though TV and radio advertising restricted to 30 and 40 mins/day respectively. Most advertising direct, in newspapers and magazines. Main advertising subjects: cars, finance, food. Advertisements for TV and radio accepted by Stichting Ether Reclame (STER). Advice on agents from the Dutch advertising association, Nederlandse Vereniging van Erkende Reclame Adviesbureaus (VEA).

Domestic economy: Broad industrial base with heavy dependence on foreign trade. Also large service sector, with public spending accounting for around 55 per cent of GDP. Growth rate, although below EC average in the past years, is now a little higher than EC average. Despite recent improvements in economic performance, unemployment and public sector deficit remain high. Current emphasis on export-led growth and monetary and fiscal restraint.

Employment: Around 80 per cent employed in private sector, mainly manufacturing and services. High levels of unemployment. After a rising unemployment trend until 1984, unemployment as percentage of labour force is now at a level of around 10 per cent. Government and union consensus over strategy for jobs involving work creation, part time jobs, training programmes and work-sharing. Restrictive policies and wage restraint reduced real earnings by an average of 1.6 per cent per annum between 1980 and 1984 before they started to revive in 1987.

External trade: Regular trade surplus since 1981 owing to increased competitiveness and relatively low levels of imports. Major exports include petroleum, gas, organic chemicals, artificial resins and plastics, foodstuffs, machinery and electronics. Petroleum and petroleum products account for over 10 per cent of total imports. Other major imports are non-electrical machinery, raw materials and transport equipment. Around 70 per cent of foreign trade is conducted with EC countries, particularly West Germany, Belgium and France. However, noticeable rise in exports (most of which are aid tied) to developing countries. Amsterdam is also major centre for trade in tobacco, diamonds and precious metals.

Agriculture: Small-scale but highly specialised and efficient, and an important part of the economy. Trend towards larger and more highly mechanised farms. Dairy farming is most important activity, involving over half the country's farmers. Horticulture also increasingly significant. Sector employs only five per cent of workforce but contributes around four per cent to GDP and accounts for nearly 15 per cent of total exports. Principal products include wheat, cheese, butter, barley, oats and meat.

Industry: Sector contributes around 30 per cent to GDP and is largely dependent on foreign markets, both for raw material imports and sales. Large finished-product industries predominate, most of which are concentrated in Randstad i.e. the western part of the country. Major industries include food processing, engineering, metals, chemicals, shipbuilding, textiles and electronics.

Leading companies: Royal Dutch Shell, Unilever NV/PLC, Philips, DSM, Nationale Nederlanden, Akzo, Fokker.

Mining: Almost totally confined to gas and oil.

Energy: Major producer and exporter of natural gas, with sufficient reserves to last well into next century (1.9 trillion cubic metres). Oil production may soon meet around 20 per cent of domestic requirements. Emphasis on developing coal, gas, solar and wind power. Development of nuclear power has been slow owing to popular opposition. Social welfare system is to some extent dependent on revenue from natural gas.

government. But Dutch state finances are not out of the woods yet. On the one hand, economic growth is expected to slow slightly in the 1990s, and this would naturally affect tax receipts. On the other hand, the easy cuts in state spending have all been made already. It will be more difficult to cut into the bone than it was to slice through the fat. In addition, the Netherlands' biggest two privatisation operations – the sale of a 33 per cent shareholding in chemicals group DSM in early 1989 and the flotation of a second 33 per cent holding in the autumn of 1989 – were both due to be completed by the end of the 1989 fiscal year, leaving the government with no comparable source of extra revenue in the years ahead.

Economic outlook

Four weeks before the September 1989 election, the semi-official Central Planning Bureau (CPB) released its forecasts for the medium-term economic outlook up to 1994 – and immediately showed that all the political parties had been slightly too optimistic when drawing up their electoral programmes. In the CPB's view, national income will rise by 2.25 per cent a year in 1990-94, and not by the 2.5 per cent annual rate foreseen by the politicians, nor by the 3 per cent rate enjoyed in 1988.

All in all, the CPB report held out the promise of respectable, but unexciting, economic growth in the years ahead. Another message was that further spectacular rises in Dutch corporate profits should not be expected, partly because rises in wage costs, which had been moderate through most of the 1980s, are likely to accelerate. And inflation, while still low by comparison with other countries, will probably run at an average rate of 1.75 per cent. Corporate investment and exports, estimated to rise by an annual 4.5 per cent and 5.25 per cent respectively, will remain important pillars of the economy, the office said.

The CPB's estimates have more than a simple academic appeal. The figures will be used by the new government to draw up a coalition agreement, and the whole edifice of commitments on spending and finances will be built around the office's forecast. The CPB's own assessment is that policy-making will not be any easier in the early 1990s than it was in the late 1980s. 'Even with a reasonably favourable economic climate in the first half of the 1990s, it will be extraordinarily difficult to bring about a realistic reconciliation of desires in the areas of state services, tax burdens, purchasing power, and employment,' the CPB document said.

The environment

Clearly, one of the new government's most urgent policy tasks would be to launch the country's national environment plan. The outlines of the plan were drawn up without too much trouble by the outgoing Lubbers government, but its financing – the issue that brought the coalition down – was left to be sorted out by the new cabinet.

The speed with which the environmental issue moved to the forefront of political debate caught the Netherlands by surprise in 1988 and 1989. If the precise origins of this new environmental awareness can be pinpointed at all, they probably lie in a startling Christmas Day address given by Queen Beatrix in 1988. The queen, departing from her tradition of touching on a number of subjects in her yuletide address to the nation, devoted her entire speech to the calamitous state of the environment. Without mincing words, she warned the Netherlands that time was running out: 'The earth is slowly dying. We human beings have become a threat to our planet,' she said, adding 'the future of creation itself is now endangered.'

These words were to usher in a period of unusual political activism on environmental issues. In the EC, for example, the Netherlands dropped its usual penchant for consensus – and even risked being hauled before the European court of justice — by insisting in early 1989 on its right to introduce tax-breaks designed to stimulate the purchase of cars with pollution-cutting catalytic converters. In the end, the 'green wave' that swept over the Netherlands also swept over Brussels: far from making the Netherlands an outcast, the EC later announced, to the satisfaction of the Dutch, that it would introduce tough US-style exhaust standards in the early 1990s. Meanwhile, the Dutch were given permission to go ahead with their tax-break scheme.

What accounts for this Dutch attachment to the environment? The answer is partly geographic. The most densely populated country in Europe, the Netherlands lies on the North Sea, reputed to be the world's most polluted sea, and serves as an outlet for the River Rhine, Europe's longest and dirtiest inland waterway. Huge manure surpluses produced by Dutch farmers, and industrial waste generated by Dutch firms, add to the problem, of course.

The national environment plan aims to cut pollution by 70 per cent over the next 20 years through the setting of stricter standards for emissions which cause acid rain, through promoting public transport and through conserving fuel and natural resources. As such, it is the most ambitious and concerted clean-up campaign ever devised by a national government. It will not come cheaply. By 1994, the total cost of the clean-up, to be borne by industry, consumers and government alike, will double to 15bn guilders from 8bn guilders in 1988. And by the year 2010, some 3.5 per cent of GNP will be devoted to pollution-cutting measures, more than double the 1988 figure of 1.5 per cent.

The government itself will come off relatively cheaply. The basic principle behind the plan is the idea that 'the polluter should pay', and by this reckoning some 80 per cent of the plan's costs will be borne by consumers, farmers, and Dutch-based corporations. The latter two categories will, of course, try to pass on as much of these extra charges as possible in the form of higher prices. Industry itself is not expected to feel any adverse effects of the planned measures. On the contrary, capital goods producers in particular hope to benefit from increased investment in environmental protection.

Foreign policy

In July 1989, US President George Bush made a whirlwind, 24-hour stopover in the Netherlands on his way home to Washington from the Western economic summit in Paris. The visit, the first ever by a US president, was meant mainly to correct this long-standing oversight.

But equally, Mr Bush's decision to go to the Netherlands served as a reward for the services rendered by the Dutch in the interests of the Western alliance. With the controversy of the early 1980s over nuclear cruise missiles more or less forgotten, the Netherlands was now highly regarded in Washington for its ability to mediate in disputes within the Western alliance.

This was particularly true of the row in mid-1989 over the modernisation of short-range nuclear missiles. Mr Lubbers and Foreign Minister Hans van den Broek were widely praised for their efforts to help bridge the gulf between West Germany on the one hand and the USA and the UK on the other.

In the EC, the Netherlands also earned a reputation as a mediator. This was particularly true when the issue at hand involved the UK, because this allowed Mr Lubbers to make use of his personal rapport with Mrs Thatcher.

Elsewhere, the only other burning issue in the field of foreign relations in mid-1989 was the matter of the Netherlands' ties to its former Dutch colony of Surinam, located on the northern coast of South America. Although formal relations were restored after a civilian government took over the reins of power from the Surinam military, The Hague and Paramaribo were still at odds over the conditions which the Netherlands wished to see fulfilled before it 'thawed' 1.6bn guilders in aid that was frozen soon after the military seized control.

COUNTRY PROFILE

Banking: System is extensive and provides a full range of services. Four main banking institutions (apart from the central bank): mortgage banks, commercial banks, savings banks and agricultural/co-operative banks.

Legal system: Civil law system incorporating French penal theory.

Membership of international organisations: ADB, Benelux, CCC, Council of Europe, DAC, EC, ECE, EIB, ELDO, EMS, ESCAP, ESRO, FAO, GATT, IAEA, IBRD, ICAC, ICAO, ICES, ICO, IDA, IDB, IEA, IFAD, IFC, IHO, ILO, IMF, IMO, INRO, INTELSAT, International Lead and Zinc Study Group, Interpol, IPU, IRC, ITC, ITU, IWC, NATO, OECD, UN, UNESCO, UPU, WEU, WHO, WIPO, WMO, WSG.

Time: GMT + 1 (GMT + 2 from late Mar to late Sep).

Climate: Temperate; warm summers and mild/cold winters. Temperatures range from about two degrees C to 25°C. Rainfall throughout year but heaviest Nov-Jan.
Clothing: Mediumweight throughout year, topcoat in winter. Raincoat useful any time.

Entry requirements: Visa not required for nationals of most European countries and several others, for visits of up to three months.
Health: No vaccination certificates required. EC nationals are covered for medical treatment.
Currency: No restrictions on import or export of currency.
Customs: Personal effects duty-free.

Air access: Regular flights by all major international airlines.
National airline: KLM (Royal Dutch).
Main international airport: Schiphol (AMS), 15 km south of Amsterdam.
International airports: Rotterdam (RTM), Maastricht (MST), Groningen (GRQ).

Surface access: Good access by road and rail from all surrounding countries, plus ferry services to all major ports. Boat trains operate to Hook of Holland from many European countries.
Main ports: Rotterdam, Hook of Holland, Scheveningen, Vlissingen, Amsterdam.

Hotels: Classified from one- to five-star by Netherlands Board of Tourism, the Royal Dutch Touring Club and the Royal Netherlands Automobile Club. Accommodation may be booked through the Netherlands Reservation Centre in Leidschendam (tel: (070)202500). Advisable to book well in advance during spring and summer. Foreign nationals must present passports before booking in, and are automatically registered with local police. Single room (five-star) costs up to about fl 250 per night, including 15 per cent service charge. Guest houses, bungalows and summer houses also available.

BUSINESS GUIDE

Restaurants: Wide variety, with many specialities often available. Service charge and VAT included in prices on menus additional tips discretionary.

Currency
Exchange rate: Netherlands is a participant in the European Monetary System. Guilder is pegged to other participants' currencies within a band of 2.25 per cent (six per cent with respect to the Italian lira and the Spanish peseta).

Foreign exchange information
Account restrictions: There are no restrictions on international payments of any kind.
Trade policy: Import licences not required for goods from most countries. Goods which do require an import licence or registration include some animal derivative and agricultural products, electrical equipment, raw iron and steel, some textiles, explosives and firearms. Also food products subject to quality and hygiene restrictions.

Credit cards: All major credit cards are accepted.

Car hire: Widely available. Addresses can be obtained from Tourist Information Offices (VVV). Rates vary from about fl 35-150 per day plus fl 0.25-1.00 per km, depending on make and model. Cheaper rates for rental periods of a week or more. Drive on the right, usually overtake on left. Speed limits: urban areas 50 kph, normal roads 80 kph, motorways 120 kph. Wearing of seat belts compulsory. Information and assistance from Royal Dutch Touring Club (ANWB).

City transport: Nationwide fares zoning system also applies to cities. Buses, trams and metro charge same fare per zone. You can buy a 'Nationale Strippenkaart' comprising 15 tickets (two for first zone and one for each additional zone) at railway stations, transport companies, post offices and some VVV offices.
Taxis: Recognised by 'taxi' signs on top. Usually boarded at ranks, at railway stations and other parts of the city, or called by telephone. Fares vary from city to city. The meter price includes a service charge.
Buses: Good services throughout cities linking tram services to provincial routes.
Metro: Rapid transit systems in Amsterdam and Rotterdam.
Trams: Good service in central regions of most large cities.

National transport: Whole country divided into zones. Fares for bus travel depend on number of zones crossed. Tickets (covering up to 45 zones) should be obtained in advance, available from railway stations, transport companies, some VVV offices and post offices. Separate tickets are required for trains, available from railway stations.
Air: Internal service linking Amsterdam with Groningen, Eindhoven, Rotterdam and Maastricht.
Road: Approximately 110,000 km of roads including 2,000 km of motorways (autowegen). Frequent bus connections between all cities. Journey times usually less than four hours.
Rail: Almost 3,000 km of track, mostly electrified. Fast regular services operated by Netherlands Railways (Nederlandse Spoorwegen).
Water: Extensive inland waterways. Scheduled boat services operate from Enkhuizen to Urk and Staveren and between the mainland and the islands in the north.

Public holidays
Fixed dates: 1 Jan, 30 Apr (Queen's Day), 5 May (Liberation Day), 25-26 Dec.
Variable dates: Good Friday, Easter Monday, Ascension Day, Whit Monday.

Working hours
Business: (Mon-Fri) 0830 – 1730.
Government: (Mon-Fri) 0830 – 1700.
Banking: (Mon-Fri) 0900 – 1600, some open Sat and on late shopping evenings.
Shops: 0830/0900 – 1730/1800, (half day closing usually Mon or Wed). Main shops open Sat and Thu/Fri evening.

Social customs: Dutch people shake hands frequently. When invited to a meal in a Dutch home it is courteous to give flowers or a small gift.

Business language and interpreting/translation: Available through the Association of International Conference Interpreters or the United Dutch Translation Office. Dutch is the official language, but English and German widely spoken and most business houses will correspond in English.

Telecommunications
Telephone: Fully automatic inland and overseas service. Telephone information given in French, English and German. Credit facilities available. Cheap rates from 1800 to 0800 and weekends. All calls can be made from public booths and post offices. Telephone booths accept 25 c, fl 1 and fl 2.5 coins, or cards, which can be bought at post offices, VVV offices and shops displaying the 'PTT-telephone card' poster. International dialling code: 31 followed by area code, 20 for Amsterdam, 10 for Rotterdam, followed by subscriber's number. For international calls from the Netherlands, the IDD access code is 09.
Telex: Available in main hotels. Country code 44 NL.
Telegram: Available at all main post offices. Telegrams can also be sent direct from telephone kiosks.

Postal service: Rates vary for both land and air mail. Following items may be sent by air without extra postage: fl 1.30 air letters, letters to all European countries, postcards and money orders.

Trade fairs: Several trade fairs and exhibitions held throughout the year. Two large exhibition centres: RAI in Amsterdam and the Jaarbeurs in Utrecht. Also smaller centres in Ahoyhal in Rotterdam and in Maastricht.

Electricity supply: 220 V AC.

Representation in capital: Algeria, Argentina, Australia, Austria, Belgium, Brazil, Bulgaria, Canada, Cape Verde Islands, Chile, China PR, Colombia, Costa Rica, Cuba, Czechoslovakia, Denmark, Egypt, El Salvador, Finland, France, German Democratic Republic, Federal Republic of Germany, Greece, Holy See, Honduras, Hungary, India, Indonesia, Iran, Iraq, Ireland, Israel, Italy, Japan, Kenya, Republic of Korea, Kuwait, Lebanon, Luxembourg, Malaysia, Mexico, Morocco, New Zealand, Nicaragua, Nigeria, Norway, Pakistan, Peru, Philippines, Poland, Portugal, Romania, Saudi Arabia, South Africa, Spain, Sudan, Suriname, Sweden, Switzerland, Tanzania, Thailand, Tunisia, Turkey, USSR, United Kingdom, USA, Uruguay, Venezuela, Yemen AR, Yugoslavia, Zaïre.

Hotels: Amsterdam
Alexander, Prinsengracht 444, 10017 KE (tel: 267721; tx: 13161).

American, Leidsekade 97, 1017 PN (tel: 245322; tx: 12545; fax: 253236).

Amstel Intercontinental, Prof. Tulpplein 1, 1018 GX (tel: 226060; tx: 11004).

Arthur Frommer, Noorderstraat 46, 1017 TV (tel: 220328; tx: 14047).

Caransa Crest, Rembrandtsplein 19, 1017 CT (tel: 229455; tx: 13342; fax: 222773).

Delphi, Apollolaan 101-105, 1077 AN (tel: 795152; tx: 16659).

Hans Brinker, Kerkstraat 136, 1017 GE (tel: 220687; tx: 12127).

Hilton, Apollolaan 138-140, 1077 BG (tel: 6034567; tx: 15186).

Hotel de l'Europe, 2-4 Nieuwe Doelenstraat, 1012 CP (tel: 234836; tx: 12081).

Museum, PC Hoofstraat 2, 1017 (tel: 831811; tx: 14275; fax: 160320).

Pulitzer, Prinsengracht 315/331 1016 GZ (tel: 228333; tx: 16508; fax: 276753).

Trianon, J.W. Brouwersstraat 3, 1071 LH (tel: 831811; tx: 14275; fax: 160320).

Victoria, Damrak 1-6, 1012 LG (tel: 234255; tx: 16625; fax: 252997).

Car hire: Amsterdam
Avis: Nassaukade 380 (tel: 836061; tx: 15669; Schiphol Airport (tel: 6041301).

Budget: Overtoom 121 (tel: 126066).

Europcar: Overtoom 51-53 (tel: 832123; tx: 17191).

Hertz: Overtoom 333 (tel: 122441; tx: 15160); Schiphol Airport (tel: 170866; tx: 11103).

Chambers of Commerce
Kamer van Koophandel en Fabrieken voor Amsterdam, Koningin Wilhelminaplein 13, 1062 HH Amsterdam (tel: 172882; tx: 18888 amtra).

Kamer van Koophandel en Fabrieken voor Midden Gelderland, Nieuwe Plein 1B, 6825 KX Arnhem (tel: 516969; tx: 45276 kkarh).

Kamer van Koophandel en Fabrieken voor Haarlem en Omstreken, Nassauplein 4-6, 2011 PG Haarlem (tel: 319017; tx: 41567 kkhlm).

Kamer van Koophandel en Fabrieken voor s'Gravenhage, Alexander Gogelweg 15, 2517 JH The Hague (tel: 795795; tx: 33003 kkgrh).

Kamer van Koophandel en Fabrieken voor Rotterdam, Beursgebouw, Coolsingel 58, 3011 AE Rotterdam (tel: 4057777; tx: 23760 kvkr).

Kamer van Koophandel en Fabrieken voor Tilburg en Omstreken, Reitseplein 1, 5037 AA Tilburg, (tel: 654524; tx: 52384 vaspa).

Kamer van Koophandel en Fabrieken voor Utrecht en Omstreken, Waterstraat 47, 3511 BW, Utrecht (tel: 331412; tx: 47730 kkutr).

Kamer van Koophandel en Fabrieken voor Noordelijk Overijssel, Weeshuisstraat 27, 8011 TZ Zwolle (tel: 218047; tx: 42281 kkzwl).

Banking
De Nederlandsche Bank (central bank), Westeinde 1, 1000 AB Amsterdam (tel: 5249111; tx: 11355).

Algemene Bank Nederland NV (commercial bank), PO Box 669, 1000 EG Amsterdam (tel: 299111; tx: 11417).

Amsterdam-Rotterdam Bank NV (commercial bank), Foppingadreef 20-22, 1000 EH Amsterdam (tel: 289393; tx: 11006).

Bank der Bondsspaarbanken, Singel 236, 1001 AR Amsterdam (tel: 5501601; tx: 15565).

De Nationale Investeringsbank NV (national investment bank), Carnegieplein 4, 2517 HJ The Hague (tel: 425425).

NCB (Nederlandse Credietbank NV) (commercial bank), Herengracht 458, 1017 CA Amsterdam (tel: 556911; tx: 14385).

Nederlandse Bankiersvereniging (Netherlands Association of Banks), PO Box 19870, 1000 GW Amsterdam (tel: 230281; tx: 16785).

NMB (Nederlandsche Middenstandsbank NV) (commercial bank), PO Box 1800, Eduard van Beinumstraat 2, 1077 XT, Amsterdam (tel: 5439111; tx: 11402).

Postbank, Haarlemmerweg 506, 1041 BL Amsterdam (tel: 5849111; tx: 10102).

Rabobank, Croeselaan 18 (commercial bank), 3521 CB Utrecht (tel: 909111; tx: 40200).

Netherlands Antilles & Aruba

Tony Thorndike

Aruba having separated from the Netherlands Antilles federation, and with the leadership of St Maarten clamouring to do the same, Curaçao has given notice that it would not support a smaller federation given the financial burdens involved. The smaller members expressed considerable alarm at the prospect. Yet, still heavily reliant on Dutch aid, the federation continues to resist Dutch demands for mutually acceptable audit procedures, making it all the less likely that more Dutch largesse will be available to bail out the warring territories.

Besides the political problems facing the Antillean federation, concern grew over economic difficulties in Saba and St Eustasius, as well as in Curaçao itself, the administrative centre of the federation, now called the 'Antilles of the Five' following Aruba's departure. In contrast, Aruba experienced a remarkable economic recovery following the collapse of the oil refining industry, and the elections of January 1989 brought to power a more radical government bent on greater redistribution of wealth. The new administration also supported independence by the agreed date, 1996, but most Arubans continue to dislike the prospect; to them 'status aparte' from Curaçao was sufficient.

Federation strains

The original federation was composed of six islands in two geographically and culturally distinct groups: the 'ABC' dry and flat limestone islands of Curaçao, Bonaire and Aruba off the Venezuelan coast, and the volcanic, northerly 'Windward Islands' of St Maarten, St Eustasius ('Statia') and Saba. The people of the ABC islands speak Papiamiento, a rich mixture of Dutch, English, Portuguese and Spanish, the others English. Curaçao and Aruba became very prosperous with the siting by Shell and Lago (Standard Oil, later Exxon) respectively of large oil refineries and oil transshipment facilities. But from much earlier times sophisticated financial and commercial services had developed, particularly in Curaçao, in which the descendants of Jewish refugee settlers from the seventeenth century played a prominent role. Not surprisingly, the two islands became a magnet to many from the Windwards (and elsewhere) whose economies relied heavily on remittances. But the advent of the tourist trade, and the shock closure of the Aruban plant in 1985 and the severe reduction of the Curaçao facility, ushered in change.

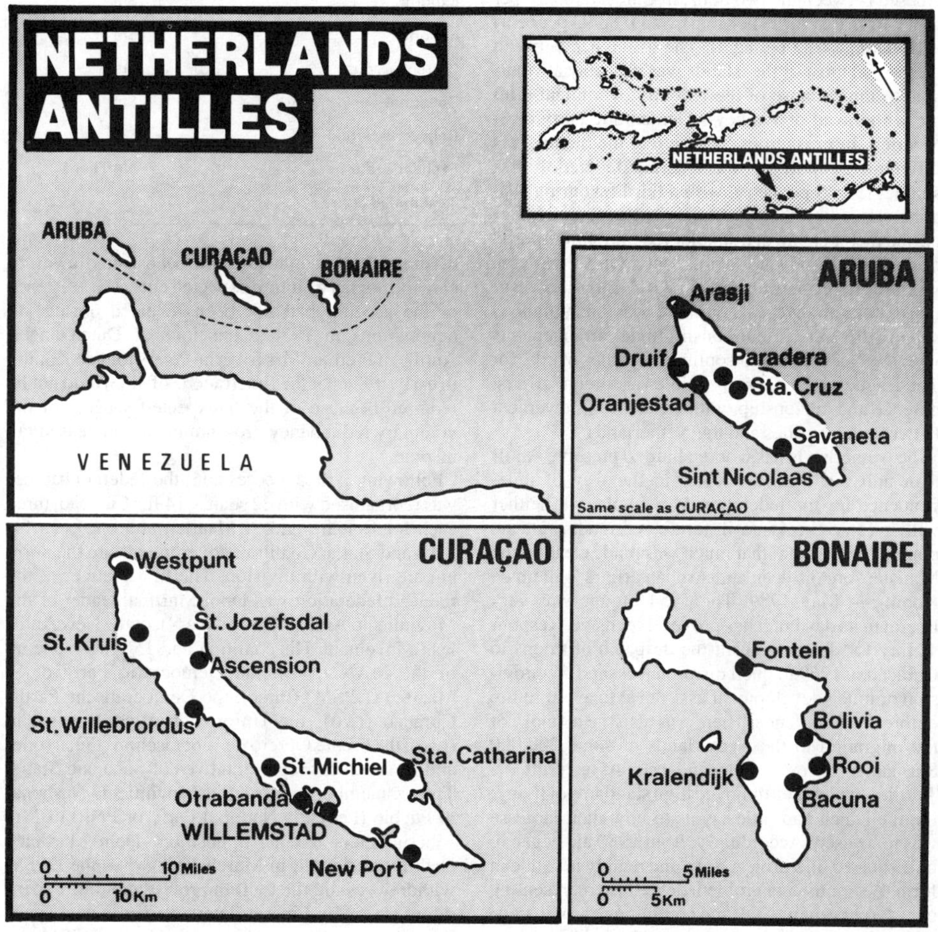

Aruba parted from the federation on 1 January 1986, becoming a separate entity within the Dutch Kingdom under the terms of the Kingdom Charter of 1954. 'Status aparte' was finally conceded by the Dutch on condition that Aruba proceeded to independence in 1996. But many of the 65,000 islanders are unsure. Neither do the majority of Antilleans want independence, even in St Maarten where the flamboyant leader Claude Wathey demands secession and sovereignty for the Dutch third of the small island. As for Curaçao, there is a desire at the very least for more autonomy, but a growing number also favour 'status aparte' if Wathey's demands are fully met. If this was granted, it would leave the Netherlands with the burden of supporting the three smallest and economically weakest members of the federation, a prospect which appeals neither to The Hague nor the three territories.

KEY FACTS

Official title: Aruba

Head of state: Queen Beatrix

Head of government: Prime Minister Nelson Oduber

Ruling party: Three-party coalition led by Movimiento Electoral di Pueblo

Capital: Oranjestad

Official Languages: Dutch, Papiamento, English

Currency: Aruba Guilder (Afl) = 100 cents

Exchange rate: Afl 1.79 per $ (Nov 1989)

Area: 193 sq km

Population: 65,000 (1987)

KEY FACTS

Official title: Netherlands Antilles

Head of state: Queen Beatrix

Head of government: Prime Minister Maria Liberia-Peters

Ruling party: Multi-party coalition led by Partido Nashonal di Pueblo

Capital: Willemstad (on Curaçao)

Official Languages: Dutch, Papiamento, English

Currency: Netherlands Antilles Guilder (NAG) = 100 cents

Exchange rate: NAG1.80 per $ (fixed)

Area: 800 sq km

Population: 204,000 (1986)

GNP per capita: $7,500 (1987)

GNP real growth: –2.5% (1987)

Inflation: 2.6% (1988)

Trade balance: –NAG349m (1987)

Foreign debt: $1.2bn (1987)

Visitor numbers: 1m (1987)

KEY INDICATORS

	Unit	1984	1985	1986	1987	1988
GDP	$m	1,171	1,287	–	–	–
Inflation	%	2.2	0.4	1.3	3.8	2.6
Exports	$m	3,760.6	1,679.4	64.4	77.2	–
Imports	$m	4,248.8	2,132.3	655.0	767.2	–
Current account	$m	135.7	369.6	102.2	–50.0	–
Debt total	$m	766	1,007	1,082	1,231	–
Debt service ratio	%	23.5	16.1	16.4	14.6	–
Exchange rate	NAfl per $	1.8	1.8	1.8	1.8	1.8

* estimated figure

An indication of public attitudes was provided by two successive surveys conducted by a team from the University of Utrecht in February and April 1988, the results of which were published a year later. It revealed a high level of support for maintaining the present 'Antilles of the Five', a sentiment that increased between the two polls, in Curaçao (63/79 per cent), Bonaire (71/89 per cent), Saba (83/93 per cent) and Statia (84/85 per cent). The difference between the two on most of the islands was probably attributable to the collapse of the federal government after the withdrawal of Wathey's Democratic Party of St Maarten. In prosperous St Maarten, the picture was different. Support for the existing federation was smaller (37/45 per cent), with a third favouring separate status like that of Aruba. The remainder supported membership of a smaller federation composed of the three Windward Islands only. Only 7 per cent wished to continue links with Curaçao. But throughout all five islands there was unambiguous support for the present constitutional arrangements with the Netherlands to continue; in other words, for The Hague to remain responsible for foreign affairs, defence and citizenship, with freedom of movement between the Antilles and the Netherlands.

The Netherlands has worked hard to preserve Antillean unity. It does not stand in the way of independence for the federation as a whole, but neither will it force the issue if future referenda indicate popular opposition: that much was made clear by the Minister for Antillean and Aruban affairs Willem de Koning in May 1989. The 1954 Statute was very liberal in its day but it has proved expensive. Responsibility for defence meant the despatch of troops to Curaçao in 1969 to quell disturbances, and the provision of much development assistance to avoid future outbreaks of violence. There was also the memory of the migration to the Netherlands of some 200,000 Surinamers prior to independence. As a result the Antilles (and Aruba) enjoyed high standards of living; but the Dutch had little say as to how their largesse was to be used. Accordingly, from 1980 aid began to be disbursed on a project basis and mainly in soft loan form. A serious debt problem developed representing over 40 per cent of GDP and continuing chronic trade and budgetary deficits made repayment difficult.

By 1988 Dutch patience ran out. The Antilles had since 1984 refused both to agree to oversight by the Dutch government audit office of the annual aid budget and to audit investigations into the most serious breakdowns in expenditure control, on the grounds that the 1954 Statute did not recognise such metropolitan interference. In November 1988 the Dutch legislature passed a motion calling upon the government to authorise such control. Budget restructuring followed involving detailed cutback plans. Aid was trimmed from US$138m for both the Antilles and Aruba in 1986 to US$82m (G170m) in 1989, government expenditure had to be cut by up to 40 per cent and public sector employment from 13,000 to 9,000, although generous redundancy funds were made available. Further encouragement was the threat of a 'solidarity tax' payable by government employees to keep superfluous colleagues in a job. In June 1988 the Antillean government requested an additional US$189m for the period 1989-92 but only a loan of US$83.3m and a grant of US$11.1m was finally granted.

The first casualty was Statia, whose revenue, mainly from the 'solidarity fund' which supports the administrations of the three smaller territories (to which the federation contributes over half, the Netherlands almost a quarter and Aruba the remainder), covered only 60 per cent of expenditure. As a result, services were dramatically reduced. Metropolitan resolve was further stiffened by the revelation that Curaçao had maintained a secret government account, worth over US$60m, into which some of the earnings of the offshore finance industry had been salted away to ease the impact of future expected cuts. The existence of the account had not been revealed during aid negotiations in 1988. Conscious of Dutch anger, Antillean Premier Mrs Liberia-Peters told de Koning in early 1989 that the first tranche of aid would not be required because of the unexpected success of the voluntary redundancy programme and increased tax income.

Following Aruba's secession, the federal chamber was reorganised with 22 seats – 14 for Curaçao, three each for Bonaire and St Maarten and one each for Saba and Statia. Coalition governments are the norm but are riven with division. The first premier of the reduced federation was Dom Martina, leader of the Movimiento Antiyas Nobo (MAN), or the New Antillean Movement. His coalition had 15 seats made up of the MAN (five), the Democratic Party of St Maarten (DPSM) (three), the Democratische Partij-Curaçao (two), the Union Patriotico Bonairiano (two), the Frente Obrero de Liberaschon (FOL) (one) and one each for representatives of Saba and Statia. The six-member conservative Movimiento Nashonal di Pueblo (People's National Party or PNP) of Mrs Liberia-Peters led the opposition. Dom Martina's coalition collapsed in March 1988 when the DPSM withdrew, as did the FOL in protest at public expenditure cuts. Mrs Liberia-Peters formed a new coalition, led by her PNP, and supported by the DPSM, the FOL, the Social Independiente – a Curaçao group expelled from the PNP following a corruption scandal in 1986 – and the parties from the three smaller islands, giving a total of 13 seats.

Her attitude to the federation is simple: maximum autonomy to its constituents and minimum subsidy from Curaçao. Balancing all these interests is a very time-consuming act leaving little opportunity to address the economic crisis. To add to the complexity, each island has its own assembly and division is commonplace. In Bonaire political deadlock became so intractable in 1989 that direct Dutch intervention was threatened, notwithstanding the constitutional niceties.

Aruba

Aruba's secession was rooted in history. Being too dry for plantation agriculture, slaves were unknown. Despite the mining of gold in the nineteenth century, Aruba always played a poor second fiddle to Curaçao, whose entrepot trade with Latin America – and its slave market – brought riches. The Lago (Exxon) refinery meant prosperity and with it demands for separation and local control of the economy. It was therefore ironic that separation coincided with the refinery's closure. Overnight 1,500 jobs were lost, as were US$140m in annual foreign exchange earnings. GDP slumped by 30 per cent, from US$545.5m in 1984 to US$372m in 1986.

The Arubans tackled the problem with gusto, constructing a large number of hotels (tripling the number of rooms to 6,000) and heavily promoting the island in North America. Excellent beaches, large casinos, a dry, breezy climate, a multilingual workforce and duty-free shopping did the trick. Stopover arrivals quadrupled in three years, reaching nearly 300,000 in 1988, necessitating airport expansion and new air services. Cruise ship business was an added bonus. Generous investment incentives provided a good deal of the money, some from Japan, and imaginative legislation led to a mushrooming of offshore financial activity. A capital-intensive industrial sector is planned on the refinery site and oil-shipment facilities are being revived. Oil prospecting also showed promise. By 1989 unemployment had fallen from 40 per cent in 1985 to a bare 4 per cent, and positive growth returned in 1987 when per capita income reached AFl.12,202 (US$6,855). Indeed, the problem became one of overheating and imports threatened to capsize the balance of payments.

Resentment against Curaçao had crystallised around the Movimiento Electoral di Pueblo (MEP) led by the charismatic Betico Croes. However, the MEP (affiliated to the Socialist International) was defeated in 1985 following the secession agreement by a coalition headed by Jan Hendrik ('Henny') Eman and his Aruban People's Party (AVP), and Croes himself was to die in a car crash only hours before the separation. The MEP returned to power in the January 1989 election, by now led by Nelson Oduber, its representation increasing from eight to 10 seats in the 21-seat legislature. Two further seats were won by smaller coalition partners. Eman's AVP won one more seat to make eight. The MEP, while acknowledging the AVP's role in restoring Aruba's fortunes after the refinery closure, charged that the benefits of the tourist and construction boom were unevenly distributed and that unemployment would reappear once all the new building was completed.

As Aruban fortunes changed for the better, the prospect of independence in 1996 no longer seemed frightening. Eman had consistently warned of the dangers and advocated a referendum on the issue. But Oduber had no such reservations. Nonetheless, popular sentiment remained ambivalent, seeming to favour a kind of theoretical independence within a Dutch commonwealth.

St Maarten

Tourism was also responsible for St Maarten's success. Once fewer than 2,000 scratched a living on the Dutch side; now some 20,000 live there, nearly half of them non-Antilleans. 1988 saw total tourist arrivals near the one million mark, and plans were unveiled in mid-1989 for a new US$200m port capable of taking eight cruise ships simultaneously. The transformation has been masterminded by Claude Wathey, a political buccaneer with a highly individualistic style. In power for over 30 years, the benefits that have accrued have disguised serious deficiencies in infrastructure and financial management. Wanting greater autonomy, he tried to force the federation's hand by first walking out of Dom Martina's government. Sensing power, he then demanded full independence. But a mission to The Hague in early 1988 yielded little as there was little public support for the proposal – as indeed the Utrecht University opinion poll made clear. His DPSM joined Mrs Liberia-Peters' coalition because of her support for greater autonomy, and because the distribution of seats gave him considerable power. Although Wathey still demanded independence by 1990, it is likely that some form of loose compromise will be reached.

Antillean economics

The oil refining industry owed its origin to the shallow water around oil-rich Venezuela which prevented access by large tankers. The Shell facility dates back to 1915, that of Lago on Aruba to 1929. In the 1970s oil transshipment terminals were built to handle African and Arab oil destined for the United States whose legislation prevented tankers of more than 60,000 tonnes from entering US ports. The building of huge refineries linked by pipelines to the US mainland sealed the fate of the Antillean industry. But whereas the Aruban plant closed, that in Curaçao was leased (to 1994) to the state-owned Petroleos de Venezuela, albeit with a sharply reduced output. President Perez of Venezuela is interested in a long-term arrangement as part of a wider relationship in which Venezuelan companies can operate in Curaçao and so access the United States market through the Caribbean Basin Initiative.

Like Aruba, Curaçao has had to come to terms with new realities. Capitalising on its trading skills, the new International Trade Centre and expanded free zone has stimulated business. Investment in the ship repairing docks has also paid dividends. Oil bunkering at Bullenbai, one of the largest facilities in the world, remains profitable. There is also a bunkering facility on Statia. But as with the rest of the group, tourism is the fastest growing sector. Although tourism was hit badly by the devaluation of the Venezuelan bolivar, the first new hotel in 17 years opened in 1988, specialising in diving holidays. Man-made beaches on the dry, rocky island are under construction and duty-free goods priced more competitively. In 1988 stopover arrivals were an estimated 160,000, still below the 1981 record of 176,269. Known to divers as one of the best sites in the world, Bonaire's share of the stopover market also increased from 30,422 in 1987 to some 35,000 in 1988. Saba and Statia also recorded higher figures. But both Curaçao and Aruba also see benefits in the offshore finance business.

Offshore initiatives

For some four decades Curaçao was synonymous with offshore finance. It came into its own as a centre as a haven for Dutch companies fleeing Nazi-occupied Netherlands, but the foundation for the modern industry, now fallen on hard times, was the extension of a US-Netherlands tax treaty to the Antilles in 1955. The offshore company was born. Conferring anonymity on their owners, such corporate vehicles (NVs) were exempt from virtually all local taxes and, more importantly, could legally be used to avoid US tax. The next boost came with the popularity for US banks and major corporations of the Eurobond market. Using the tax treaty, a shell finance company would be registered to sell the bonds, guaranteed by the parent; the shell company would then re-lend the proceeds of the bonds and receive interest, free of the usual 30 per cent US witholding tax. Overseas purchasers of the bonds would, in turn, receive interest free of the tax. Therefore borrowing costs were reduced and overseas investors received a high return. Over US$90bn was raised in this way, producing high fee income for the Antillean Government, up to US$220m annually.

The US Treasury decided to strike and in mid-1984 abolished the withholding tax on interest and dividends paid to foreigners. After abortive negotiations on a new tax treaty (the Antillean authorities refused to divulge information on suspected tax evaders) the United States revoked its treaty, and Britain soon did the same. No action was taken on existing bond issues but they will all be redeemable by 1992. Making matters worse was a new Netherlands tax regulation which boosted tax on European dividends on offshore bonds. Curaçao's unique advantages had vanished.

Since then the industry has had to rethink its future and to revise legislation to attract new business. Although a no-tax jurisdiction was rejected, tax laws now benefit mutual funds and trusts, and draft legislation is being devised to develop offshore insurance. There is relatively little offshore bank activity mainly due to the Latin American debt crisis: Venezuela and Colombia are the major customers, favouring the lack of exchange control. Future expansion will be limited until there is a legal framework for bank supervision, to avoid charges of permitting 'laundering', and better secrecy arrangements. By 1989 progress had been made for on-site bank inspections, introduced with the assistance of an inspector from the US Federal Deposit Insurance Corporation. But the lack of a no-tax jurisdiction will disadvantage Curaçao.

Aruba has been more imaginative. Its infant offshore industry is rapidly expanding, concentrating on tax-free companies – the Aruba Exempt Corporation (AEC). AECs are inexpensive (incorporation costs around US$1,100, minimum authorised capital around US$5,600, annual fee US$825) and flexible, require only one (anonymous) shareholder and one local trustee (a NV trust company) and need not file any reports. AECs cannot engage in banking or insurance. The response by Europeans and Latin Americans has been encouraging, a clear sign that entrepreneurship is alive and well in the face of adversity.

COUNTRY PROFILE

Historical profile: First European sighting 1499 by Spanish mariners; islands settled around 1527; Dutch took over 1634; Spanish and Portuguese Jews escaping from persecution contributed numerous settlers; Peter Stuyvesant, Governor 1642-46, later became Governor of New Amsterdam (which became New York); British Protectorate 1800-02; held by British again 1807-15; gold discovered in Aruba 1825 – mined until 1916; slavery completely abolished 1863; internal autonomy for Netherlands Antilles 1954. In January 1986 Aruba separated from the other islands, which became the Antilles of Five; both entities are part of the Kingdom of the Netherlands.

Political system: Executive power for external affairs and defence is exercised by the appointed governor (as representative of the sovereign of the Netherlands) assisted by an advisory council. Internal affairs are handled by the governor and a council of ministers – this council is responsible to the legislature, the Staten (22 members – Curaçao 14, Bonaire 3, St Maarten 3, Saba 1, St Eustatius 1; four-year term, elected by universal adult suffrage and proportional representation). Each island has a local administration, part of which is elected. Aruba achieved separate status from the Netherlands Antilles on 1 January 1986 as a Dutch dependency, and full independence is scheduled for 1996; in the meantime loose ties are retained with the Netherlands Antilles, and defence and foreign policy are the responsibility of the Netherlands.

Main cities/towns: Willemstad (population 130,000) is the capital of Netherlands Antilles, on the largest and by far the most populous of the five major islands that comprise Netherlands Antilles. Kralendijk is a small town on Bonaire and Philipsburg on St Maarten. Oranjestad (16,000) is the capital of Aruba.

Media
Press: Several daily newspapers published in Curaçao, including *Amigoe* and *Beurs-en-Nieuwsberichten* (Dutch) and *The News* (English). Dailies in Aruba are *Amigo di Aruba* (Papiamento), and *The News*.
Broadcasting: Two commercial channels operated by Netherlands Antilles Television Co Ltd and Leeward Broadcasting Co. Numerous radio stations include Radio Victoria (Aruba), broadcasting in English, and Radio Hoyer (Curaçao) operating on solar energy. In Aruba, television is operated by Tele-Aruba.

External trade: Trade in the Antilles of the Five is dominated by Curaçao's import and re-export of petroleum and products, though output has sharply reduced (Aruba and the other islands rely on tourism). Invisible earnings from tourism, offshore finance, transport and intercompany remittances are in most years sufficient to offset trade deficit at present. Repeal of withholding tax in the USA in 1984 affected revenues for central government and for Curaçao. Bilateral aid and grants provided by the Netherlands.
Imports: Oil (80-90 per cent of total value), machinery and transport equipment, food etc. Main sources: Venezuela, Mexico, USA, Nigeria, Netherlands, Saudi Arabia, France, UK.

Agriculture: Less than five per cent of total area is cultivated arable land. Soil is generally poor and rainfall inadequate for most crops. Small amounts of fruit and vegetables are grown for local consumption. There is little commercial fishing except on St Maarten.

Industry: Sector is dominated by petroleum refining and trans-shipment. Although still the country's main employer, the oil industry has declined dramatically owing to excess capacity in USA, new offshore US refineries and falling oil supplies from Venezuela and Middle East. After operating at well below capacity of 320,000 b/d for several years, the Shell refinery on Curaçao was sold to the government in 1985 and leased to the Venezuelan state oil company; throughput reduced by end 1987 to 195,000 b/d. The 440,000 b/d refinery on Aruba, owned by a subsidiary of Exxon, was closed in 1985 and later sold to the Aruba government for a nominal amount, depriving the island of one-third of its revenue. The trans-shipment and ship repair industries have also been affected. Manufacturing is concentrated on food processing and import substitution (paints, paper, soap, beer, chemicals). Emphasis on diversification into light export-based industries such as electronics.

Tourism: Major industry on the islands of Aruba, Curaçao and St Maarten, increasing tourist interest in Bonaire. Majority of visitors from the USA, Venezuela, Netherlands, Canada and Caribbean.

Mining: Around 70,000 tonnes per year of phosphates were mined at Curaçao but mine closed, and some 400,000 tonnes of salt recovered from Bonaire and St Maarten. Known deposits of pumice on St Eustatius.

Fiscal system: Leading Caribbean offshore finance centre. Over 30,000 registered offshore companies, mostly from the USA. Country's attraction as a tax haven was affected by changes to tax laws in USA and Netherlands.

Membership of international organisations: State associated with the EC. US Caribbean Basin Initiative beneficiary country.

Time: GMT -4.

Climate: Low levels of humidity and rainfall on Curaçao, Aruba and Bonaire where temperatures average 29°C. Temperatures are cooler on St Maarten, Saba and St Eustatius, due to the north-east trade winds, averaging 26°C and higher levels of rainfall between May and Dec.

Entry requirements
Passport: Required by all, except US and Canadian citizens, who only need proof of citizenship.
Visa: Required only for passengers from communist countries. Onward or return ticket also required. Maximum stay 14 days with possibility of extension to 90 days. N.B. regulations regarding entry are extensive and prior checking is essential.
Currency: No restrictions on import/export of foreign currency. Import/export of local currency limited to NAGI,200 or Afl 1,200. Exchange of local currency abroad can be difficult.

Health precautions
Mandatory: Yellow fever vaccination certificate required if arriving from an infected area.
Advisable: Typhoid, polio vaccination.

Air access
National airlines: Air Aruba, ALM Antillean Airlines (Antilliaanse Luchtvaart Maatschappij), and Windward Islands Airways.
Other airlines: Air BVI, Air Caribe International, Air France, Air Guadeloupe, Air Jamaica, American, Avensa, Dominicana, BWIA, Continental, Eastern, KLM, LAV, LIAT, Pan Am, Surinam Airways, TAP, VASP, Viasa.
International airports: Aruba-Reina Beatrix (AUA), 2.5 km from Oranjestad, duty-free shop, bar, restaurant, post office, car hire; Curaçao-Dr Albert Plesman/Hato (CUR), 12 km from Willemstad, duty-free shop, bar, restaurant, buffet, bank, post office, shops, hotel reservations, car hire; St Maarten-Queen Juliana (SXM), 16 km from Philipsburg, duty-free shop, bar, restaurant, shops, hotel reservations, car hire; Bonaire-Flamingo (BON), six minutes from Kralendijk.

Surface access: Weekly ferry from Curaçao to Venezuela.
Main ports: Willemstad, (Curaçao), Oranjestad (Aruba).

Hotels: Advisable to book in advance. Numerous tourist hotels on St Maarten, Aruba, Curaçao and Bonaire. Very limited accommodation on Saba, and St Eustatius. Five per cent government tax and 10-15 per cent service charge added to bill.

Car hire: Widely available at reasonable prices. International licence required.

City transport
Taxis: Usually identified by 'TX' before the licence plate or by 'Bus' for collective taxis on Curaçao. Advisable to negotiate fares for tours in advance. Tipping is discretionary.

National transport
Air: ALM operates services between the islands, and between Netherlands Antilles and Aruba. Services also provided by ABC Commuter, Windward Islands Airways and Air Aruba.
Road: Network on all islands is all-weather, though less extensive on Saba and St Eustatius.
Buses: Regular services in and around main centres. Also cars/jitney services and sightseeing tours. No regular services on smaller islands.

Public holidays: 1 Jan, Carnival Monday (Feb or Mar), Good Friday, Easter Monday, 30 Apr (Coronation Day), 1 May, Ascension, 25 Dec, 26 Dec. Flag days: Curaçao 2 Jul, Bonaire 6 Sep, St Maarten 11 Nov, St Eustatius 16 Nov, Saba 6 Dec; Aruba 18 Mar. Jewish establishments closed at the Jewish New Year and Yom Kippur.

Working hours
Government and business: (Mon-Fri) 0730 – 1200, 1330 – 1700.
Banking: (Mon-Fri) 0800 – 1200, 1400 –1600.
Shops: (Mon-Sat) 0800 – 1200, 1400 – 1800.

Telecommunications
Telephone: Dialling for Netherlands Antilles: IDD access code + 599 followed by area code (Bonaire 7, Curaçao 9, St Maarten 5) followed by subscriber's number. Dialling code for Aruba: IDD access code + 297 followed by area code 8 followed by subscriber's five-digit number. 100 per cent automatic service.
Telex: Services operated by Lands Radio Dienst (government service) and All-American Cables on Curaçao and Aruba.

Banking
Netherlands Antilles
Central bank: Bank van de Nederlandse Antillen.
Curaçao: Algemene Bank Nederland, Banco di Caribe, Banco Industrial de Venezuela, Banco Mercantil, Citco Bank Antilles, McLaughlin Bank, Heldring & Pierson, Maduro & Curiel's Bank, Pierson.
St Maarten: Algemene Bank Nederland (St Maarten), Bank of Nova Scotia, Barclays Bank, Chase Manhattan Bank, Citco Bank Antilles, Windward Islands Bank.
Bonaire: Maduro & Curiel's Bank (Bonaire).

Aruba
Central Bank: Central Bank of Aruba.
Oranjestad: Algemene Bank Nederland, Aruba Bank, Barclays Bank, Caribbean Mercantile Bank, Citibank.

Advertising: All newspapers accept advertising in English. Also advertising on commercial radio and television and outdoors available.

Electricity supply: Variable; 120/127/220 V AC at 50 cycles or 60 cycles.

Hotels: Aruba
Americana Aruba, L.G.Smith Boulevard, Oranjestad (tel: 24500; tx: 5044).

Aruba Concorde, L.G.Smith Boulevard, Palm Beach, Oranjestad (tel: 24466; tx: 2075).

Aruba Palm Beach, Palm Beach, Oranjestad (tel: 23900; tx: 529519).

Divi Divi Beach, Oranjestad (tel: 23300; tx: 5023).

Golden Tulip, 81 L.G.Smith Boulevard, Palm Beach, Oranjestad (tel: 33555; tx: 5042; fax: 23260).

Holiday Inn, Palm Beach, Oranjestad (tel: 23600; tx: 5025; fax: 825165).

Tamarijn Beach, Oranjestad (tel: 24150; tx: 5023).

Hotels: Bonaire
Bonaire Beach, P.O. Box: 34 (tel: 8448; tx: 1291).

Flamingo Beach, J.A.Abraham Boulevard (tel: 8285; tx: 1293).

Hotels: Curaçao
Curaçao Caribbean, J F Kennedy Boulevard, P.O. Box: 2133 (tel: 625000; tx: 1146; fax: 625846).

Curaçao Plaza, Willemstad, P.O. Box: 229 (tel: 612500; tx: 3324; fax: 616543).

Holiday Beach, Pater Euwensweg 31, P.O. Box: 2178 (tel: 625400; tx: 3317; fax: 624397).

Las Palmas, John F.Kennedy Boulevard, P.O. Box: 2179 (tel: 625200; tx: 1080; fax: 625962).

Princess Beach, Martin Luther King Boulevard (tel: 614944; tx: 1268; fax: 614131).

Hotels: St Maarten
Bel Air Beach, P.O. Box: 140 (tel: 3366).

Cupecoy Beach, P.O. Box: 14 (tel: 4297).

Dawn Beach, P.O. Box: 389 (tel: 2929; tx: 8085).

Maho Beach (tel: 42115; tx: 8018; fax: 43180).

Mullet Bay Sheraton, P.O. Box: 309 (tel: 42801; tx: 8019; fax: 44281).

St Maarten Beach Club, P.O. Box: 465 (tel: 3434; tx: 8046).

Business advice and information
Aruba Chamber of Commerce and Industry, PO Box 140, Zoutmanstraat 21, Oranjestad, Aruba (tel: 21566, 23423).

Aruba Tourist Board, A Schuttestraat 2, Oranjestad (tel: 23777; tx: 5006).

Aruba Trade & Industry Association, PO Box 562, Oranjestad, Aruba.

Bonaire Chamber of Commerce and Industry, P L Brionstraat, PO Box 52, Bonaire (tel: 8995).

Curaçao Chamber of Commerce and Industry, Pietermaai 21, PO Box 10, Willemstad, Curaçao (tel: 611455; tx: 1055 kkcur; fax: 615652).

Curaçao Tourist Bureau, Schouwburgeweg z/n, Willemstad (tel: 77121, 77150; tx: 1185).

Curaçao Trade & Industry Association, PO Box 49, Pietermaai, Willemstad.

St Maarten Chamber of Commerce & Industry, Frontstraat, Philipsburg (tel: 23590).

St Maarten Tourist Board, Ruyterplein, Philipsburg (tel: 22337).

Banking
Bank van de Nederlandse Antillen (central bank), Breedestraat 1, Willemstad, Curaçao (tel: 613600; tx: 1155, 1202; fax: 615004).

Algemene Bank Nederland, Nassaustraat 89, Oranjestad, Aruba (tel: 21515).

Algemene Bank Nederland, Pietermaai 17, PO Box 468, Willemstad, Curaçao (tel: 611488; tx: 1098).

Algemene Bank Nederland, Frontstraat, PO Box 295, Philipsburg, St Maarten (tel: 3505, 2504; tx: 8004).

Aruba Bank, Nassaustraat 41, Oranjestad, Aruba (tel: 21550).

Citibank, Nassaustraat 76, Oranjestad, Aruba (tel: 24666).

Maduro & Curiel's Bank, Plaza Jojo Correa 2-4, Willemstad, Curaçao (tel: 611100; tx: 1126).

Maduro & Cureil's Bank (Bonaire), Kerkweg, Kralendijk (tel: 8420, 8404; tx: 1201).

Car hire: Aruba (Oranjestad)
Avis: Kolibristraat 14, Simeon Antonio (tel: 28787; tx: 5132).

Budget: Luchtaven (tel: 28600; tx: 5136).

Europcar National: PO Box 150 (tel: 21967/24641).

Hertz: L G Smith Boulevard 142 (tel: 4545; tx: 5049).

Car hire: Aruba International Airport
Avis: (tel: 25496).

Europcar National: (tel: 25451).

Hertz: (tel: 4800 ext 364).

Car hire: Curaçao
Avis: Erieweg, Zeelandia, PO Box 3677, Willemstad (tel: 611255; tx: 1129).

Hertz: Ric Car Rental, Perseusweg 22, PO Box 3402, Willemstad (tel: 613622, 613499; tx: 1029).

New Caledonia

David Robie

(map on page xxiv)

The double assassination of Kanak independence leaders Jean-Marie Tjibaou and Yeiwene Yeiwene on 4 May 1989 threw an ominous cloud over the Matignon Accord and jeopardised a hard-won peace in New Caledonia. Gunned down at the hands of fellow Kanaks, the deaths of Kanak Socialist National Liberation Front (FLNKS) President Tjibaou, an architect of the peace agreements, and his deputy Yeiwene were particularly tragic. But their chief assassin, former pastor Djoubelly Wea, a maverick and respected political leader on the island of Ouvea, failed in his attempt to sabotage the accord.

KEY FACTS

Official title: New Caledonia

Head of state: President of France

Head of government: President of Congress, Simon Loueckhote (French High Commissioner, Bernard Grasset)

Ruling party: Rassemblement Populaire pour la Calédonie dans la République

Capital: Nouméa (on New Caledonia)

Official Languages: French

Currency: CFP Franc (CFP Fr = 100 centimes)

Exchange rate: CFP Fr113 per $ (Nov 1989)

Area: 22,125 sq km – New Caledonia Island: 16,750 sq km

Population: 145,400 (1983)

GDP per capita: CFP Fr892,669 (1985)

Labour force: 58,000

Inflation: 3.5% (1988)

Trade balance: CFP Fr500m (1988)

Foreign debt: $368m (1987)

Visitor numbers: 61,079 (1988)

Provincial elections

Instead, the provincial elections staged under the terms of the accord a month later to fill the Territorial Congress showed that support for the pact had grown. The mainstream parties gained ground at the expense of the fringe groups of both left and right.

Leaders of both the FLNKS and the anti-independence Rally for Caledonia in the Republic (RPCR) party hailed the results as a victory for the accord signed in 1988 paving the way for partial self-government and a referendum on independence in 1998.

Jacques Lafleur, leader of the RPCR, claimed his party had fared best, winning 27 seats – one seat short of an outright majority – in the new 54-seat congress and dominating the South, one of the three provincial assemblies. He pledged to work for a 'consolidation of the peace, economic development and justice' of New Caledonia. The RPCR won 21 out of the 32 seats at stake in the resource-rich South province and claimed to have the biggest political support in the territory with almost 45 per cent of the vote.

Francois Burck, leader of the Caledonia Union (UC), biggest and most influential party in the FLNKS, also welcomed the results. The FLNKS scored predicted victories in the North and Loyalty Islands provinces, but the northern win was far more decisive than expected. 'Voters gave their approval to the process launched by the Matignon accords,' Burck said. 'That's the important thing.'

The rival political groups, whose leaders Lafleur and Tjibaou had forged a truce in negotiating the accords with French Prime Minister Michel Rocard, shared 85 per cent of the seats. Two extreme right-wing parties, the National Front and Caledonian Front, which had previously been gaining support, suffered major losses. The Caledonian Front lost both its seats in congress and the National Front's share of seven seats was slashed to three.

Besides the National Front and FLNKS, which won four seats in the South, the new Oceanian Union – representing migrant workers from Wallis and Futuna – won two seats. Previously the islanders were strong supporters of the RPCR, but they began to follow a more independent political path (while still backing French rule), wanting to establish a direct dialogue with the Kanaks.

Although the Oceanian Union did not take a stand on independence, the shift in allegiance of the Wallis and Futuna Islanders is significant. It is believed that they will eventually support the Kanaks and independence.

The election results breakdown was as follows: Congress (54 seats): RPCR 27 seats, FLNKS 19, National Front 3, Oceanian Union 2, Caledonia Tomorrow 2, Anti-Neo-Colonialists 1; South Province (32 seats): RPCR 21 seats, FLNKS 4, National Front 2, Oceanian Union 2, Caledonia Tomorrow 2; North Province (15 seats): FLNKS 11, RPCR 4; Loyalty Islands (7 seats): FLNKS 4, RPCR 2, Anti-Neo-Colonialists 1.

Opposition to Matignon Accord

At the root of the crisis marking the murders of Tjibaou and Yeiwene was the accord itself. The accord has frozen the electoral rolls for the lead up to the referendum on independence in 1998; it has also embraced a development plan designed to boost poor rural and urban Kanak areas and to train Kanaks in professional, managerial and technical skills. But it gives no guarantees for independence.

While the peace plan was welcomed throughout the Pacific as a hopeful and significant step forward, several minority groups of Kanaks became bitterly opposed to it, regarding the accord as a sellout by their own leadership. Many opponents of the accord personally blamed Tjibaou and Yeiwene. While Tjibaou was alive he could be challenged over the accord, but after his death the majority of Kanaks vowed to honour his vision.

During negotiations on the pact the FLNKS had called for a 'real' vote on self-determination in five years' time, to be limited to Kanaks and first generation settlers. However, the RPCR wanted the vote put off until 1998 and kept open to all residents with French nationality.

The fiercest resistance to the accord came from Ouvea, a crescent-shaped isle 40 km long, in the Loyalty Islands, which has the least development of any part of New Caledonia. It has a population of less than 3,000 and a handful of telephones – none north

COUNTRY PROFILE

Political system: New Caledonia is an overseas territory of France. Under the Matignon Accords of July 1988, a previous regional system was replaced by direct rule from Paris until 14 July 1989 when three provincial Assemblies took effect. The Assemblies (two Kanak-dominated) will rule with limited self-government powers until 1998 when a referendum on independence will be held. The provincial Assemblies together comprise the 54-seat Territorial Congress. The High Commissioner has overall responsibility while the president of the Territorial Congress is the head of local government.

Political parties: The anti-independence party, Rally for Caledonia in the Republic (RPCR), has a majority 27 seats in the 54-seat Congress; pro-independence Kanak Socialist National Liberation Front (FLNKS) 19; National Front three; Oceanian Union two; Caledonia Tomorrow two; Anti Neo-Colonialists one.

Population: According to 1983 figures, 43 per cent of the population is made up of Melanesian Kanaks, with 38 per cent Europeans (mainly French) and the remainder mainly Wallisian and Tahitian, Indonesian and Vietnamese.

Main town: Noumea (on New Caledonia, population 62,000* in 1987).

Language: French is the official language, and should be used in documentation; English is often understood.

Media
Press: Les Nouvelles Calédoniennes is the daily newspaper on New Caledonia. *Tele 7 Jours* is a weekly TV magazine.
Broadcasting: Radio France Outre Mer (RFO).
Radio: Private local radio services operate 24 hours a day; stations include NRJ 2000, Radio Djiido, RRB.
Television: Metropolitan and international TV programmes are broadcast by the state broadcasting authorities in colour from 1600 – 2400. There are two channels.

Domestic economy: Economy almost entirely dependent on nickel extraction and processing. 1987 and 1988 have been excellent years for nickel exports because of a good international conjecture and a high dollar exchange rate. The new statute adopted in 1988 plans a development of the northern mainland and Loyalty Islands via economic, cultural and social programmes.

External trade
Exports: Main exports are processed nickel ore products, nickel ore, coffee, copra and chrome. Main destinations are France, Japan, USA, French Polynesia, Wallis and Futuna, Australia, Vanuatu and Fiji.
Imports: Main imports are food products, mineral products, machinery and transport equipment, raw materials and textiles. Main sources are France, Australia, other EC countries, USA, New Zealand and Japan.

Agriculture: Although the soil is fertile, less than 10 per cent of the land area is cultivable. Sweet potatoes, bananas, pawpaws, fruit and vegetables are produced for local consumption, but many

of the chief town of Fayaoue. When the FLNKS decided to boycott and obstruct the French presidential elections in May 1988, Ouvea militants staged a dramatic raid on the Fayaoue gendarme barracks, killing four gendarmes who resisted and seizing 27 as hostages. At dawn on 5 May 1988, French military and secret service forces launched an attack on the Ouvea cave and killed 19 Kanaks. The hostages were freed.

Faced with mounting allegations of atrocities against several of the dead Kanaks – including summary executions – Defence Minister Jean-Pierre Chevenement ordered an inquiry, saying that former Overseas Territories Minister Bernard Pons had not used 'all possible means' of negotiation with the nationalists. However, while the French Government later admitted that 'excesses' had happened, no judicial action was taken.

After the cave assault, 32 prisoners, including Djoubelly Wea, were flown to France. Wea was released after a legal battle and returned home to a hero's welcome; the other prisoners were eventually freed as part of the Matignon accord. Wea bitterly resisted the linking of the prisoners to the accord, but he was ignored by Tjibaou when he wanted to take part in the peace negotiations. The FLNKS failed to act over the Ouvea islanders' major demands, such as an international inquiry into the allegations of atrocities by the French military.

France also refused to fund the island's development projects until the barricades were demolished and weapons seized by the militants were surrendered.

Ironically, Wea and his fellow assassins hardened the resolve of the majority within the FLNKS – a coalition comprising five political parties and a trade union – to honour the pact. And the factions opposed to the accord became more isolated in spite of their serious grievances.

One fringe left-wing party, the Kanak United Liberation Front (FULK), led by former Kanak 'foreign minister' Yann-Celene Uregei, was barred from FLNKS meetings. The party, which made ambiguous statements endorsing the killings and describing the accord as a 'socialist trap', was openly accused of being behind a plot to wipe out the Kanak leadership. FULK boycotted the provincial elections but its threats to cause disruptions proved empty.

Still others among the FLNKS coalition do not trust the accord and the lack of an absolute guarantee for independence. Palika, the second largest front party, for example, accused the UC of selfishness by dominating key positions. Both the provincial presidents – Leopold Joredie (North) and Richard Kaloi (Loyalty Islands) – and three of the four vice presidents belong to the UC.

The three new provincial governments came into force on 14 July 1989: Bastille Day, when French citizens commemorated the 200th anniversary of their revolution. On 17 July two of the provincial assemblies held their first sessions in Kone (North) and We (Lifou, Loyalty Islands). Two days later the South province assembly met following the first official session of the congress in Noumea.

Elected president of the new congress, with the help of National Front votes, was Simon Loueckhote, a 32-year-old Kanak opposed to independence and a member of the RPCR.

'Somebody is here among us,' High Commissioner Bernard Grasset told deputies in the North province assembly session. 'It is Jean-Marie Tjibaou. Without him, the task wouldn't have been achieved. He would have been here at this table today.'

Tjibaou's goal was to have Kanaks trained in running offices, businesses, towns and provincial administrations; to turn political militants into modern farmers, accountants, lawyers, and teachers. He believed that for independence to succeed the Kanak community needed a pool of skilled professionals to take over key positions in the future independent state of Kanaky.

Visiting New Caledonia after the provincial assembly sessions, Overseas Territories Minister Louis le Pensec declared that the accord was in 'good health'. 'Peace is restored,' he added. 'It must be consolidated so that an incident – always possible – doesn't undo the new-found confidence.'

A monitoring committee, comprising representatives of the three parties to the Matignon accord was expected to meet in Paris in October 1989 for the first time. Committee members were likely to visit Brussels to discuss the implications for New Caledonia of Project Europe 1992.

Among critics of the accord are Nidoish Naisseline, leader of Kanak Socialist Liberation (LKS). He believes little has changed. Long a radical leader, Naisseline became in 1984 acknowledged as a moderate Kanak leader, still seeking independence but in co-operation with French political institutions. He regards the RPCR as still controlling the administration and he criticises the modification of the plan under the Matignon accord to train 400 Kanak public servants, into training for 400 'New Caledonians'.

The powers of the provincial administrations to trade and develop relations with other South Pacific states are dependent on approval from France.

Naisseline warned that if France succeeded in removing New Caledonia from the United Nations Decolonisation Committee list of colonies in defiance of resistance from the South Pacific Forum, 'then we will have no more arguments to use with militants against violence – we must take up arms.'

COUNTRY PROFILE

food requirements are imported. Exports of coffee and copra crops have increased in recent years. About one-third of the main island's land area devoted to cattle raising, chiefly on central and north-west coast. Reduction in number of head (from 120,300 in 1969, to 80,000 in 1987) has been blamed on the Land Reform Act, which redistributed land to Melanesians and provoked heavy slaughter by several cattle growers. Domestic forests supply about 50 per cent of timber demand, some reafforestation undertaken. Fishing also for local consumption (oysters, shrimps, some aquaculture).

Industry: Nickel ore processed at the Doniambo smelting plant on the outskirts of Noumea by the Soci\et\e M\etallurgique Le Nickel (SLN), a subsidiary of two French companies. Some processing and light industries established: domestic equipment, clothing, foodstuffs and beer. All sectors of industry have slumped in recent years, many by as much as 50 per cent. The tourist industry has been badly affected since 1984 by the continued unrest and occasional violent incidents. Recently small local industries have started to develop again, especially in the building trade in Noumea.

Mining: New Caledonia holds 43.7 per cent of known world nickel deposits and is the world's third largest producer. Chrome extraction resumed in 1982. Also deposits of iron ore, manganese, cobalt, lead and zinc.

Energy: Electricity and gas consumption have been dropping in recent years, from 1,156m kWh of electrical power in 1981 to around 500 kWh in 1988.

Main companies: Société Le Nickel.

Membership of international organisations: EC (associate member), Franc Zone, South Pacific Commission (Conference representation), South Pacific Forum.

Time: GMT + 10 hrs.

Climate: Hot (average temperature 26°C), with occasional tropical depressions and cyclones, from mid-Nov to mid-Apr; and cool (average temperature 23°C), with moderate rains, from mid-May to mid-Sep. Rainfall is quite irregular and can be extremely heavy. The east coast (at about 2,000mm p.a.) has twice the rainfall of the west; the wettest months are Feb and Mar.

Entry requirements
Passport: Required by all except certain French nationals.
Visa: Visas are required for most countries except European Community nationals and Japanese tourists, who may stay up to 90 days without a tourist visa.
Customs: Personal effects are allowed entry duty-free. Duty is not payable on goods of EC origin, although all imported goods are subject to a general tax, and an increasing number of goods require import licences. Importation of parrots, parakeets, pigeons, turtle-doves and non-domestic mammals is prohibited; plants and seeds require a health certificate. Export of Cagous and objects of ethnographic interest is prohibited.

Health precautions
Mandatory: Vaccination certificates for cholera and yellow fever if travelling from infected areas.
Advisable: Typhoid and paratyphoid precautions.

Representation overseas: The New Caledonia Tourist Office maintains offices in Auckland, Los Angeles, Paris, Singapore, Sydney and Tokyo.

Air access
National airline: Air Calédonie International.
Other airlines: Air New Zealand, Polynesian, Qantas, UTA.
International airports: Tontouta International (NOU), 47 km from Noumea; duty-free shop, bar, restaurant, bank. shops, car hire.

Surface access: There are regular shipping services from Australia, Europe, Japan, New Zealand and South East Asia.
Main ports: Noumea (containers).

Hotels: Tourist hotels are classified by category and size on three-star system. Hotel tax is levied, the amount varying according to classification. Details of rural or tribal lodgings in some Melanesian villages and areas are available from tourist information offices.

Credit cards: Credit cards and foreign currency are often accepted at hotels and shops.

Car hire: Self-drive car hire is available in Noumea. A current valid driving licence is required. Driving is on the right-hand side of the road.

City transport
Taxis: Taxis are available at the central square (Place des Cocotiers); radio-controlled taxis operate 24 hrs.

National transport: Sporadic unrest has occurred since late 1984, especially in rural districts. It is advisable to ascertain the latest position before travelling.
Air: Air Calédonie operates regular flights from Noumea's domestic airport, Magenta, to east and west coasts of New Caledonia island and daily flights to the Isle of Pines, Mare, Tiga, Lifou and Ouvea. Charter and tour airplanes and helicopters available.
Road: New Caledonia, the main island, has a total road network of 3,352 km, about 16 per cent

BUSINESS GUIDE

sealed in municipal areas and about 78 per cent sealed elsewhere, and a considerable length of track suitable for four-wheel drive and similar vehicles. Exercise care driving along the west or east coast, as some roads are not sealed. Some main roads are one way only, with direction of traffic changing at scheduled times.
Buses: Noumea has a bus service; regular bus services operate on the mainland.
Water: Small trading vessels sail from New Caledonia to nearby islands.

Public holidays
Fixed dates: 1 Jan, 1 May, 14 Jul, 15 Aug, 24 Sep, 2 Nov, 11 Nov, 25 Dec.
Variable dates: Good Friday, Easter Monday, Ascension Day, Whit Monday.

Working hours: Some shops and offices open half day on Sat.
Government and business: (Mon-Fri) 0730 – 1130, 1330 – 1730.
Banking: (Mon-Fri) 0745 – 1600.
Shops: (Mon-Fri) 0800 – 1100, 1400 – 1800. Half-day Sat.

Social customs: Tipping not customary. Islanders find it offensive when women sunbathe topless.

Telecommunications
Telephone: Dialling code for New Caledonia, IDD access code + 687 followed by subscriber's number (no area code necessary). International communications via the INTELSAT satellite system, and the system is 99 per cent automatic. In 1987 there were 22,000 private lines and 87 public telephones.
Telex: Telex facilities available at some hotels.
Telegram: The central telex/telegram agency is the Office des Postes et Télécommunications (PTT) on the Rue Porcheron in Noumea.

Banking
Head offices: Banque de Nouvelle-Calédonie, Banque Paribas Pacifique, Société Générale Calédonienne de Banque.
New Caledonia branches: Banque Indosuez, Banque Nationale de Paris.

Advertising: Advertising is available in the local press, on radio and TV and in cinemas.

Electricity supply: 220 V AC, with two-pin plug fittings.

Weights and measures: Metric system.

Consulates in capital: Australia, Belgium, Indonesia, Italy, New Zealand. Japan and Pays-Bas have honorary consuls.

Useful telephone numbers
Fire station: 18 Police: 17 Ambulance (Nouméa): 252100.

Hotels: Noumea
Anse Vata Motel, 19 rue Gabriel Laroque, Val Plaisance, P.O. Box: 4453 (tel: 262612).

Club Méditerranée, Anse Vata Beach, P.O. Box: 515 (tel: 261200; tx: 3077 NM CHATROY).

Escapade Island, Ilot Maitre Island, P.O. Box: 819 (tel: 285320).

Ibis Hotel, Baie des Citrons, 8 From R Laroque, P.O. Box: 819 (tel: 262055; tx: 3045 COMIN NM).

Ile de France Apartments, on Rue Boulari, Anse Vata, P.O. Box: 1604 (tel: 262422; tx: 3143 FRANLAG NM).

Le Lagon Noumea, 143 Route de l'Anse Vata, P.O. Box: 440 (tel: 261255; tx: 3017 NM LAGONOU).

Le Nouvata Beach, Anse Vata, P.O. Box: 137 (tel: 262200; tx: 3145 NOUVATA).

Le Paris, rue de Sébastopol, City, P.O. Box: 2226 (tel: 281700; tx: 3037 NM PARITEL).

Mocambo, Baie des Citrons, P.O. Box: 678 (tel: 262701; tx: 148 NM).

Paradise Park Motel, 34 rue R P Roman, Valée des Colons, P.O. Box: 234 (tel: 272541).

Surf, Rocher a la Voile, Anse Vata, P.O. Box: 4230 (tel: 286688; tx: 3118 NM).

Vouméa Village, 1 rue de Sébastopol, City (tel: 283006, 273229; tx: 3808 NM).

Car hire
AB Location Travel Car International: 21 rue Jules Ferry, Noumea (tel: 281212).

Avis: PO Box 438 (tel: 275484; tx: 3123 NM).

Europcar: PO Box 805 (tel: 262414, 262444; tx: 3083).

Hertz: 113 Route de l'Anse Vata (tel: 261822; tx: 045 NM).

Mencar: PO Box 286 (tel: 276125).

Pacific Car: 9 Rue Soissons Faubourg Blanchot (tel: 276060).

Business advice and information
Air Calédonie (tel: 272177).

Chambre de Commerce et d'Industrie, Rue de Verdun, PO Box M3, Noumea (tel: 272551; tx: 3045 COMIN NM).

Service Statistique, PO Box 823, Noumea (tel: 275481).

South Pacific Commission, PO Box D5 Cedex, Noumea (tel: 262000; tx: 3139 SOPACOM NM).

Travel information
Office Territorial du Tourisme, 25 Avenue Maréchal Foch, PO Box 688, Noumea (tel: 272632; tx: 3063 TURISM NM).

New Zealand

Ian Templeton

After five turbulent years as New Zealand's often brilliant and sometimes erratic leader, David Lange in August 1989 announced his resignation, and the parliamentary Labour Party elected his deputy Geoffrey Palmer as prime minister. Health Minister Helen Clark (39) was elected deputy prime minister, the first woman in New Zealand's political history to hold that office. Lange's action stunned the nation but the transition to a new administration went surprisingly smoothly. The former prime minister had been under intense strain in previous months as he sought to repel dissidents within his government, led by one-time Finance Minister Roger Douglas, who had been dismissed from cabinet in December 1988. The new prime minister pledged continuity and stability both in the government's policies and in its administration. In his first major speech he said that the negative mantle over New Zealand politics needed to be lifted: New Zealanders were tired of gloom, misery and despondency. But the government had presented a budget which confirmed the practice of a responsible economic strategy and at the same time recognised and developed social reforms. Initial reaction to the change at the top was highly favourable, and where the government had been trailing the National Party in opinion polls, it suddenly looked as if it might have a fighting chance of winning the election due to be held in October 1990.

Lange government in disarray

Palmer carried through a reconstruction of cabinet which had the effect of unifying the warring factions that had split the governing party. He allocated the post of attorney-general to his predecessor who would however be outside cabinet, and he offered Roger Douglas the portfolios of police and immigration. The justice portfolio which Palmer had held went to Transport Minister Bill Jeffries, and education, previously held by Lange, was accepted by Phil Goff, at 36 the youngest minister in cabinet.

The new prime minister came into politics only 10 years ago. Previously he had held the chair of constitutional law at Victoria University. Somewhat austere, in contrast to the warm and witty Lange, he aimed to project solidity and dependability, after the roller-coaster ride the country had experienced under his predecessor.

When the Labour Party swept into office in 1984, Palmer took the portfolios of attorney-general and

KEY FACTS

Official title: New Zealand

Head of state: Queen Elizabeth II

Head of government: Prime Minister Geoffrey Palmer

Ruling party: Labour Party

Capital: Wellington

Official Languages: English

Currency: New Zealand Dollar (NZ$ = 100 cents)

Exchange rate: NZ$1.70 per $ (Nov 1989)

Area: 268,676 sq km: N Island (115,690); S Island (152,356)

Population: 3.29m (1988)

GDP per capita: $9,389 (1988)

GNP real growth: 0.6%* (1988)

Labour force: 1.53m (1988)

Unemployment: 7.3% (1988)

Inflation: 6.4% (1988)

Trade balance: –$2,015m (1987/88)

Foreign debt: NZ$43bn (1987)

Visitor numbers: 855,492 (year to March 1988)

* estimated figure

justice and he played a key role in drafting the mountain of legislation the government enacted in its first term.

New Zealand seemed set for a period of consolidation and perhaps an era of more orderly politics under his rule. Lange's popularity had waned sharply in 1988-89, as unemployment deepened and divisions within his government widened into irreparable splits.

As the country's leading newspaper the *New Zealand Herald* put it, Lange had come to believe exaggerated commentaries that the entrepreneurial forces unleashed by his government were overwhelming socialist verities. 'Mr Lange in recent times could not give the leadership it needs. He has, belatedly, done the right thing'.

Lange's decision had been triggered by the vote in the caucus of the parliamentary Labour Party on 3 August to restore Roger Douglas to the cabinet on terms that were intolerable to the leader. Lange had sought a pledge of personal loyalty from those standing for the cabinet vacancies, but did not get one from Douglas. Lange's problem had been that in order to stop what he called a 'new right juggernaut', he had frequently stepped outside cabinet conventions, and antagonised many within cabinet and his caucus.

The 'Lange factor', once the government's most significant asset, had become a liability. As he left office he promised to be a 'thorn in the side of those Labour MPs who had betrayed him'.

The hard fact was that five years after the Lange government had come to power pledging policies to restore economic prosperity and rebuild the welfare state, New Zealand was still awaiting the fruits of economic reform and restructuring. The Lange government had fallen into disarray, and Lange himself had survived two challenges to his leadership, one in December 1988 and the other in June 1989.

Sydenham MP and former president of the Labour Party Jim Anderton split from the government to form the New Labour Party and recruited support from those in the Labour movement who believed the Lange government had betrayed the principles of the traditional Labour Party.

The Lange government had won a second term in 1987 in a general election that was perceived to be essentially a referendum on 'Rogernomics' – the economic policies of Finance Minister Roger Douglas designed to transform an economy that resembled, in the colourful phrase of the prime minister, a Polish shipyard. It had been weakened by excessive protectionism, deep government intervention and suppressed inflation. Decades of high tariffs, import controls and agricultural subsidies had ossified the economy. Productivity growth was among the lowest in the OECD area. The government in the previous era had propped up a large number of loss-making commercial activities in the public sector, sheltering a significant segment of the economy from competition.

It was a huge task to redirect the economy back towards the road to sustainable, non-inflationary growth. Even though the government proclaimed that its policies had a medium-term time-frame, expectations were raised that it could produce results overnight. In fact, the pursuit of macroeconomic stabilisation policies took more than three years to begin to reduce the severe inflationary pressures. In 1989 for the first time for 15 years (excluding periods of price and wage freezes) inflation was down to the OECD average. But the cost was unemployment that most New Zealanders regarded as being unacceptably high. The number out of work or in government training schemes reached more than 170,000 – and those classified as long-term unemployed, persons who have been out of work for more than six months, totalled 64,000 in May 1989, more than the total unemployed when Labour came to office.

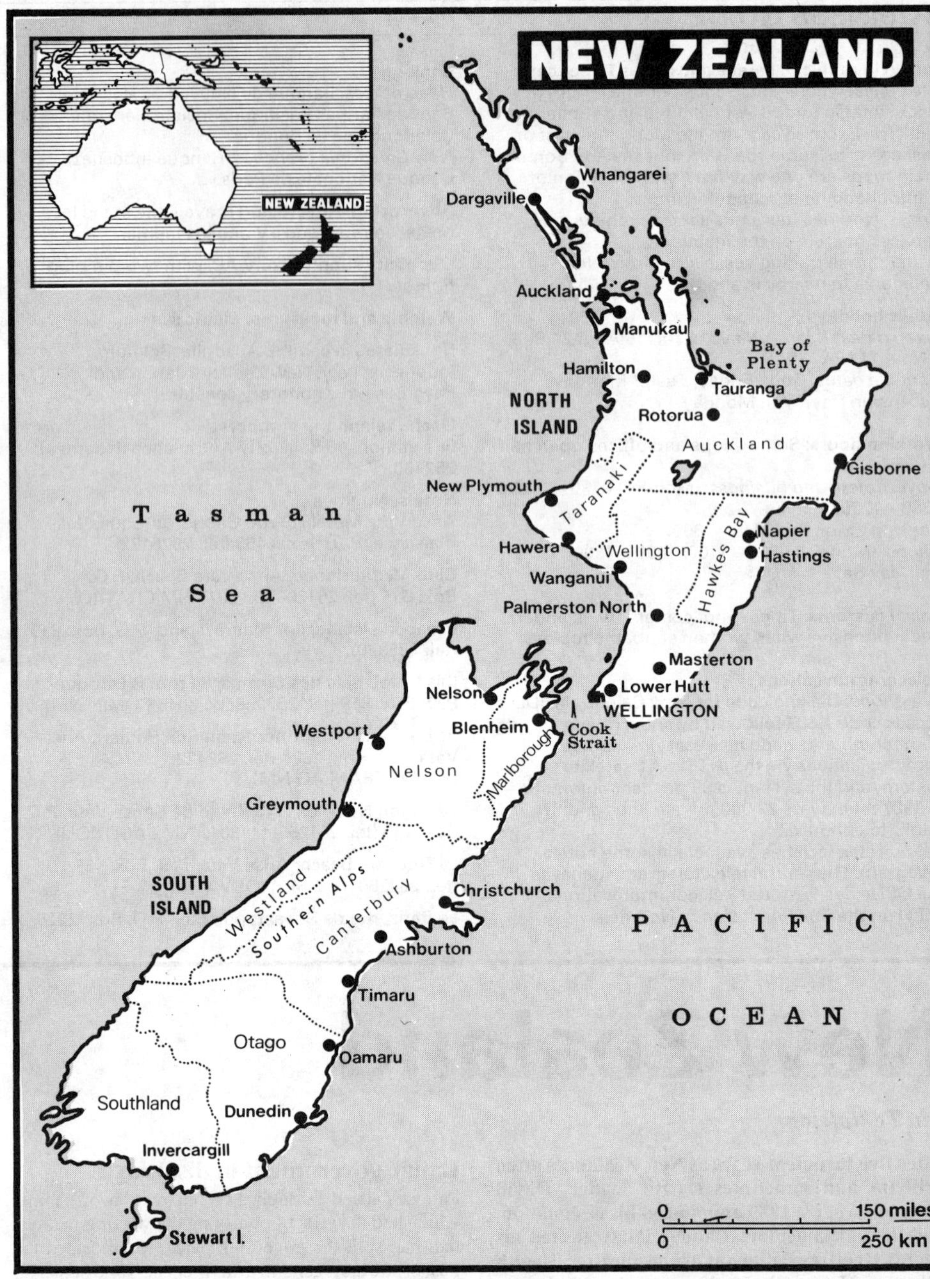

Labour's left lost its nerve as the consequences of the Rogernomic juggernaut became plain. Led by Jim Anderton it sought to push the government back to its socialist roots. Lange responded by trying to moderate the Rogernomic drive towards reform of sectors untouched during the first term from 1984 to 1987. The conflict within the Labour government steadily deepened, culminating in what political commentators described as a 'bloodbath' in the government caucus (the weekly meeting of government MPs) on 29 June. Lange survived a vote of no confidence by a margin he claimed to be between 8 and 10 votes, but others asserted to be by 27 votes to 25. Whatever the margin, it was clear that Lange did not have the confidence of a significant segment within his own government. Even the current Minister of Finance David Caygill was reported to have spoken out vigorously against the Lange leadership and called for the reinstatement of his predecessor to cabinet. A backbencher, Dr Bill Sutton, who initiated the caucus debate asserted that if there had not been some last minute defections, then Deputy Prime Minister Geoffrey Palmer would at that time have been elected leader in Lange's place, in order to reconstitute the cabinet team and restore the unity of purpose which characterised the Labour government during its first term.

The June caucus vote followed an earlier vote in December 1988 when Lange had put his leadership to the test, after in effect sacking Roger Douglas as minister of finance. He commanded 38 votes to the 15 cast for Douglas. The erosion of his support in the six months that followed reflected the sense of drift, even paralysis, that appeared to envelop the government in the wake of Douglas's departure.

The debilitating struggle between Lange and Douglas originated when the prime minister unilaterally scuttled Douglas's plan to introduce a flat rate of tax at 23c in the NZ dollar. Lange argued that people on low incomes would have to pay more tax, while rich people would pay less. But Douglas planned to compensate low-income families with a negative tax that would have guaranteed a better standard of living for families. Lange was influenced in his attitude by the feminist lobby which contended that Douglas's proposal offered no assistance to low-income single women.

In the year that followed the relationship between the prime minister and his most powerful minister deteriorated. One crisis occurred in June in the run up to the budget; another in August when the prime minister actively sought Douglas's resignation. On both occasions cabinet backed Douglas rather than Lange.

Lange was more successful in securing the dismissal of Richard Prebble who had been minister in charge of state enterprises, and was forcing through the sale of state assets, often against the opposition of the Labour Party's left wing. Prebble had been rated one of the most effective ministers in the Lange cabinet, and he did not leave the ministry without a fight, accusing Lange of an abuse of prime ministerial power. In a remarkable television appearance, Prebble charged the prime minister with irrationality, of acting contrary to cabinet decisions, and of being dictatorial. According to Prebble, the prime minister had said that the issue of state asset sales had been 'eating away at' him, that he could not sleep, and it was 'killing' him. The prime minister did not deny that he had used those phrases. He went on to remove Prebble from cabinet, and that November 1987 crisis led on directly to the confrontation with the finance minister the following month.

Douglas precipitated that confrontation with what he called a 'personal statement' to the country, in which he said a situation had developed within the government over the previous year which had gravely damaged the government, the interests of backbenchers, and the interests of every New Zealander. He said the problems which had developed between the prime minister and himself 'are tearing the government, the economy and the nation apart.' They had paralysed the government at exactly the worst point in the process of creating a modern, efficient economy out of the mess it had inherited. Douglas said the gulf that had opened up between himself and Lange was not of philosophy: it concerned the process of government, with the prime minister overturning cabinet decisions without consultation. That destroyed the trust between ministers.

Fiscal policy

If Lange had survived the departure of two of his most powerful ministers, the succession of political crises damaged business confidence, and slowed the fall in interest rates which the government had been seeking as a prerequisite to the upturn in investment it had been looking for to fuel recovery.

Lange reshuffled his cabinet, bringing David Caygill into the finance portfolio, promoting Helen Clark into the health portfolio left vacant by Caygill, and adding the responsibility of state-owned enterprises to the minister of labour, Stan Rodger. Caygill had served as an associate finance minister during the first term of the Labour government, and vowed to continue the Douglas drive to reduce inflation and eliminate the fiscal deficit. As health minister however he had backed away from the Rogernomic call for major reforms in the hospital system, and advanced less drastic measures to improve efficiency in health care. According to the Douglas faction, the health vote had risen 20 per cent in real terms since 1984, yet the waiting lists for surgical treatment had lengthened and the delivery of other health services had deteriorated. The extra funds being voted by parliament were too often being captured by the health providers, rather than by the consumers. Hospitals should be run in a more businesslike manner, according to Douglas. These issues were at the core of the conflict between the prime minister and those ministers who were concerned that the government was not getting value for money in the social policy areas.

Lange subsequently admitted that trespassing into social policy by Douglas was the real reason for his dismissal. The drive to eliminate the fiscal deficit was eroded by the refusal to contemplate moves which could put a cap on social policy spending. The consequent imperative for David Caygill was a search for other means of aligning revenue with expenditure. New tax rates had come into effect from October 1988 (the flat tax proposal of Douglas had been replaced with a compromise two-step scale of 24c in the NZ dollar on income to NZ$30,875 and 33c on income beyond that point, replacing the old four-step scale with a top rate of 48c in the NZ dollar).

Caygill had two problems in looking ahead to fiscal year 1989-90. He had to allow for the full-year implications of the October 1988 tax reductions, and he had to adjust for the fact that in the previous year a number of changes to tax provisions had brought forward tax revenue which would have fallen in subsequent years. Taken together those revenue adjustments would have meant a widening gap in the fiscal equation, moving the deficit back from 2 per cent of GDP to 4 per cent. The extra borrowing required would have propped up interest rates, and disrupted the government's medium-term strategy of lowering inflation to the level of the country's main trading partners. Caygill's attack was two-pronged. In an economic statement in March he announced that the goods and services tax would be raised from 10 per cent to 12.5 per cent from 1 July and company tax from 28 per cent to 33 per cent from 1 April. On the expenditure side, he put new restraints on departmental votes. Inevitably that produced a political backlash when it became apparent, for example, that the number of frontline police would have to be cut, at a time when public concern over increasing violent crime was widespread. The spending cuts also brought into fresh focus the difficulties in the health sector, particularly when the largest hospital board in the country had to be dismissed for exceeding statutory limits on spending and was replaced by a government-appointed commissioner.

Sale of state assets

Caygill's problems were compounded by a series of bungles in the state asset sale programme, the worst of which occurred in the case of the Bank of New Zealand, the largest of the country's trading banks. The sale had been virtually concluded in December 1988 but had been vetoed in effect by the government caucus, on the grounds that the price was too low. Subsequently it became clear that the bank's losses in the wake of the 1987 sharemarket crash had been much greater than earlier suspected (in the 1988-89 year it had to write off NZ$1.3bn in bad and doubtful debts). The government forced the resignation of Chairman Sir Ronald Brierley and the Deputy Chairman Rob Campbell. At the same time it had to mount a bail-out operation, with 30 per cent of the bank's shares being sold to Capital Markets, an off-shoot of merchant banker Fay Richwhite Ltd.

The manoeuvres over the Bank of New Zealand raised eyebrows in the international banking world, and left the government with a much reduced pay-off than if the bank had been sold in December.

Another asset sale which ran into trouble was the disposal of the Crown's interest in the Maui gas contract and the synthetic petrol plant at Motunui. In a complicated deal Petrocorp (owned by the country's biggest conglomerate Fletcher Challenge) was to pay the government NZ$585m for its stake in the Maui contract, and in turn was to be paid NZ$236m to take the Crown's 75 per cent share in the loss-making petrol plant. However, the Commerce Commission would not approve the first part of the deal, on the grounds that it gave Petrocorp too dominant a position in the wholesale gas market.

The government did succeed in completing a deal for the sale of Air New Zealand to a consortium led by Brierley Investments and including Qantas, Japan Airlines and American Airlines. That consortium beat off a bid from British Airways, which was favoured by management of Air New Zealand, concerned at the prospect that it would become a mere satellite of the Australian international airline. Some aviation industry sources believed that the NZ$660m received by the government for Air New Zealand understated its real value. However the government pressed ahead with its asset sale programme, with keen bidding expected for the vast holdings of state forests.

The imperative driving the asset sale programme had been the urgency of reducing the country's public

KEY INDICATORS

	Unit	1984	1985	1986	1987	1988
Population*	'000	3,231	3,259	3,273	3,283	3,290
Gross domestic product (GDP)	NZ$m	34,010	39,364	47,007	52,247	–
GDP per capita	NZ$	10,526	12,079	14,362	15,914	–
Consumer prices	% change	6.1	15.5	13.2	15.7	6.4
Weekly wage rates	% change	2.3	7.8	16.4	8.0	–
Manufacturing employment	1980 = 100	98	99	100	102	–
Volume of exports (March yrs)	1985=1000	90	100	98	101	105
Wool	1985=100	104	100	104	92	98
Lamb and mutton	1985=1000	91	100	94	87	81
Butter	1985=1000	77	100	83	87	94
Beef and veal	1985=1000	87	100	100	121	130
Volume of imports	1985=1000	101	100	99	109	101
Exports	NZ$m	9,584.1	11,603.2	11,225.1	12,149.7	13,388.2
Wool (June years)	NZ$m	1113.3	1475.4	1,470.5	1,540.2	1,808.0
Lamb and mutton	NZ$m	999.7	1,63.9	1,007.8	976.9	914.5
Butter (June years)	NZ$m	562.4	636.5	498.8	517.7	555.0
Beef and veal (June years)	NZ$m	634.8	952.0	800.4	1,000.3	1,053.2
Imports value cif	NZ$m	10,939.0	12,075.5	11,622.1	12,242.5	11,216.9
Balance of trade (March years)	$m	–247	87	224	531	2,015
Current account	NZ$m	–1,748	–1,411	–1,481	–1,758	–761
Exchange rate	$ per NZ$	0.58	0.50	0.52	0.59	0.66

* estimated figure

debt, which according to some calculations had reached 80 per cent of GDP. Servicing the debt was consuming 25 per cent of the annual budget. The government besides using the proceeds of asset sales to retire debt also brought external debt back onshore in order to reduce pressure on the balance of payments. The policy lowered official overseas debt from 68 per cent of GDP to just under 50 per cent, with the projection that it should be down to 36 per cent by 1992.

Trade situation

If the government had its problems on the economic front, it also had some successes. Inflation had fallen to 4 per cent, and the country's competitive edge, after a decade of being blunted by high inflation, sharpened. The international commodity boom contributed to a lift in the terms of trade, which reached their highest point since 1979. Low import flows as a result of the recession, in combination with record receipts for exports, led to a surplus on merchandise trade of NZ$2.36bn for the 12 months ended May 1989, compared with a surplus of NZ$751m for the equivalent period the previous year. High prices for dairy products and wool restored some confidence to the rural sector, while strong pulp and paper markets boosted profitability in the forest sector. Receipts from aluminium also soared as a consequence of the metal boom.

The government, while it talked of an incipient recovery led by a resurgent agricultural sector, found that the initial reaction of farmers to better prices was to reduce debt and catch up on deferred maintenance rather than to make new investments. The dairy provinces were the first to show some signs of an upturn in consumer spending, but even there the recovery looked fragile. The Dairy Board expected sales abroad to earn NZ$3bn in the 1989 calendar year, up from NZ$1.9bn the previous year. Dairy farmers had received NZ$5.30 a kilo of milkfat over the past season, up from NZ$3.60 a kilo the season before, and in the season ahead were anticipating a figure above NZ$6 a kilo. While the dairy industry, with its unified production and marketing structure, was able to provide increasing returns to farmers, the more loosely organised meat industry, inhibited by inefficient work practices, was locked into deteriorating profitability. The lamb kill fell from 39 million to 36 million, and unless returns to farmers improved sharply, the decline looked likely to continue.

The fishing industry which underwent rapid expansion in the 1980s is perhaps due for some retrenchment as stocks of some species become depleted. The kiwifruit industry, facing competition from other suppliers, has been forced to reorganise its marketing, and many producers are only marginally profitable.

While the Lange government argued that it had created a platform for sustainable economic growth, there is little evidence of a shift into activities in which the country has a comparative advantage, the process which deregulation and restructuring was supposed to encourage. Tourism was a sector earmarked for rapid expansion in the first years of the Labour government. Instead it stagnated, with hotels impeded from supplying the service expected by visitors due to out-dated union rules.

Labour market

The government was criticised for failing to deregulate the labour market, and for keeping a tight control on immigration. Meanwhile there was a continual drain of the best educated and most highly qualified New Zealanders across the Tasman to the booming economy, and higher living standards, in Australia. With hindsight, the short-term pains of restructuring would have been smaller if the government had tackled the least responsive (that is, the labour market) first, and the most responsive (the financial markets) last, according to no less an authority than the London *Economist*. Labour market rigidities were partly to blame for the sharp rise in unemployment. Reforming the wrong way round was one reason that the economy continued to struggle.

Lange reiterated his opposition to what he called the 'excesses' that would occur in the labour market if it were deregulated. In place of deregulation he deputed his third ranked minister Mike Moore to try to reach a 'compact' with the trade unions, rather similar to the 'accord' the Hawke government in Australia had with the trade unions there. Some unions were suspicious of the 'compact' concept in which the implicit trade-off for some input into government economic decision-making would be acceptance, they thought, of lower real wages.

Unions were concerned that with the government targeting its economic policy on getting inflation down to between 0 and 2 per cent, there were still more job losses to come. In the 1988-89 wage round, settlements were mostly in the 3-5 per cent range, but the forecast for the 1989-90 round was for a reduction in real wages. Hard-line monetarists in the Reserve Bank and Treasury reinforced their advice to government with studies that suggested that an increase in real wages, relative to output prices, of about 10 per cent from 1985 to 1988 contributed a reduction of 47,000 full time private sector jobs. Thus instead of the estimated 160,000 unemployed by March 1989 a constant real wage might have kept the unemployment level down to 113,000.

Maori affairs

Although across the nation in 1989 unemployment comprised about 10 per cent of the workforce, provincial cities like Whangarei and Rotorua had nearly 20 per cent of the workforce without permanent jobs. The hardest hit sector of the community is the Maori population, and that has tended to exacerbate sensitivity in race relations, already an explosive political and social issue in many regions. The Lange government sought to defuse Maori grievances by reforms in the administration of Maori affairs, and by judicial review of tribal losses of lands and fisheries which under the Treaty of Waitangi should have been retained. In leaning over backwards to be fair to the often educationally and socially deprived Maori population, the government was in danger of creating a so-called Pakeha backlash. The political climate engendered on this issue was exploited by the National Party's spokesman on Maori affairs, Winston Peters, MP for Tauranga, who in attacking what he called the 'sickly white liberalism' of the Lange government struck a chord with the redneck element in New Zealand society. Peters, himself a Maori, was also the opposition's spokesman on employment and he cleverly orchestrated these points of public concern into a broad onslaught on what he called the 'warped and anti-social' philosophy of the Labour government. In the process he out-ranked the opposition leader Jim Bolger as the most effective critic of the government, climbing rapidly up the popularity stakes as measured by opinion polls. In some quarters he was regarded as a protege of the former Prime Minister Sir Robert Muldoon, and his espousal of interventionist policies brought him into conflict with those in the opposition who rejected Muldoonism, notably the finance spokesperson Ruth Richardson. Tensions within the National Party over economic policy tended to parallel those within the Lange government, and its failure to present a united front undermined its credibility as an alternative government.

Election outlook

The National Party was heartened by opinion polls which showed support for the Lange government falling away, and at some points during the year the gap between the two parties stretched to 30 percentage points, the widest in the history of political opinion polling in the country. Labour was weakened by defections to the Anderton-led New Labour Party whose support tended to solidify around the 10 per cent mark. Opposition leader, Jim Bolger, a farmer from the King Country in the middle of the North Island, is a politician in the mould of Sir Keith Holyoake who led the country through the 1960s. Bolger's efforts to moderate between the two conflicting strands of ideology – one leaning to the interventionism of Muldoon and the other towards the Friedmanite monetarism and pure free market theory – left the impression of a sometimes ineffectual leader. The consequence was that even before the change in prime minister, National's lead in the opinion polls looked quite unstable, particularly as the level of 'undecideds' remained consistently high, ranging between 20 and 30 per cent. The Labour government pinned its hopes on a resurgence in the economy and the change in leadership to refurbish its chances in the October 1990 election. Through 1989 ministers sought to jawbone interest rates, which remained stubbornly high, impeding the investment essential to stimulate economic activity. There were some indications of a pick-up in the rural regions, but that was offset by a downturn in the Auckland metropolitan area, which was the last region to feel the impact of economic restructuring, stemming from the removal of import barriers, and the lowering of tariff protection. Manufacturing industries which had been set up during the years of total import control suddenly found themselves exposed to the chill winds of competition. The number of registered unemployed in the Auckland region was 18 per cent higher in May than in November. By contrast the number of unemployed in the South Island rural areas in May was 6.5 per cent lower than the November figures.

If the lowering of import barriers had an adverse effect on Auckland unemployment, it contributed significantly to price stability. New Zealanders now have access to goods at prices the rest of the world pays, where before importers during the era of import licensing brought in only items on which they could maximise profits, often top of the market goods.

Finance Minister David Caygill who as minister of trade and industry during the first term of the Lange government had put in place the policies to lower protectionism and open up the economy to market competition was emphatic there would be no backing off the economic reforms.

The problem for the government before August 1989 was that its internal conflicts generated so much uncertainty that the investment community was unconvinced Caygill could hold the line against a prime minister who increasingly sought to fence off the economic reformers. For his part ousted Finance Minister Roger Douglas continued to argue that the government must raise its sights. It was not good enough simply to aim at lifting New Zealand back towards the average OECD level of economic growth; New Zealand had to do better in order to make up for the ground it lost in the 1970s. Douglas continued to advocate reforms in the labour market to make it more responsive and flexible, and in the tax-benefit stucture, so that there would be greater incentives for people to move off benefits and into jobs.

Companies in trouble

While political uncertainty clouded the country's economic progress, the pessimism prevalent in the business community was compounded by a succession of company crashes, as one after another the entrepreneurial companies that flourished in the overheated climate before the October 1987 sharemarket crash toppled into receivership. The biggest crash was that of Alan Hawkins' Equiticorp whose ambitious takeovers turned out to be based more on financial sleight of hand than on sound business practice. The

Equiticorp collapse pushed receivers into selling off assets such as Equiticorp's 80 per cent stake in NZ Steel Ltd which owned the country's modern steel plant at Glenbrook (largely financed by the government). The Chinese trading corporation Minmetals lodged a bid for the steel mill, but its option to purchase lapsed after the crackdown in China, in the wake of the 4 June massacre in Tianamen Square in Beijing.

Investment across the Tasman

The sharemarket crash in conjunction with the economic recession left many New Zealand companies vulnerable, and Australian companies picked up major assets at fire sale prices. BTR Nylex bought the big industrial company Feltex from Equiticorp's receivers and Elders Resources expanded their operations in several sectors, particularly in forestry with a dominant stake in what used to be New Zealand's biggest forestry company, NZ Forest Products. Brierley Investments sold off its 52 per cent stake in Sydney-based Industrial Equity Ltd, and Chase Corporation, whose property assets in New Zealand were placed under government-appointed statutory managers, sold off Australian assets including Hanimex to various Australian companies.

The Australian influence was extended in aviation, with Qantas securing a minority stake in Air New Zealand. Meanwhile Ansett NZ which was providing Air New Zealand with stiff competition on domestic routes was told by the government that it could seek international routes where New Zealand held bilateral rights in other countries currently not being used by Air New Zealand. Ansett intensified competition on domestic routes by the introduction of the ultra-quiet BAe 146 jets.

The investment traffic across the Tasman was not all one way. The Dunedin-based Wilson Neill expanded its liquor interests, with the purchase of the Tasmanian-based brewery group Cascade. Goodman Fielder Wattie became one of the dominant food groups in the Australasian market, run by Pat Goodman out of his hometown Motueka in the South Island.

Australia and New Zealand were in any case moving towards the creation of a single market. Full free trade between the two countries was due to come into effect in 1990. During 1989 the process of lowering trade barriers accelerated with several industries like apparel, textiles and footwear emerging from quotas and other restrictions. Leading economist Len Bayliss thought the development of a single market had moved so swiftly that both countries should consider the next step: a common currency.

Defence links

The ties between the two countries were also consolidated in defence. The decision of the Reagan administration to declare the Anzus Treaty 'non-operative' between the United States and New Zealand, as a result of the Lange government's decision to refuse access to warships that might be carrying nuclear weapons, led the New Zealand defence forces to rely heavily on their links with the Australians in defence exercises and other areas of co-operation.

Out of that environment emerged a joint project for the construction of an Anzac class frigate in Australian shipyards. The plan was that New Zealand would buy, or lease, four of the frigates, which would be designed for operations in the Pacific, equipped for anti-submarine detection, able to operate at long range, and be fuel efficient. Two consortiums, one offering a Dutch design, and the other a German design, were competing for the contract, and a decision was due in August 1989, on the successful tender. The Lange government insisted that its final decision would depend on the price, with the prime minister declaring that the Australians 'were making us an

COUNTRY PROFILE

Political system: New Zealand is an independent parliamentary democracy and member of the Commonwealth. The head of state is the British sovereign, represented by a Governor-general who acts on the advice of the Cabinet. The prime minister heads the Cabinet. The 97-member unicameral House of Representatives is elected every three years by universal adult suffrage: four seats are reserved for Maoris.

Political parties: Ruling party: Labour Party (58 parliamentary seats in Sep 1987 election), National Party (39 seats), Social Credit Political League (0), New Zealand Party (0).

Population: Stood at 3.35m in 1988. 81.3 per cent of the population are of European origin, with Maoris representing another 12 per cent. Pacific Islanders are 2.9 per cent and mixed European/Maori/Islanders number 4.3 per cent. 74 per cent of the population live on North Island. 23.8 per cent of New Zealanders are under 15 years.

Main cities/towns: Auckland, North Island (population 829,000); Wellington, North Island (pop 329,697); Christchurch, South Island (pop 299,373); Dunedin, South Island (pop 107,000); Hamilton, North Island (pop 103,000); Palmerston North, North Island (pop 67,405); Napier/Hastings, North Island (107,059).

Language: English and Maori are the official languages.

Media
Press: One national newspaper, the *National Business Review* published Monday to Friday. Eight morning and 25 evening papers are published daily in the main towns, Sunday papers published in Auckland and Wellington. A few of the dailies published in main centres are distributed throughout the surrounding provinces. Major weekly publications are the two national weekly newspapers, *Listener*, the *TV and Radio Journal*, *Truth*, and *The Women's Weekly*.
Broadcasting: The public broadcasting system is run by the Broadcasting Corporation of New Zealand. *Radio:* Radio New Zealand provides 37 commercial community stations, three commercial ZM popular music stations, 26 non-commercial stations broadcasting in a national network and four serious music stations broadcasting in a concert programme network. There are also 15 private commercial radio stations and four private non-commercial stations. *Television:* Television New Zealand provides two colour television channels throughout the country, with about one-third of the programming produced in New Zealand. Broadcasts are estimated to reach 90 per cent of the population, and there are commercial broadcasts on weekdays. TV3 begins transmission in late 1989.

Domestic economy: The economy is dependent on agricultural products, which are subject to an increasing degree of processing for export income. Import licensing to protect domestic industry has now been largely dismantled. New Zealand has suffered balance of payments deficits owing to its heavy dependence on imported oil, low worldwide demand for dairy products and EC import restrictions. Since coming to power in 1984, however, the Labor Government has attempted to reverse the interventionist tendencies of its predecessor. It has cut back on export support and import quotas, and has reduced agricultural subsidies and industrial assistance, while deregulating the financial sector (and, increasingly, also the manufacturing sector). This has resulted in improved productivity while some state trading organisations are actually making profits after years of losses. Growth has been hampered, however, by debt servicing costs. By the end of 1987 the country owed more than $43bn abroad, with a debt servicing ratio of 67 per cent (Bank of New Zealand estimates). The persistent fiscal deficit was turned into surplus in the fiscal year to March 1988 by the sale of state assets. This pushed up interest rates, strengthened the dollar, discouraging exports and investment.

Employment: The normal working week is 40 hours, with overtime available in many occupations and an average weekly earning rate of NZ$411.58 (1987). Workers by law must belong to a trade union, and are entitled to three to four weeks annual leave with pay.

External trade: Policy is to substitute tariffs for import licensing as a main form of industry protection.
Exports: Main exports are meat (19.3% of 1987 total), manufactured goods (17.5%), other foodstuffs including fruit and vegetables (16.9%), wool (13.4%), dairy produce (12%) and others (14.5%). Main destinations are Australia, Japan, USA, UK and the EC.
Imports: Main imports are machinery and transport equipment (40.3% of the 1987 total), manufactures (26%), raw materials (22.3%) food (6.2%) and metals (5.2%). Main sources were Japan, Australia, USA, and UK.

Agriculture: Agriculture, fishing and forestry account for eight per cent of GDP and employ 10.4 per cent of the work force. Of the total land area, about three per cent is cultivated arable, 50 per cent permanent pasture and 25 per cent forest, woodland and plantations. Pastoral products (meat, dairy products and wool) account for over 45 per cent of exports. New Zealand is the world's second largest exporter of wool, and its lamb and mutton processing industry accounts for 20 per cent of export earnings. Over 67m sheep are kept (1987), mainly on family-owned and operated farms. Dairy farming is a main activity, and 85 per cent of milk production is used for dairy products such as butter, cheese, preserved milk and casein. Deer and goat farming are increasingly important. Wheat production is sufficient to meet national demand, and other crops include barley, maize and fresh vegetables. Horticultural production has increased; kiwi fruit accounts for over two per cent of export earnings, and other products include apples, pears, stone and berry fruits and citrus and sub-tropical fruits. Plantation forests cover over 1m ha. Logging of large areas planted in the 1920s and 30s yields about 10 cu metres p.a., and 45,000 ha p.a. are being planted with Radiata pine (which matures in 20-30 years), Douglas fir and Corsican pine.

Industry: Manufacturing accounts for 24 per cent of GDP and over 23 per cent of export earnings, and employs 22 per cent of the workforce. Manufacturing industries rely on agricultural products (food processing) or natural resources (forest products and aluminium smelting). About 10 per cent of manufacturing production is exported. Large-scale operations have been established for pulp and paper, fertiliser, oil refining, aluminium smelting, steelworks, methanol, synthetic petrol and ammonia/urea. Most manufacturing activity is small-scale production for the domestic market of goods such as food, tranport equipment, machinery, wood products, clothing, toilet requisites, pharmaceutical preparations, china and glassware, electrical goods, radio and TV sets, household appliances, hardware, rubber and plastic goods, fancy goods, sports goods, toys and electronic components. Long-term development projects plan to expand pulp and paper production alongside increased reafforestation and to develop transport fuels based on Maui natural gas, but this has been postponed due to low oil prices. The Development Finance Corporation is promoting development of tourist facilities to ensure continued expansion of the tourist industry.

offer we cannot refuse'. But peace groups in New Zealand campaigned vigorously against the frigate purchase, arguing variously that they were too costly, that the money would be better spent on hospitals and schools, and that the frigates, designed as they were to operate in conjunction with the Australian fleet, would drag the country back into a nuclear alliance. Opposition to the frigates came also from within the Labour Party, whose annual conference adopted a remit urging the government to reject the purchase. However Defence Minister Bob Tizard was confident the government would go ahead with the frigate deal. He believed a majority of ministers were in favour of the purchase, which would consolidate the alliance with Australia, provide considerable work for New Zealand engineering shops, and demonstrate to the rest of the Western alliance that despite its anti-nuclear stance, the Lange government was willing to play its part in preserving security and stability in the South Pacific.

Relations in the Pacific

While the integration of the Australian and New Zealand economies moved ahead in 1989 and defence links were consolidated, a rift between the two countries opened up when the Hawke government refused to sign the Antarctic Treaty mineral convention which New Zealand had sponsored and developed over a span of nearly a decade. Australia supported turning the Antarctic into a world heritage park, but New Zealand took the line that as a first step controls should be agreed by all the Antarctic nations.

The two countries were in closer harmony over the damaging impact of 'wall of death' drift net fishing in the Pacific Ocean. At the South Pacific Forum in Kiribati in July they mounted a diplomatic offensive against the nations (Japan and Taiwan) whose fishing fleets are depleting the stocks of fish in the southern Pacific oceans.

The New Zealand government maintained friendly relations with the island states, although those with Western Samoa were strained to a degree by New Zealand's efforts to hold down the inflow of migrants from that territory. The annual quota of migrants from Western Samoa is fixed at 1,100 but four times that number had been arriving as 'adopted' offspring of Western Samoans already resident in New Zealand. Depopulation of Niue was also a matter of concern, with more Niueans living in New Zealand than in the island of their origin. The Lange government put pressure on the government of Sir Robert Rex to use the annual flow of aid funds running at nearly NZ$12m a year for the development of the economy rather than to sustain an inflated bureaucracy.

During 1988-89 Japan became New Zealand's biggest export market, taking more than NZ$2.3bn worth of exports. That was due in part to the commodity boom, with high prices being received in particular for aluminium, paper and pulp, wood products, meat, wool and dairy products. New Zealand sales of fish and horticultural products to Japan were also substantial.

Exports of aluminium ingot from the Bluff aluminium smelter owned 79.5 per cent by Comalco and 20 per cent by Sumitomo, as well as of fabricated aluminium products, reached more than NZ$800m in the June year. For Comalco NZ (fully owned by Comalco Pty of Australia), it was a bonanza. Profits for the company reached NZ$173m, or 38 per cent on funds invested. Comalco talked of expanding the Bluff smelter, adding a fourth potline, and upgrading the technology of the existing potlines, at a cost of NZ$500m, to expand output by at least 100,000 tonnes on top of the 267,000 tonnes produced in 1988. However, that expansion would depend on whether Comalco could reach agreement with the state-owned Electricorp on a long-term power contract. For some time there has been public controversy over the price

COUNTRY PROFILE

Mining: Recoverable reserves of coal and lignites are variously estimated at between 4,000m and 6,000m tonnes, enough for over 400 years. Coal production is around 2.5m tonnes a year, but is planned to quadruple by 1995 in order to supply new steelworks and thermal power stations. Coal is also seen as a future export. Other minerals present include sulphur, clay, ironsand, limestone, dolomite, titanium, gold and silver.

Hydrocarbons: Two main natural gas fields, Kapuni and Maui, yield about 228m cu metres/month. The offshore Maui gas field, one of the largest in the world, with reserves of over 230b cu metres*, has an estimated life of 35 years. Gas production is used as CNG, LPG, generating station fuel and in the petroleum refining process, and supplies a number of downstream industries, which will include a synthetic petrol plant at Motunui. The oil price collapse in 1985-86 severely limited the plant's viability, however. Crude oil exports to Australia commenced in 1985.

Energy: Hydroelectric power generated on numerous rivers and lakes provided 72 per cent of all electricity generated in 1986, with geothermal power providing another five per cent (a large geothermal plant at Wairakei).

Financial markets: Stock exchanges are at Auckland, Christchurch, Dunedin and Wellington. In 1985 a computerised quotation system was introduced, effectively linking them into one floor. Banking has been opened up to international competition, but domestic demand for credit remains weak.

Membership of international organisations: ANZUS, ASPAC, Asian Development Bank, CCC, Colombo Plan, Commonwealth, DAC, ESCAP, FAO, GATT, IAEA, IBRD, ICAO, ICO, IDA, IEA, IFAD, IFC, IHO, ILO, IMF, IMO, INTELSAT, INTERPOL, IPU, ISO, ITU, OECD, South Pacific Agreement on Regional Trade and Economic Co-operation, South Pacific Commission, South Pacific Forum, SPF, UN, UNESCO, UPU, WHO, WIPO, WMO, World Bank, WSG.

Time: GMT + 12 hrs.

Climate: Temperate and moist, with an average temperature of 12°C, although the north is considerably warmer. Temperatures range from 15-25°C in summer (Dec-Feb) and 7-11°C in winter (Jun-Aug). Extremes of temperature, from -9°C to 30°C, occur in the mountainous interior of South Island. Rainfall is plentiful, ranging from 60-150cm p.a.

Entry requirements
Passport: Required by all.
Visa: Entry visas are not required by UK citizens or nationals of certain countries (German FR, Japan, USA) with onward passage for short stays. Otherwise, entry visas should be obtained before arrival.
Customs: Personal effects are allowed duty-free. A wide range of imports have been subject to licensing, mainly goods which are made in New Zealand, such as clothing, pulp and paper products, whiteware and various consumer goods. Import Control Regulations are administered by the Department of Trade and Industry.
Income tax: New Zealand has double taxation agreements with a number of countries. There are concessional rates for some foreigners with special skills working for periods of 6-24 months.

Health precautions
Mandatory: Agricultural regulations require visitors to declare any animal, vegetable or fruits products or plants. Visitors arriving from countries suffering from certain diseases affecting livestock and plants may have items of clothing and produce disinfected.

Representation overseas: Adelaide, Apia, Athens, Baghdad, Bangkok, Beijing, Bonn, Brisbane, Brussels, Canberra, Delhi, Geneva, Harare, The Hague, Hong Kong, Honiara, Jakarta, Kuala Lumpur, Lima, London, Los Angeles, Manama, Manila, Melbourne, Mexico, Moscow, New Delhi, New York, Noumea, Nuku'alofa, Ottawa, Paris, Perth, Port Moresby, Rarotonga, Riyadh, Rome, San Francisco, Santiago, Seoul, Singapore, Suva, Sydney, Tehran, Tokyo, Vancouver, Vienna, Washington.

Air access
National airline: Air New Zealand.
Other airlines: Aerolineas Argentinas, Air Pacific, Air Rarotonga, Ansett New Zealand Ltd, British Airways, Canadian, Cathay Pacific, Continental, Garuda Indonesia, Hawaiian, JAL, Lufthansa, Polynesian, Qantas, Singapore, Thai Airways, United, UTA.
Tax: An airport tax of NZ$2 is levied on departure.
International airports: Auckland International, Mangere (AKL), 22.5 km from Auckland, with duty-free shop, bar, restaurant, bank, hotel reservations, post office, shops, car hire; Christchurch International (CHC), 10 km from Christchurch, with duty-free shop, bar, restaurant, buffet, bank, hotel reservations, post office, shops, car hire; Wellington International, Rongotai (WLG), 8 km from Wellington, with duty-free shop, bar, restaurant, buffet, bank, hotel reservations, post office, shops, car hire.

Surface access
Main ports: Auckland (containers), Dunedin, Lyttelton (containers), Tauranga, Wellington (containers) Port Chalmers (containers), Picton, Opua.

Hotels: Motel, serviced-unit accommodation is widespread. Neither a service charge nor tipping is customary. Advance booking is advisable for major hotels in urban centres.

Credit cards: All major credit cards are accepted.

Car hire: Self-drive car hire is available in main centres. Visitors may drive using Australian, British, Canadian, US or International driving licences. Driving is on the left-hand side of the road. Parking can be a problem in larger cities.

City transport
Taxis: Taxis may be hired from ranks or by telephone 24 hrs/day, although there is an extra charge for telephone booking. Fares are generally charged per km, but rates vary throughout the country and are generally higher at night and on weekends. Tipping is not customary.

National transport
Air: Internal air services are well developed, with scheduled services linking main centres and serving some 30 airports. Air New Zealand operates the majority of regular domestic services, and other operators include Mount Cook Airlines, Newmans Airways, Eagle Airways, Wairarapa Airlines, and Bell-Air; Ansett New Zealand in 1987 started operating between Auckland, Wellington and Christchurch. Charter services are also available.
Road: The 96,000 km road network includes over 11,000 km of state highways. The main routes are surfaced, and unsealed roads are generally well-maintained.
Buses: Luxury coach services link the main centres. Advance booking is advisable, especially during the main holiday periods Dec-Feb and Easter.
Rail: The government-owned rail service operates on 4,500 km of track. Express services link the main centres on both islands.
Water: Ro-ro rail and car ferries operate a regular service between Wellington and Picton several times a day. Advance booking is advisable, especially during the main holiday periods Dec-Feb and Easter.

Comalco pays for its supplies of electricity, with Finance Minister David Caygill describing as 'bum contracts' the agreements signed by the previous government. Comalco offered to buy into the underground Manapouri power station, but this did not appear to be acceptable to the Lange government, which had not finalised its plans for restructuring the electrical industry.

Lange reforms

As part of the programme of microeconomic reform, the government employed teams of international consultants to review the electrical industry and advise whether, and how far, it should be deregulated. The generation and transmission system, run as a national network, is owned and operated by the newly corporatised Electricorp, formerly the New Zealand Electricity Department, while at the retail level electricity is sold by independent and elected power boards or by departments of territorial local authorities. Electricorp has become one of the country's biggest businesses, yielding a profit of NZ$332m in the March 1989 year, and returning over NZ$1.6bn to the government in debt and interest payments, taxation and dividends. Its performance reflected the success of the government's corporatisation programme, with trading departments which during the 1980s consumed NZ$20bn in capital funds without returning a dividend to the taxpayer now becoming cash cows for the government. Coalcorp, for example, took over the nationalised coal mines which made losses for 21 years out of the previous 22; but in the two years following corporatisation, it returned profits from which taxation and dividends were paid. One of the biggest departments, the New Zealand Post Office, was split up into three corporations, Telecom, NZ Post and Postbank, the last of which was sold to the ANZ Bank for NZ$660m. The government decided against complete deregulation of the postal service, in return for a commitment by NZ Post to maintain a universal postal service, and it was taken off the list of state assets to be sold. NZ Post made profits of over NZ$100m in aggregate in 1987-89. Telecom, whose monopoly of telecommunication services is being removed gradually was expected to be one of the first state corporations to be privatised. Another major state asset lined up for sale in 1990 is the timber in the huge state forests.

The Lange government, its reformist zeal undiminished, pushed through parliament legislation which will change the structure of local government and if it fulfills the objectives of Local Government Minister Dr Michael Bassett will deliver more efficient services to ratepayers through rationalisation of the diversity of territorial and ad hoc authorities. Another reformist thrust was directed towards changes in the country's antiquated liquor laws. A beginning was made in seeking greater efficiency in the country's port system by transferring the commercial operations of harbour boards to newly formed port companies. Similarly the country's international airports, previously administered by local bodies, were transferred to airport companies, and the airways system, previously administered by the civil aviation division of the ministry of transport, is now operated by an independent airways corporation.

Not all the government's plans for reforms proceeded smoothly. Its pledge to enact a Bill of Rights, first enunciated in its 1984 election policy, was quietly buried. Reform of the electoral system recommended by a Royal Commission, and embracing the introduction of proportional representation, extension of the parliamentary term from three to four years, and enlargement of the House of Representatives, was rejected by the government caucus.

The Lange government also had trouble in meeting its promises to redress the grievances of the Maori people under the 1840 Treaty of Waitangi, which according to some interpretations gave sovereignty in perpetuity to the Maori tribes of their ancestral lands and fisheries. In its efforts to defuse Maori radicalism, the government aroused a white backlash, which left a serious question mark on the planning for the celebrations in 1990 of New Zealand's sesquicentenary. There was concern that the Maori people might stand back from the celebrations or boycott them completely or worse still that Maori radicals could stage protests to attract world attention to their grievances.

BUSINESS GUIDE

Public holidays: In addition to statutory holidays, provincial anniversaries are also observed by a one-day holiday in each province. Mid-Dec to mid-Feb is the summer holiday season during which the majority of New Zealanders take most of their annual leave.
Fixed dates: 1 Jan, 6 Feb (Waitangi Day), 25 Apr, 25-26 Dec.
Variable dates: Good Friday, Easter Monday, Birthday of HM the Queen, Labour Day.

Working hours
Business: (Mon-Fri) 0830 – 1630.
Government: (Mon-Fri) 0800 – 1630.
Banking: (Mon-Fri) 0900 – 1630.
Shops: (Mon-Fri) 0900 – 1830; (Sat) 0900 – 1300.

Social customs: Tipping is not particularly common and is generally reserved for occasions when a special service has been rendered.

Telecommunications: Overseas links for the extensive telecommunications network are via undersea cable and satellite.
Telephone: Dialling code for New Zealand, IDD access code + 64 followed by area code (9 for Auckland, 3 for Christchurch, 4 for Wellington) followed by subscriber's number. In 1987, there were 640 telephones per 1,000 people, with over 97 per cent automatic service. 24.5 per cent of telephones are in Auckland, 13.15 per cent in Wellington and 9.95 per cent in Christchurch. Telephone dials feature the numbers in reverse order to normal custom. Thus the '0' appears in the place of the '1' and the '1' in place of the '0'.
Telex: International telex facilities are available at chief post offices and some hotels.
Telegram: Telegram service is available from post offices and telegrams may be sent by telephone, although this is expensive.
Other: Data transmission services are available e.g. fax transmission.

Banking
Central bank: Reserve Bank of New Zealand.
Main centre: Wellington.
Trading banks: Australia and New Zealand Banking Group, Bank of New Zealand, National Bank of New Zealand, Taranaki Savings Bank, Westpac Banking Corp.
Other banks: Barclays Bank International, Chase Manhattan Bank, Indosuez New Zealand, Lombard New Zealand, Marac Corporation.

Advertising: Advertising is available on commercial radio and TV, in the press, in cinemas and outdoors on buses and inside telephone booths. Direct mail advertising services are operated by firms in Wellington and Auckland. Various Acts regulate advertising relating to drugs, cosmetics, dentifrices, medical instruments, etc.

Electricity supply: 230 V AC, with three-pin flat plug fittings and bayonet-type light sockets, although most hotels supply 110 V AC sockets for razors.

Weights and measures: Metric system.

Representation in capital: Australia, Belgium, Canada, Chile, China, Cook Islands, Czechoslovakia, Egypt, Fiji, France, Germany (Federal Republic), Holy See, India, Indonesia, Israel, Italy, Japan, Korea, Malaysia, Netherlands, Papua New Guinea, Peru, Philippines, Poland, Romania, Singapore, Sweden, Switzerland, Thailand, UK, USSR, USA, Western Samoa, Yugoslavia.

Hotels: Wellington
Autolodge, Corner Willis & Dixon Streets, P.O. Box: 27267 (tel: 851304; tx: 3977 NZ).

Iona Towers, P.O. Box: 2013 (tel: 850404; tx: 30126 NZ).

James Cook, The Terrace, Private Bag 147 (tel: 725865; tx: 3871 NZ; fax: 731875).

Melksham Towers Motel, P.O. Box: 10047 (tel: 851569; tx: 31446; fax: 733892).

Sharella Motor Inn, 20 Glenmore Street (tel: 723823; tx: 3450 NZ; fax: 723887).

Shaw Savill Lodge, Kemp Street, Kilbirnie (tel: 872189; tx: 3489).

St George, Willis & Boulcott Streets, P.O. Box: 11328 (tel: 739139; tx: 3674 NZ; fax: 8445).

Terrace Regency, Private Bag, Courtenay Place Post Office (345 The Terrace) (tel: 859829; tx: 30126 Ref.TERENCY).

Travelodge Wellington, 40-44 Oriental Parade, P.O. Box: 9208 (tel: 857799; tx: 30058; fax: 852936).

Wellington Parkroyal, 360 Oriental Parade, P.O. Box: 9078 (tel: 859949; tx: 30106; fax: 845443).

Business advice and information
Auckland Chamber of Commerce, PO Box 47, 2 Courthouse Lane, Auckland (tel: 31969; tx: 2326).

British-New Zealand Trade Council, Level 3, Eagle Technology House, 150-154 Willis Street, Wellington (tel: 852145; tx: 31255; fax: 853510).

Canterbury Chamber of Commerce, PO Box 187, 159 Oxford Terrace, Christchurch (tel: 64992; tx: 4586).

Department of Trade and Industry, Bowen State Building, Bowen Street, Wellington 1. Postal address: Private Bag, Wellington.

Development Finance Corporation of New Zealand, Development Finance Centre, Corner Grey and Featherstone Streets, PO Box 3090, Wellington (tel: 737081; tx: 31351 NZ).

Export-Import Corporation of New Zealand, Willeston Street, PO Box 11332, Wellington (tel: 738110; tx: 31081).

Industries Development Commission, PO Box 27046, Wellington (tel: 850469).

New Zealand Bureau of Importers and Exporters, Parnell House, 470 Parnell Road, Parnell, Auckland 1. Postal Address: PO Box 9248, Newmarket, Auckland (tel: 771273).

New Zealand Chambers of Commerce, PO Box 11043, Wellington (tel: 723376; tx: 3714).

New Zealand Manufacturers' Federation (Inc), 3–9 Church Street, PO Box 11543, Wellington.

New Zealand-United Kingdom Chamber of Commerce and Industry, Suite 615, 162-168 Regent Street, London W1R5TB, United Kingdom.

Reserve Bank of New Zealand, PO Box 2498, Wellington (tel: 772029; tx: 3368).

Nicaragua

John Burstein

The end of February 1990 saw a dramatic change in Nicaragua's political complexion as the US-backed opposition grouping, 'UNO', gained a landslide victory in the presidential elections. The leader of the 'UNO' alliance, Mrs Violeta Barrios de Chamorro won 55 per cent of the votes, against 41.5 per cent for former President Daniel Ortega. As this review went to press, it remained to be seen whether Mrs Chamorro would secure the necessary two-thirds majority in the Assembly to allow her party to see through its planned constitutional changes.

As the Sandinista revolution embarked on its second decade, change was already in the air. The country had been racked in recent years by a proxy war with the United States, tense relations with its Central American neighbours, deep internal political divisions, major economic mismanagement and a hurricane and drought. More than a tenth of the population had fled the country altogether. Yet the much-expected counterrevolution still appeared more remote than ever. Indeed the Sandinista Front was thought to have a better-than-even chance to win the February 1990 elections. Washington appeared more open to a political accommodation than at any time in the last eight years. And the Sandinistas – having received a cautionary message regarding future aid from Moscow – seemed set on moderating their revolutionary programme.

Nicaragua is the largest, and most sparcely-populated, of the Central American countries. The Pacific plains produce sugar and cotton; the central highlands, coffee, and the extensive tropical rainforest region of the Atlantic Coast still contains precious hardwoods, though it is rapidly being converted to pasture lands. Despite ample natural resources, social convulsion and a radical economic experiment have led the Nicaraguan economy to plummet in recent years, now leaving it arguably the poorest of the Central American countries. Yet Nicaragua remains a great political rallying point for the hemisphere, or, depending on point of view, its great *bête noir*.

Nicaragua probably suffered more than it gained from a special relationship with the United States since the last century. Singled out initially as the continent's best transshipment point because of its large internal lake and accessible river system, the country was held briefly, in 1855, by William Walker, a US adventurer, or *filibustero*. The US Marines occupied the country intermittently from 1912 through 1933, when they left in place Anastasio Somoza to rule the country as head of the National Guard. One of Somoza's first acts was to have assassinated a rebel who had held out against the US occupation, Augusto Cesar Sandino.

The Somoza family – the father and the two sons who succeeded him – came to control some 23 per cent of Nicaragua's arable lands, as well as sugar and other industrial installations, and transportation and retail operations throughout Nicaragua, while maintaining the favour of Nicaragua's traditional elite, heavily represented in the Conservative party. The Somozas came to dominate the Liberal party. During the 1950s and 1960s, Nicaragua, participating in the burgeoning Central American Common Market, realised annually 4 or 5 per cent real growth.

In 1972, in response to an earthquake which devastated Managua, millions of dollars of international aid poured into the country. Public outrage sparked by the diversion of much of this money to Somoza family funds led to wide middle- and entrepreneurial-class

disillusionment. With the January 1978 murder of Somoza's most influential critic, *La Prensa* newspaper editor Pedro Joaquin Chamorro, opposition political leaders – later coalescing into the group called Los Doce (The Twelve) – opened communication with the Sandinista Front of National Liberation (FSLN), which had been leading an ineffectual but persistent guerrilla movement against Somoza since 1961.

A popular uprising militarily led by the FSLN, and with broad political support, forced Somoza to flee the country in July 1979. The Sandinista revolution had received international support from Venezuela (under the presidency of Carlos Andres Perez) and Costa Rica. US President Jimmy Carter diplomatically worked for the end of the Somoza dictatorship but did not support the FSLN-led movement.

Relations with the United States

Early efforts by the United States to encourage the new government (including a 1979-80 package of $75m in aid to the private sector) were quickly replaced by a policy of confrontation with Managua as Ronald Reagan assumed the presidency (1981-89). The United States began to block loan prospects from the World Bank and Inter-American Development Bank in 1982. In 1983 Washington cut Nicaragua's sugar quota by 90 per cent and in May 1985 it imposed a trade embargo, effective to the present.

By early 1982 the CIA had already sent military

KEY FACTS

Official title: Republic of Nicaragua

Head of state: President Violeta Chamorro (from April 1990)

Ruling party: UNO coalition (from April 1990)

Capital: Managua

Official Languages: Spanish (Miskito and English also spoken)

Currency: New Córdoba (C) = 100 centavos

Exchange rate: C21,263 per $ (Nov 1989)

Area: 147,950 sq km

Population: 3.62m (1988)

GDP per capita: $819 (1988)

GDP real growth: –8.0% (1988)

Inflation: 33,000% (1988)

Trade balance: –$611.5m (1988 merchandise)

Foreign debt: $7.3bn (1987)

advisers to form the Nicaraguan counterrevolutionary, or contra forces – which continue to receive US aid in their bases in Honduras. In March 1984 the CIA – under cover of the contras – mined Nicaraguan harbours. And in May 1985 the United States imposed the trade embargo on Nicaragua. The contras reached an apex of power when their military leader (ex-National Guardsman Enrique Bermudez) was joined by highly-respected disenchanted politicians, like Arturo Cruz and Alfredo Cesar – both presidents of the Central Bank under the Sandinistas. The war created over 60,000 casualties (including some 28,000 deaths) and over $1.4b worth of damage before the contras and the government signed a cease-fire in the town of Sapoa on 23 March 1988.

Efforts to overthrow the Sandinista-dominated government appear to have been doomed once the five Central American presidents signed in August 1987 the Central American peace plan – sometimes called the 'Arias Plan' after Costa Rican President Oscar Arias, or the Esquipulas II accord, because of its origin in an earlier meeting in Esquipulas, Guatemala. Through subsequent Central American summits, and exercise of the good offices of Venezuelan President Carlos Andres Perez and President Arias, most contending parties (from former contra leaders to official US government policy) have agreed to support open elections in Nicaragua, with full rights of participation by the opposition, including those who were in arms.

KEY INDICATORS

	Unit	1984	1985	1986	1987	1988
External debt	$m	5,113	5,691	6,205	7,291	–
Debt service ratio	$m	14.7	13.2	10.9	–	–
Population	m	3.16	3.30	3.38	3.50	3.62
GDP	(1988 $m)	2,388a	3,281	3,248	3,225	2,967
GDP per capita (1988 figures)	$	755	994	961	921	820
Inflation	%	35.5	219	681	912	–
GDP real growth	%	–4.7	–5.9	–1.0	–0.7	–8.0
Agricultural production	1980=100	93.0	90.8	84.7	78.9	80.9
Food production	1980=100	77.5	76.6	73.4	66.6	65.2
Coffee production	'000 tonnes	50	50	43.0	37.0	43.0
Cotton production	'000 tonnes	85	69	44.0	28.0	44.0
Sugar production	'000 tonnes	2,545	2,831	2,810	2,575	1,932
Exports	$m	385.7	301.5	247.2	299.9	235.7
Cotton	$m	133.8	97.0	45.2	46.4	–
Meat	$m	17.6	10.9	5.3	12.1	–
Coffee	$m	122.4	120.9	109.6	134.2	–
Imports	$m	848.4	964.3	856.8	922.6	–
Crude petroleum	$m	97.1	112.8	86.4	71.2	–
Balance of trade	$m	–413.9	–498.7	–479.3	–622.7	–
Current account	$m	–639.3	–759.4	–693.0	–	–
Exchange rate	C per $	0.91	0.04	0.1	0.1	269.73

* estimated figure

The Sandinistas in power

Called *los muchachos* (the boys) in their early days in power, the Sandinista forces were largely young men and women from urban middle-class backgrounds inspired to take up arms against Somoza through participation in the left-wing student movement or the Catholic Church's liberation theological tradition of the 1960s-70s. Some of the founders and *comandantes* of the FSLN came of age in the Nicaraguan Socialist party (pro-Moscow) or received training in Moscow itself. But three Nicaraguan rebel factions pursued independent strategies (choosing to target peasant or, conversely, urban sectors) until, in March 1979, Cuba's Fidel Castro convinced them to coalesce in a common front. The *Frente* has continued to function as a composite of the three factions under the evenly-distributed nine-member directorate. However, the *tercerista* faction – which advocated a 'third-way' revolutionary approach in alliance with bourgeois sectors – gained prominence under the leadership of Daniel Ortega (now president) and his brother Humberto (now minister of defense). Tomas Borge, the minister of interior, was an FSLN founder and head of the maoist-oriented 'Prolonged War' faction. Jaime Wheelock, now agriculture minister, was a leader of the 'Proletarian Tendency'.

Sandinista bases of power

The Sandinistas, represented by Daniel Ortega (aged 34 at the time), took unquestioned control of the governing junta when, in April 1980, Violeta Chamorro (widow of and successor to the assassinated *La Prensa* editor) and a second non-Sandinista felt squeezed out of power. In 1984 Ortega and his vice-presidential candidate, Sergio Ramirez, won elections widely hailed as clean, but incomplete because significant opposition sectors withdrew following an obstructionist strategy. The Sandinistas also gained 61 of the 96 seats of the Constitutional Assembly, which later was converted into the National Assembly. (Six parties share the remaining 35 seats.)

'Mass organisations' – of women, rural workers, youth, etc. – have diminished in membership and importance in recent years, as have the neighbourhood committees, Comites de Defensa Sandinista (CDS), charged with vigilante functions by Sandinista opponents. The FSLN has continued to rely on its labour organisation, the Confederacion Sandinista de Trabajadores (CST, Sandinista Workers' Confederation), while it outlawed strikes as part of a wartime state of emergency (lifted in January 1988). The middle-sized farmers' organisation, Union Nicaraguense de Agricultores y Ganaderos (UNAG), is perhaps the most active revolutionary sectoral grouping, with considerable influence on Sandinista decision-making.

The Sandinistas also rely heavily on the armed forces as their institutional base of support. Indeed the military bears the name Ejercito Popular Sandinista (EPS), or Sandinista Popular Army. Having come to power through force of arms and then overcome the US-contra military challenge to Sandinista rule, the army's loyalty to the nation and to the FSLN itself has become difficult to distinguish. But its effect on Nicaraguan society may be measured by the fact that since the time of Somoza the armed forces have grown from 6,000 to some 65,000 active forces (and another 80,000 reserves). Since establishing conscription in 1983, the EPS has inducted over 150,000 youths for two-year tours of duty. The Nicaraguan army is now the largest and best equipped in Central America, though its air force is limited and considerably inferior to that of Honduras.

Political opening

In February 1989 the government announced that general elections would occur one year later, to choose the president, vice-president, all National Assembly seats, municipal government and two 45-member Regional Councils in the northern and southern Atlantic Coast autonomous regions. (The date was advanced by nine months as part of a diplomatic agreement with other Central American leaders and the internal opposition.)

A flurry of political activity ensued as opposition parties initiated common actions and carried out mass organising efforts to test the political opening. By the end of the registration period there were 21 legally constituted parties, including the FSLN. The result of years of factional infighting, the 21 may be gathering under the following categories: conservatives, liberals, social christians, social democrats and marxists. Knowing it was their only chance victory, 14 of the major parties came together in June 1989 to form the Union of Nicaraguan Opposition (UNO), agreeing to enter the elections with a single platform and a single slate of candidates.

Polls during the spring and summer of 1989 showed that approximately 30 per cent of the electorate supported the FSLN, another 30 per cent supported opposition parties, and 40 per cent remained undecided. Even recognising that electoral laws prohibit use of state resources by the governing party, the FSLN – with some 14,000 full members and another 17,000 'aspirants' – is far better organised and equipped than its rivals. Though the social christian parties boycotted the 1984 election, many consider them to be the second largest political force in the country. The Liberal Independent Party (PLI), headed by Virgilio Godoy, and the conservative parties closely follow.

Labour and business organisations play an active political role in Nicaragua. The four opposition trade union federations – from communists to AFL-CIO grantees – recently formed the umbrella organisation Congreso Permanente de los Trabajadores (CPT, Permanent Workers' Congress), which has come to represent the progressive wing of the UNO. On the other hand, the Consejo Superior de la Empresa Privada (COSEP, Superior Council of Private Enterprise) has long been the country's most forceful articulator of capitalist interests.

The country's largest-distribution newspaper, *La Prensa* and the Catholic Church leadership have also been crucial to the anti-Sandinista political movement. Under the direction of Violeta Chamorro and her brother-in-law Jaime Chamorro, *La Prensa* has propounded a radical criticism of the government, thus falling victim to censorship and even periods of closure by the government during the war years. This made *La Prensa* a symbol of the struggle for civil liberty for many Nicaraguans as well as in the United States, which government has always generously funded the newspaper. Catholic Church radio and print journalism, too, suffered censorship and closure periodically during the state of emergency (1982-88). Feuding between the extremely influential Church and the government has focused on a coterie of liberation-theologian clergy who have remained in high government positions (including Foreign Minister Fr.

Miguel d'Escoto) despite Vatican demands that the priests resign. Nicaragua's chief prelate Miguel Obando y Bravo – made a cardinal by Pope John Paul in 1986 – is an ardent critic of the Sandinistas (having also opposed Somoza in the 1970s) and is sometimes said to be the only Nicaraguan who might amass a larger following than that of Daniel Ortega.

Relations with Soviet bloc

Economic relations with the Soviet bloc were initiated in 1980 through various accords which, in addition to opening military aid, provided access to donations – mostly foodstuffs – and concessionary trade credits. Total Soviet military aid since 1980 is estimated to be worth about $1bn; in 1987 the Soviet bloc extended to Nicaragua $550m in economic aid and another $515m in military aid, according to US government sources. The USSR has supplied nearly all Nicaragua's petroleum needs since 1986 when Mexico and Venezuela stopped their shipments because of previous unpaid bills. Approximately 40 per cent of Nicaragua's trade is with the Soviet bloc, or Council for Mutual Economic Assistance (COMECON).

In May 1989, Soviet leader Mikhail Gorbachev wrote President Bush that the USSR would cut its supply of weapons to Nicaragua, although US intelligence reports made public in August purported that increased deliveries from East Germany and Cuba had left total inflows undiminished. After Cuba – which has provided assistance in sugar production, health care and to the military – Bulgaria is the Soviet ally most active in Nicaragua. Sofia helped in a port project in Bluefields and in construction of a canned-foods complex, the Agroindustrial Enterprise of the Sebaco Valley, which began production in 1989.

Economic policy and results

The Sandinistas came to power nationalising Somoza family holdings, as well as the banks and foreign trade. They created state farms and promoted agricultural co-operatives – the two coming to occupy some 40 per cent of land in cultivation. And they greatly expanded the state bureaucracy. Social services rose dramatically: health and education programmes represented 32 per cent of national output in 1982 resulting in the halving of both illiteracy and infant mortality rates. In 1987, however, expansion of the statist economy ended and, with a fundamental redirection of economic policy in 1988-89, the public sector was actually reduced and the beginnings of privatisation have become apparent.

The revolutionary transformation of the system stopped well short of Nicaragua's becoming a centrally-planned economy. In fact the private sector has maintained control over some 60 per cent of production. Even early efforts by the Sandinistas to develop state farms were reversed by the mid-1980s and land being redistributed went to individual holders. But with its closely-held control of fiscal and spending policy, the government has thoroughly dominated the Nicaraguan economy.

Following three years of growth after the revolution, the Nicaraguan economy has shrunk each year since 1983. A 1989 mission by international economists led by the Swedish Government found that per capita output fell by one-fourth since 1980, and private consumption dropped by almost 70 per cent. With export crops (coffee, cotton, meat, bananas) down 40 per cent and the manufacturing sector down 84 per cent, Nicaragua's imports are now four times larger than its exports (worth $252m in 1988). The 1988 foreign trade deficit, of $579m, actually equalled one-third of the total national income. In addition, the government's budget deficits in recent years have averaged a quarter of the entire national economy (approximately $1bn in 1988). In October 1988 the economy was rocked by Hurricane Joan, hitting the Atlantic Coast and the central region and causing some $800m damage, and again by severe drought during the first half of 1989. The gross domestic product is expected to decline by 8 per cent during 1989.

Nicaragua's economic collapse must be attributed to several factors in addition to government mismanagement. The US-directed contras waged economic war, targetting everything from agricultural co-operatives to bridges and electrical installations, causing up to $3bn worth of damage. In addition to the negative investment climate created by the war, most Nicaraguan entrepreneurs chose to decapitalise their holdings as part of their opposition to Sandinista economic policy. The US trade embargo and blocking of international loans further hindered economic activity, with spare parts for transport and manufacture becoming difficult and expensive to obtain. And the government was obliged to invest over 50 per cent of the national budget in defence, as well as direct a great part of the productive population into the war effort.

The Sandinistas have consistently claimed to reward the productive, or 'loyal' bourgeoisie. The end of confiscations as part of the land reform programme came in 1987 – though as late as July 1989 three politically-active large landholders lost their farms by extraordinary government proclamation. In January 1989 the government initiated a dialogue, or *concertacion*, with organisations of the private sector, such as the coffee and cotton-producers and dairy farmers. In the hopes of sparking new investment, the government promised to pay hard currency for export products and stabilise interest rates. This extended the neo-liberal austerity measures begun in June 1988 and amplified in the first months of 1989. The government has virtually eliminated preferential exchange rates and in June 1989 slashed the exchange value of the córdoba by 61 per cent, with mini-devaluations to follow regularly. It has also lifted most subsidies, including those on basic foods, and cut its own budget through a 30 per cent reduction, or *compactacion*, of the public sector, including lay-offs of

COUNTRY PROFILE

Political system: Executive power is vested in the president who is directly elected for a six-year term and is assisted by a vice-president (also directly elected) and an appointed cabinet. Legislative power is vested in the unicameral National Assembly, currently consisting of 96 members elected by proportional representation, although a new constitution promulgated in January 1987 provides for a decrease in numbers to 90. Until April 1990, the ruling party was the Sandinista National Liberation Front (FSLN).

Main cities/towns: Managua, Chinandega, Matagalpa, Masaya, Granada, León, Bluefields.

Media
Press: Several daily newspapers include *Barricada* (circulation 95,000), *La Prensa* (75,000) and *El Nuevo Diario* (45,000), all based in Managua. There are also several weeklies and periodicals.
Radio: There are over 50 commercial, cultural, government and religious radio stations.
Television: Sistema Sandinista de Televisión (SSTV) runs two television channels.

External trade: Heavy dependence on a few basic agricultural exports makes the balance of payments vulnerable to world commodity price shifts. The performance of the external sector has, moreover, been seriously undermined in recent years by US-supported war waged by anti-government 'contra' guerrillas and by US trade sanctions.
Exports: 1984: Coffee (30% of total value), cotton (34%), meat (4%), sugar, bananas. Main destinations: Japan (typically 18-22% of total value), EC (18-22%), USA (15%), CACM (7-8%), Eastern Europe (5-6%).
Imports: Raw materials and intermediate products (typically 30-35% of total value), capital goods (25-30%), fuels and oils (18-22%), non-durables (15-18%). Main sources: CACM (typically 15% of total value), USA (13-14%), EC (10-13%), Eastern Europe (12%).

Agriculture: Sector forms mainstay of economy, contributing 25 per cent to GDP and employing 45 per cent of the workforce. The principal export crops are coffee and cotton, which combined account for around 60 per cent of total exports. Meat, and to a lesser extent, sugar are the other main agricultural exports. The main food crops are maize, rice, beans and sorghum. Production of timber, tobacco, sugar-cane and rubber is geared towards local agro-industries. Commercial fishing of the Atlantic coast (mainly for crustaceans) is an important source of protein. The government is actively engaged in the agricultural sector, though most of the cultivated land (estimated at 60 per cent of total arable land) is still farmed by private small-holders. A major programme of land distribution is under way, involving the establishment of co-operatives for food crops and large estates for export crops. Despite significant agrarian reform, food production has not kept up with demand, owing mainly to poor weather, war damage and shortages of vital inputs.

Industry: Manufacturing sector contributes 25 per cent to GDP and employs around 11 per cent of the workforce. Manufacturing is centred on food processing, especially sugar refining (annual output over 200,000 tonnes), beverages, meat processing and cooking oils. Other agro-allied industries are based on textiles, tobacco, soluble coffee, cocoa, rubber and timber. There are a few large-scale industries including a cement works, chemical plant, metals industry and 650,000 tonne/year capacity petroleum refinery. The industrial sector is predominantly state-controlled with production heavily geared towards meeting defence and domestic consumption demands. Since the revolution, industrial performance has been hampered by contra guerrilla attacks, shortages of imported machinery and raw materials, shortages of foreign exchange and lack of skilled manpower.

Mining: Sector contributes less than one per cent to GDP and is confined to the extraction of gold, silver and salt. Workable deposits of tungsten, lead and zinc also exist in northern Nicaragua.

Energy: Fuel wood is the major source of energy, accounting for 50 per cent of total energy consumption. Almost all the country's petroleum requirements (12,500 b/d) are imported, mainly from the USSR (although the proportion supplied was cut from about 90 per cent to 50 per cent in 1987) with smaller amounts from Libya, Algeria and Iran. Possible oil and gas deposits located off Pacific and Caribbean coasts. Emphasis on developing hydroelectric and geothermal potential with construction of a second geothermal plant at Momotombo, and hydroelectric plants at Asturias and Malacatoya.

Membership of international organisations: CACM, CEMA (Observer), FAO, G-77, GATT, IADB, IAEA, ICAC, ICAO, ICO, IDA, IFC, IFAD, ILO, IMF, IMO, INTELSAT, Inter-American Development Bank, Interpol, IPU, IRC, ISO, ITU, NAM, OAS, ODECA, Organisation of American States, PAHO, SELA, UN, UNESCO, UPEB, UPU, WFTU, WHO, WMO, World Bank, WTO.

army personnel and Interior Ministry police.

It appeared from first half 1989 statistics that the country's extraordinary hyperinflation – which surged to between 33,000 and 36,000 per cent in 1988 – was coming under control, with monthly averages under 20 per cent. According to Vice-President Sergio Ramirez, the government hopes to contain it at between 900 and 1,000 per cent in 1989. Not surprisingly, Merrill Lynch Capital Markets reported in August 1989 that Nicaragua's loans were selling at 2-4 US cents for the dollar.

Outlook

As the country prepared for February 1990 general elections, the nature of its government was under tremendous domestic and international scrutiny. The question has been raised consistently whether the revolutionary government could be removed from office through the ballot. Sandinista leaders had given various replies consistent with the interpretation that the FSLN would leave the government if it loses the presidency, but would be unlikely to relinquish control of the military, known as the 'Popular Sandinista Army'. The opposition – depending on whether it maintains its coalition – might minimally be expected to fare well in, and even capture control of, the National Assembly.

Following the election result, it appears that the majority of Nicaraguan contras presently based in Honduras will be demobilised. It remains to be seen whether a significant number of the many expatriates and refugees – some 250,000 in Costa Rica, 100,000 in Honduras and another 100,000 in the United States – will return while conditions remain critical.

Nicaragua, is now likely to move toward normalisation of relations with the other Central American countries and the United States. In 1990-91 it will gain some access to foreign exchange and short-term trade credits from international lending institutions as well as from a $4.3bn UN plan for Central American regional development, largely funded by European countries.

Mrs Chamorro's surprise election victory caught the US Administration off-guard. As President Bush called for a peaceful transition to domestic rule, State Department officials were putting together hurried plans for a subsantial aid package. Mrs Chamorro will need all the help she can get, not only to restore Nicaragua's economic fortunes, but also to keep a grip on UNO's potentially discordant coalition.

BUSINESS GUIDE

Time: GMT -6.

Climate: Tropical with high humidity; cooler and drier at higher altitudes. Dry season from Dec to May; temperatures reach 30°C to 35°C in Managua. Rainy season from May to Nov; temperatures up to 27°C to 30°C in Managua.

Entry requirements
Passport: Required by all.
Visa: Required by all except nationals of Central American and most Western European countries (excluding Austria, France, German FR, Italy and Portugal). An onward or return ticket is also required. Visa regulations vary frequently; check with embassy/consulate.
Currency: Import/export of local currency limited to C5000. Import/export of foreign exchange unrestricted. Tourists have to exchange $60 at official rate at airport or border entry point. Check current regulations with embassy/consulate.

Health precautions
Mandatory: None.
Advisable: Typhoid, polio vaccinations. Malaria risk exists in most areas – prophylaxis recommended. Water precautions should be taken.

Representation overseas: Addis Ababa, Algiers, Beijing, Berlin, Bogotá, Bonn, Brasília, Brussels, Budapest, Buenos Aires, Caracas, Guatemala City, Hanoi, Harare, Havana, The Hague, Lima, London, Madrid, Malabo, Maputo, Mexico City, Montevideo, Moscow, New Delhi, Ottawa, Panama City, Paris, Prague, Rome, San José, San Salvador, Santo Domingo, Sofia, Stockholm, Tegucigalpa, Tehran, Tokyo, Tripoli, Vatican City, Vienna, Warsaw, Washington.

Air access
National airline: Aeronica (Aerolíneas Nicaragüenses).
Other airlines: Aeroflot, Copa, Cubana, Iberia, LACSA, Sahsa.
International airport: Managua-Augusto César Sandino (MGA), 9 km from city; duty-free shop, bar, restaurant, buffet, post office, shops (restricted hours in some instances), banks.

Surface access
Main ports: Bluefields, Corinto, Puerto Cabezas, Puerto Sandino, San Juan de Sur, Puerto Arlen Siu.
Overland access: Pan-American Highway runs from Honduras, through Managua, to Costa Rica.

Hotels: Availability limited but there are good hotels in Managua, the main coastal towns and along the Pan-American Highway. Bills are subject to 10 per cent sales tax, and must usually be paid in dollars. 10 per cent tip is usual.

Car hire: Foreign licence acceptable for short stays (up to 30 days). Due to poor public transport, often the best way to get around in Managua.

City transport
Taxis: Fares by negotiation in advance of journey; tipping not necessary.

National transport
Air: Aeronica runs domestic service, centred on Managua. Major towns are served but flights are often subject to cancellation.
Road: Main centres on Pacific side of the country are linked by paved roads. There is only one major road to the Caribbean side and this stops, before the coast, at Rama.
Buses: Services are regular and connect main centres served by the road system (e.g. Managua-Rama, Managua-León, Chinandega, Corinto). City buses are cheap and crowded.
Rail: Several services a day link Managua to the port of Corinto, via Leon; and Managua to Granada on lake Nicaragua.
Water: A boat service links Rama and Bluefields port on the Caribbean coast.

Public holidays: 1 Jan, Holy Thursday, Good Friday, 1 May, 19 Jul, 1 Aug, 10 Aug (Managua only), 14 Sep, 15 Sep, 8 Dec, 25 Dec.

Working hours
Government and business: (Mon-Fri) 0900 – 1730; (Sat) 0900 – 1300.
Banking: (Mon-Fri) 0800 – 1530; (Sat) 0800 – 1200.

Telecommunications
Telephone: Dialling code for Nicaragua: IDD access code + 505 followed by area code (Managua 2) followed by subscriber's number. Few public telephones – mainly in hotels and bars.
Telex: Facilities available at some hotels and main post offices.

Banking
Banking centre: Managua.
Central bank: Banco Central de Nicaragua.
Commercial banks (head offices): Banco Caley-Dagnall, Banco de América, Banco de Centroamérica, Banco de Crédito Popular, Banco Immobiliario, Banco Nacional de Desarrollo, Banco Nacional de Nicaragua, Banco Nicaragüense, Corporatión Nicaragüense de Inversiones.
Foreign branches: Bank of America, Bank of London & South America, Citibank.

Advertising: Available on radio and in the press, cinemas and posters.

Electricity supply: 110 V AC, 60 cycles.

Representation in capital: Argentina, Brazil, Bulgaria, Chile, China PR, Colombia, Costa Rica, Cuba, Czechoslovakia, Dominican Republic, Ecuador, El Salvador, France, German DR, German FR, Ghana, Guatemala, Honduras, Hungary, Iran, Italy, Japan, Kampuchea, Korea DPR, Libya, Mexico, Mongolia, Panama, Peru, Poland, Spain, Sweden, UK, USA, USSR, Uruguay, Vatican City, Venezuela, Vietnam, Yugoslavia.

Hotels: Managua
Camino Real Managua, Carretera Norte, P.O. Box: 118 (tel: 31380/5; tx: 1403; fax: 31385).

Inter-Continental Managua, 101 Octava Calle SO, P.O. Box: 3278 (tel: 23531; tx: 3751054; fax: 25208).

Las Mercedes, P.O. Box: 4655 (tel: 31710/4).

Car hire
Managua
Budget: Estatua Montoya 1C Abajo 1C Sur (tel: 24827, 22819, 66226).

Business advice and information
Aeroli\01neas Nicaragüenses (AERONICA), Contiguo Aeropuerto Internacional August C Sandino, Apdo 3688, Managua, JR (tel: 31801).

Banco Central de Nicaragua, Apartado Postal 2252 and 2253, Managua (tel: 50500, 50518; tx: 1056, 1391).

Cámara de Comercio Americana de Nicaragua, Km 3\55, Carretera Sur, Apdo 2720, Managua, JR (tel: 662486; tx: 1255).

Cámara Nacional de Comercio de Nicaragua, Apdo 1345, Fte a Edificio de Lotería Popular, Managua (tel: 70412).

Cámara de Industrias de Nicaragua, Apdo 1436, Managua.

Cámara Oficial Española de Comercio de Nicaragua, Iglesia El Carmen, \55 c al Sur, Apdo 4103, Managua, JR (tel: 25403; tx: 1396).

Dirección General de Promoción de Exportaciones, Km 6, Carretera a Masaya, Apdo 2412, Managua, JR (tel: 70154; tx: 2410).

Empresa Nicaragüense de Promoción de Exportaciones, Apdo 1449, Managua.

Instituto Nicaragüense de Turismo, Avenida Bolivar Sur, Apdo 122, Managua.

Ministry of Economy, Industry and Commerce, Km 6, Carretera Masaya, Apdo 2412, Managua, JR (tel: 70116; tx: 2410).

Ministry of Finance, Palacio de Héroes y Mártires de la Revolución, Apdo 78, Managua, JR (tel: 26881; tx: 1213).

Niger

Susi Pollak

Niger has started to describe itself as 'the development society' and its sole political party – created in May 1989 – is the Mouvement National de Societe de Developpement (MNSD). The president, General Ali Saibou, was unanimously elected chairman of the MNSD; he also heads the Supreme Council of National Orientation (CSON) which is the highest organ of state. Presidential elections with Saibou as sole candidate were scheduled for December 1989. An elected national assembly is to meet for the first time in 1990 as the culmination of the constitution-making process which has taken the best part of two years.

The assembly will consist of civilians, but Saibou has taken great trouble to explain that it must not be seen as some half-way house to civilian rule: the new arrangements are intended to be permanent, and the soldiers have no intention of returning to their barracks. Their presence on the side of civilians is 'indispensable and obligatory'. 'This does not mean that I will not hand over power to the civilians,' Saibou explained, 'but whether there is a civilian president or a military president, there will always be cabinet members, prefects and sub-prefects who are members of the military.'

When Saibou says 'always' this is what he means. The MNSD guarantees free expression of ideological opinions and affiliations: 'I will therefore not enter the endless quarrel between the champions of a sole party and the advocates of the multiparty system. We respect the political choice of other people; after all we do not reject the idea of a permanent political debate in this country.' In practice he has already adopted policies with a left-wing flavour, which are bound to appeal to the great majority in a poor country. For instance, he has promised health care to all who need it regardless of their means; hospital fees and the price of medicines have been cut.

The MNSD's charter also echoes populist ideas. It rejects all forms of economic domination; it emphasises national sovereignty over agricultural and mineral wealth which must be exploited principally for the benefit of Niger's people. Its sentiments sound like a polite rejection of the reforms demanded by the World Bank in return for the $80m structural adjustment loan agreed in 1987 with the government. In fact, the World Bank withheld the bulk of that money because it considered the pace of Niger's economic reforms to be too slow. The structural adjustment programme (SAP) included more stringent cuts in public expenditure, the removal of price controls in order to attract foreign investment, and the privatisation of parastatals. The government did in fact abolish a number of loss-making public companies, including Niger Air and the Caisse Nationale de Credit Agricole (CNCA). Others, including the Development Bank of Niger, were being restructured.

Economy

But the full implementation of the SAP may have been considered politically too dangerous by the government, which relies heavily on the loyalty of its civil servants. Those made redundant have received compensation ranging between CFA3.5m ($11,000) and CFA7.5m ($23,000) to allow them to start their own businesses. Although the 1988–89 budget followed SAP policy in reducing public expenditure by 2.6 per cent, it increased public sector wages by 6 per cent.

Saibou was lucky that his first year in office, with its constitutional and political changes, coincided with a period of exceptionally favourable weather. The 1988–89 cereal harvest (sorghum and millet) was estimated at 1.98m tonnes, against an annual consumption of 1.77m tonnes. (A second off-season crop of 80,000 tonnes in 1989 was forecast.) For the first time in several years no emergency food aid had to be requested; instead, granaries were being built. The government urgently called on commercial banks to devise a scheme for marketing the season's produce through the Niger Foodstuffs Produce Company (OPVN) and the Niger Groundnut Marketing Company (Sonara). The OPVN, which is on the SAP restructuring list, has long been responsible for providing farmers with guaranteed minimum prices and ensuring nationwide distribution; in 1989 it was given the additional task of administering emergency stockpiles.

Niger's arable land amounts to only 12 per cent of the total, and the country cannot look to agriculture for its economic salvation. The government is encouraging market gardening – Galmi onions are now an important cash crop – and it exports 'Morocco' leather: goatskins from Maradi. For the time being, uranium will remain the mainstay of the economy. Income from 'yellow cake', which formed 40 per cent of budgetary receipts in 1979, has declined by two-thirds over the decade. Production at the Akouta mining company stabilised at about 2,960 tonnes a

KEY FACTS

Official title: Republic of Niger

Head of state: President, Colonel Ali Saibou (acting, Nov 1987)

Ruling party: Mouvement National de la Société de Développement (MNSD)

Capital: Niamey

Official Languages: French

Currency: CFA Franc = 100 centimes

Exchange rate: Franc CFA316 per $ (Nov 1989)

Area: 1,267,000 sq km

Population: 6.8m (1987)

GNP per capita: $318 (1987)

GNP real growth: 6.6% (1986)

Inflation: –6.7% (1987)

Trade balance: –$158.6m (1987)

Foreign debt: $1.7bn (1987)

year. In 1987 uranium exports were worth about CFA80bn – about 75 per cent of total foreign exchange earnings. France's Cogema takes about 2,000 tonnes and pays double the world spot price, as a hidden form of aid for a friendly ex-colony. But Cogema itself is looking to cost-cutting and may try to negotiate lower prices for the 1990 production.

A less favourable Cogema contract would be a serious blow to the economy; debt servicing, which consumed about 22 per cent of export earnings in 1988, would probably be the first to suffer. In 1988 external indebtedness was estimated at $1.5bn, much of it incurred during uranium boom years. Some of the debt was recently converted into local currency to be used internally for development. This was the first such transaction in the CFA zone, an American initiative to which Paris was basically opposed: because of the FFr/CFA convertibility, members of the zone are only supposed to create money against external assets.

Niger needs World Bank support if it is to diversify away from uranium, and Saibou is aware of this. He has referred to the gradual liberalisation and privatisation of the economy, promising that the recovery programme would be implemented and the investment code revised. But then he spoke of the benefits which the manufacturing sector would derive from the establishment of a new parastatal. Investment in this sector during 1984–89 was officially put at CFA6bn, with the creation of 1,400 jobs.

Niger's other minerals, including gold, received some attention in 1989. Coal was being extracted, and a salt mining company was established in 1989 in Teguidon N'tessoum (Agadez Department) under the joint ownership of the government and Belgian interests. But this seemed likely adversely to affect the traditional family-sized salt miners, whose caravans travel into neighbouring countries and across the desert.

KEY INDICATORS

	Unit	1983	1984	1985	1986	1987
Population	m	6.02	6.20	6.39	6.59	6.80
Gross domestic product (GDP)	CFA bn	677.3	638.3	647.1	643.4	649.9
GDP per capita	CFA '000	112.5	102.9	101.3	97.6	95.6
Inflation	%	–2.5	8.4	–0.9	–3.2	–6.7
GDP growth	%	–0.7	–14.8	6.0	6.6	–
Agricultural production	1979-81=100	99.9	81.7	101.8	104.6	97.0
Food production	1979-81=100	99.9	81.6	101.8	104.6	97.0
Exports of goods and services	$m	387.5	348.1	250.7	330.0	320.7
Imports of goods and services	$m	538.0	462.0	437.9	479.5	479.4
Balance of trade	$m	–150.5	–113.9	–187.2	–149.3	–158.7
Current account	$m	–61.8	1.2	–57.2	–5.8	–66.9
Total external debt	$m	915	949	1,202	1,447	1,679
Debt service ratio	%	18.3	17.9	24.9	25.3	33.5
Exchange rate	CFA per $	381.1	437.9	449.3	346.3	300.5

COUNTRY PROFILE

Political system: Since April 1974 political power has been held by the Supreme Military Council. Ministerial portfolios are held by civilian and military members of the council of ministers appointed by the president. The National Development Council was formed as a consultative body in 1983 and requested in 1988 to draw up a new constitution. Niger gained independence from France in 1960.

Political parties: Only legal political party is Mouvement National de la Société de Développement (MNSD).

Population: Niger's population of 6.8m (1987 estimate) lives mostly in the southern and more fertile part of the country.

Main cities/towns: Niamey (population 450,000), Zinder, Tahoua, Maradi Agadez, Birni n'Konni. Arlit, in the north-west, is a new town built for the local uranium mining industry.

Language: French is the official language, and the universal language for documentation; others in daily use include Hausa, Djerma, Tamacheq, Fulani and Arabic.

Media
Press: News bulletins include *Le Sahel* (daily) and *Sahel Hebdo* (weekly) both in French.
Radio: The government operates *La Voix du Sahel* radio service in French and several African languages.
Television: Télé-Sahel is broadcast four days per week.

External trade: Regular trade and current account deficits, partly financed by aid flows and foreign borrowing.
Exports: Uranium concentrates, live animals, vegetables, hides and skins, cotton (raw, yarn, fabrics), animal products. Main destinations: France, Japan, Nigeria, Italy, German FR, Côte d'Ivoire, Benin, UK, USA, Burkina Faso.
Imports: Capital goods, petroleum and products, food, textiles, paper etc, tobacco etc, rubber and products, pharmaceuticals. Main sources: France, Côte d'Ivoire, German FR, Italy, UK, USA, Algeria, Nigeria, Japan, China PR, Netherlands, Senegal.

Agriculture: Subsistence farming and stock rearing contributes around 50 per cent to GDP and employs nearly 90 per cent of the population. Less than three per cent of the total land area is cultivated. Hides and skins and cotton account for around 20 per cent of export earnings. Production of the two principal food crops, millet and sorghum, totalled 1.78m tonnes in 1986 (estimated 1.74m tonnes in 1987). Other food crops include rice (50,000 tonnes in 1986, estimated 75,000 tonnes in 1987), cowpeas and green beans. Recurrent drought and desertification have hindered efforts to increase food production by the government which announced a 350,000 tonnes food shortfall in 1987-88. The drought has also had a serious impact on livestock rearing, and a policy to promote animal husbandry has been announced by the government.

Industry: Sector is small-scale, contributing less than two per cent to GDP. Manufacturing is concentrated on the processing of agricultural commodities such as sugar refining, brewing, cotton ginning, tanning and flour/rice milling. Other activities include small-scale production of cement and metals, textiles, plastics, soft drinks and construction materials. Industrial development is handicapped by the shortage of capital and skilled labour and by the country's weak infrastructure.

Mining: Sector accounts for around seven per cent of GDP. Since early 1970s uranium has been the country's principal export. Proven reserves total 280,000 tonnes with extraction undertaken mainly at two opencast mines, at Arlit and Akouta. Production has stagnated in recent years at 3,000-3,200 tonnes (3,108 tonnes in 1986) after peaking at 4,360 tonnes in 1981. This stagnation has been caused by a world slump in demand for uranium, which is expected to continue for several years. The major export markets are France, Japan, Spain, German FR and Egypt. Interest in gold-mining is picking up. Other minerals exploited are the tin-bearing ore cassiterite, phosphates, molybdenum and coal. Also known reserves of iron-ore. Government mineral exploration programmes are in progress to assess the potential of new minerals such as copper, manganese, lithium.

Energy: Heavy dependence on imported petroleum. A small oilfield was discovered by Elf in 1983. Demand for electricity has more than doubled over last 10 years owing to expansion of uranium industry. Domestic generation (mainly thermal) meets about 50 per cent of local needs, the deficit being imported from Nigeria. Hydro potential is estimated at 250 mW.

Membership of international organisations: African Development Bank, African Groundnut Council, African States Associated with the EC (Lomé Convention), Arab Bank for Economic Community of West African States, Economic Commission for Africa, Economic Development in Africa (recipient), FAO, french Monetary Area, GATT, IAEA, IDA, IFAD, ILO, IMF, Islamic Conference, Islamic Development Bank, OAU, Permanent Inter-State Committee on Drought Control in the Sahel, UN, West African Economic Community, WHO, WIPO, World Bank.

Time: GMT +1.

Climate: Very hot with temperatures ranging between 28°C and 44°C. Rain mainly in the south between Jun and Sep. Frequent Sahara dust-storms between Nov and Jan. Dry season from Oct to May.

Entry requirements
Passport: Required by all, except nationals of certain African countries with national ID card.
Visa: Required by all except nationals of EC countries (excepting Republic of Ireland and Greece), Scandinavian countries and certain African countries. Visitors must report to police on arrival.
Currency: No restriction of import on foreign currency subject to declaration. Declaration forms must be retained and all transactions recorded. Local currency import is unlimited. Export of local currency limited to CFA 25,000 (French nationals exempt).

Health precautions
Mandatory: Yellow fever and cholera vaccination certificate required by all.
Advisable: Typhoid and polio vaccination. Malaria prophylaxis should be taken as risk exists throughout the country. Water precautions should be taken. Personal medicines can be difficult to obtain.

BUSINESS GUIDE

Representation overseas: Abidjan, Addis Ababa, Algiers, Beijing, Bonn, Brussels, Cairo, Cotonou, Jeddah, Kuwait City, Lagos, Moscow, Ottawa, Paris, Tripoli, Washington.

Air access
National airline: Air Niger (Niger is also a minority shareholder in Air Afrique).
Other airlines: Air Algérie, Air Burkina, Ethiopian, Sabena, UTA.
International airport: Niamey International (Code: NIM), 12 km south east of city; bar, currency exchange, post office, shops, car hire, hotel courtesy coaches.

Surface access
Overland access: The Nigerian border was reopened and road access is now possible. Access from Algeria, Mali and Burkino Faso is difficult, but a surfaced road connects Benin with Niamey. Access from Chad is restricted. Ferries on the Niger River coming from Mali are dependent on the water level.

Hotels: Advisable to book in advance. Good hotels in Niamey. 10 per cent service charge usually added to bills.

Electricity supply: 220/380 V AC, 50 cycles.

Car hire: Self-drive or chauffeur-driven cars available in Niamey. Drivers compulsory outside the capital. International driving licence required. Petrol and spares in short supply.

City transport
Taxis: Fixed rates apply for long-distance and urban services. Taxis in Niamey are cheap and widely available. Tipping is optional.

National transport
Air: Air Niger operates regular services from Niamey to Agadez, Tahoua, Maradi, Zinder. Charter flights are available in Niamey.
Road: Main highways link Niamey with Zinder, Tahoua and Arlit but road travel can be difficult and petrol is not always available. Best months for road travel Dec-Mar. Visitors must report to police on arrival at main centres.
Buses: Services operate from Niamey to Zinder, Agadez, and other towns.

Public holidays
Fixed dates: 1 Jan, 15 Apr (Armed Forces' Day), 1 May, 3 Aug (Independence), 18 Dec (Proclamation of the Republic), 24-26 Dec.
Variable dates: Easter Monday, Eid al-Fitr, Eid al-Adha (Tabaski), Prophet's Birthday.

Working hours
Government and business: (Mon-Fri) 0730 – 1230 and 1530 – 1830.
Banking: (Mon-Fri) 0800 – 1100 and 1600 – 1700.
Shops: (Mon-Fri) 0800 – 1200, 1600 – 1900; (Sat) mornings and evenings only.

Telecommunications
Telephone: Dialling code for Niger: IDD access code + 227 followed by subscriber's number. (There are no area codes). Facilities mainly in Niamey. Good international links via Paris. Calls should be made via 'L'inter Radio' Service.
Telex: Services available from Chief Telegraph Office , Niamey, and some major hotels.

Banking
Central bank: Banque Centrale des États de l'Afrique de l'Ouest.
Main centre: Niamey.
Head offices: Banque Arabe Libyenne Nigerienne pour le Commerce Extérieur et le Développement, Banque de Développement de la République du Niger, Banque Internationale pour le Commerce et l'Industrie du Niger.
Foreign branches: Banque Internationale pour l'Afrique Occidentale, Caisse Centrale de Coopération Economique, Citibank.

Representation in capital: Algeria, Belgium, Benin, China PR, Egypt, France, German FR, Nigeria, Pakistan, USA, USSR.

Hotels: Niamey
Gaweye Sofitel, Place Kennedy, P.O. Box: 11008 (tel: 723400, 722710; tx: 5367).

Grand, P.O. Box: 471 (tel: 732641; tx: 5239).

PLM Tenere, Boulevard de la Liberté, P.O. Box: 10734 (tel: 732020; tx: 5330).

Car hire: Niamey
Agence Trans-Cap: (tel: 2324).

Hertz (NigerCar): PO Box 715 (tel: 732331; tx: 5327).

Other useful addresses
Air Niger, PO Box 865, Niamey (tel: 733899, 733840).

Banque Centrale des États de L'Afrique de L'Ouest, BP 3108, Avenue des Barachois, Dakar, Senegal (tel: 21-16-15, 23-10-42; tx: BCEAO 21.815 SG, 21.817 SG).

Caisse Centrale de Coopération Economique (CCCE), PO Box 212, Niamey (tel: 723393; tx: 5298).

Caisse Nationale de Crédit Agricole (CNCA), PO Box 201, Niamey (tel: 725303).

Chambre de Commerce, d'Agriculture et d'Industrie du Niger, PO Box 209, Niamey (tel: 732210, 735155; tx: 5242).

Chambre de Commerce et d'Agriculture de Zinder, PO Box 83, Zinder (tel: 570078).

Conseil National de Développement, c/o Ministry of Planning, Niamey (tel: 722233; tx: 5214).

Direction des Statistiques, c/o Ministry of Planning, Niamey (tel: 7227799).

Office Nationale des Ressources Minères (Onarem), PO Box 210, Niamey (tel: 723935; tx: 5300).

Office du Tourisme du Niger, PO Box 617, Niamey.

Société Nationale de Commerce et de Production du Niger, PO Box 615, Niamey.

Société Nigérienne de Produits Pétroliers (Sonidep), PO Box 2735, Niamey (tel: 733335; tx: 5343).

Société Nigerienne pour le Développement du Tourisme et l'Hôtellerie (Nigertour), PO Box 781, Niamey.

Syndicat des Commerçants, Importateurs et Exportateurs du Niger, PO Box 535, Niamey.

Nigeria

Elizabeth Obadina

Intense politicking and dashed living standards were the hallmarks of 1989; a year in which President Ibrahim Babangida had to muster his considerable and formidable political skill to keep his military government's key, twin programmes on course. The Transitional Programme to full civilian rule in 1992 and the Structural Adjustment Programme of economic reforms (SAP) survived some of the worst civil and student unrest witnessed in Nigeria in recent times.

Politics

Violent anti-SAP street riots unhinged normal daily life throughout southern Nigeria in late May and early June, just weeks after the government had legalised civilian politics and days after it had completed a major nationwide revision of voter's lists. Despite fears that the riots would derail the political process, President Babangida kept to the timetable mapped out for the transfer of political power from the military to the civilians and went ahead with the registration procedures for two political parties, the Social Democratic Party, SDP, and the National Republican Convention, NRC, slated to fight local government elections in 1990.

On 3 May 1989 President Babingida lifted a five-year-old ban on civilian politics and opened the road to Nigeria's Third Republic.

When the army seized power on the last day of 1983 political parties were outlawed and many leading politicians imprisoned, largely for the embezzlement of public funds. By 1989 most of those politicians had been released from jail but were banned from ever holding political office again. In April President Babangida strengthened the ban with the Prohibition Amendment Decree 9 of 1989 which prescribed jail terms of up to five years and a quarter of a million naira fine for old politicians trying to resurface in Third Republic politics. The search was on for 'new-breed' politicians untainted by the politics of the past.

Also banned from political activity during the tran-

KEY FACTS

Official title: Federal Republic of Nigeria

Head of state: President General Ibrahim Badamasi Babangida

Capital: Abuja

Official Languages: English (Hausa, Yoruba, Igbo widely spoken)

Currency: Naira (N) = 100 kobo

Exchange rate: N7.36 per $ (Nov 1989)

Area: 923,768 sq km

Population: 105.0m (1988)

GDP per capita: $287 (1987)

GDP real growth: 4.2% (1988)

Inflation: 3.2% (1988)

Trade balance: $2,419m (1987)

Foreign debt: $28.7m (1987)

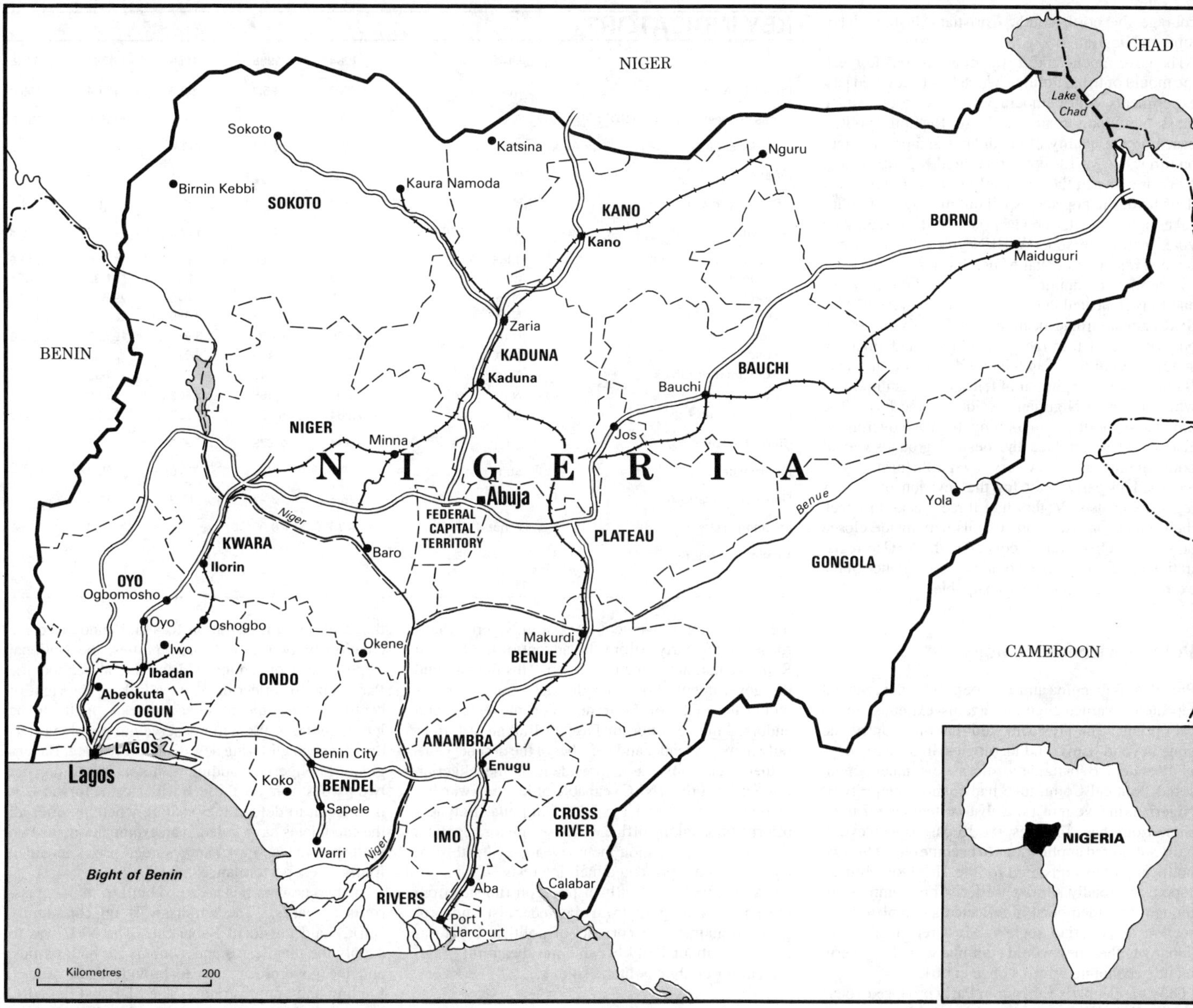

sitional period were current and former top civil servants including the heads of the National Electoral Commission (NEC), the National Population Commission (NPC), the Code of Conduct Bureau, the Directorate of Food, Roads and Rural Infrastructure (DFFRI), the National Revenue Mobilisation Committee and MAMSER, a government body set up to inculcate democratic political principles among Nigerians.

The chairman of the NPC, Alhaji Shehu Musa, a former secretary to the federal government of former President Shehu Shagari, was particularly hard hit by the ban as he had been confidently tipped as a presidential candidate for 1992.

Decree 9 dealt a blow to old politicians who had unashamedly been pulling the strings of embryonic political clubs and associations which had taken root in the activities of the Constituent Assembly. Under the chairmanship of retired Justice Anthony Aniagolu the Constituent Assembly wound up its revision of Nigeria's 1979 constitution in early March. In January it had been forced to lower its quorum from 301 of the 567 members to 201 in order to complete its work on time and in late 1988 the military government had stepped in to a bitterly divisive argument between christians and muslims over the future status of Islamic Sharia law and had ruled the debate beyond the Assembly's jurisdiction. By these two strategies the government kept to its transitional timetable.

Whether Constituent Assembly members served to fashion Nigeria's new constitution or served to stir the crucible of Third Republic politics is debatable. Justice Aniagolu confidently declared at the closing session of the Assembly: 'It shows that we Nigerians can disagree and agree to be together'. But the old schisms in Nigeria's body politic were there, albeit much papered over.

Nigeria's First Republic (1960–66) had pitted southern politicians against the entrenched northern political establishment. During the Second Republic charismatic elder statesmen, Chief Obafemi Awolowo (Awo), Dr Nnamdi Azikwe (Zik) and Malam Aminu Kano contested the presidency heading broadly welfarist parties. They split the vote opposing the northern establishment. The old North backed the National Party of Nigeria (NPN) lead by a relative unknown, Alhaji Shehu Shagari, and the NPN won. By the end of 1983 Awoists, some Zik loyalists and most northern radicals had formed a loose alliance, the Progressive People's Alliance (PPA) against the NPN.

When the Constituent Assembly began its business two opposing political, and wholly illegal, tendencies transpired: the federalists and the progressives, reflecting the interrupted political developments of 1983. The federalist camp rapidly split as wealthy political mandarins, especially from Bendel State and other ethnic minority regions, sought to differentiate their interest in a modern capitalist state from the traditional northern constituency. From this beginning the Nigerian National Congress (NNC) and the Liberal Convention (LC) evolved. The Liberals quickly came to be dubbed 'the millionaires' party'.

The progressive camp was in disarray. Chief Awolowo had died in 1987 leaving a leadership vacuum in the vast Yoruba heartland of western Nigeria. Eventually the People's Solidarity Party (PSP) picked up Awo's leadership mantle and counted among their prominent members Chief Awolowo's widow, Hannah, and his youngest daughter, Dr Tokhunbo Dosunmo. Some Awoists however decamped into the People's Front of Nigeria (PFN). The socialist followers of the late Aminu Kano melted into many camps while the venerable, 86-year-old Zik saw his Igbo constituency divide between the federalists and the progressives, continuing a trend already evident in the early 1980s.

When President Babangida lifted the ban on politics he reaffirmed his government's belief that a two-party democracy would 'encourage cohesion in a land of pluralistic interests such as ours, whilst at the same time guaranteeing an avenue for alternative choice'. Critics, including former head of state, Olusegun Obasanjo, said the two-party state would simply encourage the north–south, Christian-Muslim dichotomy in Nigeria.

The government said it was most anxious to break the mould of tribal politics which had bedevilled past experiments with democracy. It tried to encourage mass participation in the Third Republic politics based on the quality of candidates and programmes on offer. MAMSER concerts toured Nigeria in early 1989 drawing in thousands of young people to hear their favourite pop stars spell out messages of political enlightenment. 'You be prisoner, IF you sell your vote. You be prisoner WHEN you sell your vote,' warned reggae superstar Majek Fashek.

A nationwide opinion poll conducted during April and May indicated that to a certain extent the MAMSER campaign had been successful. Six out of ten Nigerians said that the sex, religion and financial status of a political candidate didn't matter to them. But a revealing eight out of ten said it was important where the next Nigerian president came from. The innovative political marketing by the government also failed to produce the desired groundswell of political enthusiasm. A very low public turnout was recorded in early May for the revision of voter's registration lists. Voters had three weeks to check their names on lists, and schools nationwide closed down as teachers were seconded to the NEC as registration officials. No one could claim ignorance of the exercise but the apathy was palpable.

KEY INDICATORS

	Unit	1984	1985	1986	1987	1988
Population	m	93.3	95.7	98.2	101.4	105.0
Gross domestic product (GDP)	N bn	70.0	78.8	79.7	110.6	137.0
GDP per capita	N	750.3	823.4	811.6	1,090.7	1,304.8
Inflation	%	24.9	5.5	5.4	10.2	3.2
GDP real growth	%	−6.7	13.4	3.2	1.8	4.2
Agricultural output	1979-81=100	112.8	121.3	130.9	123.1	125.8
Manufacturing output	1985=100	83.4	100	96.1	128.5	142.6
Oil production	m b/d	1.39	1.50	1.47	1.32	1.37
Exports	N m	9,139	11,721	9,048	29,578	33,682*
Oil	N m	8,841	10,891	8,369	28,209	29,293
Imports	N m	4,485	5,537	5,975	15,698	22,700
Food	N m	1,052	1,199	802*	1,573*	–
Consumer durables	N m	632	247	302	839	–
Capital goods	N m	2,307	2,486	2,422	7,568	–
Balance of trade	$m	2,984	5,735	2,535	3,523	2,419
Current account	$m	114	2,623	371	−71	−1,013
External debt	$bn	18.7	19.5	24.5	28.7	–
Debt service ratio	%	25.5	30.7	17.9	10.0	–
Exchange rate	N per $	0.76	0.89	1.27	4.02	4.54

* estimated figure

Politics without ideology

President Babangida and members of his government repeatedly warned Nigerians against extremist political options. The president said that the country had gone beyond 'isms' and ideologies. In the early days of President Babangida's rule a government-sponsored Political Committee had found a majority of Nigerians in favour of a socialist welfare state. Taking a cue from these findings, the 13 parties who eventually submitted applications to become one of the two political parties registered for the Third Republic all espoused broadly similar welfarist programmes, although their membership reflected a complete cross section of Nigerian society. Most remained vague about whether they would continue with the government's economic reforms once in office.

Only the People's Liberation Party promised voters a radically different Nigeria, without traditional rulers and without the government's Structural Adjustment Programme of economic reforms (SAP) which had the backing of the IMF. The PLP succeeded Aminu Kano's People's Redemption Party (PRP) of the Second Republic which had held power in Kano and Kaduna States. The PLP was launched by Alhaji Balarabe Musa, the former PRP governor of Kaduna State who had been impeached by the NPN opposition. His reputation was clean but he was arrested and detained for contravening the ban on former politicians. In October he was released on bail.

The PLP built their activities around the trial of Balarabe Musa and never in fact filed papers for registration. There was no other ideologically distinct party among the 40 or so associations who indicated their existence in the days after the ban on politics was lifted although some novel ideas emerged like the 'Black But Beautiful' party's pledge to scrap the naira currency and replace it with the American dollar or the British pound. The government's prohibition on extremism was interpreted to mean that communism was not on offer for Nigerians.

The political choice before Nigerians boiled down to character credibility. Who would be the 'new-breed' politician most likely to deliver the welfare promises made?

The Nigerian Labour Party filed papers for registration and became the first political party sponsored by the trade union movement to exist in Nigeria. Plans for a similar party before the advent of the Second Republic had not come to fruition. This time around the government shook out the rancorous elements within the ranks of the trade union movement and endorsed the selection of Paschal Bafyau, a former railwaymen's leader and member of the defunct Constituent Assembly, as a president of the Nigerian Labour Congress (NLC) capable of keeping warring factions united under the NLC umbrella. Bafyau's priority upon taking office was to secure for Nigeria's five million trade union members and sympathisers, a party which represented their interests.

As a party with a radically distinct constituency from the parties emerging from the federalist and progressive mainstream currents of political organisation, the Labour Party became the 'beautiful bride' for Third Republic political suitors.

Two new parties

Speaking after the 13 parties had lodged their applications for registration as political parties with the National Electoral Commission in mid-July President Babangida noted that, 'In my observations of the power structures of the political associations so far, it appears that the financial element of power dominates all the associations'. He then went on to express the hope that the 13 parties would 'unite as the case may be in order to domesticate political conflict along programme lines'.

The stage was set for a summer of feverish political machinations both to convince the 2,065 NEC verification officials that the parties had fulfilled the stringent requirements laid down for a truly national political party and to forge mergers between aspirant parties jockeying for power.

But the politicking came to nought. The NEC found no cause for cheer when they investigated the thirteen hopefuls. They had rigged their membership figures, they had allowed those President Babangida termed 'moneybags' to turn the parties to their own selfish ends and had begun surrounding themselves with 'bodyguards', resurrecting the spectre of political thuggery which had bedevilled the First and Second Republics. On 7 October President Babangida dissolved all thirteen aspirant parties, refunded their fifty thousand naira registration fees and announced the creation of two entirely new parties, the National Republican Convention and the Social Democratic Party, one to reflect interest, 'a little to the right of centre' and the other to reflect interests, 'a little to the left of centre'. 'We will not serve our people yesterday's food in glittering new dishes,' President Babangida told Nigerians adding of his failure to register two parties from the thirteen which came forward, 'it is difficult to defend the position whereby, after all the candidates have failed in an examination, the two with the highest marks are awarded prizes intended for the "best performance".'

The two new parties are to be built from the grassroots upwards. 'The emphasis is on community spirit,' said President Babangida. The NEC was to supply the manifestos and symbols for both parties, and the government was to provide them offices, funding and administrators. The SDP and the NRC were given identical constitutions. President Babangida hoped that his bold political engineering would lay the foundations for political stability rendering future military intervention unnecessary.

Civil and student unrest

Nigeria's stringent austerity measures called for a degree of belt-tightening never experienced before. Prices of basic food commodities rocketed beyond the pockets of ordinary people. Inflation, running at over 60 per cent for general items, increased to over 80 per cent in the case of food.

Students were in the forefront of demands that SAP should be ended in compliance with the original two-year plan which ended in 1988. In November 1988 the banned National Association of Nigerian Students issued a six-month ultimatum to government to end SAP and institute various university reforms. The deadline expired without being met in May 1989.

The result was devastating. Students began anti-SAP protests in Bendel State in mid-May. In Lagos students had been protesting about curricular issues at the state university and teachers' colleges which had brought anti-riot police on to campuses. The discontent simmered. At one stage nearly 100 students were arrested and charged with the murder of an anti-riot policeman whom newspapers reported alive and well.

On the last day of May students of the University of Lagos joined the demonstrations against SAP and the protests against heavy handling by police of their colleagues at Lagos state institutions.

The protests spread like wildfire, involving youth far beyond the university campuses. Children spilled out of schools and unemployed youth from the motor parks joined the demonstrations. Life in Lagos came to a standstill as cars and public transport disappeared from the streets and pedestrians struggled home clutching green leaves as a mark of solidarity with the rioters. The police withdrew to guard strategic locations and the army came on to the streets to keep the peace around their barracks. Lagosians had never seen anything like it before. The riots left scores of dead, many allegedly suffocated in overcrowded police custody.

Few major towns in southern Nigeria escaped unscathed in the riots which continued to spread though Bendel State during June.

The government took stern measures against the students. Known student leaders were arrested and six leading universities closed down until April 1990. They were the universities of Ibadan, Benin and Lagos (all federal institutions) and the state universities of Bendel, Imo and Lagos. Other institutions which took part in the riots were shut for shorter periods of time, but no northern establishments were affected. Education Minister Jibril Aminu was accused by southerners of bias, but in August northern traditional rulers led by the influential Sultan of Sokoto joined with their southern counterparts and virtually every other leader of public opinion to plead with the government to re-open the universities, which they did in early November.

Economics and politics

There is a Yoruba proverb which translates 'When the stomach is empty no other word is welcomed'. The challenge of the next local government elections is to convince the electorate that the recent years of hardship are both at an end and to be rewarded.

Most emerging political parties veered clear of making a populist promise to abandon SAP. It was clear to all that any aspiring political party which declared opposition to SAP would not be registered, and besides none of them had substantially better ideas for economic reform. Chief Emmanuel Iwuanyanwu, chairman of the NNC, ingenuously remarked on his party's position of staying with SAP that, 'perhaps if another name is used people won't see it negatively'.

The military government removed the rights to free health care and free education from the recommendations of the Constituent Assembly saying that such issues were not justiceable, thereby removing a potentially rich source of revenue for lawyers going to court in the Third Republic to enforce constitutional rights which the country cannot fund.

The military government continued the quest for new, non-oil, sources of revenue. After the riots substantial increases in electricity prices were announced, in fulfilment of a condition laid down by the World Bank. Adroitly, more civil unrest was avoided by applying the higher prices to big consumers of electricity only, principally industrial users. Nevertheless, Nigerians were likely to demand some proof of improving living standards as a precondition to accepting further austerity measures under any government, civilian or military.

The Babangida factor

Central to the success of the government steering a steady course through 1989 was the figure of President Babangida. Nigeria's military has ruled for al-

COUNTRY PROFILE

Political system: Military government with political power vested in the 28-member Armed Forces Ruling Council, headed by Major General Babangida. Return to civilian rule planned for 1992. Nigeria gained independence from UK in 1960. Politics in Nigeria have been historically determined by the tensions between the three main tribal groups: the Muslim Hausas in the north, the Christian Igbos in the east and the Christian/Moslem Yorubas in the west; it was this, for example, which caused the Biafran war of the 1960s, an attempt at Igbo secession. The Constituent Assembly currently preparing for the planned restoration of civilian rule in 1992 is considering a federalist constitution designed to defuse these ethnic tensions. Major-General Ibrahim Babangida came to power in late 1985, in a military coup widely supported by intellectuals, the press, some former politicians and the business community. A return to full civilian rule in 1992 was announced in 1987, starting with local elections. In February 1989 President Babangida appointed a smaller, reconstituted Armed Forces Ruling Council, thus strengthening his position within the military regime in the run-up to the transfer of power to civilian rule in 1992.

Political parties: Under current circumstances there are no formally active political parties in Nigeria.

Population: There have been no reliable population counts since the early 1960s. Estimated population in 1987 was 106.7 million. The Hausas, Igbos and Yorubas comprise the overwhelming majority of the population.

Main cities/towns: Lagos (population probably around 5m) has been the capital and is the main port. The new federal capital is Abuja in the middle of the country. Capitals of the 21 states are Kaduna, Kano, Jos, Sokoto, Maiduguri, Ilorin, Ibadan, Benin City, Port Harcourt, Calabar, Enugu, Bauchi, Minna, Makurdi, Abeokuta, Akure, Ikeja, Owerri, Yola, Katsina and Uyo.

Language: English is the official language, and is used in business and public life. African languages include Hausa, Yoruba, Igbo.

Media

Press: Around 15 major national dailies, and 15 major national weeklies published in English. Principal dailies are *Daily Times, National Concord, New Nigerian, The Guardian, The Vanguard, Tribune, Daily Star* and *Daily Sketch.*
Radio: Four regional radio services each in English and appropriate local languages from stations based in Lagos, Ibadan, Kaduna and Enugu. External services in English, French, Hausa, Arabic, German, Swahili. In addition to the national service operated by the Federal Radio Corporation of Nigeria, each state runs its own commercial station. There are some 6m receivers.
Television: National TV network operated by the Nigerian Television Authority. The service covers virtually the whole of the country from NTA stations in each state. Several states also run their own stations. There are about 500,000 TV sets.

External trade: The trade surplus formerly produced by oil exports was sharply reduced because of world market conditions. Central Bank figures for 1987 show a trade surplus of N12.4bn (compared to N2.9bn in 1986), the result of a substantial rise in exports.
Exports: Petroleum and derivatives (93 per cent of total value in 1987), cocoa, timber, manufactured goods. Main destinations: USA (37% of total value in 1987, according to CBN figures for first half of 1987), France (31%), German FR (10%), UK (3.2%), Italy, Netherlands, Spain.
Imports: Raw materials (37.5 per cent of total value in 1986/87), machinery, spare parts (25.2%), general merchandise (21.3%), food (9.7%), capital goods/aircraft and shipping vessels (1.6%). Main sources: UK (14.4% of total value in 1987), France (10.3%), German FR (9.6%), USA (8.7%), Italy, Japan, Brazil.

Agriculture: Sector has suffered a relative decline because of the dominance of oil in the economy, but it is still the main area of employment and receives much official encouragement. It contributes about 26 per cent to GDP (1988) and five per cent to exports. Plantations, sometimes owned by or in partnership with multinational corporations, are gaining ground in producing raw materials for company use (e.g. grain for breweries), but the sector is still dominated by traditional smallholders who raise subsistence crops such as sorghum, maize, cassava, yams, millet, rice and increasing quantities of wheat. The importation of rice and wheat was outlawed in 1987. Irrigation schemes, higher producer prices, the expansion of credit and improvements in the rural infrastructure are beginning to show results. Cash crops include cocoa production (up from 110,000 tonnes in 1986-87 to 130,000 tonnes in 1986-87 to 475,000 tonnes in 1987-88), coffee, cotton, rubber and palm kernels. Livestock farming is important, and poultry farming is rapidly increasing. Arable potential has been put at 25 per cent of total area, of which about 12 per cent is cultivated.

Industry: The sector developed rapidly during 1970s, but slowed in the mid 1980s, contributing around 12 per cent to GDP in 1985. Development largely centred on import-substitution of consumer goods, vehicle assembly (production down by 41 per cent in 1987), iron and steel and construction. Recent emphasis has shifted towards more low-cost integrated high value-added industries which rely on local rather than imported raw materials and capital goods, and on shifting production away from Lagos. Although it has benefited since 1985 from lower exchange rates, industry is vulnerable because of the high proportion of imports among its inputs. Manufacturing is currently operating at 40 per cent of capacity. Beverages, textiles, cigarettes, food processing and soaps continue to account for over 60 per cent of total manufacturing output. Manufacturing contributes some nine per cent of GDP. Main heavy industries are the Delta and Ajaokuta steel complexes and the petroleum refineries at Kaduna, Warri and Port Harcourt.

Mining: Extraction of solid minerals confined to coal (307,000 tonnes in 1987, which is used chiefly by Nigeria Cement Co and for electricity generation), tin (1,450 tonnes p.a.) and columbite (120 tonnes p.a.). The 1985 tin market collapse has badly affected all metals. Exploitation of iron-ore deposits at Itakpe is, however, going ahead with the long-term aim of supplying the requirements of the national steel industry. Deposits of uranium, lead, zinc, tungsten and gold are not yet exploited.

Hydrocarbons: The petroleum sector is the mainstay of the economy, contributing 20 per cent to GDP, 75 per cent to government revenues and accounting for virtually all foreign exchange earnings. Total reserves are estimated at about 20bn barrels (16bn proved, 3.2bn inferred), sufficient at current rates of extraction to last until 2020. Nigeria now ranks as the sixth largest producer in OPEC and by far the largest in Africa. But exports fell by 13.2 per cent in 1986 and by a further 12.7 per cent in 1987, and output from a peak of 2.4m b/d in 1979 to as little as 1.1m b/d in January 1988, although production fluctuates and sometimes exceeds Nigeria's OPEC quota of 1.3m b/d. The Niger Delta, with 78 of the 142 oil fields, is the main producing area. Oil extraction and exploration are undertaken by the state-owned Nigerian National Petroleum Corporation (NNPC) in partnership with 10 major international petroleum conglomerates. Petrol shortages in Nigeria are due to frequent breakdowns of the three refineries in Warri, Kaduna and Eleme near

most two-thirds of the country's post colonial history. With the banning of more experienced former politicians the military is arguably the most experienced political force in the country and Lagos wags have it that there is already a two-party system in the country: the civilians and the military. Among the military, General Ibrahim Babangida was described as a survivor instrumental in every coup staged or foiled since 1976.

In early February 1989 a radio relay of a military band concert sparked off coup rumours. It was said that even the military were finding SAP hard to bear. Later that month when the whole country was expecting a significant cabinet reshuffle to free key actors for the political race ahead, President Babangida surprised everyone by dissolving the entire Armed Forces Ruling Council. He reconstituted it a week later trimmed from 29 members to 19 to reflect 'routine military postings' and to make the ruling body more manageable. He also established an Armed Forces Consultative Assembly to give voice to the opinions of the lower ranks.

With his military constituency assured he went on to announce the long-awaited cabinet reshuffle. No 1992 presidential candidates were released for the political race as forecast but new entries to the cabinet were Minister of Industry Air Vice-Marshal Mohammed Yahaya, Minister of Power and Steel Air Vice-Marshal Nura Imam, Minister of Trade S. J. Ukpanah and Minister for Youth, Sports and Culture I.O. Graham-Douglas. Dropped from the cabinet were Dr Kalu Idika Kalu, from transport, Air Vice-Marshal A. Ishaye from special duties, Air Vice-Marshal Bayo Lawal from social development, youth, sports and culture and Air Vice-Marshal Anthony Okpere from aviation. Five ministers were redeployed but there were no changes in 15 ministries, reflecting the president's desire for cabinet stability both at federal and state level during the transitional process.

Two months later, on 3 May, President Babangida signed into law an AFRC amended constitution which left the contentious issue of Sharia Law applying to personal issues and only willing Muslims. He received the report of the Revenue Mobilisation Committee and created 149 new local government areas making a national total of 449 and bringing governance and revenue allocation closer to the grass roots. Finally he ordered the NEC to release the guidelines for the registration of political parties and with the lifting of the ban on political activity he wished Nigerians an exciting time ahead. It was a masterpiece of political right-footing worthy of his nickname in the Nigerian press, 'Maradonna'. Some said the timing was to create a favourable impression in Britain which he visited a few days later. Others said that with everyone so busy politicking there was no inclination to start coup-making while the president was away. Whatever the reason, it was perfect timing which confounded sceptics expecting the transitional programme to be delayed as a result of dilly-dallying by committees and the Constituent Assembly.

Political observers are never quite sure which way the wind from Dodan Barracks is blowing. Early in the year a shadow group called the Committee of Patriots (COP) with a Dodan Barracks address, appeared to be making underground political waves at the Constituent Assembly. 300 of the 567 assembly members were reported to be card-carrying COP members, presumably with an open eye to spotting the party that might win. But in April the government, through its number two, Vice-Admiral Augustus Aikhomu, severed all links with the group.

A party of local government chairmen elected in 1987 on a non-partisan ticket seemed another arrangement to curry favour with a government on the look-out for newbreed political aspirants. The Republican Party was born. But the rug was pulled from under its feet when in July all 449 local governments were dissolved and replaced by sole administrators to eliminate the 'incumbency factor'.

Human rights

President Babangida came to power in 1986 with a promise to curb the human rights abuses of his predecessors, Major Generals Muhammadu Buhari and Tunde Idiagbon. However, by the end of July 1989 the leading newsmagazine *Newswatch* was carrying a cover depicting a military 'Maradonna' hiding behind Decree 2 and dribbling a football marked Freedom.

As problems became more complex the government increasingly turned to the State Security (Detention of Persons) Decree 2 of 1984 to silence its critics. The Decree provides for six months renewable detention without charge and its operations lie outside the control of Nigeria's judiciary. It has been roundly condemned by civil liberties activists including the Nigerian Bar Association (NBA) and Nobel prize winner, Wole Soyinka.

The Civil Liberties Organisation, a Lagos-based human rights group, said that on 19 July to their knowledge there were 67 detainees in just three Lagos prisons being held under Decree 2. They included students, journalists and lawyers. In August it was revealed that many of those arrested during the May riots were also being held under Decree Two. The president of the NBA Mr Aka-Bashorun said that the majority of those in detention constitute political prisoners although that category of prisoner does not officially exist in Nigeria.

Most celebrated among the Decree 2 detainees was human rights lawyer, and persistent thorn in the government's side, Chief Gani Fawehinmi who was arrested on 17 June along with Dr Tai Solarin and 86-year-old Pa Michael Imoudu, a veteran trade union activist and an architect of Nigerian independence, while trying to hold an 'Alternatives to SAP' conference in Fawehinmi's home. The two old men were released after several hours but Fawehinmi was flown out of Lagos and confined in Gashua, a remote town in Borno State. On his forty-fifth day in detention without charge a High Court Judge ruled his arrest illegal and awarded him 10,000 naira damages but said that his court lacked the jurisdiction to challenge his continued imprisonment under Decree 2.

Alhaji Balarabe Musa faced a similar predicament when a court allowed him free on bail only for him to be immediately re-detained under Decree 2.

Fawehinmi's 'Alternatives to SAP' conference was aborted, as was a similar conference convened a month later by the National Association of Seadogs (Pyrates Confraternity) an organisation involving Wole Soyinka. Police surrounded the Abeokuta venue of the event and prevented participants from entering. Soyinka afterwards called the action, 'another indication of a government in the state of advanced paranoia'. He called on the government to pre-empt the 'language of the streets' by promoting the 'language of rational discourse'. He said that the government's early commitment to open debate was constantly being eroded by 'acts of harassment, intimidation, arbitrary arrests and detentions under the amorphous Decree 2'.

In October in the wake of the registration of the two new political parties, the military government reviewed the cases of many Decree 2 detainees, releasing Chief Fawehinmi and Alhaji Balarabe Musa on bail and freeing others from detention.

Added to the clampdown on free discussion of alternatives to the government's economic austerity programme was the marginalisation of Nigeria's intellectual elite in the political process. All university lecturers were banned from participating in politics and the first casualty of the ban was Dr Bala Usman, who was retired from his post in the northern Ahmadu Bello University, Zaria for his open and illegal participation in the PLP. The prospects for critical participation in the government's key programmes of economic reform and transition to civil rule wavered during 1989.

Outlook

The future success of the government's twin political and economic programmes depends largely on the behaviour of the politicians aspiring to rule and some signs of economic recovery.

President Babangida sounded a cautionary note to politicians in July when he said 'Why is it that it is only our political parties that appear to be natural grounds for the idle and the illiterate?' Through testing aspirant politicians on their understanding of the new Constitution the military government hoped to weed out such dead wood from the political process. Many politicians themselves privately admitted to fears that if they were to fail Nigeria this time the military might come back with a vengeance and ruin any future prospects for civilian rule.

For the first time in 1990 Nigeria was to experience power-sharing between the military at federal level and civilian governors at state level. The extent to which the ship of state would be able to steer a steady course from the local government elections onwards also depended on the government's economic policies bringing some relief to ordinary Nigerians. Any government under threat of food riots and further street violence would find it difficult to rule through normal democratic practices.

International relations

1989 saw the vigorous pursuit of Nigeria's new policy of economic diplomacy. External Affairs Minister Major General Ike Nwachukwu said in March: 'There has been a deliberate effort to move away from the over preoccupation by earlier administrations on political issues to those which will support the rejuvenation of our national economy'.

Nigeria tabled a proposition before the 1989 summit of the Organisation of African Unity (OAU) for the establishment of an African Common Market and in June the long-awaited trade liberalisation programme between the 16 nations of the Economic Community of West African States (ECOWAS) was scheduled to begin fully operating in January 1990 with the lifting of restrictions on the free movement of manufactured goods within the community.

Three of Nigeria's neighbours, Benin, Equatorial Guinea and Niger benefited from Nigerian aid and joint ventures. Relations with Chad cooled when the Chadian authorities paraded alleged Nigerian drug traffickers before diplomats but hopes lifted towards the end of 1989 that Chad too would benefit from Nigerian aid. A major contract for oil exploration in Ghana went to a subsidiary of the Nigerian National Petroleum Corporation (NNPC). The establishment of a joint trade commission with Sierra Leone, which like Nigeria was implementing IMF approved SAP economic reforms, was important. However, despite these promising developments the challenge of presenting a united market to the other major trading blocs, particularly the newly unified Europe of 1992, remained daunting.

In Southern Africa Nigeria signed a major trading and mutual co-operation pact with Zimbabwe which was cemented by a state visit by President and Mrs Babangida to Harare in July. The government launched the Namibian Solidarity Fund to raise 100m naira within Nigeria to assist SWAPO during the electoral transition period to full independence in Namibia. It also despatched a contingent of Nigerian policemen to participate in the United Nations handover of the territory.

South–South co-operation continued during 1989 with Nigerian trade delegations visiting Brazil, Argentina, Indonesia, Singapore, Zaire, Zimbabwe, Guinea Bissau, Liberia, Sierra Leone, Cape Verde, Togo, Angola, Zambia, Equatorial Guinea, Libya, Egypt and Benin.

Relations with Britain continued to improve with British Prime Minister Margaret Thatcher making a stopover visit to Nigeria and President and Mrs Babangida being hosted by the Queen to a state visit to Britain in May. A similar state visit to France had to be cancelled because of Nigeria's civil unrest. But the groundwork underpinning both visits was to attract foreign investors to Nigeria.

The touchy issue of renewing diplomatic ties with Israel was revived when President Babangida and Israeli President Chaim Herzog met during the funeral of Emperor Hirohito in Japan in February. It was the first official contact between the two countries since 1973 when links between the two countries were severed. However, in March Nigeria denied that a secret government delegation had visited Jerusalem as claimed by an Agence France Press report.

During 1989 Nigeria successfully vied for the presidency of the 44th session of the United Nations against her neighbour Cameroon and Egypt. Retired Major General Joseph Garba was appointed to the post. In October career diplomat Emeka Anyaoku successfully challenged former Australian prime minister Malcom Fraser for the post of Secretary-General of the Commonwealth.

COUNTRY PROFILE

Port Harcourt. A second refinery at Eleme was under construction; its completion would bring refining capacity to 445,000 b/d, more than enough to meet domestic needs, saving much foreign exchange now spent on foreign refining. The second Eleme refinery is linked to Phase II of the petrochemical scheme, funding for which is in doubt. There are vast reserves of LNG, but the implementation of plans to export the gas, costed at $2bn to $5bn, will depend on the state of the international energy market.

Energy: Expansion of industry is based on the utilisation of country's oil and gas wealth. Domestic consumption of oil currently accounts for about 20 per cent of total production. Development of Nigeria's hydro-electricity potential has been curtailed by drought and funding difficulties. Total installed capacity is 3,000 mW.

Membership of international organisations: African Development Bank, African Groundnut Council, African States Associated with the EC, Arab Bank for Economic Development in Africa (recipient status), Commonwealth, Economic Commission for Africa, Economic Community of West African States, FAO, GATT, IDA, IFAD, IFC, ILO, IMF, International Cocoa Organisation, Niger Basin Authority, OAU, OPEC, UN, WHO, WIPO, World Bank.

Time: GMT +1.

Climate: Ranges from tropical on the coast to sub-tropical in the northern reaches. Dry season from Nov-Mar with the dry harmattan wind blowing through the country during several days in Jan and Feb. The rainy season is from Aug to Oct with rainfall and humidity (80-100 per cent) highest in the south. Temperatures remain fairly constant throughout the year reaching 30-32°C in the coastal areas while in the north can reach reach 35-40°C but can drop sharply at night, especially from Nov to Feb.

Entry requirements
Passport: Required by all.
Visa: Required by most visitors. Visas must be obtained in advance. A return or onward ticket is also required. Regulations regarding visas, transit etc should be checked carefully before travelling as they are liable to change at short notice.
Prohibited entry: South African nationals and visitors with a South African entry visa in their passport.
Currency: Import/export of more than N20 in local currency prohibited. All visitors must exchange the equivalent of $100 into local currency on arrival. Foreign currency import unlimited but must be declared on arrival – export limited to amount declared. Regulations may change at short notice. Credit cards are not widely used. Banks charge a heavy commission for changing travellers' cheques.

Health precautions
Mandatory: Yellow fever vaccination certificate, and cholera, if arriving from an infected area.
Advisable: Cholera, typhoid and polio vaccination, malaria prophylaxis. Drinking water should be boiled or filtered (water precautions are essential outside of main towns).

Representation overseas: Abidjan, Accra, Addis Ababa, Algiers, Ankara, Athens, Baghdad, Bamako, Bangui, Banjul, Beijing, Beirut, Belgrade, Berlin, Berne, Bissau, Bonn, Brasilia, Brazzaville, Brussels, Bucharest, Cairo, Canberra, Conakry, Cotonou, Dakar, Dar es Salaam, Dublin, Freetown, Gaborone, The Hague, Havana, Hong Kong, Islamabad, Jakarta, Jeddah, Kampala, Khartoum, Kingston, Kinshasa, Kuwait City, Libreville, Lisbon, Lomé, London, Luanda, Lusaka, Madrid, Malabo, Manila, Maputo, Mogadishu, Monrovia, Moscow, Nairobi, N'Djamena, New Delhi, Niamey, Nouakchott, Ottawa, Ouagadougou, Paris, Port of Spain, Prague, Rabat, Rio de Janeiro, Rome, Stockholm, Tehran, Tokyo, Tripoli, Vienna, Warsaw, Washington, Yaoundé.

Air access
National airline: Nigeria Airways.
Other airlines: Aeroflot, Air Afrique, Air Gabon, Air Zaire, Alitalia, Balkan, British Airways, Cameroon, Egyptair, Ethiopian, Ghana Airways, Iberia, Interflug, KLM, Lufthansa, MEA, Pam Am, Sabena, Saudia, Swissair, UTA, Varig.
Tax: Airport tax of N5 on departure.
International airports: Abuja (Code: ABV), 35 km from city; Kano-Malam-Aminu (Code: KAN), 8 km from city, duty-free shop, restaurant, bank, post office, shops, car hire; Lagos-Murtala Mohammed (Code: LOS), 22 km from city, duty-free shop, restaurant, bar, buffet, bank, post office, car hire; Port Harcourt (Code: PHC), 24 km from city.

Surface access
Main ports: Apapa (Lagos), Port Harcourt, Calabar and the Delta Port complex including Warri, Sapele and Koko.
Overland access: Good roads linking Niger (Maradi, Zinder, Agadez, Niamey) to Kano, and from Benin; all-weather roads from Cameroon (Maroua, Mokolo) and Chad (N'Djamena). Southern road from Cameroon (Mamfe) to Enugu is not generally recommended.

Hotels: Wide range available, though difficult to obtain in Lagos and very expensive – around N100 per night for bed and breakfast in main Lagos hotels, N50 per night in state capitals. Bills must be paid in foreign currency and a deposit in advance is required (this can be as high as N300). Advisable to book in advance, though a confirmed reservation does not guarantee a room will be available on arrival. Most main hotels are air-conditioned.

Car hire: Available in most of the large towns through the main hotels. International driving licence and two passport sized photographs required; chauffeur-driven services generally recommended.

City transport
Taxis: Widely available in Lagos and other main towns. The traditional taxis are usually yellow Peugeots in Lagos, other colours elsewhere. Also numerous cars belonging to 'car hire' companies. Fare and tip should be agreed before starting journey. Daily hire available for around N80-100.

National transport
Air: Nigeria Airways service Lagos, Benin City, Calabar, Enugu, Ibadan, Jos, Kaduna, Kano, Maiduguri, Port Harcourt, Sokoto, Yola. It is advisable to book seats well in advance. Shuttle services by Kabo Air between Lagos and Kano or Kaduna, and by Okada Air between Lagos and Port Harcourt. Air charter facilities available from Aero Contractors, Pan-African Airlines and Delta Air Charter.
Road: National road system of 124,000 km links all main centres. Principal main roads connect Lagos and Port Harcourt in the south with Kano and Katsina in the north. The motorway running from Lagos-Ibadan is often congested. Some secondary roads can become impassable during rainy season.
Buses: Coach services scheduled include: Kaduna-Jos; Lagos-Umuahia; etc.
Rail: Cheap, but slow. Two classes; some restaurant cars; buffet facilities also available; some air-conditioning. Major services: Lagos-Kano Express (via Ibadan and Minna) branches run to Baro, Kaura Namoda and Nguru, Plateau Express (Lagos-Jos); Port Harcourt-Kano with branch to Jos and Maiduguri.
Sea: Ferry services on the Benue and Niger rivers, and along the southern coast.

Public holidays
Fixed dates: 1 Jan, 1 May, 1 Oct (National Day), 25 Dec, 26 Dec.
Variable dates: Good Friday, Easter Monday, Eid-el-Fitr, Eid-el-Kabir, Prophet's Birthday (Eid-el-Maulud).

Working hours
Business: (Mon-Fri) 0800 – 1230 and 1400 – 1630. Some offices also (Sat) 0800 – 1230.
Government: (Mon-Fri) 0730 – 1530, some states also (Sat) 0800 – 1300.
Banking: (Mon) 0800 – 1500, (Tue-Fri) 0800 – 1300.
Shops: (Mon-Fri) 0800 – 1730, (Sat) 0800 – 1630.

Telecommunications: International telephone, telegram and telex facilities are available at NITEL's offices in Lagos and other large cities.
Telephone: Automatic telephone and telex service operated by Nigerian Telecommunications Ltd (NITEL). Number of public telephones steadily increasing in Lagos, but the internal network is heavily overloaded and long distance calls to some areas are virtually impossible. Dialling code for Nigeria: IDD access code + 234 followed by area code (Lagos 1, Ibadan 22, Kano 64).
Telex: Facilities are available in main centres but service can be erratic.

Banking
Central bank: Central Bank of Nigeria.
Main centre: Lagos. African Banking Consortium Merchant Bank, African Continental Bank, Allied Bank of Nigeria, Bank of Credit & Commerce (Nigeria), Bankers Trust International (Nigeria), Chase Merchant Bank Nigeria, Commerce Bank Nigeria (Crédit Lyonnaise), Federal Mortgage Bank of Nigeria, Federal Savings Bank Nigeria, First Bank of Nigeria, First City Merchant Bank, Grindlays Merchant Bank, ICON, Indo-Nigerian Merchant Bank, Habib Nigeria Bank, International Bank for West Africa, International Merchant Bank (Nigeria), Merchant Banking Corporation Nigeria, NAL Merchant Bank, National Bank of Nigeria, Nigeria Merchant Bank, Nigerian American Merchant Bank, Nigeria-Arab Bank, Nigerian Bank for Commerce & Industry,

The economy staggers to its feet, but is by no means cured

Richard Synge

As Nigeria's Structural Adjustment Programme (SAP) entered its fourth full year in 1990, Nigeria still seemed to be reacting unwillingly to the medicine prescribed by its creditors. In many ways addicted to the habits which the SAP was supposed to eliminate, the Nigerian economy was proving a recalcitrant patient. As a result, the pundits began to stretch their forecasts of an effective economic recovery into the second half of the 1990s.

The intractability of many of Nigeria's economic problems continued to defy the efforts of those designated to resolve them. Although the SAP continued to be a central tenet of President Babangida's programme, the gap between promise and performance remained as wide as ever, creating a troublesome political burden of widespread dissatisfaction. In 1989 renewed and intensified rioting in a number of key cities exposed the government's failure to carry the populace with it in the reform process.

But despite the domestic worries created by high inflation, rising unemployment and growing social and political problems, the overall outlook for the economy in 1990 was considerably brighter than at any time since the early 1980s. There was a good prospect of international balance of payments support, in contrast to the first three years of the SAP when only the World Bank's 'seed money' materialised, and there were the additional benefits of significantly improved oil earnings in 1989 and a substantial boost in non-oil earnings (which jumped 254 per cent to hit $931m in 1988). These factors helped improve Nigeria's international standing as it contemplated a return to the commercial borrowing market for a number of key projects.

Preliminary figures for the first half of 1989 showed the healthiest balance of payments picture in a decade. Between May and June 1989 the monthly foreign exchange inflow, mainly from oil, jumped from $367m to $714m. Economists anticipated a significant rebuilding of the country's foreign exchange reserves over the rest of the year.

Difficulties still abound in the balance of payments, given the heavy burden of the country's $30,000m national debt. In 1989 the looming finance gap would have been $9,400m without rescheduling, but new London and Paris Club agreements have now deferred some $8,300m of it. This is an important element of relief, although servicing costs remain uncomfortably high. Even after rescheduling, the government was still committed to pay around $2,500m in 1989 in moratorium interest, interest on arrears and fees. How much of Nigeria's export earnings this was likely to consume could still not be determined at the time of writing because of continued volatility in the oil market.

Even with windfall improvements in foreign exchange flows, there was little cause for celebration about any change in Nigeria's economic fortunes. The restructuring exercise seems set to pursue its harsh monetarist logic in all parts of the Nigerian economy for a number of years to come.

Having failed to mount a sustainable return to growth in 1988 with a 'reflationary' budget, the government adopted more modest targets in 1989, with a tight reining in of credit, of government borrowing and of reliance upon oil earnings. In its commitments to the IMF in 1989 the government attempted to cut the budget deficit from 12 per cent of GDP in 1988 to no more than 8.3 per cent of GDP, and to hold credit growth to below 6.4 per cent as opposed to the previous year's level of 34.5 per cent. It was also committed to using oil sold at above $16 per barrel to rebuild the foreign exchange reserves, which slumped below $400m at the end of 1988.

The private sector felt the pinch of reduced allocations of foreign exchange under a new daily auction system, which in most of 1989 was releasing only $150m a month to cover the nation's import needs, down by a third from the levels available the previous year. This restrictive policy seemed likely to be eased in 1990 as the government rebuilt its battered reserves. An apparent priority in 1989 was to halt the rapid decline in the value of the naira, after a sharp fall in the official rate at the beginning of the year.

A battery of Central Bank of Nigeria (CBN) regulations in mid-1989 seemed designed to dry up any remaining hoards of naira, as a crude but effective means of restraining the currency's fall in value. The recessionary impact of these moves created howls of pain from the commercial and industrial sectors, but the government and the CBN seemed determined to conquer the spectre of high inflation which had shaken their confidence in the first half of the year. A report from the Manufacturers' Association of Nigeria (MAN) for first-half 1989 said capacity utilisation in industry had fallen to 31 per cent from 40 per cent a year before.

There was little chance of an import boom when most medium-term projections pointed to continuing financing gaps of between $2,000m–3,000m a year between 1989 and 1992. The situation called for sustained and substantial international support for Nigeria. The alternative could be a damaging break with its creditors at some point in the future. The debt burden had not been lightened but merely stretched over 10–20 years. Debt/equity swaps had so far made little impact.

To muster the scale of international support needed, serious consideration was being given to the convening of a World Bank consultative group meeting to focus attention on Nigeria's short and long-term economic and financial problems. Although the outlook for the oil market at the time of the conference would determine how much money donors pledge, the event would be important both in deciding the future direction of the SAP and in bringing the wider social and developmental issues under the spotlight.

Japan and the UK led the way in 1989 on co-financing of the World Bank's $500m Trade and Investment Policy Loan (TIPL), to help relieve a yawning $1,100m finance gap for the year. The African Development Bank followed suit with a clutch of loan approvals in mid-1989, while the USA, France and West Germany promised small symbolic contributions.

The price demanded for this new and wide international support was a continued bill of health from IMF, under the $625m standby agreement approved in February 1989. The Fund was looking for further movement on evolving a flexible exchange rate policy, improved 'commercialisation' of the public sector, increased public sector tariffs (especially for fuel and power) and a reduction of inflation in 1989 from its high 40 per cent level of the previous year. As these were all declared government objectives, few observers anticipated any difficulties with these conditions in principle; the only likely sticking points with the IMF could be over the pace of actual implementation and the trend of the naira's exchange rate.

Following the persistent 'round-tripping' of official foreign exchange funds into the 'autonomous' market, the introduction of a daily auction/allocation system in January 1989 was seen as a move in the right direction towards a freer market, but it failed to dampen the price of 'back door funds'. Critics added that the daily operations also tended to create a system not unlike that operated by the CBN before the SAP came into being in 1986. Since all banks were guaranteed a slice of the foreign exchange on offer, bankers were seen as vulnerable to corrupt pressures to favour some clients above others.

Privatisation gets moving

The planners of the SAP worked to deliver two new stimuli for greater private involvement in the economy: the 1988 Privatisation Decree and the Industrial Policy statement of January 1989. The one spelt out a huge list of federal government divestitures over the next three to five years, while the other completely rewrote the Indigenisation Decrees of the 1970s

Balance of payments 1984–88 ($M)

	1984	1985	1986	1987	1988*
Exports	11,891	12,566	6,784	7,732	7,117
Imports	8,882	8,279	6,744	5,774	5,600
Trade balance	3,009	4,287	40	1,958	1,517
Services	–3,743	–4,059	–3,484	–2,888	963,015
Transfers (net)	–332	–260	–178	–134	–150
Current account balance	–1,066	–32	–3,622	–1,064	–1,648
Capital account balance	1,433	310	–3,735	–2,272	–1,632
FINANCING:					
Effect of reschedulings (net)	–492	300	7,682	6,192	–1,309
Accumulation of arrears (net)	857	767	1,617	305	4,312
Rescheduling of arrears	–257	–1,049	–2,793	–3,240	**
Movement of reserves (– = increase)	–475	–296	851	78	277

* provisional
** to be calculated following Paris and London Club reschedulings

Source: Central Bank of Nigeria (CBN) and IMF staff estimates.

allowing up to 100 per cent foreign ownership in almost any new enterprise.

The proposed revisions in favour of foreign investors went further than any observers anticipated. But there was little optimism about the prospects for an investment invasion, largely because the existing investor community were denied the advantages offered to newcomers, and there were few newcomers prepared to confront the risks of setting up in Nigeria, when better opportunities existed for them elsewhere.

In the course of 1989, share applications were invited from the Nigerian public for government holdings in a number of large companies: in flour milling, oil product marketing and insurance. The new Technical Committee for Privatisation and Commercialisation (TCPC) reported a high level of over-subscription, a good indication that the privatisation process had a future. But, typically in Nigeria, the verification and approval process was a long-drawn-out affair. Thus it seemed likely to take several years to sell off the full list of government interests targeted in the 1988 decree: 67 firms to be sold off in full; and 43 for partial privatisation.

Privatisation was intended primarily to attract domestic Nigerian investors, although in due course it might also provide opportunities for foreigners. In 1989 the government was asking foreign investors to focus on brand new projects. Any new investment had almost by definition to be based on locally-available raw materials. The sectors attracting attention therefore include farming, horticulture, textiles, leather goods, timber, paper and packaging and plastics (in anticipation of new petrochemicals investment by the government). Investors from elsewhere in the Third World were already moving in to take up key positions in the emerging Nigerian export sector as well as in the planned petrochemicals and plastics downstream industry. Horizon Fibres, a group with Singapore-based management and experience of Indonesia, in two years committed over $26.7m to installation of plants to produce fibres, filament yarns, packaging film and other polypropylene by-products, mainly supplied from Warri oil refinery.

For the existing big investors such as Unilever, Leventis, ICI, Blue Circle, SCOA, John Holt, Cadbury, Nestle or Patterson Zochonis, the growth picture remained clouded by the difficulties faced in their everyday operations, and in particular their restricted access to foreign exchange.

Acquisition of existing assets was emerging as a new growth area for business, bringing a new boost to the hitherto restricted market. The rigidities of the capital market that evolved under the influence of the Nigerian Enterprises Promotion Decrees of 1972 and 1977 were giving way to a new flexibility induced both by privatisation and the new industrial policy. The market was always dominated by foreign-affiliated groups, with the shareholders locked into their maximum holdings. Nigerian Stock Exchange (NSE) director-general Hayford Alile now saw more room for expansion of the market, and eventually a full internationalisation of the NSE.

Most of the big groups were looking hard at diversification of their business in order to survive, and here the new investment provisions were seen as helpful. UAC of Nigeria chairman Ernest Shonekan said he was delighted that companies like his own could now be treated as fully Nigerian for the purposes of new investment. His counterpart at John Holt, Garba Ja' Abdulkadir, foresaw forming joint ventures with foreign firms in order to use their technical expertise.

As an earnest of good intent, the much-heralded Industrial Development Co-ordination Committee (IDCC) went into business in Abuja, with a brief to review all new investment applications within one month, and to grant all the necessary pre-investment approvals, industry incentives, expatriate quotas and technical agreement fees in as short a time as possible.

Unlike many African countries, Nigeria already had a large domestic private sector able to perform many of the key roles normally performed by international capital. There was however less confidence in the ability of the domestic private sector to move towards replacing the public sector in key and strategic areas of the economy and its infrastructure. Here there was also concern that the privatisation exercise would in many cases leave the government as the single-largest shareholder even if it did not have a formal majority. This might militate against efficient management practices, and could cause the privatisation programme to slow up before it got fully under way, some experts warned.

Another view on the investment prospects was that the government had not yet started to market the attractions of Nigeria in an aggressive enough manner. Most visitors to Nigeria were easily put off by the obvious difficulties of operation, in overcoming bureaucratic hurdles and the poor communications. Several reports advised the government to attend to these matters as an urgent priority.

Areas of great opportunity

The largest new investments in the near future would be those in the oil sector, where borrowing for projects could be isolated from the conditions governing other Nigerian debts. The new Nigerian LNG Company, for example, was looking to borrow $1,600m over 1990–95 to realise its project for the export of 4m tonnes a year of liquified natural gas.

First on the market, however, would be the $900m Oso petroleum condensates field, being developed by Nigerian National Petroleum Corporation (NNPC) and Mobil Corporation. Japanese and British export credits were being sought, alongside World Bank and International Finance Corporation (IFC) funds.

A competing energy-sector project, requiring investment of around $800m, was an ethylene-based petrochemicals industry to be built at Alasa Eleme, Rivers state. Japanese firms were expected to take small equity stakes and NNPC hoped to go to the financial markets with the project during 1990.

The logic of the SAP required that Nigeria's economic future continued to depend largely on exports. The first experience of the programme showed that the greatest export potential lay in the energy sector, much as it did before, and in the export of local raw materials. What was missing from this scenario was any guarantee of an appropriate and profitable industrialisation process – a vital social necessity in view of the unacceptably high levels of unemployment prevailing in the country.

The next phase of the SAP will need to provide clear evidence that there is a positive economic and social pay-off from Nigeria's hardships of recent years. The groundwork has been laid. What the country now needs is the funds to invest in a substantial reconstruction.

GDP in 1987 by sector

(N '000m at constant 1984 prices)

Sector	GDP 1987	%	%rise
Agriculture	17.92	24.3	–0.4
Livestock	6.01	8.1	1.5
Forestry	2.29	3.1	0.0
Fishing	1.29	1.7	3.2
Crude petroleum	9.94	13.5	–8.9
Mining, quarrying	0.21	0.3	5.0
Manufacturing	4.65	6.3	9.9
Utilities	0.41	0.6	5.1
Building, construction	0.92	1.2	2.2
Transport	3.21	4.3	3.5
Communications	0.10	0.1	0.0
Wholesale, retail trade	19.75	26.8	5.0
Hotels, restaurants	0.21	0.3	5.0
Finance, insurance	1.55	2.1	10.7
Real estate, business serv ices	0.22	0.3	4.8
Housing	2.43	3.3	3.0
Government services	2.66	3.6	1.1
TOTAL	**73.77**	**100.0**	**1.2**

Source: Central Bank of Nigeria Annual Report 1987.

Projections of balance of payments 1989–92 ($M)

	1989	1990	1991	1992
Exports	6,914	7,352	7,800	8,668
Imports	5,814	6,235	6,407	6,638
Trade balance	1,100	1,117	1,393	2,030
Services	–3,182	–3,278	–3,356	–3,408
Transfers (net)	–150	–165	–182	–200
Current account balance	–2,232	–2,326	–2,145	–1,578
Capital account balance	–1,732	–730	423	397
Overall balance	**–2,962**	**–1,903**	**–1,749**	**–1,265**
FINANCING:				
Net effect of existing reschedulings	–1,052	–1,857	–986	–1,426
Arrears (net)	–347	–	–	–
Rescheduling of arrears	–4,813	–	–	–
New rescheduling	8,269	867	87	26
Reserves	–200	–100	–100	–100
Balance of payments support loans*	1,105	950	–	–
Residual financing gap**	–	–2,043	–2,748	–2,765

* existing commitments

** gap before any balance of payments loans that may become available

Source: estimates based on Nigerian government data.

Nigerian Industrial Development Bank, Nigeria International Bank, Savannah Bank of Nigeria, Société Générale Bank of Nigeria, Union Bank of Nigeria, United Bank for Africa. Also: Bank of the North (HQ: Kano), Cooperative Bank (HQ: Ibadan), Cooperative & Commercial Bank (HQ: Enugu), Kaduna Cooperative Bank (HQ: Kabuna), Kano Cooperative Bank (HQ: Kano), Lobi Bank Nigeria (HQ: Makurdi), Mercantile Bank of Nigeria (HQ: Calabar), New Nigerian Bank (HQ: Benin City), Nigerian Agricultural & Cooperative Bank (HQ: Kaduna), Owena Bank Nigeria (HQ: Apapa), Pan African Bank (HQ: Port Harcourt), Progress Bank of Nigeria (HQ: Owerri), Wema Bank (HQ: Ebute Metta).

Electricity supply: 230 V AC, 50 cycles.

Representation in capital: Algeria, Angola, Argentina, Australia, Austria, Belgium, Benin, Brazil, Bulgaria, Burkina Faso, Cameroon, Canada, Central African Republic, Chad, Chile, China PR, Colombia, Cuba, Czechoslovakia, Denmark, Egypt, Equatorial Guinea, Ethiopia, Finland, France, Gabon, Gambia, German DR, German FR, Ghana, Greece, Guinea, Haiti, Hungary, India, Indonesia, Iran, Iraq, Ireland, Italy, Côte d'Ivoire, Japan, Kenya, Korea PDR, Rep. of Korea, Lebanon, Liberia, Malaysia, Mauritania, Morocco, Netherlands, Niger, Norway, Pakistan, Philippines, Poland, Portugal, Romania, Saudi Arabia, Senegal, Sierra Leone, Somalia, Spain, Sudan, Sweden, Switzerland, Syria, Tanzania, Thailand, Togo, Trinidad and Tobago, Turkey, UK, USA, USSR, Uruguay, Vatican, Venezuela, Yugoslavia, Zaïre, Zambia, Zimbabwe.

Hotels: Abeokuta
Ogun State, Ibara, P.O. Box: 30 (tel: 231574; tx: 24670).

Hotels: Abuja
Nicon Noga Hilton, P.O. Box: 200 (tel: 5231811, 5231830; tx: 71504; fax: 5231839).

Hotels: Benin City
Bendel, P.O. Box: 1954 (tel: 6382; tx: 41123).

Motel Benin Plaza, 1 Reservation Road, P.O. Box: 1152 (tel: 242680).

Hotels: Calabar
Government Guest House, P.O. Box: 1056 (tel: 113).

Metropolitan, Murtala Mohammed Way, P.O. Box: 1071 (tel: 220911).

Hotels: Enugu
Enugu Guest House, P.O. Box: 289.

Phoenix, P.O. Box: 1130 (tel: 2074).

Presidential, Independence Layout, P.O. Box: 1096 (tel: 252050; tx: 51121).

Hotels: Ibadan
Catering Guest House, Jericho Reservation (tel: 414237).

Green Springs, Old Ife Road, P.O. Box: 159 (tel: 70010/1; tx: 31140).

Lafia, Moor Plantation, P.O. Box: 5353 (tel: 416472; tx: 31175).

Premier Mokola Hill, Mokola, P.O. Box: 1206 (tel: 400340/8; tx: 31174).

Hotels: Jos
Hill Station, Tudun Wada Road, P.O. Box: 72 (tel: 55301; tx: 81130).

Plateau, P.O. Box: 2170 (tel: 55740; tx: 81118).

Hotels: Kaduna
Durbar, Independence Way, P.O. Box: 2218 (tel: 201100/8; tx: 71134).

Hamdala, Wharf Road, P.O. Box: 311 (tel: 211207; tx: 71163).

Hotels: Kano
Central, Bompai Road, P.O. Box: 3023 (tel: 601120/606238; tx: 77151).

Daula, 150 Murtala Mohammed Way, P.O. Box: 3228 (tel: 5311; tx: 77241).

Hotels: Lagos and district
Bristol, 8 Martin Street, P.O. Box: 1088 (tel: 661201; tx: 21144).

Durbar, Badagry Expressway, P.O. Box: 12815 (tel: 880201,881131,884111; tx: 21100).

Eko Holiday Inn, Victoria Island, P.O. Box: 12724 (tel: 615000; tx: 22650).

Excelsior, 3/15 Ede Street, Apapa, P.O. Box: 1167 (tel: 803680/4; tx: 22447).

Federal Palace (and Fed.Palace suites), Ahmadu Belle Way, Victoria Island, P.O. Box: 1000 (tel: 610030; tx: 21432).

Ikeja Palace, Kingsway Road, Ikeja, P.O. Box: 7487 (tel: 900880).

Ikoyi, Kingsway Road, Ikoyi, P.O. Box: 895 (tel: 603200; tx: 22632).

Lagos Airport, Obafemi Awolowo Way, Ikeja, P.O. Box: 21041 (tel: 901001/5; tx: 26203).

Lagos Hilton (not Hilton International), P.O. Box: 2199 (tel: 960604, 964947; tx: 26329).

Mainland, 2/4 Murtala Mohammet Way, P.O. Box: 2158 (tel: 841101; tx: 21595).

Maryland, Ikeja, P.O. Box: 34 (tel: 934465).

New World, P.O. Box: 7621 (tel: 900920/1/2; tx: 26516).

Sheraton, 30 Airport Road, Ikeja, P.O. Box: 21189 (tel: 900930; tx: 27202; fax: 900930).

Trade Fair, Lagos/Badagry Express Road (tel: 880787; tx: 22632).

Hotels: Maiduguri
Lake Chad, P.O. Box: 368 (tel: 232794; tx: 82121).

Hotels: Port Harcourt
Presidential, Aba Road, P.O. Box: 5141 (tel: 335655; tx: 61182).

Hotels: Sokoto
Sokoto, Gasau Road, P.O. Box: 2199 (tel: 232412; tx: 73112).

Hotels: Warri
River Valley, Effurun, P.O. Box: 275 (tel: 233295).

Car hire: Lagos and district
Dilvy (Nigeria) Ltd: 17A Modele Road, Surulere (with a branch at the Bristol Hotel).

Executive: Ikeja Airport (tel: 934184).

Hertz: 9-23 Atitebi Street (tel: 835290; tx: 21383).

Intra: 11 Martin Street (tel: 634884).

Nigerian Rent-a-Car: Apapa Road, Inganmu (tel: 836336).

Safedrive: 19 Adam St Yaba, PO Box 3932 (tel: 844615).

UTC Motors: PMB 1005, Lagos (tel: 803720).

Banking
Central Bank of Nigeria, Tinubu Square, PO Box 12194, Lagos (tel: 660100; tx: 21350 cebank).

First Bank of Nigeria, 35 Marina, PO Box 5216, Lagos (tel: 665900/20; tx: 21231 banker).

First City Merchant Bank, Primrose Tower, 17A Tinubu Street, PO Box 9117, Lagos (tel: 665944/52; tx: 22912, 22913 fecity).

IBWA – Afribank, 94 Broad Street, Lagos (tel: 664135, 662301, 663608; tx: 21345, 22382, 22383 afribank).

ICON (merchant bankers), NIDB House, 63/71 Broad Street, PO Box 12689, Lagos (tel: 661812, 660103, 662607; tx: 21437, 21166, 22430).

International Merchant Bank (Nigeria), 1 Akin Adesola Street, Victoria Island, PO Box 12028, Lagos (tel: 612204,613000; tx: 21169).

NAL Merchant Bank, NAL Towers, 20 Marina, PO Box 2432, Lagos (tel: 600420/9, 600850/9; tx: 215055, 22941 accepto and 23677, 23732 nal).

Nigerian Bank for Commerce and Industry, 26 Idejo Street, Victoria Island, PO Box 4424, Lagos (tel: 614071, 617665; tx: 21917 nibacind).

Savannah Bank of Nigeria, 62/66 Broad Street, PO Box 22317, Lagos (tel: 600470/9, 600970/9).

Union Bank of Nigeria, 40 Marina, Lagos (tel: 665439, 665441; tx: 21222).

United Bank for Africa, 97/105 Broad Street, PO Box 22406, Lagos (tel: 667410, 667510, 667610; tx: 21241, 21580 mibank).

Other useful addresses
DHL International, 1 Sumbo Jibowu Street, Ikeja, Lagos (tel: 681106).

Economic Community of West African States (ECOWAS), 6 King George V Road, Lagos.

Enugu Chamber of Commerce and Industry, Trade Fair Complex, Abakaliki Road, PMB 734, Enugu (tel: 330575).

Federal Ministry of Finance, PMB 12591, Ikoyi, Lagos (tel: 681908).

Federal Ministry of Industries, PMB 12614, Ikoyi, Lagos (tel: 682607).

Federal Ministry of Internal Affairs, PMB 12595, Ikoyi, Lagos (tel: 683694).

Federal Radio Corporation, PO box 12504, Lagos (tel: 603010; tx: 21484).

Federal Secretariat, PMB 12514, Ikoyi, Lagos (tel: 684451).

Ibadan Chamber of Commerce and Industry, PMB 5168, Ibadan (tel: 710021).

Kano Chamber of Commerce and Industry, 138 Ibrahim Taiwo Road, PMB 10, Kano (tel: 629940, 620165).

Lagos Chamber of Commerce and Industry, Commerce House, 1 Idowu Taylor Street, Victoria Island, PMB 109, Lagos (tel: 613898, 613911, 610533).

Manufacturers' Association of Nigeria, PO Box 3835, Lagos.

National Oil & Chemical Marketing Co, PMB 2052, Lagos.

National Science & Technology Development Agency, PO Box 12695, Lagos.

Nigeria Airways, PO Box 1024, Ikeja, Lagos (tel: 933631, 900476; tx: 22646).

Nigerian Association of Chambers of Commerce, Industry, Mines & Agriculture, 15A Ikorodu Road, Maryland, PMB 12816, Lagos (tel: 964727, 964737; tx: 21368).

Nigerian Industrial Development Bank, PO Box 2357, Broad Street, Lagos. (In particular encourages foreign capital, personnel).

Nigerian Institute of Management, PO Box 2557, Lagos.

Nigerian National Petroleum Corp, PO Box 12701, Ikoyi, Lagos (tel: 603100; tx: 21226).

Nigerian Television Authority, PO Box 12036, Lagos (tel: 613793).

Nigerian Tourist Board, Tafawa Balewa Square, PO Box 2944, Lagos (tel: 632240).

Union of Importers & Exporters, PO Box 115, Ibadan.

Niue

T. R. Hughes

(map on page xxiv)

At the end of the 1980s the tiny Polynesian community of Niue was facing a crisis of confidence. Despite persistent attempts by New Zealand to prop up the economy, Niueans were departing at such a rate from the 256-sq km coral atoll that the residents left behind – probably less than 2,000 – were outnumbered by five to one by fellow islanders attracted to the bright lights of Auckland where they enjoyed full rights. Niue, independent in association with New Zealand, was not really viable. The big headache for the island's premier, Sir Robert Rex, a veteran Pacific elder statesman, was how to deploy New Zealand aid money so as stop the drain in human resources.

Constitutional changes

The islanders resented complaints from some Wellington quarters that donor money had been squandered through lethargy or inefficiency and were sensitive to threats that the aid lifeline would be cut unless more energetic efforts were made to revive the island. In what was seen as one of the withdrawal symptoms, the legislative assembly signalled in mid-1989 that it wanted to loosen constitutional ties with New Zealand to give it more freedom of action.

It approved 37 changes to the constitution drawn up in 1974 when the island was accorded self-government and intended to put them to a referendum early in 1990. The changes included localising the New Zealand-dominated public service commission and replacing the New Zealand governor-general, as head of state, with a Niuean, probably Sir Robert who was expected to retire early from the premiership.

New Zealand aid

Ever since it was annexed by New Zealand in 1901, Niue has shown little interest in going it alone. Compared with half a century ago, however, when it was a prosperous community of self-reliant smallholders, the island has seen repeated attempts to create a modern business base end in failure.

New Zealand handles external affairs and defence, and also provides about two-thirds of the budget. Some New Zealand taxpayers said this is excessive, particularly as the atoll has no strategic or economic value. It would be a lot simpler, they suggested, to move the islanders and their belongings to the outskirts of a small New Zealand country town and give them an annual grant.

That did not find favour with a joint Niue-New Zealand review group chaired by Sir James Stewart in 1986 which recommended that the community should be left intact even if it meant a radical rethinking of policies by everyone concerned. The key, said the review, was to raise the standard of living for the islanders to that of their compatriots in New Zealand. The review gave birth to the Niue Concerted Action Plan (NCAP) in 1988 in which Wellington agreed to provide NS$33m over three years to revitalise and maintain Niue as a 'living community' by simply providing jobs rather than by trying to persuade the islanders to achieve full self-reliance.

The NCAP suggested taking a new look at agriculture, tourism, and light industry; urgently reviewing Niue's sole air carrier link; guaranteeing a direct shipping service; getting the Niue government to complete a comprehensive review of land titles; and stepping up social welfare benefits.

New Zealand, struggling in its own economic straitjacket, warned that it was a last attempt to settle the depopulation problem. One disagreement was over the bloated public service. Wellington would like to see it cut by a half to two-thirds and the saving in wages put into private enterprise, a move wholeheartedly supported by the island's own small business group. Over the years the aid money has helped to create a new social class of bureaucrats, well-paid and privileged. Sir Robert's government said defensively that recruiting to the public service was one way of ensuring that people stayed and that there were not enough jobs in the private sector to take up the slack if they were laid off. Medical and dental services, as well as plumbing and electrical work, are free, and it is not difficult to borrow government equipment for nothing. The public servants also have access to cheap housing loans.

Communications

Niue's economy is largely dependent on agriculture and large amounts of aid have already gone into the primary-product sector with its 'all or nothing' mentality – crops are abandoned if prices drop. Although the land can support a variety of fruit and vegetables, the community is exposed to the caprices of communications. There is no real harbour. Ships have to anchor off the capital of Alofi and cargo is transferred by barge. Niue relies heavily on trade with New Zealand and shipping problems are a major disincentive to farmers. Tourists have a tough time getting there and there is a shortage of good beaches. Compounding the problem, Niue lost its direct regular air link with New Zealand in 1989 because Wellington's civil aviation authorities had grounded Air Nauru. This effectively isolated the atoll, dampening down exports and also emigration – Niueans began the exodus to New Zealand when the island's airport was opened in 1971. If Air Nauru were to fail to return, one solution being considered by Wellington was to extend the runway and have an aircraft calling there as a stopover on a main route.

Outlook

Politically, the pressure was mounting on Sir Robert who had been leader of government since 1966 and premier since 1974. In April 1989, the New Zealand auditor general issued a blunt report calling the Rex cabinet negligent with aid money and accusing it of failing to protect the interests of taxpayers.

This sparked off an unprecedented vote of no-confidence from the opposition led by Young Vivian, the man regarded by many as premier material. Although he lost by 13-7, Vivian claimed a moral victory and many saw it as a sign of Sir Robert's waning influence.

Three MPs crossed the floor to support Vivian who heads Niue's first political party, the Niue Peoples' Action Party, and even Sir Robert's number two, Frank Lui, hinted that he would join the opposition if asked to take over as premier. The seven MPs who voted against Sir Robert represent more than half the island's population.

Barring a new constitutional post, Sir Robert was thought unlikely to stand again in the 1990 election although he said this once before. With a political time-bomb ticking away in the shape of a possible cut-off of New Zealand aid he might not have much choice.

KEY FACTS

Official title: Niue

Head of state: Queen Elizabeth II

Head of government: Premier Sir Robert Rex

Capital: Alofi

Official Languages: English, Niuean

Currency: New Zealand Dollar (NZ$ = 100 cents)

Exchange rate: NZ$1.70 per $ (Nov 1989)

Area: 258 sq km

Population: 2,532 (1986). 10,000 Niueans are resident in New Zealand

Labour force: 600 (1988)

Visitor numbers: 600-800 p.a.

COUNTRY PROFILE

Political system: Niue is self governing in free association with New Zealand. Under the Niue constitution, New Zealand remains responsible for defence and foreign affairs and is ready to provide necessary economic and administrative assistance, and Niueans are New Zealand citizens. The island is governed by a 20-member Legislative Assembly - 14 elected from village constituencies and six from the common roll. The four-member executive cabinet is headed by a premier elected by the Legislative Assembly. Village affairs are handled by 14 village councils elected to three-year terms in conjunction with the Community Affairs Department.

Political parties: In March 1987 elections, the parliamentary group around Sir Robert Rex, the Premier, was returned to office.

Population: The population is mainly of Polynesian descent, with some New Zealand elements. Considerable migration to New Zealand, as people from Niue are New Zealand citizens.

Main town: Alofi (population 900 in 1987).

Language: English is widely understood, but a Polynesian dialect similar to Tongan and Samoan is spoken.

Media
Press: Tohi Tala Niue is published weekly in English and Niuean by the government's information office.
Broadcasting: Radio Sunshine broadcasts are transmitted six days/week during limited hours. Television Niue operates Mon-Fri, 1730 – 2200.

COUNTRY PROFILE

Domestic economy: The economy is mainly agricultural, with inhabitants earning subsistence wages on government relief work and raising vegetables, fruit, pigs and poultry for their own consumption. Industry is limited to processing agricultural production, although some small building and joinery operations have been started. Sale of postage stamps and remittances from Niuean workers overseas are important sources of revenue. Niue requires annual grant-in-aid from New Zealand to balance its budget. Under the Niue Concerted Action Plan, New Zealand aid totals NZ$33m over 1988-91 to help the economy become self-sufficient.

Employment: More Niueans work overseas, particularly in New Zealand, than on the island. Opportunities for paid work exist only within the government services, small industries, retail outlets and hotels. Apart from the local Public Service Association, there are no industrial unions.

External trade: There is free trade between Niue and New Zealand.
Exports: Main exports are sports goods, passion fruit, limes, handicrafts, coconut cream and honey. Main destinations are New Zealand, Cook Islands and Fiji.
Imports: Main imports are foodstuffs, fuels and oils, machinery, beverages. New Zealand typically supplies 65 per cent of Niue's imports. Other main sources are Fiji, Japan, Western Samoa with the balance coming from China, USA, Australia and UK.

Agriculture: Development of agriculture is hindered by the limited amount of fertile or cultivable land, lack of surface water and susceptibility to hurricane damage. Alienation of land is forbidden, but the government may grant a lease of land for a maximum term of 60 years. About 78.5 per cent of the land area is available for agriculture. Main subsistence crops are limes, passion fruit, coconuts, taro and cassava. Plantations of coconut palms provide copra for export, and the government has encouraged increased planting of lime trees and passion fruit vines. There is some bee-keeping and livestock raising (pigs, poultry); goats have recently been introduced on a trial basis. About 20 per cent of the land area is forest with millable timber, and logging serves local demand.

Industry: The Office of Economic Affairs is responsible for planning and financing productive ventures relating to agriculture, tourism and industry. Niue Handicrafts handles production and marketing of fine ware plaited from pandanus and coconut palm leaves, and other organisations operate privately. Small-scale industries include fruit processing; lime and passion fruit juice production; honey extraction and bottling; saw milling; joinery, furniture and handicrafts.

Legal system: There is a High Court under the control of a chief justice, with right of appeal to the New Zealand Supreme Court. The Land Court has jurisdiction over land disputes.

Membership of international organisations: ESCAP (associate member), South Pacific Commission (Conference representation), South Pacific Forum, Forum Secretariat, Tourism Council of the South Pacific (TCSP).

Time: GMT -11 hrs.

Climate: Subtropical and damp, with temperature 25°C to 30°C and average rainfall of nearly 200 cm p.a.

Surface access: There are no port facilities. Ships anchor off Alofi and barges transfer cargo.

National transport
Road: There are approximately 128 km of all-weather road and 96 km bush track negotiable by heavy trucks and four-wheel drive vehicles. A 60 km road circles the island, and roads link main centres.

Telecommunications
Telephone: Dialling for Niue is via international operator. Telephone system is double-line, with overseas communications via radio link.
Telex: Telex services started in 1986.

Banking: Westpac Bank offers full banking facilities.

Representation in capital: New Zealand representative.

Business advice and information
Office of Economic Affairs, PO Box 42, Alofi.

Travel information
Tourism Section, Office of Economic Affairs, PO Box 42, Alofi.

Hotels: Alofi
Esthers Village Holiday Motel, Avatele Village, P.O. Box: 107 (tel: Home: 825 Motel: 139; tx: NF 67013 (Kuso)).

Hotels: Alofi Niue
Damianas Holiday Motel, Avatele Village, P.O. Box: 119 (tel: Home: 70V Motel: 16; tx: NF67013(Mrs M Kingi)).

Hotels: Alofi South
Niue, Amanau, P.O. Box: 80 (tel: 91, 92; tx: NF 67005).

Norfolk Island

T. R. Hughes

(map on page xxiv)

History vies with scenery in attracting visitors to the 34.5 sq km former penal colony basking in the Tasman Sea. Every summer the 2,300 resident population – of which about a third, known as 'islanders', are descendants of the Bounty mutineers – can be swelled by up to 1,200 visitors at any one time. But tourism appeared to be faltering in mid-1989 after several flourishing years. The island's economy is heavily influenced by events on mainland Australia, and to a lesser degree by New Zealand, and is feeling the draught from the dearer money policies in both countries. Even the celebrations in 1988 marking the bicentennial of the first European settlement (and, in 1989, the 200th Bounty mutiny anniversary) failed to arrest the decline.

The people of the independently-governed Australian territory are proud of the island's transmutation from a first convict settlement to a modern-day capital exporting economy displaying a high degree of affluence in per capita terms. But storm warnings were hoisted as the Canberra policy of pushing up interest rates and pushing down the dollar to squeeze domestic demand had a knock-on effect on local business. Profit margins have narrowed and savings are down.

While an optimistic assessment is that the economy is still running at a reasonable level even with the yearly growth rate cut back from 4 per cent to 2.5 per cent, the government admits the trend is 'flat'. And it is likely to continue that way unless the local government can rethink its strategy.

Overheated economy

A worrying aspect is that the steady and profitable growth of tourism may have lulled residents into what can best be described as 'consumer over-confidence' in an overheated economy. Many locals, who do not pay income tax on earnings within the island, now see their living standards threatened by sharply higher interest repayments on bank borrowings made in the good times.

Norfolk's inflation since 1983 fuelled an explosion in property estate prices. The market has now gone into reverse gear and is estimated to be dropping by 20 per cent a year.

Tourism, which accounts for about half of the local jobs, began to supplement subsistence farming, grazing and fishing as the islanders' main livelihood in the mid-1960s and received a strong boost in 1983 when the Australian Government upgraded the island's airport (owned by Australia) at a cost of A$7m to cater for medium-sized jet aircraft. The number of visitors, mainly from Australia, average 29,000 a year – compared with 900 about 25 years ago – and help to close

KEY FACTS

Official title: Norfolk Island

Head of state: Queen Elizabeth II

Head of government: Administrator: B.H. MacDonald; President of the Assembly: David E. Buffett

Capital: Kingston

Official Languages: English, Norfolk dialect

Currency: Australian Dollar (A$ = 100 cents)

Exchange rate: A$1.26 per $ (Nov 1989)

Area: 34.5 sq km

Population: 2,367 (1988)

Visitor numbers: 28,300 (1987/88)

the trade gap by injecting some A$10m a year into the economy.

Norfolk is expected to pay its way under the terms of its association with Australia which provides technical aid and expertise in various fields, such as health, but normally gives no other cash help. The local community is almost wholly dependent on imported foodstuffs. Primary industries are limited and locally-raised poultry and beef falls short of demand especially with the fluctuation in the number of visitors. Farming on any real commercial level is limited by the sub-divison of land. Fishing in the surrounding ocean is good but is inhibited by the lack of harbour facilities.

The government runs on an annual budget of around A$5.7m and revenue is derived mainly from postage stamps, customs duty and liquor sales. The re-exports of watches, cameras, clothes, radios and jewellery, imported with low duty and sold to tourists, also produce a useful income.

Sales of these re-export goods, however, were hit by Canberra's introduction in 1988 of a A$400 duty-free limit on items that tourists could take back to the mainland. Many shops hurriedly changed their product lines to avoid being left with high-priced goods on their shelves. The government said this was having 'an important adverse influence' on revenue and promised to urge Canberra either to exempt the island from the duty or increase the limit.

Strategic value

Norfolk Island, 1,676 km from Sydney, continues to be of strategic value as it has been since Captain James Cook landed there in 1744 to exploit the tall, distinctive pines for masts and the native flax for sails. A bicentennial anniversary expedition dived on the shipwreck of the flagship of the Australian First Fleet, *HMS Sirius*, wrecked off Norfolk in 1790, seeking to show the importance of the island to Britain's then colonial ambitions in the Pacific.

In 1788, the New South Wales colony's first governor, Captain Arthur Phillip, sent 23 settlers, including 15 convicts, to the island. Over the years more convicts and settlers arrived, clearing dense rain forests and pines from the central and southern regions for farming. By 1797 the population had reached 1,188. But the island could not sustain the new community and the British authorities transferred them to Tasmania. By 1814, the island had been abandoned. 11 years' later, more convicts and soldiers were landed to begin the second settlement, excluding free settlers.

Meanwhile, the Pitcairn descendants of the Bounty mutineers had been looking for a new home after their expanding community began to run out of land on the five sq km island. Queen Victoria granted them the former penal colony and they replaced the convicts in 1856, an event now celebrated as Bounty Day.

The bicentennial celebrations in 1988 revived the touchy subject of the island's aspirations. Technically, Norfolk is an Australian external territory even though it is self-governing. But the residents dispute the view that they are part of the Commonwealth of Australia. One legal opinion is that the island is still a colony and entitled to self-determination. It is an argument has been going for about 100 years between the residents and the mainland government and has led to Royal Commissions, petitions to the Crown and applications to the United Nations. There are no signs, however, of an imminent solution.

COUNTRY PROFILE

Political system: A former penal colony, Norfolk Island has an Administrator appointed by Australia, but an increasing number of local matters are dealt with by the elected Legislative Assembly, subject to the Administrator's consent. The position vis-à-vis internal self-government is subject to review.

Political parties: There are no organised political parties.

Population: Polynesian, Australian and New Zealand (2,367 in 1988).

Main cities/towns: Burnt Pine, Kingston.

Language: English, the offical language, is spoken in business circles; otherwise, the local Norfolk (Polynesian) dialect.

Domestic economy: Norfolk Island's economy is based largely on its tourist industry, catering mainly to visitors from Australia and New Zealand. It receives no direct grants or aid from Australia. Australia owns and maintains the airport, restores certain historic buildings and provides certain technical services. In addition to importing most requirements, Norfolk Island has developed a re-export industry geared to its tourist industry. In 1987-88 visitors numbered 28,297.

Employment: Immigration is strictly controlled, and tourism has absorbed any increase in the labour force. There is virtually no unemployment recorded.

External trade
Exports: Value of exports in 1986-87 was A$3.6m. Main exports are re-exports, Norfolk pine seeds, Kentia palm seeds, avocados and local pottery. Main destinations are Australia, other Pacific islands, New Zealand.
Imports: Value of imports in 1987-88 was A$23.5m. Main imports are fuel, food, etc., and luxury goods, including alcoholic beverages (typically, 4.4% of total imports), building material (4.1%), photographic goods (3.5%), jewellery (3.2%), footwear (2.2%) and clothing (11.3%). Main sources are Australia, other Pacific islands and New Zealand.

Agriculture: The 11-12 per cent of total land considered cultivable is volcanic and fertile. Production is constrained by poor terrain, porous soil, a low water table and extremely fragmented holdings. Main crops are cereals, vegetables and fruit, including avocados for export. There is some cattle and poultry raising. Norfolk Island and Kentia pine seeds are an important export, and some hardwood afforestation is being undertaken. Lack of a harbour restricts fisheries development, and catches serve local consumption only.

Industry: Aside from the tourist industry, manufacturing has been small-scale and spasmodic.

Time: GMT + 11.5 hrs.

Climate: Temperatures can range from 11-27°C, with an average rainfall of 1,320 mm/year.

Entry requirements
Visa: Visas are not required by citizens of Australia and New Zealand with proof of onward passage.

Air access
National airline: Norfolk Island Airways.
Other airlines: Air New Zealand, East-West Airlines, Qantas.
International airport: Norfolk (NLK).

Surface access: Ships anchor offshore.

Car hire: Arrangements may be made locally for hiring cars and motorcycles.

National transport
Road: The entire road network of about 200 km is now passable to vehicles.
Buses: There are bus services and tour buses available.

Telecommunications
Telephone: Dialling code for Norfolk Island, IDD access code + 672 followed by area code 3 for all parts of the island, followed by subscriber's number.
Telex and telefax: Telex and facsimile facilities are available.

Banking
Central Bank: Reserve Bank of Australia. *Norfolk Island branches:* Commonwealth Bank of Australia, Westpac Banking Corp.

Hotels: Norfolk Island
Cascade Gardens Apartments, New Cascade Road, P.O. Box: 266, 2899 (tel: 2741; tx: 32019; fax: 2785).

Castaway, Taylors Road, P.O. Box: 15, 2899 (tel: 2625; tx: 32019; fax: 2785).

Hibiscus Motels, Taylors Road, Burnt Pine (tel: 2325).

Hillcrest Gardens, Taylors Road, P.O. Box: 215 (tel: 2255; tx: 2899; fax: 67232907).

Islander Lodge, P.O. Box: 169 (tel: 2114).

Norfolk White Heron, Douglas Drive (tel: 2377).

Panorama Garden Apartments, Middlegate Road, Kingston, P.O. Box: 266, 2899 (tel: 2625; tx: 76632019; fax: 2785).

Paradise, Quality Rd, 2819 Kingston (tel: 2077).

Polynesian Licensed Motel, New Cascade Road (tel: 2309; fax: 3040).

Seaview Apartments, Taylor's Road (tel: 3033).

South Pacific, Taylors Road, Burnt Pine 2899, P.O. Box: 215 (tel: 2166; tx: 32016; fax: 2907).

Sundowner Central Apartments, P.O. Box: 2785.

The Colonial, Queen Elizabeth Avenue, 2899 (tel: 2177; fax: 67232907).

Villas Apartments, P.O. Box: 10,2899 (tel: 2177).

Car hire: Aloha Rent-a-Car: PO Box 387, 2899 (tel: 2510; tx: 766 32006; fax: 3064).

Borry's: Taylors Road, Burnt Pine (tel: 2114; tx: NV32008 BORRYS; fax: 3014).

FitzPatrick's Rental Cars: (tel: 2195/2198; tx: 76632002; fax: 2197).

Island Rent-A-Car: Taylor Road, PO Box 100, 2899 (tel: 2656; fax: 2663).

Martins Rentals: Taylors Road, PO Box 73, 2899 (tel: 2021; tx: NV32023).

Business advice and information
Norfolk Island Chamber of Commerce, PO Box 370.

Travel information
Norfolk Island Government Tourist Bureau, Burnt Pine (tel: 2147; tx: 32010).

Northern Marianas

T. R. Hughes

(map on page xxiv)

The Northern Marianas recorded rapid economic growth over the past few years as investors began to realise the benefits of its newly-found status as a self-governing commonwealth of the USA. The tiny state was part of the Trust Territory of the Pacific Islands (TTPI) administered by the United States under a United Nations mandate since the end of the Second World War. Three-quarters of the Marianan voters chose to retain their close links with the USA in 1985 but the commonwealth, similar to that of Puerto Rico, was not implemented until November 1986 when the trusteeship was terminated by the UN Trusteeship Council. At the same time the Federated States of Micronesia and the Marshall Islands, both partners in the TTPI, became independent in free association with the USA.

The commonwealth status gives the 21,700 Marianans, many of whom have Japanese ancestry, internal self-government with a wide range of civil and political rights in the USA. It also allows for duty-free and quota-free exports of locally-produced goods to the US market together with the use of alien workers in the manufacturing process and freedom to determine a minimum wage – all of which have done wonders for the economy.

Foreign workers

Thousands of foreign workers have been allowed to live there temporarily to fill jobs created by the boom in tourism and other industries but, unlike the Marianans, they are not entitled to US citizenship. The foreign workforce jumped by 50 per cent in 1986–87 swelling the total to about 9,000 and sharpening a social dilemma. Employers are compelled to recruit them because the Marianans prefer to work for the government which pays nearly twice as much. In 1986, the average going rate for a government job was $7,944 a year against $4,043 in the private sector. So far this disparity has not provoked any problems beyond stirring up the home-grown complaint that the foreigners repatriate too much of their income to the detriment of the local economy.

Land ownership is restricted to people of direct Northern Marianan descent until the year 2011 but land may be leased by US citizens and third-country nationals for commercial purposes. The government encourages free market conditions in which the private sector operates with a minimum of restraint. Foreign investment is sought from all sources and not just from the USA.

Political, governmental and financial ties are with the USA but there are strong economic links with Japan, specifically in tourism which accounts for more than a third of the gross national product and is growing steadily. Visitor arrivals were 142,284 in 1985, 157,201 in 1986 and 186,203 in 1987. About 80 per cent of visitors are package-tour Japanese in the middle to lower income bracket attracted to Saipan partly because it is cheaper than Hawaii or Europe. Catering as they do for the Japanese and with a heavy involvement of Japanese capital, the hotel trade depends on the economic climate in Tokyo as much as anything. The hotels are among the best and biggest in the world.

A by-product of the tourist boom is the flourishing construction industry with a host of new hotels and shops appearing in the past few years. There is confidence that as long as tourism remains strong, the construction growth will continue. Commercial activity also centres on light industry with more than 1,600 firms on the books although on a comparatively small scale.

Much hope is pinned on a fledgling textile industry seeking to draw on garment makers' expertise and inputs from Hong Kong, Korea and Taiwan to produce goods locally so as to escape US quota restrictions. Prospects for the textile trade are good as long as it can retain this concession.

The boom has much improved the government's coffers. Internal revenues were up 20 per cent to $52m in 1987 with another $14m coming from US government grants. The US Government has promised the state $228m for capital development, government operations and special programmes over a seven-year period from 1986. But the trade imbalance is worrying. Imports are climbing to meet the demands of tourism and building work while exports are declining, despite the textile contribution.

In an assessment of prospects, the Bank of Hawaii warns that any economic diversification or import-substitution projects would have to 'concede to economies of scale constraints'. The added cost of importing raw materials or component parts must also be considered. The Bank suggests there is potential in marine resources such as fisheries, marine minerals, ocean thermal energy conversion and geothermal energy – so far relatively unexploited.

As commercial and industrial development outpaces the infrastructure, the government plans major investments over the next few years to modernise and expand it, recognising that lack of basic necessities could rein in the state's growth. The focus is on electric power, water and sewage works, ports, public buildings and roads. Saipan's external communications, however, are moderately good. The airport is modern and can accommodate the biggest jets; there is a direct air service to Japan. Telecommunication links are among the best in the Pacific.

The sovereignty issue

Politically, the old sore of constitutional relations between the Northern Marianas and Washington bubbled up again in the 1989 election campaign. Saipan and the federal government have been at odds over internal sovereignty with increasing acerbity since the covenant came into effect in 1986.

Some prominent voices in Saipan believe the commonwealth has full control of domestic affairs without any reference to Washington, merely yielding responsibility for defence and foreign affairs to the federal government. They argue that the territorial clause of the US constitution does not apply.

The Washington representative, Froilan C. Tenorio, who was the Democratic Party's nominee for governor in the 1989 elections, called for court action to get a definitive ruling (unlike some other Pacific territories, the Washington job although elected is not a congressional seat). But no one, including the vocal Tenorio, wanted to end the commonwealth status and by mid-year as the election campaign rolled on there were clear signs of olive branches in the offing. Any changes in the political spectrum were more likely to result from local factors than differences on internal sovereignty. Party labels in Saipan have little in common with those on the US mainland.

KEY FACTS

Official title: Commonwealth of the Northern Mariana Islands

Head of state: President of the USA

Head of government: Governor Pedro P. Tenorio (elections Nov 1989)

Capital: Saipan

Official Languages: English (Chamorro, Carolinian and Japanese also spoken)

Currency: US Dollar ($ = 100 cents)

Area: 471 sq km (16 islands) – Saipan (122); Tinian (101)

Population: 42,000 (1989), of which some 23,000 are indigenous

Visitor numbers: 233,300 (1988)

COUNTRY PROFILE

Political system: The Northern Marianas Islands, a former Japanese colony, was part of the Trust Territory of the Pacific Islands, administered by the US under a 1947 mandate. The seat of the Trust Territory Government of the Pacific Islands was located on Saipan until 1987 when the US Trusteeship in Micronesia was phased out. The Northern Marianas was administratively separated from the Trust Territory in 1975, having voted to become a self-governing US Commonwealth. Residents became US citizens on 3 Nov 1986, when President Reagan signed a proclamation declaring that the Trusteeship Agreement no longer applied to the Northern Marianas. The constitution provided for an executive governor and a bicameral legislature consisting of a 15-member House of Representatives (elected every two years) and a nine-member Senate (with staggered terms), all elected by popular vote.

Population: Total population is about 42,000 of which locals account for 23,000. 85 per cent of the population live on Saipan.

Main town: Saipan (population 21,000* in 1989).

Language: English, Chamorro and Carolinian are the official languages. Japanese is spoken in many shops and hotels.

Media

Press: Weekly papers – *Marianas Variety, Marianas Review* and *Marianas Focus.*

Radio: Broadcasts by KSAI Radio (religious station), KYOI Radio and KCNM (both commercial).

Television: TV transmission by Saipan Cable TV.

Domestic economy: The Northern Marianas government agreed a seven-year commonwealth covenant with USA in Jul 1985 which provided for

COUNTRY PROFILE

$228m US aid. It aimed to promote economic development reducing the islands' dependence on US assistance. The people are mainly employed by the government. There are also many subsistence farmers and fishermen, although some crops and livestock produced on a commercial scale. The main export is garments. The small tourist industry is growing; total visitor figure was 230,000 (1988) of which majority of visitors came from Japan. This represents a 25 per cent increase in tourist numbers over 1987. Japanese construction company, Shimizu, plans to build a $400m resort on Saipan.

Employment: The tourist industry employs 45 per cent of the labour force; although garment industries are also important, they employ mainly Philippine, South Korean, Thai and Chinese workers. Local government provides most other paid employment for local residents.

External trade
Imports: Main imports are food, construction materials, fuel and cars.

Agriculture: Cultivatable land is rich and volcanic. Vegetables such as coconuts, breadfruit, tomatoes, melons and cucumbers widely grown on smallholdings. Some vegetables, fish, beef, pork, fruit and milk have been exported.

Industry: Aside from tourism, industrial activity consists of textile assembly and small-scale fish processing and handicrafts manufacture.

Legal system: There is a US district court (headed by a federal judge) and a Commonwealth Superior Court and Supreme Court are located on Saipan.

Membership of international organisations: Association of Pacific Island Legislatures, Council of Offshore Territories, Pacific Basin Development Council, South Pacific Commission, World Health Organisation.

BUSINESS GUIDE

Time: GMT + 10 hrs.

Climate: The climate is tropical marine.

Entry requirements

Passport: Passports required by all except US citizens with proof of citizenship.

Visa: Non-US citizens with onward passage and a valid visa for next destination may be given tourist entry permits for stays of less than 30 days on arrival.
Health: Vaccination certificates required for cholera or yellow fever if travelling from an infected area.

Representation overseas: The Marianas Visitors' Bureau is represented in Tokyo. There is also an elected representative in Washington DC, although not accredited to Congress.

Air access

Main airlines: Continental Air Micronesia, Freedom Air Passenger, Guam-Marianas Air, JAL, Northwest Orient.

International airport: Saipan International (SPN); duty-free shops, bar, restaurant, currency exchange, handicraft shop and car hire.

Surface access

Main ports: Saipan.

City transport

Taxis: Taxi service available on Saipan.

Buses: Tour bus service is available.

National transport

Air: Guam Marianos Air, Continental Air Micronesia and Freedom Air operate services from Saipan International to Tinian. Also services to Rota.

Road: Roads are good on the main islands.

Water: Sea links maintained between the islands.

Working hours

Government: (Mon-Fri) 0730 – 1130, 1230 – 1630.

Banking: (Mon-Thu) 0900 – 1500, (Fri) 1000 – 1800.

Shops: (Mon-Sat) 0800 – 2100, (Sun) 0800 – 1800.

Telecommunications

Telephone: Dialling code for Northern Marianas, IDD access code + 670, followed by area code (Rota 532, Tinian 433, followed by subscriber's number; Saipan 322 if followed by a 4-digit number commencing with 9 or 3, Saipan 234 if followed by any other 4-digit number).

Telex and telefax: Telex service can be used 0730 – 1930. A facsimile service is also available.

Banking

Head office: Bank of Saipan.

Northern Marianas branches: Bank of Hawaii, Bank of Guam, First Savings and Loan, Marine Merchant Bank, Union Bank.

BUSINESS DIRECTORY

Hotels: Rota Island
Blue Peninsula, Song Song Village, CM 96951, P.O. Box: 539.

Rota Coconut Village, P.O. Box: 855 (tel: 5323448; tx: 5323449).

Rota Pau Pau, Song Song Village, 96951, P.O. Box: 503 (tel: 5323561; fax: 5323562).

Hotels: Saipan
Chalan Kanoa Beach Club, P.O. Box: 356 96950 (tel: 7829; tx: 717; fax: 2346534).

Coral Ocean Point Country Club, Agingan Point, MP 96950, P.O. Box: 1160 (tel: 2347000).

Diamond Hotel, Susupe, Saipan MP 96950, P.O. Box: 66 (tel: 23459006; tx: 762 DIAHTL SPN; fax: 2345909).

Hafadai Beach, 96950, P.O. Box: 338 (tel: 6495/8; tx: 783616).

Hyatt Regency Saipan, on Micro Beach, 96950, P.O. Box: 87 CHRB (tel: 234-1234; tx: 659 HYATT SAIPAN; fax: 234-7745).

Marianas, (at Navy Hill and at Marpi), P.O. Box: 527 (tel: 9708/3324; tx: 608; fax: 3223057).

Nikko Saipan, P.O. Box: 152, 96950 (tel: 322-3311/3312; tx: 760702; fax: 322-3144).

Pacific Islands Club, MP96950, P.O. Box: 2370 (tel: 2347976).

Saipan Beach, P.O. Box: 1029,96950 (tel: 2346412; tx: 694; fax: 2347064).

Saipan Grand, Susupe Beach, P.O. Box: 369 (tel: 23466013; tx: 628; fax: 2348007).

Saipan Ocean View Hotel, Garapan, Saipan MP969950, P.O. Box: 799 (tel: 23489002; fax: 2347176).

Hotels: Tinian

Meitetsu & Fleming, P.O. Box: 68 (tel: 4333232; fax: 4333022).

Orinesia, San Jose Village.

Tinian Centre, San Jose Village, P.O. Box: 103.

Car hire: on Saipan

Dollar: PO Box 569 (tel: 234-9001/3447).

Hertz: International Airport, PO Box HR (tel: 234-8336; tx: 724).

National: PO Box 562 (tel: 234-7259/6434; tx: 657).

Nitto Saipan Inc: PO Box 910, 98950 (tel: 234-3548; fax: 2343473).

Pacific Rent-A-Car: PO Box 706, 96950 (234-7141; fax: 2347955).

Sunn Inn Rent-A-Car: PO Box 920, 96950 (tel: 234-6639/3232/3233).

Thrifty Car Rental: PO Box 478, 96950 (tel: 234-8356).

Toyota Rent-a-Car: PO Box 267 (tel: 234-59118; tx: 613).

Rota Travel Bureau, PO Box 603 (tel: 5323561; fax: 5323562).

Business advice and information

Saipan Chamber of Commerce, PO Box 806, Saipan, MP 96950 (tel: 2346132 or 2347058; fax: (670) 2347151).

Travel information

Marianas Visitors Bureau, PO Box 861, MP 96950 (tel: 234-8325; tx: 676 MVB CM SPN; fax: (670) 2343596).

Norway

Fay Gjester

Norwegian voters, preparing to elect a new parliament in the early autumn of 1989, found themselves being bombarded with somewhat conflicting messages about the state of the country's economy. Following three and a half years of tough austerity policies imposed by the minority Labour government, unemployment had reached a post-Second World War record, and even people with jobs were feeling poor. Concern at the high jobless figures, and apparent stagnation in some sectors – particularly the property market – was naturally exploited in the election campaigns of the non-socialist opposition parties. They claimed that Norway needed a change of government to get things moving again.

On the other hand, the unpopular squeeze had achieved many of its aims: the rate of inflation was down to 4.8 per cent – lower than in most of the countries Norway competes with; the mainland trade balance (excluding oil and gas) was showing a marked improvement, with imports falling, and exports rising; two years of tight curbs on pay increases had led to a 400 per cent increase in corporate profits; turnover and prices on the stock exchange were setting new records; interest rates were on the way down, from the peaks reached at the height of the economic crisis; independent analysts were reporting that the recession had bottomed out, and that most economic indicators were now pointing upwards again. Labour argued that its policies were working, and that a divided opposition (comprising five different parties) would not be able to form the kind of strong government that the country still needed.

Economic trends

Labour politicians agreed with the opposition that the time had come to administer some stimulus to the economy. In election speeches and media interviews, they stressed that the success of their policies had paved the way for this. Finance Minister Gunnar Berge promised 'limited' tax concessions, in the budget for 1990, which would be designed to stimulate investment and economic activity generally – since 'we have now come so far in improving the basic balance of Norway's economy'. Prime Minister Gro Harlem Brundtland pledged to reduce a special tax on investment – with effect from 1 November 1989, because a delay until the new year might lead businessmen to defer badly-needed new investments.

Most of the opposition parties promised even more vigorous pump priming, in the form of reduced taxes, if they should form the next government. In the event, the Labour government was displaced by a Conservative led coalition, as the two major parties conceded support to those at the extemes of the political spectrum. The non-socialist parties continued to control 84 seats but the increased support for the Progress Party, which rose from 2 to 22 seats, thwarted Labour's chances. Divisions within the coalition suggest that Norway risks a period of ineffectual government.

The change in government was welcome news to most Norwegians, who had seen their once high living standards deteriorate sharply since the 1985-86 oil price collapse.

The change from the boom years of the early 1980s had been dramatic. By autumn 1989, house and apartment prices had fallen steeply from the peak that they reached around mid-1987. But they were still too high for many prospective buyers and the housing

market was stagnant, with thousands of new units standing empty and virtually unsaleable, all over the country.

The glut of commercial property was even more marked. Office blocks and shopping complexes, started when prices were soaring, came onto the market just as demand for space was shrinking, because of the recession. Building contractors accounted for 13 per cent of the Norwegian firms declared bankrupt in the first half of 1989. Bankruptcies in the six

KEY FACTS

Official title: The Kingdom of Norway

Head of state: King Olav V

Head of government: Prime Minister Jan P Syse

Ruling party: Coalition of Conservative Party, Christian Democratic Party and Centre Party

Capital: Oslo

Official Languages: Norwegian

Currency: Norwegian Krone (NKr) = 100 ore

Exchange rate: NKr6.76 per $ (Dec 1989)

Area: 323,985 sq km

Population: 4.2m (1988)

GDP per capita: NKr141,476 (1988)

GDP real growth: 1.1% (1988)

Labour force: 2.2m (1988)

Unemployment: 3.2% (1988)

Inflation: 6.7% (1988)

Trade balance: –NKr6.8bn (1988)

Foreign debt: NKr99.0bn (1988)

months affected 2,393 companies – an increase of 32 per cent on January-June 1988. The sector hit hardest was retail trade, with 29 per cent of the total.

Advertising expenditure, a key economic indicator, had also been tumbling. After peaking in 1987, at NKr10.13bn, it dropped to NKr9.85bn in 1988 and was expected to fall still further, to around NKr9.5bn, in 1989. The advertising agencies' association said operating revenues in the first half of 1989 were 22 per cent down on a year earlier. Many newspapers had to reduce page numbers, and lay off workers.

Car sales dropped so steeply, in 1988–89, that some car dealers went out of business, and importers had to trim their payrolls. New car registrations in the first seven months of 1989 were 30 per cent down on the same period of 1988 – itself a bad year for the trade. Customers who would previously have replaced their old car with a new model made do with second hand. One dealer said he was selling about five second-hand vehicles for each new one.

Unemployment had been rising rapidly and steadily since about the middle of 1988; by mid-August 1989 it had reached a post-war record of 93,894 – 4.3 per cent of the labour force. That might be low by EC standards, but Norwegians regarded it as unacceptably high. Until recently, they had believed that serious unemployment was something which only happened in other countries. In addition to the registered unemployed, around 40,000 people were kept off the market by public works schemes – so that the 'real' number of unemployed, at mid-August, was around 130,000.

Particularly worrying was the fact that more and more of those on labour exchange rolls were long-term unemployed – many of them people in their late teens and early twenties, who had been hunting work unsuccessfully ever since leaving school. The 'make work' schemes launched by the authorities were unable to keep pace with their rising numbers. That threatened to create serious social tensions, as fit young men, living off social security handouts, wandered the streets with nothing to do. During the summer of 1989, in a part of south Norway where the problem was particularly acute, gangs of self-styled 'patriotic' unemployed youth began attacking immigrants. They claimed that the latter were pushing them to the back of the queue for jobs and subsidised housing.

When Labour came to power, early in 1986, it adopted a rather generous attitude to third-world economic immigrants claiming to be 'asylum seekers'. Large numbers of those who failed to qualify for political refugee status were allowed to remain anyway, on 'humanitarian' grounds. The word got around and between summer 1986 and the autumn of 1989 many thousands of them arrived, swelling social service outlays.

The unemployment figures would have been even higher without the country's flexible social security system. This allowed older workers, with litle prospect of finding new jobs, to drop out of the labour market and claim disability pensions to tide them over the last few years of what would normally have been their working lives (67 is the pension age – for both men and women). Many of those made redundant by the contraction of the shipbuilding and offshore construction industry chose this solution.

The number of Norwegians drawing disability pensions soared by 60,000 between 1980 and autumn 1989. A Conservative Party critic of the system, MP Harald Ellefsen, pointed out that this increase was five times as high as in Switzerland, taking population differences into account. 'There is nothing to indicate that health in Norway is so much worse than in Switzerland,' he added drily.

Economists have warned that a continued increase in the number of disability pensioners, coupled with the rise in old-age pensioners that is due anyway,

KEY INDICATORS

	Unit	1984	1985	1986	1987	1988
Population	'000	4,141	4,159	4,178	4,199	4,200
Gross domestic product (GDP)	NKr bn	452.5	500.2	513.7	562.9	594.2
GDP per capita	NKr	109,273	120,269	122,953	134,056	141,476
Inflation	%	6.3	5.7	7.2	8.7	6.7
GDP real growth	%	5.7	5.3	4.2	0.9	1.1
Unemployment	%	3.2	2.6	2.0	2.1	3.2
Industrial production	1985=100	96	100	103	111	116
Manufacturing	1980=100	104	106	108	110	–
Investment goods	1980=100	136	174	246	–	–
Consumer goods	1980=100	102	104	104	107	–
Intermediate goods	1980=100	101	103	103	108	–
Exports	NKr bn	154.0	170.7	133.8	144.5	143.9
Oil/gas	NKr bn	78.3	85.0	55.6	56.4	49.1
Petroleum products	NKr bn	4.4	5.1	3.4	3.7	2.9
Ships	NKr bn	5.8	11.1	11.1	10.0	5.5
Aluminium	NKr bn	8.9	8.2	7.6	9.4	12.6
Paper	NKr bn	4.4	5.0	4.9	5.4	5.8
Fish	NKr bn	6.2	7.0	8.1	9.5	10.1
Imports	NKr bn	113.1	132.6	150.1	152.0	150.7
EEC	%	45.7	47.4	58.5	75.5	–
CMEA	%	3.6	2.8	1.8	3.0*	–
Balance of trade	NKr bn	40.9	38.1	–16.3	–7.5	–6.8
Current account	$m	2,919	3,052	–4,545	–4,151	–3,678
Exchange rate	NKr per $	8.2	8.6	7.4	6.7	6.5

* estimated figure

owing to demographic trends, may put an intolerable strain on funding of the state social security system.

The collapse of world oil prices in 1985-86 is generally given most of the blame for Norway's economic problems in the latter half of the 1980s. In fact this is only part of the story. In an economy which had become heavily dependent on oil and gas, the sudden drop in petroleum revenues *did* hit both the state budget and the balance of payments hard. It forced the resignation – in May 1986 – of the recently re-elected, Conservative-led coalition, and obliged the new minority Labour government to embark on the austerity programme which so effectively punctured Norway's boom.

But there was trouble on the horizon even before oil prices started to slide. A credit-led surge in private consumption, encouraged by the government of the day, was getting out of hand by mid-1985, feeding inflation and sucking in a rising tide of imports. High oil earnings cushioned the problem, however, and the above-mentioned coalition, then seeking re-election, felt the time was not quite ripe to stop the party. This meant that when Labour Finance Minister Gunnar Berge took over, and ordered 'full speed astern', the jolt was much harder than perhaps it need have been.

Visitors to Norway, before the recession, used to ask the natives how they could afford their impressively high living standards – the designer clothes, the week-end shopping trips to London and Paris, the marinas full of glistening sailing boats and day cruisers, the roads clogged with large new cars, the busy pubs and restaurants. An honest answer would have been: 'Much of it is on borrowed money'.

Much of it was. The Norwegian tax system encouraged borrowing, and penalised saving, by making all interest payments deductible – even interest on hire-purchase agreements. The country's credit institutions were given greater freedom to lend, in the first half of the 1980s, as the non-socialist coalition gradually lifted many of the restrictions that had previously affected the credit market.

The result was that both borrowers and lenders went overboard. Norwegian banks lent so liberally and incautiously, to both individuals and corporate customers, that for the past few years they have had to write off billions of kroner in bad debts.

Den Norske Creditbank, once Norway's largest commercial bank (this title is now held by Christiania Bank), paid no dividend for 1986, 1987 or 1988. Some smaller banks, facing virtual ruin, were forced, in the late 1980s, to merge with stronger partners. All had to slim down payrolls, close branches, and increase prices for their services.

Personal borrowers were hit two ways. The Labour government initiated a series of changes to the tax system, making credit-financed consumption increasingly less advantageous for personal-income-tax payers. In addition, during the first years of the crisis, there was a steep rise in interest rates.

From autumn 1988 to autumn 1989 there was a nominal decline in interest rates, reflecting a series of cuts in the central bank's overnight lending rate. But with inflation slowing, the real cost of borrowing remained high.

By autumn 1989, ministers were citing the low inflation rate as one of many indications that the government was getting on top of the problems it inherited when it took office in 1986. At 4.8 per cent, it was the lowest in the Nordic area, and lower than the rate prevailing in most of the countries with which Norway competes.

Another positive development was the marked improvement in mainland Norway's trade balance (that is, excluding oil and gas). In July 1989, the deficit on non-oil trade was only NKr1.1bn – the lowest for any single month since December 1974. With oil and gas included, the country had a surplus of NKr4.6bn that month. Over the first seven months of 1989, imports of goods were worth NKr76.7bn (2 per cent down on a year earlier) and exports NKr102.4bn (30 per cent up). The export total included NKr41bn worth of oil and gas, but improved prices for oil, and rising oil output, did not alone account for the high export total.

Sales of 'traditional' Norwegian products such as light metals, paper and pulp and fish had also been doing well. Moreover, two successive years of tough curbs on pay increases had led to a 400 per cent rise in corporate profits – from a total of NKr5bn, in 1987, to an estimated NKr20 bn, in 1989. Not everybody was going bankrupt.

Early in September 1989, the official Central Bureau of Statistics published its quarterly analysis of the economic outlook. This concluded that the recession had bottomed out; production was on its way up again, and the decline in private consumption and investment had halted. The rise in the number of jobless was slowing, but unemployment remained a serious problem. Despite slower growth in some of Norway's main markets, Norwegian export industry had continued to achieve rising sales. Moreover, private consumption – having fallen continuously from fourth-quarter 1987 to early 1989, now appeared to be stabilising. It actually showed a 1.5 per cent rise, in second-quarter 1989, compared with the same period of 1988. Investment in industry also showed signs of recovering.

About a week earlier, Christiania Bank had issued a similarly upbeat survey of the economic situation, with optimistic forecasts of the way the main economic indicators were likely to move between late 1989 and 1993. The bank warned, though, that the predicted drop in unemployment during the period (to around 80,000-90,000 by 1993-94, including people on public works schemes) could turn into a rise if pay increases were allowed to exceed 4.5-5 per cent per year, in the two to three years from 1990, or if the government elected in September should pursue an excessively expansive fiscal policy.

The stock exchange

Foreign investors appeared to share the optimistic view of Norway's prospects. Heavy buying by foreigners of shares in Norwegian companies, through the spring and summer of 1989, was widely believed to be the main reason for the steady rise in both share prices and trading volumes on the Oslo stock exchange. By the first week in September, turnover for the year to date had already passed NKr60.2bn – until then the record (set in 1987) for a full year's trading. By comparison, the figure for the whole of 1988 (in the wake of the October 1987 stock market crash) was only NKr31.5bn. The Oslo *Aftenposten* said the event was not just the result of high prices, it also reflected a steep increase in the number of shares traded – to over 485m by end-August, compared with 210.5m in the whole of 1987, the previous record year.

The busy market pushed prices to record highs – by early September the all-share index was 53 per cent up on its end-1988 level. According to the Oslo business newspaper *Dagens Naeringsliv* this rise in value was exceeded, worldwide, only by the Vienna stock exchange (where the all-share index climbed by 90 per cent, over the same period).

Industry

Paper and pulp and metals, two of Norway's main land-based export industries, continued operating at virtually full capacity in the first half of 1989. Among leading industrial concerns, Elkem (ferroalloys, aluminium, silicon metal) quadrupled pre-tax profits in first-half 1989, compared with a year earlier, reflecting strong demand and high prices. But the half year report warned that results in the second half of 1989 would be affected by weaker ferro silicon prices. Aluminium prices began moving downwards in the first half of the year, and this trend helped trim the second quarter net profits of Norsk Hydro, the industrial and energy group, compared with first-quarter figures. Nevertheless, Hydro Aluminium (created in 1986 by the merger of ASV – formerly state-owned – and the aluminium division of Norsk Hydro) is going ahead with plans to boost capacity at its Ardal smelter by some 20,000 tonnes a year, in the early 1990s, to 620,000 tonnes. At the start of 1989, Hydro Aluminium was already Western Europe's largest producer of the metal, and Norway as a whole, with capacity of over 860,000 tonnes a year, had overtaken West Germany as the continent's dominant producer of primary aluminium. This was part of a shift of world capacity to regions with cheap power.

Another Norwegian industry that benefits from cheap energy is paper and pulp. The leading Norwegian concern in this sector, Norske Skog, announced early in 1989 an agreement to merge with Tofte Industrier and Follum Fabrikker, two pulp and paper firms with which it had previously had close ownership links. The merger between the three was due to be implemented with effect from the 1989 accounting year, provided it was approved by the Norwegian authorities. The concern thus created has a labour force of 5,300 and a combined annual production of 1.2m tonnes of paper and 370,000 tonnes of market pulp. Sales for 1989 were expected to reach about NKr8bn. Late in August, Norske Skog concluded an agreement of intent with Kosmos, a shipping and investment group, to buy the latter's paper factory, Saugbrugsforeningen.

Energy – oil and gas

Shortly before it recessed in summer 1989, Norway's Storting (parliament) approved a government scheme that will make an extra 3 TWh/yr of power available to industry, under ten-year contracts with Statkraft, the state electricity utility, at a price well below the cost of putting new electricity into the Norwegian grid. The plan was partly designed to create a more

COUNTRY PROFILE

Political system: Constitutional monarchy, with legislative power vested in unicameral Storting (parliament) currently comprising 157 members elected every four years by system of proportional representation. A change in the constitution approved in May 1988 will increase the number of representatives to 165 with effect from the 1989 general election. Executive power (nominally held by the Monarch) exercised by Council of State, which is led by the Prime Minister and is responsible to Parliament.

Main cities/towns: Oslo (population 453,730), Bergen, Trondheim, Stavanger, Kristiansand and Drammen.

Media
Press: Activity concentrated in south-eastern part of country, especially around Oslo where majority of papers are published. However, local papers tend to dominate each particular region.
Dailies: Mostly regional but *Aftenposten* (morning circulation 259,231; evening circulation 181,015), *Verdens Gang* (333,698), *Dagbladet* (198,937) and *Dagens Naeringsliv*, all published in Oslo, have national circulation. Important regional papers are *Bergens Tidende* (Bergen, 100,608), *Adresseavisen* (Trondheim, 87,718) and *Stavanger Aftenblad* (Stavanger, 65,803). No Sunday papers.
Weeklies: Large number of general and special interest magazines. Those aimed at women and the teens/twenties market have the largest circulations.
Business: Large number covering all aspects of Norwegian trade and industry, but most tend to have comparatively small, specialised readership. Includes *\17konomisk Rapport* (fortnightly) and *Farmand* (weekly).
Broadcasting: Of Norway's 1.5m households, approximately 1.45m have one or more radios and about 1.4m have one or more TV sets. New sets purchased annually total 500,000 radios and 180,000 TVs. All nationwide radio and TV programmes of domestic origin are broadcast by the Norwegian Broadcasting Corporation (NRK). A growing number of local FM stations have been operating since 1982, when NRK lost its monopoly. Since 1 May 1988 these stations have been allowed to braodcast advertisements, but only after securing a licence from the local authority ('kommune'). By end June, some 110 stations had been granted such licences, and the number was expected to rise considerably before the end of 1988. A US armed forces radio station broadcasts in the Oslo area, serving American forces stationed at NATO's northern headquarters. Many cabled foreign and satellite stations broadcast commercials. A growing number of households have their own 'dish' antennas, enabling them to pick up satellite broadcasts directly.
Advertising: All usual media available, although NRK still carries no advertising. Some politicians favour allowing NRK to carry commercials on a planned second TV channel. A decision on the issue could come late in 1988 or early in 1989. Total advertising spend (1987) NKr10.13bn, with cars, office equipment and travel the 'top three' categories, as regards total expenditure. Ban on advertising for tobacco and all types of alcohol except the very weakest kind of beer.

Domestic economy: Mixed economy. North Sea oil and gas still the largest single foreign currency earner, although its share of total export earnings has fallen steeply since the 1986 oil price collapse. Revenue from oil and gas continued to decline in 1987-88, but the 1987 payments deficit was smaller than in 1986, because of a marked rise in traditional exports, and a relatively modest rise in imports. 1988 showed a further improvement. Living standards remain among highest in Europe, despite belt-tightening measures since the minority Labour government's accession in Spring 1986. Current policies still dominated by need to stimulate exports and curb domestic demand, including public sector demand. However, the increase in unemployment caused a readjustment in this policy during 1982. A new temporary law regulating wage rises and dividends took effect on 22 April 1988. A temporary law took effect from 1 January 1989 and will remain in force until the end of the year. It is expected to hold the average overall wage increase at around 5 per cent. This is well below what it would have been if free bargaining had been allowed.

Employment: A third of the workforce employed in the public, social and private service sector, excluding banking and financial services, and also excluding the wholesale and retail trade and hotel/restaurant business, which employs about 18 per cent. Manufacturing industry (excluding mining, and construction), also employs about 16 per cent, and agriculture, forestry and fisheries, six per cent. Unemployment remains low by OECD standards. Average wage costs are high, but the 1988 wage and dividend regulation law could slightly narrow the gap between wage levels in Norway and in the countries with which it competes.

External trade: Norway depends heavily on foreign trade. Exports of oil and gas accounted for 33 per cent of total exports in 1988. Fish, paper, aluminium and ships also are important exports. Imports of road vehicles dropped by 29 per cent on a year earlier. Imports of second-hand ships rose by 238 per cent, and accounted for 10 per cent of total imports in 1988. This reflects the return of ships to Norwegian registration, following the establishment of the Norweigian International Ship Registry (NIS), which allows employment of cheaper foreign crews. For the same reason, exports of second-hand ships were down on a year earlier, because fewer shipowners were 'flagging out'. Other important imports are:

flexible domestic electricity market, but its main aim was to encourage industry – particularly power-intensive industry – to invest in new capacity.

Predictably, industrialists welcomed the new deal. Environmentalists and economists were less enthusiastic. The former said it was mainly designed to boost domestic power usage, in order to justify the construction of gas-fired power plants. Two prominent economists claimed it would represent a state subsidy to industry of around NKr130m/TWh.

For several years, Norwegian electricity output has exceeded demand – the consequence, some argue, of excessive hydro development in the 1970s and early 1980s, based on unrealistically high demand forecasts. In 1989 the supply glut was particularly acute, owing to heavy rain and snowfalls and the exceptionally mild winter of 1988–89. Exports in the first half of the year grew to 6.8 TWh – about five times the figure for the same period of 1988. Their value, however, merely trebled, to NKr350m, indicating the low price prevailing on the spot electricity market.

While Norwegian power-intensive industry enjoys electricity tariffs ranging from around NKr0.15/kWh to as little as NKr0.05/kWh, prices to households have been rising steadily over recent years, reflecting the high cost of servicing the huge, long-term loans raised to finance hydro-power plant construction. The average price paid by households in 1988 was over NKr0.40/kWh, including electricity tax and value-added tax. This has led many home-owners to switch to oil and paraffin for house-heating – thus further aggravating the supply glut. The government's budget for 1990 (due to be tabled in October 1989) was expected to include an 'environmental' tax on fossil fuels which might slow this trend.

Gas-fired power plants are far less capital-intensive than hydro, and this is the argument advanced by the government, and Norway's power-intensive industry, for building such plants in the second half of the 1990s. A scheme to pipe gas from Norway's offshore fields to Sweden, starting in 1995, could well depend on a decision to build a gas-fired generating plant in eastern Norway by then. Initial Swedish demand for Norwegian gas might not be enough to justify construction of the cross-country pipeline. A gas-fired plant in eastern Norway could boost throughput to the total – at least 5 bn cubic metres a year – believed necessary to make the line pay its way.

The reserves of gas on Norway's continental shelf far exceed those of oil. With oil being extracted at an ever-rising rate, opening up new markets for gas is a high priority goal. This explains official enthusiasm for the cross-country gas pipeline, despite its apparently shaky economics.

Outlook

In the run-up to the 11 September 1989 general elections, it was widely feared that the new Storting would be unable to mobilise a majority in favour of any party or coalition able to form a stable government. Opinion polls showed the country's two largest parties – Labour and the Conservatives – losing ground heavily to the two groups at either end of the political spectrum - the rightist Progress Party, and the radical Socialist Left Party (SV). At one point, Progress (which secured only two seats in the 1981 elections) seemed to be running neck-and-neck with the Conservative Party, until then indisputably Norway's second largest party. It managed to recruit voters from both Labour and Conservatives. Deserters from the latter were attracted by the party's liberalistic economic policies, its hard line on law and order, welfare state 'scroungers' and third-world immigrants claiming 'asylum'.

The immigration issue, in particular, won over many floating Labour voters. While politicians of the established parties initially avoided the topic, for fear of being labelled 'racist', Progress leader Carl Hagen repeatedly promised to stop the influx of 'fake' refugees.

SV won support – mainly at Labour's expense – partly because of its strong environmentalist platform and partly because many people who normally favoured Labour were tired of the middle of the road policies which the party had been obliged to pursue during its three-and-a-half years as a minority government.

Joker in the pack was the small Liberal Party, unrepresented in the Storting for the past four years. Following an increase in the number of MPs by eight to 165, and a revision of the voting system, aimed mainly at strengthening the position of the smaller parties, the party was expected to resume its place as part of the non-socialist team. This would then number five parties, compared with only four in the old Storting: Conservative, Progress, Christian Democrat and Centre (farmers).

These five looked set to secure a majority, between them – though few pollsters would predict just how many seats each would secure. But while all five agreed that a change of government was desirable, they did not agree on much else. The Centre party, representing rural interests, was bitterly opposed to Norwegian membership of the EC, warmly endorsed by the Conservatives. No party but Progress would accept all of that group's far-right ideas. The Centre and Christian Democrat parties favoured more cautious economic policies (less generous tax concessions) than the Conservatives.

There was every indication that the election would be followed by a prolonged period of inter-party negotiation and manoeuvering – probably with the minority Labour government staying in office, on a 'caretaker' basis, while the non-socialists sorted themselves out.

COUNTRY PROFILE

manufactures of metals, machinery, and ready-made clothing/textiles. Major trading partners are Sweden, West Germany and the UK.

Agriculture: Sector contributes around 1.6 per cent to GDP. Majority of farms consist of smallholdings. All farmers benefit from large subsidies and are protected by import controls. Grain and fodder are main lowland crops; mountain farms mainly raise livestock and grow fodder. Some dairy produce is exported and there is self-sufficiency in meat, milk, cheese, butter, fish and potatoes. Forestry often combined with farming. Fishing is a significant industry in the northern and western regions; fish, including farmed fish, is an important export. Mackerel, cod and capelin are main species caught, but catches have been falling, because of over-fishing. Fish farms product mainly salmon and trout, although some are experimenting with other varieties, such as turbot.

Industry: Sector contributes around 15 per cent to GDP and accounts for 33 per cent of exports. Production centred on traditional industries, which are based on local raw materials or exploitation of cheap bydro power, and essentially export-orientated industries such as chemicals, oil and gas, metals, forestry products and electronics. An important sector is the offshore-related engineering industry – mainly comprising shipyards which switched to building oil platforms and oil platform equipment during the prolonged world shipbuilding slump. The government's desire to maintain employment in this industry often significantly affects policy decisions concerning petroleum exploration and development.

Mining: Activity confined to small-scale mining of iron ore, copper, titanium, coal (on Spitsbergen), zinc, lead and pyrites. Most of these ores and concentrates are exported. Mining continues to contract, causing many problems in areas where there is no alternative employment.

Energy: Major European oil and gas producer and exporter. Norway's share of output from Norwegian and Anglo-Norwegian fields (including Ekofisk, Frigg, Statfjord, Oseberg, Gullfaks, Murchison and Valhall) reached 86.4m tonnes of oil equivalent in 1988. Including pipeline transportation, this represented a gross production value of NKr51.8bn, corresponding to about nine per cent of the GDP. Oil production has been rising, as new fields come on stream, despite government-enforced curbs on the rate of increase, aimed at helping OPEC support world price levels. Also ample power supplies (virtually all hydro), some of which is exported (to Sweden and Denmark). Plans to build gas-fired power plants.

Banking: Well developed and efficient system headed by the central bank (Norges Bank). Around 30 commercial banks, of which Den norske Creditbank, Christiania Bank og Kreditkasse and Bergen Bank are the largest. Also over 300 savings banks and a small number of foreign bank subsidiaries. Almost all banks had heavy losses in 1987-88 in the wake of runaway credit book in 1984-86. In 1987 alone losses amounted to NKr6bn, around 1 per cent of assets, and the losses are likely to continue at least until 1990.

Legal system: Mixture of customary law, civil law and common law traditions. Constitution adopted 1814, modified 1884, now under review.

Membership of international organisations: ADB, CCC, Council of Europe, DAC, EC, EFTA, ESRO (observer), FAO, GATT, IAEA, IBRD, ICAC, ICAO, ICES, ICO, IDA, IEA (associate member), IFAD, IFC, IHO, ILO, IMF, IMO, INTELSAT, INTERPOL, International Lead and Zinc Study Group, IPU, ITU, IWC, NATO, Nordic Council, OECD, UN, UNESCO, UPU, WHO, WIPO, WMO, WSG.

Time: GMT + 1 (GMT + 2 from late Mar to late Sep).

Climate: Temperate in west, colder inland and in north. Average temperatures range from about 17°C down to about -10°C. Lowest recorded temperature is Karasjok with -51°C. The winters are cold and dark with snow and ice Nov-Mar. Summers warm Jun-Aug. Rain throughout year. West coast wet. Jan and Feb coldest months. North of the Arctic circle the sun shines day and night for part of the summer, with a corresponding period of total darkness in the winter.
Clothing: Mediumweight all year, heavy topcoat and overshoes for winter. Most buildings centrally heated so indoor clothing not too heavy. Some restaurants require that men wear a jacket and tie.

Entry requirements: Visas not required by nationals of most European countries and USA. Nationals from non-Nordic countries must have valid passport.
Health: Vaccination certificates not normally required.
Currency: No limit on currency taken in. Up to NKr5000 in Norwegian currency may be taken out.
Customs: Personal effects duty-free, plus duty-free allowance.

Air access: Regular direct flights by most major international airlines.
National airline: Scandinavian Airlines System (SAS), jointly owned with Sweden and Denmark. Private- and government-owned.
Main international airport: Oslo served by Fornebu (FBU), 10 km from city and Gardermoen (GEN), 51 km from city.
International airports: Bergen (BGO) and Stavanger (SVG).

Surface access: Frequent ferry services from many European ports. Also road and rail access from Denmark, Sweden and Finland.
Main ports: Bergen, Stavanger, Kristiansand, Oslo.

Hotels: Wide range available in most towns, but no official rating system in operation. Advisable to book well in advance, especially during tourist season. Private accommodation can also be obtained through local tourist offices or accommodation offices in central railway stations. Luxury single room, including bath and breakfast, costs at least NKr800 per night. Service charge of 15 per cent included in bill, but tipping also expected.

Restaurants: Good selection in larger towns. Seafood is Norwegian speciality. Service charge added to bill, but five per cent tip is usual.

Currency
Exchange rate: Norwegian krone is pegged to a basket of currencies, including US dollar, Canadian dollar and Japanese yen. Fluctuations in rates within basket kept within certain limits.

Foreign exchange information
Account restrictions: Non-resident accounts permitted, but nature must be confirmed in writing to local bank holding the account.
Trade policy: Norway is export-oriented country, so few restrictions. Foreign currency for payment of most goods purchased abroad is subject to import licence.

Credit cards: All usual credit cards accepted.

Car hire: Available at airports and in major towns. Rate for medium-sized car (covering 800 km) about NKr3000. For travel between towns public transport tends to be quicker, and much cheaper. Studded or winter tyres recommended during winter. Very strict laws against drinking and driving. 50 kph speed limit in built-up areas, 80 kph on highways. Seatbelts compulsory.

City transport
Taxis: Available in most cities. Can be obtained at ranks or by telephone (Oslo 388090, Bergen 90 0990, Stavanger 526040). Telephone numbers of taxi stands are listed in the directory under 'Drosjer'. Rates in Oslo about NKr16 plus NKr6 per km. Meters are compulsory. It is not expected that the tip will be more than the small change.

National transport
Air: Efficient services operated by SAS and others, linking Alta, Alesund, Andenes, Bardufoss, Bergen, Bodo\18, Farsund, Kirkenes, Kristiansand, Kristiansund, Lakselv, Molde, Oslo, Roros, Skien, Stavanger, Tromso and Trondheim. Aircraft also available for charter.
Road: Approximately 86,800 km of roads, including 500 km of motorway. Several roads in mountainous regions are open only during summer. Good bus service, but journeys tend to be long owing to distances between main towns.
Rail: Over 4,200 km of lines operated by Norwegian State Railways. Main lines radiate from Oslo. Good service, but advisable to make seat and sleeper reservations well in advance through State Railways Travel Bureau or other travel agencies.
Water: Regular efficient motor ship services visiting all major ports. Also numerous local ferry, hydrofoil and catamaran services.

Public holidays
Fixed dates: 1 Jan, 1 May, 17 May (National Day), 24-26 Dec.
Variable dates: Maundy Thursday, Good Friday, Easter Monday, Ascension Day and Whit Monday.

Working hours
Business: (Mon-Fri) 0830 – 1600.
Banking: (Mon-Fri) 0815 – 1530, (Thu) 0815 – 1700.
Shops: (Mon-Fri) 0830 – 1700, except (Thu) 0830 – 1900, (Sat) 0830 – 1400.

Social customs: Punctuality is expected. Send flowers to hostess when visiting a Norwegian home for the first time. Shake hands on meeting. Business lunches are rare; for business dinners Norwegian men often bring their wives. The main meal of the day is generally taken at home at 1700 hours, though people will expect to eat later if invited out.

Telecommunications
Telephone: Automatic system offering direct dialling to all foreign cities. Cheap off-peak rates in evening and at weekends. International dialling code: 47 followed by 2 for Oslo, 5 for Bergen, 7 for Trondheim and 4 for Stavanger.
Telex and telefax: Further information available from telegraph authorities in Oslo (tel: 488990) and from Telegrafbygningen near the city park in Bergen. Country code 56N.
Telegram: Available from post offices, or can be dictated over the telephone.

Trade fairs: Several international events held annually, most at Sjølyst in Oslo, Forum in Stavanger and Info-Rama in Sandvika. All major sectors of industry are covered, but emphasis is on oil, shipping and fishing. Most exhibitions are organised by Norges Varemesse (the Norwegian Trade Fair Foundation) based at Sjølyst. International Fisheries Fair, Aug, Trondheim; Offshore Northern Seas Technology, Aug, Stavanger.

Electricity supply: 220 V AC.

Representation in capital: Argentina, Austria, Belgium, Brazil, Bulgaria, Canada, Chile, China PR, Colombia, Cuba, Czechoslovakia, Denmark, Ecuador, Egypt, Finland, France, Democratic Republic of Germany, Federal Republic of Germany, Greece, Hungary, Iceland, India, Indonesia, Iran, Israel, Italy, Japan, Democratic People's Republic of Korea, Republic of Korea, Mexico, Netherlands, Panama, Poland, Portugal, Romania, Spain, Sweden, Switzerland, Turkey, United Kingdom, USA, USSR, Venezuela, Yugoslavia.

Hotels: Oslo
Akershus (a ship in the harbour), Akershuskaia, N-0150 (tel: 428660; tx: 76858).

Anker, 55 Storgate 0182 (tel: 114005; tx: 72704; fax: 110136).

Bellevue, Nesbru N-1360, P.O. Box: 10 (tel: 785060).

Bristol, Kristian 4 gt 7 (tel: 415840; tx: 71668).

Continental, Stortingsgt 24-26 (tel: 419060; tx: 71012; fax: 429689).

Europa, St Olavsgate 31 (tel: 209990; tx: 71512; fax: 112727).

Grand, 31 Karl Johansgate N-0159 (tel: 429390; tx: 71683; fax: 421225).

Helsfyr, 108 Stromsveien (tel: 654110; tx: 76776).

Imi, 4 Staffeldsgate, P.O. Box: 6830 (tel: 205330; tx: 78142; fax: 111749).

Munch, Munchsgate 5, N-1065 (tel: 424275; tx: 74096).

Ritz, Fr Stangsgt 3, N-0272 Oslo 2 (tel: 443960; tx: 19668).

SAS Park Royal, N-1324 Lysaker (tel: 120011; tx: 78745).

SAS Scandinavia, Holbergs Gate 30, N-0166 (tel: 113000; tx: 79090; fax: 113000 EXT. 400).

Sara, Biskop Gunnerus' Gate 3, P.O. Box: 768 (tel: 429410; tx: 71342; fax: 424622).

Triangel, Holbergs Plass 1 (tel: 208855; tx: 19413).

Car hire: Oslo
Avis: Munkedamsveien 27 (tel: 410060; tx: 76815, 76342).

Budget: (tel: 331709; tx: 72074).

Hertz: Fornebu A/P (tel: 533647; tx: 78429).

Inter-Rent: Postboks 62, 1473 Skarer (tel: 700400; tx: 19139).

Chambers of Commerce
Bergens Handelskammer, Olav Kyrresgt 11, N-5000 Bergen (tel: 316569).

Kristiansands Handelskammer, Østromdg 39, N-4600 Kristiansand 5 (tel: 24370).

Oslo Handelskammer, Drammensveien 30, Oslo 2 (tel: 557400).

Stavanger Handelskammer, Kongsgaten 10, N-4000 Stavanger (tel: 536035).

Trondheim Handelskammer, Dronningensgt 12, N-7000 Trondheim (tel: 520625).

Banking
Norges Bank (Bank of Norway) (central bank), Bankplassen 2, PO Box 1179, Sentrum, 0107 Oslo 1 (tel: 316000; tx: 71369).

ABC Bank (Union Bank of Norway), Kirkegaten 14-16-18, PO Box 1172, Sentrum, N-0107 Oslo 1 (tel: 319050; tx: 19470).

Bergen Bank, Torgalmenning 2, PO Box 826, N-5002 Bergen (tel: 211000; tx: 42018) and Kirkegaten 23, PO Box 1170, N-0153 Oslo 1 (tel: 400550; tx: 71069).

Bergens Skillingsbank AS, Rådstuplass 4, N-5001 Bergen (tel: 210700; tx: 42082).

Christiania Bank og Kreditkasse, PO Box 1166 Sentrum 0107 Oslo 1 (tel: 485000; tx: 71043).

Den Norske Creditbank, Kirkegaten 21, N-0107 Oslo 1 (tel: 481050; tx: 78175).

Fokus Bank A/S, Vestre Rosten 77, PO Box 6090, 7003 Trondheim; (tel: 882011; tx: 55050).

Rogalandsbanken AS, Haakon VII Gate 8, PO Box 209, N-4001 Stavanger (tel: 526080; tx: 33023).

Principal newspapers
Adresseavisen, PO Box 6070, 7003 Trondheim (tel: 884411; tx: 55220).

Aftenposten, Akersgt 51, PO Box 1178 Sentrum 0107, Oslo 1 (tel: 205060; tx: 71230).

Bergens Tidende, PO Box 875, 5001 Bergen (tel: 214500; tx: 42026).

Dagbladet, PO Box 1184, Sentrum 0107, Oslo 1 (tel: 202090; tx: 11020).

Stavanger Aftenbild, PO Box 229, 4001 Stavanger (tel: 501600; tx: 33114).

Verdens Gang, PO Box 1185, Sentrum 0107, Oslo 1 (tel: 114040; tx: 71306).

Other useful addresses
AS Norsk Telegrambyraø (Norwegian news agency), Holbergsgt 1, Oslo 1 (tel: 201670; tx: 11046).

Ministry of Finance, Akersgt 42, PO Box 8008 Dep, 0030 Oslo 1 (tel: 349090; tx: 21444).

Ministry of Industry, Akersgt 42, PO Box 8014 Dep, 0030 Oslo 1 (tel: 349090).

Norges Eksportrød (Export Council of Norway), Drammensveien 40, N-0255 Oslo 2 (tel: 437700; tx: 78532).

Oman

Tim Owen

The cessation of hostilities between Iraq and Iran removed a long-standing anxiety for Oman, particularly as regards the Strait of Hormuz, where the main shipping lanes pass through Omani territorial waters. Oman's policy of maintaining a good and balanced relationship with both contestants undoubtedly paid off. In 1989, Sultan Qaboos went on an extensive tour of visits to Arab countries, as well as France, Spain and the UK. In spite of low oil prices in late 1988 and early 1989, the economy was maintained on a steady course and Oman's credit rating remained high.

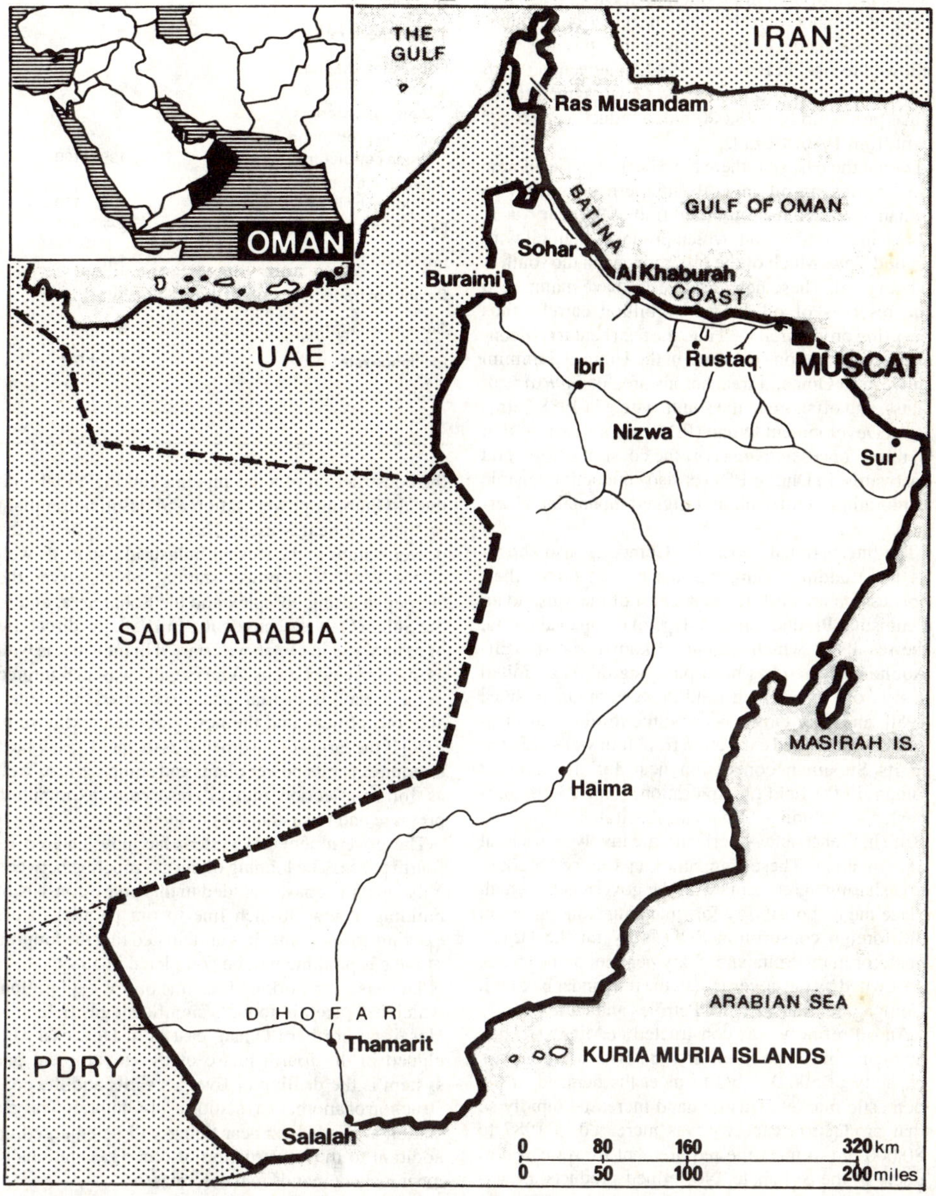

Emergence from a century of isolation

The Sultanate of Oman, strategically positioned on the flank of the Indian Ocean and Arabian Gulf trade routes, has a long maritime tradition. It also has a history of internal turbulence and has been a prey to foreign aggression and occupation ranging from the Persians to the Portuguese. Following a 150-year occupation of Muscat and the coastal areas by the Portuguese, the Omanis drove out the Portuguese, pursued them across the Indian Ocean and expelled them from their East African possessions. They established an Omani empire based on Zanzibar, which reached its zenith in the first half of the nineteenth century under the leadership of Sultan Sayyid Said bin Sultan (Said the Great). On his death in 1856, the succession was disputed between two of his sons, as a result of which the East African possessions were separated from Oman under the rule of one of the contestants. The loss of the East African possessions and their revenues, coinciding with the advent of the steamship, which lost Muscat its role as a haven and a port of replenishment, reduced Oman to a state of penury, which lasted for a century.

In 1964, oil was discovered in commercial quantities in the Sultanate and went into production in 1967. Following the ousting of his father, Said bin Taimour, in 1970, Qaboos bin Said acceded as Sultan. It was from this date that the remarkable renaissance and development of Oman was started and the Sultanate re-emerged as a power which has achieved international recognition. When visiting Oman today for the first time, visitors should appreciate that 19 years ago modern development consisted of one hospital, three boys' schools of primary standard and 8 km of paved road.

Economy largely dependent on oil

It was the discovery of oil in commercially viable quantities in 1964 which sparked development in the Sultanate, which really got under way in 1970. Since then the economy of the country has been overwhelmingly based on the production and export of oil. Oman was fortunate, in that during the early years of oil production in the 1970s the world price of oil rocketed to heights previously not envisaged, and remained there until 1986. Since then the fluctuations in the price of oil have been considerable, with their inevitable impact on the economy of the country.

Due to the fragmented nature of the geology of Oman, oilfields are small but numerous throughout the interior of the Sultanate. Production costs are therefore high compared with some other oil-producing countries in the region. Again, for climatic and geographical reasons, all oil production has to be fed into the main pipeline which passes underground through the Hajar mountains to the terminal at Mina al-Fahal on the north coast near Muscat. Lengthy pipelines have therefore had to be constructed throughout the interior. However, Oman does have one important advantage over the other Gulf States, which are dependent on passage through the Strait of Hormuz for the export of their oil. This has been particularly important over the last eight years, in that its oil terminal at Mina al Fahal lies outside the Strait of Hormuz in the Gulf of Oman.

Oman is not a member of OPEC or OAPEC, but has always kept the price of its oil in line with that of OPEC. It is not subject to quotas in the volume of its

KEY FACTS

Official title: Sultanate of Oman
Head of state: Sultan Qaboos bin Said
Capital: Muscat
Official Languages: Arabic
Currency: Rial (RO) = 1,000 baiza
Exchange rate: RO0.38 per $ (Nov 1989)
Area: 320,000 sq km (including Kuria Muria Islands)
Population: 1.38m (1988)
GNP per capita: $5,780 (1987)
GNP real growth: –1.6% (1986/87)
Inflation: 0.0% (1987)
Oil reserves: 4bn* barrels (1988)
Trade balance: $439m (1988)
Foreign debt: $3,812m (1987)

* estimated figure

oil production, which has worked to its advantage in the past, when by increasing production Oman was able to compensate for the fall in price. Now, however, in line with several other non-OPEC producers of oil, Oman does voluntarily reduce the volume of oil production, as necessary, to help maintain the world price of oil. Oil production in Oman reached a peak of 600,000 barrels a day (b/d) in 1986. Reductions were made in three stages in 1988, but then allowed to rise. In February 1989, in agreement with other non-OPEC oil producers, Oman agreed to reduce production by 28,000 b/d, a reduction of 5 per cent from 1 April 1989.

During the past year there have been significant new discoveries of oil in both southern and northern Oman, but the most important find of the year was at Rajaa in central Oman, which produces light oil with a good flow. Much of the oil produced in the south is a heavy oil. These new discoveries have maintained the reserves of oil at over 4 billion barrels, thus ensuring production well into the next century. Meanwhile, exploration continues in the Umm al Sammim in western Oman, a treacherous area of *sabhka* (salt flats), and offshore in the south, where in 1988 Petroleum Development Oman (PDO) started exploration in three concession areas off the coast of Dhofar and eastwards. In Dhofar PDO is also conducting seismic exploration in difficult and rugged mountainous terrain.

The International Petroleum Company, also known as the Lundin Group, has discovered oil in their concession area off the west coast of the Musandam Peninsula. Production will depend on an evaluation, the results of which should be known shortly. Elf-Aquitaine Oman has been pumping oil on a limited scale from its Sahmah field in western Oman since 1980, and Oxy Oman became the third producer in 1984, with limited extraction from four wells at Safah in its Sunainah concession near Ibri in northern Oman. In the field of exploration, some 23 oil companies – including Canadian, Australian, Japanese, British, French and American – are involved in a total of eight areas. These companies have signed production sharing agreements with the government, which place the responsibility for financing exploration on the foreign consortium. PDO is by far the largest operator in the Sultanate. Sixty per cent of the shares are owned by the government, the remainder by Shell, Compagnie Francaise des Petroles and Partex.

An oil refinery was constructed at Mina al-Fahal, and came into operation in 1982 with a production capacity of 500,000 b/d to meet the demand of the domestic market. This demand increased rapidly so that production capacity was increased in 1987 to 80,000 b/d to meet the requirement for refined products in the Sultanate. No refined products are exported.

Natural gas: power for industry

Oman has large reserves of natural gas, which are playing an increasing role in the development of the country and its industries. Demand has been increasing rapidly and output has increased correspondingly. Production almost doubled in 1987 compared with the previous year, and in 1988 reserves of associated and non-associated gas were estimated to be 9.3 trillion cu ft.

The natural gas system is wholly owned by the State and is operated by PDO. In order to increase reserves to meet industrial demand well into the next century, the government has financed a PDO gas exploration programme which started in 1984. The programme has resulted in confirmed discoveries of major quantities of gas, notably at Yibal in the Khuff formation below the earlier developed oilfield, where significant quantities of gas were found as well as, coincidentally, an oil reservoir.

The government gas system was started in 1978 with the construction of a gas pipeline linking Yibal with the al-Ghubrah power station and desalination plant on the coast near Muscat. Gas is also supplied to the power station in the Rusayl Industrial Estate, as well as the Wadi Jizzi power station, which serves the copper smelting complex and the Sohar area, and to the Sultan Qaboos University. In 1988, the direct use of natural gas in Oman was started in industry as well as in the power stations. A gas supply was inaugurated in the Rusayl Industrial Estate to be used by a vegetable oil refinery and the newly-opened textile mill. PDO is the largest consumer of associated gas, which it requires for its power plants and equipment, as well as for re-injection into oil reservoirs to maintain pressure and to sustain oil production.

The government gas system has now entered its fourth phase. The looping of the gas pipeline between Yibal and the coast included in this phase consists of building a new 36-inch line to run parallel to the existing 30-inch line. It is anticipated that the looping of the gas pipeline will be completed in 1989.

PDO has a rig working full-time on gas exploration, which has recently found a significant gas field at al Husain in northern Oman, east of al Huwaisah. Included in the fourth phase of the government gas system is the drilling of five gas wells and the construction of another gas station at al-Ghubrah. The gas station at Wadi Jizzi near Sohar will be enlarged, in addition to the construction of other gas stations to meet local demands in these areas.

KEY INDICATORS

	Unit	1984	1985	1986	1987	1988
Population*	'000	1,190	1,250	1,290	1,330	1,380
Gross domestic product (GDP)	RO m	3,046.7	3,456.6	3,187*	–	–
GDP per capita	RO	2,561	2,771	2,470*	–	–
Time deposit rate	%	9.03	9.04	8.33	7.48	7.57
Price index changes						
Consumer prices	%	–7.9	–1.1	6.4	–	–
Building materials	%	–10.7	–10.8	2.1	–	–
Industrial production						
Crude petroleum	1985=100	84.8	100.0	110.7	111.4	116.3
Volume of exports						
Crude petroleum	1985=100	81.9	100.0	113.8	119.6	–
Exports	$m	3,932	4,360	2,671	3,168	3,174
Imports	$m	2,748	3,153	2,398	1,822	2,735
Balance of trade	$m	1,184	1,207	273	1,346	439
Current account	$m	300	265	–1,017	851	–
Exchange rate	$ per RO	2.90	2.90	2.60	2.60	2.60

* estimated figure

Emphasis on agriculture

Historically, the economy of Oman was based on agriculture and fishing, and was largely a subsistence economy. Indeed, although the economy is now based on the production of oil and gas, the majority of the population still work on the land or are engaged in fishing, in spite of a drift into the urban areas for higher wages when development got under way. In 1967, the year that oil first went into production, agriculture represented 34 per cent of GDP, but by 1988 it had fallen to 3 per cent.

The government has always been aware from the start of the development programme that in the long term agriculture must play an important role in the economy of the country. Every effort has been made to limit the drift from the land by providing incentives to farmers in the form of subsidies and other assistance. The success of the government's policy is now beginning to become apparent. 1988 was proclaimed by Sultan Qaboos to be the year of agriculture, when a particular drive was to be made to encourage farming by means of a number of different schemes. This proved to be so successful that it was decided to extend the programme through 1989.

Oman has basically an arid climate, but its size – it is the second largest country on the Arabian Peninsula – and topographical variations produce considerable local contrasts in climatic conditions. The population of rather less than two million is unevenly distributed as a result, the most densely populated areas being in the north and the extreme south, with large areas of virtually uninhabited desert in between. Northern Oman has a climate similar to the Gulf, with a long and very hot summer which is practically rainless, and a brief mild winter when any rain is most likely to fall. Rainfall is sporadic, with considerable annual variations, and often comes in the form of violent local storms. The average annual rainfall along the north coast of Oman is about 4 ins. Inland there is a considerably greater rainfall over the Hajar range of mountains, which run parallel to the coast. Along the Batinah coastal plain in the north of Oman there is a belt of intensive cultivation based on groundwater irrigation. Dates and limes are grown in large quantities, as well as other semi-tropical fruits and vegetables such as tomatoes, onions and aubergines and small quantities of tobacco for local consumption. On the high mountain plateau of the Jebel Akhdar in the Hajar range, peaches, apricots and walnuts are cultivated as well as temperate climate vegetables. The coastal plain and seaward facing slopes of the mountains in Dhofar Province in the extreme south have, by contrast, a monsoon climate, with regular rains falling between June and September. Coconuts, bananas, papayas and other tropical fruits and vegetables are grown intensively on the coastal plain, whilst the mountain slopes behind are a cattle rearing region.

The government subsidises farmers by providing fertilisers, seeds, pesticides and farm machinery at low cost, as well as free advice. The policy is to increase output from existing farmland by more efficient farming methods and higher quality produce. At the same time, efforts are being made to increase the area of land under cultivation. The main limiting factor is lack of fresh water. Whilst water for irrigation along the coastal plain is produced from wells, in the interior it is obtained by the ancient *falaj* system, the origins of which go back nearly a thousand years. The system is based on tapping the water table in the mountains and leading the water by subterranean man-made channels, often 10 km or more in length, to the plains where it is used for irrigation and domestic purposes. Many *falajs* have fallen into disuse over the years for various reasons, usually due to blockages along the channels, so the government is engaged in a drive to repair them and thus bring more land into cultivation. The latest experiment to conserve water is the construction of recharge dams, started in 1985, which has already proved to be successful. Dams are constructed across wadi beds to hold back floodwaters after storms in the mountains, which would otherwise have to run to waste in the sea or the desert. The water sinks into the ground to recharge the subterranean aquifers, from which water is drawn for irrigation purposes.

The government has the long-term aim of making Oman self-sufficient in food production. To this end, a Public Authority for Marketing Agricultural Produce (PAMAP) was established in 1981 to provide a market outlet for Omani farmers. The organisation is proving to be particularly successful. In the same year, an Agriculture and Fisheries Bank was established to the same end.

Rich fisheries ripe for development

The Sultanate of Oman has a 17,000 km-long coastline extending from the Strait of Hormuz in the north to the border with the People's Democratic Republic of Yemen in the south. Fishing has always been a traditional occupation of the inhabitants of the coastal regions, but it is only in recent years that the potentialities of the rich offshore fishing grounds have been fully appreciated. Extensive surveys have been carried out in recent years to assess these potentialities, and it has been estimated that the annual yield could be as high as 300,000 tonnes, which would be sustainable without exhausting the growth and reproductive capacities of the rich variety of fish in these waters. The recently opened Marine Science and Fisheries Centre near Muscat has identified more than 150 species of fish in Omani waters.

In the early days of development, many fishermen migrated to the urban areas for the same reasons as the farmers. Much has been done by the government to encourage fishermen to remain in their traditional occupation. In 1976, the Fishermen's Encouragement Fund was introduced, whereby fishermen can apply for a grant which on an average covers two-thirds of the cost of a fibreglass fishing boat and an outboard engine. This has proved to be highly successful. Large sums have been invested in constructing refrigeration and ice plants, landing sites and harbours, refrigerated lorries, and on development studies. A new Fisheries Company of Oman has just been launched with a share capital of RO10m, into which the Oman National Fishery Company has been absorbed. A Korean Company has had a concession since 1977 in the deeper offshore waters. It employs Omani fishermen on its trawlers, and 38 per cent of its catch goes to the government as a form of royalty. Korea is the biggest importer of Omani fish, whilst other major importers include Saudi Arabia, the UAE, Japan and the other Gulf States.

Light industry is developing fast

One of the major long-term aims for the development of modern Oman has been the expansion of light industry, which was non-existent before 1970. It is the intention to evolve a national economy based on private enterprise, free from monopolistic practices, and preferably using indigenous materials as far as possible. Considering that Omanis had no previous experience of developing and running light industry, it is remarkable, after a hesitant start, how much they have achieved in less than 20 years.

By the beginning of 1976, at the start of the first Five-Year Plan, there were only ten industries in production. By the end of 1987, the contribution of the manufacturing sector in the economy had risen to RO119.1m from RO2.1m in 1975, representing 3.71 per cent of GDP, and employing a labour force of 25,000. In the first half of 1988, 114 industries with an investment of nearly RO7m were registered. The government provides incentives to local industry through tax reliefs, customs exemptions, tariff protection, interest-free loans and provision of planned and serviced industrial plots for the establishment of factories and workshops. The first phase of a 100 ha fully-serviced industrial estate comprising 125 plots was completed at Rusayl near Muscat in 1982. Factories already on the site comprise a wide variety of industry ranging from cement production and a marble factory to tile production, tea-packaging, air-conditioners and carbonated drinks. An aluminium plant on the estate with a 6,000-tonne annual capacity was opened in 1988. Four more industrial estates in other regions of the country are due to be opened in the near future, two of them at Sur and Raysut in 1990.

The Oman Chamber of Commerce (OCCI) was set up in 1983 to encourage industrial and commercial development in the Sultanate. All business organisations in Oman are expected to be members. With its headquarters in the capital, the OCCI has six other branches in the country. A Stock Exchange was opened in the middle of 1989.

Minerals, a potentiality

The Sultanate of Oman is rich in mineral deposits, particularly in the mountainous regions of northern Oman. It is known that copper ore was being mined 3,000 years ago and shipped to the Sumerian empire

COUNTRY PROFILE

Political system: Absolute monarchy where executive power, wielded by Council of Ministers, is derived from the Sultan. A 55-member State Consultative Council was formed in 1981 to increase participation of the Omani people.

Population: Most of the population (approx 1.4m in 1988) lives on the Batinah Coastal Plain, where 40 per cent of the farming land is found. Urban populations, however, are growing at some 17 per cent a year (compared with a national increase of 4.2 per cent), because of a drift away from agriculture. The Ibadhi sect of Islam is dominant.

Main cities/towns: Capital/commercial area of Muscat/Ruwi/Muttrah/Qurum (population estimated 280,000), Nizwa (10,000), Salalah (10,000), other main towns are Sohar, Rustaq and Sur.

Language: Arabic. English is often used in business circles. Some Swahili.

Media
Press: There are two English-language newspapers: the *Oman Daily Observer* (circulation 8,775), and the *Weekly Times of Oman* (12,500); and several Arabic language papers.
Radio: Broadcasts 19\55 hrs/day in Arabic, 14 hrs/day in English.
Television: Broadcasts eight hrs/day, with details of English programmes given in English programme on Oman radio. There were 110,000 sets* in 1988.

Domestic economy: Until oil production began in 1967, the economy was based on subsistence agriculture and fishing. Under the Second Five-Year Plan (to 1985) crude oil output was raised to 450,000 b/d, and emphasis placed on diversification of the economy, improving infrastructure and encouraging private sector activity. Some projects, including the Salalah sewage scheme, Wadi Dayqah dam near Qurayat and extensions to the port at Mina Qaboos, carried forward into the Third Plan (1986-90), which continues to emphasise private-sector import substitution industries, expansion of social services and improvement of agriculture and fisheries. The government, whose budget determines the level of domestic activity, relies on oil for 86 per cent of its revenue, but it is weighed down by heavy defence spending which accounted for 36 per cent of government expenditure in 1987. Saudi Arabia, Kuwait and the United Arab Emirates have been sources of aid, and the formation of the GCC made these countries even more important sources of finance. In 1987-88 the economy started to suffer the effects of increasingly difficult conditions in the Gulf, and a number of major intrastructural projects had to be postponed.

Employment: The first official census was undertaken in 1987. The total labour force is estimated at 410,000, of which 60 per cent (about 295,000) are expatriates, mainly Indians and Pakistanis working in construction, trade and agriculture; but their numbers are declining as construction activity slowed. In some sectors expatriate employment was being discouraged. Local labour force is about 120,000, with 22.5 per cent in the private sector, 16 per cent in the public sector and the remainder in traditional occupations. There is an overall shortfall of 8,000 technically qualified workers from within the indigenous population. Fishing employs 10 per cent of workforce, and 70 per cent of population relies on agriculture for livelihood, mostly at subsistence level.

External trade
Exports: Petroleum accounts for 97.0 per cent of all exports (1985) and re-exports for another 2.0 per cent. Main destinations for oil are Japan (66.3% of 1985 sales), Republic of Korea (14.1%), Thailand (6.7%), India (3.8%), Singapore (2.9%), Taiwan (2.7%), France (1.1%).
Imports: Main imports are machinery and transport equipment (39.2% of 1985 imports), other manufactures (20.7%), food and live animals (10.7%), chemicals (3.7%), beverages and tobacco (1.5%), fuels and lubricants (1.5%) and other crude materials (1.3%). Main sources are United Arab Emirates (19.8% of the 1985 total), Japan (18.9%), UK (15.4%), German FR (7.3%), USA (5.3%), France (3.5%), Netherlands (2.9%), and India (2.1%).

Agriculture: Agriculture and fisheries accounted for 3 per cent of GDP in 1988, compared with 34 per cent in 1967, when oil first went into production. Main crops include dates, alfalfa, lucerne, wheat, mangoes, limes and bananas, with market gardening of tomatoes, cabbages, eggplant, okra and cucumbers. Irrigation studies show potential arable land to be twice that presently under cultivation. Fishing, a traditional industry, has been identified as a development area, with government investment in onshore processing and refrigeration plants, harbours and repair yards.

Industry: Industrial activity is limited by small domestic market and lack of skilled manpower, and manufacturing accounted for only 6.4 per cent of GDP in 1986. Import substitution projects include two cement works (Oman Cement Company and Mina Raysut Cement Company) and the Dhofar Cattlefield Company (replacing cattle feed imports from India). Other important industrial plants are an oil refinery, an aluminium factory and a copper smelter. The 1986-90 plan aims to encourage small and medium-sized industry; currently there are factories producing spark plugs, paper bags, electric wire and cable. An industrial estate is established at Rusayl, near Muscat, and others were due to open in 1990 at Sur and Raysut. A stock exchange was opened in 1989.

at the northern end of the Gulf. Now the same deposits are being mined again, in the Wadi Jizzi near Sohar. Three mines are in production, and the ore is refined at a neighbouring smelter and exported as copper cathodes. Over 15,000 tonnes of cathodes were exported in 1987. Chromite is also being extracted from the same mines, and is being exported at a rate of 7,000 tonnes annually. Further deposits of copper and chromite have been discovered and their commercial potentialities are being assessed.

Much remains to be done by detailed geological mapping, drilling and evaluation to assess the potentialities of manganese, nickel, iron and coal which are known to exist in the Sultanate. Many of these deposits are in remote mountainous regions, where there are at present few road communications. The mineral potentialities of Oman therefore are still to be fully explored and assessed.

Outlook

The economy of Oman is still 95 per cent dependent on the production of oil and gas, and recent discoveries have increased reserves. Revenues from oil production have been extended well into the next century. Nevertheless, the government has always been aware of the need to diversify the economy and has made great efforts in this respect with considerable success. Fluctuations in the world price of oil in the last three years have, if anything, emphasised the need to diversify the economy. Agriculture, fisheries and light industry offer the best avenues for maximising the diversification of the economy. The exploitation of mineral resources might also provide another contribution, but this as yet is uncertain.

In the autumn of 1988, financial planning for the intermediate term was reassessed and certain policy adjustments were adopted. Amongst other things, the planning price for oil was reduced provisionally from $15.00 to $12.50 a barrel, and selected development projects were moved from the 1989-90 period to the period starting in 1991. Other measures taken include accelerated plans for the privatisation of government assets to mobilise resources and encourage private initiative, renewed use of reserve funds, and the sale of government securities in the domestic market.

This revised policy stance is reflected in the 1989 budget. Government oil revenue is forecast to be RO904.4m, as compared with RO1,065.3m in the 1988 budget. On the other hand, total government outlay, including all spending, lending and equity transactions, is slightly higher than in the 1988 budget at RO1,617.6m, as compared with RO1,602.3m in 1988. Thus the budget deficit expected for 1989 is larger, at RO407.6m, than the budgeted deficit of RO252m in 1988. Since $12.50 a barrel is a conservative target, it is reasonable to assume that the actual outcome for the 1989 deficit may well be better than the budget figure suggests. Now that skilled young people are emerging from the top end of the education system, Omanisation is being pressed ahead, so that Omanis replace foreigners in skilled and semi-skilled jobs in all walks of life, therefore achieving considerable savings in the economy.

The economy of the Sultanate has always been prudently handled – the finances are sound and credit is good. Future development plans are well thought out and all expenditure on development is based on being necessary and in the best interests of the country. Prestige projects have been eschewed. The future political and economic outlook of the Sultanate is both stable and promising.

COUNTRY PROFILE

Mining: Mineral deposits of copper manganese, lead, iron, zinc, chrome, asbestos, phosphates, coal, gold, silver and nickel exist, many of them in inaccessible areas. Copper and chromite is mined near Sohar by Oman Mining Company (OMC), which is 99 per cent government-owned. Output, starting in 1984, reached 15,500 tonnes in 1987. Saudi Arabia is expected to be main market for anticipated 16,000-20,000 tonnes p.a. cathode copper production. Major prospecting operations are in progress for other minerals.

Hydrocarbons: Oil accounted for 50 per cent of GDP in 1987 and 97 per cent of exports, and production was increased to compensate for falling oil prices – from 280,000 b/d in 1980 to 565,000 b/d in 1987. Oil reserves in 1989 of an estimated 4,000m barrels would last well into the next century at current rates of production, and exploration was under way throughout the country and offshore, with some significant finds in 1989. Petroleum Development Oman (PDO), in which the government has a 60 per cent stake and Shell Oil a 34 per cent stake, produces 96 per cent of crude oil output and operates the only refinery (Mina al-Fahal). Oman is not a member of OPEC nor OAPEC, but has come under pressure to comply with OPEC policy and has so far complied. There are large reserves of natural gas, estimated in 1988 to be 9.3 trillion cubic feet, and gas has played a significant role in industrial development.

Energy: Natural gas is used as energy substitute to reduce domestic demand for oil, mainly by power plants, copper smelters and cement plants. Proven reserves are 8 trillion cubic feet, and a 10 year gas exploration and development plan announced in 1984 aims to link all the main towns into a national gas grid. Electricity production increased by 50 per cent between 1983 and 1986, and major projects further to enlarge capacity include three new power stations at Ras Sawadi, Manah, Wadi Jizzi, and Raysut.

Banking: Central banking authority is Central Bank of Oman, which has licensed 25 banks, nine of them local, 13 foreign and three specialised (Oman Development Bank, Oman Housing Bank and Oman Bank for Agriculture and Fisheries). Foreign banks enjoy same privileges and rights as local banks.

Legal system: The Authority for Settlement of Commercial Disputes handles all business disputes, maintaining close connection with the courts and relying on basic Islamic principles. All documents submitted to the Authority must be in Arabic and hearings are conducted in Arabic.

Membership of international organisations: Arab League, FAO, G-77, GCC, ICAO, IDA, IFAD, IFC, IMF, IMO, INTERPOL, INTELSAT, Islamic Conference Organisation, Islamic Development Bank, ITU, IUCE, NAM, UN, UNESCO, UPU, WFTU, WHO, WMO, World Bank.

BUSINESS GUIDE

Time: GMT +4 hrs.

Climate: Interior region largely hot and dry, except at high altitudes in north east. Coastal area has high temperatures (47°C) and high humidity (85-90 per cent) in summer and moderate winters (16-32°C). Rainfall low and irregular, except in Salalah and the southern province of Dhofar, where there are light monsoon rains from Jun to Sep-Oct.

Entry requirements
Passport: Required by all.
Visa: Not required by nationals of Bahrain, Kuwait, Saudi Arabia, United Arab Emirates and Qatar. Impossible to enter Oman without prior arrangement for admission. Business travellers require a no objection certificate (NOC), for which an application must be filed in Oman by a local sponsor. Bona fide first-time business visitors should apply through consulates or representatives abroad (letter from company explaining purpose of visit and letter of support from home Chamber of Commerce must accompany application). If entering Oman by road from the United Arab Emirates, a road permit must also be obtained in advance. Visitors should ensure that their passport has at least one clean right-hand page.
Prohibited entry: Holders of Israeli passports, passports in any way indicating travel to or from Israel and passports with Libyan visas are not allowed entry.
Currency: There are no currency control regulations, though capital transfers abroad exceeding RO20,000 require approval.
Customs: Personal effects, most goods from GCC countries and imports for Petroleum Development Oman (PDO) or the Government are duty free. Goods of Israeli origin and imports from Israel, South Africa or companies boycotted by the Arab League are prohibited. Restrictions on import of arms, ammunition, certain motor vehicles, and dangerous drugs. Alcohol is prohibited. Video cassettes are temporarily impounded for police examination. All imports require a certificate of origin and a commercial invoice, which must be certified by an Omani consulate or representative abroad. Documents should be in both Arabic and English.
Income tax: There is no individual income tax, but companies with 'permanent establishment' in Oman are liable to tax on income arising there.

Health precautions
Mandatory: Vaccination certificates against cholera and yellow fever if travelling from infected areas.
Advisable: TAB vaccination and anti-malaria precautions.

Representation overseas: Abu Dhabi, Aden, Algiers, Amman, Ankara, Baghdad, Beijing, Bonn, Brunei, Cairo, Damascus, Djibouti, Doha, Islamabad, Jeddah, Khartoum, Kuala Lumpur, Kuwait City, London, Manama, Mogadishu, New Delhi, Paris, Rabat, Riyadh, Rome, Sana'a, Seoul, Tehran, Tokyo, Tunis, Vienna, Washington DC.

Air access
National airlines: Gulf Air (GF) (jointly with Bahrain, Qatar and the United Arab Emirates). Oman Aviation Services (WY), Seeb International Airport, PO Box 1058, Muscat (tel: 519237; tx: 5424).
Other airlines: Air India, Air Lanka, Air Tanzania, Biman Bangladesh, British Airways, Egyptair, KLM, Kuwait Airways, Lufthansa, MEA, Oman Aviation, PIA, Royal Jordanian, Saudia, Sudan Airways, TWA, Thai Airways International, UTA.
Tax: Airport tax of RO3 is levied on departure.
International airport: Seeb International Airport (Code: MCT), 38 km from Muscat, with duty-free shops and a restaurant.

Surface access
Main ports: Mina Qaboos (containers), Mina al-Fahal (oil terminal), Mina Raysut (containers).
Overland access: The border with Yemen PDR is closed, and although there are road links with the UAE, special permission is required for entry (see above, *Entry requirements*).

Hotels: There are a number of very good international hotels. It is advisable, though not generally essential, to book in advance. A service charge of 20-25 per cent and a municipal tax of 5 per cent is usual. Dhofar coast beaches are main tourist area.

BUSINESS GUIDE

Electricity supply: 220/240 V AC, plug fittings two or three-pin round or three-pin flat types.

Credit cards: Major credit cards and travellers' cheques are accepted. Local banks do not cash personal cheques.

Car hire: International driving licence necessary, and considered valid for three months. Check licence rules in advance. Speed limits are 100 kph and driving is on the right-hand side of the road. Most hotels have self-drive car hire facilities.

City transport
Taxis: Taxis should have scale of charges, but fares should be negotiated in advance (there are no meters). Tipping is not usual. Typical fare from airport to Muscat/Ruwi is RO12 (1987) or RO3 for short journey within the capital area.
Buses: An urban bus service operates in Muscat, but is not recommended for visitors.

National transport
Air: Oman Aviation Services operate daily scheduled flights to Salalah Airport and daily flights to Fahud and Marmul on charter to Petroleum Development (Oman) (PDO). There are also flights to the Musandam Peninsula, Buraimi, Sur and the island of Masirah. All flights should be booked well in advance and confirmed on the day before travelling.
Road: 780 km highway links Dhofar with the north and main centres are connected by asphalt roads. Surfaced road network comprises 3,830 km; graded road network 18,280 km.
Buses: National services twice daily to Salalah by luxury coach from the bus stations at Ruwi and the Muttrah Corniche. Local services in the capital area. Contact the Oman National Transport Company, PO Box 620, Muscat (tel: 590046, 590603; tx: 5018 transerv).

Public holidays
Variable dates: Lailat al-Mi'raj, Eid al-Fitr, Eid al-Adha, Hijra New Year's Day, Prophet's Birthday, National Day and Official Birthday of Sultan Qaboos bin Said (two consecutive days in November).

Working hours: Hours liable to vary during month of Ramadan.
Business: (Sat-Wed) 0830 – 1300, 1600 – 1800; (Thu) 0800 –1300.
Government: (Sat-Thu) 0730 – 1430.
Banking: (Sat-Wed) 0800 – 1200; (Thu) 0800 – 1130.

Social customs: It is discourteous to eat, drink or smoke in front of Muslims in daylight hours during Ramadan. Women should wear skirts of reasonable length and avoid sleeveless dresses. Men should wear trousers and shirt to work and in public (never shorts or shirtless). Swimwear should be worn only at hotel pools and on the beach. It is polite to accept the refreshments customarily offered to visitors.

Telecommunications
Telephone: Dialling code for Oman: IDD access code + 968 followed by subscriber's number (no area code required). 45,500 lines by end of 1986. National improvement programme continues. Oman has recently changed practically all telephone numbers.
Telex: Available at good hotels. Fully automatic Arabic and English telex network covers the Capital Area and Salalah.
Telegram: Available through the National Telephone Service, Omantel.
Other: Mobile telephone system to cover the Capital Area, Batinah coast, Salalah and parts of the interior under development.

Banking
Central bank: Central Bank of Oman.
Main centre: Muttrah Business District.
Head offices: (Capital Area) Al Bank Al Ahli Al Omani, Bank of Muscat, Bank of Oman, Bahrain and Kuwait, Bank of Oman & the Gulf, Commercial Bank of Oman, National Bank of Oman, Oman International Bank, Oman Arab Bank, Union Bank of Oman, Oman Bank for Agriculture and Fisheries, Oman Development Bank, Oman Housing Bank.
Oman branches: (Capital Area) National Bank of Abu Dhabi, Bank of Baroda, Bank of Credit & Commerce International, Bank Melli Iran, Citibank, Grindlays Bank, Habib Bank AG Zurich, Habib Bank, Bank Saderat Iran, Banque du Liban et d'Outre-Mer, Banque Paribas, British Bank of the Middle East, Standard Chartered Bank.

Advertising: Advertising of alcohol and suggestive or erotic advertising is prohibited in accordance with Islamic law. There is no commercial radio or TV at the moment, but English-language newspapers all accept advertising.

Weights and measures: Metric.

Representation in capital: Algeria, Austria (consulate), Bangladesh, Belgium (consulate), China PR, Cyprus (consulate), Denmark (consulate), Egypt, France, German FR, Greece (consulate), India, Iran, Iraq, Italy, Japan, Jordan, Republic of Korea, Kuwait, Malaysia, Mexico (consulate), Morocco, Netherlands, Norway (consulate), Pakistan, Portugal (consulate), Qatar, Saudi Arabia, Somalia, Sudan, Switzerland, Tunisia, Turkey, UAE, UK, USA, Yemen AR.

BUSINESS DIRECTORY

Useful telephone numbers
Capital area
Police: 560099
International operator: 195
International enquiries: 197
Directory enquiries: 198
Operator: 190

Hotels: Capital Area
Al Bustan Palace, Muttrah (on beach), P.O. Box: 8998 (tel: 799666, 40200; tx: 5477; fax: 799600).

Al Falaj, Ruwi, P.O. Box: 5031 (tel: 702311; tx: 3229 hotel).

Al Nahdha, Muttrah, P.O. Box: 8561 (tel: 712385, 714196; tx: 5610 hand).

Gulf, Ruwi (Qurum, on beach), P.O. Box: 4455 (tel: 560100; tx: 5416 gulfotel; fax: 560650).

Holiday Inn, Seeb, P.O. Box: 2185 (tel: 697123; tx: 5644 holinn; fax: 697789).

Inter-Continental, Muttrah (on beach), P.O. Box: 7398 (tel: 600500; tx: 5491 ihcmct; fax: 600012).

Khasab, Ruwi, P.O. Box: 5286 (tel: 660267; tx: 8006 mdckhb).

Mina, Corniche Roundabout, Ruwi, P.O. Box: 3504 (tel: 711828; tx: 5350 minotel).

Muttrah, Ruwi (Muttrah High Street), P.O. Box: 4525 (tel: 714403, 798401; tx: 5226 mutrotel).

Qurum Beach House, Ruwi, P.O. Box: 5148 (tel: 564070; tx: 5441 qurumbh).

Ruwi Novotel, Ruwi, P.O. Box: 5195 (tel: 704244; tx: 3456 ruwiotel; fax: 704248).

Seeb Novotel, Seeb Airport, P.O. Box: 1069 (tel: 510300; tx: 5199 seebnovo; fax: 510055).

Sheraton, Ruwi, P.O. Box: 6260 (tel: 799899; tx: 3353/7/8; fax: 795791).

Hotels: Nizwa
Nizwa Motel, P.O. Box: 26812 (tel: 410500; tx: 8001 nizhotel).

Hotels: Salalah
Holiday Inn, P.O. Box: 18870 (tel: 235333; tx: 7638 holisal; fax: 235137).

Redan, P.O. Box: 18957 (tel: 292266; tx: 7708 rydhan).

Hotels: Sohar
Sohar, P.O. Box: 12959 (tel: 840058; tx: 3456 ruwiotel).

Car hire: Muscat/Ruwi
Avis: PO Box 3833, Ruwi (tel: 706526, 701488, 705457; tx: 3264).

Budget: Ruwi Novotel, PO Box 889, Muscat (Tel: 794721/3; tx: 3389, 2456).

Europcar: Muttrah Business District, PO Box 6275, Ruwi (tel: 700190/1; tx: 3045 asha).

Rent a Car: PO Box 4455, Ruwi (tel: 560100, ext 816; tx: 5416).

Business advice and information
Ministry of Commerce and Industry, PO Box 550, Muscat (tel: 799500; tx: 5351, 3665 wizara; fax: 794238, 794239).

Ministry of Finance and Economy, PO Box 506, Muscat (tel: 738201).

Ministry of Information, PO Box 600, Muscat (tel: 603888; tx: 5265 inform).

Ministry of Petroleum and Minerals, Sultan Qaboos Street, Al Khuwair, PO Box 551, Muscat (tel: 603333; tx: 5280 plande).

Musandam Development Committee, PO Box 5286, Ruwi (tel: 704599; tx: 3103 mdcmct).

Regional Development Committee, PO Box 6255, Muscat (tel: 563669; tx: 5204).

Oman Chamber of Commerce and Industry, Mutrah Business District, PO Box 4400, Ruwi, Muscat (tel: 707694; tx: 3389 alghurfa).

Oman Development Bank, PO Box 309, Muscat (tel: 738021, 738025; tx: 5179 odebe).

Petroleum Development (Oman) Ltd, Mina al Fahal, PO Box 81, Muscat (tel: 67811, 678111; tx: 5212 petro).

Banking
Central Bank of Oman, PO Box 4161, Ruwi (tel: 702222; tx: 3794 markzi on; fax: 702253).

Al Bank Al Ahli Al Omani (SAO), PO Box 3134 Ruwi (tel: 703044; tx: 3450 bkahlion; fax: 793536).

Bank of Muscat SAO, PO Box 6326, Al Burj Street, Ruwi (tel: 701769; tx: 3708 iddikhar on; fax: 707806).

Bank of Oman, Bahrain & Kuwait SAO, PO Box 4708 Ruwi (tel: 701528; tx: 3290 obk on; fax: 705607).

Bank of Oman and The Gulf (SAO), PO Box 4175 Ruwi, Greater Muttrah (tel: 703443, 703846; tx: 3403 bnkhalij on).

Commercial Bank of Oman Ltd, PO Box 4696 Ruwi, Corniche Road (tel: 793226; tx: 3275 combnkho on).

National Bank of Oman Ltd SAO, PO Box 3751 Ruwi, (tel: 708894; tx: 3281 nbo on; fax: 707781).

Oman Arab Bank SAO, PO Box 5010 Ruwi (tel: 700161; tx: 3285 arombnk on; fax: 797736).

Oman International Bank, PO Box 4216 Ruwi (tel: 793450; tx: 3364 ominbank on; fax: 707328).

Travel information
Flight information, Seeb International Airport (tel: 519223, 519584).

Oman Automobile Association, PO Box 7776, Muscat (tel: 706322; tx: 3129).

Oman Aviation Services, Seeb International Airport, PO Box 1058 Seeb (tel: (flight info in Ruwi) 705098, 705140).

Pakistan

Anthony Hyman

A new government along with an essentially new pattern of national politics headed by a younger generation of leaders emerged in Pakistan. These major political developments all came from the crash of a plane carrying President Zia-ul-Haq, together with the army's most senior generals, in August 1988. It marked the end of an era, and the hopeful rebuilding of democracy on firmer foundations. In a strenuous series of foreign trips through 1989, Benazir Bhutto presented an attractive image of Pakistan which earned respect in the West and in Asian capitals. But a serious economic crisis faced the Bhutto government, and underlying political tensions threatened to upset the country's surface stability.

Many in the country believed the air crash which killed the military dictator who had dominated Pakistani politics ever since 1977 was deliberate sabotage rather than an accident, as officially given out, but theories abounded as to who was responsible. The most favoured explanations were a plot within the military, and/or the 'hidden hand' of various foreign agencies – the CIA, KGB and Afghanistan's KHAD all being cited. However it occurred, the death of Zia-ul-Haq had important and rapid political consequences.

In spite of early scepticism that the army would not consent to give up power and would never permit free and fair elections on a party basis, that is precisely what happened. Vigorous campaigning by rival politicians and parties, with lavish campaign pledges to the poor, was followed by orderly elections to national and provincial assemblies, held in November 1988. Many old-established politicians lost out to a younger breed which had come up in the previous decade under Zia-ul-Haq. Though it was not the predicted landslide, Benazir Bhutto's Pakistan People's Party (PPP) swept Sind, and won millions of votes in the Punjab, emerging as the biggest overall party and the only truly national party, with 92 out of 205 seats in the national assembly, short of an absolute majority.

After two tense weeks of recriminations and horse-trading, Benazir was finally called in by President Ghulam Ishaq Khan to form a government. The country's first woman prime minister took office amid great hopes and much international goodwill. It was clear, though, that daunting problems and challenges faced the country's economy and society as a whole.

Political developments

The same PPP government which had started in such high hopes at the end of 1988 looked much more vulnerable by the summer of 1989. The public's honeymoon with the youthful, glamorous leader was fading fast. Economic woes and past mismanagement required determination, special skills and above all time to solve. Chronic law and order problems were common in many areas of Sind province. In Karachi, the country's largest city, port and main industrial centre, the ethnic tensions and violence had resumed in 1989, along with other cities in Sind, reinforcing the existing trend of industrialists to set up new factories in the Punjab.

Benazir Bhutto's rise to power after the general election was only made possible after political alliances patched together with leaders of rival parties in three out of the four provinces of Pakistan. Vital regional alliances made by the ruling party with smaller locally based parties had come unstuck, in North West Frontier Province and the main urban areas of Sind, with ANP (Awami National Party) and MQM (Mohajir Qaumi Mahaz) respectively. Relations had sharply deteriorated with leaders of MQM, the new party of the Urdu-speaking *mohajirs* or migrants from India which won control of Karachi and Hyderabad cities in Sind.

In the third province, Baluchistan, a tactical alliance was wisely abandoned, after proving unworkable. After a 50-day crisis when the newly-elected Baluchistan Assembly was dissolved, the PPP government began to show itself conscious of the need to tread softly in what had been known since the 1970s as Pakistan's problem province. With the appointment as chief minister of the veteran Baluch politician Nawab Akbar Bugti, there were grounds for hoping that the dangerous, longstanding power vacuum in Baluchistan might at last be over.

It used to be the conventional wisdom that nobody could be prime minister in Islamabad and not also control the Punjab, the biggest and wealthiest province by far. This notion has been stood on its head. Instead of the central government promptly collapsing, it is coping so far with a hostile Punjab. But the political polarisation between Punjab and Sind is clearly a factor for instability.

Even well-wishers of the government found distinctly worrying the inexperience of many of the PPP appointees. The new cabinet, enlarged to a record 41 ministers, gave many signs of inability or sheer incompetence to deal with pressing matters. There was continuity from previous regimes of bureaucrats at the helm of the key ministries of finance and foreign affairs.

KEY FACTS

Official title: Islamic Republic of Pakistan
Head of state: President Ghulam Ishaq Khan
Head of government: Prime Minister Benazir Bhutto
Capital: Islamabad
Official Languages: Urdu (English, Punjabi, Pushto, Sindhi and Balucki also spoken)
Currency: Rupee (Rp = 100 paisa)
Exchange rate: Rp20.9 per $ (Nov 1989)
Area: 803,943 sq km
Population: 105.41m (1988)
GDP per capita: $361.5 (1988)
GDP real growth: 5.8% (1988)
Labour force: 30.5m (1988)
Inflation: 8.8% (1988)
Trade balance: –$2,607m (1988)
Foreign debt: $16.289bn (1987)
Visitor numbers: 142,000 (1987)

In addition, charges of nepotism were raised – even within PPP ranks – over two family appointments announced in March 1989. The most important was the decision to make Benazir's mother deputy prime minister. Begum Nusrat Bhutto, the Iranian-born widow of Zulfikar Ali Bhutto, was a dignified symbol for the Pakistan People's Party in opposition, but even her admirers admit she had limited political abilities. Consternation at this decision was aggravated by the announcement that Benazir's father-in-law, rich Karachi businessman Hakim Ali Zardari, would become chairman of the influential public accounts committee of the parliament. By convention, it had been a post usually reserved for an opposition MP.

Family loyalties are accepted in Pakistani politics, but by such actions the PPP risked looking more like a family business than a vital, democratic political party. Ever since the trial and hanging of her father, ex-Prime Minister Z.A. Bhutto in 1979, Benazir tried hard to identify the cause of democracy in Pakistan with herself and the party whose leadership she inherited. For millions of Pakistanis, the identification of the two is, indeed, a self-evident fact. For many more, though, the government's actions will have to convince them.

On ideology, Benazir Bhutto took care to distinguish between socialism and social democracy. It is the second which she says the PPP favours, qualifying it as social democracy of an Islamic type. By many sharp observers, the PPP government is classified not so much as social democratic as populist. It is not only left-wing critics who see the party as dominated by big feudal landlords and businessmen, with vested interests in the status quo. This hardly adds up to a formula for radical changes.

New firmer measures in 1989 to tackle Pakistan's acute drugs problem were warmly greeted. Much larger sums of money were made available for treatment of drug addicts, estimated at 1.9m by the Pakistan Narcotics Board, mostly in the 20-30 age group. Heroin addiction grew remarkably from 1980, along with heroin production and smuggling abroad. For the first time, a Pakistani government showed itself willing to arrest some of the richest and most influential 'drug barons'. But there were legitimate doubts that only drug criminals were being targeted, with the PPP's political rivals in North West Frontier Province, notably ex-Governor General Fazle Haq, also arrested.

On the political scene, PPP manoeuvres to unseat Benazir's bitter rival Nawaz Sharif as leader of the Islamic Democratic Alliance (IDA) and chief minister of the Punjab failed miserably. The PPP had claimed that dozens of IDA members and independents in the assembly were about to join them, and at first tended to dismiss Nawaz Sharif and the IDA as a temporary nuisance which would soon go away. But this forceful young politician, one of the richest industrialists in Pakistan, evidently has staying power.

Nawaz Sharif is himself a major beneficiary of the Zia years, and a loud apologist for its record. Claiming to represent 'Ziaism' rather as Benazir does 'Bhuttoism', Nawaz Sharif is a thorn in the side of the government's strategy. If this Centre-Punjab confrontation continues, it could well have a destabilising effect. Yet the IDA itself does have grave internal weaknesses. The opposition alliance is a loose patchwork of disgruntled interests, held together by little more than an appetite for getting political power.

The Punjab, with almost two-thirds of Pakistan's 107m people, has to be reckoned with. As entrepreneurs Punjabis are found in all areas of the economy, while in the armed forces they have established an even greater dominance. Nawaz Sharif personifies Punjabi nationalism in a strident form, appealing to popular prejudice against a Sindhi ruler, as well as exploiting for all it is worth the conservative Muslim backlash against a woman as ruler. Another strand of Punjabi nationalism is a fierce hostility to India. Nawaz Sharif's line has been that the PPP government is unreliable and soft on Pakistan's national interests.

KEY INDICATORS

	Unit	1984	1985	1986	1987	1988
Population	m	93.29	96.18	99.16	102.24	105.41
Gross domestic product (GDP)	Rp bn	421.25	482.11	547.13	608.15	685.87
GDP per capita	Rp	4,515	5.013	5,518	5,948	6,507
Share prices	1985=100	102.5	100.0	103.8	128.9	150.1
Wholesale prices	1985=100	97.2	100	104.9	113.6	124.7
Consumer prices	1985=100	94.7	100	103.5	108.4	117.9
Manufacturing production	1985=100	92.6	100	107.3	115.0	–
Energy production						
Coal	m tonnes	2.24	2.20	6.17	–	–
Crude petroleum	m barrels	4.88	9.53	15.06	–	–
Natural gas	bn cu metres	9.5	10.3	11.1	–	–
Volume of exports	1985=100	79.8	100	155.6	142.3	142.6
Cotton	1985=100	20	100	150	114	144
Volume of imports	1985=100	87.8	100	100.5	100.9	106.2
Exports	RP$m	35,9941	43,645	56,336	72,583	81,348
Cotton	$m	1,637	6,712	7,918	8,024	13,055
Rice	$m	4,692	4,334	5,034	5,224	5,991
Imports	$m	82,038	93,793	89,297	101,310	118,681
Crude petroleum	$m	12,605	14,141	7,077	8,638	7,138
Balance of trade	$m	–3,753	–3,230	–2,780	–2,316	–2,607
Current account	$m	–1,185	1,080	–645	–562	–1,403
External debt	$m	12,178	13,375	14,815	16,289	–
Debt service ratio	%	14.6	18.2	16.4	17.5	–
Exchange rate	Rp per $	14.05	15.93	16.65	17.40	18.00

* estimated figure

From crisis to crisis

The government's first budget, presented in June 1989, was well received by the IMF as signalling a seriousness of economic intent lacking under the previous regime. The budget had three goals: cutting the budget deficit, controlling inflation (currently running at around 20 per cent, instead of the official 11 per cent) and improving inadequate infrastructure so as to help the poor cope. Budget receipts for 1990 were set at Rs136.8bn ($6.5bn), of which Rs9.6bn was raised from new taxes and unpopular price increases for state-provided services. Privatisation of state-owned businesses, including banks and 30 per cent of the profitable airline PIA, was expected to raise at least $1bn.

Expenditure was set at Rs152.3bn, with the large defence element of the budget being Rs51.75bn (in real terms down to Rs46.05bn).

Minister for Finance Ihsanul Haq Piracha declared that 'the deficit of the budget will be reduced, in a balanced manner, through expenditure restraints and moderate resources mobilisation.'

There have been gratifying successes in some areas of Pakistan's economy. Agricultural output as well as industrial growth made good progress through the 1980s, with GDP growth rate of 5-6 per cent in 1989. But the general picture is rather bleak, even alarming. The Bhutto administration inherited a bankrupt treasury, along with many financial pressures. Defence spending accounts for 40 per cent of total revenues. Debt servicing takes up a full 27 per cent of state revenues – as much as total state spending on general administration and social services.

The two most pressing economic problems were the budget deficit and the current account imbalance. Pakistan's budget deficit for the 1989 fiscal year (ending 30 June) was a projected Rs40.8bn ($2.17bn), equivalent to 5.3 per cent of GDP. Unofficial estimates place it much higher, and in the previous year the budget deficit actually exceeded 8 per cent of GDP, at Rs58.6bn

The previous government had attempted in 1988 to raise badly needed additional revenues by widening the tax net. It led to a well organised agitation by the business lobby. Tax evasion is on a massive scale, with a 'black' or underground economy estimated in 1986 at some $10bn – equivalent to a startling one-third of official GDP.

There are severe financial constraints on any ambitious expansion of facilities for improved health, wider access to education, better housing and creation of jobs. It had been hoped that the new PPP administration would act quickly to improve labour conditions, by formulating a new labour policy with teeth, helped by the new labour minister and former trade union leader Malik Mukhtar Awan. There is still no comprehensive labour policy for the country, which partly explains the violence and deaths which commonly occur during industrial disputes in textile mills and other manufacturing enterprises. A harsh contract-labour system is in operation in Pakistan's textile industry, and near-slavery conditions apply for scores of thousands of labourers in Pakistan's brick kilns.

The new PPP government was faced with what some describe as the worst economic crisis yet faced by Pakistan. So serious was the shortage of finance that the administration had to resort to heavy borrowing from banks at the end of 1988, simply to pay salaries of civil servants.

From soon after the Soviet invasion of Afghanistan, it became easy for Pakistan to get large-scale international aid, from the Pakistan Aid Consortium of 10 industrialised countries plus international institutions. They pledged nearly $3,100m for 1989-90, more than the government had requested. This tide of aid, together with regular remittances from Pakistanis working abroad in the Gulf and elsewhere, enabled the government to cope with a negative balance of trade. However, foreign debt rose alarmingly under General Zia's regime, from $3.5bn to $12.5bn from 1977 to 1988. As for debt servicing, the relevant figures rose from $310m annually to $1.2bn.

Apart from extensive military aid, a large USAID $465m agreement, signed in June 1989, went to help finance overdue primary education programmes to increase literacy, for expansion of higher education and $100m for private sector housing construction. It was part of a ten-year aid package worth $4,000m, approved during Benazir Bhutto's visit to the USA, after she had assured the US Congress that Pakistan did not possess, nor intend to make, a nuclear weapon.

The government's hands appeared to be tied by agreements entered into in 1988 by the caretaker government under President Ishaq Khan. It had agreed to borrow $1bn over a three-year period from the IMF, in return for accepting stringent terms recommended for years by the World Bank as well as the IMF. Conditions for the IMF standby loan included the ending of generous subsidies on agricultural and other products, big increases in pegged electricity prices and other utility charges, together with liberalisation of imports. These measures would not only increase the rate of inflation, but certainly fuel public unrest. However, the IMF agreed in early 1989 to defer their implementation.

King cotton

Pakistan's balance of trade is regularly in deficit. In 1988-89 the value of exports was $4,645m and imports $6,700m, with a net deficit of $2,355m. The estimated deficit for 1989-90, though, was considerably lower than for recent years, placed at $1,730m. The government's hopes for resolving the serious shortage of foreign exchange are pinned on boosting exports of cotton, rice, wheat and other agricultural products in surplus. Agriculture is the mainstay of Pakistan's export strategy, as well as the backbone of the economy.

The biggest foreign exchange earner for Pakistan is the cotton industry. Exports of raw cotton and cotton goods (mainly to Asia) benefited by intensive farming methods and better quality seeds, giving a better-than-average yield. High international prices were being paid for record crops of cotton from 1985.

The target set of 8.2m bales for 1988-89 was reached, in spite of damage done in the Punjab by drought and flooding late in 1988. This compared well with the previous year's record of 8.6bn bales. However, cotton lost 21 per cent of its unit value between July 1988 and March 1989, according to the finance ministry.

Pakistan's cotton industry was shaken in 1988 by allegations of corruption and mismanagement. Private exporters and officials of the state-owned Cotton Export Corp each alleged that large sums of money were being made illegally in underhand dealings. The PPP government began an investigation into the scandal, and tried to block gaps in export procedures which helped siphon off money to overseas bank accounts.

In spite of the better-than-anticipated cotton crop, the rate of growth in agriculture as a whole was expected to decline to under 3 per cent in 1989-90, instead of 4 to 5 per cent in earlier years. After cotton, rice is the most important staple export item. The volume of rice exported in 1988-89 was 915,000 tonnes, of which 190,000 was top-grade basmati rice, to a total value of $313m. With a bigger crop and higher international prices, value of rice exports in 1989-90 is expected to rise to $430m.

A sugar surplus in 1989 resulted in Trading Corporation of Pakistan contracting to supply Iran with 300,000 tonnes of sugar, worth $127m.

Pakistan will face difficulties in holding on to its share of the world markets in cotton, cotton yarn, textiles and rice, let alone expanding it, as hoped by the government. Exporters face stiff competition, especially from new producers. Value-added goods accounting for an increased share of Pakistan's exports in the 1980s included textiles and clothing, carpets, leather goods and fish products.

The thriving cotton textile industry dominates Pak-

COUNTRY PROFILE

Political system: After nearly eight years of martial law, parliamentary democracy with a civilian prime minister was reintroduced in December 1985. But former military dictator and President Zia Ul Haq remained pre-eminent – and showed it in May 1988 when he sacked Prime Minister Mohammed Khan Junejo, dissolved Parliament and appointed a caretaker government without a prime minister. Elections are supposed to be held within 90 days, but sceptics doubted whether they would be held, especially since Zia declared his intention of turning Pakistan into a completely Islamic state. The influence of the military remains strong. The President is advised by a three-member military council and a larger advisory council. New general elections on a party basis were finally allowed in November 1988, after President Zia ul Haq lost his life in an air crash. Benazir Bhutto's Pakistan People's Party (PPP) won the elections overall, with a landslide in Sind, and formed the new government.

Political parties: The PPP government formed coalitions with smaller parties in three of Pakistan's four provinces, but two of these fragile provincial governments had collapsed by April 1989 – that with the ANP in the North West Frontier Province and with the MQM in Sind. The biggest and most important province, Punjab, remains in the control of the Islamic Democratic Alliance (IDA) led by Nawaz Sharif, Benazir Bhutto's bitter rival. Besides the parties in the IDA, there are dozens of regional parties claiming to be national.

Population: Pakistan is the world's ninth most populous country. Its population increased from 36m in 1951 to 105.7m in 1988. The present rate of growth (3.1 per cent) is planned to be reduced to 2.7 per cent per annum by the year 2000, when the population is projected to reach 150m. Population density ranges from 230 per square kilometre in Punjab to 12.5 per square kilometre in Baluchistan. Seventy per cent live in rural areas. During the period 1951-81, the urban population rose by 4.7 per cent annually compared with a 2.6 per cent increase in the rural population. The urban population may reach 56.7m by the year 2000. About 44.5 per cent of the present population is below 15 years of age. Over 90 per cent are Muslims. There are an estimated 3m Afghan refugees living mostly in camps on the western borders.

Main cities/towns: Islamabad (population 204,000 in 1981 – last reliable census), Karachi (pop 5,208,000), Lahore (pop 2,953,000), Faisalabad (pop 1,104,000), Rawalpindi (pop 795,000) and Hyderabad (751,000). In 1987, Lahore was estimated to have over 4m inhabitants, Karachi 8m, Rawalpindi 800,000 and Islamabad 400,000.

Language: The offical language is Urdu; English is quite commonly spoken in business circles.

Media

Press: There are 105 Urdu daily newspapers, published on a semi-national basis, and 14 English-language papers including the *Pakistan Times*.

Radio: There are 16 radio stations; the national service broadcasts in numerous languages and on several channels, with an external service in 17 languages. There are over one million radios in use.

Television: There are five TV stations broadcasting at least 10 hrs/day. There are around 1.5m TVs in use (1987).

Domestic economy: Pakistan's economy is dependent on agriculture, and has suffered set-backs from warfare, political upheaval, poor weather conditions reducing crops, rapid population growth and the of 3m refugees from Afghanistan. Although still agrarian in character, the economy of Pakistan has undergone significant transformation in terms of growth, patterns of production and consumption, employment structure, composition and direction of foreign trade, and relative roles of public and private sectors. The share of agriculture in GDP has declined from 53 per cent (1949-50) to 26 per cent (1987-88), with a corresponding rise in the share of industry from 9.6 per cent to 25 per cent. Over the last few years, Pakistan has managed a self sustained broad-based growth rate of about 6 per cent per annum. GDP grew by 6.2 per cent in 1987-88, comprising 2.7 per cent in agriculture and 10 per cent in manufacturing. A relatively slow growth in agriculture is attributed to unfavourable weather. The investment (16 per cent of GNP) and savings (12 per cent of GNP) levels in Pakistan, are low in absolute terms as well as in comparison with the levels attained by other countries. Over the past few years, Pakistan's current account deficit has averaged 3.5 per cent of GNP and has financed over 20 per cent of its total investment. Remittances from abroad, which have been a major source of foreign exchange, slumped to US$2.1bn in 1987-88, thus adding to the balance of payment difficulties. The Seventh Five-Year Plan (1988-1993), currently under implementation, emphasises the twin objectives of achieving efficient growth of output on the one hand and improving the quality of life on the other. The government encourages the private sector, and welcomes foreign private investment. For this purpose export processing zones have been established, the procedure for investment has been simplified, the list of specific industries requiring sanction has been considerably reduced, and policy and incentives for foreign investment in Pakistan have been announced. Inflation since 1980-81 has averaged 6.1 per cent per annum, including 6.3 per cent increase registered in 1987-88.

Employment: The official unemployment rate of 3.5 per cent is considered to be on the low side even by the Planning Ministry. Although unemployment eased somewhat in the mid-1970s, owing to out-migration to the Middle East, it is now reasserting itself as a major problem. Under-employment is high. Roughly 26 per cent of the employed labour force is made up of unpaid family helpers in rural areas. Only 27 per cent of the labour force are wage-employees. In the early 1980s 10 per cent of the labour force worked abroad, mainly in the Middle East, but since then the world oil glut and the completion of labour-intensive construction projects, have forced many to return home. Labour force in 1987-88 was estimated at around 31m, out of which 29.9m (96.5 per cent) was employed and the remaining 3.5 per cent unemployed. Of the employed labour force in 1987-88, about 48.8 per cent was engaged in agriculture, 14.3 per cent in manufacturing and mining, 6.1 per cent in construction, and the remaining 30.8 per cent in other activities.

istan's private corporate sector. The trust of the business community in the government was still clearly lacking in 1989, because Benazir Bhutto was identified with her father's nationalisation programme of the 1970s, and the economic and political chaos which followed. But there were plans to open a third stock exchange in Islamabad, (after Karachi and Lahore). In spite of a very low domestic savings level, the Stock Exchange listings of companies was up from 412 to 431 in 1989.

Push for energy expansion

The government has a bold plan to triple power generation capacity in 10 years, to reach 22,745 mW. It is made necessary partly by Pakistan's rapid growth in population. Present inability to satisfy rapidly growing domestic demand for electricity handicaps industrial production in Karachi and other centres. Because of sharp limitations on state financial resources, private sector as well as foreign investment by build-operate schemes are favoured. Pakistan's Seventh Five-Year Plan (1988-93) has a private element of 2,300 mW, which could rise by 50 per cent in the Eighth Plan.

Build-operate schemes were being funded by the World Bank, USAID, Asian Development Bank, Export-Import Bank of Japan, the UK, Italy, and Canada. The initiative differs from earlier ones in Turkey by giving a big role to the public sector and multilateral agencies. Its success hinges on tariff rates to be paid by Pakistan's Water and Power Development Authority, whose function it will be to buy and distribute electricity produced by private plants.

Increased use of plentiful coal resources is certain if the ambitious targets are to be met. Coal currently accounts for only 6 per cent of Pakistan's power capacity, with proven reserves (albeit of low-grade coal) of 460m tonnes. Oil and gas development, together with hydro-electric resources, cannot produce more than two-thirds of planned capacity by the year 2000. Pakistan's oil production rose to 45,000 barrels a day in 1988, with considerable activity in oil exploration on and offshore, with imported petroleum accounting for 12.3 per cent of total imports in 1988-89.

Foreign policy

Foreign policy was dominated by the continuing war in Afghanistan, in spite of completion of Soviet troops' withdrawal by February 1989. The presence of three million Afghan refugees in Pakistan was felt both as a heavy burden and a danger to political stability, especially in North West Frontier. The refugees still enjoyed substantial sympathy, but had become scapegoats for the spread of drugs and the so-called 'Kalashnikov culture', the new availability of automatic guns and trend towards greater violence in the country.

Pakistan's own role in the war was highlighted from February 1989, with the holding of a *'Shura'* or assembly of delegates of the seven Afghan parties established in Pakistani exile in Rawalpindi, and a large-scale offensive by Afghan guerrillas upon Jalalabad. It proved a costly disaster for the uncoordinated and lightly armed guerrillas.

The new prime minister took this failure as the opportunity to show some degree of independence of Pakistan's armed forces, by dismissing General Hamid Gul, director general of the ISI (Inter Services Intelligence), the shadowy military agency which was built up under Zia-ul-Haq's regime.

Benazir Bhutto was given credit in some Pakistani circles for seeking some form of political compromise in Kabul, and trying to convince the USA that the war had no prospect of conclusion, and that arms shipments should be stopped. But there were also many critics who felt she was allowing Pakistan to be manipulated for cynical US interests.

For the long term, it is Pakistan's relations with India which may matter most. These steadily improved, though unevenly, in 1989, with an evident effort by Rajiv Gandhi and Benazir Bhutto to clear some of the ill will and mistrust between the two countries. There was a renewed attempt to settle the undeclared state of war on the northernmost disputed frontiers between India and Pakistan, on the Siachen glacier in the little Pamirs.

Rapprochement with India was the reason why Pakistan announced it would rejoin the Commonwealth in 1989. While Zia-ul-Haq remained president, India had said it would veto Pakistan's application. The decision to quit this organisation had been taken, ironically, by Benazir's father Z.A. Bhutto, after the 1971 war which led to Bangladesh being created from East Pakistan.

COUNTRY PROFILE

External trade
Exports: Pakistan's economy has gradually opened up to the rest of the world. Its total foreign trade in 1987-88 was roughly 30 per cent of GDP, with exports and imports sharing 11.5 per cent and 18.2 per cent respectively. In volume, exports totalled US$4.4bn in 1988. Of the total exports in 1987-88, raw cotton accounted for 13.7%, cotton yarn and thread 12.5%, cotton fabrics (10.9%), rice 8.2%, leather 6.4%, carpets and rugs 5.7%, fish and fish preparations 2.8% and sports goods 1.5%. Likewise the main destinations were Japan (11.3%) USA (11.0%), Germany (7%), UK (6.8%), Italy (5.8%), Saudi Arabia (5.0%), Hong Kong (3.5%), France (3.2%) and Dubai (3.0%). In terms of economic categories, the primary commodities constituted 28% of total exports and semi and manufactured goods 20% and 52% respectively.
Imports: Imports were US$6.9bn in 1988, with machinery and transport equipment accounting for 20.9 per cent, followed by mineral fuels and lubricants (15.5%), manufactured goods, animal and vegetable fats and oil (8.1%), food and live animals (6.9%). Main sources are Japan (15.2% of 1987-88 total), USA (11.2%), Germany (7.9%), Kuwait (7.2%), UK (6.9%), Saudi Arabia (5.1%), China (3.5%), Malaysia (3.4%), Italy (2.4%) and the Netherlands (1.9%). In terms of economic categories, 36% consisted of capital goods, 49% industrial raw materials (42% consumer goods) and 15% consumer goods.

Agriculture: Agriculture employs 48.8 per cent of the labour force and accounts for 26.2 per cent of GDP (1987-88). Pakistan has an agricultural economy with a network of canals irrigating an area of about 16.1m hectares. Total cropped area (1987-88) stood at 20.8m hectares. About 26 per cent of the land area is cultivated arable. River and ground water are plentiful, and irrigation has been an essential element. Except in times of bad drought or pest infestation, as occurred in 1983-84, Pakistan is self-sufficient in wheat production. Other main food crops include rice, maize, barley, millet, sorghum, sugar-cane, tobacco, groundnuts, pulses, rapeseed, sesame, potatoes, onions, mangoes and citrus fruit. Cotton is the main cash crop (Rs10.8bn in export earnings in 1987-88) and the largest foreign exchange earner, with productivity rising as greater use is made of insecticides. Rice is also an important export (Rs6.4bn in 1987-88). Livestock raising of goats, sheep, cattle, buffaloes, pigs, donkeys, horses, mules, camels and poultry accounts for 7.4 per cent of GDP, providing meat, eggs, dairy products, leather and wool, as well as draught power and fertiliser for cultivation. Seven per cent of the total land area is in pasture. There are grazing difficulties in border regions, where Afghan refugees have brought 3m* head of cattle with them. Fishing contributes to foreign exchange earnings and output should increase with the rehabilitation of Karachi Fish Harbour. Forests cover 3.17m hectares (four per cent of the land area) but contribute little to the economy. Most logging is for fuel wood. For the encouragement of agriculture, incentives such as price support, supply of basic inputs at subsidised rates, supply of certified seeds and provision of extension services and credit on easy terms have been major policy instruments.

Industry: Manufactured and semi-manufactured goods account for 72 per cent of Pakistan's exports (1987-88). Manufacturing employs 14 per cent of the labour force and accounts for 17.5 per cent of GDP (1987-88), depending on the agricultural sector for most of its raw materials: cotton for weaving, spinning and processing industries, leather and wool for handicrafts and carpet weaving. The manufacturing sector has been dominated by public sector industries. These include large-scale cotton cloth mills, and it will continue to develop in steel, fertiliser, cement and oil refining; there is, however, a growing emphasis on private sector development (about 85 per cent of manufacturing output takes place in the private sector). Refinery products include lubricating oil, LPG, naphtha, kerosene, diesel oil and furnace oil, with petrochemical capacity for benzene, toluene and xylene. The textile industry is the leading sector, although cotton cloth production has been declining steeply in recent years, while cotton yarn increases. Other main industries are sugar refining and chemicals (Pakistan has exported two sugar plants to Indonesia and Bangladesh). There are several private sector tractor factories under construction. Other light industrial activities include food processing, sports goods, surgical instruments and paper. Construction accounts for 4.3 per cent of GDP (1987-88). Tourism has been slow to develop, hampered by lack of facilities and infrastructure.

Mining: Mining accounts for 0.53 per cent of GDP (1987-88). Currently about 67 per cent of the mining sector's output is contributed by coal, crude oil and natural gas. There are proven reserves of 187m tonnes coal. There are plans to increase coal production from deposits at Lakhra, Sind province. Proven reserves of natural gas are 441.13bn cu metres (1987), and most production is from the Sui field in Baluchistan. In 1987-88, domestic oil production remained at 16.3m barrels, mainly from the Meyal field, with increasing amounts from Dabhi. Offshore oil exploration intensified in late 1985. Limestone, rock salt, gypsum, silica sand, china clay and chromite deposits are mined; there are development plans for phosphate from Kakul and gold, silver and copper from Saindak, Baluchistan. Other minerals include iron ore, gem stones, magnetite, barites, antimony, bauxite, celestite, dolomite, fireclay, fluorite, fuller's earth, marble and soapstone. Extraction and exploration for minerals is frequently hampered by difficult terrain, technical problems or insufficient investment.

Energy: Pakistan meets about 30 per cent of its commercial energy needs through imports. In 1987-88, oil fulfilled 40% of energy requirements, gas (37%), hydroelectric power (13%), coal (9%), nuclear power (0.5%) and LPG (0.4%). Domestic oil consumption is still growing fast. Domestic oil output accounts for 23 per cent of total oil requirements. The proven reserves of gas and coal are enough to meet the country's requirements for another 30 years.

COUNTRY PROFILE

Energy development plans, including those for a new dam with a 1,760 mW generating plant at Kalabagh on the Indus and a 900 mW nuclear power station at Chasma, have been delayed by political controversy, with the three smaller provinces complaining it will help Punjab but hurt them. The generation and distribution of electricity is the sole responsibility of the public sector. However, to supplement the public sector's efforts towards eliminating electricity shortages, the private sector is now becoming involved and a Private Sector Energy Development Fund has been established with help from the World Bank.

Financial markets: Karachi Stock Exchange, Lahore Stock Exchange. Islamic banking is being developed.

Membership of international organisations: Asian Development Bank, CCC, Colombo Plan, Economic Cooperation Organisation, EC trade and co-operation agreement, ESCAP, FAO, GATT, G-77, IAEA, IBRD, ICAC, ICAO, IDA, IFAD, IFC, IHO, ILO, IMF, IMO, INTELSAT, INTERPOL, Islamic Conference Organisation, Islamic Development Bank, IPU, IRC, ITU, IWC, NAM, OIC, South Asian Association for Regional Cooperation, UN, UNESCO, UPU, WFTU, WHO, WIPO, WMO, World Bank, WSG, WTO.

BUSINESS GUIDE

Time: GMT + 5 hrs.

Climate: The terrain ranges from mountainous to desert, and the climate varies accordingly. It is cool in the mountains and foothills, with rain in summer and snow in winter, to hot in summer and cool in winter in plateau regions, with some rain in winter, to year-round heat in the Indus Valley, with extreme heat and dry winds in summer.

Entry requirements
Passport: Required by all except certain seamen and holders of certain UN documents.
Visa: British citizens require entry visas as of 1 Jan, 1987. Please check other details with embassies. No permits are now required for travel up the Karakoram Highway via Khunjerab Pass to Kashgar in China. Chinese visas available in Islamabad. Nationals of India, Afghanistan, Bangladesh and Iran require visas, which should be obtained before arrival. Foreigners require permits for approved treks in restricted zones (the districts of Chitral, Gilgit and Skardu). Permits are required for all mountaineering expeditions, which must also be accompanied by a Pakistani Liaison Officer.
Prohibited entry: Israeli and white South African nationals are not permitted entry.
Currency: The amount of domestic currency which may be imported or exported is limited to Rp 100 per person. No more than Rp 500 will be reconverted to foreign currency on departure. Foreign exchange receipts should be retained. Large sums imported may be re-exported only if covered by Repatriation Guarantee.
Customs: Personal effects are allowed duty-free. Visitors are not permitted to import alcoholic beverages in any quantity. Motor vehicles may be imported duty-free for a period of up to three months. Certain items may only be imported from tied sources under credits, barter trade or the US PL-480 food aid programme or by public sector agencies. Exports to South Africa, Israel and Taiwan are prohibited. Export of certain antiques may require a permit.

Health precautions
Mandatory: Vaccination certificates are required for cholera or yellow fever if travelling from an infected area.
Advisable: Non-compliance with health regulations may involve quarantine. Water should be boiled before drinking.

Representation overseas: Abu Dhabi, Accra, Addis Ababa, Aden, Algiers, Amman, Ankara, Athens, Baghdad, Bangkok, Beijing, Beirut, Belgrade, Berlin, Berne, Bonn, Brasilia, Brussels, Bucharest, Buenos Aires, Cairo, Canberra, Colombo, Copenhagen, Dakar, Damascus, Dar es Salaam, Dhaka, Doha, The Hague, Harare, Jakarta, Jeddah, Kabul, Kathmandu, Khartoum, Kinshasa, Kuala Lumpur, Kuwait City, Lagos, Lisbon, London, Madrid, Malé, Manama, Manila, Maputo, Mexico City, Mogadishu, Moscow, Muscat, Niamey, Nairobi, New Delhi, Nouakchott, Ottawa, Paris, Port Louis, Prague, Pyongyang, Rabat, Rangoon, Rome, Sana'a, Sofia, Stockholm, Tehran, Tripoli, Tokyo, Tunis, Vienna, Warsaw, Washington.

Air access
National airline: PIA Pakistan International Airlines.
Other airlines: Aeroflot, Air France, Air Lanka, Alia, Biman Bangladesh, British Airways, CAAC, Egyptair, Emirates, Gulf Air, Indian, Interflug, Iranair, Iraqi Airways, JAL, Jamahiriya Libyan Arab, KLM, Kuwait Airways, Lufthansa, MAS, Royal Jordanian, Royal Nepal, Pan Am, Philippine, Saudia, Singapore, Swissair, Syrian Arab, Tarom, Thai International, THY-Turkish, Yemenia.
International airport: Karachi Civil (KHI), 16 km from Karachi; duty-free shop, bar, restaurant, buffet, bank, hotel reservations, post office, shops; Lahore (LHE), 11 km from Lahore; restaurant, bank, post office, shops, car hire; Peshawar (PEW), 16 km from Peshawar; Islamabad International (ISB), 8 km from Islamabad; restaurant, car hire.

Surface access: Roads run from Afghanistan (via Qandhar-Chaman-Quetta or Kabul-Peshawar), India (Amritsar-Lahore) and China (the 805 km Karakoram Highway serving Sinkiang Province-Islamabad and Rawalpindi via the Khunjerab Pass). Railway track runs from Iran (Zahedan-Naukundi-Quetta). The route from China was open to foreigners, but routes from Afghanistan were closed in 1987 by hostilities.
Main ports: Karachi, Bin Qasim.

Hotels: A 10 per cent service charge is usually added to bill; further tipping optional.

Car hire: Self-drive and chauffeur-driven car hire is available, and minibuses may be hired. Driving is on the left-hand side of the road.

City transport
Taxis: Taxis may be hailed in the street. Metered taxis are painted black and yellow, although the meter may not be used. Tipping normally 10 per cent.

National transport
Air: PIA operates domestic services to main centres.
Road: About 35 per cent of the 101,300 km road network is classified 'high type'. The 805 km Karakoram Highway permits entry to Northern Areas otherwise accessible only by jeep or goat tracks. Avoid travelling at night on mountain roads in northern areas.
Buses: Ordinary services are generally crowded but express coach services are available between main centres (hourly between Lahore and Rawalpindi). Seats should be booked in advance. The minibuses operating on numerous routes are generally recommended.
Rail: There is an extensive network but slow and somewhat dilapidated with over 8,850 km of track. The main route runs Karachi-Lahore-Rawalpindi-Peshawar, with three classes. Certain services include air-conditioning, restaurant cars, sleeping cars, ice containers and 'ladies only' accommodation. Advance booking is generally advisable, and is essential for some services.

Public holidays
Fixed dates: 23 Mar (Pakistan Day), 1 May, 14 Aug (Independence Day), 6 Sep, 11 Sep, 9 Nov, 25 Dec.
Variable dates: Eid al-Fitr, Eid al-Adha, Ashura, Eid-i-Milad-un-Nabi.

Working hours
Government and business: (Sun-Wed) 0900 – 1300; (Thu) 0900 – 1100.
Banking: (Sun-Wed) 0900 – 1330; (Thu) 0900 – 1100.
Shops: (Sat-Thu) 0800/0900 – 1800/1900.

Social customs: In the northern areas of Baltistan (Skardu, etc) Gilgit, Hunza and the Chitral Valley, do not photograph women, military installations, bridges or airports; in the Kalash Valleys, women may be photographed with their permission, and will usually pose for payment.

Telecommunications: Satellite links are maintained with many countries.
Telephone: Dialling code for Pakistan, IDD access code + 92 followed by area code (51 for Islamabad, 21 for Karachi, 42 for Lahore, 51 for Rawalpindi) followed by subscriber's number. There were 600,000* telephones in 1987. In northern areas, there are a number of public telephone offices, with both the Post Office and the Special Communications Organisation maintaining telephone and radio networks.
Telex: The Central Telegraph Office in Karachi provides 24-hr telex service. Telex facilities are usually available in good hotels.
Telegram: Telegraph offices in main centres are open 0900 – 1600; the Central Telegraph Office in Karachi, Lahore and Rawalpindi provide 24-hr service.

Banking: In July 1985 Pakistan adopted Islamic banking practices which forbid payment of interest. Profit-and-loss (PLS) accounts instead divide the net profit (or loss) of a bank among its depositors in proportion to their savings.
Central bank: State Bank of Pakistan.
Main centre: Karachi.
Head offices: Agricultural Development Bank (Islamabad), Allied Bank of Pakistan, Federal Bank of Co-operatives (Lahore), Habib Bank, Industrial Development Bank of Pakistan, Lahore Commercial Bank, Muslim Commercial Bank, National Bank of Pakistan, Punjab Provincial Co-operative Bank, State Bank of Pakistan, United Bank.
Pakistani branches: Algemene Bank Nederland, American Express International Banking Corp, Bank of America, Bank of Credit & Commerce International SA Luxembourg, Bank of Oman, Bank of Tokyo, Banque de l'Indochine et de Suez, Banque Indosuez, Chartered Bank, Citibank, Dubai Bank, European Asian Bank, Grindlays Bank, Middle East Bank, Rupali Bank, Union Bank of Middle East, Hong kong and Shanghai Bank, Chase Manhattan, ANZ Bank, Deutsche Bank.

Advertising: Advertising is available in the press, in cinemas and on commercial radio and television. Varied sites are available for outdoor advertising, but population coverage is not considered high. Advertising of alcohol is prohibited. Use of direct mail is limited. Nudity is not acceptable.

Electricity supply: 220-240 V AC, with two or three-pin round plug fittings.

BUSINESS GUIDE

Weights and measures: Metric system (local units are also in use).

Representation in capital: Afghanistan, Algeria, Argentina, Australia, Austria, Bangladesh, Belgium, Brazil, Bulgaria, Burma, Canada, Chile, China PR, Czechoslovakia, Denmark, Egypt, France, German DR, German FR, Ghana, Greece, Hungary, India, Indonesia, Iran, Iraq, Italy, Japan, Jordan, Kenya, Korea PDR, Republic of Korea, Kuwait, Lebanon, Libya, Malaysia, Mauritius, Morocco, Nepal, Netherlands, Nigeria, Oman, Philippines, Poland, Portugal, Qatar, Romania, Saudi Arabia, Somalia, Spain, Sri Lanka, Sudan, Sweden, Switzerland, Syria, Thailand, Tunisia, Turkey, UAE, UK, USA, USSR, Vatican, Vietnam, Yemen Arab Republic, Yugoslavia.

BUSINESS DIRECTORY

Useful telephone numbers
Karachi: Police: 222222/224400; Fire: 74891; Ambulance: 73259/70600; International calls: 0102; Calls to India, Bangladesh, China PR, USSR: 102; To check on booked call: 0104

Islamabad Police: 23333; Fire: 27222; International call: 109; To check on booked call: 103

All places Directory enquiries: 17

Hotels: Hyderabad (area code 221)
Fataz Thandi Sarak (tel: 26223).

Hotels: Islamabad (area code 51)
Holiday Inn Islamabad, Aga Khan Road, Shalimar 5, P.O. Box: 1251 (tel: 826121/135; tx: 5740/5612; fax: 820648).

Islamabad, on Municipal Rd, Rama 6, P.O. Box: 1373 (tel: 827311; tx: 5643).

Hotels: Karachi
Airport, Karachi Airport, 11 Stargate Road (tel: 480141/5).

Beach Luxury, Moulvi Tamizuddin Khan Road, New Queens Rd 2 (tel: 551031/7; tx: 23899 AVARI PK).

Holiday Inn Karachi, 9 Abdullah Haroon Road, P.O. Box: 10444 (tel: 520111; tx: 25466, 28035; fax: 511610).

Jabees, Abdullah Haroon Road, 0301 (tel: 512011/5; tx: 24325).

Karachi Sheraton, Club Road, P.O. Box: 3918 (tel: 521021; tx: 25255 ASHER PK).

Mehran, Shah-reh-E-Faisal (tel: 515061; tx: 23616).

Metropole, Club Road (tel: 512051; tx: 24329).

Midway House, Karachi Airport, Stargate Road (tel: 480371/5; tx: 25860).

Pearl Continental, P.O. Box: 8513 (tel: 515021; tx: 23617).

Plaza International, 18/2 Civil Lines, Dr David Pota Road, P.O. Box: 8538 (tel: 520351; tx: 25706 PLAZA PK).

Ramada Renaissance – Avari Towers, Fatima Jinnah Road, P.O. Box: 15503/4 (tel: 525261; tx: 24400).

Taj Mahal, Shahrah-e-Faisal, 0408 (tel: 520211; tx: 24267).

United, David Pota Road, Saddar (tel: 515010/4; tx: 24755).

Hotels: Lahore (area code 42)
Faletti's, Egerton Road, P.O. Box: 144 (tel: 303660; tx: 44469).
Hilton International Lahore, 87 Sharah-e-Quaid-e-Azam (tel: 69971; tx: 44678).

International, Shahrah-e-Quaid-e-Azam (tel: 870281/7; tx: 44230).

Pearl Continental, Shahrah-e-Quaid-e-Azam (tel: 69931/67931; tx: 44877).

Hotels: Peshawar
Deans's, Islamia Road (tel: 79781/3).

Pearl Continental, Khyber Road, P.O. Box: 197 (tel: 76361; tx: 52389).

Hotels: Rawalpindi
Flashman's, 17/22 The Mall (tel: 581480; tx: 5620).

Gatmells Motel, Jail Road (tel: 581648/65123).

Pearl Continental, The Mall, P.O. Box: 211 (tel: 66011; tx: 5736).

Shalimar, off the Mall, P.O. Box: 93 (tel: 62901/9; tx: 5764).

Silver Grill, 8 The Mall, P.O. Box: 266 (tel: 64729).

Car hire: Islamabad
Avis: 7 Rahim Plaza, Murree Road (tel: 827614; tx: 5769/5836).

Hertz: 4 Block E, School Road, F-6/3; (tel: 826930/31; tx: 5652).

Car hire: Karachi
Akbar Tours: Hotel Intercontinental, Club Road (tel: 514689; cable: RENTACAR).

Avis: International Airport, 14 Services Club Mereweather Road (tel: 526295; tx: 5769/5836).

Car hire: Lahore
Avis: International Airport, 122-a Tufail Road (tel: 374429; tx: 5769/5836).

House of Travel: 14A Asif Chambers, Davis Road (tel: 305921, 301446; cable: HOT).

Car hire: Rawalpindi
Avis: 53/9 Haider Road (tel: 68879; tx: 5769/5836).

Sitara Travel Consultants: 25 Shalimar Plaza, off The Mall, PO Box 63 (tel: 64750, 66272; tx: 5751).

Business advice and information
All Pakistan Newspapers Society, 3/F 32 Farid Chambers, Abdullah Haroon Road, Karachi 3.

Federation of Pakistan Chambers of Commerce and Industry, Street 28, Block 5, Shahrea Firdousi, Clifton, Karachi 6 (tel: 534621, 532198, 534616; tx: 25370; cable: FEDCOMERC).

Investment Corporation of Pakistan, National Bank Building, Chundrigar Road, PO Box 5410, Karachi 2 (tel: 225861).

Karachi Chamber of Commerce and Industry, PO Box 4158, Aiwan-e-Tijarat Road, Karachi (tel: 226091/5).

Lahore Chamber of Commerce and Industry, PO Box 597, 11 Racecourse Road, Lahore (tel: 305538; tx: 4388).

National Bank of Pakistan, Chundrigar Road, Karachi, PO Box 4937 (tel: 226780/9).

Rawalpindi Chamber of Commerce and Industry, Chamber House, 108 Adamjee Road, Rawalpindi (tel: 62098, 67598).

State Bank of Pakistan, PO Box 4456, 1.1. Chundrigar Road, Karachi (tel: 234141/10).

Trading Corp. of Pakistan, Press Trust Building, 1.1. Chundrigar Road, Karachi (tel: 210515, 210519).

Travel information
Automobile Association of West Pakistan, 8 Multan Road, PO Box 76, Lahore.

Karachi Automobile Association (KAA), Standard Insurance House, 1 Chundrigar Road, Karachi 0226 (tel: 232173).

Pakistan Ministry of Culture and Tourism, 13-T/U, Comm Area, F-7/2, Islamabad (tel: 27023; cable: MINTOUR).

Pakistan Tourism Development Corp, House 2, Street 61, F-7/4, Islamabad (tel: 26327; tx: 5620; cable: PAKTOURS).

Palau

T. R. Hughes

(map on page xxiv)

Palauans went to the polls in late 1988 to choose a new president, vice-president and a national congress, fervently hoping the elections would see an end to the trauma of the past few years with the deaths of two presidents by gunshot wounds (one was assassinated and the other committed suicide), arson, political intimidation, allegations of corruption and the financial repercussions of a bankrupt power project. All left scars on the republic as it continued the quest for a new political relationship with the United States.

There were signs in mid-1989 that the new Bush administration and Palau leaders were moving to break the deadlock in the on-again off-again Compact of Free Association with the United States which is the key to Palau's future. At stake, however, was Palau's unique anti-nuclear constitution.

Palau is the last remaining component of the Trust Territory of the Pacific Islands administered by the USA on behalf of the UN Trusteeship Council. Its former bedfellows, the Federated States of Micronesia (FSM), Marshall Islands and the Northern Maria-

KEY FACTS

Official title: Republic of Palau

Head of state: President Ngiratkel Etpison

Head of government: President Ngiratkel Etpison

Capital: Koror

Official Languages: English, Palauan

Currency: US Dollar ($ = 100 cents)

Area: 380 sq km

Population: 14,800* (1989)

Visitor numbers: 22,675 (1988)

* estimated figure

nas, have long since graduated to new relationships with the USA. When Palau is able to complete its domestic ratification process which, in effect, means ironing out both internally and with the USA the ambiguities between its constitution and the compact (or contract) then it can get on with the job of running its own affairs.

Compact stalemate

The compact includes a provision for the United States defence of the strategic Western Pacific archipelago for 50 years, with access for its nuclear-armed and powered ships. But Washington refuses to endorse it until Palau lifts its nuclear ban and the Palau Supreme Court refuses to validate any referenda (there have been four on the constitution and six on the compact since 1980) unless at least 75 per cent of the population vote for it. The national vote has come very close several times.

In the November 1988 elections, the pro-compact Ta Belau Party retained control of the country's top post when Ngiratkel Etpison, a state governor and one of the republic's most successful businessmen, scraped home with a 37-vote majority in a seven-handed presidential race. Etpison takes over the reins held by President Lazarus E. Salii who shot himself four months before the poll. Etpison's marginal victory over the reformist and compact-critical elements in Palau politics was tempered by a landslide win by his vice-president, Senator Kuniwo Nakamura, of the opposing Coalition for Open, Honest and Just Government. Unlike the US system, the Palau president and vice-president do not run on a single ticket.

The Coalition was formed ostensibly to protect local civil liberties and human rights. It campaigned for a special prosecutor and auditor to look into political violence and corruption, and endorsed the ongoing investigation by the US watchdog agency, the General Accounting Office (GAO), into Palau's many financial problems. Futhermore, like Bush, Etpison faces a hostile congress.

Under such pressure, it was no surprise that he adopted a conciliatory tone and set up a 26-member commission reflecting a cross-section of Palauan politics, including the anti-nuclear faction, to consider the compact options. In a brisk opening shot, the commission said that any changes would have to be made to the compact not the constitution – a solution that did not appeal to Washington.

The early months of 1989 saw a flurry of activity on the compact front amid strong hints that another referendum was in the offing. The Bush administration paved the way by introducing legislation yet again in congress to implement the compact through a package of subsidiary agreements that would give the embryo sovereign state a chance to straighten out its finances.

The USA and Palau signed the agreements in Guam in May 1989. The deal helps the tiny republic with a population of 14,800 to pay off the massive $48m debt incurred in the power project fiasco. It also provides $9.3m, spread over six years after the compact enters into force, for a special prosecutor, public auditor, a programme for preventing and treating drug and alcohol abuse, plus a new prison. And it tops up the $10m which the USA has already given for the new hospital under construction.

Economic development

Overburdened with the compact troubles, Palau is missing out on economic development. The government runs on an annual budget of around $40.9m derived mainly from US subsidies and local taxes. Foreign investment is at a low ebb (there were 13 foreign business applications in 1988 but only 7 were approved) and there is barely any growth. One area of potential development is tourism. There were 22,675 arrivals in 1988 against 16,639 in 1987 with a preponderance of Japanese visitors. But hotel accommodation is limited.

Other development possibilities lie in aquaculture and fisheries, light industry for exports and import substitution, and agriculture. It is not an easy place to do business: transportation services are expensive and often inadequate, and the commercial infrastructure is, as yet, not well endowed. Any improvement in Palau's financial climate really rests on solving its political problems.

COUNTRY PROFILE

Political system: The Republic of Palau was a Japanese colony and, in 1947, became part of the Trust Territory of the Pacific Islands administered by the USA under UN trusteeship mandate. The Compact of Free Association agreed between the United States and Palau in 1986 gives the islands independence subject to defence being the responsibility of the United States for 50 years. Successive referenda in Palau, however, failed to produce a 75 per cent majority in favour. In 1987, Palau voted to amend the constitution to allow approval of the Compact by a simple majority and, in a subsequent plebiscite, 73 per cent of the votes were cast for the Compact. But the Palau Supreme Court ruled this invalid in August 1988. The court also ruled there were inconsistencies between the Compact and Palau's nuclear-free constitution.

Population: Population growth rate is low by Pacific standards, at 1.8 per cent, but this is partly due to emigration. 22 per cent of Palauans live abroad.

Main town: Koror.

Media
Press: Pacific Daily News.
Radio: Radio broadcasts are transmitted at least 18 hrs/day in English and Palauan. *Television:* TV coverage is fairly widespread, with some cable relay.

External trade
Exports: Main exports are coconut oil, copra, trochus and fish, also some beef, fruit and vegetables. Main destinations are US, Japan and Guam.
Imports: Main imports are food and beverages, machinery, construction materials, textiles and fuel. Main sources are USA, Japan, Taiwan and Australia.

Agriculture: Subsistence farming of coconuts, taro, breadfruit, bananas, yams, sweet potatoes and vegetables, with pig and poultry raising, is main occupation. Commercial farming practised where climate and soils are favourable. Land parceled into an estimated 20,000 holdings; a Land Commission maintains register to provide security of land tenure for Palauan citizens. Fishing is mainstay of economy, supplying principal source of protein and export revenues.

Industry: Small-scale industries include handicrafts, fish processing, bottling, bakeries and boat building. A new electric power plant was inaugurated at Aimeliik in Oct 1986. Hotel projects are planned on Koror, Peleliu and Angaur. The Government is issuing bonds in an effort to raise further development funding.

Membership of international organisations: Association of Pacific Island Legislatures, ESCAP, (associate member), South Pacific Commission.

Time: GMT +11.

Climate: Warm and humid, temperatures 23-30°C, humidity around 80 per cent. Rainfall (variable, minimum 250 mm/year), can occur in downpours. Hurricanes are possible.

Entry requirements
Passport: Required by all except US citizens with proof of citizenship.
Visa: Not required by US citizens with proof of adequate funds. Entry permits granted to all others with proof of onward passage and adequate funds for stays of less than 30 days. Special regulations may apply to some non-tourist destinations. Extension fee is $50 for 30 days.
Health: Precautions against tetanus, typhoid and paratyphoid advisable.

Air access
Airlines serving: Air Micronesia.
International airport: Koror (ROR).

Hotels: There are hotels ranging from 100-room international resort type to small guesthouses in Koror, Peliliu and Angaur.

National transport
Air: Paradise Air flies twice daily between the main centres. Charter services available.
Road: Outside administrative areas, the road network, which is being extended, may consist of tracks not passable to ordinary vehicles. Ngiwal has a new road system which links up with Imekang dock. A new east-west road is under construction from Ngaremlengui to Melekeok; power lines are to follow.
Water: Frequency of inter-island services governed by weather conditions; trips to inhabited islands at least every 30-45 days.

Public holidays
Fixed dates: 1 Jan (New Year's Day), 15 Mar (Youth Day), 5 May (Senior Citizens' Day), 9 July (Constitution Day), 4 Sep (Labour Day), 24 Oct (United Nations Day), Thanksgiving Day, 25 Dec (Christmas Day).

Working hours
Government and business: (Mon-Fri) 0730 – 1630.

Telecommunications
Telephone: Dialling code for Palau, IDD access code + 680 followed by subscriber's number.

Banking: Bank of Guam, Bank of Hawaii, Bank of Palau.

Other useful addresses
Office of the President, PO Box 100, Koror (tel: 96940).

Palau/Hawaii Liaison Office, 1441 Kapiolani Blvd, Suite 1120, Honolulu, Hawaii 96814 (tel: 941-0988/89).

Palau/Guam Liaison Office, ITC Bldg, Suite 615, PO Box 9457, Tamuning, Guam 96911 (tel: 646-9281/81).

Palau Visitors' Bureau, PO Box 100, Koror, Republic of Palau 96940 (tel: 489).

Washington Representative for Palau, Hall of the States, Suite 308, 444 North Capitol St, NW Washington DC 20001, USA (tel: (202)624-7793).

Hotels: Koror
DW Motel, P.O. Box: 738 (tel: 641).

Grace, P.O. Box: 6020 (tel: 1506; fax: 1505).

New Koror, P.O. Box: 339 (tel: 1582; fax: 1582/3).

Nikko Palau, P.O. Box: 310 (tel: 486).

Palau, P.O. Box: 64 (tel: 1703/4).

Palau Pacific Resort, P.O. Box: 308 (tel: 600; tx: 728-8920; fax: 1605).

West Motels, P.O. Box: 280 (tel: 1551; fax: 628-9-1783).

Panama

Luis Kaffman

One hour after midnight on 20 December 1989, US troops launched an assault on the headquarters of the Panamanian defence forces, and other key positions in the country, under the code name operation 'just cause'. The ostensible objectives of the invasion were 'to protect American lives', 'to assist the democratically elected government', 'to seize and arrest an indicted drug trafficker' and 'to defend the integrity of US rights under the Canal Treaties'.

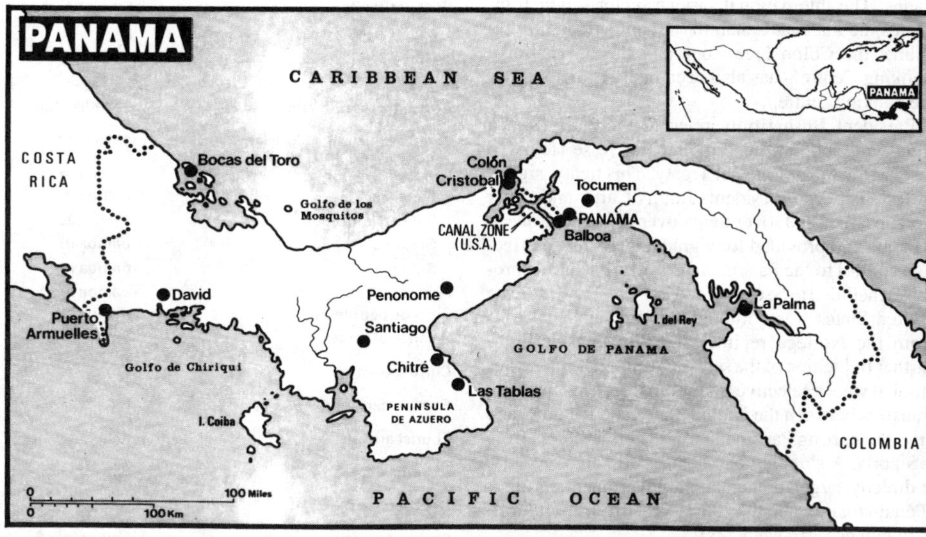

The US forces met fiercer resistance than they had anticipated. The fighting lasted five days. The principal objective – the arrest of Noriega – proved elusive as he went into hiding, emerging subsequently on 24 December as a fugitive in the Vatican embassy in Panama City. On 3 January, Noriega gave himself up to the US forces and was quickly flown to Miami to be indicted on drug charges.

Noriega had named Francisco Rodriguez (an old school friend of his) as the new president from 1 September 1989. Rodriguez was not even a candidate at the May 7 presidential elections, which were declared null and void when international observers reported a landslide victory for the opposition ADOC list headed by Guillermo Endara. The dramatic political crisis which started in June 1987 – when Noriega's second-in-command Colonel Roberto Diaz gave a detailed account of the Panamanian Defence Forces' mafia-style involvement in international drug trafficking, assassinations and electoral fraud – had led to a US economic boycott and an economic collapse: Panama's GDP contracted by about 22 per cent in 1988 and a further 5 per cent in 1989.

Noriega's regime was repudiated by about 70 per cent of Panamanians and became internationally isolated – except for the 'anti-imperialist solidarity' expressed by Cuba, Nicaragua and Libya. But the Panamanian military had neutralised domestic dissent through brutal repression, and the opposition of most Latin American countries to military intervention had previously discouraged the US government from embarking on such a course.

The US Government had consistently underestimated Noriega's political ability and resolve. The financial strangling of Panama since mid-1987 caused great hardship and economic dislocation, but neither this nor military intimidation – before the invasion the United States had sent over 2,000 combat troops with armoured vehicles to Panama to reinforce the 10,000 already based in the country to protect the Panama Canal – was sufficient to persuade the Panamanian Defence Forces to dump Noriega and sue for peace.

This is not wholly surprising. The PDF had been acting for many years as an institutionalised mafia, benefiting handsomely from their control of the Panamanian state (the sale of privileges and profits from domestic racketeering has been supplemented by payments from the Colombian Medellín cartel of drug traffickers for protecting their cocaine shipments and providing them with a safe operating base). President George Bush had stressed that the PDF has an important role to play in a democratic Panama, but Noriega proved more persuasive with his message that they stood to lose their privileges and could even be disbanded and prosecuted if they failed to remain united under his command. The United States diverted aircraft carriers towards Panama and conducted a series of military manoeuvres designed to keep the PDF's nerves on edge. However, Noriega's description of these moves as growls of a paper tiger proved correct until 20 December. Military intervention against the 15,000-strong PDF was known to be a very risky operation. Whilst it lacked the heavy weapons necessary to resist a frontal assault there was always a risk of 'guerrilla resistance' plus the real threat of sabotage to the Canal. Also, the 41,000 US citizens living in Panama were highly vulnerable, although ironically it was the murder of a US citizen that seemed to provoke the invasion. Finally, the United States could not afford the risk of intervening militarily in Panama without the political support of the Latin American countries, and such support was not only most unlikely to be forthcoming but the invasion was a setback for US-Latin American relations.

The Organisation of American States (OAS) met on 17-18 May 1989 to discuss the Panamanian crisis, and it then condemned the Noriega regime's decision to annul the elections by a 20-2 vote, with seven abstentions and only Panama and Nicaragua in opposition. It also agreed to dispatch a mission headed by OAS Secretary-General Joao Baena Soares to seek an early agreement which would ensure an early transfer of power in line with the wishes of the Panamanian people. However, when its mediating efforts failed, the OAS could not agree on what to do next.

In a new meeting in August 1989 the US Government tried to push the OAS to issue an unequivocal declaration calling for the removal of General Noriega, but the Latin American countries baulked and called instead for more negotiations to establish a transitional government in Panama.

The ADOC opposition alliance was dismayed by the OAS climbdown. They rejected new elections and said they were only prepared to negotiate on procedural arrangements for the transfer of power to Guillermo Endara, the man they – as well as the Catholic church and a group of foreign observers headed by former US President Jimmy Carter – claim was elected president by a three-to-one majority on 7 May 1989. However, the Noriega-dominated Council of State, a government policy co-ordinating board which included ministers and heads of the main state enterprises belonging to the ruling Revolutionary Democratic Party (PRD), insisted that the May elections were null and void and said new elections would require amendments to the electoral code and an end to US economic and military pressures. These elections would only take place after six months of the end of the US boycott against Panama.

The US economic sanctions against the Noriega regime included the elimination of Panama's US sugar quota, its exclusion from the preferential export system under the Caribbean Basin Initiative, the cancellation of all economic and military aid, the freezing of about $100m deposited in US banks by the Banco Nacional de Panamá and an order to US companies operating in Panama to suspend the payment of taxes to the authorities. President Bush announced on 7 April 1989 the extension of these sanctions for another year, and on 31 August he suspended diplomatic relations with Panama. These measures proved insufficient to remove General Noriega from power, despite their devastating effect on the Panamanian economy. Real GDP fell by about 25 per cent in 1988 due to plummeting demand and severe cash short-

KEY FACTS

Official title: Republic of Panama
Head of state: President Guillermo Endara
Capital: Panama City
Official Languages: Spanish
Currency: Balboa (B/) = 100 cents
Exchange rate: B/1.00 per $ (fixed)
Area: 77,082 sq km

Population: 2.32m (1988)
GDP per capita: $2,229 (1988)
GDP real growth: –25% (1988)
Inflation: 0.3% (1988)
Trade balance: –$177m (1988)
Foreign debt: $5.3bn (1987)

ages, and there are indications that it contracted again by as much as 10 per cent in 1989. Government revenue fell by 44 per cent in 1988, manufacturing industry was operating at less than half of capacity, and falls in output of over 35 per cent were recorded by some sectors of industry and the distributive trades. The international service sector – which includes the Panama Canal, the Trans-Isthmian oil pipeline, the Colon Free Zone and the International Banking Centre – has also been badly hit, especially the banking centre.

President Bush firmly rejected the calls from a right-wing lobby headed by Senator Jesse Helms to rescind the Panama Canal Treaty. This treaty, signed in 1977 between Presidents Jimmy Carter and Omar Torrijos, returned sovereignty over the Canal Zone to Panama and provided for a gradual transfer of effective control to the Panamanian Government in a process scheduled to be completed in 1999. However, the United States was withholding Panama Canal tolls from the Noriega regime, and was considering a further tightening of the screws. Among the measures studied were a selective trade embargo, a halt on bank transfers between the United States and Panama, and an order barring Panamanian-registered vessels from US ports. A shipping ban is most effective, because it directly targets Panamanian government earnings (Panama earns about $70m a year from its flag-of-convenience arrangements).

The Panamanian business sector was desperate to get rid of the military dictatorship, but had strong misgivings about US tactics that were driving them to ruin. Private investment was paralysed and, with domestic demand continuing to decline, many firms were on the edge of bankruptcy. Public investment was also in the doldrums. The plans to build Centropuerto, a major container transshipment centre, had been put on ice. According to the feasibility study prepared by the US consultants PRC Engineering, the required investment is $324m ($218m for infrastructure and $106m for machinery and equipment), but without access to multilateral finance and with most official financial sources cut off, Panama had no capacity to develop a project of this magnitude.

Gross fixed investment in 1988 fell to just over a quarter of the levels of 1987, while private consumption was down by a fifth and government consumption fell by over 15 per cent. Real exports contracted by 18.6 per cent, but the contribution of the external sector to GDP was positive as a result of the virtual halving of real imports.

Inflation was only 0.3 per cent in 1988 as a result of the sharp contraction in demand, and prices remained stable in 1989 thanks to measures to further restrict demand (wages of government employees were reduced by between 3 per cent and 10 per cent) and substitute imported foods with local produce. The country's external payments position has been severely weakened by the political crisis. International reserves were virtually non-existent in early 1988, when a desperate situation was allegedly solved thanks to loans from Libya and Medellin drug dealers. The reserves position has been improving slowly since then. It is estimated that imports fell by some 43 per cent in 1988 to only $1.77bn, while export revenues fell by 29 per cent to $1.8bn. This produced a small trade surplus – a most unusual phenomenon in Panama – which coupled with the suspension of foreign debt servicing led to a current account surplus in 1988. Interest arrears reached $370m in 1988, while principal arrears exceeded $1bn by the end of that year. The World Bank suspended all disbursements to Panama in May 1989 and put it on its non-accrual list, and in early July 1989 the IMF followed suit declaring Panama ineligible for further borrowing due to arrears of $161m in its dealings with the Fund. Arrears continued to build up in 1989.

According to the latest World Bank figures, Panama's external debt totalled $5,324m at the end of 1987, 3.4 per cent more than in 1986. Public medium- and long-term debt rose 4.2 per cent to $3,722m, while short-term debt had risen by 24 per cent to $1,256m. The total debt is estimated to have increased by a further $470m in 1988 – including interest arrears – reaching $5.8bn.

KEY INDICATORS

	Unit	1984	1985	1986	1987	1988
Population	m	2.1	2.2	2.2	2.3	2.3
GDP (1985 prices)	balboa m	4,680	4,901	5,066	5,188	–
GDP per capita	$	2,229	2,228	2,303	2,256	–
Inflation	%	1.6	1.0	–	1.0	0.3
GDP real growth	%	–	4.7	3.4	2.4	–
Agricultural production	1979-81=100	108.5	112.7	117.1	120.5	113.6
Food production	1975=100	98.0	100.5	101.9	102.4	93.5
Exports	balboa m	277.0	335.4	349.6	356.7	279.8
Bananas	balboa m	74.7	78.1	69.9	85.7	76.5
Refined petroleum	1980=100	7	32	–	–	–
Shrimps	balboa m	46.7	59.8	68.0	65.4	51.8
Sugar	balboa m	33.3	27.3	21.0	17.0	6.2
Imports	balboa m	1,276.3	1,238.8	1,103.8	–	673.4
Crude petroleum	balboa m	288.6	236.4	106.5	174.3	93.9
Balance of trade	$m	–823	–757	–535	–596	–177
Current account	$m	218	286	366	264	737
Foreign exchange	$m	216	85	68	78	73
Tourist arrivals	'000	302	–	–	–	–
External debt	$ m	4,368	4,774	4,935	5,324	–
Debt service ratio	%	8.2	6.5	7.6	6.5	–
Exchange rate	balboa per $	1.00	1.00	1.00	1.00	1.00

COUNTRY PROFILE

Political system: Constitutional republic of nine provinces and an autonomous Indian reservation. Executive power is exercised by a directly-elected president, two vice-presidents and an appointed cabinet. Legislative power is vested in the 67-member unicameral Legislative Assembly.

Main cities/towns: Panama City (population 590,000), Colón, David.

Media
Press: Main US dailies and a few European newspapers can be found in Panama City. Principal Panamanian dailies are *The Star and Herald* (English), *La República* (evening, circulation 5,000), *Crítica*, *La Prensa* (45,000) and *La Estrella de Panamá* (17,000).
Broadcasting: Around 82 AM and 31 FM radio stations and six TV stations, mainly commercial. All TV and radio broadcasting in Spanish. In vicinity of Canal Zone US armed forces radio and TV stations broadcast a variety of programmes in English.

External trade: Regular trade deficit partly offset by invisible earnings – provided by the Colón Free Zone, the Canal, offshore financial services and more recently the transisthmian oil pipeline. Debt rescheduling and World Bank and IMF loans have helped improve overall external position.
Exports: 1985: Bananas (23% of total value), shrimps (18%),raw sugar (8%), refined petroleum (6%), coffee, clothing. Main destinations (1985): USA (74% of total value), Costa Rica (5%), FRG (5%), Belgium (4%), Italy (2%).
Imports: 1985: Crude petroleum (19% of total value), manufactures (25-28%), machinery and transport equipment (20-22%), chemicals (10-12%), food (8-10%). Main sources (1985): USA (31% of total value), Mexico (8%), Ecuador (6%), Venezuela (6%), Japan (9%). NB: main suppliers to the Colón Free Zone (an autonomous institution within Panama) are Japan (typically 20-24% of total value), USA (14-16%), Taiwan (10-12%), Hong Kong (9-11%). The main items passing through the duty free zone are electronics, watches, pharmaceuticals, toiletries, clothing and jewellery.

Agriculture: Around 60 per cent of total land area given over to agriculture, of which 16 per cent is under cultivation and the remainder is natural pasture and forest. Although generating two-thirds of the country's visible exports, and employing about 25 per cent of workforce, agriculture's share of GDP has fallen to around 10 per cent. Principal cash/export crop is bananas (60 per cent of output derives from the Chiriquí Land Company, a subsidiary of United Brands Inc), followed by sugar-cane and raw coffee. Production of staple foodstuffs (rice, meat, potatoes, maize and beans) is insufficient to meet domestic demand – hence imports of foodstuffs have steadily increased. Fishing, particularly for deep-sea shrimps, has increased in importance since completion of new port and fishing terminal at Vacamonte. Panama now ranks as the world's third largest exporter of shrimps (mainly to US). Livestock raising has also grown in importance, and small quantities of meat are exported.

Industry: Sector is small-scale and geared towards domestic consumption – contributes around 10% to GDP. Activity is largely centred in Panama City and Colón, and includes food processing, textiles and clothing, footwear and leather goods, chemicals, plastics, paper, beverages, cigarettes, construction materials and petroleum products from the Las Minas refinery near Colón (capacity 100,000 b/d). Emphasis on encouraging foreign investment in labour-intensive, light assembly, export-based industries. Privatisation of some state assets in progress.

COUNTRY PROFILE

Mining: Activity limited to the extraction of limestone, clays and sea salt. Vast deposits of copper at Cerro Colorado, Petaquilla and Río Tinto Zinc, although plans to develop mines have been postponed owing to poor world trading conditions. Also coal deposits in provinces of Colón and Chiriquí, and known reserves of gold, silver and manganese.

Energy: Heavy dependence on imported oil. Some discoveries of oil and gas have been made, but not in commercially viable quantities. Rapid development of hydroelectric potential, estimated at 2,500 mW. La Estrella, Los Valles and La Fortuna plants supply around 90 per cent of country's electricity requirements. Plans to construct new hydroelectric dam at Changuinola, to be operational by 1990.

Services: Sector is mainstay of economy, with banking contributing over 70 per cent to GDP. Revenues from Panama Canal amount to around $680m annually, though passages through canal steadily declining due to opening of trans-Panamanian oil pipeline (capacity 830,000 b/d). When functioning at full capacity the new pipeline, which links Puerto Armuelles (Pacific Ocean) with Chiriquí Grande (Atlantic Ocean), will contribute $75m annually. Colón Free Zone offers duty-free storage and redistribution to over 1000 foreign companies. It is expected to become Latin America's leading free-trading centre. Earnings from merchant ships registered under Panama's flag of Open Registry are also significant. It is estimated that Panama's merchant fleet is the second largest in the world – 44m gross tonnes of ships registered under Panama's flag in 1985.

Banking: Most dynamic sector of economy, especially since tax and currency controls were liberalised in early 1970s. The assets of some 120 international banks with operations in Panama fell from $36.6bn to $27.8bn during 1987 and continued to fall in 1988.

Membership of international organisations: FAO, G-77, IADB, IAEA, ICAO, ICO, IDA, IFAD, IFC, ILO, IMF, IMO, Inter-American Development Bank, IRC, INTELSAT, IWC (Wheat), Interpol, ITU, IWC (Whaling), NAM, Organisation of American States, PAHO, SELA, UN, UNESCO, UPEB, UPU, WFTU, WHO, WMO, World Bank, WTO.

BUSINESS GUIDE

Time: GMT -5.

Climate: Tropical with little seasonal variation; temperature averages 21°C to 32°C in coastal areas. Rainy season from Apr to Dec (heaviest Oct-Nov); driest months Jan-Apr.

Entry requirements
Passport: Required by all.
Visa: Required by all except nationals of UK, Austria, Spain, Switzerland, FRG, Honduras, Costa Rica and Dominican Republic. All other nationals (except citizens of communist countries, Arab countries, India and Pakistan) require both visa and tourist card. Onward tickets and exit permit if stay longer than 30 days.
Currency: No restrictions on import/export of foreign exchange or Panamanian currency. Visitors must have a minimum of $300. Local currency exists only as coins and is interchangeable with US currency of same denomination. International credit cards widely accepted.

Health precautions
Mandatory: Cholera vaccination certificate if arriving from an infected area. Yellow fever vaccination certificate may be required for visits to certain regions.
Advisable: Yellow fever, cholera, typhoid and polio vaccinations. Malaria risk exists in rural areas – prophylaxis recommended. Water precautions should be taken, especially outside cities.

Representation overseas: Ascunción, Athens, Beirut, Belgrade, Belize City, Berlin, Bogotá, Bonn, Brasilia, Brussels, Buenos Aires, Cairo, Caracas, Damascus, Havana, Kingston, La Paz, Lima, Lisbon, London, Madrid, Managua, Mexico City, Monaco, Montevideo, New Delhi, Paris, Port-au-Prince, Quito, Rome, San José, San Salvador, Santiago, Santo Domingo, Seoul, Stockholm, Taipei, Tegucigalpa, Tel Aviv, Tokyo, Vatican City, Vienna, Washington.

Air access
National airline: Air Panamá Internacional.
Other airlines: Aerolineas Argentinas, Aeronica, Aeroperú, Avensa, Avianca, Copa, Cubana, Eastern, Ecuatoriana, Iberia, KLM, LACSA, Lan-Chile, Lloyd Aéreo Boliviano, Pan Am, Sahsa, Sociedad Aeronica de Medellín (SAM), Taca International, Varig.
Tax: Airport tax of $20 is levied on all passengers, except those in transit staying for less than nine hours.
International airport: Panama City-General Omar Torrijos (PTY), 26 km from city; duty-free shop, bar, restaurant, buffet, bank (restricted hours), post office, shops, hotel reservations, car hire.

Surface access
Main ports: Balboa (Pacific), Cristóbal (Atlantic).
Overland access: The Pan-American Highway links Panama with Costa Rica; a rail link runs between Puerto Armuelles and David. Routes to Colombia are not generally considered passable – except for intrepid tourists.

Hotels: Wide variety available. Advisable to book in advance, particularly between Dec and May. 10 per cent government surcharge on bills. Major credit cards accepted. Tipping 10 per cent (not compulsory).

Car hire: Available in main towns and at the airport. International licence required. After 90 days a local permit is required.

City transport
Taxis: Large (*grandes*) and small (*chicos*) taxis available. Can be ordered by telephone; fixed fare system varies by size of taxi and number of 'zones' traversed. Negotiate fares beforehand. Tipping 10 per cent.

National transport
Air: Frequent internal flights by Copa/Adsa, and others link main centres e.g Panama City with David (generally Mon-Sat only) and Colón (Mon-Fri).
Road: Main means of transportation. Panama section of Pan-American Highway connects Chepo and Panama City with Costa Rican border. Trans-Isthmian highway links Panama City and Colón.
Buses: City buses are cheap but crowded. Regional buses link most towns. Ticabus run modern air-conditioned service to main centres (e.g. Ciudad David-Panama: typical journey time 6 hrs; 10 buses/day). Advisable to book in advance.
Rail: Ferrocarril de Panamá run five trains daily between Panama City and Colón (around 1.5 hrs; single class; air-conditioning extra).

Public holidays: 1 Jan, 9 Jan (National Mourning), Shrove Tuesday (Carnival), Good Friday, 1 May, 15 Aug (Foundation of Panama City; Panama City only), 11 Oct (Anniversary of the Revolution) , 3 Nov (Separation from Colombia), 5 Nov (Colón only), 10 Nov (Los Santos), 28 Nov (Independence from Spain), 8 Dec (Mothers' Day), 25 Dec.

Working hours
Business: (Mon-Fri) 0800 – 1200, 1400 – 1700; (Sat) 0800 – 1200.
Government: (Mon-Fri) 0900 – 1700.
Banking: (Mon-Sat) 0800 – 1300.
Shops: (Mon-Sat) 0800 – 1200, 1400 – 1800/1900.

Telecommunications
Telephone: Automatic exchanges in Panama City and Colón. Inter-continental communications by satellite available through Pan-American Earth Satellite station. The local company is Intercomsa. Dialling code for Panama: IDD access code + 507 followed by the subscriber's number. No area code needed.
Telex: Facilities available at main hotels, also at the Tropical Radio Telegraphic Co's public booth at General Omar Torrijos Airport and at offices of All-American Cables.
Telegram: Radio telegraph services provided by TRT Communications and by Intel.

Postal service: All mail must be addressed 'Republic of Panama' or RP. All parcels must be sent from Central Post Office in Panama City.

Banking
Banking centre: Panama City.
Central bank: Banco Nacional de Panama.

Trade fairs: International Commercial Exposition, Mar, Panama City.

Electricity supply: 110 V AC, 60 cycles (domestic), 220 V AC (industrial).

Representation in capital: Angola, Argentina, Bolivia, Brazil, Chile, Colombia, Costa Rica, Cuba, Dominican Rep., Ecuador, Egypt, El Salvador, France, German FR, Haiti, Honduras, India, Israel, Italy, Jamaica, Japan, Rep. of Korea, Libya, Mexico, Nicaragua, Peru, Poland, Spain, Switzerland, Taiwan, UK, USA, Uruguay, Vatican, Venezuela, Vietnam, Yugoslavia.

Hotels: Colón
Washington Hyatt, P.O. Box: 482 (tel: 47-7501).

Hotels: Panama City
El Ejecutivo (The Executive), Calle 52 Aquilino de la Guardia (tel: 64-3333, 64-3464; tx: 3282056; fax: 691944).

Gran Soloy, Avenida Perú, P.O. Box: 3385 (tel: 27-1133; tx: 2509).

Granada, Avenida Eusebio A. Morales, P.O. Box: 8457 (tel: 64-4900; tx: 2699; fax: 64-0930).

Holiday Inn, Punta Paitilla, P.O. Box: 1807 (tel: 69-1122; tx: 2957; fax: 691645).

Marriott, Caesar Park, Via Israel, P.O. Box: 64248 (tel: 26-4077; tx: 2905; fax: 26-4262).

Riande Continental, Via España, P.O. Box: 8475 (tel: 64-4900; tx: 2699).

Riande Continental-Airport, Tocumen (tel: 20-3333; tx: 3683292).

Business advice and information
Air Panamá, Avenida Justo Arosemena y Calle 39, Apartado 8612, Panama City.

Asociación Panameña de Radiodifusión SA, Avenida 11 y Calle 28, Apartado 1795, Panama City (tel: 25-0160; tx: 2622).

Banco Nacional de Panamá, Vía España, Apartado 5220, Panama City (tel: 69-2111; tx: 2136).

Cámara de Comercio Industria y Agricultura de Panamá, Avenida Cuba y Ecuador, 33A - 18, Panama City (tel: 25-0833, 25-1158; tx: 2434).

Central Post Office, Plaza Catedral, Calle 6, Panama City.

Colón Free Zone, Avenida Roosevelt, Apartado 1118, Colón (tel: 45-3011, 41-5486).

COPA Airways, Avenida Justo Arosemena y Calle 39, Apartado 8612, Panama City.

Edificio Las Vagas (translator service), El Cangrejo, Panama City (tel: 23-0834).

Papua New Guinea

Kevin Rafferty

Australians, who might be expected to know about such things, have sometimes dubbed Papua New Guinea as 'the Next Lucky Country'. This accolade, coming from citizens of 'the Lucky Country', at first sight seems enviously accurate. Papua New Guinea (or PNG as it is often affectionately known) is a vast treasure trove or Aladdin's cave of riches – timber, minerals galore, probably rich reserves of oil and gas, fish in its teeming waters – all in a land nearly half a million sq km in area, with only 3.6m people to share the riches around. But during 1989, Prime Minister Rabbie Namaliu must have wondered many times whether it was all a dream or a nightmare.

It has been a period of great turbulence for PNG. A large number of its troops protested over their pay in February 1989 and went on the rampage through the streets of the capital Port Moresby. A cabinet minister was murdered by gunmen in the middle of the year. In March hundreds of tribesmen went on a 'peace march' to parliament, which turned into a riot and in September there were 'serious riots and looting' in Lae, the country's second city. The giant Bougainville copper mine in the east of the country had to be closed for months because of sabotage attacks from protesting local landowners. There was also trouble of various kinds at the Ok Tedi gold and copper mine in the jungly mountains in the west of the country: it was hit by a landslide, by industrial disruption and by the demands of an MP that it should be closed because of environmental concerns. It seemed that the Next Lucky Country was having its share of the 'luck' of the Australians in developing its great resources.

There can be no doubt about the wealth of the resources. The two mines in the east and west of the country are already open. Ok Tedi produces almost 70,000 metric tonnes of ore a day, yielding 170,000 metric tonnes of copper annually, plus more than 350,000 ounces of gold and a similar amount of silver each year. The Bougainville copper mine, which also produces gold, is among the world's 10 biggest open cut mines. The country is a land of potentially booming and undiscovered mineral resources.

At Lihir, on the eastern fringes of the country north-west of Bougainville, there are an estimated 25m ounces of gold reserves. At Misima Island in the south-east and at Porgera in the Enga highlands there is more gold. In Iagifu-Hedinia, again in the highlands of the main island, there are prospects of oil and gas reserves which would provide as much revenue as the two huge mines of Bougainville and Ok Tedi combined. Geologists and explorers say there are hosts of other minerals dotted about in other parts of the country still waiting to be plucked out of the earth.

This is part of the Pacific 'rim of fire' stretching from Indonesia east and south through the Solomon Islands, Vanuatu, Fiji and New Zealand. But there are two penalties for playing with the fire to extract the resources. One is the nature of the terrain. Equally difficult – and more unpredictable – is the nature of the politics and personalities involved in extracting the treasure from the treasure trove.

It is difficult to exaggerate the remoteness and wildness of much of the land of PNG. 'Rugged' or 'wilderness' are too civilised and pastoral to describe the landscape. Australian soldiers on 'death marches' either as prisoners or trying to capture the Japanese armies in the Second World War knew the horror and terror of a land that combines mountain and jungle. Even today there is no road through from the capital to the second city of Lae. Communications are by air or by ship. And most of the mineral discoveries so far have been made in some of the remotest, bleakest parts of the country.

The Ok Tedi mine is located in the Star mountains. There, before the excavators came, jungle tribesmen might see each other's bush fires a matter of miles away but never ever meet. Nearby, the border with the Indonesian province, Irian Jaya, has still not been demarcated. The excavation of the mine involved taking the top off a mountain hidden in the jungle, and is one of the world's greatest engineering achievements. Even today it is a magnificent sight – when you can see anything at all: for on most days rain and mist impair the view.

The Bougainville mine lies on the edge of a cliff top just beyond a smoking volcano. Until recently the only way in to Porgera, site of one of the world's largest gold deposits, was by helicopter – or after a long foot-slog through the jungle. Lihir is part of a collapsed and half-submerged volcano. And the Iagifu-Hedinia deposit is also only accessible by helicopter. That these finds have been discovered at all is thanks to the work of brave explorers with more than a slice of romantic foolhardiness in their hearts, modern equipment close to the cutting edge of technology – and a chunk or two of good luck. The search for oil, for example, began in PNG more than a decade before the Second World War on the theory that the highlands of Papua looked like those of Iran; unfortunately with rain forests and mountains thrown in to make the search that little bit tougher.

If all the finds live up to their potential, PNG, which is already an important copper producer, could rank as one of the biggest gold producers in the world as well as being an oil and gas exporter. The finds could transform the country's export earnings, not to speak of the government budget.

But '*if*' is always a big word. In spite of its small population, PNG is a complicated country. PNG has a civilisation going back 40,000 years and the people are almost all of Melanesian stock. But their division into tribes physically remote and separated from each other has produced a modern country where people speak more than 700 languages. It is common to find inhabitants from neighbouring villages, – neighbouring, that is, as the crow flies – who can't understand each other's languages. So pidgin has been developed as a lingua franca to enable communication.

There is a lack of a class or caste structure among these peoples. Even the chiefship system has disappeared. But, perversely, this adds to the problems, especially when villagers have moved rapidly from a

KEY FACTS

Official title: Papua New Guinea

Head of state: Queen Elizabeth II

Head of government: Prime Minister Rabbie Namaliu

Ruling party: Pangu Party

Capital: Port Moresby

Official Languages: English (Hiri Motu, Melanesian Pidgin widely spoken)

Currency: Kina (K = 100 toea)

Exchange rate: K0.86 per $ (Nov 1989)

Area: 461,691 sq km

Population: 3.56m (1988)

GDP per capita: $721 (1987)

GDP real growth: 3.9%* (1988)

Inflation: 5.4% (1988)

Trade balance: –$1,198m (1988)

Foreign debt: $2.7bn (1987)

* estimated figure

virtual stone age to the twentieth century, where they are expected to co-operate in a parliamentary system of government based on the Westminster model.

PNG has more or less played by the rules. It has had no military coup or other upheaval of that kind. But its government has always been a turbulent business, particularly in the past few years since 'the father of the nation', Michael Somare, yielded power. Somare took PNG first to self government, then to independence in 1975. In the 1980s, the political restlessness increased, with parties and independent 'big fellas' squabbling for spoils and trying to topple governments without waiting for the electorate – almost to prove their own standing.

Namaliu, in his early 40s, quiet and unassuming in comparison to his predecessors, took over as prime minister in July 1988 after Paias Wingti's five-party coalition was overthrown in a no-confidence motion. The new prime minister's tragedy has been that not only has he had to fight to keep together fractious parties in a coalition with his Pangu Pati, but he's also had to face increasing unrest outside parliament among the diverse elements that make up such a wonderful but wild country.

In February 1989 1,000 out of the 3,500 defence force personnel went on the rampage through Port Moresby to protest against their pay. Armed only with sticks and stones they broke windows at Parliament House and smashed cars before police used tear gas to disperse them. Prime Minister Namaliu denied that the government had 'caved in' to the troops and said that their huge pay rises of up to 70 per cent had already been accepted by the government. Administrative mistakes had meant that the first round of increases were lower than they should have been.

Many of the other troubles came because of disputes between the central government and the provinces or because independent spirits tried to take the law into their own hands. At Bougainville a militant band of disaffected landowners caused problems throughout 1989. The mine was closed for 113 days until September, then opened for nine hours before gunshots were fired at a bus carrying mine employees and explosives brought down an electricity power pylon. It was promptly shut again. The PNG armed forces were given orders to 'shoot to kill' to quell violence. Namaliu said in April that the landowners were being supplied with weapons from foreign sources. 'This is truly a sad event in the nation's history,' the prime minister declared, adding 'those responsible for bringing in and using such weapons can expect no sympathy or mercy.'

In Lae, the riots were partly an expression of tension between the provincial government and Port Moresby. The PNG Government suspended the Morobe provincial government after what it called 'mass lawlessness' and also imposed a dusk to dawn curfew. The provincial premier protested that the suspension was 'politically motivated'.

At Ok Tedi, opposition MP Parry Zeipi, a former environment minister, sought to close the mine after the government had granted permission for the dumping of up to 150,000 tonnes of waste every day into the Fly river. Zeipi declared 'lives need to be balanced with the needs of economic development'. The alternative to dumping the waste was to build a tailings dam which would cost $380m, expensive enough to threaten the mine's viability.

An associated problem of developing the resources, which has flared up time after time, is that the initial capital investment of developing the minerals is so huge and the technology so complicated that only big foreign mining companies have the resources available to take the risk. Most PNG governments have welcomed foreign investment, but they differed in how open their arms have been to receive it. On balance, Namaliu has been more level-headed and welcoming, but has suffered from other forces inside the country demanding their share of the cake.

Foreign involvement in the Bougainville mine, 53 per cent owned by CRA, the Australian affiliate of Britain's RTZ mining group, has been the cause of a major argument, provoking provincial resentment and demands for a bigger share of the cake. All of the big developments have strong foreign partners putting up the cash. Ok Tedi is owned 30 per cent by Australia's Broken Hill Proprietary, 30 per cent by Amoco of the United States and 20 per cent by the PNG Government. The other prospects are led by big Australian or US companies.

If the projects were all to come off, and yield the expected reserves and income, PNG is on the brink of bonanza. Paul Pora, the finance minister, boasted, 'We'll be self-reliant by the late 1990s. We won't be seeking aid. We may be investing in Australia and elsewhere. We might be a donor country to the rest of the Pacific.' But there is that famous *if*. Mineral resources and higher prices have lifted the economy since 1985, taking growth to almost 5 per cent a year against a minimal 0.5 per cent in the early 1980s and only 2 per cent in the 1970s, both rates below the 2.6 per cent annual population increase. Per capita income has still not reached $800 because of the years of slow growth.

Strip the economy of its minerals and PNG is left with a largely agricultural-based economy, depending heavily on what central bank Governor Sir Henry To Robert called the 'four choppy Cs' – coffee, cocoa, copra and copper. In recent years palm oil has been added with some success. The expanding forestry business has suffered from mismanagement. Other aspects of the economy have more than a share of problems: the civil service is bloated, labour skills are comparatively low and wage costs comparatively high, even compared with Australia. The government needs to spend much more on opening up physical communications and on education and health. It has struggled with some success to keep its deficit under control and to bring inflation down to under 5 per cent from much higher levels in the early 1980s.

But it badly needs the mineral boom. According to the World Bank, mining will bring K400m a year to government revenues by the mid-1990s, and oil and gas will add another K250m. This is about 25 per cent of PNG's total current GDP and nearly 70 per cent of current government spending. PNG needs to pray that its luck will run for it and that its truculent power brokers will see that it is in everyone's interest to pull together and not tear each other apart.

KEY INDICATORS

	Unit	1984	1985	1986	1987	1988
Population*	'000	3,430	3,330	3,440	3,480	3,560
Gross domestic product (GDP)	K m	2,134	2,284	2,468	2,764	–
GDP per capita	K	622	686	717	794	–
Consumer prices	% increase	7.3	3.7	5.5	3.3	5.4
Volume of exports						
Copper concentrate	1985=100	98.4	100	105.3	119.6	135.8
Coffee	1985=100	121.7	100	131.1	158.3	110.7
Cocoa beans	1985=100	110.2	100	100	112.1	120.6
Copra & coconut oil	1985=100	93.6	100	92.7	88.4	78.3
Exports	K m	822.13	926.2	1,000.8	1,134.8	1,253.0
Copper concentrate	K m	296.5	311.3	357.2	472.5	493.7
Coffee	K m	110.8	118.0	208.5	129.5	113.9
Cocoa beans	K m	67.00	62.5	56.4	56.6	44.4
Copra and coconut oil	K m	88.6	57.10	20.50	29.8	37.8
Gold	K m	181	175	397	–	–
Imports cif	K m	971.4	1,006.0	1,037.4	1,106.3	1,194.0
Balance of trade	$	–962.8	–874.8	–928.8	–1,209.3	–1,198.8
Current account	$	–321.9	–154.6	–105.3	–325.2	–161.1
External debt	$m	2,062	2,276	2,396	2,711	–
Debt service ratio	%	13.7	12.4	9.2	13.0	–
Exchange rate	$ per K	1.12	1.00	1.03	1.10	1.15

* estimated figure

COUNTRY PROFILE

Political system: Papua New Guinea became independent in 1975 having previously been administered by Australia under a United Nations mandate. It has a 109-member unicameral parliament elected for five years by universal suffrage but the government has changed several times between general elections on the floor of parliament. This happened in July 1988 when Rabbie Namaliu took over as the country's fourth prime minister after Paias Wingti lost a vote of confidence. The prime minister normally heads a government that is an uneasy coalition of small unstable parties and independents, all prone to be fractured or bought off. The governor general represents the head of state, the British monarch.

Political parties: In July 1988 the Pangu Pati took over the leadership of the new coalition government. Other partners in the coalition movement are the Melanesian Alliance, the People's Action Party, the Papua Party, the League for National Advancement, plus most of the National Party. Other parties are the People's Democratic Movement (led by Wingti), Papua Besena, United Party, People's Progress Party. There are nine parties and 27 independents in parliament and expected to be 2,000 candidates at the the next election.

COUNTRY PROFILE

Population: Population 3.6m (1988 estimate). Mainly of Papuan, Melanesian, Polynesian and Micronesian descent, although there is a sizeable minority of Australians and some Europeans. The population is scattered, with highest concentrations in the Highland regions. Growth rate of 2.6 per cent per annum means that over half the population is aged under 20.

Main cities/towns: Port Moresby (population 132,000 in 1987), Lae (72,800), Madang (23,600), Goroka (25,000), Rabaul.

Language: English is the offical language, but Pidgin and Hiri Motu are also found. 700 dialects in use.

Media
Press: Niugini News and *Papua New Guinea Post Courier* (Papuan dailies), some English-language newspapers.
Radio: National Broadcasting Commission transmits in English, Pidgin and certain other local languages. Services from Australia are also received.

Television: E M TV is the only television network and is privately owned by PNG and Australian businessmen.

Domestic economy: Economic development is hindered by geography and inaccessible terrain, which makes resources difficult and expensive to extract, impedes communications and limits agriculture to coastal sites and upland plateaux. The economy is largely dependent on export of commodities subject to wide fluctuations in market prices: copper, coffee, copra and cocoa. Two big mines are the Bougainville copper mine and Ok Tedi gold and copper mine more than 2,000 metres above sea level when mining started. Australian aid accounts for almost 30 per cent of budget receipts. In 1987 aid from Australia was $234m and was expected to decline to $216m in 1988. The government introduced austerity measures and a five-year development plan in 1985. It hopes to increase tax revenues and to make state industries more productive. Since 1985 the economy has enjoyed real annual growth rates of 4.8 per cent and an underlying inflation rate of less than five per cent.

Employment: Wages are relatively high but unemployment and consequent unrest has become a big problem in the major towns where gangs of unemployed youths, known as 'rascals' roam.

External trade
Exports: Exports were valued at $1,000m in 1987. Main exports are gold, copper, coffee beans, forest products, palm oil, cocoa, copra, copra oil and tea. Main destinations are German FR, Japan, Australia, UK, Spain, Netherlands, USA, China PR and New Zealand.
Imports: Imports were valued at $1,265m in 1987. Main imports are machinery and transport equipment, food and live animals, mineral fuels and lubricants, manufactured goods, chemicals and miscellaneous manufactures. Main sources are Australia, Japan, Singapore, USA, New Zealand, UK, Hong Kong and German FR.

Agriculture: Approximately five per cent of land area is cultivated arable, which is restricted by dense rain-forests and mountainous terrain. Agriculture, fishing and forestry employ over 70 per cent of the labour force, account for about 30-45 per cent of GDP and supply many of the main exports. Coconuts, coffee, cocoa, palm oil, rubber and tea are grown as cash crops on plantations, occupying about one-third of those employed in agriculture. Processing, quality control and pricing for main crops are the concern of the Coffee Marketing Board and the Copra Marketing Board, which operate stabilisation funds for these products, and for cocoa. Ground-clearing work on a 5,750-ha palm oil and cocoa project funded by World Bank and the Commonwealth Development Corporation began in 1985, and extraction of oil is expected to begin in late 1988. Smallholdings produce not only 70 per cent of all coffee for export, but also subsistence crops of yams, sago, cassava, bananas, pineapples, vegetables, sweet potatoes, tea, natural rubber, peanuts, sorghum and rice, with some raising of pigs, cattle, goats and poultry. Timber production is increasing, with permits granted for logging 1m ha of forest (of a total 40m ha).

Industry: Most industry is based on agriculture, including the production of palm and coconut oil, sugar processing, brewing and meat processing. Main activities include boat-building, steel fabrication and manufacture of cement, paper products, soap, matches, chemicals, paint, sawn timber, furniture, plywood, bottles and cigarettes. There is a small tourist industry.

Mining: Most mineral resources are difficult and costly to extract, but are currently the country's best hope for the future. Since 1981, copper ore has been mined at Bougainville copper mine, now one of the world's largest producers. PNG became an important gold producer with the opening of Ok Tedi mine (owned 30 per cent each by BHP and Amoco, 20 per cent each by the PNG government and a group of three West German companies). In 1988 the gold cap on the mine was reduced and copper production came on stream. The mine produced almost 600,000 ounces of gold in 1988 and will be producing 150,000 tonnes of copper metal in 1989. Other copper/gold deposits are to be worked in several areas, the most promising being the Frieda River project. There have also been plans to develop a chromite mine in Madang province, and there is natural gas in the Sepik valley.

Hydrocarbons: The US Geological Survey has found huge underwater sediment-filled basins that could be holding oil deposits near Papua New Guinea and the Solomon Islands. There is active exploration for oil both on- and offshore.

Energy: The only large-scale hydroelectric power plant is on the Ramu River. The government is to run a high-voltage power line between the Ramu River plant and Port Moresby across 450 km of mountainous jungle in order to meet projected demand.

Membership of international organisations: ANRPC, ASEAN (observer), Asian Development Bank CIPEC, Colombo Plan, Commonwealth, ESCAP, FAO, GATT, G-77, IBRD, ICAO, IDA, IFAD, IFC, ILO, IMF, IMO, INTELSAT, INTERPOL, ITU, Lomé Convention (EC), South Pacific Bureau for Economic Co-operation (SPEC), South Pacific Forum (participating member), SPC, UN, UNESCO, UPU, WHO, WMO, World Bank.

BUSINESS GUIDE

Time: GMT + 10 hrs.

Climate: Hot and wet, with temperatures ranging from 23-30°C. Rainfall varies considerably from region to region (up to 4,600 mm/year in some places) and falls mainly from Nov-Mar.

Entry requirements
Passport: Required by all except certain seamen.
Visa: Visas are required by all. Visa conditions are liable to change and should be checked before travelling.
Currency: Foreign exchange is controlled, and only available through authorised dealers. No restrictions on importing Kinas; export limited to K200.
Customs: The export of items of ethnographic interest is banned.

Health precautions
Mandatory: Vaccination certificates required for cholera or yellow fever if travelling from an infected area, otherwise traveller is liable to quarantine. Typhoid and cholera vaccinations are advised generally.
Advisable: Malaria is endemic outside Port Moresby and precautions must be taken.

Representation overseas: Brussels, Canberra, Jakarta, London, Manila, Singapore, Suva, Sydney, Tokyo, Washington DC, Wellington (elsewhere contact Australian diplomatic representatives).

Air access
National airline: Air Niugini.
Other airlines: Douglas Airways, Qantas, Talair.
Tax: Airport Tax of NGK10 is levied on international departures.
International airport: Jackson Field (POM), 8 km from Port Moresby; duty-free shop, bar, buffet, bank, hotel reservations, shops and car hire.

Surface access: Cruise ships call, and passenger accommodation is sometimes available on cargo ships.
Main ports: Port Moresby (on New Guinea) due for major expansion, Lae (New Guinea), Aitape, Alotau, Bialla (oil palm wharf), Kavieng, Kieta, Kimbe, Lorengau, Madang, Oro Bay, Rabaul, Samarai, Vanimo (timber wharf), Wewak.

Hotels: In addition to Western-style hotels, the Tourist Board operates a scheme of village-style guest houses run by nationals. Tipping is not usual.

Car hire: International driving licence required, with driving on the left-hand side of the road.

City transport
Taxis: Metered taxi service is available in main centres, but scarce and expensive. Negotiate fares wherever possible.

National transport: The government at times declares a state of emergency in Port Moresby because of the increasing problem of violent crime, and travellers should make enquiries about curfew provisions. Much domestic travel is by air.
Air: Air Niugini, Talair, Bougair and Douglas Airways operate flights to the hundreds of smaller air strips in remote locations. Helicopter and light aircraft charter is available from some 20 small companies.
Road: The 4,900 km road system is being extended. Few road networks connect the various provinces because of the mountainous terrain.
Buses: PMVs (public motor vehicles), usually light buses or covered trucks, operate within and between main centres from bus shelters in towns (or they can be hailed elsewhere).
Water: Sea and river transport are available.

Public holidays
Fixed dates: 1 Jan, 23 Jul, 16 Sep, 25-26 Dec.
Variable dates: Good Friday/Easter Monday, Birthday of HM the Queen.

BUSINESS GUIDE

Working hours
Government and business: (Mon-Fri) 0800 – 1200, 1300 – 1700
Banking: (Mon-Thu) 0900 – 1400; (Fri) 0900 – 1700.
Shops: (Mon-Fri) 0900 – 1630/1700; (Sat) 0900 – 1200. Markets open all daylight hours.

Social customs: Tipping is discouraged.

Telecommunications: Direct dialling via a satellite receiving station in Port Moresby, which permits direct contact with Pacific nations at present lacking an undersea cable link, including New Zealand, Fiji, New Caledonia and the Solomon Islands, as well as Hong Kong, Tokyo, Singapore, New York and Auckland.
Telephone: Dialling code for Papua New Guinea, IDD access code + 675 followed by subscriber's number (no area code is required). There were 330,000* telephones in 1987.
Telex and telegram: Facilities are readily available.

Banking
Central bank: Bank of Papua New Guinea.
Main centre: Port Moresby.
Head offices: Australia and New Zealand Banking Group (PNG), Agriculture Bank, Bank of Papua New Guinea, Bank of South Pacific, Niugini-Lloyds International Bank, Papua New Guinea Development Bank, PNG Banking Corp, Westpac Bank – PNG.
Merchant Bank: Resources and Investment Finance Ltd.

Electricity supply: 240/415 V AC, 50 cycles; plugs are three-pin Australian type.

Weights and measures: Metric system.

Representation in capital: Australia, China, France, German FR, Indonesia, Japan, Republic of Korea, Malaysia, New Zealand, Philippines, UK, USA, Vatican.

BUSINESS DIRECTORY

Hotels: Angoram
Angoram, P.O. Box: 35 (tel: 521589 ext 14; tx: 86134 NE).

Kunai, P.O. Box: 62.

Hotels: Banz, Western Highlands Province
Kunai, P.O. Box: 16 (tel: 562245).

Hotels: Buka
Buka Luman Guest House, Buka Passage, P.O. Box: 251 (tel: 108).

Hotels: Bulolo, Morobe Province
Pine Lodge, P.O. Box: 90 (tel: 445220; tx: 82707; fax: 445284).

Hotels: Collingwood Bay
Waijuga Guest House Cottages, Wanigela.

Hotels: Goroka
Minogere Lodge, P.O. Box: 2 (tel: 722299).

Hotels: Goroka, Eastern Highlands Province
Bird of Paradise, P.O. Box: 12 (tel: 721144; tx: 72628).

Lantern Lodge, P.O. Box: 450 (tel: 721776).

Hotels: Kainantu, Eastern Highlands Province
Kainantu Lodge, P.O. Box: 31 (tel: 771021; tx: 77632 GOROKA NE).

Hotels: Kavieng, New Ireland Province
Kavieng, P.O. Box: 4 (tel: 942199).

Hotels: Kieta, North Solomons Province
Davara, Toniva Beach, P.O. Box: 241 (tel: 956175; tx: 95852).

Hotels: Kundiawa, Chimbu Province
Chimbu Lodge, P.O. Box: 191 (tel: 751144; tx: 75651).

Kundiawa, P.O. Box: 12 (tel: 751033).

Hotels: Lae, Morobe Province
Baubling Haus, Cassowary Road, P.O. Box: 1055 (tel: 424412).

Cecil, P.O. Box: 12 (tel: 423674).

Huon Gulf Motel, Markham Road, P.O. Box: 612 (tel: 424844; tx: 44187).

Klinkii Lodge, P.O. Box: 1982 (tel: 421281).

Lae Lodge, 4th Street, P.O. Box: 2774 (tel: 422000; tx: 42473).

Melanesian, P.O. Box: 756 (tel: 423744; tx: 44187).

Hotels: Madang
Coastwatchers Motel, Coralita Street, P.O. Box: 324 (tel: 822684; fax: 822716).

Jais Aben Resort, Nagada Harbour, 15 km North of Madang Centre, P.O. Box: 105 (tel: 823311; tx: 82716).

Madang Resort Hotel, Coastwatchers Avenue, P.O. Box: 111 (tel: 822655; tx: 82707; fax: 823543).

Plantation, North Coast Road, P.O. Box: 302 (tel: 823176).

Smuggler's Inn, Modilon Road, P.O. Box: 303 (tel: 822744; tx: 82722).

Hotels: Maprik
Maprik Wakken, P.O. Box: 104 (tel: 891221).

Hotels: Mendi, Southern Highlands Province
Mendi, P.O. Box: 108 (tel: 591188).

Hotels: Minj, Western Highlands Province
Minj, P.O. Box: 13 (tel: 565538; tx: 52012).

Hotels: Mount Hagen, Western Highlands Province
Hagen Park Motel, Kupin Terrace, P.O. Box: 81 (tel: 521388; tx: 52056; fax: 522282).

Highlander, P.O. Box: 34 (tel: 521355).

Karawari Lodge, P.O. Box: 371 (tel: 521438; tx: 52012).

Kimininga, P.O. Box: 408 (tel: 521865).

Plumes and Arrows Inn, Mount Hagen Airport, P.O. Box: 86 (tel: 551555; tx: 52088; fax: 551546).

Hotels: Popondetta, Northern Province
Lamington, P.O. Box: 27 (tel: 297152).

Hotels: Port Moresby, National Capital District
Boroko, Okari Street, Boroko, P.O. Box: 1033 (tel: 256677).

Davara, Ela Beach Road, P.O. Box: 799 (tel: 212100; tx: 23236; fax: 212434).

Gateway, Morea Tobo Road, Boroko, P.O. Box: 1215 (tel: 253855; tx: 23082).

Islander, Waigani Drive, Wards Rd and Hohola, Boroko, P.O. Box: 1981 (tel: 255955; tx: 22288).

Papua, Cnr Musgrave & Douglas Streets, P.O. Box: 92 (tel: 212622; tx: 22353).

Port Moresby Travelodge, Hunter Street, P.O. Box: 1661 (tel: 212266; tx: 22248; fax: 217534).

Hotels: Rabaul, East New Britain Province
Hamamas, Mango Avenue, P.O. Box: 139 (tel: 921999; tx: 92959; fax: 921927).

Kaivuna, Mango Avenue, P.O. Box: 395 (tel: 921766; tx: 92982).

Kulau Lodge, North Coast Road, P.O. Box: 359 (tel: 922115; tx: 92930; fax: 922450).

Rabaul Travelodge, Mango Avenue, P.O. Box: 449 (tel: 922111; tx: 92975).

Hotels: Sogeri
Kokoda Trail, Boroko, P.O. Box: 5014 (tel: 282256).

Hotels: Tapini
Tapini, Central District, P.O. Box: 19 (tel: Tapini 298039).

Hotels: Wabag, Enga Province
Wabag Lodge, P.O. Box: 2 (tel: 571122).

Hotels: Wau, Morobe Province
Wau Lodge, P.O. Box: 41 (tel: 446233; tx: 44402).

Hotels: Weam
Bensbach Wildlife Lodge, Mount Hagen, P.O. Box: 930 (tel: 521438; tx: 52012).

Hotels: Wewak, East Sepik Province
Sepik International Beach, P.O. Box: 152 (tel: 862388; tx: 86119; fax: 862701).

Sepik International Motel, Hill Street, P.O. Box: 496 (tel: 862422; tx: 86143).

Wewak, P.O. Box: 20 (tel: 862155; fax: 862701).

Hotels: Wuvulu
Alaba Lodge, Wewak, East Sepik Province, P.O. Box: 437.

Car hire
Avis: Jacksons Airport, PO Box 1533, Port Moresby (tel: 258299; tx: 22279), Goroka (tel: 721084); Mount Hagen (tel: 551350); Lae (tel: 422722); Rabaul (tel: 921131); Wewak (tel: 862041; tx: 86143).

Budget: Port Moresby (tel: 254514; tx: 22115); Rabaul (tel: 921008); Lae (tel: 422069).

Travelodge Hotel Car Service: Port Moresby (tel: 212266).

Business advice and information
Agriculture Bank of Papua New Guinea (formerly the Papua New Ginuea Development Bank), PO Box 6310, Goroko.

Bank of Papua New Guinea, PO Box 121, Douglas Street, Port Moresby (tel: 212999; tx: NE22128).

Department of Industrial Development, PO Box 5644, Goroko (tel: 272286).

Investment Corp of Papua New Guinea, PO Box 155, Port Moresby (tel: 212855).

Lae Chamber of Commerce and Industry, PO Box 265, Lae.

National Investment Development Authority, PO Box 5053, Goroko (tel: 258777; tx: 72123).

National Statistical Office, Post Office, Ward's Strip (tel: 271782, 271705).

Papua New Guinea Associated Chambers of Commerce and Industry, PO Box 1621, Port Moresby (tel: 422340).

Port Moresby Chamber of Commerce, PO Box 1764, Port Moresby (tel: 214587).

Travel information
Air Niugini, PO Box 7186, Boroko (tel: 273200).

Melanesian Tourist Services, PO Box 707, Madang (tel: 822766; tx: 82707; fax: 823543).

National Tourist, Authority, Marea House, Port Moresby (tel: 272160/3).

Paraguay

Andrew Nickson

The military coup which overthrew the regime of General Alfredo Stroessner on 2 February 1989 marked a major turning-point in the modern history of Paraguay. Stroessner had ruled the country in a highly autocratic fashion ever since 1954, making him the longest-serving president in Paraguayan history. In recent years his increasing identification with the extremist *militante* faction within the ruling Partido Colorado had alienated many of his former supporters, including the private sector, the United States Government and large sectors of the Colorado Party itself. The coup was hatched in January 1989 when the *militantes* attempted to retire a number of senior army generals as part of their plan to groom Stroessner's eldest son, Air Force Colonel Gustavo Stroessner, as a presidential successor. It was led by General Andres Rodriguez, head of the powerful First Army Corps, who had provided Stroessner's main support within the armed forces throughout most of his regime. Rodriguez has promised to democratise Paraguay and to introduce a neo-liberal economic growth strategy. The major obstacle that he will face in pursuing both objectives will come from two deep-rooted legacies of the Stroessner regime – political intolerance and institutionalised corruption.

Economic developments

Economic performance during 1988 improved for the second year running as a result of higher agricultural production and improved export prices. According to official Central Bank figures, GNP rose by 5.9 per cent in 1988, compared with an increase of 4.2 per cent in 1987 and a zero increase in 1986 when a severe drought reduced agricultural output. However, these figures were questioned by foreign government sources. According to the US embassy, GNP rose by only 4.0 per cent in 1988. The official inflation rate for 1988 was 17 per cent, although the independent research body, CDE, recorded a rate of 32 per cent.

Agricultural production rose by 9.0 per cent as a result of a combination of increased production and higher world prices for major export crops. The production of raw cotton, the second major export crop, rose dramatically from a low of 243,834 tonnes in 1987 to an all-time record of 550,000 tonnes in 1988 as a result of higher producer prices. Wheat production also reached a record 318,000 tonnes. Meanwhile the volume of production of soya, the major agricultural export, failed to respond to higher world prices and remained at 1,150,000 tonnes, largely due to a drought during the first quarter of the year. Domestic food crops such as maize, mandioca, sweet potato and beans were also adversely affected by the drought. Sugar-cane production fell to only 260,000 tonnes as a result of frost during 14 days in the year. Flooding of the River Paraguay during the winter months (June-September) had an adverse impact on cattle production.

Industrial production rose by 5.3 per cent, compared with a rise of 3.5 per cent in 1987. The inflow of capital associated with large-scale bi-national hydroelectric projects was the lowest for 13 years. The Itaipu project (with Brazil) was almost completed, while the recently commenced Yacyreta project (with Argentina) displayed a very low Paraguayan involvement. In December 1988 two major loans were signed with the World Bank and the Inter-American Development Bank which guaranteed the continuation of work on Yacyreta, which is scheduled to come on stream by 1993.

The foreign trade sector continued to improve during 1988. Officially recorded exports increased by 44 per cent to $509m, due to higher world prices, higher export volumes and higher minimum export prices *(aforos)*. Recorded imports (fob) fell by 4 per cent to $495m, resulting in a small trade surplus of $14m, compared with a deficit of $164m in 1987. Nevertheless, these flows grossly understate the true extent of the foreign trade sector due to the extensive role of contraband in the Paraguayan economy. Non-registered trade was estimated by the Central Bank to be equivalent to the value of officially registered trade. The multiple exchange rate system which exchanged export earnings at G550 to the dollar (compared with a free market rate which fluctuated between G900-G1,000 during the year) further increased the incentive to under-register exports.

Despite the improved trade balance, the overall balance of payments deteriorated sharply in 1988, especially during the second half of the year. A number of state companies represented a heavy drain on the nation's domestic and external finances. These

KEY FACTS

Official title: Republic of Paraguay

Head of state: President Andrés Rodríguez

Ruling party: Partido Colorado (Asociación Nacional Repúblicana)

Capital: Asunción

Official Languages: Spanish (Guaraní also spoken)

Currency: Guaraní (G)

Exchange rate: G1,185 per $ (Nov 1989)

Area: 406,752 sq km

Population: 4.04m (1988)

GDP per capita: $1,503 (1988)

GDP real growth: 6.2% (1988)

Inflation: 23.4% (1988)

Trade balance: $68.6m (1988)

Foreign debt: $2.0bn (1988)

companies, in particular the state steel company, ACEPAR, and the state cement company, INC, had borrowed heavily from commercial banks in the early 1980s to finance ill-advised turnkey projects. A combination of overcapacity, mismanagement and rampant corruption led to a growing requirement for access to cheap domestic credit and foreign exchange in order to avoid their financial collapse. Their absorption of highly subsidised credit from the state banking system starved the private sector of loan finance, while the multiple exchange rate system subsidised public sector debt service payments ($1=G320 compared to the market rate of $1=G1,000) at a heavy cost to private importers.

The foreign debt burden has increased markedly in recent years, reaching 53 per cent of GNP by 1987. Total outstanding foreign debt reached $2,002m by December 1988, almost all of which was owed by the public sector. The principal creditors were multilateral development agencies (34 per cent), commercial banks (25 per cent), bilateral donors (15 per cent) and the Bank of Brazil (26 per cent). Of the total shortfall in debt service payments by the end of 1988, 55 per cent was accounted for by ACEPAR and 21 per cent by INC.

By the second half of 1988, heavy loan repayment obligations to foreign commercial banks, together with growing financial deficits among these state corporations, led to a sudden and rapid decline in the level of foreign reserves. In December an internal Central Bank study was leaked to the press which revealed the seriousness of the crisis in graphic detail and warned that foreign reserves would be exhausted within a matter of months unless there was a sharp devaluation and immediate renegotiation of debt service obligations. No action was taken.

KEY INDICATORS

	Unit	1984	1985	1986	1987	1988
Population	m	3.6	3.7	3.8	3.9	4.0
GDP	G bn	1,070.4	1,393.9	1,833.8	2,493.6	3,432.9
GDP per capita	$	1,219	854	933	957	858
Inflation	%	20.4	25.2	31.7	32.0	–
GDP real growth	%	3.1	4.0	–	4.3	6.2
Agricultural production	1979-81=100	120.3	137.1	119.4	126.6	152.0
Food production	1979-81=100	104.7	111.2	97.9	107.5	117.1
Sugar-cane	'000 tonnes	1,700	2,330	2,158	3,188	3,382
Exports	G m	76,537	96,708	92,128	208,613	333,827
Sawn timber	G m	5,004	3,072	7,246	15,003	–
Cotton	G m	30,737	45,281	30,528	59,723	–
Soya beans	G m	21,928	32,134	14,252	67,802	–
Vegetable oils	G m	4,447	4,278	3,373	5,601	–
Imports	G m	135,743	195,727	275,330	373,407	354,167
Machinery & transport	%	38.0	32.1	32.1	29.4	–
Fuels & lubricants	%	26.8	25.9	19.0	19.9	–
Balance of trade	$m	–287.8	–191.5	–162.4	–97.2	68.6
Current account	$m	–317.4	–225.5	–358.9	–133.4	–125.0
Tourist arrivals	'000	291.9	262.7	370.7	303.2	–
External debt	$m	1,461	1,791	2,040	2,447	–
Debt service ratio	%	17.7	23.9	25.5	21.3	–
Exchange rate	G per $	201.0	306.7	339.2	550.0	550.0

* estimated figure

Economic reform plan

The severe financial crisis at the beginning of 1989, with a public sector deficit in excess of 10 per cent of GNP and foreign exchange reserves down to only $143m, undoubtedly contributed to marshalling support for the coup which ousted Stroessner. The new government of General Andres Rodriguez announced an economic recovery plan shortly after taking office in February. The plan was designed to overcome many of the structural problems associated with the economic policies of the Stroessner regime in a manner which strengthened the role of the private sector in the economy. The former Ambassador in Britain and newly-appointed Minister of Industry and Commerce, Antonio Zuccolillo, soon came to personify this neo-liberal growth strategy.

The following important measures were introduced as part of the plan: the replacement of the multiple exchange rate system by a single exchange rate, the ending of cheap credit subsidy to public sector companies, the elimination of export tax and abolition of the *aforo* system and an increase in tax exemptions on domestic and foreign investment. The monopoly of the state company, APAL, on trade in spirits and alcohol was abolished and credit limits imposed on lending to the public sector. Debt service obligations to the Brazilian Government and commercial banks were rescheduled and plans announced to privatise ACEPAR and INC, which have a combined foreign debt of $840m. In the longer term, the government stated its intention to reform Paraguay's anachronistic economic and financial system through measures such as the establishment of a state audit commission, the introduction of personal income tax and VAT and the creation of a professional civil service.

Economic performance improved somewhat during the first half of 1989, encouraged in part by the policy measures introduced in February. Cotton production in 1989 reached a new record, estimated at 600,000 tonnes, compared with 550,000 tonnes in 1988. Wheat production, however, was seriously affected by widespread smuggling of imported flour from Argentina. Officially recorded exports rocketed following the introduction of a single exchange rate and abolition of the *aforo* system. Recorded exports for January-June 1989 were $535m, $26m higher than the total for the whole of 1988. Imports for the same period were 94 per cent higher than for January-June 1988. There was a marked reduction in the rate of inflation which fell to 15 per cent for the first half of 1989. However, an average price rise of 25 per cent in water, electricity, telephone and public transport announced in June 1989, coupled with a 15 per cent increase in the minimum wage rate, will fuel inflationary pressure during the second half of 1989.

There was also a marked improvement in the foreign debt situation. In April a total of $436m owed to Brazil was renegotiated over 13 years with a five-year grace period. In June Finance Minister Enzo Debernardi announced that Paraguay would soon receive $213m of unpaid royalty payments from Itaipu Binacional which had accumulated since March 1985. By June 1989, foreign exchange reserves had risen to $220m. However, plans to privatise state companies received a setback in July when the bulk of the Colorado Party opposed the stated intention of Industry Minister Zuccolillo to sell the state cement company, INC, at a knock-down price of $100m.

Industrial production was severely hit during the first half of 1989 by an unprecedented increase in contraband as a result of hyperinflation and repeated devaluations in Argentina and Brazil. A sizeable proportion of cotton and soya exports were believed to have originated in Argentina and Brazil for re-export through Paraguay. The government policy of reducing import tariffs to a standard 10 per cent on all border traffic as part of its neo-liberal strategy aggravated the situation by legalising much former contraband trade. In June the head of the industrialists' organisation, UIP, called on the government to declare a national emergency and for the armed forces to repress contraband. However, despite the pleas of the industrialists, the government remains wedded ideologically to a free market philosophy and pragmatically to the need to hold down living costs through cheap imports at a time of price hikes in basic services. As a result of the economic problems in neighbouring Argentina and Brazil, the tourist industry suffered one of its worst seasons for a decade during the winter months of 1989 due to the reduction in foreign visitors.

Politics

Following the military coup in February 1989, General Andres Rodriguez surprised observers by initiating democracy in Paraguay. Opposition political parties were recognised, draconian anti-subversive laws repealed, relations with the Catholic church improved and media freedom promoted by the re-opening of the leading daily newspaper, *ABC Color*, and Radio Nanduti, both of which had been closed by Stroessner. A noticeable improvement in respect for personal liberties took place and several leading *militantes* were detained under accusation of torture and corruption. In the following months, the press investigated cases of rampant corruption by former ministers in the Stroessner regime and reported on individual cases of disappearances of political detainees going back many years.

On 1 May 1989, exactly three months after deposing Stroessner, General Andres Rodriguez was elected president of Paraguay as the official candidate of the Colorado Party. After considerable internal wrangling, the opposition parties agreed to participate in the election in exchange for official recognition, freedom of association and access to the media, despite the absence of a clean electoral register and the deeply-ingrained tradition of ballot-rigging inherited from the past. Despite some irregularities, Rodriguez' well-organised campaign and convincing electoral victory re-inforced the fact that the reputation of both the Colorado Party and the armed forces had emerged largely unscathed from their long association with the Stroessner regime. Rodriguez received 73 per cent of the presidential vote, while the Colorado Party received 69 per cent of the vote for senators and deputies in Congress. The Partido Liberal Radical Autentico, which was not officially recognised during the Stroessner years, was the only opposition party to register a substantial vote. Its presidential candidate, Domingo Laino, received 21 per cent of the vote, while its candidates won 20 per cent in elections to Congress. However, the remaining opposition parties polled surprisingly low votes. The social democrat Febrerista Party scored only 1.1 per cent, while the Christian Democrats performed even worse at 0.7 per cent of the presidential vote. The three factions of the Liberal Party which had collaborated as a 'loyal opposition' in Congress under the Stroessner regime, together received only 1.5 per cent of the vote.

By August 1989, President Rodriguez faced growing divisions within the Colorado Party. The old party stalwarts, known as the *tradicionalistas*, who had fallen out with Stroessner in 1987, resented the ascendancy of the *ex-contestarios*, Colorados, who had been forced into exile back in 1959, and who increasingly have the ear of President Rodriguez. Despite the arrest or exile of its leading members, it became clear that the *militantes* are not a spent force and still retain a nationwide organisation led by Martin Chiola. There were signs that, in order to counter the *ex-contestarios*, the *tradicionalistas* were prepared to allow the *militantes* to re-enter the Colorado fold.

Meanwhile, the opposition parties remain wedded to a highly personal and outdated style of politics. They are unlikely to provide a credible alternative to the Colorado Party in the run-up to the municipal elections in November 1990. A political vacuum is emerging as new movements for social change have arisen in the new democratic atmosphere which remain largely independent of opposition parties. In both rural and urban areas, popular organisations have mushroomed in order to defend families from land evictions. In May 1989 the state Rural Welfare Institute, IBR, recognised the existence of 28 such associations or movements in rural areas. In mid-August 1989 more than 60 trade union and peasant bodies attended the founding congress of an independent Trade Union Congress, Central Unitaria de Trabajadores. The increasing official reluctance to pursue prosecution against high-ranking military officials and civilians accused of corruption, together with the dropping of charges against former torturers, has also served to dent the initial popularity of the Rodriguez government. There has been considerable discussion in the media concerning the eventual legalisation of the Paraguayan Communist Party, with several leading members of the Colorado Party voicing their support.

International relations have improved markedly since the coup. The Bush administration has given strong support to the Rodriguez government and is likely to re-establish Paraguay's trade benefits under the Generalised System of Preferences which were withdrawn by the Reagan administration in 1987. Relations with Argentina, which deteriorated sharply during the last years of the Stroessner regime, have improved and General Rodriguez attended the inauguration of President Menem.

The major issue on the political agenda during 1990 is likely to be the question of land reform. With the exhaustion of virgin forest land available for colonisation, the pressure of population growth coupled with the expansion of commercial forms of agriculture has led to a dramatic increase in the number of landless families, now estimated at around 60,000. The introduction of democracy since February 1989 has given expression to growing conflict between land-hungry peasant communities and large landowners. Between February-May 1989 alone, 19 land invasions totalling 70,000 hectares were registered, involving 7,227 families. In all but one case, the squatters were forcibly evicted by the army, suggesting a hardening of attitude in government policy in comparison with that of the Stroessner regime. However, a National Council for Rural Development was set up in June 1989 with the task of formulating a strategy to defuse this potentially explosive social problem. Although peasant organisations are not represented on this body, there are signs that it will recommend a far-ranging and long-overdue land reform programme. However, the implementation of such a programme is likely to meet with resolute opposition from Paraguay's powerful land-owning elite.

COUNTRY PROFILE

Political system: Executive power rests with the president (elected by direct vote for a five-year term) who appoints a council of ministers. The legislature is the National Congress, made up of the Senate (36 members) and the Chamber of Deputies (72 members). The term of both chambers is normally five years and representation is decided by the vote in the presidential elections – the largest share carrying with it two-thirds of the seats.

Main cities/towns: Asunción (population 700,000), Puerto del Este, Pedro Juan Caballero, Encarnación, Villarrica, Concepción.

Media
Press: Main daily newspapers, based in Asunción, are *ABC Color, El Diario, Ultima Hora* (circulation 45,000), *Hoy* (40,000). USA dailies are usually on sale in the capital. A few periodicals *(La Opinión* and *Sendero)* are also published.
Broadcasting: There are over 30 government and commercial radio stations and two television channels. All broadcasts are in Spanish.

External trade: Heavy dependence on agricultural exports, especially cotton fibre and soya. Trading affected by EC restrictions on meat imports, Argentina's restrictions on wood imports and fluctuating world commodity prices. The 1987 trade deficit was down by 40 per cent to $164m.
Exports: 1988: Raw cotton (40% of total value), soya (36%), meat and hides (8%), timber (3%), coconut oil (2%), essential oils (2%), tobacco (1%). Main destinations (1988): Brazil (22% of total value), Argentina (17%), Netherlands (12%), Switzerland (10%), Chile (5%), German FR (4%), Panama (4%), USA (4%), Italy (3%), Spain (3%), Uruguay (2%), France (1%).
Imports: 1985: Machinery and transport equipment (9% of total value), mineral fuels etc (20%), beverages etc (8%), chemicals (5%), wheat and other food products (2%), agricultural equipment, iron, steel manufactures etc. Main sources (1986): Brazil (32% of total value), Argentina (14%), Algeria (7%), USA (14%), FRG (6%), Japan (6%), UK, Spain, Switzerland, France, Netherlands, Uruguay.

Agriculture: Sector contributes 25 per cent to GDP and accounts for 90 per cent of exports. Principal export crops are cotton, soya and vegetable oils. Sawn timber and, to a lesser extent, meat products, fruit, vegetables, hides, skins are also important export earners, though less so than in the 1970s. Cane sugar, wheat, tobacco and various new specialist crops for industrial use are expanding as more land comes under cultivation. Near self-sufficiency achieved in staple foods (rice, maize, wheat, beans). Agricultural production fluctuates from year to year owing to climatic conditions (both flooding and drought) and widespread smuggling (particularly livestock and soya). Approximately five per cent of total land area is arable land or under permanent crops; pasture 35 per cent; 50 per cent woodland/forest. The Chaco region (the western side of the country which makes up 60 per cent of the total area) is covered by scrub woodland, grass etc. The eastern region of the country is generally fertile; flooding in the flatter areas around the main rivers generally occurs every year.

Industry: Manufacturing contributes around 16 per cent to GDP and is centred on the processing of agricultural products (particularly textiles, cotton yarn and food processing). The country is self-sufficient in cement and possesses an oil refinery (capacity 10,000 b/d) and steel works. A new cement plant, which will quadruple capacity, is under contract, while a 150,000 tonne/year steelworks started operations at Villa Hayes in 1986. Industry and construction sectors operating well below capacity.

Mining: Activity is confined to the extraction of limestone and salt. Also known deposits of iron-ore, manganese and copper. Extraction of these uneconomic at present due to transport difficulties. The government has introduced financial incentives to promote exploration for petroleum and uranium.

Energy: Self-sufficient in electric energy with supplies coming mainly from the 190 mW Acaray hydroelectric plant, and the massive 12,600 mW Itaipú hydroelectric plant on the river Paraná. This joint venture with Brazil commenced operation in 1984 and is expected to become fully operational in 1991. 50 per cent of the energy generated belongs to Paraguay, most of which will be exported to Brazil. Exports of electricity to Argentina from Yacyretá scheme had been due to begin 1987 but bilateral disputes have delayed work on the project so scheduled commencement date is Nov 1993. Wood remains the main source of fuel. Oil requirements of 10,500 b/d are met through imports.

Membership of international organisations: CCC, FAO, G-77, IADB, IAEA, ICAO, ICO, IDA, IDB, IFC, IFAD, ILO, IMF, INTELSAT, Interpol, IPU, IRC, ITU, Latin American Integration Association, Organisation of American States, SELA, UN, UNESCO, UPU, WHO, WMO, World Bank, WSG.

Time: GMT -4 (GMT -3 from Oct to Mar).

Climate: Sub-tropical. Summer from Jan to Mar is hot with maximum temperatures ranging from 25°C to 40°C. Winter from Jun to Sep is mild with average temperatures 10C to 20C, although daily range from 5°C to 30°C is likely. Rain falls all year round but is heaviest from Oct to Apr.

Entry requirements
Passport: Required by all except tourist travellers from some neighbouring countries.
Visa: Required by all except nationals of EC countries (excluding Greece, Republic of Ireland), USA, Canada, Austria, Israel, Scandinavian countries, Switzerland and most South American countries.
Prohibited entry: Visitors with USSR, Cuban or various Central American entry stamps in their passports may be refused entry.

BUSINESS GUIDE

Currency: No restriction on import/export of foreign or local currency.
Tax: All visitors, except children, required to pay an airport tax of $3 on departure.

Health precautions

Mandatory: Yellow fever vaccination certificate required if arriving from an infected area.

Advisable: Typhoid and polio vaccination. Malaria prophylaxis recommended for visits to certain rural areas. Water precautions should be taken outside Asunción.

Representation overseas: Beirut, Bogotá, Bonn, Brasília, Brussels, Buenos Aires, Caracas, Geneva, La Paz, Lima, London, Madrid, Mexico City, Montevideo, Quito, Paris, Pretoria, Rome, San Salvador, Santiago, Seoul, Taipei, Tokyo, Vatican City, Washington.

Air access

National airline: LAP (Líneas Aéreas Paraguayas).

Other airlines: Aerolíneas Argentinas, Eastern, Iberia, Ladeco, Lloyd Aero Boliviana, Pluna, Varig.

International airport: Asunción-Silvio Pettirossi/Campo Grande (ASU), 15 km from city; duty-free shop, restaurant, currency exchange. All visitors, except children, required to purchase a tourist card for $3.

Surface access
Main ports: Asunción (on river Paraguay), approximately 1,500 km from the sea; in suitable conditions ships can reach Concepción, 290 km upstream; smaller vessels travel another 950 km after that. The Paraná river is also navigable via Encarnación and Puerto Presidente Stroessner.

Overland access: Paved road from Brazil (Rio de Janeiro-São Paulo-Asunción; length 1,700 km). Also paved roads from Argentina (Buenos Aires-Sante Fé-Asunción, length 1,450 km; and Buenos Aires-Encarnación). Buses on these routes are fast and fairly comfortable. Routes also from Bolivia and ferry service on Paraná river from Argentina. Regular train runs from Buenos Aires to Posadas (Argentina) from where a ferry or bridge over the river Paraná allows connection with Encarnación–Asunción railway. River and rail journeys are very slow – buses quicker.

Hotels: Accommodation limited outside Asunción. Rooms should be booked well in advance, especially for the tourist season in Asunción (Jul-Aug) when accommodation is in short supply. If service charge not included in bill, 10 per cent tip is usual.

Car hire: Foreign or international licence acceptable. Chauffeur and self-drive cars are available at reasonable rates.

City transport

Taxis: In Asunción metered with minimum fare system; can be hired on time basis; around 10 per cent tip is optional.

Trams: Lower fuel costs said to be one reason for resumption of some services (using ex-Belgian vehicles). Two routes operate.

National transport

Air: LAP, TAM (Transportes Aéreas Militares) and LATN (Líneas Aéreas de Transporte Nacional) operate scheduled services to most parts of the country. Also planes can be chartered and seats booked on air taxis for many destinations. Flights are frequently affected by weather.

Road: Around 10 per cent of total network is surfaced. Main routes from Asunción to Encarnación and to Puerto Presidente Stroessner. The Trans-Chaco Highway runs to the Bolivian border, but is paved only half way. Some unsurfaced roads closed in bad weather conditions; service stations etc may be widely spaced. However, roads serving main centres are in good condition.

Buses: Bus/minibus service in Asunción. For longer distances advance booking advisable; frequent express (usually comfortable) services link major towns (e.g. Asunción – Encarnación: six hrs; Asunción – Puerto Presidente Stroessner: five hrs).

Rail: Main route Asunción – Villarrica – Encarnación. Services can be unreliable with long journey times, and has sometimes been out of action for months at a time.

Water: Asunción – Concepción service not frequent but generally recommended if time (approximately 24 hrs) available (with Flota Mercante del Estado).

Public holidays: 1 Jan, 3 Feb (San Blas), 1 Mar (Heroes' Day), Maundy Thursday, Good Friday, 1 May, 14-15 May (National Independence), Corpus Christi, 12 Jun (Chaco Peace), 15 Aug (Foundation of Asunción), 25 Aug (Constitution Day), 29 Sep (Battle of Boquerón), 12 Oct (Columbus Day), 1 Nov (All Saints'), 8 Dec (Immaculate Conception), 25 Dec.

Working hours

Business: (Mon-Fri) 0700 – 1130 and 1500 – 1730/1900; (Sat) 0700 – 1130.

Government: (Mon-Fri) 0700 – 1200; (Sat) 0700 – 1100.

Banking: (Mon-Fri) 0730 – 1100.

Shops: (Mon-Sat) 0730 – 1130, 1500 – 2000.

Telecommunications

Telephone: Dialling code for Paraguay: IDD access code + 595 followed by area code (Asunción 21) followed by subscriber's number. Service generally reliable.

Telex: Good service. Available at main hotels and at ANTELCO offices (Alberdi y General Díaz (tel: 44001)).

Telegram: Services widely available.

Postal service: Letters should be registered if important. Service using post box (Casilla) system is reasonably reliable.

Banking

Banking centre: Asunción.

Central bank: Banco Central de Paraguay.

Commercial banks (head offices): Banco Alemán Transatlántico, Banco Comercial Paraguayo (Bancopar), Banco Continental, Banco de Asunción, Banco de la Nación Argentina, Banco do Brasil, Banco Exterior, Banco Exterior de España, Banco General, Banco Holandes Unido, Banco de Inversiones del Paraguay, Banco Nacional de Fomento, Banco Nacional de Trabajadores, Banco Paraná, Banco Real del Paraguay, Banco Sudameris Internacional, Banco Sudameris Paraguay, Banco Unión SA, First National City Bank of New York, Lloyds Bank International.

Electricity supply: 220 V AC, 50 cycles.

Representation in capital: Argentina, Belgium, Bolivia, Brazil, Chile, China, Ecuador, El Salvador, France, German FR, Guatemala, Israel, Italy, Japan, Republic of Korea, Malta, Mexico, Panama, Peru, South Africa, Spain, Switzerland, Taiwan, UK, USA, Uruguay, Vatican City, Venezuela.

Hotels: Asunción

Chaco, Caballero y Mcal, Estigarribia, P.O. Box: 1878 (tel: 92066/9; tx: 281; fax: 444223).

Excelsior, Chile 980 y Manduvira, P.O. Box: 2863 (tel: 95632/6; tx: 5192; fax: 96748).

Gran Hotel del Paraguay, De la Residenta y Triunvirato (tel: 20851/2; tx: 286).

Guaraní, Plaza Independencia, P.O. Box: 1053 (tel: 91131/6; tx: 917).

Husa, 15 de Agosto 420 y Estrella (tel: 93760/1/3).

Itá Enramada, Cacique Lambaré, P.O. Box: 1878 (tel: 33041/9; tx: 309; fax: 444223).

Plaza, Eligio Ayala y Paraguari (tel: 444196; fax: 862).

Hotels: Encarnación

Cecilia, Calle Estrados Unidos y 25 de Mayo (tel: 205924; tx: 5251).

Gran Hotel Paraná, Caballero y 25 de Mayo (tel: 44543/6; tx: 269).

Novotel, Ruta 1, K.M.361, Villa Quitena (tel: 4131,; tx: 232).

Car hire: Asunción

Europcar National: Yegros 501 (tel: 91379/45890; tx: 367).

Hertz: Dr Eusebio Ayala Km 4.5 (tel: 503921/8; tx: 124).

Car hire: International airport:

Hertz: (tel: 22012/14, 206195/9).

Business advice and information

Agencia Publicitaria Visión, 25 de Mayo, 966, Asunción (tel: 24796).

Banco Central del Paraguay, Pablo VI y Sargento Mareco, Barrio Santo Domingo, Asunción (tel: 68015/60; tx: 134, 205, 5166/7/8, 5176).

Banco Nacional de Fomento, Independencia Nacional, 25 de Mayo y Cerro Corá (tel: 44440).

Banco de Asunción, Palma y 14 de Mayo, Asunción (tel: 93191; tx: 132).

Banco Sudameris Paraguay, Independencia Nacional y Cerro Corá, Asunción (tel: 44172/3; tx: 120).

Cámara y Bolsa de Comercio, Estrella 540, Asunción (tel: 47312\46622).

Dirección Nacional de Turismo, Alberdi/Oliva, Asunción.

Ministry of Agriculture and Livestock, Presidente Franco, Asunción (tel: 385; tx: 324).

Ministry of Finance, Chile 128 esq Palmas, Asunción.

Ministry of Foreign Affairs, Palacio de Gobierno, Asunción (tx: 111).

Ministry of Industry and Commerce, Avenida España 323, Asunción (tel: 204693; tx: 259).

Ministry of Public Works and Communications, Olivia y Alberdi, Asunción (tel: 41530; tx: 162).

National Electricity Board (Administración Nacional de Electricidad – ANDE), Padre Cardozo 360, Asunción (tel: 22713/19; tx: 22727/29).

Water Authority (Corporation de Obras Sanitarias Carposana), José Berges: e/Brasil y San José, Asunción (tel: 25001/3).

Peru

Jane Monahan

Whoever were to win Peru's presidential elections, which were due to be held on the second Sunday in April 1990, and after municipal elections were held in November 1989, they would find that there is a consensus among politicians on at least two counts.

First, a majority of political parties wanted the country to mend fences with the IMF, the World Bank and other US-influenced banks. It was realised that without foreign investment and foreign exchange Peru would not be able to take its economy out of hyperinflation and a deep recession. Most of the country's manufacturing industries, and the oil industry, were dependent on foreign investment and foreign exchange to buy essential imports such as the equipment needed for the exploration and exploitation of oil in the eastern jungle.

Secondly, all the democratic parties wanted a firm line to be taken against the country's terrorist organisations – the ten-year-old maoist Shining Path guerrillas and the right-wing death squad, the Rodrigo Franco Command, which only made its appearance for the first time in Peru in July 1988.

To reach a consensus on these two matters before the elections were held, President Alan Garcia of the American Popular Revolutionary Alliance (APRA) changed five ministers in his cabinet in May 1989 (the ministers of finance, defence, energy and mines, fisheries and health). Luis Alberto Sanchez, a widely respected 88-year-old conservative elder statesman of APRA, was named the new prime minister.

The new finance minister, Cesar Vasquez Bazan, called for 'a truce' between businessmen, industrialists and trade unionists shortly after taking office. Bazan, Sanchez and the president have said that their principal economic objective is to tackle inflation in a rational way, by cutting state subsidies, lifting price controls for certain food products and using the increase in Peru's foreign reserves caused by a collapse in imports in 1989 to finance 'selective growth' in agriculture, industry and construction. All of this, the government has said, is going to be achieved without 'shocks'.

In April 1989 President Garcia said disposable reserves stood at $750m, $300m more than at the beginning of the year, because of the decline in imports. However, economic analysts believe the size of Peru's reserves has declined since then. Meanwhile, the president, whose popularity was rated at 96 per cent in opinion polls when he was elected in 1985, now has a popularity rating of only 6 per cent, according to opinion polls published in June 1989.

The main reason for the change in the electorate's preferences is hyperinflation. In the year to the end of April 1989, the rate of inflation in Peru was 5,149 per cent and wages in June 1989 were estimated to be worth only half of what they were in June 1988. Most workers blame President Garcia for this development.

Another indication of the economic crisis was that GDP growth declined 7.1 per cent according to official figures in 1988 (and by 9 per cent according to opposition politicians and Confiep, the association of privately-owned businesses). Confiep has also forecast a 10 per cent decline in GDP growth in 1989. The country, indeed, was in one of its most severe recessions for many years.

The political repercussions of the government's programme to tackle inflation are this: if it has some success, APRA, whose candidate is Luis Alva Castro – the party's secretary-general in the presidential elections, as President Garcia is not allowed by the

Constitution to run for a second term – may still make a recovery and do reasonably well. Alternatively, if hyperinflation continues, this will play into the hands of Peru's right-wing candidate, the 53-year-old author-cum-politician Mario Vargas Llosa, who heads the FREDEMO movement, as this could sharpen the argument that a vote for the APRA or for the Left could be a vote for economic chaos.

Mario Vargas Llosa became the right-wing movement's firm candidate in July 1989. A month earlier he had resigned from this position when it was learnt that Accion Popular, the party of the two times former president of Peru, Fernando Balaunde Terry, wanted to run a separate candidate from FREDEMO in the municipal elections. The decision was confusing for the electorate, as Accion Popular was still planning to run with FREDEMO in the presidential elections. It was also strongly opposed by FREDEMO's other main political grouping, the Partido Popular Cristiano, led by Luis Bedoya Reyes.

In these circumstances, Vargas Llosa, acting as a moderator, said he could no longer be FREDEMO's presidential candidate unless Accion Popular and the Popular Christian Party ran together in the municipals and patched up their differences. To drive his point home, Vargas Llosa then packed his bags and left for a trip to Europe.

The two parties reached an agreement in July. It would have been foolish for them to have done otherwise. Opinion polls in June showed Vargas Llosa leading the field over his rivals. He was the only candidate who had started active campaigning. Al-

KEY FACTS

Official title: Republic of Peru
Head of state: President Alan García Pérez
Ruling party: Alianza Popular Revolucionaria Americana (APRA)
Capital: Lima
Official Languages: Spanish (Quechua, Aymara also spoken)
Currency: Inti (IN) = 1,000 soles
Exchange rate: IN5,606 per $ (Nov 1989)

Area: 1,285,216 sq km
Population: 21.3m (1988)
GDP per capita: $1,503 (1988)
GDP real growth: –7.1% (1988)
Inflation: 1,191% (1988)
Trade balance: $122m (1988)
Foreign debt: $18.1bn (1987)

fonso Barrantes, a former mayor of Lima, and the most popular candidate of the Left, was running second in opinion polls.

However, it was still uncertain whether the United Left coalition would overcome their differences and agree on Barrantes as their candidate. In January 1989 the most moderate group in the coalition walked out of the United Left Congress and tried to drive a wedge between its group and the radical Left, PUM, convinced that the PUM's presence (it supports armed violence in certain circumstances) could be a provocation to the army. Apart from the moderate faction, the other groups in the United Left coalition accept Barrantes as their candidate.

Vargas Llosa's weakness in the elections could be his sophisticated urban image. Barrantes, on the other hand, is a humble lawyer from Cajamarca, and he has more appeal in the countryside as a man of the people. However, Vargas Llosa may count on strong financial backing from Peruvian businessmen and from abroad, as well as the support of Peru's television stations, as these have a right-wing bias.

In spite of the sense of drift in Peru caused by the increasing withdrawal of President Garcia from all but the essentials of political life, and by the country's economic crisis, analysts consider that the possibility of a military coup is remote. In January President Garcia averted a crisis in civilian-military relations with a timely pay rise and a commitment to give the army new helicopters to help them fight the Shining Path guerrillas.

On top of this development the army is divided. Many soldiers are on the left of the political spectrum and they supported the reforms introduced by the Garcia government at the beginning of its mandate aimed at improving the incomes of poorer Peruvians. They also supported the government's controversial decision to nationalise the banks, to establish less greater control over credit, taken in July 1987. (Significantly, many bishops in Peru's Catholic Church and many Catholic priests also supported this.)

So it would seem unlikely that any but the most right-wing sectors of the armed forces would support a coup, as this could increase divisions in the army. In addition, a military coup in Peru, when most of the country's neighbours have introduced (or are in the process of introducing) a democracy would suggest that it would get little Latin American, or international support.

President Garcia also made another concession to the army in May when the new defence minister, General Julio Velazquez, was appointed, replacing a civilian. The retired general was given a much freer hand in the campaign against the Shining Path guerrillas, and in June 37 out of a total of 181 Peruvian provinces were classified as emergency zones, meaning that the army had overall jurisdiction in them, and access by journalists and the Red Cross was restricted.

Under General Velazquez' supervision, the armed forces have struck the Shining Path guerrillas hard. On 5 June the number three in the guerrilla movement, David Orozco Pello, alias Camarada Merino, was killed with 15 other rebels. (The number two in the guerrilla movement, Osman Morote, was captured by the army in June 1988 and the leader of the organisation, Abimael Guzman, has not been seen for years and is believed to be dead.)

At the end of June a Lima newspaper, *La Republica*, reported that up to 900 rebels had been killed in various military operations in Tingo Mario and Huanuco provinces. The newspaper said the army had used helicopters in these operations, gunning down guerrillas from the air.

The crackdown came after 16 soldiers died and a further 10 were seriously injured when they were ambushed by Shining Path guerrillas in Aguaytia near Tingo Maria. In Lima, a busload of troops on their way to the government palace was blown up early in June and seven more soldiers died. This attack was also believed to have been perpetrated by the maoist guerrillas.

KEY INDICATORS

	Unit	1984	1985	1986	1987	1988
Population	m	19.2	19.7	20.2	20.7	21.3
GDP	inti bn	59,865	157,977	370,778*	590,326*	4,115,088*
GDP per capita	$	867	611	1,233	1,693	1,500
Inflation	%	109.9	163	60.7	85.9	1,191*
GDP real growth	%	4.8	2.4	11.3	7.8	–7.1
Agricultural production	% real gwth	11.8	3.7	6.2	5.0	5.7*
Mining	% real gwth	4.8	4.3	–4.5	–2.9	–18.7
Manufacturing	% real gwth	5.5	4.9	16.8	13.7	–14.3
Construction	% real gwth	0.8	–10.5	21.3	15.5	–5.4
Oil production	'000 b/d	183.0	189.0	178.4	154.5	–
Exports	$m	2,979	2,531	2,661	2,547	2,695
Fishmeal	$m	137	118	206	224	364
Coffee	$m	126	151	262	240	–
Copper	$m	442	475	450	559	607
Refined silver	$m	227	139	107	92	60
Lead	$m	233	203	172	257	–
Zinc	$m	341	267	246	250	261
Petroleum & derivatives	$m	618	645	232	273	166
Imports fob	$m	1,843	1,686	2,424	2,967	2,567
Balance of trade	$m	1,007	1,172	–65	–411	122
Current account	$m	–223	135	–1,080	–1,477	–1,128
Tourist arrivals	'000	207	264	233*	–	–
Foreign debt	$bn	13.3	13.8	14.0	17.7	–
External debt	$m	13,159	14,190	15,956	18,058	–
Debt service ratio	%	15.9	17.8	14.4	12.5	–
Exchange rate	soles per $	3467	10.98*	13.95	16.84	128.83

* estimated figure

Having started their operations in Ayacucho, in the south of Peru, the Shining Path movement, which is said to have between 3,000 and 6,000 members, established its base in the Upper Huallaga. However, in 1989 the movement also began penetrating the Huancayo Valley, a rich agricultural region in the Andes, known as Peru's breadbasket because of its strategic importance as a food supplier to Lima, the capital. When this happened, the government feared that guerrilla claims that they were progressively encircling the capital were becoming true. But even if the guerrillas fail to establish a second base in Huancayo, they may still be able to survive for several years in the Upper Huallaga, as this area has virtually become a liberated zone where the guerrillas can train recruits, heal their wounded and re-supply with food and arms.

The government estimates that in the 10 years since the Shining Path's first act of violence in May 1980, 15,000 Peruvians have been killed in the terrorist struggle, the majority of them peasants; and the Shining Path has caused $15bn in damage – most of this caused by blowing up electricity pylons and installations.

Quite apart from the problems of law and order caused by the terrorism, if the army and police cannot guarantee the conditions that will make an election possible in the countryside because of the violence of Sendero guerrillas, then the rural vote will be greatly reduced. This would be a disadvantage to the Left, given the strength of the Left in peasant unions.

Adding to the political uncertainty, in July 1988 a right-wing death squad, the Rodrigo Franco Command, made its first appearance, claiming credit for the assassination of a lawyer who was representing the captured rebel leader of the Shining Path movement, Osman Morote.

The Rodrigo Franco Command is named after an official in APRA who was killed by the maoist guerrillas in 1987. Some suspect, though there has been no firm evidence brought forward to support it, that the Rodrigo group was set up by APRA.

The death squad is also believed to have been responsible for the killings of three labour leaders belonging to the Peruvian mine workers' federation on 13 February 1989 and on 15 April 1989. The emergence of the group is seen by analysts as further evidence of the breakdown in law and order – in addition to the bombings, kidnappings and assassinations of local government officials (mayors and town councilmen) and peasant leaders in the countryside by Shining Path guerrillas.

The economy

Reducing the high rate of inflation became a priority of government economic policy in 1989. But the problem is not an easy one to solve, as Peru's inflation has many causes. For instance, the growth of 430 per cent in the main M2 money supply aggregates in 1988 was far less than the 1,722 per cent rise in inflation that year. So it is not true to say that the chief causes of inflation are the budget deficit (estimated in 1989 to be equivalent to 15 per cent of GDP), state subsidies and state credits for the public sector. Although these factors play a part in fuelling inflation, economic analysts think that other causes are more important, such as a shortage of basic food products and of industrial raw materials, and the scarcity of foreign exchange, which has coincided with delays in the flow of exports because of strikes. In addition, political uncertainty has encouraged inflationary expectations, and speculation on the black market, or the Calle Ocona market, as it is called, has also put pressure on prices.

In these circumstances the government has been tackling inflation in certain specific areas. It has been limiting public expenditure by cutting the subsidies on petroleum and petroleum products distributed by Petroperu, the state oil company, as well as on basic food products like rice, wheat and fertilisers, distributed by the Empresa Nacional de Comercializacion de Insumos (ENCI), the state marketing monopoly.

As a consequence of this action, the prices of goods that were controlled at the beginning of the Garcia government are being allowed to rise. In January 1989 petrol prices were increased 61 per cent, and rice, milk, cooking oil and sugar prices were increased 60 per cent on average. At the same time, the government allowed the tariffs of basic services like transport, telephones, electricity and water to go up. So now the prices of both goods and services more accurately reflect the real costs of production, and the prices recorded on the Consumer Price Index.

The government has also begun cutting capital expenditure, particularly for public works. Combined, these policies have resulted in a contraction of public sector credit. In the third quarter of 1988, credits for state-owned companies, which had been increasing in the first six months of 1988 at an annual rate of 1,192 per cent, fell back to an annual rate of 735 per cent.

Private sector credit also declined in 1988. It dropped 66.6 per cent compared with 1987. The government has tried to stimulate savings by allowing 20 per cent average monthly increases on the interest rates of bank loans in 1988 and 17 per cent monthly increases in the rates for term-deposits, but these incentives were wiped out by high inflation rates. Total Inti deposits held by the banks in 1988 fell 80 per cent compared with 1987.

The withdrawal of bank deposits also reflected uncertainty following the nationalisation of banks in 1987. Indeed, total M2 money supply shrank by a huge 57.8 per cent in real terms in 1988, which is what sparked the recession in the manufacturing and construction industries and an estimated drop in GDP growth of 7.1 per cent.

The three exchange rates that exist in Peru – the official one, the bank rate and the 'free' or black market rate – are gradually being phased out, with periodic devaluations of the Inti on the official rate, which is where rates have been most controlled up to now.

In January 1989 the Inti on the official rate was devalued 40 per cent, and the official rate applying for exporters was raised from Intis 920 to the dollar in February to Intis 1,400 in May. In addition, in June the new finance minister started a policy of mini–devaluations of the Inti, of 1 per cent a day.

A further reflection of a new pragmatic approach by the Garcia government is that in June 1989 it proposed privatising 32 out of a total of 240 state-controlled companies. The companies offered are involved in milk production, cement, coal, paper, fishing, chemicals, industrial spirits and wood. The National Tobacco Company, which is 100 per cent state-owned, is also up for sale.

The purpose of the privatisation is not only to reduce the state deficit, but also to diversify the shareholdings of state-owned companies. The latter was a reason given by the government when it nationalised the privately-owned banks in July 1987. It was believed then, as now, that employees should be given a chance to own some of the shares in the enterprises they work for.

One of the conditions of the privatisation scheme is that the companies that are sold to private owners should show profits in a matter of three years; otherwise they could devolve to the state again. In each of the 32 companies being offered, the state is going to retain a 20 per cent stake.

The main uncertainty of the scheme is that only 13 of the companies on the list were operational in June 1989, while 13 had been inactive for some time, and six were in the process of being liquidated. Furthermore, many of the companies on the list had been offered for sale six months before, and no-one – either in Peru or abroad – showed any interest in buying them.

The government's anti-inflationary policies began to have some effect on prices in 1989 – in the first six months of the year Lima's Consumer Price Index rose 601 per cent compared with an increase of 6,000 per cent on the same Index in the first six months of 1988. Nevertheless, while government figures showed a rise in inflation in May of 28.6 per cent, and in June of 23 per cent, the private sector estimated that inflation rose 31.8 per cent in May and 35 per cent in June.

There was also still no sign in mid-1989 that the government's policies were having an effect on the recession. Provisional estimates suggest that industrial production overall in the nine months up to the end of June 1989 was declining at an annual rate of 40 per cent. Meanwhile, in the main cities, factories have been closing and workers have been laid off or put on short-time. Alberto Ramirez, the secretary-general of the biggest trades union in the country, the communist Central General de Trabajadores del Peru, claimed in mid-1989 that the recession had put one million workers, out of a total workforce of 12 million, out of a job. In addition, Peru has an unemployment rate of around 30 per cent and under-employment is around 60 per cent.

The National Industrial Association, which represents privately-owned manufacturing industries, estimated that the output of capital goods industries fell 64.3 per cent in the first six months of 1989 compared with the same period in 1988. And the construction industry, which boomed in 1986 and 1987, declined 44 per cent in the first 11 months of 1988 compared with January–November 1987, because of budget cuts for public works, a decline in orders, and a scarcity and increase in the price of cement. The construction industry, no less than other industries, is also susceptible to the electricity cuts caused by Shining Path guerrilla attacks on Peru's electricity pylons.

But the main reason for the country's recession is that, because of hyperinflation, real incomes have fallen drastically and Peruvians are buying less. Indeed, the shopping spree that was the result of a mixture of wage increases and price controls during 1985–87, ground to a halt in 1988. Retail sales in Lima declined 21 per cent in the first 11 months of 1988 compared with January–November 1987, and in the month of November 1988 alone they declined 58 per cent compared with the same month the previous year.

To compensate for the officially authorised price increases the government has started allowing wages to go up for unionised and non-unionised labour. The monthly minimum wage was increased in January 1989. Public sector wages rose 77 per cent on average between January–30 June 1989. This increase compares with an average increase of 30 per cent in private sector wages during the same six months. Nevertheless, neither public sector nor private sector wage rises are keeping up with inflation and real wages declined 50 per cent in the year to June 1989. The possibility of automatically adjusting wages to keep up with inflation is now at the centre of disagreements between the government and the trades unions. As a result of wage disputes, the number of working hours lost through strikes in the first nine months of 1988, compared with the same period in 1987, shot up 160 per cent, according to Confiep.

The strikes have affected up to half a million workers in public administration, public transport and banks. Another extremely damaging strike in terms of its effect on the economy (because it stopped exports and therefore reduced Peru's foreign exchange earnings) was a strike in the mines.

The 80,000 miners employed by the state-owned Centromin Peru company, which owns all the mines in central Peru including the Cerro de Pasco mines, stopped work for five months in 1988. Indeed, this strike was one of the reasons why international copper and zinc prices increased so much in 1988. Peru is one of the world's most important producers of both of these minerals.

In April 1989 miners stopped work again, but this time throughout the country. The strike was not about wages, but about the killing of three union leaders in February and April, allegedly by the right-wing death squad, the Rodrigo Franco Command. According to Centromin Peru, this 72-hour general strike caused losses of $4m a day. Centromin estimates that as a result of strikes, and terrorist attacks, the company lost over $35m between 1980 and 1988, of which $12m was lost in 1988 alone.

In the light of the strikes, mineral production dropped 30 per cent in Peru in 1988 between January and November compared with the same 11 months in 1987. However, the international prices of those minerals that are exported by Peru – copper, zinc, silver, lead, gold – rose by an average of 33 per cent.

The problems in the petroleum industry are different. Petroleum production declined 16.6 per cent from January–November 1988 compared with the same period in 1987, but this was because of a shortage of foreign exchange, and bureacratic problems, which delayed the buying of drilling and other essential equipment.

Petroleum sales have also fallen in Peru because of the recession. However, this decline has been compensated by the government allowing increases in the prices of petrol, and of kerosene and butane gas.

Petroperu, the state oil company, and the US oil company Occidental Petroleum, announced an important oil discovery north of Pucallpa, in the 36th jungle block, in 1989. According to preliminary estimates, the reserves discovered amount to at least 600m barrels of high-quality oil – an amount that on its own is sufficient to cover Peru's oil needs for about 10 years. Occidental Petroleum, which has already drilled three wells in the 36th jungle block, said in April 1989 that it was going to invest $150m more in drilling five more wells. The company is prepared to set world drilling records by going underground to depths of 9,000 metres.

In May 1989 it was reported that negotiations had begun between Petroperu and Mobil Oil of the United States with a view to Mobil getting a contract to explore a 4m hectare block near the source of the Huallaga River.

An agreement was reached earlier in the year with Ecopetrol, Columbia's state petroleum company, to explore a 3–4m hectare area on the border between the two countries. Ecopetrol is going to provide the technicians and equipment for the exploration. In return, Peru is going to let Columbia refine 30,000–40,000 barrels a day of its oil in Peru's coastal refineries at La Pampilla and Talara.

In the last agricultural season, which lasted from August 1988 to July 1989, farming was disrupted by many adverse developments, such as a shortage of credit, a dry climate which affected several sierra regions out of season, a decline in food sales because of a drop in consumption and a scarcity of pesticides and fertilisers.

In terms of calendar years, farming output grew between January–November 1988 by 8.2 per cent compared with January–November 1987. However, in the 1988/89 season the total area sown to crops declined significantly compared with the 1987/88 season. The harvests of coffee and cotton in the 1988/89 season were reported to be good, however.

It is a reflection of the shortage of certain food

products that in May 1989 the US government provided Peru with 68m tonnes of wheat for immediate distribution to needy families in the sierra. At the same time, the Garcia government started lifting import controls for certain foods. For instance, in 1989 it allowed 30,000 tonnes of rice to be imported.

Meanwhile, a great number of privately-run rural development projects are stopping because many European volunteers have left following guerrilla attacks. Concern for the safety of agronomists, engineers and rural doctors has risen sharply after Sendero Luminoso rebels killed two Frenchmen in a remote Andean village agro-industrial project in December 1988, and declared that foreigners would be targets of their insurgency in future.

By March 1989, 15 Europeans from various social-assistance programmes had left the country, and a further 70 rural volunteers had moved to safer provincial capitals like Cuzco, Puno and Arequipa. The United Nations-affiliated Potato Centre is also reportedly planning to move part of its operations to neighbouring Ecuador.

Unlike agriculture, fishing and fishmeal production is on a healthy footing. In 1988 fish catches trebled because of a big increase in the anchovy catch following a lifting of the bans on anchovy fishing in October. Meanwhile, 1988 fishmeal exports were worth an estimated $420m, an 83 per cent increase on fishmeal sales in 1987. After fishmeal prices fell to a low of $620 a tonne in June 1988, they ended the year at $500 a tonne. In 1989 Peru again became the world's number one producer of fishmeal.

Foreign investment is also forthcoming for the fisheries sector, unlike other areas of the Peruvian economy. For instance, according to trade officials, one project involving a foreign investment of $20m was in the pipeline for June 1989. In March 1989 the EC provided IMARPE, Peru's Maritime Institute, with $2m for various fisheries and ecological projects.

From registering a decifit of $411m in the balance of its trade in 1987, Peru achieved a $122m surplus in 1988 because of the high international prices of its mineral exports and a significant decline in imports.

According to the Ministry of Trade, in the first 11 months of 1988 imports dropped 5.5 per cent overall, clearly demonstrating the effects of the recession. Imports of consumer goods (which ceased being luxury consumer goods long ago) fell 26 per cent during January–November. Capital goods imports dropped 13 per cent. Even food imports declined 0.8 per cent. However, industrial raw materials and imports of intermediate goods and inputs rose 21.3 per cent.

Exports for the whole of 1988 totalled $2.6bn, the same as in 1987, which was remarkable in view of the large number of strikes. It is a further reflection of the importance of international prices for Peru's overall export performance.

Another positive aspect of Peru's balance of trade is that while traditional exports (coffee, sugar, fish, oil, copper, zinc, silver, gold) stayed at the same level as in 1987, totalling $1.9bn, non-traditional exports (mainly textiles, machinery and chemicals) increased by $40m compared with 1987 and totalled $756m.

There have also been improvements in Peru's balance of payments. The deficit in the current account went down from $943m between January–November 1987 to $871m between January–November 1988. In the whole of 1987 the current account deficit was $1.4bn.

In 1988 it also seemed as if Peru was making some progress in raising more foreign loans. Long-term loans for the public sector rose from $117m in 1987 to $458m in 1988, 86 per cent of which came from foreign governments. Peru also obtained $35m in import credits in 1988 from Spain, France and Russia, and another $50m from the Andean Reserve Fund.

Thanks to the increase in exports and a decline in imports, Peru's gross reserve position stood at $1.1bn

COUNTRY PROFILE

Political system: Executive power rests with the president (elected for a five-year term by universal adult suffrage). President governs with assistance of an appointed council of ministers. The legislature is a bicameral Congress, also elected for a five-year term by universal adult suffrage,comprising a Senate (60 members elected by region plus former constitutional presidents) and a Chamber of Deputies (180 members elected by proportional representation).

Main cities/towns: Lima, Arequipa, Cuzco, Callao, Trujillo, Iquitos.

Media
Press: 10 daily newspapers in Lima, of which most influential are *El Comercio* (circulation 150,000), *La República* (114,000), *Expreso* (123,000)and *Ojo* (180,000). Local dailies are published in most main regional centres. A number of periodicals are published, mainly in Lima, including *The Lima Times* (weekly, in English, 8,000). Main US and European periodicals available at Librerías ABC in Lima. *Radio:* Nearly 300 radio stations including government-run network and numerous commercial services.Radio de Pacífico (640 AM, 102.1 FM) has an English language news and music programme 'Night Line' (Mon-Fri) 2030 – 2100. *Television:* Eight TV stations in Lima. Empresa de Cine, Radio y Televisión Peruana SA (RTP) operates some 27 TV channels. All telecasts are in Spanish.

External trade: Imports declined 5.5 per cent overall between 1 January 1988 and 30 November 1988 reflecting a big drop in consumption and a slowing down of most industrial activity. In contrast, Peru's exports during the same period grew 7.8 per cent reflecting a 33 per cent average increase in the international prices of the minerals that Peru exports (mainly copper and zinc).
Exports: Manufactured goods and other non-traditional products (24% of total value in 1986), copper (17%), petroleum and derivatives (9%), zinc (9%), lead (6%), fishmeal (8%), coffee (5%), refined silver (4%). Main destinations: USA (30% of total value in 1986), Japan (10%), USSR (5%), German FR (5%), UK (4%), Netherlands (4%), Belgium and Luxembourg (4%), Italy (3%), Colombia (3%), Taiwan (2%), Republic of Korea (2%).
Imports: Intermediate goods (43% of total value in 1986), capital goods (34%), consumer goods (8%). Main sources: USA (17% of total value in 1986), Japan (6%), German FR (6%), Argentina (5%), Brazil (4%), Spain (2%), Canada (2%), France (1%).

Agriculture: Less than three per cent of land area is devoted to arable land and permanent crops. Sector, however, employs 40 per cent of population and contributed 15 per cent to GDP in 1986. Subsistence farming predominates, productivity is low due to drainage and salinity problems but has been increasing steadily since 1985. Animal husbandry (sheep, poultry, cavies and cattle) important in southern regions. Main food crops are rice, maize, sorghum, potatoes and fruit. Principal cash crops are cotton (tanguis), sugar-cane, coffee and soya. Agricultural production fluctuates widely, owing to changeable climatic conditions and the effect of subsidised imports. Regular need to import wheat, edible oils and dairy products.

Fishing: Peru's importance as a leading exporter of fishmeal declined during the 1970s and early 1980s because of ecological factors and over-fishing. However in 1989 Peru was again expected to recover its position as the world's leading fishmeal producer and exporter, after a return of anchovy resources in sufficient quantities along the coast, and a lifting of bans on chanicy fishing in October 1988. Fishmeal exports in 1988 were worth an estimated $420m, an 83 per cent increase on 1987.

Industry: Manufacturing and construction contributed 23 per cent to GDP in 1986. Manufacturing activity, centred in Lima and Callao, comprises food processing, beverages, chemicals, petro-chemicals, rubber, plastics, basic metallurgy, metal products, cement, textiles, footwear, paper products, machinery and motor vehicle assembly. Industry entered a deep recession in 1988 reflecting a collapse in demand due to hyperinflation and a shortage of credits and of foreign exchange. Textile output, which accounts for about 16 per cent of all manufacturing output, was about the only industry that stayed buoyant. The food processing, rubber, paper, chemicals, beverages, footwear, machinery and car assembly industries were all depressed. Sideperú, the state-owned steel company, is the largest single industrial entity.

Mining: Extensive mineral reserves including 13 metallic and 25 non-metallic minerals, of which most important are copper, silver, lead, zinc, gold and iron-ore. Main active regions in Sierra and on southern coast. Although employing under two per cent of work-force, share of mining in GDP increased from seven per cent in 1970 to 10 per cent in 1985, owing to increased production capacity. Mineral exports (excluding oil) account for around 40 per cent of export earnings. In 1988 the international prices of Peru's main mineral exports – copper, zinc, silver, lead and gold – rose an average of 33 per cent. This helped compensate for five months of strikes in Peru's main mines in the sierra. Around 75 per cent of total mineral production is generated by Mineroperú, Centromín, Hierroperú (all state-owned) and the international consortium Southern Peru Copper Corporation. New mining projects include the Bayovar phosphoric mine, expansion of the Corbriza mine, the Tintaya copper mine and the Aurifera Cayara gold mine.

Hydrocarbons: Exploitation of oil reserves by state agency, Petroperú, is important determinant of country's economic growth. Petroleum and products generally account for over 50 per cent of domestic energy consumption and 20 per cent of export earnings. Existing oilfields (of which most productive in the north-east jungle region) contain estimated reserves of 11,000m barrels, sufficient for 18 years' production at current rates (170,000 b/d). Petroleum refining undertaken at La Pampilla and Talara, which are linked by pipeline to main onshore fields. Production of gas minimal, though total estimated reserves in the region of 450m cu ft.

Energy: Self-sufficient in oil, with a small (50,000 b/d) surplus for export. Total electricity generating capacity 3,400 mW, of which 60 per cent hydroelectric. Estimated that nearly 70 per cent of population without electricity. New hydroelectric projects include plants in Huancavelica, Yuncan, Sheque and Carhuaquero.

Membership of international organisations: Andean Group, AIOEC, ASSIMER, CCC, CIPEC, Economic Commission for Latin America, FAO, G-77, GATT, IADB, IAEA, IATP, ICAO, ICO, IDA, IFAD, IFC, ILO, IMF, Inter-American Development Bank, Interpol, INTELSAT, International Lead and Zinc Study Group, ISO, ITU, IWC (Wheat), Latin American Integration Association, NAM, Organisation of American States, PAHO, SELA, UN, UPU, WFTU, WHO, WMO, World Bank, WSG, WTO.

on 7 February 1989 according to the central bank. However, the central bank emphasised that net reserves, or those reserves that are available after all obligations with foreign banks are met, were in fact negative on 7 February by $303m. The central bank pointed out that Peru's net reserve position has been negative since January 1988. The central bank's analysis seems to contrast with President Garcia's declaration made in April 1989 that Peru had *disposable* reserves then of $750m. The reason for the discrepancy is that it is not clear, from the president's remark, how much of the country's reserves are free and how much are committed, for instance as a guarantee for loans already received.

What is clear is that Peru's reserve position has not improved enough for the country to overcome its debt crisis. Total foreign debt at the end of June 1989 amounted to $21bn according to ministry of economy officials. The arrears in interest payments at that time totalled $5bn. This $5bn on its own is equivalent to about two years of Peru's total export earnings.

But Peru's foreign currency problems are also aggravated by the continuing rows between the Garcia government and the major US-influenced international banks. After falling out with the IMF and the World Bank at the beginning of his mandate in 1986 and 1987, on 6 March 1989 the Inter-American Development Bank (IADB) announced the suspension of all further credit to Peru because of delays by the Garcia government in servicing a debt payment of $69.4m that fell due on 28 February.

Peru's foreign ministry officials, in their statement on why the country had stopped servicing IADB debt, claimed that it was not because Peru lacked foreign exchange, or investment plans, but because the amount of funds being provided by the IADB to Peru in 1988 had been less than the amount of funds Peru had paid the bank. The officials added that in 1986 Peru had cancelled payments of $86.1m to the IADB, of $97.4m in 1987, and of $69.1m in 1988, but that in spite of this the international bank was not prepared to extend fresh credits to Peru, and had cancelled an investment project that had already begun. The Peruvian government said that it would spend the money it owed the IADB completing a hydro-electric power station at Carhuaquero in the Andes.

Relations between the APRA government and major US companies are also tense. A dispute with the US oil multinational Enron Corporation, whose subsidiary Belco Petroleum and its installation in Tacna were nationalised in August 1985, has come to a head. In December 1988 nine American insurance companies led by the American International Group (AIG) decided to support Enron Corporations's demands for compensation of $200m for the nationalisation of Belco's Tacna assets. However, the Peruvian government, while not ruling out the possibility of paying compensation altogether, has made it clear that it considers that the installations formerly owned by Belco are not worth more than about $60m.

AIG has said it is prepared to ask for the application of the Hickenlooper Amendment against Peru if the $200m compensation is not paid to its client. The application of this amendment would mean a complete freeze on US loans to Peru. However, in Peru's present circumstances such a development would be more of a blow to the government's prestige than to its finances. There is already an almost complete freeze on US loans to Peru, by both the US government and the US private sector.

Another important conflict, casting a shadow over Peru-US relations, is that between the government and the US company American Smelting (or ASARCO), which owns the Southern Peru Copper Corporation (SPCC). In April, Peru's former mines and energy minister, José Carrasco, said that he had given the central bank instructions to start a law suit against ASARCO for breaking Peru's exchange controls. The former minister claimed SPCC had illegally remitted $71m abroad in 1988. However, a criticism of SPCC and other privately-owned mines in Peru is that the government has been fixing the exchange for mineral exports at a lower rate than the international rates. It is possibly significant that the dispute with SPCC has not been publicly mentioned by the new mines and energy minister, Mario Semame, who succeeded Carrasco in May.

Although Peru's relations with US-influenced international banks and US companies are strained, this has not prevented the Garcia administration from reaching agreements with other organisations and countries on at least part of its foreign debt.

In April 1989 the economy minister succeeded in renegotiating $10m of debt with private commercial banks, with Lloyds underwriters acting as intermediaries. The operation included a write-off of part of the debt, and an agreement, for the first time ever in Peru's case, that throughout the time the debt is being restructured (i.e. seven years) Peru will not have to pay any interest on the debt at all. The equivalent of part of the $10m debt is also going to be reinvested by the banks with purchases of Peruvian textiles and of other products.

Earlier in the year, in February 1989, Peru also arranged the cancellation of $5m of debt with Hungary by agreeing to give Hungary products in lieu of foreign exchange during 1989 and 1990. At the same time, Hungary agreed to buy Peruvian goods up to a total of $10m over the same two years.

A similar agreement was struck with Czechoslovakia involving $2m of debt owed to that country. Peru is going to repay this with non-traditional exports such as textiles. Czechoslovakia, for its part, has agreed to buy the equivalent of $4m of Peruvian goods in 1989 and 1990.

But the area where Peru's trade is growing most is not in the West or in Eastern Europe but in the Far East. In 1988 Peru ran up a $295m surplus in its trade with Japan (a 46.4 per cent increase on its trade with that country in 1987). Peru also had a surplus in its trade with Taiwan throughout the period 1980–88. Taiwan is Peru's second most important trade partner in the Far East after Japan.

BUSINESS GUIDE

Time: GMT -5.

Climate: Great variations according to region. Coast, including Lima, is arid with most of the very low rainfall occurring between May and Sep. Average temperature is around 19°C although summer daytime temperatures can reach 25°C to 30°C (Oct - Apr). Humidity is high throughout the year. In the Sierra, average annual temperature is 15°C with high daily ranges of up to 20°C; wet season is from Oct to Apr. East of the Andes are the sub-tropical uplands (Montaña) and the tropical Amazonian forests.

Entry requirements
Passport: Required by all.
Visa: Required by all except for tourist visits by nationals of EC countries, USA, Canada, Japan, Scandinavian countries, Switzerland, Austria, Bolivia, Brazil, Uruguay and certain others.
Currency: Severe restrictions on the import/export of foreign currency for businessmen. Local currency may be reconverted to foreign currency on production of exchange receipts. International credit cards accepted in main centres.

Health precautions
Mandatory: Yellow fever vaccination certificate required if arriving from infected area.
Advisable: Yellow fever vaccination recommended (essential for visits to some rural areas). Typhoid, polio, tetanus vaccinations advisable. Malaria risk exists in some rural areas – prophylaxis recommended. Water precautions should be taken – it is advisable to drink only bottled water.

Representation overseas: Algiers, Asunción, Beijing, Belgrade, Berlin, Berne, Bogotá, Bonn, Brasília, Brussels, Bucharest, Budapest, Buenos Aires, Cairo, Canberra, Caracas, Copenhagen, Guatemala City, The Hague, Havana, Helsinki, Kingston, La Paz, Lisbon, London, Madrid, Managua, Manila, Mexico City, Montevideo, Moscow, New Delhi, Ottawa, Panama City, Paris, Port-au-Prince, Prague, Quito, Rome, San José, San Salvador, Santiago, Santo Domingo, Sofia, Stockholm, Tegucigalpa, Tel Aviv, Tokyo, Vatican City, Vienna, Warsaw, Washington.

Air access
National airline: Aeroperú.
Other airlines: Aeroflot, Aerolíneas Argentinas, Air France, Air Panamá, Alitalia, Avianca, Canadian Airlines International, Cubana, Eastern, Ecuatoriana, Faucett, Iberia, KLM, Lan-Chile, LAP, Lloyd Aéreo Boliviana, Lufthansa, Pluna, Varig, Viasa.
International airport: Lima - Jorge Chávez International (LIM), 16 km north-west of city; duty-free shop, bar, restaurant, bank, post office, shops, hotel reservations, baggage storage, car hire.

Surface access
Main ports: Callao, Pacasmayo, Paita, Salaverry, San Juan, Matarani, Chimbote.
Overland access: Possible from Ecuador, Chile and Bolivia via the Pan-American Highway.

Hotels: In main centres classified by stars (maximum 5) according to available facilities. In smaller towns the best accommodation is often the government-run Hoteles Turistas. All hotel bills include 15-21 per cent service charge and tourism tax. An additional five per cent tip is usual. N.B. Visitors arriving in Lima are well advised to inform their hotel of their arrival flight number and time. Most major hotels operate a free courtesy coach service to Jorge Chávez airport and will meet arriving guests.

Car hire: Major international companies operate in Lima and other main centres. Chauffeur and self-drive cars available. International licence preferred and credit cards essential. Cost includes basic insurance cover. Traffic is congested in Lima.

City transport
Taxis: Quickest means of travel in main cities. Can be hailed in the street or found on ranks. Taxis from recognised ranks (*estaciones*), for example in front of hotels and at airport, usually charge 50 per cent more. *Colectivos* (communal taxis) carrying up to six persons follow set routes into the city suburbs. In Lima they leave from La Colmena on both sides of the Plaza San Martín and charge officially-set fares. Advisable to agree fare for other taxis in advance as they do not have meters. 50 per cent surcharge after 2400 hrs and on public holidays.

National transport
Air: Aeroperú operates regular services between Lima and all main towns. Due to weather conditions flights are frequently delayed or cancelled. Essential to reconfirm bookings as flights often over-booked. Unlimited travel tickets are available.

BUSINESS GUIDE

Road: Major links include: Pan-American Highway, paved over most of the distance, runs north to south along the coast from the Ecuador border to the Chile border (with a north-east arm into the Sierra, through Arequipa and on to the Bolivian frontier). The Trans-Andean Highway runs from Lima to Pucallpa, via La Oroya and Huanuco. The Central Highway connects Lima with La Oroya, Huancayo, Huancavelica, Ayacucho, Cuzco, Puno, (linking with the Pan-American Highway spur from Arequipa). Highways between the coast and mountain regions are often blocked by landslides in rainy season (Dec-Apr).
Buses: Cheap but fairly uncomfortable services available on Pan-American Highway north to Ecuador, south to Chile and on the highway to Callejon de Huaylas in northern Andes. Yellow city buses and mini-buses connect Lima with Callao and the residential suburbs.
Rail: Regular rail services between Lima and La Oroya with branches to Cerro de Pasco, Huancayo and Huancavelica. The Southern Railway of Peru operates between Arequipa-Puno (on Lake Titicaca) with one weekly connection (Wed) by steamer across the Lake to Bolivia. Also regular rail connections from Puno to Cuzco. A short line runs from Tacna to Arica in Chile. Railways have separate summer and winter schedules.

Public holidays
Fixed dates: 1 Jan, Maundy Thursday, Good Friday, 1 May, 24 Jun, 29 Jun, 28-29 Jul (Independence), 30 Aug, 8 Oct, 1 Nov, 8 Dec, 25 Dec.

Working hours
Business: (Mon-Fri) 0830 – 1700, (Sat) mornings. Many regional variations occur.
Government: Apr-Dec, (Mon-Fri) 0800 – 1600. Jan-Mar, (Mon-Fri) 0745 – 1330.
Banking: Apr-Dec, (Mon-Fri) 0845 – 1245. Jan-Mar, (Mon-Fri) 0830 – 1130. NB: Some banks open in the afternoons and on Sat mornings. Banks closed 27 May, 30 Jun and 31 Dec.
Shops: (Mon-Sat) 1000 – 2000. Most close for 1-2 hours in afternoon. Some closed Sat afternoons. Many regional variations occur.

Telecommunications
Telephone: Direct international and long-distance dialling available from all major hotels (dial 108). Dialling code for Peru: IDD access code + 51 followed by area code (Arequipa 54, Lima 14) followed by subscriber's number.
Telex: Empresa Nacional de Telecommunications (ENTEL) operate some 90 public telex booths in Lima and 48 in other cities. These close evenings and during holidays. Facilities in Bolívar, Crillon and Sheraton hotels. Permanent service at Avenida Bolivia 347, downtown Lima. Country telex code: PE.

Postal service: Main post office, near Plaza de Armas and Government Palace, opens from 0800 – 1800 (Mon-Sat), and 0800 – 1200 (Sun).

Banking
Banking centre: Lima. Many of the country's commercial banks have been nationalised, under legislation introduced in 1987.
Central bank: Banco Central de Reserva del Peru.
Commercial banks (head offices): Banco Agrario del Perú, Banco Amazónico, Banco Central de Reserva del Perú, Banco Central Hipotecario del Perú, Banco Continental, Banco de Crédito del Perú, Banco de Desarrollo-Bandesco, Banco de la Nación, Banco de la Vivienda del Perú, Banco de Lima, Banco de Sur del Perú, Banco Industrial del Perú, Banco Internacional del Perú-Interbanc, Banco Latino, Banco Mercantil, Banco Minero del Perú, Banco Nor-Perú, Banco Peruano de los Constructores, Banco Popular del Perú, Banco Regional del Norte, Banco Regional del Sur Medio y Callao, Banco Wiese, Caja de Ahorros de Lima, Corporación Financiera de Desarrollo.
Foreign branches: Arlabank-Arab Latin American Bank, Banco Central (Madrid), Banco Exterior de Los Andes y España (Extebandes), Bank of America, Bank of London and South America, Bank of Tokyo, Citibank, Chase Manhattan.

Trade fairs: Various local fairs throughout the year. Pacific International Trade Fair held each Nov in Lima.

Electricity supply: Generally 220 V AC, 60 cycles. Exceptions include Arequipa (220 V AC, 50 cycles) and Iquitos (110 V AC, 60 cycles).

Representation in capital: Algeria, Angola, Argentina, Austria, Belgium, Bolivia, Brazil, Bulgaria, Canada, Chile, China PR, Colombia, Costa Rica, Cuba, Czechoslovakia, Dominican Rep., Ecuador, Egypt, El Salvador, Finland, France, German DR, German FR, Guatemala, Haiti, Honduras, Hungary, India, Israel, Italy, Japan, Republic of Korea, Mexico, Morocco, New Zealand, Nicaragua, Panama, Paraguay, Philippines, Poland, Portugal, Romania, Saudi Arabia, Spain, Sweden, Switzerland, USSR, UK, USA, Uruguay, Vatican, Venezuela, Yugoslavia.

Hotels: Cuzco
El Dorado Inn, Avenida Sol 395, P.O. Box: 961 (tel: 283112; tx: 52039).

Libertador, San Agustín 400 (tel: 23-1961; tx: 52043).

Picoaga, Santa Teresa 344 (tel: 5393; tx: 52041).

Savoy, Sol 954 (tel: 22-4322; tx: 52002).

Hotels: Lima
Cesar's, S/n Av. La Paz y Diez Canseco, Miraflores, P.O. Box: 5172 (tel: 44-1212; tx: 21348).

Country Club, Los Eucaliptos, San Isidro, P.O. Box: 1875 (tel: 443614/443610; tx: 21219; fax: 441981).

Crillón, 589 Av. N de Pierola, P.O. Box: 2981 (tel: 28-3290; tx: 25274; fax: 280682).

El Condado, 465 Alcanfores, Miraflores (tel: 44-3614/3610; tx: 21219; fax: 441981).

El Pardo Hotel, S/n Av. J. Pardo e Independencia, Miraflores (tel: 47-0283/0558; tx: 21160; fax: 442171).

Garden Hotel, 450 R. Rivera Navarrete, San Isidro (tel: 42-1771/3520/1734; tx: 21239).

Gran Bolívar, Plaza San Martín, P.O. Box: 864 (tel: 27-6400; tx: 20338; fax: 338626).

Lima Sheraton, Paseo de la República 170, P.O. Box: 588 (tel: 32-9050/8676; tx: 20286).

Pueblo Inn, Km 11 Central Highway (outside Lima), P.O. Box: 5516 (tel: 35-0777/1338/6363; tx: 25402; fax: 355354).

Car hire: Lima
Avis: Shopping Arcade Sheraton Hotel, Paseo de La República Ave 170 (tel: 32-71-46, 32-72-45, 32-90-50 ext 4151; tx: 25093).

Budget: Av La Paz 522 (tel: 44-45-46, 44-07-60).

Europcar National: Apartado 10351 (tel: 23-25-26; tx: 20494).

Hertz: Jr Ocoña 262 (tel: 28-84-78, 28-63-30; tx: 25652).

Car hire: Lima International Airport Avis: (tel: 52-4774).

Budget: (tel: 52-8706).

Europcar National: (tel: 52-3426).

Dollar: (tel: 52-6741).

Hertz: (tel: 51-8189; tx: 25652).

Car hire: Lima International Airport
Avis: (tel: 52-4774).

Budget: (tel: 52-8706).

Dollar: (tel: 52-6741).

Hertz: (tel: 51-8189; tx: 25652).

Nacional: (tel: 52-3426).

Business advice and information
ADEX (export association), Las Flores 326, Lima 27 (tel: 40-4929).

Andean Group, Avenida Paseo de la República, Casilla Postal 3237, Lima.

Banco Central de Reserva del Peru, Miró Quesada, Lima (tel: 276250; tx: 20169, 25437).

Cámara de Comercio de Lima, Avenida Abancay 291, 2nd Floor, Lima (tel: 28-7770).

COFIDE (State Development Council), Avenida Garcilaso de la Vega 1456, Lima.

Confederación Nacional de Comerciantes (CONADO) (National Federation of Commerce), Avenida Abancay 210, Lima (tel: 27-3528, 28-6026).

CONITE (National Commission for Investments and Foreign Technology), Avenida Abancay 500, Piso 6 (MEF), Lima 1.

CORPAC (Corporación Peruana de Aeropuertos y Aviacíon Comercial), Aeropuerto Internacional Jorge Chavez, Avenida Faucett (S/N, Callao (tel: 52-9570).

DHL Worldwide Courier, Avenida La Marina 2469, San Miguel (tel: 52-5559; tx: 25061 PE DHL).

Empresa Minera del Centro del Perú SA (CENTROMIN), Cdra 21, Avenida Javier Prado, San Borja, Lima (tel: 35-5467).

ENAPU SA (National Port Company), Avenida Guardia Chalaca S/N, Callao (tel: 29-9210).

Federal Reserve Board (BCR), Jr A M Quesada 441, Lima 1.

FOPEX (promotion fund for non-traditional exports), Avenida José Pardo 275, Lima 18 (tel: 61-7094).

FOPTUR (tourist promotion), Av Rep de Panamá (tel: 41-9778, 52-4416).

HIERROPERU (State Iron Company of Peru), Avenida Paseo de la República 3587, Lima (tel: 41-0636).

International Translation Service, Avenida Arequipa 3200, San Isidro, PO Box 6046, Lima (tel: 41-1396; tx: 20240).

Lima Times, Pasage Los Pinos 156, Miraflores (tel: 47-2552).

MINEROPERU (State Mining Company of Peru), Avenida Bernardo Monteagudo Orrantia 222, Magdalena del Mar, Lima (tel: 62-0740).

Ministry of Economy and Finance, Cuadra Avda Abancay 5, Lima (tel: 289590; tx: 20187).

Ministry of Fisheries, Avda Javier Prado Este 2465, San Luis, Lima (tel: 362630; tx: 21058).

Ministry of Health, Avenida Salaverry 8, Jes\01us María, Lima (tel: 32-3535; tx: 20433).

Ministry of Housing, Domingo Cueto 120, Jesús María, Lima (tel: 71-6070; tx: 25158).

Ministry of Industry, Commerce, Tourism and Integration, Calle 1 Oeste, Corpac, San Isidro, Lima 27 (tel: 407120; tx: 20194).

Ministry of Transport and Communications, Avenida 28 de Julio 800, Lima 1 (tel: 23-7800; tx: 25211).

PETROPERU (State Petroleum Company), Paseo de la República 3361, San Isidro, Lima (tel: 42-5000).

Skyway SA (international courier), Centro Com. Camino Real, 1103, PO Box 2552, Lima 100 (tel: 40-2353, 22-9225, 41-6725; tx: 21418 PE Batanero).

Sociedad de Industrias (Society of Industries), Los Laureles 365, San Isidro, Lima (tel: 40-8700).

Philippines

Vaudine England

December 1989 saw the sixth coup attempt against Mrs Aquino's government in just over three years. Rebel troops under the leadership of Colonel Gregorio Honasan proved more resilient than anticipated. President Aquino sought US assistance in overcoming the coup and declared a state of emergency giving her additional economic powers. Rumours of a further 'phase 2' coup persisted to the end of the year, although the economic effects of the failed coup proved less harmful than anticipated.

In early 1990 President Aquino was still in charge; in many respects that was her major achievement. Yet her rule is perceived to be tenuous and attempts at consolidation brought with them other problems. Perhaps Mrs Aquino's period in power is a salutary lesson to idealists of all persuasion in the country as traditional sources of friction persist: corruption, human rights abuses, the communist insurgency, vast-scale poverty and continued indebtedness. Progress during 1989 has, however, been seen in economic growth figures, negotiations over future foreign funding, relations with the Soviet Union, and simply the persistence of the forms of democratic government.

The state of the nation

Mrs Aquino set the tone for the next year when she gave a State of the Nation address to Congress in July 1988. She was facing an armed forces which, after its violent restlessness of the preceding year, had quietly softened its opposition to her presidency. Congress itself was, if anything, becoming more restless and somewhat wary of her lack of clarity toward key policies. The presidential address was upbeat and optimistic, deftly laying blame for most problems at the feet of her detractors. She outlined ambitious development plans, detailed government victories against communism and blamed foreign creditors for hampering growth. Referring to land reform and the future of US bases in the Philippines only in passing, her message focused on economic growth. She cited a growth rate of nearly 6 per cent adding that 1.6m jobs had been created in the last year with wages rising more than 20 per cent. By 1992, Mrs Aquino plans to provide 90 per cent of the population with clean water, concrete roads in villages and proper school classrooms for all school children. She claimed Filipinos were witnessing the year that the back of the communist insurgency was broken. But the country had 'one foot manacled' due to its foreign debt of more than £28bn, which cost the country more than 40 per cent of its national budget to service it.

But Mrs Aquino's message omitted any guidance to Congress on how it should legislate on the debt problem, what the country's stand on the bases should be, and how corruption was to be stopped. Her stirring phrases about national unity were greeted with less than the usual excitement. The 40-minute speech was broken by applause only twice, and the standing ovation at the end lasted a mere 30 seconds.

Beyond the rhetoric, it is true that some issues bedevilling her first year in office were abating. The military had received a pay rise, and housing for the troops in military camps had been improved. Mrs Aquino's tougher stance against the communist insurgency had pleased the soldiers and she had sided with the troops in the face of mounting allegations of human rights abuses. The lack of coups since the 28 August 1977 attempt does not mean that the under-

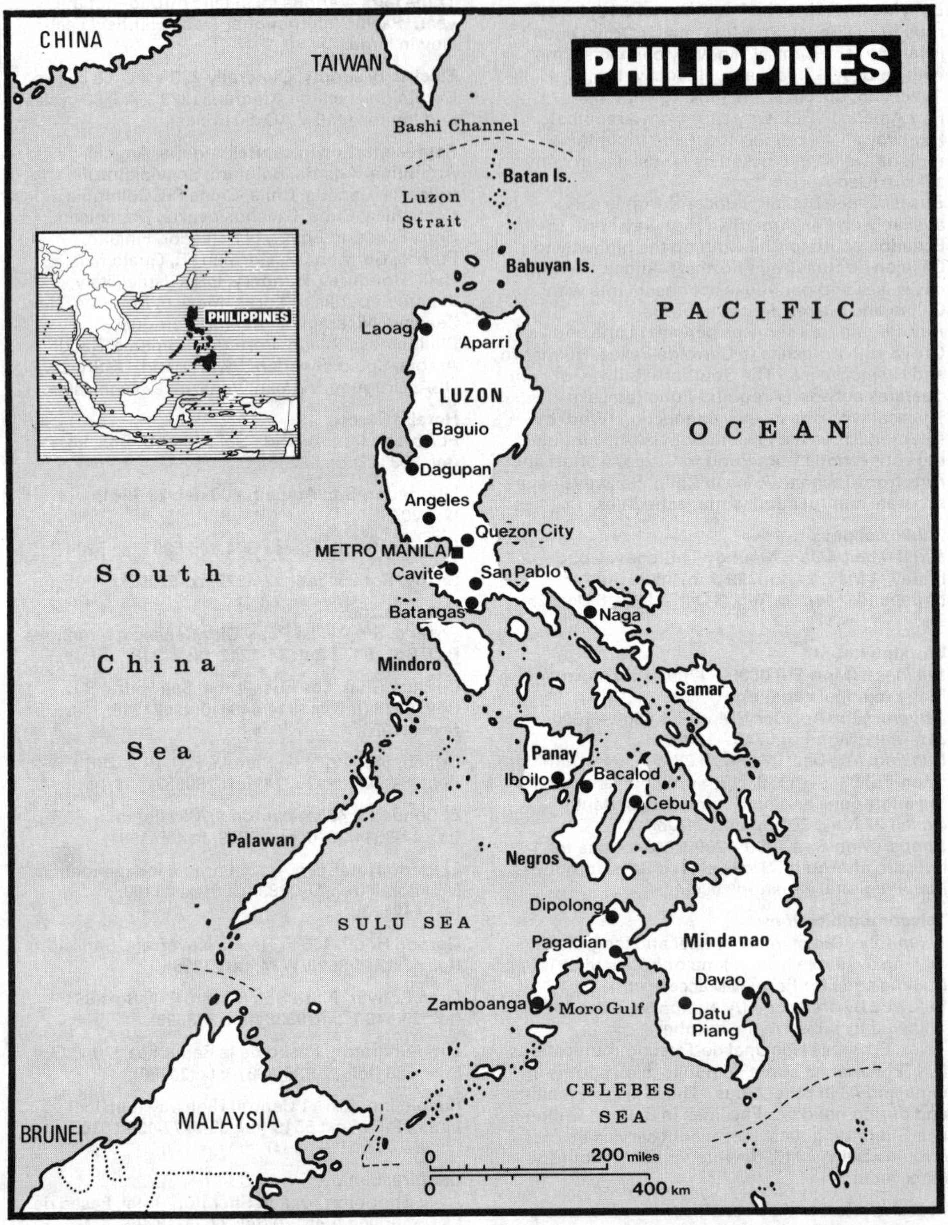

KEY FACTS

Official title: Republic of the Philippines
Head of state: President Corazon Aquino
Ruling party: Lakas ng Bansa
Capital: Manila (on Luzon)
Official Languages: Pilipino (based on Tagalog); English widely spoken
Currency: Peso (P = 100 centavos)
Exchange rate: P21.1 per $ (Nov 1989)
Area: 300,439 sq km (7,107 islands)
Population: 60m (1988)

GDP per capita: $668* (1988)
GDP real growth: 6.1%* (1988)
Labour force: 21.2m (1988)
Unemployment: 8.5% (1988)
Inflation: 9.0% (1988)
Trade balance: –$1,085m* (1988)
Foreign debt: $29.9bn (1988)
Visitor numbers: 1,043,114 (1988)

* estimated figure

lying communication gap which persists between the president and the army has been solved, however, and the military remains one of the strongest power bases in the country's politics.

Mrs Aquino's personal image was becoming more vulnerable too. The often overt sympathy shown her as she rose to power early in 1986 has since changed to a more critical stance. Part of this shift has been attributed to Mrs Aquino herself who, journalists say, has created a distance between herself and her public. She now rarely holds press conferences or interviews, criticising the 'insults of short-term journalists' in her State of the Nation address.

Despite the impressive growth figures, there is a mixed message on the economy as well. Economists note that what the country is witnessing is an artificial growth in the economy, fuelled by consumer spending on imported goods following implementation of an import liberalisation programme. It is only the urban middle class which is able to fuel the spending boom, leaving the peasantry in the same kind of poverty it has suffered for decades. While GDP grew 5.7 per cent in 1987 and by 6.1 per cent in 1988, the birth rate was spiralling too. A leaked report from the World Bank stated in July 1988 that the population of about 60m is growing by about 1.5m every year. Even a decade of fairly rapid and sustained growth will likely not keep pace with the growth in available labour. (Nationwide population control is virtually ruled out because of the predominance of the Catholic church in the country.) While growth is predicted to be over 6 per cent for 1988, the crucial agriculture sector seemed to have stalled early in the year. About 70 per cent of the population lives in rural areas. Investments increased two to three times over the previous year but, perhaps because of unrealistic expectations, the hoped-for explosion of investment interest did not come about. An unwillingness or inability to transform the euphoria of Mrs Aquino's accession to office into focused and effective economic policy took part of the blame for this.

KEY INDICATORS

	Unit	1984	1985	1986	1987	1988
Population*	'000	53,351	54,668	56,000	57,360	58,720
Gross domestic product (GDP)	P bn	540.47	612.68	627.13	708.37	827.41
GDP per capita	P	10,130	11,207	11,199	12,350	14,091
Unemployment rate	%	5.5	6.1	11.1	12.3	–
Wholesale prices	1985=100	84.6	100	98.4	107.4	121.8
Consumer prices	1985=100	81.2	100	100.8	104.6	113.7
Volume of exports	1985=100	98.5	100	116.6	124.2	138.2
Centrifugal sugar	1985=100	167	100	39	29	25
Copra and coconut oil	1980=100	64	67	124	111	–
Copper concentrates	1985=100	131	100	105	105	131
Wood	1985=100	130	100	85	76	77
Volume of imports	1985=100	111.1	100	119.7	156.2	191.3
Exports	P m	88,339	85,283	97,375	116,237	148,151
Centrifugal sugar	P m	4,471	3,501	1,754	1,229	1,261
Coconut products	P m	11,728	8,480	9,415	11,441	12,096
Copper	P m	1,876	1,565	1,820	2,254	4,534
Wood	P m	4,294	3,469	3,886	4,773	5,378
Imports	P m	107,670	101,518	107,324	147,875	184,246
Balance of trade	$m	–679	–482	–202	–1,017	–1,085
Current account	$m	–1,268	–18	971	–498	–373
External debt	$m	24,358	26,190	28,853	29,962	–
Debt service ratio	%	13.7	15.4	22.4	22.7	–
Exchange rate	P per $	16.70	18.61	20.4	20.6	21.1

* estimated figure

Bases accord is reached

Seven months of tough and often acrimonious negotiations ended with a deal regarding terms over the US bases in the Philippines, in October 1988. The agreement referred to compensation from the USA for usage of the bases up until 1991 – the term of the current lease. The thorny question of whether that lease will be renewed and if so at what price, remains to be tackled in 1990. The October deal requires the USA to provide the Philippines with $481m in both 1990 and 1991 in return for continued use of Subic Bay and Clarke Air Base, the most important US military facilities in the Asia–Pacific. The Philippines had been demanding much more than this. In the midst of a re-assessment of Filipino attitudes toward their traditional benefactors (and former colonial rulers) Manila had first asked for a more than six-fold increase in US aid, together with restrictions on military operations. But the deal did increase compensation from the $180m originally specified in the lease and Washington relaxed the rules governing the use of money, allowing Manila to use some of it in payment of its foreign debt.

Domestic criticism of the bases deal was readily forthcoming. Philippine politicians and nationalist groups decried the agreement as a sell-out, saying the USA had achieved the better terms. One group called the deal a case of high treason. It took Mrs Aquino a few days, but she did publicly support the agreement as 'a definite and substantial improvement' on the old one. At the same time, her speech-writer, Teodoro Benigno noted the deal was little different from what the USA had been offering at the start of the negotiations. Some senators said the deal could be in violation of the constitution's nuclear clause, as it allows US ships and planes to cross Philippine territory without reporting any nuclear armament. A presidential aide denied the claim.

USA calls the Marcoses to account

In October 1988, the USA indicted former President Ferdinand Marcos and his wife Imelda on criminal charges for fraudulently acquiring US property worth hundreds of millions of dollars. The indictment was wide-ranging. It said the Marcoses embezzled millions of dollars from the Philippine Government to purchase American real estate. They are also accused of defrauding US banks of almost $200m. The approval for an indictment came after Mr Marcos, ill and in exile in Hawaii, failed to respond to a take-it-or-leave-it deal that would keep him from possible gaol: in return for staying out of gaol, Mr Marcos would have had to hand back all his ill-gotten gains. But the deadline on the offer passed with no comment from the Marcos home. The decision to proceed against the Marcoses in court was not easy for the then US President Ronald Reagan because of his personal friendship with Mr Marcos, and also because it raised questions about continued US willingness to shelter the Marcoses in exile.

Even more difficult, however, is the situation facing the Aquino Government. It would like to charge the Marcoses with a wide range of crimes but to do so would mean the banished couple would almost certainly have to return to the Philippines to stand trial. And Mrs Aquino still does not want to see the people she believes are responsible for killing her husband, Benigno Aquino, back on Filipino soil. The problem is a constitutional provision which disallows the prosecution of anyone in absentia. Yet the hesitancy marking this issue is holding up investigations into the billions presumed to have been hidden in Swiss bank accounts. The Swiss say they cannot allow Philippine investigators into Swiss bank accounts, until charges are laid; Philippine investigators say they cannot lay charges until they see the Swiss accounts. The government has already filed 39 separate civil suits against Mr Marcos, seeking a total of $96.7bn in damages, but with civil suits, defendants do not have an automatic right to be present.

Behind the mass of legal detail lies a more profound problem: that of the Marcos image. There remained, among a small but vocal group in the country, an enduring mystique surrounding the frail form of Mr Marcos. Even the mention of a Marcos return caused panic in Manila; it was enough to put the military on alert, send the cabinet into emergency session and spark protest rallies. Marcos supporters continue to postpone the burial of his mother, who died in May 1989, in the hope that her son would be allowed to come home for the funeral. The death of Mr Marcos in September 1989 altered this situation.

Mrs Aquino's first 1000 days

By November 1988, Mrs Aquino was celebrating her first 1,000 days in office with carefully orchestrated events at Malacanang Palace. The only failure she admitted to was her slowness in responding to the right-wing threat from the military which has tried five times to unseat her. Her speech on this occasion was remarkable for the number of issues untouched. While the establishment of democratic forms of government was remembered, there was little time spent on every other persistent problem. Her speech seemed aimed at countering a growth tide of criticism, from businessmen, church leaders, human rights activists and news organisations.

As Commander-in-chief of the armed forces, President Aquino gives a speech at the military graduation ceremony every year at the country's top military academy. When she first tried to do this in 1987, a bomb went off underneath the platform where she was due to speak. In February 1989, however, the event went off peacefully, allowing Mrs Aquino to call first for talks with her armed foes on left and right, then to promise victory by arms if talks failed. She recognised that there were wounds in Philippine society, as she put it, that often seemed too deep to be healed. But she called on Filipinos not to pass on the legacy of narrow causes and partisan hatred to the next generation. Mrs Aquino's continuing need to

COUNTRY PROFILE

Political system: The form of government is a presidential system with separation of powers. Under the constitution, executive power is vested in the president and cabinet, legislative power with the elected National Assembly (a Senate and a House of Representatives) and judicial power with the Supreme Court and lower courts. Local government is vested in 13 regions, with provincial, city and municipal councils. In the presidential elections of February 1986 then President Ferdinand Marcos (in power from 1965) claimed to have defeated his challenger Corazon Aquino. But the election was so blatantly rigged that a popular revolt began and spread even to the army. Marcos and associates fled the country and Aquino took over. A plebiscite in February 1987 ratified a new constitution with Aquino as president and congressional elections in May 1987 confirmed her popular support. The rule of the central government is not altogether easy, notably because of a guerilla rebellion led by the banned communist New People's Army and because of unrest in the south, centred around Mindanao, by dissatisfied Muslims. Disgruntlement within the armed forces has led to a number of unsuccessful coup attempts led by lower ranking army officers.

Political parties: Presidential Party (Lakas ng Bansa, comprising the PDP-Laban, Lakas ng Bayan, National Union of Christian Democrats, Liberal Party, Bandila); UNIDO Grand Alliance for Democracy; and the Partido ng Bayan, Kilusang Bagong Lipunan.

Population: The Philippines is the only Christian majority country in Asia. About 85 per cent of the 58.7m people (1988 estimate) are baptised as Roman Catholics but there is a strong Muslim presence especially on Mindanao.

Main cities/towns: Metropolitan Manila, (on Luzon), including Manila, Quezon City, Makati, Pasay City, Caloocan City and Pasig (population 6.9m); Cebu (on Visayas), pop 557,000; Davao (on Mindanao), pop 725,000; Iloilo (on Panay), pop 269,000; Iligan (on Mindanao), pop 199,000.

Language: Pilipino (based on Tagalog) is the national language, but practically all business and politics is conducted in English. Spanish is less commonly spoken than previously.

Media
Press: There are 20 daily newspapers, 11 of which are in English. Of the three English national daily newspapers, the largest is *Manila Bulletin.* Daily business newspapers and periodicals cover most fields, with trade journals in architecture, building, construction, engineering, transport, agriculture, mining, trade and commerce, finance, law, medicine, education and other fields.
Radio: There are 286 commercial and non-commercial broadcasting radio stations, with some 3.6m radio sets.
Television: Manila has five commercial TV channels and the provinces are served by relay stations, with 2.2m TV sets in 1985.

Domestic economy: At the time of Marcos' departure the economy was in a huge mess. Economic growth was negative; the country was heavily in debt it could not repay. Marcos and his cronies were alleged to have salted billions of dollars illegally from the country; capital flight by richer middle class Filipinos was a major problem. By 1988 things were better, but per capita income was still only about US$600 a year. Foreign debt still totalled more than $28bn and the cabinet was split on key questions of economic policy. The Philippines is still largely an agricultural economy, dependent on commmodity earnings. Major industries are in the hands of the state because they were taken over from Marcos' cronies. Nevertheless economic growth in 1986 was positive after two negative years and in 1987 rose to more than five per cent with a target of between six and seven per cent set for 1988. Coconut prices trebled, manufacturing expanded 7.4 per cent, construction was up 16 per cent, local and foreign investment grew by nearly 20 per cent. Textiles were the fastest growing export industry, with garment sales totalling $1bn. The economy is boosted and balance of payments gap closed through remittances of about $1.5bn a year from Filipinos working abroad mainly in the Middle East, Hong Kong and throughout South East Asia. Altogether 175,000 women were employed abroad, a number so high that in 1988 the Aquino government imposed constraints on their employment with the aim of preventing exploitation of Filipinas. The US military also provides $180m a year for the rent of two bases. American forces have been in the Philippines for 80 years but in 1988 there was renewed argument about renewal of the leases due in 1991.

Employment: 671 firms closed in 1987. The Filipino workforce is 23.3m. Unemployment is 11.2 per cent but large sections of the population especially in rural areas are underemployed.

External trade: All trade with South Africa is prohibited, and trade with communist countries is channelled through state agencies. There are considerable difficulties with foreign cash transactions, and geographical trade patterns have become highly erratic.
Exports: Merchandise exports grew 7.8 per cent in 1987. Total exports were valued at $5.4bn in 1987 and rose to about $7bn in 1988 helped by better prices for some commodities, notably coconuts. Exports have been helped by the devaluation of the peso to around 21 to the US dollar. Main exports are electrical and electronic goods, clothing, coconut oil, copper, fish, gold and bananas. Main destinations are the USA, Japan, Singapore, Hong Kong and the UK.
Imports: Total imports were valued at $6bn in 1987 and rose to about $8bn in 1988. Main imports have declined steeply in recent years, due to the currency squeeze. Main requirements are mineral fuels and lubricants, electrical and electronic components, chemicals, electrical machinery, other machinery, food and beverages, textile yarns, iron and steel and transport equipment. Main sources are the USA, Japan, Taiwan, Hong Kong, Saudi Arabia, Malaysia, German FR, Republic of Korea, Indonesia and Singapore.

Agriculture: Agriculture (including forestry and fishing) accounts for 27.9 per cent of GDP (1987), provides about 40 per cent of export earnings and employs about 44 per cent of the labour force. Some 35 per cent of the total land area is used to cultivate food crops, mostly on smallholdings. About one-third of the population depends on coconuts, the major export crop. Other commercial crops include sugar-cane, hemp, bananas, coffee, tobacco, peanuts and various fruits. Rice and maize production is sufficient to meet domestic demand, and other crops include sweet potatoes, cassava, plantains, pineapples, mangoes and cocoa. Livestock reared for local consumption include carabaos, cattle, goats, pigs and poultry. Woodland and forests cover 51.0 per cent of the land area and contain an estimated 1,454m cu metres of hardwood, but exports of logs were phased out to assist the local timber processing industries. The reafforestation programme is seriously behind schedule but is being accelerated. There is a substantial offshore and deep-water fishing industry, and fish-farming is widely practised. Annual fisheries production is about 1.5m tonnes of tuna, milk fish, scad, anchovies, sardines, molluscs and crustaceans. In 1988 President Aquino agreed a land reform programme whereby some 2.8m hectares will be distributed between two million landless farmers over the next 10 years.

Industry: Manufacturing expanded 7.4 per cent in 1987. It accounts for 24.6 per cent of GDP (1986) and is based on processing primary products and import substitution, with some assembly of electronics products for foreign companies. Recent austerity measures have jeopardised major industrial developments, and emphasis, under agreements on debt rescheduling with the IMF, World Bank and international banks, has shifted towards production for export. The industrial scene is complicated because there are almost 300 companies owned or controlled by the government employing about 140,000 people. Of these 50 are agricultural companies, 61 in industry proper and 180 are in the service sector. But by May 1988 only 11 concerns had been privatised. Disagreements within the government were holding up further sell-offs. Food and beverage processing is the main manufacturing activity, including sugar, meat, fruit and vegetable processing, fish and shrimp, soft drinks, alcoholic beverages and cigarettes. Major development projects have included a petrochemical complex, a pulp and paper plant, an integrated iron and steel mill, a copper smelter, a phosphate fertiliser plant, an integrated coconut products project, cement industry expansion and diesel engine manufacturing. Main light industrial products, which are often produced from imported materials or components, are cotton and textiles, vehicles, chemicals, machine tools and electrical and consumer goods such as refrigerators, radios, TVs, freezers, air-conditioning equipment, sewing machines and watches. Tourism, formerly an important industry, has suffered steep declines in income since 1983.

Mining: The Philippines, with copper reserves estimated at 3,603m tonnes, is Asia's leading producer, although copper production has declined sharply in recent years; the country is also a leading Asian producer of gold. Coal production was 1.13m tonnes in 1986. Silver, nickel, chromium, manganese, zinc, mercury, sand, gravel and rock asphalt are also mined. Other metal and mineral resources include iron, molybdenum, lead, platinum, palladium, cadmium, cobalt, uranium, phosphate, guano, sulphur, pyrites, limestone, shale, gypsum, clay, kaolin, feldspar and silica sand.

Hydrocarbons: The Philippines covers about 10 per cent of its energy requirements from indigenous petroleum production.

Energy: Geothermal power is being developed and currently contributes 19 per cent of the country's energy requirements. A nuclear power plant has been built on the Bataan peninsula, near Manila, but has proved highly controversial and has been put in mothballs.

Financial markets: Two stock exchanges operate in Metro Manila.

Membership of international organisations: ASEAN, ASPAC, Asian Development Bank, CCC, Colombo Plan, EC Trade and Co-operation Agreement, ESCAP, FAO, GATT, G-77, IAEA, IBRD, ICAO, IDA, IFAD, ILO, IMF, INTELSAT, INTERPOL, IPU, IRC, ISO, ITU, UN, UNESCO, UPU, WFTU, WHO, WIPO, WMU. World Bank, WTO.

know who to trust in the military was underscored by her reappointment, also in February, of an old friend and loyal supporter, Lieutenant-General Rafael Ileto, to a job in her government. He was defence minister when Senator Enrile lost the job in November 1986, but was retired in January 1988 in favour of the now powerful General Fidel Ramos. Ileto is now Mrs Aquino's national security adviser, and may be a counter-balance for Mrs Aquino in the face of General Ramos' growing power and possible presidential ambitions.

Vice-President leaves cabinet

In the middle of August 1988, Vice-President Salvador Laurel made formal what everyone knew already, namely that he was not part of the cabinet under Mrs Aquino. Ever since Laurel and Aquino jointly won power in early 1986, Laurel felt himself to be left out of any important decision-making. He constantly accused Mrs Aquino of breaking promises she made to him before they formed a government. And while he presented a more liberal and conciliatory face during voting times, there was no doubt that Laurel's own approach to government policy was more conservative that Mrs Aquino's. Mr Laurel's resignation from the cabinet meant he retained his largely ceremonial job of vice-president, leaving him next in line for the presidency if anything should happen to Mrs Aquino.

Mr Laurel's formal departure from the Aquino cabinet left him free to annouce, before the end of August, the formation of a new alliance of right-wing politicians, aimed at the replacement of Mrs Aquino. Most of the several dozen men who appeared with Mr Laurel to inaugurate the Unity for National Action in Manila, were members of the cabinet of Ferdinand Marcos. Notable among them was former defence minister and now Senator Juan Ponce Enrile, former senator Arturo Tolentino and the Filipino Moslem leader Abul Khyer Alonto. This new-clothes-for-old alliance was the beginning of a revival of the Nacionalista Party. This party was born in 1906 to fight for reforms in pre-independence days, and has split and been reunited many times in its history. By March 1989, a reunification ceremony was again being held, this time between the United Nationalist Democratic Organisation (UNIDO) and the Union for National Action. The aim was to draw on the name of the oldest Filipino political party to fight for power. If gained, the party's platform would see the ushering in of federal government, denunciation of graft and corruption and advocation of nationalist industrialisation. However, the Grand Alliance for Democracy (GAD), another right-of-centre grouping, has not joined the revived party, even though Senator Enrile, a GAD patron, is also supportive of the revival of the Nacionalista Party.

Lined up against these forces remain the three political parties which support Mrs Aquino: the Laban Ng Demokratikong Pilipino (Struggle of Philippine Democrats), the Liberal Party and the PDP-Laban.

Mrs Aquino's term of office expires in 1992 and she has said she will not run again. It is this eventuality that those hoping for a political revival of the right await. Yet the public display of unity between such weather-beaten politicians as Enrile and Laurel seems almost as improbable as the Laurel–Aquino double-act. Neither Enrile nor Laurel are men happy with second place. Of the two, Enrile can claim to more consistency: Laurel, like his father before him, has changed sides regularly in the messy world of Philippine politics. But Enrile and Laurel have also expressed views on key security issues and attitudes which are at stark variance with one another.

The extent to which Marcos still provides a focus for the grievances of the country's right-wing politicians, along with Laurel's own position in the political wilderness, was highlighted when Laurel paid a visit to the dying Marcos at his hospital bed in Hawaii in February 1989. Little came of the trip, but Laurel's allegiances – which had once been wholeheartedly pro-Marcos – had turned another circle.

With politicking seemingly taking up more time than the formulation of economic policy, new opinion surveys show congress to be suffering what one columnist calls a crisis of confidence. The fact that only three of its 200 members come from disadvantaged sectors of society, helps explain how the land reform ideals of the Aquino administration finally turned into yet another damp squib. Land reform is arguably the most important plant of any programme for the alleviation of poverty. The Land Reform Bill is designed to redistribute about 2.8m hectares of agricultural land to more than 2m farmers over 10 years. The plan is for the government to buy the land off large landowners, and sell it to the peasantry at 'affordable' prices. But the compromise bill that was eventually passed early in 1988 includes no specific funding allocation.

The government's approach toward privatisation seems similarly bogged down, even though it is not a landlord-dominated congress which is causing the trouble. The government intends to generate much-needed cash by selling off many of the national corporations which have been a drain on administration for years. The highest-profile of the attempted sell-offs is that of the huge Nonoc Mining and Industrial Corporation. But it remains unsold after two public bids, open negotiations with three potential buyers on a non-bid basis, and then the breakdown of what were supposed to be final talks with one of the buyers, Australian businessman Alan Bond.

Moscow made diplomatic headway with the strongest ally of the USA in south-east Asia in March 1989, when the Philippines signed an agreement with the Soviet Union to expand trade and allow repair access for the Soviet fishing fleet. The deal came after a process of talks over the new year, which have included a visit to the Soviet Union – for the first time since diplomatic relations were established in 1976 – of Foreign Secretary Raul Manglapus, and a visit to Manila by Soviet Foreign Minister Eduard Shevardnadze.

NPA still in action

A few high-profile arrests, the discovery of secret files, and the government is proclaiming the incipient demise of the country's communist insurgency. But the New Peoples' Army (NPA) can just as easily claim a kind of victory: like Mrs Aquino, the communists are still at large. Early in 1989, the communist leadership including recently-escaped Romulo Kintanar issued an uncompromising statement, which declared that it was reasonable to aim for total victory within the next 10 years. If 10 years was not enough, the document went on, then the NPA remained ready to fight for as long as it took. The communist movement has concentrated more on political progress, expanding their political machine, and less on military actions. But there is no perceptible sign of the level of violence subsiding in either the towns or countryside. This is by no means all due to NPA action. The still-growing bands of armed, right-wing vigilante groups are just as convinced of their right to kill. Villages formerly under communist control are now ruled by self-appointed vigilantes. The killing of an American who fought in Vietnam, Colonel James 'Nick' Rowe, in April 1989, was linked to his work in so-called 'low-intensity conflict' activities, such as arming the vigilantes. Yet the last formal stage of Mrs Aquino's democratisation programme (the holding of *barangay* or village-level elections in March 1989) with only nine killings, was adjudged to mark a newly peaceful level of voter interest.

The documents and computer discs seized by the government in 1988 reveal a similarly mixed picture of the state of the communist movement. On the one hand, the confidential reports show that communists have infiltrated various sectors of society, including the church, the press and local government. The insurgents have a complex financial structure including stock holdings and representatives around the world. They claim a presence in 11,392 of 27 per cent of the country's barrios, and considerably more advanced community health programmes than those of the government. The discs also reveal, however, that support for them from the middle class has diminished with the peaceful accession to power of Mrs Aquino, and the decline of student radicalism. Just as telling, however, is the military's admission in January 1989, that it had subcontracted the decoding of the discs to a private contractor who was working on them in his spare time.

Perhaps the military's more significant act was the apprehension in July 1989 of the top communist leader Satur Ocampo and his wife. Mr Ocampo is a member of the Politburo of the Communist Party of the Philippines and has served as chief spokesman for the National Democratic Front (NDF), the insurgents' political and international umbrella organisation. His wife Carolina Malay, also a ranking Communist Party leader, was his fellow negotiator for the NDF in failed peace talks with Mrs Aquino's government early in 1987. Five days before his capture, Mr Ocampo had promised a unilateral cease-fire if the government pledged to remove US bases from the Philippines by 1991. His arrest also followed the massacre by NPA guerillas of 27 members of a Protestant community in southern Mindanao province in June 1989. Mr Ocampo had publicly apologised for the incident, thought to be the worst of its kind throughout the 20-year insurgency.

While the government continued in 1989 to gain access to old money (from the Marcoses for example), it also sought new money from old problems. The government decided in December 1988 to sue the US firm, Westinghouse Electric, claiming it paid bribes which benefited Mr Marcos, to procure the contract for a nuclear power station. The plant eventually cost $2.1bn but has never operated. Westinghouse is seeking arbitration on the dispute in Geneva, while various Filipino officials are alternately seeking a court settlement, or an out-of-court settlement liable to go into millions of dollars.

New money arrives – or does it?

By July 1989, it looked as if the Aquino government's hope for easier foreign financing had come true. For months, there had been talk of a mini-Marshall Plan for the country, led by the USA, funded substantially by Japan, and including a range of other donors. The theory was to provide Manila with extra funding to help it consolidate the fledgling democracy. But talk of the plan became tangled up with talk about the US bases. The USA denies a direct connection, but it seemed clear that the aid initiative would deprive Filipinos of the upper hand if a threat to national sovereignty were to be seen in future base negotiations. More to the point, Japan's contribution would allow that country to contribute to the so-called defence of South-East Asia, despite its non-military constitution. The end result was a promise of $3.5bn but it remained unclear how much of this would be new money, above aid commitments already in train on a bilateral basis. Japan pledged the most, at a special conference in Tokyo in July 1989 to launch the initiative. Japan's contribution of $100m in aid for development projects, for example, comes on top of $900m in grants and loans already promised.

On the home front, the plan was immediately assailed by nationalist economists such as Solita Monsod. The argument against the Philippine Aid Plan

(PAP) is played out largely in terms of nationalist rhetoric. Mrs Aquino was finally forced to admit that only $1.5bn is truly 'new' money but her former economics secretary Mrs Monsod said it was just $250m.

Outlook

As a pay-off for allowing the bases to stay at least until 1991, and for bringing democratic forms to the Philippines, the gains of the PAP would seem to lie more in the hands of the donors than in Manila. But the problem for Mrs Aquino is that she will have critics whatever she does. Perhaps nowhere else in South-East Asia is the populace so articulate, and so fond of participating in political debate. Her government's agreement in April 1989 to a new IMF austerity drive in return for a $1.3bn aid package, was assailed by nationalists left and right. The spiral of indebtedness remains as hard to break as does the spiral of violence in the country.

The Aquino government plans to enter 1990 with a stronger budget than that of 1989. Yet of the $10.6bn budget, one third will go toward debt servicing. A 20 per cent rise in government revenue is predicted for 1990, as tighter tax and customs regimens take root. A deficit of $360m is projected, down from the $695m of 1989.

BUSINESS GUIDE

Time: GMT + 8 hrs.

Climate: Tropical, with seasonal variations in rainfall. In the northern part of the country, there is a rainy season from May-Nov, during which typhoons may occur, causing loss of power supply and flooding. Rainfall is evenly distributed in the south. Annual mean temperatures vary between 31-33°C, with humidity from 65-90 per cent. In Manila, Apr and May are the hottest months and Nov-Feb is the cool season.

Entry requirements
Passport: Required by all except holders of certain documents (UN, Hong Kong, USA armed forces).
Visa: Nationals of most countries with assured onward passage do not require visas for short stays. Otherwise, visas should be obtained before arrival. Visitors who have stayed in the country for 60 days must obtain an emigration clearance certificate before departure. Travel regulations are liable to alter at short notice.
Currency: There is no limit on the amount of foreign currency which may be imported, but amounts in excess of $3,000 must be declared. Foreign currency exports are limited to amounts declared on entry. No more than P500 may be imported.
Customs: Personal effects are allowed duty-free. Visitors may import motorcycles and boats duty-free for stays of up to one month; longer stays require a bond guaranteeing re-export. Importation of arms, explosives, pornography, seditious publications and gambling equipment are prohibited.

Health precautions
Mandatory: Vaccination certificates required for yellow fever if travelling from an infected area and, sometimes, cholera if arriving by sea.
Advisable: Anti-malaria precautions and vaccinations against tetanus and typhoid.

Representation overseas: Algiers, Baghdad, Bangkok, Beijing, Belgrade, Berne, Bonn, Brasilia, Brussels, Bucharest, Buenos Aires, Cairo, Canberra, Colombo, Dakar, Dhaka, The Hague, Hanoi, Islamabad, Jakarta, Jeddah, Kuala Lumpur, Kuwait City, Lagos, Libreville, London, Madrid, Mexico City, Moscow, Nairobi, New Delhi, Ottawa, Paris, Port Moresby, Rabat, Rangoon, Rome, Seoul, Singapore, Stockholm, Tehran, Tel Aviv, Tokyo, Tripoli, Vatican City, Vienna, Vientiane, Washington DC, Wellington.

Air access
National airline: Philippine (PAL).
Other airlines: Air France, Air Niugini, British Airways, CAAC, Cathay Pacific, China, Air Micronesia, Egyptair, Garuda, Gulf Air, Hang Khong Vietnam, JAL, KLM, Korean Air, Kuwait Airways, Lufthansa, MAS, Northwest, PIA, Qantas, Royal Brunei, Saudia, Singapore, Swissair, Thai Airways International, United.
International airport: Manila International (MIA), 12 km from Manila; duty-free shop, bar, restaurant, buffet, bank, car hire. Cebu International Airport, 6km from Cebu city.

Surface access: Several shipping lines operate cargo and containerised freight services.
Main ports: Manila, Cebu, Davao, Iloilo, Zamboanga, Cagayan de Oro.

Hotels: A service charge of 10 per cent and a government tax of 10 per cent are usually added to hotel bills.

Credit cards: Major credit cards such as Diners, Visa and American Express are accepted.

Car hire: Self-drive and chauffeur-driven car hire is available. International driving licences are acceptable. Driving is on the right-hand side of the road. Local driving habits make traffic conditions extremely difficult.

City transport
Taxis: Metered taxis are available, and are preferable to negotiating pre-arranged fares. Tipping is not customary.
Buses: Numerous inexpensive bus and jeep-type 'jeepney' services operate in and around main centres, but they can be crowded, and knowledge of the area is recommended before travelling by bus. Outside Manila there are several local and long-distance bus services. Some air-conditioned services are available.
Rail: In Metro Manila, the mass transport system is complemented by a 15 km elevated Light Rail Transport (LRT) system.

National transport
Air: Philippine Airlines operates domestic services between Manila and 50 towns throughout the country.
Road: 164,000 km of roads (44 per cent surfaced), mainly confined to coastal areas, with 68,600 km of highways, 60 per cent of which is surfaced. The Maharlika highway runs from Luzon to Mindanao, with connecting ferry services.
Rail: The Luzon railway system consists of 474 km of track (Manila-Legaspi) and 266 km (Manila-San Fernando). The Panay railway, 117 km. Some services on Luzon have sleeping accommodation, restaurant cars and air-conditioning.
Water: Inter-island services are operated by several companies, some with air-conditioned cabins and dining rooms. There are 622 public and 188 private ports, many serving coastal shipping traffic.

Public holidays
Fixed dates: 1 Jan, 9 Apr (*Araw Ng Kagitingan* or Hero's Day), 1 May (labor Day), 12 Jun (Independence Day), 1 Nov (All Saints Day), 30 Nov (Bonifacio Day), 25 Dec (Christmas Day), 30 Dec (Rizal Day), 31 Dec.
Variable dates: Maundy Thu, Good Fri, National Heroes day.

Working hours: Working hours vary. Some banks and offices open for a half day on Sat and, in the Manila area, many shops open for a half day on Sun.
Business: (Mon-Fri) 0800 – 1200/1300, 1300/1400 – 1700; (Sat) 0830 – 1200.
Government: (Mon-Fri) 0730 – 1130, 1230 – 1630 or 0800 – 1200, 1300 – 1700.
Banking: (Mon-Fri) 0900 – 1600.
Shops: (Mon-Sat) 0900 – 1200, 1400 – 1930.

Telecommunications: International telecommunications are via satellite communications, telex lines, microwave relay stations and undersea cable.
Telephone: Dialling code for the Philippines, IDD access code + 63 followed by area code (2 for Manila) followed by subscriber's number. There are over 890,000 registered telephones (1987), 76 per cent of them in Metro Manila, with automatic service.
Telex: Public telex booths are operated by Eastern Telecommunications Philippines Inc, Globe-Mackay Cable and Radio Corp and RCA Communications Inc. Telex facilities are also available at good hotels.
Telegram: Telegram service is available.

Banking: Central Bank of the Philippines is the banking authority. There are 32 commercial banks, four foreign banks (The Chartered Bank, Hongkong and Shanghai Banking Corp, Citibank, and Bank of America), three government-specialised banks (Land Bank of the Philippines, Philippine Amanah Bank and the Development Bank of the Philippines). As of October 1986, there were a total of 116 thrift banks.

Advertising: Advertising is available in the press, radio, TV, cinemas and via direct mail.

Trade fairs: International Association of Business Communicators Asia Pacific Conference (Manila Jan), Energy '90 (Manila Jan), 11th World Congress of Cardiology (Manila Feb), 1st Asian Photo Congress '90 (Manila Jun), Corporate Gifts Fair (Manila Jun). For details of further exhibitions and conventions, contact the Philippine International Convention Center.

Electricity supply: 220 or 110 V AC, 60 cycles with flat and round two-pin plug fittings and screw-type lamp sockets.

Weights and measures: Metric system, with some local units still in use.

Representation in capital: Argentina, Australia, Austria, Belgium, Brazil, Bulgaria, Burma, Canada, Chile, China PR, Colombia, Cuba, Czechoslovakia, Denmark, Egypt, Finland, France, Gabon, German DR, German FR, Greece, India, Indonesia, Iran, Iraq, Israel, Italy, Japan, Republic of Korea, Libya, Malaysia, Malta, Mexico, Nepal, Netherlands, New Zealand, Nigeria, Norway, Pakistan, Panama, Papua New Guinea, Peru, Romania, Saudi Arabia, Singapore, Spain, Sri Lanka, Sweden, Switzerland, Thailand, UK, USA, USSR, Vatican, Vietnam, Yugoslavia.

Hotels: Baguio City, Benguet, Luzon
Hyatt Terraces Baguio, South Drive, P.O. Box: 219 (tel: 5670, 5780; tx: 27202; fax: 8153395).

Hotels: Calatagan, Batangas, Luzon
Punta Baluarte Inter-Continental, P.O. Box: 731, 4001 (tel: Ma'la 884164/8159711; tx: 23314; fax: 8171330).

Hotels: Cebu
Magellan International, Gorordo Avenue, Lahug, P.O. Box: 6401 (tel: 74613/9; tx: 24729).

Montebello Villa, Banilad, 6401 (tel: 85021/31; tx: 66030).

Rajah, Fuente Osmena Street, P.O. Box: 584,6401 (tel: 96231).

Hotels: Davao City, Davao de Sur, Mindanao
Davao Insular Inter-Continental, Lanang, 9501, P.O. Box: 144 (tel: 76051/61; tx: 48209; fax: 62959).

Hotels: Manila
Admiral, 2138 Roxas Boulevard, P.O. Box: 7155 (tel: 572081/94; tx: 74240488; fax: 5210711).

BUSINESS DIRECTORY

Bayview Prince, Roxas Boulevard and UN Avenue, P.O. Box: 771,2801 (tel: 503061; tx: 22311; fax: 632503061).

Century-Park Sheraton, Corner Vito Cruz and M Adriatico Streets, Malate, P.O. Box: 117 (tel: 5221011; tx: 40650; fax: 5213413).

Holiday Inn Manila, 3001 Roxas Boulevard, Pasay City, 3129 (tel: 597961/80; tx: 63487; fax: 5223985).

Hyatt Regency Manila, 2702 Roxas Boulevard, Pasay City (tel: 8312611; tx: 63462; fax: 8179742).

Inter-Continental Manila, Hotel Drive, Ayala Avenue, Makati, P.O. Box: 731 (tel: 8159711; tx: RCA 23314/63597).

Mandarin Oriental, Makati Avenue, Makati, P.O. Box: 1038 (tel: 8163601; tx: 63756; fax: 8172472).

Manila, Rizal Park on Manila Bay, P.O. Box: 307 (tel: 470011; tx: 40537; fax: 471124).

Manila Garden, 4th Quadrant, Makati Commercial Centre, Makati (tel: 8104101; tx: 45883; fax: 8171862).

Manila Hilton International, United Nations Avenue, Ermita, P.O. Box: 4430 (tel: 573711; tx: 63387).

Manila Midtown, Corner Pedro Gil and M Adriatico Streets, Ermita, P.O. Box: 4252 (tel: 573911; tx: 27797).

Manila Peninsula, Corner Ayala and Makati Avenues (tel: 8193456; tx: 22507; fax: 8154825).

Manila Royal, Carlos Palanca Street, Santa Cruz, P.O. Box: 2999 (tel: 488521; tx: 40334).

Silahis International, 1990 Roxas Boulevard (tel: 573811; tx: 63163; fax: 502526).

Westin Philippine Plaza, Roxas Boulevard, P.O. Box: 1146 (tel: 832-0701; tx: 40443 FILPLAZA; fax: 8323485).

Car hire: Manila
Avis: Airport, Mia Road, Pasay (tel: 832-2088).

Europcar: 108 Jupiter Street, Bel-Air, Makati (tel: 8188667; tx: 66496 CIFAR PN).

Hertz: China Banking Corp Building, Corner Peseo de Roxas and Villar Streets, Makati, Metro Manila, (tel: 8172761; tex: 64075).

Le Car: 583 Shaw Boulevard, Mandaluyong (tel: 799306).

Qualitrans: 3 Chapel Road, Pasay (tel: 8323208/3010).

Business advice and information
American Chamber of Commerce, 2nd Floor, Corinthian Plaza, Paseo de Roxas, Legaspi Village, Makati, Metro Manila (tel: 8187911; tx: 45181; cable: AMCHAMBCON).

Asian Development Bank, 2330 Roxas Boulevard, Metro Manila (tel: 8344444).

Bankers Association of the Philippines, 12th Floor, Philbanking Corp Building, Port Area, Manila.

Board of Investments, Industry and Investments Building, 385 Sen Gil Puyat Avenue, Makati, Metro Manila.

Central Bank of the Philippines, A Mabini corner Vito Cruz, Malata, Metro Manila (tel: 507051; tx: 63343).

Chamber of Agriculture and Natural Resources, 3rd Floor, Rico House, 126 Amorsolo Street, Legaspi Village, Makati, Manila.

Chamber of International Trade, Room 914, L & S Building No. 2, 1515 Roxas Boulevard, Ermita, Metro Manila (tel: 591372/3).

Development Bank of the Philippines, DBP Building, corner Makati and Buendia Avenues, Mataki, Metro Manila (tel: 8189511).

European Chamber of Commerce, 3rd Floor, Electra House Building, 115-117 Esteban Street, Legaspi Village, Makati, Metro Manila.

Ministry of Finance, Finance Building, Agrifina Circle, Manila.

National Economic Development Authority, NEDA Building, Amber Avenue, Pasig, Metro Manila (tel: 6735031; tx: 29058; cable: NEDAPHIL).

Philippine Chamber of Commerce and Industries, CCP Building, Magellanes Drive, Intramuros, Manila 2801 (tel: 403082).

Philippine Convention Bureau, 4th Floor, Legaspi Towers, 300 Roxas Boulevard, Manila (tel: 575031; tx: 40604).

Philippine International Trading Corp, Tordesillas Street, Salcedo Village, PO Box 1056 MCC, Metro Manila (tel: 8189801; tx: 63745). Private Development Corp of the Philippines, PDCP Building, Ayala Avenue, Makati, Metro Manila (Tel: 850686).

Trade and Industry Information Centre, Ministry of Trade and Industry, Trade and Industry Building, 361 Sen Gil Puyat Avenue, Makati, Metro Manila (tel: 8185701; tx: 14830; cable: MTI).

Pitcairn Island

(map on page xxiv)

COUNTRY PROFILE

Political system: Pitcairn Island is a British Crown Colony. The British High Commissioner in New Zealand is concurrently the Governor of Pitcairn Island. He is traditionally represented on the island by the Government Adviser/Education Officer, a teacher on secondment from New Zealand. Internal matters are dealt with by a 10-member local council headed by an elected magistrate.

Domestic economy: Native islanders, who comprise about 75 per cent of the population, mainly farm, raise livestock and fish for their own consumption, and trade handicrafts. Stamp sales and investment income account for 90-95 per cent of foreign exchange earnings. Foreign exchange deficits are made up from capital reserves and investments.

External trade: Revenue for 1987-88 was NZ$922,701; expenditure for 1987-88 NZ$799,722.

Agriculture: The island has highly fertile volcanic soil and rainfall is adequate. Main crops include taro, coconuts, beans, sugar-cane, bananas, yams and citrus fruits. There is some goat and poultry raising.

Energy: Electricity is produced by diesel generator.

Legal system: The magistrate exercises judicial power on the island.

Membership of international organisations: Lomé Convention (EC), South Pacific Commission (Conference representation).

Time: GMT -9 hrs.

Climate: Subtropical, with temperatures ranging from 13-30˚C, averaging 17-19˚C in Aug and 20-28˚C in Feb. Rainfall varies, but can exceed 2,000 mm/year.

Surface access: There are very few scheduled shipping services, with up to four regular supply and mail ships p.a. from New Zealand. In 1988, 52 ships and 21 yachts stopped at Pitcairn Island.
Main ports: There are no port facilities, and vessels anchor or heave to off the island. Bad weather can prevent contact.

National transport
Road: There are no surfaced roads, and in 1987 vehicles in use included 10 conventional motor cycles, 16 three-wheelers, one four-wheeled motorcycle, one motor car and three tractors. A bulldozer was parachuted on to the island in 1983.
Water: In Bounty Bay, on the north side of the island, there is a slip for launching the islanders' longboats. The jetty, built in 1977, was recently damaged, but has been repaired by local labour at a cost of 303,000 sterling.

Telecommunications: External contacts are by SSB, wireless telegraphy and radio telephone.

Telephone: The island has a party-line telephone service with 28 telephones, including two public telephones (1984).
Other: Radio: Radio officer Tom Christian (call sign VR6TC) is a keen radio amateur and can be reached on the following frequencies: 0630 – 0730 Mon on Z 14180 KHz; 1700 – 1800 Tue on Z 28950 KHz; 2200 – 2300 Tue and 1630 – 1730 Fri on Z 21350 KHz. Other licensed amateurs are Karl Young VR6KY, Kay Brown VR6KB, Betty Christian VR6YL, Irma Christian VR6ID and Meralda Warren VR6MW.

Weights and measures: Metric system.

KEY FACTS

Official title: Pitcairn Islands

Head of state: Queen Elizabeth II

Head of government: British High Commissioner in New Zealand

Capital: Adamstown

Official Languages: English, Pitcairn dialect

Currency: New Zealand Dollar (NZ$ = 100 cents)

Exchange rate: NZ$1.70 per $ (Nov 1989)

Area: 27 sq km (4 islands)

Population: 52 (1988), all on Pitcairn Island

Poland

Dr George Blazyca

In a remarkable year the pace of political reform quickened in Poland. In rapid succession round-table talks were held with the 'opposition', an historic compromise was reached, Solidarity was legalised and a general election was held in which Solidarity's victory was complete and surpassed all expectations. If the agreed election formula guaranteed the ruling parties a majority in parliament there was little doubt that this was a poor reflection of the mood in the country. Nonetheless a major step towards full democratisation was taken and the international community signalled its intention to aid the reform process with some ready cash and, more important, the promise of full normalisation of financial relations. These positive developments on the political scene nationally and internationally were continually blighted, however, by an economy on the point of collapse and the threat of social breakdown. The key problem for the period ahead was how to manage growing social tension alongside an economic restructuring and modernisation that might take some time before it brought widespread and tangible benefits.

Fundamental political changes

The contours of the political map were totally reshaped during the early part of 1989. The changes were as radical as they were unexpected and brought about by an increasing awareness on the part of the reforming wing of the Polish leadership, headed by General Jaruzelski, that social tensions, in 1988, were growing to explosive proportions. Two serious outbursts of industrial unrest, one in May 1988 followed by another in September, convinced the authorities that a political settlement with the still illegal 'opposition' was crucial to the success of the economic reform that was clearly running into the ground.

The September 1988 strike wave was brought to an end with the help of the Solidarity leader Lech Walesa when the authorities agreed, in return, to open 'round table' talks with the opposition. These talks would be wide ranging and cover all aspects of Polish social, political and economic life. But when striking workers returned to work the authorities stalled. Seemingly interminable 'talks about talks' filled most of the rest of the year, creating an impasse that was not broken until November 1988 when Lech Walesa appeared in a live television debate with the official trade union leader and politburo member, Alfred Miodowicz. Walesa's impressive performance gave him a famous victory over Miodowicz and helped to bring back to centre stage the question of the round table.

With inflation and shortages threatening to get out of control and society becoming increasingly restive it was clear that the authorities would have to act. The exhausted state of the economy meant of course that Poland's leaders could offer society no material relief; political reform became the number one item on the agenda. At last in February 1989 the round table did get under way and for the next six weeks the government-party coalition was locked in heated debate with Solidarity over the wholesale reconstruction of Polish economic and political life.

The talks were fraught with difficulties and took place against a background of severe industrial unrest that many observers believed to have been partly stimulated by the official trade unions and the party. A certain amount of tension in the air was designed to impress on Solidarity the urgency of finding a compromise agreement before a social breakdown threatened more drastic measures. Eventually, by early April, enough agreement was hammered out to support a grand concluding ceremony attended by Lech Walesa and the chief government-party negotiator, General Czeslaw Kiszczak.

Although Solidarity by no means won every point it pressed, the scale of the concessions made by the authorities was stunning. First, after more than seven years of a clandestine existence Solidarity would be legalised. Second, Solidarity would be able to contest 35 per cent of seats in the lower house of parliament in the June 1989 general election. It was also agreed that by the time of the next election in 1993 all parliamentary seats should be contested in an open competition. Third, a new upper house, or senate, would be created with 100 members whose election would be completely free even in June 1989. A state presidency, with sweeping powers (including the right to dissolve parliament) would be created, the post-holder to be elected by a simple majority of both houses of parliament.

The Solidarity side was less successful in its attempts to trim the party's nomenklatura, the system by which key posts throughout the state administration and the economy were awarded, through party patronage, to those considered to be politically reliable. This party control of the state apparatus was an issue that was to

KEY FACTS

Official title: The Republic of Poland

Head of state: General Wojciech Jaruzelski

Head of government: Prime Minister Tadeusz Mazowiecki

Ruling party: Solidarity, Polish United Workers' Party (PZPR), Peasants' Party (ZSL) and Democratic Party (SD)

Capital: Warsaw

Official Languages: Polish

Currency: Zloty (Zl) = 100 groszy

Exchange rate: Zl 5,978 per $ (Dec 1989)

Area: 312,683 sq km

Population: 37.8m (1988)

GDP per capita: $1,363* (NMP, 1988)

GDP real growth: 4.7% (NMP, 1988)

Labour force: 17.1m (1988)

Inflation: 60% (1988)

Trade balance: $1,089m (1988, hard currency)

Foreign debt: $39.2bn (end 1988, hard currency)

* estimated figure

dog Solidarity when, later in 1989, it faced the question of whether to participate in government. The nomenklatura question was a major piece of unfinished business and looked likely to be the greatest single blockage in the way of further economic and political reform. It was bound to be a central subject in future disputes between the authorities and the opposition.

Solidarity's election victory

The round table agreement allotted 299 of the Sejm's 460 seats to the established political parties (the Polish Communist Party – PZPR, the Peasants' Party – ZSL and the Democratic Party – SD) with the remaining 161 made available to the opposition around Solidarity and any other independent candidates. The authorities entered the race to parliament on the calculated gamble that the hastily called vote would disadvantage the opposition which lacked a national election machine. Moreover, to tip the balance even more in their favour the authorities had set up an extraordinarily complicated voting procedure. This included considerable numbers of bogus 'independents' (the best known was the former government press spokesman, Jerzy Urban) whose names appeared on ballot papers that did not declare political affiliation, and where voters were expected to delete the names of those not chosen. In addition, the round table gave the authorities the right to run a 'national list' of 35 candidates who would stand unopposed and be elected provided they captured 50 per cent of the vote.

Solidarity, aware that it was involved in an election struggle where a number of cards had been stacked against it, believed that it would capture at most around 115 seats in the lower house and 70 in the upper chamber. This would nonetheless have given the organisation a respectable presence in parliament. The authorities however had miscalculated the public mood and grossly underestimated the strength of Solidarity support. When the votes were counted it became clear that Solidarity was facing a landslide victory that gave it all 161 of the seats available to it in one house and 99 out of 100 in the senate.

Solidarity's clean sweep was matched by a humiliating result for the party made all the worse by the fact that, even on its uncontested national list, it carelessly managed to lose all but two candidates. In late July the party conducted its first election post-mortem amidst much recrimination that the misguided reforms of General Jaruzelski were to blame. Party hardliners, – the betony, or concretes, as they were known in Poland – made little secret of their anger. In the subsequent reshuffle of the party leadership Mieczyslaw Rakowski stepped in as new first secretary as General Jaruzelski moved over to the state presidency and, ominously for the reform process, two well-known hardliners, Janusz Kubasiewicz and former planning commission chief, Manfred Gorywoda, were elevated to the politburo.

If the momentum of political reform in Poland continued the party would face the prospect of utter defeat at the polls in 1993. In the second half of 1989 the question of how to build within four years an electable Communist Party must have been uppermost in the mind of many an apparatchik.

But Solidarity's stunning election victory also caused problems for the opposition. It had to decide how to respond to invitations from General Jaruzelski to participate in government. The idea found favour with a section of the movement represented by Adam Michnik and Jacek Kuron whose position was neatly summarised in the title of an article written by Michnik, 'Your president, our prime minister'. If General Jaruzelski was to take the presidency (which he did by a nail-biting margin of just one vote on 20 July 1989) the Michnik view was that Solidarity's parliamentary leader, Professor Bronislaw Gemerek, should lead a government even if the key defence and interior ministry posts had to go to the party.

KEY INDICATORS

	Unit	1984	1985	1986	1987	1988
Population	'000	36,914	37,203	37,460	37,660	37,860
Net material product (NMP)	Zl bn	7,182	8,586	10,697	14,013	23,475*
NMP per capita	$	1,719	1,569	1,629	1,376	1,363*
Inflation	%	15.1	15.1	17.7	25.2	60.0
NMP real growth	%	5.6	3.4	4.9	2.0	4.7
Agricultural production	%	3.1	2.6	7.0	–5.4	–2.2
Crop output	%	8.6	–2.6	9.0	–6.5	–1.0
Industrial production	1985=100	96.4	100.0	104.2	107.5	113.5
Mining	1980=100	100.9	100.4	101.2	105.0	–
Manufacturing	1980=100	94.9	98.8	103.2	106.1	–
Hard coal output	m tonnes	192	192	192	193	193
Exports	$m	11,654	10,945	11,926	12,026	13,846
USSR	%	29.6	28.4	27.6	24.8	24.5
German FR	%	8.5	8.7	9.4	–	–
Czechoslovakia	%	5.6	6.2	6.3	–	–
Imports	$m	10.995	10.598	11,459	11,236	12,757
USSR	%	36.4	34.4	32.5	–	–
German FR	%	7.4	9.0	9.8	–	–
German DR	%	6.3	6.1	5.9	–	–
Balance of trade	$m	659	347	467	790	1,089
Current account	$m	–1,083	–982	–1,106	–379	–107
Foreign debt	$bn	26.8	29.3	33.5	39.2	39.2
Exchange rate	Zl per $	113.2	147.1	175.3	265.1	430.1

* estimated figure

Walesa and Gemerek, however, were unmoved by this and argued that joining or forming a government in the immediate post election period would be mistaken.

Solidarity would end up, according to Walesa, as no more than decoration on a system that, because of the party's pervasive influence through the nomenklatura, was essentially unchanged. A Solidarity government would have little power so long as the party filled the state machinery with its nominees. Indeed, it could be expected that the party apparatus would do its utmost to sabotage and discredit any Solidarity administration. No doubt beyond this very real fear of taking responsibility without power also lay another, the fear of the strains that would be imposed on the movement by taking charge of economic policy at a time of deepening economic crisis.

In the event, Tadeusz Mazowiecki, a senior Solidarity official became prime minister in August 1989. General Jaruzelski retained the presidency.

During 1989 the Polish economy showed every sign of running out of control. A misguided approach towards prices policy during 1988 simply set off, as some observers had warned might happen, a series of inflationary wage demands. Before long the economy was held in the grip of a ferocious price–wage spiral. The inflation rate rocketed upwards from a relatively modest 15 per cent per annum in 1984-85 to 60 per cent in 1988 and still upwards to a 1989 rate that looked set to reach well over 100 per cent. In the first half of 1989 the monthly inflation rate was 8.3 per cent, almost twice that of the same period a year earlier (4.4 per cent). This was combined with deepening chaos on the domestic market as shortages were reported on a scale, said some observers, approaching that of the very worst period of 1981. Little wonder that social tensions were running high in 1989.

Meat supply crisis

To make matters even worse one of the most acutely felt shortages in 1989 was that of meat and meat products. For the first time since meat rationing was introduced at the beginning of the 1980s the government admitted in June 1989 that it could not guarantee a meat supply large enough to satisfy even rationed demand.

A food crisis had been brewing for some time, signs of which were evident even in October 1988 when the Rakowski government took office. The new government claimed to be well aware of the difficulties and declared agriculture to be one of its priority areas. One of the first things it did was to raise and backdate the purchase prices paid by the state to farmers for agricultural produce. But this had little noticeable effect on farmers' willingness to sell produce to the state. Purchase prices were raised again a few months later in March and then again in April 1989 but with equally little impact. Indeed the major consequence of higher purchase prices alongside virtually frozen retail prices was felt by the state's budget where the deficit grew astronomically in the first months of 1989. By the end of May 1989 that deficit stood at a colossal Zl3,231bn while the plan for the whole of 1989 had, in theory, set a limit to the deficit of Zl1,000bn. Any hope of effective agricultural policy innovation ceased in the spring of 1989 as the round table talks got under way and the food economy slipped further towards chaos.

From the time of its coming to power the preferred solution to these problems of the Rakowski administration was the so-called 'marketisation' of the food economy. This was to have a number of components including price liberalisation, the withdrawal of rationing and a freeze on subsidies. What the policy lacked, however, was a convincing recipe of measures that would genuinely liberalise and open to competition the state's stranglehold on produce purchase, food processing and supply to agriculture of industrial parts. The 'food marketisation' policy was little more than a repeat of previous 'reforms' that had done nothing except raise prices and boost inflation.

One of the leading critics of the policy in the summer months of 1989 was former national bank chairman and politburo member Professor Wladyslaw Baka. Baka had been tipped as likely successor to Rakowski as prime minister but this chance slipped away, at least temporarily, as his criticisms of 'food economy marketisation' grew. Despite widespread objections the outgoing Rakowski administration decided to press ahead with the policy from 1 August 1989. From the time of the round table a policy paralysis had gripped the Rakowski government so it was somewhat surprising to see, on the eve of its departure, such radical action on food prices.

Policy innovations

If the Rakowski government lost direction during the political upheavals of 1989 it was a different story during its first few months in power. Then it appeared to have a dynamism that contrasted sharply with previous administrations. Rakowski's ministerial appointments were, for Poland, exciting. A youthful and enthusiastic Dominik Jastrzebski was given the foreign trade ministry, while a private entrepreneur, Mieczyslaw Wilczek, went to industry. The deputy premier responsible for the economy, Ireneusz Sekula, was also generally highly regarded. Moreover, several potentially important economic acts passed through parliament when the government still had some energy in late 1988. But it is also clear on looking back that the Rakowski team was unusually good, at least in its early days, at self promotion.

Nonetheless some of the early policy might prove durable and highly useful if it could be lifted from the statute books and introduced in practice. This holds particularly for the highly publicised Bill of December 1988 which guaranteed all sectors of the economy (state, co-operative and private) equality in terms of access to materials and foreign exchange, and in matters of taxation and fiscal privilege. At the same time the authorities declared that all enterprises should be free (with only very few exceptions) to develop new lines of business as and where it seemed profitable to do so. Where the previous presumption as regards new economic initiatives had been to allow nothing in the absence of express permission from one ministry or another, in the new system the presumption was to be turned upside down, allowing anything unless expressly prohibited.

Unfortunately, during 1989 there was precious little evidence that these new rights were being exploited and the real economy continued to operate much as it always had.

There was also little sign of there being any practical consequence of another of the December 1988 legislative Acts, one which reduced the rank of the planning commission to that of a central planning office with mainly strategic responsibilities.

COUNTRY PROFILE

Political system: Socialist state, in which supreme organ of power is the parliament (Sejm) with 460 members in lower house and 100 members in upper house (Senate) directly elected for four years. Parliament is composed of a lower and upper house and chooses the president for a six-year term. The president has power to dissolve parliament and also nominates the prime minister. Supreme executive power is vested in Council of Ministers which is responsible to the Sejm. Until 1989, the leading political organisation was the Polish United Workers' Party (PZPR). After the 'Round Table' Agreement of April 1989, 299 of the Sejm's 460 seats were allocated to the established parties (PZPR – Polish Communist Party, Peasants' Party and Democratic Party) with the remaining 161 available to opposition parties. The June 1989 election gave the Solidarity Party all 161 available seats in the parliament, and 99 of 100 in the Senate. Next elections are in 1993. Solidarity provided the prime minister. Interior and Defence portfolios were retained by Communist Party. In 1990 the Communist Party was reconstituted.

Main cities/towns: Warsaw (population 1,671,400), Lodz, Cracow, Wroclaw, Poznan, Gdansk, Szczecin, Katowice and Bydgoszcz.

Media
Press: Most newspapers and periodicals are published by PZPR, or state agencies, though there is now an 'opposition' press as well as a Catholic press.
Dailies: Around 100 newspapers published, mostly in Warsaw, with a total circulation of over 10m. *Trybuna Ludu* (circulation 602,100) is the party daily and *Zycie Warszawy* (337,000) is published daily in Warsaw.
Weeklies: Around 2,900 periodicals, of which one of the most influential is the weekly *Polityka* (400,000). Sports magazines, TV/radio guides and youth magazines also have sizeable readerships. The *Warsaw Voice* is published weekly in English.
Business: Several commercial papers published by Polish Chamber of Foreign Trade, and Foreign Trade Publicity and Publishing Enterprise (AGPOL). Technical papers are published by Naczelna Organizacja Techniczna (NOT), Wydawnictwa Komunikacyji and Lacznosci. Most influential periodicals are *Zycie Gospodarcze* (economic weekly), *Rynki Zagraniczne* (3 per week; foreign trade), *Veto* (consumers' weekly), *Handel Zagraniczny* (monthly; foreign trade) and *Gazeta Bankowa* (bankers' weekly).
Broadcasting: All TV and radio programmes broadcast by Polskie Radio i Telewizja (Polish Radio and Television).
Radio: Over 10m radio licences current. Four programmes are transmitted nationally and there is a foreign service which broadcasts in most European languages.
Television: Approximately 9.8m TV licences issued. Programmes broadcast on two channels.
Advertising: Two state agencies, Polish Advertising Agency (PAR) which serves domestic advertisers, and AGPOL which operates on behalf of foreign clients. Advertisements in English possible through papers such as *Warsaw Voice.* Almost 50 per cent of advertising expenditure by foreign firms is in the form of exhibitions (organised by AGPOL), a further 30 per cent on advertising in the press, and the remainder mostly on posters. A number of privately-owned or co-operative advertising and business consultancies have recently appeared.

Domestic economy: A 'planned' economy characterised by large industrial and agricultural sectors, but during 1982-86 period when economic reform was in progress the economy was neither centrally planned nor subject to extensive market forces. After a decade of high import-led growth economic performance was undermined by high levels of inflation, heavy debt burden, labour unrest and contraction in export demand. Positive rates of economic growth 1983-85, with incomes rising faster than inflation, more recently some evidence of slowdown. The 1986-90 plan aims for self-sufficiency in foodstuffs, expansion of new export industries and real growth in investment and consumption.

Employment: Approximately 36 per cent of the labour force in the socialised sector employed in industry, nine per cent in construction and 23 per cent in services. Private sector employment is almost entirely in the agricultural sector, which employs an estimated 4.4m workers. The non-agricultural private sector has grown rapidly in recent years, to just over 1m in 1986.

External trade: Over the 1980s a reduction of imports has helped reverse regular balance of trade deficits. Expansion of exports still constrained by lack of credit and, to some extent, uncompetitiveness of products, but some improvement in export earnings in 1986-87. Traditional exports largely from agricultural and extractive industries, particularly coal, copper, foodstuffs and wood products. Exports of electrical engineering products, chemicals and motor vehicles also important. Imports consist mainly of oil, gas, iron ore, zinc concentrates, cereals, industrial machinery and equipment. The USSR is by far the largest trading partner, accounting on average for around 30 per cent of total trade turnover. Major OECD trading partners include Germany FR, UK and Austria.

Agriculture: Sector contributes around 15 per cent to national income and accounts for over four per cent of exports. Although co-operative and state farming important, just over 76 per cent of agricultural land is privately owned and farmed. Production is concentrated on livestock farming (dairy and pigs), potatoes, cereals and sugar-beet. Current emphasis on encouraging private sector farming, expansion of grain production and growth in meat production for export and home market.

Industry: Sector forms mainstay of economy, contributing 49 per cent to NMP and accounting for over 77 per cent of exports. Heavy export-based industries dominate, such as shipbuilding, metallurgy (particularly steel), chemicals, motor vehicles and cement. Also recent growth of electronics and light industries. Food processing and forestry industries also significant. Production is state-controlled, though small business is generally privately-owned, and some 700 foreign-owned companies (Polonian firms) and about 200 joint ventures produce goods locally. Industrial policy orientated, with limited success over 1982-87, towards expansion of consumer goods and export-based production, the development of a broader industrial base to support needs of agriculture and the more efficient use of materials, energy and labour.

Mining: Rich mineral resources including largest deposits of copper ore in Europe and substantial deposits of coal, zinc-lead ores, sulphur and salt. Also lesser deposits of nickel and other precious metals. Coal is most important mineral resource, accounting for nine per cent of total exports and meeting 80 per cent of local industries' fuel needs.

Energy: Coal makes significant contribution to energy needs. Domestic supplies of oil are negligible. Thus strong dependence on imported oil, mostly from the USSR. Gas production, which currently meets around 50 per cent of local needs, has been declining steadily. Gas consumption is however increasing due to the completion of Kobrin-Brest-Warsaw pipeline.

Banking: The National Bank (Narodowy Bank Polski) is in overall control of the banking system and provides a full range of banking services, but four banks specialised in foreign trade relations also exist. The provision of credit and banking services for foreign trade is made by the Bank Handlowy w Warszawie SA. Bank Rozwoju Eksportu, responsible for export promotion, began operations in 1987 and also supervises new system of fortnightly currency auctions introduced in 1987. Other major banks include the Bank Polska Kasa Opieki SA, the official savings bank, and the cooperative bank Gospodarki Zywnosciowej whose main business is farm and food industry finance. A network of nine commercial banks began operating in 1989.

Legal system: Mixture of continental civil law, contract law and communist legal theory.

Membership of international organisations: CCC, CMEA, FAO, GATT, IAEA, ICAO, ICES, IHO, ILO, IMF, IMO, Indochina Truce Commission, IPU, International Lead and Zinc Study Group, ISO, ITC, ITU, Korea Truce Commission, UN, UNESCO, UPU, Warsaw Pact, WFTU, WHO, WIPO, WMO.

On the other hand the third piece of December 1988 legislation was more important. This involved a comprehensive rewriting of the rules governing the establishment of joint ventures. The aim was to produce a framework for JVs that would be at least as competitive as that offered by any other socialist country. Tax privileges were considerably extended and foreign exchange restrictions much relaxed. All new JVs would be eligible for a three year tax holiday extendable in special circumstances for a further three years (previously two years had been the norm). Firms' requirement to sell a portion of hard currency earnings to the state at the official exchange rate was reduced from 25 per cent to 15 per cent. A new office, the Agencja d/s Inwestycji Zagranicznych, was established to oversee all aspects of the JV process, including the granting of licences, and to take on board the more general task of selling Poland as a worthwhile location for Western investment.

One other policy innovation of key significance was enacted before the political events of 1989 paralysed the government. This was the decision to 'legalise' the black market. On 15 March 1989 a number of branches of the PKO bank began to buy and sell hard currency at rates indistinguishable from those of the genuine black market. A small number of private exchange bureaux also opened for business. This currency liberalisation, long urged by the IMF, was a major step towards strengthening the zloty. It was unfortunate however that the authorities should have allowed their macro-economic management to slip so badly in 1989, generating immense inflationary pressures which could do nothing but weaken confidence in the zloty.

When trading on the new 'grey' market began in March 1989 the bank's average exchange rate was Zl3,000 per $, some 5.2 times the official rate of Zl571 per $. By mid July 1989 the bank's average rate reached Zl5,050 per $, and had crossed two psychologically significant thresholds (Zl4,000 and Zl5,000). Meanwhile the official rate was depreciated to Zl836,06 per $ and the ratio of grey to official exchange rates increased to a factor 6. People were clearly trying to escape from zloty and turning to the dollar as a safer anti-inflationary bet.

Inflation to 'stagflation'

It became clear during 1989 that the economy was moving into a new phase where rapidly accelerating inflation was combining with decelerating output growth – 'stagflation' was settling on Poland. In June 1989, while prices were 91.3 per cent higher than a year earlier, earnings were a colossal 145 per cent greater. Meanwhile industrial production in the first half of 1989 was only 1 per cent up on the same 1988 period. The output slow-down (from a highly respectable 5.4 per cent growth in 1988) was widely attributed to difficulties with intermediate supplies, particularly those imported from the Socialist bloc.

The decline in the flow of supplies of raw materials, materials and fuel amounted to 2.3 per cent in the first half of 1989 but the drop in intermediate supplies imported from the rouble area amounted to 6.3 per cent. This was only partly compensated by a 6.6 per cent increase in intermediate supplies from the West, which payment difficulties would clearly make hard to sustain. With supplies down and production growth slipping it was little wonder that exports came under pressure too. In the first half of 1989 these increased by only 1.5 per cent.

Difficulties with production were bound to lead to problems with the trade balance and the country's capacity to service debts. Commercial banks in the West had a whiff of this when in April 1989 the Poles suddenly and unexpectedly announced that they would not be honouring some $20m imminently due for payment. Though this problem was later resolved the banks were told to expect a sharp deterioration in

BUSINESS GUIDE

Time: GMT + 1 (GMT + 2 from late Mar to late Sep).

Climate: Temperate, with warm summers and cold winters. Temperature can reach 25°C from Jun to Aug, falling to an average of 10°C Sept to Oct. Most snow from Nov to Mar when temp can fall below -10°C.
Clothing: Lightweights from Jun to Aug, medium to heavyweight for rest of year, plus heavy topcoat in winter.

Entry requirements: Visas (valid for up to 90 days) required by all except CMEA nationals. Transit visas are valid for 48 hours. N.B. regulations are liable to change at short notice. Polish consulate or travel agent should be consulted prior to departure.
Health: Vaccination certificates not normally required.
Currency: No Polish currency may be taken into or out of the country. No limit on foreign currency taken in, but this amount must be declared and not exceeded by amount taken out. Visitors must exchange $15 into Polish currency per day of visit at the official exchange rate. Beyond this minimum requirement, leading banks will buy dollars from visitors at the 'black market' rate.
Customs: Customs and currency declaration must be completed and presented at border on entry.

Air access: Regular direct flights operated by most major international airlines.
National airline: Polish – LOT (Polskie Linie Lotnicze).
International airport: Warsaw (WAW) at Okecie, 10 km from city centre.

Surface access: Several sea, rail and road routes from neighbouring CMEA countries. From Western Europe main road and rail routes are via German FR and German DR and there is access by ferry from Sweden, Finland and Denmark.
Main ports: Gdansk, Szczecin, Gdynia and Swinoujscie.

Hotels: Mostly belong to ORBIS hotel chain and are classified one- to four-star and Lux (Intercontinental, Holiday Inn, Novotel). Accommodation can be scarce in all main towns, so it is advisable to book at least six weeks in advance. In an emergency a very large travel agency or airline may be able to provide a hotel room. Travellers' cheques must be endorsed. Bills include 10-15 per cent service charge; tipping around 10 per cent is customary. N.B. avoid drinking the tap water. *Hard currency (Pewex) shops:* Almost all large hotels have a Pewex kiosk, selling alcohol, tobacco, toiletries, food and a wide range of other articles for hard currency. In addition Pewex supports specialised branches throughout Warsaw and other large towns selling all manner of goods from chocolate to motor components exclusively for convertible currency.

Credit cards: All major credit cards are accepted.

Car hire: Minimum age 21 years. Available in all main towns through ORBIS. Rates vary, depending on size of car, from about $15.00 to $30.00 per day plus $0.15 to $0.30 per km, not including fuel and insurance. International driving licence and insurance cover required. Minimum renting period is 24 hours. Speed limits: built-up areas 60 kph, normal roads 90 kph, motorways 110 kph.

City transport
Taxis: Metered taxis available in all main towns. Customary to wait at ranks. Tipping usually 10 per cent of fare. Taxi drivers may insist on payment in hard currency.
Buses and trams: Good bus services in all towns, also trams in some. Tickets can be bought at newspaper kiosks and used indiscriminately.
Metro: Network of about 90 km began construction in 1983.

National transport
Air: LOT operates regular services connecting Warsaw, Gdansk, Cracow, Poznan, Wroclaw, Szczecin, Rzeszow, Katowice, Koszalin, Bydgoszcz, Zielona, Gora, Slupsk and Czestochowa.
Road: Approximately 154,000 km surfaced roads of which 80 per cent are main roads. Extensive bus and coach services operated by Polish Motor Communications (PKS) – state enterprise which also organises goods transport by road.
Rail: Approximately 30,000 km of track, with about 33 per cent electrified. Regular services, operated by Polish State Railways (PKP), connect major towns. Intercity express trains are inexpensive and usually very reliable.
Water: About 4,000 km of navigable inland waterways, including about 400 km of canals.

Public holidays
Fixed dates: 1 Jan, 1 May, 22 Jul (National Day), 15 Aug (Assumption), 1 Nov (All Saints'), 11 Nov (Independence), 25 and 26 Dec.
Variable dates: Easter Monday, Corpus Christi.

Working hours
Business: (Mon-Sat) 0800 – 1530.
Government: (Mon-Sat) 0800/0900 – 1500/1600.
Banking: (Mon-Sat) 0900 – 1200.
Shops: (Mon-Sat) usually 1100 – 1900, but food shops often 0600/0700 – 1800/1900. N.B. companies and shops, other than food shops, close on 'Free Saturdays' which vary from business to business, but usually three per month (one for shops).

Social customs: Organisations do not stop for lunch in the middle of the day. The main meal 'obiad' is taken from 1500.

Business language and interpreting/translation: Available through Polish Travel Office – ORBIS.

Telecommunications
Telephone: Rates for calls vary with distance. Cheap rate on long-distance calls between 1600 and 0600. International dialling code: 48 followed by 22 for Warsaw (6-digit numbers) and 2 for Warsaw (7-digit numbers).
Telex: Available at main ORBIS hotels, Foreign Trade Enterprises and, 24 hours a day, at URZAD POCZTOWY in Warsaw. Country code 63 PL.
Telegram: Available in post offices and via telephone.

Trade fairs: Most important is Poznan International Trade Fair, held annually in mid-June. Poznan Fairs Authority also organises specialised international fairs for goods not covered by main fair. Other international events also held regularly in various cities. International Book Fair, May, Warsaw; International Trade Fair of Medical Equipment (Salmed), Apr, Poznan; International Fair in Textiles (Interfashion), Lodz.

Electricity supply: 220 V AC.

Representation in capital: Afghanistan, Albania, Algeria, Argentina, Australia, Austria, Bangladesh, Belgium, Brazil, Bulgaria, Canada, China PR, Colombia, Cuba, Czechoslovakia, Denmark, Egypt, Finland, France, German Democratic Republic, Federal Republic of Germany, Greece, Hungary, India, Indonesia, Iran, Iraq, Israel, Italy, Japan, Democratic People's Republic of Korea, Laos, Libya, Malaysia, Mexico, Mongolia, Morocco, Netherlands, Nicaragua, Nigeria, Norway, Pakistan, Palestine, Peru, Portugal, Romania, Spain, Sweden, Switzerland, Syria, Thailand, Tunisia, Turkey, USSR, United Kingdom, USA, Uruguay, Vatican City, Venezuela, Vietnam, Yugoslavia, Zaïre.

the current account of the balance of payments in 1989. Stronger signals of the scale of the likely deterioration appeared when, at the end of June 1989, the hard currency trade surplus was reported to be $447.7m. This was well below what would have been expected if the economy was on course for an annual surplus of $1.2bn, the optimistic target of the annual plan.

Western aid on the way

Just as the economy looked set to take a further dive in the middle of 1989 the international community signalled its willingness to support, through a range of carefully prepared measures, the Polish reform process. The ball was set rolling in April 1989 when President Bush announced, in very general outline, a series of aid proposals. During his subsequent visit to Poland, three months later in July, the aid package was firmed up into a promise of $100m to support Polish private enterprise plus another $15m to help clean up the Krakow environment. To this was added a general commitment to reschedule official debt.

Other Western leaders tended towards a broadly similar approach in offering help to Poland. In June 1989 General Jaruzelski visited Mrs Thatcher at Chequers and was told that apart from a willingness to argue Poland's case at the Paris Club of official creditors she would also be setting up a fund worth £25m over 5 years to support Polish management education. A few days later while President Mitterand paid an official visit to Poland he made a similar debt-rescheduling commitment topped with new short-term credits worth $23m and a promise of a further medium-term credit of $77m once Poland reached an agreement with the IMF.

In addition to the aid outlined above the West responded promptly to requests for food, supplies of which started to reach Poland in August 1989. The European Community also reached a five-year trade agreement with Poland cutting some EC tariffs on Polish foodstuff exports and phasing out most EC quotas on Polish industrial goods.

The fairly uniform approach of Western countries towards Poland in 1989 was characterised by two main features. First, relatively small cash sums were promised (certainly much smaller than Solidarity's grandiose appeal for $10bn) and tended to be tied to some aspect of the reform process and the introduction of market mechanisms. Second, there was a general willingness within the Paris Club to break the deadlock that had hitherto existed over official debt rescheduling. Previously the Paris Club had insisted that Poland agree an adjustment programme with the IMF as a precondition to official debt rescheduling, but by mid-1989 its position on this had softened. Leading official creditors were saying instead that it would be enough for Poland to be involved in serious discussions with the IMF over an adjustment programme.

Some attempt was made by the outgoing government of Mieczyslaw Rakowski to get a dialogue with the IMF off the ground on the basis of a carefully prepared three-year programme for 1989-92. This offered a severely deflationary economic policy, cuts in consumption and in the budget deficit, as well as further economic liberalisation. Though the IMF had misgivings over certain aspects of policy, notably wage – price indexation, the programme was widely considered to offer the best basis yet for negotiation of an adjustment programme.

But it faced several hurdles. One was the fact that given the changing composition of the government the status of the programme was, at least in August 1989, unclear. Another problem was that the plan would have to win the backing of the new parliament where a rough ride could be expected. Despite these qualifications it was widely expected in mid-1989 that Poland would draw a first tranche of between $100m and $200m from the IMF before the year was out. It seemed likely too that World Bank assistance for food and agricultural projects worth around $300m would begin to flow at around the same time.

Outlook

In mid-1989 the Polish political scene had reached an unusually volatile stage. It was certainly a healthy development that political differences were coming much more into the open and parliament looked as if it would, for the first time since 1980-81, become a forum for genuine debate. However there was also a very real danger that a power vacuum would develop where no side is able to govern effectively. Governments without a strong base and the authority to act are apt to be short-lived. It was possible that Poland might be about to experience a period of instability where coalitions rise and fall and where the short ten-month tenure of Prime Minister Rakowski was simply a taste of things to come. This would be bad enough but set against the economic crisis that was unfolding in 1989 it could, for Poland, be disastrous.

A much more clearly positive development of 1989 was the generally supportive attitude of the West which held out the prospect of full normalisation of financial relations. If General Jaruzelski and the party reformers could stay on top and further develop the historic compromise of 1989 then internal political forces might not end up in 'self-destruct'. It seemed likely that the West might find that it was worth paying just a bit more to help keep Polish reform on course.

BUSINESS DIRECTORY

Hotels: Warsaw
Europejski Orbis, 13 Krakowskie Przelmeicie Street, 00071 (tel: 265051; tx: 813615).

Forum Warsaw, Nowogrodzka Street 24/26 00511 (tel: 210271; tx: 812521).

Grand Orbis, 28 Krucza Street 00522 (tel: 294051; tx: 813422).

MDM, Pl Konstytucji 1, 00647 (tel: 216211; tx: 814871).

Metropol, 99A Marszalkowska, 00693 (in city centre) (tel: 294001; tx: 813627).

Novotel Orbis, Ulica 1 Sierpnia 1, 02134 (4km from airport) (tel: 464051; tx: 812525).

Solec Orbis, 1 Zagorna Street, 00441 (tel: 259241; tx: 814676).

Victoria Inter-continental, 11 Krolewska Street, 00065 (facing Victoria Square) (tel: 278011; tx: 812516).

Warszawa, Pl Powstancow Warszawy 91, 00039 (tel: 269421; tx: 814456).

Car hire: Warsaw
Europcar: Parking House, ul Nowogrodzka St 27 (tel: 293875; tx: 813722).

Hertz: Hotel Intercontinental, 11 Królewska Street (tel: 274185; tx: 812516).

Chambers of Commerce
Polish Chamber of Foreign Trade, 00950 Warsaw, ul Trebacka 4 (tel: 260221; tx: 814361).

Banking
Bank Gospodarki Zywnosciowej (bank of food economy), Warsaw, ul Swietokrzyska 12.

Bank Handlowy w Warszawie SA, 00950 Warsaw, ul Chalubinskiego 8, Box 129 (tel: 303000; tx: 814811).

Bank Polska Kasa Opieki SA (guardian bank), 00950 Warsaw, ul Traugutta 7/9, Box 1008 (tel: 269211; tx: 813441).

Narodowy Bank Polski (national bank), 00950 Warsaw, ul Swietokrzyska 11/21 (tel: 200321; tx: 814681

Principal newspapers
Polityka, 00-182 Warsaw, ul Dubois 9 (tel: 6353491; tx: 812546).

Trybuna Luda, 02-015 Warsaw, Pl Starynkiewicza 7 (tel: 285271).

Zycie Warsawy, 00-624 Warsaw, ul Marszalkowska 3/5 (tel: 25289; tx: 814507).

Other useful addresses

AGPOL (foreign trade publicity and publishing enterprise), ul Kierbedzia 4, 00957 Warsaw, Box 7.

Central Statistical Office, Al Niepodleglosci 208, 00925 Warsaw (tel: 252431; tx: 814581).

Eximpol, 00950 Warsaw, Stawki 2, Box 810.

LOT (Warsaw office of the Polish Airlines), ul Warynskiego 9 (tel: 217021; tx: 8143278).

Ministry of Finance, 00-916 Warsaw, ul Swietokrzyska 12 (tel: 200311; tx: 815592).

Ministry of Foreign Economic Relations, 00-489 Warsaw, ul Wiejska 10 (tel: 210331; tx: 814501).

Ministry of Industry, 00-926 Warsaw, ul Wspólna 4 (tel: 210351; tx: 814261).

Mundial, 00957 Warsaw, Czerniakowska 58, Box 23.

NOT (technical interpreter service), Czackiego 3/5, Warsaw (tel: 267461).

Orbis (Tourist Board), 00028, Warsaw, ul Bracka 16 (tel: 260271; tx: 814761).

Polcargo (cargo experts and supervisors), Zeromskiego 32, Box 223, 81963 Gdynia (tel: 213921/957; tx: 0542472).

Polcomex, 00061 Warsaw, Marszalkowska 140, Box 478 (tel: 266810; tx: 813452).

Poliglot, 00950 Warsaw ul Stawki 2, Box 40.

Polish Chartering Agents (Polfracht), ul Pulaskiego 8, Box 206, 81368 Gdynia (tel: 214991; tx: 054201 and 054331).

Polska Agencja Interpress (Polish Agency Interpress), 00585 Warsaw, ul Bagatela 12 (tel: 282221; tx: 814775).

Polska Agencja Prasowa (Polish press agency), 00950 Warsaw, Al Jerozolimskie 7 (tel: 280001; tx: 813480).

Polskie Koleje Panstwowe (state railways), 00928 Warsaw, ul Chalubinskiego 4-6 (tel: 244300).

Polskie Linie Lotnicze (Polish Airlines - LOT), 00096 Warsaw, 17 Stycznia 39 (tel: 461251; tx: 813552).

Polskie Linie Oceaniczne (Polish Ocean Lines), ul 10 Lutego 24, 81364 Gdynia (tel: 201901).

Poznan International Fairs, 60734 Poznan, ul Glogowska 14 (tel: 61221; tx: 0415210).

Supreme Customs Office, ul Swietokrzyska 12, 00950 Warsaw.

Timex, 00193 Warsaw, ul Stawki 2, Box 268.

The Warsaw Voice (English language weekly), 00-585 Warsaw, ul Bagatela 12 (tel: 211328; tx: 814775).

Transactor, 00950 Warsaw, ul Stawki 2, Box 276.

WCT-NOT (foreign co-operation dept), ul Swietokrzyska 14a, 00950 Warsaw, Box 1004.

Portugal

Peter Wise

Prime Minister Aníbal Cavaco Silva's centre-right government was making steady progress, as the 1980s closed, with a reform programme aimed at bridging the development gap separating Portugal from its richer European Community partners. Investment in modernisation was producing economic growth well above the European average and obstacles to progress built into Portuguese society were being gradually removed. More restrictive monetary policies were being implemented to cool the economy after three years of exceptional expansion produced renewed inflationary pressures in 1989. But growth was expected to remain strong. Cavaco Silva's Social Democrats (PSD) looked likely to recover from a mid-term term slump in popularity and win re-election in 1991. But they may fall short of regaining an overall parliamentary majority.

Inherited problems

The reform programme is designed to overcome deep-rooted problems inherited from 48 years of right-wing dictatorship under the Salazar-Caetano regime and from more than a decade of political instability following the almost bloodless military coup that restored democracy in 1974. Dictator António d'Oliveira Salazar, fearing the rise of an educated industrial class as a threat to his power, strove to keep Portugal a poor, rural society, devoutly Roman Catholic and isolated from foreign influences. Industry was starved of investment. Long colonial wars in Africa drained resources and eventually led to the overthrow of the regime by discontented military officers.

In 1974, the 'Revolution of the Carnations' restored political freedoms, ended police repression and brought independence to the former African colonies of Angola, the Cape Verde Islands, Guinea-Bissau, Mozambique and São Tomé e Principe. But ensuing political turmoil led to the collapse of 17 governments in 12 years, severely shaking investors' confidence. The loss of the colonies deprived Portugal of captive markets and cheap raw materials. Coupled with the oil shock of the early 1970s, these developments took a heavy toll on the economy. A boom–slump cycle developed, marked by overheated phases of soaring inflation and massive current-account deficits that twice forced governments to resort to the International Monetary Fund. The IMF backed stabilisation programmes involving austerity measures that resulted in two deep recessions but succeeded in bringing Portugal's external accounts back under control. In the wake of the revolution, sweeping nationalisations also brought one-third of national output under state control. But they were motivated by political rather than business reasons and have since proved a debilitating burden on the economy.

Portugal began to emerge from its last recession in 1985 as the international economic climate, including a weaker dollar, falling oil prices and lower international interest rates, grew more favourable. At the same time, Cavaco Silva was elected prime minister of a minority government on a reform platform focusing on free enterprise and privatisation. Amid rapid economic growth, the PSD won a landslide re-election victory in July 1987, becoming the first single party to win an overall majority in parliament since the 1974 revolution. This added unprecedented political stability to favourable economic conditions. In addition, Portugal joined the European Community

KEY FACTS

- **Official title:** The Portuguese Republic
- **Head of state:** President Mário Soares
- **Head of government:** Prime Minister Aníbal Cavaco Silva
- **Ruling party:** Social Democratic Party (PSD)
- **Capital:** Lisbon
- **Official Languages:** Portuguese
- **Currency:** Portuguese Escudo (Esc) = 100 centavos
- **Exchange rate:** Esc159 per $ (Dec 1989)

- **Area:** 92,072 sq km
- **Population:** 10.41m (1988)
- **GDP per capita:** $3,465 (1987)
- **GDP real growth:** 4.7% (1987)
- **Labour force:** 4.5m (1988)
- **Unemployment:** 7.9% (1988)
- **Inflation:** 9.6% (1988)
- **Trade balance:** –$5.1bn (1988)
- **Foreign debt:** $18.3bn (1987)

with Spain in January 1986. Efforts to modernise in preparation for the single European market planned for 1992, the availability of massive EC development funds and the dismantling of trade barriers as a result of Community membership greatly stimulated economic growth.

In this climate, the Portuguese economy has grown at double the average European rate since 1985, unemployment has been halved to below 6 per cent and inflation has fallen from 30 to 12 per cent. Growth at 1-1.5 per cent above the EC average is forecast over the medium term. Even at this rapid rate of expansion, time is short to overcome the cumulative effect of Portugal's inherited problems – which have left it the poorest country in western Europe – and to prepare for competition within post-1992 Europe. The challenge of rapid modernisation faces a backward industrial sector where more than two-thirds of companies have less than 10 workers and only 0.5 per cent of firms employ more than 500 people. Farming, where the workforce is three times the European average but productivity three times lower, has an even greater distance to cover. The problems include a predominance of tiny family plots; primitive methods; crops unsuited to the soil and climate; and the old age and low literacy rate of farm workers.

Lack of investment has also left Portugal badly in need of basic infrastructures to enable the economy to function efficiently and to attract investment. Massive spending on roads, bridges, ports, airports, telecommunications and other infrastructures, funded by extensive EC grants, is one of the main forces behind current growth. But the country is starting from very low down on the European development scale and a great deal of catching up lies ahead. On a less tangible level, bureaucracy and red tape that deter initiative and hamper development have become almost an intrinsic part of Portuguese officialdom.

A major structural problem resulting from these legacies of the past is the budget deficit. Although the general government deficit has been almost halved since 1985 as a percentage of GDP, it still represented close to 8 per cent in 1989. Financing the deficit absorbs savings that would otherwise be available for productive investment in the private sector. Crude monetary controls have had to be imposed to ensure that the bank liquidity needed to finance the deficit is not released into the general economy. The deficit also stimulates domestic demand. The government aimed to reduce the deficit to 5 per cent of GDP by 1990. But it had very little room for manoeuvre. Spending on public sector wages and debt-servicing account for 80 per cent of budget expenditure. And capital spending is expected to increase by 12 per cent a year up to 1993 to meet Portugal's share of the cost of EC funded projects and to improve the underfunded health and education services.

KEY INDICATORS

	Unit	1984	1985	1986	1987	1988
Population	m	10.09	10.16	10.21	10.25	10.41
Gross domestic product (GDP)	Esc bn	2,806	3,536	4,419	5,115	–
GDP per capita	Esc '000	278	348	432	499	–
Inflation	%	28.9	19.3	11.7	9.4	9.6
GDP real growth	%	-1.6	3.3	4.3	4.7	–
Unemployment	'000	381	385	382	320	262
Agricultural production	1979-81=100	101.8	106.4	105.0	113.4	107.0
Industrial production	1985=100	96	100	106	110	116
Manufacturing	1980=100	105.0	117.3	123.6	128.4	–
Exports	$bn	5.2	5.7	7.2	9.2	10.7
UK	%	15.4	14.6	14.3	14.0	–
German FR	%	13.7	13.8	14.7	15.4	–
France	%	13.9	12.7	15.2	15.8	–
Imports	$bn	7.2	7.1	8.8	12.5	15.8
Import prices	1985=100	79	100	110	–	–
Balance of trade	$bn	–2.0	–1.5	–1.6	–3.4	–5.1
Current account	$m	–514	410	1,144	640	–629
Total foreign debt	$bn	14.9	16.6	16.3	18.3	–
Exchange rate	Esc per $	146.4	170.4	149.6	140.9	144.0

* estimated figure

Strategy for growth

A key objective of Cavaco Silva is 'real and nominal convergence' with the rest of the EC – in other words establishing a stable economic position in the mainstream of the Community. Real convergence requires bringing living standards, wages, per capita GDP, industrial and agricultural production, education, health and other social services up to a level approaching that in other EC countries. Nominal convergence implies bringing Portugal's most wayward areas of economic performance – chiefly inflation, the current-account deficit, and escudo devaluation – down to average Community rates. The two goals are not easily compatible as the investment effort required for modernisation heats up the economy, placing inflation, external accounts and exchange policy under increased pressure. This makes it essential that growth is steady, to improve Portugal's productive and competitive capacity, but sufficiently controlled, so as not to produce an overheated, inflationary climate that would require harsh austerity measures to cool down.

Government policy consists of a broad framework for economic development supported by an extensive programme of sectoral reforms. Medium-term macro-economic strategy revolves around the 'Programme for the structural adjustment of the foreign deficit and unemployment' (PCEDED). The eight-year plan was launched in March 1987 and revised in July 1989. The main objectives of the revised programme are: modernisation and increased productivity through the continuation of a strong investment effort; reduced inflation; and reduction of public sector financing as a percentage of GDP.

To keep Portugal on this narrow path to development, the revised PCEDED sets out the following goals for 1990-92:

• GDP growth at an annual average of 1.5 to 2 percentage points above the EC average; currently implying growth of approximately 4 per cent a year;
• investment growth at double the rate of GDP expansion
• restructuring of productive sectors and strong increases in productivity to reduce both the imbalance between global demand and domestic supply and Portugal's vulnerability to adverse external conditions;
• the gradual reduction of inflation to within 2 or 3 percentage points of the EC average by 1992;
• gradual reduction of the deficit of the enlarged public sector – state industries and the administration – from 9 per cent to 6 per cent of GDP by 1992;
• halting and reversing the growth of the public debt as a percentage of GDP.

Government policy aimed at supporting the modernisation and restructuring of Portugal's productive sectors and creating the conditions for greater productivity and business efficiency should involve the following:
• privatisations and management contracts to improve the efficiency of companies currently within the public sector and to reduce the amount of credit and subsidies to the state-owned sector;
• improving infrastructures;
• modernising and reducing the bureaucracy of the public administration;
• providing incentives and grants and creating a favourable economic climate to encourage productive investment
• encouraging foreign investment;
• expanding professional training and retraining and strengthening social security payments to cushion the effects of restructuring on the labour market;
• optimising the effects of Portugal's EC membership and supporting fuller integration of the Community.

The government considers the PCEDED guidelines essential for channelling investment into overcoming the deficiencies of the productive sector. Portugal has one of the most open economies in Europe – imports and exports accounted for 77.7 per cent of GDP in 1988 – and this openness will be reinforced by the single European market. Investment needs to be directed into expanding and improving production to ensure the gradual reduction of the gap between global demand and domestic output. The main threats to fulfilment of the PCEDED objectives include possible difficulties in maintaining wage restraint that will put pressure on the programme's goal of keeping wage increases below productivity gains; the stimulative effect of EC grants on domestic demand and government spending; and the vulnerability of Portugal's small, open economy where external factors such as the dollar, oil prices and interest rates can have considerably more impact on results than government policy.

Progress with reforms

One of the most far-reaching reforms the Cavaco Silva government achieved in 1989 was a revision of the Socialist-inspired constitution of 1976. This required agreement with the opposition Socialists to ensure the necessary two-thirds majority in parliament. The revision, completed in June 1989, stripped the constitution of ideological references, striking out marxist-oriented clauses that described Portugal as 'a society in transition to Socialism' and defined the post-revolutionary nationalisations as 'irreversible'. The changes include a reduction in the number of parliamentary deputies from 250 to 235 after the next general election; establishing the referendum as a legitimate means of consulting voters and permitting private television broadcasting.

The chief alteration cleared the way for companies nationalised following the 1974 revolution to be fully privatised. Previously, the government could not sell more than a minority stake of up to 49 per cent of the capital of state-owned companies. To show their political will to privatise, the PSD had pressed ahead with minority sales of state-owned stock before the constitutional revision. Stock market floatations of 49 per cent of the successful brewers Unicer in April 1989 and of Banco Totta e Açores, Portugal's fourth largest bank, three months later, were both heavily oversubscribed. A 49 per cent share in both the Alianca and Tranquilidade insurance companies was to be sold later in 1989. Following the constitutional revision, the government began drawing up basic legislation setting out conditions for 100 per cent privatisations. Banco Portugues do Atlantico, a leading state-owned bank, and the brewers Centralcer were announced as the first companies to be 100 per cent privatised. But it seemed that these sales might be made in stages because the government was expected to restrict privatisations to 49 per cent for some time. Cavaco Silva made clear that only minority stakes would be put on the market in the medium term to keep control of privatised companies in Portuguese hands.

The government believes Portuguese investors cannot absorb the capital of more than four or five companies over the next two years. By limiting denationalisations to 49 per cent, it will be able keep control until the Portuguese market is ready to purchase more. The success of the BTA sale, which was 4.5 times oversubscribed, indicated that there was more demand in the market than the government calculated. One argument against delaying full privatisations is the need to reduce the public debt. More importantly, most public sector companies are badly in need of private management to put them on a competitive basis before 1992. Cavaco Silva has suggested one solution may be for the state to retain a majority share in the companies but give up the right to appoint the chairman.

A second important reform was the implementation of a package of new labour laws from mid-1989. Employers had been pressing for change since strongly pro-worker legislation was introduced following the 1974 revolution. Those laws made it virtually impossible to dismiss individual workers except in clear cases of misconduct. Collective dismissals and layoffs required government permission and involved such complicated processes that they were rarely used. An earlier attempt by the Cavaco Silva government to liberalise labour legislation resulted in a 24-hour general strike in early 1988 and the proposals were subsequently watered down. However, the resulting legislation fell short of many employers' expectations – a fact indicated by the relatively low level of union opposition to the new laws. The legislation makes it possible to dismiss individual employees for economic as well as disciplinary reasons and removes the need for government approval of collective dismissals and layoffs. It also limits the use of short-term work contracts, which had become widespread under the previous laws, to strictly short-term situations such as seasonal work and construction projects. Generally, Portuguese workers, whose wage levels are by far the lowest in the EC, are likely to remain among the best protected in Europe until their pay, unemployment and other social security benefits begin to approach average EC levels.

Sweeping tax reforms were also introduced from January 1989. The new tax code replaces a complex cluster of separate taxes with a single tax on corporations, known as the IRC, and a single tax on individuals, the IRS. The changes have produced a modern, integrated tax system. But the effective corporate tax rate of 40.15 per cent is nevertheless relatively high. In addition, withholding tax on dividends remitted from Portugal has been increased from 12 per cent to 25 per cent, or to 20 per cent for the 12 countries that have tax treaties with Portugal. This lack of an extensive network of tax treaties means that companies from countries including the USA, Japan and the Netherlands cannot repatriate profits in a tax- efficient way.

Portugal's two small stock exchanges in Lisbon and Oporto began to revive in late 1989 following a deep depression that set in with the 1987 crash. Foreign investors, attracted by good investment opportunities and solid economic growth prospects, led the return to the market. Reforms planned for the two exchanges will improve the efficiency and flexibility of the market, modernising trading mechanisms and providing investors with better company information. The reforms, due to be introduced from January 1990, involve: privatising both exchanges; transferring supervision from the finance ministry to an independent commission; creating a second market for smaller and developing companies; stricter regulations on company reporting and information; creating a new organisation for stock registration that will unite the separate Lisbon and Oporto systems; and introducing computerised trading, a move that will help unify prices on the two exchanges.

Some of the most advanced anti-pollution laws in Europe, setting down even more stringent standards

COUNTRY PROFILE

Political system: Republic with legislative power vested in the President, the Assembly of the Republic and the government. The President, who is directly elected for five years, appoints a Prime Minister, and, on his recommendation, the rest of the government. The principal organ of executive power within the government is the Council of Ministers which is responsible to the Assembly of the Republic. The Assembly comprises 250 members directly elected for four years.

Main cities/towns: Lisbon (population 2,124,000 in 1986), Oporto, Setúbal, Coimbra, and Braga.

Media

Dailies: 15 major dailies. Most influential published in Lisbon: *Diário Popular* (circulation 62,242), *Correio da Manhã* (78,000), *Diário de Noticias* (58,898), and Oporto: *Jornal de Noticias and O Comércio do Porto* (53,934).

Weeklies: Eight major weeklies; most popular tend to be general news magazines, such as *Expresso* (113,500), *Tempo* (72,000), *Tal & Qual* (58,500), *Semanário* (73,700) and *O Jornal* (54,000), and women's magazines such as *Cronica Feminina* and *Maria* (369,000). Economic weekly *Jornal do Comércio.* Majority of weeklies are published in Lisbon.

Business: Rather limited number of publications available, primarily monthly, mainly concerned with agriculture, engineering and automobile industry.

Radio: Approximately 2.3m licensed radios in use. There are four home service networks provided by nine radio stations, all of which have merged with the national broadcasting company Radiodifusao Portuguesa EP (RDP), which also provides an international service. There is also Radio Renascenca which is run by the Catholic Church.

Television: Over 1.5m TV sets in use. All broadcasting done by state-owned Radiotelevisão Portuguesa EP(RTP). Programme 1 (VHF) received by over 90 per cent of population, Programme II (UHF) received by about 60 per cent. Programmes also transmitted to Madeira and the Azores. Portugal has no cable TV but is experimenting with videotext.

Advertising: TV is the most popular medium of advertising consumer goods, followed by radio, newspaper and magazine advertisements. Other methods used are cinema shorts, posters, direct mail and door-to door canvassing. There are about 50 advertising agencies, mainly specialising in radio and TV work and newspapers. Market research firms also available. Total advertising spend (1987) Esc28,450m, mostly on food, automotives, and alcoholic beverages.

Domestic economy: Small mixed economy with heavy dependence on foreign trade, less than hitherto on agriculture. Tourism and export-orientated manufacturing sector also of growing importance. Rapid industrialisation during 1970s curtailed by international recession. IMF-backed stabilisation package introduced 1983-85 in effort to reduce external debt growth, cut inflation rate and improve trade balance. Current emphasis on modernisation, rehabilitation of public enterprises and export-led growth. Joining EC brought access to funds for agriculture and infrastructure development. Improved economic performance in 1986 and 1987, when highest growth rate of all OECD countries was registered (five per cent).

Employment: Around 20 per cent of labour force engaged in agriculture, 35 per cent in industry and 45 per cent in services. Unemployment around 6.7 per cent and real wages have risen by around 5.8 per cent during period 1986-88.

External Sector: In Jan 1986 Portugal became a member of EC. EC entry should mean rapid economic growth in the long term, but considerable competitive pressure in early years. Significant reduction in traditionally large current account deficit due to lower debt servicing and increased earnings from exports of manufactured goods, tourism and emigrants' remittances. Principal visible exports are textiles and clothing, wood and cork, paper pulp, wines, footwear, chemicals and electrical equipment. Principal visible imports include petroleum and other fuels, cereals, vehicles and components, raw materials, metals, cotton and machinery. Major trading partners are W Germany (15%), Spain (12%), France (11%), Italy (9%), UK (8% share imports) and the USA (5%).

Agriculture: Sector contributes around nine per cent to GDP and together with forestry and fishing accounts for around eight per cent of exports. Small-holders predominate, particularly in the north, and productivity remains low by Western European standards despite mild climate and natural resources such as forests. Recurrent droughts, fragmentation of the land, poor-quality soil and outdated methods of production continue to handicap government efforts at agrarian reform. Current policy aimed at encouraging private investment, co-operatives and integration into EC. 85 per cent of farm produce unable to compete effectively with other EC produce. These will be excluded from Common Agricultural Policy for first five years, and included on gradual basis over following five years. Remaining 15 per cent subject to 7-year transition period. Programme for the Development of Portuguese Agriculture (PEDAP) has been set up with EC funds and Portugal may also benefit from EC agricultural fund.

than the EC, were also to be implemented in Portugal in late 1989. Following legislation covering noise, toxic waste and solid waste products, three other fundamental laws were for the first time to define clear rules on water pollution, air pollution and the evaluation of the environmental impact of major projects. At the same time the government is moving ahead with reforms to the the education, health and judiciary systems and was to restructure the public administration pay system from October 1989 in a move to improve efficiency.

Cooling an overheated economy

Inflation began to rise again towards the end of 1988 after falling for three consecutive years. The upward trend grew more marked during the first half of 1989 as the annual average rate rose from 9.5 per cent to 12.6 per cent. The current account also registered a negative result in 1988 for the first time in three years, changing from a surplus of $444m in 1987 to a deficit of $654m. These figures strengthened earlier warning signals that the economy was in danger of overheating. The government introduced an anti-deflationary package in April 1989, tightening credit restrictions, including those on consumer purchases. But investment growth and its effect on private consumption will more than compensate for tighter monetary and budgetary policies. This should produce steady GDP growth of 3.5-4 per cent over the medium term. EC aid, much of it channelled into investment incentives, is a determining factor in rising investment. EC financial transfers for 1988-92 are expected to total $7.8bn, representing 4.8 per cent of GDP. Foreign investment, representing 8.8 per cent of the total in 1988 at $656m, is virtually doubling each year.

Reviving a national prices and wages pact will be essential to cutting inflation without sharply reducing overall growth. An accord between unions, employers and government played an important role in reducing inflation and producing real wage gains in 1986 and 1987. But it collapsed in 1988 as inflation rose considerably above the government forecasts that wage rises were geared to. A pact will be more difficult to negotiate in the current climate of strong economic growth and high employment. But the government has made a clear policy option to focus anti-inflationary efforts on wage restraint rather than stringent monetary restrictions, which would have severe repercussions on employment and the modernisation effort.

Export growth should remain buoyant at 6-7 per cent over the medium term and import growth was expected to drop from the soaring rates of 1986-89 as domestic demand slowed. High import growth pushed the trade deficit up to $5.1bn in 1988 from $1.7bn two years earlier. The current-account deficit was expected to grow to $1.5bn in 1989. Prime lending rates climbed steadily from September 1988 to approximately 20 per cent – a real rate of 7-8 per cent – in mid-1989 and are likely to stabilise at this level over the near future. Moves towards further liberalisation of financial markets will produce upward pressure on interest rates over the medium term. The principal measure in this area, expected to be phased in from the end of 1989, is the transition from the physical control of credit through the fixing of separate lending limits on each bank to a more indirect system, mainly regulated by central bank management of bank reserves. Portugal will maintain a gradual 'crawling-peg' devaluation of the escudo as part of an exchange rate policy aimed at full participation in the European Monetary System from 1992 or as soon afterwards as possible.

Government popularity dips

Several factors contributed to an erosion of support for the government that led to a marked mid-term dip in popularity in 1989. Reforms such as labour legislation and the new tax code – which resulted in some double taxation in 1989 – were then at their most costly phase in terms of public support. State employees had to bear the brunt of government efforts to control inflation through wage restraint. A series of public sector strikes and protests resulted, involving not only workers in traditionally militant sectors such as transport but also teachers, doctors, medical, tax and bank workers. Reform of public services including health, education and the judiciary also brought the government into conflict with the professions. Media campaigns suggesting alleged irregularities – either personal or within their ministries – by at least two ministers also damaged the government's image, although independent inquiries had attributed no wrong-doing to the ministers involved by mid-1989.

The swing away from the government became clear in the European Parliamentary elections of June 1989, when the PSD's vote fell by 4.7 per cent to 32.7 per cent, while the Socialists, running in a coalition with the small Democratic Renewal Party (PRD), increased their joint vote by 1.6 per cent to 28.5 per cent, in comparison with the 1987 Euro-elections. The PSD was most heavily penalised by abstention, which rose from 27.4 per cent to 48.7 per cent – the biggest in any Portuguese election since the 1974 revolution. A large sector of the electorate appeared to be expressing disenchantment with the government without being sufficiently motivated by the opposition to change their vote.

Government popularity was again to be put to the test in local government elections in December 1989, where the PSD were to face an unprecedented coalition of Socialists and Communists in the capital. The Socialist Party leader, Jorge Sampaio, who succeeded Vitor Constancio on his resignation in 1988, was to run for the post of Lisbon mayor. The coalition was an indication of how opposition parties and the labour movement are responding to the PSD's domination of politics by overcoming past rivalries and increasing co-operation. However, it is unlikely that the Socialist–Communist pact in Lisbon will develop into any kind of understanding between the two parties for the legislative elections in 1991.

President Mario Soares, a former Socialist leader, is virtually assured of re-election if he decides to stand in the presidential election that is scheduled to precede the 1991 legislative ballot by a few months. Soares has proved amenable to the government and Cavaco Silva has no strong motive to oppose his re-election. But the PSD is unlikely to switch to supporting Soares, whose election they opposed. This leaves them with the politically weak choices of backing an almost certain loser or staying out of the race altogether.

Outlook

The effect of these negative factors on the PSD's medium-term political prospects should not be exaggerated, however. Cavaco Silva has four years of strong growth behind him and the economic outlook continues relatively bright. By 1991, many recent reforms, such as the new tax code and privatisation, will be paying more positive dividends and other more popular reforms, including private television broadcasting will be implemented. In the likely event of Cavaco Silva failing to regain an overall majority, he will be forced to reach some kind of understanding with the conservative Centre Social Democrats (CDS). This will reintroduce the instability factor of minority governments and coalitions to Portugal. But integration into the single European market, the openness of the Portuguese economy and the evolution of the main political parties will tend to make Portugal's political and economic stability increasingly less vulnerable to such changes.

COUNTRY PROFILE

Industry: Although contributing over 40 per cent to GDP, industry is under-developed and dependent on imported energy and materials. State-run companies predominate, particularly in the home-based sector. Production has been sluggish since the period of rapid industrialisation in the 1960s and 1970s, though there has been an expansion in 1985-87 in the export-based industries such as footwear, textiles, chemicals, base metallurgical products and machine components. Other important industries include processed cork, paper, cement, fertilisers, steel, motor vehicles and glassware. Investment in manufacturing industries increased in 1986 and 1987, and a new system of regional cash incentives was introduced. Programme for the Development of Portuguese Industry (PEDIP) was introduced, with EC funding of 500m ECUs over five years plus contributions from the Social and Regional Funds. PEDIP's aim is to strengthen the industrial infrastructure.

Mining: Though there is considerable mineral wealth, deposits are scattered and not easily exploitable on a large scale. Small-scale mining of tin, copper, tungsten concentrates, marble, stone and iron pyrites. Also known reserves of uranium. Recent growth in development of resources following fall in revenues from former overseas territories.

Energy: Portugal lacks natural resources so is heavily dependent on imported fuels, particularly oil. The National Energy Plan emphasises increased conservation and shift to oil substitutes such as coal, hydro-electricity, wood, biomass, solar and wind power. Hydro-electricity contributes around 50 per cent of electricity generated.

Banking: System headed by Bank of Portugal (central bank). Main commercial banks, which are all state-owned, are: Banco Português do Atlantico, Banco Pinto & Sotto Mayor, Banco Espirito Santo & Comercial de Lisboa, Banco Nacional Ultramarino, Banco Borges & Irmão, Banco Totta & Açores, União de Bancos Portugueses and Banco Fonsecas & Burnay. Main foreign banks include Crédit Franco-Portuguais, Lloyds Bank and Manufacturers Hanover Trust. Four Portuguese and nine foreign private banks were by 1987 authorised to operate: Banco Português de Investimento, Banco Comercial Português, Banco de Comércio e Indústria, Banco Internacional de Crédito; Banco do Brasil, Chase Manhattan Bank, Citibank, Banque Nationale de Paris, Société Générale de Banque and Barclays Bank International. Prominent among specialised banks are Caixa Geral de Depositos (savings bank) and Banco de Fomento Nacional (medium and long-term credit for industry). Deregulation in progress.

Legal system: Civil law system. Constitution adopted 1976, revised 1982 and 1989.

Membership of international organisations: CCC, Council of Europe, EC, EFTA, FAO, GATT, IAEA, IATP, IBRD, ICAC, ICAO, ICES, ICO, IDB, IEA, IFAD, IFC, IHO, ILO, IMF, IMO, INTELSAT, INTERPOL, IOOC, IRC, ISO, ITU, IWC, NATO, OECD, UN, UNESCO, UPU, WHO, WIPO, WMO, WSG.

BUSINESS GUIDE

Time: GMT (GMT + 1 from late Mar to late Sep).

Climate: May-Oct dry and warm, Nov-Apr cool with rain in north, mild in south (though often wet and windy Jan-Mar). Temperatures vary between about 8°C and 28°C.
Clothing: Mediumweight and raincoat Nov-Feb, lightweight for summer.

Entry requirements: Visas not required by nationals of European countries, plus a few others. NB: regulations liable to change at short notice. Portuguese Consul or travel agent should be consulted prior to departure.
Health: Vaccination certificates not normally required. In main towns water is drinkable, but visitors advised to drink bottled water.
Currency: No limit on import of foreign currency, but amount should be declared on entry.
Customs: Personal effects duty-free, plus small duty-free allowance.

Air access: Regular direct flights from most major international airports.
National airline: TAP - Air Portugal.
Main international airport: Lisbon (LIS) at Portela, 7 km from centre of capital.
International airports: Porto (OPO), Faro (FAO), Funchal (FNC) on Maderia and Santa Maria (SMA) in the Azores.

Surface access: Car and passenger ferry services either direct or via Spain. Several road and rail routes through Spain, mostly via Madrid.
Main ports: Lisbon, Leixões, Setúbal, Aveiro, Sines and Funchal (Madeira).

Hotels: Available to various standards throughout country, classified from one-star to five-star. Single five-star room with breakfast costs about Esc23,000 per night, including 10 per cent service charge. A tip is also expected. Accommodation should be booked well in advance, especially during holiday season, when all hotels become very busy. N.B. drinking water is not purified outside main towns.

Restaurants: Numerous and varied. Service charge usually included in bill. Small tip usually given.

Currency
Exchange rate: Escudo is floating.

Foreign exchange information
Account restrictions: No restrictions on non-resident savings deposits in foreign currency. Other accounts subject to authorisation by Bank of Portugal.
Trade policy: Importers and exporters must register their operations beforehand at the Direcção-Geral do Comércio. Progressive reduction of import duties for EC products by 1992.

Credit cards: All usual credit cards are accepted.

Car hire: Self-drive and chauffeur-driven cars available throughout country. A medium-sized self-drive car costs about Esc83,000 per week (covering about 800 km). International driving licence or full national licence required, as well as international insurance Green Card. Motoring information available from 'Automovel Clube de Portugal' in Lisbon. The wearing of seat belts is compulsory outside city areas.

City transport
Taxis: Lisbon taxis are green and black. Relatively cheap but efficient service, costing about Esc74 plus Esc39 per kilometre in capital. Tip of 15 per cent expected. Taxis may be scarce during rush-hour.
Buses and trams: Good services in main centres, such as Lisbon and Oporto. Lisbon also has a small metro network.

National transport
Air: TAP operates scheduled flights between Lisbon and Oporto, Faro, Madeira and the Azores.
Road: Over 52,000 km of roads, including about 22,000 km of main or national roads. Regular coach services link major towns.
Rail: Nationalised railway company 'Caminhos de Ferro Portugueses, EP'(CP) own about 3,500 km of track, of which 500 km are electrified. Services between main centres – Oporto and Lisbon, Lisbon and Faro – are good and include express trains with restaurant cars; seats on these should be booked in advance, but rail services on other lines are poor.
Water: About 800 km of inland waterways which are only rarely used. Some coastal shipping, including services to Madeira and the Azores.

Public holidays
Fixed dates: 1 Jan, 25 Apr, 1 May, 10 Jun, 13 Jun (Lisbon only), 24 Jun (Oporto only), 15 Aug, 5 Oct, 1 Nov, 1 Dec, 8 Dec and 25 Dec.
Variable dates: Carnival, Good Friday, Easter Day, Corpus Christi.

Working hours
Business: (Mon-Fri) 0900 – 1300 and 1500 – 1900, half day on Sat.
Government: (Mon-Fri) 0930 – 1200 and 1430 – 1800, close 1730 on Mon and Tue.
Banking: (Mon-Fri) 0830 – 1145 and 1300 – 1445.
Shops: (Mon-Fri) 0900 – 1300 and 1500 – 1900, half day on Sat.

Business language and interpreting/translation: French and English fairly widely spoken in business circles, but knowledge of Portuguese useful when dealing with smaller companies. However information on translation services for business visitors can usually be obtained from Chambers of Commerce.

Telecommunications
Telephone: Government-owned system, offering direct dialling to most parts of the country and several other parts of the world. International dialling code: 351 followed by 1 for Lisbon.
Telex: Available in major hotels and from main post offices in Lisbon and Oporto. Facilities also available on Madeira and Azores. Country code 404 P.
Telegram: Can be sent from post offices or via telephone.

Trade fairs: The main international fair (FIL) is held annually in Lisbon. Other specialist fairs are held regularly in Lisbon, Oporto and several other towns. National Agricultural Show, Jun, Santarem; Congress of Opthalmology, May, Lisbon; European Opthalmological Congress, Jun, Lisbon.

Electricity supply: 220V AC.

Representation in capital: Algeria, Angola, Argentina, Australia, Austria, Belgium, Bolivia, Brazil, Bulgaria, Canada, Cape Verde Islands, Chile, China PR, Colombia, Costa Rica, Cuba, Czechoslovakia, Denmark, Egypt, Finland, France, German Democratic Republic, Federal Republic of Germany, Greece, Guinea-Bissau, Holy See, Hungary, India, Iraq, Ireland, Israel, Italy, Japan, Democratic People's Republic of Korea, Republic of Korea, Libya, Mexico, Morocco, Mozambique, Netherlands, Nigeria, Norway, Pakistan, Panama, Peru, Poland, Romania, Sao Tomé and Príncipé, South Africa, Spain, Sweden, Switzerland, Turkey, USSR, United Kingdom, USA, Uruguay, Venezuela, Yugoslavia, Zaïre.

Hotels: Lisbon
Altis, Rua Castilho 11, P.O. Box: 1200 (tel: 524206, 522496; tx: 42520; fax: 548696).

Avenida Palace, Rua 1 Dezembro 123 (tel: 360151; tx: 12815).

Borges, Rue Garrett 108 (tel: 361951; tx: 15825).

Diplomatico, Rua Castilho 74 (tel: 562041; tx: 13713).

Eduardo VII, Avenida Fontes Pereira de Melo 5 (tel: 530141; tx: 18340; fax: 533879).

Fenix, Praça Marques de Pombal 8 (tel: 535121; tx: 12170).

Flamingo, Rua Castilho 41 (tel: 532191/2/3/4; tx: 14736; fax: 532191).

Meridien, Rua Castilho 149 (tel: 690900; tx: 64315; fax: 693231).

Novotel, Avenida José Malhoa 1642 (tel: 7266022; tx: 40114; fax: 7266496).

Penta, Avenida dos Combatentes (tel: 724054; tx: 18437).

Ritz, Rua R. da Fonseca 88 (tel: 692020; tx: 12589; fax: 691783).

Roma, Avenida da Roma 33, P.O. Box: 1700 (tel: 767761; tx: 16586).

Sheraton, Rua Latino Coelho 1, P.O. Box: 1097 (tel: 575757; tx: 12772; fax: 547164).

Tivoli, Avenida da Liberdade (tel: 530181; tx: 12588; fax: 579461).

Chambers of Commerce
Associação Comercial e Industrial do Funchal, Avenida Arriaga 41, 9000 Funchal, Madeira (tel: 3013718; tx: 72293).

Associação Comercial do Porto, Palacio da Bolsa, Porto 4000 (tel: 24497, 311448; tx: 22159).

Camara de Comercio de Ponta Delgada, Rua dos Mercadores 68, 9500 Ponta Delgada, Azores (tel: 22427; tx: 82118).

Camara de Comercio Portuguesa, Palacio do Comercio, Rua das Portas de Santo Antao, Lisbon 1200 (tel: 327179, 327289, 323472; tx: 13441).

Banking
Banco de Portugal (central bank), Rua do Comercio 148, 1100 Lisbon (tel: 362931; tx: 16554).

Banco Borges e Irmão (commercial bank), Rua Sá da Bandeira 20, 4001 Oporto (tel: 324517; tx: 26899).

Banco de Fomento Nacional (National Investment Bank), Avenida Casal Ribeiro 59, 1000 Lisbon (tel: 562021; tx: 64752).

Banco de Comércio e Indústria, Rua Tenente Valadim 290, 4102 Oporto (tel: 695671; tx: 28606).

Banco Fonsecas & Burnay (commercial bank), Rua do Comércio 132, 1106 Lisbon (tel: 874801; tx: 12210).

Banco Internacional de Crédito, Avenida Fontes Pereira de Melo 27, 1000 Lisbon (tel: 527135; tx: 62353 banicr).

Banco Português do Atlântico (commercial bank), Palácio Atlântico, Praça D. João 1, 4001 Oporto (tel: 381494; tx: 22720).

Banco Português de Investimento, Rua Tenente Valadim 284, 4100 Oporto (tel: 699951; tx: 26887).

Banco Pinto & Sotto Mayor (commercial bank), Av Fontes P de Melo 7-5\28, 1000 Lisbon (tel: 542078; tx: 12516).

Banco Espirito Santo e Comercial de Lisboa (commercial bank), Avenida da Liberdade 195, 1200 Lisbon (tel: 578005; tx: 12191).

Banco Nacional Ultramarino (commercial bank), Rua Augusta 24, 1111 Lisbon (tel: 369981; tx: 13305).

Banco Totta & Açores (commercial bank), Rua Aurea 88, 1100 Lisbon (tel: 369421; tx: 12266).

Caixa Geral de Depósitos (savings bank), Largo do Calhariz, 1109 Lisbon (tel: 371881; tx: 12621).

União de Bancos Portugueses (commercial bank), 80 Praça D João 1, Oporto (tel: 20961; tx: 26061).

Puerto Rico

A special correspondent

Governor Rafael Hernández-Colón, re-elected in November 1988 by a slim majority, surprised virtually everyone at his inauguration (2 January 1989) by announcing he would seek an agreement with the United States Government to hold a plebiscite on Puerto Rico's future political status. Although he ran on a platform downplaying status, his announcement immediately brought it back to the forefront. Subsequently, pro-Commonwealth Hernández-Colón of the Popular Democratic Party (PDP), pro-Statehood Carlos Romero-Barcelo of the New Progressive Party (NPP) (who replaced defeated Baltasar Corrada del Rio as party president) and pro-Independence Ruben Berrios of the Puerto Rico Independence Party (PIP) entered into negotiations with US Senator J Bennett Johnston, democrat of Louisiana (who, as chairman of the Senate Energy and Natural Resources Committee, has oversight of Puerto Rico) to devise the formula under which the plebiscite would be conducted, tentatively in June 1991. The status debate has eclipsed government attention from more pressing social and economic issues, including educational reform, crime and drugs, infrastructural expansion and streamlining the bureaucracy.

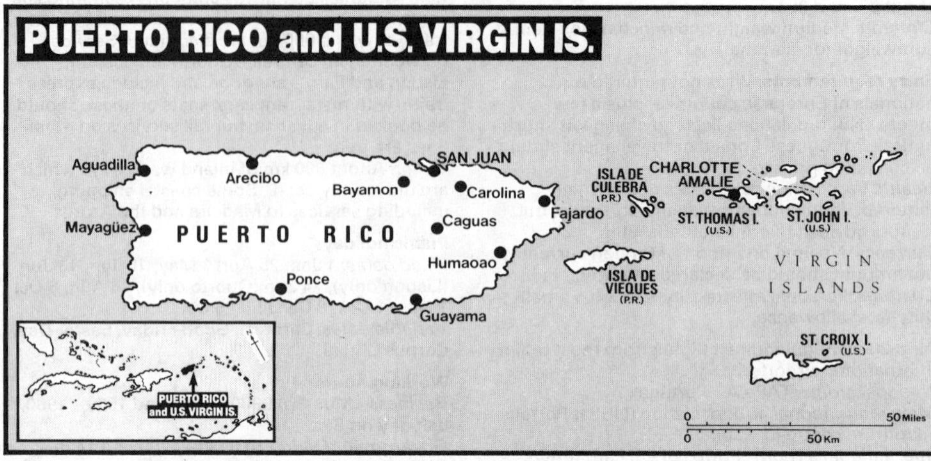

The status debate has the PDP seeking enhancement of the current Commonwealth, while the NPP wants Puerto Rico's admission as the 51st state of the Union and the PIP wants an independent republic. President George Bush has expressed his preference for statehood on several occasions. Discussions could continue for several months in both houses of the US Congress. The PDP wants greater local autonomy within permanent union with the United States. The NPP requests guarantees for an economic transition into statehood as well as protection for Puerto Rico's predominant Spanish language. The PIP also wants financial guarantees to smooth a transition and a friendship treaty with eventual removal of US military bases from Puerto Rico.

The requests of all three parties have caused quite a stir in Washington. In the not too extensive colonial history of the United States, all territories, with the exception of the Philippines, have become states of the Union.

As proposed, the plebiscite would be self-executing, requiring the United States to implement the results. Assuming a simple majority (50 per cent plus one vote) is not reached in the plebiscite, a runoff election between the two leading options would be required.

Much of the debate in Washington, carried live on Puerto Rico's television stations but virtually ignored by US news media, has raised considerable anxiety among Puerto Ricans regarding their US citizenship, access to the United States, current benefits to individuals (Social Security, veterans' pensions and food checks) and Section 939 (a US tax incentive which is the backbone of Puerto Rico's industrial base and banking sector). Conversely, some others warn that statehood could create a Northern Ireland-type situation for the United States with a militant pro-independence minority remaining in the new state.

While the debate continues and the legislation is being formulated, it is too early to forecast the outcome of the plebiscite. However, statehood appears to be gaining ground and some observers are even predicting it will receive a 60 per cent majority. In the meantime, some US and foreign investors are taking a wait-and-see attitude, which will certainly begin to take its toll. There has also been increased home and small business investment in Florida by Puerto Ricans, uneasy over the future as well as crime and the high cost of living. While currently at a low level, there has also been an upsurge in terrorist activity in the form of bombings of US government agencies and some private US companies.

Crime and drugs

Crime and drugs continue to plague Hernández-Colón as well as an expansion in bureaucratic malaise and continued decline in the infrastructure. Puerto Rico remains a major transit point for the entry of narcotics into the United States and Europe. Increased crime on the island is related to the trafficking and local drug consumption. The government bureaucracy, already bloated and employing 250,000, continues to expand as services deteriorate further. Treasury Secretary Juan Augusto-Alicea resigned his post after five years in frustration over his inability to affect change in a government preoccupied with status politics. Hernández-Colón, although accepting his resignation with regret, described Augusto as idealistic and utopian. Infrastructural deficiencies increased with periodic power outages, water rationing and chronic traffic congestion. Plans to increase electricity rates by 15 per cent to finance new power plants have met strong tripartisan opposition in the Legislature and warnings that they will fuel the inflationary spiral. The Hernández-Colón administration is studying proposals to privatise some government agencies like the Telephone Company to raise $1bn for an Infrastructural Fund to improve the situation.

Relations between the US Navy and residents of the off-island of Vieques again became strained as the Navy sought to oust squatters from a buffer zone outside a major training area. Residents also protested against the Navy's failure to get US defence contractors to deliver on a 1983 pledge to create and maintain factory jobs on the island.

Economic development

The Puerto Rican economy continued to show steady growth (4.9 per cent) in 1989, fuelled by continuing expansion of construction and tourism but considerable concern is being expressed that investment will become inhibited as the status debate creates more caution. Unemployment remains at a six-year low (15 per cent) and federal transfers from the US Government continue at about $6bn per annum.

Uncertainty over the application of an increased US minimum wage to Puerto Rico abated in 1989 as the US Congress failed to challenge President Bush who, on taking office, voiced his vehement opposition to it.

Transportation remains strong in spite of the Eastern Airlines strike because of the dominance of American Airlines, which now controls more than 60 per cent of the market. Although a massive terminal was erected for Eastern in 1988, as it retrenched foreign carriers were to be moved to the Eastern terminal as American expanded in the original terminal. Because of the American Airlines hub, Luis Muñoz Marin International Airport has become the tenth busiest US airport. Traffic through the hub has increased tremendously because of the Eastern strike, as tourists bound for the US Virgin Islands, the Dominican Republic, St Maarten, Antigua and elsewhere seek alternate routes. The number of cruise ship passengers visiting San Juan rose 15 per cent over last year to 767,000 tourists. That number is projected by the Puerto Rico

KEY FACTS

Official title: Commonwealth of Puerto Rico
Head of state: President of the USA
Head of government: Governor Rafael Hernández Colón
Ruling party: Popular Democratic Party
Capital: San Juan
Official Languages: Spanish and English
Currency: US Dollar ($) = 100 cents

Area: 8,897 sq km
Population: 3.3m (1988)
GDP per capita: $7,762 (1988)
GDP real growth: 5.2% (1988)
Inflation: 2.8% (1988)
Trade balance: $2.0bn (1988)
Aid flows: $4.0bn (1988; USA)

Ports Authority to reach 936,000 in 1992 and to exceed 2m by the year 2000. Fourteen cruise ships are now home berthed in San Juan and each weekend sees thousands of cruise ship passengers arriving in San Juan from North America and Europe.

936/Caribbean Basin Initiative

The Puerto Rico Caribbean Development programme yielded disappointing results in 1989 for many of the same reasons that have hampered it from the beginning. Although the US Treasury Department gave the go-ahead for major loans of low-cost 936 funds to Barbados and Jamaica, few new manufacturing joint ventures creating permanent employment developed. Only four CBI countries have signed Tax Information Exchange Agreements (TIEA) – Barbados, Dominica, Grenada and Jamaica. Four others – Costa Rica, the Dominican Republic, St Lucia and Trinidad and Tobago – have signed but not ratified. A TIEA is required for accessing 936 funds from Puerto Rico. At mid-1989 the number of projects were 61 in 11 countries valued at $163.5m. The fact that half of them are in the Dominican Republic has resulted in frustration and resentment in other countries.

The much touted CBI Loan Fund failed to materialise and private banks are reluctant to make loans to the debt-strapped countries without guarantees. Because they do not generate permanent employment, as capital-intensive projects (like one for $51m to Air Jamaica) increase it is considered unlikely that the Caribbean and Central American countries (which backed Puerto Rico's efforts to save Section 936) and their supporters in Washington (both in the Bush Administration and Congress) will accept these as the fulfilment of Puerto Rico's pledge to stimulate regional economic development. Many observers believe that without massive improvement to regional infrastructure and transportation, it is going to be difficult for Puerto Rico to deliver on its commitment.

In the Tax Reform Act of 1986, the US Congress made the US Virgin Islands eligible for Section 936 benefits. With a limited industrial sector, the US Virgin Islands did not create its own 936 mechanisms but instead has been granted eligibility to access Puerto Rico's 936 funds. Massive capital-intensive projects in the neighbouring territory are being proposed for financing from Puerto Rico. Critics of the expansion of Puerto Rico's Caribbean Development programme cite the US Virgin Islands' high per capita income and warn that such loans will irritate less developed Caribbean countries and raise questions in Washington about how the US Virgin Islands got involved. Some observers believe the move has been deliberately orchestrated by the Hernández-Colón administration in recognition of the considerable clout Virgin Islands Delegate to Congress Ron de Lugo exercises in Washington. De Lugo chairs the House Interior and Insular Affairs Subcommittee which will review the Puerto Rico status plebiscite bill.

KEY INDICATORS

	Unit	1984	1985	1986	1987	1988
Population	m	3.3	3.3	3.3	3.3	3.3
GDP	$m	18,622	19,683	21,270	23,706	25,567
GDP per capita	$	5,698	6.000	6,481	7,214	7,755
GDP real growth	%	6.6	2.2	7.0	7.6	5.2
Unemployment rate	%	22.0	21.4	20.5	17.7	15.9
Manufacturing output	%	12.1	6.8	7.0	13.1	7.3
Tourist arrivals	'000	1,496	1,545	1,573	1,871	2,077
Exports	$m	9,146	10,544	11,571	12,068	13,141
USA	$m	7,561	9,060	10,100	10,516	11,568
Imports	$m	9,528	10,112	10,099	10,724	11,869
USA	$m	5,457	5,946	6,116	6,977	7,928

* estimated figure

COUNTRY PROFILE

Historical profile: First European sighting by Columbus 1493; member of expedition, Juan Ponce de Leon given permission to settle island - named by him San Juan; more healthy, more easily defended site chosen and capital transferred to Puerto Rico (Rich Port) 1521; subsequently names reversed; strategically important to exploration, settlement and invasion of region; Charter of autonomy granted by Spain in 1897; troops from USA landed (in course of Spanish-American War) 1898; island taken over by USA 1899; inhabitants became citizens of USA 1917; first elected governor 1948; commonwealth in voluntary association with USA 1952.

Political system: Detailed laws governing status and relationship of the Commonwealth of Puerto Rico with the United States cover among other aspects: military conscription, tax and trade, social security, citizenship, constitutional changes, internal autonomy. Executive power rests with the governor (elected by direct vote for a four-year term) who appoints an executive consisting of secretaries of government departments. The legislature is the bicameral Legislative Assembly: the Senate (27 members – 16 from the eight districts, 11 on wider franchise) and the House of Representatives (51 members – 40 from constituencies, 11 on wider franchise). The term for both Houses is four years and representation of minority parties is constitutionally guaranteed – from a quarter to one-third of seats in each chamber.

Main cities/towns: San Juan (population over 1m), Aguadilla, Arecibo, Bayarnón, Caguas, Carolina, Mayagüez, Ponce and Yuaynabo.

Media
Press: Four main dailies, of which the largest are *El Nuevo Día, El Mundo, El Vocero de Puerto Rico* and *San Juan Star* (English). Weeklies include *Caribbean Business.*
Broadcasting: Around 103 national commercial radio stations and 10 television stations. US armed forces also operate an English-speaking radio station and three TV channels.

External trade: Regular trade surplus since 1985 owing to increase in exports of manufactured goods. Balance of payments is, however, still reliant on US federal aid flows.
Exports: Chemicals and allied products (35-40% of total value), non-electric machinery (10-15%), electric machinery (5-10%), textiles and clothing (5-7%), refined petroleum (2-5%), food and beverages, metals, drugs and pharmaceuticals. Main destinations: USA (80-90% of total value), US Virgin Islands (1.2%), others (10%).
Imports: Petroleum and derivatives, foodstuffs, machinery and transport equipment, textiles, chemical products. Main sources: USA (55-65% of total value), US Virgin Islands (.03%), others (33%).

Agriculture: Sector is small-scale, contributing around two per cent to GDP. Dairy and livestock farming of increasing importance. Main cash crops include sugar-cane, pineapples, coffee, oranges, other fruits and tobacco. Around 50 per cent of food requirements met by imports.

Industry: Sector forms mainstay of economy, contributing around 40 per cent to GDP and accounting for 98 per cent of exports. Production is centred on food processing, textiles, petro-chemicals, rum distilling, pharmaceuticals, metal fabrication and assembly of electrical/electronic components. Most of the assembly industries are US-owned and are heavily dependent on the US market. Industrial development remains focal point of government planning, with emphasis on expanding the island's free trade zone and encouraging foreign investment in import substitution.

Tourism: Important sector with annual earnings of over $1bn. Number of regular visitors around two million, of which 75 per cent come from USA.

Mining: Activity in this area extremely small – production centred on non-metals such as stone, sand, salt and clay. Reserves of copper, nickel, cobalt, iron, chromium, lead, gold and silver (not quantified).

Energy: Dependence on imported energy fuels, mainly from Venezuela and Netherlands Antilles. Petroleum exploration under way off coastline.

Time: GMT -4.

Climate: Tropical with extremes of heat tempered by constant sea winds. Average temperature ranges between 24°C and 30°C. Occasional hurricanes.

Entry requirements
Passport: Required by all except nationals of USA and Canada.
Visa: Required by all except nationals of USA and Canada.
Currency: No restrictions on import/export of any currency to the value of $10,000. Higher amounts must be declared.

Health precautions
Mandatory: None.

Air access
Airlines serving: Aeroportal, Aero Virgin Islands, Air BVI, Air Caribe Int, Air France, Air Guadeloupe, ALM, American, British Airways, BWIA, Delta, Delta, Dominicana, Eastern Metro, Iberia, LACSA, LIAT, Lufthansa, Mexicana, TWA, Viasa, Vieques Air Link.
International airport: Luis Muñoz Marin (SJU), 8 km from San Juan; duty-free shop, bar, restaurant, bank, post office, shops, hotel reservations, car hire.

Surface access
Main ports: Ponce, Mayagüez, San Juan.

Hotels: Most of the first-class hotels are in Condado or Isla Verde. 15 per cent tip usual.

Car hire: Advisable to book through airline well in advance. Foreign licence acceptable.

City transport
Taxis: Within San Juan: metered - small charge for taxis ordered by telephone. Also shared taxis (*públicos*) which have yellow number plates and run to all parts of the island. 15 per cent tip usual.

National transport
Air: American Eagle, Eastern Metro Express and Dorado Wings link San Juan, Ponce and Mayaguez.
Road: Modern highways link all main centres.
Buses: Regular bus (*guagua*) service operates in San Juan from centre terminal at Plaza Colón. Buses are scarce after 2100.

Public holidays: 1 Jan, 6 Jan (Epiphany), 11 Jan (De Hostos's Birthday), Martin Luther King Day (Mon in mid-Jan), Washington's Birthday (Mon in mid-Feb), 22 Mar (Emancipation Day), Good Friday, 16 Apr (De Diego's Birthday), 29 May (Memorial Day), 4 Jul (Independence Day), 17 Jul (Muñoz Rivera's Birthday), 25 Jul (Constitution Day), 27 Jul (Barbosa's Birthday), Labour Day (first Mon in Sep), 9 Oct (Columbus Day), 11 Nov (Veterans' Day), 19 Nov (Discovery Day), Thanksgiving (Thu in late Nov), 25 Dec.

Working hours
Government and business: 0800 – 1630. (Exceptions: Sat, Sun closed.)
Banking: 0900 – 1430. (Exceptions: some banks 0900 – 1600; some banks open Sat; all banks closed Sun.)

Telecommunications: Direct dialling and telex facilities available at all main hotels. Dialling code for Puerto Rico: IDD access code + 1.809 followed by subscriber's seven-digit number.

Electricity supply: 120 V AC.

Hotels: San Juan
Carib Inn, Loiza Station, P.O. Box: 12112 (tel: 791-3535; tx: 3450165).

Caribe Hilton, P.O. Box: 1872 (tel: 721-0303; tx: 2733; fax: 7210303 EXT 1725).

Condado Beach, P.O. Box: 41226 (tel: 721-6090; tx: 2433; fax: 7225062).

Condado Plaza, 999 Ashford Avenue, P.O. Box: 1270 (tel: 721-1000; tx: 3450183).

El Convento, Calle Cristo 100, P.O. Box: 1048 (tel: 723-9020; tx: 3453199; fax: 7212877).

El San Juan, P.O. Box: 2872 (tel: 791-1000; tx: 3252585; fax: 7912856).

Excelsior, 801 Ponce de Leon (tel: 721-7400; tx: 3450187; fax: 7230068).

Howard Johnson Nabori Lodge, 1369 Ashford Avenue (tel: 721-7300; tx: 3252822; fax: 7236151).

Hyatt Dorado Beach, Dorado (tel: 796-1600; tx: 3450196).

Pierre, 105 De Diego, Santurce (tel: 721-1200; tx: 0216).

Ramada San Juan, 1045 Ashford Avenue, 00907 (tel: 724-5657; tx: 9660; fax: 7228230).

Regency, 1005 Ashford Avenue, P.O. Box: 4484 (tel: 721-0505; tx: 3450079).

Business advice and information
Chamber of Commerce of Puerto Rico, PO Box 3789, Chamber of Commerce Building, 100 Tetuan Street, San Juan 00904 (tel: 721-6060).

Government of Puerto Rico Economic Development Administration, PO Box 2350, San Juan 00936 (tel: 758-4747).

Puerto Rico Manufacturers' Association, Suite 404, 07 Midtown Condominium, Ponce de Leon Ave, Hato Rey, San Juan 00918.

Puerto Rico Ports Authority, PO Box 2829, San Juan (tel: 723-2260; tx: 9464).

Puerto Rico Tourism Company, 301 San Justo Street, PO Box 4435, Old San Juan Station 00905 (tel: 721-2400; tx: RCA 2376, ITT 1058; fax: 725-4417).

San Juan Convention Bureau, Ashford Avenue 1110, Santurce, 00907 (tel: 725-2110; tx: 3252351).

Banking
Banco Central, Av Ponce de Leon 221, Hato Rey, San Juan (tel: 753-2500; tx: 3090).

Banco Comercial de Mayagüez, Mayagüez, 00708 (tel: 834-3717; tx: 3859898).

Banco de Ponce, Plaza Degetau, Ponce 00731 (tel: 842-8000; tx: ITT 3450064).

Government Development Bank for Puerto Rico, PO Box 42001, San Juan, 00940-2001 (tel: 726-2525; tx: 3857265).

Qatar

Teresa English

Qatar is undergoing a period of social and political change prompted by the prospect of new industrial development using the abundant gas supplies from the North Field. In its traditionally stable and cautious fashion, the emirate is reshaping at the highest levels to cope with the challenges of the 1990s. The Gulf War cease-fire in mid-1988 has brought greater security to the region, but Qatar has not prospered as much as neighbouring UAE from the post-cease-fire optimism. Oil, the economic mainstay, accounting for 90 per cent of annual income, has proved unstable in the 1980s and has prompted a re-think of economic policies instigating a review of priorities.

From dependence on nomadic livestock farming, pearls and fishing, Qatar has used its oil revenues since the 1950s to set up the infrastructure of a modern developed state. The last 30 years have been a period of rapid change. Qatar now has a per capita gross domestic product of some $16,000 for its 350,000 population, a large proportion of whom are expatriates. Oil has enabled the country to introduce a reasonably high standard of living: education, health and social standards have been improved considerably. Qatar, like other Gulf states, is increasingly aware of its dependency on an unpredictable oil market and its new priority is to increase non-oil income. This will not be an easy process.

Budget deficit continues

A further deficit budget was announced for 1989-90 in April after a series of difficult years. In 1986 no official budget was announced for the year. Budgets will now begin on 1 April, previously they ran to the Islamic year. The three previous budgets were overspent by about 20 per cent. The government hopes to cut the deficit for 1989-90 by about 7.5 per cent to QR5,648m, against revenues which are predicted to fall by nearly 8 per cent to QR5,835m. Capital spending has been cut drastically by 43 per cent to QR1,513m. Only priority projects, like the vitally needed power and water schemes which went to tender in 1989, will be allocated funds. Current spending is increased slightly to QR9,969m from QR9,827m. The largest fall in capital spending comes from public works and construction which is cut from QR1,507m to QR906m. Economic services falls to QR412m from QR627m. Social and health services fall to QR63m from QR227m. Education and youth spending is cut to QR1,513m from QR2,616m.

After a tough year for oil earnings in 1988, Qatar looked set to enjoy higher earnings for 1989. Long-term forecasts also predicted improvements. As a result, the increased oil earnings should ease some of the restraints within Qatar's economy. Export earnings were forecast to rise by about 23 per cent to around $1,950m, according to Washington-based Petroleum Finance Company. This assumes a continued output of 330,000m barrels a day (b/d) and an average price of $15.85 a barrel. It also depends on prices continuing steadily to the end of the year. In 1988, Qatar exported an estimated $1,580m at an output of 319,000m b/d with an average price of $13.75 a barrel. OPEC itself has been predicting a better year for its 13 members. In mid-1989 it said that demand for OPEC crude would be about 4.5 per cent higher than in 1988, with prices about 17.2 per cent up on the 1988 average at around $16.20 a barrel. Officially, Qatar sticks closely to quotas and prices. Its production is usually fairly close to quota. However, in 1988 the country began to come under pressure from crude term customers to adopt more flexible oil pricing

KEY FACTS

Official title: State of Qatar
Head of state: Shaikh Khalifa bin Hamad al-Thani
Capital: Doha
Official Languages: Arabic
Currency: Rial (QR) = 100 dirhams
Exchange rate: QR3.64 per $ (Nov 1989)
Area: 11,437 sq km
Population: 350,000 (1988)
GDP per capita: $15,645 (1988)
GDP real growth: 0.4% (1987)
Inflation: 1.6% (1987)
Trade balance: $619m (1988)
Foreign debt: $619m* (1987)

* estimated figure

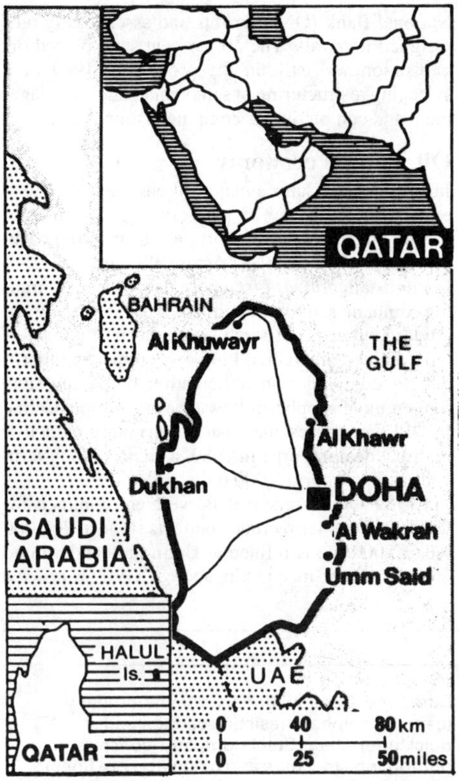

KEY INDICATORS

	Unit	1984	1985	1986	1987	1988
Population	'000	280	300	310	330	350
Gross domestic product (GDP)	$m	6,870	6,153	5,017	5,354	5,476
GDP per capita	$	24,535.7	20,510.0	16,183.9	16,224.2	15,645.7
Consumer prices	1985=100	98.1	100.0	101.6	–	–
Oil production	'000 b/d	325.3	290.1	313.6	291.4	319.4
Natural gas production	m cubic m	5,930	5,460	5,800	5,610	6,470
Exports	$m	4,513	3,098	1,849	1,985	1,869*
Petroleum	$m	4,386	3,068	1,720	1,829*	1,709*
Imports	$m	1,162	1,139	1,099	1,134	1,250*
Balance of trade	$m	3,351	1,959	750	851	619
Current account	$m	830	549	–189	–132	–317*
Exchange rate	QR per $	3.64	3.64	3.64	3.64	3.64

* estimated figure

policies, largely because of the slumping market. Oil industry observers say that a flexible pricing policy has been in existence since May 1987. As Qatar's output is fairly small, it does not have the same bargaining power as larger producers like Saudi Arabia.

A new and youthful cabinet

1989 saw Qatar going through its most important political changes since independence in 1971. In July 1989, a major government reshuffle was announced by the Amir Shaikh Khalifa Bin Hamad al-Thani. Seven of Qatar's most senior ministers who had held power since independence were replaced. Eleven much younger newcomers were appointed to the 16-strong cabinet. It was a response to demands for a younger and more energetic approach to meet the challenges of the 1990s. Political power is still firmly under the control of the al-Thani family, who are represented in the most important portfolios. The Amir, in charge since 1972, remains prime minister.

Analysts also say the changes appear to have been at the urging of the 39-year old Heir Apparent and Minister of Defence, Commander-in-Chief of the armed forces Shaikh Hamad Bin Khalifa al-Thani, who is taking a firmer hold of the reigns of government. For example, the new foreign affairs minister is Abdullah Bin Khalifah al Attiyah, former advisor to the Heir Apparent. Shaikh Hamad's more vigorous role is still seen in Doha as a preparatory one. The key portfolio of finance and petroleum remains with one of the Amir's sons, Shaikh Abdel Aziz Bin Khalifa al-Thani. Education and the interior, two portfolios formerly held by senior members of the al-Thani family, have been given to much younger men. The Amir's third son, Abdulaziz Bin Khalifah al-Thani, has taken on the interior ministry, which the government hopes will be at the forefront of new planning initiatives and will begin to establish a tourist industry in Qatar. He was formerly deputy commandant of the armed forces and is well known for the development of sport in Qatar. Education passes from Shaikh Mohammad Bin Hamad al-Thani to Abdel-Aziz Abdullah Turki, previously the under secretary. The economy and trade portfolio, vacant for some time, has been filled, an indication that the new government will place greater emphasis on economic management. The new minister is former chief of police, Shaikh Hamad Bin Jassim Hamad al-Thani. The new justice minister is Shaikh Ahmad Bin Seif al-Thani, former ambassador in London. Shaikh Ahmad was active in the development of the Gulf Co-Operation Council. Health goes to Shaikh Khaled Bin Mohammad Bin Ali al-Thani. Municipal affairs teams up with agriculture under newly-appointed Shaikh Hamad Bin Jassion Bin Jabr al-Thani. The industry ministry, split from agriculture and teamed with public works, has been raised to new prominence to manage Qatar's new industrial development schemes. The post has gone to Ahmad Mohammad Ali al-Subaie, already prominent in industrial circles as chairman of the Qatar Steel Company (QASCO) and head of the aluminium steering committee. He is also a director of Qatar National Cement Company (QNCC) and of Qatar Public Telecommunications Corporation (Q-Tel). The information portfolio has been linked to a culture brief under Shaikh Hamad Bin Suhaim al-Thani. Hamad had been acting foreign affairs minister since his father, the previous incumbent, died in 1978. Former information minister, Isa Ghanem al-Kuwairi, the Amir's influential chief advisor, has taken a new portfolio, Amiri Diwan Affairs.

New appointments amongst the smaller ministries represent some of the other major families. The new communications and transport minister is Abdullah Bin Saleh al-Manna, the former ambassador to Saudi Arabia and North Yemen, influential in GCC circles. Electricity and water goes to Mubarak Ali al-Khater, formerly on the Supreme Council for Youth Welfare at the justice ministry. Khater will have to tackle Qatar's severe power and water shortages which are a constraint on new development plans. Labour, social affairs and housing goes to Abdel Rahman Saad al-Dirhem. Foreign affairs goes to Abdullah Bin Khalifah al Attiya. Three new under secretaries were appointed at the start of September. Shaikh Mohamed Bin Khalifah al-Thani, the Amir's fourth son who obtained a BSc in 1987, was appointed under secretary to the finance and petroleum ministry. The post had been vacant for several years. Ahmad Abdullah al Mahmoud, former ambassador to Washington, is now foreign affairs under secretary. Abdullah Hussain Salatt, a senior official at QGPC is now under secretary at the ministry of transport and communications.

Linked to the cabinet changes, a major move towards reorganising economic resources came in May 1989 when the Heir Apparent Shaikh Hamad set up the Supreme Council for Planning to co-ordinate social and economic development plans and to maximise the country's economic resources. The council has been briefed to design, implement and monitor all development plans. It will also propose planning legislation and reorganise government administration and services and will co-ordinate relations with other GCC countries. Framework for the council, including a new secretariat, was due to be set up by the end of 1989. Secretary-General Saleh Abu Dawood Al-Mohannadi said in June that the council would start with the development of public administration as a whole, and follow it up with restructuring of various government ministries, departments and public corporations. The government changes appear to be reflected by restructuring at all levels throughout Qatar's industrial and public sector organisations. These changes are expected to have visible effects in the next few years. Jaber Al Marri replaced Shaikh Rashid Bin Owedidah al-Thani as managing director of Qatar General Petroleum Company (QGPC) in January 1988, and Qatar Steel Company came under direct Qatari control when the Japanese management contract ran out in 1989. Qatar National Bank, 50 per cent state owned, completed a major reorganisation of senior management and by mid-1989 was contemplating the same for middle and lower management. In addition, QGPC took over responsibility from the finance and petroleum ministry for all oil exploration and term contracts.

Economy

One striking sign of the changes in Qatar's economic and political thinking was the government's debut in the international financial markets to raise two loans in 1989. The borrowing comes after several years of financial constraints and budget deficits. It has been interpreted as a positive sign that the government is willing to adopt a more flexible approach to economic management.

The first of these was a $400m bullet loan to QGPC arranged in January 1989 to pay for work on the North Field: much of the cash was expected to go to the major international contractors, rather than into the local economy. The $200m Euroloan in mid-1989 raised eyebrows in financial circles when it was announced that London-based Chase Investment Bank had been mandated to raise the loan before the QGPC deal had even been signed. The government would not comment on how the loan would be used. The second loan was welcomed by some commercial observers in Doha as a sign that the government was prepared to relax its traditionally cautious financial stance. Some suggested that the money would go towards new and vitally needed projects, or to finance the budget deficit. Others, that it would be used to reduce a massive backlog of government payments due to contractors, which had severely restrained the local economy, resulting in a slowdown on salaries.

The government has for many years been the prime source of work for contractors, who have been shaken by the payment delays and virtual slowdown on most government projects, apart from the priority areas of defence and the North Field. Contractors' difficulties have had a knock-on effect throughout the economy.

By mid-1989, an estimated QR1,000m ($275m) backlog was owed to contractors. The government's solution was to continue with its austerity programme, calling a halt to any new project spending, except for essential work. The new loan follows heavy government borrowing from local banks to finance the 1989 spending programme, to pay salaries and utility bills, rather than contractors. Three local banks were understood to have provided QR200m ($55m) and a fourth to have provided large ad hoc amounts. In addition, one of the state owned utilities was understood to have lent QR200m, according to the London-based *Middle East Economic Digest*. However, by mid-1989, the project slowdown had improved the situation. Payment delays were reduced to around six to eight months. Construction firms in Doha said in late August that the new government had promised to reduce payment delays to 120 days.

Banks are pressing for changes in monetary policy and financial administration which would encourage more Qatari money to stay in the local banking system. Banks say that lack of sophisticated borrowing and treasury activity has held commercial banking back. Many would like to see the government raising funds through treasury instruments. The Qatar Monetary Agency (QMA – Central Bank) deposit rates and lending rates are fixed at 7 and 9.5 per cent respectively, which has meant that large amounts of private money has left the country in search of US prime rates, which stood at around 11.5 per cent in May 1989. Local banks also want interest rates to be raised, as in Kuwait and Bahrain. In August 1989, QMA announced its first reserve requirements changes for several years. The legal reserve requirement for commercial banks was raised from 3 per cent to 4.5 per cent. The reserve for foreign currency savings and fixed interest accounts that have to be placed with QMA has been raised from 4 per cent to 5 per cent. All four of the conventional commercial banks did well in 1988 with profits up in all except the Commercial Bank of Qatar, due to depreciation costs of setting up a new headquarters. The Qatar National Bank (QNB) profits and assets easily outweighed its rivals. The 50 per cent state owned organisation was still in the process in 1989 of a sweeping restructuring of senior and middle management and completing its computerisation system.

Oil and the economy

Japan remains Qatar's major oil customer. In 1988, exports increased by 33 per cent to $1,040m from $788m in 1987. Gas exports to Japan dropped to $79.5m in 1988 from $81.6m in 1987. In July 1989, London daily *Lloyds List* reported that Japanese trading companies, the major outlet for Qatar's evergreen crude contracts, are steadily increasing their long-term direct deal crude oil import contracts with producing companies in anticipation of the possible tightening of supplies following a production decline by OPEC countries. Idemitsu Kosan is the most active in direct deal imports. In 1989, it signed a long-term contract to import 35,000 b/d.

In 1989, QGPC renewed its evergreen contracts for the year. Total term crude contracts stood at 269,000 b/d; 40,000 b/d is refined at Umm Said and around 20,000 b/d is refined in Singapore. QGPC proposed a

COUNTRY PROFILE

Political system: Qatar has been an independent and sovereign state since 1971. Power rests with the Emir, assisted by the supreme executive body, the Council of Ministers, and a 30-member advisory council.

Population: About a quarter of the estimated 300,000 residents (1987) are Qataris, the rest being largely foreign workers. 80 per cent of the population live in or around Doha.

Main cities/towns: Doha (population estimated 215,000), Umm Said, al-Khor, Dukhan, Wakrah, Madinet al-Shamal.

Language: Arabic is the official language, but English is widely spoken in business circles.

Media
Dailies: Main daily newspapers are *Al Arrayah* (Arabic), *Al-Arab* (Arabic, circulation 25,000) and *The Gulf Times* (English, 10,000). Numerous weeklies and monthlies in Arabic and English.
Broadcasting: Local reception for some services from neighbouring states is quite good. Radio broadcasts 19 hours a day in Arabic and 19 hours a day in English, plus three hours a day folklore and Urdu programmes and one hour in French. Transmission from six transmitters on medium wave, short wave and stereo.
Television: Broadcasts nine hours a day weekdays and 14 hours on Fri (plus extra morning transmissions in summer), with programmes on Channel 37 in English. There were 74,000 sets in 1987.

Domestic economy: Poorly endowed with natural resources apart from oil and gas. Qatar hopes to reduce its heavy dependence on oil exports and develop markets for gas and manufactured goods, mainly in the East and the Third World in general. Suffering from reductions in oil revenues, the government, whose treasury is the driving force behind the economy, enforced cost-cutting measures in 1983, resulting in postponement of some non-essential projects and a significant drop in imports. The 1986 budget was never published at all, so great were the uncertainties at the time. There was, however, a big improvement in prospects after 1986, with oil prices rising and a more optimistic situation in the Gulf by mid-1988. A small current account surplus was recorded in 1987. The volume of activity in the construction sector was expected to increase with the North Field gas development project which was finalised in May 1987; construction and engineering contracts are to go to local companies. Law (3) of 1985 bans non-Qataris from owning any trade, or importing, contracting, working as trade agents or engaging in industry or agriculture, although nationals of GCC states and certain companies are exempt. Foreigners may participate up to 49 per cent in large industrial or agricultural projects and technology-intensive projects essential for economic development.

Employment: Immigrant workers, some of whom have been resident for many years, comprised 75 per cent of the population in 1987. In recent years, the government has taken steps to review the manpower situation, resulting in a reduction in the overall workforce and in the resident population.

External trade
Exports: Exports are dominated by crude oil, which represented 90.0 per cent of all sales in 1987. Main destinations are Japan (60.2% of 1985 crude sales), France (9.5%), Singapore (4.0%), Netherlands Antilles (2.3%), Australia (2.2%) and Netherlands (2.2%).
Imports: Main imports are machinery and transport equipment (36.6% of the 1985 total), manufactured goods (19.7%), other manufactures (14.8%), food and live animals (13.9%) and chemicals (5.0%). Main sources are Japan (17.0% of the 1986 total), UK (16.5%), German FR (9.6%), USA (5.4%), Italy (5.3%) and France (5.2%).

Agriculture: Agricultural production is limited by scarcity of water and unfavourable terrain, and contributes only one per cent to GDP. Major emphasis is placed on educating the population in agricultural techniques, experimenting with new methods of cultivation and developing better marketing structures. One project considered vital to long-term productivity is experimental cultivation of crops on sand using solar energy and sea water. Vegetable production meets local demand in winter and a high proportion of it in summer. Other crops include fruit, dates and fodder crops. Fish catches meet 85 per cent of demand.

Industry: Heavy industry, often developed as state-owned joint ventures, received the bulk of government investment in industry, and now contributes about nine per cent of national revenues, despite shortages of fuel and feedstock. Major companies include: Qatar Fertiliser Company (ammonia and urea), Qatar Fertiliser Factory (organic fertilisers), Qatar Flour Mills Company (flour and bran), Qatar National Plastics Company, Qatar National Cement Company (cement, sulphate-resisting cement and quicklime), Qatar Steel Company (concrete reinforcing steel bars and oxide iron ore fines), and Qatar Petrochemical Company (ethylene, low-density polythylene and sulphur). The government now offers incentives to encourage private sector development of light industry. New industries include the production of intermediate building materials (cement, concrete, moulded aluminium, marble tiles and paving stone), food processing, freezing and packaging, paper products, batteries, paint, plastics, detergents, lubricants, household utensils and furniture.

Hydrocarbons: Oil accounts for about 75 per cent of GDP, 80 per cent of government revenue and 90 per cent of export earnings in a typical year. Oil reserves are relatively small and will be exhausted by the end of the century at present rates, but at present production is limited by OPEC quota systems. All oil operations are nationalised under the Qatar General Petroleum Company (QGPC). The much-delayed first phase of the development of the offshore gas deposit known as the North Field, estimated – at 300-400 trillion cu ft – to be one of the largest single gas deposits in the world, was expected to come onstream in 1992, and would meet local industries' demand for fuel with a maximum output of 800m cu ft, rising later to 2.4bn cu ft per day. The second phase would turn 1.2 bn cu ft of gas per day into liquefied natural gas (LNG), of which some 40-50,000 b/d would be for export (subject to a long-term sales agreement with Japan); but this was unikely to materialise until the late 1990s.

Energy: Heavy industries have been designed to use gas as an alternative to oil as fuel, but reductions in oil production have affected supplies of associated gas. An oil production level of 450,000 b/d is necessary to keep local industries supplied with gas until the North Field comes on stream, but at times actual production fell as low as 100,000 b/d. Solar energy is being developed in conjunction with desalination. The major Al Wusail power and water station project has been postponed.

Membership of international organisations: Arab League, FAO, G-77, GATT, GCC, IAEA, ICAO, IFAD, ILO, IMF, IMO, INTELSAT, INTERPOL, Islamic Conference Organisation, Islamic Development Bank, ITU, NAM, OAPEC, OPEC, UN, UNESCO, UPU, WHO, WIPO, WMO, World Bank.

5 per cent increase in its January 1989 price formula for term contract customers – Oman monthly assessment price plus 10 cents for Dukhan and plus 5 cents for Marine. This was continued into March 1989, leaving actual prices for Qatar crudes at $15.2 and 14.92 a barrel in January and February 1989 for Dukhan and $15.15 and $14.87 for Marine in the same period, against an average price of $13.63 for Dukhan and $13.38 for Marine in 1988.

The North Field

Qatar is pinning its economic, social and political hopes on the successful development of the North Field Gas Scheme which is expected to come on stream by 1991.

The North Field, discovered in 1972, was the largest discovery of offshore non-associated gas. It lies in relatively shallow water to the north east of Qatar and covers approximately 5,900 sq km. The field's proven reserves are put at a minimum of 100m cu ft and probable reserves are estimated at 300m cu ft. The field could have a productive life of between 115 and 200 years, compared to around 40 years for known oil resources. In addition to associated gas, the North Field gives Qatar about 4.5 per cent of the world's total gas reserves. The North Field is Qatar's main hope of weaning itself off oil dependency.

By mid-1989 all 23 major development packages of the field were awarded and work is well under way on phase one. Qatari planners are turning their attention to the next phases of the field's development. Phase one will set up the field's infrastructure and cater for domestic requirements. About 300m cu ft of gas a day will be surplus during phase one. A number of projects to utilise the gas are under discussion. Later stages will involve the full exploitation of the field, and set up a variety of downstream and industrial projects. A further phase will set up exports of around 6m tonnes of LNG a year. Once phase one of the North Field development programme is complete, it will have a production capacity of 8,000 cu ft of gas a day, 270,000 tonnes a year of propane, 190,000 tonnes a year of butane, 110,000 tonnes a year of natural gasoline, 730,000 tonnes a year of light naptha and 330,000 tonnes a year of heavy naphtha; around 400 cu ft a day of dry gas will meet local demand. The government is paying for the first phase by the sale of 40,000 b/d of crude set aside from government reserves and by entering the international markets for a $400m loan.

In May 1987, Qatar General Petroleum Company (QGPC) appointed US Bechtel Corporation and France's Technip as managing consultants for the project. By mid-1988, both companies had moved staff to Doha to supervise the construction work. Total cost of phase one was estimated at $1,300m. The first phase of the project centres on a cluster of six production platforms 80 km offshore. These comprise two wellhead and utility platforms to provide power, air and water; riser/treatment platforms to separate and dry the gas and accommodation and flare platforms. After separation, the gas and liquid condensates will be piped ashore to Ras Laffan or Umm Said, where pipelines will take them 130 km to a new natural gas liquids (NGL) processing plant, NGL-3. The condensates will be fractionated into light naphtha for export and heavy naphtha for use by the National Oil Distribution Company (Nodco). Raw NGL from the plant will be turned into ethane-rich gas, propane, butane and natural gasoline at the NGL plant.

Bidding for all of the North Field contracts was intense; the more technical and complicated jobs have gone to well known outside contractors. By June 1989, all of the 23 major engineering and construction packages had been awarded. A total of 240 purchase orders had been sent out for building materials and equipment supplies for the major North Field contractors. In August 1988, work started on a two-year contract to sink the first of the field's 16 wells, after the successful completion of two wellhead platforms in Package One by National Petroleum Construction Company of Abu Dhabi. By March 1989, 65,000 feet had already been drilled. Drilling is expected to be completed and production tested by July 1990 with commissioning starting by December 1990 or early in 1991. Work on the six production platforms, a major component of the project, was well under way by first quarter 1989 and has been shared between several firms. The US firm Brown and Root have the $56m Package Two to supply and will fabricate jackets and piles for the production platforms and supply $24.6m worth of work on Package 19 for offshore installation of jackets and topsides. Semabang of Singapore had the $40m contract to build the accommodation platforms. The US firm McDermott has the $25.6m contract to provide the topsides for the utilities platforms with local Qatar National Navigation & Transport Company. Belleli Saudi Heavy Industries has the $53m contract to build the treatment and riser topsides. Local firm Mansal Offshore has the $15m package to provide the accommodation platform to support the offshore hook-up and commissioning. Italy's Saipem has the lucrative onshore and offshore pipeline work worth a total $101m. Another Italian firm, Fochi, won the $79m contract to build NGL-3, and the $7m storage tank contract. Local contractors gained only a small share of the contracts on the North Field. The most successful contractor with three packages was Gulf Housing Corporation which won $11m worth of work. But according to QGPC sources, local firms stand a good chance of picking up subcontracts and supply orders from the main contractor.

Exports from the North Field

Qatar is hoping that sales of surplus North Field gas will provide a boost to its economy in the 1990s. In 1984 the government set up the Qatar Liquefied Gas Company (Qatargas) to develop and operate a liquefaction plant and market the liquefied gas overseas. The UK's British Petroleum Company, France's Total CFP and Japan's Marubeni each took out a 7.5 per cent shareholding with the government as main shareholder. In May 1989, another Japanese company, Mitsui Corporation, took out a 7.5 per cent stake in Qatargas, also with the brief of helping to find export partners. Qatargas meanwhile is busily organising customers to buy surplus gas from the North Field once it has fulfilled domestic requirements and the new industries that are planned. Talks had already begun as far back as 1986 when the Indian Government submitted a formal offer to buy 3.5 to 5m tonnes a year of LNG from Qatar in the mid-1990s. Qatargas has already proposed a project for the initial production and export of 4m tonnes of LNG, with two 2m tonne capacity gas trains which would be capable of expansion. In addition, the project would require seven 125,000 cu metre LNG tankers to deliver 4m tonnes a year of gas to potential markets.

Another industrial scheme being considered is the possibility of building a 350 km undersea pipeline from Qatar to Jebel Ali in Dubai which would transport about 800m cu ft of gas a day. Dubai is trying to attract more industrial and manufacturing companies to Jebel Ali and needs an assured power supply.

Oil – still an economic mainstay

Qatar's first oil well was spudded in 1938 in Dukhan and oil is still the mainstay of the country's economy. Offshore exploration began in earnest in 1952. By 1988, proven oil reserves were estimated at around 4,000m barrels. Qatar's huge Dukhan field is the only onshore reservoir, with an estimated 2,000m barrels. It has a production capacity of around 250,000 b/d, although present output is well below this level.

Offshore reserves are concentrated in three fields forming a triangle to the east of the Qatari peninsula. Bul Hanine is the largest with recoverable reserves put at 1,100m barrels. Idd Al Shargi, discovered in 1960, has around 350m barrels, and Maydan Mahzam has about 830m recoverable barrels. Combined production is put at about 320,000 b/d, but currently the three produce about half this amount. Qatar has to keep oil production at a relatively high level whatever the market climate to maintain the flow of associated gas which powers the Umm Said industrial complex, at least until the North Field gas becomes available. The Bunduq Field covers both Qatari and Abu Dhabi's territorial waters and output was expected to reach 30,000 b/d during 1988.

With the gradual lessening of restraints imposed by low oil prices, Qatar is moving ahead with plans to refurbish its oldest offshore oil platforms and improve recovery techniques. The UK's Ewbank International Consultants won a two-year design and consultancy contract in the second quarter of 1989. The programme, which concentrates on the topside pipework and living quarters, is aimed at increasing the platform's working life by about 30 years. Refurbishment is due to start in 1990 for completion in 1994: QGPC is doing most of the front-end design work.

In February 1989, a new oil product storage and distribution centre at Abu Hamour, just outside Doha, came on stream to supply the local market. It was built by South Korea's Hyundai Engineering & Construction Company, with a pipeline to Umm Said refining complex, equipped with a 430 cu metre an hour pumping station. Work on two other pipelines, linking Umm Said refineries to the oil export terminal in the nearby port, was completed in mid-1988.

Qatar exports about 46,500 b/d of refined products, nearly 75 per cent of the output of its two refineries. The first refinery was built in the 1960s and today has a 12,000 b/d capacity. A much larger one was commissioned in 1983 with a 50,000 b/d capacity, exporting about 46,500 b/d. They produce all types of petroleum products including light gasoil, fuel oil and gasoline. The National Oil Distributing Company (Nodco), which runs the refineries, is currently working on expansion plans, including installing a new isomerisation unit at the larger refinery as part of plans to produce lead-free gasoline. Plans also include upgrading or expanding the crude distillation unit and the naphtha and kerosene hydro-treating plant to make more efficient use of feedstock. The project is estimated to cost around $70m, and is expected to be given high priority by the Qatar Government. It should increase the level of income by 20 per cent. Output at the Umm Said refining complex in 1988 increased by 15 per cent to 12.1m barrels or 33,150 b/d, compared with 1987, when about 7.6m barrels, or 20,800 b/d, were exported. This allowed Qatar to establish a strategic reserve equivalent to two months' consumption. Nodco's sales increased by 18 per cent to QR650m ($178.6m).

On the exploration front, no new oil reserves have been found in Qatar for some time. The UK's British Petroleum Corporation announced in 1989 that results from its exploratory drilling in Qatar had proved unsatisfactory, despite positive seismic studies a few years earlier. However, two other companies were actively exploring. France's Elf Aquitaine was awarded its first offshore exploration licence in Qatar in January 1989. Under a five-year agreement, Elf will explore the 2,800 sq km block 6 permit in the highly productive zone of the east coast. Elf says it will undertake a 5,000 km seismic survey and drill five exploration wells in the next five years. The US firm Amoco Corporation began the second phase of its exploration programme. Amoco signed a 25-year production sharing agreement in 1986. Three exploration wells sunk in 1988 were dry. Four new wells will be drilled in block two, the largest onshore block covering all of the Qatar peninsula, by mid-1990.

Industry takes the lead

Qatar's industrial base is concentrated at Umm Said, 30 km from Doha; its activities account for roughly 10 per cent of annual revenue. Development has concentrated on downstream activities. It also has some basic industries which were set up in the 1970s: the Qatar Steel Company (QASCO), the Qatar Fertiliser Company (QAFCO), Qatar Petrochemical Company (QAPCO) and the Qatar National Cement Company (QNCC). Light and medium scale industries have also been built up in the past decade; for example, furniture and food processing. Qatar wants to expand its industrial base, using North Field gas to increase non-oil export revenue. Industry has been given a higher profile and increased responsibility with the appointment of a new industry minister in July 1989. Previously the ministry concentrated on monitoring the performance of established industries and organising loans for small and medium scale ventures. In August 1989, the ministry took over the responsibilities of the industrial development technical centre (IDTC) which handled Qatar's industrial planning. Plans now reach far into the late 1990s. When Umm Said is fully utilised, the government hopes to create a second industrial base at Ras Qirtas, an onshore deposit of non-associated gas in the north east, close to Ras Laffan where North Field gas will be brought ashore. The new industrial development area would have industries such as ammonia, urea and methanol NGL and LNG production; a port; marine terminals; roads and public services. However, such a scheme would cost an immense amount to develop and would be dependent on attracting joint venture partners.

Several projects to utilise the North Field's gas are under discussion. Strong world demand for aluminium, which rose to record levels in 1989 for the third successive year, fuelled Qatar's plans to develop its own aluminium industry. Demand in non-socialist countries has been put at 13,637,000 tonnes in 1987, and is forecast to increase to 14,650,000 tonnes by 1990. The Doha-based Gulf Organisation of Industrial Consulting (GOIC) prepared feasibility studies for the scheme in 1987 and reviewed proposals from several international firms. However, the government shelved the plans until the first phase development of the North Field got under way. In 1988, a UK group known as the London Consortium put forward plans to build a 240,000 tonne-a-year smelter which would use 100-115m cu ft of gas a day. In November 1988, the consortium signed an agreement with Qatar's executive aluminium committee to set up the Doha Aluminium Company (Dohal). The consortium had originally proposed to build the smelter in the UAE's Umm Al Qaiwain, but transferred to Qatar because of the abundant supplies of gas. Plans for setting up the smelter were, therefore, very advanced and the consortium hoped to set up the project more quickly than would otherwise be possible. The project, expected to cost an estimated $1,200m, was to be self-financing with pre-sold 12-year offtake agreements with Dohal's shareholders. The company had a planned equity of $300m. The 'A' group of shareholders, instrumental in setting up the project, were to take a $60m shareholding. Lloyds International Corporation of New York was to lead a syndicated loan of up to $500m. Included in the plans was a 500-MW power station and a 25m gallon a day desalination plant; commodities urgently needed in Qatar. Dohal would also sell water and electricity in return for gas. However, by mid-1989 no formal contract had been signed and at least four other international groups emerged to bid for the smelter project. These included Norway's Norsk Hydro, a European consortium led by the UK's Davy McKee, France's Péchiney, and West Germany's Ferrostaal. Qatar is poised to enter the

BUSINESS GUIDE

Time: GMT +3 hrs.

Climate: Desert climate with extremely hot and humid summers, when temperatures can reach 44°C from Jul to Sep, and mild winters with occasional rainfall.

Entry requirements
Passport: Required by all except certain seamen.
Visa: Not required by nationals of some Arab countries and those born and resident in the UK, for stays of up to 30 days. Others must obtain visas from Qatar representatives overseas before arrival, if staying for more than 72 hours. In some cases a 'No Objection Certificate' (NOC) applies instead of a visa; check well before departure.
Prohibited entry: Nationals of Israel and South Africa are prohibited entry, as are holders of passports with evidence of travel in these countries.
Currency: There are no exchange restrictions.
Customs: Personal effects are duty-free. Import of pork and pork products, goods from Israel or South Africa, cultured pearls, and obscene or seditious literature is forbidden. The Israeli boycott is firmly enforced. Foodstuffs must be labelled in Arabic. Importers must register with the Controller of Companies and appear on the Chamber of Commerce's register of importers. Certain goods (firearms, ammunition, drugs and alcohol) may only be imported under licence.
Income tax: There is no personal income tax.

Health precautions
Mandatory: Vaccination certificates against yellow fever and cholera if travelling from infected areas.
Advisable: TAB vaccination.

Representation overseas: Abu Dhabi, Algiers, Amman, Baghdad, Beirut, Bombay, Bonn, Brussels, Cairo, Damascus, Dhaka, Geneva, Islamabad, Jeddah, Khartoum, Kuwait City, London, Madrid, Mogadishu, Muscat, New Delhi, New York, Nouakchott, Paris, Rabat, Tehran, Tokyo, Tripoli, Tunis, Vienna, Washington DC.

Air access
National airline: Gulf Air (jointly owned with Bahrain, Oman and UAE).
Other airlines: Air France, Air India, Alyemda, Balkan, Biman Bangladesh, British Airways, Egyptair, Iran Air, KLM, Kuwait Airways, MEA, PIA, Royal Jordanian, Saudia, Sudan Airways, Syrian Arab, TWA, Yemenia.
International airport: Doha International (Code: DOH), 6.5 km from city, with restaurant, buffet, bank, hotel reservations, shops, car hire.

Surface access
Main ports: Umm Said (bulk ammonia, grain, fertilisers, tankers), Doha (containers), Halul Island (tankers).
Overland access: It is possible to enter by road from Saudi Arabia and the UAE.

Hotels: It is not generally essential to book hotels in advance. A service charge of 10 per cent and a five per cent government tax are usually added to the bill.

Electricity supply: 220/240 V AC, with three-pin flat plug fittings most common.

Credit cards: Major credit cards are accepted.

Car hire: If hiring for more than seven days, it is necessary to obtain a 90-day local licence – international or foreign licences are not acceptable. For this, a foreign or international licence, a letter from a local sponsor and passport must be produced within a week of arriving, and a test on road signs may be required. Third-party insurance is compulsory. Speed limits are 60 kph in cities and 100 kph on highways. Driving is on the right-hand side of the road.

City transport
Taxis: Taxis have black-on-yellow number plates with metered fares. They can be hired on a time basis, with a set hourly rate.

National transport: There is a good, extensive road network, but no rail or bus services.

Public holidays
Fixed dates: 22 Feb (Anniversary of the Accession of the Amir), 3 Sep (Independence Day).
Variable dates: Eid al-Fitr, Eid al-Adha, Hijra New Year's Day.

Working hours: Fri is the weekly holiday, and oil company offices are also closed on Thu afternoons. Hours may be restricted during Ramadan or very hot weather.
Business: 0700/0730 – 1200/1300, some also 1530/1600 – 1800 (closed Fri).
Government: (Sat-Thu) 0600 – 1300.
Banking: 0730 – 1100 (Thu 0730 – 1030, closed Fri).

Social customs: It is not advisable to wear shorts or sleeveless dresses in public. Avoid eating, drinking or smoking in front of Muslims during daylight hours in Ramadan; displaying the soles of your feet; and using the term 'Mohammedan'. Under some circumstances, photography may be impolite or illegal, so exercise caution.

Telecommunications: International communications are via satellite, microwave radio (to Saudi Arabia) and submarine cable links (with Bahrain and the UAE). Qatar is building a ground station as part of the Arabsat system.
Telephone: Dialling code for Qatar: IDD access code + 974 followed by subscriber's number (no area code is necessary). International direct dialling. There were 114,000 telephones in 1987. Car telephone services are available.
Telex and telegram: Available 0600 – 2300 at Cable & Wireless office, and at good hotels.
Other: Several private courier services operate in Doha.

Banking
Central bank: Qatar Monetary Agency.
Main centre: Doha.
Head offices: Alahli Bank of Qatar, Commercial Bank of Qatar, Doha Bank, Qatar Islamic Bank, Qatar National Bank.
Branches: Arab Bank, Bank of Oman, Banque Paribas, British Bank of the Middle East, Standard Chartered Bank, Citibank, Grindlays Bank, United Bank, Bank Saderat Iran, United Bank Ltd.

Advertising: Publicity material for general distribution should preferably be in Arabic, although correspondence and technical literature is acceptable in English. Advertising is available in the press, on commercial TV and in cinemas. Outdoor advertising and direct mail are available subject to prevailing regulations. There are several advertising agencies in Doha. The promotion of cigarettes or any tobacco products is banned.

Weights and measures: Metric system; other weights are still in use, however.

Representation in capital: Algeria, Bangladesh, Egypt, France, Gabon, German FR, India, Iran, Iraq, Japan, Jordan, Rep of Korea, Kuwait, Lebanon, Mauritania, Morocco, Oman, Pakistan, Saudi Arabia, Somalia, Sudan, Syria, Tunisia, Turkey, UK, USA, Yemen AR.

aluminium market at a time when other Gulf countries are in an expansionary mood. Cheap gas supplies, low energy costs and the low amount of labour required gives them a comparative advantage for producing aluminium. Péchiney, for example, is active in new aluminium ventures in Bahrain, Saudi Arabia and also in Iraq, whilst the UAE is also working on expansion plans. Many of the schemes will be operational before Qatar's plant gets under way which may mean a tougher market for Qatar's aluminium.

Apart from the aluminium smelter, several other industrial ventures are on the drawing board and the government appears to be open to offers from foreign joint ventures. They include an expansion of the petrochemicals and fertiliser plants. QAPCO is also studying plans to build a 1,500 tonnes a day ammonia plant, according to Doha commercial sources. The plant would cost over $200m. QAPCO is 75 per cent owned by QGPC and 25 per cent by Norway's Norsk Hydro. Norsk's management agreement with QGPC will be reviewed by 1990. The company wants to continue the association and has expressed an interest in the ammonia project and a proposed methanol plant. QAFCO produced 728,000 tonnes of ammonia and 780,000 tonnes of urea in 1988 – an increase of around 6 per cent. Most is exported to China and India, although the company is looking hopefully at the Far East. Operating profits for 1988 were QR181m. QGPC says it is looking at various methods of expanding ethylene and polyethelene production and is talking to companies from Europe, Japan and the USA. The US firm Phillips Petroleum Corporation was considering setting up a joint venture to build a 450,000 tonne a year ethylene production complex to export ethylene and polyethylene. QGPC was also considering proposals to expand polyethylene production at QAPCO. The company has an ethylene surplus at present. It produced 262,687 tonnes in 1987. France's Technip was also undertaking a study for QAPCO on a high density polyethylene plant.

Australia's Pennant Holdings headed a consortium to set up a $300m ferro-alloy production plant to produce new materials for the steel industry and for export. The 230,000 tonne a year factory will produce ferro manganese, silico manganese and ferro chrome. It will also have a 180 MW power station on site. QASCO, a company which has run at a loss since it began production in 1978, made its first profit in 1988 and said that projections for 1989 were similarly optimistic. It attributes its results to the 15-20 per cent increase in steel bar prices and the fall in steel sales to the Gulf by Japanese, Brazilian and Korean producers. The company does not publish financial results, but in 1987 it introduced a wide-ranging cost cutting programme to halt heavy losses. Qatar took sole management of the plant in January 1989 after the management agreement with main shareholder Japan's Kobe Steel ran out. Production in 1988 was 6 per cent up at 533,000 tonnes of steel bars; well above the original design capacity of 350,000 tonnes a year. By mid-1989, the company said it had no major production expansion plans.

Qatar National Cement Company's (QNCC) production was 7 per cent up in 1988 to 449,800 tonnes, although actual sales reached only 324,590 tonnes, mainly due to the lack of new construction projects. The company satisfies about 90 per cent of local demand and says it has no major development plans for 1990.

Foreign policy

Qatar's small size limits its influence on the world political stage. However, as a regional player its membership of the major Gulf and Arab organisations, the Gulf Co-operation Council, the Arab Monetary fund and OPEC, lend Qatar a much louder voice in international affairs. Qatar's consistent foreign policy tends to follow along the same lines as its more influential neighbour Saudi Arabia. It supports the traditional Gulf and Arab causes and regularly speaks out on the Palestinian issue; in common with the other Gulf states, Qatar was among the first to recognise the recently declared independent Palestinian state.

The Gulf cease-fire has eased tensions brought about by eight years on the fringes of a major conflict. Like the UAE, Qatar is keen to improve trading relations with the rest of the world. The high proportion of Arab and Asian expatriates ensures that relations with those countries are continuous. There was a noticeable increase in contact with Pakistan, as Pakistani firms tried to attract investment in housing and industry. Relations with its major trading partners in Japan and Western Europe remain cordial. Qatar is likely to increase contact with these countries as it attempts to attract joint industrial ventures. The country is also hoping to benefit from the large increase in tourism to the Gulf since the cease-fire and has sent several tourism missions to Europe, particularly West Germany.

Qatar continued to improve relations with the Eastern Bloc in 1989, establishing links with Poland and Hungary for example, following the establishment of ties with the two major communist states, the Soviet Union and China, in 1988. At the same time, relations with the USA became slightly cooler over an arms sales dispute.

Outlook

1989 saw three hopeful signs for Qatar's handling of its social, political and economic affairs in the next decade. Qatar is a stable country and the al-Thani family has not relinquished its control. However, the first major cabinet change brought an injection of new and youthful talents which promised a more modern and energetic approach to government. At the same time, the decision to set up a centralised planning body to co-ordinate resources suggests that economic decisions might be more efficient in future. The government's decision to finally borrow to finance its spending and debts after a long period of extreme financial caution has raised the hopes of business people in Qatar.

Qatar has considerable hydrocarbon resources. It suffered from the falling oil revenues of the latter part of the 1980s and will continue to remain dependent on oil for many years. Towards the end of 1989, the prospect of abundant gas supplies from the North Field had already begun to stimulate interest in Qatar. By 1991, when the gas begins to come on stream, the pace of change will increase. However, Qatar's hydrocarbon resources will need careful, efficient management and considerable investment if the country is to benefit fully.

BUSINESS DIRECTORY

Useful telephone numbers
Emergency (all services): 999
International operator: 150
Directory enquiries: 180
International enquiries: 190

Hotels: Doha
Gulf, P.O. Box: 1911 (tel: 432432; tx: 4250 glfhtl; fax: 418784).

New Capital, Mushairib Street, P.O. Box: 833 (tel: 445452; tx: 4741 newcap).

New Doha Palace, P.O. Box: 710 (tel: 26131/4; tx: 4263 newdph).

Oasis, Ras Abboud Road, P.O. Box: 717 (tel: 424424; tx: 4212 oasis).

Qatar International, P.O. Box: 1188 (tel: 321761; tx: 4648 habib).

Qatar Palace, Al-Asmakh Street, P.O. Box: 1101 (tel: 321515; tx: 4649 qphotl).

Ramada Renaissance, P.O. Box: 1768 (tel: 417417; tx: 4664 ramda; fax: 410941).

Sheraton, P.O. Box: 6000 (tel: 833833; tx: 5000 dosher; fax: 832323 EXT 4422).

Shezan, P.O. Box: 5616 (tel: 865225; tx: 4460 shezan).

Sofitel Doha Palace, Musheirib Street, P.O. Box: 7566 (tel: 435222; tx: 5151 softel; fax: 439186).

Car hire: Doha
Ahmed Transport Co: PO Box 2376 (tel: 428896; tx:4436).

Al Muftah Rent-A-Car: PO Box 1316 (tel: 328100, 426649; tx: 4056 rntcar).

Avis: PO Box 2515 (tel: 447766; tx: 4892 engulf).

Budget: Rayyan Road, PO Box 7552 (tel: 419500; tx: 4544 budget).

Europcar (Mannai Car Rental): Rayyan Road, PO Box 76 (tel: 442402; tx: 4208).

Business advice and information
Ministry of Industry and Agriculture, PO Box 1966, Doha (tel: 412525; tx: 4096 sinzra).

Ministry of Economy and Commerce, PO Box 1968, Doha (tel: 434888; tx: 4488 ecom).

Ministry of Finance and Petroleum, PO Box 83, Doha (tel: 414436; tx: 4233 qatfin, 4315 dohmfp; fax: 413617).

Qatar Chamber of Commerce, PO Box 402, Doha (tel: 423677, 425131; tx: 4078 tijara; fax: 324338).

Qatar Fertiliser Company, PO Box 50001, Umm Said (tel: 770252; tx: 4215 qafco).

Qatar General Petroleum Corporation, PO Box 3212, Doha (tel: 491491; tx: 4343 petcor).

Qatar Petrochemical Company, PO Box 756, Doha (tel: 321105; tx: 4361 qapco).

Qatar Steel Company Ltd, PO Box 10090, Umm Said (tel: 770011, 321000; tx: 4606 qasco).

Banking
Qatar Monetary Agency (central bank), PO Box 1234, Doha (tel: 413456; tx: 4335 QATMON DH; fax: 413650).

Al Ahli Bank of Qatar, PO Box 248, Doha (tel: 436633; tx: 5126 ALAHLIDH; fax: 414014).

Commercial Bank of Qatar, PO Box 3232, Doha (tel: 431010; tx: 4351 tejari dh; fax: 438182).

Doha Bank, PO Box 3818, Doha (tel: 435444; tx: 4534 dohbnk dh; fax: 416631).

Qatar Islamic Bank, PO Box 559, Doha (tel: 438000; tx: 5177 islami dh; fax: 412700).

Qatar National Bank, PO Box 1002, Doha (tel: 413511; tx: 4636 qbkgm dh; fax: 413753).

Travel information
Department of Tourism and Antiquities, PO Box 2777, Doha (tel: 327286; tx: 4229 infor).

Doha International Airport information (tel: 438111).

Gulf Air reservations (tel: 455444).

Ministry of Information, PO Box 1836, Doha (tel: 831333; tx: 4040 teevee, 4229 infor).

Romania

George Fodor

1989 reached a dramatic conclusion for Romania, with the trial and execution on 25 December of President Ceausescu and his wife, Elena. In stark contrast to the peaceful democratisation of Poland, Hungary and Czechoslovakia, Romania witnessed a bitter, bloody revolution as pro-Ceausescu forces – notably the hated *Securitatae* (secret police) – fought a last ditch battle to retain power. Various estimates were produced for the death toll, but in early 1990 no accurate figures were available.

Romania's revolution was sparked off, perhaps predictably, by the tensions among the ethnic Hungarian minority in the north-western region of Transylvania. When Hungarian preacher Laszlo Tibor spoke out in Timisoara in mid-December against the excesses of the Ceausescu regime, the authorities found out that not only the Hungarian community, but also the Romanians rallied to his defence. The government's efforts to crush this open challenge to its authority resulted in pitched street battles and heavy civilian casualties.

Revolutionary fever soon spread throughout the country. On 20 December, the day after his (early) return from an official visit to Iran, President Ceausescu discovered that a government-organised 'popular' rally in Bucharest turned in to an open expression of opposition. Attempts to quell the growing resistance resulted in further bloodshed. Ceausescu's attempts to flee were thwarted, and after a summary televised trial he and his wife were executed by firing squad.

Pending general elections in April 1990, the provisional government of the 'National Salvation Front' is headed by President Iou Iliescu and Prime Minister Petre Roman. A confused attempt to outlaw the Communist Party in mid-January 1990 was later reversed although the death penalty and the hated Securitatae were quickly abolished.

On 12 March 1989 President Ceausescu announced that Romania had finished repaying its foreign debt. Meetings were held throughout the country to hail the event and a law was passed by the National Assembly to ban the taking of further foreign credits in the future. It was stressed throughout the celebrations that for the first time in its history Romania owed nothing to anyone and was in a position to resist all foreign pressure or influence. This was indeed largely the case and the Romanian leadership made full use of the freedom, which indebted countries do not have, to reject all criticism from abroad and follow its own course. That course, as preparations for the November 1989 communist party congress showed, was to carry on with the same policies as before, notwithstanding what was happening in other Soviet bloc countries.

Debt repayment

It had been widely expected that the repayment of Romania's foreign debt in hard currency would be completed by the end of 1989. The date was probably brought forward and the announcement made in March to make a political point. In February, at the concluding meeting of the Vienna Conference on European Security and Co-operation, relations between Romania and the West became very strained as Bucharest announced that it would sign the final common declaration but it would not abide by some of its crucial provisions concerning human rights. Romania was told in very strong terms that its position was unacceptable. The announcement shortly afterwards that the debt had been paid off was a signal to the West that it had no real leverage over Bucharest, and to domestic public opinion not to expect Western pressure to be effective.

It was later suggested that Ceausescu's announcement had been premature, as one or more repayment instalments were still outstanding. Whether true or not, the fact remained that unlike other countries which got heavily in to debt in the 1970s, Romania was now free from the economic and political constraints which they have to bear. Over the seven years spent in efforts to repay the debt, imports were cut to the bone and it is generally assumed that Romanian industry is extremely short of spare parts and in great need of modernisation. But at the same time a considerable export capacity has been built up, only partly at the expense of the domestic market, so that the country now commands an annual foreign currency revenue which could be as high as $2bn. 1989 was the year when detailed plans for the next five years and long-term plans for the next 20 were to be drawn up and submitted to the fourteenth congress of the ruling communist party. The government was in an ideal position to consider the global economic situation and establish new priorities.

Early in July the draft of the main policy document of the November party congress was published. The so-called Congress Theses proved to be to a great extent an ideological manifesto setting out President Ceausescu's opposition to the attempts going on in other countries of the Soviet bloc to reform the communist system. Only one of its 19 chapters dealt with economic development, giving a very broad outline of the leadership's further intentions. It gave, however, a clear indication that no radical change was contemplated in economic strategy. The one bright spot was a promise that by 1995 the housing problem in urban areas would be solved. This was a realistic prospect : housing construction on a large scale had been going on throughout the past ten years, some of it as a complement to industrial developments which eventually had to be cancelled. This resulted in places in a surplus of housing and in quite a few Romanian

KEY FACTS

Official title: The Socialist Republic of Romania

Head of state: Ion Iliescu (acting President)

Head of government: Petre Roman (acting Prime Minister)

Ruling party: Council for National Salvation appointed late Dec 1989; elections planned for April 1990.

Capital: Bucharest

Official Languages: Romanian

Currency: Leu (plural Lei) = 100 bani

Exchange rate: Lei 8.77 per $ (Dec 1989)

Area: 237,500 sq km

Population: 23.1m (1988)

GDP per capita: Lei 34,900 (1987)

GDP real growth: 3.6% (1987)

Labour force: 10.7m (1987)

Trade balance: $1.9bn (1986)

Foreign debt: $8.0bn (1986)

towns the previously grave housing shortage came to an end. But the rest of the programme made depressing reading. The bulk of investment in the next five-year plan was to go to what was described as the peak industrial sectors, which include the giant chemical and steel industries that have already drained so much of the country's resources. There was no mention in the Theses of any boost to the consumer goods industry, to compensate for the massive shortages experienced on the domestic market in the 1980s. After a decade of severe food shortages investment in agriculture was to be only 15 per cent of the total investment. Industrial production was to increase by 30 per cent until 1995, but the average wage by less than 6 per cent. This was an increase of 1 per cent per annum in a country with significant hidden inflation. All in all it looked like a programme aimed at carrying on with the industrial build-up as originally envisaged in the 1970s, coupled with a milder version of the austerity of the 1980s. It is probably true to say that the main consideration of the Romanian leadership when it decided to repay the foreign debt was to regain the freedom to carry on with this programme rather than admit that something was wrong with it. This would have been inevitable had they taken the course of countries like Hungary, Poland or Yugoslavia.

KEY INDICATORS

	Unit	1984	1985	1986	1987	1988
Population	m	22.7	22.7	22.8	22.9	23.1
Gross national product (GNP)	lei bn	843.2	873.6	917.5	771.8	–
GNP per capita	lei	37,145	38,485	39,547	34,900	–
GNP real growth	%	7.2	5.5	6.1	4.6	3.3
Consumer prices	1985=100	100.4	100.0	99.9	–	–
Private consumption	lei bn	453.2	490.7	509.8	–	–
Labour productivity	% growth	7.1	5.0	–	–	–
Average earnings index	1980=100	122.6	123.3	127.6	135.6	–
Agricultural production	% growth	13.1	–5.7	14.8	–1.92	–6.7
Industrial production	1985=100	94.2	100.0	106.4	–	–
Exports	$bn	12.6	12.2	12.5	–	–
Machinery & equipment	$bn	4.0	3.6	4.6	–	–
Fuel, minerals, metals	$bn	3.5	3.6	2.8	–	–
Industrial consumer goods	$bn	2.0	2.1	2.1	–	–
Imports	$bn	10.3	10.4	10.6	–	–
Machinery & equipment	$bn	2.6	2.6	3.1	–	–
Fuels, minerals, metals	$bn	5.4	5.6	5.2	–	–
Balance of trade	$bn	2.3	1.4	1.9	–	–
Current account	$m	1,719	1,239	1,489	–	–
External debt	$bn	7.2	6.6	8.0	–	–
Exchange rate	lei per $	21.3	17.1	16.2	14.6	14.3

* estimated figure

Technology and trade

There were a number of major problems which this strategy failed to address. One of them was the imbalance between oversized industries – the petro-chemical industry had grown to provide one-third of the country's total industrial output – and inadequate energy and raw material resources. This was one of the main factors that precipitated the debt crisis in the first place and subsequently caused the chronic energy crisis experienced throughout the 1980s. There was a major imbalance between the needs of the growing urban population and the poor state of agriculture. A fresh problem was the urgent need for the import of advanced technology to resupply existing industry and expand it in new directions. As the Romanian leadership had decided not to resort to foreign credits any more, and a law had been passed to ban foreign credits which had no parallel in Europe except in Albania, straight imports of technology were out. In the Theses for the party congress it was announced that Mrs Elena Ceausescu, the president's wife and the country's head scientist, had been put in charge of a research programme that would solve the problem with the country's own means.

Another unanswered question was what shape Romania's foreign economic relations were now to take. The Romanian government had started an advertising campaign in the West to persuade companies to make use of its existing production capacities in joint ventures. Romania built huge industrial capacities with the money borrowed in the 1970s and now that the debt has been repaid would like to put them to use. The trouble is that the money was often invested wrongly, so that many of these newly built industries produce goods which are obsolete, expensive and difficult to sell. Their inability to bring a proper return on investment caused Romania's 1982 debt crisis in the first place. Now, 10 to 15 years later they would probably appear even less attractive to foreign investors.

Difficulties in trade relations with the West have forced Romania back into the Soviet bloc and especially into increased trade and co-operation with Russia. But economic co-operation within the bloc is hampered by growing disagreements among member states. Some of them are demanding radical reforms in mutual economic relations, such as a convertible currency and direct market-to-market links, while others, including Romania, want to maintain central planning and state-to-state trade agreements. Differences have been so acute that some meetings of CMEA's policy-making bodies have had to be postponed and there has been open talk about groups of member-countries going their separate ways. Romania, together with East Germany, emerged as the most militant among the anti-reform group.

The last of the true Bolsheviks

In economic as well as political terms, Ceausescu had become the standard-bearer of conservatism in the Soviet bloc. A speech in June 1989 to the communist party's central committee contained a complete exposition of his version of socialism. In the economy he stands for the abolition of all private ownership. He maintains that large-scale units in both industry and agriculture are not only preferable to small- and medium-sized ones, but represent the end-result of a law of economic development. In social policies he is egalitarian and stands for maintaining a narrow range of salaries and eliminating other sources of income. In political terms he is strictly for the one-party system and against any separation between party and state. This is the way socialism was practised in the USSR at the time of Stalin.

In 1988-89 Romania took an increasingly active position in the rift developing inside the Warsaw Pact concerning reform. Up to a point Ceausescu was continuing his traditional strategy, in dealings with the Warsaw Pact, of raising issues which other member-states support but are reluctant to raise themselves. In 1989 the most vocal condemnations of the idea of a multi-party system, of the market economy, or of President Bush's openings to the Eastern bloc, came from Bucharest. There was, however, no significant deterioration in relations with the Soviet Union. In June 1988 President Ceausescu went on an official visit to Moscow, taking along his wife Elena who sat next to him during the talks with Gorbachev. It was the first time the Kremlin had acknowledged her position as second-in-command in the political leadership. The talks revealed considerable differences between the two sides and Gorbachev, in an obvious reference to Romania, said in a luncheon speech that the international reputation of socialism could be spoiled by the failure of one country alone. But the new line of the Kremlin in relations with allies is to express opinions and possibly advice, but not to interfere in the way leaders of other countries conduct their affairs. This line prevailed in relations with Bucharest as well.

Relations with Hungary

Relations with Hungary, on the other hand, went through phases of great drama and hostility. In August 1988 the newly elected Hungarian party leader Karoly Grosz met Ceausescu in the Romanian town of Arad, near the common border. He returned with a programme for normalising relations between the two countries step by step. A number of bilateral committees which had been inactive for years were to be revived. Bucharest was to allow a team of Hungarian observers to look into the so-called rural systemisation plan and see to what extent it affected Hungarian villages. More high-level meetings were to be held. Nothing of this was implemented. The major development over the year that followed was the emergence in Hungary of a mass of 20-30,000 refugees from Romania, partly travellers who decided not to return, partly defectors who crossed the border illegally. The vast majority were ethnic Hungarians, members of the 2m-strong minority who lived mostly in Romania's north-western region of Transylvania. Some 10 per cent were ethnic Romanians trying to escape to the West. The refugees set up their own organisations in Hungary to denounce the Ceausescu regime, and took part in various political demonstrations which provoked angry responses from Bucharest. The climax came after the rehabilitation and reburial of Imre Nagy, the leader of the 1956 Hungarian revolution, when the Romanian media charged that in Hungary communism was being abandoned and fascism allowed to return. It is believed that shortly afterwards Romania tried to raise its objections to political developments in Hungary at a Warsaw Pact summit conference held in July 1989 in Bucharest, but the issue was ruled out of order. The leaders of the two countries met at the end of the conference and had an extremely heated exchange. Hungarian officials started talking about a military threat, alleging that Romania was in the process of building its own short-range missile.

Western debacle

The controversy concerning Romania's reservations against the Final Document of the CSCE Conference in Vienna marked the beginning of a series of attacks on Romania by Western governments. In the framework of a new procedure agreed in Vienna, the Bucharest government was challenged by Britain to provide information about a number of dissidents whose whereabouts were unkown. A group of Western governments led by Sweden called on the UN Committee for Human Rights to start an investigation on Romania. Their motion was carried at the annual session of the Committee in Geneva, partly because several Warsaw Pact countries did not take part in the vote. The need to take a collective attitude against the Romanian government was raised in the European Community and as a first step negotiations for an EEC-Romania agreement were frozen. In a series of diplomatic incidents Western diplomats in Bucharest trying to keep in touch with dissidents were prevented by the police from contacting them. The ambassadors of France, West Germany and the USA were for short periods recalled from Bucharest as a diplomatic signal of discontent. At the same time international protests continued against the projected resettlement of peasants in agro-industrial centres as part of the so-called rural systemisation plan. In mid-1989 the future of the plan was not clear. On the one hand only a few villages were known to have been demolished and all mention of demolitions had been dropped from official pronouncements. This prompted some observers to suggest that the idea of eliminating thousands of villages, envisaged in the original plan, might have been dropped. But many others believed that the entire programme, including village demolitions, was merely waiting for funds which would now be forthcoming, and had not been abandoned.

Political opposition

Throughout the 1980s dissent in Romania consisted mostly in individual gestures that came from people moved by a combination of courage and exasperation. But in 1989, in the space of just a few months, there were four different instances in which several people as a group confronted the regime. In January three leading journalists and a printer were arrested, after they had tried to launch an anti-Ceausescu manifesto. In March six formerly prominent personalities in the Romanian communist party wrote a letter to Ceausescu. Later in March seven well-known writers wrote jointly to the President of the Writers' Union protesting against the persecution of some of their dissident colleagues. Finally, in April, a letter addressed to the president by Romania's best known dissident, Doina Cornea, carried the signatures of five other people, some of them industrial workers and participants in the workers' riots which had occurred in the town of Brasov in November 1987. One of the main tasks of the security police, that of preventing collective action and keeping dissent at the level of individuals, had obviously failed. With the emergence of several groups representing different milieux the outlines of a comprehensive critique of the regime started taking shape. The letter of the six former communist party dignitaries was the most far-reaching analysis of the situation in the country. Among the six signatories there were two former members of the party politburo, a former foreign minister, and a former ambassador to the USA and the United Nations. They argued in their letter that major areas of government policy such as labour regulations which force employees to work on Sundays, the plan that involved the destruction of villages and the resettlement of their inhabitants, and the decree that forbade Romanian citizens to have contacts with foreigners, were unconstitutional. The constitution was virtually suspended, they said, and there was no legal system in force. They pointed out the grave damage done to the country by its deteriorating international position, and by the massive emigration of Germans, Hungarians and Jews. The letter contained three major demands, described as a first step : the restoration of constitutional guarantees for the rights of citizens, a clear-cut statement by Ceausescu to the effect that the village resettlement plan had been abandoned, and a decision to stop all exports of food. The signatories did not press broader economic demands because, as they put it, the existing record showed that the government was incompetent and inept to govern. Finding ways to overcome the present impasse, the letter said, was to be the object of further dialogue after the immediate demands had been met.

Outlook

The demands were not met and a dialogue never took place. The six were immediately put under house arrest and great pressure was exerted on each of them to recant. At the same time the stepson of one of the signatories, a high-ranking official in the Romanian foreign ministry, already under arrest, was accused of spying. This marked the beginning of a public campaign that stressed the need for vigilance against hostile foreign influences and Romanians who yield to them. The theme was repeatedly taken up by Ceausescu himself in the following months, and found its way in to the Theses for the party congress in a statement to the effect that people who have betrayed socialism must be brought to justice. This was a new and ominous development which suggested that political trials might again be held in Romania after a pause of more than 30 years. The Ceausescu regime enjoyed considerable popularity during its first decade in power and continued to claim thereafter that there was no opposition and that it had the support of the entire people. Until 1989 the regime avoided putting political opponents on trial and preferred to use other means of repression against them. Their mail was stopped, their telephones disconnected, they were sacked from their jobs, some forced to move to other towns, and trials, if held, were held in secret and mostly on trumped-up economic charges such as black-marketeering. This approach left little evidence of political repression and maintained the facade that there was no political opposition. By 1989 official propaganda had abandoned this claim and spoke of the presence of hostile elements. It seemed that the next development in Romania might consist in one or several political trials surrounded by noisy hate-mongering campaigns.

COUNTRY PROFILE

Political system: 150-member 'Council for National Salvation' appointed late December 1989. The council has an 11-member executive bureau, with only the acting President (Ion Iliescu) and the acting Prime Minister (Petre Roman) represented, with a loose opposition group 'Social Dialogue' and the Army also represented. Elections planned for April 1990.

Main cities/towns: Bucharest (population 1.96m), Constanta, Iasi, Timisoara, Cluj, Galati, Craiova, Braila, Oradea, Brasov and Ploiesti.

Media
Press: Highly regionalised. Several publications are produced in minority languages such as Hungarian, German and Serbian.
Dailies: Most important national daily is *Scinteia* (circulation 1,400,000), the organ of the RCP Central Committee. Regional newspapers are published jointly by RCP District Committees and District People's Councils.
Periodicals: Several special interest and business publications, mostly produced by various state councils.
Broadcasting: Controlled by Radioteleviziunea Romana.
Radio: Approximately 3.2m receivers in use. Three domestic channels operated by Radiodifuziunea Romana, plus foreign broadcasts in 13 languages.
Television: Approximately 4m sets in use. Service is centred in Bucharest, with daily broadcasts operated by Televiziunea Romana.
Advertising: Advertisements can be placed through State Public Relations Agency (PUBLICOM), in Bucharest.

Domestic economy: Centrally planned economy with large industrial and agricultural sectors. High levels of export-led growth during 1970s. Economic growth affected in recent years by hard currency debt burden, import restrictions and energy problems. The 1986-90 five-year plan targets annual economic growth at eight per cent, with annual growth in investment and domestic demand at two per cent.

Employment: Around 70 per cent workforce employed in industry and other non-agricultural sectors. Emphasis on increased productivity through linking of wages to performance.

External trade: All foreign trade carried out by state enterprises, mostly through foreign trade organisations. Regular trade and current account surplus since 1981 due to reduction in imports and debt rescheduling. Principal exports include machinery and equipment, fuel, mineral raw materials, metals and industrial consumer goods. Convertible currency exports targeted to rise 9 per cent per annum over period 1986-90. Main imports are fuel, raw materials, electrical engineering and power equipment, machinery and chemicals. About 50 per cent of trade conducted with CMEA, principally USSR. Major OECD trading partners include West Germany, Italy and France.

Agriculture: Sector accounts for 16 per cent of national income and nine per cent of exports. Approximately 90 per cent of agricultural land is socialised, either into co-operatives or state enterprises. Large-scale irrigation programe under way, and farm land is being developed in Danube delta. Important agricultural produce includes grapes (leading European producer), wheat, maize, sugar-beet, potatoes, plums, apples and meat. Output targeted to increase by six per cent per annum over period 1986-90.

Industry: Sector accounts for around 62 per cent of National Income and is main determinant of economic growth. Production is state-owned and largely export-orientated. Main industries include metallurgy, chemicals, motor vehicles, machine building, petroleum products and timber processing. Also important: food processing and light manufacturing.

Mining: Mineral deposits include coal, lignite, iron ore, bauxite, manganese and smaller quantities of gold, zinc, uranium, tin and copper. In recent years, considerable problems in mining sector as deposits declined so became increasingly difficult to provide industry with raw materials. However, Romania finds it increasingly worthwhile to exploit domestic resources instead of relying on imports, even if initial cost is high.

COUNTRY PROFILE

Energy: Significant domestic production of oil, gas and coal. Oil used to be major strength of Romanian economy, but its supply is now causing major difficulties. Gas is the most important domestic fuel resource, though output is now declining. Aiming to be energy self-sufficient by 1990. Also rapid development of hydro-electricity and nuclear power. Winter energy rationing in effect.

Banking: State-controlled system with all finance overseen by National Bank. All banking operations with foreign countries carried out through the Romanian Bank for Foreign Trade. Investment Bank provides capital funds to state enterprises and State Savings Bank operates a limited giro. Personal cheque system not in use.

Legal system: Mixture of civil law system and Communist legal theory.

Membership of international organisations: CCC, CMEA, FAO, G77, GATT, IAEA, IBRD, ICAO, ICJ, IFAD, ILO, IMF, IMO, Interpol, IPU, ITC, ITU, UN, UNESCO, UPU, Warsaw Pact, WFTU, WHO, WMO, WTO, WIPO.

Time: GMT + 2 (GMT + 3 from late Mar to late Sep).

Climate: Temperate - continental with long hot summers and cold winters; average temperature varies between about -2°C and 22°C. Frosty with good deal of ice and snow in winter.
Clothing: Mediumweight plus heavy topcoat and overshoes for winter. In principal hotels and most Government offices heating adequate. Lightweight and light raincoat for summer.

Entry requirements: Visas required by all except nationals of CMEA countries and a few others. Usually issued on short-term basis and renewed as required.
Health: Vaccination certificates not normally required but typhoid innoculation advisable. The water is safe to drink.
Currency: No restrictions on movement of foreign currency. Non-CMEA visitors must exchange a minimum of $10 per day of stay into Romanian currency unless in an exemption category. Romanian currency may not be taken into or out of the country.
Customs: Strict inspection of luggage on arrival and departure. Personal effects duty-free. No limit on amount of alcohol and tobacco taken in, provided it is for visitor's own consumption.

Air access: Frequent services from all East European capitals plus flights from many other major international airports.
National airline: TAROM.
Main international airport: Bucharest served by Otopeni (OTP) 16 km from capital and Baneasa (BBU) 9 km from capital.

Surface access: Good rail connections with all neighbouring countries. Road links with USSR, Hungary, Yugoslavia and Bulgaria, and from Western Europe via Vienna and Budapest.
Main ports: Constanta, Sulina and Mangalia on the Black Sea; Tulcea, Galati, Braila, Giurgiu, and Turnu-Severin on the Danube.

Hotels: Classified as de luxe, A and B. Rates in the capital vary from about $35 to $80 per night for de luxe single room. Accommodation outside Bucharest is generally cheaper. Advisable to purchase pre-paid vouchers for accommodation through travel agents, as a confirmed reservation, if not pre-paid, is not a guarantee of accommodation.

Restaurants: Main meals at hotels or at few Bucharest restaurants of similar standard. Tax on bills – sometimes called service charge – is not paid to staff; optional tip 10 per cent.

Currency
Exchange rate: Two rates of exchange – an official rate for foreign trade and a more favourable rate for visitors and tourists.

Foreign exchange information: Import and export of Romanian currency is prohibited. Import and export of foreign currency unrestricted. Most import and export transactions handled by Foreign Trade Organisations. Customs duties on imports. All imports and exports subject to licensing.

Credit cards: Credit cards not widely used but are accepted in most major hotels.

Car hire: Self-drive and chauffeur-driven cars available through National Tourist Office (ONT). Petrol is obtained in exchange for coupons available from ONT, hotels or the Romanian Automobile Club. International or national driving licence required. Drive on the right, overtake on the left.

City transport: Public transport services in principal towns. Taxis available and can be hailed in the street; a 10 per cent tip is usual. Limited metro service in Bucharest.

National transport
Air: Relatively cheap services operated by TAROM connecting Bucharest with Arad, Bacau, Baia Mare, Caransebes, Cluj, Constanta, Craiova, Iasi, Oradea, Satu Mare, Suceava, Sibiu, Timisoara, Tirgu Mures and Tulcea. Charter service available from LAR.
Road: Approximately 15,000 km of roads of various standards. Main cities connected by good roads. Bus service not widely used.
Buses: Local services to most towns and villages.
Rail: Efficient service between principal towns operated by Caile Ferate Romane (CFR). About 11,200 km of track of which 30 per cent is electrified. All long distance trains are provided with sleeping cars.
Water: The volume of freight transported by water increased with the opening of 64 km canal link between Danube and Black Sea.

Public holidays: 1 Jan, 1 May, 23 Aug (National Day). Day after each date is also a public holiday unless it is a weekend in which case the day before may be taken.

Working hours
Government and business: (Mon-Fri) 0800 – 1600 (Sat) 0800 – 1230, lunch usually 1230 – 1300.
Banking: (Mon-Fri) 0900 – 1200 and 1300 – 1500, (Sat) 0900 – 1200.

Social customs: Accepting hospitality from and giving social invitations to officials is normal. Such entertainment usually takes place in restaurants and hotels.

Business language and interpreting/translation: Available from National Tourist Office.

Telecommunications
Telephone: Public call boxes operated by 1 and 3 lei coins. International calls can be made from main hotels and post offices. International dialling code: 40.
Telex and telegram: Available from main hotels and post offices. Country code 65 R.

Postal service: International deliveries should be sent by air mail to avoid delays.

Trade fairs: Important method of promotion for foreign businesses. Main international event is Bucharest International Trade Fair, organised by Bucharest International Fair Authority and held every October. Several small specialised fairs are also held throughout the year. Information on these available from State Public Relations Agency (PUBLICOM).

Electricity supply: 220 V AC.

Representation in capital: Albania, Algeria, Argentina, Austria, Bangladesh, Belgium, Brazil, Bulgaria, Burundi, Canada, Chile, China PR, Colombia, Congo, Costa Rica, Cuba, Czechoslovakia, Denmark, Ecuador, Egypt, Finland, France, Gabon, German Democratic Republic, Federal Republic of Germany, Greece, Guinea, Hungary, India, Indonesia, Iran, Iraq, Israel, Italy, Japan, Jordan, Democratic People's Republic of Korea, Lebanon, Liberia, Libya, Malaysia, Mauritania, Mexico, Mongolia, Morocco, Netherlands, Nigeria, Pakistan, Peru, Philippines, Poland, Spain, Sudan, Sweden, Switzerland, Syria, Thailand, Tunisia, Turkey, USSR, United Kingdom, USA, Uruguay, Venezuela, Vietnam, Yugoslavia, Zaïre, Zimbabwe.

Hotels: Bucharest
Ambassador, Boulevard Magheru 10, 70161 (tel: 159080; tx: 11160).

Athenee Palace, Episcopiei Street 1-3 (in city centre), 70144 (tel: 140899; tx: 11162).

Continental, Calea Victorei 56, 70104 (tel: 166596; tx: 10380).

Dorbanti, Calea Dorobanti 5-9, 70186 (tel: 110860).

Flora, Boulevard Poligrafiei 3, 71556 (tel: 184640; tx: 86410977).

Inter-Continental, Boulevard Nicolae Balcescu 4, 79242 (tel: 140400; tx: 11541; fax: 136470).

Lido, Boulevard Magheru 5, 70161 (tel: 144930; tx: 11161).

Modern, Republicii Boulevard 44, 70334 (tel: 164320; tx: 11409).

Negoiul, 13 Decembrie Street 16, 70118 (tel: 155250).

Nord, Calea Grinitei 143 (near to North Station), 78102 (tel: 506081; tx: 11164).

Parc, Poligrafiei Boulevard 3, 71556 (tel: 180950; tx: 11768).

Union, 13 Decembrie Street 11, 70116 (tel: 132640).

Business advice and information
Chamber of Commerce of the Socialist Republic of Romania, Boulevard Nicolae Balcescu 22, Bucharest 70122 (tel: 139883; tx: 11374).

Ministry of Foreign Trade, 14 Boulevard Republicii, Bucharest (tel: 334050).

Banking
Banca de Investitii (investment bank), Bucharest, Doamnei Street 4 (tel: 134640).

Banca Nationala a RSR (national bank), Bucharest, Lipscani Street 25 (tel: 130410; tx: 11136).

Banca Romana de Comert Exterior (foreign trade bank), Bucharest 4, Calea Victoriei 22-24 (tel: 138010; tx: 11235).

Casa de Economii si Consemnatiuni (savings and consignation bank), Bucharest, Calea Victoriei 13 (tel: 154810).

Principal newspapers
Romanian Foreign Trade, Bucharest, Chamber of Commerce and Industry, Bd N Bălcescu 22.

Scînteia, 71341 Bucharest, Piata Scînteii 1.

Other useful addresses
Agerpres (news agency), Piata Scienteii 1, Bucharest (tel: 182030).

Anuarul Statistic al Republicii Socialiste Romania (general statistics), Directia Centrala de Statistica, St Stavropoleos 6, Bucharest (tel: 158200).

Bucharest International Fair Authority, Bucharest, Boulevard Ba\06lcescu 22 (tel: 147535; tx: 11374).

Caile Ferate Romane (railways board), Bucharest 7, Boulevard Dinicu Golescu 38 (tel: 184020).

Ministry of Finance, Bucharest, Str Doamnei 8 (tel: 132000).

Rwanda

Diana Hubbard

Presidential and legislative elections in December 1988 saw the anticipated re-election of President Juvénal for a further five-year term and the formation of a new 70-member Conseil National de Développement (CND) – the national parliament – with some new faces. The authorities were mightily relieved when the majority of the 60,000-odd refugees from Burundi, who had fled the massacres in their own country in August 1988, decided to return home voluntarily. With the highest population density in Africa – an average 225 inhabitants per sq km, rising to 800 in some areas – the sudden large influx could have created intolerable pressures, particularly when the ethnic dimension was also taken into account. Despite the high quality of its arabica coffee, depressed world prices point to a drop in export revenue in 1989. External donors, however, continue to be impressed by Rwanda's political stability and the World Bank announced two substantial loans during the year, along with smaller commitments from a wide variety of other donors.

President Habyarimana re-elected

In a very slight relaxation of the autocratic form of government found throughout most of the continent, 140 candidates were allowed to stand at the elections on 26 December for 70 seats in the CND although they all had to belong to the sole political party, the Mouvement Revolutionnaire National pour le Développement (MRND). The results saw 26 newcomers elected, but of the total 70 deputies, all but three came from the majority Hutu ethnic group, with just two representatives of the Tutsi, who comprise about 14 per cent of the population, and one Twa. The elections passed off smoothly, although one of the nine female deputies – Dancille Mukarushma – was eventually disqualified and her election declared void in April 1989, because of electoral fraud. Juvénal Habyarimana, the sole candidate at the presidential elections, was returned with 99.98 per cent of the vote on 19 December and has now embarked on his third five-year term, although he has, in fact, been in power since 1973.

Burundian refugees return

Although the problem of the Burundian refugees more or less took care of itself during the year, in February 1989 Habyarimana set up a special commission to examine the situation of Rwandan refugees in neighbouring countries, and particularly in Uganda. The government in Kampala puts the number of Rwandans living in Uganda as high as 200,000 and, although it would like to see the return home of many, it has agreed to look favourably on applications for naturalisation. The refugees are mainly Tutsi, who fled during the 1959 uprising (prior to independence in 1962) when the majority Hutu turned on the minority Tutsi, who were, at that time, dominant in the public, political and economic sectors. Since then, however, and particularly under the pragmatic leadership of Habyarimana, Rwanda has managed to avoid the massive bloodbaths which have occurred on two occasions in recent years in neighbouring Burundi because of ethnic tensions between the Hutu and the Tutsi.

Coffee earnings squeezed

External economic forces, however, continue to frustrate economic growth. The 1988 coffee harvest was estimated at 42,000 tonnes – equivalent to 700,000 bags – but export revenue will have been depressed both by world market prices and by the quotas imposed by the International Coffee Organisation (ICO), which permitted Rwanda exports of less than 500,000 bags for the 1988-89 season. Coffee exports in 1988 totalled 29,930 tonnes and earned an estimated RwFr7.3bn, compared with RwFr8.35bn in 1987. The cultivation of coffee has expanded rapidly over the last 25 years, rising from 8,000 tonnes in 1964 to

KEY INDICATORS

	Unit	1984	1985	1986	1987	1988
Population	m	5.9	6.1	6.3	6.5	6.8
Gross domestic product (GDP)	Frw bn	159.1	173.6	168.9	168.5	–
GDP per capita	$	266	277	301	319	–
Inflation	%	5.4	1.7	–1.1	4.1	2.9
GDP real growth	%	–4.2	4.4	4.5	–	–
Agricultural production	1979-81=100	101.4	114.7	105.8	101.2	102.5
Food production	1979-81=100	100.7	113.5	103.2	95.9	97.4
Coffee production	'000 tonnes	43.8	48.0	47.9	53.8	42.0*
Tea production	'000 tonnes	8.6	11.3	10.3	11.6	8.0*
Exports	Frw bn	14.5	13.2	16.5	8.9	8.3
Coffee	Frw bn	9.4	9.4	13.2	7.4	6.5
Tin	Frw m	938	808	155	47	–
Tea	Frw m	2,668	1,713	1,713	971	1,082
Imports	Frw bn	27.9	30.2	30.6	28.0	28.3
Petroleum	Frw m	4.4	4.4	4.2	3.7	3.6
Balance of trade	$m	–54.9	–93.2	–75.1	–145.5	–160.7
Current account	$m	–41.6	–64.1	–69.2	–134.3	–119.0
Total external debt	$m	291.4	354.8	440.6	583.4	–
Debt service ratio	%	5.3	8.9	7.5	11.3	–
Exchange rate	Frw per $	100.2	101.3	87.6	79.7	76.4

* estimated figure

KEY FACTS

Official title: Republic of Rwanda

Head of state: President Major General Juvénal Habyarimana

Ruling party: Mouvement Révolutionnaire National pour le Développement (MRND).

Capital: Kigali

Official Languages: Kinyarwanda, French (Kiswahili widely spoken)

Currency: Rwanda Franc (Frw)

Exchange rate: Frw80.2 per $ (Nov 1989)

Area: 26,338 sq km

Population: 6.8m (1988)

GDP per capita: $319 (1987)

GDP real growth: 4.5% (1986)

Inflation: 2.9% (1988)

Trade balance: –$160.7m (1988)

Foreign debt: $583m (1987)

42,000 tonnes in 1988. The number of coffee bushes planted has risen from 31m to 94m and the number of coffee farmers from 285,000 to 682,000. Reduced earnings from coffee and failure to meet customs duty targets were two factors behind a decision announced by Finance Minister Benoit Ntigulirwa in June to cut planned expenditure in the 1989 budget. The original budget had provided for expenditure of just over $324m, but by mid-year revenue projections had fallen to $284m. Precise details of the cuts were not revealed but it was announced that the one major aim was to reduce the deficit on the balance of payments, which in 1988 totalled $31.75m. Total outstanding public debt – both internal and external – was $932.5m at the end of 1988.

Long-term economic growth, however, will get a considerable boost from two loans announced in 1989 by the International Development Association (IDA), the World Bank's soft loan arm. $20m will be used to promote agricultural development and $32m will go to a $66.2m project aimed primarily at setting up a municipal administration in the capital Kigali and improving the functioning of the communes – the smaller town councils – in other urban areas. The strengthened administration is to introduce a property tax, to help create a tax base, and the national mortgage bank will be restructured with the aim of improving housing finance services.

Other donors during the year tended to concentrate on specific infrastructure projects, with Japan providing 3bn yen for construction of the Mukungwa 2 hydro-electric power station. The viability of the project was seriously questioned by the World Bank, which maintained that there were considerable potential hydrological and environmental problems attached to the scheme. With a capacity of just 3.8 MW, Mukungwa 2 will represent construction costs of $5,300 per kilowatt, compared with the average for the continent of just $2,000.

In February 1989, the Arab Bank for Economic Development in Africa (BADEA) announced a $9.5m to part finance construction of a road between Gitarama and Mukamira and, in July, Habyarimana inaugurated the new 116 km tarred road from Kayonza – Kagitumba. This will run from the east of the country towards Uganda, crossing through the Akagera national park. The government is hoping that it will be used by some of the many new foreign tourists expected to visit Rwanda following the publicity attached to the film *Gorillas in the Mist*, which portrayed the career and murder in Rwanda in 1985 of the American naturalist Dian Fossey, who gained fame through her commitment to the protection of Rwanda's mountain gorillas.

COUNTRY PROFILE

Political system: Republic with executive power vested in the president (elected for five years) assisted by a council of ministers whom he appoints. Legislative power is exercised by the president in conjunction with the National Development Council, whose members are elected by universal suffrage. Gained independence in 1962. Formerly Belgian trust territory of Rwanda-Urundi.

Political parties: Mouvement Révolutionnaire National pour le Développement (MRND) is the only political party and is headed by President Major-General Juvénal Habyarimana. Habyarimana came to power in a military coup in 1973 and has offered himself for re-election three times in the intervening years; once in 1978 when the country returned to civilian rule, again in 1983 and again in 1988. On each occasion he was the sole candidate. The presidential election in December 1983 was followed immediately by elections to the National Assembly where voters were given a choice between various candidates but all approved by the MRND. In June 1988 the MRND again nominated Habyarimana as the sole candidate at the presidential elections that were to be held in December 1988.

Population: Rwanda's population of 6.64m (1987 estimate) comprises two major groups, the Hutu and the Tutsi. The two communities maintain quite distinct cultural and social traditions.

Main cities/towns: Kigali, Butare, Gisenyi, Ruhengeri, Cyangugu.

Language: French is the usual business language, while Kinyarwanda and Kiswahili are widely spoken. English is sometimes spoken as a business language.

Media
Press: Government publishes the *Daily Bulletin*, along with magazines in French and Kinyarwanda. Also influential church newspaper, *Kinyamateka*. There is also the weekly *Imvaho*, published in Kinyarwanda.
Radio: Two AM and five FM radio stations broadcast daily in various languages including Kinyarwanda, Swahili, French, German, English and Amharic.
Television: There is no television service as yet.

External trade: Reliance on commodity exports. Regular trade and current account deficits, which worsened after 1984, were partly funded by aid inflows. Export potential reduced by transport problems.
Exports: Coffee (typically 70-80 per cent of total value), tea (10-13%), tungsten and tin-ores and concentrates (5%), quinquina, pyrethrum, cut flowers, fruit and vegetables. Main destinations: UK (typically 40-45% of total value), Switerland (18-20%), German FR (10-12%), USA (8-10%).
Imports: Food (typically 12-15 per cent of total value), machinery and tools (13-14%), transport equipment (11-13%), clothing (10-12%), mineral fuels and lubricants (8-10%), construction materials, semi-manufactures. Main sources: Kenya (typically 20-25% of total value), Belgium/Luxembourg (13-15%), Japan (10-13%), France (10-12%), China (8-9%).

Agriculture: Contribution to GDP has fallen to around 42 per cent – though the sector remains the major employer and principal export earner. The main food crops are plantain, sweet potatoes, beans, cassava, sorghum and maize. Crop yields have fluctuated due to drought, soil erosion and underinvestment, but rebounded in 1985-86. Coffee is by far the most important export crop, mainly high quality arabica. Other export crops include tea (also of high quality), pyrethrum extract and quinquina. Approximately 35 per cent of the land area is cultivated arable land, 20 per cent pasture and 11 per cent forest; tree planting programmes under way to combat deforestation.

Industry: Manufacturing sector contributes about 18 per cent to GDP. Industries include brewing, food processing, cigarette production, soaps, plastics, tin smelting and textiles. Growth of the sector is limited by the small domestic market, transport difficulties and irregular supply of imported fuels and raw materials which comprise 77 per cent of inputs.

Mining: Activity confined to the extraction of cassiterite (known reserves 90,000 tonnes), since 1985 on an artisanal scale only.

Energy: Subsistence farming sector reliant on wood and charcoal. Urban areas and industry energy requirements are met by hydroelectricity – over 50 per cent of which is generated by the Mururu power station in Zaire. Another plant, being jointly developed with Burundi and Zaïre, is expected to be completed before 1990. Plans are also underway to exploit natural methane gas under Lake Kivu – possibly reserves of 50,000 cu m. Two important power stations in Rwanda are the Ntaruka station and Mukungwa 1.

Membership of international organisations: African Development Bank, African States Associated with the EC (Lomé Convention), Arab Bank for Economic Development in Africa (recipient), Economic Commission for Africa, FAO, GATT, IDA, IFAD, IFC, ILO, IMF, OAU, UN, WHO, WIPO, World Bank.

BUSINESS GUIDE

Time: GMT +2.

Climate: Warm, tempered by altitude. Rainfall is low and is concentrated in two seasons from mid-Jan to mid-May, and mid-Oct to mid-Dec. Average temperatures in Kigali range from 12-14°C at night to 28-32°C during the day. Cooler in highland areas.

Entry requirements
Passport: Required by all.
Visa: Required by all. Special rules for Burundi nationals.
Currency: Import and export of local currency limited to a maximum RwFr 5,000. Import of foreign currency unlimited but amounts should be declared on arrival, and export is allowed up to amount declared.

Health precautions
Mandatory: Yellow fever and cholera vaccination certificate required if arriving from an infected area.
Advisable: Yellow fever, typhoid and polio vaccinations. Malaria prophylaxis recommended as risk exists throughout the country. Water precautions should be taken. AIDS risk.

Representation overseas: Abidjan, Addis Ababa, Beijing, Bonn, Brussels, Bujumbura, Cairo, Dar es Salaam, Kampala, Kinshasa, Moscow, Nairobi, Ottawa, Paris, Tokyo, Tripoli, Washington.

Air access
National airline: Air Rwanda.
Other airlines: Aeroflot, Air Burundi, Air France, Cameroon, Ethiopian, Kenya Airways, Sabena.
International airport: Kigali-Kanombe (Code: KGL), 12 km east of city; duty-free shop, bar, currency exchange, post office, shops, coach, taxi service.

Surface access
Overland access: Roads from Zaïre, Tanzania, Uganda and Burundi. Buses run from the Tanzanian border at Rusumo.

Hotels: Tend to be expensive in Kigali; cheaper in Butare, Gisenyi and Ruhengeri. Advisable to book in advance.

Electricity supply: 220 V AC.

Car hire: Limited service available in Kigali. International driving licence required. All weather roads are sparse and in poor condition.

City transport
Taxis: Available in main towns. Fares should be agreed at start of journey. Tipping not necessary.

National transport
Air: Air Rwanda operates internal services between Kigali, Kamembe and Butare. Chartered air taxis can be taken to Gisenyi, Ruhengeri, Nemba and Gabiro; these tend to be expensive.
Road: Roads, apart from the paved main routes (including Kigali to Butare), are poor with many being impassable in bad weather.
Buses: Services are available: e.g. Kigali-Ruhengeri (3 hrs), Ruhengeri-Gisenyi (2 hrs), and Cyangugu-Butare (6 hrs). Tend to be unreliable.

BUSINESS GUIDE

Water: Services run between Gisenyi and Cyangugu, on Lake Kivu.

Public holidays
Fixed dates: 1 Jan, 28 Jan (Democracy Day), 1 May, 1 Jul, 5 Jul, 1 Aug (Harvest Festival), 15 Aug, 8 Sep (Culture Day), 25 Sep, 26 Oct, 1 Nov, 25 Dec.
Variable dates: Easter Monday, Ascension, Whit Monday.

Working hours
Government and business: (Mon-Fri) 0700 – 1200, 1400 – 1700.
Banking: (Mon-Fri) 0700 – 1200, 1400 – 1800; (Sat) 0700 – 1300.
Shops: Dawn to dusk.

Telecommunications
Telephone: Dialling code for Rwanda: IDD access code + 250 followed by subscriber's number. (There are no area codes). Local service largely automatic.
Telex: Facilities available in Kigali and at major hotels.

Banking
Central bank: Banque Nationale du Rwanda.

Representation in capital: Belgium, Burundi, China PR, Egypt, France, German FR, Korea PDR, Tanzania, Uganda, USA, USSR, Vatican, Zaïre.

Hotels: Kigali
Kiyovu, P.O. Box: 1331 (tel: 5108).

Umubano Meridien, P.O. Box: 874 (tel: 2176; tx: 561; fax: 83361).

des Diplomates, P.O. Box: 269 (tel: 75111/2).

des Mille Collines, P.O. Box: 1322 (tel: 76530).

Car hire: Kigali
Agence Solhard: PO Box 335 (tel: 5660).

Europcar: BP 493 (tel: 72983; tx: 22689).

Rwandamotor: PO Box 448 (tel: 5294).

Siarwanda: PO Box 448 (tel: 452).

Other useful addresses
Agence Rwandaise de Presse (ARP), 27 Avenue du Commerce, BP 83, Kigali (tel: 5665; tx; 557).

Air Rwanda, BP 808, Kigali (tx: 554).

Banque Nationale du Rwanda, BP 531, Kigali (tel: 74282, 75249; tx: 508, 589 rwanda bank).

Chambre de Commerce et d'Industrie du Rwanda, PO Box 319, Kigali (tel: 29072319; tx: 22662 CCIR RW).

Economat Général (tobacco exports), PO Box 45, Ruhengeri.

L'Institut des Sciences Agronomiques du Rwanda, PO Box 138, Butare.

Office des Cafés, PO Box 104, Kigali (tel: 5277; tx: 513).

Office des Cultures Industrielles du Rwanda (OCIR), PO Box 104, Kigali (tel: 5600, 5004, 5277, 5398; tx: KGL 13).

Office du Pyrèthre au Rwanda, PO Box 79, Ruhengeri.

Office du Thé, PO Box 1344, Kigali.

Office Rwandais du Tourisme et des Parcs Nationaux, PO Box 905, Kigali (tel: 6512/5; tx: RWATRAVEL 42).

St Kitts Nevis

Tony Thorndike

The March 1989 general election, as expected, returned the People's Action Movement to power but prompted a demand for secession by some Nevisians. It also triggered a change of leadership of the opposition in St Kitts. Economically, the twin-island state continued its pattern of slow but steady growth, but suffered increasingly serious labour shortages.

Now one of the few plantation-based sugar producers left in the Commonwealth Caribbean, St Kitts' economy complements that of its near neighbour, Nevis, which is based on peasant agriculture. But political relations between the two parts of the state are becoming more tense as Nevis experiences relative prosperity as a result of a rapidly developing tourist industry and the opportunity for greater autonomy. Indeed, although only two miles separates the two islands at their nearest point, relations between them have for long been uneasy.

Political murmurs

Dr Kennedy Simmonds once again led his party, the People's Action Movement (PAM), to victory in the March 1989 general election, winning six of the 11 seats. Since 1980, PAM has ruled as the senior partner in a coalition with the Nevis Reformation Party (NRP), led by lawyer Simeon Daniels. But this election saw the NRP lose one of the three Nevis seats to the Concerned Citizens' Movement (CCM). The CCM advocates secession from St Kitts, just as the NRP used to do before it shared power with the PAM. The once-omnipotent St Kitts Labour Party (SKLP) again won only two seats. Faced with this third consecutive disaster, and his second personal defeat, Lee Moore resigned the leadership which passed to newcomer Denzil Douglas, like Prime Minister Simmonds a medical doctor, with lawyer Henry Browne as deputy leader.

The SKLP was the political offspring of the St Kitts Trades and Labour Union, founded by sugar worker Robert Bradshaw in 1932. It led the struggles against the hated plantation owners, but after all sugar lands were nationalised in 1975 and Bradshaw's death three years later, the party lost much of its appeal. Bradshaw's experiences with the plantocracy tended to make him intolerant of opposition. There was no love lost between the SKLP and Nevis as Bradshaw insisted that all resources be divided on a per capita basis which, since Nevis had only a population of about 12,000, meant relatively little for the 'ward' island. This formula was followed by Lee Moore, his successor, which only served to further alienate Nevisians so far as the SKLP was concerned.

PAM, founded in 1966 to represent the small middle class and Labour dissidents, also experienced Bradshaw's antagonism: to him Simmonds and PAM cofounder Billy Herbert were agents of the planters. Matters worsened when PAM supported the Anguillans in their (eventually successful) quest for secession from what was a three-island state, and then a similiar quest by the Nevisians. Independence was held up because of this issue but the breakthrough eventually came with the victory of the PAM-NRP alliance in the momentous 1980 election. All talk of secession ended with Daniels named as Finance Minister and the opening of negotiations for a quasi-federal state, which would permit Nevis its own Assembly, guarantee a generous share of national resources and the right to secede at any time, a privilege denied to St Kitts. Once agreed, independence followed in 1983.

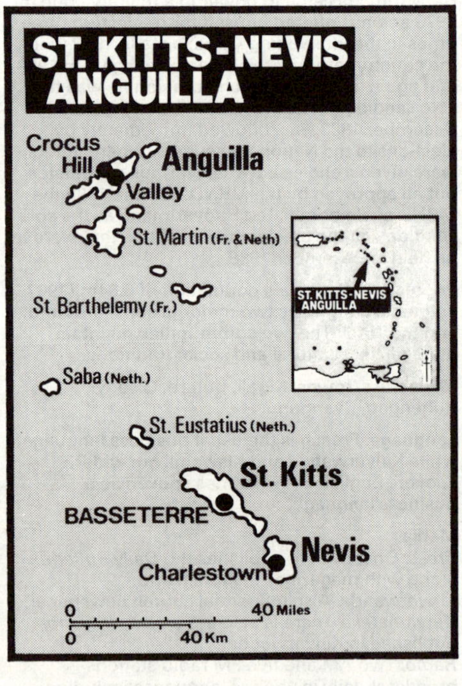

KEY FACTS

Official title: St Christopher and Nevis

Head of state: Queen Elizabeth II

Head of government: Prime Minister Dr Kennedy Simmonds

Ruling party: Coalition: People's Action Movement/Nevis Reformation Party

Capital: Basseterre

Official Languages: English

Currency: East Caribbean Dollar (EC$) = 100 cents

Exchange rate: EC$2.69 per $ (fixed)

Area: 269 sq km

Population: 46,500 (1987)

GDP per capita: $2,119 (1987)

GDP real growth: 4.0% (1988)

Inflation: 1.0% (1988)

Trade balance: EC$64.5m (1987)

The 1984 election saw PAM consolidate its position. Although the alliance was continued, PAM now had an absolute majority. Under pressure, in particular from minister Micheal Powell, who is regarded as the cabinet heavyweight, the NRP leaders were given nominal posts. Not surprisingly, all three NRP members withdrew to their home base. A successful businessman with a finger in numerous pies, Powell had little time for the 'incessant whining' of the Nevisians. The CCM then took up the secessionist torch. It won a seat in the Nevis elections in December 1987 and gained momentum when two other groups merged with it to form a united opposition to Premier Daniels. Bolstered by its good showing in the general election, the CCM called for separation. But the central government adopted a 'do-nothing' stance, knowing the parlous economy of Nevis and the likelihood that few countries would wish to support the CCM's call.

The Nevis issue has injected a certain degree of political uncertainty. As for the Labour party, the election of Douglas will mean a new start and a loosening of the ties that bind it to the declining sugar sector and the union. However, most of the electorate associate the PAM with a new prosperity and PAM will consequently be hard to beat. After all, it survived the wide publicity given to allegations in 1987 that the fast-talking Billy Herbert had misused his diplomatic position as Ambassador to the United States by laundering drug funds, presumably through the variety of offshore companies in which he had an interest in his Anguillan base. He denied all charges but resigned his post.

Economic diversification

It is ironic that when the government made strenuous attempts to diversify away from sugar towards manufacturing and tourism in the early 1980s there was scant success; but when investors were finally attracted they were met with labour shortages, especially of skilled people. The shortage affected economic growth which fell 4 per cent in 1988. Output increased in all sectors except sugar.

Sugar still accounts for over half of exports (54 per cent in 1988). But production fell from 28,940 tonnes in 1986 to 25,256 tonnes in 1988, well below the 30,000-tonne target. Labour shortages left over 60,000 tonnes of cane rotting (equal to about 5,800 tonnes of sugar), despite the introduction of two West German mechanical harvesters. There was a similiar experience in 1987 when 40,000 tonnes were abandoned. Stocks were used to supply a buoyant UK market, but depreciation of the pound sterling against the US dollar meant a 6 per cent decrease in export earnings. Ironically, market prospects are good: the US quota was increased to 8,128 tonnes and construction began on a EC$8m ethanol plant in conjunction with three US investors. Nevertheless, some sugar land will be converted to other agricultural use under the land leasing programme, the government having finally acquired the title to the nationalised lands by borrowing EC$60.1m from the Social Security Fund.

The manufacturing sector expanded, and once again the electronics and data-processing industries were the pace-setters. This was not surprising as they are directed to the US market and not to the region, where trade conditions remained sluggish. The garment factories experienced both market access and labour problems, prompting a move towards high-quality operations permitting more competitive wages and conditions. Construction boomed, the centrepiece being the work on the south-east peninsular road and associated developments, a US$27m project funded largely by US aid. This is opening up one of the greatest areas of tourist potential. St Kitts, unlike Nevis, was to an extent bypassed by tourists because of its volcanic nature and consequent shortage of beaches. But the impetus provided by the extensive Frigate Bay development, adjoining the peninsular, and the provision of direct air services through the twice-expanded Golden Rock airport, has transformed prospects. Frigate Bay boasts a casino, a championship golf course, several high-class hotels and a popular condominium development.

However, greater promotion is needed as stopover arrivals rose only by 4.9 per cent in 1988 to 66,608, especially since three large hotels are under construction (two on Nevis, one on St Kitts). The number of cruise ship and yacht passengers rose by an astonishing 70.6 per cent from 31,438 to 53,645, but these tourists do not spend as much. Nearly all in this category landed on St Kitts as that has the only deep-water harbour, and to cater for them an attractive tourist centre opened in the Kittitian capital Basseterre in 1988. Nevis has good beaches but suffers from access problems. Not only is the small harbour at the capital Charlestown shallow, but the little airfield at Newcastle can accept only small aircraft. While large facilities are being built, especially on Pinney's beach, the emphasis is on luxury 'great house' plantation facilities.

Future plans

A five-year development plan was launched in October 1987. It was preceded by the establishment of the Investment Promotion Agency to screen investment proposals. The plan calls for greater mechanisation of the sugar industry and electricity generation from bagasse (sugarcane residue), together costing EC$16m. Agricultural diversification will include more cattle rearing to provide high quality meat for the tourist industry, and the resuscitation of cotton and copra in Nevis. Employment in manufacturing and tourism is planned to increase to 1,500 and 2,000 respectively, and tourist arrivals to 104,000 by 1990, to earn some EC$50m. But other than the successful completion by the Rothschilds group of a EC$3.5m distillery, which makes a remarkable product, 'the Baron's CSR', a distinctive cane spirit using pure mountain water, it is hard to see how these targets can be achieved as the skilled labour shortage grows more acute. Suggestions to import up to 100 Dominicans were rejected as politically unacceptable; in any event, they would bring few non-agricultural skills with them.

Budgetary prudence

Simmonds announced in the January 1989 budget that recurrent expenditure would be EC$74.6m (US$27.6m) and revenue EC$86.9m (US$32.2m). It was hoped that the projected surplus would be achieved, unlike the experience in 1987 when virtually all the surplus was swallowed up with a 14 per cent pay increase for civil servants. Capital expenditure in 1988 was low at EC$17.8m (US$6.6m) because of heavy loan repayments to local banks. The external debt at the end of 1988 stood at US$29.6m and debt servicing consumed 4 per cent of recurrent expenditure. Domestic debt was estimated to be US$37m, the result of accumulated deficits incurred in the past and one which is a millstone around the neck of an enterprising government.

COUNTRY PROFILE

Political system: Independent state within the Commonwealth with the British monarch as head of state represented by a governor-general who exercises executive power. The legislature is the National Assemby comprising 11 members elected for a five-year term (eight for St Kitts, three for Nevis) plus three appointed members. The cabinet headed by a prime minister is collectively responsible to the National Assembly.

Main cities/towns: Basseterre on island of St Kitts (capital, population 16,400), Charlestown on Nevis island (1,700).

Media
Press: The St Kitts-Nevis Trades and Labour Union publishes *The Labour Spokesman* twice weekly and PAM publishes the weekly *Democrat.*
Broadcasting: Government-owned commercial radio and television station (ZIZ); a cable TV system has been installed. Also regional religious broadcasts operated by Radio Paradise and TBN Television.

External trade: Balance of trade deficit, largely due to need for massive food imports (as much as 70 per cent of requirements), partly offset by earnings from tourism and remittances from workers abroad.
Exports: Sugar and molasses (54% of total value), clothing and shoes (14%), sea island cotton, copra oil, electronic goods. Main destinations: USA, UK.
Imports: Food (21% of total value), machinery and transport equipment, fuels and chemicals. Main sources: UK, USA, Trinidad and Tobago.

Agriculture: Although sector's contribution to GDP has declined to around 12 per cent, the economy remains largely dependent on the cultivation of sugar-cane for export. About 80 per cent of available arable land on St Kitts is used for sugar growing, and the government is trying to improve productivity. At the same time, diversification into food crops is also being encouraged to reduce dependence on imports. Main cash crop on Nevis is cotton, although most of the island's population are engaged in subsistence farming. Inshore fishing is a traditional occupation and a significant source of protein. Owing to poor pasturage, the livestock sector remains under-developed.

Industry: Manufacturing sector has expanded rapidly in recent years, with the diversification from sugar processing to light industries such as electronics, clothing and furniture. Government policy is directed towards the promotion of labour-intensive, export-orientated industries, although a labour shortage developed from 1986. However, the sugar industry remains the islands' single most important economic concern.
Tourism: Expanding sector aided by government investment in improving port facilities at Basseterre, the construction of many new hotels and improved air connections with other Caribbean islands and the USA.

Membership of international organisations: CARICOM, Economic Commission for Latin America (associate), IMF, OECS, Organisation of American States, state associated with the EC, UN, WHO, World Bank.

Time: GMT -4.

Climate: Tropical tempered by trade winds, with an annual mean temperature of 27°C. Dec-Apr are the driest months. Rain can occur throughout the year, although generally wetter from May to Oct.

Entry requirements
Passport: Not required by citizens of Canada, UK and USA subject to proof of identity, maximum length of stay six months, onward tickets required.
Visa: Required by all except nationals of Commonwealth, Scandinavian and most EC countries, USA, Switzerland and certain other countries.

BUSINESS GUIDE

Currency: No restriction on import/export of foreign currency. Local currency imports should be declared; export allowed up to the declared amount.

Health precautions
Mandatory: Vaccination certificates for yellow fever and cholera required when travelling from infected areas.
Advisable: Typhoid, polio vaccinations. Water precautions.

Air access
National airline: St Kitts Nevis is a shareholder in LIAT.
Other airlines: Air BVI, BWIA, Eastern, Windward Islands Airways.
International airport: Golden Rock (SKB), 5 km from town (Basseterre); duty-free shop, restaurant, hotel reservations.

Surface access
Main ports: Basseterre (St Kitts) has a deep-water harbour, Charlestown (Nevis).

Hotels: Advisable to book in advance. A seven per cent room tax is added to bills and 10 per cent service charge usual.

Car hire: National licence required in order to obtain visitor's temporary licence. Traffic drives on the left.

City transport
Taxis: Serve both islands; standard fare system applies, 10 per cent tip usual.

National transport
Air: Scheduled daily flights between the two islands.
Road: 300 km of roads. Main routes cover perimeters of both islands.
Buses: Regular but unscheduled bus services are operated.
Water: The government operates daily commercial boat service between St Kitts and Nevis, except on Thursdays and Sundays.

Public holidays: 1 Jan (Carnival Day), Good Friday, Easter Monday, Labour Day (first Mon in May), Whit Monday, August Bank Holiday (first Mon in Aug), 19 Sep (Independence), 25-26 Dec.

Working hours
Business: (Mon-Fri) 0800 – 1200, 1300 – 1600/1630. N.B. businesses generally close (Thu) afternoons, and open (Sat) 0800 – 1600.
Government: (Mon-Fri) 0800 – 1200, 1300 – 1600/1630.
Banking: (Mon-Fri) 0800 – 1200/1300. Also (Fri) 1500 – 1700.

Telecommunications
Telephone: Dialling code for St Kitts: IDD access code + 1-809-465 followed by subscriber's four-digit number. Dialling code for Nevis: IDD access code + 1-809-469 followed by subscriber's four-digit number. 100 per cent automatic service.
Telex: SKANTEL public telex booths available in Charlestown and Basseterre. Telex facilities also available at main hotels.

Banking
Central bank: East Caribbean Central Bank (ECCB).
Other banks: Bank of Nevis, Bank of Nova Scotia, Barclays Bank International, Development Bank of St Kitts Nevis, Nevis Co-operative Bank, Royal Bank of Canada, St Kitts Nevis Anguilla National Bank.

Electricity supply: 230 V AC, 60 cycles. (Some hotel supplies are at 110 V AC).

Hotels: Basseterre, St Kitts
Fairview Inn, P.O. Box: 212 (tel: 465-2472; tx: 6834; fax: 9651056).

Fort Thomas, Basseterre, Fortlands, P.O. Box: 407 (tel: 465-2695; tx: 6816).

Golden Lemon, Dieppe Bay (tel: 465-7260).

Ocean Terrace Inn, P.O. Box: 65 (tel: 465-2754; tx: 6821).

Hotels: Charlestown, Nevis
Rest Haven, P.O. Box: 209 (tel: 469-5208; tx: 5208).

Zetland Plantation, P.O. Box: 448 (tel: 469-5454; tx: 6855).

Hotels: Frigate Bay, St Kitts
Frigate Bay Beach, P.O. Box: 137 (tel: 847-9445; tx: 6884).

Royal St Kitts, P.O. Box: 406 (tel: 465-8651; tx: 6819).

Hotels: Nevis
Montpelier Plantation Inn, P.O. Box: 474 (tel: 469-5462; tx: 6818; fax: 1932).

Hotels: Newcastle, Nevis
Nisbet Plantation Inn, Beachlands (tel: 469-5325).

Car hire: Basseterre
Budget: The Circus (tel: 465-3377).

Caines Car Rental: Prince's Street (tel: 465-2366).

Delisle Walwyn & Co, Liverpool Row (tel: 465-2631; tx: 6835).

Holiday Car Rental: South Independence Square Street (tel: 465-2616).

Sunshine Agency: Cayon Street (tel: 465-2193).

TDC: Fort Street (tel: 465-2511).

Car hire: Charlestown
Budget: (tel: 469-5390).

Powell's Rental: Bath Bridge (tel: 469-5300).

Skeete's Rental: Newcastle (tel: 469-5458).

TDC Rentals: Main Street (tel: 469-5430).

Business advice and information
Investment Promotion Agency, Bay Road, Basseterre (tel: 465-4106).

Ministry of Trade, Industry & Development, PO Box 186, Basseterre (tel: 465-2521; tx: 6820).

St Kitts Nevis Chamber of Industry and Commerce, PO Box 332, Basseterre (tel: 465-2980; tx: 6822 HRSFDSKB).

St Kitts Nevis Hotel Association, PO Box 65, Basseterre (tel: 465-2754; tx: 6807).

St Kitts and Nevis Manufacturers' Association, PO Box 392, Basseterre (tel: 465-6226).

St Kitts Nevis Tourist Board, Church Street, PO Box 132, Basseterre (tel: 465-2620/4040).

Banking
East Caribbean Central Bank (central bank), PO Box 89, Basseterre (tel: 465-2537; tx: 6828).

Barclays Bank, PO Box 42, Basseterre (tel: 465-2449; tx: 6823).

Development Bank of St Kitts Nevis, Basseterre (tel: 465-2288).

East Caribbean Central Bank, PO Box 89, Basseterre (tel: 465-2537; tx: 6828).

St Kitts Nevis Anguilla National Bank, Church Street, Basseterre (tel: 465-2204).

St Lucia

Tony Thorndike

The banana industry occupied centre stage both economically and politically. Improved production enabled a strong contribution to economic growth, and the tourist and construction sectors also did well. Although manufacturing was not so dynamic a sector before, the outlook is very favourable, especially when the ambitious capital development programme is taken into account.

Prime Minister John Compton's United Workers' Party government continues to preside over an increasingly prosperous economy where investment in both the private and public sectors is growing. But his vision of a politically united Eastern Caribbean is not likely to be realised despite his strenuous efforts.

Prosperity

From being billed 'the sick man of the Caribbean', St Lucia's economy has more than recovered. A diversification programme away from the dominant banana industry and agriculture generally, a sector particularly vulnerable to adverse weather and externally-determined market conditions, has resulted in the largest manufacturing sector in the Commonwealth Eastern Caribbean and a healthy tourist industry. The credit for the turnround must in large part go to Compton, who has dominated St Lucian politics

KEY FACTS

Official title: St Lucia

Head of state: Queen Elizabeth II

Head of government: Prime Minister John Compton

Ruling party: United Workers' Party

Capital: Castries

Official Languages: English

Currency: East Caribbean Dollar (EC$) = 100 cents

Exchange rate: EC$2.69 per $ (fixed)

Area: 616 sq km

Population: 142,400 (1987)

GDP per capita: $1,400 (1987)

GDP real growth: 5.0% (1988)

Inflation: 1.0% (1988)

Trade balance: –$82.28m (1988)

Foreign debt: $48m (1986)

Visitor numbers: 214,500 (1988)

for nearly 25 years, although favourable weather conditions played their part.

The economy reached its nadir in 1982, largely as a result of a deep leadership division in the then St Lucia Labour Party government. The split involved the trade union movement, and a series of mainly politically motivated strikes forced some companies to seek more stable environments. Following Compton's return to power in the May 1982 election, peace gradually – and painfully – returned. The banana industry was the first to recover and productivity on model farms sharply increased. Unfortunately, the 1987 crop fell victim to heavy rains, as did part of that of 1988 due to the effects of Hurricane Gilbert. But in spite of this, 1988 saw a record production of 136,690 tonnes, the previous record of 114,800 tonnes set in 1986 being easily surpassed. This was a rise of 47 per cent over 1987. Export volume reached 128,000 tonnes, 50 per cent up, and earnings increased by some 52 per cent to EC$176.9m (US$65.6m). In its early 1988 budget the government imposed a controversial 5 per cent levy to help pay for feeder roads, but this was halved in June because of the devaluation of sterling against the US dollar and a consequent fall in the green market price in Britain, the sole market. Other agricultural products, notably copra, fared poorly.

Manufacturing was sluggish because of relatively depressed conditions in the regional CARICOM market. But the reciprocal removal of levies by St Lucia and its main customers, particularly Trinidad and Tobago, brought improvement. To take advantage of investment opportunities under the US Caribbean Basin Initiative and the Puerto Rico '936' programme, a Tax Information Exchange Treaty was signed with the United States in 1987. The National Development Council reported in May that exports from the sector in 1988 totalled EC$80m (US$29.6m) and that there were over 200 enterprises involved in such industries as clothing, furniture, electronic goods, food products and sports goods. The United States was the main market (worth EC$50m in 1988, compared to only EC$3m 10 years ago) and clothing accounted for 76 per cent of US imports from the island. The 11 garment factories employed more than 3,000 and another 600 worked in five electronics plants owned by North American, Asian and European companies. Data entry industry was expected to expand following the investment by Cable and Wireless in a satellite telecommunications centre. Expansion plans for the four industrial zones (two are to be free trade areas) will, the Centre estimated, create nearly 2,000 extra jobs.

Construction was the most dynamic sector during 1988, spurred by large infrastructural projects in and around the capital, Castries. Major hotel building work was completed in the Rodney Bay area to the north of the island, adding a much needed extra 300 rooms. The trend in the tourist industry, besides upward, showed a swing back to St Lucia's original main market, Western Europe. Stopover tourist arrivals totalled 132,973, an increase on the 1987 figure of 117,577; hotel occupancy averaged a healthy 75 per cent over the entire year. The proportion of North Americans fell sharply as cheaper destinations in Latin America and Mexico were heavily promoted. As flights from the United States decreased, so those from Europe increased. This trend was welcomed as European tourists generally stay longer and spend more. Cruise ship excursionists fell slightly to 83,550.

Development outlook

Capital expenditure and widespread infrastructural development was the hallmark of Compton's budgetary plans, announced in April 1989. A 58 per cent rise from EC$97.4m (US$36m) to EC$153.8m (US$57m) was complemented by a 14 per cent rise in recurrent spending from EC$190m (US$70.4m) to EC$216.5m (US$80.2m). A surplus of EC$39.8m was planned, and that, together with an estimated EC$45.7m in grants and aid and EC$61m from loans, would fund all but EC$10m of the capital programme. Confidence was high that the gap would be filled because of St Lucia's economic performance: inflation at 1 per cent (1987: 7 per cent), unemployment below 20 per cent and growth estimated at just over 6 per cent.

Planned capital work included better water supplies from a dam in the Roseau valley funded by a Canadian-led consortium, a new US$60m power station, the long awaited rebuilding of the West Coast main road and the exploitation of geothermal power at the Soufriere 'walk-in' volcano, always popular with visitors.

Compton has for long been an advocate of regional co-operation and in May 1987 he, together with Prime Minister Mitchell of St Vincent, proposed a union of the Commonwealth Eastern Caribbean islands. The economic benefits were especially stressed but during discussion it became apparent that only the Windward group were interested. It was eventually agreed in June 1989 that referenda would be held the next year on the question in the four islands, St Lucia, St Vincent, Dominica and Grenada.

Banana politics

Despite the apparent calm, political passions are never far from the surface. The St Lucia Labour Party (SLLP) only narrowly lost the 1987 elections, with eight seats to nine for the United Workers' Party (UWP). There were two elections because after the first, in April, Compton declared that the margin was too small for effective government. Another was called but the result was the same. Then a SLLP member was enticed to cross the floor with the promise of ministerial office. Since then SLLP leader Julian Hunte, a successful businessman who reorganised the party after its disastrous split and expelled the extreme left which had caused most of the damage, left no opportunity to regain the initiative against the UWP.

The SLLP's defeat was due to the intervention of a third party, the SLLP breakaway Progressive Labour Party (PLP), led by George Odlum. Odlum, a radical, was the dynamic former SLLP foreign minister during its term of government between 1979 and 1982, and noted for his Libyan and Cuban sympathies. It was he who had challenged the leadership and, when finally defeated, formed the breakaway party. The PLP won a seat in 1982 (the SLLP two) but the breakdown in relations was complete.

The struggle between the UWP and the SLLP moved into the banana industry when the SLLP orchestrated protests against the production levy, in association with the St Lucia Banana Growers' Association (SLBGA). Compton accused the Association's board, and especially its openly pro-Labour chairman Rupert Gajardar, of 'economic sabotage' and unacceptable political activity, especially as the implications for the industry of the single internal European market demanded close cabinet attention. He successfully steered a bill through the legislature removing the board and placing the industry under direct government control for an initial period of one year. There were other reasons for Compton's drastic move. One was pricing policy. Compton wanted low prices in order to build up the SLBGA's reserves in preparation for 1992, whereas the board wanted higher levels to encourage quality. Another were debts owed by farmers, allegedly pro-UWP, who refused to repay. While both sides claimed to be acting in the interests of democratic control of the industry, their only common ground may be the ousted chairman's summing-up: 'Whoever controls the banana industry and has the support of the banana growers has the power in St Lucia'.

COUNTRY PROFILE

Political system: Independent state within the Commonwealth with the British monarch as head of state, represented by the governor-general who exercises executive power. He appoints as prime minister (from the House of Assembly) the person likely to command majority support there and, on his advice, members of the cabinet. The legislature is bicameral: the appointed Senate (six members on the prime minister's advice, three on that of the leader of the opposition and two chosen by the governor-general acting after external consultation), and the House of Assembly (17 members elected by single-member constituencies for a five-year term).

Main cities/towns: Castries (population 51,500), Vieux Fort, Soufriere.

Media
Press: Main newspapers are *The Voice* (twice per week), *The Star, The Crusader* and *The EC News* (all weekly).
Broadcasting: One private and one government-owned radio service plus two commercial television services. Broadcasts in Creole, English and French.

External trade: Regular trade deficits are offset by invisible earnings from tourism, workers' remittances and aid flows.
Exports: Bananas (58% of total value in 1988), clothing (20%), paper & board (10%), electronics (6%), coconut oil, cocoa, beans, beverages, electronic goods. Main destinations: UK (60 per cent of total value in 1986), CARICOM and USA.
Imports: Manufactured goods (30-35% of total value), foodstuffs (18-20%), fuels (9-10%), machinery and transport equipment (18-20%). Main sources: UK (20-25% of total value), USA (20-25%), Trinidad and Tobago, Canada.

Agriculture: Mainstay of economy, contributing 20 per cent to GDP and employing 30 per cent of workforce. Main export crop is bananas, production of which has only recently recovered from hurricane damage in 1980 and received a further setback when a tropical storm hit the area in late 1986. Other cash crops are coconuts and cocoa. Self-sufficiency aimed at in the production of tomatoes, onions, carrots. A number of agricultural diversification projects are underway including the planting of tropical fruits and vegetables in the fertile Roseau valley for export to Europe. Over 50 per cent of the total area is cultivated arable land.

COUNTRY PROFILE

Industry: Rapid expansion of light manufacturing and agro-industries since 1970s. Main activities are cigarettes and tobacco products, tyre re-treading, concrete blocks and quarry products, textiles and clothing, furniture, bedding, soft drinks, rum, other alcoholic products, animal feed, brewing (Heineken), plastics, electronic assembly, vehicle batteries, fruit juices. There is also a petroleum storage and trans-shipment terminal constructed by the US company Amerada Hess. Emphasis is on expanding facilities at the Vieux Fort free zone and improving the island's infrastructure. Tourism sector is growing in importance with the total number of arrivals, mainly from USA and Britain, rising to more than 200,000. Estimated that around 50 per cent of total foreign exchange earnings derived from tourism; in 1987 it produced over $78m.

Energy: Attempts to reduce dependence on imported fuel include construction of a dam in Roseau and a geothermal project in Soufrière.

Membership of international organisations: CARICOM, Caribbean Development Bank, Commonwealth, FAO, GATT (de facto), ILO, OECS, Organisation of American States, state associated with the EC, UN, World Bank.

BUSINESS GUIDE

Time: GMT -4.

Climate: Mean annual temperature of 26°C. The island is cooled by the north east trade winds. Driest from Jan to Apr. The rainy season is between Jul and Oct.

Entry requirements
Passport: Required by all except Canadian and USA citizens and British subjects with proof of identity and assured onward passage staying not longer than six months.
Visa: Not required by nationals of Commonwealth, most EC, and Scandinavian countries, USA, Switzerland and certain other countries for any length of stay. For visits under 14 days visas not required by most nationals (except those from communist countries and Haiti).
Currency: No restrictions on local currency provided export does not exceed import. No restrictions on import/export of foreign currency. Visitors must satisfy immigration that they have sufficient for length of stay.

Health precautions
Mandatory: Yellow fever vaccination certificate required if arriving from an infected area.
Advisable: Typhoid, polio vaccination.

Representation overseas: London, Ottawa, Washington.

Air access
National airline: St Lucia is a shareholder in the regional airline LIAT.
Other airlines: Air Canada, Air Antilles, Air Martinique, American, British Airways, BWIA, Helen Air.
Tax: Airport service charge EC$20 payable on departure (EC$10 if proceeding to most Caribbean destinations).
International airports: Hewanorra (UVF), 64 km from Castries, duty-free shop, bar, restaurant, shops, car hire, tourist information outlet; Vigie (SLU), 3km from Castries, bar, restaurant, car hire.

Surface access
Main ports: Castries, Vieux Fort, Soufrière.

Hotels: Bills include eight per cent tax and usually 10 per cent service charge.

Car hire: Available in Castries, Vieux Fort and Soufrière, and through hotels. National or international licence acceptable. Traffic drives on the left.

City transport
Taxis: Relatively cheap and widely available. Fixed-rate system but advisable to negotiate fares in advance, especially for long journeys. Tips are not expected.

National transport
Road: All centres are served by the well-maintained road network. Main roads constitute over half of 800 km network.
Buses: Unscheduled local basic service by independent drivers; also boats ply to various destinations.

Public holidays: 1 Jan, 2 Jan, 22 Feb (Independence), Carnival (Feb or Mar, one and a half days immediately before Ash Wednesday), Good Friday, Easter Monday, 1 May, Whit Monday, Corpus Christi, Emancipation Day (first Mon in Aug), Thanksgiving (first Mon in Oct), 13 Dec (National Day), 25 Dec, 26 Dec.

Working hours
Government and business: (Mon-Fri) 0800 – 1230, 1330 – 1630.
Banking: (Mon-Thu) 0800 – 1300; (Fri) 0800 – 1200 and 1500 – 1700.
Shops: (Mon-Fri) 0800 – 1200, 1300 – 1600; (Sat) mornings only.

Telecommunications
Telephone: Dialling code for St Lucia: IDD access code + 1.809.45 followed by subscriber's five-digit number. Direct international dialling now in operation.
Telex: Fully automatic telex, international telegraph and leased circuit services available. Public telex booth available from Cable and Wireless in Castries.

Banking: Bank of Nova Scotia, Barclays Bank, Canadian Imperial Bank of Commerce, National Commercial Bank of St Lucia, Royal Bank of Canada, St Lucia Co-operative Bank.

Electricity supply: 220 V AC, 50 cycles.

Hotels: Cap Estate
Club St Lucia – Smugglers Village, P.O. Box: 915 (tel: 20551, 20519; tx: 6228; fax: 20281).

Le Sport, Caribue Beach, Anse du Cap, P.O. Box: 437 (tel: 28551; tx: 6330; fax: 27419).

Hotels: Castries
Couples, Vigie Beach, P.O. Box: 190 (tel: 24211; tx: 27419).

Cunard La Toc, P.O. Box: 399 (tel: 23081/9; tx: 6320; fax: 21012).

Green Parrot, The Morne, P.O. Box: 648 (tel: 23399, 23167/8).

Vigie Beach Hotel, Vigie Beach, P.O. Box: 395 (tel: 25211; tx: 6219).

Hotels: Choc Bay
Halcyon Beach Club, Choc Beach, P.O. Box: 388 (tel: 25331; tx: 6343; fax: 25434).

Hotels: La Brelotte Bay
Bois d'Orange Holiday Village, P.O. Box: 1741 (tel: 28213, 20021).

East Winds Inn, P.O. Box: 193 (tel: 28212).

Windjammer Landing Villa, P.O. Box: 1504 (tel: 21041; fax: 24115).

Hotels: Marigot
Marigot, Marigot Bay, P.O. Box: 101 (tel: 34357, 34246; fax: 34303).

Hotels: Reduit
St Lucian, P.O. Box: 512 (tel: 8351/5; tx: 6326; fax: 28331).

Hotels: Soufriere
Anse Chastanet, P.O. Box: 216 (tel: 47553, 47354; fax: 47097).

Hotels: Vieux Fort
Club Mediterranée (tel: 46547).

Car hire: Castries
ABC Cars Ltd: PO Box 93 (tel: 26445).

Avis: Vide Bouteille PO Box 1010 (tel: 24554, 22700, 22202; tx: 6363; fax: 31536).

Budget: PO Box 986 (tel: 20233, 28021; tx: 6303 SUMOL); Marisule (8021).

Colly's Car Rental: PO Box 62 (tel: 22303).

St Vincent

Tony Thorndike

A remarkable election victory by the incumbent National Democratic Party resulted in a one-party parliament. While it was a clear endorsement of James 'Son' Mitchell's leadership, there were fears that an 'elected dictatorship' could result.

St Vincent and the Grenadines, to give the state its full name, experienced mixed economic fortunes in both 1987 and 1988. Real growth in both years barely reached 3 per cent on the strength of a recovery in the banana industry and considerable construction activity. This moderate growth was not enough to cut already high levels of unemployment and reduce excess liquidity in the banking system. But a very effective election campaign by the governing party glossed over the record, stressing prudent financial management and the need for a continuing structural adjustment programme, and promoting land reform.

Sweeping victory

On a voter turnout of over 71 per cent, the National Democratic Party (NDP) swept to victory with 66 per cent of the votes, 29,038 against 13,308 for the opposition St Vincent Labour Party (SVLP). In the last election, in 1984, it attracted 51.4 per cent. The SVLP's share of the vote dropped from 41.4 per cent in the 1984 election to 30.4. All 15 seats were taken by the NDP. Two small socialist parties also competed, each with two candidates: the Movement for National Unity (MNU) and the United People's Movement (UPM), having 1,031 (2.3 per cent) and 466 votes (1.1 per cent) respectively. In the outgoing parliament the NDP had held 10 out of 13 seats (another two seats were created through a constitutional amendment) to the SVLP's three. Fears were expressed, even by some NDP supporters, that government legislation and performance could go unchallenged.

The NDP first came to power in 1984. Its victory finally marked the end of a prolonged period of political instability. SVLP leader Milton Cato retired, to be succeeded by his former deputy Hudson Tannis, but he died in an air crash in 1986. Vincent Beache was elected leader in preference to union leader Burns Bonadie, and attempted to rebuild the demoralised party. Bonadie later joined the NDP. A man of gentle character, Beache was no match for the NDP leader, former agronomist James 'Son' Mitchell, a nationalist with firm views on development.

Mitchell, although a professed democrat, had since 1987 influenced the state radio to favour the NDP. Most of the press also supported his administration and he and the party benefited from generous donations from the business community, especially Hans Neuman, the Caracas-based owner of the 'royal' island of Mustique. Not surprisingly, there was little media coverage or discussion of SVLP policies, only denigration. Worse, the SVLP in its campaign missed the point. It said little about unemployment or the situation of the poorer sectors of society, rather it promised further tax relief (for the more well off), an aim the NDP insisted was impossible. The NDP on the other hand championed local food production (a favourite Mitchell theme) and land reform in favour of the many peasant farmers. In short, Beache's 'People's Agenda' was not credible. Beache himself was defeated by a political newcomer, Jonathan Peters, the former ambassador to the United Nations, although he did defeat Ralph Gonsalves, the torch-bearer of the MNU, a breakaway from the UPM.

1990 will see a referendum on the question of political union of the four Windward islands – St Vincent, St Lucia, Dominica and Grenada. The original proposal in May 1987, which Mitchell firmly endorsed, included Antigua, Montserrat and St Kitts-Nevis. However, the leaderships of those three islands declined to participate.

Management and reform

Underlying Mitchell's success was not only political stability and leadership but also firm economic management. His incoming government inherited a sorry legacy of deficits, confusion and poor morale in the public service, and a moribund economy. St Vincent's image in the international community, especially among aid donors, was poor.

Since 1984 a number of loss-making public enterprises have been closed, greater efficiency levels achieved and aid negotiated from a variety of sources. Donors included Britain and the European Community – who together helped finance a new hospital and high school in the capital, Kingstown, and paid for an engineering feasibility study for an airport on Bequia, Mitchell's home island in the Grenadines – the World Bank, Canada, France and Japan. Emphasis was put on agriculture (notably for the Crop Diversification Programme), fisheries and forestry projects. Budgetary surpluses became the norm, and 1988'89 was no exception. The July 1988 budget substantially raised tax thresholds and set recurrent expenditure at EC$123.6m (US$45.8m) compared to EC$112.7m in 1987'88. Revenue was projected to be EC$135m (US$50m), producing a surplus of EC$11.4m to be directed to the EC$99.7m capital expenditure programme. A surplus of EC$10m was achieved in 1987'88. Despite pay awards associated with the Administrative Reform Programme, the projected surplus will be achieved. Another plus was Mitchell's success during 1987'88 in negotiating soft loans and grants worth some EC$80m to fund the capital budget. But there were two concerns. External debt reached US$67m, or 60 per cent of GDP, a good deal of which had been inherited in 1984. The other was the near doubling of the trade deficit to EC$63.4m.

Agriculture the key

As a predominantly agricultural state, St Vincent's economic fortunes are often adversely affected by events beyond its control, such as inclement weather and world prices. Storm damage severely reduced banana production in 1987 but the industry's fortunes were restored in 1988 when 64,887 tonnes were cut, a 75 per cent rise. But adverse prices reduced arrowroot production by 15.8 per cent to 148 tonnes. In order to encourage diversification into arrowroot, prices to farmers were increased. But world market conditions for the commodity, the source of natural starch, are uncertain due to the availability of corn-based substitutes and strong competition from Brazil. However, plans were made to penetrate the Jamaican market. Government intervention was also apparent in the land reform programme, which stresses food production. But heavy rainfall and depressed market conditions in the main Trinidad and Tobago export market hampered progress.

Manufacturing activity was, as in earlier years, stagnant. Inadequate infrastructure and transport problems were the main reasons, but the completion of the Diamond Industrial Estate in 1989 provided much needed space. Strong demand for beer and tennis rackets boosted exports. The electronic and switchgear plant also did well. One industry, apparel, is home-based, the women workers being very low paid. Work continued on the Cumberland hydroelectric project; inaugurated by the Labour government, it should supply the bulk of St Vincent's electricity when completed.

The tourist industry also remained static, in sharp comparison with other Caribbean destinations. Poor air services from North America did not improve matters and the numbers of US tourists fell. Stopover arrivals in 1988 were virtually the same as the 45,990 who came in 1987. Air excursionists to the Grenadines, mainly from Martinique, contributed another 13,000 visitors. A worrying statistic was a reduction in yacht arrivals, as the Grenadines tourist industry is particularly geared to that sector. Tourism will never be a major industry despite the beauty of the Grenadines. St Vincent, the 'mainland', has by virtue of its volcanic origin only one white sand beach, while such Grenadine islands as Mustique, Union, Petit Nevis, Palm and Canouan have strictly limited facilities and access, are very expensive and hold no appeal for large-scale package tour operators; some are even private fiefs, a legacy of the past which, under Mitchell, will never return.

KEY FACTS

Official title: St Vincent and the Grenadines

Head of state: Queen Elizabeth II

Head of government: Prime Minister James Mitchell

Ruling party: New Democratic Party

Capital: Kingstown

Official Languages: English

Currency: East Caribbean Dollar (EC$) = 100 cents

Exchange rate: EC$2.69 per $ (fixed)

Area: 388 sq km

Population: 112,400 (1987)

GDP per capita: $1,210 (1987)

GDP real growth: 3.0% (1988)

Inflation: 2.0% (1988)

Trade balance: –$99.27m (1988)

Foreign debt: EC$67.0m (1988)

Visitor numbers: 127,755 (1987 including cruises)

COUNTRY PROFILE

Political system: Independent state within the Commonwealth with the British monarch as head of state represented by a governor-general who exercises executive authority and is appointed on the prime minister's advice. The governor-general appoints six senators to the legislature, the House of Assembly (four on the prime minister's advice and two on the advice of the leader of the opposition); the other 15 representatives in the House are elected for five years by universal adult suffrage.

Main cities/towns: Kingstown (population 25,000), Georgetown, Barrouallie, Chateaubelair. Seven of the Grenadines are inhabited, including Bequia, Canouan, Mustique and Union.

Media
Press: Weeklies include *The Vincentian, The Star, The New Times, The Unity* and *The Justice.*
Broadcasting: Government-owned radio station and two private cable television services, using US satellite programmes.

External trade: Heavy dependence on banana exports and aid flows.
Exports: Bananas (30-35 per cent of total value), vegetables, flour, arrowroot starch, coconuts, sweet potatoes, tannias, eddoes, coconut oil. Main destinations: UK (75 per cent of total value), Trinidad and Tobago, Barbados, USA, Canada.
Imports: Foodstuffs (25-30% of total value), manufactured goods (19-22%), machinery and transport equipment (14-16%) mineral fuels (8-10%). Main sources: USA (30-33 per cent of total value), UK, Trinidad and Tobago, Canada, Japan.

Agriculture: Sector is mainstay of economy, though its contribution to GDP has fallen from 40 per cent in 1960 to 18 per cent. Around half of the total area is arable land with only a small percentage unused. Bananas (grown mainly on small farms under auspices of St Vincent Banana Growers Association) are the main export crop. Banana shipments, mainly to the UK, have fallen steadily since mid-1970s due to slump in world demand and storm damage. St Vincent is the world's leading producer of arrowroot and an exporter of coconut oil. Carrots, and plantains are also important cash crops. The main food crops are sweet potatoes, tannias, yams, vegetables, and various fruits. Offshore fishing production has increased rapidly.

Industry: Sector employs around eight per cent of the workforce and contributes 10 per cent to GDP. Activity is primarily based on agricultural processing. Units in operation include those producing: cigarettes, tobacco products, coconut oil, textiles and clothing, soft drinks, fruit juices, milk, beer, rum, furniture, arrowroot starch, tyre re-treading, concrete blocks and quarry products. A flour mill, serving all the Windward Islands, banana box factory and yacht building yard are the main export industries.

Membership of international organisations: CARICOM, Commonwealth, GATT (de facto), state associated with the EC, IMF, UN, World Bank.

Time: GMT -4.

Climate: Tropical, tempered by trade winds, with temperature range 18°C to 32°C. High levels of rainfall between May and Nov, especially in the north.

Entry requirements
Passport: Required by all except nationals of Canada, USA and UK for visits not exceeding six months.
Visa: Not required as length of stay established by immigration authority on arrival.
Currency: Declaration usually required on arrival. Export must not exceed import.

Health precautions
Mandatory: Yellow fever vaccination certificate required if arriving from an infected area.

Air access
National airline: St Vincent is a shareholder in LIAT.
Other airlines: Air Martinique.
International airport: Arnos Vale (SVD), 3 km from Kingstown.

Surface access
Main port: Kingstown.

Hotels: Wide range of good hotels available at reasonable prices – except in the privately-owned islands of Palm, Mustique and Petit St Vincent. Five per cent tax is added to room rates. 10 per cent tip is usual if service charge not included on bill.

Car hire: International or temporary Vincentian licence required. Traffic drives on the left.

City transport
Taxis: Usually ordered by telephone; fixed-rate system, 15 per cent tip is usual.

National transport
Air: There are landing strips (suitable for light planes) on Mustique, Canouan and Union island – several companies operate flights between the islands.
Road: Total network is over 1,000 km – although surface varies, main/key routes are generally of good surface.

Public holidays: 1 Jan, 22 Jan (Discovery Day), Good Friday, Easter Monday, Labour Day, Whit Monday, Carnival (first Mon and Tues in Jul), first Mon in Aug, 27 Oct (Independence), 25 Dec, 26 Dec.

Telecommunications
Telephone: Dialling code for St Vincent: IDD access code + 1.809.45 followed by subscriber's five-digit number. 100 per cent automatic service.
Telex: Automatic telex and telegraph facilities available from Cable and Wireless. Modern computerised telex service provided for subscribers.

Electricity supply: 230 V AC, 50 cycles.

Hotels
Heron, P.O. Box: 226 (tel: 71631; tx: 7557).

Sunset Shores, Villa, P.O. Box: 849 (tel: 84411; tx: 7557; fax: 74800).

Business advice and information
Barclays Bank, PO Box 604, Halifax Street, Kingstown (tel: 456-1706; tx: 7520).

Department of Tourism, PO Box 834, Halifax Street, Kingstown (tel: 457-1502).

San Marino

David Lane

The flurry of activity connected with local elections for the Republic's nine 'castles' in June 1989 barely rippled the tranquil surface of a domestic scene which San Marinese officials described as quiet. With the 'historic compromise' coalition of Communists and Christian Democrats seeking endorsement from the electorate, the republic's two largest parties, from opposite ends of the political spectrum, were encouraged to avoid polemic.

The results clearly rewarded the stability, style and work of the Grand and General Council central government formed after the general elections of May 1988 when the Christian Democrats won 44.1 per cent of the vote and the Communists 28.7 per cent. In those elections the Socialists took 11.1 per cent and the Unitary Socialists 13.6 per cent.

Local elections

When the votes were counted on 4 June 1989, the Christian Democrats were satisfied with a result giving them 47.1 per cent of the poll, which was half a point better than they achieved in 1984's 'castle' elections. The Communists had less cause for rejoicing with a 26.9 per cent share, though this was a substantial 4.7 per cent improvement on 1984. The Socialists' 14.4 per cent was 3.3 per cent higher than their 1988 vote but was still one quarter of a point below 1984's achievement. San Marino's Unitary Socialists were clear losers, the 11.7 per cent share in 1989's 'castle' elections keeping them on the downward slope; they won 14.6 per cent of the vote in 1984.

The Christian Democrats won their highest share (58.4 per cent) in one of the smallest 'castles', Chiesanuova, though they also did well in the large San Marino Citta 'castle', winning 52.4 per cent of the vote. The weakest turnout of Christian Democrat voters was in Fiorentino (29.8 per cent), the medium to small 'castle' where the Communists achieved their best result (38.8 per cent). San Marino's Communists were weakest in Domagnano (18.1 per cent), the 'castle' where the Unitary Socialists obtained their best result (15.8 per cent).

With 89 per cent of eligible San Marinesi going to the polling booth, June 1989's 'castle' elections represented a reliable public judgement of the Congresso di Stato central government. At July 1989 this comprised 10 members, six Christian Democrats and four Communists. The key posts of foreign affairs and finance and economic planning were held by Christian Democrats while a Communist headed home affairs.

Flourishing economy

San Marino's 'historic compromise' government was helped by good economic conditions in 1988 and 1989. In contrast to neighbouring Italy, unemployment was not a serious problem during the 1980s and statistics showed that at the end of the decade, it was even less so. From 750 jobless at the end of 1986 the figure declined to 650 at the end of 1988. At the end of March 1989 the number of unemployed San Marinesi was 580, equal to 5.2 per cent of the working population. Though female unemployment was 9.3 per cent, for men the problem was virtually non-existent. With 2.2 per cent unemployment, San Marino could reasonably claim full male employment.

Shortage of appropriate work was proportionately greater among San Marino's professionals or aspirant professionals. Nearly 50 of the 580 out-of-work at the end of March 1989 were university graduates or of equivalent educational attainment. And a further 200 had high school diplomas. No less than nine doctors and surgeons were among the San Marinesi looking for jobs.

In common with Italy, unemployment was concentrated among the young. Nearly 15 per cent of March 1989's unemployed were under 20 and a further 53 per cent were between 20 and 30 years old. Yet expansion of employment in San Marino seemed likely to give most of them good hopes of work. The total number of jobs rose steadily from 10,560 in December 1985 to 11,330 at the end of 1988, and increased further to 11,680 at the end of March 1989. About 21 per cent of the Republic's workforce were self-employed in 1989 compared to 25 per cent in 1985.

In addition to the lower rate of unemployment, the increase in hours lost through strike action in the first quarter of 1989 was also an indicator of the flourishing economy. After two very peaceful years with strike action less than one tenth of 1986's level, in the January-March period of 1989 nearly 72,000 hours were lost to strikes, one third higher than in the whole of 1986. Virtually all the strikes were by workers seeking better conditions in new labour contracts.

Another indicator of the health of San Marino's economy was automobile registrations. In 1988 these returned to the 2000 mark last reached in 1981, and 1770 of these were new vehicles compared to 1560 in the previous year. San Marinesi continued to visit car showrooms in the first quarter of 1989 and the figures pointed to a record year, with 620 new automobiles being registered compared to 590 in the same period of 1988. A point of interest was the republic's high level of purchases of non-Italian cars. The Fiat Group's share of its home market was 59.9 per cent. In San Marino, the Italian group's market share was 41.6 per cent in 1988.

Member of Council of Europe

In 1988 and 1989 San Marino appeared to give most attention to exploiting the favourable period which its economy was enjoying. Politics were quiet at home, and foreign policy was in a reflective mood without the major journeys by San Marino's leaders abroad or important foreign visitors to the republic which characterised the mid-1980s.

Officials pointed, however, to a significant step which was taken in November 1988 when the Republic of San Marino became a member of the Council of Europe. San Marino, which with its adhesion to this European body signed the European Convention on Human Rights, was accorded a warm welcome as a nation which has always strenuously defended the idea of liberty and equality. Speaking on the occasion, the republic's secretary of state for foreign affairs emphasised that San Marino attached particular importance to its entry into the Council of Europe.

KEY FACTS

Official title: The Republic of San Marino

Head of state: Two Captains-Regent elected for six months at a time

Ruling party: Coalition of San Marino Communist Party and San Marino Christian Democratic Party

Capital: San Marino

Official Languages: Italian

Currency: San Marino Lira (SML) and Italian Lira

Exchange rate: Lire 1,360 per $ (Nov 1989)

Area: 60.5 sq km

Population: 22,543 (1988)

GDP per capita: $7,001 (1985)

Visitor numbers: 2.9m (1988)

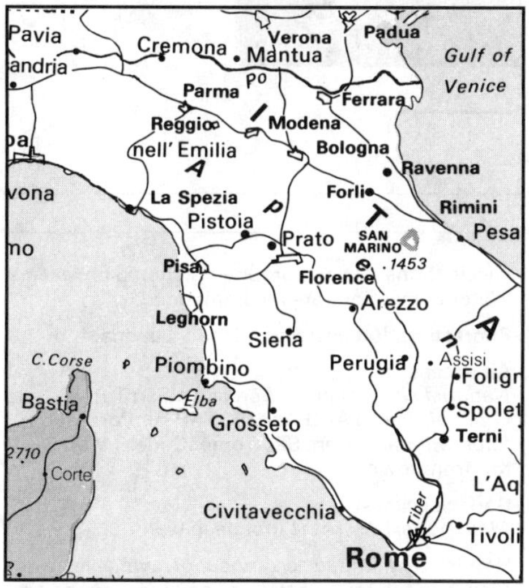

COUNTRY PROFILE

Political system: Republic with legislative power vested in unicameral Grand and General Council, with 60 members elected every five years. Council elects two members every six months to act as Captains-Regent (joint Heads of State and Government) and a ten-member Congress of State which exercises executive power. Country divided into nine 'Castles', each governed by a Castle-Captain.

Main cities/towns: San Marino (population 4,178).

Media

Press: Several periodicals, mostly organs of the political parties. Daily insert in Italian newspaper.
Broadcasting: Broadcasts from Radiotelevisione Italiana are received. Daily local information bulletins from private TV and radio stations in Italy.

Domestic economy: Primarily based on tourism (which accounts for 60 per cent of revenue) and agriculture. Sale of postage stamps to foreign collectors also an important source of income, as are more traditional industries such as quarrying for building stone, although emphasis is currently being placed more on light manufacturing industries.

External trade: Customs and currency union with Italy. Major exports are wine, textiles, furniture, stone, stamps, coins and ceramics. Import dependence on Italy.

Legal system: Based on civil law system with Italian law influences.

Membership of international organisations: Council of Europe, ICJ, International Institute for Unification of Private Laws, International Relief Union, IRC, ITU, UNESCO, UPU, WFTU, WHO, WTO, NAM, (observer).

Time: GMT + 1 (GMT + 2 from late Mar to late Sep).

Climate: Mediterranean, with warm summers and dry, cold winters. Temperatures can range from about 30°C to below 0°C.
Clothing: Lightweight for summer, mediumweight and topcoat for winter.

Entry requirements: No regulations in operation.

Air access

Main international airport: Rimini (RMI) 27 km from San Marino, or Bologna (BLQ) 135 km from San Marino.

Surface access: Regular bus service along highway between Rimini and capital. Funicular operates between Borgo Maggiore and capital.

Hotels: Approximately 30 hotels, of various standards.

National transport: No railways. Limited bus service operating from Italy. Good internal services.

Public holidays

Fixed dates: 1 Jan, 6 Jan, 5 Feb, 19 Mar, 25 Mar, 1 Apr, 1 May, 29 Jun, 28 Jul, 15 Aug, 3 Sep, 1 Oct, 1 Nov, 2 Nov, 8 Dec, 24-26 Dec and 31 Dec.
Variable dates: Good Friday, Easter Monday, Ascension Day, Corpus Christi.

Telecommunications: Approximately 15,500 telephones in use. International dialling code: 39 549. There are no area codes. Telex country code: 505 1, SO.

Business advice and information: There is no chamber of commerce in San Marino. The following government offices are useful sources of information:

Office for Industry, Handicrafts and Trade, Palazzo Mercuri (tel: 992745, 991385).

Secretariat of State for Finance and the Budget, Palazzo Begni (tel: 992345; tx: 330).

Travel information

Ufficio di Stato per il Turismo (tourist board), Palazzo del Turismo 47031 (tel: 992101/5).

Hotels: San Marino

Grand, Via Antonio Onofri 31, 47031 (tel: 992400; tx: 0505555; fax: 992274).

Titano, Contrada del Collegio 21, 47031 (tel: 991007; tx: 505444).

São Tomé and Príncipe

Susi Pollak

During an overseas tour in 1989 São Tomé's President Manuel Pinto da Costa predicted, not without apprehension, that 1990 would be a turning-point in his country's political life. There would be important political consultations, a referendum and elections: this was all to the good. But there was also the possibility of social unrest, caused by the tough measures imposed by the International Monetary Fund (IMF) on his small nation. After four years of economic liberalisation there was need to review the situation. Improvements could be effected by co-ordinating the contributions made by the various national and multilateral aid agencies whose projects did not always complement each other, he said.

While the Portuguese ambassador in São Tomé described the government's economic reforms as 'brave, pragmatic and responsible', the people most affected by the stringent measures – particularly the redundant civil servants – were more critical of structural adjustment. A quarter of all public employees lost their jobs while hundreds of foreign experts flooded in. In the space of one year, the IMF imposed three devaluations of the local currency, the dobra, and was demanding an end to food subsidies. An agricultural labourer earned Db140 a day while rice cost Db100 a kilo. There was a lot of public resentment.

Economy Minister Teotone Torres resigned in November 1988 over the World Bank's austerity measures. After two years of negotiations with the IMF over a $1.9m facility no agreement had been reached. But in mid-1989 the World Bank signalled that São Tomé's grievances were receiving attention: an IDA credit of $5m was announced for an $8m project designed to increase income and employment opportunities and improve social services. An autonomous agency, the social and infrastructure fund, would be established to this end. Included in the services to be improved were health care, education and low-cost housing.

KEY FACTS

Official title: Democratic Republic of São Tomé and Príncipe

Head of state: President Dr Manuel Pinto da Costa

Head of government: Prime Minister Celestino Rocha da Costa

Ruling party: Movimento de Libertação de São Tomé e Príncipe (MLSTP)

Capital: São Tomé

Official Languages: Portuguese

Currency: Dobra (Db) = 100 centavos

Exchange rate: Db107 per $ (Nov 1989)

Area: 964 sq km

Population: 111,000 (1988)

GDP per capita: $280 (1987)

GDP real growth: 5.1% (1988)

Trade balance: –$10.0m (1988)

Foreign debt: $90m (1988)

Education in particular is in need of a boost. 47 per cent of the population are under 15, and there are not enough teachers for children of school age, two-thirds of whom actually attend classes. There is a shortage of skilled personnel; not all of the students sent abroad for further studies return after graduation. The foreign experts are clearly needed, at least for the time being, and must be credited with some of the economic progress made.

The volume of the traditional agricultural exports rose by 33 per cent to 7,532 tonnes in 1988. Cocoa production that year was 4,900 tonnes (compared with 3,500 tonnes in 1987); copra 2,100 tonnes; coffee 31 tonnes; palm oil 79 tonnes. But the collapse of the world cocoa market underlined even more than before the need for diversification. There are two immediate options for this: food production (including fishing) and tourism.

São Tomé still imports 90 per cent of its food, for historical as well as geographical reasons: the colonial system of cocoa plantations left little land on the small islands for the cultivation of other crops and discouraged the development of peasant farming. Government policy and world food programme schemes are designed to demonstrate the desirability of local food production, which will also contribute to the success of the other opportunity for growth: tourism.

There is one international hotel – the 50-room Swiss-run Miramar – which already accepts group bookings from operators like International Air Services (IAS) and Swissair. IAS, an Irish group, owns 73 per cent of Equatorial Airlines and uses São Tomé as a base for its freight operations and aircraft maintenance; it also has the management contract for Miramar. Swissair flights are being planned from Frankfurt via Douala, and TAP hopes to inaugurate direct weekly flights to Portugal.

These prospects will break the islands' geographic isolation. Indeed, interest in São Tomé has widened since the Geneva donor conference in 1989, when generous commitments were made: Italy pledged $16m over two years for agricultural development, West Germany DM3m for improving harbour facilities, while Spain is to contribute 1,500 tonnes of wheat and 200 housing units. The Japanese agreed to 323m yen for fishery development. But 1989 was officially described in São Tomé as 'the year of co-operation with France', which promised FFr50m in aid. Indeed, France and Portugal, the former colonial power, are both vying for São Tomé's friendship, and some of da Costa's ministers fear that their colleagues are trying to drag the country back into Portugal's orbit. But this competition between suitors has its advantages: when the Portuguese started work on a 436m escudos television service for São Tomé, the French announced FFr4m in grants for media development.

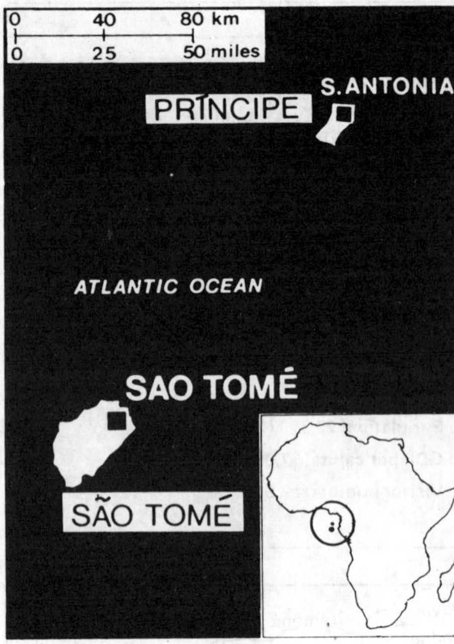

COUNTRY PROFILE

Political system: The MLSTP (Movimento de Libertação de São Tomé e Príncipe) is the only legal party, but according to the revised constitution (1987) candidates presented by 'recognised organisations' and independents are also allowed to stand in the four-yearly legislative elections. The president is elected separately at four-yearly intervals by direct universal suffrage from two candidates chosen for the purpose by the MLSTP. As head of state, the president has executive power and is answerable to the National Assembly. He has sole responsibility for defence, security and foreign affairs, and he appoints a prime minister who has to lay the government's programme before the National Assembly for approval. Gained independence from Portugal in 1975.

Political parties: The MLSTP is the only legal political party.

Main cities/towns: São Tomé is the main population centre.

Language: Portuguese is the official language; Fôrro is widely spoken.

Media
Press: Two weekly newspapers – *Revolucão* (official organ of Ministry of Information) and *Diario da Republica*.
Radio: Approximately 25,000 receivers in use. Home service broadcast in Portuguese by Radio Nacional de São Tomé e Principe.
Television: There is no television service.

External trade: Balance of payments position depends on cocoa exports. The main trading partner is France. Current account deficit is widening because of growing debt burden.
Exports: Cocoa (85-90 per cent of total), coffee, copra, palm kernels. Main destinations: EC countries.
Imports: Fuel, capital goods, food. In 1987 food imports of 7,256 tonnes (part of the international aid effort) amounted to about 50 per cent of total consumption. Main sources: UN, France, Portugal and Japan. Eastern Europe and Angola remain important.

Agriculture: Based on cocoa production. (The islands were the world's largest cocoa producer at the beginning of the century). Prior to independence, production was in the hands of the Portuguese, but output slumped drastically after their departure in 1975, when 70 per cent of the plantations was nationalised. There is a programme of returning cocoa growing to the private sector, and government efforts to increase production are being intensified. The second largest export crop is coffee; other cash crops are copra, palm kernels, cinnamon, pepper, breadfruit. Priority is currently being given to the diversification of food crops in an effort to reduce the large import bill. Fishing is being encouraged for local consumption and possible future export.

Industry: Is limited to small-scale manufacturing concerns such as soap, soft drinks, timber processing, palm oil, bricks, textiles.

Energy: All petroleum products are imported. Exploration for oil reserves in the early 1970s was halted due to technical difficulties, even though some oil was found. Installed electricity generating capacity is currently around 9 mW.

Membership of international organisations: African Development Bank, African States Associated with the EC (Lomé Convention), Arab Bank for Economic Development in Africa (recipient), CEEAC, Economic Commission for Africa, GATT, IDA, IFAD, IMF, OAU, UN, WHO, World Bank.

Time: GMT.

Climate: Equatorial with high temperatures and humidity. Average temperatures remain fairly constant throughout the year with a daily range from 20°C to 32°C. Driest month July, wettest March.

Entry requirements
Passport: Required by all.
Visa: In general required by all. Application should be made well in advance of visit as there are often considerable delays before visa is issued.

Health precautions
Mandatory: Yellow fever vaccination certificate required if travelling from infected area or if visit is longer than two weeks.
Advisable: Yellow fever, cholera, typhoid, polio vaccinations. Malaria prophylais should be taken. Water precautions are recommended.

Representation overseas: Lisbon, Luanda.

Air access
National airline: Linhas Aéreas de São Tomé.
Other airlines: TAAG Angola, TAP-Air Portugal.
International airport: São Tomé (Code: TMS), 6 Km from town.

Surface access
Main ports: São Tomé (not deep water).

Hotels: Very limited accomodation available mainly in the form of state owned inns; no major hotels.

National transport
Air: Restricted service links the two islands.
Road: Only about 300 km of roads, of which about two-thirds are asphalted, but network is being improved.
Buses: Frequent, efficient bus service.

Public holidays
Fixed dates: 1 Jan, 1 May, 10 Jun, 15 Aug, 5 Oct, 1 Nov, 1 Dec, 25 Dec.
Variable dates: Shrove Tuesday, Good Friday, Easter Monday, Corpus Christi.

Working hours
Banking: (Mon-Fri) 0730 – 1130.

Telecommunications
Telephone: Dialling code for São Tomé and Príncipe: IDD access code + 239 followed by subscriber's number. (There are no area codes). Local telephone telegraph service described as good but international facilities are poor. 45 per cent automatic service.
Telex: Available in hotels, and in capital. Telex country code: ST.

Banking
Main centre: São Tomé.
Head offices: Banco Nacional de São Tomé & Príncipe, Caixo de Crédito.
Foreign branch: Banco Nacional Ultramarino.

Electricity supply: 220 V AC.

Representation in capital: China PR, Korea PDR, Portugal.

Saudi Arabia

A special correspondent

Saudi Arabia, like all the Arabian Gulf monarchies, was immensely relieved by the August 1988 cease-fire between Iran and Iraq. Saudi Arabia's relations with Iran, however, remained glacial in early 1989, and Iran boycotted the summer 1989 pilgrimage to Makkah and Median for the second year running. Saudi relations with Iraq seemed more edgy after the cease-fire than during the war itself, but Saudi diplomacy scored a real success in the spring of 1989 by promoting Egypt's re-admission to most Arab organisations. Within Saudi Arabia, business and economic confidence improved steadily in 1988–89 from the low point reached in 1986. But Saudi oil revenues and spending on imports, while starting to recover, remained far below the levels of the early 1980s. The drive to diversify industrial production, which had slowed down during the mid-1980s, was set for another move forward in the early 1990s, with major chemical projects on the horizon and the prospect of tangible results from the various offset investment programmes linked to arms purchases.

The cease-fire in the Iran–Iraq war had not been translated into anything like a true peace agreement by autumn 1989. The Iranian regime continued to express its strong hostility to what it termed the 'American Islam' of Saudi Arabia, and exacerbated the tension by supporting a mujaheddin group in Afghanistan which was strongly opposed to the Saudi-sponsored mujaheddin alliance.

While the ending of the immediate military danger to Gulf shipping and Saudi installations on the Gulf coast as a result of the cease-fire was clearly welcome to Saudi Arabia, the ending of the war also created the prospect of a divergence between Saudi Arabia and Iraq over future oil policy. During 1989, Iraq was moving to increase its oil production capacity in order to return to its very strong pre-war position within OPEC. Iraq's interest in maximising its oil revenues may well conflict with Saudi Arabia's long-term oil policy. Saudi Arabia also showed signs of unease about Iraq's political and military ambitions in the Arab world after the end of the war. In March 1989, King Fahd found it necessary to sign a non-aggression pact with Iraq during a visit to Baghdad; an unexpected development in view of the close ties which had existed between Iraq and Saudi Arabia during the war.

In other parts of the Arab world, Saudi diplomacy was more positive. King Fahd's visit to Cairo in March 1989 marked a major step forward in the slow process of re-admitting Egypt to the various Arab organisations from which it was expelled in 1979 following the Camp David agreements. Saudi Arabia had been keen to see Egypt re-admitted, but had to proceed with caution because of the strong resistance to the idea in certain more hard-line Arab states. Saudi Arabia's next objective will be the holding of an Arab summit involving both Egypt and the hard-line states such as Syria.

Economy

The Saudi Arabian Monetary Agency claimed that total GDP in Saudi Arabia rose by 3.2 per cent during 1988. There was particularly strong growth in agricultural GDP (+10.8 per cent) and in the industrial sector (+4.7 per cent). The considerable growth in imports during 1988 (up from SR75.3bn to SR81.6bn) was also an indication of increased activity in the Saudi economy. The impression of livelier business conditions was supported by the evidence of marketing surveys for consumer and electrical products conducted in the Kingdom during early 1989.

The upturn in confidence in the business community has apparently encouraged Saudi Arabia's planners to revive a long-standing policy of increasing the role of the Saudi private sector in the economy. The General Principles for the 1990–95 Development Plan which were formally approved at the start of 1989 called for private capital to participate more energetically in investment in Saudi Arabia. The private sector was also scheduled to become more involved in managing the services now provided by the state. A conference of academic and planning ministry experts in Saudi Arabia in January 1989 discussed possible long-term privatisation of urban and road maintenance, libraries and sports facilities. Corporations which were identified as candidates for immediate privatisation included the oil refining and oil marketing organisations, the national telephone organisation and the state airline Saudia. It was also expected that the private shareholding in the petrochemicals group SABIC would be raised from the 1989 level of 30 per cent. But specific plans for privatisation were still not available by August 1989.

It must be emphasised that, despite the signs of growing confidence, economic activity in Saudi Arabia is well down from the levels of the boom years of high oil revenues. In particular, the volume of government spending is still low. The 1989 state budget allocated a total of SR141bn for government spending during 1989; exactly the same amount as in 1988. To put this in context, government spending in the peak year of the oil boom (1981–82) amounted to SR288bn.

The 1989 state budget gave very little information about the proposed breakdown of government spending during 1989. But Saudi officials have generally stressed that the major infrastructural projects are now more or less complete, and that the overwhelming priority is to develop human resources; i.e. the trained Saudi manpower to operate the infrastructure and to staff the new industrial projects. In the January 1988 state budget for the calendar year 1988, education received 15.4 per cent of total government spending allocations. This proportion will tend to increase,

KEY FACTS

Official title: Kingdom of Saudi Arabia

Head of state: The Custodian of the Two Holy Mosques, King Fahd bin Abdul Aziz

Capital: Riyadh

Official Languages: Arabic

Currency: Riyal (SR) = 100 halalas

Exchange rate: SR3.75 per $ (Nov 1989)

Area: 2,149,690 sq km (approximately)

Population: 14.0m (1988)

GDP per capita: $5,535 (1988)

GDP real growth: 0.0% (1987)

Inflation: 1% (1988)

Oil reserves: 255bn barrels (1988)

Trade balance: $1,952 (1988)

as the government pursues its drive to reduce dependence on immigrant labour.

It should be noted that most estimates suggested that Saudi Arabia was still heavily dependent on immigrants, and plans to reduce the immigrant population appeared to have had little effect by mid-1989. The total number of immigrants in Saudi Arabia was perhaps as high as 3.5–4m. The main hope for achieving at least some reduction in the number of immigrants would be to increase the number of trained Saudis able to assume technical and managerial positions.

The second major priority for government spending was certainly the defence sector. In the 1988 state budget, defence was allocated 36 per cent of total spending. In view of the large arms contracts concluded during 1988 and 1989 with Britain, France and the United States, it was hard to imagine that the defence sector's share of spending would fall in coming years, especially in view of known Saudi plans to buy new battle tanks and to create a submarine fleet.

Apart from education and defence, the overwhelming priority of Saudi planners continued to be the containment of costs. The General Principles of the 1990–95 Development Plan stressed this priority on several occasions, saying that the government would aim to reduce the scope and extent of state services in areas where they exceeded reasonable levels. The government was also to limit future specifications for both development projects and operational projects. More positively, the Plan would tackle the problem area of inadequate maintenance of government projects, by adopting a new formula to evaluate project costs not only in terms of the cost of construction but also in terms of operation, maintenance, administration and repair throughout the life of the project.

Turning to the revenue side of the government's accounts, hard information remained difficult to come by. The 1989 state budget projected that total state revenues would rise from SR105.3bn in 1988 to SR116bn in 1989, reducing the planned budget deficit to be financed through domestic borrowing from SR35bn to SR25bn. Since no recent information was given for actual government revenues from oil, it is only possible to say that the generally higher level of world per-barrel crude oil prices during much of 1989 gave good reason to hope that the government would achieve its objective of higher revenues for 1989 as a whole.

The commitment to greater efficiency in government spending was clearly shown by the work of the Saudi Arabian health ministry during 1988–89. The ministry has in effect stopped the programme of constructing large hospitals, apart from completing a large complex of hospitals in Riyadh and a few large hospitals started in the first half of the 1980s. Instead, the ministry has been constructing a number of small (30–50 bed) hospitals in the smaller towns and more remote areas, with the aim of extending its coverage to new and previously unserved areas. The ministry was also concentrating on expanding the work of the primary health centres, which would act in future as the initial point of contact between patients and the health system and would maintain files on all families in Saudi Arabia. The enormous task of registering families and opening files was more or less complete in the summer of 1989. The new system should lead to far more efficient use of hospitals and treatment facilities, as well as a big improvement in the quality of information and statistics on illness and mortality.

KEY INDICATORS

	Unit	1984	1985	1986	1987	1988
Population	'000	11,130	11,590	12,070	13,610	14,020
Gross national product (GNP)	$m	112,807	95,135	78,887	74,418	77,599
GDP per capita	$m	10,135.4	8,208.4	6,535.8	5,467.9	5,534.9
Consumer prices	1985=100	103.4	100.0	97.0	96.1	97.1
Crude petroleum production	'000 b/d	4,079.1	3,175.0	4,784.2	3,975.2	5,086.3
Volume of exports						
Crude petroleum	1980=100	34.6	41.2	32.0	40.2	–
Refined petroleum	1980=100	93.0	85.0*	92.0*	90.0	–
Exports	$m	37,522	26,908	19,480	22,590	23,736
Crude petroleum	$m	34,243	24,180	16,975	19,271	20,500
Imports	$m	33,696	23,622	19,112	20,110	21,784
Balance of trade	$m	3,826	3,246	368	2,480	1,952
Current account	$m	–18,400	–12,932	–11,935	–9,570	–9,683*
Exchange rate	$ per SR	0.29	0.28	0.27	0.27	0.27

* estimated figure

Foreign trade

Saudi Arabia's imports slumped dramatically after the fall in oil revenues began to bite in 1983. But figures for 1987 and 1988 show a steady increase from the low point reached in 1986. From their peak in 1983 at SR135.4bn, Saudi Arabia's imports of goods fell to SR70.8bn in 1986. In 1987, the value of imports rose to SR75.3bn, and there was a further increase to SR81.6bn in 1988.

Saudi exports were also on a rising trend recovering from SR72bn in 1986 to SR84.6bn in 1987 and SR88.9bn in 1988. As a result, Saudi Arabia has had useful trade surpluses, amounting to SR7.3bn in 1988. But the very large Saudi deficit on services and invisibles continued to produce a large current account deficit. The latest available figures showed that this deficit was SR38bn in 1986.

Exports were still dominated by revenues from crude oil, which were SR75.7bn in 1988. But the successful start-up of the first generation of SABIC's petrochemical plants in 1985–86 at a time of strong world demand created a measure of diversification. Exports of petrochemicals and plastics accounted for 11.4 per cent of the total value of Saudi exports in 1988. Although the world market for chemicals may become tougher in the early 1990s, SABIC's ambitious plans for new projects will help to reduce further the dependence on crude oil. On a much smaller scale, Saudi industrial plants were starting to register successes in export markets. In 1989, these plants won orders to supply, for example, electrical cable to Iraq, air-conditioning units to the People's Republic of China and cement to neighbouring Arab and African countries.

Among suppliers, the United States held first place in 1988, with 16.2 per cent of the Saudi market, followed by Japan with 16 per cent. This reversed the 1987 positions, when Japan had a small lead over the USA. The United Kingdom retained third place among Saudi suppliers in 1988, with 7.3 per cent of the Saudi market. West Germany remained in fourth place in 1988 (7.2 per cent), followed by Italy (6.5 per cent), France (5.2 per cent) and South Korea (4.7 per cent).

All those involved in exporting to Saudi Arabia agree that the market has become extremely tough, and that Saudi customers are more price sensitive than ever before. Local Saudi factories were also taking an increasing share of the market, and were especially strong in building materials and certain foodstuffs. That said, imports of consumer products appeared to be holding up well, with continuing market opportunities for high quality products which could fit in with an increasingly sophisticated life-style and take advantage of the trend towards supermarket complexes and shopping malls in major cities.

Imports of private cars rose from about 52,000 units in 1986 to 75,000 units in 1987, still far below the 172,000 units imported in 1983. Saudi car buyers are now more careful about maintenance and servicing than in the past, and expect their cars to last in the tough climate. In consequence, Japanese cars with their excellent reputation for reliability continue to dominate the Saudi market.

Saudi Arabia is a major market for pharmaceuticals, with imports rising from SR1.25bn in 1986 to SR1.63bn in 1987. A large pharmaceuticals factory was due to open in Central Saudi Arabia in 1990 with the aim of meeting up to 40 per cent of Saudi demand for basic generic medicines and antibiotics.

Purchases of defence equipment were clearly likely to have a growing influence on the pattern of Saudi imports, because of the large size of the arms contracts recently negotiated. The very large contract for the supply of Tornado aircraft obtained by British Aerospace in 1985 and expanded in 1988 was a major factor enabling the United Kingdom to rise from fifth place in 1985 to third place among Saudi suppliers in 1988. The major deliveries under this contract would be made during the 1990s: of the total of 110 Tornado aircraft ordered, only 24 were delivered by May 1989. UK exports during the next few years will be boosted by the 60 training aircraft, more than 80 Westland helicopters and 6 minesweeping vessels which were ordered via British Aerospace in July 1988.

French exports will benefit in coming years from the impact of the £1.6bn agreement signed in June 1989 for the supply of two helicopter-carrying frigates and large numbers of Mistral surface-to-air missiles. Since at least part of the payment for this contract was to be made in the form of Saudi petrochemical and agricultural products, the deal would also help to diversify Saudi exports away from crude oil.

Oil

Saudi Arabia's oil production and oil revenues recovered strongly in the period 1987–89 from the low levels of 1986 (when Saudi Arabia suddenly and briefly increased its production so as to force down world crude prices and re-establish discipline among OPEC oil producers). During 1988, Saudi Aramco, the extraction company responsible for practically all Saudi crude oil production, averaged production of 4.93m barrels a day (b/d), well above the company's average of 3.99m b/d in 1987. Although weaker per-barrel prices meant that Saudi Arabia's export revenues from oil fell slightly in 1988 to SR75.7bn, compared with SR76.5bn in 1987, these revenues were still comfortably above the SR66.9bn earned from oil exports in the difficult year of 1986.

During the first two thirds of 1989, world demand for crude oil was strong by the standards of recent years, and average OPEC prices were frequently above the organisation's target of $18/barrel. In the face of this improvement, Saudi Arabia had to make some delicate choices about its long-term strategy within OPEC. Many in Saudi Arabia naturally welcomed the recovery in revenues caused by higher average prices for crude oil. But the Saudi Oil Minister Hisham Nazer was perhaps the leading supporter within OPEC of the view that the long-term interests of oil-producing states are best served by *not* allowing average per barrel prices to rise too fast. A period of high prices for crude oil could cause a long-term shift away from oil as an energy source in the industrial world. Such a shift would not suit Saudi Arabia, with its massive reserves: proved Saudi reserves were recalculated during 1988 to be about 252bn barrels, with a possible further 315bn barrels available.

As a consequence, Saudi Arabia generally opposes any prolonged period of prices much above $18/barrel. But the Kingdom is also keenly anxious to avoid a collapse of production discipline within OPEC which would remove, perhaps for ever, the organisation's ability to influence energy markets in the long term. At the June 1989 OPEC meeting, therefore, Saudi Arabia avoided a head-on confrontation with states such as Iraq which seek lower total oil production and an immediate rise in prices per barrel, so as to maximise their revenues in the short term. Although the Saudi oil minister was known to consider that there were good reasons to increase the total OPEC production quota to 20m b/d or even more, Saudi Arabia went along with the OPEC decision to limit the total quota to 19.5m b/d during the second half of 1989.

Looking to the longer term, Saudi Arabia seems confident that its oil production will rise considerably in the future, and is taking steps to prepare for this scenario. During 1989, Saudi Aramco moved ahead with preparations to expand the capacity of the East–West oil pipeline across the Arabian peninsula from the present level of 3.6m b/d to perhaps as much as 4.8m b/d. Since the capacity of this expanded pipeline

COUNTRY PROFILE

Political system: Absolute monarchy with King Fahd bin Abdul Aziz, the Custodian of the Two Holy Mosques, heading the government as Prime Minister and General Commander of the Armed Forces; and the Crown Prince, Abdullah bin Abdul Aziz, as First Deputy Prime Minister and Commander of the National Guard. Executive power rests with the council of ministers. Local government is administered through general municipal councils, district councils and tribal and village councils. A national consultative council has been formed and is believed to be preparing a draft for a constitution.

Main cities/towns: Riyadh (population estimated 1.5m), Jeddah, Makkah, Medina, Hofuf, Taif, Yanbu, Abha, Dammam, Al-Khobar.

Language: The official language is Arabic; English is widely spoken, especially in business circles.

Media
Press: Leading Arabic dailies are *Okaz* (circulation 75,000), *Al-Riyadh* (140,000 Sat-Thu), *Al-Jazirah* (90,030), *Al-Youm* (25,000), *Al-Madinah* (46,370), and the English dailies are *Saudi Gazette, Arab News, Riyadh Daily* and *Asharq Al-Awsat* (published daily in London). There are a number of magazines and periodicals, including *Saudi Economic Survey*, weekly in English. *Um-Al Qura* is the official weekly newspaper issued by the Saudi Government.
Radio: Saudi Arabian Broadcasting stations at Jeddah, Riyadh and Dammam transmit in Arabic and English. Aramco Radio station at Dhahran broadcasts in English. Approximately 2.9m recievers (1988).
Television: State TV services in Riyadh, Jeddah, Medina, Dammam and Qassem broadcast regional coverage eight hrs per day. Aramco TV broadcasts from Dhahran. Kuwait TV can be received by about 50 per cent of population and the Eastern Province is within range of Bahrain and Qatar radio broadcasts, and Bahrain, Qatar and UAE TV broadcasts. Some commercial stations. There were 2.7m* sets in 1988.

Domestic economy: The international oil glut reduced Saudi Arabia's oil revenue from $110bn in 1981 to $25bn* in 1986, and the resulting cuts and delays in government spending affected all sectors of the economy. After 15 years of development planning characterised by major infrastructural projects, the Fourth Five-Year Development Plan (1985-1990) stressed operational efficiency, reducing dependence on oil revenues, reducing the number of unskilled workers (and dependence on foreign labour) and encouraging the private sector to become more involved in economic development as the government's role is reduced.

Employment: There were in the 1980s some 3m foreigners living in Saudi Arabia, but their numbers dwindled by up to 700,000 a year as their employment contracts expired. It has proved difficult to fill the less highly-skilled jobs the foreigners leave behind.

External trade
Exports: Crude oil represented 94.3 per cent* of exports in 1987. Main destinations are Japan (23.3% of 1986 exports), USA (17.6%), France (9.4%), Italy (8.1%), Republic of Korea, UK, German FR, Brazil, Singapore.
Imports: Main imports are machinery and equipment (23.9% of 1984 total), foodstuffs (15.8%), transport equipment (13.4%), metals and metal products (11.9%), minerals and chemicals (9.8%), and textiles and clothing (8.1%). Main sources are USA (16.2% of 1988 imports), Japan (16.0%), UK (7.3%), German FR (7.2%), Italy (6.5%), France (5.2%), Republic of Korea (4.7%).

Agriculture: Agricultural development projects helped the country achieve self-sufficiency in wheat, eggs, some dairy products and vegetables, and although agriculture still only contributes three per cent of GDP it accounts for 25 per cent of employment. Development of water desalination plants is crucial to future development. Wheat production rose from 130,000 tonnes in 1980 to 2.8m tonnes in 1988.

Industry: Major industrial development began in the late 1970s, when Saudi Basic Industries Corporation (SABIC) was established. SABIC co-operated with a number of international companies to set up the first generation of industrial plants to produce petrochemicals, fertilizers and steel. Industrial development was concentrated at Jubail and Yanbu, although most major cities have a designated zone for industrial activities. Basic petrochemical feedstocks (ethane and methane) are very competitively priced in relation to European, US and Japanese producers, and export markets are being developed. By 1986 most of the SABIC first generation plants were on stream, and output of petrochemicals, fertilisers and steel rose from 8.5m tonnes in 1987 to 9.2m tonnes in 1988. Petrochemicals and plastics accounted for 11.4 per cent of the total value of exports in 1988. Other industries include cement production, building materials, food and beverage processing, textiles, furniture and household appliances. Increasingly, the authorities are looking to the private sector for further industrial development funding.

Minerals and mining: Extensive mineral exploration has been carried out and some 700 deposits have been found in the Kingdom. Among those minerals found are gold, silver, copper, lead, zinc, iron, phosphates, beryl, fluorite, magnesium, salt and sulphur and certain radioactive minerals. During the 1980s the government has promoted a policy of exploration of some of these minerals, including iron ore from the Wadi Sawawin in the north-west; gold, and its associated minerals, from Mahd adh Dhabab north of Al-Madinah; copper from the Zaid region about 350 km north-east of Jeddah; and bauxite from between Ha'il and Qassim in the north-central area. Not only are minerals to be found onshore but also in the Red Sea, where the Saudi-Sudan Red Sea Exploitation Commission has found silver, zinc and copper in substantial quantities.

Hydrocarbons: Saudi Arabia is the world's largest exporter of oil, with proven reserves of 166.6bn barrels (1988). During the 1980s, within the framework of OPEC, Saudi Arabia limited oil production to maintain price stability. However, in 1986 it briefly increased production so as to force down world crude prices and re-establish discipline among OPEC oil producers. In 1987-1989 oil production and oil revenues recovered strongly. During 1989 Saudi Arabia's OPEC quota was 4.524m barrels a day (b/d). There are export refineries, each with a capacity of 250,000 b/d, at Jubail, Yanbu and, at Rabegh on the Red Sea coast, a larger one was under construction in 1989. Natural gas deposits amount to 3,500bn cu metres (1988). Following a successful drilling programme in the Eastern Province to secure enough non-associated gas supply for the new national gas grid and petrochemical plants, there is no immediate prospect of further drilling, except in the Khuff Field, where gas is not tied to petroleum production.

Energy: $11.6bn has been set aside for new power projects designed to provide reliable electricity services for all population growth centres.

Banking: During the 1980s there was significant expansion in the banking sector and by 1985 there were more than 600 bank branches around the Kingdom, belonging to a total of ten commercial banks in the Kingdom. The end of the economic boom in the early 1980s had posed certain challenges for the banks. Many institutions had lent large sums against little or no collateral. Islamic strictures on the payment of interest were invoked by defaulting borrowers to avoid settling outstanding loans. Some of the Kingdom's most notable private companies were forced to reschedule payments or go into liquidation. The level of provisioning against possible loan losses undermined bank profits for several years. Banks embarked on a strategy of cost cutting and refined their services to improve underlying profitability. Several international banks sold holdings in local joint venture operations and withdrew entirely from the market. However, by 1988 the worst of the debt problems were over for the majority of banks. The Saudi Arabian Monetary Agency (SAMA) acts as a central bank and is steadily introducing modern regulation to the financial community. Islamic banking rules are in force.

Legal system: Shariah, the system based on Islamic laws, is the dominant force in the Saudi legal system. However, commercial courts have been set up to take a more active role in implementing a rather more sophisticated legal system to deal with the complexities of the business world.

Membership of international organisations: Arab League, CCC, FAO, G-77, GATT, GCC, IAEA, ICAO, IDA, IFAD, IFC, ILO, IMF, IMO, INTELSAT, International Maritime Satellite Organisation, INTERPOL, Islamic Conference Organisation, Islamic Development Bank, ITU, IWC, NAM, OAPEC, OPEC, UN, UNESCO, UPU, WHO, WMO, World Bank.

could even exceed Saudi Arabia's total volume of crude exports at present, this project only appears to be justified if total crude production rises sharply once the expansion is completed at the end of 1992.

Saudi Aramco resumed activity during 1989 on the development of its offshore oil fields after a lapse of some years. Development of the offshore Safaniya oil field received the go-ahead. A small on-shore oil find was also reported south of Riyadh, in an area with no previous history of discoveries.

As well as preparing for increased crude production in the longer term, the Saudi oil ministry made significant changes to its distribution and marketing strategy during 1988–89. The major element in the new strategy was the move to control downstream refining and distribution operations outside Saudi Arabia itself, a move which followed the successful example set by Kuwait. For Saudi Arabia, the move downstream took the form of an agreement in June 1988 to buy a 50 per cent stake in the refining and marketing assets of the American oil giant Texaco in eastern and southern parts of the United States, including refineries in Louisiana, Texas and Delaware, together with about 1,500 owned/leased service stations and a distributor network of 10,000 stations. This first downstream investment by Saudi Arabia began operations in January 1989 under the name Star Oil. It meant that Saudi Arabia was directly involved in the third biggest supplier of refined petroleum in the eastern United States.

The Saudi oil minister pointed out that Star Oil would benefit the United States as well as Saudi Arabia, because it would ensure security of supply and would create greater efficiency through integration of supply, refining and marketing operations. He also made it clear that Saudi Arabia would seek similar agreements in other parts of the world. Saudi planners apparently consider that their presence in downstream operations in the United States itself could greatly strengthen the position of Saudi Arabia if the US Government was to move towards imposing customs duties on imported oil.

The second element in Saudi Arabia's distribution strategy during 1989 was the creation of the Saudi Marketing and Refining Company, which was intended to manage all the refining, distributing and marketing activities within Saudi Arabia which were previously managed by the oil ministry's subsidiary organisation Petromin. The new company was intended to take a more profit-oriented approach to marketing than Petromin, and would apparently be able to close down unprofitable activities. For foreign buyers of refined products, the main impact was that the new company was to take over Petromin's holdings in the two 250,000 b/d export refineries in operation at Jubail and Yanbu and the third, rather larger, export refinery being constructed at Rabegh on the Red Sea coast. Already in mid-1989, the new company was holding spot tenders for refined products in addition to the term contracts traditionally offered by Petromin.

Industry

Saudi Arabia began its move into industry in the late 1970s, when the large statal corporation Saudi Basic Industries Corporation (SABIC) co-operated with a number of major international companies to set up the first generation of Saudi Arabia's fertiliser, petrochemical and steel plants. Most of these plants came on stream in 1985–86, and the petrochemical plants were well placed to profit from the strong world demand for most petrochemicals in 1987–89. As a result, SABIC's total output of fertilisers, petrochemicals and steel rose from 8.5m tonnes in 1987 to 9.2m tonnes in 1988, and the company's profits rose from just over SR1bn in 1987 to SR3.2bn in 1988. Saudi Arabia now has about 5 per cent of total world capacity of petrochemicals.

Apart from the large SABIC-backed plants, most Saudi factories are rather small. Many of the factories producing construction materials suffered from the collapse of the construction sector in 1983–85, and had to reduce or suspend production. But some companies such as the Saudi Cable company emerged as successful world class producers, able to win orders all over the world.

The government's strategy is to develop the range and volume of SABIC's petrochemical and steel output, through construction of a second generation of large plants. These plants will be able to provide basic materials for new downstream private industries. As a further sign of its commitment to industry, at least two of the government's large weapons purchases during 1985–89 have included 'off-set' investment packages, under which the foreign suppliers of weapons were scheduled to arrange investment of a proportion of their contract receipts in industrial projects in Saudi Arabia, especially in high technology projects.

SABIC's planned second generation of factories will be a major source of business in the 1990s. Contracts were awarded in mid-1989 to expand the SABIC direct reduction iron plant and design work was well advanced for an expansion of the steel plant. In the petrochemical sector, new ethylene crackers are scheduled to be installed in the early 1990s at the Yanpet and Petrokemya plants, each producing 500,000 tonnes/year of ethylene. Significantly, the crackers will be able to use either naphtha or ethane feedstock, which will reduce dependence on the natural gas feedstocks obtained from crude oil production which are currently in short supply in Saudi Arabia. Looking ahead, SABIC has targeted propylene derivatives as a priority area, and is considering a polypropylene unit at either Jubail or Yanbu. SABIC has also commissioned designs to expand its PVC production by 50 per cent to 300,000 tonnes/year.

Turning to fertilisers, SABIC plans to complete its first phosphatic fertiliser plant by the end of 1991. This plant will use Saudi urea and ammonia together with imported phosphate rock to produce a range of fertilisers. The SABIC affiliate SAFCO invited bids in mid-1989 for a 1.2m tonne/year urea and ammonia plant, to replace SAFCO's present obsolete plant. SABIC also resumed study during 1989 of a proposal to expand methanol ouput at its Al-Razi plant from 600,000 tonnes/year to 1.2m tonnes/year.

Outside the scope of SABIC's projects, the Saudi private sector has generally been rather slow to proceed with petrochemicals projects. One reason often given to explain this reluctance is that SABIC's large plants have already pre-empted much of the available methane and ethane feedstock available in the Kingdom. But the Saudi Venture Capital Group, set up in 1986, completed by early 1989 its evaluation of a paraxylene plant, to be located in Jubail, which would be the first basic petrochemicals plant undertaken in Saudi Arabia without SABIC's involvement. Another private sector group, Xenel, was looking in mid-1989 at setting up a 350,000 tonne/year polypropylene plant in Yanbu, separate from SABIC's plans to produce polypropylene.

A private sector plant to produce 45,000 tonnes/year of titanium dioxide is scheduled to start production in 1991.

Outside the chemicals sector, a very important test of the private sector's capacity to create large projects will come in the early 1990s with the implementation of the planned 214,000 tonne/year aluminium smelter at Yanbu, which was being promoted in 1989 by the Alujain corporation with assistance from Bechtel of the USA and Pechiney of France. This project only seems to be competitive if it receives some form of guaranteed access to low-cost power over a fairly long period. But, if successful, the plant could represent a major leap forward by the private sector in Saudi Arabia, as well as a significant step in the process of diversification away from oil.

More industrial projects are being planned by the National Industrialisation Corporation (NIC), an industrial holding company set up in 1984–85. These plants include a wire-drawing plant, a factory to make X-ray films, a car battery factory, a newsprint factory to serve the entire Gulf region, and a plant for disposable glass bottles. NIC is also considering a float glass plant, which might produce as much as 100,000 tonnes/year.

Looking further ahead, the offset investment programmes linked to Saudi Arabia's arms purchases are seen by Saudi planners as a vital element in the move towards acquiring leading-edge technology in Saudi Arabia. The investment programme linked to the US Peace Shield defence programme has produced four definite projects, all of which are still in an early stage. These are the $230m Advanced Electronics Centre, involving Westinghouse; the $122m Al-Salam Aircraft Modification Centre; the $53m Aircraft Accessories and Components Centre and the $20m Systems Engineering Centre.

The off-set programme linked to Britain's sale of Tornado aircraft and other equipment is potentially even bigger than the Peace Shield programme, since the British programme has an investment target of £1bn. Specific projects under discussion in the British programme by mid-1989 included a missile engineering plant and possible involvement in the Yanbu aluminium smelter, as well as a factory for polyethylene yarn and a national network of computer training centres. British officials were confident that talks between leading British and Saudi companies would produce other worthwhile projects which would cover a wide range of industrial sectors.

Agriculture

Despite the obstacles presented by the harsh climate and desert terrain of much of the Kingdom, Saudi planners are firmly committed to the development of agriculture, as a key element in the strategy of economic diversification away from oil. In most of Saudi Arabia, intensive modern agriculture has to rely on centre pivot irrigation, bringing water up from underground aquifers (water reservoirs). Only in the mountainous and rather more rainy south-west is it possible to trap surface water with dams. In Bisha in the south-west one of the largest dams in the Middle East was being constructed during 1989 with storage capacity of 325m cubic metres of water.

Mainly as result of centre pivot irrigation, the total area used for agriculture in Saudi Arabia grew rapidly from 150,000 hectares in 1975 to 2.9m hectares in 1988. The generous incentives given to agriculture have produced some astonishing successes. In particular, the production of wheat rose from 130,000 tonnes a year in 1980 to 2.8m tonnes in 1988 and about 3m tonnes in 1989. This level of production was far above Saudi Arabia's annual consumption of about 800,000 tonnes/year, and most of the surplus had to be sold or given away in aid. In the dairy sector, production of milk was officially put at 457,000 tonnes in 1986, and Saudi Arabia was self sufficient in many dairy products. Production of broiler chickens amounted to well over 50 per cent of local demand by 1987, and production of eggs, which had reached about 138,000 tonnes in 1987, was sufficient to allow for export of a surplus to neighbouring countries.

Cultivation of fruit and vegetables also expanded, using both traditional methods and sophisticated greenhouse techniques. After a period of decline caused by labour shortages in rural areas, production of dates recovered to about 500,000 tonnes in 1987. Many surplus dates are now sent abroad as food aid, but the agriculture ministry plans to develop local packaging plants so as to increase commercial exports. During 1989, the ministry's research centres in south-west Saudi Arabia targeted citrus fruits as an area of potential in Najran, and exotic fruit such as mango, guava, pineapple and pawpaw in the neighbouring area of Jizan.

But these impressive achievements have raised some dilemmas for the Kingdom. In particular, the wheat support programme in its present form is expensive to operate, and results in very large surpluses. In 1989, Saudi growers of wheat received a guaranteed support price from the government of SR2,000 per tonne in the case of small farmers and SR1,500 per tonne in the case of the six largest farming enterprises. Both prices were far in excess of the prices the government could obtain from sales of the surplus abroad. Wheat farmers were also able to claim subsidies for much of their equipment and fertiliser. The 666,000 tonnes of wheat exported in the first four months of 1989 therefore represented a considerable net loss to the Saudi exchequer. To make matters worse, the intensive use of underground aquifers for centre pivot irrigation led to the depletion of many of the aquifers to a level which the government's own 1985–90 Development Plan described as 'critical'. However, Saudi officials rejected the estimates of some foreign observers that the life of the underground aquifers at present rates of depletion could be less than 20 years.

One solution favoured by the government is to encourage the largest agricultural enterprises to diversify away from wheat to barley. Demand for barley for animal feed is strong in Saudi Arabia, with 4.9m tonnes of barley imported in 1987, mainly from EC countries. But progress towards self-sufficiency is slow, with barley output only reaching about 90,000 tonnes in 1987. Meanwhile, wheat production which had originally been targeted to stabilise at about 2.4m tonnes per year in 1988 and 1989 continued to rise to around 3m tonnes in 1989.

The dairy sector was also beginning to see problems of over-capacity in the late 1980s. Dairy farmers complained that their milk, most of which is made into a form of yoghurt called laban, was often unable to compete on price with products based on cheap imported powdered milk. Saudi fresh milk was also too expensive by world standards to allow for competitive large-scale production of cheese and other products.

But, despite some reservations, the agricultural sector will continue to be an area of opportunity, as technology develops to increase the options open to Saudi farmers. Already, Saudi vegetable production uses some of the most advanced hot climate techniques anywhere in the world. While the government may amend some aspects of its subsidy policy in coming years, the overall commitment to agriculture remains as strong as ever.

Banking

Saudi banks were recovering in 1988 and 1989 from a painful crisis which began with the collapse of activity in the construction sector in 1983-84. This collapse left many Saudi companies thoroughly overstretched financially. All the major Saudi banks faced a sudden accumulation of bad and doubtful debts as a result. At the same time, the down-turn in the volume of imports reduced the profits available from letters of credit, a traditional staple of the Saudi banking sector.

The scale of the difficulties can be illustrated by the Saudi British Bank, which had to make provisions in its 1988 accounts of SR301.8m for possible loan losses. With these provisions, the bank claimed to have provided for 90 per cent of its doubtful loan portfolio, but this was achieved at the cost of a declared loss of SR239m in 1988.

Saudi banks were aided in their adjustment to the crisis by support from the Saudi Arabian Monetary Agency, which supervises the banking sector. In the case of the Saudi Cairo Bank, the most obviously fragile of the leading banks, the Agency intervened publicly in June 1988, ordering the government's Public Investment Fund to subscribe SR300m to the bank so as to more or less double its capital. In 1989, the Agency deposited SR1,000m with the bank as a further support. The Bank had to show a loss of SR70m in 1988, but hoped that its new management could achieve a turn-round.

Among other leading banks, profits in 1988 were generally a strong improvement on 1987. As an example, the Riyadh Bank was able to raise net profits from SR155.4m in 1987 to SR264.5m in 1988, and the Saudi American Bank increased profits from SR90.1m in 1987 to SR235.2m in 1988. Most other banks saw similar growth, and results for the first half of 1989 showed further improvements.

But the continued need for heavy provisioning will certainly depress profits in future. The largest of all the Saudi banks, the National Commercial Bank, announced that it intended to assign all its 1988 profits of SR965.2m to general provisions, so as to cover 24 per cent of loans and advances.

As part of the readjustment, most of the leading banks in Saudi Arabia strengthened their capital base in 1987–89. The Saudi British Bank achieved this by issuing new shares to the public. Most other banks made transfers from reserves.

Assuming that the problems of the mid-1980s are now on the way to a solution, the banking sector in Saudi Arabia will offer some attractive prospects. A lot of Saudi money is still circulating outside the banking system, and could be attracted into the banks by improved services such as those offered by on-line Automatic Telling Machines. Such a widening of the customer base would help to reduce the dependence of Saudi banks on a small number of trading families as their main source of business. The cost of deposits to Saudi banks is low by most standards, as a significant (though declining) proportion of deposits are made free of interest. One bank particularly well-placed to compete for small customer deposits is the new Al-Rajhi Banking and Investment Corporation, the former Al-Rajhi money changing network, which was incorporated as a bank in March 1988 and by mid-1989 had a network of 288 branches.

But the lending business of the Saudi banks still needs clear legal and administrative guidelines, since neither Saudi law nor Saudi culture accept the legitimacy of charging interest on bank loans, and recovery of bad loans is extremely difficult. To improve the situation, the Saudi Arabian Monetary Agency set up a special committee in 1987 to hear cases involving disputes about bank loans. By the end of 1989, the committee was winning praise for the speed with which it heard cases and issued decisions to settle disputes. But the committee is not a law court, and its powers to enforce its solutions are limited.

The lack of clear regulations on lending means that Saudi banks are understandably cautious about plans to issue a specifically Saudi Arabian credit card. Indeed, those Saudi banks which currently market non-Saudi credit cards through their Saudi branches stress that they have to be very cautious about granting credit limits.

There was strong growth in 1988 in the volume of share trading involving shares of leading Saudi companies. This trading generates business for the banks, since it is done by matching bids and offers through the major banks. Trading volume rose by 22 per cent during 1988. But there were no immediate plans to reopen the Saudi Stock Exchange, whose brief experimental opening in 1987 lasted only a few weeks.

It is still not clear how far the Saudi banks will be involved in the Monetary Agency's borrowing campaign, launched early in 1988 to finance the state budget deficits planned for 1988 and 1989. Although the Agency announced that its new bonds would be marketed to the major banks, it seemed that the banks were reluctant to take on the bonds, which were mainly placed with the government's financial institutions. But in July 1989 one of the public institutions which was a major purchaser of the new bonds, the Public Investment Fund, announced that it had agreed with a syndicate of 11 Saudi and Gulf banks to borrow $660m over five years, apparently to help with the purchase of the Agency bonds. To this extent, therefore, the Saudi banks appeared to have become involved in financing the state deficit at one remove.

BUSINESS GUIDE

Time: GMT +3 hrs.

Climate: Desert climate. Jeddah is hot and humid, with temperatures from 18 to 38°C and 90 per cent humidity in summer. Riyadh's climate is more continental, hotter than Jeddah in summer (to 49°C) and colder in winter (can be near freezing), with occasional heavy rainstorms from Nov to Apr. The Eastern Province climate is similar to other Gulf States, cooler in winter than Riyadh and with high humidity and dust storms in summer.

Entry requirements

Passport: Required by all except certain seamen and pilgrims (with passes). The unexpired duration of passport must extend at least six months beyond the proposed stay.

Visa: Entry requirements are subject to periodic modification. Transit without visa may be allowed for bona fide passengers in transit through Jeddah airport. For entry, visas are required by all except nationals of GCC states and must be obtained in advance. Visa applications must be accompanied by a letter of invitation from a Saudi sponsor together with a visa number issued by the Ministry of Foreign Affairs and a No Objection Certificate. No certificate of religion required, but Judaism and atheism are not accepted as stated beliefs. Visitors with residence permits (as opposed to temporary visas) may also require an exit visa. Visitors should not surrender their passports to their Saudi sponsors without valid reasons.

Prohibited entry: Israeli nationals or holders of passports with Israeli visas are denied entry and transit. Travellers who arrive obviously inebriated are liable to arrest or deportation.

Currency: No currency restrictions.

Customs: Import of anything with alcoholic content, certain foodstuffs (such as pork), pornography, censored literature and controlled substances (drugs) is strictly prohibited. Prescription drugs should be carried only in small quantities in original containers. Dogs are banned with the exception of guard dogs, hunting dogs and seeing-eye dogs. Duty is chargeable on many imported items, starting at 12 per cent but rising to 20 per cent for goods normally manufactured in the Kingdom. No duty on samples of low value.

Income tax: Foreign nationals' wages and salaries are not taxed.

Health precautions

Mandatory: Certificates of vaccination against cholera and yellow fever is required if travelling from infected area.

BUSINESS GUIDE

Advisable: Vaccination for hepatitis, polio, typhoid and tetanus, and anti-malaria precautions.

Representation overseas: Abu Dhabi, Accra, Addis Ababa, Aden, Algiers, Amman, Ankara, Athens, Baghdad, Bamako, Bangkok, Beirut, Berne, Bonn, Brasilia, Brussels, Buenos Aires, Canberra, Caracas, Conakry, Copenhagen, Dacca, Dakar, Damascus, Dhaka, Djibouti, Doha, Fort Lamy, The Hague, Islamabad, Jakarta, Kabul, Kampala, Khartoum, Kuala Lumpur, Kuwait City, Lagos, Lisbon, London, Madrid, Manama, Manila, Mogadishu, Muscat, Nairobi, New Delhi, Niamey, Nouakchott, Ottawa, Paris, Rabat, Rome, Sana'a, Seoul, Singapore City, Stockholm, Taipei, Tehran, Tokyo, Tripoli, Tunis, Vienna, Washington DC, Yaoundé.

Air access: Direct flights are available from all three international airports.
National airline: Saudia (Saudi Arabian Airlines).
Other airlines: Air Afrique, Air Djibouti, Air France, Air India, Air Lanka, Air Mauritius, Alitalia, Alyemda, Austrian, Biman Bangladesh, British Airways, Cathay Pacific, China Airlines, Cyprus Airways, Egyptair, Emirates, Ethiopian, Garuda, Gulf Air, Iraqi Airways, JAL, Kenya Airways, Korean Air, KLM, Kuwait Airways, Libyan Arab, Lufthansa, MAS, MEA, Nigeria Airways, Olympic, Pan Am, Philippine, PIA, Royal Air Maroc, Royal Jordanian, Sabena, SAS, Singapore, Somali, Sudan Airways, Swissair, Syrian Arab, Thai International, THY Turkish, Tunis Air, Yemenia.
International airports: Dhahran International (Code: DHA), 8 km from Al Khobar, with hotel, buffet, bank, post office, shops and car hire; Jeddah, King Abdul Aziz International (Code: JED), 25 km from city, with hotel, restaurant, bank, post office, shops, car hire and special pilgrimage facilities; and Riyadh King Khaled International (Code: RUH), 35 km from city, with ultra-modern facilities including mosque and air cargo terminal.

Surface access
Main ports: Dammam (containers, grain terminal, ro-ro), Jeddah (bulk cement and grain facilities, containers, ro-ro tankers), Jizan (bulk cement, containers), Jubail (containers, dry and liquid bulk, tankers), Yanbu (bulk cement, containers) and Riyadh (dry dock).
Overland access: Roads from all countries sharing a land border with Saudi Arabia, as well as Bahrain via the new causeway (toll of SR20 for cars).

Hotels: There are many good hotels in the Kingdom. Alcoholic drinks are strictly prohibited.

Electricity supply: 110 or 220 V AC, 60 cycles, with 2-pin European-type plugs and both bayonet and screw light fittings in use. 50 cycles is still found in some remote areas; 380 V AC, 60 cycles is used by industry.

Credit cards: All major credit cards are accepted.

Car hire: Available at airports and main hotels. Driving licence required. Valid licences from most countries will be accepted by car hire companies. Women not allowed to drive. Driving on the right-hand side of road at maximum 110 kph on motorways and 40 kph in city traffic. Insurance claims not legally enforceable unless a police certification of the damage is obtained. Chauffeur-driven service is usually recommended.

City transport
Taxis: All taxis are orange and should have visible meter and taxi number in addition to normal registration. Although drivers are supposed to use the meters, it is best to agree the fare in advance. Taxi drivers do not expect a tip. *Saudi limousines:* White American air-conditioned saloons and white London taxis with clearly identifiable 'limousine' logos are operated by a number of companies within the cities and especially to and from airports. These limousines have virtually replaced ordinary taxis. Fixed fares for specific journeys (average SR50) are prominently displayed at airports and available from drivers.

National transport: Non-Muslims may not travel to the holy cities of Medina and Makkah.
Air: 23 domestic airports, including Al-Julaidan (Code: MED), 12 km from Medina, with a restaurant. Saudia operates a comprehensive schedule of domestic flights between Jeddah, Riyadh and Dhahran and other major centres. Always confirm flight bookings 24 hours before take-off, especially during the annual pilgrimage (Hajj), when pilgrims come to Saudi Arabia from all over the world.
Road: There is a 1,800 km network of main and secondary highways, with a total of 82,380 km asphalted roads (1988).
Buses: Saudi Arabian Public Transport Company (SAPTCO) operates bus services within three major urban areas, airport bus services and inter-city bus services between major centres.
Rail: Daily rail service links Riyadh and Dammam, with refreshments and air-conditioning available. Passengers and freight to double in 1985-90.

Public holidays: Public holidays for the public sector are all day Thursday and Friday. For the private sector it is mainly Thursday afternoon and all day Friday. There are no official public holidays in Saudi Arabia but business activity stops for Eid al-Fitr, the Hajj and Al-Hijra.

Working hours: Friday is usual weekly holiday, although many shops remain open. Private sector working hours may be shortened by two hours or more a day during Ramadan.
Business: 0830 – 1200, 1630 – 1930 in Riyadh; 0900 – 1330, 1630 – 2000 in Jeddah; 0730 – 1200, 1430 – 1730 in Eastern Province (closed Thu pm and Fri).
Government: 0730 – 1430 (closed Thu, Fri).
Banking: (Sat-Wed) 0830 – 1200, 1700 – 1900; (Thu) 0830 –1130.

Social customs: In public places, women should wear long skirts, long sleeves and high collars and men, shirts and long trousers. Unless addressing members of the Royal Family normal Western forms of address and greeting are usual. At business meetings it is not uncommon for several people to be present. If coffee is served it is courteous to accept it. Cups will generally be refilled automatically unless the cup is shaken from side to side as it is returned to the server. Refrain from taking photographs without permission. It is prohibited to photograph airports, military installations and certain other buildings.

Telecommunications
Telephone: Dialling code for Saudi Arabia: IDD access code + 966 followed by area code (3 for Al-Khobar/Dammam/Dhahran, 2 for Jeddah/Makkah/Taif, 1 for Riyadh) followed by subscriber's number. Telephone network has 1.6m phones (expected to reach two million by 1990), and is serviced by satellite and microwave systems as well as underground cables.
Telex: Available at most hotels and PTT offices.

Delivery services
Post: There is no home delivery service: all mail is posted to post office boxes. Along with automatic mail sorting, five-digit postal (zip) codes are being introduced to replace PO Box numbers. Until the changeover is complete, some addresses will be expressed as PO Box number and others as town names plus postal codes. Air mail post takes about five days to Europe. Surface mail takes three to five months.
Courier: SNAS-DHL operates door-to-door shipment of documents and small parcels, including collection, transport, customs clearance and delivery, with overnight service to Europe and 48-hr service to the USA. The post office also offers a courier service (Mumtaz).

Banking: Islamic banking rules are in force.
Central bank: Saudi Arabian Monetary Agency (SAMA).
Main centres: Jeddah and Riyadh.
Head offices: (Jeddah) Bank Al Jazira, Islamic Development Bank, Islamic Investment Bank, National Commercial Bank, Saudi Cairo Bank. *(Riyadh)* Al Bank Al Saudi Al Fransi, Al Bank Al Saudi Al Hollandi, Al-Bank Al-Saudi Al-Tejari Al-Muttahed (United Saudi Commercial Bank), The Arab Investment Company SAA, Arab National Bank, Riyadh Bank, Saudi American Bank, Saudi British Bank, Saudi Industrial Development Fund, The Saudi Investment Bank. *(Al-Khobar)* Arab Petroleum Investments Corporation (Apicorp).
Branches: Banque Indosuez, Citibank, Foreign Commercial Bank Switzerland, Korea Exchange Bank.

Advertising: Restrictions on tobacco advertising and medicinal products. There is no public cinema. Outdoor advertising subject to local controls. Direct mail advertising is available.

Trade fairs: During the 1980s trade fairs and exhibitions have grown rapidly in number and importance. The main centres are Riyadh, Jeddah and (from 1988) Dhahran.

Representation in capital: Algeria, Argentina, Australia, Austria, Bahrain, Bangladesh, Belgium, Brazil, Burkina Faso, Burundi, Cameroon, Canada, Chad, China PR, Denmark, Djibouti, Egypt, Ethiopia, Finland, France, Gabon, Gambia, Germany FR, Ghana, Greece, Guinea, India, Indonesia, Iran, Iraq, Ireland, Italy, Japan, Jordan, Kenya, Korea, Kuwait, Lebanon, Libya, Malaysia, Mali, Malta, Mauritania, Mexico, Morocco, Nepal, Netherlands, New Zealand, Niger, Nigeria, Norway, Oman, Pakistan, Philippines, Portugal, Qatar, Rwanda, Senegal, Sierra Leone, Singapore, Somalia, Spain, Sri Lanka, Sudan, Sweden, Switzerland, Syria, Taiwan, Thailand, Tunisia, Turkey, Uganda, UAE, Uruguay, UK, USA, Venezuela, Yemen AR, Yemen PDR, Zaïre.

BUSINESS DIRECTORY

Useful telephone numbers
Emergency police: 999; Ambulance: 997; Traffic accidents: 993; Fire: 998; Directory enquiries: 905

Hotels: Riyadh (area code 1)
Al Batha, Batha Street, P.O. Box: 17199 (tel: 405-2000; tx: 403132/3).

Al Carawan, 11452, P.O. Box: 6882 (tel: 403-8855, 401-2205; tx: 402146 carawan).

Al Khozama, Oleya Road, 11491, P.O. Box: 4148 (tel: 465-4650; tx: 400100 khoz; fax: 464-8576).

Al Mansour, Old Airport Road, 11481, P.O. Box: 962 (tel: 479-4288; tx: 404873 mansho).

Al Muraba'a, King Fahad Bin Abdul Aziz Street, 11415, P.O. Box: 18055 (tel: 405-7177; tx: 402186 obaid; fax: 4059055).

Al Mutlaq Novotel, Old Airport Road, 11481, P.O. Box: 3525 (tel: 476-2193/3483; tx: 405266 homtq; fax: 4780696).

Al Seteen Palace, 11415 Seteen Road, 11365, P.O. Box: 18329 (tel: 477-3006; tx: 404282 seteen).

Hyatt Regency Riyadh, Redec Plaza, 11415, P.O. Box: 18006 (tel: 479-1234; tx: 402963 hyat; fax: 477-5373).

Minhal Sofitel, Old Airport Road, 11484, P.O. Box: 17058 (tel: 478-2500; tx: 403088 mnal; fax: 4772819).

Oleya House Residence, Oleya Street, 11491, P.O. Box: 10835 (tel: 464-2010; tx: 400536 hamam).

BUSINESS DIRECTORY

Riyadh Inter-Continental, Mathar Street, 11481, P.O. Box: 3636 (tel: 465-5000; tx: 401076 ichhot; fax: 4657833).

Riyadh Marriott, Mathar Street, 11464, P.O. Box: 16294 (tel: 477-9300; tx: 400983 marrsj; fax: 4779300 EXT 77054).

Riyadh Palace, Prince Fahd bin Abdul Aziz Street, 11461, P.O. Box: 2691 (tel: 405-4444; tx: 400312 rph; fax: 4054444 EXT 2161).

Riyadh Sheraton, 11441, P.O. Box: 1975 (tel: 454-3300; tx: 401415 sher; fax: 454-1889).

Sahara, 11473 King Khaled International Airport, P.O. Box: 12120 (tel: 220-4500; tx: 407400 airhtl; fax: 2204505).

Salahedin, King Abdul Aziz Road, 11691, P.O. Box: 85396 (tel: 454-4888; tx: 405888 saltel).

White Palace, Old Airport Road, 11482, P.O. Box: 8145 (tel: 478-7800; tx: 405793 wplctl).

Zahret Alshark, Airport Road, 11481, P.O. Box: 3616 (tel: 403-8800; tx: 405590 zahtel; fax: 4038800 EXT 713).

Hotels: Jeddah (area code 2)
Al Amoudi, 21473, P.O. Box: 12496 (tel: 660-5611/3062; tx: 604109 monira).

Al Attas, King Fahad Street, 21441, P.O. Box: 1299 (tel: 642-0211; tx: 601158 atas).

Al Attas Oasis, P.O. Box: 1789 (tel: 642-0400; tx: 601158).

Al Haramain Palace, Madaris Street (tel: 642-6055/1894; tx: 603122, 601609).

Al Harithy, Palestine Road, P.O. Box: 7584 (tel: 667-0520; tx: 600687 dalia).

Al Salam, King Khaled Street & Makkah Road, 21452, P.O. Box: 6582 (tel: 631-4000; tx: 601327; fax: 6324194).

Albilad Mövenpick, Al Corniche Highway, P.O. Box: 6788 (tel: 682-8282,691-0111; tx: 603010 bilad; fax: 691-2208).

Alhamra Sofitel, Palestine Road, Al Hamra, P.O. Box: 7375 (tel: 660-2000; tx: 600749 sofi; fax: 660-4145).

Holiday Inn Al Fau, Al Hamra, P.O. Box: 10924 (tel: 661-1000; tx: 600755 holi; fax: 660-6326).

Hyatt Regency, Medina Road, 21482, P.O. Box: 8483 (tel: 651-9800; tx: 602688 hyat; fax: 651-6260).

Jeddah Kaki, off Old Airport Road, P.O. Box: 2559 (tel: 631-2201; tx: 601739 khtl; fax: 6311350).

Jeddah Marriott, 8 King Fahad Street, at Palestine Road, 21442, P.O. Box: 6448 (tel: 671-4000; tx: 605135 mariot; fax: 6715990).

Jeddah Sheraton, Jeddah 21424, P.O. Box: 14315 (tel: 699-2212; tx: 607229 JEDSHE SJ; fax: 699-2660).

Kandara Palace, Old Airport Road, P.O. Box: 473 (tel: 631-2177/2944/2833; tx: 401095).

Khojah Residence Center, Old Airport Road, 21474, P.O. Box: 16955 (tel: 631-8000; tx: 605127 amkcnt).

Red Sea Palace, King Abdul Aziz Street, 21421, P.O. Box: 824 (tel: 642-8555; tx: 601014 rsph; fax: 642-2395).

Sands, Hamra Street, P.O. Box: 7030 (tel: 669-2020; tx: 600789 sands; fax: 6609971).

Business advice and information
Correspondence to government departments (in Riyadh unless otherwise indicated) should be addressed simply to the ministry in question, with no reference to PO Box or street address.

Al-Qassim Chamber of Commerce and Industry, PO Box 444, Burayda (tel: 324-2887; tx: 301060 sinaia).

Central Department of Statistics, Ministry of Finance and National Economy, Riyadh (tel: 29120; tx: 401021).

Central Planning Organisation, Ministry of Planning, Riyadh.

Council of Saudi Chambers of Commerce and Industry, PO Box 16683, Riyadh 11474 (tel: 405-3200; tx: 405808; fax: 4024747).

Dammam Chamber of Commerce, PO Box 719, Dammam (tel: 834-7454); tx: 802584; fax: 8335755).

Dhahran International Expo, PO Box 7519, Dammam 31472 (tel: 833-7900; tx: 803113; fax: 8338010).

Director-General of Mineral Resources, Ministry of Petroleum and Minerals, Jeddah.

Islamic Development Bank (International Headquarters), PO Box 5925, Jeddah (tel: 643-3994/5; tx: 601137, 601407).

Jeddah Broadcasting Service, Broadcasting Station, Jeddah.

Jeddah Chamber of Commerce and Industry, PO Box 1264, Jeddah (tel: 647-1100; tx: 603783 al ghurfa; fax: 6484603).

Jeddah TV Service, Ministry of Information Building, Jeddah.

Makkah Chamber of Commerce and Industry, PO Box 1086, Al-Ghazza Street, Makkah (tel: 574-5775; tx: 540271 chamec; fax: 5741200).

Medina Chamber of Commerce and Industry, PO Box 443, Medina (tel: 822-5190; tx: 570009; fax: 8268965).

Ministry of Commerce, PO Box 1774, Airport Road, Riyadh (Companies Registration, Committee for Investigation of Commercial Disputes, Hotels, Standards).

Ministry of Defence and Aviation, Riyadh (Saudia, Jeddah).

Ministry of Finance and National Economy, Riyadh (Customs, Income Tax, National Computer Centre, Statistics) (tel: 405-0080; tx: 401021).

Ministry of Industry and Electricity, PO Box 5729, Riyadh 1432 (Committee for the Investment of Foreign Capital, Industrial Studies and Development Centre) (tel: 477-2722; tx: 401154).

Ministry of Information, Nasiriyah Street, Riyadh (tel: 401-4440; tx: 401461).

Ministry of Interior, Riyadh (Public Security Passports and Nationality, Residence Permits) (tx: 404416).

Ministry of Labour, Riyadh (General Organisation for Social Insurance, Training, Work permits) (tx: 401043).

Ministry of Petroleum and Minerals, Riyadh (PETROMIN, Director-General of Mineral Resources) (tel: 478-1661; tx: 401058).

Ministry of Planning, PO Box 358, Riyadh 11411 (Central Planning Organisation) (tx: 402851).

National Industrialisation Co, PO Box 26707, Riyadh 11496 (tel: 476-7166; tx: 406662; fax: 4770898).

Organisation of the Islamic Conference, Secretariat-General, Kilo 6, Makkah Road, PO Box 178, Jeddah.

PETROMIN (General Petroleum & Mineral Organisation), PO Box 757, Riyadh 11421 (tel: 478-7777; tx: 401490).

Riyadh Chamber of Commerce and Industry, PO Box 596, Riyadh (tel: 404-0044; tx: 404110).

Riyadh Exhibition Co Ltd, PO Box 56010, Riyadh 11554 (tel: 454-1448; tx: 406359; fax: 4544846).

Royal Commission for Jubail and Yanbu, PO Box 5964, Riyadh 11432 (tel: 479-4445; tx: 401386).

Sabic (Saudi Basic Industries Corp), PO Box 5101, Riyadh 11422 (tel: 406-9900, 401-2033; tx: 401177 sabic).

Saudi Arabian Oil Co, PO Box 5282, Dhahran 31311 (tel: 875-4915; tx: 8012220 asao); Riyadh (tel: 464-1055; tx: 401084).

Saudi Consulting House, PO Box 1267, Riyadh; PO Box 1001, Jeddah.

Saudi Expo (Saudi Corporation for International Exhibitions and Fairs), PO Box 1252, Jeddah 21431 (tel: 651-1212; tx: 603908; fax: 6530869).

Saudi Industrial Development Fund, PO Box 4143, Riyadh (tel: 477-4002, 403-3200; tx: 402583).

Saudi Television, Yarmuk Street, Al Badia'a Riyadh.

Taif Chamber of Commerce and Industry, PO Box 1005, Taif (tel: 746-4624; tx: 751009).

The Co-operation Council of The Arab States of the Gulf (GCC), PO Box 7153, Riyadh 11462 (tel: 476-0000; tx: 404830 tawini).

Banking
Saudi Arabian Monetary Authority (SAMA) (central bank), PO Box 2992, Riyadh 11169 (tel: 463-3000; tx: 401734 markazi sj; fax: 4411384); PO Box 394, Jeddah (tel: 631-4740/4757; tx: 601511/011/094).

Al Bank Al Saudi Al Fransi, PO Box 1290, Riyadh 11431 (tel: 476-0284, 476-0288; tx: 401428 safria); PO Box 1, Jeddah 21411 (tel: 660-5863; tx: 603879 sfgm).

Al Bank Al Saudi Al Hollandi, PO Box 1467, Riyadh 11431 (tel: 401-0288; tx: 405178 bshr sj; fax: 4031104).

Arab National Bank, PO Box 56921, Riyadh 11564 (tel: 402-9000; tx: 402660 arna sj).

Bank Al Jazira, PO Box 116, Jeddah 21411 (tel: 642-4201; tx: 601036 jazira).

Banque du Caire, PO Box 2848, Riyadh 11461 (tel: 402-8003; tx: 401051 bancair).

National Commercial Bank, PO Box 3555, Jeddah 21421 (tel: 644-6644/7880; tx: 603249/250, 605571 ncbh).

Riyadh Bank, PO Box 229, Riyadh 11411 (tel: 402-4011; tx: 401061 riyadex).

Saudi American Bank, PO Box 833, Riyadh 11421 (tel: 477-4770; tx: 400195 samba sj; fax: 4774770 (Ext 200).

Saudi British Bank, PO Box 9084, Riyadh 11413 (tel: 405-0677; tx: 402349 sabb sj; fax: 4050660).

Saudi Cairo Bank, Al Faiha Building, Medina Road, PO Box 11222, Jeddah 21453 (tel: 651-6432; tx: 600205 scho, 602190).

The Saudi Investment Bank, PO Box 3533, Riyadh 11481 (tel: 477-8433; tx: 401170 saibrh sj; fax: 4776781).

United Saudi Commercial Bank, PO Box 25895, Riyadh 11476 (tel: 478-4200; tx: 405461 saucom sj; fax: 4783197).

Travel information
Saudi Arabian Airlines (Saudia), PO Box 620, Jeddah (tel: 686-0000, 632-3333; tx: 601007 sj).

Senegal

Kaye Whiteman

After a lull in which the wounds of the 1988 election crisis were nursed, the beginning of 1989 saw a quickening of the political pace, as the opposition parties returned to the offensive. However, April 1989 saw a dangerous new and violent turn to the situation, with the outbreak of a crisis with Mauritania, which involved rioting in the two countries, and a mass two-way exodus of populations, as well as massacres. This may have had some short-term political side-effects in rallying opposition parties to a national cause, but the year also saw a disturbing, and most un-Senegalese, rise of xenophobia.

Senegal in 1988–89 seemed to be going through a time of travail. The crisis and state of emergency which followed the February 1988 elections was the worst moment yet of the period in office of President Abdou Diouf, who came to power in 1980. Arguably, the political, social and economic crisis of 1989 constituted the most serious period the country has had to face since independence 30 years ago. Most prognostications were gloomy, in spite of Diouf's personal resilience and international standing. The gap between Senegal's aspirations to be a pillar of democracy and cultural leadership, and its inadequate resource base and deeply imbalanced economy sometimes seems too great.

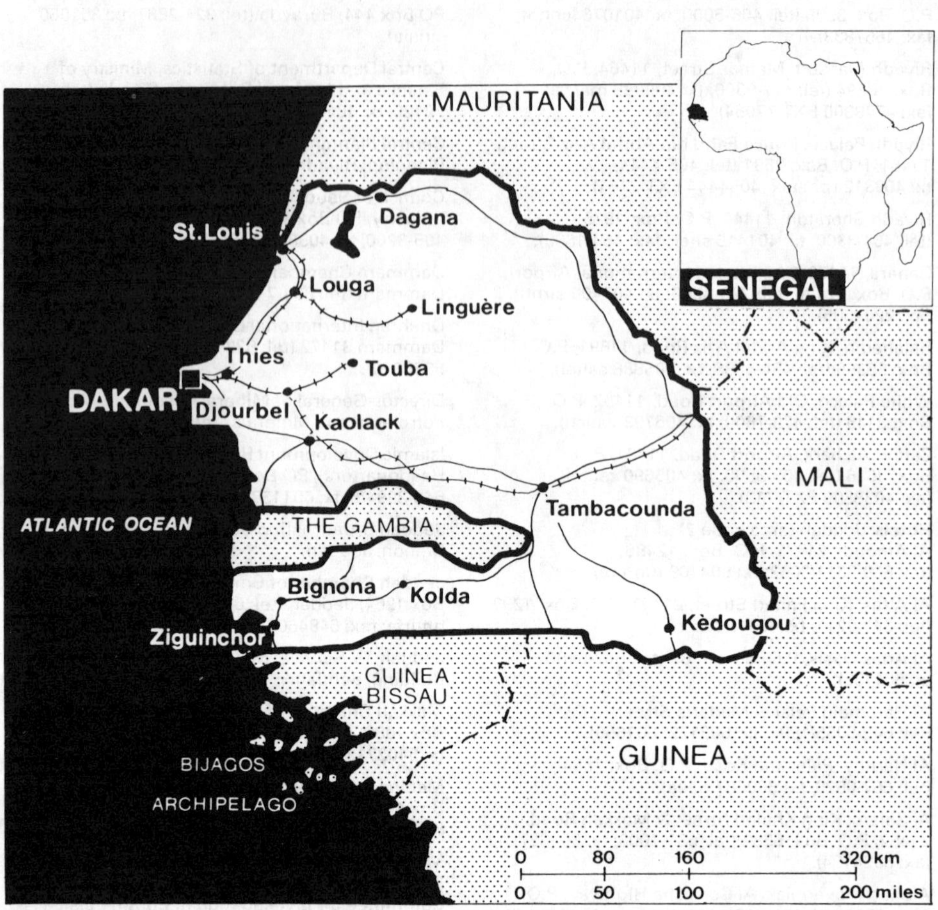

Political activity

The post-election deal of May 1988 between President Diouf and Maître Abdoulaye Wade, the leader of the main opposition party, the Parti Democratique Senegalais (PDS) came unstuck fairly rapidly as the round-table talks became deadlocked. It was generally believed that the tough stance of Diouf, which effectively blocked the talks, was dictated by hardliners in the ruling Parti Socialiste (PS). Wade, however, had made his own contribution to the breakdown, by putting forward exaggerated and unrealistic demands. In fact, the breakdown of the talks had less political impact than might have been expected, indicating that, in treating Wade indulgently after his trial in May and securing his agreement to talks on national unity (and, incidentally, his *de facto* acceptance of Diouf's election as president which was implied), the regime had substantially regained the initiative.

The next hurdle which Diouf had to face was the 'rentree', the beginning of the school year in October 1988. There were dire predictions of trouble, in view of the fact that the previous academic year (1987–88) had been written off as an 'annee blanche' (an empty year). In fact, following certain undertakings on expenditure in the education sector which went some way to meeting some of the students' demands on conditions, (for which French financial assistance was conveniently provided) pupils and students returned to the classrooms. The year saw the occasional hiccup, such as sporadic student strikes, as well as two short-lived university teachers' strikes in February and March 1989, but it proved possible to continue the year and hold examinations at the end of it, which in the circumstances was rated as a kind of achievement. It resulted from the basic instinct for survival among both students and teachers, rather than from a sudden rush of support for the government or the ruling party, who still correctly identified the education sector as a major power base for the opposition parties.

In spite of these signs of a slight defusion of tension, the impression that the Diouf regime was both beleaguered and accident-prone continued. In December 1988 there was a mysterious explosion in central Dakar, and later in the month two soldiers were killed and several others wounded in an ambush in the troubled region of Casamance. The government attributed the ambush to the Movement of Democratic Casamance Forces (MDFC), a hitherto little-known grouping. Reinforcements had been sent into the region in December partly because the end of the year is always a sensitive, being the time of Diola festivals, during which there has sometimes been unrest in the past. Then, in February 1989, there was a mysterious affair of tracts in Dakar, in which prominent personalities were accused of borrowing large amounts, running into millions of francs, from Senegalese banks, which had not been paid back. The affair came at a time of unease in the banks themselves, because of a 'restructuring' reform.

March 1989 was a month of considerable activity, with an extraordinary congress of the PS, the biggest in its history, being followed by the triumphal return of Maître Wade, from an extended overseas stay. This was followed by what appeared to be an abortive attempt to force through a national government. The PS party congress was seen as carrying through, much further than previously, Diouf's attempt to clear out the old guard of the party, sometimes known as the barons, which he had begun after the 1983 elections, and which he had continued to strive for. In December 1988, he had secured the fall of a prominent party baron, Dr Daouda Sow, who was dethroned as president of the national assembly, to be replaced by the elderly and more amenable Abdoul Aziz Ndaw. At the party congress itself, the rank and file were stunned by the provisional suspension of the party's

KEY FACTS

Official title: Republic of Senegal	**Area**: 196,192 sq km
Head of state: President Abdou Diouf	**Population**: 7.0m (1988)
Ruling party: Parti Socialiste	**GNP per capita**: $510 (1987)
Capital: Dakar	**GNP real growth**: 1.3% (1986/87)
Official Languages: French	**Inflation**: 6.4% (1986)
Currency: Franc CFA = 100 centimes	**Trade balance**: –$447m (1987)
Exchange rate: Franc CFA316 per $ (Nov 1989)	**Foreign debt**: $3.7bn (1987)

main institutions, the political bureau and the central committee, to be replaced by a 12-man executive committee charged with renewal of the party, until the first quarter of 1990.

The favourable effect of such renovation was upstaged however by Wade's return, which brought out the huge crowds chanting 'Sopi' (–change' in the Wolof language) which had marked the post-election emergency. The possibility of renewed political tension did not materialise, however, since Wade cancelled a mass meeting intended to call on Diouf to dissolve his government and form a coalition, saying that he was to see the president, following discussions in France with Diouf's powerful right-hand man Jean Collin, at which such a coalition had been agreed. However, it emerged that there was to be no meeting, no coalition had been agreed, and further, that communication between Wade and Collin had been via a bizarre intermediary, the so-called 'Ayatollah of Kaolack', Ahmet Niasse, a politically dubious religious leader who has veered between alliance with Libya and running errands for the Diouf government. He substantiated Collin's position, which meant that Wade was left looking foolish. Indeed, Wade's lack of political *savoir-faire* was one of the best cards held by the government, and led to increasing friction among the opposition parties.

At the beginning of April 1989, Diouf offered a further olive branch to the opposition in the form of a commitment to reform the electoral law. This was one of the main grievances of the opposition, which claimed that loopholes in the existing law were used by the ruling party to rig the 1983 and 1988 elections. At the same time, it was announced that the municipal elections, due to have been held in November 1989, were postponed for one year, which further defused inter-party tension, as an electoral contest in the prevailing atmosphere might have led to new frictions and violence. By the end of March however, Diouf and Collin were becoming increasingly worried about the army, as was seen by the dismissal from the army and confining to barracks of the former Armed Forces Chief of Staff, General Tavares da Souza. At the same time, news was leaked of the detention for questioning of four prominent army officers: Col. Oumar Ndiaye, Col. Didier Bampassy, Major Gabar Diop, and Lt. Col. Francis Gomis. They had been, until April 1988, respectively Quartermaster-General, and commanders of the Commandoes, the Armoured Squadron and the Paratroop regiment. At the same time as General da Souza was replaced and sent to Bonn as ambassador, all four were removed. An official statement in April said that all had been released, but would face an inquiry headed by the new head of the armed forces, General Mansour Seck. These problems were soon submerged, however, in the crisis in relations with Mauritania.

KEY INDICATORS

	Unit	1983	1984	1985	1986	1987
Population	m	6.21	6.38	6.57	6.77	6.97
Gross domestic product (GDP)	CFA bn	939.5	1,015.5	1,152.0	1,295.3	1,419.0
GDP per capita	Franc CFA	151,288	159,169	175.3	191.3	203.6
Inflation	%	11.7	11.7	13.0	6.4	–4.3
Agricultural production	1979-81=100	88.3	106.0	125.8	131.5	144.9
Food production	1979-81=100	88.0	104.9	126.1	132.0	145.3
Industrial production	1985=100	99.5	97.6	100.4	107.0	109.8
Exports	CFA bn	235.5	277.0	252.5	202.2	161.2*
Imports	CFA bn	341.5	374.7	324.3	258.0	291.5*
Balance of trade	$m	–311.4	–221.2	–293.1	–261.6	–447.0
Current account	$m	–287.9	–272.3	–317.0	–415.0	–316.0
Total external debt	$m	1,894	2,000	2,442	3,014	3,695
Debt service ratio	%	5.5	8.7	10.2	19.7	21.2
Exchange rate	CFA per $	381.1	437.9	449.3	346.3	300.5

* estimated figure

Relations with Mauritania

Problems with Senegal's northern neighbour, Mauritania, had been apparent for a little while; in December 1988, for example, there was a row about Mauritanian livestock, and in January–February 1989, a mini-economic war broke out in which some Mauritanian exports were banned from Senegal, which provoked retaliation. But it was in April that the serious crisis began with an incident on Diawara island in the Senegal River, when two Senegalese peasants were killed by Mauritanian border guards in a clash over grazing rights. The incident brought to the surface latent hostility to Mauritanians of whom there were up to 300,000 in Senegal. In the face of apparent Mauritanian indifference to the deaths, there

COUNTRY PROFILE

Political system: Legislative power is vested in the unicameral 120-member National Assembly. Executive power is held by the president and his chosen council of ministers. Senegal gained independence from France in 1960.

Political parties: Parti Socialiste is the ruling party; others include Parti Démocratique Sénégalais, Mouvement Républicain Sénégalais and Parti Africain de l'Indépendance.

Population: The population of 7.0m (1988 estimate), which grows by 2.9 per cent a year comprises a large number of ethnic groups. Over 85 per cent of the population are Muslim and about five per cent Christian (mainly Roman Catholic).

Main cities/towns: Dakar (population 850,000), Thiès (125,000), Kaolack (110,000), St Louis (95,000), Ziguinchor (65,000).

Language: French is the official language; Ouolof, Serer, Banbala, Diola, Peul and Toucouleur are also spoken. In business, it is essential to speak French.

Media
Press: Only national daily newspaper, *Le Soleil*, controlled by the government. A number of periodicals are published.
Radio: Two state-run radio networks broadcasting in French, Portuguese, Arabic, English and six African languages.
Television: ORTS, the state radio and TV authority, broadcasts news and educational TV programmes. About 1,500 TV sets.

External trade: Regular substantial trade and current account deficits. The total external debt in 1987 was $3.4bn, leading to a sharp rise in the debt service burden. There is a continued reliance on aid. With donor pledges of $1.8bn for 1987-88 – 1989-90, Senegal enjoys one of the world's largest per capita aid inflows.
Exports: Total value in 1987 was $741m, consisting of groundnuts, petroleum products, phosphates, fish. Main destinations: France, UK, Côte d'Ivoire, Netherlands, Italy.
Imports: Total value in 1987 was $1,052m, consisting of semi-finished goods, food, capital goods. Main sources: France, Nigeria , USA, Spain, German FR.

Agriculture: Sector employs 70 per cent of the workforce. Basic food crops include millet, rice, maize and beans. Output has failed to keep pace with population growth, but it responded in 1985-86 to government incentives. Estimated that 40 per cent of farm land taken up by groundnut cultivation. Output of groundnuts in 1987-88 was 963,000 tonnes, up by 15 per cent on the previous year. Emphasis on diversifying agricultural output with rice, sugar, millet and cotton being grown with some success. Also significant expansion of horticultural sector and development of marine fishing (tuna, oysters, prawns, lobsters). The completion of the Manantali Dam on Senegal River, will allow the irrigation of some 120,000 ha of land, but at the prohibitive cost of $10,000 a hectare.

Industry: Sector contributes around 16.5 per cent to GDP and is centred around the Cap Vert peninsula near Dakar. Light industries such as textiles, food processing and beverages predominate. The only heavy export industries are an oil refinery at Dakar-Mbao, a sulphuric/phosphoric acid plant at Darou Khoudou and a fertiliser complex (Industries Chimiques du Senegal) at Mbao. State participation in industry is confined to the key sectors such as energy and construction, but production is heavily subsidised and protected by tariffs. Around 60 per cent of manufacturing companies are privately owned. French firms are active in most sectors. The Government plans to boost industry's sluggish growth rate, which has averaged two per cent a year over the past decade, and to encourage tourism, which is developing rapidly.

Mining: Mainly extraction of phosphates from open mines near Thiès. Workable deposits around 130m tonnes. Production (around 1.3m tonnes per annum) is mainly for export, although it is also important source of supply for new fertiliser complex at Mbao. Titanium, zircon and rutile are mined along the south coast of Cap Vert. Some 330m tonnes of iron-ore deposits have been located at Falémé in the east of the country near the Mali border, but their extraction seems uneconomic at current market prices. Prospecting for uranium, copper and gold is underway.

Energy: Virtually all commercial energy requirements are imported. Government's 10-year energy plan aims to substitute 50 per cent of imported oil by local products, including oil/gas from the Dome Flore, peat deposits from Niayés and expansion of hydro-power from the Senegal and Gambia rivers.

was a rising tide of attacks on Mauritanian shops. It was in part fuelled by economic hardship, but awareness of the poor treatment accorded to blacks in Mauritania also played its part. The troubles reached Dakar before the end of the month, and on 23–24 April there was a serious rampage/riot in which not just Mauritanian but other shops were looted, vehicles were burnt and police stoned. It was in danger of turning into a new 'Sopi' riot, and gendarmes were brought in to help restore order. The only fatalities, however, were one Mauritanian trader and three Senegalese looters.

When reports of the rampage reached Mauritania, the crisis took a more dramatic turn, as revenge attacks took place on Senegalese in Nouakchott, Nouadhibou and other major towns, in which the security forces apparently stood by without intervening. Around 400 people were reported killed, and when reports of this, with varied atrocity stories, came back to Dakar, revenge attacks took place on Mauritanians, in which maybe 70 were killed before the authorities restored order. By now most of the Mauritanians in Senegal had fled, either northwards or into the Gambia and Guinea-Bissau, though many sought refuge in mosques, or in a special camp at the Dakar trade fair site. An airlift of some 60,000 Mauritanian refugees was organised with international assistance. The river frontier was virtually closed, the ferries non-operational. At the same time, maybe 30,000 Senegalese citizens were expelled from Mauritania, and a vituperative media war was developing. Mediation bids were proferred by Mahgreb countries, and by the OAU chairman, President Traore of Mali.

Throughout May 1989 the situation deteriorated, amid rumours that Mauritania was planning to expel all its black citizens, and, despite a certain embarrassment, the francophone summit took place in Dakar late that month amid rumours of troops massing on both sides of the frontier. Mediation bids by the French at the time failed to achieve anything, and the Mauritanian president failed to attend the meeting. During June and July, in spite of a meeting between the presidents of the two countries at the ECOWAS summit in Burkina, relations worsened, and there were growing reports of tension along the Senegal River itself, as refugee camps grew up composed of those the Senegalese claimed were black Mauritanians, particularly from the Toucouleur ethnic group that is predominant in the valley, whom the Mauritanians now seemed to be saying were Senegalese. These groups began to carry out raids over the river.

In Senegal itself there was initial criticism of the government for reacting too mildly, but soon government and opposition parties were outdoing each other in tough talk. This had a useful effect on relations between the parties, especially as opposition groups were given space in public media to air their views. At the same time the Party of Independence and Work (PIT) of Ameth Dantsoho broke its alliance with the PDS, and began talks with the government. Although not directly related to the Mauritanian quarrel, there was certainly a changed climate. To some it was even alarming, especially when the normally cautious Khalif-General of the Tijjaniya brotherhood, El Hadj Abdoul Aziz Sy in June 1989, reproached the government for passivity and called on his followers, strong among the Toucouleur, for direct action. By the end of July, there were said to be 30,000 refugees along the river, and there was open talk of the possibility of war. In mid-August, just as a mediation mission set up by the OAU arrived in the area, the Mauritanians expelled the Senegalese ambassador with 48 hours notice, for circulating 'insulting material', and in return the Senegalese broke diplomatic relations with only 24 hours notice.

Relations with two other neighbouring countries were by now also deteriorating: at the beginning of August 1989, President Diouf cancelled a visit to Guinea-Bissau, following the Bissau Government's rejection of a ruling by an international arbitration tribunal in Geneva in Senegal's favour in a border dispute. In mid-August, after announcing that it would be pulling its troops out of the Gambia, President Diouf announced that the Senegambian confederation had been 'frozen'. Defence Minister Medoune Fall, visiting Banjul, said that the pull-out related to the worsening relations with Mauritania. There was no doubt, however, that the Senegalese had been irritated for some time at the lack of progress in the economic aspects of the union. It was reported from Dakar that Diouf was annoyed at a request from Jawara for the presidency of the confederation to rotate, instead of the present system in which the Gambia is a permanent junior partner.

Economy

In the background to these tensions, both internal and international, was Senegal's continuing economic crisis. Despite the political problems, the economic recovery programme in place since 1979 continued, with new allocations from the IMF. In November 1988, an Enhanced Structural Adjustment Facility (ESAF) of nearly 150m SDRs was awarded, with the first 29.78m SDRs being transferred immediately. This meant an intensification of adjustment programmes, and a continuing high level of debt servicing. An ensuing rescheduling exercise in January 1989 helped here, although growth expectations for 1989 were said to be less encouraging than in the previous two years, especially in the agricultural sector. The budget for 1989–90 was forecast to balance at CFAfr507bn, an increase of 3.7 per cent over the previous year. Measures for privatisation of uneconomic state enterprises have not been making much headway, for lack of suitable takers, but new incentives are planned to attract both local and foreign investors. A new three-year investment programme was launched in April 1989.

BUSINESS GUIDE

Membership of international organisations: African Development Bank, African Groundnut Council, African States Associated with the EC (Lomé Convention), Arab Bank for Economic Development in Africa (recipient), Economic Commission for Africa, ECOWAS, FAO, French Monetary Area, GATT, IAEA, IDA, IFAD, IFC, ILO, IMF, Islamic Conference, Islamic Development Bank, OAU, UN, West African Economic Community, WHO, WIPO, World Bank.

Time: GMT.

Climate: Tropical in south (Casamanca), more temperate in north. During dry season (Dec-May) temperatures range from 18°C to 30°C. Jun-Nov is generally hot and humid with frequent thunderstorms. Temperatures and rainfall higher in land than on the coast.

Entry requirements
Passport: Required by all.
Visa: Required by all except nationals of Denmark, Italy, France, German FR and certain African countries. Return or onward travel ticket necessary for visa approval.
Prohibited entry: Nationals of South Africa.
Currency: Local currency import is unlimited, export to countries outside the franc zone is limited to around franc CFA 250,000. All foreign currency must be declared on arrival and departure. Amount taken out must not exceed amount brought in. Visitors should ensure that their travellers' cheques are made negotiable in Senegal.

Health precautions
Mandatory: Yellow fever vaccination certificate required by all.
Advisable: Cholera, typhoid and polio vaccinations. Malaria prophylaxis should be taken as risk exists throughout the country. Water precautions advisable.

Representation overseas: Abidjan, Addis Ababa, Algiers, Baghdad, Beijing, Berne, Bissau, Bonn, Brasilia, Brussels, Cairo, Conakry, Jeddah, Kuwait City, Lagos, Libreville, London, Moscow, Ottawa, Paris, Rabat, Rome, Stockholm, Tokyo, Tunis, Vatican City, Washington, Yaoundé.

Air access
National airline: Air Sénégal (Senegal is also a minority shareholder in Air Afrique).
Other airlines: Aeroflot, Air Algérie, Air France, Air Gabon, Air Mauritanie, Air Zaïre, Alitalia, Ethiopian, Ghana Airways, Iberia, Linhas Aéreas de Guiné-Bissau, Nigeria Airways, Royal Air Maroc, Sabena, Saudia, Swissair, Tap Air Portugal, Transportes Aereos de Cape Verde, Tunis Air.
Tax: Airport tax is added to price of tickets bought in Senegal.
International airport: Dakar-Yoff (Code: DKR), 17 km from city; duty-free shop, bar, restaurant, buffet, bank, post office, shops, car hire, taxis.

Surface access
Main ports: Dakar.

Overland access: Principal road routes are from Gambia, Mali, Mauritania – those from Guinea and Guinea-Bissau are not generally recommended. Twice weekly rail services between Dakar and Bamako (Mali) via Kaolack and Tambacounda.

Hotels: Air-conditioned hotels widely available in Dakar, though can be expensive. Advisable to book in advance, particularly between Dec-May. Hotel bills usually include service charges and local tax. Tipping therefore optional.

Electricity supply: 127/220 V AC, 50 cycles.

Credit cards: Major credit cards accepted in hotels, by airlines and some shops.

Car hire: International or national driving licence, insurance and car registration document (Carte Grise). Vehicles coming from the right always have right of way.

City transport

Taxis: Good service in Dakar, where most taxis are licensed and fitted with meters. Tipping is optional.

Buses: Regular flat-fare service operated by SOTRAC (large green and yellow public buses).

National transport

Air: Air Senegal links Dakar with all the main towns. Also possible to charter small aircraft, e.g. from Africair Service, at Yoff Airport. Gambia Air Shuttle operates a twice-daily service between Dakar and Banjul in The Gambia.

Road: Tarred roads mainly near coast; inland areas served by roads of variable quality. Main highways Dakar to Saint-Louis, Rosso, Djourbel, Joal, Kaolack and Ziguinchor.

Buses: Coach services Dakar-Ziguinchor; Tambacounda-Ziguinchor; Tambacounda-Gaoual have been operated subject to demand.

Rail: Links Dakar with the Mali border via Tambacounda to the east, and with St Louis and Linuère to the north-east.

BUSINESS GUIDE

Public holidays
Fixed dates: 1 Jan, 1 Feb (Confederation Day), 4 Apr (National Day), 1 May, 15 Aug (Assumption), 1 Nov (All Saints'), 25 Dec.
Variable dates: Good Friday, Easter Monday, Ascension, Whit Monday, Eid al-Fitr (Korité), Eid al-Adha (Tabaski), Ashura (Tamkharit), Prophet's Birthday (Maouloud).

Working hours
Government and business: (Mon-Fri) 0800/0900 – 1200, 1500 – 1800; (Sat) 0800/0900 – 1200.
Banking: (Mon-Fri) 0800 – 1115, 1430 – 1600.
Shops: 0800 – 1200, 1430 – 1800.

Telecommunications
Telephone: Dialling code for Senegal: IDD access code + 221 followed by subscriber's number. (There are no area codes). Dial 16 for operator service.
Telex: Public telex facilities available at Hotel Teranga, Dakar and at main post offices.
Telegram: Telegrams can be sent from main post offices.

Banking
Central bank: Banque Centrale des États de l'Afrique de l'Ouest.
Main centre: Dakar. *Head offices:* Bank of Credit and Commerce International, Banque Internationale pour l'Afrique Occidentale, Banque Internationale pour le Commerce et l'Industrie du Sénégal, Banque National de Développement du Senegal, Banque Senegalo-Kuweitienne, Citibank, Caisse Nationale d'Epargne, Société Générale de Banques au Sénégal, Union Sénégalaise de Banque pour le Commerce et l'Industrie.

Trade fairs: International general trade fair is held in Dakar in late November every two years.

Representation in capital: Algeria, Argentina, Austria, Bangladesh, Belgium, Brazil, Bulgaria, Canada, Cape Verde, China PR, Czechoslovakia, Egypt, Ethiopia, France, Gabon, Gambia, German FR, Guinea, Guinea-Bissau, Haiti, India, Indonesia, Iran, Italy, Côte d'Ivoire, Japan, Korea PDR, Rep. of Korea, Kuwait, Lebanon, Mali, Mauritania, Mexico, Morocco, Netherlands, Niger, Nigeria, Pakistan, Philippines, Poland, Portugal, Qatar, Romania, Saudi Arabia, Somalia, Spain, Syria, Switzerland, Thailand, Tunisia, Turkey, UK, USA, USSR, Vatican, Zaïre, Zimbabwe.

Hotels: Dakar
Club Méditerranée Almadies, Pointe des Aladies, P.O. Box: 3236 (tel: 221-00).

Hacienda, Route de Cambérène, P.O. Box: 315 (tel: 22-07-80; tx: 291).

Indépendance, Place de l'Indépendance, P.O. Box: 221 (tel: 23-10-19; tx: 454).

Meridien Diarama, (near Airport), P.O. Box: 8092 (tel: 23-10-05; tx: 682).

N'Gor (Meridien), (near Airport), P.O. Box: 8092 (tel: 23-10-05; tx: 31516).

Novotel Dakar, Avenue du Barachois, P.O. Box: 2073 (tel: 21-88-49, 23-10-90; tx: 21645).

Savanna Pullman, Pointe Bernard Petite Corniche, P.O. Box: 1015 (tel: 23-10-44; tx: 21634).

Teranga (Sofitel), Rue Colbert, P.O. Box: 3380 (tel: 23-10-44; tx: 21634).

Hotels: Saly
Novotel, Saly Portudal, P.O. Box: 70 (tel: 21-03-86; tx: 777111).

Savana Koumba, Sali Portudal, P.O. Box: 88 (tel: 57-11-12; tx: 77128).

Car hire: Dakar
Avis (Eurafricars): PO Box 3249 (tel: 22-33-33; tx: 414).

Europcar: PO Box 3115 (tel: 22-74-42, 22-72-53; tx: 51667, 51214).

Hertz: 64 Rue Felix Faure, PO Box 2302 (tel: 21-56-23, 22-20-16; tx: 51655).

Banking
Bank of Credit and Commerce International, 4 Avenue Roume, Place de l'Indépendance, PO Box 2060, Dakar (tel: 22-72-62/3; tx: 3318).

Banque Internationale pour l'Afrique Occidentale, Po Box 129, Dakar.

Banque Internationale pour le Commerce et l'Industrie, Po box 392, Dakar.

Citibank, 2 Place de l'Indépendance, PO Box 3391, Dakar (tel: 21-29-81; tx: 296).

Société Générale de Banques au Sénégal, 19 Avenue Roume, PO Box 323, Dakar (tel: 23-10-60; tx: 534).

Union Sénégalaise de Banque pour le Commerce et l'Industrie, 17 Boulevard Pinet-Laprade, PO Box 56, Dakar (tel: 23-10-08; tx: 506).

Other useful addresses
Dakar: Africair Service, Yoff Airport (tel: 20-09-87; tx: 664 piper).

Agence Havas Afrique, 29 Boulevard Pinet-Laprade, PO Box 503.

Air Afrique, PO Box 3132 (tel: 21-53-66).

Air Sénégal, PO Box 8010 (tel: 20-09-13; tx: 449).

Chambre de Commerce, d'Industrie et d'Artisanat, Place de l'Indépendence, PO Box 118 (tel: 21-71-89, 21-64-78; tx: 3112).

DHL (Courier), 22 Rue Carnot, PO Box 3554 (tel: 22-13-93, 22-13-94).

Direction de la Statistique, PO Box 116 (tel: 23-08-81).

Foire Internationale de Dakar (SOFIDAK), Route de l'Aéroport, PO Box 3329 (tel: 23-10-11; tx: 430).

Ministère du Tourisme, PO Box 4049 (tel: 22-22-26).

Port Autonome de Dakar, 35 Boulevard de la Libération (tel: 22-45-45, 22-74-21; tx: 406).

Société Nationale d'Études et de Promotion Industrielle, PO Box 100.

Société Nationale des Télécommunications du Sénégal (SONATEL), 6 Rue Wagane Diouf, PO Box 62 (tel: 23-10-23, 21-42-42; tx: 1296, 982 sonatel).

Seychelles

Gillian Cronje

In June 1989 President France Albert René was returned to office unopposed for a third and last term, under the country's constitution. Although Seychelles is a one-party state, his popularity appears to be genuine. Under his leadership the Seychelles islands have become a welfare state. There is a free health service, free education and a housing policy which seeks to provide a home for each family. Per capita income is $3,300 a year. But the benefits and prosperity could be threatened by the inroads made on the budget by increasingly heavy foreign and domestic debt servicing costs: SR121.5m in 1987; SR145.2m in 1988; SR147m (projected) in 1989.

René insists that his government will not abandon socialism, but it seeks foreign investment. It has recently lowered duties on imported technology, cut taxes and reduced interest rates to encourage private enterprise. It derives most of its income from providing a playground for the international jet-set. Upmarket tourism earns it 70 per cent of the foreign exchange, contributes 40 per cent to the GDP and employs 30 per cent of the workforce. The government has set out to court the very rich who can afford to admire its rare fauna and flora from the luxury hotels of Mahé and Praslin.

Forty per cent of the land surface has been turned into nature reserves. The UN has declared Aldabra atoll a world heritage site: it is the home of some 150,000 tortoises, members of the last surviving land species. Aldabra is too remote for tourists, but the more accessible Curieuse Island has started to breed tortoises with financial help from Britain's Royal Zoological Society. Praslin, the second-largest island, is another UN world heritage site because of its rare coco-de-mer palms.

Tourism

The government is careful to preserve these assets and the beauty of the beaches. Would-be property devel-

KEY FACTS

Official title: Republic of Seychelles
Head of state: President France Albert René
Ruling party: Seychelles People's Progressive Front
Capital: Victoria
Official Languages: Creole, French and English
Currency: Seychelle Rupee (SR) = 100 cents
Exchange rate: SR5.56 per $ (Nov 1989)

Area: 453 sq km
Population: 66,000 (1987)
GDP per capita: $3,723 (1987)
GNP real growth: 2.5% (1987)
Inflation: 1.7% (1988)
Trade balance: –$113m (1988)

opers come up against stringent planning laws: no hotel may be higher than the highest palm tree. There is a ceiling on the number of tourists: 100,000 is the maximum annual number envisaged. The 1988 figure was 77,000 – 7.5 per cent up on the previous year – and the 1989 estimate is 91,000. The visitors come mainly from Britain, France, Italy, West Germany and South Africa, but Japan has become increasingly important since Air Seychelles started its Far Eastern flights.

The airline was formed in 1983 with a single aircraft. By 1989 it operated five flights a week, including one from Singapore. The new Boeing 767 which recently joined the fleet is intended for a new direct Japanese route, as well as allowing an additional weekly service to Europe. It is capable of flying non-stop for 11,900 km, which means it can offer highly competitive long-distance flights to Europe or the Far East without the inconvenience of refuelling stops.

The Seychelles' other sources of income are the export of copra and cinnamon, fishing and light industry. In 1988 some 220,000 tonnes of tuna were caught, mainly by vessels from Spain, South Korea and Japan which pay $4.8m a year in licence fees. There are plans for an indigenous fleet of 10 purse-seiners, the first of which was scheduled to be launched in 1989. Meanwhile the Seychelles Fishing Authority leases its newly expanded port facility to foreign vessels. A new tuna-canning factory, Conserverie de l'Ocean Indien (70 per cent government-owned) exported SR50m worth of fish in the first nine months of 1988.

Seychelles is the site of a new £10m BBC relay station which started broadcasting in September 1988. The BBC's 25-year lease adds to foreign exchange earnings, as does the $10m annual rent paid by the US Air Force for its satellite tracking station on Mahé. The contract with the Americans comes up for renewal in 1990, and René's socialist government has no intention of terminating the arrangement. The country's non-aligned status shows itself in its arrangements for military training: naval personnel receive instruction from 30 Soviet advisers; India is training army officers; North Korea a militia; and the helicopter pilots receive instruction from French personnel. The country's forces number 8,000.

Seychelles has become the most active member of the Indian Ocean Commission, which also has Mauritius, Comores, Reunion and Madagascar as members. The government hopes for shared maritime patrols to strengthen regional security, and there are talks on economic co-operation, particularly on the pooling of imports, to secure better prices from suppliers. There are already some joint ventures with Mauritius, for instance a paint factory and a video and television assembly plant. But the Indian Ocean Commission's basic objective remains the creation of a regional zone of peace.

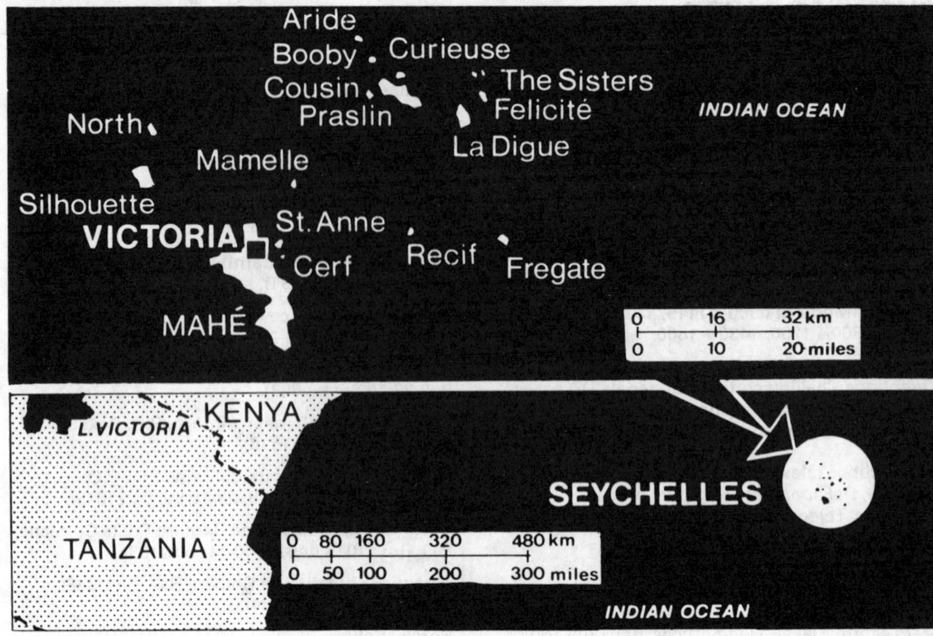

KEY INDICATORS

	Unit	1984	1985	1986	1987	1988
Population	'000	64	65	66	66	–
Gross domestic product (GDP)	Rupees m	1,068	1,205	1,284	1,376	–
GDP per capita	Rupees	16,687	18,538	19,454	20,848	–
Inflation	%	4.1	0.8	0.3	2.6	1.7
Exports	Rupees m	181.5	199.7	113.7	124.1	169.2
Copra	Rupees m	8.2	7.6	3.8	3.2	2.8
Re-exports	Rupees m	160.1	177.8	99.8	88.1	95.3
Imports	Rupees m	618.7	704.7	652.0	633.9	823.0
Balance of trade	$m	–69.0	–79.5	–84.9	–88.2	–112.5
Current account	$m	–13.3	–19.2	–33.3	–21.1	–22.4
Total external debt	$m	61.0	79.0	106.9	119.2	–
Debt service ratio	%	3.6	5.0	6.9	–	–
Exchange rate	Rupees per $	7.06	7.13	6.18	5.60	5.38

* estimated figure

COUNTRY PROFILE

Political system: Executive power rests with the president, elected for five years. A council of ministers is appointed by the president, as are two members of the unicameral National Assembly, which exercises legislative power; the other 23 members are directly elected for four years. Gained independence from UK in 1976.

Political parties: The Seychelles People's Progressive Front is the only legal political party. There are, however, nominated representatives in Parliament from small islands without permanent populations.

Population: The population of 66,000 (1987) is largely of Creole and European origin. Practically the whole population is Christian, with 90 per cent belonging to the Roman Catholic faith.

Main cities/towns: Victoria, on Mahé island (population 30,000).

Language: Creole, French and English are the official languages in Seychelles.

Media
Press: One daily newspaper, *Nation*, which is government owned. A monthly SPPF paper, *The People*. A few periodicals are also published.
Broadcasting: Radio Television Seychelles (RTS) run radio and television services in Creole, English and French. Reception is good on Mahé and the other main islands.

External trade: Visible trade deficit is partially offset by earnings from tourism which is the main foreign exchange earner, plus foreign aid and investment, and rental from US satellite tracking station.
Exports: Copra (27 per cent of total value excluding re-exports in 1986), fish (39%), cinnamon (17%); also re-exports of fuel.
Imports: Fuels (18 per cent of total value in 1986), machinery and transport equipment (27%), food and live animals (15%), basic manufactures (19%), chemicals (6%). Main sources: UK (16% of total value in 1986), France (16%), South Africa (13%).

Agriculture: Contributes about eight per cent to GDP. Overall importance of sector to economy has declined with the expansion of the tourist industry. Less than 10 per cent of land area is cultivable, much of which is given over to copra

COUNTRY PROFILE

and cinnamon – the islands' major cash/export crop. Small quantities of vanilla, tea and limes are also exported. Crops grown for local consumption include tropical fruits, cassava, sweet potatoes, yams, sugar-cane, bananas, tea and vegetables; but the country's staple food crop, rice, has to be imported. Government reforms have included setting up 10 state farms and six smaller co-operatives, a reafforestation project and new marketing structures. The fishing industry is also being expanded as part of the government policy of economic diversification, and foreign companies are encouraged to become involved.

Industry: Small-scale manufacturing sector contributing under 10 per cent of GDP. Main activities include production of canned tuna, soft drinks, juices, jams, beer, cigarettes, paints, assembling of television sets and processing of cinnamon and coconuts. Tourism employs a third of the workforce and generates up to 50 per cent of GDP. The Government hopes to attract 100,000 tourists by 1988, though the number of tourists declined slightly in 1986 to 66,782 from 72,542 in 1985. In 1987 it rose to 71,626.

Mining: There is no mining on any large scale, but some granite is quarried for construction. Offshore surveys have indicated the presence of certain metals on the seabed, but there are no definite plans for exploitation.

Energy: Some gas resources have been found offshore but as yet have not been exploited. Annual generation of electricity amounts to around 70 gWh.

Membership of international organisations: African Development Bank, African States in Association with the EC (Lomé Convention), Commonwealth, Economic Commission for Africa, FAO, GATT, ICAD, IFC, ILO, IMF, IMO, Indian Ocean Commission, OAU, UN, UNESCO, UPU, WHO, WIPO, WMO, World Bank.

BUSINESS GUIDE

Time: GMT +4.

Climate: Tropical and humid. Average temperature 24°C to 32°C throughout year. Hottest months Dec to May; wettest from Dec to Mar; cooler from Jun to Nov. Islands lie outside cyclone belt.

Entry requirements

Passport: Required by all.

Visa: Generally not required. A visitor's pass is issued on arrival.

Health: Cholera vaccination certificate required if arriving from Africa or India.

Currency: Unlimited import/export of foreign or local currency.

Representation overseas: London, Nairobi, Paris, Rome, Stockholm, Tokyo, Washington.

Air access

National airline: Air Seychelles.

Other airlines: Aeroflot, Air France, British Airways, Kenya Airways, Lesotho Airways.

International airport: Seychelles-Pointe Larue International (Code: SEZ), Mahé Island, 10 km from Victoria; duty-free shop, bar, restaurant, buffet, bank, shops, car hire.

Surface access

Main port: Victoria.

Hotels: Good standard and widely available. All the large hotels in Mahé are on the beach. Advisable to book and confirm reservation in advance, particularly at Christmas and during August. Government trades tax of five per cent is added to bill, and usually also a service charge. Tipping optional.

Electricity supply: 240 V AC, 50 cycles.

Credit cards: Major credit cards widely accepted.

Car hire: Available on Mahé and Praslin. Advance reservation advisable. Foreign or international driving licence required.

City transport

Taxis: Available on Mahé and Praslin. Fixed charges in operation; tipping is optional.

National transport

Air: Air Seychelles operates regular services from Mahé to Praslin, Fregate, Bird and Denis islands. Airstrips exist on most of the outlying islands.

Road: Mahé and Praslin are the only islands with appreciable surfaced roads. Gravel roads elsewhere.

Buses: Cheap, regular bus service (the SPTC) operates to most areas of Mahé from Victoria. Services also operate on Praslin and La Digue.

Water: Regular ferry services connect Mahé, Praslin, La Digue and other islands.

Public holidays

Fixed dates: 1-2 Jan, 1 May, 5 Jun (Liberation Day), 29 Jun (Independence), 15 Aug (Assumption), 1 Nov (All Saints'/La Digue Festival), 8 Dec (Immaculate Conception), 25 Dec.

Variable dates: Good Friday, Easter Saturday, Corpus Christi.

Working hours

Government and business: (Mon-Fri) 0800 – 1200, 1300 – 1600.

Banking: (Mon-Fri) 0830 – 1300; (Sat) 0800 – 1200.

Shops: (Mon-Fri) 0800 – 1200, 1330 – 1700; (Sat) 0800 –1200; some open Sun morning.

Telecommunications

Telephone: Dialling code for Seychelles: IDD access code + 248 followed by subscriber's number. There are no area codes. Direct dialling from Seychelles to most countries. As well as from hotels, calls can be made 24-hours a day from SEYTELS/Cable and Wireless, Francis Rachel Street, Victoria, Mahé.

Telex and telegram: Facilities available at some hotels. 24-hour telex service from Cable and Wireless, PO Box 4, Francis Rachel Street, Victoria, Mahé.

Banking

Central bank: Central Bank of Seychelles.

Head offices: Development Bank of Seychelles, Seychelles Savings Bank.

Foreign branches: Bank of Baroda, Bank of Credit & Commerce International, Banque Française Commerciale, Barclays Bank, Habib Bank, Standard Chartered Bank.

Representation in capital: Belgium, China PR, Cuba, Cyprus, Denmark, France, Germany FR, India, Madagascar, Mauritius, Netherlands, Sweden, UK, USA, USSR.

BUSINESS DIRECTORY

Hotels: Mahé

Auberge Club des Seychelles, Bel Ombre, P.O. Box: 550 (tel: 47550; tx: 2237).

Barbarons Beach (Meridien), P.O. Box: 626 (tel: 78253; tx: 2258; fax: 78484).

Beau Vallon Bay, P.O. Box: 550 (tel: 47141; tx: 2237).

Coral Strand, Beau Vallon, P.O. Box: 400 (tel: 47036; tx: 2215).

Fisherman's Cove (Meridien), Beau Vallon, P.O. Box: 35 (tel: 47252; tx: 2296; fax: 47540).

Northolme, Glacis, P.O. Box: 333 (tel: 23222; tx: 2355).

Reef, Anse aux Pins, P.O. Box: 388 (tel: 76251; tx: 2231).

Sheraton, Port Glaudi, P.O. Box: 540 (tel: 78451; tx: 2245; fax: 71571).

Vacoa Village Resort, Beau Vallon, P.O. Box: 550 (tel: 47141; tx: 2237).

Vista do Mar, Glacis, P.O. Box: 622 (tel: 47351/2; tx: 2318).

Car hire: Victoria

Avis: PO Box 224 (tel: 24511, 24542; tx: 2281).

Europcar: PO Box 559 (tel: 23303, 47018; tx: 2263).

Hertz: Barbarons Beach Hotel (tel: 22447; tx: 2310).

Ideal: PO Box 72, Victoria (tel: 44280; tx: 2399 ideal).

Jean's: (tel: 22278).

Tropicar: (tel: 23838, 41382).

Union Vale: (tel: 47052).

Other useful addresses

Air Seychelles, PO Box 386, Victoria House, Victoria (tel: 21547; tx: 2337,2289).

Barclays Bank, Independence Avenue, PO Box 167, Victoria.

Central Bank of Seychelles (CBS), Independence Avenue, PO Box 701, Victoria (tel: 21580; tx: 2301).

Development Bank of Seychelles, PO Box 217, Victoria.

National Travel Agency, Kingsgate House, PO Box 611, Victoria (tel: 21180; tx: 2356,2384).

RTS Radio, PO Box 321, Union Vale, Victoria (tel: 24220; tx: 2320).

RTS TV, PO Box 321, Hermitage, Mah\e (tel: 24220; tx: 2320).

Seychelles Chamber of Commerce & Industry, PO Box 214, Victoria (tel: 23812).

Seychelles Marketing Board, PO Box 516, Victoria (tel: 24444; tx: 2368,2377).

Seychelles Savings Bank, Independence House, PO Box 531, Victoria.

Seychelles Tourist Office, PO Box 92, Victoria, Mahé (tel: 22881).

Sierra Leone

Stephen Riley

President Joseph Momoh's government faced major economic problems in late 1989 after the externally-supported economic emergency programme was abandoned. It was seeking further sources of external support, and was hoping to resume links with the IMF. The expected investment in deep mining diamonds was delayed, and major exports fell dramatically. Despite the hike in consumer prices caused by the June budget, the regime remained politically stable, although there were some signs of dissent.

Budget indicates economic decline

The 1989–90 budget of Finance Minister Hassan Kanu gave an indication of Sierra Leone's economic woes. The government retreated from its implementation of the ambitious adjustment programme agreed with the IMF in 1986, and, as a consequence, it was in difficulties with the major financial institutions. It was declared ineligible to receive further funding from the World Bank, and has substantial arrears with the IMF. Though there were some contacts between the Fund and the government in early 1989, there is a long way to go before the rift can be healed.

The June budget did contain some items of which the IMF would approve. It doubled the price of petrol to Le110, consequently increasing other consumer prices. It claimed that inflation was falling from the 60 per cent of the previous year, but few believed this. Kanu also sought to improve the finance ministry's control over general government expenditure, and reduce the independent spending power of ministries. Unlike the centralising rule of Siaka Stevens, President Joseph Momoh delegates a great deal to his ministers, and, as a result, government expenditure rocketed beyond agreed budgetary limits. Kanu also intervened in the payment of civil service wages, by establishing payment terms to pay salaries direct and to cut out payments to 'die men' (or imaginary, dead or redundant staff).

Recorded exports were worth $78m in the first three-quarters of 1988, a fall of 13 per cent on the corresponding period the previous year. There was a major fall in official diamond exports, but coffee and cocoa exports also went down. Coffee exports fell from 3,600 tonnes to 1,600 tonnes, while cocoa exports fell from 8,800 tonnes to 5,600 tonnes. The fall in agricultural exports was attributable in part to smuggling across the border into Liberia, and partly to a decision by farmers to opt out of cash crop production for food production. Low world market prices also contributed. The only positive sign was the increase in the export of rutile (or titanium dioxide), with exports of 108,000 tonnes in the first nine months of 1988 compared to 89,000 tonnes in the same period in 1987. Rutile contributed $42m in value, although the trade gap widened.

For the period July 1988 to March 1989 the recorded trade gap was $56m, an increase on the $20m of the previous year, as recorded imports at $134m were up 20 per cent on the last corresponding period. Kanu's budget did not disguise his disquiet at this deteriorating performance, though there were few measures to alleviate it. In the June presidential address to parliament, President Momoh announced the ending of the state of economic emergency. He was very mild in his description: 'the results of this exercise have been very mixed'. In fact, the economic emergency was a complete failure, and had little effect upon the growing smuggling and parallel market activity which underminded the government's pursuit of an economic revival strategy.

1989 saw greater economic hardship for many in Sierra Leone. Rice, the staple food, was often only available through political intermediaries at a high price. Its market price for small amounts reflected the considerable speculation involved. The government's economic difficulties led to the late payment of salaries. Salary increases were agreed – such as those for university staff – but payment was delayed or postponed altogether. Inflation and hoarding of essential commodities was also a problem, as was the continuing shortage of the national currency, the Leone. Withdrawals from the banks were limited due to shortages of notes, which most thought had been smuggled out of the country to Liberia and Guinea as part of the growing parallel market in diamonds, gold and agricultural produce. The relatively low market price for petrol and kerosine also led to its extensive smuggling over the borders. Momoh's government has also done little to counteract these growing problems.

Few developments in mining

1989 did not see many new developments in Sierra Leone's important mining sector. In 1988 there was considerable optimism that the rise in world diamond prices would lead to a local reward when the long-delayed project to deep-mine diamonds down the kimberlite pipes in the Kono region would be started. The project was costed by the Swedish mining company Otukumpu, and interested parties included, initially, De Beers and the Saudi Fund for Development. However, interest waned as legal problems over the ownership of the rights and assets of the official part-government owned mining company emerged. One legal disputant was Jamil Said Mohammed, a formerly influential Afro-Lebanese businessman who left Sierra Leone in the early years of the Momoh government. The windfall which the kimberlite project represents has been postponed.

Official export production of diamonds fell dramatically during the year. This was partly due to the poor quality of the plant at the mines in Yengema, but many suspect that extensive smuggling also took place. Officially exported diamonds fell from 141,000 carats in 1987–88 to 4,800 carats in 1988–89, according to the 1989 budget speech of Finance Minister Hassan Kanu. There is evidence, however, that there was a large increase in the private mining of gold and diamonds, particularly in the Northern and Eastern Provinces. Harry Winston, a New York diamond magnate, opened a new gold mining operation near Bo in 1989, and there were a series of ad hoc mining operations, some with important political backers. The problem for Momoh's government is that unless the gold and diamonds are exported through official channels, revenue is lost.

One interesting feature of 1988 and 1989 was the re-emergence of an apparently Israeli-backed mining company with interests in gold and diamond mining. The firm is N. R. Scipa, a company with a high profile as some of its offices are in the presidential villa of the former president, Siaka Stevens. Like the case of Liat, the company once headed by Shabtai Kalmanowitch, Scipa is alleged to be the front-runner for the revival of significant Israeli diplomatic and commercial interests in Sierra Leone. N. R. Guer, the Israeli managing director of Scipa, has denied this interpretation of his company's role, but there is no doubt that it is engaged in extensive ad hoc mining activities. It is also suggested that Scipa has gained important political allies through its other role as a rice importer at a time of shortages, hoarding and speculation. It imported 20,000 tonnes during the first part of 1989,

KEY FACTS

Official title: Republic of Sierra Leone
Head of state: President Joseph Saidu Momoh
Ruling party: All People's Congress (APC)
Capital: Freetown
Official Languages: English
Currency: Leone (Le) = 100 cents
Exchange rate: Le62.5 per $ (Nov 1989)

Area: 72,325 sq km
Population: 4.0m (1988)
GNP per capita: $300 (1987)
GNP real growth: 0.4% (1987)
Inflation: 179% (1987)
Trade balance: $24.8m (1987)
Foreign debt: $659m (1987)

with much of it being distributed to the politically influential.

Little political change

Some have argued that Scipa's emergence represents the politics of the Middle East again intruding into Sierra Leonean life, as it is a challenge to the role of the many Lebanese and Afro-Lebanese entrepreneurs operating in trade and mining. Although Momoh's government is pro-Western, it is always on the lookout for a new source of external finance. The government has also sought to improve its relations with Iran, and also recognised the new state of Palestine in early 1989. The new ambassador, Saad al Gerwaji, is playing a prominent role, while Iran is supplying oil for the Freetown refinery and promoting itself through cultural events and links with the politically influential surrounding Momoh. However, the government is merely searching for external financial support, and support that will have less politically destabilising effects than the imposition of IMF conditionalities.

Domestically the government is stable after the shock-waves of the alleged coup attempt of 1987. The coup conspirators, convicted of treason, continue to remain in Freetown's Pademba Road prison under tight military security, and their appeals continue. The prospects for former first vice-president Francis Minah, and the others implicated, do not look good however. The quashing of the alleged coup attempt gave Momoh much greater authority within the political system. He is becoming more of a contrast with his predecessor, Siaka Stevens, as he is a delegator who has given the cabinet much autonomy.

A 'kitchen cabinet' has grown up around the president with influential members of the circle including Dr A. K. Turay, the minister for presidential affairs, and Mansa Samura, head of protocol in the president's office. Criticism of the two vice presidents as weak and ineffectual has led to proposals to re-create the post of prime minister and to reform the presidential office. There are a number of front-runners for the premiership, some of whom share ethnic loyalties with Momoh. They include Bambay Kamara, the Inspector-General of Police who is also responsible for government's distribution of rice, Hassan Kanu the finance minister, and Dr Abdullai Conteh, the Attorney-General. However, as with much in Sierra Leone politics in the 1980s, there has been drift and delay and no decisions on the premiership have been taken.

In January 1989 the 10th national delegates conference of the ruling All People's Congress (APC) was finally held after delay caused by internal divisions. The conference sought to revive the party, which faced much criticism for its poor organisation and incompetence. It may also have to face a real challenge, as in mid-1989 the flags and slogans of the former Sierra Leone's Peoples Party – now banned under the one party state – were again seen in the Eastern province towns. The support for the former opposition party was related to the growing economic hardship being suffered in the provinces. The APC delegates, meeting in Makeni near Momoh's home town, stressed the need for accountability, honesty and responsibility in government, and passed resolutions supporting the government's Green Revolution programme. This has made little progress since it was started in 1986. The conference also suggested that the constitution should be revised, particularly where it relates to senior officials.

The attempt to revive the ruling party indicated that the president is aware of the potential political challenges to him. Although the military under Mohammed Tarrawallie is supportive of Momoh, and is generously supplied with subsidised rice, new uniforms, equipment and a new barracks building programme, there is always the possibility of a coup attempt by non-coopted junior officers. There was some unorganised protest activity against the June price increases, but the trade union movement has not stirred itself. Momoh remains firmly in charge, but the regime's economic difficulties could always spark off a major challenge to his regime. Unless there is a major investment in deep mining diamonds – with an expected return of $750m over 10 years – or a new agreement with the IMF in return for further credit, the government's economic concerns could lead to increased political dissent.

KEY INDICATORS

	Unit	1983	1984	1985	1986	1987
Population	m	3.49	3.57	3.66	3.75	3.85
Gross domestic product (GDP)	Le m	1,876	2,729	4,310	6,353	29,143
GDP per capita	Le	538	764	1,178	1,694	7,570
Inflation	%	68.4	66.6	76.6	80.9	178.7
GDP real growth	%	1.0	–0.5	0.0	–2.4	0.4
Agricultural production	1979-81=100	110.9	99.6	101.0	112.7	111.0
Food production	1979-81=100	103.9	101.5	111.6	109.5	108.1
Crop production	1979-81=100	110.9	97.9	98.5	110.6	107.0
Livestock	1979-81=100	110.9	116.0	124.1	135.1	135.1
Diamond production	'000 carats	352	265	250*	272	360
Bauxite production	'000 tonnes	785	885	1,001*	1,390	1,379
Exports	Le m	201.9	334.7	649.0	1,951.7	4,668.9
Diamonds	Le m	66.2	100.6	122.0	446.9	1,072.9
Cocoa beans	Le m	22.8	58.7	50.5	193.5	653.4
Imports	Le m	286.9	394.3	788.7	2,021.3	4,210.5
Petroleum	Le m	87.9	154.0	166.1	200.7	503.7
Balance of trade	$m	–25.9	–17.1	2.4	14.6	24.8
Current account	$m	–17.5	–22.8	19.3	128.9	–24.3
Total external debt	$m	531	437	540	595	659
Debt service ratio	%	7.0	10.8	5.8	9.7	–
Exchange rate	Le per $	1.68	2.51	4.73	8.40	30.77

* estimated figure

COUNTRY PROFILE

Political system: Executive power rests with the president, who is elected (for seven years and not more than two consecutive terms) by the delegates at the conference of the All People's Congress (APC). The president appoints a first and second vice-president. The legislature is the unicameral House of Representatives, with 60 or more elected members (approved by the APC), up to seven presidential appointees, a speaker and a deputy speaker. Gained independence from the UK in April 1961, and became a republic in 1971.

Political parties: The All People's Congress (APC) is the only legal political party. It was formed by Siaka Stevens in 1960, shortly before independence in 1961, as a rival to the Sierra Leone People's Party (SLPP). The APC came to power in 1968, after a short period of military rule. In the 1973 and 1977 elections, the APC was able to maintain its support, often due to unfair means. In 1978 a one-party state constitution was passed and the remaining SLPP members joined the ruling party.

Population: The population of 4m (1988 estimate) is growing at the rate of 2.6 per cent a year. The principal religions are Islam and Christianity. The number of foreign businessmen (Lebanese and Indians) is declining.

Main cities/towns: Freetown (population 470,000), Bo (39,000), Kenema (31,000), Makeni (26,000), Koidu-Sefadu.

Language: English is the official language, and the main medium for business; major languages are Krio, Mende and Temne.

Media

Press: New Citizen and *New Shaft* published weekly. Ministry of Information and Broadcasting publishes *Sierra Leone Trade Journal* quarterly. Various other periodicals are also published.

Broadcasting: One commercial radio and one commercial TV station (mainly received in Freetown area). Broadcasts in English, French and local languages.

External trade: Trade balance was in deficit from 1978 throughout 1980s, but considerable improvements were recorded in the late 1980s. The first half of 1987 showed a small surplus due to higher exports. The IMF suspended its stand-by facility in April 1988. The current account deficit, which is usually offset by aid inflows, is burdened with heavy debt servicing.
Exports: Diamonds (19 per cent of the 1985 total), rutile 20%), bauxite (15%), cocoa (8%), coffee (3%). Main destinations: UK (typically 33%), Netherlands (14%) USA (13%), other EC (16%), German FR (10%).
Imports: Fuel (typically 36 per cent of total value), foodstuffs (20%), machinery and transport equipment (18%) and manufactured goods (13%). Main sources: Nigeria (typically 18%), UK (16%), German FR (11%), Japan (5%), Netherlands (5%).

Agriculture: Contributes around 32 per cent to GDP and employs 72 per cent of labour force. Area under cultivation is limited by traditional land tenure system and is mostly in hands of smallholders engaged in subsistence farming. Major cash export crops are cocoa, coffee, palm kernels and ginger. Rice is the country's staple foodstuff, but despite government efforts towards self-sufficiency, imports have risen. Other food crops include maize, cassava, sweet potatoes and sorghum. Production has been hampered by a lack of incentives, low producer prices and a poor marketing/distribution system. The government's Green Revolution Programme, launched in 1986, aims to counter these problems; producer prices have been substantially increased. The fishing sector offers considerable potential and is currently being expanded. Afforestation programmes are under way, and rubber plantations are being revitalised.

Manufacturing: Contributes seven per cent to GDP and is mostly limited to light manufacturing of consumer goods such as cigarettes, alcoholic beverages, plastic footwear, nails, paint and confectionery. Emphasis placed on import substitution industries, but attempts to establish heavy industry, such as oil refining, have only met with limited success. Expansion limited by weak local demand, power shortages, foreign exchange shortages and low investment.

Mining: Most important sector is diamond extraction, although earnings have fallen due to depletion of reserves, illegal digging and smuggling. Plans to raise output with development of an underground mine with Kimberlite diamond pipes. Rutile is mined near Bonthe, bauxite at Mokanji Hill in Southern Province and gold near Magburaka.

Energy: Sierra Leone imports all its oil requirements. Several companies have carried out exploratory surveys for domestic supplies. Installed electrical capacity approximately 100 mW (mostly oil-fired thermal power stations). A $300m hydro-electric project is planned on the Seli river, in an attempt to reduce oil imports, but financing remains uncertain.

Membership of international organisations:
African Development Bank, African States Associated with the EC (Lomé Convention), Arab Bank for Economic Development in Africa (recipient), Commonwealth, Economic Commission for Africa, Economic Community of West African States, GATT, IDA, IFAD, IFC, ILO, IMF, Islamic Conference, OAU, UN, WHO, World Bank.

Time: GMT.

Climate: Tropical, with high humidity and rainfall. Dry season Nov-Apr, wet season rest of year. Average temperature range of 21-32°C remains fairly constant throughout the year.

Entry requirements
Passport: Required by all.
Visa: Required by all except nationals of Liberia, Republic of Guinea and holders of re-entry permits. (Note: visitors are advised to check regulations carefully before travelling.)
Prohibited entry: White nationals of South Africa.
Currency: Import/export of local currency limited to Le20. No limit on foreign currency brought into country, but amount must be declared on entry and not be exceeded by amount taken out. Visitors must exchange at least $100 worth of hard currency into local currency.
Customs: All visitors must complete a customs declaration form for presentation on entering country. Import of goods from South Africa prohibited.

Health precautions
Mandatory: Yellow fever vaccincation certificate required.
Advisable: Cholera, typhoid and polio vaccinations. Malaria risks exists throughout the country. Water precautions should be taken.

Representation overseas: Addis Ababa, Banjul, Beijing, Bonn, Brussels, Buenos Aires, Conakry, Havana, Jeddah, Lagos, London, Monrovia, Moscow, Paris, Rome, Washington.

Air access
National airline: Sierra Leone.
Other airlines: Aeroflot, British Airways, Ghana Airways, KLM,Nigeria Airways, UTA.
Tax: Airport tax of Le15 levied on departure.
International airport: Freetown-Lungi International (Code: FNA), 24 km from city (other side of wide river with helicopter and ferry links with central Freetown and Lumley Beach area); bar, currency exchange, post office, shops.

Surface access
Main ports: Freetown.
Overland access: Road: Possible from Guinea and Liberia but routes not generally recommended.

Hotels: Available in Freetown, especially at Lumley Beach, within easy taxi access of the centre of Freetown. Limited availability outside capital. Credit cards accepted only in major hotels. A service charge usually included in bill.

Electricity supply: 230/240 V AC, 50 cycles. Voltage fluctuation and power cuts occur.

Car hire: Car hire is available at relatively high rates. International driving licence required.

City transport
Taxis: Available in main towns; cannot be summoned by phone; fares by negotiation; tipping is not usual.

National transport
Road: Main roads include Freetown-Makeni-Kabala and Freetown-Bo-Kenema; other roads can be difficult particularly in the rainy season.
Buses: Buses in Freetown tend to be overcrowded and unreliable. Regular service Freetown-Kambia, Freetown-Pendembu.

Public holidays
Fixed dates: 1 Jan, 27 Apr (Independence Day), 25 Dec, 26 Dec.
Variable dates: Good Friday, Easter Monday, Eid al-Fitr, Eid al-Adha, Prophet's Birthday (Moulid an Nabi).

Working hours
Government: (Mon-Fri) 0800 – 1230, 1330 – 1645; close 1500 on Fri.
Banking: (Mon-Thu) 0800 – 1330; (Fri) 0800 – 1400.
Shops and business: (Mon-Fri) 0800 – 1200, 1400 – 1630; shops open Sat.

Telecommunications
Telephone: Dialling for Sierra Leone: IDD access code + 232 followed by area code (22 for Freetown), followed by subscriber's number. Service can be unreliable. International connections can be difficult, except at the Sierra Leone External Telecommunications Office (SLET), where it may be necessary to wait for a time.
Telex: Telex facilities available in Freetown at the SLET Office.

Banking
Central bank: Bank of Sierra Leone.
Other banks: Standard Chartered Bank (Sierra Leone), Sierra Leone Commercial Bank, Bank of Credit and Commerce International, Barclays Bank of Sierra Leone, International Bank of Trade and Industry.

Representation in capital: China PR, Cuba, Egypt, France, Gambia, German FR, Guinea, Italy, Côte d'Ivoire, Korea PDR, Rep. of Korea, Lebanon, Liberia, Nigeria, UK, USA, USSR. The PLO representative has ambassadorial status.

Hotels: Freetown
Bintumani, Lumley Beach, P.O. Box: 655 (tel: 37019/34; tx: 3316).

Brookfields, New England, P.O. Box: 1193 (tel: 41860; tx: 3424 bfh).

Cape Sierra, Lumley Beach, P.O. Box: 54 (tel: 37270; tx: 3367).

Paramount, Tower Hill, P.O. Box: 574 (tel: 24531, 24533; tx: 3257).

Sofitel Mammy Yoko, Lumley Beach, P.O. Box: 563 (tel: 37444, 37459; tx: 3416).

Banking
Bank of Sierra Leone, Siaka Stevens Street, PO Box 30, Freetown (tel: 26501; tx: 3232).

Barclays Bank of Sierra Leone, 25/27 Siaka Stevens Street, Freetown (tel: 22501).

Sierra Leone Commercial Bank, Siaka Stevens Street, Freetown (tel: 25264).

Standard Chartered Bank (Sierra Leone), 9/12 Lightfoot Boston Street, Freetown (tel: 25760, 25021).

Other useful addresses
Central Statistics Office, Tower Hill, Freetown (tel: 23897,24267).

Chamber of Commerce of Sierra Leone, Guma Building, Lamina Sankoh Street, PO Box 502, Freetown (tel: 26305).

Europcar, PO Box 23, (tel: 37444; tx: 3416).

Ministry of Information and Broadcasting, Yeuyi Building, Brookfields, Freetown.

Ministry of Tourism and Culture, Wallace Johnson Street, Freetown (tel: 26345,24776).

Ministry of Trade and Industry, Ministerial Building, George Street, (tel: 25211; tx: 3218).

National Trading Co, Howe Street, Freetown (tel: 23986,26179; tx: 3340).

PD Media, 49 Wilkinson Road, Freetown.

Sierra Leone External Telecommunications (SLET) Office, Wallace Johnson Street, Freetown (tel: 22801,24591).

Sierra Leone Ports Authority, PO Box 386, Freetown (tx: 3262).

Sierra Leone Produce Marketing Board, Cline Town, Freetown (tel: 50431; tx: 3211).

Singapore

A special correspondent

The Singapore government of Prime Minister Lee Kuan Yew could be forgiven a feeling of smugness. Few governments in the world could boast that in freely contested general elections, they would win all but one out of 81 seats in parliament. Yet in September 1988, Lee's People's Action Party did just that. Nor could many regimes point to such a consistent record of outstanding economic achievement. The real economic growth rate of around ten per cent in 1988 out-stripped even the other Asian 'tigers', let alone the developed countries. Yet Lee's government still acts as if all these achievements are fragile, as if it is under imminent threat of subversive conspiracy, and as if such hard-won prosperity could easily be lost in a moment of careless political liberalisation.

The 1988 elections

This government edginess had showed itself in the way it had stacked the decks for the elections. The PAP had been consistently in power since self-government in 1959. Only a shock by-election defeat in 1981, and the election of two opposition MPs in the 1984 elections had broken its monopoly of parliamentary seats. But in 1988, there were two worries. There was a growing sense of popular discontent with the authoritarian style of the PAP government. The PAP had been engaged in a bruising series of encounters both with the domestic opposition, some of whom it had locked up as parties to an alleged marxist conspiracy, and with the regional media, whose reporting of its actions it regarded as meddling in Singapore's internal business. The second worry was the advent of 220,000 new young voters. Lacking the older generation's experience of the bad old days, it was feared these young might frivolously vote for the opposition. So anti-government candidates found themselves facing two new hurdles, other than the traditional bias of the pro-government media. For the first time 13 'group' constituencies were introduced. Parties had to field a team of three candidates. One at least must come from a minority race, to ensure representation from non-Chinese communities in parliament. Although the move was presented as designed to ensure that Singapore's delicate ethnic mix was reflected in the composition of parliament, it also had the effect of penalising the young and poorly organised opposition parties, who found it hard to nominate three credible candidates. Similarly, a new scheme designed to ensure that MPs retain grass-roots links also militated against the opposition. MPs in some constituencies will now also be chairmen of local councils, with direct administrative responsibility for local services. PAP candidates were naturally better placed both to prove a record of administrative competence, and to allay any fears of central government discrimination against particular constituencies.

Government victory

Even given these handicaps, the opposition's showing in the elections was not impressive. The PAP's share of the poll fell one per cent, to 61.8 per cent. Because of Singapore's British style first-past-the-post system, this left only one opposition MP, the Singapore Democratic Party leader Chiam See Tong. Two other Worker's Party candidates, Francis Seow and Lee Siew Choh, exercised their right under a 1984 amendment to take up non-voting 'non-constituency' seats. But Francis Seow was then disqualified, after being tried and convicted *in absentia* for tax evasion. Mr

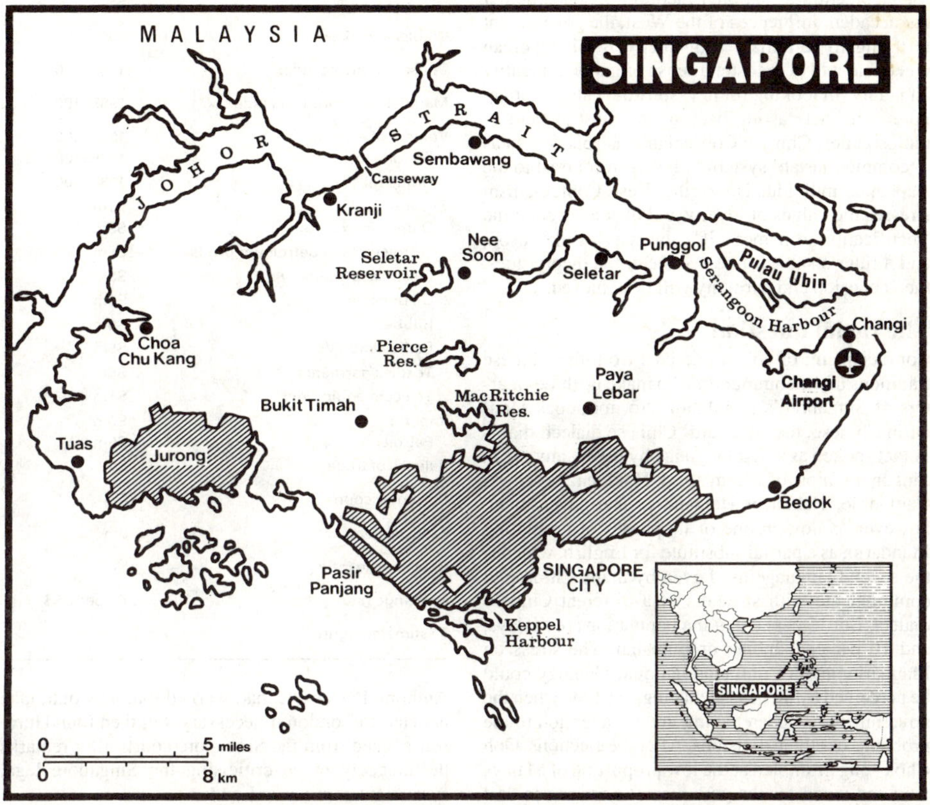

Seow was in New York, for heart treatment, he said, while the government alleged he was in fact living it up and staying away to avoid facing the music.

The government's comprehensive victory was taken as a popular mandate for the two big issues in the campaign – the harsh paternalism of government style, and a proposed constitutional change beefing up the powers of Singapore's hitherto largely ceremonial presidency. The new, elected executive president would have a power of veto over the spending of Singapore's foreign exchange reserves, and over senior appointments in public service. Lee Kuan Yew has said that he hopes to exert an influence over Singapore politics 'even as he is lowered into the grave'. So it was widely assumed that he saw himself as stepping into this role of custodian of the national nest-egg. He would become like former British prime minister Harold Macmillan, later Lord Stockton, railing against his successors' selling off the family silver. But he would have powers to stop it. However, it now seems Lee has no immediate designs on the presidency.

Succession question looms

This has to be seen in the context of Lee Kuan Yew's expressed desire to stand back from the premiership, as the 1980s ended, in his sixty-sixth year. There are two contenders for the succession. The man formally in line is the first deputy prime minister, Goh Chok-tong. But Lee took the unusual step of publicly criticising Goh during the election campaign, for his stiff public manner, and alleged lack of toughness. A longer-term heir presumptive is Lee's own son, Lee Hsien-loong, known as 'B.G', for his army rank 'Brigadier General', and more scurrilously, for his family connections – 'Baby God'. Lee Hsien-loong's out-

KEY FACTS

Official title: Republic of Singapore
Head of state: President Wee Kim Wee
Head of government: Prime Minister Lee Kuan Yew
Ruling party: People's Action Party
Capital: Singapore City
Official Languages: English, Mandarin Chinese, Malay, Tamil
Currency: Singapore Dollar (S$ = 100 cents)
Exchange rate: S$1.96 per $ (Nov 1989)
Area: 625.6 sq km
Population: 2.65m (1988)
GDP per capita: S$18,131 (1988)
GDP real growth: 9.0% (1988)
Labour force: 1.25m (1988)
Unemployment: 3.3% (1988)
Inflation: 1.5% (1988)
Trade balance: –$2,345m (1988)
Foreign debt: $4,491m (1987)
Visitor numbers: 3.68m (1987)

spokenness belies any Goh-like lack of fibre, and he is wheeled out now as government spokesman to the foreign media, even on issues like internal security which do not seem to fall within his brief. B.G. Lee kept his portfolio at trade and industry in a largely un-shuffled cabinet. But he also took on responsibility for inculcating a 'national ideology'. Concerned at the decadent influences of the West, the government is trying to formulate a distinctively Singaporean creed that will help foster a sense of national identity and unity. In looking for new spiritual values, Singapore is in fact casting back to the most ancient of philosophies, Chinese Confucianism. This is seen as a 'communitarian' system of belief in contrast to the slaphappy individualism of the West. Confucianism stresses the values of loyalty and obedience, and the 'rectification of names'. If a son will act like a son, and a ruler's subject like a subject, the theory goes, then cohesion and harmony will be achieved.

The racial question

Confucianism, of course, derives from the Chinese traditions of the community who make up three-quarters of Singapore's population. So, too, does Mandarin Chinese, the 'standard' Chinese dialect, that is in fact spoken as a first language by hardly anyone in Singapore, most of whom speak Hokkien, Teochew, Cantonese, Hakka or Hainanese. The government, however, is now in one of its phases of promoting Mandarin, as a partial substitute for English, which is the second language used even by many Chinese to communicate with speakers of a different Chinese dialect. But 15 per cent of the population are Malays, and 10 per cent Indian or Eurasian. The stress on Mandarin, and a Confucian national ideology could be perceived by them as menacing, at a time when the government has been paying greater attention to the problems of ethnic divisions. After the elections, Goh Chok-tong grumbled at the low proportion of Malays who had voted for the government, seeming to cast aspersions on their patriotism.

In February 1989, the government made a move widely seen as designed to stop this alleged disaffection of the non-Chinese community leading to Malay-dominated electoral constituencies. It fixed strict racial quotas for the allocation of flats in government-controlled Housing Development Board blocks, where 86 per cent of Singaporeans live. No more than a quarter of apartments in any given block can now be allocated to Malays. The decision alienated some Malays, who pointed to the religious imperatives facing their islamic majority, who feel required to live in easy reach of food-stores and restaurants selling halal food, and within earshot of the dawn call to prayer from the local mosque.

Opposition dealt with severely

With two leading lights of the legal opposition – Francis Seow and Worker's Party leader J.B. Jeyeretnam – barred from parliament for criminal offences, the government still uses the full extent of its powers under the law against perceived political threats. At times it seemed that when that legal power was not enough, it simply changed the law. Thus when J.B. Jeyeretnam had an appeal against legal disbarment upheld by the Privy Council in London in December 1988, the right to Privy Council Appeal was removed in such cases. Similarly, convictions under the Internal Security Act are no longer subject to judicial review in London. Under the ISA, those considered a threat to national security can be detained without trial indefinitely, subject to periodic renewals of the detention orders. Of the 22 arrested under the ISA from May 1987 for their part in an alleged marxist conspiracy, only two remained in jail by mid-1989, the Church social worker Vincent Cheng, and the lawyer Teo Soh Lung. On 19 June, both their detentions were extended by a year. Teo's British counsel,

KEY INDICATORS

	Unit	1984	1985	1986	1987	1988
Population	'000	2,529.1	2,558.0	2,586.2	2,610	2,650
GDP (current market prices)	S$m	40,048.0	38,924	38,217	42,038	48,046
GDP per capita	S$	15,835	15,217	14,777	16,107	18,131
Wholesale prices	1985=100	102.3	100	84.9	91.3	89.6
Consumer prices index	1985=100	99.5	100	98.6	99.1	100.6
Manufactured products index	1985=100	107.9	100	108.6	127.5	150.9
Volume of exports	1985=100	100	100	114	135	180
Petroleum	1985=100	93.7	100	125.3	116.0	129.7
Crude rubber	1985=100	117.5	100	90.4	94.7	118.4
Exports	S$m	51,340.0	50,178.8	48,985.5	60,278	79,088
Domestic exports	S$m	33,051.2	32,575.5	32,062.4	–	–
Petroleum and petroleum prods	S$m	15,785	16,209	13,225	12,076	11,842
Electrical machinery	S$m	4,832.2	4,205.1	4,755.1	–	–
Oil bunkers	S$m	2,994.3	2,858.0	2,214.1	–	–
Rubber	S$m	2,135	1,492	1,335	1,835	2,315
Radios and TVs	S$m	1,319.7	1,368.9	1,562.8	–	–
Textiles/garments	S$m	940.2	913.6	1,119.4	–	–
Telecom equipment	S$m	804.5	759.5	799.0	–	–
Imports	S$m	61,134	57,819	55,543	68,416	88,241
Petroleum	S$m	16,951	17,020	10,996	12,516	12,413
Balance of trade	$m	–4,071	–2,829	–2,065	–2,636	–2,345
Current account	$m	–385	–4	542	553	1,660
External debt	$m	3,290	3,406	3,875	4,491	–
Debt service ratio	%	1.0	2.4	1.4	1.4	–
Exchange rate	S$ per US$	2.1331	2.2002	2.1774	2.1060	2.0124

* estimated figure

Anthony Lester QC, had warned that he would take her case to London if necessary, but then found himself banned from the Singapore courts after remarks he allegedly made criticising the Singapore legal system at a seminar in London.

In May, the government released the longest serving ISA detainee, Chia Thye Poh, regarded by organisations like Amnesty International as one of the world's longest serving political prisoners. Chia was freed without meeting the government's long-held condition, that he admitted his links with the Communist Party of Malaya, and renounced violence. But his release was not unconditional. He is to be confined to the tiny resort island of Sentosa, and subject to restrictions on whom he meets, and on engaging in political activities. Still, at a time when Lee Hsien-loong was on a visit to the United States, Chia's release did look suspiciously like a very rare sop from the Singapore government to international opinion.

Meanwhile, government spokesmen carried on lengthy exchanges in the pages of regional news magazines both with the exiled Francis Seow, and with Singapore's former president, Devan Nair, now accused by them of being a philandering alcoholic. The attitude to the media itself did not ease. The magazine *Asiaweek*, which had its circulation in Singapore slashed in 1987 for its 'meddlesome' reporting, was in 1988 allowed to sell more copies. But the other Hong Kong-based regional publications *The Asian Wall Street Journal*, and the *Far Eastern Economic Review* remain subject to punitive limitations on sales in Singapore.

Foreign policy

The Singapore government's banning of Anthony Lester led to a frosty spell in relations with London, which complained. But a similar wrangle with the United States following the May 1988 expulsion of a US diplomat blew over, though Singapore remains suspicious of US support for 'democratisation' in Asia, following the reforms in the Philippines, South Korea and Taiwan. Nevertheless, in August 1989 it was announced that Singapore was to be used as a more important US military base, a significant boon to American strategic requirements as the US base at Subic Bay in the Philippines faces an uncertain future. Relations with neighbouring Malaysia also improved, as a long drawn-out negotiation over the supply of water and natural gas was concluded in June 1988, and as the two countries announced in early 1989 that they would resume joint military exercises on land for the first time since Singapore left the Federation of Malaysia in 1965. Within its regional grouping, ASEAN, Singapore took one of the hardest lines against Vietnam and the government it backs in Cambodia. Progress towards resolution of the Cambodian problem poses the question of whether ASEAN will again find an issue that is so capable of uniting it politically. Even in 1989, with Thailand, under Prime Minister Chatichai, moving from the most intransigent of ASEAN anti-Vietnamese postures to among the most active in searching for compromise, Singapore found itself more on a limb in holding out against any concession to Vietnam.

The Cambodian issue, indirectly, threw up another diplomatic conundrum for Singapore. Partly because of its wish to play a leading intermediary role on the issue, Indonesia in February announced that it would start talks on unfreezing relations with China. Singapore has always said it would follow Indonesia in opening diplomatic relations with China. This is a nuisance, because Singapore's current position, of having full diplomatic relations with neither China nor Taiwan, enables it to enjoy very close relations with both, and had put it in a good position to capitalise on the 1997 reversion of Hong Kong to China, as a natural intermediary between Taiwan and the mainland. In this context, it was surprising that the Singapore government's reaction to the 4 June Beijing massacre was among the most forceful. Outrage and grief at the death of unarmed Chinese outweighed any considerations of diplomatic caution.

The economy

With foreign trade totalling three times its GDP, Singapore enjoyed the buoyant world markets of

1988, now clearly emerged as the boom year in the current cycle. Exports grew 37 per cent in US dollar terms, while imports rose 35 per cent, to feed a consumption boom fuelled by very strong domestic demand, as well as the processing, re-exporting industries which are Singapore's mainstays. With per capita foreign exchange reserves among the highest in the world, negligible foreign debt, low inflation and full employment, the Singapore economy presented a picture of rude health. However, its dependence on foreign trade, and on the US market, which takes about a quarter of its exports, in particular, made some slow-down inevitable, and 1989 GDP growth looked unlikely to exceed a still robust 7 per cent, before a further slow-down in 1990.

Foreign trade

The reliance on the US market made the passage of the 'Super 301' provisions of the August 1988 US 'Omnibus Trade and Competitiveness Bill' of particular concern. Singapore is less vulnerable to protectionist retaliation than, say, Taiwan and South Korea, because of its liberal trade policies. It does, however, face a number of anti-dumping investigations in the USA, in colour television tubes, ball bearings and steel pipes among other products. Moreover, like the other 'tigers', Singapore does face pressure to see the local currency appreciate faster against the US dollar, and so far it has managed to lag behind the South Korean won, and the New Taiwan dollar. This helped the 1988 boom in electronics exports. Singapore has become the world's largest producer of computer disc drives and disc drive parts, which in 1988 were second in importance only to refined oil products among Singapore's exports. Similarly, integrated circuits, telecommunications equipment, radio receivers and other electronics products also had a good year. But the notoriously cyclical nature of the electronics industry, and especially the glut in the disc drive market, made the 1988 successes appear unrepeatable in 1989. Continued modest expansion of the US and world economies, however, was expected to delay any sharp downturn until 1990. By then, rising wage costs, and the introduction of a new consumption tax, as well as continuing appreciation of the Singapore dollar, looked certain to have eroded Singapore's international competitiveness. A race is on with the other Newly Industrialising Economies to secure market share in non-American markets, like Japan, south-east Asia, the communist bloc, and, especially, the European Community, where fear of the implications of the 1992 integrated market is another spur to market diversification. In the meantime, the government will resort to its traditional tactic for riding out the troughs in the global economic cycle – stimulating a construction boom.

Tourism and construction

In fact, the construction industry was by early 1989 already enjoying one of its frenetic phases. The 1984-85 trough came at a time of massive government infrastructural spending, notably on the spanking new underground Mass Rapid Transit project. This time, it is the private sector which is the principal source of construction spending. The government still has MRT contracts to let and pay for, as well as a number of highway projects and a plan for a road link with Sentosa island, but the Housing Development Board has fewer new blocks to build. This slack over the next two years will be taken up by a number of massive commercial developments, including the UOB and Republic office blocks, and the Marina South project, for various parks and leisure facilities. After what looked at the time like chronic over-building in the hotel sector in the mid-1980s, a renewed tourism surge may also lead to a fresh scramble to construct more hotel rooms. Singapore has doubled its hotel room capacity to about 24,000 in just a few years. But now, tourist arrivals are growing by more than 10 per cent a year, and by the end of 1988, occupancy levels had topped 80 per cent. The Tourism Promotion Board, taking a conservative average growth rate of 5 per cent a year, estimated that Singapore will have a shortfall of 8,500 hotel rooms by the year 2000.

Labour shortage

One of the problems facing the construction industry is Singapore's acute labour shortage, which is especially severe at the top and bottom ends of the labour market. Builders have relied heavily on immigrant labour, particularly from Malaysia, Sri Lanka and Thailand. The government is not keen to encourage this, seeing the influx as potentially socially divisive, and wanting the construction and other labour-intensive industries to mechanise and upgrade. Hence, in 1989 there was a tightening up on illegal immigration, and culprits were caned with the rattan, in the rough

COUNTRY PROFILE

Political system: Singapore, a former British colony, achieved internal self-government in 1959. It became a state of the Federation of Malaysia in 1963, but seceded in 1965. Its 79-member unicameral legislature (parliament) is elected to a five-year term by universal and compulsory adult suffrage, and elects a president, who serves as head of state. The president appoints as prime minister the member of parliament who commands the confidence of the majority of MPs. Cabinet ministers are appointed by the president on the advice of the prime minister. Lee Kuan Yew became prime minister in 1959. In the 1988 elections the system of single-member constituencies for parliament was changed. A constitutional amendment had been passed for 'group representation constituencies' to be formed. Each of these has three MPs, one of whom must be non-Chinese. Altogether 13 group constituencies have been formed, leaving another 42 as single-member seats. This raises the total number of MPs to 81. The government also plans to draft legislation so that the country's president should be popularly elected and should have stronger executive powers.

Political parties: There are some 20 registered political parties in Singapore. Of the 81 seats in parliament, the ruling PAP holds 80 seats and the Singapore Democratic Party one seat. One non-voting MP from the Workers Party also takes his seat in parliament, as provided by a 1984 amendment.

Population: Singapore has one of the world's highest population densities, at 4,072 per sq km in 1987. Chinese are 76.3 per cent of the population; with Malays, 15 per cent and Indians, 6.4 per cent. Population growth has slowed rapidly since the 1960s, and the government is now attempting to boost growth with financial incentives to parents. Population estimated to be 2.65m in 1988.

Language: English is the main business language and is almost universally understood. Other official languages are Mandarin, Malay and Tamil.

Media
Press: Newspapers and magazines are published only under government licence. In 1984, the three major newspaper groups were merged into Singapore Press holdings, which is effectively a monopoly, controlling six daily newspapers, a regional magazine and a book distribution network. There are two English-language newspapers, the morning *Straits Times* (Singapore circulation 218,000) and the *Business Times* (Singapore circulation 15,000). The two main Chinese dailies are *Lian He Zao Bao* (United Morning News) and *Lin He Wan Bao* (United Evening News). *Berita Harian* (in Malay) and *Tamil Murasu* (in Tamil) have smaller circulations. International editions of leading foreign newspapers are also available. According to the government 3,700 foreign publications are circulated in Singapore. But in the last few years, tough government legislation has led to several foreign newspapers and magazines having their Singapore circulation forcibly restricted by the government. Publications affected included *Time*, the *Asian Wall Street Journal*, *Asiaweek* and the *Far Eastern Economic Review*. As of mid-1988 only *Journal* was cut from 5,000 to 400 copies a day. The three newspapers were restricted for failing to publish in full letters from the government and the *Far Eastern Economic Review* was accused of interfering in domestic politics. There are numerous trade publications, including *Singapore Business, Building Materials and Equipment SE Asia, The Manufacturer, Petroleum News, Shipping 'N' Shipbuilder* and *The Economic Bulletin* published by the Singapore International Chamber of Commerce.
Radio: The Singapore Broadcasting Corporation broadcasts radio programmes in English, Chinese, Malay and Tamil from 0600 – 2400. Rediffusion (Singapore) Ltd operates a wired broadcasting service, with advertising facilities in each of the four languages.

Television: The Singapore Broadcasting Corporation broadcasts TV programmes daily in English, Chinese, Malay and Tamil on Channels 5, 8 and 12, most in colour; reception of Channels 3 and 10 from Malaysia is usually possible. Many hotel TVs are linked to the teletext system, which operates 0700 – 2300. Commercial cable services are available. In 1984 there was a TV for every five persons.

Domestic economy: After high growth averaging nine per cent a year in the 1960s and 1970s, in 1985 Singapore ran into a slump. By 1989 the economy was back on the fast track. The 1988 growth rate was 11 per cent after being 8.8 per cent in 1987 and a paltry 1.8 per cent in 1986. The impressive performance of the 1960s and 1970s was founded on a flourishing port and entrepôt business, and on a strong international banking role made possible by its high standard of telecommunications and its first-class hotels, as well as its tax incentives and its relaxed rules on repatriation of profits and capital. Since the 1970s the country has been moving consciously away from the traditional low-cost manufacturing and toward high-margin activities such as electronics manufacture and chemicals and petrochemicals, involving the use of skilled and high-waged labour. In the early 1980s the government deliberately encouraged high wage increases to push the economy away from low paid, low-skilled jobs and towards higher productivity. The move partially backfired and high wages were partly responsible for the slump. It has also spent vast sums on schools, hospitals and other infrastructure amenities. But the economy depends on external business for over 65 per cent of its GDP, and the problems of the mid-1980s focussed attention on the inflexibility of its unusally interventionist government; for example, a severe glut of new office and hotel space was largely due to excessive incentives to builders and developers. Construction continued its decline even as the overall economy grew quickly. The main reason was that the government had used construction to pull out of the slump. With per capita of more than US$7,000 Singapore disputes with Hong Kong third richest Asia Pacific country after Brunei and Japan.

justice typical of Singapore. This led to a damaging row with ASEAN partner Thailand. Singapore has also increased the levy payable by employers on imported labour, with effect from 1 July 1989, from S$220 to S$250 per month. But in affluent Singapore, it is hard to find residents willing to take on low-paid construction jobs, and the shortage is likely to become a noticeable bottleneck in economic growth.

Perhaps even more seriously, Singapore is witnessing significant emigration from its professional classes. An estimated 3-4,000 Singaporeans are emigrating every year, mostly middle-class, English-speaking professionals. Australia is the favourite destination, followed by Canada. The trend seems to puzzle the government, in the light of Singapore's prosperity and social and political stability. But English-speakers fear their children may be handicapped by the growing stress on Mandarin, and an indefinable quality of life present in the wide open spaces and political free-for-alls of Canada and Australia seems to be missing in clean and green, but cramped and claustrophobic, Singapore.

The government is trying to make up for some of the lost talent by coaxing jittery Hong Kong Chinese who look forward to the 1997 reversion to Chinese sovereignty with less than equanimity. In June 1989, after the Beijing massacre, Singapore announced an easing of the terms on which it will give residents' permits to Hong Kong citizens. The consulate was at once besieged, and had to close its doors, and accept only postal applications.

Industry to internationalise?

Facing a labour shortage at home, a rising cost base and fierce international competition, while remaining by and large extremely liquid, it is surprising that Singapore's industries have been so slow to invest overseas. Indeed, the only sizeable 'multinational' in Singapore is the national flag carrier, Singapore Airlines, which, clearly, has little choice. As so often in its economic policy, the government is likely to show the way, with a gradual programme of overseas expansion by some of the state-controlled industrial groups. Apart from neighbouring Malaysia and Indonesia, the obvious relocation targets for the labour-intensive industries are the Philippines and, if the political climate eases, mainland China. In both markets, Singapore lags behind Taiwan as an investor. In part this is a result of Singapore's conscious 'high cost' policy of the early 1980s, designed to promote industrial growth not through internationalisation, but enforced automation and product diversification into the more capital-intensive, labour-extensive industries. However, not all industries are entirely susceptible to this approach. Ship-repairers in particular are now looking offshore for joint venture opportunities.

Outlook

The domestic cost pressures aside, Singapore seems well placed to weather the anticipated slow-down in world trade in 1990. It has domestic inflation under control, is in an extremely sound external financial position, and is ahead of its immediate competitors in moving up-market into the information technology and other 'sunrise' industries. The period up to 1993, by when the next election has to be held, gives also a good opportunity to manage an orderly transition of power to what Lee Kuan Yew calls the 'second echelon' leaders, the youngish technocrats with whom he has stacked his cabinet. So political stability should also be maintained, barring clumsy handling of the ethnic tensions simmering below Singapore's serene surface, and unless the government's siege mentality when faced with domestic or international criticism alienates too many of the educated young the government so badly needs if its economic strategy is to work.

COUNTRY PROFILE

Employment: The rate of unemployment has traditionally been low. Relative to neighbouring countries, the workforce has a high proportion of skilled workers, with graduates accounting for about 15 per cent of those entering in 1987 and technicians or skilled operatives for 65 per cent. About 100,000 workers, nine per cent of the labour force, are foreign, mainly from Malaysia but also from the Philippines and Thailand and as far afield as Sri Lanka. The government originally aimed to phase out foreign labour by 1991, but amended its policy in March 1987 to allow a revolving pool of foreigners after 1992.

External trade: Most imported goods are not subject to quota restrictions or licences and in most instances are duty-free. Out of 2,200 items in the trading classification, only eight per cent are subject to import duties. Singapore's entrepôt function has shifted from exporting goods from the South-east Asian region to becoming a gateway for goods into the region. Its traditional role as a trading centre for agricultural commodities is declining. While all exports are technically subject to control there are restrictions only on a few categories.

Exports: Total exports in 1988 were S$79.1bn. Main exports are machinery and transport equipment, petroleum products, manufactured goods, chemicals, food, beverages and tobacco and raw materials. Main destinations are USA, Malaysia, Japan, Hong Kong, Thailand, Australia, German FR, UK and China PR.

Imports: Total imports in 1988 were S$88.2bn. Main imports are machinery and equipment, mineral fuels, manufactured goods, foodstuffs, chemicals and crude materials. Main sources are Japan, USA, Malaysia, China PR, Taiwan, Kuwait, Iran and Saudi Arabia.

Domestic export: Domestic exports are growing somewhat faster than re-exports, and now account for about two-thirds of the total. Particularly important are computers and other electronic products, office equipment and plastics; but on the other hand sales of oil rigs, ships and certain low-margin, labour intensive goods have been declining.

Agriculture: Agriculture and fishing employ 0.8 per cent of the labour force and account for about 0.5 per cent of GDP (1987). Small-scale farming supplies vegetables, eggs and meat products for domestic demand. Apart from eggs, poultry and pork, most food is imported. Orchids and aquarium fish have become valuable exports. A special fishing port has been established at the Jurong Industrial Estate as part of efforts to modernise the fishing industry.

Industry: Manufacturing industry is the biggest contributor to gross domestic product, accounting for more than a quarter of output and of employment. The second biggest sector as a contributor to GDP is financial and business services with about 29 per cent. The island republic has used its position as an important crossroads of world shipping and communications markets and has also built upon its educated mobile labour force. Shipbuilding and repairing and its offshoot of oil rig building, oil refining and air support services use Singapore's strategic location; newer industries like textiles and garments, electrical and electronic manufacture and even food and beverages processing and chemicals take advantage of high labour skills. Printing and publishing has been built up trying to use the two advantages. Singapore has the world's third largest petroleum refining capacity and five refineries account for about a third of manufacturing output. More than 300 companies associated with the oil industry have set up regional bases in Singapore. But oil and shipbuilding and associated industries have suffered with the decline in the world oil price and the petroleum industry declined seven per cent in 1987. The textiles industry was important at the start but as Singapore's labour costs have gone up, it has been surpassed by electronics and electrical industries. These have continued to grow with impressive rates of 25-30 per cent in recent years, inspite of the wage increases. This is because Singapore has managed to continue to move upmarket. More than 200 firms are producing consumer products from radios and tape recorders, calculators and clocks to electronic components like printed circuit boards, electronic electrolytic and ceramic capacitors and transformers. The government would like to see Singapore become more knowledge intensive and a brain centre for the region, but the industry is handicapped by the small market size and the fact that most manufacturing investment, especially in high tech areas, is done by foreign concerns. Foreign investment slumped in the mid-1980s but picked up again with the economy. Japanese and US companies were in the forefront. Tourism, or tapping through traffic from Europe and the US to Australia as well as the growing Japanese market continues to be important. Singapore gets more than 3m tourist visitors a year but optimistic estimates led to excessive hotel-building in the mid-1980s. By the late 1980s, many Singapore hotels were still finding it difficult, occupancy was still only 60-70 per cent.

Energy: Electricity is supplied by three thermal power stations with a total generating capacity of 2,741mW. A fourth, rated at 750 mW is being built on Pulau Seraya. 60 per cent of natural gas production is consumed domestically.

Financial markets: Singapore is an important financial centre and pioneered the Asian dollar market in 1968 as part of its plans to be a sophisticated international financial centre. Since then financial activities expanded and more than 130 banks, most of them foreign, have banking operations. Another 50 have representative offices and there are about 60 merchant banks. The Asian dollar market has grown and is worth several hundred billion dollars with almost 200 Asian currency units operating in the country. The authorities have consistently pioneered diversification and in 1984 Singapore International Monetary Exchange (SIMEX) started trading with a mutual offset link-up with the Chicago Mercantile Exchange. Among the contracts offered are gold futures, Japanese yen/US dollar currency deals, three-month interest rate Asian dollar deposits, German mark/US dollar currency deals, sterling/US dollar currency contracts; US Treasury bond futures contracts; and Nikkei Stock average futures. The government has also been keen to attract fund management and insurance business to Singapore as part of an increasingly sophisticated market, but critics grumble that the government sometimes forces the pace too hard rather than let growth come naturally. The Singapore stock market is small but has attracted worldwide interest. To help widen and deepen the mark an unlisted securities market was set up.

COUNTRY PROFILE

Legal system: Laws of the country are based on the British judicial system. The Supreme Court is divided into the High Court, the Court of Appeal and the Court of Criminal Appeal. The final court of appeal is the UK Judiciary Committee of the Privy Council, although in 1988-89 right of appeal to the Privy Council was withdrawn for some offences. There are 17 district hearing courts, nine magistrates courts, three mention Courts, one juvenile court and one coroner's court. The Shariah is a religious court with jurisdiction over Muslim law and domestic proceedings between Muslim parties.

Membership of international organisations: ASEAN, Asian Development Bank, Colombo Plan, Commonwealth, EC trade and co-operation agreement, ESCAP, FC, GATT, IAEA, IBRD, IDA, IFAD, ILO, IMF, SPF, UN, UPU, WHO, World Bank.

BUSINESS GUIDE

Time: GMT + 8 hrs.

Climate: Tropical, with abundant rainfall throughout the year and a monsoon season in Nov-Jan. The wettest season is Oct-Mar. Humidity is high, and temperatures vary little from season to season, ranging from an average maximum of 31°C by day to 25°C by night.

Entry requirements
Passport: Passport required by all with few exceptions.
Visa: Visas are not required by Commonwealth citizens and nationals of certain other countries (Eire, Liechtenstein, Monaco, Netherlands, San Marino and Switzerland), nor by passport holders of most other nations for stays of up to 14 days. Entry may be refused to male visitors whose hair reaches below the collar or extends over the ears or eyebrows.
Health: Vaccination certificates required for yellow fever or cholera if travelling from an infected area.
Currency: There are no restrictions on the importation or exportation of currency.
Customs: Personal effects are allowed duty-free. Importation of goods of South African origin is prohibited.

Representation overseas: Bandar Seri Begawan, Bangkok, Bonn, Brussels, Cairo, Canberra, Geneva, Jakarta, Jeddah, Kuala Lumpur, London, Madrid, Manila, Moscow, New Delhi, Paris, Rangoon, Riyadh, Stockholm, Tokyo, Vienna, Washington DC, Wellington.

Air access
National airline: Singapore (SIA).
Other airlines: Aeroflot, Air Canada, Air-India, Air Lanka, Air Mauritius, Air New Zealand, Air Niugini, Alitalia, Air Seychelles, Biman Bangladesh, British Airways, CAAC, Cathay Pacific, China, CSA, Finnair, Garuda, Indian, Interflug, JAL, JAT, KLM, Korean Air, LOT-Polish, MAS, Northwest, Olympic Airways, Philippine, PIA, Qantas, Royal Brunei, Royal Jordanian, Royal Nepal, Sabena, SAS, Saudia, Swissair, Tarom, Thai International, THY-Turkish, Tradewinds Pte Ltd, United, UTA. Charter helicopter services to Malaysia and Indonesia are available.
International airport: Changi (SIN), 20 km from Singapore; duty-free shop, bar, restaurant, buffet, bank, hotel reservations, post office, shops, car hire. Upgrading in progress.

Surface access: There are excellent sea links with other countries. Road transport arrives via the causeway from Malaysia, with express bus services from Kuala Lumpur and Johore Bahru. There are rail services to Kuala Lumpur and Bangkok.
Main ports: Singapore (containers). More than 30,000 vessels docked in 1986, handling 48m tonnes of cargo (excluding bulk mineral oils).

Hotels: Most international-class hotels have shopping arcades, bars and swimming pools. Tipping is prohibited in hotels. A three per cent tax and a 10 per cent service charge are generally added to the hotel bill.

Credit cards: All major credit cards, such as Visa, American Express, Diners Club and Mastercard, are widely accepted.

Car hire: An international driving licence is required for car hire. Driving is on the left-hand side of the road. Coupons for use of the public car parks managed by the Urban Redevelopment Authority (URA) or Housing & Development Board (HBD) can be purchased at post offices, URA parking kiosks and some gas/petrol stations. Car hire companies are listed in the Yellow Pages of the telephone directory.

City transport
Taxis: Metered taxis are widely available. Before undertaking a journey, make sure the driver knows the destination, that the meter is flagged and that the fare reads S$1.60, (the minimum). There is an additional charge of S$1 for luggage, a 50 per cent surcharge of the metered fare from 2400 – 0600, a S$3.00 surcharge for taxis travelling from (but not to) Changi Airport and a surcharge of S$0.50 for each adult in excess of two passengers. Taxis can also be hired at about S$25 per hour. Radio taxis are available, but a S$1.00 booking fee is charged.
Metro: A Mass Transit System started up in Dec 1987. It will eventually consist of two lines running north/south and east/west, with 42 stations (15 of them underground).

National transport
Road: In 1986, the road network comprised 2,686 km of roads, including 77km of expressways. Vehicular access to the Central Business District (CBD) is restricted and there are charges for vehicles entering the area at certain times. A causeway links the Singapore road system with that of Malaysia across the Straits of Johore.
Buses: Timetables for the extensive and inexpensive bus network are widely available at news stands. Exact fares, which should be displayed on a signboard in the front of the bus, must be tendered. *Trishaws*: Bicycles with sidecars (trishaws) are less common than they were, but can be hired for tours. There is no standard fare structure.
Water: Regular ferry services from the World Trade Centre operate to some of the islands; others may be reached by charter boat. Cruise boats offering tours of the harbour and southern islands operate from Clifford Pier.

Public holidays: Owing to its multi-ethnic composition, Singapore celebrates a wide range of world religious festivals and holidays in addition to the public holidays listed below. Many festivals are based on a lunar calendar, while the dates of some are only finalised at the last minute. Check with the Singapore Tourist Promotion Board for exact dates and locations affected. When a holiday falls on a Sun, the following Mon is a public holiday.
Fixed dates: 1 Jan, 1 May, 9 Aug (National Day), 25 Dec.
Variable dates: Chinese New Year, Good Friday, Vesak Day, Hari Raya Puasa (Eid al-Fitr), Hari Raya Haji (Eid al-Adha), Deepavali. Actual dates of Muslim festivals may vary. During the Chinese New Year, many Chinese firms close for an entire week.

Working hours
Business: (Mon-Fri) 0900 – 1300, 1400 – 1700; (Sat) 0900 – 1200.
Government: (Mon-Fri) 0800 – 1300, 1400 – 1700; (Sat) 0800 – 1300.
Banking: (Mon-Fri) 1000 – 1500; (Sat) 0930 – 1130.
Shops: (Mon-Sat) 0900 – 1800. Many shops open on Sun and many in the 'shopping centres' are open until 2100.

Social customs: Smoking is prohibited in many public places, including taxis, public transport, lifts/elevators, cinemas, theatres and government offices, with fines of up to S$500 imposed on offenders. There are also fines for jaywalking and littering, and the use of narcotic drugs is strictly prohibited. Long hair is regarded unfavourably, and male visitors are advised to have their hair cut if it reaches below the top of the shirt collar. Tipping is discouraged.

Telecommunications: Telecommunication Authority of Singapore (TAS) provides international telephone, telex and telegram services via satellite, submarine cable, HF, VHF and UHF. Visa and Mastercharge credit cards may be used to pay for TAS services at its airport service counters.
Telephone: Dialling code for Singapore, IDD access code + 65 followed by subscriber's number (no area code required). Except for calls from public telephones, which are widely available in shopping centres and hotels, local calls are free. IDD is available to over 160 destinations. There were 348 telephones per 1,000 inhabitants in 1987.
Telex: Public telex facilities are available at Telecoms Customer Services centres. Incoming telex calls are accepted only in reply to an outgoing call. Telex facilities are also available at good hotels.
Telegram: Telegrams may be sent from any post office, Telecoms Customer Services Centres or good hotels.
Other: TAS provides Datel (data transmission), Teleconference, Telefax (international and local facsimile), Telebox (electronic mail), Telepac (computer access), Maritime Satellite Communications, Prisnet (private switching network) services, as well as leased circuits, private user networks and common-interest group networks.

Banking: Three types of banks are licensed to operate: full licence banks, restricted licence banks and offshore licence banks. (*Indicates banks with Asian Currency Unit or ACU licences). Monetary authority: Monetary Authority of Singapore. *Head offices, full banks:* Bank of Singapore, Chung Khiaw Bank (80 per cent owned by UOB), *DBS Bank, Far Eastern Bank, Four Seas Communications Bank (99 per cent owned by OCBC), *Industrial Bank of Singapore, Lee Wah Bank, *Overseas Chinese Banking Corporation, *Overseas Union Bank, Post Office Savings Bank, *Tat Lee Bank, *United Overseas Bank.

Advertising: Advertising is available on commercial radio, TV and in cinemas, and via direct mail and house-to-house distribution of samples. Outdoor advertising, especially short-lived posters, is widely used.

Electricity supply: 230-250 V AC, 50 cycles for domestic use and 400 V AC, 50 cycles for industry, with three-pin (square) plug fittings and bayonet-type lamp sockets. Most hotels also have 110 V AC supply.

Weights and measures: Metric system (Imperial and local units still used).

Representation in capital: Argentina, Australia, Austria, Bangladesh, Belgium, Brazil, Brunei, Bulgaria, Burma, Canada, Chile, Cyprus, Czechoslovakia, Denmark, Egypt, El Salvador, Fiji, Finland, France, German DR, German FR, India, Indonesia, Iran, Iraq, Ireland, Israel, Italy, Japan, Republic of Korea, Malaysia, Mexico, Netherlands, New Zealand, Norway, Pakistan, Philippines, Poland, Romania, Saudi Arabia, Somalia, Spain, Sri Lanka, Sweden, Switzerland, Thailand, UK, USA, USSR, Vatican, Yugoslavia.

BUSINESS DIRECTORY

Hotels
Amara, 165 Tanjong Pagar Road, 0208 (tel: 224-4488; tx: 55887; fax: 2243910).

Asia, 37 Scotts Road, 0922 (tel: 737-8388; tx: 24313; fax: 7333563).

Boulevard, 200 Orchard Boulevard, 1024 (tel: 737-2911; tx: 21771; fax: 7378449).

Cairnhill, 19 Cairnhill Circle, 0922 (tel: 734-6622; tx: 26742; fax: 2355598).

Carlton, 76 Bras Basah Road, 0718 (tel: 338-8333; tx: 42076; fax: 3396866).

Century Park Sheraton, 16 Nassim Hill, 1025, or PO Box 89,Tanglin Post Office (tel: 7321222; tx: 21817; fax: 7322222).

Cockpit International, Oxley Rise and Penang Road, 0923, P.O. Box: 528 (tel: 737-9111; tx: 21366).

Crown Prince, 270 Orchard Road, 0923 (tel: 732-1111; tx: 22819; fax: 7327018).

Dai-Ichi, 81 Anson Road, 0207 (tel: 224-1133; tx: 40163; fax: 2220749).

Dynasty Singapore, 320 Orchard Road,0923 or PO B.176,Killiney Rd Post Off.9123 (tel: 734-9900; tx: 36633; fax: 7335251).

Equatorial, 429 Bukit Timah Rd, 1025 (tel: 732-0431; tx: 21578; fax: 7379426).

Excelsior, 5 Coleman Street, 0617 (tel: 338-7733; tx: 20678; fax: 3393847).

Furama, 10 Eu Tong Sen Street, 0105 (tel: 533-3888; tx: 28592; fax: 5341489).

Garden, 14 Balmoral Road, 1025 (tel: 235-3344; tx: 50999; fax: 2359730).

Glass, 317 Outram Road & Havelock Road 0316 (tel: 733-0188; tx: 50141; fax: 7330989).

Goodwood Park, 22 Scotts Road, 0922 (tel: 737-7411; tx: 24377; fax: 7328558).

Grand Central, 22 Cavenagh Road, at Orchard Road, 0922 (tel: 737-9944; tx: 24389).

Hilton International Singapore, 581 Orchard Road, 9123, P.O. Box: 115 (tel: 737-2233; tx: 21491; fax: 7322917).

Holiday Inn Park View, 11 Cavenagh Road, 0922 (tel: 733-8333; tx: 55420; fax: 7344593).

Hyatt Regency Singapore, 10-12 Scotts Road, 0922 (tel: 733-1188; tx: 24415; fax: 7321696).

Imperial, 1 Jalan Rumbia, 0923 (tel: 737-1666; tx: 21654; fax: 7374761).

King's, Havelock Road, 0316 (tel: 733-0011; tx: 21931; fax: 7325764).

Ladyhill, 1 Ladyhill Road, 1025 (tel: 737-2111; tx: 23157; fax: 7374606).

Lion City, 15 Tanjong Katong Road, 1543 (tel: 744-8111; tx: 21789; fax: 7444050).

Mandarin Singapore, 333 Orchard Road, 0923 or PO Box 135 (tel: 737-4411; tx: 21528; fax: 7322361).

Marco Polo, 247 Tanglin Road, 1024 (tel: 474-7141; tx: 21476; fax: 4710521).

Marina Mandarin, Marina Square, 6 Raffles Boulevard, 0103 (tel: 338-3388; tx: 22299; fax: 3394977).

Meridien Changi-Singapore, 1 Netheravon Road, off Upper Changi Road, 1750 (tel: 542-7700; tx: 36042; fax: 5425295).

Ming Court, 1 Tanglin Road, 1024 (tel: 737-1133; tx: 21488; fax: 7330242).

Miramar, 401 Havelock Road, 0316, P.O. Box: 2768 (tel: 733-0222; tx: 24709; fax: 7334027).

New Otani, 177a River Valley Road, 0617 (tel: 338-3333; tx: 20299; fax: 3392854).

Novotel Orchid Inn, 214 Dunearn Road, 1129 (tel: 250-3322; tx: 21756; fax: 2509292).

Orchard, 442 Orchard Road, 0923 (tel: 734-7766; tx: 35228; fax: 7325061).

Orchard Singapore, 100 Orchard Road, 0923 (tel: 733-8855; tx: 50163; fax: 7327886).

Oriental, Marina Square, 6 Raffles Boulevard, 01-200, 0103 (tel: 338-0066; tx: 29117; fax: 3399537).

Pan Pacific, Marina Square, 6 Raffles Boulevard, 01-300, 0103 (tel: 3368111; tx: 38821; fax: 3391861).

Paramount, 25 Marine Parade Road, 1544 (tel: 344-5577; tx: 22234; fax: 4474131).

Pavilion Inter-Continental Singapore, 1 Cuscaden Road, 1024 (tel: 733-8888; tx: 37248; fax: 7328838).

Peninsula, 3 Coleman Street, 0617, P.O. Box: 3751 (tel: 337-2200; tx: 21169; fax: 3393580).

Phoenix, Orchard and Somerset Road, P.O. Box: 263,0923 (tel: 737-8666; tx: 23718; fax: 7322024).

Plaza, 7500A Beach Road, 04-301, The Plaza, 0719 (tel: 298-0011; tx: 22150; fax: 2963600).

President Merlin, 181 Kitchener Road, 0820 (tel: 295-0122; tx: 35581; fax: 2960719).

Raffles, 1-3 Beach Road, 0718 (tel: 337-8041; tx: 21586; fax: 3397650).

Regent, No 1 Cuscaden Road, 1024 (tel: 7338888; tx: 37248; fax: 7328838).

River View, 382 Havelock Road, 0316 (tel: 732-9922; tx: 55454; fax: 7321034).

Royal, 36 Newton Road, 1130 (tel: 253-4411; tx: 21644; fax: 2538668).

Royal Holiday Inn, 25 Scotts Road, 0922 (tel: 737-7966; tx: 21818; fax: 7376646).

Sea View, 26 Amber Close, 1543 (tel: 345-2222; tx: 21555; fax: 3451741).

Shangri-La, 22 Orange Grove Road, 1025 (tel: 737-3644; tx: 21505; fax: 733-7220/733-1029).

Sheraton Towers, 39 Scotts Road, 0922 (tel: 737-6888; tx: 37750; fax: 7371072).

Supreme, 15 Kramat Road, 0922 (tel: 737-8333; tx: 33283).

Tai-Pan Ramada Singapore, 101 Victoria Street, 0718 (tel: 336-0811; tx: 21151; fax: 3397019).

Westin Plaza and Westin Stamford, 2 Stamford Road, 0617 (tel: 338-8585; tx: 22206; fax: 3382862).

York Hotel, 21 Mount Elizabeth, 0922 (tel: 737-0511; tx: 21683; fax: 7321217).

Business advice and information
American Business Council, 10-12 Shaw House, 354 Orchard Road, 0923 (tel: 235-0077; tx: RS 50296).

British Businessmen's Association, 3rd Floor, Inchcape House, 450-452 Alexandra Road, 0511 (tel: 475-4192; tx: RS 21400)).

Customs & Excise Department, Customs House, Maxwell Road (tel: 222-3511).

Economic Development Board, 250 North Bridge Road, 24-00 Raffles City Tower (tel: 336-2288; tx: RS 26233).

Export Credit Insurance Corporation of Singapore Ltd, 10 Shenton Way, 17-03 MAS Building, 0207 (tel: 220-8344; tx: RS 21524; fax: 224-2887).

Housing and Development Board, National Development Building, Maxwell Road, 0106 (tel: 225-1212/4444; tx: RS 22020).

Japanese Chamber of Commerce, 79 Robinson Road, 24-04 CPF Building, 0106 (tel: 221-0541/0665; fax: 225-6197).

Jurong Town Corporation, Jurong Town Hall, Jurong Town, 2260 (tel: 560-0056; tx: RS 35733).

Ministry of National Development, National Development Building, Maxwell Road, 0106 (tel: 222-1211; tx: RS 34369).

National Productivity Board, 55 Cuppage Road, 08-16 Cuppage Centre, 0922 (tel: 734-5534; tx: RS 36047).

Registry of Companies and Businesses, Rooms 06-6, 6th Floor, Colombo Court, 0617 (tel: 336-1203; tx: RS 28395).

Singapore Chinese Chamber of Commerce Foundation, 47 Hill Street, 0617 (tel: 337-8381; tx: 33714).

Singapore Ministry of Trade and Industry, Suite 201, 2nd Floor, World Trade Centre, Telok Blangah Road (tel: 225-9911; tx: RS 28617).

Singapore Federation of Chambers of Commerce & Industry, 03-01 Chinese Chamber of Commerce Building, Hill Street, 0617 (tel: 338-9761; tx: 26228).

Banking
Monetary Authority of Singapore, 10 Shenton Way, 0207 (tel: 225-5577; tx: 21382 ORCHID RS; fax: 229-9491; cable: MONETARY).

Head offices, full banks Bank of Singapore, Tong Eng Building, 101 Cecil Street 01-02, 0106 (tel: 223-9266; tx: 27149 BOS LTD RS).

Chung Khiaw Bank, 10 Anson Road, 01-01 International Plaza, 0207 (tel: 222-8622; tx: 22027).

Development Bank of Singapore, DBS Building, 6 Shenton Way, 0106 (tel: 220-1111; tx: 24455; fax: 221-1306).

Far Eastern Bank, 156 Cecil Street, Far Eastern Bank Building, PO Box 2950, 0106 (tel: 221-9055; tx: 23029).

Four Seas Communications Bank, 19-25 Cecil Street, 0104 (tel: 224-9898; tx: 23670).

Industrial and Commercial Bank, ICB Building, 2 Shenton Way, 0106 (tel: 221-1711; tx: 21112).

International Bank of Singapore, 01-02 Overseas Union House, 50 Raffles Place, 0104 (tel: 223-4488; tx: 23579).

Lee Wah Bank, 01-093 Robina House, 1 Shenton Way, 0106 (tel: 225-8844; tx: 21529).

Oversea-Chinese Banking Corporation, OCBC Centre, 65 Chulia Street, 0104 (tel: 535-7222; tx: 21209).

Overseas Union Bank, OUB Centre, 1 Raffles Place, 0104 (tel: 533-8686; tx: 24475 RS; fax: 225-2082; cable: OVERSUNION).

Post Office Savings Bank, POSB Centre, 73 Bras Basah Road, 0218 (tel: 339-3333; tx: RS 25450; fax: 339-4739).

Tat Lee Bank, Tat Lee Building, 63 Market Street, 0104 (tel: 533-9292; tx: 26767; fax: 533-1043).

United Overseas Bank, 01-00 UOB Building, 1 Bonham Street, Raffles Place, 0104 (tel: 533-9898; tx: 21539).

Car hire
Avis: PO Box 40, Orchard Point, Bukit Timah Road (tel: 737-9477; tx: RS 50000); airport (tel: 542-8833).

Europcar: Tong Nam Building, 73 Bukit Timah Road, (tel: 338-8444; tx: RS 28789 NATCAR).

Hertz: 19, Tanglin Road, 0120 (tel: 734-4646; tx: 5112); airport (tel: 545-8228).

Taxis
Taxis Jurong Port/Pier Taxi Service (tel: 261-8175, 265-1802).

NTUC Comfort Taxi Service - Sin Ming Drive Control Room (tel: 452-5555); Jurong Control Room (tel: 264-0288/0329).

Solomon Islands

Kevin Rafferty

(map on page xxiv)

KEY FACTS

Official title: Solomon Islands
Head of state: Queen Elizabeth II
Head of government: Prime Minister Solomon Mamaloni
Ruling party: Alliance Party
Capital: Honiara (on Guadalcanal)
Official Languages: English
Currency: Solomon Islands Dollar (SI$ = 100 cents)
Exchange rate: SI$2.38 per $ (Nov 1989)

Area: 27,556 sq km: Guadalcanal: 5,302 sq km
Population: 295,000* (1988)
GDP per capita: SI$790 (1988)
GDP real growth: 8.0% (1988)
Inflation: 16.6% (1988)
Trade balance: –$21.6m (1988)
Foreign debt: $88.7m (1987)
Visitor numbers: 10,679 (1988)

* estimated figure

Solomon Islands was shaken up in 1989 by the return to power of the mercurial Solomon Mamaloni, who quickly offered reminders of his reputation for being unpredictable. The newly elected prime minister declared his intention of replacing expatriates, who, he claimed, were 'members of the Freemason society'; he attacked the whole government framework as 'untidy, disrespectful'; and he quickly clashed with the governor of the country's central bank, an expatriate who has taken Solomons' citizenship.

Mamaloni's new position as prime minister is stronger than that of any other previous leader of the country. At the election in February 1989 his People's Alliance Party won 14 of the seats in the 39-member parliament, far ahead of the six seats achieved by the United Party led by outgoing Prime Minister Ezekial Alebua. When they saw the tide of popular support in Mamaloni's favour, a number of members elected as independents joined his party, giving him 21 supporters in parliament, effectively the first time that a government has had an overall majority of its own party.

Hitherto, Solomons' governments have been uneasy alliances built on the shifting sands of powerful ambitious individual personalities. Instability was the order of the day simply because of the distrust between the principal players, many of them anxious for a bigger share of the spoils and quite happy to be disloyal and bring down a government they were part of if they could get a better share in its successor. The game of changing parties was helped merrily along by the high turnover of members of parliament at election time, the number of ambitious personalities and the consequent loose discipline among members of the same so-called 'party', not to speak of a lack of proper political programmes. Prime ministers were then elected by a mixed barter system, in which support at the election, personal popularity and willingness to horse-trade the ministerial jobs all played a part.

With his election success, Mamaloni was thus in a new commanding position. His choice of cabinet reinforced his own prestige. The only minister with experience of politics is Foreign Minister Sir Baddeley Devesi, who was the country's governor general until he retired in 1988. Mamaloni's ministers include four civil servants, four teachers, a male nurse, a mechanic and a policeman. Six of the cabinet had no experience of politics until they were elected in February 1989, and another five became MPs for the first time in 1984. No wonder that the prime minister claimed that he was spending much of his time instructing his neophyte ministers what they should do. Attempts of the remaining members of parliament to form a grand opposition alliance fell apart, partly because of rivalry for the title of leader of the opposition.

Economic situation

The new prime minister immediately announced a suspension of all aid, whether bilateral or multilateral, to his country. He declared that his government was dedicated to free enterprise and to private enterprise and wanted trade rather than aid. The 46-year-old chain-smoking Mamaloni who loves to play cricket 'because it is a game for gentlemen like me' seemed determined to antagonise key people in the Solomons. Some of his other opening statements raised more than a few eyebrows. He attacked the civil service and his fellow politicians for being lazy, inefficient and self-seeking. He also questioned the efficiency of some foreign aid projects and the integrity of workers supplied under foreign aid.

Most surprising, Mamaloni locked horns with the country's central bank and notably its governor Tony Hughes. In April the finance ministry put out an angry statement that the country's economic state was much worse than the bank had admitted. It accused the central bank of an act of 'monumental deception'. What was surprising about this was that for years the bank has been sounding a clarion call demanding urgent government action to curb the government deficit and inflation. On the surface, the concerns of Mamaloni and Hughes were the same, and they might have been expected to join forces. Instead, personality conflicts inaugurated by the prime minister threatened effective joint action to put the economy right.

In its annual report for 1988, the central bank recorded that the Solomons' economic growth rate at 8 per cent was the best for many years, thanks to rises in both volume and prices of Solomons' exports. But the bank was also quick to point out the worrying signs. Inflation was running at 17 per cent. The government deficit was large and growing and had to be financed by overseas borrowing. The foreign exchange reserves stood at SI$39m, but would not have been this high except for economic assistance from the European Community.

More worrying for the long term is the rapid growth in the Solomons' population. Between 1976 and 1986 the population grew by a massive 45 per cent. By the end of 1988 there were more than 305,000 Solomon Islanders. In spite of the improved economic growth rate, there has been virtually no growth of employment over the past four years. 'The fundamental problem facing the Solomon Islands' economy in the 1990s has not changed,' the central bank states. 'It is that the rate of population growth is outrunning the ability of the formal monetary economy to provide jobs and incomes and the capacity of the government to provide social and administrative services. Moreover, no foreseeable combination of investible resources will substantially change this trend by itself... if the underlying balance is to change there will have to be a marked decline in population growth rates over the next five to ten years.'

KEY INDICATORS

	Unit	1984	1985	1986	1987	1988
Population	'000	260	270	280	290	295
Gross domestic product (GDP)	SI$m	220.8	228.7	223.7	213.9	230.8
GDP per capita	SI$	849	871	894	750	790
Agricultural production						
Copra	1985=100	146	144	109	93	–
Rice	1985=100	50	42	17	–	–
Cocoa	1985=100	494	486	575	776	–
Palm oil & kernel	1985=100	143	146	107	87	–
Logs(timber)	1985=100	142	127	158	101	–
Fish	1985=100	157	136	193	130	–
Exports	SI$m	118.56	103.87	114.86	128.30	170.56
Fish	SI$m	25.09	27.66	46.89	42.97	66.50
Timber (logs)	SI$m	28.74	23.71	33.95	35.07	38.79
Copra	SI$m	32.20	23.47	5.95	10.26	15.66
Palm oil	SI$m	17.13	12.38	5.61	6.90	12.34
Imports cif	SI$m	100.61	123.20	125.19	161.93	–
Balance of trade	$m	25.9	0.7	–0.9	–3.3	–21.6
Current account	$m	5.3	–22.9	–9.1	–4.4	–16.1
External debt	$m	46.4	67.8	74.2	88.7	–
Debt service ratio	%	0.8	2.6	–	–	–
Exchange rate	SI$ per $	1.27	1.48	1.74	2.00	2.08

* estimated figure

Export potential

As the bank and other outside economists are quick to point out, the Solomons also has happier prospects – if the population does not explode out of control. Tuna fishing – in which the Solomons has a joint venture deal with the Japanese – forestry, palm oil and copra all offer good prospects of providing jobs and export earnings. If world gold prices increase, then the small gold mine on Guadalcanal may become a real gold mine for the economy (as well as a reminder of the name of the country, so-called by early European explorers because they thought it was the site of the fabled King Solomon's Mines). Tourism could also be developed as a source of income. The country has its share of unspoilt South Pacific beaches, and its waters are a ready exploration ground for scuba divers and other water-going sports.

The Solomons is close to Australia and New Zealand. The problem is that the country is small and off the main air routes round the world. With the advent of the new generation of long-range aircraft capable of flying 8,000 miles without refuelling, the Solomons risks being isolated from mainstream communications. In the 1970s Boeing 707 and VC-10 aircraft used to make stops in places like Fiji. Now air transport planners are saying that the long-range jets may by-pass advanced tourism spots like Hawaii. Small island countries like the Solomons have to try harder to get their own place in the tourist sun, and their share of the tourism dollar.

Barriers to economic development

This leads on to a wider point of the country's vulnerability. Unlike other Pacific island states the Solomons has a relatively homogeneous population 93 per cent of whom are Melanesian. Even so, in 1989 small-scale troubles broke out between the minority Polynesians and Melanesians near the capital. It was probably an early warning signal of bored youth.

The country's inhabitants live in small and widely dispersed settlements along the narrow coastlines. Honiara, the capital, with about 10 per cent of the population, is the only urban centre. The mountainous and rugged terrain means that little of the extensive land area is suitable for mechanised agriculture. But the limiting factor is not really a shortage of arable land.

As an International Monetary Fund study points out there are plenty of barriers to economic development: 'Village life, the extended family system and a custom of sharing income among relatives exercised a pervasive influence on economic activity. While there are several large agricultural plantations jointly owned by the government and foreign investors, 87 per cent of land is communally owned. Investment in agriculture is impeded by the land tenure system and unresolved claims over land.'

The system of government has further handicapped the development process. Devolution to the provincial government has, according to the IMF, proved 'to be slow and expensive.' The fragile coalitions in the central government have not helped in working out a coherent development policy.

The central bank in its annual report pointed out that the main components for economic growth were in place, and the financial system would support a bigger and more diversified monetary economy. But it urged flexibility in bringing the various components of the Solomon Islands together. 'The economy is an intricate web of activities by a wide variety of actors, large and small. It does not respond well to orders or exhortations, but it will respond to clear price signals, fair rules, durable incentives and credible opportunities, to do more, and do it better. That is the economic environment the authorities must provide.'

Evidence in 1989 was that the economy would grow by 5 per cent or more. This was largely due to the private sector responding to new opportunities. The main question was whether the government could put its own house in order. But new Prime Minister Mamaloni seemed to be headed in the direction of exhortation and even hectoring and condemnation rather than following the policy urged by the bank.

COUNTRY PROFILE

Political system: The Solomon Islands, which became an independent state (a constitutional monarchy) and member of the Commonwealth in 1978, is a parliamentary democracy with a unicameral, 38-member parliament elected to a four-year term by universal suffrage (over age 18). The British Crown is represented by a governor general appointed by the parliament. Executive power rests with a prime minister elected by and from the members of parliament and the 15-member cabinet of ministers he appoints. The seven provinces are each administered by elected provincial assemblies and Honiara, by a town council.

Political parties: Solomon Islands United Party (four seats in the 1988 elections); People's Alliance Party (22 seats), National Democratic Party (one). There are a number of independents. Political parties are still in a process of development with party allegiances fluid and forming around individual leaders. This is helped by a high turnover of MPs with about half sitting members being rejected at election time.

Population: 94 per cent of the population is of Melanesian origin (1987), with four per cent Polynesians (living mainly on the outlying islands), 1.4 per cent Micronesians and a scattering of Europeans and Chinese. The population growth rate is high at around 3.5 per cent per annum, and nearly 60 per cent are aged 20 or below. Total population 297,000 (end 1987 estimate).

Main cities/towns: Honiara (on Guadalcanal, population 26,000 in 1987), Gizo (on Gizo Island), Auki (on Malaita).

Language: There are around 100 local dialects in addition to English, the offical language, and *Pijin* English, the most effective street medium.

Media
Press: The Solomon Star, The Nius and *Solomon Toktok* are the weekly newspapers.
Radio: Solomon Islands Broadcasting Corporation produces radio programmes in English and Pidgin, transmitting local and overseas programmes 118 hrs per week from Honiara, Gizo (on New Georgia), Santa Cruz and Malaita.
Television: No television broadcast; video cinema and three film cinemas in Honiara and Gizo (on New Georgia).

Domestic economy: Based on agriculture, with 90 per cent of population in rural areas. Much real national income derived from non-market, subsistence production of food crops, fish, meat, houses, fuel, canoes and certain services. The subsistence economy contributes about 38 per cent of GDP, and growth in this sector is one objective of government policy. Overall, the economy has suffered from the high cost of imported fuel and low worldwide prices for its agricultural products. World Bank estimates GNP at US$510 per head, but in 1988 the Solomons Central Bank suspended national income estimates, saying that the quality of data was unreliable. The country has relied heavily on foreign aid from UK, Japan, Australia and New Zealand and multilateral aid from World Bank, IMF and UN agencies such as WHO, ILO and UNESCO. Payments from the EC Stabex Fund, $25m in 1987, have helped offset a large trade deficit. Development projects under the 1985-1989 Plan include improvements to the road network and to Henderson Airport and rural development. Government priority on investment in agriculture, livestock, agro-industrial industries, added value manufacturing for import substitution, export assembly industries and tourism.

Employment: In 1987, 10 per cent of the population was in organised employment. There are 14 trade unions, the largest being the Solomon Islands National Union of Workers, the Solomon Islands Public Servants' Association and the Government Workers' Union.

External trade
Exports: SI $128.3m (FOB) 1987. Main exports are logs and timber (29% of 1987 exports), fish (41%), copra (8%), cocoa (7.4%), palm oil and kernels (5.9%). Main destinations are Japan (35.6% of 1987 total), UK (13.8%), Thailand (12.2%), Republic of Korea (6.1%), German FR (4.2%), Australia (3.7%), Netherlands (2%), Sweden (1%), Singapore (0.9%).
Imports: SI $165.7 (cif) 1987. Main imports are machinery and transport equipment (29.1% of 1987 imports), manufactures classified by materials (20.5%), food (15%), mineral fuels and lubricants (14.7%), miscellaneous manufactures (9.2%), chemicals (6.2%), beverages and tobacco (3.1%). Main sources are Australia (41.4% of 1987 total), Japan (19.1%), Singapore (9.2%), New Zealand (7.9%), Papua New Guinea (7.9%), UK (4.4%), China PR (3.5%), USA (2.9%), Hong Kong (2.4%).

Agriculture: Agriculture normally accounts for over 50 per cent of GDP, most exports and about one-third of paid employment. The 1986-87 crop was, however, devastated by cyclones. About 25-30 per cent of total land area is suitable for intensive, non-traditional agriculture, mainly on Guadalcanal; the islands are self-sufficient in rice, beef and vegetables. Copra and cocoa are produced for market on smallholdings and private plantations; an oil palm plantation was commenced in 1971 to produce for local industry. The Government Shareholding Agency, in association with major plantations, is encouraging new coconut and cocoa planting and extension of the oil palm plantations. Large-scale, mechanised rice growing covers all domestic consumption, but interest in the crop is declining as export markets disappear. Forests contain some 170,000 hectares of exploitable land having 10.4m cu metres of commercial timber, with timber resources expected to last 10-15 years at current logging rates. Further expansion of logging is only licensed in conjunction with provision for reafforestation. The National Livestock Development Authority (NLDA) administers a $5.5m programme to develop cattle, goat, pig and poultry raising. Commercial fishing and fish processing, mainly of skipjack tuna, developed by Solomon Taiyo Fishery Co, a long-term (to 1992) joint venture between government and a Japanese company based at Tulagi (in the Florida Islands) and at Noro (on New Georgia). Fish exports are expected to account for one-third of total export earnings in 1988. Domestic seafood demand served by small, local operations.

COUNTRY PROFILE

Industry: Manufacturing activities include palm oil milling; rice milling; fish smoking, canning and freezing; saw milling; copra drying; food processing (biscuits, bread, tobacco, snack foods and soft drinks); production of nails, detergents and soaps, wood and rattan furniture, fibreglass articles, boats, clothing, handicrafts, shell jewellery, buttons. Most timber is exported as logs, but an increasing proportion is being sawn; the government hopes to develop wood processing where species enable profitable marketing of sawn timber and veneers. Many engineering companies are foreign-owned. New industrial estates are being built at Guadalcanal and at Honiara. There is strong investor interest in tourism.

Mining: Panning of alluvial gold produces some 50-100 kg p.a. Amoco Minerals Solomons has been actively prospecting for gold deposits suitable for large-scale mining and have decided to open a small mine on Gold Ridge. Undeveloped mineral resources include small deposits of copper, lead, zinc, silver, nickel, cobalt, bauxite, phosphates and asbestos.

Hydrocarbons: The government is investigating possible development of offshore petroleum resources indicated by the US Geological Survey's discovery of huge, undersea sediment-filled basins that could hold oil deposits near the Solomon Islands, Tonga and Papua New Guinea. New petroleum storage is being built at Honiara.

Energy: A new 6 mW hydro-electric scheme is under consideration at Komarindi. Other schemes include a new power station at Noro, estimated cost of $7.2m, and re-equipping the Lungga power station at a cost of $5.5m.

Main companies: Solomon Taiyo Ltd, Lever Solomons Ltd, Solomon Islands Plantations Ltd, National Fisheries Development Corp, Pacific Timbers (SI) Ltd.

Legal system: The UK Statutes of General Application to 1961 administered through the government's court system. Local courts presided over by elders of local communities, who deal with matters of custom, by-law, etc. Customary Land Appeal Courts hear land appeals from Local Courts.

Membership of international organisations: Asian Development Bank, Commonwealth, ESCAP, GATT (de facto), G-77, IBRD, IDA, IFAD, IFC, IMF, Lomé Convention (EC), South Pacific Commission (participating member), SPEC, SPF, UN, UPU, WHO, World Bank.

BUSINESS GUIDE

Time: GMT + 11 hrs.

Climate: Warm and humid, equatorial with average temperatures from 22°C (mountainous areas inland) to 28°C (coastal areas). Rainfall averages about 3500 mm p.a., but varies greatly according to location and mostly falls Nov-Apr, when cyclones may occur as well.

Entry requirements
Passport: Required by all except certain seamen.
Visa: Visas required by all except citizens of UK, Commonwealth countries, the US and some European countries. Travellers with onward passage and adequate funds may obtain a visitor's permit for up to two months on arrival.
Currency: No restriction on the importation of currency, but travellers are advised to declare any foreign currency on arrival, and may only export up to the same amount of that currency. Notes and coins to the value of SI$250 may be taken out of the country without exchange control approval.
Customs: Personal effects (including an allowance of alcoholic beverages and tobacco) allowed duty-free. Import licences required for spirits, wines, tobacco, cigars, cigarettes, firearms, ammunition, animals, seeds, soil and plant material.
Income tax: Non-residential individuals and companies are liable for tax only on income which accrues in or is derived from Solomon Islands.

Health precautions
Mandatory: Vaccination certificates required for yellow fever if travelling from an infected zone.
Advisable: Anti-malaria precautions throughout the visit and for six weeks after departure.

Air access
National airline: Solomon Airlines Ltd.
Other airlines: Air Niugini, Air Pacific, Qantas, Singapore.
Tax: An airport tax of SI$20 is levied on departure.
International airport: Henderson International (HIR), 13 km from Honiara; buffet and car hire (being upgraded).

Surface access: Regular shipping links with Australia, New Zealand, Hong Kong, Japan, the UK and Europe.
Main ports: Honiara (to be upgraded).

Hotels: Visitors are advised to book well in advance. Hotel tax of five per cent added to bill. In addition to Honiara's three hotels, there are resorts, guesthouses and government resthouses of varying standards and quality scattered throughout the islands.

Car hire: Car hire available; driving licences with four months' unexpired duration are acceptable.

City transport
Taxis: Taxis available in Honiara. Advisable to check the fare before hiring a taxi.

National transport
Air: Solair provides inter-island transport from Henderson Airport to most main islands and towns. Charter services available. 25 airfields throughout the islands with more planned.
Road: Honiara (on Guadalcanal), Auki (on Malaita) and Gizo (on New Georgia) served by 100 km of roads, about half of which are tar-sealed. 1,700 km of rural roads and tracks mainly coral or gravel-surfaced; terrain can be difficult.
Buses: Bus services operate in and around Honiara.
Water: Inter-island shipping services operated by private companies, missions and the government provide main means of transport. Boats used to transport cargo also carrying passengers in varying degrees of comfort; some trading centres have wharves suitable for small vessels and some are being enlarged.

Public holidays
Fixed dates: 1 Jan, 7 Jul (Independence Day), 25 and 26 Dec.
Variable dates: Good Friday to Easter Monday, Whit Monday, Birthday of HM the Queen.

Working hours: Many shops open Sat afternoons and Sun; Chinese stores often open at other times.
Business: (Mon-Fri) 0800 – 1200, 1300 – 1630; (Sat) 0800 – 1200.
Government: (Mon-Fri) 0800 – 1200, 1300 – 1630.
Banking: 0830 – 1500 (Mon-Fri).
Shops: (Mon-Fri) 0800 – 1200, 1330 – 1700; (Sat) 0800 – 1200.

Social customs: Tipping is not customary, and visitors are strongly advised to refrain from the practice. Women should avoid wearing shorts and make sure their legs are adequately covered to avoid giving offence. The social structure of the Solomon Islands is extremely complex, with traditions, culture and even language varying from island to island and among villages on the same island.

Telecommunications
Telephone: Dialling code for Solomon Islands, IDD access code + 677 followed by subscriber's number (no area code required). Automatic telephone service available in Honiara, Auki and Tulagi, with about 3200* subscribers in 1987 (90 per cent in Honiara). Radio links between most islands, international services available 24 hrs via SOLTEL (Solomon Islands International Telecommunications Ltd) satellite and cable links. Specialised systems such as private leased circuits and data links can be provided.
Telex: Public telex facilities available Mon-Fri 0700 – 1800, Sat 0800 – 1200, Sun 0900 – 1100 and public holidays at SOLTEL offices on Mendana Ave, Honiara (tx: 66301).

Postal service: There is no local delivery system.

Banking: The Central Bank of Solomon Islands is the monetary authority responsible for issuing and managing international exchange and for guiding overall banking and lending policy.
Main centre: Honiara.
Head offices: Development Bank of the Solomon Islands, National Bank of the Solomon Islands.
Solomon Islands branches: ANZ Banking Corp, Westpac Bank.

Advertising: Newspapers, radio and cinemas accept advertising.

Electricity supply: 230/415 V AC with flat three-pin plug fittings and bayonet-type sockets, using the standards and wiring regulations of the Standards Association of Australia. Availability confined to Honiara and five other urban centres, but to be increased under the present plan.

Weights and measures: Metric system (gradual conversion from imperial system).

Representation in capital: Australia, New Zealand, Japan, Papua New Guinea, Singapore, Taiwan, UK, USA.

Useful telephone numbers
Police and fire: 22999; Ambulance: 22200; Marine emergency: 21535; Emergencies outside Honiara: 111; Directory enquiries: 101; Overseas operator: 102; Shipping and time: 107; Operator assistance: 100; Customs: 22301; Immigration: 22243.

Hotels: Honiara, Guadalcanal
Hibiscus (town centre), P.O. Box: 268 (tel: 21205; tx: 66409).

Honiara (2 km from Honiara), Prince Phillip Highways, P.O. Box: 4 (tel: 21737/40; tx: 66347 HQ; fax: 23592).

Mendana, P.O. Box: 384 (tel: 20071).

Tambea Village Tourist Resort, Honiara, Tambea Village, P.O. Box: 506 (tel: 23639; tx: 66338).

Car hire: On Guadalcanal
Avis (Solair World Travel): PO Box 333, Honiara (tel: 20031; tx: HQ 66312).

Business advice and information
Central Bank of Solomon Islands, PO Box 634, Honiara (tel: 21791/3; tx: 66320 BOKOLO HQ).

Development Bank of Solomon Islands, PO Box 760, Honiara (tel: 21595; tx: 66350).

Ministry of Commerce and Primary Industries, PO Box G26.

Ministry of Finance and Economic Planning, PO Box 26.

Ministry of Foreign Affairs and Trade Relations, PO Box G10.

Ministry of Tourism and Aviation, PO Box G20.

Solomon Islands Chamber of Commerce, PO Box 64, Honiara, Guadalcanal.

Trading Co (Solomons) Ltd, Mendana Avenue, PO Box 114, Honiara, Guadalcanal (tel: 22588; tx: 66313: cable: TRADE).

Travel information
Solomon Islands Tourist Authority, Honiara, Guadalcanal (tel: 22442; cable: TOURIST).

Somalia

Lindsey Hilsum

Somalia's ageing president, Mohamed Siad Barre, was under siege in 1989. The war in the north of the country continued sporadically, and insecurity spread east and south. A section of the Ogadeni clan, the president's traditional allies, withdrew their allegiance, and his support base is narrowing down to his own Marehan clan and his immediate family. The war disrupted livestock exports, and the price of basic commodities shot up.

The war

The Somali army was unable to rout the Somali National Movement, after the rebel assault on Hargeisa in May 1988. Young boys pulled off the streets of Mogadishu and herded to the front with little training, have neither the will nor the ability to fight a hit-and-run guerrilla force. Visitors to Hargeisa, Burao, Berbera and Erigavo report that much of the gunfire heard at night is the result of young soldiers playing with their weapons or harassing people on the streets.

In January 1989, President Siad Barre announced a programme of 'reconstruction and reconciliation' for the north. He released some prisoners of the Issaq clan, who form the support base of the SNM, and who were picked up during the initial stages of the conflict. But the 200,000 or so Issaqs who fled to Ethiopia last June remained in refugee camps, and the Issaqs did not return to Hargeisa. Several thousand went to Mogadishu, swelling the capital's slums.

The government appealed for aid, and the United Nations pledged $19m to reconstruct the north. However, most aid agencies and Western donors were reluctant to give the government money to reconstruct buildings and water sources which it destroyed itself in bombing raids.

The SNM does not appear to administer consolidated 'liberated zones', but vehicles travelling along northern roads are prone to ambushes by the SNM or clan militias. The government's control does not extend beyond the main towns.

The government is trying to beat the Issaq-dominated SNM by exploiting clan conflicts. Automatic weapons have been dished out to clans which traditionally oppose the Issaq. As a result, old scores are being settled with bullets, and many observers believe that Somalia is fragmenting into ancient tribal groupings, with little influence from the centre. The proliferation of weapons is a potential destabilising force throughout the region, as Somalis travel and trade in all neighbouring countries.

The government's most controversial move was to arm the Ogadeni refugees who fled from Ethiopia ten years ago, and who are settled in camps in the north. They oppose the Issaq and have joined the army in fighting the SNM. The policy brought the government into conflict with the United Nations High Commission for Refugees, which refused to feed refugees carrying guns.

Relations between the UNHCR and the government were already strained, because UNHCR announced in late 1988 that, following the agreement between Ethiopia and Somalia, the refugees should go home and UNHCR would withdraw from Somalia. After discussions with the government, which regards the annual $33m aid for refugees as a major economic support, UNHCR agreed to delay the start of the withdrawal until 1990, and proceed gradually.

International relations

Somalia's international image has been damaged by persistent reports of human rights abuses. In January 1989, the British Foreign Secretary, Sir Geoffrey Howe, speaking in neighbouring Djibouti, expressed his concern and mentioned reports that chemical weapons had been imported from Libya. Britain and the United States froze aid commitments. An Australian non-governmental organisation, Community Aid Abroad, pulled out of Somalia and published a first-hand account of atrocities in Erigavo, including the massacre of 120 nomads in mid-1988. The atrocities continued: on 16 March 1989 the army is reported to have carried out another massacre of over 200 Issaqs in Erigavo.

Amnesty International visited Somalia in July 1989, but no Western journalists have been allowed in since before May 1988.

At the beginning of 1989, the government embarked on a 'diplomatic offensive'. Ministers were despatched around the world, including to Britain and the United States, to explain Somali government policy and deny reports of human rights abuses. Britain and America restored some aid, but non-governmental agencies are scaling down their programmes. In May 1989, a year after expatriate personnel were evacuated from Hargeisa, aid agencies withdrew from the refugee camps at Borama because of insecurity.

Although the Americans are still Somalia's main backers, they have reneged on the original agreement. In the early 1980s, the Americans provided $25-30m annual military assistance in return for naval facilities at the Red Sea port of Berbera. In 1988, they provided only $5m.

Relations between Ethiopia and Somalia remain steady, following the 1988 accord, which formally brought to an end a decade of hostilities. Although Issaqs in the Ethiopian refugee camps express support for the SNM, there is no evidence that the Ethiopian government is arming the rebels.

Political alliances

In March 1989, a group of between 200 and 600 Ogadeni soldiers mutinied at the southern military base of Kismayo. Some felt betrayed because the government relinquished its claim to the Ogaden, as part of the accord with Ethiopia. Others were angry about the dismissal of the defence minister, Major General Mohammed Adan Nur, himself an Ogadeni, who had been vying for power with President Siad Barre's son-in-law, General Morgan. General Morgan has a reputation for ruthlessness and unusual tastes: when he lived in Hargeisa he kept two pet cheetahs in a cage, and was building himself a replica of the White House. He was made deputy minister of defence in 1989 and has more power than the newly-

appointed civilian minister of defence.

The mutiny has not spread throughout the Ogadenis in the army, but by July 1989 the rebel soldiers were still controlling the area between Kismayo and Afmadu. Ogadenis hold the balance of power between the president's Marehan clan, and the Issaqs. If more desert the government cause, President Siad Barre's grasp on power will become shakier.

Political instability and war began affecting Mogadishu seriously in mid-1989. In July, troops shot at Muslims who were apparently protesting about the arrest of some immams. At least 23 people were killed, and President Siad Barre went on the radio to appeal to the security forces to 'stop shooting at random'.

Two political reshuffles promoted close members of President Siad Barre's family. However, the president's half-brother, Abdel-Rahman Jama Barre, was removed from the ministry of finance and the portfolio was returned to Mohamed Shaikh Osman. The perennial question of who will follow the geriatric leader when he dies remains unanswered.

KEY FACTS

Official title: Somali Democratic Republic

Head of state: President Mohamed Siad Barre

Ruling party: Somali Revolutionary Socialist Party (SRSP)

Capital: Mogadishu

Official Languages: Somali, Arabic

Currency: Somali Shilling (SoSh) = 100 centesimi

Exchange rate: SoSh408 per $ (Nov 1989)

Area: 738,000 sq km

Population: 5.7m (1987)

GNP per capita: $290 (1987)

GNP real growth: 12.4% (1987)

Inflation: 125% (early 1989)

Trade balance: –$382m (1987)

Foreign debt: $2.5bn (1987)

In June 1989, the question of turning Somalia into a multi-party state was mooted. The president professed to support the idea, but organised a demonstration of Mogadishu residents clamouring for continuing one-party rule. The multi-party idea was widely regarded as a publicity stunt, and dropped.

KEY INDICATORS

	Unit	1983	1984	1985	1986	1987
Population	m	5.08	5.23	5.38	5.55	5.71
Gross domestic product (GDP)	SoSh bn	35.2	62.2	87.4	119.0	170.7
GDP real growth	%	–14.0	9.9	10.0	4.2	12.4
GDP per capita	SoSh bn	6,929	11,893	16,245	21,444	29,895
Inflation	%	36.2	91.1	37.8	35.8	28.1
Agricultural production	1979-81=100	103.9	106.6	116.4	125.1	124.1
Livestock production	1979-81=100	103.5	105.2	114.3	123.6	121.8
Exports (fob)	SoSh m	1,717	1,096	3,756	6,372	10,900
Live animals	SoSh m	1,244	514	2,604	4,420	7,301
Imports (cif)	SoSh m	5,509	2,172	4,425	20,474	13,914
Balance of trade	$m	–263.6	–410.9	–240.2	–294.5	–381.5
Current account	$m	–141.6	–132	–96.6	–87.7	–100.2
Total external debt	$m	1,425	1,598	2,070	2,308	3,534
Debt service ratio	%	10.0	21.5	4.0	9.2	–
Exchange rate (official)	SoSh per $	15.8	20.0	39.5	72.0	105.2

* estimated figure

The economy

Most Somalis live outside the formal economy. As pastoralists, they rely on their camels, cattle and goats for subsistence. Their bartering and trading keep the country going, but this informal economy is not measured. Weather conditions were largely good in 1988 and 1989, so the bulk of the population was scarcely affected by the economic problems reflected in official statistics.

However, a rid increase in food prices is making life hard for the growing numbers of people who live in Mogadishu. Commodities such as sugar are exorbitantly expensive, and, at times, unavailable. Despite a 100 per cent wage increase for civil servants, salaries are so low that government employees must moonlight to survive. Mogadishu residents reported an increase in robbery and violent crime in mid-1989.

Somalia's economy is suffering because donors are reluctant to increase aid, and the war eats up money. The main export is livestock, but the war has prevented herders from getting their animals to the port of Berbera. Livestock exports reached an all-time low in 1988, but picked up a little in 1989 as some animals were exported from Mogadishu and others from Bossasso, a port built with Italian aid.

Somalia's two other main exports are bananas and mhyrr, and total export earnings are equivalent to only one-quarter of imports. Illegal trade in frankincense, mhyrr and ivory is relatively significant, but unquantifiable.

In 1987, the government, alarmed by the slide in the value of the shilling, caused by a system of currency auctions, broke an agreement with the IMF. The IMF cancelled credits, and donor countries suspended aid commitments. In 1988, Somalia should have received $137.7m, but only $21m was delivered.

The government started to patch up its quarrel with the IMF in mid-1988. In 1989, foreign currency became available to importers on a non-discriminatory allocation basis. The government lifted price controls, devalued the shilling and increased interest rates.

In January 1989, the government allowed private participation in foreign trade and banking. Two months later, the IMF agreed a $260m standby credit. In June 1989, the US government released $15.1m in balance of payments support pledged in 1987.

Somalia's biggest development project is the $500m Baadere Dam, intended to irrigate the medium-potential Juba Valley and provide hydro-electric power for the capital. Although construction was scheduled to begin in mid-1989, experts predict a two- or three-year delay as the environmental and social impact is assessed. About 100,000 refugees and 20,000 farmers will have to be moved from the area to make way for flooding.

Amoco and Chevron suspended oil exploration during 1988, because their main drilling sites were in the conflict zone. Petron and Conoco signed agreements to prospect in central and eastern Somalia. The plan to build the largest oil refinery in Africa on a site in Mogadishu, with Middle Eastern backing, remains on paper.

Although international donors responded positively to the government's economic reforms, private investment remained sluggish. No significant upturn in the economy can be expected until the civil war ends.

COUNTRY PROFILE

Political system: Executive power vested in president who is elected every seven years. Legislative power is held by the 177-member People's Assembly. Gained independence in 1960. Became the Somali Democratic Republic in 1969.

Political parties: The Somali Revolutionary Socialist Party is the sole legal party.

Population: The 5.8m inhabitants (1987 estimate) are predominantly Muslim, with a small Christian element.

Main cities/towns: Mogadishu (population 1m), Berbera, Hargeisa, Kismayu, Merca.

Language: The official languages are Somali and Arabic; English is often used for business, however.

Media
Press: All newspapers are government-controlled. The only daily is *Xiddigta Oktobar* (in Somali). Weekly papers include *Heegan* (English) and *Horseed* (Italian and Arabic).
Radio: The Somali Broadcasting Service, Radio Hargeisa and Radio Mogadishu broadcast in Somali, Italian, English, Arabic, Swahili and Amharic.
Television: Experimental television service operates in Mogadishu area for 2-3 hrs/day; there are plans to extend the service.

External trade: At the end of 1987 the foreign debt was $1.8bn. During 1987 the IMF suspended its stand-by facility and a structural adjustment loan. The poor balance of trade reflects the continuing need for the large-scale import of fuels and food. An estimated $400m in remittances from migrant Somali workers is contributed annually to the current account.
Exports: Bananas, livestock, myhrr. Main destinations: Italy, Saudi Arabia, Yemen Arab Republic.
Imports: Petroleum, food, machinery and transport equipment, construction materials. Main sources: Italy, Saudi Arabia, USA, UK.

Agriculture: Mostly a pastoral form of production. Most important sector of economy. Only 1 per cent of the land surface is used for agriculture, although nine per cent is potentially arable. Livestock raising, mainly by nomads, still suffers from the six-year-old ban imposed by Saudi Arabia on African animals. Nomads (a very small sector) are being encouraged to set up permanent agricultural or fishing communities. Subsistence farmers grow maize and sorghum; wheat and rice are imported. The most important cash crops are bananas, cotton, and frankincense (of which Somalia is the world's largest producer). Irrigation projects are being discussed with aid donors.

Industry: Small industrial sector continues to be dominated by state-owned agro-industries, although some parastatals have recently been hived off, and the government is encouraging foreign investment in small-scale manufacturing and processing industries. Main industries at present include meat and fish processing, sugar refining, fruit and vegetable canning, textiles and leather goods. Some factories are idle because foreign exchange shortages have cut off foreign inputs; but a cement factory is under construction at Berbera. Emphasis on development of value added export-based light industries and import substitution.

COUNTRY PROFILE

Mining: Reserves of iron, uranium, beryl and columbite have been found, but as yet only salt and gypsum are extracted commercially.

Energy: Potential oil and gas reserves have been found near Mogadishu and in the north, but the country still relies on imports for most of its fuel needs. Electricity generation (all thermal) amounts to about 100 gwh.

Membership of international organisations: African Development Bank, African States Associated with EC (Lomé Convention), Arab Common Market, Arab Fund for Economic and Social Development, Arab League, FAO, IDA, IFAD, IFC, IGGAD, ILO, IMF, Islamic Conference, Islamic Development Bank, UN, WHO, WIPO, World Bank.

BUSINESS GUIDE

Time: GMT +3.

Climate: Tropical. Humid on coast, drier in north. Average temperatures 27°C to 32°C throughout year, but can reach 42°C on coast. Dry seasons from Jan-Feb and Aug-Sep. Rainy seasons from Mar-Jun and Oct-Dec.

Entry requirements

Passport: Required by all.

Visa: Required by all.

Currency: Import/export of only small amounts of local currency (SoSh200) allowed. Unlimited import of foreign currency, which must be declared on a form for which a small charge may be made. Currency transactions should be recorded at each exchange. Export of foreign currency limited to amount declared on arrival.

Health precautions

Mandatory: Yellow fever and cholera vaccination certificates usually required.

Advisable: Cholera, yellow fever, typhoid and polio vaccinations. Malaria prophylaxis should be taken as risk exists throughout the country (two types of prophylaxis now recommended). Water precautions are essential. Risk of meningitis in north of country.

Representation overseas: Abu Dhabi, Aden, Algiers, Ankara, Baghdad, Belgrade, Berlin, Bonn, Brussels, Cairo, Dakar, Damascus, Dar es Salaam, Djibouti, Doha, Islamabad, Jeddah, Kampala, Khartoum, Kuwait City, Lagos, London, Lusaka, Maputo, Moscow, Muscat, Nairobi, New Delhi, Ottawa, Paris, Rome, Sana'a, Stockholm, Tehran, Tripoli.

Air access

National airline: Somali.

Other airlines: Aeroflot, Air Djibouti, Alyemda, Ethiopian Airlines, Kenya Airways, Saudia.

International airport: Mogadishu International (Code: MGQ), 6 km from city.

Surface access

Main ports: Berbera, Kismayu, Merca, Mogadishu.

Overland access: No railway system in Somalia. Road links with Kenya in the south and Djibouti in the north. Four-wheel drive vehicles recommended. Roads being upgraded.

Hotels: Available in all principal towns. Advisable to book in advance. Service charge of 10 per cent added to bills.

Electricity supply: 220V AC, 50 cycles.

Car hire: Available in Mogadishu. Four-wheel drive vehicles available from Marill (Somalia) Limited.

City transport

Taxis: Fares by negotitation; tipping is not usual. Can be hired on a time basis.

National transport

Air: Somali Airlines operate domestic services to main centres.

Road: Travel may be restricted outside Mogadishu. Local enquiries should be made. Good roads from Mogadishu to Kismayu (via Merca) and Baidoa in the southern part of the country, and to Hargeisa and Berbera in the north. Most other routes are mainly tracks and gravel roads. Traffic moves on the right of the road. Travel to Hargeisa and Berbera is restricted because of insecurity.

Buses: Network outside Mogadishu is restricted, with few and irregular bus services.

Sea: Coastal shipping of both freight and passengers is extensive.

Public holidays

Fixed dates: 1 Jan, 1 May, 26 Jun (Independence), 1 Jul (National Day), 21-22 Oct (Anniversary of 1969 Revolution).

Variable dates: Eid al-Fitr, Eid al-Adha, Islamic New Year, Prophet's Birthday (Maulid an-Nabi).

Working hours

Business: (Sat-Thu) 0800 – 1230, 1630 – 1900.

Government: (Sat-Thu) 0800 – 1400.

Banking: (Sat-Thu) 0800 – 1130.

Shops: (Sat-Thu) 0900 – 1300, 1600 – 2000.

Telecommunications

Telephone: Dialling code for Somalia: IDD access code + 252 followed by area code (Mogadishu 1), followed by subscriber's number.

Telex and telegram: Facilities available from main post office opposite Hotel Juba in Mogadishu.

Banking

Central bank: Central Bank of Somalia.

Main centre: Mogadishu.

Other banks: Commercial and Savings Bank of Somalia, Somali Development Bank.

Representation in capital: Algeria, Bulgaria, China PR, Czechoslovakia, Egypt, France, German DR, German FR, India, Iran, Iraq, Italy, Kenya, Korea PDR, Kuwait, Libya, Malta, Nigeria, Pakistan, Saudi Arabia, Syria, Turkey, UK, USA, USSR, Yemen AR, Yemen PDR, Yugoslavia.

BUSINESS DIRECTORY

Hotels: Mogadishu

Al Uruba.

Croce del Sud, P.O. Box: 91 (tel: 20250; tx: 745).

Gouled.

Juba.

Mukka-al-Mukkarama.

Taleh.

Other useful addresses

Agricultural Development Corporation, PO Box 930, Mogadishu.

Central Bank of Somalia, PO Box 11, Corso Somalia 55, Mogadishu (tel: 725; tx: 604).

Chamber of Commerce, Industry and Agriculture, PO Box 27, Mogadishu (tel: 3209, 3122; tx: 632).

Commercial and Savings Banks of Somalia, PO Box 203, Mogadishu (tx: 602).

Livestock Development Agency of Somalia, PO Box 1759, Mogadishu.

National Agency of Foreign Trade, PO Box 6021, Mogadishu.

National Petroleum Agency of Somalia, PO Box 573, Mogadishu.

Somali Airlines, Medina Road, PO Box 726, Mogadishu (tel: 80487; tx: 619).

Somali Broadcasting Service, Ministry of Information and National Guidance, Private Bag, Mogadishu (tel: 2455).

Somali Development Bank, PO Box 79, Mohgadishu (tel: 21800; tx: 635).

Statistical Department, PO Box 1742, Mogadishu (tel: 80385; tx: 715).

South Africa

Anthony Robinson

The release of Nelson Mandela, the leader of the Africa National Congress (ANC) from prison on 11 February 1990 overshadowed all other events in South Africa's recent history. As this review went to press, it was difficult to assess fully the implications of the new president's policies; however, Mandela's release suggested that de Klerk's reforms could be more than cosmetic.

After one of the most closely contested general elections since the Afrikaner-dominated National Party swept to power in 1948 South Africa's crucial white electorate returned the party to power with a sharply reduced majority in nationwide polling on 6 September 1989. The election was marred by bloodshed in several black townships, especially in the Cape Town area, and marked again by widespread poll boycotts in the Coloured (mixed race) and Indian communities. Unlike the still unenfranchised black majority both communities were given the vote, for racially segregated separate houses of parliament, in the deeply flawed 1984 constitution. The reform, repeal and possible replacement of this constitution is likely to form the leitmotiv of the five-year mandate received by President Frederik Willem de Klerk. Before the election he spoke of convening a 'great Indaba' of representatives of all races to seek a negotiated settlement of the country's bloody political stalemate.

The mould of South African politics was bent but not broken after a general election campaign in which, for the first time since 1948, the mood of the electorate and rapidly changing circumstances offered the possibility of a hung parliament and an end to the long hegemony of the National Party. As the dust cleared after polling day however it became clear that the government party had retained its hold on power. But the party saw its majority whittled down from 123 of the 166 contested seats in the old parliament to only 93 in the new.

Squeezed from both left and right, and unsettled by a prolonged leadership crisis which culminated in the sulky resignation of President P. W. Botha only three weeks before the poll took place, the total was still comfortably above the 84 seats for the party to continue to govern alone with the barest majority. It was also well above the 39 seats won by the pro-apartheid Conservative Party, which retained its status as the official opposition, and the 33 seats won by the 'liberal-left' Democratic Party. The latter was born as a coalition of three former 'reformist' parties only six months before the poll.

Even so, it was a close run thing by South African white electoral standards and saw the National Party's support dip below 50 per cent of the total vote for the first time since the 1950s. The ruling party's main consolation was that the break-away Conservative Party failed to fulfil its own confident pre-poll predictions of massive gains at the ruling party's expense. Although it fell short of its target the right wing party managed to raise its total seats from 22 to 39 and broke out of its Transvaal home base for the first time by gaining a handful of seats both in the Cape and the Orange Free State. The Conservatives also gained several parliamentary constituencies in Pretoria, seat of government and heartland of the largely Afrikaner bureaucracy for the first time and are in a position to breathe down the National Party's neck in several marginal constituencies, especially in the Transvaal.

But the net result of the party's limited gains was to reduce its credibility as a possible alternative government while the unexpectedly strong showing of the Democrats indicated that the bulk of the white electorate, the half which voted for the National Party and the over 20 per cent which voted for the Democrats, was more willing to put its faith in still vaguely defined 'reform' than in the Conservative party's promise to turn back the clock to the classic apartheid of Dr Hendrik Verwoerd in the 1960s.

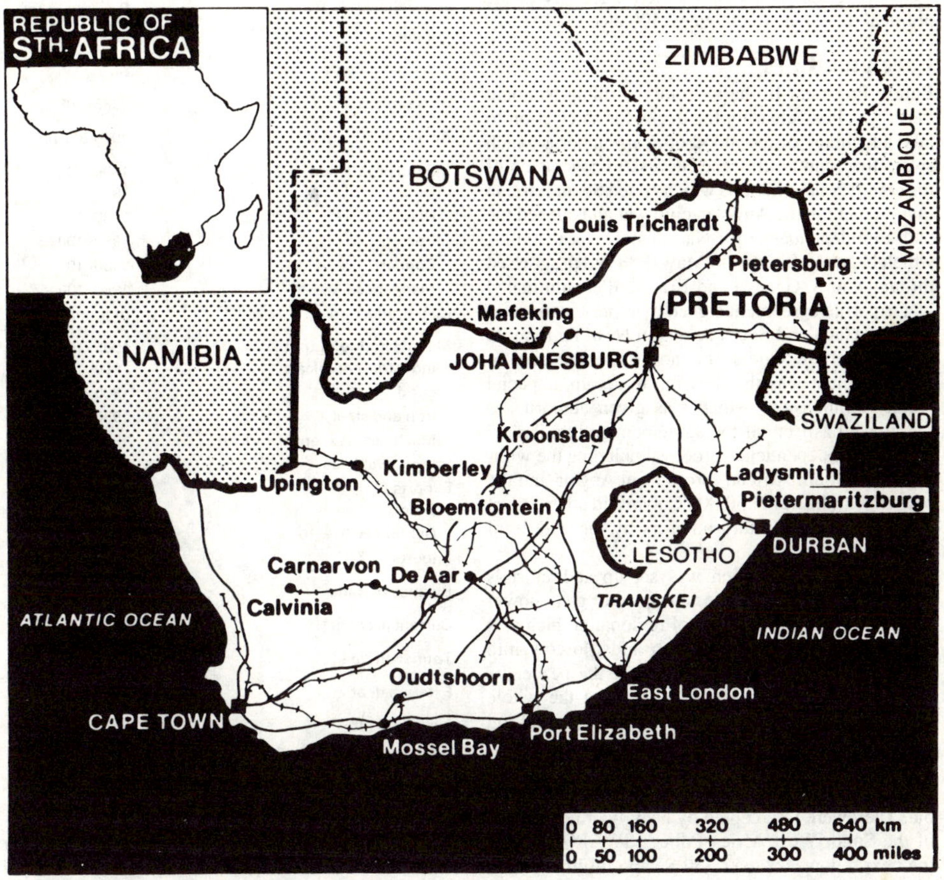

That at least was how President de Klerk chose to interpret the election results which he described as 'a clear and resounding mandate' for pushing ahead with vaguely defined plans to 'create a new South Africa' within the next five years.

The months leading up to the election saw unprecedented flux and change in South Africa's domestic politics against the background of progress towards peace in Angola and Mozambique and independence for Namibia. The ruling National Party's original intention to go for general elections early in 1989 to take advantage of the improved international situation, but before the economy felt the full impact of rising inflation and a declining gold price, were frustrated by President Botha's 'minor stroke' on 17 January. This unexpected event put the timetable back by six months so that the economy was clearly in trouble during the vital run-up to the election. Pressure on the balance of payments in particular forced the authorities to raise politically sensitive mortgage and other interest rates to stifle growth.

The buoyant economy, which registered 3.2 per cent growth in 1988, helped to defuse political tensions but fuelled inflation and sucked in imports at an unsustainable rate, given the cut-off from foreign funds and

KEY FACTS

Official title: Republic of South Africa

Head of state: State President Frederik Willem De Klerk

Ruling party: National Party of South Africa

Capital: Cape Town (legislative); Pretoria (administrative); Bloemfontein (judicial)

Official Languages: Afrikaans, English (Xhosa, Zulu, Sesotho widely spoken)

Currency: Rand (R) = 100 cents

Exchange rate: R2.65 per $ (Nov 1989 – commercial); R3.91 (financial)

Area: 1,221,037 sq km

Population: 33.8m (1988)

GDP per capita: $2,574 (1988)

GDP real growth: 3.2% (1988)

Inflation: 12.8% (1988)

Trade balance: $5.2bn (1988)

Foreign debt: $21.2bn (1987)

heavy debt repayment obligations. Growth for 1989 was expected to be around 2 per cent at most, although the increasingly dynamic parallel or 'black' economy, in both senses of the world, continued to ensure that overall economic activity and unemployment remained at a much higher level than the depressed 1985-86 period which contributed massively to the political unrest in which over 2,500 died.

Foreign affairs

Before President Botha's stroke the ruling National Party was busy clearing the decks for an election to take advantage of the mood of near-euphoria which swept the country after the successful conclusion of the complex negotiations which ended South African involvement in the Angolan civil war and opened the path to independence for Namibia. The peace talks, which began in London in May 1988 and progressed through Cairo, Geneva and Brazzaville, culminated in a tripartite agreement signed by representatives of Angola, Cuba and South Africa in New York on 22 December 1988. One of the most important parts of the agreement was the decision to set up a special 'Joint Commission' which was given authority to monitor all aspects of the agreement and the linked Namibian independence process, including the withdrawal of South African forces from Angola and the phased withdrawal of 50,000 Cuban and other Soviet bloc forces and military advisers by the middle of 1991.

The Joint Commission was set up at Pretoria's request and underlined the new spirit of great power co-operation in the solution of regional problems. It effectively brought Washington and Moscow into Southern Africa as co-guarantors of the peace. Its value was proved in April 1989 when the whole carefully negotiated peace package risked falling apart as some 2,000 armed guerrillas of the South West Africa Peoples Organisation (Swapo) slipped illegally across the frontier from Angola into Namibia. They were intercepted by heavily armed patrols of the South West Africa Police (SWAPOL). Over the next few weeks over 330 Swapo infiltrators and 30 police were killed in a series of bitter skirmishes in the Ovamboland war zone. Armed units continued to come over the border while Swapo leaders in exile were variously either denying that any infiltration had taken place or claiming that returning guerrillas had no hostile intention and were merely looking for the nearest United Nations Transition Assistance Group (UNTAG) forces in order to give themselves up.

A week after the first incursion all five members of the Joint Commission met in emergency session at Mount Etjo, 250 km from the Namibian capital, Windhoek, to discuss the crisis. There it soon became clear that neither the Russians nor the Cubans were amused at Swapo's tactics. Shortly thereafter Swapo forces were repatriated under UN auspices beyond the Angolan 16th parallel, at least 150 km from the Namibian border. The independence process, culminating in elections to a constituent assembly on 1 November 1989, was put back on course.

The new prospects for regional peace opened up by the linked Angolan and Namibian negotiations were accompanied by agreement on two important economic development projects in which South Africa is closely involved. These are the $2bn Lesotho Highland water scheme and the R950m Sua Pan Soda Ash project in Botswana in which AECI, a joint venture between Anglo American Corporation and Britain's ICI, has a major stake. Several other black states, including Zambia, Mozambique and several francophone black African countries have also shown greater interest in improving political and economic links with South Africa in recent months.

President Kenneth Kaunda of Zambia, presiding over a rapidly disintegrating economy and facing food riots and other dangerous challenges to the stability of his regime, made clear his desire for closer relations both by his public recognition, at a royal function in Swaziland, of South Africa's legitimate role in Africa and his desire for talks with Mr F. W. de Klerk. Recognition of Pretoria's potentially vital role in economic recovery also played a major part in reviving the rapprochement between Mozambique and South Africa which faltered dramatically after the March 1984 Nkomati accord failed to end clandestine South African support for the Renamo rebels. Closer ties were underlined when Mr de Klerk flew to Maputo at the end of July 1989 for top level talks with President Joaquim Chissano only days before the Frelimo Party congress. The latter paved the way for further co-operation by quietly dropping Marxist–Leninist goals from its party programme, thus legitimising the significant shift back towards more pragmatic economic, social and foreign policies which was under way before the death of President Samora Machel in October 1986 but which has accelerated since.

In a significant gesture towards Mr de Klerk personally President Chissano said that he now accepted Pretoria's assurances that it was no longer supporting Renamo, while reserving judgement on the activities of clandestine supporters within South Africa and abroad. Both sides agree that no dramatic improvement could be expected in the Mozambican situation until the civil war ended and the security situation was brought back to normal. Pretoria is still committed to restoring power from the Cabora Bassa dam, put out of action by Renamo's destruction of the power lines. But the immediate priority appears to be securing and improving communications with Maputo and clearing the southern part of the country of bandits before gradually extending the secure area further north. Meanwhile Pretoria has called on Renamo to come to the negotiating table to end the civil war through internal reconciliation – the same policy being urged on UNITA and the MPLA in Angola.

It is too early to state with confidence that black Africa is ready to reverse decades of political opposition to closer economic ties: key countries like Zimbabwe are still awaiting clearer signs of domestic political progress. But there is certainly great awareness in Pretoria that the key to re-acquiring international respectability lies through Africa and that the future of regional economic as well as political co-operation hinges on the evolution of South Africa's domestic politics. Meanwhile South Africa's trade with black Africa continues to grow apace as the cheap rand, combined with products acclimatised for African conditions and available through relatively quick, cheap and reliable transport links ensures a major competitive advantage for South African-made goods and services.

On a more strategic level senior executives from the electricity corporation Eskom are continuing with their long-term aim of participating in a regional power grid which would link South Africa with its neighbours and stretch as far north as the Zaïre copper belt. Given South Africa's importance as a consumer as well as producer of electric power, Eskom's long-term aim is to encourage countries like Swaziland, Botswana and Zimbabwe to build power stations on their extensive but under-utilised coal reserves for sale to South Africa, thus increasing the interdependence of the region and providing income for increased multi-lateral trade in a region which after decades of war and economic collapse now faces improved prospects for peace and economic recovery.

Domestic politics

Meanwhile prospects for substantial domestic politi-

KEY INDICATORS

	Unit	1984	1985	1986	1987	1988
Population	m	30.9	31.6	32.3	33.0	33.8
Gross domestic product (GDP)	R bn	105.8	124.2	145.6	168.5	200.1
GDP per capita	R	3,424	3,930	4,508	5,106	5,920
Inflation	%	11.5	16.3	18.7	16.1	12.8
GDP real growth	%	5.1	–0.8	2.2	0.1	3.2
Share prices	1985=100	94	100	128	188	148
Agricultural output	1979-81=100	89	95	97	102	104
Mining production	1985=100	100	100	96	92	93
Gold	1980=100	101.8	99.8	94.5	88.2	–
Diamonds	1980=100	100.8	103.8	103.0	95.1	–
Iron ore	1980=100	92.4	93.3	89.1	77.4	–
Copper	1980=100	100.6	96.9	96.7	109.8	–
Manganese	1980=100	55.8	63.6	66.0	51.0	–
Coal production	m tonnes	159	173	175	–	178
Manufacturing output	1985=100	105	100	100	104	110
Industrial chemicals	1980=100	131.6	130.8	136.9	132.0	–
Food	1980=100	133.0	138.0	139.5	145.6	–
Iron and steel	1980=100	148.6	155.6	170.1	165.0	–
Machinery (except electrical)	1980=100	124.4	99.9	99.8	95.3	–
Motor vehicles	1980=100	110.3	75.0	70.9	81.8	–
Exports	R bn	25.4	36.7	42.1	43.8	49.2
Gold	R bn	11.7	15.5	16.7	17.8	19.3
Diamonds	R bn	1.4	2.2	–	–	–
Imports	R bn	21.7	23.0	26.9	28.7	39.5
Balance of trade	$bn	2.2	5.8	7.2	7.2	5.2
Current account	$m	–1,588	2,622	3,152	3,002	1,272
Tourist arrivals	'000	792	728	645	703	–
External debt	$bn	24.3	23.5	22.6	22.6	21.2
Exchange rate	R per $	1.4	2.2	2.3	2.0	2.3

* estimated figure

cal change in South Africa appear far more dramatic than they have for 41 years. President Botha's stroke injected a new dimension into the political calculus. The news threw the ruling National Party into turmoil. For years it had been forced to put up with the increasingly imperious and autocratic style of a president given sweeping executive powers by the 1984 constitution. But it had no way of getting rid of him, until on 2 February 1989 Mr Botha made his fatal mistake.

On that day he announced his surprise decision to resign as leader of the National Party. He apparently failed to take sufficiently into account the fact that leadership of the National Party was the font from which all his other powers ultimately flowed. His plan was to lighten his own burdens of office by leaving party and governmental matters to the new party chairman. His intention was to concentrate on a statesmanlike role above party aimed principally at trying to persuade 'moderate' black leaders to take part in the government's proposed national council and other bodies. Until now all such government offers have been rejected by key players like Zulu Chief Mangosuthu Buthelezi, black mayors, councillors, businessmen, community leaders and others. Aware of the minimum demands of their constituent power bases black leaders continue to call for the abolition of the tri-cameral parliament, the release of jailed leaders, the lifting of bans on organisations and other conditions like the lifting of the state of emergency before even considering negotiations about an agenda.

Instead of lifting the state of emergency re-imposed in June 1986 the government re-imposed it virtually unchanged for another 12 months when the annually reviewed regulations came up for renewal in mid-June 1989. Once again the government cited the persistence of what it calls 'a revolutionary climate' just below the surface for renewing the tough package of personal and media restrictions. It also claimed that the emergency regulations were needed to ensure a relatively peaceful run-up to the elections. Prior to re-imposing the emergency, however, the government showed unexpected flexibility when it commuted the death sentence of several convicted political prisoners, including the so-called Sharpeville Six, and released several hundred detainees after they went on hunger strike to demand that they either be charged or released. By mid-year most of the 30,000 people, mainly blacks, who had been detained for varying periods under the emergency or security laws had been released, several hundred of them under tight restrictions designed to prevent them functioning as opposition leaders and organisers.

On the negative side of the ledger, however, the murder of Dr David Webster, a well-known university lecturer and human rights activist by unknown gunmen in early May added a new victim to the long list of unsolved crimes stretching back decades. The inability of the police force to trace such killers strongly suggests the existence of a clandestine hit-squad targeted at left wing 'radicals', church, trade union, community and other organisations actively involved in the struggle against apartheid.

President Botha

Meanwhile President Botha's plan to lighten the burdens of his office while retaining ultimate control might have worked in a limited way had he been able to secure the election of his protégé, the relatively youthful Finance Minister Barend du Plessis, as the new leader of the party. Instead the party caucus, on the second ballot, elected the 53-year-old Transvaal party boss Frederik Willem de Klerk as the new party leader by a narrow 69 to 61 majority. It was the signal for the start of a three-month long power struggle which severely damaged the electoral prospects of the National Party. It formally ended in a victory for Mr

COUNTRY PROFILE

Political system: Republic with legislative power vested in tricameral parliament, with one house for each racial group (excluding blacks). House of Assembly (178 members) represents whites, House of Representatives (85 members) represents people of mixed race, and House of Delegates (45 members) represents Indians. Most members are directly elected by their own electorate; the rest are either appointed by the State President or elected by other members, for five years. Executive power is vested in the State President, on advice from ministers' councils (one for each house), acting with a cabinet which he appoints. System of apartheid operating throughout country.

Political parties: House of Assembly (white chamber): the ruling party is the National Party of South Africa. The Conservative Party is the official parliamentary opposition. The other party is the newly-formed Democratic Party. Coloured chamber: Labour Party, Freedom Party and United Democratic Party. Indian chamber: Solidarity, National People's Party, Progressive Independent Party and People's Party of South Africa. Extra-parliamentary groups include the African National Congress, Pan Africanist Congress and United Democratic Front (these three movements are banned in South Africa), the Herstigte Nasionale Party (reconstituted National Party), the Afrikaner Weerstandsbeweging and Inkatha.

Population: 14 per cent of South Africa's population of 36.8m (1988 estimate, including the so-called black 'homelands') are classified as white, 11 per cent as 'Coloured' and Asian, and 74 per cent as Black. Population densities are especially high in the urban centres and in some 'homelands'.

Main cities/towns: Pretoria, Johannesburg, Cape Town, Bloemfontein, Durban, Pinetown, Port Elizabeth, Kimberley, Pietermaritzburg, East London, Welkom, Vereeniging.

Language: Afrikaans and English are the official languages; Xhosa, Zulu and Sotho are widely used African languages.

Media
Press: Main daily newspapers include *Business Day* (English), *The Star* (English), *Die Beeld* (Afrikaans), *The Argus* (English), *The Citizen, Cape Times, Sowetan.* Largest Sunday papers are *Sunday Times, Rapport* (Afrikaans). Weeklies include *Financial Mail, Finance Week, Farmer's Weekly, Finansies en Tegniek* (Afrikaans).
Radio: Commercial radio services broadcast throughout country in various languages.
Television: With the exception of the M-Net, television is a state-influenced monopoly (SA Broadcasting Corporation) comprising four networks: TV 1 – English and Afrikaans – commercial advertising, since 1978; TV 2 and TV 3 – various African languages – introduced 1983, commercial advertising; TV4 – entertainment and sport – English and Afrikaans. M-Net – independent network introduced March 1986 – popular, with about 300,000 subscribers.

External trade: The 1987 R6bn current account surplus shrank to R3bn in 1988 as higher growth and spending drew in more imports while gold exports remained relatively static. In January-March 1988 the first quarterly balance of payment deficit in more than three years was recorded. For political reasons, capital inflow cannot be taken for granted.
Exports: Gold (38 per cent of total value in 1988); structure of merchandise exports: food, beverages, base metals (23%), precious metals and stones (14%), tobacco (4%). Main destinations: To protect itself against the possibility of sanctions, the South African government no longer lists its trading partners.
Imports: (1988) Structure of merchandise imports: machinery and transport (46%), chemical products (11%). Main sources: To protect itself against the possibility of sanctions, the South African government no longer lists its trading partners.

Agriculture: Although the relative importance of agriculture has declined – currently accounting for only 5-6 per cent of GDP – the country remains a net food exporter. Output has roughly kept pace with population growth, though imports of wheat grains were needed during drought years to meet demand. Sector is highly diversified, and a wide variety of food crops are successfully grown. Agricultural land, like residential areas, is racially zoned. Main crops include maize, sugar, citrus fruits, sorghum and groundnuts. Livestock farming also important.

Industry: Manufacturing has developed into the most important sector of the economy, and currently contributes about 25 per cent to GDP. Expansion mainly due to increased domestic demand and policy of import substitution. Output dominated by engineering and metal products, especially steel which is to be further developed with the South African Iron and Steel Corporation's (ISCOR) current programme of expansion. Other major growth areas have been automobile production and the chemical industry, which currently accounts for 20 per cent of all manufacturing. Although food and tobacco processing remain of great importance, their share of total output has fallen significantly. The slow growth in manufacturing output in recent years is largely attributable to the fall in world demand, low levels of capacity utilisation, reduced investment and disinvestment by foreign companies, and labour problems. The government is attempting to alleviate some of these with a programme of industrial decentralisation, offering various benefits to industries willing to move to the so-called 'black homeland' areas. Expenditure on consumer durables increased very strongly in 1988.

Mining: Sector accounts for about 13 per cent of GDP and a large proportion of export earnings. Foreign exchange earnings from coal exports increased by 19 per cent in 1988. Gold production declined from 1984 to 1987 but increased in 1988. More than 60 other minerals are produced and these include chromium, diamonds, uranium, manganese, titanium, vanadium, zirconium and fluorspar.

Energy: Based on vast coal reserves (estimated at 120bn tonnes) which provides 97 per cent of locally generated electricity and about 20 per cent of liquid fuel requirements – after oil-from-coal conversion by South African Coal, Oil and Gas Corporation (SASOL). However, the virtual absence of indigenous oil reserves means the country is still heavily dependent on imports of petroleum products, particularly diesel, although SASOL has somewhat reduced this dependence by expanding its oil-from-coal conversion programme. The bulk supply electricity company, ESKOM, has an installed generating capacity of about 33GW. At present, South Africa has only one nuclear plant – Koeberg, near Cape Town, rated at 1,840 MW. Hydro-electric capacity is approximately 3,000 MW.

Membership of international organisations: GATT, IDA, IFC, IMF, IWC, UN, WHO, WIPO, World Bank.

de Klerk on 6 April when President Botha announced that a general election, which he had earlier opposed in a televised interview, would take place on 6 September and hinted that he would not seek re-election as state president. This left the way clear for Mr de Klerk to take over the combined roles of head of party, government and state in October – provided the National Party retained its pre-eminent role in white politics and continued to dominate the 88-strong presidential electoral college which appoints the new president after general elections.

A new leadership for the National Party and the imminence of elections had a galvanising effect on black and white politics throughout the year. The combination of progress towards Namibian independence and the emergence of a new leader from a younger and presumed 'post-apartheid generation' put Mr de Klerk at the receiving end of invitations from several European leaders. With an election looming the internationally inexperienced leader took up the offers with alacrity. For 10 days at the end of June he talked with senior European politicians, bankers and businessmen and gave them the general outlines of the National Party's new policy under his leadership. He made clear that while black/white power sharing negotiations would be on the agenda after the elections the party remained wedded to what sounded like an essentially neo-apartheid vision of politics based on ethnic groups.

Among his principal hosts were Mrs Margaret Thatcher, the British Prime Minister, the principle Western opponent of sanctions and disinvestment, West German Chancellor Helmut Kohl, Portuguese Premier Anibal Cavaco Silva and the veteran Italian politician Giulio Andreotti, who became the premier of Italy's forty-ninth post-war government six weeks later. It was originally expected that the new leader would crown his European tour with a pre-election visit to the United States for talks with President Bush, senior officials and bankers. Opposition to the visit by 100 congressmen and anti-apartheid and pro-sanctions lobbies led to the visit being cancelled, ostensibly at Pretoria's request, citing pressure of the election campaign.

On his return from Europe Mr de Klerk launched his party's election campaign at a special meeting of the Federal Congress in Pretoria. After years of seeking to be all things to all whites Mr de Klerk made clear that in these crucial elections the National Party intended to fight the Conservative Party by telling voters that its policies would lead to economic ruin and foreign hostility on an unprecedented scale. The other leg of its strategy was to woo the 70 per cent of white voters who, with varying degrees of enthusiasm, support the idea of greater democracy and political power sharing between all sectors of South Africa's ethnically and culturally diverse society.

The main novelty of the white election campaign was the launch of the new Democratic Party at an American-style congress in Johannesburg in early April 1989. The Democratic Party is essentially an amalgamation of the three principal white 'liberal' opposition groups who between them could hope to attract the votes of between 20 to 25 per cent of the electorate. The mainstay of the new party in financial and organisational terms is the former Progressive Federal Party (PFP), which went down to a humiliating defeat in the May 1987 whites-only elections when it was replaced as the official opposition by the right wing Conservative Party, led by Dr Andries Treurnicht. Its other components are the former National Democratic Movement (NDM) led by Wynand Malan, a dissident Afrikaner who left the National Party two years ago, and the Independent Party, which was formed in 1987 essentially as a vehicle for the political ambitions of Dr Denis Worrall, the former Nationalist senator and ambassador to London.

The emergence of the Democratic Party, which appeared to be well-funded and stiffened by a much tougher breed of younger politicians like Tony Leon, the former PFP Johannesburg City leader, underlined the way in which the long-ruling National Party faced an unprecedented challenge from both left and right at the September elections. The Democrats went into the election hoping that the Nationalists would lose enough seats on both right and left to leave a hung parliament in which no single party would have a majority. At that point, Democrats told their electorate, it would offer to form a ruling coalition with those Nationalists ready for a quick start to black/white power sharing negotiations and the abolition of remaining apartheid laws like the Population Registration Act, the Group Areas Act and the Separate Amenities Act.

On the opposite end of the political spectrum the Conservative Party went into the elections with an impressive local and parliamentary power base and determined to translate its voting strength into firm seats. The Conservatives gained only 22 seats after polling nearly nearly 27 per cent of the white vote at the May 1987 general elections thanks to the first past the post voting system. But the heavy concentration of Conservative votes in a relatively few Transvaal and Free State constituencies left them breathing down the necks of sitting Nationalist MPs in dozens of marginal seats vulnerable to a relatively small further swing. Such a swing occurred 15 months later at the October 1988 municipal elections when the Conservative Party raised its share of the white vote to over 33 per cent. That gave it control over more than 100 small towns and rural *platteland dorps* in wide areas of the Transvaal and the Orange Free State, as well as a stronghold in several blue collar working class areas on the reef and in Pretoria and Johannesburg.

Both the major opposition parties, like the ruling Nationalists, had their weaknesses. Having been in power without interruption since 1948 the National Party has acquired an arrogant image and been tainted with several corruption scandals, both in the homelands and the Republic, involving ministers, MPs, senior officials and their families. As for the new Democratic Party, it went into the election without a clear leader as none of the leaders of its constituent parts would agree to stand down for the other. Thinly disguised personal ambitions and rivalries masked ideological differences between various factions of the new party.

Right wing voters meanwhile were embarrassed by the alleged sexual and drinking habits of Eugene Terre'blanche, leader of the paramilitary Afrikaner Resistance Movement (AWB) whose highly publicised escapades with a female journalist shocked many of the more puritanical bible-punching Afrikaners who until then had worshipped the blue-eyed leader with his nazi-style oratory and penchant for nazi-style uniforms, symbols and fancy dress.

Away from the fringes of Afrikanerdom, the mainstream Conservative Party supporters were also given something to think about by the black consumer boycotts which badly hit the local economies of Conservative controlled towns like Boksburg and Carletonville. Blacks, now more prosperous, more mobile and more conscious of their rights and dignities, simply refused to shop at stores in the openly bigoted white towns. Instead they took black-owned taxis to nearby shopping centres rather than accept the attempt of Conservative town councils to re-introduce petty apartheid by putting up 'whites only' signs in parks and public places.

But the most dramatic and unpredictable event in a year of surprises took place in early July when President Botha invited Nelson Mandela, the jailed leader of the African National Congress (ANC), to the Tuynhuys presidential complex in Cape Town for 45 minutes of tea and sympathy. The government had long been looking for a way of releasing its most distinguished prisoner, a man seen as a hero, martyr and symbol of the anti-apartheid struggle by millions of South Africans and countless more in the wider world. He had been transferred from Robben Island to Pollsmoor prison near Cape Town as far back as 1982. In 1988 he was taken from the prison to a luxury clinic to be treated for tuberculosis. On his discharge from the Constantiaberg clinic he was taken to the Victor Verster Prison in Paarl, 50 km from Cape Town. There he was not placed in a cell but given a former prison warder's bungalow, complete with swimming pool and garden, in the grounds of the heavily-guarded prison complex.

It was at this bungalow that, ten days after meeting President Botha, Mr Mandela was permitted to meet five of his old friends and co-accused at the 1964 Rivonia treason trial: Walter Sisulu, Ahmed Kathrada, Elias Motsoaledi, Raymond Mhlaba and Andrew Mahlengi. A few days later he entertained his wife, Mrs Winnie Mandela, and close family at a reunion to celebrate his 71st birthday.

The cabinet had agreed in principle to top level meetings with Mr Mandela but the timing of the meeting with President Botha appeared to take many in the cabinet, including Mr de Klerk, by surprise. It was widely interpreted as another personal initiative by a brooding and resentful president anxious to establish his place in the history books as the man who got black/white negotiations on the road, and to make life difficult for his successor at the same time. Certainly such a gesture, with its dramatic impact on black expectations and conservative white fears, was not expected less than two months before such a crucial election.

It was a gesture which exploded like a grenade into pre-electoral white and black politics. The Democratic Party welcomed the meeting, partly because it made it much more difficult for the National Party to accuse the Democrats of being soft on the ANC. The Conservative Party seized the opportunity to tell its supporters that here was further proof of the National Party's willingness to sell out the white man and negotiate with 'terrorists.' But the most dramatic impact was on black politics where leaders of the so-called 'mass democratic movement', a loose amalgamation of trade union, community and church groups formed after the virtual banning of the United Democratic Front (UDF) in February 1988, initially reacted in disbelief to the news. Early statements reflected concern that the government might have extracted concessions out of Mr Mandela. Privately several younger and more radical black leaders voiced concern that the veteran leader, once freed, might be willing to adopt a more flexible position in any future political negotiations than the traditional ANC demand for the straight transfer of power to the black majority in a unitary state. In a subsequent statement Mr Mandela made clear that his much-quoted 'support for peaceful development in South Africa' did not imply any change in his long-held view that this could only be achieved through the participation of the mass democratic movement in general and the ANC in particular.

Whatever the finer doctrinal points, the fundamental point made by the Botha–Mandela meeting was that the timetable for future negotiations, even the preliminary 'talks about talks', had suddenly moved a lot closer to reality after decades of vague discussion. It raised expectations that getting such talks off the ground would be a major task of whatever new government emerged from the general elections.

Whether future developments justify such optimism remains to be seen. But a major first step towards an apparent re-alignment of black political forces to prepare for such an eventuality appeared to have been signalled in April when Mr Mandela sent a personal letter to Zulu Chief Mangosuthu Buthelezi, president of the Inkatha movement. In it Mr Mandela called for a restoration of the good relations which existed between the ANC and Inkatha and their respective leaderships in the 1970s. This was the signal which made it possible for Inkatha and leaders of the Mass Democratic Movement to meet in June and July in Durban to try and bring about an end to the violence in Natal. Over 2,000 people are believed to have been killed and tens of thousands either injured or made homeless in over three years of fighting in the townships around Durban and Pietermaritzburg which, at some levels, is seen as a proxy struggle for land, hearts and minds between more traditional Inkatha-supporting chiefs and UDF-supporting comrades.

The economy.

While unpredictable political developments made 1989 an astonishing year on the formal political front, economic and financial developments were less spectacular. Government's inability to curb its expenditure, boosted by politically inspired pay rises for civil servants before the elections, coincided with a steady decline in the gold price from around $500 per ounce in 1987 to just over $375 per ounce by end-July 1989 after falling to just over $350 in June. Gold still accounts for over 40 per cent of South Africa's hard currency income and the lower gold price heavily impacted both on the currency and on an industry which in the short space of four years has been transformed from one of the world's cheapest gold producers to one of the dearest. Higher black wages and low productivity gains in a generally inflationary environment have placed the gold mines in a cost-price scissor grip. As a result nearly a third of the gold mines are now marginal or loss-making. Management responded by ordering mass retrenchments at Gencor and other mining houses. Management has opted to close specific loss-making shafts and leave unprofitable ore in the ground rather than risk wholesale pit closures.

Heavy investment also continues in mine mechanisation to raise labour productivity. The industry is rapidly changing from its traditional dependence on a cheap, migrant workforce towards reliance on a smaller, better paid and more highly skilled labour force, with major social effects both inside South Africa and the region as single sex migrant hostels are gradually phased out. They will eventually be replaced by owner-occupied family homes for the next generation of skilled black miners whose career prospects have been opened up by abolition of racial job restrictions.

Fortunately for the balance of payments, however, the declining gold price coincided with strong demand at higher world prices for most other mineral exports including chrome, nickel, vanadium, coal and platinum while good rains for the second year running meant bumper maize and other export crops. Half the 11m tonne maize crop last year was available for export while higher world prices for sugar, wool and other agricultural commodities boosted farm exports to a record R4bn.

Higher overall exports showed up in better than expected trade figures for the first half of 1989 when the surplus rose to R5.32bn from R4.13bn in first half 1988. Exports rose to a record R27.37bn while imports rose to R22.05bn. The continuing high level of imports reflected the recovery of private domestic fixed investment which rose by 16 per cent in 1988 after five years of sharp decline. But import growth was expected to fall over the second half as the impact of 20 per cent prime and mortgage rates and government hire purchase and other credit restrictions bit deeper into consumer pockets already ravaged by inflation which touched 16 per cent and was still rising at mid-year.

As part of a series of measures aimed at cooling an economy which recorded an unsustainable 7 per cent rise in Gross Domestic Expenditure in 1988, the monetary authorities introduced a package of interest rate hikes, forced loans and hire purchase restrictions in May 1989. Finance Minister Barend du Plessis, announced a one point rise in bank rate to 17 per cent, a 1.5 per cent rise in overnight funds for the banking system, a R750m forced loan on company profits and a 5 to 10 per cent rise in minimum hire purchase deposits. The commercial bank prime rate and politically sensitive mortgage rate rose to 20 per cent in the wake of the new package.

Despite reluctance to squeeze the economy in the run-up to the election the authorities had no choice given the need to maintain a strong current account surplus in order to service the $21bn foreign debt and cope with the cut-off in foreign funding since August 1985 when the money markets were closed after US banks tried to call in their loans. This sparked off a payments crisis which forced the authorities to declare a partial debt moratorium and re-introduce the two-tier financial and commercial rand system. Over $14bn of the total $24bn debt was frozen under the terms of the so-called debt-standstill but some $6bn has since been repaid and over $3.5bn has been converted from frozen short term to longer-term debt, re-payable over 10 years. Bankers, having twice rescheduled the debt since 1985, are awaiting Pretoria's formal request for another re-scheduling agreement. This is expected to be signed before the current three-year agreement, signed in March 1987, expires in June 1990.

Anti-apartheid groups in the United States and Europe have made it clear that they will campaign against creditor banks who, on purely technical grounds, are otherwise perfectly willing to roll-over performing loans which are regularly serviced and carry a higher than market coupon as a sweetener. South African bankers meanwhile hope that political improvements at home and abroad will eventually allow South Africa to re-open its blocked credit lines and permit the economy to expand at the 5 to 6 per cent annually of which it is capable, and which it needs if unemployment is to be absorbed and living standards raised.

This is not an argument accepted by the various sanctions and disinvestment lobbies who continue to argue that Pretoria will only abandon apartheid if forced to do so by economic pressure. Such pressures remain strong in parts of Europe and particularly the United States where the South African question is closely tied in with internal political manoeuvring apparently related in only a tenuous way to changing realities on the ground in South Africa.

Under pressure from such lobbies, and legislative changes such as the 1987 Rangel Amendment which scrapped the US double taxation agreement with Pretoria, American businesses continued to disinvest throughout 1989 as did several British and other European companies. The biggest single disinvestment was by Mobil oil which sold its $400m refinery and downstream distribution interests for a reputed $150m to the Gencor mining, industrial, paper and energy group. Gencor is controlled by the Afrikaner financial and insurance group Sanlam which welcomed the purchase as a cheap addition to its rapidly expanding energy division.

Mobil was one of six household name US corporations to leave in the first half of 1989, the others being Hewlett-Packard, NCR, Control Data, IMS International and RJR Nabisco. The UK-controlled BP also announced the sale of its South African coal interests although that, like several other disinvestments, was part of a global re-structuring. After Mobil's departure only 131 US companies remained in South Africa. They include companies like Palmolive, Johnson and Johnson, Kelloggs and Caltex, all of whom now have to pay tax twice on their South African profits thanks to the Rangel amendment.

Partly as a result of disinvestment and sanctions, South African companies have acquired first class assets on the cheap while South African trade has shifted increasingly towards Asia, the 'Pacific Rim' countries, Latin America – and even Eastern Europe where Hungary and Poland are shortly expected to open trade missions, which reflects both improved relations with the Soviet bloc generally in the aftermath of the Angolan agreements and the communist world's desperate need for greater trade and new markets. It is just one more example of South Africa's capacity to surprise.

BUSINESS GUIDE

Time: GMT +2.

Climate: Sub-tropical, with regional variations. In Cape coastal area summers, from Dec to Mar, are warm and wet with temperature averaging 25-28°C; average winter temperature 17-20°C. In Pretoria and other highveld areas winters are warm and dry, while most rain falls during the hot summer. On the Natal coast humidity can be high during the summer, while the winter is drier with temperatures around 22-24°C. Most rain falls during the summer months.

Entry requirements

Passport: Required by all.
Visa: In general required by all – except UK, German FR and Irish passport holders, business or holiday visitors from Switzerland and Liechtenstein. (Note: there are certain regulations on entry, transit, re-entry etc which should be checked before travelling. Visitors intending to travel to Lesotho, Swaziland, Zimbabwe, Botswana, Bophuthatswana, Venda or Ciskei require re-entry visas before they will be allowed back into the country; allow at least a week. Also, holders of passports showing evidence of a visit to South Africa are liable to be refused entry to several other African countries.)
Currency: Import and export of local currency is limited to R200. Unlimited amount of foreign currency may be imported although amount must be declared, and re-export is limited to declared amount. Visitors must have sufficient funds to finance their visit and, if not in possession of a return ticket, enough money to buy ticket home.
Customs: Licences required for certain imports, including consumer goods; all items need certificates of origin and value.

Health precautions

Mandatory: Yellow fever and cholera vaccination certificates required if travelling from infected or endemic area.

Advisable: Malaria risk exists throughout the year in certain areas of Northern Transvaal, Eastern Lowveld and Northern Natal – prophylaxis should be taken for visits to these areas. Water precautions should be taken in rural areas.

Representation overseas: Athens, Berne, Bonn, Brasilia, Brussels, Buenos Aires, Canberra, Copenhagen, Guatemala City, The Hague, Helsinki, Hong Kong, Lilongwe, Lisbon, London, Luxembourg, Madrid, Monte Carlo, Montevideo, Oslo, Ottawa, Paris, Rome, Reykjavik, Santiago, Stockholm, Taipei, Tel Aviv, Tokyo, Vienna, Washington.

BUSINESS GUIDE

Air access
National airline: South African Airways.
Other airlines: Air Botswana, Air Malawi, Air Mauritius, Air Zimbabwe, Alitalia, British Airways, Comair, El Al, Iberia, KLM, LAM-Linhas Aéreas de Moçambique, Lesotho Airways, Lufthansa, Olympic Airways, Royal Swazi Air, Sabena, Swissair, TAP-Air Portugal, Transkei Airways, UTA, Varig, Zambia Airways.
Main international airport: Johannesburg-Jan Smuts (Code: JNB), 24 km from city; duty-free shop, bar, restaurant, buffet, bank, post office, shops, car hire, airport bus.
International airports: Cape Town-D.F.Malan (Code: CPT), 14 km from city; duty-free shop, car hire, bank, bar and restaurant. Durban-Louis Botha (Code: DUR), 16 km from city; duty-free shop, car hire, bank, bar and restaurant.

Surface access
Main ports: Cape Town, Durban, East London, Port Elizabeth. Richards Bay and Saldanha Bay are major bulk cargo ports.
Overland access: Road: Possible from Botswana, Zimbabwe, Namibia, Swaziland (travellers are generally advised to check regulations and conditions regarding entry by road with The Automobile Association of South Africa). Rail: Services from Mozambique, Zimbabwe, Namibia, Botswana.

Hotels: Wide choice available in main commercial centres. Advisable to make reservations well in advance, especially during Dec/Jan and around Easter. Hotels classified from one to five-star. All star-rated hotels are, in principle, open to all people regardless of race, but many smaller ones, especially in outlying areas, still segregate. 10 per cent service charge usually added to bill.

Electricity supply: Usually 220/230 V AC – but 220/250 V in Port Elizabeth and 250 V in Pretoria.

Car hire: Self-drive and chauffeur-driven cars widely available. An international driving licence is required unless visitor's national licence carries the photograph and signature of the holder. Driving is on the left. Speed limits: built-up areas 60 kph; country roads 100 kph; declared freeways and some main roads 120 kph. Heavy fines for speeding.

City transport
Taxis: Widely available in all towns. Cannot be hailed in the street and must be called from a rank. Fares within the city areas depend on distance and time, while longer distance fares are lower and are obtained on a quotation basis. 10 per cent tip is usual.
Buses: Services are operated in all main centres. Various pre-paid ticket schemes are in use. Some services are infrequent outside rush hours.

National transport
Air: South African Airways operates frequent flights to all major centres; independent operators provide scheduled and charter flights to other towns.
Road: Extensive network of tarred roads, including over 52,000 km of tarred roads linking main centres. Also, over 130,000 km of untarred roads – some of the remoter sections can become impassable in wet weather.
Buses: Inter-city services are operated by South African Transport Services, Greyhound, Citiliner and private companies.
Rail: Network of some 24,000 km with good services throughout the country. Reservations for express trains should be made well in advance. Two classes available, although visitors are advised to travel first class. Most long distance mainline trains have restaurant cars and all have sleeping accommodation (couchettes operated in both first and second class; bedding is charged). Named services include: Blue Train, three times a week (Pretoria-Johannesburg-Cape Town; sleeping accommodation, restaurant cars, air-conditioning, suites, staterooms available); Trans-Karoo, daily (Cape Town-Johannesburg); Trans Orange, once a week (Durban-Cape Town); Trans Natal, daily (Durban-Johannesburg).

Public holidays
Fixed dates: 1 Jan, 6 Apr (Founder's Day), 1 May (Workers' Day), 31 May (Republic Day), 10 Oct (Heroes' Day), 16 Dec (Day of the Vow), 25 Dec, 26 Dec.
Variable dates: Good Friday, Easter Monday, Ascension Day, Family Day.

Working hours
Government and business: (Mon-Fri) 0730/0830 – 1600/1700. Some businesses have extended hours.
Banking: (Mon-Fri) 0900 – 1530; (Sat) 0830 – 1100. Some banks have extended hours.
Shops: (Mon-Fri) 0830 – 1700; (Sat) 0830 – 1300. Most shops are closed on Sunday.

Telecommunications
Telephone: Dialling code for South Africa: prefix + 27 followed by area code (Cape Town 21, Durban 31, Johannesburg 11, Pretoria 12, East London 431) followed by subscriber's number. IDD access code from South Africa: 091. IDD available to most countries. Total number of telephones, 4.4 million; 96 per cent automatic service, of which 32% are in Johannesburg, 13% in Cape Town, 10% in Durban and 10% in Pretoria.
Telex: Automatic exchanges at main centres. Over 31,000 telex subscribers recorded; automatic service to numerous countries including: Australia, Brazil, Canada, German FR, Hong Kong, Japan, New Zealand, UK, USA, much of Western Europe; public facilities available at main Post Offices in Cape Town, Durban, Pretoria, Johannesburg (hours: 0830 – 1700; exceptions: Sat 0830 – 1200; Sun closed).

Banking
Central bank: South African Reserve Bank, 370 Church Street, PO Box 427, Pretoria (tel: 3133911; tx: 322136, 320455, 322137).
Main centre: Johannesburg.
Head offices: Banco Borges & Irmao, Banco de Credito Commercial e Industrial, Bank Julius Bär, Bank Leumi le Israel, Bank of Lisbon & South Africa, Bank of Tokyo, Banque Worms, Bayerische Hypotheken- und Wechsel-Bank, Bayerische Landesbank Girozentrale, Bayerische Landesbank International, Bayerische Vereinsbank, Berliner Handels-Gesellschaft Frankfurter Bank, Chase Manhattan Overseas Corporation, Commerzbank AG, Compagnie de Banque et de Crédit SA, Credit Commercial de France, Credit Lyonnais, Credit Suisse, Deutsche Bank AG, Dresdner Bank AG, First National Bank, International Bank for Swaziland, Kleinwort Benson, Kredietbank NV, Kredietbank SA, Rea Brothers (Guernsey Ltd), Société Générale, Swiss Bank Corporation, Union Bank of Switzerland. N.B. Foreign bank representation is subject to change.

Representation in capital: Argentina, Australia, Austria, Belgium, Bolivia, Brazil, Canada, Chile, China PR, Denmark, Finland, France, German FR, Greece, Guatemala (Cape Town), Iceland, Israel, Italy, Japan, Malawi, Mozambique, Netherlands, Norway (Cape Town), Paraguay, Portugal, Spain, Sweden, Swaziland, Switzerland, Taiwan, UK, USA, Zimbabwe.

Hotels: Cape Town
Alphen, Constantia 7848, P.O. Box: 35 (tel: 741011; tx: 526195).

Cape Sun, Strand Street, 8001, P.O. Box: 4532 (tel: 238844; tx: 522453).

Capetonian, Pier Place, Heerengracht 8001, P.O. Box: 8012 (tel: 211150; tx: 520000; fax: 252215).

Carlton Heights, 88 Queen Victoria Street, 8001 (tel: 231260).

Claridges International, 47 Main Road, Granpoint, 8000, P.O. Box: 251 (tel: 441171; tx: 527860).

De Waal Sun, Mill Street Gardens, 8000, P.O. Box: 2793 (tel: 451311; tx: 520653).

Holiday Inn, Eastern Boulevard, Melbourne Road, 8000, P.O. Box: 2979 (tel: 474060; tx: 520576).

Holiday Inn, Belville, Cross Street, 7535, P.O. Box: 233 (tel: 978111; tx: 520097; fax: 978111).

Metropole, 38 Long Street, 8001, P.O. Box: 3086 (tel: 236370; tx: 520295).

Mount Nelson, 76 Orange Street, 8000, P.O. Box: 2608 (tel: 231000; tx: 527804; fax: 247472).

Newlands Sun, Private Bag (Newlands), 7700 (tel: 611105; tx: 520686).

President, Beach Road, Sea Point, P.O. Box: 62 (tel: 441121; tx: 5717973).

St George's, St George's Street, 8000, P.O. Box: 5616 (tel: 214190808; tx: 52533).

Town House, 60 Corporation Street, P.O. Box: 5053 (tel: 457050; tx: 520890; fax: 453891).

Hotels: Johannesburg
Carlton, Main Street, P.O. Box: 7709 (tel: 3318911; tx: 486130; fax: 3313555).

Down Town Holiday Inn, 88 Plein Street, 2000, P.O. Box: 11026 (tel: 281770; tx: 484092).

Holiday Inn Jan Smuts, P.O. Box: 388 (tel: 9751121; tx: 749534; fax: 9755846).

Johannesburg Sun & Towers, P.O. Box: 535 (tel: 297011; tx: 482327).

Johannesburger, Corner Twist and Wolmarans Street, P.O. Box: 23566 (tel: 7253753; tx: 422314).

Rand International, 290 Bree Street, P.O. Box: 4235 (tel: 292724; tx: 485349).

Sandton Sun, Fifth Street, Sandhurst, P.O. Box: 784902 (tel: 7838701; tx: 430338).

Victoria, 25 Plein Street, P.O. Box: 2463 (tel: 281530; tx: 485963).

Hotels: Pretoria
Arcadia, 515 Proes Street (tel: 269311/6; tx: 320496).

Boulevard, 186 Struben Street, 0001, P.O. Box: 425 (tel: 264806; tx: 322145).

Burgerspark, Cnr. van der Walt/Minnaar Street, P.O. Box: 2301 (tel: 3227500; tx: 322525).

Other useful addresses
Bureau for Information, Private Bag X745, Pretoria 0001 (tel: 3253160).

Cape Town Chamber of Commerce, PO Box 204, Union Castle Building, St Georges Street, Cape Town 8000 (tel: 232323; tx: 527566).

Central Statistical Service, Steyn's Building, 274 Schoeman Street, Pretoria 0002, or Private Bag X44, Pretoria 0001 (tel: 3108911).

Chamber of Mines of South Africa, PO Box 61809, Marshalltown, Johannesburg 2107 (tel: 8388211; tx: 487057).

Development Bank of Southern Africa, PO Box 1234, Midrand (tel: 3133911; tx: 422666).

Industrial Finance Corporation of South Africa, PO Box 8575, Johannesburg 2000 (tel: 8363536).

Industrial Development Corporation of South Africa, 19 Fredman Drive, PO Box 784055 Sandton 2146 (tel: 8831600; tx: 427201).

Sasol 1, PO Box 1, Sasolburg 9570 (tel: 6889111; tx: 487938).

Sasol 2, PO Box 1, Secunda 2302 (tel: 401111; tx: 482702).

South African Reserve Bank, 370 Church Street, PO Box 427, Pretoria (tel: 3133911; tx: 322136, 320455, 322137).

Spain

Tom Burns

At the end of October 1989 Spanish Prime Minister Felipe González was returned with a reduced majority for a third successive four-year term in office. His Partido Socialista Obrero Español (PSOE), the Spanish Socialist Workers Party, gained 39.5 per cent of the vote to obtain 176 seats in the 350-member congress of the Cortes (parliament), down from the 184 seats that the party won in 1986. The result once more gave Mr González, albeit by a hair's breath, an overall majority and will allow him to govern without recourse to pacts with smaller political groupings such as the Basque and the Catalan nationalists. The polls reaffirmed his overall leadership and the nationwide strength of his party. Second past the post was the conservative Partido Popular (PP) which picked up 25.8 per cent of the vote for 106 congress seats, one more than in the 1986 poll.

'Spain in progress'

González campaigned on his record and that amounted to a 'You've never has it so good' message. The Spanish economy was at full steam and growing faster than anywhere else in Europe; foreign investment was pouring into the country; jobs were being created at the rate of 400,000 a year; welfare spending, from the health service to the schools system had never been so lavish. The campaign slogan neatly summed it up saying, 'Spain in progress'.

Continued voter support for the Socialists had already been anticipated in the European elections of June 1989 when González was virtually the only one of the Community premiers to derive satisfaction from the Euro poll. After seven years in office, González, 47 years old, was able to maintain a grip on the electorate that mirrored that of Mrs Margaret Thatcher. His particular blend of magic has sheer professional political skills, experience and pragmatism enveloped by a mantle of charisma, idealism and innate decency that he acquired more than 20 years ago as a clandestine opponent of the Franco dictatorship. The mantle is fraying at the edges but is still very much there. Although he poses now virtually as an elder statesman, his hair is greying and his face jowly, González continues to be known nationwide simply as Felipe.

The restatement of the premier's electoral strength and of Spain's overall political stability came at a key period in the nation's surging economic boom for it emerged last year that careful financial stewardship was more necessary than ever. The pitfalls and the traps of strong growth began to crowd the Spanish horizon and González needs at present all the political muscle he can muster in order to control an economy that is hovering close to overheating point.

EMS entry

González took a major gamble in 1989 with his decision to bring the peseta into the exchange rate mechanism of the European Monetary System (EMS). Joining the mechanism at pesetas 65 to the D-Mark just before he hosted the Madrid European Community summit in June, González ignored advice from many quarters that urged caution; rising prices, falling tourism receipts and a widening trade gap amounted to a scenario that appeared less than propitious for EMS membership albeit with a generous 6 per cent fluctuation band on the lines of the Italian lira.

González nevertheless rushed in where the likes of Mrs Thatcher feared to tread because he firmly believed that EMS discipline was precisely what the Spanish economy required. The move, for all its risks, was a clear signal that the period of complete reliance by his government on monetary measures to steer the economy was at an end.

By the standards of any politician anywhere, EMS entry was a courageous decision. In part it was a reflection of González's own political confidence. Formerly it had been the Bank of Spain's job to raise interest rates every time the economy needed cooling, but now González appeared willing to take fiscal measures and introduce spending cuts in order to maintain on an even keel the momentum of Spain's growth. It reflected also the Spanish premier's unequivocal support for European unity in general and monetary union in particular.

The Euro polls gave González 40 per cent of the national vote which was the same proportion the Socialist Party had gained when European Parliament elections had been held in 1987. The centre-right opposition lost ground and the Communist Party, which had hoped to reap the benefit of trade union discontent against the government, advanced only marginally.

The returns, which were to be repeated in broad terms in the general elections four months later, were all the more remarkable for González given that the centre-right had worked hard to regain the political initiative with a series of alliances and a new leadership, and that the governing Socialist Party had been under fire for months from the left wing of the party and from its 'fraternal' labour arm, the Socialist-led Union General de Trabajadores (UGT).

Relations with unions

If EMS membership was the single most significant economic development in Spain since the country joined the EC in 1986, the open quarrel between the Socialist government and the unions marked a no less far-reaching watershed on the political front. During his second term in office, González moved towards a confrontation with trade unions. He endured battle scars but he appeared to have emerged triumphant from the collision.

Trouble between PSOE and the UGT had been brewing for a good three years. The union case, as expressed by its veteran leader Nicolas Redondo, was that González and the Socialist Party administration were implementing a conservative economic policy and had reneged on party principles that had to do with income redistribution and a fairer society. Echoing the British Fabians, Redondo accused the government of entering into the 'aristocratic embrace'.

Both the party and the union had marched more or less evenly in step ever since they had been founded at the end of the last century by Pablo Iglesias, the prophet-like grandfather of Spanish socialism. The UGT traditionally helped to frame the PSOE's electoral manifestoes and all party members were under an obligation to join the union. The close historical ties have now taken on the appearance of a fast receding memory and the break between the PSOE and the UGT has introduced a wholly new element into Spanish socialism and, by extension, into the domestic political scene.

The break with the unions effectively allowed González to pursue, with greater emphasis than ever, the pragmatic social democracy that he has consistently preached. The tactical victory that the premier scored over the labour movement cleared the way for his adoption of tough economic policy decisions.

The issue came to a crunch in December 1987 when Redondo put an end to a longstanding rivalry between the UGT and the Communist-led trade union Comisiones Obreras (CCOO) and engineered a joint strike platform against the government. The two unions organised a 24-hour national stoppage that effectively brought the country to a standstill in what was the biggest show of force by organised labour in Spain for nigh-on half a century.

The strike slogan was a broad demand for a *giro social*, a catchword that meant literally a 'social about-turn' and which amounted in practice to an economic U-turn by González. UGT and CCOO demanded radically increased government spending to improve pensions, unemployment benefits and the wages of public sector employees.

From the December strike highpoint, the fortunes of the unions began to decline. They committed what, with hindsight, appeared a colossal tactical mistake by refusing a government offer in February of a Pta334bn ($2.8bn) package aimed at compensating public sector salaries and increasing welfare, on the grounds that it fell short of the Pta477bn demand that they had tabled.

In the event the government withdrew its offer and implemented a number of expenditure cuts in defence and in education that had already been approved by parliament in order to allocate the considerably lower sum of Pta197bn to the elderly, the jobless and the lower-paid.

On the heels of the reduced package Finance Minister Carlos Solchaga unveiled the first of a succession

KEY FACTS

Official title: The Kingdom of Spain

Head of state: King Juan Carlos I

Head of government: Prime Minister Felipe González Márquez

Ruling party: Partido Socialista Obrero Español (Socialist Workers' Party)

Capital: Madrid

Official Languages: Spanish

Currency: Peseta (Pta) = 100 centimos

Exchange rate: Pta114 per $ (Dec 1989)

Area: 504,782 sq km

Population: 39.1m (1988)

GDP per capita: $4,236 (1986)

GDP real growth: 5.0% (1988)

Labour force: 10.4m (1985)

Unemployment: 22.0% (1987)

Inflation: 4.8% (1988)

Trade balance: –$18.0bn (1988)

Foreign debt: $28.9bn (1985)

of restrictive fiscal plans to dampen demand, to rein in imports and to curb inflation. Solchaga axed Pta115bn-worth of budgeted public expenditure in defence, public works and transport and communications.

González, for his part, went to parliament to explain the breakdown of the negotiations with the unions and to earn support for his policies. He received backing from his own party, where the socialist ranks stood firm, and also from the centre-right oppostion and, for good measure, he obtained the go-ahead for drafting anti-strike legislation.

The unions committed a second error of judgement when they flatly refused, in tripartite talks with the employers and the labour ministry, a Pta200bn government aid programme aimed to subsidise companies hiring women, youth and the long-term unemployed. The catch was that the programme applied only to short, six-month work contracts and this touched a raw nerve with both the UGT and the CCOO. The government could nevertheless plausibly argue that the unions were impeding job creation and thus undermine still further the support that they had enjoyed in December.

A series of wearing public-sector strikes in the railways and in municipal bus companies tarnished the overall image of the unions still more and by 1 May Redondo and the CCOO leader Antonio Gutiérrez had a desultory turn-out for the May Day rallies that they had jointly organised in the expectation of repeating the mobilising effectiveness they had displayed just six months earlier in December.

In the June Euro poll the UGT leadership announced that for the first time ever they would not aid the PSOE campaign nor even endorse its candidates. As at the time of the December joint strike platform with the CCOO, the boycott decision by the erstwhile fraternal union heaped pain and anguish on the PSOE but the suffering was short-lived.

The UGT leadership, which alternatively backed abstention or a vote for the Communist Party ticket, was unable to translate its confrontational mood into a determined electoral protest against the government. González and the PSOE romped home in the polls with the same ease that they had shown in the days when they still enjoyed the confidence of Redondo and his union colleagues.

Thus a key lesson was drawn from the PSOE-UGT divorce that dominated so much of Spain's political scenario last year. The bottom line was that after standing up successfully to the unions, González gained considerable room to manoeuvre. Having weathered the confrontation he was able to embark unhampered on the programme of fiscal and credit-squeezing measures that his finance team had identified as the correct course for cooling the economy.

At different moments during the González premiership there has been talk of what is known as the 'British example', and this is taken to mean the collapse of the British Labour Party government in the second half of the 1970s due to pressure from the Trades Union Congress. The Spanish prime minister was tempted early in the year to go to the country, calling a snap election on the issue of whether parliament or the unions ruled the country but he opted instead for a policy of brinkmanship that paid full dividends. The British precedent served its purpose.

Opposition performance

Comparisons with the record of former British prime minister James Callaghan more than a decade ago may however be misplaced in part because the frame-

KEY INDICATORS

	Unit	1984	1985	1986	1987	1988
Population	'000	38,369	38,561	38,668	38,830	39,050
Gross domestic product (GDP)	Ptabn	25,111	27,889	31,948	35,715	39,618
GDP per capita	Pta '000	4,201	4,339	5,927	–	–
Inflation	%	11.3	8.8	8.8	5.2	4.8
GDP real growth	%	1.8	2.3	3.3	5.5	5.0
Unemployment	'000	2,475.2	2,642.0	2,758.7	2,924.1	2,852.2
Share prices	1985=100	161.7	204.1	415.7	597.7	–
Agriculture production	1979-81=100	114.1	110.6	108.7	121.8	114.9
Industrial production	1985=100	92.6	100.0	100.9	101.8	104.8
Manufacturing	1980=100	90.1	93.2	96.0	98.0	–
Tourism earnings	$bn	7.6	8.1	8.0	8.9	–
Exports	$bn	22.7	23.7	26.8	33.6	39.7
France	$bn	3.4	3.7	4.9	–	–
German FR	$bn	2.2	2.3	3.2	–	–
UK	$bn	2.1	2.1	2.4	–	–
USA	$bn	2.2	2.4	2.5	–	–
Imports	$bn	27.0	27.8	33.3	46.5	57.7
USA	$bn	3.4	3.3	3.5	–	–
France	$bn	2.3	2.8	4.1	–	–
German FR	$bn	2.4	3.2	5.3	–	–
UK	$bn	1.9	1.9	2.7	–	–
Balance of trade	$bn	–4.3	–4.2	–6.5	–13.0	–18.0
Current account	$bn	2.0	2.8	4.0	–0.2	–3.8
Foreign exchange	$m	11,448	10,490	13,805	29,287	35,401
Exchange rate	Pta per $	160.8	170.0	140.0	123.5	116.5

* estimated figure

work binding González's political options is one of an economic boom rather than a recession and in part, also, because it has become clearer than ever in recent months that the Spanish premier does not face a credible political alternative on the centre-right.

The centre-right did, to its credit, try hard. Last year the strongman of the Spanish conservatives, Manuel Fraga, returned from a temporary and self-imposed retirement to run the Spanish right and he recruited what on paper looked like a promising candidate to head the party's ticket in the mid-year Euro polls. The hand-picked MEP leader was Marcelino Oreja, a former foreign minister and lately the secretary general of the Council for Europe.

The new Conservative candidate's chief asset was his experience, his impeccable democratic credentials (this has often been a problem for the authoritarian Spanish right) and his moderate, Christian Democrat-based political beliefs. With Oreja as deputy party leader, Fraga conducted a thorough-going spring cleaning of the Conservative Party that included a change of name from Alianze Popular to Partido Popular (PP) and which managed to attract a number of seasoned politicians who had served, as Oreja had done, in centre-right administrations a decade ago during Spain's exemplary transition from Francoism to democracy.

Most important of all the revamped PP managed to bring about a tactical alliance with the third force in Spanish politics, the minority Liberal group called the Centro Democrático y Social (CDS) that is led by the architect of the 1970s political transition, former prime minister Adolfo Suárez.

The deal between the PP and the CDS consisted in joining forces at regional government and city hall level in order to eject provincial administrations and town councils that were run by the PSOE. The plan was effectively a blueprint and a trial run for a later, grand centre-right alliance that aimed to take on González and the Socialists in national polls.

The strategy, however, backfired miserably. The one eye-catching success that the centre-right could demonstrate was its capture of the Madrid city hall where the combined forces of the PP and the CDS councillors managed to replace the sitting, minority Socialist mayor with a close Suárez associate. Voters up and down the country were nevertheless not impressed. Supporters of each party understood that their respective party principles had been compromised by the newly-forged links and both the PP and the CDS lost ground in the European elections.

González's decision at the begining of September to call elections on 29 October caused the Conservative Party to cast about urgently for a new leader to replace the lacklustre Oreja. Fraga refused to run for national office in part because he was concentrating on regional elections scheduled for December in his home state of Galicia and in part because he remained convinced that the revamped right required a new leader.

The politician eventually picked by Fraga to lead the PP in the national polls was 36-year-old Jose Maria Aznar who was at the time chief minister of Castilla-León region of the North Central plain. Little known outside his own constituency Aznar was handicapped from the start by the extremely short notice he had been given to fight the national campaign. That he headed the PP campaign at all was a sombre indicator of how low the conservative fortunes had fallen. Aznar came across as a lightweight and never threatened González.

Suárez, the former premier who had dazzled so many Spaniards during the heady days of the political transition, posed an even lesser threat to the Socialist premier. He campaigned on a series of populist promises – a three-month military service in place of the present 12 months, free school text books, unemployment benefits for first job seekers – that failed to impress a sceptical public. The 'progressive centre' star of the CDS had waned considerably and Suárez himself seemed to lack the fire of former times.

The root problem of the centre-right was that González, with the success of his orthodox, monetarist policies and his wide appeal to the middle ground of the electorate had stolen their ground from them. There was all too little room for the liberal CDS and the conservative PP to manoeuvre. Suárez took on the appearance of a yesterday's man and Aznar was, if at all, a man for some time in the future.

The only credible platform from which to attack the PSOE was on the left of the political spectrum and it was no suprise that the communist party, campaigning as Izquierda Unida (IU), United Left, managed to significantly increase its support in October and gain some 10 per cent of the votes cast. That level of support is similar to the one that the Spanish communists enjoyed in the late 1970s and it is not likely to rise much further.

IU took up the trade union theme that González and the PSOE had betrayed socialism and had become corrupt in power and made some electoral mileage as a result. The communists were however handicapped by PSOE taunts that it was to them and to their leader González that Poles and Hungarians were turning for advice on how to abandon centralist planning and how to embrace social democracy and a market economy. 1989, when eyes were trained on the disintegrating Soviet bloc, was not a vintage year for a West European communist party.

The key reading of the Spanish political status quo is that the PSOE, having captured the electoral centre in the 1982 elections which, with a landslide win, gave González his first term in office, remains the arbiter of national politics. The party's strength is based on strong and enduring leadership in the person of the prime minster and on its nationwide presence. In the Basque country and in Catalonia, where politics are on the whole dominated by the local nationalist parties, neither the centre-right nor the communist left count for anything. The only national party that counts in either of these two areas is the PSOE; the socialists have a power-sharing agreement in the Basque country with the Basque nationalists and they form the main opposition group in Catalonia to the Catalan nationalists.

Economy

Long isolated from the rest of Europe and from the developed world, and complexed as a result, Spain has in recent years gained a new found confidence through its membership of the European Community. EC entry occurred in 1986 with González at the helm and in 1988 the prime minister increased his, and Spain's international prestige with the smooth-running Spanish presidency of the EC in the first half of the year.

That the June Madrid summit succeeded in extracting a measure of flexibility from Mrs Thatcher on the issue of monetary union was a tribute to the Spanish leader's determined search for a consensus. His stature enhanced, at home and abroad, González was able to concentrate on the complexity of an overheated economy.

The root problem was that while Spain's Gross Domestic Product (GDP) was growing in 1989 at a handsome 4.5-5 per cent, domestic demand was increasing by 7 per cent. Exports grew at 6.5 per cent in the first half of the year and this was less than half the growth of imports. Although a sizeable proportion of the import volume represented capital goods equipment, thus indicating the continued retooling of Spanish industry, there was little doubt that avid domestic consumers were taking full advantage of Spain's lowering tarrif barriers.

The current-account deficit was forecast to top $11bn, a rise of more thatn 300 per cent. The deficit estimate was an ominous one for it represented 3.1 per cent of GDP and it effectively meant that Spain was beginning to accumulate foreign debt. It began to look last year, for the first time since the Spanish economy began to take off halfway through the decade, as though the all-important foreign-exchange cushion was beginning to lose its stuffing.

Reserves leapt by $1.6bn in July, the third highest monthly climb on record, to reach $46.15bn. With such a figure Spain can, with a degree of comfort, cover trade deficits for two years and but there was a nagging doubt as to the reliability of the reserves for a major proportion of the foreign exchange flowing into Spain was earmarked for speculative portfolio investments.

Spain's impressive ability to attract reserves in the past three years could in fact have introduced a measure of complacency over the widening trade deficits during the same period. The $46bn reserves figure meant that Spain could live with a current account deficit that represented 2 per cent of the GDP. The problem is that Spain cannot so happily live with a deficit that is expected to total $11bn and which represents 3.1 per cent of the GDP.

One sure indicator that the economy urgently required cooling came halfway throught the year when the July consumer price index shot up by 1.6 per cent, the highest monthly rise in more than three years, to bring the accumulated 1989 inflation up to 4.5 per cent and the year-on-year total to 7.4 per cent. The government had originally hoped to keep inflation to 6 per cent, marginally up on the 1988 figures and a far cry from confident forecasts made 18 months previously which had estimated that in 1989 inflation would drop to 3 per cent.

What really caused concern was that the shock inflation figures came on the heels of credit-squeezing measures which meant, for example, that prime borrowers were already paying banks 15 per cent for funds. By July the finance ministry was estimated to have raised the total amount of money taken out of circulation since the beginning of the year to $10bn.

Independent economists reckoned nevertheless that a further $10bn had to be pulled out of the money system if demand was going to be effectively brought into line with what the country was actually producing. By autumn it was clear that González, bereft of further monetary ammunition, had no choice other than to introduce a stringent 1990 budget. Major public spending cuts would have to be accompanied by a squeeze on hire purchases, an increase in the VAT rate on certain products and even by a hike in income tax.

These growing economic concerns lay behind González's decison to call an election at the end of October 1989, a good eight months before his term ran out in June this year. The prime minister, his economic team and, not least, the international business community which throughout last year viewed Spain as the focus of a sustained investment drive were all betting nevertheless on the continued good fundamentals that were propping up the economy.

The governing underlying perception was that Spain lay in the 'sun belt' European zone and was set to grow, as a result, at a higher rate than the EC average. The sun belt syndrome embraced at least three key principles: that Spain would attract a higher proportion of high tech industrial investment than elsewhere in Europe; that Spain was in the midst of an agricultural revolution that would increasingly provide the European market with early harvested, and with subtropical, fruits and vegetables of the kind substituting imports from, for example, Florida or California; and finally that Spain was earmarked not to be the package-tour catchment area of European tourism but, rather, the retirement zone for growing contingents of northern Europeans who will increasingly cluster in fellow-national expatriate communities on the *costas* under the shadow of a satellite TV disc and within driving distance of a golf course.

Investment and banking

At a domestic business level last year keynote reforms continued to direct Spain along the path of growing integration in a Europe without frontiers. The stock markets in Madrid, Barcelona, Bilbao and Valencia underwent a considerable overhaul with the introduction of new guidelines that ushered in broking and market-making companies in the place of the age-old, *agentes de cambio y bolsa*, who had controlled floor operations for fixed commissions. The Spanish markets' 'mini bang' also introduced a computer-based, continual market aimed at eventually substituting the existing open cry system that was a restrictive characteristic of the domestic stock exchanges.

The downside in the banking sector was the failure of a proposed merger between Banco Español de Credito, Banesto, with Banco Central which would have created a financial institution to rival that of Banco Bilbao Vizcaya, BBV, and would, addition-

COUNTRY PROFILE

Political system: Parliamentary monarchy, with legislative power vested in the Cortes Generales (divided into the Congress of Deputies and a Senate). Also seven autonomous regions, whose governments are elected every four years.

Main cities/towns: Madrid (population 4.7m), Barcelona, Valencia, Seville, Zaragoza, Bilbao, Málaga.

Media
Press: Only 34 per cent of the adult population reads a newspaper every day.
Dailies: Published regionally, though those published in Madrid are available in most parts of the country. Principal dailies include *El País* (circulation 25,000), *Diario 16* (136,687), *YA* (75,447) and *ABC* (Madrid, 247228) and *La Vanguardia* (195,088) and *El Periódico* (Barcelona, 153,735).
Weeklies: Several general and special interest magazines, plus news magazines such as *Blanco y Negro* and *Sábado Gráfico*. Major business magazines *Actualidad Económica* (30,659), *Dinero* and *Mercado* (25,000).
Broadcasting: Co-ordinated by Radiotelevisión Española (RTVE). Approximately 10.5m radios and 10m TVs in use. Radio programmes are broadcast by Radio Nacional de España (RNE) and several commercial and independent stations. All TV programmes are broadcast by Television Española (two channels). Private commercial TV imminent.
Advertising: All usual media are available. TV and hoardings tend to be most popular method, though radio, newspaper and cinema advertising are also widespread. Information, including current rates, appears in quarterly magazine *Guia de los Medios.* Total advertising spend (1987) Pta540bn. Major advertising products: food, automotives. Newspapers and TV are the major advertising media.

Domestic economy: Mixed economy with large agricultural and industrial sectors. Also important: tourism and banking industries. Wide regional disparities of wealth and large 'black economy'. High rates of economic growth during 1970s curtailed by world recession and rising inflation. Introduction of austerity measures allowed for export-led growth, but only at cost of higher unemployment. Economic policy based on modernisation of industry and integration into EC, and economy is now the fastest growing in the EC.

Employment: Of the total workforce around 14 per cent are employed in agriculture, 31 per cent in industry including construction and 46 per cent in service sector. World recession and decline of traditional heavy industries pushed up unemployment rate to over 21 per cent in 1986. Situation exacerbated by retrenchment in public sector and deflationary domestic policies.

External trade: Recent growth in visible exports owing partly to depreciation of peseta led to reduction in trade deficit and improvement in terms of trade but deficit rose again sharply in 1986 and 1987. Regular balance of payments deficit meanwhile offset by recovery in private sector capital inflows and rise in earnings from tourism. Major exports include fruit and vegetables, metals, motor vehicles, machinery and transport equipment. Spain sends half its exports to the EC. Import growth has been sluggish due to stagnating domestic demand. Main imports include foodstuffs, energy products, raw materials, basic manufactures and machinery. Principal trading partners are France, Germany FR, USA and UK.

Agriculture: Formerly mainstay of the economy; currently accounts for around 5.5 per cent of GDP. Production has fluctuated in recent years due to droughts and heavy flooding, dropping 6.7 per cent in first year of EC membership (1986) compared with two good preceding years. Near self-sufficiency in wheat, barley, sugar and poultry. Major exporter of wine, fruit and vegetables. EC entry likely to result in modernisation of sector. At present over half of farming plots are smaller than 10 hectares.

Industry: Heavy industry expanded rapidly during the 1970s, particularly steel, shipbuilding, chemicals, motor vehicles and machine tools. Also important: food processing, textiles and electrical goods. Production has since fallen in most sectors. Emphasis on restructuring industry and promoting hi-tech industries – most important are steel, shipbuilding, textiles, footwear, automobile parts and home appliances. Public sector involvement in industry is largely confined to the operations of the state holding company, Instituto Nacional de Industria (INI). The most active industrial region is Catalonia, with Barcelona as its focal point. EC entry has meant changes to the way industry financed.

Mining: Extensive mineral deposits including iron ore, mercury, uranium, wolfram, lead and zinc. Also large reserves of coal and lignite, production of which has risen due to increased domestic demand.

Energy: Gradual reduction in dependence on imported oil and increased utilisation of domestic fuel sources, particularly coal. By stepping up oil exploration programme Spain hopes to satisfy 15 per cent of its total oil requirements from domestic sources by 1990.

Banking: System of private and savings banks supervised by the Banco de Espana. Commercial banking dominated by eight highly profitable banks (Central, Banesto, Hispano Americano, Bilbao, Vizcaya, Santander, Popular, Exterior).

Legal system: Civil law system with regional applications.

Membership of international organisations:
Andean Pact (observer), ASSIMER, CCC, Council of Europe, EC, ESO, FAO, GATT, IAEA, IBRD, ICAC, ICAO, ICES, ICO, IDA, IDB, IEA, IFAD, IFC, IHO, ILO, IMF, IMO, INTELSAT, International Lead and Zinc Study Group, INTERPOL, IOOC, IPU, ITC, ITU, IWC, NATO, OAS (observer), OECD, UN, UNESCO, UPU, WHO, WIPO, WMO, WSG, WTO.

ally, have sparked off a merger process right across the Spainish business board by bringing together interests controlled by the two banks that ranged from insurance to petroleum and from construction to agribusiness.

BBV, the resultant of Banco Bilbao and Banco Vizcaya, was able, in contrast, to consolidate not only the multiple fingers that each of the banks had in Spain's corporate pie but also a domestic economy-of-scale strategy.

The banking upside lay with the initiatives taken by somewhat smaller Spanish banks to build bridges with European institutions in preparation for the 1992 liberalisation. Banco de Santander which acquired 10 per cent of the Royal Bank of Scotland in a share swap deal was a trailblazer on this particular path and the example was followed by Banco Hispano Americano which took up an option to acquire 5 per cent of West Germany's Commerzbank.

Elsewhere the trend pointed to greater-than-ever penetration by foreign capital. At year's end Volvo, Fiat, DAF and Daimler-Benz were all bidding for Enaso, Spain's state-owned truck manufacturer. Enaso's proprietor, the public sector holding company Instituto Nacional de Industria (INI), was determined to marry the company off to a multinational group.

The Enaso sale mirrored that of the car manufacturer SEAT three years earlier to Volkswagen and it came on the heels of a major investment programme by the West German giant that heralded the consolidation of Spain as a major auto producer.

The growing openness of Spain's public sector had been foreshadowed in the first half of last year by the minority floatation of Repsol, the state energy company, whose shares were listed in New York as well as in Madrid. On the private sector front there was a different kind of revolution going on. A bevy of institutions were offering financial leverage services, acquiring companies and arranging mergers. Technology parks were launched in places such as Bilbao in the North and in Malaga on the Mediterranean coast as the existing ones near Madrid and outside Barcelona found that demand for space far outstripped the available ground on offer. Spain, bolstered by its political stability, continued to ride the crest of an economic boom wave.

In the midst of it all foreign investment even permeated the Spanish media scene with Italy's Silvio Berlusconi and France's Canal Plus acquiring stakes in Spain's new commercial television channels and with the ubiquitous Mr Rupert Murdoch investing heavily into Gruppo Zeta, a household name in Spanish newspaper and magazine publishing.

BUSINESS GUIDE

Time: GMT + 1 (GMT + 2 from late Mar to late Sep); Then Canaries GMT (GMT + 1 during summer).

Climate: Very varied. Mainland is temperate in the north, hot, dry summers in central and southern regions. On central plateau weather very cold Dec-Mar. Northern coastal region has mild winters and warm summers with slightly less sunshine than the rest of Spain. Rainfall Jan-Apr. Summer storms can be violent. Temperatures range from one degree C to 35°C.
Clothing: Mediumweight for winter, plus topcoat in central and northern areas. Lightweight for summer.

Entry requirements: Visa not required by nationals of most European and American countries, and several others, for visits up to three months. N.B. regulations liable to change at short notice. Travel agent or Spanish consulate should be consulted prior to departure.
Health: Vaccination certificates not normally required. Tap water not always safe to drink outside cities, so bottled water advisable.
Currency: No restriction on amount of Spanish or foreign currency that may be taken into the country. Up to Pta100,000 in Spanish banknotes and the equivalent of Pta500,000 in foreign currency may be exported. (Greater amounts may be re-exported provided such amounts were declared on entry to the country.)
Customs: Personal effects duty-free. Duty-free goods facilities at main international airports.

Air access: Regular flights by all major international airlines. Very busy during holiday season.
National airline: Iberia.
Main international airport: Barajas (MAD) 16 km from Madrid, Barcelona (BCN) 15 km from city and Málaga (AGP) 10 km from city.
International airports: Alicante (ALC), Bilbao (BIO), Almeria (LEI), Valencia (VLC), Santiago (SCQ).

Surface access: Good road and rail access from France. Regular ferry and shipping services from several countries.
Main port: Barcelona, Valencia, Alicante, Málaga, Algeciras, Cádiz, La Coruña, Bilbao.

Hotels: Classified from one- to five-star, plus a 'Grand De Luxe' category (pensions/hostels classified from one- to three-star). Also paradores (national tourist inns) increasingly popular. Daily rate for a single room, in a major city, ranges from about Pta 6,000 – 10,000 depending on category and season. Rates in other parts of the country tend to be slightly cheaper. Accommodation should be booked well in advance, especially during holiday season. N.B. term 'Residencia' denotes establishments without dining-room facilities.

Restaurants: Numerous and varied. Even modest restaurants offer very substantial meals. Service charge usually included in bill, but small tips customary.

Currency
Exchange rate: Peseta floats freely against all foreign currencies. Spain is a member of the EMS.

Foreign exchange information
Account restrictions: Non-resident accounts may be convertible or non-convertible.
Trade policy: Tariff reductions associated with membership of EC. Quotas imposed on some goods. Spanish import duties being replaced by EC Common External Tariff. Proceeds of exports must be surrendered in return for local currency. Credit may be obtained to finance exports at favourable rates of interest.

Credit cards: All major credit cards are accepted.

Car hire: Available in most large towns. International driving licence, or foreign licence translated into Spanish and authorised by Spanish consulate, required. Third-party insurance compulsory, and a Green Card is advisable. Details of hire from local tourist offices. Drive on the right. Speed limits: towns 60kph, national highways 100kph, motorways 120kph, others 90kph. Traffic coming from right generally has priority. Seat belts must be worn in front seats.

City transport
Taxis: Available in most major cities; all metered. Tend to have a distinct colour in each city. Fares vary with location, generally starting at Pta 60-80 for local journeys. Tipping about 10 per cent.

National transport: Based on radial routes centred on Madrid. Often very busy during holiday season.
Air: Frequent services from Madrid to all major centres operated by Iberia and Aviación y Comercio (Aviaco).
Road: Good roads connecting all main towns. Network of over 150,000 km, including 2,000 km of motorways (usually toll) mostly confined to coastal regions. Regular bus and coach services.
Rail: Approximately 14,408 km of track, of which about 11,500 km is operated by Red Nacional de los Ferrocarriles Españoles (RENFE), and the rest (narrow gauge) by Ferrocarriles Españoles de Vía Estrecha (FEVE).
Water: Regular steamer and hydrofoil services operated by Compañía Transmediterránea connect Balearic Islands with Barcelona, Valencia and Alicante. Also weekly ferry service to Las Palmas (Canary Islands) from Barcelona.

Public holidays
Fixed dates: 1 Jan, 6 Jan (Reyes), 19 Mar (San José), 1 May, 15 Aug (Assumption), 12 Oct (National Day), 1 Nov (All Saints'), 6 Dec (Constitution Day), 8 Dec (Immaculate Conception), 25 Dec.
Variable dates: Holy Thursday, Good Friday, Corpus Christi. Each town and region also has its own fiesta, eg Madrid in May and Nov, Barcelona in Jun and Sep, and Valencia in Jan. Dates of these also vary.

Working hours
Government and business: Vary considerably from region to region and according to time of year. In Madrid: (Mon-Fri) generally 0900 – 1330 and 1500 – 1800; except Jul and Aug, 0830 – 1430 (1400 on Fri) with only skeleton staff remaining during afternoon.
Banking: (Mon-Fri) 0900 – 1400, (Sat) 0900 – 1300.

Social customs: Lunch 1430, dinner 2130. Spaniards hospitable towards business visitors but do not usually invite them to their homes.

Business language and interpreting/translation: Available in all major towns. Working knowledge of Spanish great advantage to business visitor.

Telecommunications
Telephone: Automatic system in operation throughout most of country. Rates vary from province to province. For international calls from public phone booths dial 0744. International dialling code: 34 followed by 1 for Madrid.
Telex: Available in main post offices and larger hotels. Country code 52E.
Telegram: Available in post offices and by telephone.

Postal service: Letters may be sent *poste restante* care of main post offices. These should be addressed: Addressee, Lista de Correos, Name of village etc.

Trade fairs: Many specialised fairs and exhibitions held in various parts of the country. Important method of promotion. Goods may be imported temporarily, with an ATA carnet, then later either fully imported or returned to country of origin.

BUSINESS DIRECTORY

Electricity supply: 220 V AC (occasionally 110-125V AC in older buildings).

Representation in capital: Algeria, Angola, Argentina, Australia, Austria, Belgium, Bolivia, Brazil, Bulgaria, Cameroon, Canada, Chile, China PR, Colombia, Costa Rica, Cuba, Czechoslovakia, Denmark, Dominican Republic, Ecuador, Egypt, El Salvador, Equatorial Guinea, Finland, France, Gabon, German Democratic Republic, Federal Republic of Germany, Greece, Guatemala, Haiti, Holy See, Honduras, Hungary, India, Indonesia, Iran, Iraq, Ireland, Israel, Italy, Ivory Coast, Japan, Jordan, Republic of Korea, Kuwait, Lebanon, Libya, Malaysia, Mauritania, Mexico, Morocco, Netherlands, Nicaragua, Nigeria, Norway, Pakistan, Panama, Paraguay, Peru, Philippines, Poland, Portugal, Qatar, Romania, Saudi Arabia, South Africa, Sweden, Switzerland, Syria, Thailand, Tunisia, Turkey, USSR, United Arab Emirates, United Kingdom, USA, Uruguay, Venezuela, Yugoslavia, Zaïre.

Hotels: Barcelona
Avenida Palace, Gran Via 605, 08007 (tel: 301-9600; tx: 54734; fax: 3181234).

Avenida Victoria, Avenida Pedralbes 16, 0803 (tel: 204-2754; tx: 98302; fax: 2042766).

Diplomatic, Pau Claris 122 (tel: 317-3100; tx: 54701).

Gran Sarria, Avenida de Sarria 50, 08029 (tel: 239-1109; tx: 51638).

Presidente, Diagonal 570, 08021 (tel: 200-2111; tx: 52180; fax: 2002266).

Ritz, Gran Via 668, 08010 (tel: 318-5200; tx: 52739; fax: 3180148).

Hotels: Bilbao
Aranzazu, Rodrigues Arias 66 (tel: 441-3100; tx: 32164).

Carlton, Plaza Fererico Mayua 2, 48009 (tel: 416-2200; tx: 32233; fax: 4164628).

Gran Ercilla, Ercilla 37/39, 11 (tel: 443-8800; tx: 32449; fax: 4439395).

Villa de Bilbao, Gran Via Lopez de Haro 87, 11 (tel: 441-6000; tx: 32164).

Hotels: Madrid
Alameda (airport), Avenida de Logron\08o 100, Barajas-Madrid 28042 (tel: 7474800, 2239868; tx: 43809; fax: 7478928).

Arosa, Gran Via 29 (tel: 2321600; tx: 43618).

Carlton, Paseo Delicias 28 (tel: 2397100; tx: 44571).

Castellana, Paseo de la Castellana (tel: 4100200; tx: 27686; fax: 4195853).

Cuzco, Paseo de la Castellana 133, Madrid, P.O. Box: 16 (tel: 4560600; tx: 22464).

Emperador Plaza (reservations Office), Princesa 40, P.O. Box: 8 (tel: 2472800; tx: 46261; fax: 2472817).

Eurobuilding, Padre Damián 23 (tel: 4577800; tx: 22548; fax: 4579729).

Florida Norte, Paseo de la Florida 5 (tel: 2416190; tx: 23675; fax: 2477833).

Gran Colón, 119 Avenida Doctor Esquerdo (tel: 2735900; tx: 22984).

Los Galgos Sol, Claudio Coello 139 (tel: 2626600; tx: 43957; fax: 2617662).

Meliá, Calle Princesa 27 (tel: 2418200; tx: 22537).

Meliá Castilla, 43 Capitan Haya (tel: 5712211; tx: 23142; fax: 5712210).

Miguel Angel, Miguel Angel 29/31 (tel: 4420022; tx: 44235; fax: 4425320).

Palace, Plaza de la Cortes 7 (tel: 4297551; tx: 22272; fax: 4298266).

The Ritz, Plaza de la Lealtad 5 (tel: 5212857; tx: 439986; fax: 5328776).

Villa Magna, Paseo de la Castellana 22 (tel: 5641236; tx: 27738; fax: 2753158).

Wellington, Velazquez 8 (tel: 2755200; tx: 22700).

Car hire: Madrid
Avis: Gran Via 60 (tel: 2472048; tx: 22410).

Budget: (tel: 4575968; tx: 45223, 48538).

Europcar: General Yague, Nr 6 Bis, Madrid 20 (tel: 5971500; tx: 43087).

Hertz: Gran Via 88 (tel: 2485805; tx: 27379).

Chambers of Commerce
Cámara Oficial de Comercio, Industria y Navegación de Barcelona, Avenida Diagonal 45, 08006 Barcelona (tel: 219-1300; tx: 54713).

Cámara Oficial de Comercio, Industria y Navegación de Bilbao, Alameda Recalde 50, 48008 Bilbao (tel: 444-4054; tx: 4445891, 32182).

Cámara Oficial de Comercio, Industria y Navegación de Cádiz, Ahumada 2, 11004 Cádiz (tel: 323-050; tx: 76013).

Cámara Oficial de Comercio, Industria y Navegación de Ibiza, Historiador J Clatés 4, Ibiza, Balearic Islands (tel: 301-492; tx: 69679).

Cámara Oficial de Comercio, Industria y Navegación de Las Palmas de Gran Canaria, León y Castillo 24, Las Palmas, Canary Islands (tel: 371-000, 361-130; tx: 95317).

Cámara Oficial de Comercio e Industria de Madrid, Huertas 13, 28012 Madrid 12 (tel: 429-3193; tx: 27307, 23261).

Cámara Oficial de Comercio, Industria y Navegación de Málaga, Bolsa 1, Málaga (tel: 211-673; tx: 77368).

Cámara Oficial de Comercio e Industria de Zaragoza, Don Jaime 1 18, Zaragoza (tel: 295-900; tx: 58702).

Cámara Oficial de Comercio, Industria y Navegación de Sevilla, Plaza de la Contratación 8, 41004 Sevilla (tel: 211-204; tx: 72407).

Cámara Oficial de Comercio, Industria y Navegación de Valencia, Poeta Querol 15, 46002 Valencia (tel: 351-1301; tx: 62518).

Banking
Banco de España (central bank). Alcalá 50, 28014 Madrid (tel: 446-9055; tx: 49461).

Banco Atlántico SA, Diagonal 407 bis, Barcelona (tel: 237-1240; tx: 52267).

Banco Central, Calle de Alcalá, 28014 Madrid (tel: 222-2991; tx: 42003).

Banco de Bilbao, Vizcaya Plaza de San Nicolas 4, 48005, Bilbao (tel: 424-4620).

Banco de la Exportacón SA, Barcas 10, Valencia 2 (tel: 351-7862; tx: 62676).

Banco de Sabadell, Placa Sant Roc 20, 08201 Sabadell (tel: 726-2100; tx: 52847).

Banco Español de Credito (Banesto), Paeo de la Castellana 7, Madrid (tel: 338-1000; tx: 27410).

Banco Exterior de España, Carrera de San Jerónimo 36, Madrid (tel: 429-4477; tx: 22033).

Banco Hispano Americano, Plaza de Canalejas 1, Madrid (tel: 522-4660).

Banco Internacional de Comercio, José Ortega y Gasset 56, 28006 Madrid (tel: 402-8362; tx: 23914).

Banco Popular Español, Velázquez 34, 28001 Madrid (tel: 435-3620; tx: 44351).

Banco Santander, Apdo de Correos 00045, Santander (tel: 221-200; tx: 35833).

Caja de Pensiones para la Vejez y de Ahorros de Cataluña y Baleares, Avda Diagonal 621-629, 08028 Barcelona (tel: 330-3000; tx: 52623; fx; 3300480).

Confederación Española de Cajas de Ahorros (confederation of Spanish savings banks), Alcalá 27 Madrid 14 (tel: 232-7810).

Consejo Superior Bancario (central committee of Spanish banking), José Abascal 57, Madrid 9 (tel: 441-0611).

La Caixa de Barcelona (savings bank), Avenida Diagonal 530, 08006 Barcelona (tel: 201-6666; tx: 97214).

Principal newspapers
ABC, Serrano 61, 28006 Madrid (tel: 435-3100).

Actualidad Economica, Recoletos 1, 7\28, 28001, Madrid (tel: 431-0917; tx: 49455).

Diario 16, San Romualdo 26, 28037 Madrid (tel: 754-4066; tx: 45934).

El País, Miguel Yuste 40, 28037 Madrid (tel; 754-3800; tx: 42187).

El Periódico, Comte D'Urgell 71-73 08011 Barcelona (tel: 323-4212; tx: 50624).

Mercado, Hernández de Tejada 10, 28027 Madrid (tel: 268-3300; tx: 47466).

Ya, Mateo Inurria 15, 28036 Madrid (tel: 259-2800).

Other useful addresses
Agencia EFE, SA (news agency), Espronceda 32, Apartado 1112, 28003 Madrid (tel: 441-5599).

Bolsa de Comercio de Valencia (stock exchange), Pascual y Genis, 19/46001 Valencia (tel: 352-1487; tx: 62880).

Bolsa de Madrid (stock exchange), Palacio de la Bolsa, Plaza de la Lealtad 1 (tel: 232-8484).

Central de Reservas de los Paradores de España, calle Velázquez 25, 28001 Madrid (tel: 435-9700/9744/9768/9814; tx: 44607 rrpp).

Confederación Española de Organizaciones Empresariales (Spanish confederation of employers' organisations), Diego de León 50, 28006 Madrid (tel: 262-4410; tx: 46527).

Instituto Nacional de la Seguridad Social, Subdirección General de Relaciones Internacionales, Padre Damian 4, 28036 Madrid. (tel: 450-1900).

Kompass Register España, Avenida del General Perón 26, Madrid 20 (tel: 455-9753).

Ministry of Economy, Finance and Trade, Alcalá 9, Madrid 14 (tel: 232-6124; tx: 48387).

Ministry of Industry and Energy, Paseo de la Castellana 160, Madrid 16 (tel: 458-8010; tx: 42112).

Tur España, Secretaria Gral de Turismo, Maria de Molina 50, 28001 Madrid (tel: 411-4014).

Sri Lanka

Nalin Wijesekera

Sri Lanka's first Executive President Junius Jayawardene's ten year political reign will go down in history as four years of unprecedented growth 'matching West Germany's post war development', while his last six years in power was besmirched with a bloody political *blitzkrieg* leading to an economic bender that sent shock waves through the country. Yet in his last encounter with the foreign media before stepping down from the presidency, his wily mind went in top gear to artfully respond, 'in other countries like Burma, Malaysia, Pakistan, Bangladesh, Philippines they have violence and bloodshed calling for elections; Sri Lanka is the only country in the world where there is violence by some people who oppose elections being held'.

After the election at the end of 1988 of Ranasinghe Premadasa as president, and the general election that brought back the ruling United National Party (UNP) into the seat of power under the new populist image of Premadasa, the air seemed no longer haunted with a fear-psychosis, and even those timorous migratory birds called tourists came back to the island's exotic haunts of ancient cities and burnished beaches. President Premadasa successfully set up a dialogue with the warring Tamil terrorist group – the Liberation Tigers of Tamil Eelam (LTTE). Schools, closed for so long, re-opened and universities began to stir with scholastic activity. Colombo traffic began limping about again at its sclerotic pace – virtually free of the unofficial curfews clamped down sporadically by southern terrorists.

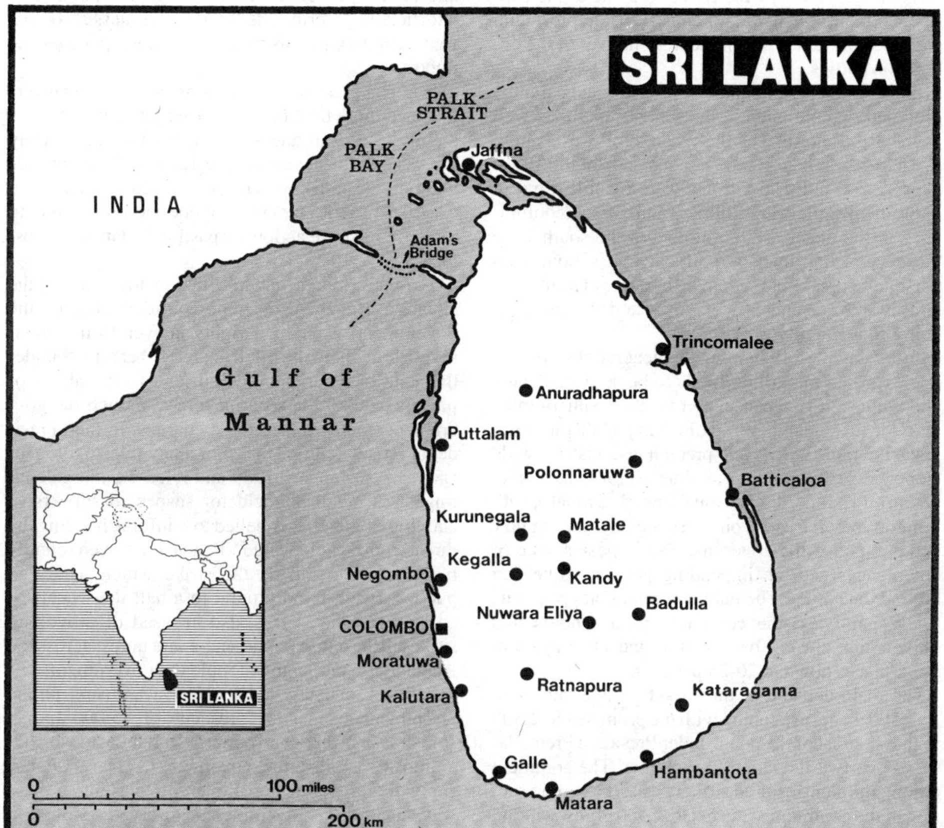

History of unrest

The 1971 insurrection activated by the Janatha Vimukthi Peramuna (JVP) – the same militants who were in 1989 paralysing the island republic – was indeed a bad dream. The July 1983 ethnic riots were cruel and ugly because little was done by the government to quell them. The July 1987 disturbances after the signing of the Indo-Lanka Peace Accord were contained, though for three days Colombo and some of the country's main cities were under siege, looking like a typical Third World scene. Here, it was not that government failed to do anything; there was nothing it could do. And nothing drains public confidence as much as the image of political impotence.

By the last quarter of 1988 militant anti-government forces were attacking President Jaywardene's exhausted government – which was facing two critical elections – on many fronts. Assorted weapons of economic dislocation created a state of paralysis. The weapons included protests, demonstrations, work stoppages; rumours, selective killings, death threats; crippling strategic services – transport, fuel distribution, the port and banking; and small formation guerilla attacks on vital static targets like police stations and army cantonments. This harassing multi-front war strained the regime and its resources to the limit.

Election success

Within weeks of Mr Premadasa's convincing win over Mme Bandaranaike, both at the presidential election (polling 50.43 per cent of votes cast against 44.95 per cent) and the general election (polling 50.71 per cent against 31.85 per cent), he was able to throw her party into a state of disarray, and turn the tide of popular favour, so widely discredited when his predecessor Jayawardene led the ruling party. In fact Premadasa turned his apparent handicaps to his own advantage. For instance, his modest origins, once bruited about as an insuperable obstacle to reaching the nation's highest seat, were adroitly turned around to give him a boost as the 'Messiah of the Masses'.

To consolidate his populist image, the new president offered the 'poorest of the poor' in Sri Lanka his 'Janasaviya' (People's Strength) Poverty Alleviation Programme (PAP) – costing Rs10bn for the first phase – which is today the government's priority lead project, harnessing virtually every ministry, and husbanding the private sector and social organisations. It is not meant to be a state handout, but will help over 7m people in 30,000 village hamlets earning less than Rs750 a month, to purchase Rs1,358 worth of essential commodities monthly and Rs1,042 a month to be invested in a fixed deposit which will generate Rs25,000 in two years, which would be used to set up a productive business. Some economists scoff at Janasaviya as a piece of 'political scheming', whose financing semed likely to compel government to print money to the tune of Rs11bn in 1989. It would fuel inflation, they say, from the 14 per cent of early 1989 to 20 per cent and would push the debt ratio to 29 per cent. But then Mahatma Gandhi's 'Antyodaya' programme of a similar nature in India met with similar opposition. The Premadasa government also launched a free midday meal for schoolchildren, costing Rs2,500m, to alleviate malnutrition.

In early June 1989 it seemed that the redoubtable

KEY FACTS

Official title: Democratic Socialist Republic of Sri Lanka

Head of state: President Ranasinghe Premadasa

Head of government: Prime Minister D.B. Wijetunge

Ruling party: United National Party

Capital: Colombo (commercial); Sri Jayawardenapura (administration)

Official Languages: Sinhala, Tamil and English

Currency: Rupee (Rs = 100 cents)

Exchange rate: Rs39.5 per $ (Nov 1989)

Area: 65,610 sq km

Population: 16.59m (1988)

GNP per capita: $423* (1988)

GNP real growth: 3.5%* (1988)

Unemployment: 12.0% (1986)

Inflation: 14.0% (1988)

Trade balance: –$544.2m (1988)

Foreign debt: $4.7bn (1987)

Visitor numbers: 230,106 (1985)

* estimated figure

new President Premadasa had managed to bring the warring Liberation Tigers of Tamil Eelam (LTTE) to the conference table, without the usual preconditions for dialogue. The modalities of the peace talks were finalised. A ministerial level delegation began a more detailed round of talks – with the promise of the Tamil militant leadership of the LTTE to come to the conference table – that was to consist of the president, cabinet representatives and representatives of the opposition parties.

But behind this semblance of progress and calm the bloodletting continued in the Sinhala south, as well as the Tamil north. The daily death toll was as high as 15 to 20, and during the last six years of the 1980s over 12,500 political killings were recorded, but in 1989 the Tamils were killing Tamils for predominance in the north, while murders in the south were more vendettas than of political motivation, with personal scores being settled in the name of militancy, and law enforcement officers suspected of 'fingering' being conveniently 'rubbedout'.

In the dawn of 1989 after the general election, no one expected the killing fields to be activated anew and with such vengeance, and for the southern militants of the Janatha Vimukthi Party (JVP) to gatecrash the trade unions and precipitate an island-wide stoppage of the public omnibus sector. The strike, begun on 12 June, dragged on for nearly a month, with ominous repercussions on every sector of the country and compelled the government to impose a state of emergency, with an impending general strike just around the corner. The public omnibus stoppage virtually paralysed the economy, with schools and universities closing once again, and attendance at work places down to 20-25 per cent.

The main demand of the strikers was a minimum Rs2,500 wage, in keeping with the promised Rs2,500 to the 'poorest of the poor' under President Premadasa's Poverty Alleviation Programme. The argument among militant-inspired strikers was: if a certain section of the community can get Rs2,500 gratis, without working, why shouldn't the working majority also be entitled to the same benefit? It was in this context that the President hinted at 'a foreign hand' adroitly fuelling the disturbances that ripped across the island leaving in their wake over 200 dead in 2 1/2 months of bloody political turmoil.

It was in this bleak scenario that Sri Lanka, once a peaceful, exotic, paradise isle, was gasping for economic sustenance, fuelled by a credit squeeze that struck at the heart of the liberalised economy, and the 'immediate stoppage' of commercial banks being allowed to open letters of credit on some 50 odd items unofficially described as 'luxuries', unless a 100 per cent cash deposit was made in advance to cover the import.

Sri Lankan security forces using emergency powers detained more than 600 suspected left-wing rebels – mainly in central and southern Sri Lanka. Security forces, able to detain people for indefinite periods without trial under the state of emergency, were given orders to 'shoot on sight' anyone inciting people to violence or strike action, or pasting inflamatory posters.

The opposition, meanwhile, taking advantage of the situation, brought a vote of no-confidence against the government, as human rights lawyer Prins Gunasekera brought to the forefront fears that arrests under the state of emergency could lead to the abuse of powers held by the security forces, though the government stoutly denied these charges. In fact, President Premadasa lifted the emergency rule, in force for six years, in January 1989 and released a large contingent of detainees held for suspected subversive activities, but was compelled to reinforce it, citing the threat from the left-wing JVP, who have been roundly blamed for organising the strike which paralysed public transport and ground to a halt the wheels of commerce. The JVP is also accused of provoking strikes in the entire public sector and unrest across the island in its campaign to topple the government.

Political analysts said the JVP was using public resentment against the presence of 45,000 Indian soldiers in northern and eastern Sri Lanka, to whip up animosity against the government. 'The JVP has shrewdly turned the anti-Indian protests into a mass wave against the government,' observed a Colombo watcher. The strained relations between Colombo and New Delhi due to the continued presence in the island of the Indian Peace Keeping Force (IPKF) prompted Premadasa to decide against representing Sri Lanka at the Foreign Ministers' Conference of the South Asian Association for Regional Co-operation (SAARC), and it seemed that he might not host the Summit scheduled for late 1989, if 'even one member of the IPKF remained on Sri Lankan territory.' The LTTE, the strongest Tamil militant group in the island, the JVP and all the opposition – except for the now New Delhi-backed Eelam People's Revolutionary Liberation Front (EPRLF) who head the Provincial Council of the temporarily merged north and east and the Eelam Revolutionary organisation with 13 members in parliament – categorically stated that they wanted the IPKF out of Sri Lanka immediately. Further fuelling the problems in the Tamil-speaking northern and eastern provinces was the news that the EPRLF had kidnapped young boys and girls, getting them trained with IPKF assistance, in order to have a strong people's army to fight the LTTE and the Lankan security forces and to set up a State of Eelam – which demand they dropped when they decided to join the democratic process and contest the provincial and general elections.

India has had her fair share of machismo in the Maldives, Nepal and Sri Lanka and both politically and diplomatically needed to pull back from her stance of 'brown imperialism' – especially as 1989 was election year for Rajiv Gandhi, and such jingoism by Congress (I) was being exploited by the majority of opposition parties in India. Gandhi, of course, consented to speak with Colombo for a phased withdrawal of the IPKF by the end of 1989, but had refused to pull the forces out by the deadline date of 29 July – which marked the second anniversary of the IPKF's presence in Sri Lanka. By that date it was apparent that the most forceful way to keep the IPKF in Sri Lanka would be to fuel problems in the island's southern, central and western provinces. This was successfully achieved. With the limited numerical strength of the Sri Lanka security forces, it seemed virtually impossible for them to wage two battles – one in the north and east and the other in the southern, central and western parts of the island. By mid-1989, it was very apparent that India's small neighbours like Nepal, Bhutan and Sri Lanka too want their share of self-assertiveness, sovereignty and territorial integrity.

Economic performance

Economic indicators revealed that the Colombo Consumers' Price Index, considered a key reflector of the country's inflation rate, had risen to 14 per cent in 1988 – almost twice the increase for the previous year. Balance of payments too showed a further deterioration in 1988, with the estimated trade deficit at SDR 564m – a 9 per cent increase from the 1987 deficit. Expenditure on imports rose by 4 per cent from Rs61,102m in 1987 to Rs71,276m in 1988, as against the value of exports increasing marginally by less than 2 per cent from Rs41,097m in 1987 to Rs46,764m in 1988. The current account deficit deteriorated to Rs13,841m in 1988 compared with the Rs10,093m deficit in the previous year. The total number of commercial banks operating in Sri Lanka has risen to 25 and the number of indigenous banks to six with the opening of the Seylan Trust Bank, while remittances by Sri Lankans employed in the Middle East during the first half of 1988 rose to Rs4960m.

The Premadasa government has given the go ahead to broadbase ownership of 25 state-owned commercial undertakings, identified by the Presidential Commission on Privatisation – now styled the 'Peopleisation' process. This was an area the World Bank was keen the government should commit itself to and with the first government-owned undertaking, United Motors, already under the hammer of the 'Peopleisation' initiative, strong aid flows from the US Agencies and the ADB are well on the cards. However, to the chagrin of the World Bank, who

KEY INDICATORS

	Unit	1984	1985	1986	1987	1988
Population	'000	15,599	15,840	16,100	16,400	16,590
Gross domestic product (GDP)	Rs m	153,746	162,375	179,474	196,723	223,010
GDP per capita	Rs	9,856	10,251	9,918	11,995	13,442
Wholesale prices	1985=100	117.9	100	97.0	110.0	129.6
Consumer prices	1985=100	98.5	100	108.0	116.3	132.6
Wages (minimum)	1985=100	91.5	100	105.3	110.5	138.3
Volume of exports	1985=100	97	100	107	108	112
Tea	1985=100	103	100	105	102	111
Rubber	1985=100	105	100	92	88	82
Coconut products	1985=100	52	100	104	75	37
Volume of imports	1985=100	104.6	100	113.8	116.5	119.3
Exports	RS$m	37,347	36,207	34,072	41,133	46,928
Tea	Rs m	15,764	12,003	9,253	10,654	12,299
Rubber	Rs m	3,301	2,566	2,622	2,929	3,706
Coconut products	Rs m	2,118	3,093	2,389	2,140	1,539
Imports	Rs m	41,541	54,049	54,559	60,528	71,200
Balance of trade	$m	–237.1	–522.6	–555.7	–472.2	–544.2
Current account	$m	0.9	–418.5	–417.1	–326.1	–403.7
External debt	$m	2,993	3,538	4,063	4,723	–
Debt service ratio	%	9.6	11.9	14.6	16.1	–
Exchange rate	Rs per $	25.44	27.16	28.02	29.45	31.81

* estimated figure

spent some US$70,000 on various studies and surveys pertaining to the privatisation of Telecommunications, government has decided not to go ahead with it, but has successfully offered foreign companies like Mitsubishi Motor Corporation shares in United Motors.

Meanwhile the Governor of the Central Bank, Dr Neville Karunatilleke, in a sharp rebuttal of certain statements by opposition leaders that the World Bank has frowned on the government's Poverty Alleviation Programme (PAP), said, 'the World Bank now proposes to get involved in this programme and we look forward to its active participation for its second phase, which deals with the investment component. The Bank has already worked out preliminaries on a 'Janasaviya'-related project called 'Work Plan on Employment and Poverty Alleviation' and the World Bank will make an extensive study of how best to assist the investment programme.'

The staggering Rs53bn budget deficit for 1989-90 glaring in the face of Lanka's overheated economy, is being financed predominantly by foreign loans totalling Rs18,000m, foreign outright grants of Rs5600m and domestic non-inflationary borrowings of Rs17,000m, together with borrowings from the local banking system to the tune of Rs395m.

With the violence in the north, east and southern segments of the country continuing, there has been an enhanced pressure on limited agricultural land, which has in turn marginalised the small farmer, the share cropper and the daily wage labourer – who form the backbone of Sri Lanka's agricultural society.

Despite the political turmoil due to terrorism and adverse publicity in the international media, 1988-89 proved to be an exceptional year for the Greater Colombo Economic Commission (GCEC) – the auth-

COUNTRY PROFILE

Political system: The small teardrop-shaped island of Sri Lanka is currently going through political turmoil because of racial conflicts between the Sinhalese majority and minority Tamils. Under the 1987 agreement with India the central government had to accept an Indian peace-keeping force (50,000 strong) and to devolve some powers to provincial governments in the north and east of the island where the Tamils are strong.Sri Lanka is governed by a 168-member National State Assembly, which is elected by universal suffrage, and by an executive president. The president appoints a prime minister and cabinet. Local authority is represented by 24 district councils in nine provinces and the new Pradesiya Sabas (councils based on local administrative divisions).

Political parties: *Ruling Party:* United National Party; Sri Lanka Freedom Party, Sri Lanka Mahajana Party, Sri Lanka Communist Party, Ceylon Workers' Congress, Tamil United Liberation Front, Lanka Sama Samaja Party, Nava Lanka Sama Samaja Party, United Lanka Janatha Party.

Population: About 74 per cent of the estimated 16.5m population are Sinhalese, 18 per cent are Tamils and the rest mainly Moors and Malays. Sinhalese are predominantly Theravada Buddhists, and Tamils are Hindus, while Moors and Malays are mainly Muslims.

Main cities/towns: Colombo (population 700,000* in 1987), Dehiwala Mount Lavinia (205,000), Jaffna (151,000), Kandy (122,000), Galle (100,000), Negombo (80,000), Kurunegala (40,000), Newara Eliya (25,000) and Trincomalee (52,000).

Language: Sinhalese, English and Tamil are the offical languages. English is widely spoken, particularly in business circles. Literacy, at 88 per cent, is among the highest rates in South Asia.

Media
Press: The main English-language dailies are the *Daily News*, the *Evening Observer*, the *Sun* and the *Island. Sunday Times* is published weekly. There are in addition at least eight daily newspapers published each in Sinhalese and Tamil. Business: The Ceylon Chamber of Commerce issues the monthly *Ceylon Commerce*, the Ceylon National Chamber of Industries publishes the quarterly *Industrial Ceylon* and the Trade and Shipping Information Service of the Ministry of Trade and Shipping publishes the quarterly *Business Lanka*, and *Expo News* fortnightly.
Radio: The Commercial Broadcasting Service of the Sri Lanka Broadcasting Corp transmits three national radio services, in English, Sinhalese and Tamil, and overseas Radio Ceylon has services in five languages. There were about 800,000 radios in Sri Lanka in 1987.

Television: Two TV channels, ITN and Rupavahini, broadcast from 0500 – 2345 daily in all three languages. English language news is broadcast at 2100 on Rupavahini and at 2130 on ITN. The CCIR system is used with PAL colour on 625 lines.

Domestic economy: Sri Lanka in 1988 was in the doldrums. Economic growth had averaged 6.2 per cent during 1977-82 but slumped thereafter as ethnic violence and weather both hit a mainly agricultural economy. Real growth was only 2 per cent in 1987. Drought hurt the agricultural backbone, offset only partially by better industrial growth, mainly in the private sector. Racial violence also frightened away tourists, another important contributor to growth and foreign exchange earnings. But there were improvements in port and insurance earnings. Remittances from overseas workers also rose in rupee terms in 1987. Sri Lanka had a considerable current account deficit in 1987 of $395.7m, up from 1986. In the I988 budget the government targeted privatisation of uneconomic state enterprises, better control and monitoring of expenditure and promised further liberalising measures. Continuing development projects include the Mahaweli irrigation and hydroelectric power plant programme. The Greater Colombo Economic Commission (GCEC) administers the Investment Promotion Zone programme, which offers generous tax concessions in order to attract foreign investment. Business investment outside these zones is the responsibility of the Foreign Investment Advisory Committee.

Employment: The labour force of 5.8m is about 34.6 per cent of the population (between the ages of 15-64). Agriculture has been declining but still employs a little over half of the labour force. Until the 1980s, most of the increase in employment went into the services sector. More recently industry has been taking more workers and accounts for more than 15 per cent. The unemployment rate is officially around 18 per cent, but taking underemployment into account a more realistic figure would be 30 per cent or more given the tourist slump.

External trade
Exports: Total exports in 1988 were Rs46,764m against Rs41,097m (1987). Total exports in 1987 were over $1.4bn, only a slight rise in dollar terms. In rupee terms, the rise was a much higher 21 per cent. Textiles and clothing exports have overtaken tea and in 1987 rose by 34 per cent. Tea export earnings rose 15 per cent and minor agricultural products rose 14 per cent except for coconut products which declined in 1987, hit by drought. Other exports include rubber and gems. Main destinations are the USA, UK, Egypt, German FR, Japan, Iraq and Saudi Arabia.
Imports: Total imports (1988) were Rs71,276m against Rs61,102m (1987). Total imports in 1987 were $2.1bn. Imports of intermediate goods rose, while imports of capital goods fell seven per cent and consumer-goods imports also fell three per cent. Major imports include petroleum, machinery and transport equipment, food and beverages, fertilisers and chemicals. Main sources of supply are Japan, Saudi Arabia, Iran, USA, German FR and the UK.

Agriculture: Agriculture accounts for 27 per cent of GDP, 53 per cent of exports and employs 50 per cent of the labour force. This sector includes huge, largely state-owned plantations of tea, rubber and coconuts, and smaller holdings where rice, sugar-cane, cassava, sweet potatoes, soyabeans, other vegetables, cashew nuts, cocoa, castor, spices, chillies, onions and other crops are produced, sometimes at virtually subsistence level. Plantation crops provide most export earnings, although output has declined in recent years. Sri Lanka is now self-sufficient in rice, the main food crop, but in 1987 because of drought, production fell 18 per cent. About 44 per cent of the land area is cultivated arable. In addition to extending cultivation through the Mahaweli irrigation project, a major development project to replant and redevelop neglected tea and rubber plantations is starting up and a sugar-cane plantation is to be developed in an undeveloped south-eastern district. Livestock raised include buffaloes, goats, pigs, sheep and poultry. About 38 per cent of the land area is forested, and forestry accounts for 1.8 per cent of GDP, providing timber for local demand. Fishing accounts for 1.8 per cent of GDP, producing prawns, shrimps, lobsters, crabs and beche-de-mer for export, and adding considerably to the local diet.

Industry: Industrial production increased by six per cent in 1988, against eight per cent in 1987. Manufacturing accounts for about 15 per cent of GDP and is expected to rise. About one-third of manufacturing output is based on raw materials from the agricultural sector. Public sector industry declined one per cent in 1988 but the private sector grew 11 per cent. State-owned industries include petroleum products, chemicals, textiles, leather goods, steel, rubber goods, shipping, cement, tiling, ceramics, kaolin, mineral sands, bricks, foodstuffs, paper, hardboard and plywood. Private sector industries include furniture, plastic products, metal products and machinery. Tourism was a growth industry until racial violence increased. Visitor arrivals dropped from more than 400,000 in 1982 to about 160,000 in 1988.

Mining: Mining and quarrying accounts for 2.3 per cent of GDP (1985). Sri Lanka is rich in minerals such as ilmenite, plumbago, graphite, dolomite, kaolin, rutile, feldspar, quartz, mica, monazite, apatite, industrial clays and limestone. A new plant for processing phosphate deposits is planned. Precious and semi-precious stones, such as sapphires, rubies, catseyes, alexandrites, aquamarines, garnets, tourmalines, zircons, topaz, spinels, amethysts and moonstones provide some export income.

Energy: Over two-thirds of all power requirements are met by imported oil. Hydroelectric generation supplies about 85 per cent of electricity requirements, and this proportion is likely to grow.

Membership of international organisations: ANRPC, Asian Development Bank, CCC, Colombo Plan, Commonwealth, EC Trade and Co-operation Agreement, ESCAP, FAO, GATT, G-77, IAEA, IBRD, ICAO, IFAD, IFC, ILO, IMF, IMO, INTELSAT, INTERPOL, IPU, IRC, ITU, NAM, SARC, South Asian Association for Regional Cooperation, UN, UNESCO, UPU, WFTU, WHO, WIPU, WMO, World Bank, WTO.

ority for the country's Export Processing Zones (EPZ). Foreign investment flows to the country through the EPZs increased significantly, with 31 projects being approved at an estimated investment of Rs1,713m against Rs812m in 1987, while employment at the Zones stood at 55,662 persons. Said President Premadasa, 'with the success of the EPZ at Katunayake in terms of investment and employment creation, we hope to further develop the Biyagama EPZ and create another at Koggala for the south'. Meanwhile, the Business Aquisition Act has been repealed giving greater confidence to the private investor, and a single umbrella body is being established to promote and co-ordinate foreign investment and operations of the Greater Colombo Economic Commission (GCEC) and the Foreign Investment Advisory Committee (FIAC). According to Tulie Cooray, Senior Manager (Appraisals), GCEC, 'it would be advantageous to have one authority for approval of all foreign investment ventures with the line ministry overlooking the project, and this policy package should be under one piece of legislation. After all the GCEC is performing extremely well with 12 projects having been approved up to April 1989 and our target for the year is around 40 projects.'

FIAC Director Kumar Fernando commented, 'in spite of unsettled conditions the number of investments sought for and approved rose significantly to 518 projects up to June 1988. These projects envisage an investment of Rs21,575m and have the potential to employ 64,873 persons. The 298 projects currently in operation and the 28 projects under implementation will yield a total investment of Rs22,355m by the end of 1989.'

In the face of adverse circumstances the economy recovered considerably in the agricultural sector from the drought-hit production levels experienced in 1987. Though tea production increased by 6 per cent to a record 227m kg, a major problem facing the industry is that wages erode 70 per cent of production costs and sales prices have not been reaching viable levels. A rational adjustment of discontinuing the payment of six days' wages irrespective of whether work is available or not, and paying estate labour in terms of actual work done, has been successfully negotiated by Plantations Minister Gamini Dissanayake. Meanwhile, rubber production rose marginally to reach 125.5m kg, but it was in the garments and textile export sector, for the third year running, that the highest foreign exchange earnings were recorded, with Rs13,651.68m against Rs.12,383.36m in 1987. However, coconut production declined substantially by 16 per cent, while paddy production, which suffered during the previous two years due to unfavourable weather and unsettled political conditions, recovered sharply to register a 16 per cent growth and an output level of 2.5m tonnes in 1988. Earnings from minor agricultural exports rose to Rs2561m from Rs1714m in 1987. Industrial output increased by 6 per cent with private sector industries showing an upturn of 11 per cent and those in the public sector declining by 1 per cent, due to production shortfalls in a number of state corporations in 1988-89.

However, Sri Lanka's trade deficit which widened steeply in 1988 showed a welcome improvement during the first quarter of 1989, according to government export-import figures.

The trade deficit which stood at Rs6,409m (SDR 152m) in the first three months of 1988 dropped to Rs6,243m (SDR 143m) in the corresponding period of 1989. The trade deficit at the end of 1988 rose to Rs24,512m from Rs20,000m in 1987. Total income from exports during the first quarter of 1989 stood at Rs11,871m (SDR 272m) as against imports worth Rs18,113m (SDR415m). The corresponding figures for the first three months of 1988 were exports Rs10,483m (SDR248m) and imports Rs16,893m (SDR 400m).

According to these statistics sharp increases in coconut and garment exports contributed to this improvement. The coconut exports rose from Rs209m in the first quarter of 1988 to Rs444m in the same period of 1989. Similarly garment exports rose from Rs3,359m to Rs3,981m. During the same period, however, tea exports had dropped to Rs2,612m from Rs2,664m and rubber from Rs950m to Rs916m.

Another cause for concern was the sudden drop in the export of gems from Rs573m to Rs514m. On the imports front, the pre-budget rush to import cars and motorcycles resulted in the country spending Rs813m on these two items during the first three months of 1989.

The tourist industry, which had been experiencing a downward trend since 1983, showed signs of recovery by mid-1988. However, the deterioration in the security situation in the last quarter put a damper on the optimism within the industry. With the opening of the newest five-star property in the Colombo Hilton, total tourist arrivals in 1988 eased off at 182,662, remaining roughly at the same level as in the previous year, but a significant 41,338 visitors came to Sri Lanka in the first three months of 1989, with gross earnings estimated at Rs2500m compared to Rs2415m in 1987.

With the completion of the One Hundred Thousand Houses Programme, and the coming to an end of the Million Houses Programme, President Premadasa announced the even more ambitious 1.5 Million Houses Programme for commencement in 1990 to meet his promise of 'shelter for all by the year 2000'. According to the new Housing & Construction Minister, Sirisena Cooray, 'Sri Lanka's bold step in launching the Million Houses Programme with a support based strategy has yielded extraordinary results in the development of the poorer groups of society. It has gone far beyond mere production of housing units. It has provided a base for the development of the urban community and the consolidation of rural societies throughout the island'. In fact, by the end of 1989 over 650 model villages were due to be in occupation, with an average 80 units in each village.

Outlook

Looking at the rising militancy in Sri Lanka, it seems that there is new sub-culture taking root in the island – whose fraternity comprises mostly the young, educated unemployed, who cannot be found jobs if production is at a standstill due to the destabilisation process brought about by terrorism. In fact production would not only generate employment but would cushion the cost of living that has been spiralling due to intermittent work stoppages. The work of these young people was their concession to the necessities. Their out-of-work identities are contra-establishment. In the context of the overall socio-economic scenario it is peace and normality that Sri Lanka is crying out for.

BUSINESS GUIDE

Time: GMT + 5.5 hrs.

Climate: Colombo and the south-west experience monsoon rains May-Aug; likely temperature range is 22-31°C with average rainfall of 2,240 mm p.a. The north-east experiences monsoon rains Nov-Feb; lower temperatures (down to 10°C) occur inland at higher altitudes, with average rainfall of 1,000-1,500 mm p.a.

Entry requirements
Passport: Passports valid for at least three months from the date of issue of visa are required by all except certain seamen.
Visa: Visas are not required by bona fide tourists who are nationals of 38 countries. Business visitors generally need to obtain a visa before arrival. Travel regulations are liable to change at short notice, and enquiries should be made before departure. Travel to areas in the north and east may be difficult.

Currency: Foreign currency must be declared on arrival and may be exchanged only at banks or authorised dealers. Visitors should have money for the number of days they intend to stay, at the rate of US$15 pr day. In addition, they should have a return ticket, or the moeny for it. Certain foreign currencies, including Indian and Pakistani, may not be exchanged. Unspent Rs may be reconverted on departure. It is advisable to retain foreign exchange, hotel and shopping receipts.
Customs: Personal effects are allowed duty-free, but must be re-exported on departure. Importation of firearms, ammunition, explosives, dangerous weapons, drugs and pornography is strictly prohibited. Export of antiques, rare books, palm leaf manuscripts, rare anthropological material and any wild animal, bird or reptile or part thereof is restricted, if not prohibited.

Health precautions
Mandatory: Vaccination certificates required for cholera if travelling from an infected zone, or from or via Bangladesh, Pakistan, India, Burma, Thailand or the Far East. Certificates for yellow fever are required if travelling from Africa, the Caribbean, or the northern areas of South America.
Advisable: Anti-malaria precautions and TAB and poliomyelitis vaccinations. Dengue fever is an additional hazard. Water should be boiled and filtered before drinking.

Representation overseas: Abu Dhabi, Baghdad, Bangkok, Beijing, Belgrade, Bombay, Bonn, Brussels, Burma, Cairo, Canada, Copenhagen, Dhaka, Dusseldorf, Geneva, Islamabad, Jakarta, Karachi, Kuala Lumpur, Kuwait, London, Maldives, Malé, Manila, Moscow, Nairobi, New Delhi, Ottawa, Paris, Prague, Rangoon, Riyadh, Rome, Singapore, Stockholm, Sydney, Tokyo, Washington.

Air access
National airline: Air Lanka.
Other airlines: Aeroflot, Balkan, Emirates, Gulf Air, Indian, KLM, Kuwait Airways, PIA, Saudia, Singapore, Thai International, UTA.
Tax: Airport tax of Rs350 is levied on departure.
International airport: Colombo International, Katunayake (CMB), 30 km from Colombo, with duty-free shop, bar, restaurant, buffet, bank, hotel reservations, post office, car hire and rail link to Colombo (journey time 1.25 hrs).

Surface access: There is a railway service linking Colombo with ferry boat services between Talaimannar and Rameswaram. Ferry services may be suspended during bad weather.
Main ports: Colombo (containers, LASH, ro-ro), Trincomalee, Galle. Jaffna was not frequently used in 1988-89.

Hotels: A 10 per cent service charge is added to hotel bills.

Credit cards: International credit cards such as American Express, Visa and Mastercard are widely accepted by hotels and shops. No surcharge should be levied on their use.

BUSINESS GUIDE

Car hire: Self-drive and chauffeur-driven car hire is available. Mini-buses and coaches are also available. National or international driving licenceBUSINESS GUIDE

must be presented for local endorsement (on weekdays only). Chauffeur-driven cars are generally recommended (and cheaper). Driving is on the left-hand side of the road.

City transport
Taxis: Metered taxis usually have yellow tops and red numbers on a white plate. Although mini-cabs are generally accurately metered, inaccurate meters or excessive charging should be reported to the police. A five per cent tip is usual.

National transport: *Note:* Severe internal disruption has resulted in recent years from civil unrest. It is essential to ascertain the security situation before travelling in the north, the east or the extreme south of the country.
Air: Upali Travel and Air Travel Ltd operate domestic services to several destinations; charter planes and helicopters are also available. There were no flights to Jaffna in 1987.
Road: The extensive road network has 27,000 km of road, 19,000 km of which is surfaced.
Buses: Express services, some air-conditioned, are available to all main centres, and may be booked in advance; local services are generally crowded.
Rail: Around 1,800 km of track is in regular use. Service to Jaffna has been discontinued. Regular services link Colombo with all other main centres, with air-conditioning, dining cars and first-class accommodation available on some services.

Public holidays: The full moon (Poya) day of each month is a public holiday.
Fixed dates: 4 Feb (National Day), 12 Apr (day prior to the Sinhala and Tamil New Year Day), 13 Apr (Sinhala and Tamil New Year Day), 1 May, 22 May (National Heroes' Day), 30 Jun (Bank Holiday), 25 Dec, 31 Dec (Bank Holiday).
Variable dates: Duruthu Full Moon Poya Day (Jan), Thai Pongal Day (Jan), Navam Full Moon Poya Day (Feb), Maha Sivarathri Day (Mar), Medin Full Moon Poya Day (Mar), Bak Full Moon Poya Day (Apr), Good Friday, Full Moon Poya Day (May), Eid al-Fitr, Wesak Full Moon Poya Day (May, 2 days), Poson Full Moon Poya Day (June), Eid al-Adha, Esala Full Moon Poya Day (Jul), Nikini Full Moon Poya Day (Aug), Binara Full Moon Poya Day (Sep), Wap Full Moon Poya Day (Oct), Birthday of the Prophet, Deepavali Festival Day (Nov), Il Full Moon Poya Day (Nov), Unduvap Full Moon Poya Day (Dec).

Working hours
Business: (Mon-Fri) 0830/0900 – 1615/1700. (Sat) (some) 0830/0900 – 1300.
Government: (Mon-Fri) 0830 – 1615.
Banking: (Mon) 0900 – 1300; (Tue-Fri) 0900 – 1330.
Shops: (Mon-Fri) 0800/0900 – 1700.

Social customs: Alcoholic drinks are not served in hotels or restaurants on Poya (full moon) days. Footwear and head-gear should be removed before entering Buddhist shrines; photographing statues of the Buddha is acceptable but not posing beside them; a yellow-robed Buddhist *bhikku* should not be asked to pose for photographs nor should visitors attempt to shake hands with him; discreet dress in public places is appreciated.

Telecommunications: An earth satellite station and the SE-WE-ME submarine cable provide communications with a wide range of countries.
Telephone: Dialling code for Sri Lanka, IDD access code + 94 followed by area code (1 for Colombo Central, 9 for Galle, 8 for Kandy, 30 for Katunayake International Airport) followed by subscriber's number. When dialling from within Sri Lanka, the area code is prefixed by 0 (i.e. 01 for Colombo Central). There is some domestic direct dialling, but trunk calls must be booked for some areas. There are some 90,000 telephone lines in existence. Public telephones are available in hotels, stores, post offices and telephone booths.
Telex and telefax: There is a public telex service at the General Post Office. Facsimile facilities also available.
Telegram: International telegraph service is available at the Central Telegraph Office in Colombo.

Postal service: The main Colombo post office is open 24 hours a day, 365 days a year.

Banking
Central bank: Central Bank of Sri Lanka.
Main centre: Colombo.
Head Offices: Bank of Ceylon, Central Bank of Ceylon, Commercial Bank of Ceylon, Hatton National Bank, National Savings Bank, People's Bank.
Foreign branches: Algemene Bank Nederland, American Express International Banking Corp, Amsterdam-Rotterdam Bank, Bank of America NT & SA, Bank of Credit & Commerce International (Overseas), Bank of Oman, Bankers Trust Co, Banque Indosuez, Chartered Bank, Citibank, Dubai Bank, European Asian Bank, Grindlays Bank, Habib Bank, Hongkong & Shanghai Banking Corp, Indian Bank, Indian Overseas Bank, Middle East Bank, Overseas Trust Bank, State Bank of India, Union Bank of the Middle East.

Advertising: Press, cinema and commercial TV and radio available. Outdoor advertising is widespread; direct mail advertising must conform to code of practice.

Electricity supply: 230-240 V AC, 50 cycles.

Weights and measures: Metric system (local units also in use).

Representation in capital: Australia, Austria, Bangladesh, Belgium, Brazil, Bulgaria, Burma, Canada, China PR, Cuba, Czechoslovakia, Denmark, Egypt, Finland, France, German DR, German FR, India, Indonesia, Iran, Iraq, Italy, Japan, Republic of Korea, Libya, Malaysia, Maldives, Netherlands, Pakistan, Philippines, Poland, Romania, Sweden, Switzerland, Thailand, UK, USA, USSR, Vatican, Yugoslavia.

Useful telephone numbers: Colombo Emergency: 33333 Fire and ambulance: 22222 Accident service: 93184 Directory enquiries: 161

Telephone information International calls: 100 Trunk calls: 101 Speaking clock: 104 Phonograms: 133 Tourist police: 26941/2 Railway information: 35838 Bus information: 28081 General post office: 26203, 25834 Flight information Air Lanka: 546175 (24 hrs) Other airlines: 030 2281 (24 hrs)

Hotels: Colombo
Airport Garden, 234 Negombo Road, Seeduwa (tel: 030-3771; tx: 22430).

Ceylinco, 69 Janadhipathi Mawatha, 1 (tel: 20431).

Ceylon Inter-Continental, 48 Janadhipathi Mawatha, 1, P.O. Box: 408 (tel: 21221; tx: 21188; fax: 547326).

Galadari Meridien Colombo, 64 Lotus Road, Echelon Square 1, P.O. Box: 282 (tel: 544544; tx: 22334/22335; fax: 549875).

Galle Face, 2 Kollupitiya Road, 3, P.O. Box: 63 (tel: 541010/6; tx: 21281; fax: 541072).

Holiday Inn, 30 Sir Mohamed Macan Markar Mawatha, 3 (tel: 22001/9; tx: 21200; fax: 547977).

Lanka Oberoi, 77-83 Steuart Place, P.O. Box: 252,3 (tel: 20001; tx: 21201; fax: 549280).

Mount Lavinia, 104 Hotel Road (tel: 715221/4; tx: 21294; fax: 715228).

Pegasus Reef, Santa Maria Mawatha, Wattala (tel: 530205; tx: 21168).

Ramada Renaissance, 115 Sir Chittampalam A, Gardiner Mawatha, 2 (tel: 544200; tx: 22386; fax: 549184).

Ranmuthu, 112 Galle Road, 3 (tel: 33986/7/8/9; tx: 21839).

Taj Samudra, 25 Galle Face Centre Road, 3 (tel: 546622; tx: 21729; fax: 546348).

Taprobane Hotel, 2 York Street, P.O. Box: 152,1 (tel: 20391/3, 548734; tx: 21557).

World Trade & Tourist Centre, 237 Galle Road (tel: 83372, 83345; tx: 21501).

Hotels: Kandy
Chalet, 32 Rajaphilla Road (tel: 24353).

Citadel, 124 Srimath Kuda Ratwatta Mawatha (tel: 25314/5; tx: 21228,21389; fax: 547087).

Hill Top, 200/201 Bahirawakanda, Peradeniya Road (tel: 24162; tx: 21788).

Mahaweli Reach, 35 Siyambalagastenne Road, P.O. Box: 78 (tel: 32062/3).

Suisse, 30 Sangarajah Mawatha, P.O. Box: 13 (tel: 32083, 22672, 22637; tx: 21135).

Thilanka, 3 Sangamitta Mawatha (tel: 25496/7; tx: 22144).

Topaz, Anniewatte (tel: 24150; tx: 21121).

Business advice and information
Central Bank of Sri Lanka, PO Box 590, 34-36 Janadhipathi Mawatha, Colombo 1 (tel: 21191, 27325, 29938, 20863/4, 22094, 22918; tx: 21176, 21290, 21627).

Ceylon Chamber of Commerce, 127 Lower Chatham Street, PO Box 274, Colombo 1 (tel: 22084, 36444, 21745/6/7; tx: 1193).

Colombo Plan (for Co-operative Economic and Social Development in Asia and the Pacific), 12 Melbourne Avenue, PO Box 596, Colombo.

Department of Commerce, Raksaana Mandiraya, Vauxhall Street, Colombo 2 (tel: 36114; tx: 21908 comece ce; cable: COMMERCE).

Export Promotion Council of Ceylon, 5 Charlemont Road, Colombo.

Federation of Chambers of Commerce and Industry, Ground Floor, People's Bank, PO Box 2015, 220 Deans Road, Colombo 10 (tel: 599530; tx: 22186 WILCO attn FCCISL).

Foreign Investment Advisory Council (authority for non-Free Trade zone projects), International Economic Co-operation Division, Ministry of Finance and Planning, 1st Floor, Room 120, Galle Face Secretariat, PO Box 1689, Colombo 1 (tel: 28367, 26286).

Greater Colombo Economic Commission (authority for Free Trade zone projects), 14 Sir Baron Jayatillaka Mawatha, PO Box 1768, Colombo 1 (tel: 34403/4/5, 35407/8/9; tx: 21332 ECONCOM CE).

National Chamber of Commerce of Sri Lanka, 2nd Floor, YMBA Building, Main Street, PO Box 1375, Colombo 1 (tel: 25271, 547412).

National Chamber of Industries, 1st Floor, 20 Galle Face Courts 2, Colombo 3 (tel: 29038).

Sri Lanka Chamber of Small Industries, 430, R.A. De Mel Mawatha, Colombo 3 (tel: 573464).

Sri Lanka Export Credit Insurance Corporation, 278 Union Place, Colomob 2, (tel: 547508/9).

Sri Lanka Export Development Board, 310 Galle Road, Colombo 3 (tel: 573044, 573042; tx: 21457).

Sri Lanka Importers', Exporters' & Manufacturers' Association, 55 De La Salle Street, Colombo 15.

Sri Lanka State Trading (Consolidated Exports) Corp, 68/70 York Street, PO Box 263, Colombo (tel: 25167/8).

Sudan

Suzanne Cronje

General Omar Hasan al-Bashir's military coup of 30 June 1989 came within one week of the previous administration's 1989–90 budget. This predicted a record deficit of S£13.2bn (£1.9bn or $2.93bn): foreign aid was expected to cover less than half of this shortfall; foreign debt stood at $13bn; and the debt service ratio was 110 per cent of export earnings. Inflation was running at 80 per cent a year, and the rise in the price of food and other commodities was causing public protests and at times even riots. The ousted government of Prime Minister al-Mahdi could clearly be accused of economic mismanagement.

In his early statements al-Bashir said the army had taken power because of the increasing deterioration of the economic, political and social situation. He promised action to tackle the danger of economic collapse: prices would be controlled, bank deposits frozen. The previous administration's finance minister, Omar Nour al-Dayem, had openly admitted only the previous week that corruption and embezzlement had 'grown to pose a danger to our values, society and financial and economic conditions'; the new military government merely repeated what he had said. It established special courts to try profiteers, and it embarked on an anti-corruption drive. Hoarders of scarce goods – usually members of the old administration – were exposed with much publicity.

Money dealers were given one week to sell all their foreign exchange at set rates; unfortunately the banks soon ran short of local currency and were forced to issue unpopular promissory notes. There were large-scale arrests of trade unionists, intellectuals and even army officers, none of whom had ever been suspected of profiteering. Essential commodities remained scarce. The new government's early measures soon came to be regarded as hasty improvisations, especially as there were no signs that it was devising a new economic policy. Instead, it imposed a new 'defence tax' on all workers. During its first month in office it announced nationwide conscription and the imminent arrival of new arms supplies. Al-Bashir, who had come to power declaring: 'our primary goal is peace', appeared to be preparing for an indefinite continuation of the six-year-old civil war.

Civil war

The fighting in the south was the main reason why this large and potentially rich country was at the brink of economic collapse. In 1988 alone 250,000 people died of starvation caused by the war. The fighting had halted major development projects such as the construction of the Jonglei Canal and the exploitation of important oil reserves. It had also destroyed an important part of the agricultural sector, disrupted transport and other public services, and consumed vast amounts of military supplies. An end to the conflict was clearly essential for economic reconstruction, a fact accepted by the al-Mahdi government. On 30 June, the day of the coup, it was scheduled to sign a document which would have effectively abolished the Muslim sharia law: al-Bashir's military takeover was designed to stop this from happening, and had very little to do with the economic mismanagement of which al-Mahdi stood accused.

The imposition of sharia law is the principal bone of contention in the civil war between the Muslim north and the non-Muslim south. The main issue agreed during lengthy negotiations in 1988–89 between the Khartoum government and the leader of the Sudan People's Liberation Army (SPLA), Colonel John Garang, was the suspension of sharia laws pending the holding of a constitutional conference in September 1989. This conference would have asked all parties to accept a new, secular constitution which would have guaranteed all parts of the country equitable access to political power.

The fundamentalist National Islamic Front (NIF) would have opposed the proposals; it had resigned some months earlier from the ruling coaltion in protest at the discussions which were being held with Garang. But the NIF's national strength would probably have been insufficient to stop the new constitution from going through. The NIF had only 55 parliamentary seats against the 168 of the two main coalition partners, al-Mahdi's Umma and the Democratic Union Party (DUP). In April 1989, after a report by al-Mahdi on the progress of the peace talks with

KEY FACTS

Official title: Republic of Sudan

Head of state: Chairman of Revolutionary Command Council for National Salvation; Lt Gen. Omar Hasan Ahmad al-Bashir

Ruling party: All political parties banned in June 1989

Capital: Khartoum

Official Languages: Arabic

Currency: Sudanese Pound (S£) = 100 piastres

Exchange rate: S£4.50 per $ (Nov 1989)

Area: 2,505,813 sq km

Population: 23.1m (1987)

GNP per capita: $482 (1987)

GNP real growth: 3.2% (1987)

Inflation: 24.4% (1986)

Trade balance: –$430m (1987)

Foreign debt: $11.1bn (1987)

the SPLA, parliament voted by 128 to 23 for the suspension of the Islamic laws.

The attempt by northern fundamentalists to impose sharia law on the nation and turn Sudan into an Islamic republic is equated by most southerners with the 'Arabist domination and exploitation' by Khartoum of the south. This is why the SPLA is opposed to any special links between Sudan and its northern neighbours, and in the course of the 1988–89 negotiations Garang asked that the government abrogate its defence treaties with Libya and Egypt. Although al-Mahdi's government was reluctant to do this at first, it finally agreed. But from his first announcement of the military takeover, al-Bashir stressed that his revolution had 'a pan-Arabist orientation'. He and his 15-man Revolutionary Command Council (RCC) would seek union with Egypt and Libya, he said. When Egypt recognised the new government within hours of al-Bashir's coup, the south was convinced that the whole affair was an Arabist plot.

The SPLA therefore seemed scarcely surprised when al-Bashir declared the outcome of the peace talks with the al-Mahdi government to be null and void. The newly promoted General offered the SPLA an amnesty and tried to summon Garang to Khartoum for fresh talks. Garang ignored the invitation, and remained impervious for some weeks to the increasingly less peremptory RCC appeals for a meeting. This finally took place in August 1989, but not before Garang had added a new dimension to the SPLA's peace terms – the RCC's immediate resignation. This was to be followed by the establishment of a broad-based interim government which would include the SPLA and representatives from the political parties and other groups; the merging of the government army and his troops into a single non-regional and non-sectarian force; and finally the holding of a constitutional conference. The chances of any early agreement with the RCC on these terms were nil: the new government's fundamentalist make-up was incompatible with the SPLA demand for a secular state. Bashir denies any personal links with the NIF, but the membership of both the RCC and the government consists overwhelmingly of Islamic fundamentalists and NIF sympathisers.

KEY INDICATORS

	Unit	1983	1984	1985	1986	1987
Population	m	20.5	21.2	21.8	22.5	23.1
Gross domestic product (GDP)	S£m	9,178	11,311	13,913	22,009	31,157
GDP per capita	S£	447	533	638	978	1,349
Inflation	%	30.5	34.1	45.3	24.4	–
GDP real growth	%	0.4	–2.8	–11.2	5.9	3.2
Agricultural production	1979-81=100	104.7	98.9	115.6	113.0	99.9
Food production	1979-81=100	101.1	94.1	112.7	112.2	97.0
Exports	S£m	810.6	817.3	844.7	833.2	1,497.1
Cotton	S£m	395.9	405.0	374.3	366.7	–
Gum arabic	S£m	73.6	66.4	66.0	141.7	267.1
Groundnuts	S£m	16.4	26.8	23.2	2.5	10.0
Sesame	S£m	70.0	96.1	97.8	58.9	134.8
Imports	S£m	1,760.8	1,490.8	1,739.1	2,402.2	2612.9
Petroleum	S£m	451.3*409.1	298.7	257.6	–	
Balance of trade	$m	–188.9	–80.8	–134.8	–306.9	–429.7
Current account	$m	–219.5	25.4	151.7	–20.6	–232.7
Total external debt	$m	7,445	8,466	8,929	9,568	11,126
Debt service ratio	%	7.9	6.3	9.0	24.7	6.8
Exchange rate	S£ per $	1.3	1.3	2.3	2.5	2.8

* estimated figure

Economic reconstruction

The RCC may eventually be forced by economic circumstances to agree to the terms negotiated by its predecessors with the SPLA. If it does not, foreign aid is likely to dwindle. Even before the coup some donor countries made it clear that their future support might depend on a negotiated end to the conflict. The government of the Netherlands stopped the $40m it had committed for 1989 because of the continued fighting. US assistance has been intermittently suspended for some years, partly because of the war, partly because of Sudan's debt repayment arrears. In 1988 Sudan's unpaid arrears to the IMF alone amounted to some $800m, and the government agreed to a four-year reconstruction programme on approved IMF lines. Its implementation was to start in October 1988, but the August–September floods – the worst in Sudan's recorded history – intervened, and implementation had to be postponed for a few months. But the IMF was widely unpopular in Sudan, and several previous attempts to follow its recommendations met with strong public resistance. The anti-IMF riots of 1985 developed into a revolt which led to the overthrow of the Nimeiri regime. The RCC will therefore need political and economic skills as well as wide public support if it is to implement the four-year programme.

The major World Bank structural adjustment loan which Sudan needs to reverse its economic position depends on the successful execution of this four-year plan which, among other changes, demanded another devaluation of the Sudanese pound. In October 1988 the al-Mahdi government made a gesture towards this requirement by introducing a dual exchange rate. Essential imports could be paid for at the rate of S£4.45 per $1, while a floating rate (S£11.45 to $1 at the time) was to be used for non-essentials (about 30 per cent of all imports) and remittances.

This did not satisfy the IMF conditions laid down, and the World Bank did not relax its pressure on Khartoum. The only regular support from the World Bank and its agencies to reach Khartoum in the 1988–89 financial year was an agricultural credit of $20m from the IDA, to help increase food production in Kassala. Ironically, the World Bank release announcing this loan was issued in November 1988 – two months after the floods – but still quoted research showing 'that severe floods are not likely to happen in the coming decade'. In May 1989 the World Bank annouced a $75m contribution by the IDA to a $300m recovery programme to help repair the damage caused by the 1988 floods which left 2m people homeless and destroyed their roads and crops.

The floods also damaged Sudan's main export, cotton. As a result, the 1989 harvest was expected to be down by 20 per cent of the 780,400 bales of the previous year. The urban drift has led to labour shortages in the vast Gezira cotton scheme, and there was a controversial official suggestion that some of the 1.5m displaced southerners who live in Khartoum's shanty towns might be resettled near the agricultural schemes to supply the essential labour. The second most important export crop is groundnuts, production of which has increased.

The growing of sorghum was Sudan's most spectacular success. Mechanised farming, good rains and a new strain led to a record harvest of 4m tonnes in 1986–87, and the 1988–89 crop was expected to be 6m tonnes, against an average annual consumption of 2.4m tonnes. Some of the sorghum is exported to Saudi Arabia, but earnings from this commodity do not cover the foreign exchange required to import wheat. Bread consumption has been rising at the rate of 10 per cent a year. In March 1989 the government arranged with Saudi Arabia for the delivery of 110,000 tonnes of wheat, and at the same time it reached an agreement with the American aid agency, USAID, to exchange 25,000 tonnes of sorghum for wheat on the basis of ruling world prices. The USA supplies Sudan with about 70 per cent of its wheat requirements, and the government has tried in vain to persuade the urban public to use sorghum rather than wheat as a staple food. In 1989 Khartoum was reported to face a 28 per cent deficit in its wheat flour requirements, and the government claimed that wheat imports cost the country $214m annually, with an additional $111m in food subsidies. An attempt at the end of 1988 to raise bread prices led to serious riots. In the rural areas sorghum is still accepted as the traditional staple, and in the starving south it would have been even more acceptable, but transport problems prevented it from being sent.

The difficulty with all agricultural commodities is inefficient marketing and inadequate transport. The port and the railways run at less than 20 per cent capacity. Sudan Railways – with a staff of 33,000 the country's largest employer – is undergoing a much needed three-year overhaul. Costed at $160m, the work is being financed by a number of Western donors and includes an IDA credit of $35m, authorised in 1987–88 by the World Bank. Most of the money will go on replacing faulty sleepers on the 4,752 km of track and on rehabilitating rolling stock. But some $27m is being made available specifically to allow a reduction of the workforce by 9,000 – a move which might have serious political effects.

Phase III of the Port Sudan modernisation programme which would increase its annual capacity to 11m tonnes was due to start in 1989 at the cost of more than $40m, but after the coup there were reports that the work may be delayed. Phase II was completed in 1986 and increased the annual capacity to 8m tonnes at the cost of $25m. In July 1989 foreign shippers complained about a 750 per cent increase in harbour dues in Port Sudan; this meant that the rate for a call by a 17,000 dwt ship had gone up from $16,000 to $120,000 without prior warning.

Fortunately many of Sudan's donors continued to finance important projects. In May 1989 the Islamic Development Bank ratified a $4.1m loan for agricultural development in Kordofan, and Japan granted $10.4m to increase agricultural production. The Japanese also continued to finance the repair of the telephone and sewage systems in Khartoum. Britain, which has provided some budgetary assistance, contributed $23.3m to the railway recovery programme. But total aid continued to diminish, and while there are no reliable statistics for 1987–89, the downward trend shown by the 1985 and 1986 figures ($1,223m and $1,000m respectively) is indicative.

Apart from trying to reassure donor countries and persuade them to restore their previous levels of aid, there was one obvious way in which the RCC hoped to generate in the short term some of the much needed foreign exchange: reactivate promising development projects. Within days of the coup Chevron, the American oil company, was invited to return. Chevron has already spent $1bn exploring and developing the rich Unity oil field in the south, and formed joint ventures with the government and other oil companies. In 1984 it was forced to withdraw because of SPLA attacks on its operations. A further estimated $1bn is required for infrastructural work, including roads and airfields. International contracts for the laying of a 1,400 km pipeline between the field and the Red Sea coast near Port Sudan had to be cancelled by Khartoum in 1986. This scheme was particularly opposed in the south because it would practically guarantee that only the north would benefit from any oil exports. An alternative southern pipeline to Mombasa in Kenya was suggested by some of Chevron's competitors, but Garang was not prepared to discuss schemes which might suggest that the SPLA was seeking secession.

Even without export arrangements, the return of Chevron would quickly improve the country's foreign exchange position. The Unity field, with estimated recoverable reserves of 250m barrels, could be brought into production within months and save Sudan some $150m a year on oil imports. Large tracts of land near the Chevron wells, which are believed to contain even bigger petroleum reserves, are awaiting exploration. Total confirmed reserves in Sudan are 2,000m barrels, a quarter of which is recoverable. The snag is that no foreign company will move into the south because of the security risks: there was no guarantee in 1989 that the cease-fire, which came into operation during the later stages of the al-Mahdi/SPLA negotiations, would hold after the al-Bashir coup. The Sudanese army – which had suffered severe setbacks before the truce – was in no position to make credible offers of protection.

A smaller but safer mineral venture took off in March 1989, when the first gold was poured in the Aberketeib mine in the Red Sea hills. The mine was out of production for 20 years, but the old tailings were found to have a high gold content: 7.4 grams per tonne, on the average. The tailings are being processed at the rate of 90 tonnes a day, and it was expected that 5,000 troy ounces of gold would be recovered by mid-1990. Aberketeib is run by an Irish-based concern, Kenmara Resources, which has a 49 per cent share in the project; the Sudanese company, Central Desert Mining, has the majority holding. Two other companies were already working old mine dumps in the area: Gbeit mine, which started production in 1987 and is a joint venture between the state-owned Sudan Mining Company (SMC) and the London-based Greenwich resources; and the Hassai project, operated jointly between the SMC and French interests, which went into operation in 1988. A new gold mine at Anab was planned to start up in 1992.

Jonglei and Egypt

Security considerations have prevented completion of the Jonglei canal project south of Malakal, abandoned in 1984. This was intended to increase the supply of water for Egypt by creating a shortcut in the White Nile. This resource is regarded as imperative by the economic planners in Cairo if Egypt is to irrigate enough land to keep up with its population growth and keep the level in the Aswan dam sufficiently high to generate enough power. It is likely that these considerations motivate Egypt in all its dealings with Sudan, and that this explains the quick recognition by Cairo of the new RCC government. However, Egypt is fully aware that a political settlement with the SPLA is essential if the Jonglei canal is to be completed and – despite al-Bashir's initial position – is putting heavy pressure on Khartoum to reach agreement with Garang.

The SPLA's 1984 attacks on the Jonglei contractors were designed to put a permanent end to the work. Southern objections to the scheme centre on the fact that the ecological dislocation which would result from changing the course of the river would disrupt their economy and traditional way of life for the benefit of the northern 'Arabists', in both Cairo and Khartoum. In June 1989 – two months after the SPLA's capture of the Jonglei provincial capital, the garrison town of Bor – Sudan and Egypt agreed to pay $17m in compensation to the French contractors, with the cost shared equally by the two countries.

Towards the end of 1989 the immediate future seemed to be bleak. There were increasing signs that the RCC regime was planning to divide the country so that it could establish a theocratic Islamic republic in the north. The SPLA accused Khartoum of having secret plans to this effect, and when al-Bashir denied it, Garang claimed that al-Bashir himself had advocated partition 'at least three times in the last forty days, and this is enough reason for disquiet'. If any attempt was made to put such plans into operation, he said, the SPLA would mobilise the trade unions, members of the now outlawed parties, and even parts of al-Bashir's own army to overthrow the regime in Khartoum. This was no idle threat since a similar combination of forces had managed to topple the Nimeiri regime in 1984. But if Garang were to carry out his threat and succeed, he would merely drive the fundamentalists into a revolt which would probably enjoy external support. With Egyptian President Hosni Mubarak the new chairman of the Organisation of African Unity, there was a growing belief that only Cairo could solve Sudan's problems.

COUNTRY PROFILE

Political system: Multi-party parliamentary system introduced in 1986, comprising a five-member supreme council, a 301-member National Assembly and a 20-member cabinet. Sudan gained independence in 1956. Formerly a condominium of the UK and Egypt.

Political parties: In 1987 the National Islamic Front, until then in opposition, joined the ruling coalition between Umma and the Democratic Unionist Party. The Sudanese Communist Party and the South Sudan Political Alliance remained in opposition. Outside parliament, the Sudan People's Liberation Movement leads an armed struggle against the government.

Population: The population, estimated in 1987 at 23.7m, is growing at around 2.9 per cent per annum. About 70 per cent live in rural areas, 20 per cent in urban areas and 10 per cent are nomadic. From the mid-1980s, the civil war and natural disasters drove large numbers to the towns, particularly Khartoum.

Main cities/towns: Khartoum (swollen by up to 2m refugees to an estimated 4m, including Khartoum North), Omdurman, Port Sudan, Wad Medani, Juba, Kassala, El Obeid, Atbara.

Language: Arabic is the official language and is generally used in business; English is also widely used, however. African languages include Nilotic and Nilo-Hamitic.

Media

Press: Each of the major political parties has its own newspaper. Main national dailies are *El-Ayam, El-Sahafa, Sudan Times and New Independent.* Also monthly English language magazine, *Sudan Now*.

Radio: National service broadcasting in numerous languages, including Arabic and English.

Television: Single TV channel broadcasts 35 hrs/week, chiefly in Arabic with a weekly news review in English.

External trade: The 1986 deficit on the trade balance was S#1.57bn. The debt ratio (total debt outstanding over exports of goods and services) was 1,428.6 at the end of 1987, the second highest in Africa. Despite workers' remittances and reductions in imports, the balance of payments deficit worsened to $17.5bn, largely because of reduced aid inflows. IMF facilities were withdrawn in 1986.

Exports: Total value in 1988 estimated at $500m: cotton and cotton products (44 per cent of total value in 1986), gum arabic (17%), sesame (7%). Main destinations: Saudi Arabia, Italy, Japan, France.

Imports: Total value in 1988 estimated at $1,500m, and consisting of petroleum (11 per cent of total value in 1986), manufactured goods, machinery and equipment, chemicals, foodstuffs, transport equipment and textiles. Main sources: USA (17% of total value in 1985), Saudi Arabia (11.6%), UK (9.5%), German FR (8%), France (6%), Italy (5%).

Agriculture: Mainstay of economy, providing over 90 per cent of total export earnings and employing around 80 per cent of workforce. Estimated that only eight per cent of total area is cultivated and under one per cent is irrigated. The traditional farming areas are semi-arid and used for livestock rearing, while export crops are grown in the irrigated areas, mostly in the Gezira area between the Blue and White Niles. The principal food crops are sorghum (durra), cassava, millet, wheat and beans. Major cash/export crops are cotton, groundnuts, dates, sesame, sugar-cane (entirely consumed in Sudan, with imports), and gum arabic from the country's forests (world's leading supplier). Output of both food and cash crops has been severely affected by drought, locusts, and the worst floods this century in 1988. Major irrigation schemes such as the Jonglei Canal have been held back due to funding difficulties and the civil war. Sudan possesses vast fresh-water and marine fishing potential, but the Nile fisheries are barely exploited; fishing on the Red Sea coastline is being encouraged with government assistance.

Manufacturing: Together with construction contributes around 10 per cent to GDP and employs 4 per cent of the workforce. Activity is centred on the processing of agricultural products, particularly textiles, sugar, oilseeds, flour and footwear. Inadequate infrastructure, shortages of imported inputs, power cuts and lack of skilled manpower have thwarted government efforts at developing large-scale industries. A 1987 government plan envisages the privatisation of some industries and the reconstruction and expansion of others, including tanning, food processing and cement manufacture.

Mining: Chromite is extracted from the Ingessana Hills, near the Ethiopian border. Gold and gypsum are mined in the Red Sea Hills. There has been an upsurge in gold prospecting. Other mineral deposits (few of which are found in sufficient quantities to warrant commercial extraction) include zinc, lead, copper, talc, coal, nickel and tin.

COUNTRY PROFILE

Energy: Besides fuel wood and imported petroleum, hydroelectricity is the main source of energy. Power cuts are regular occurrences, particularly during the summer. Oil discoveries in the south-west were due to come on stream in late 1986, and were scheduled to reach 50,000 b/d, much of which would be exported via a 1400-km pipeline from the production area near Bentui to Port Sudan. However, work was abandoned in 1985 after the site was attacked in the civil war. Total oil reserves are estimated at over 250m barrels.

Membership of international organisations: African Development Bank, African States Associated with EC (Lomé Convention), Arab Common Market, Arab Fund for Economic and Social Development, Arab League, FAO, IAEA, IFAD, IDA, ILO, IMF, Islamic Conference, Islamic Development, OAU, UN, WHO, WIPO, World Bank.

BUSINESS GUIDE

Time: GMT +2.

Climate: Tropical in the south, hot and dry in the north. In Khartoum hottest month May (26-42°C), coldest Jan (16-32°C).

Entry requirements
Passport: Required by all.
Visa: Required by all except transit visitors, Egyptian and Tanzanian residents. Travellers are required to register with police headquarters within three days of arrival and must obtain special permit to travel anywhere south, west or east of Khartoum.
Prohibited entry: Nationals of Israel and South Africa, and holders of passports with Israeli or South African visas.
Currency: No restrictions on import/export of foreign exchange, though all currency must be declared. Import/export of local currency is prohibited. From August 1989 new foreign currency restrictions mean that possession of foreign currency bank notes and/or travellers cheques without an official declaration is considered illegal. Hotels and such establishments are therefore not allowed to exchange foreign currency for the settlement of bills unless supported by a valid declaration.
Customs: Alcohol is strictly forbidden, and travellers arriving with alcoholic beverages are liable to immediate arrest.

Health precautions
Mandatory: Certificates of yellow fever and cholera vaccination required if travelling from infected areas.
Advisable: Yellow fever, cholera, typhoid and polio vaccinations. Malaria prophylaxis recommended as risk exists throughout the country. Water precautions are essential.

Representation overseas: Abidjan, Abu Dhabi, Addis Ababa, Algiers, Amman, Bangui, Beijing, Bonn, Brussels, Bucharest, Cairo, Dar es Salaam, Islamabad, Jeddah, Kampala, Kuwait City, Kinshasa, Lagos, London, Moscow, Muscat, Nairobi, N'Djamena, New Delhi, Paris, Rabat, Rome, Tokyo, Tripoli, Tunis, Washington DC.

Air access
National airline: Sudan Airways.
Other airlines: Aeroflot, Air France, Balkan, British Airways, Egyptair, Ethiopian, Gulf Air, Iraqi Airways, Libyan Arab, KLM, Kenya Airways, Kuwait Airways, Lufthansa, MEA, Saudia, Yemenia.
Tax: Airport tax of LS30 on international departures.
International airport: Khartoum (Code: KRT), 4.5 km from city; duty-free shop, bar, buffet and restaurant.

Surface access
Main ports: Port Sudan.
Overland access: From Egypt via Wadi Halfa. Access from other surrounding countries not practicable.

Hotels: Accommodation can be difficult to obtain outside Khartoum and Port Sudan. Advisable to book in advance. Service charge of 10 per cent is usual. Hotel bills subject to 10 per cent sales tax.

Electricity supply: 240 V AC.

Car hire: Available in main centres but expensive. National or international driving licence required.

City transport
Taxis: Easily available in Khartoum. Can be hailed, or taken from ranks. Official tariffs, but fares still subject to negotiation in advance of journey.

National transport
Air: Sudan Airways operates regular service between Khartoum, Port Sudan and El Obeid. Flights to the south have been suspended because of the civil war. Small air taxi companies fly from Khartoum to main towns.
Road: Main tarred roads are the 1186km route from port Sudan to Khartoum, from Port Sudan to Kassala and on to Shavak, from Khartoum to Sennar and on to Malakal. Another 800 km of tarred roads exist, but the rest of the country's 48,000 km network is of very variable quality.
Buses: Scheduled coach services include Khartoum-Kosti, Khartoum-Omdurman, Khartoum-El Fasher, Juba-Nimule, Juba-Faradge.
Rail: 5500 km network links Khartoum with Port Sudan, Kassala, Wau, Nyala and Wadi Halfa, but not Juba. Three-class service with sleeping, dining and air-conditioned cars available. Services can be very slow.
Water: Ferries and steamers ply the Nile between Khartoum and Juba and on the Sobat and Jur rivers.

Public holidays
Fixed dates: 1 Jan (Independence), 6 Apr (Uprising Day), 21 Oct (1964 Revolution), 25 Dec.
Variable dates: Sham el Naseem, Isra and Merag, Eid al-Fitr, Eid al Adha, Islamic New Year, Mouloud.

Working hours
Government and business: (Sat-Thu) 0800/0830 – 1330/1400.
Banking: (Sat-Thu) 0830 – 1200.
Shops: (Sat-Thu) 0800 – 1330, 1730 – 2000.

Telecommunications
Telephone: Direct dialling to Sudan from Europe and USA – but frequent delays. Dialling code for Sudan: IDD access code + 249 followed by area code (11 for Khartoum) followed by subscriber's number.
Telex: Telex services available through main post offices. Telex numbers have become five-digit, and all previous three-digit numbers are now prefixed by 22.
Telegram: Telegraph service available at Khartoum main post office 24 hours.

Banking
Central bank: Bank of Sudan.
Main centre: Khartoum.
Head offices: Agricultural Bank of Sudan, Bank of Khartoum, El Nilein Bank, Tadamun Islamic Bank, Faisal Islamic Bank, Industrial Bank of Sudan, National Bank of Sudan, Sudan Commercial Bank, Sudanese Estates Bank, Sudanese International Bank, Unity Bank.
Sudan branches: Bank of Oman, BCCI, Citibank, Habib Bank, Middle East Bank, National Bank of Abu Dhabi.

Representation in capital: Algeria, Belgium, Bulgaria, Central African Republic, Chad, China PR, Czechoslovakia, Denmark, Egypt, Ethiopia, France, German DR, German FR, Greece, Hungary, India, Italy, Japan, Jordan, Korea DPR, Rep. of Korea, Kuwait, Lebanon, Libya, Morocco, Netherlands, Niger, Nigeria, Oman, Pakistan, Poland, Qatar, Romania, Saudi Arabia, Somalia, Spain, Switzerland, Syria, Tanzania, Turkey, Uganda, UAE, UK, USA, USSR, Vatican, Yemen AR, Yemen PDR, Yugoslavia, Zaïre.

BUSINESS DIRECTORY

Hotels: Juba
Juba (tel: 2811, 3310).

Hotels: Khartoum
Araak (city centre), P.O. Box: 1957 (tel: 74826; tx: 251).

Friendship Palace, Baladia Park, North Khartoum, P.O. Box: 148 (tel: 78204; tx: 26017).

Grand (city centre), P.O. Box: 316 (tel: 77756; tx: 22436).

Green Village (in suburbs), P.O. Box: 2366 (tel: 73949; tx: 407).

Hilton International Khartoum, P.O. Box: 1910 (tel: 74100/78930; tx: 22250; fax: 75793).

Meridien Khartoum, Qasr Avenue, P.O. Box: 1716 (tel: 75970/79066; tx: 22974).

Sudan (city centre), En Nil Avenue, P.O. Box: 1845 (tel: 76407; tx: 594).

Hotels: Port Sudan
Red Sea, P.O. Box: 105 (tel: 2813/4).

Car hire: Khartoum
Auto Rent: PO Box 816, Khartoum (tel: 81098; tx: 22345).

Nile Automobile: USA Embassy Building (tel: 75452).

Banking
Bank of Sudan, PO Box 313, Khartoum (tel: 78064; tx: 22352).

Bank of Khartoum, Sharia el Gamhoria, PO Box 1008, Khartoum (tel: 80137, 70198; tx: 22181 bkhof).

Sudan Commercial Bank, PO Box 1116, Khartoum (tel: 71468, 76344; tx: 22242/52).

Other useful addresses
Arab Bank for Economic Development in Africa (headquarters), PO Box 2640, Khartoum.

Ministry of Finance & Economic Planning, PO Box 700, Khartoum (tel: 77255).

Sudan Air Transport, PO Box 253, Khartoum (tel: 70325).

Sudan Chamber of Commerce, PO Box 81, Khartoum.

Sudan Development Corp, PO Box 710, Khartoum (tel: 79536; tx: 22427).

Sudan Exhibitions and Fairs Corporation, PO Box 2366, Khartoum (tel: 77702; tx: 22407).

Sudan National Broadcasting Corporation, PO Box 1094, Omdurman (tel: 512100).

Sudan News Agency, PO Box 1506, Khartoum (tel: 75771; tx: 22418).

Sudan Railways, PO Box 65, Atbara (tel: 2000; tx: 40001).

Sudanese Tourism Corp, PO Box 2424, Baladiya Avenue, Khartoum (tel: 74664, 74266).

Suriname

Tony Thorndike

Despite the return to civilian rule after the November 1987 general elections, peace has not returned to this troubled land, nor is it clear to what extent the military holds the reins of power. The question of the full resumption of Netherlands' aid causes extra concern as the economy remains sluggish.

The future is uncertain for what was once one of the richest Caribbean territories. Ethnic divisions among the beleagured 350,000 Surinamers remain and two small radical political groups which found favour during the seven years of military rule continue to exercise influence. The economy generally, and the critically important bauxite sector in particular, continues to suffer from the effects of military mismanagement and from the continuing Brunswijk insurrection in the near-empty hinterland. Although peace talks have a reasonable chance of success, devaluation and recourse to the IMF appear inevitable.

Civilian rule?

A referendum on a new constitution in September 1987 and a general election two months later officially restored civilian government. There was no doubting the relief felt by the great majority of the people to whom military rule signified oppression and fear. The new government was installed in January 1988, composed of the Front for Democracy and Development (Het Front voor Democratie en Ontwikkeling) which had won 40 of the 41 seats it contested. The military-backed National Democratic Party won only two. The remaining six seats in the 51-seat National Assembly went to a Javanese party, Pendawa Lima (four) and the radical Agricultural Workers and Peasants Party (PALU) (two). These six seats represented interior constituencies suffering from civil war, well away from the coastal strip where 95 per cent of the population live. Since the Front secured a two-thirds majority it was entitled to nominate the president. This it duly did, the mantle of office falling on Ramsewak Shankar, once an agriculture minister before the February 1980 military takeover.

The way seemed clear for the army, led by Lieutenant-Colonel Desi Bouterse, to return to its barracks at Fort Zeelandia in the capital, Paramaribo. But to believe this would be to fly in the face of reality. Not only does the army profess a 'vanguard' role in national development, a role which is open to wide interpretation, but it also exploits deliberate ambiguities in the constitution. In theory, ultimate authority rests with the 51-seat National Assembly in conjunction with an executive presidency. But the Assembly and president are advised by the National State Council, on which the army is represented. Its powers are regulated by 'organic law' and are far from precise. It was not until January 1989 that Council appointments were announced. Seven seats were to go to the Front, two to trade unions, one to employers, one to the military (and allocated to a marxist) and up to three for parties sympathetic to the military. Similar ambiguity surrounds the powers of the five-person all-military Security Council. The insurrection led by former bodyguard of Bouterse, Ronnie Brunswijk, and the consequent need for security, has also enhanced military influence. In addition, the NDP has taken advantage of post-election widespread disillusionment with the Front, which was expected to achieve the impossible.

The military takeover by 16 NCOs led by the then Sergeant Bouterse in 1980 was largely unplanned.

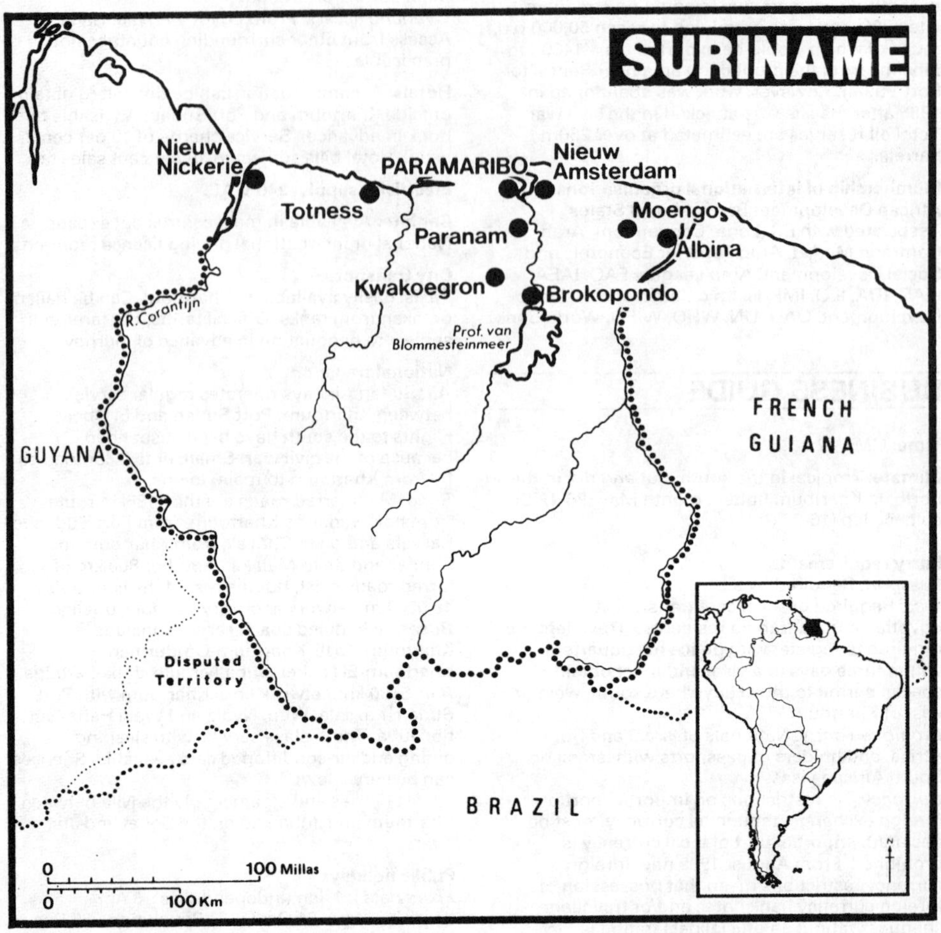

Initially popular, it capitalised on a succession of weak coalition governments and a marked lack of national unity. Although to a large extent disunity was attributable to ethnic division, part of the blame could be levelled at the impact of an extraordinarily generous aid package. Granted by the Netherlands at independence in 1975, it was worth $2.7bn over 12 years. Coupled with profitable rice and, above all, bauxite industries, it meant that little effort was needed to ensure prosperity. But all was not well, Surinamese nationalists deplored the extent of the country's dependency and penetration by foreign capital.

The basis of Surinamese politics was *apanjaht*, or voting according to ethnic background. The three main parties, all right-of-centre, therefore represented the dominant ethnic groups, namely the Hindustanis or Indo-Surinamese, themselves divided into Hindus and Moslems; the Creoles, or Afro-Surinamese; and those of Javanese descent. The respective parties were the Progressive Reform Party (VHP) led by wealthy businessman Jaggernath Lachmon, the Suriname National Party (NPS) led by the last pre-military prime minister Henk Arron, and the Javanese Farmers Party (KPTI) of Willy Soemita. The Front was composed of these three and far from the military experience jolting the populace out of *apanjaht*, the results of the 1987 election again reflected the size of each ethnic community: 16 to the VHP, 14 for the NPS and 10 for

KEY FACTS

Official title: Republic of Suriname
Head of state: President Ramsewak Shankar
Ruling party: Three-party coalition: the Front for Democracy and Development
Capital: Paramaribo
Official Languages: Dutch
Currency: Suriname Guilder (SG) = 100 cents
Exchange rate: SG1.78 per $ (Nov 1989)
Area: 163,265 sq km

Population: 359,000* (1988)
GDP per capita: $3,420 (1988)
GDP real growth: 2.2% (1987)
Inflation: 30% (1988)
Trade balance: $119.3m (1986)
Foreign debt: $66m (1987)

* estimated figure

the KPTI. All three leaders became ministers (Arron was appointed vice president), although Soemita's appointment was controversial since he had been convicted of corruption while a minister in the 1970s. The old order, it seemed, lived on. But the experience of military rule would never be forgotten as it involved a massacre, the loss of Dutch aid and an intermittent flirtation with socialism, Cuba and Libya.

Bouterse's government was characterised by a lack of political direction and vacillation, but eventually a philosophy of socialist nationalism emerged, strongly influenced by a small Marxist-Leninist pro-Cuba group, the Revolutionary Workers' Party (RVP), two members of which, Chas Mijnals and Badrasaen Sital, were among the 16 NCOs who had seized power. The pro-Libya PALU was also influential. PALU's philosophy is one of nationalism, co-operative self-help and the promotion of a peasant-based economy stressing self-sufficiency. The two factions were favoured alternately and mutual suspicions were rife.

There was never direct military rule, rather the NCO *junta* ruled through a succession of prime ministers and councils of ministers. The RVP assumed prominence in late 1981 and, as the regime became more authoritarian, so increasing numbers of exiles left to join the 200,000 or so Surinamers who had left for the Netherlands immediately prior to independence. Protests, especially orchestrated by the active trade union movement, became more strident. In December 1982 Bouterse and his allies struck, killing 15 leading opponents. An outraged Dutch government suspended the aid programme, but the effect was muted as spending could continue, using over $200m worth of reserves. Cuba became a valued political ally but the Embassy rather overplayed its hand and suffered expulsion in the aftermath of the US-led invasion of Grenada in October 1983. PALU was delighted as it saw Cuba as another coloniser. One of its leaders, Errol Alibux, assumed the premiership, but as the economy sharply deteriorated, he was replaced. Faced with deteriorating support, the military established the cell-based 'vanguard' February 25th Movement and attempted popular mobilisation. The effort was in vain; by March 1984 the political pendulum swung to the right as gradually the business sector assumed influence and the three traditional parties began to participate in government. The February 25th Movement became the NDP and an orthodox mass-based party. But even here ethnicity showed: the military, as in neighbouring Guyana, were predominantly Creole and party leaders were not above making racist remarks during the election campaign.

KEY INDICATORS

	Unit	1984	1985	1986	1987	1988
Population	'000	395	400*	395*	397*	359*
GDP	SG	1,878	1,980	1,750	1,820*	2,192*
GDP real growth	%	–1.9	1.1	1.2	–8.1	2.2
Inflation	%	3.6	7.0*	10.7	20.5	7.3
Exports fob	$m	374.1	336.1	337.1	339.8	358.4
Imports fob	$m	391.6	309.5	304.1	274.3	239.0
Trade balance	$m	–17.5	26.6	32.0	64.5	119.3
Current account	$m	–78.7	–11.6	–22.4	74.8	63.2
External debt	$m	59	57	57	66	–
Debt service ratio	%	13.6	12.3	8.8	18.2	–
Exchange rate	SG per $	1.785	1.785	1.785	1.785	1.785

* estimated figure

Illusive peace

The insurrection led by Brunswijk began in July 1986 and focused attention on a near-forgotten and marginalised segment of Surinamese society, the 'bosnegers' or Maroons, descendants of slaves who escaped into the interior. Brunswijk, a bosneger in his late twenties, was a member of one of the five clans. Claiming discrimination, he deserted the army and began a 'Robin Hood' campaign, giving proceeds of bank and other robberies to his kinfolk. After his home village was razed by the army he travelled to Amsterdam to gain support from the exile movement. On returning, well-organised attacks began mainly in the east and south of the country where most of the bauxite plants are. The rebels, by now termed the 'Jungle Commando', attacked the international airport at Zanderij, occupied the important bauxite town of Albina and even reached the outskirts of Paramaribo. Crucial power lines to the Paradam alumina smelter were also destroyed, further crippling the depressed economy. The army responded by attacking bosneger villages and some 500 villagers were killed, prompting highly critical reports by the United Nations human rights investigator Amos Wako and by Amnesty International. Large numbers of refugees, about 12,000 in all, fled into neighbouring French Guiana.

The rebellion weakened in 1987 in the face of army offensives and lack of supplies. Neither was there much support in the urban areas for the rebellion, the low status of the bosnegers being a factor. Peace talks began and a series of cease-fires came and went, involving church mediators, some of the bosneger *granmans* (chiefs) and the former Ombudsman, Denise de Hart. The atmosphere was not helped by the arrest in December 1988 of Stanley Rensch, the leader of the Suriname human rights organisation, Mooiwana 86 (so named in memory of a particularly notorious massacre of bosnegers by the army) on his arrival from the United States.

A major stumbling block was the refusal by the army to negotiate directly with the rebels. However, attitudes appeared to soften when President Shankar announced in early 1989 that an amnesty would also be negotiated covering actions (and atrocities) undertaken by both sides in the war. But that angered the rebels: to them an amnesty would benefit the army and several Jungle Commando leaders demanded judicial enquiries into the killing of their kinfolk. Brunswijk experienced particular difficulty in restraining the more militant bosnegers and it was not long before different rebel commanders began to act independently. Paradoxically, the government found that it had to bolster Brunswijk's position in order to bring peace to the war-torn interior. By April the threat of renewed fighting became very real and some rebels talked about attacking Paramaribo. Fighting indeed resumed the next month. Important bridges were damaged and a rebel group led by the militant Barabbas captured the Afobaka Dam and its power station, which he threatened to destroy. This would have resulted in a total loss of power in Paramaribo and disastrous flooding.

Although Brunswijk ordered Barabbas to withdraw, the occupation prompted an emergency purchase of $2m worth of modern arms by the army and the establishment by the Assembly of a seven-person commission headed by Assembly chairman Lachmon to negotiate with Brunswijk. Meeting in French Guiana, an outline agreement was hammered out in early June. The amnesty law had been approved at the beginning of the month but Brunswijk insisted that it was not to be made effective until an agreed date and only after certain other conditions were met. The agreement required ratification by the Assembly and it was clear that if this was not done then hostilities would be specifically directed at economic and civilian targets, a grim prospect indeed for the battered economy.

The Dutch aid issue

The major priority of the new government was the resumption of Netherlands' aid, about $760m of which was still outstanding. After a series of meetings, interrupted by a number of diplomatic upsets, an outline agreement was reached in May 1988. Disagreements remained over the timing of disbursement but a visit by Overseas Development Minister Piet Bukman in August resulted in a NG100m (US$47m) 'bridging' aid over two years. In addition there would be payment of NG200m annually from 1990 to 1996, the same level as before the 1980 coup, making NG1.5bn (US$816m) in total. This was granted on the strict condition that disbursememts were to be project based and that on no account would any funds go to the army.

Notwithstanding the agreement, nothing was disbursed for several months. Diplomatic relations again soured when Bouterse refused to meet a Dutch parliamentary delegation in late 1988. The Netherlands government also protested at the arrest by military police of Rensch (he was released a week later) and the detention of a Dutch diplomat at the scene. The overall sensitivity of relations between the two countries does not augur well for the future.

Relations with Brazil, on the other hand, are more positive. Brazil made known its anger at the strong Cuban presence during the revolutionary phase of military rule and negotiated a series of arms deals in place of those allegedly offered by Havana. Loans and lines of credit totalling some $34m were granted in 1987 and 1988. But suspicions that narcotic smugglers were using Suriname as a transit point led Brazil to put considerable pressure on Bouterse. The message sank home and in April President Sarney made the first official visit by a Brazilian head of state to Suriname. Greater co-operation against trafficking was agreed and Brazilian technical assistance in establishing free trade zones and agricultural improvement was promised. In addition, a considerable number of projects were identified which Brazil was prepared to support, including the important Kalalebo Dam industrial project.

Economic reconstruction

The Kalalebo project, if realised, will supply all the country's electricity needs and play a critical role in diversifying the economy away from reliance on bauxite-alumina. But the expected cost is over $1bn and is regarded as being too ambitious by multinational aid agencies. In any event, the immediate priority of the civilian government has been to rebuild the economy after the damage caused by the Brunswijk revolt, conservatively estimated at $100m.

The bauxite-alumina industry was the worse hit. It could not have come at a worse time as the industry was already suffering from the prolonged depression in aluminium prices that began in 1979. The two companies, Shell (Billiton) and the Alcan-owned Suralco (Surinamese Aluminium Company) suffered sharp contraction; worse, production levies were imposed by an increasingly desperate government. In 1986 combined losses exceeded $75m; nonetheless they offered to invest $70m and $85m respectively in return for the removal of the levies and a special exchange rate to take account of the unofficial devaluation from the official SG1.8/US$ to parallel market rates which on occasions have reached SG14. A deal was struck and work went ahead on the reopening of the Moengo mines, closed after rebel attacks, and feasibility studies began on proposed new workings.

The voracious needs of the alumina smelters meant that as metal prices rose in 1987 bauxite had to be imported to ease shortages. There is no doubt that confidence in the industry has been restored although production remains below that achieved in earlier years. Export earnings rose from $240m in 1987 to over $350m in 1988, and workers earn about three times the national average of $4,500 per year. There is, however, an increased risk that industrial unrest, for which the sector has an unfortunate reputation, will return with prosperity. PALU activists wrested control of part of the trade union movement in the industry in early 1989.

Rice output also expanded and targets were exceeded, boosted by new polders in the west of the country, healthy demand from the Middle East and an agreement that farmers could retain 35 per cent of foreign currency earnings, which in 1988 exceeded $6m. The state-owned banana company, Surland, also recorded gains. Production reached 40,000 tonnes in 1988, the British market being particularly important. The feasibility of producing exotic fruit for the European Community is being actively explored in conjunction with Fyffes. Plans to privatise the company are at an advanced stage. Timber production, however, remains sluggish despite 85 per cent of the country being forested. The largest company, Bruynzeel, has exploited new regional markets with its plywood and prefabricated houses. Other sawmills produce hardwood, most of which is exported to the United States. Manufacturing remains limited, constrained by a small market and relatively high pay. The national oil company, Staatsolie, produced about 3,000 barrels per day but the planned construction of a $35m refinery to produce diesel oils and propane gas was temporarily deferred due to continuing depressed oil prices and excessive refinery capacity in the region. The intention is to save about $40m per year by halving oil imports worth $40m.

Oil imports consume about a third of export earnings necessitating severe exchange controls. As a result the trade imbalance has been turned around from a negative $95.5m in 1983 to a positive $30m in 1987. Fortunately there are extensive family remittances from the Netherlands and without these many of the 33 per cent unemployed would face extreme privation. The high unemployment rate (only 13 per cent in 1980) would have been higher but for the expulsion of numerous Guianese, mainly illegal, immigrants. It is the result of heavy job losses not only from the bauxite industry (2,000 lost) but also from government. The public sector as a whole accounts for some 42 per cent of employment and more jobs will go as a result of continuing falls in public expenditure. In the 1988 budget, for instance, this was reduced by a substantial SG272m to SG997m. Formal devaluation is likely but resisted; nevertheless, rates of SG5 to the dollar and SG2.5 to the Dutch guilder were applied to consumer goods in June 1989. A request for a loan of at least $35m from the IMF remains under consideration. Given the fiscal measures taken so far, and the resumption of Dutch aid at least in part, the request will most probably be favourably received. But the economy as a whole has much ground to regain.

COUNTRY PROFILE

Political system: Became independent of The Netherlands in 1975. Under military rule from 1980 to 1987. Parliamentary democracy returned with the November 1987 general election and the election by the Assembly of the president in January 1988.

Main cities/towns: Paramaribo (population 150,000), Nieuw Nickerie, Albina, Moengo.

Media
Press: Principal dailies include *De Ware Tijd* (a government newspaper), and *De West*. They and a number of periodicals appear irregularly due to shortage of newsprint.
Radio: Two government-run stations in Paramaribo: Stichting Radio-omroep Suriname (SRS) and Radio Suriname International (RSI).
Television: Government-owned commercial TV service by STVS and ATV (mainly Dutch and English)

External trade: Imports reduced by 70 per cent since 1981, to offset falling export revenues. Balance of payments is reliant on very limited foreign aid, remittances from expatriate workers and earnings from rice, bauxite and its derivatives.
Exports: Bauxite, alumina, aluminium (typically 75-80% of total value), rice, bananas, wood, shrimps. Main destinations: Netherlands (24% of total value in 1987), Norway (22%), USA (20%), Japan (13%).
Imports: Raw materials and semi-manufactures (37% of total value), refined petroleum products (28%), machinery and transport equipment (12%). Main sources: USA (30% of total value in 1988), Trinidad and Tobago (23%), Netherlands (10%), Brazil (9%), Japan.

Agriculture: Sector contributes around 10 per cent to GDP and employs about 14 per cent of the workforce. Permanent crop and arable land accounts for less than 0.5 per cent of total land area, and is concentrated mostly along the coastal plain. Self-sufficiency is nevertheless achieved in most basic foodstuffs. The staple food crop and most important agricultural export is rice, the farming of which is highly mechanised. Other major crops include palm oil, coconuts, bananas, sugar, citrus fruits and coffee. Commercial fishing, particularly for shrimps, is growing in importance and currently accounts for around seven per cent of export earnings. Around 80 per cent of Suriname is covered by forest, but only two per cent is commercially exploited as access is limited. The timber industry is, however, an important sector of the economy.

Industry: The largest single activity is the processing of alumina and aluminium/ingots, which is undertaken by Suralco (a subsidiary of the US conglomerate Alcoa) and Billiton Maatschappij (part of the Royal Dutch/Shell Group). Production of alumina/aluminium has fallen in recent years due to a slump in export demand, labour problems and civil war. Other industrial activities centre on the processing of agricultural produce (particularly timber) and the manufacture of construction materials. Until 1984 there was a steady growth in light manufacturing, especially clothing, soaps, footwear and plastics but this was limited by energy shortages and relatively high wages.

Mining: Economy is primarily based on the mining of bauxite. Ranked as world's seventh largest producer with exports of bauxite and derivatives accounting for 75-80 per cent of country's total export earnings. Annual production of bauxite, however, declined from 6.9m tonnes in 1973 to 2.5m tonnes in 1987 and in 1987 Suralco closed all its plants including the refinery following damage by anti-government rebels and by suspended miners. Billiton also announced in late 1987 that it was halting exports of bauxite from Suriname. Exports of bauxite, aluminium and alumina fell sharply in 1987, but 1988 saw a reversal of fortune due to reconstruction and high metal prices. Other commercially viable minerals as yet unexploited include iron-ore, copper, nickel, platinum and kaolin. Small quantities of gold are mined.

Hydrocarbons: Increasing amounts of oil (2,500 b/d in 1987) are extracted. Continued exploration underway in the Saramacca oil region is expected to double proven reserves to 350m barrels.

Energy: Total installed electrical generating capacity estimated at 415 mW. Electricity for the aluminium industry is supplied by the Afobakka hydroelectric power station. Construction of a 500 mW hydroelectric plant on the Kabalebo river in west Suriname has been delayed.

Membership of international organisations: Economic Commission for Latin America, FAO, G-77, GATT, IBA, ICA, ICAO, IDB, IFAD, ILO, IMF, IMO, INTERPOL, ITU, NAM, Organisation of American States, PAHO, SELA, UN, UNESCO, UPU, WHO, WIPO, WMO, World Bank.

Time: GMT -3.

Climate: Tropical but cooled by trade winds. Rain throughout the year but heaviest from Nov-Jan and from Apr-Jul. Average daily temperature remains fairly constant throughout the year at 27°C; daily range from 22°C to 35°C from May to Oct; slightly lower temperatures from Nov to Apr.

Entry requirements
Passport: Required by all.
Visa: Most visitors do not require a visa, but temporary residence is granted for only three days after which registration with the Immigration Office is required. Onward tickets required by all.
Currency: Foreign currency over the equivalent of SG5,000 should be declared on arrival, and the equivalent of SG500 must be exchanged at the official rate. On departure SG5,000 in foreign currency and SG100 in local currency does not need to be declared. Any foreign currency over this amount must be covered by the declaration made on arrival.

Health precautions
Mandatory: Yellow fever vaccination certificate required if arriving from an infected area.
Advisable: Yellow fever, typhoid and polio vaccinations. Malaria prophylaxis recommended for visits outside Paramaribo and some coastal areas. Water precautions should be taken.

Representation overseas: Brasília, Brussels, Caracas, Cayenne, Georgetown, Munich, The Hague, Washington.

Air access
National airline: Suriname Airways (SLM).
Other airlines: ALM, Cruzeiro, KLM.
International airport: Paramaribo-Zanderij (PBM), 45 km from city; restaurant, bar, bank, car hire, post office, hotel reservations.

Surface access
Main ports: Paramaribo.
Overland access: Coastal roads link Paramaribo with borders of Guyana (at Nieuw Nickerie) and French Guiana (at Albina). Some bus services operate from Cayenne to Paramaribo, but they tend to be slow and irregular.

BUSINESS GUIDE

Hotels: Limited availability in Paramaribo and Nieuw Nickerie – advance booking essential. Service charge of 10 per cent is usual.

Car hire: Available in Paramaribo through main hotels and Tourist Information Office, Waterkant 8. International driving licences accepted.

City transport
Taxis: Fares by negotiation in advance of journey; tipping is not usual.

National transport
Air: SLM operates a service from Paramaribo to Nieuw Nickerie. Charter flights are also available.
Road: Coastal towns are linked by road from Guyana border (Nieuw Nickerie), through Paramaribo, to French Guiana border (Albina).
Buses: Paramaribo and most towns have a local bus service. Bus routes link coastal towns but service is irregular and tends to be crowded.
Rail: Twice-weekly service from Onverwacht to Brownsweg by narrow gauge railway.
Water: Major form of local transport, centred on Paramaribo. Service operated by SMS. Local enquiries recommended.

Public holidays: 1 Jan, 25 Feb (Revolution Day), March (Phagwa), Good Friday, Easter Monday, 1 May, 7 May (Id al-Fitr), 1 Jul (National Union Day), 25 Nov (Independence Day), 25 Dec, 26 Dec.

Working hours
Business: (Mon-Fri) 0730 – 1700; (Sat) 0700 – 1230.
Government: (Mon-Fri) 0700 – 1400; (Sat) 0700 – 1130.
Banking: (Mon-Fri) 0800 – 1400; (Sat) 0800 – 1100/1200.
Shops: (Mon-Sat) 0700/0730 – 1300, 1600/1700 – 1800/2000.

Telecommunications
Telephone: Dialling code for Suriname: IDD access code + 597 followed by subscriber's five-digit number. No area code needed. International direct dialling from Suriname available.
Telex and telegram: Facilities available at the Government Telegraph Service, Gravenstraat 33, Paramaribo. Outside the capital telegrams may be sent from some post offices.

Banking
Banking centre: Paramaribo.
Central bank: Centrale Bank van Suriname.
Commercial banks (head offices): Algemene Bank Nederland, Handels-Krediet & Industriebank (Hakrinbank), De Surinaamsche Bank.

Electricity supply: 110/127 V and/or 220 V AC, 60 cycles.

Representation in capital: Brazil, China PR, France, Guyana, Indonesia, Republic of Korea, Libya, Netherlands, USSR, USA, Venezuela.

Hotels: Paramaribo
Ambassador, Dr S. Redmondstraat (tel: 77555; tx: 295 hotam).

Krasnapolsky, Domineestraat 39, P.O. Box: 1866 (tel: 74050, 75050; tx: 245 krasna).

Stardust, Leonsberg, P.O. Box: 914 (tel: 51959/51544; tx: 375).

Torarica, Rietbergplein 1, P.O. Box: 1514 (tel: 71500; tx: 167 surtor).

Business advice and information
Algemene Bank Nederland, Kerkplein 1, Paramaribo (tel: 71555).

Centrale Bank van Suriname, PO Box 1801, Waterkant 20, Paramaribo (tel: 73741; tx: 152, 211 (Foreign Dept).

Dienst Toerisme (tourist service), Grote Combeweg 99, Paramaribo.

Hakrinbank, Dr S Redmondstraat 13, Paramaribo (tel: 77722).

Suriname Tourism and Catering Department, PO Box 656, Waterkant 8, Paramaribo (tel: 71163, 78421; tx: 292 surair).

Suriname Chamber of Commerce & Industry, PO Box 149, Dr J de Mirandastraat, Paramaribo (tel: 73527/7; tx: 375).

Stichting Planbureau Suriname, PO Box 172, Dr S Redmonstraat 110, Paramaribo (tx: 170).

Suriname Airways, PO Box 2029, Paramaribo (tel: 73939; tx: 292).

Surinaamsche Bank, Gravenstraat 26-28, Paramaribo (tel: 71100).

Verenging Surinaams Bedrijfsleven (trade and industry association), Prins Hendrikstraat 18, PO Box 111, Paramaribo (tel: 75286/7).

Swaziland

Gillian Cronje

Swaziland is anxiously following developments in South Africa. It is more prosperous than almost all of South Africa's other neigbours, with the possible exception of Botswana. But this prosperity depends to a large extent on its proximity to South Africa: some of its wealth is undoubtedly derived from the strategems Pretoria has adopted to evade boycotts and sanctions. At the same time Swaziland is trying to forge more modern political institutions and a more independent foreign policy stance. Gradual though this evolution has been, it is beginning to attract adverse reactions in Pretoria.

Land

The government's policies of encouraging private enterprise and industrial development were briefly upset in 1988, with the passing of a controversial land decree. This effectively allowed the minister of justice to intervene in any land or property transaction. When this evoked outraged protests from foreign investors who suggested they might have to reconsider their plans in Swaziland, it was suggested that the publication of the law had been improperly influenced by the desire of some powerful personages to gain the upper hand in a deal involving some large farms.

Bankers, who were holding land as collateral, were particularly concerned; so were their clients who feared that their bank loans would be recalled. King Mswati is an absolute monarch whose word is law. To put the matter right he stated publicly that the 'misunderstanding' which had arisen over the recent decree 'has been taken care of'. This was followed by a government statement acknowledging 'the importance attached to the security of land title by investors and financial institutions'. The matter was being reviewed and the public would be notified in due course. Meanwhile people were requested 'to refrain from making further undue public announcements on the matter'.

Sugar is the country's main foreign exchange earner, but according to the Swaziland Sugar Association, production levels in 1988–89 declined for the second consecutive season. Export volumes were down by 16.1 per cent in the first 10 months of 1988, while values declined by 2 per cent during the same period to Emalangeni 192.5m (about $76m). Fortunately, domestic sales took up some of the slack due to the establishment of some new food processing industries, particularly a cold drinks plant in Matsaph and a jam factory in Malkerns.

Good progress was made with government policies aiming at food self-sufficiency. Maize production rose by 14 per cent to 98,180 tonnes in 1988 and it is expected that 1989 will show a similar degree of growth. The improvement is being achieved through better irrigation, made possible by a $8.3m IFAD scheme which involves the cement-lining of irrig-

KEY FACTS

Official title: Kingdom of Swaziland

Head of state: King Mswati III

Capital: Mbabane (Lobamba to become legislative capital)

Official Languages: English, SiSwati

Currency: Lilangeni – plural Emalangeni (E) = 100 cents

Exchange rate: E2.65 per $ (Nov 1989)

Area: 17,363 sq km

Population: 713,000 (1987)

GNP per capita: $773 (1987)

GNP real growth: –0.9% (1986/87)

Inflation: 12.4% (1987)

Trade balance: $26.8m (1987)

Foreign debt: $293m (1987)

ation canals. This also allows farmers to grow two vegetable crops a year, with the result that the yield of market gardens went up by 40 per cent in 1988. Smallholders now grow a great variety of higher-value crops, some of which are being sold to Mozambique. While Maputo's war-torn economy is unlikely to become a major customer for Swaziland's vegetables in the immediate future, the demand is expected to soar as peace and the tourists return, and the export potential is good. Meanwhile a domestic wholesale market has been established by the National Agricultural Marketing Board. It is the first such venture in Swaziland, where farmers are accustomed to retail their produce. But the market provides the growers with an assured outlet, and it is being increasingly used.

'Made in Swaziland'

The use by South Africa of Swaziland as a base for sanctions busting is a matter of record and must be taken into consideration in any analysis of this small country's trade figures. For instance, US imports from Swaziland rose by 116 per cent to $24m in 1987, a boost which was largely the result of re-routing South African exports. There was a congressional enquiry into this issue, and it was revealed that 'Made in Swaziland' goods, like garments, owed very little to Swazi labour. The main local input consisted of packaging the goods in boxes from Johannesburg. This situation is said to have slightly improved, but there is still a great deal of Swazi resentment at South Africa's subterfuges in its search for 'safe' export havens.

The situation is slightly different in the case of Coca-Cola, which moved from Durban to Swaziland in 1987, after American anti-apartheid activists threatened to boycott the parent factory in the USA. Coca-Cola sold its South African bottling plants to local interests, but the syrup production now takes place in Swaziland, employing some 100 people. Coca-Cola syrup has become Swaziland's second largest export, after sugar. With revenues some 40 per cent above initial expectation, it accounts for about a fifth of total foreign exchange earnings.

Usutu and South Africa

The complex relationship between the two countries' economies was illustrated by the story of SAPPI, the South African paper giant. In 1988 the British firm Cortaulds sold its Swazi and South African wood pulp industries to SAPPI for $403m. The deal included a share in the Usutu kraft pulp mill in Swaziland and 25,000 hectares of Swazi forest. Usutu provides about 13 per cent of the world's kraft softwood pulp exports and is Swaziland's largest employer of labour. Its acquisition was of strategic importance to SAPPI because water shortages make it difficult to run paper mills in South Africa itself. SAPPI wanted a controlling interest in Usutu, but Cortaulds owned only 50 per cent of the mill – the Commonwealth Development Corporation (CDC) owned the other half.

SAPPI and Cortaulds approached the CDC, which agreed to sell them 60 per cent of its half (i.e. 30 per cent of the total), so that SAPPI would have a total of 80 per cent of the shares. But there was strong opposition in Swaziland to allowing the transfer of Usutu to South African interests, and the government in Mbabane stepped in. The CDC part of the transaction was cancelled. It is not known how CDC was persuaded to forego the profitable sale, but the company has other important interests in Swaziland to consider – among other things, it is involved in sugar, fruit and fruit juice production and agricultural management training. The Swazis themselves bought some of CDC's Usutu shares.

A new venture, Swazi Paper Mills (SPM), recycles local waste cardboard and turns it into high quality tissue paper. Established in 1988, the $12m plant expected to employ 550 people by the end of 1989, and to have an output of 7,000 tonnes a year. SPM might eventually process some of the local wood pulp. The volume of Swaziland's pulp exports declined in 1987, but favourable world prices meant that its value ($29m) was up 27.6 per cent over the previous year.

New ventures

Swaziland has slightly reduced its dependence on South Africa by buying for R800,000 the fixed assets of the South African Transport Services (SATS). The transaction involved mainly buildings and workshops: SATS continues to provide road services, and the trains running through Swaziland are still firmly under South African control. But the government felt that an essential step had been taken to allow the country to develop its own road industry.

Foreign exchange earnings from the mining sector rose by 18 per cent in 1988 to $9.2m. The Havelock asbestos mine, one of the country's large employers, announced that it would remain in production for the foreseeable future because of improved demand for the mineral. Fortunately, a newly discovered ore body was found to have a rich vein of asbestos. The mine had been operating at a loss and required government-guaranteed loans in 1988, a year when production declined by 14 per cent. Asbestos exports earned some $2.3m, a 15 per cent decline compared to 1987. But diamond sales rose a surprising 77 per cent, and there is a great deal of prospecting by foreign companies for precious minerals.

Exports in 1987–88 totalled E795.4m ($316m), 25 per cent higher than in 1986–87, and the trade balance registered a surplus of E54.6m ($22.4m). The outflow of dividends and invested earnings almost doubled during the same period, from E68.9m to E130.8m – an indication of the extent of foreign interests in this small country. The secrecy surrounding fiscal affairs was underlined by reports in 1989 that Prime Minister Sotsha Dlamini would not allow the budget to be debated in parliament 'for security reasons'.

Royal politics

Soon afterwards the king fired Dlamini for unspecified acts of disobedience. The dismissal came in the wake of an industrial dispute. Sotsha Dlamini was replaced by an accountant, Obed Dlamini (no relation) who will act as prime minister 'while I search for the right man to fill the post', the king announced.

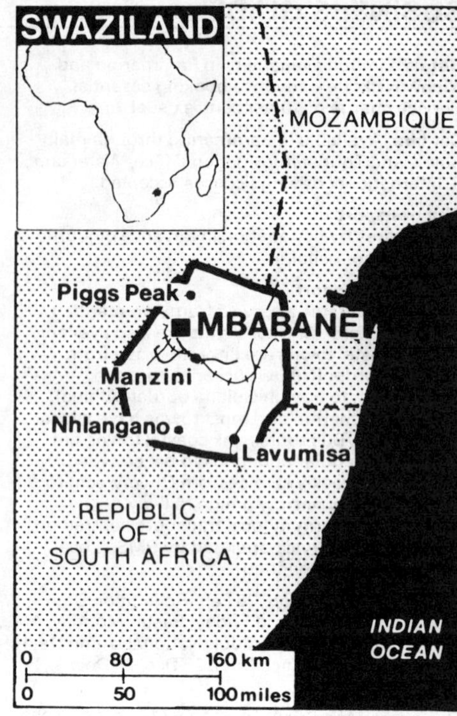

King Mswati – who turned 21 in April 1989, when he assumed full executive powers – is increasingly taking charge of politics, and it is thought that this is leading to a slight distancing in relations with Pretoria. At the end of 1988 Swazi diplomats joined their colleagues from other African countries for the first time when they walked out of an international conference addressed by a South African minister, a fact which was noted and resented in Pretoria. On the eve of the Non-Aligned Movement summit in 1989, new Prime Minister Obed Dlamini announced that his government was considering closer links with the countries of the Eastern bloc 'to allow us to become more effective in the Non-Aligned Movement'.

KEY INDICATORS

	Unit	1983	1984	1985	1986	1987
Population	'000	623	644	665	689	713
Gross domestic product (GDP)	E m	579.6	655.1	742.2	956.0	1,102.3
GDP per capita	E	930	1,017	1,116	1,388	1,546
Inflation	%	11.0	13.0	19.8	11.8	12.4
Agricultural production	1979-81=100	110.4	113.0	110.1	126.2	124.4
Food production	1979-81=100	110.4	113.4	110.2	128.0	126.1
Sugar production	'000 tonnes	403	429	402	537	461
Exports	Em	338.5	340.5	385.2	606.0	632.0
Sugar	Em	121.9	140.6	139.6	245.3	223.7
Wood pulp	Em	54.4	66.8	88.7	108.7	129.6
Imports	Em	612.8	642.7	707.1	798.9	876.5
Balance of trade	$m	–159.7	–140.4	–94.7	–18.0	26.8
Current account	$m	–91.1	–67.3	–23.8	24.0	44.8
Total external debt	$m	222.9	176.8	210.0	243.0	292.9
Debt service ratio	%	4.4	5.0	8.6	7.0	6.1
Exchange rate	E per $	1.1	1.4	2.2	2.3	2.0

* estimated figure

COUNTRY PROFILE

Political system: Executive power is vested in the monarch and exercised through a cabinet of ministers (all appointed by the monarch). Parliament is elected by an electoral college and consists of the Senate (20 members) and the House of Assembly (50 members). Gained independence from the UK in 1968.

Political parties: There are no political parties.

Population: The population was estimated at 725,000 in 1988. Some 60 per cent are Christian, the remainder being animists.

Main cities/towns: Mbabane (population 40,000), Manzini.

Language: English and Siswati are the official languages.

Media
Press: Two local daily papers, *The Times of Swaziland* and *The Swazi Observer*.
Radio: Programmes broadcast by Swaziland Broadcasting and Information Service, Swaziland Commercial Radio Ltd and Trans World Radio. Broadcasts Siswati and English.
Television: Government-owned Swaziland Television Broadcasting Corporation transmits colour services in English. Some programmes in Siswati.

External trade: Trade balance showed improvements in mid-1980s, but could deteriorate as a result of US sugar import quotas. The 1988 debt ratio of 58.3 (ratio of debt to current account exports) is among the lowest in Africa.
Exports: Sugar, woodpulp, canned/fresh fruit. Main destinations: South Africa, UK.
Imports: Capital goods, manufactures, fuels, chemicals, food. Main source: Southern African Customs Union.

Agriculture: Commercial farming is centred on sugar (major export earner), wood pulp (timber plantations cover six per cent of the total land area), citrus (grown on large estates), pineapples, tobacco and cotton. Main food crops are maize, beans, groundnuts and sorghum. About half of the working population are engaged in subsistence farming, mainly on small plots. Agriculture suffers from competition with cheap South African imports, but it recovered somewhat in 1985-86 and 1986-87. Livestock accounts for six per cent of GDP.

Manufacturing: Traditionally centred on sugar refining, fruit canning and woodpulp processing. Starting with textile production, the modern industrial sector is growing rapidly, with the South African market its main outlet. The government is promoting foreign investment and is attracting interest from South Africa; it is, however, concerned about excessive dependence on Pretoria. Tourism has stagnated recently.

Mining: Importance to the economy has fallen considerably, and the sector now contributes only three per cent to GDP. The future of the 50-year-old Havelock asbestos mine is in the balance. Some diamonds (mainly industrial grades) are extracted for export. Coal is mined at Mpaka and there are plans to exploit the 1m tonnes of known semi-anthracite reserves. About half the coal output is used domestically; the rest is exported to Kenya, Mozambique and the Republic of Korea.

Energy: No known oil or gas reserves but considerable deposits of coal (at least 300m tonnes of which are economically recoverable). Swaziland imports its oil needs from South Africa, mostly via Mozambique.Although still heavily dependent on South Africa for electricity supplies, imports have been reduced dramatically with the completion of the Lupholohlo-Ezulwini hydro-electric project in 1985.

Membership of international organisations: African Development Bank, African States in Association with the EC (Lomé Convention), Arab Bank for Economic Development in Africa (recipient), Commonwealth, Economic Commission for Africa, FAO, G-77, GATT, IDA, IFAD, IFC, ILO, IMF, OAU, Southern African Customs Union, Southern African Development Co-ordination Conference, UN, UNESCO, WHO, World Bank.

Time: GMT +2.

Climate: Semi-temperate in Mbabane and the rest of the highveld; drier in Manzini. Range from about 7-10°C (with occasional frost) during Apr-Sep, to 20-30°C during Aug-Jan. Wettest months Dec and Mar.

Entry requirements
Passport: Required by all.
Visa: Not required by UK and USA nationals and nationals of a few European, Commonwealth and African countries.

Health precautions
Mandatory: Yellow fever within five days of visiting infected area.
Advisable: Cholera, anti-malaria. Water precautions essential. NB: Health checks on overland access have been abolished.

Representation overseas: London, Maputo, Nairobi, Washington.

Air access
National airline: Royal Swazi National Airways.
Other airlines: Air Botswana, Air Zimbabwe, Comair, LAM, Lesotho Airways, Magnum, Zambia Airways.
International airport: Manzini-Matsapa (Code: MTS), 8 km from Manzini, 40 km from Mbabane; refreshments, currency exchange, car hire. NB: No direct intercontinental flights to Swaziland. Regular services via South Africa, Kenya, Tanzania, Zambia, Mozambique.

Surface access
Overland access: Road: Tarred roads from Johannesburg and Durban; and from Mozambique (Siteki-Lomahasha road). Coach services from Maputo and Johannesburg are scheduled. NB: Opening hours of border posts vary but the main border post with South Africa at Oshoek/Ngwenya is open to 2200 hrs daily.

Hotels: No official rating system. Accommodation is fairly scarce, especially during national holidays, so rooms should be booked well in advance. Bills generally include service charge, but 10 per cent tip also usual. A 10 per cent government tax is added to the room rates.

Electricity supply: 230 V AC, 50 cycles; 15 amp/round pin.

Car hire: Self-drive cars available at airport. National driving licence required; drive on left. General speed limit of 80 kph.

City transport
Taxis: Scarce. Best to order from hotel. A tip is usual.

National transport
Road: Approximately 3,000 km of roads, including 700 km of tarred roads connecting most main towns. Extensive road improvement schemes under way.
Rail: Network of about 320 km, connected to both Mozambique and South African railway systems, operated by Swaziland Railway Board; to be expanded. Used for freight transport only. South African railheads servicing the country are at Gollel, Pietretief, Breyten and Komatip\08oort.

Public holidays
Fixed dates: 1 Jan, 19 Apr (King's Birthday), 25 Apr (National Flag Day), 6 Sep (Independence), 24 Oct (United Nations Day), 25 Dec, 26 Dec.
Variable dates: Incwala, Good Friday, Easter Monday, Ascension Day, Commonwealth Day, Umhlanga.

Working hours
Government and business: (Mon-Fri) 0800 – 1300, 1400 – 1700; (Sat, business only) 0815 or 0830 – 1230.
Banking: (Mon-Fri) 0830 – 1300/1430; (Sat) 0830 – 1100.

Telecommunications
Telephone: Dialling code for Swaziland: IDD access code + 268. (There are no area codes).
Telex: Telex facilities available in main towns; country code: WD.

Banking
Central bank: Central Bank of Swaziland.
Head offices: Barclays Bank of Swaziland, Standard Bank Swaziland, Swaziland Development & Savings Bank.
Foreign branches: Bank of Credit & Commerce International.

Representation in capital: Israel, Republic of Korea, Mozambique, Taiwan, UK, USA.

Hotels: Ezulwini Valley
Ezulwini Sun Cabanas, P.O. Box: 123 (tel: 61201; tx: 2147).

Lugogo Sun Cabanas, Private Bag (tel: 61101; tx: 2058).

Royal Swazi Sun, Private Bag (tel: 61001; tx: 2014; fax: 61606).

Hotels: Manzini
The George V, P.O. Box: 51 (tel: 52061; tx: 2071).

Uncle Charlie, Ngwanest Mzimnene Riverside, P.O. Box: 48 (tel: 52297).

Hotels: Mbabane
Mountain Inn, Princess Drive, P.O. Box: 223 (tel: 42781; tx: 2135; fax: 45393).

Swazi Inn, Malagwane Hill, Manzini Road, P.O. Box: 121 (tel: 42235; tx: 2135; fax: 45393).

Car hire: Mbabane
Car Hire Services: (tel: 42672; tx: 2051).

Hertz: (tel: 43486; tx: 2111).

Car hire: Matsapa airport
Avis: (tel: 52734; tx: 2281).

Hertz: (tel: 84393; tx: 2085).

Other useful addresses
Central Bank of Swaziland, PO Box 546, Mbabne (tel: 43221; tx: 2029, 2108).

Central Statistics Office, PO Box 456, Mbabane (tel: 42151/4; tx: 2109).

Federation of Swaziland Employers, PO Box 386, Manzini (tel: 22768).

Ministry of Commerce, Industry, Mines and Tourism, PO Box 451, Mbabane (tel: 43201).

Ministry of Works, Power and Communications, PO Box 58, Mbabane (tel: 42321).

Royal Swazi National Airways Corporation, Mbabane (tel; 43744).

Swaziland Chamber of Commerce and Industry, PO Box 462, Mbabane (tel: 52889).

Swaziland Citrus Board, PO Box 343, Mbabane (tel: 43547).

Swaziland Commercial Board, PO Box 509, Mbabane (tel: 42930).

Swaziland Cotton Board, PO Box 230, Manzini (tel: 52775).

Swaziland Industrial Development Company, PO Box 866, Mbabane (tel: 43391; tx: 2052; fax: 43391).

Sweden

Sarah Webb

Sweden faced exciting times. The ruling Social Democrats, who had spent most of the past half century in power creating one of the world's most advanced welfare states, were leaning more towards market socialism as the 1980s closed. At the same time, they had come under considerable pressure to make environmental issues a priority even though it recognised that placing the environment first would have a particularly pronounced effect on the future of Swedish industry and the strength of the economy.

KEY FACTS

Official title: The Kingdom of Sweden
Head of state: King Carl XVI Gustaf
Head of government: Prime Minister Ingvar Carlsson
Ruling party: The Social Democratic Party
Capital: Stockholm
Official Languages: Swedish
Currency: Swedish Krona (SKr) = 100 ore
Exchange rate: SKr6.34 per $ (Dec 1989)
Area: 448,964 sq km
Population: 8.4m (1988)
GDP per capita: $21,289 (1988)
GDP real growth: 2.0% (1988)
Labour force: 4.39m (1987)
Unemployment: 1.6% (1988)
Inflation: 5.8% (1988)
Trade balance: SKr24.9bn (1988)
Foreign debt: $19.3bn (1987)

Changes in ruling-party policy

Some of the most exciting changes have taken place within the Social Democratic Party. The finance minister, Mr Kjell-Olof Feldt, who has been responsible for the recovery of the Swedish economy during the 1980s and who has spear-headed many of the reforms and the deregulation of the financial markets, sparked off a political debate which challenged some of the party's basic principles.

Since the Social Democrats were re-elected in September 1988, they have removed foreign currency exchange rules, outlined an impending reform of the tax system, and reduced subsidies to the textile industry and farmers. For his pains, the media has compared Feldt with the Soviet president Mikhail Gorbachev, primarily because of his efforts to give socialism a new direction. However, Feldt's proposals have met with opposition from the party stalwarts such as Sten Andersson, the foreign minister, and the trade union movement which has traditionally had a close alliance with the party.

Feldt extols the virtues of the market economy with its decentralised economic decision-making, competition in free markets and free choice of consumer goods, and sees it as an attractive alternative to the planned economy. He acknowledges the contribution made to the welfare state of a tax system which claims half the wages of the average worker, but says that such a state of affairs cannot continue. One of his major coups was to propose a complete reform of the Swedish tax system which (with taxes accounting for 55.8 per cent of GNP putting Sweden at the top of the world league when it comes to tax) was well overdue for change.

Much of Swedish industry is controlled not by the state but lies in private hands. Big companies like Volvo (cars, trucks), Ericsson (telecommuniations), Stora (forestry), and Electrolux (white goods) help to drive the economy while transport, communications and health care come under the state's control.

The building up of an economy which is heavily dependent on the export of manufactured goods has helped to give Swedes one of the world's highest living standards. The catch is that individual Swedes have traditionally paid through the nose for these services because of the very high taxes and as a result have little left in their pockets to spend afterwards.

The row within the party over the direction in future for social democracy left Prime Minister Ingvar Carlsson with the difficult task of reassuring party members that the government would maintain a general welfare policy while trying to reconcile the two diverging wings. Since taking over at the helm in 1986 after the assassination of Prime Minister Olof Palme, Carlsson has adopted a 'hands-off' approach to government, giving Feldt a free rein when it comes to economic policy and, furthermore, making a point of promoting women to senior or ministerial posts within his government in order to add some weight to the party's stated aim of equality.

Sweden's Social Democrats have long had a reputation for being committed to a strong economy. They favour full employment, high levels of income, and equality. While they are standing back from applying to join the EC, they are aware of the changes taking place and recognise the need to harmonise from the outskirts in preparation for 1992 and the internal market. Sweden stops short of applying for membership because it says it does not wish to compromise its neutrality.

Environmental and social issues

In their manifesto for the 1990s, the Social Democrats have taken environmental issues to heart, not only in terms of non-nuclear policy but with plans to restructure industry to eliminate many stressful, dull and monotonous jobs. In August 1989, a government-appointed committee recommended that Swedish workers should get six weeks' vacation and have more flexible working hours. Most salaried employees work a 7.5 hour day, five days a week. They also get a minimum five weeks' vacation and twelve months' parental leave with full pay to take care of their babies in the early months.

The committee favoured the idea of raising the minimum annual vacation to six weeks and increasing parental leave, which can be taken by either parent, to 18 months.

It also said that workers should have more flexibility in choosing when to work and when to take their holidays since many small businesses and shops close

KEY INDICATORS

	Unit	1984	1985	1986	1987	1988
Population	'000	8,331	8,358	8,387	8,370	8,440
Gross domestic product (GDP)	SKrbn	798.58	860.88	931.78	1,005.23	1,096.05
GDP per capita	$	11,549	11,977	15,647	19,063	21,289
Inflation	%	8.0	7.4	4.2	4.2	5.8
GDP real growth	%	4.0	2.1	1.1	2.4	2.0
Unemployment	%	3.1	2.8	2.7	1.9	1.6
Share prices	1985=100	107	100	163	195	218
Wages (hourly earnings)	1985=100	91.7	100.0	107.2	115.1	123.1
Retail sales	1980=100	146	157	172	180	–
Agricultural production	1979-81=100	108.22	108.10	109.99	107.28	111.41
Industrial production	1980=100	107	109	110	114	118
Manufacturing	1980=100	107	110	110	119	119
Wood pulp	'000 tonnes	9,293	9,123	8,922	9,420	–
Shipbuilding	'000 tonnes	358	224	–	–	–
Exports	SKr bn	241.81	260.50	265.10	281.33	304.19
EEC	%	49.0	46.8	57.2	–	–
USA	%	9.5	11.6	6.9	–	–
Imports	SKr bn	218.57	244.61	232.48	257.39	279.26
EEC	%	53.0	56.0	57.2	–	–
USA	%	8.3	8.4	7.8	–	–
Balance of trade	SKr bn	23.24	15.89	32.62	23.94	24.93
Current account	$m	246	–1,608	62	–1,247	–2,549
Public foreign debt	$bn	15.1	17.8	18.5	19.3	–
Exchange rate	SKr per $	8.3	8.6	7.1	6.3	6.1

* estimated figure

down for the month of July, forcing employees to take most of their vacation time then.

However, there is a cost to such reforms. The committee added that a shorter working day would lead to a 25 per cent loss of production within a few years. Swedish industry already suffers from a chronic shortage of labour at present because unemployment is so low, and measures aimed at encouraging foreigners, such as Danes and Norwegians, to come and work in Sweden have not met with much success.

The environmental issues being considered are sensitive ones. At the election in 1988, the Green Party entered parliament for the first time. It was assumed that the Greens would play a key role in pushing for environmental reforms. However, it seems the party is far too small to exert any real power, but has the effect instead of keeping the Social Democrats on their toes when it comes to green issues.

Back in 1980, Sweden chose to phase out nuclear power – which currently accounts for 50 per cent of the country's electricity production – by the year 2010. The first reactors are due to be phased out by the mid-1990s. However, the wisdom of abandoning nuclear power altogether is now being questioned since it will entail much higher costs for those industries such as forestry and chemicals which consume a lot of electricity. Other measures aimed at reducing pollution (from chemical emissions) will probably mean extra costs for industry too. The unions have voiced their opposition to the proposals because they worry that the higher costs will be met by job cuts and make their industrial employers less competitive – a serious threat to their high living standards.

The main points that the Social Democrats have pinpointed for the 1990s are greater freedom of choice in the workplace while maintaining full employment – one of the party's sacred cows. They see the need for strong economic growth, and the introduction of measures to combat inflation and tackle the inefficiencies of the public sector (where one in three employees work).

Economic performance

The biggest problem facing the Swedish economy is wage-push inflation. Pay increases in 1989 were set to reach around 9 per cent and are likely to be around 7.5 per cent in 1990 while price increases were expected to be between 6.7 and 7 per cent in 1989 and 1990. A report by one of the main banks suggested that once the tax reforms are introduced in 1991, these will have an inflationary effect because of the rise in indirect taxation on goods and services.

Meanwhile, the bank expected economic growth to be around 1.9 per cent in 1989 and 1.1 per cent in 1990. The balance of payments deficit was expected to deteriorate further from a forecast SKr22bn in 1989 to SKr25bn in 1990.

While the opposition parties criticise the government's policies when it comes to bringing down inflation and the fact that wages have risen faster than in competitor countries, thereby wiping out the advantages gained in the 1982 devaluation, they pay scant regard to the Social Democrats' successes. Unemployment has come down from a post-war peak of 3.5 per cent (in 1983) to 1.6 per cent in 1988. Economic policies, particularly wage policies, are usually based on a high degree of social and political consensus and reflect the government's commitment to equality and full employment.

Sweden's economic performance has improved considerably in recent years: the fiscal and external imbalances of the early 1980s have been virtually eliminated; the economy's competitiveness, and the balance of payments, improved after the devaluations of 1981 and 1982, the better terms of trade in 1985 and 1986, and the progressive reduction in price differentials between Sweden and its main trading partners. Increased exports and better current account performance reduced the external debt burden.

A recent OECD report warned that Sweden's high inflation rate and weak economic growth should inspire the government to introduce structural reforms to make the economy more competitive. It warned that high wage settlements and the focus of Sweden's export mix on basic products will tend to limit growth prospects over the longer term.

At the same time, it suggested that shortages of skilled labour and various regulatory factors could curtail the development of more dynamic industries which would otherwise play an important role in Sweden in the future. Sweden is not a country noted for its small entrepreneurs – in fact the emphasis traditionally has been very much on the big multinational companies.

Some of these worries may be addressed by a comprehensive tax reform which is going to take place in 1991. For example, lower marginal tax rates could provide more incentive for people to work longer hours and acquire labour skills, while simultaneously helping to hold down new wage demands. In the past, the double taxation of distributed profits and moves to encourage self-financing of investment have tended to discourage the creation of new enterprises.

On the wage front, the OECD observed that the recent move away from a centralised negotiationg structure and the tendency towards negotiating at branch or industry level has been accompanied by

higher settlements. The OECD forecast that gross domestic product would grow at an inflation-adjusted rate of 1.8 per cent in 1989 before slowing to 1.4 per cent growth in 1990. GDP was up 2.8 per cent in 1988.

Unemployment was expected to hold steady at 1.6 per cent in 1989 before rising to 1.9 per cent in 1990 while the current account deficit was forecast to widen to Skr17.3bn in 1989 and SKr21.5bn in 1990 – from 13.6bn in 1988.

By mid-1989, the government's plans to tackle inflation had met with very little success. In May, Feldt proposed plans for compulsory savings to cool off the overheated economy. This move represented quite a U-turn since Feldt had originally suggested increasing valued-added taxes and employers' payroll taxes to dampen the economy. The plan for higher value-added taxes met with considerable opposition because it was seen as a way of eroding the pay increases which had been won by the unions only a short time before.

Compulsory savings mean that private citizens and companies must pay 4 per cent out of their annual tax bills into savings funds which are closed for 16 months (from September 1989 to the end of 1990). The funds will be released from April 1992.

The government also proposed an environment tax of 1.5 per cent on companies, and higher taxes on oil and electricity to encourage energy saving, as well as a 10 per cent duty to be charged on investment in new building in the overheated areas of Sweden.

Financial markets

While Feldt and Mr Bengt Dennis, the governor of the central bank, have together introduced sweeping reforms in the financial markets, the biggest step came in early 1989 with the abolition of foreign currency restrictions on investments at home and abroad. The step was part of the the government's programme to eliminate all currency restrictions, thereby opening Sweden up to foreign competition and bringing it into line with the rest of Europe where controls on capital movement are part and parcel of the creation of the internal market.

Compared with other countries, Sweden has been rather slow when it comes to dismantling its foreign exchange controls which date from the Second World War. Prominent economists have pressed for their abolition during the 1980s but Sweden's economic problems during the 1970s (with a large budget deficit and current account deficit) provided a handy excuse to keep the status quo.

With the recovery of the economy, Sweden could afford to pay much closer attention to what was happening elsewhere in the world. In 1986, the central bank proposed allowing direct investments by Swedish companies overseas so that they were no longer forced to finance these investments abroad. Swedes were also allowed to buy foreign equities but the system worked on a pool basis whereby there was a limited amount of foreign equity available to Swedish investors – so depending on the demand, investors would pay an additional (but variable) premium.

The changes introduced in 1989 mean that foreign investors can now buy Swedish domestic bonds, something which the government and central bank were wary of for many years because of fears that this would allow unwelcome intervention in financial policy. In addition, all restrictions were removed on the purchase of shares in foreign companies and the sale of shares abroad in Swedish companies.

Foreign shares have to be deposited with a Swedish bank or negotiated through a Swedish broker – in order to help clamp down on tax dodgers. The Riksbank also removed restrictions on company investment in property overseas. Previously companies needed government permits in order to invest in real estate abroad.

COUNTRY PROFILE

Political system: Constitutional monarchy, with legislative power vested in unicameral Riksdag (349 members directly elected for three years). Executive power exercised by the Cabinet (Regeringen) which is led by the Prime Minister and is responsible to Parliament.

Main cities/towns: Stockholm (population 1.4m), Gothenburg, Malmö and Uppsala.

Media
Dailies: Major medium for political discussion as many are owned or run by political parties or trades unions. Most influential include *Dagens Nyheter* (liberal, independent; circulation weekdays 406,726, Sundays 513,258), *Expressen* (liberal; Sundays 695,985), *Svenska Dagbladet* (conservative; weekdays 227,050, Sundays 233,718) and *Aftonbladet* (Social Democratic; weekdays 387,503, Sundays 479,772), all Stockholm-based but available nationally; also the Gothenburg-based *Göteborgs-Posten* (liberal; weekdays 283,310, Sundays 316,528).
Business: Main publications: *Affärsvärlden* and *Veckans Affärer* (both weekly) and the daily *Dagens Industri.* Some business publications are in English.
Broadcasting: Programmes broadcast by Swedish Broadcasting Corporation (Sveriges Radio AB) under licence from the State are non-commercial.
Radio: Approximately 3.3m receivers in use. Sveriges Radio broadcasts three national channels and there is also a network of 24 independent local stations. Also overseas service broadcast in six languages, including English.
Television: Approximately 3.2m receivers in use. Sveriges Radio transmits two channels. Latest figures suggest over 300,000 homes are cabled. Satellite channels only available in a local test area, although rapid expansion is projected. TV3 is a commercial station broadcasting via cable to the Nordic countries.
Advertising: Modern and of a high standard, but tends to be expensive. All media available except radio and state TV. Total advertising spend SKr3829m (1984). Key categories include automotive, food, financial, fuel and household appliances. Promotion of alcohol is not allowed (except for light beers), while tobacco advertising is restricted to the press only.

Domestic economy: Well-developed mixed economy strongly dependent on international trade. Stable socio-political environment combined with effective utilisation of local resources has helped maintain high living standards despite effects of recession. Major constraints to growth include heavy dependence on imported oil, large budget deficit and relatively high rates of inflation although counter-inflationary and stabilisation policies reduced both the deficit and inflation rate in 1986.

Employment: Unemployment remains relatively low owing partly to government support measures but partly because of low birth rates in previous decades which are squeezing the jobs market. Broad industrial consensus on maintaining wage levels.

External trade: Significant improvement in balance of payments in 1986 due to increase in exports and reduction in capital outflows. Export earnings account for about 30 per cent of GDP, and over 50 per cent of industrial sales. Major exports include manufactured goods, forestry products, machinery and transport equipment and shipping services. Imports dominated by crude oil and petroleum products, machinery and equipment, chemicals and manufactured goods. The EC is major trading partner, with West Germany, UK and Denmark accounting for about 35 per cent of trade turnover. Nordic countries account for a further 20 per cent and the USA approximately 10 per cent.

Agriculture: Sector is highly efficient though it contributes less than four per cent to GDP. Farming concentrated in southern regions, where livestock farming predominates. Around 80 per cent self-sufficiency in basic foodstuffs. Large forestry sector supplying raw materials to industry and for export.

Industry: Sector contributes over 20 per cent to GDP and is based largely on indigenous resources (iron ore, timber and water-power). Small-scale privately owned firms predominate. Major industries include food processing, textiles, motor vehicles, chemicals, shipbuilding, iron and steel, electrial and electronic equipment and forestry products. Production is export-based and import-dependent.

Leading companies: Volvo, Electrolux, ASEA, Saab-Scania, Ericsson, SCA, SAS, Nobel Industries, Stora.

Mining: Rich in mineral deposits, the most important of which are iron ore (5-6 per cent of world production), zinc, lead, copper, silver and pyrites. Also large deposits of uranium, exploitation of which has been held back by environmental and political objections.

Energy: High levels of energy consumption (5.9 tonnes of oil equivalent per capita in 1984), around 70 per cent of which is met by imports. Energy policy aimed at reducing import dependence to 40 per cent by 1990 – in 1970 oil accounted for 75 per cent of energy supply and in 1986 for 48.4 per cent, and expanding hydro-electric power and other indigenous energy resources while phasing out nuclear power by 2010. Oil substitution programme under way, aimed at increasing imports of coal and natural gas. Gas imports from Denmark began in 1985 to south and south-west Sweden. Proposals to import natural gas from both Norway and USSR via Finland. In 1986 nuclear power accounted for 50 per cent of electricity production and hydropower for 45 per cent. According to a 1980 referendum Sweden is committed to phasing out nuclear power by the year 2010. Provided that alternative energy forms can be found Sweden is planning to phase out the first of the country's 12 reactors during 1993-95 and the second during 1994-96. Energy consumption 1986: hydroelectric (26%), nuclear (30%), other domestic (8%), coal and coke (5%), petroleum products (31%).

Banking: System is dominated by four commercial banks – Skandinaviska Enskilda Banken, Svenska Handelsbanken and Götabanken (all privately owned), and state-owned Post-och Kreditbanken. Around 25 commercial banks, around 160 savings banks and over 400 co-operative banks.

Legal system: Civil law system influenced by customary law. New constitution adopted 1975.

Membership of international organisations: ADB, CCC, Council of Europe, DAC, EC, EFTA, ESRO, FAO, GATT, IAEA, IBRD, ICAC, ICAO, ICES, ICO, IDA, IDB, IEA, IFAD, IFC, IHO, ILO, IMF, IMO, INTELSAT, Interpol, International Lead and Zinc Study Group, IPU, ISO, ITU, IWC, Nordic Council, OECD, UN, UNESCO, UPU, WHO, WIPO, WMO, WSG.

BUSINESS GUIDE

Time: GMT + 1 (GMT + 2 from late Mar to late Sep).

Climate: Mild summers, cold winters. Large north-south variations with north colder and drier.

Very cold from Nov to Apr.
Clothing: Mediumweight throughout year, heavy topcoat for winter and preferably snowboots.

Entry requirements: Visa not required by nationals of most European countries and several others. Passports not required by Scandinavian nationals. N.B. Regulations liable to change at short notice. Travel agent or Swedish Embassy should be consulted prior to departure.
Health: Vaccination certificates not required unless travelling from an infected area.
Currency: Up to SKr6,000 in Swedish currency may be taken in or out. No limit on foreign currency.
Customs: Personal effects duty-free.

Air access: Regular flights by all major international airlines. Swedish airport tax included in cost of ticket.
National airlines: AB Aerotransport (ABA). Also Scandinavian Airlines System (SAS) which serves all of Scandinavia (three-sevenths owned by ABA).
Main international airport: Stockholm (STO) served by Arlanda (ARN) 41 km from capital.
International airports: Gothenburg (GOT) and Malmö (MMA).

Surface access: Several ferry links with northern Europe. Rail and road links with Norway and Finland.
Main ports: Gothenburg, Norrköping, Hëlsingborg, Malmö.

Hotels: No official rating system in operation. Shortage of accommodation in major cities so reservations should be made well in advance. Booking agencies available in Stockholm and Malmö. Single room in Stockholm costs up to SKr850 per night. Rates in other areas tend to be slightly cheaper.

Currency
Exchange rate: The Swedish krona is related to a trade-weighted basket of 15 major currencies. Only minor fluctuations around a base are allowed, with the Central Bank intervening to counteract larger imbalances.

Foreign exchange information
Account restrictions: Non-residents may hold accounts in krona for payments to or from residents. Foreign currency non-resident accounts are credited or debited for payments made under an exchange control permit.
Trade policy: Certain iron and steel products, textiles and certain goods from Eastern Europe have import restrictions.

Credit cards: American Express, Diners, Eurocard and Visa are all accepted.

Car hire: Available at all airports and stations in main towns. Medium-sized car costs about SKr1,761 per week. Speed limits: urban areas 50 kph, normal roads 70 or 90 kph, motorways 90 or 110 kph. Strict enforcement of drink-driving, speeding and other laws. Information and assistance from automobile club – Motormannens Riksforbund/Kungliga Automobile Klubben (M/KAK).

City transport
Taxis: Available in all major towns. Cost about SKr10 plus SKr3 per km. Can be called by telephone (in Stockholm dial 150000) at an additional charge. Waiting time charged at SKr120 per hour.
Buses, trams and underground: Available in Stockholm at a flat fare of SK6 (two coupons). Books of 20 coupons available at Press Agency news stands for SKr35.

National transport
Air: Daily flights connecting all main towns.
Road: Network of over 100,000 km, including about 1,000 km of motorways. Efficient bus service, mainly controlled by the State Railways (Statens Jarnvagar). Services integrated with rail service.
Rail: Network of over 11,000 km, about 60 per cent electrified, operated by State Railways. Generally efficient service. Seats on express services must be booked in advance.
Water: Over 2,000 km of navigable inland waterways. Ferry service operated by State Railways.

Public holidays
Fixed dates: Day preceding holiday often taken as half-day holiday. 1 Jan, 6 Jan (Twelfth Day), 1 May, 25 and 26 Dec.
Variable dates: Good Friday, Easter Monday, Ascension Day, Whit Monday, Midsummer's Day, All Saints' Day.

Working hours
Business: (Mon-Fri) 0830 – 1700 (often closed one hour earlier in summer).
Banking: (Mon-Fri) 0930 – 1500 (larger branches open longer).
Shops: (Mon-Fri) 0900 – 1800 (closed 1400 or 1600 on Sat).

Social customs: It is usual to take flowers when being entertained in a Swedish home for the first time. Punctuality is important.

Business language and interpreting/translation: Available through major Chambers of Commerce.

Telecommunications
Telephone: Widely used, especially long-distance. Direct dialling throughout country and to most parts of the world. Shops displaying 'TELE' or 'TELEBUTIK' sign offer cheap international phone services (also telex and telefax). International dialling code: 46 followed by 8 for Stockholm and 31 for Gothenburg.
Telex: Available in some major hotels and at telegraph offices. Country code 54 S.
Telegram: Can be sent by telephone or from telegraph offices (telexpedition). Also accepted by a few post offices.

Postal service: Good service. Stamps available at most newspaper kiosks and tobacconists. Main post office in Stockholm opens daily from 0800 – 2400.

Trade fairs: Several international events, mostly held in Stockholm, Gothenburg and Malmö. Wide range of products covered, but heavy emphasis on boats (Gothenburg late Jan/early Feb, Stockholm late Feb/Mar), International Fashion Fair (Feb, Stockholm), International Furniture Fair (Feb, Stockholm), International Exhibition for the Pulp Paperboard and Conversion Industries (May, Stockholm), Scanautomatic Hydraulics Pneumatics and Automotive Trade Fair '90 (Nov, Gothenburg).

Electricity supply: Electricity supply 220 V AC.

Representation in capital: Albania, Algeria, Angola, Argentina, Australia, Austria, Bangladesh, Belgium, Bolivia, Botswana, Brazil, Bulgaria, Canada, Chile, China PR, Colombia, Cuba, Czechoslovakia, Denmark, Ecuador, Egypt, Ethiopia, Finland, France, German Democratic Republic, Federal Republic of Germany, Greece, Guatemala, Guinea-Bissau, Hungary, Iceland, India, Indonesia, Iran, Iraq, Ireland, Israel, Italy, Japan, Kenya, Democratic People's Republic of Korea, Republic of Korea, Laos, Lebanon, Libya, Malaysia, Mexico, Morocco, Mozambique, Netherlands, Nicaragua, Nigeria, Norway, Pakistan, Panama, Peru, Philippines, Poland, Portugal, Romania, Saudi Arabia, Senegal, Somalia, South Africa, Spain, Sri Lanka, Switzerland, Tanzania, Thailand, Tunisia, Turkey, USSR, United Kingdom, USA, Uruguay, Venezuela, Vietnam, Yugoslavia, Zambia, Zimbabwe.

Hotels: Stockholm
Anglais, Humlegardsgatan 23 (tel: 249900; tx: 19475; fax: 211629).

City, Slojdgatan 7, Hoetorget (tel: 222240; tx: 12487; fax: 208224).

Eden, Sturegatan 10 (tel: 223160; tx: 10570; fax: 6608067).

Flamingo, Hotellgatan 11, Solna (tel: 830800; tx: 10060; fax: 839814).

Fly (near Bromma Airport), Brommaplan (tel: 262620).

Grand, P O Box 16424 (tel: 221020; tx: 19500; fax: 218688).

Kungs Hatt, P.O. Box: 476 (tel: 180140; tx: 19488).

Malmen, Gotgatan 49-51 (tel: 226080; tx: 19489; fax: 411148).

Oden, Karlbergensvagen 24 (tel: 349340; tx: 17151; fax: 322299).

Palace, St Erikgatan 115 (tel: 241220; tx: 19877; fax: 302329).

Sara Amaranten, Kungsholmsgatan 31, P.O. Box: 8054 (tel: 541060; tx: 17498; fax: 526248).

Sergel Plaza, Brunkebergstorg 9 (tel: 226600; tx: 16700; fax: 215070).

Sheraton, Tegelbacken 6 (tel: 142600,3348484; tx: 17750; fax: 217026).

Strand, Nybrokajen 9 (tel: 222900; tx: 10504).

Terminus, Vasagatan 20 (tel: 222640; tx: 11749; fax: 248295).

Car hire: Stockholm
Avis: Arlanda Airport (tel: 61885; tx: 13492).

Budget: (tel: 576800; tx: 15089).

Europcar: Box 45038 10430 Stockholm (tel: 7445200; tx: 10109).

Hertz: Arlanda Airport (tel: 60555; tx: 17254).

Inter-Rent: Box 7545 (tel: 240280; tx: 11565).

Chambers of Commerce
Handelskammaren for Gavleborgs, Kopparbergs och Uppsala lan, Drottninggatan 25, 80104 Gavle (tel: 105430).

Handelskammaren i Göteborg-Vastsvenska Handelskammaren, PO Box 5253, 40225 Göteborg (tel: 835900; tx: 27430).

Stockholms Handelskammare, PO Box 16050, Vastra Trädgårdsgatan 9, 10322 Stockholm (tel: 231200; tx: 15638).

Sydsvenska Handelskammaren, Malmö Börshus, Skeppsbron 2,21120 Malmö (tel: 73550; tx: 33388).

Banking
Sveriges Riksbank (central bank), Box 16283, 10325 Stockholm (tel: 7870000; tx: 19150).

Götabanken, Sveavägen 14, 10377 Stockholm (tel: 7904000; tx: 19420) and Hamngatan 16, 40509 Gothenburg (tel: 625000; tx: 21750).

Nordbanken, Regeringsgatan 38, PO Box 7133, 10387 Stockholm (tel: 7963000; tx; 17616; fax: 213482).

Post-och Kreditbanken (PKbanken), Hamngatan 12, 10571 Stockholm (tel: 7818000; tx: 19310).

Skandinaviska Enskilda Banken, Kungsträdgårdsgatan 8, 10640 Stockholm (tel: 7635000; tx: 11000).

Svenska Bankforeningen (Swedish bankers' association), Regeringsgatan 42, Box 7603, 10394 Stockholm (tel: 243300).

Svenska Handelsbanken (commercial bank), Blasieholmstorg 11, 10670 Stockholm (tel: 7011000; tx: 11090).

Svenska Sparbanksforeningen (Swedish savings banks association), Drottninggatan 29, Box 16426, 10327 Stockholm (tel: 572000).

SwedBank (commercial bank), Brunkebergstorg 8, 10534 Stockholm (tel: 222320; tx: 12826).

Principal newspapers
Affärsvärlden, Stockholm.

Aftonbladet, Vattugt 12, 105 18 Stockholm (tel: 7880000; tx: 17138).

Dagens Nyheter, 105 15 Stockholm (tel: 7381000; tx: 10450).

Expressen, Gjörwellsgt 30, 105 16 Stockholm (tel: 7383000; tx: 17480).

Göteborgs-Posten, Polhemsplatsen 5, 405 02 Göteborg (tel: 624000; tx: 2505).

Svenska Dagbladet, Rålambsvägen 7, 105 17 Stockholm (tel: 135000; tx: 17400).

Other useful addresses
Federation of Commercial Agents of Sweden, Hantverkargatan 46, 11221 Stockholm (tel: 540975).

Federation of Commercial Agents of Sweden, Western Division, Box 36059, 40013 Gothenburg, (tel: 192045; tx: 27080).

Federation of Swedish Industries, Storgatan 19, 11485 Stockholm (tel: 7838000).

Federation of Swedish Wholesalers and Importers, Grevgatan 34, Box 5512, 11485 Stockholm (tel: 635280).

Handels Arbetsgivareorg (HAO) (commercial employers' confederation), Box 1720, 11187 Stockholm (tel: 243080).

Handelsdepartementet (Ministry of Trade), 10333 Stockholm (tel: 7631000).

Landsorganisationen (LO) (Swedish trade union), 10553 Stockholm (tel: 7962500).

Ministry of Finance, Rödbodgt 6, 10333 Stockholm (tel: 7631000; tx: 11741).

Ministry of Industry, Fredsgt 8, 10333 Stockholm (tel: 7631000; tx: 14180).

SACO/SR (confederation of professional associations), Box 2206, 10315 Stockholm (tel: 225200).

Stockholm Stock Exchange, Kallagrand 2, Box 1256, 11182 Stockholm (tel: 143160; tx: 13551).

Stockholm Technical Fair (Stockholmsmassan AB), Alvsjo, 125 80 Stockholm (tel: 7491100).

Svenska Arbetsgivareforeningen (employers' confederation), Sodra Blasieholmshammen 4A, 10330 Stockholm (tel: 7626000; tx: 19923).

Sveriges Exportrad (Swedish export council), PO Box 5513, 11785 Stockholm (tel: 7838500).

Sveriges Turistrad (tourist board), Hamngatan 27, Box 7473, 10392 Stockholm (tel: 789200; tx: 17076).

Swedish Automobile Association (Motormannens Riksforbund), Sturegatan 32, 11436 Stockholm 5 (tel: 7823800; tx: 19827).

The Swedish Institute, Box 7434, 103 91 Stockholm (tel: 7892000; tx: 10025).

Swedish Trade Fair Foundation (Svenska Mässan Stiftelse), Ska\20negatan 26, Box 5222, 40224 Gothenburg (tel: 120000; tx: 20600).

TCO (central organisation of salaried employees), Box 5252, 10245 Stockholm (tel: 221760).

Tidningarnas Telegrambyra (news agency), Kungsholmstorg 5, 10512 Stockholm (tel: 132600; tx: 19167).

Trafik Sakerhetsverket (National Swedish Safety Office) (tel: 024378000).

Switzerland

Andrew Lycett

The fate of Mrs Elisabeth Kopp, the hard-line former justice and police minister, overrode other political and economic developments in Switzerland in 1988-89. If everything had gone according to plan, the woman described as Switzerland's Margaret Thatcher would in 1990 have become the country's first female head of state. In December 1988, with none-too-subtle timing, the federal assembly had elected her vice-president, and therefore automatically president-elect for 1990.

But only five days after her appointment, she was forced to resign, after admitting she had telephoned her husband, Hans, on 27 October 1988 to warn him that a Zurich-based, Lebanese-owned company, of which he was deputy chairman, was about to be named in an official investigation into drug money laundering. The Kopp affair subsequently became Switzerland's biggest political scandal this century.

As her successor, her party – the Radicals – and subsequently the national assembly chose Kaspar Villiger, a cigar-chomping bicycle manufacturer from near Lucerne. This did not go down well in Zurich for it meant that the country's industrial and financial capital was without a member of the seven-person council of ministers, or cabinet, for the first time since the Swiss Confederation was formed in 1848.

Mrs Kopp's position as justice minister was taken by Defence Minister Arnold Koller, who stepped up anti-drugs surveillance and suggested he might establish a federal anti-drugs brigade. This news came as Switzerland was being forced to come to terms with the fact that not only is it a dumping ground for hot money, it has also become one of the drugs centres of Europe. During 1989 police in Zurich and Berne used tear-gas and batons to break up demonstrations of young people protesting against lack of cheap housing. The country's social problems were only exacerbated by the news from the World Health Organisation in April 1989 that – per capita – Switzerland has the highest number of AIDS victims in Europe.

Role in the world

The Kopp affair raised difficult and wide-ranging questions about Switzerland's role in the world. Is it to continue as a bolt-hole where all types of business can be conducted and any kind of assets banked, no questions asked? In recent years Switzerland's strict banking secrecy laws have been under fire – and have yielded slightly – to incursions by the US Securities Exchange Commission (SEC). In January 1989 the Swiss authorities agreed to co-operate with the SEC in its investigations into alleged insider dealing surrounding the takeover in November of the US company, Triangle Industries, by Péchiney SA, a French state-owned aluminium business. (The fact that purchasers of Triangle shares were linked to President Mitterand similarly threatened the political establishment in France.)

The Kopp affair was no mere financial scandal, however. It was alleged that, for the last decade, not only had Switzerland connived at the laundering of Middle East drugs and arms money in its banks, but also its so-called 'Lebanese connection' had been monitored and used by the Swiss secret service and the CIA. The secret network was said to have played a part in Irangate – channelling funds from illegal arms sales to Iran to Contra rebels in Nicaragua. The reputation of the federal prosecutor, Rudolph Gerber, was reported to be as much on the line as Mrs Kopp's.

The scandal touched a disturbing xenophobic strain

KEY FACTS

Official title: The Swiss Confederation

Head of state: President Arnold Koller (for 1990 only)

Ruling party: Coalition of Radical Democratic Party, Social Democratic Party, Christian Domocratic People's Party and Swiss People's Party

Capital: Berne

Official Languages: German, French and Italian

Currency: Swiss Franc (SFr) = 100 centimes

Exchange rate: SFr1.60 per $ (Dec 1989)

Area: 41,293 sq km

Population: 6.5m (1988)

GDP per capita: $37,361 (1988)

GDP real growth: 3.2% (1988)

Labour force: 2.9m (1984)

Unemployment: 0.7% (1988)

Inflation: 1.9% (1988)

Trade balance: $10.9bn (1988)

in the Swiss. As well as anti-Arabism, some observers noted an element of anti-semitism. The Kopps are Jewish. Hans W. Kopp is a fashionable company lawyer with a seat on the board of some forty companies. One of them, where he was deputy chairman, was Shakarchi Trading, a gold and foreign currency trader, which – like many similar businesses – moved to Zurich from the Lebanon in the 1970s.

The company was named in an investigation in early November 1988 by the public prosecutor in the Southern canton of Ticino, where, back in February 1987, the Lebanese connection was rumbled, following the seizure of 100 kg of heroin and morphine. The trail led to Zurich, where Lebanese couriers were alleged to have laundered over half a million dollars of drug money through the three main Swiss banks. The cash was brought into Switzerland, via Turkey and Bulgaria, and carried in suitcases laden with notes valued up to $200,000. Bank certificates of deposit were then used to purchase gold from Shakarchi, though there was no suggestion Kopp knew about this side of the operation. $30m worth of Californian drug money was also said to have been cleaned this way. The banks protested that, although some of the defendants in this case kept accounts with them, large-scale stashing away of criminal assets is impossible, as well as illegal. As the president of one of the Big Three put it, 'Unlike the American banks, we know our clients personally'. Unfortunately, the clubiness – the secretive, personal nature – of Swiss business and political relationships was just what was under scrutiny.

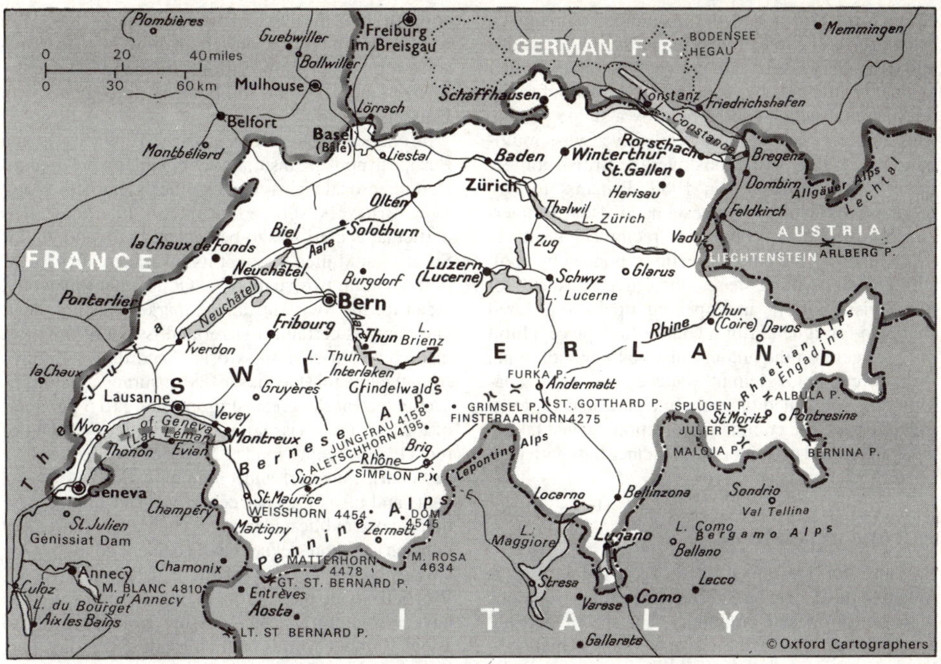

Banking and finance

In April 1989 Switzerland's Federal Banking Commission officially cleared the big Swiss banks of having done anything illegal in the scandal which became known as Koppgate. However the Commission emphasised the need to tighten its control over Switzerland's international banknote trading business.

To conduct banknote business, banks now have to obtain clearance which the Commission only grants if it is satisfied about the 'internal rules and allocation of personnel sufficient to guarantee control'. Predictably the banks were unhappy with this ruling. The banks were also unhappy about proposed legislation – designed to prevent Swiss financial markets being exploited by organised crime – which would require them to check systematically the origin of funds credited to their customers' accounts. According to Robert Studer, president of the Union Bank of Switzerland, the authorities were 'forcing us to regard every one of our customers as a potential criminal and to look into the origins of all larger sums moved via bank accounts'.

These measures coincided with another assault on the banks' autonomy: proposed stricter rules from the Federal Banking Commission over the way banks use hidden reserves to disguise poor earnings or losses. A survey by Commission showed that, although 65 banks had made losses in the 1987 – the year of the stock market crash – 46 had actually reported earnings gains in their annual reports. Light was thrown on the reality of the situation from London where Union Bank of Switzerland, Swiss Bank Corporation and Credit Suisse were all forced to admit they had made losses on their London operations. UBS took the steepest losses through its London securities subsidiary, Philips and Drew, which lost £81m ($144m) in the year ending March 1988. However, there was no suggestion of this in UBS's annual report for the year.

In the past Swiss law allowed banks to build up and run down hidden reserves without disclosing the fluctuations in their published accounts. The banks had not been required to publish consolidated accounts either. By the end of the 1980s both these practices were under fire, not least because the Swiss stock exchange realises that greater disclosure is essential if Switzerland is to compete in the global capital markets. In 1988 securities turnover in Zurich, Geneva and Basel fell for the first time in ten years – dropping by 12 per cent to SFr850bn.

In response to competition on the world financial markets, the banks themselves had lobbied in 1988 for the reduction of stamp duties on financial transactions, including trading in securities and money market instruments. By mid-1989, however, they were up in arms as the the government attempted to claw back some of the revenues it had thus lost by proposing a new stamp duty on the banks' fiduciary business. The banks claimed this could lead foreign investors to withdraw funds under management.

The closed world of Swiss finance was under further attack during 1989 following pressure from inside and out for the relaxation of the country's protectionist takeover rules. In the past Swiss companies have been able to protect themselves against takeovers by refusing to register a bidder's shares. Foreign shareholders have been particularly discriminated against in this way. However, local takeover merchants such as Tito Tettamenti have also suffered from the restrictions – in his case, by having his bids for the engin-

KEY INDICATORS

	Unit	1984	1985	1986	1987	1988
Population	'000	6,440	6,470	6,504	6,550	6,510
Gross domestic product (GDP)	SFrbn	214.1	228.0	243.4	254.7	355.1
GDP per capita	$	14,147	14,325	20,791	26,098	37,361
Inflation	%	3.0	3.4	0.8	1.4	1.9
GDP real growth	%	2.6	3.3	2.8	2.0	3.2
Unemployment	%	1.1	1.0	0.8	0.8	0.7
Share prices	1985=100	76.9	100.0	121.8	126.1	119.0
Interest rate (official disc.)	%	4.0	4.0	4.0	2.5	3.5
Industrial production	% growth	2.1	6.2	4.0	0.0	6.4
Textiles	% growth	6.0	3.0	2.3	–	–
Chemicals	% growth	6.0	6.0	2.0	–	–
Machinery	% growth	–1.0	7.0	7.9	–	–
Watches and jewellery	% growth	7.0	15.0	11.9	–	–
Paper	% growth	4.0	2.0	8.1	–	–
Food and beverages	% growth	1.0	1.0	1.2	–	–
Exports	SFr bn	55.6	66.7	67.0	67.5	74.0
EEC	%	50.3	50.2	54.8	–	–
USA	%	9.8	10.3	9.5	–	–
CMEA	%	3.0	3.0	3.0	–	–
Imports	SFr bn	69.2	74.8	73.5	75.2	82.6
Foreign exchange	$m	14,706	17,463	21,334	27,162	24,045
Balance of trade	$bn	–1.1	–1.6	–5.0	–5.4	10.9
Current account	$bn	6.2	6.0	4.7	6.3	8.3
Exchange rate	SFr per $	2.35	2.46	1.80	1.49	1.46

* estimated figure

eering firm Sulzer and the insurance company La Suisse rejected. Some Swiss businesses, like Sandoz and Ciba-Geigy, sought to protect themselves further against hostile bids by reducing the amount of registered capital which can be held by a single person. This embarrassed the more acquisitive Swiss multinationals. When Nestlé took over the British confectionery firm Rowntree in 1988, Britain's foreign secretary Sir Geoffrey Howe demanded that Switzerland allow British companies reciprocal rights to make takeover bids there. Nestlé responded by abolishing its non-voting shares (the usual repository for non-Swiss money) and opening up its registered shares to foreigners. But it continues to place a limit of 3 per cent on the amount of stock any individual investor can hold. Helmut Maucher, Nestlé's managing director, stated that this rule is designed not so much to give the company protection against takeovers but to ward off raiders who finance bids with junk bonds and short-term borrowings.

European relations

Pressure for liberalisation of such regulations is linked to the debate on Switzerland's future relations with the European Community. The spectre of 1992 – and the arrival of the European single market – hangs over many of the political developments in the country. Swiss opposition to the EC is based on fears that closer relations may force it to compromise its unique federal political system, abandon its neutrality, remove protection on its agriculture and accept free movement of labour. However, initial Swiss concern about the threat of the single market has given way to pragmatic debate about how best to reach some form of accommodation with the EC. Currently Switzerland has some 130 bilateral agreements with Brussels.

In one such agreement in early 1989, Swiss and European non-life insurance companies were guaranteed reciprocal rights to each other's markets. However Brussels showed it meant business: this particular accord was to be declared null and void if Switzerland failed to adapt its insurance laws to EC norms within twelve months. Quite what this would mean for the Swiss insurance market was not clear. One thing is certain: Switzerland is one of the most insured countries in the world. According to figures from the federal statistics office, insurance premiums accounted for the greatest proportion (16.5 per cent) of the average Swiss household's budget – ahead of rent (13.8 per cent), food (12.5 per cent), leisure (12.1 per cent), transport and communications (10.3 per cent) and taxes (9.3 per cent). Most insurance payments go to public and private schemes for retirement pensions, and on sickness and accident cover.

Many Swiss are wary of EC legislation, however. Worried about damage to their environment, they have resolutely resisted pressure from Brussels to allow 40-tonne lorries on their roads. Swiss law bars vehicles above 28 tonnes. To appease its powerful neighbours, Switzerland offered to build a new combined road-rail route which would carry heavy trucks through its countryside on railway wagons.

The environment has been an important issue in domestic politics. In late 1988 the States Council, or upper house of the Swiss national assembly, voted to scrap long-debated plans to build a nuclear power plant at Kaiseragst, near Basel. This meant that compensation totalling SFr350m would have to be paid to the consortium of power companies which hoped to construct the facility. It has been reported that, without the plant, Switzerland could experience a shortfall of electricity by the turn of the century. The association of Swiss power stations has forecast a shortfall of some 7.5m kW by the year 2004.

Sandoz, the Basel-based chemicals company, subsequently threatened that it might be forced to site a proposed new factory costing SFr170m in another country if environmental protesters continued to resist the project. Ciba-Geigy similarly experienced problems over its plans to build a waste disposal facility. The Sandoz plant was to have produced chemical ingredients for Sandimmun, one of the company's most successful products – a drug widely used in heart surgery, which has worldwide sales of SFr400m. Not linked to environmental concerns, Sandoz also plans to build a new biotechnology laboratory in the United States.

As a predominantly export-oriented industry, chemical and pharmaceutical companies continue to act as driving forces in the economy. An unusual decline in the value of the Swiss franc (down 10 per cent against the dollar during 1988, and still falling during 1989) helped boost exports in this sector by 8.5 per cent to SFr15.86bn – roughly 20 per cent of total exports – in 1988. This pushed GDP up by 3 per cent during the year, compared with 2.3 per cent in 1987. In 1989 this growth is expected to slip back to around 2.5 per cent.

COUNTRY PROFILE

Political system: Federal state composed of 26 cantons and half-cantons, with bicameral parliament (federal assembly) consisting of a Council of States or upper house (46 members – two from each canton, one from each half-canton) and a National Council or lower house (200 members elected for four years). Executive power exercised by seven-member Federal Council, elected by the federal assembly, representing the four government parties, and headed by an annually elected president. Major issues frequently decided by referendum. In December 1988 the federal assembly elected Jean-Pascal Delamuraz as President for 1989.

Main cities/towns: Berne (population 136,292), Zurich, Basle, Geneva, Lausanne.

Media
Press: Decentralised press owing to regional variations in language and culture. Large number of publications with relatively small circulations.
Dailies: About 120 regional newspapers (75 per cent printed in German, 20 per cent in French with total circulation of only 2m. Of international repute is *Neue Zürcher Zeitung* (circulation 145,735), published in Zürich. Most popular are *Blick* (382,275) and *Tages Anzeiger Zürich* (256,767), both published in Zürich.
Weeklies: Weltwoche and several general and special interest magazines, including large number imported from Germany, France and Italy.
Business: As well as *Neue Zürcher Zeitung* (daily) there is *Finanz und Wirtschaft* (twice weekly, 32,197). Two major business magazines *Bilanz* (55,300) and *Politik und Wirtschaft* (both monthly).
Broadcasting: Controlled by Swiss Broadcasting Corporation (SBC), under licence from the Federal Government. Actual transmission is controlled by Swiss postal and telecommunications service.
Radio: Approximately 2.5m receivers in use. There are three SBC programmes broadcast in French and German, two in Italian and also some broadcasts in Romansch as well as some private stations. Swiss Radio International broadcasts in nine languages.
Television: Over 2m receivers in use. Each linguistic region has its own TV channel, and also receives other regions' broadcasts. Satellite, cable and TV and teletext available.
Advertising: Expensive due to language variations. TV advertising restricted to specific times, and there is no advertising on Swiss state radio. Newspapers, cinemas and direct mail are widely used, but poster advertising is confined to selected sites. Information from Union Suisse d'Agences-Conseils en Publicité, in Zurich. Total advertising spend (1984) SFr2043m. Major subjects cars, banks and tobacco; main advertising media newspapers and magazines.

Domestic economy: Switzerland is one of the richest countries in the world, with a well developed manufacturing sector as well as a highly skilled labour force, an important tourist industry and leading banking and insurance sectors. However, Switzerland is almost completely dependent on imported raw materials and is vulnerable to world economic developments.

Employment: Around 54 per cent of labour force is employed in service sector, 35 per cent in industry and seven per cent in agriculture. Heavy reliance on foreign labour. Unemployment rate is lowest of any OECD country, despite large number of redundancies in traditional industries. Wage inflation has decelerated sharply since 1983.

External trade: Regular trade deficit counterbalanced by net earnings from tourism and service sector. Principal exports include electrical/non-electrical machinery, chemical/pharmaceutical products, clocks and watches and metals. Imports consist mainly of fuels and lubricants, industrial machinery, metals, motor vehicles, chemicals, jewellery, foodstuffs and clothing. Major trading partners are the FRG (taking 22 per cent of exports and supplying over 30 per cent of imports), France, Italy, UK and USA.

Agriculture: State-subsidised sector employing 6 per cent of active population, with activity concentrated on dairy farming. Production sufficient to satisfy around 50 per cent of domestic food requirements. Major crops include wheat and spelt, potatoes, sugar-beet, grapes and apples. One of the largest tractor densities in the world. Farmers have use of large and expensive equipment through machinery syndicates.

Industry: Developed export-based manufacturing sector with emphasis on production of finished goods. Traditional industries include machines, tools, chemicals and pharmaceuticals, textiles, watch-making, food processing, chemicals and engineering. Emphasis on increased specialisation and development of high-technology products.

Leading companies: Nestlé, Ciba-Geigy, Brown Boveri, Migros.

Energy: Water power is prime energy source, though heavy, but declining, dependence on imported oil and gas. Rapid development of nuclear power (40 per cent of electric energy), despite local opposition.

Banking: Zürich ranks as one of the leading international banking centres. Five main commercial banks – Swiss Bank Corporation (1986 balance sheet SFr 137,828m), Union Bank of Switzerland (SFr 152,162m), Crédit Suisse (SFr 103,741m), Swiss Volksbank and Bank Leu, 29 cantonal banks and numerous foreign banks operating. Banking system characterised by strict secrecy laws.

Legal system: Civil law system marginally influenced by customary law. Constitution adopted 1874, has been amended several times.

Membership of international organisations: ADB, CCC, Council of Europe, DAC, EFTA, ELDO (Observer), ESTO, FAO, GATT, IAEA, ICAC, ICAO, ICO, IDB, IEA, IFAD, ILO, IMO, INTELSAT, INTERPOL, IPU, ITU, IWC, OECD, UN (permanent observer), UNESCO, UPU, World Confederation of Labour, WFTU, WHO, WIPO, WMO, WSG, WTO.

Trade gap

The lower value of the Swiss currency did not dent imports. In August 1989 Switzerland experienced record monthly imports of SFr7.07bn contributing to a foreign trade deficit for the month of SFr1.44bn. The trade gap for 1989 was expected to exceed the previous high of SFr11.25bn recorded in 1980. In 1988 the trade deficit was SFr8.3bn. However there was an overall current account balance of SFr9.1bn – the result of Switzerland's surplus on services and the annual return on its foreign investments.

Nevertheless, high import penetration helped contribute to a widely acknowledged overheating of the economy. Despite virtual full employment, wages rose by 3 per cent in 1988 and a figure of 4 per cent was expected in 1989. Prices were reported to have been rising faster than in West Germany. This led to pressure on the Swiss National Bank (SNB) to rein in the money supply and to maintain price stability. Marcus Lusser, president of the Bank, stated publicly that the weakness of the currency was only temporary. Nevertheless the SNB committed itself to restricting monetary growth to 2 per cent in 1989. As part of its attack on inflation, which was expected to reach an unprecedented 3 per cent within a year, it raised interest rates in April 1989 for the third time since December.

The leading Swiss industrial sector (ahead even of chemicals and pharmaceuticals) continued to be machinery. In 1988 the 200 members of the Swiss Machine Builders Association reported record new orders of SFr21bn – 8.1 per cent above the figure for 1987. Exports of the industry as a whole rose by 8.7 per cent to SFr34bn, the bulk going to western Europe and particularly West Germany.

However, manufacturers were concerned about the rising costs of imported raw materials and about the problems of recruiting adequate skilled and semi-skilled labour. They attempted to tackle these problems in a traditional Swiss way – by increasing expenditure on research and development. Swiss companies spend 2.9 per cent of GDP on research and development, one of the highest figures in the world. In the chemicals industry, the major companies spend around 10 per cent of their sales revenues on research and development.

One sector where expenditure on R and D has paid off is the watch industry, where in 1988 exports were up by 16.4 per cent, topping the SFr5bn mark for the first time ever. Having dropped below 50m in the early 1980s, the number of watches exported by the end of the decade exceeded 70m and was on its way back to the record high of 90m in 1974.

Ongoing discussion about Switzerland's links with the outside world, particularly with the European Community, forced planners to look more closely at Swiss agriculture – as the 1980s closed, the most protected on the continent. The Swiss government guaranteed a price of SFr105 per quintal of wheat – more than double the price in Europe and five times that available in the United States. Berne contributed SFr6bn to keeping its 120,000 farmers in business. But they managed to provide only 65 per cent of the country's food needs – the lowest self-sufficiency level in the world bar Hong Kong. Consumers and retailers were beginning to lobby against agricultural policies which lead to the unacceptably high cost of groceries and other goods in the shops.

Outlook

Again these developments must be seen against the wider debate about how far Switzerland adapts its well-established customs and policies to meet the challenges of the 1990s. How can the Swiss grow closer to Europe, but not be part of it? Luckily they can call upon well-established reserves of pragmatism. As Jean-Pascal Delamuraz, the 1989 President of the Confederation, put it in a debate on Europe in the federal assembly, 'We must be active and dynamic. That is the only chance Switzerland has of retaining its autonomy and not just being towed along in the wake of European development.'

BUSINESS GUIDE

Time: GMT + 1 (GMT + 2 from late Mar to late Sep).

Climate: Temperate; large variations with altitude. Temperatures range from about -1°C to 18°C. Generally warm summers and cold winters.
Clothing: Mediumweight throughout year. Topcoat and lined boots for winter.

Entry requirements: Visa not required by nationals of W Germany, Austria, Belgium, France Luxembourg, Netherlands, Portugal, Italy, Spain, UK and Greece and several others. N.B: Strict laws apply to business visits. Advisable to check with Swiss Embassy regarding regulations.
Health: Vaccination certificates not usually required. Medical insurance advisable as treatment is expensive.
Currency: No restrictions on Swiss and other currencies. Importation of coins of gold and platinum are liable to turnover-tax.
Customs: Personal effects, and gifts up to value of SFr100, duty-free.

Air access: Regular flights by all major international airlines. Reservations should be made in advance during holiday season.
National airline: Swissair.
International airports: Zürich (ZRH), 12 km from city; Geneva (GVA), Berne, Basle (BSL) and Lugano. Zürich and Geneva airports directly linked to national rail system.

Surface access: Good road and rail links with all surrounding countries. Advisable to book for rail travel beforehand.

Hotels: High standard throughout country. Classified by Swiss Hotel Association from one- to five-star. Single bedroom (five-star) costs about SFr160 per night, including 15 per cent service charge.

Restaurants: Fairly expensive but usually good quality. Service charge included in bill.

Currency
Exchange rate: Swiss franc has been floating since 1973. Switzerland does not officially try to maintain exchange rates for Swiss franc within margins, but sometimes does intervene.

Foreign exchange information
Account restrictions: Non-resident accounts allowed, but certain criteria must be met.
Trade policy: No exchange controls over imports or exports. Apart from agricultural produce and some chemicals, there are few restrictions on imports. There are some 22 free ports or zones, where goods may be stored for an unlimited period duty-free.

Credit cards: All major credit cards are accepted.

Car hire: Self-drive and chauffeur-driven cars available in all main towns. Rate for self-drive car about SFr200-300 per day plus SFr0.50-2.00 per km, depending on size of car. Also available at weekly and monthly rates. Valid national or international driving licence required, and insurance is compulsory. Speed limits, built-up areas 50kph, normal roads 80kph and motorways 120kph. Information from The Touring Club Suisse or the Automobile Club Suisse. For road information tel:163.

City transport
Taxis: Widely available but do not ply for hire. Fare generally about SFr4.00 plus SFr2.00 per km. Charge usually includes tip.
Buses and trams: Good services in major towns. Tickets should be bought in advance from vending machines. Multi-journey tickets also available. Flat fare up to five stops.

National transport
Air: Several daily flights linking Zurich, Geneva, Basle, Lugano and Berne, but only Zurich-Geneva route is widely used.
Road: Network of about 70,000 km, including 1,000 km of motorways. Good quality but tend to be slow due to terrain and volume of traffic.
Rail: Over 5,000 km of track, practically all electrified, of which about 60 per cent is operated by Swiss Federal Railways (Schweizerische Bundesbahnen),and the rest by about 120 small private companies. Good services with journeys between major towns rarely exceeding two or three hours.

Public holidays
Fixed dates: 1 Jan, 2 Jan, 1 May, 1 Aug (afternoon) 24, 25 and 31 Dec.
Variable dates: Good Friday, Easter Monday, Ascension Day and Whit Monday. There are other local holidays.

Working hours
Business: (Mon-Fri) 0800 – 1200, 1400 – 1800.
Government: (Mon-Fri) 0730 – 1145, 1330 – 1800 or 0800 – 1230, 1315 – 1730.
Banking: Regional variations but generally (Mon-Fri) 0800 – 1200, 1330 – 1630.
Shops: 0800 – 1215, 1330 – 1800 (close earlier on Sat).

Social customs: If being entertained at someone's home give the hostess flowers or chocolates.

Business language and interpreting/translation: Available from the Internationale Vereinigung dipl. Dolmetscher and various associations. Can be difficult to obtain at short notice. Rates fixed by International Association of Interpreters. Though English often spoken by business people, a knowledge of French and German is advantageous.

Telecommunications
Telephone: Excellent service. Subscriber trunk dialling in operation throughout country. International dialling code: 41 followed by 31 for Berne, and 1 for Zürich. Cheap rate at certain times for internal calls over 20 km.
Telex: Available at main post offices in larger towns and at hotels. Country code 45 CH.
Telegram: Efficient service, but seldom used within country.

Postal service: Reliable. Use of four-figure postal code obligatory.

Trade fairs: Several fairs of international significance held throughout the country. Also three official national fairs.

Electricity supply: 220V AC.

Representation in capital: Algeria, Argentina, Australia, Austria, Belgium, Brazil, Bulgaria, Burundi, Canada, Chile, China PR, Colombia, Costa Rica, Cuba, Czechoslovakia, Denmark, Ecuador, Egypt, Finland, France, German Democratic Republic, Federal Republic of Germany, Ghana, Greece, Guinea, Holy See, Hungary, India, Indonesia, Iran, Iraq, Ireland, Israel, Italy, Ivory Coast, Japan, Jordan, Democratic People's Republic of Korea, Republic of Korea, Lebanon, Libya, Liechtenstein, Luxembourg, Malaysia, Mexico, Monaco, Morocco, Netherlands, Nigeria, Norway, Pakistan,

BUSINESS DIRECTORY

Peru, Philippines, Poland, Portugal, Romania, Rwanda, Saudi Arabia, Senegal, South Africa, Spain, Sweden, Thailand, Tunisia, Turkey, USSR, United Kingdom, USA, Uruguay, Venezuela, Yemen Arab Republic, Yugoslavia, Zaïre.

Hotels: Berne
Bellevue Palace, Kochergasse 3-5 (tel: 224581; tx: 911524; fax: 224743).

Bern, Zeughausgasse 9 (tel: 211021; tx: 911555; fax: 211147).

Kreuz, Zeughausgasse 41 (tel: 221162; tx: 912365).

Metropole, Zeughausgasse 26-28 (tel: 225021; tx: 912807).

Savoy, Neuengasse 26 (tel: 224405; tx: 911863; fax: 211978).

Schweizerhof, Schweizerhoflaube 11 (tel: 224501; tx: 911782; fax: 224501).

Touring (near railway), Zieglestrasse 66 (tel: 458666; tx: 33356).

Zu/03/rich, Neumühlequai 42, 8001 Zurich (tel: 3636363; tx: 817587; fax: 3636015).

Hotels: Geneva
Beau Rivage, Quai du Mont Blanc 13 (tel: 310221; tx: 23362; fax: 867841).

Metropole, Général-Guisan 34 (tel: 211344; tx: 421550; fax: 211350).

Président, Quai Wilson 47 (tel: 7311000; tx: 22780; fax: 7312206).

Hotels: Zurich
Baur an Lac, Talstrasse 1 (tel: 221-1650; tx: 813567; fax: 2118139).

Dolder Grand, Kurhausstrasse 65 (tel: 251-6231; tx: 8164169; fax: 2518829).

Zurich, Neumühlequai 42 (tel: 363-6363; tx: 817587; fax: 3636015).

Car hire: Zürich
Avis: Gartenhofstrasse 17 (tel: 242-20-40; tx: 812094).

Budget: Lindenstr 33 (tel: 383-17-47).

Europcar: Badenerstrasse 812 (tel: 437-21-21; tx: 822287).

Hertz: Lagerstrasse 33 (tel: 242-84-84; tx: 812184).

Chambers of Commerce
Basler Handelskammer, St Albangraben 8, CH-4051 Basel (tel: 23-18-88).

Berner Handelskammer, Gutenbergstrasse 1, CH-3001 Berne (tel: 26-17-11).

Chambre de Commerce et d'Industrie de Genéve, Boulevard du Théâtre 4, CH-1211 Geneva 11 (tel: 21-53-33).

Chambre Fribourgeoise du Commerce et de l'Industrie, Rue de la Banque 1, CH-1700 Fribourg (tel: 22-56-55).

Chambre Vaudoise du Commerce et de l'Industrie, Avenue d'Ouchy 47, CH-1000 Lausanne 13 (tel: 27-72-91; tx: 25405).

Handelskammer-Kaufmännisches Directorium, Gallusstrasse 16, CH-9001 St Gallen (tel: 23-15-15).

Zürcher Handelskammer, Bleicherweg 5, CH-8001 Zürich (tel: 221-07-42, 2210742).

Banking
Banque Nationale Suisse (central bank), Börsenstrasse 15, CH-8022 Zürich (tel: 221-37-50; tx: 812400).

Banca della Svizzera Italiana, 2 Via Magatti, 6901 Lugano (tel: 58-71-11; tx: 841020).

Bank Leu, Bahnhofstrasse 32, CH-8001 Zürich (tel: 219-11-11; tx: 812174).

Crédit Suisse, Paradeplatz 8, CH-8021 Zürich (tel: 215-11-11; tx: 812412).

Swiss Bank Corporation, Aeschenvorstadt 1/Gartenstrasse 9, Basel (tel: 20-20-20; tx: 962334 BVB CH).

Swiss Bankers' Association, Aeschenplatz 4, Postfach 4182, CH-4002 Basel (tel: 23-58-88; tx: 963248).

Swiss Volksbank, Weltpoststr 5, 3015 Berne (tel: 32-81-11; tx: 912486).

Union Bank of Switzerland, Bahnhofstrasse 45, CH-8000 Zürich (tel; 234-11-11; tx: 822802).

Zürcher Kantonalbank, Bahnhofstrasse, PO Box 4039, 8022 Zürich (tel: 41-12-20; tx: 812140; fx: 4112111525).

Principal newspapers
Bilanz, Edenstr 20, 8021 Zürich (tel: 207-72-21; tx: 816765).

Blick, Dufourstr 23, 8021 Zürich (tel: 259-62-62; tx: 817300).

Finanz und Wirtschaft, Bäckerstr 7, Postfach, 8021 Zürich (tx: 812577).

Neue Zürcher Zeitung, Falkenstr 11, Postfach, 8021, Zürich (tel: 258-11-11; tx: 817099).

Tages Anzeiger Zürich-Werdstr 21, 8021 Zürich (tel: 248-44-11; tx: 812236).

Weltwoche, Edenstr 20, 8021 Zürich (tel: 207-73-11; tx: 815503).

Other useful addresses
Basel Stock Exchange, Aeschenplatz 7, 4002 Basel (tel: 23-06-66; tx: 962524).

Federal Department of Finance, Bundesgasse 3, 3003 Berne Bernerhof (tel: 661-11; tx: 912868).

Federal Department of Public Economy, Bundeshaus-Ost, 3003 Berne (tel: 61-21-11; tx: 912889).

Federal Office for Industry, Crafts and Labour, Bundesgasse 8, CH-3003 Berne (tel: 61-29-44).

Fédération Suisse des Agences de Voyages, Postfach, Hardstrasse 316, CH-8027 Zurich (tel: 42-64-42).

Federation of Swiss Employers' Organisations, Florastrasse 44, CH-8034 Zürich (tel: 363-52-85).

Swiss Federal Tax Administration, Eidgenössische Steuerverwaltung, Eigerstrasse 65, CH-3003 Berne (tel: 61-71-12).

Swiss Federation of Commerce and Industry, Börsenstrasse 26, CH-8022 Zürich (tel: 221-27-07; tx: 813294).

Swiss Lawyers' Federation, Lavaterstr 83, CH-8027 Zürich (tel: 202-56-50).

Swiss National Tourist Office, Bellariastrasse 38, CH-8027 Zürich (tel: 202-37-37; tx: 815391).

Swiss News Agency, Langgasstrasse 7, CH-3012 Berne (tel: 24-44-61; tx: 33733).

Union Suisse d'Agences-Conseils en Publicité (advertising), Kurfürstenstr 80, CH-8002 Zürich (tel: 202-65-40).

Zürich Stock Exchange, Bleicherweg 5, CH-8021 Zürich (tel: 229-21-11; tx: 813065).

Syria

A special correspondent

The year 1989 was an uncomfortable one for Hafiz al-Assad, who was in his 19th year as president of Syria. In Lebanon he faced the 'war of liberation' launched by General Michel Aoun, backed to the hilt by an Iraq seeking revenge against its old adversary. On the regional level, Syria found itself increasingly isolated within an Arab world that was grouping itself into formal politico-economic blocs. In the Arab-Israeli peace process, Assad was upstaged by PLO Chairman Yasser Arafat whom he had long tried to undermine. Arafat's peace initiative, and the *intifada* in the occupied territories meant that Syria was pushed to one side in the search for a settlement. On the economic front, the oil sector continued to give grounds for cheer, but there were still some deep-rooted economic problems.

Iraq punishes Syria via Lebanon

It was only to be expected that once Iraq's hands were freed by the Gulf cease-fire agreement, which took effect on 20 August 1988, it would seek to inflict punishment on Syria which had supported Iran during the war. The two countries, ruled by rival wings of the Ba'ath Party, have long pursued a vendetta that has turned into a personalised feud between Assad

KEY FACTS

Official title: Syrian Arab Republic

Head of state: President Hafiz al-Assad

Head of government: Prime Minister Mahmoud Zubi

Capital: Damascus

Official Languages: Arabic

Currency: Syrian Pound (Syr£) = 100 piastres

Exchange rate: Syr£21.0 per $ (Nov 1989)

Area: 185,180 sq km (plus 1,295 sq km Israeli-controlled area)

Population: 11.34m (1988)

GDP real growth: –9.3% (1987)

GNP per capita: $1,820 (1987)

Inflation: 59% (1987)

Trade balance: –$869m (1987)

Foreign debt: $4.0bn* (1988)

* estimated figure

and Iraq's President Saddam Hussain. The cease-fire meant that one of Syria's main justifications for maintaining the alliance with Iran – that it could use its influence with Tehran to help bring about peace in the Gulf – was no longer valid. For Syria, however, the Iranian card was too valuable to sacrifice readily. Even after the Gulf cease-fire the Iranian connection continued to give Syria additional room for manoeuvre in the Middle Eastern arena, and a vital consideration was Iran's influence within Lebanon and the need for Damascus to maintain a working relationship with it there.

Lebanon proved to be a convenient arena for Saddam Hussain in which to prosecute his campaign against Syria. Damascus regards Lebanon as being of crucial strategic importance. It is a possible staging point for an attack by Israel, and Syria has no wish to see it partitioned into confessional statelets which could prove a security threat and also stir up confessional strife in Syria itself. Syria's direct and costly involvement in Lebanon started in 1976 when it sent its troops there in support of the Maronites, to counter the threat of a PLO-leftist victory. Over time it came to ally itself with various Muslim factions. But the years of involvement in Lebanon failed to bring about the long-term constitutional changes that Syria seeks, and the unity of the country, which Syria claims it wishes to preserve, was thrown into doubt. It was probable that were the Syrians to withdraw, Lebanon would collapse into anarchy. It could not be denied, however, that the behaviour of Syrian troops in Lebanon had left a lot to be desired, and had aroused the resentment of various sections of the population.

The protracted crisis Syria faced in Lebanon in 1989 could be traced back to the fiasco of the presidential election the previous summer. Syria had hoped to see a successor elected to Amin Gemayel who would be sympathetic to the views of Damascus, and ready to start the process of reform. After a fruitless search by the Syrians for a compromise candidate acceptable both to themselves and to at least some important sections of the Maronites (the group that will ultimately have to cede some of their traditional privileges in any programme of reform), there was surprise when Syria came out in support of the seventy-eight year-old pro-Syrian former president Sulaiman Franjiya, who was anathema to the Lebanese Forces (LF) and other Maronite quarters. A first attempt to hold the ballot on 18 August failed when Christian deputies stayed away and a quorum could not be achieved. Before the second attempt to stage a ballot, Syria dropped its support for Franjiya in favour of Mikhael Daher, a lawyer and parliamentary deputy from a Syrian-controlled area of northern Lebanon. Syria was supported by the USA in its promotion of Daher, but his candidature was vehemently criticised by Christian militias and by the Lebanese Army. Despite talks between President Gemayel and President Assad on its eve, the second ballot, set for 22 September, failed like the first when a quorum could not be achieved.

On 22 September 1988, Gemayel's term ended, and his final act as president was to issue a decree appointing the army commander, General Michel Aoun, as prime minister at the head of a military government. The Muslims saw the appointment of Aoun as a constitutional coup, and the three Muslim generals also named as members of Aoun's government refused to serve in it. The government of Salim al-Hoss declared that it was the legitimate government, and in this was supported by Syria. By the end of September, Iraq was reported to be supplying the LF with arms, and the following month Iraq was the first country to come out in public support of Aoun. To Syria's consternation, it was followed in this by Libya, Damascus' long-time supporter.

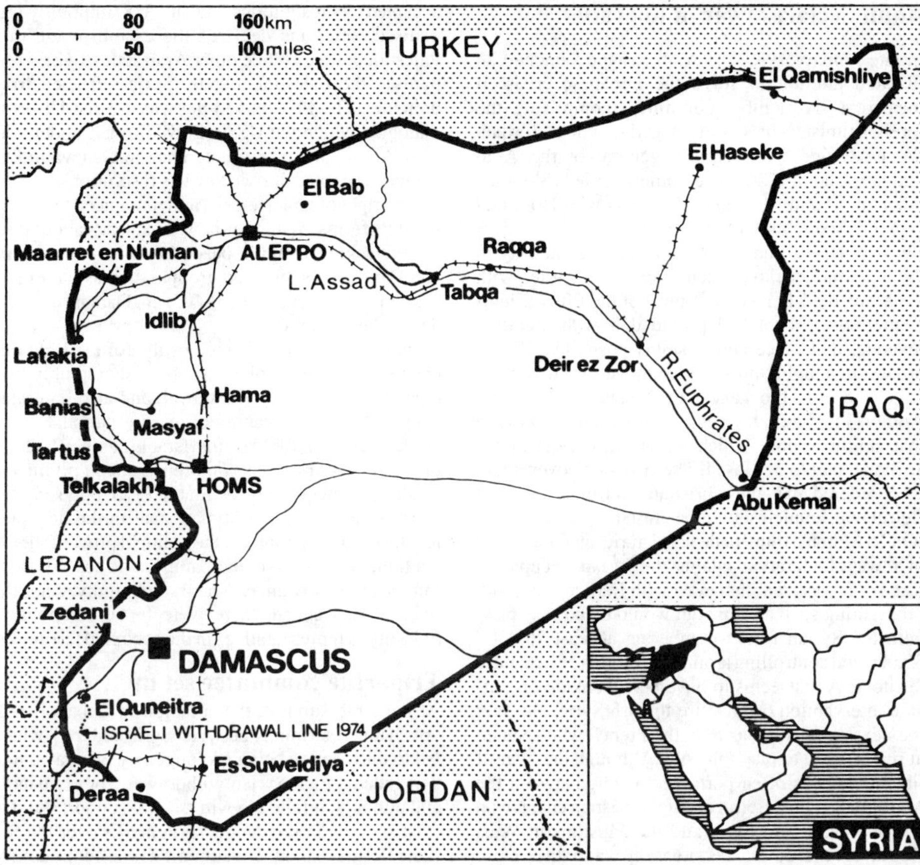

Syria brokers Amal-Hizbollah accord

In early 1989, there was not only mounting tension between the two governments and their supporters, but also renewed fighting between the rival Shi'a groups – Amal, which has been backed by Syria, and Hizbollah which has been sponsored and financed by Iran. The two organisations had long been at each other's throats in the contest for influence. In spring 1988, there were major clashes between them, which resulted in Amal making gains in the south of Lebanon, and Hizbollah seizing much of the southern suburbs of Beirut. Following a cease-fire, Syria had sent troops into the southern suburbs. There were further battles in the southern suburbs and other areas of west Beirut in November 1988, and the following January these widened to duels in southern Lebanon. In the meantime, the PLO's moderate line had led to a split between it and Hizbollah, and to an improvement in relations with Amal, which in 1985–88 was sporadically engaged in the 'camps war' against the PLO in Lebanon. Top-ranking Syrian and Iranian politicians were involved in moves to end the clashes between Amal and Hizbollah, and on 30 January 1989 a comprehensive peace settlement was announced. Syria gained in that Hizbollah agreed that security in Beirut's southern suburbs should be the responsibility of Syrian forces. Syria gave Hizbollah the right to a presence in the south, although Amal was granted the right to control security there.

KEY INDICATORS

	Unit	1983	1984	1985	1986	1987
Population	'000	9,610	9,930	10,270	10,610	10,970
Gross domestic product (GDP)	Syr£m	73,291	75,342	83,225	100,300	126,325
GDP per capita	Syr£	7,626	7,587	8,104	9,453	11,515
Wholesale prices	1985=100	83.4	90.8	100.0	140.5	206.1
Consumer prices	1985=100	78.1	85.3	100.0	136.1	217.0
Industrial production	1985=100	102	102	100	108	123
Volume of exports						
Crude petroleum	1985=100	114	107	100	96	101
Cotton	1985=100	117	141	100	108	66
Exports	Syr£m	7,548	7,275	6,437	5,199	5,312
Crude petroleum	Syr£m	4,131	3,608	3,119	1,319	5,047
Cotton	Syr£m	675	1,076	546	391	877
Imports	Syr£m	17,828	16,154	15,570	10,611	9,761
Balance of trade	$m	−2,106	−1,853	−1,877	−1,326	−869
Current account	$m	−844	−794	−860	−530	−187
Exchange rate	Syr£ per $	3.92	3.92	3.92	3.92	11.22

* estimated figure

Aoun declares 'war of liberation'

In late January 1989, the Arab League launched an ill-fated attempt to solve the Lebanese crisis. It formed a six-member committee comprising the foreign ministers of Kuwait, Algeria, Tunisia, Jordan, the UAE and the secretary general of the Arab League, Chedli Klibi. The committee held separate talks on 30 and 31 January in Tunis with Hoss and General Aoun, but with little result.

In February, Aoun moved to assert his authority in the Christian enclave when after fierce battles with the army, the LF gave up control of the fifth basin of Beirut port, which had provided it with lucrative revenue, and of certain customs posts. On 27 February, in a highly provocative move, Aoun began a naval and aerial blockage of the Muslim militia ports. And on 11 March, he went further when he ordered that Beirut airport, which was within range of his artillery, should be closed. The Hoss-led government then ordered that the Christian-controlled ports of Beirut and Jounieh should be closed.

On 15 March, Aoun made his historic announcement of a 'war of liberation' against the Syrian 'occupation forces'. It was clear that even with Iraqi military aid, prosecuting such a campaign would be no easy task, with the Syrian forces numbering at least 25,000–35,000 and controlling around two-thirds of Lebanese territory. Aoun seems to have been banking on outside intervention or pressure to curb Syria, but over the next months it was only the French who seemed at times ready to play this role. Although the 'war of liberation' was often portrayed as one between the Maronites in East Beirut and the Christian enclave on the one hand, and Syria and the Muslims in West Beirut on the other, the situation was rather more complex than this. Aoun's stand against the Syrians struck a sympathetic chord among a not inconsiderable number of West Beirutis, while on the Christian side the unity of the Maronite camp could not be taken for granted, as Aoun's earlier clashes with the LF, and criticisms he had come in for from the Maronite clergy, showed. And Syria still had its supporters among the Maronites, in the shape of the followers of Franjiya in northern Lebanon.

Syrian forces played a direct role in the artillery bombardments which raged in the months following Aoun's launching of his war. The bombardments were among the fiercest since the civil war started in 1975, and caused many civilian casualties. The Syrian media tried to play down the Syrian role, however, referring to the military activities of Syria's Muslim allies rather than directly to those of Syria.

Arab summit supports Arafat

The Lebanese crisis was a major preoccupation of the extraordinary Arab summit which started in Casablanca on 23 May 1989. This summit underlined the weak position in which Syria found itself in the Arab arena. One notable decision of the summit was the restoration of Egypt to full membership of the Arab League and its organisations after a ten-year rift. Syria, which had in any case softened its stand on Egypt in the preceding months, acquiesced in this. Egypt's readmission lent its considerable weight to the group of 'moderate' countries which were fully behind the PLO'S peace offensive.

The summit also clarified its position on the Middle East peace process, and supported the decisions taken by the Palestine National Council (PNC) at its 12-15 November 1988 meeting in Algiers. The initiative launched at the PNC, and the *intifada* in the occupied territories, meant that it was PLO Chairman Yasser Arafat and the Palestinians who were the focus of international peace moves in 1989. Despite recognition of Syria's key role in any eventual settlement, it was pushed to the sidelines. In Algiers, the PNC had accepted UN resolutions 242 and 338, implicitly recognised Israel and declared the establishment of a Palestinian state. While other Arab states swiftly recognised the Palestinian state, Syria refrained from doing so. The PNC paved the way for the US-PLO dialogue, which the USA allowed to get under way after Arafat, at a press conference in Geneva in December, accepted the two key UN resolutions, spelled out the right of Israel to exist and renounced terrorism in all its forms. While Syria was cautious about publicly attacking Arafat's peace initiative, anti-Arafat and Damascus-sponsored groups, such as the PFLP-General Command (PFLP-GC) in Damascus, condemned it as treason.

The summit supported the position of the PLO that elections in the occupied territories should only take place after an Israeli withdrawal, and also reiterated support for the convening of an international peace conference. On 13 May the Israeli cabinet had approved a peace plan that was seen as uncompromising – ruling out negotiations with the PLO, opposing the establishment of a Palestinian state in the occupied territories, and stating that the political process should continue on the basis of the Camp David accords, with limited autonomy under Israeli sovereignty. Once the *intifada* was ended, there were to be elections of Palestinian representatives in the occupied territories.

Tripartite committee set up

At the Arab summit, the situation in Lebanon provoked much acrimony between Syria and Iraq, which led to the meeting being prolonged by two days. Iraq demanded a total Syrian withdrawal from Lebanon, and although it failed to win general support it did at least succeed in raising question marks over the Syrian role in Lebanon. The summit's final communiqué did not mention Syria, and called for continued Arab action to arrive at a solution which restored Lebanon's independence and security and preserved its Arab character, sovereignty and territorial integrity. A high-level tripartite committee was set up to replace the previous Arab League committee. It comprised King Hasan of Morocco, King Fahd of Saudi Arabia and President Bendjedid of Algeria. This committee was charged with making the necessary contacts to call the members of the Lebanese Chamber of Deputies to a meeting to discuss the charter of political reforms, hold presidential elections and form a government of national unity within a period of six months – close to what Syria had long been trying to achieve.

While the tripartite committee sought international support for its efforts in Washington, London, France and the Soviet Union, the crisis intensified. The Syrian navy was deployed for the first time on 2 July to prevent weapon shipments reaching the ports in the Christian enclave, and Syria said that before the blockade could be lifted the medium-range missiles which Iraq had allegedly supplied to Aoun should be removed. The Soviet-made Frog-7 missiles would be capable of hitting Damascus if fired from mountains in the Christian enclave. The Christians for their part refused to allow Beirut airport to reopen as long as the blockade went on.

The Soviet Union raised its profile over Lebanon when Soviet First Deputy Foreign Minister Alexander Bessmertnyk made a late June–early July visit to Baghdad and Damascus, although any positive results were short-lived. During Soviet President Mikhail Gorbachev's visit to France of 4-5 July 1989, a final communiqué called for an independent united Lebanon and a halt to arms supplies as well as for an immediate cease-fire. The Syrian media particularly welcomed the call for a halting of arms supplies, and claimed the communiqué was in essence identical to the Syrian viewpoint.

The US attitude to Syria was somewhat vacillatory. In an interview broadcast on 26 June 1989, Assistant Secretary of State Lawrence Eagleburger declared that Syria 'has to back off' from Lebanon. However, in a statement to the Senate foreign affairs committee six weeks earlier, he had said that withdrawal of Syrian troops from Lebanon would worsen the situation because they imposed restraint on the warring parties. The US ambassador in Damascus, Edward Djerdjian, said on 3 July 'we realise Syria has an important role to play in any political settlement in Lebanon', and said that the Lebanese must agree on elections and reforms and create a central government that could extend its authority over all Lebanese territory. The ambassador also said that a solution would require withdrawal of all foreign forces in Lebanon, including Israeli and Syrian troops.

The welter of international concern over Lebanon had no effect on the ground, with Aoun insisting that a Syrian withdrawal must precede any settlement of the Lebanese crisis. At the end of July, the tripartite committee frankly admitted it had failed to achieve the cease-fire and political reforms it had been seeking. Although the chances of the tripartite committee's succeeding had always looked slim, the Arab initiative had been seen as the last hope perhaps, and its failure was a blow. It seemed it would be followed by a fight to the death.

Hostage issue resurfaces

A further twist was added to the situation in Lebanon when, in late July 1989, Israeli troops penetrated southern Lebanon and kidnapped Sheikh Abdul Karim Obeid, a leader of Hizbollah. The intention was to use him to bargain for the release of three Israeli soldiers and possibly for Western hostages held in Lebanon, most of whom were assumed to be held by groups linked to Hizbollah. Syria was thrust to centre stage by this because many of the hostages are assumed to be held in areas supposedly under its control, and because of its influence with Iran and with Muslim groups in Lebanon. Syria has in the past been eager to show its ability to cooperate over the hostages, and to take credit for releases. Thus when an Indian-US hostage Mr Singh was released in Beirut in October 1988, he was handed over to the US ambassador in Damascus, and US-Syrian relations were said to have been improved by Syria's role in the release.

The abduction of Obeid was followed by the release in Lebanon of a video tape that was claimed to be of the hanging of one of the hostages – US Colonel William Higgins – and there were threats that another US hostage, Joseph Cicippio, would be killed. With the USA unable to take effective military action with guaranteed success, it and its European allies appealed to Syria to do all possible to ensure the safety of the surviving hostages.

It seems to have been intervention by the newly elected Iranian President Hashemi Rafsanjani that resulted in the death threat against Cicippio being lifted. The changes in the Iranian leadership following the death of Ayatollah Khomeini on 3 June 1989, and the election of Hashemi Rafsanjani as president on 28 July, were of crucial interest to Syria as they had implications for future Iranian policy in Lebanon and for Tehran's relations with Syria and other Arab states. Rafsanjani said, in a spoken message to the White House on 4 August that was notably conciliatory in tone, that there were solutions to the problems of Lebanon and the hostages. Ali Mohtashami, the interior minister, who was replaced in Rafsanjani's first cabinet – much to the chagrin of radicals in the Iranian parliament – had been seen as the key link with Hizbollah in Lebanon. It was he who when ambassador to Damascus played an important part in setting up the organisation. Soon after Rafsanjani took power in August 1989, it became apparent that he was shifting from the policy of supporting Hizbollah only to also supporting Amal, and it was reported that the aid to Amal would include weapons. It was also reported that Iran had promised shipments of oil to Lebanese Muslim militia leaders.

Syria's 'council of war'

The furore stirred up by the abduction of Obeid was overshadowed in mid-August by ever more vicious fighting between the two sides in Lebanon. On 13 August, Syria and its allies launched a ground attack on the strategically vital position of Souk al Gharb. Although the attack was repulsed, its significance lay in the fact that it was the first ground offensive of the five-month-old war, and that Syria had shown itself prepared to make inroads into the Christian enclave. With the situation in Lebanon becoming graver by the hour, the UN Security Council called for a cease-fire on 15 August. Aoun expressed unconditional support for this, but said it must be followed by talks about a Syrian withdrawal from Lebanon. Hoss also agreed to the cease-fire, as did Syria, but there was no impact on the ground. On 16 August there was a 'war council' in Damascus of the Syrian and Iranian foreign ministers and leaders of several pro-Syrian Lebanese militias, including Walid Junblatt, Druse leader, Nabih Berri, leaders of Hizbollah and the heads of two radical Palestinian factions – Abu Musa's dissident Fatah movement, and the Popular Front for the Liberation of Palestine General Command (PFLP-GC). The meeting declared that Hizbollah would participate in the fighting in Lebanon, and that there were no longer any 'red lines' that could not be crossed. Cynics said that the formation of the coalition of forces in Damascus was intended to give Syria a 'cover' should it cross the red lines in Lebanon.

Syria left out of ACC

Syria's growing regional isolation was underlined by the fact that it was excluded from the Arab Cooperation Council (ACC), a regional grouping that was inaugurated in Baghdad on 16 February 1989 by the heads of state of Egypt, Jordan, Iraq and North Yemen. Although the grouping publicly emphasised its economic rather than political function, it had undoubted political dimensions, and formalised the Cairo-Baghdad-Amman axis. It also appeared to be a counterbalance to the Gulf Cooperation Council (GCC).

Mid-February also saw the creation of the five-member Arab Maghreb Union (AMU). Libya, which has in the past been one of Syria's closest Arab allies, seemed to be adopting a pragmatic stance by entering the Maghreb grouping. Syria was anxious about Qadafi's support for Aoun and his rapprochement with Iraq and with the mainstream PLO and its chairman Yasser Arafat. In late March, Assad made a surprise visit to Libya, Algeria and Tunisia, and Syria supported Libya in warning against possible US actions against the alleged Libyan chemical weapons plant. Some observers believed that Syria and Saudi Arabia were drawn closer together as a result of the changes on the regional political board. Syria appeared to continue to receive financial aid from Saudi Arabia, after the expiry of aid provided through the ten-year Baghdad summit aid agreement of 1978. One factor drawing them together is shared suspicions about Iraqi ambitions in the Arab political arena, particularly in the post-Gulf War era.

Relations with USSR

There was continuing speculation in 1989 about strains in Syria's relations with the Soviet Union. Syria has been the Soviet Union's closest ally in the Middle East since the signing of a friendship and cooperation treaty in October 1980. Over the years, Syria has received generous arms supplies from the Soviet Union, and one reason for the strain was said to be Syria's mounting debt to the Soviet Union, which has been estimated as high as $15,000m, although Soviet sources in Damascus have put it at $4,000m. The Soviet Union has for some years been unhappy about Assad's role in splitting the Palestinian movement, and has tried to encourage a reconciliation with the mainstream PLO. In 1989, it urged him not to undermine Arafat's peace initiative. The Soviet initiatives over Lebanon in 1989 showed a desire to be more involved in the Middle East. Nor has Syria been happy over increasing Soviet and Eastern bloc economic links with Israel in 1989, which were seen as paving the way for a full resumption of diplomatic relations.

When Soviet Foreign Minister Eduard Shevardnadze visited Damascus in February 1989, he was reported to have tried to persuade the Syrians to abandon their cherished aim of reaching strategic parity with Israel, and to concentrate, rather, on having the ability to defend their legitimate interests. However, Israeli military sources claimed that Syria was still pursuing a policy of achieving strategic parity, and repeated their allegations that it has missiles capable of carrying chemical warheads.

The Syrian air force was said to have received Soviet SU-24 ground attack jets in 1989. But during his visit to Damascus in March 1989, Soviet Defence Minister Dimitri Yazov was reported to have refused Syrian requests for the updated version of the MiG-29. Western intelligence sources said in 1989 that the number of Soviet advisers in Syria had shrunk from around 6,000 in 1982 to some 2,400.

Economic problems persist

Compared with many other countries in the Middle East, Syria is endowed with a good resource base; it has water for irrigation from rivers, notably the Euphrates, relatively generous rainfall, oil and phosphates, and a traditionally entrepreneurial population. Because of its strategic position as a 'frontline' state in the Arab confrontation with Israel, and because of its political clout, the Arab oil states have been willing to make it generous (though declining) subventions over the years. Yet the economy has been in difficulties throughout the 1980s. One reason for this is the heavy burden of defence spending. Thus in the 1988 budget, £Syr13,320m or 45 per cent of current spending went on defence. Mismanagement and corruption have also taken their toll on the economy, and periodic well-publicised campaigns launched from the top and involving the sackings of top officials, including on occasion ministers, have done little to improve matters. The large state industries have often proved unwieldy and inefficient; Syria has yet to follow the example of its major backer the USSR and undertake a meaningful process of perestroika (economic restructuring). One reason for the resistance to change is that there are deeply entrenched political interests in the status quo. One indication of the pressures to which the Syrian economy has been subjected is the fact that the government has in recent years found its coffers almost bare at times. IMF figures show that Syria's total reserves minus gold stood at only $10m in mid-1986. By the end of 1987, they had risen to $2,323m – the highest end of year figure for reserves since 1984.

An additional challenge for Syria is the growth rate of its population, which is one of the highest in the world. According to the World Bank, the population growth rate increased from 3.4 per cent a year in 1965-80 to 3.6 per cent a year in 1980-87 and will increase further to 3.7 per cent a year in 1987-2000. The World Bank predicts that on current trends, the size of the Syrian population will grow from 11m in 1987 to 18m in 2000 and 37m in 2025.

Syria's economic problems are reflected in trends in gross domestic product (GDP). GDP grew by 9.5 per cent in 1981, but in the years since has shown either small or negative growth. At constant 1980 prices, GDP fell by 9.3 per cent in 1987 according to provisional government figures, after falling 1.2 per cent the previous year. With population increasing at a fast pace, GDP per head has shown a relentless drop. At constant 1980 prices, it dropped progressively from £Syr6,216 in 1982 to £Syr4,775 in 1987. In 1987 alone, per capita GDP fell by 12.2 per cent. Western estimates prepared in mid-1989 put GDP growth in 1988 at 1.5 per cent, and predicted that there would be growth of 2 per cent in 1989. The same economists had earlier predicted that GDP would grow by 6 per cent in 1989, but reduced this figure when the economy was hit by drought and by production problems at the new Omar oilfield.

For a number of years before 1988 there was a succession of austerity budgets, with tight controls on spending. In the 1987 budget, there was a cut in total spending of 4.9 per cent. The 23.6 per cent increase in total spending allowed for in the 1988 budget, to £Syr51,545m, therefore seemed generous in comparison. Current expenditure in the 1988 budget was increased by 22.6 per cent to £Syr29,665m and investment spending by 25 per cent to £Syr21,880m. But taking inflation into account, the nominal increase in spending in 1988 still represented a substantial drop in real terms.

Inflation at high levels

According to the IMF, consumer prices rose 60 per cent in 1987 and 36 per cent in 1986, and unofficial estimates in 1989 estimated inflation as at least 100 per cent. The high rate of inflation negates the value to employees of government pay increases. Effective from June 1989, the government awarded a 25 per cent pay increase to public employees, including the army and the police, and raised pensions by 15 per cent. But these increases had to be set against rises of 50 per cent in both fuel prices and customs duties. As well as a drop in the purchasing power of their wage packets, Syrian consumers have also had to contend with periodic shortages of goods, not only of consumer durables, but also of such basics as coffee, tea, tissue paper, fruit and gas. A new system of customs tariffs introduced in 1989 was intended to try to improve the supply of basic goods. The new system gives preference to 'necessary imports', with lower duties on commodity foodstuffs including tea, rice, sugar and vegetable oil, as well as medicine, spare parts and certain agricultural and industrial inputs. Duties on luxury goods were increased.

The black economy has come to play a vital role in the 1980s, with widespread smuggling from Lebanon assuring a supply not only of consumer and luxury goods, but also of essential raw materials. Although there have been drives against smuggling from time to time, it persists and indeed provides a sort of economic safety valve. In 1986 there was a drive against black-market currency dealers after the value of the Syrian pound plunged, but dealers do still operate, albeit in a discreet fashion, through family and personal contacts. A decree of September 1986 laid down stiff penalties for currency violations, and smugglers of currency face imprisonment for up to 25 years. People who fail to repatriate funds earned abroad from business in Syria can also be penalised. Such legislation added to the nervousness of the private sector, with businessmen who had previously operated in a 'grey area' in which they were able to bend the rules afraid of falling foul of the sometimes unpredictable authorities.

Current account position improves

Despite its economic woes, Syria has seen its current account position improve since 1985. In the first half of the 1980s the external payments position had deteriorated, with a current account deficit of $860m in 1985 compared with a surplus of $959m in 1979. One reason for the deterioration was a steady decline in Arab aid, and in remittances from Syrians working abroad. In 1981, net official transfers stood at $1,819m, but by 1987 they had fallen to $696m. Net private transfers dropped from $901m in 1979 to $222m in 1987. In 1986 the current account deficit decreased to $520m, and in 1987, according to provisional figures, it shrank to $171m. It was estimated that the current account went into surplus, to the tune of $110m in 1988, and an increase in the surplus to $200-300m rising to $500m in 1990 was predicted.

A major factor in the improvement has been a reduction in the trade deficit. Measures to curb imports since the early 1980s have had a marked effect, and the value of imports declined progressively from $4,116m in 1984 to $2,492m in 1987 and to an estimated $2,000m in 1988. Exports rose from $1,037m in 1986 to $1,356m in 1987, and are estimated to have been $1,600m in 1988. In early 1988, the government devalued the official rate for the Syrian pound from $1=£Syr3.925 (the level since 1976) to $1=£Syr11.20, and this has helped the downward pressure on the volume of imports.

Unlike many developing countries, Syria is not burdened with a huge foreign debt, although World Bank debt figures do not include the massive debt to the Soviet Union, estimated at up to $15,000m. According to the World Bank, public and publicly-guaranteed foreign debt outstanding and disbursed at the end of 1987 stood at $3,648m, a 17 per cent increase from a year earlier. The debt service ratio was 14.9 per cent, compared with 15.6 per cent in 1986. The ratio has grown markedly – in the first half of the 1980s it hovered between 9.5 and 10.8 per cent. It was estimated that the debt rose to $3,950m at the end of 1988, with a rise in the debt service ratio to 17.2 per cent.

Private and foreign investment sought

The government has tried to encourage private investment in recent years through allowing joint government-private sector companies in which the government takes a less than 51 per cent stake. Pioneered in tourism, such companies have since proliferated in the agricultural sector. The state has also tried to encourage foreign investment, in accordance with its foreign investment laws of 1971. In 1982, an office was established in the planning ministry with the purpose of facilitating foreign investment procedures, and in early 1989 it was announced that the government planned to set up an official investment office under the direct authority of the prime minister.

Oil finds bring benefits

The one sector of the economy that has seen much Western involvement in the 1980s is hydrocarbons. The discovery of high-grade oil in the Deir ez-Zor region by a consortium of foreign companies, which was revealed in 1984, has transformed the oil sector and made it by far the most dynamic sector of the economy. The discovery of the Al-Thayyem field was made by a partnership of Pecten, Royal Dutch Shell and Deminex. The Al-Ward and Al-Asharah fields were later found in the Deir ez-Zor concession. Commercial production at Al-Thayyem started in September 1986, and subsequently Brown & Root carried out work to increase production capacity to 120,000 barrels a day (b/d) through linking the Al-Ward and Al-Asharah fields into the Al-Thayyem pipeline system.

In early 1988, the Pecten consortium made another promising discovery in the Al-Thayyem concession – the Omar field. Omar was brought on stream at full capacity of 100,000 b/d in early 1989, but working of the field at full production stressed it and led to a drastic loss in pressure. By July 1989, output had fallen to only 29,000 b/d. Production from the other Deir ez-Zor fields was around 150,000 b/d, and that from Syria's traditional oil fields in the north east 160,000 b/d or so. Total output was thus around 300,000 b/d. In order to remedy the problem at Omar, it was decided that a water injection system costing around $100m should be installed, and companies attended a pre-tender meeting in July 1989. It was thought that if the injection system were installed on schedule, production from the Omar field would be restored to 100,000 b/d in late 1990 or early 1991.

Before the problems at Omar, Syrian officials had hoped that oil output would rise to 400,000 b/d by the end of 1989. In early 1989, Brown & Root signed a contract to develop two oil fields – Tayani in the Deir ez-Zor concession, and Tanak, the first field to be developed in the Al-Sham concession, east of the Deir ez-Zor area and exploited by the Pecten consortium. The work was expected to be completed by the end of 1989, and to result in production of about 20,000 b/d from each of the fields. The Pecten consortium had meanwhile given up its exploration rights to the Deir ez-Zor concession, and these were taken over by France's Société Nationale Elf Aquitaine. Another field in Al-Sham said to be of particular interest was Sinjan.

In mid-1989, it became clear that the Syrian Petroleum Company (SPC) was introducing a new relationship with contractors in the development of oil fields. Whereas previously foreign companies had been given turnkey contracts, SPC was now to take on the role of main contractor, letting contracts for consultancy and equipment supplies to foreign firms. This formula was first applied to the Izba field in the Deir ez-Zor concession, with the American company Fluor Daniel being appointed to carry out designs and supervise construction of production facilities for about 25,000 b/d of crude oil.

The discoveries in the Deir ez-Zor region have fuelled the interest of other oil companies. Among the companies that signed exploration agreements between 1987 and the first half of 1989 were the British companies Arco British (formerly Tricentrol) and British Petroleum; the US companies Pentagon Petroleum, Union Oil of California (Unocal) Corporation, Occidental Corporation and Enron Corporation, and the French concerns Total (Compagnie Française des Pétroles). Further companies were known to be negotiating with the Syrians over possible concession agreements.

The development of the Deir ez-Zor concession has greatly reduced Syria's need for concessionally-

COUNTRY PROFILE

Political system: Executive power rests with the president, who governs with an appointed Council of Ministers led by the Prime Minister. Legislative power rests with the 195-member unicameral People's Council.

Population: The Alawite (Shi'a) Muslim community dominates the political structure, though it represents only around 12 per cent of the population. Most others are Sunnis. Population growth rate is running at the high level of 3.5 per cent per annum.

Main cities/towns: Damascus (population estimated 2.8m in 1988), Aleppo (2.1m), Homs (860,000), Hasakeh (700,000), Latakia (584,000), Tartus (475,000), Deir-ez-Zor (430,000).

Language: The official language is Arabic; English and French are sometimes spoken in business circles.

Media
Radio: Service in Arabic, French, English, German and certain other languages. 2.5m* radios (1988).
Television: Estimated 1.3m sets (1988).

Domestic economy: The economy is highly vulnerable to the vagaries of external aid from Iran, Saudi Arabia, Libya and the Arab Fund for Economic and Social Development, with its finances bled by the defence expenditure which maintains Syria's presence in Lebanon. Military spending has channelled funds away from development projects, and shortage of foreign exchange has stifled economic activity, leading to a shortage of raw materials and spare parts, and to domestic rumblings of discontent. Heavy indebtedness to the USSR and Eastern Europe. In August 1987, however, measures were introduced to encourage private savings and investment in an effort to attract much-needed foreign currency.

Employment: About a quarter of the workforce is engaged in agriculture.

External trade: Trade in certain items, such as phosphates, is conducted under countertrade arrangements.
Exports: Main exports are petroleum and petroleum products (51.8% of 1987 total), yarn and woven fabric (12.9%), cotton (5.8%), phosphates (4.0%), live animals (2.7%) and wool (0.4%). Main destinations are Italy (22.6% of 1985 total), Romania (18.5%), USSR (11.8%), France (11.5%), Spain (6.7%), UK (5.2%), Greece (1.9%).
Imports: Main imports are petroleum and petroleum products (19.1% of 1987 total), machinery and transport equipment (25.1%), iron and steel (7.0%), cereals (6.3%), textiles (5.7%) and medical products (2.0%). Main sources are Iran (31.8% of 1985 total), German FR (7.2%), Italy (7.0%), Libya (5.0%), France (4.8%), USSR (3.3%), Japan (3.3%).

Agriculture: Agriculture contributes about 17 per cent to GDP and employs about 25 per cent of the labour force (1987). Agricultural land is mainly privately owned. Main crops include cotton, wheat, barley, millet, maize, lentils, sugar beet, grapes, tobacco, sesame, figs, pomegranates, apricots, apples, onions, tomatoes, potatoes, melons, citrus fruit and chick peas. Wool is also an important product. Serious problems with an ambitious irrigation system have induced the government to opt instead for intensification of farming in the rain-fed areas.

Industry: Industrial production including mining contributes about 16.5 per cent of GDP and employs about 18 per cent of the labour force. Fertiliser plant at Homs converts local naptha to ammonia and urea. Other industries include cement, soap, glass, food processing, iron and steel, TVs, footwear, leather goods and brassware. The textile industry is the oldest-established, producing cotton yarn and fabrics, silk fabrics, knitwear and hosiery. Sugar processing, an important activity, is mainly conducted by state-owned enterprises.

Minerals and mining: 62 per cent of phosphate mined is exported, with 10 per cent used at the Homs fertiliser factory.

Hydrocarbons: The discovery of high-grade oil in the Deir ez-Zor region has transformed the oil sector. The Al-Thayyem field was revealed in 1984, and the Al-Ward, Al-Asharah and Omar fields, in the following years. Originally, oil production, mainly from Suwaidiyah, Karachuk and Jubaissiah in the north east, was largely of heavy crude, which must be mixed with imported light crude for use in domestic refineries. Production of light, low-sulphur oil began in the 1986 in the Deir-ez-Zor region. Total output in mid-1989 was running at 300,000 barrels a day (b/d). Natural gas reserves were estimated in 1986 to be 104bn cubic metres, but only a small percentage is utilised. Oil refineries at Homs and Banais.

priced oil from Iran, which it supported in the Gulf war. From 1982 onwards, with the oil pipeline from Iraq to the Syrian coastline closed for political reasons, yearly deals were reached with Iran for the supply of oil. Typically, 1m tonnes would be supplied free to the Syrian army and the rest – 5m tonnes in 1985 – at a discount. The agreements were disrupted by Syria's mounting oil debt to Iran, which was reported to be at least $1,500m in mid-1986. A new one-year agreement allowing for the supply of 1m tonnes of free oil was reached in April 1988, but in mid-June 1989 the Syrian oil minister said that no Iranian oil had been received since the end of 1988.

The production of oil from the Deir ez-Zor region has improved Syria's oil trade balance. In 1983, Syria was a net importer of oil in terms of value, with an oil trade deficit of £Syr123m. However, in terms of quantity it had a surplus of 900,000 tonnes. In 1986, imports dropped from 5.4m tonnes the previous year to 2.7m tonnes, rising in 1987 to 3.3m tonnes. Exports dropped from 7.4m tonnes in 1985 to 6.3m tonnes in 1986, rising to 6.6m tonnes the following year. The oil surplus was 3.3m tonnes in 1987, a slight decrease from the 3.6m tonne figure seen in 1986.

Cotton output up

Syria's main cash crop is cotton. The 1987-88 crop was hit by drought, falling from 126,000 tonnes of ginned cotton in 1986-87 to 96,489 tonnes. The Aleppo-based Cotton Bureau put production in 1988-89 at 109,126 tonnes. The cotton area was increased in 1988-89 to 154,170 ha from 128,688 ha the previous year, and for 1989-90 there were plans to increase it to 165,175ha which was expected would lead to output of 115,000 tonnes. Syria's record cotton crop was in 1983-84, when 193,760 tonnes of ginned cotton was produced.

Winter rains in 1988/89 were around 40 per cent below average, and the drought badly hit the wheat crop which was expected to be 50 per cent down on the previous year. The drought also led to water shortages in the cities and affected the level of water in the Euphrates river, which reduced electricity output from the Euphrates dam.

Tourism holds promise

Tourism is one sector that has great potential, but that has been held back by the political situation in the region and also by a somewhat ambivalent attitude on the part of the government to the prospect of an influx of large numbers of foreigners. There has been a growth in specialised tours from Europe in recent years, interested mainly in the rich historic heritage of the country. Private sector–government joint ventures in tourism, with the government taking a minority stake, have been permitted for some years. There has been some investment by other Arab countries or by rich expatriate Syrians in the tourist sector, and in December 1988 the Abu Dhabi Fund for Arab Economic Development (ADFAED) announced that it planned to invest in agricultural and tourist projects, and to set up a joint investment company with local institutions and businessmen.

In 1987, the number of tourists dropped by 13 per cent to 493,400. While the number of Arab tourists stayed virtually constant at 266,600 or 54 per cent of the total, the number of non-Arab tourists dropped by 25 per cent to 226,800. A major reason for this was a decline in the number of Iranian tourists. As part of the alliance between Syria and Iran, and because of an arrangement whereby Iranian visitors could visit Syria in part payment for Iranian oil supplies, there had been a sharp increase in the 1980s in the number of Iranian visitors. In 1986, the number peaked at 174,000, but in 1987 there was a 33 per cent drop to 117,500.

In 1989, a tourism project aimed primarily at Iranian pilgrims was unveiled. It consists of two hotels and other facilities such as a shopping centre, clinic and cafeteria and parking lots, and will cater for visitors to the Saydeh Zeinab shrine near Damascus. The project was initiated by the Saydeh Zeinab Company for Tourism and Pilgrimage, a joint venture of influential local businessman Saeb Nahas, together with Arab and Iranian businessmen. Companies were invited to prequalify for the project by 15 August 1989.

BUSINESS GUIDE

Time: GMT +2 hrs (GMT +3 hrs Mar-Oct).

Climate: Hot dry summers (30-35°C) May to Sep and fairly cool winters (5-15°C) on the coast. Temperatures inland are more extreme.

Entry requirements
Passport: Required by all except Arab League nationals.
Visa: Required by all except nationals of certain Middle Eastern countries and must be acquired in advance. Journalists must report to the Ministry of Information within 24 hours of arrival. All travellers intending to stay longer than 15 days must report to the Direction d'Immigration des Passeports et de la Nationalité within 15 days.
Prohibited entry: Nationals of Israel and holders of passports with evidence of travel in Israel are denied entry.
Currency: Import of local currency is limited to Syr£200. No limit on foreign currency, but it must be declared on entry. All foreign visitors are obliged to change $100 or equivalent when entering the country.
Customs: Imports controlled by licensing, and importation of certain goods produced domestically and certain non-essential luxuries is forbidden. Most basic commodities are imported by government agencies only. Trade with South Africa and Israel prohibited. Duty-free shop, known as Gota, near the Central Bank in Damascus, stocks cigarettes, liquor, perfumes, etc for sale to foreigners on production of passport, with payment in US or Lebanese currency only. Transit visitors (visas valid for up to three days) are not liable for duty.

Weights and measures: Metric system (local units also in use).

Health precautions
Mandatory: Certificate of vaccination against yellow fever if travelling from infected area.
Advisable: TAB and cholera vaccinations, anti-malaria precautions.

Representation overseas: Abu Dhabi, Algiers, Amman, Ankara, Athens, Baghdad, Beijing, Berlin, Belgrade, Bonn, Brasilia, Brussels, Bucharest, Budapest, Buenos Aires, Caracas, Conakry, Dakar, Dar es Salaam, Doha, Havana, Islamabad, Jakarta, Jeddah, Khartoum, Kuwait City, Lagos, Madrid, Mogadishu, Moscow, New Delhi, Nicosia, Paris, Prague, Rabat, Rome, Sana'a, Santiago, Sofia, Tehran, Tokyo, Tripoli, Tunis, Washington DC, Warsaw.

Air access
National airline: Syrian Arab.
Other airlines: Aeroflot, Air Algérie, Air France, Alitalia, Alyemda, Austrian, Balkan, CSA, Cyprus Airways, Emirates, Gulf Air, Interflug, Iran Air, JAT, KLM, Kuwait Airways, Libyan Arab, LOT, Lufthansa, MALÉV, MEA, Olympic, PIA, Royal Jordanian, Saudia, Sudan Airways, Swissair, Tarom, THY Turkish, Tunis Air, Yemenia.
International airport: Damascus International (Code: DAM), 30 km from Damascus, with duty-free shop.

Surface access
Main ports: Banias, Latakia (containers, ro-ro), Tartus. Schedules are liable to change at short notice.
Overland access: International roads run from Iraq, Jordan, Lebanon and Turkey, with a scheduled coach service Baghdad-Damascus. Rail services link Iraq (Haleb-Mosul), Jordan (Amman-Damascus), Lebanon (Beirut-Haleb) and Turkey.

Hotels: Rooms are in short supply, and it is essential to book in advance. At first-class and international hotels, it will be necessary to pay in foreign currency (Arab nationals and resident foreigners exempted).

Electricity supply: Damascus, 115/200 or 220/380 V AC; Aleppo, 115 V AC.

Credit cards: Widely but not by any means universally accepted, American Express being the most familiar.

Car hire: Private cars are rarely available, but taxis are reasonably priced.

City transport
Taxis: Fares by negotiation. Drivers do not expect a tip. Besides city services, long-haul 'service' taxis travel between Damascus and Aleppo.

National transport
Air: Syrian Arab Airlines operate domestic services between main centres. Important airport at Aleppo.
Road: The 20,160 km road network has some 16,000 km of surfaced roads linking main centres.
Buses: Regular long-distance bus services between main centres.
Rail: Two classes of rail service available, with restaurant car and air conditioning in some cases. 2,000 km network (325 km narrow gauge).

Public holidays
Fixed dates: 1 Jan, 8 Mar (National Day), 25 Dec (Christmas).
Variable dates: Eid al-Fitr, Eid al-Adha, Hijra New Year's Day, Ashura Day, Prophet's Birthday.

Social customs: Do not smoke or drink in public during Ramadan. Islamic customs should be respected, but alcohol is available to visitors. Do not photograph anything remotely connected with the armed forces, including radio transmission aerials, and remember that peasants may regard cameras with suspicion.

Telecommunications
Telephone: Dialling code for Syria: IDD access code + 963 followed by area code (11 for Damascus, 21 for Aleppo) followed by subscriber's number. There were 500,000* telephones in 1988, with 90 per cent automatic service (often erratic in rural areas).
Telex and telegram: There are telephone, telex and telegraph facilities at the main telegraph office in Damascus.
Trade fairs: International Fair of Damascus, Kouwatli Street, Damascus (tel: 229853/840/914; tx: 411393 SY).

Business advice and information
Aleppo Chamber of Commerce & Industry, Al-Moutanabbi Street, PO Box 1261, Aleppo (tel: 38236/7; tx: 331012).

Arab Advertising Organisation, PO Box 2842, Damascus.

Central Bank of Syria, 29 Ayar Square, Damascus (tel: 224800; tx: 411007).

Cotton Marketing Organisation, BP 729, Rue Bab al-araj, Aleppo (tel: 238486; tx: 331210).

Damascus Chamber of Commerce, Mou'awiah Street, Damascus (tel: 113399, 118339; tx: 411326).

Director General of the Damascus International Fair, Kouwatli Street, Damascus (tel: 229853/840/914; tx: 411393; cable: MARAD).

Taiwan

Bob King

If 1988 was the year in which Taipei cautiously peered out through a crack in the door at China and at greater participation by its citizens in the running of government, 1989 was the year in which that same door blew wide open. During 1989, Taiwan greatly expanded the scope of contacts between its citizens and the mainland, which were inaugurated in late 1987 with permission for Taiwan residents with close kin in China to return there for family reunions. Interestingly, the openings to China were not, as of mid-1989, significantly affected by Beijing's violent response to increasing calls for greater democracy and freedom there. Taipei apparently believed that its interests would be best served by maintaining its dialogue with the mainland, however horrendous the communist regime's actions against its own citizens appeared to the rest of the world.

Liberalisation moves

Decisions by the Taipei authorities, beginning late in 1988 and gathering steam in the new year, allowed virtually anyone with any interest at all in China to visit there. At the same time, Taipei inaugurated a programme whereby certain mainland residents could visit Taiwan for more than purely familial visits.

By May 1989, dozens of mainland residents had made the trek via Hong Kong across the Taiwan Straits: relatives of ageing scholars, doctoral candidates studying in the USA, and others. At the same time, though, more and more Taiwan residents were being allowed, tacitly if not explicitly, to visit the mainland for a variety of purposes.

The most startling of these liberalisations toward China was the government's decision in April to send a delegation led by Ms Shirley Kuo, minister of finance and thus a member of the cabinet, to attend the annual meeting of the Asian Development Bank in Peking from 4 to 6 May.

The Taipei government had for years maintained a staunch attitude of no contacts, no negotiations, and no compromise toward the Beijing regime, which it regarded as composed of 'rebels and bandits' who had usurped the Nationalist government's avowed position as the ruler of all China. Just a few years before, for instance, it would have been unthinkable for a member of the nationalist government to acknowledge an official from Beijing, even in a third country.

Ms Kuo's delegation to the ADB gathering, however, appeared to have shattered that rigid stance by journeying to Beijing, no matter what semantic contortions the government's propagandists went through to prove that her presence at the bank meeting did not constitute 'contact' with the 'rebel' regime.

The government took great pains to insist that the group was in Beijing only to deal with bank business. It would not, the authorities stressed, 'deal with' the communists. The Taiwan delegation had to attend the meeting, analysts said, or else their country would lose credibility in the international community.

One China, two governments

Splitting hairs is, however, a time-honoured Chinese tradition. The Taiwan-born Ms Kuo and her delegation were, in their daring visit to Beijing, simply taking one step further Taiwan's most recent policy of detente with the nationalists' arch-enemy of four decades. They were taking into the 'enemy' camp the latest brain-child of younger, more progressive Taiwanese planners: the idea of 'one China, two governments'.

Roughly translated, that meant a Chinese confederation, under which both the nationalists on Taiwan and the communists on the mainland would co-exist internationally as separate-but-equal governments, with all the rights and privileges pertaining thereto.

Taiwan was seeking such an arrangement with China for a number of reasons. First, it had seen its international position being eroded year after year as nation after nation recognised Beijing and dropped ties with Taipei. Taiwan, in turn, realised that its survival as a *de facto* independent entity – and perhaps even its economic well-being – depended on at least substantive recognition by the major developed nations.

Second, Taipei saw in accommodation with China a chance to lower tensions across the Straits of Taiwan. Such a detente would allow both sides to concentrate on the immediate problems facing their respective economies and societies without having to worry about external threats.

Finally, Taipei saw increased contact between the two sides as a way of acquainting China with an alternative to the rigid central planning and political controls that had dogged China's development ever since the communist victory in the civil war that nominally ended in 1949. After all, eventual unification still remained the explicit goal of both Taiwan and the mainland, and one way to achieve this would be to break down the prejudices of decades on both sides through increased contacts.

Despite Beijing's charges during the summer of 1989 that Taipei had somehow 'instigated' Chinese students and workers to stage a 'counter-revolution', there were few signs that China really meant to set back the clock in its improving relations with Taiwan. One clear indicator of this was Beijing's release after eight days of a Taiwan journalist it had arrested on charges of aiding and harbouring a major student leader, Wang Dan, after the bloody crackdown.

Challenges for planners

Taipei was, however, undertaking this radical opening to China at a time of other unprecedented changes at home – in the economic, financial, and social fields – that presented planners with unparalleled challenges on all fronts. These changes, and the social upheavals that accompanied them, meant that the government was having to tackle long-neglected but important problems in a variety of areas, all at the same time.

For instance, since 1986, the Taiwan government

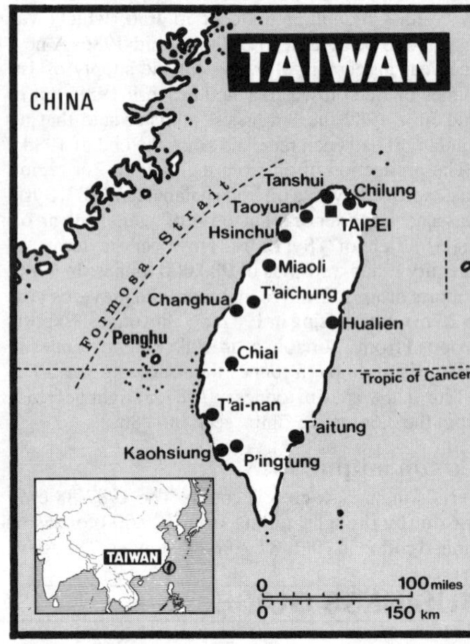

had increasingly acceded to requests by the USA that it allow its currency to appreciate to levels the USA believed to be consistent with 'economic realities', levels that, coincidentally, made American goods more competitive in Taiwan. Washington had also successfully prevailed upon Taipei to beat down numerous tariff and non-tariff barriers to imports – again, with the idea of increasing US sales to the island and thereby slashing Taiwan's massive, and previously-growing, trade surplus with the USA.

The results of this programme were, from the US side, a major success. Taiwan in 1988 managed to reduce its surplus with the USA, its major trading partner and, coincidentally, chief political ally, by 14 per cent, from $16bn in 1987 to $13.8bn in 1988. If one were to include $2.9bn in gold purchases through American dealers which Taiwan's Central Bank undertook that year, supposedly to diversify its foreign-exchange holdings, the drop in the surplus was even larger: 32 per cent, to $10.9bn.

From the Taiwan side, though, the changes brought as many new problems as they solved. For one thing, the rise in the value of the Taiwan dollar strained the resources of many small and medium-size exporters, almost all of whom had in the past quoted prices in

KEY FACTS

Official title: Republic of China
Head of state: President Lee Teng Hui
Head of government: Head of the Executive Yuan, Li Huan
Ruling party: Kuomintang (Nationalist Party)
Capital: Taipei
Official Languages: Mandarin Chinese (Taiwanese, Hakka also spoken); English
Currency: New Taiwanese Dollar (NT$ = 100 cents)
Exchange rate: NT$25.8 per $ (Nov 1989)

Area: 35,961 sq km
Population: 19.8m (1988)
GDP per capita: $5,696 (1988)
GDP real growth: 7.03% (1988)
Labour force: 8.18m (1987)
Unemployment: 2% (1987)
Inflation: 5.0% (1988)
Trade balance: $10.76bn (1988)
Foreign debt: $19.8bn (1987)
Visitor numbers: 1.76m (1987)

US currency. The appreciation, which totalled almost 40 per cent since late 1985, sent major buyers scurrying to other nations in the region – such as China, Indonesia, Malaysia, Thailand, and the Philippines – where costs were lower and the currencies more stable against the US dollar. Taiwanese companies that lacked the financial foundation or the sophistication to follow the trend and set up manufacturing operations in these countries simply lost out on the business, and began to shut up shop in ever-greater numbers.

At the same time, Taipei's greater willingness to open its market to goods from other nations was starting to spell an end to the protected environment for local producers that had allowed many of them over the years to corner the market in various goods and services. In 1987, for instance, the government began a drastic programme that in two years cut import duties on roughly 3,500 items by an average of 50 per cent. That programme continued with the planned inclusion of another 400 items by mid-1989. At the same time, bans on imports of such 'luxury' goods as foreign-made automobiles were lifted; as from mid-1989, foreign cars had captured half of the local market, sharply cutting into the sales of local assemblers.

Unions legalised

Local companies faced possibly an even greater problem with the government's ending of a ban on unions and strikes. Suddenly, employers found the country crowded with dissatisfied workers, many of whom only vaguely understood that the bosses were already strapped by higher costs and the strong Taiwan dollar. Actions by labour increased dramatically during 1988 and 1989, including walkouts by transportation workers as well as by employees in the private sector.

During May 1989, for instance, Taiwan saw its first 'legal' strike in decades when thousands of workers at a massive synthetic fibre plant, which supplied about one-fourth of the island's demand, walked off the job to protest against the dismissal of three labour activists, allegedly because of their union activities. The shut-down cost the company, Far Eastern Textiles, about $1.1m a day in lost revenues, and panicked downstream manufacturers of yarn and fabric who relied on the facility for their raw materials.

Action on environment

Environmental issues also came to the fore in 1989 after years of neglect by planners intent on maintaining huge annual increases in GNP. An action by villagers living near the massive Linyuan Industrial Zone in southern Taiwan – which supplied as much as 40 per cent of the island's petrochemical feedstocks, on which basic industries such as textiles and footwear depended – shut down the 18 facilities there for almost two weeks. In fact, the villagers held the government to ransom during this period, until the authorities agreed to pay an unprecedented NT $1.27bn (US$49m) in compensation for damage to fishing grounds by poorly-treated waste water from the plants.

The success of that protest – involving clearly illegal actions by the villagers, that were nevertheless tolerated by the government, which feared an even worse calamity should it resort to force to remove the blockage – encouraged residents of other areas around the island to try their luck as well. In fact, some environmentalists, along with government officials, deplored the development, saying that the resolution – which was in essence a government agreement to pay the villagers compensation in return for the right of industry to continue polluting – had set the stage for further actions in the future that would do little to solve the basic problems.

KEY INDICATORS

	Unit	1984	1985	1986	1987	1988
Population	'000	19.013	19,558	19,455	19,673	19,800
Gross domestic product (GDP)	NT$bn	2,255.1	2,357.1	2,701.8	3,013.9	3,225.65
GDP per capita	NT$	119,489	123,177	139,576	154,154	162,912
Wholesale prices	1981=100	99.1	96.6	91.8	88.7	–
Consumer prices	1981=100	104.1	103.9	102.4	102.9	108.0
Exports	$bn	30.46	30.72	39.80	43.54	61.30
Imports	$bn	21.96	20.10	24.14	34.51	50.54
Balance of trade	$bn	8.50	10.62	15.62	19.03	10.76
Current account	$bn	7.0	9.4	16.22	18.06	–
External debt	$m	9,557	9,019	12,638	19,847	–
Debt service ratio	%	26.9	26.4	14.1	14.3	–
Exchange Rate	NT$ per $	39.50	39.85	37.84	31.85	28.6

* estimated figure

Excess liquidity

Finally, there was the problem of too much money circulating; money that had been in many instances too easily obtained. Coupled with this were unreasonable expectations of acquiring more money, without much effort. This situation, of course, stemmed from government policies of the previous two decades which strove to show to the world, and to the Taiwanese public, that the nationalists' programme of development was far superior to that of the communists on the mainland, who had promised so much and delivered so little.

While praiseworthy from a developmental viewpoint – Taiwan's per-capita GNP, for instance, roughly doubled between 1984 and 1989 to a projected US$6,870 in 1989 – the accent on material success, often at the expense of other, more traditional values, affected the citizenry's attitudes to an extent that was likely to have profound effects on future development.

Analysts noted from around 1988, for instance, the increasing notion among the public that wealth that was not accumulated rapidly and with a minimum of effort was not worth the trouble of striving for. As a result, a significant skewing of the economy became evident: productivity fell seriously, for instance, on days when the winning numbers in Hong Kong's Mark Six lottery and other technically illegal games of chance were announced; workers were simply too excited at the possibility of windfalls to pay attention to their jobs. Absenteeism was reported high on the days preceding and following the announcements: punters were out placing their bets before the numbers were drawn, and out celebrating or consoling themselves afterwards.

At the same time, Taiwanese, with too much money on hand and few places to invest it, turned to the stock market with a fervour unmatched elsewhere in the region. During June 1989, for instance, the Taiwan market, with only 177 companies listed, was nevertheless worth US$211bn, with daily turnover running as high as US$4bn. Both figures exceeded those of the Hong Kong and Singapore exchanges combined, a continuation of a phenomenon that economists considered unnatural.

This phenomenon of stock-market investments on a record scale was matched only by the rise of underground investment houses which promised the Taiwan public a return on their investments of as much as 50 per cent per annum. Many of the operators of these schemes were themselves speculating in the Taipei exchange, essentially with borrowed money. Although such operations remained illegal in mid-1989, and at least one major house had collapsed, leaving billions in debt to investors and other creditors, the government was still hedging on whether it would move strongly against them.

Critics claimed that some government officials might be colluding with the illegal operations, and were thus profiting from their operations. But more sober analysts offered another explanation: should the government move too fast and too severely against the underground houses, with so much public money at stake, the economy might be badly hurt by a lack of public confidence.

Industrial investment

This public fascination with the fast buck on extremely risky terms contrasted sharply with a public unwillingness to re-invest in industry. This reluctance had in fact been a major headache for government planners during the 1980s: while the government was full of blueprints for Taiwan's continued economic development, the private sector was less than forthcoming in following the plan.

As a result, significant re-investment in industry, apart from a few industrialists tied closely to the government, did not rise as economic planners had hoped. Instead, many small- and medium-size businesses were content to continue their traditional operations using traditional methods, or else to move part or all of their manufacturing offshore to other parts of the region with lower labour costs and more stable currencies.

During 1988-89, such offshore investments gathered steam. Exact accounting was not possible, given the government decision during 1987 to allow each resident of Taiwan to remit abroad up to the equivalent of US$5m a year for whatever purpose he or she intended.

But Taiwan's investment commission, after comparing notes with its counterparts elsewhere in the region, came up with $110m for the Philippines, $307m for Malaysia, $913m for Indonesia, and $842m for Thailand. (Figures for these last two countries included capital invested by Thai and Indonesian partners in joint ventures with Taiwanese firms).

Estimates for China were harder to come by, because Taiwan still officially forbade its citizens to invest directly there. But local press reports, quoting Taiwan government sources, placed 1988 investments well above US$500m. They added that for Xiamen alone, Taiwan entrepreneurs invested US$34m during the first quarter of 1989, double the level during the same period a year earlier.

Outlook

It seemed likely that those investments would continue to increase, if China returned to a more rational road that would increase Taiwanese confidence. China was a natural source of the inexpensive manpower sorely needed by Taiwanese firms that were forced by their lack of capital and expertise to stick with low-end, labour-intensive products instead of moving upmarket to newer, more sophisticated goods. At the same time, many Taiwanese companies saw China as a natural market for their products, especially given the common culture and language shared across the Taiwan Straits.

Worrying as the shortcomings of the system were, however, planners were confident that many of them could be overcome in time. The government, for instance, set a growth target for the economy of seven per cent for 1989 – about the same as the performance during the previous year – and all indications at mid-year were that if the target was not met, it would not be missed by much.

At the same time, the government finally began in earnest 14 major public projects – including a mass rapid transit system for Taipei, a major upgrading of the telecommunications system, expansion of railways and harbours, and energy projects – for which the government has budgeted US$32.5bn over the following decade. While several hundred million dollars of that $32.5bn would be used for procurement of foreign equipment and services, the bulk would be pumped back into the domestic economy through construction and plant contracts.

There were also encouraging signs that domestic consumption would begin to account for a vastly increased share of GNP. For example, as margins on exports sales shrunk over the previous few years because of the currency's appreciation and higher costs of labour, more and more businessmen started serving the sharply increased local taste for imported goods. Products that were especially popular included foreign-made cars, despite retail prices that were as much as three times what they would be abroad; home appliances, such as dishwashers, washing machines, refrigerators, and televisions, of which sales during 1988 increased by 40 per cent; and fashion items, such as designer footwear, clothing, and accessories.

At the same time, Taiwan, and, by extension, its economy, was looking outward into areas other than the China mainland. Scholars and politicians were in mid-1989 still splitting semantic hairs over the so-called 'flexible diplomacy' initiated at the top levels of government early in the year. But most would agree that in essence Taiwan was moving quickly to re-establish substantive international connections that had earlier been severed as the country clung to its position that Taipei alone was the legitimate ruler of all China.

Top government and business leaders said, in fact, that Taiwan's goal was nothing less than full and equitable membership of the United Nations, a membership it lost when China was admitted in 1971. To that end, Taiwan sought to participate in a number of international bodies, and among the first it targeted was GATT, which China was also seeking to join.

Before its overtures could begin to make an impact, however, Taipei had first to inform the world, in a substantive manner, just what and where Taiwan was. Thus, the government set up a $1.1bn development fund to be made available to developing nations in the region and elsewhere that would include soft loans, credit guarantees, and the like, to cover worthy projects. It also encouraged serious investments abroad – for instance, a US$1.3bn petrochemical project in the USA by the Formosa Plastics Group, one of the world's three largest producers of PVC – as well as co-operative ventures with existing companies around the world.

Taiwan also revised its banking laws to allow Taiwanese banks to privatise to some extent, and to set up international branches wherever it made economic sense. That would, according to most bankers in the country, mean that there would within a year be branches of the 'big three' foreign-exchange banks – Changhua Commercial, Hua Nan Commercial, and First Commercial – as well as anyone else who cared to play, serving the banking needs of the international community in money centres such as New York, London, Frankfurt and Hong Kong. At the same time, financial insiders predicted that other regulations would be quickly amended to allow Taiwan's citizens to avail themselves more fully of securities-investment opportunities around the globe.

In short, Taiwan's planners had clearly decided to free the island of nearly 40 years of restraints, and were determined to move up to date in the shortest time possible. This brought serious strain on the system in a number of areas, and it was by no means clear how the authorities planned to relieve this strain to achieve their ambitious programme for Taiwan's 20m people, and, by intended extension, the mainland's teeming 1.1bn.

COUNTRY PROFILE

Political system: The head of state is the president, who is elected to a six-year term by the National Assembly, which included about 951 members in 1987. The president governs through a 322-member Legislative Yuan and appoints a premier to head the Executive Yuan, subject to the consent of the Legislative Yuan. The 77-member Provincial Assembly and its administrative body control domestic affairs; local government is vested in county and municipal authorities. Major opposition party is the Democratic Progressive Party. 1988 saw the death of President Chiang-Ching-Kuo and the end of rule by the Chiang family from 1928. The new president was Taiwan-born Lee Teng-Hui. In 1989 the country held its first significant general election in its 40-year history.

Population: 19.8m was the mid-1988 estimate. Growth is about 1.2 per cent per annum.

Main cities/towns: Taipei (population 2.69m in Mar 1989), Chilung, Chia-I, Hsinchu, Kaohsiung, Taichung, Tainan.

Media
Press: English-language dailies include the *China Post* (morning) and *China News* (evening). *Industry of Free China*, a monthly bilingual (English/Chinese) publication of the Executive Yuan's Council for Economic Planning and Development, publishes extensive economic statistics and articles about current economic performance. Bilingual statistical material is also published by the Department of Statistics, Ministry of Economic Affairs (*Taiwan Industrial Production Statistics Monthly*) and by the Directorate-General of the Budget, Accounting and Statistics Executive Yuan (*Statistical Yearbook, Social Indicators of the Republic of China*). Also available is *Monthly Key Economic Indicators of Taiwan the Republic of China.*
Radio: There are 179 broadcasting radio stations, with services generally in Chinese.
Television: There are three TV networks with multiple channels. There were 4.45m colour TVs in 1986, reaching 94.42 per cent of all homes.

Domestic economy: Taiwan is a newly industrialised country with labour-intensive exports driving the economy. Manufacturing emphasis has shifted in recent years away from textiles and towards capital-intensive, high technology industries with low energy requirements. A policy of developing export-orientated manufacturing (as opposed to import substitution) has resulted in an economy largely dependent on exports. The government offers generous tax and credit incentives, particularly for productive and export-orientated enterprises. Recently the government has liberalised capital and foreign exchange markets. Residents can now purchase shares abroad. There has been considerable investment in infrastructure, which is well developed along the west coast, and which has been largely domestically financed. A vigorous recovery from the economic slump of 1985-86 has helped to revive such imperilled projects as the completion of the round-the-island rail network, the construction of three new highways and the construction of additional facilities at Suao and Taichung harbours. The laws on foreign investment are gradually being eased.

Employment: Between 1965 and 1987 the proportion of agricultural workers declined from 47 per cent to 15 per cent, as industrial employment rose from 22 per cent to 42.7 per cent and services from 31 per cent to 42 per cent. Unemployment in 1987 was around two per cent, down from 2.7 per cent in 1986.

External trade: Taiwan is heavily dependent on foreign trade, and suffered badly in 1985-86; first from the strength of the US dollar, to which the Taiwan dollar is linked, and later from market shrinkage and protectionism in the US. Nevertheless, in 1987 exports rose following the depreciation of the NT$ against the Japanese Yen. Taiwan's consistent heavy trade surpluses have amassed an enormous foreign currency reserve of an estimated $75bn (second only to Japan) of which 42 per cent is invested in banks abroad. Trade relations with China PR are offically banned but are in fact flourishing, amounting to an estimated $1.2bn in 1987, with a significant black market in operation.
Exports: Exports grew by 34.5 per cent in 1987 to $54bn. In 1988 exports were about $60bn. Main exports are electrical machinery, textile products, metal products and machinery (9.9 per cent), plastic products, wood products, transport equipment, and refined petroleum products. Main destinations are USA (44% of 1987 total), EC (15%), Japan (13%), Hong Kong (13%), Asia (10%), Canada (3%) and Middle East (3%).
Imports: Imports grew by 43 per cent to $35bn in 1987. In 1988 imports were about $50bn. Main imports are electrical machinery, chemicals, chemical products, machine tools, crude oil, food and drinks, transport equipment, refined petroleum products, paper and pulp, maize, soya beans, raw cotton, lumber, and wheat. Main sources are Japan (34%), USA (22%), Europe (15%), Asia (12%), Middle East (8%) and Canada (2%).

Agriculture: About 25 per cent of total land area is devoted to agriculture, with two or three crops per year. The sector benefits from a favourable climate. Rice is the main crop but consumption is falling as attention shifts toward vegetables, sugar-cane, sweet potatoes and bananas; government efforts to diversify production have favoured soya beans, groundnuts, tea, wheat and processed fish. Industrial crops include cotton, hemp and jute. About half the land area is wooded, and forestry products include sawn timber, plywood, paper and fuel for local use.

COUNTRY PROFILE

The timber industry is limited by inaccessibility and poor quality of much of the forestry resources and by an official policy of conserving supplies. The deep-water fishing industry is large and expanding, supplemented by aquaculture. Agriculture's share of GDP has been sharply reduced since the 1950s when it was the cornerstone of the economy. The sector now employs just under 17 per cent of the population, against 56 per cent in 1952. This decrease is due partly to limited agricultural resources and partly to rapid development of the industrial sector.

Industry: In 1988 41.5 per cent of the workforce were employed in the industrial sector. Manufacturing industry contributes 43.5 per cent of GDP and 85.4 per cent of export revenues (1987), and has shifted since the 1970s towards high value-added activities such as electronics (now the biggest single export product) and toward light industries which supply plastic goods, synthetic yarns, motor spares, and plywood. Successful heavy industries include seven motor companies, 300 steelworks, and major shipbuilding yards run mainly by the China Shipbuilding Corporation. Taiwan is self-sufficient in petrochemicals. The government has defined 127 industrial sites, of which 59 have special infrastructural facilities. There are three export processing zones at Taichung, Kaohsiung and Nantze (the largest), and a science park has been built in Hsinchu, south west of Taipei. Industrial production has been slowing since the early 1980s, when up to 16 per cent annual growth was recorded, and is susceptible to fluctuations in export markets. Foreign investment has been deterred by a lack of adequate patent laws and by tight regulations on local content.

Mining: Mining accounts for less than one per cent of GDP and there are few resources present. Coal, marble and dolomite are the principal resources, plus small amounts of limestone, gold, silver, copper and salt. Coal reserves are becoming increasingly uneconomic, and provided only 2.6 per cent of energy requirements in 1987. Crude petroleum and some natural gas is also produced, but Taiwan is heavily dependent on imported oil; bottled gas is more common than natural gas.

Energy: In 1987, over 80 per cent of energy resources were imported, including all crude oil. Lack of energy resources has made nuclear power increasingly important. In 1987 it provided 48.5 per cent of electricity requirements. There are three nuclear power stations (comprising six reactors), with a fourth planned. Hydro-electric power provides 10.8 per cent of requirements and thermal stations another 40.6 per cent (1987). Total installed capacity exceeds 15,000 mW (1987), reaching about 26,700 mW by 1989.

Main companies: BES Engineering Corp, China Petroleum Development Corp, China Shipbuilding Corp, China Steel Corp, Chinese Petroleum Corp, Kaohsuing Ammonium Sulfate Corp, Ret-Ser Engineering Agency, Taiwan Aluminium Corp, Taiwan Chung Hsing Paper Corp, Taiwan Fertilizer Co, Taiwan Power Co, Taiwan Metal Mining Corp, Taiwan Machinery Manufacturing Corp, Taiwan Salt Works, Taiwan Sugar Corp, Taiwan Tobacco & Wine Monopoly Bureau, Tang Eng Iron Works Co, Nan Ya Plastics Corp, Formosa Plastics Corp, Formosa Chemicals & Fibre Corp, Tatung Co, Far Eastern Textiles Ltd, Yue Loong Motor Co, President Enterprises Corp, Matsushita Electric Taiwan Co, San Yang Industry Co, Hualon Corp, Taiwan Cement Corp, Chung Shing Textile Co, Sampo Corp, Teco Electric & Machinery Co, Chi Mei Industrial Co, Wei-Chuan Co Inc, Asia Cement Corp, Ford Lio Ho Motor Co, China Rebar Co, Yuen Foong Yu Paper MFG. Co, Shinkong Synthetic Fibres Corp, Formosa Taffeta Co, Tai Yuen Textile Co, Sanyo Electric (Taiwan) Co, Chia Hsin Flour Feed & Vegetable Oil Corp, China Motor Co, China General Plastics Corp, Tainan Spinning Co, Pacific Electric Wire & Cable Co, USI Far East Corp.

Financial markets: Taiwan's stock exchange was opened to indirect foreign investment in 1984, though the International Investment Trust Company. Bank loans are still the preferred means of raising capital. In 1987 the rules on investing abroad were also relaxed. The Securities and Exchange Committee opened its doors to new securities firms after 15 years. There are currently 28 licensed securities companies on the stock market.

Membership of international organisations: Asian Development Bank.

BUSINESS GUIDE

Time: GMT + 8 hrs.

Climate: Subtropical with temperatures ranging from 33°C in Jul-Aug to 12°C in Jan-Feb. Average rainfall is 2,500 mm/yr, with typhoons from May-Oct and occasional snow in the mountains in Jan-Feb.

Entry requirements
Passport: Required by all except certain seamen.
Visa: Tourist, transit or entry visas are required by all, and may be obtained at any Republic of China Embassy or Consulate. Visas cannot be obtained on arrival. From countries without diplomatic relations with Taiwan, detailed enquiries should be made to the nearest Taiwanese information or trade service. Applications made in Korea for a visa to enter Taiwan are often subject to delays. Visitors and residents must register with the Foreign Affairs Police, although hotels will usually handle this formality.
Currency: Travellers leaving Taiwan may not take out more than US$5,000 cash or the equivalent in any other foreign currency unless it was declared on arrival in the country. Import and export of local currency is restricted to NT$8,000 and 20 coins.
Customs: Personal effects are duty-free. Narcotics, gambling aids (including mahjong sets), firearms and explosives are prohibited. Articles produced in China PR, Korea PR, Albania, Romania, USSR or Cuba may be held in bond and reclaimed on departure, as they are prohibited. Gold and silver ornaments must be declared. Import of certain items is controlled, such as military supplies, national defence and communications equipment and poisonous chemicals. Most permitted imports require an import licence.
Income tax: Residents (not including short-term visitors) and non-residents are taxed on all income derived from Taiwan. There are no full tax treaties except with Singapore.

Health precautions
Mandatory: Vaccination certificates or cholera if travelling from an infected area.

Representation overseas: Amman, Asuncion, Athens, Bandar Seri Begawan, Bangkok, Bogota, Bonn, Brussels, Copenhagen, Costa Rica, Dhaka, Dubia, Frankfurt, Funafuti, Guatemala City, The Hague, Geneva, Hong Kong, Honiara, Jakata, Jeddah, Johannesburg, Kingstown, Kuala Lumpur, Lilongwe, Lima, London, Luxembourg, Manama, Manila, Madrid, Mbabane, Montreal, Montevideo, Muscat, Nauru, Nicosia, Nuku'alofa (Tonga), Osaka, Oslo, Panama City, Paris, Port-au-Prince, Pretoria, Rio de Janeiro, Riyadh, Rome, San José, Santo Domingo, Santiago, San Salvador, Seoul, Singapore, Stockholm, Sydney, Tegucigalpa, Tokyo, Tripoli, Tuvalu, Vatican City, Vienna, Washington, Wellington.

Air access
National airlines: China Airlines, Far Eastern Air Transport.
Other airlines: Cathay Pacific, Delta, Garuda Airways, Japan Asia Airways, KLM, Korean Air, MAS, Northwest, Philippine, Royal Brunei, Singapore, South African Airways, Thai International, United.
Tax: An airport tax of NT$300 is levied on departure.
International airport: Chiang Kai Shek International, Taoyuan (TPE), 40 km from Taipei, with duty-free shop, bar, restaurant, buffet, bank, post office, hotel reservations and shops; Kaohsiung International (KHH).

Surface access
Main ports: Kaohsiung (containers), Keelung (including Suao), Hualien,(main ports) Taichung.

Hotels: It is advisable to book hotel rooms in advance. Room facilities usually include TVs and refrigerators. Larger hotels will arrange transport to/from the airport. A 10 per cent service charge is added to the bill. Reasonably priced accommodation is available at Japanese-style hot springs resorts in the mountains.

Car hire: Self-drive car hire is available, although chauffeur-driven cars are recommended due to traffic conditions. An international driving licence is required, and driving is on the right-hand side of the road.

City transport
Taxis: Metered taxis are available in Taipei, and fares are metered by kilometres and delay time. Have the destination (and the return address) written in Chinese for the taxi driver's reference. Tipping is not an established practice, though it is becoming more usual. Specialist taxis operated (in the manner of buses) on certain routes, but these are not recommended.

National transport
Air: Domestic air services operated by China Airlines, Far East Air Transport and five other carriers connect most of the main cities.
Road: The road network covers 20,000 km (1987), most of it surfaced. A good highway links the main centres between Keelung and Kaohsiung, and is being extended. Bad terrain and one-way systems can make road travel difficult outside urban centres.
Buses: Extensive bus services, covering coastal, cross-island and inland areas. Express coach services link Taipei, Kaohsiung and other main centres. Advance booking is recommended. Local bus services are crowded during rush hours. Destinations are clearly marked in English at urban bus stations.
Rail: The railway extends the whole length of Taiwan, mainly along the west coast, with the main electrified route serving Taipei-Kaohsiung. Rail services are good, with air-conditioned express trains linking main centres. Urban train stations have destinations clearly marked in English. An underground rail system and a Rapid Mass Transit System are under construction in Taipei.
Water: There are ferry services from Kaohsiung to the Pescadores Islands, from Taitung to the Lanyu and Green Islands.

Public holidays
Fixed dates: 1 Jan (Founding Day of the Republic of China), 29 Mar (Youth Day), 5 Apr (Tomb-sweeping Day), 10 Oct (Double Tenth Day), 31 Oct (Chiang Kai Shek's Birthday), 25 Oct (Taiwan Retrocession Day), 12 Nov (Birthday of Dr Sun Yat-sen), 25 Dec (Constitution Day).
Variable dates: Chinese New Year, Dragon Boat Festival, Mid-Autumn Festival.

Working hours
Government and business: (Mon-Fri) 0830 – 1230, 1330 – 1730; (Sat) 0830 – 1230.

BUSINESS GUIDE

Banking: (Mon-Fri) 0900 – 1530; (Sat) 0900 – 1200.
Shops: 0900 – 2100 (department stores 1030 – 2130).

Social customs: The topic of mainland China is generally not discussed. Although the country is known as Taiwan (or sometimes Formosa), it should only be referred to as the Republic of China.

Telecommunications: Two satellite communication earth stations provide direct international radio circuits with the rest of the world.
Telephone: Dialling code for Taiwan, IDD access code + 886 followed by area code (7 for Kaohsiung, 2 for Taipei) followed by subscriber's number. The telephone system has 100 per cent automatic service. In 1987 there were 6.95m lines.
Telex: Telex facilities are often available at good hotels.
Telegram: Telegram service is available 24 hrs/day at the Telecommunications Office, Hang Chow S. Road, Taipei and at good hotels.

Banking: An offshore banking centre was set up in 1984.
Central bank: Central Bank of China.
Commercial banks: Bank of Communications, Bank of Taiwan, Central Trust of China, Chang Hwa Commercial Bank, City Bank of Taipei, City Bank of Kaoshiung, Co-operative Bank of Taiwan, Export-Import Bank of China, Farmers' Bank of China, First Commercial Bank, Hua Nan Commercial Bank, International Commercial Bank of China, Land Bank of Taiwan, Medium Business Bank of Taiwan, Overseas-Chinese Commercial Banking Corporation, Shanghai Commercial and Savings Bank, United World Chinese Commercial Bank.
Taiwanese branches: American Express International Banking Corporation, Amsterdam-Rotterdam Bank NV, Bangkok Bank, Banque National de Paris, Banque Paribas, Bank of America, Bankers Trust Company, Chase Manhattan Bank, Chemical Bank, Citibank, Continental Illinois National Bank and Trust Co. of Chicago, Crédit Lyonnais, Nippon Dai-ichi Kangyo Bank, European Asian Bank, Development Bank of Singapore, First Interstate Bank, The First National Bank of Boston, Grindlays Bank, Hollandsche Bank-Unie, Hongkong and Shanghai Bank, International Bank of Singapore, Irving Trust Company, Lloyds Bank International, Manufacturers Hanover Trust, Metropolitan Bank and Trust Co, Morgan Guaranty Trust Co of NY, Rainier National Bank, Royal Bank of Canada, Standard Chartered Bank, Toronto-Dominion Bank.

Advertising: Advertising is available in the press, in cinemas and via commercial radio and TV.

Trade fairs: The new Exhibition Hall of the Taipei World Trade Centre, with 99,100 square metres of exhibition space, opened in 1986. The Exhibition Hall is designed for both short-term and long-term product displays. On the ground floor is a 26,000 square metres exhibition area, the site of 20-plus TWTC-sponsored international trade shows a year. Trade fairs for 1990 are: Taipei Elec (Mar), Taipei International Food Industry Show (Mar), Taipei International Gift & Stationery Spring Show (Mar), Taipei International Industrial Machinery Show (Apr), Taipei International Sporting Goods Show (Apr), Taipei International Cycle Show (Apr), Taipei International Footwear & Leather Goods Show (May), Taipei International Auto Show (May), Taipei International Hardware, Building Materials & Houseware Show (May), Instrument '90 (Jun), Computex Taipei (Jun), Taipei International OEM/Subcontracting Show (Jun), Taipei International Hi-Technology Show (Jun), Taipei Import fair (Aug-Sep), Taipei International Jewellery & Timepiece Show (Sep), Taipei International Toy Show (Sep), Taipei International Electronics Show (Oct), Taipei Fashion Week (Oct), Taipei International Gift & Stationery Autumn Show (Nov), Taipei International Furniture Show (Nov). The International Trade Mart, located on floors 2 through 7, is an extensive permanent exhibition area for both import and export products.

Electricity supply: 110 V AC, 60 cycles.

Weights and measures: Metric system (some Chinese units in use).

Representation in capital: Costa Rica, Dominican Republic, El Salvador, Guatemala, Haiti, Honduras, Rep of Korea, Nauru, Nicaragua, Panama, Paraguay, Saudi Arabia, South Africa, Swaziland, Vatican.

BUSINESS DIRECTORY

Useful telephone numbers
Fire: 119; Police: 110; Ambulance: 721-6315; International calls: 100
Directory enquiries: Chinese language: 104 (long-distance: 105); English language: 311-6796

Hotels: Kaohsiung
Ambassador, 202 Min Sheng 2nd Road (tel: 211-5211; tx: 72105; fax: 2811115).

Holiday Garden, 279 Liu Ho 2nd Road (tel: 241-0121; tx: 81948; fax: 2512000).

Kingdom, 42 Wu-Fu 4th Road, 800 (tel: 551-8211; tx: 81938; fax: 5210403).

Hotels: Taipei
Ambassador, 63 Chung Shan North Road, Section 2 (tel: 551-1111; tx: 11184; fax: 5617883/5315215).

Asiaworld Plaza, 100 Tun Hwa North Road (tel: 715-0077; tx: 26299; fax: 7134148).

Brother, 255 Nanking East Road, Section 3, 10566 (tel: 712-3456; tx: 25977, 28930; fax: 7173334, 7175857).

Fortune Dragon, 172 Chung Hsiao East Road, Section 4 (tel: 772-2121; tx: 19794; fax: 7210302).

Gloria, 369 Lisen North Road, 104, P.O. Box: 17172 (tel: 581-8111; tx: 11192; fax: 5815811).

Golden China, 306 Sung Chiang Road (tel: 521-5151; tx: 19550; fax: 5312914).

Grand, 1 Chung Shan North Road, Section 4 (tel: 596-5565; tx: 11646; fax: 5948243).

Hilton International, 38 Chung-Hsiao West Road, Section 1 (tel: 311-5151; tx: 11699; fax: 3319944).

Howard Plaza, 160 Jen Ai Road, Section 3, P.O. Box: 24-733 (tel: 7002323; tx: 10702; fax: 7052803, 7000729).

Imperial, 600 Lin Shen North Road, 104 (tel: 596-5111/3333; tx: 11382, 11730; fax: 5927506).

Lai Lai Sheraton, 12 Chung Hsiao East Road, Section 1 (tel: 321-5511; tx: 23939; fax: 3944240).

Majestic, 2 Min Chuan East Road (tel: 581-7111; tx: 21828; fax: 5623248).

Mandarin, 166 Tun Hwa North Road, 10590 (tel: 712-1201; tx: 11386; fax: 7122122).

Miramar, 420 Min Chuan East Road (tel: 5053456; tx: 19788; fax: 5029173).

New Asia, 139 Chung Shan North Road, Section 2 (tel: 5117181; tx: 23394; fax: 5224204).

Paradise, 24 Hsi Ning South Road, P.O. Box: 70212 (tel: 314-2122; tx: 26972; fax: 3147873).

President, 9 Teh Hwei Street, 10469 (tel: 595-1251; tx: 11269; fax: 5913677).

Rebar Crown, 32 Nanking East Road, Section 5 (tel: 763-5656; tx: 14207; fax: 7679347).

Ritz Taipei, 155 Min Chuan East Road (tel: 597-1234; tx: 27345; fax: 5969223).

Riverview, 77 Huan Ho South Road, Section (tel: 311-3131/311-3688; tx: 11609; fax: 3613737).

Royal, 37-1 Chung Shan North Road, section 2 (tel: 542-3266; tx: 23915; fax: 5434897).

Santos, 439 Cheng Teh Road (tel: 596-3111; tx: 27155; fax: 5983120).

Taipei Fortuna, 122 Chung Shan North Road, Section 2 (tel: 563-1111; tx: 21578; fax: 5619777).

The Regent, 41 Chung San North Road Section 2 (opening late 1989).

United, 200 Kuang-Fu South Road (tel: 773-1515; tx: 11679; fax: 7412789).

Business advice and information
Board of Foreign Trade, MOEA, 1 Hukou St, Taipei (tel: 3510271; tx: 11434 BOFT; fax: 3513603).

China External Trade Development Council, 4-8th Fl, International Trade Building, 333 Keelung Road, Sec 1, Taipei 10548 (tel: 738-2345; tx: 21676 CETRA; fax: 7576653).

Chinese National Association of Industry & Commerce, 13th Fl, 390 Fu Hsing S Rd, Sec 1, Taipei (tel: 707-0111; fax: 7017601).

Chinese National Federation of Industries, 12/F, 390 Fu Hsin South Road, Section 1, Taipei (tel: 703-3500; tx: 14565 CNFI).

Council for Economic Planning and Development, Executive Yuan, 9/F, 87 Nanking East Road, Section 2, Taipei (tel: 551-3522; tx: 11385).

Directorate-General of Budgets, Accounting & Statistics, Executive Yuan, 1 Chung Hsiao East Road, Section 1, Taipei.

General Chamber of Commerce of the Republic of China, 6th Floor, 390 Fu Hsin South Road, Section 1, Taipei (tel: 708-0350; tx: 13396 JUNGTAI).

Industrial Development and Investment Centre, MOEA, 10th Fl, 7 Roosevelt Rd, Sec 1, Taipei (tel: 3947213; tx: 10634 INVEST; fax: 3926835).

Industry of Free China, 9th Floor, 87 Nanking East Road, Section 2, Taipei (tel: 543-5988).

Taipei Chamber of Commerce, 6/F, 72 Nanking East Road, Section 2, Taipei (tel: 531-8217).

Taipei World Trade Centre, 5 Hsinyi Road, Section 5, Taipei (tel: 725-1111; tx: 28094 TPWTC; fax: 3513603).

Banking
Central Bank of China, 2 Roosevelt Road, Section 1, Taipei (tel: 393-6161; tx: 21532).

Travel information
Republic of China Tourism Bureau, 9th Floor, 280 Chung Hsiao East Road, Section 4, PO Box 1490, Taipei (tel: 721-8541; tx: 26408 ROCTB; cable: ROCTB).

Taiwan Visitor's Association, 111 Minchuan East Road, 5/F, Taipei (tel: 594-3261; fax: 5943265).

Tourist Service Centre at the CKS International Airport (tel: 03-383-4631/2).

Tourist Information Hot Line (tel: 717-3737).

Tanzania

Jane Carroll

President Mwinyi continued in 1989 to press forward quietly but firmly with his measures for liberalising the economy, having to satisfy two virtually opposing groups: on the one hand the international donors on whom economic recovery largely depended, and who wanted reforms to be speeded up; on the other, the still powerful 'old guard' of politicians who bitterly opposed what they saw as the abandonment of socialist ideals. His ultimate objectives were closer to those of the donor community, but his views on the desirable pace of change were very different from theirs. Since he took office in November 1985 he showed great skill in steering his reform programme, backed by such key figures in government as Finance Minister Cleopa Msuya, and by a strong body of opinion in official and business circles. Despite periodic hitches caused by disagreement with the International Monetary Fund (IMF) in particular, as well as the chronic problems connected with weather and fluctuating commodity prices, and although export earnings continued to disappoint, overall economic growth in 1988 and 1989 was better than it had been for many years.

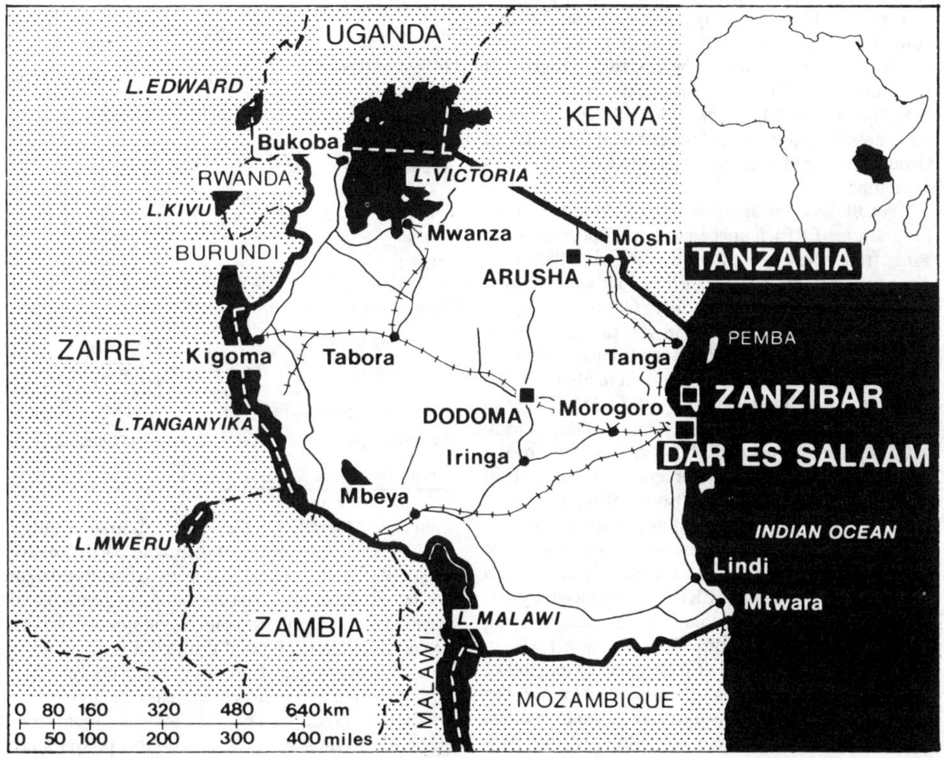

Discontent in Zanzibar

The political situation in Zanzibar continued to threaten the 25-year-old-union between islands and mainland. Simmering discontent, which had first surfaced in 1984 and erupted into violence in early 1988, received a new stimulus in August 1988 when Seif Shariff Hamad, who had earlier in the year been dismissed from his post as chief minister for an alleged coup plot, circulated an open letter demanding a referendum on separation from the mainland. In September restrictions were placed on press criticism of the government. Three months later a dissident movement, Zanzibar Democratic Alliance, was formed in exile in Sweden, its leaders repeating the demand for a referendum and calling for multiparty democracy in Zanzibar. In March 1989 Julius Nyerere, as chairman of CCM (Revolutionary Party of Tanzania), toured the islands and declared that potential agitators must be detained. This was possibly designed to help Mwinyi in a tricky political situation in his home area. Factors behind the discontent were resentment at the perceived dominance by the mainland, the Muslim population's fear of the ascendancy of Christian, and specifically Roman Catholic, influence within the CCM, and mutual distrust between those of Arab and those of African descent.

Secessionist sentiments were fuelled by the islands' economic crisis, with the collapse of close exports and a disappointing level of foreign aid and investment, despite a vigorous programme of liberalisation. Early in 1989 the government ended the state monopoly on the marketing of cloves and other export crops and relaxed restrictions on private sector trading in general.

With one wary eye on the islands, Mwinyi kept the other on the balance of power in central government. He deftly strengthened his own power base in a cabinet reshuffle in March 1989, aimed at boosting the reformist camp. A new planning commission was set up, under the leadership of Kighoma Malima, a diehard socialist like most of the five other members. The commission's role was expected to be marginal in the new era of liberalisation and encouragement of private enterprise, and it appeared to be a means of removing from key cabinet posts some of the opponents of reform, shunting them into a siding with a deceptive title.

An earlier move by Mwinyi was his first important change to the armed forces' leadership, in September 1988. He appointed General Ernest Mwita Kiaro to replace the retiring commander of the armed forces, in a bold and potentially risky move aimed at reinforcing his control of the army, traditionally an important part of Nyerere's power base. Kiaro was said to be unpopular with officers; on the other hand, he was from Mara, home region of both Nyerere and of Prime Minister Joseph Warioba, so perhaps this was yet another of Mwinyi's well-tuned operations.

Economic reforms

Central to the economic reforms was exchange rate management. The government continued in 1988–89 to push the shilling downwards at the fastest speed possible without upsetting political stability. In November 1988, in what Mwinyi described as a 'bitter but inevitable' measure, the currency was devalued from \$1=TSh98 to \$1=TSh120. This, together with agreement on further reforms of crop marketing and other measures, was enough for the IMF to release the second tranche of the three-year structural adjustment facility (SAF). In turn, this unblocked funds from other sources, including International Development Association (IDA) credits totalling about \$143m, mostly for financing imports, and \$39m from the EC. Bilateral donors followed suit and in December 1988 the Paris Club of Tanzania's main creditors agreed to reschedule part of its external debt, which by early 1989 was estimated at \$5,100m. Debt servicing costs in 1988 were more than 50 per cent of export earnings, but Tanzania was believed to have made no servicing payments on bilateral debt in 1986, 1987 and probably also 1988.

By May 1989 the shilling had fallen to \$1=TSh138. The 1989–90 budget, presented in June, devalued it further, to \$1=TSh145. Other proposals were targeted in particular at cutting public spending. Total spending was to rise by 19 per cent in absolute terms, to TSh141,248m, of which 30 per cent was to come from foreign loans and grants. Recurrent expenditure was to go up by 31 per cent to TSh118,552m, but the development budget was cut in absolute terms by 20

KEY FACTS

Official title: United Republic of Tanzania
Head of state: President Ali Hassan Mwinyi
Ruling party: Revolutionary Party of Tanzania (CCM)
Capital: Dodoma
Official Languages: Kiswahili, English
Currency: Shilling (TSh) = 100 senti
Exchange rate: Tsh150 per \$ (Nov 1989)
Area: 945,087 sq km
Population: 23.2m (1988)
GDP per capita: \$153 (1987)
GDP real growth: 3.9% (1987)
Inflation: 35% (1988)
Trade balance: –\$860m (1988)
Foreign debt: \$5.1bn (1989)

per cent to TSh22,696m. There were increases in education charges and in various licences, the duty on alcohol went up by 25 per cent and that on cigarettes by 50 per cent, the airport tax was doubled and six new road toll stations were introduced. Minimum wages were raised by 26 per cent. The budget contained no new incentives for either production or investment and it was coolly received by the IMF and other donors, particularly because they had wanted a much larger devaluation. The hoped-for Enhanced Structural Adjustment Facility (ESAF), which the government wanted to replace the SAF, was not immediately forthcoming and the planned Consultative Group meeting of donors was postponed from July to September.

Nevertheless, the three-year Economic Recovery Programme (ERP), launched in June 1986, had borne some fruit. Although export earnings in 1988 were disappointing, mainly because of low world coffee prices, Gross Domestic Product (GDP) grew by an estimated 4.1 per cent in 1988–89, the best performance since the mid-1970s. The annual consumer price inflation rate was pulled down to about 28 per cent in the first quarter of 1989, from 35 per cent in 1988 – although it was expected to have moved back to 35 per cent by the end of 1989, as a result of the devaluation. A slight improvement was discernible in the financial performance of some of the country's more than 400 parastatals and as a result of encouragement offered to the private sector, the number of small enterprises registered rose from 4,782 in 1986 to 5,240 by 1988. The ERP was replaced by the 1988–93 Five-Year Development Plan, delayed from 1988 and launched in April 1989. The target was an annual GDP growth rate of 6 per cent by the end of the Plan's period and total investment required was $1,300m, of which transport and communications were allocated 23.8 per cent, agriculture 18.5 per cent and industry 10.1 per cent. Export earnings were scheduled to rise from $388m in 1987–88 to $681m in 1992–93 and imports from $1,200m to $1,450m.

KEY INDICATORS

	Unit	1983	1984	1985	1986	1987
Population	m	20.41	21.06	21.73	22.46	23.22
Gross domestic product (GDP)	TSh bn	70.5	88.9	120.6	161.9	227.9
GDP per capita	TSh	3,454	4,221	5,545	7,208	9,815
Inflation	%	27.0	35.4	34.0	32.4	30.0
GDP real growth	%	–2.4	3.4	2.6	3.6	3.9
Agricultural production	1979-81=100	103.2	107.0	110.6	114.4	117.5
Coffee (cash crop)	'000 tonnes	48.6	56.0	56.0	54.0*	49.0*
Seed cotton (cash crop)	'000 tonnes	140.4	155.0	200.0	218.0	246.0
Exports	TSh m	4,138	5,761	4,960	11,116	18,512
Coffee	TSh m	1,449	2,216	1,993	5,465	5,793
Cotton	TSh m	689	713	414	1,009	2,832
Sisal	TSh m	145	150	94	176	322
Imports	TSh m	8,877	12,956	17,962	27,778	59,340
Petroleum	TSh m	1,945	2,393	3,911	3,651	6,954
Balance of trade	$m	–333.5	–385.7	–532.6	–566.0	–801.9
Current account	$m	–196.0	–214.5	–147.4	–174.3	–265.4
External debt	$m	3,405	3,473	3,879	4,066	4,335
Debt service ratio	%	11.7	12.7	15.5	17.0	18.5
Exchange rate	TSh per $	11.1	15.3	17.5	32.7	64.3

* estimated figure

Export crops

At least one-third of export earnings came from coffee. Low world prices in 1988–89 and the threat of further decline following the failure of International Coffee Organisation (ICO) talks in 1989 were a major blow to Tanzania's export hopes, although arabicas, which make up more than 75 per cent of its coffee, were less badly affected by price falls than were robustas. Output for 1988–89 was forecast at 49,000 tonnes compared with 47,000 tonnes the previous year and a record 67,300 tonnes in 1980–81. The ICO quota for the year, after several adjustments, was 646,000 bags (of 60kg each). Exports to quota and non-quota countries usually total about 800,000 bags. In October 1988 the government and a syndicate of international banks, lead-managed by Bankers Trust Company, signed a line-of-credit agreement for pre-export finance for the coffee sector. The agreement was first signed a year earlier, for $50m, and the renewed agreement was for $40m. The main purpose was to speed up the Coffee Authority of Tanzania's payment to farmers.

The 1988–89 cotton crop was forecast at 500,000 bales, compared with 450,000 bales purchased the previous season. Major renovation schemes aimed particularly at dilapidated ginneries and at transport bottlenecks, and funded by the Netherlands, the UK and the European Investment Bank (EIB), began to have the desired effect, although large stocks of cotton were still lying unginned at the villages. The tea sector showed signs of optimism in 1989, despite weak world prices. Devaluation of the shilling and changes in government policies encouraged new investment plans. More than half of production (17,000 tonnes of made tea in 1988) comes from the estates of Brooke Bond Tanzania, George Williamson Tanzania and Mufindi Tea Company, all in the southern highlands. The government is fully or partly privatising several of the estates that were nationalised in the 1970s. A Tanzanian Tea Research Institute, aimed particularly at increasing yields and quality, is being set up in Mufindi, funded through a levy on all tea sales. Local consumption of tea accounts for 25 per cent of output and is slowly increasing.

An IDA credit of about $25m was approved in mid-1989 to finance a project to double production of both cashew nuts and coconuts within 10 years. Only about 23,000 tonnes of cashew nuts were produced in 1988, compared with 145,000 tonnes a year in the early 1970s, and Tanzania has been unable to benefit fully from the high world prices now prevailing.

Cloves (mostly grown on Pemba) provide about 80 per cent of Zanzibar's foreign exchange earnings. In 1988–89 they earned probably less than $10m, compared with $15m the previous year, mainly because of depressed world prices and the difficulty of finding markets to replace Indonesia which once took the bulk of the output but had ceased to be an importer. Production in 1988–89 was believed to have risen to 6,000 tonnes, from the dismal 1,900 tonnes recorded two years earlier, but there were probably stocks of at least 8,000 tonnes.

Heavy rains in April 1989 caused severe flooding in seven regions right across the country, destroying food stocks and standing crops, and disrupting road and rail links. The movement of food from surplus to deficit areas was hampered by transport problems and the continued relaxing of controls on prices and marketing could make it even more difficult to maintain a balance of food supplies throughout the country. The National Milling Corporation, once the principal reprobate in the vast ranks of inefficient and loss-making parastatals, handled only about 25 per cent of grain sales in 1989, having been stripped of its monopoly of marketing in favour of the reinstated rural co-operatives.

Diversification of export crops and the export of processed rather than raw agricultural commodities were urgent priorities in 1989. Small quantities of fresh fruit and vegetables were beginning to be air-freighted to the UK and the Middle East but the trade was hindered by the lack of air cargo space, fierce competition from more experienced exporters, including Kenya, and lack of expertise in this demanding market. Nevertheless the government saw further potential in it and in 1989 was investigating the use of refrigerated ships for carrying some items such as pineapples. Studies were also in progress for setting up export-oriented plants for canning fruit and fruit juice, probably as joint ventures with foreign firms. Exports of honey from Tabora to the UK were resumed on a small scale in 1989, after a break of several years. They were expected to grow sharply, with strong demand in Europe for Tanzania's high quality honey and also for beeswax.

Industrial recovery

The industrial sector showed signs in 1989 of continuing its painfully slow recovery from the downward slide in output estimated at 2.4 per cent a year from 1980 to 1987. This was largely thanks to the better performance of a small number of individual industries. Output of cement in 1989 was expected to be 700,000 tonnes, giving an exportable surplus of 70,000 tonnes, following completion of rehabilitation of the country's three plants. The Tanga Fertiliser plant re-opened in April 1989 after a nine month shutdown while it was awaiting funds for essential repairs. The 30,000 tonnes-a-year bitumen plant at the Dar es Salaam oil refinery, due to start operating in 1990, should have a surplus for export. A small range of manufactured goods were exported to other African countries, and paper from Southern Paper Mills' plant at Mufindi was also sold to the Far East. A significant success was scored in 1988 by an electrical firm, NEM, which exported wall lamps and fluorescent light fittings worth $148,000 to Sweden and West Germany: helped by the shilling's devaluation, for the first time Tanzania was able to compete with Far Eastern and European suppliers.

Finance – possibly as much as $700m is required – had still not been found for the Kilwa ammonia/urea fertiliser project based on the offshore gas field near Songo Songo island in the south east. A large export market needed to be found for the scheme to be viable. Prospecting for oil by several international companies, both on and offshore, still produced no positive results but Shell Petroleum Development Tanzania in early 1989 signed a new four-year agreement, under which it was to spend $8m exploring offshore near Mafia island. It had already spent $15m over the previous three years.

The Minjingu phosphate mine in Arusha region was hit by the temporary closure of the Tanga fertiliser plant which it supplies, and in 1988 output was below 18,000 tonnes, compared with a capacity of 100,000 tonnes. Controls were relaxed on private sector mining in late 1988, with the aim of reducing the amount of illegally mined gold, believed to be worth about $3.9m a month. An Arab-owned firm, Dar Tadine al-Umma, in February 1989 was able to renew the gold-mining agreement first signed in 1984, despite widespread criticism of certain aspects of the agreement's implementation. Mwadui Diamond Mining Company in Shinyanga region reported expansion plans and said its reserves would last until the late 1990s.

The transport sector was especially successful in attracting aid from donors, who see its rehabilitation as one of the main priorities for Tanzania's economic recovery. The Tanzania–Zambia Railway Authority (Tazara) line's renovation is one of the main projects under SADCC's programmes, because of its importance not only to Tanzanian trade but also to that of Zambia, Malawi and Burundi. In February 1989 Tazara held a donors' meeting at which it asked for $123m in new aid. The Dar es Salaam port modernisation programme resulted in markedly improved efficiency from 1987 onwards, enabling the port to compete with Mombasa. The new container terminal went into service in 1989 and an inland container depot was being developed at Ubungo, 15 km away. The $150m project is due for completion in the early 1990s, when throughput capacity will be about 7m tonnes of cargo a year. Early in 1989 donors gave a favourable response to a World Bank report recommending a $750m programme for upgrading the road network over 8–10 years.

The question of tourism development brings into relief the policy debate within government and party. For entrenched socialists, it remained an unacceptable means of achieving economic recovery; others saw it as the most promising potential source of foreign exchange. In 1988 and 1989 the government signed agreements with several international firms to fund and renovate a number of Tanzania Tourist Corporation (TTCs) hotels and lodges. In late 1988 a deal was signed with the French hotel group Accor to take a shareholding in and to co-manage several TTC hotels and lodges. This drew strong criticism from some circles, opposed to the idea of foreign ownership in principle and resentful at the suggestion that Tanzanians were incapable of managing their own hotels. Dar es Salaam's run-down Kilimanjaro Hotel was being renovated in 1989 by a Yugoslavian firm under a $13.5m contract, and will be managed by Sheraton Corporation of the USA. The Zanzibar Government in December 1988 signed an agreement with the Aga Khan Fund for Economic Development for two new hotels and the development of tourist sites. The island's airport runway was being extended in 1989 by Cogefar of Italy so that long-haul aircraft could be handled.

Gross foreign exchange revenue from tourism was $31m in 1987–88 and was expected to exceed $40m in 1988–89.

Merchandise exports were worth an estimated $390m in 1988, compared with an original projection of $424m, and imports costs about $1,150m. Forecast figures for 1989 were $450m for exports and £1,250m for imports, which would produce a current account deficit on balance of payments of $359m, up from an estimated $300m in 1988.

In March 1989 the foreign exchange retention scheme was altered so that proportion of sales revenue which exporters of non-traditional items could keep was 35 per cent, instead of within a 10–50 per cent range as previously. They could use half of it for importing essential inputs or spare parts, and the remainder for importing certain specified goods for local sale. The Open General Licensing (OGL) scheme, introduced early in 1988, was extended in 1989: the amount of foreign exchange allocated monthly to private importers for items on an approved list was increased from $5m to $15m and the ceiling on a single allocation was raised from $200,000 to $500,000. The monthly allocation was to rise to $20m by the end of 1989 and to $30m by the end of 1990.

Outlook

In the run-up to the presidential elections at the end of 1990, it was clear that Mwinyi would have to become even more cautious and skilful. Donors were insisting on accelerated reforms, but if this resulted in damaging social programmes and the population's general well-being, his chances of a second term in office might be put in jeopardy. The much lower than expected population growth rate revealed in the 1988 census was welcome news: an estimated 2.8 per cent a year over 10 years compared with 3.2 per cent in the previous decade, implying a small but still significant growth in per capita income in 1988 and 1989.

A new and long-awaited investment code, which was expected to be published before the end of 1989, planned to open virtually all sectors to foreign private investment, with special incentives such as lower duties and sales tax on imported capital goods, and a cut in corporate tax from 50 to 45 per cent.

A large question mark still hung over what would happen in 1992 with the end of Nyerere's term as chairman of CCM: some believed that there should be a return to a single leader at the helm, filling the dual role of president and party chairman, in order to settle the rather uneasy relations between government and party, which very roughly represented the two sides of the policy debate.

COUNTRY PROFILE

Political system: Executive power rests with the president and the executive of the ruling Chama Cha Mapinduzi (the Revolutionary Party of Tanzania). Legislative power is vested in the 297 member National Assembly. Both the president and the National Assembly are elected for five-year terms. The president appoints ministers; the prime minister of the Union and the president of Zanzibar are both vice-presidents. The United Republic of Tanzania was formed in 1964 following the union of Tanganyika and Zanzibar.

Political parties: Chama Cha Mapinduzi (CCM) is the only legal political party.

Population: 98 per cent of the 23.17m population (1988 estimate) are of African or Arab (Zanzibar) origin, with the remainder mainly from the Indian sub-continent. With a growth rate of about 2.8 per cent a year, it will attain about 30m by the year 2000. Only 7.5 per cent live in urban areas.

Main cities/towns: Dar es Salaam (population 1,300,000), Zanzibar (270,000), Mwanza (171,000), Dodoma (capital, 160,000), Tanga (143,000), Arusha (88,000).

Language: The official languages are Kiswahili and English (the language most used in business).

Media
Press: Daily newspapers include the government-owned *Daily News* (English) and *Uhuru* (Swahili), the official organ of CCM. Both are published in Dar es Salaam. Weekly newspapers include *Sunday News, Business Times on Fridays* (English) and *Mzalendo* (Swahili). A number of periodicals are also available: *Tantravel Magazine* (trade travel journal) and *Mfanyakazi Magazine*.
Radio: Radio Tanzania broadcasts internal service in Kiswahili and external services in English, Afrikaans and other South African languages. Radio Tanzania Zanzibar broadcasts on three wave-lengths in Kiswahili.
Television: No TV service on the mainland; the only broadcasts are in Zanzibar where colour service is transmitted by Television Zanzibar.

External trade: Regular trade deficit mainly due to low volume of several major export crops and to low world prices. But the revival in agricultural exports after 1986 generated some optimism in the late 1980s. The ratio of debt to current account was a high 677.1 in 1987. The government is now emphasising diversification of exports as a means of boosting export earnings.
Exports: Coffee (32 per cent of total value in 1987), cotton (13%), minerals (6.3%), tea (5%), cloves, sisal, diamonds. Main destinations: German FR (23.5% of total value in 1985), UK (16.8%), Indonesia (6.9%), Singapore (6.4%), Netherlands (6.1%).
Imports: Petroleum (13 per cent of total value in 1986, main single category), machinery and transport equipment, metals, food, beverages and tobacco. Main sources: UK, Japan, German FR, Italy, Netherlands, USA.

Agriculture: Mainstay of economy, contributing almost 50 per cent to GDP and up to 80 per cent of export earnings, as well as employing over 90 per cent of labour force. Weak world demand and distribution problems have tended to turn farmers away from cash to subsistence crops (maize, cassava, bananas and plantains). Poor pricing policies and late payments to farmers by state crop authorities have also turned farmers against subsistence crops. But favourable weather conditions have produced good food and cash crop harvests in 1986 and 1987, when there was a 200,000-tonne surplus of maize, some of which was exported. The 1988 maize harvest was forecast at nearly 3m tonnes. In the 1987-88 growing season, cotton was up by 15 per cent, tea by 21 per cent, and sugar by 8 per cent. Other cash crops include cashew nuts, sisal, and cloves from Zanzibar.

Manufacturing: Accounts for approximately 6 per cent of GDP, down from 10 per cent in 1982, and is dominated by government-owned concerns engaged in the processing of local minerals and agricultural raw materials for local consumption. Also paper and pulp, cement, textiles and some light engineering. Emphasis placed on the development of import substitution industries, particularly of consumer goods, promotion of small companies and streamlining of state-owned companies. There is also now emphasis on export-oriented industries. Growth has been restricted by a lack of foreign exchange needed for the import of raw materials, spare parts and fuel, and capacity utilisation is as low as 45 per cent. Foreign aid is now aimed at rehabilitation of existing industries, the majority of which are still operating at less than 30 per cent of capacity. A fertiliser plant was planned at Kilwa, and a new pulp and paper mill at Mufundi started in 1987 to export small amounts of paper.

Mining: Accounts for less than 1 per cent of GDP, 85 per cent of which comes from diamond sales. In most sectors output has fallen drastically. Gold production has also declined considerably, but is set to improve with large-scale foreign investment in old mines at Geita and Lupa. There is also a large phosphate mine at Minjingu, supplying the fertiliser plant at Tanga.

Energy: Heavy dependence on imports of oil and oil products, which may account for up to two-thirds of total foreign exchange earnings, and can be in short supply due to a lack of foreign exchange reserves. Several companies are prospecting for on and offshore oilfields. As yet these have met with no success, but gas reserves (41bn cu metres) have been discovered at Songo-Songo island and a much larger field at Kimbiji. Total installed electricity generating capacity is currently about 439 MW, of which about 70 per cent is hydro-electric. Development plans are concentrated on further utilisation of the country's hydro-electric potential, which is being developed with the aid of foreign investment. New plants are being built at Mtera and Masasi, and the Ubungo plant which supplies Dar es Salaam is being upgraded. A coal-fired power station at Kiwira was due for completion in 1988.

Membership of international organisations: African Development Bank, African States Associated with the EC (Lomé Convention), Arab Bank for Economic Development in Africa (recipient), Commonwealth, Economic Commission for Africa, FAO, GATT, IAEA, IDA,IFAD, ILO, IMF, OAU, PTA, Southern African Development Co-ordination Conference, UN, WHO, WIPO, World Bank.

BUSINESS GUIDE

Time: GMT +3.

Climate: Tropical, with variations according to altitude. Rainy seasons Apr-May and Nov-Dec. Warmer on coast, cooler in upland areas. Temperatures range from 23°C to 30°C. Best time to visit Jun-Sep.

Entry requirements
Passport: Required by all.
Visa: Required by all except nationals of certain Commonwealth countries, Republic of Ireland and Scandinavian countries. A visitors' pass must be held – it is advisable to obtain this before travel to the country. An onward or return ticket is also required.
Prohibited entry: South African passport holders.
Currency: Import of local currency prohibited and export limited to Shs 200. Import of foreign currency is unlimited but amount must be declared. The declaration form must be recorded at each currency exchange. Export of unchanged foreign currency allowed if declaration form is produced. Every visitor is required to change at least $50 or equivalent on arrival at airport. Any undeclared currency may be confiscated. US$20 departure tax must be paid in US Dollar bills.

Health precautions
Mandatory: Yellow fever vaccination certificate required if travelling from an endemic zone. Yellow fever and cholera vaccination certificates may be required by visitors arriving from certain countries.
Advisable: Yellow fever, cholera, typhoid and polio vaccinations. Malaria prophylaxis recommended. Water precautions should be taken – bilharzia risk exists. AIDS risk.

Representation overseas: Addis Ababa, Beijing, Bonn, Brussels, Bujumbura, Cairo, Conakry, Geneva, The Hague, Harare, Kigali, Khartoum, Kinshasa, Lagos, London, Luanda, Lusaka, Maputo, Moscow, New Delhi, Ottawa, Paris, Rome, Stockholm, Tokyo, Washington DC.

Air access
National airline: Air Tanzania.
Other airlines: Aeroflot, Air Comores, Air France, Air Zimbabwe, British Airways, Egyptair, Ethiopian, Gulf Air, Kenya Airways, KLM, LAM-Linhas Aéreas de Moçambique, Lufthansa, Royal Swazi National Airways, Sabena, Swissair, Air Uganda, Zambia Airways.
Tax: Airport tax of US$20 on departure, payable in US Dollar bills (no other form of payment accepted).
International airports: Dar es Salaam (Code: DAR), 13 km from city, duty-free shop, restaurant, bar, bank, shops, post office; Kilimanjaro International (Code: JRO), 50 km from Arusha (between Arusha and Moshi), bar, restaurant, post office, shops.

Surface access
Main ports: Dar es Salaam, Mtwara, Tanga. Steamer services are operated by Société Nationale des Chemins de Fer Zairois (SNCZ) on Lake Tanganyika, and Tanzania Railways Corporation on Lakes Tanganyika, Victoria and Malawi. These services connect with ports in Zaïre, Zambia, Uganda, Malawi and Burundi.
Road: Access from Zambia via surfaced road between Lusaka and Dar es Salaam. Also roads from Uganda, Rwanda (very poor quality), Kenya and Malawi.
Rail: Inefficient service from Zambia operated by Tanzania-Zambia Railway Authority (TAZARA). No through services between Zambia and Tanzania.

Hotels: Accommodation tends to be expensive and can be difficult to obtain, especially in Dar es Salaam, so reservations should be made well in advance and confirmation obtained. Bills must be settled with foreign exchange – the production of an exchange certificate is not enough.

Electricity supply: 230 V AC, 50 cycles.

Credit cards: Use restricted to a few shops and hotels, mainly in Dar es Salaam. Hotels mostly accept only American Express and Diners'.

Car hire: Expensive and very limited availablity of self-drive cars, mainly chauffeur-driven.

City transport
Taxis: Available in main towns. Fixed rates for journeys within Dar es Salaam from taxis at hotels. Fares in any other taxis are by negotiation and should be fixed before journey. Tip of 10 per cent usual.
Buses: Services operate within Dar es Salaam. Flat fare system operates. Generally unreliable and overcrowded and quite unsuitable for visitors.

National transport
Air: Air Tanzania Corporation operates services between Dar es Salaam, Kilimanjaro, Mwanza, several other, small airports, Zanzibar and Pemba. Charter companies operate to isolated airfields, national parks and numerous towns.
Road: Over 35,000 km of roads, of which about 2,500 km are surfaced. All-weather roads connect major centres, but minor roads are liable to be impassable in the rainy season except to four-wheel drive vehicles. The road from Dar es Salaam to Moshi has deteriorated badly, but the road to Zambia is generally in adequate condition. There is a new road from Songea to Makambako, and one from Mwanza to Musoma is being built.
Buses: Inexpensive services link most centres. Routes include: Dar es Salaam-Songea; Dodoma-Moshi; Lindi-Tunduma; Lindi-Mtwara. Services are unreliable.
Rail: Operated by Tanzania Railway Corporation, except for the Tanzania to Zambia line which is run by the TAZARA (see Surface Access above). Routes from Dar es Salaam include those to Kigoma on Lake Tanganyika, Mwanza on Lake Victoria, Moshi, Arusha and Mbeya.
Water: Regular ferry link from Dar es Salaam to Zanzibar and Pemba Islands. Services on Lakes Victoria and Tanganyika (may be crowded).

Public holidays
Fixed dates: 1 Jan, 12 Jan (Zanzibar Revolution Day), 5 Feb (CCM Day), 26 Apr (Union Day), 1 May, 25 Jul (Peasants' Day), 9 Dec (Independence/Republic Day), 25 Dec.
Variable dates: Good Friday, Easter Monday, Eid al-Fitr, Eid al-Adha, Islamic New Year, Prophet's Birthday.

Working hours
Government and business: (Mon-Fri) 0730/0800 – 1430/1500; (Sat) 0730/0800 –1200/1230.
Banking: (Mon-Fri) 0830/0900 – 1200/1230; (Sat) 0830/0900 –1100/1130.

Social customs: Muslim sensitivities should be respected in dress. Visitors should only use cameras in private settings and tourist resorts.

Telecommunications
Telephone: Dialling code for Tanzania: IDD access code + 255, followed by area code (Arusha 57, Dar es Salaam 51, Kilimanjaro Airport 575, Moshi 55, Tanga 53, Zanzibar 54) followed by subscriber's number. No IDD from Tanzania.
Telex: Public telex facilities available at Kilimanjaro Hotel or the Telegraph Office in Dar es Salaam. Country code: TZ.

Banking
Central bank: Bank of Tanzania.
Main Centre: Dar es Salaam. *Head Offices:* National Bank of Commerce, Tanzania Housing Bank, Tanzania Investment Bank, Co-operative Rural Development Bank.

Trade fairs: There is an International Trade Fair every year in July in Dar es Salaam.

Representation in capital: Albania, Algeria, Angola, Australia, Belgium, Botswana, Bulgaria, Burundi, Canada, China PR, Cuba, Czechoslovakia, Denmark, Egypt, Finland, France, German DR, German FR, Guinea, Hungary, India, Indonesia, Iran, Iraq, Italy, Japan, Korea PDR, Madagascar, Mexico, Mozambique, Netherlands, Nigeria, Norway, Pakistan, Poland, Romania, Rwanda, Somalia, Spain, Sudan, Sweden, Switzerland, Syria, Turkey, UK, USA, USSR, Vatican, Vietnam, Yemen AR, Yugoslavia, Zaïre, Zambia, Zimbabwe.

Hotels: Dar es Salaam
Embassy, P.O. Box: 3152 (tel: 30006; tx: 4570, 4166).

Kilimanjaro, Kivukoni Front, P.O. Box: 9574 (tel: 21281; tx: 41021).

Motel Agip, P.O. Box: 529 (tel: 23511/5; tx: 41276).

New Africa, Maktaba Street, P.O. Box: 9314 (tel: 29611; tx: 81040).

Twiga, Samora Avenue, P.O. Box: 1194 (tel: 22561; tx: 41658).

Other useful addresses
Bank of Tanzania, PO Box 2939, 10 Mirambo Street, Dar es Salaam (tel: 21291; tx: 41024 BENKUU TZ).

Cashew Nut Authority of Tanzania, PO Box 533, Mtwara (tel: 2310; tx: 56134 korosho).

Coffee Authority of Tanzania, PO Box 732, Moshi (tel: 4011).

Dar es Salaam Chamber of Commerce, PO Box 41, Dar es Salaam (tel: 21893).

Economic Development Commission, Dar es Salaam.

National Development Corp of Tanzania, PO Box 2669, Dar es Salaam (tx: 41068).

Radio Tanzania, PO Box 9191, Dar es Salaam (tel: 63342; tx: 41201).

State Mining Corporation, PO Box 4958, Dar es Salaam (tel: 28781; tx: 41354 stamico).

Tanzania Tourist Corporation, IPS Building, PO Box 2485, Dar es Salaam (tel: 27671; tx: 41061).

Television Zanzibar, PO Box 314, Zanzibar (tel: 32816).

Thailand

Nicholas Cumming-Bruce

Major-General Chatichai Choonhavan could reflect with some satisfaction on a first year as prime minister which saw his political base consolidated and the country enjoying an economic boom with higher-than-expected growth. It seemed that Chatichai, contrary to initial expectations, might be able to serve a full term as premier; and Thailand, under a government wedded to pragmatic pursuit of business, now nursed ambitions of becoming the hub of economic activity in mainland South East Asia.

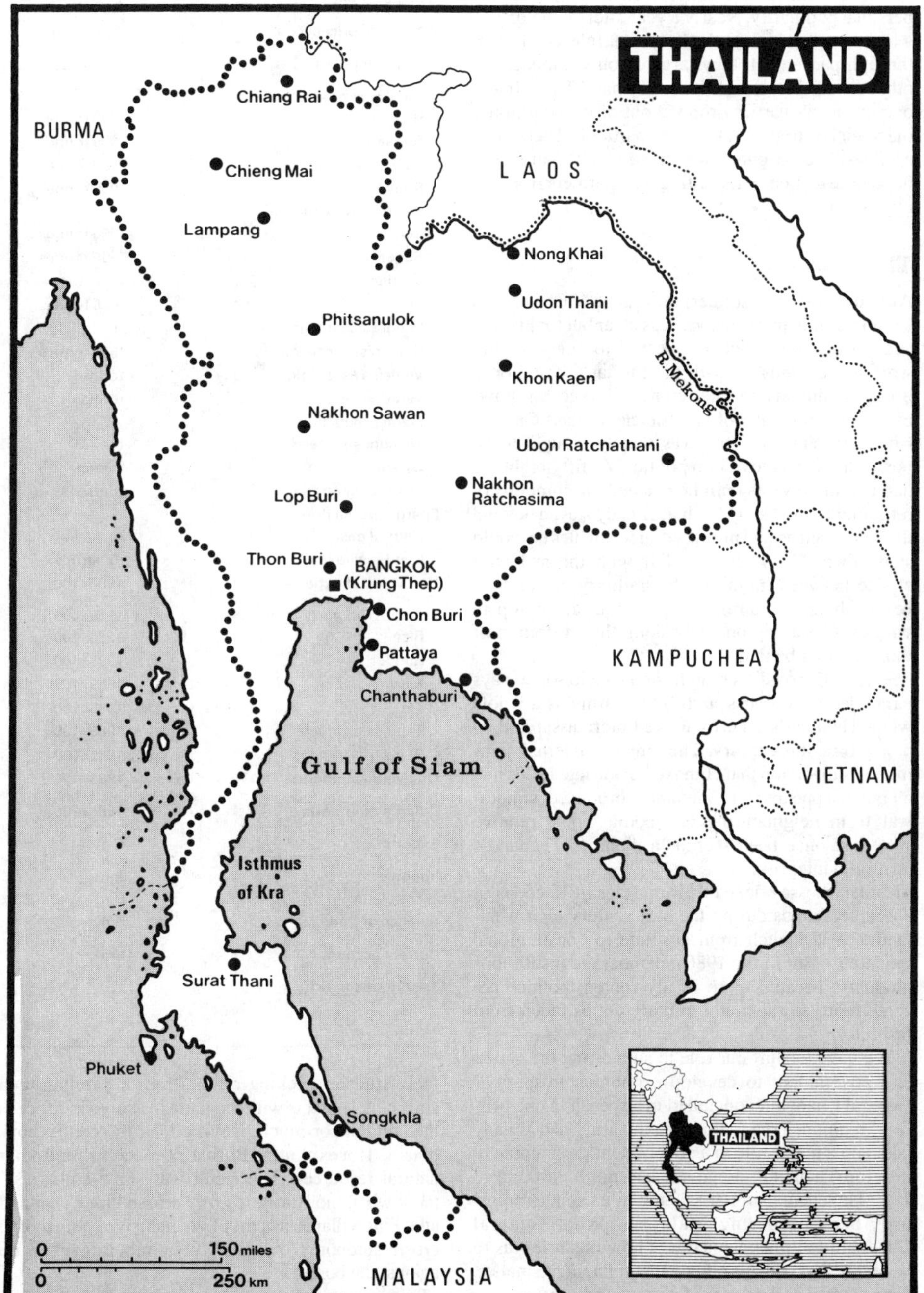

Chatichai's successes

When Chatichai took over in August 1988 as the first prime minister to be an elected member of parliament in a decade, few gave him much hope of surviving more than perhaps six months. His predecessor, Prem Tinsulanonda, came to the premiership direct from command of the army and kept it eight years with the help of backing there and from Thailand's revered monarchy. Chatichai, although a minister in the previous government and for many years a leader of the Chart Thai party, had no obvious power base. Chatichai, thought to be seventy years old, appeared to be a political flyweight, a wealthy businessman-socialite better known for riding a motorcycle and a love of discos than for powers of leadership.

But Chatichai proved a better communicator than Prem and established a more positive public image. His line up of ministers looked suspiciously like a cabal of political godfathers including many with reputations for corruption and he was warned of the dangers for the national economy of cronyism in placing major contracts. Yet Chatichai also appointed a team of young intellectuals and technocrats – among them his left-leaning academic son Kraisak – to be his personal advisers, and acted outside purely party or business interests.

In a move that earned public plaudits the government banned all logging in January 1989 in a bid to check the dangerously fast pace of deforestation. Chatichai took the initiative in the face of anguished cries from a logging industry with powerful connections in government, even in the cabinet, after disastrous flooding and loss of life in the south exposed the social and economic costs of this environmental destruction.

His government supported a number of other reforms in parliament. Constitutional amendments would transfer the presidency of parliament from the mainly appointed and military-dominated senate to the house of representatives in 1991. MPs had long argued for this measure to strengthen democratic processes in a country where frequent coups reflected the political pre-eminence of the army. The lower house also unanimously passed a social security bill setting up a fund that would provide compensation for

KEY FACTS

Official title: Kingdom of Thailand

Head of state: King Bhumibol Adulyadej (Rama IX)

Head of government: Prime Minister General Chatichai Choonhavan

Ruling party: Chart Thai Party leads coalition

Capital: Bangkok

Official Languages: Thai (English and Chinese also spoken)

Currency: Baht (B = 100 satang)

Exchange rate: B25.5 per $ (Nov 1989)

Area: 514,000 sq km

Population: 54.5m (1988)

GDP per capita: $841 (1987)

GDP real growth: 8.4% (1987)

Labour force: 27.91m (1986)

Inflation: 3.8% (1988)

Trade balance: –$2,074m (1988)

Foreign debt: $20.7bn (1987)

Visitor numbers: 4.2m (1988)

sickness and injury. The bill was welcomed by labour as putting Thailand on course for a more comprehensive system of social security that would eventually cover old age pensions and unemployment benefits.

Chatichai came under criticism for a haphazard and piecemeal approach that left his government without clear priorities but this did not inhibit a growth in his personal popularity. Nearly a year after taking office, when the opposition pushed a 'no confidence' motion through parliament, they targeted only ministers of other parties in the coalition, not Chart Thai's. If any one of the six parties dropped out of the coalition – the Social Action and Democrat parties had been seen as possible drop-outs – it seemed Chart Thai would have little difficulty recruiting new partners from the opposition.

The army

Although a former soldier, the extent of Chatichai's support within the army was less clear but for his first year in office Chatichai appeared to enjoy a solid working relationship with the military, based on a policy of mutual non-interference. His success, however, was something of an obstacle to plans General Chaovalit Yongchaiyuth was widely thought to be pondering for a move into politics. At fifty-eight, he had two more years until he reached the compulsory retirement age of sixty but had already stayed beyond the date when he had promised subordinates he would step down. Chaovalit was left with the awkward choice between giving up his military power base before there was an opening for him at the top in politics, or staying on and risking the frustration of those waiting behind.

General Chaovalit's search for a role in some ways paralleled a similiar search by the army as a whole when Thailand's security looked more assured than at any time this century. The threat of internal communist subversion that climaxed in the late 1970s had all but disappeared and Vietnam's promised withdrawal from neighbouring Cambodia would remove what had once been seen as a threat to Thailand's territorial integrity.

Military bosses flexed their muscles in 15 coups or attempted coups during the half century after Thailand's 1932 switch from absolute to constitutional monarchy. But in the 1980s, democratic institutions gradually became more firmly rooted, politics became more pluralist and military coups faded from fashion.

'We have to shift our role to supporting the nation and government to develop economic prosperity,' General Chaovalit concluded in a speech to one officers' reunion. True to this idea the army had already taken on a major role in a government programme to develop Thailand's impoverished north east and in 1989 launched a similar scheme in its southernmost provinces bordering Malaysia. What General Chaovalit left unexplained was how this role was to be reconciled with his other idea for having a smaller, better armed and more professional military.

A golden land

In the meantime, however, the military found an outlet for some of its energies in foreign policy towards Thailand's immediate neighbours. In January 1989 General Chaovalit put forward the notion of creating a 'Suwannaphume' or Golden Land in South East Asia with Thailand at its centre. Thailand, in effect, would guide neighbouring Burma and the countries of Indochina towards creating an area of stability and economic prosperity. Thailand's more advanced state of development would make it the natural focal point of this Suwannaphume.

To this end, the army assumed the lead role in relations with Burma, deciding the treatment of Burmese students seeking refuge from the ruling junta and using contacts with the junta in Rangoon to act as the conduit for much, if not all, of the concessions obtained for exploiting Burma's timber, fish and other natural resources. The result was some embarrassment when the Burmese army crossed into Thailand attacking villages as part of its intensive campaign to crush autonomy-seeking Karen rebels backed up against the border.

The army also took a major, if less exclusive, role in opening up relations with Laos, helped by close contact between General Chaovalit and the Laotian army chief General Sisowath, and played an important part, if more behind-the-scenes, in developing contacts with Cambodia's Vietnamese-backed government in Phnom Penh.

KEY INDICATORS

	Unit	1984	1985	1986	1987	1988
Population	m	50.71	51.68	52.65	53.60	54.54
Gross domestic product (GDP)	B bn	988.9	1,041.4	1,098.4	1,223.2	–
GDP per capita	B	19,501	20,151	20,862	22,821	–
Wholesale prices	1985=100	100	100	99.6	105.5	114.2
Consumer prices	1985=100	97.6	100.0	101.8	104.4	108.4
Agricultural production						
Rice	'000 tonnes	19,549	19,905	20,264	18,868	–
Sugar-cane	'000 tonnes	23,869	25,055	24,093	24,450	–
Tapioca root	'000 tonnes	20,000	19,263	15,255	19,550	–
Rubber	'000 tonnes	629	722	790	840	–
Industrial production						
Cement	'000 tonnes	8,240.0	7,915.7	7,913.6	–	–
Sugar	'000 tonnes	2,431.0	2,293.6	2,607.1	–	–
Textiles	m sq metres	1,959.7	2,072.4	–	–	–
Tin	'000 tonnes	111.3	86.0	–	–	–
Integrated circuits	m	783	406	580	758	–
Galvanised iron sheet	'000 tonnes	153.8	128.7	140.9	–	–
Vehicles assembled	tonnes	111,037	82,106	74,155	–	–
Tyres	tonnes	44,442	31,564	33,239	–	–
Mineral production						
Tin concentrates	tonnes	29,979	22,995	23,299	–	–
Gypsum	tonnes	1,110,726	1,273,459	1,665,557	–	–
Lead concentrates	tonnes	39,204	46,245	61,885	–	–
Energy production						
Natural gas	m cu feet	85,508	132,272	127,765	–	–
Condensate	m barrels	3.01	5.20	5.21	–	–
Unit value of exports	1980=100	97.3	101.2	102.9	110.0	–
Volume of exports	1985=100	93.1	100	116.1	136.7	–
Rice	1985=100	113.6	100	111.4	109.4	125.3
Rubber	1985=100	85.8	100	110.3	128.4	135.9
Maize	1985=100	113.0	100	144.3	59.3	43.7
Tin	1985=100	102.7*100	105.0	76.4	72.8	
Tapioca products	1985=100	92.7	100	89.1	87.6	114.6
Sugar	1985=100	71.7	100	113.7	117.5	107.6
Unit value of imports	1985=100	91.3	100	94.1	100.3	–
Volume of imports	1985=100	107.0	100	100	130.6	–
Exports	B m	175,237	193,366	233,383	301,453	403,570
Imports	B m	245,155	251,169	241,358	334,209	513,114
Petroleum	B m	35,035	38,526	21,939	26,248	21,889
Balance of trade	$m	–1898	–1332	388	–424	–2074
Current account	$m	–2,109	–1,537	247	–365	–1,671

* estimated figure

Foreign relations

In what was the most pronounced change instituted by the Chatichai administration, the prime minister decided it was time to establish a direct relationship with leaders in Phnom Penh, despite Thailand's close links with the UN-recognised Cambodian resistance coalition. Moreover, Thailand had previously supported the policies of isolating Phnom Penh which had been pursued by its allies in the West and by its partners in the Association of South East Asian Nations. To their amazement and in some instances, anger, Chatichai agreed to have direct talks in Bangkok in January with Phnom Penh Prime Minister Hun Sen.

In a sense Chatichai was simply being consistent. Among his first statements on taking office was a determination to turn the battlefields of Indochina into markets for Thailand. The initiative was welcomed by those who had been critical of Thailand's previous policy as too rigid. But the handling of Hun Sen's visit, something of a watershed in Thai foreign policy, was clumsy and amateurish. The timing of the visit, shortly before a second round of Cambodian peace talks in Jakarta handed Hun Sen a diplomatic coup that clearly strengthened his hand and correspondingly weakened that of the resistance coalition. As confusing for Thailand's regional allies as the sudden change was the uncertainty over who was making Thailand's foreign policy. Prime Minister Chatichai's new policy on Cambodia had been decided with scant reference to the foreign ministry. Comments by his advisers were at clear odds with views of veteran Foreign Minister Siddhi Savetsila and his senior officials. When Prime Minister Chatichai flew to Tokyo for the funeral of Emperor Hirohito he took the opportunity to present US President George Bush with an aide memoire inviting American support and participation in his policy on which Siddhi had not even been consulted. When senior ministers from Phnom Penh came to Bangkok later for a conference on Indochina opened by Chatichai, foreign ministry officials tried to block their appearance.

An inevitable result was tension between Siddhi and his officials and Chatichai's advisers and speculation as to whether the foreign minister, who leads the Social Action Party, would remain in the coalition. After friction between his advisers and other ministries, Chatichai reined in their public speaking, emphasising that their task was to advise, not to formulate policy.

The effect on Thailand's foreign relations seemed, however, likely to prove more long lasting. China's disapproval was clearly communicated when Deng Xiaoping lectured Chatichai on the matter during a March visit to Beijing. Some ASEAN countries were also irritated and pondered whether the new policy, coupled with the talk of Suwannaphume, signalled a weakening Thai commitment to ASEAN.

Economic performance

Whatever doubts some in government entertained about this approach to Cambodia, there was no argument that the moves towards a settlement of the 10-year old Cambodian conflict and Vietnam's efforts to normalise ties with the West enhanced Thailand's attractions as a location for investment. Indeed, Chatichai's sole observation on China's brutal suppression of demonstrations was to instruct the cabinet to prepare for extra investment that would come Thailand's way as a result. Hong Kong's deepening anxieties about its future were expected similarly to raise Thailand's investment prospects.

Such expectations reflected the degree to which Thailand appeared to have caught the crest of an economic wave. To the steadily rising number of investors examining opportunities in the region, Thailand offered a domestic market of 55m people second only to Indonesia's within ASEAN, a relatively low-cost workforce, a stable political environment and a competitive range of investment incentives. On top of that it emerged as ASEAN's strongest performing economy.

Thailand turned in its best economic results for two

COUNTRY PROFILE

Political system: Thailand is a constitutional monarchy governed by a bicameral National Assembly consisting of a 347-member House of Representatives elected to a four-year term and a smaller Senate appointed to a six-year term by the sovereign and traditionally dominated by the military. As head of state, the sovereign appoints a prime minister and cabinet of about 44 ministers, and is advised by a privy council. Local government is vested in 73 provinces, with locally-elected councils operating at all levels. There have been several constitutional changes in the 1980s. Last general elections were held in July 1988. In August General Chatichai Choonhavan was appointed prime minister.

Political parties: Ruling party: Chart Thai party (87 seats following Jul 1988 elections). Other parties in government: Social Action Party (54 seats), the Democrats (48 seats) and Rassadorn, or 'Citizen' party which has close ties with the army (21 seats). There are another 15 political parties.

Population: In 1988 population was 54.5m. The population was growing in 1987 only 1.6 per cent per annum (compared with 3.2 per cent in the 1960s). 90 per cent of the population are ethnic Thais, with eight per cent Chinese, and 0.7 per cent Malays. Over 40 per cent are under 15 years of age, and 82 per cent live in rural areas.

Main cities/towns: Bangkok Metropolis (population 7.0m* in 1987).

Language: Thai is the national language; Chinese, English and sometimes French are spoken in business circles.

Media
Press: The Bangkok Post (morning) and *The Nation* (morning) are the main English-language newspapers. Various international publications, such as the *New York Times, Newsweek, The Economist* and *Asiaweek* are sold by newsagents. Thai-language papers include *Thai Rath, Siam Rath, Khao Panich* and *Daily News, Dao Siam.* Chinese-language papers include *Sing Sian Yit Pao, Universal Daily News* and *Sirinakorn, Thai Shang Yic Pao*. There are three main English-language business journals: *The Investor* (Industrial Finance Corp of Thailand), *Business in Thailand* (Business Information and Research Company) and *Business Review,* a supplement of *The Nation. Industry,* a bi-monthly, is produced for the Association of Thai Industries.
Radio: There are over 200 heavily commercial AM radio stations, FM radio services and an educational network. External services are broadcast in several languages, including English, French, Malay, Lao, Khmer, Chinese, Vietnamese, Japanese and Burmese.

Television: Bangkok has five colour TV channels: Thai Television Co, Army Television, Bangkok Broadcasting Co and Bangkok Entertainment Co and an educational TV channel.

Domestic economy: In 1988 the Thai economy grew by 10.3 per cent, a recent year record. The economy is traditionally agricultural, although industry has grown rapidly. Textile production, construction materials, consumer goods and trade in imported products are important sectors, with mainly foreign ownership of certain others, including petrochemicals, motor vehicles, tyres and detergents. Successive balance of payments deficits have been financed by foreign borrowing, increasing exports and with receipts from tourism and overseas workers' remittances. External debt services were budgeted at 24.7 per cent of government expenditure in 1987. The current Five-Year Development Plan (the Sixth National Economic and Social Development Plan 1987-1991) aims at the development of tourism and exports, efficiency in developing and improving production structure and to raise the quality level of basic services and distribution of income and prosperity to rural regional areas. Foreign aid and loans have provided funds for energy sector projects, port development, new highways, electrification, family planning, housing and labour-intensive and export-orientated industries. The Eastern Seaboard Development Programme, inaugurated in 1982, is a long-term project for a complete industrial region, involving petrochemical and heavy industry plant as well as new deep-sea ports at Map-Ta-Phut and Laem Chabang. Large influxes of refugees act as a drain on resources and place severe burdens on the local economy in certain areas. There were 250,000 Kampucheans, 80,000 Laotians and 7,000 Vietnamese.

Employment: The labour force of 27.91m (1986) is growing at 2.8 per cent per annum, enough to swamp the growing GDP growth rate. Official unemployment rates are low (two per cent in 1986), but there is considerable under-employment and seasonal unemployment. In 1986, 85,662 Thai workers were employed abroad, mainly on construction sites in the Middle East. The government's Labour Management Organisation actively encourages Thais to seek work overseas. The Eastern Seaboard Programme is expected to generate 300,000 new jobs by 1992.

External trade: Thailand's economy is strongly dependent on its foreign trade especially with the USA, and it has proved unusually susceptible to fluctuations in the US market. The former emphasis on import substitution has now given way to export promotion. In 1987 balance of payment surplus stood at B18,000m.
Exports: In 1987 export value was B291,000m, up 26 per cent, volume up 15 per cent. Main exports are rice, tapioca, rubber, precious stones, integrated circuits, canned food, textiles, sugar, and tin. Main destinations are USA, Japan, Singapore, Netherlands, German FR, Malaysia, Hong Kong, China PR and UK (3.2 per cent).
Imports: In 1987 imports totalled B337,000m, up 37 per cent, mainly due to growth in investment and the economic recovery. Main imports are fuels and lubricants, non-electrical machinery, chemicals electrical machinery, electrical appliances, food and beverages, vehicles and parts, fertilisers and pesticides. Main sources are Japan, USA, Singapore, German FR, Malaysia, UK, Republic of Korea.

Agriculture: Agriculture accounted for 17 per cent of GDP in 1988. In 1986 it accounted for 22.4 per cent of GDP, 36.9 per cent of exports and employed 65.6 per cent of the workforce. Thailand is known as the rice bowl of Asia and is one of the world's leading net exporters of food. To date, production generally has been increased by expansion of planted acreage, rather than productivity improvements such as irrigation or use of fertilisers. Yield per paddy is one of the lowest in SE Asia. Drought affected crops in 1987 but the weather was exceptionally good in 1988. Major crops include rice, sugar, tapioca, maize, rubber, kenaf, cotton, jute, tobacco, fruit (especially pineapple), beans, oilseeds and coffee. Livestock raised includes pigs, cattle, sheep and poultry. Buffalo, oxen, horses and elephants are used as draught animals. Crocodiles are farmed for their skins. Fish is an important source of protein and foreign exchange, with freshwater fish farming widely practised in flooded paddy fields. Forests cover about 25 per cent of total land area, with a further 18 per cent subject of a reafforestation programme. Forest reserves are considerably depleted, partly due to land clearance for cultivation. Export of most logs and sawn wood is now banned.

Industry: Manufacturing accounted for 21.1 per cent of GDP (1986) and 35.4 per cent of exports, employing 8.9 per cent of the workforce. Most industries are private sector and most registered factories are small undertakings, but there is a range of medium and large concerns. Main industrial products include processed food, precious stones and jewellery, cement, sugar, refined oil, textiles, assembled vehicles and parts, paint, steel, paper, pharmaceuticals, galvanised iron sheet, plastics (including artificial flowers), electronics, electrical appliances, glass, tin ingots, condensed milk, tin plate, detergent, hydrochloric acid, caustic soda. The government is actively developing heavy industries associated with natural gas. A gas separation plant with associated LNG depot and marine terminal was recently completed in Rayong province. There are plans for a petrochemical complex along with plants for soda ash, fertilisers, potash and sponge iron. There are also plans for oil refinery expansion and a zinc smelter. Thailand has 14 assembly plants producing about 22,000 cars, with 45 per cent of the spare parts made locally and has ambitions for exports. Local processing of wood has been encouraged and exports of plywood, veneer, parquet, furniture, household

decades in 1988 when real GNP growth surged 11.1 per cent and initial forecasts that this would slow to about 7.5 per cent in 1989 were progressively raised by commercial banks to 9 and perhaps 10 per cent. Manufacturing led the way with 12.4 per cent growth but even agriculture, recovering from drought in 1987, managed a remarkable 8.6 per cent rise in output.

A consistently strong rise in exports, up 36 per cent in 1988, reflected one of Thailand's greatest economic strengths: the diversity of its products and the diversity of its markets. Exports of traditional products showed a marked increase, helped by both high crop production and export prices for rice, expected to reach a record 5m tonnes in 1989, as well as for rubber and to a lesser extent maize. Yet overall sales of the eight main traditional items, which also included tin, tapioca and sugar, fell behind the higher value-added manufactures, particularly textiles, food products and gems and jewellery.

Thailand's main trading partners were the USA and Japan but dependency on the USA was only some 18 per cent, well below the level of countries like South Korea and Taiwan, and a significant share of its exports was spread among developing countries. With a market share in the USA much lower than those of Asian NIC's (newly industrialised countries), Thailand was less exposed to protectionist pressures. It, however, was one of eight countries put on a watch list by the USA as a possible target of retaliatory measures after November because of enduring dissatisfaction over the lack of protection for intellectual property under Thailand's patent and trademark laws.

Thailand's main trade concern, however, was not over exports but the rate at which its economic boom was sucking in imports. Thailand's trade deficit almost tripled in 1988 to $3.4bn and worried finance ministry and central bank officials were contemplating the prospect of it rising to $5bn by the end of 1989.

To Finance Minister Pramual Sabhavasu this situation was 'critical', but it generated little sense of alarm among the business community. Economists emphasised that capital goods and raw materials for industry, much of it export-oriented, were mainly responsible for the fast rise in imports and the imbalance could start to correct itself as factories came on stream and the pace of expansion slowed. In the meantime there were other mitigating factors.

Thailand was still reaping the benefits of its 1987 Year of Tourism in steadily rising numbers of visitors, more than 4m of them in 1988, who brought in a net income of more than $3bn. That was expected to rise to around the $5bn mark over the next two years.

This still left a current account deficit of $1.7bn in 1988 and forecasts by some bankers that it might rise by two-thirds or more in 1989 to about 7 per cent of GDP. The signs were that this would be more than offset by foreign investment. The Board of Investment gave its blessing to 180 projects planning investments of some $550m in the first half of 1989, increases of 175 and 290 per cent respectively over the same period of 1988. Foreign capital inflows in

COUNTRY PROFILE

utensils and paper products are increasing. The tourist industry has become the single most important foreign exchange earner, with the majority of tourists from Malaysia, Japan, UK, USA and German FR. 1987 was a special visit Thailand year with celebrations extended to July 1988. 3.5m tourists were recorded in 1987.

Mining: Mining accounts for 1.5 per cent of GDP (1986) and provides less than one per cent of employment (1985). Thailand is the world's third largest tin exporter, but has suffered since mid-1985 from depressed markets which followed the collapse of the International Tin Agreement. It has now lifted its 20-year-old ban on the export of tin concentrates, but has been unable to revive the industry. Other minerals present include ores of lead, zinc, tungsten, copper, antimony, barite and iron, fluorite, wolfram, limestone, gypsum, potash, rock salt, silica sand and precious stones, especially sapphires. Lignite reserves of 180-600m tonnes are being mined, with output of 5.5m tonnes in 1986.

Hydrocarbons: Significant offshore natural gas reserves of 170-453m cu metres have been discovered in the Gulf of Thailand. The Erawan field, which began production in 1981, produces over 5.6m cu metres/day, and the Satun, Kapgong, Pladang, Baanpot and Platong fields have more than doubled production. There are also onshore natural gas deposits on the Korat Plateau. Oil production at the Sirikit field in the north central plain began in 1984 and was yielding 25,000 b/d by early 1986, covering about a third of Thailand's energy requirements.

Energy: Thailand is attempting to reduce its dependence on imported oil by developing use of indigenous natural gas, lignite and hydroelectric resources. Geothermal resources have been discovered in Chiang Mai, Mae Hong Son, Chiang Rai and Phrae provinces.

Main companies: Esso Standard Thailand Ltd, Siam Cement Co Ltd, Toyota Motors (Thailand) Ltd, Boon Tawd Brewery Co Ltd, Siam Motors Co Ltd, East Asiatic Co (Thailand) Ltd, Siam City Cement Co Ltd, Luckytex (Thailand) Ltd, Italian Thai Development Corp, Thai Central Chemical Co, Saha-Union Corp Ltd, Lever Brothers (Thailand) Ltd, Mitsubishi Co (Thailand) Ltd, Berli Jucker Co Ltd, Sittipol Motors Co Ltd, United Flour Mill Co Ltd, Union Textile Ind Corp, Construction Materials Mktg Co, Borneo Co (Thailand) Ltd, Saha Pathanapibul Co Ltd, The Serm Suk Co Ltd, Louis T Leonowens & Co, Colgate-Palmolive (Thailand) Ltd, Thai Tinplate Manufacturing Co, Kamal Sukosol Co Ltd, Firestone Tyre & Rubber (Thailand), Jalaprathan Cement Co, Thai Durable Textile Co.

Financial markets: Securities Exchange of Thailand, Bangkok. Before the global stockmarket crash of 1987 the Bangkok Exchange was doing well with average volume of $60m – at the end of the year it was down to $5.6m. Consists of five domestic funds totalling $100m, the Thailand Fund $30m started in October 1986 and the Bangkok Fund started in 1985 at $10m.

Membership of international organisations: ANRPC, ASEAN, ASPAC, Asian Development Bank, Association of Tin Producing Countries, CCC, Colombo Plan, EC trade and co-operation agreement, ESCAP, FAO, GATT, G-77, IAEA, IBRD, ICAO, IDA, IFAD, IFC, IHO, ILO, IMF, IMO, INRO, INTELSAT, INTERPOL, IPU, IRC, ITC, ITU, UN, UNESCO, UPU, WHO, WMO, World Bank, WTO..

BUSINESS GUIDE

Time: GMT + 7 hrs.

Climate: Tropical savanna in the north and tropical monsoon in the south, with a hot season Feb-Apr, a rainy season May-Oct and a cool season Nov-Jan. Average temperature is 28°C, and in Bangkok temperatures range from 35°C (Apr) to 17°C (Dec).

Entry requirements
Passport: Required by all except certain seamen.
Visa: Visas are not required by most tourists from non-communist countries with onward passage for stays of up to 15 days/14 nights. Otherwise, visas must be obtained before arrival. New Zealand passport holders may stay in the country for up to 90 days without a visa.
Currency: There is no limit on the amount of foreign currency which may be imported, although it should be declared on arrival. Foreign currency in excess of $2,000 may be exported only if declared at entry. No more than B500 may be exported and only BT2,000 imported. Foreign currency should be exchanged only by authorised banks and dealers.
Customs: Personal effects are allowed duty-free. Narcotic drugs, pornographic material, imports from South Africa, firearms and certain luxury goods are forbidden. Importation of many goods produced locally may be restricted. Export of images or statues of Buddha is only allowed with a certificate from the Department of Fine Arts. Antiques and archaeologically valuable items may only be exported with a certificate from the Department of Fine Arts. Articles exceeding B10,000 in value require a Certificate of Exportation.
Income tax: Travellers entering on a Business Visa must produce a Tax Clearance Certificate from the Revenue Department on departure.

Health precautions
Mandatory: Vaccination certificates for yellow fever or cholera if travelling from an infected area.
Advisable: Vaccinations for typhoid, paratyphoid and cholera and anti-malaria precautions. Water should be boiled and filtered before drinking. Take anti-rabies precautions if bitten by an animal.

Representation overseas: Ankara, Baghdad, Beijing, Belgrade, Berne, Bonn, Brasilia, Brussels, Buenos Aires, Cairo, Canberra, Colombo, Copenhagen, Dakar, The Hague, Hanoi, Islamabad, Jakarta, Jeddah, Kathmandu, Kuala Lumpur, Kuwait City, Lagos, London, Madrid, Manila, Mexico City, Moscow, Nairobi, New Delhi, Ottawa, Paris, Rangoon, Rome, Seoul, Singapore, Stockholm, Tehran, Tokyo, Vienna, Vientiane, Warsaw, Washington, Wellington.

Air access
National airline: Thai Airways International.
Other airlines: Aeroflot, Air France, Air India, Air Lanka, Alia, Alitalia, All Nippon Airways, Bangladesh Biman, British Airways, Burma Airways, CAAC, Canadian, Cathay Pacific, China, CSA, Dragonair, Egyptair, Finnair, Garuda, Gulf Air, Hang Khong Vietnam, Indian, Interflug, Iraqi Airways, JAL, JAT, KLM, Korean Air, Kuwait Airways, Lao Aviation, Lauda Air, LOT-Polish, Lufthansa, MAS, Northwest, Olympic, Philippine, PIA, Qantas, Royal Brunei, Royal Jordanian, Royal Nepal, Sabena, SAS, Saudia, SIA, Singapore, Swissair, Tarom, THY-Turkish, United, Tradewinds Pte Ltd, UTA.
Tax: Airport taxes are levied on departure.
International airports: Don Muang (BKK), 25 km from Bangkok; duty-free shop, bar, restaurant, buffet, bank, post office, hotel reservations; Chiang Mai International (CNX); and Phuket International (HKT) (being upgraded to international standard). Hat Yai International is now open. Photography at airports is forbidden.

Surface access: There are three weekly rail services between Singapore and Bangkok via Kuala Lumpur, Ipoh, Butterworth and Haadyai (journey time 48 hrs). Road access is possible from Malaysia.
Main ports: Bangkok, Sattahip.

Hotels: Most top hotels have good facilities for meetings and can arrange secretarial services if notified in advance. A 10 per cent service charge and 11 per cent tax are added to hotel bills, and it is customary to give small tips for good service. Book well ahead during summer months and Christmas/New Year.

the first half of 1989 alone reached more than $1.5bn. About a quarter of this was estimated to have been invested on the Thai Stock Exchange, helped by the creation of a number of foreign mutual funds investing in Thailand in the previous two years and by registration of Bangkok's stock market as an authorised exchange in Japan and the UK. The influx left a comfortable balance of payments surplus of $1.6bn in 1988.

Strained infrastructure

The main immediate constraint expected to brake Thailand's development was the strain on infrastructure. The capital, Bangkok, epitomised many of the problems. Traffic was approaching a standstill and a rush of high rise construction threatened to make it worse; international telecommunications lagged far behind the standard of other ASEAN countries like Malaysia and Singapore; and Bangkok port, Thailand's only major commercial port, continued to be plagued by congestion. Bangkok's problems only emphasised the need for better road and rail links particularly between the eastern seaboard – one of the main areas of industrial growth – and other parts of the country, which would bypass the capital.

There was now a huge crop of infrastructure projects pending that bankers said would need investment of more than $30bn over the coming six years, much of it in the first three. They included a $1.5bn expressway serving Bangkok, three other stretches of highway priced at some $950m, an overhead transport system for Bangkok that would cost more than $1.6bn, electricity generating schemes costing close to $4bn, two new ports at Laem Chabang and Mab Ta Pud on the eastern seaboard and substantial investment in telephones and transport. The problem was how they would be financed.

Fiscal policy

One of the benefits of Thailand's boom was an abrupt turnaround in the state of government finances, burdened throughout much of the 1980s with runaway deficits. Revenues in 1988 jumped 27 per cent to $9.7bn and for the first time in 10 years gave the government a small surplus over expenditures of $9.6bn, which included some $900m in repayment of loan principal. The improvement allowed the government to draw up a budget for 1990 that would raise overall spending by 17 per cent to some $13bn. It proposed to raise investment spending more than 50 per cent to account for some $3bn of the total.

The government, however, remained committed to a ceiling on new foreign borrowing of $1bn to $1.2bn a year which forced it to work out other methods of bringing in the additional capital needed to bridge a hefty savings-investment gap. As a start the finance ministry in mid-1989 raised the ceiling on deposit rates and offered tax incentives to small depositors and companies setting aside capital for reinvestment. It also embarked on negotiations to raise the number of foreign bank branches, from 14, in the hope that they would also attract foreign funds and help develop the local capital market.

Foreign banks had been queueing for years to open in Thailand but negotiations with a group of nine Japanese, European, Canadian and Australian banks appeared to run into difficulty over conditions of entry. In addition to bringing substantial capital with them, Thai authorities reportedly wanted each to take over one from a collection of 11 finance companies saved from failure in the mid-1980s by a costly government bail-out but still in difficulty. Some of the targeted foreign banks appeared distinctly uninterested.

The scheme was not viewed with much enthusiasm by Thai bankers either. An expanded foreign bank presence had been under consideration for some years but there was uneasiness over the conditions on which they would be allowed to compete. Thai bankers were also quick to warn that an influx of foreign institutions would aggravate a problem of limited availability of qualified Thai managers and staff.

Privatisation

Another cornerstone of the government's strategy to get round financial constraints was a programme of privatising state enterprises. The 61 state enterprises would need investment of some $15bn between 1989 and 1993 on top of what the government would be spending on utilities and infrastructure, Finance Minister Pramual explained. The intention was not so much to sell off these enterprises as to augment their capital by offering shares to the public.

After years of government talk, the finance ministry's plans appeared to single out the three biggest and most profitable enterprises, the national airline Thai International, the Electricity Generating Authority of Thailand and the Telephone Organisation of Thailand for some form of privatisation. It seemed that TOT would be listed on the stock exchange, THAI partially so, while EGAT would simply subcontract some 30bn bahts worth ($1.2bn) of planned projects to the private sector.

By mid-1989, however, little had happened. The managements of THAI and EGAT in particular proved hostile to such measures. The government response in EGAT's case was the peremptory dismissal of its board of directors, sparking a strike by angry EGAT workers. In the end the new board announced by the government was withdrawn and a compromise team assembled. THAI's privatisation, which had supporters and critics within the cabinet, promised to be more difficult to resolve.

Outlook

In the longer term it seemed that physical infrastructure and finance might not be the only constraints on development. Thailand's ambition was to move progressively into higher technology industries en route to achieving NIC status. But to achieve this, Thai planners warned government that it should attempt to bring about a more equitable regional distribution of investment, check the ever-widening disparities in wealth and develop human resources to meet the demand for skills. Enrolment in secondary education is among the lowest of all the ASEAN countries. Unemployment among graduates was a perennial problem yet industry remained short of engineers and middle management and banks complained about the fierce competition for qualified Thai executives. Its problem, in short, was the perhaps enviable one of keeping up with the opportunities for development.

BUSINESS GUIDE

Credit cards: Major credit cards, such as Access, American Express, Diners Club, Visa and Mastercharge, are accepted by main hotels and shops frequented by travellers.

Car hire: Self-drive and chauffeur-driven car hire is available in Bangkok, Pattaya, Hat Yai, Phuket and Chiang Mai. An international driving licence is required and driving is on the left-hand side of the road.

City transport
Taxis: Taxis have yellow number plates and, although they are metered, fares should be agreed in advance. Taxi drivers rarely understand English and it is best to have the name and address of one's destination written in Thai to show to the driver. Air-conditioned limousine services provided by main hotels are more expensive than ordinary taxis. Tipping is not usual.

National transport: Considerable investment has gone into improving the country's transport facilities. In remote areas, however, conditions are still uncertain, and banditry occurs in the north-west of the country.
Air: Thai Airways and Bangkok Airways operate domestic services to main centres.
Road: There are 45,838 km of national and provincial roads, with another 106,000 of rural roads and many modern highways. Major roads are generally metalled.
Buses: Long-distance (air-conditioned) express coaches operate between main centres; local services are not generally recommended.
Rail: The 4,440 km railway network links all main centres except Phuket, with three classes of coach on express routes. A new 134 km link has recently been added to connect centres on the Eastern Seaboard. Rail services are generally recommended: the system is equipped with modern rolling stock, including air-conditioned coaches, sleeping accommodation and restaurant cars on main express services.
Water: There are 1,110-1,600 km of navigable inland waterway, depending on the season. Various types of ferries and passenger/cargo boats operate on rivers and in coastal areas.

Public holidays
Fixed dates: 1 Jan, 6 Apr (Chakri Day), 13-15 Apr (Songkran – Water Festival), 1 May (Labour Day), 5 May (Coronation Day), 12 Aug (Queen's Birthday), 23 Oct (Chulalongkorn Day), 5 Dec (King's Birthday), 10 Dec (Constitution Day), 31 Dec.
Variable dates: Makha Bucha (Feb/Mar), Ploughing Ceremony (May), Visakha Bucha Day (Apr), Asalaha Bucha Day (Jul), Khao Phansa (day following Asalaha Bucha), Loi Krathong (Festival of Lights, Nov).

Working hours
Business: (Mon-Fri) 0830 – 1700. (Sat) 0830 – 1200.
Government: (Mon-Fri) 0830 – 1630.
Banking: (Mon-Fri) 0830 – 1530.

Social customs: Dress should be modest for visiting temples and royal palaces. Remove shoes before entering temples, Thai homes or mosques. It is considered rude to point your foot at someone. Public displays of affection between men and women are frowned upon.

Telecommunications
Telephone: Dialling code for Thailand, IDD access code + 66 followed by area code (2 for Bangkok) followed by subscriber's number. Telephone numbers throughout Bangkok are being changed as the system is revamped. A directory enquiry service is available. It is advisable to ask hotel operators to dial required numbers when calls are difficult to obtain. Public international telephone facilities are available at the Central General post Office. It was hoped to double the number of telephones to 1m between 1985 and 1988. Submarine cable links with Singapore and Malaysia.
Telex: Telex services are available at the Central GPO, and at hotels (at extra cost).
Fax: Facilities are now operated by the Communications Authority of Thailand.
Telegram: Cable and telegram services are available at major hotels, any telegraph office or the Central GPO.

BUSINESS GUIDE

Banking
Central bank: Bank of Thailand.
Main centre: Bangkok.
Government-owned banks: Bank of Agricultureand Agricultural Co-operatives, Bank of Thailand, Government Housing Bank, Government Savings Bank, Industrial Finance Corp of Thailand, Krung Thai Bank, Small Industries Finance Office, Thai Development Bank.
Head offices: Asia Trust Bank, Bangkok Bank, Bank of Commerce, Bangkok First Investment & Trust, Bangkok First Tokai Co, Bangkok Metropolitan Bank, Bank of Asia, Bank of Ayudhya, First Bangkok City Bank, Laem Thong Bank, Siam City Bank, Siam Commercial Bank, Thai Danu Bank, Thai Farmers Bank, Thai Military Bank, Union Bank of Bangkok, Wang Lee Bank.
Thai branches: Bank of America, Bank of California, Bank of Canton, Bank of Tokyo, Banque de l'Indochine et de Suez, Banque Nationale de Paris, Bharat Overseas Bank, Chartered Bank, Chase Manhattan Bank, Continental Illinois National Bank & Trust Co of Chicago, European Asian Bank, Export-Import Bank of Japan, Four Seas Communications Bank, Hongkong and Shanghai Banking Corporation, International Commercial Bank of China, Mercantile Bank, Mitsui Bank, United Malayan Banking Corporation.

Advertising: Advertising is available in the press, on commercial radio and TV, in cinemas and outdoors. Static and mobile loudspeakers are widely used. Tobacco advertising is banned, and TV advertising of spirits is permitted only after 2200 hours.

Trade fairs: International Construction and Building Machinery, Equipment and Material and Electrical Technology Exhibition (Jan), International Exhibition on Fish Farming, Fishery and Refrigeration Technology and Equipment (Mar), 4th Asian International Food Processing and Packaging Technology Show and Conference (Sep).

Electricity supply: 220 V AC, 50 cycles for domestic use, with plug fittings having two round or flat pins and both bayonet and screw-type light sockets; 220 V and 380 V AC three-phase for industrial use.

Weights and measures: Metric system (local units also in use).

Representation in capital: Argentina, Australia, Austria, Bangladesh, Belgium, Brazil, Brunei, Bulgaria, Burma, Canada, Chile, China PR, Colombia, Czechoslovakia, Denmark, Dominican Republic, Egypt, Finland, France, German FR, Hungary, Greece, Iceland, India, Indonesia, Iran, Iraq, Ireland, Israel, Italy, Japan, Jordan, Republic of Korea, Laos, Malaysia, Nepal, Netherlands, New Zealand, Norway, Pakistan, Philippines, Poland, Portugal, Romania, Saudi Arabia, Senegal, Singapore, Spain, Sri Lanka, Sweden, Switzerland, Turkey, UK, USA, USSR, Vatican, Vietnam, Yugoslavia.

BUSINESS DIRECTORY

Useful telephone numbers
Directory (Bangkok): 13
Directory (provinces): 183
International calls: 100

Hotels: Bangkok
Airport Hotel, 333 Choe Thanon Wudthakas, Don Muang Airport, 10210 (tel: 5661020; tx: 87424; fax: 5661941).

Ambassador, 171 Thanon Sukhumvit (tel: 2540444/2550444; tx: 82910; fax: 2534123).

Asia, 296 Thanon Phya Thai, 10400 (tel: 2150808; tx: 81177; fax: 2154360).

Bangkok Palace, 1091-336 Thanon Petchburi Tadmai, 10400 (tel: 2530550; tx: 84278/80; fax: 2530056).

Dusit Thani, Thanon Phra Ram 6, Silom, 10500 (tel: 2331130; tx: 81170, 81027; fax: 2366400).

Erawan Hotel, 494 Thanon Rajadamri, 10500 (tel: 2529100; tx: 82189).

First, 2 Thanon Petchburi, P.O. Box: 1144 (tel: 2525011; tx: 82027).

Hilton International, Nai Lert Park, 2 Thanon Witthayu, 10500 (tel: 2530123; tx: 72206; fax: 2536509).

Hyatt Central Plaza, 1695 Thanon Phaholyothin, Bangkaen, 10900, P.O. Box: 10-182 (tel: 5411234; tx: 20173; fax: 5411087).

Impala, 9 Soi 24, Thanon Sukhumvit, 10110 (tel: 2590053; tx: 84056, 21146; fax: 2588747).

Imperial, Thanon Witthayu, 10500 (tel: 2540111; tx: 84418).

Indra Regent, Thanon Rajaprarob (tel: 2511111; tx: 82723).

Mandarin, 662 Thanon Phra Ram 4, Samyan 5, 10500 (tel: 2334980; tx: 87689, 84131).

Manhattan, 15 Thanon Sukhumvit, 11 (tel: 2527141; tx: 87272).

Menam, 2074 Thanon Jaroen Krung, Yannawa, 10120 (tel: 2890352/3; tx: 21098/87423; fax: 2911048).

Montien, 54 Thanon Surawongse, P.O. Box: 1098,10500 (tel: 2348060; tx: 81160; fax: 2365219).

Narai, 222 Thanon Silom, P.O. Box: 908 (tel: 2333350; tx: 81175).

New Peninsular, 295/3 Thanon Surawongse, 10500 (tel: 2343910/6; tx: 84079; fax: 2365526).

Oriental, 48 Oriental Ave, 10500 (tel: 2360400, 2348621/9; tx: 82997; fax: 622236.).

President, 135/26 Thanon Gaysorn, 10500 (tel: 2520444; tx: 81194; fax: 2537565).

Rajah Hotel, 18 Thanon Sukhumvit, Soi 4 (Soi Nana Tai) (tel: 2525102/9; tx: 87385).

Rama Gardens, 9/9 Thanon Vibhavadi Rangsit, Bangkaen, 10210 (tel: 5795400; tx: 84250; fax: 5611025).

Regent, 155 Thanon Rajadamri, 10500 (tel: 2516127; tx: 20004; fax: 2539195).

Royal Orchid Sheraton, 2 Captain Bush Lane, Siphya Road, 10500 (tel: 2345599; tx: 84491; fax: 2368320).

Shangri-La,, 89 Soi Wat Suan Plu, Thanon Jaroenkrung 10500 (tel: 2367777; tx: 84265; fax: 2368579).

Siam Inter-Continental, Sarapatum Palace Property, Thanon Phra Ram 1 (tel: 2530355/7; tx: 81155; fax: 2532275).

Tawana Ramada, 80 Thanon Surawongse, 10500, P.O. Box: 1742 (tel: 2360361; tx: 81167; fax: 2363738).

Windsor, Soi 20 Thanon Sukhumvit (tel: 2580160/5; tx: 82081).

Hotels: Chiangmai
Orchid Chiang Mai, 100-102 Thanon Huay Kaew (tel: 222099; tx: 4337; fax: 221625).

Rincome, 301 Thanon Huey Kaew, 50000, P.O. Box: 103 (tel: 221044; tx: 49314; fax: 221915).

Hotels: Pattaya, Cholburi
Asia, Pattaya Beach (tel: 418602/6; tx: 85902; fax: 423496).

Grand Palace, No 1 Sunset Strip, Pattaya Beach (tel: 419901/3; tx: 85917).

Merlin Resort, Pattaya Beach (tel: 418755/9; tx: 85905).

Montien, Pattaya Beach, 20260 (tel: 428155/6; tx: 85906; fax: 423155).

Novotel Tropicana, 45 North Pattaya Road (tel: 8645/8; tx: 85910).

Palace, Beach Road (tel: 428319; tx: 75904).

Royal Cliff Beach, Pattaya Beach (tel: 418513/5; tx: 85907; fax: 428511).

Siam Bayshore, Pattaya Beach (tel: 428677/81; tx: 20119; fax: 2253173).

Siam Bayview, Beach Road (tel: 428728/9; tx: 82820; fax: 2253173).

Wong Amat, Pattaya Beach (tel: 422118/19/20; tx: 85908; fax: 428599).

Car hire
Avis: 10/1 N Sathorn Road (Downtown) (tel: 2330397; tx: 72118 TH).

Hertz: 987 Ploenchit Road, Bangkok 10500 (tel: 25362514; tx: 87190; cable: HERTZ).

King Car Rent: 18/1 Sathorn Nua Road, Bangrak, Bangkok (tel: 2334514).

Business advice and information
American Chamber of Commerce, 7/F Kian-Nguan Building, 140 Thanon Withayn, Bangkok (tel: 2519266, 2511605; tx: 82778; cable: AMCHAM).

Association of Thai Industries, 294/14 Thanon Samsen, Tambol Dusit, Bangkok.

Board of Investment, Thanon Vibhavadee Rangsit (tel: 2701400).

Board of Trade of Thailand, 150 Thanon Rajabophit, Bangkok (tel: 2219350; tx: 84309).

British Chamber of Commerce, Room 206, Bangkok Insurance Building, 302 Thanon Silom (tel: 2341140/69).

Economic and Social Commission for Asia and the Pacific (ESCAP), United Nations Building, Bangkok (tel: 2829608).

Export Service Centre, Department of Commercial Relations, Ministry of Commerce, 22-77 Thanon Rachadaphisek-Ladprao, Bangkok 10900 (tel: 5131905).

General Post Office, 1160 Thanon Jaroenkrung, Bangkok 10501 (tel: 2331050).

Port Authority of Thailand, Thanon Sunthomkosa, (tel: 2490362).

Thai Chamber of Commerce, 150 Thanon Rajabophit, Phra Nakhom, Bangkok (tel: 2216532/4; tx: 72093).

Banking: Bank of Thailand (central bank), 273 Bang Khunprom, Bangkok (tel: 2823322; tx: 72012/3).

Travel information
Royal Automobile Association of Thailand, 151, Soi Aphaisongkram, Phaholyothin, 10900, Bangkok (tel: 5112230/1).

Tourism Authority of Thailand, Thanon Ratchadamnoen Nok, Bangkok (tel: 2821143/7; tx: 72059 TAT BKKTH).

Thai Airways International Ltd, 6 Thanon Larn Luang, Bangkok (tel: 2800090, reservations 2800700) (internal travel).

Thai Airways International Ltd., 89 Vibhavadi Rangsit, Bangkok (tel: 5130121, reservations 2333810; tx: 82359 Thainter TH).

Togo

Kaye Whiteman

The dominance of President Gnassingbé Eyadéma over Togolese political life became more total than ever. The fall of the once powerful Justice Minister Kpotivi Lacle confirmed this. Meanwhile, after a lean period in which precepts of structural adjustment were fairly rigorously applied, the economy appeared to be modestly on the up-turn.

After a difficult patch in the mid-1980s, when he faced a domestic security problem which culminated in the 'invasion' of September 1986, President Eyadéma has in the late 1980s seemed to be in total control. The fall of Thomas Sankara in Burkina Faso in 1987, and policy changes in Ghana and Benin, meant that the Togo president no longer felt that he was surrounded by hostile regimes. Opposition in exile was muted, and any evidence of internal dissent was non-existent. Thus Eyadéma was increasingly able to make longer journeys overseas: in the first part of 1989 he not only attended the funeral of Emperor Hirohito in Japan in February, the francophone summit in Dakar in May, and the celebrations of the Bicentenary of the French Revolution in Paris in July, but also made a longer visit to both Japan and China in April. The president was also active in the affairs of the Economic Community of West African States (ECOWAS), where he is seen as a 'founding father'.

Domestic politics

If Togo has any domestic politics other than the activities of the ruling single-party, the Rassemblement du Peuple Togolais (RPT), it is the game of ministerial snakes and ladders, especially for southern politicians, since all depends on the favour of a northern president. The nearer to the top, and the more apparently powerful, the more precarious is likely to be the tenure. Thus it was with Kpotivi Djidjogbe Lacle, who for years, as minister of the interior, was held to be the second most powerful man in the country and number two of the regime, and who in 1989 came to grief. First signs of his disfavour came when he was shuffled to the ministry of justice in 1987. Then, in March 1989, he was summarily dismissed. The reason, it was indicated, was because charges were pending against him, and it was necessary to lift his ministerial immunity, so that he could go to court like an 'ordinary citizen'. The case against him soon became clear when an action was brought against him by the public prosecutor involving falsification of tickets for the national lottery to the value of CFA15.5m (about £288,000). The former minister's brother and several officials were also implicated. Lacle was released from custody (after a court appearance in which he was booed by a hostile crowd) to await prosecution.

Subsequently a demonstration was organised in the middle of Lomé by the Togolese Trade Union Confederation (CNTT), in support of what was thought to be a crackdown on corruption in public life. Further evidence of this came in June 1989, when Head of Security and Commander of the Togo Armed Forces Lt. Col. Seyi Memene was placed under arrest having been implicated in defrauding the army in a series of contracts. The scandal first became public in an independent weekly newspaper in neighbouring Benin, the *Tam-Tam Express*, which reported in June that President Eyadéma had disclosed the fraud at a special meeting of senior officers on 6 May 1989.

Economy

After the difficulties of 1987 in which growth figures were very disappointing, the first figures for 1988 seemed more promising. The estimated growth of GDP per capita of 3.5 per cent in real terms (the same figure was 1.4 per cent in 1987), was put down in part to improved world prices for phosphate rock, as well as a spectacular increased production of almost 30 per cent in a year from the phosphate mines. The product from the agricultural sector also improved in money terms, although not in quantity of production. Likewise there was an increase of 14 per cent in manufacturing production, though this was from historically low levels that occurred during the transition to privatisation. Togo is one of the few African countries where the beginnings of a privatisation programme has got off the ground.

Buoyed, perhaps, by evidence of improvement, however slight, the 1989 budget showed a 5.5 per cent increase in expenditure by CFA2.800m to CFA92,500m, although tax receipts in the previous year were below target. The major increase was in the wage bill which was scheduled to rise by 12 per cent. Nonetheless the budget was said to reflect policies of 'austerity and rigour', in force since 1983. The budget was expected to be a balanced one, and after its approval, a delegation from the International Monetary Fund (IMF) said that Togo had taken courageous economic measures in conformity with IMF requirements.

International benevolence towards Togo was seen in June, when an important rescheduling of Togo's official debt took place at the Paris Club. It was stated that the favourable terms accorded were in accordance with the decisions of the G7 summit in Toronto in 1988. This follows an earlier rescheduling in 1988 at the same time as Togo received a new structural adjustment facility from the IMF. Continuing international approval also kept up aid flows, especially from France and the EC. In June 1989, for example, the latter accorded 16.6m Ecus for maintaining road infrastructure. Togo is notable for having had many of its roads built with aid from the European Development Fund.

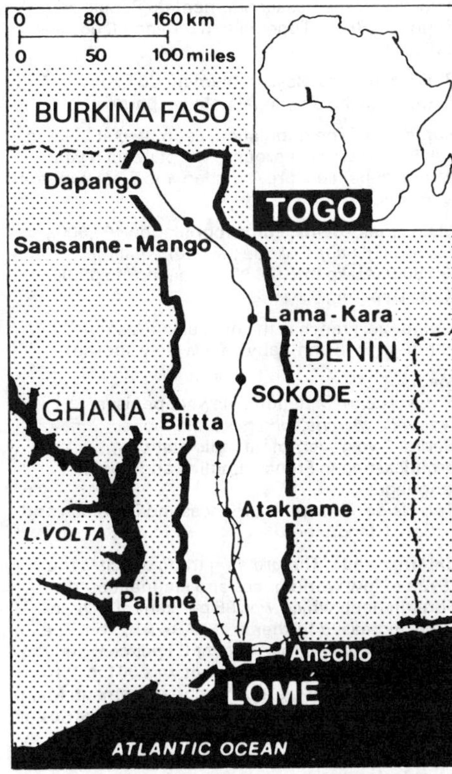

KEY FACTS

Official title: Republic of Togo
Head of state: President Gen. Gnassingbé Eyadèma
Ruling party: Rassemblement du Peuple Togolais
Capital: Lomé
Official Languages: French
Currency: Franc CFA = 100 centimes
Exchange rate: Franc CFA316 per $ (Nov 1989)
Area: 56,600 sq km
Population: 3.3m (1988)
GDP real growth: 1.4% (1987)
GNP per capita: $300 (1987)
Inflation: –0.1% (1988)

COUNTRY PROFILE

Political system: Executive power rests with the president, who appoints and leads a council of ministers. Legislative power is vested in the 77-member National Assembly. Candidates are selected by the Rassemblement du Peuple Togolais (RPT). Togo gained independence in 1960.

Political parties: Rassemblement du Peuple Togalais is the only legal political party.

Population: The population, estimated at 3.3m in 1987, is growing at around two per cent a year. About 25 per cent are Christian and 8 per cent are Muslims.

Main cities/towns: Lomé (population 400,000), Mango (76,800), Sokodé (36,000), Kpalimé (30,000), Atakpamé (26,000), Anécho (15,000), Kara, Dapaong.

Language: French is the official language, although Ewe and Kabyè are widely spoken.

Media
Press: The main national newspaper is *la Nouvelle Marche*, which is state owned.
Radio: Two government radio stations broadcasting in French, English and African languages.
Television: TV station broadcasting in French and national languages.

External trade: The growing trade deficit is the result of low world prices and stagnant output. Foreign debt, which was 95 per cent of GDP in 1986, increased further in 1987, to $1.27bn. The debt service ratio rose to over 33 per cent and the ratio of debt to current account was 301 in 1987.
Exports: Phosphates, cocoa, coffee, cement. Main destinations: Netherlands, France, EC countries in general, Côte d'Ivoire, Yugoslavia, Ghana.
Imports: Petroleum and petroleum products, machinery, construction materials, metals, transport equipment. Main sources: France, Netherlands, UK, Japan, German FR.

Agriculture: Sector is mainstay of economy, contributing 25 per cent to GDP in 1986 and employing over two-thirds of the workforce. Self-sufficiency in basic foodstuffs is generally attained. The majority of farmers are smallholders who raise stock and grow maize, millet, yams, cassava, sorghum and rice. The production of cocoa and coffee, the main export crops, is stagnating and the 1987 cotton crop was expected to be 5 per cent down on 1986.

Industry: Sector is small-scale but expanding, contributing around 10 per cent to GDP. Activity is centred on the processing of agricultural commodities and production of phosphoric acid, fertilisers and cement. Also beverages, footwear, textiles and plastics. The service sector, which focuses on transport and tourism, generates about 50 per cent of GDP.

Mining: Activity is centred on phosphate mining, which contributes nine per cent to GDP and is the pricipal export earner. Reserves estimated at over 100 m tonnes, mainly located on Lake Togo. Phosphate production has steadily fallen from 2.9m tonnes in 1979 to around 2.4m tonnes in 1986 and 2.1m tonnes in 1987. Proceeds from phosphate exports are down by 20 per cent. Limestone and marble are quarried, and there are known reserves of iron-ore, bauxite, dolomite and chromite.

Energy: Heavy dependence on imported fuel, although exploration is in progress for oil. Crude petroleum from Nigeria is refined locally. The joint hydroelectric power project with Benin has been in full operation since 1988.

Membership of international organisations: African Development Bank, African States in Association with the EC (Lomé Convention), Arab Bank for Economic Development in Africa (recipient), Economic Commission for Africa, Economic Community of West African States, FAO, French Monetary Area, GATT, IDA, IFAD, IFC, ILO, IMF, International Cocoa Organisation, OAU, UN, WHO, WIPO, World Bank.

BUSINESS GUIDE

Time: GMT.

Climate: Tropical Mean annual temperature 28°C. Dryer in the north. Two rainy seasons between Apr-Jun and Sep-Oct.

Entry requirements
Passport: Required by all except nationals of France and certain (mainly African) countries. NB: Joint passports not accepted.
Visa: Required by all except nationals of most West European countries (up to 90 days) USA and all Commonwealth countries. Multiple entry visas and extensions can be obtained within the country.
Prohibited entry: Nationals of South Africa.
Currency: No restriction on import of foreign currency. However export of foreign currency must not exceed amount declared on arrival. No more than franc CFA 1m may be taken in and no more than franc CFA 25,000 exported. Major credit cards accepted.

Health precautions: Vaccination certificate for yellow fever required if arriving from infected area; anti-malarials advisable. Drink bottled water.

Representation overseas: Accra, Beijing, Bonn, Brussels, Harare, Havana, Kinshasa, Lagos, Libreville, London, Moscow, Ottawa, Paris, Tel-Aviv, Tripoli, Washington.

Air access
National airline: Air Togo.
Other airlines: Aeroflot, Air Afrique, Air Burkina, Air Gabon, Air Zaïre, Ethiopian Airways, KLM, Sabena, Nigeria Airways, UTA.
International airport: Lomé (Code: LFW), 4 km from city; duty-free shop, bar, restaurant, buffet, bank, post office, shops, car hire.

Surface access
Main ports: Lomé.
Road: Well surfaced coastal road connecting Accra (Ghana) and Lagos (Nigeria) runs through Lomé. Also road access from Cotonou (Benin) and Ouaga Fougou (Burkina Faso).

Hotels: High standard hotels in Lomé, which should be booked well in advance. 10 per cent tip is usual.

Electricity supply: 220 V AC, 50 cycles.

Car hire: International driving licence required.

City transport
Taxis: Readily available in Lomé, while shared taxis ply to other main towns; tipping is not usual.

National transport
Air: Air Togo flies between Lomé, Sokodé, Mango, Lama-Kara, Niamtougou and Dapango.
Road: Main surfaced roads run to the borders of neighbouring countries; other roads may not be passable in the rainy season. Main route north is called 'Highway of Unity'.
Rail: Main lines run from Lomé to Blitta (midway between Atakpamé and Sokodé), and to Kpalimé.
Sea: Coastal ferries visit southern coastal ports.

Public holidays
Fixed dates: 1 Jan, 13 Jan (National Liberation Day), 24 Jan (Sarakawa Day), 24 Apr (Victory Day), 27 Apr (Independence), 1 May (Labour Day), 21 Jun, 15 Aug (Assumption), 24 Sep (Anniversary of failed attack on Lomé), 1 Nov (All Saints'), 25 Dec.
Variable dates: Ascension Day, Whit Monday, Eid al-Adha (Tabaski).

Working hours
Government and business: (Mon-Fri) 0700 – 1200, 1430 – 1730.
Banking: (Mon-Fri) 0730 – 1130, 1430 – 1600.
Shops: (Mon-Fri) 0800 – 1200, 1430 – 1730; (Sat) 0730 –1230.

Telecommunications
Telephone: Dialling code for Togo: IDD access code + 228 followed by subscriber's number. There are no area codes. International calls effective. Phone connections are best in Lomé, more difficult elsewhere.
Telex: Public facilities available. Messages are cheaper from main Post Office in Lomé, but no Sunday service.
Telegram: Reliable international service. Calls to France and West African countries are cheaper.

Banking
Central bank: Banque Centrale des Etats de l'Afrique de l'Ouest.
Main centre: Lomé.
Head offices: Banque Arabe Libyenne Togolaise du Commerce Extérieur, Union Togolaise de Banque, Banque Togolaise pour le Commerce et l'Industrie, Banque Internationale pour l'Afrique Occidentale, Bank of Credit and Commerce International, Ghana Commercial Bank, Banque Togolaise de Développement, Caisse Nationale de Crédit Agricole.

Representation in capital: Belgium, Brazil, China PR, Egypt, France, Gabon, German FR, Ghana, Israel, Korea PDR, Libya, Nigeria, Tunisia, USA, USSR, Zaïre.

Hotels: Lomé
2 Fevrier Sofitel, P.O. Box: 131 (tel: 21-00-03; tx: 5347; fax: 216266).

De la Paix Pullman, Route d'Aného, P.O. Box: 3452 (tel: 21-52-97; tx: 5252; fax: 212302).

Le Benin, Avenue Gen de Gaulle, P.O. Box: 128 (tel: 24-85; tx: 5264).

Miramar, P.O. Box: 502 (tel: 21-61-41).

PLM Azur Sarakawa, P.O. Box: 2232 (tel: 21-65-90; tx: 5354; fax: 217180).

Tropicana, P.O. Box: 2724 (tel: 21-34-04; tx: 5269).

Car hire: Lomé
Budget: PO Box 1542 (tel: 21-79-31; tx: 5244).

Hertz: Lomé airport (tel: 21-50-52; tx: 5208).

Other useful addresses
Banque Centrale des Etats de l'Afrique de l'Ouest, Avenue de Sarakawa, PO Box 120, Lomé.

Chamber of Commerce, Agriculture & Industry of Togo, BP 360, Lomé (tel: 21-20-65).

Direction de la Statistique, PO Box 118, Lomé (tel: 27-06-62).

Direction des Professions Touristiques, BP 1289, Lomé (tel: 21-56-62, 21-43-13; tx: 5007).

National Tourist Board, BP 1177, Lomé.

Office des Produits Agricoles du Togo, BP 1334, Lomé (agency dealing with marketing, export, development) (tel: 21-44-71; tx: 5220).

Tokelau

(map on page xxiv)

KEY FACTS

Official title: Tokelau

Head of state: Queen Elizabeth II

Head of government: Administrator, H.H. Francis

Capital: Each atoll has its own administrative centre

Official Languages: Tokelauan (English also spoken)

Currency: New Zealand Dollar (NZ$ = 100 cents); also W Samoan Tala

Exchange rate: NZ$1.70 per $ (Nov 1989)

Area: 12 sq km (3 coral atolls: Nukunonu, Fakaofo, Atafu)

Population: 1,703 (1985)

COUNTRY PROFILE

Political system: Tokelau is a non-self-governing territory of New Zealand. It is administered by a Wellington-based Administrator of Tokelau, appointed by the New Zealand Ministry of Foreign Affairs, most of whose powers are delegated to the Official Secretary who heads the Office for Tokelau Affairs in Apia, Western Samoa. Each atoll has an Administration Officer, who acts in a supervisory capacity over public servants, and is governed by a Council of Elders, variously composed of family heads, elders and elected officals. *Faipule* (headmen) elected on each atoll to a three-year term administer the law, preside over the court and represent the atoll in its dealings with the Administrator. The three *Faipule* act as joint chairmen of the General *Fono*, which draws up an annual budget, and travel overseas to represent Tokelau. Each atoll also elects a *pulenuku* (village mayor) responsible for maintenance of order, sanitation, cleanliness, water supplies and the inspection of plantations.

Political parties: There are no organised political parties.

Population: The 1,703 residents (1985) are mainly Polynesians, with close links to Western Samoa. Practically all are Christian.

Main cities/towns: Atafu village (population 589 in 1986), Nukunonu village (pop 416) and Fakaofo village (pop 622).

Language: Tokelauan is the offical language, but English is also spoken.

Media

Broadcasting: Radio stations on each atoll broadcast traffic and weather reports.

Domestic economy: The economy is based on communal subsistence agriculture and fishing. The atolls' size, isolation and lack of land-based resources allow little scope for economic development. Copra, postage stamps, souvenir coins, handicrafts and remittances from migrant workers are the principal sources of foreign exchange. Grants from New Zealand account for about 80 per cent of expenditure. Funding from the New Zealand bilateral aid programme, the UN Development Programme (UNDP), the South Pacific Commission, the ILO and other international agencies has been the main source of development assistance. Since 1982 the General *Fono* has collected a tax on the salaries of public servants unavailable for communal service (called the Community Services Levy) in order to subsidise copra and handicrafts producers, provide honoraria to members of island councils and supplement village projects. Tokelau joined with Tuvalu, Western Samoa and Niue in negotiations with the American Tuna Boat Association for a joint access agreement to the four countries' exclusive economic zones.

Employment: Jobs created by public works and agriculture programmes are rotated among workers so as to distribute the income more widely throughout the community.

Agriculture: The soil is thin and infertile, and the land does not rise more than five metres above sea level. Rainfall is erratic, and crops are subject to drought and storm damage. Main subsistence crops are coconut and breadfruit, supplemented by pulaka, ta'amu, pandanus, bananas, pawpaw, with experimental crops of cucumbers, tomatoes, beans, cabbage and water melon. A coconut replanting project commenced in 1980 continues and copra production subsidies have recently resulted in increased exports. Livestock raised now includes pigs, poultry and goats. Fishing for tuna, bonito, trevally and mullet supplies a main source of protein, although Apia-built fishing boats supplied by UNDP now need major repair.

Industry: Copra production, woodwork and manufacture of woven and plaited goods such as hats, mats, bags and fans are the only industries of significance.

Legal system: The High Court of Niue has civil and criminal jurisdiction under the Tokelan Amendment Act of 1970, and the High Court of New Zealand has concurrent jurisdiction in some cases, as well as original and appellate jurisdiction. There is provision for a Tokelauan Commissioner for each of the three atolls (currently the *Faipule*), with jurisdiction in limited civil proceedings and criminal offences. The legislative and judicial systems are based on the Tokelau Act of 1948 and its amendments, under the provisions of which a variety of British colonial regulations continue in force, and work is under way to compile and translate a set of laws applicable in Tokelau. Unless expressly extended, New Zealand statute law does not apply in Tokelau.

Membership of international organisations: South Pacific Commission (Conference representation).

Time: GMT -11 hrs; EDT -6 hrs.

Climate: The average mean temperature is 28°C; warmest in May and coolest in Jul. Rainfall is heavy but irregular. Severe tropical storms are rare, but possible.

Surface access: There are scheduled charter services from Apia, Western Samoa. Otherwise, there is no regular shipping service.

Main ports: Vessels anchor off the main settlements, as there are no harbour facilities, and purpose-built aluminium boats provide transport to concrete wharves and beaches, often losing substantial amounts of cargo in getting through the reef. A chartered vessel sails from Apia 10-12 times a year.

Hotels: No provisions for tourists.

National transport

Air: Airstrips are planned for each atoll but none are built yet.

Road: There are no roads, trucks, buses or cars.

Water: Official charter services between atolls are available, and Tokelauans are not permitted to make unscheduled inter-atoll voyages because the risk of missing landfall is too great. Dinghies with outboard motors provide the main means of local transport.

Telecommunications: Telecommunications services are being upgraded under a programme funded by the New Zealand Post Office, the UNDP and Tokelau's General *Fono*.

Telephone: Radio-telephone has provided the main international link.

Banking: The nearest commercial banking services are in Apia, Western Samoa, although savings facilities under the control of the administrative officer have been set up on each atoll.

Business advice and information

Office for Tokelau Affairs, PO Box 865, Apia, Western Samoa (tel: 20822/3; cable: TOKALANI APIA).

Tonga

T. R. Hughes

(map on page xxiv)

The ancient Polynesian kingdom, probably the most feudal of the Pacific nations, was slowly getting to grips with the modern age as the 1980s closed. But the crucial question was whether it could change fast enough politically as well as economically. The resource-scarce islands depended on injections of foreign aid and remittances from Tongans overseas. Unemployment was high. Agriculture was still the backbone of the country and the government recognised the need to diversify as speedily as possible. Though outwardly stable, Tonga and its seventy-two year old absolute monarch King Taufa'ahau Tupou IV faced pressing problems in a society where the divisions between the haves and the have-nots were becoming more clearly visible. A relatively small number of nobles in the 100,000-strong nation held vast tracts of the land and also had an equal voice with commoners in the legislative assembly. But the commoners wanted more political clout. In a muscle-flexing demonstration of their growing discontent, reformist MPs held up the government's budget in June 1989.

KEY FACTS

Official title: Kingdom of Tonga

Head of state: King Taufa'ahau Tupou IV

Head of government: Prince Fatafehi Tu'ipelehake

Capital: Nuku'alofa (on Tongatapu)

Official Languages: Tongan (English widely spoken)

Currency: Tongan Dollar or Pa'anga (T$ = 100 seniti)

Exchange rate: T$1.26 per $ (Nov 1989)

Area: 748 sq km (170 islands)

Population: 94,535 (1986) – Tongatapu: 63,614; Vava'u: 15,170

Trade balance: –$53.8m (1985/86)

Foreign debt: $33.3m (1986)

Visitor numbers: 39,550 (1987)

Foreign policy

Externally, the Pacific's last monarchy was adept at playing off superpower rivalries. Tonga's foreign policy, described by critics as sitting on the fence, was undoubtedly pragmatic. The king was in Tahiti in 1987 to seek French aid and Crown Prince Tupouto'a, the foreign minister, risked irritating friends in the region by making three trips to the Soviet Union looking for advanced technology and raw materials.

But the government placed Tonga firmly in the Western orbit — if, indeed, there was any real doubt — by signing a treaty of Amity, Commerce and Navigation with the United States in July 1988. The concord renewed a century-old treaty signed by President Cleveland and King George Tupou I. Apart from the mutual backslapping, the treaty meant that Tonga wanted to stay under the US nuclear umbrella.

Both states had refused to sign the South Pacific Nuclear Free Zone Treaty and the July pact guaranteed transit of nuclear-armed vessels in the Tonga archipelago. To criticism that Tonga had 'sold out', the government contended that Tongan foreign policy, albeit with a justifiable element of self-interest, was directed at trying to act as a stabilising influence in the Pacific.

Agriculture and industry

As a nation that was almost exclusively agricultural just over a decade before, Tonga in 1989 was making steady progress in a switch of emphasis into light industry. But agriculture still employed nearly half the workforce mainly in traditional plantation farming. Tonga's first farm census published in November 1988 disclosed that nearly two-thirds of the kingdom's 14.366 households were producing agricultural crops solely for their own use and just 10 per cent were raising crops primarily for sale. The land, said to be as fertile as anywhere, supports coconut, vanilla and bananas for exports and subsistence crops including tomatoes, watermelons, cassava, taro, citrus, breadfruit and capsicums.

The farm sector is vulnerable to low prices and oversupply, particularly as most of Tonga's produce can be matched by other Pacific island countries. New Zealand, once a big buyer of tomatoes and capsicums, now had its own hothouse industry. The government recognised that it needed to be more flexible in introducing new crops and modernising processing methods. Top priority was given to developing light industry and the government's hopes were reflected in the thriving Small Industries Centre in Nuku'alofa.

An incentive to prospective investors and businessmen was a 2m pa'anga programme, launched jointly by the government and the Tonga Development Bank, to encourage small businesses to take a positive role in the country's national development.

Announcing the plan in May 1989, the government admitted it had to rethink its policies to attract the right kind of industries —'otherwise we face problems', warned an official of the ministry of labour, commerce and industries. New training facilities were needed as the plentiful supply of trained labour of the past had dried up.

Vital to the plan's success was the need to create 'a will for both public and private involvement'. Concentrating on producing more industrial goods both for export and for meeting local demands — and so providing more jobs — was seen by the government as the best and perhaps the only way of tackling a potentially explosive situation that might arise when the rising population runs out of the agricultural land. Government planners were banking on industrial development outdistancing agriculture and livestock-raising in the early 1990s.

The need for business training at middle management level was also acute. This was made painfully evident when the king revealed in his speech from the throne at the opening of parliament in May 1989 that the Tonga Commodities Board had run into heavy debt. The board was rapped for 'trying to load its debts on to others'.

Development projects

For a nation steeped in tradition, King Taufa'ahau was in the vanguard of the world's leaders in encouraging high-technology as the answer to his country's problems. A Norwegian firm, Kvaerner Brug A\S, was contracted to build a wave-powered electricity generating plant that could conceivably set the pattern for other micro-states. The T$8m, 2mW plant on the rocky shoreline at Makeke, Tongatapu, was based on a British-designed turbine and, although the output would be small compared with conventional generating stations, the Norwegians hoped to prove that it could be made financially viable.

The king saw this as only one example of cost-cutting innovations. He was known to be keen to experi-

KEY INDICATORS

	Unit	1982	1983	1984	1985	1986
Population	'000	93.0	95.0*	96.4	97.3*	97.9*
Gross domestic product (GDP)	T$m	71.02	76.60	78.0*	80.0*	80.0
GDP per capita	T$	870	912	809	822	817
Exports	T$m	4.29	6.45	10.00	7.78	–
Coconut oil products	T$m	1.29	2.12	5.01	2.74	–
Copra	T$m	–	–	–	–	–
Desiccated coconut	T$m	0.24	0.46	0.76	0.53	–
Bananas	T$m	0.09	0.30	0.86	0.82	–
Vanilla beans	T$m	0.63	0.61	1.08	0.40	–
Imports	T$m	41.20	41.66	46.61	58.93	–
Balance of trade	$m	–28.0	–28.0	–30.1	51.15	–
Balance of payments	$m	3.0	9.0	5.0*	–2.0	–
Exchange rate	T$ per $	0.98	0.91	1.20	1.43	–

* estimated figure

ment with wind-powered electric generators and aquaculture farms producing inexpensive protein from shrimps and shellfish.

Tonga also looked for a useful income from the regional fisheries treaty with the US — it was the last of the Pacific Island states to sign in 1989. But tourism, still in its infancy, was regarded as the number two foreign exchange earner and the sector was being given a facelift to attract more visitors. The main runway at Fua'amotu international airport was being extended by another 600 metres at a cost of T$6m to take wide-bodied jets, together with a new terminal building, new roads and much-improved airport equipment. The main donors for these projects were Australia, Japan and the European Community.

The overall spending on development, financed by foreign aid and loans to government, rose in 1989-90 from T$41m to T$75m but the government decided not to embark on any new schemes before agreeing on the next 10-year plan. The budget earmarked some T$43m for recurrent expenditure, up 23 per cent on 1988-89. A key source of overseas income is the large number of expatriate Tongans living in New Zealand, Australia and the USA whose remittances run into millions of dollars. The Church of the Latter Day Saints has a strong following in Tonga, partly for the educational opportunities it offers members, and Tongans can be found in Salt Lake City or attending the Mormon university in Hawaii.

AIDS education

Nuku'alofa was the focal point of an intensive anti-AIDS campaign after one student returned home from the university and was diagnosed as the victim of AIDS, believed to have been contracted through a blood transfusion at a Honolulu hospital. Another similar victim died. The government's education campaign drew on MPs, church and village leaders to plug the message that chastity was more effective than condoms. So it was no wonder that Tongans, members of a strictly Victorian society, were stunned to learn in April 1989 that a sixteen year old former high school male student was the first to become infected from a contact within the kingdom.

COUNTRY PROFILE

Political system: Tonga is governed by a hereditary monarch who appoints and chairs a privy council consisting of seven life ministers, including an executive prime minister, and the governors of Ha'apai and Vava'u. A unicameral legislative council meets annually.

Political parties: There are no organised political parties.

Population: The population of around 100,000 is of Polynesian descent, and about 25,000 live in the capital Nuku'alofa. Most of the 170 islands are uninhabited. The Wesleyan church is the major religion.

Main cities/towns: Nuku'alofa (on Tongatapu, population 28,000 in 1986), Pangai (on Lifuka), Neiafu (on Vava'u).

Language: English is used in education, and for administrative purposes; otherwise, Tongan is usual.

Media
Press: The Tonga Chronicle and *Tonga Times* are published weekly in Tongan and English editions. *Matangi Tonga* and *Tonga Today* are bi-monthly general interest magazines. *What's On In Tonga*, a monthly tourist magazine available free at hotels, motels and major stores.
Radio: Tonga Broadcasting Commission administers radio services heard over a wide geographical area in Tongan and English, with much programming devoted to local interests, culture, etc. There are two radio stations: AM (Tongan) and FM (English).

Television: A private TV station, ASTL-TV3, makes limited broadcasts, and the Government of Fiji is studying proposals to set up a service serving Fiji, Western Samoa and Tonga via satellite from a ground station to be built in Fiji. There are also plans for an internal television service using equipment offered by two US church groups. Cable style, small-scale TV service ('Vision of Pilolevu') covers Nuku'alofa area.

Domestic economy: Tonga is a small, open economy dependent on export of agricultural products, especially coconut products, foreign aid and private remittances from abroad to offset its regular trade deficit. Official development aid is about 19 per cent of GNP at around $13.6m. The tourist industry is a major factor, and a 10-year tourist development plan aims to provide more hotel accommodation and reduce dependence on the cruise business. Total visitors numbered some 15,000 in 1986-87. Other areas of possible development include offshore oil, fish and vegetable canneries and coconut-based industries. Offshore banking has been established, and is growing rapidly.

External trade
Exports: Total exports were valued at $5m in 1987. Main exports are coconut oil products, vanilla beans, desiccated coconut and bananas. Main destinations are New Zealand, Australia, Fiji, East Samoa, USA and Sweden.
Imports: Total imports were valued at $44m in 1987. Main imports are food and live animals, manufactured goods classified by materials, fuels and lubricants, machinery and transport equipment, other manufactures, inedible raw materials and food and beverages. Main sources are New Zealand, Australia, Japan, the USA, Fiji, Spain and Singapore.

Agriculture: Agriculture accounts for 50 per cent of GDP (1987) and employs 47 per cent of the labour force. The soil is generally fertile, but production can suffer from hurricane damage. All land is held by the Crown, and every male Tongan is entitled to a smallholding (alienation of land is forbidden). Families raise their own livestock (pigs and poultry) and subsistence crops of manioc, yams, breadfruit, watermelons, tomatoes, cassava, oranges and capsicum. Coconut, vanilla and bananas are produced for export. Fishing provides additional food and occasional export revenue. A new $4m fisheries harbour with a fish market equipped with storage freezers is expected to boost fisheries sector. Old coconut tree trunks now fulfil up to 25 per cent of timber needs.

Industry: Efforts to develop industry have led to approval of over 100 projects since 1978. Annual industrial production is now about $10m p.a. Coconut products supply most export revenue: coconut oil, desiccated coconut and, until 1982, copra. Manufacturing accounts for 8 per cent of GDP (1987) and employs 5 per cent of the labour force. The wide range of products includes shoes, saddles, footballs, knitwear, wooden toys, corrugated iron, plastic piping, bicycle assembly, wire netting, paper, paint, biscuits and processed milk, pulp and passion-fruit processing dumper truck bodies and mini-excavators. A joint venture between Tongan and W. German concerns has formed Marine Tonga to manufacture 10-metre epoxy veneer-moulded yachts. A mini-industrial estate has been established in Nuku'alofa, where most of the more advanced products are made.

Mining: The US Geological Survey has found huge, undersea sediment-filled basins that could hold oil deposits near Tonga, the Solomon Islands and Papua New Guinea. German and Soviet researchers have discovered copper and zinc deposits off Tonga.

Membership of international organisations: Asian Development Bank, Commonwealth, ESCAP, FAO, GATT (de facto), IFAD, IMF, ITU, Lomé Convention (EC), Overseas Private Investment Corporation, South Pacific Bureau for Economic Co-operation, South Pacific Commission (Conference representation), South Pacific Forum, UNESCO, UPU, WHO.

Time: GMT + 13 hrs.

Climate: From May-Nov, temperatures relatively cool (11-29C) and humidity low; Dec-Apr is the wet season, with temperatures reaching 32°C and high humidity. Average rainfall 1,700mm p.a., but varies from place to place.

Entry requirements
Passport: Required by all.
Visa: Visas not required by citizens of Commonwealth and EC countries. When required visas must be obtained before arrival. Most visitors with onward passage and sufficient funds may be admitted for up to 30 days; longer stays require special authorisation.
Currency: All outward remittances of Tongan currency require approval; all proceeds of exports from Tonga must return to Tonga.
Customs: Visitors should not attempt to bring arms or drugs into the kingdom. Otherwise, no restrictions on personal goods. Imports from some countries may require an import licence or be subject to temporary control. Quarantine is required for all imported live animals and plants.

Health precautions
Mandatory: Vaccination certificates for cholera or yellow fever if travelling from an infected area.

Representation overseas: Hong Kong, London, San Francisco, Sydney.

Air access
National airline: Friendly Island Airways.
Other airlines: Air New Zealand, Air Pacific, Hawaiian, Polynesian.
International airport: Fua'amotu International, Tongatapu (TBU), 22 km from Nuku'alofa; duty-free shop and car hire.

Surface access: No regular passenger services to the kingdom, but berths may be available on cruise ships visiting Nuku'alofa and Vava'u.
Main ports: Nuku'alofa (on Tongatapu), Neiafu (on Vava'u), Pangai (on Lifuka), Ha'apai.

Hotels: Information regarding various types of tourist accommodation is available in Nuku'alofa and throughout the islands can be obtained from the Tonga Visitors Bureau.

Credit cards: Visa is accepted.

Car hire: Self-drive or chauffeur-driven cars are available. International and national driving licence must be presented to obtain local driving licence. Speed limits of 40 km/hr in towns and 65 km/hr in country areas are enforced.

City transport
Taxis: Private taxis are for hire. Fares should be agreed before undertaking a journey.

National transport: No public transport, shipping or air services operate on, to, within or out of Tonga on Sundays.
Air: From Tongatapu, Friendly Island Airways operate internal flights to 'Eua, Ha'apai, Vava'u and the two Niuas.
Road: Total road network about 400 km, of which 190 km are on Tongatapu. Some 80-90 per cent of total bituminised. Horses used extensively for transport.
Buses: Buses (or trucks) serve all parts of Tongatapu from Nuku'alofa.

BUSINESS GUIDE

Water: Various shipping lines operate between the islands. No advance bookings or inter-island ferry services as schedules frequently change at short notice due to weather or other local commitments. Charter yachts available.

Public holidays
Fixed dates: 1 Jan (New Year's Day), 4 May (HRH The Crown Prince's Birthday), 4 Jun (Independence Day), 4 Jul (HM The King's Birthday), 4 Nov (Constitution Day), 4 Dec (Tupou I Day), 25 and 26 Dec (Christmas Day & Boxing Day).
Variable dates: Good Friday, Easter Monday, Anzac Day.

Working hours: Sunday is widely observed as a day of rest, with work, sports and transport services forbidden.
Business: (Mon-Fri) 0800 – 1700; (Sat) 0800 – 1200.
Government: (Mon-Fri) 0830 – 1630.
Banking: (Mon-Fri) 0930 – 1530. (Sat) 0930 – 1130.

Social customs: Tipping is not customary. Tongans appreciate modesty in dress. Bikinis are acceptable on the beach but not in public places. It is an offence to appear in public without a shirt. Drunkenness is frowned upon; alcohol consumption may be restricted.

Telecommunications
Telephone: Dialling code for Tonga, IDD access code + 676 followed by subscriber's number (no area code necessary). Telephone system fully automatic with 24-hr international communications via Cable and Wireless satellite earth station and 2,765 telephones (1985).
Telex and telefax: Telex service is available at Cable and Wireless. Facsimile service also in operation.
Telegram: Telegrams via Cable and Wireless.

Banking
Main centre: Nuku'alofa.
Head offices: Bank of Tonga, Tonga Development Bank.

Advertising: Radio advertising is accepted in Tongan and English.

Electricity supply: 240 V AC.

Weights and measures: Metric system.

Embassies and consulates in capital: Australia, France, German FR, Republic of Korea, Nauru, New Zealand, Spain, Sweden, Taiwan, UK.

Hotels: Neiafu, Vava'u
Paradise International, P.O. Box: 11 (tel: 70211; tx: 66282; fax: 70184).

Hotels: Nuku'alofa, Tongatapu
Captain Cook's Vacation Apartments, Vuna Road (1.5 km from town centre), P.O. Box: 838 (tel: 23617; tx: 66231).

Friendly Islander, (3 km from town centre), P.O. Box: 142 (tel: 219000; tx: 66208 TF).

International Dateline, Waterfront Vuna Rd, (0.5 km from town centre), P.O. Box: 39 (tel: 23411; tx: 66223; fax: 23410).

Ramanlal (town centre), P.O. Box: 74 (tel: 23218; tx: 66025; fax: 833).

Car hire: Nuku'alofa
Avis: Burns Philip, Salote Road (tel: 21720; tx: 66220).

Five Star Enterprises: (tel: 21429; tx: 66275).

Friendly Islander Rental Cars: PO Box 142 (tel: 21900).

Gateway Rental Cars: PO Box 956 (tel: 21215; cable: GATEWAY RENTAL)

Nuku'alofa Rental (Nuku'alofa Motors): Taufa'ahau Road, Fanga.

Business advice and information
Department of Labour, Commerce and Industries, Salote Road, PO Box 110, Nuku'alofa (tel: 21888).

Travel information
Friendly Island Airways Ltd, PO Box 100, Nuku'alofa (tel: 21866; tx: 66228).

Tonga Visitors Bureau, PO Box 37, Nuku'alofa (tel: 21733, 23507; tx: 66269 PRIMO TS; fax: 22129; cable: TOURBUREAU).

Trinidad and Tobago

Tony Thorndike

Continuing economic crisis and deepening political division has resulted in widespread pessimism. However, the rate of decline has decelerated for the first time in six years and there is hope that positive growth will be restored, particularly since the twin-island's external debts were successfully rescheduled and two agreements reached with the International Monetary Fund. Other grounds for optimism are a positive trade balance and reductions both in state sector employment and the losses of state sector enterprises.

What limited progress there is, however, is not being achieved without pain. A 15.3 per cent devaluation to TT$4.25/US$ in August 1988, and a controversial 10 per cent cut in public sector salaries as part of a restrictive budget announced in January 1989, caused widespread hardship. Unemployment exceeds 25 per cent compared to less than 10 per cent eight years ago and private sector investment remains disappointingly low despite tax and other incentives. Predictably the trade union movement demanded the withdrawal of the budget and called for industrial action. But fewer than expected workers responded, probably because of the threat of unemployment. The key to decline is the fall in the oil price. As oil and oil-based products have constituted over 80 per cent of exports (and 25 per cent of GDP and 30 per cent of government revenue) the impact has been severe. Politically, the unity expressed in the establishment of the National Alliance for Reconstruction coalition and its subsequent victory in the December 1986 elections has all but gone with the founding of the United National Congress in April 1989 by mainly Indo-Trinidadian dissidents.

Out of the trough

The 1989 budget was more severe than many expected. But it reflected the reality of economic decline. Economic growth had plunged to a negative 7.4 per cent in 1987 although it recovered to a negative 4.0 per cent in 1988. GDP at current prices stood at TT$15.5bn compared to TT$19.1bn in 1984, and total debt reached US$1.8bn, 60 per cent (US$1.08bn) of which was external. The 1988 budget had cut government expenditure and more of the same followed in the wake of the August devaluation as revenue declined by some 17 per cent, mostly due to reduced business activity and rising unemployment. Revenue had been TT$6.7bn in 1985 but only TT$4.8bn in 1988. The August austerity programme halved the TT$2,091m capital budget for 1988 and slashed TT$235.2m from the TT$5,147.1m recurrent budget. Recurrent expenditure was cut again in the January budget to TT$4,627m, but some latitude was given to capital spending which was set at TT$1,324.6m. A recurrent surplus of TT$479.9m was expected.

There were also tax changes, the underlying philosophy being a gradual emphasis on indirect taxation. A subsequent White Paper published in May 1989 forecast the introduction in 1990 of Value Added Tax of between 15 and 20 per cent to replace a variety of indirect taxes. Although top marginal income tax rates were reduced from 50 to 45 per cent, and the number of bands cut from 11 to 4, virtually all allowances were removed, so few gained. Corporation tax was reduced from 49.5 to 45 per cent in order to encourage investment. But there was widespread doubt that investment would grow: Prime Minister Robinson himself admitted that only TT$136m worth of investment and 869 jobs could be identified in the manufacturing sector during 1989. Conscious of this, the May White Paper gave notice that the top income tax rate will fall to 35 per cent, as will corporation tax.

There is no doubt that the budget was in tune with IMF strictures; indeed, most of the measures featured in the government's letter of intent to the Fund in November 1988 in support of its application for a

KEY FACTS

Official title: Republic of Trinidad and Tobago

Head of state: President Noor Mohammed Hassanali

Head of government: Prime Minister A. N. R. Robinson

Ruling party: National Alliance for Reconstruction (NAR)

Capital: Port of Spain

Official Languages: English

Currency: Trinidad & Tobago Dollar (TT$) = 100 cents

Exchange rate: TT$4.23 per $ (Nov 1989)

Area: 5,128 sq km

Population: 1.2m (1988)

GDP per capita: TT$13,090 (1987)

GDP real growth: –4.0% (1988)

Inflation: 7.7% (1988)

Trade balance: $25m (1988)

Foreign debt: $1,080m (1988)

standby facility of TT$547m (US$128.7m) over the period October 1988 to December 1989, which was granted with commendable speed in February 1989. Besides promising to reduce public expenditure from 7 to 4 per cent of GDP, the letter of intent included a voluntary redundancy scheme for at least 9,600 civil servants (the government wage bill was TT$2bn per year for 64,000 employees, some 45 per cent of recurrent expenditure) and the complete liberalisation of import policy by 1990. The standby arrangement came on the heels of a successful bid for a loan of TT$470m (US$110.5m) under the Fund's Compensatory Financing Facility to make up for the shortfall in expected export earnings. In the period June 1987-June 1988, crude oil earnings fell from US$990m in the previous 12 months to US$790m.

A major component of the IMF 'medicine' (although Robinson consistently disclaimed that government policy was linked to Fund preferences) was the sharp cutback in subsidies to loss-making public sector enterprises and, where possible, privatisation. Some 70 enterprises had been created by the previous government and few made money; by 1989 their number had been reduced to 29. Only TT$186m were to be transferred to these parastatal enterprises in 1989, but this is bound to be exceeded. Some had been created to bypass public sector bureaucracy and slothfulness, for which the Trinidadian civil service was justly known; hardly a firm basis for profitable operation. Some were forced into receivership or merged with more viable enterprises. Others, however, such as the Caroni sugar company, faced very adverse world prices. In Caroni's case, production plans were scaled down from 100,000 tonnes to 75,000 tonnes by 1993; exceptionally favourable weather conditions enabled 97,000 tonnes worth TT$125m to be produced in 1989. But that was little when set against accumulated post-tax losses since 1985 of TT$1.2bn.

The Iron and Steel Company of Trinidad and Tobago (ISCOTT), the centrepiece of the Point Lisas industrial complex, has perhaps the worst economic record. It has been a notorious loss-maker since it was built in the 1970s just as world steel prices fell. Although losses fell from TT$230m in 1987 to TT$124m in 1989, accumulated losses exceed TT$1.6bn. After a series of management failures, a West German concern, Hamburger Stahlwerke (HSW), was awarded a management contract. Negotiations for a lease involving an option for HSW to purchase after five years collapsed in the face of widespread political objection to foreign control of what was seen as a vital national asset. As a result, the plant was leased for 10 years in May 1989 to Ispat, an Indian steel company.

ISCOTT's debt highlights the general debt problem. About US$575m of external debt is attributable to the parastatal enterprises, creating severe servicing difficulties. Worse, all but US$500m of the total US$1.8bn was due for payment by 1991. Reserves had reached a low minus TT$217m in August 1988 and were barely above zero three months later (at their peak in 1981 they were over TT$8bn, then a huge US$3.3bn). Rescheduling therefore became a pressing priority, made possible by the IMF agreements. Talks in Paris and London in late 1988 resulted first in a payment moratorium; successful negotiations followed in February with the Paris Club. With repayment pressure reduced, the reserves began to climb. Loans totalling TT$4.8bn were expected over the next three years from eight international agencies, including the IMF and the World Bank, the latter following the reversal in December 1988 of the 1983 'graduation' which made Trinidad and Tobago ineligible for concessionary finance because of the then high per capita income. Another source of finance was accessed with the signature of a tax information agreement with the United States, enabling investment of Puerto Rico '936' funds and tax concessions for US companies holding conventions in the country

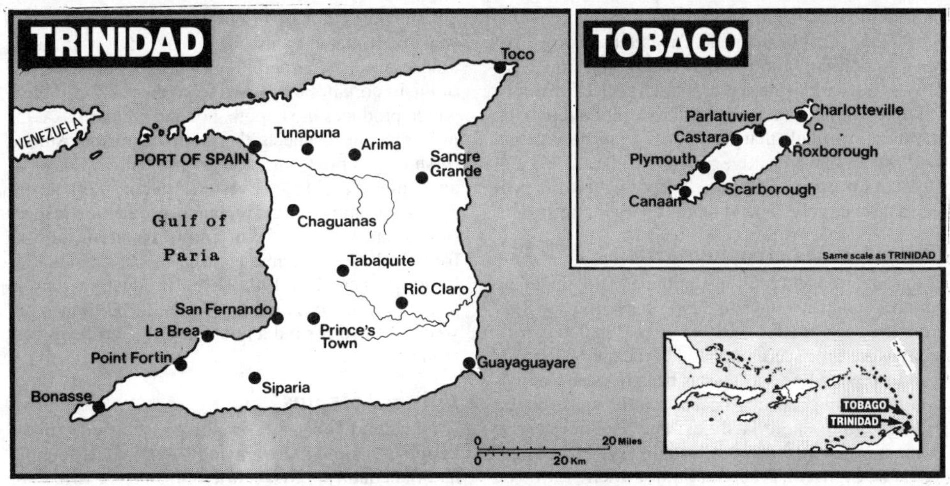

or wishing to establish foreign sales corporations. The confidence of the agencies was further boosted by a cabinet reshuffle in early 1989 and the appointment of the respected Selby Wilson as minister of finance, Prime Minister Robinson relinquishing that onerous portfolio.

There were further grounds for optimism with the success of the government's diversification campaign away from overreliance on the energy sector. Food production rose 19 per cent in 1989, food imports falling by virtually the same amount. Tourism receipts also rose following intensive marketing of Tobago's facilities and better air services, and exports from the non-oil and non-chemical manufacturing sectors showed an astonishing 52 per cent increase in 1988. Exports as a whole in 1988 were worth TT$5,388m, producing a trade surplus of TT$1,009m (US$25m). Oil exports as a percentage of total exports fell to 62 per cent.

It would appear that Trinidad and Tobago touched the bottom of the economic cycle in late 1988. But plans for economic growth, first laid out in July 1988, entitled 'Restructuring for Economic Independence', are predicated on over-ambitious assumptions about local and foreign investment. Spanning seven years, the plan envisages the creation of 100,000 jobs by 1995, a surplus on the government's current account of TT$700m a year and a 7 per cent increase in exports. Capital investment was to be some TT$30bn during this period, the government only contributing 30 per cent. The remainder was to come from foreign investment – a sharp reversal of the policy of econ-

KEY INDICATORS

	Unit	1984	1985	1986	1987	1988
Population	m	1.2	1.2	1.2	1.2	1.2
GDP	TT$m	20,136*	18,921	17,300	15,933	–
GDP per capita	TT$	17,238	15,767	14,417	13,090	–
Inflation	%	13.4	7.6	7.7	12.0	7.7
GDP real growth	%	–5.9	–4.4	–3.3	–7.4	–4.0
Agricultural production	1979-81=100	64.0	92.6	77.5	68.9	78.5
Sugar	'000 tonnes	879	1,002	1,165	1,104*	1,100*
Coffee	'000 tonnes	0.9	2.1	1.3	1.8	1.0
Cocoa	'000 tonnes	1.6	1.3	1.4	1.4	2.0*
Citrus fruit	'000 tonnes	3.3	6.1	4.3	3.0	4.0*
Industrial production	1977=100	111.0	106.6	108.0	–	–
Food processing	1977=100	130.7	119.6	139.2	–	–
Textiles	1977=100	60.3	46.6	58.4	–	–
Chemicals	1977=100	193.4	243.7	286.0	–	–
Assembly industries	1977=100	253.6	223.7	331.6	–	–
Crude oil production	'000 b/d	169.5	175.6	168.9	143.3	149
Exports	TT$m	5,216.2	5,247.1	4,988.6	5,264.6	5,424*
Sugar	TT$m	60.6	52.7	91	–	–
Machinery & transport	TT$m	73	–	–	–	–
Mineral fuels	TT$m	4,169	4,192	3,515	3,740	–
Chemicals	TT$m	565	677	–	742	–
Imports	TT$m	4,605.9	3,739.0	4,860.2	4,387.5	4,310.3
Mineral fuels	TT$m	35	124	0	–	–
Manufactured goods	TT$m	988	758	–	1,200	–
Machinery & transport	TT$m	1,385	1,135	–	–	–
Balance of trade	$m	405.9	756.1	153.6	349.6	–
Current account	$m	–552.5	–90.3	–441.7	–247.1	–
External debt	$m	1,222	1,448	1,858	1,801	–
Debt service ratio	%	7.2	9.7	17.8	–	–
Exchange rate	TT$ per $	2.4	2.45	3.6	3.6	3.84

* estimated figure

omic nationalism pursued by the previous government – and local investment. The latter has proved to be disappointingly low while foreign investors have proved difficult to attract, despite the selective removal of restrictions under the Aliens Landholding Act. It is therefore likely that growth will again be negative in 1990 but that a positive profile will be achieved in 1991. As over the past three decades, recovery will ultimately largely depend on the energy industry.

Oil, gas and petrochemicals

Oil production and refining continue to dominate the economy but revenue had nearly halved in 1988-89 from the peak in 1981. In 1978 some 230,000 barrels a day were produced; only 154,000 came to the surface in 1988. More serious has been the sharp reduction in refining capacity because of the construction of large US offshore facilities. The Shell refinery at Point Fortin was nationalised in 1974, renamed TRINTOC; that of Texaco at Point-a-Pierre followed in 1985, becoming TRINTOPEC. TRINTOPEC also acquired a 49.9 per cent holding in the Tesoro plant. After much delay, integration of the two companies began in 1989. Plans to upgrade the former Texaco plant to produce higher grade oils were announced in late 1988, costing an initial US$87m, although the total cost will exceed US$410m. When combined, the state oil company will be responsible for about 28 per cent of extraction, Amoco for over half (Amoco is obliged to refine a portion locally). Most extraction (75 per cent) is offshore. Until recently, there has been little exploration, notwithstanding the probability that proven reserves will last only until the turn of the century. Perhaps because of this uncomfortable thought, considerable interest centres on the discovery of what is thought to be a major oil reservoir off the south coast by TRINMAR, the offshore consortium owned by the two state oil companies and Texaco.

Natural gas production remains broadly constant at about 7,500bn cubic metres, mostly exported to the United States in frozen liquefied form. Proven reserves are estimated to last for at least 45 years: that is indeed fortunate as natural gas has become the basis of the important Point Lisas petrochemical complex, which produces urea, methanol and ammonia. A 75-mile pipeline connects the complex to the Amoco offshore gas platforms. Joint ventures involving ICI and Union Carbide have resulted in considerable expansion in ammonia and methanol production respectively, and a US$5.5m plant to manufacture formaldehyde concentrate (used in plastics and adhesives) opened in 1988. Urea production for the fertiliser industry is rising to over 1,200 tonnes per year, destined for India and Western Europe in particular.

Political strains

Trinidad and Tobago was for long associated with the People's National Movement (PNM) and its founder and the country's first prime minister, Dr Eric Williams. Williams died in 1981 but PNM rule continued to December 1986 when the National Alliance for Reconstruction (NAR) won a stunning election victory. Proclaiming a multi-class and multi-ethnic 'One Love Rainbow', the NAR won 33 of the 36 seats.

The scale of the victory expressed popular frustration at the deeply-rooted inefficient public sector with its amoeba-like bureaucracy, political drift and economic mismanagement. Under Williams, and to a lesser extent under his successor George Chambers, corruption became a scandal and planning a mere theory, for as the oil flowed, to use Williams' words, 'money is no object'. Most of the huge profit windfall from oil in the 1970s was frittered away. The PNM was able not only to use money to gain favour but also to exploit a divided opposition. One important factor is the ethnic vote: the Afro-Trinidadians, who comprise some 43 per cent of the population, generally supported the PNM, whereas the Indo-Trinidadians, mainly Urdu and Gujerat-speaking, followed the United Labour Front (ULF) led by veteran politician Basdeo Panday. The left-of-centre ULF, as its name suggested, had close links with the union movement, especially that concerned with the Indian sugar workers. But it never fulfilled its multi-ethnic promise, despite its association with the militant Afro-dominated Oilfield Workers' Trade Union. The PNM could also count on the votes of the Indian middle class.

The victory of the NAR also expressed popular anger at economic decline. Many had become accustomed to the good life and believed that a change of government would restore both national and personal fortunes. Expectations were therefore very high. Not surprisingly, many welcomed the gradual coming together of the opposition and few questioned the opposition's ability to fulfil expectations, let alone question its unity. But the opposition were only united on an anti-PNM platform. There were four constituents: the ULF, the Tobago-based Democratic Action Congress (DAC) led by Arthur Robinson, the middle class right wing Organisation for National Reconstruction created by Karl Hudson-Phillips (like Robinson an ex-PNM minister sacked by Williams) and the small intellectual-based Tapia House Movement of economist Lloyd Best. They united a bare nine months before the election and went on to capture 66 per cent of the votes cast. Chambers lost his seat and retired, his place taken by Patrick Manning, who had the thankless task of reorganising the morale-shattered party.

However, the cracks soon appeared. The first to go was the Tapia group and increasingly bitter arguments in the cabinet became public. The main issues were Robinson's alleged authoritarian attitude and his insistence on austerity programmes which the ULF faction found very difficult to sell to their union co-patriots. Ethnic-based accusations that the ULF faction were only interested in promoting the interests of the Indian minority made a deteriorating situation worse. In fact, Panday (deputy Prime Minister and External Affairs minister) was the one leader most

COUNTRY PROFILE

Political system: Executive power is exercised by the prime minister (appointed by the president) and other members of the cabinet. The legislature is bicameral: the Senate (31 members, all appointed by the president, nine chosen by himself, 16 on the prime minister's advice and six on the advice of the leader of the opposition) and the House of Representatives (36 members; elected by universal adult suffrage; five-year term). Under the constitution adopted 1 Aug 1976 the president is the constitutional head of state, elected by an electoral college made up of members of both houses of parliament. The separate Tobago House of Assembly is made up of 18 members (15 elected, three appointed by the majority party; four-year term).

Main cities/towns: Port of Spain, San Fernando, Arima; Scarborough is the main town on the island of Tobago.

Media
Press: Two major dailies, the *Trinidad Guardian* and *Trinidad and Tobago Express* and two major evening papers, the *Evening News and The Sun.*
Radio: One national and one commercial radio station. *Television:* Commercial TV operated by state-owned Trinidad and Tobago Television Company (TTT).

External trade: Trade balance remains heavily dependent upon oil exports. Imports were reduced sharply in 1987 and 1988 by tighter foreign exchange controls and reduced domestic demand.
Exports: Petroleum and products (77% of total value in 1985), fertilisers (4-6%), sugar (1%), chemicals (5-6%). Main destinations: USA (63% of total value in 1985), EC (13%), CARICOM (8%), UK (5%).
Imports: Consumer goods (27-30% of total value), food (15-18%), intermediate goods (34-36%), capital goods (30-33%). Main sources: USA (38% of total value in 1985), EC (17%), UK (10%), Saudi Arabia, Japan (10%), Canada (7%), CARICOM.

Agriculture: Sector has been overshadowed by growth of the petroleum industry and currently contributes under three per cent to GDP and employs less than 12 per cent of workforce. Due to the recession the former pattern of importing food has been partially offset by increased production. Almost 25 per cent of food requirements are imported. Farming of the major cash crops (sugar, coffee, cocoa and citrus fruits) has slumped owing to labour shortages, diseases and falling export demand. The fishing sector is small-scale, but an important local source of food. Nearly half the land area is forested, with production of hardwood nearly sufficient to meet local needs.

Industry: Manufacturing sector contributes around seven per cent to GDP and construction around 12 per cent. Development since 1970s centred on heavy export-orientated industries which are geared towards maximising the country's energy resources. Main industries so far developed are the state-owned iron and steel works and a number of downstream chemicals plants at the Point Lisas industrial estate. Smaller scale industries include car assembly, concrete production, food processing, rum distilling, paper production, radio, TV and vehicle assembly, paper products, textiles and asphalt (major world supplier). Industrial production has been sluggish due to falling export demand, foreign exchange shortages and low levels of productivity. Exports of industrial products to the region are increasing. Industry plays only a small role in Trinidad, contributing less than two per cent to GDP.

Tourism: Earnings from tourism form the cornerstone of the Tobago economy.

Hydrocarbons: The petroleum sector is the mainstay of the economy, contributing around 25 per cent to GDP, 30 per cent of government revenue and 62-65 per cent of exports. Although oil production has declined over the last five years from a peak of 230,000 b/d in 1978 to 160,900 b/d in 1987, the country remains the largest producer in the Caribbean. Amoco is the largest operator with 50 per cent of the country's output from its fields off Guayaguayare. Proven reserves of oil amount to only 700m barrels, sufficient to last until 1995 at current rate of extraction. The two main oil refineries, at Pointe-à-Pierre and Point Fortin, are primarily engaged in exporting low-grade fuel oil. Both plants are currently operating well below capacity. Production of gas, largely from the Amoco oilfields, has been steadily rising. Proven reserves are in the region of 16tn cu ft, sufficient to last for 100 years. Pipelines link gas-fields with the Point Lisas industrial estate.

Membership of international organisations: CARICOM, Commonwealth, Economic Commission for Latin America, FAO, GATT, ILO, IMF, UN, WHO, Inter-American Development Bank, Organisation of American States, World Bank.

committed to racial unity. Despite these divisions, the NAR could still show it had public support: it won the local government elections in September 1987 against the PNM and captured 11 of the 12 seats in the Tobago House of Assembly in November 1988, although the former DAC in Tobago has little opposition.

The first casualty was the outspoken Works Minister John Humphrey who was fired in late 1987. Not long afterwards Robinson dismissed Panday and others who opposed government policy. Panday and the dissidents formed 'Club 88' and were formally expelled from the NAR at the September 1988 congress as a result. The group went on to establish the United National Congress at the end of April 1989. Although it commands six seats in the House of Assembly, Panday has declined to supplant Manning as the official leader of the opposition. The new party has much ground to cover, as a public opinion poll conducted in May 1989 suggested that if a general election were to be held, 24 per cent would support the NAR, 26 per cent the PNM and only 8 per cent the UNC. On that showing, the UNC would hold the balance of power as it would be assured of the seats in the Indian-dominated 'sugar belt'. Ominously for Robinson, only 7 per cent of those polled expressed satisfaction with his performance as prime minister.

Another party may also emerge in 1989-90, a socialist offspring of the Committee for Labour Solidarity (CLS). The CLS is the political arm of the Council of Progressive Trade Unions, with which the Oilfield Workers' Trade Union is closely associated. It is the radical of the two trade union groupings, the other being the Trinidad and Tobago Labour Congress, for long associated with the PNM. The era of the 'One Love Rainbow' is well and truly dead.

BUSINESS GUIDE

Time: GMT -4.

Climate: Tropical and particularly humid during rainy season from Jun-Dec. Mean temperatures are 13°C at night and 29°C during the day.

Entry requirements
Passport: Required by all and must be valid for at least six months from date of arrival.
Visa: Not required by citizens of most Commonwealth countries, nor from tourists from USA and certain West European and Latin American countries, staying for limited periods – generally 2-3 months.
Prohibited entry: Nationals of Taiwan and South Africa.
Currency: Travellers must not import or export more than $200 of local currency.

Health precautions
Mandatory: Yellow fever vaccination certificate if arriving from infected area.
Advisable: Polio, typhoid vaccinations and precautions with drinking water.

Representation overseas: Brasília, Bridgetown, Brussels, Caracas, Geneva, Georgetown, Kingston, Lagos, London, New Delhi, New York, Ottawa, Washington.

Air access
National airlines: BWIA International (Trinidad & Tobago Airways Corp).
Other airlines: Aeropostal, Air Canada, Air India, ALM, American, British Airways, Caricargo, Cruzeiro Brazilian, Guyana Airways, KLM, LAV, LIAT, Pan Am.
Tax: Departure tax of TT$50 for all except direct transit passengers and children five years or under.
International airport: Port of Spain-Piarco (POS), 25 km from city; duty-free shop, bar, restaurant, bank, post office, shops, car hire.

Surface access
Main ports: (Trinidad) Chaguaramas, Point Lisas, Port of Spain; (Tobago) Scarborough.

Hotels: Limited range available in Trinidad, generally expensive, though less so in Tobago. 10 per cent tip usual. Book well in advance if arriving during carnival time.

Car hire: National driving licences of most countries accepted for a period of three months from arrival. Insurance required. Cars driven on left and maximum speed of 110 kph on highways.

City transport
Taxis: Routes with standard fares operated by passenger cars bearing 'H' registration plates and two-coloured Maxi Taxis (yellow stripe in Port of Spain). Negotiate fares for regular taxis in advance. Limousine service available at airport.

National transport
Air: There are flights approximately every half hour operated by BWIA between Port of Spain and Tobago.
Road: Extensive network of around 8,000 km. Major highways run north-south and east-west. Traffic jams are common.
Buses: Cheap and generally crowded. Plenty of 'road taxis' and 'maxi taxis'.
Water: Regular ferries sail Port of Spain to Scarborough.

Public holidays: 1 Jan, Good Friday, Easter Monday, Whit Monday, Corpus Christi, Eid al-Fitr, 19 Jun, 1 Aug (Emancipation Day), 31 Aug (Independence), 24 Sep (Republic Day), Divali, 25 Dec, 26 Dec. Little business is conducted at Carnival time (Feb or Mar).

Working hours
Business: (Mon-Fri) 0800 – 1630.
Government: (Mon-Fri) 0800 – 1630.
Banking: (Mon-Thu) 0900 – 1400; (Fri) 0900 – 1200, 1500 – 1700.
Shops: (Mon-Fri) 0800 – 1630; (Sat) 0800 – 1200.

Telecommunications
Telephone: Totally automatic service. Dialling code for Trinidad and Tobago: IDD access code + 1-809 followed by subscriber's seven-digit number. No area codes required. System currently being improved.
Telex: Good International services in Port of Spain from Textel House and Hilton Hotel.

Postal service: Airmail to Europe takes one to two weeks.

Banking
Banking centre: Port of Spain.
Central bank: Central Bank of Trinidad and Tobago.
Commercial banks (head offices): Agricultural Development Bank, Bank of Commerce, Trinidad and Tobago, Bank of Nova Scotia Trinidad and Tobago, Merchant Bank of Trinidad and Tobago, National Commercial Bank of Trinidad and Tobago, Republic Bank, Royal Bank of Trinidad and Tobago, Trinidad Co-operative, United Bank, Workers Bank of Trinidad and Tobago.

Electricity supply: 115 and 230 V AC, 60 cycles.

Representation in capital: Argentina, Brazil, Canada, China, Colombia, France, German FR, India, Jamaica, Japan, Netherlands, New Zealand, Nigeria, Portugal, UK, USA, Venezuela.

Hotels: Piarco Airport, Trinidad
Belair International Airport (tel: 664-4771/3; fax: 6251150).

Hotels: Port of Spain
Chaconia Inn, 106 Saddle Road, Maraval (tel: 628-8603/5; tx: 22645).

Hilton International, P.O. Box: 442 (tel: 624-3211; tx: 22319; fax: 6243211 EXT 6133).

Holiday Inn, Wrightson Road, P.O. Box: 1017 (tel: 625-3361/8; tx: 22456 holin; fax: 625-4166).

Normandie, 10 Nook Avenue, St Anns, P.O. Box: 851 (tel: 624-1181; tx: 22639; fax: 6270779).

Car hire
Auto Rentals: corner Richmond and Tragarete Road, Port of Spain (tel: 623-3063).

Auto Rentals: Piarco Airport (tel: 664-3905).

Baird's Car Rentals: Tobago (tel: 639-2528).

Berk's Car Rentals: 8-10 Purcell street, San Fernando (tel: 652-3880/3275).

Hertz: Hilton Hotel, Port of Spain (tel: 624-3316); Piarco Airport (tel: 664-4771/3).

Singh's Auto Rentals: Piarco Airport and 7-9 Wrightson Road, Port of Spain (tel: 625-7265).

Business advice and information
Businessmen's Association of Trinidad and Tobago, PO Box 322, Time Plaza, Room 10, 28 Henry Street, Port of Spain (tel: 623-4568).

Caribbean Employers' Confederation, 43 Dundonald Street, Port of Spain (tel: 625-4723).

Caribbean Industrial Research Institute, O'Meara Industrial Estate, Arima (tel: 662-7161/4; fax: 6634180).

Export Development Corporation, Export House, 17 Richmond Street, PO Box 582, Port of Spain (tel: 623-6022/3; tx: 22732 export).

Industrial Development Corporation, 10-12 Independence Square,PO Box 949, Port of Spain (tel: 623-7291/6, 623-7289; tx: 22255).

Management Development Centre, Room 212, Salvatoria Building, PO Box 1301, Port of Spain (tel: 623-4951/3).

Ministry of Industry and Enterprise, Riverside Plaza, Port of Spain (tel: 623-2931/4).

Reinsurance Company of Trinidad and Tobago, Trinre House, 52 Jerningham Avenue, Belmont, PO Box 1087, Port of Spain (tel: 623-6194/6602; tx: 22811 trinre; fax: 624-4021).

Small Business Association of Trinidad and Tobago, Third Floor, MPU Building, 3 Besson Street, Port of Spain (tel: 624-3666).

Shipping Association of Trinidad and Tobago, Room 12a, 64-66 South Quay, Port of Spain (tel: 623-8570).

Trinidad and Tobago Chamber of Industry and Commerce, PO Box 499, Corner Frederick and Hart Street, Port of Spain (tel: 623-1561).

Trinidad and Tobago Development Finance Co Ltd, PO Box 187, 8-10 Cipriani Boulevard, Port of Spain (tel: 623-4665/7, 625-4666/8; tx: 22856 dfc; fax: 624-3563).

Trinidad and Tobago Manufacturers' Association, 155 Tragarete Road, Port of Spain (tel: 628-8342/5).

Trinidad & Tobago Tourist Board, 122-124 Frederick Street, PO Box 222, Port of Spain (tel: 623-1932/3; tx: 22318 tttb; fax: 623-3848); Jerningham Street, Scarborough, Tobago (tel: 639-2125, 639-3566).

Banking
Central Bank of Trinidad and Tobago (central bank), PO Box 1250, Independence Square, Port of Spain (tel: 625-4835/4921/5028; tx: 22386, 22532).

Agricultural Development Bank, 86 Duke Street, Port of Spain (tel: 623-6262).

Tunisia

Jon Marks

After a year spent establishing his credentials as Tunisia's undisputed leader following the 7 November 1987 overthrow of Habib Bourguiba, President Zine el-Abidine Ben Ali consolidated during his second year in power. Major challenges remained, notably claims by the main Islamic opposition grouping to legitimate political party status – which the government resisted – and the need for further reform in the ruling Rassemblement Constitutionel Democratique (RCD) and bureaucracy. Following strenuous efforts to stabilise the economic situation in the wake of the 1986 balance of payments crisis the economy remained on an even keel, although there were fears that drought and other factors beyond the government's control would add to the pressure on limited resources, leading to an upsurge of tensions in society. In order to build a more solid base for future development, the government pushed ahead with its liberal economic strategy, promising further reforms and, eventually, a convertible dinar.

In the year following the July 1988 RCD congress, a major date in Ben Ali's calendar of reform, the president showed his determination to open up a political system which had been stifled during the Bourguiba years. By continuing his policy of giving amnesty to former opponents and liberalising laws covering activity in the public domain, Ben Ali maintained the high level of support Bourguiba's overthrow had generated.

Following the persistent crises of the later Bourguiba years this was widely welcomed within Tunisia, where even the Islamists – Bourguiba's last and most intractable foes – were apparently prepared to give Ben Ali the benefit of the doubt. The re-establishment of a political consensus was also welcomed abroad by Tunisia's allies, creditors and potential investors who feared that an upsurge of political troubles could damage the country's economic prospects.

With agriculture and oil production in decline, and tourism becoming the main source of foreign exchange, the economy's future prosperity is now tied to encouraging outside investment, with the aim of making Tunisia a service centre for the southern Mediterranean and, exploiting its geographical position, possibly an offshore base for manufacturers targeting their production at the European Community (EC), as well as Arab and African markets.

1989 elections

The strength of Ben Ali's position in the country was underlined by presidential elections held on 2 April 1989. These legitimised Ben Ali's status as head of state. No opposition party dared to offer an alternative candidate – with even Islamists saying that they recognised that for now Ben Ali was the only viable leader for Tunisia – and 'Zine' polled 99.27 per cent of the vote.

Ben Ali's personal prestige was highlighted by the lack of a convincing protest vote. This was not the case in contested legislative elections, also held on 2 April, when 489 candidates competed for 141 seats in an enlarged chamber of deputies.

Among their number were dozens of independents representing the Islamic movement. As the former Mouvement de Tendance Islamique (MTI), led by the country's best-known Islamist leaders, Rachid Ghannouchi and Abdelfattah Mourou, and now known as *Harakat* (Movement) and subsequently *Hizb Nahda* (the Renaissance Party), had not been recognised as an official political party it could not contest elections under its own name. Nevertheless in the majority of constituencies 'independent' lists made up of Nahda members were entered. This strategy forced concessions from the government, which agreed to give 'independents' television time, an advantage previously only offered to recognised parties. This showed Ben Ali's pluralist impulse, but also underlined the Islamists' strength and political sophistication – the strategy of fielding 'independent' candidates had effectively outflanked a government still committed to blocking their advance, albeit by peaceful means rather than the repression of 1986–87.

The results of the legislatives were revealing. As expected the RCD drew on its traditions of strong organisation and localised power bases established during its past incarnations as the Neo-Destour and Parti Socialiste Destourien. According to official figures, the RCD won 80.48 per cent of the vote and, aided by an electoral system which favours majority parties, took all 141 seats in the chamber of deputies.

This did not necessarily represent a victory for Ben Ali, who had staked his prestige on forming a viable multi-party democracy, dominated by the RCD but with acceptable, rival parties providing a form of 'loyal opposition' in parliament. This was not to be. The election results showed that for all Ben Ali's commitment to reform, the RCD was not yet ready to give up power in the country. Even before the campaign formally opened on 19 March, there were complaints of irregularities in procedures to register candidates – a traditional problem in Tunisian elections.

Among the opposition, only the largest secular party, the Mouvement des Democrates Socialistes (MDS) and Islamist candidates polled more than 1 per cent of the vote. The MDS, led by the respected former minister and long-time opponent of Bourguiba, Ahmed Mestiri, managed to poll only 3.76 per cent. The independents, on the other hand, polled 13 per cent of the national vote, and much more in their strongholds such as Gafsa, Tunis and Ben Arous.

This figure underlined the strength of the Islamic movement: and the vote was possibly larger still as local observers said there was some malpractice working in favour of the RCD – if certainly not as much as in past elections. To balance the equation there were also isolated reports of violence during the campaign involving radical Islamists.

The elections showed that for all Ben Ali's efforts to promote an opposition acceptable to the liberal RCD establishment, it was Nahda that represented the main challenge, which would have to be accommodated if the tensions of the Bourguiba years were not to re-emerge.

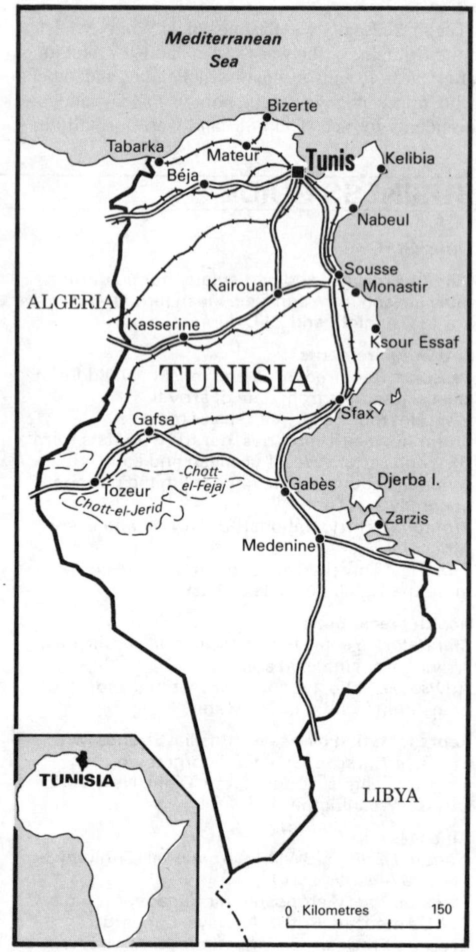

After a year-and-a-half of grace, political conflict was re-emerging – albeit at a far lower level than before – with the first reports of serious violence on Tunis University campus, and in colleges and lycees elsewhere. The violence involved Islamists and leftists fighting for better conditions and control of student bodies (in battles where the RCD was largely absent due to lack of student members). There were also renewed protests on campus over such 'Islamic' issues as prayer rooms and arabisation.

KEY FACTS

Official title: Republic of Tunisia

Head of state: President Zine el-Abidine Ben Ali

Ruling party: Rassemblement Constitutionnel Démocratique

Capital: Tunis

Official Languages: Arabic (French widely spoken)

Currency: Tunisian Dinar (TD) = 1,000 millimes

Exchange rate: TD0.96 per $ (Nov 1989)

Area: 164,150 sq km

Population: 7.8m (1988)

GDP per capita: $1,286 (1988)

GDP real growth: 1.5% (1988)

Inflation: 6.3% (1988)

Trade balance: –$1,090m (1988)

Foreign debt: $6.9bn (1987)

Ben Ali's response to these challenges was to stick to his by now tried and trusted formula for promoting reconcilliation in the country: further amnesties, overtures to opposition groups and reform of the RCD and administration. On 11 April 1989 he reshuffled his government and a day later announced a restructured RCD politburo. Into it came more technocrats and politicians who had past associations with opposition groups or were known to favour dialogue. The new government thus included former opponents such as Health Minister Dali Jazy (previously an MDS member) and Mohammed Charfi, who gave up the chair of the Ligue Tunisienne pour la Defence des Droits de l'Homme (LTDH) to become education and science minister.

Also associated with promoting dialogue in the new cabinet were Youth and Sports Minister Hamouda Ben Slama (who before joining the government in 1988 had acted as a negotiator in Ben Ali's talks with the MTI), Culture and Information Minister Habib Boulares and Social Affairs Minister Moncer Rouissi. Ben Slama and Boulares were also appointed politburo members, underlining a will to introduce pluralist tendencies into sometimes resistant party structures. Other new members in an enlarged nine-member politburo were Interior Minister Chedli Neffati (who replaced Habib Ammar) and Planning and Finance Minister Mohamed Ghannouchi, a key advocate of economic liberalism in the politburo, who joined the influential central bank governor Ismail Khelil, already a member. Neffati, like newly-appointed Defence Minister Abdallah Kallel and Foreign Affairs Minister Abdelhamid Escheikh, represents another element in the Ben Ali administration: those senior officials who, like the president, have a background in the interior ministry or military establishment.

The extent of changes at the top is shown by the fact that the only members of the last Bourguiba administration left in the government were Ben Ali, Prime Minister Hedi Baccouche, Khelil and Ghannouchi. However, as the elections showed, lower down the ranks of the RCD, change had been much less drastic.

Ben Ali also pushed ahead with his policy of encouraging national 'reconciliation'. On 27 June 1989 legislation was passed for a long-awaited amnesty which would restore civil rights to 5,416 Tunisians of varying political shades. Included on the list were followers of Salah Ben Youssef (who was murdered in Frankfurt in 1961) and former minister Ahmed Ben Salah. Ex-premier Mohamed Mzali was not included; nor were Mansour Skhiri and a few other senior aides of Bourguiba, as those charged with corruption and embezzlement were not eligible.

Possibly most significant, members of Nahda were included. This opened the way for the party to be recognised. In early June Nahda's application to be recognised as a political party was rejected. Officially this was because much of its leadership had served prison terms (albeit during the Bourguiba years) making them ineligible to participate in party politics. The decision was not unexpected: although it was clear that to maintain a viable dialogue with the opposition the leadership must talk to Nahda, the RCD leadership (including, it seems, the president) remained acutely sensitive to what it saw as Nahda's claims to have a monopoly of religious authority.

Although the general amnesty quashed the sentences of the Nahda leadership, opening the way to let Nahda into formal politics, there were clear signs that the attitudes of government and Islamists were hardening. In a major speech on 25 July Ben Ali argued that even in a multi-party system, too many parties were too much of a good thing. 'We do not believe in the usefulness of a plethora of parties,' Ben Ali said. 'We do not think this serves the interest of the nation as it means a waste of energy and creates antagonism.' This seemed to be directed at not only Hizb Nahda

KEY INDICATORS

	Unit	1984	1985	1986	1987	1988
Population	m	7.0	7.3	7.5	7.6	7.8
Gross domestic product (GDP)	TD m	6,240	6,910	7,004	7,970	8,624
GDP per capita	TD	891	946	934	1,048	1,106
Inflation	%	8.4	8.0	5.8	7.2	6.3
GDP real growth	%	5.7	5.7	–1.6	13.9	1.5
Agricultural production	1979-81=100	104.4	133.5	115.1	134.8	110.5
Food production	1975=100	104.4	133.9	115.1	135.1	110.4
Industrial production	1980=100	118.3	127.7	118.4	–	–
Phosphate production	tonnes m	5.4	4.6	5.7	5.8	6.7
Crude petroleum production	'1000 b/d	110.0	113.0	107.5	105.1	103.3
Exports	TDm	1,399	1,443	1,388	1,771	2,055
Crude petroleum	TDm	584	573	315	382	323
Phosphates	TDm	142	164	203	243	264
Olive oil	TDm	57	43	54	66	71
Imports	TDm	2,509	2,287	2,308	2,510	3,167
Balance of trade	$m	–1,152	–866	–935	–728	–1,090
Current account	$m	–770	–587	–618	–60	212
Total external debt	$m	4,095	4,915	5,881	6,909	–
Debt service ratio	%	20.6	22.3	26.8	24.1	–
Exchange rate	TD per $	0.78	0.83	0.79	0.83	0.86

* estimated figure

but also other elements of an opposition which seemed without direction during a critical period.

Ben Ali also seemed to be losing patience: to underline the president's determination that parties should play by the existing rules, a troublesome, but otherwise insignificant opponent, Jalloul Azzouna, was imprisoned for a year in August 1989 for 'disparaging the head of state' in a press statement. Azzouna's problems leading an unauthorised splinter group of the already tiny Parti de l'Unite Populaire (PUP) were widely reported and his conviction was described as the first political trial of the Ben Ali era. It seemed to show the parameters for political activity.

Despite unconfirmed rumours of an outbreak of infighting within the leadership, Ben Ali's main political problems seemed to centre on the lack of a credible opposition, as a liberal complement to liberal economic policies. With even the MDS, whose appeal was much greater when Tunisians were forced to react to Bourguiba's intolerance, offering little real opposition, that is proving hard to find. Ben Ali's hopes for the development of a controlled democracy were dealt another blow when Mestiri – one of the few figures of any weight in the legal opposition and, as such, an essential player if the current model of democracy was to be seen to work – refused appeals by his party and the president to stay on to lead the MDS following the election defeat.

More serious, rising frustrations among the Islamic opposition at the government's refusal to authorise Nahda were being expressed in increasingly strong language. In a communique issued in early September Nahda took off its kid gloves to criticise a 'regression' in political life and especially alleged abuses of power against those expressing their 'religious' convictions. If this trend continues further trouble could follow. Despite its relatively moderate line, Nahda was being placed in a position where it might have to challenge the government's position, if only to reconcile its more radical supporters.

Economic pressures

Following a period of stabilisation in 1986–88, two years of drought have increased economic pressures, despite the strong showing of other sectors, most notably tourism, where an influx of Libyans in 1988 – benefiting from the two countries' rapprochement – is calculated to have added up to $500m to the balance of payments. Analysis of economic performance in 1988–89 shows that the economy has become much more resiliant than in 1986 when IMF support was sought, and that Tunisia could weather the storm.

Drought, affecting the country's most productive regions, was a major reason why gross domestic product (GDP) failed to meet projections, posing a serious threat to efforts to reform the economy. Planning and Finance Minister Mohamed Ghannouchi and other officials announced revised projections for economic performance when the extent of problems became clear in mid-1989.

GDP, which grew by only 1.5 per cent in 1988, was now projected at 3.5 per cent in 1989, down from an original figure of 5.5 per cent. Meanwhile, the current account deficit was set to rise from a projected TD285m to TD415m, double the 1988 level. These revisions were hardly surprising as food imports were 40 per cent above target in the first five months, maintaining the trend towards higher trade deficits in 1989. Also causing concern was the estimated TD5.4bn debt, with the rising value of the dollar (in which currency about 60 per cent of Tunisia's debt is denominated) pushing up interest payments. It is calculated that the dollar's strengthening will add TD100m to interest payments in 1989, when debt service is estimated at TD1,155m. According to the central bank, the debt service ratio is about 33 per cent.

International support

In order to overcome these difficulties Tunisia again mobilised substantial quantities of multilateral and bilateral support. The World Bank approved a $84m loan to back agricultural reforms, supporting efforts to stimulate private sector farming, as well as reforms to pricing and marketing mechanisms. The Bank also approved a $95m loan for primary and secondary education and vocational training, and $5.5m to back oil exploration with the purchase of equipment, technical assistance and training.

The EC also announced new aid, including funds from the European Investment Bank which approved a ECU17m ($18m) loan for Banque Nationale de Developpement Touristique (BNDT) for on-lending for hotel and other tourism investments, drawn from the ECU131m third EC financial protocol. BNDT in mid-1988 had been allocated a ECU10m EIB loan which was quickly used.

Among other supporters, the African Development Bank announced a $75m structural adjustment loan and funds for company restructuring, also supported by the World Bank with a loan worth $130m. Arab agencies were also active.

Leading European trading partners continued to finance their rivalry with credits. Reflecting its concerted campaign to replace France as Tunisia's leading supplier, Italian trade figures showed a steady increase. In 1988 Italian exports to Tunisia rose to IL604.9m from IL516.6m in 1987 and IL486.1m in 1986. Tunisian exports to Italy rose to IL625.2m, from IL514.4m in 1987 and IL457.7m in 1986.

France responded to this challenge: following a visit to Tunis by President Mitterrand in June 1989, accords offering credits worth more than Fr1bn were signed on 21 August during a visit by the French Minister of State for the Economy, Budget and Finance Pierre Beregovoy. These were offered at very favourable terms, with interest set at 1.5 per cent (down from 2 per cent or 2.5 per cent in previous official lines) on loans repayable over 33 years (previously 20–25 years), with 10 years' grace.

Towards an open economy

In order to overcome domestic problems the government showed its clear commitment to pushing ahead with liberal reforms, most notably by acting to ease the burden on subsidies on the state budget and the Caisse Generale de Compensation (CGC), which in 1989 it was calculated would pay out an unprecedented TD400m to support food prices. Subsidies were cut in August 1989 – and further reforms were planned – balanced by new mechanisms to protect the poor.

These involved increased prices for most bread and basic foods. In a move previously used in 1986 to avoid a replay of the 1983–84 bread riots, the cost of a large loaf was held at the same level, but its weight was reduced by 100 grams to 500 grams. Subsidy cuts were seen as essential. In his 25 July speech Ben Ali said that subsidies absorbed a quarter of recurrent state spending, 'sufficient to build 150 secondary schools or eight dams a year'. According to national economy ministry data, the problem would get worse: in the period 1984–88 subsidies channelled through CGC cost TD960m, but in 1989–93 were estimated to rise to TD2.7bn. This was mainly due to the high cost of cereal's subsidies, forecast to cost TD243.7m in 1989 – 85.5 per cent of the total and 45 per cent higher than in 1988.

Major changes were planned to stimulate growth and investment, including further tax reforms following the introduction of a value added tax on 1 July 1988, which gradually began to have positive effects. This was to be followed by a profits tax, the Impot sur les Benefices, and a unified income tax, the Impot General sur le Revenu (IGR), both to be presented in 1989.

Fiscal reforms were intended to ease the tax burden on those paying very high rates of tax, which under the current regime does not necessarily mean the highly paid. Under consideration was a reduction of the maximum level of income tax to 40 per cent.

More controversial would be reforms to the law governing employment, which employers argue can be very restrictive but which trade unionists say is essential to protect already vulnerable workers from further privations. The government, supported by the World Bank, had decided to act on labour law and wage bargaining in order to make Tunisia even more attractive to potential investors. However, it was likely to move with caution – putting in place safety nets for those whose jobs would be in jeopardy as companies were 'restructured' – to avoid a new round of labour disputes, polarising the country and damaging the Ben Ali consensus.

For the longer term more radical measures still are envisaged, including convertibility for the dinar. For some liberals and business people convertibility is seen as the acid test of the government's liberal

COUNTRY PROFILE

Political system: Legislative power is vested in the 136-member unicameral National Assembly. Executive power is held by the president, who as head of government appoints the prime minister and council of ministers. Gained independence in 1956. Formerly a French Protectorate.

Political parties: Rassemblement Constitutionnel Démocratique (RCD) is the new name given in February 1988 to the ruling Parti Socialiste Destourien (PSD) of former President Habib Bourguiba. It occupied all the seats in the National Assembly. The centre-left Mouvement des Démocrates-Socialistes (MDS), the Parti Communiste, and the centre Mouvement de l'Unité Populaire were the only legal opposition parties until September 1988 when the president announced the legalisation of the liberal Parti Social pour le Progres (PSP) and the leftist Rassemblement Socialiste Progressiste (RSP). Mouvement de Tendance Islamique (MTI), the main Islamic party, remained illegal up to the end of 1988, though there were hopes that it would be recognised, perhaps under a new name.

Population: The population of 7.6m (1988 estimate) is growing at an annual rate of 2.7 per cent, and 40 per cent are aged under 15. Islamic observance is strong.

Main cities/towns: Tunis (population 1.7m), Sfax (600,000), Kairouan (450,000), Bizerte (400,000), Sousse (340,000).

Language: The official language is Arabic, but French is widely spoken and is acceptable for business purposes.

Media
Press: National dailies include (in French) *Le Renouveau, La Presse, Le Temps* and (in Arabic) *Es Sabah, Al Houria.* Also several periodicals and provincial newspapers are published.
Broadcasting: Nationwide radio and television service operated by Radiodiffusion Télévision Tunisienne; radio in Arabic, French and Italian, TV in Arabic and French. There are no commercial broadcasting stations. There are 742,000 radios (1985) and 705,000 TV sets.

External trade: Regular trade deficit, which is partly offset by earnings from tourism and remittances by emigrants. Current account deficit offset by net inflows of capital. External debt has risen during the 1980s to more than 50 per cent of GDP. Heavy government subsidies have added to the problems. An attempt in 1984 to remove food subsidies caused rioting and had to be retracted. But gradual subsidy cuts were subsequently implemented.
Exports: Textiles and leather (30.3 per cent of total value in 1986), petroleum and products (23.8%), food products (12.2%), industrial goods (7.2%). Main destinations: France (25.3% of 1985 total), Italy (16.1%), German FR (11.1%), USA (6.6%), Spain (6.4%).
Imports: Foodstuffs (15.6 per cent of total value in 1986), capital goods (27%), consumption goods (26.5%), raw and semi-finished materials (20.6%), energy (10.3%). Main sources: France (typically 25-28% of total value), Italy (14-16%), German FR (10-12%), USA (7-9%), Spain (4-6%), UK (2-3%).

Agriculture: Sector's share of GDP dropped from 56 per cent in 1960 to 13.5 per cent in 1986, and now employs under 30 per cent of workforce. Although production has increased since 1970s, imports of food and feedstock have shot up and currently meet 40 per cent of domestic requirements. Rural depopulation, an inequitable land tenure system, soil erosion, over-grazing and low producer prices remain the major constraints to development. Main crops are wheat, barley, maize, sorghum, olives, dates and citrus fruit. Around 55 per cent of the total land area is cultivated. Only a relatively small area is irrigated, though several large-scale irrigation and drainage projects are under way. By 1986 there were 15 dams with a total 945m cu metres capacity in existence, with another three (259m cu metres) under construction. Water provision, especially drinking water, will be a major problem in coming years. Fishing and poultry farming are growing in importance.

Manufacturing: Sector contributes 18 per cent to GDP and is based primarily on processing domestic raw materials, notably agricultural commodities and minerals. Major industries include textiles, foodstuffs, steel, cement, phosphate processing and downstream chemicals. Manufacturing has been a major catalyst to overall economic growth, with output rising by 8-9 per cent a year between 1983-86. Foreign investment is being actively sought in vehicle assembly, engineering and electrical industries, and these industries are growing particularly rapidly. Particular emphasis is placed on export-orientated industries and on encouraging the decentralisation of manufacturing away from Tunis and the main coastal towns.

Mining: Tunisia is the world's sixth largest source of phosphates (5.8m tonnes in 1986), though the quality of the rock mined is poor. Extraction (largely in Metlaoui and Gafsa) is geared increasingly towards local phosphate processing rather than exporting it in a raw state. Chemicals industries are being developed. Other important minerals mined include lead (3,000 tonnes mined in the coastal region), zinc (8,000 tonnes), iron-ore (310,000 tonnes), potash, barytes (13,000 tonnes), salt (440,000 tonnes), fluorspar (35,000 tonnes) and uranium.

Hydrocarbons: Oil reserves estimated at 1.65bn barrels. The main producing fields are El-Borma and Ashfat and Tazarka in the Gulf of Gabes. Other small discoveries have been made offshore in the Gulfs of Hammamet and Gabes, but are largely uncommercial. Weak prices, stagnant production (around 5m tonnes p.a.) and an increase in local consumption will result in Tunisia becoming a net energy importer by the early 1990s. Gas reserves put at 88 bn cu metres. Production of natural gas has fallen to around 373m cu metres p.a. The country receives gas as royalties, however, from the Algerian Trans-Med gas pipeline, which crosses 360 km of Tunisian territory. A new liquified petroleum gas plant is being built for the El Borma field.

Energy: Lack of sufficient refining capacity requires Tunisia to import substantial quantities of petroleum products. Emphasis on expanding electricity production and distribution (up to 3.75bn kWh in 1986), and stepping up oil and gas exploration programme.

Membership of international organisations: African Development Bank, Arab Fund for Economic and Social Development, Arab League, FAO, GATT, IDA, IFAD, IFC, ILO, IMF, Islamic Conference, Islamic Development Bank, OAPEC, UN, WHO, WIPO, World Bank.

commitment. Supporters say that capital flight would not prove a problem as exchange controls would remain. The government is more cautious, arguing that while convertibility could give Tunisia an edge over its rivals for investment, it would also mean further devaluation and pressure on reserves which for now Tunisia can ill afford. Officially, convertibility is on the agenda, but for the next five-year plan, which runs from 1992–96.

Maghreb developments

Arguing that convertibility must wait for the optimum conditions is Khelil, a leading exponent of liberal economic theory in the regime. However, there are other arguments. Baccouche – more often associated with caution and even *étatiste policies* – is understood to have argued that it may not be necessary to wait for the optimum conditions, but to go ahead with convertibility as one way of speeding Maghreb unity.

Baccouche's argument is based on the assertion that to accelerate moves to bring North African states together, convertibility could be a necessity. This underlines Tunisia's commitment to Maghreb unity – expressed in its membership of the five-nation Arab Maghreb Union (UMA).

Even when Tunisia's ambivalent relationship with Libya is taken into account, the pace of developments at a regional level accelerated sharply over 1989. Among major development was a decision to go ahead with several industrial projects with Maghreb partners – including the 7 November oilfield, to be operated jointly with Libya – and agreements to regulate bilateral trade. Advances in conventional trade were reported. According to provisional figures, Banque Tuniso-Libyenne (a joint institution revived as a result of the two states' rapprochement) had transacted TD33m in business by the end of 1988, mainly during the latter months of the year. The influx of Libyans in 1988, unlikely to be repeated on such a scale in 1989, was credited with supporting the balance of payments during a drought year.

As remarkable was the growth in trade outside the authorities' direct control. By mid-1989 it was clear throughout the Maghreb, and especially on Tunisia's borders, that efforts should be made to crack down on uncontrolled cross-border trade. A checkpoint was set up within Libya's borders to stop the export of subsidised goods which include tea, coffee, cheese and cooking oils. Cross-border trade had in some cases got out of control: in Tunisian markets, for example, olive oil from Turkey, Spain and even Algeria was available at prices cheaper than local brands. It was oil imported into Libya for sale at subsidised rates resold by travellers who could get a higher price (and dinars to spend on cheaper goods) in Tunisia.

The olive oil saga showed one facet of Tunisia's cross-border trade. The government wanted to promote others – not least trade with Europe as the EC prepared for 1992 and the single market. To this end further incentives for investment would be introduced over the next few years, while the regime would monitor its domestic opponents carefully to insure that a replay of the troubles in Bourguiba's last years did not emerge to undermine further its attempts at economic recovery.

BUSINESS GUIDE

Time: GMT +1.

Climate: Northern coastal area has Mediterranean climate with warm, rainy winters (Dec-Mar) and hot summers. Southern and inland area is hot and arid. Temperature in Tunis ranges from 6-14°C in Jan to 21-33°C in Aug.

Entry requirements
Passport: Required by all.
Visa: Generally not required for visits of less than 90 days. Exceptions: Belgium, Luxembourg, Netherlands, Australia. Rules are complex; check with embassy.
Prohibited entry: Nationals of Israel and South Africa.
Currency: Import/export of local currency prohibited. No restrictions on amount of foreign currency which can be imported, although amounts over TD500 should be declared to enable re-export. Note that reconversion of dinars into foreign exchange is limited to 30 per cent of any foreign currency previously converted, or TD100, whichever is the greater. All currency forms should be retained.

Health precautions
Mandatory: Yellow fever vaccination certificate required if arriving from an infected area.
Advisable: Water precautions should be taken outside main towns.

Representation overseas: Abu Dhabi, Algiers, Amman, Ankara, Athens, Baghdad, Beijing, Beirut, Belgrade, Berlin, Berne, Bonn, Brussels, Cairo, Dakar, Damascus, Doha, The Hague, Islamabad, Jeddah, Kinshasa, Kuwait City, Libreville, Lomé, London, Madrid, Moscow, Muscat, New Delhi, Ottawa, Paris, Prague, Rabat, Rome, Sana'a, Stockholm, Tripoli, Vienna, Warsaw, Washington.

Air access
National airline: Tunis Air.
Other airlines: Aeroflot, Air Algérie, Air France, Alitalia, Austrian, Balkan, British Airways, CSA, Egyptair, Iberia, Interflug, Iraqi Airways, JAT, KLM, Kuwait Airways, Libyan Arab, LOT, Lufthansa, MEA, Royal Air Maroc, Royal Jordanian, Sabena, Saudia, Swissair, Syrian Arab, THY Turkish.
International airports: Tunis-Carthage (Code: TUN), 8 km from city, duty-free shop, bar, buffet, restaurant, currency exchange, post office, car hire; Djerba-Melita (Code: DJE), 7 km from city, duty-free shop, bar, buffet, restaurant, bank, hotel reservations, post office, shops, car hire (all restricted hours); Monastir-Skanès (Code: MIR), 8 km from city, duty-free shop, bar, buffet, restaurant, bank, hotel reservations, post office, shops, car hire (all restricted hours); Sfax-el Maou (Code: SFA), 5 km from city; Tozeur-Nefta (Code: TOE), 10 km from city, duty-free shops, bar, restaurant.

Surface access
Main ports: Bizerte (oil), Sfax, Gabès, Sousse (goods), Tunis-La Goulette (passengers), and Rades (a newly built container port). Regular shipping services operate to Tunis from France, Italy and Malta.
Overland access: By road and rail from Algeria and Libya.

Hotels: Classified into four categories with special rates applying to deluxe hotels. Government hotel tax and service charge are added to bill.

Electricity supply: Electricity supply is a mixture of 220 V AC (in tourist resorts) and 110 V AC.

Car hire: International or national driving licence required. Traffic drives on the right. 100 km/h speed limit. Permission must be obtained to drive in Saharan areas.

City transport
Taxis: Available in all main towns and fairly easy to obtain. Fares are metered and are generally cheap.
Buses: Extensive services operate in main towns.

National transport
Air: Tunis Air operates regular domestic services linking Tunis with Djerba, Monastir, and Tozeur. The air taxi company, Tunisavia, also operates weekly flights to Sfax and some other towns.
Road: 18,260-km network linking all main towns. Long distance taxis *(louages)* operate between all main towns. These are considered the fastest method of road transport.
Buses: Extensive services; long-distance routes include: Tunis-Kelibia; Tunis-Kairouan; Sousse-Sfax; Sousse-Gafsa; Sfax-Gabès-Djerba.
Rail: A 2,000-km network links the main towns. Two classes with some additional 'grand confort' accommodation; refreshment facilities usual, air-conditioning in 'grand confort' and some first class cars; many routes are served by rail cars. Services include: Tunis-Nabeul/Sfax (frequent, average journey time 3.75 hrs); Tunis-Bizerte (several rail cars/day; 1.75 hrs); Tunis-Ghardimaou (several services/day; express 3-3.75 hrs; rail car 3.75-4 hrs); Tunis-Gabès; Tunis-Béja; Sfax-Gabès; Sfax-Tozeur.
Water: Regular daily ferries from Sfax-Iles Kerkenna and Jorf-Djerba Island.

Public holidays
Fixed dates: 1 Jan, 18 Jan, 20 Mar (Independence), 21 Mar (Youth Day), 9 Apr (Martyrs' Day), 1 May (Labour Day), 1 Jun, 25 Jul (Republic Day), 3 Aug, 13 Aug (Women's Day), 3 Sep, 15 Oct (Evacuation Day).
Variable dates: Aid el-Seghir (2 days, Eid al-Fitr), Aid el-Kebir (2 days, Eid al-Adha), Islamic New Year, Prophet's Birthday.,

Telecommunications
Telephone: Internal automatic service covers nearly all of the country. 24-hour service operates for international calls with direct dialling available to many countries. There were 217,000 lines in 1986. Dialling code for Tunisia: prefix + 216 followed by area code (1 for Tunis) followed by subscriber's number. IDD access code from Tunisia: 00.
Telex: International facilities available from Central Post Office, Tunis and major hotels. 3,082 lines in 1986.
Telegram: 24-hour service at Central Post Office, Tunis.

Representation in capital: Algeria, Austria, Belgium, Brazil, Bulgaria, Canada, China PR, Cuba, Czechoslovakia, Egypt, France, German DR, German FR, Ghana, Greece, Iraq, Italy, Ivory Coast, Japan, Jordan, Korea PDR, Rep. of Korea, Kuwait, Lebanon, Libya, Mauritania, Morocco, Netherlands, Oman, Pakistan, Poland, Portugal, Qatar, Romania, Saudi Arabia, Senegal, Spain, Sudan, Sweden, Switzerland, Syria, UAE, UK, USA, USSR, Vietnam, Yugoslavia, Zaïre.

Hotels: Tunis
Africa (Meridien), Avenue H Bourguiba (tel: 347-477; tx: 15536).

Du Lac, 2 Rue Mont Calmé (tel: 258-322; tx: 12277).

El Mechtel, Boulevard Ouled Haffouz (tel: 783-200; tx: 14611).

Hilton, Avenue Salambo, P.O. Box: 165 (tel: 282-000; tx: 15372; fax: 782208).

Iben Khaldoun, Tunis 1002 (tel: 892-211; tx: 13923; fax: 780273).

International, 49 Avenue H Bourguiba (tel: 254-855; tx: 15198).

Maison Dorée, 6 Bis Rue de Hollande (tel: 240-632; tx: 13168).

Majestic, 36 Avenue de Paris (tel: 242-848/9; tx: 15156).

Other useful addresses
Agence de Promotion des Investissements, 7 rue de l'Arabie Saoudite, Tunis.

Banque Centrale de Tunisie, 8 rue de la Monnaie, Tunis (tel: 259-977, 254000; tx: 12375, 12322, 12138, 13308).

Turkey

Jim Bodgener

Turkey's Prime Minister Turgut Ozal faced a gruelling autumn, it was clearly apparent in late summer 1989. With his ruling Motherland Party's (ANAP's) electoral standing at an all-time low, and with inflation sliding towards three digits, how long the tubby premier could fend off calls for early general elections was uncertain – particularly with the impending selection of a replacement for President Kenan Evren in November.

It would be the most testing time yet since the turbulent end to the 1970s and the 1980 military coup for the still youthful parliamentary tradition in Turkey, not to mention the structural adjustment programme set in train under IMF tutelage in 1980 and adumbrated by the two ANAP governments since the return to civilian rule in 1983. While money supply was at record levels, the economy was contracting so fast that some predictions were economic growth would fall back to 2.5 per cent for the year, compared with the previous slump in 1988 to 3.4 per cent.

Economy

Ratings in June opinion polls of around 15 per cent nationally were a sorry pass for Ozal, who had presided over perhaps the most far-reaching structural economic changes in the Turkish economy in such a short span since the founding of the republic in the 1920s. No longer was the dominant economic mentality one of insularity and hostility towards integration with the world economy; indeed, the export drive aided by tourism had produced a $1,500m current-account surplus in 1988, the first in a long series of deficits since 1973.

Then the surplus was unhealthy in that it was dependent on remittances from *gasterbeiters* (expatriate workers) in West Germany and other European countries; now the main worry of the financial authorities was how to insulate the surplus' inflationary implications if monetarised for the domestic economy.

Tourism was the success story of the Turkish economy in 1988 – earning $2,200m from 4.1m visitors, whose numbers had doubled since 1983. That offset a decline in the rate of growth of exports, by 15 per cent to $11.6bn compared with the 36 per cent increase in 1987.

But on the other hand, the current account surplus owed as much to a sharper fall-off in the rate of increase of imports, by only 5 per cent to total $14.3bn. This was evidence that the surplus masked deep industrial and commercial distress in the domestic economy from high inflation, which peaked at around 87 per cent in November 1988. This meant the government had failed to curb public-sector spending sufficiently, and that the private sector had shouldered the brunt of the consequent economic deterioration.

Local elections

However Ozal, in the first quarter of the year, appeared to carry on regardless, arousing widespread concern that he had lost touch with reality. But had he not realised the government's true predicament, it was brought home suddenly and forcefully in ANAP's disastrous showing in end-March local elections.

Here ANAP only managed to scrape up a dismal 22 per cent of the vote, compared with the 28 per cent won by the main opposition Social Democratic Populist Party (SDPP), and the surge into second place with 25 per cent of the third-largest parliamentary grouping, the True Path Party (TPP) led by Ozal's arch-rival, former premier Suleyman Demirel. He was ousted in the 1980 coup while Ozal was kept on as economic supremo by the military.

To ANAP's chagrin, it lost all three main cities to the SHP. Ankara was expected to go, Izmir was finely balanced, but the loss of what was previously regarded as an impregnable bastion in Istanbul under the redoubtable mayor Bedrettin Dalan was a stunning blow. Whether or not the Turkish brand of democracy was strong enough to stand the stresses and strains of a party which was in power at the periphery being in opposition at the centre was now to be one of the main themes of Turkish politics.

True to form, Ozal lost no time in belittling the local election result, defiantly ruling out general elections. At the same time, he instituted a sweeping cabinet reshuffle in which his brother, Yusuf Bozkurt Ozal, was dropped as state minister in charge of the economy, along with his nephew, Husnu Dogan, as minister of agriculture. This was to freshen up the government's image, jaded by accusations that Ozal had retreated from the second half of 1988 onwards

KEY FACTS

Official title: The Republic of Turkey
Head of state: President Turgat Özal
Head of government: Prime Minister Yildirim Akbulut
Ruling party: The Motherland Party
Capital: Ankara
Official Languages: Turkish
Currency: Turkish Lira (TL)
Exchange rate: TL2,292 per $ (Dec 1989)

Area: 779,452 sq km
Population: 54.1m (1988)
GNP per capita: $1,342 (1988)
GNP real growth: 3.4% (1988)
Labour force: 19.0m (1988)
Unemployment: 14.0% (1988)
Inflation: 75.4% (1988)
Trade balance: –$1,800m (1988)
Foreign debt: $37.6bn (1988)

into an ever more distanced rule surrounded by advisers and close family becoming increasingly Ottoman in its appurtenances.

The new team was largely of unknown faces. By mid-year it had yet to prove itself amid charges that too many heads were confusing formulation of clear policies. Top bureaucrats like Rusdu Saracoglu were becoming increasingly irritated too with political interference and vacillation in sound economic management.

High inflation

And hard behind the government was the spectre of inflation. Despite strict austerity measures introduced following the local elections, in June and July, traditionally months of low prices due to the harvests, the gradual decline since the November peak inexorably turned around in a climb back upwards to 74 per cent, from hovering at around 62 per cent in the spring.

The main reasons were the big increases in food prices because of the lower harvest due to drought, and price hikes introduced by State Economic Enterprises (SEEs). But in July the government followed a May cave-in to workers' pay demands with massive handouts to civils servants and farmers – to compensate them for income lost to inflation – auguring badly for prices in the autumn as production contracted.

A prime engine of inflation was budget shortfalls. Although the consolidated budget deficit declined by 20.4 per cent to TL922bn in the first half of 1989 compared with January-June 1988, the state's cash deficit actually increased by 1,594 per cent to total TL1.8tn over the same period. First half budget revenues rose by 77.2 per cent to total TL12.9tn and expenditure by 63.9 per cent to total TL13.9tn. To blame was major increase in domestic borrowing, which over the above period rose substantially by 294 per cent to total TL4.6tn. For all the austerity measures, and criticism from international organisations like the OECD, the IMF and the World Bank, high domestic borrowing remained the main domestic source of revenue for the government.

According to Finance and Customs Minister Ekrem Pakdemirli, the government expected to finance the hand outs from sound resources – part of an expected increase in tax revenues of TL1.3tn in 1989 would be channelled towards the white collar salary increases, he said.

The treasury had a battle plan ready – simply to issue bonds and bills to the tune of TL 1.5tn, in order to soak up the excess liquidity. To achieve this, however, in the open market without coercion as pledged, it would have to offer attractive interest rates well above inflation – which in turn might force the banks to increase their rates as well to keep in step.

Increasing rates to keep in step with treasury bonds might attract enough liquidity to ease banks over an uncomfortable period in the autumn when one-year deposits contracted at 85 per cent a year previously matured, but that would only be deferring the problem into 1990.

If interest rates were skewed in this way by government borrowing, it was likely to have an adverse effect on the capital markets, particularly stocks and shares, which were beginning to stage a recovery from the doldrums of 1988. Similarly, the privatisation programme would suffer from the dampening of investors' appetites for alternative savings instruments other than government bills and bonds and deposits. Already in late July, the rally on the stock exchange was fast fading.

Adding to the strain on the financial organs of government was the decision to make the treasury stand on its own in future, and not come running cap in hand to the central bank whenever its own resources were inadequate. The treasury repaid all its debts to the central bank in July, amounting to around TL1.3tn – during the first half of the year its borrowings from the central bank had increased by 20 per cent to total TL984bn.

KEY INDICATORS

	Unit	1984	1985	1986	1987	1988
Population	m	48.3	49.3	50.3	51.4	52.4
GNP real growth	%	6.0	5.1	8.1	7.4	3.4
GNP per capita	$	1,037	1,080	1,159	1,325	1,342
Inflation	%	48.4	45.0	34.6	38.9	75.4
Unemployment	%	16.5	16.5	15.5	15.2	14.0
Agricultural production	% growth	0.1	3.1	4.1	0.6	4.1
Wheat	m tonnes	17.2	17.0	19.0	18.9	20.5
Sugar-beet	m tonnes	11.1	9.8	10.7	12.7	11.0
Industrial production	% growth	8.8	6.6	11.1	9.7	7.4
Manufacturing	% growth	9.3	5.7	10.5	10.1	18.4
Cement	% growth	15.7	11.7	13.7	9.8	3.1
Motor vehicles	% growth	28.6	10.2	35.9	30.7	12.7
Mining	% growth	18.2	17.5	15.3	1.1	–13.4
Energy	% growth	6.0	9.5	15.7	11.6	12.0
Oil	% growth	–4.5	0.7	12.3	10.2	–2.5
Electricity	% growth	12.2	12.2	16.7	11.7	8.2
Exports	$m	7,389	8,255	7,583	10,322	11,846
Agricultural products	$m	1,749	1,719	1,886	1,853	2,341
Manufactures	$m	4,337	5,348	5,324	8,065	8,943
Minerals	$m	240	244	247	272	377
Imports	$m	10,331	11,230	10,664	13,551	13,646
Oil and oil products	$m	3,639	3,611	2,008	2,923	2,778
Chemicals	$m	1,210	1,294	1,422	1,937	1,984
Iron and steel	$m	859	1,059	1,028	1,537	1,655
Machinery	$m	1,599	1,552	2,304	2,452	2,400
Balance of trade	$m	–2,942	–2,975	–3,081	–3,229	–1,800
Current account	$m	–1,407	–1,030	–1,528	–979	1,500
Workers' remittances	$m	1,807	1,774	1,696	2,102	1,844
Total foreign debt	$bn	21.2	25.3	31.2	38.3	37.6
Exchange rate	lire per $	366.7	522.0	674.5	857.2	1,422.3

* estimated figure

Foreign trade and tourism

There was more gloom too on the external trade front. The current account in January-April still registered a $211m surplus compared with a $425m deficit in the same period of 1988. But this fell compared with the previous month's surplus of $255m – and ominously, tourism revenues declined on an annual basis by 11 per cent to total $375m, admittedly still only one month into the important beach tourism season from April to October.

The government had targeted tourism revenues to increase to around $3,000m in 1989 from about $2,200m in 1988. According to Tourism Minister Ilhan Akuzum, however, tourist numbers were still expected to increase by around 10-15 per cent on the 4.1m visiting in 1988. The slowdown was blamed on poor marketing, over-pricing, and transport difficulties, including a lack of charter flights.

Unfortunately, in 1989 the sector also had to contend with a depressed season generally for Mediterranean resorts, with higher interest rates and leaner pockets in most of Europe keeping people closer to home. This had an impact on Britons especially, and combined with bad publicity over human rights and sanitation. Whereas their numbers doubled to around 450,000 in 1988, it seemed doubtful that 1989's tally would even level-peg.

The rate of increase in accommodation construction had outpaced that in tourists, so there were smaller slices of business all round. Earlier on in the season, there was a flood of advertisements in the Turkish press of cut-price offers by hoteliers frantically trying to fill space.

However, the gap left by British visitors was filled by increases in the staples of the Turkish trade, from West Germany, France and Austria and Switzerland. And running counter to expectations was an encouraging upsurge from Scandinavia, where human rights criticism of Turkey in the past had been most vociferous. Clearly, Turkey's image in Europe despite the poor British press was gradually changing for the better from the horror stereotypes of films like 'Midnight Express' and 'Dark Holiday'.

The outlook for the remainder of the year appeared to be better, however, according to the Aegean Association of Tour Operators. This was partly because many hoteliers and tour operators had slashed their prices.

The foreign trade deficit fell by 32 per cent in the first four months of 1989 compared with the same period of 1988, but economists said this masked the domestic recession. Exports level-pegged over the period at $3,812m, but imports were down by 5 per cent to $4,363m, symptomatic of production cutbacks due to demand slumps. The government revised its 1989 targets downwards by about $2bn for imports to $14,500m, and by around $1bn for exports to $11,500m.

Political situation

While generally in the economy there were encouraging reforms, the danger was that these might be smothered by short-term measures ensuing from Prime Minister Turgut Ozal's difficult political predicament.

All this was mounting pressure on Ozal to move decisively. However, even in early August the premier remained as enigmatic as ever about his intentions. Opposition parties were clamouring in the heat of summer for early general elections so that a fresh parliament with a truly popular mandate could elect a replacement for President Kenan.

Internally, ANAP at times appeared to be in danger of falling apart, though Ozal held the ring as the central mediating force in the party between its rival factions. But his leadership bond had been weakened by serious bickering following the local election catastrophe. Enfeebled political leadership gave way to a lack of dialogue and poor co-ordination between the central bank, treasury, and State Planning Organisation. Top bureaucrats were meddling overmuch in politics beyond their remits, alleged other officials.

How ready the main opposition parties might be for government, however, was a moot point – in terms of experience in office, the TPP was ahead of the SHP. The latter anyway was riven with differences between the party executive controlled by its moderate, pragmatic secretary general, Mr Deniz Baykal, and left-wing deputies. The left-wingers came largely from the east, where the continuing official rejection of Kurdish ethnic aspirations widened party divisions.

Mr Demirel on the other hand was fiercely battling to wrest back leadership of the traditional right lost in the 1980 military coup. He had not let up in strident campaigning since the local elections, concentrating on the cities to build on his strongholds in the countryside. His attempts to persuade the business community that he was a changed man from his last premiership during the troubled end of the 1970s were gaining ground, notably in the forum of the influential Turkish Businessmen's and Industrialists' Association (TUSIAD).

At least thus far, the military did not appear to be stirring in their barracks. Towards the end of July, Prime Minister Turgut Ozal made a round of visits to senior military officers, to smooth any ruffled feathers at his surprise appointment of General Dogan Gures as Land Forces Commander. The appointment was yet another example of the degree to which the Turkish military had become professionalised since the return to civilian government in 1983, into a service arm of the civilian authorities.

In addition, the army's reputation had been badly impaired by its apparent failure to curb an offensive since the spring by guerrillas of the separatist, marxist Kurdish Workers' Party (PKK) in the south-east.

Foreign affairs

Abroad, even the crisis with Bulgaria over its driving out of more than 200,000 refugees since the spring was not enough to distract domestic criticism of Ozal in late summer. The upswelling of protest that erupted in May in Bulgaria by ethnic Turks against Sofia's four-year Bulgarisation assimilation campaign had been building over the past year, encouraged by the example of glasnost across the border in the Soviet Union.

However, the Bulgarian government clearly overreached itself when the expulsion of the protest movements' leaders into Turkey was followed by an outpouring of Bulgarian ethnic Turks voting with their feet. While economic dislocation threatened in Bulgaria as key workers left for the border crossing, even Soviet mediation attempts had not solved the impasse between Sofia and Ankara, which was demanding a formal immigration agreement permitting a more orderly egress.

Warming bilateral relations with the big bear to the north, however, were hardly dented in the spring by the arrival in Trabzon airport of an advanced Mig-29 fighter with a fugitive Soviet airman. While Turkey's NATO allies fumed with indignation, the aircraft was handed back without letting US specialists get a fleeting glimpse of the plane – though the pilot later went on to America.

Steadily increasing trade underpinned the growing friendliness with Moscow. Major credit agreements were reached, and total bilateral trade was expected to increase to $1,000m for the year on the back of a Soviet natural gas import programme.

Turkey's links within NATO had been under strain anyway, Turkish objections to the country's flanking and expendable role in the so-called central front strategy being largely responsible for the latter's jettisoning. As regards the debate over a response to the Warsaw Pact's zero option offer, Turkey inclined naturally as a front-line state to the West German position of bending an ear.

By August, there had been no major breakthrough in the two main regional issues, relations with Greece and moves towards a settlement on the divided island of Cyprus. Though a subject of some interest in the Turkish press, the descent of Ozal's counterpart Andreas Papandreou from heart problems through the scandal of his love life and the corruption revelations involving his senior ministers towards electoral defeat in June was more than enough to put on hold the Davos process of reconciliation initiated in Switzerland in early 1988. However, Ozal and Constantine Mitsotakis, the new Greek premier, both expressed a desire to carry on the dialogue.

On Cyprus, Turkey continued its unwavering support for the northern Turkish-Cypriot administration headed by its leader Rauf Denktash in the UN sponsored talks with his Greek Cypriot counterpart, Georges Vassiliou. But although the talks continued slowly, the fundamental differences between the two sides on power-sharing and equality in any future bi-communal government appeared to be as wide as ever.

With its neighbours to the south, Turkey continued to tread as wary a path as ever, keeping to its policy of 'active neutrality' between Iraq and Iran even though the Gulf War had ended in summer 1987. Trade seemed likely to pick up most with the former; yet with the death of Ayatollah Khomeini it seemed trade with Iran might not be so beset with the uncertainties of previous prickly relations between the secular Turkish republic and the Iranian theocracy. Deep mutual suspicion continued to underlie relations with Syria.

Outlook

Thus Turkey headed towards the 1990s with many fundamental questions unresolved – the durability of the structural adjustment programme, whether or not the economy could shrug off perennially high inflation, and externally, whether or not the government could keep to the letter and spirit of the republic's founding father, Mustafa Kemal Ataturk, in his famous dictum, 'Peace at home, peace in the world.' Perhaps the most pressing was whether Prime Minister Turgut Ozal could hold together his creation, ANAP, with all its internal fissiparous tendencies, against the external buffeting of electoral disenchantment – and perhaps sooner than expected, a period in the wilderness of opposition.

COUNTRY PROFILE

Political system: Single chamber Grand National Assembly of 450 members elected every five years. Out of 67 provinces state of emergency still in force in nine in early 1988. New system of local government established in 1982. In 1987 the electorate voted by referendum in favour of permitting leaders of abolished parties to re-enter political life.

Main cities/towns: Ankara (population 3.3m), Istanbul, Izmir, Adana, Konya, Mersin, Bursa.

Media
Dailies: Five with circulation above 500,000. Most influential include *Hürriyet* (circulation 600,000), *Tercüman* (150,000), *Milliyet* (300,000) and *Günaydin* (300,000). Also several hundred regional papers. *The Turkish Daily News* is the main English-language publication.
Weeklies: Nokta is the most significant.
Broadcasting: Approximately 6.m radio receivers and 6.9m TVs in use. Service controlled by Turkiye Radyo Televizyon Kurumu (TRT). National, local, external and educational radio services and about 63 hours per week of TV are broadcast. A second TV channel opened in Ankara in 1986 and there are plans to extend it nationwide.
Advertising: Local code of practice/ethics in operation. Newspapers account for 45 per cent of advertising revenue, broadcasting 29 per cent (commercials allowed only at specified times) and cinemas 15 per cent.

Domestic economy: High levels of economic growth during 1970s based on well-developed agricultural sector and rapidly expanding manufacturing sector. Stabilisation programme introduced 1980 owing to rising imported energy costs, hyper-inflation and growing debt burden. Aim for 1989 is to curb growth further (1988, 3.4 per cent), aiming for an inflation rate of 50-60 per cent (from 75 per cent). Underlying structural problems remain, despite relative success of massive sales drive abroad and domestic monetary austerity. Tourism expanding fast, but faced trough in 1989 from general Mediterranean recession.

Employment: Agricultural sector employs 56 per cent of the workforce, industry 14 per cent, public and personal services 30 per cent. High rates of both unemployment and under-employment. Approximately 1.6m nationals working abroad, mainly in German FR. Workers' remittances important source of foreign exchange. Real earnings falling because of inflation.

External trade: Rapid growth in exports since 1980, particularly textiles, metals, foodstuffs and agricultural products but exports to Middle East have fallen sharply since drop in oil prices in 1986. Import growth slower by comparison owing to import and foreign exchange restrictions. Major imports include oil, machinery and chemicals. Principal trading partners include W Germany, Iraq, Libya, Iran, Italy, USA, France and UK.

Agriculture: Mainstay of economy accounting for around 20 per cent of GNP and 25 per cent of exports. Declining in relative importance, but with strong growth potential. Country runs steady food balance of payments surplus despite liberalisation of food imports since 1984. Self-sufficiency in basic foodstuffs although only a third of arable land under cultivation. Sector continues to suffer from lack of irrigation and low levels of productivity. Wheat output was down in 1989 because of drought. Food processing growing rapidly and agricultural products continue to provide bulk of export revenue. Principal crops include wheat, cotton, fruit and nuts, tobacco and sugar-beet.

Industry: Sector contributes 29 per cent to GNP. Dominated by large state-owned industries (State Economic Enterprises), mainly engaged in textiles, food processing, metals and motor vehicles. Rapid development of light industry and growth in exports of manufactured goods. High levels of industrial growth since mid-1970s despite low levels of capital investment and plant utilisation. Privatisation programme under way.

COUNTRY PROFILE

Main holding companies: Sabanci, Koc, Cukurova, Enka, Dogus, Kutlutas, Tekfen, Anadolu, Transturk.

Mining: Substantial mineral reserves, including copper, zinc, lead, iron ore, coal and lignite. Deposits of borax, wolfram and chromite are internationally significant. Etibank controls 60 per cent of all mining activity.

Energy: Heavy dependence on oil and gas imports despite efforts to develop local energy resources and stepping up of energy conservation programme. Extensive oil exploration under way to bolster falling indigenous oil production (2.6m tonnes). Rapid expansion of hydro-electricity capacity, including development of the Ataturk dam. Also plans to develop nuclear energy as prime source of electricity.

Banking: The state-owned central bank is Türkiye Cumhuriyeti Merkez Bankasi, which has its head office in Ankara and sub-offices in a few major cities. The state-owned TC Ziraat Bankasi acts as agent for the central bank in places where the central bank does not have an office. Ziraat Bankasi is the largest completely state-owned commercial bank, but Türkiye Is Bankasi, in which the state has major holding, is larger. The largest banks in the private sector are Akbank and Yapi ve Kredi. There are close links between banks and industry – several banks are owned by industrial holdings or specialise in particular sectors. However, the specialisation of such banks as Sekerbank (sugar), Etibank (minerals), Pamukbank (cotton) is steadily declining as they go more and more into general commercial banking.

Legal system: Derived from Swiss civil, German commercial and Italian penal codes during 1920s/30s; subsequently modified.

Membership of international organisations: ASSIMER, CCC, Council of Europe, EC (associate member), Economic Co-operation Organisation, ECOSOC, FAO, GATT, IAEA, IBRD, ICAC, ICAO, IDA, IDB, IEA, IFAD, IFC, IHO, ILO, IMF, IMO, INTELSAT, Interpol, IOOC, IPU, ITC, ITU, NATO, OECD, OIC, UN, UNESCO, UPU, WHO, WIPO, WMO, WSG, WTO.

Time: GMT + 2 (GMT + 3 from late Mar to late Sep).

Climate: Extreme regional variations. Apr-Oct warm and dry in north. Southern coast has higher temperatures and humidity. North can be cold Nov-Mar, south usually milder. Temperatures range from -40°C to 43°C.
Clothing: Lightweight May-Sep, tropicals in southern areas. Mediumweight and light topcoat Sep-Nov and Mar-Apr. Warm clothing for Nov-Feb.

Entry requirements: Visa required by all except nationals of some European countries and a few others.
Health: Cholera certificate required if travelling from an infected area. Anti-malaria and anti-cholera precautions advisable. Not usual to drink tap water, but hotels and restaurants supply spring water which is safe to drink, except at times of epidemic when all water and milk should be boiled.
Currency: No restrictions on amounts of foreign currency taken in; large amounts should be declared and endorsed by authorities. Not more than $1,000 worth of Turkish currency may be brought in or taken out. Exchange slips should be retained.
Customs: Personal effects allowed in duty-free. Gifts up to the value of (German FR) DM500 may be taken out duty-free. Advisable to produce invoices and foreign currency exchange slips to cover value of purchases. Export of antiquities and some antiques prohibited.

Air access: Frequent direct flights from most European cities to Istanbul; limited flight access to Ankara. Antalya, Dalaman and Izmir airports cover only international charter flights.
National airline: Turk Hava Yollari (THY).
International airports: Istanbul-Atatürk (IST), 24 km from city; Ankara Esenboga (ESB), 35 km from city; Izmir-Adnan Menderes (IZM), 25 km from city.

Surface access: Express rail services from Munich, Vienna, Venice, Belgrade, Baghdad and Tehran Coach services from Austria, France, West Germany and Switzerland, as well as a number of countries in the Middle East.
Main ports: Mersin, Iskenderun, Istanbul, Izmir, Samsun, Trabzon.

Hotels: Classified into five categories – de luxe and first to fourth class. Prices vary and many hotels reduce their rates between mid-Oct and mid-Apr. Service charge of 15 per cent usually added and tipping is extra. 12 per cent VAT is also added. Advance reservations are advisable. Tap water is drinkable in major hotels.

Restaurants: Both in hotels and outside as going to restaurants is a national pastime. Bill usually includes service charge, but optional extra tip for good service of up to 10 per cent.

Currency
Exchange rate: Turkish lira floats via rates determined daily by Central Bank of Turkey dealing room based on value of US dollar. All banks free to establish own rates within margins set by Central Bank.

Foreign exchange information
Account restrictions: Blocked accounts in operation. Non-residents and residents can maintain accounts in Turkey and use these accounts freely.
Trade policy: Recent liberalisation of import controls. Around 77 per cent of imported goods are now allowed free entry (subject to duties).

Credit cards: Access, Diners Club, VISA, American Express and Eurocard credit cards accepted in larger hotels, restaurants and shops only.

Car hire: Available at main hotels, airports and travel agents. International driving licence preferred, but most foreign licences accepted.

City transport: Metered taxis available in major towns and cities. Also available are the much cheaper 'Dolmus' taxis, which have fixed routes and carry up to eight passengers. Tipping not customary. Bus tickets must be bought in advance.

National transport
Air: Turkish Airlines operate regular services between Istanbul, Izmir, Ankara and other major towns.
Water: Steamship service between Istanbul and most major coastal towns.
Road and rail: Most major cities and towns are linked by regular rail and coach services.

Public holidays: 1 Jan, 23 Apr, 19 May, 30 Aug, 29 Oct. Seker Bayrami and Kurban Bayrami (Muslim holidays) are also celebrated. These last several days and their dates are fixed by sightings of the moon.

Working hours
Government and business: (Mon-Fri) 0830 – 1230 and 1330 – 1730.
Banking: (Mon-Fri) 0830 – 1200 and 1330 – 1700.
Shops: (Mon-Sat) 0900 – 1300 and 1400 – 1900. Many food shops open on Sun.

Social customs: Hospitality very important. Turkey is a Muslim country and religion plays an important part in Turkish life. Practically all business entertaining conducted in restaurants and clubs.

Business language and interpreting/translation: Available from Turk Argus Ajans, Istanbul (tel:144-16-34, 149-84-66); Universal, Istanbul (tel: 167-24-88); Modern Tercume Burosu, Ankara (tel: 117-81-22, 118-14-70); and many other sources.

Telecommunications
Telephone: Approximately 3.8m phones, around half of which are in Istanbul or Ankara. International dialling code: 90 followed by 1 for Istanbul, 4 for Ankara. Public phones operated by discs. Direct dialling available. For long distance calls first dial 9, and for overseas calls 99.
Telex: Available in major hotels and certain main post offices in Ankara and Istanbul. Country code:607TR.

Postal service: Main post offices in major cities are open 24 hours a day.

Trade fairs: International Trade Fair held annually (Aug-Sep) at Izmir.

Electricity supply: 220 V AC (110 V in parts of Istanbul).

Representation in capital: Afghanistan, Albania, Algeria, Argentina, Australia, Austria, Bangladesh, Belgium, Brazil, Bulgaria, Canada, Chile, China PR, Czechoslovakia, Denmark, Egypt, Ethiopia, Finland, France, German Democratic Republic, Federal Republic of Germany, Greece, Holy See, Hungary, India, Indonesia, Iran, Iraq, Israel, Italy, Japan, Jordan, Republic of Korea, Kuwait, Lebanon, Libya, Malaysia, Mexico, Morocco, Netherlands, Norway, Pakistan, Poland, Portugal, Romania, Saudi Arabia, Somalia, Spain, Sweden, Switzerland, Syria, Thailand, Tunisia, Turkish Republic of Northern Cyprus, USSR, United Kingdom, USA, Venezuela, Yugoslavia.

Hotels: Ankara
Bulvar Palas, Ataturk Bulvari 141 (tel: 175020; tx: 42613).

Buyuk Ankara, Ataturk Bulvari 183 (tel: 1256655; tx: 42398; fax: 1255070).

Dedeman, Bulklum Sok 1, Akay (tel: 171100; tx: 42408).

Etap Mola, Ataturk Bulvari 80 (tel: 339065; tx: 42294).

Kent, Mithatpasa Caddesi 4 (tel: 1355050; tx: 42424; fax: 1344657).

Stad, Istiklal Caddesi 20, Ulus (tel: 3104848; tx: 42248).

Hotels: Istanbul
Buyuk Tarabya, Kefelikoy Caddesi Tarabya (10 km from city) (tel: 6210100; tx: 26203).

Cinar (near Ataturk Airport), Fener Mevkii, Yesilkoy (tel: 732910; tx: 28861; fax: 5735701).

Dedeman, Yildiz Posta Caddesi 50, Esentepe (tel: 1728800; tx: 22238).

Dilson, Siraselviler Caddesi 49, Taksim (tel: 1521305; tx: 25689).

Divan, Cumhuriyet Caddesi 2, Sisli (tel: 1314100; tx: 22402; fax: 1314100).

Etap, Mesrutiyet Caddesi, Tepebasi (tel: 1448880; tx: 24345).

Etap Marmara, Taksim Meydani (tel: 1448850; tx: 24137).

Hilton, Cumhuriyet Caddesi, Harbiye (tel: 1314646/50/60/70; tx: 27027; fax: 1404165).

Macka, Eytam Caddesi 35, Tesvikiye (tel: 1343200; tx: 23114).

Pera Palas, Mesrutiyet Caddesi 98-100, Tapebasi (tel: 1514560; tx: 24152).

Sheraton, Taksim Parki (tel: 131212160; tx: 22729).

Turban Carlton, Koybasi Caddesi 119, Yenikoy (on shore of the Bosphorus) (tel: 1621020; tx: 26260).

BUSINESS DIRECTORY

Car hire: Istanbul
Avis: Hilton Hotel (tel: 114-87-752; tx: 38144).

Budget: (tel: 149-57-14, 145-07-66; tx; 25908).

Europcar: Cumhuriyet Caddesi 47/2, Taksim (tel: 150-88-88; tx: 24175).

Hertz: Airport of Atatürk (tel: 573-59-87, 573-29-20; tx: 26019).

Chambers of Commerce
Adana Chamber of Commerce, Abidinpasa Caddesi, Adana (tel: 1-24-04 or 1-37-79).

Ankara Chamber of Commerce, Sehit Tegmen Kalmaz Caddesi 20, Ankara (tel: 310-41-45 or 324-32-60; tx: 42029).

Istanbul and Marmara Region Maritime Chamber of Commerce (tel: 143-54-95).

Izmir Chamber of Commerce, Ataturk Caddesi 126, Izmir (tel: 14-43-55 or 12-25-76; tx: 52331).

Banking
Türkiye Cumhuriyet Merkez Bankasi (central bank), Ataturk Bulvari 40, Ulus, Ankara (tel: 312-52-00-09; tx: 42993).

Akbank, Meclisi Mebusan Caddesi 65/69, Findikli, Istanbul (tel: 152-65-04; tx: 24139).

Garanti Bankasi, Mele Caddesi 40, Taksim, Istanbul (tel: 149-18-39; tx: 24538).

Iktisat Bankasi, Buyukdere Caddesi 165, Esentepe, Istanbul (tel: 172-70-00; tx: 26021; fx: 1747028).

Ottoman Bank, Voyvoda Caddesi 35-37, 80835 Karaköy, Istanbul (tel: 152-30-00).

TC Ziraat Bankasi, Ataturk Bulvari 42, Ulus, Ankara (tel: 311-11-85; tx: 44217).

Türk Ekonomi Bankasi, Istiklal Caddesi 284, Odakule 80050, Beyoglu, Istanbul (tel: 151-21-21; tx: 25358 tebu; fx: 1496568).

Türkiye Bankalar Birligi (bankers' association), Mithatpasa Caddesi 12, Yenisehir, Ankara (tel: 131-66-78).

Türkiye Emlak Bankasi (including Anadolu Bankasi), Büyükdere Caddesi 45 Maslak, 80670 Istanbul (tel: 176-16-20; tx: 27780).

Türkiye Halk Bankasi, Ilkiz Sokak 1, Sihhiye, Ankara (tel: 231-75-00; tx: 42004).

Türkiye Is Bankasi, Abdi Ipekçi Caddesi 75, 80412 Maçka, Istanbul (tel: 133-03-60; tx: 31000 isex).

Yapi ve Kredi Bankasi, Istiklal Caddesi 285, 80050 Beyoglu, Istanbul (tel: 144-28-63; tx: 24166 yapm; fax: 1448996).

Principal newspapers
Günaydin, Alayköskü Cad, Cagaloglu, Istanbul (tel: 528-50-00; tx: 22284).

Hürriyet, Babiali Cad 15-17, Cagaloglu, Istanbul (tel: 512-00-00; tx: 22249).

Milliyet, Nuriosmaniye Cad 65, Istanbul (tel: 127-00-34; tx: 23403).

Tercuman, Londra Asfalti Tercümen Tesisleri, Topkapi, Istanbul (tel: 582-12-12; tx: 22253).

Turkish Daily News, Tunus Cad 49/7, 06680 Kavaklidere, Ankara (tel: 128-29-56).

Yanki, Konur Sok 27/7, Ankara (tel: 118-55-24; tx: 42218).

Other useful addresses
Borsa Komiserligi (stock exchange), Menkul Kiymetler ve Kambiyo Borsasi, Rihtim Caddesi 245, 80030 Karakoy, Istanbul (tel: 152-48-00; tx: 22748 imkb; fax: 143-72-43).

Free Zones Directorate, State Planning Organisation, Necatibey Caddesi 108, Ankara (tel: 229-00-71; tx: 42110).

Ministry of Culture and Tourism, Gazi Mustafe Kemal Bulvari 33, Ankara (tel: 229-29-30).

Ministry of Finance, Ankara (tx: 42285).

Ministry of Industry and Commerce, Ankara (tx: 42598).

Ministry of Tourism, Gazi Mustafa Kemal Bul 33, Demirtepe, Ankara (tel: 231-73-90; tx: 46892).

Modern Tercume Burosu (translation service), Karanfil Sokak 21/4, Yenisehir, Ankara (tel: 117-81-22).

State Institute of Statistics, Necatibey Caddesi 114, Ankara (tel: 125-63-30).

Turban Travel Avency, Atatürk Bulvari 169, Atayurt Han, Bakanliklar, Ankara (tel: 118-29-19; tx: 42614).

Turk Argus Ajansi (translation service), Lamartin Caddesi 32/4 Taksim, Istanbul (tel: 144-16-34).

Turk Haberler Ajansi (news agency), Turkocagi Caddesi 1/4, Cagaloglu, Istanbul (tel: 528-50-60).

Turkiye Isveren Sendikalari Konfederasyonu (Turkish confederation of employers' unions), Mesrutiyet Caddesi 1/4-5, Bakanlikar, Ankara (tel: 118-32-17).

Turk Snayicileri ve Isadamlari Dernegi (Turkish industrialists' and businessmen's association), Cumhuriyet Caddesi, 233/9-10 Harbiye, Istanbul (tel: 114-01-205).

Turks and Caicos Islands

Tony Thorndike

Political corruption is now well in the past in this British Dependency although the fight against the drug traffickers continues. Investor confidence is growing as the tourist and offshore finance sectors expand.

The Turks and Caicos Islands (TCI) are composed of two island groups at the end of the Bahamian chain, the Caicos Islands to the west and the Turks to the east. Only a handful of the 48 low-lying islands are populated, their main natural asset being some 230 miles of often stunning beaches and coral reefs. They were favourite haunts of the early explorers and pirates; much excitement was generated when the remains of what is thought to be one of Columbus' flagships, the 'Pinta', was discovered on the Molasses reef.

Budgetary turnround

A very welcome development was that British budgetary aid ceased in the 1989-90 financial year, two years ahead of schedule. Budgetary aid involves close scrutiny and approval of expenditure by the British Development Division (BDD) in Barbados and inevitable financial restraints. It constituted nearly 22 per cent of recurrent expenditure in 1986-87 but in 1988 the TCI and the British agreed on a three-year aid package providing \$3.5m (the US dollar is the currency of the territory) annually for budgetary support. Any surplus could be transferred to the Retainable Surpluses Fund for approved capital projects. In addition there was a \$1.5m technical assistance programme. This incentive, plus sharply rising revenues from the fast-expanding offshore finance centre on the 'capital' island of Grand Turk, rising tourist arrivals and more efficient collection of customs duties had the desired effect. Estimates had to be revised following a review in December 1988 when it was concluded that budgetary support could be dispensed with. Indeed, it was predicted that local revenue would increase by 50 per cent by 1992-93 to over \$30m. Although careful planning directed by the BDD was partially responsible, a good deal of the credit must go to the government headed by Oswald Skippings.

The July 1988 budget called for \$17.1m in recurrent expenditure and \$5.2m in capital expenditure. \$1.084m of British budgetary support was agreed. Revenue was forecast to be \$16m, up from the \$12.7m of 1987. Import duties would account for nearly half of the revenue, and salaries for 52 per cent of spending. These targets were for the most part met and a surplus of \$237,000 resulted. The 1989 budget was expected to boost recurrent expenditure to over \$20m with a similar surplus.

Economic buoyancy

Tourism and associated construction continued to be the main source of economic activity. Following vigorous promotion, there were 47,079 arrivals in 1988 compared to 36,647 in 1987, a 28 per cent rise, and some 60,000 are forecast for 1989. Providentiales, or Provo, is the major destination. It hosts the TCI's most popular resort, Club Med, which will face stiff competition when the Sheraton and Ramada develop-

KEY FACTS

Official title: Turks and Caicos Islands
Head of state: Queen Elizabeth II
Head of government: Governor Michael Bradley; Chief Minister Oswald Skippings
Ruling party: People's Democratic Movement
Capital: Cockburn Town (Grand Turk)
Official Languages: English
Currency: US Dollar (\$) = 100 cents
Area: 430 sq km
Population: 12,500 (1988)
GNP per capita: \$3,500* (1988)
Trade balance: –\$23.9m (1984/85)
Aid flows: \$2.5m* (1988/89)
Visitor numbers: 47,079 (1988)

* estimated figure

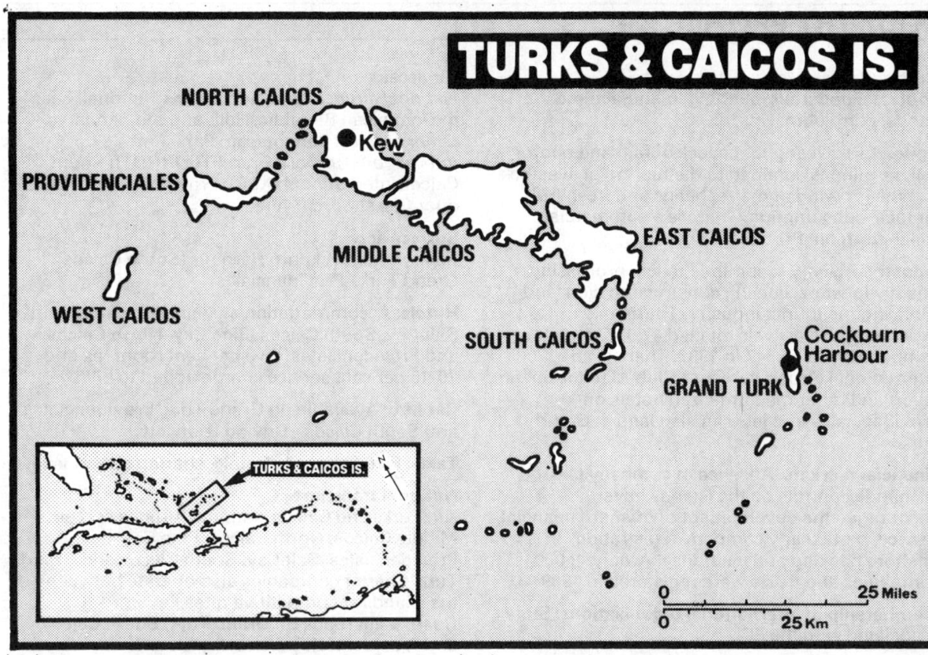

ments (480 rooms in total) open in 1990. In contrast, hotel development on South Caicos stopped some time ago due to financial difficulties, and facilities on North Caicos remain limited by choice, appealing to those who are prepared to pay for luxurious solitude. Grand Turk has much to appeal to the visitor, especially the Grand Turk diving wall. Diversification of gateways to the islands away from Miami, 575 miles away to the north-west, is a high priority. Plans for twin-centre holidays with Puerto Plata in the Dominican Republic are at an advanced stage and scheduled air services are planned between the TCI and Puerto Rico.

Traditional industry was salt-raking, founded as early as 1678 by Bermudian settlers. Disused salinas, relics of the once-thriving industry, are to be seen on Grand Turk, South Caicos and Salt Cay. Salt Cay was the last centre until work ceased in 1962. The island was saved from virtual abandonment by the government work-aid programme, but recent investment by a Florida-based water pressure group will reopen the salt pans in 1989. Fishing is the only traditional occupation since the flat, dry limestone-based islands render agriculture impractical. Conch and lobster are the main and most profitable catches but resources of both are over-exploited. Although the lobster catch increased by 15 per cent in 1987-88 after three years of decline, that of conch fell by 7.5 per cent for the second successive year. Fish resources are, by contrast, under-exploited and various projects aimed at improving catches and techniques are under way.

Active surveillance of fishing grounds was stepped up in 1988 with the provision of a patrol launch and an aircraft by Britain and the United States respectively. Both are also used to detect drug traffickers who find the scattered islands highly attractive to transship cargoes destined for the US mainland. Co-operation with the United States in combatting drug smuggling – compared with the recent past – is both extensive and constant. In late 1989 a new agreement was signed by Britain, on behalf of the TCI, and the United States, stipulating the access limits to TCI waters and airspace by US coastguard and other federal agencies.

An emerging offshore centre

The TCI laid the foundation for what is now a growing offshore industry with the Companies Act in 1981. Growth was slow because of bad publicity surrounding drug-related scandals centred on a number of ministers. However, by 1989 over 8,000 companies were registered, yielding fee income of over $1.8m annually; the industry as a whole contributing over 11 per cent of GDP. Professional services are efficient and cheap and full confidentiality is assured. Companies can be formed in 24 hours, are not subject to capital restrictions and not liable for income or corporation taxes. There are two types, ordinary and exempt. While the former requires the filing of annual returns naming directors and shareholders, the latter need not. But an exempt company, which can be incorporated for about $1,000, must trade mainly outside the islands although it can maintain a local office. Exempt companies must have a local representative but management may be anywhere in the world, as can annual meetings. About 70 per cent are used as personal asset holding vehicles, while 10 per cent are insurance concerns.

The British Government commissioned a feasibility study of the industry by Coopers and Lybrand and their recommendations were the basis of new legislation due to be effective in 1989-90. The proposed regulatory framework encompassing trusts, companies and insurance will be simple, flexible and low cost. One important feature will be the facilitation of both the return of capital through redemption of shares and the merger of Turks and Caicos companies with foreign companies, a useful provision for companies formed and managed in less stable jurisdictions, such as Panama.

Two growth areas were pinpointed by the report: banking and insurance. Banking operations were hampered by outdated legislation which, for instance, forbad the lending of more than 25 per cent of paid up capital to any one borrower. As a result, most offshore transactions are limited except for Eurocurrency transactions which are efficiently conducted. Great care is taken to prevent money laundering: no large cash deposits are accepted and all new customers are vetted. The only private investment bank with a physical presence, the Turks and Caicos Banking Company, is the primary institution servicing the offshore centre. Pending the new legislation, which will include the establishment of a new Superintendent of Banking, there has been an unofficial moratorium on new licences. The insurance legislation, based on that operating in the Cayman Islands, will also create a Superintendent post. It will simplify procedures and protect all parties against irregular practices. Flexibility will be a central feature, especially of risk assessment in the credit life reinsurance sub-sector. This accounts for 55 per cent of existing business, whereas captive insurance accounts for only 33 per cent, in contrast to Bermuda and the Caymans.

Political stability

The March 1988 election marked the end of a short but traumatic period in the TCI's political history. The People's Democratic Movement (PDM) swept to victory with 11 of the 13 seats. The Progressive National Party (PNP) retained only two seats, none going to the PNP breakaway, the New Democratic Alliance. The PDM was the first modern political party in the TCI, although founded as late as 1975. Radical in tone, it won power but lost to the PNP in both the 1980 and 1984 polls.

The TCI gained international notoriety when in March 1985 Chief Minister Norman Saunders was arrested in Miami with two other ministers and a businessman on drug running charges. Subsequent scandals eventually led the British Government to institute an enquiry led by the eminent British lawyer Louis Blom-Cooper, QC. His report unveiled an appalling picture of corruption and intimidation; another revealed other misdeeds over the disposal of Crown Land by a British official. The British Government were embarrassed, especially as it was revealed that the British Governor, who was in office for most of the period in question, had been totally inadequate. Following the release of Blom-Cooper's report the constitution was suspended, the government dismissed and an interim administration established headed by Governor Christopher Turner, known for a 'no nonsense' approach. Thereafter a Constitutional Commission was appointed which reported in March 1987. It recommended multi-member constituencies to encourage voting on a personality rather than a party basis, but the results proved otherwise.

Given the trauma of these events, it will probably be some time before the PNP can regain public support. In the meantime Saunders, released in 1987, directs the party from his South Caicos base through his wife, the nominal leader. As for the new government, it assumed power under a more restricted constitution than before. The PDM is less radical than it was, but it expects enhanced autonomy to follow in the fullness of time. But independence, once a central aim of the PDM, is not on the agenda.

COUNTRY PROFILE

Political system: British Caribbean dependency with appointed governor and 13-member executive council. The governor is responsible for external affairs, defence and internal security. Most other executive decisions are taken by the council (consisting of the chief secretary, attorney-general, financial secretary, chief minister and four ministers from the legislature, with the governor as chairman). The legislature comprises 13 elected members plus a speaker, the chief secretary, the attorney-general, the financial secretary, the chief minister and three nominated members.

Main cities/towns: Cockburn Town on the island of Grand Turk (population 2,500), Cockburn Harbour on South Caicos (1,000). Six out of some 30 small islands are inhabited.

Media
Press: No dailies. Periodicals published include *The Turks and Caicos News* and *Conch News* (both weekly).
Radio: Small government-owned radio station operated by Radio Turks and Caicos (RTC).
Television: Cable television services are expanding.

External trade: Large trade deficit is partly offset by aid flows from UK and earnings from tourism and offshore finance.

COUNTRY PROFILE

Exports: Lobster, conch, fish meat, shells.
Imports: Foodstuffs, clothing, manufactured goods, petroleum.

Agriculture: Owing to the lack of rain and poor soil, farming is confined to the rearing of livestock and the growing of maize, beans and fresh fruit for local consumption. There is a commercial conch farm on Providenciales.

Industry: Activity is confined to fish processing (mainly lobsters, conch), construction work and servicing the tourist industry. Tourism is expanding and arrivals totalled 47,079 in 1988 compared with 36,647 in 1987. There is an international airport serving a Club-Mediterranean resort and other medium-sized hotels on Providenciales and jets can also land at Grand Turk.

Financial markets: Absence of company tax and exchange controls on the islands have encouraged the development of offshore financial service, particularly company registration, offshore banking and insurance. Around 8,000 companies and trusts were registered in 1989.

Membership of international organisations: State associated with the EC.

BUSINESS GUIDE

Time: GMT -5 (GMT -4 from Apr to Oct).

Climate: Tropical, tempered by trade winds. Winter nights sometimes cool, summers are hot. Mean temperature range from 25°C to 29°C.

Entry requirements
Passport: Required by all. A return ticket is also required for visitors.
Visa: Not required by travellers with onward passage assured (except nationals of communist countries).
Currency: No exchange control.

Health precautions
Mandatory: Yellow fever vaccination certificate required if arriving from an infected area.
Advisable: Typhoid/polio vaccinations. Water precautions.

Air access
National airline: Turks and Caicos National. Flies to Dominican Republic, Haiti and the Bahamas.
Other airlines: Bahamasair, Pan Am.
International airport: Grand Turk (GDT); South Caicos International (XSC); Providenciales (PDS), duty-free shop, car-hire.

Surface access
Main ports: Cockburn Harbour (South Caicos), Grand Turk, Providenciales.

Hotels: Accommodation available on Grand Turk, Salt Cay, South Caicos, Pine Cay, North Caicos and Providenciales. Five per cent room tax and 10-15 per cent service charge added to bill.

Car hire: Available on Grand Turk, Providenciales and South Caicos. Driving is on left.

Taxis: Fixed prices to hotels, sharing not unusual.

National transport
Air: Turks and Caicos National Airlines serves Middle Caicos, North Caicos, Pine Cay, Providenciales, Salt Cay, South Caicos and Grand Turk. Charter companies also operate between the islands at competitive rates.
Road: Main roads on Grand Turk, South Caicos and Providenciales are surfaced.

Public holidays: 1 Jan, Good Friday, Easter Monday, 13 Mar (Commonwealth Day, 22 May (South Caicos Regatta Day), Queen's Birthday (Mon in mid-Jun), Emancipation Day (first Mon in Aug), Columbus Day, Human Rights Day, 25 Dec, 26 Dec.

Working hours
Government and business: (Mon-Fri) 0800/0830 – 1300, 1400 – 1630; (Sat) 0800 – 1200.
Banking: (Mon-Fri) 0900 – 1300, 1400 – 1630; (Sat) 0800 – 1200.

Telecommunications
Telephone: Dialling code for Turks and Caicos Islands: IDD access code + 1-809-946 followed by subscriber's four-digit number.
Telex and telefax: Facilities operated by Cable and Wireless.

Banking
Grand Turk Bank of Nova Scotia, Barclays Bank, Turks and Caicos Banking Company.
Providenciales Bank of Nova Scotia, Barclays Bank.

Electricity supply: 110 V AC.

Hotels: Grand Turk
Guinep Lodge (tel: 2977; tx: 8220).

Island Reef, P.O. Box: 10 (tel: 2055; tx: 8252).

Kittina, P.O. Box: 42 (tel: 2232; tx: 8220).

Salt Raker Inn, P.O. Box: 1 (tel: 2260).

Hotels: North Caicos
Pelican Beach (tel: 4692).

Prospect of Whitby, P.O. Box: 21 (tel: 4512116; fax: 4513245).

Hotels: Providenciales Island
Club Méditerranée Turkoise (tel: 4491/5; tx: 4495).

Erebus (tel: 946-4343; tx: 8436; fax: 9464700).

Island Princess, The Bight (tel: 4260; tx: 8432).

Mariner Hotel (tel: 4343; fax: 4700).

Third Turtle Inn (tel: 4230; tx: 8453; fax: 4560).

Turtle Cove Yacht & Tennis Resort (tel: 4203; tx: 8455).

Hotels: South Caicos
Harbour View (tel: 3251).

Car hire: Providenciales
Budget: Butterfield Square (tel: 4214).

Provo Car Rental: (tel: 4224, 4387).

Provo Rent-a-Car: (tel: 4404).

Business advice and information
Cable and Wireless Public Telex Booth, Mercury House, Queen's Street, Grand Turk.

Chamber of Commerce, PO Box 148, Grand Turk (tel: 2043).

Development Board, Front Street, Grand Turk (tel: 2058).

Governor's Office (tel: 2308/9).

Main Government Offices, Grand Turk (tel: 2801).

Tuvalu

T. R. Hughes

(map on page xxiv)

Although one of the poorest Pacific countries, Tuvalu entered its second decade of independence in October 1988 in a cheerful mood, confident that the years ahead would see an all-round improvement in its economic and social state of health. The first 10 years were not easy. Tuvalu's 8,500 people are dispersed over 750,000 sq km in the Central Pacific with a total landmass of only 26 sq km. The nation is not only tiny, widely-scattered and short of land but is resource-poor. Despite its isolation, the islanders have a natural affinity with other Polynesian communities, especially neighbouring Samoa, and it was no great surprise when, as the Ellice entity of the British colony of the Gilbert and Ellice Islands, they were separated from their Micronesian partners three years before independence.

Tuvalu can chalk up substantial progress in health, education and the economy in its stocktaking exercise after the first 10 years of independence. The Princess Margaret hospital in Funafuti, opened in 1978, has been upgraded, new dispensaries are being built on all the nine islands and a family planning programme launched. Each island now has its own primary school and a new University of the South Pacific centre was opened in 1988. The physical infrastructure is much improved with the construction of a deep-sea wharf in Funafuti, the capital, and the installation in 1989 of a satellite earth station (Funafuti was introduced to telephones only five years ago). There are regular air services to Fiji and Kiribati.

Civil service

Going it alone, Tuvalu has demonstrated quite a good record of public administration. A British Government adviser recently commented after reviewing the public service that 'considering the difficult birth pangs of the civil service, the geographical disposition of the environment and the range of essential functions needed, the overall picture is indeed encouraging'.

The government has encouraged the expansion of small enterprises with long-term, low-interest loans. Existing businesses range from small construction firms to bakeries and chicken farms. Apart from a few traditional root crops grown in specially-prepared pits, pandanus and coconut palms, the scope for agriculture appears limited. The soil is nutrient-poor and low-lying, leaving it vulnerable to salt spray and susceptible to encroachment from the sea.

KEY FACTS

Official title: Tuvalu

Head of state: Queen Elizabeth II

Head of government: Prime Minister Tomasi Puapua

Capital: Funafuti

Official Languages: Tuvaluan (Tuvalu Samoan), English

Currency: Australian Dollar (A$ = 100 cents)

Exchange rate: A$1.26 per $ (Nov 1989)

Area: 26 sq km (9 coral atolls)

Population: 8,500 (1988)

The sale of postage stamps overseas continues to be a major source of foreign exchange. They are marketed by a British company which employs more than 50 people in a philatelic bureau in the capital and bring in around A$1m a year. Remittances from Tuvaluan seamen working aboard foreign vessels also produce vital income. The Tuvalu Maritime School trains young men to work as seamen on foreign ships. The government also receives fees from fishing rights granted to other countries.

Pioneer aid plan

Tuvalu is also the beneficiary of a new concept of aid-giving which has gone a long way towards helping it solve its acute financial problems. It persuaded aid donors to set up a A$27m trust fund in mid-1987 from which it can draw the accrued interest at regular intervals to make up its budget deficit. The initial contributions came from Britain (A$8.5m), New Zealand (A$8.2m), Australia (A$8m), with A$1.6m from the Tuvaluans themselves. The fund was topped up with A$700,000 from Japan and A$30,000 from South Korea. The Tuvalu Trust Fund is now a going concern and is expected to boost the budget for 25 years or longer to replace British aid which was tapering off, as is usual, after independence. Westpac, the Australian bank, took control of the fund in September 1987. The bank's investment strategy is to maintain the fund's growth at a rate greater than inflation in Australia where most of the money is invested.

The fund did not escape the October 1987 stockmarket crash and Tuvalu's portfolio losses were reportedly put at A$3m but, by the middle of 1988, the portfolio had recovered better than expected. Tuvalu has made two withdrawals from the fund and, in what could be seen as a promising omen, passed up an opportunity for a third withdrawal.

This once-for-all aid handout runs counter to all established principles of public accounting. But it is being watched keenly by aid experts who see it as a possible blueprint for other small, less developed countries with financial headaches. Best of all it endows Tuvalu with genuine, no-strings-attached political independence giving the tiny state the ability to pick and choose its own friends: 'we can now try to manage ourselves without forever going begging,' says Henry Naisali, the Tuvalu finance minister who thought up the idea.

The fund will at least give Tuvalu a breathing space to work out further development projects. Prime Minister Dr Puapua, said in Britain in 1988, when he went there for the launching of Tuvalu's new inter-island passenger and cargo ship, *Nivaga II*, that his government was looking hard at tourism and fisheries.

The A$10m ship was a gift from Britain to replace the original *Nivaga* (the Sea Goddess) which was the 'lifeline' of the islands for a decade. It sailed into Funafuti just in time for the independence anniversary festivities. As an inter-island workhorse, it has a vital role in Tuvalu's daily life by bringing in food, getting Tuvaluans from one atoll to another and taking cargoes to overseas markets. Built in Lowestoft, England, by the 111-year-old firm of Richards (Shipbuilders) the 1,200-tonne vessel carries 144 passengers instead of its predecessor's 58.

Tuvalu is aggressively looking for trade opportunities abroad. Benefiting from the new Pacific Forum Line shipping service linking it with the Marshall Islands, it negotiated in July 1989 export orders to that country for sweet potatoes, honey, jams, soap and copra. Tuvalu is also defensively alert to regional developments that might affect its vital fisheries interests. It joined other Pacific countries in 1987 in signing a regional fisheries pact with the United States but it has refused to allow the Russians to fish for tuna within its exclusive economic zone.

COUNTRY PROFILE

Political system: British sovereign is head of state, represented by a Governor-General. Executive power exercised by Prime Minister and cabinet. The 12-member unicameral parliament, which elects the Prime Minister, is elected to a four-year term by universal suffrage. Local administration organised through elected island councils. The Governor-General's power to veto government measures was abolished under a constitutional amendment announced in May 1986.

Political parties: There are no political parties, although there is an opposition group within the parliament.

Population: Tuvalu's population is Polynesian in origin. About a quarter of the 8,300 inhabitants live on Funafuti.

Language: English is spoken, as well as the Tuvaluan Polynesian dialect.

Media
Press: The Tuvalu government publishes the *Tuvalu Echoes* fortnightly.
Broadcasting: Tuvalu Broadcasting Services transmits daily radio programmes in Tuvaluan with English news broadcasts. There are about 4,000 receivers.

Domestic economy: A small, subsistence economy dependent on sales of postage stamps and remittances from overseas workers to pay for imports. Annual budgetary aid from UK comprises 25 per cent of government spending. GDP the smallest of any independent state. Tuvalu is conducting a capital development programme to build infrastructure, and planned to spend $A 9.5m in 1987-88 on new projects. In 1987, a trust fund of approximately A$27m, donated by New Zealand, Australia, the UK, Japan and South Korea, was set up to produce income to support the budget.

Employment: The main employers are government, Co-operative Society and Philatelic Bureau. Many Tuvaluans work overseas, as seamen and in the Nauru phosphate industry.

External trade: In 1983, imports were double GDP and 20 times exports. No exact figures are issued.
Exports: Copra is a major export. There is a worldwide trade in Tuvalu postage stamps. Some fish also exported. Coconut oil is exported to New Zealand.
Imports: Main imports are food, textiles, metal manufacturers and petroleum products. Main sources are Fiji, Australia, New Zealand.

Agriculture: Infertile soil, variable rainfall, crops liable to hurricane damage. Copra is the only export crop. Family smallholdings produce subsistence crops of pulaka, taro and other vegetables, bananas and coconuts. Some livestock (pigs, poultry, goats). Fishing and exploitation of the sea are the mainstays of the economy, serving mainly local consumption. Potential to increase income by negotiating fisheries agreements.

Industry: Small industry (baking, construction, boat building, coconut oil mill, soap making, etc.) serves local needs, some handicrafts exported. Since signing a contract with a UK firm in 1980 to advise on postage stamp issue and sell stamps, export earnings from stamps has become largest single foreign exchange earner. No developed tourist industry owing to Tuvalu's remote location, infrequent flights and lack of amenities. Work started in 1987 on a university centre at Funafuti, as part of the University of the South Pacific.

Mining: Tuvalu has no known mineral resources, but the UN Law of the Sea has given it an exclusive economic zone of 12,949 sq km. In 1985, Tuvalu, along with the Cook Islands and Kiribati, signed a tentative agreement to undertake joint exploration for ocean floor minerals with Japan.

Membership of international organisations: Commonwealth (special member not attending Heads of Government meetings), ESCAP, GATT (de facto), Lomé Convention (EC), South Pacific Bureau for Economic Co-operation, South Pacific Commission (participating member), South Pacific Forum, UN, UPU, WHO.

Time: GMT + 12 hrs.

Climate: Hot and humid, temperatures 26-32°C. Rainfall varies considerably, up to 3,000 mm/year, falling most heavily from Nov-Feb.

Entry requirements
Passport: Required by all except certain seamen.
Visa: Citizens of UK and colonies and nationals of certain European and Commonwealth countries with proof of onward passage or sufficient funds do not require a visa for stays of up to one month.
Currency: Tuvaluan coinage (up to one Tuvaluan dollar, equals 100 cents, on a par with the A$) used alongside Australian currency. Foreign exchange regulations may be enforced.
Customs: Personal effects allowed duty-free. Quarantine regulations for plants and animals, inadvisable to carry fruit or plant material. Certain goods may be subject to regulation or import licensing, such as arms, fireworks, drugs, some secondhand goods, motorcycles, jewellery.

Health precautions
Mandatory: Vaccination certificates for cholera and yellow fever required if travelling from an infected zone.

Representation overseas: Suva.

Air access
Other airlines: Airline of the Marshall Islands, Fiji Air. Book well in advance.
International airport: Funafuti International (FUN).

Surface access: Shipping services operate from Fiji, Australia and New Zealand.
Main ports: Funafuti.

Hotels: Reservations at the one hotel should be made well in advance. Visitors may be asked to share rooms when there are accommodation shortages.

National transport
Road: Funafuti has some impacted-coral roads; elsewhere there are tracks. A limited number of vehicles including some minibuses.
Water: An inter-island service available which calls at Suva at times.

Public holidays
Fixed dates: 1 Jan, 1 and 2 Oct, 25 and 26 Dec.
Variable dates: Commonwealth Day, Good Friday, Easter Monday, Birthday of HM the Queen, National Children's Day (Aug), Prince of Wales' Birthday.

Telecommunications
Telephone: Dialling code for Tuvalu via the international operator.
Telex: Telex service available via the post office in Funafuti.
Telegram: Telegram service available via the post office in Funafuti.

Advertising: Radio and press advertisements accepted by the Broadcasting and Information Officer in Funafuti.

Electricity supply: 240 V AC (on island of Funafuti only).

Weights and measures: Imperial system (metric units allowed in some instances).

Business advice and information
Broadcasting and Information Office, Vaiaku Funafuti.

Business Development Advisory Board, PO Box 9, Funafuti (tel: 850; tx: 4800).

National Bank of Tuvalu, PO Box 13, Funafuti (tel: 802, 803; tx: 4802).

Uganda

Catharine Watson

By mid-1988, President Yoweri Museveni had been in power for two and a half years, and for his reformist National Resistance Movement (NRM) government the honeymoon was over. Insecurity continued, as did corruption and inefficiency in government. Even more worrying, dishonesty and 'soft-living' began to involve close Museveni aides. At the top of the NRM there was a loss of energy and purpose. Ordinary Ugandans, meanwhile, grew bitter about the austerity that accompanied the structural adjustment of the economy. There were food shortages across the north, crime waves due to army deserters, and rebel activity in the hitherto peaceful West Nile and Ruwenzori regions. But 1988–89 still gave good reason to be optimistic. Museveni's honesty and commitment to democratisation was never in doubt. The economy grew by over 7 per cent. There were free and peaceful elections for a new legislature. Thousands of rebels were absorbed into the government army, the NRA.

Economy: inflation down

Backed by the IMF and the big donors, Uganda began major economic reforms in May 1987. The first year, however, was a failure: the $307m of aid pledged by the Consultative Group was slow to come, the government had to print money, and inflation exploded to 20 per cent a month.

In 1988–89, however, economic management and aid flows were better. The government kept a tight rein on expenditure and money supply, and inflation fell to 6 per cent a month. Recognising Uganda's efforts to meet IMF targets, the Consultative Group pledged $550m in October 1988. There was also aid from Libya, China and elsewhere.

In his budget speech on 3 July, Minister of Finance Crispus Kiyonga said inflation would have declined further but that coffee revenue was much less than expected – 5.6bn shillings instead of 18.9bn. There were also unforeseen expenses: the famine in West Nile, continued war and elections.

Uganda's foreign debt in 1989 stood at $1.6bn, up from $1.2bn when the NRM took power in January 1986. The debt service ratio is up from 50 per cent to 70 per cent, due mostly to reduced earnings from coffee.

Few countries were likely to be as badly affected by the suspension of the economic clauses of the International Coffee Agreement in July 1989: 96 per cent of Uganda's foreign exchange came from coffee exports, and 95 per cent of that from robusta coffee. Robusta fetched less than the more valuable arabica.

Uganda was also badly placed to exploit a free market in coffee. Its Coffee Marketing Board lacked dynamism and was unlikely to come up with a keen marketing policy. Uganda was unable to switch to growing more arabica as there was little new land at the required high altitude. Improving the quality of its robusta was a long-term project.

In 1986 Uganda was earning about $400m a year from coffee. In 1989 it was likely to earn less than $150m. The irony was that Uganda produced in 1989 more coffee than for years. The factories destroyed by war in central Uganda were all repaired. Peasants responded to a new system of prompt payments by coffee unions. Coffee smuggling declined. In 1987–88 Uganda produced 160,000 tonnes of coffee; in 1988–89 it produced 187,000 tonnes. It produced twice as much as its quota of 2.1m bags, some of the balance going on barter. Barter had a mixed record in

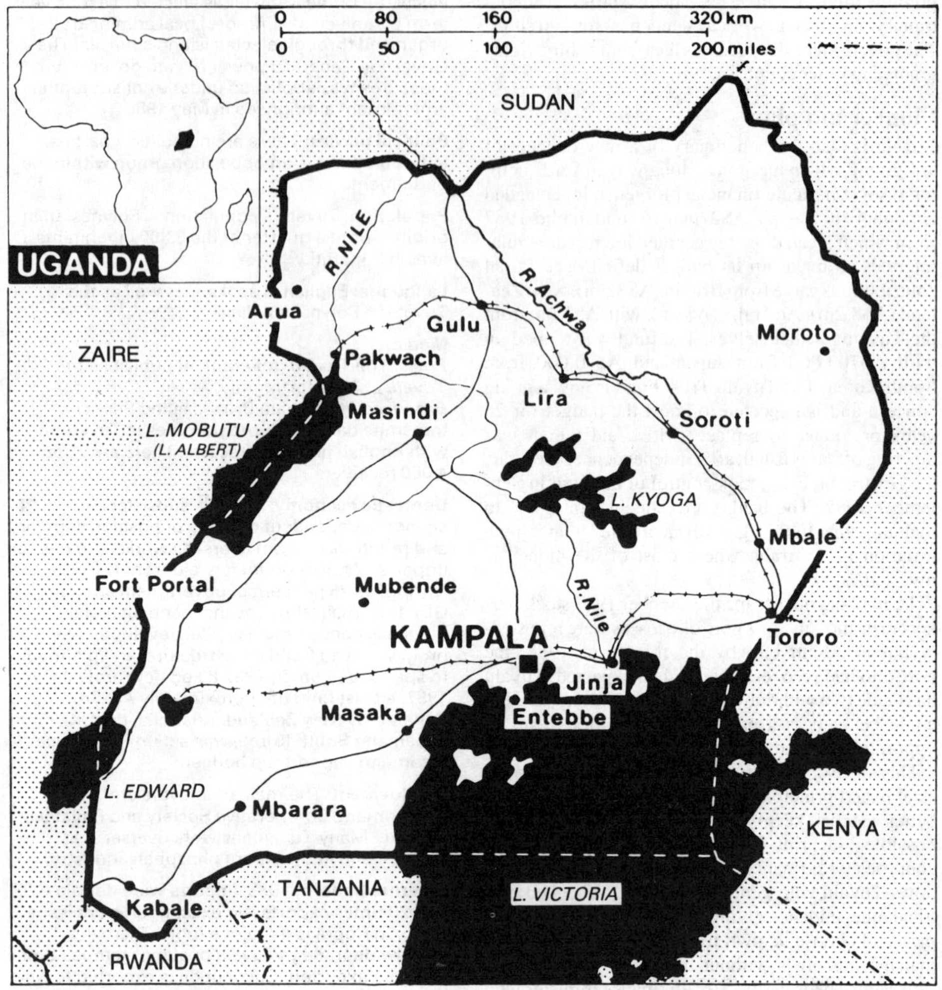

Uganda but the NRM remained committed to it. One large contract worth $12.5m was signed in May. In the third such deal, Ugandan coffee, beans and leather were to be traded for Cuban sugar, salt and pharmaceuticals.

The full implications of the coffee crisis were not clear. Minister Kiyonga said the 1989–90 budget was calculated on the rock bottom price of $0.60 a pound and did not need reworking. A worst-case scenario was that Uganda might default on its foreign debt. Certainly, coffee's poor prospects made it doubly urgent to revive crops like cotton, tobacco and tea and to find new ones. Much hope rested on out-of-season vegetables and tropical fruits that could be flown to wealthy markets in Europe and the Middle East. USAID gave $12.5m to promote horticultural exports; and the revitalisation of Uganda Airlines hinged on horticulture becoming big business. A key blockage was the cold store at Entebbe Airport: in a state of dilapidation, it needed to be rehabilitated first.

Cotton production remained moribund, due to insecurity in cotton-growing areas and to the low producer price for the labour intensive and unpopular crop. Production of tea and tobacco, however, was climbing back up, though in 1989 both were affected by drought.

KEY FACTS

Official title: Republic of Uganda

Head of state: Yoweri Museveni

Capital: Kampala

Official Languages: English, Kiswahili (Luganda, Luo widely spoken)

Currency: New Ugandan Shilling (USh) = 100 cents

Exchange rate: USh339 per $ (Nov 1989)

Area: 236,036 sq km

Population: 17.2m (1988)

GDP real growth: 5.0% (1988)

GNP per capita: $260 (1987)

Inflation: 60% (1988)

Trade balance: –$316m (1988)

Foreign debt: $1.6bn (1989)

Tea production rose to about 3,000 tonnes from 2,500 in 1987–88. Mechanical harvesters were in place on the biggest estates to alleviate the labour shortage. Tobacco registered its best crop since 1972, with almost 3,000 tonnes of flue-cured and 1,060 tonnes of fire-cured. Cigarettes are a major source of tax.

Economists said that most of the 7.2 per cent economic growth in 1988–89 came from agriculture. All food crops increased, except maize.

Industrial recovery was slower, due to lack of foreign exchange and managerial and technical skills, difficulties in raising local cover, problems with power and water, and ownership disputes (many factories were previously owned by Asians and were now managed by government). Key revenue-raising industries like soft drinks had preferential access to foreign exchange, but there was not enough to go around. Nevertheless, manufacturing, including food processing, was said to have grown 40 per cent, though much of this was recovery rather than new growth.

Privatisation and Liberalisation

Privatisation was a crucial issue. In principle, the government believed that loss-making parastatals should be sold and that the 7,000 or so Asian properties it managed would be better maintained if they were returned to their original owners or sold to new ones who would invest in them. In practice, however, privatisation was complex and sensitive. Who were the original owners? Where were the Ugandan entrepreneurs to buy the factories and estates? The value of the Asian properties far exceeded the money in circulation; where were the shillings to come from? Would foreigners snap up the properties? USAID and the World Bank said the lacklustre state sector was a brake on growth. The government was committed to 'settling the Asian question': it returned over 200 properties to the Aga Khan foundation in July 1989 and also earmarked 20 (out of 112) parastatals to be sold. But the process was painfully slow, a fact Museveni often complained about.

Steps were also being taken to maintain a realistic exchange rate and to liberalise the economy. Donors talked enthusiastically about the improved climate for the private sector – although a new more welcoming foreign investment act had still to be written.

There were four devaluations since May 1987, raising the official rate for the shilling from 60 to the dollar to 200. A special import programme, begun in June 1989, was disbursing $40m to businessmen at a rate of 400 shillings to the dollar. It was hoped that this would mop-up excess liquidity and curb the parallel market where dollars cost about 550–600 shillings. Donors pointed optimistically to the fact that the unofficial rate was only two-and-a-half times the official rate. In 1986–87 it was 11 times.

State monopolies on tea and various other products were ended, and private exporters could sell any commodity except coffee. It was thought that even the Coffee Marketing Board's monopoly might soon be over. Since March 1989 private exporters have been allowed to retain 100 per cent of the foreign exchange they earned as long as they brought in imports of equivalent value. There were no controls on the shilling price at which they resold them.

In general, signs of economic growth abounded, particularly in Kampala – repaired roads, more cars, painted buildings, new restaurants. British Airways and Air France were looking into resuming flights, BA after a halt of 15 years. Some parastatals made all staff reapply for their jobs, to flush out poorly qualified staff hired under past political pressure, and started to grow. Last year Uganda Post and Telecommunications doubled the number of international calls passing through its system, licensed over 200 private fax machines, and computerised its accounts with bills going out by the middle of the following month. After the disastrous crash in Rome in October 1988, which left 33 dead, Uganda Airlines began a major restructuring. Performance quickly improved: flights between Entebbe and Nairobi were by mid-1989 90 per cent full, from 10–20 per cent in early 1989.

But many parastatals were still stuck in the morass and confusion inherited from previous regimes: 50 per cent of water connections and 60 per cent of electricity connections in Kampala were illegal. Rehabilitation of Owen Falls Dam progressed slowly, and electricity production fell by 23 per cent in 1988–89. Plans to build a second hydroelectric dam at Murchison Falls in the middle of a game park were opposed by environmentalists.

Tourism – a hope

Tourism remained a hope of the government. An interesting circuit existed in Western Uganda, taking in chimpanzees, pygmies, the Ruwenzori Mountains, gorillas and Queen Elizabeth National Park.

The Sheraton Corporation planned to rehabilitate and take over the management of the lodge at Queen Elizabeth; the Aga Khan expressed interest in other lodges in other parks. But all this was wishful thinking if the government did not control poaching and encroachment. The chimps and gorillas were losing their habitat; the elephants in the northern parks were being slaughtered by a mixture of rebels, soldiers and bandits.

Austerity

For ordinary people, however, salaries had never been lower compared to the cost of living. This gave rise to a wave of strikes, especially in parastatals. A strike at Makerere University in May 1989 caused panic in the NRM as it could have spread to the civil service. A professor earned about $10 a month; a kilo of sugar cost $1, a kilo of meat $1.20, a box of 8 candles $1.50. In the countryside peasants complained they never saw essential commodities.

But the high cost of education was the biggest crisis for ordinary Ugandans. Primary schools in Kampala routinely charged $100 a year per child. Across the country parents withdrew children, particularly girls, from school for lack of money.

Under the structural adjustment programme, government spending on education and health fell, but a further problem was that officials stole much of the little that was given. In 1988 numerous headmasters and district education officers were dismissed or arrested. In Mbale the District Education Officer was arrested for having 'diverted' 216 iron sheets and 100 bags of cement meant for a school.

Politics

Corruption was also a key political issue. To keep its credibility the NRM needed to stamp on it. The difficulty arose when key NRM figures, perceiving that their term in power might be short, began to 'eat' as well. This appeared to be the trend in late 1989.

The NRM had, however, made genuine efforts to control corruption. The government newspaper, *New Vision*, was allowed a free hand to expose it. The Public Accounts Committee resumed work for the first time since the Amin years. The inspectorate of government, a new investigative office, was set up to halt dubious deals and payments. Many government officers were dismissed, including in mid–1989, 8 permanent secretaries and about 100 senior accounting officers. Museveni was pushing for a 'leadership code' under which all prominent figures would declare their assets; at a press conference in July 1989 he expressed his irritation that it was taking so long to become law.

The biggest political achievement of 1989 undoubtedly was the elections for the National Resistance Council, the legislature, in February and March. Slightly unconventional, the rules allowed no campaigning and particularly no references to the old political parties. Grassroots 'resistance committees' elected an electoral college at county level: members of the electoral college then queued behind the candidates. Critics argued that this was undemocratic and that there should have been a system of one person–one vote.

The government replied that its method was cheaper, quicker and prevented fraud. Queuing was immensely popular with ordinary people, who remembered how ballot boxes were destroyed, switched and stolen in the 1980 elections.

The elections were peaceful, with only Gulu district unable to participate due to rebels. Fourteen serving ministers were not elected, while several prominent Amin and Obote figures were – proof that the government was not fixing the results. The old political parties did not fare well. Candidates were chosen largely on the basis of whether they would bring benefits to their counties. Plenty of corrupt and dubious individuals were elected, however, and there was much surreptitious campaigning. Museveni's attitude seemed to be: 'That's too bad, but that's democracy for you. Maybe they'll choose better next time.'

After the election the NRC consisted of 38 'bushmen' (longstanding NRM-ers), 10 representatives of the NRA, 20 presidential nominees, a woman from each of the 34 districts, 149 county representatives, and 3 worker and 5 youth representatives. NRM supporters no longer predominate, and the NRC is no longer a mere rubber stamp. 48 ministers and 27 deputy ministers were drawn from the new NRC: care was taken that they reflected the country's ethnic and religious make-up.

With the elections, Museveni fulfilled his promise to hold elections within four years of coming to power. Less progress was made towards a new constitution. In late 1989 a 21-member commission was only just begining to sit.

In March 1989 Amnesty International published a large report on Uganda, which praised the government for human rights improvements, but criticised it for continuing to detain in prison over 3,000 men, women and children picked up in counter-insurgency sweeps in the north and east. It also alleged that in northern Uganda in late 1989, NRA soldiers had murdered numerous civilians, burning some inside their huts. Subsequently, about 600 detainees were released and a government enquiry set up to investigate the atrocities in the north. Several NRA soldiers and militiamen were also executed in the north after being found guilty of harming civilians. Just as the government took these steps, however, disaster struck: in July 1989 over 60 young men died after the NRA locked them in a railway wagon in eastern Uganda. The NRA's top commander, Major General Salim Saleh said it was the 'worst thing' he'd ever seen in the NRA and that 'someone would have to die for it': 14 NRA officers and men were promptly arrested, but easterners were not mollified. Discipline had declined in the NRA with the incorporation of tens of thousands of rebels and soldiers from the UNLA, the national army of the Obote and Okello regimes. The NRA expanded from about 14,000 fighters in 1986 to at least 50–60,000 in 1989: some observers believed the NRA to be 100,000 strong. Many NRA commanders were unhappy about this rapid growth, but it was inevitable if the NRM was to draw rebels out of the bush with promises of jobs in the army.

The war in northern Uganda appeared almost over. An offensive in April, May and June 1989 netted over 500 guns. Rebels died and surrendered in their hundreds. Joseph Kony, leader of the Holy Spirit Army, was still at large, but many of his commanders had come in. In eastern Uganda small groups of rebels-cum-bandits were still ambushing vehicles, but they too were weakened by desertions and the hatred of the local people. In July ex-Obote and Amin soldiers based in Sudan and Zaire attempted to start a war in West Nile. They failed, but the fact that they could even briefly disrupt life around several key towns indicated they had some support. For much of 1989 West Nile was cut off from the rest of Uganda due to Holy Spirit attacks on convoys. This, together with a food shortage, led to frustration which rebels were quick to exploit. In western Uganda, members of the Ruwenzururu movement killed local officials. More bandits than political rebels, they appeared to resent the crackdown on coffee smuggling. In July there was talk of guerrilla activity in Mpigi, a Baganda district where Museveni fought part of his bush war. The guerrillas were said to be disappointed that Museveni had not restored the Baganda king. The talk proved to be rumours: no guerrilla camps were found.

Karimojong cattle-raiders continued to plunder eastern Uganda, causing intense fury among their prime victims, the Itesot, who were left with almost no livestock. A commission of NRC representatives and army officers hoped to get to the roots of the raiding. Curbing the raiding would be the NRA's key concern when the rebellions died down. A government agency created to improve conditions within Karamoja and thus decrease the motive to raid was proving corrupt and inefficient.

Internationally, Uganda's profile rose. Museveni visited France, the UK, the USA, China. Uganda was visited by Danielle Mitterand, Sir Geoffrey Howe, Christopher Patten, Jimmy Carter, Maureen Reagan, Jerry Rawlings, Mohamed Ali and others.

Regionally, however, problems continued. In January, former president Idi Amin left his exile in Saudi Arabia and flew to Zaire. There was little doubt that he hoped to encourage rebels along Uganda's western frontier. In March, Kenya accused Uganda of bombing Lokichoggio, a small town in northwest Kenya on the supply route into SPLA–controlled southern Sudan. The government stayed calm about Amin's presence in Zaire, but denied involvement in the Lokichoggio affair. Uganda did not have the aircraft to carry out such a raid nor did it have a motive. In April, Uganda expelled the Zaire ambassador 'subversive' activities. Well-placed sources said that Kenya and Zaire had been attempting to destablise the Museveni regime.

With Rwanda, there were talks to resolve the 30-year-old problem of the Rwandese refugees. In March, Uganda received about 50,000 refugees from the fighting in southern Sudan. Relations were cordial with Sudan, although the NRM/A did not hide its sympathy for the SPLA.

Health and environment

AIDS continued to mount, with over 8,000 confirmed cases by September 1989. Museveni wanted an anti-AIDS law banning early marriages and extra-marital sex, but a precipitous decline in the incidence of sexually transmitted diseases indicated that Ugandans had already curbed their sexual activity. TB and herpes epidemics reflected the high-seropositivity rate in the south and southwest.

Floods and landslides in western Uganda and drought in the south and west were symptoms of rapid deforestation and swamp drainage. A ministry of environmental protection was set up but it was weak and under-resourced. The rehabilitation of Uganda's power grid would do much to stop deforestation. Uganda, green and fertile compared to its neighbours, could not afford complacency about the environment.

Outlook

The key challenge for 1990 was to cope with the loss of revenue from coffee and to curb corruption: good plans were on the drawing board but they could falter for lack of funds or if officials were more interested in building their personal businesses. The wars should peter out, but there will be continued ambushes by deserters and bandits. President Museveni will need all his skills to explain to the people why the austerity measures are necessary.

COUNTRY PROFILE

Political system: Government comprising members of National Resistance Movement, UPM, Democratic Party, Uganda People's Congress, Conservative Party and two smaller guerrilla organisations. The National Resistance Council (268 members) is led by an executive chairman, the President of the Republic, and is the country's sole legislative body.

Population: The population, estimated at 15.5m in 1987, is growing at an annual rate of 2.8 per cent. 45 per cent are aged under 15. Christianity is the dominant religion.

Main cities/towns: Kampala (population 592,200 in 1986), Jinja (256,900), Masaka (764,700), Mbale (655,400), Kabarole (655,800). The population is concentrated in southern Uganda in a belt from east to west with only 8 per cent living in towns of any size. The north is sparsely populated.

Language: English is the official language and the universal language for documentation. The most widely spoken African languages include Luganda and Kiswahili. The latter can be used in the National Resistance Council.

Media
Press: Daily newspapers include *New Vision* (government-owned), and *The Star* (in English); and *Munno, Ngabo* and *Taifa Empya* (in Luganda). Several weeklies, periodicals and trade papers are produced in English and Luganda.
Radio: Radio Uganda operates a service in English, Kiswahili and other local languages. 21.8 sets per thousand of the population.
Television: A government television channel transmits over a radius of 320 km from Kampala. More relay stations are being built to extend coverage. 5.5 sets per thousand of the population.

External trade: Increased coffee exports. Much of the trade is barter and difficult to quantify. Since 1988 private companies are also allowed to enter into barter deals. At the end of 1987, the government had concluded bilateral barter deals worth $550m with Algeria, Cuba, Czechoslovakia, German DR, Italy, Libya, Korea DPR, Rwanda, Somalia, Sudan, German FR and Yugoslavia. The heavy debt burden remains a strain on the economy.
Exports: Coffee (96 per cent of total value in 1989), tea, hides and skins, timber.
Imports: Capital goods (19 per cent of total value in 1986), manufactured goods, foodstuffs, fuels.

Agriculture: Contributes 70 per cent to GDP and employs about 90 per cent of the labour force. Coffee is the main cash crop. Rehabilitation of coffee holdings has been the main stimulus to economic growth in recent years. Cotton was once an important cash crop, but has been badly hit by war. Sugar is grown on the small number of estate farms, and tea is produced both on estates and small-holdings, although output remains low. The main food crops are plantains (matoke), cassava, millet, maize, sweet potatoes, beans and cereals. 20-25 per cent of total land area is arable, 20 per cent pasture and 12 per cent forest. Agricultural development hampered by shortages of vital inputs and damage caused by civil war.

Manufacturing: Sector's contribution to GDP has fallen from 12 per cent in 1970 to under 4 per cent. The main industries are processing of agricultural commodities, textiles, cement, soap, steel, metal products and brewing. Most of these are operating well below capacity, but output has increased in the beverage industry, soap factories and others receiving preferential access to foreign exchange inputs. This has been mainly due to shortages of imported materials, spares and fuel, inadequate infrastructure and a lack of skilled manpower. Rehabilitation of the sugar, textiles, paper and steel industries is under way, mainly funded by international aid agencies. The Lugazi sugar factory has been completed.

Mining: Uganda has deposits of copper, cobalt and iron ore, as well as less viable fields of tungsten, beryl, columbo-tantalite, gold, bismuth, tin, limestone and phosphates. Copper was Uganda's third most important export during the 1970s, but output from the Kilembe mine was halted in 1980. Proposals to reopen it will hang on the rather doubtful construction of a new smelter to replace the plant at Jinja. The government is setting up purchasing centres for gold, although the overvalued exchange rate means that miners make more if they sell it on the black market.

Energy: Uganda relies on imported oil for 50 per cent of commercial energy needs, the balance being provided by hydroelectricity. Total installed generating capacity is about 154 MW, mainly from the Owen Falls Station. Considerable amounts of electricity are exported to western Kenya. Generating capacity is currently being increased in anticipation of an upturn in industrial demand; Owen Falls being upgraded to 210 MW.

Membership of international organisations: African Development Bank, African States in Association with the EC (Lomé Convention), Arab Bank for Economic Development in Africa (recipient), Commonwealth, Economic Commission for Africa, GATT, ICAO, IDA, IFAD, IFC, FAO, ILO, IMF, Islamic Conference, Islamic Development Bank, Non-Aligned Movement, OAU, UN, WHO, WIPO, World Bank, WTO.

BUSINESS GUIDE

Time: GMT +3.

Climate: Equatorial, tempered by altitude. Temperatures are fairly constant throughout the year, hottest months Dec-Feb, with day-time range (in Kampala) of 27°C to 29°C compared with an annual average of 26°C (night-time average 16°C). Heaviest rainfall occurs Mar-May, Oct-Nov; Apr is wettest month (average fall for month 175 mm).

Entry requirements
Passport: Required by all.
Visa: Required by all except nationals of Britain, Kenya, Tanzania, Rwanda, Zaïre and Sudan. A return or onward ticket is also required. Regulations alter frequently.
Prohibited entry: Nationals of South Africa.
Currency: Import/export of local currency prohibited. No restrictions on import of foreign currency, but amount must be declared on arrival. The requirement to change a set amount of foreign exchange during a visit has been suspended to encourage tourism.

Health precautions
Mandatory: Yellow fever and cholera vaccination certificates required by all.
Advisable: Typhoid, polio vaccinations. Malaria prophylaxis should be taken as risk exists throughout the country. Water precautions are necessary; bilharzia risk is present. Mild form of dysentery is common. AIDS risk.

BUSINESS GUIDE

Representation overseas: Addis Ababa, Beijing, Bonn, Brussels, Cairo, Copenhagen, Dar es Salaam, Khartoum, Kigali, Kinshasa, London, Lusaka, Moscow, Nairobi, New Delhi, Ottawa, Paris, Riyadh, Rome, Tripoli, Washington.

Air access
National airline: Uganda.
Other airlines: Aeroflot, Air Rwanda, Air Tanzania, Ethiopian, Kenya Airways, Sabena.
Tax: Airport tax of $20 is levied on departure.
International airport: Entebbe (Code: EBB), 36 km from Kampala; duty-free shop, restaurant, bank post office, hotel reservations, car hire.

Surface access
Water access: Water access is possible via Lake Victoria, from Mwanza (Tanzania) and Kisumu (Kenya).
Overland access: Good road access from Kenya. Links with Rwanda being improved. It is possible to travel by train from Nairobi (Kenya) to Kampala via a connection at Tororo; can be slow and uncomfortable.

Hotels: Variable standard, but available in all main centres. Should be booked in advance. Foreign visitors are required by law to settle hotel bills in foreign currency.

Electricity supply: 240 V AC, 50 cycles.

Car hire: Valid international driving licence required. Hire cars are now available, mainly with driver, from a number of rental firms, and through independent taxi drivers.

City transport
Taxis: Available at airport and in Kampala at hotels, the railway station and near major office blocks.

National transport
Air: Uganda Airlines operates services to main centres (including Arua and Kasese) from Entebbe. Charter services are also available.
Road: Network of 27,540 km. Main towns are connected by 1,280 km of tarmac roads. These and other roads are of variable quality.
Buses: Regular services scheduled include Entebbe-Kampala (journey-time: 1 hr); also services to/from airport. Services between most main centres but tend to be crowded.
Rail: Two routes are operated: Tororo-Kampala-Kasese and Kampala-Tororo. The line north of Tororo has been closed for two years due to rebel attacks. Daily services are scheduled on these routes, but they can be slow and irregular.

Public holidays
Fixed dates: 1 Jan, 26 Jan (National Resistance Day), 6 Feb, 1 May (Labour Day), 9 Oct (Independence), 25 Dec, 26 Dec.
Variable dates: Good Friday, Easter Monday, Eid al-Fitr.

Working hours
Government and business: (Mon-Fri) 0830 – 1230, 1400 – 1700.
Banking: (Mon-Fri) 0830 – 1400. Bureaux de change open Sat and Sun.
Shops: (Mon-Sat) 0830 – 1230, 1400 – 1700.

Telecommunications
Telephone: Dialling code for Uganda: IDD access code + 256 followed by area code (Kampala 41, Entebbe 42) followed by subscriber's number. Service is good for calls abroad but can be unreliable within Uganda.
Telex and telefax: Facilities available at central post offices in Kampala, Jinja and Mbale from 0800 – 1600; also at some hotels.

Banking
Central bank: Bank of Uganda.
Head offices: Bank of Baroda (Uganda), Barclays Bank of Uganda, Centenary Rural Development Bank, Co-operative Bank, East African Development Bank, Gold Trust Bank, Grindlays Bank (Uganda), Libyan Arab Uganda Bank, Nile Bank, Standard Chartered Bank Uganda, Uganda Commercial Bank, Uganda Development Bank.

Representation in capital: Algeria, Burundi, China PR, Cuba, Egypt, France, German DR, German FR, Ghana, India, Iraq, Italy, Kenya, Korea PDR, Rep. of Korea, Libya, Nigeria, Rwanda, Somalia, Sudan, UK, USA, USSR, Vatican, Yugoslavia, Zaïre.

BUSINESS DIRECTORY

Hotels: Entebbe
Lake Victoria, P.O. Box: 15 (tel: 20645/452).

Hotels: Jinja
Crested Crane, P.O. Box: 444 (tel: 21514; tx: 61150).

Hotels: Kampala
Equatoria, William Street, P.O. Box: 5100 (tel: 259595; tx: 61150).

Fairway, P.O. Box: 4595 (tel: 259571/5; tx: 61396).

Imperial, Buganda Road, P.O. Box: 7088 (tel: 259800; tx: 61169).

Kampala International, Nile Avenue, P.O. Box: 7041 (tel: 258022; tx: 61034).

Lion, Namirembe Road, P.O. Box: 6751 (tel: 243687, 243490; tx: 61322 agase).

Lunar, Gaba Road, P.O. Box: 16211 (tel: 268196; tx: 61108 lunar).

Sheraton, Nile Avenue, P.O. Box: 7041 (tel: 244590; tx: 61517; fax: 256696).

Silver Springs, Portbell Road, P.O. Box: 734 (tel: 261301; tx: 61150).

Speke, Nile Avenue, P.O. Box: 7036 (tel: 254553; tx: 61150).

Car hire: Kampala
Hertz: Diamond Trust Building, PO Box 5914 (tel: 235395, 233738, 268179; tx: 61153).

Sunshine Tours and Travel: PO Box 5011, Kampala (tel: 253255; tx: 61196).

Uganda Tours and Travel: PO Box 7211, Kampala (tel: 243822).

Other useful addresses
Bank of Uganda, PO Box 7120, 37-43 Kampala Road, Kampala (tel: 258441; tx: 61059).

Central Tender Board (responsible for administering most government- and aid-funded tenders), PO Box 3925, Kampala.

Coffee Marketing Board, PO Box 7154, Kampala.

East African Development Bank, 4 Nile Avenue, PO Box 7128, Kampala (tel: 230021, 259761; tx: 61074 ugair).

Lint Marketing Board, PO Box 7018, Kampala.

Mbale Chamber of Commerce, PO Box 396, Mbale.

Ministry of Agriculture and Forestry, PO Box 102, Entebbe (tel: 20752).

Ministry of Animal Industry and Fisheries, PO Box 7003, Wandegeya, Kampala (tel: 257351, 543964; tx: 61266).

Ministry of Commerce, PO Box 7000, Parliament Avenue, Kampala (tel: 233561, 245027).

Ministry of Co-operatives and Marketing, Parliament Avenue, PO Box 7103, Kampala (tel: 232665).

Ministry of Industry and Technology, Parliament Avenue, PO Box 7125, Kampala (tel: 243833).

Ministry of Planning and Economic Department, PO Box 7086, Uganda House, 10 Kampala Road, Kampala (tel: 254081; tx: 61117).

National Chamber of Commerce and Industry, 17/19 Jinja Road, PO Box 3809, Kampala (tel: 258791).

National Housing and Construction Corporation, PO Box 659, Kampala.

National Mining Corporation, Kampala.

Statistics Department, PO Box 7086, Kampala (tel: 254081).

Radio Uganda, Nile Avenue, PO Box 2038, Kampala.

Uganda Advisory Board of Trade, PO Box 6877, Kampala (tel: 233311; tx: 61085).

Uganda Airlines Corporation, Airways House, 1 Kimathi Avenue, PO Box 5740, Kampala (tel: 232990; tx: 61239 ugair).

Uganda Commercial Bank, 12 Kampala Road, PO Box 973, Kampala (tel: 234710, 258012; tx: 61073).

Uganda Development Bank, 4 Parliament Avenue, PO Box 7210, Kampala (tel: 230257, 230706; tx: 61143).

Uganda Development Corporation Ltd, PO Box 442, Kampala (tel: 234281; tx: 61069).

Uganda Export Promotion Council, Kampala.

Uganda News Agency (UNA), c/o Ministry of Information and Broadcasting, PO Box 7142, Kampala (tel: 254461; tx: 61188).

Uganda Railways Corporation, PO Box 7150, Nasser Road, Kampala (tel: 254961; tx: 61111).

Uganda Tea Authority, PO Box 4161, Kampala.

Uganda Television Service, PO Box 4260, Kampala.

Uganda Tourist Development Corporation, Hotel Equatoria Building, 37/39 William Street, PO Box 7211, Kampala (tel: 234230, 234822; tx: 61150 tourism).

United Arab Emirates

Teresa English

The UAE faces the 1990s in a much stronger position after emerging from economic and regional uncertainty in the 1980s. Key challenges facing the emirates in the next 10 years will be to continue to diversify the economy, to maintain the internal political balance that unites seven independent states, and to maximise the potential of its vast oil reserves.

The effects of the Gulf cease-fire in August 1988 have made themselves felt. A new sense of security is apparent in the emirates, which spent eight long years on the fringes of major hostilities. Oil, a major revenue earner, has stabilised since the dramatic slump in the mid-1980s and the emirates, in particular Abu Dhabi, can take comfort from the optimistic predictions for oil markets in the 1990s.

The UAE has a better reputation for economic and financial management than some of its Gulf neighbours. The aftershock of the oil slump meant introducing sweeping cutbacks and reorganisation to cope with the sudden loss of oil revenues. It has resulted in a rationalisation and streamlining of the major government departments and industries, allowing a trimmer economy to emerge.

The federation has survived the difficulties attached to 30 years of rapid oil-led growth. Abu Dhabi and Dubai are the main contributors to the federal budget, and the falling oil revenues have meant that smaller states, such as Ras Al Khaimah, Ajman, Umm Al Qaiwain, and Fujairah, have seen reduced federal spending on infrastructure in late 1980s. Nevertheless, the smaller emirates have pushed ahead with plans to build up light industry and attract tourists.

Umm Al Qaiwain suffered a major disappointment in 1989 when the team of investors who planned to set up a new aluminium plant in the emirate, pulled out. They offered the scheme to neighbouring Qatar, attracted by readily available gas supplies. However, in the end another group won the smelter project.

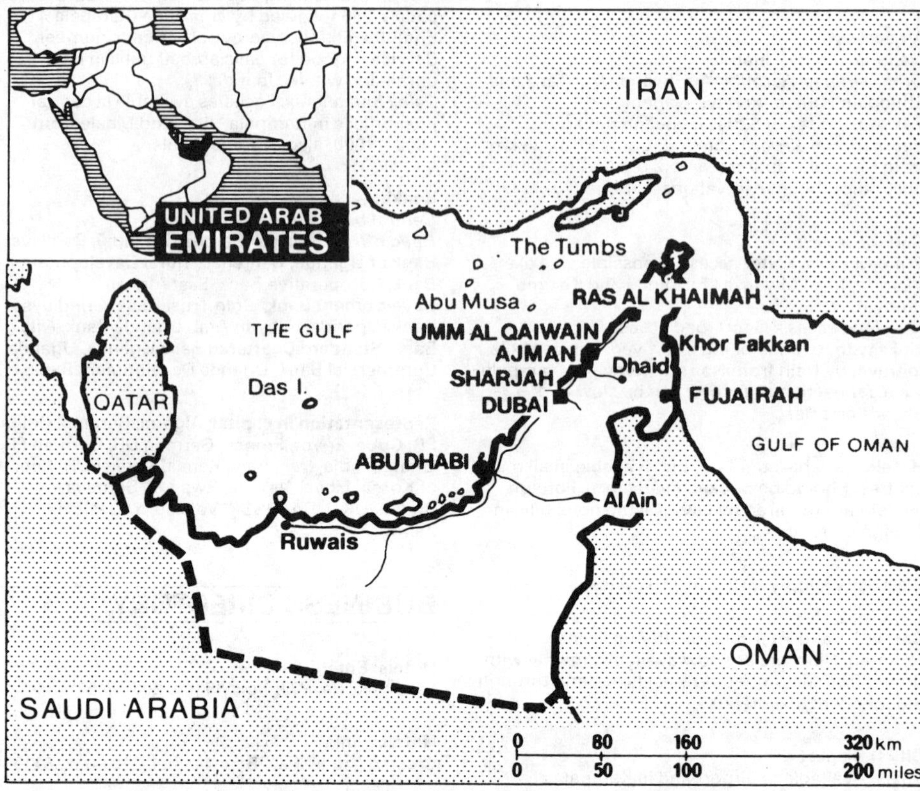

Economy

Diversification has become an economic watchword in the Gulf and UAE is no exception, but the fall in oil revenues have slowed the flow of funds necessary to develop alternative domestic industries. However, the UAE appears to be making some progress. Oil and gas exports fell in 1988 by 4.5 per cent but this was offset by a rise in other exports and re-exports.

The squeeze that falling oil prices have put on the federal budget which has been running a deficit for some years, became apparent when the government announced in mid-1989 that the draft UAE budget for 1989 would have to be cut to reduce the proposed deficit of $900m. Budgets begin on 1 January each year, but have been late for five years, 1985–89. Ministries and government departments are instructed not to exceed the previous year's allowances. The 1989 budget was due in the last quarter of 1989.

The budget relies heavily on contributions from the main oil producers Abu Dhabi and Dubai which have decreased in recent years as the emirates suffered financial pressures of their own. Nearly 75 per cent of the federal budget is taken up by recurrent spending. President Shaikh Zayed still provides funds for the Northern Emirates from time to time.

The deficit increased by 5.5 per cent in 1988 on gross revenues and expenditure Dh11,500m ($3,133m) in 1988, according to a preliminary report from the Central Bank released in 1988. The increase was attributed to a 22 per cent fall in the total value added from the oil sector, which stood at Dh15,200m. Other revenues rose by 49 per cent to Dh6,100m. Gross expenditure was cut by 3.2 per cent to Dh34,100m.

Gross domestic product (GDP) fell by 1.2 per cent to Dh87,550m in 1988. Oil's contribution to GDP fell to 33 per cent in 1988, from 36.5 per cent in 1987. Imports increased 19 per cent to Dh31,000m, causing the trade surplus to fall sharply in 1988 by 26 per cent to Dh14,000m. The USA accounted for 10 per cent of imports, while 63 per cent came from European and industrialised countries.

The 1989 draft budget stood at $1,410m, against expected revenues of $2,510m, according to Minister of State for Finance and Industry Ahmad Humaid al-Tayer. However, by October 1989 it became apparent that oil revenues for the year would be much higher than could have been predicted at the start of the year, increasing from an average yearly forecast of $13.80 a barrel in January to $16.20 a barrel by September, even with the higher production levels in the latter half of the year.

Oil markets and OPEC

The UAE has waged a long-standing battle to get its share of the OPEC production quotas increased. One of the UAE's main arguments was that the quotas should be shared out in line with production capacity and reserves. In the peak year of 1977, the UAE was producing a daily average of 1,998,700 barrels per day (b/d).

Abu Dhabi dominates the UAE oil scene, producing two-thirds of the total UAE hydrocarbon production, which is controlled by ADNOC. ADNOC has the main shareholding in the three main oil field operating companies: ADCO, Zakum and Abu Dhabi Marine Operating Company (ADMA–OPCO). ADNOC also has the majority control of the Emirate's two gas companies ADGAS and GASCO and refineries at Umm Al Nar and Ruwais.

Dubai continues to charge market related prices, while Abu Dhabi tends to base its prices on OPEC pricing formulas. Aware of the short life span of its

KEY FACTS

Official title: United Arab Emirates

Head of state: President HH Shaikh Zayed bin Sultan al Nahyan (ruler of Abu Dhabi)

Head of government: Vice-president HH Shaikh Rashid bin Said al Maktoum (ruler of Dubai)

Capital: Abu Dhabi

Official Languages: Arabic

Currency: Dirham (Dh) = 100 fils

Exchange rate: Dh3.67 per $ (Nov 1989)

Area: 90,559 sq km

Population: 1.5m (1988)

GDP per capita: $15,527 (1988)

GDP real growth: –2.1% (1988)

Inflation: 6%* (1987)

Oil reserves: 98.1bn barrels (1988)

* estimated figure

oil reserves, Dubai has continued to pump close to its maximum of 400,000 b/d.

Towards the end of 1987, the UAE produced a new estimate of its hydrocarbon reserves which nearly trebled previous predictions for proven oil reserves. Reserves were put at 97,200m barrels, against 33,000m in 1986. If this level were proven, UAE would have the largest reservoir in the world after Saudi Arabia and Iraq, accounting for about one seventh of proven global reserves. Natural gas reserves were also raised to 183,545,000m cu feet. These claims underpin the UAE's demand for a higher quota. Throughout 1988, UAE's Minister for Petroleum and Mineral Resources Dr Manaa Bin Said al-Otaibah pushed for a 1.5m b/d quota to match its large reserves.

The UAE's quota at the end of 1988 was 948,000 b/d. In 1988 the quota was increased to 988,000, to 1.480m and eventually to 1.094m when OPEC agreed to raise the ceilings in response to strong winter demand. The emirates continued to produce well over quota throughout 1989 and had reached 859,000 b/d over quota by September to total 1.96m b/d.

However, oil industry observers in Abu Dhabi are at last more optimistic about Abu Dhabi's OPEC future. Predictions of strong oil demand in the 1990s and beyond have bolstered OPEC confidence. The stronger prices throughout 1989, despite higher production ceilings, have made OPEC producers more flexible on the quota issue, recognising the need for a restructuring of the system.

This was amply illustrated in the autumn 1989 OPEC conference in Geneva when Iran put forward a new proposal for settling quotas. It proposed to increase the quota proportions for countries like the UAE, Kuwait, Gabon and Ecuador and raising the overall ceiling to 21.5m b/d. The scheme would upgrade the UAE's quota to 1.35m b/d. However, it was strongly opposed by some of the smaller countries, such as Libya and Nigeria, which would lose out under the plan. The plan has been welcomed in Abu Dhabi as a first step towards settling the quota question.

The turbulent oil years of the mid-1980s led UAE to reorganise and restructure its oil management in mid-1988, aimed at cutting some of its bureaucracy. The department of petroleum was abolished and a new higher council for petroleum was created. ADNOC'S board was replaced by an 11-member petroleum council and Sohal Fares al-Mazrui was appointed as new ADNOC general secretary and secretary-general to the higher council, chaired by President Shaikh Zayed.

The move was designed to allow the industry to respond more readily to the government's decisions and to improve relations between ADNOC and the foreign equity partners. The changes continued into 1989 when Mazrui was also appointed head of the Abu Dhabi Company for Oil Distribution (ADNOC–FOD). Two offshore oil operators, the Zakum Development Company (Zadco) and Umm al-Dalkh Development Company (Udeco), were merged.

During 1987, and 1988 oil exploration and development spending were put on hold and only essential maintenance and operations projects were carried out. By mid-1989, improved oil revenues were making themselves felt. At least four major projects were given the go-ahead and plans for several more were being dusted down.

Abu Dhabi is planning to increase its refining capacity at home for export and for home use. Domestic consumption of refined products is about 80,000 b/d. Abu Dhabi has two refineries at Umm Al Nar and Ruwais with a combined design capacity of 180,000 b/d. The expansion of Abu Dhabi's Umm Al Nar and Ruwais refineries was approved. Sulphur production was also being restructured with facilities at Ruwais and Das Island being overhauled and expanded.

Expansion in Abu Dhabi's refining and gas export capacity was prompted by an increased demand for oil and gas from Japan, Abu Dhabi's largest customer. Many Japanese companies hold interests in Abu Dhabi operations. In 1989, two Japanese oil companies Nippon Oil Company and Showa Oil Company set up liaison offices in the emirate, joining Mitsui, Cosmo Oil Company and Idemitsu.

KEY INDICATORS

	Unit	1984	1985	1986	1987	1988
Population	'000	1,290	1,350	1,400	1,450	1,500
Gross domestic product (GDP)	$m	27,676	27,023	21,602	23,699	23,291
GDP per capita	$	21,454	20,017	15,430	16,344	15,527
Crude oil production	'000 b/d	1,069	1,057	1,309	1,418	1,510
Volume of exports						
Crude petroleum	1985=100	100.4	100.0	124.8	128.4	134.5
Exports	$m	15,990	14,756	10,129	12,765	12,250
Petroleum	$m	12,978	11,842	7,453	8,665*	7,352*
Imports	$m	6,571	6,537	6,806	7,080	8,522
Balance of trade	$m	9,419	8,219	3,323	5,685	3,728
Current account	$m	7,464	6,946	1,852	3,812*	2,023*
Exchange rate	Dh per $	3.67	3.67	3.67	3.67	3.67

* estimated figure

Abu Dhabi Gas Liquefaction Company (ADGAS), which is owned by ADNOC, Mitsui, BP and CFP Total, exported a record 2.4m tonnes of LNG to Japan from Das Island in 1988, up from 2.28m tonnes in 1987. The Das Island plant sells all of its output to Tokyo Electric Power Company (Tepco). Tepco's 20-year contract with ADGAS runs out in 1997. It is one of the world's largest gas plants and has 11 gas streams feeding two identical process trains.

In October 1989, local press reports said that a plan to build a further liquefaction train to increase the output at Das Island was given approval and the US consultants, Arthur D. Little was appointed to undertake feasibility studies.

In 1989, the second phase of the Upper Zakum field was commissioned. Over 100 production and water injection wells were planned to increase production from the current level of 350,000 b/d to 500,000 b/d within two years. The offshore Umm al-Anbar structure in the West Mubarraz oil field began producing 8,000 b/d in February 1989.

Like other Gulf countries Abu Dhabi is intensifying its interests in downstream operations; a policy that was continued in 1989. The International Petroleum Investment Company (IPIC) took out an option to increase its stake in Spanish refinery operator CEPSA. Its stake had increased from 10 to 12 per cent by mid-1989. IPIC said it would continue to search for additional downstream investments, particularly in the Far East.

The Abu Dhabi Investment Authority, which invests most of the personal fortune of the ruling Al Nahyan family, also increased its investments in downstream refineries, with an increase in its holding in French oil major Total to around 9 per cent. In 1989, local press reports said that the Authority was holding talks with Japanese trade officials about taking out a shareholding in a major Japanese refinery. ADIA suggested that the reports were premature.

Investment continued to be an important source of income for Abu Dhabi. Bankers estimated that before 1995 the emirate could find itself in the same position as Kuwait where investment income contributes more to the economy than oil revenue. This trend was behind the success of organisations such as National Investment and Securities Corporation (Niscorp) which showed a profit in its first year of operations. Niscorp has set its sights on the small to medium investor and increased its ties with the Far East during 1989.

Trade, Dubai's traditional activity, cushioned the emirate during the poor oil years. Dubai has used its oil revenues from the 400,000 barrels of oil a day that it produces to lay the foundations of a modern Gulf state. However, Dubai's planners are increasingly aware that unlike Abu Dhabi which has reserves that will last more than 100 years, Dubai has only about 30 years of oil wealth left. Oil exploration and maximisation of existing oil revenues have become a priority for the emirate. Consultants say they expect to see oil maintenance and exploration projects intensify.

In October 1989, Dubai Petroleum Company (DPC), the state-owned organisation that operates the emirates' four onshore producing fields, granted a new exploration concession to a group led by Dubai Exploration Onshore (DEOL). The concession is south of the city and the first exploration well was due to be spudded in January 1990.

The Fateh field, Dubai's oldest and largest field, has been continuously developed since its discovery in 1966. Oil from the second largest southwest Fateh is piped to Fateh for processing, storage and export. Dubai Petroleum Company (DPC) began tendering for the installation of a third central compression plant in the field. The platform will be connected to the field's existing facilities. It includes a bridge link to the field's second platform and will allow additional gas injection compressors to be installed later.

A new development well was spudded in September 1989 at Dubai's condensate and gas producing onshore Margham field, operated by Arco Dubai and the UK's BP. In October 1989, BP announced that its one third share in the Margham concession was included in a worldwide package of hydrocarbon assets which it was selling to the US Oryx Energy for $1,300m. The sale was due to go through in early 1990.

Dubai has plans to set up the emirate's first ever refinery, in the Jebel Ali freezone. The estimated $900m project is being set up by an Indian joint venture led by Reliance Industries. Reliance is putting together financing proposals from a variety of Gulf investors, among them the Abu Dhabi Investment Authority.

Dubai Natural Gas Company (Dugas) built a new gas line from the onshore Margham field to the Dugas liquefaction plant at Jebel Ali to process the gas produced by Arco and BP. Abu Dhabi's Star Energy opened a privately-owned products terminal at Jebel Ali with a storage capacity of 2m barrels. Consultants in Dubai saw plenty of opportunities in the oil and gas sector. Dubai's producing fields are old and require constant maintenance. Added to this the emirate is keen to keep up exploration activity in the hope of extending its reserves.

Sharjah produces about 10,000 b/d of crude, and 60,000 b/d of condensate and about 27m cu feet a day of gas. Sharjah also produces some LPG and supplies natural gas direct to Dubai and to the Abu Dhabi-owned Emirates General Petroleum Corporation (EGPC). The emirate began supplying EGPC about 200m cu feet of gas a day in 1982, dispute arose over payment levels which is still unsettled.

The UAE plans a census in 1990, the first since 1985. It will be used as the basis for plans to develop the seven emirates into the 1990s and beyond. It should reveal the changes that have taken place in the emirate since the mid-1980s and the rate of growth of UAE's 1.6m population.

The cease-fire appeared to be a turning point for the UAE. Up to August 1988, the emirates had struggled nervously through eight years on the sidelines of a major conflict, which had several times threatened UAE's security and its investment potential. A new sense of confidence has returned to the emirates, borne up by the improved markets for oil and aluminium – Dubai's largest non-oil export – and by the new security in the region which has proved a boost to trade in the UAE.

Abu Dhabi's size guarantees it a major say in the emirates' affairs and its ruler Shaikh Zayed is president of the federation set up in 1971 following the termination of the UK's treaties with the seven emirates on what was known as the Trucial coast.

The UAE takes an assertive role in regional issues and is an active member of regional organisations such as the GCC. The UAE has made a supportive stand towards the Palestinian issue and has attempted to act as a mediator between Iran and Iraq. The emirates have good foreign relations with Europe, including traditional strong links with the UK, and with the USA and Japan. During the 1980s the emirates have tried to expand links with less traditional trading partners. Talks with the Soviet Union began in 1985 and more recently China which is becoming an increasingly important trading partner. A new aviation agreement was signed with China in 1989. Eastern bloc countries have sent several trade delegations to the emirates during 1989, including Poland, Romania and Hungary.

Development in Abu Dhabi

Abu Dhabi has commissioned a comprehensive masterplan study to guide the development of the emirates for the next 20 years. The plan will look at population, employment, industry and landuse and study the co-ordination of the emirate's resources.

In April 1989, Under-secretary for Planning Shaikh Said Bin Zayed al-Nahyan said that the emirate proposed to spend Dh14,800m on development planning and infrastructure projects over the next 10 years. Power and water schemes came top of the list. The first phase of the massive Taweelah power plant was well under way by 1989, and tenders were due for the main construction work on phase two, with Belgium's Tractebel and West Germany's Fischner competing. Spreading power resources to rural areas is also a priority.

Abu Dhabi is also planning further work on hospitals, roads and bridges, expecting eventually to extend the city beyond the Maqta bridge. The expansion would include building other bridges to Hoderiay and Sadiyat islands and developing the infrastructure of these islands.

Abu Dhabi's man-made island, eventually planned to be a leisure and recreational facility for Abu Dhabi, went into the second construction phase. Also in the recreational sector, Abu Dhabi National Hotels company said it planned to expand its existing hotels and acquire new ones abroad, with North Africa, particularly Morocco and Egypt, appearing to be target areas.

Al Ain's new airport will be operational in the early 1990s. The university project is moving slowly ahead with consultants drawing up plans for a medical faculty at Al Ain. The city is also overhauling its electrical distribution network to link into the Taweelah project.

Development in Dubai

Now that the immediate security threat of the Gulf War is over, Dubai's planners are thinking of the 1990s and beyond, so that the decline in oil revenues a few decades hence can be comfortably managed.

In October 1989, the Dubai Commerce and Tourism Promotion Board opened, headed by executives with considerable experience in Hong Kong. The board was given the mandate to co-ordinate Dubai's efforts to diversify its economy away from oil. The Board includes representatives from Dubai's major commercial sectors and businesses. The group was told to

COUNTRY PROFILE

Political system: The seven emirates making up the UAE – Abu Dhabi, Dubai, Sharjah, Ras Al-Khaimah, Fujairah, Umm al-Qaiwain and Ajman – formed their present federation in 1971. Highest government authority is vested in the Supreme Council, which consists of the Rulers of the seven emirates and which elects the President; he in turn appoints the Council of Ministers, which exercises executive authority. The 40 members of the Federal National Council are appointed by the Rulers. The individual emirates have retained a great degree of autonomy, and all powers not specifically reserved for the federal government belong to them. Abu Dhabi is the largest emirate, and its Ruler, Shaikh Zayed bin Sultan al-Nahayan, is President. Ruler of Dubai, Shaikh Rashid bin Said al-Maktoum, is Prime Minister and Vice President.

Political parties: There are no political parties.

Population: Abu Dhabi (population estimated 420,000 in 1987), Dubai (419,104), Sharjah (268,722), Ras Al-Khaimah (116,470), Ajman (64,318), Fujairah (54,425), Umm al-Qaiwain (29,229).

Language: The official language is Arabic; English is generally understood in business circles and elsewhere.

Media
Press: Magazines, weeklies and dailies are published in Arabic and English.
Radio: Extensive national services in Arabic, also in English, French and Urdu, with shortwave world service broadcasts to North America, the Far East and Europe.
Television: Services throughout UAE, with TV stations in Abu Dhabi, Dubai, and Sharjah. 280,000* sets in 1987.

Domestic economy: The country's infrastructure was built up in a period of rapid growth in the 1970s, which halted with the fall in oil prices and the government's adoption of tight fiscal policies. Existing projects are being completed although the government budget for development has been cut. In 1987 with rises in the oil price, however, the economy began to pick up and GDP had grown by nine per cent by the end of the year. Abu Dhabi is financially dominant, providing almost 75 per cent of federal revenues and grants to smaller emirates.

Employment: Unemployment is negligible and the workforce (550,000 in 1986) is dominated by single, male foreign workers, 70 per cent of whom are unskilled.

External trade: Dubai is the main trade centre, and accounts for over 50 per cent of UAE's imports (1987).
Exports: Main exports are crude oil (75% of total) natural gas (10%), and certain re-exports, mainly aluminium, from Dubai. Main destinations are Japan (49.4% of 1985 total), Singapore (3.9%), USA (3.7%), France (3.2%), Oman (2.7%).
Imports: Main import requirements are machinery and transport equipment (35% of total), manufactures classified by materials (22%), food and live animals (10%), fuels and lubricants (7%), and chemicals (6%). Main sources are UK (14.8% of total value in 1987), Japan (13.4%*), USA (13.2%), German FR (6.9%), France (4.9%), Italy (4.5%).

Agriculture: Harsh climate and sandy soil make self-sufficiency in food production unlikely, and, although agriculture contributes only one per cent to GDP, this sector receives special priority for strategic reasons. Locally grown tomatoes, cucumbers, cabbages, aubergines and other vegetables are now available, and demand for poultry, livestock, eggs and dairy products is increasingly met by local production. Overall local production accounts for less than 30 per cent of needs but roughage and forage crop production now supplies 80 per cent of local market. Water supplies and distribution are a constant source of problems, and salinity also depletes productivity. Ras Al-Khaimah which has greater rainfall, has the best developed agricultural sector.

Industry: Oil and gas are the main industries (mainly in Abu Dhabi and Dubai), along with aluminium, steel, chemicals and petroleum products. Gas has become increasingly important in terms of industrial development and a new LPG plant was opened at Jebel Ali in 1986. The main heavy industries in existence or under development are related to hydrocarbons and located in Jebel Ali Free Zone (Dubai, 100 sq km), and the industrial city at Ruwais (Abu Dhabi), where there are free trade zones. The Dubai Aluminium Company is producing up to 150,000 tonnes a year of cast metal. Construction materials predominate in Ras Al-Khaimah. Manufacturing trebled its contribution to GDP between 1980 and 1986 with emphasis on building materials, plastics, soft drinks and other consumer goods. High labour costs and the small domestic market limit private sector industry.

Minerals and mining: Oil and gas are the only natural resources of economic importance. The commercial viability of exploiting known deposits of chromite, nickel, copper, platinum, bauxite, magnesium and asbestos remains doubtful.

Hydrocarbons: Exploration activity and development of new oil and gas fields have slowed down. Production, export and processing of hydrocarbons generate 35 per cent of economic activity (1986, as opposed to 64 per cent in 1980), and provide about 80 per cent of government revenues. Abu Dhabi produces two-thirds of the country's crude oil and almost all of its natural gas. Dubai accounts for around 30 per cent of output (1987) and Ras Al-Khaimah 0.5 per cent. At current rates of production, oil reserves are estimated at 33bn barrels, enough to last for 75 years. Significant new finds have been made at Sajaa (Sharjah), providing gas feed stocks. Some reinjection of gas into oil fields.

Stock exchange: Plans for a federal stock exchange have been postponed. A private venture stock exchange for GCC nationals opened in Umm al-Qaiwain in late 1984, but has seen little activity.

Membership of international organisations: Arab League, CCC, FAO, G-77, GATT, GCC, IAEA, ICAO, IDA, IFAD, IFC, ILO, IMF, IMO, INTELSAT, INTERPOL, Islamic Conference Organisation, Islamic Development Bank, ITU, NAM, OAPEC, OPEC, UN, UNESCO, UPU, WHO, WIPO, World Bank, WTO.

stimulate increased investment in Dubai in manufacturing, processing, distributing, trading, transport, finance and other service industries and to further the development of the Emirate as a tourist destination. It will draw up ideas for enhancing the city as a business and tourist centre. Its functions are also to promote and publicise Dubai abroad and it will be opening representative offices in target markets in Europe, the USA, and the Far East.

Dubai hopes to build on its reputation as a centre for regional business and government investment. One of the most important innovations for Dubai is the Jebel Ali Freezone industrial base. The freezone was born just before the Gulf War broke out and has managed to expand during the war years. By 1989, prospects had revived and more than 250 companies had joined the freezone, up from 180 at the end of 1988 and just 25 three years earlier. The freezone is building on Dubai's reputation as the traditional supplier to Iran and attracting many new distribution and warehousing customers including those who were interested in sidestepping political difficulties by trading with Iran through a third party. These firms were encouraged by the large reconstruction spending programmes that Iran was drawing up.

The freezone is trying to attract heavy industrial ventures, as well as light industry and distribution operations. In mid-1989, the freezone's chairman Sultan Bin Sulayem said that he expected the capital invested in the freezone, which stood at $600m in 1989, to increase to $1,000m in 1990. This is dependent on attracting several new heavy industry projects to the zone; for example, Reliance Industries wants to set up Jebel Ali's first oil refinery which could supply India National Oil Company. Financing proposals were being put together in October 1989. ADIA the government's investment arm, had shown an interest in the project in line with its current policy of increasing downstream investment.

The freezone has also had some success in attracting distribution companies like Sony, and Philips of the Netherlands, which set up in the zone in 1989. Dubai Natural Gas Company (Dugas) built a new gas line from the onshore Margham field to the Dugas liquefaction plant at Jebel Ali to process the gas produced by Arco and BP. Abu Dhabi's Star Energy opened a privately owned products terminal with a storage capacity of 2m barrels.

Traditionally Iran is Dubai's top re-export partner. During the Gulf War, Iranian merchants based in Dubai kept Iran supplied with nearly all its basic commodities. The peak year was 1987 when re-exports reached Dh1,306m and total trade $360.7m. In 1988, this had reduced to just Dh692m.

Traders were disappointed when Iranian currency restrictions led to a 60 per cent drop in re-exports to Iran in the first half of 1988. After the cease-fire, traders bought in large stocks in anticipation of a boom in trade with Iran, which took a long time to arrive because the Iranian Government laid down stringent trade rules. However, by 1989 currency restrictions had been eased and re-exports recovered enough to attract the attention of international firms interested in taking part in Iran's lucrative reconstruction plans. By October 1989, Dubai's trade had fully recovered. In the first half of 1989, total re-exports increased by 21 per cent to Dh3,228.5m. Re-exports to Iran increased to Dh813.1m.

Aluminium is Dubai's leading export and the rising prices for aluminium in 1989 boosted export revenues. In 1989, Dubai Aluminium Company (Dubal) embarked on a major upgrading programme to increase its hot metal capacity by around 40 per cent. The development includes a new pot line installation which will increase hot metal capacity by 235,000 tonnes a year, against an original design capacity of 135,000 tonnes. Dubai has been working on a continuous programme to improve operations which have already included projects in power generation, anode manufacturing, casting process computer systems, and cell technology.

The smelter development's improvements are due to be fully operational by 1991. The upgrading is a timely one for Dubai, since the new operation will be ready well before other Gulf smelters and upgrading, prompted by better world prices, will be ready.

Dubai is also turning new attention to the Far East as a trading partner to take advantage of its geographical proximity to the area. UAE banks have built close working relationships with the Far East and Abu Dhabi is increasing its investments there.

Japan is one of the largest trade partners throughout the emirates. Several new Japanese oil companies have set up new offices in Abu Dhabi in 1989 and Abu Dhabi is planning to buy into Japanese refining operations. Japan's large projected demand for gas in the 1990s has also attracted Abu Dhabi's interest.

As part of plans to develop Dubai as a hub for the region's transport and commerce, Dubai's municipality is pressing ahead with several road, landscaping and power projects, including a new 400MW power and water plant at AL Mamzar. Two of the emirate's largest park projects are out to tender and beach development is under way. The municipality is also pressing ahead with road and creek crossing projects to ease the heavy burden of traffic in the area.

Transport

All of the emirates ports have reported improved results since the cease-fire. Abu Dhabi has appointed consultants from West Germany to redevelop Mina Zayed and attract more shipping to the port. Port Rashid is studying the reorganisation and upgrading of its facilities to cope with record levels of cargo. Jebel Ali port brought in crane equipment in 1989 as capacity expanded when the Sealand/Norasia consortium began operating from the port in early 1989. Both Rashid and Fujairah experienced an increase in the level of sea and air cargo during 1989. Sharjah port also restarted a link to Bandar Abbas in Iran during the year. A $70m cargo terminal is under way in Dubai to increase the airport's ability to handle the large volumes of cargo to the Far East.

Tourism is also on the up-and-up, a relatively new sector given a considerable boost by the Gulf cease-fire. The leisure facilities and comparatively relaxed attitudes for an Islamic country, and the exceptional array of merchandise, draws a large number of intra-Gulf visitors each year. The prospect of winter holidays in the sun is now drawing visitors from Europe and hotels are being upgraded.

The expected increase in traffic to the region has prompted both Dubai's four-year-old Emirates airline and Gulf Air, which is part-owned by Abu Dhabi, to go ahead with massive investments in new aircraft. In October 1989 Emirates announced it would be buying an additional five new aircraft, to double its cargo and passenger handling capacity in the early 1990s. Bahrain-based Gulf Air announced plans to buy 12 new airbus A320s.

Outlook

The federation has withstood many tests during the 1980s and seems set to survive into the 1990s. The Gulf War cease-fire has given the emirates a new lease of life and the UAE is in a good position to benefit from the return of optimism to the area. The uncertainty of the oil markets in the mid-1980s has strengthened the emirates' determination to continue economic diversification. However, this process will require careful husbanding and forward planning. The death of Deputy Prime Minister Shaikh Hamdan in October 1989 was a timely reminder that the management of government will gradually pass into younger hands as the decade progresses.

BUSINESS GUIDE

Time: GMT +4 hrs.

Climate: Dry, sub-tropical weather with exceptionally hot summers and high humidity near the coast, where temperatures can reach 45°C, and mild winters (10-30°C).

Entry requirements
Passport: Required by all except certain seamen.
Visa: Required by all except nationals of certain neighbouring states and UK passport holders. Must be obtained in advance of arrival, unless being met at the airport by a sponsor. Visas are valid for all seven emirates.
Prohibited entry: Nationals of Israel and South Africa, and holders of passports with evidence of travel to Israel.
Currency: There are no currency control regulations. Import of South African or Israeli currency is prohibited.
Customs: Personal effects are duty-free. Trade with Israel and South Africa prohibited, as are goods originating from either country. Imports of pork products, alcoholic beverages, firearms and ammunition require a special permit. Most goods may be imported freely, but importers must hold a general import licence permitting trade only in the categories of goods specified in the licence.

Weights and measures: Metric system (imperial system and local units also used).

Health precautions
Mandatory: Certificates of vaccination for cholera and yellow fever if travelling from infected area.
Advisable: Cholera vaccination and anti-malaria precautions.

Representation overseas: Algiers, Amman, Ankara, Baghdad, Beijing, Beirut, Bonn, Brussels, Cairo, Dacca, Damascus, Geneva, Islamabad, Khartoum, Kuwait City, London, Madrid, Mogadishu, Moscow, Muscat, New Delhi, Paris, Rabat, Riyadh, Rome, Sana'a, Seoul, Tehran, Tokyo, Tripoli, Tunis, Vienna, Washington DC.

Air access
National airlines: Gulf Air (owned jointly with Bahrain, Oman and Qatar); Emirates (Dubai).
Other airlines: Aeroflot, Air Djibouti, Air France, Air India, Air Lanka, Air Tanzania, Alitalia, Alyemda, Ariana Afghan, Balkan, Biman Bangladesh, British Airways, CAAC, Cathay Pacific, CSA, Cyprus Airways, Egyptair, Emirates, Ethiopian, Garuda, Interflug, Iran Air, Iraqi Airways, Japan, JAT, Kenya Airways, KLM, Kuwait Airways, LOT, Lufthansa, MALÉV, MAS, MEA, Olympic Airways, Philippine, PIA, Royal Air Maroc, Royal Brunei, Royal Jordanian, Royal Nepal, Sabena, Saudia, Singapore, Somali, South African Airways, Sudan Airways, Swissair, Syrian Arab, Tarom, Tunis Air, Turk Hava Yollari, TWA, Uganda, Yemenia, ZAS Airline of Egypt.
International airports: Nadia, New Abu Dhabi International (Code: AUH), 40 km from city, with duty-free shop, buffet, bank, shops, bar; Dubai International (Code: DXB), 4 km from city, with duty-free shop, bar, buffet, bank, hotel reservations, post office, shops, car hire (new terminal under construction); Sharjah International (Code: SHJ), 10 km from city, with duty-free shop, bar, buffet, restaurant, bank (restricted hours), hotel reservations; Ras Al-Khaimah International; Fujairah's new airport is open and an international airport at Al-Ain is under construction.

Surface access
Main ports: Abu Dhabi: Mina Zayed (containers, ro-ro), Abu Al Bu Khoosh (tankers), Arzannah Island (tankers), Das Island (tankers), Jebel Dhanna/Ruwais (bulk fertilisers, LPG tankers), Mubarras Island (tankers), Zirku Island (tankers); Dubai: Mina Rashid (containers), Fateh (tankers), Mina Jebel Ali (containers, ro-ro); Fujairah: Port of Fujairah (containers); Ras Al-Khaimah: Mina Saqr (bulk cargo, ro-ro);

Sharjah: Mina Khalid (containers, ro-ro), Mubarek (tankers), Khor Fakkan (bulk, ro-ro), Umm Al-Qaiwain (bulk ore, containers, tankers). Gas handling facilities are planned at Mina Khalid (Sharjah).

Hotels: Excellent standards in Abu Dhabi and Dubai, and rooms are generally in adequate supply. Advance booking is always advisable, however. Major credit cards are acceptable.

Car hire: Chauffeur-driven car hire is available. International licences are not acceptable. Driving is on the right-hand side of the road, with speed limits of 60 kph in towns and 80-100 kph elsewhere.

City transport
Taxis: Metered taxis are available in Abu Dhabi. It is advisable to negotiate fares in advance in Dubai as taxis are not metered. Fixed fares are available for pre-paid journeys from Dubai airport to the city (Dh 30 in 1988). Dubai to Abu Dhabi costs Dh 250-250 by special taxi.

National transport
Air: Scheduled services are infrequent. Small light aircraft and helicopters may be chartered.
Road: Good, surfaced roads along the coast link all emirates. There are numerous oases, the best-known being Al-Ain and Liwa in Abu Dhabi and Dhaid in Sharjah, and small fishing villages in Dubai, Sharjah, Ras Al-Khaimah and along the east coast. The motorway between Abu Dhabi and Al-Ain is being upgraded.

Public holidays
Fixed dates: 1 Jan, 6 Aug (Accession of the Ruler of Abu Dhabi), 2 Dec (National Day).
Variable dates: Laitat al-Mi'raj, Eid al-Fitr, Waqfa, Eid al-Adha, Hijra New Year's Day, Prophet's Birthday.

Working hours: Working hours vary between the emirates, and change from summer to winter.
Government and business: 0700/0800 – 1300/1400, sometimes also 1600 – 1900/1930 (Thu 0730 – 1100/1200, closed Fri).
Banking: 0800 – 1200 (Thu 0800 – 1100 in Abu Dhabi, Thu 0800 – 1200 in northern emirates, closed Fri). Some banks open 1630 – 1830 (not Thu or Fri).

Social customs: Religious customs and practices must be respected. There are very strict controls on the sale of alcohol, which is banned in Sharjah. Avoid: sitting with the soles of the feet facing anyone; eating, drinking or smoking in front of Muslims in daylight hours during Ramadan. Photographers should be careful, especially when photographing women, and should obtain permission before taking pictures of installations such as bridges, oil installations, etc.

Telecommunications: The UAE subscribes to the Arabsat satellite communications system, and is upgrading the entire telecommunications service with digital technology.
Telephone: Dialling code for UAE: IDD access code + 971 followed by area code (2 for Abu Dhabi, 4 for Dubai, 70 for Fujairah, 77 for Ras al Khaimah, 6 for Ajman, Sharjah and Umm al Qaiwain) followed by subscriber's number. Dubai telephone numbers which started 42 or 43 now start 52 and 53. There were 300,000* lines in 1988; approximately 34% are in Dubai, 30% in Abu Dhabi and 13% in Sharjah. Telephone calls within each emirate are free. Direct dialling to over 110 countries is available. IDD access code for international calls from UAE is 00 followed by the country code etc.
Telex: There are public telex facilities in Abu Dhabi and Dubai at Etisalat and main post offices. 6,600* lines in 1988.
Telegram: There are telegraph facilities in main centres.
Other: Telefax services are available at Etisalat offices in all main centres.

Banking: Governing authority is UAE Central Bank. Banks have been affected by the economic downturn, and many are carrying heavy bad debt burdens. Since July 1987 the charging of compound interest has been acknowledged by the state.
Central bank: Central Bank of the United Arab Emirates.
Head offices: (Abu Dhabi) Abu Dhabi Commercial Bank, Abu Dhabi Fund for Arab Economic Development, Abu Dhabi Investment Authority, Abu Dhabi Investment Company, Arab Bank for Investment & Foreign Trade, National Bank of Abu Dhabi, United Arab Emirates Central Bank, United Arab Bank. *(Dubai)* Bank of Oman, Commercial Bank of Dubai, Emirates Bank International, Middle East Bank, Dubai Islamic Bank, National Bank of Dubai, Union Bank of the Middle East. *(Ajman)* First Gulf Bank. *(Fujairah)* National Bank of Fujairah. *(Ras Al-Khaimah)* National Bank of Ras Al-Khaimah. *(Sharjah)* Bank of Sharjah, Investment Bank for Trade & Finance, National Bank of Sharjah. *(Umm al-Qaiwain)* National Bank of Umm Al-Qaiwain.
Foreign branches: (Abu Dhabi) Algemene Bank Nederland, Arab African International Bank, Arab Bank, Arab Monetary Fund, Banca Commerciale Italiana, Bank Saderat Iran, Bank Melli Iran, Bank of Baroda, Bank of Tokyo, BNP, Standard Chartered Bank, Crédit Lyonnais, Crédit Suisse, El Nilein Bank, Indosuez, Citibank, First Interstate Bank of California, First National Bank of Chicago, Grindlays Bank, Habib Bank, Habib Bank Al Zurich, Janata Bank, National Bank of Bahrain, National Bank of Oman, Rafidain Bank, Union Bank of Switzerland, United Bank, United Arab Bank. *(Dubai)* Algemene Bank Nederland, Amro Bank, Arab African International Bank, Arab Bank, Bank of Baroda, Bank Saderat Iran, Banque de l'Indochine et de Suez, Banque Paribas, Banque Bruxelles Lambert, Banque du Liban et d'Outre-Mer, Banque Libanaise pour le Commerce, Barclays Bank International, Banque du Caire, Bank of the Arab Coast, British Bank of the Middle East, Chemical Bank, Standard Chartered Bank, Citibank, Creditanstalt, Grindlays Bank, Habib Bank AG, Janata Bank, Lloyds Bank International, Natwest Investment Bank, Dryx Merchant Bank, Royal Bank of Canada, United Bank, Wardley Middle East Ltd. *(Fujairah)* Arab Bank, Bank of Baroda, Bank Melli Iran, Bank of Credit & Commerce International, Bank Saderat Iran, British Bank of the Middle East, Habib Bank, United Bank. *(Ras Al-Khaimah)* Arab Bank, Bank Melli Iran, Bank of Baroda, Bank Saderat Iran, Banque du Caire, Banque Libanaise pour le Commerce, British Bank of the Middle East, Standard Chartered Bank, Citibank, Grindlays Bank, Habib Bank, Janata Bank, Middle East Bank, United Bank. *(Sharjah)* Algemene Bank Nederland, Arab Bank, Bank Melli Iran, Bank of Baroda, Bank Saderat Iran, Banque du Caire, Banque du Liban et d'Outre-Mer, Banque Libanaise pour le Commerce, Barclays Bank International, British Bank of the Middle East, Standard Chartered Bank, Citibank, First National Bank of Chicago, Grindlays Bank, Habib Bank AG Zurich, Janata Bank. *(Umm Al-Qaiwain)* British Bank of the Middle East, Habib Bank, United Bank.

Advertising: Over 50 per cent of the population are non-Arabic speaking and rely on English-language press, radio and TV. Advertising is also available in public cinemas, outdoors, by negotiation with local municipalities and by direct mail in Arabic and English. There are strict regulations controlling 'indecency' of male and female models in advertising. Tobacco advertising is banned in Dubai; alcohol prohibited in Sharjah.

Trade fairs: The Dubai International Trade Centre has a full programme of exhibitions and fairs of regional significance. Exhibitions and fairs are also held in Abu Dhabi and Sharjah.

Electricity supply: 240/415 V AC (Abu Dhabi) and 220/380 V AC (Northern Emirates), with three-pin round or flat type plug fittings.

Representation in capital: Algeria, Argentina, Australia, Austria, Bangladesh, Belgium, Brazil, Canada, China PR, Denmark, Egypt, France, German FR, India, Indonesia, Iran, Iraq, Italy, Japan, Jordan, Kenya, Rep of Korea, Kuwait, Lebanon, Malaysia, Mali, Mauritania, Morocco, Netherlands, Pakistan, Philippines, Qatar, Saudi Arabia, Somalia, Spain, Sri Lanka, Sudan, Sweden, Switzerland, Syria, Tunisia, Turkey, UK, USA, USSR, Yemen AR, Yemen PDR.

Useful telephone numbers
Directory enquiries: 180 Operator: 100 Call enquiries: 160 Call bookings: 150

Hotels: Abu Dhabi (area code 02) & Al Ain (03)
Abu Dhabi Inter-Continental, Al Khalidiya Street, P.O. Box: 4171 (tel: 666888; tx: 23160; fax: 369153).

Al Ain Inter-Continental, Al Ain, P.O. Box: 16031 (tel: 654654; tx: 34034; fax: 641096).

Al Ain Palace, Abu Dhabi Corniche, P.O. Box: 33 (tel: 322377; tx: 22227 enotel; fax: 329713).

Centre (Novotel), Sheikh Hamadan Street, P.O. Box: 7136 (tel: 333555; tx: 22488 ctrnov; fax: 343633).

Corniche Residence, Sheikh Hamadan Street, P.O. Box: 279 (tel: 211200; tx: 23780; fax: 244696).

Gulf, P.O. Box: 3766 (tel: 448777; tx: 22904; fax: 446537).

Hilton International Abu Dhabi, Corniche Ave, P.O. Box: 877 (tel: 661900; tx: 22212; fax: 669696).

Hilton International Al Ain, Al Ain, P.O. Box: 1333 (tel: 641410/3; tx: 33505; fax: 654958).

Holiday Inn Abu Dhabi, P.O. Box: 3541 (tel: 335335; tx: 23030 holinn; fax: 335766).

Khalida Palace, Corniche Street, P.O. Box: 4010 (tel: 362470; tx: 22506 khp).

Meridien Abu Dhabi, P.O. Box: 6066 (tel: 776666; tx: 23794, 22233; fax: 727221,729315).

Nihal, Sheikh Hamadan Street, P.O. Box: 3789 (tel: 829900; tx: 23659/60 nihalem).

Omar Al Khayam, Corniche Street, P.O. Box: 152 (tel: 722000; tx: 23459 khayam).

Ramada Abu Dhabi (at airport), P.O. Box: 3766 (tel: 377260; tx: 48333; fax: 421033).

Ramada Dhafra, Jebel Dhannah, Ruwais, P.O. Box: 11828 (tel: (052) 71600; tx: 52201 abu dhabi; fax: 5271002).

Sheraton Abu Dhabi (on beach), P.O. Box: 640 (tel: 773333; tx: 23453 sherad; fax: 725149).

Zakher, Sheikh Hamadan Street, P.O. Box: 932 (tel: 327900; tx: 22392 zakher).

Hotels: Ajman (area code 06)
Ajman Beach, Corniche Road, P.O. Box: 874 (tel: 423333; tx: 69519).

Hotels: Dubai (area code 04)
Al Khaleej, P.O. Box: 10559 (tel: 211144; tx: 48629 kleej; fax: 237140).

Al Khaleej Palace, P.O. Box: 3720 (tel: 231000; tx: 48388 kleej; fax: 211293).

Ambassador, P.O. Box: 3226 (tel: 531000; tx: 45467 amhtl; fax: 434751).

Astoria, P.O. Box: 457 (tel: 434000; tx: 45817).

Bustan (at airport), P.O. Box: 1533 (tel: 221261; tx: 45460).

Carlton Tower, Binyas Street, P.O. Box: 1955 (tel: 227111; tx: 46410 cartel).

Chicago Beach, P.O. Box: 11416 (tel: 480000; tx: 47490 cbhtl; fax: 482273).

BUSINESS DIRECTORY

Claridge, P.O. Box: 1833 (tel: 227141; tx: 45875 clatel).

Dubai International, Airport Road, P.O. Box: 10001 (tel: 245111; tx: 47333 dihtl; fax: 246438).

Dubai Marine, Khalid Bin Walid Street, Al Raffa (in suburbs), P.O. Box: 5182 (tel: 434140; tx: 46772 dumar).

Dubai Metropolitan, Dubai-Abu Dhabi Road, P.O. Box: 4988 (tel: 440000; tx: 46999 mthtl; fax: 441146).

Dubai Plaza, P.O. Box: 7646 (tel: 459000; tx: 47978 dupla; fax: 497390).

Excelsior, Al Mateena Street, Deira, P.O. Box: 3548 (tel: 665222; tx: 46990 excel).

Hatta Fort (at Hatta, in mountains), P.O. Box: 9277 (tel: (085) 23211; tx: 47999 hatta).

Hilton, P.O. Box: 927 (tel: 370000; tx: 46670/71; fax: 371383).

Hyatt Regency, Deira, P.O. Box: 5588 (tel: 238000; tx: 47555 hyatt; fax: 211868).

Inter-Continental, Bin Yass Street, Deira, P.O. Box: 476 (tel: 227171; tx: 45779; fax: 284777).

Jebel Ali, PO Box 9255, Dubai (tel: 35252; tx: 48000 jebel; fax: (084)35543).

Phoenicia, Deira, P.O. Box: 4467 (tel: 227191; tx: 45853 fnicia; fax: 221629).

Ramada, P.O. Box: 7979 (tel: 421010; tx: 48333; fax: 421033).

Riviera, Bin Yass Street, Deira, P.O. Box: 1388 (tel: 222131; tx: 46505 riviera; fax: 211220).

Sheraton, Deira, P.O. Box: 4250 (tel: 281111; tx: 46710; fax: 213468).

Hotels: Fujairah (area code 070)
Beach Motel, Fujimrah Beach, P.O. Box: 283 (tel: 22001; tx: 89013 motel).

Hilton (on private beach), P.O. Box: 231 (tel: 22411; tx: 89018; fax: 26541).

Hotels: Khorfakkan (area code 070)
Oceanic, P.O. Box: 10444 (tel: 85111; tx: 89089 holinn).

Hotels: Ras Al Khaimah (area code 077)
Al Nakheel, P.O. Box: 5333 (tel: 22822; tx: 99261).

Ras Al Khaimah, P.O. Box: 56 (tel: 52999; tx: 99113 hotels; fax: 52990).

Hotels: Sharjah (area code 06)
Aladin, P.O. Box: 1859 (tel: 350111; tx: 68594).

Beach (by lake), P.O. Box: 5977 (tel: 358311; tx: 68558).

Carlton Sharjah, Al Khan, P.O. Box: 1198 (tel: 523711; tx: 68012 cartel; fax: 374962).

Continental, P.O. Box: 3527 (tel: 37111; tx: 68123 contel).

Coral Beach, Corniche Al Moutazah, P.O. Box: 5524 (tel: 351011; tx: 68533 em).

Federal, Al Wahada Street, P.O. Box: 5143 (tel: 354106; tx: 68189).

Holiday Inn, P.O. Box: 5802 (tel: 357357; tx: 68305 holinn).

Marbella Club (on lagoon), P.O. Box: 5017 (tel: 357123; tx: 68281 em; fax: 382050).

Palace, P.O. Box: 5734 (tel: 353555; tx: 68588).

Sharjah Airport, P.O. Box: 2088 (tel: 581100; tx: 68238 aphotel).

Sharjah Beach, Al Khan (at airport), P.O. Box: 6015 (tel: 356566; tx: 68213 novtel).

Sharjah Grand, Sultan Al Awal Street, P.O. Box: 6059 (tel: 356557; tx: 68204 homer).

Sheba, King Faisal Street, P.O. Box: 486 (tel: 22522; tx: 68053).

Summerland Motel, P.O. Box: 1083 (tel: 354321; tx: 68079).

Hotels: Umm al-Qaiwain (area code 06)
Umm al-Qaiwain Beach, P.O. Box: 158 (tel: 666647/8; tx: 47472).

Car hire: Abu Dhabi
Abu Dhabi Car Sales & Rental Co: PO Box 2554 (tel: 772317).

Al Borg Rent-A-Car: PO Box 4250 (tel: 770353, 779644).

Avis: PO Box 3237 (tel: 323760; tx: 22669); Airport (tel: 757180; tx: 22669).

Budget: Al Nasr Street, PO Box 3292 (tel: 334200; tx: 23494); Airport (757178);

City Car Rental: PO Box 7700 (tel: 771188).

Eisa Car Hire: PO Box 294 (tel: 324832, 335297; tx: 324832).

Emco Rent-A-Car: PO Box 6761 (tel: 338933/4).

Europcar: PO Box 4399 (tel: 772590, 701181; tx:22343, 24176); Airport (tel: 757183).

Safetycar Rent-A-Car: PO Box 861 (tel: 330187, 335687; tx: 22795).

Star Rent-A-Car: PO Box 2945 (tel: 323971; tx: 22990).

Sunny Rent-A-Car: PO Box 6363 (tel: 770867, 773342; tx: 23914).

Car hire: Dubai
Avis: PO Box 6891 (tel: 282121; tx: 47717).

Budget: PO Box 8323 (tel: 668200; tx: 68720).

Europcar: PO Box 2533 (tel: 434221/2; tx: 47012).

International Rent-A-Car: PO Box 10990 (tel: 662233; tx: 46799).

Car hire: Fujairah
Avis: PO Box 231 (tel: 22021; tx: 47717).

Car hire: Sharjah
Al Boorj Rent-A-Car: PO Box 67 (tel: 357884).

Budget: King Faisal Street, PO Box 2200 (tel: 373600; tx: 68720 BUDGET EM).

UAE Rent a Car: PO Box 6318, Port Khan Road (tel: 350437; tx: 68058).

Business advice and information
Abu Dhabi Chamber of Commerce & Industry, PO Box 662, Abu Dhabi (tel: 214000; tx: 22449; fax: 215867).

Abu Dhabi Company for Onshore Oil Operations (ADCO), PO Box 270, Abu Dhabi (tel: 366100).

Abu Dhabi Gas Liquefaction Co Ltd, PO Box 3500, Abu Dhabi (tel: 333888).

Abu Dhabi National Oil Co (ADNOC), PO Box 898, Abu Dhabi (tel: 366000; tx: 22215).

Ajman Chamber of Commerce & Industry, PO Box 622, Ajman (tel: 422177).

Arab Monetary Fund (headquarters), PO Box 2818, Abu Dhabi (tel: 328500).

Department of Information, Dubai Municipality, PO Box 67, Dubai (tel: 221141).

Dubai Chamber of Commerce & Industry, Bin Yass Street, PO Box 1457, Dubai (tel: 221181; tx: 45997).

Dubai International Trade Centre, PO Box 9292, Dubai (tel: 372200; tx: 47474 ditc; fax: 373493).

Federation of Chambers of Commerce & Industry, PO Box 3014, Abu Dhabi (tel: 214144).

Jebel Ali Free Zone Authority, PO Box 3258, Dubai (tel: Jebel Ali (084)56578; tx: 47398 paja; fax: (084)56093).

Gulf Arab Marketing & Exhibition Co (GAME), PO Box 610, Abu Dhabi (tel: 376900; tx: 23896; fax: 378894).

Ministry of Economy & Commerce, PO Box 901, Abu Dhabi (tel: 215455; tx: 22897 ecocom; fax: 215339); PO Box 3625, Dubai (tel: 284151; fax: 225685); Sharjah (tel: 541100); Ras Al-Khaimah (tel: 23344); Fujairah (tel: 24677).

Ministry of Finance & Industry, PO Box 433, Abu Dhabi (tel: 726000; tx: 22937).

Ministry of Information & Culture, PO Box 17, Abu Dhabi (tel: 463100); PO Box 5053, Dubai (tel: 326000; tx: 22283).

Ministry of Petroleum & Mineral Resources, PO Box 59, Abu Dhabi (tel: 651810).

Port Rashid Authority, PO Box 2149, Dubai (tel: 451545; tx: 47530).

Ras Al-Khaimah Chamber of Commerce, Industry & Agriculture, PO Box 87, Ras al-Khaimah (tel: 33511; tx: 99140).

Sharjah Chamber of Commerce & Industry, PO Box 580, Sharjah (tel: 541444; tx: 68205).

Sharjah Port Authority, PO Box 510, Sharjah (tel: 541666; tx: 68138).

Umm al-Qaiwain Chamber of Commerce and Industry, PO Box 436, Umm al-Qaiwain (tel: 666915/895; 69714).

Banking
Central Bank of the United Arab Emirates (central bank), PO Box 854, Abu Dhabi (tel: 652220; tx: 22330 markzi em; fax: 668483).

Abu Dhabi Commercial Bank, Al Salam Street, PO Box 939, Abu Dhabi (tel: 720000; tx: 22244 adcbcr em; fax: 776499).

Arab Bank for Investment & Foreign Trade, PO Box 6733, Abu Dhabi (tel: 721900; tx: 22455 arbift em; fax: 723264).

Bank of Oman Ltd, PO Box 858, Abu Dhabi (tel: 332300; tx: 22251 omanbk em; fax: 345935).

Commercial Bank of Dubai Ltd, PO Box 2466, Abu Dhabi (tel: 345700; tx: 22507 trbnk em; fax: 334126).

Emirates Bank International, PO Box 2923, Dubai (tel: 281181; tx: 47160 ebiho em; fax: 221005).

Middle East Bank, PO Box 5547, Dubai (tel: 220121; tx: 46074 mebnk em; fax: 213481).

National Bank of Abu Dhabi, PO Box 4, Abu Dhabi (tel: 335262; tx: 22266 masrip em; fax: 336078).

National Bank of Dubai, PO Box 777, Dubai (tel: 222241; tx: 45421 natnal em; fax: 215939).

National Bank of Fujairah, PO Box 887 (tel: 24518; tx: 89050 bankfj em; fax: 24516).

National Bank of Ras Al Khaimah, PO Box 5300 (tel: 21127; tx: 99109 nbrak em; fax: 23238).

National Bank of Sharjah, PO Box 4, Sharjah (tel: 547745; tx: 68085 natbnk em; fax: 543483).

National Bank of Umm Al Qawain, PO Box 800 (tel: 665225; tx: 69733 nbuaq em; fax: 665440).

Travel information
Emirates, PO Box 686, Dubai (tx: 45728).

Gulf Air, Hamdan St/Airport Road, Abu Dhabi (tel: 332600).

United Kingdom

Roland Hill

The year of Margaret Thatcher's tenth anniversary in 10 Downing Street was also the year when Britain had the highest inflation rate among the big seven industrial economies, the largest current account deficit as a percentage of GDP, and the slowest growth. Britain was also demoted to sixth place in all the tables in the OECD Half Yearly Economic Outlook (July 1989), having been overtaken even by Italy. Nigel Lawson, the former chancellor of the exchequer, stated the obvious when he said that the British was like economies of the other developed countries, 'only more so'. Lawson consoled himself and the British by putting on a brave face and looking on the bright side: the economy was still strong, with investment growing twice as fast as consumption, and British investment in the 1980s growing faster than that of all other major countries. Also, one third of all the new jobs in the European Community were created in the United Kingdom in 1988. As mortgage interest and prices are included in the British inflation rate (8.3 per cent), it was actually two points lower in comparison with other countries. However, there was slower growth in Britain's producer prices, but raw materials are less important than labour costs, which were accelerating.

Rising inflation and wage costs

The disastrous effects of the chancellor's 1988 Budget in which he had loosened the economic reins were still being felt in mid-1989. The British people enjoyed a spending spree, living largely on credit cards; debts increased, savings were at an all time low. Then Lawson applied the monetary brakes by increasing interest rates to nearly 14 per cent. This particularly upset house buyers with mortgages to pay off, mainly Tory voters, with corresponding political effect on the popularity of the government.

The British continued to indulge their old habit of paying themselves more than they earned. Average wage settlements moved up from 6 to 8 per cent – 8.8 per cent in the case of the striking railwaymen and BBC employees, though these two were regarded as exceptional cases. With inflation expected to have reached its peak these settlements were kept near it, an improvement, some said, for instance on 1986 when the inflation rate was less than 3 per cent, but wage settlements about 7.5 per cent. Average earnings were expected to top 10 per cent by the end of 1989. Chancellor Lawson's interest-rate medicine had its effect. It slowed down demand. House prices fell in London and the expensive south east as well as the East Anglia region, though not in other parts of the country. Because of the high interest rates house sales had fallen by a fifth compared with the previous year. Consumer spending was squeezed by reducing the amount of mortgage lending that merely released cash for other spending. And with fewer people moving house, spending on carpets, furniture and DIY products also declined.

The labour market remained tight. The unemployment rate continued to fall (6.3 per cent in June 1989). With the supply of young workers diminishing, it seemed that employers might be more reluctant to lay off people as demand slowed. The trade-weighted exchange rate of the pound slid no less than 7 per cent in 1988-89 which threatened to push up import prices – if sterling, shadowing the Deutsche Mark, were to reach DM 3.00 or less interest rates might have to be raised again. The pressure on the pound was partly eased by the monetary policy of the United States, but the trade deficit would cause difficulties for the pound and increase pressures for Britain to join the EMS, resistance to which came mainly from the prime minister although the chancellor, too, had made joining dependent on a lowering of inflation by at least half. For the foreseeable future interest rates looked likely to stay high. Rising inflation and wage costs limited the government's spending plans in departments with the biggest demands, especially in their reforms for health and social security, education and transport. Getting inflation down became a question of life and death for the Thatcher government. Everything depends on it.

The sudden resignation of Nigel Lawson as Chancellor in November highlighted the divisions within Mrs Thatcher's cabinet and the failure of conservative policies to get to grips with the economy. Speculation centred on whether Mr Lawson was getting out while the going was good – or whether he was simply unable to agree with Mrs Thatcher.

When figures for the enormous pay increases for company bosses were published – 176 per cent (£386,000) for Lord King, the chairman of British Airways, 220 per cent (£408,558) for Sir Edward du Cann of Lonrho, 73 per cent (£382,800) for Sir Jeffrey Sterling of P & O – Mrs Thatcher was said to have been 'appalled'. She herself persisted in setting a lonely example by refusing to take up her full pay, thus foregoing £100,000 over her ten years in office. Being 'appalled', however, is not really in accordance with her free market philosophy. Businessmen and trade unionists alike will always pay themselves what they think they are worth. What she used to emphasise, rightly, was that people should be free to increase their pay by collective bargaining as long as they earned it by productivity, otherwise they would price themselves out of the market. That was evidently the fate of those young men who earned high salaries after the 'big bang', the deregulation in the London Stock Exchange, and were sacked in large numbers as the market contracted in 1988-89.

Slump or signs of defeat?

Was all this a mere mid-term slump of the Thatcher government or the first sign of a coming electoral defeat in 1991 or 1992 when the Conservative mandate expires? Both interpretations were possible after what was a bad year economically and politically for the government. A series of labour disputes by dockers, railwaymen, tube and bus drivers, BBC-, passport office- and local authority employees, and engineering and building workers was widely called the 'summer of discontent' in allusion to the 'winter of discontent' which in 1978-79 proved the downfall of the Labour government.

But to say that was more of a political dig. Labour unrest in 1988-89 was very different. The historic defeat of Arthur Scargill, who led the miners' strike, a major political challenge to the Tory government in 1984-85, allowed the industry to be thoroughly overhauled. The nationalised British Coal was able to report in July 1989 that it was back in the black for the first time in twenty years. Production had doubled – it is now the best in Europe. Colliery costs had been cut by a third, with less than half the workforce. Intense competition in the energy market held coal prices down. Had they risen in line with inflation, revenue (£498m) would have been a billion pounds. Absenteeism, the bane of the mining industry, was lower than at any time, and the industry had the lowest number of disputes for 13 years. Restoring coal to profitability was a triumph for Thatcherism and pleasing for the taxpayer.

When the railwaymen's strike was ended, the question was asked whether a similar miracle could be worked for British Rail. Its chairman, Sir Robert Reed, had to face heavy public criticism for not having foreseen and forestalled this particular dispute, expressed in a series of one-day strikes with the lowest cost to the employees and maximum cost to the railway users. The suffering public sympathised with the badly paid railwaymen despite the fact that, as a watchdog committee reported, BR's service had reached an all-time low in punctuality, efficiency and cleanliness. A new man with proven experience in industrial efficiency was clearly needed to run British Rail, long ago the country's pride and world famous.

The railway strike highlighted a gap in Mrs Thatcher's popular labour relations reform. It was not enough to remove legal immunity from trade unions which call a strike, without first holding a properly conducted secret ballot of members concerned and securing a majority vote for their action. In the railway and dock strikes the arbitration procedures were used, court action sought. The dockers' strike was found to be illegal, but the dockers walked out nevertheless; in the case of the tube drivers the court found the ballot paper to be confusing and the strike illegal, yet the men came out all the same; in the case of the railwaymen a court action at first stopped the strike but an Appeal Court reversed the previous judgement.

The lesson for the government was that in the case of unofficial strikes existing legislation was useless and there was also a feeling that new laws would not

KEY FACTS

Official title: The United Kingdom of Great Britain and Northern Ireland

Head of state: Queen Elizabeth II

Head of government: Prime Minister Margaret Thatcher

Ruling party: Conservative Party

Capital: London

Official Languages: English (English and Welsh in Wales)

Currency: Pound Sterling (£) = 100 pence

Exchange rate: £0.63 per $ (Dec 1989)

Area: 244,103 sq km

Population: 57.1m (1988)

GDP per capita: £7,937 (1988)

GDP real growth: 2.6% (1988)

Labour force: 27.1m (1984)

Unemployment: 8.4% (1988)

Inflation: 4.8% (1988)

Trade balance: –£24,937m (1988)

help much, as they could be regarded as an arbitrary interference in the relationship between union and union members which ought to be by voluntary agreement, and this the Thatcher government wanted to retain. There was also the fear that a law which made officials responsible for their members' actions would only drive union members back into militant union control.

Politicised unions

The example of the German Federal Republic as a model for industrial relations is often held up by Labour Party supporters for imitation. But they do not mention that workers in the public service unions in the FRG are not permitted to strike. It is a condition of employment in all essential public service jobs from teachers to train drivers that a no-strike clause must be accepted and that, due to the special history of Germany, political membership of an extremist organisation like the Communist Party or neo-nazi groups is not permitted. In return for this understanding, employees get job security, index-linked pay and generous pension benefits. In Britain, and because of longer and unbroken traditions for more than 150 years, trades unionists have secured for themselves greater political rights, indeed, become politicised to such an extent that their real purpose – proper job training, improvement of working conditions, pension agreements – has almost become a matter of lesser concern. In Britain there is a confrontational relationship between the unions and any Conservative government because of their close relationship with the Labour Party. For the Thatcher government this is a double dilemma because the unions are by nature its political enemies, even though many of their members vote Conservative. But they do not have the same interests as for instance the less political German unions who collaborate with the government, whatever its political colour, in sharing out the national cake.

Another basic difference between the German and British union movements is that in Germany two disastrous post-war inflations have entered deeply into the national psyche with regard to the short-sightedness of inflationary wage settlements. Economic illiteracy in Britain, however, still suggest to many workers and their leaders that wage increases act not as a self-defeating spiral but as a means to get the better of inflation. Besides, in their confrontational attitudes ('them' and 'us') they believe that they would be victimised by the bosses as they used to be in the nineteenth century – disadvantages of being burdened with a great and unbroken tradition.

In Britain, unlike any other Western democracy, the unions were first in the field. They were the pioneers who founded the Labour Party as their parliamentary representation. The Labour-affiliated unions are the party's paymasters and exercise a five-to-one majority influence over its members by the bizarre voting system at the annual party conferences. Party policy is decided by block votes cast on behalf of millions of union members who may not even have been consulted on particular matters. The Labour leader Neil Kinnock would dearly like, if he could, to sever this traditional connection with the 'brethren', since the old custom practised by Labour politicians like Harold Wilson, to use these left-wing union allies to get into power but then ditch and govern without them, is no longer possible. Today the Labour Party and the left-wing unions stand for different things. And the Labour leader is anxious to present an image that would enable him to win the next election. Moreover, modern union leaders have long ceased to be the docile Labour supporters they used to be a generation ago.

The wave of disputes in 1989 was mainly in the public sector industries culminating with the protracted dispute with the ambulance drivers, where it seemed the government had seriously misjudged the drivers' resolve and the degree of public sympathy and support for their case. The private sector had already been through a similar phase in the early 1980s when, in the 'health slimming process' of the Thatcher government, many companies pushed through flexibility deals against the threat of unemployment and bankruptcies. These flexibility disputes also distinguished the 1989 disputes from previous ones. The dock strike, similar to the former typesetters' strike in Fleet Street, concerned the preservation of jobs in a declining industry. It was a protest against the abolition of the National Dock Labour Scheme, one of those restrictive practices by which the power monopoly of the unions could insist on the preservation of jobs regardless of work availability, the advance of modern technology and the laws of economics. The train drivers' strike was a protest against the decentralisation of pay bargaining, the tube strike partly about driver-only trains.

Transport – a key question

These strikes also showed the key importance of transport problems for the Thatcher government. Crippling congestion on the roads and sub-standard public transport services made this one of the major challenges for Mrs Thatcher. There was a series of railway disasters, the worst in December 1988, when a rush-hour train crashed near Clapham Junction, south London, and left 33 people dead and 113 injured, with the subsequent enquiry finding the cause to be a faulty signal circuit and slack maintenance work. This was followed in March 1989 by two rail crashes, at Purley in Surrey and in Glasgow, with five people killed in the first and two in the second.

Road traffic is likely to double in the foreseeable future. Many motorways, especially London's M25 ringroad, carry far more vehicles than originally planned. Traffic speed in most of London has dropped below 8mph during the day. The cost to businesses caused by traffic jams has been estimated by the CBI at £15bn a year. Railways are increasingly overladen, and in London rush-hours close to capacity, with demand increasing. In 1989 the underground in the central London area was so dangerously overburdened that London Transport considered raising fares to deter people from using the tube. The denationalisation of British Rail and London Transport, the privatisation of the municipal buses, the decision whether to introduce tolls to price many cars out of city centres and devise ways to tempt private enterprise to fund new motorways, light railway systems and new underground railways looked like major tasks for Mrs Thatcher's remaining span. The crash of an American Boeing at Lockerbie, Scotland, caused by a terrorist bomb in December 1988 highlighted airport security (along with airport delays) as another sensitive topic.

KEY INDICATORS

	Unit	1984	1985	1986	1987	1988
Population	m	56.5	56.6	56.7	56.9	57.1
Gross domestic product (GDP)	£bn	323.0	354.0	378.9	414.2	453.3
GDP per capita	£	5,717	6,254	6,683	7,279	7,937
Consumer prices	1985=100	94.3	100.0	103.4	107.8	113
Retail prices	%	5.0	6.1	4.7	4.1	4
GDP real growth	%	1.8	3.7	2.8	4.3	2.6
Share prices	1985=100	81.0	100.0	124.1	163.8	147.6
Unemployment	%	11.8	11.8	11.7	10.6	8.4
Wages (monthly earnings)	1985=100	89.8	100.0	108.0	116.7	126.5
Industrial production	1985=100	94.9	100.0	102.2	105.8	109.6
Manufacturing	1980=100	101	104	105	111	119
Metals	1980=100	108.4	112.9	–	–	–
Chemicals	1980=100	113.9	118.2	–	–	–
Engineering	1980=100	99.2	103.7	–	–	–
Food, beverages, tobacco	1980=100	102.1	101.7	–	–	–
Textiles	1980=100	97.9	101.8	–	–	–
Construction	1980=100	83	100	114	118	–
Retail sales	1980=100	110.7	115.3	121.0	126.0	–
Exports	£m	70,373	78,263	72,812	79,849	81,476
Food, beverages, tobacco	£m	4,693	4,999	5,478	–	–
Basic materials	£m	1,989	2,143	1,941	–	–
Fuels	£m	15,367	16,739	8,683	–	–
Manufactured goods	£m	46,668	52,530	54,596	–	–
Imports	£m	78,967	84,790	86,067	94,026	106,413
Food, beverages, tobacco	£m	8,936	9,281	10,059	–	–
Basic materials	£m	5,420	5,397	4,623	–	–
Fuels	£m	10,193	10,311	6,294	–	–
Manufactured goods	£m	52,886	58,381	62,833	–	–
Balance of trade	£m	–8,594	–6,527	–13,255	–14,177	–24,937
Current account	$m	2,608	4,765	158	–6,172	–26,015
Foreign exchange	$bn	6.97	9.74	14.89	38.56	41.12
Exchange rate	$ per £	1.34	1.30	1.47	1.64	1.78

* estimated figure

Tackling the 'sacred cow'

The government's major programme of reform for the National Health Service introduced in February 1989 encountered strong resistance from the doctors' lobby, nurses and midwives. Three stages were planned, the second and third in 1990 and 1991, and their purpose was to offer patients a new deal with more choice, higher standards and a better quality health service. If implemented they would enable Britain's state medicine to continue to be open to all, regardless of income, and to be financed mainly out of general taxation, true to the Conservatives' promise 'the Health Service is safe in our hands', which was much attacked and greatly ridiculed by opponents. The system has been burdened by enormous problems of accountability, long waiting lists for operations where surgery is regarded as not immediately essential, and with some operations virtually only obtainable privately. The gigantic system has become Britain's 'sacred cow', a popular institution regarded with national pride because it was thirty years ago a world leader and can still command great devotion and selfless service from many doctors, nurses and ancillary staff. This explains the great resistance to change, in addition to the traditional British dislike for changing things that have proved their worth even if everybody realises that change is needed. Under the Conservatives NHS gross expenditure increased from £8bn (1978-79) to £26bn (1989-90), an increase of 40 per cent in real terms. The expenditure for 1988-89 was equivalent to around £35 a week for an average family of four, compared with about £11 in 1978-79. NHS hospital staff now treat 8.5m patients a year, 1.5m more than ten years ago. The idea of the reforms is to delegate as much power as possible to the local level, involving greater flexibility in setting pay and conditions of employees, and financial incentives to make better use of hospital assets. While remaining within the NHS, some 320 of the largest hospitals would be able to apply for a new self-governing status, take full responsibility for their affairs, earn revenue for the services they provide and thus have an incentive to attract patients by offering the service patients want. To reduce waiting lists and help cut the long hours worked by junior doctors, 100 new consultants' posts would be created by 1991. To help family doctors improve their services, large GP practices would be able to run their own budgets and obtain a range of services direct from hospitals. GPs would also be encouraged to compete for patients by offering better services, and it would be easier for patients to choose and change doctors.

The annual bill for drugs alone, prescribed under the NHS, is £2bn, more than the cost of the doctors issuing the prescriptions. To ensure quality of service, a value-for-money system of medical auditing would be more rigorously applied. The criticisms which the proposals encountered were concentrated on the possibility that GPs with limited budgets might be forced to deny patients certain treatments or drugs; also that 'market-orientated medicine' would contribute less to raising the level of care for the chronically ill and elderly, because that sector does not bring in money and constitutes merely a growing burden for the community. In the 40th anniversary year of the National Health Service the Tories have started on its most significant reform – its success or failure will be a determining factor in the next general election.

Thatcher in the doldrums?

The year which saw on 4 May 1989 the tenth anniversary celebrations of Margaret Thatcher's entry into 10 Downing Street also produced considerable political reverses notably, Mr Lawson's resignation. In March 1989, in a by-election in the middle-class constituency of Richmond, North Yorkshire (where Labour had no chance), the Conservatives suffered their biggest drop in support since they came into office, with their majority reduced from 19,000 to 2,000, and the small remainder of Dr David Owen's Social Democrats (SDP) and Paddy Ashdown's social and liberal democrats (SLD) winning between them the most votes. Labour held its majority in the by-election of Pontypridd, Wales, and won Vale of Glamorgan from the Conservatives with a 12 per cent swing. But by-elections are traditionally unreliable indicators of the political mood of the country because they are usually occasions for letting off steam, allowing people to vote with their hearts rather than with their heads. That is why they are generally a feast for third party support which, when general elections dawn, melts away again to one or the other of the two main parties. Britain's two-party system was widely back in favour due to the recovery of the Labour Party and the inability of Paddy Ashdown's Social and Liberal Democrats to rise and to Dr David Owen announcing the partial demise of his Social Democrats as a national political force by deciding to operate in future on a 'guerrilla' basis.

Letting off steam was certainly also the point about the elections to the European Parliament in June 1989 because no one in Britain can take the Strasbourg Assembly seriously, either as a parliament or as a power factor, least of all Mrs Thatcher. That was a tactical mistake because the opposition did take it seriously, if only as a mini-election test which it was in a way, being the first national election since 1987. The result was not so bad for the Tories as BBC television predicted – it is becoming a sort of stunt habit for the corporation to make wild predictions – nor was it as bad as in mid-term by-elections under previous Tory governments when they nevertheless won the general election that followed. However, Labour triumphantly polled 40.1 per cent and captured 13 of the previous 45 Tory Euro-seats, with the Conservatives polling 34.7 per cent. There was an amazing breakthrough of the Greens (14.9 per cent), but the poor Democrats and Nationalists only got 6.2 and 3.4 per cent respectively.

If the British had voted by proportional representation, their 78 Euro-seats would have been divided into 32 Labour, 28 Tory, 12 Green, 4 SLD, 2 Nationalists. Bad for the Tories was not so much that their Scottish and Welsh representation was wiped out or only three of their 32 MEPs just held on to their seats north of the river Trent, but that the Tory-vote slide was actually bigger in their strong areas south of the Trent. The British result corresponded to that of the other eleven EC member states: a victory for the Left and the Green Party (although their 1.3m votes produced not a single seat under the British majority voting system, though they practically took the place of the third party). The poll, with 34 per cent, was better than in previous Euro-elections but still the lowest in Europe.

There were bitter recriminations in the Conservative Party: its campaign was muddled, wrong-footed, too negatively emphasised, for example Mrs Thatcher's question: 'Do you want socialism to come back through the European backdoor?' To which the Labour leader Neil Kinnock could only reply that he was intending to come in through the front door. On Europe the Conservative campaign merely reflected

COUNTRY PROFILE

Political system: Monarchy, with sovereign as head of state. Legislative power vested in bicameral parliament consisting of House of Commons (650 members directly elected by universal adult suffrage for up to five years) and House of Lords (composed of hereditary and life peers). Executive power is vested in the cabinet, led by the prime minister, which is responsible to the House of Commons.

Main cities/towns: London (population 6.8m), Birmingham, Glasgow, Leeds, Sheffield, Liverpool, Bradford, Manchester, Edinburgh, Bristol, Belfast, Cardiff and Newcastle.

Media
Press: National and regional press, dominated by nine dailies with combined circulation of over 14m. The Press Council (a voluntary body under an independent chairman) responsible for maintaining general standards and dealing with complaints.
Dailies: Most influential: *The Times* (circulation 450,626), *Financial Times* (286,774), *The Guardian* (470,023) and *Daily Telegraph* (1,138,673). *The Independent* (375,317) emerged in 1987. Most popular are *The Sun* (4,146,644), *Daily Mirror* (3,082,215), *Daily Express* (1,679,438), *The Star* (28,544) and *Daily Mail* (1,792,701).
Weeklies: Several Sunday newspapers – most influential: *The Sunday Times* (1,314,504), *The Observer* (749,644) and *Sunday Telegraph* (716,044). Most popular: *News of the World* (5,213,901), *Sunday Mirror* (2,778935) and *The People* (2,932,472). Most widely read weekly magazines include *Radio Times* (BBC radio and TV programme listings), *TV Times* (IBA radio and TV programme listings), *Woman's Weekly, Woman and Woman's Own.* Also vast number of general and special interest magazines.
Business: Large number covering all aspects of business, often with international circulation including *Financial Times* (daily), *The Economist* and *Investors Chronicle* (both weekly).
Broadcasting: Controlled by the non-commercial British Broadcasting Corporation (BBC), which is financed by TV licence fees; and Independent Broadcasting Authority (IBA) which has authority over Independent Local Radio (ILR), Independent Television (ITV) and the Channel Four Television Co., which are all financed primarily by sale of advertising time.
Radio: Over 53m receivers in use. BBC operates four national networks and 32 local radio stations. Also over 40 commercial (ILR) stations in operation.
Television: Approximately 23m receivers in use. Four national TV networks in operation - BBC1 and BBC2 controlled by BBC; and ITV and Channel Four (in Wales 'Awdurdod Sianel Pedwar Cymrw' – S4C). ITV and Channel Four programmes are produced by 15 regional TV companies. It is estimated that over 200,000 homes are cabled, mainly to older low capacity systems. Wideband cable and satellite systems are expected to grow slowly.
Advertising: Highly developed industry covering all usual media. Television advertising is particularly important, with both ITV and Channel Four accepting advertisements (limited to a daily average of six minutes per hour). Advertising on ILR also effective (limited to nine minutes per hour). Practically all newspapers and magazines carry advertising in some form, and high circulation figures make this a particularly useful form of promotion. Short films in cinemas, and posters – on billboards and public transport – are also widely used. The Advertising Standards Authority acts as an independent watchdog, maintaining general standards. Total advertising spend #4,055m (1984). Key categories include food, household equipment, financial leisure and automotive.

Domestic economy: Large and open mixed economy characterised by export-orientated manufacturing sector, energy self-sufficiency and growing service industry. Mining and agriculture also important. Economic growth has been sluggish despite increased offshore oil and gas earnings and reductions in inflation. Unemployment has steadily increased and real living standards have fallen in all regions except the south. Government policy remains essentially monetarist with particular emphasis on increased liberalisation, cuts in public sector spending and reduction in labour costs.

Employment: Around 32 per cent of the workforce is employed in manufacturing and construction, three per cent in agriculture and 64 per cent in services. Unemployment remains high, particularly in areas with the greatest dependence on traditional industries. Special employment training and relief schemes have failed to reverse the underlying trend. Over the last year real wages, particularly in the public sector, have lagged behind inflation.

External trade: Traditional deficit on non-oil visible trade offset by invisible earnings (which account for 30 per cent of total export earnings). However, recent rise in imported manufactured goods, capital outflows and slump in non-oil exports significantly reduced current account surplus. Major European exporter of oil, aerospace products, chemicals, motor vehicles, metals, finished textiles and machinery. Also major importer of agricultural products, raw materials, semi-manufactures and most recently finished manufactures. Major trading partners include German FR, USA, France and the Netherlands.

Agriculture: Highly efficient sector which contributes around three per cent to GDP and meets about 75 per cent of national food requirements. Also important European exporter of agricultural produce, fertilisers and foodstuffs. Farm holdings are largely owner-occupied and production is concentrated on livestock rearing and growing of wheat, barley, oats, potatoes, oilseed and sugar-beet. Horticulture and trawler fishing also important.

Industry: Sector forms mainstay of economy contributing over 25 per cent to GDP and accounting for around 65 per cent of visible exports. Production has remained sluggish, particularly in construction and 'heavy' industries. Also gradual decline in international competitiveness. Principal industries include metal manufacturing, chemicals, engineering, motor vehicles, aircraft, textiles, telecommunications, shipbuilding, electronics, food processing and other agro-business. Current emphasis on privatisation, rehabilitation and development of high-tech industries.

Mining: Major European coal producer with estimated operating reserves of over 7bn tonnes. Also significant producer of zinc, lead, limestone, sand and tin. Other minerals include small-scale deposits of silver, copper, gold, iron ore and potash.

Energy: Exploitation of oil and gas from continental shelf under North Sea enabled self-sufficiency in energy. Around 25 offshore fields producing crude oil of which Brent and Forties among largest in Europe. Recoverable reserves estimated at between 1.6 and 2.7bn tonnes. Water and coal resources sufficient for domestic and industrial requirements. Current emphasis on developing nuclear power which accounts for 18 per cent of electricity generated. British Gas was privatised in 1986, British Petroleum (BP) in 1987; the privatisation of the electricity supply industry is also planned.

Banking: Highly developed system with the City of London acting as major international financial centre. Monetary policy and Exchange Equalisation Account managed by Bank of England (central bank). Also four main clearing banks (Lloyds, National Westminster, Barclays and Midland), National Girobank, Trustee Savings Bank, National Savings Bank, numerous merchant banks and some 540 overseas banks and financial institutions.

Legal system: Common law tradition with early Roman and modern continental influences. No written constitution.

Membership of international organisations: ADB, CCC, CENTO, Colombo Plan, Council of Europe, DAC, EC, ELDO, ESRO, FAO, GATT, IAEA, IBRD, ICAC, ICAO, ICES, ICO, IDA, IDB, IEA, IFAD, IFC, IHO, ILO, International Lead and Zinc Study Group, IMF, IMO, INTELSAT, Interpol, IOOC, IPU, IRC, IRO, ISO, ITC, ITU, IWC, NATO, OECD, UN, UPU, WEU, WHO, WIPO, WMO, WSG.

the differences of emphasis whether to join the other eleven in building up the Community by pretending to think less about national interests or fighting for national interests, (as the others do) as well as greater accountability of the European Commission. The vote, in so far as Conservatives stayed at home, was less a victory for Labour than an expression of widespread dissatisfaction with Thatcherism, especially with inflation and interest rates, but above all for apparently not knowing their own mind in Europe. The first major success of the Green Party was to respond to a series of scandals concerning drinking water and food production. They made the protest character of the Euro-vote quite clear. Many of the third party voters seemed to have gone over to them. Discontented, however, as the average Tory voter might be, he would hardly vote in a general election for a party that is against nuclear arms, nuclear energy production, economic growth, and for the cancellation of the Channel Tunnel project and the European agricultural policy.

Maggie and the Greens

The new factor in British politics – environmental concern – was highlighted by the prime minister's speech to the Royal Society, the oldest and leading institution of British scientists, in September 1988 in which, as a trained scientist herself, she stressed the importance, apart from all utilitarian considerations, of basic science and the need to take seriously the evidence that man's activities were harming the global environment. The value of the work of scientists could not be measured by economic output, she said, but that was no argument for lack of careful management in the way specific projects were conducted in an age of enormous technical change. Man had unwittingly begun a massive experiment with the system of the planet itself, and she gave examples of three changes in atmospheric chemistry that must cause grave concern: the increased levels of carbon dioxide and methane producing the 'greenhouse effect', chlorofluorocarbons acting in the same way and also depleting the ozone layer, and industrial emissions of nitrogen and sulphur oxides producing acid rain.

The speech was influential, and led to charges that the prime minister had become a late convert to environmental issues for political reasons – though, if true, that seems to be as good a reason as any. An international conference in London on the ozone layer in March 1989 was a direct result. The rise in public consciousness of environmental pollution could be said to be a direct result of the rise in living standards under the Thatcher government. There was a great scare about salmonella-infected eggs and poultry, largely due to feeding methods in battery farming. Millions of eggs and chickens had to be destroyed, and the high-profile Junior Health Minister Edwina Currie resigned after an unguarded remark on television, taken as critical of the industry. She was generally thought to have been sacrificed to the powerful farming lobby, strong supporters of the Conservatives. In parliament the government got into trouble over its timetable for bringing water purity levels up to EC standards.

With the government under pressure on these and other political fronts and shown to have slipped on many, the Labour Party advanced in the opinion polls, and in early 1990 was 8 per cent ahead, unthinkable a year before. If Labour could keep this up throughout the remainder of the Thatcher mandate, even the enormous advantage of the government's 100 seat majority in the House of Commons might be in danger. What must have been worrying for Labour was the fact that in its opinion poll advances it had as yet not made a significant dent in the middle-class vote of southern and south east England, Britain's fastest growing area which, as the potential party of the future, it must win over. The general opinion seemed to be that Mrs Thatcher was liked less and less, and that voting for her again was temporarily in suspense.

The Iron Lady has suddenly become vulnerable and fallible, and Mr Kinnock, a shining knight in new armour, seemed to many a less frightening prospect as a future prime minister. He was more confident than before. He had purged his party of its wild men and their ideas which was a considerable achievement for one who not long before had agreed with them. But the most difficult task was to appear credible not only to his own supporters but to a whole new Thatcherite generation that could not be less interested in Labour's traditional politics and concerns. Kinnock managed to move the Labour Party away from its election-losing unilateralist stance on nuclear weapons, though the Transport Workers Union remained true to the old beliefs.

Labour on the advance

But the accusation of a typical Labour Party fudge, wanting to be everything to all men, remained due to the divergent pulls. It was not much of a change of policy to pretend that one wanted to keep and was prepared to use a weapon which a Labour prime minister would put into cold storage (i.e. the Strategic Arms Limitation Talks), thus relying on a deterrent which would no longer deter. Mr Kinnock got very angry with people who asked him whether he, the former anti-nuclear protester, would if he had to, press the nuclear button. None of his opponents would answer that embarrassing question either, he said. But the point about deterrence is not whether he or they would press the nuclear button, but whether the enemy believes that he might, and also perhaps whether the British voters believe that he would.

The Labour Party's revised 90-page programme, *Confronting the Challenge*, the result of more than two years' work by many committees, contained accordingly some good, some bad and a lot of boring and irrelevant points. It discarded, apart from the old unilateralist policy, two other ideological 'sacred cows': extensive nationalisation and high taxation. In almost Thatcherite language the Labour leader declared: 'We must not spend more than we can afford. We must live within our means. We want to work with, not against the market.' The costs of the programme were not shown. Taxation for the richest was to be raised from 40 to 59 per cent, general taxation to be between 20 and 50 per cent, plus 9 per cent for social security; there would be a minimum wage which would be half the average wage, a 9 per cent charge for share and capital earnings, a limitation of the tax benefits introduced by the Conservatives, and increases in child benefit. The Labour Party promised support for NATO 'as long as it exists', since the dissolution of the Atlantic as well as the Warsaw Pact remains its final aim.

On European policy Labour spoke with the voice of a convert who has relapsed and reconverted, relapsed again and reconverted. Labour had new confidence, as had the British Trades Union Congress, that the Social Charter of the French socialist president of the European Commission, Monsieur Delors, would be of great advantage to the British Left. The reason was largely that continental politicians and their public, accustomed to progressive social policies, but not to unions with the political clout of their British counterparts, could not begin to understand what Mrs Thatcher's fuss about the European Commission's political aspirations was all about. However, the Labour Party was careful enough not to appear too idealistically 'European'. It sensed that the prime minister's standing up for sovereign rights and against the EC's superstate postures was soundly based in the nation's political instincts.

The cabinet strategy

The importance of Britain's policy in Europe in the run-up to the completion of the single European market in 1992 was the key to the cabinet changes which were announced in July 1989 and which were expected to see the Thatcher government into the next election. The circumstances in which these changes took place recalled Macmillan's 'night of the long knives' in July 1962, but while Macmillan sacked seven cabinet ministers, Mrs Thatcher made do – relatively bloodlessly – with only two, though four left for their own reasons, among them Industry Secretary Lord Young and Defence Minister George Younger.

The biggest surprise was Sir Geoffrey Howe's replacement by Mr John Major, the 46-year-old former chief secretary at the treasury, with no experience in foreign affairs but proven as a monetarist and in keeping government departments' spending in check, which was the reason for his promotion not only as a preferred Thatcher heir-apparent but also as someone with her instinctive approach to European policy. However, Mr Major's tenure at the foreign office was short-lived. Following Mr Lawson's resignation, he was appointed chancellor, with Mr Douglas Hurd moving from the Home Office to become foreign secretary, with Mr Waddington replacing Mr Hurd at the home office. Under Sir Geoffrey Howe the foreign office had gone, as far as the prime minister was concerned, too 'native' in regard to Brussels affairs. Major, unlike Sir Geoffrey Howe, is a 'European' of a younger generation, not an enthusiast who, like Mrs Thatcher, accepts British EC membership soberly as a fact of life and as a matter of national interest. What disagreements there were between Mrs Thatcher and Howe concerned not major issues but rather the question of 'when' Britain would join the European Monetary System (EMS) and the Exchange Rate Mechanism (ERM), which will not be before inflation had at least been halved. With Major and his number two, Francis Maude, Mrs Thatcher would have the Foreign Office machine working more pliantly in the way she wanted it to function.

Mrs Thatcher intended keeping the great strategic questions of British foreign policy in her hands: how far had the so-called Anglo-American 'special relationship' been affected by President Bush's siding with Bonn's delaying tactics rather than with British urgency in the matter of modernising short-range nuclear weapons? Mrs Thatcher inclined to the belief that the Americans would sooner or later discover that the British were their best and most reliable friends in Europe. For the moment, however, they were relatively isolated since at the Madrid Summit they willy-nilly joined the other eleven in going ahead with the first stages of greater European unity, and the NATO disagreements seemed to have been merely postponed to a later date. This might change again when the question of a successor for the Lance rocket might produce renewed German resistance to weapons, the destructive power of which would hit them more than any of their allies. Basically, Americans, British and French were agreed on the need for maintaining NATO's strength and treating Chancellor Kohl with understanding for his internal political difficulties. But the danger of a red–green coalition in Bonn could bring the French and British closer who, much as they dislike each other, can in case of need always fall back on the old 'Entente Cordiale'.

On East-West relations and after a spectacular London visit of President Gorbachev, with British 'Gorbymania' falling somewhat short of the German variety, the British were inclined to be careful. Mrs Thatcher, having been the first to discover that one could do business with the Soviet leader, seemed to find that he was trying to do too much and not systematically enough, as though he, as a new Peter the Great, had to 'Westernise' the USSR in a desperate hurry. In London he seemed to be regarded as a pragmatist which in the Thatcher perspective is no longer a recipe for success.

Realpolitik for Hong Kong

Sir Geoffrey Howe, now deputy prime minister and leader of the House of Commons, had to end his much praised six years in the foreign office unpleasantly, first because he was plainly unwilling to leave and had to be 'persuaded' by Mrs Thatcher with a number of sweeteners, second because the Anglo-Chinese Agreement he had reached regarding the due return of Hong Kong to China in 1997 looked very different after the bloody civil war-like events of Tianamen Square in Peking in the early summer of 1989. Mrs Thatcher told the House of Commons that there was no question of giving the right of abode to 3.25m Hong Kong Chinese who are British passport holders. Sir Geoffrey Howe's disagreeable last task was having to explain that to jeering and demonstrating Hong Kong crowds. Unsatisfactorily he proposed to organise international assistance in the event of a mass exodus. This kind of traditional foreign office *Realpolitik* led to attacks on the government by friends and enemies alike, to charges of betrayal of the Hong Kong Chinese as well as of appeasement of dictators of which the foreign office had not been guilty for over fifty years.

The British dilemma in Hong Kong was compounded by the unfortunate 36,000 Vietnamese 'boat people' who had landed in the overcrowded British colony and were to be persuaded even with money bribes to return voluntarily to Vietnam. The British claimed that they were 'economic emigrants' rather than 'political refugees', a distinction not popular with the Americans who had done a lot for Vietnam's refugees and made that quite clear to the British. Moreover, people were reminded of the shameful actions in 1945, when British troops in Austria forcibly sent tens of thousands of Cossacks, white Russians and Croatians, women and children among them, back to Stalin and Tito and thus to their certain death, because they were unable to cope with the millions of surrendering prisoners of war and were glad to have fewer mouths to feed.

Mrs Thatcher's cabinet changes were intended to serve the better presentation of policies which her government would stick to come what may, and in the hope that they would eventually be found to be more acceptable. Important apart from the foreign office changes was the move of Kenneth Baker from education to become chairman of the Conservative Party. Having introduced the Thatcherite school reform, he is regarded as the prototype of the ever-smiling and smooth presenter of policies who now has to inspire new confidence in the flagging Conservative ranks. If successful, he would be another contender for the leadership. So would, under the same assumptions, be the new minister for the environment, 45-year-old Chris Patten, who took the place of Nicholas Ridley (now again in charge of trade and industry). Ridley was disastrously misplaced because he appeared to lack understanding for environmental issues which seemed to be distasteful to him. He also lacked the talent, an essential prerequisite in this hot seat, to suffer idealists, journalists and generally fools gladly, and as a chain-smoker was also a bad advertisement for environmental protection. Patten looked likely to please the environmental lobby better as a former 'wet' or 'leftist' Conservative, as more liberal-minded Thatcher critics used to be called before they were either converted to Thatcherism or left the cabinet. His main task would be to deal with the Greens and to counter the attraction they might present to dissatisfied conservatives.

By appointing the effective and persuasive Cecil Parkinson, another of her favourites, to be minister of transport, Mrs Thatcher had clearly recognised the key political importance of that post for the British future although it soon became apparent that Mr Parkinson was making heavy weather of the job. In this connection the full Channel Tunnel operation was in progress by the middle of 1989 with eleven tunnel-boring machines working on the project and the aim of finishing by the end of 1992. The cost of the project went up to £5.3bn, with Mrs Thatcher again insisting that no government money would be available multilaterally. However, with higher compensation paid in France to owners of land for the new high-speed links to and from the tunnel entrance and exit, the British with their less efficient railways and lagging behind in the planning anyway were feeling to be disadvantaged.

Environmental protesters in Kent succeeded in getting the rail connection London to Folkestone to run 25 miles, a third of the journey, underground so as to avoid ruining the beautiful countryside of Kent more than necessary. This will cost an additional £500m, and the building of the British rail link will not be completed until the late 1990s. So the journey from London to Folkestone will take half an hour longer, which means three hours for the London to Brussels run and 2 hours 45 minutes for London to Paris. The protest movement extended to south London which will be greatly inconvenienced by the required modernisation of the rail link. The transport problems in connection with the Channel Tunnel seemed to be far greater than the project itself.

No hooligans

And as though the year did not have enough political and economic setbacks, 95 people, more than half of them under the age of 20, were crushed to death and 175 injured during a football match at the Hillsborough stadium in Sheffield in April 1989. The cause for once was not football hooligans but the very fences designed to stop them invading the pitch, which trapped the fans in an overcrowded area. An exhaustive enquiry conducted by Lord Justice Taylor found the police to be mainly to blame because they had been unprepared for the thousands of fans arriving at the main gates, many without tickets. They opened the gates to let in the people who, by their sheer weight of numbers, pressed upon one of the overcrowded stands. By concentrating on averting the threat of public order outside the football ground an imbalance was created between the need to quell a minority of drunkards and troublemakers and the need to safeguard the security and comfort of the majority. The lack of communication between the police and the organisers of the football match, as well as the lack of first-aid facilities, were contributing factors and confirmed public criticism of football clubs for their resistance to introducing changes for improved safety on the terraces. Hooliganism remained associated with football matches and had by mid-1989 led only to the controversial government proposal to introduce identity cards. This goes against the grain with the British public who are not used to the idea of identity cards in general as people are in continental Europe, where they are not regarded as an infringement of personal freedom. But the clubs themselves had not come forward with any proposals for viable measures to combat the problem. Inhuman conditions on the stands with their inadequate toilet facilities contributed to brutalising Britain's most popular sport.

Church and state

England's national church can always be relied upon to cater for more than just the nation's spiritual needs. The Bishop of Durham contributed his almost annual entertainment by saying that the resurrection of Jesus Christ was not a physical event, and the Archbishop of Canterbury got into the news because Portia, a pig owned by him, won first prize at the Kent Agricultural Show. The House of Commons made unusual use of its right to run the national church by rejecting the General Synod's measure which would have allowed the ordination of remarried divorcees. This confronted the two primates of Canterbury and York with the question whether they should accept defeat or consider renouncing the gold fetters of the state that provide them with much national prestige though they do little to fill the churches. The rebuff was administered to the church by a small group of anglican MPs, feeling strongly that clergymen who preach and administer indissoluble Christian marriage should set a 'good example' by not being divorced themselves. Whether the MPs were acting constitutionally in teaching their church Christian morals which is not any longer the House of Commons' business, is another question.

The bishops had at first justified the Synod's proposal by pointing to the shortage of priests and saying that only a few hard cases would be involved. Then it turned out that no fewer than 240 potential ordinands would be affected. It was said that it would be unchristian and pharisaical to deny the poor sinners forgiveness. But then some of the faithful grumbled and pointed out to their bishops that sins were only forgiven if they were repented. The remarried priests, however, surely felt themselves to be rightly married and were not sorry at all. And besides, they asked, what about those clergymen who had renounced divorce to marry someone they loved, out of regard for their calling and the indissolubility of Christian marriage? To Charles Moore, the editor of *The Spectator*, the proposal to ordain divorced priests seemed equivalent to ordaining pork butchers as rabbis.

But the Anglican Church was not alone in responding to the idiosyncratic religious needs of British people. The Presbyterian Church, a calvinist fundamentalist sect of some 7,000 members, did not hesitate to suspend even its most illustrious brother, Lord Chancellor Mackay. His sin was to have taken part twice in funerals of Roman Catholic judges who were his friends and to have attended a requiem mass which in John Knox's Scotland was still tantamount to idolatry and blasphemy. The lord chancellor preferred to leave the church in which he and his family had been brought up rather than promise that he would not do it again.

Tolerance for blasphemers?

A blasphemy of a different kind according to Islamic law was committed by the Anglo-Indian author Mr Salman Rushdie in his novel *The Satanic Verses*, which Muslims found deeply insulting to their religion and its founder, the Prophet. The Ayatollah Khomeini declared (before his death) that the author and all those involved in the book's publication were sentenced to death and hellfire, and killing them would be a religious act (as at one time killing Protestants was for loyal Roman Catholics). There were anti-Rushdie riots and killings all over the Islamic world. Contracts worth $1m dollars were offered for Rushdie's life. The Muslim community in Britain demonstrated in their thousands, burned his book publicly, and demanded changes in the ancient British blasphemy laws, but this was rejected on the grounds that it was not legally feasible in a pluralist society. Mr Rushdie, who had to go into hiding and live under police protection, in danger of his life from fanatics, made what appeared to be a partial 'apology' for what he called 'genuine distress' caused to 'sincere Muslims'.

It was pointed out with some irony that Mr Rushdie, whose freedom to publish what he writes was defended everywhere, had himself been a vocal supporter of legislation against racial discrimination which involves censorship on the very grounds upon which the withdrawal of his novel has been urged, namely that the statement of certain views is liable to provoke violence and cause racial hatred. Muslims in Britain are aware that living in the country involves accepting its laws of religious tolerance which they do not share. Under British law they can expect protection of their liberty but not protection of their religion, just as Christians have to accept that living in a pluralist society gives them no right to special religious protection, if one excepts the above mentioned freak-case of the Church of England still under state control.

TV ban for IRA

1988 and 1989 were notable for various commemorations going back to the First and Second World Wars. Sadly it was also the twentieth anniversary of the arrival of British troops in Northern Ireland, then welcomed by the Roman Catholic Community as their liberators. But this was soon to change. The troops' presence was exploited for propaganda purposes by the IRA which had been caught unawares by the success of the civil rights movement. However, it soon infiltrated its ranks and mounted a series of attacks designed to break Northern Ireland's link with Great Britain.

In 1970, when the 'provisional' IRA broke with the 'official' IRA, the nature of the violence in Northern Ireland changed substantially from a public order problem to a terrorist problem. A further split in 1974 led to the emergence of the Irish National Liberation Army (INLA) which has been responsible for particularly brutal murders and terrorist attacks and internal feuds. Terrorism continues, but the level of violence has been reduced from 467 dead in 1972 to an annual average of 90 dead in the 1980s, not counting, of course, the immeasurable tragedies afflicting thousands of injured and ruined lives.

The Anglo-Irish Agreement concluded in 1985 continued four years later to function more or less as a token of the accord of both governments that no change in the (British) status of Northern Ireland would come about without the consent of the majority of its people, that there was no such wish at present but that, if a majority expressed such a wish in the future, both governments would support it. The intergovernmental conference met regularly, and cross-border co-operation between the police and the army was extensive. The British government imposed a ban on television and radio interviews with members of Sinn Fein, the political wing of the IRA, and IRA spokesmen as well as a curb in Northern Ireland on the right not to be challenged in court if the accused remained silent. What frustrates any political advance is the conflicting aspirations of the 60 per cent Protestant majority and the 40 per cent Catholic minority. The Protestants regard themselves as British, want to belong to the United Kingdom and are opposed to a united Ireland; the Catholics regard themselves as Irish and want a united Ireland. Northern Ireland is clearly one of the world's many insoluble problems. The only hope must be fora *modus vivendi*, which exists for example in the continental economically prospering regions such as in the South Tyrol between Italians and Austrians or among Belgium's Flemish and Walloon ethnic groups. The terrorist outrages conceal how much is being done to rebuild what has been destroyed, to improve Ulster's viability, and by way of reconciliation in a land in which the great majority of both traditions want to live at peace with each other. The cause of the IRA was given a considerable boost with the release in late 1989 of the 'Guildford Four' who, it appeared, had been wrongly jailed in 1974 for the bombing of a public house in Guildford. This unprecedented reversal, and the doubts it raised about British impartiality, lead to further pressure for the release of other convicted terrorists – notably the 'Birmingham Six'.

Evolution not revolution

The French celebrations on the two-hundredth anniversary of the Storming of the Bastille on 14 July 1989 had a memorable response from Britain where people had the good – or according to some, bad – fortune to have been saved from the havoc caused by the French Revolution on the continent. Mrs Thatcher, in the tradition of Burke, did not question President Mitterand's right to national celebrations; what she did criticise was the linked French claim that '*liberté, egalité and fraternité*' were an unmixed blessing for mankind. 'Pardon me', she said with topical allusions, 'let us not forget the terror; then came Napoleon in order to unite Europe by force. The only universally valid lesson is rather that people have rights which no state can take away from them and which nowadays ought to be respected by all governments.' And she added with pride that the human rights proclaimed at the beginning of the revolution were not invented by the French. 'We had our Magna Carta, then there were the Greeks. ...' She forgot about Judeo-Christian ideas and much else, as historians were quick to point out, but she was right in stressing the continuing difference between the revolutionary approach and the British preference for evolution. Instead of one clean and bloody sweep the British prefer their revolutions to be unbloody, and consequently were landed with many more and smaller ones.

BUSINESS GUIDE

Time: GMT (GMT + 1 from late Mar to late Oct).

Climate: Temperate, with warm summers and cold winters. Rainfall throughout year, heaviest Oct to Mar. Average temperatures range from about 5°C to 15°C.
Clothing: Mediumweight and raincoat throughout year, topcoat for winter.

Entry requirements: Passport (valid until at least two months after intended departure date) required by all visitors. Visas required by nationals of CMEA countries; all Asian countries except Bahrain, Israel, Japan, Kuwait, South Korea, Qatar, Turkey and UAE; all African countries except Algeria, Ivory Coast, Morocco, Niger, Tunisia and Republic of South Africa; Cuba and Argentina.
Health: Vaccination certificates not normally required.
Currency: No restrictions on movement of UK or foreign currencies.
Customs: Personal effects duty free (other than alcoholic drink, tobacco products, perfume), plus small duty-free allowance.

Air access: Regular direct flights operated by all major international airlines.
National airline: British Airways.
International airports: London Heathrow (LHR) 24 km from capital; London Gatwick (LGW); Birmingham International (BHX); Prestwick (PIK); and Luton International (LTN).

Surface access: Regular ferry and hovercraft connections from the continent.
Main ports: London, Liverpool, Grimsby, Southampton, Milford Haven, Teesport, Hartlepool, Dover and Felixstowe.

Hotels: Classified from one- to five-star by AA and RAC (automobile clubs). Rating system in Northern Ireland – A star, A, B star, B, C and D. Single room (five-star) generally costs £60 to £75 per night. Prices usually include 10-15 per cent service charge, but tipping also expected.

Currency
Exchange rate: The pound floats, with occasional intervention by the Bank of England during unstable market conditions.

Foreign exchange information
Account restrictions: There are no blocked or non-resident accounts.
Trade policy: There are no exchange controls over imports or exports. Imports of certain items such as food products, livestock and textiles are subject to quota or other restrictions.

Credit cards: All major credit cards accepted.

Car hire: Widely available at airports and in main towns, with all major international hire firms represented. International driving licence or full national licence required. Drive on left. Speed limits: motorways/dual carriageways 70 mph (113 kph), normal roads 60 mph (97 kph) and built-up areas 30 or 40 mph (48 or 64 kph).

City transport
Taxis: Available in all major cities, and can be hailed in the street, at ranks, or be called by telephone. Taxis may charge extra – over and above metered charge – for number of passengers, large items of luggage, journeys at night, at weekends and journeys over 8 km. Tipping (about 10 per cent) expected. For black cab in London, tel: 253-5000 or 272-0272.
Buses and underground: Operated in London by London Regional Transport (LRT). Extensive network linking all parts of the capital. Good bus services also available in all other major towns. Glasgow, Liverpool and Newcastle also have reliable metro services.

National transport
Air: Most major cities linked by regular flights to 21 main commercial airports provided by seven domestic operators. Not widely used, with exception of London-Scotland services, due to expense and relatively short distances between principal commercial centres.
Road: Extensive network of about 370,000 km, including 2,800 km of motorway. Regular bus and coach services, mainly operated by state-controlled National Bus Co. and Scottish Bus Group, and various municipal and private companies.
Rail: Relatively expensive first and second class service. Network of about 18,400 km, of which 4,000 km is electrified. All principal towns are connected by regular inter-city services, operated by the British Railways Board and, in Northern Ireland, the Northern Ireland Railways Co.
Water: Over 3,000 km of navigable inland waterways, under control of the British Waterways Board.

Public holidays
Fixed dates: 1 Jan, 25 and 26 Dec.
Variable dates: Good Friday, Easter Monday, May Day (first Mon in May), Spring Bank Holiday (last Mon in May) and Summer Bank Holiday (last Mon in Aug). N.B.: Some variations in Scotland and Northern Ireland.

Working hours
Government and business: (Mon-Fri) usually 0900 – 1700.
Banking: (Mon-Fri) 0930 – 1530, some banks open Sat morning and there are variations in hours in Scotland and Northern Ireland.
Shops: (Mon-Sat) generally 0900 – 1730, usually early closing one day per week.

Business language and interpreting/translation: Available from The Institute of Linguists (tel: (01)359-7445), the Guild of Guide Lecturers (tel: (01)839-7438); or see 'Translators and Interpretors' section of 'Yellow Pages' telephone directory.

BUSINESS GUIDE

Telecommunications: British Telecom, although no longer having statutory monopoly to provide telecommunications, continue to operate vast majority of services.
Telephone: Direct dialling available on all local and overeas calls. Calls to emergency services are free (dial 999 and ask for police, fire brigade or ambulance). Operator service dial 100. Public phones are found at railway stations, in the streets and in hotels and restaurants. A growing number are operated by telephone cards, which can be purchased at post offices. For international telephone enquiries dial 153. International dialling code: 44 followed by 1 for London.
Telex: Full automatic service with direct dialling available to most countries. Available in all major hotels and there is a public telex facility at Electra House, Victoria Embankment, London.
Telegram: Can be sent via telephone or from any post office. Telegrams are transmitted through computer-controlled Telegram Retransmission Centre, in London, to 81 other countries.

Postal service: The Post Office has a monopoly on the collection and delivery of letters, as well as providing other specialist high-speed delivery services.

Trade fairs: Several annual events of international significance, such as the Motor Show, the Boat Show, the Ideal Home Exhibition and many others covering all aspects of trade and industry. Principal exhibition venues are Earl's Court and Olympia in London, the National Exhibition Centre (NEC) in Birmingham, and the G-Mex Centre in Manchester. Further information available from UK Trade Fairs Dept. of the Department of Trade and Industry, in London.

Electricity supply: 240 V AC.

Representation in capital: Afghanistan, Algeria, Antigua and Barbuda, Argentina, Australia, Austria, Bahamas, Bahrain, Bangladesh, Barbados, Belgium, Belize, Bolivia, Botswana, Brazil, Brunei, Bulgaria, Burma, Cameroon, Canada, Chile, China PR, Colombia, Costa Rica, Cuba, Cyprus, Czechoslovakia, Denmark, Ecuador, Egypt, El Salvador, Ethiopia, Fiji, Finland, France, Gabon, Gambia, German Democratic Republic, Federal Republic of Germany, Ghana, Greece, Grenada, Guatemala, Guyana, Holy See, Honduras, Hungary, Iceland, India, Indonesia, Iran, Iraq, Ireland, Israel, Italy, Ivory Coast, Jamaica, Japan, Jordan, Kenya, Republic of Korea, Kuwait, Lebanon, Lesotho, Liberia, Libya, Luxembourg, Malawi, Malaysia, Malta, Mauritius, Mexico, Mongolia, Morocco, Mozambique, Nepal, Netherlands, New Zealand, Nicaragua, Nigeria, Norway, Oman, Pakistan, Panama, Papua New Guinea, Paraguay, Peru, Philippines, Poland, Portugal, Qatar, Romania, Saint Kitts & Nevis, Saint Lucia, Saint Vincent and the Grenadines, Saudi Arabia, Senegal, Seychelles, Sierra Leone, Singapore, Somalia, South Africa, Spain, Sri Lanka, Sudan, Swaziland, Sweden, Switzerland, Syria, Tanzania, Thailand, Togo, Tonga, Trinidad and Tobago, Tunisia, Turkey, Uganda, USSR, United Arab Emirates, USA, Uruguay, Venezuela, Vietnam, Yemen Arab Republic, Yemen People's Democratic Republic, Yugoslavia, Zaïre, Zambia, Zimbabwe.

Hotels: Birmingham
Midland, New Street (tel: 643 2601; tx: 338419; fax: 6435075).

Plough & Harrow, Hagley Road, Edgbaston (tel: 454 4111; tx: 338074; fax: 4541868).

Royal Angus Thistle, St. Chads, Qeensway (tel: 236 4211; tx: 336889; fax: 2332195).

Strathallan Thistle, Hagley Road, Edgbaston (tel: 455 9777; tx: 336680; fax: 4549432).

Hotels: Glasgow
Holiday Inn, Anderston Cross, Argyle Street (tel: 226 5577; tx: 776355; fax: 2219202).

Stakis Grosvenor, Grosvenor Terrace, Great Western Road (tel: 339 8811; tx: 776247; fax: 3340710).

The Albany, Bothwell Street (tel: 248 2656; tx: 77440; fax: 2218986).

The MacDonald Thistle, Eastwood Toll, Giffnock (tel: 638 2225; tx: 779138; fax: 6386231).

Tinto Firs Thistle, Kilmarnock Road (tel: 637 2353; tx: 778329; fax: 6331340).

Hotels: London
Belgravia-Sheraton, 20 Chesham Place (tel: 235 6040; tx: 919020; fax: 2596243).

Britannia Inter-Continental, Grosvenor Square (tel: 629 9400; tx: 23941).

Brown's, Albemarle Street (tel: 493 6020; tx: 28686; fax: 4939381).

Cavendish, Jermyn Street (tel: 930 2111; tx: 263187; fax: 8392125).

Charing Cross, Strand (tel: 839 7282; tx: 261101; fax: 8393933).

Chesterfield, 35 Charles Street (tel: 491 2622; tx: 269394).

Churchill, Portman Square (tel: 486 5800; tx: 264831; fax: 9350431).

Claridge's, Brook Street (tel: 629 8860; tx: 21872).

Cumberland, Marble Arch (tel: 262 1234; tx: 22215; fax: 7244621).

Dorchester, Park Lane (tel: 629 8888; tx: 887704; fax: 4090114).

Dukes, 35 St James's Place (tel: 491 4840; tx: 28283; fax: 4931264).

Gloucester, 4-18 Harrington Gardens (tel: 373 6030; tx: 917505; fax: 3730409).

Grosvenor, 101 Buckingham Palace Road (tel: 834 9494; tx: 916006; fax: 6301978).

Grosvenor House, Park Lane (tel: 499 6363; tx: 24871; fax: 4933341).

Hilton International, 179-199 Holland Park Avenue (tel: 603 3355; tx: 919763; fax: 6039397).

Holiday Inn, 134 George Street (tel: 723 1277; tx: 27983; fax: 4020666).

Hyatt Carlton Tower, Cadogan Place (tel: 235 5411; tx: 21944; fax: 2456570).

Inn on the Park, Park Lane (tel: 499 0888; tx: 22771; fax: 4931895).

Inter-Continental, 1 Hamilton Place, Hyde Park Corner (tel: 409 3131; tx: 25853).

London Hilton, 22 Park Lane (tel: 493 8000; tx: 24873; fax: 4934957).

London Marriott, Grosvenor Square (tel: 493 1232; tx: 268101; fax: 4913201).

Mayfair Inter-Continental, Stratton Street Mayfair (tel: 629 7777; tx: 262526; fax: 6291459).

Park Lane, Piccadilly (tel: 499 6321; tx: 21533).

Ritz, Piccadilly (tel: 493 8181; tx: 267200; fax: 4932687).

Royal Garden, Kensington High Street (tel: 937 8000; tx: 263151; fax: 9384532).

Russell, Russell Square (tel: 837 6470; tx: 24615; fax: 8372857).

The Portman, 22 Portman Square (tel: 486 5844; tx: 261526; fax: 9350537).

The Savoy, The Strand (tel: 836 4343; tx: 24234).

Waldorf, Aldwych (tel: 836 2400; tx: 24574; fax: 8367244).

Westbury, New Bond Street (tel: 629 7755; tx: 24378; fax: 4951163).

Hotels: Manchester
Piccadilly, Piccadilly Plaza M60 1QR (tel: 236 8414; tx: 668765; fax: 2281568).

Post House, Palatine Road, Northenden (tel: 998 7090; tx: 669248; fax: 9460139).

The Grand, Aytoun Street (tel: 236 9559; tx: 667580; fax: 4362340).

The Portland Thistle, Portland Street, Piccadilly Gardens (tel: 228 3400; tx: 669157; fax: 2286347).

Chambers of Commerce
Association of British Chambers of Commerce, Sovereign House, 212 Shaftesbury Avenue, London WC2H 8EW (tel: 240-5831).

Birmingham Chamber of Industry and Commerce, 75 Harborne Road, Birmingham B15 3DH (tel: 454-6171; tx: 338024 BIRCOM G).

Cardiff Chamber of Commerce and Industry, 101-108 The Exchange, Mount Stuart Square, Cardiff CF1 6RD (tel: 481-648; tx: 497492).

Edinburgh Chamber of Commerce and Manufacturers, 3 Randolph Crescent, Edinburgh EH3 7UD (tel: 225-5851; tx: 72465 CHAMCO G).

Glasgow Chamber of Commerce, 30 George Square, Glasgow G2 1EQ (tel: 204-2121; tx: 777967 CHAMCOM G).

Leeds Chamber of Commerce and Industry, Commerce House, 2 St Alban's Place, Leeds LS2 8HZ (tel: 430-491; tx: 55293 CHACOM G).

London Chamber of Commerce and Industry, 69 Cannon Street, London EC4N 5AB (tel: 248-4444; tx: 888941 LCCI G).

Manchester Chamber of Commerce and Industry, 56 Oxford Street, Manchester M60 7HJ (tel: 236-3210; tx: 667822 CHACOM G).

Sheffield Chamber of Commerce and Manufacturers (Incorporated), Commerce House, 33 Earl Street, Sheffield S1 3FX (tel: 730-114; tx: 547676 CHAMCO g).

Banking
Bank of England (central bank), Threadneadle Street, London EC2R 8AH (tel: 601-4444; tx: 885001).

Bank of Scotland, The Mound, Edinburgh EH1 1YZ (tel: 442-7777; tx: 72275).

Barclays Bank, 54 Lombard Street, London EC3P 3AH (tel: 626-1567; tx: 887591).

British Bankers' Association, 10 Lombard Street, London EC3V 9EL (tel: 623-4001; tx: 888364).

British Overseas and Commonwealth Banks Association, 99 Bishopsgate, London EC2P 2LA (tel: 636-2366).

Institute of Bankers, 10 Lombard Street, London EC3V 9AP (tel: 623-3531).

Institute of Bankers in Scotland, 20 Rutland Square, Edinburgh EH1 2BB (tel: 229-9869).

Lloyds Bank, 71 Lombard Street, London EC3P 3BS (tel: 626-1500; tx: 888301).

Midland Bank, Poultry, London EC2P 2BX (tel: 260-8000; tx: 8811822).

National Westminster Bank, 41 Lothbury, London EC2P 2BP (tel: 726-1000; tx: 888388).

Royal Bank of Scotland, 42 St Andrew Square, Edinburgh EH2 2YB (tel: 556-8555; tx: 72230).

Standard Chartered Bank, 38 Bishopsgate, London EC2N 4DE (tel: 280-7500; tx: 885951).

Trustee Savings Bank, 25 Milk Street, London EC2V 8LU (tel: 606-7070; tx; 8812487)

United States of America

Susan Traill

The USA voted for continuity in the 1988 presidential and congressional elections, electing President George Bush and a Republican administration counterbalanced by a Democratic controlled Congress. Their uneasy relationship opened with a protracted political ethics war. Although a truce was called in summer 1989, the underlying causes were not resolved. The large budget and external deficits remained intractable, but the economy enjoyed strong growth in 1988 with a relatively weak dollar. In 1989 the dollar was remarkably strong and growth slowed much less rapidly than expected. Detente ruled foreign political relations, but trade relations took a tougher turn with implementation of the 1988 Trade Act.

President Ronald Reagan ended his eight-year tenancy of the White House in January 1989. Reagan won the hearts of Americans through his tax cuts and tough policies on crime and foreign relations. It was felt that he had re-established US prestige in the eyes of the world after the disastrous Carter years, a notion confirmed when he signed the Intermediate Nuclear Forces (INF) treaty with Soviet Premier Mikhail Gorbachev in Washington DC in December 1987. The ensuing detente unfroze decades of public misgivings towards the Soviet Union at a pace that alarmed Vice-President Bush, who stressed caution.

For seven of the Reagan years, the US economy was in expansion following the recession of 1980–81. The public went on a massive spending spree and employment rose to meet their aspirations for more money. However, Reagan's policies were building up serious long-term problems. Tax cuts, which favoured the rich, and increased defence spending produced huge budget deficits. Social programmes were slashed and greater social responsibilities pushed onto state and local budgets where no deficits were allowed. The gap between rich and poor widened. House prices soared under pressure of demand from the newly rich whilst federally funded low-cost housing construction was phased out. The homeless multiplied in the inner cities. As cocaine snorting lost favour among young stockbrokers in the mid-1980s, so the new Latin American drug kings found an alternative market in the ghettoes, where drugs could bridge the gap to wealth. Violent crime increased.

These were the more visible failures of Reaganomics and are the issues of greatest public concern under the Bush administration. But equally important are the structural changes in the economy. From a slight trade surplus in 1981, the USA now has a chronic trade deficit caused by an overvalued dollar. The budget deficit lies behind this, since government borrowing pushed up interest rates attracting a massive influx of foreign capital. The USA has thus made itself dependent on foreign confidence in the currency, which necessitates strong exchange and interest rates, a self-perpetuating situation unless the budget deficit is cured. The main cure would be a major tax increase and neither Republicans nor Democrats want to be responsible for such an unpopular move.

Reagan was avowedly conservative and his main legacy in this respect is his alteration of the Supreme Court majority from liberal to conservative. Following Reagan's appointment of Anthony Kennedy in mid-1988, the court had five conservative judges to four liberals. It is likely to become more strongly conservative as more liberals retire. Supreme Court decisions play an important role in the setting and implementation of policy, since the court is the ultimate arbiter of the constitutionality of federal and state laws.

Campaigns and elections

The Bush election campaign against Governor Michael Dukakis of Massachusetts was described by some observers as the dirtiest presidential campaign ever. The Republicans started knowing that the boom economy with the lowest unemployment in 14 years would be their strongest ticket to win. They therefore set out to discredit Dukakis personally, portraying him as inexperienced in national and international affairs, soft on defence and on crime and likely to raise taxes. Dukakis did not rise to the challenge, particularly failing to present a strong and resourceful personality in television interviews and debates. The inexperience argument seemed vindicated by the contrast between Dukakis and his running mate Lloyd Bentsen, chairman of the powerful Senate Finance Committee. Bentsen presented cohesive policies and a tough image in his television debate against Bush's running mate, the vacuous Dan Quayle. Quite late in the campaign, Bush promised 'a kinder, gentler America' and 'no new taxes', without saying how he would accomplish either. He also made the war on drugs a top priority, but so had Dukakis from the outset.

The media complained of a lacklustre campaign with both candidates so close on issues that it was difficult to tell them apart. Indeed, many reckoned that the campaign had ended when the Reverend Jesse Jackson lost the Democratic Party nomination. Jackson was the only candidate on either side with the charisma to rival that of the departing Reagan.

Dukakis handled the Jackson camp with insensitivity by not informing them of his choice of running mate until the Democratic Convention. Jackson seemed unwilling to admit defeat. He secured an unprecedented debate at the convention on amendments to the election platform. Although they were rejected, the amendments were almost guaranteed to alarm the electorate: higher taxes, an end to nuclear weapons and the right to Palestinian self-determination. As the price of his support for Dukakis, Jackson demanded more places for his supporters on the Democratic National Committee and the introduction of proportional representation (PR) in the primaries for the 1992 presidential election. Jackson would have had a larger share of the vote under PR and this change could deepen racial divisions in the Democratic Party in 1992 if Jackson runs again, as it seems likely he will.

The vote in both the presidential and congressional elections reflected the electorate's desire for continuity: Republicans to run the executive, Democrats to control Congress while playing the role of moderator. Bush won a decisive victory with only 10 states opposing him. The Democrats won an extra Senate seat giving them 55 to the Republicans' 45 seats. They also won five additional seats in the House of Representatives, giving them 260 seats to the Republicans' 175 seats. The Democrats have held a large majority in the House for over 35 years. About 98 per cent of incumbents are re-elected to the House every two years. Of the few defeated in November 1988, two (one Democrat and one Republican) were the subject of ethics inquiries. 12 states held gubernatorial and 45 states held legislative elections at the same time as the presidential and congressional elections. The Democrats won one additional governorship giving them 28 states to the Republicans' 22, but by April 1989 they held 30 governorships. 41 states presented their voters with 230 measures to be adopted by direct vote. The most controversial was Maryland's law for stricter rules on handgun purchases, which passed despite costly lobbying against it by the National Rifle Association.

The three largest cities of New York, Los Angeles and Chicago all have Democratic mayors. All three faced elections in 1989. New York elected its first black mayor, David Dinkins, in November 1989. Los Angeles re-elected its black mayor, Tom Bradley. Black politicians were divided in Chicago, however, and lost the Democratic nomination and the election to Richard Daley, son and namesake of one of Chicago's most controversial mayors. Jesse Jackson began to earn himself the reputation of a divisive force when he backed a minor black candidate in this election against the acting black mayor. His support for David Dinkins in the New York primaries was regarded by many as a liability more than an asset because of his antagonism of the Jewish community there. Jackson might run for mayor of Washington DC in 1990 if drug-taking charges stick against the incumbent Marion Barry.

In 1991 the congressional map will be redrawn in accordance with the shifts of population recorded in the 1990 census. New York city, New York state, California, Chicago, Houston and Miami claimed that their population, and therefore their share of federal funds and representation, were underestimated in the 1980 census. Re-districting in most of these areas would help the Democrats rather than the Republicans. The top job in the census bureau was still awaiting a Bush nominee in September 1989.

Bush versus Congress

President Bush has taken a more conciliatory line with the Democratic majority in Congress than his predecessor. Reagan did not have to cope with complete Democratic domination of Congress and the inhibiting effect of this on his policies until the Democrats won control of the Senate in November 1986.

Although Bush has sought concensus with Congress,

KEY FACTS

Official title: United States of America
Head of state: President George Bush
Ruling party: Republican Party
Capital: Washington DC
Official Languages: English
Currency: US Dollar ($) = 100 cents

Area: 9.3m sq km
Population: 246.3m (1988)
GNP real growth: 3.9% (1988)
Unemployment: 5.4% (1988)
Inflation: 4.0% (1988)
Trade balance: –$126.78bn (1988)

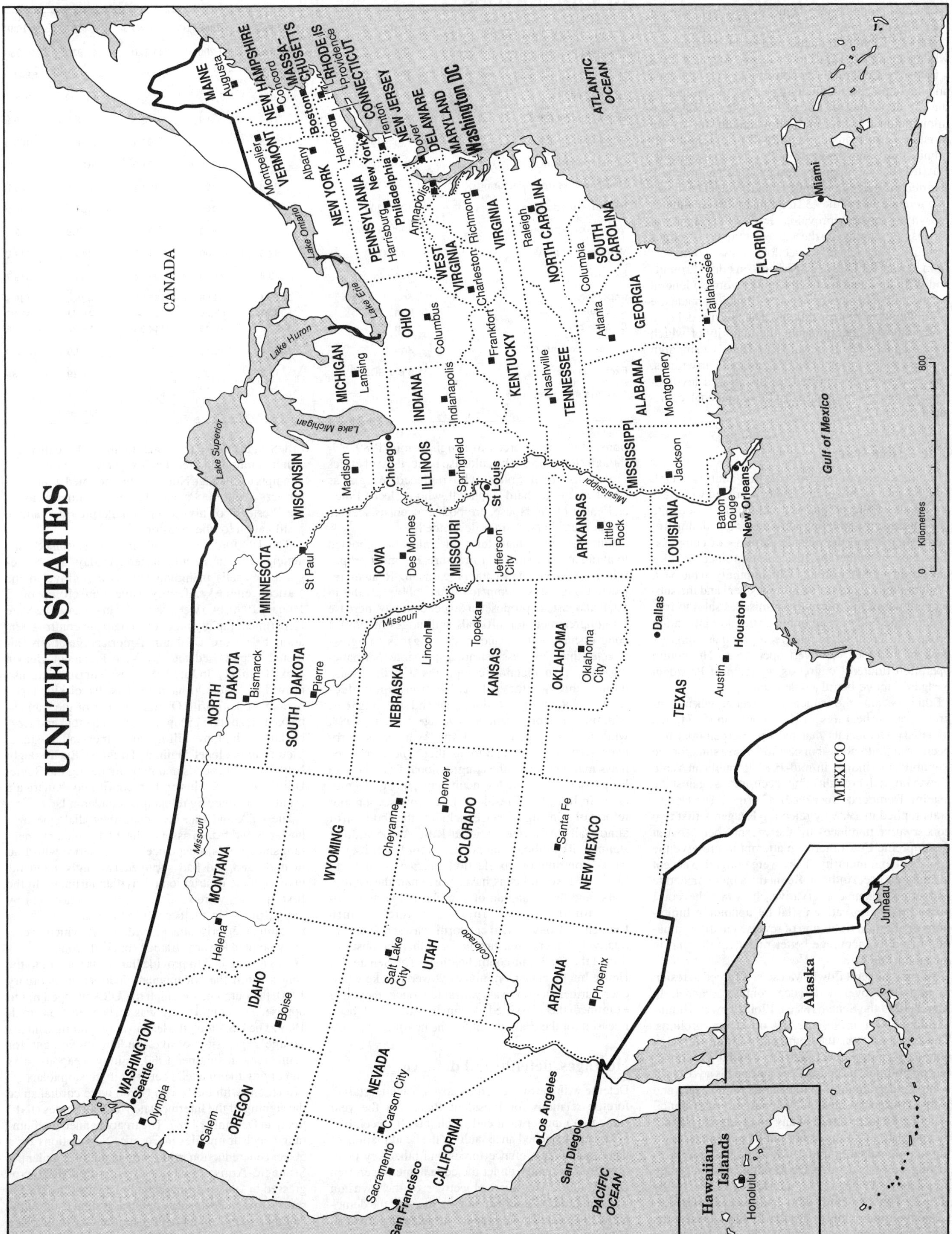
UNITED STATES
CANADA
MEXICO
ATLANTIC OCEAN
PACIFIC OCEAN
Gulf of Mexico
Lake Superior
Lake Michigan
Lake Huron
Lake Erie
Lake Ontario
Mississippi
Missouri
Colorado
MAINE
Augusta
NEW HAMPSHIRE
Concord
VERMONT
Montpelier
MASSA-CHUSETTS
Boston
RHODE IS
Providence
CONNECTICUT
Hartford
NEW YORK
Albany
New York
NEW JERSEY
Trenton
PENNSYLVANIA
Harrisburg
Philadelphia
DELAWARE
Dover
MARYLAND
Annapolis
Washington DC
WEST VIRGINIA
Charleston
VIRGINIA
Richmond
NORTH CAROLINA
Raleigh
SOUTH CAROLINA
Columbia
GEORGIA
Atlanta
FLORIDA
Tallahassee
Miami
OHIO
Columbus
MICHIGAN
Lansing
INDIANA
Indianapolis
KENTUCKY
Frankfort
TENNESSEE
Nashville
ALABAMA
Montgomery
MISSISSIPPI
Jackson
LOUISIANA
Baton Rouge
New Orleans
ILLINOIS
Springfield
Chicago
WISCONSIN
Madison
MINNESOTA
St Paul
IOWA
Des Moines
MISSOURI
Jefferson City
St Louis
ARKANSAS
Little Rock
NORTH DAKOTA
Bismarck
SOUTH DAKOTA
Pierre
NEBRASKA
Lincoln
KANSAS
Topeka
OKLAHOMA
Oklahoma City
TEXAS
Austin
Dallas
Houston
MONTANA
Helena
WYOMING
Cheyenne
COLORADO
Denver
NEW MEXICO
Santa Fe
IDAHO
Boise
UTAH
Salt Lake City
ARIZONA
Phoenix
WASHINGTON
Olympia
Seattle
OREGON
Salem
NEVADA
Carson City
CALIFORNIA
Sacramento
San Francisco
Los Angeles
San Diego
Alaska
Juneau
Hawaiian Islands
Honolulu
0
Kilometres
800

appearances can be deceptive. In the 1990 budget bargaining, for example, he neatly shifted blame for spending cuts onto Congress by setting an overall target of $10bn for reductions in social programmes while leaving the details to Congress. Any new taxes will also be Congress's responsibility. This approach and the unprecedentedly long process of nominating the cabinet and other key officials left the Bush administration in a virtual no-policy situation for several months. Bush blamed Congress for holding up his nominations and therefore policy pronouncements. But the Senate majority leader, George Mitchell, retorted in September 1989 that the president had still to nominate 200 of the 432 administration candidates for whom Senate approval is required. The approval process is lengthy in the current climate of ethics scrutiny. The Senate's rejection of two nominees, John Tower for Defense Secretary on ethics grounds and William Lucas for Civil Rights Attorney General on account of his inexperience in litigation, made the administration more cautious. The Senate delayed ambassadorial appointments, the majority of which were handed out as rewards to Bush's campaign workers and contributors to Republican Party funds. One nominee was rejected for his alleged involvement in the Housing and Urban Development Department scandal.

KEY INDICATORS

	Unit	1984	1985	1986	1987	1988
Population	'000	237,000	239,280	241,620	243,930	246,330
GDP	$bn	3,724.8	3,974.2	4,205.4	4,497.2	4,847.3
GDP per capita	$	15,811	16,593	17,361	18,209	19,678
Prime interest rates	%	12.0	9.9	8.4	8.2	9.3
Wholesale prices	1985=100	100.5	100	97.1	98.6	103.6
Consumer prices	1985=100	96.6	100	101.9	105.7	109.9
Hourly wages (manufacturing)	1985=100	96.4	100	102.1	104.0	106.8
Industrial production	1985=100	97.8	100	99.4	104.3	110.2
Volume of exports	1985=100	101.9	100	99.0	111.9	132.0
Volume of imports	1985=100	92.0	100	110.5	113.4	117.7
Exports	$bn	219.9	215.9	224.0	250.9	319.7
Imports cif	$bn	346.4	352.5	382.3	424.4	459.5
Petroleum	$bn	59.21	52.36	37.64	44.71	41.81
Balance of trade	$bn	–112.51	–122.15	–144.54	–158.88	–126.78
Current account	$bn	–107.14	–115.16	–138.5	–143.09	–126.18
Exchange rate	$ per SDR	1.03	1.02	1.17	1.29	1.34

* estimated figure

The ethics war

The ethics war resulting from the Iran-contra scandal, which broke in November 1986, outraged Congress and made public probity a watchword. But despite laws limiting the lobbying activities of ex-administration officials and the outside earnings of congressmen, they are vulnerable to accusation since the rules have been regularly broken with impunity in the past. With the Iranian arms-for-hostages deal and the subsequent use of the money from missiles sales to Iran to fund the Nicaraguan contras, the administration broke a different set of rules requiring that Congress be kept informed of covert operations. The contra operation flouted Congress's rejection of President Reagan's above-board aid demand.

Ethics became the vehicle for revenge when Congress's public hearings into Iran-contra in 1987 failed to satisfy Democrats that the true perpetrators had been brought to book, but cast too many slurs for the Republicans' liking. In mid-1987 Republican Whip Newt Gingrich brought ethics accusations against the leading Democrat, Speaker Jim Wright. The Democrats replied in 1988 by rejecting Reagan's first two conservative nominees for the vacant liberal seat at the Supreme Court in a vain attempt to preserve the court's liberal majority. They were obliged to accept his third choice, Anthony Kennedy, a more moderate and ethics-clean conservative. In 1989 the court proved its conservative metal by upholding limitations of abortion rights and of reverse job discrimination (used to encourage better racial equilibrium in the labour force).

Attorney General Edwin Meese was forced to resign in mid-1988 over a defence contract scandal. In March 1989 the Senate rejected John Tower's nomination as Defense Secretary on ethics grounds. Tower's report on the Iran-contra affair, although nominally independent, set the official version of responsibilities. It accused Reagan and his advisers of a 'misguided attempt' to trade arms for hostages and blamed the contra deal on National Security Council officials. It cleared Bush of any involvement. Neither the majority in Congress nor public opinion, according to polls taken in mid-1989, agree with this. Following Tower's demise, the Republicans stepped up pressure on Wright and on the Deputy Leader of the House, Tony Coelho, who had successfully persuaded business lobby groups to fund Democrats more than Republicans in the 1986 and 1988 elections. Both were forced to resign in mid-1989. Although Newt Gingrich could also be accused of ethics violations, Tom Foley, who replaced Wright as Speaker, and Richard Gephardt, who replaced Foley as Leader of the House, are both concensus seekers and peacemakers. A truce descended.

Both sides agree that ethics rules must be tightened to avoid another such war in future, but most proposals have been unacceptably partisan. Reagan rejected Congress's attempt to tighten lobbying rules in 1988 and instead proposed a 50 per cent pay increase for congressmen, top officials and judges in 1989 if congressmen would abandon speaking fees. Congress rejected this. Congressmen are responsible for financing their own election campaigns and they raise money through the political action committees (PACs) of lobby groups, donations and speaking fees. Winning senators spent an average $3m in 1988, while the losers spent only $1.4m. 'War chests' carry over from one campaign to another. These and free mass mailings under the congressional frank help to keep House seats safe for incumbents, mostly Democrats. In July 1989 President Bush proposed a major reform of campaign finance including the abolition of almost all PACs (the National Rifle Association, a staunch Republican supporter, would survive), a larger contribution to election finance from party funds (the Republicans have more than the Democrats) and the scrapping of war chests and the congressional frank. In the same month Lieutenant-Colonel Oliver North was sentenced to relatively lenient punishment for aiding the obstruction of the 1987 Iran-contra hearings. The Senate and House Intelligence Committees vowed to take whatever further action was warranted once they had examined the classified documents that had been released for the trial but not for the hearings.

Hostages, detente and defence

Detente with the Soviet Union ensured a quiet start to foreign relations for President Bush, but the year came to a more critical end when, on 20 December, US troops launched an assault on the headquarters of the Panamanian defence forces and other key positions in the country under the code named operation 'Just Cause'. The overt objectives of the invasion were 'to protect American lives', 'to assist the democratically elected government', 'to seize and arrest an indicted drug trafficker' and 'to defend the integrity of US rights under the Canal treaties.' In reality, the administration had reached exasperation point in its attempts to dislodge Noriega; this, coupled with nervousness about the Panama Canal and a desire to rid the President of his 'wimp' image provided ample justification for the invasion.

The US forces met with fiercer resistance than anticipated. The fighting lasted five days and Noriega, after spells in hiding and as a fugitive in the Vatican embassy, finally gave himself up on 3 January 1990, two weeks after the invasion. The State Department rapidly mounted a damage control operation to restore US-Latin American relations, although a proposed tour by Vice President Quayle looked unlikely. In July 1989 President Bush had also proved well able to handle crisis. Israel kidnapped Sheikh Abdel Karim Obeid, a leader of the Iranian-backed Hezbollah group in south Lebanon, and Hezbollah replied by killing an American hostage, Lieutenant-Colonel William Higgins. Bush sought help from old friends and new, including the Soviet Union, avoided bluster but confirmed America's traditional policy by refusing to condemn Israel.

The USA and Iran continued their dialogue over hostages and frozen assets. This had gained momentum since Iran's acceptance of a ceasefire with Iraq in mid-1988 and had survived two ugly incidents involving the destruction of civilian airliners. In the first, Iranians were killed apparently by accident by the US navy; in the second, Americans died in an unclaimed terrorist attack, evidence of which eventually pointed towards Iranian- backed groups.

Congress agreed to provide humanitarian aid to the contras until the Nicaraguan elections in February 1990, the outcome of which the USA pledged not to oppose. The contra army was disbanded from mid-1989. The USA took the lead in condemning China's savage repression of the student movement for democracy in summer 1989, halting weapons sales and urging the financial community to stop loans.

Relations with the Soviet Union were cordial since the signing of the Intermediate Nuclear Forces (INF) treaty in December 1987. The treaty banned medium-range nuclear missiles in Europe. Negotiations for a 50 per cent reduction in long-range missiles under the Strategic Arms Reduction Treaty (START) progressed in 1988 but finally ran up against the USA's refusal to include its space defence system in the talks. Another round of START negotiations took place with little effect in July 1989. President Bush was not keen to rush into summitry with Soviet leader Mikhail Gorbachev on account of START. In December 1988 Gorbachev upstaged him by announcing unilateral reductions in conventional forces in a speech to the United Nations in New York. Gorbachev took the limelight as international peacemaker while Bush was cast in the role of doubtful follower. NATO allies, led by West Germany, turned against the planned modernisation of short-range nuclear weapons which would have upgraded them almost to medium-range weapons. After months of silence, Bush regained the initiative at the NATO summit in May 1989 by calling for a treaty with the Warsaw Pact on conventional force reductions as a prerequisite to any further nuclear arms limits. The modernisation of short-range missiles was shelved. In September 1989 Secretary of State James Baker and Soviet Foreign Minister Edward Shevardnadze discussed tentative agenda items for a future summit. This included Shevardnadze's proposal for chemical weapons control, scientific and economic cooperation, the problem of Soviet Jews emigrating to the USA (an embarrassment to the USA now that the doors are finally open), regional conflicts such as the Middle East and global issues such as the environment. They signed a protocol on mutual inspection of chemical weapons plants. President Bush announced after the meeting that he would hold a summit meeting with Gorbachev in spring or summer 1990 and that nuclear arms control would be discussed. Refinement of the Strategic Defense Initiative (SDI) and President Bush's decision to increase land-based nuclear missiles hold out some hope for START in the long term, since clear

COUNTRY PROFILE

Political system: The US is a federal republic comprising 50 states, the District of Columbia and overseas territories. The president, elected to a four-year term by an electoral college of representatives elected from each state, wields executive power. Legislative power is vested in a bicameral Congress consisting of a 435-member House of Representatives elected every two years and a 100-member Senate, one-third of whose members are elected every six years. Each of the 50 states is headed by an executive governor, with legislative power vested in a bicameral state legislature.

Political parties: Republican Party; Democratic Party.

Population: America is the fourth largest nation in the world, and has a varied social and ethnic mix. Population growth is low at around 0.8 per cent per annum.

Main cities/towns: New York (population 7.1m in 1985), Los Angeles (3m), Chicago (3m). The political and administrative centre is Washington DC. Other cities with populations over 0.5m are: Atlanta, Baltimore, Boston, Buffalo, Cincinnati, Cleveland, Dallas, Denver, Detroit, Houston, Kansas City, Memphis, Miami, Milwaukee, Minneapolis, New Orleans, Philadelphia, Phoenix, Pittsburgh, San Antonio, San Diego, San Francisco, San José, St Louis, Seattle.

Media

Press: The only national newspaper is the tabloid *USA Today*. The *Wall Street Journal*, a financial and business daily, is published in four regional editions. *The New York Times*, the *Washington Post* and the *Los Angeles Times* are distributed nationally in main centres.

Magazines: In addition to numerous specialist consumer journals, *Time*, *Newsweek*, *Forbes*, *Fortune* and *Business Week* are general interest and business news magazines with national circulation, some of them appearing in regional editions.

Radio: There are more than 5,000 private radio stations.

Television: Well over 1,000 TV stations in the largely commercial broadcasting system. Largest TV networks are American Broadcasting Company (ABC), Columbia Broadcasting System (CBS) and National Broadcasting Company (NBC). Public TV stations within the Public Broadcasting System (PBS) are funded by membership subscription, corporate sponsorship and government credit through the Corporation for Public Broadcasting.

Domestic economy: The economy is well developed, highly industrialised and rich, catering to a huge domestic market and exporting machinery and electrical goods, transport equipment, chemicals and food. Since 1982, recovery from recession has come especially from heavy government spending on defence (producing massive budget deficits to which there is no end in sight), and also from personal consumption, which has been financed by rapid income growth, aided by tax cuts in 1981, plus a fall in the saving ratio from 6.2 per cent in 1982 to five per cent in 1983. The rapid pace of US growth relative to the rest of the world and the strength of the US dollar from 1982, especially in 1982-85, has led to a growing external trade deficit.

Employment: There have been major shifts of the workforce from older industrial cities of the north-east to the south and the Pacific coast, where the cost of living is lower and high-technology industries are located, and of retired people to communities in the warmer south.

External trade: The steep rise in imports in recent years has given rise to trade restrictions affecting imports of Japanese automobiles, steel, motorcycles, foodstuffs and textiles. There are restrictions on trade with South Africa.

Exports: Main exports are machinery (33.6% of 1987 total), transport equipment (18.2%), chemicals and related products (12.7%), agricultural products (11.7%), food production (5.9%). Main destinations: Canada (23% of total value in 1987), Japan (11%), Mexico (6%), UK (5.6%), and German FR (4.6%).

Imports: Main imports are industrial supplies (27.4% of total value in 1988), capital goods (22.8%), consumer goods (21.6%), vehicles and parts (19.7%), petroleum products (8.8%) and food and beverages (5.6%). Main suppliers: Japan (20.1% of total value in 1988), Canada (18.9%), German FR (5.9%), Taiwan (5.6%) and Mexico (5.2%).

Agriculture: Agriculture accounts for 2.1 per cent of GDP (1987) and provides 2.9 per cent of employment. Although small in relation to the overall economy, the agricultural sector accounts for half world corn production and over 20 per cent of world grain output. The US is the world's largest agricultural exporter, and exports account for about 25 per cent of farmers' receipts. Highly-capitalised farming techniques produce dairy produce, potatoes, fruit, vegetables and poultry for urban markets in the north-eastern states; wheat, barley, maize, oats, soya beans, fodder crops, pigs and cattle in the mid-west and central plains; cotton, tobacco, peanuts, citrus fruits, rice and sugar-cane in the south; cattle and sheep in the central and western states; apples, berries and nuts in the Pacific north-west; and vines, apples, citrus fruits, peaches, tomatoes, olives, cotton and rice in California. Price support programmes operate for most main agricultural commodities, via direct purchase of some products and via non-recourse loans for crops. Income support programmes operate for producers of certain crops, but the general level of assistance to farmers has been declining in the mid-1980s. Forestry is widespread and about half domestic timber needs are met by Oregon and Washington, with the south-east producing increasing quantities of softwood for pulp. Fishing industries have developed on both the Pacific and Atlantic coasts.

Industry: Industrial production growth rate for 1988 was 5 per cent. In 1985, manufacturing industry accounted for 20.1 per cent of GDP and provided 18.4 per cent of employment, although its importance is declining relative to service industries such as transport, communications, finance and other services, which account for 58 per cent of GDP and 55 per cent of employment. In 1984, the US accounted for 17 per cent of manufactured goods exported by the main industrial countries. There has been a shift in the manufacturing sector away from 'smokestack' industries, such as cars, primary metals and heavy machinery, toward high technology industries, such as aerospace, communications equipment, electronic components and computers.

Mining: Mineral resources include abundant coal and lignite, and ores of iron, copper, nickel and other metals. Proven oil reserves of 26.9bn barrels (1986) would last eight years at 1985 rates of production, but there are vast reserves currently classified as probable or possible. In 1988, 2,975m barrels of crude oil were produced and 4,913m barrels consumed. Proven natural gas reserves were 187.2 trillion cu ft at end-1988, sufficient to last 11.2 years at 1988 production levels. Coal reserves amount to 24 per cent of world reserves of bituminous coal and anthracite and 31 per cent of sub-bituminous coal and lignite resources. Coal production for 1988 was 861.9m tonnes.

Energy: Energy conservation programmes have resulted in a net decline in crude oil imports, and energy requirements per unit of GDP are expected to continue to fall. Oil accounted for 52 per cent of energy consumption in 1984, gas for 24 per cent, coal for 10 per cent and electricity for 14 per cent. Privately-owned local monopolies regulated by state public utility commissions account for 80 per cent of electricity sales. Over 70 proposed nuclear plants were cancelled in the 1980s, reflecting massive cost overruns due to high interest rates, regulatory delays, changes in design requirements during construction and poor cost and quality control – not to mention increasing reservations among the public about nuclear safety. Coal inputs for electricity generation have risen and are expected to continue to grow.

Financial markets: The New York Stock Exchange is the largest of over a dozen stock exchanges. The Chicago Board of Trade is the largest of 11 commodity and financial futures exchanges. The National Association of Securities Dealers Automated Quotation System (NASDAQ) is a computer-based over-the-counter market in securities.

Legal system: Exporters to the US face the risk of litigation if products cause injury or damage, and should investigate the question of product liability. The US system of strict liability for defective products and the liberal interpretation of 'defective products' by juries, combined with the lawyers' contingent fee system, has led to substantial awards for damages and steep insurance premiums.

Membership of international organisations: African Development Fund, Asian Development Bank, Colombo Plan, FAO, GATT, Inter-American Development Bank, IAEA, IDA, IFAD, IFC, IMF, IMO, OECD, South Pacific Commission, WHO, WIPO, WMO, World Bank.

policies on these issues are essential before reductions can be negotiated. Under pressure from Congress, the Defense Department agreed in autumn 1988 to give more attention to land-based SDI equipment, the beginning of some equivalence with existing systems. The administration abandoned the notion of a fully protective SDI barrier, a dream that had undermined the incentive for reduction of long-range missiles. Bush considered two SDI systems in 1989, one to protect military bases against Soviet attack and the second to protect cities against 'accidental' attack. He decided to deploy 50 multiple warhead MX land-based missiles transportable by rail and 250–500 single warhead Midgetman missiles transportable by road.

The Stealth bomber, designed to be invisible to radar, made its runway debut in July 1989, but the programme will be slowed with a $1bn cut in 1990, part of the $3.7bn defence budget cuts. Development of the V-22 attack aircraft was cancelled. NASA's shuttle programme resumed in August 1989 with the successful launch of Columbia, the first shuttle flight since the Challenger explosion in 1986. The Pentagon used Columbia to launch a new spy satellite but will give NASA only seven more payloads before switching exclusively to unmanned Titan rockets.

Economic boom

1988 was the seventh year of economic expansion, the longest upward trend of the business cycle in a peacetime period. Slower growth did not set in until the second half of 1989, later than expected and a 'softer landing' than many feared. Real GNP growth was 3.9 per cent in 1988, up from 3.4 per cent in 1987. Growth is expected to be 2.8 per cent in 1989. Growth was export-led in 1988 with the volume of total exports increasing by 18 per cent. In contrast, imports rose by 8.7 per cent. In 1989 the gap is expected to narrow somewhat with exports increasing by 9.9 per cent and imports by 6.7 per cent. Private consumption rose by 2.8 per cent in real terms in 1988 and is expected to increase by 2.6 per cent in 1989. It is by far the largest component of GNP, accounting for $3,227.5bn or 66 per cent in 1988. Inflation was 4.2 per cent in 1988 (up from 3.8 per cent in 1987) but is likely to step up pace to 5.2 per cent in 1989 on account of continued consumer spending, combined with slower domestic output. The increase in industrial production was a robust 5.7 per cent in 1988 but is due to ease to 3.2 per cent in 1989.

Investment growth mirrored that of industrial production with a 5.6 per cent increase in 1988 and 3.2 per cent in 1989. However, these figures bely the shortage of fixed capital in US industry. Fixed assets per employee have not increased since 1983, because of fuller employment and faster capital depreciation. In 1988 industry reached almost full capacity utilisation, a factor that contributed to the slowing of the economy in the second half of 1989. Unemployment fell to 5.4 per cent in 1988 from 6.1 per cent in 1987 and is not expected to increase significantly in the short-term. Average labour costs rose by 4.2 per cent in 1988 and by 5 per cent in 1989.

By August 1989 stock market prices on the Dow Jones Industrial Index had recovered to almost pre-1987 crash levels. However, the volume of shares available fell due to mergers and acquisitions (which cancelled $300bn in shares in 1988) and the new trend for public companies to go private as an anti-leveraged buyout (LBO) mechanism and to avoid the short-term profit constraints imposed by institutional investors. These constraints also make companies vulnerable to raids. Firms bought $40bn of their own shares in 1988, while slightly less than that was offered in new stock. Company profits improved by 33 per cent in 1988 and dividends paid out in 1989 increased by 10 per cent.

High external deficit

Canada is the USA's main trading partner and accounted for $84.4bn of imports and $73.5bn of exports in 1988. The US trade deficit with Canada narrowed by just under $1bn to $10.7bn in 1988, while significant improvements were recorded with the EC ($15.8bn deficit, $11.7bn less than in 1987) and Japan ($52.6bn deficit, $3.8bn less than in 1987) despite increased imports from these countries. Main traded goods include motor vehicles, machinery, chemicals and food. Paper and pulp are also major imports from Canada.

The total trade balance improved in 1988 leaving a deficit of $127bn against $153bn in 1987. The trade deficit is expected to fall again in 1989 to $109bn. However, the balance of payments deficit is expected to remain stable at about $129bn in 1989 owing to the heavier external debt service burden. The improvement in the 1988 current account deficit, which fell to $127bn from $154bn in 1987, was due to the rapid rise in exports boosted by the decline in the dollar. But the dollar strengthened from mid-1988 and remained remarkably strong in 1989 due to tight monetary policy by the federal reserve and the unexpectedly sharp decline in domestic credit demand which slowed monetary growth in the first half of the year. Prime lending rates strengthened throughout 1988 to a quarterly average of 10.18 per cent in the fourth quarter, supporting the strengthening of the dollar. First and second quarter averages in 1989 of 11.17 per cent and 11.36 per cent continued the trend. In addition, the savings rate improved from 3.2 per cent of disposable income in 1987 to 4.2 per cent in 1988 and 5.4 per cent in the first quarter of 1989. The external deficit has been allowed to remain high to foster non-inflationary growth without resorting to a drastic cut in the budget deficit or raising taxes. The current account had a small surplus of $6.4bn in 1981 but President Reagan's tax cuts and increased defence spending produced soaring budget deficits and public borrowing pushed up interest rates. The exchange rate subsequently strengthened as foreign capital, especially Japanese, was drawn in. This undermined exports and boosted imports.

As debt has built up (to $2,129bn at the end of the first quarter 1989), the USA is increasingly dependent on maintaining attractive interest and exchange rates to prevent large-scale selling of the dollar abroad. The abolition of exchange controls in other countries has increased the volume of dollars held by funds managed by the private sector. The greater internationalisation of the European financial market in the 1990s will reinforce this trend. Despite the vulnerability of the dollar to foreign interests, other OECD governments are unlikely to allow any sudden deterioration in its value because of the damaging effect this would have on their exports and growth. Most of these governments would like to see a gradual depreciation of the dollar, which would cut the external deficit and increasingly heavy debt service burden. The OECD has also urged the USA to cut its budget deficit and raise taxes.

Protectionism

Increased indebtedness, the persistence of the external deficit and a rapid increase in foreign direct investment of over 20 per cent a year in recent years led Congress to adopt the 1988 Trade Act. This mandated retaliatory action against barriers to US exports deemed to be unfair, and the Exon-Florio amendment allowing the president to block foreign takeovers if these endanger national security. The administration blocked a Japanese takeover of defence contractor Fairchild Industries in 1987, although Fairchild is French controlled.

Total foreign direct investment reached $328.9bn in 1988, surpassing US direct investment overseas, at $326.9bn, for the first time. Most foreign investment in the USA has gone into the chemicals industry, followed by insurance, banking and services. Takeovers have become increasingly large. The main investors in terms of asset value are the UK, Japan, Canada, Australia, France and the Netherlands. US overseas investment is mainly in manufacturing, petroleum industries, finance and insurance. Main recipients are Canada, UK and West Germany. Official US statistics tend to understate foreign assets, thus making the debt appear worse than it is.

The administration is unlikely to take hasty unilateral action on trade, as evidenced by former Trade Representative Clayton Yeutter's rejection of the US Rice Millers Association's demand for action against Japan in late 1988. Unilateral action could contravene GATT rules but threatened action is a useful bargaining counter. In July 1989 GATT partners accepted the principle of 'tariffication' put forward by Yeutter in his new role as Agriculture Secretary. This involves converting a wide range of import restrictions into a tariff equivalent which can be compared internationally and for which a reduction schedule can be established. The idea, based on the 1988 US–Japan beef agreement, offers the most solid solution yet to increased protectionism by the major western economies. The new Trade Representative Carla Hills set about drawing up a general list of unfair practice offenders in 1989 and a 'super-301' list of countries for priority negotiations. South Korea and Taiwan lobbied relentlessly in Washington and made concessions to stay off the priority list. The super-301 list named Japan, India and Brazil for action, put eight other countries on a general watchlist and 17 others on a copyright and patent watchlist. Japan refused to take part in super-301 negotiations and a group called the structural impediments initiative (SII) was set up instead. When talks began in September 1989, however, Japan had its own list of US structural impediments to US exports – especially the high level of domestic consumption, lack of private and public savings and the obsession with short-term corporate results. The Japanese attitude hardened in 1989 with the greater influence of the socialists in national politics. They pointed out that Japan had accepted 62 voluntary restraint limits on its exports to the USA while imposing only 13 such restrictions on US imports.

The EC's aim to create a single market for its 12 member countries by 1992 gradually gained a more favourable response from US government and business circles in 1989, especially after the EC Commission agreed not to demand reciprocal access to the US financial market in return for nondiscrimination against American banks operating in the EC. The commission also seemed ready to allow US observer status on EC standardisation committees. Administration officials declared that a strong and united Western Europe was essential to counterbalance reforms and the influence of Mikhail Gorbachev in Eastern Europe. The Western Economic Summit in Paris in July 1989 agreed that the EC Commission should co-ordinate emergency aid to newly democratic Poland, including $100 m from the USA. Several trade problems remain, however, including reciprocal farm product bans and a proposal for minimum European content in television programmes. EC labelling and testing standards for telecommunications products were included in Carla Hills' unfair practices list.

Congress ratified the US–Canada Free Trade Act in 1988. Canadian exports that will benefit from the agreement include hydroelectricity as utilities in the north-east USA decommission oil-fired and nuclear power stations. Canadian officials and environmentalists are concerned, however, that the agreement will force them to raise restrictions on plans for drawing water from the Great Lakes and reversing rivers in Alaska and Canada to supply agriculture in the Midwest.

The scheme to bail out Mexico from its $100bn debt failed to produce the blueprint for Latin American debt for which Treasury Secretary Nicholas Brady had hoped. By mid-1989 commercial banks demonstrated their preference to quit rather than produce new money to sweeten the debt-for-bonds swap proposed by Brady. The Third World applauded Brady's call for easing the debt terms of middle-income developing countries, however, a policy which has been broadly followed by the IMF and World Bank.

Budget acrobatics

President Bush's first major task on taking over the White House was to present a 1990 budget with a maximum deficit of $100bn, the ceiling set by the Gramm-Rudman law. Gramm-Rudman allows a $10bn leeway so that the effective deficit can be $110bn for 1990. The budget deficit built up in the Reagan years from $58bn in 1981 to $212bn in 1985 because of Reagan's increase in defence spending (which rose 50 per cent during this time) and his 1981 tax cuts. In 1985 Congress passed legislation to reduce the budget deficit to zero by 1991. The law, known as Gramm-Rudman, was revised in 1987 with new targets and an extension of the period to 1993. Deficits for 1991 and 1992 are $64bn and $28bn respectively. The limit in 1989 was $136bn, but the actual deficit was $170bn. Gramm-Rudman targets proved impossible to meet, despite the revision, with the result that more and more expenditure was pushed off-budget. Off-budget debt was estimated at $700bn in mid-1989. It is more costly to the taxpayer since the money raised does not qualify for low government interest rates. Gramm-Rudman targets apply to estimates, not to real expenditure, so that supplemental budgets can be passed once the fiscal year has begun. A Supreme Court ruling in 1987 against co-ordination between the administration's Office of Management and Budget (OMB) and the Congressional Budget Office (CBO) means that the OMB's estimates are the only ones that count, however inaccurate they may be. These conditions have made nonsense of the law and make another revision a strong likelihood.

Bush's 1990 budget demonstrated masterful acrobatics. It took Congress three days to discover that his education budget represented a cut in spending, not an increase as he had claimed in his budget speech. Budget Director Richard Darman had altered the basis of comparison between 1989 and 1990 for most budget sections (defence was the exception) to one of straight cash. Previous budgets had always taken inflation into account. A major 1990 expense, the $50bn bail-out of the savings and loans associations (thrifts) was shifted off-budget to the tune of $30bn, the remaining $20bn carried back to the 1989 budget. $850 m in farm subsidies were also carried back to 1989. This contravened the two-year budget agreement for 1988 and 1989 which banned any mid-term increases in the budget, but was accepted by Congress as the only means of meeting the Gramm-Rudman target. Congress completed its approval of 1990 budget appropriations in summer 1989 but discussion of details continued, especially regarding the new money to be spent on the drugs war and Bush's controversial proposed cut in capital gains tax from 28 per cent to 15 per cent. The OMB claimed that this would net an additional annual $4bn in government revenue from increased investment. Congressmen conceded that there might be a gain in the first year because of the sale of assets held back in anticipation of the new cut, but insisted that the government would lose substantial revenue thereafter. Opponents pointed out that the cut would favour the rich and encourage them to make arrangements to declare assets rather than income for tax purposes since the new rate on capital gains would be well below the top income tax rate of 28 per cent. This would lose the government additional revenue.

No new taxes

It is virtually impossible for the USA to substantially reduce the budget deficit without raising taxes. President Reagan shifted a larger share of spending onto the states. By the late 1980s, federal spending had been pared back to slightly over half that of the states, a reversal of the 1970s situation. There is now little scope left for cutting non-defence programmes. Part of the appeal of a conventional arms reduction treaty with the Soviet Union is that it would rapidly allow for large savings in the defence budget. However, such an agreement will be hard to reach because of the massive imbalance of forces in the Warsaw Pact's favour.

Federal tax measures in the Reagan years restricted locally funded social programmes. The 1986 Tax Reform Act cut tax exemptions on gifts to charities, which traditionally carry out a large part of social care, thus reducing their source of funds and ability to operate. Corporate charity contributions now go more towards education in which companies have a vested interest. The special tax credit for low income housing construction expired in 1989. This was offered after the Housing and Urban Development Department (HUD) ceased its public housing construction programme. The credit helped states, cities and charities take over this responsibility. The states now have to raise taxes or lower standards. Many have increased taxes but others have taken the second route. For example, federal officials have urged some states to raise the poverty limit to allow more needy people to benefit from Medicaid, the health insurance programme for the poor. Proposals to reduce the number of police and fire stations to balance city budgets have met with public outcry. The devolving of more responsibilities and thus more regulatory powers onto the states has led to a proliferation of different laws and agencies. Business operating conditions are therefore more complex and time-consuming, especially with regard to environmental and safety standards.

President Bush had said 'No new taxes,' but at the start of his administration, Budget Director Richard Darman spelt out a list of tax equivalents. A petrol tax could be construed as a 'user fee' if nominally used for road building, tobacco and alcohol taxes, too, if spent on related medical care. Income tax applied to untaxed categories of fringe benefits would be a 'definitional change". An oil import fee would be a tariff. Apart from such euphemisms, taxes due to be phased out could be retained, including for example, taxes on gifts and legacies and the 3 per cent excise duty on telephone services. Another measure discussed in 1989 was the proposal to remove the tax deductibility of interest on large loans to discourage LBOs. The treasury is concerned at the high level of debt taken on by companies and at the escalating number of large mergers and acquisitions. But critics pointed out that it would be difficult to distinguish between loans for this purpose and others. Companies built up huge amounts of debt in the 1980s, following the federal requirement to consolidate all majority-owned subsidiaries for tax purposes. This gave major manufacturing and retailing companies greatly enlarged cashflow, through the inclusion of financial subsidiaries on which to borrow. It also made them more attractive to hostile bids. Thus they built up an even higher debt to ward off potential takeovers. From 1984 to 1988 net worth increased by 8 per cent while debt rose by 45 per cent giving a debt/equity ratio of 44 per cent. Return on total physical assets was 9.4 per cent in 1988.

Kinder, gentler America

President Bush made action against drugs and crime a top priority. His $7.8bn programme, announced in September 1989, emphasised law enforcement and included aid for the Andean drug-producing countries of Colombia, Bolivia and Peru to curb the supply problem at source. The programme allocates $716 m more than the original 1990 allowance, the additional funds to be cut from other programmes. In May 1989 Bush introduced tougher penalties for violent crime including the federal death penalty for murder and espionage, a doubling of the minimum sentence to 10 years for crimes with semiautomatic weapons, a ban on gun ownership for drug offenders and violent criminals and a reduction of plea bargaining. Bush is against gun control but banned the importation of semiautomatic rifles following a mass murder in California in January 1989. In early 1989, 37 states had the death penalty but this continues to be a controversial issue. Critics of Bush's get-tough policies point to the already burgeoning prison population, and the length of time and cost of building new prisons. 1984 sentencing guidelines and the abolition of the federal parole board were intended to produce a tougher and fairer response to crime, but in practice they have tended to make judges more lenient in sentencing.

Poverty, homelessness and lack of access to health care have increased as the gap between rich and poor has widened. Some 20 m Americans were living below the poverty line in 1987. Estimates of the number of homeless range up to 3 m. The 1990 census will attempt to count them accurately. 35m adults do not have health insurance and do not qualify for Medicaid. As the decade closes, there is a greater desire to tackle poverty if only to cut down on surging street crime and violence and the eyesore of homelessness. With a buoyant labour market, some states have introduced workfare programmes to help unemployed single mothers on welfare find jobs. In June 1989 the Senate approved a $10bn five-year childcare programme providing subsidies to childcare centres and tax credits to low income families. The aim of the Bill is not only to enable the parents to work but to help reduce the relatively high rate of infant mortality and the build-up of health problems from an early age. The Head Start preschool programme was given an additional $400 m in the 1990 federal budget for similar reasons; this programme improved infant care as well as laying the basis for sounder educational achievement in later years. 23m Americans are illiterate and, with an increasingly tight labour market, companies are thought to spend some $25bn a year on remedial education for prospective employees. Some large companies allocate part of their employees' time towards teaching local schoolchildren. In Boston and some other cities, local companies have offered jobs and training in return for higher school standards. Companies increased their payments for employee and pensioner health insurance by 30 per cent in 1988. Despite their complaints, their contribution to health costs is unlikely to fall in the foreseeable future. Congress agreed a $2.7bn cut in the government's Medicare health insurance programme for the elderly in 1990 although costs were expected to rise by $12bn. The pensioners' contribution to Medicare costs was raised from 20 per cent to 25 per cent in 1989, but it was not raised again for 1990. A new programme of pensioner insurance cover for chronic illness was agreed in principle by Congress in April 1989, with funding to be raised from a graduated surtax on the elderly.

A cleaner environment is another item on which companies will have to spend more. Almost the entire cost of measures under President Bush's Clean Air Bill, introduced in June 1989, will be borne by the private sector. The government will hold talks with individual industries to limit the production of toxic chemicals. Electricity generating companies will foot a $3.8bn bill for reducing sulphur dioxide emissions by 10 m tonnes by the year 2000. They will also be asked to match a $2.5bn government subsidy for the development of clean coal technology. Car manufacturers will be responsible for reducing nitrogen oxide emissions by 2m tonnes a year and for producing at least 1m cars capable of using methanol only by 1997. Bakeries, furniture manufacturers, drycleaners, paintmakers and others are called on to reduce hydrocarbons emissions by 45 per cent by the year 2005. The oil industry and the federal budget both suffered from the Exxon Valdez disaster in March 1989. When the ship ran aground spilling 11m gallons of oil into Prince William Sound, Congress shelved plans to allow oil drilling at the nearby Arctic National Wildlife Reserve, a measure that would have increased federal tax revenue by $2.1bn in 1990. Exxon had spent $1bn on trying to clean up the Sound by September 1989 with only partial success.

The Department of Energy's nuclear weapons production programme faced problems when investigations into plant safety, ordered by the Department following the 1986 Chernobyl disaster in the Soviet Union, revealed poor management and possible rule violations in 1988–89. Tritium plants operated by Du Pont at Savannah River, Georgia, were closed following a neutron surge in August 1988. Du Pont was accused of poor management and the plants remained closed a year later. The FBI and Environmental Protection Agency (EPA) opened investigations in mid-1989 into possible safety violations by another private contractor, Rockwell International, at the Rocky Flats, Colorado plutonium plant, only 16 miles north of heavily populated Denver. The Energy Department was also accused of lax management.

Oil, manufacturing and property

Crude oil production was 2.98bn barrels in 1988, down from 3.05bn barrels in 1987. Lower oil prices from the mid-1980s, which undermined the economies of Texas and Louisiana, have also reduced exploration activities. Reserves fell to 26.9bn barrels in 1986, or eight years of production, from a peak of 39bn barrels in 1970. The newest oilfields, in Alaska, are expected to decline from 1991. Oil exports fell by 2.2 per cent to $4.4bn in 1988. Imports declined by 8.4 per cent to $39bn due to more fuel efficient vehicles and factories, the continued decline of heavy industry and the gradual phasing out of oil-fired power stations. Coal-fired generating accounted in 1989 for about 57 per cent of electricity production, while nuclear accounts for 18 per cent, natural gas for 11 per cent, hydropower for 10 per cent and oil for 5 per cent.

Manufacturing industry, encompassing a wide field of products, accounts for about 19 per cent of GNP. Total manufacturing output rose by 8 per cent in 1988, sales increased by 9.3 per cent, inventories by 7.6 per cent and producer prices by 4.7 per cent. Wholesale and retail sales rose by 7.1 per cent each. Wholesale stocks increased by 9.6 per cent while retail inventories rose by 6.9 per cent. The main manufacturing sectors in order of value added are transportation equipment (including motor vehicles), electrical and electronic goods, food, machinery (excluding electrical), chemicals, printing and publishing, metal goods, textiles and clothing, paper, instruments and precision goods. From 1985 manufactured goods regained some of the share of world exports lost due to the dollar appreciation in the early 1980s. However, reconquest of the home market was only beginning at the end of the decade. Americans, on a long-extended shopping spree, have become used to buying cheap everyday items from the Far East and luxury goods from Europe. European cars are a price-competitive status symbol, while Japanese and South Korean cars provide excellent value for money. But Detroit's auto giants have produced new designs to compete with the influx and they have captured a much faster growing market – that for light trucks bought by families as a fuel-efficient alternative to cars. Fears that the USA would lose its world lead in computers to the Japanese proved unfounded on account of more rapid innovation and greater reliability of US hardware. The European invasion of the US chemicals industry gave European firms the top four slots in the world league in terms of sales in 1988. However, they were outshone in terms of overall performance by Dow Chemical and in terms of net profit and productivity by Du Pont.

Completed construction of new buildings increased by 10.7 per cent in 1988, but housing starts fell by 8.2 per cent and the value of total construction contracts declined by 1.9 per cent. New mortgages also fell significantly due to the inability of prospective home buyers to raise sufficient capital. House prices spiralled in the 1980s, especially in the north-east and south-west, and by mid-decade were already out of reach for 50 per cent of first-time buyers. The collapse of oil prices in Texas led to massive defaults on mortgages. The Texan property market had soared due to the influx of new light industry to the Dallas-Fort Worth area. The defaults were the last straw for some 500 savings and loans associations (thrifts) which filed for bankruptcy by the end of 1988. By June 1989 Arizona was badly affected and the number of thrift bankruptcies had risen to 600. Loan interest rates were subject to fixed ceilings until 1986, whereas savings rates were variable. Thrifts experienced increasing difficulty in matching their costs to this fixed income in the 1970s, but deregulation, phased in from 1980, put them in direct competition with the banks. More heavily capitalised and diversified, banks are better able to withstand loan defaults. Even so, all major banks in Texas had either been rescued or bought out following the real estate collapse. Banks suffered from mortgage defaults elsewhere in the south and in the north-east in 1988. They may have to increase loss provisions further as the government asks them to take on responsibility for part of the thrift business. The government's rescue operation will certainly cost more than the $50bn budgeted in 1989–90, and it will have to handle the sale of foreclosed property with caution not to depress the market further.

Agriculture

The long drought of 1988 saved the government $7bn in farm subsidies on the 1989 budget because scarcity pushed up crop prices. In August 1988 Congress approved $3.9bn in emergency aid for farmers who lost more than 35 per cent of their harvest with a limit of $100,000 per farm. Farms with an annual income of $2m and more were excluded. Dairy farmers received an extra 50 cents a cwt subsidy for milk from April to June 1989 to help cover the extra cost of feed. A general 50 cents/cwt cut in the milk subsidy planned to take effect from January 1989 was cancelled. Federal farm support fell from $25.7bn in 1986 to $10.7bn in 1990 ($13.9bn in 1989). Grains and dairy products absorb most of this, while sugar and peanuts also receive support. At the end of the 1988–89 season, wheat stocks were down to 16.8m tonnes, maize to 46.5m tonnes and soyabeans to 3.4m tonnes. Prices and farm incomes had risen. Farm debt fell to $140bn in 1988 from $183bn in 1983.

Outlook

GNP growth is expected to slow to 1.7 per cent in 1990. Inflation should ease to 4 per cent on account of this and the legacy of tight monetary control in 1989. Interest rates and the dollar will weaken. With exports continuing to grow at a faster pace than imports, albeit more slowly than in 1989, and with the dollar depreciation, the trade deficit should fall to $86bn. This is enough to reduce the balance of payments deficit to $109bn despite an increase in the invisibles deficit (including the debt service) to $23bn. A revival of growth from 1991, along with a strengthening of the dollar, higher interest rates and slightly higher inflation, will be investment-led resulting in a deterioration in the balance of payments deficit. Employment levels are expected to remain high in 1990 and 1991.

Congress may vote some new taxes in 1990, the most likely contender being the petrol tax, but with reluctance because of the midterm elections in November 1990. It may decide to go for a revision of Gramm-Rudman instead, since similar gymnastics to those of the 1990 budget package will be necessary to meet the 1991 budget deficit target. The impact of the petrol tax, if approved, would be slight. Although the Gramm-Rudman target was nominally met for 1990, the real deficit is likely to be $140bn. The OMB forecasts a $94bn deficit for 1991, $22bn above target.

The ethics war could revive in 1990 once Congress has had the chance to examine the evidence of the North trial and when Vice-Admiral John Poindexter comes to trial. North's boss has claimed that he destroyed documents to protect ex-President Reagan. Supreme Court rulings are likely to build up an increasingly conservative body of legislation, undermining the civil rights gains of the 1970s. The summit between President Bush and Soviet leader Mikhail Gorbachev in 1990 could make progress on nuclear arms reductions.

BUSINESS GUIDE

Time: Pacific Standard Time = GMT -8 hrs (-7 hrs Apr-Oct); Eastern Standard Time = GMT -5 hrs; Central Standard Time = GMT -6 hrs; Mountain Standard Time = GMT -7 hrs.

Entry requirements

Passport: Required by all with few exceptions.
Visa: Extensive regulations govern visa requirements, and travellers should make detailed enquiries as far in advance of their intended visit as possible about the visa appropriate to their circumstances and the method of applying for such visas.
Currency: Amounts of $5,000 or more in currency or other negotiable form should be declared.
Customs: Personal effects are allowed duty-free, including up to $100 in gifts every six months. Import restrictions apply to dairy products, fish, chocolate crumb confectionery, wheat products, sugar, oil products, textiles from low-cost countries, firearms, protected wildlife and other goods. Importation of goods from Cuba and North Korea is prohibited. Special regulations regarding quality safety or technical requirements apply to certain goods. Would-be exporters to the USA should comply with the relevant US standards.
Income tax: Under certain circumstances, business visitors may be liable for payment of federal social security and unemployment taxes, and liability to taxes may be incurred under state or municipal laws, which vary.

Health precautions

Mandatory: No vaccination certificates are required, unless visitor has been in certain areas within the previous six months.
Advisable: Medical fees are extremely high, and visitors are advised to arrange adequate health insurance before arrival.

Hotels: Major hotels have toll-free telephone numbers (with an 800 area code) for reservations. Unless a deposit has been paid, a hotel room will often not be held after 1700/1800, even when the hotel is notified of late arrival. Check-out times vary, from 1000 to 1300, and short extensions can be arranged.

BUSINESS GUIDE

Visitors may be charged for overstaying check-out time without making arrangements. Most good hotels have restaurant facilities, bars, free parking and swimming pools, and many provide courtesy transport or an airport bus service.

Credit cards: Considerable use is made of credit cards, and it is advisable to carry an internationally recognised card such as American Express, Master Card or Visa. A credit card is essential for car hire.

Car hire: Car hire is widely available in major cities. A valid overseas or international driving licence and an international credit card are required. Other methods of payment may not be accepted. Driving is on the right-hand side of the road. The speed limit is 55 miles/hr (88 kph) on the interstate highway system.

City transport
Taxis: Taxi fares vary in different cities. Most cities have severe parking problems in downtown areas, and it is most convenient to travel by taxi.

National transport
Air: A highly developed network of airline services connect most towns of importance. Fare systems have recently been deregulated, leading to sharp competition (and some insolvencies). Major savings can be made by shopping with care.
Road: High-speed motorways (intra- and interstate freeways and highways) connect main centres.
Buses: There is a wide network of air-conditioned long-distance buses linking all major cities, but smaller cities and rural areas are generally not well served by public transport.
Rail: 62,155 km of railways link 500 stations, and most long-distance trains are air-conditioned and equipped with dining cars, bar cars and sleeping accommodation. Much of the national network is in severe need of new equipment however, and in terms of time and cost, rail travel compares poorly to air travel on most inter-city routes.

Public holidays: Legal and public holidays are fixed by state legislation and vary considerably from state to state. Most offices remain closed on the Friday following Thanksgiving.
Fixed dates: 1 Jan, 4 Jul (Independence Day), 11 Nov (Veterans' Day), 25 Dec.
Variable dates: Martin Luther King Day (Jan), Washington's Birthday (Feb), Memorial Day (last Mon in May), Labor Day (first Mon in Sep), Columbus Day (second Mon in Oct), Thanksgiving Day (Thu in late Nov).

Working hours: Working hours vary considerably.
Business: (Mon-Fri) 0900 – 1700.
Government: (Mon-Fri) 0830 – 1730.
Banking: (Mon-Fri) 0900 – 1500.
Shops: (Mon-Sat) 0930 – 1800; Sun 1200 – 1700.

Telecommunications
Telephone: Dialling code for USA: IDD access code + 1 followed by area code followed by subscriber's number. For international calls, the IDD access number from the US is 1 (followed by country code followed by area code followed by local number). IDD is available from most public telephones, otherwise dial 0 (for Operator). When dialling from public telephones, you must pay for the first 3 minutes on long distance and international calls (the operator will tell you when and how much money to deposit). When dialling from hotel telephones, first dial the local hotel access code, which should be posted near the telephone in the hotel room, then the rest of the call numbers as usual. Ringing tone lasts two seconds, followed by a four-second silence; number unobtainable is indicated by a recorded message; number engaged is indicated by rapid pips.

Postal service: General delivery: Mail directed c/o General Delivery, Main Post Office must be collected in person by the addressee. Proof of identity is required. It should be marked 'Hold for 30 days' and the return address should be clearly indicated. Post office opening hours: (Mon-Fri) 0800 – 1730/1800; Sat 0800 – 1200. Self service facilities for stamps are often available 24 hrs/day.

Banking: Banking regulation is overseen by numerous federal and state authorities with overlapping jurisdictions, and is thus extremely complex. US law prohibits inter-state commercial banking, and raises tough barriers between banking and securities business, but there were proposals in 1987 which were expected to soften these restrictions. Regulatory agencies include the Federal Reserve, the Federal Deposit Insurance Corp (FDIC), the Securities and Exchange Commission, the Comptroller of the Currency, the Department of Justice and state bank departments. Depositors in banks or savings and loan associations which are members of the FDIC or Federal Savings and Loan Insurance Corporation have their deposits guaranteed to a limit of $100,000 by the government's system of deposit insurance.

Advertising: Press, trade press, television, radio and direct mail advertising are all widely available. Costs are high, and it is not uncommon for manufacturers and distributors to share promotion expenses. Videocassettes used to demonstrate products should conform to the US line standard, which is NTSC 525 lines. The standard format is the 0.75 inch U-matic cassette.

Trade fairs: Trade fairs and exhibitions featuring a wide range of industries and products are held in cities throughout the country. There are good fair facilities in many centres. Some events are always in the same city, on an annual, or six-monthly basis, while others move from city to city.

Electricity supply: 110-120 V AC, 60 cycles single phase, with flat two-point plug fittings and screw-type light sockets. It is advisable to purchase any adaptors which may be required before travelling to the USA.

Weights and measures: Units of measurement used in the US are in general the same as the imperial system, but the old winchester bushel and wine gallon are still in use. Wine gallon = 0.83268 imperial gallon; bushel = 0.969 imperial bushel; 1 pint = 16 fl oz. A short hundredweight contains 100 lb and a short ton contains 2,000 lb. Converstion to metric system is taking place very slowly and on a voluntary basis, with the US Metric Board co-ordinating the process.

Useful business addresses
US and Foreign Commercial Service Office, Department of Commerce, 14th Street and Constitution Avenue, NW, Washington DC 20230

Securities and Exchange Commission, 450 5th Street, NW, Washington DC 20549

Chamber of Commerce of the USA, 1615 H Street, NW, Washington DC 20062

Uruguay

Mike Rose

As confirmation of its democratic heritage, Uruguay went to the polls in December 1989 to vote for a new president and congress, its second free election this decade and ample proof that the 12 years of military rule which ended in 1985 was an aberration unlikely to be repeated in a hurry.

With the vexed question of the military amnesty settled by a referendum accepted by the majority of Uruguayans, electioneering became quite intense, and by September 1989 it appeared increasingly likely that continuity would be ensured with a Colorado party victory. However, a new factor in the political equation was what some commentators saw as the terminal decline of the once-influential left-wing coalition, the Broad Front, making the elections look even more like a two-horse race between the Colorados and the Blancos (National Party). The victory of Ernesto Lacalle of the Blancos promised a consolidation of centre-right government in Uruguay for the indefinite future, a possibility likely to leave large numbers of metropolitan left-wing voters disillusioned. There seemed little likelihood this would lead to the insurrectionary situation of the late 1960s and early 1970s, culminating in the military coup. But the vacuum caused by the splintering of the left would engender new tensions affecting the country's political and economic life well into the future.

The economy, likewise, seemed to be destined for a period of uncertainty, although there were some promising signs on the horizon.

By late August, the trade balance for 1989 was running at nearly $250m, $50m over the forecast for the whole year. And the Central Bank reported that its net reserves increased by $16m in June alone and by $56m over the first half of the year. However, it was thought doubtful that the government would meet

KEY FACTS

Official title: Oriental Republic of Uruguay	**Area:** 176,215 sq km
Head of state: President Luis Ernesto Lacalle	**Population:** 3.06m (1987)
Ruling party: Partido Blanco	**GDP per capita:** $2,989 (1988)
Capital: Montevideo	**GDP real growth:** 0.5% (1988)
Official Languages: Spanish	**Inflation:** 62.6% (1988)
Currency: New Peso (N$) = 100 centésimos	**Trade balance:** $292.3m (1988)
Exchange rate: N$730 per $ (Nov 1989)	**Foreign debt:** $6.33bn (1988)

its target of containing inflation within the 60-65 per cent range or its growth target of 4 per cent. Inflation was expected to end the year near the 70 per cent mark and gross domestic product to grow by less than 2 per cent.

Government amnesty vote triumph

The most pressing and potentially divisive political issue facing the country was settled in April 1989, when the government of Julio Maria Sanguinetti triumphed in the referendum over the military amnesty issue. Ever since President Sanguinetti pushed through Congress in December 1986 the bill amnestying military officers accused of human rights abuses during the 12 years of military rule, the issue had overwhelmingly dominated the political agenda. Against Sanguinetti's advice that it was best forgotten, a large, mainly left-wing bloc set up a pro-referendum commission which secured the signatures of well over the constitutional requirement of one-quarter of the electorate necessary to hold a referendum. The government became increasingly concerned, afraid that the process would upset the armed forces and could even lead to their intervention and a possible military coup. Official propaganda over the two years of the campaign openly reflected this concern. On the eve of the referendum, Sanguinetti told the nation that ratification of the amnesty law would be 'the final step in the country's transition to democracy. The alternative would be a step backwards in the establishment of peace within the country.' The pro-referendum lobby, although lacking the government's extensive propaganda machinery, nevertheless argued persuasively that a democracy which places the military above the law is no democracy at all. The result was a government victory of 58 per cent to 42 per cent against repealing the amnesty. In the capital, Montevideo, the majority voted against the amnesty but this was not sufficient to offset government support in the deeply conservative countryside. The losers promised to respect the result and Sanguinetti said unity had overcome disagreement and Uruguay could look forward to a great future. With the major hurdle overcome, all energies would be devoted to the forthcoming election.

Political continuity assured

There was little doubt that continuity rather than change in government policies would be the order of the day following last November's general elections, no matter which of the two principal, and centre-right, parties won. The well-oiled party machinery of the ruling Colorados was expected to assure them victory over the Blanco (National) party, although both were expected to maintain substantial representation in Congress, probably at the expense of the troubled Broad Front coalition. Under the complex Uruguayan electoral system, political parties can divide into factions, each of which nominates its own presidential candidate who stands in the national election. The candidate who receives the most votes from the party whose combined vote is the largest, wins the presidency.

Within the Colorados, Jorge Pacheco Areco and Jorge Batlle had been chosen as presidential candidates, with the former ahead in the opinion polls. The Blancos, whose highly popular leader Wilson Ferreira Aldunate had died of cancer in 1988, had three presidential hopefuls, conservative Senator Luis Alberto Lacalle (the most popular, according to opinion polls), Carlos Julio Pereyra and Senator Alberto Zumaran. At one time, the Blancos were seen as more nationalistic than the Colorados, and promised to take a much tougher line over issues like the foreign debt and foreign investment. Recently, however, their policies have moderated. The bulk of the party supported Sanguinetti over the military amnesty issue and there was little to differentiate Blanco and Colorado policy over major political and economic issues. For this reason, little was likely to change in government policy after the elections, whichever party was victorious.

The big question in Uruguayan politics is over the future role of the left. The Broad Front coalition was a major force, both within and outside Congress, and boasted considerable support in the capital. However, increasing splits within its ranks, exacerbated by a bitter debate over the military amnesty question, culminated in two moderate parties withdrawing in March 1989. The Christian Democrats and the People's Government Party left to set up another leftist bloc in which, according to a spokesman, 'only Marxists-Leninists will have no place'. This loss was seen by many observers as heralding a major decline in the Broad Front itself, and the left in general. There were some who saw the organisation behind the pro-

KEY INDICATORS

	Unit	1984	1985	1986	1987	1988
Population	m	3.0	3.01	3.03	3.04	3.06
GDP	N$bn	294.36	528.15	981.1	1,753.86	2,855.32
GDP per capita	$m	1,761	1,744	2,086	2,465	2,989
Inflation	%	55.3	72.2	76.4	63.5	62.2
GDP real growth	%	−1.5	0.3	7.5	5.9	0.5
Agricultural production	1974-81=100	105.1	106.2	109.3	111.0	110.6
Food production	1979-81=100	100.88	103.4	102.3	103.1	102.4
Exports	$m	933.8	909.0	1,087.8	1,189.2	1,404.5
Wool	$m	162.8	163.8	201.6	244.7	348.6
Meat	$m	145.9	117.9	198.9	129.3	152.0
Hides	$m	91.6	61.3	74.7	83.9	104.6
Imports	$m	776.7	707.7	870.0	1,141.9	1,157.2
Crude petroleum	$m	260.1	223.5	140.3	154.6	112.2
Balance of trade	$m	192.4	178.2	273.3	102.4	292.3
Current account	$m	−129.1	−120.1	67.2	−131.2	33.8
Tourist arrivals	'000	855	787	–	–	–
External debt	$m	3,271	3,919	3,908	4,235	–
Debt service ratio	%	29.8	30.0	21.1	24.4	–
Exchange rate	N$ per $	56.1	101.4	152.0	226.7	359.4

* estimated figure

referendum campaign as a sign that a new grassroots politics was emerging, free from the traditional political divisions. However, it was hard to see such a movement consolidating itself, except within the ranks of the left. And with such disarray, it seemed unlikely that a viable challenge to the centre-right would emerge in the near future.

Economic prospects unremarkable

Despite the much-lauded regional integration pact between Uruguay and its two giant neighbours, Brazil and Argentina, little of substance appears to have been achieved in giving the Uruguayan economy the sort of boost its three million inhabitants are unable to provide. At their summit meeting in August 1989, Presidents Sanguinetti, Sarney and Menem reiterated their commitment to full integration, but it remains to be seen how far this will go beyond rhetoric, especially considering the dire state of the Argentine economy.

With the trade balance for the first half of 1989 registering a healthy surplus of $240m and an increase of $56m in reserves over the same period, government economists were far from pessimistic over economic prospects for 1989 and beyond. It was nevertheless hard to see how their 4 per cent growth target would be achieved, especially as inflation was expected to end the year around the 70 per cent mark. The main inflationary factors towards the end of 1989 were an influx of Argentine capital fleeing economic uncertainty, swelling the money supply and making interest rate increases necessary; a summer drought pushing up the prices of agricultural production; increased world oil prices and another substantial public sector deficit. Another factor deterring economic growth from exceeding 2 per cent during 1989 was the government's determination not to let real wages increase by more than 1.5 per cent, further dampening down economic activity. If this proves to be the case, it will have been the second year running in which the economy behaved disappointingly. In 1988 real GDP growth was 0.5 per cent, a dismal effort compared to the previous two years. The central government deficit for 1988, at 54,000m pesos, was equivalent to 2 per cent of GDP, up from the previous year's 1.3 per cent. The aim was to reduce this deficit to 1 per cent of GDP in 1989 and also to cut the overall public sector deficit from its unsatisfactory 1988 level of 4.3 per cent.

Unemployment down

Unemployment for the first three months of 1989, at 8.17 per cent, was the lowest figure for this quarter for three years. The government hailed the constant fall in the unemployment rate since May 1988 as a great success for its policies. In the last quarter of 1988, unemployment fell to 8.89 per cent, down from 9.35 per cent the year before, and in the December to February quarter, it fell to one of the lowest rates in recent times, 7.8 per cent. Part of this was attributable to the success of the construction industry, which grew 3.9 per cent in 1988, making it one of the few success stories of the year. Much of this growth came from the building boom in the tourist centres of Maldonado and Punta del Este. The momentum was expected to decline somewhat in 1989 due to a downturn in tourism in troubled Argentina and spending cuts in the public sector. One bright spot on the horizon was a planned modernisation of ports and highways, to be funded partly by the World Bank and Japan's Eximbank.

In contrast, the manufacturing industry had a bad year in 1988, declining 3.7 per cent, after growing a much-needed 11 per cent the previous year. For example, machinery and equipment output declined 7.7 per cent; chemicals, petrochemicals 7.6 per cent and textiles, clothing and leather 8.6 per cent. Both domestic and external demand for manufactures was sluggish.

Drought hits agriculture, energy

Agriculture and the energy sector were both hit by a drought, affecting large parts of the north and north east of the country during the first half of 1989, although it failed to have quite the impact some rice farmers and cattle ranchers forecast. Indeed, the rice harvest was said to be considerably up on 1988, although of inferior quality. However, maize, soya and sorghum output registered sharp falls (of 52 per cent, 29 per cent and 48 per cent respectively). The cattle industry, crucial for foreign exchange earnings, claimed it was in dire straits, forcing the government to introduce special measures in March last year. These extended to 90 the number of days for ranchers to pay slaughterhouses and also offered the latter compensatory credits. Cattle ranchers were allowed to defer social security payments and wealth tax dues were reduced by 20 per cent. Offsetting the drought toll was an extremely high slaughter rate, with 650,000 head being killed during the first four months of 1989, 50 per cent more than in the same period in 1988. The government also announced a plan to cut 5 per cent off export tariffs on preserved meats and to subsidise its production by $40 a tonne. It also opened up credit lines of up to eight years' repayment for the replacement of herds. Notwithstanding, the Rural Federation estimated losses would exceed $200m. The energy sector was badly affected, due to the failure of hydroelectric plants. At the largest, Salto Grande, only two of 14 turbines were operating in March due to the low level of the Uruguay river. Rationing affected towns and cities throughout the country, with television transmission limited to four hours nightly.

Foreign debt increases

Uruguay's foreign debt was heading towards the $7bn mark at the end of 1989, having reached $6.33bn by the end of 1988 (up 8.4 per cent in the year). Private borrowing abroad increased more rapidly than state borrowing, bringing it to over $2bn by the end of 1988. But a repetition of the record trade figures for 1988, was going to be difficult to achieve in the light of Argentine austerity measures and Brazil's continuing economic difficulties. The value of exports in 1988 reached a record level of $1,405m, mainly thanks to sales of farm produce, notably soyabeans, barley and wool, the latter increasing 36 per cent over the previous year. China was the major wool buyer, followed by the USSR, USA, West Germany and the UK.

COUNTRY PROFILE

Political system: Executive power vested in the president, who is directly elected for a period of five years. The president is assisted by a vice-president and an appointed council of ministers. Legislative power is vested in a bicameral Congress, comprising a 30-member Senate and a 99-member Chamber of Deputies.

Main cities/towns: Montevideo, Paysandú, Salto.

Media
Press: Most important dailies are published in Montevideo. These include: *El Día, La Mañana, El País, La Hora, La República* and *Gaceta Comercial* (morning); *El Diario* and *Ultimas Noticias* (evening). Several regional newspapers are available in main centres outside Montevideo. The main Buenos Aires papers can also be found. There are numerous periodicals and a few trade publications.
Radio: There are approximately 40 radio stations in Montevideo and over 60 outside the capital.
Television: Four TV stations transmit from Montevideo and 16 others serve towns outside the capital.

External trade: The trade balance in the early 1980s suffered from falling export earnings from traditional goods, such as meat and wool, combined with increased cost of oil imports. However since 1983 declining levels of imports have caused a surplus on the trade account. The trend of falling meat and wool exports was reversed in early 1986.
Exports: Wool and products (21% of total value in 1987), hides and skins (16%), meat and products (12%), textiles etc, vegetable products, chemicals, shoes, ceramics, glassware, transport equipment, machinery, electrical energy. Main destinations: Brazil (30% of total value in 1986), EC (30%), USA (8-13%), Argentina (8-10%), German FR (5-8%), Netherlands (3-4%), Iran (5-12%), Italy, UK, France, Spain, USSR, Paraguay, Chile, Israel, Japan, Greece.
Imports: Crude petroleum (17% of total value), machinery and transport equipment (10-14%), chemicals (10-12%), base metals etc (8-10%), food etc (6-8%), plastics etc, textiles etc, skins, hides etc, paper products. Main sources: Brazil (12-16% of total value), Argentina (10-12%), USA (7-10%), German FR (5-9%), Venezuela (5-7%), Iran (9-15%), Nigeria (10-15%), Kuwait, Italy, UK, Netherlands, Japan, France.

Agriculture: Mainstay of economy, contributing 11 per cent to GDP and accounting for around 85 per cent of export earnings. Sector is also important supplier of raw materials (sugar, oilseeds etc) to industry. Livestock rearing forms the basis of the sector with cattle and sheep being produced for domestic consumption and for exports (as meat, wool, hides and skins). Milk, poultry and pigs are largely produced for the home market but exports of dairy products are increasing in importance. Principal crops are wheat (mainly grown on mixed farms), rice (the main export crop, grown almost entirely in the north-east), sugar cane and beet, maize, barley, sorghum, linseed, sunflower seed, vegetables (mainly grown by small-holders) and citrus fruits (mainly oranges and tangerines). Commercial fishing is being strongly promoted by the government as an important source of foreign exchange. Fish exports have risen from under 3,000 tonnes in 1974 to over 50,000 tonnes in 1984.

Industry: Manufacturing and construction combined contribute around 25 per cent to GDP and employ 26 per cent of the workforce. Principal industries are textiles and clothing, food processing (meat, sugar, beverages, fish), hides and leather, cement, petroleum refining, chemicals, metallurgy, rubber, paper, electrical machinery, vehicle assembly. Emphasis is on development of export-orientated light industries and modernisation of traditional industries. Industrial production rose sharply in 1986.

Mining: Sector is undeveloped, except for marble and construction materials. Known deposits of iron-ore, gold, manganese, copper, zinc, lead and low-grade coal. Intermittent oil exploration is carried out but results have been disappointing so far.

BUSINESS GUIDE

Energy: Dependence on imported petroleum, and fuel wood. 40,000 b/d oil refinery at Montevideo. Electricity production is entirely hydro-electric. The 1,800 mW hydropower plant at Salto Grande (built with Argentina), the 300 mW Palmar plant, and two plants on the Río Negro are sufficient to meet local demand and provide surplus for export.

Membership of international organisations: Economic Commission for Latin America, FAO, GATT, ILO, IMF, Inter-American Development Bank, Latin American Integration Association, Organisation of American States, trade & co-operation agreement with the EC, WHO, World Bank.

Time: GMT -3 (GMT -2 from Dec to Feb).

Climate: Temperate with mild summers and winters. Summer lasts from Nov to Mar; warm with temperatures 21°C to 30°C. Winter from Jun to Sep tends to be cool and rainy with strong winds; average temperatures 10°C to 15°C, occasionally falling to freezing-point at night.

Entry requirements
Passport: Required by all.
Visa: Required by all except nationals of EC countries, USA, Canada, Japan, Scandinavian countries, Switzerland, most Latin American countries and certain others.
Currency: No restrictions on import/export of local or foreign currency.

Health precautions
Mandatory: none
Advisable: Typhoid, polio vaccinations recommended. Water precautions should be taken, especially outside Montevideo.

Representation overseas: Asunción, Athens, Beirut, Belgrade, Berlin, Berne, Bogotá, Bonn, Brasilia, Brussels, Bucharest, Budapest, Buenos Aires, Cairo, Canberra, Cape Town, Caracas, Guatemala City, The Hague, Havana, Jeddah, La Paz, Libreville, Lima, Lisbon, London, Madrid, Managua, Mexico City, Monaco, Moscow, Quito, Panama City, Paris, Prague, Rome, San José, San Salvador, Santiago, Santo Domingo, Seoul, Sofia, Stockholm, Taipei, Tegucigalpa, Tehran, Tel Aviv, Tokyo, Warsaw, Vatican City, Vienna, Washington.

Air access
National airline: PLUNA (Primeras Líneas Uruguayas de Navegación Aérea).
Other airlines: Aerolíneas Argentinas, Air France, Avianca, Cruzeiro, Iberia, Lan-Chile, LAP, Lufthansa, KLM, Pan Am, SAS, Varig.
International airport: Montevideo-Carrasco International (MVD), 21 km from city; duty-free shop, bar, restaurant, buffet, bank, post office, shops.

Surface access
Main port: Montevideo.
Overland access: A rail link runs from Brazil – services operate from São Paulo to Montevideo. Also rail services from Argentina via Salto. Several roads run from the borders with Argentina and Brazil. Entry from Brazil can be made via the Pan-American and Coastal Highways. Several bus services link Montevideo with Pôrto Alegre, São Paulo and Rio de Janeiro. Entry from Argentina is possible at Fray Bentos, Paysandú and the Salto Grande dam. Regular buses link Montevideo with Buenos Aires. (N.B. A quicker route is by ferry or hydrofoil from Buenos Aires to Colonia across the Río de la Plata, then by coach to Montevideo. There is also a ferry service direct to Montevideo from Buenos Aires).

Hotels: Graded into four classes by the National Tourism Bureau – de luxe, 1, 2A, 2B. Should be booked in advance during tourist season. 20 per cent value added tax on hotel bills. Service charge normally included – if not, usually 10 per cent tip.

Car hire: International driving licence must be accompanied by two photographs; traffic conditions within Montevideo can be difficult and chauffeur-driven cars often recommended. Driving permit for 90 days can be obtained from Montevideo town hall.

City transport
Taxis: Widely available in towns and from airports. Can be hailed. Fares are metered, with higher charges for extra passengers and at night; can be hired on a time basis, in which case fare should be negotiated in advance. 10 per cent tip is usual.

National transport
Air: TAMU and PLUNA operate regular internal flights from Montevideo to the more distant towns (eg Artigas, Melo, Rivera, Paysandú and Salto).
Road: Total network of 50,000 km, of which 90 per cent are all-weather and 20 per cent are motorway, giving good connection between major towns.
Buses: Montevideo served by numerous bus and trolley-bus services. Good services in most other towns. ONDA, CITA and COT run fast and frequent lines, connecting most towns over all the country (e.g routes include: Montevideo-Punta del Este, 2 hrs; Montevideo-Paysandú, 6 hrs).
Rail: System connects main cities. Tends to be slower but cheaper than the bus services. Routes include: Montevideo – Paysandú – Salto – Artigas.
Water: Some Buenos Aires-Montevideo ferries call at Colonia; there are various other services on the rivers Plate, Uruguay and Negro to suit local needs.

Public holidays: 1 Jan, 6 Jan, Carnival, Easter, 19 Apr, 1 May, 18 May, 19 Jun, 18 Jul, 25 Aug, 12 Oct, 2 Nov, 25 Dec.

Working hours
Business: (Mon-Fri) 0830 – 1200, 1430 – 1830/1900. (Some business offices open Sat mornings).
Government: From mid-Mar to mid-Dec: (Mon-Fri) 1200 – 1900. From mid-Dec to mid-Mar: (Mon-Fri) 0730 – 1330.
Banking: (Mon-Fri) 1300 – 1700 in Montevideo; summer variations may apply in other areas.

Telecommunications
Telephone: Dialling code for Uruguay: IDD access code + 598 followed by area code (Montevideo 2) followed by subscriber's number.
Telex: Facilities available in most commercial offices, banks and hotels.
Telegram: Service widely available throughout country.

Banking
Banking centre: Montevideo.
Central bank: Banco Central del Uruguay.
Commercial banks (head offices): Banco Comercial, Banco de Crédito, Banco Exterior, Banco Hipotecario del Uruguay, Banco de la República Oriental del Uruguay, Banco de Montevideo, Banco de Seguros del Estado, Banco Financiero Sudamericano, Banco la Caja Oberra, Banco Panamericano Uruguay, Banco Pan de Azucar, Banco Real del Uruguay, Hapoalim (Latin America), Casa Bancaria, Unión de Bancos del Uruguay.
Foreign branches: American Express Bank, Banco de la Nación Argentina, Banco do Brasil, Banco Francés e Italiano para la América del Sud, Banco Holandés Unido, Bank of America, Bank of London and South America, Banque Nationale pour le Commerce et l'Industrie, Banco de Santander, Banco Sudameris, Citibank, Discount Bank (Overseas), Dresdner Bank, First National Bank of Boston.

Electricity supply: 220 V AC, 50 cycles.

Representation in capital: Argentina, Australia, Austria, Belgium, Bolivia, Brazil, Bulgaria, Canada, Chile, Colombia, Costa Rica, Czechoslovakia, Denmark, Dominican Republic, Ecuador, Egypt, El Salvador, Finland, France, German DR, German FR, Greece, Guatemala, Finland, Honduras, Hungary, Haiti, Israel, India, Indonesia, Italy, Japan, Republic of Korea, Lebanon, Malta, Mexico, Netherlands, Nicaragua, Pakistan, Panama, Paraguay, Peru, Philippines, Poland, Portugal, Romania, Senegal, South Africa, Spain, Sweden, Switzerland, Taiwan, Thailand, Turkey, UK, USA, USSR, Vatican, Venezuela, Yugoslavia.

BUSINESS DIRECTORY

Hotels: Montevideo
Columbia Palace, Rambla Rep. de Francia 473 (tel: 960001; tx: 22524; fax: 960009).

Internacional, Colonia 823 (tel: 920001/2; tx: 22295).

Lancaster, Plaza Cagancha 1334 (tel: 984621; tx: 822).

Victoria Plaza, Plaza Independencia 759 (tel: 920237; tx: 22037; fax: 921628).

Car hire: Montevideo
Autorent: Yaguaron 1683 (tel: 90-87-78).

Avis: Rbla República de Mexico, Montevideo (tel: 50-46-02; tx: 22337).

Europcar National: Mercedes 802 (tel: 98-19-48; tx: 6282).

Hertz: Colonia 813 (tel: 90-79-57, 98-57-56; tx: 22484).

Car hire: International Airport
Avis: (tel: 50-22-61 ext 209).

Europcar National: (tel: 51-23-29).

Hertz: (tel: 51-00-01).

Business advice and information
Aero Consultora Uruguaya, Florida 1280/202, Montevideo (tel: 98-73-12; tx: 747).

Asociación de Importadores y Mayoristas de Almacén, Ed de la Bolsa de Comercio, Rincón 454, Montevideo.

Banco Central del Uruguay, Av Juan P Fabini esq, Florida, Montevideo (tel: 91-71-17, 91-67-47, 91-65-40, 98-57-42, 98-77-66; tx: 6659, 6939).

Banco de la República Oriental del Uruguay (Central Bank Foreign Trade Department), Cerrito 351, Montevideo (tel: 95-41-71).

Cámara de Industrias del Uruguay, Brig. Gral Lavalleja, Montevideo (tel: 98-11-30; tx: 6505).

Cámara Nacional de Comercio, Ed de la Bolsa de Comercio, Misiones 1400, Montevideo (tel: 96-12-77, 96-12-43; tx: 6996 bolsa).

CENCI (Centro de Estadísticas Nacionales y Comercio Internacional del Uruguay) Misiones 1361, Montevideo (tel: 95-45-78).

Latin American Integration Association, Cebollatí 1461, Casilla de Correo 577, Montevideo (tel: 40-11-21/28; tx: 326944).

Ministerio de Turismo, Brig. Gral Lavalleja 1409, Montevideo (tel: 91-43-40).

Oficina de Planeamiento Coordinación y Difusión, Edificio Libertad.

PLUNA Airlines, Colonia 1021, Montevideo (tel: 98-35-16).

USSR

Margie Lindsay

On 7 February 1990, the Soviet Communist Party agreed to end its 72-year monopoly of political power, at the same time endorsing President Mikhail Gorbachev's plans for an increased executive role for the state president. These positive political steps contrasted sharply with the dreary litany of riots, strikes, man-made and natural disasters, ethnic and nationalistic tensions and violence that had characterised 1989. Whether the political reforms will succeed in ending the turmoil – and if so when – is an open question.

Economy

The common theme which unites all these diverse happenings is the poor state of the Soviet economy. Most of the ethnic tension is motivated by economic problems: Central Asian republics have a lower living standard and average wage – sometimes almost three to four times below those in the more industrially and economically advanced Baltic republics; Central Asian unemployment is also high compared with other parts of the USSR.

Similarly, the calls for more autonomy and even secession – particularly by the three Baltic republics of Estonia, Latvia and Lithuania – are primarily a result of growing frustration with the slowness of reform throughout the Soviet empire and a consequence of a failed economic theory which has produced shortages, low living standards (compared to the developed West) and a mind-numbing sense of futility for many young people wishing to escape the closed circle of Soviet life. The Baltics, for one, think they could do much better by themselves, unencumbered by a nonconvertible rouble and other remnants of central planning.

Shortages of basic consumer goods, rising inflation and perceived differentials between state workers and the budding co-operative (private) sector have aggravated ethnic tensions in many republics. Many of the riots began because of economic advantages (real or imagined) of one ethnic group over another – or fears that one group is better off than another.

Strikes – which must have set alarm bells ringing throughout the Kremlin hierarchy – were called by coal miners not because of political disagreements with Moscow, but (initially) because of the failure of the system to provide basic commodities and decent living conditions for workers.

The impact of a string of natural disasters – like the Armenian earthquake in early 1989 – was aggravated by poor construction of buildings; infringements of construction regulations; an inability of the state to offer financial and medical assistance quickly and efficiently; as well as the general inertia of the economy to cope with a major hiccough.

The man-made disasters – such as the liquid petroleum gas pipeline explosion in Siberia and the string of fires aboard nuclear submarines in summer 1989 – highlighted the technological backwardness and overall inefficiency of the economic system.

Although many blame perestroika (restructuring) for the growing economic chaos and social tensions, the Gorbachev reforms are in many ways merely trying to cope with a situation brought about over 70 years of central planning. The muzzle which effectively kept the issues of nationality and annexation silent, too, has been lifted in the era of glasnost (openness). This new-found freedom to say exactly what is true (within some limits), rather than accept the less-than-accurate words of the party apparatus, has contributed to the economic, ethnic, national and social tensions in the country.

Reform attempts are further exacerbated by ideological taboos (the two biggest still concern ownership of property, the private sector and agriculture), which constrain the more radical approach that some of Mr Gorbachev's advisers are recommending.

The extent of the economic problems confronting the country became all too obvious in the second half of 1989 during the first session of the (partially) freely-elected Supreme Soviet (parliament). Parliamentarians were told by the prime minister, Nikolai Ryzhkov, that the country had borrowed too much and that the debt-service ratio topped 25 per cent of export earnings. The finance minister, Valentin Pavlov, said sweeping tax reforms were needed, as well as drastic cuts in investment spending.

The first-half 1989 figures showed a continuation of the decline: too many enterprises still operated with a shortage of capital; the increasing use of bank credits to cover budget deficits and the granting of deferments on loans helped to boost inflation and restrict the potential for providing credit. Every fifth rouble in short-term bank credits issued – accounting for R63bn – had no material backing.

Money emission into circulation in the first half of 1989 was one-third more than in the same period in 1988. Unemployment for the first time was admitted and was seen to be rising, along with inflation and wages, while productivity and output fell: every tenth enterprise in the non-food consumer sector failed to fulfil production targets and agricultural supplies were equally poor.

By June 1989, government debt was put at around R330bn – about 10 per cent of GNP – and rising. Mr Pavlov said the debt would be manageable only if the annual budget deficit ran at less than 2 or 3 per cent of GNP. Getting the budget deficit under control would be one of his main priorities in the short term. Others would include the promised tax reform, anti-inflation measures (linking wage increases to price rises), a new investment system forcing state enterprises to finance capital investment from their own savings or loans and a substantial increase in interest rates as well as a greater differentiation in rates, helping to create a capital market (a long term aim of perestroika).

Getting it right before 1991

In many ways 1990 is a last chance at getting perestroika right. The yearly plan, as drafted by the State Planning Committee (Gosplan), aims to halt the worsening of the country's economic position and diffuse social tensions as the next five-year-plan period (1991-96) approaches. Getting demand and supply right on the consumer market is the key to this. Primarily, the government needs to find a way of soaking up the estimated R70-150bn of 'hot money' – roubles the population has, but cannot spend – the equivalent of one- to two-fifths of all turnover of goods.

So far attempts to boost output of consumer goods production has met with little success. A spending spree in the West – using an estimated R10bn of hard currency equivalent – may help curb consumer appetites in the short term for such consumer items as razor blades, soap and soap powder, women's clothing, video recorders and other scarce goods, but could equally raise expectations even higher. The problem is that the country cannot buy itself out of the crisis; it simply does not have the hard currency needed to satisfy consumers and modernise and restructure its industrial base at the same time. With the fall in world oil prices, and Soviet sales of oil and gas hit, too, by shortages at home and problems in the production sphere, revenues from this vital hard-currency earner will be difficult to absorb.

Growing debt

Already by mid-1989 Soviet debt was beginning to worry some, although Western bankers were not yet sounding warning bells. Bank for International Settlements figures showed big increases in Soviet borrowing from commercial banks in both the final quarter of 1988 and in the first three months of 1989, without any let-up expected. Soviet bank debt outstanding increased from \$33.1bn in September 1988 to \$36.8bn in December 1988 and stood at \$38.6bn at the end of March 1989. Soviet deposits with Western banks during the first quarter of 1989 dropped to \$14.85bn from \$15.3bn at the end of 1988. This was worrying, given the prospects for Soviet hard-currency exports of goods other than raw materials (like oil and gas), which looked unlikely to increase significantly in the short-term. The prime minister went on record as saying the USSR needed to borrow at least \$14bn in 1989 alone to service hard currency foreign debt. More radical economists, such as Nikolai Shmelev, a member of parliament, thinks the Soviet Union needs to borrow \$30bn over 1990-91 to avoid ruin.

Feeding the millions

A lot of money could be saved if Soviet agriculture could work properly. Reforms in this area – together with the political hot potato of wholesale and retail price reform – have been tricky. The successful con-

KEY FACTS

Official title: The Union of Soviet Socialist Republics

Head of state: President Mikhail Gorbachev

Head of government: Prime Minister Nikolai Ryzhkov

Ruling party: Communist Party of the Soviet Union; General Secretary: Mikhail Gorbachev

Capital: Moscow

Official Languages: Russian

Currency: Rouble (R) = 100 kopeks

Exchange rate: R0.61 per \$ (Dec 1989)

Area: 22,402,200 sq km

Population: 287.5m (1988)

GDP real growth: 4.4% (NMP, 1988)

GNP real growth: 5.0% (1988)

Labour force: 130.8m (1987)

Trade balance: –2.725bn (with West, 1988, excluding gold)

Foreign debt: \$21.5bn (1988, hard currency)

clusion of the special Central Committee plenum on agriculture, held in March 1989, saw some of the main points of agricultural reform agreed. These included the leasing of land to farmers for periods between five to 50 years (and possibly more); allowing the establishment of family farms on land outside the existing collective farms or in place of such farms; and the dismantling of at least some of the inefficient and money-hungry state farms and collectives. The 'super ministry' Gosagroprom, set up only a couple of years before, was chopped into smaller bits. Some price liberalisations were agreed for potatoes, fruit and vegetables, beginning 1 January 1990.

Ironically, despite being the second largest producer of wheat and the third largest producer of meat in the

world, the Soviet Union finds it impossible to feed itself adequately. It spend on average over $15bn a year on importing food – 15-25 per cent of the entire Soviet import bill. In the first half of 1989, the Soviet Union imported 38m tonnes of grain (an increase of 19 per cent over the same period in 1988); 306,000 tonnes of meat and meat products; 159,000 tonnes of butter (up 700 per cent on 1988); 515,000 tonnes of vegetable oil (up 20 per cent); 9,000 tonnes of fish (up 100 per cent); 36,000 tonnes of granulated coffee (up 47 per cent); and 100,000 tonnes of tea (up 25 per cent). The problems in the agriculture sector are not primarily of stagnate production but of selection of products and pricing.

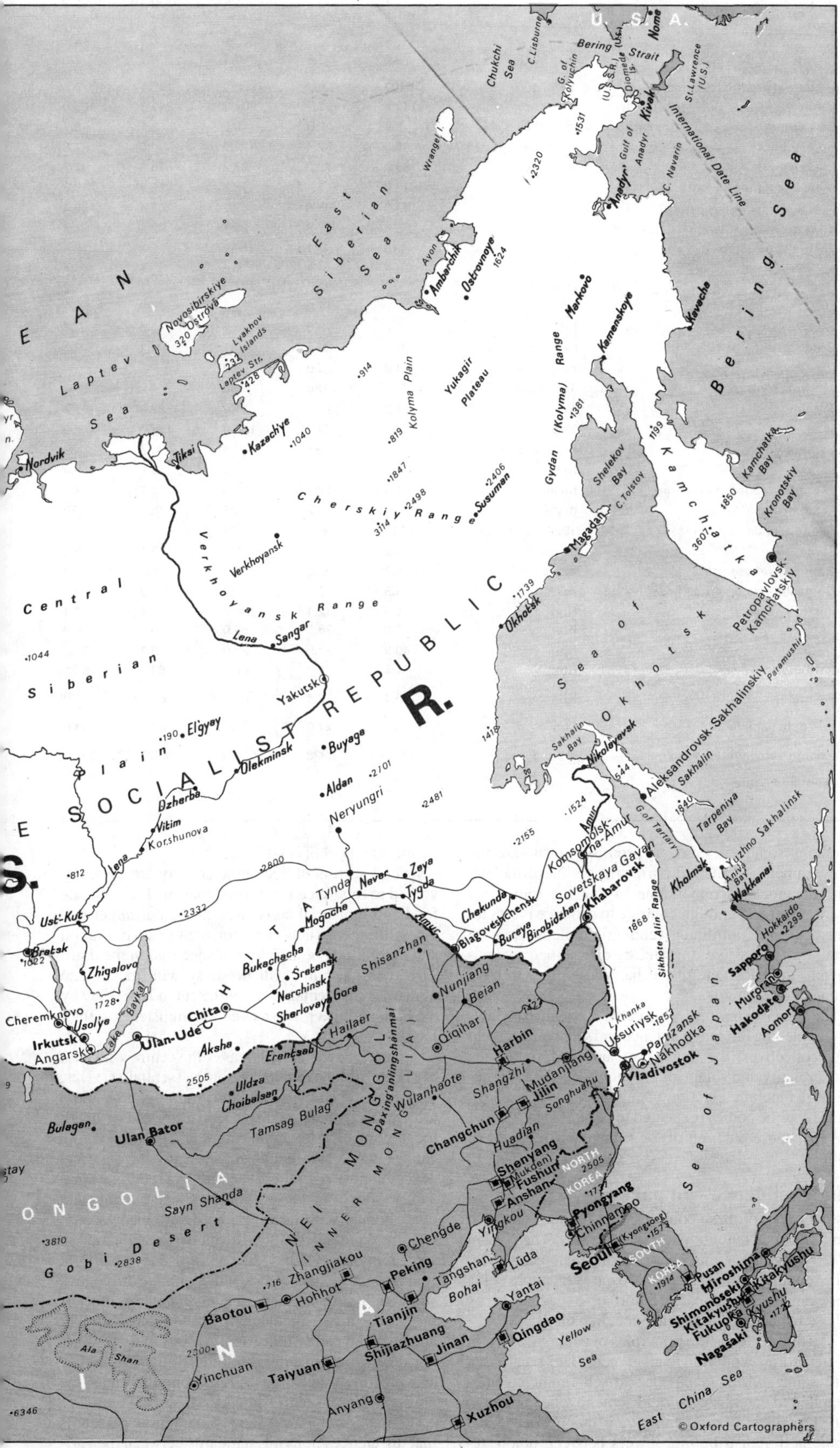

One way, the government reasons, to cut the import bill, boost production incentives and feed the population is to pay farmers partially in hard currency for food produced over and above the state orders. The aim is 'to stimulate the expansion of the production of needed cultures [in other words, get people to work harder]... and reduce imports'.

With the possibility of creating a free market (and pricing) for food unlikely for some time – and particularly unwise while shortages of meat and other major items continue – the authorities had little leeway in finding a way of paying farmers properly for output.

The 'experiment' sets relatively easy targets for collective farms for high quality grain production. The hard currency paid for over-production will, it is hoped, not only increase production and cut imports, but help farmers buy Western machinery, food processing equipment and other items, including consumer goods, for themselves.

The plan will make critics of the present import policy happy. They maintain that it is primarily the pricing system which does not provide enough incentives to farmers to produce wheat of a satisfactory quality. Most wheat produced at present is low-feed-value grain and when used in livestock feed mix creates serious problems in that sector, too. Official statistics show that 70 per cent of the feed mix for cattle is not balanced adequately in protein, vitamins and minerals. This means that feeding costs are higher and livestock productivity lower than they should be.

Not all the problems with farming are to do with production itself. There are many problems connected with the sector, not least of all the storage, transportation and distribution of foodstuffs. Experts estimate that up to 30 per cent or more of food produced is lost between field and table. Therefore, modernisation and expansion of the food processing sector is a priority investment and development area for the government.

Many foreign companies are involved, too, in helping the Soviet food sector become more efficient. Indeed, the West German Government has made substantial D-Mark credit facilities available for purchases of German machinery and equipment for the sector. Several major projects are under negotiation and some have been agreed, for example, the building with Italian help of the largest plant in the USSR to produce long-life milk and baby food.

Perestroika at home

Agricultural problems are only part of the overall economic malaise affecting the Soviet Union. One of the main points of perestroika is the streamlining of the economy and this means modernising industry. In a country where heavy industry has played – and is still playing – a major role, this is not easy. The lack of hard currency available for investment – regardless of any Western restrictions on the import of technology – to modernise obsolete industries is only one of the problems.

In general Soviet enterprises, although asked to bear a large part of the burden of restructuring, are ill-equipped to cope with the new tasks placed on them by the dictates of self-financing and self-management. Under the new rules for industry many things are to happen – in theory. The reality, however, is far from the hope.

Soviet enterprises are expected to make enough money from sales to fund research and development, new projects and products, pay taxes, buy better and more equipment (usually from the West), as well as develop exports particularly in the hard currency areas. They are expected to do this when almost every part of the enterprise is still under fairly tight central control. Prices are centrally set; the state controls, through monopolies, the supply and distribution of almost all inputs and outputs of enterprises. Only in enterprises where management is willing to take a risk and use the new rules to benefit the factory is perestroika producing the results Mr Gorbachev wants.

In general, however, most enterprises continue to perform badly. The main reasons for this are: low quality and no choice of inputs, together with uncertain and erratic deliveries; poor technology for production, using antiquated machinery and excessive manual labour (some 60m are estimated to be 'over-employed'); little or no co-ordination between production and consumer demand (state orders still account for almost 100 per cent of most enterprise output); a general imbalance in the economy which central planning has failed to balance between supply and demand for raw materials and finished products; and poor management (caused in part by the promotion of 'party' men rather than those with the right skills).

The serious problems a real overhaul of the industrial sector would throw up – large-scale overmanning, retraining, redundancy and the consequences of bankruptcy – are unlikely to happen in the short term, unless there is a dramatic (and radical) shift in the balance of power. The worsening economic crisis may push the leadership into more radical solutions. However, if Soviet planners, politicians and party members are watching the rest of eastern Europe, they will be wary of implementing the kinds of changes needed to shake up Soviet industry.

Ethnic unrest

While Moscow and the party continued in the late 1980s to struggle with these and other questions, several parts of the country were growing restive. In Armenia, there were demands for an autonomous region, Nagorno-Karabakh, to be transferred from Azerbaydzhan to Armenia. Massacres of Armenians in Sumgiat and Nagorno-Karabakh led to direct intervention from Moscow, which took over control of the region.

In Georgia there were demonstrations – some ending in violence and killings – for independence. The autonomous region of Abkhazia saw clashes between Georgians and Abkhazis, resulting in deaths. Uzbekistan saw attacks on Meskhetian Turks, forcibly resettled there by Stalin during the Second World War. Clashes there between Uzbeks and the Soviet army led to deaths.

Disorders in Alma Ata, the capital of Kazakhstan, followed changes in the party leadership, imposed by Moscow. Deaths resulted from attacks by Uzbeks on other nationalities. Moldavia saw national demonstrations against Slavicisation of its Latin culture.

Most vocal – and visible – of all disturbances have been seen in the three Baltic republics of Estonia, Latvia and Lithuania where calls for independence have grown following the admission by the Soviet Government that Stalin, illegally and against the will of the Baltic people, annexed the three states in 1939.

Glasnost has allowed these problems to be more or less freely discussed in the Soviet press – and reported world-wide. Perestroika is frustrating attempts at holding the empire together. The way to diffuse the situation was first seen to be the almost free elections to the Supreme Soviet, where the first session caused the average Soviet citizen to sit up and take notice. Not only were the discussions frank – even the sacred KGB came in for some tough questioning from the deputies – but the parliament was given the chance to help solve some of the economic and ethnic strife.

An Appeal to the Soviet People passed by the Supreme Soviet at the end of its first session in 1989, contains some revolutionary promises that it – and the government – will find hard to keep. Nevertheless, the document is remarkable in that the semi-tame parliament had enough courage to produce the statement. Fear of imminent economic collapse and the country-wide strikes initiated by disgruntled coal miners – the backbone of the party – appears to have concentrated minds.

KEY INDICATORS

	Unit	1984	1985	1986	1987	1988
Population	m	273.8	276.3	278.8	281.7	287.5
Net material product (NMP)	Rbn	558.6	568.7	576.0	585.8	611.0
NMP per capita growth	%	1.7	2.2	2.5	0.5	3.3
NMP real growth	%	2.7	3.7	4.1	2.3	4.4
Average monthly wages	% growth	2.5	2.8	2.7	3.7	7.0
Agricultural production	1975=100	107	107	112	–	–
Grain	m tonnes	172.6	191.7	201.1	211.3	195.0
Raw cotton	m tonnes	8.6	8.8	8.2	8.1	8.7
Sugar beet	m tonnes	85.0	82.4	79.3	90.0	87.8
Potatoes	m tonnes	85.0	73.0	87.2	75.9	62.7
Meat (slaughter weight)	m tonnes	17.0	17.1	18.0	18.6	19.3
Milk	m tonnes	97.9	98.6	102.2	103.4	106.4
Eggs	bn	76.5	77.3	80.7	82.1	84.6
Industrial production	1980=100	116	121	127	–	–
Electricity	bn kwh	–	3.5	4	4	2
Coal	m tonnes	–	2.0	3	1	2
Oil	m tonnes	–	–3.0	3	2	–
Gas	bn cu m	–	9.0	7	6	6
Crude steel	m tonnes	154	155	161	162	163
Plastics	m tonnes	4.8	5.0	5.3	5.5	–
Machine tools	1980=100	123	138	150	–	–
Turbines	m kw	21.3	21.6	21.9	22.5	–
Cars	'000	1,300	1,300	1,300	1,300	1.300
Cement	m tonnes	130	131	135	137	139
Paper	m tonnes	5.9	6.0	6.2	6.2	6.3
Fabrics (not man-made fibres)	bn sq m	11.8	12.1	12.3	12.7	13.1
Leather shoes	m pairs	764	787	801	805	820
Canned foods	1980=100	112	117	129	–	–
TV sets	m	9.0	9.4	9.4	9.1	9.6
Domestic refrigerators	m	5.7	5.9	5.9	6.0	6.2
Domestic washing machines	m	4.5	5.1	5.4	5.8	6.1
Total foreign trade	Rbn	139.8	142.1	130.9	128.8	132.1
CMEA	Rbn	72.8	78.1	80.0	79.6	78.8
OECD	Rbn	40.9	37.9	29.0	28.1	31.0
Total exports	Rbn	74.4	72.7	68.3	68.1	67.1
Total imports	Rbn	65.4	69.4	62.6	60.7	65.0
Gross debt liabilities	$bn	26.4	27.0	36.0	36.5	38.0
Exchange rate (annual average)	R per $	0.818	0.838	0.703	0.627	0.611

* estimated figure

The appeal commits the Supreme Soviet to pass laws guaranteeing full independence and self-management for collectives; the transfer of land to farmers; the sovereignty and financial autonomy of union republics; democratic election of representatives to local, and republic bodies (expected to be held in late spring 1990); and protection of the rights and freedoms of all.

Independence – in part

Most tricky of these could prove to be the sovereign and financial autonomy of the republics. The present draft programme on economic self-management for the republics goes some way towards trying to solve grievances, but stops far short of expectations by those bent on full independence, like the Baltic.

The draft extends the principles of economic reform to territorial economic management. This means that the same self-management and self-financing rules applied to enterprises should somehow be applied to the republics, strengthening their independence and rationally redistributing powers between union, republic and local official bodies.

Several sectors of economic activity are to be devolved to republican and local control. They include food and industrial consumer goods, consumer services, the construction sector, nature conservation and other kinds of activity not specified in the draft. This reorganisation will do away with many state ministries that remained after the cut in June 1989 by the prime minister, when over 100 ministers lost their jobs.

The draft, however, still leaves the centre with a lot of control. Management of some essential services, such as local transport, health, training, education, tourism and other 'social' industries would be under local control. However, in practice little would change, since local authorities already have a great deal of control over these sectors. The reorganisation and personnel of these ministries, departments and other bodies will, however, be determined locally, and this could lead to significant de-Russification in some republics.

Despite this, the centre will continue to control strategic industries, such as fuel and energy, metallurgy, chemical, medical, microbiological, timber and pulp and paper industries, geology and prospecting, defence and machine building. Rail, maritime, air and pipeline transport will also continue under the draft to be in the centre's hands.

The main problem, if the programme is passed, is that differences in living standards between the various regions could increase. Even if only the major parts of the draft are implemented, the more prosperous republics – especially the Baltic ones – would stand to benefit more than average, while the poorer republics of Central Asia would do worse than average.

Foreign affairs

While problems multiplied within the Soviet system and criticism of Mr Gorbachev's policies continued, he appeared to find a more receptive audience outside the Soviet Union. Although he had some problems during a trip to China in the first half of 1989, when he finally healed the rift between the two communist superpowers, he must have found it difficult to try to ignore the extraordinary events happening literally outside his meeting rooms. The students occupying Tiananmen Square, he must have felt, could far too easily be imagined in Moscow's own Red Square.

Nevertheless, Mr Gorbachev's low-profile trip managed not only to mend a lot of broken political fences between the two countries, but also helped to boost trade prospects. Cuts in troop deployments along the 4,500-mile Sino-Soviet border will help stem the flow of funds from the already deficit national budget. Fewer soldiers could also mean more trade. Normalised relations between the two are expected to help expand trade.

In particular the Soviet Union wants more consumer goods from China in exchange for Soviet raw materials from Siberia and the Far East region. Chinese labour and investment could also play a part in developing these two important economic areas – if the USSR relaxes immigration controls. At the time of his May 1989 trip, over 45,000 Chinese specialists were already at work in various parts of the Soviet Far East. As a result of the bilateral agreements, another 4,000 to 6,000 people are expected to cross the border for work in the Soviet Union.

Mr Gorbachev sees Asia as a whole as an important region. The Soviet Union itself is considered not only a European nation, but an Asian one, too. In general Asian nations are developing at a much faster rate than his own and he would like to see at least the Asian parts of the Soviet Union benefiting from the same kind of economic progress seen in the region.

Relations with such economic successes as South Korea made progress and the Soviets were expected to normalise relations with South Korea in 1990. The South Koreans have expressed interest in investing in the Soviet Union, probably through joint venture companies, and are likely to be involved in helping the Soviets set up a free economic trade zone in Nakhodka.

The Soviets also hope to attract investment from Japanese companies. Joint development of such areas of the Soviet Union as the three offshore oil and natural gas fields around Sakhalin Island, appear more likely – particularly once the long-running territorial dispute over the Kurile Islands is settled. Mr Gorbachev's desire to open up the Far East still further to foreign companies was shown by an announcement in August 1988, when several hitherto 'closed' areas along the border near Japan were opened. The Japanese, however, remain cautious about involvement in Soviet projects.

Western Europe and EC

While Mr Gorbachev may doubt the sincerity of devotion of many of his people at home, unqualified endorsement for his policies tends to greet his appearances in western Europe. During trips to West Germany and France in 1989, Mr Gorbachev inspired the kind of support that a US president would envy – and did. Indeed, the US president, George Bush, failed to make an equally strong impact in his forays into eastern Europe in the same year.

Nevertheless, the bridge-building between east and west Europe via the Soviet connection is important for Mr Gorbachev's perestroika plans. He needs not only Western government support for his policies, he needs money. European Community backing for reforms in Poland and Hungary has given those countries an economic carrot to continue. The Soviet Union would like closer links with the EC, as well as with the larger developing world.

The exclusion of the Soviet Union from major economic summit meetings is justified from a Western point of view, but Mr Gorbachev would like to take part. He wants his country to be a leading economic power, not just a military threat. The USA and other Western nations belonging to the Group of Seven (Canada, Italy, France, West Germany, Britain and Japan) responded coolly to overtures made in a letter from Mr Gorbachev sent to the Paris summit meeting in July 1989, asking to participate in the Western economic system. Their polite rebuff, however, highlights the isolation of the Soviet Union in the world economic system. It is still not a member of the International Monetary Fund (IMF) or of GATT (the generalised agreement on trade and tariffs).

Western companies are wary, too, of investing where their governments are still cautiously optimistic. With the threat of military tension almost gone in Europe at least (as perceived by the west Europeans), investment opportunities should bloom.

The more sceptical US businessmen, however, proved to be just as susceptible to Mr Gorbachev's charm as west Europeans. Despite dire predictions by the CIA that the Soviet Union is heading rapidly towards economic collapse, US businessmen are flocking to Moscow in an attempt to find a way of cracking the market which provides goods (badly) and services (rarely) to over 200m people. They are finding it a hard slog, not least because of a lack of hard currency, Western business and management techniques and the non-convertible rouble.

While American businessmen may find progress slow, the US Government appears to have had more success in getting the Soviet Union to sign on the dotted line. In fact, the Soviet desire to cut its military spending is helping speed detente – something which the US military establishment is still none too eager to accept at face value. After a rather sluggish start in its policy formation towards the Soviet Union, the Bush Administration came up with its own ideas.

Nevertheless, there are still many within the US Government who worry more about what happens if Mr Gorbachev fails than if he succeeds. 'We don't know whether that's a permanent change [on the hands-off policy towards eastern Europe affairs] that's going to last or whether it's a temporary aberration. And as long as there's a possibility it's an aberration, we have to be careful', said the US secretary of defence, Richard Cheney, in a television interview in the USA. 'I don't think you can take a system, run it for 70 years as a community dictatorship, stamp out all of the spirit of initiative and entrepreneurship in that society and then overnight or even within a short period of time, say, a few years, come back and create an efficient economy. I have yet to see any evidence of any improvement in the economic performance of the Soviet Union.'

Although Mr Cheney has a point, the US Congress appears more favourable in its reaction to the dramatic changes that Mr Gorbachev has made, or is trying to make. There were signs in 1989 that the USA might change its trade law in order to grant the Soviet Union most favoured nation (MFN) status. The author of the present law which links US tariff treatment of Eastern bloc countries to their treatment of emigrants recommended that the USSR be granted a two-year waiver. Former Congressman Charles Vanik, who co-sponsored the Jackson-Vanik amendment to the 1974 US Export Administration Act, thinks Soviet policies on the emigration of Jews, if it continues to be liberalised, should be rewarded by a dropping of the ban on MFN.

Mr Bush has been more cautious. He wants the Soviet Union to 'work with the West in positive, practical – not merely rhetorical – steps towards diplomatic solutions to regional disputes around the world' and to work together with the USA to combat 'global problems', such as international drug smuggling and environmental pollution.

In at least one area, Mr Bush's ideas made progress in 1989. Soviet relations with the Middle East countries improved. Economic ties with Kuwait, the UAE, Saudi Arabia, Iran and Oman improved. Even links with Israel were being developed and there were hopes that diplomatic relations with more Middle East states would be established in 1990.

Eastern Europe and CMEA

Closer to home, however, things are not so happy. Although political relationships with eastern Europe may not be hostile, there are growing trade problems, highlighted in the Soviet-led and dominated trading club, CMEA. The Soviet Union, along with some of the more reform-minded in eastern Europe – in particular Hungary and Poland – want to see major changes in the way trade between the countries is carried out.

The Soviets are understood to be keen on ending the barter arrangements under which Soviet raw materials are traded for east European goods, often of inferior quality to those available on hard currency markets. The reforms in Soviet industry and foreign trade also mean that USSR enterprises have more say in who they trade with, and many are expected to prefer Western partners to Eastern ones, if deliveries and quality do not meet expectations.

Increasing trade

The problems besetting CMEA relationships are a natural outcome of the changes taking place in Soviet foreign trade organisation. Here, at least, perestroika has made some definite advances. For example, the joint venture law, liberalised in 1989 in order to attract more foreign investment, attracted many Western companies. At 1 July 1989, 80 per cent of the 680 joint ventures registered in the Soviet Union involved Western partners.

Plans to create free trading zones in the Soviet Union are proceeding slowly, but could help attract Western companies producing the high-technology goods Soviet industry needs but cannot yet afford.

The ability on a limited scale of Soviet enterprises with foreign trading rights to retain at least a portion of the hard currency earned from exports, and the promise of hard currency auctions (where enterprises without hard currency can buy it), will help further de-monopolise foreign trade and encourage more direct contact with Western companies.

The rouble's non-convertibility

One indicator of success for perestroika will be the rouble. Soviet economists realise that the role of money in the economy needs to be enlarged and that a system of supplying material resources, based on realistic economic principles, including a real exchange rate, is needed. The rouble, if given enough time, may emerge at the end of perestroika as a hero of the new economic system. Converting the rouble on the world money exchanges, however, still appears to be a long way off. Some Soviet economists, when pressed to guess a date, will hazard the end of the 1990s, but such a move depends on getting things right at home first – and that could take quite a long time.

Meanwhile, there are several schemes being floated to introduce partial convertibility. At the same time, the government has committed itself to a devaluation in 1990 of the rouble by 50 per cent for commercial and trade transactions only. This could have an impact on future joint ventures which rely on the Soviet market for the majority of sales.

Out of order, comes chaos

The comfortable monopolies in foreign trade, under which the West was able to buy and sell with little or no risk, have all but ended. Out of the established state order has come quite a lot of chaos for both sides as Soviet enterprises struggle to cope with new rules and accounting procedures and Western businessmen at-

COUNTRY PROFILE

Political system: Socialist state, comprising 15 national republics dominated by the Russian republic. The Communist Party holds control through a Politburo consisting of 12 full members and eight candidate (non-voting) members. The Politburo chooses the party leader and is itself appointed by the Central Committee, answerable to the party membership of 19m. The Government is the Council of Ministers, headed by the Prime Minister. In March 1989 a new Congress of National Deputies was elected and from its members a smaller Supreme Soviet (parliament) set up which sits in permanent session. A similar system of organisation is planned in all the union and autonomous republics; election of these deputies is scheduled for end-1989.

Main cities/towns: Moscow (population 8.8m), Leningrad, Kiev, Tashkent, Baku, Khar'kov, Gorki, Novosibirsk, Minsk, Kuybyshev, Sverdlovsk, Tbilisi, Odessa, Chelyabinsk, Donets, Yerevan, Omsk.

Media
Press: Approximately 8,000 dailies in some 55 languages. Combined circulation of 20m. Most influential dailies *Pravda* (9,664,000) and *Izvestia* (10,500,000). Principal news agency, Tass. Also more than 5,000 journals and periodicals in some 25 languages.
Broadcasting: Eight main national radio stations and extensive local network. Foreign service broadcasts in 70 languages and totals over 1,000 hours daily. Six central TV stations broadcasting some 40 hours each day. Also 130 major TV centres and over 2,000 small stations. Around 100 main towns have two channels broadcasting in Russian and local language. Six channels in Moscow and three in Leningrad. Approximately 87.4m TVs in use and 245m radios (January 1988).
Advertising: Commercial advertising is expanding quickly. Limited television and radio advertising is available to Western companies; billboards and illuminated sign-space can be bought in most major cities. Most newspapers are beginning to accept Western advertising. Baltic republics are the most advanced in this area. Many co-operatives have been set up in advertising services and are expanding at a good rate, but the main foreign trade organisation dealing with advertising remains Vneshtorgreklama.

Domestic economy: The centrally planned economy is in a state of reform. Due to changes, industrial output is difficult to assess as indicators have changed and statistics are unreliable at present. Main emphasis is on self-financing and self-management, with end-goal being as little as possible central and state interference in production plans and sales. Wholesale and retail price reforms delayed, so little room for manoeuvre at enterprise level as monopoly suppliers and buyers still prevail. Continuing shortages with acute supply problems in some products, including foodstuffs, clothing and light household goods. Emphasis being placed on increased production of consumer goods for population. State intends to invest heavily in expansion and modernisation of food processing and packaging, light industry (mainly consumer goods) and modernisation including cutting energy and raw materials. Environmental issues expected to become more important.

Employment: Industry is the largest employer in socialised sector, followed by agriculture and construction. Officially unemployment does not exist, but a leaked confidential Soviet document indicates that in 1984 there were an estimated 500,000 long-term unemployed. This will probably increase due to reforms tackling overmanning. Labour productivity increasing at average annual rate of 2.3 per cent pa (3.8 per cent in 1986). Emphasis on improving quality and skills of the labourforce.

External trade: USSR plays only a small part in world trade: share of USSR in world export trade is just over three per cent compared to 11.3 per cent for the USA. Majority of trade (65 per cent) is with socialist countries. Under reforms, increasing and developing commercial ties with capitalist countries is a priority. Biggest item in Soviet imports from the West is grain, in second place comes machinery and transport equipment. Imports of some consumer goods have increased. All enterprises and organisations now have the right of access to foreign markets, but the right is subject to registration; imports must be financed by exports. Export and import licensing is required for all enterprises directly involved in foreign trade. Licences are issued by relevant ministries and a fee is payable. Documents regulating licensing of foreign trade transactions are published in Ekonomicheskaya Gazeta (newspaper), Vneshnyaya Torgovlya (magazine) and the MVES (foreign trade ministry) Bulletin of Foreign Commercial Information. Special office to deal with enquiries is at MVES (tel: 2441533 or 2441335). Introduction of new customs tariff, to be written up in 1989, with preparation for transitional phase beginning on 1 Jan 1991. New foreign exchange rate for foreign trade to be introduced. New rate is devaluation which is to be carried out in two phases beginning 1 Jan 1990 when for trade and commercial transactions only the rouble will be devalued by 50 per cent. Further adjustments expected in 1990 with new exchange rate to come into effect that year. Special economic zones are to be set up in the Far East. Despite reforms, ministries manage the export of fuels and raw materials and import of agricultural goods, raw materials, a good proportion of intermediate goods and equipment for new projects, accounting for 81 per cent of exports to OECD countries and around 50 per cent of Soviet imports.

Agriculture: World's largest producer of several crops, including cereals, potatoes, sugar beet, oilseeds and cotton, although poor soil and climate mean only 10 per cent of land area is suitable for cultivation. Large-scale mechanical state farms (sovkhozy) account for 53 per cent of arable land, smaller collective farms (kolkhozy) account for 44 per cent. Production hampered by climatic problems, rural migration, inferior equipment and poor storage and distribution system. Private plots account for less than 4 per cent of total arable acreage. Area sown to various crops reduced slightly in recent years to leave more land fallow within rotation of crops. Land area used is 2,227,200,000 hectares, of which majority is forest. Total area sown is 211.5m ha of which non-food crops (115.2m ha) account for majority of land. Decree of March 1989 abolished Gosagroprom (State Agro-Industrial Commission) and established State Food and Procurements Commission. Bulk of responsibility of production to be transferred to republicans, local or farm level.

Industry: 29.1 per cent of labour force was employed in industry in 1987. Output indices are gross or global which means intermediate goods are counted in rather than netted out as in the West. Fuel and raw materials use is wasteful. No inflationary factors are shown so real output conceals hidden inflation. Official figures are also distorted by price factors. Data on capital stock tends to give clearer picture. Heavy industry is still politically powerful. Stated intentions for new investment priorities have had little effect on capital stock. Major production bottlenecks include steel, construction outputs (like cement) and consumer and light industry products (ie, TV sets, robots, computers). Emphasis in 1990s to be on light industry, modernisation and computerisation. More emphasis also placed on individual enterprise and factory production decisions, but no real freedom on supply, sales or pricing yet introduced.

Mining: World's largest producer of iron ore, asbestos, manganese ore, nickel, chromite, platinum group metals and potassium salts. Second largest producer of gold, lead and phosphate ores. Vast reserves, but extraction has been held back due to rising production costs, labour shortages and a shortage of technology. Estimated to have 30 per cent of world iron ore reserve and 20 per cent of many other minerals. Most production for own use. Major foreign exchange earners include gold and diamonds. Significant quantities of iron ore, chromium, nickel, asbestos and fertiliser materials also exported. Also large deposits of antimony, beryllium, cadmium, mercury, molybdenum, tin and vanadium plus workable deposits of all rare earth metals. Large-scale investment in the sector is set to improve extraction and processing techniques, while reducing wastage and controlling production costs.

Energy: Largest gas reserves (about 40 per cent of world total) and probably the largest oil reserves in the world. Oil production accounts for over 20 per cent of total world output. Oil exports increasing (currently account for up to 60 per cent of foreign exchange earnings), although gas is planned to take over as major earner with exports increasing at 15 per cent per annum. USSR currently ranks as world's largest gas producer - 686 bn cubic metres in 1986. Coal reserves also vast, but output consistently falls short of production targets due to the exhaustion of seams in traditional mining areas and the lack of investment in new areas. Second largest generator of electricity in the world with an output of about 1705bn kWh, of which about 75 per cent is produced thermally. Nuclear power is being reassessed and hydro-electric generation is being increased. Oil-fired stations are gradually being replaced by gas or coal. Wasteful energy use is a major defect of the economy.

Legal system: Civil law system as modified by communist legal theory. Fundamental review, codification and systematisation of laws undertaken as part of legal reform. Courts should become more important and be used to bolster democracy. Also play role in contract law as self-management and financial autonomy develop (i.e. arbitration, employment, company rights).

Membership of international organisations: CMEA, ESCAP, Geneva Disarmament Conference, IAEA, IBEC, ICAC, ICAO, ICCAT, ICCO, ICES, ILO, IMO, International Lead and Zinc Study Group, INRO, IPU, ISO, ITC, ITU, IWC – International Whaling Commission, IWC – International Wheat Council, UN, UNESCO, UPU, Warsaw Pact, WFTU, WHO, WIPO, WMO, WTO. Applied for observer status at Gatt; expected to approach IMF/World Bank for membership.

tempt to find the right people, many of whom will increasingly be found outside Moscow. Creative thinking will be needed by Western businessmen eager to start – or continue – doing business in the Soviet Union. Soviet enterprises will be coming under increased pressure to produce more goods for the home market, but at the same time told to export. They will need partners interested in long-term projects, like joint ventures, co-productions and other co-operations, as well as Western partners that can help bring new technology and machinery, as well as know-how, including managerial skills and marketing ability, to the partnership.

While this happens in trading relations, the Soviet people are also struggling with new ideas in politics. The comfortable one-party state is not quite as firm as it was when Mr Gorbachev took over in 1985. Although he now has netted for himself both key jobs – general secretary of the communist party and president of the country – there are more changes expected over the next few years. Mr Gorbachev wants to reform the communist party, not necessarily destroy it. He assumes it will retain its leading role, but that it will manage to discharge its duties in a more democratic way. That may be increasingly challenged as the east Europeans move towards multi-party democracy. He will also have to listen to increased criticism of the unfairness of the one-party state from his own republics, many of which are unhappy about membership without a full say in the way the government is run and those content to stay as part of the union, but anxious to re-draw borders set under Stalin.

All these pressures will combine to force changes on the party. Not least among them will continue to be the pressure exerted by perestroika itself, which encourages decentralisation and challenges the party's right to decide at every level on the running of the economy, including the appointments of enterprise managements as well as the control of economic levers.

Along with the pressure for change will be some strong tendencies to hold back reform. Not everyone is happy with perestroika and glasnost. Many older party (and non-party) members have been traumatised by the revelations that heroes of the revolution, like Stalin, have feet of clay and are now denounced as criminals or enemies of the country. Likewise the rehabilitation of those whose names were never mentioned in textbooks (and erased from official photographs) must cause confusion among both the old and young. It is also costing the state a small fortune to re-write and reprint many textbooks.

The public's ability to absorb these tremendous changes has so far not caused too many problems. However, it may shake their confidence in the state's ability to know what it is doing.

Perestroika and glasnost both challenge the party's monopoly not only on power, but on truth and the interpretation of events both historical and presently taking place. These potential conflicts between governed and the governing could explode, particularly if the economic decline is not halted.

Filling a vacuum

At present, Mr Gorbachev is desperately trying to fill a vacuum left by the partial withdrawal of central planning and the end of total communist party domination of all spheres of life. The future is uncertain and highly volatile. Already in 1989 there were hints that the communist party's future was anything but certain. The semi-free elections showed that given the choice, people were not at all happy about many of the party members wielding authority; they showed their displeasure by crossing out names on ballot papers, often where there was no choice of candidates. This led to a major shake-up in the party hierarchy and may have also strengthened Mr Gorbachev's hand for advocating even more democracy within the party. His policy has up to now been to change the party and streamline the economy, not throw either away in favour of completely different systems. His arguments in favour of this policy should now fall on more sympathetic ears. After the major changes in personnel and cuts in the Central Committee, Politburo and major party leaders in crucial cities, such as Leningrad and Kiev, Mr Gorbachev's power base appears more firm.

At the same time, his reform of the party and government has presented him and the Politburo with a more-or-less uncontrollable Supreme Soviet, which cannot now be relied upon merely to rubber-stamp. Its first sessions in 1989 showed it had teeth and could bite. It will be more difficult for the parliamentarians, however, to get to grips with – and take part responsibility for – dragging the economy out of the current mess.

Outlook

Over the 1990s, and particularly in the final year of the current plan period, 1990, Mr Gorbachev will need to find answers to some outstanding questions. Not least of these is how to tackle price reform without setting off uncontrolled inflation and further aggravating an economy of shortages. He will also have to find – with the help of the Supreme Soviet – a way of holding the crumbling empire together.

The new constitutional draft promised for 1990 needs to give the more nationalistic republics (like the Baltics) some reasonable incentives for staying inside the Soviet Union, without resorting to forced membership and stifling economic constraints. At the same time, the country needs to get to grips with its own North-South divide where the prosperous northern areas continue to grow while the poor, mainly Asian republics are helped to bring up their standard of living to that of the rest of the country. This will mean a new approach to taxation, government responsibility and public spending as well as jettisoning some ideological hang-ups that still remain, in particular those on ownership and agriculture.

Warnings over the steady deterioration of the economy are not to be ignored, particularly when they come from leading members of the government such as Leonid Abalkin, chairman of the State Commission for Economic Reform. Inefficient industries, he said, need to shed 12-15m workers by the end of the century. Unemployment, however, is already a serious problem in Central Asia and among young people aged 18-29. Food queues and rationing cannot end until the budget deficit is cut and it is already considered to be growing out of control. At the same time more money is needed, preferably hard currency, to stimulate industry and fulfil consumer demand.

Mr Gorbachev and his supporters may be able to use these negative points as a means to push through more radical reforms and break the stranglehold of party bureaucracy over many measures. The Supreme Soviet will have to stop being merely a 'talking shop', too. At the same time there remains the real risk that opponents to the reform process, after five years of visible failures, may try to call a halt to the process.

BUSINESS GUIDE

Time: USSR divided into 12 time zones. Moscow (also Kiev, Leningrad and Riga): GMT + 3 (GMT + 4 from 27 Mar to 24 Sep).

Climate: Sub-arctic to sub-tropical. Moscow – warm summers and long, cold winters.

Entry requirements: Entry and exit visas are required by all visitors except CMEA nationals, and must be obtained in advance.
Health: Visitors from Asia, South America and Africa may require an International Certificate of Vaccination against cholera. Anti-malaria precautions advisable. Water precautions recommended.
Currency: Import/export of roubles is prohibited. No limit to amount of foreign currency taken in, but must be registered with customs and not be exceeded by the amount taken out.
Customs: Small amount of personal goods duty-free.

Air access: Direct flights from several European airports. Aeroflot operate flights from over 90 countries.
National airline: Aeroflot.
International airports: Moscow-Sherenetyevo International (SVO) 29 km from city centre, Moscow-Vnukovo (VKO) 29 km from city, Moscow-Domodedovo (DME), 40 km from city; Leningrad (LED) 17 km, Kiev (IEV) 38 km, Irkutsk (IRK) 7 km.

Surface access: Border crossings from: Finland, Poland, Turkey, Iran, China, Czechoslovakia, Hungary and Romania.
Main ports: Arkhangelsk, Baku, Batumi, Berdyansk, Ilichevsk, Ilia, Izmail, Klaipeda, Krasnovodsk, Leningrad, Mezen, Murmansk, Novorossiysk, Odessa, Onega, Poti, Reni, Riga, Sevastopol, Sochi, Sukhumi, Tallin, Tuapse, Ventspils, Vladivostok, Vyborg, Yalta, Zhdanov.

Hotels: Accommodation difficult to obtain in Moscow at short notice. Bookings should be made through Intourist prior to departure. First class and tourist class available, with all prices fixed by Intourist. Main hotels have foreign currency restaurants and bars. Tipping increasingly common, typically 10 per cent.

Restaurants: Difficult to find top class restaurants. Selection of food is limited to one or two main dishes. Foreign currency readily accepted. Co-operatives now opening throughout country. Bookings recommended; prices vary but food and service usually better than state sector.

Foreign exchange information
Trade policy: Foreign trade is a state monopoly. Most trade with West conducted on barter basis or in hard currency. Customs duties levied on imports under a double column tariff. Payment is usually made in cash against documents or documents against letters of credit.

Credit cards: Diners' Club, American Express, Bank of America, Carte Blanche, Eurocard and AIT credit cards are accepted at main hotels.

Car hire: Available in major towns. International driving licence required with Russian translation of details. Self-drive costs around R10 per day plus R0.10 per km, excluding petrol and insurance. Chauffeur-driven cost around R8-10 per hr. Notification of route to be taken should be given if travelling outside main cities.

City transport
Taxis: May be identified by checkerboard on side of car. Green light at top right hand corner of windscreen indicates availability. Can be hired at taxi ranks or by telephoning 2250000 or 2270040. Average charge around R0.20 plus R0.20 per km. Also possible to hail private (unofficial) cars by raising arm in street. Prices negotiable and usually higher than state-run cars, but availability plentiful although service patchy.
Metro: Flat fare service in Moscow connecting some 120 stations. Excellent service.

National transport
Air: Extensive internal service operated by Aeroflot.
Road: Major highways connecting Moscow with Kiev, Leningrad, Minsk, Riga and Warsaw. Secondary roads often untarred. In total over 1.5m km of roads.
Rail: Network extends over 140,000 km, almost all of which is electrified. Cheap and efficient service to all major towns. Sleepers should be booked well in advance.
Water: Approximately 145,000 km of inland waterways, of which largest River Volga. Number of inland ports and canals.

Public holidays: 1 Jan, 8 Mar (Woman's Day), 1-2 May, 9 May (Victory Day), 7 Oct (Constitution Day), 7-8 Nov (October Revolution). If holiday falls on Tue it is customary to declare Mon as a holiday and the preceding Sat as a normal working day.

Working hours
Business: (Mon-Fri) 0900 – 1800 (appointments best between 0900 – 1000).
Shops: (Mon) 0800 – 1900, (Tue-Sat) 0800 – 2100.

Social customs: Customary to take a small gift on business/social visit.

Business language and interpreting/translation: Available from Intourist and Sovincentr.

Telecommunications
Telephone: Local calls may be made free of charge from hotels. Cost of call from public kiosk R0.20. Telephone directories rarely available. Numbers of official organisations and enterprises can be obtained at hotels or street information kiosks (spravochnoe). Direct dialling available, but service is slow. Best conducted through hotel or central post office. Operator numbers: international calls (English-speaking operator) 8196, general enquiries (Moscow area) 09, police 02. International dialling code: 7 followed by 095 for Moscow. International, direct dial phone boxes now in operation in Moscow, use Western credit cards or phone cards purchased for hard currency.
Telex: Efficient and relatively cheap service. Codes and ciphers must not be used. Country code: 64 SU. Also available through joint venture with Rank Xerox are photocopying and facsimile services – only in Moscow at present.
Telegram: Available at most major hotels and post offices.

Postal service: Air mail facilities available. Surface mail often slow.

Banking: State bank for Foreign Economic Relations, responsible for all banking transactions connected with foreign currency/trade, including export credit guarantees and the issuing of travellers' cheques. Foreign banks represented in Moscow include: Amro Bank, Banca Commerciale Italiana, Banca Nationale del Lavoro, Banco Central Madrid, Banco di Napoli, Banco di Roma, Banco Exterior de Espana, Banco Hispano Americano, Bank-Melli-Iran, Bank of America, Bank of Scotland, Banque Paribas, Banque Nationale de Paris, Barclays Bank, Bergen Bank, Chase Manhattan Bank, Cic-Union Europeene, International et Cie, Commerzbank Aktiengesellschaft, Creditanstalt, Credit Lyonnais, Credito Italiano, Deutsche Bank, Dresdner Bank, Generale Bank (Brussels), Kansallis-Osake-Pankki, Lloyds Bank, Midland Bank, Monte dei Paschi di Siena, Morgan Grenfell and Co, Moscow Narodny Bank, National Westminster Bank, Okobank, Ost-West Handels Bank, Postipankki, Pk Banken, Privatbanken, Skandinaviska Enskilda Banken, Skopbank, Société Générale, Societe Generale de Banque, State Bank of India, Svenska Handelsbanken, Swedbank, Union Bank of Finland, Union Bank of Switzerland, Westdeutsche Landesbank Girozentrale.

Trade fairs: Numerous international trade fairs and exhibitions held by Foreign Trade Corporation, Expocentre. Other trade fairs organised from time to time by non-Soviet exhibition specialists.

Electricity supply: 220 V AC.

Representation in capital: Afghanistan, Algeria, Angola, Argentina, Australia, Austria, Bangladesh, Belgium, Benin, Bolivia, Brazil, Bulgaria, Burkina, Burma, Burundi, Cameroon, Canada, Cape Verde, China PR, Colombia, Congo, Costa Rica, Côte d'Ivorie, Cuba, Cyprus, Czechoslovakia, Denmark, Ecuador, Egypt, Equatorial Guinea, Ethiopia, Finland, France, Gabon, German Democratic Republic, Federal Republic of Germany, Ghana, Greece, Guinea, Guinea-Bissau, Guyana, Hungary, Iceland, India, Indonesia, Iran, Iraq, Ireland, Italy, Jamaica, Japan, Jordan, Kampuchea, Kenya, Democratic People's Republic of Korea, Kuwait, Laos, Lebanon, Libya, Luxembourg, Madagascar, Malaysia, Mali, Malta, Mauritania, Mexico, Mongolia, Morocco, Mozambique, Nepal, Netherlands, New Zealand, Nicaragua, Niger, Nigeria, Norway, Oman, Pakistan, Palestine, Papua New Guinea, Peru, Philippines, Poland, Portugal, Romania, Rwanda, Senegal, Sierra Leone, Singapore, Somalia, Spain, Sri Lanka, Sudan, Sweden, Switzerland, Syria, Tanzania, Thailand, Togo, Tunisia, Turkey, Uganda, United Arab Emirates, United Kingdom, USA, Uruguay, Venezuela, Vietnam, Yemen Arab Republic, People's Democratic Republic of Yemen, Yugoslavia, Zaïre, Zambia, Zimbabwe.

Hotels: Leningrad
Astoria, 39 Herzen Street (tel: 191100).

Europeiskaya, 1/7 Brodsky Street (tel: 2103295/86).

Moskva, 2 Alexander Nevsky Square (tel: 2742051).

Pulkovskaya, Peochad Pobedy 1 (tel: 2645122).

Hotels: Moscow
Belgrad, 5 Smolenskaya Street (tel: 2487848).

Berlin, 3 Zhdanov Street (tel: 2256901).

Bukharest, 1 Ylitsa Balchag (32km from Vnukovo airport) (tel: 2331005).

Central Hotels Information Centre, Kompositorskaya Ulitsa 25/5, Moscow.

International, Krasnopresnenskaya Nab 12 (tel: 2537729).

Intourist, 3/5 Gorky Street (tel: 2034007).

Kosmos, Mir Avenue 150 (tel: 2170789).

Leningradskaya, 21/40 Kalanchevskaya Ulitsa (36km from Vnukovo airport) (tel: 2255730).

Metropol, 1 Marksa Prospekt (tel: 2256677).

Minsk, 22 Ulitsa Gorkovo (33 km from Vnukovo airport) (tel: 2991448).

Moskva, 7 Prospekt Marksa (tel: 2922121).

Mozhaisky, 165 Mozhaisky Hwy (tel: 4473434).

National, 1/14 Marksa Prospekt (tel: 2036539).

Ostankino, 29 Botanicheskaya Ulitsa (opposite Botanical Gardens) (tel: 2195411).

Peking, 1/7 Bolshaya Sadovaya Ulitsa (tel: 2538347).

Rossiya, 6 Razin Street (opposite the Kremlin) (tel: 2981442).

Soyuz, 12 Levoberezhnaya Ulitsa (tel: 4583240).

Ukraina, 2/1 Kutuzovsky Prospekt (tel: 2433020/21).

Business advice and information
USSR Chamber of Commerce and Industry (Promishlennosti SSSR), Ulitsa Kuibysheva 6, Moscow (tel: 2210811; tx: 411126).

USSR Chamber of Commerce Consultancy Centre, same as above (tel: 9253529; tx: 411431).

Expocentr, 1a Sokolnicheskiy Val, Moscow 107113 (tel: 2687083; tx: 411185) (Responsibility for organising, on a commercial basis, international and foreign exhibitions and symposia in USSR).

Sovicentr, Krasnopresnenskaya Nab 12, Moscow 123610 (tel: 2566303; tx: 411486).

Banking
Agroprombank of the USSR, Neglinnaga 12, Moscow (Tel: 9215119).

Promstroibank (All-Union capital investment bank), Moscow, Tverskoybul 13.

Sherbank of the USSR, Neglinnaya 12, Moscow. (tel: 9214182).

State Bank of the USSR (Gosbank) (central bank), Neglinnaya 12, Moscow (tel: 9231870; tx: 411283).

USSR Bank for Foreign Economic Relations (Vneskeronombank) Ulitsa Plyushchikha 37, Moscow 119121 (tel: 2466780; tx: 411174).

Zhilsotsbank of the USSR, Daev Lane 13, Moscow. (tel: 2071567).

Principal newspapers
Izvestiya, Moscow, Pushkinskaya Pl 5 (tel: 2099100).

Pravda, Moscow, ul Pravdy 24 (tel: 2575311).

Other useful addresses
Aeroflot, Moscow (flight enquiries) (tel: 1568002 [international]; 1550922 [domestic]).

Agentstvo Pechati Novosti (news agency), Moscow, Zubovsky bul 4 (tel: 2012424; tx: 411323).

All-Union Central Council of Trade Unions, 117119 Moscow, Leninsky Prospekt 42.

Autoexport (vehicles, spares and garage equipment), Volkhonka Ulitsa 14, Moscow 119902 (tel: 2028535; tx: 411125).

Chief State Sanitary Inspector, Vodkovskiy Pereulok 18-20, Moscow.

Customs House, Komsomolskaya Ploshchad la, Moscow (tel: 2084584).

Exportkhleb (agriculture and most foreign trade corporations), 32/34 Smolenskaya Sennaya Ploshchad, Moscow 121200 (tel: 2444701, 2441247; tx: 411145/7).

Exportles (timber and timber products), 32/34 Smolenskaya Sennaya Ploshchad, Moscow 121200 (tel: 2447432, 2447440; tx: 411229).

Ingosstrakh (insurance organisation), Ulitsa Pyatnitskaya 12, Moscow (tel: 2332070; tx: 411144).

Intourist, Prospekt Marksa 16, 103009 Moscow (tel: 2036962; tx: 411211).

Litsenzintorg (licensing), Minskaya Ulitsa 11, Moscow 121108 (tel: 1451111; tx: 411246).

Metallurgimport (mining), Ulitsa Arkhitektora Vlasova 33, Moscow 117393 (tel: 1280775, 1283720; tx: 411388/9).

Ministry of the Electronic Industry, Moscow, Kitaisky proyezd 7 (tel: 2206600).

Ministry of Finance, Moscow, ul Kuybysheva 9 (tel: 2989101).

Ministry of Foreign Economic Relations, Moscow, Smolenskaya-Sennaya pl 32/34 (tel: 2441947; tx: 411441, 411451).

Ministry of the Gas Industry, Moscow, ul Stroitelei 8, kor 1 (tel: 1331300).

Ministry of Light Industry, Moscow, prospekt Kalinina 29 (tel: 2914710).

Promsyrioimport (metal products), Ulitsa Chaikovskovo 13, Moscow 121814 (tel: 2030577, 2030646, 2030595; tx: 411151/2, 411112).

US Virgin Islands

A special correspondent

As 1989 progressed in the US Virgin Islands, two subjects dominated the territory: political relations with the United States and how rapid overdevelopment is to be controlled.

Relations with the USA

The US Virgin Islands was purchased by the United States from Denmark in 1916 with formal transfer taking place in 1917. It has been an unincorporated territory of the United States ever since. Although enjoying a measure of self-government, the territory has no constitution and is governed by an Organic Act of Congress enacted in 1935 and revised in 1954. Four attempts to adopt a constitution have failed, the last two rejected by the territory's voters and the two earlier ones not acted upon by Washington.

Out of the impasse in territorial-federal relations evolved the Legislature's Select Committee on Status and Federal Relations chaired by St Thomas Senator Lorraine Berry. A referendum in 1982 established that Virgin Island voters considered the definition of federal relations should come before any further effort at adopting a constitution. Senator Berry's committee in turn created the Virgin Islands Status Commission, of which she became co-chair with University of the Virgin Islands Professor Marilyn Krigger. The Status Commission laid the ground work for a referendum in November 1989 which asked voters to choose between three broad categories of relationship with the United States: closer union, continued status quo or greater autonomy/independence. A runoff election was to be held if the winning category did not obtain 50 per cent plus one vote. The outcome, unlike with Puerto Rico's plebiscite, would not be binding on the US Congress.

The US Virgin Islands is a fragmented community of minorities. Residents from Puerto Rico, elsewhere in the Caribbean and the United States outnumber the natives in a population of only 110,000. The largest group is from the former British Eastern Caribbean islands (downislanders), about 15 per cent are Puerto Ricans (mainly on St Croix) and about 12 per cent are continental Americans (mostly white). In addition, Crucians (the natives of St Croix) have different priorities and attitudes from those of St Thomians and St Johnians. Rapid influx during the boom decade of the 1960s led to the population more than tripling, but there are indications of a contraction in immigration from the Caribbean; caused by tighter US immigration policy but also because of the high cost of living which has made the US Virgin Islands less attractive than before. Migration of US mainlanders, on the other hand, has accelerated. A recent CBS *Sixty Minutes* programme highlighted several hundred Texans who have moved to St Thomas and St John following the severe recession in that state. The programme showed their impact on the small insular societies. All of these factors contribute to making it extremely difficult to reach consensus, because politicians cater to each minority group and its interests. It is not easy to predict the outcome of the referendum – natives are upset with the territory's relations with the United States while continental residents and some former downislanders tend to support the status quo or closer relations with the United States. A growing minority supports independence and there is a longing by some for integration into an as-yet-to-be-defined Caribbean union, since US Virgin Islands residents, particularly youth, see nearby former British colonies (like St Kitts and Nevis and Antigua and Barbuda) having acquired independence with smaller populations and resources. Among independence supporters is former gubernatorial candidate Adelbert Bryan, who received 36 per cent of the vote in 1986 and forced Governor Alexander A Farrelly into a runoff. Governor Farrelly himself told *The Economist* that Washington would be surprised at the pro-independence vote in the referendum, but others dismiss his prediction as exaggerated and/or an effort to send a message to the US Government, from which he encounters growing difficulty extracting federal funding.

What are the principal irritants in US Virgin Island relations with the United States? First and foremost is immigration. With an area of only 132 square kilometres, population density is among the highest in the western hemisphere, particularly on St Thomas. The territory has traditionally been a way station for entry into the United States by West Indians, and a growing number of Canadians, Europeans and Latin Americans – attracted by the tourism industry – find jobs and stay. Other areas of concern are control over customs collections and more equitable sharing in federal programmes. But opponents of greater local control point to a record of poor management, a bloated and inefficient bureaucracy and corruption in a number of agencies. They question how a government with such a history of chronic mismanagement could ever assume greater responsibilities.

The economy

The economy continued to experience growth, with tourism dominating it more than ever before. While tourist arrivals in 1988 reached all time highs (1.9m – a 7 per cent increase over 1987), warning signals began to appear in 1989. While some attribute contractions in arrivals in early 1989 to the impact of the Eastern Airlines strike, others point to a declining attractiveness of the US Virgin Islands as a destination. The character of the territory's tourism is changing radically, with cruise ship passengers now surpassing overnight visitors by two to one. But the one-third overnight visitors spent two-thirds of total annual tourist expenditure ($676m). St Thomas and St John are experiencing severe congestion and increases in crime and anti-social behaviour. The *Miami Herald* compared a cruise ship visit to St Thomas with getting stuck on Interstate 95 in Florida at rush hour.

Hess Oil's St Croix refinery no longer provides the massive employment it once did because of operating on reduced capacity and greater automation. Efforts by the Virgin Islands Government to acquire the former Martin Marietta alumina processing plant became bogged down because of the involvement of the controversial Marc Rich, wanted by the United States on tax fraud charges. The property was eventually purchased by Clarendon Ltd. The territory's high wage scale and competition from other Caribbean states make it unlikely that efforts to diversify the economy into light industry will be very successful and therefore tourism will continue to predominate.

Overdevelopment

The predominance of the tourist sector will, in turn, aggravate the other pressing problem facing the territory: overdevelopment. Great concern over where the US Virgin Islands is going and what it will be like in

KEY FACTS

Official title: Virgin Islands of the United States

Head of state: President of the USA

Head of government: Governor Alexander A. Farrelly

Ruling party: Democratic Party of the Virgin Islands

Capital: Charlotte Amalie (on St Thomas)

Official Languages: English

Currency: US Dollar ($) = 100 cents

Area: 354.8 sq km

Population: 110,000 (1988)

GNP per capita: $9,724* (1986)

GNP real growth: 6.2% (1986)

Trade balance: –$521m (1986)

Visitor numbers: 1.9m (1988)

* estimated figure

the future became pronounced in 1988 when the West Indian Company (a Danish multinational conglomerate which enjoys extra-territorial rights in St Thomas under the 1916 Treaty of Sale with the USA) dredged and filled in 7.5 square acres of the beautiful harbour of Charlotte Amalie. It claimed the land for itself and future resort and commercial development under agreements with several previous governors. Outraged St Thomians formed the Save the Long Bay Coalition to fight the action and took the case all the way to the US Supreme Court and the United Nations Decolonisation Committee, charging that the company is usurping land that rightfully belongs to the people of the US Virgin Islands. While US courts were indifferent to the petition of the natives, the United Nations was sympathetic and supportive. To add insult to injury, the company has now unveiled plans for a massive expansion of its land which would turn the harbour of Charlotte Amalie into a gigantic company-owned and -operated marina with shopping malls, cruise ship docks, hotels, etc. It is expected to face stiff opposition in obtaining permits.

Another organisation, VI 2000, wants a moratorium on future building until a full-scale development plan can be adopted. While some segments of the business community share the organisation's concerns, others continue to advocate expansion of the tourist-based economy as the answer to the territory's perennial need for more revenues.

Meanwhile the chronic infrastructural problems which have plagued the territory for decades continue to worsen, with frequent power outages, deteriorating water quality and garbage problems, temporary shutdowns of two new hospitals, severe traffic congestion and woefully inadequate airport facilities (the St Thomas terminal was built by the US Navy in the Second World War). Further complicating this dire situation is rising crime, caused in part by an increase in narcotics trafficking and use in the territory, especially of crack. The narcotics epidemic has involved the US Virgin Islands in a drug scandal in Washington DC where Charles Lewis, a St Thomian former aide to Mayor Marion Barry, has been convicted of cocaine possession and use. Lewis subsequently turned prosecution witness against Barry in exchange for leniency.

While the problems in America's paradise are going to require decades to resolve, long-term observers of the US Virgin Islands are impressed by the efforts of a broad coalition of young, middle-aged and older residents, both natives and newcomers, to force politicians and the community at large to be more conscious of their obligations to future generations who will inherit some of the most beautiful islands in the world. Whether the coalition will succeed remains to be seen, but they certainly offer hope that is nowhere else in sight.

COUNTRY PROFILE

Historical profile: Discovered by Columbus 1493; settled by French and English mid-17th century but French left; changed hands between English, French, Spanish, Knights of Malta until Danes took over – St Thomas 1666, St Croix 1733; sold to USA ($25m) March 1917.

Political system: Unincorporated territory of USA under its full jurisdiction, electing one non-voting member to the US House of Representatives. Executive authority is exercised by the governor (elected for a four-year term by popular vote) who makes other executive appointments with the concurrence of the legislature. The unicameral legislature has 15 members elected by popular vote for a two-year term; its measures are subject to the governor's approval.

Main cities/towns: Charlotte Amalie on island of St Thomas (capital), Christiansted and Frederiksted on St Croix. The third major island and the least populous is St John (Cruz Bay).

Media
Press: The two major dailies are the *Virgin Islands Daily News* and *St Croix Avis*. Both are independent.
Broadcasting: Cable and television station and several commercial radio stations operating.

External trade: Heavy reliance on imports and aid from the USA.
Exports: Petroleum products, alumina, chemicals, clocks and watches, meat, ethanol. Main destinations: USA, Puerto Rico.
Imports: Crude oil, machinery and transport equipment, semi-manufactures (particularly clock, watch parts), food, chemicals etc, consumer goods (particularly servicing tourist trade). Main sources: USA, Japan, Nigeria, UAE.

Agriculture: Poor quality of the soil and lack of rain precludes large-scale cultivation. Small quantities of sorghum, fruit and vegetables are produced on St Croix and St Thomas. Leaves from the bay tree forest on St John are used in making bay rum and cattle are raised on St Croix for meat export to Puerto Rico. The waters around the islands are mainly used for game rather than commercial fishing: some farming, however, is undertaken.

Industry: Manufacturing is small-scale and export-based (mainly to USA). The main activities are rum distilling (3-4m proof gallons per year), watch/clock assembly, ethanol refining, woollen textiles and garments. The largest single employer is the 728,000 b/d Hess oil refinery on St Croix, but production at the plant (the world's biggest) has slumped in recent years.

Tourism: Sector forms mainstay of the economy, employing 30 per cent of the workforce. The emphasis is on cruise ship business (mainly from USA) and the sale of duty-free products.

BUSINESS GUIDE

Time: GMT -4.

Climate: Tropical with a mean annual temperature of 26°C. Low levels of humidity.

Entry requirements
Passport: Not required by citizens of the US or Canadian nationals with an ID card.
Visa: Required by all except US citizens or Canadian nationals with proof of ID. Certain exceptions applied; check before departure.
Currency: The US Dollar circulates. Import/export over $5,000 must be declared.

Health precautions
Mandatory: Yellow fever vaccination certificate if arriving from infected area.
Advisable: Health insurance strongly advised. Precautions with water in rural areas. Typhoid/polio vaccinations.

Air access
Airlines serving: Aero Virgin Islands, Air BVI, Air Caribe Int, Air Guadeloupe, American, Coastal Air Transport, Eastern, LIAT, Midway, Pan Am, Sunaire, TWA, Vieques Air Link, Virgin Air, Windward Islands Airways.
International airports: St Croix-Alexander Hamilton (STX), 14.5 km from Christiansted, car rental; St Thomas-Cyril E. King (STT), 4 km from Charlotte Amalie, duty-free shop, bar, restaurant, bank, shops, car hire.

Surface access
Main ports: Charlotte Amalie (St Thomas), Christiansted (St Croix), Frederiksted (St Croix), South Shore cargo port (St Croix).

Hotels: Advisable to book in advance, especially in winter months. There is a 7.5 per cent hotel tax. 15 per cent tip is usual.

Car hire: Wide selection of cars available. National licence required. Daily cars about $35. Traffic drives on the left.

City transport
Taxis: Widely available; fixed-rate system applies but is not always strictly adhered to. Higher charges for extra passengers, luggage and at night.

National transport
Air: There are frequent services between St Thomas and St Croix (by Aero VI, Eastern Metro, American Eagle, VI Seaplane Shuttle) plus air-taxi and flying boat services covering the area. Helicopter service.
Road: Over 800 km, spread over the islands. Surfaces are good.
Buses: Public service on all main routes and group tours available.
Water: Regular ferry service between St Thomas and St John and the British Virgin Islands.

Public holidays: 1 Jan, 6 Jan (Three Kings' Day), Martin Luther King Day (Mon in mid-Jan), Washington's Birthday (Mon mid-Feb), Holy Thursday, Good Friday, Easter Monday, 31 Mar (Transfer Day), Children's Carnival Parade (late Apr), Memorial Day (late May), Organic Act Day (mid Jun), 3 Jul (Emancipation Day), 4 Jul (Independence Day), last Mon in Jul, first Mon in Sep, Columbus Day (Mon around 10 Oct), Hurricane Thanksgiving Day (Mon around 18 Oct), 1 Nov (Liberty Day), 11 Nov (Veterans' Day), 25 Dec, 26 Dec.

Working hours
Government and business: (Mon-Fri) 0800 – 1700.
Banking: (Mon-Fri) 0900 – 1430; also Fri 1530 – 1700.

Telecommunications
Telephone: Modern, 100 per cent automatic service in operation. Dialling code for US Virgin Islands: IDD access code + 1-809 followed by subscriber's seven-digit number.
Telex and telefax: International services are available.

Postal service: Air Mail to US and Europe takes under one week.

Banking: Banco Popular de Puerto Rico, Bank of Nova Scotia, Barclays Bank International, Chase Manhattan Bank, Citibank, First Federal Savings, First Pennsylvania Bank.

Electricity supply: 110/120 V AC, 60 cycles.

Hotels: St Croix
Anchor Inn, 58A King Street, Christiansted (tel: 773-4000; tx: 3471111).

BUSINESS DIRECTORY

Divi St Croix, Christiansted, P.O. Box: Z (tel: 773-9700; tx: 3471038).

Mill Harbour Apartments, Golden Rock Estate, Christiansted (tel: 773-3800).

Queen's Quarter, Christiansted, P.O. Box: 770 (tel: 778-3784; tx: 3475020).

Hotels: St John
Caneel Bay, P.O. Box: 720 (tel: 776-6111; tx: 3470019).

Virgin Grand Beach, Great Cruz Bay (tel: 776-7171; tx: 3474093; fax: 7753858).

Hotels: St Thomas
Bluebeard's Castle, Charlotte Amalie, P.O. Box: 7480 (tel: 774-1600; tx: 3470023; fax: 7745134).

Frenchman's Reef, P.O. Box: 7100 (tel: 776-8500; tx: 3470080; fax: 7768500).

Island Beachcomber, Lindbergh Beach, P.O. Box: 1618 (tel: 774-5250; tx: 3677310; fax: 7745250).

Limetree Beach, P.O. Box: 7307 (tel: 776-4770; tx: 3470032; fax: 3208).

Ramada Yacht Haven, P.O. Box: 7970 (tel: 774-9700; tx: 3470072).

Stouffer Grand Beach, Smith Bay (tel: 775-1510; tx: 990142).

Virgin Isle, P.O. Box: 3188 (tel: 774-1500; tx: 3470105).

Windward Passage, Charlotte Amalie, P.O. Box: 640 (tel: 774-5200; tx: 3470126; fax: 7741231).

Car hire: St Croix
Avis: Seaplane Port (tel: 773-4377; tx: 347103).

Europcar National: 1 King Cross Street, Christiansted (tel: 773-2485).

Car hire: Cyril E. King Airport, St Thomas
Avis: (tel: 774-1468).

Hertz: (tel: 774-1879).

Business advice and information
Department of Economic Development and Agriculture, PO Box 6400, St Thomas 00801 (tel: 774-8784).

Industrial Development Commission, PO Box 3499, St Croix (tel: 773-6499); PO Box 6400, St Thomas (tel: 774-8784).

St Croix Chamber of Commerce, 16AA Church Street, Christiansted, St Croix (tel: 773-1435).

St Thomas/St John Chamber of Commerce, PO Box 324, St Thomas Apothecary Hall, St Thomas (tel: 776-0100).

Vanuatu

Kevin Rafferty

(map on page xxiv)

Vanuatu, the most surprising and enterprising of all the Pacific island states, was anxiously awaiting for general elections in 1991 to see whether the electorate would confirm the rule of Father Walter Lini and his way of running things under his Vanuaaku Pati. The last 18 months of the 1980s saw a series of crazily adventurous challenges to Lini's rule from inside his own party, then an attempted parliamentary coup led by the president. But Lini stayed on top. The perpetrators of the coup were first jailed, then the convictions were overturned because, the Appeal Court judge claimed, the trial judge had made serious errors.

Lini, an Anglican priest, has been prime minister of the country ever since its independence from Britain and France in 1980. He has pursued an adventurous policy and seemed to rejoice in a radical reputation. Vanuatu is the only South Pacific member of the Non-Aligned Movement, and has established diplomatic relations with the Soviet Union and, even more daringly, with Libya. It also caused less consternation by granting fishing rights to Soviet trawlers.

Coup attempt

But Lini was not radical enough for Barak Sope, a younger member of the ruling party once considered the prime minister's heir apparent. When Prime Minister Lini suffered a stroke, Sope claimed he was not fit enough to rule. He tried several times in several different ways to capture power. First he and his supporters marched on the capital Port Vila complaining of injustices in the use of land, probably the most touchy and emotional question in Melanesia. He challenged Lini within the party and failed. He tried to overthrow him in parliament, then he formed an alliance with his uncle, President George Sokomanu.

As far as public political postures were concerned, Sokomanu and Sope seemed unlikely partners. The president had previously criticised Lini for being autocratic and for his own radical policies. The alliance between the uncle and nephew proved that blood is thicker than water and that possibly Sope was not as radical as he seemed, though he had benefited from having some of his assistants trained in Libya. Sokomanu dissolved parliament and appointed Sope prime minister instead of Lini. Unfortunately for them, Prime Minister Lini still had the allegiance of the small Vanuatu police force – the country has no standing army – who arrested the two plotters and later brought them to trial on treason charges.

Early in 1989 Sokomanu was found guilty of incitement to mutiny and sentenced to six years in jail. Sope and Maxime Carlot, former leader of the opposition Union of Moderate Parties (UMP), were each sentenced for five years and another former member of parliament Willie Jimmy was given a two-year sentence. Sokomanu had been stripped of the presidency by the country's electoral college. Justice Gordon Ward overruled a two-man jury of 'assessors' who had found the four, plus three other defendants, not guilty of the mutiny charges. 'For a few days in December parliamentary democracy in this country hung by a delicate thread. The fact the thread did not snap owes very little or nothing to you,' declared the trial judge in the country's Supreme Court.

But in April an Appeal Court overturned the convictions of all four men. The panel of three judges found errors so numerous and serious that they prevented a fair trial. In a fair trial, the appeal judges declared, the four politicians would have been acquitted. Chief Justice Geoffrey Martin of Tonga said that after listening to the evidence at the trial the court had been left with a 'very real doubt, not just a lurking doubt' about the verdict. This left Sope free to challenge Lini either in the party or in the country. Inside the Vanuaaku Pati Lini had sufficiently recovered from his stroke to consolidate his lead. He was re-elected party president in June 1989. This means the battlefield will be transferred to the electorate next year.

Traditions

The botched and unsuccessful attempt to overthrow Lini was well in keeping with Vanuatu's peculiar, sometimes hilarious traditions. This is a country after all that was ruled for almost 75 years by an Anglo-French condominium or 'pandemonium' as it was jokingly and more accurately described. The colonial powers set up two sets of everything, from customs and immigration to law courts and even simple things like stamps. For an island country of only 120,000 people (150,000 by 1988), it was excessive. It culminated in a 'bow and arrow struggle' against Lini at independence, in which the French rulers did not quite openly support the rebels and the British refused openly to put them down. It was left to the newly independent government to summon Melanesian brotherly help from Papua New Guinea to put down the rebellion.

Vanuatu is still a country where Melanesian tribal customs are strong and twentieth-century economy sits oddly. Some tribes follow strange cargo cult worship, numbering the Duke of Edinburgh among their gods. It is a place where a woman's worth is still calculated in pigs, where the colourful pidgin Bismala language is still spoken. In the local language it is common to hear the word 'bugger-up', an everyday expression meaning broken.

Economy

Such a country has found it hard making the transition

KEY FACTS

Official title: Republic of Vanuatu

Head of state: President Frederick Karlomuana Timakata

Head of government: Prime Minister Father Walter Lini

Ruling party: Vanuaaku Pati

Capital: Port Vila (on Efate)

Official Languages: Bislama (Ni-Vanuatu Pidgin), English, French

Currency: Vatu (V = 100 centimes)

Exchange rate: V118 per $ (Nov 1989)

Area: 11,880 sq km (82 islands)

Population: 150,000* (1988)

GDP real growth: 0.7% (1987)

Inflation: 8.7% (1988)

Trade balance: –$42.5m (1988)

Foreign debt: $134.6m (1987)

Visitor numbers: 17,544* (1988)

* estimated figure

to modern life. Per capita income is about $800, higher than the average for the Pacific, and taking Vanuatu well outside the ranks of poor developing countries. But the figures can be questioned as arbitary. They are inflated by the high earnings of the few expatriates. Many ni-Vanuatu are involved in subsistence farming. That is not to say that they go hungry. One of the country's problems is that the land is rich enough for the people to make an easy living without having to work really hard. It is estimated that 80 per cent of the population live on subsistence agriculture, though its contribution to gross domestic product is only 25 per cent.

There is plenty of scope for modern economic development. The main cash crop is copra. Cocoa and beef production are important but still have to be properly developed. Fish and forestry exploitation has been minimal. Manufacturing is mainly agro-based and accounts for a mere 5 per cent of GDP. Plantations, which can also be opened up, have declined in importance over the last two decades as trees have aged, the independence struggle discouraged investment and the ni-Vanuatu largely preferred to work for themselves in subsistence agriculture rather than be employees on a big plantation. Further development of the modern economy depends on sorting out land rights. At independence, land granted to European settlers reverted to its customary owners. In practice this has meant many disputes between rival claimants. There is a backlog of cases ploughing their way slowly through the courts. The government has the power to lease disputed land for up to 75 years, but has only exercised this sparingly.

Government's policy has been to encourage export-oriented industry rather than import substitution, except for the few products used in local raw materials. But there is still a lot more to be done. Exports are worth about 15 per cent of GDP in a good year and dominated by copra, worth 70-75 per cent of the total.

KEY INDICATORS

	Unit	1984	1985	1986	1987	1988
Population*	'000	130.0*	135.0	140.2	140.0	150.0
Consumer prices % Change	1980=100	5.5	1.0	4.8	14.8	–
Exports	V m	4395	3205	1526	1908	–
Copra	V m	2,644	1,312	450	699	–
Re-exports	V m	1174	1283	584	440	–
Cocoa	V m	134	153	196	207	–
Beef	V m	142	198	130	–	–
Timber and logs	V m	148	136	63	–	–
Imports cif	V m	6631	7462	6105	7476	–
Balance of trade	$m	–18.9	–33.6	–38.0	–43.4	–42.5
Current account	$m	19.3	1.3	–2.8	–5.3	2.7
External debt	$m	88.3	108.9	120.0	134.6	–
Debt service ratio	%	0.6	0.7	1.3	–	–
Exchange rate	V per $	99.23	106.03	106.08	109.85	104.43

* estimated figure

On the other hand, imports are 40 per cent of GDP, of which food and basic consumer goods make up 40 per cent of the total. This means that to fill the gap Vanuatu is dependent on aid and earnings from tourism and from its offshore financial centre. Establishment of the centre, which today numbers more than 100 banks and financial institutions, proved that Lini was a genuine radical in the old-fashioned sense of the word, not a mere Leftist. People who know the prime minister say some of his postures are the product of a genuinely flexible and open mind, others are for public relations effect rather than a real Leftist commitment. He is, after all, an Anglican priest. His fishing deal with the Russians gave Vanuatu valuable earnings and did not lead to any Soviet influence onshore inside the country.

As the events of the late 1980s showed, the country is highly vulnerable to external circumstances. Its agriculture was badly mauled by cyclones in 1985 and 1987. The balance of payments was affected by lower world copra prices. Then tourism income was disrupted by a combination of factors including a downturn in the Australian economy, the main source of tourists, the appreciation of the Australian dollar and the end of the national airline after its management contract with a foreign airline was terminated.

COUNTRY PROFILE

Political system: The 46-member parliament is elected to a four-year term by universal adult suffrage, and elects the president, the head of state. It also elects an executive prime minister who appoints the cabinet. Local authority is partly exercised by local councils.

Political parties: *Ruling party: Vanuaaku Pati* (40 seats in parliament, following 1987 and 1988 elections); Tan Union (six).

Population: The great majority of the population are Melanesian in origin, with another 5 per cent of European descent. 80 per cent of the population are Christian, but animism is still in evidence, and the cargo cult remains on Tanna Island. There have been localised secessionist movements in Santo, Malekula, Ambrym, Aoba, Pentecost and Maewo.

Main cities/towns: Port Vila (on Efate, population 15,000 in 1986), Santo (6,000).

Language: The offical languages are Bislama (the local pidgin dialect), English and French.

Media

Press: The Vanuatu Weekly is a government-run newspaper.

Radio: Radio Vanuatu programmes are broadcast 16 hrs/day in Bislama, English and French.

Domestic economy: Vanuatu is a small, agricultural economy whose main export, copra, is a commodity subject to wide fluctuations in price on the world market, although steps have been taken to reduce dependence on it by developing exports of cocoa, timber, beef and fish. Most islanders are farmers and fishermen, and about 82 per cent of the population lives in rural areas. Earnings from tourism and the tax-free financial centre help offset trade deficits and provide employment.

External trade

Exports: Exports have declined steeply since reaching their 1984 peak and were $18m in 1987. Main exports are copra (typically, 60% of total), fish re-exports (17%), beef (3.0%), and cocoa (3.0%). Main destinations are Netherlands (37%), Belgium (22%), Japan (13%), France (10%), Singapore (5%), and New Caledonia (1%).

Imports: Imports were worth $72m in 1987. Main imports are machinery and transport equipment (typically 23% of total requirements), food and live animals (20%), basic manufactures (15%), other manufactures (15%), fuels and lubricants (10%), and chemicals (7.1%). Main sources are Australia (35% of 1984 requirements), Japan (13%), New Zealand (10%), France (9%), Hong Kong (5%), New Caledonia (4%), and Singapore (4%).

Agriculture: The soil is generally fertile and rainfall adequate, though crops can be subject to cyclone damage. Subsistence crops of coconuts, taro, bananas, sweet potatoes, fruit and vegetables are grown on smallholdings, alongside pig and poultry raising. Smallholdings and plantations produce crops destined for export (copra and cocoa), but much plantation area needs replanting. Cattle-raising and forestry are becoming increasingly important foreign exchange earners. Operations by the South Pacific Fishing Co, a joint Japanese/Vanuatu venture, engaged in fish exports, were handed over to the national government in August 1986. Access only by sea to exploitable forests has limited timber production. An experimental project to seed reefs with trochus raised in hatcheries is under-way (the shells are collected and sold as buttons).

Industry: Main industries include copra processing, meat canning, fish processing, soft drinks bottling, furniture making, metalwork and handicrafts for the growing tourist market. Although there is a concern about the potential effect of tourism on native culture, the government plans to develop the industry. Estimated revenue from tourism in 1986 was $14.3m, when 23,500 visitors arrived from Australia (over 63% of total), New Caledonia (20%), UK (10%) and New Zealand (8%). Japan has played an important part in helping to improve regional commercial centres for transporting and distributing agricultural products and other goods, providing investment for wharves on Tanna and Malakula Islands.

Mining: Production of manganese from the large-scale mine on Efate has ceased. There has been prospecting for nickel, copper and bauxite. The US Geological Survey has found undersea platinum and copper deposits off the coasts of Vanuatu and Kiribati.

Membership of international organisations: Asian Development Bank, Commonwealth, ESCAP, FAO, G-77, IBRD, ICAO, IDA, IFC, IMF, ITU, NAM, South Pacific Bureau for Economic Co-operation, Pacific Islands Association of Commerce, South Pacific Commission, South Pacific Forum, World Bank, UN, UNO, WHO, WMO.

BUSINESS GUIDE

Time: GMT + 11 hrs. Summer time +1 hr.

Climate: Temperatures can range from 16-33°C, and rainfall varies from 1,200-2,000 mm/year. Cyclones may occur from Dec-Apr.

Entry requirements
Passport: Required by all.
Visa: Visas are not required by citizens of the UK and colonies and nationals of France, German FR, Japan, the USA and other countries who have assured onward passage. Regulations govern the length of stay permitted for different purposes.
Prohibited entry: Anyone whose demeanour is not considered acceptable is prohibited entry.

Health precautions
Mandatory: Vaccination certificates for yellow fever or cholera if travelling from an infected area.
Advisable: Anti-malaria precautions.

Air access
National airline: Air Vanuatu.
Other airlines: Air Calédonie, Air Melanesie, Air Niugini, Air Pacific, Norfolk Island Air.
Tax: Airport tax of VT 1,500 is levied on departure.
International airport: Bauerfield (VLI), 6.5 km from Port Vila (on Efate); duty-free shop, buffet, currency exchange, hotel reservations, post office, shops, car hire. Currently being upgraded.

Surface access
Main ports: Port Vila, Luganville (Santo).

Hotels: A 10 per cent tax is added to hotel bills. There are three luxury hotels now operating.

Car hire: Car hire is available in Port Vila and Luganville. International, French and UK licences are acceptable.

City transport
Taxis: Taxi service is available in Port Vila and Luganville.
Buses: Buses serve the whole of Port Vila.

National transport: Tours by minibus, airplane, car and boat are available.
Air: Air Melanésie operates inter-island services to 16 destinations from Bauerfield, Port Vila Santo and charter services are available. Dovair also provides services to outer islands of Vanuatu.
Road: There are some 150 km of surfaced road on Efate, and 100 km on Espiritu Santo, which are passable in dry weather.
Water: Sea links between the islands are generally good.

Public holidays
Fixed dates: 1 Jan (New Year's Day), 5 Mar (Custom Chiefs' Day), 1 May (Labour Day), 30 Jul (Independence Day) 15 Aug (Assumption Day), 5 Oct (Constitution Day), 29 Nov (Unity Day), 25 & 26 Dec (Christmas Day & Boxing Day).
Variable dates: Ascension Day, Easter.

Working hours
Government and business: (Mon-Fri) 0730 – 1700, with two-hour midday break.
Banking: (Mon-Fri) 0830 – 1530.

Social customs: Tipping is considered contrary to Melanesian customs of hospitality.

Telecommunications: Domestic and international telecommunications are operated by Telecom Vanuatu. Formed in 1989, this is jointly owned by the government of Vanuatu and British and French telecommunications companies.
Telephone: Dialling code for Vanuatu, IDD access code + 678 followed by subscriber's number (no area code is necessary).
Telex: 24-hr telex service is available.

Banking: Vanuatu's development as an offshore banking centre has attracted some 100 banks, of which a few trade locally.
Central bank: Central Bank of Vanuatu.
Head offices: Vanuatu Co-operative Savings Bank, Vanuatu Development Bank.
Vanuatu subsidiaries: ANZ (Vanuatu), Banque Indosuez Vanuatu.
Vanuatu branch: Westpac.

Electricity supply: 230 V AC, 50 cycles.

Weights and measures: Metric system.

Hotels: Luganville, Espiritu Santo
Bokisa Island Resort, P.O. Box: 261 (tel: (03)855).

Jaranmoli Guest Bungalows, P.O. Box: 238 (tel: (03)857, 712).

New Look Motel, P.O. Box: 114 (tel: (03)440).

Relais Boughainville, P.O. Box: 149 (tel: (03)615).

Hotels: Port Vila, Efate
Coral Apartments, P.O. Box: 95 (tel: 3569).

Erakor Island Resort, P.O. Box: 24 (tel: 2983; tx: 1069).

Hide-Away Island, P.O. Box: 875 (tel: 3930, 2963; tx: 1069; fax: 3867).

Iririki Island, P.O. Box: 230 (tel: 3388; tx: 1080; fax: 3880).

Kalfabun Guest House and Back Packers, P.O. Box: 494 (tel: 3277).

Le Lagon Pacific Resort, P.O. Box: 86 (tel: 2313; tx: 1032; fax: 3817).

Manuro Paradise,, East Efate,, P.O. Box: 1009 (tel: 2378).

Marina Motel, Kumul Highway, P.O. Box: 681 (tel: 2566; tx: 1111 VANTEX NH).

Olympic, Kumul Highway, P.O. Box: 709 (tel: 2464; tx: 1090; fax: 2953).

Pacific Lagoon Apartments, P.O. Box: 827 (tel: 3860).

Port Vila Inter-Continental Inn, Tassiriki Park, P.O. Box: 215 (tel: 2040; tx: 1042; fax: 3340).

Rossi, P.O. Box: 11 (tel: 2528; tx: 1022).

Solaise, rue de Picardie, P.O. Box: 810 (tel: 2150; tx: 1065 NH; fax: 2678).

Takara Resort, North Efate,, P.O. Box: 947 (tel: 3576).

Vila Chaumiers, Teouma Road, P.O. Box: 400 (tel: 2866; tx: 1067).

Whitesands Country Club, Rentabau, P.O. Box: 906 (tel: 2090).

Hotels: Tanna
Tanna Beach Resort, Isangel, P.O. Box: 27 (tel: 3510).

Whitegrass Bungalows, Lenakel, P.O. Box: 5 (tel: 2288).

Car hire
Budget: Port Vila (tel: 3170).

Discount Rentals: Port Vila (tel: 3242).

Hertz: Port Vila (tel: 2244; tx: 1019; fax: (678)3241). Airport (tel: 2016).

Business advice and information
Central Bank of Vanuatu, PO Box 271, Port Vila (tel: 3333; tx: 1049; fax: 2037).

Ministry of Trade, Commerce, Co-operative Industry and Energy, Port Vila (tel: 3979).

Vanuatu Chamber of Commerce, PO Box 189, Port Vila (tel: 3255; tx: 1107 CHACOM NH).

Travel information
National Tourism Office, PO Box 31, Port Vila.

Tourist Information Bureau, PO Box 209, Port Vila (tel: 2515, 2685; tx: 1102; fax: 3889).

Tour Vanuatu (Tourism Office), PO Box 409, Port Vila.

Vatican City

David Lane

The celebrations of Pope John Paul II's first decade as head of the Church of Rome in October 1988 were a moment for reflection. What had been achieved and what kind of Papacy had the Polish Holy Father created were the questions which most Vatican watchers tried to answer.

A conflicting picture emerged, rigid conservatism contrasting with Pope John Paul II's strong human appeal. During 1988 and 1989 the Pope continued in style and content the pattern designed in the earlier years of the Papacy. His Polishness remained a firm reference point, though his journeys marked him as an international figure. With his uncompromising views, John Paul II often seemed to seek conflict during his first 10 years as Pope, and controversy shadowed him throughout his eleventh year.

The role of women and the attitude of the Church to the female sex was one issue raising emotional temperatures. 'Mulieris Dignitatem' (Dignity of Women), the document published in September 1988, condemned sexual discrimination but underlined the differences between the sexes and their different roles. By closing the door on ordination for women, Karol Wojtyla disappointed many of the Church of Rome's more progressive followers, not least those in the USA who had been calling for a change.

In his Apostolic Letter the Pope attempted to place the Church at woman's side in the cause of freedom from male dominion, stating the principle of equality before God and the consequent condition that woman cannot become an object of male dominion and possession. Pope John Paul II gave two reasons for denying the extension of equality to ordination. First, Jesus called only men as his Apostles and, second, in the Eucharist the priest acts in the person of Christ.

The growing role of women in the Anglican Church contrasts sharply and widens the gap with Rome.

KEY FACTS

Official title: The State of the Vatican City
Head of state: Pope John Paul II
Official Languages: Italian
Currency: Italian Lira
Exchange rate: Lire 1,304 per $ (Dec 1989)
Area: 0.44 sq km

Pope John Paul II reacted predictably in 1988 to the real equality sought by many Anglicans. Unyielding preaching held a firm line of Roman doctrine on women in the Church, attempting to deny the existence of independent Protestant churches, as though the Reformation had never happened. Some interpreted the Pope's announcement in June 1989 in Copenhagen that Martin Luther's excommunication was no longer effective as further evidence of Roman ecclesiastical expansionism.

Conservative about women, during 1988 and 1989 the Vatican under Karol Wojtyla continued to crusade against abortion and contraception. Some cyclists participating in the 'Giro d'Italia' race in April 1989 wore vests with slogans campaigning against Italy's abortion law. But the biggest boost to the Church's anti-abortion crusade was given by the US Supreme Court's ruling in July 1989.

All involved in abortion earn the Roman Church's condemnation and automatic excommunication. Teaching on contraception is almost as severe. The Vatican holds that contraception is the world's great sickness and that ignoring the Church's doctrine is a denial of God. AIDS education spots on Italian television, which won widespread approval, were criticised by the Vatican. But its teaching that abstinence is the only ecclesiastically acceptable prophylactic drew comments from Italian politicians comparing the Vatican to the ostrich.

Unsurprisingly, Pope John Paul II's views on woman's role in the Church, abortion, contraception and AIDS precautions earned him few followers during his Scandinavian journey in June 1989. The visits to Norway, Iceland, Finland, Denmark and Sweden were in a low key. A somewhat louder note was heard when the Pope spoke to the European Parliament in Strasbourg in October 1988 and Ulster Protestant Ian Paisley marked the Papal visit with a vigorous demonstration.

Pope John Paul II had enjoyed an easier trip to Southern Africa in September 1988, though bad weather forced an unplanned landing at Johannesburg in the Republic of South Africa. In Harare he had severely attacked racialism and called for the release of the imprisoned nationalist leader Nelson Mandela. When he visited Zambia and Malawi in May 1989 he again spoke strongly against racial discrimination. He also took the opportunity to complain about the sins of irregular marriage, divorce, contraception, abortion and excessive emphasis on sex.

Eastern Europe is much nearer to home and to the Pope's personal experience. Developments in the Communist bloc have brought changes in the Vatican's relations with Hungary and Czechoslovakia, and a visit to Russia is closer. The Pope's Polishness and the attention the Vatican dedicates to his country of birth was underlined when Solidarity leader Lech Walesa was received in the Holy See in April 1989.

That Poland held the cause of a major matter of contention between Jews and the Roman Catholic Church was attributed to continuing anti-semitism in Poland and the Vatican. Jews had been given the Roman Church's word that the Carmelite community which had annexed part of the Auschwitz concentration camp would be moved before 22 July 1989. This date passed, with the Church's promise revealed hollow. Tacit Papal approval of the Carmelites' presence and apparent Vatican policy of distorting the history of suffering in World War II was a continuing affront to Jews during 1988 and 1989.

The Church's bivalence on the extremism of Muslim fundamentalists and the welcome accorded to Yasser Arafat by the Vatican in December 1988 helped reinforce Jewish reservations about the Vatican's position.

In addition to taking stands on issues in distant countries, Vatican diplomacy was kept busy in 1988 and 1989 on a matter concerning the state's immediate neighbour. But in March 1989, after four years of uncertainty, clerical power was defeated when Italy's Constitutional Court ruled that alternative teaching to the weekly hour of religious instruction in state schools, agreed in 1985's Revised Lateran Treaty, is not obligatory. This was widely seen as a resounding victory for lay forces.

That the Vatican can no longer assume that its vision of the world and its rulings will automatically be accepted in Italy was underlined in May 1989 when a group of Italian theologists wrote critically about the Curia's regressive tendencies. At the beginning of the year a much stronger attack had been launched by theologists in Germany. Such declarations point to widespread disaffection with an intransigent Vatican, seemingly careless in its use of power and out of touch with church members.

However, providing convincing spiritual leadership and winning over dissenters were not the only problems facing the Vatican as Karol Wojtyla started his second decade as Pope. A budget deficit was forecast for 1989 with revenues of L73bn falling L102bn short of expenses. But though the 'Obolo di San Pietro', which is collected on 29 June and realised L69bn in 1988, covers a large part of the Vatican's annual deficits, additional funding is needed. Keeping the Vatican's battalions deployed, whether through the campaigning of Vatican Radio and the printed word of the 'Osservatore Romano' or the Papal Nunzio outposts throughout the world, is a costly business.

COUNTRY PROFILE

Political system: Overall jurisdiction vested in the Pope, elected for life by conclave of members of the Sacred College of Cardinals. The Pope appoints a Pontifical Commission, headed by a President, to oversee general administration.

Media
Press: Only daily newspaper is *L'Osservatore Romano* which covers religious matters plus some general news items. Also weekly editions published in several languages. Official bulletin of Vatican City is *Acta Apostolicae Sedis* (circulation 6,000), published monthly.
Broadcasting: Vatican Radio broadcasts 24 hours a day in 35 languages, providing a link between the Vatican City and Catholics throughout the world. Broadcasting centre at Santa Maria di Galeria, about 20 km from Vatican, has diplomatic privileges similar to foreign embassies.

Domestic economy: The main sources of income are the Istituto per le Opere di Religione (Vatican Bank, which collects money from Vatican residents), 'Peter's pence' (voluntary contributions to the Catholic Church) and interests on investments.

Legal system: Code of Canon Law devised 1917. New code with many revisions ratified and introduced 1983.

Membership of international organisations: IAEA, INTELSAT, ITU, IWC, UPU, WIPO, WTO, FAO (observer), OAS, UN, UNESCO.

Time: GMT + 1 (GMT + 2 during summer).

Climate: Mediterranean, with hot summers and mild winters. Temperatures range from about 30°C down to 4°C.
Clothing: Lightweight for summer, mediumweight and light topcoat for winter.

Entry requirements: No formal regulations.

Air access: Heliport used by Vatican officials and visiting dignitaries.
Main international airport: Rome served by Leonardo da Vinci (Fiumicino) (FCO) 35 km from Vatican City.

Surface access: By road or rail through Rome.

National transport: Railway line covers 862m before leaving Vatican City.

Telecommunications: International telephone dialling code: 3966982, followed by subscriber's number. Telex country code 504 VA.

Representation in capital: Argentina, Australia, Austria, Belgium, Bolivia, Brazil, Canada, Central African Republic, Chile, Colombia, Costa Rica, Cuba, Denmark, Dominican Republic, Ecuador, Egypt, El Salvador, Finland, France, Federal Republic of Germany, Greece, Guatemala, Haiti, Honduras, Indonesia, Iran, Ireland, Italy, Ivory Coast, Japan, Republic of Korea, Lebanon, Liberia, Lithuania, Luxembourg, Monaco, Netherlands, Nicaragua, Panama, Paraguay, Peru, Philippines, Poland, Portugal, San Marino, Senegal, Spain, Sweden, Taiwan, Turkey, United Kingdom, USA, Uruguay, Venezuela, Yugoslavia, Zaïre.

Principal newspapers
Acta Apostolicae Sedis, The Secretariat of State, Palazzo Apostolico 00120 Cittá del Vaticano (tel: 6983693).

L'Osservatore Romano, Via del Pellegrino, 00120 Cittá del Vaticano (tel: 6983404; tx; 2021).

Other useful addresses
Acta Apostolicae Sedis (The Secretariate of State), 00120 Città del Vaticano (tel: 6983693)

Agenzia Internazionale Fides (AIF), Palazzo di Propaganda Fide, Via di Propaganda lc, 00187 Rome (tel: 6792414).

Annuario Pontificio, Palazzo Apostolico, 00120 Città del Vaticano (tel: 6983064).

Istituto per le Opere di Religione (IOR), 00120 Città del Vaticano (tx: 2012).

Ministry of Finance and Housing, PO Box 31, Port Vila (tel: 2951; tx: 1040).

Ministry of Trade, Co-operatives, Energy and Industry, Port Vila (tx: 1040).

Radio Vaticana, Piazza Pia 3, 00193 Roma (tel: 6983551; tx: 613423).

Statistical Yearbook of the Church, Via della Tipografia, 00120 (tel: 6984834)

Venezuela

Peter J. West

Faced by a balance of payments crisis, newly-elected president Carlos Andrés Pérez launched a stiff IMF-backed structural adjustment package after taking office in February 1989. The about-turn in economic policy was accompanied by an unprecedented wave of violent rioting, followed by more co-ordinated opposition from organised labour. The stabilisation measures led to record inflation and a slump in activity in 1989, but also contributed to a sharp improvement in the balance of payments current account and in the fiscal deficit. At the same time, the public was transfixed and investor confidence shaken by a scandal-ridden investigation into a multi-million dollar fraud in the allocation of preferential rate foreign exchange during the previous government of Jaime Lusinchi. Although unexpectedly orthodox at home, Pérez channelled his populist energies into pressing for a reduction in Latin America's external debt burden, seizing on this issue to promote his role as a Third World statesman. Having restored a measure of macroeconomic stability, the challenge in the early 1990s will be to press forward with the structural reforms needed to achieve sustained export-led growth and a more equitable distribution of income. This will require strong political leadership, in order to prevent interest-group pressures against change from derailing the reform programme.

Unsustainable imbalances

From the standpoint of macroeconomic stability, the final year in office of the Lusinchi administration was little short of an unmitigated disaster. Financial imbalances which had emerged earlier were allowed to balloon, and practically no corrective action was taken, in part out of fear of upsetting the ruling Acción Democrática (AD) party's chances in the December 1988 general election. In the event, AD's candidate, Carlos Andrés Pérez, won a comfortable victory in the presidential race by taking 52.9 per cent of the vote as against the 40.4 per cent polled by his major rival, Eduardo Fernández of the COPEI christian democratic opposition, although his party lost its absolute majority in congress. But the cost in delaying potentially unpopular but unavoidable adjustment measures was very dear.

On the external front, the current account deficit widened dramatically and international reserves plummeted. Oil exports in 1988 dropped by $900m to $8.2bn due to a 17 per cent decline in the average price to $13.5 per barrel, which was only partly compensated for by an 8.5 per cent increase in the volume of shipments to 1,650,000 barrels a day (b/d). Although non-oil exports rose by almost 30 per cent to $2bn, this was entirely due to a sharp increase in non-monetary gold shipments and the beneficial effect of higher world prices on primary aluminium sales. Other non-oil exports put in a disappointing performance, highlighting the failure of economic policy to foster a reduction in dependency on oil revenues. In addition, imports rose by 31 per cent to $11.6bn, and this coupled with weaker exports pushed the trade balance into deficit for the first time in 10 years. Furthermore, the services deficit expanded substantially due to higher freight and insurance charges on imports and greater interest payments on external debt. Under the influence of all these pressures, the current account shortfall in 1988 soared to a massive $4.5bn, equivalent to 7.9 per cent of GDP, compared with $1.3bn or 2.6 per cent of GDP in 1987.

Although amortisation payments remained at the high level of $2.1bn, capital account inflows increased somewhat. This was in part due to the flotation of a number of bond issues in international capital markets. Nonetheless, foreign exchange reserves were under heavy pressure throughout the year and in gross terms (including the assets held by the state investment fund, FIV, as well as the central bank) declined by $4bn. Moreover, the underlying loss in reserves was worse than this as, in order to make the situation look a little better, a total of $850m was obtained from the Bank for International Settlements and the Bank of Spain in the form of short-term credits. In addition, short-term trade loans continued to build up and by the end of the year amounted to about $6bn. These obligations were to become a major headache for the Pérez administration in 1989.

This untenable balance of payments position was directly related to poor economic policies. Instead of adjusting to the renewed fall in oil revenues, the Lusinchi government did exactly the opposite. Boosted by higher state expenditure, the overall public sector deficit was allowed to expand to 7.8 per cent of GDP from 5.3 per cent in 1987, putting pressure on demand and prices. At the same time, interest rates were maintained at artificially low levels by government controls, thereby discouraging financial saving and creating serious distortions in financial markets. The controlled market exchange rate was not changed and became increasingly overvalued in real terms as domestic inflation was again much higher than in major trading partner countries. Amortisation payments on medium and long-term external debt continued to be made but, beyond the rather modest amounts acquired through bond flotations, little was obtained in the way of new finance. It was only in December that the government finally announced a suspension of public sector principal payments to commercial banks. Within this environment, it became ever more apparent that drastic corrective action would be required immediately after if not before the general election. Anticipating large future price in-

KEY FACTS

Official title: Republic of Venezuela
Head of state: President Carlos Andrés Pérez
Ruling party: Acción Democrática
Capital: Caracas
Official Languages: Spanish
Currency: Bolivar (B) = 100 céntimos
Exchange rate: B42.9 per $ (Nov 1989)
Area: 916,490 sq km

Population: 18.8m* (1988)
GDP per capita: $4,544 (1988)
GDP real growth: 5.7% (1988)
Inflation: 35.5% (1988)
Trade balance: –$1,347m (1988)
Foreign debt: $37bn (1987)
* estimated figure

creases, consumers engaged in a round of pre-emptive buying, particularly of durable goods. Private demand was hence further boosted by a fall in the savings rate.

The expansion in domestic demand resulting from these policies, coupled with the fixed controlled market exchange rate, were the principal factors behind the surge in imports and the poor performance of non-oil exports other than gold and aluminium. In addition, stockbuilding stemming from widespread expectations of a major devaluation contributed to the growth in imports, and it would also seem that there was a substantial degree of overinvoicing of products imported at the controlled market exchange rate so as to acquire dollars which could be sold at the much higher free market rate.

Higher domestic demand led to an extension of the recovery in activity which had begun in 1985 into a fourth consecutive year in 1988. Indeed, non-oil GDP growth accelerated to 5 per cent from 4.2 per cent in 1987 and open unemployment fell to 7 per cent in the second half of 1988 from 8.5 per cent a year earlier. In addition, although much higher than originally programmed, consumer price inflation fell moderately to 35.5 per cent in 1988 as against 40.3 per cent the year before. Although these short-term favourable results may have contributed to AD's success in the presidential election, they were based on extremely shaky foundations. The cost of poor macroeconomic management was soon to become all too apparent.

KEY INDICATORS

	Unit	1984	1985	1986	1987	1988
Population	m	16.9	17.3	17.8	18.3	18.8
GDP	B bn	409.5	464.6	493.8	719.4	924.4
GDP per capita	B	24,302	26,826	27,755	39,377	–
Inflation	%	12.3	11.4	11.5	28.2	29.5
GDP real growth	%	–1.4	1.3	6.8	3.0	5.7
Unemployment	%	13.4	12.1	10.3	8.5	–
Agricultural prod real growth	%	0.8	8.3	8.3	4.1	4.3
Crude oil production	m bls	620.6	613.6	631.8	631.3	667.9
Manufacturing prod real growth	%	4.6	4.8	7.1	2.5	4.4
Cement	'000 tonnes	4,783	5,121	5,875	–	–
Crude steel	'000 tonnes	2,511	2,723	3,031	–	–
Motor vehicles	'000 units	110.0	116.0	143.0	–	–
Aluminium	'000 tonnes	377	407	418	427	–
Fertilisers	'000 tonnes	759	650	868	1,014	–
Iron-ore production	'000 tonnes	13,116	14,449	16,249	17,007	17,600
Gold production	tons	1.1	1.0	1.0	1.0	1.0
Exports	$m	15,841	14,660	9,122	10,567	10,234
Petroleum & Products	%	92	90	83	86	85
Imports	$m	7,260	7,530	7,862	8,832	11,581
Balance of trade	$m	8,581	7,130	1,260	1,735	–1,347
Current account	$m	4,569	3,668	–1,471	–1,125	–4,692
Foreign debt	$bn	36.1	34.2	33.2	35.0	–
External debt	$m	36,457	34,693	34,709	36,519	–
Debt service ratio	%	14.6	14.0	28.7	22.4	–
Exchange rate	B per $	7.5	7.5	14.5	14.5	14.5

* estimated figure

Structural adjustment and reform

Force of circumstance more than anything else explains the conversion of Carlos Andrés Pérez, one of the region's best known populists, to orthodox economic policies on his assumption of power in February 1989. By this time, readily available foreign exchange reserves were almost exhausted and growing shortages of many basic consumer goods were indicative of the strength of latent inflationary pressures. To his credit, Pérez was sufficiently pragmatic to see the need for a sea change in policy, and filled the top economic posts in his administration with able people, including the experienced banker Pedro Tinoco as president of the central bank and the young Yale economics graduate Miguel Rodríguez as planning minister. He also had the political courage to implement harsh remedies in the face of strong opposition from all quarters, including from within his own party.

Although a gradualist approach to adjustment has considerable theoretical merits, in practice it had become an excuse for long periods of inaction during the Herrera (1979-84) and Lusinchi (1984-89) administrations. Sensing that it would be best to take unpopular measures at the beginning of his administration rather than later on, Pérez's initial steps bore all the hallmarks of a shock stabilisation programme. Perhaps the most radical move was the scrapping of the unwieldly multi-tiered exchange rate regime, in place since early 1983, and its replacement by a single, floating rate system. As the new unified rate started trading close to the previous, heavily depreciated, free market rate, this implied a large devaluation for most commercial transactions, with the average rate for imports increasing by about 120 per cent. In Venezuela a large devaluation of this nature not only encourages the production of import substitutes and exports, but also results in a shift of resources from the private to the public sector by raising oil revenues in local currency terms. It was hence expected to make a major contribution to the lowering of the fiscal deficit.

Accompanying the major overhaul in exchange rate arrangements, price controls were lifted on all but a small basket of basic necessities, interest rates more than tripled and substantial increases were decreed for a wide range of public sector goods and services, including petroleum products, electricity, aluminium, steel and petrochemicals. The size of the initial price hikes was often dramatic: many consumer staples rose by over 100 per cent, the most widely-used low octane petrol was lifted by 96 per cent, and electricity tariffs went up by an initial 65 per cent for business users and 35 per cent for residential users, followed by regular monthly adjustments. Further large price increases were planned for petroleum products on an annual basis starting in early 1990, to bring them to levels prevailing in international markets by 1992. Likewise, the large subsidy on domestic fertiliser prices is to be removed in stages by 1992.

Following the initial stabilisation measures, a start was made in implementing structural reforms, required to raise economic efficiency and enhance growth potential over the medium term. Following much acrimonious debate, a trade liberalisation programme was unveiled in May 1989. Designed to boost international competitiveness and improve the allocation of resources, if fully implemented the new policy will represent a fundamental shift towards a more outward-oriented development model, away from the inward-looking import substitution strategy pursued previously. Focused on lowering import barriers, the programme encompasses the removal of most quantitative restrictions and tariff exemptions, the reduction and simplification of tariff rates and the abolition of import monopolies. The initial steps included the elimination of specific import duties based on weight or volume and the reduction in the maximum *ad-valorem* rate from 135 per cent to 80 per cent for consumer products and 50 per cent for raw materials, intermediate goods and capital equipment. Future annual adjustments are planned to bring the top rate down to 20 per cent by March 1993. Quantitative restrictions in the form of prohibitions and licences have been halved to 25 per cent of import categories for manufactured products, and will be cut to 5 per cent by September 1990. Restrictions for agricultural products will be phased out beginning in 1991.

The opening up of the economy to foreign competition has hence been designed to take place over a period of four years, in order to give domestic producers time to adapt to the new circumstances. Confirming its commitment to a more liberal trade regime, the country also applied to join the General Agreement on Tariffs and Trade (GATT). Following the initiation of its trade liberalisation programme, the government intended to begin reforms in the fiscal and financial areas, including the introduction of a value-added tax, improvements in the income tax system, a restructuring of the state enterprise sector, and a revamping of the banking law.

In order to alleviate the social costs of adjustment and protect vulnerable groups, a number of social measures were taken. The minimum wage was raised by 54 per cent, the earnings of general government employees lifted according to a sliding scale, varying from 30 per cent for low-income earners to 5 per cent for the higher paid, and an across-the-board flat rate increase decreed for private sector and other public sector workers. In addition, a temporary firing freeze was implemented. Provision was made in the government budget for subsidies for a number of basic consumer goods, a special employment plan for 1989 and 1990 designed to generate 170,000 new jobs, and a nutritional assistance programme for critically poor families. Furthermore, in a concession which will mainly benefit the middle class, a special subsidy was approved to protect most existing mortgage holders from the sharp increase in market interest rates.

Pérez goes to IMF

Pérez's willingness to seek financial assistance from the IMF, an institution he had frequently criticised in the past for its imposition of draconian economic policy conditions on developing countries, highlighted the fundamental difference in orientation of his new government from that of his first administration during the free-spending, oil boom years of the mid-1970s. Coinciding with the eruption of violent rioting on the streets of Caracas and other major cities, a letter of intent was signed with the IMF at the end of February 1989. This paved the way for a first credit tranche drawdown shortly afterwards amounting to $445m, which was the country's first ever use of Fund resources.

Subsequently, at the end of June a much larger operation known as an Extended Fund Facility (EFF) was approved, under which the country may borrow up to $4.6bn over the three-year period until mid-1992. In order to qualify for this arrangement, the authorities submitted another letter of intent to the Fund in which they committed themselves to a number of quite stringent macroeconomic targets. Amongst these, the fiscal deficit was programmed to be halved to 4 per cent of GDP in 1989, lowered to zero in 1990 and converted into a moderate surplus during the following two years. The aim was to make room for a considerable increase in private sector investment within the context of a viable balance of payments, leading to a rapid recovery in non-oil GDP growth to 6 per cent per year by 1992 and a reduction in inflation to an annual rate of only 10 per cent by that time.

As part of the strategy of taking advantage of the country's large but previously untapped borrowing potential from multilateral organisations, the government also sought assistance from the World Bank. This bore fruit in June 1989 when the Bank agreed to provide $755m in the form of two quick-disbursing

COUNTRY PROFILE

Political system: Executive power rests with the president (elected by universal suffrage for a five-year non-renewable term), who appoints a council of Ministers. The legislature is the bicameral Congress: the Senate (two members for each state, two for the federal district; plus, as life members, ex-presidents of the republic and a number of other senators whose presence is as representatives of minorities – 55 seats), and the Chamber of Representatives (currently 213 members) – number set at one for every 50,000 population with a minimum of two per state and one for every territory. Term for both chambers is five years.

Main cities/towns: Caracas (population about 4m), Maracaibo, Barquisimeto, Valencia, Maracay, Puerto Cabello, Ciudad Bolívar, Ciudad Guayana, Puerto La Cruz/Barcelona.

Media
Press: There are approximately 50 newspapers, all with low circulation in proportion to the population. The major Caracas papers include *El Nacional, El Universal, El Diario, El Mundo, Ultimas Noticas, Economía Hoy, Daily Journal* (English). A number of regional dailies are published in main centres. Also numerous periodicals.
Radio: Radio Nacional operates the state broadcasting service. There are also two cultural and over 140 commercial stations. All broadcasts are in Spanish. *Television:* Three government and three private TV stations broadcast in Spanish.

External trade: Visible trade balance is heavily reliant on petroleum export earnings. The trade balance has hence deteriorated seriously as a result of lower world oil prices and, with imports rising strongly, fell into deficit in 1988. The growing external debt became a problem in the 1980s, but a $21bn refinancing deal was signed in 1986 and subsequently modified in 1987. In mid-1989 Venezuela requested debt relief under the Brady Plan.
Exports: Crude oil (68% of total value in 1986), petroleum products (30-33%), aluminium (3-4%), iron and steel (1-2%), coffee, cocoa. Main destinations: USA (45% of total value in 1986), Netherlands Antilles (18-25%), Canada (4-9%), German FR (6%), Italy (4-6%), Cuba (4%), Japan (3%), UK, Jamaica, Brazil, Netherlands, Spain.
Imports: Machinery and transport equipment (40-45% of total value), semi-manufactures (12-15%), chemicals (12-15%), foodstuffs (10-12%). Main sources: USA (46% of total value in 1986), Japan (7%), German FR (7%), Italy (5-6%), Brazil (5-6%), Canada, France, UK, Spain.

Agriculture: Sector contributes six per cent to GDP and employs around 14 per cent of the workforce. Although about 20 per cent of the land area is under cultivation or used as pasture, Venezuela was importing until recently as much as 50 per cent of its food requirements (particularly cereals – wheat, maize, sorghum – vegetable oils, and dairy products). The main arable centres are Acarigua, El Tigre, Maracay, Valencia and Barquisimeto. There has been little investment in modern farm technology, inefficient marketing and farm management, unremunerative prices (until 1984), and little irrigation. The major crops are rice, bananas, maize, sorghum, sugar-cane, coffee (the main export crop)and cotton. Tropical fruits, cassava, beans, groundnuts and other vegetables are staple crops for small farmers. Poultry and pig-farming are of growing importance with small quantities of meat exported. Beef production has, however, slumped due to the smuggling of cattle to Colombia (where prices are higher), and cheap imports. The forestry sector remains undeveloped. A 10-year agricultural programme (1984-94) is underway, involving the improvement and irrigation of 350,000 ha of existing agricultural land and the use of about 1m new ha for cultivation. The programme aims to increase output of cereals, sugar and oilseeds (to reduce dependence on imports), and promote crop diversification and has already achieved a decrease in food imports. Financial measures have been introduced such as the extention of credit, higher prices and subsidies on vital inputs. The fishing sector has seen good growth since 1984, due mainly to a threefold increase in the tuna catch (making Venezuela the world's fourth largest producer).

Industry: Non-petroleum manufacturing and construction combined contribute 21 per cent to GDP and employ around 26 per cent of the workforce. Industrial development since 1960s has been based on the utilisation of the country's natural resources, particularly iron-ore, bauxite, natural gas and oil. Heavy industries that have arisen with the intention of using local materials as inputs include refining of aluminium (an increasingly significant export), petrochemicals (ammonia, sulphuric acid, fertilisers, plastics etc), cement, and steel production. Import-dependent industries include motor vehicle assembly, tyres, rubber, pharmaceuticals, electrical goods and machinery. The traditional home market industries are beverages, textiles, food processing, ceramics and paper/pulp. Approximately 44 per cent of the total value of manufacturing output derives from the state sector. Major state enterprises include Sidor (steel), Venalum and Alcasa (aluminium), Pequiven (petrochemicals). Manufacturing production (which is highly concentrated, with around 10 per cent of all firms accounting for 75 per cent of output) benefited from the devaluation of the bolivar and grew rapidly in 1986, but the pace of expansion slackened in 1987 and 1988, in part due to shortages of imported components. Vehicle assembly recovered sharply in 1986 as demand increased in anticipation of higher prices, but plummeted in 1989 as devaluation led to large price increases. Activity in the chemicals, textiles, electrical equipment, machinery and food processing sectors has also showed signs of improvement. High public sector investment has helped pull construction out of a massive slump, but the recovery has been constrained by lack of mortgage finance for private housebuilders.

Mining: Vast and largely undeveloped mineral wealth including deposits of iron-ore, bauxite, coal, gold, diamonds, zinc, copper, lead, silver, manganese, titanium, nickel, marble, sulphur, phosphates, mercury and uranium. The sector contributes less than one per cent to GDP. The major mineral exploited is iron-ore (reserves 2,800m tonnes, 80 per cent high-grade). The largest deposits are located at Cerro Bolívar and San Isidro. Production which had declined from over 25m tonnes in the mid-seventies to under 10m tonnes in 1983, has increased to 17m tonnes. Estimated reserves of bauxite at Los Pijiguaos amount to some 4bn tonnes of high grade ore. Production of around 2-3m tonnes a year is geared towards meeting the requirements of the Interalumina refinery. Production of gold has steadily risen, while diamond production has fallen; it is estimated that more than 50 per cent of domestic output of gold and diamonds is smuggled out of the country without tax. However, in 1987 the Central Bank began purchasing gold from producers at the free market exchange rate, giving rise to the start of official gold export. Plans to develop extensive coal reserves in Zuliamstates and Tachira.

Hydrocarbons: Petroleum industry is mainstay of economy, accounting for around 80 per cent of export income and around 50 per cent of government revenues. Sector contributes around 15 per cent to GDP. Venezuela is the world's ninth largest oil producer, and the fourth largest exporter. Proven reserves (excluding the Orinoco belt) were 58.5bn barrels at end 1988, sufficient at current rates of extraction to last 90 years. Current oil production (1.7m b/d in 1988 against capacity of 2.7m b/d) derives from three major sedimentary basins: the Maracaibo basin, the Eastern Venezuelan basin and the Apure-Barinas basin. Exploration is focused on the Orinoco heavy oil belt, the Monagas region and the area south-west of Lake Maracaibo. The extraction, refining and export of oil is undertaken by the state-owned company PDVSA and its subsidiaries. There are six operating refineries which absorb around 50 per cent of crude production. Crude is also processed in the refinery on Curaçao and a number of half-owned refineries in USA and Europe. Emphasis is currently on more flexibility, increasing the share of gasoline and other light fractions refined. Large natural gas reserves (estimated at 1,650bn cu m) in eastern Venezuela are currently being exploited. Annual production of natural gas has remained at around 17bn cu m during the early 1980s. Exploitation of gas reserves is now a priority. A new LNG complex was completed in 1985, and an 800 km gas pipeline is under construction.

Energy: Self-sufficient in oil and natural gas. Emphasis on developing hydroelectric power which contributes 32 per cent of electrical generating capacity and which is planned to replace thermal electricity as the principal energy source by 1989. Major projects include a 1,600 mW hydro-plant on the Uribante and Caparo rivers and the 10,300 mW Guri dam scheme (inaugurated in 1986).

Membership of international organisations: Andean Group, Economic Commission for Latin America, FAO, ILO, IMF, Inter-American Development Bank, Latin American Integration Association, OPEC, Organisation of American States, UN, WHO, World Bank.

policy loans: one for $402m to support the first phase of the structural adjustment programme and the other for $353m, linked to the changes in trade policy. These were the first loans the World Bank had approved for Venezuela since 1974, after which the country became ineligible for financial help as a result of the oil-induced increase in per capita income. They were expected to be followed by other policy loans, including one related to financial sector reform. In addition, it was intended to step up borrowing from the Inter-American Development Bank for infrastructure and other projects.

Both the IMF and World Bank loans provide for the allocation of resources to support debt and debt service reduction under the new initiative to tackle the developing country debt problem launched by US Treasury Secretary Nicholas Brady. The IMF agreed that 25 per cent of the EFF (around $1.2bn) could be set aside for debt reduction and that an additional $700m would be provided to back arrangements with commercial banks designed to lower debt service. Likewise, the World Bank approved the use of 25 per cent of its loans in transactions to reduce debt principal. It was expected that these resources would be augmented by a contribution from Japan. The idea was that the funds obtained in this way would be employed to guarantee the repayment of principal and part of the interest on new instruments issued to replace existing debt and carrying a lower face value or interest rate. Wringing concessions from commercial bank creditors was nonetheless expected to be a difficult and lengthy process.

Championing debt relief

Although turning his back on the interventionist and expansionary policies which characterised his first administration (1974-79), Pérez took up where he had left off in terms of an active, pro-Third World foreign policy. In this he was motivated not only by a conviction that the resolution of Venezuela's economic troubles requires changes at the international level, particularly as regards foreign debt and oil, but also by a deeply-rooted ambition to be recognised as a major world statesman. Given that Venezuela is Latin America's fourth largest debtor, with total obligations of about $34bn at the end of 1988, it is hardly surprising that he focused on the external debt issue. Demonstrating his customary enthusiasm and energy, he untiringly promoted the cause of Latin American debt relief, arguing that social and political stability in the region is threatened by never-ending economic misery, and that lifting the debt burden is a precondition for restoring growth.

Pérez also took up the task of encouraging a greater degree of internal unity within OPEC, of which Venezuela is a founding member. To this end, he called for a meeting of the heads of state of the organisation to reassert its role in the world oil market. In accordance with these aims, his oil minister, Celestino Armas, adopted a conciliatory stance in negotiations over the thorny issue of condensates, or extra-light crudes, which do not form part of OPEC quotas. Using rather liberal definitions, Venezuela had built up its production of condensates over the 1983-88 period as a way of raising export volume without technically violating its quota. However, this became one of the many divisive issues in the organisation, so in the interest of unity the country accepted the recommendation of a technical committee to reclassify 120,000 b/d of its condensates as crude and to lower production accordingly. To avoid a sudden decline in revenues, it was agreed that the reduction would be phased in gradually as the overall quota was increased; thus when the country's production ceiling was lifted by 88,000 b/d to 1,724,000 b/d in July 1989, 40,000 b/d of condensates were reclassified as crudes.

Closer to home, an effort was initiated to achieve a rapprochement with neighbouring Colombia over a long-standing dispute concerning rights to territorial waters in the Gulf of Venezuela, as well as to tackle other issues relating to the 1,400 mile long border between the two countries. Border problems include drug-trafficking, guerrilla attacks, cattle rustling and contraband trade. Reflecting a new spirit of co-operation, after a naval confrontation was narrowly avoided in the Gulf of Venezuela in late 1987, Pérez met with his Colombian counterpart, Vigilio Barco, on the border bridge at Urena. The meeting resulted in the designation of three commissions, the main one headed by former Spanish Prime Minister Adolfo Suarez, to study the delimitation question and other aspects of bilateral relations. Meetings got under way immediately and progress was reported to be encouraging.

As regards Central America, Pérez took a leading role in denouncing the rigging and subsequent annulment of the May 1989 election in Panama by military henchman General Noriega. Venezuela called for a special meeting of the Organisation of American States in Washington, and offered Noriega asylum if he were to step down from power voluntarily. The General angrily rejected these overtures, but Venezuela's standing in Washington apparently benefited from the anti-Noriega, pro-democracy line towards Central America. This helped cement a friendly relationship between the Bush and Pérez governments, which Caracas hoped to turn to its advantage in its negotiations with commercial banks for debt relief.

However, the talks with the bank advisory committee did not get off to a good start. Venezuela's initial proposal for a 50 per cent cut in principal or in interest payments received an immediate rebuff from the advisory committee on the grounds that it was 'unilateral, involuntary and retroactive'. There was a clear feeling among many bankers that Venezuela neither deserved nor needed concerted debt reduction, and that it should use the enhancement resources available from official sources to support market-oriented or voluntary buy-back and exchange offers. The Venezuelans were equally adamant in their request for a binding reduction package along the lines of the arrangement agreed to in principle between the banks and Mexico, and counted on support from the IMF, World Bank and US Treasury to become one of the first beneficiaries of the Brady initiative. The accumulation of interest arrears was without doubt a problematic factor, considered by the Venezuelans as the only means of quickly acquiring new financing which the banks were reluctant to provide, but viewed by the banks as an unacceptable disruption in debt servicing. The scene was hence set for tough and drawn-out discussions.

Economy in tailspin

Although the structural adjustment programme aims at restoring a sustainable rate of economic growth in a low inflationary environment, its initial impact was diametrically the opposite of this due to the size of the pre-existing imbalances. Under the impact of devaluation, the easing of price controls, public sector price adjustments and higher wages, the annual rate of increase in the cost-of-living index shot up to a record 103.4 per cent in May 1989. However, after large initial rises, the monthly pace of advance in the index had slowed down to 2-3 per cent by mid-year, compared with a peak of 21.3 per cent in March. This more moderate pace of monthly increase was expected to continue in the second half but, given the large rises in the first half, inflation for the whole year undoubtedly will have been around twice as high as the previous record of 40.3 per cent established in 1987.

The acceleration in inflation far outstripped the average increase in earnings, giving rise to a sharp downturn in real income and demand. The tightening in fiscal and monetary policies also had a negative impact on demand, as did the abrupt halt to anticipatory buying and the rundown in accumulated inventories. The positive impact of devaluation, in the form of expenditure switching from imports to domestic goods and higher non-oil exports, did little to counteract these negative forces in the short run. This was partly because the non-oil export sector still accounts for only a small proportion of total output, and also due to the fact that it takes time to shift resources into new areas of activity. As a result, the economy entered a deep recession in 1989, characterised by dramatic declines in sales and production in cyclical industries such as motor vehicles, and near-total paralysis in residential construction. When the adjustment programme was launched, government planners predicted that real non-oil GDP would stagnate rather than decline, but by mid-year they were talking of a reduction of around 5 per cent, while private analysts were forecasting an even steeper fall. It hence seemed likely that 1989 would see the most serious slump in output in recent history, exceeding the previous record drop in GDP of 5.6 per cent which occurred in 1983. Thus, much of the gains from the previous recovery in activity were wiped out in one fell swoop, demonstrating the ephemeral nature of growth in the context of financial instability.

The collapse in activity contributed to a major improvement in the current account of the balance of payments, by giving rise to sharply lower import demand and making room for higher non-oil exports. In addition, imports fell as a result of a drawdown in stocks, while oil exports benefited from stronger world prices. The medium and long-term capital account position was also better, due to borrowing from the IMF and World Bank coupled with the suspension of most amortisation payments to commercial banks whilst negotiations took place over a new financial package. Nonetheless, reserves remained under pressure as a consequence of the need to make repayment of the large amount of short-term trade credit which came due in 1989. In effect, much of the large increase in imports which took place in the profligate final year of the Lusinchi administration had to be paid for the year after. Hence, despite the accumulation of interest arrears with commercial banks totalling around $500m, by mid-year central bank international reserves had risen only marginally to $6.9bn (including $3.4bn of gold) from the $6.5bn to which they had fallen at the end of 1988.

Riots and protests

In late February and early March of 1989, well before the full force of the adjustment package had been felt, a wave of violent rioting broke out which turned areas of Caracas and other major cities into a battleground. The disturbances were touched off by a hike in bus fares, and the immediate background to them was a period of growing shortages of basic necessities such as coffee, bread and corn flour. In addition, there was obviously great apprehension about the impact of the impending austerity measures, but at a deeper level the uprising reflected the resentment among a broad section of society about the way in which the country's vast oil income had been squandered as a result of corruption, capital flight and irresponsible policies. At first there was a festive air amongst the demonstrators, as middle class people joined the poor in looting expensive electrical appliances as well as basic necessities. However, the situation quickly got out of control, and the authorities had to take tough action, including the deployment of a large number of army troops, the temporary suspension of constitutional guarantees and the imposition of a curfew. Serious gun battles occurred during the clampdown, as the security forces moved into normally off-limit areas in the slums clinging to the hillsides around

Caracas. Before order was restored, 300 were killed, according to official estimates, with some independent observers putting the death toll much higher.

Following this largely spontaneous outburst, the government faced more co-ordinated protests against its policies from all sides. Realising that a large devaluation was imminent, business interests pushed hard for outstanding letters of credit to be recognised at the old exchange rates so as to avoid exchange losses. This was vehemently resisted by the authorities, as it would have resulted in such large subsidy payments that a correction in the fiscal deficit would have been impossible. In the end an arrangement was worked out to split the losses roughly equally between private sector importers and the government. This did not satisfy some, who took the issue to the Supreme Court but with little hope of success.

More importantly, the organised trade union movement mounted a serious challenge to the austerity programme, which was particularly significant given that the largest union federation, known as the Confederación de Trabajadores de Venezuela (CTV), is controlled by Pérez's AD party. As part of its campaign for pro-labour changes in policies, including further across-the-board wage increases, the CTV organised a one-day national strike, the first such stoppage since 1958 when a number of general strikes took place to support the establishment of democracy rather than for economic reasons. With the government refusing to give in to its demands, the CTV lent its support to a new labour law originally promoted by ex-president Rafael Caldera, founder of the major opposition party COPEI. This was expected to be approved by congress in the latter part of 1989, but only after many of its most controversial clauses, such as wage indexation, had been watered down through amendments. Nonetheless, the CTV did sign a consensus-building document with the government and the private sector business organisation Fedecámaras, suggesting that it did not want to rock the boat too much.

In addition to complaints from business and labour, Pérez faced criticism from within his own party over the adjustment programme. This was partly motivated by the fear that certain elements of the programme would curtail opportunities for patronage and influence-peddling, by limiting the discretionary powers of bureaucrats. It was opposition from within AD that blocked an attempt to reform the organisational structure of the state development corporation for the Guayana region, CVG, which controls the country's basic metal industries. There was also great concern that the austerity measures would damage AD's prospects in the municipal and first ever gubernatorial elections scheduled for December 1989.

Corruption scandal

Investor confidence, vital to the ultimate success of the adjustment programme, was badly shaken by the way in which the initial investigations were conducted into fraud in the allocation of preferential dollars in the Lusinchi administration. The investigations arose from well-founded allegations that, in return for bribes and commissions, the old exchange control office, known as RECADI, allotted hundreds of millions of cheap dollars to phantom companies or to importers who overstated the value of the products they purchased abroad. This gave rise to massive profits for those involved, as the dollars fraudulently obtained from RECADI at controlled exchange rates could be sold in the free market at premium prices.

Although former ministers and the private secretary (and alleged lover) of ex-president Lusinchi were questioned by congress, the criminal proceedings led by Judge Luis Guillermo La Riva focused on the private sector. A wave of arrest warrants was issued against foreign businessmen, including the president and other top executives of the local Ford subsidiary; others were barred from leaving the country and the president of a large domestic grain company was jailed. The scandal took on an international dimension when it was revealed that an arrest warrant was also out for the newly-appointed Economy Minister of Argentina, Néstor Rapanelli. Fearing that they would be unjustly arrested in what was termed judicial terrorism, many foreign executives temporarily left the country. Although nobody can receive a bribe unless someone else pays it, it looked as though the attacks on the private sector were intentionally designed to divert attention away from the guilty politicians and government bureaucrats.

Petrochemicals expansion

Despite the stringent financial environment, the groundwork has been laid for an ambitious expansion programme for the country's petrochemical industry. The aim is to more than triple capacity to some 10m tonnes per year by 1995 at a cost of around $5bn. Joint ventures involving state petrochemicals producer Pequiven, along with both foreign and private domestic companies, will account for around 60 per cent of the planned investments. Given Venezuela's evident comparative advantages in petrochemicals production, foreign multinational companies have shown considerable interest in the programme, and are already involved in a variety of projects. The industry, which was only recently returned to a sound financial footing under the tutelage of state oil company PDVSA, thus looks set on becoming an important element in the country's drive to diversify export revenues.

In a related move, the Pérez administration put an end to further expansion of PDVSA's downstream investments abroad, arguing that the present overseas refining capacity of 600,000 b/d is adequate, and that resources should now be focused on expanding domestic output. To this end, tentative plans were announced to seek foreign participation in the construction of a new $1.3bn refinery with a capacity of 180,000 b/d. If these plans materialise, they will represent a major change in policy towards foreign investment by Pérez, under whose first government the oil industry was nationalised.

However, despite the fact that state aluminium producers Venalum and Alcasa completed their respective projects to expand their existing smelters, serious doubts arose about the feasibility of the plan to raise industrial capacity to 2m tonnes per year by the year 2000. Foreign interest was not encouraged by a political scandal over alleged corruption in the sector, nor by stories of managerial inefficiency in existing companies. Nonetheless, the country is clearly suited to aluminium production in light of its abundant supplies of cheap energy, and expansion was expected to continue but perhaps at a slower pace than originally envisaged.

Outlook

With a degree of macroeconomic stability restored after the traumatic events of 1989, the major challenges in the early 1990s will be to restore economic growth and achieve a more equitable distribution of income. For this to occur, the authorities will not only have to maintain financial discipline but also proceed with the structural reforms required to raise productivity and economic efficiency. Apart from changes in the trade, fiscal and financial areas, a major effort will have to be made to improve the country's transportation and ports system in order to permit non-oil exports to grow rapidly. With the passing of the immediate crisis confronted in 1989, the temptation for the Pérez government to give way to interest-group pressures and grant unwarranted concessions will rise. Pérez's resolve in pushing through structural reform and modernisation will determine whether the 1989 adjustment measures really represented a clean break with the past, or were merely a response to a crisis, after which the country will slip back into the same pattern of behaviour as in the past.

BUSINESS GUIDE

Time: GMT -4.

Climate: Tropical with variations according to altitude. Dry season from Dec to Apr, with mean temperature in Caracas 19°C, rising to 28°C during the day; nights are cool. Rainy season from May to Nov, with mean daytime temperature in Caracas 28°C.

Entry requirements
Passport: Required by all.
Visa: Not required if entering by air for nationals of USA, Canada and most Western European countries – tourist landing card is sufficient, subject to assured onward passage. Otherwise tourist transit ordinary visa required. As well as by certain visitors, ordinary visa is required by those wishing to stay indefinitely and involves some elaborate formalities and a charge of B20 on entry, and B400 plus airport tax on departure. Also, ordinary visa holders require a certificate of clearance from income tax authorities on departure.
Currency: No restrictions on the amount of foreign or local currency or traveller's cheques which may be imported or exported.

Health precautions
Mandatory: None (yellow fever and cholera vaccination certificates have sometimes been required. Check latest regulations).
Advisable: Yellow fever, typhoid, polio vaccinations. Malaria prophylaxis recommended for visits to some rural areas. Water precautions should be taken outside main towns.

Representation overseas: Addis Ababa, Algiers, Ankara, Asunción, Athens, Baghdad, Beijing, Beirut, Belgrade, Berlin, Berne, Bogotá, Bonn, Brasília, Bridgetown, Brussels, Bucharest, Budapest, Buenos Aires, Cairo, Canberra, Colombo, Copenhagen, Damascus, Georgetown, Guatemala City, The Hague, Harare, Havana, Helsinki, Jakarta, Jeddah, Kathmandu, Kingston, Kuwait City, Lagos, La Paz, Libreville, Lima, Lisbon, London, Madrid, Mexico City, Managua, Moscow, New Delhi, Oslo, Ottawa, Panama City, Paris, Port-au-Prince, Port of Spain, Prague, Quito, Rome, St George's (Grenada), San José, San Salvador, Santiago, Santo Domingo, Sofia, Stockholm, Tegucigalpa, Tehran, Tel Aviv, Tokyo, Tripoli, Vatican City, Warsaw, Washington.

Air access
National airline: Viasa (Venezolana Internacional de Aviación), Avensa, Aeropostal, Aerotuy.
Other airlines: Aerolíneas Argentinas, Aeroperú, Air France, Air Panamá International, Alitalia, ALM, American, Avensa, Avianca, British Airways, BWIA, Dominicana de Aviación, Eastern, Ecuatoriana, Iberia, KLM, Lan-Chile, LIAT, Lloyd Aéro Boliviano, Lacsa, LAV, Lufthansa, Pan Am, Swissair, TAP-Air Portugal, Varig.
Tax: Airport tax of B150 payable on departure.
International airports: Caracas-Simón Bolívar (Maiquetía) (CCS), 21 km from city, duty-free shop, bar, restaurant, buffet, bank, post office, shops, car hire; Maracaibo-La Chinita (MAR), 16 km from city, restaurant, car hire.

BUSINESS GUIDE

Surface access
Main ports: La Guaira, Maracaibo, Puerto Cabello, Puerto Ordaz.
Overland access: Possible from Colombia by Caribbean Coastal Highway, or by Pan-American Highway via San Cristobal. The only road from Brazil (via Santa Elena de Uairen) is very rough and is difficult in the rainy season. No direct access from Guyana.

Hotels: Good standard in Caracas and main centres. Graded into classes by Tourism Department on a one to five star basis. Booking in advance is essential. Usually service charge is not added, and there are no seasonal variation of rates. There is a 10 per cent tourist tax.

Car hire: Available in main towns and at airports. National or International licence accepted. A credit card is required. Rates are expensive. Insurance cover is recommended.

City transport
Taxis: In Caracas taxis are metered; meter can be turned off at passenger's request and fare negotiated. Tips are only necessary for metered fares. Higher fares are charged for late night journeys. Shared taxis *(Por Puestos)* widely used. Outside Caracas fares are by negotiation and tend to be expensive.
Buses: City services are cheap, and crowded in rush hours. Services have deteriorated in recent years. Shared taxis are main form of public transport in main towns.
Metro: First stage of the Caracas metro opened in 1983. Fast, comfortable services are operated, linking with Metrobus services.

National transport
Air: Best means of internal travel. LAV (Linea Aéropostal Venezolana), Aeronares del Centro and Avensa operate domestic flights serving most major towns. Over-booking is common and it is advisable to arrive at the airport well before minimum check-in-time. Cancellations and schedule changes are also likely to occur. Unlimited travel tickets are available.
Road: Roads between major towns are built to a high standard. Pan-American Highway runs from Caracas via Valencia and Barquisimeto to the Colombian border; other main highways include: Valencia-Puerto Cabello; Coro-La Ceiba; Caracas-Ciudad Bolívar.
Buses: Services are efficient with frequent journeys between major cities (e.g. Caracas-Maracaibo: journey time approximately 10 hrs.) It is advisable to book in advance.
Rail: Very limited (Barquisimeto-Puerto Cabello; around four trains/day) – eventual expansion is proposed.

Public holidays: 1 Jan, Carnival Monday, Shrove Tuesday, Maundy Thursday, Good Friday, 19 Apr (Act of Independence), 1 May, 24 Jun (Battle of Carabobo), 5 Jul (Independence), 24 Jul (Bolivar's Birthday), 12 Oct (Columbus Day), 24 Dec, 25 Dec.

Working hours
Business: (Mon-Fri) 0800 – 1800 (with long midday break).
Government: (Mon-Fri) range from 0730 – 1530 to 0930 – 1730.
Banking: (Mon-Fri) 0830 – 1130, 1400 – 1630.
Shops: (Mon-Sat) 0900 – 1300, 1500 – 1900.

Telecommunications
Telephone: Dialling code for Venezuela: IDD access code + 58 followed by area code (Caracas 2) followed by subscriber's number. Calls out of the country are often subject to delay and cost more than calls in.
Telex: Public telex system is widely available at hotels and post offices.
Telegram: Service is very slow and often takes longer than airmail letters. Telex or telephone is more advisable.

Banking
Central bank: Banco Central de Venezuela.
Main centre: Main centre: Caracas.
Head offices: Banco Andino, Banco Barinas, Banco Capital, Banco del Caroni, Banco del Orinoco, Extebandes, Banco Federal, Banco Financiero, Banco Principal, Banco Popular, Banco Internacional, Banco Caracas, Banco Consolidado, Banco de Desarrollo Agropecuario, Banco de Fomento Regional Coro, Banco de Fomento Regional Guayana, Banco de Fomento Regional Los Andes, Banco Construcción, Banco de Lara, Banco Latino-Americano de Venezuela, Banco del Caribe, Banco de Maracaibo, Banco de Occidente, Banco del Orinico, Banco Exterior, Banco Hipotecario Unido, Banco Industrial de Venezuela, Banco Italo Venezolano, Banco la Guaira Internacional, Banco Latino, Banco Mercantil Agricola, Banco Metropolitano, Banco Nororiental de Venezuela, Banco Occidental de Descuento, Banco Provincial de Venezuela, Banco República, Banco Royal Venezolana, Banco Unión, Banco Venezolano de Crédito.

Advertising: Widely used in the press, television and radio, cinema and on posters. Tobacco and alcohol advertisements are banned from television and radio. Major areas of coverage are: radio 95 per cent and TV 85 per cent (of population).

Electricity supply: 110 V AC, 60 cycles.

Representation in capital: Algeria, Argentina, Austria, Barbados, Belgium, Bolivia, Brazil, Bulgaria, Canada, Chile, China, Colombia, Costa Rica, Czechoslovakia, Denmark, Dominican Republic, Ecuador, Egypt, El Salvador, Finland, France, Gabon, German DR, German FR, Greece, Grenada, Guatemala, Guyana, Haiti, Honduras, Hungary, India, Indonesia, Iran, Iraq, Israel, Italy, Jamaica, Japan, Republic of Korea, Lebanon, Libya, Mexico, Netherlands, Nicaragua, Nigeria, Norway, Panama, Paraguay, Peru, Poland, Portugal, Romania, Saudi Arabia, Spain, Suriname, Sweden, Switzerland, Syria, Trinidad and Tobago, Turkey, UK, USA, USSR, Vatican City, Yugoslavia.

BUSINESS DIRECTORY

Hotels: Barquisimeto
Hilton, Carrera 5, Nuevo Segovia, P.O. Box: 614 (tel: 536-022; tx: 51464; fax: 539601).

Hotels: Caracas
Avila, Avenida Washington, San Bernadino, P.O. Box: 629 (tel: 515155; tx: 21637).

Centro Ciudad Tamanaco (CCT), P.O. Box: 1060 (tel: 926122; tx: 21991).

Crillón, Avenue Libertador, Con Acacias (tel: 714411; tx: 21203).

Eurobuilding, La Guairita, Chuao (tel: 920403, 951133; fx: 922069).

Hilton, Avenida Libertador/Sur 25, P.O. Box: 6380 (tel: 571-2322/574-2111; tx: 21171; fax: 5750024).

Tamanaco Inter-Continental, Apartado 467, Avenue Las Mercedes (tel: 924522; tx: 23260; fax: 914555).

Tampa, Avenida Solano López, P.O. Box: 51094 (tel: 723771; tx: 24403).

Hotels: Litoral Central

Macuto Sheraton, La Guaira, P.O. Box: 65 (tel: 944300; tx: 31165; fax: (031) 944318).

Meliá Caribe, La Guaira, P.O. Box: 285 (tel: 945555; tx: 31192; fax: 941509).

Hotels: Ciudad Guyana
Inter-Continental Guayana, Puerto Ordaz, P.O. Box: 293 (tel: 222244; tx: 86238).

Hotels: Maracaibo
Inter-Continental del Lago, Avenida El Milagro, P.O. Box: 90 (tel: 912022; tx: 62309; fax: 914551).

Hotels: Valencia
Inter-Continental Valencia, P.O. Box: 762 (tel: 211033; tx: 41342).

Car hire: Caracas
Avis: Sabana Grande Ave Casanova (tel: 718614, 711771, 711737; tx: 28146).

Budget: Av Luis Roche, entra 5a y 6a Transversal Qta Los Irunes, 50 Altamira (tel: 2834333).

Europcar National: Avenida Francisco Miranda-La Castellana (tel: 9863222).

Hertz: Reservations, PO Box 50577 (tel: 715332, 719441; tx: 21283).

Car hire: Caracas International Airport
Avis: (tel: (031)551190).

Hertz: Centro Comercial Plazuela Calle 4/Calle 61 (tel: (061)72616, 71669).

Business advice and information
Asociación Nacional de Comerciantes e Industriales, Plaza Panteón Norte 1, Apdo 33, Caracas.

Banco Central de Venezuela, Ave Urdaneta, Esquina de Carmelitas, Caracas 1010 (tel: 829811/60; tx: 28250; fax: 811437).

Cámara de Comercio de Caracas, Avenida Este 2 No 215, Los Caobos, Caracas (tel: 57123222; tx: 24652).

Cámara de Comercio de Maracaibo, Apartado 414, Edificio Cámara de Comercio, Piso 2, Avenida El Milagro No 86-78, Maracaibo (tel: 061.913255; tx: 61294).

Cámara de Puerto Cabello, Apartado 52, Edificio Cámara de Comercio, Urb Cumboto Norte, Puerto Cabello (tel: 042.69980; tx: 42482).

Consejo Nacional de Comercio y los Servicios, Ed Cámara de Comercio de Caracas, Af Este 2, Los Caobos, Caracas.

Consejo Venezolana de la Industria, Ed Cámara de Industriales, esq. de Puente Anauco, Caracas.

Cordiplan, Palacio Blanco, Avenida Urdaneta, Miraflores, Caracas.

Corporación de Turismo de Venezuela, Centro Capriles, Plaza Venezuela, Caracas.

Federación Venezolana de Casmara y Asociaciones de Comercio y Producción, Ed Federación 5\28 piso, Avenida El Empalme, Urb el Bosque, Apdo 2568, Caracas.

PDVSA (Petróleos de Venezuela), Avenida Liberator, La Campina, Apartado 169, Caracas 1010-A (tel: 7081111).

Superintendencia de Inversiones Extranjeras, CCC Tamanaco, Nivel C-2 Mezzanina 2, Oficina 4-M Apartado 64582 (tx: 2434).

Unión Patronal Venezolana de Comercio, Ed General Urdaneta, 2\28 piso, Marrón a Pelota, Apdo 6578, Caracas.

Veneconomia SRL, Apartado 76.634, Caracas 1070-A (tel: 574511).

Vietnam

Michael Williams

The process of reform in Vietnam – known as *doi moi* – appeared by late 1989 to have run into serious difficulties. A meeting of the Communist Party central committee in August 1989 made it clear that the leadership would not countenance political reform of any sort. The tone set by that meeting was so negative about the prospects of political reform, that it appeared to seriously undermine the chances of economic reform in the country. In April 1989 Vietnam announced that it would be withdrawing all its military forces from Cambodia by the end of September bringing to an end its ten-year involvement in the conflict. But an international conference convened in Paris in August 1989 failed to produce a genuine comprehensive settlement of the Cambodian conflict and continued to see Vietnam denounced by the main supporters of the Cambodian resistance – China, the United States and the non-communist countries of the Association of Southeast Asian Nations (ASEAN). The absence of an international settlement meant that Vietnam's military disengagement was unlikely to see its long sought after reintegration into the international community. There is little doubt that this will have harmful effects on the country's efforts to promote economic reform.

Perestroika maybe; no glasnost

Ever since the sixth congress of the Vietnamese Communist Party in December 1986, the country has committed itself to a programme of economic reform. In taking this course the party seemed to be acknowledging both Vietnam's own dire economic predicament and to be taking note from the example set by other socialist countries who had pioneered the path of economic reform and closer contact with the capitalist world. Before this date Vietnam had extremely limited political and economic contacts with the West since the communist takeover in the south in 1975, and even more so since its 1979 intervention in Cambodia. In response to this isolation Vietnam aligned itself more closely with the Soviet Union, becoming a member of CMEA in 1978. But although Vietnam has made some headway with economic reform since 1986, even attracting a limited amount of foreign investment, and has repeatedly said it desired to expand relations with non-socialist countries, the whole nature of the reform process was thrown into question by a meeting of the Communist Party's central committee in Ho Chi Minh City in August 1989.

A foretaste of the extraordinary hardline that emerged from that meeting had been obvious for some months. In February 1989 the party General-Secretary Nguyen Van Linh warned the country's media against publishing reports that caused a loss of confidence in the Communist Party or the government. In 1987 and 1988 Linh himself had written several critical articles in the press, which are believed to have caused concern to more conservative elements in the leadership. But by 1989, Linh also appears to have become concerned that critical comment in the press was undermining the prestige and credibility of the Communist Party. At the same time Linh delivered the first of several warnings that there could be no question of political pluralism developing in Vietnam and that the Communist Party intended to keep its monopoly of political power.

The concern of the leadership that economic reform and the more relaxed atmosphere that has been ob-

KEY FACTS

Official title: Socialist Republic of Vietnam
Head of state: President Vo Chi Cong
Ruling party: Communist Party of Vietnam
Capital: Hanoi
Official Languages: Vietnamese
Currency: New Dong (D = 100 xu)
Exchange rate: D4,476 per $ (Nov 1989)
Area: 329,556 sq km
Population: 65.2m (1988)
GNP per capita: $150* (1986)
Labour force: 34m (1988)
Inflation: 700%* (1988)
Trade balance: –$1,200m* (1986)
Foreign debt: $10bn (1987), of which $1.5bn in hard currency

* estimated figure

vious since 1986 might get out of hand increased as a result of a series of student demonstrations in May and June 1989. Although largely concerned with demands for better housing and educational facilities, it is more than probable that the students were influenced by the wave of student demonstrations in China at the time. There had also been protests in late 1988 in Ho Chi Minh City by students protesting at the ideological domination of marxism-leninism in their studies. In August 1989 the party newspaper, *Nhan Dan* (People's Daily) said that a minority of extremist students were active in the country. Evidence that Vietnam's leaders, like their counterparts in China, were growing concerned about Western influences, was also clear in July when new restrictions were brought in to guard against the growing use of video recorders to show Western films.

The growing conservatism of leadership statements was first obvious in official comments on developments in other socialist countries. Despite the fact that China and Vietnam have been at loggerheads for more than a decade over the Cambodian conflict, there was no criticism in the Vietnamese media of the bloody suppression in June 1989 of the democracy movement in Beijing. The Vietnamese media simply carried the Chinese authorities' version of events. Conservatives within the Vietnamese leadership must already have been drawing the conclusion that the June events in Beijing were an inevitable consequence of socialist countries embracing reform. More striking and strident were the comments that began appearing from July on developments towards greater democracy and political pluralism in Poland and Hungary. While the Soviet Union, Vietnam's main ally, has been extremely cautious in its comments about Poland and Hungary, the Vietnamese media have denounced the reforms in these countries as bourgeois liberalisation and as representing an attack by imperialist forces against socialism.

It was against this background that the communist leadership met in Ho Chi Minh City on 15 August 1989 to address what it called 'urgent ideological problems'. The communique published at the end of the meeting categorically rejected any possibility of political reform in Vietnam, and also denounced such measures in other socialist countries as being tantamount to the restoration of capitalism. In a closing speech at the conference, the party leader, Nguyen Van Linh, went out of his way to threaten 'severe punishment' for those who organised or incited civil unrest. He also admitted that there had been ideological confusion within the ranks of the Communist Party as a result of recent events in other socialist countries and suggested that in some cases party cadres had organised demonstrations.

But not only was the August 1989 meeting a rejection of political reform. The Cold War rhetoric reserved for denouncing the United States and 'imperialist forces' generally did not sound like a country opening its doors to economic co-operation with the capitalist world and looked as if it would harm the prospects for economic reform. Indeed the call in the official communique after the August meeting to strengthen the forces of international socialism, only underlined the growing isolation of the Vietnamese Communist Party from the profound changes under way within the communist world itself. Having closely aligned themselves with the Soviet Union and eastern Europe for years, Vietnam's leaders seemed at pains to distance themselves from their hitherto closest allies.

Indications that the conservatives within the Vietnamese leadership were strengthening their hand had been clear as early as March 1989, when a politburo meeting failed to produce personnel changes in fa-

COUNTRY PROFILE

Political system: Under the 1980 constitution executive power is exercised by a council of ministers under a prime minister which is responsible to the legislative National Assembly, itself elected to a five-year term by universal adult suffrage. A small Council of State represents the presidency. Membership of the council of ministers generally coincides with that of the Politburo and Secretariat of the Communist Party of Vietnam. Local government is vested in elected provincial, municipal and district people's councils.

Political parties: The Communist Party dominates the political life of Vietnam. Other parties are the Democratic and the Socialist Parties.

Population: The population of around 64m (1988) is growing at about 2.2 per cent p.a., and is expected to reach 85m by 2000. The workforce in 1988 totalled 26m.

Main cities/towns: Hanoi (population 1.75m* in 1985), Ho Chi Minh City (pop 3.5m* in 1985), Da Nang, Haiphong.

Language: Official language is Vietnamese; French is spoken in official circles, and some English spoken in business circles, especially in the south.

Media
Radio: The national radio service broadcasts in Vietnamese, with external service in numerous languages.

Television: There are TV stations in main cities, with over 2m sets in 1982.

Domestic economy: Vietnam's economy is based on a highly centralised communist five-year plan system. This has brought moderate success in repairing the ravages of three decades of war on infrastructure, forests and farmland. But in 1987 and 1988 shortfalls in the communiist system were exposed. Drought fuelled acute food shortages and nearly three million people in the north were close to starvation. Population growth of 2 per cent a year only increases the strains on the economy. So does the fact that Vietnam has 1.3m men under arms, the third largest standing army in the world. More than 30 per cent of total income goes on defence spending. By the mid 1980s the government began to liberalise in a desperate attempt to encourage new economic resources. The dong was devalued by 500 per cent in November 1986 and again in December 1987. In 1986 the Politburo allowed enterprises to raise their own capital to encourage foreign involvement and to be responsible for their own profits and losses. One result was the resurgence of blackmarkets which drove up prices. In 1988 there were new investment laws to attract overseas capital. The law prohibits nationalisation of foreign concerns or joint ventures. To expedite matters, regional investment boards are allowed to grant licences. New Prime Minister Muoi said that Vietnam sought to attract investment from both socialist and capitalist countries. Neither the World Bank nor the ADB have published any economic growth rate figures for the last few years.

Employment: The unemployment rate in 1988 was put at 20 per cent. About 6m people in a workforce of 32m are without full-time jobs.

External trade: All trade is through state agencies. Japan, Singapore, Hong Kong, Thailand, France and Australia all have significant trading links with Vietnam, but usually on a cash or barter basis.
Exports: Total exports in 1987 rose eight per cent in value to $881m. Main exports are coal, minerals (e.g. zinc up 7.6%, chromium), handicrafts and knitted wear (74%), rubber (14%), food and beverages (tea, coffee up 13%, canned and fresh fruit, coconuts and eggs), jute carpet (8.5%), furniture etc. (e.g. embroidery, rattan, bamboo) and footwear. Main destinations are USSR, Hong Kong, Japan, Czechoslovakia, Singapore, Poland, Bulgaria, Hungary, France, Belgium/Luxembourg and other EC countries.
Imports: Total imports in 1987 rose 14.2 per cent in value to $2,409 million. Main imports are machinery and transport equipment, mineral fuel, fertilisers, food (grain, particularly rice, maize, meat, sugar), textiles, (e.g. wool, cotton fabrics), pharmaceuticals. Main sources are USSR, Japan, Hong Kong, Singapore, Czechoslovakia, France, USA, Poland and Italy.

Agriculture: Agriculture accounts for 70 per cent of employment, with 15-18 per cent of the total land area used as cultivated arable. In the south especially, climate and soils are ideal for rice production. Considerable losses can be sustained from typhoons and flooding and drought. In the south, 60 per cent of the land is privately farmed; in the north, 95 per cent of farms have been turned into substantial collectives. A new contract system on the land has spurred a marked improvement in agricultural production in recent years, but in 1987 output was down two per cent to 17.6 million tonnes. In 1988 there was a severe food shortage and the government had to introduce emergency measures. It cut rice rations, called on farmers to produce fast-growing vegetables and called for food aid of 108,000 tonnes. Ambitious plans include increased use of fertilisers, development of irrigation systems and resettlement of small farmers. Vietnam is by no means self-sufficient in rice. Other main food crops include sweet potatoes, cassava, maize, sorghum, pulses, groundnuts, sugar-cane, fruit and vegetables; coffee, cotton, cashews, coconuts, tea, tobacco, jute, natural rubber, flax and opium are grown for the domestic and export markets. Livestock raised includes pigs, buffaloes, cattle, sheep, goats, horses and poultry. Marine and river fisheries are an important source of protein, and shrimps and other seafood are exported. Marine fisheries account for some 70 per cent of the catch. About 50 per cent of the total land area is forest, and forestry is developing, with 12-15 per cent of the removed volume of timber classified as industrial wood.

Industry: Most industries, particularly heavy industries, are located in the north, and have been adversely affected by recent conflicts with China. There has been encouraging growth, however, since 1985 and output was up 6.7 per cent in 1987. Main industries are state-owned, and include food processing, synthetic yarns and fabrics, textiles, engineering, cement, fertilisers, glass, rubber products, chemicals, paper and steel. Consumer goods have grown less rapidly than basic products.

Mining: Minerals, mainly located in the north, include coal (reserves of 20,000m tonnes), phosphates, salt, tin, copper, chromium, wolfram, lead, zinc, silver, nickel, gold, bauxite, iron ore, manganese, titanium, apatite and limestone.

Hydrocarbons: There has been exploration for oil and gas in the Mekong Delta and on the Nam Bo continental shelf, and exploitable reserves have been located. Electricity generation is still very inadequate.

Membership of international organisations: Asian Development Bank, Colombo Plan, CMEA, ESCAP, FAO, G-77, IAEA, IBRD, ICAO, IDA, IFAD, IFC, ILO, INTELSAT, IMF, IMO, IRC, ITU, International Bank for Economic Co-operation, International Investment Bank, NAM, UN, UNDP, UNESCO, UNICEF, UPU, WFTU, WHO, WIPO, WMO, World Bank, WTO.

vour of reformers. A subsequent statement noted that the meeting had 'rejected appeals for political changes'. A special party conference scheduled for March and expected to make major changes in the leadership did not take place. Speculation was intense in Vietnam that the growing conservative tone of the leadership reflected the influence of veteran old guard leader, Le Duc Tho, who was due to retire from the politburo in December 1989, but still wields considerable influence in the Communist Party. Some observers have suggested that he is opposed to the leadership of Nguyen Van Linh and is trying to manoeuvre his younger brother, Mai Chi Tho, currently minister of the interior, into a more powerful position with a possible eye on the future leadership of the party.

Halting economic reforms

In the initial period after the sixth congress of the Communist Party in December 1986, considerable progress was made in reforming Vietnam's economy. However, this has to be measured against the almost total stagnation that had existed in the previous ten years. To all intents and purposes the economy only survived through massive injections of Soviet aid, currently running at US$2.5bn a year. It is clear that pressure from the Soviet Union, as well as Vietnam's own dire economic predicament, were the main motivating factors behind the reform.

Despite apparent reservations by conservative sections of the leadership, especially with regard to the social consequences, restructuring or *doi moi*, proceeded apace in 1988. The aim of the reforms was to reorient investment toward agriculture, light industry and exports, rely more on the market place and break out of the international embargo placed on Vietnam because of its military presence in Cambodia. But while a small number of people benefited handsomely from the resurgence of private enterprise, the vast majority of the population, estimated at 65m, has yet to see an improvement in their impoverished living standards. The pace of reform was brutally interrupted in 1988 when a combination of bad weather and inept economic management took the country to the brink of disaster. Hyperinflation, unofficially estimated at 700 per cent, ravaged the country and food shortages left millions hungry. On top of this the accelerating reforms brought turmoil and convulsions to the primitive economy. State enterprises, deprived of subsidies, stagnated while the unshackled private sector threatened to run out of control.

By early 1989, however, the economy was beginning to show signs of recovery. Food prices and currency rates steadied after their dizzying spiral in 1988. A new competitive environment was obvious, fuelled by a surge in imports. Foreign investors visited Vietnam in 1989 on a scale not seen since the takeover in the south in 1975. In the countryside, where 80 per cent of the population lives, conditions have improved considerably since early 1988, when some 7m people were close to starvation for several months. Food production in 1988 finally totalled 19m tonnes of paddy, considerably up from an abnormally low 17.6m tonnes in 1987. Prospects of reaching 1989's target of 20m tonnes looked reasonable.

The most impressive reforms in 1988-89 were in the financial sector. A single exchange rate was introduced that accurately reflected the purchasing power of the dong against the dollar. In addition there was an almost free market in gold. A floating exchange rate of 4,500 dong to the dollar was set in March 1989, replacing a dozen or more official rates. The rate will be adjusted periodically to keep it close to the black market rate for dollars, which by June 1989 had fallen to 5,000 dong. In the recent past, the black market rate has often hovered around ten times the official rate.

But restructuring or *doi moi* alone cannot overcome the Vietnamese economy's multitude of ills, which include a pre-industrial structure, lack of capital, a shortage of qualified personnel, inadequate institutions and a stifling bureaucracy. The plan for 1989 sets further liberalisation as a priority, but the government has shown signs of worry at the social consequences of full-blooded reform. In the absence of greater discipline of the state sector, the targets for 1989 of 10 per cent growth in industrial production and 7.5 per cent in GNP looked over-ambitious. The government is also disappointed with the response of foreign companies to the liberal investment code it introduced in 1987. But Vietnam's inherent problems of a dilapidated infrastructure, continuing high inflation, foreign debt totalling some US$9bn and poor international credit rating has slowed down potential investment. Apart from production-sharing oil exploration deals signed with four international groups, including BP and Shell, foreign investment agreed in 1988 totalled only $130m for projects, with a total investment of $190m. If the oil deals are included the figure rises to $360m.

A key element in Vietnam's economic recovery programme has been the negotiations it has been conducting with the IMF. The main aim of the negotiations has been to agree on a short-term economic stabilisation plan that would clear the way for IMF assistance in removing the country's arrears to the organisation, which in turn would give Vietnam access to IMF funds for development. At the beginning of 1989 those arrears stood at US$135m. In the first six months of 1989 Vietnam resumed interest payments on the arrears, for the first time since 1985, paying out some $10m. Many of the policies recommended by the IMF have been implemented, but officials say IMF calls for even more stringent economic policies are inappropriate and need to be negotiated.

Foreign relations

Vietnam's foreign relations remained bedevilled by the Cambodian conflict. The announcement in April 1989 by Hanoi that it would withdraw all its forces by the end of September led to expectations in Vietnam that there would be a breakthrough in relations with the West, ASEAN and China. But the failure to secure an internal settlement between the warring Cambodian factions prevented any dramatic improvement in relations with the outside world.

Relations with China did show a modest improvement in 1988-89. Border trade along their mutual border was resumed and plans are in hand for upgrading roads near the border. In January 1989 a Vietnamese deputy foreign minister visited Beijing for talks on improving relations, but the Cambodia conflict as well as rival claims to the disputed Spratley Islands in the South China Sea prevented any real progress in bilateral relations.

There were no signs either of any dramatic improvement in relations with the US. Vietnam has long expressed the wish to see diplomatic relations established, but Washington has consistently said this is impossible as long as the conflict continues in Cambodia and the problem of US servicemen missing in action (MIAs) remains. In July 1989 the two countries did announce that they had reached agreement on the resettlement in the USA of thousands of Vietnamese who worked for the former Saigon regime. The first group of 3,000 were expected to leave by the end of 1989.

An unexpected visitor to Vietnam in April 1989 was a British foreign office minister, Lord Glenarthur. His visit was the first by a British minister since 1975. The visit marked the growing concern of the British government at the continuing large numbers of Vietnamese 'boat-people' reaching Hong Kong. It was reported subsequently that the two countries had reached agreement on the mandatory repatriation of considerable numbers of Vietnamese boat-people back to Hong Kong. Opposition from the United States, however, delayed implementation of the agreement.

In April 1989 the Vietnamese leader Nguyen Van Linh visited the Soviet Union for talks with President Gorbachev. The visit came one month before the Soviet leader's historic trip to China and indicated Vietnam's concern that its interests might be sacrificed at the Beijing summit. There were also signs of Vietnamese nervousness that Soviet economic aid would be substantially reduced in the next five year plan from 1990-95.

BUSINESS GUIDE

Time: GMT + 7 hrs.

Climate: The summer climate is warm and sultry, with monsoon rains in the south during Apr-May; winters are cool and dry. Typhoons occur on the eastern plains and foothills in spring and autumn. The likely temperature range for Hanoi is from 20°C in Jan to 26-33°C in Jun-Aug; for Ho Chi Minh City, from 25-35°C in Apr-May to 21-30°C in Nov-Jan.

Entry requirements
Passport: Required by all.
Visa: Entry and exit visas are required by all, and application should be made well in advance of any visit planned. Special permits are required for travel in the countryside. It is possible to extend visas once in Vietnam.
Currency: The Bank of Foreign Trade of Vietnam handles all legitimate foreign exchange transactions. Widespread black market in operation. Import and export of local currency is prohibited.

Health precautions
Mandatory: Vaccination certificates are required for cholera.
Advisable: Anti-malaria precautions.

Representation overseas: Addis Ababa, Aden, Algiers, Antananarivo, Baghdad, Bamako, Bangkok, Beijing, Belgrade, Berlin, Bonn, Bucharest, Budapest, Cairo, Canberra, Colombo, Conakry, Dakar, Damascus, Dar es Salaam, Dhaka, Havana, Jakarta, Kabul, Kuala Lumpur, London, Luanda, Managua, Manila, Maputo, Mexico City, New Delhi, Oslo, Ottawa, Paris, Phnom Penh, Prague, Pyongyang, Rangoon, Rome, Sofia, Stockholm, Tirana, Tokyo, Tripoli, Tunis, Ulan Bator, Vientiane, Warsaw.

Air access
National airline: Hang Khong Vietnam.
Other airlines: Aeroflot, Air France, Garuda, Lao Aviation, Interflug, Philippine, Thai Airways.
International airports: Noi Bai (HAN), 45 km from Hanoi; Ho Chi Minh City (SGN), 7 km from Ho Chi Minh City.

BUSINESS GUIDE

Surface access: Status of overland routes (e.g. Hanoi-Guangzhou) should be checked, as passage has not always been practicable.
Main ports: Da Nang, Haiphong, Ho Chi Minh City.

Hotels: Visitors may have no choice in accommodation, which is often officially allocated. Hotel bills must be settled in US dollars. Travellers' cheques may not be accepted, or may be subject to a surcharge.

Car hire: Car transport is normally arranged by Vietnam Tourism. Car hire bills must be settled in US dollars. International driving licence is required.

City transport
Taxis: Hire is normally arranged through official channels and can be obtained on time-basis. Some non-official vehicles are available in Ho Chi Minh City although they should be used with discretion and care. 'Discovery' trips are discouraged and destination difficulties are likely.

National transport: Travel by foreign visitors has been restricted, and visitors should check the current position on the scope of movement allowed.
Air: Hang Khong Vietnam operates domestic services.
Road: There are over 4,000 km of main roads in the 347,200 km road network.
Rail: There are two-class rail services between main centres, including Hanoi-Ho Chi Minh City, Hanoi-Haiphong, Hanoi-Lao Cai, Loc Ninh-Ho Chi Minh City-My Tho and Hanoi-Lang Son. Access to rail services is difficult or impossible for foreigners.

Public holidays
Fixed dates: 30 Apr (Unification Day), 2 Sep (National Day).

Working hours
Government and business: (Mon-Sat) 0800 – 1200, 1300 – 1630.
Banking: (Mon-Fri) 0800 – 1630; (Sat) 0800 – 1200.

Telecommunications
Telephone: Dialling for Vietnam is via the International Operator and, in practice, it may not be possible to get through. International calls from Vietnam are slow.
Telex: Available through central post office in Hanoi or Ho Chi Minh City.

Banking
Main centre: Hanoi.
Head offices: Bank of Agriculture, Bank of Foreign Trade of Socialist Republic of Vietnam, Bank of Investment and Reconstruction, State Bank of the Socialist Republic of Vietnam.

Weights and measures: Metric system.

Representation in capital: Albania, Algeria, Australia, Belgium, Bulgaria, Burma, China PR, Cuba, Cyprus, Czechoslovakia, Egypt, Finland, France, German DR, German FR, Hungary, India, Indonesia, Iraq, Italy, Japan, Kampuchea, Korea PDR, Laos, Malaysia, Mexico, Mongolia, Netherlands, Philippines, Poland, Portugal, Sweden, Switzerland, Thailand, Turkey, UK, USSR, Yugoslavia.

Business advice and information
Chamber of Commerce of the Socialist Republic of Vietnam, 33 Ba Trieu Street, Hanoi (tel: 529862); 69 Dong Khoi Street, Ho Chi Minh City (tel: 20101/2).

Foreign Trade Bank of Vietnam, 7 Le Lai Street, Hanoi (tel: 52831; tx: 4504).

Vietnam National Foreign Trade Corp (TRANSAF), 46 Ngo Quyen, Hanoi.

Travel information
Vietnam Tourism, 54 Nguyen Du Street, Hanoi (tel: 52986; tx: 296).

Hotels: Hanoi
Dan Chu.

Hao Binh.

Thang Loi.

Thong Nhat.

Hotels: Ho Chi Minh City
Cuu Long (formerly Majestic), 1 Dong Khoi St, Quani (tel: 95515; tx: 275276).

Doc Lap (formerly Caravelle), 19 Pl Lam Son, Tu Do Street (tel: 93704; tx: 8259).

Wallis and Futuna

(map on page xxiv)

COUNTRY PROFILE

Political system: Wallis and Futuna is an Overseas Territory of France, administered by an appointed Administrator (Administrateur Supérieur) and represented in the French parliament by a deputy and a senator. Local affairs are conducted by an elected Territorial Assembly and traditional councils of rulers and leaders.

Main town: Mata Utu (on Wallis).

Domestic economy: The economy is based on subsistence agriculture and fishing, with some copra exports. Remittances from migrant workers and grants from France are the other sources of income.

External trade
Exports: Copra is the only export, averaging 100 tonnes per year, but none was exported in 1987 and 1988.
Imports: Main imports are food (about 33 per cent of total), raw materials, fuel, machinery and transport equipment. Main sources are France, Australia, New Zealand and New Caledonia.

Agriculture: The soil of the main islands is volcanic and rainfall is adequate. The main subsistence crops are coconuts, ignamo, taro, cassava, bananas, breadfruit, yams and pineapple. There is some pig, poultry and cattle raising, and tuna fishing for local consumption. Timber is logged for local consumption, and some pine reafforestation has been undertaken.

Industry: Industrial activity is limited to very small-scale copra processing and handicrafts.

Membership of international organisations: Franc Zone, South Pacific Commission (Conference representation), Territory Associated with the EC.

Time: GMT + 12 hrs.

Climate: Hot and humid, although Jul-Aug can be dry and cool.

Air access
Other airlines: Air Calédonie International, UTA.
International airport: Wallis Island (WLS). Most flights arrive via Noumea. A flight from Noumea to Tahiti and return, via Wallis, started on 9 June 1989.

Surface access
Main ports: Wallis Island.

Hotels: There is a small guesthouse and three hotels on Wallis (20 rooms).

National transport
Air: Air Calédonie International operates an inter-island service between Wallis and Futuna islands (three planes weekly).
Road: There are surfaced roads in Mata Utu, and a road network links main centres on Wallis. Car hire is available, and minibus services operate on Wallis. There is a running-track all around Futuna.
Water: One boat monthly from Noumea (New Caledonia) to Wallis and Futuna.

Telecommunications: Limited radio link is maintained. Communications by satellite from August 1989.
Telephone: Available via International Operator.

Banking: Banque Indosuez.

Weights and measures: Metric system.

KEY FACTS

Official title: Wallis and Futuna

Head of state: President of France

Head of government: Administrator, M. le Préfet Roger Dumec

Ruling party: Rassemblement pour la République, Mouvement des Radicaux de Gauche

Capital: Mata Utu (on Wallis)

Official Languages: Wallisian, French

Currency: CFP Franc (CFP Fr = 100 centimes)

Exchange rate: CFP Fr113 per $ (Nov 1989)

Area: 274 sq km – Wallis (or Uvea): 6,000 hectares; Futuna: 6,400

Population: 12,500* (1985)

BUSINESS DIRECTORY

Travel information
Air Calédonie International, Magenta Airport, PO Box 212, New Caledonia (tel: 272177; tx: 112 AIRCAL NM).

Hotels: Mata Utu
Moana Hou, P.O. Box: 136 (tel: 72-21-35).

Teone, P.O. Box: 36 (tel: 72-29-19).

Hotels: Wallis
Lomipeau chez Paola, P.O. Box: 84 (tel: 72278/33; tx: 076 WF).

Western Samoa

Kevin Rafferty

(map on page xxiv)

Western Samoa's vulnerabilities and dependence on its neighbour New Zealand were exposed in 1989 when the Wellington government suggested a cut in Samoa's annual immigration quota of 1,100. The move brought an outcry in Apia. The then Deputy Prime Minister Geoffrey Palmer (who later became David Lange's successor as prime minister) flew to Samoa to try to get agreement. But he had to retire hurt saying 'I came to consult and that is all I have done.' Later the New Zealand government started a crackdown on an estimated 6,000 visitors from various Pacific island countries who had overstayed their permits in New Zealand.

It is easy to understand the New Zealand concern. Its Employment Minister Phil Goff said with unemployment at a record 170,000 New Zealand could not afford to allow 1,100 people free entry. But it is equally easy to understand Western Samoa's outrage both on political and economic grounds. The quota had been agreed back in 1982 after the two countries signed a deal stripping almost 100,000 Samoans of New Zealand citizenship that the Privy Council had said was theirs as of right.

A huge protest march was launched in Samoa against Wellington's plans. Former Attorney General Misa Foni Retzlaff spoke for local opinion when he said that New Zealand had conned Samoans out of their rights in 1982 and was trying to do it again. Acting Prime Minister Patu Afaese suggested to the visiting Palmer that New Zealand should put its own house in order first.

KEY FACTS

Official title: Independent State of Western Samoa

Head of state: Malietoa Tanumafili II

Head of government: Prime Minister Tofilau eti Alesana

Ruling party: Human Rights Protection Party

Capital: Apia (on Upolu)

Official Languages: Samoan, English

Currency: Tala or Western Samoan Dollar (WS$ = 100 senes, or cents)

Exchange rate: WS$2.30 per $ (Nov 1989)

Area: 2,840 sq km (9 islands) – Savai'i (1,708); Upolu (1,118)

Population: 163,000 (1988)

GNP per capita: $580 (1988)

Inflation: 8.4% (1988)

Trade balance: –$53.4m (1988)

Foreign debt: $80.2m (1987)

Visitor numbers: 60,000 (1986)

Economy

Remittances sent by expatriate Samoans mainly in New Zealand, but also in American Samoa, Hawaii and the mainland United States, help to keep afloat an economy that has been consistently under pressure for almost two decades. As an International Monetary Fund report records 'GDP per capita, which is at present special drawing rights (SDR) 500 (approx US$630 in 1989) annually, increased little in real terms over the past two decades. Economic growth was constrained mainly by the slow growth of output in the plantation sector. As a result, exports declined steadily in relation to GDP and their coverage of imports fell from four-fifths to only one-fifth.'

To get the economy into balance, the government of the mid-1980s imposed virtually a siege economy. Arrears in external payments were having a cumulative negative impact, affecting the country's ability to attract capital and service debts. The tough measures had an effect. Commercial arrears, largely because of oil imports, were repaid. Arrears on government and government-guaranteed debt were partly repaid and partly rescheduled. But it was a very hard struggle, and the domestic economy was not helped by the decline in world prices of coconut products which saw exports halved from the 1984 value of almost $20m. Official figures show that the external current account has continued to register surpluses, thanks to the remittances, some earnings from tourism, helped along by lower prices of petroleum imports.

IMF figures show the trade deficit continuing at around SDRs35m a year (Tala 96m), but a current account in surplus from 1983 onwards and in credit to the tune of SDRs9.4m in 1986 and 7.5m in 1987. The debt service ratio was reduced from 18.2 per cent in 1984 to 12.5 per cent in 1987 with external debt falling slightly from 80 per cent of GDP to 77.7 per cent. Thanks to the improved external position, inflation measured by the Consumer Price Index was reduced sharply from 24 per cent in 1983 to less than 3 per cent in 1986 though it rose thereafter.

Development prospects

The way forward for the country's economy is difficult. Coconuts and cocoa account for two-thirds of agricultural output, but production has been hit by disease and the ageing of trees. Various attempts have been made at rehabilitation, but replanting has not been very effective, either at commercial or at smallholder level. Production of bananas, once a major export, ceased. Other crops like passion fruit, coffee and mangoes which might be exportable are grown almost entirely for domestic consumption. To make progress and improve agriculture, the country would probably have to change the land tenure system – but this would hit at the traditional social and political system.

Western Samoan villages normally consist of 40 to 50 households presided over by a council of chiefs. Land allocations and distribution of earnings are at the discretion of the chief, so that the system acts as a disincentive for family financed capital improvement. The political system is also on a *matai* (chiefly) suffrage, in which the *matai* representing family units, can vote and stand as candidates in all but two of the constituencies. The total number of people eligible to vote is about 21,000 out of a total population of more than 160,000. Politicians have promised to look at introducing universal suffrage, but have not gone further than this.

There are prospects of the development of the non-agricultural sector, but these have been handicapped by shortage of skills and good managers. Unlike other Pacific islanders the Western Samoans have largely neglected the resources of the seas around them. The government is pinning its greatest hopes on expansion of tourism and would like to see 200,000 visitors a year by the early 1990s, more than double present rates. The country is physically very beautiful. Compared with other Pacific islands its air links are good, but this is a relative matter. If Hawaii is fearing that its future is under threat from the new long-range Jumbo jets, Samoa will have to fight hard for its tourists.

KEY INDICATORS

	Unit	1984	1985	1986	1987	1988
Population*	'000	159	162	161	162	163
Exports	WS$m	36.8	66.7	23.5	25.0	31.4
Cocoa beans	WS$m	2.39	2.35	2.26	2.62	1.26
Coconut oil	WS$m	20.78	15.62	6.54	8.73	11.69
Re-exports	WS$m	961	3,782	1,158	1,822	1,666
Imports	WS$m	93.29	115.07	105.38	131.01	157.30
Balance of trade	$m	–27.2	–30.45	–32.34	–44.02	–53.40
Current account	$m	0.66	1.78	7.58	7.33	3.88
External debt	$m	74.3	75.2	74.8	80.2	–
Debt service ratio	%	9.7	11.1	10.9	6.6	–
Exchange rate	$ per WS$	0.54	0.45	0.45	0.47	0.48

* estimated figure

COUNTRY PROFILE

Political system: The head of state, now hereditary but to be elected in future, appoints an executive prime minister who must have majority support of the 47-member legislative assembly. Executive power rests with the prime minister and his appointed cabinet.

Political parties: Human Rights Protection Party (25 seats in the legislative assembly, following Feb 1988 elections); Samoa National Development Party (22 seats).

Population: The population is of Polynesian origin, and is growing at a low 0.5 per cent per annum.

Main town: Apia (on Upolu, population 32,196 in 1987).

Language: Samoan is the official language, but English is widely spoken.

Media
Press: Observer, Samoa Times, Samoa Weekly, South Seas Star.
Radio: 24-hour radio broadcasts in Samoan and English. Some external services are received.
Television: The government of Fiji is studying proposals to set up a TV service for Fiji, Western Samoa and Tonga via satellite from a ground station to be built in Fiji. It is possible to pick up TV broadcasts from American Samoa.

Domestic economy: Small, subsistence economy based on agriculture with a few developed resources and a large population, dependent on coconuts and cocoa, remittances from migrant workers and aid from New Zealand. Industrial development limited by small domestic market and remote location. A small but growing tourist industry; Apia has become an international air staging-post for Pacific routes. A 10-year tourist development plan to 1993 designed to expand industry to around 100,000 visitors p.a. In 1987 visitors numbered 45,000. Earnings are about $10m a year. In 1985, Japan agreed to give 433m yen in grant aid to help improve port facilities. In 1988 offshore banking centre was approved by parliament.

External trade: Western Samoa is one of 32 developing nations designated in 1985 to receive exemption from US competitive trade requirements in accord with the US Generalised System of Preferences.
Exports: Exports in 1987 were valued at $11m. Main exports are coconut oil, cocoa, taro, timber, veneer, bananas, canned coconut cream, other food products, handicrafts and textiles. Main destinations are New Zealand (30.3%), Australia (21.37%), German FR (21.3%), USA (9.3%), American Samoa (8%) and Netherlands (2.5%).
Imports: Imports in 1987 were valued at $66m. Main imports are food and live animals, machinery and transport equipment, fuel, manufactured goods, textiles, beverages and tobacco, iron and steel, paper. Main sources are New Zealand (31%), Australia (20%), other Pacific states (14.9%), USA (5%), UK (1.9%).

Agriculture: Agriculture accounts for about 50 per cent of GDP and employs two-thirds of the population. Fertile volcanic soil, adequate rainfall. Export crops of coconuts, cocoa, bananas, taro and taamu produced on plantations and smallholdings. The latter also produce subsistence crops of yams, taro, breadfruit, maize, vegetables and fruit; also major replanting of coconut plantations (for copra). Other produce under development include macadamia nuts, annatto (dye), timber and cattle. Some development of fisheries with refrigeration facilities. Some timber exported.

Industry: Small-scale manufacturing and industry has expanded. The government's industrial area of Vaitele (on Upolu) houses a brewery, a cigarette factory and a match factory. Other industries include copra processing, food processing, light engineering, wood-working and manufacture of coconut oil, paint, concrete and construction materials, bottled gases, leather goods, corned beef and garments. Samoa Forest Products Ltd has upgraded plant and equipment at its Asau sawmill. Pacific Electrical Products is to assemble electronic printed circuit boards and receivers. In 1987, 45,000 tourists visited the country, contributing $10m. 40-45 per cent came from Pacific area (mainly Australia) and 22 per cent from USA. Some hotel construction is in progress, with government encouragement.

Energy: The Electric Power Corporation has recently constructed a wood-fuel power station with $12.8m in soft loans from Kuwait, Saudi Arabia and New Zealand. There are three hydro-electric plants with a fourth on the way.

Main companies: Western Samoa Trust Estates, Western Samoa Breweries, Samoa Coconut Products, Rothmans Tobacco.

Membership of international organisations: Asian Development Bank, Commonwealth, ESCAP, FAO, G-77, IBRD, IDA, IFAD, IFC, IMF, Lomé Convention (EC), South Pacific Bureau for Economic Co-operation, South Pacific Commission (participating member), South Pacific Forum, UN, UNESCO, WHO, World Bank.

Time: GMT -11 hrs.

Climate: Temperatures 24-30°C (hottest in Mar) and high humidity. Rainy season Nov-Apr, rainfall at least 5,000 mm/year, heaviest in Jan.

Entry requirements
Passport: Passport or Certificate of Identity with photograph required by all.
Visa: Entry permits issued for stays of up to 30 days for visitors with proof of onward passage and sufficient funds. Visas required for longer stays.
Currency: Import and export of local currency is prohibited. No limits on foreign currency.
Customs: Personal effects allowed duty-free. Firearms, ammunition, explosives, drugs and pornography are prohibited. Plants and animals subject to approval from the Department of Agriculture and Forestry.

Health precautions
Mandatory: Vaccination certificates for yellow fever if travelling from an infected area.

Representation overseas: Wellington. There is a Western Samoa Consulate in Auckland, a Samoan mission to the United Nations in New York and a Western Samoan Commission in Brussels.

Air access
National airline: Polynesian.
Other airlines: Air New Zealand, Air Pacific, Hawaiian, Samoan Air.
Tax: Airport tax of US$20 on departure.
International airport: Faleolo International (APW), 35 km from Apia.

Surface access: Ferry services operate to American Samoa; shipping services to New Zealand, Australia, Japan and other Pacific islands, as well as Europe and USA.
Main ports: Apia, Asua.

Hotels: In addition to Apia's hotels, there are hotels in rural areas: Aggie Grey's Hotel (Apia), Harbour Light (Apia), Olivia's Casual Accommodation (Apia), Safua Hotel, Taffy's Paradise Inn (Savaii), Tusitala Hotel, Valentine's Parkers (Apia), Vaisala Hotel.

Credit cards: Major cards (American Express, Visa) are accepted.

Car hire: International or national driving licences required.

City transport
Taxis: Taxi service is available in Apia.

National transport
Air: Polynesian Airlines operates regular services between Maota (near Salelolga), Asau and Savai'i, and a service to Fagali'i Airport (on Upolu).
Road: The extensive road network is being improved. Traffic drives on the left.
Buses: Scheduled bus services operate in and around Apia; other services in most islands.
Water: Ferry services operate between Apia and Savai'i.

Public holidays
Fixed dates: 1 and 2 Jan, 25 Apr, 1-3 June, 25 and 26 Dec.
Variable dates: Good Friday, Easter Monday, Monday after Whit Sunday (Oct).

Working hours
Government: (Mon-Fri) 0800 – 1200, 1300 – 1630.
Banking: (Mon-Fri) 0830 – 1500.
Shops: (Mon-Fri) 0800 – 1200, 1330 – 1630; (Sat) 0800 – 1230.

Social customs: Tipping is uncommon.

Telecommunications: The Telecommunications Corporation of Western Samoa formed in 1985 a 10-year plan for development of a comprehensive communications network prepared by the International Telecommunications Union and the South Pacific Bureau for Economic Co-operation (SPEC). Telecommunication facilities have improved with completion of an earth satellite station.
Telephone: Dialling code for Western Samoa, IDD access code + 685 followed by subscriber's number (no area code necessary). Over 7,000 phones in 1987, with automatic service in Apia.
Telex: Automatic telex exchange provides telex service.
Telegram: Telegram service available.

Postal service: Even air mail takes several weeks.

Banking
Central bank: Central Bank of Samoa. *Head offices:* Bank of Western Samoa, Development Bank of Western Samoa, Pacific Commercial Bank.

Advertising: Most weekly newspapers accept advertising; facilities available on Radio 2AP. Cinemas do not offer advertising facilities.

Electricity supply: 230 V AC.

Weights and measures: Imperial system, with metric systems in use.

Representation in capital: Australia, China PR, New Zealand.

Hotels: Apia, Upolu
Aggie Grey's, P.O. Box: 67 (tel: 22880; tx: 257).

Samoan Hideaway Beach Resort, P.O. Box: 1191.

Tiafau, Sogi Street, P.O. Box: 34 (tel: 21141; tx: 262).

Tusitala, P.O. Box: 101 (tel: 21122; tx: 226).

Hotels: Savai'i
Safua, Saleologa, P.O. Box: 5002 (tel: 24262).

Business advice and information
Bank of Western Samoa, Beach Road, PO Box 187, Apia (tel: 22422; tx: 258).

Central Bank of Samoa, Private Bag, Apia (tel: 24100; tx: 200 Sx).

Chamber of Commerce and Industry, Nelson's Building, Beach Road, PO Box 2014, Apia (tel: 21728).

Trade Information Centre, Nelson's Building, Beach Road, Apia.

Travel information
Western Samoa Visitors' Bureau, PO Box 862, Apia (tel: 20878; tx: 220A PTRGRMS SX; cable: MALO; fax: 20886).

Yemen Arab Republic

Tim Owen

In early 1989 a new, ostensibly economic bloc of Arab countries, the Arab Co-operation Council (ACC), was set up by Iraq, Egypt, Jordan and the YAR. What economic benefits the YAR expects to derive from this is far from clear. The oil industry is now established and its benefits are beginning to accrue. The YAR's relations with the People's Democratic Republic of Yemen (PDRY) may be expected to improve should current negotiations, still in their early stages, between the regime in Aden and the adherents of ex-President Ali Nasser Mohamed, now in the YAR, achieve success.

A sense of political isolation

Having achieved a large measure of internal stability after years of political turbulence and convulsions since the revolution in 1962, which overthrew the Imamate, the YAR has recently been experiencing a sense of isolation in the Arab world. In spite of a long-standing agreement with the PDRY on the principle of unification of the two Yemens, relations with the PDRY remain ambivalent, whilst the YAR is deeply aware of being overshadowed to the north by its large and wealthy neighbour, Saudi Arabia, which, through its economic aid, wields considerable influence over the policies of any government in the YAR. Approaches to the Gulf States to join the Arab Gulf Co-operation Council (AGCC) were politely rebuffed. It is almost certainly against this background that the YAR joined the ACC. It can hardly be on the grounds of getting economic assistance from the other member States, as the one thing they all have in common is their considerable economic problems. In other respects, the YAR is physically separated from the other member States by considerable distances.

International relations

Since the revolution the YAR has attempted to maintain a position of neutrality in international relations. With a marxist regime to the south and a staunchly anti-communist regime to the north, the YAR has found it necessary to keep in with both sides of the ideological spectrum. In practice this means that the government has accepted economic and military assistance from both east and west. On occasion Saudi Arabia has shown its displeasure when it has considered that the YAR has been leaning too far towards the east by reducing its economic aid, only to restore it when it has seen that a proper balance has been restored. Membership of the ACC has not inhibited the YAR in its relations with other Arab States. In early 1989 agreement was reached with Syria to increase economic co-operation, and further agreements on joint investment in agriculture and oil have been achieved by the Libyan–Yemeni Higher Committee.

Following the fall of the Imamate in 1962 and the British withdrawal from Aden and the Protectorates in 1967, it was assumed that the two parts of the Yemen, artificially divided by colonial rule, would unite as a matter of course. This has turned out to be very far from the case. The main obstacle has been a basic difference in ideology, of marxism in the south and laissez-faire liberalism in the north. At present there is no sign of these basic differences disappearing. Nevertheless, despite these differences and two spells of open hostility since 1967, as well as the convulsions in the PDRY of early 1986, continuing efforts have been made and some success achieved in creating unified action where this has been feasible and to their mutual advantage. An example of this has been the setting up of a joint company to prospect for oil in border regions where oil has been discovered on both sides of the frontier. In 1983 a Supreme Yemen Council was established under the joint chairmanship of the Presidents of the two Yemens, which met alternately in Sana'a and Aden. This was disrupted by the overthrow of President Ali Nasser Mohamed of the PDRY in 1986, but was resuscitated in 1988 when the two respective Presidents signed an agreement reactivating the Supreme Yemen Council, and committing both countries to joint development of the potentially oil-rich border zone. A mutual point of friction, which has yet to be resolved, is the presence on the YAR side of the border of camps containing the supporters of ex-President Ali Nasser Mohamed, which are tolerated by the YAR, and represent a constant threat to the internal political stability of the PDRY; while on the other side of the border the PDRY has given shelter to the National Democratic Front (NDF), a coalition of left-wing groups opposed to President Ali Abdullah Saleh of the YAR, which has led an insurgency group in the southern areas of the YAR. Ultimate unification of the two Yemens is a long way off yet.

Oil: economic key to the future

The search for oil in the YAR began many years ago, but it was not until 1984 that oil was first struck in commercial quantities by the Hunt Oil Company of the USA in the Marib region of eastern YAR. Soon after this discovery, a Ministry of Petroleum and Mineral Resources was established to supervise the exploration for, and production of oil, gas and other minerals. The Alif field near the historic site of Marib is the largest field discovered so far, and has estimated reserves of 500m barrels. Further discoveries in the same area are the Azal field, discovered in 1987, and the Asaad Al-Kamil field in 1988. Other oilfields, the commercial viability of which has yet to be assessed, have been discovered in the Marib/Jawf region. A pipeline, which started operating in November 1987, has been constructed which links the Alif and Azal fields with an export terminal at Ras Issa on the Red Sea coast near Salif. The pipeline's current capacity is 225,000 barrels a day (b/d), and could be expanded to 400,000 b/d. Production at present is 200,000 b/d and is not likely to be increased in the near future, as the government has agreed not to raise production in the first half of 1989 in solidarity with the efforts of OPEC and some non-OPEC producers to strengthen the market price of oil.

The oil that has been discovered so far is a high-quality light oil with a low sulphur content and a high ratio of associated gas. This bodes well for using gas

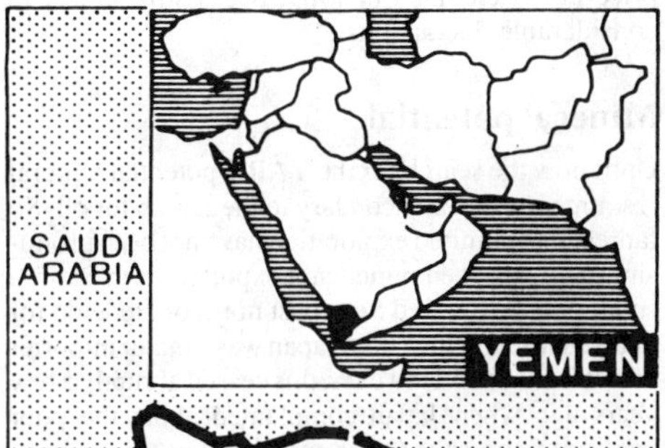

for power generation and industry, as considerable reserves of non-associated gas have also been discovered. The third Five-Year Development Plan (1987-91) includes a project for the construction of a Liquefied Petroleum Gas plant and a natural gas pipeline connecting fields in the Marib/Jawf area with Sana'a, Hodeida and the Red Sea coast, as well as an oil refinery with a 60,000 b/d capacity. Conservative estimates put the YAR's reserves of gas at 20 trillion cu ft, but potentially three or four times as much might be found. The YAR's economic future might well then lie rather with gas than with oil.

The government's share of crude oil output, in agreement with foreign oil companies, is administered by the Yemen Crude Oil Marketing Company (YCOMC). Owing to the considerable fluctuations in the world price of oil since 1986, it has been difficult

KEY FACTS

Official title: Yemen Arab Republic
Head of state: President Ali Abdullah Saleh
Capital: Sana'a
Official Languages: Arabic
Currency: Rial (YR) = 100 fils
Exchange rate: YR9.70 per $ (Nov 1989)

Area: 195,000 sq km
Population: 9.3m (1988)
GNP per capita: $576 (1987)
GNP real growth: 2.4% (1986/87)
Trade balance: –$1,141.2m (1987)
Foreign debt: $3.2bn (1987)

KEY INDICATORS

	Unit	1983	1984	1985	1986	1987
Population	'000	6,510	6,690	6,890	7,090	7,310
Gross domestic product (GDP)	YR m	21,870	24,756	30,969	38,389	43,559
GDP per capita	YR	3,359	3,700	4,495	5,414	5,959
Time deposit rate	%	9.5	9.5	9.5	9.5	9.5
Consumer prices	1980=100	114	128	144	172	–
Exports	YR m	122.6	84.1	96.7	74.1	491.8
Coffee	YR m	2.5	1.7	10.6	–	57.4
Hides & skins	YR m	5.5	4.0	10.4	0.1	47.0
Imports	YR m	7,293	8,333	9,573	11,155	–
Balance of trade	$m	–1,755.8	–1,392.6	–1,070.7	–780.5	–1,141.2
Current account	$m	–559.0	–300.6	–287.2	–125.3	–452.2
Exchange rate	YR per $	4.58	5.35	7.36	9.64	10.34

* estimated figure

to forecast future revenues from oil production. If the official price of $18 a barrel, as set by OPEC in 1987, had been maintained, the YAR's revenues from oil production would have been $600m annually. However, at one stage since then the price fell to $10.13 a barrel, reducing earnings to $335m. Fortunately the price has risen again in 1989, but is still subject to considerable fluctuations.

Mineral potential

Until now the search for the YAR's potential mineral resources has been secondary to the search for oil. So far results of limited exploration have not been promising. Salt has been mined and exported from Salif, a small port on the Red Sea coast north of Hodeida for many years, and for a time Japan was a major importer of salt from Salif. However, this ceased abruptly a few years ago when Japan complained about the poor quality of the salt. Other markets have since been sought. A limited amount of marble is quarried, and a marble factory started operations in 1982. There are deposits, yet to be assessed and exploited, of copper and zinc near Taiz in the south, and deposits of zinc, lead, iron, gold, silver, uranium and sulphur are known to exist. There are also some deposits of coal.

Important role for agriculture

The YAR was the Arabia Felix of Roman times and the centre of the Himyaritic civilisation. It is fortunate in having the most salubrious climate and heaviest rainfall of the Arabian Peninsula. Throughout the ages, agriculture has been the basis of its economy. It is the most densely populated of all the countries of the Peninsula, with a total population of over 9m, one million of whom are working abroad at any one time, and its people are overwhelmingly agricultural.

Geographically the country can be divided into four zones. Along the Red Sea coast is the Tihama Plain, a hot and humid region with a low rainfall of under 25cm a year. Largely desert and semi-desert, it is crossed by several large wadis with both surface and subterranean water, which makes irrigated cultivation possible. The main products of the Tihama are dates, cotton and animal hides. To the east of the Tihama, the slopes of the mountains which rise to the high plateau of the interior have heavier annual rainfall, which allows for the cultivation of coffee, qat and semi-tropical fruits and vegetables. On the high plateau, which lies between 3,000 and 4,000 metres, temperate climate grains such as sorghum, millet and wheat as well as fruit and vegetables are cultivated, mainly on terraces by means of irrigation and natural rainfall. To the east of the high plateau, the mountains slope away to the deserts of the Empty Quarter in the interior of the Arabian Peninsula. These have a drier climate and are mostly open rangelands with cultivation in the wadi beds.

Although decreasing as a result of oil production and the steady development of industry, agriculture still accounts for nearly 30 per cent of GNP, and employs three-quarters of the national workforce. The YAR is a country of smallholdings, the size of which varies from one to 100 ha, and there are about 700,000 families working on the land. Much of the cultivation is on terraced hillsides and so does not lend itself to mechanised farming. Labour costs are high, as many of the men left their farms to work for higher wages in the neighbouring oil-rich States, leaving in many cases the women and children to work the land. In the last two years many have returned from abroad – jobs having dried up as the economies of neighbouring States have gone into recession – but instead of returning to the land they find employment in the urban areas. As a result of these distortions, many farmers have switched from the cultivation of coffee, which was exported and earned foreign exchange, to growing qat, a mild narcotic used by Yemenis, which favours the same climatic conditions as coffee, but produces more than one crop a year and is therefore more profitable than coffee. The difficulty is that qat is grown entirely for home consumption and therefore in no way benefits the economy of the country. It also, incidentally, has a detrimental effect on the work output of the average Yemeni male. However, there are some positive aspects to the current situation in agriculture. Yemenis working abroad have been sending home remittances, which have partly been used for consumer spending, but have also been invested in fruit and vegetable farming, as the government has created the necessary incentives by the imposition of restrictions on agricultural imports. Such farms use modern labour-saving equipment and water-conserving irrigation methods. A great variety of fruit and vegetables are now being cultivated on various schemes throughout the country. On one 600 ha farm near Taiz in the south, a region which has the highest rainfall in the country, production includes bananas, mangoes, guavas, maize, sesame, onions, coffee and tobacco. On another 100 ha scheme bananas are being cultivated, whilst in the Marib region a citrus farm has an annual production of 10,000 tonnes.

Ever since the revolution of 1962, successive governments have given priority to the development of agriculture in the YAR, but plans have consistently fallen behind schedule largely due to the chronic instability in the country, only recently overcome. Agriculture ranks high in the current Five-Year Plan and several schemes are in hand, most with foreign assistance, from organisations such as the International Development Association (IDA), the Kuwait Fund for Arab Economic Development (KFAED) and the Abu Dhabi Fund for Arab Economic Development (ADFAED). Amongst these is a large development scheme on the Tihama Coastal Plain where the Tihama Development Authority (TDA) is operating. This is developing a number of irrigation projects based on a number of large wadis which cross the Tihama from the mountains to the Red Sea, together with an integrated rural development operation. Other irrigation schemes are in progress in the Wadi Raysan basin, the Wadi Rima and the Wadi Jawf. Now that the Consultative Council have ratified the necessary loan agreements, the Northern Agricultural Development Authority (NADA), which includes the Governorates of Sana'a, Hajja and Saada, is about to start operations. The Authority will receive a total of $52.9m, divided amongst the government ($18.4m), the IDA ($17.6m), AFESD ($13.5m) and a grant from the UN Development Programme (UNDP) of $3.4m.

One of the most spectacular schemes is the construction of the new Marib Dam, completed in 1986, 3 km upstream from the site of the ancient Marib dam of antiquity. The new dam has been funded by the ADFAED at a cost of $90m. It is 763 metres long and 40 metres high, with a capacity of 390m cu metres. It stores rainwater derived from the heavy rains that fall over the mountain chain of the interior. Initially the dam will provide irrigation water for 9,000 ha in a region adjacent to the sands of the Empty Quarter. A proposed network of irrigation works associated with the dam is expected to double the area under irrigation. The dam will also provide electricity for neighbouring rural centres of population, and is expected eventually to boost farming income by about 15 per cent. A second stage of the project was due to be completed in 1989.

Livestock breeding has always been an important part of traditional agriculture in the YAR. In 1989 there was estimated to be 950,000 cattle, 2m sheep, over 2m goats and 57,000 camels in the country. Badly needed veterinary services are now being established throughout the country as animal diseases are rife. In 1976 a Livestock Development Association was set up and established several modern fattening farms which provide credits for smallholders of livestock and poultry.

The YAR has rich fisheries in the Red Sea, which have yet to be developed. A start has been made with aid from the World Bank, Denmark and State funds. A new fishing port was being constructed at Hodeida. In 1982 work was started on the construction of harbour, storage and processing facilities at Khawbah, Mokha and Hodeida. The potential annual catch is estimated to total 26,500 tonnes, including shrimp and lobster. Record catches of 19,300 tonnes were made in 1978 and 1979, but then fell to 12,200 tonnes in 1983, but have now recovered to near the 1978 figure. It is hoped to export fish to Saudi Arabia, Egypt and Sudan.

A modest expansion of industry

Larger industrial enterprises in the YAR are either owned by the government or are in mixed ownership, whilst private enterprise is confined to light industry. Amongst the major industries, textile manufacture is one of the most successful. The cotton is grown locally, mostly on the Tihama Plain, and the government-owned Yemeni General Cotton Company has monopoly rights. The largest textile mill was built by the Chinese in Sana'a, but most textile processing is concentrated in Hodeida.

There are two cement plants at Amran and Bajil built by Soviet companies. The Amran plant is to be expanded to produce 500,000 tonnes in a first phase with the intention of ultimately increasing capacity output to one million tonnes a year. Japan may provide the finance for this project. The USSR made a commitment to increase the annual capacity output of the Bajil plant from 300,000 tonnes a year to one million tonnes. An agreement was signed in 1987 between the government and a Japanese company for the construction of a cement plant at Mafraq, on which work was expected to start in 1989. Japan signed a $166m soft loan agreement for this scheme.

There is a factory producing aluminium goods at Taiz in the south, a fish canning factory in Hodeida, as well as a cigarette factory and other small enterprises. The privately owned Yemen Fertilisers Company is planning to construct an ammonia and nitrogenous fertiliser plant at a cost of $120m using associated gas from the Marib/Jawf oilfields. A feasibility study is in progress for the establishment of local production of galvanised iron. Industrial estates are being set up in Sana'a, Hodeida and Taiz. A joint venture of local private companies, with a capital of $15m, is planning to build a factory for the manufacture of fluorescent lamps and electrical fittings near Sana'a. There is also a private initiative to establish a company with a capital of $5.1m to build a factory for the manufacture of household appliances such as freezers, cookers and water heaters.

Progress in social development

Until 1962, when the Imamate was overthrown, the North Yemen was one of the most isolated and backward countries in the world. Although it has a homogeneous population, the YAR is divided in the religious sense. Sixty per cent of the people are Sunni Muslims of the Shafei school, as they are known locally, whilst 40 per cent are Zaidis of the Shia branch of Islam. The Imams were Zaidis, and in the political sense the Zaidis were always dominant in Yemeni society and continue to be so. Nevertheless the religious divide is not the source of much internal friction. However, the fact that the population of the PDRY are Shafeis, as are the population of the southern regions of the YAR, has had some bearing on political unrest in these regions. A census held in 1986 put the population of the YAR at 9m, but as there are a million or more North Yemenis working outside the country at any one time, it is difficult to assess the accuracy of the figure.

Advances in health services had to start from a very low base; until 1962 there were only rudimentary services and the mortality rate was very high, particularly amongst the children. In 1965 the life expectancy was 37 years, and in 1985 it was only 50. This is considerably lower than the average life expectancy in similar low income countries. The mortality rate for children under five years of age is still at 210 per thousand. Only Afghanistan and Somalia in the region have a higher rate. By 1982 there were 28 hospitals with 3,529 beds and 130 health centres, as well as numerous pharmacies and rural health units. Since then new hospitals have been built in Dhamar, Saada and al-Baidha. A new 55-bed hospital at Hajja, funded by the Saudis, is to be built by North Korea after an original contract fell through. Ten new health centres were due for completion in 1988. An immunisation programme has been in progress for several years, covering 3,520 vaccination sites and involving 30,000 personnel. The aim is to reduce the incidence of measles, whooping cough, tetanus and diphtheria. The programme is being assisted with Arab aid as well as the IDA, WHO, USAID and UNICEF.

Under the Imamate there were a few modern schools in addition to the Koranic schools. By 1985 the number of modern schools had risen to 5000, which included 94 secondary schools, 17 training colleges and 20 specialised colleges. There were at that time a total of 570,000 pupils in the schools and 3,000 college students. There is a university in Sana'a with 5,200 students and a further 2,300 students or more at Arab universities abroad.

The financial situation improves

Statistically the YAR is one of the poorest countries in the world, but clearly this fact is belied by appearances. In the particular case of the YAR, statistics should be treated with some degree of scepticism. Two main factors serve to distort the financial situation. A significant proportion of the country's earnings are in the form of remittances and money brought home by emigrant workers in the wealthy Gulf States, although this source of income is dwindling. The rural

COUNTRY PROFILE

Political system: Since the dissolution of the constitution in the military coup of 1974, legislative power has rested with the president and a mainly civilian consultative council of ministers helped by a civilian prime minister. The General People's Congress comprises 300 appointed and 700 elected representatives.

Main cities/towns: Sana'a (population estimated 510,000 in 1987), Hodeidah (148,000) and Taiz (200,000).

Language: The official language is Arabic; English is often understood in business circles.

Media
Press: Dailies are *Al-Jumhuriya* and *Al-Thaura*. There are eight weekly newspapers and periodicals.
Radio: Three commercial stations broadcast in Arabic.
Television: There is one national network with some advertising. English-language news at 1930. 800,000 sets in 1988.

Domestic economy: Falling remittances from Yemeni expatriate workers in the oil-rich states, reductions in Arab economic assistance (due to the Iran-Iraq war) and drought have all reduced the rate of economic growth over the last seven years. Large-scale development projects commenced in the 1970s and financed by expatriate workers' remittances and Arab aid created large deficits, leading to government cutbacks in the early 1980s. The exploitation of oil deposits in the late 1980s has, however, created a basis for continued development, and economic prospects are much improved for industry. Large volume of goods smuggled across the Saudi Arabian border makes indirect taxes difficult to impose.

Employment: A vast number of Yemenis work abroad (1.4m* in 1988), mostly as immigrant labour in Saudi Arabia and other Arab states where salaries are high. A large proportion of domestic labour is unskilled. Returning expatriate workers tend not to enter the domestic labour market, leading to high wage rates for skilled and semi-skilled workers and a dependence on foreign labour.

External trade: Import controls were imposed in 1983 and non-essential imports banned in 1984 in an effort to control trade deficit. Yemen AR is one of the 32 developing nations designated in 1985 to receive exemption from US competitive trade requirements in accord with the Generalized System of Preferences.
Exports: Exports are minimal, and shrinking; main products are biscuits (66 per cent of total), hides and skins, coffee, potatoes, cotton yarn, fruit and fish. Main destinations are USA (41.4% in 1985), Yemen PDR (14.3%), Saudi Arabia (8.3%), Netherlands (6.9%), France (3.5%) and UK (2.6%).
Imports: Main imports are sugar products (11.5 per cent), vehicles, petroleum products, foodstuffs, wood, machinery and metal products. Main sources are Italy (9.7% of 1985 total), Saudi Arabia (8.8%), Japan (8.7%), UK (8.3%), Netherlands (6.2%), German FR (5.9%), France (5.2%) and USA (3.3%).

Agriculture: Agriculture contributes 36 per cent of GDP and this is expected to increase in the following years as more agricultural projects get off the ground. Main crops are sorghum, millet, cotton, coffee, vegetables, fruits and qat (a mildly narcotic leaf). There are three main agricultural projects in YAR: the Southern Upland Rural Development Project, the Tihama Development Project and the Central Highlands Agricultural Project. Private sector trading companies have now invested in agriculture in Tihama and Mareb, concentrating on bananas and citrus fruits. Periodic drought and manpower shortages remain a serious problem however. Fishing has considerable improvement potential.

Industry: Import-substitution industries are being encouraged by the government to reduce the volume of imports. The private sector is taking the lead in this field, which currently employs five per cent of the labour force, and local businessmen are now involved in many light industrial projects producing handicrafts, paint, plastic products, mattresses, soap, cosmetics, perfumes, aerosol insecticides, textiles, garments, cigarettes, foods and soft drinks. There are more projects in the pipeline such as dry-cell batteries, glass, table salt and fertilizers. Three industrial estates are in existence. The Bajil and Amran cement plants produce about 750,000 tonnes p.a. There are plans to build another, producing 1m tonnes p.a., at Mafraq (near Mocha). Meanwhile work continues on the restoration of the old city at Sana'a.

Minerals and mining: Salt is mined at Salif, where deposits total 25m tonnes, and gypsum and marble are extracted. Mineral exploration is increasing, and gold, silver, copper and nickel is being explored for.

Hydrocarbons: In 1984 the first major oil field, the Alif field was discovered in the Marib area in eastern YAR. It had estimated reserves of 500m barrels of high-quality, light oil and a high ratio of associated gas. Subsequently, in the same area, the Azal field was discovered in 1987 and the Asad al-Kamil field in 1988. A refinery was built, and a pipeline linking these fields with the export terminal at Ras Issa on the Red Sea near Salif began operation ion 1987, with an initial capacity of 225,000 barrels a day (b/d). In 1989, oil production was at 200,000 b/d. Conservative estimates of reserves of gas were 20 trillion cubic feet in 1989, including considerable quantities of non-associated gas, which could be used for power generation and industry. The Third Five Year Development Plan (1987/91) includes an LPG plant and a natural gas pipeline connecting the gas fields in the Marib region with Sana'a, Hodeida and the Red Sea coast.

Energy: Expansion of electricity supply has been a major priority of the Development Plans, and installed capacity more than doubled between 1978 and 1981. A 160 MW power station is being built at Mocha, along with other diesel and thermal plants. The grid is under development, and may eventually be linked with Yemen PDR.

Membership of international organisations: Arab League, FAO, G-77, ICAO, IDA, IFAD, IFC, ILO, IMF, IMO, INTELSAT, INTERPOL, Islamic Conference Organisation, Islamic Development Bank, ITU, NAM, UN, UNESCO, UPU, WFTU, WHO, WIPO, WMO, World Bank.

population in particular have been traditionally distrustful of banking and therefore hoard their earnings at home and spend them on consumer goods, virtually all imported. Taxes are difficult to collect owing to the geographical nature of the country, inadequate communications and the remoteness of many of the villages. The urban population is still a minority in the YAR. Smuggling is the other factor which distorts the economic situation. This occurs on a large scale across the northern border with Saudi Arabia. This is an area where the central government still has minimal control: inhabited by powerful tribes, considerably influenced by the Saudis, and whose traditional occupation is smuggling. In short, the Yemenis are not as poor as they statistically should be.

Since the revolution, the finances of the YAR have been run on chronic budget deficits covered by foreign loans, aid and development grants and workers' remittances. In 1982 the government deficit amounted to 25 per cent of GDP and it was forced to introduce an austerity budget in 1983. Imports of consumer goods rose steadily for reasons already explained, whilst exports, which traditionally consisted for the most part of coffee and cotton fell, as farmers switched from the cultivation of coffee to qat, which was more profitable but only sold on the home market, and cotton was used for domestic textile production. As a result, severe import restrictions were imposed in 1986.

By maintaining a policy of political neutrality in world affairs, the YAR has succeeded in cultivating friendly relations with countries on both sides of the ideological divide as well as those who are not ideologically committed. As a result, over the years it has received considerable economic aid from a wide variety of sources including the USSR, the Arab countries (particularly Saudi Arabia, Kuwait and the UAE), China and the Western European countries, as well as the USA. The USSR has modernised the port of Hodeida, whilst China financed much of the basic road system of the country. In 1987 China agreed to provide aid amounting to $47m for road and dam construction. Much aid has also come from such international agencies as the World Bank, IDA, the Arab Fund for Economic and Social Development (AFESD), ADFAED and UNDP.

Now that oil has gone into production and has started to play an important role in the exports of the YAR, the financial situation is confidently expected to alter in favour of the YAR. The Budget for 1989 was approved by President Saleh in March. It allows for an estimated deficit $486m, which is about the same as for 1987, but greater than that for 1988. However, if past experience is anything to go by, official budgets are not an accurate guide to figures actually achieved.

The outlook is promising

Subject to certain provisos, the outlook for the YAR is favourable. Oil revenues are likely to increase as more oilfields are discovered. Internal political stability is being maintained after years of turmoil, and external relations are good. The potential cloud on the horizon is the unstable situation in the PDRY. Upheavals there are invariably reflected in the YAR, and major differences between the two countries remain to be settled. Provided that these and the internal problems of the PDRY can be resolved, the YAR will experience steady progress, but unification of the two Yemens is still a long way off.

BUSINESS GUIDE

Time: GMT +3 hrs.

Climate: The semi-desert coastal plain known as the Tihama is hot, humid and dusty. The highlands, which are agreeable in summer but cold in winter, enjoy most of the unreliable rainfall (Mar-Apr and Jul-Sep).

Entry requirements
Passport: Required by all.
Visa: Required by all with few exceptions. Exit visas are needed for stays of more than seven days.
Prohibited entry: South African or Israeli nationals or holders of passports with Israeli visas are forbidden entry.
Currency: Not all currencies can be exchanged inside the country; it is essential to check before arrival.
Customs: Personal effects may be imported free of duty. No alcohol allowed. All goods from Israel are prohibited, and permitted imports require import licences.

Health precautions
Mandatory: Certificate of vaccination against yellow fever if travelling from infected areas.
Advisable: Anti-cholera and anti-malaria precautions (malaria has been endemic in Tihama). Avoid stagnant water and avoid drinking untreated water. Bilharzia is widespread and gastric upsets common.

Representation overseas: Abu Dhabi, Addis Ababa, Algiers, Amman, Baghdad, Beijing, Beirut, Berlin, Bonn, Brussels, Cairo, Damascus, Djibouti, Doha, Geneva, The Hague, Islamabad, Jeddah, Khartoum, Kuwait City, London, Mogadishu, Moscow, Muscat, New Delhi, New York, Paris, Prague, Rabat, Rome, Tokyo, Tripoli, Tunis, Washington.

Air access
National airline: Yemenia (Yemen Airways).
Other airlines: Aeroflot, Air Djiouti, Air France, Alyemda, Egyptair, Ethiopian, Gulf Air, Iraqi Airways, Kuwait Airways, Lufthansa, PIA, Royal Jordanian, Saudia, Somali, Syrian Arab, Sudan Airways.
Tax: Airport tax of YR60 is payable on international departures, and YR15 on internal flights.
International airports: Sana'a International, El Rahaba (Code: SAH), 13 km from Sana'a, with duty-free shop (no alcohol), restaurant, buffet, bank; Al Ganad International (Code: TAI), 22 km from Taiz, with restaurant; and Hodeidah. A fourth airport is under construction at Denaem.

Surface access
Main port: Hodeidah is the main port, with other ports at Mocha and Salif. An oil terminal is expected to be sited near Salif.
Overland access: Via tarmac road from Jizan in Saudi Arabia or Aden.

Hotels: Sana'a has three first-class hotels; Ta'izz and Hodeidah Ma'rib have one each. It is advisable to book in advance. The major hotels have good restaurants and all major credit cards are acceptable.

Electricity supply: Generally 220 V AC, with two-pin plug fittings.

Car hire: Available in Sana'a.

City transport
Taxis: Fare by negotiation, and minimum charge system in cities. A fixed fare of YR100 is charged between Sana'a airport and the city centre.

National transport: Internal travel may be affected by local night-time curfews and military check points.
Air: Yemenia has regular scheduled services linking Sana'a, Hodeidah, Taiz and Marib.
Road: 15,600 km of road, of which 1,300 km are main roads and 830 km are secondary roads. There are metalled roads between all major cities.
Buses: There are scheduled services between Sana'a and all major centres.

Public holidays
Fixed dates: 26 Sep (Revolution Day).
Variable dates: Eid al-Fitr, Eid al-Adha, Hijra New Year's Day, Prophet's Birthday.

Social customs: Islamic culture and customs are strictly observed, but visitors are allowed to drink alcohol in hotels or private homes. Business is commonly conducted while chewing qat, a mild narcotic. Foreigners are not obliged to join in, if they attend these afternoon sessions.

Telecommunications
Telephone: Dialling code for Yemen AR: IDD access code + 967 followed by area code (3 for Hodeidah, 2 for Sana'a, 4 for Taiz) followed by subscriber's number. There is a telephone directory in Arabic. For help ask the telephone operator at your hotel or ring 18 (English spoken). Yemen AR is building an earth station as part of the Arabsat communications programme.
Telex: Telex facilities at major hotels.
Telegram: Cable & Wireless offices at Sana'a (PO Box 168, tx: 2235), Hodeidah (tx: 5501/2) and Taiz (tx: 8801/2).

Banking
Central bank: Central Bank of Yemen.
Main centre: Sana'a.
Head offices: International Bank of Yemen, Yemen Bank for Reconstruction & Development, Co-operative and Agricultural Credit Bank, Industrial Bank of Yemen.
Branches: Arab Bank, Bank of Credit and Commerce International, Banque Nationale de Paris, Rafidain Bank, United Bank.

Advertising: Advertisements must not offend Islamic laws. In addition to the press, radio and TV, some 10 cinemas permit advertising.

Weights and measures: Metric system.

Representation in capital: Albania, Algeria, China PR, Czechoslovakia, Ethiopia, France, German DR, German FR, India, Iran, Iraq, Italy, Japan, Jordan, Korea PDR, Kuwait, Lebanon, Libya, Morocco, Netherlands, Oman, Pakistan, Romania, Saudi Arabia, Somalia, Sudan, Syria, Turkey, UAE, UK, USA, USSR.

Hotels: Sana'a
Hadda Ramada, N Hadda Road, P.O. Box: 999 (tel: 215212; tx: 2227).

Sam City, 5 November Street, P.O. Box: 520 (tel: 270751; tx: 2301).

Sheraton, P.O. Box: 2467 (tel: 237500/9; tx: 2222, 2644 shsan; fax: 251521).

Taj Sheba, Ali Abdolmoghni Street, P.O. Box: 773 (tel: 272372; tx: 2324 sheba; fax: 274129).

Business advice and information
Central Bank of Yemen, PO Box 59, Sana'a (tel: 75219; tx: 2210, 2280).

Central Planning Organisation, PO Box 175, Sana'a (tel: 250101-8; tx: 2266).

Foreign Trade Corporation, PO Box 77, Sana'a (tel: 72058; tx: 2262).

General Post Office, Liberation (Tahreer) Square, (tel: 71401/2).

Hodeidah Chamber of Commerce, PO Box 3370, Azzoubairi Street, Hodeidah (tel: 217401; tx: 5610).

Middle East Association, 33 Bury Street, London SW17 6AX (tel: 839-2137/8)

Ministry of Economy, Supply & Trade, PO Box 1704, Sana'a (tel: 202471; tx: 2360).

Ministry of Oil and Mineral Resources, PO Box 81, Sana'a (tel: 202313; tx: 2257 YOMIN YE, 2757 YOMIN YE).

Yemen PDR

Tim Owen

To the surprise of many, the PDRY has remained politically quiet since the bloodbath of early 1986. Nevertheless, the basic instability remains. The regime is sustained both politically and economically by the USSR and it is only this which prevents trouble breaking out again. The country is economically in a mess, despite the efforts of the regime to increase trade with the West and obtain economic assistance. However, with its marxist regime and background of internal turbulence, the PDRY is looked at askance by Western businessmen and governments. The regime also fears a diminution of economic and political backing by the USSR after the withdrawal from Afghanistan and the USSR's new policies towards its foreign protégés. Devastating floods did extensive damage in early 1989.

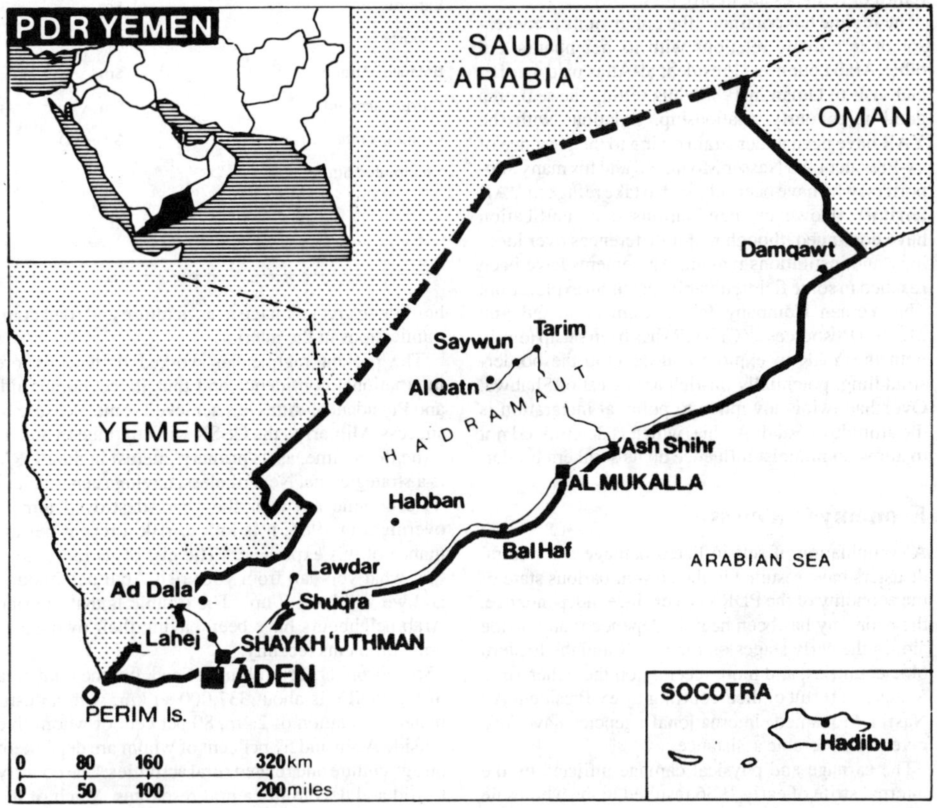

Instability is inherent

Aden came to be occupied by the British in the 1840s at much the same time as Hong Kong, and for much the same reasons. In both cases there was nothing there except a few small fishing villages, but both have excellent natural harbours and strategic positions. Originally, the hinterland of Aden was only taken over as a buffer against the hostile Turks, who occupied the Yemen until the end of the First World War. Further to the east, the Hadramaut and its environs only became a British Protectorate in the 1930s, as otherwise this region represented a dangerous political void. Little was done to develop the Protectorates during the British occupation and all economic activity was concentrated in Aden. A few years before the British withdrawal in 1967, a federation of Aden and the Protectorates was established which collapsed as soon as the British withdrew. It was a highly artificial set-up which never could have worked, composed as it was of highly politically-conscious and relatively sophisticated Adenis on the one hand, and intensely tribal and uneducated inhabitants of the Protectorates on the other. Political power was invested in the tribal leaders, whom the Adenis despised.

When the British withdrew in 1967, a civil war was raging between the Nasser-supported Front for the Liberation of South Yemen (FLOSY) and the rather more left-wing National Liberation Front (NLF) from which the NLF emerged victorious. The British withdrawal, with all the economic benefits that the British presence brought to Aden, coinciding with the closure of the Suez Canal as a result of the conflict in Palestine – which had a devastating effect on the port of Aden – left the emerging State economically bereft. The Western nations were disinterested, so it was not surprising that the regime looked to the USSR and the Eastern bloc for assistance. With an eye to the strategic importance of Aden, the USSR responded with alacrity, providing both military and economic assistance. The People's Democratic Republic of Aden (PDRY), as it was ultimately named, went through a series of internal convulsions over the succeeding years, from which a marxist State eventually emerged. It also got itself involved in periodic bouts of hostilities with neighbouring Yemen Arab Republic (YAR) and the Sultanate of Oman.

In the PDRY, marxism is only a veneer over deeply ingrained tribalism and tribal factions. Power is concentrated in the Politburo, where there is intense rivalry between individuals representing the various tribal factions. Shoot-outs at Politburo meetings mark the culmination of these power struggles. It was one of these periodic heavily personalised conflicts which led to the bloodbath of January 1986. In 1980, as a result of losing one of these power struggles, President Abdul-Fatah Ismail, a hard-line pro-Soviet marxist, resigned and went into exile in Moscow.He was replaced by Ali Nasser Mohamed, a pragmatist, who did much to improve relations with neighbouring States and the economy. In 1985 Abdul-Fatah Ismail was permitted by the USSR to return to the PDRY, a decision which the USSR was subsequently to regret deeply, and which showed an abysmal lack of understanding of internal politics in the PDRY. The outcome of this decision was inevitable. Abdul-Fatah Ismail, with the backing of his supporters, was re-admitted to the Politburo, where a coup against the President and his supporters was organised. The plot went badly wrong; the President got wind of it, and took pre-emptive action. In the ensuing shoot-out at the Politburo meeting, at which the President failed to appear, Abdul-Fatah Ismail and his chief supporters were killed. However, their tribal supporters, already moving on Aden, engaged the supporters of President Ali Nasser Mohamed and defeated them in a particularly bloody battle in Aden. The President fled to the YAR, where he remains, leaving the curious situation of the victorious side bereft of its leaders. This is basically the situation that exists today, with a weak leadership of second- rank supporters of Abdul-Fatah Ismail and some more pragmatic personalities, none of whom are strong enough to emerge on top. The current President, Haidar Abu Bakr al-Attas, was Prime Minister to Ali Nasser Mohamed and happened to be out of the country at the time of the bloodbath, and immediately hastened to Moscow where he enlisted the support of the Soviets.

Support for Ali Nasser Mohamed was strong in Aden, where most of the intelligentsia were to be found. Large numbers of them fled to North Yemen, where they remain. The PDRY ever since has suffered from a severe lack of administrators and skilled personnel. For several years the regime has stubbornly

KEY FACTS

Official title: People's Democratic Republic of Yemen

Head of state: President Haider Abu Bakr al Attas

Capital: Aden

Official Languages: Arabic

Currency: Yemeni Dinar (YD) = 1,000 fils

Exchange rate: YD0.34 per $ (Nov 1989)

Area: 332,968 sq km (boundaries ill-defined)

Population: 2.48m* (1988)

GNP per capita: $420 (1987)

GNP real growth: -1.2% (1986/87)

Foreign debt: $1,548m (1987)

* estimated figure

refused to negotiate with Ali Nasser Mohamed and his supporters, who are in camps just inside the YAR in large numbers, and poised to take advantage of any crisis which might arise south of the border. Now, at last, efforts at reconciliation have been reported and negotiations over the release of prisoners and the return of refugees are in progress.

The regime has continued the policy of its predecessor of improving relations with its neighbours and other Arab countries. President al-Attas paid an official visit to Oman in 1988, thus setting the seal on a good-neighbourly relationship. Relations with the YAR have been under strain owing to the presence of ex-President Ali Nasser Mohamed and his many supporters, who have been allowed to take refuge in YAR territory. However, negotiations over unification have continued, though major differences over ideology and institutions remain. Agreements have been reached in some fields, notably that of oil exploration. The Yemen Company for Investment in Oil and Mineral Resources (YCIOMR) has been set up jointly with the YAR to explore and develop the border-straddling, potentially oil-rich area west of Shabwa. Overshadowing any move to political integration is the attitude of Saudi Arabia, which is determined not to allow communist influence on its southern border.

KEY INDICATORS

	Unit	1984	1985	1986	1987	1988
Population	'000	2,230	2,290	2,360	2,440	2,510
Consumer prices	1985=100	95.2	100.0	–	–	–
Exports	$m	30.7	42.6	30.4	70.9	82.2
Imports	$m	734.2	623.6	447.9	456.9	597.9
Balance of trade	$m	–703.5	–581.1	–417.5	–385.9	–515.6
Current account	$m	–253.0	–222.4	–175.7	–129.7	–383.0
Exchange rate	$ per YD	2.90	2.90	2.90	2.90	–

* estimated figure

Economy in a mess

A combination of self-inflicted damage and natural disasters have resulted in the current parlous state of the economy of the PDRY. Ever since independence, the economy has been heavily dependent on outside aid; in the early stages on the USSR and the Eastern bloc countries, and more recently on the richer Arab States as a result of their courting by ex-President Ali Nasser Mohamed. International agencies have also given considerable assistance.

The carnage and physical damage inflicted by the internal strife of early 1986 resulted in the wholesale flight of foreign experts, as well as most of those with skills and expertise in the various Ministries and administration. Foreign governments, not least that of the USSR, were deeply disturbed by these events. Disastrous floods in 1982, and again in March 1989, have done great damage to agriculture and road communications in the interior.

The present regime has worked hard to restore international confidence and thus its economic aid, and President al-Attas has achieved some measure of success. Militarily, the USSR will continue to underwrite the regime, as its primary interest in the PDRY is a strategic one. Nevertheless, having seen much of its economic aid over the years destroyed almost overnight in 1986, it is loth to risk a repeat performance of this experience. The gist of its message is 'get what you can from your rich Arab neighbours, and we will back it up'. The trouble is that the rich Arab neighbours have been having their own economic problems recently.

Some frontiers are still undefined, but the total area of the PDRY is about 337,000 sq km, with an estimated population of 2.4m, 80 per cent of whom live outside Aden and 57 per cent of whom are dependent on agriculture and related rural activities. The country is arid and devoid of natural resources. Much of the interior is mountainous, and only about one per cent of the land is under cultivation, mostly confined to the Lahej and Abyan areas in the west and the Wadi Hadramaut in the east. Efforts have been made to develop the country's rich fisheries potential and oil has been discovered in limited quantities and has recently gone into production.

The PDRY has a chronic balance of payments deficit and has always relied on outside aid to finance its planned expenditure. Following the original development plan, largely organised with Soviet assistance, there have been three successive Five-Year Plans. The first, from 1974 to 1979, was extended by a year and was quite successful with increasing external funding. Priority was given to fishing, irrigation, agriculture, industrial development and minerals. The second plan was delayed owing to hostilities with neighbouring YAR and was dependent on 70 per cent external assistance. Emphasis was on industry, agriculture and fisheries as well as health and communications. Foreign financing was provided by the Eastern bloc, Arab countries and international agencies such as the International Development Association (IDA) and the UN. In early 1982 extensive flooding inundated agricultural areas and caused major damage to roads, bridges and factories. Five hundred people were killed and 250,000 rendered homeless. Then in January 1986, major destruction and damage were caused by intense fighting in the Aden area and the oil refinery was damaged. All this was a major setback to the second Five-Year Plan and

COUNTRY PROFILE

Political system: Executive power rests with the chairman of the Presidium, which is appointed by the Supreme People's Council, a unicameral legislature with 111 elected members. President Ali Nasser Mohammed was deposed in January 1986 in an armed revolt; a major political reshuffle ensued, but the new administration maintained its essentially conciliatory line toward Yemen AR and the eventual reunification of the two states is still a long-term goal.

Main cities/towns: Aden (population 250,000) and Al Mukalla (50,000*).

Media
Press: All newspapers are government-run. Three weeklies and two dailies (total circulation 49,000). There are also three magazines published.
Radio: One radio station and some 111,000 radios.
Television: One TV station and some 38,000 sets.

Domestic economy: During the years immediately following independence in 1967 all major businesses and concerns were nationalised. The economy is largely based on agriculture and light industry and has traditionally run a budget deficit. Almost half its foreign debt is owed to the USSR and Eastern bloc countries. Since 1981 measures have been taken to encourage foreign joint-venture capital investment. Much of the 1986-7 budget went to repairing buildings damaged in the Jan-Feb 1986 fighting. However, the prospect of crude oil revenues has brought some grounds for optimism.

Employment: 49 per cent in agriculture, 23 per cent in services, seven per cent in commerce and four per cent in construction.

External trade: Accurate foreign trade figures are not available.
Exports: Main exports are refined petroleum, cotton, hides/skins, dried/salted fish and coffee. Main destinations in 1985 were Italy (36.0%), Japan (12.4%), Yemen AR (11.6%), France (11.2%), Sudan (8.3%).
Imports: Main imports are oil and foodstuffs, with recent increases in the import value of machinery and transport equipment. Iran supplies most of the crude oil for the refinery at Little Aden, and other imports come mainly from the UK, Australia, France, Japan, Kuwait, United Arab Emirates, China PR and Italy.

Agriculture: About 0.3 per cent of land area is cultivated, mainly in fertile silted valleys, although potential arable estimated to be one per cent of total. Main cash crops are cotton, coffee, qat and tobacco. Food crops include millet, sesame, dates, wheat and barley. Much farming is on subsistence level, employing traditional irrigation techniques, and accurate assessment of yields is not possible.

Industry: Industry contributes about 25 per cent of GDP, with manufacturing accounting for 10 per cent. An oil refinery at Little Aden, originally owned by BP, was taken over by the state in 1977, and in recent years the Soviet Union has helped finance an oil pipeline and terminal at Bir Ali. Development of industry has been promoted through joint ventures, and recent projects have included setting up light industries to produce clothing and shoes, processed foods and electronic goods, cigarettes and salt. YD20m has been allocated to promote tourism through a joint company formed with Yemen AR (tourist revenue in 1980 was $4m).

Petroleum: The discovery of modest reserves of oil offshore (about 1bn barrels) of the Hadhramaut in the east of the country created interest in other areas, notably the Shabwa (bordering on Yemen AR and the subject of an agreement with the USSR), the northern Hadhramaut and onshore and offshore around Aden. Production is running at about 10,000 b/d. Oil development is carried out by the Petroleum and Minerals Board.

Energy: The Hiswa power station is still incomplete after 13 years in construction. Plans for 1987-88 included electrifying five towns east of Mukalla.

Membership of international organisations: Arab League, FAO, G-77, GATT, ICAO, IDA, IFAD, ILO, IMF, IMO, Islamic Conference Organisation, Islamic Development Bank, ITU, NAM, UN, UNESCO, UPU, WFTU, WHO, WMO, World Bank, WTO.

left the economy badly battered. The third Five-Year Plan was delayed by these events and was to be launched in 1988. How this will work out remains to be seen. In 1986 virtually all expatriates, comprising advisers and technicians, fled the country, as well as large numbers of refugees, including most of the more highly qualified workers. Work on all major projects came to a halt, and there is now a great labour shortage. Development expenditure in the current Five-Year Plan has been set at $1.7bn, a 30 per cent increase on the previous plan. Of this, $793m is to go to industry, $399m to services, and $286m to social services. The floods of early 1989 have caused major economic dislocation, which will adversely affect the current plan. Two thousand homes have been destroyed and 50,000 people rendered homeless. Damage has been particularly severe in the Wadi Hadramaut, where date palms have been uprooted and other agriculture devastated. Roads and bridges have been destroyed.

Setbacks for industry and agriculture

The major contributor to the economy in the industrial field is the oil refinery at Little Aden. Originally owned by BP, it was nationalised in 1977. Entirely dependent on imports of crude oil until very recently, it has been subject to the vicissitudes of events in the outside world. It has never fully recovered from the effects of the closure of the Suez Canal in 1967. With a capacity of 170,000 barrels a day (b/d) it was down to a processed production of 14,000 b/d in 1977. It then climbed to 40 per cent of capacity, but received a further setback when it was damaged in the fighting of 1986. In 1982 an agreement was signed with Iran for the supply of crude oil for the refinery, but this was severely curtailed in 1986 by Iraqi bombing of Iranian oil installations. At the present time, crude oil is being supplied by Kuwait, Iraq and the USSR together with some of the PDRY's own oil.

Non-oil industrial production remains limited. Foreign companies are being encouraged to enter into joint ventures by such incentives as freedom to remit hard currency. So far 15 such schemes have been established, mostly involving expatriate Yemeni capital. The major enterprises in this sector are salt extraction, soap production and the bottling of mineral water. The Chinese have helped set up a textile plant.

Foreign trade is being boosted

One of the objectives of the current Five-Year Plan is the further development of foreign trade. The UK is second among the PDRY's top suppliers, and the biggest from amongst the OECD countries, with 18 per cent of the market share. It exports chemical materials and products, electrical machinery and equipment to the PDRY. In return it imports petroleum products and hides. BP is amongst oil companies engaged in oil exploration in the PDRY. UK companies, and especially consultants, have a good reputation in the PDRY. Foreigners working in the PDRY are permitted to transfer abroad up to 70 per cent of their taxed earnings.

Provided there are no further internal convulsions, the economy of the PDRY will gradually improve. It is one of the poorest countries in the world. The most urgent need is to settle the problem of ex-President Ali Nasser Mohamed and his supporters, now outside the country, but badly needed back in the PDRY to help the development of the country. Foreign capital on any scale will only be attracted back in when foreigners are assured that political stability has really been achieved in the country.

BUSINESS GUIDE

Time: GMT +3 hrs.

Climate: Coastal region has humid monsoon climate and the mountains are cooler. Summers hot and humid (to 54°C) with very little rain. Temperatures from Oct to Apr average 20-30°C and from May to Sep 30-40°C.

Entry requirements
Passport: Required by all.
Visa: Required by all except nationals of some Middle Eastern countries.
Prohibited entry: Nationals of Israel and South Africa and holders of passports with visas from these countries.
Currency: Credit cards are not recognised. All currency to be declared on entry and any spent must be accounted for on departure. Only YDS may be exported in local currency. Body searches possible.
Customs: Personal effects duty free. Trade with Israel and South Africa prohibited.

Health precautions
Mandatory: Vaccination certificates against cholera if travelling from infected area.
Advisable: Anti-malaria precautions.

Representation overseas: Addis Ababa, Algiers, Baghdad, Beijing, Beirut, Berlin, Dar es Salaam, Djibouti, Havana, Jeddah, Khartoum, Kuwait City, London, Mogadishu, Moscow, New Delhi, Paris, Tehran, Tripoli.

Air access
National airline: Alyemda.
Other airlines: Aeroflot, Air Djibouti, Air France, Ethiopian, Kuwait Airways, MEA, Saudia, Yemenia.
International airport: Khormaksar International (Code: ADE), 11 km from Aden.

Surface access: From Taiz in Yemen AR (currently closed to foreigners).
Main port: Aden (containers).

Hotels: Book hotels well in advance, particularly if visiting near the time of religious festivals.

City transport
Taxis: Taxis are available. Tipping is not customary.

National transport
Air: Alyemda operates a domestic service.
Road: Limited length of surfaced roads serving centres, about 8,300 km in 1987.
Buses: Service linking Aden with Al Mukalla (via Zinjibar, Shugra).

Public holidays
Fixed dates: 8 Mar (Women's Day), 1 May (Labour Day), 22 June (Corrective Move), 26 Sep (Revolution Day), 14 Oct (National Day), 30 Nov (Independence Day).
Variable dates: Lailat al-Mi'raj, Eid al-Fitr, Eid al-Adha, Hijra New Year's Day, Prophet's Birthday.

Working hours
Business: 0800 – 1230, 1500 –1800 (Sat-Thu).
Government: 0700 – 1400 (Sat-Thu).
Banking: 0800 – 1200/1230 (Sat-Thu).

Social customs: Alcohol and cigarettes are available. Religious customs and traditions must be respected. Do not carry cameras outside Aden.

Telecommunications
Telephone: There is no direct international dialling to Yemen PDR. International calls from within the country are usually booked in advance.
Telex: Some large firms, airlines and a few hotels have their own telex facilities. There are public call offices in Aden, at Steamer Point and in Khormaksar.
Telegram: There is a telegraph office at Steamer Point (tel: 22223; open 24 hours), and another at Crater.

Postal service: Use air mail for international correspondence. Domestic and surface deliveries are liable to severe disruption at present.

Banking
Central bank: Bank of Yemen.

Representation in capital: Afghanistan, Algeria, Bulgaria, China, Cuba, Czechoslovakia, Egypt, Ethiopia, France, German DR, German FR, Hungary, India, Iran, Iraq, Italy, Japan, Korea PDR, Kuwait, Libya, Oman, Pakistan, Romania, Saudi Arabia, Somalia, UK, USSR, Vietnam.

Banking
Bank of Yemen (central bank), PO Box 452, Crater, Aden (tel: 51814; tx: 242, 279).

Travel information
Alyemda, PO Box 6006, Alyemda Building, Khormaksar Civil Airport, Aden (tx: 2269).

Hotels: Aden
Aden Rock (26 September Hotel), Steamer Point (on ocean front), P.O. Box: 1151 (tel: 22266/67).

Chalet Complex, Khormaksar (tel: 31301).

Crescent, Tawahi (tel: 23471/2).

Gold Mohur, Tawahi (tel: 24171/7; tx: 2312 mohur).

Pullman, Khormaksar, P.O. Box: 6111 (tel: 42911; tx: 2319).

Yugoslavia

Christopher Cviic

Yugoslavia entered a period of high political turbulence in 1988-89. A dramatic deterioration occurred in the relations between the country's eastern half (Serbia, with its provinces of Kosovo and Vojvodina; Macedonia; and Montenegro) on one side, and its western half (Croatia and Slovenia) on the other. Bosnia and Hercegovina continued to hover uneasily in the middle, but increasingly leaning towards the western side and thus earning the popular description of 'Yugoslavia's duty-free shop'. The split was reflected in the country's ruling communist party (official name since 1952: the League of Communists of Yugoslavia). The work of the party's top organs, the 165-member central committee and the 23-member presidium, was completely paralysed by deep ideological clashes with strong nationalist overtones. Political conflicts continued to push economic issues into the background.

Opposition from the hardline coalition in the east led by Serbia threatened to undermine the reform programme of the strongly market-oriented new federal government in Belgrade. In mid-summer 1989, as inflation looked set to reach the annual 1,000 per cent rate (it had reached 790 per cent by the end of July), widespread social unrest was expected. There was speculation about a Serbian-inspired coup, ostensibly to save the country from becoming another Lebanon, but in reality aimed at recentralising it under the dominance of the Serbs, Yugoslavia's largest nation (36 per cent of the total population).

As internal divisions spilt over into foreign policy, there was a slackening of the drive for closer links with Western Europe, which had been gathering momentum since the mid-1980s. Slovenia and Croatia came under attack from Belgrade for developing close co-operation with their western neighbours and western Europe in general. Yugoslavia's diplomacy strove to maintain the country's prominent position within the non-aligned movement, whose chairman it had become in the autumn of 1988.

Crisis of the system

The worsening internal situation affected Yugoslavia's intricate political system which continued to function, but with increasing difficulty. On 15 May Janez Drnovsek, a little-known Slovene economist aged 39, took over from Raif Dizdarevic as president of Yugoslavia's collective state presidency, the eleventh since Tito's death in 1980. Under Yugoslavia's rotating system of government, which Tito introduced in the last decade of his life, each of the eight members of the state presidency, representing the six federal republics (Bosnia and Hercegovina, Croatia, Macedonia, Montenegro, Serbia and Slovenia) and the autonomous provinces (Kosovo and Vojvodina), serves a one-year term as head of state in strict rotation and then reverts to being an ordinary member. Until recently, the presidency had a ninth member – the current president of the League of Communists of Yugoslavia sitting *ex officio* – who was not included in the annual rotation. But under the newly amended constitution (see below), the party is no longer officially represented on the presidency.

Like the country's federal system as a whole, the collective presidency is meant to reflect Yugoslavia's multinational character and to emphasise the equality of its constituent parts. More specifically, Tito saw the collective presidency as a means of averting a power struggle after his death and the eventual predominance of a single individual or, even more important, a single nation. Since Tito, the president has changed every year. The vice-president has always stepped up to the presidency, as the current vice-president, Borisav Jovic, Serbia's member on the presidency, is expected to do in 1990. In Yugoslavia's highly volatile political situation, with the ruling party unable to function as a united body at the top, the state presidency has assumed an ever greater significance. According to the Yugoslav constitution, it is, for example, 'the supreme body in charge of the administration and command of the armed forces' and also it directs foreign policy. It has the power to introduce a state of emergency, a very important prerogative. But, under the newly amended constitution, members of the state presidency are no longer allowed to sit on the Yugoslav party's central committee.

KEY FACTS

Official title: The Socialist Federal Republic of Yugoslavia

Head of state: President Janez Drnovsek

Head of government: Premier Ante Markovic

Ruling party: The League of Communists of Yugoslavia

Capital: Belgrade

Official Languages: Serbo-Croat, Slovenian, Macedonian, Albanian, Hungarian

Currency: Dinar (D)

Exchange rate: D90,953 per $ (Dec 1989)

Area: 255,804 sq km

Population: 23.6m (1988)

GDP real growth: 3.6% (NMP, 1986)

GNP per capita: $2,355 (1986, NMP)

Labour force: 10.1m (1984)

Unemployment: 14.1% (1988)

Inflation: 194% (1988)

Trade balance: $779m (1988)

Previously, members of the presidency were elected by the assembly of their respective republic or autonomous province for a five-year term that can be extended once, after a nomination process conducted by the Socialist Alliance, the party's umbrella organisation. This procedure, previously a formality, is no longer so. Slovenia was the first to decide to give its own representative on the presidency a broader democratic mandate in line with the process of democratisation taking place in that republic since the mid-1980s. A referendum was held on 2 April throughout Slovenia, in which the little known Drnovsek defeated a well-known party establishment figure, the 63-year-old Marko Bulc. In his campaign, Drnovsek had advocated a market economy, cuts in public spending, closer ties with western Europe and political pluralism with the end of the one-party state and the party's control of economic decision making.

Serbia did not follow Slovenia's example: its presidency member, Borisav Jovic, was elected by the republic's assembly in April, unopposed. In Croatia, also in April, Stipe Suvar, the Yugoslav party's president for the 1988-89 period, was elected by the republic's assembly in a contest against another candidate. The province of Vojvodina elected its member, Dragutin Zelonovic, a university professor, also through a two-man contest in the assembly. The Kosovo assembly elected its member, Riza Sapunxhiu, unopposed. The remaining three federal units – Bosnia and Hercegovina, Macedonia and Montenegro – elected their representatives at public referenda: Nenad Bucin (Montenegro) in April; Vasil Tupurkovski (Macedonia) in June; and Bogic Bogicevic (Bosnia and Hercegovina) in July. The referenda in both Bosnia and Macedonia were held after the original votes in the assembly produced a stalemate. In Bosnia, this occurred after a favourite, a well-known university professor, was accused of 'improper contacts with foreign powers'. He withdrew, but neither of the two remaining candidates obtained the required majority. Hence the decision to hold the referendum.

Within the ruling party, the change-over at the end of June 1989 saw Stipe Suvar hand over the party's presidency to Milan Pancevski, recently elected president of the party in Macedonia. But his election was for six months only – till the party's congress scheduled to be held in December 1989, ahead of the normal date in 1990. The decision to hold the congress earlier was preceded by an acrimonious debate between Serbia and its allies on one side and Slovenia and Croatia on the other. Serbia had been demanding the holding of an extraordinary (i.e. emergency) congress for some time on the grounds that the country's critical state required the party to decide on far-reaching changes of the entire system left behind by Tito.

This demand was bitterly opposed by the two western republics. They saw it as an attempt by Serbia to achieve its aim of a recentralised Yugoslavia under Serbian leadership by first of all capturing the control of the party organs at the congress and then using the congress as a mandate for sweeping changes of the federal system that would drastically reduce the powers of the republics and autonomous provinces in favour of the federal centre in Belgrade. These suspicions were confirmed by Serbia's attempts to change the weighted system of representation at the congress, which was presently in force, to a one-member, one-vote system, which would give Serbia, the largest republic with most party members, an enormous preponderance over the others. The Yugoslav party had just over 2m members in 1989, of which the Belgrade party organisation alone had 200,000 (10 per cent). Slovenia as a whole had only 105,000 party members, about 5 per cent below Slovenia's proportion of Yugoslavia's population (8.3 per cent).

In the end, a solution was worked out under which most of the elements of the old weighted system would be preserved during the preparations for the next congress which would also – in line with old procedures – be preceded by republican or provincial congresses if individual republics or provinces so wished. But the compromise did not go far enough for Slovenia. At the April 1989 meeting of the central committee of the Yugoslav party in Belgrade, the Slovene members voted against it *en bloc*. Their fear, openly expressed, was that more not fewer safeguards were needed to prevent a Serbian *diktat* at the congress, given Serbia's dominance in the whole of the eastern half of the country as well as in the largely Serbian-officered Yugoslav army. In June, the Slovene party presidium announced that Slovenia's party would not submit to any *diktat* at the congress and might instead declare its own 'organisation independence', and split off from the federal party. There were also strong voices urging Slovenia's boycott of the congress, but those were rejected by the party leader, Milan Kucan. Controversy about where the congress was to be held continued right into the summer with suggestions that, rather than in Belgrade, it should be held in Sarajevo, the capital of Bosnia, or a city in Croatia or even in Ljubljana. But Serbia continued to insist that the place to hold the congress was Belgrade. Bitter controversies, which erupted at the central committee meeting at the end of July called to discuss the crisis in intra-national relations within Yugoslavia, led to a closer alignment of Bosnia with Croatia and Slovenia, in this and other questions, but this only emphasised the polarisation in the party.

KEY INDICATORS

	Unit	1984	1985	1986	1987	1988
Population	m	23.1	23.2	23.3	23.4	23.6
Net material product (NMP)	dinar bn	6,653	11,951	25,083	–	–
NMP per capita	$	2,661	2,716	2,355	–	–
NMP real growth	$	2.3	0.5	3.6	–1.0	–2.0
Inflation	%	54.7	72.4	90	121	194
Retail sales	1980=100	381	651	1,284	3,000*	–
Unemployment	'000	975	1,040	1,087	1,081	1,132
Labour productivity	% growth	2.7	–0.7	–2.2	–2.0	–
Agricultural production	1979-81=100	109	100	113	103	103
Industrial production	1985=100	97	100	104	104	103
Investment goods	1980=100	109	114	115	113	–
Consumer goods	1980=100	112	114	120	118	–
Intermediate goods	1980=100	112	114	119	117	–
Exports	$bn	10.1	10.6	11.1	11.4	12.8
Imports	$bn	10.9	11.2	11.8	11.3	12.0
Balance of trade	$m	–789	–588	–702	83	779
Current account	$m	478	833	1,100	1,249	2,489
Foreign debt	$bn	20.2	19.9	20.6	–	–
Exchange rate	dinar per $	152.8	270.2	379.2	737.0	2,522.6

* estimated figure

Constitutional reform

Most of the extraordinary bitterness within the party in Yugoslavia was caused by the problem of the Kosovo province and the Serbian leaders' handling of it within the framework of Yugoslavia's constitutional reform which had been agreed to by all on the grounds that the last 1974 Tito constitution was both unwieldly and out of date. While there was a consensus (at least in principle) ever in favour of a market-oriented reform in all the republics since the process of constitutional reform started formally in 1987, there never was a corresponding broad agreement on the future shape of the federal system. Serbia stressed from the start its view that what was needed was a stronger, more united and, by implication, a more centralised Yugoslavia. Behind this lay the even stronger wish for the re-integration of the provinces of Kosovo and Vojvodina into Serbia proper. Under the 1974 constitution, the two had become *de facto* mini-republics in line with the then Tito policy of giving more scope to the Albanians in Kosovo (90 per cent of the total population there now) and the Hungarians and the other non-Serbs in Vojvodina (46 per cent of the total population of the province). And so there was, right from the start of the process of constitutional revision in 1987, a difference here: for most Serbs a stronger Serbia within a more centralised Yugoslavia was the main demand with everything else to follow; for most non-Serbs the retention of the existing federal structure guaranteeing the equality of all, Serbs and non-Serbs, was a *conditio sine qua non*. In the event, a constitutional reform resolved nothing and, in some ways, made matters even worse.

On 25 November 1988 the Yugoslav federal assembly in Belgrade adopted 39 constitutional amendments affecting 135 out of the 406 articles of the 1974 constitution. The core of the amendments dealt with the economy. Ten amendments of Chapter One of the 1974 constitution were aimed at removing legal obstacles to the passage of new laws regulating economic reform after January 1989 and allowing foreign ownership of Yugoslav enterprises and banks in several free-trade zones (mostly in tourist areas); foreigners with a majority share-holding would be allowed to manage and lay off workers in joint-venture enterprises. The amendments cleared the way for Yugoslav citizens to become shareholders in Yugoslav firms and for firms to have more freedom in their own decision-making. Workers' councils, which at least in theory represent the interests of the workers and their firm, were given a greater say in appointing and dismissing their managers and in running day-to-day operations. Limitations on wages and salaries were abolished, with wages to be based solely on a firm's profits. The amendments provided for state assistance (subsidies) only in cases of clear national interest (steel, energy and arms industries). Workers were guaranteed the right to strike – until then strikes had been tolerated without being either legal or illegal. The amendments also increased the so called 'land maximum' – the amount of land a private individual may own – from 10 to 30 hectares.

The amendments relating to political reform included new provisions for the election of public officials based on multiple candidacies and secret balloting. The process of separating the party from the state also found expression in the amendments (the party had already initiated the process at its central committee meeting in October 1988). One of the amendments stipulated that the president of the party should no longer be an *ex officio* member of the collective state presidency of Yugoslavia. The federal assembly was given the right to dismiss the whole of the state presidency or individual members of it. The amendments also stipulated that the council of the federation, a 170-member body of retired senior politicians and other public figures serving as advisers to the state presidency, was to be abolished in 1991. A compromise was reached on the original amendment relating to the financing of the Yugoslav army: instead of supplementing the federal budget allocation for the army with income tax, a Slovene proposal was accepted to raise a sales tax on some consumer and luxury items. The amendments altered little in the sphere of the relations between the federation on one side and the republics and provinces on the other. But one amendment, dealing with the constitutional relationship between Serbia and its two autonomous provinces, Kosovo and Vojvodina, alluded to the need for the 'consistent expression of Serbia's constitutional position'. The Serbs took this as the green light from the rest of Yugoslavia for action to extend the Serbian government's control over the two provinces. But, as subsequent events showed, strong reservations over any serious reduction in the autonomy of the two provinces remained in non-Serbian parts of Yugoslavia, though nobody was prepared to confront Serbia head-on on this issue.

The Milosevic offensive

The Serbian authorities, for their part, attempted to reassure doubters by insisting that nothing drastic was being planned – only a relatively minor constitutional adjustment.

These reassurances carried little conviction in the light of a political campaign to regain control of the provinces by, if necessary, overthrowing their leaderships by pressure and intimidation – a clear departure from the rules of post-Tito political behaviour – which was already under way by the time the amendments were being voted upon in the federal assembly in November. A prolonged campaign against Vojvodina's 'autonomist' leadership orchestrated by the Belgrade press, under the full control of Mr Slobodan Milosevic, Serbia's party president since 1986 and the leader of the populist movement to make Serbia a 'proper republic just like any other', culminated in a two-day siege by a large crowd, including groups from Kosovo and Belgrade, of the Vojvodina party and government headquarters in Novi Sad, the province's capital, on 9 and 10 October. Vojvodina's leadership resigned *en bloc* and was quickly replaced by pro-Milosevic men. A wholesale purge of 'autonomists' was set in train.

The federal party showed its displeasure with the pro-Milosevic coup in Novi Sad (and a similar but failed one at the same time in Titograd, the capital of Montenegro, Yugoslavia's poorest and smallest republic, which many Serbs believe should belong to Serbia) at a central committee in Belgrade in the second half of October. Dusan Ckrebic, Milosevic's right-hand man in Serbia, was voted off the party's 23-man presidium at a secret vote of confidence. The indignant Milosevic promptly announced that the Serbian party would not allow Ckrebic to resign. A pro-Milosevic Macedonian, Vasil Tupurkovski, attacked the 'unprincipled coalition' of anti-Milosevic forces. But, after a brief lull, the Milosevic camp scored an important victory in Montenegro in January, when the local leadership fell following a large-scale demonstration of more than 100,000 people. This political upheaval gave Milosevic another vote in the federal bodies – that of Montenegro.

In Kosovo, there was a large mass demonstration in November in which many workers took part against the proposed constitutional changes in the Serbian republic that would give the Serbian government greater control over the provinces. But the demonstration was ignored and, under strong pressure from Belgrade, a number of prominent Kosovo figures, including the party president, Kaqusha Jashari, and

COUNTRY PROFILE

Political system: Socialist federation of six republics (Serbia, Croatia, Bosnia-Hercegovina, Macedonia, Slovenia, Montenegro) and two autonomous provinces (Vojvodina, Kosovo) each represented in the federal collective state presidency, which is elected by the bi-cameral Assembly every five years. Posts of president and vice-president rotate annually. The assembly also elects the Federal Executive Council, the administrative branch of the government, every four years. Political power is exercised by the League of Communists.

Main cities/towns: Belgrade (population 1.5m; Serbia), Zagreb (Croatia), Nis\06 (Serbia), Skopje (Macedonia), Sarajevo (Bosnia-Hercegovina), Llubljana (Slovenia), Novi Sad (Vojvodina), Titograd (Montenegro), Pris\09tina (Kosovo), Rijeka (Croatia), Split (Croatia), Maribor (Slovenia).

Media
Dailies: There are some 30 dailies mainly printed in Serbo-Croat, Slovenian and Macedonian. Most significant are the SAWPY paper *Borba* (Zagreb and Belgrade editions; circulation 34,780), *Politika* (Belgrade, 202,875) and *Vjesnik* (Zagreb, 68,519).
Weeklies: Most significant are *Nin* (Belgrade, 140,000) and *Danas* (Zagreb).
Business: Large assortment of business magazines and technical journals, many of which are published in French, German and English. Main business papers are *Privredni pregled* (Belgrade/daily, 14,000), *Prirednì vjesnik* (Zagreb/weekly), *It Novine* (Belgrade/weekly), *Informator* (Zagreb/twice monthly) and *Ekonomska politika* (Belgrade).
Broadcasting: Approximately 3m radio receivers and 4m TVs in use. Home and foreign services operated by 10 organisations under the auspices of Jugoslovenska Radio-Televizija (JRT). Radio Yugoslavia broadcasts daily in 10 foreign languages on mw.
Advertising: All principal media available through the state advertising agencies, such as Jugoslaviapublic, Ozeha and Interpublic or directly.

Domestic economy: Socialist market economy with emphasis on self-management. Boom years 1976-79 when economy grew at average annual rate of 6 per cent. IMF-aided austerity programme was first adopted in 1983 to combat rising inflation and liquidity problems but by the end of 1987 foreign debt was equivalent to 10 per cent of GDP and inflation was approaching 200 per cent and further austerity measures were put into effect in 1988. In an attempt to revitalise the economy export capacity is being expanded with increased worker participation.

Employment: Around 98 per cent of workforce employed in the social sector, mainly in manufacturing and service industries. Real incomes continue to lag behind inflation. Unemployment seven per cent (1986).

External trade: Heavy dependence on imported mineral fuels from USSR and other countries. Other important imports are base metals, chemicals, manufactures and foodstuffs, around 50 per cent of which come from OECD countries. Exports are based on basic manufacturing industries and agricultural produce, and are increasingly directed towards OECD countries. Trade with other CMEA countries has fallen, though is still of great importance, and overall trade deficit has widened since 1985.

Agriculture: Accounts for approximately 14 per cent of GDP. About 85 per cent of arable land is owned by private farmers. Emphasis on specialised agro-industrial 'kombinats'. Main crops: wheat, maize, sugar-beet and potatoes. Extensive forests, vineyards, orchards and livestock.

Industry: Accounts for about 43 per cent of GDP. Diversified industrial base, largely centred in the north. Major industries include: shipbuilding, metals, chemicals, engineering and other manufactures.

Mining: Accounts for about 2.5 per cent of GDP. Diverse resources, such as coal, iron ore, non-ferrous metal ores, copper, bauxite and small reservices of oil and gas.

Energy: Heavy dependence on imports of oil and gas despite energy conservation and increased utilisation of local resources. Hydroelectric power accounts for 33 per cent of electricity generation, thermal 58 per cent and nuclear 4 per cent. Gas production rose by 20 per cent in 1985 with expansion of Molve field. Plans to revitalise coal industry to contribute to future electricity generation.

Banking: Three-tier system co-ordinated by the Narodna Banka Jugoslavije. In addition to the Narodna Banka (national bank), there are national banks for each of the six republics and two autonomous provinces, as well as commercial banks, which operate competitively. After 1977 the banking system was substantially re-organised.

Legal system: Mixture of civil law system and communist legal theory.

Membership of international organisations: ASSIMER, CCC, CEMA, FAO, G-77, GATT, IAEA, IBA, IBRD, ICAC, ICAO, IDA, IDB, IFAD, IFC, IHO, ILO, IMF, IMO, INTELSAT, International Lead and Zinc Study Group, Interpol, IPU, ITC, ITU, NAM, OECD, UN, UNESCO, UPU, WHO, WIPO, WMO, WTO.

the former president, Azem Vllasi, still on Kosovo's provincial party committee, were removed from their posts. Widespread industrial unrest throughout the province in February included a hunger strike by Albanian miners of the important Trepca lead and zinc mining complex. Their action was in protest against the proposed constitutional changes and also against the appointment of pro-Milosevic figures to leading positions – notably that of Rahman Morina, former police chief, as party president. On 26 February the federal authorities proclaimed so-called 'special measures', a partial state of emergency. The strike ended when the authorities seemed to be giving in, with the three pro-Milosevic men offering their resignation. But within a day, after the miners had finished their hunger strike, the resignations were withdrawn. On 23 March, Kosovo's assembly, surrounded by tanks, voted to adopt constitutional amendments giving Serbia control over the two provinces' police, courts, economic planning and foreign relations. The amendments also gave the Serbian assembly the right to make further changes in the province's status without recourse to their assemblies. Only a handful of deputies voted against. In early March, several weeks before the vote, 44 ethnic Albanians, including political and industrial leaders, (former party leader, Azem Vllasi, among them) were arrested and charged with stirring up unrest. The vote in the Kosovo assembly led to violent demonstrations in which 24 people, two of them policemen, were killed. On 28 March the Serbian assembly solemnly adopted the constitutional amendments giving it virtually direct control over the provinces. In the wake of the unrest, 238 prominent ethnic Albanians, including a number of university professors, were sent into 'isolation', a form of preventive internment. Many of them had signed the so-called 'Petition of 215' in favour of the restoration of human and political rights in Kosovo.

The events of February and March marked the end of the constitutional battle over the status of Serbia and its provinces, which had ended with Slobodan Milosevic's victory. But the violent suppression of ethnic Albanian dissent in the province led to open public protests in the western half of the country, notably Slovenia. A protest meeting against the crackdown in Kosovo held in Ljubljana on 28 February was attended by members of the party leadership and the opposition. It provoked a strong reaction in Belgrade, including a decision by the Serbian Writers' Union to break off relations with the Slovene writers and the stepping up of the unofficial but widespread boycott of Slovene goods in Serbia. In Slovenia the Kosovo events produced a strong public demand for constitutional changes emphasising Slovenia's autonomous status within Yugoslavia. In August the Slovene authorities published proposed amendments to the constitution of Slovenia which emphasised its right to self-determination including that of secession. A further legal change sought to uphold the principle that only Slovenia's assembly should have the right to proclaim a state of emergency in Slovenia in peacetime. These moves reflected Slovene fears shared by many non-Serbs in Yugoslavia that Slobodan Milosevic, having consolidated his hold over Serbia and Yugoslavia's eastern half, was bent on gaining power over the whole of it, an impression reinforced by his statement in a speech in Vojvodina in April after the completion of the constitutional reform that 'Serbia would not be content to rest on its laurels' and his defiant speech, on the occasion of the six hundredth anniversary on 28 June of the battle of Kosovo where the Turks defeated the Serbs, in which he foresaw 'further battles, perhaps requiring the use of arms'.

In Croatia and Bosnia and Hercegovina, these words were taken as a warning of new strife likely to be caused by Belgrade's attempts to mobilise the Serbs in Bosnia (32 per cent of the total population) and those in Croatia (11.6 per cent). In Croatia there were disturbances in July in Knin on the occasion of the celebration of the Kosovo anniversary there by the local Croatian Serbs with groups from Serbia present. More serious disturbances took place on Croatia's Adriatic coast, mostly caused by Serbian tourists apparently seeking to provoke the local population by chauvinistic behaviour. Similar incidents suggesting the possibility of communal violence on a considerable scale occurred in the summer of 1989 in Bosnia and Hercegovina. In Croatia and Bosnia as well as in Slovenia these incidents were seen as part of a deliberate campaign designed to manufacture an excuse for the proclamation of a state of emergency that could be used by Milosevic to push through a recentralisation of the country under his own leadership as an executive president within the framework of a new constitution.

The economy

Having survived an attempt by Croatia and Slovenia in May 1988 to vote him out of power, Branko Mikulic, the federal prime minister since 1986, eventually tendered his resignation on 30 December 1988, a year and a half before the expiry of his term of office, universally criticised and unlamented by anybody. The ostensible reason for Mikulic's resignation was the failure of federal assembly on 28 December to approve an austerity measure that would have provisionally limited public spending in 1989 in line with demands for such cuts by the International Monetary Fund. Also the assembly had failed to accept the government's general budget proposals for 1989. In his defence Mikulic said the goverment could not be blamed for 'unveiling the mistakes and shortcomings of the past and for failing to solve them in two and a half years. He said that the real decisions were not made by the federal government but 'in 7,000-8,000 places'.

But it was clear that Mikulic had lost the support of the top party leaders when Stefan Korosec, the party's executive secretary, said on 12 December that the government's economic reforms had come too late and that its anti-inflation programme had 'collapsed completely', with the inflation rate in 1988 twice that in 1977 (170 per cent). Mikulic had also been under pressure as part of a campaign against corruption and nepotism in his own native republic of Bosnia and Hercegovina.

On 19 January 1989 the state presidency voted by a narrow majority to appoint Ante Markovic, 64, a former president and prime minister of Croatia and one of the country's most successful industrial managers, as Mikulic's successor. On 16 March the new prime minister and his smaller, 19-member cabinet, was endorsed by the federal assembly. It included a number of new faces but also from the Mikulic cabinet Budimir Loncar, the experienced foreign minister, and General Veljko Kadijevic, the defence minister. But presumably to appease Serbia, whose (unsuccessful) candidate for the job of prime minister

BUSINESS GUIDE

Time: GMT + 1 (GMT + 2 from late Mar to late Sep).

Climate: Mediterranean on Adriatic coast with warm summers and mild winters; northern mountain areas have cold winters with heavy snow. Average summer temperature in Belgrade 22°C, in winter 0°C.
Clothing: Lightweight and raincoat for summer, mediumweight and topcoat for winter.

Entry requirements: Visas not required by nationals of most European countries. Entry-exit visas available at frontier for visits up to three months, transit visas for stays up to seven days. All visitors must register with authorities within 48 hours of arrival (usually undertaken at hotel).
Health: Vaccination certificates not required.
Currency: No limit to amount of foreign currency taken into country. In any calendar year no more than D1,000,000, in notes of D100,000 or below, may be taken in or out of the country, on the first visit; and not more than D500,000 on subsequent visits.
Customs: Various personal articles and goods allowed duty-free.

Air access: Regular flights by all major international airlines.
National airline: Jugoslovenski Aero Transport (JAT).
Main international airport: Belgrade (BEG) – 20 km from city centre.
International airports: Zagreb (ZAG), Ljubljana, Dubrovnik-Cilipi, Ohrid, Split, Skopje, Pula, Titograd, Sarajevo, Osijek, Tivat, Maribor, Rijeka, Portoroz, Zadar.

Surface access: Border crossings from: Italy, Austria, Hungary, Romania, Bulgaria, Greece, Albania.
Main ports: Rijeka, Sibenik, Split, Kardeljevo, Dubrovnik, Bar, Pula, Koper.

Hotels: Advisable to book in advance. Hotels are classified into five categories: L (extra), A, B, C and D; boarding houses into three. Daily rate for single room varies, with category, from $35 to $150, not including a 10-20 per cent service charge. Visitors must also pay a residential tax, which varies between regions.

Restaurants: A wide variety with many serving national and regional food. Service charge generally not included on bills and tip of up to 10 per cent should be added.

Car hire: Cars can be hired in most main towns through Yugoslav travel agencies. There is a speed limit of 120 kph on motorways, and 60 kph in built-up areas. Drive on the right and give way to traffic from the right unless clearly marked otherwise. Seat belts are compulsory in front seats.

City transport
Taxis: Good service operating in all main cities. All taxis are metered, but there is no basic charge. A 10 per cent tip is usual.
Buses and trams: Most city centres are served by trams, and the suburbs by buses. The service is generally cheap and regular.

had been Borisav Jovic, Markovic gave the key post of interior minister to General Petar Gracanin, former chief of staff of the Yugoslav army and more recently president of Serbia. (That post, vacated by Gracanin, was filled in April by Slobodan Milosevic. Jovic became Serbia's representative on the Yugoslav state presidency and was due, under the rules of rotation, to become its president in May 1990.)

In his speech to the assembly on 16 March, Markovic once again referred to his idea of a 'new socialism' for Yugoslavia that he had first mentioned in a remarkably frank and outspoken live broadcast on television on 28 January. He had said that all socialisms were in a state of crisis and none of them had been able to solve the problems either of efficiency or of political democracy. He further said that the worldwide failure of socialism was due to dogmas and delusions, chiefly about the nature of the market identified with capitalism although it was 'one of the main achievements of civilisation' which 'cannot be negated without serious consequences'. Markovic made it clear that in contrast to his predecessor he would adopt an oblique approach to the struggle against inflation, which in any case was a symptom of a deeper malaise.

Under Markovic's management, a number of initiatives were taken immediately. At the end of July, the government claimed that 77 per cent of all imports had been liberalised; the proportion of the federal budget expenditure of the estimated gross domestic product was reduced from 11.7 per cent in 1988 to 7.14 per cent in 1989; industrial output rose by three per cent in the first half of 1989 compared with the same period in 1988. Exports to hard-currency countries grew by 27.6 per cent in the January-June period of 1989; the liquidity of the country improved with hard-currency reserves reaching the figure of $4bn by the end of July. But the government failed to achieve several of its key objectives: resistance from primarily the eastern half of the country led to the abandonment of the plan to abolish promissory notes, one of the main sources of internal credit and inflation in Yugoslavia (the estimated internal debt is around $30bn). This, in the words of the respected weekly *Ekonomska Politika* in Belgrade, devalued the effects of the government's restrictive credit and monetary policy. Its other important failure was over the priority in payments. Despite the government's strenuous opposition, firms continued to give priority to the payment of wages to their employees before their obligations to third parties. This undermined the main criterion for judging a firm set by the government – that a firm was doing well as long as it was able to meet its obligations towards third parties. The government also retreated – in the face of strong opposition from Serbia – from its planned cut of the surplus in trade with the Soviet Union (estimated at over $2bn and considered one of the important sources of inflation under which the National Bank pays off the Yugoslav firms owed money by the Soviet partners as though payment has been received from there). Last, but by no means least, the rapid growth of inflation – 31.3 per cent in July over the previous month corresponding to an annual rate of 2,525 per cent – made the prime minister vulnerable to the objection, once again voiced from Serbia, that, by not fighting inflation head-on, he was favouring the richer republics which could bear inflation more easily than the poorer ones.

By mid-summer of 1989 it was clear that the Milosevic camp was beginning to gun for Markovic. An article in the Milosevic-controlled *Politika Ekspres* in Belgrade published an open attack on Markovic, calling his government 'the government of discord'. Fortunately for the prime mininster, the ever-more numerous strikes had by mid-1989 not yet turned into a nationwide wave of unrest as in Poland in the past. So the future of the government remained open with much pointing to the likelihood of failure, perhaps resulting in the federal assembly refusing to extend Markovic's mandate beyond May 1990. On the other hand, Markovic's popularity with Western bankers and governments, where he was known and respected, as well as taken seriously as a sincere free marketeer remained an important source of his strength.

Foreign policy

Yugoslavia's image suffered as a result of the internal squabbles and, especially, the repression in Kosovo. American-Yugoslav relations came under some strain over that issue: rather conspicuously President Bush omitted to visit Yugoslavia during his trip to Eastern Europe in July. Congress passed several motions on the internal situation in Yugoslavia criticising its human rights record – notably in Kosovo. In April, the European parliament passed a resolution criticising Yugoslavia's policy in Kosovo. At the end of May, a delegation of the European parliament visited Yugoslavia, including Kosovo, to gather material about the situation there. The delegation criticised inadequate arrangements on its return, noting particularly its inability to meet the Kosovo deputies who had voted against, or abstained from voting on, the proposed constitutional changes in the republic of Serbia in March. It was noted in Yugoslavia that the prominent Croation dissident, Dobroslav Paraga, was received in Congress during a visit to Washington in May as well as by West Germany's president, Richard von Weizsäcker, when eventually Paraga was granted permission to leave the country to attend a conference abroad.

However, the Soviet Union though officially keeping out of the internal conflicts in Yugoslavia showed clear sympathy for Serbia's stance over Kosovo in a number of reports published in *Krasnaya Zvezda* the army paper, and in *Pravda* and *Izvestia* in the summer of 1989. A similar attitude was displayed by the hardline east European countries such as Czechoslovakia, East Germany and Bulgaria. Yugoslavia found itself involved in a continuing controversy with Greece over Macedonia, while its polemics with Albania over Kosovo became more and more shrill.

As a result of foreign criticism, Yugoslav diplomacy came under attack from Belgrade as 'not doing enough to represent the Kosovo situation in a true light to the world'. This was widely seen in Yugoslavia as an attack on Budimir Loncar, the foreign minister and by nationality a Croat. But in August Zivorad Kovacevic, a Serb and former mayor of Belgrade, was withdrawn from the post of ambassador to the United States because he had been appointed by Milosevic's predecessor in Serbia and had since fallen foul of Milosevic by entertaining Slovenia's president, Janez Stanovnik, in November 1988 despite the fact that Stanovnik had openly criticised Milosevic's policy while in the United States.

Outlook

Despite some modest gains achieved by the Markovic government, which could become more impressive if he were allowed to go on with his wide-ranging reforms (particularly the reform of the banking system to free it from the undue influence of local politicians) the outlook for economic reform remained bleak. The reason was lack of agreement between the mainly reformist northern republics and the anti-reformist alliance led by Serbia. It seemed likely that increasing political uncertainty about the future shape of the regime, including the possibility of deepening conflict leading to some form of military rule, would continue to have an adverse effect on direct foreign investment under the newly liberalised laws on joint ventures and other forms of co-operation.

BUSINESS GUIDE

National transport

Air: Regular air services linking main towns, plus three charter operations. JAT operate regular and charter flights. Fares are low but demand is heavy on some routes, so advance reservation is advisable.

Road: There are some 120,000 km of roads, including 756 km of motorway. Most important roads are: the main inland route through Ljubljana, Zagreb, Belgrade, Nis and Skopje to the Greek border; along the Adriatic coast from Rijeka to Ulcinj, and inland to Titograd. An extensive network of express buses links main centres. Advanced seat reservation is recommended.

Buses: Cheap and efficient coach services to most towns.

Rail: There are some 9,250 km of track, of which over a third is electrified. International express and fast trains to all main towns. Sleepers should be booked well in advance. Service is operated by Zajednica Jugoslovenskih Zeleznica. Reduced fares for foreign visitors are available.

Water: Well established inland waterways system, based on the Danube, Sava, Tisa, Drava and Begej. The Yugoslav Maritime Agency, Jadrolinija, operates express ferryboat services along the Adriatic coast.

Public holidays

Fixed dates: In all Yugoslavia: 1 and 2 Jan, 1 and 2 May, 4 Jul (Veterans' Day), 29 and 30 Nov (Day of the Republic). Holidays celebrated only in Republic concerned: Serbia, 7 Jul; Montenegro, 13 Jul; Slovenia, 27 Apr, 22 Jul and 1 Nov; Croatia, 27 Jul; Macedonia, 11 Oct and 2 Aug; Bosnia-Hercegovina, 27 Jul and 25 Nov.

Variable dates: Though not official holidays, Easter and Christmas are widely celebrated in Slovenia and Croatia.

Working hours

Business: (Mon-Fri) 0700 – 1430.

Government: (Mon-Fri) 0730 – 1530, except (Wed) 0730 – 1730.

Banking: (Mon-Fri) 0730 – 1900.

Shops: (Mon-Fri) 0800 – 1200 and 1700 – 2000, (Sat) 0800 – 1500, but many shops open throughout day.

Business language and interpreting/translation: No regular nationwide interpreter service, but a translation service can usually be arranged through hotels, tourist offices or local enterprises.

BUSINESS GUIDE

Telecommunications
Telephone: International dialling code: 38 followed by 11 for Belgrade, 41 for Zagreb. Direct dialling is available. Approximately 95 telephones per '000 inhabitants. Operator numbers: international calls 901, internal long distance 900, telegrams 96. For IDD access dial 99.
Telex: Service available in most large hotels and at main post offices. Messages can be handed in at telegram counters. Country code 62YU.
Telegram: Main post offices operate ordinary and express rates.

Trade fairs: International Technical Fair (May) and International Motor Show (Apr) held in Belgrade. Also International Agricultural Fairs in Novi Sad. Principal exhibition centres are Belgrade, Zagreb, Sarajevo, Skopje, Kranj, Novi Sad, Leskovac and Ljuibljana.

Electricity supply: 220 V AC.

Representation in capital: Afghanistan, Albania, Algeria, Angola, Argentina, Australia, Austria, Bangladesh, Belgium, Bolivia, Brazil, Bulgaria, Burma, Canada, China PR, Colombia, Cuba, Cyprus, Czechoslovakia, Denmark, Ecuador, Egypt, Ethiopia, Finland, France, Gabon, German Democratic Republic, Federal Republic of Germany, Ghana, Greece, Guinea, Holy See, Hungary, India, Indonesia, Iran, Iraq, Italy, Japan, Jordan, Kampuchea, Democratic People's Republic of Korea, Kuwait, Lebanon, Libya, Malaysia, Mali, Mexico, Mongolia, Morocco, Netherlands, Nicaragua, Nigeria, Norway, Pakistan, Panama, Peru, Philippines, Poland, Portugal, Romania, Somalia, Spain, Sri Lanka, Sudan, Sweden, Switzerland, Syria, Thailand, Tunisia, Turkey, USSR, United Kingdom, USA, Uruguay, Vatican, Venezuela, Vietnam, Zaïre, Zambia, Zimbabwe.

BUSINESS DIRECTORY

Hotels: Belgrade
Balkan, Prizenska 2, 11000 (tel: 687466; tx: 72224).

Excelsior, Kneza Milosa 5, 11000 (tel: 331381; tx: 12299).

Inter-Continental, Vladimira Popovica 10, 11070 (15 minutes from airport) (tel: 138708; tx: 12009; fax: 131402).

Jugoslavija, 3 Bulevar E. Kardelja, 11181 (facing the river Danube) (tel: 600222; tx: 11777).

Majestic, Obilicev Venac 28, 11000 (tel: 636022; tx: 11345; fax: 630793).

Metropol, Bulevar Revolucije 69, 11000 (tel: 330911; tx: 11364; fax: 332991).

Moskva, Terazijie, 11000 (tel: 686255; tx: 11505).

National, Novi Beograd, 11070 (tel: 693509; tx: 11774).

Park, Njegoseva 4, 11000 (tel: 334722; tx: 12888).

Prag, Narodnog Fronta 27, 11000 (tel: 687355; tx: 12494).

Slavija, Svetog Save 1, 11000 (business centre) (tel: 450842; tx: 11545).

Srbija, Ustanicka 127, 11000 (tel: 413255; tx: 12455).

Sumadija, Sumadijski Trg 8, 11030 (tel: 554255; tx: 12400).

Union, Kosovska 11, 11000 (tel: 187036; tx: 12955).

Hotels: Zagreb
Inter-Continental, Zagrab 41000, Krsnjavoga 1 (tel: 443411; tx: 21732; fax: 444431).

Palace, Zagreb 41000, Strossmayerov Trg 10 (tel: 275611; tx: 21526).

Car hire
Avis-Autotehna: Bulevar Revolucije 94, 11000 Belgrade (tel: 433-323; tx: 11713).

Inter rent-a-car rental system, Kidriceva 13a, 61000 Ljubljana (tel: 213-840; tx: 32170).

Kompas Hertz rent-a-car, Celovska c.206 61000, Ljubljana (tel: 575-311; tx: 31425).

Unistours Europcar rent-a-car, Djure Djakovica bb 71000, Sarajevo (tel: 511-331; tx: 41891).

Banking
Beobanka-Beogradska Osnovna Banka (Belgrade Basic Bank), 11001 Belgrade, Zeleni Venac 16, Box 55 (tel: 629-455; tx: 11802).

Investiciona Banka Titograd, 81000 Titograd, Bulevar revolucije 1 (tel: 814-2922; tx: 61118; banka).

Jugobanka-Udruzena Banka Beograd (united bank), 11000 Belgrade, 7 Jula 19-21 (tel: 630-022; tx: 11145, 11302, 11398).

Ljubljanska Banka-Zdruzena Banka, 6100 Ljubljana, Trg revolucije 2 (tel: 612-5551; tx: 31256 lblj).

Narodna Banka Jugoslavije (central bank), 11000 Belgrade, Bulevar Revolucije 15, Box 1010 (tel: 332-001; tx: 72000).

Privredna Banka Sarajevo, 71000 Sarajevo, Obala vojvode Stepe 19 (tel: 213-144; tx: 41280).

Stopanska Banka, 91000 Skopje, 11 oktombri 7 (tel: 912-35111; tx: 51749 sbank).

Udruzena Banka Hrvatske, 41000 Zagreb, Ksaver 208 (tel: 431222; tx: 22560).

Udruzenje Banka Jugoslavije (association of Yugoslav banks), 11001 Belgrade, Masarikova 5/IX, Box 715 (tel: 684-797; tx: 11767 for bank for international economic co-operation).

Udruzena Beogradska Banka (associated Belgrade bank), 11001 Belgrade, Knez Mihailova 2, Box 955 (tel: 624-455; tx: 11712).

Udruzena Kosovska Banka, 38001 Pristina (tel: 341-11).

Vojvodjanska Banka, 21001 Novi Sad, Bulevar Marsala Tita 14 (tel: 215-7222; tx: 14129 vojba).

Principal newspapers
Borba, Belgrade, Trg Marksa 1 Engelsa 7 (tel: 334-531; tx: 11104).

Ekonomska Politika, 11000 Belgrade, Trg Marksa 1 Engelsa 7 (tel: 335-355; tx: 11410).

Informatur, 41000 Zagreb, Masarykova 1, PO Box 974 (tel: 429-333; tx: 21264).

Nin, Belgrade, Makedonska 29.

Politika, 11000 Belgrade, Makedonska 29.

Privredni Pregled, 11000 Belgrade, M Birjuzova 3 (tel: 182-888; tx: 11509).

Privredni Vjesnik, 41000 Zagreb, Rooseveltov Trg 2 (tel: 443422; tx: 21524).

Vjesnik, 41000 Zagreb, Avenija Bratstva 1 Jedinstva 4 (tel: 515-555; tx: 21121).

Other useful addresses
Adria Airways (domestic and European international flights), 61001 Ljublijana, Kuzmiceva 7 (tel: 313-366; tx: 31268).

Air Yugoslavia (international flights), 11000 Belgrade, Mose Pijade 1/III (tel: 325-635; tx: 12411).

Aviogenex (flights within Europe), 11000 Belgrade, Sava Centar-Milentija Popovica 9 (tel: 603-198, 139-840; tx: 11711).

Federal Chamber of Economy, 11000 Belgrade, Terazije 23 Box 1003 (tel: 339-461; tx: 11638).

Federal Secretariat for Energy and Industry, 11070 Belgrade, Omladinskih brigada 1 (tel: 338-281; tx: 11448).

Federal Secretariat for Finance, 11070 Belgrade, Omladinskih brigada 1 (tel: 338281; tx: 11448).

Federal Secretariat for Foreign Economic Relations, 11070 Belgrade, Omladinskih brigada 1 (tel: 195444; tx: 12110).

Federal Statistical Office, 11000 Belgrade, Kneza Milosa 20, Box 203 (tel: 681-999; tx: 11317).

Interpublic (advertising agency), Meduliceva 2, Box 315, 41000 Zagreb (tel: 423-222; tx: 21662).

Jugoslavijapublic (advertising agency), Knez Mihajlova 10, Box 447, 11000 Belgrade (tel: 633-266; tx: 11125).

Jugoslovenski Aerotransport (Yugoslav Airlines), 11000 Belgrade, Box 749, Bircaninova 1-III (tel: 683-299; tx: 11401).

Jugoslovenska Pomorska Agencija (shipping agency), Belgrade, Bulevar Lenjina 165A, Box 210 (tel: 130-004).

Novinska Agencija Tanjug (news agency), 11001 Belgrade, Obilicev Venac 2, Box 439 (tel: 332-231; tx: 11220).

Prevodilacki Centar (translation agency), Palmoticeva 22, 41000 Zagreb (tel: 431-459).

Savezni Komitet za saobracaj i veze (committee for transport and communications), 11070 Belgrade, Bulevar Avnoja 104 (tel: 602-555; tx: 11062).

Statisci godisnjak Jugoslavije (foreign trade annual of the SFR Jugoslavia), Savenzi Zavod za Statistiku, Belgrade, Kneza Milosa 20.

Turisticki Savez Jugoslavije (tourist association), 11001 Belgrade, Mose Pijade 8, Box 595 (tel: 339-780; tx: 11863).

Uoruzenje strucnik prevodilaca SR SRBIJE (translation agency), 11000 Belgrade, Kicevska 9 (tel: 444-2997).

Vece saveza sindikata Jugoslavije (confederation of trades unions), 11000 Belgrade, trg Marksa i Engelsa 5 (tel: 330-481; tx: 11121).

Zajednica jugoslovenskih zeleznica (community of Yugoslav railways), 11001 Belgrade, Nemanjina 6, Box 563 (tel: 688-722; tx: 12495).

Zaïre

Susan Traill

Zaïre concluded a two-year agreement with the IMF in June 1989 which led to the unblocking of aid by the World Bank, the EC, USA and France. A major row with Belgium over debt rescheduling was still not resolved by summer 1989, although a Belgian bank loan enabled Zaïre to pay off its arrears to the IMF, thus paving the way for the new agreement. Copper prices improved in late 1988 and mining company Gecamines ordered a copper refinery that will give Zaïre a greater measure of independence from Belgium from 1991.

President Mobutu Sese Seko scored a diplomatic triumph in June 1989 by hosting, in his home town of Gbadolite, the signing of a cease-fire between Angola and UNITA in the presence of 18 African heads of state. The ceasefire did not hold, but Zaïre's strategic importance in the region was confirmed. Mobutu was therefore welcomed in Washington by the Bush Administration during his June–July visit there. The Administration pledged $60m in aid for 1989. The US Congress Foreign Relations Committee for Africa was not enthusiastic, however, accusing Mobutu of corruption and violation of human rights.

Mobutu visited President Francois Mitterrand in February 1989 at the height of his quarrel with Belgium. Although France later leased Mobutu a Concorde to return to Zaïre from his villa in Nice, and in June 1989 unblocked FF60m in aid for 1989, the French government denied that it had any ambitions to take over Belgium's role as the leading Western influence in Zaïre.

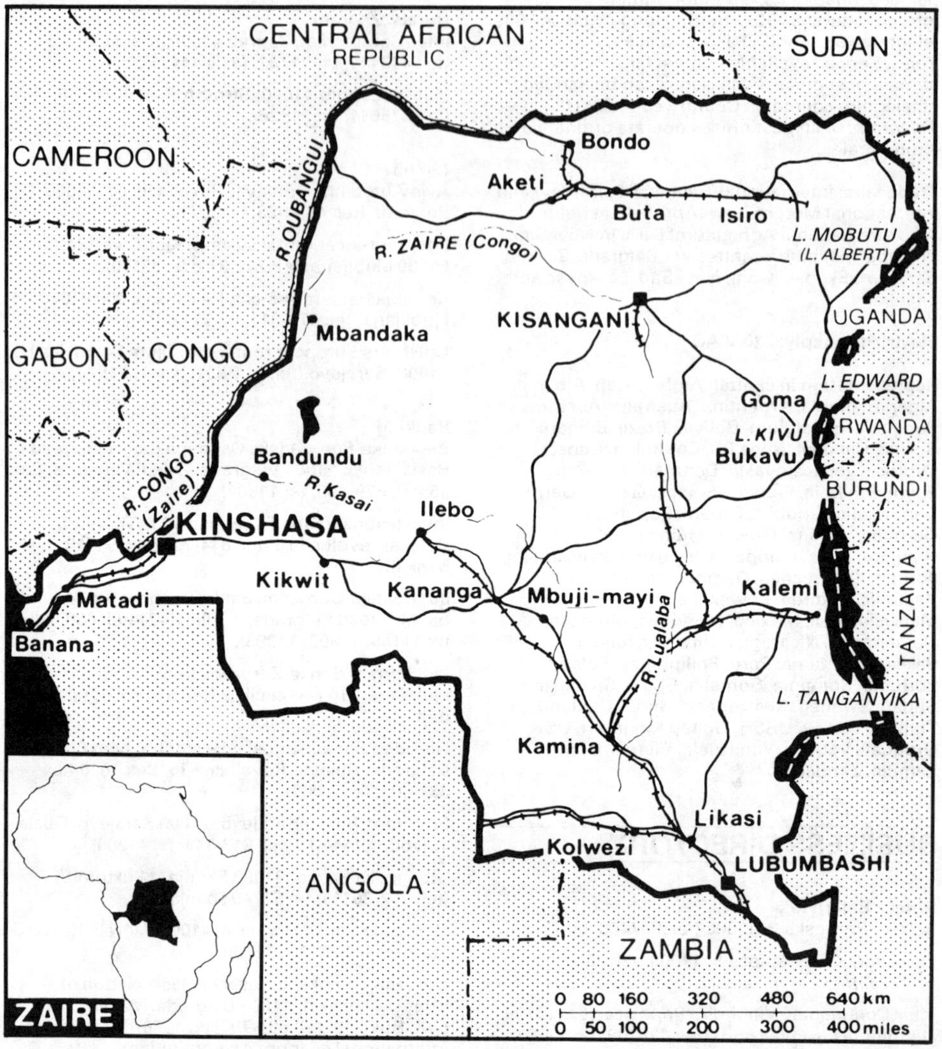

Media-generated animosity

The row with Belgium began in November 1988 following Prime Minister Wilfried Martens' visit to Zaïre, during which he offered to ease the debt repayment burden by one-third of the BF5bn in government-to-government loans by rescheduling over 25 years with 14 years' grace the BF17bn in government guaranteed commercial debt. At the Belgium–Zaïre joint commission held in Brussels in October 1988, Belgium had pledged BF20bn for its new five-year aid programme due to begin in 1989. Zaïre welcomed these moves at first, but following a spate of Belgian press articles on corruption in Zaïre and anger from the Socialist members of the Belgian coalition at Martens' debt offer, Zaïre refused any easing of the debt burden or to accept new aid from Belgium.

The quarrel escalated in December, with Zaïre accusing Belgium of trying to foment a coup against Mobutu. The government ordered its 15,000 nationals living in Belgium to sell their assets by year-end, 5,000 students to return home, state companies to move their offices to other European capitals and to transfer their bank accounts from the two major Zaïrean banks, Banque Commerciale du Zaïre (BCZ) and Union Zairoise des Banques (UZB), both partly owned by Belgian banks, to three government-owned banks. President Mobutu said he would not set foot in Belgium again and gave his properties there to the Zaïrean state. Belgian Development Co-operation Minister Andre Geens and Foreign Affairs Minister Leo Tindemans visited Zaïre to try to patch up the differences. The debt offer remained on the table. The Belgian government also distanced itself from agricultural company Cotoni's attempted injunction on the sale of the presidential properties pending compensation for the 1973 nationalisation of its operations in Zaïre.

In January 1989, the ruling party, Mouvement Populaire de la Revolution (MPR), decided to suspend debt payments to Belgium, declared that Zaïre would end copper supplies to Belgian company Metallurgie Hoboken-Overpelt (MHO), which refines some 200,000 tonnes a year for Gecamines, and demanded a renegotiation of the 1960 independence treaty with Belgium. Belgium suspended new aid and over the next few months withdrew aid workers, partly as a security measure. When Belgium refused to renegotiate the independence settlement, Zaïre threatened to take the matter to the International Court of Justice at the Hague. The government also confirmed its order to state companies to withdraw from Belgium and remove their bank accounts from BCZ and UZB. King Hasan of Morocco offered to mediate. After a few false starts, the two sides met in Rabat in May 1989 and agreed to work out the agenda for a minis-

KEY FACTS

Official title: Republic of Zaire

Head of state: President Marshal Mobuto Sese Seko

Ruling party: Mouvement Populaire de la Révolution (MPR)

Capital: Kinshasa

Official Languages: French

Currency: Zaire (Z) = 100 makuta = 10,000 sengi

Exchange rate: Z441 per $ (Nov 1989)

Area: 2,345,409 sq km

Population: 33.5m (1988)

GDP per capita: $90 (1987)

GDP real growth: 2.6% (1987)

Inflation: 100% (1988)

Trade balance: $563m (1988)

Foreign debt: $7.1bn (1988)

terial conference. Just prior to this meeting, at the Francophone Summit in Dakar, Belgium had proclaimed a 10-year debt moratorium for sub-Saharan countries, including Zaïre, and debt rescheduling over 25 years with 14 years' grace for government guaranteed commercial debt.

Some observers, including the dissident Union pour la Démocratie et le Progrès Social (UDPS), saw the hiatus as a smoke screen to cover up and refocus internal opposition to a new round of IMF austerity. Others saw a cyclical trend, namely that when copper prices rise, Zaïre likes to blow the trumpet of its independence from Belgian aid and influence. But the loan from Banque Belgolaise, BCZ's Belgian shareholder, to pay off IMF arrears was significant and Belgium will be a key player in any general rescheduling talks with Paris and London Club donors later in 1989. The haranguing could lead to a softening of terms, a well tried and often successful Mobutu tactic.

KEY INDICATORS

	Unit	1984	1985	1986	1987	1988
Population	m	30.0	31.0	31.5	32.5	33.5
Gross domestic product (GDP)	Z m	99,723	147,263	203,416	326,946	–
GDP per capita	Z	3,324	4,750	6,477	10,060	–
Inflation	%	52.2	23.8	44.4	78.7	–
GDP real growth	%	2.7	2.5	2.7	2.6	–
Agricultural production	1979-81=100	111.9	115.0	118.9	119.6	120.2
Food production	1979-81=100	111.6	114.5	118.5	119.1	119.8
Crop production	1979-81=100	112.9	116.0	120.2	120.8	121.3
Livestock production	1979-81=100	104.8	106.8	108.7	110.5	112.3
Mining production						
Copper	'000 tonnes	500.4	502.2	507.6	472.2	438.6
Zinc	'000 tonnes	131.7	86.6	79.6	–	61.0
Cobalt	'000 tonnes	9.1	10.8	14.3	11.9	10.0
Diamond	m carats	18.5	19.6	23.3	–	24.0*
Gold	'000 kg	3.5	1.9	2.2	4.4	2.6
Oil	'000 barrels	11,704	12,226	11,857	–	–
Exports	Z m	36,300	47,372	65,559	109,532	209,598
Diamonds	Z m	6,385	9,263	13,026	19,053	24,979
Coffee	Z m	7,207	8,482	19,260	12,771	8,250
Imports	Z m	24,745	39,478	52,176	84,998	142,698
Balance of trade	$m	742	666	561	349	563
Total external debt	$m	5,066	5,885	6,932	8,630	–
Debt service ratio	%	15.2	16.9	13.8	12.8	–
Exchange rate	Z per $	36.1	49.9	59.6	112.4	187.1

* estimated figure

Austerity provokes unrest

An 80 per cent rise in bus fares in February 1989 sparked student unrest in Kinshasa and the closure of the university for about a month. There were also student riots in the copper capital Lumbumbashi, where UDPS claimed 52 were killed by police. The government increased bus fares by another 40 per cent in May and tripled Kinshasa–Brazzaville ferry fares to meet IMF demands to cut subsidies.

Tougher government control over public funds was displayed in the sacking of 34 heads of state companies in 1988 for holding funds abroad, the suspension also in 1988 of 24 zonal commissioners in Kinshasa and 12 other local commissioners in Bas Zaïre, Bandundu and Equateur for raising unauthorised local taxes, and in 1989 the dismissal of six transport ministry officials for irregularities and waste of funds.

A Permanent Consultative Council for Development (CCPD) was set up in April 1989 to represent the views and problems of different sections of the population. Local and municipal elections were held in May 1989, but voting in the eastern province of Kivu was suspended for security reasons. Rwanda, Burundi and Zaïre had signed an agreement in October 1988 regarding refugees and political dissidents in the border areas, but the situation remained unstable. Increased northern border incursions by the Uganda-based Parti de Libération Congolaise led to the breaking off of diplomatic relations with Uganda in May 1989. The government had reorganised the army into eight territorial divisions in August 1988 to tighten security. President Mobutu also appointed the popular General Nsinga Boyengo Mosambaye as defence minister in December 1988, some reports said in order to purge the army of potential coup-makers. In May 1989, Mobutu took back the defence portfolio which he had held continuously since taking power in 1965 except for this brief interlude. A major cabinet reshuffle in November 1988 brought Kengo wa Dondo back as prime minister. Kengo was responsible for implementing IMF reforms in 1983-86.

IMF agreement means more austerity

The IMF agreed to provide SDR204m in 1989-90, SDR116.4m in 1989 and SDR87.3m in 1990. The agreement followed the payment of SDR100.8m in arrears by Zaïre with a $120m loan supplied by Banque Belgolaise. The IMF demanded that the government reduce inflation, raise domestic interest rates and prices, cut government subsidies and the money supply and devalue the zaire. The government had already gone a long way towards meeting these goals before the agreement was signed on 9 June 1989. Petrol prices were doubled in February (leading to the bus fare increases and riots). The oil companies had wanted a quadrupling of prices to overcome the lack of hard currency for imports. In January 1989, the zaire was devalued by 8.2 per cent against the US dollar, following a 9 per cent devaluation in December and 10 per cent devaluation in October 1988. Foreign exchange availability from the Banque du Zaïre was increased from $12m to $30m a month from January 1989. The result of these two measures was to narrow the differential between the official and black market exchange rates from 45 per cent at end-1988 to 12 per cent at end-April 1989 and thus cut down black market operations. Inflation was reduced from 12 per cent a month to 2 per cent a month in the first half of 1989. Interest rates remained at about 60 per cent.

In spring 1989, President Mobutu insisted that no additional expenditure to that already written into the budget would be allowed and that all export of funds must be authorised. State oil company Petro-Zaïre would be restructured to make it more competitive. However, he also approved a 50 per cent increase in public sector wages, at the same time removing 7,400 non-existent staff from the payroll.

The 1989 budget adopted in February allocated Z350bn for expenditure and Z329bn for receipts, giving a deficit of Z21bn. The debt service was set at Z87bn. The corresponding figures for 1988 were Z153bn, Z148bn, Z16bn and Z40bn. With inflation of 85-100 per cent in 1988, real 1989 budget expenditure increased by 14-24 per cent and receipts by 11-20 per cent. The deficit and debt service were expected to improve in real terms. The 1989 budget was revised upwards from its original December 1988 version on account of the substantial increase in copper prices. In April 1989, the government announced that first quarter 1989 revenues had exceeded targets by 20 per cent.

The total debt reached $7.1bn at end-1988, boosted by an 80 per cent fall in the value of the zaire against the US dollar. The World Bank complained in October 1988 that the debt service should have been $820m in 1988, but only $280m had been budgeted. In mid-1988, Zaïre announced that it would only make repayments to countries that provided a net inflow of capital.

Aid programmes resume

Following the 1989 IMF agreement, the World Bank disbursed the second tranche ($82m) of its 1987 $149m structural adjustment loan. It also approved a $75m IDA loan for the $280m programme to upgrade the Voie Nationale transport system operated by ONATRA and SNCZ, vital to Zaïre's copper exports. The scheme involves the supply of new railway track, rolling stock and signalling equipment, new port loading equipment and barges. It is co-financed by France (FF220m for rolling stock and FF40m for signalling equipment), ABD (UA30m for signalling equipment), Belgium $25m), the EC and West Germany. World Bank conditions for the IDA loan included an increase in rates by ONATRA and SNCZ and government agreement not to invest in new aircraft for loss-making state airline Air Zaïre.

The EC unblocked an Ecu30m credit for import support for agricultural machinery and inputs and industrial equipment. The World Bank was considering loans of $165m for the 1988-90 energy programme and $105m for social sector schemes, including health, education, family planning and nutrition. France allocated FF131m for the energy programme in 1988. In the health sector, USAID provided $3.6m for primary rural health and family planning in September 1988. An anti-AIDS campaign was launched in October 1988 with $8.1m from IDA, $2m from the EC, $1.7m from the USA, UNICEF and Belgium providing $1.5m each, Italy $1.2m, and UNDP and WHO $1m each. WHO estimated that AIDS affected 6 per cent of the urban population.

In November 1988, the World Bank was reportedly considering loans of $125m for agriculture, especially to reduce food imports which account for 20 per cent of total food consumption. An IDA credit of SDR9m was accorded to support agricultural and rural policy formulation and co-ordination between the three ministries responsible: agriculture, rural development and land management and environment.

UNDP credits of $5.1m were earmarked for agriculture in March 1989, including $3.2m for improvement of statistical data on agriculture and $1.8m for the national seeds office to strengthen storage and distribution. UNDP also provided $2.5m for food production and rural energy in September 1988.
French aid to the industrial sector included FF22m for a brewery expansion scheme in August 1988, FF11.6m for the pharmaceuticals company Biokin in October 1988, and FF25m for machinery modernisation at the Utexafrica textiles consortium in May 1989. In contrast to this tight aid programming, donor countries, multilateral agencies and the Zaïre authorities also showed a renewed desire to launch grandiose schemes. President Mobutu was apparently considering building a magnificent new cathedral in Kinshasa or Gbadolite in 1989. Belgium is building a costly hydropower station that will mainly supply the presidential palace in Gbadolite. In May 1989, ADB lent $5m to Egypt for a feasibility study to link the Egyptian and Zaïrean electricity grids. Zaïre's huge Inga power station is still under-utilised, despite the spectacular but unnecessary 1,725 km Inga–Shaba power line – responsible for about one-fifth of Zaïre's total debt. For the Egyptian scheme, Zaïre would build another gigantic power line, this time to the north, to connect with the Sudanese grid to which Egypt is linked. Sudan would also have to extend its power lines through the war zone to the Zaïre border. Other expansive ideas for Inga include extending power lines to Nigeria or to South Africa via Angola. As with the Inga–Shaba line, it seems likely that local populations along the way would remain bereft of electricity. Swedish–Swiss firm Asea Brown Boveri (ABB), which took on a 30-month SKr100m management contract for the Inga–Shaba line in autumn 1988, showed a greater sense of realism when it also set up a countertrade deal to swap ABB locomotives for used railway track.

Copper more brightly burnished

Copper company Gecamines has stabilised production at around 500,000 tonnes a year. Following Zambia's lead, Gecamines raised its prices on the London Metal Exchange in November 1988. In February 1989, Gecamines awarded West German firm Kloeckner a $30m contract to build a 100,000 tpa refinery in Shaba to upgrade its copper exports and to rely less on Belgian company MHO, its main contract refinery at present. MHO began to look for other copper suppliers following the deal and Zaïre's ban on copper supplies to Belgium. The 1989 refining contract held good, however. The Kloeckner refinery, which will come on stream in 1991, is financed by ADB as part of the World Bank-backed 1986-90 Gecamines investment programme, which the mining company estimates could cost up to $900m.

Mining company Sominki produced 625 kg of gold in 1988, up from 549 kg in 1987, while Mines d'Or de Kilo-Moto output fell to 260 kg from 438 kg in 1987. Diamond production was 10m carats in the first quarter 1989. Prices for both gold and diamonds strengthened in the first seven months of 1988. Diamond exports fell 13 per cent in volume over that period but rose 18 per cent in value, while gold exports were down by 10 per cent in volume with a 6.6 per cent increase in value.

In November 1988, Zaïre and Zambia agreed to raise cobalt prices by 12 per cent to $8.40/lb. Since the ending of the price war between the two countries in 1986, prices have more than doubled. Zaïre produces 10,000 tonnes a year of cobalt and Zambia 4,000 tonnes, giving them a 75 per cent share of non-Communist output. Oil production was 31,200 barrels a day (b/d) in 1988. A re-evaluation of exploration data produced by Texaco and Esso for the Cuvette basin in the early 1980s was less promising than hoped, and the government did not issue new exploration licences in early 1989 as planned. Zaïre's coffee quota was 1.1m bags for 1988-89.

Outlook

The economic prospects look relatively bright for 1989-90. 1989 inflation will be lower than in 1988, thus easing pressure on the currency and containing the upward drift of the debt burden. Increased availability of foreign exchange in 1989 should help boost private sector activity, both production and investment. Increased receipts from copper exports in 1989 should allow Zaïre to keep to the terms of the 1989-90 IMF agreement and thus ensure a reasonable inflow of funds in 1990. There could even be a reduction of the budget deficit below target level by the year-end if President Mobutu keeps to his pronouncement of no expenditure beyond budget limits. This would earn him high marks with the IMF and World Bank and other donors. The raising of public sector wages in 1989 should help to ease the local opposition to austerity.

With the expected slowdown in the industrialised economies in 1990, copper prices could slacken and Zaïre may find it harder to keep to IMF terms, especially later in the year. Major investment programmes will bear fruit in the medium to long term and should provide lasting benefits, but in the meantime, Zaïre desperately needs to reschedule and reduce its debt burden. The government's conduct vis-a-vis Belgium is hardly encouraging to other donors. But the Group of Seven summit in Toronto in 1988 agreed on the principle of substantial debt concessions to sub-Saharan Africa. If Zaïre resolves its dispute with Belgium, it will certainly receive generous rescheduling from Brussels and this could open the door to a general agreement with the Paris Club. If it insists on rewriting the 1960 independence treaty, such agreement seems out of the question.

COUNTRY PROFILE

Political system: Executive power rests with the president, who has a seven-year mandate. He also heads the executive council (cabinet) but may appoint a first state commissioner (prime minister). Cabinet ministers have the title of state commissioners and ministries are known as departments. The Legislative Council (parliament) comprises 210 people's commissioners (MPs), elected for a five-year term by universal suffrage. However, the central committee of the ruling MPR wields much greater political power than either of the councils. Gained independence in 1960.

Political parties: Mouvement Populaire de la Révolution is the only legal political party.

Population: The population, estimated at 34m in 1988, comprises over 200 ethnic groups. There are some 100,000 Europeans in the country, more than half being of Belgian nationality.

Main cities/towns: Kinshasa, Matadi, Bandundu, Mbandaka, Kisangani, Bukavu, Lubumbashi, Mbuji-Mayi, Kananga.

Language: French is the official language and is the universal medium for business. Among the many African languages, Kiswahili, Lingala, Tshiluba and Kikongo are the most prominent.

Media
Press: The two main newspapers are *Elima* and *Salongo* – both are government-owned. Also a number of periodicals, mostly published in Kinshasa.
Radio: Domestic radio service broadcast in French and other major languages by government-owned La Voix du Zaïre. Educational programmes broadcast by Radio Candip.
Television: Limited TV service broadcast by government-owned commercial station Zaïre Television.

External trade: Narrow export base, concentrated mainly on minerals, with some agricultural cash crops, has made Zaïre's balance of trade very susceptible to the vagaries of world commodity markets. Regular trade surpluses offset by net outflow on services account, due to huge debt service burden, repatriation of profits and high transport costs.
Exports: Copper (typically, 25-30 per cent of total value), crude petroleum (15-18%), diamonds (15-20%), cobalt (9-12%), coffee (8-10%). Main destinations: Belgium (typically 30-35% of total value), USA (25-30%), German FR (10-12%), Italy (6-8%).
Imports: Manufactured goods (typically, 35-40 per cent of total), food (15-20%), raw materials (10-15%). Main sources: Belgium (20-25% of total value), France (10%), German FR (10%), USA (8%).

Agriculture: Contributes about 29 per cent of GDP, only about half of which is derived from subsistence farming even though this provides the livelihood of 70 per cent of the population. Despite enormous agricultural potential, sector has been handicapped above all by transport problems, occasional drought in the east of Kivu, smuggling and inflexible pricing policies. Main food crops are cassava, maize, rice, plantain. Production is insufficient to meet demand, and poor transport restricts supplies to the urban areas. Although once self-sufficient in basic foodstuffs, Zaïre now imports over 125,000 tonnes of food annually. Main cash/export crops are plantation grown coffee, cocoa, oil palm, rubber, tea, cotton, sugar and tobacco. Output of these crops has steadily declined since 1970s.

Industry: Manufacturing contributes only about 9 per cent to GDP (compared with 13 per cent in early 1960s). Approximately three-quarters of production is centred around Kinshasa or in Shaba province, owing to the availability of electricity and adequate transport facilities in these areas. Output is geared towards the domestic market and is mainly concentrated on brewing, food processing, textiles, consumer goods, construction industry inputs and transport equipment. Mismanagement and shortages of spare parts and materials have lead to cutbacks in production with few firms operating at even half capacity. The government is attempting to increase production by encouraging foreign investment and offering substantial tax incentives. It is also sponsoring forestry development with multilateral assistance.

Mining: Zaïre is the world's biggest producer of cobalt (10,000 tonnes in 1987); the second largest producer of industrial diamonds (23.5m carats in 1987); and the fourth largest copper producer in the Western world, but output decreasing (507,600 tonnes in 1986, 470,000 tonnes in 1987). It also mines large quantities of zinc. With the exception of diamonds, these minerals have been hit by weak world demand in recent years. In 1987, minerals accounted for over 80 per cent of export receipts, worth about $1.5bn. There is also small-scale mining of cadium, cassiterite and wolframite. Activity is concentrated in the copper-rich Shaba province. Diamonds are mined on a commmercial scale by the Miba Company at Bakwanga in Kasai Oriental, but artisanal diggers also account for almost half the total output.

COUNTRY PROFILE

Hydrocarbons: Some 140m barrels of oil reserves, located off the Atlantic coastline and in the Zaïre River estuary. Production was running at 31,000 b/d in 1988.

Energy: The huge river Zaïre provides enormous hydro-electricity potential. Major plant at Inga, near the port of Matadi at the mouth of the river, which supplies the copper mines in Shaba, and produces two thirds of the country's 2,400 mW installed capacity. Energy in rural areas is mainly derived from charcoal and wood. Zaïre imports most of its oil requirements, due to the inability of its own oil refinery to process the very heavy domestic crude.

Membership of international organisations: African Development Bank, African States Associated with the EC (Lomé Convention), Arab Bank for Economic Development in Africa (recipient), Economic Commission for Africa, Economic Community of Central African States (CEEAC), Economic Community of the Countries of the Great Lakes (CEPGL), FAO, GATT, IDA, IFAD, IFC, ILO, IMF, International Coffee Organisation, International Tin Council, OAU, UN, WHO, WIPO, World Bank.

BUSINESS GUIDE

Time: West (Kinshasa, Mbandaka): GMT +1; east (including Shaba province): GMT +2.

Climate: Great variability. Lowlands (western region): hot and humid. Temperatures range from 27°C to 30°C, maximum likely 32°C. Plateau (central): likely temperature range 18°C to 20°C, maximum likely 25°C. Eastern Zaïre has Mediterranean type climate. Kinshasa: rainy season Oct to May.

Entry requirements
Passport: Required by all.
Visa: Required by all except certain nationals of Congo or re-entry permit holders. Business visitors require a letter of invitation from a Zaïre company. Travel regulations should be studied carefully before visit.
Currency: Import/export of local currency is prohibited. Foreign currency import is unlimited but amounts must be declared; foreign currency export is limited to the declared amount. Currency declaration forms must be kept and all currency exchanges should be recorded.

Health precautions
Mandatory: Yellow fever vaccination certificate is required if arriving from an infected area.
Advisable: Cholera, typhoid, polio and yellow fever vaccinations recommended. Malaria prophylaxis should be taken as risk exists throughout the country. Water precautions should be taken; bilharzia is endemic. AIDS risk.

Representation overseas: Abidjan, Abu Dhabi, Accra, Addis Ababa, Algiers, Athens, Bangui, Beijing, Belgrade, Berlin, Berne, Bonn, Brasilia, Brazzaville, Brussels, Bucharest, Buenos Aires, Bujumbura, Cairo, Conakry, Cotonou, Dakar, Dar es Salaam, The Hague, Harare, Jeddah, Kampala, Khartoum, Kigali, Kuwait City, Lagos, Libreville, Lisbon, Lomé, London, Luanda, Lusaka, Luxembourg-Ville, Madrid, Maputo, Monrovia, Moscow, Nairobi, Nouakchott, New Delhi, N'Djamena, Ottawa, Paris, Rabat, Rome, Stockholm, Tehran, Tel Aviv, Tokyo, Tripoli, Tunis, Vatican, Vienna, Warsaw, Washington, Yaoundé.

Air access
National airline: Air Zaïre.
Other airlines: Air Afrique, Air Gabon, Cameroon, Ethiopian, Lufthansa, Sabena, Scibe Airlift, Swissair, TAAG-Angola, TAP-Air Portugal, UTA.
International airports: Kinshasa-N'djili (Code: FIH), 25 km from city; Lubumbashi-Luano (Code: FBM), 6.5 km from city.

Surface access
Main ports: Matadi; Kinshasa is main inland river port and the ferry crossing point from Brazzaville.
Overland access: There are some roads, mostly in very bad state of repair, from neighbouring countries. Rail links connect with Tanzania, and with southern African countries via Zambia.

Hotels: Several major hotels in main cities. Most tend to be expensive and are often heavily booked, so reservations should be made well in advance – if possible through an international airline. Service charge is usually added to bill, further tipping is optional.

Electricity supply: 220 V AC.

Car hire: Self-drive cars available in Kinshasa and at airport. A deposit (usually at least Z8000) is required unless an acceptable credit card can be produced. International driving licence required. Traffic drives on the right.

City transport
Taxis: Available in main towns. No metered taxis, so fare should be negotiated in advance. Tipping is optional.

National transport
Air: Air Zaïre flies between Kinshasa (N'djili Airport) and about 40 airports. Scribe Air also provides reliable domestic services, as well as Lubumbashi and Kinshasa to Brussels. Charter facilities are available from Kinshasa.
Road: Over 150,000 km of roads, but outside main population centres these tend to be of poor quality, except for main routes linking Kinshasa with Matadi and Kikwit, and Lubumbashi with Kolwezi and Zambian border.
Buses: Very irregular, crowded and infrequent service.
Rail: Network of over 5,000 km operated by Société Nationale des Chemins de Fer Zairois (SNCZ). Four classes – travelling 'deluxe' and first class generally advised; sleeping (couchette) and dining cars available; some railcar services; air-conditioning not available. Main routes include Kinshasa-Matadi, Lubumbashi-Kamina-Ilebo and Kamina-Kalemie. Local advice may be advisable when planning timing etc of journeys.
Water: Almost 14,000 km of inland waterways, with regular passenger services on all major rivers and lakes. Main routes include Kinshasa-Ilebo and Kinshasa-Kisangani. Advisable to travel luxury or first class.

Public holidays
Fixed dates: 1 Jan, 4 Jan, 1 May, 20 May, 24 Jun, 30 Jun, 1 Aug, 14 Oct, 27 Oct, 17 Nov, 24 Nov, 25 Dec.

Working hours
Business: (Mon-Fri) 0730 – 1230, 1430 – 1700; (Sat) 0730 – 1200.
Government: (Mon-Fri) 0730 – 1500.
Banking: (Mon-Fri) 0800 – 1130.

Telecommunications
Telephone: Dialling code for Zaïre: IDD access code + 243 followed by area code (12 for Kinshasa, 2 for Lubumbashi) followed by subscriber's number. It is often very difficult to get through. Internal service is generally unreliable.
Telex: Telex and telegraph services available at main Post Offices in Kinshasa and Lubumbashi. Telex country code: ZR.

Banking
Central bank: Banque du Zaïre.
Main offices: Banque Commerciale Zaïroise, Banque de Kinshasa, Banque de Paris et des Pays-Bas (Zaïre), Banque du Peuple, Banque Grindlay Internationale au Zaïre, Banque Internationale pour l'Afrique au Zaïre, Barclays Bank Zaïre, Citibank (Zaïre), Union Zaïroise de Banques.

Representation in capital: Algeria, Angola, Argentina, Austria, Belgium, Benin, Brazil, Bulgaria, Burundi, Cameroon, Canada, Central African Republic, Chad, China PR, Congo, Côte d'Ivoire, Czechoslovakia, Egypt, Ethiopia, France, Gabon, German DR, German FR, Greece, Guinea, India, Iran, Israel, Italy, Japan, Kenya, Korea PDR, Rep. of Korea, Liberia, Libya, Mauritania, Morocco, Netherlands, Nigeria, Pakistan, Poland, Portugal, Romania, Rwanda, Spain, Sudan, Sweden, Switzerland, Tanzania, Togo, Tunisia, Turkey, Uganda, UK, USA, USSR, Vatican, Yugoslavia, Zambia.

BUSINESS DIRECTORY

Hotels: Kinshasa
Inter-Continental, Avenue de Batetela, P.O. Box: 9535 (tel: 27-284/355/895; tx: 21212; fax: 27005).

Memling Palace, Avenue Republique du Chad, P.O. Box: 68 (tel: 23-260; tx: 21654).

Okapi, P.O. Box: 8697 (tel: 80-020; tx: 21564).

Hotels: Kisangani
Zaire Palace, Avenue de l'Eglise (tel: 2105/6; tx: 21032).

Hotels: Lubumbashi
Park, Avenue du Kassaii, P.O. Box: 112 (tel: 22-35-21/23; tx: 311).

Sheraton Karavia, Route du Golf, P.O. Box: 4701 (tel: 22-45-15/12; tx: 41049; fax: 225011).

Car hire: Kinshasa
Avis: Hotel Intercontinental, BP 16198 (tel: 28-005, 28-076, 28-114; tx: 21212).

Europcar: Boulevard du 30 Juin (tel: 25-966; tx: 41039).

Hertz: 11 Avenue des Aviateurs (tel: 23-322; tx: 21191).

Banking
Banque du Zaïre, Boulevard Colonel Tshiatshi au Nord, PO Box 2697, Kinshasa (tel: 30-681; tx: 227, 388).

Banque Commerciale Zaïroise, Boulevard du 30 Juin, PO Box 2798, Kinshasa (tel: 23-772; tx: 21127).

Banque de Kinshasa, 1 Place du Marché, PO Box 8033, Kinshasa (tel: 26-361; tx: 21304).

Banque du Peuple, Boulevard du 30 Juin, PO Box 400, Kinshasa (tel: 25-161/6; tx: 21108).

Other useful addresses
Air Zaïre, PO Box 8552, Kinshasa (tel: 24-985; tx: 21313).

Association Nationale des Entreprises du Zaïre, PO Box 7247, Kinshasa (tel: 22-565, 22-286).

Chambre de Commerce et d'Industrie de Kinshasa, PO Box 7247, Kinshasa (tel: 24-623).

DHL Courier, Kinshasa (tel: 24-253).

Foire Internationale de Kinshasa, PO Box 1397, Kinshasa. (annual international fair in July).

Office National de Tourisme, Boulevard du 30 Juin, PO Box 9502, Kinshasa.

SNCZ Railways, PO Box 597, Kinshasa.

Zambia

Anthony Hawkins

During the first half of 1989, Zambia resumed the search for a new agreement with the International Monetary Fund (IMF), against the backdrop of a deteriorating economy as copper prices weakened and agricultural output fell following the February floods. Industrial and mining production was interrupted by electricity loadshedding necessitated by the fire at the Kafue Gorge hydroelectric power station while exports were constrained by transport bottlenecks on the Tazara railway to Tanzania. Initial hopes that President Kenneth Kaunda would act decisively to solve the country's proliferating economic problems after yet another overwhelming victory in the October 1988 elections in which he received 93 per cent of the votes, proved misplaced as the president continued to insist that international creditors accept a Zambian-style economic recovery programme. But as the economic slide continued, a number of bold reform measures were adopted despite mounting criticism of the government. Political events were dominated by the discovery in October 1988 of an alleged plot against the president which resulted in two Cabinet reshuffles within a six-month period as Dr Kaunda surrounded himself with loyal supporters.

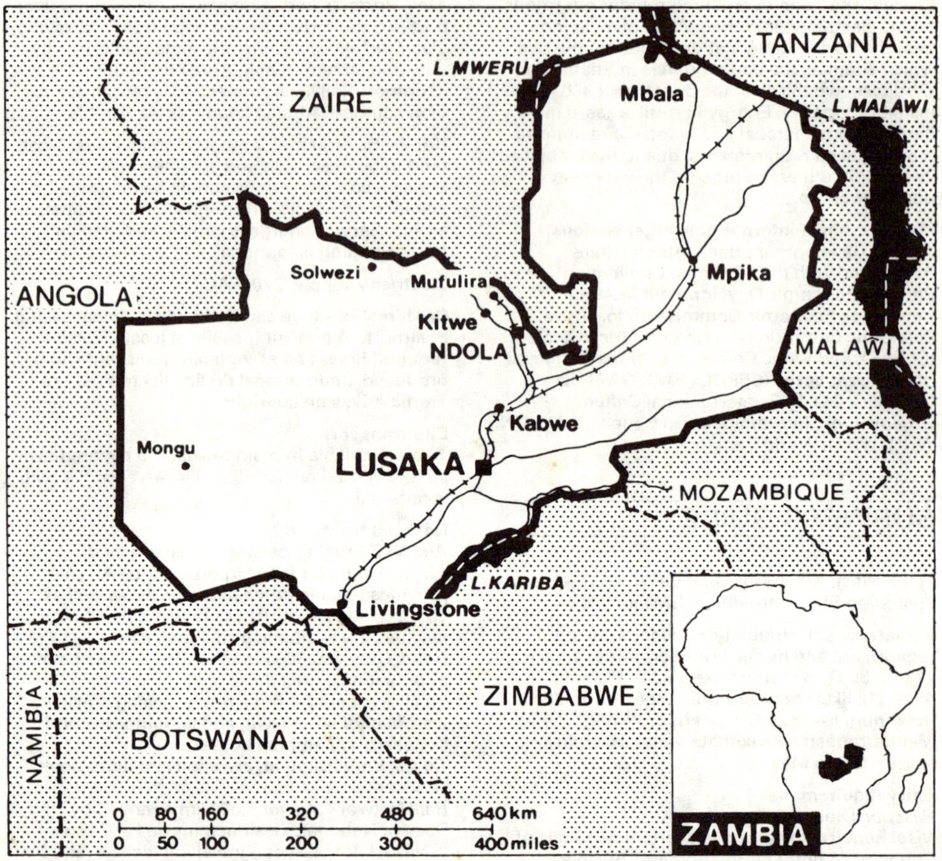

Foreign relations

Beset with seemingly intractable domestic political and economic problems, President Kaunda sought relief in the international sphere, turning his attention to foreign relations where he achieved success both in launching the Angolan peace initiative and in arranging his August 1989 meeting with Mr F. W. de Klerk, President P. W. Botha's successor as leader of the ruling Nationalist Party in South Africa. Zambia's role as a peacemaker in the region helped the president to shore up his own domestic position despite mounting public criticism of economic mismanagement and corruption.

Another foreign success for the beleaguered Zambian leader was the ending of the border dispute with Zaïre over Kaputa in the northern province. Zaïre occupied Kaputa in 1980 setting up customs and immigration posts some 20 to 30 km within Zambian-claimed territory. The area has known deposits of strategic minerals including platinum, cobalt, copper and tungsten.

President Kaunda's pivotal role in regional affairs was heightened by mounting concern within Zambia over the conduct of African National Congress (ANC) guerrillas and politicians within Zambia itself. Five people died in shooting incidents involving ANC members in April and May 1989, two of them Zambian women who were murdered by ANC members, one of whom subsequently committed suicide. Following protests by the government, the ANC leadership promised to disarm some of its members in Zambia, but there were signs of an uneasy relationship developing as some ANC members questioned the wisdom of bilateral talks with Mr De Klerk.

Zambia also contributed to the peace initiative in neighbouring Mozambique following a renewed spate of cross-border incursions by Mozambique National Resistance (MNR) rebels, fighting the Frelimo government during April and May 1989. Zambian civilians were killed and Lusaka responded by ordering raids into Mozambique claiming that more than 100 rebels were killed in such clashes, while at least 20 Zambians died at the hands of the MNR.

Relations with Britain, soured in 1986 by disagreement over economic sanctions against Pretoria, improved in the wake of the Namibia settlement and the January 1989 visit by junior Foreign Office Minister, Mrs Lynda Chalker marked a rapprochement between the two countries after three uneasy years.

Political developments

But the president's foreign policy achievements could not divert attention from a deteriorating domestic political climate, which had its roots in economic decline. As the government edged cautiously towards a new agreement with the IMF and the World Bank so it was forced to adopt unpalatable and politically unpopular policies designed to cut the budget deficit and stabilise the worsening balance of payments situation. Predictably such policies were attacked by radical politicians, student leaders and trade unionists.

Growing political unease surfaced ahead of the October 1988 elections with the arrest of six army officers, including the former army commander and ambassador to West Germany, Christon Tembo, and three civilians accused of plotting a coup against President Kaunda. It was subsequently reported that some 40 other people had been arrested including businessmen and politicians. The ruling United National Independence Party (UNIP) cleverly exploited the alleged coup during the elections, arguing successfully that Dr Kaunda was the only person capable of leading Zambia through the difficult period that lay ahead. President Kaunda, who was unopposed in the polls, won a sixth term while the fact that only four

KEY FACTS

Official title: Republic of Zambia

Head of state: President Kenneth David Kaunda

Ruling party: United National Independence Party

Capital: Lusaka

Official Languages: English

Currency: Kwacha (K) = 100 ngwee

Exchange rate: K18.2 per $ (Nov 1989)

Area: 752,614 sq km

Population: 7.5m (1988)

GDP real growth: 2.7% (1988)

GNP per capita: $250 (1987)

Inflation: 55.7% (1988)

Trade balance: $497m (1988)

Foreign debt: $6.4bn (1987)

ministers in an unpopular administration lost their parliamentary seats reflected a successful campaign by UNIP.

However, the reduced turnout – 56 per cent of Zambia's 2.6m registered voters went to the polls compared with 66 per cent in 1983 – suggested that many dissatisfied Zambians had abstained, while on the Copperbelt itself Kaunda's support was lower than elsewhere in the country. This was a warning sign, given the potential political clout of the mine-workers.

Significant political developments included the return to parliament of the former secretary-general of UNIP, Humphrey Mulemba, seen by some as a possible successor to Dr Kaunda, and the appointment of the president's son Wesai as junior minister, sparking speculation and criticism that Dr Kaunda was grooming a member of his family to succeed him.

A slimline cabinet of 16 ministers – compared with 21 previously – was appointed including 5 newcomers, while 10 former ministers were axed. The new cabinet was shortlived and in March 1989, Kebby Musokotwane was demoted from the post of prime minister to minister for general education, youth and sports. He was replaced by Home Affairs Minister General Kingsley Chinkuli only four months after the cabinet had been reshuffled in November 1988.

KEY INDICATORS

	Unit	1984	1985	1986	1987	1988
Population	m	6.44	6.73	7.28	7.56	7.53
Gross domestic product (GDP)	Kwacha m	4,931	7,072	12,963	19,632	22,495
GDP real growth	%	−0.3	1.6	0.5	0.2	2.7
Inflation	%	20.1	37.4	51.6	43.1	55.7
Agricultural production	1979-81=100	106.2	115.4	114.0	113.2	126.6
Food production	1979-81=100	104.2	114.5	112.7	113.6	126.1
Mining production	1985=100	108.1	100.0	96.6	–	–
Industrial Production	1985=100	90.0	100.0	95.8	–	–
Exports	Kwacha m	1,183.6	2,451.4	3,743.4	8,058.4	9,791.3
Copper	Kwacha m	1,031.2	1,960.6	3,644.9	6,854.5	8,299.3
Imports	Kwacha m	1,329.2	1,932.1	4,828.1	7,423.9	6,861.2
Balance of trade	$m	280	226	175	261	497
Current account	$m	−153	−398	−300	−141	−174
Total External debt	$m	3,847	4,641	5,625	6,400	–
Debt service ratio	%	11.8	10.9	19.1	13.5	–
Exchange rate	Kwacha per $	1.79	2.71	7.30	8.9	8.2

* estimated figure

Political and industrial unrest

Political unrest erupted at the Lusaka campus of the University of Zambia in April 1989 when students stoned cars during a protest against government proposals to introduce accommodation and tuition fees at polytechnics and the university itself. The Copperbelt at Kitwe was subsequently closed while some 200 teacher training college students were arrested and fined.

Riots broke out on the Copperbelt in July 1989 in protest against the withdrawal of subsidies on maize meal and the consequential 25 per cent rise in its price and against the 60 per cent petrol price hike.

Meanwhile, the Anti-Corruption Commission (ACC) reported a 20 per cent rise in the number of complaints against public officials during 1988 when the figure reached 1,300, while the auditor-general revealed that misappropriation and over-spending by government departments totalled K125m.

Further evidence of dissatisfaction with the Kaunda administration was the increase in the number of industrial disputes forcing the government to classify more occupations as essential. Strikes are illegal in classified occupations and workers who defied the law were threatened with fines or dismissal. Despite this, strike action proliferated with bank officials, air traffic controllers, school-teachers, colliery workers and industrial labourers all taking industrial action during the first half of 1989 primarily in protest against escalating living expenses and falling real earnings.

Economy

After a marginal decline of 0.2 per cent in 1987, real GDP grew strongly in 1988 rising 2.7 per cent – the country's best performance since 1981, though still below the rate of population growth. Agriculture, with output growth of 6.4 per cent was the lead sector, though manufacturing continued its strong performance with growth of more than 5 per cent for the second successive year. Mining output grew only 2.7 per cent despite the buoyant copper price which averaged 117 US cents a lb during the year, 44 per cent higher than the 1987 figure of 81 cents. Despite this, copper exports rose only 21 per cent to $930m, reflecting falling production volumes, down 14 per cent to only 417,000 tonnes in 1988.

Zambia's Fourth National Development Plan (FNDP), launched in January 1989, projected real GDP growth of 3 per cent annually in the four-year period to 1993–94. Other ambitious plan targets included a reduction in the budget deficit to 2 per cent of GDP from a forecast percentage in 1989, with inflation falling to 20 per cent from 60-70 per cent while population growth was expected to decline significantly from 3.5 per cent, partly reflecting the severe impact of AIDS in Zambia.

The Plan favours rapid agricultural growth of 5.5 per cent a year, while manufacturing will expand at 3.1 per cent and mining at 2.5 per cent annually. Investment is targeted at $2.3 bn of which almost 80 per cent is for economic services. Capital spending will be dominated by the public sector with parastatals accounting for 44 per cent, central government 29 per cent and the private sector 27 per cent.

Two dominant priorities in the FNDP were the emphasis on a switch from consumption to investment – in 1988 gross fixed capital formation accounted for only 7 per cent of GDP – and measures to strengthen the balance of payments by reducing food imports and developing new non-traditional exports.

The Plan was immediately criticised for focusing more on targets than policies, but welcomed for its commitment to reduced subsidies and market-influenced interest and exchange rates. At the same time, it seemed clear that some of the growth targets were optimistic especially if Zambia were to be forced into implementing an IMF/World Bank-sponsored structural adjustment programme in 1990 as seemed increasingly probable.

Foreign trade

The current account position improved very substantially between 1986 and 1988 from a deficit of $300m to an estimated $90m though this was largely the result of higher copper prices and tight import and exchange controls. After rising 28 per cent in 1988, export growth was forecast to flatten in 1989 in response to reduced export volumes, resulting from both production problems at the copper mines themselves and mounting transport difficulties on the Tazara railway. But with copper prices holding up better than forecast, exports of more than $1.2bn – a rise of at least 10 per cent – were in prospect. The performance of non-traditional exports was uneven, declining from $105m in 1986 to $77m in 1987 before rebounding to $150m in 1988. Further growth is anticipated in 1989–90 in response to the devaluation of the Kwacha.

Import growth was expected to accelerate despite the two major devaluations in the Kwacha in November 1988 and June 1989, growing at around 16 per cent, reflecting food imports necessitated by the floods early in 1989 and increased imports of spares for the mines. With capital flows remaining sluggish Zambia was unable to expand imports at the rate necessary to prevent shortages of spares and raw materials from interrupting its recovery programme. During 1989, it became increasingly obvious that far-reaching reforms were needed to re-establish the donor and investor confidence necessary to ensure the major improvement in import capacity which is a prerequisite for sustained economic expansion.

Prospects for economic recovery continued to be jeopardised too by Zambia's deteriorating external debt situation. Early in 1989 the government revealed that its foreign debt had risen to $6.5bn with a debt-service ratio of more than 60 per cent – the fifth worst in Sub-Saharan Africa after Mozambique, Sudan, Guinea-Bissau and Somalia.

The growth rate of the country's foreign debt accelerated after Zambia broke with the IMF and World Bank in May 1987, unilaterally imposing a 10 per cent ceiling on foreign debt-service payments. As a result, arrears to the IMF and World Bank alone have risen to an estimated $800m, while at the end of 1987 its borrowings from the Fund at $960m made it Africa's largest IMF debtor. The FNDP included the assertion that the debt ceiling strategy was to be maintained, though bankers warned that there can be no new money for Zambia until the ceiling is lifted and Zambia's debt profile restructured.

Western donors, anxious to encourage Zambia's return to the IMF fold, raised the possibility of new money being made available in addition to funds to finance the repayment of arrears. At the same time, the government's willingness to adopt harsh austerity measures despite growing public criticism of its economic management record, fuelled hopes that the basis of a new IMF-sponsored recovery programme might be achieved by early 1990.

Economic reforms

During 1988–89, President Kaunda demonstrated greater economic realism, though many of his government's economic forecasts were criticised by business leaders and bankers as unrealistically optimistic. The reform programme began to emerge shortly after the elections with the 25 per cent devaluation of the Kwacha to 10 kwacha to the US dollar. At the same time, interest rate ceilings were raised to 25 per cent from 20 per cent and a reduction in maize subsidy was promised.

In the 1988–89 budget presented in November 1988, Finance Minister, Gibson Chigaga provided for a sharp reduction in the budget deficit from K2.75bn (9.5 per cent of GDP) to K1.9bn (5 per cent of GDP). This was to be achieved through a 43 per cent growth in government revenue while spending was projected to increase only 18 per cent. By mid-1989, it was evident that several of the budget targets were unduly optimistic. Mr Chigaga's forecast of 2.9 per cent growth was replaced by unofficial estimates of GDP expansion of less than 1 per cent, while his hopes of containing inflation to a maximum of 35 per cent were misplaced given the rise in the official inflation rate to 70 per cent by mid-1989 with unofficial estimates putting it closer to 100 per cent.

A two-tier approach to maize prices was adopted in an effort to reconcile subsidy reduction with social justice and avoid political controversy. A system of maize meal coupons was introduced in January 1989 allowing people to purchase their basic requirements of the staple at subsidised prices. At the same time, the market price of maize meal was trebled and economists estimated that if the new coupon system could be made to work, the maize subsidy should decline to K360m from K500m.

A further batch of reforms was announced in June 1989 including a second devaluation of the kwacha, this time of 37.5 per cent taking the rate to K16 to the US dollar. Price controls were abolished, to the relief of industry, though maize meal was excluded. But President Kaunda also announced salary and pension increases of 50 per cent designed to defuse adverse reaction to the new measures, heightening expectations of even faster inflation in the latter half of 1989.

Industry and agriculture

Government ministers hope that the worst is past in the copper mining industry, forecasting that the 14 per cent output decline in 1988 will be reversed with production recovering to 465,000 tonnes in 1989 from 417,000 tonnes. The FNDP envisaged further output expansion to 553,000 tonnes by 1993 but these targets seem unlikely to be reached without a major improvement in the availability both of foreign exchange and technical personnel. Finance Minister Chigaga blamed the shortage of spare parts for the fall in production in 1988 but this continued into 1989 when production was forecast to fall a further 70,000 tonnes to around 350,000 tonnes.

Reduced production in recent years may extend the industry's life by a few years but by the turn of the century production is likely to be falling sharply with devastating consequences for Zambia's balance of payments unless new exports are developed. It has been estimated that at an annual output rate of 550,000 tonnes annually, the bulk of the country's exploitable ore reserves will be exhausted by the year 2000.

Torrential rains and flooding resulted in a sharp fall in maize production in 1989 to an estimated 800,000 tonnes compared with 1.35m tonnes the previous year. Normally this would be adequate to meet domestic demand but with an estimated 1m people requiring food because their crops were flooded, imports of maize and other foods were likely in 1989–90 before the new harvest. Road haulage operators warned that with spares and tyres in short supply and with only 400 of the 600 trucks in working order crop losses were likely due to inadequate collection capacity.

The FNDP projected maize self-sufficiency with output of 1.9m tonnes a year of which 1.6m would be marketed by 1993, while the area planted out to wheat would double from 11,000 hectares to 21,000 ha. which would also ensure self-sufficiency. Cotton output will more than double to 115m kgs, tobacco production will double to 10.5m kg and cashewnut

COUNTRY PROFILE

Political system: The United National Independence Party's Central Committee is the principal policy-making body. Legislative power is vested in the 135-member National Assembly, 125 members of which are elected every five years. Zambia gained independence from UK in 1964.

Political parties: The United National Independence Party is the sole legal political party.

Population: Zambia's population of 7.3m (1988 estimate) is growing at a high 3.8 per cent a year, and this is likely to cause some strain to the country's economy in future years. Largest concentrations of people are in the northern Copperbelt towns. The drift to these urban centres is continuing despite the recent downturn in the copper industry.

Main cities/towns: Lusaka (population 538,000 according to the 1980 census), Kitwe (315,000), Ndola (282,000), Mufulira (150,000), Chingola (146,000), Kabwe (144,000), Luanshya (132,000), Livingstone (72,000), Chililabombwe (62,000), Kalulushi (59,000).

Language: English is the official language and the usual medium for business; African languages include Nyanja, Bemba, Tonga, Lozi, Kaonde, Lunda and Luvale.

Media
Press: Two national daily newspapers, the UNIP-owned *Times of Zambia* (circulation 55,000) and government-owned *Zambia Daily Mail* (40,000). Also several periodicals, mostly published in Lusaka.
Radio: Over 150,000 radio receivers in use. Domestic service, in English and several local languages, provided by government-controlled Zambia National Broadcasting Corporation (five channels).
Television: Over 80,000 TV sets in use. Service broadcast in English by Zambia National Broadcasting Corporation (a parastatal) in Lusaka, Livingstone and Copperbelt towns.

External trade: Strong reliance on copper exports has left the economy vulnerable to world market conditions, but the upgrading of the Tazara railway which carries most of the exports should reduce costs. Despite strict import cuts, trade surpluses were being eroded by capital outflows and service payments, and the current account was in deficit. In May 1987 Zambia abandoned IMF austerity measures and allowed arrears to the IMF and World Bank to grow substantially. Foreign debt payments were restricted to less than 10 per cent of foreign exchange earnings.
Exports: Copper and cobalt (combined value over 90 per cent of total), zinc, lead. Main destinations: Japan, China, Italy, France, USA.
Imports: Machinery and transport equipment, crude petroleum, basic manufactures, chemicals. Main sources: UK, Saudi Arabia, South Africa, Japan, USA, German FR, Zimbabwe.

Agriculture: Contributes 17 per cent to GDP (1986) and employs around half the workforce, largely on a subsistence basis. There are vital shortages of input, but production has improved in recent years because of higher producer prices, higher spending on agricultural development and irrigation schemes. Tobacco sales in 1987 were 128m kg up in volume over 1986. Cash/export crops also include seed cotton, sunflower and groundnuts. Main food crops are maize, rice, sorghum, millet, soya beans, wheat. A bumper maize crop in 1986 helped to tide the country over the 1987 drought, which caused long-term damage to the livestock industry.

Manufacturing: Contributes 20 per cent to GDP (1985) and employs 13 per cent of the total labour force. Over one-third of output consists of processed foods and beverages. Other major products include textiles, and chemical and metal products. Production is largely for the domestic market. Some 70 per cent of output takes place under the aegis of the state-run Industrial Development Corporation. The major constraint on the sector as a whole has been the reliance on imports for raw materials, spares and machinery, which has made the sector's performance subject to the availability of foreign exchange. Problems exacerbated by port delays and operating problems on the railways. Government policy has shifted towards the development of non-traditional exports and import substitution, including increased production of inputs for the expanding agricultural sector.

Mining: Copper mining is the key sector of the economy, accounting for virtually all foreign exchange earnings. Output has been declining from 495,000 tonnes in 1986 to 470,000 tonnes in 1987; it was expected to be only 375,000 tonnes in 1988. The 1984-89 mine rehabilitation programme has helped replace equipment and raise local skills, but foreign exchange is needed for some essential spare parts. The state-controlled ZCCM, which operates the world's second biggest open cast copper mine, showed net profits of $45.3m in 1987-88 (compared to a $68.4m loss in 1986-87). Zambia is the world's second producer of cobalt after Zaïre; output was 4,565 tonnes in 1986. Other mining includes lead and zinc, both very depressed – down by some 10,000 tonnes to 29,284 tonnes in 1986.

Energy: Almost 70 per cent of total domestic energy needs are met by hydro-electricity, mostly from plants at Kafue Gorge, Kariba North and Livingstone, and some electricity is exported to neighbouring countries. Consequently, oil imports have been kept fairly modest at around 800,000 tonnes a year. A 2,700 km pipeline from Tanzania supplies oil to a refinery at Ndola, but it is badly in need of repairs and was running in 1987 at only half capacity.

Membership of international organisations: African Development Bank, African States Associated with the EC (Lomé Convention), Arab Bank for Economic Development in Africa (recipient), Commonwealth, Economic Commission for Africa, FAO, GATT, IAEA, IDA, IFAD, ILO, IMF, OAU, Preferential Trade Area, Southern African Development Co-ordination Conference, UN, WHO, WIPO, World Bank, WTO.

output rise from 20 tonnes to 3,000 tonnes as trees reach maturity.

Industrial expansion was estimated to slow from 5.3 per cent in 1988 to 3 per cent in 1989, rising modestly to 3.5 per cent in 1990. The FNDP identified more than 100 projects due to be undertaken by the private sector, including flat glass production, expansion in the footwear industry, pharmaceuticals, vehicle radiators, water pumps, irrigation equipment and textiles. The government believed that industrialists were responding to appeals to intensify the search for domestic rather than imported inputs. Despite this, industrial production remained a hostage to foreign exchange availability with plant shutdowns and widespread under-utilisation of installed capacity. Economic difficulties were exacerbated during 1989 by the major fire at the Kafue Gorge hydroelectric station in March. It was hoped to product 300 megawatts of electricity from the plant – one third of its 900 MW installed capacity – by late 1989, but Kafue is unlikely to return to capacity production until 1991. The Zambia Electricity Supply Commission (Zesco) was able to increase generation at the Kariba North and Victoria Falls power stations and import some electricity from both Zaïre and Zimbabwe, but electricity supplies still fell by a third. Loadshedding became necessary with adverse repercussions on industry and mining.

To make matters worse, the Indeni Oil Refinery was closed after a fire in April 1989 and supplies of premium petrol were reduced. Both fires were blamed on the shortage of foreign exchange which precluded adequate maintenance and modernisation, though engineers said human error was a major contributory factor in the Kafue disaster. The FNDP provided for investment of K1.3bn in the energy sector of which some K750m was earmarked for electricity development including the rehabilitation of the Kafue and Victoria Falls hydroelectric stations.

Zambia's export drive and balance of payments situation continued to be threatened by transport bottlenecks. The acute shortage of rolling stock on the Tazara railway to Dar es Salaam forced Zambia to re-route its copper export traffic through the longer and more costly route via Zimbabwe to the Mozambique port of Beira. The Tazara line normally handles 35,000 tonnes of Zambian copper exports monthly, but in May–June 1989 this dropped to an average of 27,000 tonnes. Tazara officials said the system required 3,200 wagons and 90 locomotives to operate efficiently but in mid-1989 it had only 39 locos and 1,800 wagons at its disposal. Donors were asked to speed up the delivery of promised locomotives and spares but Tazara officials warned that a further decline in tonnages was likely in the latter half of 1989.

Outlook

The 1990s will be an extremely difficult period for Zambia. The failure of the 'self-help' economic recovery plan adopted after the break with the IMF in May 1987 is widely acknowledged, and the bold reforms announced in 1988–89 represented a major shift in strategy as the Kaunda government sought a rapprochement with the international donor community. But even if this is achieved, economic recovery will be a slow process especially as the government, aware of its loss of popular support and of the need to ensure that it takes the people along with it, will be reluctant to embrace radical reforms at least until there are concrete signs that the policies already adopted are succeeding.

With the massive overhang of debt arrears and the reluctance of foreign investors and banks who have burned their fingers in the past to commit new funds, dependence on aid will increase. Given the economic conditionalities attached to aid programmes President Kaunda will have to tread a delicate path between maintaining the confidence of the donor community on the one hand and retaining popular support from his electorate on the other.

On the positive side, progress towards peace in Namibia, Angola and Mozambique will improve transport links with beneficial implications for Zambia's foreign trade. But while the FNDP goals are eminently sensible, several of the targets look to be unattainable within the plan period unless Zambian economic policymakers achieve a far greater measure of successful policy implementation and co-ordination than in the recent past.

On the negative side, copper prices are likely to weaken from 1990 and as Zambia approaches the year 2000 its export base will increasingly erode unless far greater attention is paid to developing nontraditional exports. In addition, there is the depressing prospect of a worsening AIDS epidemic during the 1990s which could have far-reaching economic as well as social and humanitarian implications. Taken together, all these influences suggest that Zambia is heading for a hard landing during the 1990s and this prospect poses a formidable challenge to the donor community whose ingenuity will be taxed to the full in arriving at a debt-restructuring formula that will enable Zambia to grow out of its current difficulties.

BUSINESS GUIDE

Time: GMT +2.

Climate: Tropical, cool on high plateaux. Three seasons: May-Sep, cool, dry, temperatures range from 15°C to 27°C; Sep-Nov, hot, dry, temperatures range from 23°C to 29°C, Dec-Apr, hot, rainy season, temperatures range from 27°C to 32°C.

Entry requirements
Passport: Required by all.
Visa: Required by all except nationals of Commonwealth countries, the Irish Republic, Sweden, Romania and Yugoslavia. Visitors must have an onward ticket, at time of arrival. Special tourist visas available at border posts, without prior application necessary for most nationalities provided that applicants are a group of tourists with prior arrangements made through a tour operator in Zambia. Individual tourists should apply for a visa well in advance.
Currency: No limit on foreign currency import, but amount must be declared and not be exceeded by amount taken out. Up to K10 in Zambian currency may be imported or exported. US$20 departure tax must be paid in US Dollar bills.

Health precautions
Mandatory: Yellow fever and cholera vaccination certificates required if travelling from infected area.
Advisable: Cholera, yellow fever and typhoid and polio vaccinations. Malaria prophylaxis should be taken as risk exists throughout the country. Water precautions recommended, especially in rural areas. Bilharzia may be present in some lakes and rivers. AIDS risk. There is tsetse fly in the Kafue National Park region.

Representation overseas: Addis Ababa, Beijing, Bonn, Brussels, Cairo, Canberra, Dar es Salaam, Gaborone, Harare, Kinshasa, Lagos, Lilongwe, London, Luanda, Maputo, Moscow, Nairobi, New Delhi, Ottawa, Paris, Rome, Stockholm, Tokyo, Washington.

Air access
National airline: Zambia Airways.
Other airlines: Aeroflot, Air Botswana, Air India, Air Malawi, Air Tanzania, Air Zimbabwe, British Airways, Kenya Airways, South African Airways, Royal Swazi National Airways, TAAG-Angola, UTA.
Tax: Airport tax of US$20 is levied on departure, payable in US Dollar bills (no other form of payment is accepted).
International airport: Lusaka International (Code: LUN), 25 km from city; duty-free shop, bar, restaurant, bank, shops, car hire.

Surface access
Overland access: Tarred roads from Zimbabwe, Zaïre, Tanzania, Malawi; motorists should check border post hours at Kariba and Chembe and regulations concerning their vehicles. Customs bond may be required for import of cars. The TAZARA railway links Zambia to Tanzania – Connection at Kapiri Mposhi. Ferry from Botswana (Kazungula); steamer from Kigoma, Tanzania across Lake Tanganyika to Mpulungu.

Hotels: Several good quality hotels available, but these are often fully booked so reservations should be made well in advance. Hotels are graded from one to five stars by the Hotel Board. A service charge of 10 per cent, plus 10 per cent sales tax are added to all bills. Room charges must be settled in foreign currency. Some hotels require a deposit to cover the room rate and an element for food and drink to be converted to Kwacha on arrival. Tipping not customary in Zambia.

Electricity supply: 230 V AC.

Credit cards: Not widely accepted, except by hotels and duty-free shops.

Credit cards: Credit cards are not widely accepted except by hotels and duty-free shops.

Car hire: The Zambia Tourist Board has authorised over 16 car hire firms serving mainly Lusaka, Livingstone and the Copperbelt which are the major urban and tourist centres. Car hire is usually on a chauffeur-driven basis. This is available from BIG-5 in foyer of Ridgeway Hotel, Church Road, PO Box 30666, Lusaka (tel: 218240; tx: 42510). On special request firms may offer self-drive car hire services.

City transport
Taxis: Difficult to obtain except at airport and between hotels and town centre. Generally unmetered.
Buses: The United Bus Company of Zambia and other privately-owned companies run domestic services over a number of routes.

National transport
Air: Zambia Airways runs regular flights Lusaka to Livingstone and to the Copperbelt in the north. Several flights a week from Lusaka to other centres including Mfuwe and Luangwa. Charter companies also operate but are in heavy demand. Total of around 150 airfields and airstrips.
Road: Total network of almost 40,000 km, of which about 5,000 km are main roads. Surfaced roads link main centres. During rainy season many rural roads are impassable. N.B. It is not regarded as advisable to travel by car between Lusaka and the Copperbelt after dark.
Buses: Eagle Travel runs regular coach services to numerous locations including tourist sites. Non-tourist services can be irregular and crowded.
Rail: Zambia Railways runs daily express service between Kitwe and Livingstone via Lusaka. Buffet and sleeping cars on some trains. TAZARA (Tanzania-Zambia Railway Authority) operates line to Tanzania border. Total network over 2,000 km.

BUSINESS GUIDE

Public holidays
Fixed dates: 1 Jan, 1 May, 24 Oct (Independence), 25 Dec.
Variable dates: Youth Day (Sat in Mar), Good Friday, Easter Saturday, African Freedom Day (Mon in May), Heroes' Day and Unity Day (first Mon and Tue in Jul), Farmers' Day (first Mon in Aug).

Working hours
Business: (Mon-Fri) 0800 – 1230, 1400 – 1630.
Government: (Mon-Fri) 0800 – 1300, 1400 – 1700.
Banking: (Mon-Fri) 0815 – 1245. (Exceptions: Thu 0815 – 1200; Sat 0815 – 1100.)
Shops: (Mon-Fri) 0800 – 1700; (Sat) 0800 – 1300.
N.B. Many variations outside city centres.

Telecommunications
Telephone: Direct dialling facilities available for most centres in Zambia. Domestic service is limited and expensive. Dialling code for Zambia: IDD access code + 260 followed by area code (Kitwe 2, Livingstone 3, Lusaka 1, Ndola 2.
Telex: Telex facilities available at good hotels and at public telex offices in Lusaka (24-hour service), Kitwe and Ndola (during normal business hours). Telex country code: ZA.

Banking
Central bank: Bank of Zambia.

Representation in capital: Angola, Australia, Belgium, Botswana, Brazil, Bulgaria, Canada, China PR, Cuba, Czechoslovakia, Denmark, Egypt, Finland, France, German DR, German FR, Guyana, India, Iraq, Italy, Japan, Kenya, Korea PDR, Malawi, Mozambique, Netherlands, Nigeria, Portugal, Romania, Saudi Arabia, Somalia, Sweden, Tanzania, USSR, UK, USA, Vatican, Yugoslavia, Zaïre, Zimbabwe.

Banking
Bank of Zambia (central bank), PO Box 30080, Lusaka (tel: 216529; tx: ZA 41560).

Other useful addresses
Central Statistics Office, PO Box 31908, Lusaka.

Export Board of Zambia (tel: 228106; tx: 40309).

INDECO Ltd, PO Box 31935, Lusaka (tel: 228626, 228462-70; tx: 41821).

Lima Bank, PO Box 39157, Lusaka (tel: 228073-77; tx: 43280).

Lusaka Chamber of Commerce & Industry, PO Box 844 (tel: 252369; tx: 40790).

Metal Marketing Corp of Zambia, PO Box 35570 Lusaka (tel: 228131-40; tx: 40070, 40490).

Ministry of Commerce & Industry, PO Box 31968, Lusaka (tel: 228301-40).

Ministry of Finance, PO Box 50062, Lusaka.

National Air Charters Zambia, PO Box 33650, Lusaka (tel: 229774-8).

National Commission for Development Planning, PO Box 50268, Lusaka.

National Import & Export Corp, PO Box 30282, Lusaka (tel: 228018).

Rural Development Corp of Zambia, PO Box 39157, Lusaka (tel: 213111; tx: 43280).

Zambia Airways, Haile Selassie Avenue, PO Box 30272, Lusaka (tel: 229154).

Zambia Broadcasting Corporation, PO Box 50015, Lusaka (tel; 219673; tx: 42120).

Zambia Industrial & Commercial Copper Industry Service Bureau, PO Box 22100, Kitwe.

Zambia National Tourist Board, Century House, Lusaka Square, PO Box 30017, Lusaka (tel: 229087; tx: 41780).

Zambia Railways, PO Box 80935, Kabwe (tel: 222201; tx: 8100).

Hotels: Kitwe
Edinburgh, Obote Avenue, P.O. Box: 21800 (tel: 212188; tx: 51220).

Hotels: Livingstone (near Victoria Falls)
Musi-o-Tunya Inter-Continental, P.O. Box: 60151 (tel: 21121/9; tx: 24221).

Hotels: Livingstone (town centre)
North Western, P.O. Box: 60069 (tel: 320711; tx: 24260).

Hotels: Lusaka
Andrews Motel, P.O. Box: 30475 (tel: 211532/211108; tx: 40450).

Inter-Continental, Haile Selassie Avenue, P.O. Box: 32201 (tel: 212366; tx: 41440).

Lusaka, Cairo Road, P.O. Box: 30044 (tel: 217370; tx: 41921).

Pamodzi, Church Road, P.O. Box: 35450 (tel: 212620, 890 215289; tx: 44720).

Ridgeway, Church Road, P.O. Box: 30666 (tel: 218240; tx: 42510).

Hotels: Ndola
New Savoy, Buteko Avenue, P.O. Box: 71800 (tel: 3771; tx: 30020).

Zimbabwe

Anthony Hawkins

As Zimbabwe nears the tenth anniversary of its independence in April 1990 President Robert Mugabe's ruling ZANU-PF administration faces proliferating challenges on both the political and economic fronts. When the unity agreement with Joshua Nkomo's minority ZAPU (Zimbabwe African Peoples Union) was reached at the end of 1987, President Mugabe appeared to be riding on the crest of the wave. At one stroke, he had achieved a *de facto* one-party state, bringing Mr Nkomo and his top lieutenants into a coalition administration, eliminating the dissident threat in the western province of Matabeleland.

But by the second half of 1989 the government was in serious difficulties and a new political party, the Zimbabwe Unity Movement (ZUM) led by former ZANU (Zimbabwe African National Union) secretary-general and long-time Mugabe associate, Edgar Tekere, was attracting considerable support in the eastern half of the country. Although ZANU-PF comfortably defeated ZUM in the latter's first electoral outing – the Dzivaresekwa by-election in July 1989 – the fact that the government's support slumped to 7,000 votes from 44,000 in 1985 was a clear reflection of its loss of popular support.

Crisis of expectations

On the surface ZANU's problems arose from the adverse publicity generated by the so-called 'Willowgate' scandal. The courageous editor of the Bulawayo newspaper, the *Chronicle*, Geoff Nyarota, broke the story that government ministers and members of parliament had been using their official positions to obtain new motor vehicles from the state-owned Willowvale assembly plant and reselling them at prices well in excess of the official price laid down by government. Mr Nyarota was subsequently threatened by government ministers, including Defence Minister Enos Nkala, before President Mugabe, bowing to the weight of public opinion, appointed a judicial commission, under Mr Justice Sandura to investigate the whole affair.

After some embarrassing public hearings in which government ministers admitted giving false evidence and pressuring witnesses, the Sandura Commission issued a report, scathingly critical of several cabinet members who, it said, should be convicted not just of

KEY FACTS

Official title: Republic of Zimbabwe
Head of state: President Robert Mugabe
Ruling party: Zimbabwe African National Union-Patriotic Front (ZANU-PF)
Capital: Harare
Official Languages: English, Shona, Ndebele
Currency: Zimbabwe Dollar (Z$) = 100 cents
Exchange rate: Z$2.26 per $ (Nov 1989)

Area: 391,109 sq km
Population: 8.9m (1988)
GDP real growth: 2.6% (1986)
GNP per capita: $580 (1987)
Inflation: 12.5% (1987)
Trade balance: $381m (1987)
Foreign debt: $2.5bn (1987)

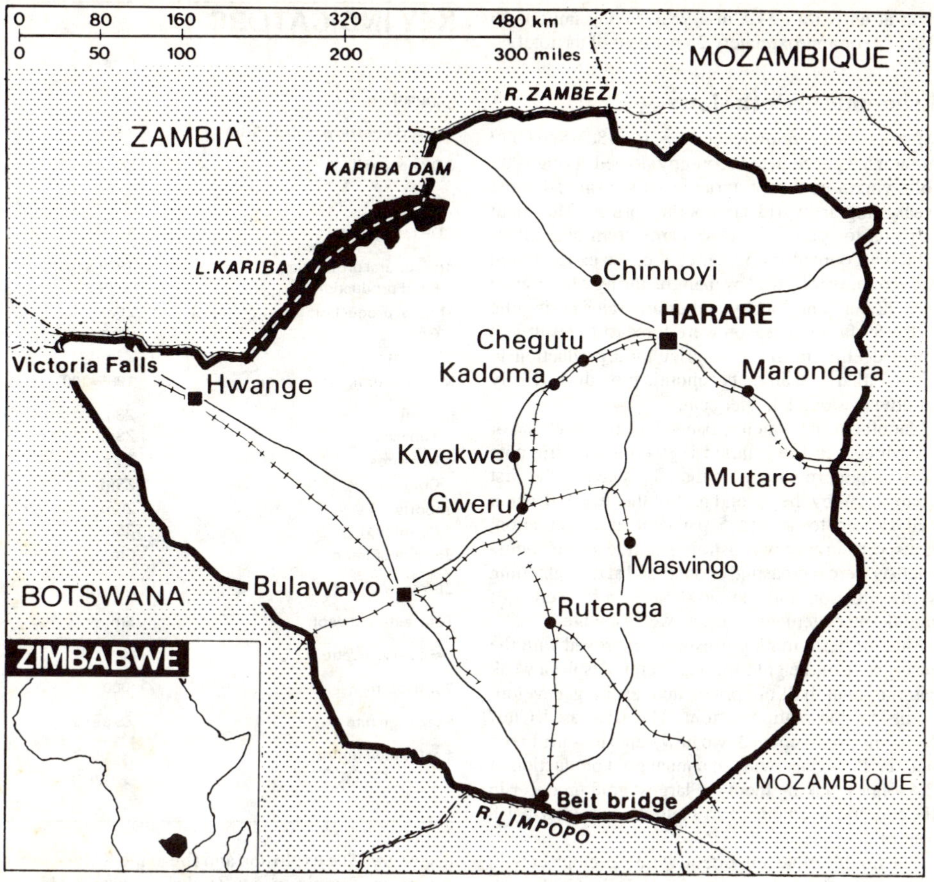

breaching price control regulations but of subornation and perjury. Five ministers, including two of the president's longest-standing colleagues, Senior Minister Maurice Nyagumbo and Mr Nkala, were forced to resign and Mr Nyagumbo subsequently took his own life.

The final stage of the drama came when Political Affairs Minister Fred Shava was fined and sentenced to a 9-month jail term, on being convicted of subornation and perjury. This was too much for the president who promptly pardoned Shava and ordered that, in future, those involved in the scandal be prosecuted only for price control offences and not for giving false evidence. This was a major miscalculation resulting in widespread criticism of the government, which previously had been acclaimed by many for its willingness to expose corruption.

Had it not been for deep-seated discontent with the government, Willowgate might well have passed off as no more than a storm in a teacup. But against the background of a deepening crisis of unfulfilled expectations over land, over stagnant living standards, falling real wages, pervasive shortages in the shops, high inflation, inadequate and deteriorating transport facilities and – above all – escalating unemployment, Willowgate assumed a far greater significance, demonstrating how members of a socialist administration were exploiting the black market opportunities created by government controls.

ZUM, formed to exploit public discontent, immediately attracted considerable 'protest vote' support especially from students, intellectuals, trade unionists and other radicals. By August 1989, ZUM had no clear-cut policy platform and it was garnering support from those opposed to the government rather than those who believed in its principles, which remained vague and ill-defined.

ZUM's public support was all the more disconcerting for the government given its refusal to allow the new party to hold public rallies and the use of the state-controlled media to defend and extol the administration. Within the government itself divisions appeared between those who wanted a return to hard-line socialism and those believing that new faces and new policies were needed. President Mugabe himself, in his April 1989 Independence Day address, signalled a more flexible, market-oriented economic strategy, welcoming foreign investment and promising to ease state controls over wages and prices. But the prolonged delays in agreeing and implementing new policies reflected conflict within the ruling party and a lack of consensus over direction and speed of change. Suggestions that socialist strategies adopted when ZANU was fighting the liberation war in the 1970s be dropped were strongly opposed by party radicals.

Land issue resurfaces

Meanwhile, after a year in the shadows, Joshua Nkomo emerged as chief protagonist of a populist land policy. At independence in 1980 the government pledged to re-settle 162,000 landless families but by 1989 only 52,000 families or 416,000 people had actually obtained new land. Some 2.7m hectares of land have been purchased amounting to only 16 per cent of the area owned by commercial farmers and Mr Nkomo insisted that the country's 4,200 white commercial farmers make more land available to ease land hunger especially in Matabeleland and the Midlands.

Mr Nkomo's unabashedly political pitch was designed to regain the political initiative particularly for his wing of the ruling party. Land remains a highly emotional issue and one which is bound to win votes in the 1990 elections. This was readily appreciated by President Mugabe himself who in August 1989 came out strongly in favour of a revolutionary land reform programme which could include amendments to the independence constitution after the reserve clauses fall away in April 1990. In particular an amendment to force farmers to sell land, at agreed prices, was under consideration, though many felt the government was using the prospect of constitutional amendments as a bargaining counter to induce farmers to sell land.

Farmers and economists warned that far-reaching land redistribution was likely to have adverse medium-term economic consequences, in two main respects. First, the commercial farmers produce more than three-quarters of the country's marketed agricultural production, including almost all the beef, virtually all the Virginia tobacco and products like cut flowers and fruit for export. Any redistribution of land from commercial to peasant farmers could have adverse implications for output, employment and especially exports. This is the familiar trade-off between equity and efficiency.

Second, under the Lancaster House Constitution agreed in 1979, land cannot be compulsorily purchased but must be acquired on a 'willing seller willing buyer' basis. Threats to amend the constitution after the ten-year safeguard clauses lapse in 1990 might be no more than a bargaining ploy, but if implemented could jeopardise the government's parallel programme of encouraging foreign investment.

The 1990 general elections scheduled for mid-year were likely to be delayed until the latter part of the year. Despite the government's loss of support, it seemed unlikely that any opposition party would seriously threaten ZANU-PF's overwhelming political domination. Much more likely was bitter infighting within the ruling party and its ZAPU partner over candidate lists, ministerial positions and future economic policies, including the handling of the land issue. Also likely was a low turnout at the polls, since if 1989 by-elections were any guide voters were more likely to stay at home than vote for opposition candidates.

Pragmatism in foreign policy

Foreign policy became more pragmatic reflecting narrowing options on the one hand and the changing Southern African environment on the other. The Namibian settlement and peace moves in both Angola and Mozambique held out the hope of a more stable regional situation and this, allied with Zimbabwe's economic difficulties, encouraged greater pragmatism in policy towards South Africa. But the rhetoric was unchanged and Zimbabwe continued to demand mandatory economic sanctions against Pretoria, notwithstanding its own continuing high level of economic dependence on the south vividly illustrated by the hiring of 20 locomotives from the South African Transport Services (SATS), and the increased number of Zimbabweans seeking visas to visit South Africa on shopping sprees. South African officials said the number of visas granted each month had risen to 16,000 by mid-1989 – double the figure for 1986.

While increasing volumes of imports and exports were diverted to two Mozambican ports of Beira and Maputo, Zimbabwe remained dependent on South Africa's transport system for upwards of 60 per cent of its foreign trade. Pretoria also remained the country's main trading partner supplying one fifth of total imports and purchasing 10 per cent of exports. Indeed, these visible trade figures seriously understate the degree of trade dependence since, when shopping trips are taken into account, it is estimated that indirect and undeclared imports from South Africa – either through Botswana or direct – were running at an extra Z$100m a year, taking imports to Z$550m or approximately 25 per cent of the total.

Economic growth slows

After a strong economic recovery in 1988 when GDP grew at 5.3 per cent, the economy slowed during 1989 following a mediocre agricultural season and some weakening in world commodity prices. The main impetus to growth in 1988 came from agriculture which accounted for 53 per cent of the expansion in the economy followed by manufacturing whose share was 22 per cent. Patchy and erratic rains during the 1988–89 farming season were forecast to result in a small decline in the real agricultural production in 1989 with the result that economic growth was likely to slow to around 3.5 per cent.

Manufacturing which expanded at an annual rate of 4.9 per cent in 1988 started 1989 on a very firm note with output growing more than 8 per cent in the first quarter, but by the second half of the year expansion had slowed to around 5 per cent as construction, transport and energy constraints started to bite. Industrialists were increasingly concerned at the tightening skills situation and the slow but steady exodus of middle-management, black as well as white.

Medium-term mining prospects improved with the resurgence of the gold-mining sector, despite a weak international bullion price, and exciting developments on the platinum front. The UK-based Cluff Minerals group opened two new deposits – the Freda and the Rebecca – with an annual gold production of 2.4 tonnes making them the largest gold producer in the country, overtaking Rio Tinto's Renco with its 1.8 tonnes.

Potentially more important were developments in platinum with the Australian-based Delta Gold group announcing plans to invest up to $200m in the Hartley project. Union Carbide is also looking into a platinum prospect as are Rio Tinto and Anglo-American who have a joint-venture possibility. A total of 57 exclusive prospecting orders were operational in mid-1989 with a further 17 pending – easily the highest-level of exploration activity since independence.

Since 1980 the value of mining production has trebled to reach almost Z$1bn in 1988 but this is almost entirely the result of higher prices on world markets and the devaluation of the Zimbabwe dollar. In US dollar terms, the value of production barely changed while output volumes in 1988 were 13 per cent below their 1976 peak. Increased exploration activity and plans to develop at least one of the platinum projects as well as Rio Tinto's Sengwa coal property, give rise to hopes that mining is poised for a return to rapid growth in the 1990s after a decade of relative stagnation.

Mining production is estimated to grow a further 15 per cent in 1989 exceeding Z$1.1bn, mainly reflecting higher prices for nickel and copper and the continuing devaluation of the Zimbabwe dollar.

A potentially-exciting development was the decision by Mobil to drill for oil in the Zambesi Valley. Geologists had long written off Zimbabwe as a potential oil producer, but Mobil believes there is a reasonable chance of discovering viable oil deposits which could have a major beneficial impact on a land-locked economy that devotes at least one-fifth of its import bill to fuel supplies.

Infrastructural problems worsen

Both the transport and electricity industries hit the headlines during the year as capacity came under severe pressure. Although Zimbabwe has an installed capacity of almost 1,900 megawatts of electricity and a maximum demand of 1,410 mW, shortages of spare parts and a breakdown at the Hwange power station meant that load-shedding and power blackouts were necessary. The explosion and fire at the Kafue power station in Zambia meant Zimbabwe was unable to draw imported supplies from its neighbour and the pressure of demand during the winter months in 1989 meant that routine maintenance could not be undertaken heightening fears of further future breakdowns.

No final decisions were made on the country's ambitious US$4bn electricity development plan involving two new turbines of 300 mW at Kariba South power complex, a new hydro-plant at Batoka Gorge on the Zambesi river and a second new hydro project close to the Victoria Falls. If all existing capacity operates to peak efficiency then no new investment would be needed until 1993, but with the thermal plants at Munyati, Harare and Bulawayo operating with ageing if not obsolete equipment, optimal efficiency is unlikely to be achieved.

A highly cost-effective and politically attractive option would be to draw power from the underutilised Cabora Bassa Dam in neighbouring Mozambique. In the past Zimbabwe has held out for electricity self-sufficiency but as the peace process in the region widens, so the possibility of greater regional co-operation in electricity grows.

In transport, regional co-operation of a different, but still highly significant, nature occurred with Zimbabwe's decision to lease 20 locomotives from South Africa. This decision was forced on the state-owned National Railways of Zimbabwe (NRZ) because in mid-1989 no less than 102 of the NRZ's 285 locomotives were obsolete, leaving 183 of which only 110 to 120 were operational at any one time.

The NRZ's problems were largely the result of inadequate foreign exchange to import the necessary spares and new equipment and a shortage of skilled maintenance personnel. The system's inability to carry all the coal traffic on offer was a major problem affecting the tobacco industry (which uses coal for curing), operations at the state-owned steel industry Zisco, and many smaller industrial concerns reliant on coal-fired furnaces.

A similar situation applied in road transport where huge investment in both buses and heavy lorries was needed to rebuild the country's ageing fleets. In 1989 170 new buses were ordered for the state-owned Zupco to ease the crisis situation which also had its roots in the shortage of foreign exchange.

If foreign exchange shortages were the root cause of infrastructural bottlenecks they were not the only one. Skilled technical personnel were increasingly difficult to obtain while in the construction industry the brick, cement and steel manufacturers were unable to satisfy domestic demand. In the case of cement, price controls were to blame with manufacturers refusing to invest in new capacity at a time when they were being forced to sell cement at a loss.

The price and wage freeze temporarily reduced inflation from 15 per cent early in 1987 to 5 per cent by mid-1988, but thereafter, as the freeze was partially dismantled, inflation regained momentum to reach 15 per cent in August 1989. Substantial food price increases, aimed at curbing subsidies, and the removal of the rent freeze were expected to push inflation to 20 per cent by the early 1990s.

Inflationary pressures were exacerbated by shortages in the economy and economists estimated that the true inflation rate was at least 5 percentage points higher than the official rate as many transactions were driven into the black market. With the failure of direct controls and the government's increased unpopularity partly explained by falling real wages, a policy of indexation was adopted with wage increases of between 11 per cent and 16 per cent awarded to public servants and much higher awards for doctors, railway artisans and others in powerful bargaining positions. Increased labour unrest was anticipated and the government introduced new emergency powers regulations in August 1989 to ban strikes in essential services.

With the approach of the 1990 elections, inflation seemed likely to accelerate still further as the government encouraged wage increases designed to offset fast-rising living costs. Economists were predicting inflation averaging 16 per cent in 1989 (more than double the 7.3 per cent for 1988) rising above 20 per cent in the first half of 1990.

KEY INDICATORS

	Unit	1983	1984	1985	1986	1987
Population	m	7.74	7.98	8.38	8.41	8.64
Gross domestic product (GDP)	Z$m	6,224	6,423	7,609	8,880	–
GDP per capita	Z$	804	805	908	1,056	–
Inflation	%	23.1	20.2	8.4	14.3	12.5
GDP real growth	%	1.3	–1.7	4.7	2.6	–3.5
Agricultural production	1979-81=100	84.7	99.6	125.9	127.2	101.9
Food production	1979-81=100	79.1	84.5	121.0	124.0	84.2
Mineral production	Z$m	470.4	546.5	589.3	–	–
Gold	Z$m	193.8	214.1	245.0	–	–
Asbestos	Z$m	69.4	80.8	98.4	–	–
Manufacturing production	1985=100	94.3	90.4	100.0	102.9	105.3
Exports	Z$m	1,025.7	1,271.0	1,795.5	2,169.4	2,371.4
Tobacco	Z$m	232.6	287.3	365.9	424.0	397.0
Asbestos	Z$m	68.6	74.0	82.7	82.7	91.6
Cotton	Z$m	73.6	115.3	149.3	130.5	121.1
Imports	Z$m	1,067.0	1,200.6	1,663.5	1,886.5	2,008.0
Petroleum	Z$m	200.2	230.0	309.2	169.9	211.3
Balance of trade	$m	84.1	184.3	200.6	311.2	381.0
Current account	$m	–460.2	–100.0	–75.7	6.8	48.0
Total external debt	$m	2,302	2,067	2,195	2,340	2,512
Debt service ratio	%	32.1	19.8	23.6	22.1	23.2
Tourist arrivals	'000	292.1	339.6	389.5	433.4	–
Exchange rate	Z$ per $	1.01	1.24	1.61	1.67	1.66

* estimated figure

Budget deficit reduced

The budget deficit rose to Z$1.1bn (11.3 per cent of GDP) in 1988–89 but Finance Minister Bernard Chidzero projected a fall to Z$989m (9 per cent of GDP) in fiscal 1990. This looks to be an ambitious target given rapid inflation and will depend on the government's ability to contain public spending well below the inflation rate as well as securing a 23 per cent rise in tax revenues. Total public spending is forecast to rise 15 per cent in the year to June 1990 during which time revenue would increase 22 per cent. For the first time since independence the government expects to finance recurrent spending without resort to borrowed funds and this move was widely applauded. There was also a significant reduction in subsidies – to Z$560m from Z$663m previously – mainly because the grant to steel was cut by Z$67m to Z$100m.

But Dr Chidzero failed to introduce any far-reaching changes leaving an unsatisfactory underlying situation. Government spending will reach 57 per cent of GDP in 1989–90with education (19 per cent of the budget), debt-service (15 per cent), defence (13.4 per cent) and subsidies (9 per cent) absorbing more than 56 per cent of total expenditure. Hopes that defence spending will decline – in relative though not absolute terms – from 1990 depend largely on progress in the Mozambique peace talks and, for the second successive year, Chidzero warned that the government was investigating ways of reducing the burden of education spending – presumably by imposing school fees.

The balance of payments remained in surplus on both current and overall account in 1988 despite a continued outflow of capital. Exports rose 18.5 per cent to a record Z$2.86bn and although imports grew more rapidly (21 per cent) the trade surplus widened to Z$708m from Z$634m in 1987. But the net deficit on invisibles grew even more rapidly reflecting interest payments abroad and rising freight and transport charges to reach Z$691m compared with Z$506m in 1987. Consequently, the current account surplus fell from Z$128m to only Z$17m but with net capital inflows of Z$163m, there was an overall surplus of Z$180m.

Although the capital account showed a surplus, there was still a net outflow once below-the-line repayments to the IMF were taken into account. However, with the IMF now having been repaid, and with the government moving to discourage disinvestment and encourage new investment by foreign companies, some modest improvement in capital account is forecast for 1989–90.

A major improvement was the further reduction in debt-service payments which declined to 27 per cent of exports in 1988 from more than 35 per cent in 1987. The debt-service ratio is forecast to fall to between 20 per cent and 25 per cent in 1990 but with new borrowings to purchase Chinese-manufactured ground-to-air missile systems and two Boeing 767 aircraft for Air Zimbabwe, the ratio is likely to level off above 20 per cent and start climbing again in the early 1990s.

Depreciation of the Zimbabwe dollar exchange rate slowed from an average of more than 10 per cent a year between 1985 and 1987 to only 4 per cent in 1988 and 1.5 per cent in the first seven months of 1989. However, as inflation accelerates, so faster depreciation is likely in 1990 with the Zimbabwe dollar forecast to fall close to the US 40 cents level by the end of that year.

Economic policy shift foreshadowed

As it became increasingly apparent that the government's socialist policies were failing to achieve their goal of 'growth with equity', President Mugabe announced plans to shift to a more market-oriented strategy with fewer direct controls. While this policy shift, unveiled in April 1989, was welcomed by investors, bankers and industrialists at home and abroad, progress towards a new policy regime was hesitant and slow.

The revised investment guidelines, released in May 1989, were the first concrete indication of a change of economic direction. These however were only a modest improvement on the original 1982 document,

COUNTRY PROFILE

Political system: Legislative power is vested in the bicameral parliament, comprising 100-member House of Assembly and 40-member Senate. Parliament is re-elected every five years. Executive power is vested in the president, elected by parliament for six years. Independence gained from UK in 1980.

Political parties: The ruling party is the Zimbabwe African National Union-Patriotic Front (ZANU-PF), which won 64 of the 79 contested seats in the 1985 elections to the House of Assembly. A merger with the Zimbabwe African People's Union (ZAPU) was effected in April 1988.

Population: Zimbabwe's population of 9.07m (1988 estimate) is growing at around 3.1 per cent a year. Main ethnic groups include the Shona, who comprise 75 per cent of the population, and the Ndebele (18 per cent).

Main cities/towns: Harare (population 700,000), Bulawayo (400,000), Gweru (80,000), Mutare (76,000), Kwekwe (65,000), Hwange (35,000).

Language: English, Shona and Ndebele are the official languages, while English is the the medium for all business. Principal African languages are Shona and Ndebele.

Media
Press: Major dailies are *The Chronicle*, published in Bulawayo, and *The Herald*, published in Harare. Weekly newspapers include *The Sunday Mail*, *The Sunday News*, *The Financial Gazette* (only wholly independent newspaper), and some regional publications. Also several weekly magazines and other periodicals available. All government-controlled except *The Financial Gazette*.
Broadcasting: Over 507,000 radio receivers and 400,000 television sets in use. Radio services in English, Shona and Ndebele, and two television channels, broadcasting for about 45 hours per week, are transmitted by the Zimbabwe Broadcasting Corporation.

External trade: Strong mineral exports and improvements in the industrial sector helped to increase the balance of trade surplus in 1987. The growth in exports was expected to continue in 1988, but the current account balance, which was greatly reduced in 1987 because of high debt service payments, is unlikely to improve substantially in the next few years, while debt servicing is expected to remain an annual Z$800m – Z$900m.
Exports: Semi-processed goods (24.8 per cent of total value in 1986), tobacco (19.5%), gold (19.0%), food (14.9%), crude materials (12.8%). Main destinations: South Africa (12.4% of total value in 1986), UK (12.3%), German FR (8.6%), Italy (5.9%), USA (5.7%), Japan (4.5%), Botswana (4.3%), Zambia (2.9%), France (1.2%).
Imports: Machinery and transport (37.9 per cent), electricity (16.1%), chemicals (15.9%), semi-processed goods (14.5%), fuels and crude materials (5.9%). Main sources: South Africa (21.4% of total value), UK (10.9%), German FR (9.9%), USA (8.3%), Japan (4.3%), Botswana (4.7%), Italy (4.2%), France (3.3%), Zambia (2.6%).

Agriculture: Sector plays a pivotal role in the economy, accounting for an important part of foreign exchange earnings and providing a large share of raw materials for the manufacturing industry. Tobacco, the main export crop, contributes about 12 per cent to GDP. Near self-sufficiency in food crops, with large-scale production of maize, wheat, soya beans and groundnuts. Also increased production of meat and milk. Production of cash crops, principally tobacco, cotton, coffee and sugar, suffered substantially less than subsistence crops from the effects of drought, but all were affected by the 1986-87 weather conditions. Small fishing industry based on Lake Kariba. Also commercial trout farms. Emphasis on land distribution and settlement, development of co-operatives and improved provision of farm machinery and fertilisers.

Manufacturing: Sector is well established and diversified. A period of decline ended in 1985 with the good harvest which boosted demand. Importance of sector can be traced back to long period of trade sanctions, when manufacturing output had to be increased dramatically to provide goods which until then had been imported. Major products, which are largely derived from the domestic mining and agriculture sectors, include food, drink, tobacco, textiles, wool, paper, chemicals, non-metallic mineral products, metals and metal products, transport equipment and construction materials. Around 75 per cent of activity is centred in Harare and Bulawayo. Heavy industry is mainly located in Redcliff, Kwekwe and Gweru. Government efforts to attract investment and encourage decentralisation of industry have been hampered by poor profitability and high taxation.

Mining: Production increased by 17 per cent in value in 1987, and expansion in both volume and value is expected to continue. The sector employs six per cent of workforce. Production, which is largely undertaken by foreign conglomerates, is both for export and local consumption. Coal – now second in importance – is currently for domestic use only. Over 40 minerals are extracted including gold (43 per cent of 1987 value), coal (13%), asbestos (12%), nickel (9%), copper and chrome ore. There is a sharp increase in gold prospecting. Higher world gold prices have boosted earnings.

Energy: Coal, of which country has 29bn tonnes of reserves, meets 50 per cent of local energy requirements, wood 25 per cent and oil and hydro-electric power 25 per cent. Current emphasis on developing local energy resources, particularly coal and hydro-power. Projects under way involve expansion of the huge Hwange coalfield, including two new thermal power stations, and the building of hydro-electric stations on the Zambezi River. Most of the country's fuel needs are imported via the Beira pipeline or through South Africa by rail.

Membership of international organisations: African Development Bank, African States Associated with the EC (Lomé Convention), Commonwealth, Economic Commission for Africa, FAO, IDA, IFAD, IFC, ILO, IMF, OAU, Preferential Trade Agreement, Southern African Development Co-ordination Conference, UN, UNICEF, WHO, World Bank.

including a willingness to sign bilateral and multilateral investment guarantee agreements and establishing an investment centre (a one-stop investment agency) to speed up and facilitate investment decisions. Financial regulations were changed to enable third parties to purchase at a discount the blocked funds of companies wishing to disinvest from Zimbabwe.

Other economic reforms included changes to the price control regulations, which were sharply criticised by businessmen on the grounds that they entrenched rather than liberalised the existing system, and a partial return to collective bargaining within parameters set by the authorities. Dr Chidzero said the government was studying the recommendations of its working party on trade liberalisation and promised further far-reaching economic changes by the end of 1989.

But many feared that the government was likely to be preoccupied with its political problems in the run-up to the elections and that economic reform was likely to remain a slow and cautious process at least until after President Mugabe obtains a fresh mandate, probably towards the end of 1990.

Whether the administration can afford to leave economic reform on the back-burner is a matter of great controversy, chiefly because of the deteriorating unemployment situation. At the end of 1988 unemployment was estimated at 1m people while employment was put at around 1,160,000. This meant that a total of 175,000 new jobs had been created since 1979 at a rate of no more than 20,000 jobs a year. With the number of secondary school-leavers projected to double from 97,000 in 1986 to 185,000 in 1989, reaching 330,000 in 1991, unemployment will reach 2m people by 1995 and exceed 3m at the turn of the century.

There are no quick solutions to this problem which could yet turn out to have its darker side in the spread of the AIDS virus, about which increasing concern was being expressed by health experts. Indeed, one respected health forecaster suggested that, in the absence of a cure for the disease, Zimbabwe's total population, currently projected to reach 13m by the year 2000 could, in fact, fall below 8m by 2010.

Assuming that this gloomy scenario can be avoided, there is little doubt that Zimbabwe faces a severe political and social crisis in the mid-1990s in the form of an unemployment rate in excess of 40 per cent of the workforce. Although this problem is widely acknowledged, government economic reforms to slow unemployment growth lacked both urgency and direction. Preoccupation with reducing the debt-service ratio stifled import capacity, economic growth and job creation during the 1984–89 period. Investment averaged only 16 per cent of GDP at a time when an investment ratio of 25 per cent was required. Some of the policy changes introduced since April 1989 should lead to faster economic growth during the 1990s provided the full reform package is implemented without further undue delay.

However, higher levels of investment and foreign exchange availability are not going to solve Zimbabwe's unemployment crisis. The problem will have to be tackled across a broad front: faster and more efficient land resettlement programmes, targeted assistance to small-scale enterprises, more emphasis on technical and vocational education, a rationalisation of the existing salary and wage structure where differentials have been squeezed to levels that discourage employment of school-leavers and under-motivation at middle and senior levels, major changes to the 'hire-and-fire' labour legislation which militates against employment growth, and effective measures to slow inflation.

Economic slowdown likely, 1989–90

Growth in the economy is forecast to slow in 1989–90 as the impetus from the excellent 1988 farming season, buoyant tobacco and commodity prices and increased export earnings, loses momentum. Given good rains during the 1989–90 farming season, GDP growth should accelerate from a forecast 3.5 per cent in 1989 to 5 per cent in 1990 though this will be accompanied by rapid inflation, a deterioration in the balance of payments and a declining exchange rate.

The government is unlikely to administer the tough and unpopular medicine necessary to stabilise the economy until after the 1990 elections, at which juncture major cuts in government spending on social services and on subsidies, higher interest rates and a reduced role for the state will be needed if a catastrophic unemployment crisis is to be avoided in the 1990s.

During the first 9 years of independence the Zimbabwe economy grew at little more than 3 per cent a year with the result that by 1988 real per capita incomes at Z$455 (at 1980 prices) were only 10 per cent above those ruling in 1979. More than 40 per cent of the growth was generated in education, public administration and health. After strong growth in the early part of the 1980s, the peasant sector's contribution to marketed farm output appeared to be levelling off at around 20 per cent while mining and manufacturing had both failed to achieve the strong growth forecast at the time of the Lancaster House settlement in 1979.

Most disappointing of all, there was a small net outflow of capital from Zimbabwe during the 1980s with few new foreign investors taking the plunge while a number of existing investors took the disinvestment route. By mid-1989, some Z$800m had been invested in 4 per cent disinvestment bonds – a measurement of the outflow of capital (much of it in the form of blocked dividend and profit remittances) from the country.

The economy's resilience and potential is unquestioned. What remains to be demonstrated is the capability of President Mugabe's administration to harness this undoubted potential before the government is overwhelmed by escalating unemployment. Considerable donor sympathy and support exists for Zimbabwe and given a more positive policy environment, the disappointments of the 1980s can be translated into vigorous self-sustaining growth in the 1990s. For this to happen, the administration needs to move rapidly to free the economy from burdensome controls and excessive government spending, exploiting the full potential of what is probably the most sophisticated private sector in Africa south of the Sahara excluding South Africa.

BUSINESS GUIDE

Time: GMT +2.

Climate: Although situated within the tropics climate is tempered by altitude and inland position. From Nov to Mar is warm and rainy with likely temperature 25-32°C. From May to Aug is a cool dry winter season with temperatures reaching 20-28°C during the day, but can fall to freezing point at night. From Aug to Oct is increasingly warm with Oct being the hottest month – up to 34°C. Low-lying areas (e.g. Victoria Falls, Zambezi Valley) are warmer all year round.

Entry requirements

Passport: Required by all.

Visa: Not required by citizens of most Commonwealth countries, EC, USA, Japan, Canada and some other countries. Visitors must have an onward or return ticket and 'sufficient' money for their stay.

Currency: No limited on foreign currency brought into country, but amount must be declared and must not be exceeded by amount taken out. Limit of Z$40 on movement of Zimbabwean currency. US$20 departure tax must be paid in US Dollar bills.

Health precautions

Mandatory: Yellow fever vaccination certificate if travelling from an infected area.

Advisable: Cholera, typhoid, polio vaccinations recommended. Malaria risk exists in Zambesi valley and other low lying areas – anti-malaria tablets should be taken. Water precautions are necessary outside main towns. Bilharzia is endemic, therefore avoid using lake or river water for drinking or swimming. AIDS risk.

Representation overseas: Addis Ababa, Algiers, Beijing, Belgrade, Berne, Bonn, Brussels, Bucharest, Dakar, Dar es Salaam, Dhaka, Berlin (GDR), Gaborone, Geneva, Havana, Lagos, Lilongwe, London, Lusaka, Maputo, Moscow, Nairobi, New Delhi, Ottawa, Paris, Stockholm, Tokyo, Washington.

Air access

National airline: Air Zimbabwe.

Other airlines: Air Botswana, Air Mauritius, Balkan, Kenya Airways, LAM-Linhas Aéreas de Moçambique, Lesotho Airways, Pakistan Airways, Qantas, Royal Swazi National Airways, South African Airways.

Tax: Departure fee of US$10 per adult payable in US dollar notes, maximum denomination $20, on leaving Zimbabwe.

International airports: Bulawayo (Code: BUQ), 24 km from city, restaurant; Harare (Code: HRE), 12 km from city, duty-free shop, bar, buffet, restaurant, bank, post office, car hire; Victoria Falls (Code: VFA) 22 km from city.

Surface access

Road: Direct route from Lusaka (Zambia) to Harare; or via Victoria Falls. Entry from South Africa at Beitbridge and from Botswana at Plumtree. The roads from Mozambique are subject to security problems. Most of the border posts are closed from 1800 to 0600 hours every day although specific hours vary.

Rail: Regular services from Zambia, via Victoria Falls, from Beira and from South Africa. Rail link with Maputo almost totally inoperative.

Hotels: Several hotels of various standards available in main cities, rated from one- to five-stars by the Tourist Board. Most of the larger ones are air-conditioned. Government imposes a bed tax per person per night, and it is usual to tip 10 per cent.

Electricity supply: 230 V AC.

BUSINESS GUIDE

Car hire: Self-drive cars are available in main cities and at Harare airport. Can be a useful method of transport as most main inter-city roads tend to be of good quality. Drive on the left. Foreign or international driving licence acceptable for short visits.

City transport

Taxis: Not usually hailed in street. Available at ranks near main hotels. A 10 per cent tip on price shown on meter is usual, or more for short journeys.

Buses: Urban services in some centres can be sporadic.

National transport

Air: Regular cheap daily flights by Air Zimbabwe between Harare and Bulawayo, Buffalo Range, Gweru, Hwange, Kariba, Masvingo and Victoria Falls. Also several private charter firms operating from Harare.

Road: Network of over 85,000 km, of which about one-quarter are classed as main or secondary roads and half are surfaced with gravel. Good roads connect major towns.

Buses: Good inter-city network operated by Express Motorways Africa Ltd, Zimbabwe Omnibus Company, plus numerous local operators. Express coach services from Harare to Bulawayo, Mutare, Kariba, Chipinge, Masvingo. Advisable to book in advance.

Rail: National Railways of Zimbabwe operate services between Harare and Gweru, Bulawayo, Victoria Falls, Mutare, Masvingo, Chinhoyi and intermediate towns (there are also certain places served by branch lines). These are 3,394 km of track. Four classes; first two generally recommended; some trains carry restaurant cars and couchette sleeping accommodation (NB: bedding is charged). Advisable to book tickets (and bedding) in advance.

Public holidays

Fixed dates: 1 Jan, 18 Apr (Independence), 1 May, 25 May (Africa Day), 11 Aug (Heroes' Day), 12 Aug (Defence Forces' Day), 25 Dec, 26 Dec.

Variable dates: Good Friday, Easter Monday.

Working hours

Government and business: (Mon-Fri) 0745/0830 – 1600/1700.

Banking: (Mon-Fri) 0830 – 1400, (Wed) 0830 – 1200; (Sat) 0830 – 1100.

Telecommunications

Telephone: Dialling code for Zimbabwe: IDD access code + 263 followed by area code (Harare 4, Bulawayo 9, Gweru 54, Mutare (Umtali) 20, Victoria Falls 13. Good international direct dialling.

Telex: Telex facilities at main hotels and central post offices. Telex country code: ZW.

Banking

Central bank: Reserve Bank of Zimbabwe.

Head offices: Agricultural Finance Corp, Bank of Credit & Commerce, Barclays Bank of Zimbabwe, Bard Discount House, Discount Company of Zimbabwe, Grindlays Industrial and Commercial Finance, Merchant Bank of Central Africa, RAL Merchant Bank, Standard Chartered Bank Zimbabwe, Standard Chartered Merchant Bank, Syfrets Merchant Bank, UDC Zimbabwe, Zimbabwe Banking Corp, Zimbabwe Development Bank.

Representation in capital: Afghanistan, Albania, Algeria, Angola, Australia, Austria, Bangladesh, Belgium, Botswana, Brazil, Bulgaria, Burundi, Canada, China PR, Cuba, Czechoslovakia, Denmark, Egypt, Ethiopia, Finland, France, German DR, German FR, Ghana, Greece, Guyana, Hungary, India, Indonesia, Iran, Iraq, Italy, Japan, Kenya, Korea PDR, Libya, Malawi, Mozambique, Netherlands, New Zealand, Nicaragua, Nigeria, Norway, Pakistan, Peru, Poland, Portugal, Romania, Senegal, Sri Lanka, Spain, Sudan, Sweden, Switzerland, Tanzania, Togo, Tunisia, Turkey, Uganda, UK, USA, USSR, Vatican, Vietnam, Yemen, Yugoslavia, Zaïre, Zambia. Palestine Liberation Organisation is represented.

BUSINESS DIRECTORY

Hotels: Bulawayo and district

Bulawayo Sun, Wilson Street and 10th Avenue, P.O. Box: 654 (tel: 60101; tx: 3242).

Cresta Churchill, P.O. Box: 9140 (tel: 41016; tx: 33051).

Holiday Inn, Milnerton Drive, Ascot, P.O. Box: AC88 (tel: 72464; tx: 3341).

Hotel Cecil, P.O. Box: 224 (tel: 60295; tx: 33023).

Selborne, Selborne Avenue, 4 Rhodes Street, P.O. Box: 219 (tel: 65741).

Hotels: Harare and district

Courtney, Selous Avenue and 8th Street, P.O. Box: 3150 (tel: 706411; tx: 22611).

Cresta Jameson, Samora Machel Avenue, P.O. Box: 2833 (tel: 794641; tx: 26483).

Cresta Oasis, 124 Baker Avenue, P.O. Box: 1541 (tel: 704217; tx: 22099).

Holiday Inn, Samora Machel Avenue, P.O. Box: 7 (tel: 795611; tx: 2076).

Meikles, Stanley Avenue, P.O. Box: 594 (tel: 795655; tx: 26063; fax: 707754).

Monomatapa, 54 Park Lane, P.O. Box: 2445 (tel: 704501; tx: 26078).

Park Lane, Samora Machel and Bodle Area, Highlands, P.O. Box: HG 192 (tel: 707631).

Sheraton, Penefather Avenue, P.O. Box: 3033 (tel: 729771; tx: 22621; fax: 796678).

Hotels: Victoria Falls

A'Zambesi River Lodge, P.O. Box: 130 (tel: 561; tx: 3076).

Makasa Sun, P.O. Box: 90 (tel: 275; tx: 3222).

Rainbow, P.O. Box: 130 (tel: 583/585; tx: 3483).

Victoria Falls, P.O. Box: 10 (tel: 4203; tx: 51651).

Car hire: Harare

Avis: 5 Samora Machel Avenue (tel: 704191; tx: 26056).

Europcar: 19 Samora Machel Avenue, PO Box 3430 (tel: 706484, 702221; tx: 4641).

Hertz: 46 Samora Machel Avenue (tel: 704915, 793701; tx: 4741).

Car hire: Harare airport
Avis: (tel: 50121).

Hertz: (tel: 50320).

Other useful addresses

Air Zimbabwe, PO Box AP1, Harare Airport (tel: 737011; tx: 40008).

Confederation of Zimbabwe Industries, PO Box 3794, Harare (tel: 739833; tx: 22073).

Express Motorways Africa Ltd, Hood/Highfield Roads, PO Box ST 102, Harare (tel: 63505; tx: 24410).

Industrial Development Corporation of Zimbabwe, PO Box 8531, Causeway, Harare (tel: 706971; tx: 24409).

National Railways of Zimbabwe, PO Box 596, Bulawayo (tel: 363111; tx: 33173).

Reserve Bank of Zimbabwe, PO Box 1283, Samora Machel Avenue, Harare (tel: 790731; tx: 6033, 6075 RES/ZIM).

Zimbabwe Broadcasting Corporation, PO Box HG 444, Harare (tel: 707222; tx: 24175).

Zimbabwe International Trade Fair, PO Box Famona, Bulawayo (tel: 64911; tx; 33273).

Zimbabwe National Chamber of Commerce, Bulawayo Branch, PO Box 1934, Bulawayo (tel: 69769).

Zimbabwe National Chamber of Commerce, Harare Branch, PO Box 1934, Harare (tel: 708611; tx: 22531).

Zimbabwe Tobacco Association, PO Box 1781, Harare (tel: 727441).

Zimbabwe Tourist Development Corporation, PO Box 8052, Causeway, Harare (tel: 793666; tx: 26087).

Index of countries & territories